ENGINEERING

ENGINEERING

DRAWING & DESIGN

3rd Edition

DAVID A. MADSEN
Faculty Emeritus, Former Department Chair Drafting Technology, Autodesk Premier Training Center
Former Member Board of Directors, American Drafting and Design Association
Clackamas Community College, Oregon City, OR

DR. JAMES FOLKESTAD
(Chapter 20) Department of Manufacturing Technology and Construction Management
Colorado State University, Fort Collins, CO

KAREN A. SCHERTZ
(Chapter 5) Former Department Chair Advanced Technology Division
Front Range Community College, Fort Collins, CO

TERENCE M. SHUMAKER
(Chapters 4, 19 and 23) Department Chair Drafting Technology, Autodesk Premier Training Center
Clackamas Community College, Oregon City, OR

CATHERINE STARK
(Chapters 22 & 25) Engineering Technician/CADD, City of Portland, OR

J. LEE TURPIN
(Chapters 9 and 10) Former Drafting Instructor/Department Chair, Vocational Counselor
Clackamas Community College, Oregon City, OR

DELMAR

THOMSON LEARNING

Australia • Canada • Mexico • Singapore • Spain • United Kingdom • United States

Engineering Drawing and Design, Third Edition
by
David A. Madsen, Dr. James Folkestad, Karen A. Schertz, Terence M. Shumaker, Catherine Stark, J. Lee Turpin

Business Unit Director:
Alar Elken

Executive Editor:
Sandy Clark

Acquisitions Editor:
Jim DeVoe

Development Editor:
John Fisher

Editorial Assistant:
Jasmine Hartman

Executive Marketing Manager:
Maura Theriault

Executive Production Manager:
Mary Ellen Black

Channel Manager:
Mary Johnson

Production Manager:
Larry Main

Production Editor:
Stacy Masucci

Art/Design Coordinator:
Mary Beth Vought

Marketing Coordinator:
Karen Smith

ISBN-13: 978-0-7668-1634-3
ISBN-10: 0-7668-1634-6

NOTICE TO THE READER

CONTENTS

Chapter 20—Solid Modeling, Animation, and Virtual Reality   677

Chapter 21—Welding Processes and Representations   693

Chapter 22—Industrial Process Piping   722

Chapter 23—Structural Drafting   765

Chapter 24—Heating, Ventilating, and Air-Conditioning (HVAC)/Pattern Development and Precision Sheet Metal Drafting 828

Chapter 25—Electrical and Electronic Schematic Drafting 883

PREFACE

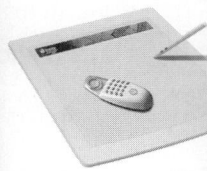

Engineering Drawing and Design is a practical, in-depth textbook that is easy to use and understand. The content may be used as presented, or the chapters may be rearranged to accommodate alternate formats for traditional or individualized instruction. *Engineering Drawing and Design* may be used for the following courses, so your students need only one text for their entire curriculum:

- THE ENGINEERING DESIGN PROCESS
- COMPUTER-AIDED DESIGN DRAFTING (CADD)
  Hardware, software, and applications.
- MECHANICAL DRAFTING
  Sketching, lettering, lines, geometric constructions, multiviews and auxiliary views, dimensioning, fasteners, springs, sectioning, working drawings, details, assemblies, parts lists, and engineering changes.
- DESCRIPTIVE GEOMETRY
  Beginning and advanced.
- MANUFACTURING PROCESSES
- WELDING PROCESSES
- GEOMETRIC TOLERANCING
  Based on ASME Y14.5M—1994.
- MECHANISMS/KINEMATICS
  Linkages, gears, cams, bearings, and seals.
- BELT AND CHAIN DRIVES
- PICTORIAL DRAWINGS
- SOLID MODELING, ANIMATION, AND VIRTUAL REALITY
- STRUCTURAL DRAFTING
- CIVIL DRAFTING
- INDUSTRIAL PIPE DRAFTING
- HEATING, VENTILATING, AND AIR CONDITIONING (HVAC)
- PRECISION PATTERN DEVELOPMENT
- SHEET METAL DRAFTING
- FLUID POWER
- ENGINEERING CHARTS AND GRAPHS
- ELECTRICAL DRAFTING
  Electrical power substation design.
- ELECTRONIC SCHEMATIC DRAFTING

MAJOR FEATURES

Engineering Drawing and Design has these important features:

- Engineering Design Applications.
- CADD throughout.
- ANSI, ASME, and related standards emphasized.
- Professional Perspectives.
- Math Applications.
- Step-by-step layout methods.
- Engineering layout techniques.
- CADD Applications.
- Practical appendices.
- Real industry problems.
- Web Site Research.

Engineering Drawing and Design provides a practical approach to drafting as related to the American National Standards Institute (ANSI) and the American Society of Mechanical Engineers (ASME) standards and common alternates that may be found in traditional industrial standards, such as the American Welding Society or the American Institute for Steel Construction. Also presented, when appropriate, are standards and codes related to specific engineering fields. One excellent and necessary foundation to engineering drawing and design and the implementation of a common approach to graphics nationwide is the emphasis of standardization in all levels of drawing and design instruction. When you become a professional, this text will go along as a valuable desk reference.

Each chapter provides realistic examples, **illustrations**, **problems**, and **related tests**. The examples illustrate recommended design presentation based on ANSI/ASME standards and other related national standards and codes with actual industrial drawings used for reinforcement. The correlated text explains drawing techniques and provides professional tips for skill development. Step-by-step layout methods provide a logical approach to setting up and completing the drawing problems.

RELATED TESTS   PROBLEMS

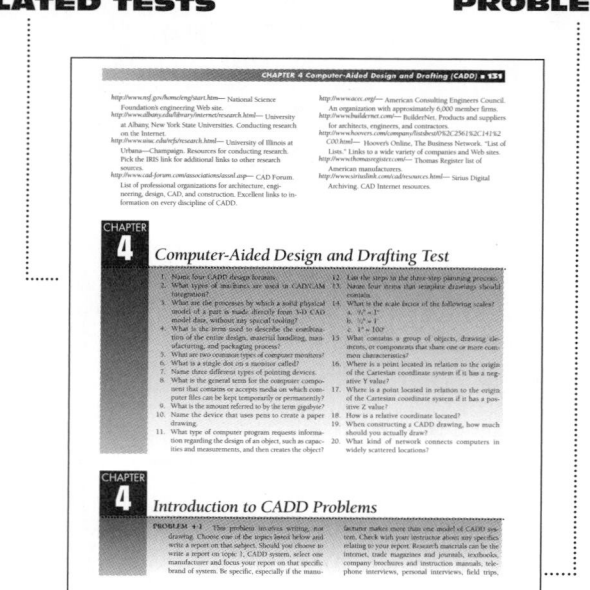

The Engineering Design Application leads the content of every chapter. This gives you an early understanding of the type of engineering project that is found in the specific design and drafting area discussed in the chapter.

ENGINEERING DESIGN APPLICATION

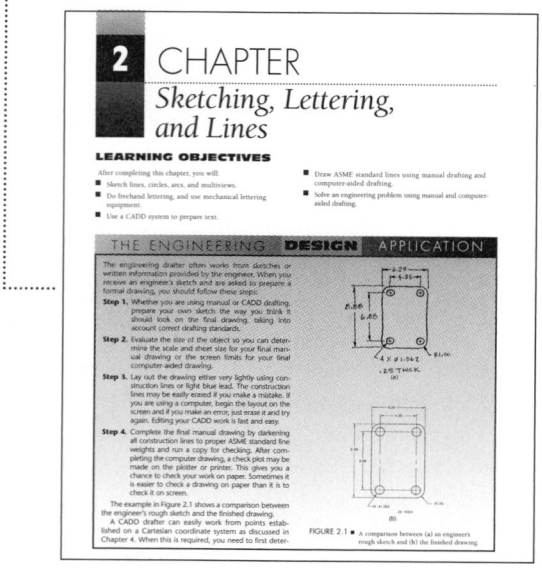

CADD Applications are provided in each chapter to illustrate how the use of CADD is streamlining the design process.

Math Applications, math problems, and practical drafting problems are contained in every chapter. These elements provide examples and instruction on how math is used in the specific discipline.

CADD APPLICATIONS

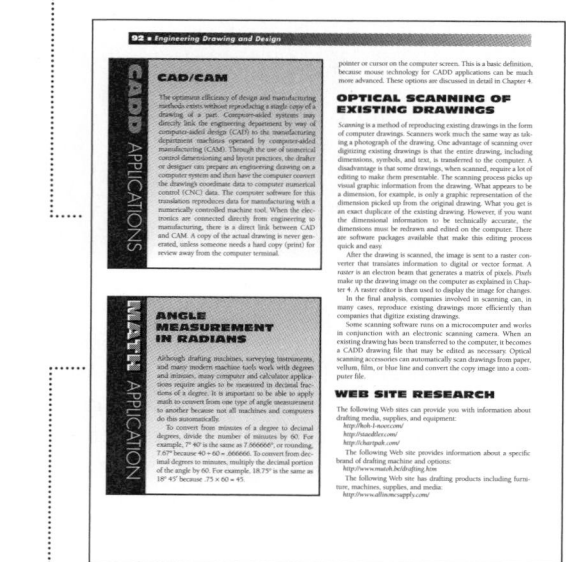

MATH APPLICATIONS

Professional Perspective, a boxed article at the end of each chapter, explains how the skills and knowledge discussed in the chapter may be applied to a real-world job-related setting.

Web Site Research is a feature that is placed at the end of every chapter, providing key Web sites where you can do additional research, find standards, and seek manufacturing information or vendor specifications related to the chapter content.

WEB SITE RESEARCH

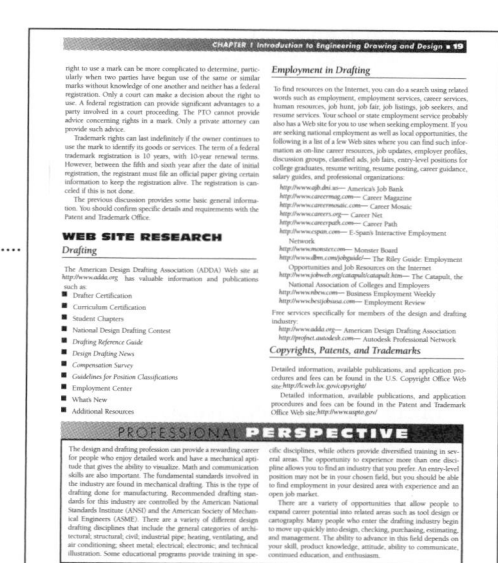

PROFESSIONAL PERSPECTIVE

The appendices contain the types of charts and information that is used daily in the engineering design and drafting environment. These appendices include common fastener types and data, fits and tolerances, metric conversion charts, tap drill charts, and other manufacturing information used on engineering drawings; a complete list of abbreviations and a comprehensive glossary.

INDUSTRIAL APPROACH TO PROBLEM SOLVING

The drafter's responsibility is to convert the engineering sketch or instructions to formal drawings. The text explains how to prepare drawings from engineering sketches by providing you with the basic guides for layout and arrangement in a knowledge-building format. One concept is learned before the next is introduced. Problem assignments are presented in order of difficulty within each chapter and throughout the text. The concepts and skills learned in one chapter are used in following chapters so that by the end of the text you have the ability to solve problems using a multitude of previously learned activities. The problems are presented as 3-D or actual industrial layouts in a manner that is consistent with the engineering environment. Early problems provide suggested layout sketches. It is not enough for you to duplicate drawings from given assignments; you must be able to think through the process of drawing development. The goals and objectives of each problem assignment are consistent with recommended evaluation criteria based on the progression of learning activities.

COMPUTER-AIDED DESIGN AND DRAFTING

Computer-aided design and drafting (CADD) is presented throughout the text. CADD topics include:

- CADD hardware.
- CADD software.
- CADD material requirements.
- Specific CADD applications.
- CADD menus and symbol libraries for specific engineering drafting applications.
- Increased productivity with CADD.
- Parametrics in CADD applications.
- The CADD environment in industry.

DESIGN PROJECTS

Chapters contain projects that allow you to practice your design knowledge and skills. These are advanced problems that require you to systematically determine a desired solution. There are challenges in manufacturing knowledge, tolerances, accuracy, and other issues related to the specific discipline.

TEAM PROJECTS

Some projects are designed to be solved in a team approach. Groups of students can develop their own team organization and establish the best course of action to create the desired solutions.

COURSE PLAN

Chapter 1 of *Engineering Drawing and Design* provides a detailed look at drafting as a profession, and includes occupations, professional organizations, occupational levels, opportunities, career requirements, seeking employment, CADD issues, workplace ethics, copyrights, patents, and trademarks.

Sketching, Lettering, Lines, and Geometric Construction

You then learn how sketching, lettering, lines, and geometric constructions may be properly used in both manual and computer-aided design in accordance with ANSI/ASME standards.

Drafting Equipment, Media, and Reproduction Methods

The study of engineering drawing and design in this text begins with equipment, materials, and reproduction for manual drafting with specific instruction on how to use tools and equipment.

Introduction to CADD

Next you received an in-depth introduction to CADD, including developing CADD skills, computer languages, drafting with the computer, using menus, CADD in industry, the CADD environment, the job market, computer drafting equipment, CADD drawing materials, and future CADD hardware and software.

Using the Engineering Design Process

This chapter provides the step-by-step use of the design process. You will see how all the aspects of the design process fit together. You will take the design of a product from engineering sketches through production, implementing CAD/CAM, parametric design, rapid prototyping, CAE, concurrent engineering, collaborative engineering, reverse engineering, team projects, and other innovative topics. Creativity and the design process are the emphasis of this chapter.

Multiviews and Auxiliary Views

A complete study of multiviews and auxiliary views, in accordance with ANSI/ASME standards, provides accurate and detailed instruction on topics such as view selection and placement, first- and third-angle projection, and viewing techniques. You are provided with a step-by-step example on how to lay out multiview and auxiliary view drawings.

Descriptive Geometry

This text provides all the descriptive geometry coverage that you need, including discussion of lines, planes, angles, slope, bearing, intersecting lines and planes, and vectors. Problems are presented as actual engineering situations. This is an easy-to-understand set of two chapters with related problems.

Manufacturing Processes

This is a complete introduction to manufacturing processes, including product development, manufacturing materials, material numbering systems, hardness and testing, casting and forging methods, design and drafting, complete machining processes, computer numerical control, computer-integrated manufacturing, machine features and drawing representations, surface texture, design of machine features, tool design, and statistical process control.

Dimensioning

This chapter is in accordance with ASME Y14.5M—1994 (R 1999) and provides complete coverage on dimensioning systems, rules, specific and general notes, tolerances, symbols, and dimensioning for CAD/CAM. You are provided with a step-by-step example showing how to lay out a fully dimensioned multiview drawing.

Fasteners and Springs, and Welding Processes

These two chapters cover the complete range of fastening devices and welding processes available. The fasteners chapter covers screw threads, thread cutting, thread forms, thread representations and notes, washers, dowels, pins, rings, keys and keyseats, rivets, and springs.

The welding chapter provides an in-depth introduction to processes, welding drawings and symbols, weld types, symbol usage, weld characteristics, weld testing, welding specifications, and prequalified welded joints.

Sections, Revolutions, and Conventional Breaks

This chapter explains in detail every type of sectioning practice available to the mechanical engineering drafter. Additionally, the chapter includes treatment of unsectioned features, conventional revolutions, and conventional breaks.

Geometric Tolerancing

This chapter provides a complete instruction to the reading and use of geometric tolerancing symbols and terms as presented in ASME Y14.5M—1994 (R 1999). It features excellent coverage on datums, feature control, basic dimensions, geometric tolerances, material condition, position tolerance, virtual condition, geometric tolerancing, and CADD.

Kinematics, Mechanisms, Belt and Chain Drives

These two chapters provide you with extensive coverage of linkage mechanisms, gears, cams, and belt and chain drives. You are given detailed information on the selection of bearings and lubricants. The use of vendors' catalog information is stressed in the design of belt and chain drive systems. Actual mechanical engineering design problems are provided for gear train and cam plate design. Working drawings include details, assemblies, and parts lists—the most extensive discussion on engineering changes available. A complete analysis on how to prepare a set of working drawings from concept to final product.

Pictorial Drawing

This chapter is a complete review of 3-D drafting techniques used in manual and computer-aided drafting. The content includes extensive discussions in isometric, dimetric, trimetric, perspective, exploded drawing, and shading methods.

Solid Modeling, Animation, and Virtual Reality

This chapter covers the modern use of solid modeling with applications from programs such as the Autodesk Mechanical Desktop, Autodesk Inventor, SolidWorks, SDRC, and others. You will see how to create automatic 2-D drawings from models. You will learn how models can be animated to function as in the real world. Virtual reality is explained and demonstrated in actual industry and architectural environments.

Engineering Fields of Study

The balance of this text provides you with chapters that offer the most complete information available in specific engineering drawing and design fields. These chapters may be used individually for complete courses. They contain comprehensive instruction, problems, and tests. These fields include:

- *Fluid power*—with complete coverage on forces, pressure, work, power, hydraulics, pneumatics, diagrams, symbols, equipment, systems, and operations. This chapter is located in the ONLINE COMPANION that can be accessed from *www.delmar.com*.

- *Industrial process piping*—with explanations and detailed examples of pipe and fittings, valves and instrumentation, pumps, tanks, and equipment, flow diagrams, piping plans and elevations, piping isometrics, and piping spools.

- *Structural drafting*—of reinforced and precast concrete, truss and panelized framing, timbers, laminated beams, steel joists and studs, prefabricated systems, structural steel, structural welding, set of structural drawings.

- *Civil drafting*—covers the complete discipline of mapping, including legal descriptions, survey terminology, and site plans. Civil drafting includes road layout, cuts and fills, and plan and profile drawings. This chapter is located in the ONLINE COMPANION that can be accessed from *www.delmar.com*.

- *HVAC pattern development, and precision sheet metal drafting*—provides complete coverage of HVAC systems, heat exchangers, HVAC symbols, single- and double-line ducted systems, working from an engineer's sketch, sections and details, common sheet metal pattern developments with step-by-step layout procedures, sheet metal intersections, bend allowances calculations, and precision sheet metal drawings.

■ *Electrical drafting and power substation design*—provides the only coverage of its type in electrical power transmission. The content is specific to electrical drafting of wiring diagrams, cable assemblies, one-line and elementary diagrams, electrical power system symbols, plot plans, bus layouts, ground layouts, conduit layouts, electrical floor plan symbols, power supply plans, and schematics.

■ *Electronic schematic drafting*—exclusively related to the electronics industry. This includes topics such as block diagrams, electronic components and symbols, engineer's sketch, component numbering, IC systems, logic diagrams, LSC schematics, SMT, electronics artwork, layers, marking, drilling, assembly drawings, photo drafting, and pictorial diagrams.

■ *Engineering charts and graphs*—with the most complete coverage on rectilinear, log and semilog scales, surface, column, pie, concurrency, alignment, organizational, production control process (SPC), distribution, and pictorial charts. The chapter includes an in-depth analysis of chart design. This chapter is located in the ONLINE COMPANION that can be accessed from *www.delmar.com.*

The text contains CADD discussion and examples, engineering layout techniques, working from engineer's sketches, professional practices, and actual industrial examples. The problem assignments are based on actual real-world products and designs.

SECTION LENGTH

Chapters are presented in individual learning segments that begin with basic concepts and build until each chapter provides complete coverage of each topic.

APPLICATIONS

Special emphasis has been placed on providing realistic problems. Problems are presented as 3-D drawings, engineering sketches, and layouts in a manner that is consistent with industry practices. Many of the problems have been supplied by the industry. Each problem solution is based on the step-by-step layout procedures provided in the chapter discussions. Problems are given in order of complexity so that you can be exposed to a variety of engineering experiences. Early problems recommended the layout to help you save time. Advanced problems require you to go through the same thinking process that a professional faces daily, including drawing scale, paper size selection, view layout, dimension placement, sectioning placement, and many other activities. Problems may be solved using manual or computer-aided drafting as determined by the individual course guidelines. All problems should be solved in accordance with recommended ANSI, ASME, or other related industry practices. Problems may not be presented in the best arrangement for a quality solution. You should always approach a problem with critical analysis based on view selection and layout, and dimension placement when dimensions are used. You should not assume that problem information is presented exactly as the intended solution. Advanced problems often require you to evaluate the accuracy of provided information. Chapter tests provide a complete coverage of each chapter and may be used for student evaluation or as study questions.

SUPPLEMENTS
e.resource

This is an educational resource that creates a truly electronic classroom. It is a CD-ROM containing tools and instructional resources that enrich your classroom and make the instructor's preparation time shorter. The elements of *e.resource* link directly to the text and tie together to provide a unified instructional system. With *e.resource* you can spend your time teaching, not preparing to teach. ISBN: 0-7668-1635-4.

Features contained in *e.resource* include:

■ **Syllabus.** Lesson plans created by chapter. You have the option of using these lesson plans with your own course information.

■ **Chapter Hints.** Objectives and teaching hints that provide the basis for a lecture outline that helps you to present concepts and material. Key points and concepts can be graphically highlighted for student retention.

■ **PowerPoint® Presentation.** These slides provide the basis for a lecture outline that helps you to present concepts and material. Key points and concepts can be graphically highlighted for student retention.

■ **Exam View Computerized Test Bank.** Over 800 questions of varying levels of difficulty are provided in true/false and multiple-choice formats so you can assess student comprehension.

■ Files for many of the drawings in the text are listed in chapter folders. You can modify these plans to create new illustrations or chapter problems, or you can make them available to students who desire to work within AutoCAD.

■ **ASME/ANSI Exercises.** Actual industry drawing files are provided that contain intentional ASME/ANSI errors. The students can correct the drawing files using CADD, or redline prints to conform to accept ASME/ANSI standards.

■ **Video and Animation Resources.** These AVI files graphically depict the execution of key concepts and commands in drafting, design, and AutoCAD and let you bring multimedia presentations into the classroom.

Workbook

We have developed a workbook to correlate with *Engineering Drawing and Design*. This workbook is correlated with the main text and contains "survival information" related to the topics covered. The plentiful problem assignments take the beginning engineering drafting student from the basics of line and lettering techniques to drawing plans. Problems may be done on CADD or manually. ISBN 0-7668-1636-2.

Solutions Manual

A solutions manual is available with answers to end-of-chapter review questions and solutions to end-of-chapter problems. Solutions are also provided for the Workbook problems. ISBN 0-7668-4336-X.

Videos

Two video sets, containing four 20-minute videos each, are available. The videos correspond to the topics addressed in the text.

■ Set #1, ISBN 0-7668-3903-6 ■ Set #2, ISBN 0-7668-3908-7

Video sets are also available on Interactive Video CD-ROM.

■ Set #1, ISBN 0-7668-3914-1 ■ Set #2, ISBN 0-7668-3915-X

Online Companion

Visit the *Engineering Drawing and Design* online companion by accessing our site at *www.delmar.com*. Additional chapter information and appendices are provided. Chapters include Fluid Power, Civil Drafting, and Engineering Charts and Graphs. The Math Instruction Appendix is also included on this site.

ACKNOWLEDGMENTS

I would like to express a special thanks to my contributing authors:

- Terence M. Shumaker, for his outstanding implementation of the introduction to CADD and related topics, industrial pipe drafting, and pictorial drawing chapters.
- J. Lee Turpin, for his experience in presenting descriptive geometry to students in a manner that is easy to follow and understand.
- Dr. James Folkestad for his ability to teach CADD modeling, animation, and virtual reality in a manner that sparks excitement among students and for his creativity in preparing instructional materials.
- Karen Schertz, for contributing the chapter Using the Engineering Design Process.
- Catherine Stark, for a professional coverage of fluid power and civil drafting.
- Steve Brown, for his chapter applications and problems. The math instruction is given in an innovative and practical manner that immediately addresses the real-world issues.
- Tim Stafford, for his extensive knowledge of electronics and his coordination of the revision for Chapter 25, Electrical and Electronic Schematic Drafting.

I would like to give special thanks and acknowledgment to the many professionals who helped formulate this work. Also, thanks to those who reviewed the manuscript in an effort to help me publish the best possible text:

William Batson
ITT Technical Institute
Dayton, Ohio

John Beauchamp,
ITT Technical Institute
Grand Rapids, Michigan

James E. Blair
Jackson State Community
College
Jackson, Tennessee

Patrick E. Connolly
Purdue University
West Lafayette, Indiana

Terry Crissey
Forest Hills School
District
Sidman, Pennsylvania

Judith Dalton
ITT Technical Institute
Spokane, Washington

Jeff Gibbs
Muskingum Technology
College
Zanesville, Ohio

Casey Guthridge
Stark State College of
Technology
Canton, Ohio

Dr. Ralph Horne
Pennsylvania College of
Technology
Williamsport, Pennsylvania

Jennifer Ingram
Pulaski Technical College
North Little Rock, Arkansas

Gregory Nelson
University of Advancing
Computer Technology
Phoenix, Arizona

Don Nimmer
ITT Technical Institute
Henderson, Nevada

James Overton
Southeast College of
Technology
Memphis, Tennessee

Mike Schnurr
Hillsborough Community
College
Tampa, Florida

Tom Singer
Sinclair Community College
Dayton, Ohio

Kevin Standiford
Arkansas State Teacher
Retirement System
Little Rock, Arkansas

Scott Thomas
Deer Valley Unified School
District
Phoenix, Arizona

The quality of this text is also enhanced by the support and contributions from industry and vendors. The list of contributors is extensive, and acknowledgment is given at each figure illustration. (Many figures appearing in Chapter 23 were reprinted from Jefferis and Madsen, *Architectural Drafting and Design,* 4th edition, © 2001 by Delmar Publishers.) The following individuals and companies gave an extraordinary amount of support:

Accugraph Corporation
Ontario, Canada

Aerojet TechSystems Company
Sacramento, California

American Institute for Design
and Drafting
Rockville, Maryland

American National Standards
Institute
New York, New York

Berol USA RapiDesign
Burbank, California

Bishop Graphics, Inc.
Westlake Village, California

CALCOMP
Anaheim, California

Chartpak
Northampton, Massachusetts

Computervision
Corporation
Bedford, Massachusetts

Consul and Mutoh, Ltd.
Anaheim, California

Hyster Company
Portland, Oregon

Doug Major
Mark Hartman
Professional Consultants

Parker-Hannifin
Cleveland, Ohio

Tom Pearce
Stanley Hydraulic Tools, A
Division of the Stanley
Works
Portland, Oregon

Jean K. Shirkoff, consultant
Portland, Oregon

T & W Systems
Huntington Beach, California

Ricardo Wilkins
Michael Spiegel
Fluid Air Components
Portland, Oregon

The quality of this edition is enhanced by the following professionals who gave many hours of dedicated work in providing reviews, edits, information, and the development of problem projects:

John Boertjens
CAD Technology Corp.
Franklin, North Carolina

Randee R. Buckle
Wendy's International, Inc.
Dublin, Ohio

Dick Button
Fastener content
Fisher Controls International,
Inc.
Marshalltown, Iowa

Kasey Cassell
The Cork Screw Project

Bill Curtis
Curtis and Associates
Portland, Oregon

Tom Durston
Applications Engineer/EDA
Consultant
Cindy Easton
Technical Communications
Manager
OrCAD, Inc.
Beaverton, Oregon

Sabine Gossart
SolidWorks Corporation
Concord, Massachusetts

David P. Hammer
Hammer Inc.
Medford, New Jersey

Chris Herford
Gary Roberts
Ball Valve Project

Richard Hertel
The Oil Pump Project

Daniel L. Hodgin
KH2A Engineering, Inc.
Portland, Oregon

Kim Lockwood
The Flash Light Project

David P. Madsen
John Melloy
Fly Tying Vice Project

Karl Mayhew
Interface Engineering
Portland, Oregon

Keith McDonald
FLIR Systems, Inc.
Portland, Oregon

Mike McIntyre
The Landing Gear Project

Vin Mehta
Hunter Fan Company
Memphis, Tennessee

John Melloy
Light Fixture Project

Jack Pitcher
The Table Vice Project

Everett Russ
PAE Consulting Engineers, Inc.
Portland, Oregon

Martin Soll
Tool Design content
GD&T Project

Leonard Sosnovske
Nevada Power
Las Vegas, Nevada

Timothy Y. Taylor
Matthew Bell
The Motor Project

Connie Willmon
Ankrom Moisan Architects
Portland, Oregon

Deborah Parker Wong
The Medialink Group
Kentfield, California

CHAPTER 1

Introduction to Engineering Drawing and Design

A Short History of Engineering Drawing

J. H. Oakey

OBJECTIVES

After completing this chapter, you will:

- Explain topics related to the history of engineering drafting.
- Identify categories and disciplines related to drafting.
- Define drafter and identify current terminology.
- Identify the professional organization that is dedicated to the advancement of design and drafting.
- Discuss the requirements for becoming a drafter.
- List and explain points to consider when seeking employment.
- Discuss issues related to computer-aided design and drafting (CADD).
- Explain workplace ethics and related issues.
- Identify topics related to copyrights, patents, and trademarks.

We are all aware of the marvelous drawings and the inventive genius of Leonardo da Vinci. It is naturally assumed that he was the originator of drafting and that after his work, all inventors and engineers carefully duplicated his designs in some form in their drawings. This is not so. History is not well documented in the realm of inventors and engineers. The only source for the detailed history of Leonardo's work is his own careful representation. His drawings were those of an artist. They were 3-D and they generally were without dimensional notations, as shown in Figure 1.1. Craftsmen worked from the 3-D representation, and each machine or device was one-of-a-kind; parts were not interchangeable. In fact, interchangeability was not realized except in special demonstrations until the development of the micrometer in the late 1800s, and even then it was not easily achieved.

Without the concept of interchangeability, accurate drawings were not necessary. Inventors, engineers, and builders worked on each product on a one-of-a-kind basis, and parts were manufactured from hand sketches or hand drawings on blackboards. Coleman Sellers, in the manufacture of fire engines, had blackboards with full-size drawings of parts. Blacksmiths formed parts and compared them to the shapes on the blackboards. Coleman Seller's son, George, recalls lying on his belly using his arms as a radius for curves as his father stood over him directing changes in the sketches until the drawings were satisfactory. Most designs used through the 1800s were accomplished by first completing a hand sketch of the object to be built. These were then converted into wooden models (3-D modeling) from which patterns were constructed. This practice was followed well into the twentieth century by some. Most of us are familiar with the stories of Henry Ford and his famous blackboards. What is news is that these were also the Henry Ford "drafting tables." Henry would sketch cars and parts three-dimensionally and have patternmakers construct full-size models in wood.

An effort to create a program to standardize drawing came when the Franklin Institute was founded in Philadelphia in 1824. It was founded "to advance the general interests" of mechanics and entrepreneurs "by extending a knowledge of mechanical science." One of the goals of the Institute was to establish a mechanical drawing school. Unfortunately, the academic program was never realized

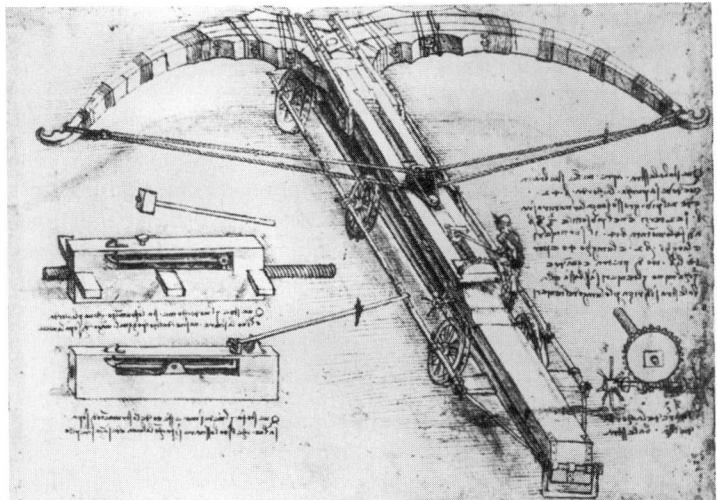

FIGURE 1.1 ■ The drawing of a design by Leonardo da Vinci.

FIGURE 1.2 ■ 3-D modeling is replacing board drafting. *Courtesy Solid Works Corporation.*

because two factions argued over whether to base it on classical academics (Latin and Greek) or on science and practical courses. Eventually the founders abandoned the academic purposes of the Institute.

Also in 1824, in upstate New York, Amos Eaton founded a school "for the purpose of instructing persons . . . in the application of science to the common purposes of life." This school has grown to be known today as Rensselaer Polytechnic Institute.

In 1862, with the coming of federal government assistance in the form of the Morrill Act, more technical schools emerged, and we can assume that mechanical drawing *slowly* became a part of the intellectual tool kits of trained mechanics, engineers, and inventors.

This can, in part, put to rest the common thought (as we look at pictures of pyramids, steam engines, and other engineering examples) that "it all started with a drawing." Certainly there were freehand sketches, cartoons, and other forms of graphic models. It is probably accurate to say that before most things were built, they were tested with a 3-D model. Just as mechanical drafting slowly took the place of hand sketches and wooden models, today computer design and drawing and 3-D computer modeling are replacing board drafting. An example is shown in Figure 1.2. The eventual utility and order of computer modeling and design remain to be established as part of the developing graphic process.

DRAFTING TECHNOLOGY

According to the *Dictionary of Occupational Titles,* published by the U.S. Department of Labor, drafting is grouped with professional, technical, and managerial occupations. This category includes occupations concerned with the theoretical and practical aspects of such fields of human endeavor as architecture; engineering; mathematics; physical sciences; social sciences; medicine and health; education; museum, library, and archival sciences; law; theology; the arts; recreation; administrative specialties; and management. Also included are occupations in support of scientists and engineers and other specialized activities such as piloting aircraft, operating radios, and directing the course of ships. Most of these occupations require substantial educational preparation, usually at the college, junior college, or technical institute level.

Men and women employed in the drafting profession are often referred to as drafters. A general definition of *drafter,* as prepared by the Career Information System, follows:

Drafters translate data and sketches of engineers, architects, and scientists into detailed drawings that are used in manufacturing and construction. Their duties may include interpreting directions given to them, making sketches, preparing drawings to scale, and specifying details. Drafters may also calculate the strength, quality, quantity, and cost of materials. They utilize various drafting tools and computer equipment, engineering practices, and math to complete drawings.

According to the *Occupational Outlook Handbook,* published by the U.S. Department of Labor and Bureau of Labor Statistics, the definition of *drafter* closely parallels that of the Career Information System with the following additional information:

Most drafters use computer-aided drafting (CAD) systems to prepare drawings. These systems employ computer work stations which create a drawing on a video screen. The drawings are stored electronically so that revisions can be made easily. These systems also permit drafters to easily and quickly prepare variations of a design. Although this equipment has become easier to operate, CAD is only a tool. Persons who produce technical drawings using CAD still function as a drafter, and need most of the knowledge of traditional drafters—relating to drafting skills and standards—as well as CAD skills.

As technology advances and the cost of systems continues to fall, it is likely that almost all drafters will use CAD systems regularly in the future. However, manual drafting may still be used in certain applications, especially in firms that produce many one-of-a-kind drawings with little repetition.

Due to advancement in the technology used to create drawings, drafters are becoming technicians. The term *drafter* is being replaced by *drafting technician, CAD technician, CADD* (computer-aided design and drafting) *technician,* and *CAD/CAM* (computer-aided manufacturing) *technician.*

OCCUPATIONS IN ARCHITECTURE AND ENGINEERING DRAFTING FIELDS

There are several types of drafting technology occupations. While drafting in general has one basic description, specific drafting areas have unique conceptual and skill characteristics. The types of drafting occupations fall into three general professional areas; architecture, engineering, and surveying. The following are specific drafting areas as defined by the *Dictionary of Occupational Titles.*

Architectural Drafter

Draws artistic architectural and structural features of any class of buildings and like structures; designs and details; confirms compliance with building codes; may specialize in planning architectural details according to structural materials used. (See Figure 1.3 (a) and (b).)

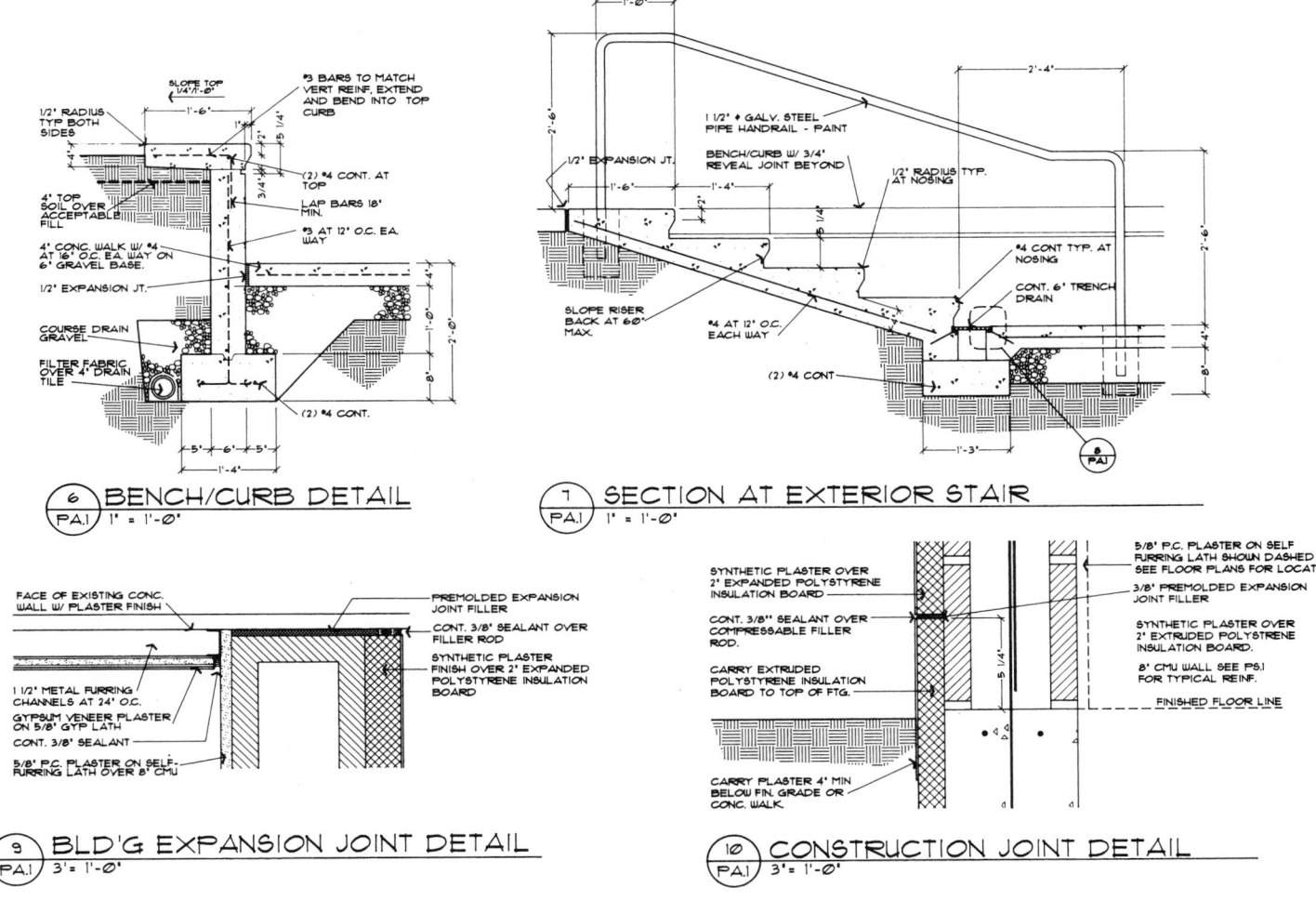

FIGURE 1.3(a) ■ Computer-generated elevations. *Courtesy Piercy & Barclay Designers, Inc.*

FIGURE 1.3(b) ■ Computer-generated architectural details. *Courtesy Soderstrom Architects PC.*

Landscape Drafter

Prepares detailed scale drawings from rough sketches or other data provided by Landscape Architect; may prepare separate detailed site plan, grading and drainage plan, lighting plan, paving plan, irrigation plan, planting plan, and drawings and detail of garden structures; may build models of proposed landscape construction and prepare colored drawings for presentation to client. (See Figure 1.4.)

Electrical Drafter

Prepares electrical-equipment working drawings and wiring diagrams used by construction and repair crews who erect, install, and repair electrical equipment and wiring in communications centers, power plants, industrial establishments, commercial or domestic buildings, or electrical distribution systems; performing duties described under Drafter. (See Figure 1.5.)

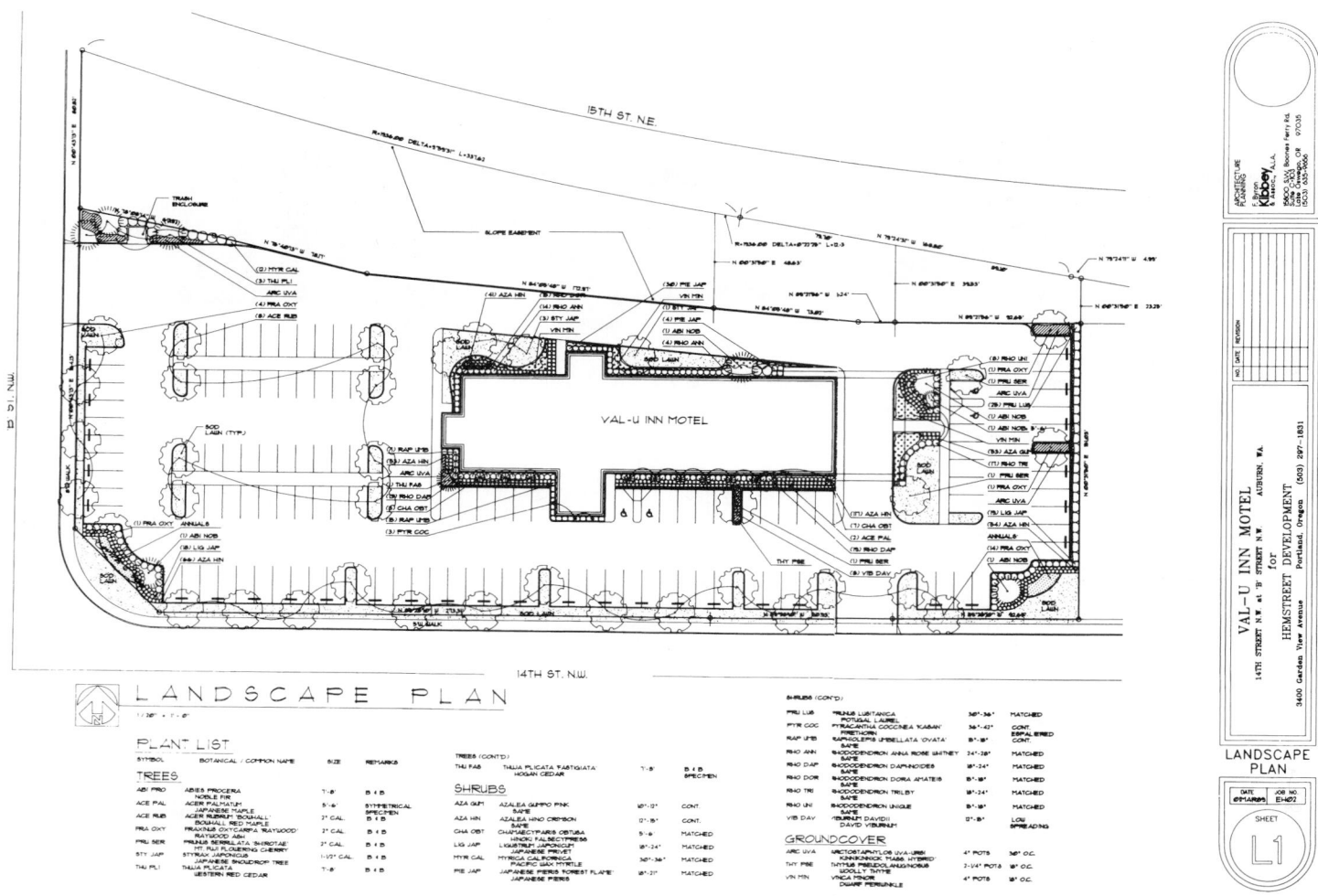

FIGURE 1.4 ■ Landscape plan. *Courtesy OTAK, Inc., for landscaping, and Kibbey & Associates, for site plan.*

Aeronautical Drafter

Specializes in preparing engineering drawings of developmental or production airplanes and missiles and ancillary equipment, including launch mechanisms and scale models of prototype aircraft, as planned by Aeronautical Engineer.

Electronic Drafter

Drafts wiring diagrams, schematics, and layout drawings used in manufacture, assembly, installation, and repair of electronic equipment such as television cameras, radio transmitters and receivers, audio-amplifiers, computers, and radiation detectors, performing duties as described under Drafter; drafts layout and detail drawings of racks, panels, and enclosures; may conduct service and interference studies and prepare maps and charts related to radio and television surveys; may be designated according to equipment drafted. (See Figure 1.6.)

Civil Drafter

This category is also known by the following titles: Drafter, Civil Engineering; Drafter, Construction; and Drafter, Engineering. Prepares detailed construction drawings, topographic profiles, and related maps and specification sheets used in planning and construction of highways, river and harbor improvements, flood control, drainage, and other civil engineering projects, performing duties as described under Drafter; plots maps and charts showing profiles and cross sections indicating relation of topographical contours and elevations to buildings, retaining walls, tunnels, overhead power lines, and other structures; drafts detailed drawings of structures and installations such as roads, culverts, fresh water supply and sewage disposal systems, dikes, wharfs, and breakwaters; computes volume of excavations and fills and prepares graphs and hauling diagrams used in earth-moving operations; may accompany survey crew in field to locate grading markers or to collect data required for revision of construction drawings; may be designated according to type of construction. (See Figure 1.7.)

Structural Drafter

Performs duties of Drafter by drawing plans and details for structures consisting of reinforced steel, concrete, masonry, wood, and other structural materials; produces plans and details of foundations, building frame, floor and roof framing, and other structural elements. (See Figure 1.8.)

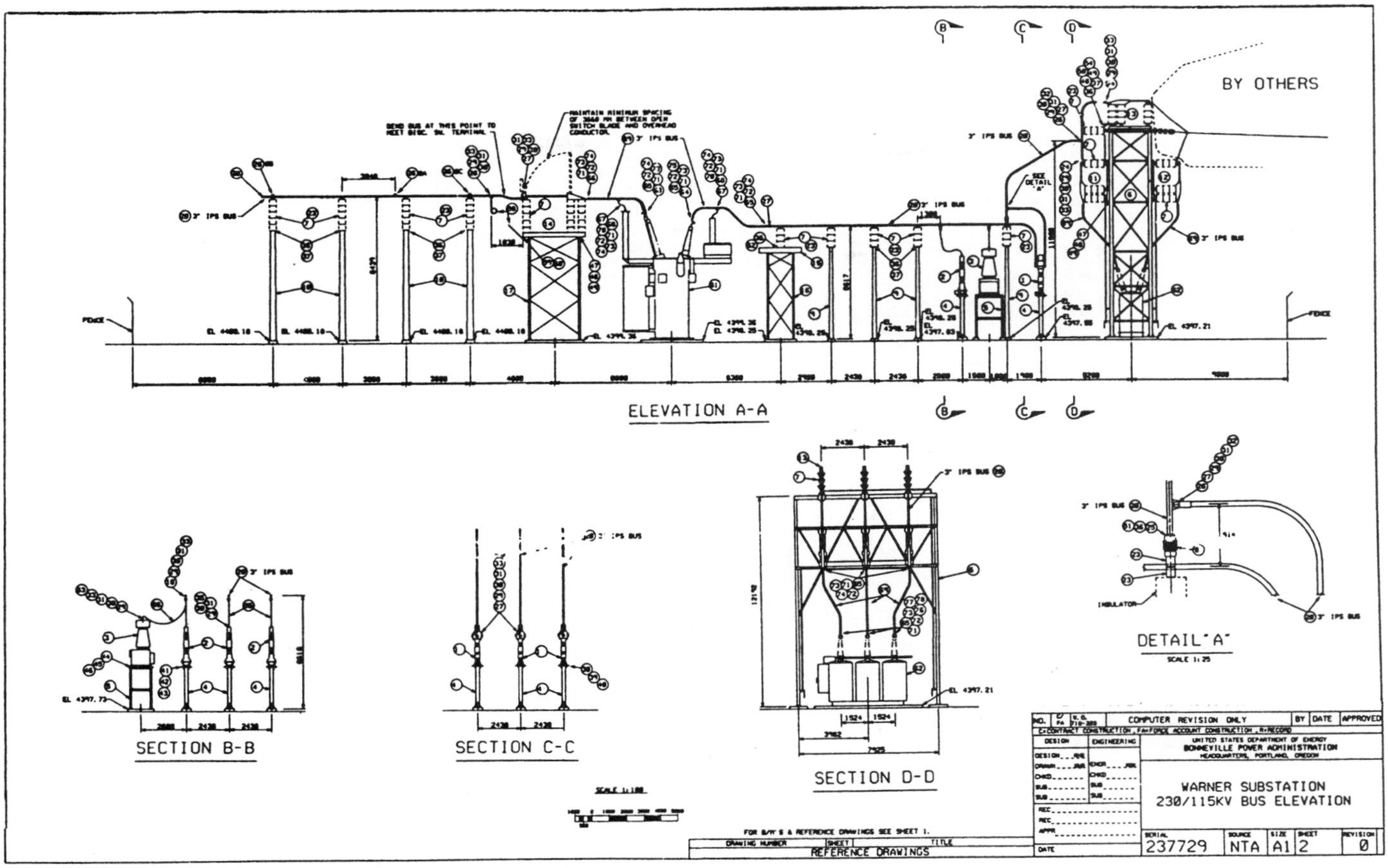

FIGURE 1.5 ■ Electrical drafting plan and elevation. *Courtesy Bonneville Power Administration.*

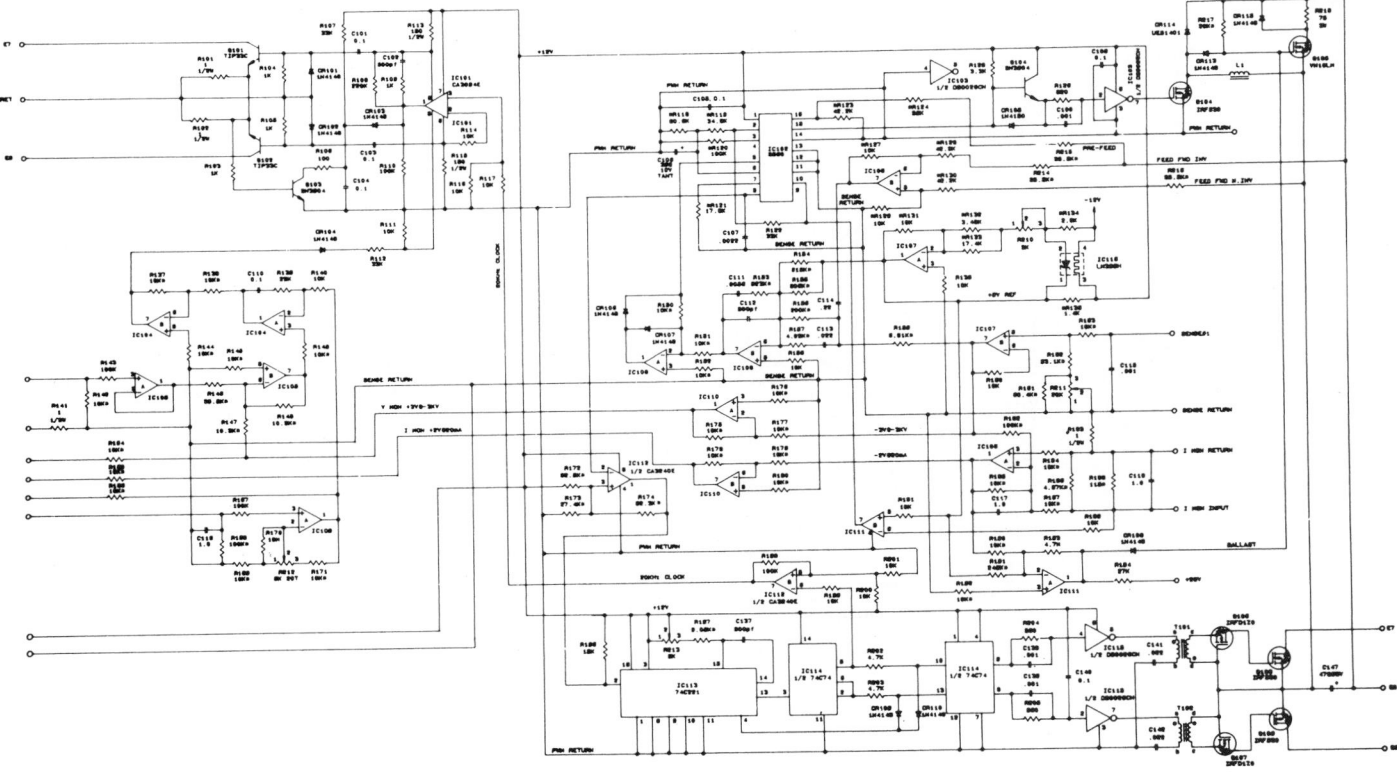

FIGURE 1.6 ■ Computer-generated electronics schematic. *Courtesy Versacad Corporation.*

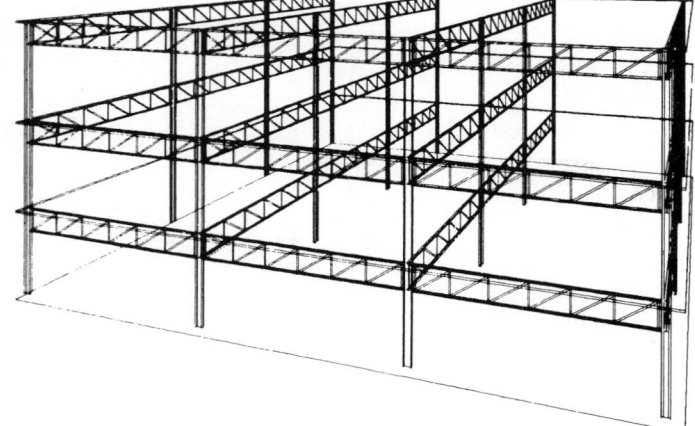

FIGURE 1.7 ■ Civil drafting, computer-generated drafting subdivision plat. *Courtesy Glads Project.*

FIGURE 1.8 ■ Computer-generated structural perspective. *Courtesy Computervision Corporation.*

Castings Drafter

Drafts detailed drawings for castings, which require special knowledge and attention to shrinkage allowances and such factors as minimum radii of fillets and rounds.

Patent Drafter

Drafts clear and accurate drawings of mechanical devices for use by a Patent Lawyer in obtaining patent rights.

Tool Design Drafter

Drafts detailed drawing plans for manufacture of tools, usually following designs and specifications indicated by a Tool Designer.

Mechanical Drafter

Drafts detailed working drawings of machinery and mechanical devices indicating dimensions and tolerances, fasteners, joining requirements, and other engineering data. Drafts multiple-view assembly and subassembly drawings as required for manufacture and repair of mechanisms; performs other duties as described under Drafter. Mechanical drafting, in general, is the core of the engineering drafting industry. (See Figure 1.9.)

Directional Survey Drafter

Plots oil or gas-well boreholes from photograpic subsurface survey recordings and other data; computes and represents diameter, depth degree, direction of inclination, location of equipment, and other dimensions and characteristics of borehole.

Geological Drafter

Draws maps, diagrams, profiles, cross sections, directional surveys, and subsurface formations to represent geological or geophysical stratigraphy locations and oil deposits; performs duties described under drafter; correlates and interprets data obtained from topographical surveys, well logs, or geophysical prospecting reports, utilizing special symbols to denote geological physical formations or oil field installations; may finish drawings in mediums and according to specifications required for reproduction by blueprinting, photographing, or other duplication methods.

Geophysical Drafter

Draws subsurface contours in rock formations from data obtained by geophysical prospecting party; plots maps and diagrams from computations based on recordings of seismograph gravity meter, magnetometer, and other petroleum prospecting instruments and from prospecting and surveying field notes.

Heating and Ventilating Drafter

Also known as Heating Ventilating and Air Conditioning (HVAC) Drafter; specializes in drawing plans for installation of heating, air-conditioning, and ventilating equipment; may calculate heat loss and heat gain for buildings for use in determining equipment specifications, following standardized procedures; may specialize in drawing plans for installation of refrigeration equipment. (See Figure 1.10.)

Plumbing Drafter

Also known as Piping Drafter; specializes in drafting plans for installation of plumbing and piping equipment for residential, commercial, and industrial installations. Commercial and industrial piping are closely related to industrial pipe drafting. (See Figure 1.11.)

Automotive Design Drafter

Designs and drafts working layouts and master drawings of automotive vehicle components, assemblies, and systems from specifications, sketches, models, prototype and verbal instructions, applying knowledge of automotive vehicle design, engineering principles, manufacturing processes and limitations, and drafting techniques and procedures, using drafting instruments and work aids; analyzes specifications, sketches, engineering drawings, ideas, and related design data to determine critical factors affecting design of components based on knowledge of previous designs and manufacturing processes and limitations; draws rough sketches and performs mathematical computations to develop design and work out detailed specifications of components; applies knowledge of mathematical formulas and physical laws; performs preliminary and advanced work in development of working layouts and final master drawings adequate for detailing parts and units of design; makes revisions to size, shape, and arrangement of parts to create practical design; confers with Automotive Engineer and others on staff to resolve design problems; specializes in design of specific type of body or chassis components, assemblies or systems such as door panels, chassis frame and supports, or braking system.

Industrial Pipe Drafter

Also known as an Oil and Gas Drafter; drafts plans and drawings for layout, construction, and operation of oil fields, refineries, and pipeline systems from field notes, rough or detailed sketches, and specifications; develops detail drawings for construction of equipment and structures, such as drilling derricks, compressor stations, gasoline plants, frame, steel, and masonry buildings, piping manifolds and pipeline systems, and for manufacture, fabrication, and assembly of machines and machine parts. (See Figure 1.12.)

Technical Illustrator

Lays-out and draws illustrations for reproduction in reference works, brochures, and technical manuals dealing with assembly, installation, operation, maintenance, and repair of machines, tools, and equipment; prepares drawings from blueprints, designs, mock-ups, and photographs by methods and techniques suited to specified reproduction process or final use, such as diazo, photo-offset, and projection transparencies, using drafting and optical equipment; lays-out and draws schematic, perspective, axonometric, orthographic, or oblique-angle views to depict function, relationship, and assembly sequence of parts and assemblies, such as gears, engines, and instruments; shades or colors drawing to emphasize details or to eliminate undesired background, using ink, crayon, airbrush, and overlays; pastes instructions and comments in position on drawing; may draw cartoons and caricatures to illustrate operation, maintenance, and safety manuals and posters. (See Figure 1.13, page 10.)

Cartographic Drafter

Draws maps of geographical areas to show natural and construction features, political boundaries, and other features;

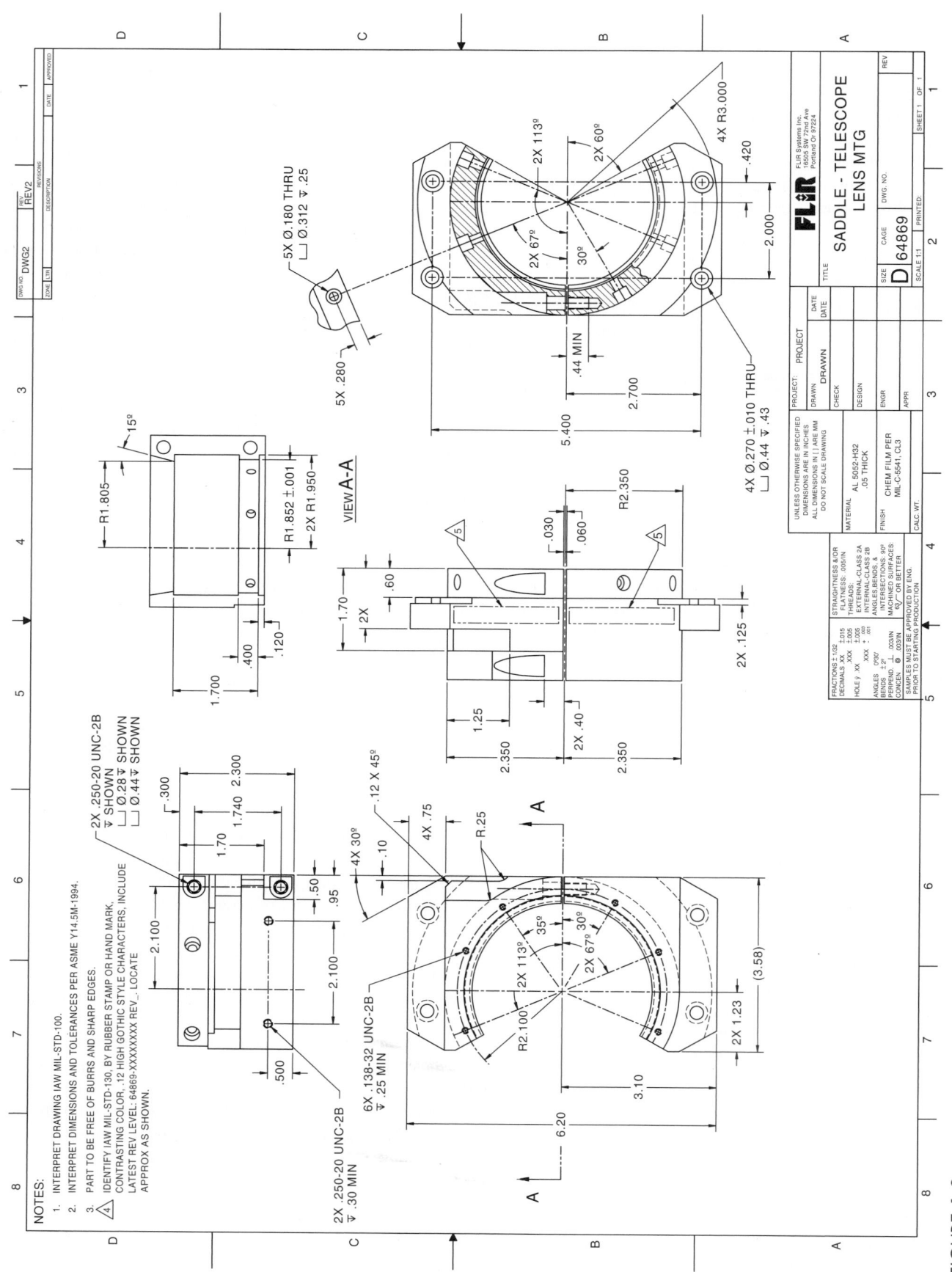

FIGURE 1.9 ■ Computer-aided mechanical drafting. *Drawing courtesy of FLIR Systems, Inc.*

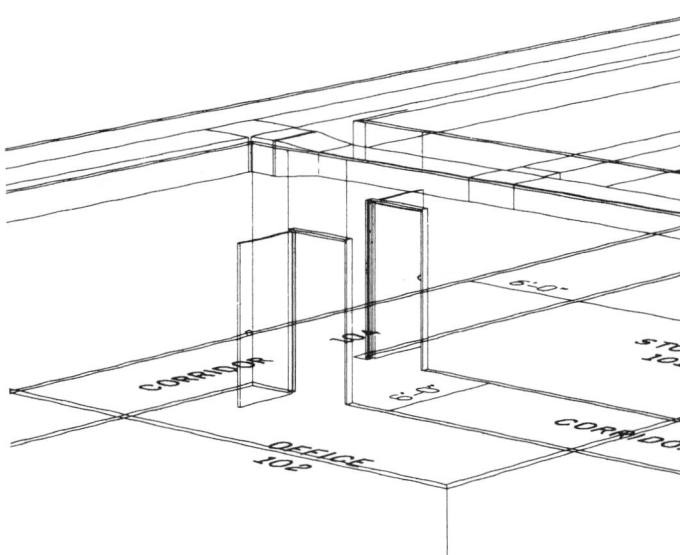

FIGURE 1.10 ■ Computer-generated HVAC pictorial. *Courtesy Computervision Corporation.*

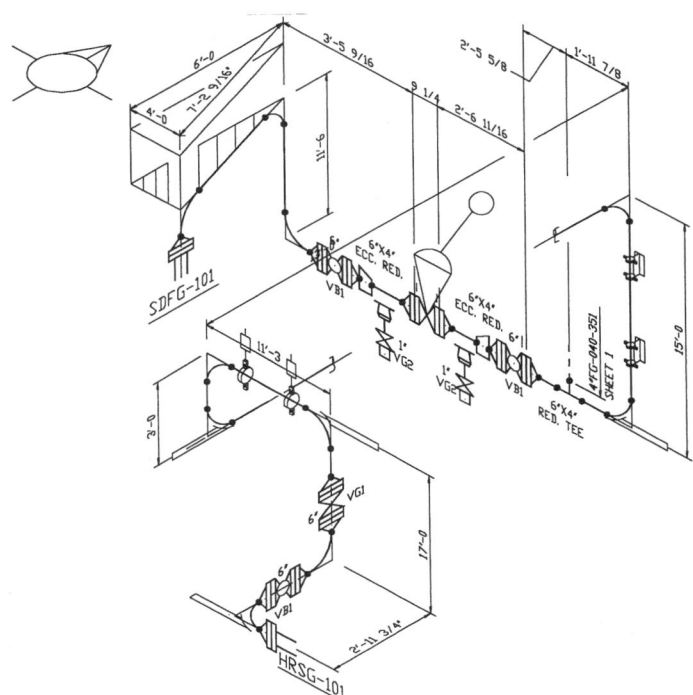

FIGURE 1.11 ■ CADD isometric piping layout. *Courtesy Autodesk, Inc.*

analyzes survey data, survey maps and photographs, computer- or automated-mapping products, and other records to determine location and names of features; studies records to establish boundaries of properties and local, national, and international areas of political, economic, social, or other significance. Geological and topographical maps are drawn by a Cartographic Drafter.

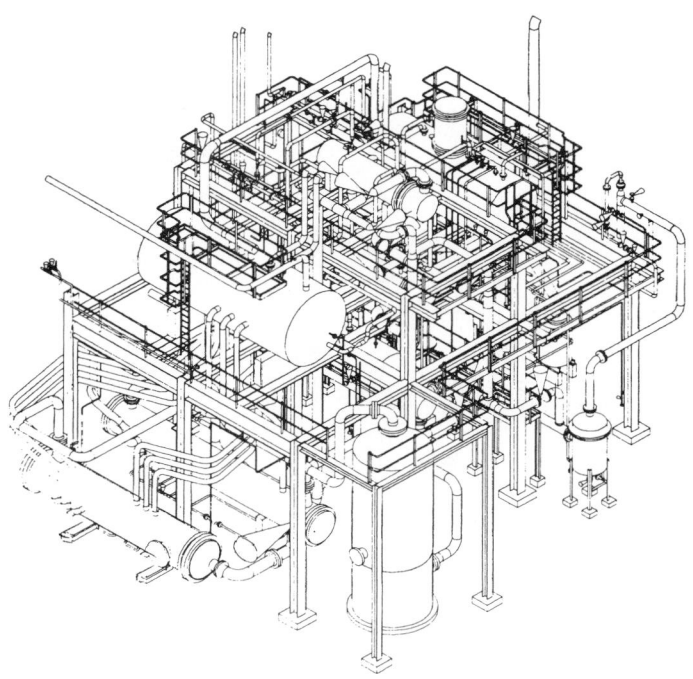

FIGURE 1.12 ■ CADD piping pictorial. *Courtesy Computervision Corporation.*

Photogrammetrist

Analyzes source data and prepares mosaic prints, contour-map profile sheets, and related cartographic materials requiring technical mastery of photogrammetric techniques and principles; prepares original maps, charts, and drawings from aerial photographs and survey data and applies standard mathematical formulas and photogrammetric techniques to identify, scale, and orient geodetic points, estimations, and other planimetric or topographic features and cartographic detail; graphically represents aerial photographic detail, such as contour points, hydrography, topography, and cultural features, using precision stereoplotting apparatus or drafting instruments; revises existing maps and charts and corrects maps in various stages of compilation; prepares rubber, plastic, or plaster three-dimensional relief models.

PROFESSIONAL ORGANIZATION

The *American Design Drafting Association (ADDA)* is a nonprofit, professional organization dedicated to the advancement of design and drafting.

The basic knowledge and skill required to do board drafting is also helpful in preparing input data for and/or operating computerized drafting systems such as digitizers, plotters, and photo composition devices.

Whatever means are used to generate drawings, persons skilled in drafting techniques are in demand. Therefore, drafting can be a rewarding and challenging career to those who can visualize and portray ideas graphically.

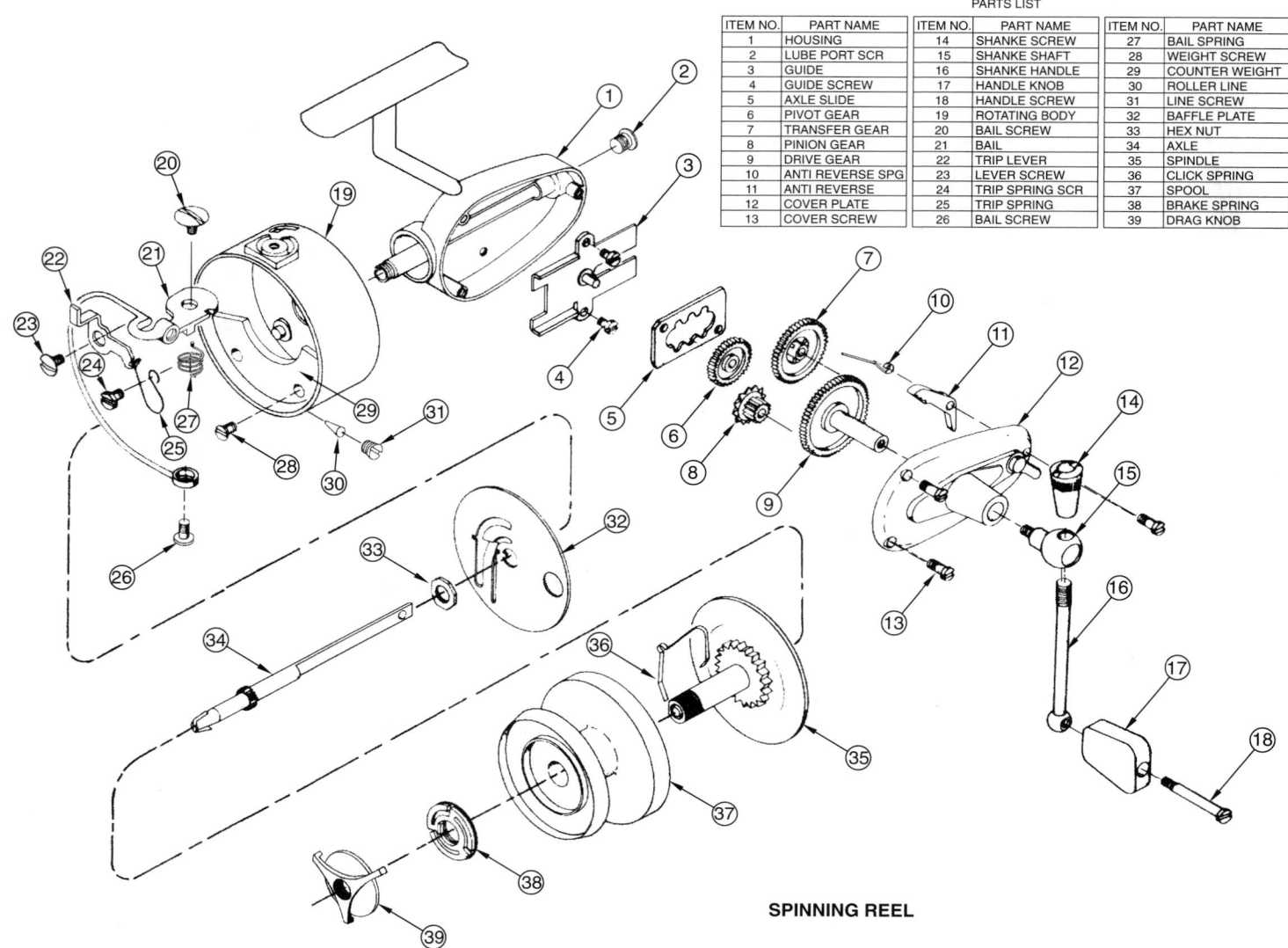

ITEM NO.	PART NAME	ITEM NO.	PART NAME	ITEM NO.	PART NAME
1	HOUSING	14	SHANKE SCREW	27	BAIL SPRING
2	LUBE PORT SCR	15	SHANKE SHAFT	28	WEIGHT SCREW
3	GUIDE	16	SHANKE HANDLE	29	COUNTER WEIGHT
4	GUIDE SCREW	17	HANDLE KNOB	30	ROLLER LINE
5	AXLE SLIDE	18	HANDLE SCREW	31	LINE SCREW
6	PIVOT GEAR	19	ROTATING BODY	32	BAFFLE PLATE
7	TRANSFER GEAR	20	BAIL SCREW	33	HEX NUT
8	PINION GEAR	21	BAIL	34	AXLE
9	DRIVE GEAR	22	TRIP LEVER	35	SPINDLE
10	ANTI REVERSE SPG	23	LEVER SCREW	36	CLICK SPRING
11	ANTI REVERSE	24	TRIP SPRING SCR	37	SPOOL
12	COVER PLATE	25	TRIP SPRING	38	BRAKE SPRING
13	COVER SCREW	26	BAIL SCREW	39	DRAG KNOB

SPINNING REEL

FIGURE 1.13 ■ Technical illustration, exploded isometric assembly.

HOW TO BECOME A DRAFTER

Persons considering drafting as a career should be able to visualize what is to be drawn and be mechanically minded. Potential drafters should be willing to work hard and learn new concepts, and be able to adjust to varying working conditions. Above all, they should be neat and accurate in their work.

Today, most industries, to fill their drafting job openings, are employing only graduates who have received specialized training in design and drafting from technical institutes, junior or community colleges and vocational schools. The specialized training includes basic and advanced drawing, mathematics (algebra, geometry, and trigonometry), physics or chemistry, English, humanities, technology courses, and courses in specific fields such as aeronautical, architectural, automotive, electrical, electronics, illustrative, mapping, mechanical, piping, structural, and sheet metal drafting. Computer-aided drafting is essential training for today's employment market. Mechanical and visual aptitude are also important.

DRAFTING OCCUPATIONAL LEVELS

There are several levels of advancement for drafters, based on educational background and practical experience. For advancement, most companies require career employees to hold a two-year college or trade-school degree; however, persons with an equivalent or larger amount of practical experience in a specific job-related area can also obtain advanced occupational levels.

Occupational levels differ slightly from one industry to another although the general classifications are accurate nationwide.

According to the *Occupational Outlook Handbook,* published by the U.S. Department of Labor and Bureau of Labor Statistics, entry-level or junior drafters do routine work under close supervision. After gaining experience, intermediate-level drafters progress to more difficult work with less supervision. They may be required to exercise more judgment and perform calculations when preparing and modifying drawings. Drafters may advance to senior drafter, designer, or supervisor. Many employers pay for continuing education, and with appropriate college degrees or experience, drafters may go on to become engineering technicians, engineers, or architects.

DRAFTING JOB OPPORTUNITIES

Drafting job opportunities, which include all possible drafting employers, fluctuate with national and local economies. This is no different from most other employment in industry or construction. Drafting is tied closely to manufacturing and construction so that a slowdown or speedup in these industries nationally affects the number of drafting jobs available. This same effect upon drafting opportunities may be experienced at the local level or with specific industries. For example, construction may be strong in one part of the country, and slow in another, so the demand for drafters in those localities is strong or slow accordingly. Other examples are demonstrated when automobile manufactures experience poor sales, and drafting opportunities decline, or when hi-tech industries expand, and more drafters are needed. Public and private indicators suggest that the demand for drafters will continue to be strong for graduates of two-year postsecondary drafting curriculums. The emphasis should be placed on drafting skills, computer-aided drafting, math, English, and oral- and written-communication skills.

The types of drafting jobs available are also controlled by local demands. A predominance of mechanical drafting jobs is found in metropolitan areas where manufacturing is strong, while in outlying areas there may be more civil or structural drafting jobs. Each local area has a need for more of one type of drafting skill than another. Also, drafting curriculums in different geographical areas usually specialize in the fields of drafting that help fill local employment needs. Some drafting programs offer a broad-based education so that graduates may have versatile employment opportunities. When selecting a school, look into curriculum, placement potential, and local demand. Talk to representatives of local industries for an evaluation of the drafting curriculum.

Opportunities for advancement for drafters are excellent, although dependent on the advancement possibilities of the specific employer. Advancement also depends on an individual's initiative, ability, product knowledge, and willingness to continue to be educated. Additional education for advancement usually includes increased levels of mathematics, preengineering, engineering, computers, and advanced drafting. Drafting has traditionally been an excellent stepping-stone to designing, engineering, and management.

SEARCHING FOR A DRAFTING POSITION

Entry-level drafting positions require that you be prepared to meet the needs and demands of industry. The previous discussions explained the recommended education level and type of training required for most job-entry opportunities. Entry into this career marketplace depends on your training and ability, and on the market demand. Some geographic areas around the country may have a greater demand than others. The amount of demand is often based on the economy of the region, the state, and the country. The job market usually changes with the economy. More jobs are commonly available when the economy is doing well. During these positive economic times, it may be easier to find a job than when the economy is down. In a poor economic environment, your training, skills, and personal presentation may make the difference in finding an employment opportunity. A two-year postsecondary degree in a college program such as Drafting Technology, Computer-Aided Drafting, or Engineering Technology can provide a big advantage when seeking a position in this industry. Programs of this type normally have a quality cross section of training in design and drafting, math, and communication skills. These programs also typically have job preparation and placement services to aid their graduates. Many of these schools have direct industry contacts that help promote hiring opportunities. Training programs also often have cooperative work experience (CWE) or internships where their students work in industry for a designated period of time while completing the degree requirements. These positions allow a company to determine if the student is a possible candidate for full-time employment, and provide the student with valuable on-the-job experience that can be included on the resume. Even if you do not go to work at the company where you do the CWE or internship, you can get a letter of recommendation for your portfolio.

When the local economy is doing extremely well and drafting job opportunities are plentiful, it may be possible to find a job with less than a two-year college degree. If you want to find entry-level employment in a job market of this type, you can take intensive training in computer-aided drafting and drafting practices. The actual amount of training required depends on how well you do and whether you can match an employer who is willing to hire with your level of training. Many people have gotten into the industry in this manner, although you would be well advised to continue schooling toward a degree while you are working.

The following are some points to consider when you are ready to seek employment:

■ Get your resume in order. Take a resume preparation course or get some help from your instructors or a career counselor. Your resume must be a quality and professional representation of you. When an employer has many resumes, the best stand out. A possible resource is the book *Resumes for Dummies*.

■ Write an application or cover letter. You can get help on this from the same people who help with your resume. The application letter should be professionally and clearly written, short and to the point, and should give reasons why you would be an asset to the company. A possible resource is the book *Cover Letter for Dummies*.

■ Prepare a portfolio. The portfolio should contain examples of school and industry drawings that you have completed. The drawings should be neatly organized and of the type that helps you target the specific industry discipline that you are seeking. For example, model your portfolio with mechanical drawings if you are interviewing with a manufacturing industry. Display architectural drawings if you are interviewing with an architect or building designer. Include your letters of recommendation from employers and instructors.

■ Register with the department, school, and state employment service. Watch the employment ads in local newspapers, and check out Internet employment sites, which are discussed later.

■ Make a realistic decision about the type of place where you want to work and the salary and benefits you realistically think you should get. Base these decisions on sound judgment. Your instructors should have this information for the local job market. Do not make salary your first issue when seeking a career position. The starting salary is often just the beginning at many companies. Consider advancement potential. A drafting technology position often is a stepping-stone to many opportunities, such as design, engineering, and management.

■ Be prepared when you get an interview. First impressions are critical. You must look your best and present yourself well. Figure 1.14 shows a job candidate making the first introduction for a possible employment opportunity. Always be on time or early. Do some research about the company, if possible. Your instructors often can provide quick information about a company. Relax as much as you can. Answer questions clearly and to the point, but with enough detail to demonstrate that you know what you are talking about. It is often unwise to talk too much. Show off your portfolio. Be prepared to take a CAD test or demonstrate your skills. A possible resource is the book *Interviewing for Dummies*.

■ Ask intelligent questions about the company, because you need to decide if you want to work there. For example, you may not want to work for a company where they have no standards, poor working conditions, and pirated software. You might prefer to work for a company where they have professional stan-

FIGURE 1.15 ■ A new employee working at a CADD position. *Courtesy Hewlett-Packard Company.*

dards, layering systems, a pleasant work environment, and advancement possibilities.

■ Respond fast to job leads. The employment marketplace is often very competitive. You need to be prepared and move quickly. Follow whatever instructions are given for you to apply. Sometimes employers want you to go in person to fill out an application; and sometimes they want you to fax or mail a resume. Either way, you can include your application letter and resume. Sometimes they want you to call for a pre-interview screening.

■ In an active economy, it is common to get more than one offer. If you get an offer from a company, take it if you have no doubts. However, if you are uncertain, ask for 24 or 48 hours to make a decision. If you get more than one offer, weigh the options carefully. There are advantages and disadvantages with every possibility. Make a list of the advantages and disadvantages with each company and carefully consider them.

■ Once you have made a decision, you need to feel good about it and move on with enthusiasm. Figure 1.15 shows a new employee working at a CADD position.

■ Send a very professional thank-you letter to the companies that were considering you. Phone the companies and send a follow-up letter where you had other offers. You never know when you might need to apply at these companies in the future, so this is an important step.

Employment Opportunities on the Internet

The Internet has become a valuable place to seek employment. There are hundreds of Web sites that are available to help you prepare for and find a job. Many Web sites allow you to post your resume for possible employers, and apply for jobs. Some employers screen applicants over the Internet. Figure 1.16 shows a person looking for a job opportunity on the Internet. The only caution is that your personal information displayed through the Internet is available for anyone to read.

FIGURE 1.14 ■ A candidate for an employment opportunity making the first introduction. You must present yourself well. First impressions are very important.

FIGURE 1.16 ■ A person looking for a job opportunity on the Internet.

DRAFTING SALARIES AND WORKING CONDITIONS

Salaries in drafting professions are comparable to salaries of other professions with equal educational requirements. Employment benefits vary according to each individual employer. However, most employers offer vacation and health insurance coverage, while others include dental, life, and disability insurance.

COMPUTERS IN DESIGN AND DRAFTING

The acronym *CAD* means *computer-aided design*, although many people refer to it as computer-aided drafting. *CADD* is *computer-aided design and drafting*. The correctness of this is not extremely important, but what is important is that CAD or CADD has revolutionized the way business is done in design and drafting for the engineering and architecture industries. The use of CADD has made the design and drafting process more accurate and faster. This has allowed for added creativity among engineers, architects, designers, and senior drafters. During the early years of CADD development, many people felt that the computer reduced the individual creativity and artistic impression of the designer and drafter. CADD has now enhanced the ability for a person to be creative by providing many new tools such as solid modeling, animation, and virtual reality. The CADD system allows the drafter to produce drawings that are accurate, very neat and legible, and matched to industry standards. Even architectural drawings, which have always had an artistic flair with lettering and line styles, can now be produced on CADD to match the quality of the finest handwork available. Plus these drawings are consistent from one person or company to the next.

Design Coordination

Software is the instructions that are stored on a disk. Software determines what you can do with a computer. CADD software programs offer a wide variety of applications that make the design and drafting

process efficient. The use of parametric and virtual design has allowed the computer program to create objects and automatically update designs based on the information you provide. *Parametric design* enables you to enter certain values or information that allows the computer to create or change drawings based on this information. An example is the design of a gear, where the computer prompts you for variables such as specific diameters, type and number of teeth, and hub and keyway specifications. The program then automatically tells you if the information is complete and accurate, provides additional data, and creates a detail drawing of the gear. With regard to computer usage, the term *virtual* refers to something that appears to have the properties of being real or actual. A *virtual design* stores your work in a file where the information remains integrated, up-to-date, and easy to manage. Working from the virtual design, you can create and modify designs in 2-D (two-dimensional) and 3-D (three-dimensional) views and automatically create sections, elevations, details, and other working drawings easily. The virtual design integrates every aspect of the design, so changes are automatically updated in all drawings, saving time and reducing the risk of errors along the way. Figure 1.17 shows the coordination taking place through the entire engineering process of design, documentation, communication, and collaboration. Engineers, architects, and designers can share their work directly across a local network or through the Intranet and Internet. *Intranet* links communication between computers within a company or organization, while *Internet* is a worldwide network of communication between computers.

Virtual Reality and Animation

Virtual reality (VR) can be used to demonstrate products or to display material on a Web site. *Virtual reality* or VR refers to a world that appears to be a real world, having many of the properties of a real or actual world. The VR world often appears and feels so real that it is almost as if it were real. This is where the computer is used to simulate environments, including inside and outside of the product or building, and including sound and touch. *Walk-through* can be characterized as a camera in a computer program that is set up like a person walking through a building, around a product or building, or through a landscape. *Fly-through* is similar, but the camera is like a helicopter flying through the area. Fly-through is generally not used to describe a tour through a building. Walk-through or fly-through is the effect of a computer-generated movie where the computer images represent the real architecture, or VR presentation where the computer images turn or move as you turn your head in the desired direction. Realistic renderings, animations, and VR are excellent tools to show the client how the building will look inside and out, or how a product operates. Design ideas can be created and changes made easily at this stage.

National CADD Standards

National CADD standards are set forth in the document titled *National Occupational Skill Standards CADD,* published by the Foundation for Industrial Modernization, 1331 Pennsylvania Avenue NW, Suite 1410 N, Washington, DC 20004. This document was developed in conjunction with business, education, labor organizations, and the U.S. Department of Education. According to

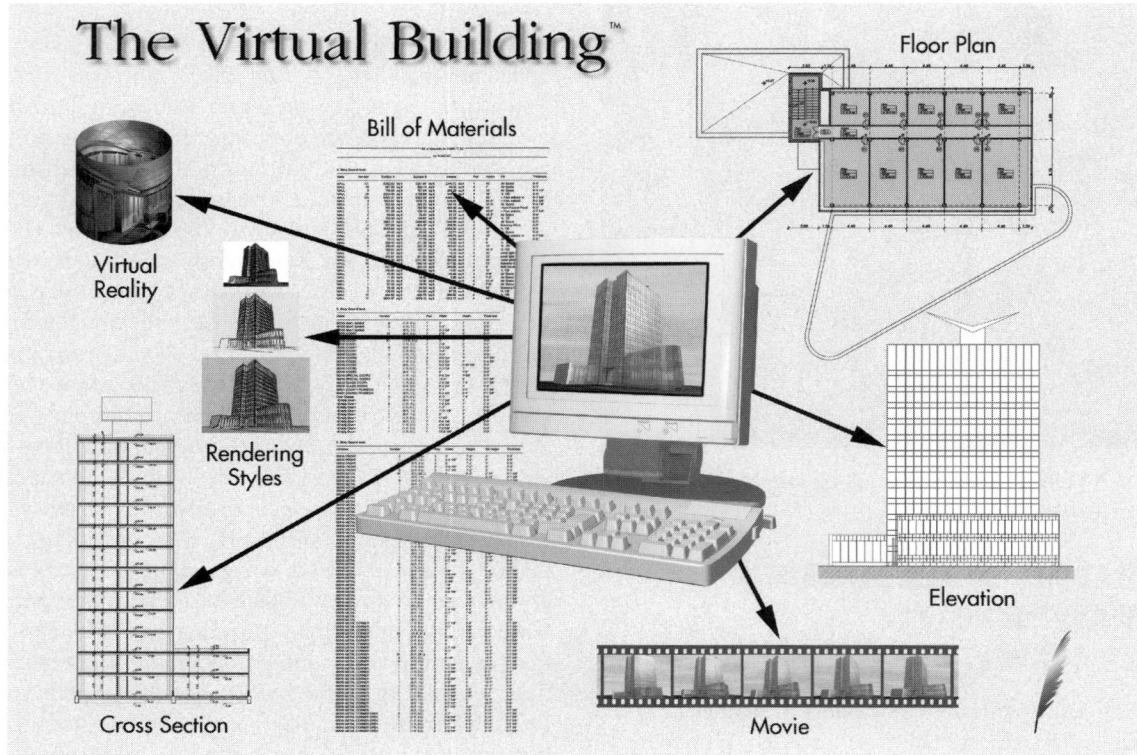

FIGURE 1.17 ■ The coordination taking place through the entire engineering process of design, documentation, communication, and collaboration. *ArchiCAD® image courtesy of Graphisoft®.*

the publication, the skill standards document represents skills that are core to all CADD disciplines, generic to all software, and entry-level. The standards can be used in the following ways:

Employers
- Criteria for hiring.
- Criteria for evaluating job performance.
- A tool for determining retraining needs.

Employees
- A list of skills needed to stay current.
- A list of skills to be tested for national certification.
- A list of core skills that can be used between disciplines.

Educators
- Identifying the skills needed in industry.
- Setting student outcomes.
- Designing curricula.
- Updating teacher materials.
- Evaluating programs.
- Guiding equipment needs.

The standards are divided into the following skill areas:

Technical skills
- Fundamental drafting skills.
- Fundamental computer skills.
- Basic CADD skills.
- Advanced CADD skills.

Academic skills
- Communication skills.
- Math skills.
- Science skills.

Employability skills
- Interpersonal skills.
- Information, systems, and technology skills.
- Thinking skills.
- Personal qualities.
- General knowledge of the industry.

The Computer

The *personal computer* or *PC* has become the primary tool in most engineering and architectural firms for the creation of drawings and calculation of engineering data. The PC that is set up for work applications is commonly referred to as a *workstation*. An engineering or architectural office might have a workstation set up for each person and one or more manual drafting tables for layout work as needed. Figure 1.18 shows an office with computer workstations and a manual drafting table.

The PC belongs to a class of computers called *microcomputers*. The most commonly used microcomputer systems for CADD applications are IBM systems and computers made by other manufacturers, which are referred to as IBM *compatibles* or *clones*. The input devices such as the mouse and digitizer, and the output devices such as the printer or plotter, are referred to as *peripherals*. Figure 1.19

FIGURE 1.18 ■ An office with computer workstations and a manual drafting table. *Courtesy Hewlett-Packard Company.*

shows a computer system being used in an architectural office. The drawing on the screen is being printed on a plotter.

WORKPLACE ETHICS

Ethics are the rules and principles that define right and wrong conduct. A *code of ethics* is a formal document that states an organization's values and the rules and principles that employees are expected to follow. In general, codes of ethics contain these main elements: be dependable, obey the laws, and be good to customers. An example of a company with a corporate code of ethics is the Lockheed Martin Corporation, recipient of the American Business Ethics Award. According to information found in the Lockheed

FIGURE 1.19 ■ A computer system being used in an architectural office. *Courtesy Hewlett-Packard Company.*

ONCE IS ALWAYS ENOUGH WITH CADD

by Karen Miller
Graphic Specialist with Tektronix, Inc.

Reusability is one of the important advantages of CADD. With CADD it is never necessary to draw anything more than once. This advantage is even enhanced by developing a CADD symbols library. Building a parts library for reusability has increased productivity, decreased development costs, and set the highest standards for quality at the Test and Measurement Documentation Group at Tektronix, Inc.

The parts library began by reusing 2-D isometric parts created by the AutoCAD illustrator and saved as blocks (symbols) in a Parts Library Directory. By pathing the named blocks back to this library directory, each block becomes accessible to any directory and drawing file. This allows the CADD illustrator to insert library symbols into any drawing simply by typing the block name.

The illustrator adds new parts to the library as a product is disassembled and illustrated. Each part is given the next available number as its block name in the library, as shown in Figure 1.20.

Originally, the AutoCAD drawings were combined with text (written using Microsoft Word) in Ventura Publisher to create technical publications. Now, the AutoCAD drawings are added to document files in Interleaf technical publishing software. The entire parts library is available to both Auto-CAD and Interleaf users (Figure 1.21).

From the point of view of cost management, the parts library has saved hundreds of hours of work. From the illustrator's view, the parts library helps improve productivity and frees time for new or complex projects.

(Continued)

CADD APPLICATIONS

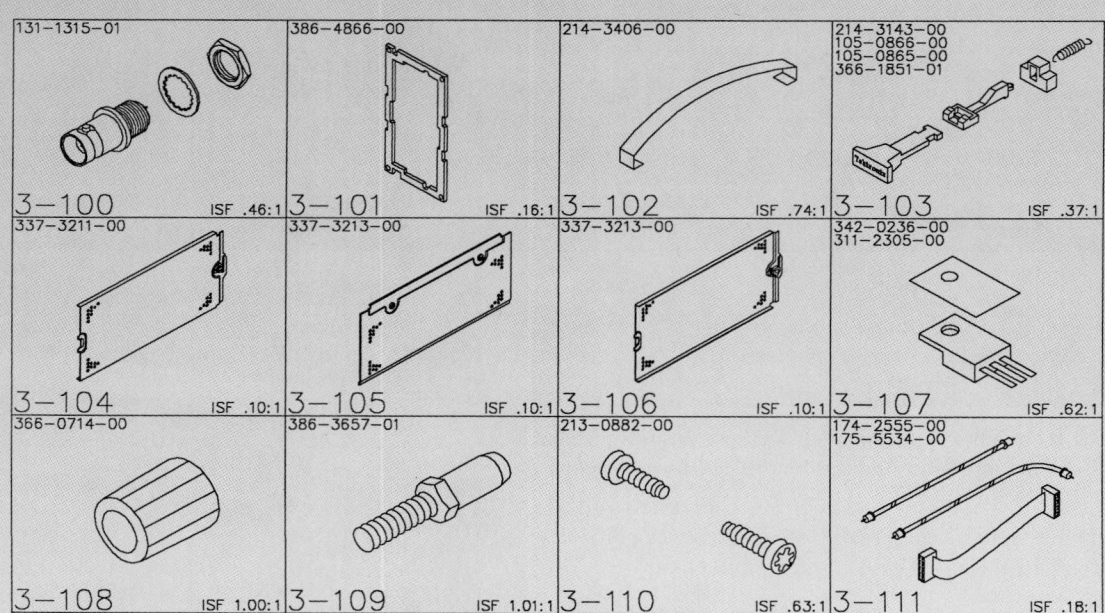

FIGURE 1.20 ■ CADD parts library. *Courtesy Karen Miller, Technical Illustrator, Tektronix, Inc.*

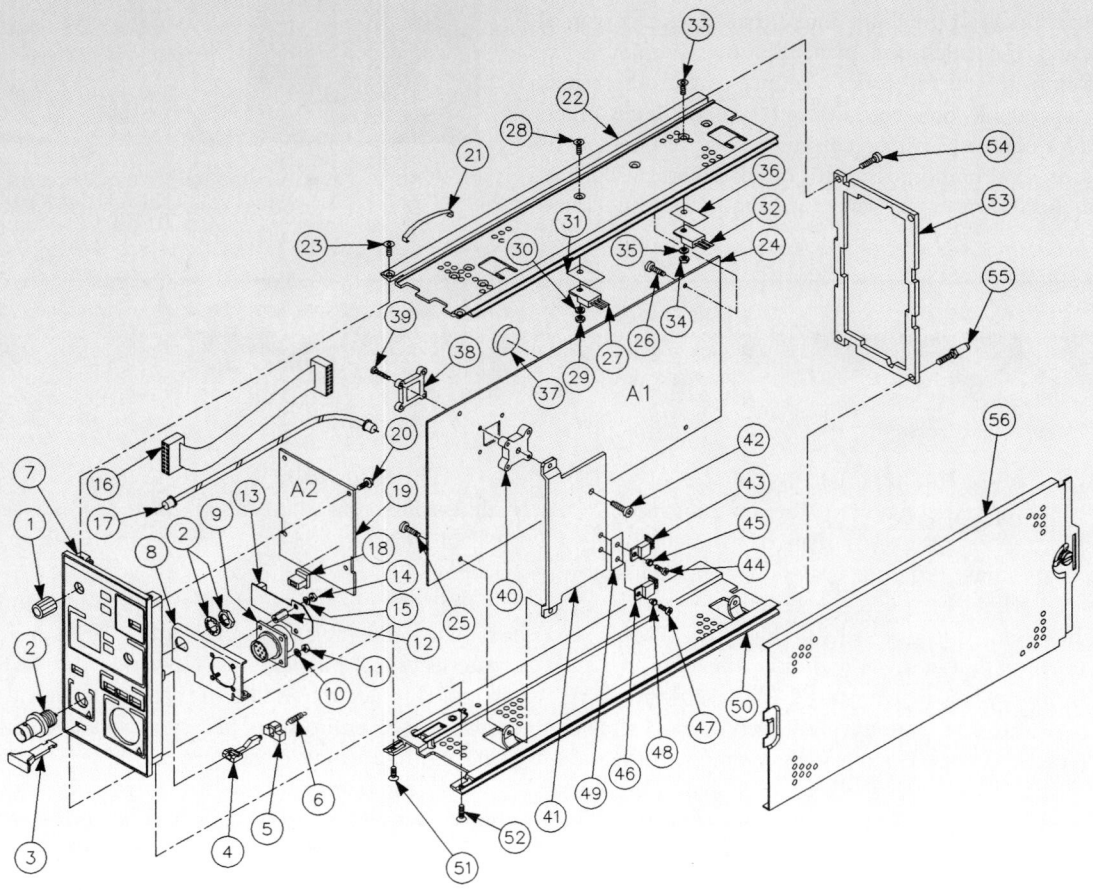

FIGURE 1.21 ■ Integrating the parts from the library for a complete exploded assembly drawing. Tektronix current probe amplifier. *Courtesy Karen Miller, Technical Illustrator, Tektronix, Inc.*

Martin Web site, Lockheed Martin aims to "set the standard" for ethical business conduct. This will be achieved through six virtues:

1. *Honesty:* to be truthful in all our endeavors; to be honest and forthright with one another and with our customers, communities, suppliers, and shareholders.

2. *Integrity:* to say what we mean, to deliver what we promise, and to stand for what is right.

3. *Respect:* to treat one another with dignity and fairness, appreciating the diversity of our workforce and the uniqueness of each employee.

4. *Trust:* to build confidence through teamwork and open, candid communication.

5. *Responsibility:* to speak up, without fear of retribution, and report concerns in the workplace, including violations of laws, regulations, and company policies, and seek clarification and guidance whenever there is doubt.

6. *Citizenship:* to obey all the laws of the United States and the other countries in which we do business and to do our part to make the communities in which we live better.

Intellectual Property Rights

The success of a company often relies on the integrity of its employees. Products are normally the result of years of research, engineering, and development. This is referred to as the intellectual property of the company. Protection of intellectual property can be critical to the success of the company in a competitive industrial economy. This is why it is very important for employees to help protect design ideas and trade secrets. Many companies manufacture their products in a strict secure and secret environment. You will often find proprietary notes on drawings that inform employees and communicate to the outside world that the information contained in the drawing is the property of the company and cannot be used by others.

COPYRIGHTS

A *copyright* is the legal rights given to authors of "original works of authorship." Copyright and patent law is established in the U.S. Constitution, article 1, section 8, which empowers the U.S. Congress "to promote the progress of science and useful arts, by securing for limited times to authors and inventors the exclusive right to their respective writings and discoveries." Copyrights control exclusively the reproduction and distribution of the work by others. In the United States, published or unpublished works that are typically copyrightable are:

■ Literary works, including computer programs and compilations.

■ Musical works, including any accompanying words.

■ Dramatic works, including any accompanying music.

■ Pantomimes and choreographic works.

■ Pictorial, graphic, and sculptural works.

■ Motion pictures and other audiovisual works.

■ Sound recordings.

■ Architectural works, and certain other intellectual works.

Copyright protection exists from the time the work is created in fixed form. *Fixed form* may not be directly observable as long as it can be communicated with the aid of a machine or device. The copyright in the work of authorship immediately becomes the property of the author who created the work. Copyright is secured automatically when the work is created, and the work is created when it is fixed in a copy or phonorecorded for the first time. *Copies* are material objects from which the work can be read or visually perceived either directly or with the aid of a machine or device. A copyright notice can be placed on visually perceptible copies. The copyright notice should have the word "Copyright," the abbreviation "Copr.," or the symbol © (® for phonorecords of sound recordings); the year of first publication; and the name of the owner of the copyright. Copyright registration is a legal formality intended to make a public record, but it is not a condition of copyright protection. A copyright claim is registered by sending a copy of unpublished work or two copies of the work as published, with a registration application and fee, to the Library of Congress, Copyright Office, Washington DC. The application forms are available from the Copyright Office upon request. The previous discussion provides some basic general guidelines. You should confirm specific details and requirements with the Copyright Office.

PATENTS

A *patent* for an invention is the grant of a property right to the inventor, issued by the Patent and Trademark Office (PTO). The term of a new patent is 20 years from the date on which the application for the patent was filed in the United States or, in special cases, from the date an earlier related application was filed, subject to the payment of maintenance fees. United States patent grants are effective only within the United States, U.S. territories, and U.S. possessions. The patent law states, in part, that any person who "invents or discovers any new and useful process, machine, manufacture, or composition of matter, or any new and useful improvement thereof, may obtain a patent," subject to the conditions and requirements of the law.

The patent law specifies that the subject matter must be "useful." The term *useful* refers to the condition that the subject matter has a useful purpose and must operate. Laws of nature, physical phenomena, and abstract ideas cannot be patented. A complete description of the actual machine or other subject matter is required to obtain a patent.

Application for a Patent

There are two types of patent applications, one is nonprovisional and the other is provisional. A *nonprovisional patent application* is for the full patent, which lasts 20 years. The *provisional patent application* is for a temporary patent that lasts for one year.

Nonprovisional Application for a Patent

According to the PTO, a nonprovisional application for a patent is made to the Assistant Commissioner for Patents and includes:

1. A written document that has a specification, and an oath or declaration.

2. A drawing in those cases in which a drawing is necessary.

3. The filing fee.

All application papers must be in the English language, or a translation into the English language is required. All application papers must be legibly written on only one side either by a typewriter or mechanical printer in permanent dark ink or its equivalent in portrait orientation on flexible, strong, smooth, nonshiny, durable, white paper. The papers must be presented in a form having sufficient clarity and contrast between the paper and the writing to permit electronic reproduction. The application papers must all be the same size—either 21.0 cm by 29.7 cm (DIN size A4) or 21.6 cm by 27.9 cm (8½ × 11 in.), with a top margin of at least 2.0 cm (¾ in.), a left-side margin of at least 2.5 cm (1 in.), a right-side margin of at least 2.0 cm (¾ in.), and a bottom margin of at least 2.0 cm (¾ in.) with no holes made in the submitted papers. It is also required that the spacing on all papers be 1½ or double spaced, and the application papers must be numbered consecutively (centrally located above or below the text) starting with page one.

All required parts of the application must be complete before the application is sent, and it is best to send all of the parts together. All applications received in the PTO are numbered in serial order, and the applicant will be informed of the application serial number and filing date by a filing receipt.

Provisional Application for a Patent

Since June 8, 1995, the PTO has offered inventors the option of filing a provisional application for patent that was designed to provide a lower-cost first patent filing in the United States and to give U.S. applicants equality with foreign applicants. Claims and oath or declaration are not required for a provisional application. Provisional application provides the means to establish an early effective filing date in a patent application and permits the term "Patent Pending" to be applied in connection with the invention. Provisional applications may not be filed for design inventions. The filing date of a provisional application is the date on which a written description of the invention, drawings if necessary, and the name of the inventor(s) are received in the PTO.

Patent Drawings

According to the *Guide for the Preparation of Patent Drawings,* published by the U.S. Department of Commerce, Patent and Trademark Office, drawings form an integral part of a patent application. The drawing must show every feature of the invention specified. Figure 1.22 shows a proper patent drawing that is used as an example in a PTO publication. There are specific requirements for the size of the sheet on which the drawing is made, the type of paper, the margins, and other details relating to creating the drawing. The reason for specifying the standards in detail is that the drawings are printed and published in a uniform style when the patent issues,

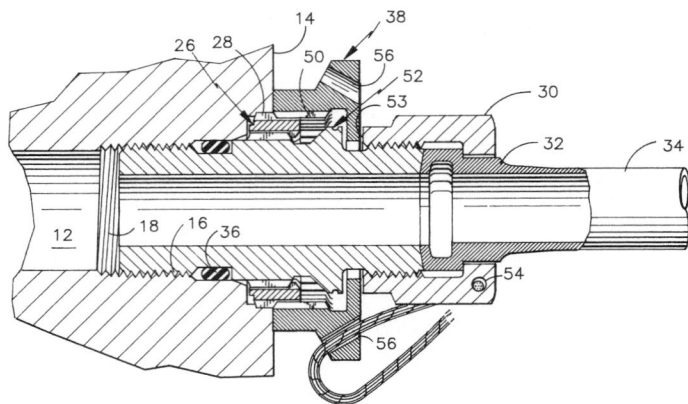

FIGURE 1.22 ■ A proper patent drawing that is used as an example in the Patent and Trademark Office publication, *Guide for the Preparation of Patent Drawings.*

and the drawings also must be understood by persons using the patent descriptions.

Drawings must be created using solid black ink lines on white media. Color drawings or photographs are accepted on rare occasions, but there is an additional petition and other specific requirements for submitting a color drawing or photograph.

The previous discussion provides some basic general guidelines. You should confirm specific details and requirements with the Patent and Trademark Office. Additional information about the preparation of patent drawings is found in Chapter 18 of this text.

TRADEMARKS

According to the PTO publication *Basic Facts about Registering a Trademark,* a *trademark* is either a word, phrase, symbol or design, or combination of words, phrases, symbols or designs, that identifies and distinguishes the source of the goods or services of one party from those of others. A *service mark* is the same as a trademark except that it identifies and distinguishes the source of a service rather than a product. Normally, a mark for goods appears on the product or on its packaging, while a service mark appears in advertising for the services. A trademark is different from a copyright or a patent; as previously explained, a copyright protects an original artistic or literary work, while a patent protects an invention.

Trademark rights start from the actual use of the mark, or the filing of a proper application to register a mark in the Patent and Trademark Office stating that the applicant has a genuine intention to use the mark in commerce regulated by the U.S. Congress. Federal registration is not required to establish rights in a mark, nor is it required to begin use of a mark. However, federal registration can secure benefits beyond the rights acquired by just using a mark. For example, the owner of a federal registration is presumed to be the owner of the mark for the goods and services specified in the registration, and to be entitled to use the mark nationwide. Generally, the first party who either uses a mark in commerce or files an application in the PTO has the ultimate right to register that mark. The PTO's authority is limited to determining the right to register. The right to use a mark can be more complicated to determine, particularly when two parties have begun use of the same or similar marks without knowledge of one another and neither has a federal registration. Only a court can make a decision about the right to use. A federal registration can provide significant advantages to a party involved in a court proceeding. The PTO cannot provide advice concerning rights in a mark. Only a private attorney can provide such advice.

Trademark rights can last indefinitely if the owner continues to use the mark to identify its goods or services. The term of a federal trademark registration is 10 years, with 10-year renewal terms. However, between the fifth and sixth year after the date of initial registration, the registrant must file an official paper giving certain information to keep the registration alive. The registration is canceled if this is not done.

The previous discussion provides some basic general information. You should confirm specific details and requirements with the Patent and Trademark Office.

WEB SITE RESEARCH
Drafting

The American Design Drafting Association (ADDA) Web site at *http://www.adda.org* has valuable information and publications such as:

■ Drafter Certification

■ Curriculum Certification

■ Student Chapters

■ National Design Drafting Contest

■ *Drafting Reference Guide*

■ *Design Drafting News*

■ *Compensation Survey*

■ *Guidelines for Position Classifications*

■ Employment Center

■ What's New

■ Additional Resources

Employment in Drafting

To find resources on the Internet, you can do a search using related words such as employment, employment services, career services, human resources, job hunt, job fair, job listings, job seekers, and resume services. Your school or state employment service probably also has a Web site for you to use when seeking employment. If you are seeking national employment as well as local opportunities, the following is a list of a few Web sites where you can find such information as on-line career resources, job updates, employer profiles, discussion groups, classified ads, job fairs, entry-level positions for college graduates, resume writing, resume posting, career guidance, salary guides, and professional organizations:

> *http://www.ajb.dni.us*—America's Job Bank
> *http://www.careermag.com*—Career Magazine
> *http://www.careermosaic.com*—Career Mosaic
> *http://www.careers.org*—Career Net
> *http://www.careerpath.com*—Career Path
> *http://www.espan.com*—E-Span's Interactive Employment Network
> *http://www.monster.com*—Monster Board
> *http://www.dbm.com/jobguide/*—The Riley Guide: Employment Opportunities and Job Resources on the Internet
> *http://www.jobweb.org/catapult/catapult.htm*—The Catapult, the National Association of Colleges and Employers
> *http://www.nbew.com*—Business Employment Weekly
> *http://www.bestjobsusa.com*—Employment Review

Free services specifically for members of the design and drafting industry:

> *http://www.adda.org*—American Design Drafting Association
> *http://profnet.autodesk.com*—Autodesk Professional Network

Copyrights, Patents, and Trademarks

Detailed information, available publications, and application procedures and fees can be found in the U.S. Copyright Office Web site: *http://lcweb.loc.gov/copyright/*

Detailed information, available publications, and application procedures and fees can be found in the Patent and Trademark Office Web site: *http://www.uspto.gov/*

PROFESSIONAL PERSPECTIVE

The design and drafting profession can provide a rewarding career for people who enjoy detailed work and have a mechanical aptitude that gives the ability to visualize. Math and communication skills are also important. The fundamental standards involved in the industry are found in mechanical drafting. This is the type of drafting done for manufacturing. Recommended drafting standards for this industry are controlled by the American National Standards Institute (ANSI) and the American Society of Mechanical Engineers (ASME). There are a variety of different design drafting disciplines that include the general categories of architectural; structural; civil; industrial pipe; heating, ventilating, and air conditioning; sheet metal; electrical; electronic; and technical illustration. Some educational programs provide training in specific disciplines, while others provide diversified training in several areas. The opportunity to experience more than one discipline allows you to find an industry that you prefer. An entry-level position may not be in your chosen field, but you should be able to find employment in your desired area with experience and an open job market.

There are a variety of opportunities that allow people to expand career potential into related areas such as tool design or cartography. Many people who enter the drafting industry begin to move up quickly into design, checking, purchasing, estimating, and management. The ability to advance in this field depends on your skill, product knowledge, attitude, ability to communicate, continued education, and enthusiasm.

CHAPTER 1

Introduction to Engineering Drawing and Design Test

DIRECTIONS

Answer the questions with short complete statements.

1. How were blackboards used for making drawings before formal drafting started being used?
2. Define drafter.
3. The types of drafting occupations fall into what three general professional areas?
4. According to the *Dictionary of Occupational Titles*, drafting is grouped with what occupations?
5. Persons who produce technical drawings using CAD still function as a _____, and need most of the knowledge of traditional drafters—relating to drafting skills and standards—as well as CAD skills.
6. What type of drafter draws artistic architectural and structural features of any class of buildings?
7. What type of drafter often prepares site plans, lighting plans, paving plans, irrigation plans, planting plans, and garden structures?
8. What type of drafter prepares electrical equipment working drawings, and communications wiring, power plants, commercial and domestic wiring drawings?
9. What type of drafter draws wiring diagrams, schematics, and layout drawings used in electronic equipment?
10. What type of drafter prepares detailed construction drawings, topographic profiles, and related maps for highways, river and harbor improvements, and drainage?
11. What type of drafter makes drawings for structures with reinforcing steel, concrete masonry, wood, and other structural materials?
12. What type of drafter creates working drawings of machinery and mechanical devices indicating dimensions and tolerances?
13. What type of drafter specializes in the plans and related drawings for the installation of heating, ventilating, and air-conditioning drawings?
14. What type of drafter creates drawings for layout, construction, and operation of oil fields, refineries, and piping systems?
15. What type of drafter lays out and draws illustrations for reproduction in reference works, brochures, and manuals?
16. Identify the organization that is dedicated to the advancement of design and drafting.
17. Give at least two titles that may be replacing the term *drafter* due to advancements in technology.
18. List at least 10 points to consider when you are ready to seek employment.
19. Identify a caution with regard to displaying your resume on the Internet.
20. Why is the Internet a valuable place to seek employment opportunities?
21. What does CAD stand for?
22. What is the term many people use to refer to CAD that is different from its intended meaning?
23. What does CADD stand for?
24. Define software.
25. Define parametric design.
26. Define virtual design.
27. Define VR.
28. What is a walk-through?
29. What is a fly-through?
30. What do the national CADD standards represent?
31. What is a PC, and what is its primary use in engineering and architecture?
32. Define ethics.
33. A company's intellectual property rights are usually the results of what?
34. Define copyright.
35. When does copyright protection exist?
36. What is a patent?
37. The patent law specifies that the subject matter must be useful. What does the term *useful* mean in this context?
38. Name the two types of patent applications.
39. What must patent drawings show?
40. Define trademark.

CHAPTER 1

Introduction to Engineering Drawing and Design Problems

DIRECTIONS

Select one or more of the following topic areas (as determined by your instructor or course guidelines) and write a 300- 500-word report on the selected topic or topics. Prepare each report using a word processor. Use double-spacing, proper grammar and spelling, and illustrative examples where appropriate. Use, but do not copy, the information found in this chapter and additional research information.

History of drafting

One or more drafting fields of your choice

Why it is never necessary to draw anything more than once with CADD

ADDA

Searching for a drafting position

Employment opportunities on the Internet

Requirements for becoming a drafter

Computers in design and drafting

Parametric design

Virtual design

VR

National CADD standards

The CADD computer workstation

Workplace ethics

Intellectual property rights

Copyrights

Patents

Trademarks

Professional perspective

Your own selected topic that relates to the content of this chapter

CHAPTER

Sketching, Lettering, and Lines

LEARNING OBJECTIVES

After completing this chapter, you will:

■ Sketch lines, circles, arcs, and multiviews.

■ Do freehand lettering, and use mechanical lettering equipment.

■ Use a CADD system to prepare text.

■ Draw ASME standard lines using manual drafting and computer-aided drafting.

■ Solve an engineering problem using manual and computer-aided drafting.

THE ENGINEERING **DESIGN** APPLICATION

The engineering drafter often works from sketches or written information provided by the engineer. When you receive an engineer's sketch and are asked to prepare a formal drawing, you should follow these steps:

Step 1. Whether you are using manual or CADD drafting, prepare your own sketch the way you think it should look on the final drawing, taking into account correct drafting standards.

Step 2. Evaluate the size of the object so you can determine the scale and sheet size for your final manual drawing or the screen limits for your final computer-aided drawing.

Step 3. Lay out the drawing either very lightly using construction lines or light blue lead. The construction lines may be easily erased if you make a mistake. If you are using a computer, begin the layout on the screen and if you make an error, just erase it and try again. Editing your CADD work is fast and easy.

Step 4. Complete the final manual drawing by darkening all construction lines to proper ASME standard line weights and run a copy for checking. After completing the computer drawing, a check plot may be made on the plotter or printer. This gives you a chance to check your work on paper. Sometimes it is easier to check a drawing on paper than it is to check it on screen.

The example in Figure 2.1 shows a comparison between the engineer's rough sketch and the finished drawing.

A CADD drafter can easily work from points established on a Cartesian coordinate system as discussed in Chapter 4. When this is required, you need to first deter-

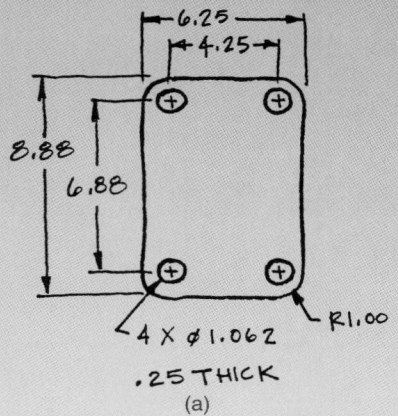

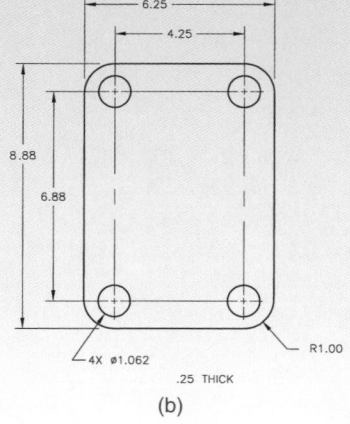

FIGURE 2.1 ■ A comparison between (**a**) an engineer's rough sketch and (**b**) the finished drawing.

mine the type of coordinate system being used: absolute, incremental, or polar. When a situation of this kind occurs, you may be given the X and Y values for each point in a chart similar to Figure 2.2.

Drawing lines between all of the listed points in this demonstration produces the drawing shown in Figure 2.3.

POINT	X	Y
1.	2.8	2.3
2.	7.7	2.3
3.	7.7	5.6
4.	5.2	5.6
5.	4.1	4.1
6.	2.8	4.1
7.	2.8	2.3

FIGURE 2.2 ■ Absolute values for X and Y point coordinates.

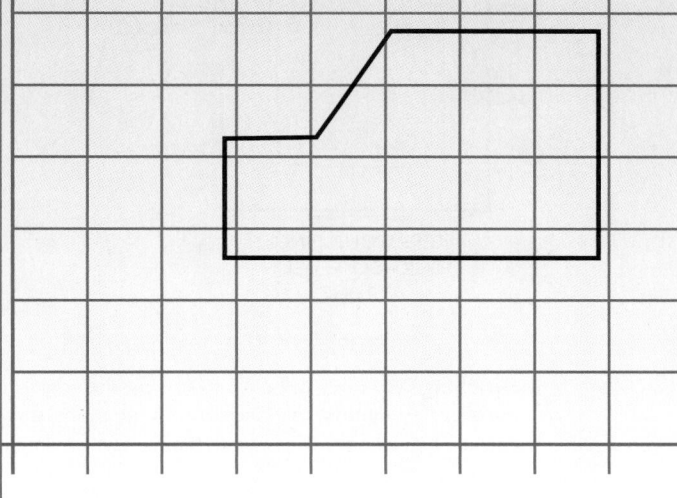

FIGURE 2.3 ■ The solution to a Cartesian coordinate system drawing problem.

SKETCHING

Sketching is freehand drawing, that is, drawing without the aid of drafting equipment. Sketching is convenient since all that is needed is paper, pencil, and an eraser. There are a number of advantages and uses for freehand sketching. Sketching is fast visual communication. The ability to make an accurate sketch quickly can often be an asset when communicating with people at work or at home. Especially when technical concepts are the topic of discussion, a sketch may be the best form of communication. Most drafters prepare a preliminary sketch to help organize thoughts and minimize errors on the final drawing. The computer operator usually prepares a sketch on graph paper to help establish the coordinates for drawing components. Some drafters use sketches to help record the stages of progress when designing, until a final design is ready for

implementation into formal drawings. A sketch can be a useful form of illustration in technical reports. Sketching is also used in job shops where one-of-a-kind products are made. In the job shop, the sketch is often used as a formal production drawing. When the drafter's assignment is to prepare working drawings for existing parts or products, the best method to gather shape and size description about the project is to make a sketch. The sketch can be used to quickly lay out dimensions of features for later transfer to a formal drawing.

The quality of a sketch depends on the intended purpose. Normally a sketch does not have to be very good quality as long as it adequately represents what you want to display. *Speed is a big key to sketching.* You normally want to prepare the sketch as fast as possible while making it as easy and clear to read as possible. Sometimes a sketch does need to have the quality of a formal presentation. The degree of quality can vary depending on the intent of the sketch. A sketch can be used as an artistic impression of a product, or as a one-time detail drawing for manufacturing purposes. However, the sketch is normally used in preliminary planning or to relate a design idea to someone very quickly. The quality of your classroom sketches depends on your course objectives. Your instructor may want quality sketches or very quick sketches that help you establish a plan for further formal drafting. You should confirm this in advance. In the professional world, your own judgment determines the nature and desired quality of the sketch.

TOOLS AND MATERIALS

Sketching equipment is not very elaborate. As mentioned, all you need is paper, pencil, and an eraser. The pencil should have a soft lead; a common number 2 pencil works fine or an automatic 0.7 or 0.9 mm pencil with F or HB lead is also good. The pencil lead should not be sharp. A dull, slightly rounded pencil point is best. Different thicknesses of line, if needed, can be drawn by changing the amount of pressure you apply to the pencil. The quality of the paper is not critical either. A good sketching paper is newsprint, although almost any kind works. Actually, paper with a surface that is not too smooth is best. Many engineering designs have been created on a napkin around a lunch table. Sketching paper should not be taped down to the table. The best sketches are made when you are able to move the paper to the most comfortable drawing position. Some people make horizontal lines better than vertical lines. If this is your situation, then move the paper so that vertical lines become horizontal. Such movement of the paper may not always be possible, so it does not hurt to keep practicing all forms of lines for best results. Graph paper is also good to use for sketching because it has grid lines that can be used as a guide for your sketch lines.

SKETCHING STRAIGHT LINES

Lines should be sketched in short, light, connected segments as shown in Figure 2.4. If you sketch one long stroke in one continuous movement, your arm tends to make the line curved rather than straight, as shown in Figure 2.5. Also, if you make a dark line, you may have to erase if you make an error, whereas if you draw a light line there often is no need to erase.

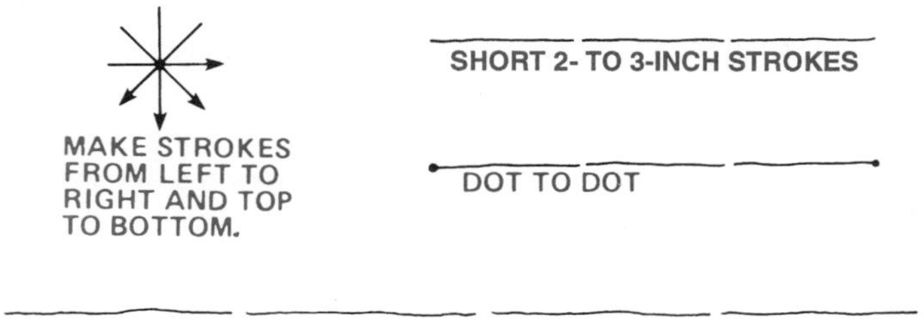

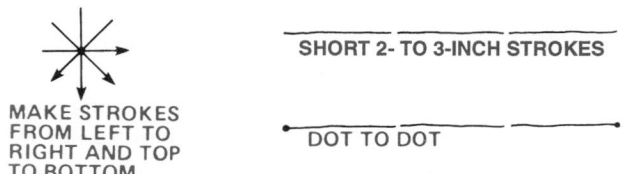

MAKE STROKES FROM LEFT TO RIGHT AND TOP TO BOTTOM.

SHORT 2- TO 3-INCH STROKES

DOT TO DOT

FIGURE 2.4 ■ Sketching short line segments.

FIGURE 2.5 ■ Long movements tend to cause a line to curve.

Following is the procedure used to sketch a horizontal straight line with the dot-to-dot method:

Step 1. Mark the starting and ending positions, as in Figure 2.6. The letters A and B are only for instruction. All you need are the points.

Step 2. Without actually touching the paper with the pencil point, make a few trail motions between the marked points to adjust the eye and hand to the anticipated line.

Step 3. Sketch very light lines between the points by stroking in short light strokes (2- to 3-in. long). Keep one eye directed toward the end point while keeping the other eye directed on the pencil point. With each stroke, an attempt should be made to correct the most obvious defects of the preceding stroke so the finished light lines are relatively straight. (See Figure 2.7.)

Step 4. Darken the finished line with a dark, distinct, uniform line directly on top of the light line. Usually the darkness can be obtained by pressing on the pencil. (See Figure 2.8.)

Very long straight lines can often be sketched by using the edge of the paper or the edge of a table as a guide. To do this, position the paper in a comfortable position with your hand placed along the edge as shown in Figure 2.9a. Extend the pencil point out to the location of the line. Next, place one of your fingers or the palm of your hand along the edge of the paper as a guide. Now, move your hand and the pencil continuously along the edge of the paper as shown in Figure 2.9b. A problem with this method is that it works best if the line is fairly close to the edge of the paper. A sketch does not have to be perfect anyway, so a little practice should be good enough.

A•   •B

FIGURE 2.6 ■ Step 1, use dots to identify both ends of a line.

A•———— ——— ———— ——•B

FIGURE 2.7 ■ Step 3, use short light strokes.

A•————————————————•B

FIGURE 2.8 ■ Step 4, darken to finish the line.

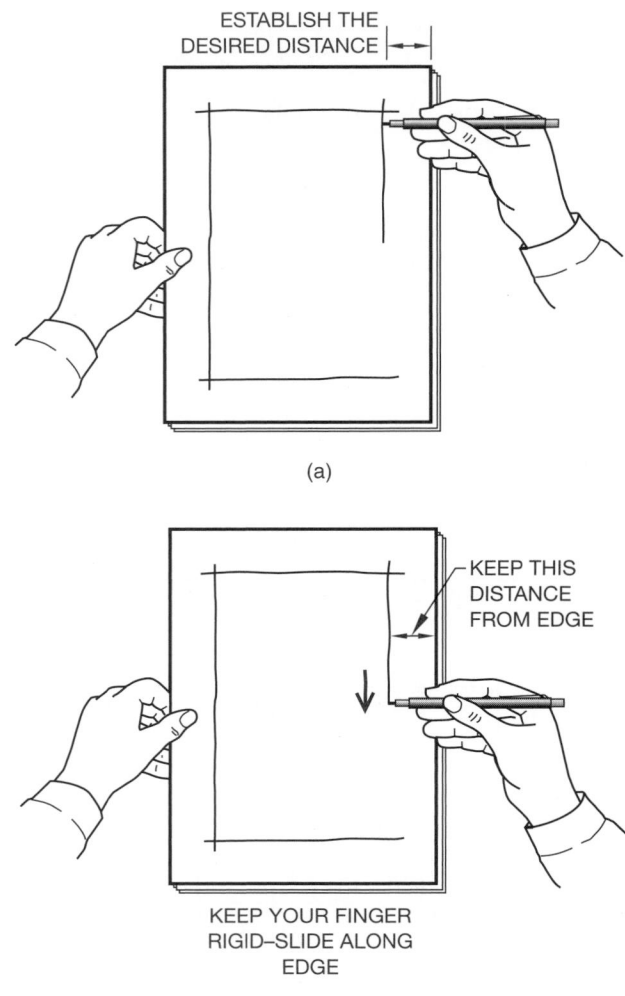

ESTABLISH THE DESIRED DISTANCE

(a)

KEEP THIS DISTANCE FROM EDGE

KEEP YOUR FINGER RIGID–SLIDE ALONG EDGE

(b)

FIGURE 2.9 ■ Sketching very long straight lines using the edge of the sheet as a guide. (**a**) Place your hand along the edge as a guide. (**b**) Move your hand and the pencil along the edge of the paper using your finger or palm as a guide to keep the pencil a constant distance from the edge.

SKETCHING CIRCULAR LINES

Figure 2.10 shows the parts of a circle. There are several sketching techniques to use when making a circle; this text explains the quick freehand method for small circles, the box method, the centerline method, the hand-compass method, and the trammel method for very large circles.

Sketching Quick Small Circles

Small circles are easy to sketch if you treat them just like drawing the letter o. You should be able to do this in two strokes by sketching a half circle on each side as shown in Figure 2.11.

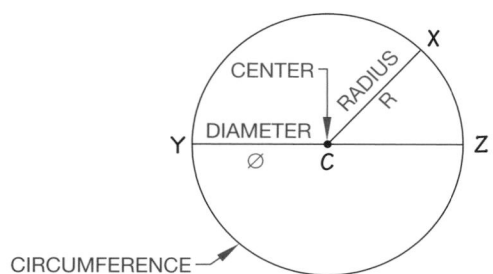

FIGURE 2.10 ■ The parts of a circle.

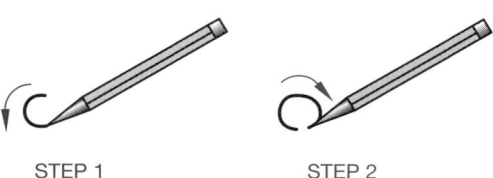

STEP 1   STEP 2

FIGURE 2.11 ■ Sketching a small circle just like drawing the letter o.

Using the Box Method

It is always faster to sketch a circle without first creating other construction guides, but it can be difficult to do so. The box method can help you by providing a square that contains the desired circle. Start this method by very lightly sketching a square box that is equal in size to the diameter of the proposed circle as shown in Figure 2.12. The very light lines are called *construction lines*. Next, sketch diagonals across the square. This establishes the center and allows you to mark the radius of the circle on the diagonals as shown in Figure 2.13. Use the sides of the square and the marks on the diagonals as a guide to sketch the circle. Create the circle by drawing arcs that are tangent to the sides of the square and go through the marks on the diagonals as shown in Figure 2.14. If you have trouble sketching the circle as dark and thick as desired, sketch it very lightly first and then go back over it to make it dark. You can easily correct very lightly sketched lines, but it is difficult to correct very dark lines. Your construction lines do not have to be erased if they are sketched very lightly.

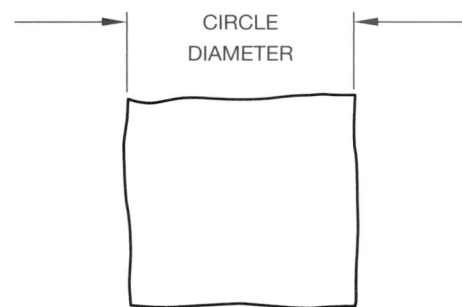

FIGURE 2.12 ■ Very lightly sketch a square box that is equal in size to the diameter of the proposed circle.

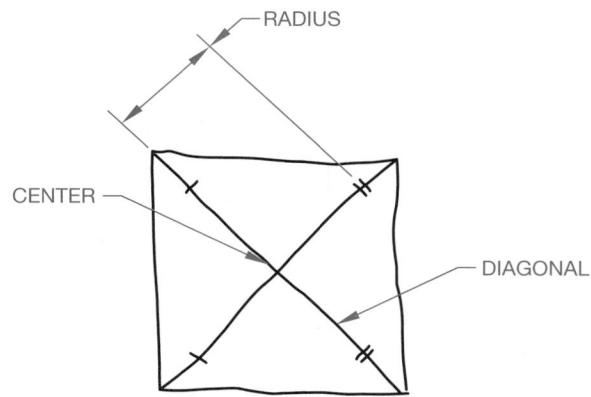

FIGURE 2.13 ■ Sketch light diagonal lines across the square, and mark the radius on the diagonals.

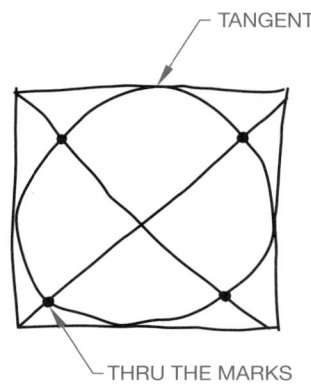

FIGURE 2.14 ■ Create the circle by sketching arcs that are tangent to the sides of the square and go through the marks on the diagonals.

Using the Centerline Method

The centerline method is similar to the box method, but without a box. This method uses very lightly sketched horizontal, vertical, and two 45° diagonal centerlines as shown in Figure 2.15. Next, mark the approximate radius of the circle on the centerlines as shown in Figure 2.16. Create the circle by drawing arcs that go through the marks on the centerlines as shown in Figure 2.17.

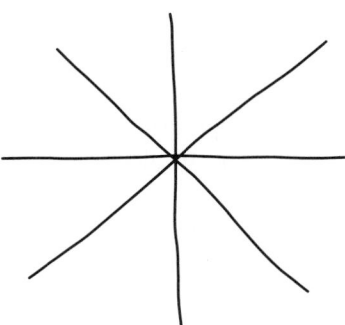

FIGURE 2.15 ■ Sketch very light horizontal, vertical, and 45° lines that meet at the center of the proposed circle.

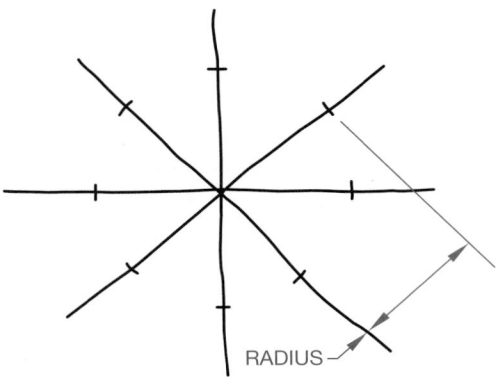

FIGURE 2.16 ■ Mark the approximate radius of the circle on the centerlines created in Figure 2.15.

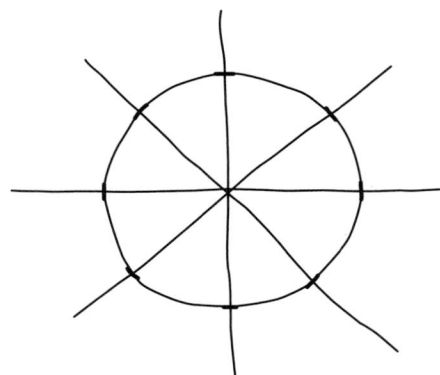

FIGURE 2.17 ■ Create the circle by sketching arcs that go through the marks on the centerlines.

Using the Hand-compass Method

The hand-compass method is a quick and fairly accurate method of sketching circles, although it is a method that takes some practice.

Step 1. Be sure that your paper is free to rotate completely around 360°. Remove anything from the table that might stop such a rotation.

Step 2. To use your hand and a pencil as a compass, place the pencil in your hand between your thumb and the upper part of your index finger so your index finger becomes the *compass point* and the pencil becomes the *compass lead*. The other end of the pencil rests in your palm as shown in Figure 2.18.

Step 3. Determine the circle radius by adjusting the distance between your index finger and the pencil point. Now, with the desired approximate radius established, place your index finger on the paper at the proposed center of the circle.

Step 4. With the desired radius established, keep your hand and pencil point in one place while rotating the paper with your other hand. Try to keep the radius steady as you rotate the paper. (See Figure 2.19.)

Step 5. You can perform step 4 very lightly and then go back and darken the circle or, if you have had a lot of practice, you may be able to draw a dark circle as you go.

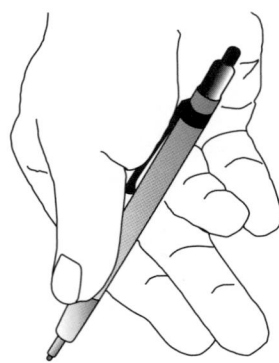

FIGURE 2.18 ■ Step 2, holding the pencil in the hand compass.

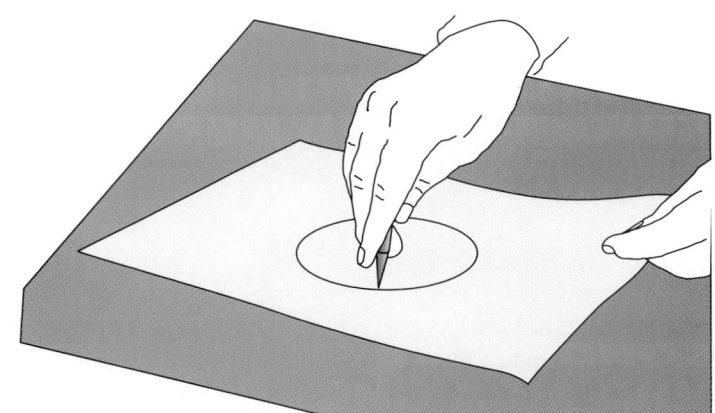

FIGURE 2.19 ■ Step 4, rotate the paper under your finger center point.

Trammel Method

The trammel method should be avoided if you are creating a quick sketch, because it takes extra time and materials to set up this technique. Also, the trammel method is intended for large to very large circles that are difficult to draw when using the other methods. The following examples demonstrate the trammel method to create a small circle. This is done to save space. Use the same principles to draw a large circle.

Step 1. Make a trammel to sketch a 6-in. diameter circle. Cut or tear a strip of paper approximately 1 in. wide and longer than the radius, 3 in. On the strip of paper, mark an approximate 3-in. radius with tick marks such as A and B in Figure 2.20.

Step 2. Sketch a straight line representing the circle radius at the place where the circle is to be located. On the sketched line, locate with a dot the center of the circle to be sketched. Use the marks on the trammel to mark the other end of the radius line as shown in Figure 2.21. With the trammel next to the sketched line, be sure point B on the trammel is aligned with the center of the circle you are about to sketch.

Step 3. Pivot the trammel at point B, making tick marks at point A as you go, as shown in Figure 2.22, until the circle is complete.

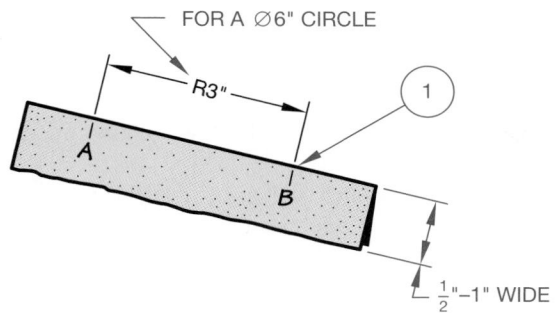

FIGURE 2.20 ■ Step 1, make a trammel.

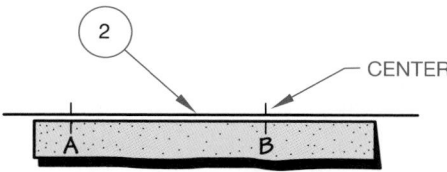

FIGURE 2.21 ■ Step 2, locate the center of the circle.

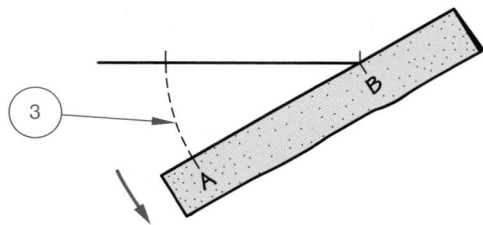

FIGURE 2.22 ■ Step 3, begin the circle construction.

Step 4. Lightly sketch the circumference over the tick marks to complete the circle, then darken it (step 5) as shown in Figure 2.23.

Another similar trammel method, generally used to sketch very large circles, is to tie a string between a pencil and a pin. The distance between the pencil and pin is the radius of the circle. Use this method when a large circle is to be sketched, since the other methods may not work as well. Workers at a construction site some-

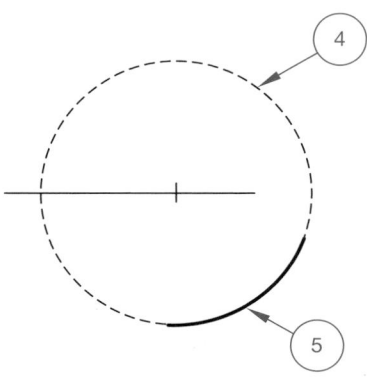

FIGURE 2.23 ■ Steps 4 and 5, darken the circle.

times use this method by tying the string to a nail and driving the nail at the center location.

SKETCHING ARCS

Sketching arcs is similar to sketching circles. An arc is part of a circle as you can see in Figure 2.24. An arc is commonly used as a rounded corner or at the end of a slot. When an arc is a rounded corner, the ends of the arc are typically tangent to adjacent lines. In short, *tangent* means that the arc touches the line at only one point and does not cross over the line as shown in Figure 2.24. An arc is generally drawn with a radius. The most comfortable way to sketch an arc is to move the paper so your hand faces the inside of the arc. Having the paper free to move helps with this practice.

One way to sketch an arc is to create a box at the corner. The box establishes the arc center and radius as shown in Figure 2.25. You can also sketch a 45° construction line from the center to the outside corner of the box, and mark the radius on the 45° line. (See Figure 2.25.) Now, sketch the arc by using the tangent points and the mark as a guide as shown in Figure 2.26. You should generally connect the straight lines to the arc after the arc is created, because it is usually easier to sketch straight lines than it is to sketch arcs.

The same technique can be used to sketch any arc. For example, a full radius arc is sketched in Figure 2.27. This arc is half of a circle, so using half of the box method or centerline method works well.

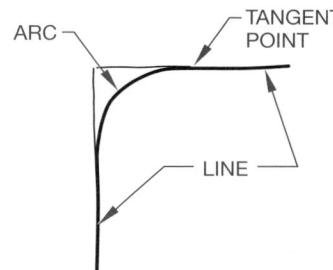

FIGURE 2.24 ■ An arc is part of a circle. This arc is used to create a rounded corner. Notice that the arc creates a smooth connection at the point of tangency with the straight lines.

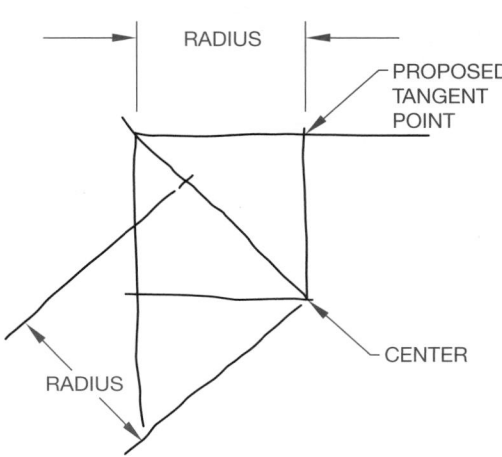

FIGURE 2.25 ■ The box establishes the center and radius of the arc. The 45° diagonal helps establish the radius.

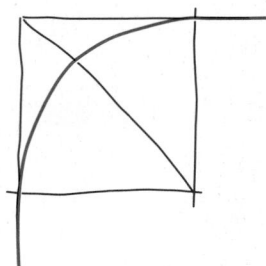

FIGURE 2.26 ■ Sketch the arc using the tangent points and mark on the diagonal as a guide for the radius.

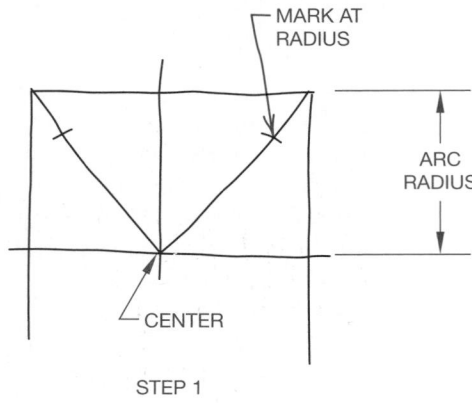

STEP 1

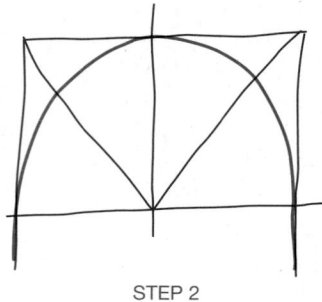

STEP 2

FIGURE 2.27 ■ Sketching a full radius arc uses the same method as sketching any arc or circle.

SKETCHING ELLIPSES

If you look directly at a coin, it represents a circle. As you rotate the coin, it takes the shape of an ellipse. Figure 2.28 shows the relationship between a circle and an ellipse, and shows the parts of an ellipse.

If you can fairly accurately sketch an ellipse without a construction line, then do it. If you need help, an ellipse can also be sketched using a box method. To start this technique, sketch a light rectangle equal in length and width to the major and minor diameters of the desired ellipse as shown in Figure 2.29a. Next, sketch crossing lines from the corners of the minor diameter to the midpoint of the major diameter sides as in Figure 2.29a. Now, using the point where the lines cross as the center, sketch the major diameter arcs. (See Figure 2.29b.) Use the midpoint of the minor diameter sides as the center to sketch the minor diameter arcs as shown in Figure 2.29b. Finally, blend in connecting arcs to fill the gaps as shown in Figure 2.29c.

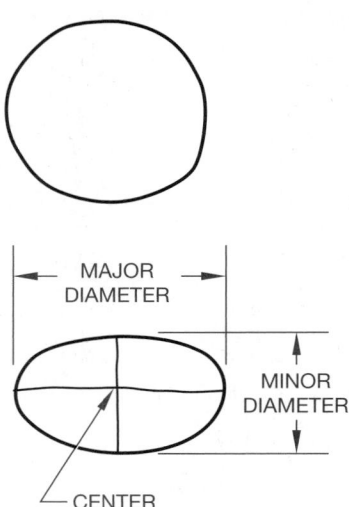

FIGURE 2.28 ■ The relationship between an ellipse and a circle.

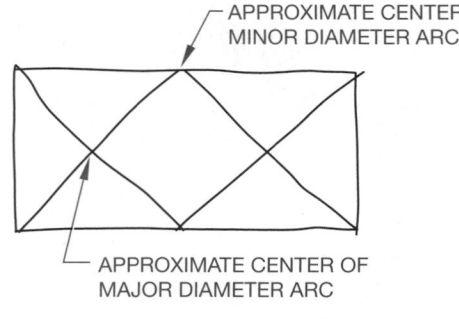

SET UP THE ELLIPSE CONSTRUCTION
(a)

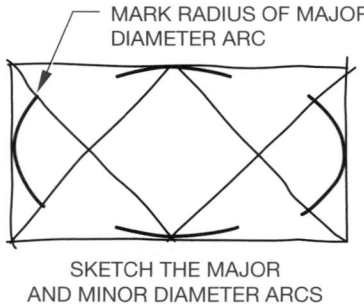

SKETCH THE MAJOR AND MINOR DIAMETER ARCS
(b)

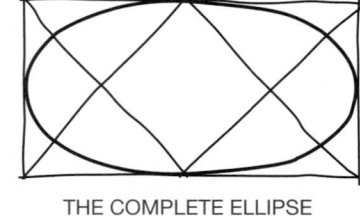

THE COMPLETE ELLIPSE
(c)

FIGURE 2.29 ■ Sketching an ellipse. (**a**) Sketch a light rectangle equal in length and width to the major and minor diameters of the desired ellipse. Sketch crossing lines from the corners of the minor diameter to the midpoint of the major diameter sides. (**b**) Use the point where the lines cross as the center to sketch the major diameter arcs. Use the midpoint of the minor diameter sides as the center to sketch the minor diameter arcs. (**c**) Blend in connecting arcs to fill the gaps.

MEASUREMENT LINES AND PROPORTIONS

When sketching objects, all the lines that make up the object are related to each other by size and direction. In order for a sketch to communicate accurately and completely, it must be drawn in the same proportion as the object. The actual size of the sketch depends on the paper size and how large you want the sketch to be. The sketch should be large enough to be clear, but the proportions of the features are more important than the size of the sketch.

Look at the lines in Figure 2.30. How long is line 1? How long is line 2? Answer these questions without measuring either line, but instead relate each line to the other. For example, line 1 could be stated as being half as long as line 2, or line 2 called twice as long as line 1. Now you know how long each line is in relationship to the other (proportion), but we do not know how long either line is in relationship to a measured scale. No scale is used for sketching, so this is not a concern. Whatever line you decide to sketch first determines the scale of the drawing. This first line sketched is called the *measurement line*. Relate all the other lines in the sketch to that first line. This is one of the secrets of making a sketch look like the object being sketched.

The second thing you must know about the relationship of the two lines in the above example is their direction and position relative to each other. For example, do they touch each other, are they parallel, perpendicular, or at some angle to each other? When you look at a line, ask yourself the following questions (for this example use the two lines given in Figure 2.31):

1. How long is the second line?
   a. same length as the first line?
   b. shorter than the first line? How much shorter?
   c. longer than the first line? How much longer?
2. In what direction and position is the second line related to the first line?

Typical answers to these questions for the lines in Figure 2.31 would be as follows:

1. The second line is about three times as long as the first line.

2. Line two touches the lower end of the first line with about a 90° angle between them.

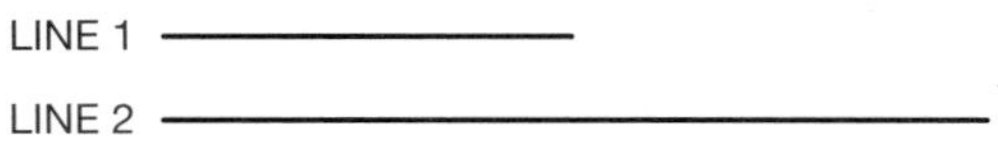

LINE 1 ───────────

LINE 2 ─────────────────

FIGURE 2.30 ■ Measurement lines.

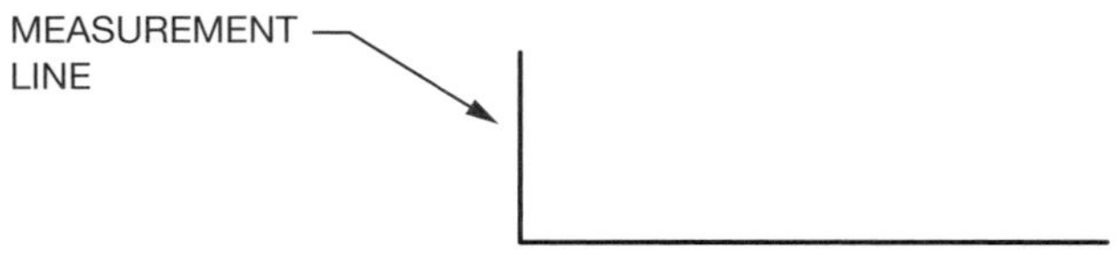

MEASUREMENT LINE ─┐

FIGURE 2.31 ■ Measurement line.

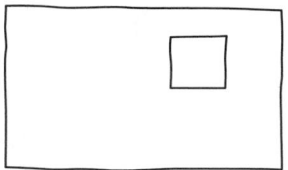

FIGURE 2.32 ■ Space proportions.

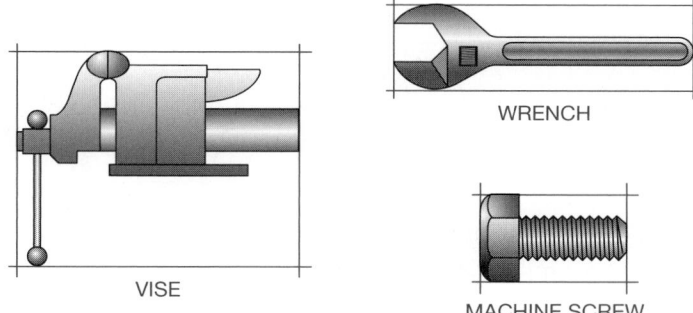

WRENCH

VISE

MACHINE SCREW

FIGURE 2.33 ■ Block technique.

Carrying this concept a step further, a third line can relate to the first line or the second line and so forth. Again, the first line drawn (measurement line) sets the scale for the whole sketch.

This idea of relationship can also apply to spaces. In Figure 2.32, the location of the square can be determined by space proportions. A typical verbal location for the square in this block might be as follows: the square is located about one-half square width from the top of the object or about two square widths from the bottom, and about one square width from the right side or about three square widths from the left side of the object. All the parts must be related to the whole object.

Introduction to the Block Technique

Any illustration of an object can be surrounded with some sort of an overall rectangle, as shown in Figure 2.33. Before starting a sketch, visualize the object to be sketched inside a rectangle in your mind. Then use the measurement-line technique with the rectangle, or block, to help you determine the shape and proportion of your sketch.

PROCEDURES IN SKETCHING

Step 1. When starting to sketch an object, visualize the object surrounded with an overall rectangle. Sketch this rectangle first with very light lines. Sketch the proper proportion with the measurement-line technique, as shown in Figure 2.34.

Step 2. Cut sections out or away using proper proportions as measured by eye, using light lines, as in Figure 2.35.

Step 3. Finish the sketch by darkening in the desired outlines for the finished sketch. (See Figure 2.36.)

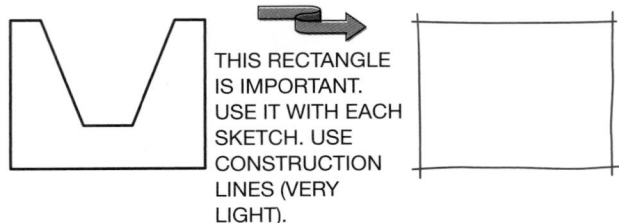

FIGURE 2.34 ■ Step 1, outline the drawing area with a block.

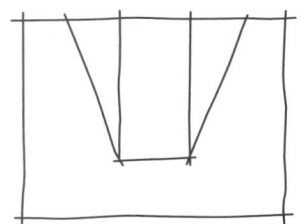

FIGURE 2.35 ■ Step 2, draw features to proper proportions.

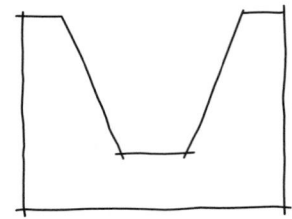

FIGURE 2.36 ■ Step 3, darken the object lines.

Sketching Irregular Shapes

By using a frame of reference or an extension of the block method, irregular shapes can be sketched easily to their correct proportions. Follow these steps to sketch the cam shown in Figure 2.37.

Step 1. Place the object in a lightly constructed box. (See Figure 2.38.)

Step 2. Draw several equally spaced horizontal and vertical lines as shown in Figure 2.39. If you are sketching an object already drawn, just draw your reference lines on top of the object's lines to establish a frame of reference. If you are sketching an object directly, you have to visualize these reference lines on the object you sketch.

Step 3. On your sketch, correctly locate a proportioned box similar to the one established on the original drawing or object, as shown in Figure 2.40.

Step 4. Using the drawn box as a frame of reference, include the grid lines in correct proportion, as seen in Figure 2.41.

Step 5. Then, using the grid, sketch the small irregular arcs and lines that match the lines of the original, as in Figure 2.42.

Step 6. Darken the outline for a complete proportioned sketch, as shown in Figure 2.43.

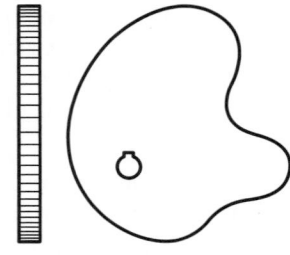

FIGURE 2.37 ■ Cam.

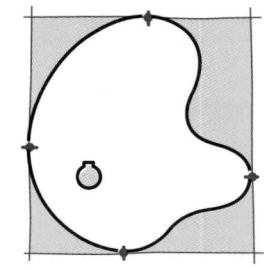

FIGURE 2.38 ■ Step 1, box the object.

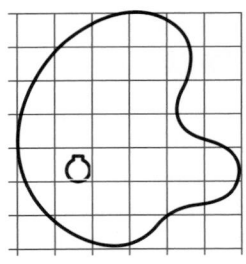

FIGURE 2.39 ■ Step 2, evenly spaced grid.

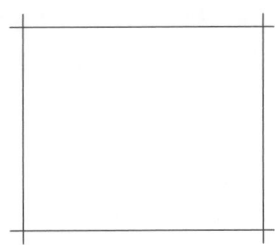

FIGURE 2.40 ■ Step 3, proportioned box.

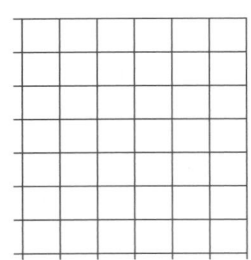

FIGURE 2.41 ■ Step 4, regular grid.

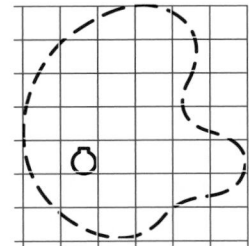

FIGURE 2.42 ■ Step 5, sketched shape using the regular grid.

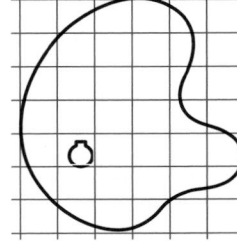

FIGURE 2.43 ■ Step 6, completely darken the outline of the object.

CREATING MULTIVIEW SKETCHES

Multiview projection is also known as orthographic projection. Multiviews are two-dimensional views of an object that are established by a line of sight that is perpendicular (90°) to the surface of the object. When making multiview sketches, a systematic order should be followed. Most drawings are in the multiview form. Learning to sketch multiview drawings will save you time when making a formal drawing. The pictoral view shows the object in a 3-D (three-dimensional) picture, while the multiview shows the object in a 2-D representation. Figure 2.44 shows an object in 3D and 2D.

Multiview Alignment

To keep your drawing in a common form, sketch the front view in the lower left portion of the paper, the top view directly above the front view, and the right-side view to the right side of the front view. (See Figure 2.44.) The views needed may differ depending on the object. Your ability to visualize between 3-D objects and 2-D views is very important in understanding how to lay out a multiview sketch. Multiview arrangement is explained in detail in Chapter 5.

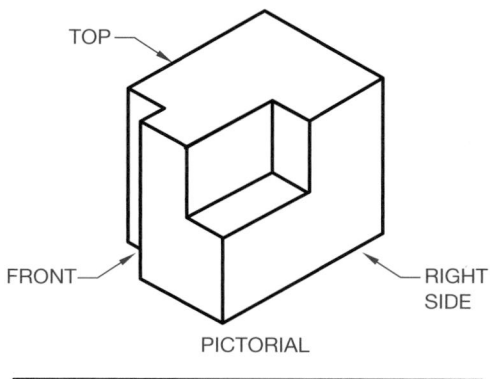

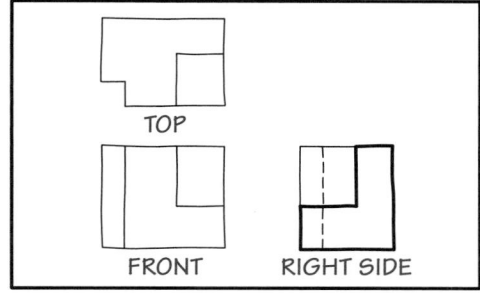

FIGURE 2.44 ■ Views of objects shown in pictorial view and multiview.

Multiview Sketching Technique

Steps in sketching:

Step 1. Sketch and align the proportional rectangles for the front, top, and right side views of the object given in Figure 2.44. Sketch a 45° line to help transfer width dimensions. The 45° line is established by projecting the width from the top view across and the width from the right-side view up until the lines intersect as shown in Figure 2.45.

Step 2. Complete the shapes by cutting out the rectangles, as shown in Figure 2.46.

Step 3. Darken the lines of the object as in Figure 2.47. Remember, keep the views aligned for ease of sketching and understanding.

Step 4. In the views where some of the features are hidden, show those features with hidden lines, which are dashed lines

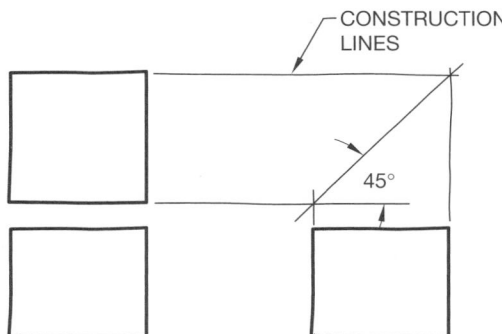

FIGURE 2.45 ■ Step 1, block out views.

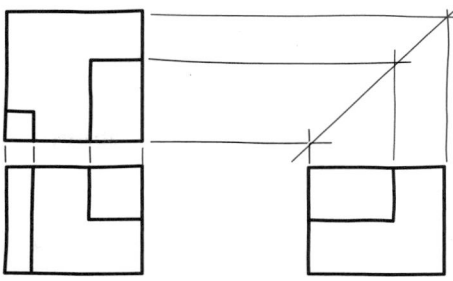

FIGURE 2.46 ■ Step 2, block out shapes.

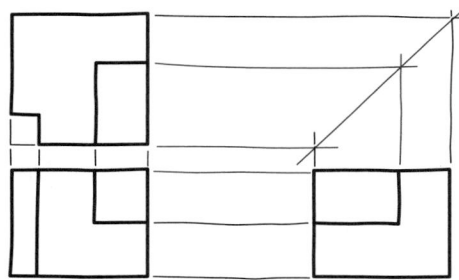

FIGURE 2.47 ■ Step 3, darken all object lines.

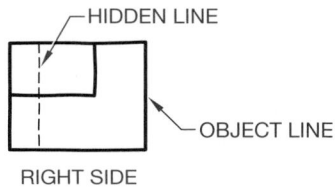

FIGURE 2.48 ■ Step 4, draw hidden features.

as shown in Figure 2.48. Start the practice of sketching object lines thick and hidden lines thin.

Creating Isometric Sketches

Isometric sketches provide a three-dimensional pictorial representation of an object. Isometric sketches are easy to create and make a very realistic exhibit of the object. The surface features or the axes of the objects are drawn at equal angles from horizontal. Isometric sketches tend to represent the objects as they appear to the eye. Isometric sketches help in the visualization of an object, because three sides of the object are sketched in a single three-dimensional view. Chapter 19 covers isometric drawings in detail.

Establishing Isometric Axes

In setting up an isometric axis, you need four beginning lines: a horizontal reference line, two 30° angular lines, and one vertical line. Draw them as very light construction lines. (See Figure 2.49.)

Step 1. Sketch a horizontal reference line (consider this the ground-level line.)

Step 2. Sketch a vertical line perpendicular to the ground line and somewhere near its center. The vertical line is used to measure height.

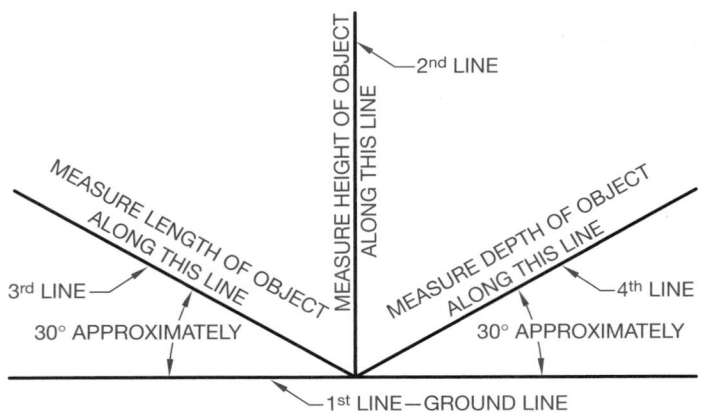

FIGURE 2.49 ■ Isometric axis.

Step 3. Sketch two 30° angular lines, each starting at the intersection of the first two lines as shown at Figure 2.49.

Making an Isometric Sketch

The steps in making an isometric sketch are as follows:

Step 1. Select an appropriate view of the object.

Step 2. Determine the best position in which to show the object.

Step 3. Begin your sketch by setting up the isometric axes. (See Figure 2.50.)

Step 4. By using the measurement-line technique, draw a rectangular box, using correct proportion, which could surround the object to be drawn. Use the object shown in Figure 2.51 for this example. Imagine the rectangular box in your mind. Begin to sketch the box by marking off the width at any convenient length as in Figure 2.52. This is your measurement line. Next estimate and mark the length and height as related to the measurement line. (See

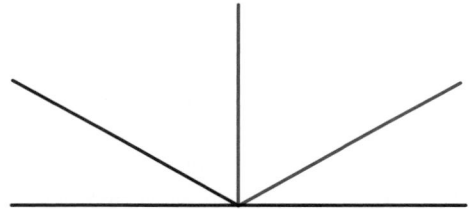

FIGURE 2.50 ■ Step 3, sketch the isometric axis.

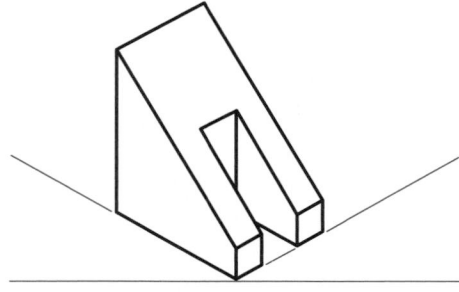

FIGURE 2.51 ■ Given object.

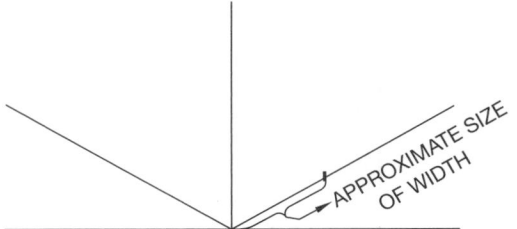

FIGURE 2.52 ■ Step 4, lay out the width.

Figure 2.53.) Sketch the three-dimensional box by using lines parallel to the original axis lines. (See Figure 2.54.) Sketching the box is the most critical part of the construction. It must be done correctly; otherwise your sketch will be out of proportion. All lines drawn in the same direction must be parallel.

Step 5. Lightly sketch in the slots, holes, insets, and other features that define the details of the object. By estimating distances on the rectangular box, the features of the object are easier to sketch in correct proportion than trying to draw them without the box. (See Figure 2.55.)

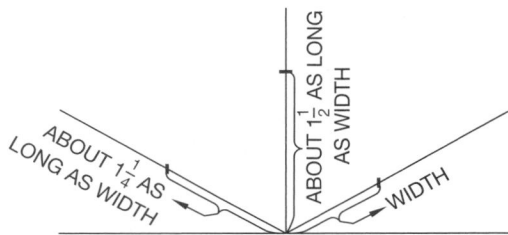

FIGURE 2.53 ■ Step 4, lay out the length and height.

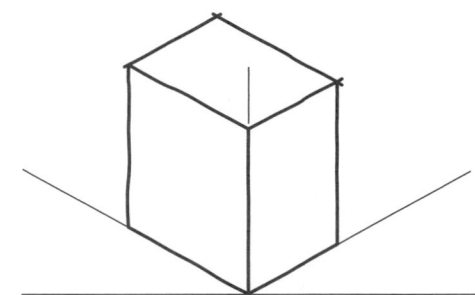

FIGURE 2.54 ■ Step 4, sketch the 3-D box.

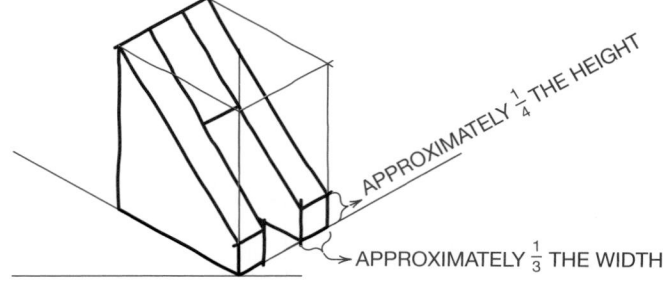

FIGURE 2.55 ■ Step 5, sketch the features.

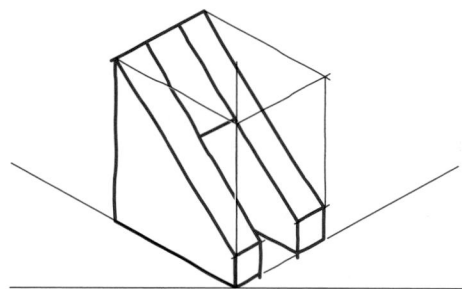

FIGURE 2.56 ■ Step 6, darken the outline.

Step 6. To finish the sketch, darken all the object lines (outlines), as in Figure 2.56. For clarity, do not show any hidden lines.

Nonisometric Lines

Isometric lines are lines that are on or parallel to the three original isometric axes lines. All other lines are nonisometric lines. Isometric lines can be measured in true length. *Nonisometric lines* appear either longer or shorter than they actually are. (See Figure 2.57.) You can measure and draw nonisometric lines by connecting their end points. You can find the end points of the nonisometric lines by measuring along isometric lines. To locate where nonisometric lines should be placed, you have to relate to an isometric line. Follow through these steps, using the object in Figure 2.58 as an example.

Step 1. Develop a proportional box, as in Figure 2.59.

Step 2. Sketch in all isometric lines, as shown in Figure 2.60.

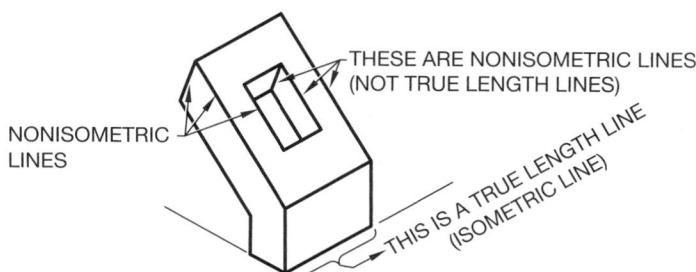

NONISOMETRIC LINES

THESE ARE NONISOMETRIC LINES (NOT TRUE LENGTH LINES)

THIS IS A TRUE LENGTH LINE (ISOMETRIC LINE)

FIGURE 2.57 ■ Nonisometric lines.

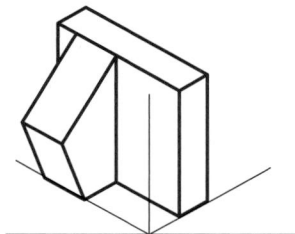

FIGURE 2.58 ■ Guide.

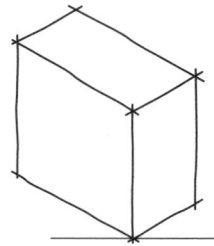

FIGURE 2.59 ■ Step 1, sketch the box.

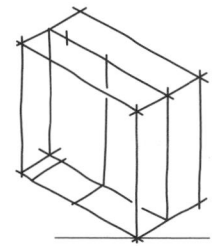

FIGURE 2.60 ■ Step 2, sketch isometric lines.

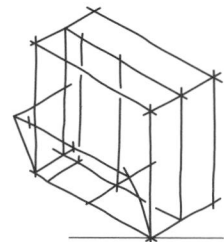

FIGURE 2.61 ■ Step 3, locate nonisometric line end points.

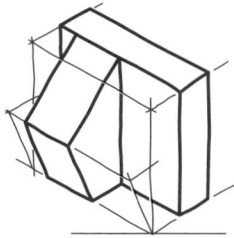

FIGURE 2.62 ■ Step 4, complete the sketch and darken all outlines.

Step 3. Locate the starting and end points for the nonisometric lines. (See Figure 2.61.)

Step 4. Sketch the nonisometric lines, as shown in Figure 2.62, by connecting the points established in step 3. Also darken all outlines.

Sketching Isometric Circles

Circles and arcs appear as ellipses in isometric views. To sketch isometric circles and arcs correctly, you need to know the relationship between circles and the faces, or planes, of an isometric cube. Depending on which face the circle is to appear, isometric circles look like one of the ellipses shown in Figure 2.63. The angle the ellipse (isometric circle) slants is determined by the surface on which the circle is to be sketched.

To practice sketching isometric circles, you need isometric surfaces to put them on. The surfaces can be found by first sketching a cube in isometric. A *cube* is a box with six equal sides. Notice, as shown in Figure 2.64, that only three of the sides can be seen in an isometric drawing.

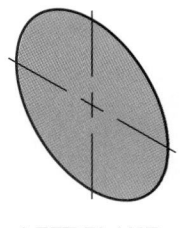 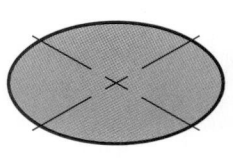 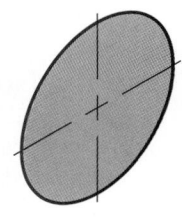

LEFT PLANE   HORIZONTAL PLANE   RIGHT PLANE

FIGURE 2.63 ■ Isometric circles.

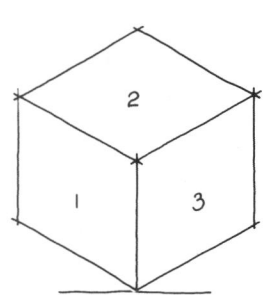

FIGURE 2.64 ■ Step 1, draw an isometric cube.

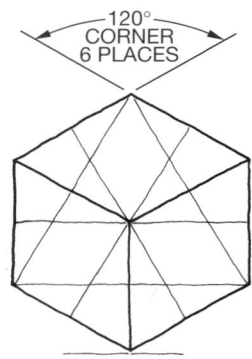

FIGURE 2.65 ■ Step 2, four-center isometric ellipse construction.

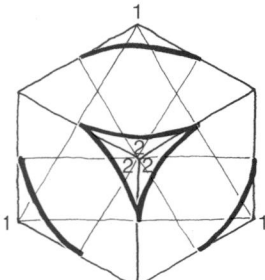

FIGURE 2.66 ■ Step 3, sketch arcs from points 1 and 2 as centers.

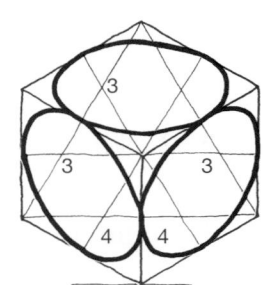

FIGURE 2.67 ■ Step 4, sketch arcs from points 3 and 4 as centers.

Four-center Method

The four-center method of sketching an isometric ellipse is easier to perform, but care must be taken to form the ellipse arcs properly so the ellipse does not look distorted.

Step 1. Draw an isometric cube similar to Figure 2.64.

Step 2. On each surface of the cube, draw line segments that connect the 120° corners to the centers of the opposite sides. (See Figure 2.65.)

Step 3. With points 1 and 2 as the centers, sketch arcs that begin and end at the centers of the opposite sides on each isometric surface. (See Figure 2.66.)

Step 4. On each isometric surface, with points 3 and 4 as the centers, complete the isometric ellipses by sketching arcs that meet the arcs sketched in step 3. (See Figure 2.67.)

Sketching Isometric Arcs

Sketching isometric arcs is similar to sketching isometric circles. First, block out the overall configuration of the object, then establish the centers of the arcs. Finally, sketch the arc shapes as shown in Figure 2.68. Remember that isometric arcs, just like isometric circles, must lie in the proper plane and have the correct shape.

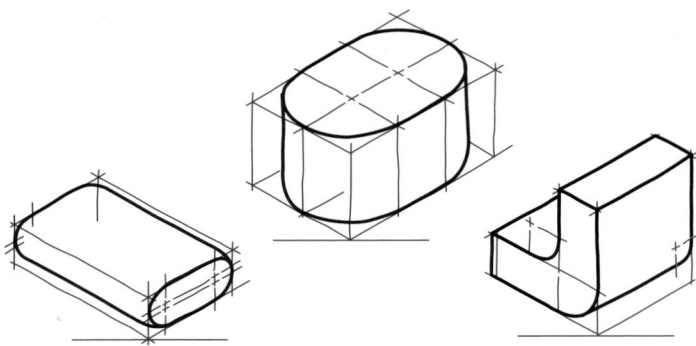

FIGURE 2.68 ■ Sketching isometric arcs.

LETTERING

Information on drawings that cannot be represented graphically by lines may be presented by lettered dimensions, notes, and titles. It is extremely important that these lettered items be exact, reliable, and entirely legible in order for the reader to have confidence in them and never have any doubt as to their meaning. This is especially important when using reproduction techniques that require a drawing to be reduced in size such as with photocopy or microfilm. Poor lettering ruins an otherwise good drawing.

SINGLE-STROKE GOTHIC LETTERING

ASME The standard for lettering was established in 1935 by the American National Standards Institute (ANSI). This standard is now conveyed by the American Society of Mechanical Engineers document ASME Y14.2M, *Line Conventions and Lettering.*

The standardized lettering format was developed as a modified form of the Gothic letter font. The term *font* refers to a complete assortment of any one size and style of letters. The simplification of the Gothic letters resulted in elements for each letter that became known as single-stroke Gothic lettering. The name sounds complex but it is not. The term *single stroke* comes from the fact that each letter is made up of a single straight or curved line element that makes it easy to draw and clear to read. There are upper- and lowercase, vertical, and inclined Gothic letters, but industry has become accustomed to using vertical uppercase letters as the standard. (See Figure 2.69.)

ABCDEFGHIJKL MNOP
QRSTUVWXYZ&
1234567890

FIGURE 2.69 ■ Vertical uppercase single-stroke Gothic letters and numbers.

OTHER LETTERING STYLES

Inclined Lettering

Some companies prefer inclined lettering. The general slant of inclined letters is 68°. One edge of the Ames Lettering Guide has a 68° slant, which may be used to help maintain the proper angle. Structural drafting is one field where slanted lettering is commonly found. Figure 2.70 shows slanted uppercase letters.

Lowercase Lettering

Occasionally, lowercase letters are used; however, they are very uncommon in mechanical drafting. Civil or map drafters use lowercase lettering for some practices. Figure 2.71 shows lowercase lettering styles.

Architectural Styles

Architectural lettering is much more varied in style than mechanical lettering; however, neatness and readability are essential. (See Figure 2.72.)

FIGURE 2.70 ■ Uppercase inclined letters and numbers.

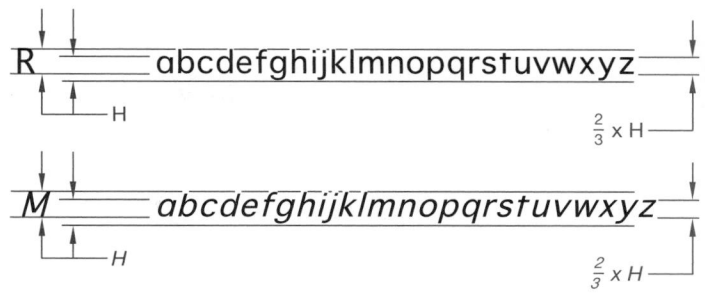

FIGURE 2.71 ■ Lowercase lettering.

FIGURE 2.72 ■ Architectural lettering.

LETTERING

Lettering with a CADD system is one of the easiest tasks associated with computer-aided drafting. It is just a matter of deciding on the style or font of lettering to use, locating the text where it is needed, and typing the desired text. Lettering is called *text* when using CADD.

The CADD drafter often rejoices when the time comes to place text and notes on the drawing because no freehand lettering is involved. The computer places text of a consistent shape and size on a drawing in any number of styles, or fonts. The FONT, STYLE, or TEXT command is one of several that can be found in a section of the menu labeled text or *text attributes*. The drafter is also able to specify the height, width, and slant angle of characters (letters and numbers). Most systems maintain a certain size of text called the *default* size that is used if the operator does not specify a value. The term *default* refers to any value that is maintained by the computer for a command or function that has variable parameters. The default text height may be ⅛ in., but you can change it if needed.

Lettering Styles

CADD systems have a variety of lettering styles, or fonts. The drafter can select the style to use simply by picking a menu command, symbol, or by typing a command at the keyboard. Figure 2.73 shows some of the styles and sizes of characters that can be used in CADD. The size and style of characters used is determined by the nature of the drawing.

Locating Text

The process of locating text has not changed. The drafter still needs to decide where to locate dimensions, notes, and parts lists, but with CADD the process of placing notes is a little more technical. Most CADD systems provide for keyboard entry of location and size coordinates, thus allowing the drafter to accurately locate text. But using the keyboard to

(Continued)

CADD APPLICATIONS

ABCDEFGHIJKLMNOPQRSTUVWXY,
abcdefghijklmnopqrstuvwxyz
1234567890

ABCDEFGHIJKLMNOPQRSTUVWXYZ
abcdefghijklmnopqrstuvwxyz
1234567890

ABCDEFGHIJKLMNOPQRSTUVWXYZ
abcdefghijklmnopqrstuvwxyz
1234567890

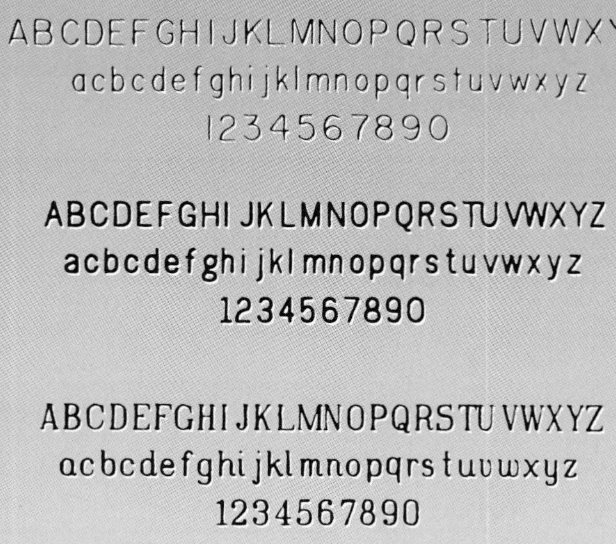

FIGURE 2.73 ■ Samples of CADD character font styles.

input text location coordinates takes time. A quicker way to select the text location is to move the screen cursor to the desired location and pick the point with the mouse.

Text is positioned by selecting one of several places on the lettering. Figure 2.74 shows an example of several points on the text that can be used for positioning purposes. Positioning and aligning text is referred to as *justification*. Not all CADD systems use all the points shown, but most systems have a command known as TEXT that is used for placing written information on the drawing. For example, AutoCAD has three commands that can be used to place text for different applications. These are TEXT, DTEXT, and MTEXT. Some systems may allow you to locate text between two

points. The computer calculates the size of each letter so the text fits in the desired space.

The first decision regarding text is to determine its height, width, and slant angle. Most CADD systems maintain a default text size that is used by the computer if the operator decides not to change it. The angle of rotation, direction, or text path is also determined by the drafter. The text path is the angle from horizontal that the text will lie on. This is also referred to as rotation angle. An example of text rotation angle is shown in Figure 2.75. Text located on a horizontal line has a direction of 0°, and text that reads from the bottom up vertically has a direction of 90°.

With the size and slant angle decided, you then need to locate it on the drawing and type the text at the keyboard. This process can occur in a couple of ways. You could be asked to first pick the text location using a pointing input device and then type the text. The second method is the reverse of the first. Type the text and it appears on the screen with the crosshairs at the point of location that you previously specified. Then *drag* the text to the proper location, press a button on the mouse, puck, or keyboard, and it is in place. A nice thing about locating notes and labels on a drawing with a CADD system is the ability to move them around instantaneously, as often as you want, without erasing holes in your drawing.

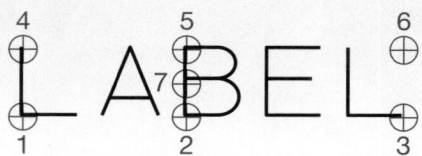

FIGURE 2.74 ■ Text can be placed on a drawing by choosing one of several location points shown here.

FIGURE 2.75 ■ Examples of text rotation angles.

LETTERING LEGIBILITY

ASME The minimum recommended lettering size on engineering drawings is .125 in. (3 mm). All dimension numerals, notes, and other lettered information should be the same height except for titles, drawing numbers, section and view letters, and other captions which are .25 in. (6 mm) high.

Either vertical or inclined lettering may be used on a drawing depending on company preference. However, only one style of let-

tering should be used on a drawing. Lettering must be dark, crisp, and opaque for the best possible reproducibility.

The composition or spacing of letters in words and between words in sentences should be such that the individual letters are uniformly spaced with approximately equal background areas. This usually requires that letters such as I, N, or S be spaced slightly farther from their adjacent letters than L, A, or W. A minimum recommended space between letters in words is approximately .06 in. (1.5 mm). The space between words in a note or sentence should be about the same as the height of the letters. The space between

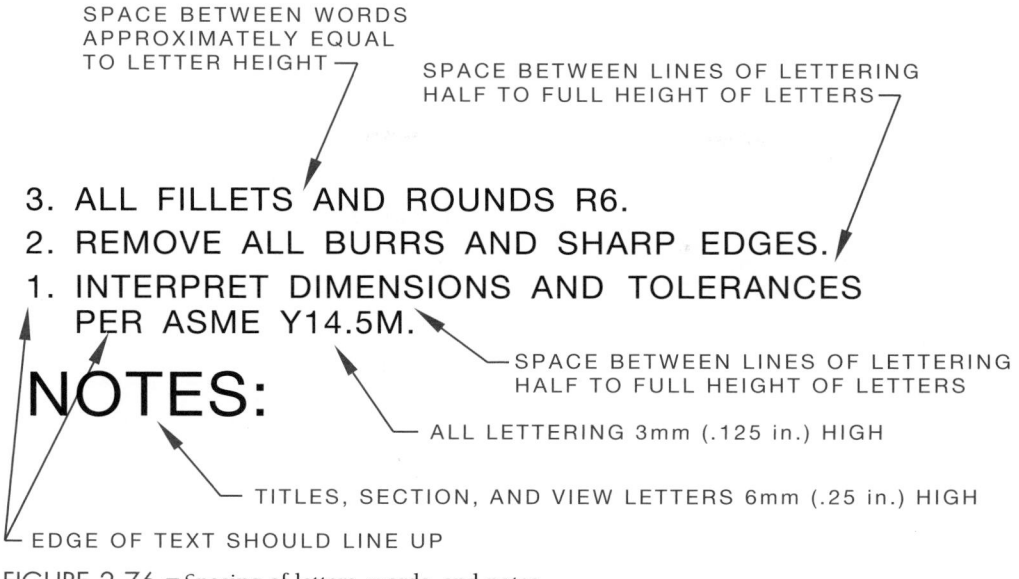

FIGURE 2.76 ■ Spacing of letters, words, and notes.

two numerals with a decimal point between them is a minimum of two-thirds of the letter height.

Notes should be lettered horizontally on the sheet. When lettering notes, sentences, or dimensions requires more than one line, the vertical space between the lines should be a minimum of one-half the height of letters. The maximum recommended space between lines of lettering is equal to the height of the letters. Some companies prefer to use the minimum space to help conserve space while other companies prefer the maximum space for clarity. (See Figure 2.76.)

Additional specific information about notes is provided in Chapter 12, Dimensioning and Tolerancing.

VERTICAL FREEHAND LETTERING

Vertical freehand lettering is the standard for mechanical drafting. The ability to perform good-quality lettering quickly is important. A common comment among employers hiring entry-level drafters is the ability to do quality lettering and line work. Although standard, not all companies require freehand lettering. Some companies allow drafters the flexibility of freehand lettering or using a template. As many companies are now changing to computer-aided drafting, traditional lettering skills may become obsolete.

Always use lightly drawn horizontal guidelines that are spaced equal to the height of the letters. Some people need vertical guidelines to help keep their letters vertical. The ability to perform quality freehand lettering requires a great deal of practice for most people.

Vertical Capital Letters

Straight Elements

Use a 0.5 mm automatic pencil for lettering. This kind of pencil does not need sharpening and H, F, or HB leads are usually easy to

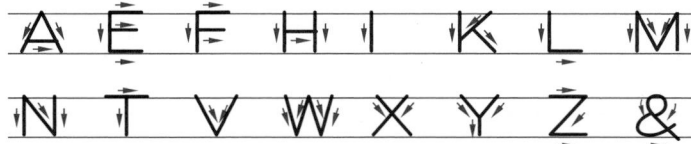

FIGURE 2.77 ■ Recommended strokes for vertical uppercase Gothic letters with straight elements.

control for effective lettering. Remember, you need to experiment with leads to determine which gives you the best results. Letters should be dark and crisp. Fuzzy letters print as fuzzy lines.

Letters composed of straight lines are shown in Figure 2.77. You should become familiar with these letter forms and the strokes needed to make them. The arrows in Figure 2.77 indicate the direction of the stroke used to form the letters. However, if these strokes are uncomfortable, you should develop your own procedure. Try the recommended strokes first. Because these letters are made of single-stroke elements, try not to let the strokes combine as the result may be curves where straight elements should be.

Horizontal guidelines should be used for all lettering at all times. Use vertical guidelines if you have difficulty keeping your letters vertical. Guidelines should be very light 6H or 4H pencil lines. Some drafters prefer to use a light-blue lead rather than a graphite lead for guidelines. Light blue will not reproduce in the diazo or photocopy processes. When lettering, protect your drawing by resting your hand on a clean protective sheet placed over your drawing. This prevents smearing and smudging, as shown in Figure 2.78.

Curved Elements

Letters that contain arcs are shown in Figure 2.79. Notice the difference in the sizes of arcs and circles used for different letters. The recommended letter elements, as with straight elements, are made up of a series of suggested strokes. These strokes when used as

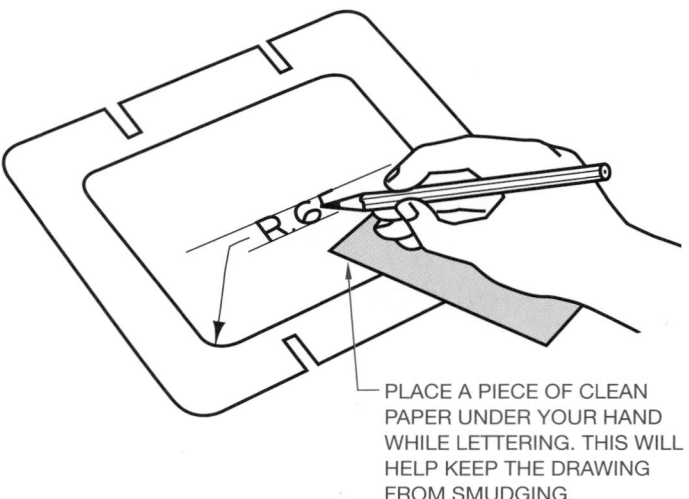

FIGURE 2.78 ■ Place clean paper under your hand when lettering.

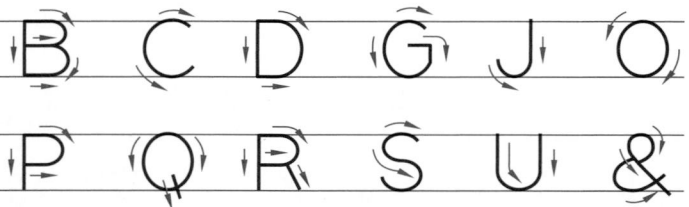

FIGURE 2.79 ■ Recommended strokes for vertical uppercase Gothic letters with curved strokes.

shown provide the best lettering results. Vertical guidelines are often an asset for the best lettering results.

Vertical Numerals and Fractions

Vertical numerals, as seen in Figure 2.80, are also made up of recommended strokes. Numerals are the same height as capital letters.

Fractions

Fractions are not as commonly used on engineering drawings as decimal inches or millimeters, but are common on architectural and structural drawings. When fractions are used on a drawing, the fraction numerals should be the same size as other numerals on the drawing. The fraction bar should be drawn in line with the direction of the dimension. For example, if all dimensions read hori-

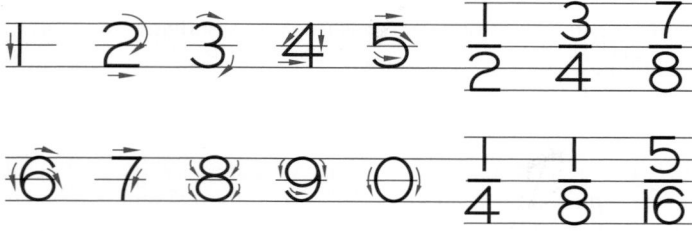

FIGURE 2.80 ■ Vertical numerals and fractions.

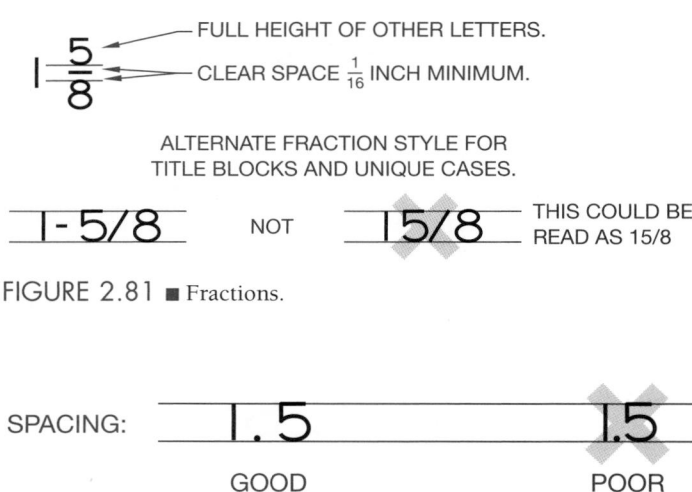

FIGURE 2.81 ■ Fractions.

FIGURE 2.82 ■ Spacing of decimal point in numerals.

zontally, all fraction bars are horizontal. The fraction's numerals should not touch the fraction bar. A space of 1.5 mm or .06 in. between the numeral and bar is suggested. The fraction bar may be diagonal in certain situations such as when used in a general note, in a drawing title, or when using CADD. (See Figure 2.81.)

Decimal Points

The placement of the decimal point in a decimal dimension is critical. If the decimal point is crowded or drawn too lightly, it may be lost and the result is an unclear dimension. Always make the decimal point dark and bold. Also space the numerals far enough to clearly provide room for the decimal point. Two-thirds the height of letters is recommended. (See Figure 2.82.)

LETTERING TECHNIQUES

Always use guidelines. Straight, even letters of consistent height look better than letters of varying heights. Even when using guidelines be sure to extend each letter directly to the guidelines. Letters that periodically extend beyond or fall short of the guidelines tend to make the words or notes irregular.

Try an H or F pencil for lettering if you have a light touch or try a 2H pencil if your touch is heavy. Use a 0.5 mm automatic pencil. Lines should be black, crisp, and sharp. All vertical lines are made perpendicular, starting at the top for each stroke; all horizontal lines are made from left to right. Balance angles for the letters A, V, W, X, and Y about a vertical guideline. Use a round form for curved letters. Be careful to not allow any space to show between the letter and the guideline.

COMPOSITION

As a rule of thumb, curved letters can be placed close together; straight letters should be placed further apart. Good lettering composition is evident when all letters in a word look as if they have the same amount of space between them. To achieve this appearance it is often necessary to shorten the horizontal strokes of open letters such as L and J. When strokes are parallel and next to each other as the WA in WALL and both Ns and I in PLANNING, they should be placed a little farther apart.

If your letters are wiggly or if you are nervous, try pressing hard to make your lines straighter. If you are pressing too hard, try to relax the pressure. Also try making each letter as rapidly as possible. This tends to eliminate wiggly letters. If your lead is too hard, wiggly letters could result; try a softer lead.

MAKING GUIDELINES

Guidelines are very lightly drawn lines equal to the height of letters in distance apart. As previously mentioned, some drafters prefer to use a light-blue lead so guidelines will not reproduce.

Ames Lettering Guide

A commonly used device for making guidelines is the versatile Ames Lettering Guide shown in Figure 2.83. With an Ames Lettering Guide it is possible to draw guidelines and sloped lines for lettering from ¹⁄₁₆ to 2 in. in height. Instructions for use are normally found with the lettering guide when purchased. The lettering guide may also be used for parallel lines, needed for such purposes as section lines, including brick, tile, and concrete block, or a music staff.

OTHER LETTERING AIDS AND GUIDELINE METHODS

Other guideline lettering aids for equidistant spacing of lines have parallel slots ranging in width from ¹⁄₁₆ to ¹⁄₄ in. These lettering guideline aids are not as complex as the Ames Lettering Guide, but they are also not as flexible.

Another method of making guidelines used by a few drafters is to place an ⅛-in. grid paper under the drawing. The lines show through the drawing sheet and guidelines need not be drawn. Be careful with this method, because lines of lettering may not be as straight as with conventional guidelines.

BASIC LETTERING CONSIDERATIONS

Always use two guidelines. Three are helpful for beginners. The center guideline is for the letter crossbars.

To adjust eye and hand coordination, first letter lightly, then darken your lettering. Doing so allows you to correct mistakes. This technique should not be necessary after you get some lettering experience.

Make all letters and numbers on the drawing at least ⅛ in. (3 mm) high, except for titles which are ¼ in. (6 mm) high. All letters should be the same height. Be consistent.

Make all lettering dark. You may have to press on the pencil to get dark letters.

When practicing lettering, practice no more than 15 minutes per day. Otherwise your hand may cramp and any further practice may be of less value. Practice every day to gain speed and neatness as you letter.

LETTERING GUIDE TEMPLATES

Some companies prefer that drafters use lettering guides so uniformity is maintained. Standard lettering guide templates are available with vertical Gothic letters and numerals ranging in height from ³⁄₃₂ to ⅜ in. (See Figure 2.84.) Lettering guides are also available in many other lettering styles, including slanted Gothic, and Microfont letters in either upper- or lowercase.

MECHANICAL LETTERING EQUIPMENT

Mechanical lettering equipment has typically been used on government projects and in civil drafting. Mechanical lettering equipment is available in kits with templates for letters and numerals in a wide range of sizes. A complete lettering equipment kit includes a scriber, templates, tracing pins, and lettering pens. Figure 2.85 shows the component parts of a lettering equipment set. Complete instructions are normally found with the lettering equipment when purchased.

FIGURE 2.83 ■ Ames Lettering Guide. *Courtesy Olson Manufacturing Company, Inc.*

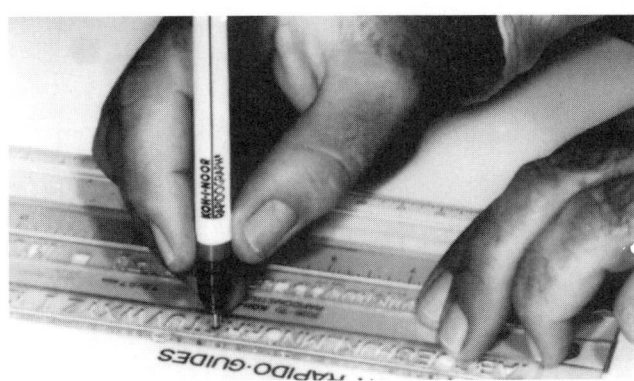

FIGURE 2.84 ■ Lettering guide template in use. *Courtesy Koh-I-Noor, Inc.*

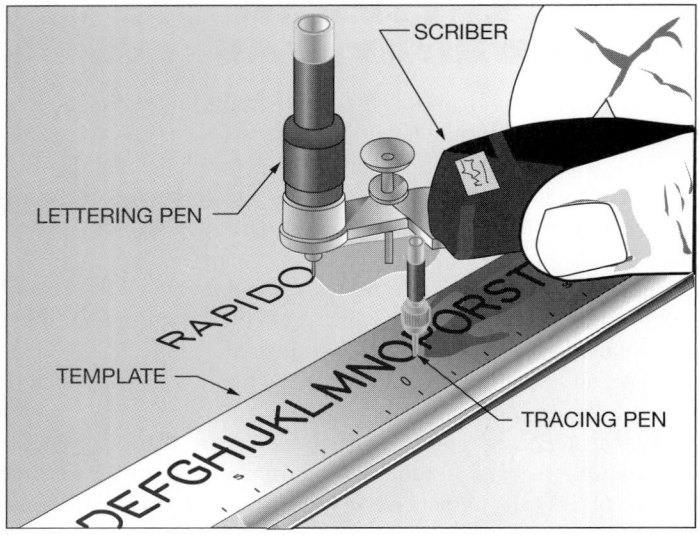

FIGURE 2.85 ■ Components of a lettering equipment set, and using the mechanical lettering equipment. *Courtesy Koh-I-Noor, Inc.*

LINES

Drafting is a graphic language using lines, symbols, and notes to describe objects to be manufactured or built. Lines on drawings must be of a quality that will easily reproduce. All lines are dark, crisp, sharp, and of the correct thickness when properly drawn. There is no variation in darkness, only a variation in thickness, known as *line contrast*. Certain lines are drawn thick so they stand out clearly from other information on the drawing. Other lines are drawn thin. Thin lines are not necessarily less important than thick lines, but they are subordinate for identification purposes.

ASME The American Society of Mechanical Engineers (ASME) recommends two line thicknesses with bold lines twice as thick as thin lines. This line standard relates to both manual and computer-aided drafting. Standard line thicknesses are 0.6 mm for thick lines and 0.3 mm for thin lines. The actual width of lines may be more or less than the recommended thickness depending on the size of the drawing or the size of the final reproduction. Drawings that meet military documentation standards require three thicknesses of lines: thick, medium, and thin. Figure 2.86 shows widths and types of lines as taken from ASME standard, *Line Conventions*

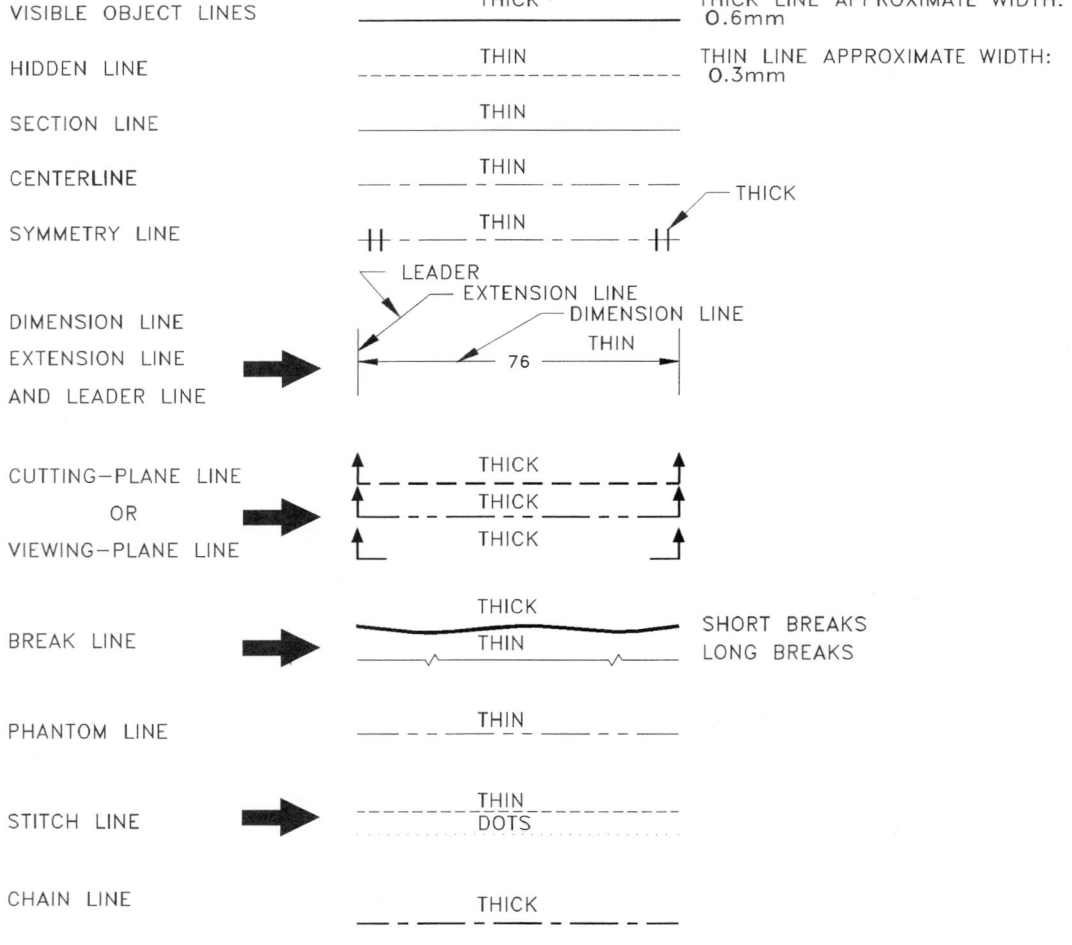

FIGURE 2.86 ■ Line conventions, width and type of lines.

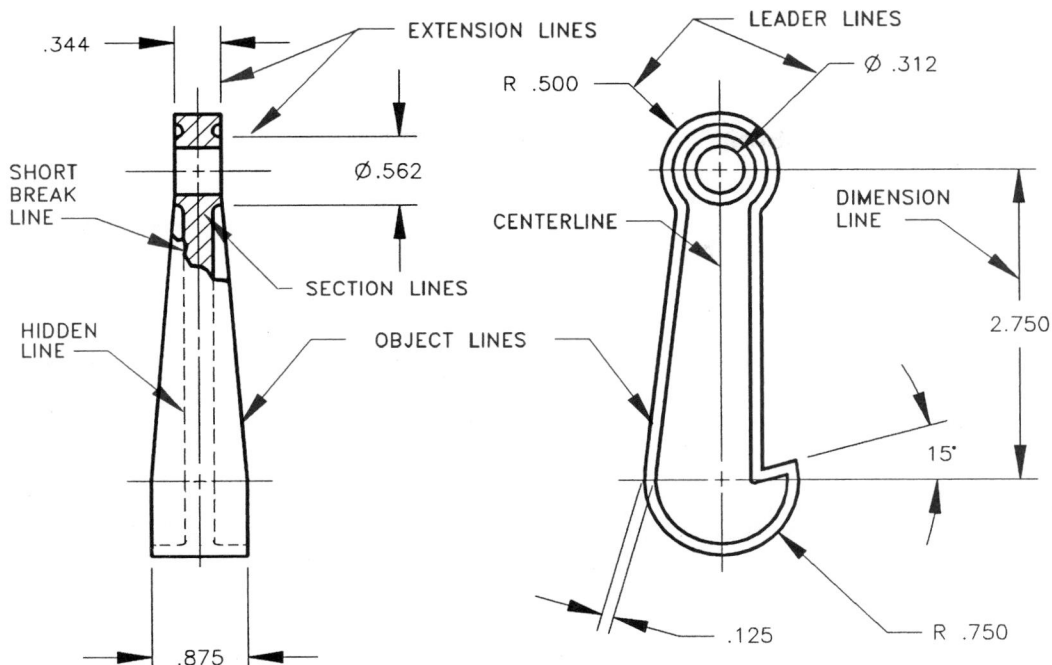

FIGURE 2.87 ■ Sample drawing with a variety of lines displayed.

and Lettering, ASME Y14.2M. Figure 2.87 shows a sample drawing using the various kinds of lines.

TYPES OF LINES

Introduction to Lines

The following discussion is an introduction to the lines that are commonly used on engineering drawings. You will use only a few of these lines as you work on the problems for this chapter. You will continue to use additional lines as you continue to learn about specific applications throughout this text. For example, the lines used in dimensioning are covered in detail in Chapter 12, Dimensioning and Tolerancing, and sectioning practices are covered in Chapter 14, Sections, Revolutions, and Conventional Breaks. You will put the use of lines to practice in every chapter where specific engineering drafting applications are fully explained.

Construction and Guidelines

Construction lines are used for laying out a drawing. Construction lines are drawn very lightly so they will not reproduce and will not be mistaken for any other line on the drawing. Construction lines are drawn with a 4H to 6H pencil and, if drawn properly, will not need to be erased. Use construction lines for all preliminary work.

Guidelines are drawn the same as construction lines and will not reproduce when properly drawn. Guidelines are used to keep lines of lettering in perfect alignment and of a constant height. For example, if lettering on a drawing is to be .125 in. high, then guidelines are drawn very lightly .125 in. apart. Guidelines must always be used for lettering. Some drafters prefer to use a light-blue lead for all guidelines and layout work. Light-blue lead will not reproduce and may be cleaner than graphite lead.

Construction lines are also commonly used in CADD for layout work. These lines are usually placed on a specific layer that can be named something such as CONSTRUCTION. This layer can be turned off or frozen so that it does not display or plot when the construction work is finished.

Object Lines

Object lines, also called visible lines or outlines, describe the visible surface or edge of the object. They are drawn as thick lines as shown in Figure 2.88. Thick lines, remember, are usually drawn at 0.6 mm or .032 in. wide. Drafters usually use a soft lead, H or F, to draw object lines properly. You need to experiment to see which lead gives you the best line quality. For inked drawings, object lines can be made with a number 2 (0.6 mm) pen. When using CADD, the recommended thickness of object lines is also 0.6 mm. This thickness is normally set in the layer that controls the specific line type. Depending on the CADD system used, the line thickness can be displayed on the screen while drawing, or it may only be represented in the final print or plot of the drawing. Some CADD programs refer to line thickness as line weight.

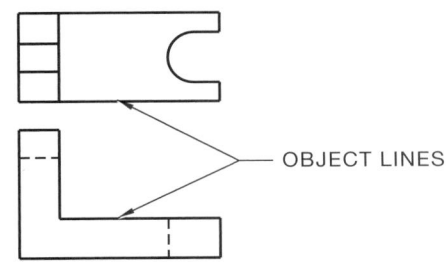

FIGURE 2.88 ■ Object lines.

Object Line Quality

To determine if your object lines are dark and crisp enough, turn the paper over and hold it up to the light. The line should have a dark, consistent density. A light table, if available, also works very well to check darkness.

You can also make a diazo or photocopy reproduction of your drawing to observe the quality. Properly drawn lines and lettering make quality prints. If the print lines are fuzzy, your original probably needs more work.

Most manual drafters use automatic pencils with a 0.7 mm lead for object lines.

Line quality is easy to achieve when using CADD. This is because the CADD program controls the way your lines look on the screen and the printer or plotter controls the way the lines look on paper. Modern engineering printers and plotters generally produce high-quality drawings.

CADD APPLICATIONS

DRAWING LINES

The basics of drawing lines with a CADD system are applicable to all systems, whether they use a puck, stylus, light pen, thumbwheels, joystick, or mouse. The term *pointers* refers to the pointing device used to select commands and digitize points. The term *digitize* or *pick* means to press the stylus to the tablet or press the proper puck or keyboard button to select, or draw, a point. Several different commands are given for one function in the following discussions, to give you an idea of the terms used in different systems.

Once you are in the drawing mode, your location on the screen is indicated by crosshairs. When you move the pointer, the crosshairs move. Unless you press a button or click the pointer, nothing is drawn on the screen. Some systems may continually display the coordinates of the crosshair location on the screen. These values change as you move the pointer. When constructing your drawing, it is best to look at the video display as you draw. You will then be able to see if your lines are straight, and if the drawing looks right.

Begin drawing a line by first selecting the proper command such as POINT, INSERT POINT, PLACE LINE, or LINE. This command is used to locate corners of the object and connect them with straight lines. Pick a point, and the coordinates of that point are displayed on the screen. Select another point and a straight line is drawn between the two. This shows as line 1 in Figure 2.89a. Each selection of the pointer after that draws a line. The steps required to draw an

object are shown in Figure 2.89a–d. There is no need to select LINE between each selected point. If you wish to break the line and move to another feature, press the [Enter] key to complete the previous command, and enter the LINE command again. This allows you to locate a new coordinate point without drawing an unwanted line.

Some systems use what is called a *rubber band line*. Once you have picked one point, a line is attached to that point and to the crosshairs and moves with the crosshairs. Selecting another point draws the first line and another is now attached to the crosshairs.

Different types of lines can be indicated either before or after picking the points. Specifying the type of line before picking points is done by selecting a command such as LINE TYPE, or selecting a layer that contains the desired line type. Any line drawn after you have entered a type will have the characteristics that you have chosen, and the type of line will change only after you select a new type of line. If you want to change an existing line to a different type, you can use an edit command such as MODIFY, CHANGE, or PROPERTIES because you are changing something that has already been drawn.

CADD systems also allow you to set line thickness for desired contrast. This is normally done by setting a line weight within the layer that is selected for a specific line. Different lines are also shown on the screen using contrasting colors for clarity.

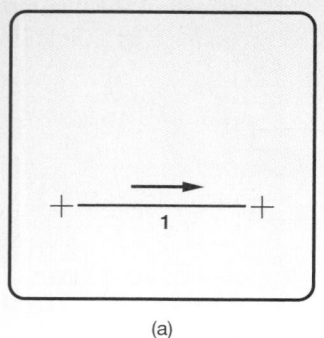
(a)

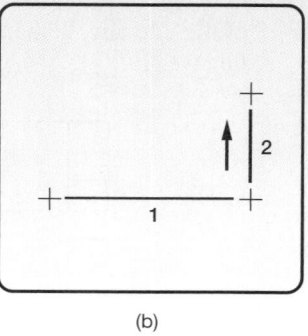
(b)

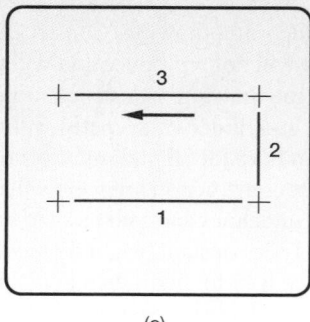
(c)

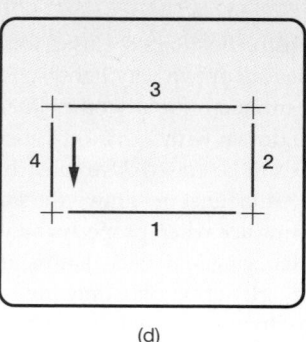
(d)

FIGURE 2.89 ■ Digitizing an object one point at a time.

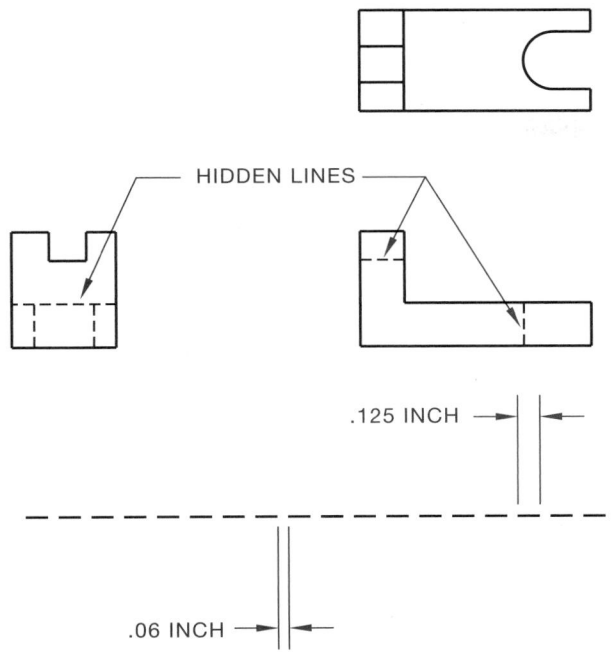

FIGURE 2.90 ■ Hidden line representation.

Hidden Lines

A *hidden line* represents an invisible edge on an object. Hidden lines are thin lines, half as thick as object lines for contrast. Figure 2.90 shows hidden lines properly draws with .125 in. dashes spaced .06 in. apart. Hidden lines, as all thin lines, can be drawn effectively with a 0.3 mm to 0.5 mm automatic drafting pencil, or a number 00 (0.3 mm) technical pen tip when inking. The best way to draw hidden lines is to draw dash lengths and spaces by eye. This takes some practice, but is the fastest way.

Hidden Line Rules

The drawings in Figure 2.91 show situations where hidden lines meet or cross object lines and other hidden lines. These situations represent rules that should be followed when possible.

Hidden lines should also be represented as previously discussed when using CADD. The dash length and spacing can be adjusted to match the desired standard when using most CADD programs. Some initial experimentation may be needed to get the correct line representation depending on the size of your drawing and other scale factors within the CADD program. You may also have some difficulty applying all of the hidden line rules, but most CADD systems should give you the desired T and L form shown in Figure 2.91. The hidden lines are normally set up on a related layer where their line type and thickness match the desired standard. They can also be displayed in a contrasting color that helps distinguish them from other lines.

Centerlines

Centerlines are used to show and locate the centers of circles and arcs, and are used to represent the center axis of a circular or sym-

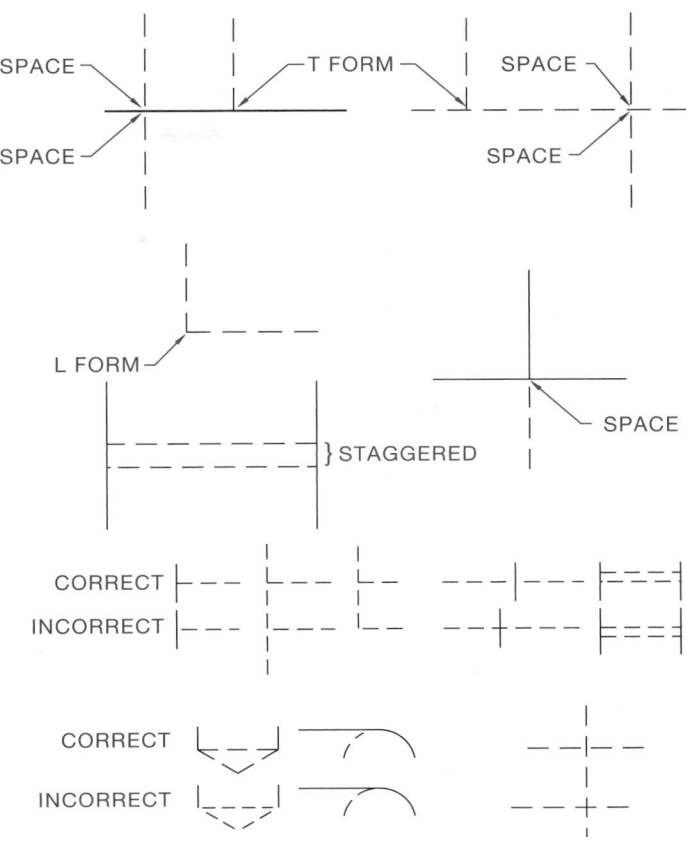

SOME OF THESE PRACTICES CAN BE DIFFICULT TO DO WITH CADD.

FIGURE 2.91 ■ Hidden line rules.

metrical form. Centerlines are thin lines on a drawing. They should be about half as thick as an object line. The long dash is about .75 to 1.50 in. The spaces between dashes are about .062 in. and the short dash about .125 in. long. The length of long lines varies for the situation and size of the drawing. Try to keep the long lengths uniform throughout the centerline. (See Figure 2.92.) Small centerline dashes should cross only at the center of a circle or arc. (See Figure 2.93.) Small circles should have centerlines as shown in Figure 2.94.

Centerlines for holes in a bolt circle may be drawn either of two ways, depending on how the holes are located, as shown in Figure 2.95. A *bolt circle* is a pattern of holes arranged in a circle.

When a centerline represents symmetry, as in the centerplane of an object, the symmetry symbol, shown in Figure 2.96, may be used if needed for clarity.

The centerline is commonly drawn with a 0.3 to 0.5 mm automatic pencil. A 0.3 mm pencil may provide best results with a soft lead and a 0.5 mm pencil with a harder lead. Remember the results should be dark, crisp, and sharp lines that are half as thick as object lines. When inking, a number 00 (0.3 mm) technical pen gives the best results.

Centerlines should also be represented as previously discussed when using CADD. The dash length and spacing can be adjusted to match the desired standard when using most CADD programs. Some initial experimentation may be needed to get the correct line

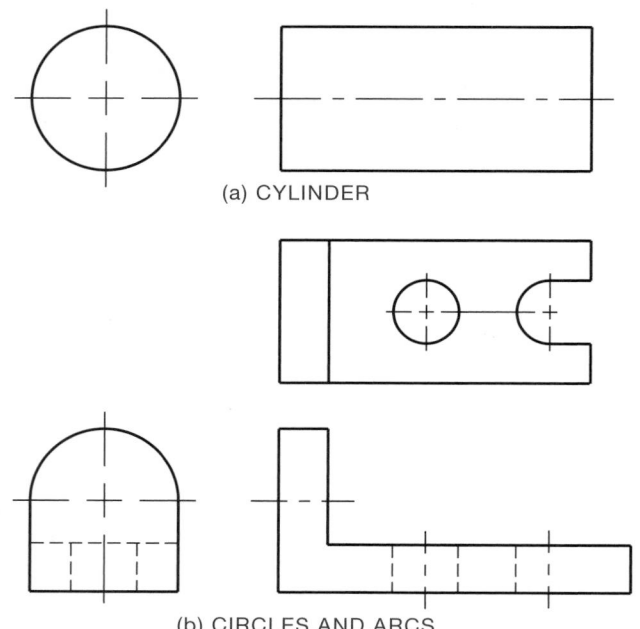

(a) CYLINDER

(b) CIRCLES AND ARCS

FIGURE 2.92 ■ Centerline representation and examples.

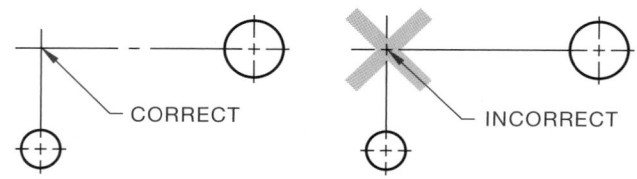

CORRECT   INCORRECT

FIGURE 2.93 ■ Centerline rules.

FIGURE 2.94 ■ Centerlines for small circles.

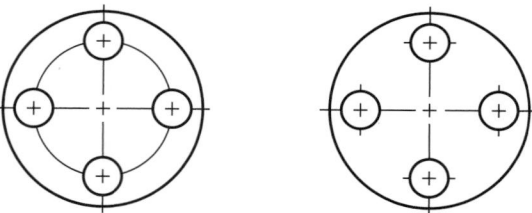

FIGURE 2.95 ■ Bolt circle centerline options.

representation depending on the size of your drawing and other scale factors within the CADD program. You may also have some difficulty applying all of the centerline rules, but most CADD systems should give you the desired results with some variation. The centerlines are normally set up on a related layer where their line type and thickness match the desired standard. They can also be displayed in a contrasting color that helps distinguish them from other lines.

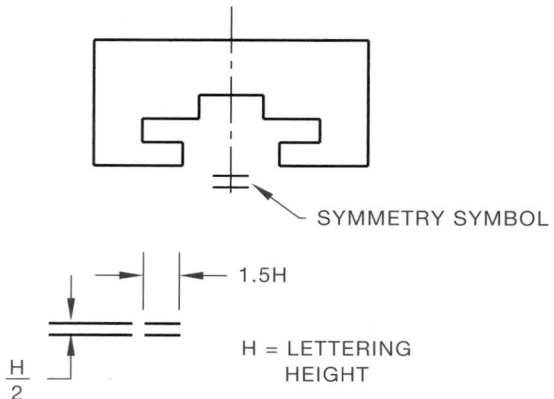

SYMMETRY SYMBOL

1.5H

$\frac{H}{2}$

H = LETTERING HEIGHT

FIGURE 2.96 ■ Symmetry symbol.

Extension Lines

Extension lines are thin lines used to establish the extent of a dimension. Extension lines begin with a short space from the object and extend to about .125 in. beyond the last dimension, as shown in Figure 2.97. Extension lines may cross object lines, centerlines, hidden lines, and other extension lines, but they may not cross dimension lines. Circular features, such as holes, are located by their centers in the view where they appear as circles. In this practice, centerlines become extension lines as shown in Figure 2.98.

Dimension Lines and Leader Lines

Dimension lines are thin lines capped on the ends with arrowheads and broken along their length to provide a space for the

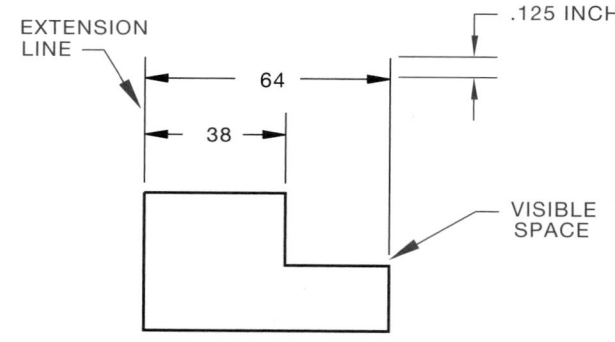

EXTENSION LINE

.125 INCH

64

38

VISIBLE SPACE

FIGURE 2.97 ■ Extension lines.

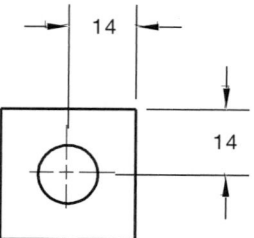

14

14

FIGURE 2.98 ■ The centerline becomes an extension line when used for dimensioning.

dimension numeral. *Dimension lines* indicate the length of the dimension. (See Figure 2.99.)

Leaders, or leader lines, are thin lines used to connect a specific note to a feature as shown in Figure 2.100. Leaders may be drawn at any angle, but 45°, 30°, or 60° lines are most common. Slopes greater than 75° or less than 15° from horizontal should be avoided. The leader has a .25 in. shoulder at one end that begins at the center of the vertical height of the lettering and an arrowhead at the other end pointing to the feature. If the leader were to continue from the point where the arrowhead touches the circle, it would intersect the center. (See Figure 2.101.)

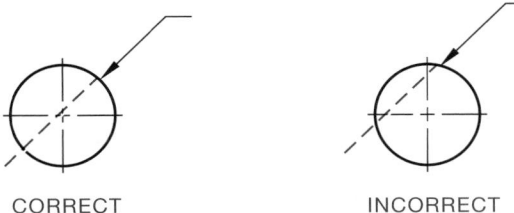

CORRECT   INCORRECT

FIGURE 2.101 ■ Circle to leader line relationship. The path of the leader should pass through the center.

Arrowheads

Arrowheads are used to terminate dimension lines and leaders. Properly drawn arrowheads should be three times as long as they are wide. All arrowheads on a drawing should be the same size. Do not use small arrowheads in small spaces. (Limited space dimensioning is covered in Chapter 12.) Some companies require that arrowheads be drawn with an arrow template, while others accept properly drawn freehand arrowheads. (See Figure 2.102.) Individual company preference dictates whether arrowheads are filled in or left open as shown.

CADD programs normally allow you to select from a variety of available arrowhead options. You should be able to match the desired ASME standard or the standard used by your company or drafting application. The CADD system allows you to set the size of arrowheads, and placement of arrowheads as related to the dimension text placement. The commonly preferred arrowhead for engineering drawings is solid-filled, but this standard can be different between offices. The solid-filled arrowhead stands out clearly on the drawing and helps identify the dimension location. Dimensions are normally set up on a specific layer and with a specific style that correlates the dimensions with the desired application.

Cutting-plane and Viewing-plane Lines

Cutting-plane lines are thick lines used to identify where a sectional view is taken. *Viewing-plane lines* are also thick and are used to

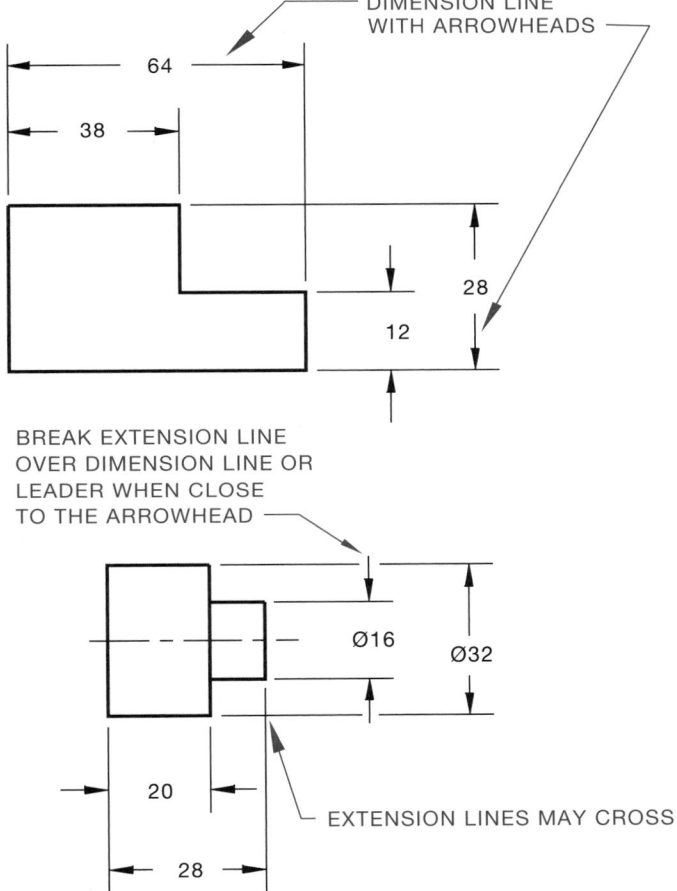

FIGURE 2.99 ■ Dimension lines.

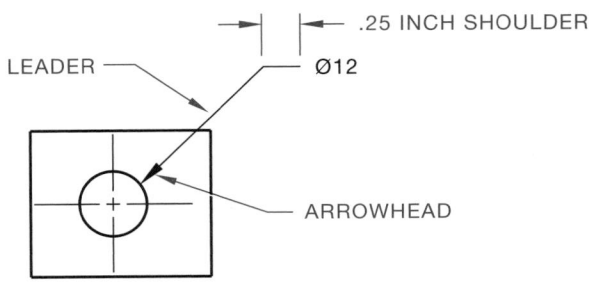

FIGURE 2.100 ■ Leader line.

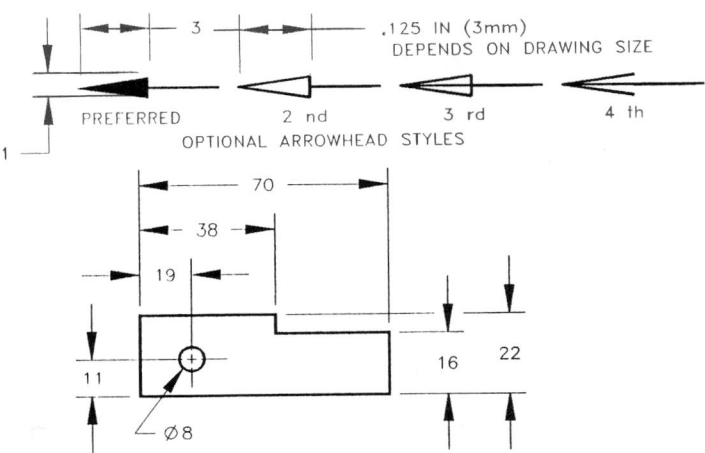

FIGURE 2.102 ■ Arrowheads.

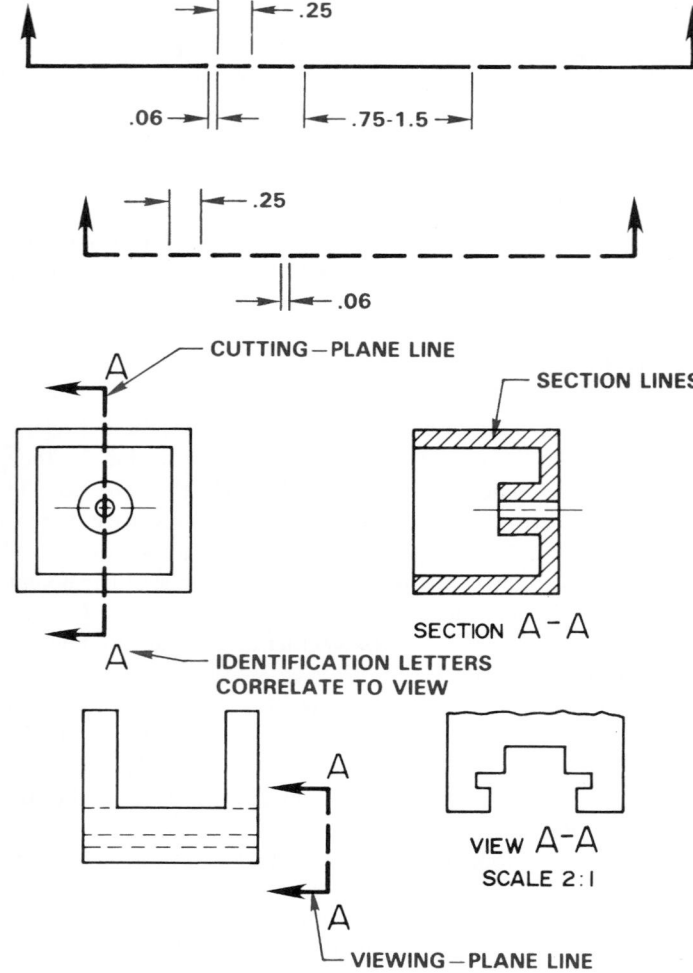

FIGURE 2.103 ■ Cutting- and viewing-plane lines.

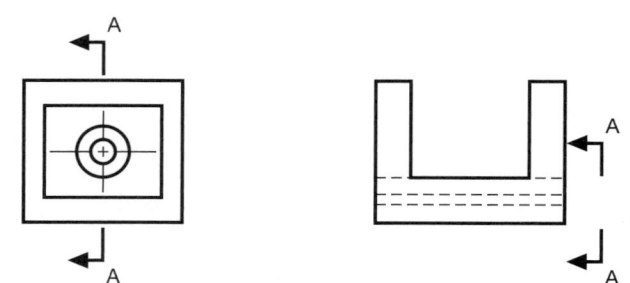

FIGURE 2.104 ■ Simplified cutting- and viewing-plane lines.

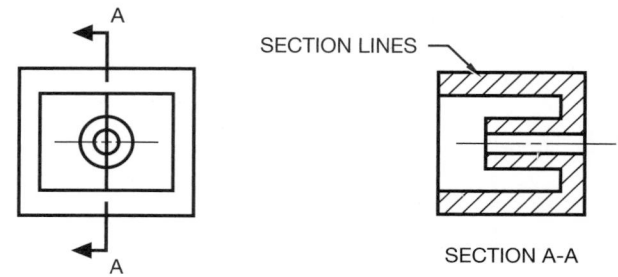

FIGURE 2.105 ■ Section lines.

identify where a view is taken for view enlargements or for partial views. Cutting-plane and viewing-plane lines are properly drawn in either of the two ways. The approximate dash and space sizes are shown in Figure 2.103. The cutting-plane line takes precedence over the centerline when used in the place of a centerline.

The scale of the view may be increased or remain the same as the view from the viewing plane, depending on the clarity of information presented. When the location of the cutting plane or viewing plane is easily understood or if the clarity of the drawing is improved, the portion of the line between the arrowheads may be omitted as shown in Figure 2.104.

Section Lines

Section lines are thin lines used in the view of a section to show where the cutting-plane line has cut through material. (See Figure 2.105.) Section lines are drawn equally spaced at 45° but may not be parallel or perpendicular to any line of the object. Any convenient angle may be used to avoid placing section lines parallel or perpendicular to other lines of the object; 30° and 60° are common. Section lines that are more than 75° or less than 15°

from horizontal should he avoided. Section lines should be drawn in opposite directions on adjacent parts. (See Figure 2.111.) For additional adjacent parts, any suitable angle may be used to make the parts appear clearly separate. The space between section lines may vary depending on the size of the object. (See Figure 2.106.) Figure 2.107 shows correct and incorrect applications of section lines. When a very large area requires section lining, you may elect to use outline section lining, as shown in Figure 2.108.

The section lines shown in Figures 2.105 through 2.108 were all drawn as general section-line symbols. General section lines can be used for any material and are specifically used for cast or malleable iron. Coded section-line symbols, as shown in

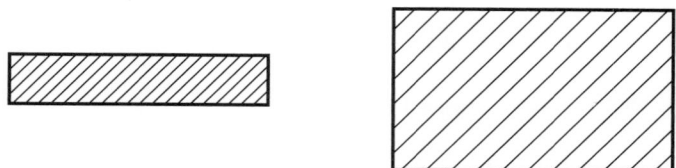

FIGURE 2.106 ■ Space between section lines.

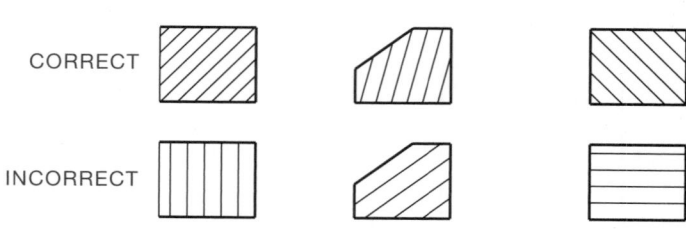

FIGURE 2.107 ■ Correct and incorrect section lines.

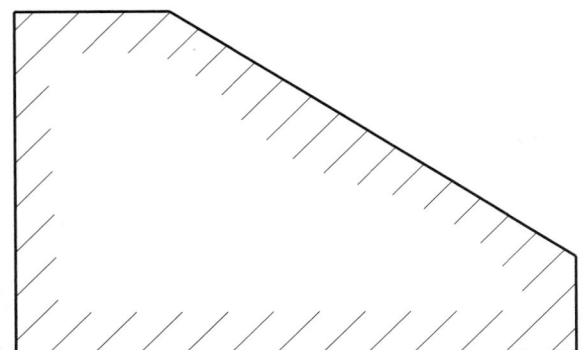

FIGURE 2.108 ■ Outline section lines.

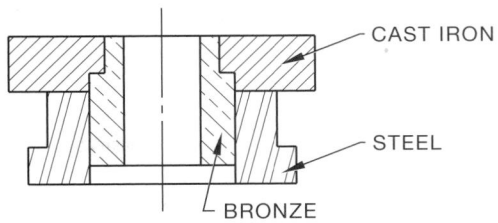

FIGURE 2.110 ■ Coded section lines in assembly.

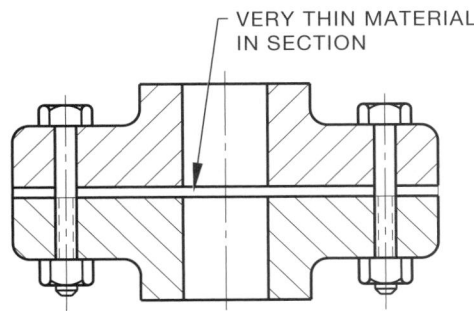

FIGURE 2.111 ■ Very thin material in section is drawn without section lines.

Figure 2.109, are not commonly used on detail drawings, as they are more difficult to draw and the drawing title block usually identifies the type of material in the part. Coded section lines may be used when the material must be clearly represented, as in the section through an assembly of parts made of different materials. (See Figure 2.110.) Very thin parts less than 4 mm thick may be shown without section lining; only the outline is shown. This option is often used for a gasket as shown in Figure 2.111.

CADD programs normally allow you to select from a variety of available section-line options. You should be able to match the desired ASME standard section-line symbol. The CADD system allows you to set the section-line type, spacing, and angle. Section lines are normally set up on a specific layer and with a specific color so they stand out clearly on the drawing. Section lines are very easy

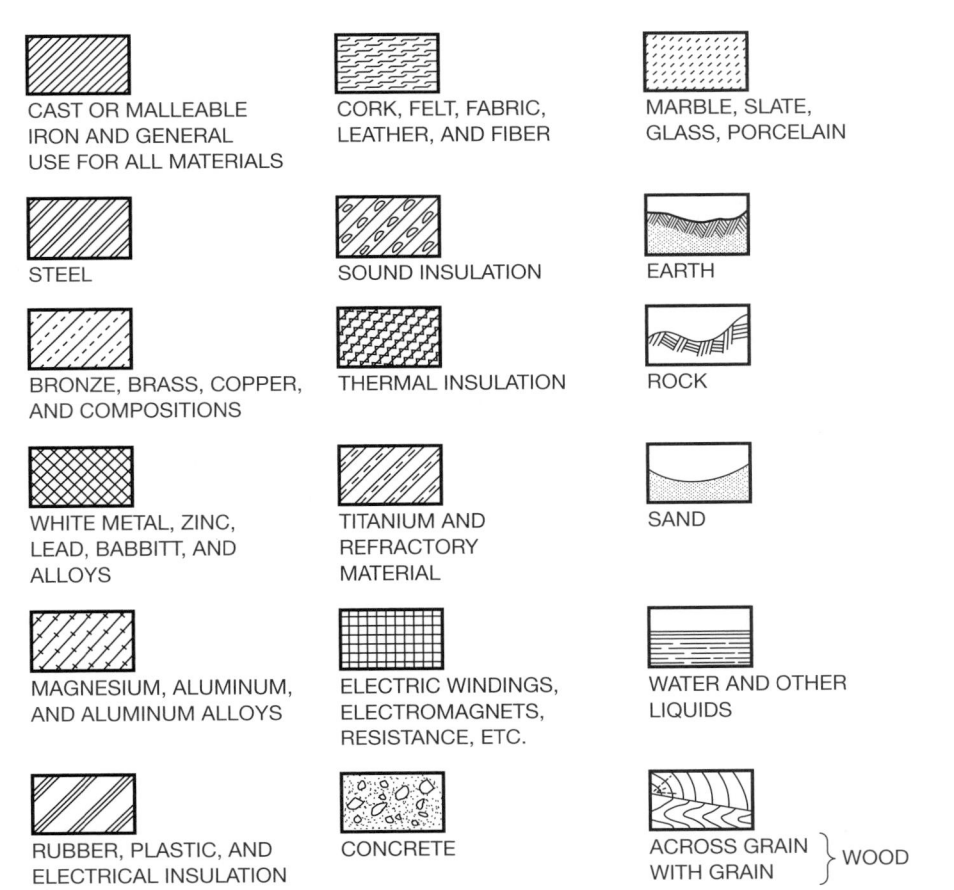

CAST OR MALLEABLE IRON AND GENERAL USE FOR ALL MATERIALS

CORK, FELT, FABRIC, LEATHER, AND FIBER

MARBLE, SLATE, GLASS, PORCELAIN

STEEL

SOUND INSULATION

EARTH

BRONZE, BRASS, COPPER, AND COMPOSITIONS

THERMAL INSULATION

ROCK

WHITE METAL, ZINC, LEAD, BABBITT, AND ALLOYS

TITANIUM AND REFRACTORY MATERIAL

SAND

MAGNESIUM, ALUMINUM, AND ALUMINUM ALLOYS

ELECTRIC WINDINGS, ELECTROMAGNETS, RESISTANCE, ETC.

WATER AND OTHER LIQUIDS

RUBBER, PLASTIC, AND ELECTRICAL INSULATION

CONCRETE

ACROSS GRAIN } WOOD
WITH GRAIN

FIGURE 2.109 ■ Coded section lines.

to draw with CADD, because all you have to do is select the object to be sectioned, or pick a point inside of the area to be sectioned. Normally, the object to be sectioned must be a closed geometric shape without any gaps in the perimeter. Some CADD systems also allow you to place section lines in an area that is not defined by an existing geometric shape. Section lines are referred to as hatch lines in some CADD programs and are drawn with a command such as HATCH.

Break Lines

There are two types of *break lines:* the short break and long break line. The thick, short break is very common on detail drawings, although the thin, long break may be used for breaks of long distances at the drafter's discretion. (See Figure 2.112.) Other conventional breaks may be used for cylindrical features as in Figure 2.113.

Phantom Lines

Phantom lines are thin lines made of one long and two short dashes alternately spaced. Phantom lines are used to identify alternate positions of moving parts, adjacent positions of related parts, repetitive details, or the contour of filleted and rounded corners. (See Figure 2.114.)

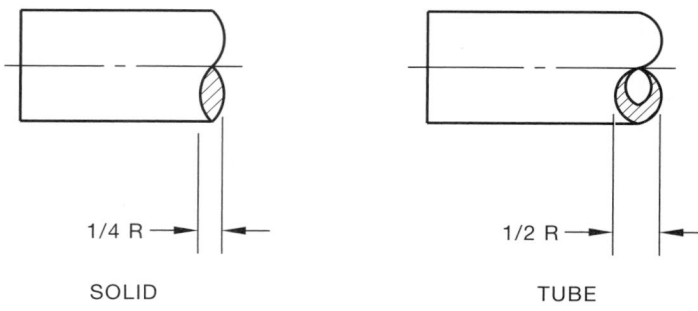

SOLID   TUBE

FIGURE 2.113 ■ Cylindrical conventional breaks.

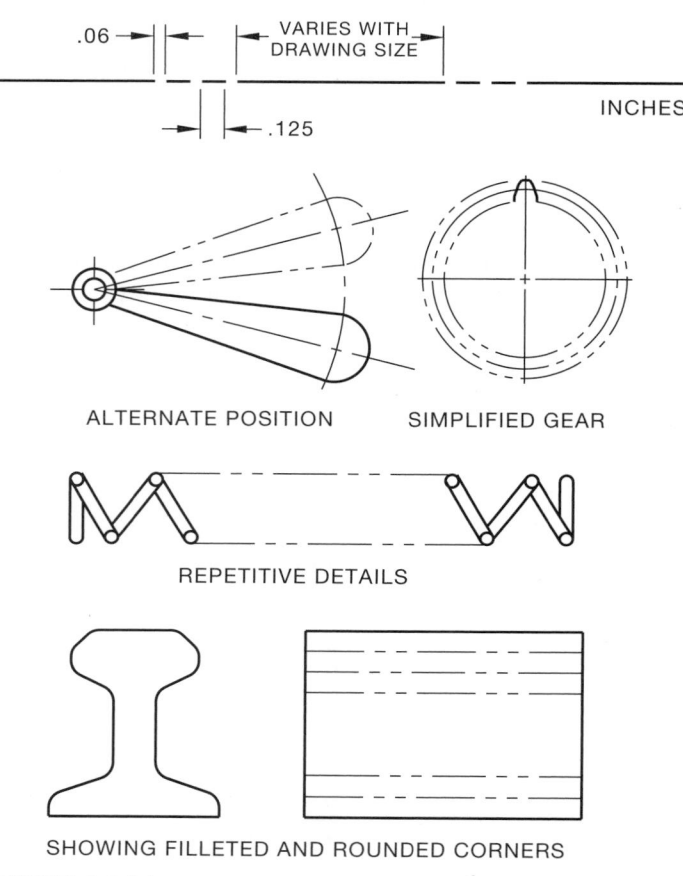

ALTERNATE POSITION   SIMPLIFIED GEAR

REPETITIVE DETAILS

SHOWING FILLETED AND ROUNDED CORNERS

FIGURE 2.114 ■ Phantom line representation and examples.

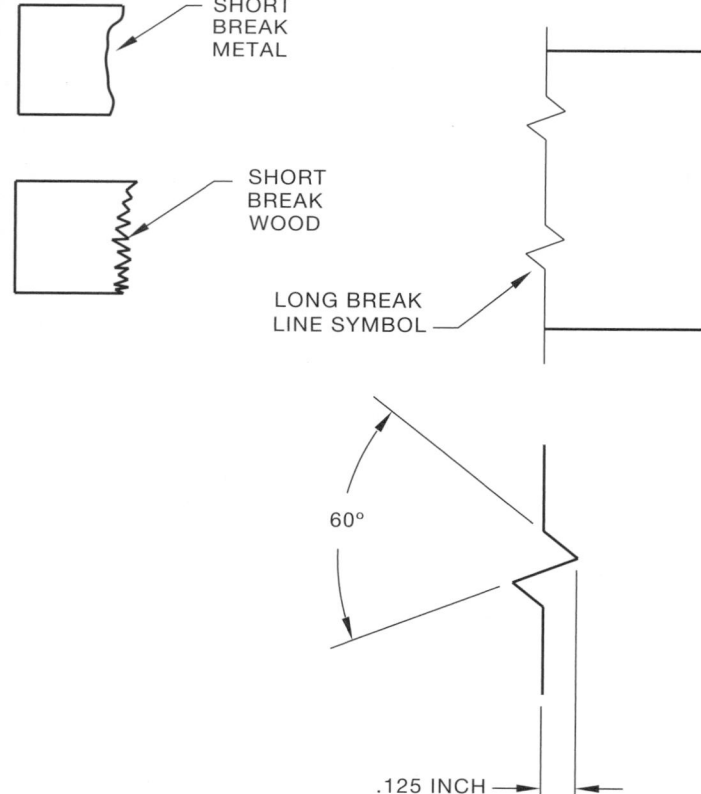

FIGURE 2.112 ■ Long and short break lines.

Chain Lines

Chain lines are thick lines of alternately spaced long and short dashes used to indicate that the portion of the surface next to the chain line receives some specified treatment. [See Figure 2.115.)

Stitch Lines

There are two types of acceptable *stitch lines.* One is drawn as thin, short dashes, the other as .016-in. diameter dots spaced .12 in. apart. They are used to indicate the location of a stitching or sewing process as shown in Figure 2.116.

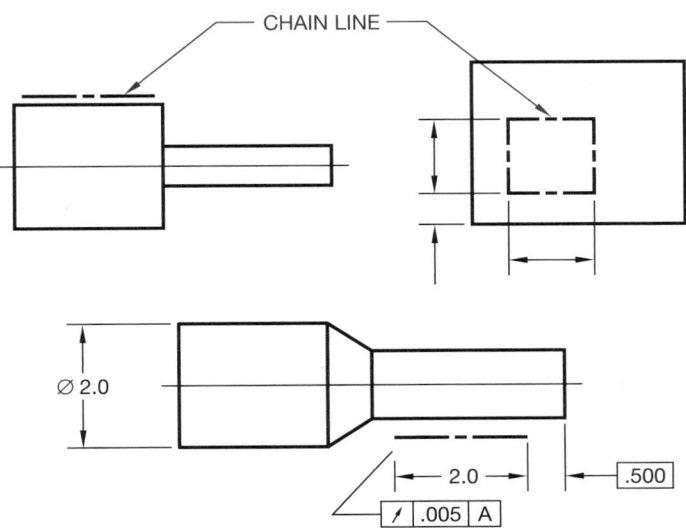

FIGURE 2.115 ■ Chain lines.

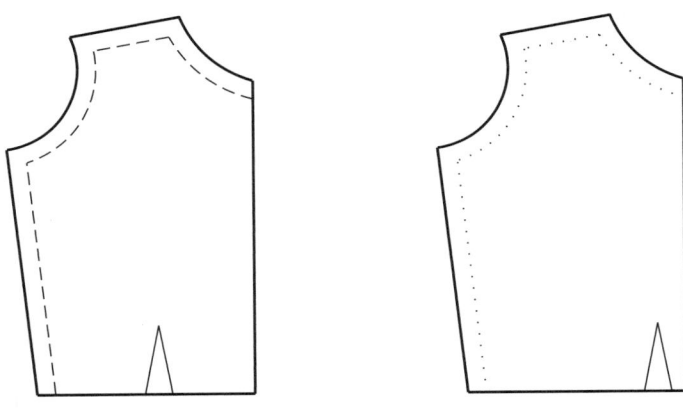

FIGURE 2.116 ■ Stitch lines.

PENCIL AND INK LINE TECHNIQUES

Pencil Techniques

There are four basic properties of correctly drawn lines: uniformity, contrast, darkness, and sharpness. *Line uniformity* means that all lines are drawn their proper thickness without variation. For example, all thick lines (object, cutting-plane) are the same degree of thickness. All thin lines (center, dimension, extension) are the same thinness. *Line contrast* is the variation that exists between different types of lines, thick and thin.

> **ASME/MIL** According to ASME standards, thick and thin are the two line thicknesses used. Military (MIL) standards recommend three line thicknesses: thick (cutting- and viewing-plane, short break, and object), medium (hidden and phantom), and thin (center, dimension, extension, leader, long break, and section).

FIGURE 2.117 ■ Draw circles and arcs first.

All lines should be the same darkness with no variation. Lines must be drawn opaque so light will not pass through. Lines are sharp when edges are clear and crisp. Fuzzy edges on lines result from the texture of the paper, or too-soft pencil lead. When using an automatic pencil, do not let the lead out too far.

Wash your hands and equipment frequently to keep the drawing clean. Try not to handle your pencil leads.

Lay out your entire drawing with construction lines before darkening any lines. A combination of pressure and pencil lead hardness makes the best lines. If you are right-handed, it is generally best to draw horizontal lines from left to right and vertical lines from bottom to top. The opposite may be true for left-handed drafters. Do not use excessive pressure. You do not have to dig trenches in the drawing board. The more pressure you use, the more difficult it is to erase. If your hand cramps while drawing or lettering, you are forcing yourself into an unnatural stroking technique. Relax! Try different methods until you find a suitable system of strokes.

Make thick lines with a 0.7 or 0.9 mm automatic pencil held vertically to establish full contact between the lead and the paper. In order to obtain lines that are dark, crisp, and sharp, you may need to go over each line several times. If light passes through your lines, they are drawn too lightly. Hold your drawing up to a light or place it on a light table to see if light passes through.

Always draw radii first, then connect the straight lines. This is also a good technique for ink drawings. (See Figure 2.117.) If you make the straight lines first, you may have trouble aligning the arcs.

Line Layout

First, draw thin horizontal lines beginning from the top of the sheet and work downward. Draw vertical thin lines next, starting on the left side of the sheet and working toward the right side for right-handed drafters. Next draw all circles and arcs. Last, draw all horizontal and vertical thick lines in the same direction as thin lines. Do all lettering when the line work is complete. Try to avoid going back over lines or lettering that has already been drawn in order to help avoid smudging.

Inking Techniques

Ink or graphite only adheres cleanly to one side of vellum. If the ink beads or skips while making a line, you are working on the wrong side of the paper. Paper with preprinted borders and title blocks have the correct side predetermined. Some manufacturers have a water mark that can be seen when the paper is held up to the light. When the water mark is readable, the paper is right side up. For

Mylar®, there is either a single or double matte surface. Matte refers to the textured surface as opposed to a glossy surface. Double matte surface may be drawn on either side. Single matte surface requires you to draw on the matte side, not the shiny, or glossy, side.

Use template lifters or templates and other equipment designed for inking that keep the edge of the tool slightly off the paper. Template lifters are plastic shapes that can be attached to the back of a template or other instrument to help keep the instrument off the drafting surface, as seen in Figure 2.118. Another option is to place a template under the one being used. (See Figure 2.119.) There are also template risers available that are long plastic strips that fit on the edges of the template to help keep it off of the drawing surface. Figure 2.120 shows how the template riser functions when attached to a template. Drafting machine scales have long been manufactured with edge relief for inking. Now, templates, triangles, and other equipment are being made with ink risers built in, as shown in Figure 2.121. If you draw with ink without taking these precautions, the ink could easily flow under the template and cause a mess.

Periodically check your technical fountain pens for leaks around the tip to prevent your hands from smearing your drawing. Also check the tip for a drop of ink before you begin a line. Have a piece of cloth or a tissue available to help keep the tip free of ink drops. Keep your pens clean and loaded with fresh ink to help keep proper ink flow for trouble free use. Also, keep a piece of paper handy for scratching your pen to start the ink flowing.

Ink your lines beginning with thin horizontal lines, and work from top to bottom and from left to right if right-handed. The technique of drawing horizontal lines top to bottom and vertical lines

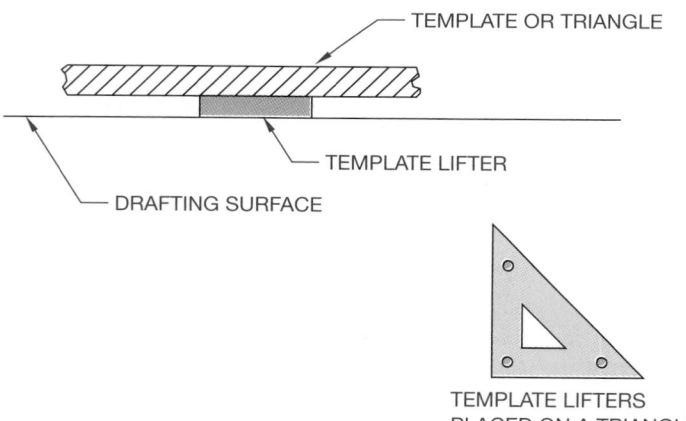

FIGURE 2.118 ■ Template lifters.

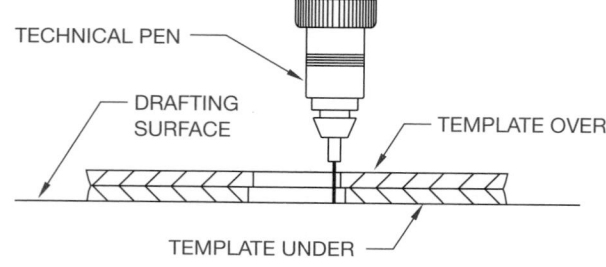

FIGURE 2.119 ■ Second template as a spacer.

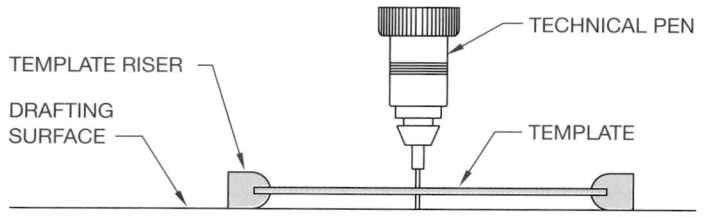

FIGURE 2.120 ■ Template risers.

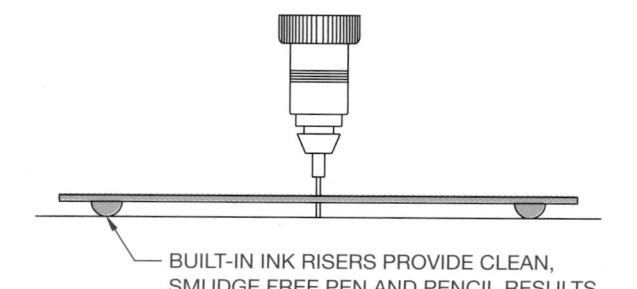

FIGURE 2.121 ■ Built-in ink risers. *Courtesy Chartpak.*

left to right allows inked lines to dry as you go along and saves valuable drafting time. Inked lines are dry when they do not appear glossy. Glossy lines are wet; keep equipment away from them. Next, ink all circles and arcs. Ink all thick horizontal and vertical lines in the same manner as thin horizontal and vertical lines. Finally, do all lettering.

When using technical pens, hold the pen perpendicular (90°) to the paper for the best results. Do not apply any pressure to the pen. Allow the pen to flow easily over the vellum or Mylar. Figure 2.122 shows proper technical pen use. If the technical pen is not held 90° to the surface, the resulting line may be fuzzy or rough on the edge. Move your technical pen at a constant speed. Do not go too fast and do not slow at the ends of lines. Following these hints will help keep your line consistent in width and the pen point from skipping.

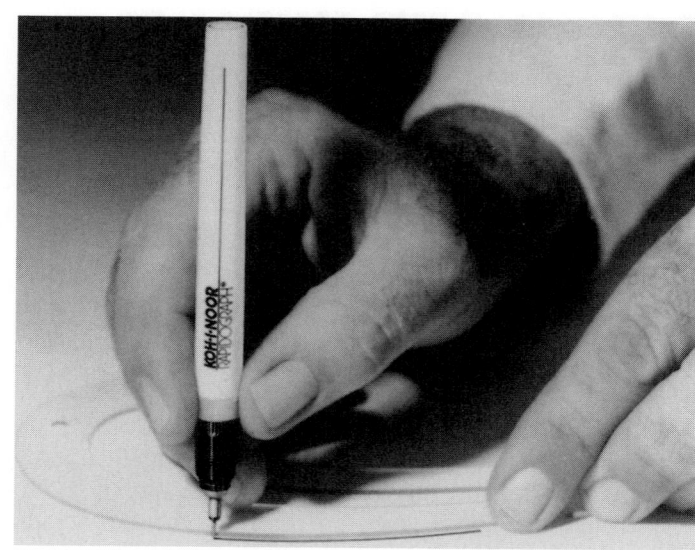

FIGURE 2.122 ■ Proper technical pen use. *Courtesy Koh-I-Noor, Inc.*

Erasing Ink

Erasing ink from vellum is possible in small areas but doing so may destroy its surface. A smooth surface cannot be inked over again satisfactorily. An electric eraser would be best used here, but overuse will burn a hole through the paper. Electric erasers are excellent tools, but be careful to touch the paper surface lightly.

To erase on polyester film, it is best to apply a little water with a felt tip or a clean cotton swab. Remove any excess water with a blotter. Allow the area to dry thoroughly. The ink remover recommended by the ink manufacturer can also be used. A polyester eraser also removes ink. Use an eraser with care so you do not destroy the matte surface, which makes further drafting in that area difficult.

Polyester Lead on Film

When pencils are used on Mylar, the recommended lead is made of polyester. The quality of a line on polyester film depends on how well the matte finish is maintained. If you use single matte film, do not try to draw on the glossy side; use polyester lead on the matte side.

Pencil Skills with Polyester Lead

■ Draw with a single line, in one direction. Tracing a line in both directions deposits a double line, and causes smearing and damage to the matte.

■ Draft with a light touch. Drafting films require up to 40 percent less pressure than other media. Smearing and embossing can be reduced with less pressure. It will take practice to use polyester lead after drawing with graphite lead.

■ The surface of your drafting board is also a factor in line quality. Use a recommended backing material on your table. Check with your local vendor.

■ Erase with a vinyl eraser. If an electric eraser is used, be very careful not to destroy the matte.

PROFESSIONAL PERSPECTIVE

Sketching

Freehand sketching is an important skill if you are a manual or a CADD drafter. Preparing a sketch before you begin a formal drawing may save many hours of work. The sketch assists you in the layout process in that it allows you to:

■ Decide how the drawing should appear when finished.

■ Decide how big to make the drawing.

■ Determine the sheet size for manual drafting or the screen limits for CADD.

■ Establish the coordinate points for the computer drawing.

A little time spent sketching and planning your work saves a lot of time in the final drafting process. Sketches are also a quick form of communication in any professional environment. You can often get your point across or communicate more effectively with a sketch.

Lettering

The appearance of manual drawings may be greatly enhanced by quality freehand lettering. An otherwise good drawing may look unprofessional with poor lettering. However, good freehand lettering does not come easily for some people; it takes a lot of practice. The only substitute for practice is an inherent talent. For some people, lettering comes naturally.

CADD lettering is a different story. Lettering on a computer is as easy as typing, and the lettering comes out fast and is perfect every time. Another exciting aspect of CADD lettering is the many styles available. There are even lettering styles available that duplicate the artistic appearance of the best freehand architectural lettering. Freehand lettering may not be important when the computer age totally takes over the drafting industry. For now, however, lettering on manual drawings is very important.

Lines

Quality line work is dependent on the opaqueness of the lines. The diazo reproduction process creates an excellent print if the light is unable to pass through the lines. If the line does not make an opaque image, you will not get a good print. Do not blame it on the print machine. You cannot make poor lines look good if they are not opaque. The job is difficult when drawing with pencil on vellum. You need the right combination of pressure, lead hardness, and skill to produce properly executed lines. This combination is different for each individual. Some people draw acceptable lines with an F lead while others are successful with a 2H lead. Also, one lead may not work well for all lines. You may need to have a few automatic pencils with different lead grades. If you are inking on polyester film, you will get good-quality lines, but you may have trouble getting used to working with the pitfalls of ink, such as smearing and error cleanup. Either manual technique takes a lot of practice.

Drafting with the computer is a different story, because all you need to do is make lines on the screen and then it is up to the plotting or printing process to reproduce a quality drawing on paper. CADD programs allow you to set line type, line thickness, and color as desired. You can also plot in different colors if you want to emphasize a part of the drawing, if a color plotter or printer is available.

WEB SITE RESEARCH

The following Web sites can provide you additional information for research or further study into topics covered in this chapter:

http://www.asme.org/asme/8.html—Find information and publications related to the American Society of Mechanical Engineers.

http://www.ansi.org—The American National Standards Institute. Information about national and international drafting standards.

http://www.adda.org—American Design Drafting Association— Drafting Reference Guide.

PULLING IT ALL TOGETHER

The drawing shown in Figure 2.123 (page 53) provides an example of lettering and line work that is commonly used in engineering drawing. The line types are labeled for your reference.

MATH APPLICATION

FRACTIONAL ARITHMETIC

The United States is one of the few countries of the world that does not commonly use a metric system of measurement. It is important for a drafter to be able to handle the arithmetic of both common fractions and decimal fractions. Determine the overall height of Figure 2.124 and the overall width of 15 of these pieces laid side to side.

The solution to overall height involves an addition of mixed numbers:

$$2\frac{1}{16}+1\frac{1}{4}+2\frac{3}{8}=2\frac{1}{16}+1\frac{4}{16}+2\frac{6}{16}=5\frac{11}{16}$$

The solution to the width of 15 pieces requires multiplication:

$$5\frac{7}{8}\times15=\frac{47}{8}\times\frac{15}{1}=\frac{705}{8}=88\frac{1}{8}"=7'4\frac{1}{8}"$$

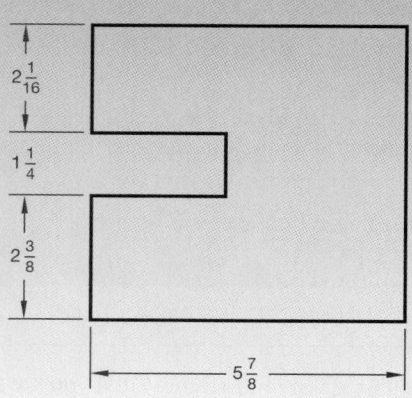

FIGURE 2.124 ■ Fractionally dimensioned part.

CHAPTER 2

Sketching Lettering, and Lines Test

DIRECTIONS

Answer the questions with short, complete statements or drawings as needed.

PART 1 • SKETCHING

1. Define sketching.
2. How are sketches useful as related to computer-aided drafting?
3. Describe the proper sketching tools.
4. When sketching, should the paper be taped down? Why or why not?
5. What kind of problem can occur if a long straight line is drawn without moving the hand?
6. What type of paper should be used for sketching?
7. Describe a method that can be used to sketch irregular shapes.
8. Define an isometric sketch.
9. What is the difference between an isometric line and a nonisometric line? An example may be used.
10. What do the use of proportions have to do with sketching techniques?
11. Why should a sketch be done quickly in most cases?
12. Why is graph paper good for sketching?
13. Briefly describe the procedure for sketching a long straight line that is fairly close to the edge of the paper.
14. How can you sketch small circles with only two pencil strokes?

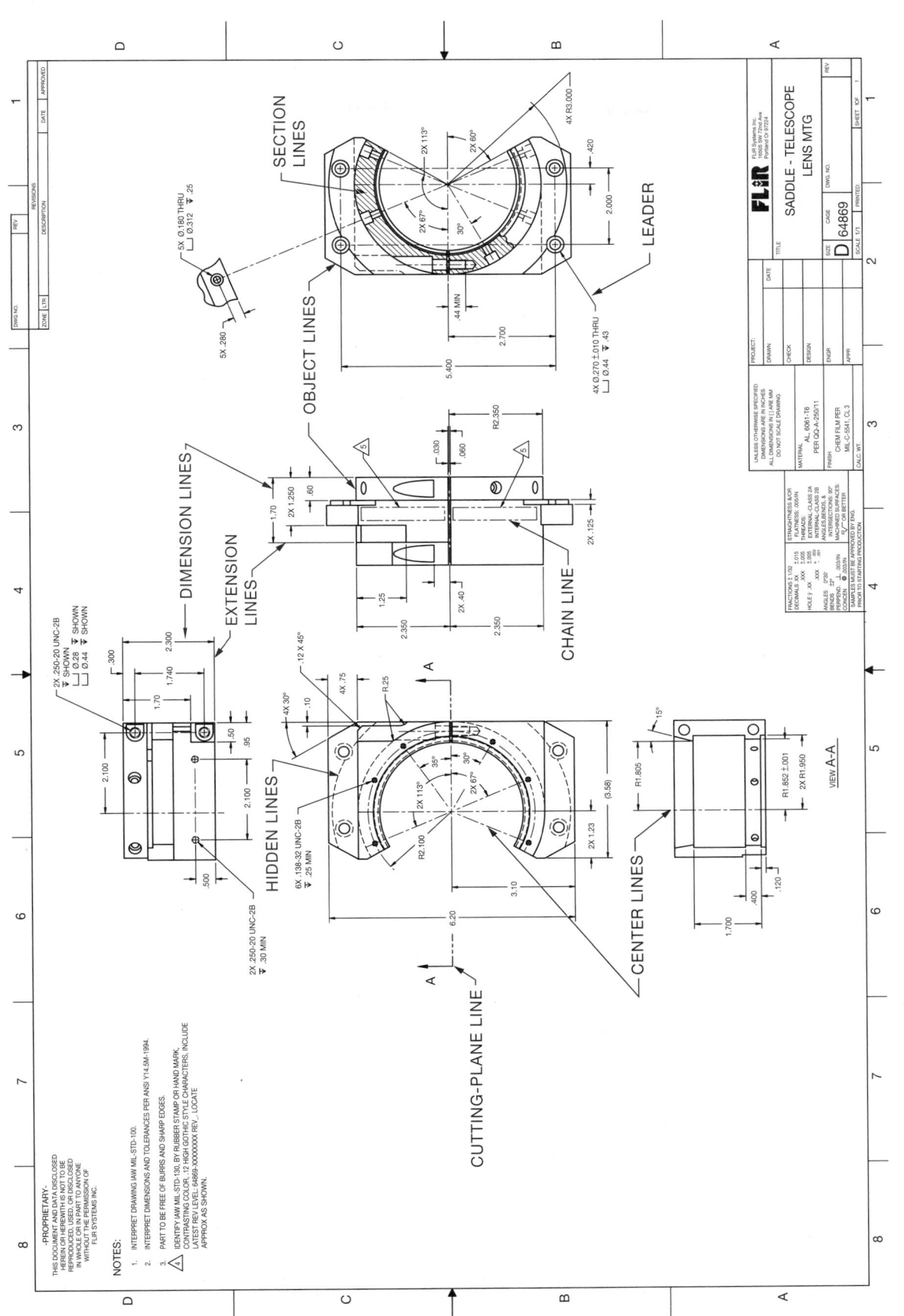

FIGURE 2.123 ■ A real-world drawing with a variety of line types identified. *Courtesy Flir Systems, Inc.*

15. Briefly explain the difference between the box and center-line methods for sketching circles.
16. Name the fastest method of sketching a circle that was described in this chapter.
17. Why is the sketching method for arcs similar to sketching circles?
18. What is the length and width of an ellipse called?
19. Define multiviews.
20. What is the difference between a multiview and an isometric view?

PART 2 • LETTERING

1. What type of lettering characters are recommended by ASME for mechanical drafting?
2. According to ASME standards, what are the minimum recommended lettering heights?
3. How should the letters within words be spaced?
4. What is the recommended minimum space between two numerals having a decimal point between them?
5. What is the recommended horizontal space between sentences?
6. Discuss the general rule that applies to the vertical spacing between lines of lettering and between individual notes on a drawing.
7. Identify the ASME document that provides the recommended lettering standards.
8. Describe guidelines.
9. Why are guidelines necessary for freehand lettering?
10. Describe the recommended leads and points used in mechanical and automatic pencils for freehand lettering.
11. Identify a method to use when lettering to help avoid smudging the drawing.
12. When lettering fractions, what is the recommended relationship of the fraction division line to the fraction numerals?
13. List two manual methods that can be used to make guidelines rapidly.
14. Why do some companies prefer the use of lettering guide templates?
15. What does the term *single stroke* mean with regard to freehand lettering?
16. What is the typical slant of inclined letters?
17. Define font.
18. When should guidelines be used for freehand lettering?

19. Give the term that refers to the spacing of letters in words and between words in sentences.
20. What is the term that refers to any value that is maintained by the computer for a command or function that has variable parameters?

PART 3 • LINES

1. Identify the ASME document that governs line standards.
2. What are construction lines and guidelines used for, and how should they be drawn?
3. Discuss line uniformity and line contrast.
4. Should there be any difference in line darkness?
5. What is the recommended thickness of object lines?
6. What do hidden lines represent on a drawing?
7. Descibe two functions that centerlines serve on a drawing.
8. Extension lines are thin lines that are used for what purpose?
9. Where should the extension lines begin in relationship to the object and end in relation to the last dimension line?
10. Describe leaders.
11. What is the correct length-to-width ratio of a properly drawn arrowhead?
12. Should arrowheads on a drawing all be the same size? Why?
13. Describe the difference between a cutting-plane and a viewing-plane line.
14. Should cutting-plane lines be drawn thick or thin?
15. Discuss the recommended spacing and angle of section lines.
16. List two uses for phantom lines.
17. What type of line is used to indicate that a portion of a surface or feature will receive a specific treatment?
18. Discuss the proper recommended technique when using a mechanical pencil with graphite lead to draw lines on vellum.
19. How will the line technique differ when using a mechanical pencil as compared to using an automatic pencil?
20. Describe the proper technique to use when drawing lines with a technical fountain pen.

PART 4 • LINE IDENTIFICATION

Given the following print (page 55), identify the lines labeled a through m.

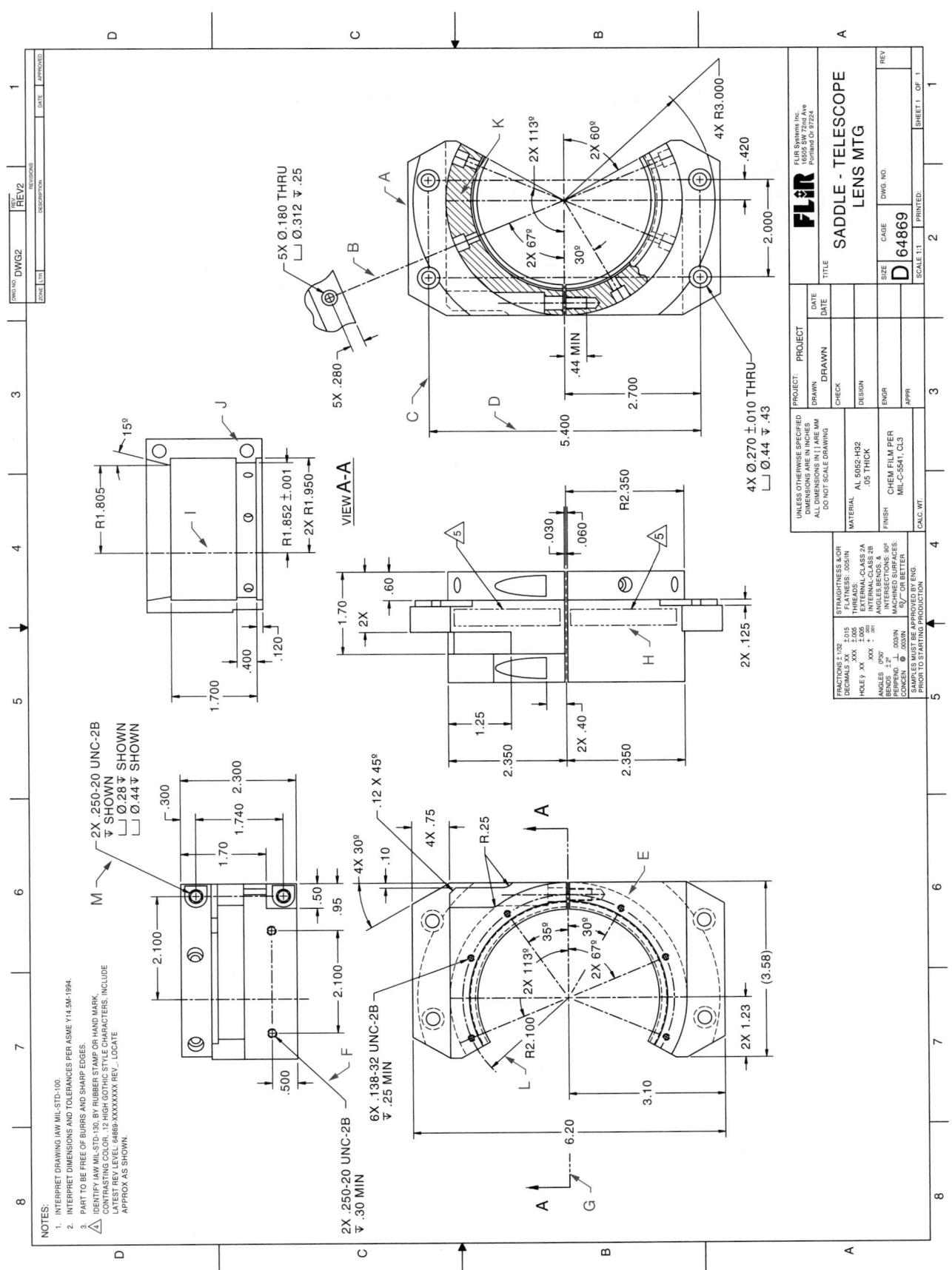

CHAPTER 2

Sketching, Lettering, and Lines Problems

DIRECTIONS

Use proper sketching materials and techniques to solve the following sketching problems, on 8½ × 11 in. bond paper or newsprint. Use very lightly sketched construction lines for all layout work. Darken the finished lines, but do not erase the layout lines.

PART 1 · SKETCHING PROBLEMS

PROBLEM 2-1 List on a separate sheet of paper the length, direction, and position of each line shown in the drawing. Remember, do not measure the lines with a scale. Example: Line 2 is the same length as line 1 and touches the top of line 1 at a 90° angle.

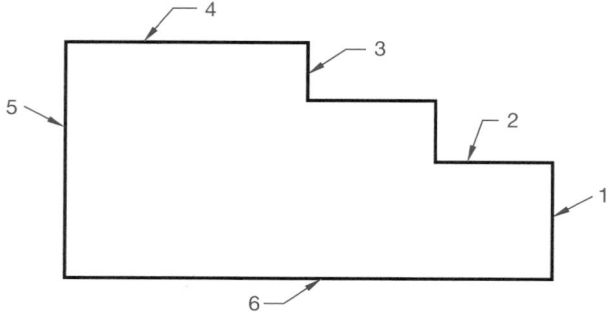

PROBLEM 2-2 Use the box, centerline, hand-compass, and trammel methods to sketch a circle with approximately a 4-in. diameter.

PROBLEM 2-3 Make a sketch of the wrench below. Use a frame of reference to make your sketch twice as big as the given sketch.

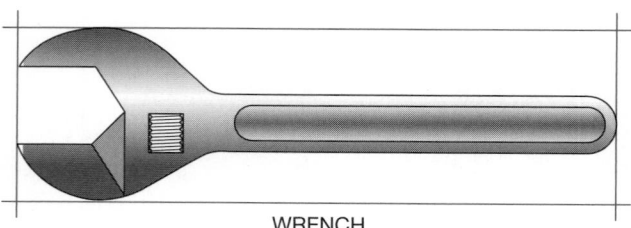

WRENCH

PROBLEM 2-4 Make a sketch of the machine screw below. Use a frame of reference to make your sketch twice as big as the given sketch.

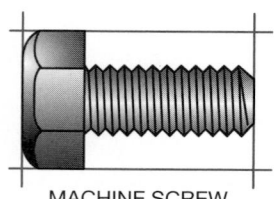

MACHINE SCREW

PROBLEM 2-5 Make a sketch of the vice below. Use a frame of reference to make your sketch twice as big as the given sketch.

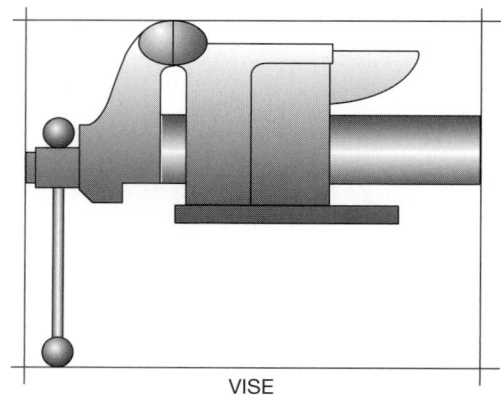

VISE

PROBLEM 2-6 Make a sketch of the patio, swimming pool, and spa below. Use a frame of reference to make your sketch twice as big as the given sketch.

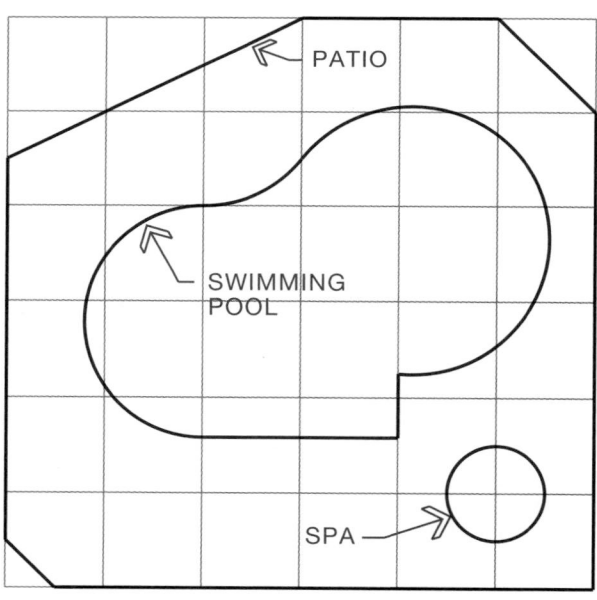

PROBLEM 2-7 Find a stapler, tape dispenser, or coffee cup and sketch a two-dimensional frontal view using the block technique. Do not measure the object. Use the measurement-line method to approximate proper proportions.

PROBLEM 2-8 Find an object with an irregular shape and sketch a two-dimensional view using the regular grid method. Sketch the correct proportions of the object without measuring.

PROBLEM 2-9 Transfer the given top and right-side views of the house below to another sheet using the sketching methods that you learned in this chapter. Make your sketch twice the size of the given example. Use projection methods to establish and sketch the missing front view.

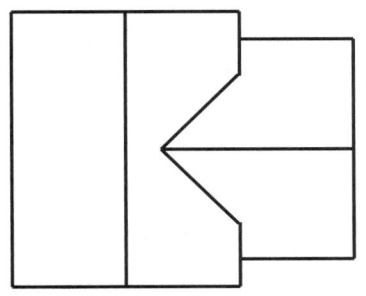

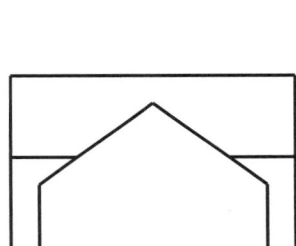

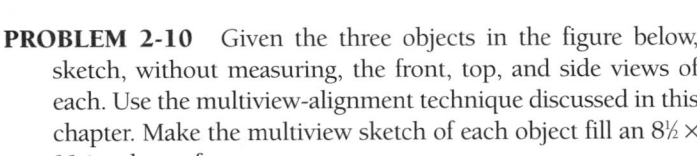

START VIEW HERE

PROBLEM 2-10 Given the three objects in the figure below, sketch, without measuring, the front, top, and side views of each. Use the multiview-alignment technique discussed in this chapter. Make the multiview sketch of each object fill an 8½ × 11 in. sheet of paper.

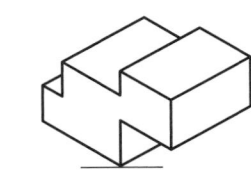

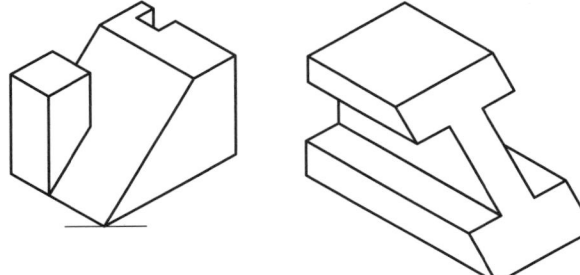

PROBLEM 2-11 Given the three objects below, sketch an isometric view of each at the adjacent location marked A. Transfer your sketch to fill an 8½ × 11 in. sheet of paper.

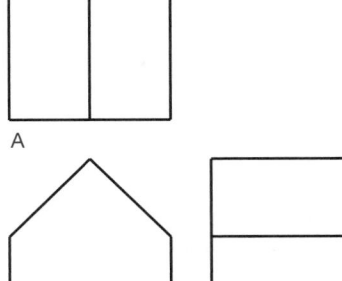

A

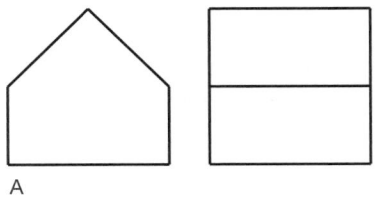

A

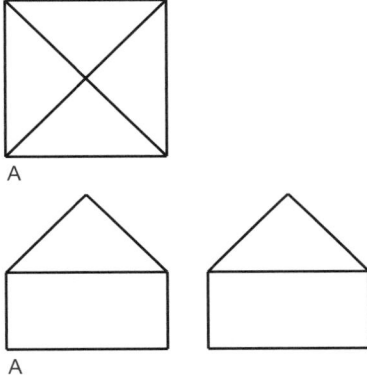

A

A

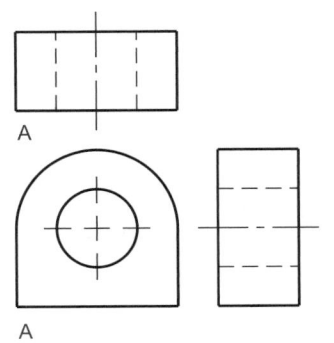

A

A

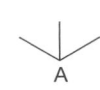

A

PROBLEM 2-12 Using the same object selected for Problem 2-7, sketch an isometric representation. Do not measure the object but use the measurement-line technique to approximate proportions.

PART 2 · LETTERING PROBLEMS

DIRECTIONS

Use vertical freehand Gothic lettering, or computer-aided drafting as required by the specific instructions for each assignment. Do all freehand lettering on an A-size drawing sheet with .125-in. letters. Space lines of lettering .125 in. apart. Use guidelines for all freehand lettering.

PROBLEM 2-13 Use vertical Gothic freehand lettering to letter the following statement:

THE QUALITY OF THE FREEHAND LETTERING GREATLY AFFECTS THE APPEARANCE OF THE ENTIRE DRAWING. DRAFTERS LETTER IN PENCIL OR INK. PROPER FREEHAND LETTERING IS DONE WITH A SOFT, SLIGHTLY ROUNDED POINT IN A MECHANICAL DRAFTING PENCIL, OR A 0.5 MM AUTOMATIC PENCIL WITH 2H, H, OR F GRADE LEAD DEPENDING UPON INDIVIDUAL PRESSURE. LETTERING IS DONE BETWEEN VERY LIGHTLY DRAWN GUIDELINES. THESE GUIDELINES ARE DRAWN PARALLEL AND SPACED EQUAL TO THE HEIGHTS OF THE LETTERS. GUIDELINES HELP TO KEEP YOUR LETTERING UNIFORM IN HEIGHT. LETTERING STYLES MAY VARY BETWEEN COMPANIES. SOME COMPANIES REQUIRE THE USE OF LETTERING DEVICES.

PROBLEM 2-14 Use vertical Gothic freehand lettering to letter the following statement:

MOST MECHANICAL DRAFTING THAT DOES NOT USE CAD DOES USE VERTICAL FREEHAND LETTERING. THE QUALITY OF THE FREEHAND LETTERING GREATLY AFFECTS THE APPEARANCE OF THE ENTIRE DRAWING. MANY MECHANICAL DRAFTING TECHNICIANS USE FREEHAND LETTERING WITH PENCIL ON VELLUM OR WITH POLYESTER LEAD ON MYLAR. LETTERING IS COMMONLY DONE WITH A SOFT, SLIGHTLY ROUNDED LEAD IN A MECHANICAL PENCIL OR A 0.5 MM LEAD IN AN AUTOMATIC PENCIL. LETTERING IS DONE BETWEEN VERY LIGHTLY DRAWN GUIDELINES. GUIDELINES ARE SPACED PARALLEL AT A DISTANCE EQUAL TO THE HEIGHT OF THE LETTERS. GUIDELINES ARE REQUIRED TO HELP KEEP ALL LETTERS THE SAME UNIFORM HEIGHT.

PROBLEM 2-15 Use vertical Gothic freehand lettering and a CADD system to letter the following notes:

1. INTERPRET DIMENSIONS AND TOLERANCES PER ASME Y14.5M-1994.
2. UNLESS OTHERWISE SPECIFIED, ALL DIMENSIONS ARE IN MILLIMETERS.
3. REMOVE ALL BURRS AND SHARP EDGES.
4. ALL FILLETS AND ROUNDS R6.
5. CASEHARDEN 62 ROCKWELL C SCALE.
6. AREAS WHERE MATERIAL HAS BEEN REMOVED SHALL HAVE SMOOTH TRANSITIONS AND BE FREE OF SCRATCHES, GRIND MARKS, AND BURNS.
7. FINISH BLACK OXIDE.
8. PART TO BE CLEAN AND FREE OF FOREIGN DEBRIS. COMPARE THE DIFFERENCE BETWEEN FREEHAND AND CADD LETTERING AS TO SPEED AND APPEARANCE.

PART 3 · LINE PROBLEMS

DIRECTIONS

1. Using the selected engineer's layout as a guide only, make an original drawing using manual or CADD as required by your course objectives. Select an appropriate scale. Draw *only* the object lines, centerlines, hidden lines, and phantom lines as appropriate for each problem. Do not dimension. Keep in mind that the engineer's sketches are rough and not meant for tracing.
2. Use construction lines (very lightly drawn) to prepare the entire drawing. When satisfied with the product, darken the drawing using proper line technique.
3. Complete the title block.
   a. The title of the drawing is given.
   b. The material the part is made of is given.
   c. The drawing number is the same as the problem number.
   d. Specify the scale.

PROBLEM 2-16 Object lines (in.)
Part Name: Plate
Material: .25-in.-thick Mild Steel

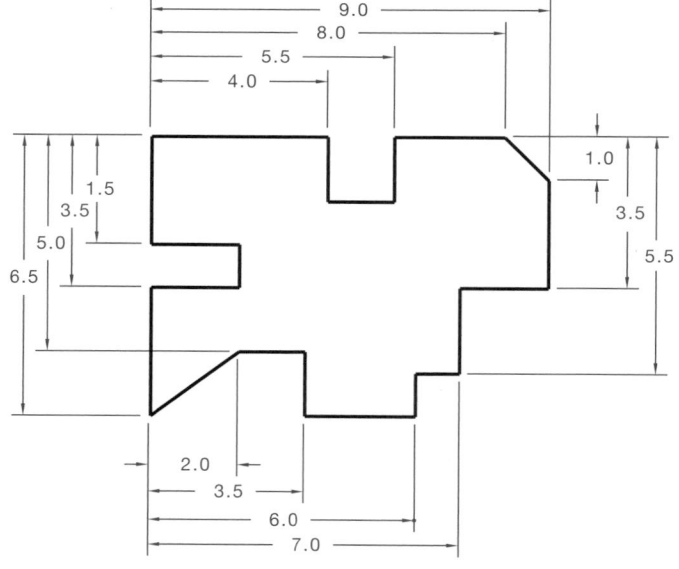

PROBLEM 2-17   Straight object lines only (in.)

Part Name: Milk Stencil

Material: .015-in.-thick wax-coated cardboard

Used as a stencil to spray paint identification on crates of milk.

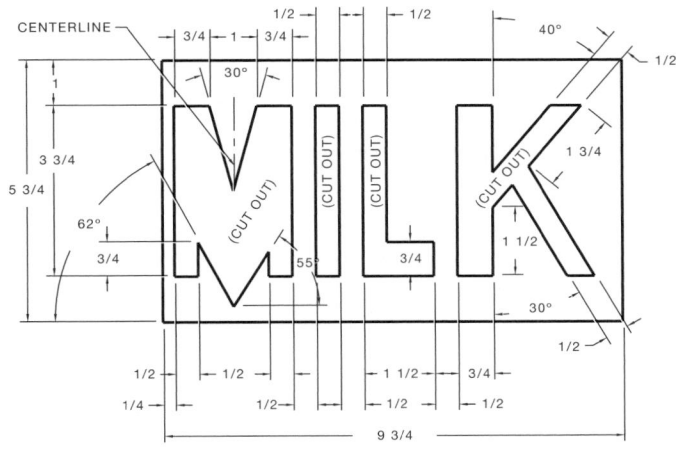

PROBLEM 2-18   Circle and arc object lines and centerlines (in.)

Part Name: Connector

Material: .25-in.-thick Steel

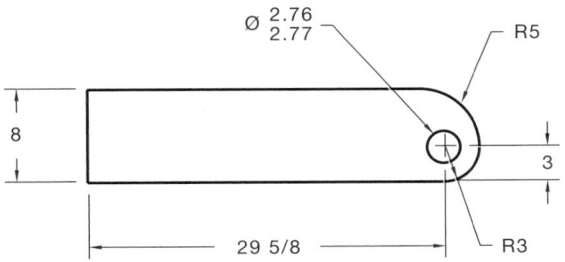

PROBLEM 2-19   Arcs, object lines, and centerlines (in.)

Part Name: Latch

Material: .25-in.-thick Mild Steel

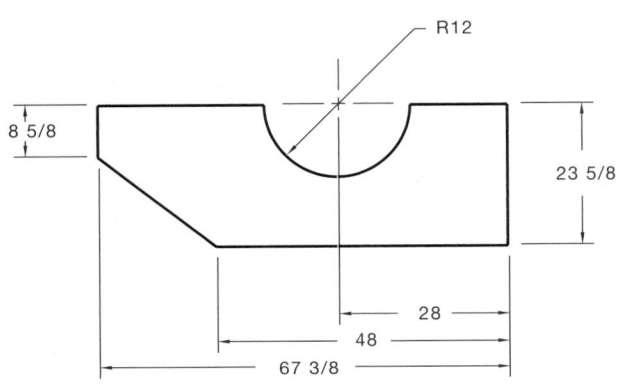

PROBLEM 2-20   Straight line and arc object lines and centerlines (metric)

Part Name: Plate

Material: 10-mm-thick HC-112

Used as a spacer to separate electronic components in a computer chasis.

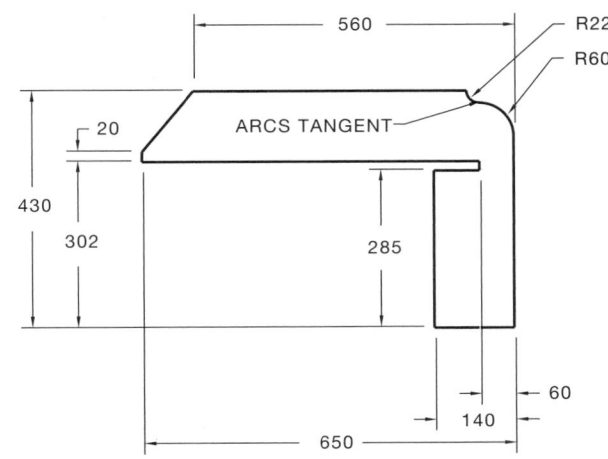

PROBLEM 2-21   Circle and arc object lines and centerlines (in.)

Part Name: Bogie Lock

Material: .25-in.-thick Mild Steel

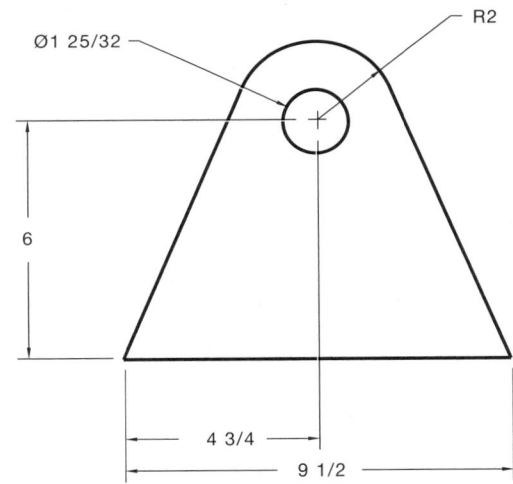

PROBLEM 2-22   Circle, object lines, and centerlines (in.)
Part Name: Stove Back
Material: .25-in.-thick Mild Steel

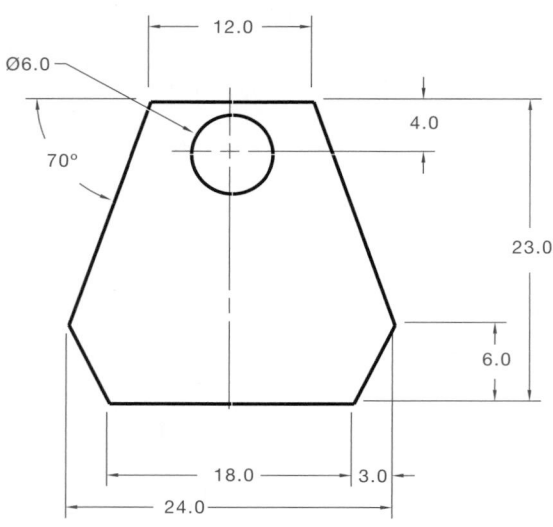

PROBLEM 2-23   Circle and arc object lines and centerlines (in.)
Part Name: Bogie Lock
Material: .25-in.-thick Mild Steel

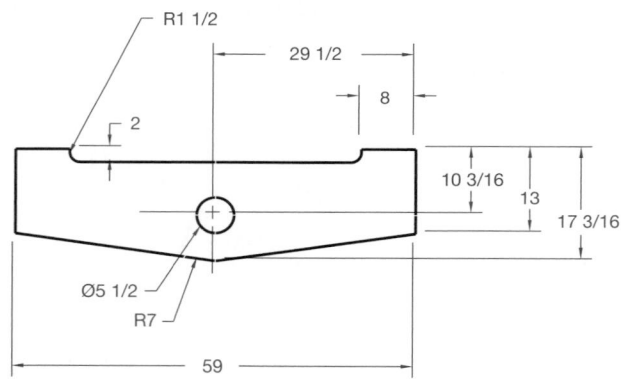

PROBLEM 2-24   Circle and arc object lines and centerlines (metric)
Part Name: T-Slot Cleaner
Material: 6-mm-thick Cold Rolled Steel

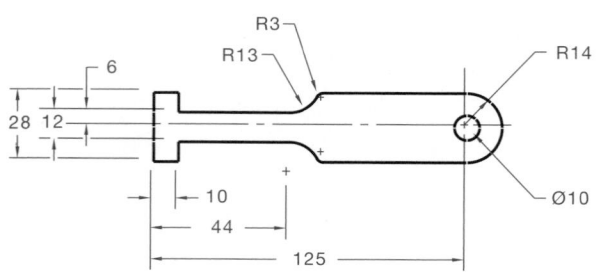

PROBLEM 2-25   Circle and arc object lines and centerlines (in.)
Part Name: T-Slot Cleaner (inch)
Material: .25-in.-thick Cold Rolled Steel

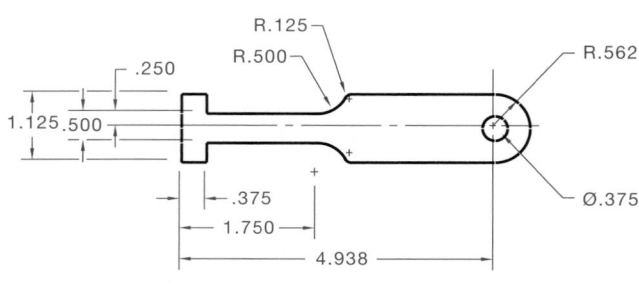

PROBLEM 2-26   Circle and arc object lines and centerlines (in.)
Part Name: Plate
Material: .125-in.-thick Aluminum

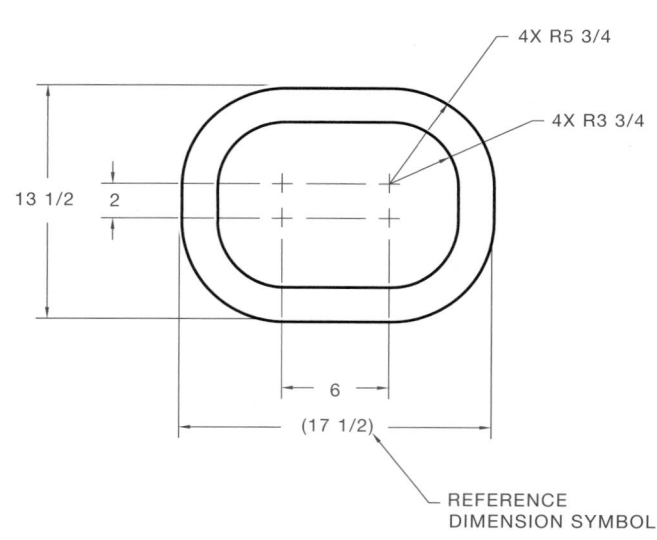

PROBLEM 2-27 **Arcs and centerlines (in.)**
Part Name: Flat Spring
Material: 26 gauge × .50 SAE 1085

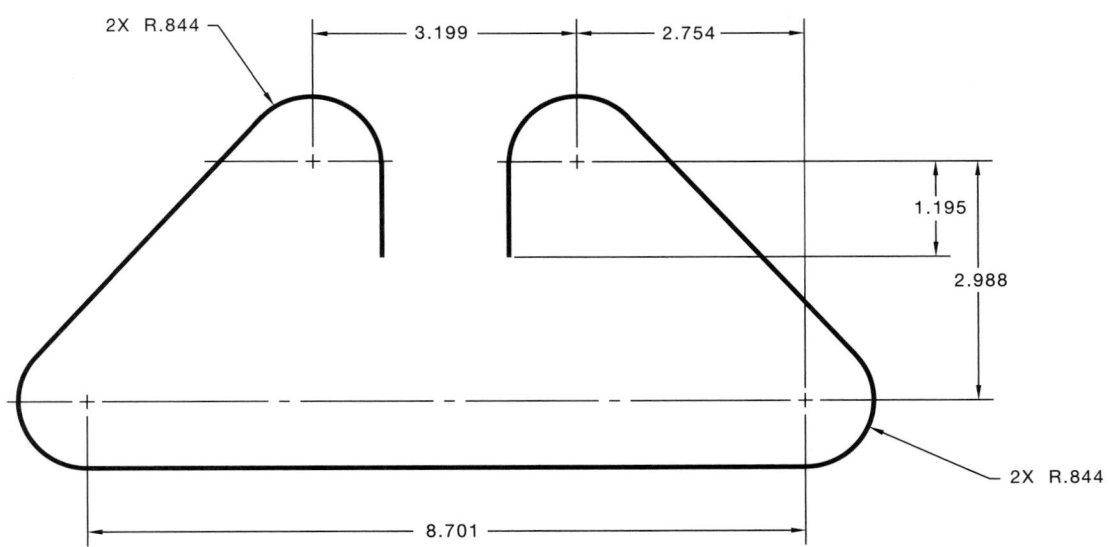

PROBLEM 2-28 **Circle and arc object lines and centerlines (in.)**
Part Name: Gasket
Material: .062-in.-thick Neoprene

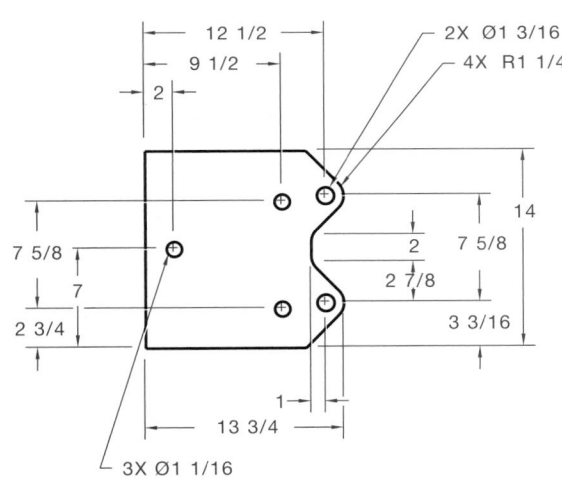

PROBLEM 2-29 **Arcs, circles, and centerlines (in.)**
Part Name: Bracket
Material: Stainless Steel

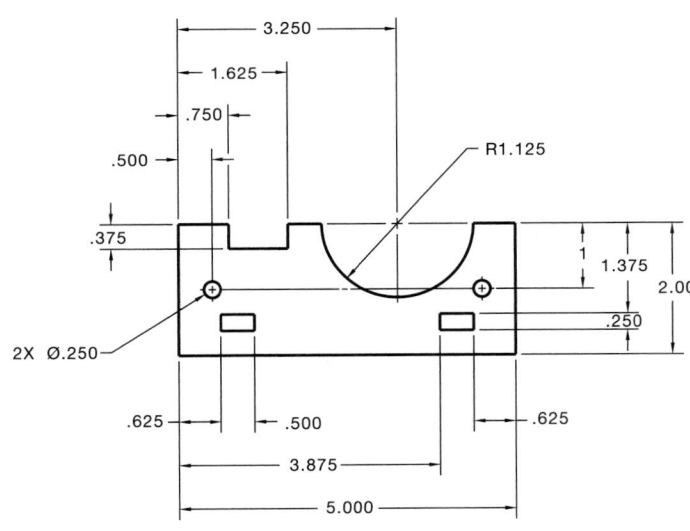

PROBLEM 2-30 Arcs, circles, and centerlines (in.)
Part Name: Gasket
Material: Brass

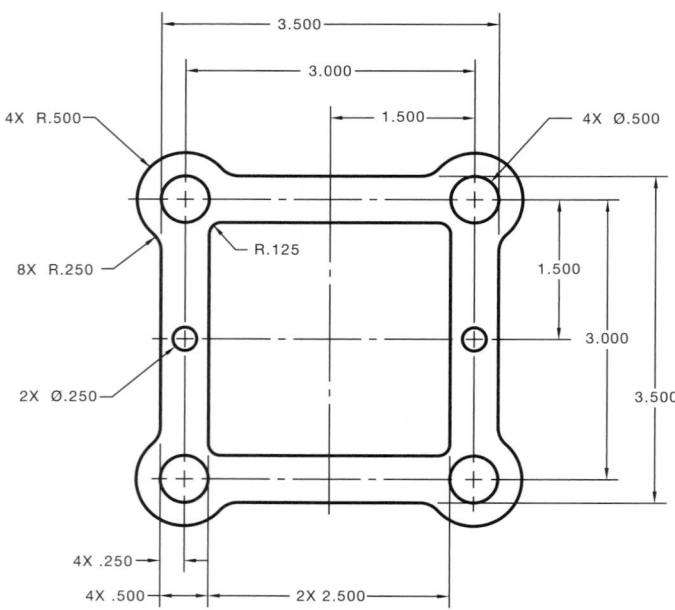

PROBLEM 2-31 Arcs and centerlines (in.)
Part Name: Clip
Material: SAE 3140

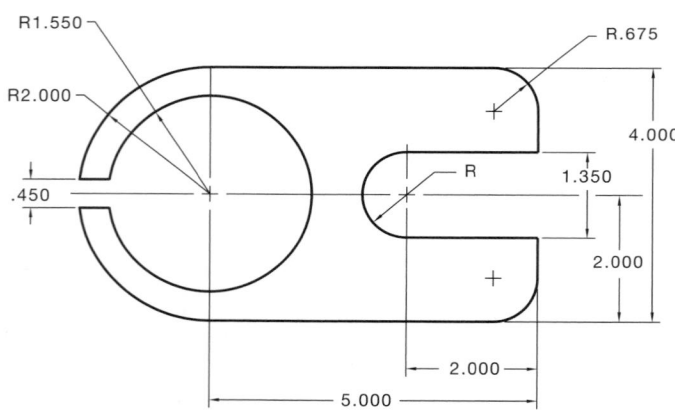

PROBLEM 2-32 Arcs, circles, and centerlines (in.)
Part Name: Bracket
Material: Mild Steel (MS)

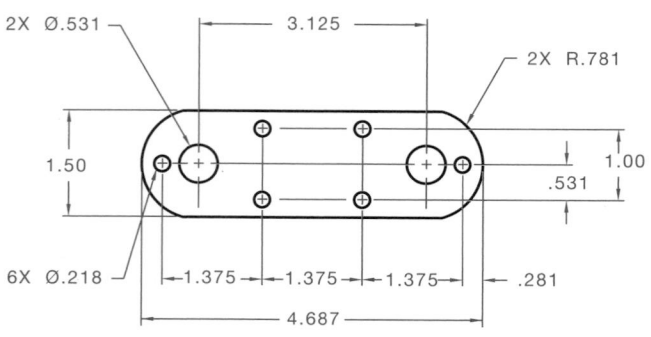

PROBLEM 2-33 Circle and arc object lines and centerlines (in.)
Part Name: Gasket
Material: .062-in.-thick Cork
Gasket for hydraulic pump.

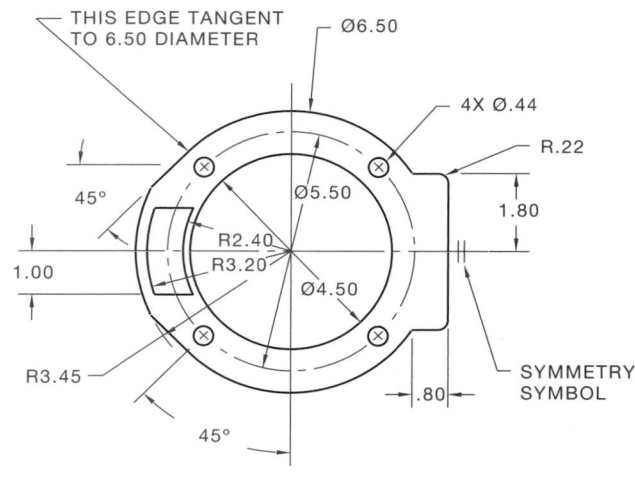

PROBLEM 2-34 Arc object lines, centerlines, phantom lines, and leader lines (in.)
Part Name: Bogie Lock
Material: .25-in.-thick Mild Steel
Connect the leader lines and place the notes on the drawing.

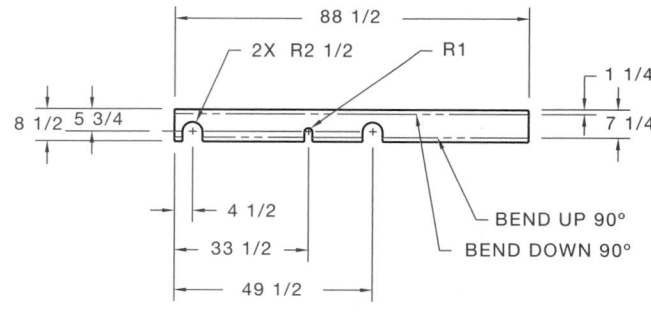

PROBLEM 2-35 Arcs, centerlines, and hidden lines (in.)
Part Name: Tube
Material: Copper

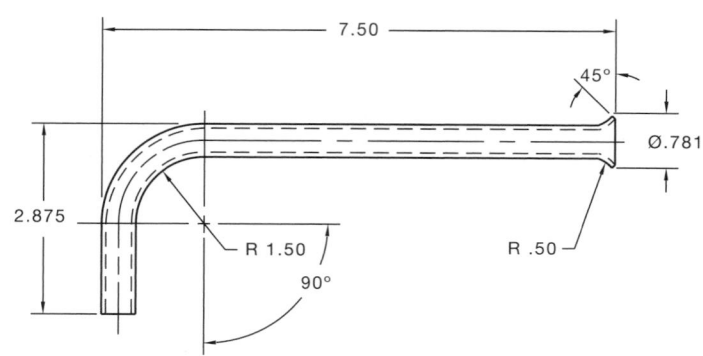

PROBLEM 2-36 Arcs, centerlines, and hidden lines (in.)
Part Name: Pin
Material: Phosphor Bronze

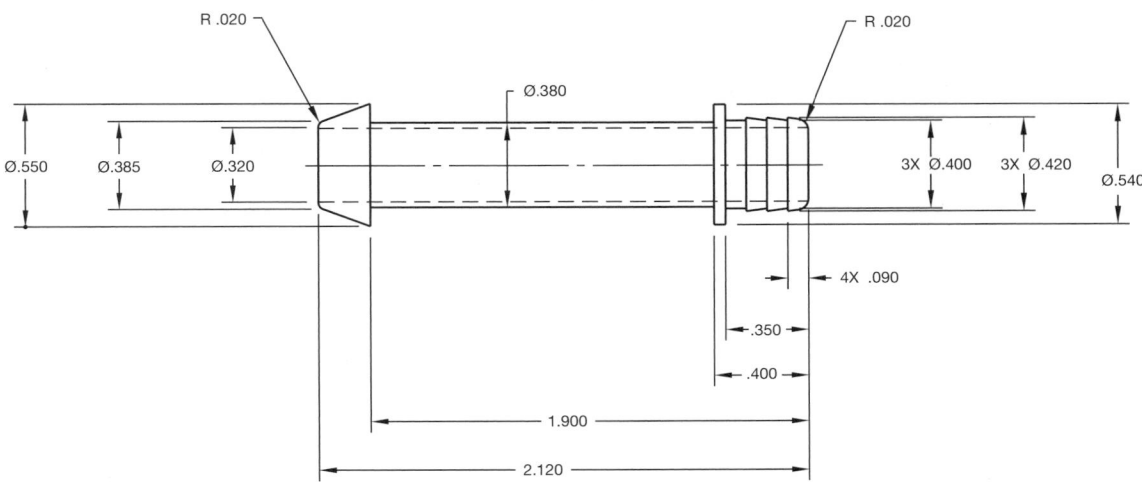

PROBLEM 2-37 Arcs, centerlines, and hidden lines (in.)
Part Name: Pivot Arm
Material: Aluminum

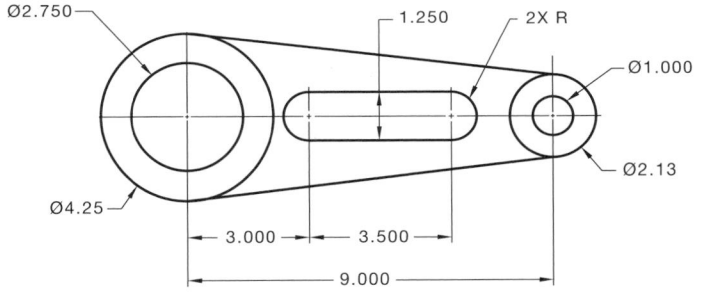

PROBLEM 2-38 Multiviews (2 views), arcs, centerlines, and hidden lines (in.)
Part Name: Bracket
Material: Cast Iron

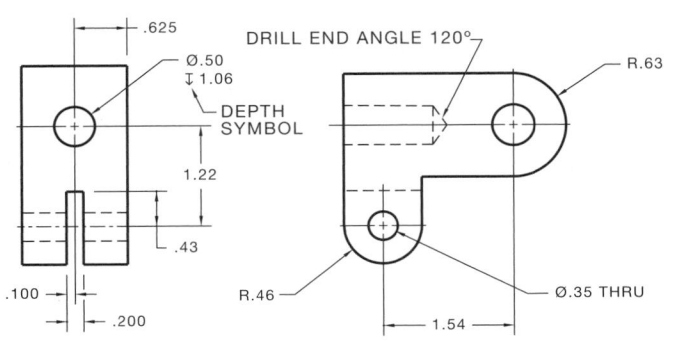

PROBLEM 2-39 Multiviews (2 views), object lines and hidden lines (in.)
Part Name: V-block
Material: 4.00-in.-thick Mild Steel

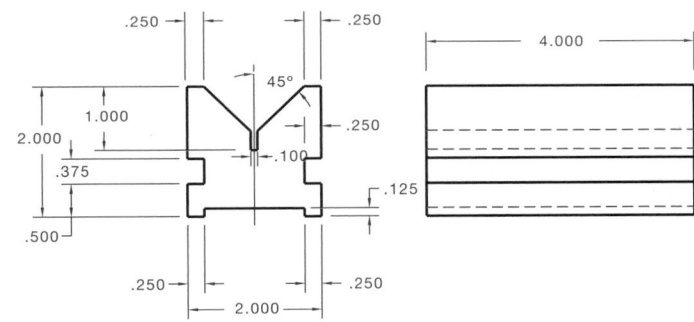

PART 4 • DRAWING WITH COORDINATES

DIRECTIONS

The problems in this section are representative of the skills required when using a computer-aided drafting system. The use of the Cartesian coordinate system is stressed because of its universal use in computer-aided drafting. Problems 2-40, 2-41, and 2-42 test this skill of locating points.

PROBLEM 2-40   You will need graph paper with 10 squares per inch for this problem. The 11 points given below each have an X and Y coordinate. Plot each of the points on the grid paper using Cartesian coordinates and connect them with a straight line. The origin of your coordinate system does not have to be the extreme lower left corner of the paper.

POINT	X	Y
1	0	1.5
2	3	1.5
3	3	0
4	7.12	0
5	7.12	2.25
6	6.18	2.25
7	6.18	.75
8	3.93	.75
9	3.93	2.25
10	0	2.25
11	0	1.5

PROBLEM 2-41   The two objects given below (objects A and B) are similar to the previous problem, but each has two views, front and right-side. Use 10-squares-to-the-inch graph paper and follow the same procedure you used in Problem 2-40. The lower left corner of each view should be the origin point for that view. Each view should have its own separate Cartesian coordinate system.

		FRONT		RIGHT SIDE	
	POINT	X	Y	X	Y
OBJECT A	1	0	0	0	0
	2	3	0	2.75	0
	3	3	3.6	2.75	3.6
	4	0	3.6	0	3.6
	5	0	0	0	0
OBJECT B	1	0	0	0	0
	2	4	0	3.8	0
	3	4	2.1	3.8	4.2
	4	3	2.1	0	4.2
	5	3	4.2	0	2.1
	6	1	4.2	0	0
					Move to:
	7	1	2.1	0	2.1
	8	0	2.1	3.8	2.1
	9	0	0		

PROBLEM 2-42   The following problem contains drawing data given in both incremental and polar coordinates. The points should be plotted and drawn on 10-squares-to-the-inch grid paper.

A. INCREMENTAL COORDINATES			B. POLAR COORDINATES		
Point	X	Y	Point	Angle	Radius
1	0	0	1	0	0
2	3	0	2	0	1.25
3	0	2	3	90	1
4	–1	0	4	0	1.25
5	0	1	5	270	1
6	–1	0	6	0	1.50
7	0	–2	7	90	2
8	–1	0	8	180	2.75
9	0	–1	9	90	1
			10	180	1.75
			11	270	3

MATH PROBLEMS

PROBLEM 2-43   For the illustration of the block shown:

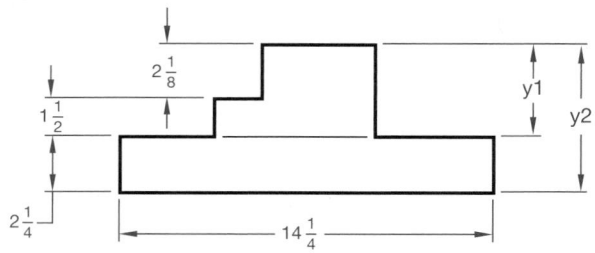

a. Determine dimension y1.
b. Determine dimension y2.
c. Determine the overall width of 12 of the blocks laid side to side.
d. Convert each of the four given dimensions of the block to decimal fractions.

PROBLEM 2-44   Twenty-five pieces of metal are needed for a job. Each piece is to be $3^{3}/_{32}''$ in length. Disregarding the cutting loss, what length of stock is needed?

PROBLEM 2-45   A piece of stock $25^{7}/_{8}''$ long is to be cut into five equal lengths. Allowing $^{1}/_{16}''$ lost per cut, what will be the length of each of the five pieces?

PROBLEM 2-46   A drawing shows a dimension of 4.1875''. Convert this decimal fraction to a common fraction.

PROBLEM 2-47   Convert each decimal fraction to a common fraction:
a. 2.5''
b. 0.125''
c. 14.4375''
d. 5.75''

Drafting Equipment, Media, and Reproduction Methods

OBJECTIVES

After completing this chapter, you will:

■ Describe and demonstrate the use of various manual drafting tools and equipment.

■ Read engineer's, architect's, and metric scales, and drafting machine verniers.

■ Discuss and use drafting media, sheet sizes, and title block information.

■ Explain common reproduction methods.

INTRODUCTION

This chapter discusses and demonstrates the type of equipment that is used for manual drafting. *Manual drafting* is the type of drafting that is done by hand with pencil or ink on a media such as paper or polyester film using drafting instruments and equipment. Manual drafting is also typically called *hand drafting*. Manual drafting is rapidly being replaced by computer-aided design and drafting (CADD) in industry. Some companies use CADD and have manual drafting tables available for use as shown in Figure 3.1. CADD equipment and the reasons for the evolution of drafting from manual to CADD are discussed in detail in Chapter 4 of this textbook. Some people say that CADD will completely take the place of manual drafting someday, while others say that manual drafting may always be used for some applications, such as one-of-a-kind products. Only the future will tell the true story; however, it is true that CADD has revolutionized the way design and drafting is done, and there are constantly new advances made in this indus-try. It does not matter if you are using manual drafting or CADD, you still need to know the basics of drafting. These basics are founded on national and industry-specific standards. As you learned in Chapter 2, the use of sketching and the proper presentation of lines and lettering are very important when using manual drafting or CADD.

DRAFTING EQUIPMENT

Drafting tools and equipment are available from a number of vendors that sell professional drafting supplies. For accuracy and long life, always purchase high-quality equipment. Local vendors can be found by looking in the yellow pages of your telephone book under headings such as: Drafting Room Equipment & Supplies, Blue-printing, Architects' Supplies, Engineering Equipment & Supplies, or Artists' Materials & Supplies.

Drafting supplies and equipment can be purchased in a kit, or items can be bought individually. Whether equipment is purchased in a kit or by the individual tool, the items that are normally needed include the following:

■ One 0.3 mm automatic drafting pencil with 4H, 2H, and H leads.

■ One 0.5 mm automatic drafting pencil with 4H, 2H, H, and F leads.

■ One 0.7 mm automatic drafting pencil with 2H, H, and F leads.

■ One 0.9 mm automatic drafting pencil with H, F, and HB leads. Drafters may elect to purchase two or more pencils and use a different grade of lead in each. Doing so reduces the need to constantly change leads. Some drafters use a light-blue lead for layout work. A combination of 0.5, 0.7, and 0.9 mm pencils and leads is good for sketching and other re-lated activities if you are doing computer-aided drafting.

■ 6-in. bow compass.

■ Dividers.

FIGURE 3.1 ■ An architectural office with computer workstations and a manual drafting table in use. *Courtesy of Hewlett-Packard Company.*

- Eraser. Select an eraser that is recommended for drafting with pencil on paper.
- Erasing shield.
- 8-in. 30°–60° triangle.
- 8-in. 45° triangle.
- Irregular curve.
- Scales:
    1. Triangular architect's scale.
    2. Triangular civil engineer's scale.
    3. Triangular metric scale.
- Drafting tape.
- Circle template (small holes).
- Circle template (large holes).
- Lettering guide (optional).
- Arrowhead template (optional).
- Sandpaper sharpening pad.
- Dusting brush.

DRAFTING FURNITURE

Tables

There is a large variety of drafting tables available, ranging from economical models to complete professional workstations. Drafting tables are generally sized by the dimensions of their tops. Standard tabletop sizes range from 24 × 36 in. to 42 × 84 in.

The features to look for in a good-quality, professional table include:

- One-hand tilt control.
- One-hand or foot height control.
- The ability to position the board vertically for use with the track drafting machine.
- An electrical outlet.
- A drawer for tools and/or drawings.

Some manufacturers ship tables with tops that are ready to draw on. Others ship tables with tops of steel or basswood that need to be covered. Basswood can be a drawing surface; however, most offices commonly cover drafting-table tops with smooth, specially designed surfaces. The material, usually vinyl, provides the proper density for effective use under normal drafting conditions. After compass or divider points have pierced the surface, the small holes close and so provide a smooth surface for continued use. Drafting tape is commonly used to adhere drawings to the tabletop, although some drafting tables have magnetized tops and use magnetic strips to attach drawings.

Chairs

Better drafting chairs have the following characteristics:
- Padded or contoured seat design.

- Height adjustment.
- Foot rest.
- Fabric that allows air to circulate.
- Sturdy construction.

DRAFTING PENCILS, LEADS, AND SHARPENERS

Mechanical Pencils or Lead Holders

The term *mechanical pencil,* also referred to as *lead holder,* is applied to a pencil that requires a piece of lead to be manually inserted; by some physical action such as twist, push, or pull, the lead is mechanically, or semiautomatically, advanced to the tip. These have been replaced by automatic pencils for most use.

Leads for mechanical pencils are bought separately. Leads are graded by hardness and designated by a number and letter, or one or two letters. The designation is usually found along or on one end of the lead. Always sharpen the end opposite the designation. Figure 3.2a shows a mechanical pencil.

Automatic Pencils

The term *automatic pencil* refers to a pencil with a lead chamber that, at the push of a button or tab, advances the lead from the chamber to the writing tip and, when a new piece of lead is needed, advances the new piece to the tip. Automatic pencils are designed to hold leads of one width so you do not need to sharpen the lead. These pencils are available in several different lead sizes. Drafters have several automatic pencils. Each pencil has a different grade of lead hardness and is used for a specific technique. (See Figure 3.2b.)

Lead Grades

The leads that you select for line work and lettering depend on the amount of pressure you apply and other technique factors. Experiment until you identify the leads that give the best line quality. Leads commonly used for thick lines range from 2H to F, while

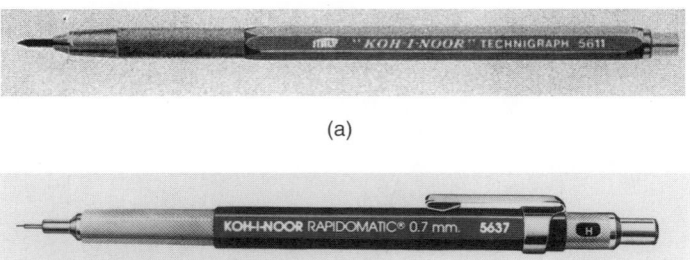

(a)

(b)

FIGURE 3.2 ■ (a) Mechanical pencil, also known as lead holder. *Courtesy Koh-I-Noor, Inc.* (b) An automatic pencil. Lead widths are 0.3, 0.5, 0.7, and 0.9 mm. *Courtesy Koh-I-Noor, Inc.*

9H 8H 7H 6H 5H 4H	3H 2H H F HB B	2B 3B 4B 5B 6B 7B
HARD	MEDIUM	SOFT
4H AND 6H ARE COMMONLY USED FOR CONSTRUCTION AND LETTERING GUIDELINES.	H AND 2H ARE COMMON LEAD GRADES USED FOR LINE WORK. H AND F ARE USED FOR LETTERING AND SKETCHING.	THESE GRADES ARE FOR ART-WORK. THEY ARE TOO SOFT TO KEEP A SHARP POINT, AND THEY SMUDGE EASILY.

FIGURE 3.3 ■ ■ The range of lead grades.

leads for thin lines range from 4H to H, depending on individual preference. Construction lines for layout and guidelines are very lightly drawn with a 6H or 4H lead. Figure 3.3 shows the different lead grades.

Polyester and Special Leads

Polyester leads, also known as *plastic leads,* are for drawing on *polyester drafting film,* often called by its trade name, *Mylar*®. Plastic leads come in grades equivalent to F, 2H, 4H, and 5H and are usually labeled with a prefix and number. Pentel™, for example, uses P1 and P2. Some companies make a combination lead for use on both vellum and polyester film.

Colored leads have special uses. Red lead is commonly used for corrections. Red prints as a black line. Blue leads can be used on the original drawing by the supervisor to tell the drafter the corrections that need to be made. Blue lead will not show on the print when the original drawing with blue lines on it is run through a diazo machine. Some drafters use light-blue lead for all layout work and guidelines because it will not reproduce on a diazo printer; however, it may show on a photocopier.

Basic Pencil Technique

Automatic pencils do not require rotation, although some drafters feel rotating the pencil makes darker lines. The automatic pencil should be held near vertical. The full surface of the lead is used when the pencil is in a vertical position. Provide enough pressure and go over each line enough times to make a line dark and crisp. Take care not to make it too thick. Figure 3.4 shows some basic pencil motions.

Sanding Block

Especially useful for sharpening compass lead, this is one of the simplest devices used for sharpening. Sandpaper stapled to a wooden paddle is called a sandpaper, or sanding, block. Plastic lead fills sandpaper rapidly so several sheets are needed as compared to graphite lead. To avoid smudging your drawing, use the sanding block away from your drawing table and dispose of the graphite carefully.

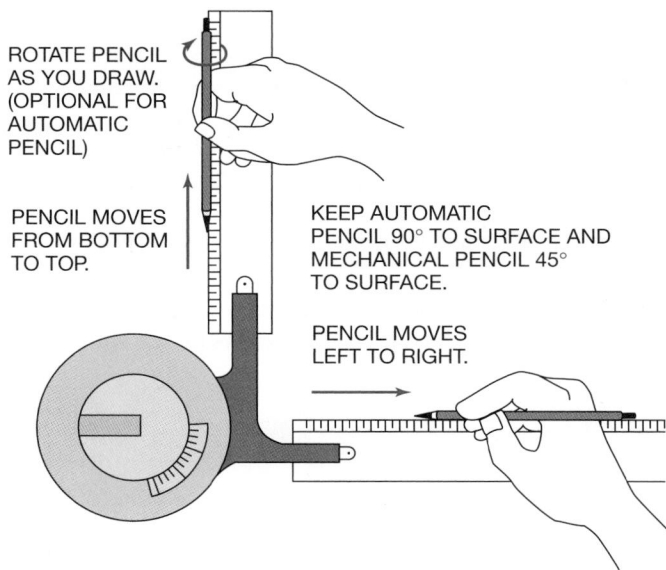

FIGURE 3.4 ■ Basic pencil motions. Move the pencil in the opposite direction if you are left-handed.

Pocket Pointer

A portable sharpener for mechanical pencils is the pocket pointer. This pointer contains blades that sharpen lead to a conical point. The pocket pointer works with either graphite or plastic lead.

Cutting Wheel Pointer

The best type of conical-point sharpener for mechanical pencil holders is a mechanical lead pointer with a tool-steel cutting wheel. Use the slots provided in the top to expose the right length of lead to get a sharp or slightly dull point. The slightly dull point is used for lettering. This sharpener works on graphite or polyester leads equally well.

TECHNICAL PENS AND ACCESSORIES
Technical Pens

Also known as technical fountain pens, technical pens have improved in quality and ability to produce excellent inked lines. These pens function on a capillary action where a needle acts as a valve to allow ink to flow from a storage cylinder through a small tube, which is designed to meter the ink so a specific line width is created. Technical pens may be purchased individually or in sets. The different tip sizes used to make various line widths range from a narrow number 6×0 (.005 in./0.13 mm) to a wide number 7 (.079 in./2 mm). Figure 3.5 shows a comparison of some of the different line widths available with technical pens.

In addition to having the advantage of a constant line width, technical pens have a reservoir that allows you to make inked lines for a long period of time before ink must be added. Technical pens may be used with templates to make circles, arcs, and symbols.

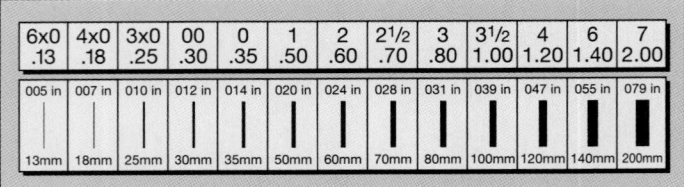

6x0 .13	4x0 .18	3x0 .25	00 .30	0 .35	1 .50	2 .60	2½ .70	3 .80	3½ 1.00	4 1.20	6 1.40	7 2.00
005 in	007 in	010 in	012 in	014 in	020 in	024 in	028 in	031 in	039 in	047 in	055 in	079 in
13mm	18mm	25mm	30mm	35mm	50mm	60mm	70mm	80mm	100mm	120mm	140mm	200mm

FIGURE 3.5 ■ Technical pen line widths. *Courtesy Koh-I-Noor, Inc.*

Compass adapters to hold technical pens are also available. Technical pen tips are designed to fit into scribers for use with lettering guides.

Pen Cleaning

Read the cleaning instructions that come with the brand of pen that you purchase. Some pens require disassembly for cleaning while others should not be taken apart. Pens should be cleaned before each filling or before being stored for a long period of time. Clean the technical pen nib, cartridge, and body separately in warm water or special cleaning solution.

Ultrasonic pen cleaners are available to clean points. Pens are placed in a tank where millions of energized microscopic bubbles, generated by ultrasonic action, carry cleaning solution into the smallest openings of the drawing point to scrub the tube inside and out.

Specially formulated pen cleaner should be used for best results. Pen cleaner can also be used to soak pen points for cleaning by hand.

Ink

Drafting inks should be opaque, or have a matte or semiflat black finish that will not reflect light. The ink should reproduce without hot spots or line variation. Drafting ink should have excellent adhesion properties for use on paper or film. Certain inks are recommended for use on film in order to avoid peeling, chipping, or cracking. Inks recommended for use in technical pens also have nonclogging characteristics. This property is especially important for use in high-speed computer-graphics plotters.

When selecting an ink, be sure to purchase one for the job you want done. First, determine how the ink will be applied, that is, from a technical pen, computer plotter, airbrush, fountain pen, calligraphy pen, or with a brush. Second, determine the surface the ink will be used on, such as paper (vellum or bond), polyester film, or acetate. Third, determine if your use requires the ink to be opaque, fast-drying, waterproof, or erasable.

ERASERS AND ACCESSORIES

The common shapes of erasers are rectangular and stick. The stick eraser works best in small areas. There are three basic types of erasers: pencil, ink, and plastic. The plastic eraser is used for plastic lead or ink on polyester film. These erasers are identified by their white or translucent color. Select an eraser that is recommended for the particular material used. When used with ink, apply very light pressure, and be very careful, because the friction developed by the speed of erasure can easily damage the drafting film surface and prevent the adhesion of ink when redrawing over the erased area. Moistening the eraser during use on polyester film helps to reduce any damage to the drawing surface, but do not do this on paper.

Erasing Tips

When erasing, the idea is to remove an unwanted line or letter. You do not want to remove the surface of the paper or polyester film. Erase only hard enough to remove the unwanted line. However, you must bear down hard enough to eliminate the line completely. If all the line does not disappear, ghosting results. A *ghost* is a line that seems to have been eliminated but still shows on a print. Lines that have been drawn so hard as to make a groove in the drawing sheet can also cause a ghost. To remove ink from vellum, use a pink or green eraser, or an electric eraser. Work the area slowly. Do not apply too much pressure or erase in one spot too long or you will go through the paper. On polyester film, use a vinyl eraser and/or a moist cotton swab. The inked line usually comes off easily, but use caution. If you destroy the matte surface of either a vellum or polyester sheet, you will not be able to redraw over the erased area.

Lead may be picked up from the drawing board surface and transmitted to the back of the drawing surface. When this happens, the drawing must be turned over and the graphite removed from the back surface of the drawing.

Electric Erasers

Professional manual drafters use electric erasers. Those with cords that plug in are best, but cordless, rechargeable units are also available. When working with an electric eraser, you do not need to use much pressure because the eraser operates at high speed. The purpose of the electric eraser is to remove unwanted lines quickly. *Use caution*; these erasers can also remove paper quickly!

Erasing Shield

Erasing shields are thin metal or plastic sheets with a number of differently shaped holes. They are used to erase small, unwanted lines or areas. For example, if you have a corner overrun, place one of the slots of the erasing shield over the area to be removed while covering the good area.

Eradicating Fluid

Eradicating fluid is primarily used with ink on film or for removal of lines from sepia (brown) prints. The eradicating fluid is most often applied with a brush or cotton swab. If in doubt about its application, follow the manufacturer's instructions; however, application is usually done by lightly moistening the area to be corrected. Then the solution is wiped with a tissue, being careful to remove all residue. Eradicating fluid is especially effective for removing aged ink lines and for erasure of large areas.

Cleaning Agents

Special eraser particles are available to help reduce smudging and to keep the drawing and your equipment clean. The particles also help float triangles, straightedges, and other drafting equipment to reduce line smudging. Use this material sparingly since too much can cause your lines to become fuzzy. Cleaning powders are not recommended for use on ink drawings or on polyester film.

Dusting Brush

Use a dusting brush to remove eraser particles from your drawing. Doing so helps reduce the possibility of smudges. Avoid using your hand to brush away eraser particles; the hand tends to cause smudges, which reduces drawing neatness.

The brush should be cleaned regularly with soap and water because it picks up graphite particles and casts a slight film over the drawing.

DRAFTING INSTRUMENTS

Kinds of Compasses

Compasses are used to draw circles and arcs. However, using a compass can be time-consuming. Use a template, whenever possible, to make circles or arcs more quickly. A compass is especially useful for large circles.

There are several basic types of compasses; however, a bow compass, shown in Fig. 3.6, is used for most drawing applications. Beam compasses consist of a bar with an adjustable needle, and a pencil or pen attachment for swinging large arcs or circles. Also available is a beam that is adaptable to the bow compass. Such an adapter works only on bow compasses that have a removable leg.

Use of Compass

Keep both the compass needle point and lead point sharp. The points are removable for easy replacement. The better compass needle points have a shoulder on them. The shoulder helps keep the point from penetrating the paper more than necessary. Compare the needle points in Figure 3.7.

The compass lead should, in most cases, be one grade softer than the lead you use for straight lines because less pressure is used on a compass than a pencil. Keep the compass lead sharp. An elliptical point is commonly used with the bevel side away from the needle leg. Keep the lead and the point equal in length. Figure 3.8 shows properly aligned and sharpened points on a compass.

Use a sandpaper block to sharpen the elliptical point. Be careful to keep the graphite residue away from your drawing and off your hands. Remove excess graphite from the point with a tissue or cloth after sharpening. Sharpen the lead often.

Some drafters prefer to use a conical point in their compass.

Protecting the Sheet during Compass Use

If you are drawing a number of circles from the same center, you will find that the compass point causes an ugly hole in your drawing sheet. Reduce the chance of making such a hole by placing a

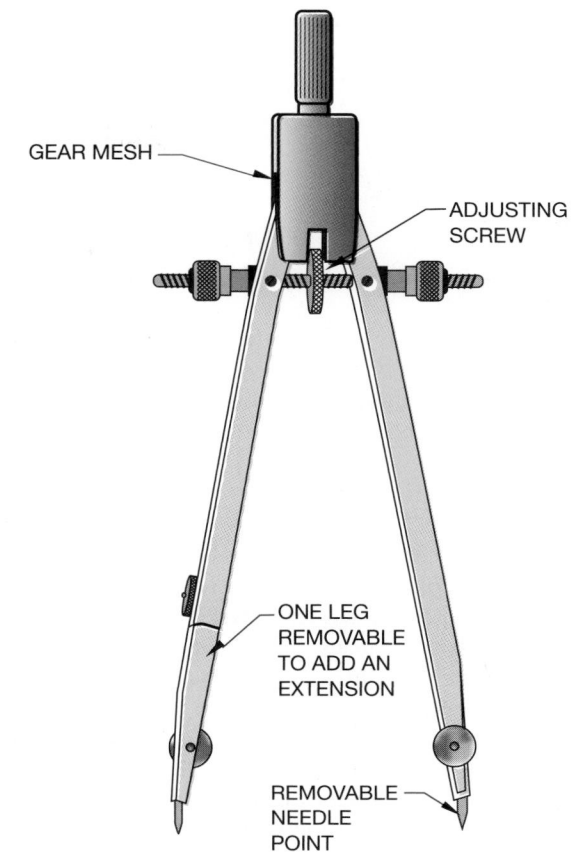

FIGURE 3.6 ■ Bow compass.

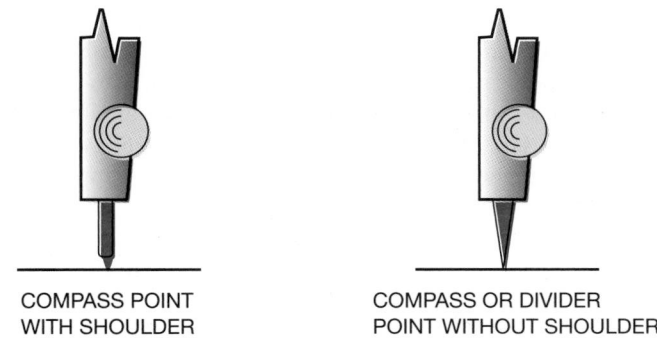

FIGURE 3.7 ■ Compass points.

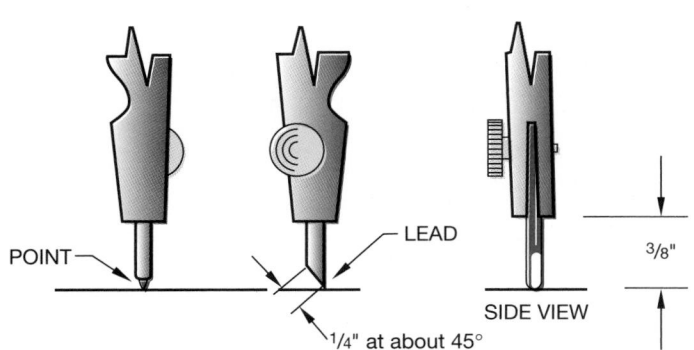

FIGURE 3.8 ■ Properly sharpened and aligned elliptical compass point.

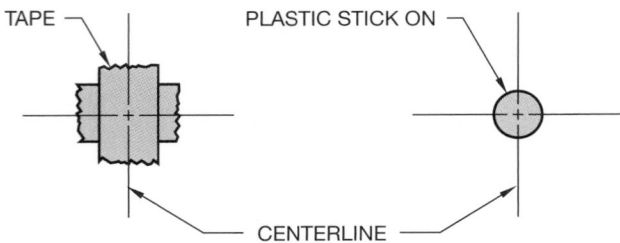

FIGURE 3.9 ■ Drawing sheet protection from compass point.

couple of pieces of drafting tape at the center point for protection. There are small plastic circles available for just this purpose. Place one at the center point, then pierce the plastic with your compass. (See Figure 3.9.)

Dividers

Dividers are used to transfer dimensions or to divide a distance into a number of equal parts.

Note: Do not try to use dividers as a compass.

Some drafters prefer to use bow dividers because the center wheel provides the ability to make fine adjustments easily. Also, the setting remains more stable than with standard friction dividers.

A good divider should not be too loose or tight. It should be easily adjustable with one hand. In fact, you should control a divider with one hand as you lay out equal increments or transfer dimensions from one feature to another. Figure 3.10 shows how the divider should be handled when used.

Proportional

Proportional dividers are used to reduce or enlarge an object without the need of mathematical calculations or scale manipulations. The center point of the divider is set at the correct point for the proportion you want. Then you measure the original size line with one side of the proportional divider and the other side automatically determines the new reduced or enlarged size.

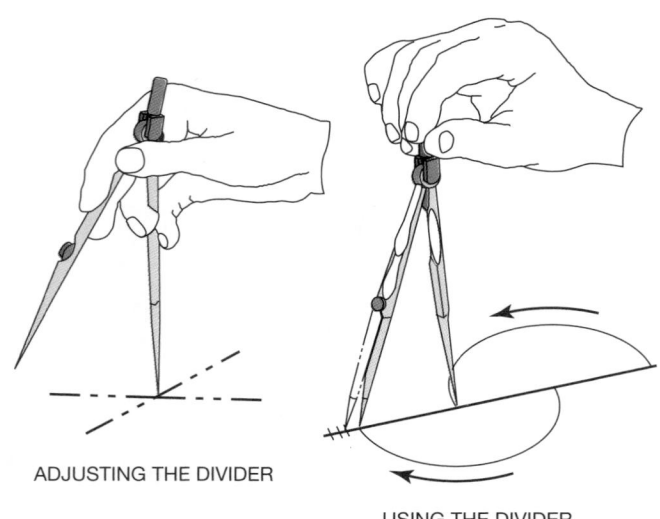

ADJUSTING THE DIVIDER

USING THE DIVIDER

FIGURE 3.10 ■ Using a divider.

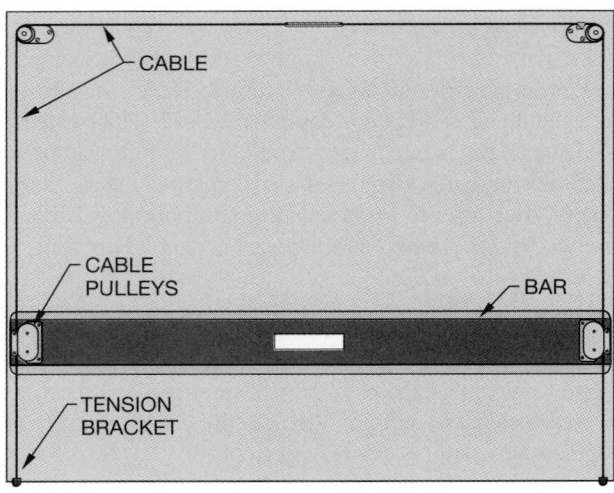

FIGURE 3.11 ■ Parallel bar.

Parallel Bar

The parallel bar slides up and down the board to allow you to draw horizontal lines. (See Figure 3.11.) Vertical lines and angles are made with triangles in conjunction with the parallel bar. The parallel bar is commonly found in architectural drafting offices because architectural drawings are frequently very large. Architects often need to draw straight lines the full length of their boards, and the parallel bar is ideal for such lines.

Triangles

There are two standard triangles. One has angles of 30°–60°–90° and is known as the 30–60 triangle. The other has angles of 45°–45°–90° and is known as the 45° triangle. Figure 3.12 shows these popular triangles.

Some drafters prefer to use triangles in place of a vertical drafting machine scale as shown in Figure 3.13. The machine protractor or the triangle can be used to make angled lines. Drafters who use parallel bars rather than drafting machines also use triangles to make vertical and angled lines.

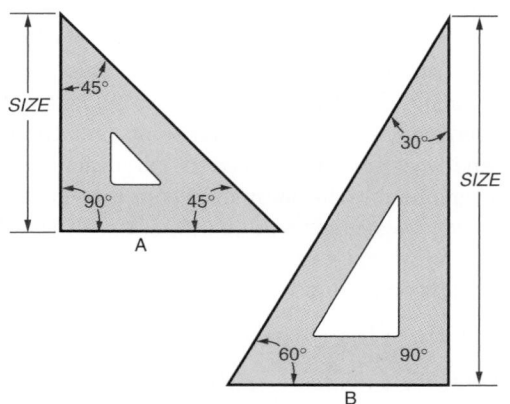

FIGURE 3.12 ■ 45° and 30°–60° triangles.

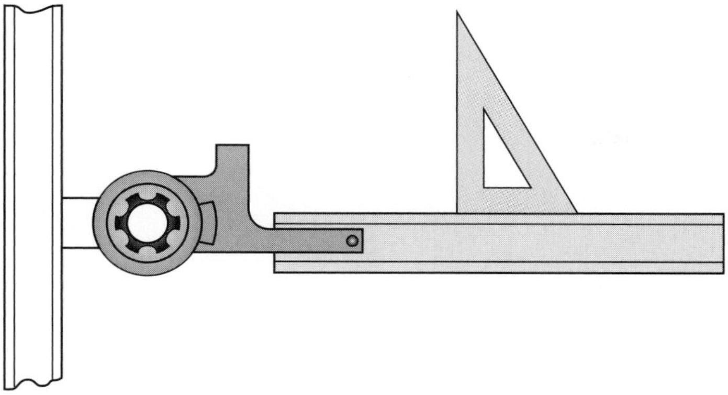

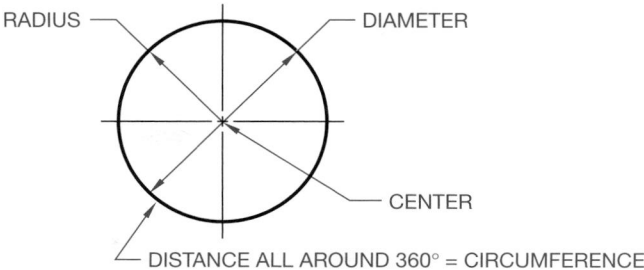

FIGURE 3.15 ■ Parts of a circle.

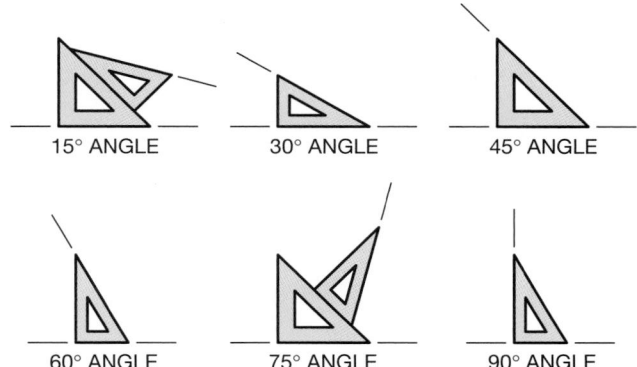

FIGURE 3.13 ■ Using a triangle with a drafting machine.

15° ANGLE   30° ANGLE   45° ANGLE

60° ANGLE   75° ANGLE   90° ANGLE

FIGURE 3.14 ■ Angles that may be made with the 30°–60° and 45° triangles individually or in combination.

Triangles may also be used as straightedges to connect points for drawing lines without the aid of a parallel bar or machine scale. Triangles are used individually or in combination to draw angled lines in 15° increments. (See Figure 3.14.) Also available are adjustable triangles with built-in protractors that are used to make angles of any degree up to a 45° angle.

Templates

Circle Templates

Circle templates are available with circles in a range of sizes beginning with 1/16 in. The circles on the template are marked with their diameters and are available in fractions, decimals, or millimeters. The parts of a circle are shown in Figure 3.15. Sample circle templates are shown in Figure 3.16. A popular template is one that has circles, hexagons, squares, and triangles.

Always use a circle template rather than a compass. Circle templates save time and are very accurate. For best results when making circles, try to keep your pencil or pen perpendicular to the paper. To obtain proper width lines with a pencil, use a 0.9 mm automatic pencil.

To use a circle template properly, first draw the centerlines of your circle. Then exactly align the dashes on the template with the centerlines as shown in Figure 3.17. Proceed to trace the outline of the circle.

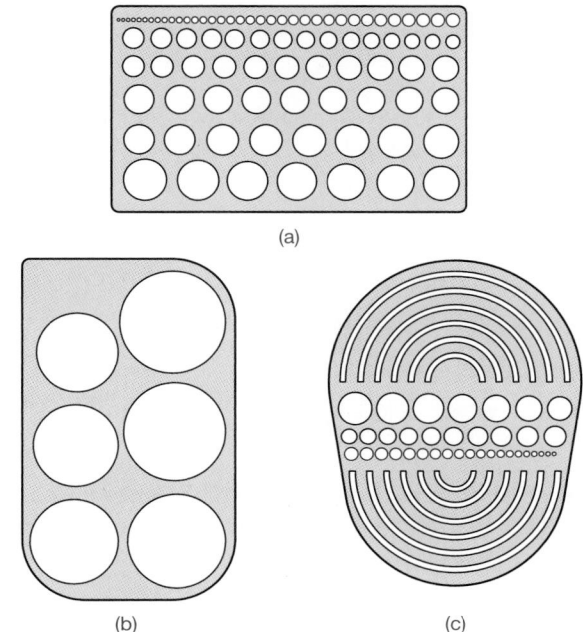

(a)

(b)   (c)

FIGURE 3.16 ■ (a) Small circles. (b) Large, full circles. (c) Large half circles.

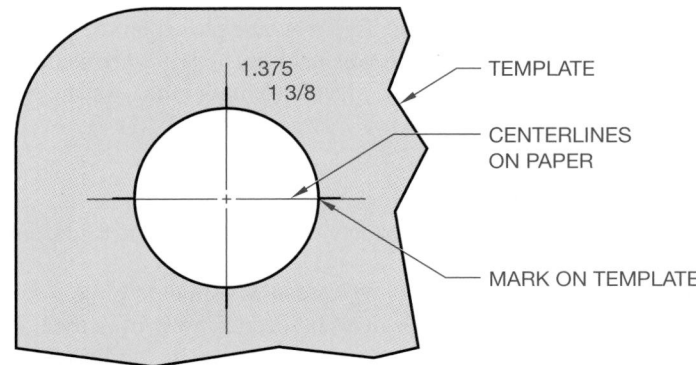

FIGURE 3.17 ■ Using a circle template to draw a circle.

To draw arcs with a circle template, use one of two methods. One method is to draw the centerlines of the arc, align the template with the proper diameter, and draw the arc. Keep in mind that the template circles are marked in diameter while the size of an arc is given in radius. Remember to divide the template size in half to find the proper arc radius. The other method of drawing arcs is to lightly

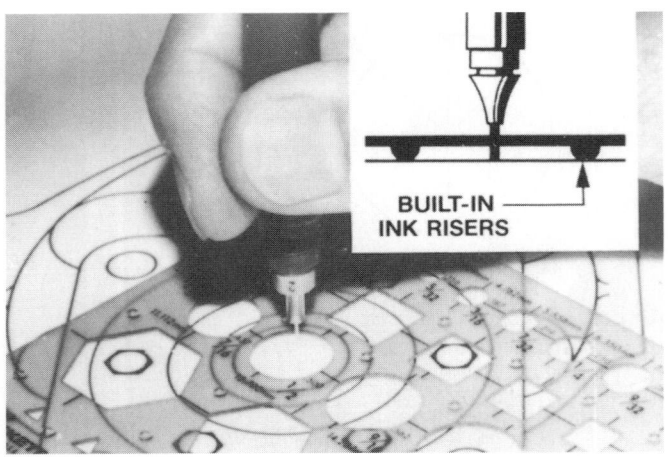

FIGURE 3.18 ■ A template with built-in risers for inking. *Courtesy Chartpak.*

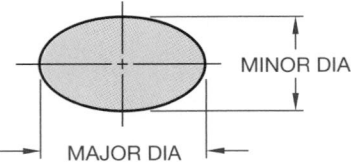

FIGURE 3.19 ■ Parts of an ellipse.

draw outside construction lines, then fill in the arc to the points of tangency. This method is demonstrated in Chapter 6, Geometric Construction. Be sure the connection between the arc and the straight line is smooth.

When using a circle template and a technical pen, keep the pen perpendicular to the paper. Some templates have risers built in to keep the template above the drawing sheet. Without this feature there is a risk of ink running under a template that is flat against the drawing. If your template does not have risers, purchase and add template lifters, use a few layers of tape placed on the underside of the template (although tape does not always work well), or place a second template with a larger circle under the template you are using. (See Figure 3.18.)

Ellipse Templates

Ellipses are circles seen at an angle. The parts of an ellipse are shown in Figure 3.19.

The type of pictorial drawing known as isometric projects the sides of objects at a 30° angle in each direction away from the horizontal. Isometric circles are ellipses aligned with the horizontal right, or left planes of an isometric box as in Figure 3.20. Isometric ellipse templates automatically position the ellipse at the proper angle of 35° 16'. (See Figure 3.21.)

Isometric drawing is discussed in detail in Chapter 19.

Use a Template

Never use a compass if you can use a template. Templates increase drafting speed and are very accurate.

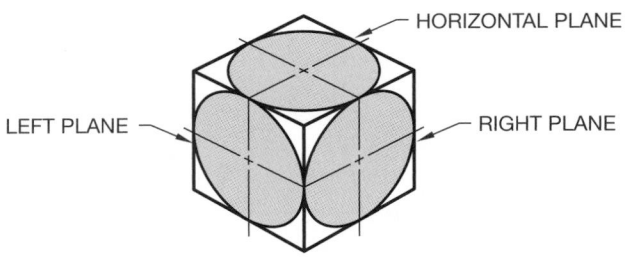

FIGURE 3.20 ■ Ellipses in isometric planes.

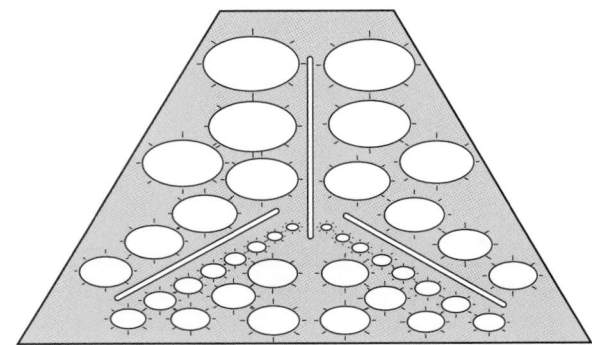

FIGURE 3.21 ■ Isometric ellipse template.

Irregular Curves

Irregular curves are commonly called *French curves.* These curves have no constant radii. An irregular curve is shown in Figure 3.22. A radius curve is composed of a radius and tangent. The radius on these curves is constant and ranges from 3 ft to 200 ft. These curves are commonly used in highway drafting. In addition to these two kinds of curves, there are available ship's curves. The curves in a set of ship's curves become progressively larger and, like French curves, have no constant radii. They are used for layout and development of ships' hulls. Flexible curves are also available that allow you to adjust to a desired curve.

FIGURE 3.22 ■ Irregular or French curve. *Courtesy The C-Thru®
Ruler Company.*

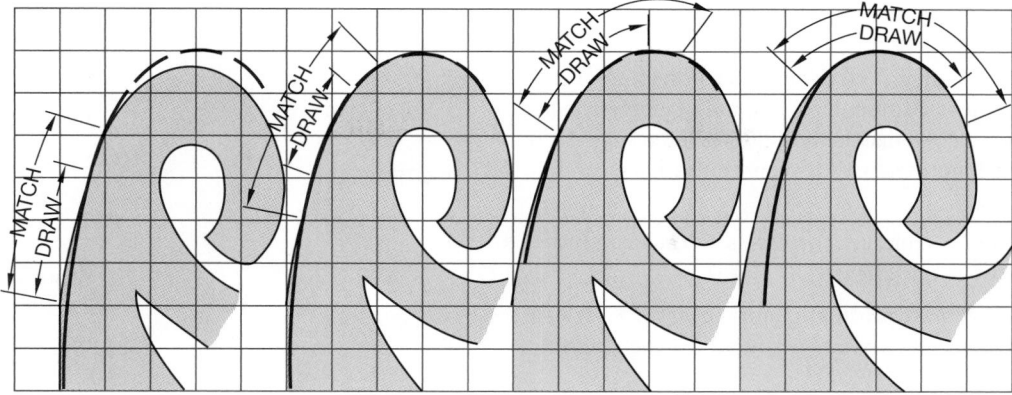

FIGURE 3.23 ■ Irregular curve development.

In order to draw an irregular curve, points on the curve are developed or plotted. The points are then connected with a construction line. Then a flexible curve is bent to fit, or a portion of an irregular curve is matched to the construction line to include at least three plotted points. Care must be taken to make the curve flow smoothly. This smooth flow of the line is accomplished by never drawing the full length of the curve, but by overlapping each successive setting of the irregular curve. (See Figure 3.23.)

DRAFTING MACHINES

Drafting machines have a vernier head allowing you to measure angles accurately to 5'. Drafting machines, for the most part, take the place of triangles and parallel bars. The drafting machine maintains a horizontal and vertical relationship between scales, which also serve as straightedges. A protractor allows the scales to be set quickly at any angle. There are two types of drafting machines, arm and track. Although both types are excellent tools, the track machine has generally replaced the arm machine in industry. A major advantage of the track machine allows the drafter to work with a board in the vertical position. A vertical drafting surface position is generally more comfortable to use than a horizontal table. For school or home use, the arm machine may be a good economical choice.

Arm Drafting Machine

The arm drafting machine is compact and less expensive than a track machine. The arm machine clamps to a table and through an elbowlike arrangement of supports, allows you to position the protractor head and scales anywhere on the board. An arm drafting machine is shown in Figure 3.24.

Track Drafting Machine

A track drafting machine has a traversing arm that moves left and right across the table and a head unit that moves up and down the traversing arm. There is a locking device for both the head and the traversing arm. The shape and placement of the controls of a track machine vary with the manufacturer, although most brands have the same operating features and procedures. Figure 3.25 shows the component parts of a track drafting machine.

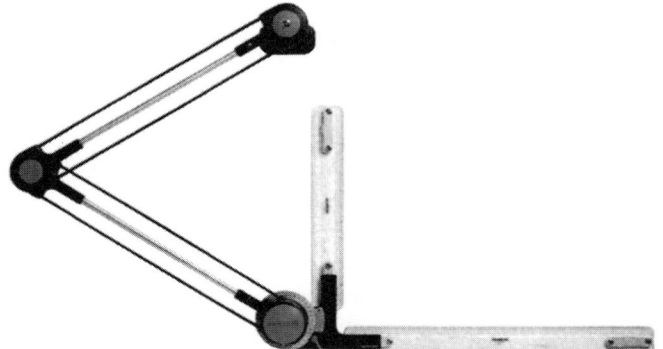

FIGURE 3.24 ■ Arm drafting machine. *Courtesy Vemco America, Inc.*

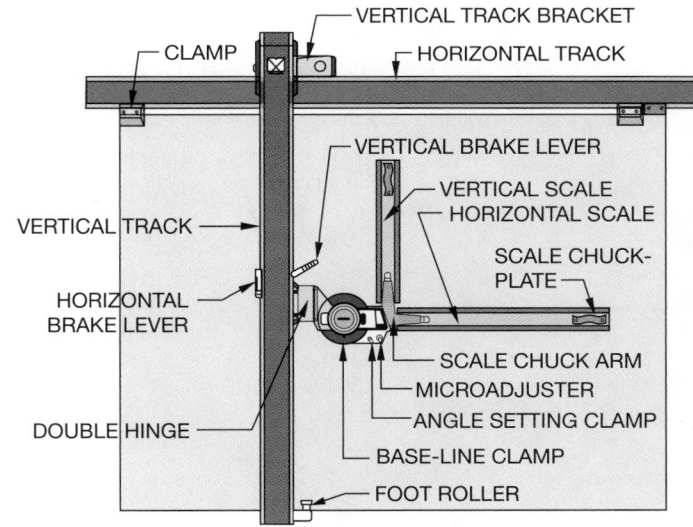

FIGURE 3.25 ■ Track drafting machine and its parts. *Courtesy Mutoh America, Inc.*

Sizes of Drafting Machine

When ordering a drafting machine, the specifications should relate to the size of the drafting board on which it is mounted. For example, a 37½ × 60 in. machine would properly fit a table of the same size.

Controls and Machine Head Operation

Drafting machine heads contain the controls for horizontal, vertical, and angular movement. Although each brand of machine contains similar features, controls may be found in different places on different brands. (See Figure 3.26.) Most machines have the following controls:

1. Baseline adjustment—releases the scales so they can move but the protractor is not affected.

2. Index control—permits automatic stops every 15°. It can also be pushed in and locked to let you adjust the machine to any angle.

3. Indexing clamp—locks the protractor at angles other than 15° increments so you can draw an accurate line without the protractor moving.

To operate the drafting machine protractor head, place your hand on the handle and, using your thumb, depress the index thumbpiece. Doing so allows the head to rotate. Each increment marked on the protractor is one degree with a label every 10°. As the vernier plate (the small scale numbered from 0 to 60) moves past the protractor, the zero on the vernier aligns with the angle that you wish to read. For example, Figure 3.27 shows a reading of 10°. As you rotate the handle, notice the head automatically locks every 15°. To move the protractor past the 15° increment, you must again depress the index thumbpiece.

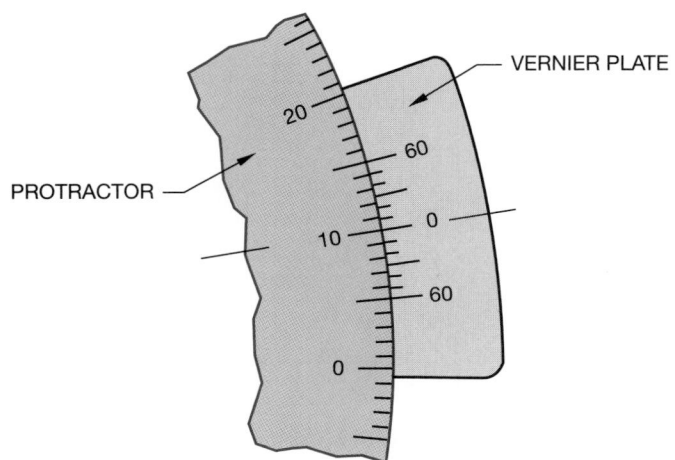

FIGURE 3.27 ■ Vernier plate and protractor showing a reading of 10°.

Having rotated the protractor head 40° clockwise, the machine is in the position shown in Figure 3.28. The vernier plate at the protractor reads 40°, which means that both the horizontal and vertical scale have moved 40° from their original position at 0° and 90° respectively. The horizontal scale reads directly from the protractor starting from 0°. The vertical scale reading begins from the 90° position. The key to measuring angles is to determine if the angle is to be measured from the horizontal or vertical starting point. See the examples in Figure 3.29.

Measuring full degree increments is easy since you simply match the zero mark on the vernier plate with a full degree mark

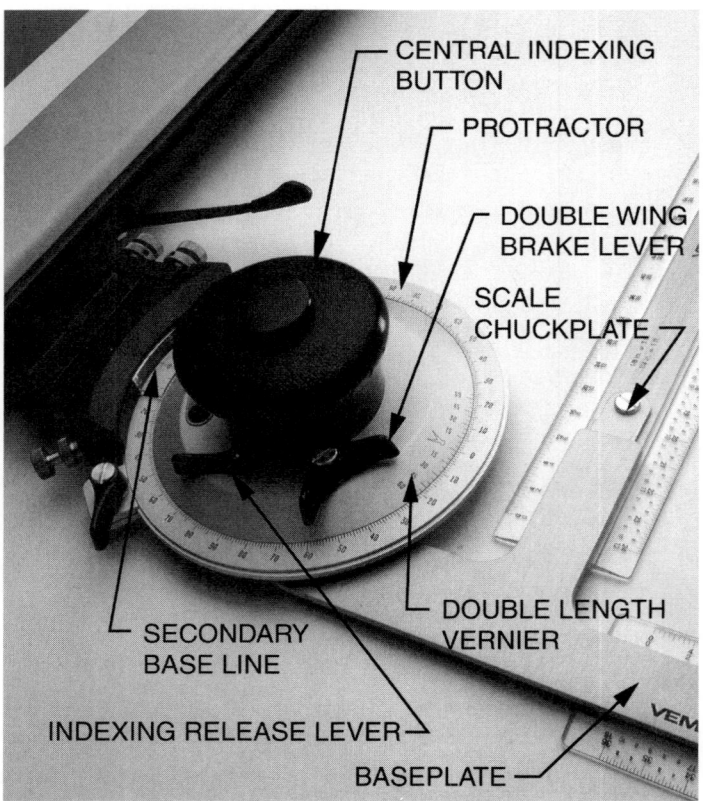

FIGURE 3.26 ■ Drafting machine head controls and parts. *Courtesy Vemco Corporation.*

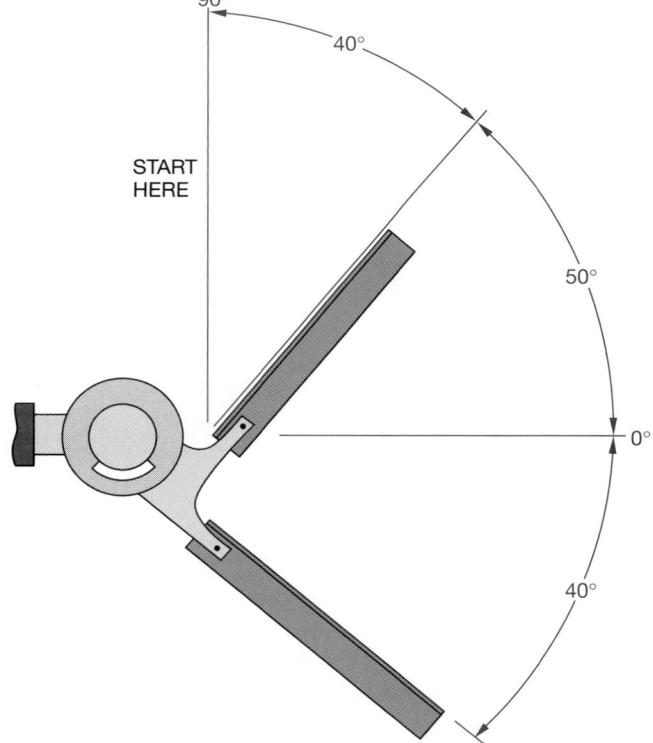

FIGURE 3.28 ■ Angle measurement with the drafting machine.

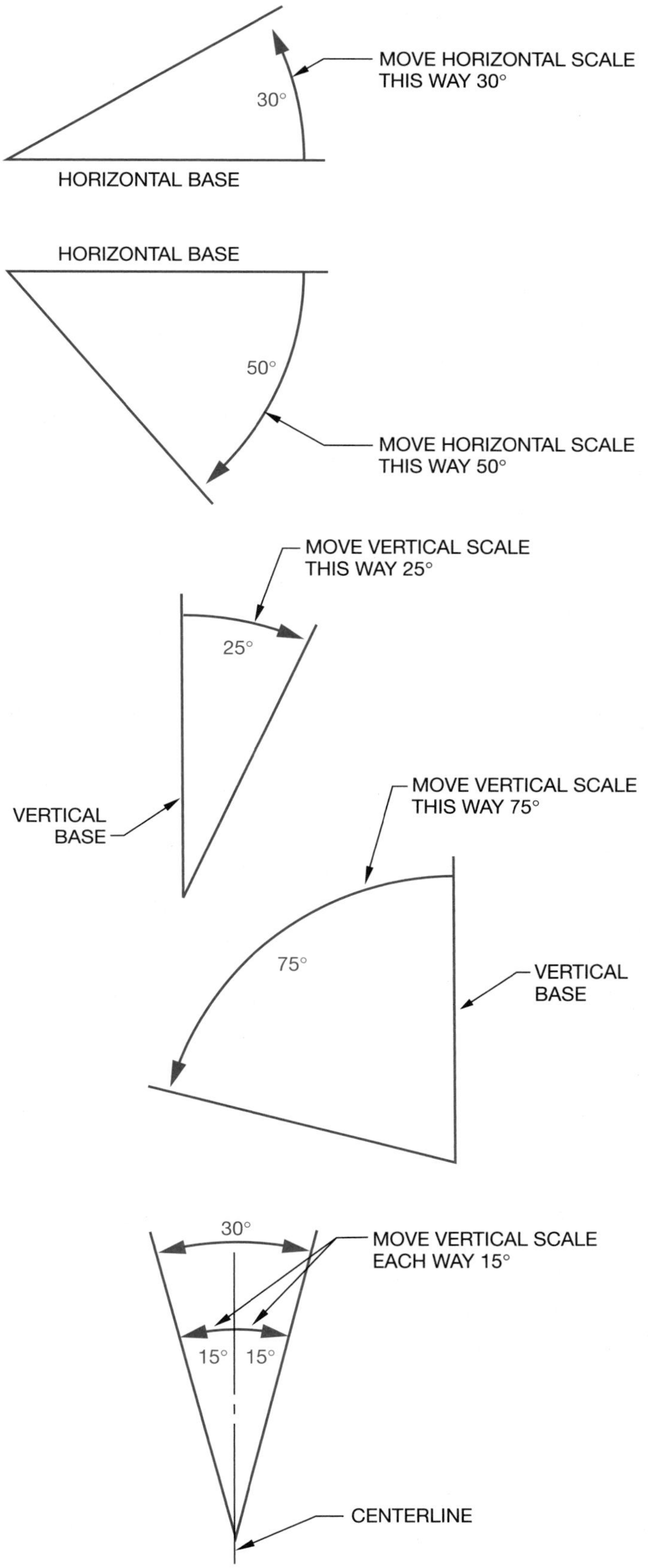

MOVE HORIZONTAL SCALE
THIS WAY 30°

30°

HORIZONTAL BASE

HORIZONTAL BASE

50°

MOVE HORIZONTAL SCALE
THIS WAY 50°

MOVE VERTICAL SCALE
THIS WAY 25°

25°

VERTICAL
BASE

MOVE VERTICAL SCALE
THIS WAY 75°

75°

VERTICAL
BASE

30°

MOVE VERTICAL SCALE
EACH WAY 15°

15° 15°

CENTERLINE

FIGURE 3.29 ■ Angle measurements from either a horizontal or vertical reference line.

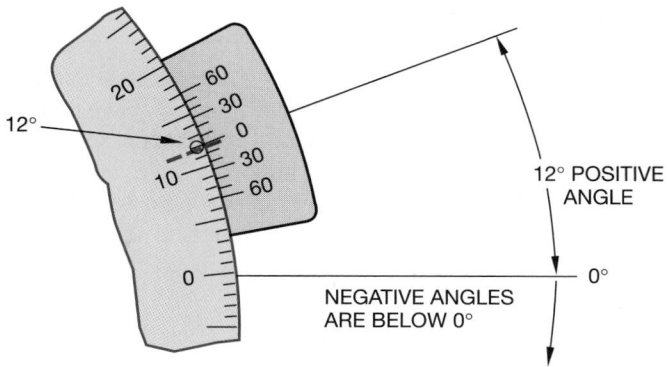

FIGURE 3.30 ■ Measuring full degrees.

on the protractor. See the reading of 12° in Figure 3.30. The vernier scale allows you to measure angles as accurately as 5' (minutes). Remember, 1 degree equals 60 minutes (1° = 60') and 1 minute equals 60 seconds (1' = 60").

Reading and Setting Angles with the Vernier

To read an angle other than a full degree, assume the vernier scale is set at a *positive angle* as shown in Figure 3.31. Each mark on the vernier scale represents 5'. First see that the angle to be read is between 7° and 8°. Then find the 5' mark on the *upper* half of the vernier (the direction in which the scale has been turned) that is most closely aligned with a full degree on the protractor. In this example, it is the 40' mark. Add the minutes to the degree just passed. The correct reading then, is 7°40'. The procedure for reading *negative angles* is the same, except you read the minute marks on the *lower* half of the vernier.

Suppose you wished to set the angle 7°40' as shown in Figure 3.31. First release the protractor brake and disengage the indexing mechanism with the thumb control. Rotate the protractor arm counterclockwise until the zero of the vernier is at 7°. Then slowly continue the rotation until the 40' mark on the upper half of the vernier aligns with the nearest degree mark on the protractor. Lock the protractor brake and draw the line. The procedure for setting negative angles is essentially the same except for turning the protractor head in a clockwise motion.

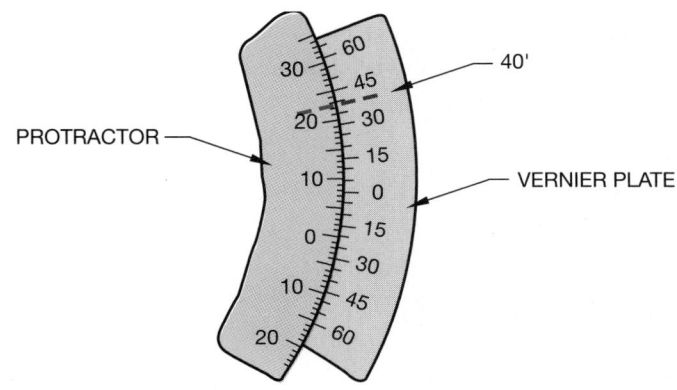

FIGURE 3.31 ■ Reading positive angles with the vernier.

Machine Setup

To insert a scale in the baseplate chuck, place the scale flat on the board and align the scale chuckplate with the baseplate chuck on the protractor head. Firmly press, but do not drive the scale chuckplate into the baseplate chuck. To remove a scale, use the scale wrench provided with the machine. Removing a scale by hand without the aid of a key could result in damage to the scale and/or machine. Refer to the operation manual provided with your drafting machine.

Scale Alignment

Before drawing with any drafting machine, the scales should be checked for alignment and, if needed, be adjusted at right angles to each other. For best results with a track drafting machine, the scales should also be aligned with respect to the horizontal track. Both operations can be accomplished by following the instructions provided with your operation manual.

SCALES

Scale Shapes

There are four basic scale shapes as shown in Figure 3.32. The two-bevel scale is also available with chuckplates for use with standard arm or track drafting machines. These machine scales have typical calibrations, and some have no scale reading for use as a straightedge alone. Drafting machine scales are purchased by designating the length needed, 12, 18, or 24 in., and the scale calibration such as metric, engineer's full scale in 10ths and half scale in 20ths, or architect's scale 1/4" = 1'–0" and 1/2" = 1'–0". Many other scales are available.

Scale Notation

The scale of a drawing is usually noted in the title block or below the view of an object that differs in scale to that given in the title block. Drawings are scaled so the object represented can be illustrated clearly on standard sizes of paper. It would be difficult, for example, to make a full-size drawing of a Boeing 747; thus, a scale that reduces the size of such a large object must be used. Machine parts are often drawn full size or even twice, four, or ten times larger than full size, depending on the actual size of the part.

The scale selected, then, depends on:

■ The actual size of the part.

■ The amount of detail to be shown.

■ The paper size selected.

■ The amount of dimensioning and notes required on the part.

 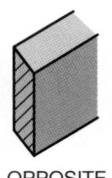 

| TWO | FOUR | OPPOSITE | TRIANGULAR |
| BEVEL | BEVEL | BEVEL | |

FIGURE 3.32 ■ Scale shapes.

Commonly used scales should be selected. Avoid scales that are not identified in the following discussions.

The following scales and their notation are frequently used on mechanical drawings.

Full scale	=	FULL or 1:1
Half scale	=	HALF or 1:2
Quarter scale	=	QUARTER or 1:4
Twice scale	=	DOUBLE or 2:1
Four times scale	=	4:1
Ten times scale	=	10:1

Some scales used on architectural drawings are noted as follows:

1/8"	=	1'–0"	1"	=	1'–0"
1/4"	=	1'–0"	1 1/2"	=	1'–0"
1/2"	=	1'–0"	3"	=	1'–0"

Some scales used in civil drafting are noted as follows:

1"	=	10'	1"	=	50'
1"	=	20'	1"	=	60'
1"	=	30'	1"	=	100'

Metric Scale

ANSI/ASME According to the American National Standards Institute (ANSI) and the American Society of Mechanical Engineers (ASME):

The commonly used SI (International System of Units) linear unit used on engineering drawings is the *millimeter*. The commonly used U.S. customary linear unit used on engineering drawings is the *decimal inch*. On drawings where all dimensions are either in inches or millimeters, individual identification of units is not required. However, the drawing shall contain a note stating: UNLESS OTHERWISE SPECIFIED, ALL DIMENSIONS ARE IN INCHES (or MILLIMETERS as applicable). Where some millimeters are shown on an inch-dimensioned drawing, the millimeter value should be followed by the abbreviation *mm*. Where some inches are shown on a millimeter-dimensioned drawing, the inch value should be followed by the abbreviation *IN*.

Metric symbols are as follows:

millimeter	=	mm
centimeter	=	cm
decimeter	=	dm
meter	=	m
dekameter	=	dam
hectometer	=	hm
kilometer	=	km

Some metric-to-metric equivalents are the following:

10 millimeters	=	1 centimeter
10 centimeters	=	1 decimeter
10 decimeters	=	1 meter
10 meters	=	1 dekameter
10 dekameters	=	1 kilometer

Some metric-to-U.S. customary equivalents are the following:

1 millimeter	=	.03937 inch
1 centimeter	=	.3937 inch

| 1 meter | = | 39.37 inches |
| 1 kilometer | = | .6214 mile |

Some U.S. customary-to-metric equivalents are the following:

1 mile	=	1.6093 kilometers	=	1609.3 meters
1 yard	=	914.4 millimeters	=	.9144 meter
1 foot	=	304.8 millimeters	=	.3048 meter
1 inch	=	25.4 millimeters	=	.0254 meter

To convert inches to millimeters, multiply inches by 25.4.

Figure 3.33 shows the common scale calibrations found on the triangular metric scale. One advantage of the metric scales is that any scale is a multiple of 10; therefore, any reductions or enlargements are easily performed. No mathematical calculations should be required when using a metric scale. Always select a direct reading scale. To avoid the possibility of error, avoid multiplying or dividing metric scales by anything but multiples of ten.

Civil Engineer's Scale

The triangular civil engineer's scale contains six scales, two on each of its sides. The civil engineer's scales are calibrated in multiples of 10. The scale margin displays the scale represented on a particular edge. The following table shows some of the many scale options available when using the civil engineer's scale. Keep in mind that *any* multiple of 10 is available with this scale.

Civil Engineer's Scale					
Divisions	Ratio	Scales Used with This Division			
10	1:1	1" = 1"	1" = 1'	1" = 10'	1" = 100'
20	1:2	1" = 2"	1" = 2'	1" = 20'	1" = 200'
30	1:3	1" = 3"	1" = 3'	1" = 30'	1" = 300'
40	1:4	1" = 4"	1" = 4'	1" = 40'	1" = 400'
50	1:5	1" = 5"	1" = 5'	1" = 50'	1" = 500'
60	1:6	1" = 6"	1" = 6'	1" = 60'	1" = 600'

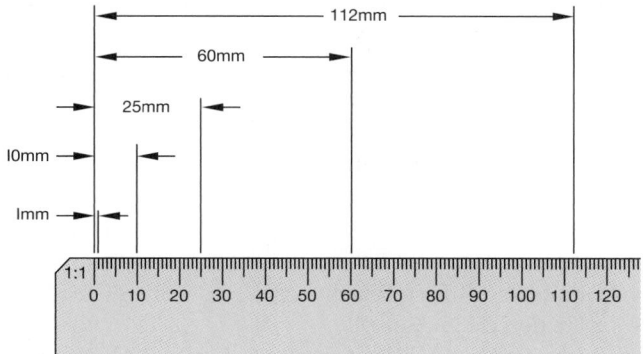

FULL SCALE = 1:1

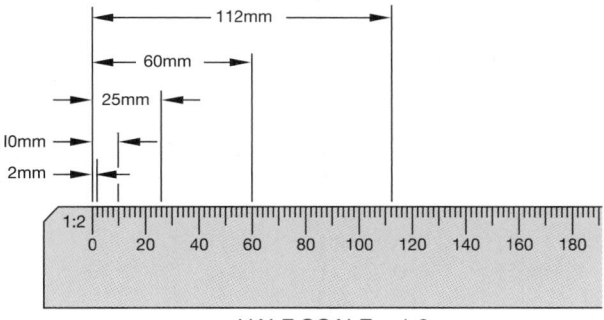

HALF SCALE = 1:2

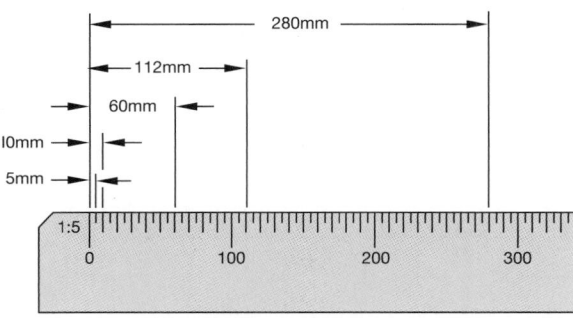

ONE FIFTH SCALE = 1:5

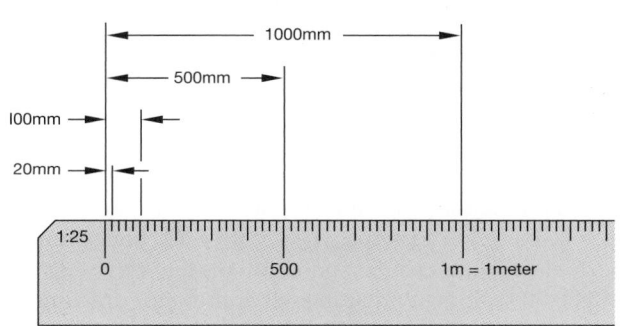

ONE TWENTY FIFTH SCALE = 1:25

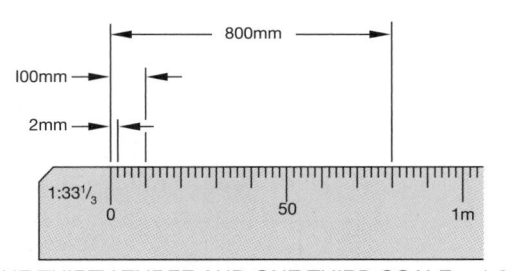

ONE THIRTY THREE AND ONE THIRD SCALE = 1:33 $\frac{1}{3}$

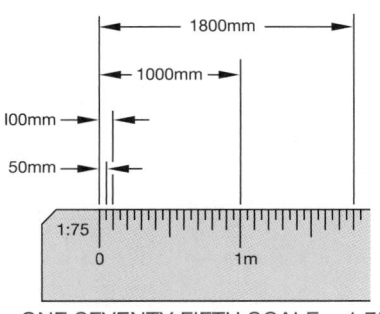

ONE SEVENTY FIFTH SCALE = 1:75

FIGURE 3.33 ■ Metric scale calibrations.

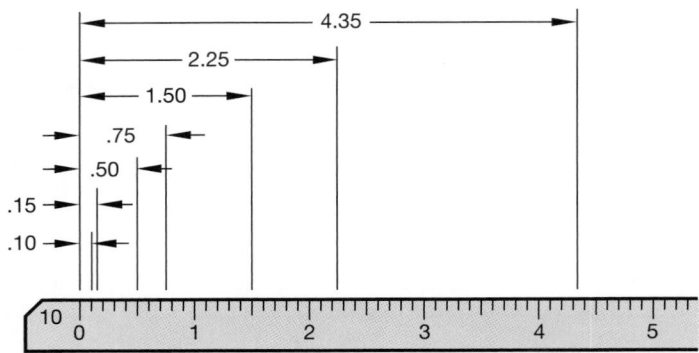

FIGURE 3.34 ■ Full engineer's decimal scale (1:1).

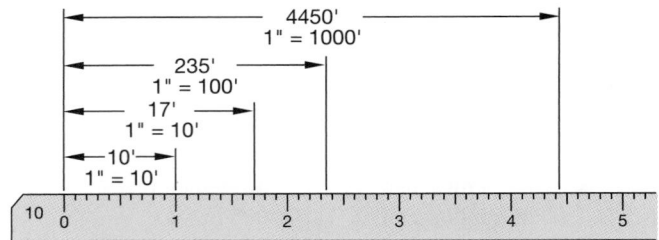

FIGURE 3.35 ■ Civil engineer's scale, units of 10.

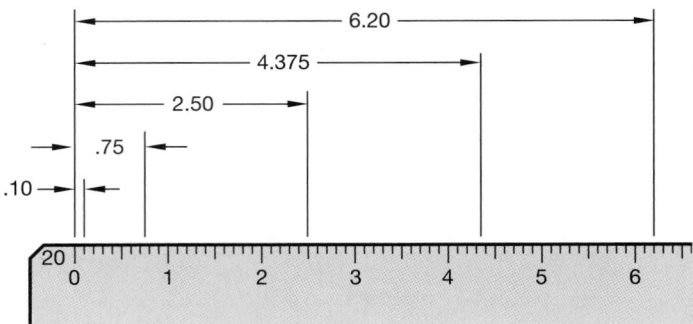

FIGURE 3.36 ■ Half scale on the engineer's scale (1:2).

The 10 scale is often used in mechanical drafting as a full, decimal-inch scale, shown in Figure 3.34. Increments of ¹⁄₁₀ (.1) in. can easily be read on the 10 scale. Readings of less than .1 in. require you to approximate the desired amount, as shown in Figure 3.34. Some scales are available that refine the increments to ¹⁄₅₀th of an inch. The 10 scale is also used in civil drafting for scales of 1" = 10' or 1" = 100' and so on. (See Figure 3.35.)

The 20 scale is commonly used in mechanical drawing to represent dimensions on a drawing at half scale (1:2). Figure 3.36 shows examples of half-scale decimal dimensions. The 20 scale is also used for scales of 1" = 2', 1" = 20', 1" = 200', as shown in Figure 3.37.

The remaining scales on the engineer's scale may be used in a similar fashion. For example, 1" = 5', 1" = 50', and so on. Figure 3.37 shows dimensions on each of the civil engineer's scales. The 50 scale is popular in civil drafting for drawing plats of subdivisions.

Architect's Scale

The triangular architect's scale contains 11 different scales. On ten of them each inch represents a foot and is subdivided into multiples of 12 parts to represent inches and fractions of an inch. The eleventh scale is the full scale with a 16 in the margin. The 16 means that each inch is divided into 16 parts and each part is equal to ¹⁄₁₆th of an inch. Look at Figure 3.38 for a comparison between the 10 engineer's scale and the 16 architect's scale. Figure 3.39 shows an example of the full architect's scale, while Figure 3.40 shows the fraction calibrations.

Look at the architect's scale examples in Figure 3.41. Note the form in which scales are expressed on a drawing. The scale notation may take the form of a word (full, half, double), or a ratio (1:1, 1:2, 2:1), or an equation of the drawing size in inches or fractions of an inch to one foot (1" = 1'–0", 3" = 1'–0", ¹⁄₄" = 1'–0"). The architect's scale commonly has scales running in both directions along an edge. Be careful when reading a scale from left to right. Do not confuse its calibrations with the scale that reads from right to left.

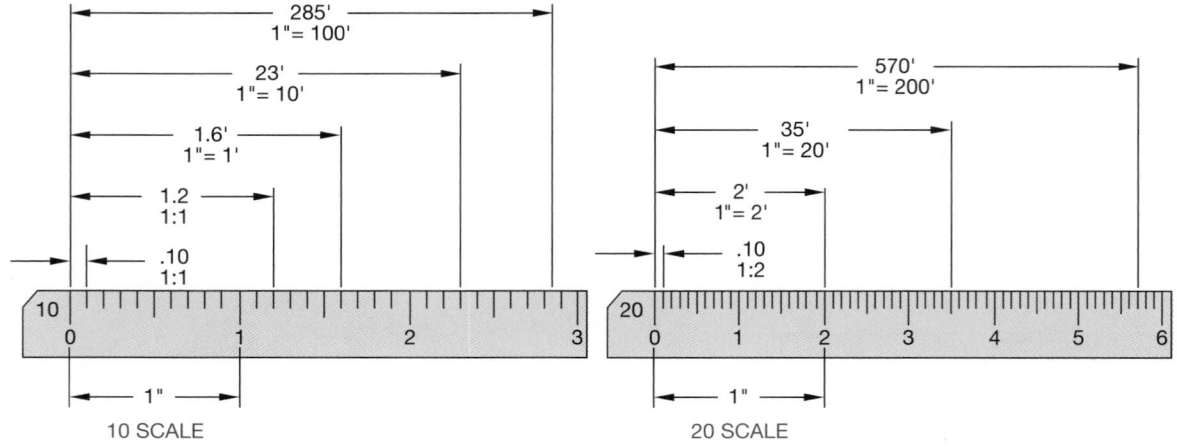

FIGURE 3.37 ■ Civil engineer's scales.

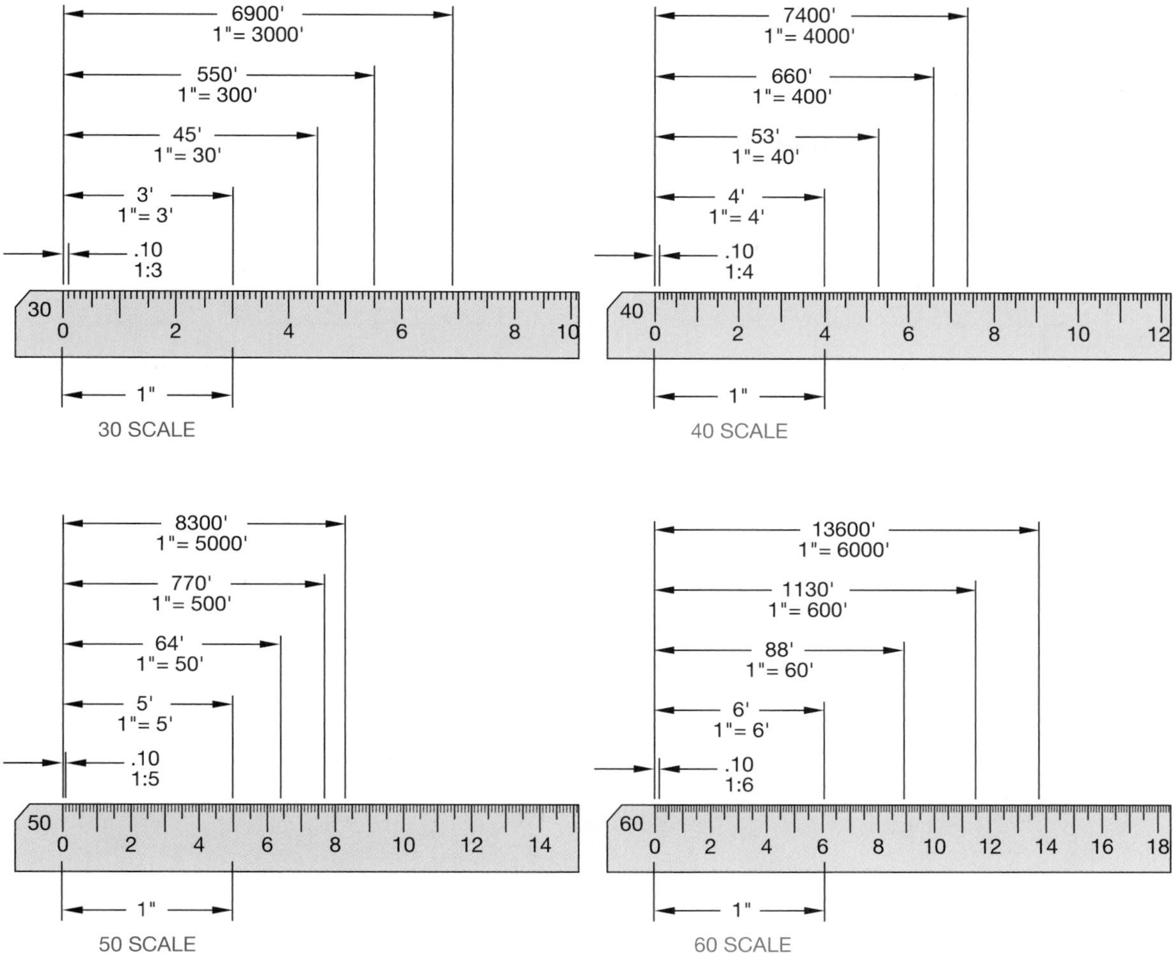

6900'
1"= 3000'
550'
1"= 300'
45'
1"= 30'
3'
1"= 3'
.10
1:3
30
0 2 4 6 8 10
1"
30 SCALE
7400'
1"= 4000'
660'
1"= 400'
53'
1"= 40'
4'
1"= 4'
.10
1:4
40
0 2 4 6 8 10 12
1"
40 SCALE
8300'
1"= 5000'
770'
1"= 500'
64'
1"= 50'
5'
1"= 5'
.10
1:5
50
0 2 4 6 8 10 12 14
1"
50 SCALE
13600'
1"= 6000'
1130'
1"= 600'
88'
1"= 60'
6'
1"= 6'
.10
1:6
60
0 2 4 6 8 10 12 14 16 18
1"
60 SCALE

FIGURE 3.37 (continued) ■ Civil engineer's scales.

ARCHITECT'S SCALE
FULL SIZE
INCH/
FOOT
DECIMAL
0
1
2

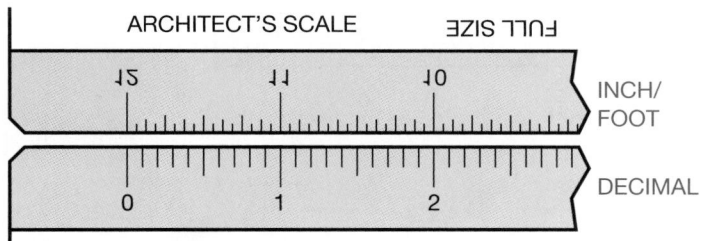

FIGURE 3.38 ■ Comparison of full engineer's scale (10) and architect's scale (16).

2 17/32
1 3/16
1/2
16
0
1
2
3

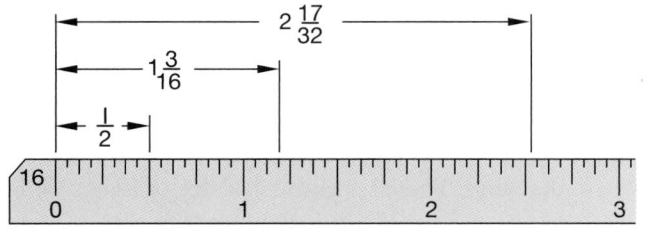

FIGURE 3.39 ■ Full (1:1), or 12″ = 1′–0″, architect's scale.

1/16
1/32
1/64
16
1/8
1/4
1/2

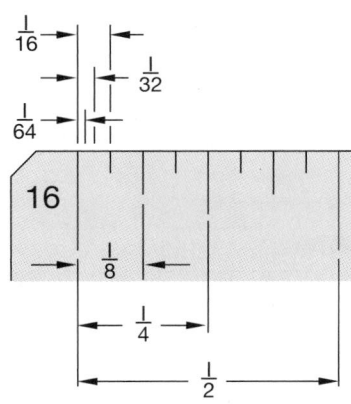

FIGURE 3.40 ■ Enlarged view of architect's (16) scale.

Mechanical Engineer's Scale

The mechanical engineer's scale is commonly used for mechanical drafting when drawings are in fractional or decimal inches. The mechanical engineer's scale typically has full-scale divisions that are divided into 1/16, 10, and 50. The 1/16 divisions are

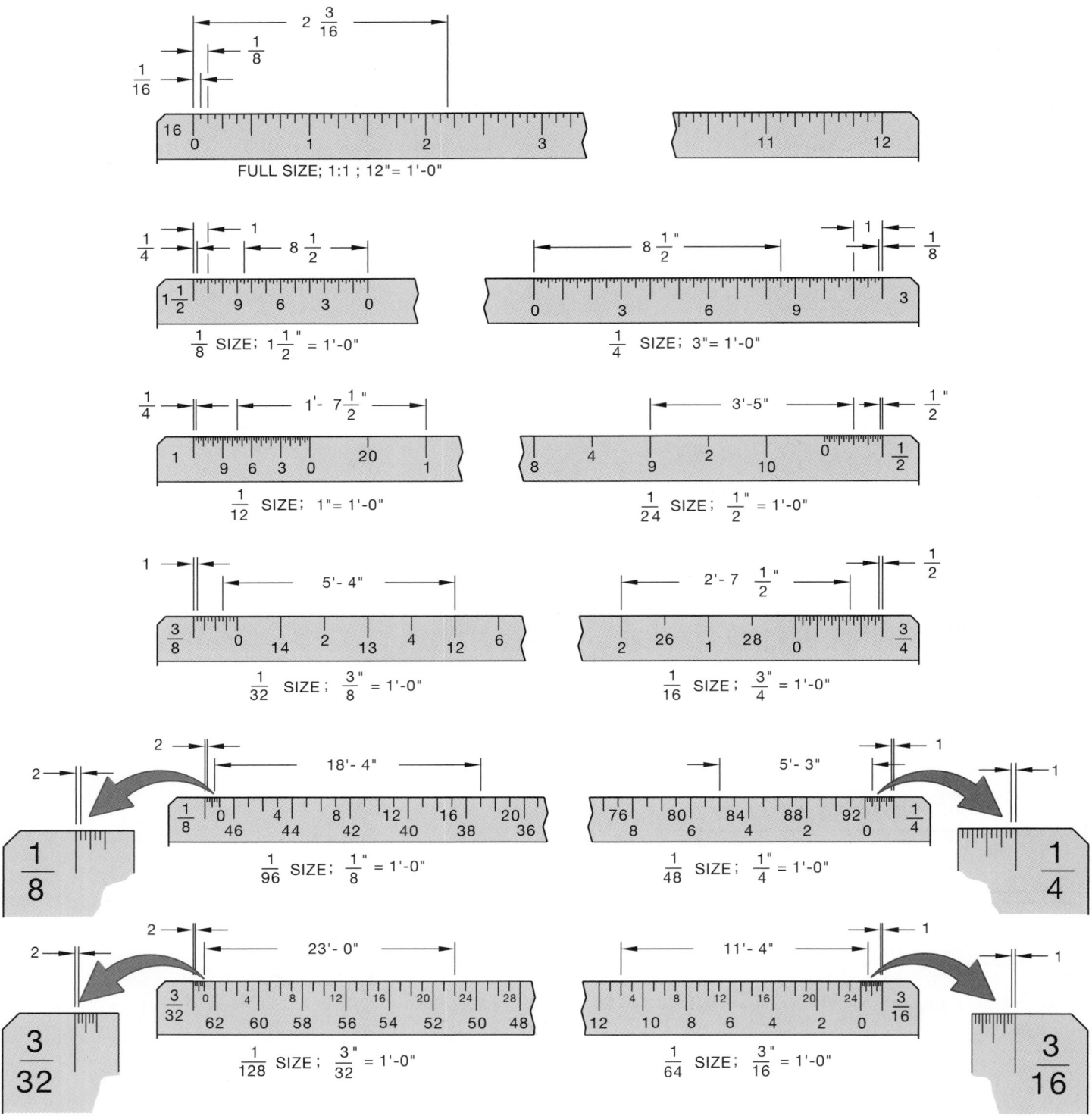

FIGURE 3.41 ■ Architect's scale examples.

just like the 16 architect's scale where there are 12 inches and each inch is divided into ¹⁄₁₆ in. increments, or sometimes ¹⁄₃₂ in. divisions. The 10 scale is the same as the 10 civil engineer's scale, where each inch is divided into 10 parts, with each division being .10 inch. The 50 scale is for scaling dimensions that require additional accuracy, because each inch has 50 divisions. This makes each increment ¹⁄₅₀ in. or .02 in. (1 ÷ 50 = .02). Fig-

ure 3.42 shows a comparison between these scales. The mechanical engineer's scale also has options for scaling down the size of drawings. These scales are half size (½ size, ½" = 1"), quarter size (¼ size, ¼" = 1"), and eighth size (⅛ size, ⅛" = 1″). These scales are shown in Figure 3.42. A drawing that is represented at full scale (1:1), half scale (1:2), and quarter scale (1:4) is shown in Figure 3.43.

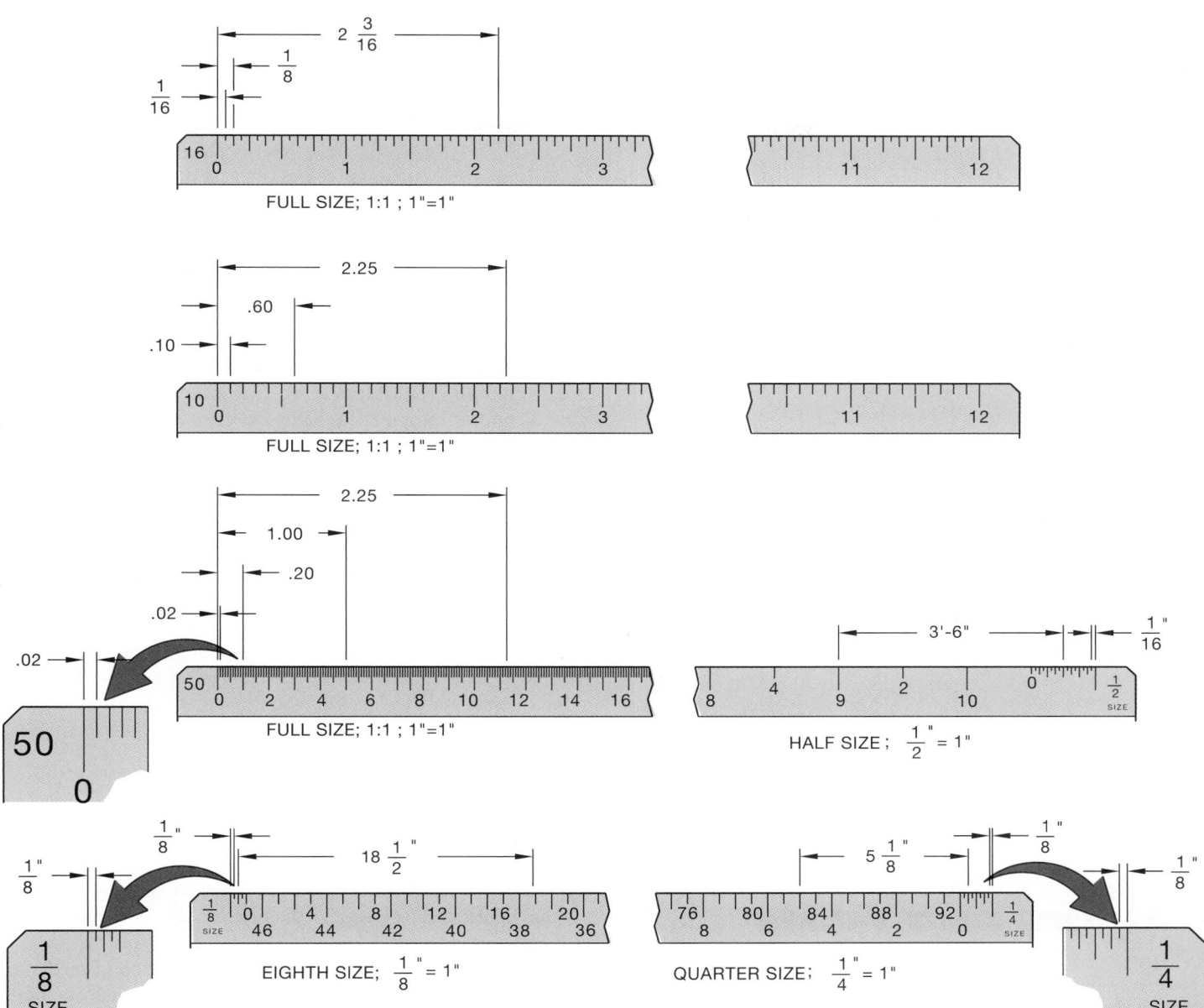

FIGURE 3.42 ■ A comparison of three types of scales.

CLEANING DRAFTING EQUIPMENT

Cleaning the drafting equipment each day helps keep your drawing clean and free of smudges. This is easily done with mild soap and water or with a soft rag or tissue. Avoid using harsh cleansers or products that are not recommended for use on plastic. It is also a good practice to clean your hands periodically to remove graphite and oil. Also, keep your hands dry.

DRAFTING MEDIA

Several factors other than cost should influence the purchase and use of drafting media. The considerations include durability, smoothness, erasability, dimensional stability, and transparency.

Durability should be considered if the original drawing will have a great deal of use. Originals may tear or wrinkle and the images become difficult to see if the drawings are used often.

Smoothness relates to how the medium accepts line work and lettering. The material should be easy to draw on so the image is dark and sharp without a great deal of effort on your part.

Erasability is important because errors need to be corrected and changes frequently made. When images are erased, ghosting should be kept to a minimum. *Ghosting* is the residue that remains when lines are difficult to remove. These unsightly ghost images are reproduced in a print. Materials that have good erasability are easy to clean up.

Dimensional stability is the quality of the media to not alter size due to the effects of atmospheric conditions such as heat, cold, and humidity. Some materials are more dimensionally stable than others.

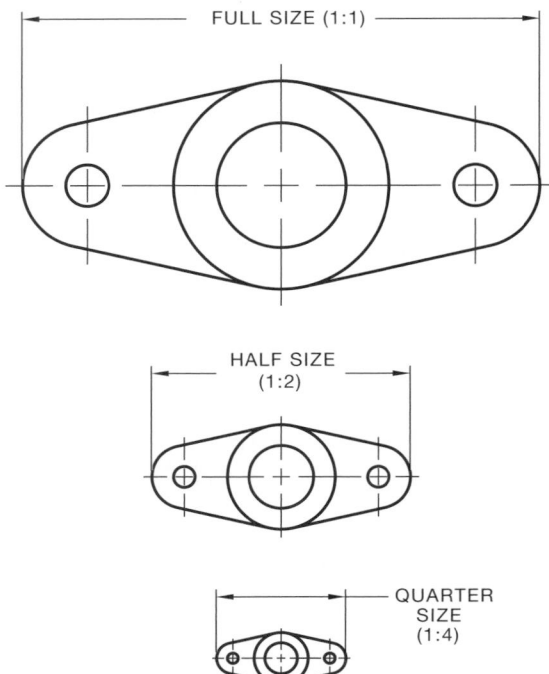

FIGURE 3.43 ■ ■ A sample drawing represented at full scale, half scale, and quarter scale.

Transparency is one of the most important characteristics of drawing media. The diazo reproduction method requires light to pass through the material. The final goal of a drawing is good reproduction, so the more transparent the material the better the reproduction, assuming that the image drawn is professional quality. Transparency is not important when using the photocopy process.

PAPERS AND FILMS
Vellum

Vellum is drafting paper that is specially designed to accept pencil or ink. Lead on vellum is the most common combination used for manual drafting. Many vendors manufacture quality vellum for drafting purposes. Each claims to have specific qualities that you should consider in the selection. Vellum is the least expensive material having good smoothness and transparency. Use vellum originals with care. Drawings made on vellum that require a great deal of use could deteriorate, as vellum is not as durable as other material. Some brands are better erased than others. Affected by humidity and other atmospheric conditions, vellum generally is not as dimensionally stable as other materials.

Polyester Film

Polyester film, also known by its brand name, Mylar, is a plastic material that offers excellent dimensional stability, erasability, transparency, and durability. Drawing on Mylar is best accom-

plished using ink or special polyester leads. Do not use regular graphite leads as they smear easily. Drawing techniques that drafters use with polyester leads are similar to graphite leads except the polyester leads are softer and feel like a crayon when used.

Mylar is available with a single or double matte surface. *Matte* is a nonglossy, slightly textured surface. The double matte film has texture on both sides so drawing can be done on either side if necessary. Single matte film is the most common in use with the non-drawing side having a slick surface.

Protect the Mylar Surface

When using Mylar, you must be very careful not to damage the matte by erasing. Erase at right angles to the direction of your lines and do not use too much pressure. Doing so helps minimize damage to the matte surface. Once the matte is destroyed and removed, the surface will not accept ink or pencil. Also, be cautious about getting moisture on the Mylar surface. Oil from your hands can cause your pen to skip across the material.

Normal handling of drawing film is bound to soil it. Inked lines applied over soiled areas do not adhere well and in time will chip off. It is always good practice to keep the film clean. Soiled areas can be cleaned effectively with special film cleaner.

Mylar is much more expensive than vellum; however, it should be considered when excellent reproductions, durability, dimensional stability, and erasability are required of original drawings.

Reproduction

The one thing most designers, engineers, architects, and drafters have in common is that their finished drawings are intended for reproduction. The goal of every professional is to produce drawings of the highest quality that give the best possible prints when they are reproduced.

Many of the factors that influence the selection of media for drafting have been discussed; however, the most important factor is reproduction. The primary combination that achieves the best reproduction is the blackest, most opaque lines or images on the most transparent base or material. Each of the materials mentioned makes good prints if the drawing is well done. If the only concern is the quality of the reproduction, ink on Mylar is the best choice. Some products have better characteristics than others. Some individuals prefer certain products. It is up to the individual or company to determine the combination that works best for their needs and budget. The question of reproduction is especially important when sepias must be made. (Sepias are second- or third-generation originals. See Sepias in this chapter.)

Look at Figure 3.44 for a magnified view of graphite on vellum, plastic lead on Mylar, and ink on Mylar. Judge for yourself which material and application provides the best reproduction. As you can see from Figure 3.44, the best reproduction is achieved with a crisp, opaque image on transparent material. If your original drawing is not good quality, it will not get better on the print.

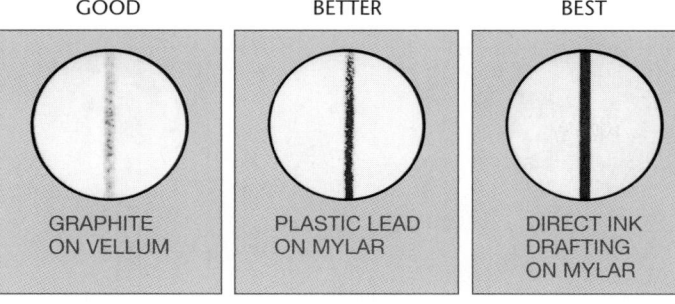

GOOD	BETTER	BEST
GRAPHITE ON VELLUM	PLASTIC LEAD ON MYLAR	DIRECT INK DRAFTING ON MYLAR

FIGURE 3.44 ■ A *magnified* comparison of graphite on vellum, plastic lead on Mylar, and ink on Mylar. *Courtesy Koh-I-Noor, Inc.*

SHEET SIZES, TITLE BLOCKS, AND BORDERS

ASME/ANSI standard sheet sizes and format are specified in the documents ANSI Y14.1—1980 (Reaffirmed 1987) *Drawing Sheet Size and Format,* and ASME Y14.1M—1992 *Metric Drawing Sheet Size and Format.*

All professional drawings have title blocks. Standards have been developed for the information put into the title block and on the surrounding sheet next to the border so the drawing is easier to read and file than drawings that do not follow a standard format.

Sheet Sizes

ANSI Y14.1 specifies sheet size specifications in inches as follows:

Size Designation	Size in Inches	
	Vertical	Horizontal
A	8½	11 (horizontal format)
	11	8½ (vertical format)
B	11	17
C	17	22
D	22	34
E	34	44
F	28	40

There are four additional size designations (G, H, J, and K); these apply to roll sizes.

The *M* in the title of the document Y14.1M means all specifications are given in metric. Standard metric drawing sheet sizes are designed as follows:

Size Designation	Size in Millimeters	
	Vertical	Horizontal
A0	841	1189
A1	594	841
A2	420	594
A3	297	420
A4	210	297

Longer lengths are referred to as elongated and extra-elongated drawing sizes. These are available in multiples of the short side of the sheet size.

Standard inch sheet sizes are shown in Figure 3.45a, and metric sheet sizes are shown in Figure 3.45b.

Zoning

Some companies use a system of numbers along the top and bottom margins and letters along the left and right margins called *zoning*. Notice in Figure 3.45a that numbered and lettered zoning begins on C-size drawing sheets, and in Figure 3.45b zoning is used on A3 sheet sizes and larger. Zoning allows the drawing to read like a road map. For example, you can refer to the location of a specific item as D4, which means that the item can be found at or near the intersection of D across and 4 up or down.

Title Blocks

Companies generally have title blocks and borders preprinted on drawing sheets to reduce drafting time and cost. Drawing sheet sizes and sheet format items such as borders, title blocks, zoning, revision columns, and general note locations have been standardized so that the same general relationship exists between engineering drawings internationally. The ANSI Y14.1 and ASME Y14.1M documents specify the exact size and location for each item found on the drawing sheet. It is recommended that standard sheet sizes and format be followed to improve readability, handling, filing, and reproduction. Each company may use a slightly different design, although the following basic information is located in approximately the same place on most engineering drawings:

1. Title block. Lower right corner.
   a. Company name.
   b. Confidential statement.
   c. Unspecified dimensions and tolerances.
   d. Sheet size.
   e. Drawing number.
   f. Part name.
   g. Material.
   h. Scale.
   i. Drafter signature.
   j. Checker signature.
   k. Engineer signature.
2. Revision column. Upper right corner, over or next to the title block.
   a. Revision symbol, number or letter.
   b. Description.
   c. Drafter.
   d. Date.
3. Border line.
   a. With or without zoning.

Figure 3.46, page 86, shows a sample title block.

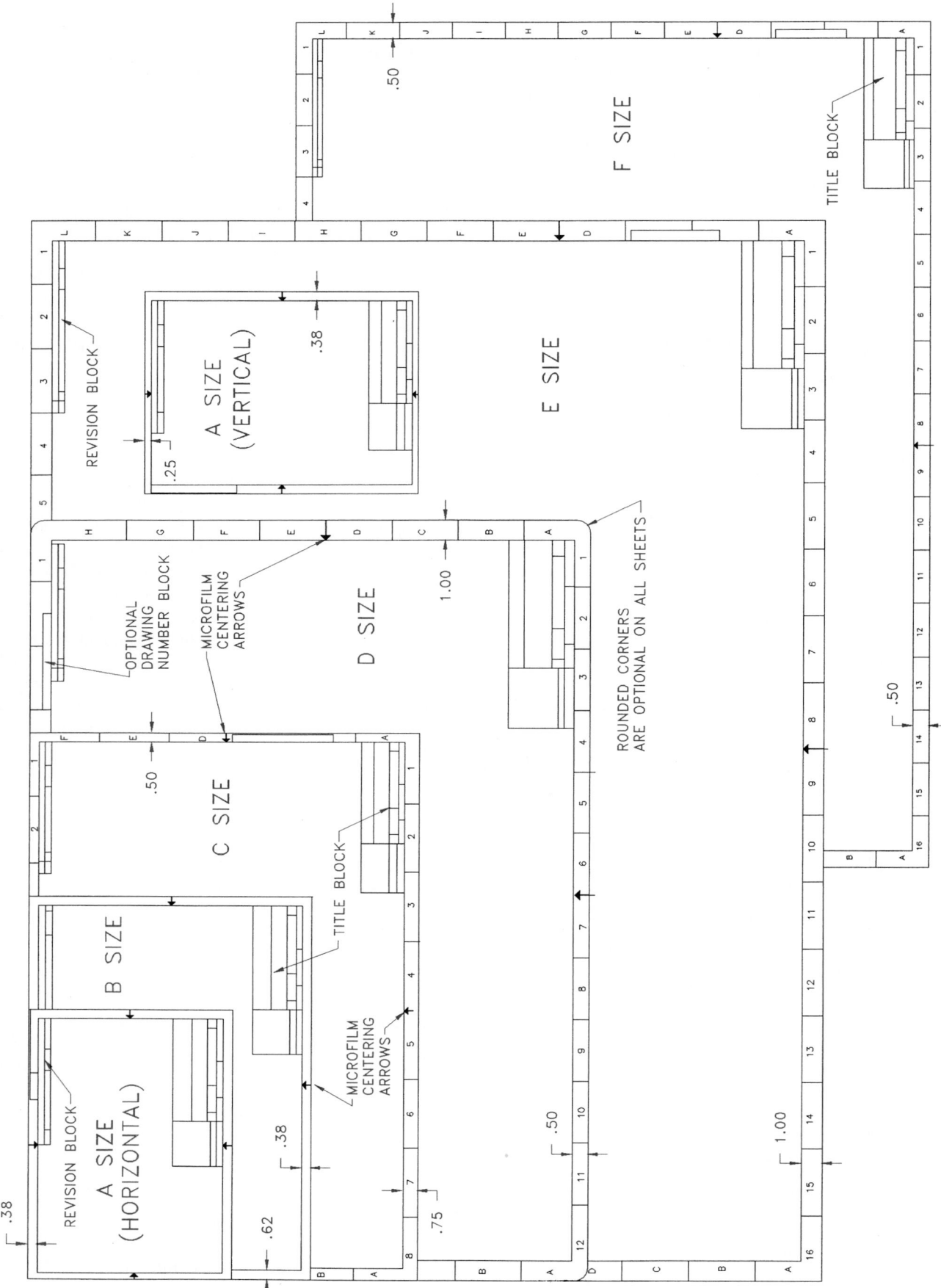

FIGURE 3.45(a) ■ Standard inch drawing sheet sizes.

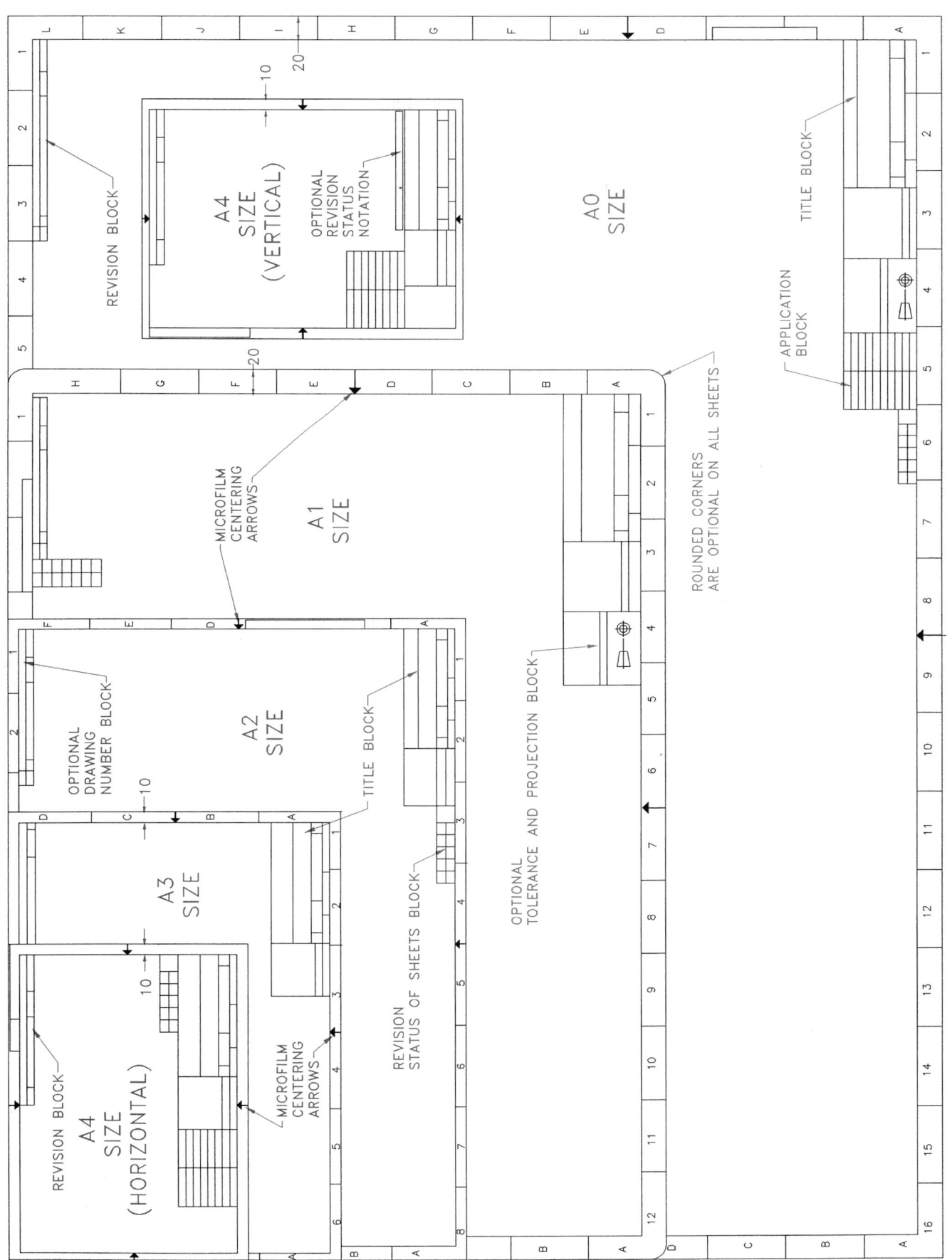

FIGURE 3.45(b) ■ Standard metric drawing sheet sizes.

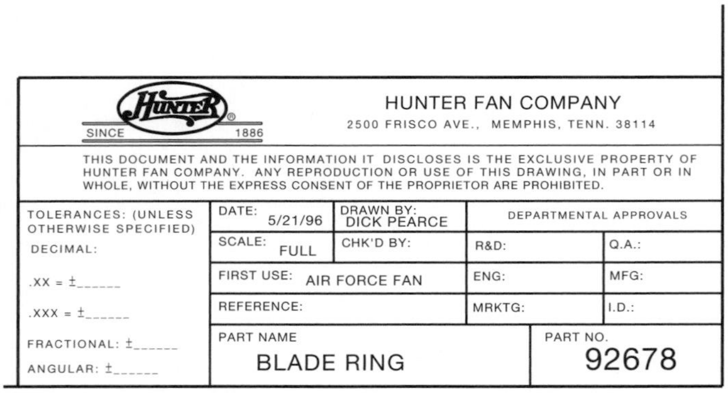

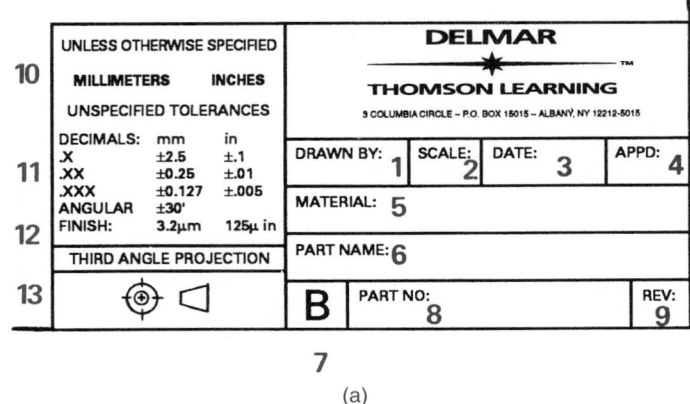

FIGURE 3.46 ■ Sample title block. *Courtesy Hunter Fan Company, Memphis, TN.*

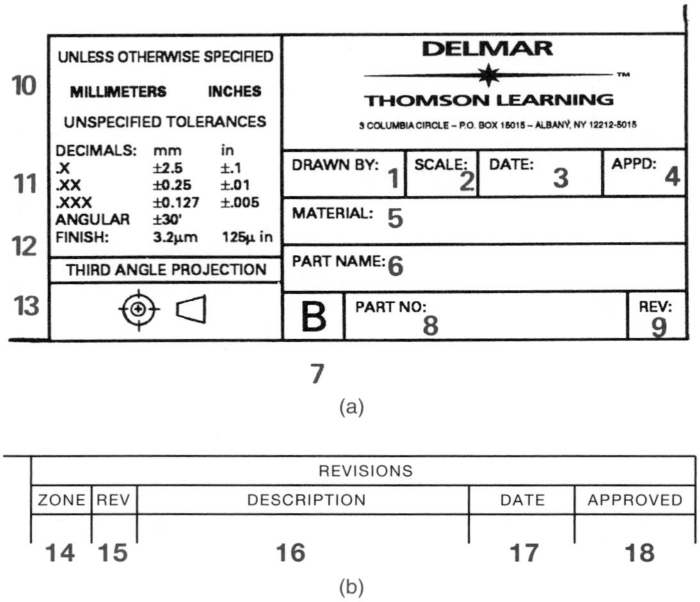

Title Block Definitions

■ *Tolerances* are discussed in detail in Chapter 10, although it is important to know that a tolerance is a given amount of acceptable variation in a size or location dimension. All dimensions have a tolerance.

■ *Millimeters* and *inches*. All dimensions are in millimeters (mm) or inches (in.) unless otherwise specified.

■ *Unless otherwise specified* means that, in general, all of the features or dimensions on a drawing have the relationship or specifications given in the title block unless a specific note or dimensional tolerance is provided at a particular location in the drawing.

■ *Unspecified tolerances* refers to any dimension on the drawing that does not have a tolerance specified. This is when the dimensional tolerance required is the same as the general tolerance shown in the title block.

■ *Revisions*. When parts are redesigned or altered for any reason, the drawing will be changed. All drawing changes are commonly documented and filed for future reference. When this happens, the documentation should be referenced on the drawing so that users can identify that a change has been made. Before any revision can be made, the drawing must be released for manufacturing.

Title and Revision Block Instructions

The title block displayed in Figure 3.47a shows most of the common elements found in industrial title blocks. The revision block, Figure 3.47b, found in this format is located in the upper right corner of the drawing sheet. Some companies use other placement.

The large numbers shown on Figure 3.47 refer to the following instructions for completing the title and revision blocks. All lettering should be .125 or .188 in. high as indicated in the instructions. The lettering style should be vertical uppercase Gothic freehand, or mechanical lettering as specified by the instructor or company standards. Computer graphic systems are to use vertical uppercase letters of a ROMAN font unless otherwise specified.

FIGURE 3.47 ■ (a) Title block elements. (b) Revision block elements.

1. DRAWN BY: .125 in. high lettering. Identify yourself by using all your initials such as DRC, DAM, JLT, unless otherwise specified.

2. SCALE: .125 in. high lettering. Examples of scales to fill in this block are FULL or 1:1, HALF or 1:2, DBL or 2:1, QTR or 1:4, NONE.

3. DATE: .125 in. high lettering. Fill in this block using the following order: day, month, and year such as 18 NOV 00, or numerically with month, day, and year such as 11/18/00.

4. APVD: .125 in. high lettering. This block is to be used by the instructor or checker who initials his or her approval of the drawing.

5. MATERIAL: .125 in. high lettering. Describe here the material used to make the part, for example, BRONZE, CAST IRON, SAE 4320.

6. PART NAME: .188 in. high lettering. Insert here the name of the part, such as COVER, HOUSING.

7. B: .188 in. high if unprinted. Gives sheet sizes shown in Figure 3.45. When preprinted borders are used, this sheet size designation is usually printed.

8. PART NO: .188 in. high lettering. Fill in the drawing number of the part. Most companies have their own part numbering system. While numbering systems differ, they are often keyed to categories such as the disposition of the drawing (casting, machining, assembly), materials used, related department within the company, or a numerical classification of the part.

9. REV: .188 in. high lettering. Fill in the revision letter of the part or drawing. A new or original drawing is — (dash) or 0 (zero). The first time a drawing is revised, the — or 0 changes to an A, for the second drawing change a B is placed here, and so on. The letters *I, O, S, X,* and *Z* are not used. When all of the available letters A through Y have been used,

double letters such as *AA* and *AB,* or *BA* and *BB* are used. Some companies use revision numbers rather than letters.

10. UNLESS OTHERWISE SPECIFIED: Unless otherwise specified, drawing dimensions will be given in millimeters or inches. When the predominant dimensions are in MIL-LIMETERS neatly blacken out INCHES; when in INCHES, blacken out MILLIMETERS.

11. UNSPECIFIED TOLERANCES: One, two, and three place decimals are established as MILLIMETERS or INCHES in a manner the same as item 10. For example, if the drawing dimensions are predominately MILLIMETERS, then INCH related tolerances are blackened out. Angular tolerances for unspecified angular dimensions are ±30'. These applications depend on company practice.

12. FINISH: In this block fill in the unspecified surface finish, for surfaces that are identified for finishing without a specific callout.

13. THIRD-ANGLE PROJECTION: The drawing assignments in this text are done using third-angle projection unless otherwise specified. A complete discussion of this topic is found in Chapter 7.

14. ZONE: This is the drawing zone where the change is located. Remember from an earlier discussion, the zoning allows the drawing to read like a road map; numbers along the top and bottom, and letters along the sides, direct you to a specific location on the drawing. If the location of the change is D4, then D4 is placed in the ZONE column for this change. The ZONE column is used only if the drawing has zoning.

15. REV: Fill in the revision letter or number in this block, such as A, B, C, etc. Succeeding letters are to be used for each Engineering Change Notice (ECN) or group of related ECNs regardless of quantity of changes on an ECN. *Note:* the REV: block in the title block must be changed to agree with the last REV letter in the revision block. Some companies use revision numbers.

16. DESCRIPTION: In this block, fill in the Engineering Change Notice (ECN) letter covering the Engineering Change Request (ECR) or group of ECRs that requires the drawing to be revised. Another option is to give a short description of the change. An Engineering Change Notice is a numbered document that provides information about the change. ECNs are discussed in Chapter 18.

17. DATE: Fill in the day, month, and year on which the ECN package is ready for release to production, such as 6 APR 00, or use month, day, and year numbers such as 4/6/00.

18. APPROVED: This column is for the initials of the person approving the change, and optional date.

Drafting revisions are completed in chronological order by adding horizontal columns and extending the vertical column lines.

The previous example showed you a typical title block and revision block. Each company has its own design and format that is similar, and most CADD programs have predefined sheet layout options with borders and title blocks. These are normally based on the ASME standard title blocks and revision blocks and can usually be customized for your own applications. The document ANSI Y14.1 gives the recommended title block and revision block dimensions shown in Figure 3.48. The title block shown in Figure 3.48 is

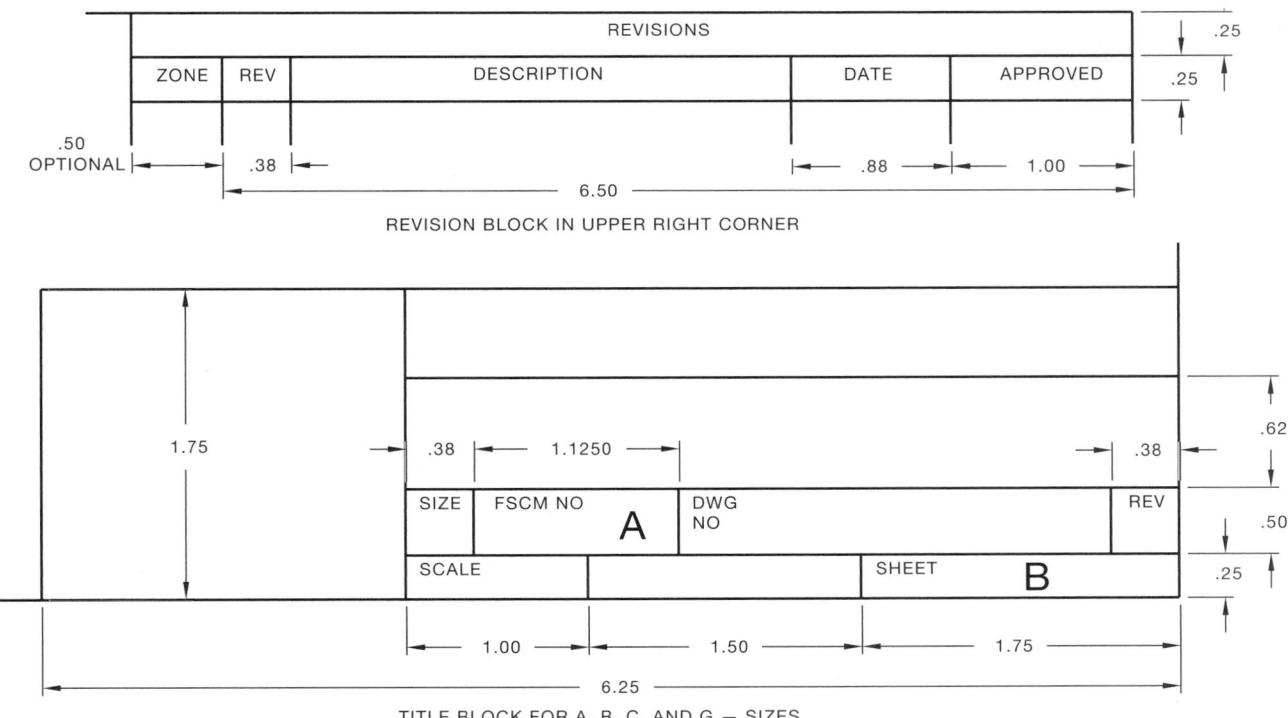

FIGURE 3.48 ■ Title block and revision block dimensions recommended by ANSI Y14.1.

the recommendation for A-, B-, C-, and G-size sheets. The recommended title block for D-, E-, F-, H-, J-, and K-size sheets is bigger, measuring 7.62 in. long and 2.50 in. high. The revision block is recommended for the upper right corner of the drawing border, but some companies place it in other locations, such as over the title block or to the left of the title block. Look again at Figure 3.48. Notice the large letters *A* and *B*. These are items found in the ANSI Y14.1 examples that are not displayed in Figure 3.47. These are described as follows:

A   This is the FSCM number, which is optional. *FSCM* stands for *Federal Supply Code for Manufacturers*. The FSCM number is a five-digit numerical code used on all drawings that produce items used by the federal government.

B   The SHEET compartment is for the identification of sheet numbers. This is used by companies that anticipate having the drawing for a particular application occupy more than one sheet. 1 OF 1 is placed here if the drawing is on one sheet. If the drawing is on two sheets, the first has 1 OF 2 and the second has 2 OF 2 in this compartment.

Notice that the ANSI title block in Figure 3.48 has DWG NO rather than PART NO, as in Figure 3.47. This is a company preference.

TAPING DOWN YOUR DRAWING

Before you tape down your drawing, be sure your hands and equipment are clean. You should always use specially designed "drafting tape" because it can be removed without tearing the drafting paper. If you are using a drafting machine, first be sure that the scales are properly adjusted. Refer to your manufacturer's instructions. Never fold the drafting original, and try not to roll it. Follow these steps:

■ Start taping down the drafting paper or film by aligning one edge with your horizontal scale or parallel bar.

■ Use short pieces of tape that are approximately 1 in. long.

■ Tape one corner, normally the top left or right corner.

■ Move your hand across the sheet toward the opposite corner, forcing the paper tight to the table, and then tape the opposite corner.

■ Tape the other top corner and then force the paper down toward the opposite corner and tape it.

The drafting paper or film should be very tightly taped to your drafting board or table. If there are any edges curling up, you can add some tape to the edges to keep them down. Be very careful when you move the drafting equipment across the table, because catching an edge of the paper can easily tear the sheet and ruin a professional drawing.

DIAZO REPRODUCTION

Diazo prints, also known as ozalid dry prints or blue-line prints, are made with a printing process that uses an ultraviolet light that passes through a translucent original drawing to expose a chemically coated paper or print material underneath. The light does not go through the dense, black lines on the original drawing. Thus the chemical coating on the paper beneath the lines remains. The print material is then exposed to ammonia vapor which activates the remaining chemical coating to produce blue, black, or brown lines on a white or clear background. The print that results is a diazo, or blue-line print, not a blueprint. The term *blueprint* is now a generic term that is used to refer to diazo prints even though they are not true blueprints.

Making a Diazo Print

Before you run a print using a complete sheet, run a test strip to determine the proper speed setting.

To make a diazo print, place the diazo material on the feedboard, coated (yellow) side up. Position the original drawing, image side up, on top of the diazo material, making sure to align the leading edges. Use diazo material that corresponds in size with your original.

Using fingertip pressure from both hands, push the original and diazo material into the machine. The light passes through the original and exposes the sensitized diazo material except where images (lines and lettering) exist on the original. When the exposed diazo material and the original drawing emerge from the printer section, remove the original and carefully feed the diazo material into the developer section, as shown in Figure 3.49.

In the developer section, the sensitized material that remains on the diazo paper activates with ammonia vapor and blue lines form, although some diazo materials make black or brown lines. Special materials are also available to make other colored lines, or to make transfer sheets and transparencies. Figure 3.49 shows the diazo print process.

Storage of Diazo Materials

Diazo materials are light-sensitive so they should always be kept in a dark place until ready for use. Long periods of exposure to room light cause the diazo chemicals to deteriorate and reduce the quality of your print. If you notice brown or blue edges around the unexposed diazo material, it is getting old. Always keep it tightly stored in the shipping package and in a dark place such as a drawer.

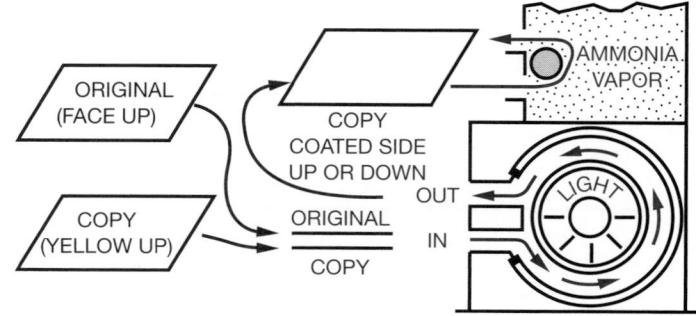

FIGURE 3.49 ■ The diazo print process.

Some people prefer to keep all diazo material in a dark refrigerator. Doing so preserves the chemical for a long time.

Sepias

Sepias are diazo materials that are used to make secondary originals. A secondary or second-generation original is actually a print of an original drawing that can be used as an original. The term *generation* refers to the number of times a copy of an original drawing is reproduced and used to make other copies. Changes can be made on the secondary original while the original drawing remains intact. The diazo process is performed with material called sepia. Sepia prints normally form dark brown lines on a clear (translucent) background; therefore, they are sometimes called brown lines. In addition to being used as a secondary original to make alterations to a drawing without changing the original, sepias are also used when originals are required at more than one company location.

Sepias are normally made with the original face down on the sensitized sepia material. The balance of the print process is the same as just described except that the sepia should be fed into the developer coated-side down. Doing so allows the coating on the sepia material to come into direct exposure with the ammonia in the development chamber. Experiment both ways and observe the quality of your reproductions. Standard diazo copies may reproduce better with the coated-side down; it depends on your machine and materials.

Sepias are made in reverse (reverse sepias) so that corrections or changes can be made on the matte side (face side) and erasures can be made on the sensitized side. Sepia materials are available in paper or polyester. The resulting paper sepia is similar to vellum, and the polyester sepia is Mylar. Drawing changes can be made on sepia paper with pencil or ink, and on sepia Mylar with polyester lead or ink. Figure 3.50 shows an illustration of the reverse sepia print process. The use of sepia reproduction is being rapidly replaced by computer-aided drafting.

Evolution in Reproduction Technology

The photocopy process and CADD printers and plotters are rapidly replacing the diazo process. The photocopy process, which is explained later, has many advantages over the diazo process. The advantages include quality reproduction in many sizes, use of most common materials, and no hazardous ammonia. The printers and plotters used with the CADD workstation create drawings on paper

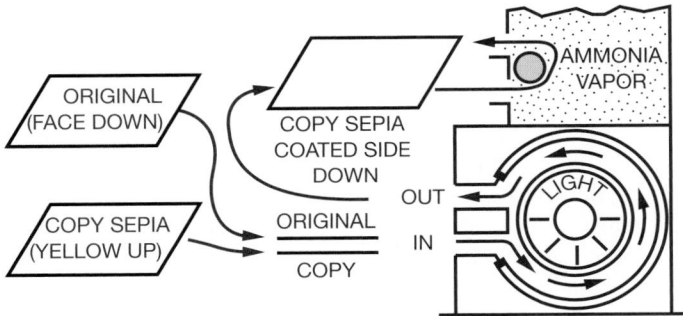

FIGURE 3.50 ■ The reverse sepia print process.

or other media, which are referred to as *hard copy*. Printers such as the laser printer or the ink-jet printer allow the company to produce quality hard copy drawings very quickly. The hard copy drawings can be printed on vellum material for further reproduction using the diazo or photocopy process. Chapter 4 of this text covers the options available with a CADD system.

SAFETY PRECAUTIONS FOR DIAZO PRINTERS

Ammonia has a strong, unmistakable odor. It may be detected, at times, while operating your diazo printer. Diazo printers are designed and built to provide safe operation from ammonia exposure. Some machines have ammonia filters while others require outside exhaust fans. When print quality begins to deteriorate and it has been determined that the print paper is in good condition, the ammonia may be old. Ammonia bottles should be changed periodically, as determined by the quality of prints. When the ammonia bottle is changed, it is generally time to change the filter also.

When handling bottles of ammonia or when replacing a bottle supplying a diazo printer, be sure to follow the manufacturer's recommendations. Specific procedures should be provided with your machine that explain safety issues, such as eye protection, ammonia contact, inhaling ammonia, ammonia exposure first aid, and ultraviolet light exposure when running the diazo machine. Keep in mind that ammonia is a hazardous waste and must be disposed of properly. Refer to your local, state, or federal Environmental Protection Agency (EPA) guidelines for proper disposal of ammonia. Several governmental agencies including the Occupational Health and Safety Agency [OSHA] and the Environmental Protection Agency [EPA] provide guidelines pertinent to identifying, classifying and disposing of hazardous materials and waste by-products. Manufacturing facilities will also generate numerous hazardous materials including paints, solvents, oils and fluids that will need to be safely handled, stored and disposed of properly according to established guidelines.

PHOTOCOPY REPRODUCTION

Photocopy printers are also known as engineering copiers. Prints can be made on bond paper, vellum, polyester film, colored paper, or other translucent materials. The reproduction capabilities also include instant print sizes ranging from 45 to 141 percent of the original size. Larger or smaller sizes are possible by enlarging or reducing in two or more steps, that is by making a second print from a print.

Almost any large original can be converted into a smaller-sized reproducible print, and then the secondary original can be used to generate diazo or photocopy prints for distribution, inclusion in manuals, or for more convenient handling. Also, a random collection of mixed-scale drawings can be enlarged or reduced and converted to one standard scale and format. Reproduction clarity is so good that halftone illustrations (photographs) and solid or fine line work have excellent resolution and density.

Engineering copiers are rapidly replacing diazo print machines in the engineering and drafting world because of their versatility and competitive cost. Also, engineering copiers do not use ammonia, which is an important consideration when purchasing a print machine.

HOW TO PROPERLY FOLD PRINTS

As you have already learned, prints may come in a variety of sizes ranging from small 8½ × 11 in. to 34 × 44 in. or larger. It is easy to file the 8½ × 11 in. size prints because standard file cabinets are designed to hold this size. There are file cabinets available called flat files that may be used to store full-size unfolded prints. However, many companies use standard file cabinets. Larger prints must be properly folded before they can be filed in a standard file cabinet. It is also important to properly fold a print if it is to be mailed.

Folding large prints is much like folding a road map. Folding is done in a pattern of bends that results in the title block and sheet identification ending up on the front. This is desirable for easy identification in the file cabinet. The proper method used to fold prints also aids in unfolding or refolding prints. Look at Figure 3.51 to see how large prints are properly folded.

MICROFILM

Microfilm is photographic reproduction on film of a drawing or other document that is highly reduced for ease in storage and sending from one place to another. When needed, equipment is available for enlargement of the microfilm to a printed copy. Special care must be taken to make the original drawing of the best possible quality. The reason for this is that during each "generation," the process of "blowback" makes the lines narrower in width and less opaque than the original. The term *generation* refers to the number of times a copy of an original drawing is reproduced and used to make other copies. For example, if an original drawing is reproduced on sepia or microfilm, and the sepia or microfilm is used to make other copies, this is referred to as a *second generation* or *secondary original*. When this process has been done four times, the drawing is called a *fourth generation*. The term *blowback* means bringing the drawing from the microfilm back to a printed copy. A true test of the original drawing's quality is the ability to maintain good reproductions through the fourth generation of reproduction.

In many companies original drawings are filed in drawers by drawing number. When a drawing is needed, the drafter finds the original, removes it, and makes a copy. This process works well although, depending on the company's size or the number of drawings generated, drawing storage often becomes a problem. Sometimes an entire room is needed for drawing storage cabinets. Another problem occurs when originals are used over and over. They often become worn and damaged, and old vellum becomes yellowed and brittle. Also, in case of a fire or other kind of destruction, originals may be lost and endless hours of drafting vanish. For these and other reasons, microfilm has been used for storage and reproduction of original drawings. ASME Y14.5M—1994 provides recommended microfilm reduction factors for different sheet sizes. These help ensure standardization.

While the microfilm storage of old drawings still exists in some companies, CADD files have replaced the use of microfilm for most modern applications. The CADD Application explains to you why this evolution has happened.

DIGITIZING EXISTING DRAWINGS

Digitizing is a method of transfering drawing information from a digitizer into the computer. The digitizing function of automatically sending commands to the computer from a tablet menu is discussed in Chapter 4. A digitizing tablet may be used to take existing drawings and digitize the information into the computer. Digitizer tablets range in size from 6 × 6 in. to 44 × 60 in. Most school and industry CAD departments use 11 × 11 in. digitizer tablets to input commands from standard and custom tablet menus. In most cases companies do not have the time to convert existing drawings to CAD, because the CAD systems are heavily used for new product drawings. Therefore, an increasing number of businesses digitize existing drawings for other companies.

While the digitizing of existing drawings still has some applications, scanning of existing drawings into CADD has become widely used because of improvements in this technology. Scanning of existing drawings is discussed next. Digitizer tablets are still used by some CADD systems; however, CADD software has rapidly evolved for use with a mouse to access commands from pull-down menus, tool bar buttons, icon menus, and dialog boxes. Mouse technology has also advanced to provide you with maximum flexibility when working at your CADD station. *Software* is the instructions that runs the computer and determines what you can do with your computer. A *mouse* is a device that allows you to control the

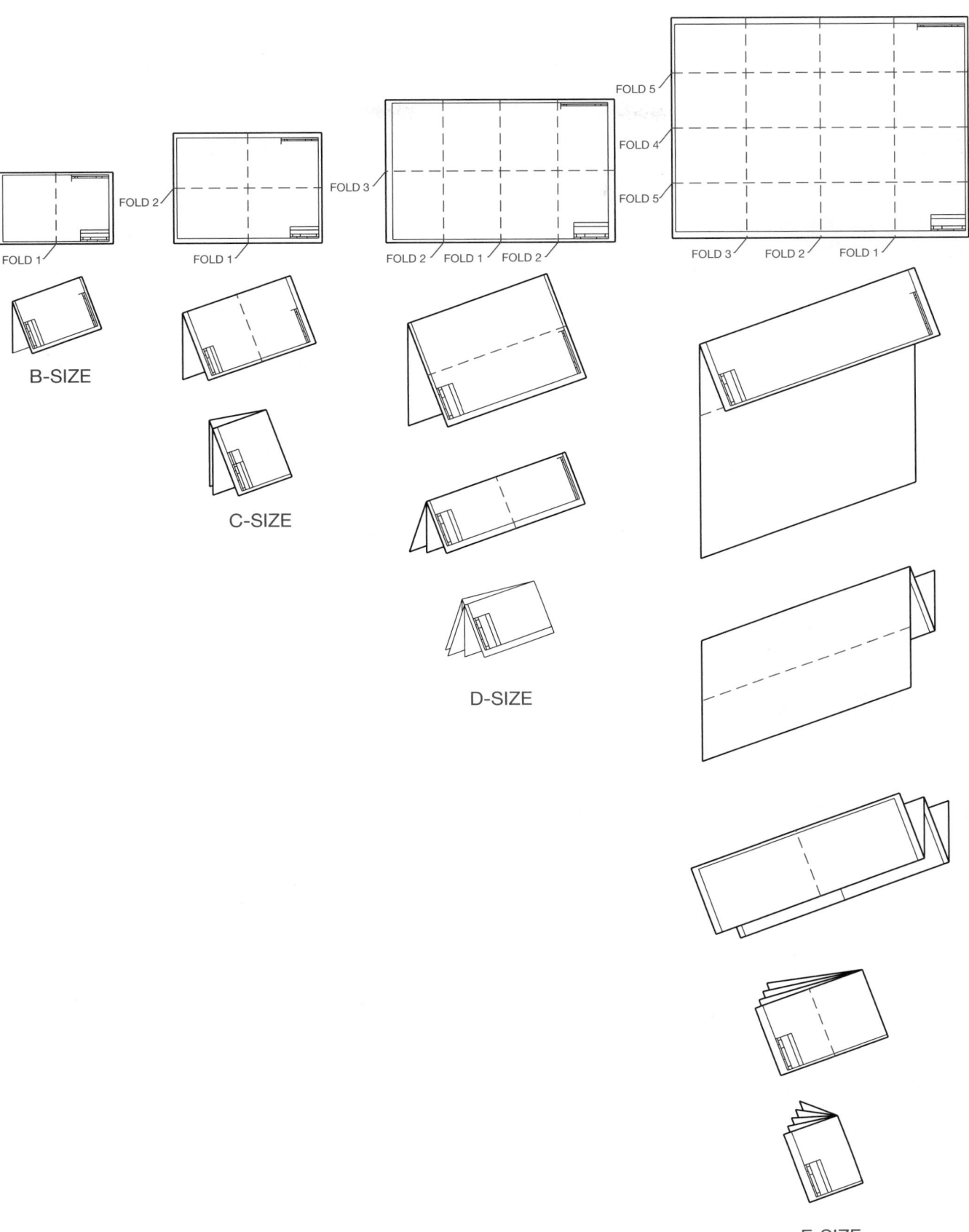

FIGURE 3.51 ■ How to properly fold B-size, C-size, D-size, and E-size prints.

CAD/CAM

The optimum efficiency of design and manufacturing methods exists without reproducing a single copy of a drawing of a part. Computer-aided systems may directly link the engineering department by way of computer-aided design (CAD) to the manufacturing department machines operated by computer-aided manufacturing (CAM). Through the use of numerical control dimensioning and layout practices, the drafter or designer can prepare an engineering drawing on a computer system and then have the computer convert the drawing's coordinate data to computer numerical control (CNC) data. The computer software for this translation reproduces data for manufacturing with a numerically controlled machine tool. When the electronics are connected directly from engineering to manufacturing, there is a direct link between CAD and CAM. A copy of the actual drawing is never generated, unless someone needs a hard copy (print) for review away from the computer terminal.

ANGLE MEASUREMENT IN RADIANS

Although drafting machines, surveying instruments, and many modern machine tools work with degrees and minutes, many computer and calculator applications require angles to be measured in decimal fractions of a degree. It is important to be able to apply math to convert from one type of angle measurement to another because not all machines and computers do this automatically.

To convert from minutes of a degree to decimal degrees, divide the number of minutes by 60. For example, 7° 40' is the same as 7.666666°, or rounding, 7.67° because 40 ÷ 60 = .666666. To convert from decimal degrees to minutes, multiply the decimal portion of the angle by 60. For example, 18.75° is the same as 18° 45′ because .75 × 60 = 45.

pointer or cursor on the computer screen. This is a basic definition, because mouse technology for CADD applications can be much more advanced. These options are discussed in detail in Chapter 4.

OPTICAL SCANNING OF EXISTING DRAWINGS

Scanning is a method of reproducing existing drawings in the form of computer drawings. Scanners work much the same way as taking a photograph of the drawing. One advantage of scanning over digitizing existing drawings is that the entire drawing, including dimensions, symbols, and text, is transferred to the computer. A disadvantage is that some drawings, when scanned, require a lot of editing to make them presentable. The scanning process picks up visual graphic information from the drawing. What appears to be a dimension, for example, is only a graphic representation of the dimension picked up from the original drawing. What you get is an exact duplicate of the existing drawing. However, if you want the dimensional information to be technically accurate, the dimensions must be redrawn and edited on the computer. There are software packages available that make this editing process quick and easy.

After the drawing is scanned, the image is sent to a raster converter that translates information to digital or vector format. A *raster* is an electron beam that generates a matrix of pixels. *Pixels* make up the drawing image on the computer as explained in Chapter 4. A raster editor is then used to display the image for changes.

In the final analysis, companies involved in scanning can, in many cases, reproduce existing drawings more efficiently than companies that digitize existing drawings.

Some scanning software runs on a microcomputer and works in conjunction with an electronic scanning camera. When an existing drawing has been transferred to the computer, it becomes a CADD drawing file that may be edited as necessary. Optical scanning accessories can automatically scan drawings from paper, vellum, film, or blue line and convert the copy image into a computer file.

WEB SITE RESEARCH

The following Web sites can provide you with information about drafting media, supplies, and equipment:

http://koh-I-noor.com/
http://staedtler.com/
http://chartpak.com/

The following Web site provides information about a specific brand of drafting machine and options:

http://www.mutoh.be/drafting.htm

The following Web site has drafting products including furniture, machines, supplies, and media:

http://www.allinonesupply.com/

CHAPTER 3

Drafting Equipment Test

DIRECTIONS

Answer the questions with short complete statements or drawings as needed.

PART 1 · MANUAL DRAFTING EQUIPMENT

1. Describe a mechanical drafting pencil.
2. Describe an automatic drafting pencil.
3. List three lead sizes available for automatic pencils.
4. Identify the lead hardness that is normally recommended for the following functions:
   Construction lines
   Thin lines
   Thick lines
5. How should the automatic pencil be held when drawing lines?
6. Identify the recommended direction that the pencil should be moved when drawing a horizontal line for a right-handed and left-handed person.
7. List three characteristics recommended for technical pen drafting inks.
8. Describe an advantage and a disadvantage of using electric erasers.
9. Name the type of compass most commonly used by professional drafters.
10. Why should templates, instead of compasses, be used whenever possible?
11. Why should a compass point with a shoulder be used whenever possible?
12. Identify two common lead points that may be used in compasses.
13. What is the advantage of placing drafting tape or plastic adhesive disks at the center of a circle before using a compass?
14. A good divider should not be too loose or too tight. Why?
15. List two divider uses.
16. List two basic types of manual drafting machines.
17. How many degrees are in the interval between the locking points for most drafting machine heads?
18. How many minutes are there in one degree, seconds in one minute?
19. What is the degree of accuracy achieved by a drafting machine vernier?
20. Why should drafting machine scales be checked for alignment daily?
21. Why should centerlines be established before drawing a template circle?
22. Describe the recommended procedure for drawing smooth irregular curves.
23. Show how the following scales are noted on a mechanical drawing:
    Full scale, half scale, quarter scale, twice scale, ten scale.
24. According to ANSI standards, identify the drawing note that indicates that all dimensions are in metric.
25. Why is it important to keep your drafting equipment and hands clean?

PART 2 · MEDIA AND REPRODUCTION

1. List five factors that influence the purchase and use of drafting materials.
2. Why is media transparency so important when reproducing copies using the diazo process?
3. Describe vellum.
4. Describe polyester film.
5. What is another name for polyester film?
6. Define matte, and describe the difference between single and double matte.
7. Which of the following combinations would yield the best reproduction: graphite on vellum, plastic lead on polyester film, or ink on polyester film?
8. What are the primary elements that will give the best reproduction?
9. Identify the standard sheet sizes as defined by ANSI for the following sheet size designations: A; B; C; D; E; A0; A1; A2; A3; A4.
10. Describe zoning.
11. Define the following title block terms: tolerances, millimeters and inches, unless otherwise specified, unspecified tolerances, and revisions.
12. Given the following title block, describe the numbered elements 1 through 13.

13. What is another name for the diazo print?
14. Is a diazo the same as a blueprint? Explain.
15. Describe how the diazo process functions.
16. List two advantages of the photocopy reproduction method over the diazo process.
17. Define the microfilm process.
18. Discuss how computerized reproduction can take place without generating a single copy.

19. Why is it important to properly fold prints?
20. Discuss the advantages of CADD file storage over microfilm.
21. Define scanning.
22. What is the advantage of scanning an existing drawing over digitizing the same drawing?

23. Define Internet.
24. Define software.
25. How can a drawing be sent from your computer to the computer at a school or company in another state?

CHAPTER

3 Drafting Equipment Problems

PART 1 · READING SCALES AND DRAFTING MACHINE VERNIERS

1. Given the following civil engineer's scale, determine the readings at A, B, C, D, and E.

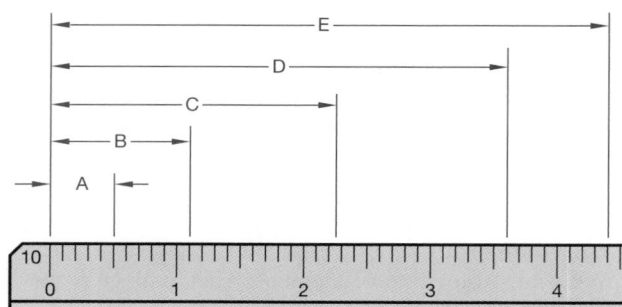

FULL SCALE = 1:1

2. Given the following civil engineer's scale, determine the readings at A, B, C, and D.

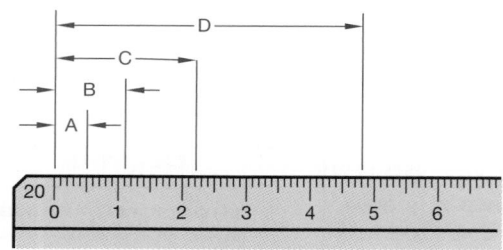

HALF SCALE = 1:2

3. Given the following architect's scale, determine the readings at A, B, C, D, E, and F.

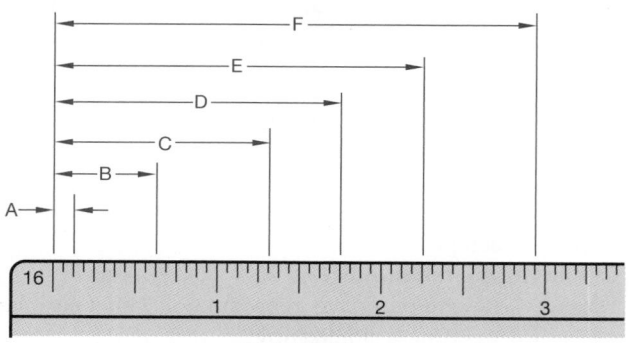

FULL SCALE = 1:1

4. Given the following metric scale, determine the readings at A, B, C, D, and E.

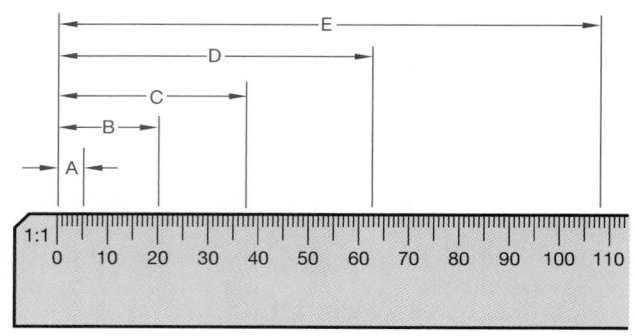

FULL SCALE = 1:1

5. Given the following metric scale, determine the readings at A, B, C, D, and E.

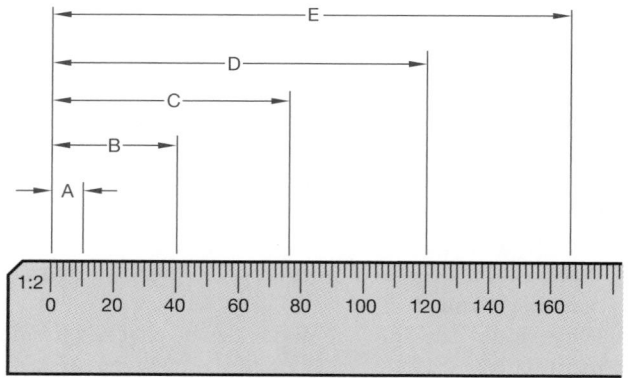

HALF SCALE = 1:2

6. Given the following architect's scale, determine the readings at A, B, C, and D.

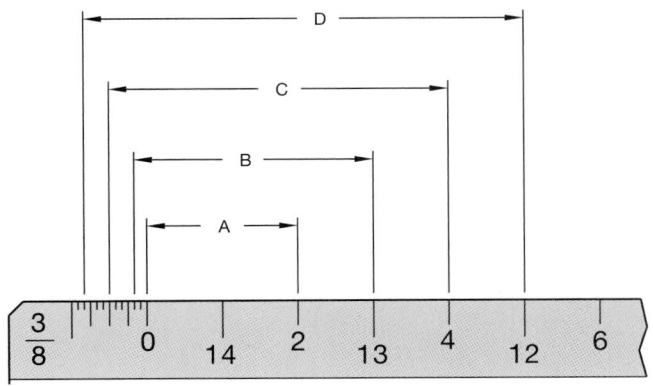

7. Given the following mechanical engineer's scale, determine the readings at A, B, C, and D.

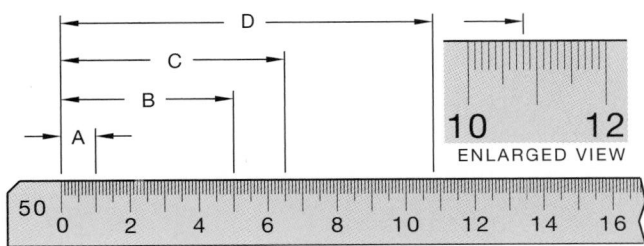

FULL SIZE; 1:1; 1"=1"

8. Given the following mechanical drawing, use the civil engineer's or mechanical engineer's 1:1 (10) scale to determine the dimensions at A, B, C (to be calculated as shown), D, E, F, and G.

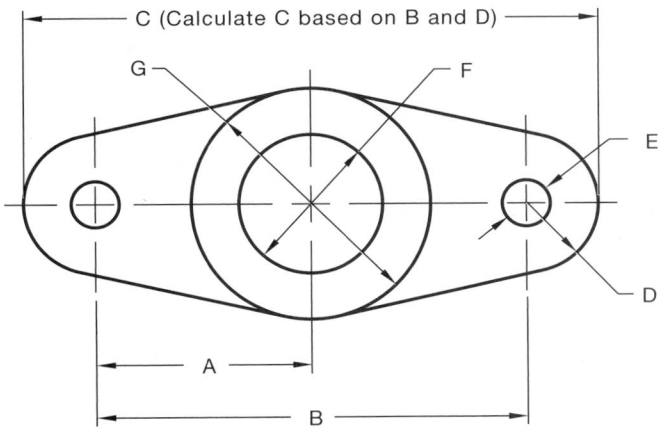

9. Given the following partial floor plan, use the architect's ¼' = 1'–0" scale to determine the dimensions at A, B, C, D, E, F, and G.

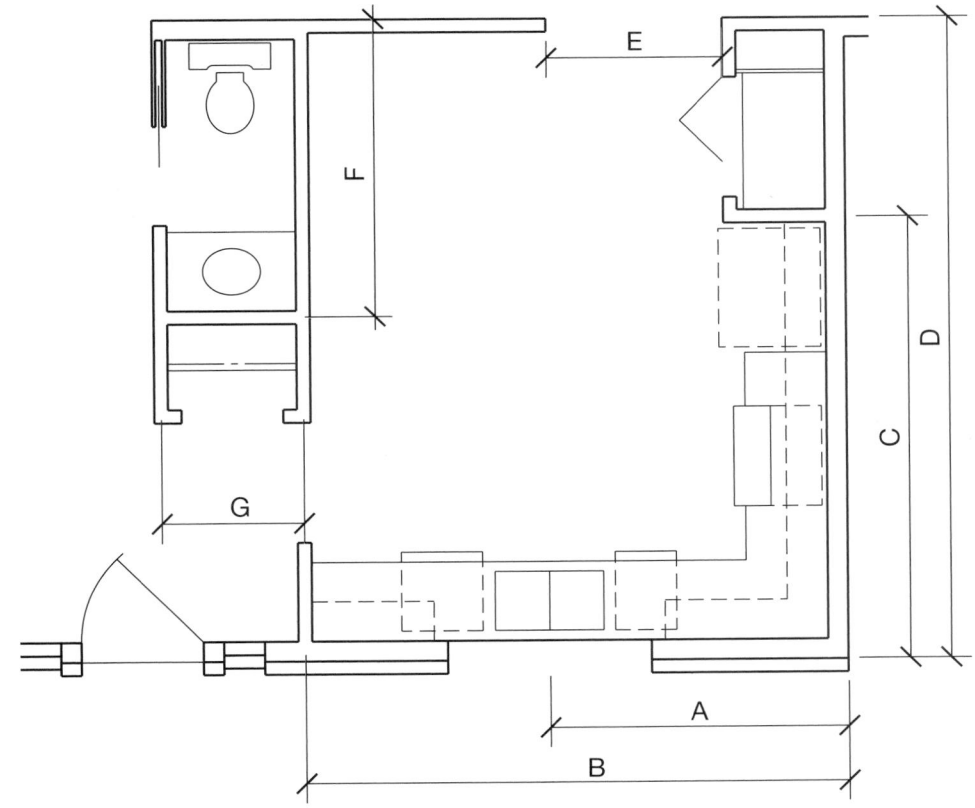

10. Given the following drafting machine protractors, determine the angular readings.

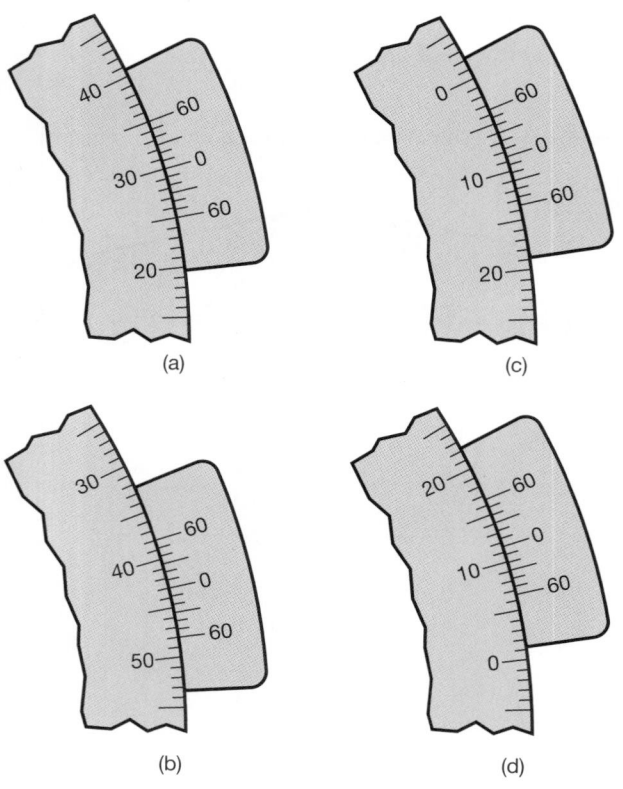

(a)

(c)

(b)

(d)

MATH PROBLEMS

Convert the following angle measurements to decimal degrees.
1. 15'
2. 7° 30'
3. 18° 5'
4. 200° 18'
5. −13° 42'

Convert the following angle measurements to degrees and minutes.
6. 60.4°
7. 9.5°
8. .27°
9. 177.8°
10. −45.1°

PART 2 · READING A TITLE BLOCK

Given the following title block, describe the numbered elements 1 through 13.

	UNLESS OTHERWISE SPECIFIED		

10 MILLIMETERS   INCHES

UNSPECIFIED TOLERANCES

DECIMALS:	mm	in
.X	±2.5	±.1
.XX	±0.25	±.01
.XXX	±0.127	±.005
ANGULAR	±30'	
FINISH:	3.2μm	125μ in

11

12

THIRD ANGLE PROJECTION

13   ⊕ ◁

DELMAR

★

THOMSON LEARNING ™

3 COLUMBIA CIRCLE – P.O. BOX 15015 – ALBANY, NY 12212-5015

DRAWN BY: **1**	SCALE: **2**	DATE: **3**	APPD: **4**

MATERIAL: **5**

PART NAME: **6**

B	PART NO: **8**	REV: **9**

7

Computer-Aided Design and Drafting (CADD)

OBJECTIVES

After completing this chapter, you will:

■ Describe CADD design formats.

■ Explain the CADD environment and its related disciplines.

■ Describe the computers and peripheral equipment that comprise the tools of CADD.

■ Describe some future directions of CADD.

■ Describe the project planning process and how it relates to all disciplines.

■ Identify the parts of the Cartesian coordinate system, and describe their function.

■ Demonstrate the process for determining drawing scale factors.

■ Describe the process of file management.

■ Identify the process used for research.

THE ENGINEERING DESIGN APPLICATION

THE PROJECT PLANNING PROCESS

The overall success of any project begins with the layout and planning stage. Proper and thorough planning is the key ingredient to ensure that the project will run efficiently, and produce drawings, designs, and products that are accurate, consistent, and well made. To that end, it is vital that you are familiar with the *drawing planning process* so that you can apply it to any type of drawing or project, in any drawing and design discipline.

When you are asked to create a design or model or to construct a new drawing, you should avoid the temptation to begin working without having a plan. Instead, take the time to sufficiently plan the drawing process using the outline given here. Because this is such an important aspect of solving any problem, or working on any type of project, it is discussed in detail in this chapter.

The Problem-Solving Process

A simple, three-step process enables you to immediately organize your thoughts, ideas, and knowledge about the specific project. Prior to beginning this process, it is critically important that you read all instructions available for the project. The process requires you (or the workgroup) to answer the following three questions about the project:

1. What do I know about the subject?

2. What do I need to know about the subject?

3. Where can I find the information I need?

The best way to answer these questions is to use *brainstorming*. In this process, individuals voice their thoughts and ideas regarding the specific topic, problem, or project at hand. Here are a few rules about working in a brainstorming session.

■ One person records all statements.

■ Place a reasonable time limit on the session, or agree to let it run until all ideas are exhausted.

■ Every statement, idea, or suggestion that is made is a good one.

■ Do not discuss or evaluate any statement or suggestion.

Return to the brainstorming list later to evaluate the items. Throw out items that the group agrees are not valid. Rank the remaining ideas in order of importance or validity. Now you are ready to conduct research and gather information required to begin and complete the project.

Research Techniques

Even though the Internet may be an inviting method of conducting research, it can also be time-consuming and nonproductive if it is not used in a logical manner. It is

(Continued)

THE ENGINEERING DESIGN APPLICATION *(continued)*

still important to know how to use traditional resources such as libraries, print media, and professional experts in the area of your study. Any research should be conducted using a simple process to find information efficiently.

- Define and list your topic, project, or problem.
- Identify key words of the topic.
- Identify all resources with which you are familiar that may provide information.
- Using the Internet, conduct quick keyword searches on your topic.
- Check with libraries for lists of professional indexes, periodicals and journals, specialized trade journals, and reference books related to specialized topics.
- Contact schools, companies, and organizations in your local area to find persons who are knowledgeable in the field of your research.
- Create a list of all potential resources.
- Prioritize the list and focus first on the most likely resource to provide the information you need.
- Begin your detailed study of the prioritized list.
- Ask questions.

Preparing a Drawing

After planning the drawing project as outlined above, establish a method for starting and completing the actual drawing. Always begin any drawing or design project by creating freehand sketches. A basic procedure to use for this drawing construction process is as follows:

1. Create a freehand sketch of the intended drawing layout showing all views, sections, details, and picto-

rials. Apply all required dimensions and notes. Using numbered items or colored pencils, label or list the components in the order they are to be drawn.
2. Try to determine if special views will be required.
3. Note the geometric shapes that must be drawn from scratch.
4. Determine if existing shapes can be edited to create new features.
5. Are there predrawn objects that can be inserted or referenced in the drawing?
6. Determine the locations and type of dimensions to be used.
7. Locate all local notes, general notes, and view titles.
8. Try to decide if using object values such as size, location, areas, etc. will assist you in drawing additional features, or provide you with useful information required for the project.
9. Determine the types of prints and plots or electronic images that are required for the project.
10. Begin work on the project

It may seem to the beginning drafter that these procedures constitute an inordinate amount of work just to plan a drawing or project. But keep in mind that if you do not take the time at the beginning of the project to plan and gather the information you need, you will still have to do so at some time during the life of the project. Therefore, is it better to begin a project from a solid foundation of plans and information, or interrupt your work at many stages of the project to find the information you need?

Make the project planning process a regular part of your work habits, and then any type of engineering design process you encounter will be easier.

WHAT IS CADD?

The term *computer-aided design and drafting (CADD)* refers to the entire spectrum of drawing with the aid of a computer, from straight lines to color animation. An immense range of artistic capabilites resides under the heading of CADD, and drafting is just one of them. Three-dimensional industrial modeling and analysis is one of the specialized areas that has developed as a result of CADD. (See Figure 4.1.)

CADD Design Formats

Productivity gains realized by the use of CADD tools are directly related to the proper use of those tools. In any discipline of drafting and design, the final format, detail, and accuracy of the drawings and design are directly related to the function of the part, model, or layout. For example, a part such as a wing-nut would not be designed and drawn using tight tolerances (see Chapter 12), whereas machine screws used in precision equipment would. In

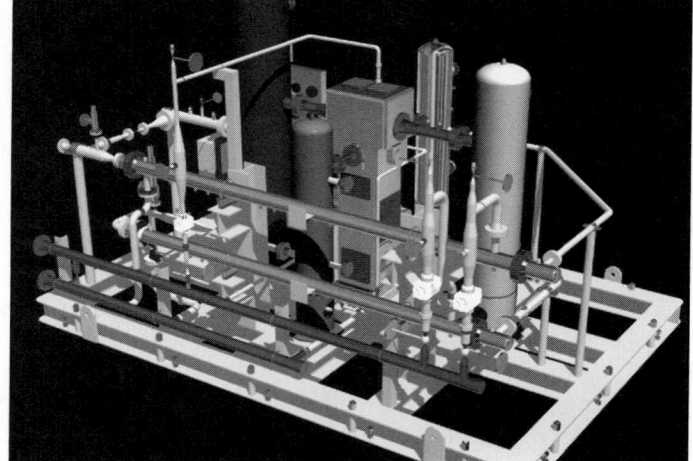

FIGURE 4.1 ■ The screen display of this piping system was generated with a 3-D modeling software package. *Courtesy Applications Development, Inc.*

addition, a simple part that is to be mass manufactured without additional testing and analysis can be drawn quickly as a 2-D multiview drawing (see Chapter 7). But a prototype design of a new component that must be analyzed and tested would best be developed as a 3-D solid model. The following sections provide a brief introduction to these and other design formats that the CADD drafter must choose in the planning stages of any project.

2-D

The abbreviation 2-D refers to *two-dimensional*: having only length and width (or width and height) dimensions. The views of the object appear in flat form and are normally rotated 90° from each other. This form of drafting is best used for objects that will not require analysis, testing, and visualization and will not be used for presentation purposes. Two-dimensional drawings are most often dimensioned and contain notes and text that describe features and details on the part, map, or plan.

3-D Wireframe

Objects initially drawn using 3-D techniques are often composed of lines connecting corners of the object. These individual points on the object are referred to as *vectors*. Technically a vector is a quantity having magnitude (length) and direction, but this can be simplified to say that a vector is a line defined by two end points. Therefore, a 3-D model must be composed of numerical values for each corner of the object. Lines connecting these corners appear as wires, hence the name *wireframe*. (See Figure 4.2.)

Three-dimensional wireframes are themselves not very useful because they are difficult to visualize. Wireframes are usually the basis for creating other types of 3-D models. Three-dimensional surface and solid models are often displayed as wireframes for the purpose of redisplaying (regenerating) the model quickly.

Three-dimensional modeling is an integral part of the design, manufacturing, and construction industry, and contributes to increased productivity in all aspects of a project. Drafters should develop a good working understanding of this skill because of the many and varied job opportunities that are available.

3-D Surface Modeling

A 3-D *surface model* is a "hollow" object on which flat planes connect all corners of the object, and the geometry is described by its

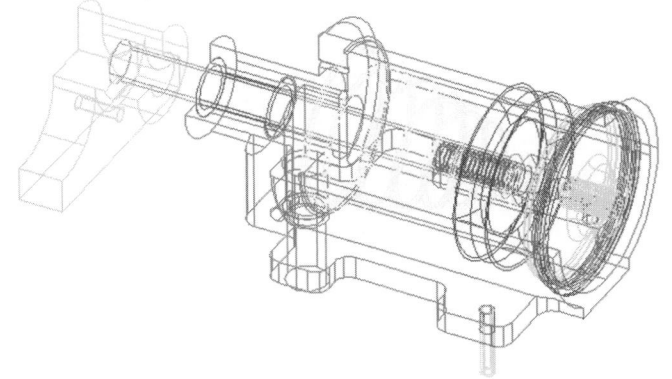

FIGURE 4.2 ■ Lines defining the edges of a model appear as wires, hence the name *wireframe*.

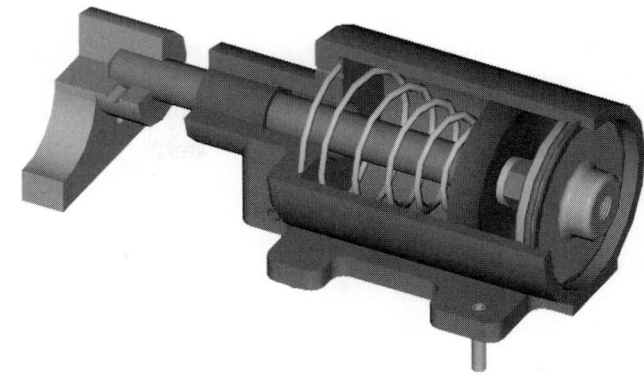

FIGURE 4.3 ■ Solid models can enhance productivity because of the ability to cut through any part of the model to create sections.

surfaces. For example, automobile body panels require the construction of accurate surface shapes. High-end surface modeling software allows the user to create surface models that can be edited, analyzed, and tested. These files can then be used in the manufacture of the part.

Basic surface modeling software is used to create realistic *presentation* models. Basic surface models appear to be solid but contain no mass property data. Surface characteristics such as color, texture, light, and shadows can be applied to the model in order to create a more realistic presentation. Surface models are also used for situations in which the presentation of the object is important, such as the shaded and rendered objects used in architectural presentations, television commercials, movies, and computer animations. In addition, this type of model is used in 3-D *worlds,* which are composite models designed for geographical information systems (GIS), video games, and virtual reality applications.

3-D Solid Modeling

This is a 3-D modeling method that accurately describes both the exterior and interior of a part or assembly. A solid model is constructed using *solid primitives* such as boxes, cones, spheres, and cylinders. These objects are combined, subtracted, and edited to create the final model. Materials can be specified for the model, which can then be submitted to testing and analysis. In addition, the solid model can be assigned surface properties and lighting for use in accurate presentations. An additional aspect of solid models that enhances productivity and usefulness is the ability to cut through any part of the model to create sections. (See Chapter 14.) This allows designers and engineers to view and edit the model while looking at interior features. (See Figure 4.3.)

The solid model can also be used to quickly create a prototype composed of plastic. This discipline is called *rapid prototyping* and is discussed later in this section.

INDUSTRY AND CADD

Computer-aided drafting systems are small enough to occupy just a few square feet of desk space. Individual contractors, designers, and architects can have one in their home or office. CADD is used in all aspects of drafting, design, and engineering, and in the disciplines of architecture, HVAC, structural, civil, process piping, landscape design, pattern making, machine design, and solid modeling.

Computer-Aided Manufacturing (CAM)

This discipline requires the use of computers to assist in the creation or modification of manufacturing control data, plans, or operations. Computers are integral to the manufacturing process. Computerized welding machines, machining centers, punch-press machines, and laser-cutting machines are commonplace. Many firms are engaged in *computer-aided design/computer-aided manufacturing (CAD/CAM)*. In a CAD/CAM system, a part is designed on the computer and transmitted directly to the computer-driven machine tools that manufacture the part. Within that process there are other computerized steps along the way.

■ The CAD program is used to create the product geometry. This can be in the form of 2-D dimensioned multiview drawings, or as 3-D models.

■ The drawing geometry is then used in the CAM program to generate instructions for the computer numerical control (CNC) machine tools. This is commonly referred to as *CAD/CAM integration.*

■ The CAM program then uses a series of commands to instruct CNC machine tools by setting up tool paths. The tool path includes the selection of specific tools to accomplish the desired operation.

■ The CAM programmer then just establishes the desired tool and tool path. The final CNC program is generated when the postprocessor is run. A *postprocessor* is an integral piece of software that converts a generic, CAM system tool path into usable CNC machine code (G-code). The *CNC program* is a sequential list of machining operations in the form of a code that is used to machine the part as needed. Figure 4.4 illustrates the CADD 3-D model, the tool path, and the G-code for the part.

■ The CNC program is then verified using the software's simulator.

■ The CNC code is created.

■ Then it is time to prove out the program on the CNC machine tool. The program is run to manufacture the desired number of parts.

Computer Numerical Control (CNC)

Also known as numerical control (NC), CNC is just one critical aspect of CAM in which a computerized controller uses motors to drive each axis of a machine such as a mill to manufacture parts in a production environment. The motors of the machine rotate based on the direction, speed, and length of time that is specified in the CNC program file. This file is created by a programmer and contains programming language such as *G-codes,* which are preparatory functions such as tool moves, and *M-codes,* which are miscellaneous functions such as tool changes and coolant settings.

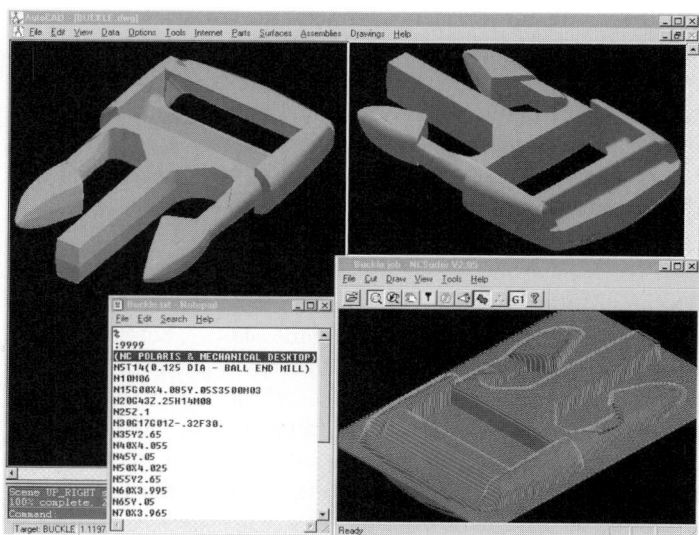

FIGURE 4.4 ■ This screen display illustrates the CADD 3-D model, the tool path, and the G-code for the part. *Courtesy NC Micro.*

Computer numerical control was a major innovation in manufacturing. It has lead to increased productivity because the consistency of the process has lowered manufacturing costs, increased product quality, and led to the development of new techniques. Persons possessing skills in CADD and CNC will find a variety of opportunities in manufacturing industries.

Rapid Prototyping (RP)

Rapid prototyping is a manufacturing process by which a solid physical model of a part is made directly from 3-D CAD model data, without any special tooling. An RP model is a physical 3-D model that can be created far more quickly than by using standard manufacturing processes. This discipline is also referred to as *stereolithography,* or 3-D printing. Rapid prototyping equipment accepts 3-D CAD files, slices the data into thin cross sections, and constructs layers from the bottom up, bonding one on top of the other, to produce physical prototypes. Computer-aided design software such as AutoCAD allows you to export an RP file from a solid model in the form of an .stl file.

A computer slices the 3-D CAD data into .005-in. thick cross-sectional planes. Each slice or "layer" is composed of closely spaced lines resembling a honeycomb. The slice is shaped like the cross section of the part. The cross sections are sent from the computer to the rapid prototyping machine, which builds the part one layer at a time. The rapid prototyping machine has a "vat" that contains a photosensitive liquid epoxy plastic, and a flat platform or starting base resting just below the surface of the liquid. (See Figure 4.5.) A laser, controlled with bidirectional motors, is positioned above the vat perpendicular to the surface of the polymer. The first layer is bonded to the platform by the heat of a thin laser beam that traces the lines of the layer onto the surface of the liquid polymer. When the first layer is completed, the platform is lowered the thickness of a layer. Additional layers are bonded on top of the first in the same manner, according to the shape of their respective cross section. This process is repeated until the prototype part is complete.

A second type of rapid prototyping called solid object 3-D printing uses an approach similar to inkjet printing. During the build process, a print head with hundreds of jets builds models by dispensing a thermoplastic material in layers. The printer can be networked to any CAD workstation and operates with the push of a few buttons. (See Figure 4.6.)

FIGURE 4.5 ■ This stereolithography system builds a physical prototype, layer by layer, using liquid epoxy plastic that is hardened by a laser to precise specifications. *Courtesy 3D Systems.*

(b)

(a)

(c)

FIGURE 4.6 ■ (a) The 3-D printer employs a print head with hundreds of jets to build models by dispensing a thermoplastic material in layers. *Courtesy 3D Systems.* (b) A designer removes a model built on a solid object printer. *Courtesy 3D Systems.* (c) Engineers built this prototype wheel from a CAD file using a solid object printer. *Courtesy 3D Systems.*

Rapid prototyping has helped to revolutionize product design and manufacture. The development of physical models can be accomplished in significantly less time when compared to traditional machining processes. Changes to a part can be made on the CAD 3-D model, then sent to the RP equipment for quick reproduction. Engineers can use these models for design verification, sales presentations, investment casting, tooling, and other manufacturing functions. Additionally, medical imaging, CAD, and RP have made it possible to quickly develop medical models.

Computer-Integrated Manufacturing (CIM)

This is the concept of an automated factory integrated by a CAD/CAM system using robotics and computer-controlled machinery from delivery of the raw material to completion and shipping of the finished product. The field of CIM incorporates the disciplines of CAD, CAM, robotics, electronics, hydraulics, pneumatics, computer programming, and process control. Computer-integrated manufacturing enables all persons within a company to access and utilize the same database that would normally be used by designers and engineers.

Computer-Aided Engineering (CAE)

The use of computers in design, analysis, and manufacturing of a product, process, or project is called *computer-aided engineering*. Computer-aided engineering can also be thought of as the umbrella discipline that includes CADD, CAD/CAM, CNC, CIM, and rapid prototyping. It includes, but is not limited to, the following topics:

■ Mechanical design and product development automation.

■ Surface and solid modeling.

■ Simulation, analysis, testing, and optimization of mechanical structures and systems.

■ Product data management (PDM) of engineering data and documents.

■ Using the Internet and other technologies to collaborate on projects.

■ Manufacturing using NC programming and RP systems and services.

■ Utilizing computers including PCs, workstations, networking, and operating systems.

Within this process, the computer and its software controls most, if not all, portions of the manufacturing. A basic CIM system may include transporting the stock material from a holding area to the machining center at which several machining functions are performed. From there, the part may be moved automatically to another station at which additional pieces are attached, then on to an inspection station, and from there to shipping or packaging.

THE CADD WORKSTATION

The electronic tools used for CADD are referred to as *hardware*. An individual CADD workstation relies on a computer for data processing, calculations, and communications with other pieces of

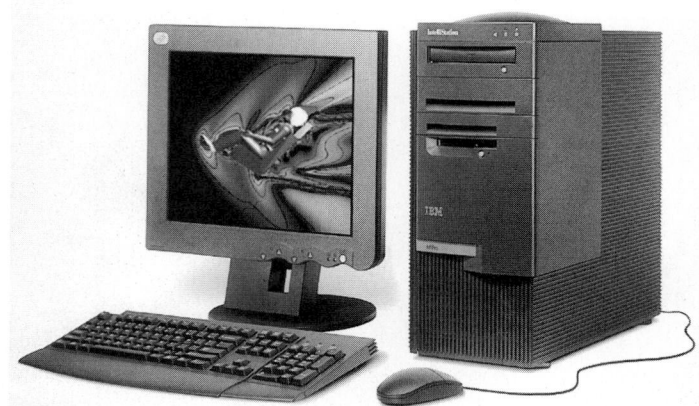

FIGURE 4.7 ■ This CADD workstation has a computer, keyboard, mouse, and flat-panel monitor. *Courtesy of International Business Machines Corportation.*

peripheral equipment. Peripheral equipment is any additional hardware item that uses the computer's files and performs functions that the computer cannot handle. A typical CADD workstation is shown in Figure 4.7.

The additional functions and services provided by the peripheral equipment fall into three categories: input, output, and storage. Input means to put information inside the computer to be acted on in some way. Input can come from the keyboard, a mouse (or other similar input device), or a digitizer. Output refers to information that is sent from the computer to a receiving point such as a monitor, a plotter, or a printer. Storage refers to disk, tape, and CD drives that allow the operator to store programs, drawing files, symbols, and data.

Computers

The computer is the heart of the CADD workstation. It contains all of the components and software that enable all of the other peripheral devices to function. Computers are changing so rapidly that it is impractical to list accurate specifications. Technological advances in all areas of hardware and software mean that schools and companies find that they must upgrade their systems, on average, every two years. The following list provides a good representation of the principal components found in a typical CADD computer.

■ Central processing unit (CPU)—The integrated circuit chip that performs all of the computer's calculations.

■ Memory—Blank chips that receive and hold programs and data while the computer is on and functioning.

■ Graphics card—The graphics on the monitor are processed separately by this card, or "board," which contains its own bank of memory chips in order to accelerate the display of complex graphics.

■ Storage devices—These are devices such as hard disk drives, floppy disk drives, optical drives, and CD-ROM drives that can be used for file and software storage. Storage devices are discussed in detail later.

■ Sound card—This card enables the use of speakers, earphones, and microphones.

A typical computer box is shown in Figure 4.7. As the size of computer components decreases and the power increases, the actual computer "box" grows ever smaller. In fact, computers are now built into many machines and tools, and future CADD systems may appear to be composed of only a monitor and input device.

Monitors

The principal display component of the CADD workstation is the monitor. Two types of monitors are in use: the CRT (cathode ray tube), and the flat-panel display. The CRT is recognizable by its deep cabinet, which houses the electron gun, whereas the flat-panel monitor's technology requires a much shallower cabinet. (See Figure 4.8a.) The flat-panel display is much more attractive because of its smaller size, lower power requirements, and ability to provide a crisp, clear image with much less glare. (See Figure 4.8b.)

The quality of the monitor's image is based on its *resolution*. The screen is composed of a gridwork of tiny squares that appear as dots and are known as pixels. The word *pixel* is an acronym for picture (pix) element (el). The resolution of the screen, or the crispness of the image, is determined by the number of pixels. Each pixel is actually a *bit,* as shown in Figure 4.9. A bit is a binary digit and can be either on or off. A bit that is receiving a signal is on, and is lit. The absence of a signal means the bit is off. A raster display is often called a bit map because it is a map composed of pixels. Low-resolution raster screens may produce lines that have jaggies, or a jagged look. This is especially true for angled lines and curved features. The lines in Figure 4.10a and 4.10b show how an image would appear on raster screens of different resolution if magnified.

The resolution of a monitor is stated in horizontal pixels by vertical pixels, such as 800×600. High-resolution monitors of 1280×1024 are common, and even higher resolutions are available. The resolution can usually be adjusted to suit the needs of the user.

The most important aspect of the monitor is how it feels to the user. Become familiar with the controls, functions, and menus that enable the user to alter the appearance of the screen image. These can be accessed via the operating system, such as adjusting the display properties in Windows. Additionally, each monitor has controls that enable the user to adjust width, height, display shape, contrast, brightness, and color. The goal is to adjust the monitor so that it provides a display that is the least tiring, and easiest to look at.

Input Devices

Keyboard

There are several methods of putting information into a computer. The most obvious is with the familiar alphanumeric keyboard (alpha-letters, numeric-numbers). Written instructions can be given to the computer from the keyboard. Graphic instructions can also be entered from the keyboard in the form of X, Y, Z coordinates, circle diameters, number of polygon sides, and curve radii.

(a)

(b)

FIGURE 4.8 ■ **(a)** These CRT monitors are recognizable by the deep cabinet, which houses the electron gun. *Courtesy of Hewlett-Packard Company.* **(b)** The flat-panel display is much smaller, has lower power requirements, and provides a crisp, clear image with much less glare. *Courtesy of Hewlett-Packard Company.*

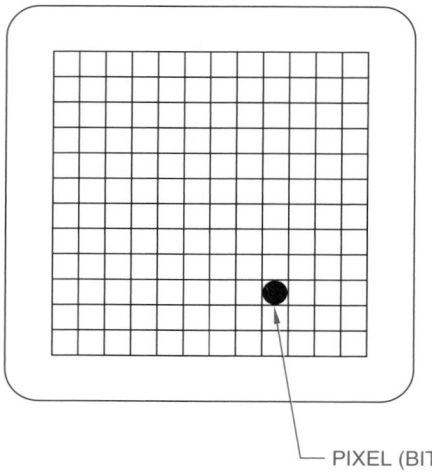

PIXEL (BIT)

FIGURE 4.9 ■ Video display screens are composed of a grid, or framework of pixels (bits), or dots, that can be turned on to display graphics or alphanumeric characters.

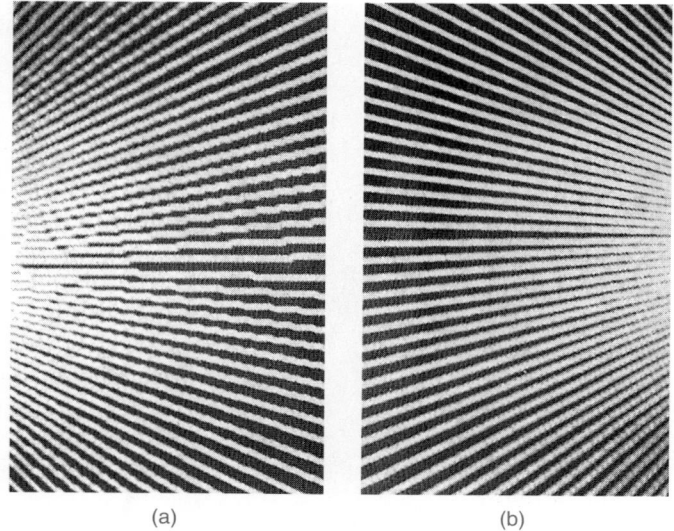

(a)   (b)

FIGURE 4.10 ■ (a) A raster display produces jagged lines called "jaggies," (b) but the jaggies can be reduced by high-resolution screens. *Courtesy Megatek.*

The layout of the computer keyboard varies between different brands of CADD systems. The standard QWERTY keys are arranged the same, and additional keys can be found along the top, left, and right sides of the standard keyboard. (See Figure 4.11.) Consider the following descriptions of some of the most common keys found on a CADD system keyboard.

Enter or Return Key. The ENTER key is usually located at the right side of the keyboard. The computer responds to a line of data entry after this key is pressed.

Escape Key. This key allows you to escape from the current command or activity in which you are engaged.

Cursor Keys. The function of these keys is to move the screen cursor. They are arranged in a group and appear as four arrows pointing in the four compass directions. They are often called arrow keys.

FIGURE 4.11 ■ A typical alphanumeric keyboard. *Courtesy Logitech.*

Home Key. This key is used in conjunction with the cursor keys. It moves the cursor to either the top left corner of the screen or to the beginning of the line. It can be found in the center of the cluster of cursor keys or at the upper left of the keyboard.

Insert Key. The insert key can be used to insert text that has been copied to the clipboard. It can also act as a toggle when inserting new text in a document. When on, typed text is inserted alongside existing text. When off, typed text replaces existing text to the right of the screen cursor.

Delete Key. This key deletes highlighted text or objects on the screen. When pressed in a text document, the character to the right of the cursor is deleted. Caution should be used when working in file management applications such as Windows Explorer.

End Key. Pressing this key sends the screen cursor to the right end of a line of text.

Page-Up Key. This key moves the cursor up one page in the current screen display.

Page-Down Key. Pressing this key moves the cursor down one page in the current screen display.

Backspace Key. If you make a mistake in typing, press this key and you can delete what you just typed. This key can also serve to move the screen cursor back along the line just typed.

Control Key. This key is similar to the shift key because it is usually pressed along with another key (or keys) to initiate a function or command within the program.

Function Keys. These 10 to 12 additional keys are located just about anywhere on the keyboard. A popular location is along the top of the keyboard. The number and grouping pattern vary by manufacturer. Function keys are used by the program to activate specific functions or commands. Their use may change in the program, or from one program to the next, hence the need for function key masks or overlays. The overlay is usually a plastic template with written labels that fits above or around the keys.

Calculator Keypad. This cluster of keys is usually referred to as the keypad and contains the numbers 0 through 9, a decimal point, arithmetic function symbols, and the Enter key.

Become intimately familiar with the keyboard you will be working with, and you will be rewarded with less wasted time searching for keys and smoother-running sessions at the computer overall.

FIGURE 4.12 ■ The mouse is the most common pointing device, having two or more buttons, and most often connected to the computer with a cable. *Courtesy IBM.*

Mouse

The most commonly used input device is the mouse. It is a *pointing device* that is used to control the movement of the screen cursor. The *screen cursor* is an on-screen symbol, usually an arrow, box, or crosshairs. The mouse (see Figure 4.12) has a small ball on the underside that rolls against internal sensors or rollers. These sensors transmit the cursor location to the screen as the mouse is rolled on a flat surface. The mouse can have two or three buttons, and may also have a small wheel on the top surface to control zooming and scrolling. The left button is normally the button used to pick an item or location on the screen. The mouse is commonly connected to the computer by a thin cable, but is also available in a wireless version.

Trackball

The trackball is merely an inverted mouse. It contains buttons similar to those on the mouse, and is available in a wide variety of shapes. The roller ball rests on the top of the device, or is mounted on the side, and is controlled by the thumb, fingers, or palm. The screen cursor moves in the direction the ball is rotated. (See Figure 4.13.)

Trackballs are often a good alternative to a mouse, especially for persons with physical problems in the hand, arm, or shoulder. The trackball requires movement of only the fingers.

Digitizer

The digitizer is often called a graphics tablet, or tablet for short. It is a low-profile piece of hardware that has a flat, smooth surface covering a gridwork of thin wires. A pointing cursor is attached to the digitizer by a wire.

The pointing cursor used with the digitizer can take two standard forms, *stylus* or *puck*. The stylus looks like a ballpoint pen. (See Figure 4.14.) The puck is usually rectangular in shape and may have up to 16 buttons on it. It may be attached to the digitizer by a cable. To locate points with the puck, you must look through a small lens containing crosshairs. (See Figure 4.14.)

When the stylus or puck is brought near the active surface of a *digitizer tablet,* its position is determined by the electronics inside the tablet. As the stylus is moved across the tablet, crosshairs on the video display screen track its movement. By pressing the stylus to

FIGURE 4.13 ■ The roller ball rests on the side of this trackball, and is controlled by the thumb. *Courtesy Logitech.*

FIGURE 4.14 ■ Digital tablet with a 16-button programmable mouse and a stylus. *Courtesy CalComp.*

the tablet at the beginning of a line, the coordinates for that point are transmitted to the computer. If the stylus is moved to the end of the line and the tablet is touched at that point, the final coordinates are then transmitted.

Digitizer tablets are also used to give instructions to a graphics system. This input method uses a menu having several graphics

FIGURE 4.15 ■ Digitizer used to establish coordinate points for drawing lines. *Courtesy GTCO Calcomp Peripherals Inc.*

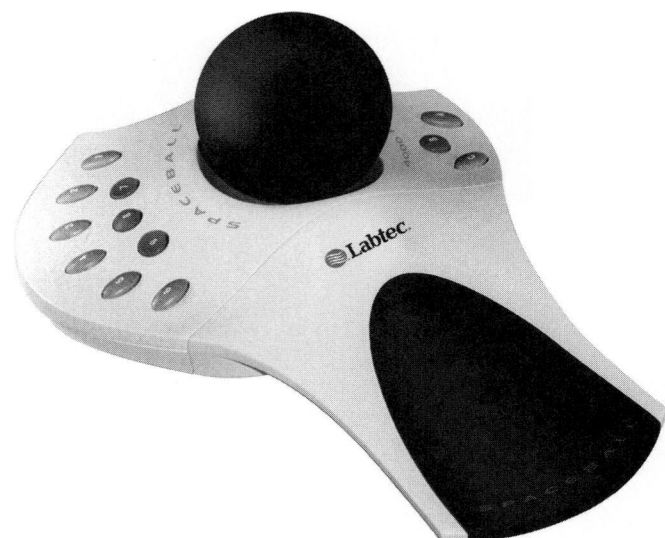

FIGURE 4.16 ■ The large ball is tilted, raised, lowered, and twisted to control the position of the object on the screen. *Courtesy Labtec.*

3-D Input Device

Through the process of *reverse engineering,* a model can be digitized with a 3-D digitizer such as the one shown in Figure 4.17. Using a stylus attached to a pivoting arm, this instrument enables you to digitize an existing object or model and input the data into a CADD system as a 3-D model.

The 3-D scanning digitizer creates a 3-D image of a model without the use of handheld input devices. (See Figure 4.18.) The computer file can be exported to a CADD program or directly to code for a CNC machine.

Data Storage Media and Devices

The computer is equipped with two or more storage devices. A *storage device* is a component that contains or accepts media on which computer files can be kept temporarily or permanently. A variety of devices are available.

Floppy Disk

The 3½" is the most common floppy disk. It is enclosed in a hard plastic case, fits in a shirt pocket, and holds 1.44 megabytes (MB) of data. The floppy disk is best used for temporary storage of files and for transferring files from one computer to another in the absence of other transfer capabilities. The 5¼" version is found in some computers and may be used for compatibility with older systems. It is much more susceptible to damage and holds far less data.

Floppy disks are magnetic media and can be ruined if placed too near magnetic materials. Any magnetic field—such as those in motors, televisions sets, and speakers—can rearrange the information stored on a disk. Temperature extremes can damage disks, so try to store them in a room with a fairly comfortable temperature range.

Many new floppy disks must be *formatted* before they can be used. The formatting process divides the disk into *sectors* and *tracks* to accommodate the storage of files. An area for the disk

commands and symbols printed on it. A command is a specific instruction that is issued to a computer. The user simply presses the stylus on the desired command and the computer system accepts it. Figure 4.15 shows a digitizer used to establish coordinate points for drawing lines.

Another method used to input graphics is to place a sketch over the digitizer surface or to work from a sketch or series of commands next to the digitizer. Here the stylus or cursor is used to establish coordinate points or geometric shapes by a touch on the active surface of the digitizer. Large-format digitizers can also be used to digitize a manual drawing into a CADD drawing.

3-D Motion Control Device

The fields of 3-D modeling, solid modeling, and animation often require more sophisticated devices that enable the user to dynamically control the presentation of the model on the screen. One such device is shown in Figure 4.16. The large ball is tilted, raised, lowered, and twisted to control the position of the object on the screen. This ability to finely control the movement of the model in 6 degrees of movement enables the user to position the object to achieve the best view for design and presentation.

FIGURE 4.17 ■ Using a stylus attached to a pivoting arm, this 3-D digitizer enables you to digitize an existing object, and input the data into a CADD system as a 3-D model. *Courtesy FARO.*

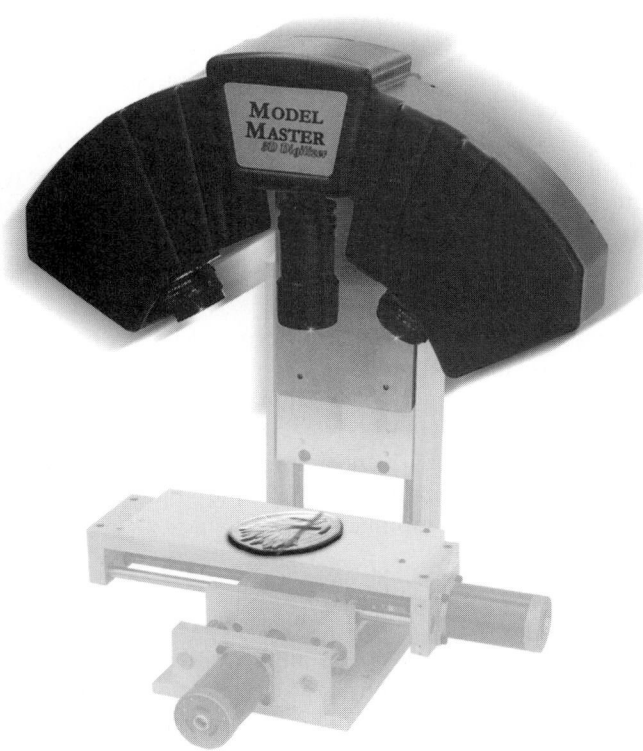

FIGURE 4.18 ■ The 3-D scanning digitizer creates a 3-D image of a model without the use of handheld input devices. *Courtesy InDepth.*

directory is also established by the drive. The sectors of a disk resemble slices of a pie, and the tracks are like concentric rings. (See Figure 4.19.)

Formatting a disk is a simple process. Refer to the Help facility in your operating system if you wish to learn more about formatting options.

Hard Disk

The *hard disk drive* is the internal storage device found in most computers. It is a sealed unit, and its principal use is to hold software that is used by the computer. In addition, working data files and

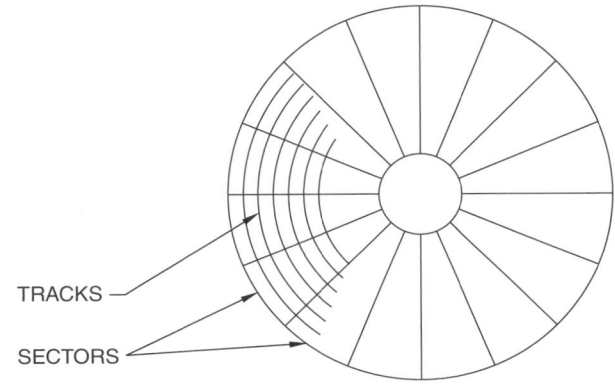

TRACKS

SECTORS

FIGURE 4.19 ■ A disk is divided into pie-shaped sectors and tracks that circle the disklike concentric rings.

some semipermanent storage are often contained on the hard disk. Permanent file storage is best suited to other forms of storage media. Storage capacities of hard disks are commonly represented in *gigabytes* (GB), which is one billion bytes of data. A *byte* is eight binary digits, or roughly one character.

An important aspect of the hard disk is the amount of empty space it contains. This space is often referred to as *disk cache* or *swap file space.* Computer programs require empty space both in the computer's memory (RAM) and on the hard drive. Programs automatically look to the hard drive to claim sectors of "workspace" while the software is active. This temporary workspace is critical for the efficient functioning of the software. Temporary files, backup files, and files that must be swapped in and out of the computer's memory are stored in this space. When the software is closed, this space is reclaimed. Therefore, it is important to be sure that your hard disk has enough empty space available when installing new software. These requirements are always given in the software installation instructions.

Optical Disk

The optical disk is commonly known as the *compact disk* or *CD.* One CD can store up to 4 GB. Some optical disks are *rewriteable,* which means you can store, erase, and replace files as needed. The thin, plastic platter of the CD is treated on one side with tiny pits and a reflective surface. Digital impulses are read by a laser and transformed into either audio or digital information.

A common optical disk drive is the *CD-ROM.* The acronym ROM stands for read-only memory. You can access data from a CD-ROM, but you cannot store additional data. A CD-ROM is important in modern computers, because CADD programs and other software are shipped on CDs.

High-Capacity Disk

These disk drives are available as internal components in the computer box or as external add-on devices to increase your storage capacity and flexibility. Common sizes are 100- and 250-MB, and 1- and 2-GB, disks. High-capacity disks are commonly used to store one or more projects or individual customer accounts. They are also used to back up the hard drive and to archive files, or to share large or multiple files with coworkers or other company sites.

Plotters and Printers

When a printed copy of a drawing is required, a plotter or printer generates it. This paper version is called *hard copy.*

Pen Plotter

On a vector graphics display screen, lines and features are drawn using X and Y coordinates. This information is stored in the computer as a display list. The display list contains all the information that the plotter needs to function. The information stored on a raster screen (bit mapped) must first be converted to a display list before it is sent to the plotter because the plotter operates in two directions only. Movement of the pen is along the X or Y axis. Some plotters allow the pen to move in both directions, while others move the paper in one direction and the pen in the other. The paper size that can be used on this plotter is determined by the width and length of the bed.

FIGURE 4.20 ∎ An architectural rendering is printed on a large format color inkjet plotter. *Courtesy of Hewlett-Packard Company.*

Inkjet Plotter

The inkjet plotter sprays tiny droplets of ink from a cartridge onto the paper to form dot-matrix images. Print quality, or *resolution,* is measured in *dots per inch* (dpi). Common quality is 600 dpi, but resolutions of 720 dpi and higher are available. Inkjet plotters are faster than pen plotters and require less maintenance. Figure 4.20 shows an architectural rendering printed on a large-format color inkjet plotter. Photorealistic renderings are best produced using a color plotter.

Inkjet plotters are available in monochrome (only one color) and color, and are available for small- and large-format drawings. Plotters are often networked within a company or office for access from several or all computers within the network.

Impact Printer

A variety of printers fall into the category of impact printer. The action of this kind of printer is to strike a ribbon and paper to produce a character or image. The dot-matrix printer hammers tiny wires in specified locations to create a character composed of dots. A matrix is just an arrangement of rows and columns (Figure 4.21). Scoreboards in stadiums around the country flash scores using a matrix of lights. Impact printers are economical and best used for check prints and drafts.

Nonimpact Printer

The nonimpact printer may be found in several different forms. The thermal printer uses tiny heat elements to burn dot-matrix characters into treated paper. Blue or black images can be produced. The inkjet printer, seen in Figure 4.20, squirts miniscule droplets of ink at the paper to produce a dot-matrix character. Some of these printers spray a continuous stream of ink at the paper.

The best-quality image produced by a nonimpact printer comes from the laser printer. The laser printer is more closely related to

FIGURE 4.21 ■ A matrix is an arrangement of rows and columns.

the electrostatic copier process than to the striking, poking, burning, and spraying printing methods. A laser draws lines on a revolving plate that is charged with a high voltage. The light of the laser causes the plate to discharge. An ink/toner will then adhere to the laser-drawn images. The ink is then bonded to the paper by pressure or heat.

THE FUTURE OF CADD

Since 1980, when CADD began to make inroads in the drafting and design industry, phenomenal changes in hardware and software have required drafters to be more flexible in accepting change. Therefore, drafters and students must be prepared to learn new tools and techniques, and be open to attending classes, seminars, and workshops on a regular basis. Although predicting the future is, at times, a dubious endeavor, there are a few aspects of technology related to CADD that can be predicted with some accuracy, and some that should be followed closely as research progresses.

Computer Size and Format

Computer size has decreased to the point where computers now reside in a host of machines, equipment, vehicles, appliances, and toys. Size will not be a deterrent to portability and mobility, and most computer-related tasks can be accomplished almost anywhere.

Computer Power and Capacity

Along with the reduction in size of computers, chips, processors, and peripheral components, has come a geometric increase in the power of computers. Approximately every 18 months computer speed, power, and memory capacity doubles. This trend will continue, and the size of computer components will continue to shrink.

In the future computer components may even reach the realm of the molecular, or even subatomic, level. Rapid advances in the two fields of chemistry and engineering have produced the discipline of *nanotechnology*. Nanotechnology is manufacturing on a molecular level where products are built one molecule or atom at a time. A *nanometer* is one billionth of a meter. This is approximately the width of three or four atoms. When, and if, nanomanufacturing is realized, it will mean that almost any type of consumer goods

can be produced using raw atoms, at very low cost without traditional labor. This would happen with the use of programmed *nanoscopic* machines and robot arms, which themselves could be self-replicating.

Although this may seem far-fetched in the first years of the second millennium, it would be advisable for you to keep abreast of developments in this area because the implications could be far-ranging for design and engineering, and for that matter, all of society.

Parametric Design and Artificial Intelligence (AI)

As CADD software tools are improved, it is evident that the function of traditionally drawing an object is occupying less of a drafter's time. Using predrawn objects and components, drafters can now "compose" a drawing by inserting these objects. Advances in parametric design enable users to input a variety of data into the computer, which then generates the drawing or model. A *parametric design* program contains a wealth of information about the design and construction of specific items, such as tanks. The program requests information such as capacities and measurements (*metric*) regarding the design (*parameters*) of the tank, hence the name *parametric*. When sufficient data has been supplied to the software, the tank design is generated, and drawings can be produced.

A somewhat related, but more abstract, field is *artificial intelligence*, or AI. The focus of AI research is to create computers and machines that can "think." This could have implications for drafters, designers, and engineers when creating designs of products or structures. Computers and software having artificial intelligence may be able to offer suggestions, gather data, suggest resources, make comparisons, or suggest alternatives when design errors are detected.

One aspect of AI is the use of *intelligent agents*. Think of this as your "assistant in a computer." You can ask questions, make requests for information, and give commands to the agent. For example, you can request, by voice commands, that the agent find all of the designs of the newest energy-efficient, double-hung windows created within the last year. You can also request that the agent select the best three designs based on efficiency, price, and durability, and then display 3-D views of each in a tiled manner on your screen so that each is visible. As you can imagine, an agent such as this would free up much more of a designer's time for making comparisons, asking questions, requesting a variety of information, and thinking more about the design of a product.

The incorporation of intelligent computers into the design process may be a great benefit to society as a whole if it means safe, accurate, environmentally sensitive designs and a time-savings in design and manufacturing.

VIRTUAL REALITY

Virtual reality (VR) is a situation in which what is seen and experienced appears real, but is not. The popular image of VR is a user who wears a helmet or goggles containing two small flat-panel screens, one for each eye. On one hand is worn an instrumented glove that senses finger movement. The virtual reality user is actually inside the model and can see his or her hand in the screen. Not only can the user move around inside the model, but he or she can also interact with it.

Virtual reality technology is a logical step in the design process. We have always lacked the ability to place ourselves inside the models we create and see them from a variety of viewpoints before they are built. A VR system gives us the capability to interact with a model of any size, from molecular to astronomical. Surgeons can learn on virtual cadavers and practice a real operation on a virtual body constructed from scanned images of the human to be operated on. Home designers can walk around inside the house, stretching, moving, and copying shapes in order to create the finished product.

Buildings can be designed and placed on virtual building sites. Clients can take "walk-through" tours of a building before it is built and make changes as they "walk." Scientists can conduct experiments on a molecular level by placing themselves inside a model of chemical compounds. Using *telerobotics,* a person can "see" through a robot's eyes while in a safe virtual environment, in order to guide a robot into a hazardous situation. These applications are possible using a variety of equipment.

The principal component of "immersive" VR is the *head-mounted display (HMD),* which may also include stereo earphones. (See Figure 4.22.) Using an HMD gives you the feeling of being totally inside the virtual world because your eyes are not distracted by outside activity, hence the term *immersive.* A second type of HMD, the *suspended display system,* is shown in Figure 4.23. It is not as heavy to wear because it is suspended by the articulated boom, which contains all of the necessary cabling and sensors.

The HMD technology is moving toward smaller, less cumbersome units. The use of lightweight eyewear, a special monitor, and sensing devices can enable the user to see in 3-D. (See Figure 4.24) An *instrumented glove* system uses cloth gloves with electrical sensors in each fingertip. Contact between any two or more digits completes a conductive path, and a complex variety of actions based on these simple "pinch" gestures can be programmed into applications. (See Figure 4.25.)

FIGURE 4.23 ■ This *suspended display system* is designed for applications such as vehicle simulation and cockpit modeling. The operator can stand or sit, with both hands free to manipulate real or virtual controls and input devices. *Courtesy Fakespace.*

FIGURE 4.24 ■ This lightweight eyewear, a special monitor, and sensing devices generate alternating images that enable the user to see in 3-D. *Courtesy Stereo-Graphics.*

A more common application of VR technology is called *through-the-window VR* (also referred to as "passive" VR). It is nothing more than manipulating a 3-D model or "world" with input from a mouse, trackball, or 3-D motion control device. This allows more than one person to see and experience the 3-D world. A variation on this is a flat-panel display with handles. It can be moved in either 3 or 6 degrees of motion. The window VR unit in Figure 4.26 is designed to allow natural, intuitive interac-

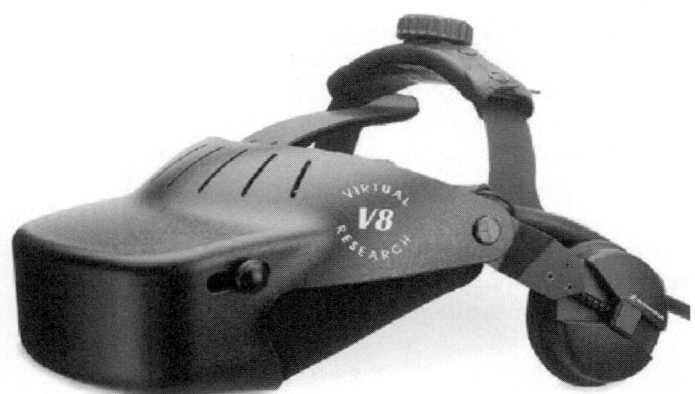

FIGURE 4.22 ■ This head-mounted display incorporates both stereo viewing with two small monitors and stereo sound using the attached earphones; it is the principal component of *immersive* VR. *Courtesy Virtual Research.*

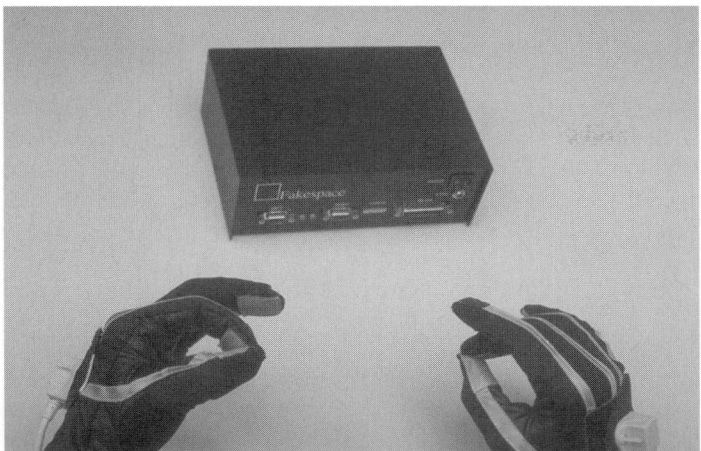

FIGURE 4.25 ■ An *instrumented glove* system uses cloth gloves with electrical sensors in each fingertip. *Courtesy Fakespace.*

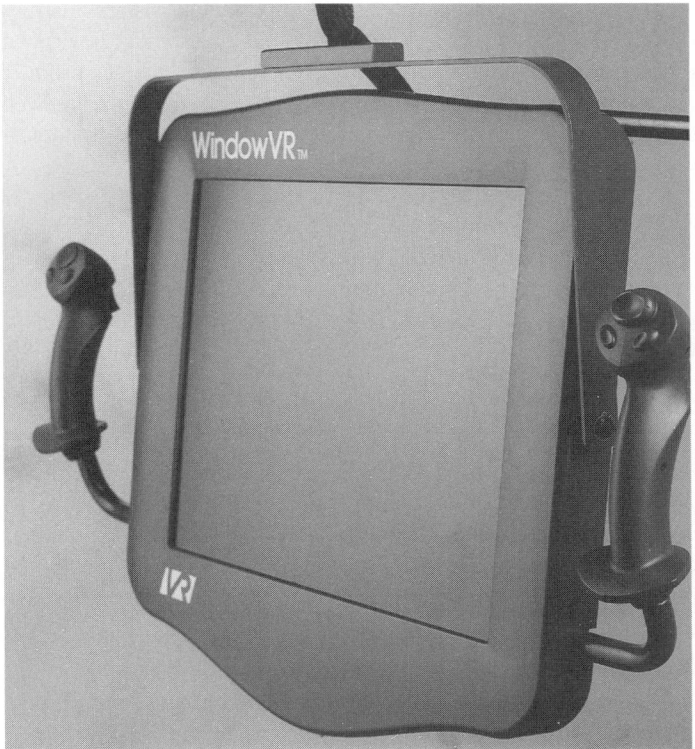

FIGURE 4.26 ■ This unit is designed to allow natural interaction with the virtual environment. Museum goers and showroom visitors can walk up, grab the handles, and instantly begin interacting. Observers can follow the action by moving beside the primary user. *Courtesy Virtual Research.*

tion with the virtual environment. Museum goers and showroom visitors can walk up, grab the handles, and instantly begin interacting. Observers can follow the action by moving beside the primary user. A variety of handle-mounted buttons can emulate keyboard keystrokes, joystick buttons, or 3-D motion control device buttons.

FIGURE 4.27 ■ This type of through-the-window VR consists of a stereoscopic viewing panel, which mounts directly over the monitor screen. The viewer wears lightweight, passive, polarized eyewear. *Courtesy StereoGraphics.*

A second type of through-the-window VR consists of a stereoscopic viewing panel that mounts directly over the monitor screen. The viewer wears lightweight, passive, polarized eyewear. The viewing panel "shutters" the images directly on the screen to generate 3-D images by users wearing the special glasses. (See Figure 4.27.) This technology also enables several persons to view the same image on the screen.

A field of tremendous opportunities is developing around the creation of virtual worlds. These worlds will be detailed 3-D models of a wide variety of subjects. Worlds will need to be constructed for many different applications. Persons who can construct realistic 3-D models will be in great demand. The fields of VR and GIS (geographic information systems) will be combined to create "intelligent" worlds from which an infinite wealth of data can be obtained while occupying the virtual world. No town or city will be without a virtual model of their municipality.

The implications for this technology are tremendous, and the only limitation for its uses is our imaginations.

GETTING STARTED WITH A CADD PROJECT

The most important and productive time you can spend working on any project or drawing is the time you use to plan it. This point cannot be overemphasized. Always plan your work carefully before you begin to use the tools required to do it.

When you plan a project and map out the directions required to complete it, you are implementing a "process" by which you can do your work. This process can be applied to any project. But when you begin the actual work of building the drawing or model, you use tools, skills, techniques, formulae, materials, and standards to complete the job. These items we refer to as the "content" of the discipline. If you begin a project by using content before process, your task will be much more difficult, take longer to complete, and may contain more mistakes. Therefore, take a few minutes to study the following procedures for planning your work.

Using the Planning Process

A simple, three-step process will enable you to immediately organize your thoughts, ideas, and knowledge about the specific project. You can use this process at the beginning of the project to map the overall goals, and also use it at any point during the project to guide you through smaller tasks and goals. The process requires you (or the workgroup) to answer the following three questions about the project:

1. What do I know about the subject?

2. What do I need to know about the subject?

3. Where can I find the information I need?

The best way to answer these questions is to use *brainstorming*. In this process, individuals voice their thoughts and ideas regarding the specific topic, problem, or project at hand. All ideas are listed by a recorder and are not discussed or criticized. This technique is a proven method to tap the knowledge, creativity, and energy of individuals and groups. Here are a few rules about working in a brainstorming session.

∎ One person records all statements.

∎ Place a reasonable time limit on the session, or agree to let it run until all ideas are exhausted.

∎ Every statement, idea, or suggestion that is made is a good one.

∎ Do not discuss or evaluate any statement or suggestion.

After the session, take a break or discuss other items. When you return to the brainstorming list, it is time to evaluate the items. Throw out any items that the group agrees are not valid. Rank the remaining ideas in order of importance or validity.

Prior to beginning this brainstorming process, it is critically important that you read all instructions available for the project. Use different colored highlighters to mark the following:

∎ Project details.

∎ Items needing further study, or those you do not understand.

∎ References to other resources or information about the project.

Now let us look at the three-step planning process questions again.

What Do I Know About the Subject?

List everything that you are sure about regarding the topic. For example, if you are working on a new drawing of a mechanical part,

write down all the things you know about the object, such as the material, machining features, tolerances, etc. Do you know the type of drawing (2-D, isometric, or 3-D) and the standards to be used? How is the object to be manufactured? Do you know basic items such as paper size and scale of the drawing, text heights and font styles, number of views, sections, and details required? You can even list specific CADD software commands and techniques that you know can be used to complete this drawing.

What Do I Need to Know About the Subject?

List anything you are not sure about. This includes things you think you know, but about which you are not sure. Do you have a question about a detail listed in the instructions? List it here. List any CADD commands or procedures that you think you may need but that you are not confident in using.

Where Can I Find the Information I Need?

Try to think of all additional resources that could assist you in completing the project. This can include textbooks; written instructions; vendor brochures and catalogs; published standards; experts in the discipline such as instructors, engineers, architects, product representatives; and other resources such as trade journals, Web sites, academic papers, government publications, and even conferences and conventions.

As stated earlier, after answering these questions using the brainstorming technique, return to your lists. Eliminate unnecessary items, and prioritize the remainder. If you make these procedures a regular part of your planning process, you will find that any project you work on will flow smoothly, take less time, and contain fewer errors.

Choosing the Drawing Format

Perhaps the most critical question to ask at the beginning of any project is "What is the function of this project?" This question is important for engineers, architects, and designers. Then a similar question must be asked by the drafter, "What will this drawing be used for?" We can break this down into discipline-specific questions as follows:

∎ *Mechanical design*—What is the function of the part? Will this drawing be used for manufacturing, inspection and testing, or presentation?

∎ *Architectural/structural*—What type of structure is required, and what is its intended use? Is the drawing to be used by contractors, for the purpose of acquiring building permits, or for presentation to the client?

∎ *Civil*—Who will use the map and what for? If the map is a property plat survey or large land survey, what is the required accuracy?

∎ *Commercial mechanical (piping, HVAC, electrical)*—What is the purpose of the building, how many people will occupy it, and how much and what types of equipment are required? Will the drawing be used for preliminary design and layout, estimating, construction, or client presentation?

It should be evident that when a drafter or designer is preparing to begin a new drawing in any of the varied disciplines, it is vital to determine the purpose of the drawing. Although engineers and designers are responsible for questions related to the overall design and materials, drafters must focus on the uses of each drawing created. A discussion of drawing formats was presented early in this chapter. Now let us look at some questions that assist the drafter in deciding on the proper format to use.

■ Will machinists use traditional methods to manufacture the part from the drawing?

■ Will CNC machines be used to manufacture the part?

■ Is the drawing to be used for 2-D presentation purposes only?

■ Is the drawing to be reduced and printed in a book for use by nontechnical persons, or in a company parts catalog or instruction manual?

■ Will the drawing be used for client presentation in order to win a contract?

■ Must the drawing be flexible so that the design can be used for presentation purposes, but also used in the testing, analysis, and manufacturing processes?

Once the purpose and function of the drawing has been decided, you can then begin the actual process of creating the drawing. Following a planned procedure for beginning a new drawing is a sure step to completing a successful project.

Drawing Setup Steps

Throughout every step of a project, you should always be applying the three-step problem-solving process. Stopping periodically to answer the three questions will always keep you aware of what information you need and where you need to get it.

After you plan the overall project and determine the number and type of drawings required, you should also have a plan indicating the progression, or chronology, of drawings. Just as every drawing begins with a single line, every project begins with a single drawing. The following drawing setup procedure is a basic outline provided to assist you in starting any type of drawing, once you have determined the type of drawing to be created. It is based on using the AutoCAD software. This process can be modified to suit specific needs.

1. Determine the width (X axis) and height (Y axis) of the real-world drawing area. This is called the drawing *limits*. For example, if you will be drawing a map that covers an area of 900 feet by 700 feet, provide for some additional space in each direction so that your drawing limits will be approximately 1,000 feet by 750 feet.

2. Determine the sheet size required for the drawing, for example, C-size or 22" × 17".

3. Determine the scale of the final plotted or printed drawing. Calculate the scale factor if it is unknown; for example, the drawing area of 1,000' × 750' will occupy a C-size using a scale of 1" = 50'. This is the same as 1" = 600".

Therefore, 600 is the scale factor. See the next section, Scaling and Scale Factors, for a detailed discussion of calculating scale factors.

4. Open the appropriate template or base drawing. Remember that the template drawing should contain objects, values, and settings that will be common to that type of drawing. This includes border and title block, title block text, general notes, text styles, dimension styles, layers, and display settings.

5. Establish layers, line types, and line weights if they are not part of the template drawing. Always adhere to applicable national and company standards when specifying line types and line weights. See Chapter 2 for information on lines.

6. Create text styles using the scale factor if they are not already saved in the template drawing. To calculate the proper AutoCAD text height for plotted text of 1/8", multiply .125" × 600". This gives you a text height of 75". Create a text style with a height of 0 to use for dimensions.

7. Create a proper dimension style using the text style with a zero height. Set all dimension sizes such as text height, arrows, ticks, extension line offsets, etc. to their actual plotted measurements. Then set the overall dimension scale to the scale factor of the drawing. Always be certain that you use a text style with a zero text height.

Following these basic steps will ensure that you always begin drawings in a consistent and accurate manner. Remember that it is most efficient to save the text and dimension styles in a template drawing so that you do not have to create them each time you begin a drawing.

SCALING AND SCALE FACTORS

Drawings created using manual techniques are laid out using a scale, whereas drawings created with CADD are drawn full scale and plotted to a specific scale. Mechanical drawings utilize scales such as 1/2" = 1", 1/4" = 1", and 1/8" = 1". Architectural drawings are representations of buildings and use scales such as 1/2" = 1', 1/4" = 1', and 1/8" = 1'. Some of the smallest scale drawings are in the civil engineering field. A *small-scale drawing* shows a large area, while a *large-scale drawing* shows a small area. These drawings represent areas of land measured in feet or miles, and some of the scales used are 1" = 10', 1" = 20', and 1" = 50'.

The most important point to remember when using AutoCAD is that you always draw full size, and plot at the proper scale. But this also means you must plan ahead and know what scale the final plotted drawing is to be. This is discussed in detail later. Drawing in AutoCAD at full scale means that if a part measures 4.35", you draw it 4.35" long. Even if the final drawing will be plotted on a B-size sheet of paper at 1/4" = 1", it is still drawn at full size in Auto-CAD. A house that measures 48'-6" along one wall must be drawn that exact size in AutoCAD. You will soon realize that it is critically important to plan your work carefully so that you set up the correct drawing limits within which to work. Establishing the correct drawing limits will enable you to draw anything full size within those limits—from a microscopic computer circuit to a map of the solar system.

The planning that should occur prior to beginning the drawing involves, but is not limited to, the following items related to scaling:

■ Size of paper for final plot or print.

■ Size of the object to be drawn.

■ Height of plotted text.

■ Type of dimensioning to be used.

Planning the scale at which a drawing is to be plotted is important because the plotted scale dramatically affects the size of drawing entities such as text and dimensions. First, determine what you have to draw and what size of paper it will be plotted on. Next, determine the scale at which the drawing should be plotted. For example, a drawing of a mechanical part must be plotted at a scale of $1/4" = 1"$. This means that when you use the plot routine in AutoCAD and use the proper scale, all entities in the drawing will be reduced to one quarter of their original size. What would happen to text in this case if it were drawn in AutoCAD .125" high? It would be reduced to one quarter of its size, which means the plotted text would be .03125" high! This is entirely too small. That is why you must plan your drawing carefully, and take the final plotted text height into consideration.

Text height and dimension text heights are directly affected by the plotting scale, and must be considered early in the planning of your drawing. Use the following procedure to create a text style and dimension style that work together to create properly sized dimension text heights.

1. Create a new text style using the font of your choice. Set the text height to zero (0).

2. Create a new dimension style. Give it a name that reflects the type of dimensioning to be used, for example, "Bilateral tolerances" or "Architectural."

3. Choose the text style you created in step 1.

4. Select the proper text height, for example, .125.

5. Set all values for dimension line, extension line, and arrowhead sizes and offsets.

6. Set the overall scale using the reciprocal value of the scale. This is explained in detail below, but the reciprocal is just the number of times smaller or larger the drawing is from the original part or location. For example, a mechanical drawing scale of $1/4" = 1"$ means that the drawing is reduced four times, or to one quarter of the original size. Therefore, the reciprocal number is 4.

7. Save the dimension style and set it current so that it can be used. Whenever you begin a new drawing, always remember to set the appropriate dimension style current.

You can easily find the reciprocal of any scale by the following math:

Mechanical
$1/4" = 1"$
$.25" = 1"$
$1/.25 = 4$

The number 4 is the reciprocal of the scale $1/4" = 1"$. We know that the part is reduced four times when it is plotted. Therefore, you must set DIMSCALE to the value of 4. Now, whenever you place a dimension on your drawing, all of its size components such as text, arrowheads, and offsets are multiplied by the DIMSCALE value of 4. This makes them four times their actual size, but they will look correct on your monitor when you draw the features of the part at their actual size.

The reciprocal is also known as the *scale factor*. The same technique is used to find the scale factors for any type of drawing. Study the examples for architectural and civil scales.

Architectural	Civil
$1/4" = 1'-0"$	$1" = 200'$
$.25" = 12"$	$1" = 2,400"$
$12/25 = 48$	$2400/1 = \textbf{2,400}$

The second important aspect of using the scale factor of the drawing is the text size. Always multiply the intended plotted text height by the scale factor to find the AutoCAD text height. For example, if you wish to have text displayed .125" high on a drawing with a scale of $1/4" = 1"$, first multiply this by the scale factor, in this case 4.

$$.125 \times 4 = \textbf{.5}$$

The value of .5 is the text height. Therefore, you should set your text height to .5 when drawing in AutoCAD. It will appear correct on the screen. When you plot the drawing at the scale of $1/4" = 1"$, the text height is multiplied by the scale to produce the plotted text height; thus, $.5 \div 1/4 = \textbf{.125}$, which is the proper text height.

This same technique is used for architectural, civil, and any other scaled drawings. Calculations for a .125" text height for the architectural and civil scales shown above are given here.

Architectural	Civil
$.125 \times 48 = \textbf{6}$	$125 \times 2,400 = \textbf{300}$

Therefore, an AutoCAD text height of 6 inches will produce .125" text when the architectural drawing is plotted correctly. Likewise, text height of 300 inches in the civil drawing produces the proper .125" text height when the drawing is plotted.

As you can see, planning your drawing regarding scales and scale factors is a critically important task and one that should become an integral part of your drawing habits.

USING LAYERS

In CADD terminology, a *layer* is a group of objects, drawing elements, or components that share one or more common characteristics. When layers are "stacked" on top of each other, they are registered perfectly. For example, an architectural drawing may contain a layer for walls, windows, plumbing, electrical, etc. A multiview mechanical drawing may contain layers for object lines, hidden lines, dimensions, section lines, etc. When all layers of a drawing are visible, the drawing is complete and all components are located accurately. (See Figure 4.28.)

The benefits of layers are many, and their use can increase drawing flexibility, productivity, and clarity.

■ Related components and information can be placed on individual layers.

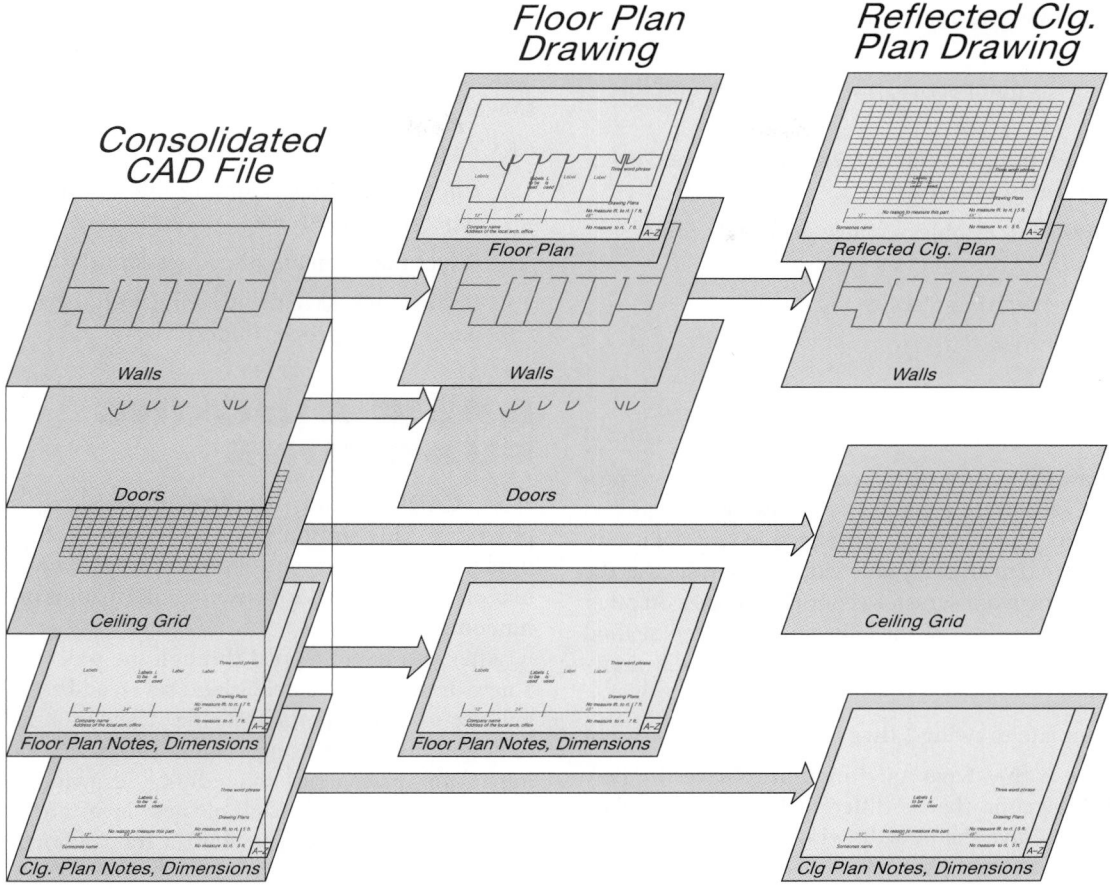

FIGURE 4.28 ■ An example of using layers to share information.

■ Drawing clarity is improved by assigning different colors, line types, and line weights to layers.

■ Layers can be turned on or off as needed to remove information from the screen.

■ Individual layers can be turned off for plotting and printing purposes. For example, the floor plan and HVAC plan can be plotted together and used by the HVAC contractor for estimating and bidding purposes.

The use of layers is one of the most basic aspects of CADD. For example, the American Institute of Architects (AIA) maintains a detailed layering standard. It is good practice to become familiar with any layering standards that are used by your school or company.

THE IMPORTANCE OF TEMPLATE DRAWINGS

A *template* is a pattern of a standard or commonly used part that is created once, then used on subsequent drawings. A collection of symbols in a symbol library could be called a template. If you create a base drawing that contains standard components, values, and settings, it is referred to as a *template drawing*.

Most schools and companies use template drawings. These template drawings save time and help produce a certain amount of consistency in the drawing process. Let us first look at what a template drawing is composed of, then at how it can be used.

Template Drawing Contents

Template drawings should be stored in a common location that is accessible to everyone who needs them. This is often on a network computer. If you maintain your own templates, be sure they are stored in at least two different locations. Template drawings can and should be updated and added to; therefore, it is important that you replace all old copies with the updated versions. Keep a variety of template drawings on file that contain settings for different drawing disciplines and scales. Template drawings can contain, but are not limited to, the following items:

■ Border and title block. Standard title block text information can be filled in.

■ One or more AutoCAD floating model space viewports.

■ Several named text styles with heights to match different scale drawings.

■ A text style with zero text height for dimensioning purposes.

■ Named dimension styles with values and settings for specific drawing scales.

- Named layers containing colors, line types, and line weights.
- Display settings for point styles, multiline styles, and line weights.
- User profiles containing display screen menu layouts, colors, fonts, and configuration.
- Drafting settings such as object snaps, grid, snap, units, and limits.
- Section patterns (hatch) and scales.
- Plot styles and settings.

Using Template Drawings

Using template drawings is a simple process, and if a standard procedure is used, it will create a productive drawing session and ensure consistency in your drawings. First be sure that your template drawings are stored in an easily accessible location such as the local hard disk or the network server. Keep backup copies on other flexible disks or optical media. Use the following three-step method to work with template drawings whenever you begin a new drawing project.

1. Open the template drawing (.dwg file).
2. Immediately use the "Save As" function to save the new drawing with a name that is different from the template name. Be sure to save the new drawing in a different location than that of the templates.
3. Begin work on the new drawing project.

Now you have a base drawing on the screen that contains many of the standard items required for any drawing. As you work, you may discover additional things that should be in your template drawings. Add them as you think of them, as this will lead to greater productivity in future work.

Using AutoCAD's Template Drawing Method

AutoCAD provides a method for saving template drawings that can be quickly opened when you begin a new drawing session. If you use this method, be aware that template files are saved, by default, in a standard location on the local hard disk drive. A *default* is a setting or value that is used automatically if you do not provide one. The default location can be changed if you are familiar with customizing AutoCAD's file managing options. Use the following method to save template files in AutoCAD using the default settings.

1. Construct the template drawing as discussed previously.
2. Save the drawing in a secure location as a .dwg file.
3. Use the **SAVEAS** command, or pick **Save As** . . . from the **File** pull-down menu.
4. Choose **AutoCAD Drawing Template File (*.dwt)** from the **Save as type:** drop-down list. Notice that the folder location automatically changes when the .dwt file type option is selected. This location can be changed in AutoCAD.

The template file is now saved as a .dwt file (drawing template). It is ready to be used whenever you begin a new drawing session. You can open the drawing just like you open any other drawing (.dwg) file in the following manner:

1. Pick **Open** . . . from the **File** menu.
2. Choose **Drawing Template File (*.dwt)** from the **Files of type:** drop-down list.
3. Select the template file you wish to use and pick **Open**.
4. Immediately save the drawing with the appropriate project name in the proper folder.

BASIC DRAWING FUNCTIONS

Most CADD software is structured along similar drawing processes. This section discusses general categories of drawing functions; therefore, you should always consult your software text or user's manual for specific information on commands and functions.

After you have stepped through the procedure for setting up a new drawing, there are at least seven additional areas that you will use to complete the drawing. They are basic geometry construction, object editing, using predrawn entities, text and dimension placement and editing, drawing display, drawing queries, and printing. Each of these areas are discussed in general terms to provide you with a framework for working efficiently on any project.

Preparing a Drawing

After planning the overall scope of the drawing project as outlined previously, you should then focus on a method for starting and completing the actual drawing. The least stressful and most productive method to use for preparing your drawing plan is to create a freehand sketch. A basic procedure to use for this drawing construction process is as follows:

1. Create a freehand sketch of the intended drawing layout. This does not have to be to scale. Sketch all views, sections, details, and pictorials that are required. Apply all required dimensions and notes. Using numbered items or colored pencils, label or list the components in the order they are to be drawn. Use a color such as red to provide notes on the sketch regarding specific commands or functions to use on certain features. Let your sketch be your roadmap for the project.
2. Try to determine if special views such as AutoCAD viewports will be required. Also decide if you will need named views that can be restored quickly for working in detailed areas of the drawing.
3. Note the geometric shapes that must be drawn from scratch.
4. Carefully study the sketch and determine what features editing existing shapes can create.

5. Are there any objects or components that have already been drawn, such as objects in a symbol library, that can be inserted or referenced into the drawing?

6. Determine the locations and type of dimensions to be used.

7. Locate all local notes, general notes, and view titles.

8. Specific commands can provide you with information about the drawing and its objects. Try to decide if using object values such as size, location, areas, etc. will assist you in drawing additional features or provide you with useful information required for the project. Commands such as these are called *inquiry* commands in AutoCAD.

9. Determine the types of prints and plots or electronic images that are required for the project. This should include paper sizes, scales, line thickness, textures, and colors. It may also include plots that require specific layers or features to be printed in bold or grayscale.

Basic Geometry Construction

In its most basic form, all art is created with just two geometric shapes: the arc and the straight line. But CADD software contains a variety of commands that enable you to draw an assortment of geometric shapes such as circles, arcs, polygons, rectangles, multiple line entities, complex curves, and ellipses.

One of your first tasks when planning the drawing with a freehand sketch is to determine which shapes can be constructed with the basic geometry commands in the software. Use your sketch to note object values such as radii, diameters, angles, and lengths. Also note the command names on your sketch if are just learning the software. An additional step you can use is to number the basic shapes in the order you will draw them.

Object Editing

A good rule to use when constructing a CADD drawing is "draw as little as possible." Think about what this implies. If you carefully plan your work, you will begin to see that, with any type of drawing, shapes and features are repeated, used in different sizes, or even appear rotated or mirrored from one another. Try to locate features on the freehand sketch that fall into this category.

After you have drawn just the "bare bones" objects in the drawing, see how many of the original features can be copied, rotated, stretched, compressed, or mirrored in some manner in order to create more objects. In this way, you actually draw just a few objects, and then use those to create new ones. This process is referred to as *editing* and can become a very productive and efficient method to use in any project. One such productive editing command is MIRROR.

Mirror

The MIRROR command can produce reverse copies of an entity or symbol by simply specifying an axis on which to reflect the object. The image can be duplicated at any distance from the original or appear to be attached to it, such as the object in Figure 4.29.

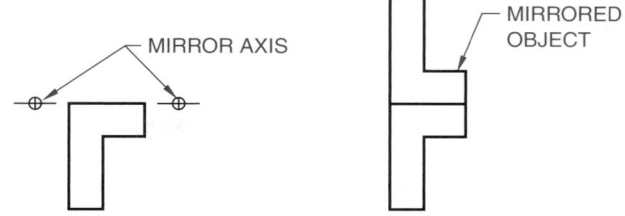

FIGURE 4.29 ■ The MIRROR command can be used to reflect a feature.

Using Predrawn Entities

A productive aspect of CADD is the ability to draw an object or component once, then save it for future use. These objects can be referred to as symbols, cells, blocks, and reference files. Regardless of the terminology, the value of the objects lies in the fact that they can be used over and over on any drawing. This saves time in drawing creation and can lead to an increase in productivity.

The most important aspect of this topic for the drafter to remember is to always be aware of the kinds of predrawn symbols that are used by your school or company. Before you begin work on any drawing, always consult available company and project standards for information on the use of symbols. AutoCAD allows you to use predrawn symbols, and entire drawings, in three different ways.

Block

When an object is saved using the BLOCK command, it is saved in the current drawing. It is always available for insertion in the current drawing, and can be copied to other drawings if needed.

Wblocked Objects

Any objects or drawing can be saved as a single entity using the WBLOCK command. The selected entities are saved as a drawing file with an insertion point. This drawing file can then be inserted in any other drawing. The same is true of any AutoCAD drawing. If you think that a drawing you create may be useful in other situations, use the BASE command to provide an insertion point, then select the entire drawing. When you insert the drawing into any other drawing, it will be placed using the insertion point specified with the BASE command.

Reference Drawings

When it is necessary to insert an entire drawing or part of a drawing into the current drawing, it can be *referenced*. This means that the reference drawing is not actually inserted in the current drawing; its name is just attached. Then, when you open the drawing containing reference files, AutoCAD finds the reference files and displays them on the screen. By default, referenced drawings do not become part of the current drawing. Therefore, it is always important to keep referenced files in the same location. This process is a bit more complex than using blocks, so it is best to consult your textbook or user's manual for detailed instructions.

As indicated previously, using predrawn entities can be a great time-saver. Therefore, always be aware of the kinds of symbols that are available in your school or company before you begin any drawing. This topic is discussed in greater detail in the upcoming section titled Creating and Using Symbols.

Text and Dimensions

Dimensions should always be placed on a drawing after the geometry is completed. Then place local notes and text. When using software such as AutoCAD, it is important to establish dimension styles and text styles before you begin a drawing. As stated previously, it is best to have dimension styles and text styles saved in template drawings so they are available for use when the drawing is opened.

Most drawings created for production purposes rely on accurate and readable dimensions. Therefore, it is imperative that you plan the style and location of dimensions before you begin the drawing. Again, it is critical that you become familiar with your school or company standards regarding dimension and text styles.

Drawing Display

Throughout the life of a drawing, you may change the display hundreds of times. This means that you may *zoom* in on an object to get closer, or *pan* around the drawing to see something that is off the screen. In addition, you may zoom in to an area that you will be working on frequently and create a *named view* that can be restored at any time. Drawings and models that require the layout of different views other than a standard orthographic layout may benefit from the use of AutoCAD *viewports*. A viewport can contain any type of view at a scale that is different from other views on the drawing. These functions are all aspects of displaying the drawing. The most common display function is the ZOOM command.

ZOOM

The ZOOM command allows the drafter to place a zoom window around an area that needs to be enlarged. After the zoom window is drawn, the entire screen is filled with the area that was inside the window. This command is useful when working in a detailed area of a drawing. Once a new window is displayed, the command can be used again and again to get even closer to the object or feature. Figure 4.30 illustrates how the ZOOM command works.

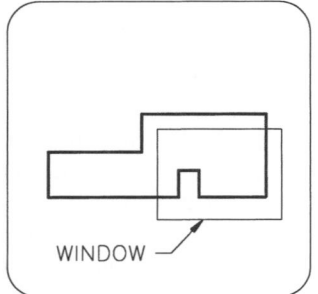 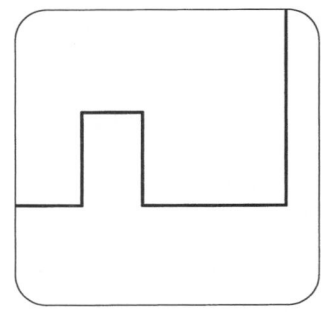

FIGURE 4.30 ■ The ZOOM command brings the object, or a feature of the object, closer.

Viewports

In the AutoCAD software, a viewport is a view through the "paper" in paper space that displays the drawing objects in model space. Basically it is a "hole" that is cut in the paper through which you can see the drawing or model. You can create as many viewports as needed, of different shapes and sizes. Most importantly, any type of view can be shown in a viewport at any scale. Since the viewport is a drawing object, it can be moved and resized. At minimum, a drawing should contain at least one viewport.

AutoCAD viewports are important objects, especially in relation to plotting the drawing on paper. Therefore, you should always include at least one viewport in all template drawings. Size it to include most of the open space inside the border and above the title block. (See Figure 4.31.) The viewport size can be quickly changed to suit the needs of the drawing.

Because the use of viewports is such an important aspect of AutoCAD drawing layout and plotting, it is important for the drafter to consult all available resources and become familiar with the use of this drawing component.

3-D Views

Display options for 3-D models are endless. If you work with 3-D models, it is important to learn the different methods for 3-D display. One benefit of a 3-D model is that it can be displayed in a variety of formats. For example, a single drawing can contain standard dimensioned orthographic views, plus one or more 3-D views of the model at different scales using different rendering (shading and coloring) techniques. As mentioned previously, several different 3-D views can be displayed at different scales on a single plotted drawing with the use of viewports.

One important aspect of 3-D display is understanding the concept of the viewpoint. The *viewpoint* is the view of the model as seen by the user looking at the screen or printed drawing. Keep in mind that the Cartesian coordinates of the model never change. We may think that the model is being rotated, but as far as the software is concerned, it is the sight of the person viewing the object, or viewpoint, that is being changed. For example, notice the two views of the truck shown in Figure 4.32. The left side view is a plan, and the coordinate location of the viewer is X=0, Y=0, Z=1. In comparison, the coordinate location of the viewer is X=4, Y=4, Z=2 in the 3-D view on the right side.

Drawing Queries

A *query* is a question. When you query the drawing for information, it may give you an answer. Typical queries involve finding the distance between two points, the area of an object, the length of a line or entity, or the diameter of a circular feature. AutoCAD refers to these functions as *inquiry* commands. An example of an inquiry command is finding the area of an object, such as a parking lot.

If you know that the project you are working on will require the calculation of areas, it is best to create the objects in such a manner that the area can be calculated quickly. Additionally, inquiry commands can be used to check your work for accuracy. It is good practice to learn how to check your drawings for accuracy. Less time required for error correction later in the project leads to greater efficiency and productivity.

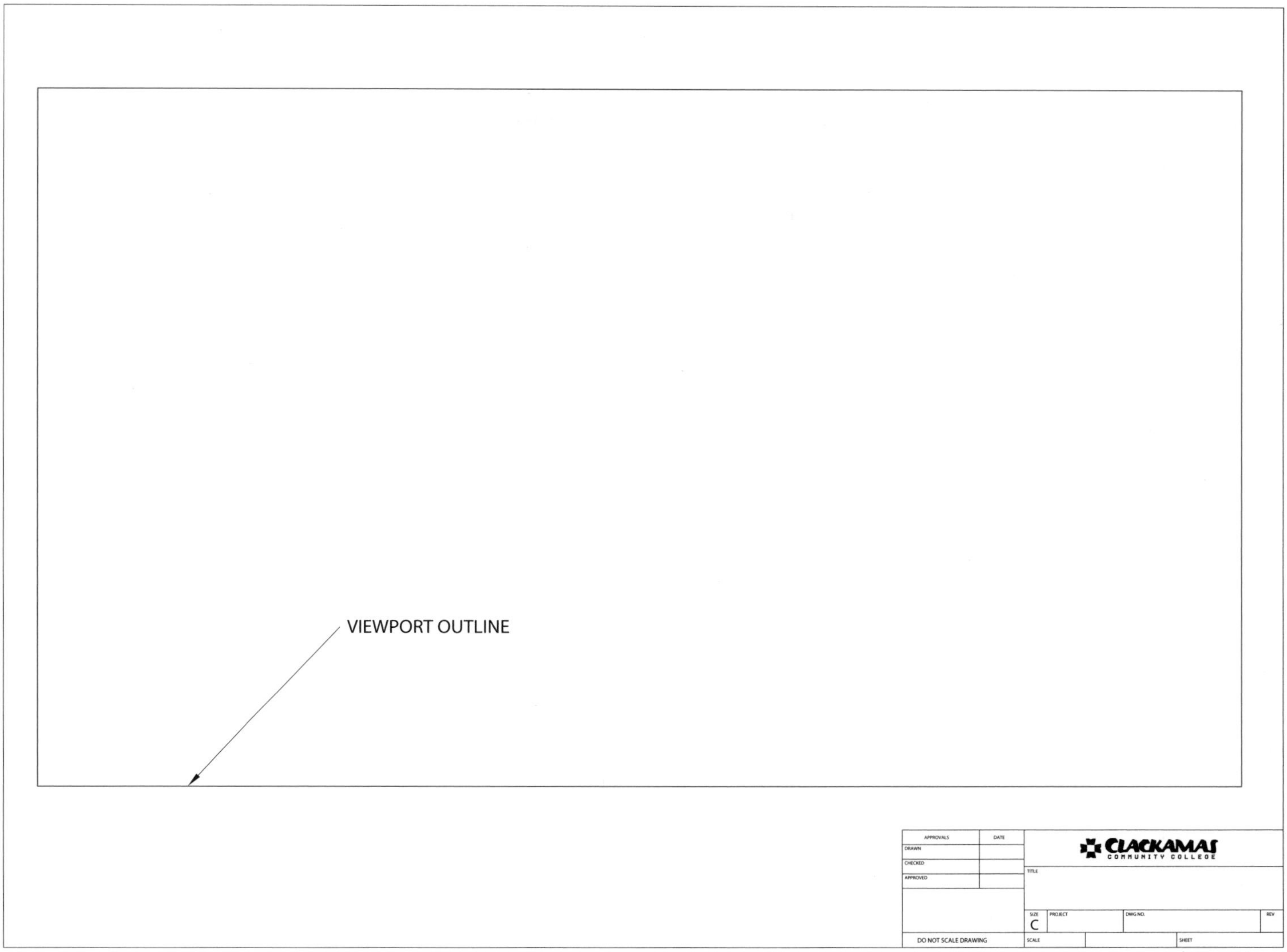

VIEWPORT OUTLINE

APPROVALS	DATE	✿ CLACKAMAS
DRAWN		COMMUNITY COLLEGE
CHECKED		TITLE
APPROVED		
		SIZE C	PROJECT	DWG.NO.	REV
DO NOT SCALE DRAWING		SCALE		SHEET

FIGURE 4.31 ■ In the AutoCAD software, a viewport is a "hole" that is cut in the paper through which you can see the drawing or model.

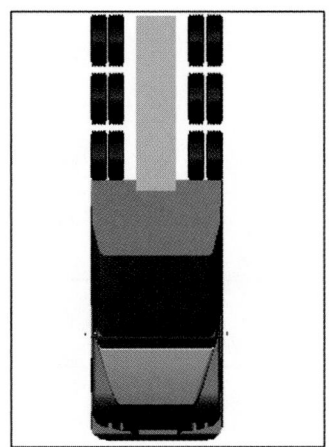 

FIGURE 4.32 ■ The left-side view is a plan, and the coordinate location of the viewer is X=0, Y=0, Z=1. The coordinate location of the viewer is X=4, Y=4, Z=2 in the 3-D view on the right side.

Printing and Plotting

The final step in the progression of a drawing or design is to put the image on a sheet of paper. This is referred to as *printing* and *plotting*. Most often the terms are used interchangeably, but printing usually refers to a small A- or B-size sheet of paper generated by a laser or inkjet printer, and not to scale. This size and type of drawing is known as a *check print*. A *plot* generally refers to a sheet of paper C-size or larger that is plotted to scale. Large-format plotters can use inkjet, laser, or ink pen technologies.

DRAWING WITH THE COMPUTER

Creating good drawings with a CADD system hinges on just a few things: your ability to learn and remember, hand-to-eye coordination between the input device and the screen, your ability to understand and visualize coordinate systems, and your typing ability, although this varies among CADD systems.

In this section we will look at how you use the above-mentioned skills to create drawings. The intention is not to discuss the techniques used with a particular brand of CADD system, but to describe the general concepts, techniques, and commands used when working with any computer drafting system. A familiarity with the basics of CADD will allow you to more readily grasp the particulars of any one system.

Cartesian Coordinate System

Absolute Coordinates

The *Cartesian coordinate system* is a rectangular coordinate system that locates a point by its distances from intersecting, perpendicular planes. Drawings done with a computer are based on the Cartesian (rectangular) coordinate system, so it is imperative that you develop a thorough understanding of it.

The two-dimensional rectangular coordinate system is illustrated in Figure 4.33. The point of intersection of the dark vertical and horizontal lines (planes) is called the *origin*. The origin has a value of zero. Values along the horizontal increase as you move to the right of the thick vertical line. This is called the *X axis*. The Y axis is the vertical line and values increase as you move upward from the origin. The values just mentioned are positive. Note in Figure 4.33 that each axis can also have negative values. Negative X values are to the left of the vertical plane, and negative Y values are below the horizontal plane. Study this figure carefully before continuing.

The best way to understand the rectangular coordinate system is to work with a piece of 10 lines to the inch graph paper. Locate the intersection of any two heavy perpendicular grid lines on your graph paper. Mark this point as the origin. Both X and Y have values of zero at the origin. At each inch mark on the horizontal line to the right of the origin, write the values 1, 2, 3, 4, etc. Do the same for the inch lines above the origin along the vertical axis line. (See Figure 4.34.) Now locate the point of X = 2, Y = 2. First count two inches to the right on the X axis, then count two inches straight up on the Y axis. This point has the value of X = 2, Y = 2. Only one point in this coordinate system can have that value. Coordinate locations are always listed with the X value given first; for example, the previous point is written 2,2.

Not all points will have even inch values. Locate the point X = 4.5, Y = 2.9 on your grid paper. This point is located directly from the origin and not from any other point. This point is written as 4.5,

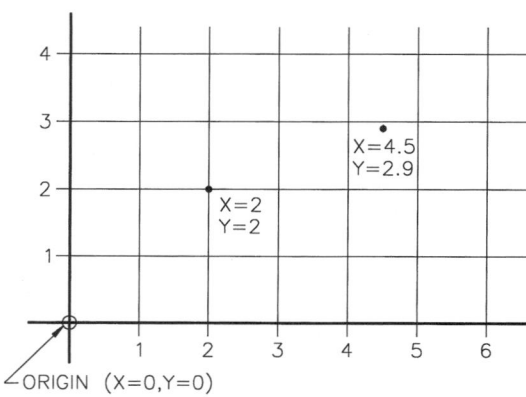

FIGURE 4.34 ■ Points located on a rectangular coordinate grid.

2.9. Count the grid squares along the X and Y axes to find this point, or use a scale and measure. Remember that every point must have at least two coordinate values. A point of just 4.5 could be any location 4.5 in. from the X plane.

Relative Coordinates

Objects can also be drawn using a moving or relative origin. In this method, each point becomes the origin and the next point is located in relation (relative) to the previous point. This method is called *relative coordinates*. If each point becomes the origin for the next point, then numbers to the left and below the origin are negative. See Figure 4.35 for an example of this method.

The relative coordinate method may be useful when using the dimensions of each line to lay out a part. It is important that you keep in mind the positive and negative aspects of this method, because each point becomes the origin of a coordinate system. For example, moving to the left of a point gives the next point a negative X value. The object in Figure 4.36 shows the relationship of points on an object constructed with relative coordinates.

When using AutoCAD, relative coordinate values must be entered at the keyboard as follows:

@2,2

This entry locates a point two units on the X axis and two units on the Y axis from the previous point.

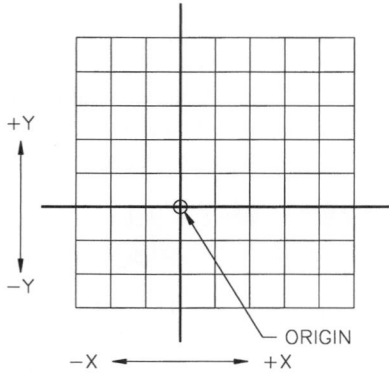

FIGURE 4.33 ■ Two-dimensional Cartesian (rectangular) coordinate system.

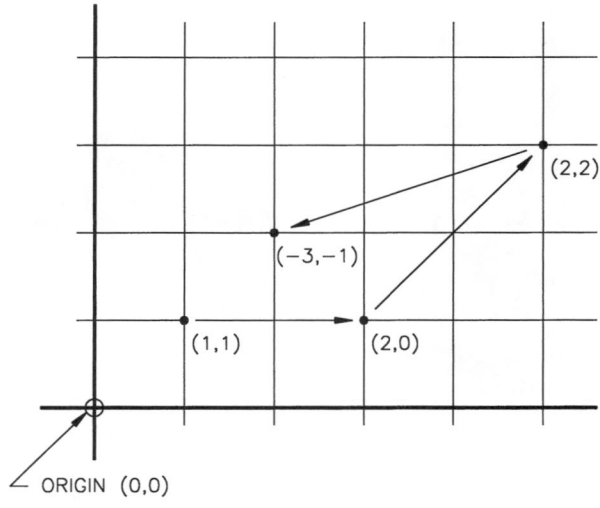

FIGURE 4.35 ■ Points located using relative coordinates.

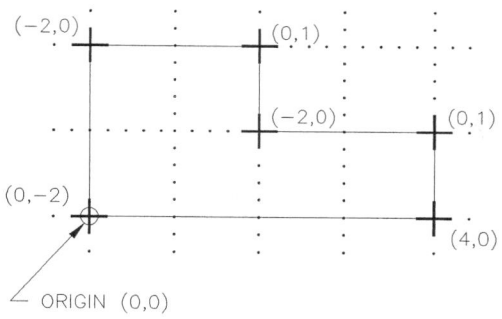

FIGURE 4.36 ■ Object drawn using relative coordinates.

Polar Coordinates

Not all points can be located properly with a rectangular coordinate system. For example, a point may be at a 45° angle from horizontal at a radius distance of 1.5 in. from a reference point. Such a location defines a point by *polar coordinates*. Note in Figure 4.37 that the horizontal line has a value of 0°. Angles are measured in a counterclockwise direction from horizontal. The radius distance is measured from a specified reference, or center, point.

Polar coordinates are relative and can be used to create a drawing if all the angles of the part are known. Figure 4.38 is the same object drawn in Figure 4.36, but the values given are polar coordinates. The first number at each corner of the object is the angle from the previous point (origin), and the second number is the radius from the origin. Study this object until you are familiar with the concept of polar coordinates.

When using AutoCAD, polar coordinate values must be entered at the keyboard as follows:

@1.5<45

This entry locates a point two units from the previous point at a 45° angle.

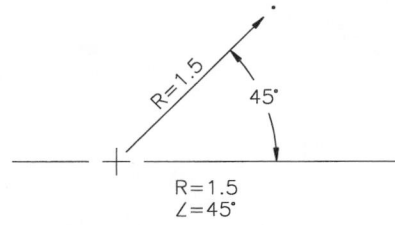

FIGURE 4.37 ■ Locating a point using polar coordinates.

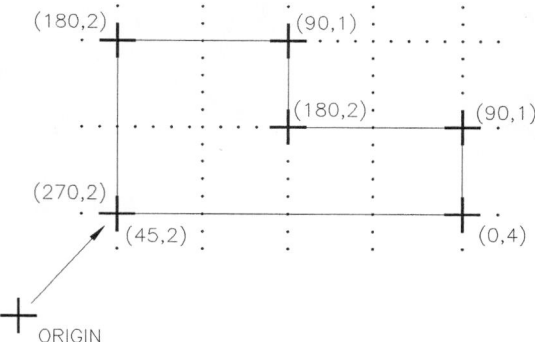

FIGURE 4.38 ■ Object drawn using polar coordinates.

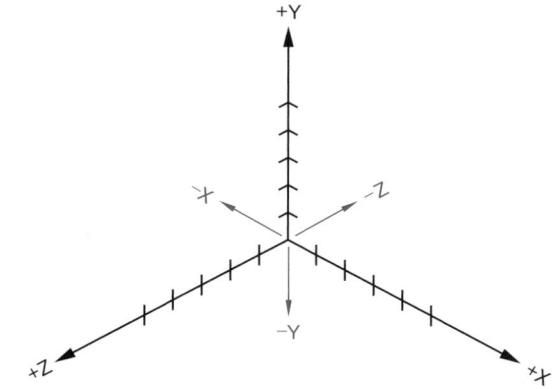

FIGURE 4.39 ■ Three-dimensional coordinate system.

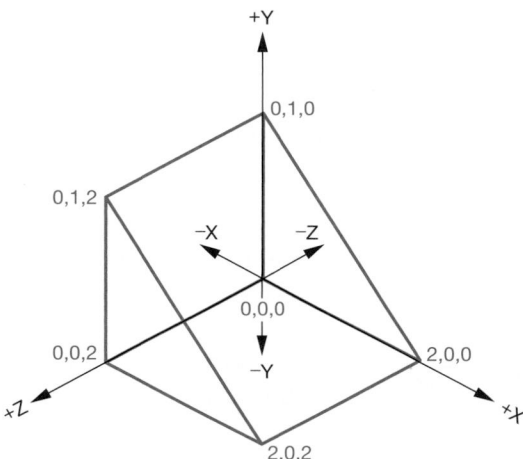

FIGURE 4.40 ■ Object plotted using 3-D coordinate system.

Three-Dimensional Coordinates

To construct a three-dimensional coordinate system, all we need do is add a third dimension, the *Z axis*. The third axis, Z, projects perpendicular from the drawing surface. This means a positive Z value projects toward the viewer from the surface of the display screen when viewing the object from the top (the "plan" view). (See Figure 4.39.) The intersection of these three axes forms the origin, which has the numerical value of X = 0, Y = 0, and Z = 0. Any point that is behind the origin on the Z axis has a negative value.

The object shown in Figure 4.40 is shown in wire frame form, and the X, Y, Z coordinate values of each point are indicated. The first number of each set is X, the second Y, and the third Z. Study the object in Figure 4.40 until you understand the technique to plot points in three dimensions (3-D).

Using Coordinates

Now try your hand with the object given in the table on this page. You are provided only the coordinate values for each point on the object. Locate each point given on the chart and draw a line as you go. Lay out a 3-D axis like the one shown in Figure 4.39. Use a scale to measure the coordinate values given and plot them as shown in the example in Figure 4.41. The solution to this exercise is found at the end of this chapter, page 133.

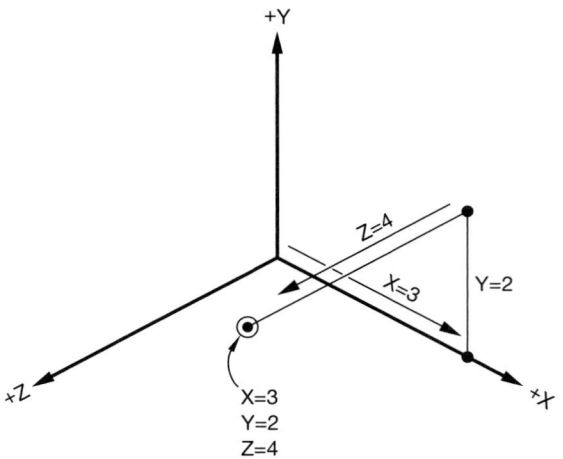

FIGURE 4.41 ■ Locating a point using 3-D coordinates.

Three-Dimensional Coordinate System

The addition of the Z axis adds depth to the object and allows the CADD professional to create a model form. A popular misconception about computer graphics is that the entire object can be shown in three dimensions (3-D) just by entering coordinates for a couple of views. This is not always true. The important thing to realize is that every point on the drawing must be defined by numerical locations, or coordinate points, for the computer to generate a 3-D form.

Three-Dimensional Coordinate Exercise			
Point	X	Y	Z
1	0	0	0
2	3	0	0
3	3	1	0
4	1	3	0
5	0	3	0
6	0	3	4
7	1	3	4
8	3	1	4
9	3	0	4
10	0	0	4
11	0	3	4
12	1	3	4
13	1	3	0
14	3	1	0
15	3	1	4
16	3	0	4
17	3	0	0
18	0	0	0
19	0	0	4

The process of entering three-dimensional data into a system is relatively simple. The principal component is your ability to visualize three dimensions and distinguish and separate any number of planes in one view. When entering data in 3-D, you should establish an origin point. (See Figure 4.40.) You can then draw using XYZ coordinates based on that origin. Programs such as AutoCAD allow you to move and rotate the origin to create new coordinate systems. You can place the origin and coordinate system on any plane of an object by using the UCS command in

AutoCAD to create a new *user coordinate system*. This allows you to draw any shape, using its actual dimensions, from any angle or viewing point you choose.

Note in Figure 4.40 that the coordinate system can accommodate negative coordinates. An object can be drawn in any or all of the four quadrants. An object can just as easily be drawn using all negative coordinates as positive ones. Most drafters work with positive coordinates unless it is more convenient to locate the origin in the center of the part or a feature.

A CADD 3-D model becomes amazingly versatile when all of its vertices have been entered. Once the computer knows the location of all of the points of an object, it can turn and rotate the object to any angle, as with the robot arms shown in Figure 4.42. And that is one of the features of CADD that makes it so appealing in engineering, design, testing, and manufacturing.

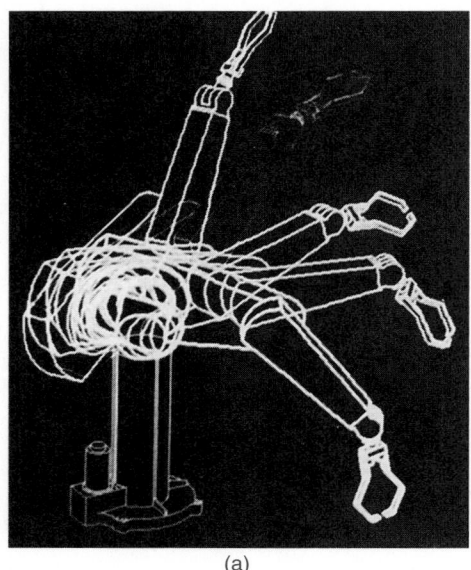

(a)

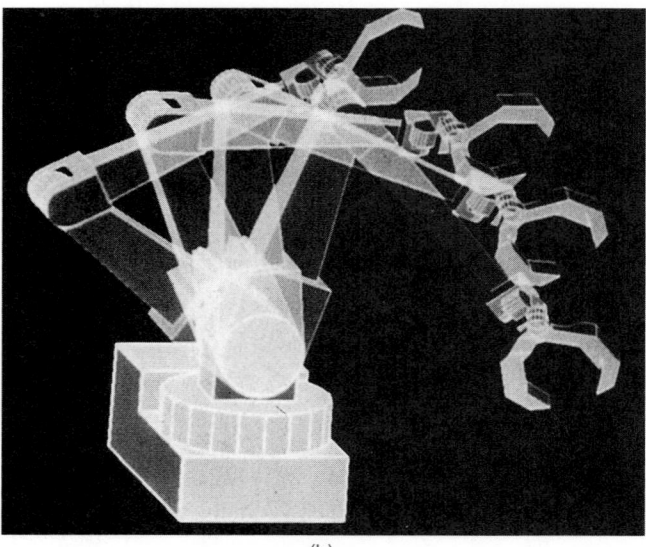

(b)

FIGURE 4.42 ■ **(a)** These robot arms can be rotated in any direction once all of the points have been entered into the computer. *Courtesy Computervision Corporation.* **(b)** Another example of robot arm movement. *Courtesy Spectragraphics Corporation.*

CREATING AND USING SYMBOLS

One of the great time-saving features of CADD technology is the ability to draw a feature once, save it, and then use it over and over without ever having to draw it again. That is why one of the first things a company does after purchasing a CADD system is to begin developing a symbol library. This is a file of symbols that can be called up, located on the menu tablet, and used on a drawing. The area for a symbol library is shown on the menu in Figure 4.43.

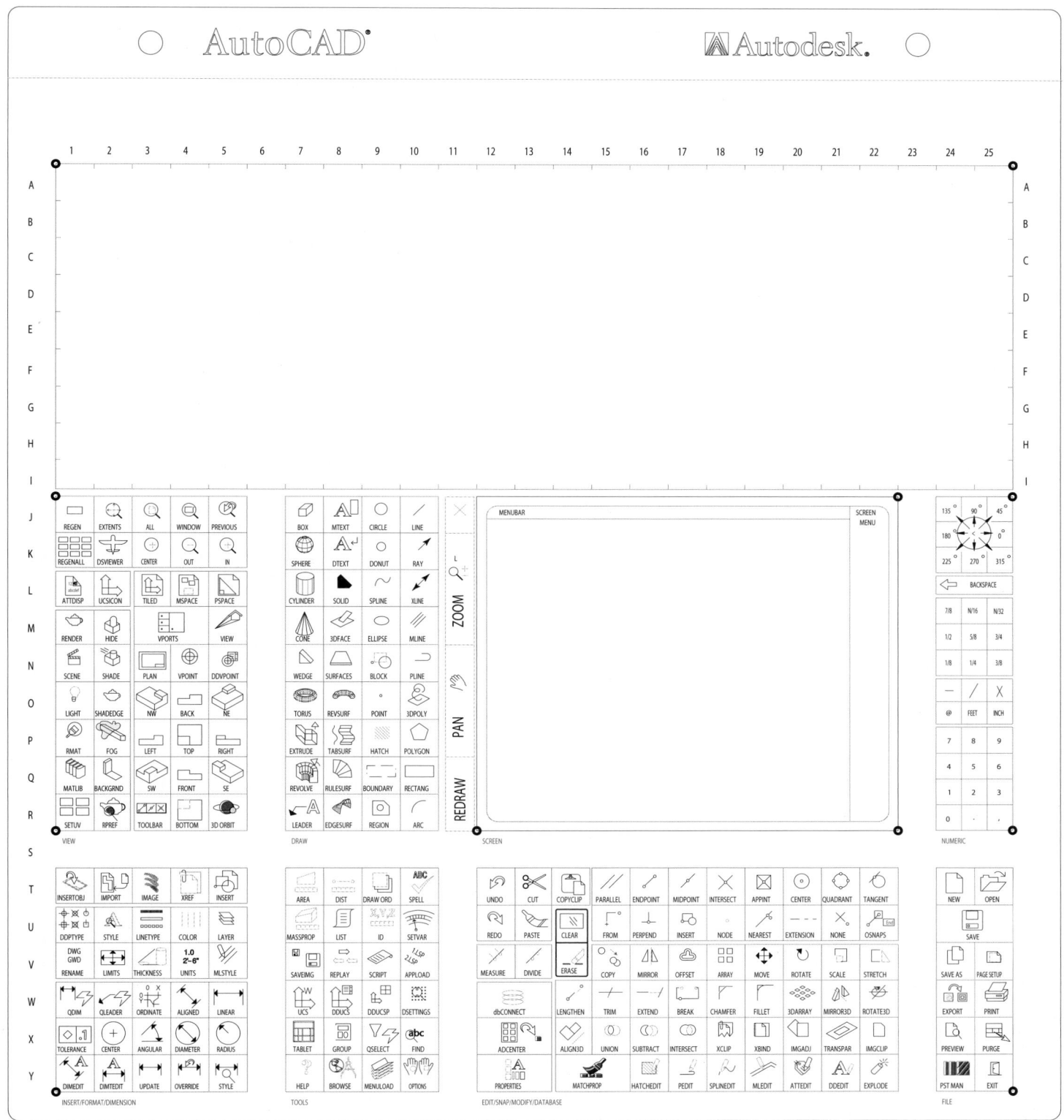

FIGURE 4.43 ■ The upper portion of this menu overlay houses the symbol library. *Courtesy Autodesk, Inc.*

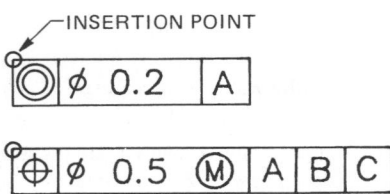

FIGURE 4.44 ■ Location or insertion points on symbols.

Drawing a Symbol

The process of creating a symbol is a simple one. Begin a drawing with the name of the symbol as the file name. When you start the actual drawing of the symbol, begin by placing the origin of your coordinate system at a spot on the symbol that would be convenient to use as a point of reference or location on a drawing. Note the symbols in Figure 4.44. They have small circles at their location points. This is the spot that you point to on your drawing to locate the symbol.

When the symbol has been drawn, it must then be stored, but in a slightly different manner from a regular drawing. It is a symbol and not a drawing, and must be stored as such. The AutoCAD command BLOCK enables you to save the object as a symbol in the current drawing. This allows the symbol to be used in ways that a regular drawing cannot be used.

The AutoCAD WBLOCK command allows you to save the selected objects as a drawing file. Although blocks and wblocked objects can be used in the same manner, keep in mind that symbols created using WBLOCK require more storage space because each symbol is a separate drawing file.

Creating a Symbol Library

After all of the required symbols have been drawn, they can be used in a variety of ways. It is important that all the symbols or blocks be located in a common area, such as a floppy disk, a directory on the hard disk drive, or a drive or folder on the network server. The simplest method is to just insert or merge them with an existing drawing, calling the symbol up by its file name. If you have the ability to customize the software's menus, the symbols can be added to a new screen or pull-down menus for easy access. In addition, they can be added to an area of the tablet menu as discussed previously. These methods of locating the symbols in menus allow them to be picked and then placed in the drawing, thus eliminating the need to type file names.

If symbols are placed on a digitizer or tablet menu, the final task would be to have all the symbols on one template plotted in a pattern that matches their layout on the template. This can then be used as an overlay or mask on the menu that would look similar to those shown in Figure 4.45.

Using Symbols

Most of the editing features that are used to revise drawings can be used to alter symbols. In addition, there are several commands that enable you to work with the symbols. The entire symbol template

SYMBOL LIBRARY: PIPING

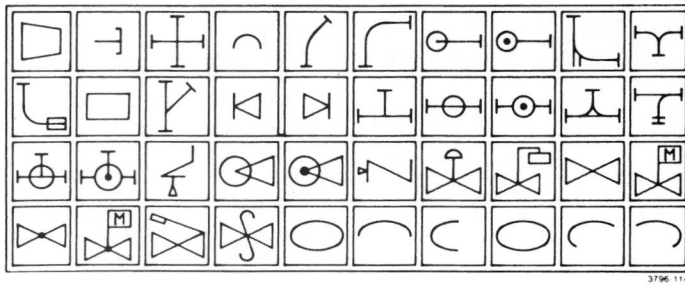

SYMBOL LIBRARY: SCHEMATIC

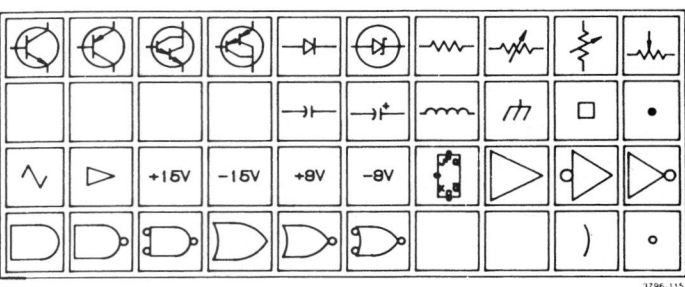

FIGURE 4.45 ■ Typical symbol template overlays.

can be selected for use by choosing a command such as MENU. Another method of choosing a specific symbol template is to have all of the template names displayed on the menu tablet. Then it is just a matter of pointing to the name of the template that is needed.

A tablet menu symbol can be selected by pressing the pointer inside the particular symbol box. The symbol can then be located on the drawing by simply pressing the pointer to the spot where it is desired. Some systems may automatically display the symbol at the crosshairs location. As the puck is moved, the symbol moves with it. When the symbol is in the desired location, simply press a button on the puck and the symbol is placed in that spot.

Symbols are fluid just like the rest of the drawing, and can be changed, moved, renamed, and copied. Commands such as ROTATE, COPY, MIRROR, RENAME, and MOVE can all be used to work with symbols.

STORING A DRAWING

The storage area for the computer drafting system is the flexible disk, optical disk, or hard disk drive. Permanent storage is best done using optical disks or magnetic tape. The disk drive does the actual storing of your drawing on the disk. It can also read the data from the disk and send it to the computer.

The act of storing a drawing that is on the screen is handled by menu commands such as SAVE and save as. After selecting the proper command, you may be given a chance to change the file name of the drawing. After that, a press of the RETURN, or ENTER, key sends the drawing on its way to the disk. Some systems eliminate the need for the operator to store the drawing because they constantly send the drawing data to the disk. This is a great feature because it removes the consequences of a computer crash, or malfunction. Should the computer malfunction during the course of work on a drawing, nothing would be lost except the command that was in progress.

How often should a CADD operator store a drawing? This could be a matter of personal preference or company policy. Practice varies, but it is a good habit to save your work every 10 to 15 minutes, just in case something happens to the computer.

FILE MANAGEMENT

The term *file management* refers to the methods and processes of creating a management system, then storing, retrieving, and maintaining the files within the guidelines of the system. From the CADD drafter's viewpoint, this may be as simple as opening a drawing file and saving it in the same location. Or the drafter may be responsible for creating a new hard disk directory tree (folders and subfolders) for new projects, and ensuring that a variety of file types are stored in their proper locations.

Be aware that a wide range of file management systems are used in the industry. The important thing for you to remember is to become familiar with the system used in your school or company. One of your goals should be to handle your work in the most efficient and productive manner possible. Proper file management procedures can help you reach this goal regardless of the kind of project you are working on.

Rather than describe a specific type of file management system here, it is more important that you are familiar with some of the aspects of file management and maintenance procedures. Proper file management should include the following:

■ Save all working files on a regular basis. When you save a file, save it to at least two different media, such as the local hard disk and a flexible or optical disk. In addition, establish automatic safeguards for saving your work on a regular basis. For example, set AutoCAD's automatic save function to save your drawing every 10 minutes.

■ Create storage media directories and folders for all files.

■ Never save data files (AutoCAD drawings, text documents, spreadsheets, etc.) to folders that contain application programs.

■ Create separate folders for each project, and subfolders for different types of files in that project.

■ Template file originals should be stored in a protected location that allows only selected individuals to edit and save these important files.

■ Separate drawing files from other types of files such as text documents, spreadsheets, and database files.

■ Maintain off-site locations for file storage and archiving.

■ Establish a regular schedule for deleting all backup, temporary, and "junk" files from your storage media.

■ If you have hard disk maintenance software on your computer, set it to *defragment* your hard disk on a regular basis. When application software is used, and files are opened

and saved, portions of the files are saved in different physical locations on the hard disk and flexible storage media. As more of this defragmentation occurs, it can begin to slow the hard drive's access to the files. Disk maintenance software can rearrange software and data on the hard disk, and reclaim space occupied by deleted files.

■ Install only licensed software on your computer system. Always report any unlicensed software to your systems manager.

■ Never install software without approval from your instructor, supervisor, or systems manager.

Network Systems

The network *server* is the computer that acts as the manager, or control center, for the network. Drives, directories, and folders for software applications and data files must be maintained on the server. Schools and companies using computer networks are faced with a much more complex file management task. In addition to maintaining all of the software and data files that are required on the local computers, a variety of applications and data backup locations must be managed on the network server. In addition, the server allows many users to be connected to peripheral equipment such as plotters, scanners, and other reproduction facilities. (See Figure 4.46.)

A network that connects the computers within an office or company is called a *local area network* (LAN). A network that connects computers in widely scattered locations is called a *wide area network* (WAN). Within these systems, there may also be electronic communication services such as e-mail and local office *intranet* systems, Internet connections, file security systems, and computer virus detection applications. Therefore, even though you may not be responsible for the management and maintenance of these systems, it is important that you know how they function. This enables both you and others who use the network to function more efficiently, and also makes the network administrator's job easier.

INTERNET-BASED CADD

The rapid proliferation of network and Internet technologies has opened doors for CADD-based design, engineering, and construction projects to be linked instantaneously through the World Wide Web. Persons working on a project can now be located at distant sites and still have access to the drawings, documents, schedules, and files they need to complete the project. This includes subcontractors, vendors, and manufacturers. Each person associated with the project is given password access to the files needed to complete his or her tasks. These files can be located in several locations around the world, and likewise can be accessed from any location.

Powerful new scheduling, database, and file management software enable projects to flow without conflict as each piece of the project is completed according to the schedule. All companies and individuals related to the project receive instant communications and documentation required to work efficiently within the project requirements. As the appropriate persons update files, new copies are transmitted to project members regardless of where they are located.

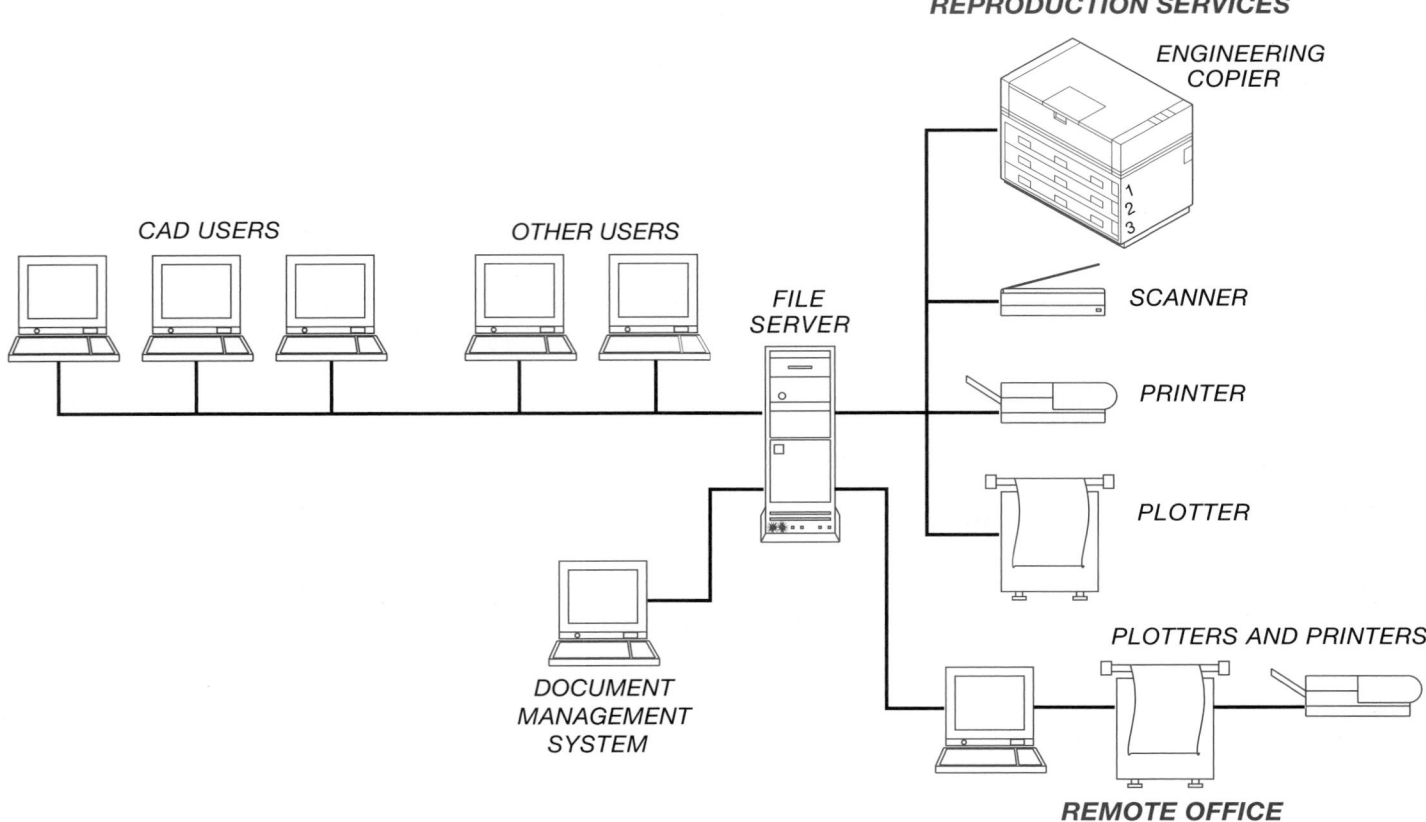

FIGURE 4.46 ■ A typical network system in an architectural office.

Web-based projects allow for the increased use of outsourcing. *Outsourcing* means sending parts of a project out to subcontractors for completion. The timely progress of a project is controlled by tracking e-mails and the scheduled completion of each piece of the project. For example, as a drawing or a portion of a drawing is completed, this information is recorded and transmitted to all team members. This signals other drawings or tasks to begin, or information to be sent. Although the entire project may be governed by a completion date, it is now based on a series of defined tasks rather than hours worked.

From the client's point of view, the Web-based collaboration can also make the project a much more open process. At any point during the life of the project, the owner can view the progress. Given this open process, the potential exists not only for increased efficiency and accuracy, but also for a greater sense of teamwork.

RESEARCH TECHNIQUES

Within the framework of the new Web-based CADD, the ability to conduct thorough and detailed research is enhanced. Having access to resources throughout the world, information on any topic is often just a few minutes away. Therefore, it is suggested that CADD students take time to study the techniques of conducting good Internet research. Several Web site addresses are provided at the end of this chapter to assist you in this effort. But do not forget that it is still important to know how to use traditional resources such as libraries, print media, and professional experts in the area of your study.

Even though the Web may be an inviting method of conducting research, it can also be time-consuming and nonproductive if it is not used in a logical manner. Any research should be conducted using a simple process that can assist you in finding information quickly and efficiently. Keep these items in mind when conducting research:

■ Define and list your topic, project, or problem.

■ Identify key words of the topic.

■ Identify all resources with which you are familiar that may provide information.

■ Using the Internet, conduct quick keyword searches on your topic. Write down the most likely Web sites and links to your topic, or copy the site locations to a text file in your word processing software. Avoid exploring sites in detail at this stage. Library Web sites often have excellent links for specific subject areas. Keep in mind that most periodical and professional journal documents on Web sites have been archived since only about the mid-1980s.

■ Check with libraries for lists of professional indexes, periodicals and journals, specialized trade journals, and reference books related to specialized topics. A wealth of information is still available in book and magazine form. Older magazine and journal articles not found on the Web are often in microfiche format in libraries.

■ Contact schools, companies, and organizations in your local area to find persons who are knowledgeable in the field of your research.

■ Create a list of all potential resources. Be sure to list enough information about the resource that you will be able to find it again quickly.

■ Prioritize the list by putting the most likely resources at the top. Focus first on the most likely resource to provide the information you need.

■ Begin your detailed study of the prioritized list.

Any project, be it a simple drawing or an entire set of construction plans, will require study, research, and testing. Regardless of the type of work you are engaged in, never forget about the rich resources and information locked within the human brain—information that is the result of untold hours of study, personal experiences, professional achievements, successes, and failures. Often the knowledge to be gained from consulting professionals is far more valuable than days of Internet and library research. Therefore, always remember the most basic and simple form of research: ask questions.

ERGONOMICS

Ergonomics is the study of a worker's relationship to physical and psychological environments. There is concern about the effect of the CADD working environment on the individual worker. Some studies have found that people should not work at a computer workstation for longer than about four hours without a break. Physical problems can develop when someone is working at a poorly designed CADD workstation; these problems can range from injury to eyestrain. The most common type of injury is referred to as *repetitive strain injury* (RSI), but it also has other

names, which include *repetitive movement injury* (RMI), *cumulative trauma disorder* (CTD), and *occupational overuse syndrome* (OOS). One of the most common effects is *carpal tunnel syndrome* (CTS), which develops when inflamed muscles trap the nerves that run through your wrists. Symptoms of CTS include tingling, numbness, and possibly loss of sensation. An RSI usually develops slowly and can affect many parts of your body. You should be concerned if you experience different degrees and types of discomfort that seem to be aggravated by computer usage.

Many factors can help reduce the possibility of RSI. These factors are outlined in the text that follows.

Ergonomic Workstations

Figure 4.47 shows an ergonomically designed workstation. In general, a workstation should be designed so that you sit with your feet flat on the floor, with your calves perpendicular to the floor and your thighs parallel to the floor. Your back should be straight, your forearms should be parallel to the floor, and your wrists should be straight. For some people, the keyboard should be either adjustable or separate from the computer to provide more flexibility. The keyboard should be positioned, and arm or wrist supports may be used to reduce elbow and wrist tension. Also, when the keys are depressed, a slight sound should be heard to assure the user that the key has made contact. Ergonomically designed keyboards are available.

The monitor is a concern. It should be 18" to 28", or approximately one arm's length, away from your head. The screen should be adjusted to 15° to 30° below your horizontal line of sight. Eyestrain and headache can be a problem with extended use. If the position of the monitor is adjustable, you can tilt or turn the screen to reduce glare from overhead or adjacent lighting. Some users have found that a small amount of background light is helpful. Monitor

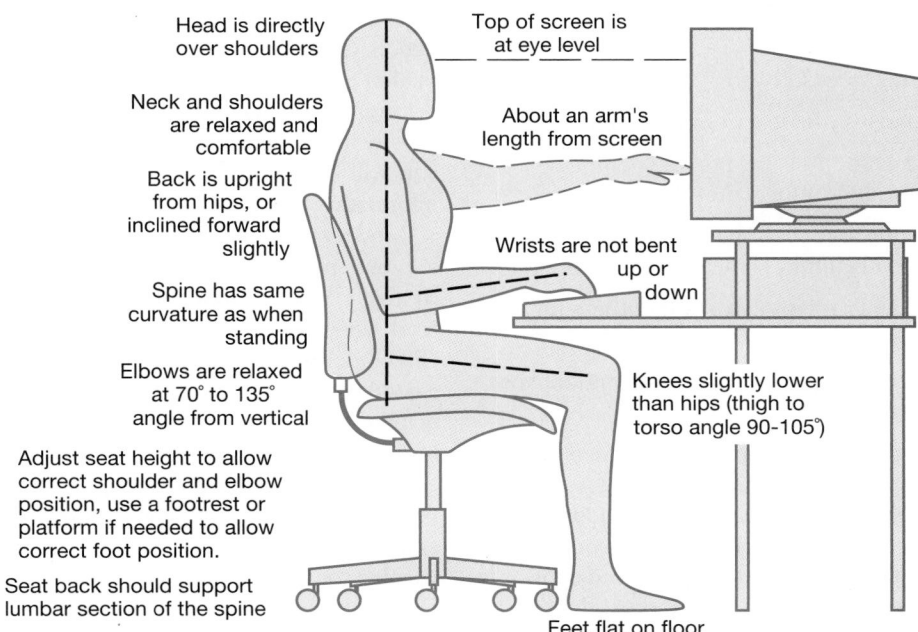

FIGURE 4.47 ■ An ergonomically designed workstation.

manufacturers offer large, flat, nonglare screens that help reduce eyestrain. Some CADD users have suggested changing screen background and text colors weekly to give variety and reduce eyestrain.

The chair should be designed for easy adjustments, to give you optimum comfort. It should be comfortably padded. Your back should be straight or up to 10° back, your feet should be flat on the floor, and your elbow-to-hand movement should be horizontal when you are using the keyboard, mouse, or digitizer. The mouse or digitizer puck should be close to the monitor so that movement is not strained and equipment use is flexible. You should not have to move a great deal to look directly over the cursor to activate commands.

Positive Work Habits

In addition to an ergonomically designed workstation, your own personal work habits can contribute to a healthy environment.

Try to concentrate on good posture until it becomes second nature. Keeping your feet flat on the floor helps improve posture. Try to keep your stress level low, because increased stress can contribute to tension, which may aggravate physical problems. Take breaks periodically to help reduce muscle fatigue and tension. You should consult with your doctor for further advice and recommendations.

Exercise

If you feel the pain and discomfort that can be associated with computer use, some stretching exercises can help. Figure 4.48 shows some exercises that can help reduce workstation-related problems. Some people have also had success with yoga, biofeedback, and massage. Again, you should consult with your doctor for further advice and recommendations.

PROFESSIONAL PERSPECTIVE

A wide variety of jobs are available for qualified CADD operators. Keep in mind that the kinds of tasks you will be engaged in may not always be traditional drafting functions. In addition to creating drawings, you may be responsible for working in some of the following areas:

- Preparing freehand sketches on the shop floor or at a job site, then converting the sketch to a finished CADD drawing.
- Digital image creation and editing.
- Text documents for reports, proposals, studies, etc.
- Incorporation of CADD drawings and images into text documents.
- Conducting research for job proposals, feasibility studies, or purchasing specifications.
- Evaluating and testing new software.
- Training staff members in the use of new software or procedures.
- Collecting vendor product information for new projects.
- Speaking on the phone and dealing personally with vendors, clients, contractors, and engineers.
- Checking drawings and designs created by others for accuracy.
- Researching computer equipment and preparing bid specifications for purchase.

The people who most often hire employees—human resource directors—agree that persons who possess a set of good general skills usually become good employees. The best jobs will be obtained by those students who have developed a good working understanding of the project planning process and can apply it to any situation. The foundation on which this process is based rests on the person's ability to communicate well verbally, apply solid

math skills (through trigonometry), write clearly, exhibit good problem-solving skills, and know how to use resources to conduct research and find information.

These general qualifications also serve as the foundation for the more specific skills of your area of study. These include a good working knowledge of drawing layout and construction techniques based on applicable standards, and a good grasp of CADD software used to construct drawings and models. In addition, those students who possess the skills needed to customize the CADD software to suit their specific needs will be in demand.

What is most important for the prospective drafter to remember is the difference between *content* and *process*. This was discussed previously in this chapter, but deserves a quick review here. *Content* applies to the details of an object, procedure, or situation. Given enough time you can find all of the pieces of information needed to complete a task, such as creating a drawing or designing a model. *Process* refers a method of doing something, usually involving a number of steps. By learning a useful process, you will find it easier to complete any task and find all of the information (content) you need.

Much traditional learning requires the student to exhibit qualities of a sponge and absorb as much content as possible. But with so much data and information surrounding us in our school and work situations, it is extremely difficult to rely on our retention of content for very long. It is initially much more beneficial to learn a good process for problem solving and project planning that can be used in any situation. Then, by using the process for any task, it becomes easier to determine what content is needed.

For these reasons, it is strongly recommended that you focus your efforts on learning and establishing good problem-solving and project-planning habits. Not only will they make the task of locating the content you need for any project easier, but they will also contribute to making all aspects of your life more efficient, productive, and relaxing.

Sitting at a computer workstation for several hours in a day can produce a great deal of muscle tension and physical discomfort. You should do these stretches throughout the day, whenever you are feeling tension (mental or physical).

As you stretch, you should breathe easily—inhale through your nose, exhale through your mouth. Do not force any stretch, do not bounce, and stop if the stretch becomes painful. The most benefit is realized if you relax, stretch slowly, and *feel* the stretch. The stretches shown here take about 2¹/₂ minutes to complete.

1. 5 seconds, 3 times:
2. 5 seconds, 3 times:
3. 5 seconds, 2 times:
4. 5 seconds, 2 times:
5. 5 seconds, each side:
6. 5 seconds, each side:
7. 5 seconds:
8. 10 seconds, each arm:
9. 10 seconds:
10. 10 seconds:
11. 10 seconds, each side:
12. 10 seconds:

FIGURE 4.48 ■ Stretches that can be used at a computer workstation to help avoid repetitive stress injury.

The Plotter

The plotter makes some noise and is best located in a separate room next to the workstation. Some companies put the plotter in a central room, with small office workstations around it. Others prefer to have plotters near the individual workstations, which may be surrounded by acoustical partition walls or partial walls.

Other Factors

The CADD working environment may be different from traditional drafting rooms in that air conditioning and ventilation should be designed to accommodate the computers and equipment. Carpets should be antistatic. Noise should be kept to a minimum. If smoking is permitted, the room should have special ventilation. Smoke or other contaminants can affect computer equipment.

Drafters who are required to do tedious, repetitive tasks on the computer could develop symptoms of fatigue. Consideration should be given to allowing CADD employees to do a variety of jobs or take periodic breaks.

WEB SITE RESEARCH

The following list of Internet Web sites provides access to information on many of the CADD-related topics discussed in this chapter. Useful Web sites that you locate should be bookmarked in your Web browser. In addition, you should create a text document that contains a list of useful Web sites with links to those sites. Separate the site names by categories. This list provides you not only with a resource that is a backup to your browser's bookmarks, but also with a file that you can add to at any time you are not working on the Internet.

http://www.caenet.com/—*Computer-Aided Engineering* magazine.

http://ecoleing.uqtr.uquebec.ca/geniedoc/gmm/productique/rpworld. htm—The World of Rapid Prototyping, University of Quebec, Three Rivers, Quebec, Canada.

http://intell-lab.engi.cf.ac.uk/Franck/RPB/RPB.html—The Rapid Prototyping Bookmark.

http://www.tandfdc.com/jnls/cim.htm—*International Journal of Computer Integrated Manufacturing.*

http://www.tii-tech.com/train.html—TII Technical Education Systems. CIM product information and useful links for industry, training, and educational sites.

http://nanozine.com/—*NanoTechnology* magazine.

http://thevrsource.com—The VR Source Newswire. Product reviews, new product listings, monthly specials, and more.

http://www.hitl.washington.edu—Human Interface Technology Labs, University of Washington, Seattle.

http://ai.about.com/compute/ai/library/weekly/aa051899.htm—About.com. Artificial intelligence topics and links.

http://websearch.about.com/internet/websearch/library/weekly/aa010899.htm—About.com. Web searching techniques.

MATH APPLICATION

One of the most important aspects of learning to operate a CADD system is the rectangular coordinate system. The polar system is one form of rectangular coordinates that uses a distance and angle for measurement. Understanding angles and how they relate to each other is useful in the solution of a variety of problems. For example, you may be given a flat sheet metal part or a property plat to draw, and the sketch contains only included angles. It is up to you to convert these included angles into polar coordinates in order to draw the object. (See Figure 4.49 below.)

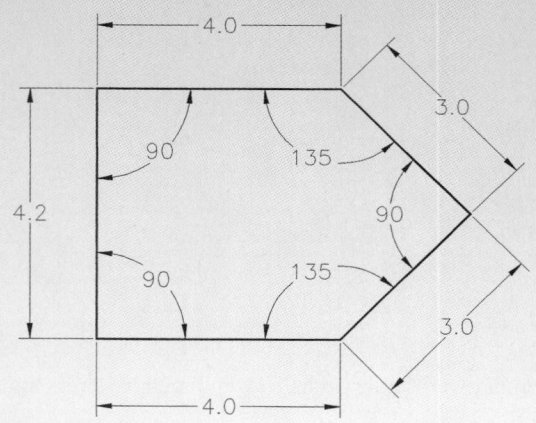

FIGURE 4.49 ■

The first line can be drawn easily by entering a length, 4", and an angle of 0°. The angle of the second line is determined by subtracting 135° from 180° (the angle of a straight line). This gives the angle of 45° measured counterclockwise from horizontal. Thus, to draw to point C, enter a length of 3" and an angle of 45°. (See Figure 4.50 below.)

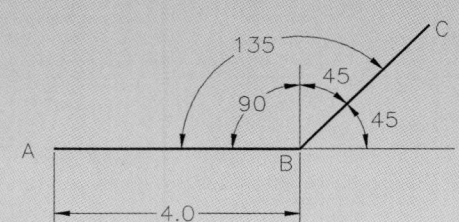

FIGURE 4.50 ■

In order to draw to point D you must remember the rule that intersecting lines create equal opposite angles. By drawing a horizontal line through point C and projecting the previous line past the object, you see that one half of the 90° angle is above the horizontal. Therefore, the angle of the line to point D is the result of subtracting 45° from 180°. To draw a line from point C to point D, enter a length of 3" and an angle of 135°.

Try to complete the remainder of the object (Figure 4.51) using the previous examples as a guide.

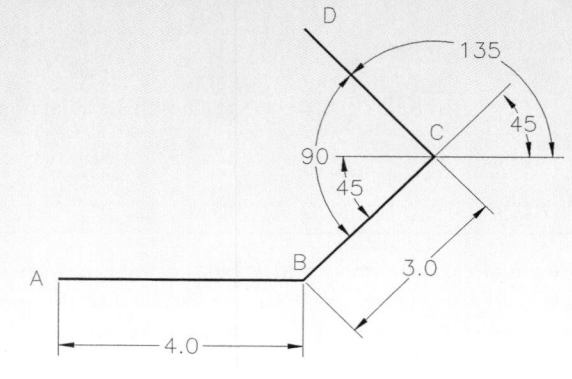

FIGURE 4.51 ■

http://www.nsf.gov/home/eng/start.htm—National Science Foundation's engineering Web site.

http://www.albany.edu/library/internet/research.html—University at Albany, New York State Universities. Conducting research on the Internet.

http://www.uiuc.edu/refs/research.html—University of Illinois at Urbana—Champaign. Resources for conducting research. Pick the IRIS link for additional links to other research sources.

http://www.cad-forum.com/associations/assnl.asp—CAD Forum. List of professional organizations for architecture, engineering, design, CAD, and construction. Excellent links to information on every discipline of CADD.

http://www.acec.org/—American Consulting Engineers Council. An organization with approximately 6,000 member firms.

http://www.buildernet.com/—BuilderNet. Products and suppliers for architects, engineers, and contractors.

http://www.hoovers.com/company/listsbest/0%2C2561%2C141%2 C00.html— Hoover's Online, The Business Network. "List of Lists." Links to a wide variety of companies and Web sites.

http://www.thomasregister.com/—Thomas Register list of American manufacturers.

http://www.siriuslink.com/cad/resources.html—Sirius Digital Archiving. CAD Internet resources.

CHAPTER 4
Computer-Aided Design and Drafting Test

1. Name four CADD design formats.
2. What types of machines are used in CAD/CAM integration?
3. What are the processes by which a solid physical model of a part is made directly from 3-D CAD model data, without any special tooling?
4. What is the term used to describe the combination of the entire design, material handling, manufacturing, and packaging process?
5. What are two common types of computer monitors?
6. What is a single dot on a monitor called?
7. Name three different types of pointing devices.
8. What is the general term for the computer component that contains or accepts media on which computer files can be kept temporarily or permanently?
9. What is the amount referred to by the term *gigabyte*?
10. Name the device that uses pens to create a paper drawing.
11. What type of computer program requests information regarding the design of an object, such as capacities and measurements, and then creates the object?
12. List the steps in the three-step planning process.
13. Name four items that template drawings should contain.
14. What is the scale factor of the following scales?
    a. $\frac{1}{2}" = 1"$
    b. $\frac{1}{2}" = 1'$
    c. $1" = 100'$
15. What contains a group of objects, drawing elements, or components that share one or more common characteristics?
16. Where is a point located in relation to the origin of the Cartesian coordinate system if it has a negative Y value?
17. Where is a point located in relation to the origin of the Cartesian coordinate system if it has a positive Z value?
18. How is a relative coordinate located?
19. When constructing a CADD drawing, how much should you actually draw?
20. What kind of network connects computers in widely scattered locations?

CHAPTER 4
Introduction to CADD Problems

PROBLEM 4-1 This problem involves writing, not drawing. Choose one of the topics listed below and write a report on that subject. Should you choose to write a report on topic 1, CADD system, select one manufacturer and focus your report on that specific brand of system. Be specific, especially if the manufacturer makes more than one model of CADD system. Check with your instructor about any specifics relating to your report. Research materials can be the internet, trade magazines and journals, textbooks, company brochures and instruction manuals, telephone interviews, personal interviews, field trips, and attendance at seminars, workshops, conferences, and conventions.

1. CADD system (choose a specific brand and model)
2. Computer
3. Input devices (handheld)
4. Digitizer, graphics tablet
5. Output devices: plotter, printer, display screen (monitor or terminal)
6. Storage devices: disk drive, optical disk drive, tape drive
7. Drawing media: paper, film, pens
8. Storage media

PROBLEM 4-2 Write a short report that discusses the differences among pen plotters, inkjet plotters, and laser printers. You may use any sources that can provide the needed information, such as the internet, textbooks, magazines, advertising brochures, demonstrations, interviews, and personal experiences. Be sure to discuss the following points in your report:

■ Type of media required (paper, pens, etc.)

■ Media size

■ Price

■ Quality of final product

■ Usefulness in your application

■ Warranties and service contracts

■ Durability

■ Reputation

■ Length of time on the market

PROBLEM 4-3 Create a specification sheet for purchasing a single CADD workstation and the software to run it. You can do this as a report, a form, or a list. The assignment should include the following pieces of equipment and software:

■ Desktop computer—specifications, speeds, and capacities change rapidly, so when researching, look for the most current specifications in the following components:

  ■ Computer memory

  ■ Processor (CPU) speed

  ■ Hard disk drive

  ■ Portable storage media device (floppy disk, optical disk, high-capacity disk, etc.)

  ■ Graphics card with memory

  ■ Ports for input devices, printers, speakers, etc.

■ Color monitor—CRT or flat-panel

■ Input device such as mouse, trackball, digitizer

■ A/B-size plotter (pen or inkjet), or D-size plotter

■ Computer operating system software

■ CADD software—current version

Keep the following things in mind when creating your specification list.

■ Contact at least three computer hardware vendors for prices.

■ Provide each vendor with the same specifications.

■ Do not specify brand names of equipment unless instructed to do so.

■ Determine the brand of CADD software you want before obtaining prices.

■ You may have to go to different vendors for the CADD software.

PROBLEM 4–4 Using existing equipment or vendor's catalogs, measure the equipment required for a CADD workstation (computer, monitor, keyboard, and pointing device), and design a workstation table. Sketch the table first, then draw it on your CADD system.

After drawing the table, create a computer lab arrangement of from 12 to 20 CADD workstations using your table design. Consider the following when designing the lab:

■ Chalkboard or marker board location

■ Glare from windows

■ Aisles and access

■ Power connections

■ Location of pen plotters or printers

■ Storage for paper, plotter pens, and supplies

■ Instructor podium or workstation

■ Projection system for instructor workstation (use overhead projector)

Remember to "draw as little as possible." Once you have drawn one workstation, it can be saved as an object in a symbol library, or just copied as many times as needed in the drawing. If your drawing is used for actual CADD lab layout purposes, keep in mind that additional information for the workstation furniture and computer components can be added to the drawing later. As you continue your CADD studies, always be cognizant of items that can be added to the drawing and its components to enhance the value of the drawing. If you are using AutoCAD, you can add *attributes* to symbols to give the drawing "intelligence." Consult your textbook or user's manual for detailed instructions.

PROBLEM 4–5 This problem involves the creation of template drawings. These drawings should be constructed early in your CADD studies and then used and added to as you progress. Refer to the discussion of template drawings in this chapter before you begin. Create a minimum of one template drawing for each sheet size and drafting discipline in which you will be working. For example, if you will be creating mechanical and civil drawings using B, C, and D sheet sizes for each, you should create six template drawings. The following items should be included in your template drawings:

■ Border and title block

■ Title block text

■ Standard layers

- Text styles for ¹⁄₈" text and 0" high text
- Dimension style
- Named views of the title block and the entire drawing

PROBLEM 4-6 For this problem you need to have access to the Internet. Use the search function of your Web browser to locate information on the following topics:

- Rapid prototyping
- Nanotechnology
- Solid modeling
- Computer-integrated manufacturing (CIM)
- Virtual reality hardware

When you locate resources and links for each topic, try to get information in the following areas:

- Schools and training facilities offering courses
- Research papers, theses, and white papers
- Research organizations
- Vendors of equipment, software, and resources

Create a text file of the list given above and include the Internet Web site addresses below each topic in the format of active links if your software supports it.

PROBLEM 4-7 Before working on this problem, you must complete problem 4–6. After you have located information for each topic in problem 4–6, choose the one that interests you the most. Write a short, informative report on the topic using word processing software. Include at least two illustrations in the report. Try to copy images from the Web and paste them in your document if they are relevant. If your word processing software supports it, include at least two active links in the report to a Web site related to the topic. One of the links should be embedded in one of the images.

When you have completed the report, send it as an e-mail attachment to your instructor. Request that the instructor test your Web site links to be sure they function properly.

PROBLEM 4–8 Use the Internet or your library's resources to find professional trade journals and magazines. Try to find at least one publication in the disciplines of CADD, mechanical engineering, architectural design, structural engineering, solid modeling, rapid prototyping, civil engineering and surveying, geographical information systems (GIS), numerical control, 3-D design, virtual reality, and nanotechnology. Keep a list of the publications that you find in each field. Add journals of other disciplines to the list as you find them.

Throughout the course of your studies, try to read one article a week from one of the journals. Choose a different journal each week.

MATH PROBLEMS

PROBLEM 4-9 Draw the following sheet metal part using CADD software. If you are using AutoCAD, the best point entry method to use is polar coordinates. Space is provided in

which you can write the coordinate values needed to draw a line to each point.

A to B_____
B to C_____
C to D_____
D to E_____
E to F_____
F to G_____
G to A_____

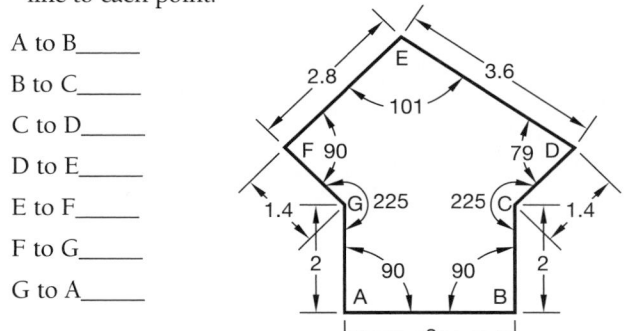

PROBLEM 4-10 The following property plat has been given to you in the form of a sketch with included angles. You are to draw the property plat full size using the dimensions given. Do not use decimal values for angles or distances. Space is provided in which you can write the coordinate values needed to draw a line to each point. Write your answers using a polar coordinate notation. For example, a line 87'–3" long, at an angle of 65° would be written @ 87'3"<65.

A to B_____
B to C_____
C to D_____
D to E_____
E to F_____
F to A_____

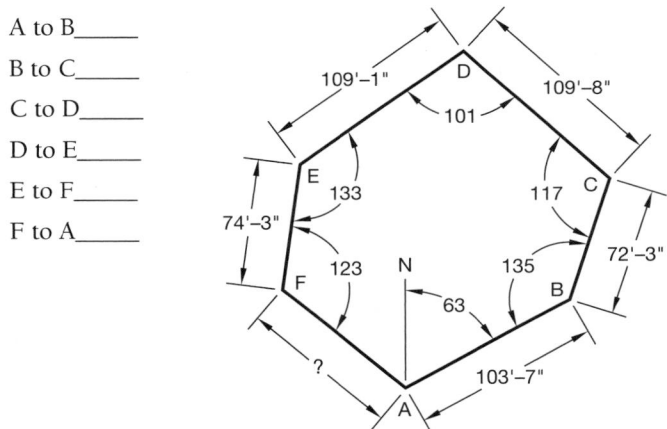

SOLUTION TO CADD DRAWING (from page 121)

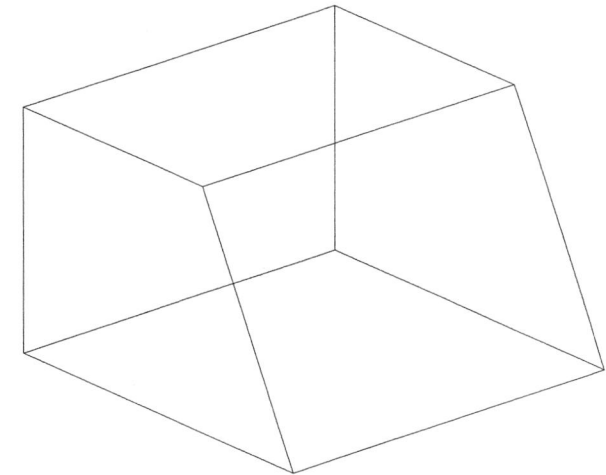

The Engineering Design Process

After completing this chapter, you will:

- Describe various factors that are changing the design process.
- Discuss the steps in a design analysis process.
- Explain the importance of creativity and innovation in the design product.
- Describe the problem-solving steps used in a design process.
- Explain the importance of concurrent engineering and teams in the development of a product.
- Define today's engineering design models.
- Explain the design review process.
- Describe design deliverables.

Today's designs are driven by cost, market trends, and customer service as well as by quality and technology. Because of these realities, engineers, designers, drafters, and technologists now need to be aware of how their designs fit into the broader function of product development. In turn, they need to understand how their function fits into the entire product development cycle. And finally, this knowledge needs to be carried one step further into the areas of sales and marketing to understand exactly what the company is trying to sell and produce, to whom and what for. These are the forces behind the change to a whole systems approach to engineering design.

As a future designer or technician in the engineering profession, it is vital to possess abilities in areas such as problem definition, resource/information acquisition, design review, problem solving, technical communication and teamwork skills, business systems, project management, and above all the ability to see the whole system. The basic premise of a whole systems approach to engineering design is the need for all people in the process to know all the necessary information and actions involved in the design process.

This sharing of knowledge, varying viewpoints, and ideas leads to better, faster, and more efficient problem solving. Accumulated knowledge and the availability of information in the engineering profession are explosive. The ability and tools to gather and archive information are available to the engineering design team more quickly and with less effort than ever before. This surge of information and its availability to everyone have brought a new model for decision making in the design process. The decisions are

THE ENGINEERING DESIGN APPLICATION

TRANSITIONS TO A SYSTEMS APPROACH: A HISTORIC POINT OF VIEW

As you begin the study of the engineering design process, it is important to reflect a moment on the past and on how in recent times design systems have changed. Traditionally an engineering process has been a collection of activities that takes one or more kinds of input and creates an output that is of value to a customer. The delivery of the product to the customer or client is the value that the process creates. This is still true, but how the design process works today is significantly different from how it worked in the past. Engineering has used a design process that goes back to Adam Smith's model for production. Mr. Smith's model focused on individual tasks in a process of designing a product.

Traditionally, engineering, testing, manufacturing, marketing, cost analysis, and so on have been components of the design cycle. Each component has functioned independently with communication occurring only when the design was passed on to the next group. Often the larger objective of customer satisfaction and product quality was lost with the individual focus on the task at hand. Revisions and engineering change notices were the rule of the day. Communication from one end of the process to the other was limited to each group's perspective. Cost of production, time to market, and the demand for quality products have driven a change from this linear design engineering process to an integrated approach that brings together a team of people from all aspects of product development. The team includes everyone in engineering, marketing, and production, as well as the customer. This whole systems approach to design is the predominant way design is occurring in engineering firms today. (See Figure 5.1.)

THE ENGINEERING **DESIGN** APPLICATION *(continued)*

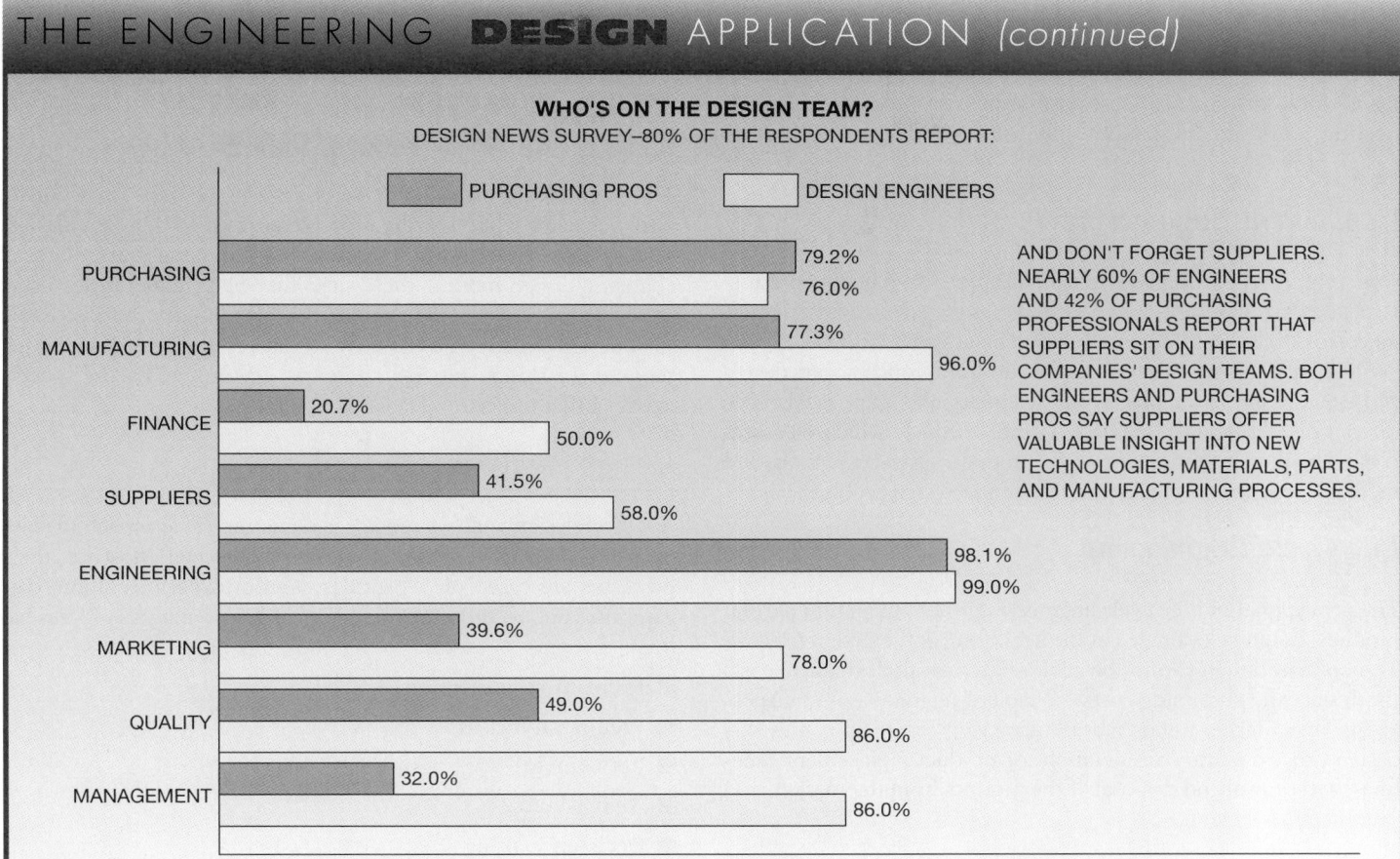

WHO'S ON THE DESIGN TEAM?
DESIGN NEWS SURVEY–80% OF THE RESPONDENTS REPORT:

PURCHASING PROS — DESIGN ENGINEERS

Category	Purchasing Pros	Design Engineers
PURCHASING	79.2%	76.0%
MANUFACTURING	77.3%	96.0%
FINANCE	20.7%	50.0%
SUPPLIERS	41.5%	58.0%
ENGINEERING	98.1%	99.0%
MARKETING	39.6%	78.0%
QUALITY	49.0%	86.0%
MANAGEMENT	32.0%	86.0%

AND DON'T FORGET SUPPLIERS. NEARLY 60% OF ENGINEERS AND 42% OF PURCHASING PROFESSIONALS REPORT THAT SUPPLIERS SIT ON THEIR COMPANIES' DESIGN TEAMS. BOTH ENGINEERS AND PURCHASING PROS SAY SUPPLIERS OFFER VALUABLE INSIGHT INTO NEW TECHNOLOGIES, MATERIALS, PARTS, AND MANUFACTURING PROCESSES.

FIGURE 5.1 ■ ■ Who is on the design team? ("Design Teams Bring Radical Change in Product Development," *Design News*, May 18, 1998.)

The design process today is influenced by the need for better, cheaper, and faster-to-market products. Behind these changes in design is global competition, communication technologies that have accelerated the speed and volumes of information available to designers. Also, a compression of time has resulted from the Information Age of Technology, which creates a need to bring designs faster to market. The need and tools are at hand to design a quality product quickly, utilizing all parties in the design process concurrently.

now made with input from everyone on a design team, not with just a few at the top passing it down the line. A person can use this shared data bank of product information for inventory planning, scheduling, purchasing, labor forecasting, or customer service. The power of all this input is a quality product that is cost-efficient and produced faster than in the past.

Traditionally designers were trained in design fundamentals, the nature of materials, the capabilities of tools and equipment, and manufacturing processes. These skills are still very important to the development of an engineering designer or drafter. Today these skills are expanded to include knowledge of market trends, safety, data acquisition and management, teaming concepts, packaging, distribution, and storage. This is a whole systems approach to design. Today's designer faces a world of complex dependencies and interrelationships that involve federal and state regulations, environmental impacts, consumer perceptions, economic/societal needs, cultural trends, and demographics impacting the whole design process. All these forces have changed the way we think about the world of engineering design

CHANGING BUSINESS PARADIGM

DESIGNER (YESTERDAY)	KNOWLEDGE WORKER (TODAY)
DETAIL WORK	INTELLECTUAL WORK
FACTOR OF PRODUCTION	KNOWLEDGE PRODUCER
QUESTION NOTHING	QUESTION EVERYTHING
DO AS YOU ARE TOLD	DETERMINE WHAT TO DO
REPETITIVE TASKS	CONTINUOUS CHALLENGES
SEGMENTED WORK	HOLISTIC WORK
DIRECT SUPERVISION	AUTONOMY

FIGURE 5.2 ■ Changing business paradigm.

and the fundamental nature of work. The change has moved from a designer assigned to a project to a knowledge-based workforce. (See Figure 5.2)

TODAY'S ENGINEERING DESIGN MODELS

Let us look at some of the systems approaches used by the engineering world to address the constantly evolving engineering design process.

Concurrent Engineering

Concurrent engineering is an integrated approach to design, production, and customer service that emphasizes the advantages of simultaneous, or concurrent, product design by utilizing individuals from various areas of the business in the up-front concept and design phase, with special emphasis on customers and their needs. The focus is on all aspects of the design simultaneously, which is accomplished by the integration of people, processes, standards, tools, and methods to achieve a quality solution to the design problem.

Life Cycle Engineering

The concept behind life cycle engineering is that the entire life of a product should be evaluated at the beginning of the design process. The optimal design cannot be achieved unless performance, costs (both start and sustaining costs), reliability, maintainability, disposability, and market trends are addressed up front. Built into this design process is the consideration for product replacement timelines, withdrawal and disposal of the product from the market, and product replacement.

Integrated Product Development (IPD)

Integrated product development is the process by which the product is designed and developed to satisfy all the conditions the product will encounter in its product life. Evaluating all the factors that will impact the design during the product life at the initial design phase increases the likelihood that unforeseen circumstances will not accumulate and damage the product's longevity in the marketplace.

Knowledge-Based Engineering (KBE)

Knowledge-based engineering is the utilization of computer models to simulate the best-known engineering processes. Typically KBE is used by creating a set of engineering data. This data is comprised of components such as CAD data, manufacturing data, tooling data, and structural information. This data is then used in an integrated manner to develop a more detailed and comprehensive design plan. The inclusion of so much information allows for better analysis and evaluation throughout the process.

Total Quality Management (TQM)

Total Quality Management is a philosophy that calls for the integration of all organizational activities to achieve the goal of serving customers. It seeks to achieve this goal by establishing process standards, maximizing production efficiency, implementing quality improvement processes, and employing integrated teams to effectively design customer-driven products. Total quality management strives to eliminate all non value-added activities in the production process and to achieve satisfaction of all its customers, both internal and external.

THE ENGINEERING DESIGN PROCESS

The engineering design process is a series of steps or actions a designer takes when solving a problem or creating a new design. Traditionally these steps were viewed as sequential but in light of changing design models, these steps may be parallel to each other or in reverse as well as sequential. It is also important to note that each design situation has its own unique set of demands and criteria, and the design process chosen is determined by the specific needs of that project.

Design Analysis

A typical design analysis process is a methodical approach to solve problems and generate new design solutions. Most of these processes are based on a logical system of experimentation, data gathering, and record keeping. A typical design analysis process has these elements:

- Problem identification.
- Preliminary ideas.
- Refinement.
- Analysis.
- Decision making.
- Implementation.

This process is often referred to as research and development (R&D). A number of factors influence product development. Beginning with research and development, a product should be designed to meet a market demand, have good quality, and be economically produced. The sequence of product development begins with an idea and results in a marketable commodity.

The Steps of Design Analysis

The different steps in design analysis are commonly started with problem identification or identifying the need. Next, the criteria, such as limitations and desirable features, need to be determined. Then exploration, research, and investigation take place, which then leads to preliminary alternative solutions. Investigation of many solutions at this stage of the process is good, but eventually one is chosen as the final design. Now, details of the design are created. The new design is then modeled, developed, and readied for production. During this phase, design testing and evaluation occur. Redesigns and improvements can be made to improve the original design at this point. The different steps of design analysis can be translated into a series of questions to be answered. (See Figure 5.3.)

CREATIVITY AND INNOVATION IN DESIGN

Creativity is very important in the world of engineering and design. Today's competitive market for products is placing a major need on

THE ENGINEERING DESIGN PROCESS

DESIGN PROCESS STEPS	QUESTIONS TO BE ANSWERED
IDENTIFY THE NEED	DO OUR CUSTOMERS HAVE A NEED, AND IF SO, WHAT ITS IT?
DEFINE THE CRITERIA FOR THE NEED	WHAT DOES OUR TEAM'S SOLUTION NEED TO DO?
EXPLORATION/RESEARCH/INVESTIGATION	HOW DO WE SOLVE THE DESIGN PROBLEM?
GENERATE ALTERNATIVE SOLUTIONS	WHAT ARE THE POSSIBILITIES?
CHOOSE A SOLUTION	WHAT IS OUR BEST POSSIBLE SOLUTION?
DETAILED DESIGN	WHAT ARE THE DETAILS OF OUR DESIGN?
MODELING/DEVELOPMENT/PRODUCTION	IS IT TURNING OUT THE WAY WE THOUGHT IT WOULD?
DESIGN TESTING AND EVALUATION	DID THE DESIGN DO WHAT OUR CUSTOMERS WANTED?
REDESIGN AND IMPROVEMENT	HOW CAN WE DO IT BETTER NEXT TIME?

FIGURE 5.3 ■ Design process steps.

the engineering design process to be more creative and innovative in order to meet the demands of faster, cheaper, and better. The only way to meet these demands is to do things differently, which is where creativity and innovation enter the picture. By definition, creativity is the ability to produce through imaginative skill, to make or bring into existence something new, to form new associations and see patterns and relationships between diverse information. Creativity's partner is innovation. Innovation can be defined as the process of transforming a creative idea into a tangible product, process, or service. Innovation is about improving the quality of a specific thing and allowing for more and better choices.

Using creativity techniques can lead to more effective problem-solving skills that will facilitate productive and satisfying designs. Creativity may be required in the following situations:

■ *New methods:* Problems in industry and construction often must be solved in areas that require original ideas or revisions of existing designs.

Creativity tools allow for new ideas to be generated.

■ *Determination:* A designer must have a lot of determination in order to keep working on the project until the problems are solved.

Creativity provides a way to look at problems from a different angle, making it easier to problem-solve.

■ *Attitude:* The designer needs a positive attitude and must realize the possibilities of modern technology.

Creativity is all about possibilities.

■ *Confidence:* A good designer must have confidence in his or her ability to solve a problem within the marketing and manufacturing or construction requirements.

Creativity contributes to confidence by providing alternative solutions.

■ *Experimentation:* Careful testing and recording of data is a key to successful design. Experimentation, prototyping, testing, and analysis are used in the design process.

Creativity is the oil for experimentation and analysis.

■ *Logic:* A designer may lose ideas and waste valuable time without a logical approach to problem solving. A step-by-step design process works when followed and documented.

Creativity generates the ideas. The designer puts them in logical order.

Principles of Creativity

There are basic principles of creativity. These principles may not be natural to a person but with practice and determination, they can be cultivated and improved upon to increase creativity. Every person has the potential to be creative. People who are looked upon as creative seem to have the ability to see the whole picture and make connections and combinations of seemingly unrelated subjects to produce new original ideas.

Let us look at the basic principles:

■ *The ability to see relationships and patterns:* This means taking existing objects and combining them in different ways for new purposes.

■ *The belief that you are creative:* Do not underestimate the power of positive thinking.

■ *The ability to look at a problem from a different point of view:* This involves changing the position or role in how a new idea or problem is approached.

■ *Playfulness and humor:* Be a dreamer; brainstorm ideas; and surround yourself with inspiring ideas.

1. Look beyond the immediate object or situation— see the Big Picture.
2. Look for potential future consequences.
3. Always ask "Why" questions.
4. Always question and remain curious.
5. Keep an interest in a wide variety of areas—the broad base of knowledge helps make connections to new ideas.
6. Keep a daily journal of ideas, thoughts, and sketches.
7. Lateral thinking: Look at the idea from a number of different directions.
8. Mindmapping: Use free word association and brainstorming. Then organize the ideas into related areas.
9. Brainstorming: Use free association of ideas from a group to generate as many ideas as possible without passing judgment on any idea presented.

FIGURE 5.4 ■ Methods for developing creativity.

- *A work environment that is flexible, open, and autonomous:* In an ideal world, an unusual idea, silly or not, is never criticized or declared a failure. It may spark a creative solution.

- *Imagery and visualization:* Daydream, sketch, and visualize the future.

- *Subconscious thoughts:* Allow an idea to simmer awhile before making judgments. Ideas keep working in the subconscious mind even while you sleep.

In thinking of ways to become more creative in designs, remember the creative individual asks more *"why"* questions as opposed to *"what"* questions. (See Figure 5.4.)

These are just a few of the ways to foster the creative process. One that works well for designers is to draw a picture of the problem instead of using words. Many times a visual representation provides a fresh perspective and insights. There are many software tools available that help in the creative process. Just remember, the bottom line is to have a mindset that nothing is set in stone.

Creativity Works Well in a Group

The world of engineering is moving to self-directed teams and concurrent team groups through which the real advantage for creative problem solving will be realized. A group of people is able to bring diverse and varied perspectives to the problem or idea at hand. A group possesses accumulated knowledge and experience that one individual cannot even begin to exhibit. A team of people is in a position to learn from each other and consider various ideas and solutions, which creates a synergy that leads to creativity. (See Figure 5.5.)

Creativity and Innovation

Creativity and innovation require a diverse, information- and interaction-rich environment. This is brought about with people who have different perspectives working together toward a common goal. This team of people requires accurate, up-to-date information and the proper tools.

Only in this kind of environment—where the organization is continuously learning about its products, services, processes, customers, technologies, industry competitors, and environment—can

FIGURE 5.5 ■ Teamwork in action.

innovation thrive. Existing tools such as CAD/CAM/CAE are good for routine problem solving but fall short in generating breakthrough concepts. To meet current market demand and customer requirements, creativity must transcend into innovative new ideas and processes.

In addition, remember that failure can be a learning experience and a necessary step in the design process. Sometimes design solutions fail. Failure prompts us to look at other solutions and promotes new ideas that lead to innovations. So when a design fails a test, think of it as an opportunity, not a dead end.

DESIGN ANALYSIS AND PROBLEM SOLVING

As the design process responds to available technology and knowledge management tools, the process for analyzing data and test results has moved more and more decisions to the designer. This is a result of more information being available. The tools that are now at our fingertips make it easy to perform very powerful test analysis exercises and to interpret the data. The interpretation and application of this data are still in the hands of the engineer and the designer, but success in today's world of engineering design requires an increased knowledge of these new software tools.

Problem-solving Steps

Problem solving is the ability to analyze and solve complex problems by using large amounts of data, changing the conditions, and recognizing how seemingly unrelated issues or events interact and affect one another. Problem solving is a process that can be defined and followed systematically.

Following are the basic steps to follow when working through a problem.

- *Identify* the problem.
- *Define* the problem.
- *Research* the cause.
- *Explore* solutions.
- *Act* on a solution.
- *Observe* the solution.
- *Evaluate* the success of the solution.

1. *Identify the problem:* Joint agreement that a problem exists is important.
2. *Define the problem:* Joint agreement on the root cause of the problem can lead to agreement on the procedures to follow for possible solutions.
3. *Explore solutions:* Brainstorming is natural for a group of people. Many brains are better than one for generating solution ideas.
4. *Act on a solution:* A team of people has more resources to get the job done.
5. *Observe the solution:* More eyes and ears are able to identify any potential problems. A coordinated standardized method of documentation is critical to this step.
6. *Evaluate the success of the solution:* A team must reach consensus on the success or failure of the solution.

FIGURE 5.6 ■ Problem solving as a team.

Depending on the success of the solution, the problem solving process may begin all over again. Once a solution is selected, give it time to work. It is easy to jump to conclusions at the first sign of failure. Sometimes it takes time to work out the kinks. After a respectable length of time, begin to critique the potential solution. Identify strength and weaknesses, determine any developing pattern such as frequency, data, voids, and so on. Based on these observations, develop a plan, chart, or spreadsheet to study any recurring patterns. Methods for measuring the solution should be designed and agreed upon before the solution is implemented.

When evaluating the success or failure of the solution, weigh the pluses and minuses by outlining the consequences of the solution. Was the problem solved? Were information needs met? Was a decision made? Was the situation resolved? Does the product satisfy the requirements as originally defined? This evaluation process determines how effectively and efficiently the problem solving process was conducted. With this evaluation, this problem-solving loop may begin again.

Problem Solving with Teams

Problem solving with a group or a team has advantages and disadvantages. Collectively the team has more knowledge than one individual. There are more people to gather the data and do the work. The diverse inputs also give a broader picture to the problem at hand. It is not a perfect world, however, and working in a team can cause people to conform when they do not really agree on the results. Also, decisions do come slower in a group. See Fig. 5.6 for some helpful tips for working in a group.

CHANGE AND THE IMPACT ON THE DESIGN PROCESS

The designer naturally embraces change as part of the design process. Every design or idea goes through an evolution before its final outcome. Today the designer and the engineering team are attempting to reduce the amount of change in the design process by taking an integrated approach to the design process at the early

stages of the design inception. By considering all aspects in the development of a product, the design team can troubleshoot in advance any potential problems that would result in the change of a design. Solid modeling and rapid product development tools aid in early identification of design flaws and eliminate the need for change down the line.

Within the engineering profession, the ability to manage change is crucial to a project's success. Although there are many variations of the engineering change process, the structure and purpose are fairly standard. The objective of any engineering change process is to incorporate design changes as quickly and accurately as possible with a minimum of disruption and cost. Changes in the engineering design process serve many functions:

1. Satisfy customer requests.
2. Improve the product.
3. Incorporate improvements in production and manufacturing.
4. Resolve design problems.
5. Integrate new technologies.

The engineering change process requires thoroughly documented entrance and exit criteria to ensure design changes are accomplished accurately and timely, as well as the maintainance of a complete history of the changes.

TRACKING DOCUMENTS: ECO/ECR

A tracking document commonly is referred to as an Engineering Change Order (ECO) or an Engineering Change Request (ECR). The style and procedure for ECO or ECR may differ from engineering firm to engineering firm, but the basic process is the same. Standard information is included on most document change forms. (See Figure 5.7.)

The engineering change process necessary to incorporate design change properly can be summarized in three steps:

1. Communicating the change.
2. Documenting the change.
3. Tracking the change.

In considering a potential change to a design, remember that change has an impact on the whole system within which the design is developed. A positive change for the designer may be a negative change for someone else who has a role in the design process. This is where a team meeting may be of value, since ideas can be considered by everyone involved.

■ Engineering drawings
■ Parts list or bill of materials
■ Production specifications
■ ECO/ECR document—problem description
■ Engineering Change Order number
■ Status of engineering change
■ Effective date for change
■ Approval list for reviewing changes

FIGURE 5.7 ■ Documents/information required for an ECO/ECR.

THE DESIGN REVIEW PROCESS

Design review and evaluation activities provide an opportunity to compare the progress of a project against a preestablished criteria. If the design fails to meet the criteria, modification activities are implemented based on an analysis of the problem. The ultimate goal is to produce an improved product through an ongoing process of evaluation. Design review and evaluation activities become part of the continuous quality process until standards and design goals are achieved. The design review is focused on the process to achieve product improvement. The general rules of a design review are tied to omissions or ambiguity in the design requirements or specifications.

The design review process usually is a process on which all parties involved in the design have an impact. In a whole systems approach to design, this would include the engineering team, customers, and other impacted departments such as production or marketing. Documentation of the data collected in a design review is critical to continuous quality. With today's technology, all data can be shared and accessed easily through intranet or Internet servers. Multiple software tools and review techniques also can be utilized for design analysis. (See Figure 5.8.)

- ■ *Design Optimization:* A systematic approach to designing a product for the optimal performance in terms of function, assembly, maintenance, ergonomics, and utilization.
- ■ *Design for Tooling:* An analysis tool that considers the engineering costs, in terms of money and design, for any tooling that will be required for the product's development and production.
- ■ *Design to Cost:* A method that utilizes cost as a design criteria, to help ensure that profit goals are met, and to help control project development costs. Initial cost estimates are derived from throughout the company, including marketing, sales, engineering, production, and finance.
- ■ *Design for Maintainability and Reusability:* The process for reviewing a product's ability to be reused in a different application and its ability to be maintained after the sale.
- ■ *Design for Manufacturability:* A method for considering the production processes required for a design.
- ■ *Finite Element Analysis (FEA):* An analytical tool used in the engineering of solid materials to determine the static and dynamic responses to specific materials under a variety of conditions.
- ■ *Reliability Analysis:* An analytical tool used for evaluating a product's reliability in terms of use and/or time.
- ■ *Designing for Safety:* An ingredient in the design process that is necessary to ensure that all safety considerations are met in the final product. Designing safety into a product up front not only helps to create a better solution, but also prevents potentially costly redesign and liability concerns later.
- ■ *Fatigue Damage Analysis and Life Prediction:* Usually in the form of a computer simulation, this analytical tool allows for formal estimation of a product's functional life span, which is crucial in terms of sales and marketing, pricing determination, and product warranty.

FIGURE 5.8 ■ Design analysis techniques. (Terry Whitney, "Design Tools for Engineering Teams—An Integrated Approach.")

As design models shift to a customer-focused business environment, customers have become an integral part of the design and development process. The customer is now involved during the entire life of the project from establishing initial criteria to monitoring and approving design and development activities. The new approach to customer involvement is keeping costs down, quality high, and, ultimately, customers satisfied.

Documentation and Specification

It is impossible to discuss the design project process without also discussing documentation and specifications. A specification is a standard by which the product or design is built to conform. Basically, specifications set guidelines for the end product. Documentation includes all the tests, changes, and final documents of a project.

The specifications are guidelines for the design and production of the part, plan, or structure. They are used by manufacturing or construction when producing the final product. In the design phase of the project, they help define the limits of the design. The management of the design process is driven by the need to meet these specifications.

Specifications can be of various types ranging from the standards set by the customer to the specific industry standards such as OSHA or ANSI. A design specification will determine the production specifications. Specifications make up part of the documentation package; other documentation for a project may include test results, inspection reports, assembly instructions, and all drawing plans or CAD files.

LIFE CYCLE OF A DESIGN PROJECT

Each design project is dictated by its own particular design elements. Factors that determine the life cycle of a design can include whether the design is for a new product, a revision of an old part, part of a larger assembly, or a stand-alone design. There usually is a basic flow of how the design process will proceed. (See Figure 5.9.)

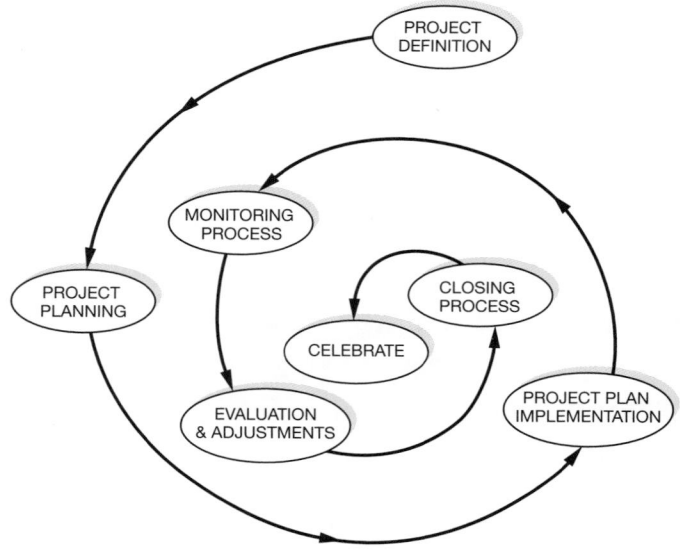

FIGURE 5.9 ■ Typical stages of a design life cycle.

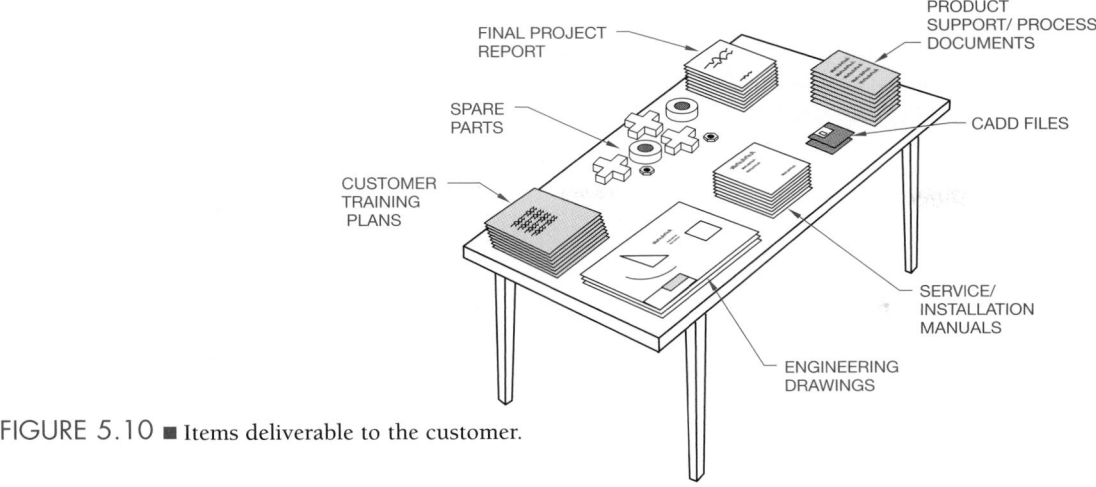

FIGURE 5.10 ■ Items deliverable to the customer.

DESIGN DELIVERABLES

As a design process comes to an end, not only is a final product produced, but many different types of documents are created to document the product, process, installation procedures, training plans, and so on. These materials are known in the industry as deliverables. Typical deliverables include such items as engineering and/or construction drawings or CAD files, prototypes, life cycle plans, installation manuals, and instruction guides. Deliverables also may include service/maintenance manuals, manufacturing process specifications, product support equipment, and materials as well as spare parts. Let us not forget the customer training plans and final project report. (See Figure 5.10.)

Deliverables are what is provided to the customer. Along with the final product and deliverables, several questions will be asked to determine the project success. The answers to the questions (see Figure 5.11) in will help foster continuous improvement by looking at the strengths and weaknesses of the project.

At the completion of a design product, the designer must ask, "Were the key elements successfully met?" Those key elements include:

1. *Time:* Was the project completed by the customer's due date?

2. *Cost:* Were the project costs within the initial budget estimates? Numerous factors can influence the cost of documenting an engineering project. Design complexity, re-use of existing designs or the use of new materials can have a substantial impact on determining the cost of a project. Another factor that is not often considered is the skill level of drafters and designers using the various tools [CAD and analysis software] needed to complete the job. Unfortunately, there are no "standard" for-

mulas for calculating the cost. This is a skill that is generally acquired through years of experience.

3. *Performance:* Does the product perform to the customer's specifications in terms of quality?

HOW THE DESIGN PROCESS HAS RESPONDED TO CHANGES IN ENGINEERING

In examining all the changes in technology and the way in which engineering and product development occurs, how the process of designing ultimately has responded to these changes cannot be overlooked. There are several truths for design that are emerging in this new world of systems engineering.

■ *Timing of the design:* Do it early in the design cycle; doing it later costs more.

■ *Reduce parts in a design:* This reduces costs throughout the whole assembly process.

■ *Standardize parts:* This reduces cost by making parts interchangeable for different designs.

■ *Keep the design simple:* The higher the tooling costs, the more complicated the design.

■ *Utilize modular designs:* these reduce cost in design time because parts can be used in other designs. Modular designs also reduce assembly time.

■ *Design with gravity in mind:* It is easier and more time-efficient to assemble from the top down.

■ *Eliminate fasteners:* This saves time and cost in assembly as well as end-of-life disassembly.

■ *Optimize part handling:* Costs go down and quality goes up with minimal proper assembly sequence.

■ *Design for easy part mating:* Easier assembly through alignment for insertion speeds up assembly.

■ *Provide nesting features in the design:* These help to show where features go.

■ *Optimize manufacturing process sequence:* This increases the speed of getting the product to market.

■ Have the deliverables met the acceptance criteria established?
■ Can the customer use the product or service as desired?
■ Did the customer accept the deliverables provided?
■ Have the "lessons learned" from the project been documented?
■ Has the project's historical information been filed for use in future project planning?
■ What techniques that might have worked well on this project were not used due to some constraints?

FIGURE 5.11 ■ Questions for continuous improvement.

PROFESSIONAL PERSPECTIVE

ONLY THE FAST SURVIVE

—Jim Leonard, Colorado Manufacturing Competitiveness, Denver, Colorado

Every morning in Africa a gazelle wakes up, knowing it must outrun the fastest lion or it will be killed. At the same time a lion wakes up, knowing it must run faster than the slowest gazelle or it will starve. It doesn't matter whether you're a gazelle or a lion, when the sun comes up you'd better be running. It's the law of the competitive jungle—only the fastest survive.

Source: A gazelle somewhere in Africa

Every product development team member has heard or read the law of the competitive jungle; many have the law stamped on their foreheads. So, why is rapid product development, a.k.a., quick time to market, so important? There are many reasons, but the main one is: *Satisfy customers' needs first and best.*

Design Tools for Engineering Teams: An Integrated Approach describes many tools and practices that product developers use to improve product quality and reduce development time. But the toolbox of processes, practices, and technologies is constantly changing. What is state-of-the-art today, will generally be mainstream in two or three years. State-of-the-art tools are rarely visible to anyone other than the most insightful industry observers because best-in-class product developers keep their new tools and practices secret as long as possible. Here is a brief list of practices and technologies that best-in-class teams are using right now.

Customer Focus

The 1990s were characterized by the increased role of "customer focus," particularly leading-edge manufacturers. That leading role will intensify in the 2000s, as articulated by Jacques Nasser, CEO of Ford, " . . . the real work at hand; getting inside the mind of the consumer to understand what he or she wants, aspires to become, and will be needed long after a purchase has been made." Product development professionals will be employing all kinds of innovative ways to understand customer needs, wants, fantasies, including living with customers, having customers as permanent members of product development teams, drilling down into and analyzing mountains of consumer data collected over the Internet.

E-business

Current application of the Internet provide a distinct competitive advantage. Although Internet applications are appearing from everywhere to improve customer service and reduce cycle times, most of the highest value-added applications are in product development. New, innovative applications will continue to appear. Every one of the practices mentioned in this brief list applies Internet technologies in some way.

Collaborative Engineering

Collaborative engineering goes beyond concurrent engineering. It is the cooperative exchange of resources, for example,

information, and ideas among a virtual team focused on an engineering-intensive project and having on overall common creative purpose. [A] virtual team is one whose members are not physically collocated, but connected by distance communication technologies such as videoconferencing, and e-mail.

Predictive Engineering, Simulation, and Virtual Prototypes

With mainstream use of 3-D solid modeling, products can now be designed, built, tested to failure, and redesigned all in digital form. The designs can be nearly optimized before time, funds, and effort are spent on hardware. Entire manufacturing plants can be simulated and optimized before the first cubic yard of concrete is poured.

Visual Engineering

3-D solid modeling also allows all team members to see and understand information in the same way. Now the marketing, finance, and other nontechnical members can see product designs without needing the skill to read 2-D engineering drawings. Realistic images of engineered products can be shown more quickly and for much less cost than building physical protoypes.

Supply Chain Integration, and Value Chain Integration

Companies and teams now coordinate the series of activities and processes that purchase, design, manufacture, and deliver products and/or services to customers. The Supply Chain or Value Chain is a network of company relationships that support materials, information, and funds flows from a firm's supplier's supplier to its customer's customer.

Design for Everything (DfX) Has New Meaning

DfX used to mean something such as "design for manufacturability, assembly, test, service, and environment." Now DfX is taking on several new functions.

- Design for supply chain: optimizes use of the distinctive competencies of all supply chain members.
- Design for postponement: Allows for incorporation of distinctive/customizable features into a product until the latest possible production step.
- Design for recycle, reuse, rebuild, and disposal: Design for environment used to mean "consider ease of recy-

PROFESSIONAL PERSPECTIVE *(continued)*

cling." In the future, expensive, high-information-content products will be designed for easy, inexpensive rebuild/refurbish, which eliminates the need for recycle or disposal for at least another life cycle. Less expensive products will be designed for convenient, environmentally acceptable disposal, which may include recycling or components made or biodegradable material.

This list is certainly not comprehensive. Most of these practices and technologies may be mainstream by the time this book is purchased and read, certainly by the early 2000s. Enlightened product development professionals constantly scan the horizon for new practices, processes, and technologies, and then intelligently apply them in their own projects. This is the only way to consistently *satisfy customers' needs first and best.*

■ *Form follows function:* Design for the user.
■ *Utilize design generations:* An initial design can be reused in the next generation of the product.

WEB SITE RESEARCH

The following Web sites can provide you with information about the engineering design process.

www.designnews.com—Trade magazine.
www.partnerwerks.com—Information on design teams.
www.sme.org—Society of Manufacturing Engineering.
www.ame.org—Association of Manufacturing Excellence.
www.manufacturingnews.com—Current articles.
http://world.std.com/-uieweb—User interface engineering.

For more information; search the Web using *whole systems engineering design* or *engineering design teams.*

CHAPTER 5

The Engineering Design Process Test

DIRECTIONS

Answer the following questions with short, complete statements.

PART 1 • ENGINEERING DESIGN APPLICATION: TRANSITIONS TO A WHOLE SYSTEMS APPROACH

1. List three items that are driving today's designs.
2. Define whole systems design.
3. List five areas that could make up an engineering design team.
4. What are the necessary skills for an engineering designer or drafter?

PART 2 • TODAY'S ENGINEERING DESIGN MODELS

5. Define concurrent engineering.
6. Define life cycle engineering.
7. Describe the process used in integrated product development (IPD).
8. List the type of data used in knowledge-based engineering (KBE).
9. Define total quality management (TQM).

PART 3 • THE ENGINEERING DESIGN PROCESS

10. List the six items in a typical design analysis process.
11. What does R&D stand for?

PART 4 • CREATIVITY AND INNOVATION IN DESIGN

12. Define creativity.
13. Define innovation.
14. Describe two situations in design in which creativity may be required.
15. List the seven basic principles of creativity.
16. Describe five methods for developing creativity in an individual.
17. What is the advantage of creative problem solving in a group?
18. What kind of environment does creativity and innovation require?

PART 5 • DESIGN ANALYSIS AND PROBLEM SOLVING

19. List the seven basic steps to follow when working through a problem.
20. What is important to do when evaluating the success or failure of a design solution?
21. List four tips for problem solving in teams.

PART 6 • CHANGE AND THE IMPACT ON THE DESIGN PROCESS

22. List two tools that are helping to eliminate changes in a design.

23. List the five functions that change serves in the design process.
24. What two things are always required on an engineering change process from the beginning to the end of the design process?

PART 7 • TRACKING DOCUMENTS: ECO/ECR

25. What does ECO stand for?
26. What does ECR stand for?
27. List four of the eight types of documents required for an ECO/ECR.
28. What three steps are necessary for a design change to be incorporated properly?

PART 8 • THE DESIGN REVIEW PROCESS

29. What is the ultimate goal of the design review process?
30. Describe the processes of a design review.
31. List two ways data for a design review may be shared.
32. Describe four of the nine design analysis techniques.
33. Describe the role of the customer in a design review.
34. Define a specification.
35. What is the purpose of a specification?

36. List two types of standards that are commonly used for specific industry standards.
37. Specifications are part of what package in the design documents?

PART 9 • LIFE CYCLE OF A DESIGN PROJECT

38. What factors determine the life cycle of a design project?
39. List the typical stages of a design life cycle.

PART 10 • DESIGN DELIVERABLES

40. Define deliverables.
41. List the typical deliverables for a design project.
42. Who receives the deliverables.
43. What are the three key elements that are reviewed at the completion of a design project?

PART 11 • HOW THE DESIGN PROCESS HAS RESPONDED TO CHANGES IN ENGINEERING

44. Describe eight out the thirteen truths for design that are emerging in the new world of systems engineering.

CHAPTER 5

The Engineering Design Process Problems

Note: This chapter is intended as a reference for engineering design processes. The concepts discussed serve as a foundation for the applied concepts of engineering drawing practices that are in the following chapters.

PROBLEM 5-1 The following topics require research and/or industrial visitations. It is recommended that you research current professional magazines, examining Internet Web sites, visit local industries, or interview professionals in the engineering field. Your reports should emphasize:

■ The link between manufacturing and engineering
■ Current technological advances
■ The design process
■ Team applications

Select *two* of the following engineering fields or as assigned by your instructor and write a 500 word report for each.

Mechanical Engineering
Structural/Architectural Engineering
Civil Engineering
Electrical/Electronic Engineering
HVAC/Sheet Metal

PROBLEM 5-2 The following topics are changing the way engineering design is currently conducted. You are to research, observe, visit industry or product vendors, or search the Internet to find ways that the following items are allowing product development to be cheaper, faster, and of higher quality. Your presentation should emphasize the following:

■ Process
■ Current technological advances
■ Team applications
■ Problem solving tools
■ Design applications for engineering

Select *three or more* of the topics listed below or as assigned by your instructor and prepare an oral presentation for each.

Rapid Product Development
Rapid Tooling
Virtual Reality
Computer-Aided Manufacturing (CAM)
Computer-Aided Engineer (CAE)
Computer-Integrated Manufacturing (CIM)
Concurrent Engineering
Self-directed Teams
Virtual Teams
Statistical Process Control (SPC)
Virtual Design Process
Design for Manufacturability (DFM)
CAD Software – 3D solid modeling
CAD Simulations
Engineering Resource Planning (ERP)
Product Data Management (PDM)
Design for Disassembly(DFD)

Geometric Construction

OBJECTIVES

After completing this chapter, you will:

- Draw parallel and perpendicular lines.
- Construct bisectors and divide lines and spaces into equal parts.
- Draw polygons.

- Draw tangencies.
- Draw ellipses.
- Solve an engineering problem by making a formal drawing with geometric constructions from an engineer's sketch or layout.

GEOMETRIC CONSTRUCTION

Machine parts and other manufactured products are made up of different geometric shapes ranging from squares and cylinders to complex irregular curves. *Geometric constructions* are methods that

can be used to draw various geometric shapes or to perform drafting tasks related to the geometry of product representation and design. The proper use of geometric constructions requires a basic understanding of plane geometry. When using geometric construction techniques, extreme accuracy and the proper use of drafting instruments are important. Pencil or compass leads should be

THE ENGINEERING DESIGN APPLICATION

You should always approach an engineering drafting problem in a systematic manner. Plan out how you propose to solve the problem. As an entry-level drafter you need to do this planning with sketches and written notes. These sketches and notes help you decide:

- The scale to use to effectively display the drawing. If the scale is too small, complex detail may be lost, but if the scale is too large, the drawing may take up too much space and cause you to spend too much time.

- What paper size or CADD drawing limits should be established. This depends on the final plotted scale of the drawing, the amount of detail, and the specifications required.

Warning! Engineering sketches are often difficult to read. This is typical because engineers do not have the time or the skills to prepare a very neat and accurate sketch. Actual engineering sketches may be out of proportion and have missing information. The dimensions frequently do not comply with ASME or other related standards. It is your responsibility to convert the engineer's communication into a formal drawing that is accurate and drawn to proper standards.

After the initial decisions about scale and paper size are made, you need to proceed with the problem in a logical manner. Refer to the engineer's sketch shown in Figure 6.1 as you follow these layout steps for either manual drafting or CADD:

Step 1. Do all preliminary manual work using construction lines so if you make an error, it is easy to erase. Begin by establishing the centers of the ⌀57 and ⌀25.5 circles and then draw the circles.

Step 2. Draw the concentric circles of ⌀71.5 and ⌀40.

Step 3. Locate and draw the 6X R7.5 arcs and then using tangencies blend the R3 radii with the outside arcs.

Step 4. Draw the 2X R7 arcs tangent to the large inside arcs.

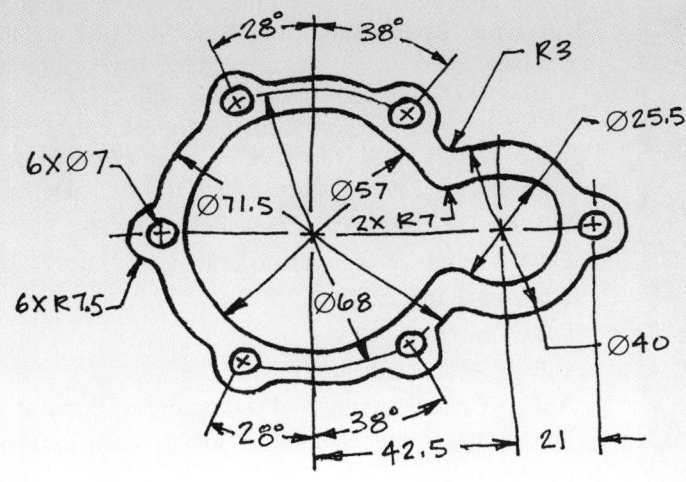

FIGURE 6.1 ■ Engineer's sketch.

(Continued)

THE ENGINEERING DESIGN APPLICATION (continued)

Step 5. Draw the centerlines for the 6X Ø7 circles and then draw the circles in place.

Step 6. Darken all thin lines and then darken all object lines. If there are arcs and straight lines, draw the arcs first and then blend the straight lines into the tangencies. The finished drawing is shown in Figure 6.2.

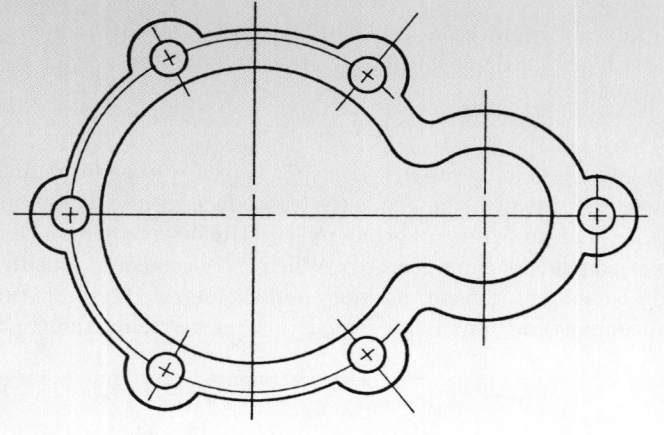

FIGURE 6.2 ■ The complete drawing (without dimensions) for engineer's sketch shown in Figure 6.1.

sharp and instruments should be in good shape. Always use very lightly drawn construction lines for all preliminary work.

When computers are used in drafting, the task of creating most geometric construction related drawings becomes much easier, although the theory behind the layout of geometric shapes and related constructions remains the same.

CHARACTERISTICS OF LINES

A straight line segment is a line of any given length, such as lines A–B shown in Figure 6.4.

A curved line may be in the form of an arc with a given center and radius or an irregular curve without a defined radius as shown in Figure 6.5.

Two or more lines may intersect at a point as in Figure 6.6. The opposite angles of two intersecting lines are equal; a = a, b = b.

Parallel lines are lines equidistant throughout their length, and if they were to extend indefinitely they would never cross. (See Figure 6.7.)

Perpendicular lines intersect at a 90° angle as shown in Figure 6.8.

GEOMETRIC SHAPES

Angles

Angles are formed by the intersection of two lines. Angles are sized in degrees (°). Components of a degree are minutes and seconds.

A LAYOUT APPROACH FOR CADD APPLICATIONS

If the engineering sketch is good quality, you can use it to plan your CADD layout, but if it does not represent your plans for making a professional drawing, then you should prepare a new sketch. Your sketch should display the drawing with dimensions so everything meets ASME or other related standards. Planning with a sketch will save you time in completing the project, because you will have a good idea of what to do before you start working with CADD. When your sketch is ready, make a form like the example given in Figure 6.3 to list all of your drawing steps. Write the command in the left column. Choose from the following list of commands and provide the information indicated.

POINT—Cartesian coordinate location

SCALE—numerical size

ORIGIN—coordinates

CIRCLE—center, diameter

ARC—center, beginning and ending angles (degrees)

POLYGON—center, radius flat to flat (RF), or radius point to point (RP)

DASHED LINE—end points

TEXT (NOTES)—(1) height, width, slant angle;(2) angle direction; (3) lower left or lower right coordinates.

Indicate the type of line in column 2, the coordinates in column 3, and additional parameters (diameter, radius, angles, etc.) in column 4.

SAMPLE FORM

Student Name _____ File Name _____
Scale _____ Date _____

① COMMAND	② LINE TYPE	③ X	Y	④ PARAMETERS
Point	Solid	0	0	
"	"	2	0	
"	"	2	1	
"	"	0	1	
Circle	Solid	1	.5	Dia. = .25
Arc	"	0	.5	R = .5, 90°- 180°

FIGURE 6.3 ■ CADD planning form.

CADD APPLICATIONS

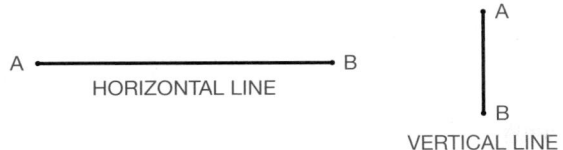

FIGURE 6.4 ■ Horizontal and vertical lines.

FIGURE 6.5 ■ Arc and irregular curve.

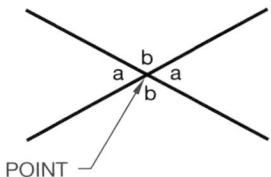

FIGURE 6.6 ■ Intersecting lines.

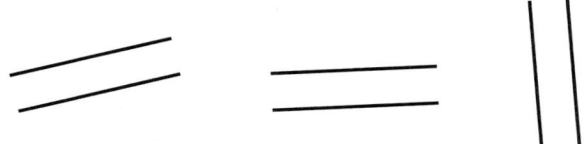

FIGURE 6.7 ■ Parallel lines.

FIGURE 6.8 ■ Perpendicular lines.

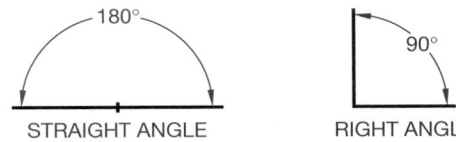

FIGURE 6.9 ■ Types of angles.

FIGURE 6.10 ■ Parts of an angle.

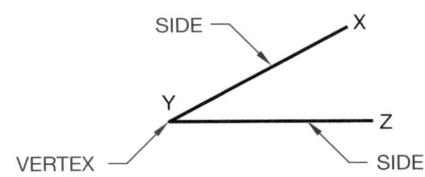

FIGURE 6.11 ■ Parts of a triangle.

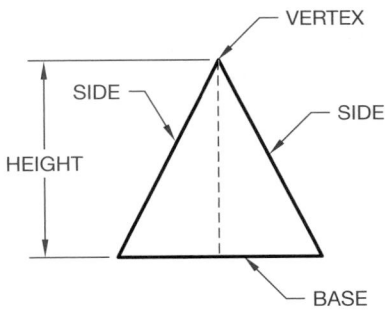

FIGURE 6.12 ■ Labeling triangles.

There are 60 minutes (') in one degree and there are 60 seconds (") in one minute. $1° = 60'$, $1' = 60"$. The four basic types of angles are shown in Figure 6.9. A straight angle equals 180° while a right angle equals 90°. An acute angle contains more than 0° but less than 90° and an obtuse angle contains more than 90° but less than 180°.

The parts of an angle are shown in Figure 6.10. An angle is labeled by giving the letters defining the line ends, and vertex, with the vertex always between the ends, such as XYZ where Y is the vertex.

Triangles

A *triangle* is a geometric figure formed by three intersecting lines creating three angles. The sum of the interior angles always equals 180°. The parts of a triangle are shown in Figure 6.11. Triangles are labeled by lettering the vertex of each angle such as ABC, or by labeling the sides abc as shown in Figure 6.12.

There are three kinds of triangles: acute, obtuse, and right as shown in Figure 6.13. Two special acute triangles are the equilateral, which has equal sides and angles, and the isosceles, which has two equal sides and angles. An isosceles triangle may also be obtuse. In a scalene triangle, no sides or angles are equal.

Right triangles have certain unique geometric characteristics. Two internal angles equal 90° when added. The side opposite the 90° angle is called the hypotenuse, as seen in Figure 6.14. A

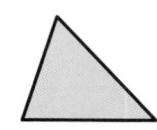

EQUILATERAL
TRIANGLE—
ALL SIDES AND
ANGLES EQUAL

ISOSCELES
TRIANGLE—
TWO SIDES
AND TWO
ANGLES EQUAL

ACUTE SCALENE
TRIANGLE—
NO EQUAL SIDES
OR ANGLES

ACUTE TRIANGLES
(NO INTERIOR ANGLE IS GREATER THAN 90°)

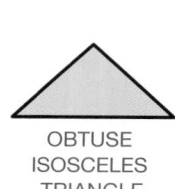
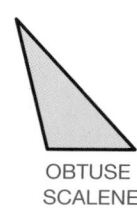
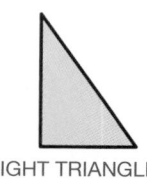

OBTUSE
ISOSCELES
TRIANGLE

OBTUSE
SCALENE
TRIANGLE

RIGHT TRIANGLE—
ONE INTERIOR
90° ANGLE

OBTUSE TRIANGLES
(ONE INTERIOR ANGLE GREATER THAN 90°)

FIGURE 6.13 ■ Types of triangles.

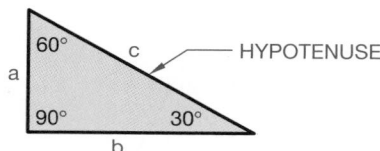

FIGURE 6.14 ■ Right triangle.

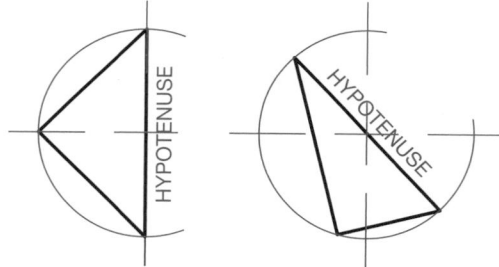

FIGURE 6.15 ■ A semicircle is formed when an arc is drawn
through the vertices of a right triangle.

semicircle (half circle) is always formed when an arc is drawn
through the vertices of a right triangle as shown in Figure 6.15.

Quadrilaterals

Quadrilaterals are four-sided polygons that can have equal or
unequal sides or interior angles. The sum of the interior angles is
360°. Quadrilaterals with parallel sides are called parallelograms.
(See Figure 6.16.)

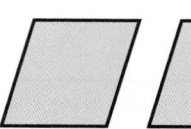

SQUARE—
EQUAL
SIDES, 90°
INTERNAL
ANGLES

RECTANGLE—
OPPOSITE SIDES
EQUAL, 90°
INTERNAL
ANGLES

RHOMBUS—
EQUAL
SIDES

RHOMBOID—
OPPOSITE SIDES
EQUAL

PARALLELOGRAMS

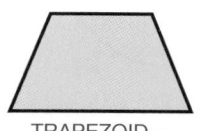
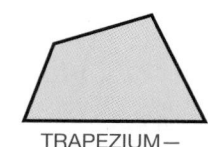

TRAPEZOID—
TWO PARALLEL SIDES

TRAPEZIUM—
NO PARALLEL SIDES

FIGURE 6.16 ■ Quadrilaterals.

Regular Polygons

Some of the most commonly drawn geometric shapes are regular
polygons. *Regular polygons* have equal sides and equal internal
angles. Polygons are closed figures with any number of sides, but
no less than three. The relationship of a circle to a regular polygon
is that a circle may be drawn to touch the corners (circumscribed)
or the sides, known as flats, (inscribed) of a regular polygon. (See
Figure 6.17.) This relationship is an advantage in constructing regular polygons. Some common regular polygons are shown in Figure 6.18.

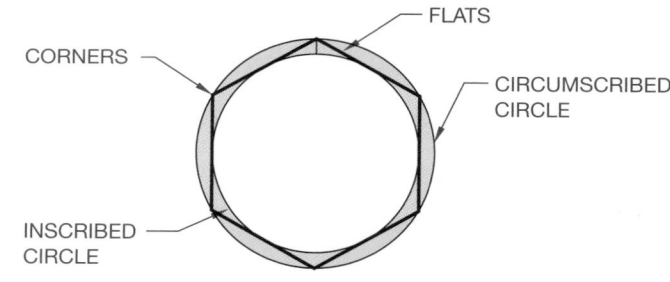

FIGURE 6.17 ■ Regular polygon.

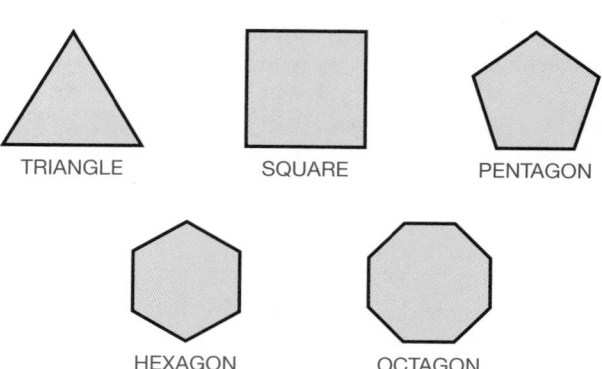

FIGURE 6.18 ■ Common regular polygons.

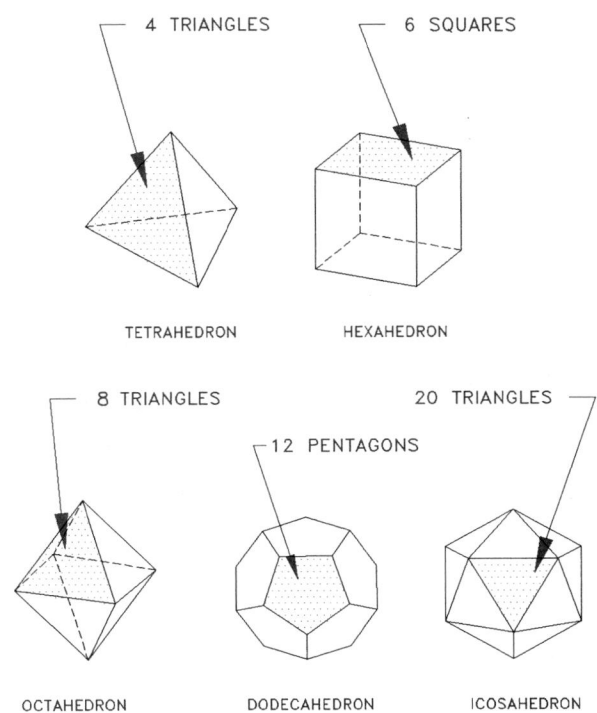

FIGURE 6.19 ■ Regular solids.

Regular Solids

Solid objects constructed of regular polygon surfaces are called regular polyhedrons. (See Figure 6.19.) A *polyhedron* is a solid formed by plane surfaces. The surfaces are referred to as *faces*.

Prisms

A *prism* is a geometric solid object with ends that are the same size and shaped polygons and sides that connect the same corresponding corners of the ends. Figure 6.20 shows a few common examples. A *right prism* has sides that meet 90° with the ends. An *oblique prism* has sides that are at an angle to the ends.

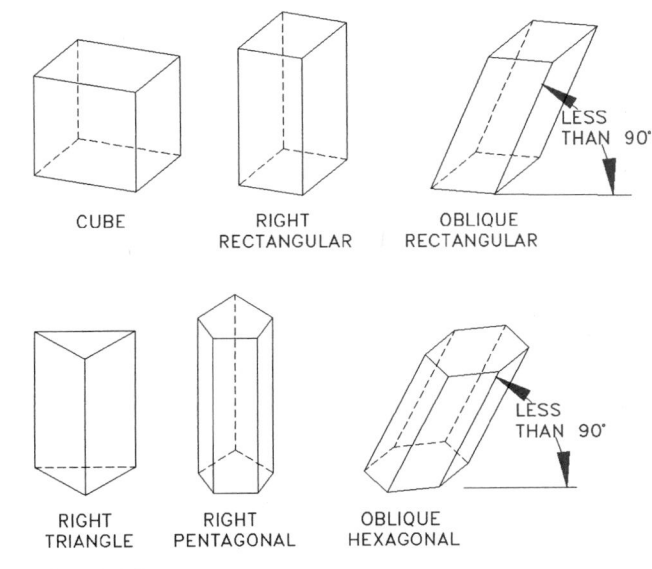

FIGURE 6.20 ■ Common prisms.

Pyramid Prisms

A *pyramid prism* has a regular polygon-shaped base and sides that meet at one point called a *vertex* as shown in Figure 6.21. A geometric solid is *truncated* when a portion is removed and a plane surface is exposed. The *axis* is an imaginary line that connects the vertex to the midpoint of the base. It may also be referred to as an imaginary line around which parts are regularly arranged.

Circles

A *circle* is a closed curve with all points along the curve at an equal distance from a point called the center. The circle has a total of 360°. The *circumference* is the distance around the circle. The *radius* is the distance from the center to the circumference. The *diameter* is the distance across the circle through the center. Figure 6.22 shows the parts of a circle and circle relationships. *Concentric circles* have the same center, and *eccentric circles* have different centers.

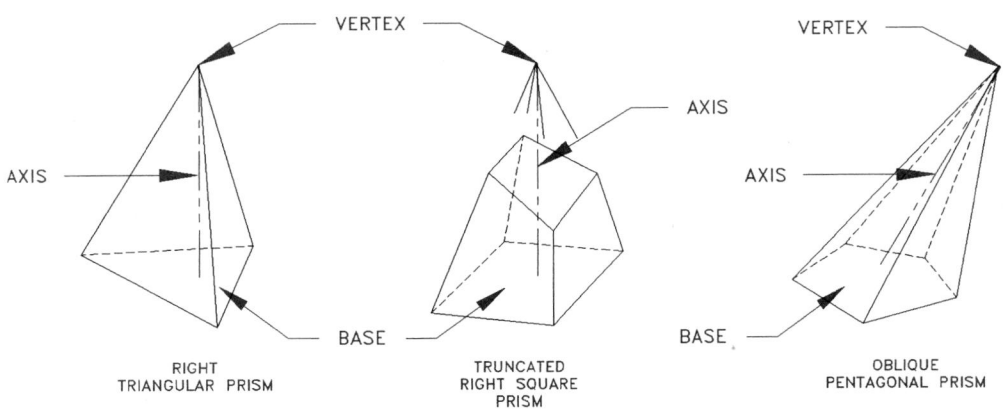

FIGURE 6.21 ■ Common pyramid prisms.

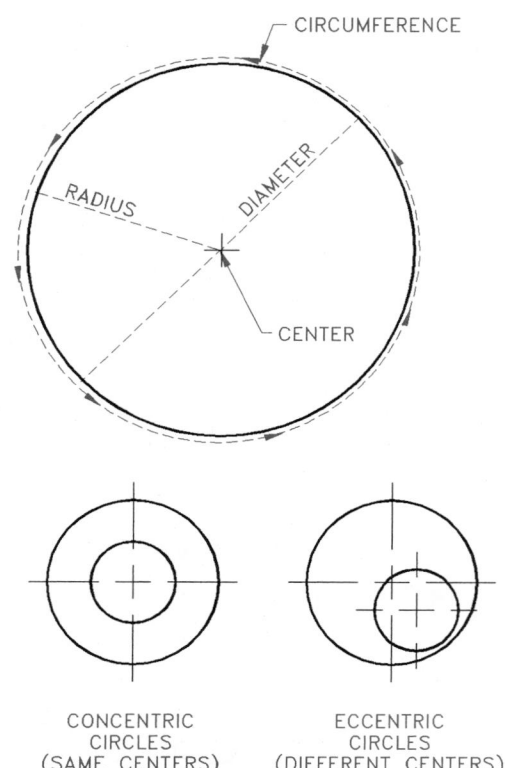

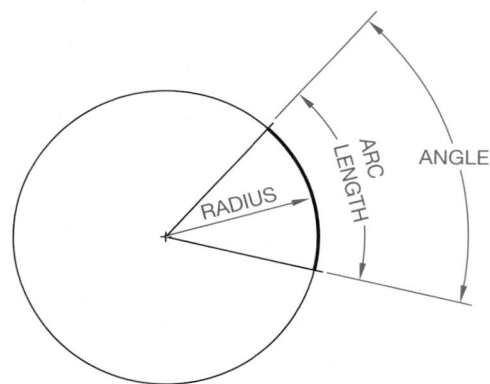

FIGURE 6.22 ■ Parts of a circle and circle characteristics.

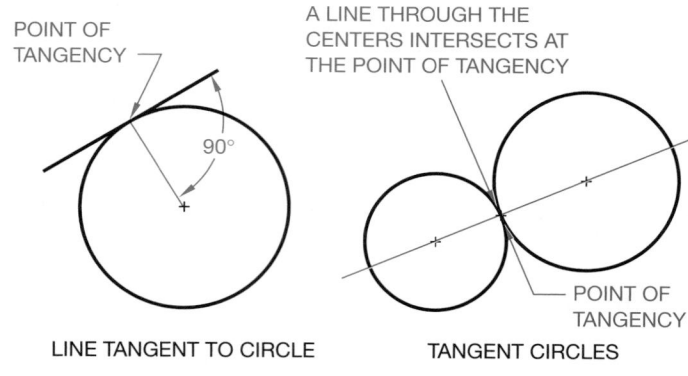

FIGURE 6.24 ■ Sphere.

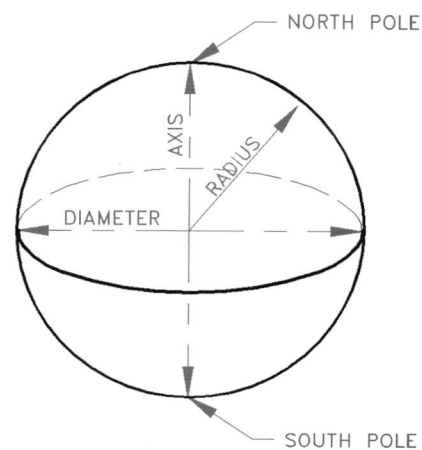

FIGURE 6.25 ■ Tangency.

An *arc* is part of the circumference of a circle. The arc may be identified by a radius, an angle, or a length (Figure 6.23).

FIGURE 6.23 ■ Arc.

Spheres

A *sphere* is the shape of a ball. Every point on the surface of a sphere is equidistant from the center. If you think of the earth as a sphere, the North and South poles are at the end of its axis as shown in Figure 6.24.

Tangents

Straight or curved lines are *tangent* to a circle or arc when the line touches the circle or arc at only one point. If a line were to connect the center of the circle or arc to the point of tangency, the tangent line and the line from the center would form a 90° angle. (See Figure 6.25.)

COMMON GEOMETRIC CONSTRUCTIONS

Parallel Lines

Parallel lines are lines evenly spaced at all points along their length and will not intersect even when extended. Figure 6.26 shows an example of parallel lines. The space between parallel lines may be any distance.

Parallel lines may be drawn horizontally, vertically, or at any angle using a drafting machine. When a drafting machine is not available, parallel lines may be drawn with the aid of standard triangles as shown in Figure 6.27. Parallel lines may be drawn with a

FIGURE 6.26 ■ Parallel lines.

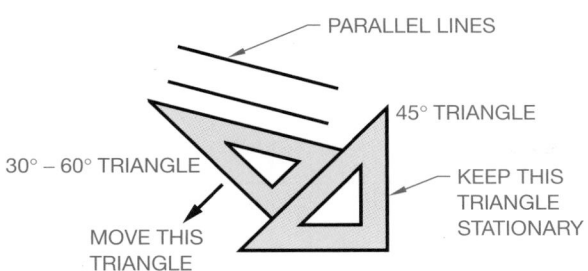

FIGURE 6.27 ■ Constructing parallel lines with triangles.

straightedge and compass when the distance between lines is established. Use the compass radius to draw arcs near the ends of the given line. The parallel line is then drawn at the points of tangency of the arcs as shown in Figure 6.28. Draw very light construction lines for all construction layout.

Parallel or concentric arcs may be drawn where the distance between arcs is equal. Establish parallel arcs by adding the radius of Arc 1 (R_1) to the distance between the arcs (X), as shown in Figure 6.29.

Concentric circles are parallel circles drawn from the same center. The distance between circles is equal, as shown in Figure 6.30.

Parallel irregular curves are drawn any given distance (R) apart by using a compass set at the given distance. A series of arcs is lightly drawn all along the given curve at the required distance. The line to be drawn parallel is then constructed with an irregular curve at the points of tangency of the construction arcs. (See Figure 6.31.)

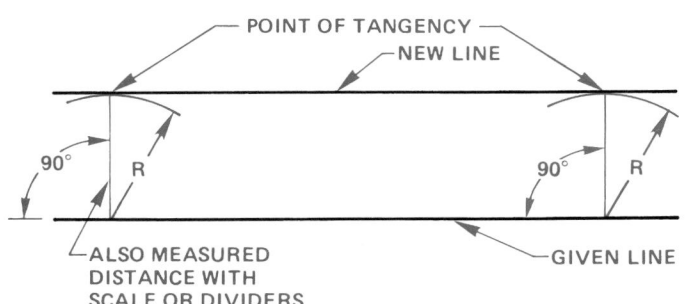

FIGURE 6.28 ■ Constructing parallel lines with a straightedge and compass.

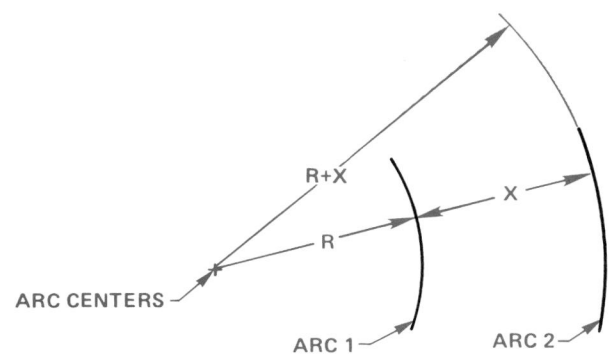

FIGURE 6.29 ■ Parallel arcs.

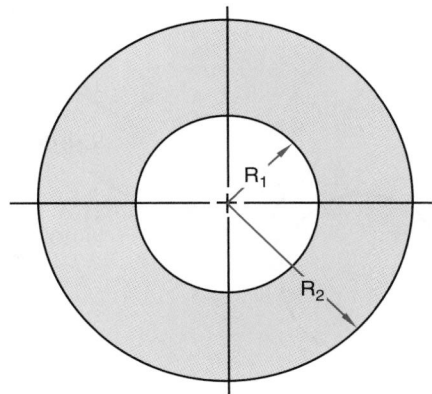

FIGURE 6.30 ■ Concentric circles.

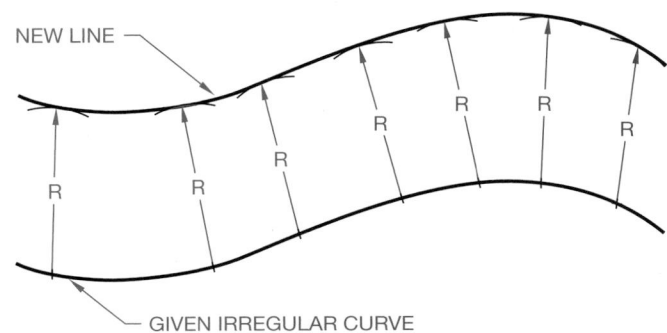

FIGURE 6.31 ■ Constructing parallel irregular curves.

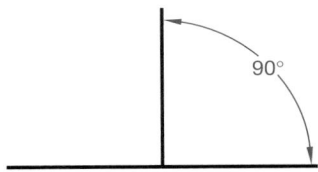

FIGURE 6.32 ■ Perpendicular lines.

Perpendicular Lines

Perpendicular lines intersect at 90°, or a right angle, as seen in Figure 6.32. Perpendicular lines are drawn with the horizontal and vertical scales of a drafting machine, even if the scales are set at an angle away from horizontal, as seen in Figure 6.33. Perpendicular lines can also be drawn with a straightedge and triangle or two triangles as shown in Figure 6.34.

Perpendicular Bisector

The perpendicular bisector of a line may be found using a straightedge and a compass.

Step 1. Set the compass radius more than halfway across the given line segment and draw two intersecting arcs from the line ends.

Step 2. Connect the points where the arcs intersect, as in Figure 6.35.

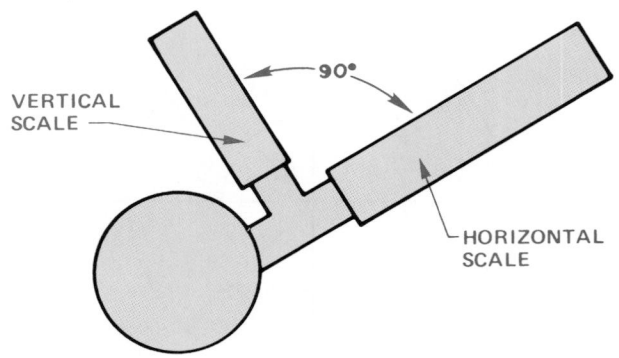

FIGURE 6.33 ■ Perpendicular drafting machine scales.

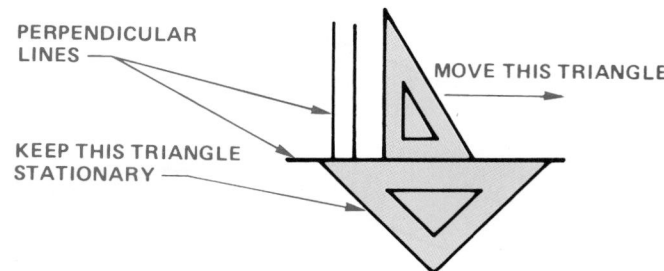

FIGURE 6.34 ■ Perpendicular lines with triangles.

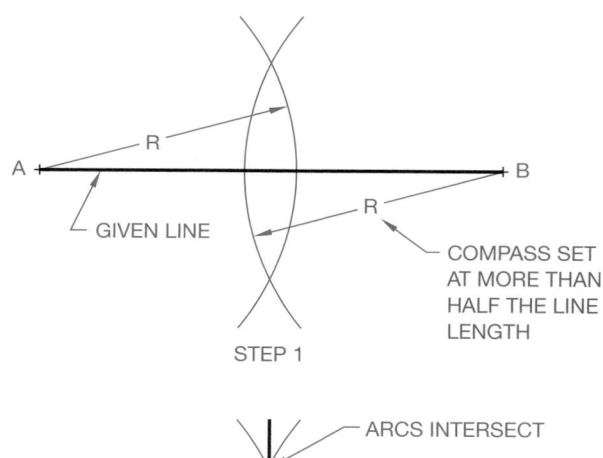

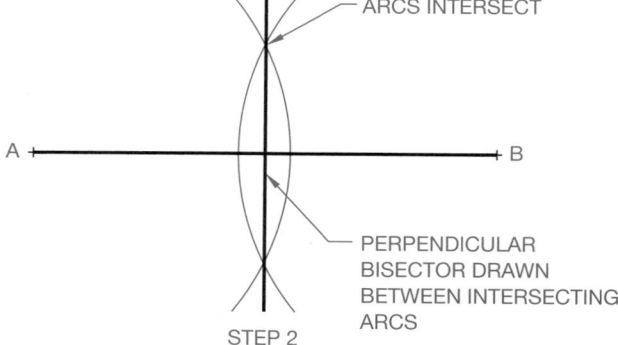

FIGURE 6.35 ■ Constructing a perpendicular bisector.

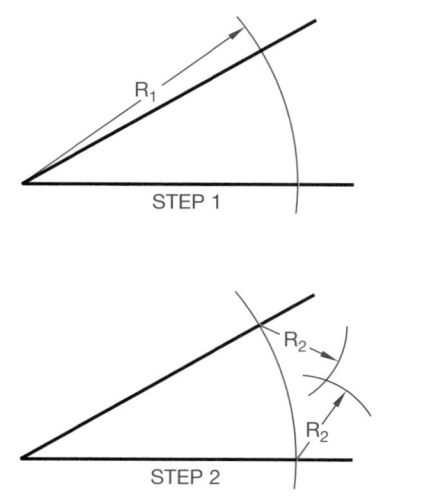

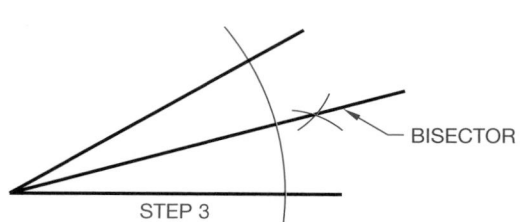

FIGURE 6.36 ■ Bisecting an angle.

Bisecting an Angle

Any given angle may be divided into two equal angles by using a compass and straightedge.

Step 1. Adjust the compass to any radius (R_1) and draw an arc as shown in Figure 6.36.

Step 2. At the points where the arc intersects the sides of the angle, draw two arcs of equal radius (R_2).

Step 3. Connect a straight line from the vertex of the angle to the point of intersection of the two arcs.

Transferring an Angle

A given angle may be transferred to a new location as shown in Figure 6.37. This method also works well for transferring a triangle.

Step 1. Given angle ABC, draw line B^1C^1 in the new location and position.

Step 2. Draw any arc, R, with its center at B on the given angle. Transfer and draw arc R with its center at point B^1.

Step 3. Using the intersection of one side of the given angle and arc R as center, set a compass distance to the opposite intersection of arc R. Then transfer the new radius, S, to the new position using the intersection of B^1C^1 and arc R as center. Connect a line from B^1 to the intersection of the two arcs, R and S, as shown in Figure 6.37.

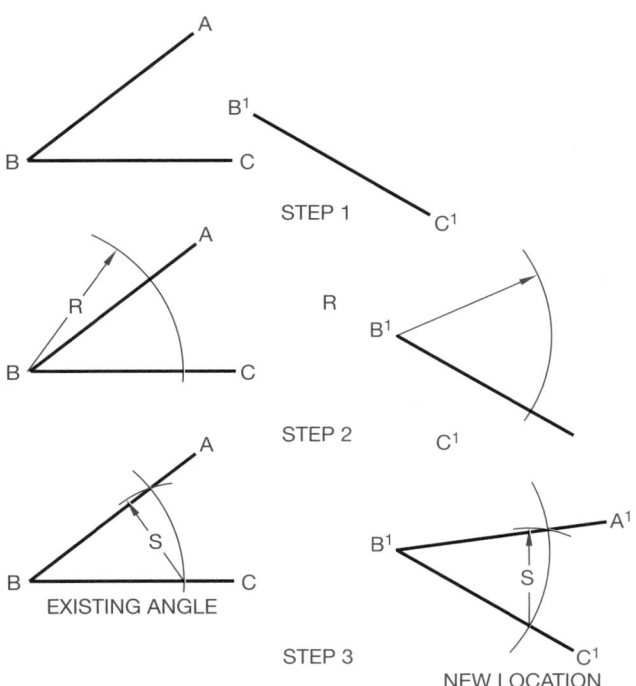

FIGURE 6.37 ■ Transferring a given angle to a new location.

Dividing a Line into Equal Parts

Any given line can be divided into any number of equal parts. Divide given line AB, shown in Figure 6.38, into eight equal parts.

Step 1. Draw a construction line at any angle from either end of line AB as shown in Figure 6.39.

Step 2. Use a scale or dividers to divide the angled construction line into eight equal parts. Use any size division that will extend down the angled line to approximately the length of the given line. (See Figure 6.40.) Using a divider is generally more accurate, but using a scale can be accurate if you are careful.

Step 3. Connect the last point (8) of the construction line to point B of the given line. Then draw lines parallel to line 8B from each of the numbered points on the construction

FIGURE 6.38 ■ Given line AB to divide into equal parts.

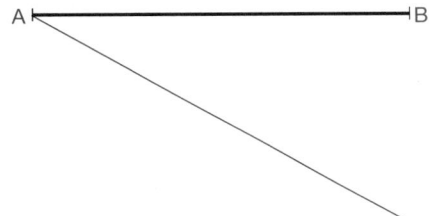

FIGURE 6.39 ■ Step 1, angled construction line.

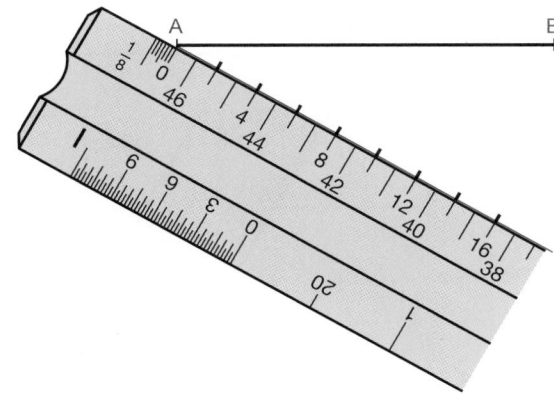

FIGURE 6.40 ■ Step 2, divide angled line into required number of equal parts. A divider is also commonly used to divide the angled line into equal parts.

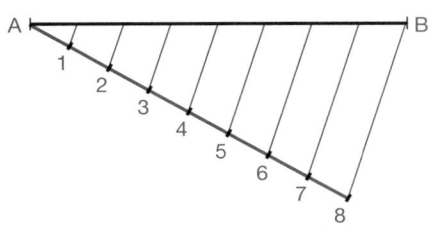

FIGURE 6.41 ■ Step 3, connect parallel line segments.

line to the given line AB. (See Figure 6.41.) You now have the given line AB, divided into eight equal parts. This same process may be used for any number of equal parts.

Dividing a Space into Equal Parts

You can divide any given space into any number of equal parts. Given the space shown in Figure 6.42, divide it into 12 equal parts.

Step 1. Between the lines that establish the given space, place a scale with the required number of increments (12) so that the zero on the scale is on either line and the 12 is on the other line. Mark each increment. (See Figure 6.43.)

Step 2. Remove the scale. Each increment you have marked has divided the given space into 12 equal spaces. Figure 6.44 shows parallel lines drawn at each mark to complete the space division.

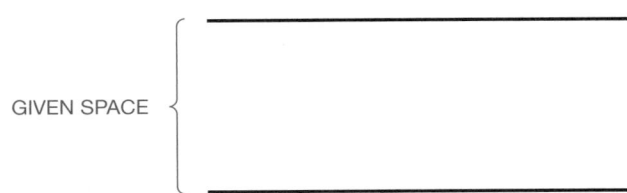

FIGURE 6.42 ■ Dividing a given space into equal parts.

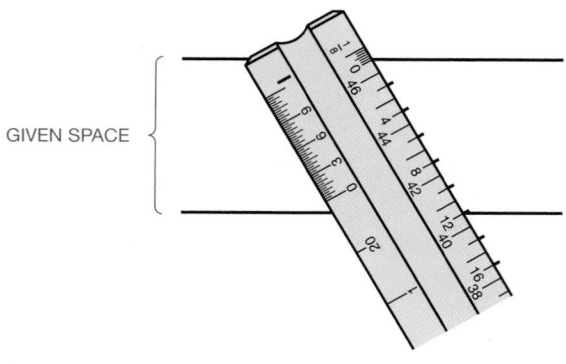

FIGURE 6.43 ■ Step 1, place a scale with required number of increments between the lines.

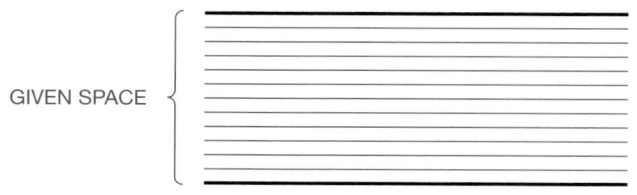

FIGURE 6.44 ■ Step 2, given space divided into equal parts.

CONSTRUCTING POLYGONS

Triangle Given Three Sides

The following technique is known as triangulation. Using given triangle sides x, y, and z in Figure 6.45, draw a triangle as shown in the following steps:

Step 1. Lay out one of the given sides. Select the side that will be the base: z has been chosen as shown in Figure 6.46. From one end of line z, strike an arc equal in length to one of the other lines, x, for example, as in Figure 6.46.

Step 2. From the other end of line z, strike an arc with a radius equal to the remaining line, y. Allow this arc to intersect

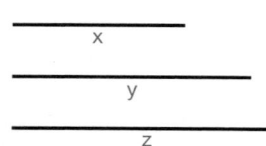

FIGURE 6.45 ■ Construct a triangle given three sides.

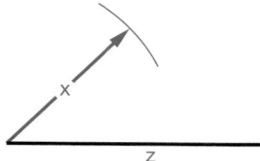

FIGURE 6.46 ■ Step 1, layout line z and swing arc x.

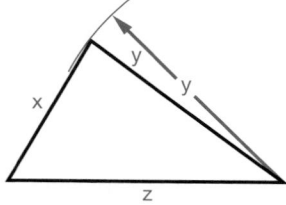

FIGURE 6.47 ■ Step 2, swing arc y.

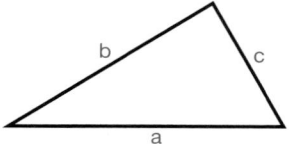

FIGURE 6.48 ■ Transfer a given triangle to a new location.

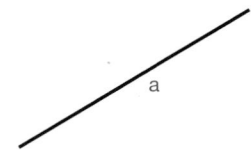

FIGURE 6.49 ■ Step 1, establish new location.

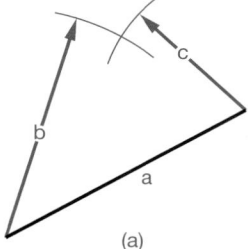

(a)

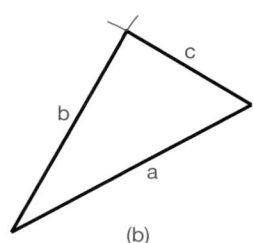

(b)

FIGURE 6.50 ■ (a) Step 2, swing arcs b and c. (b) Step 3, connect the ends of line a to intersection of arcs.

the previous arc. Where the two arcs cross, draw lines to the ends of the base line, as in Figure 6.47, to complete the triangle. This method is called *triangulation*. Triangulation is commonly used in many disciplines and especially useful in sheet metal pattern development.

Transferring a Triangle

Any given polygon may be constructed or transferred to a new location using triangulation. Given the triangle, abc, in Figure 6.48, transfer it to a new location as shown in the following steps:

Step 1. Transfer one of the sides of the triangle to the new location by measuring its length or using dividers. (See Figure 6.49.)

Step 2. From one end of line a in the new location, draw an arc equal in length to line b. From the other end of line a, draw an intersecting arc equal in length to line c. (See Figure 6.50a.)

Step 3. Connect both ends of line a to the intersection of the arcs to form the triangle in its new position as shown in Figure 6.50b.

Constructing Right Triangles

A right triangle may be drawn when the length of the two sides adjacent to the 90° angle are given as shown in Figure 6.51.

Step 1. Draw line a perpendicular to line b. (See Figure 6.52.)

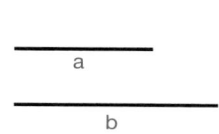

FIGURE 6.51 ■ Construct a right triangle given two sides.

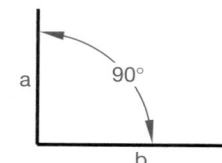

FIGURE 6.52 ■ Step 1, draw line a perpendicular to line b.

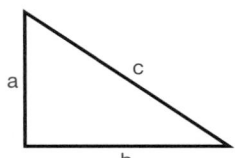

FIGURE 6.53 ■ Step 2, connect hypotenuse.

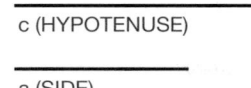

c (HYPOTENUSE)

a (SIDE)

FIGURE 6.54 ■ Construct a right triangle given a side and hypotenuse.

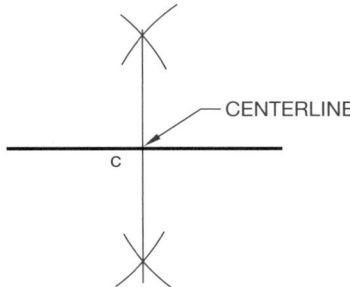

CENTERLINE c

FIGURE 6.55 ■ Step 1, draw the hypotenuse and establish its center.

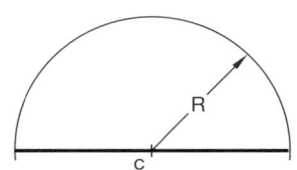

FIGURE 6.56 ■ Step 2, draw a 180° arc from the center.

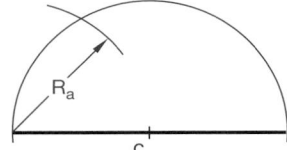

FIGURE 6.57 ■ Step 3, use radius R_a to establish the end of line a.

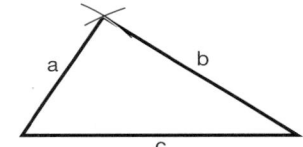

FIGURE 6.58 ■ Step 4, complete the right triangle.

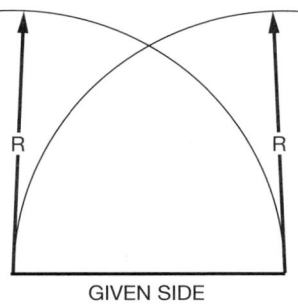

GIVEN SIDE

FIGURE 6.59 ■ Step 1, to construct an equilateral triangle given one side, swing equal arcs r.

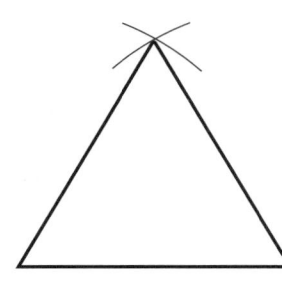

FIGURE 6.60 ■ Step 2, complete the equilateral triangle.

Step 2. Connect a line from the end of line a to the end of line b to establish the right triangle. (See Figure 6.53.)

Draw a right triangle given the length of the hypotenuse and one side as in Figure 6.54.

Step 1. Draw line c and establish its center. You can use the perpendicular bisector method to find the center. (See Figure 6.55.)

Step 2. From the center of line c, draw a 180° arc with the compass point at the line center and the compass lead at the line end. (See Figure 6.56.)

Step 3. Set the compass with a radius equal in length to the other given line, a. From one end of line c draw an arc intersecting the previous arc as shown in Figure 6.57.

Step 4. From the intersection of the two arcs in step 3, draw two lines connecting the ends of line c to complete the required right triangle. (See Figure 6.58.)

Constructing an Equilateral Triangle

An equilateral triangle may be drawn, given the length of one side, as demonstrated in the following steps.

Step 1. Draw the given side length. Then set a compass with a radius equal to the length of the given side. Use this com-

pass setting to draw intersecting lines from the ends of the given side as in Figure 6.59.

Step 2. Connect the point of intersection of the two arcs to the ends of the given side to complete the equilateral triangle as shown in Figure 6.60.

Constructing Squares

Square-head bolts or square nuts are sometimes drawn in conjunction with manufactured parts. Use geometric constructions when necessary; however, when practical, use square templates or special square-head bolt and nut templates. There are several methods that may be used to draw a square, each of which relates to the characteristics of the square shown in Figure 6.61.

Draw a square with the length of one side given as follows:

Step 1. Draw the length of the given side. Then with the drafting machine, or other appropriate method of obtaining a 90° angle, draw lines from each end equal in length to the given side. (See Figure 6.62.)

Step 2. Connect the end of sides 1 and 2 drawn in step 1 to complete the square with side 3, shown in Figure 6.63.

Draw a square with the distance across the flats or across the corners given.

Step 1. Draw a circle equal in diameter to the distance across the flats, Figure 6.64a, and a circle with a diameter equal to

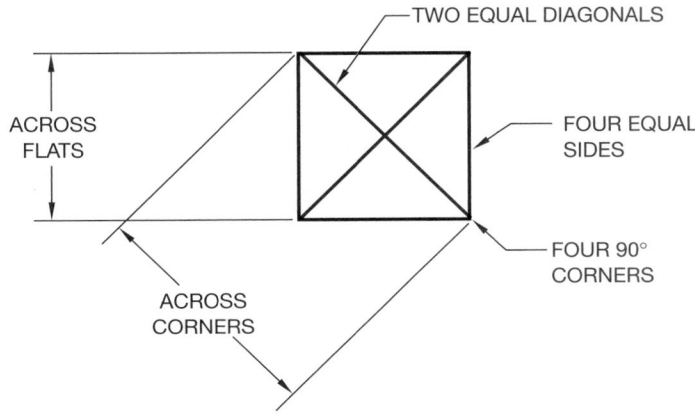

FIGURE 6.61 ■ Elements of a square.

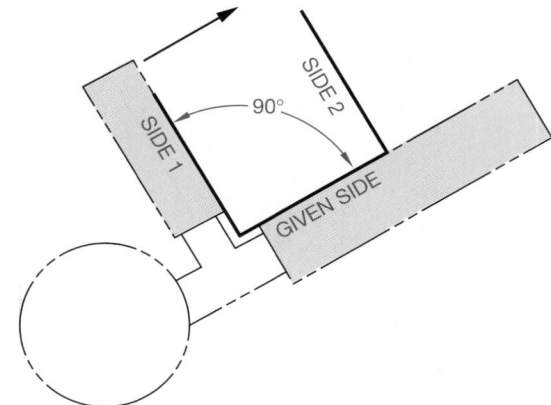

FIGURE 6.62 ■ Step 1, use the drafting machine to establish 90° angles and sides.

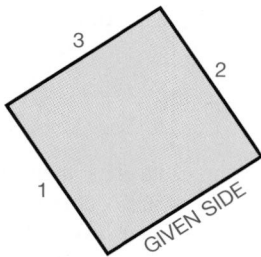

FIGURE 6.63 ■ Step 2, complete the square by drawing side 3.

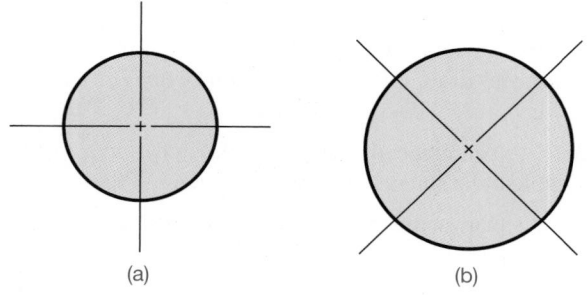

FIGURE 6.64 ■ Step 1, draw a circle to construct a square with flats or corners given.

the distance across the corners, Figure 6.64b. Notice that the position of the centers dictates the position of the squares.

Step 2. Draw 45° lines tangent to the circle and extending to the centerlines as seen in Figure 6.65a. Draw lines inside the circle connecting the intersections of the centerlines and the circle as seen in Figure 6.65b.

Constructing a Regular Hexagon

Hexagons are six-sided polygons. Each side of a regular hexagon is equal and its interior angles are 120° as shown in Figure 6.66. Hexagons are commonly used as the shape for the heads of bolts and for nuts. Generally the dimension given for the size of a hexa-

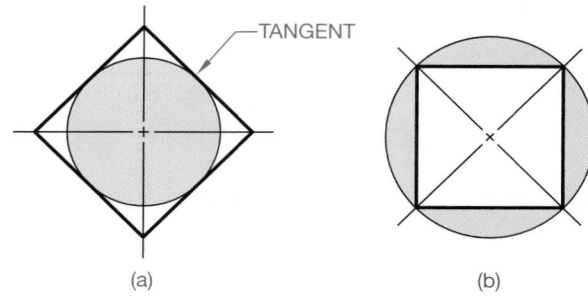

FIGURE 6.65 ■ Step 2, a circumscribed square at (a) and an inscribed square at (b).

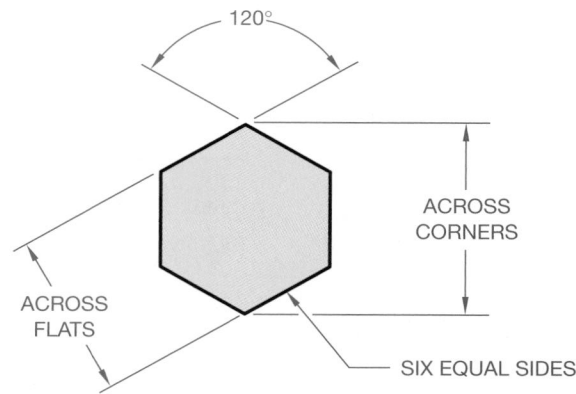

FIGURE 6.66 ■ Elements of a regular hexagon.

gon is the distance across the flats. The distances across the flats and across the corners are shown in Figure 6.66.

There are two easy geometric construction methods to draw a hexagon. One gives the distance across the flats; the other gives the distance across the corners. When practical, use a hexagon template or a special hex-head bolt, screw, or hex nut template.

Across the Flats

Given the distance across the flats, do the following to construct a hexagon:

Step 1. Lightly draw a circle with the given distance as the diameter. (See Figure 6.67.)

Step 2. Set your drafting machine at 30° from horizontal or use a 30–60 triangle. Then lightly draw lines 1 and 2 tangent to the circle as seen in Figure 6.67.

Step 3. Now set the angle at 30° the other way from horizontal and draw lines 3 and 4 tangent to the circle as shown in Figure 6.68.

Step 4. Draw lines 5 and 6 vertical and tangent to the circle and darken the six lines to complete the required hexagon as in Figure 6.69.

Across the Corners

Given the distance across the corners of a hexagon, use the following construction steps:

Step 1. Lightly draw centerlines in the desired location of the hexagon center. Then lightly draw a circle with the diam-

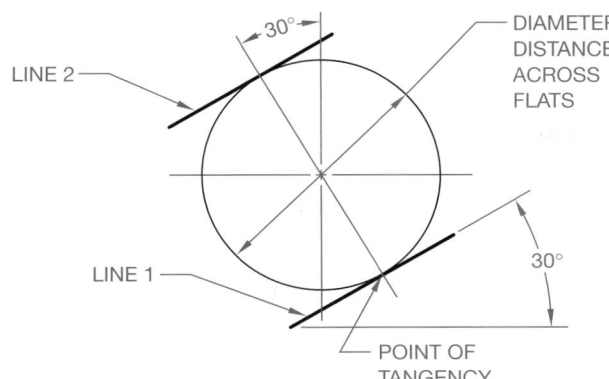

FIGURE 6.67 ■ Steps 1 and 2, constructing a hexagon.

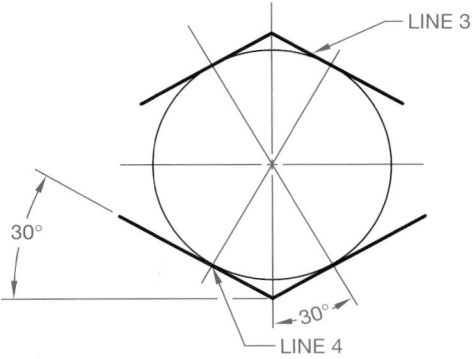

FIGURE 6.68 ■ Step 3.

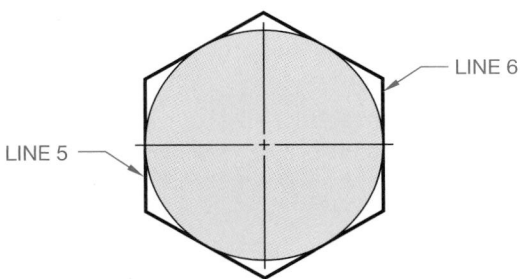

FIGURE 6.69 ■ Step 4, complete the hexagon.

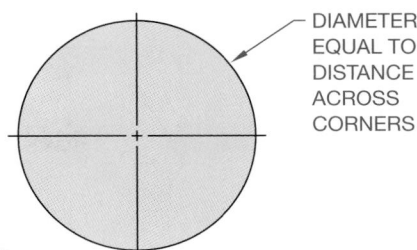

FIGURE 6.70 ■ Step 1, construct a hexagon given the distance across the corners.

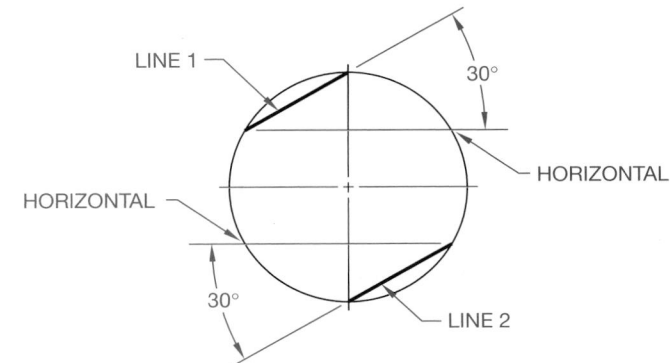

FIGURE 6.71 ■ Step 2.

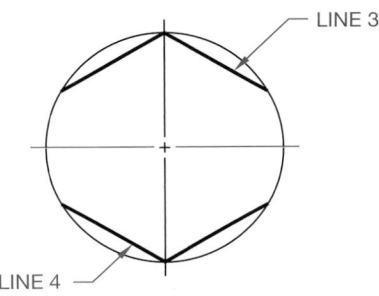

FIGURE 6.72 ■ Step 3.

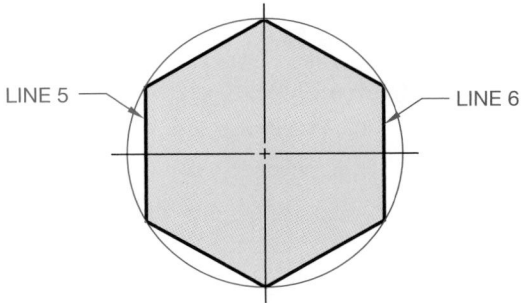

FIGURE 6.73 ■ Step 4, complete the hexagon.

eter equal to the given distance across the corners. Use construction lines as shown in Figure 6.70.

Step 2. From the location where the vertical centerline touches the circle, draw two lines, 30° from horizontal, inside the circle as shown in Figure 6.71.

Step 3. Draw lines 3 and 4 at 30° from horizontal in the other direction, as seen in Figure 6.72.

Step 4. Draw vertical lines 5 and 6 and darken the object lines to create the required hexagon shown in Figure 6.73.

The position of a hexagon may be rotated by establishing the six sides in an alternate relationship to the circle centerlines, as shown in Figure 6.74. Be sure that the included angles are 120° and each side is equal in length. Use a template when you can to draw hexagons and other shapes. Many different geometric templates are available for your use. Templates save time.

Constructing a Regular Octagon

The octagon may be drawn using the same principle applied to the construction of a hexagon. A *regular octagon* has eight equal sides. If you use a 45° line as shown in Figure 6.75, you can easily draw an octagon about the given circle with a diameter equal to the distance across the flats.

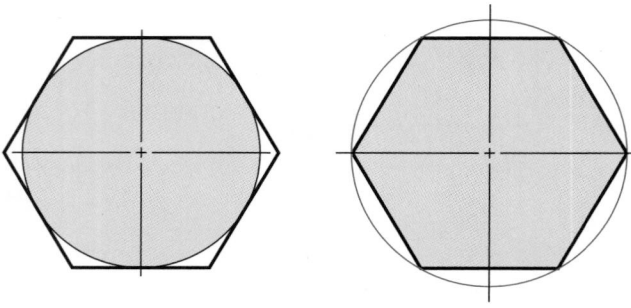

FIGURE 6.74 ■ Alternate hexagon positions.

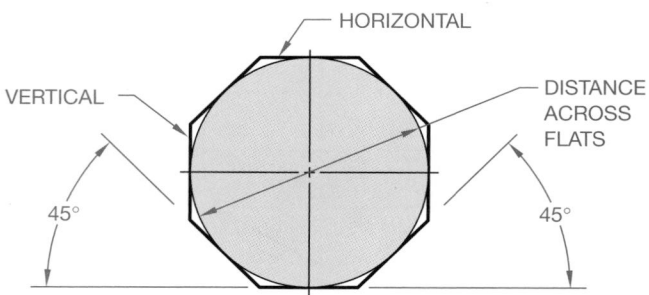

FIGURE 6.75 ■ Elements of an octagon.

Polygons

A polygon is any closed plane geometric figure with three or more sides or angles. Hexagons and octagons are polygons. You can draw any polygon (if you know the distance across the flats and the number of sides) by the same method used to draw the hexagon and octagon across the flats.

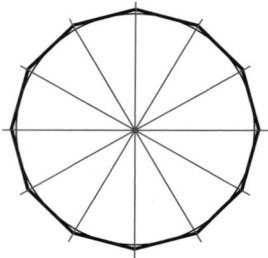

FIGURE 6.76 ■ Step 1, divide the circle into 12 parts; Step 2, connect the 12 radial lines.

There are 360° in a circle; so, if you need to draw a 12-sided polygon, divide 360° by 12 (360 ÷ 12 = 30°) to determine the central angle of each side. Since there are 30° between each side of a 12-sided polygon, divide a circle with a diameter equal to the distance across the flats into equal 30° parts as shown in Figure 6.76. Then connect the 12 radial lines with line segments that are tangent to each arc segment, as shown in Figure 6.76.

Drawing a Pentagon

A pentagon can be tricky to draw, because the angles used are not as convenient as those used with a hexagon or octagon. A pentagon has five sides. You can use a method similar to the previous examples by drawing a circle equal to the distance across the flats or corners depending on the information available. Next, divide the circle into five equal parts, with a line from the center to the circle for each part. Each part is 72° (360° ÷ 5 = 72°). Now, connect the points of tangency with the dividing lines for a pentagon that is outside of the circle, or connect where the lines intersect the circle for a pentagon inside the circle.

DRAWING GEOMETRIC SHAPES

The process used for drawing geometric shapes with CADD normally involves three steps:

1. Select the command.
2. Locate the feature.
3. Provide the needed information to draw the feature.

Circle

A circle, for example, may be drawn several different ways. After selecting the CIRCLE command, the information that you have about the circle determines the method used to draw it. The methods are shown in Figure 6.77, (a) through (d). The first method (a) requires that the center (1) and one point on the circle (2 and 3) be picked before the circle can be drawn. The second, example (b), requires that two opposite points (1 and 2) on the circumference (the diameter) be selected. The third example (c), requires that three points be picked on the circumference. The final method (d) requires

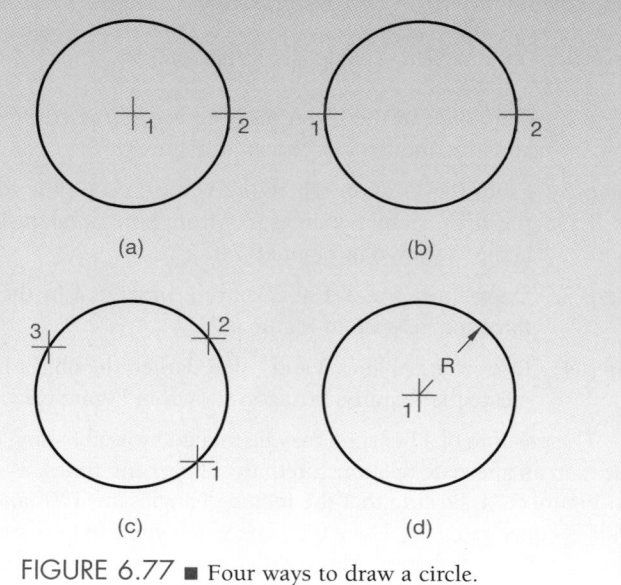

FIGURE 6.77 ■ Four ways to draw a circle.

that the center point be selected and the radius entered as a number selected from the keypad on the menu tablet, or typed on the keyboard.

Ellipse

Ellipses are quick and easy to draw on a CADD system. Most programs have a command such as ELLIPSE. In this command you are normally asked for the major and minor diameters, and the computer automatically draws an ellipse with the information.

Arc

Arcs are constructed in the same manner as circles. But arcs are only portions of circles and therefore the two end points of the arc must be established. The methods of constructing an arc are shown in Figure 6.78, (a) through (d). The fourth method (d) has two variations, each requiring numerical input. The first variation requires point number 1 (beginning point) to be selected, as well as point number 2, a location on the arc. Then a numerical value for the radius and length of the arc must be entered from the menu keypad or from the keyboard. The second variation requires the same two points, but instead of entering the radius, the center point location of the arc is entered at the keyboard.

Tangencies

Tangencies are easy to draw with the computer. Most CADD programs have a command or option such as TANGENT, which allows you to pick a circle or arc and automatically draw a line or another circle or arc tangent to the selected object.

Polygon

Any desired regular polygon is easy to draw with CADD. The command is usually POLYGON. The command prompts often begin by asking for the number of sides of the polygon. Then the command continues by asking for the center of the polygon. After you select the center, you are asked if the polygon is to be inscribed in a circle or circumscribed about a circle. This is the same decision you made when drawing a polygon manually. Now you enter the radius of the circle, and the polygon is automatically drawn. (See Figure 6.79.) POLYGON commands may also have an option that allows you to draw the desired polygon by picking the end points of an edge as shown in Figure 6.80.

Rectangle

A rectangle is one of the easiest shapes to draw. Only two points are needed. The first is one corner of the rectangle, which can be any of the four corners. Once you pick this corner, the direction location of the second point has already been determined. The second point is the corner of the rectangle opposite the first corner that was picked. The

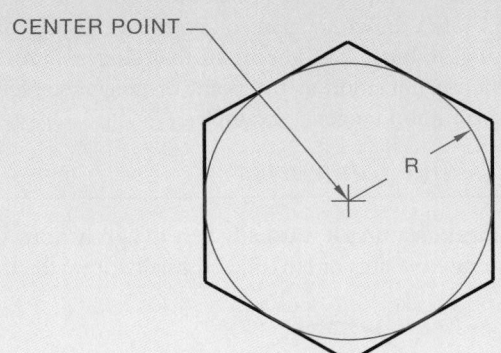

FIGURE 6.79 ■ Drawing a polygon by picking the center and entering the radius.

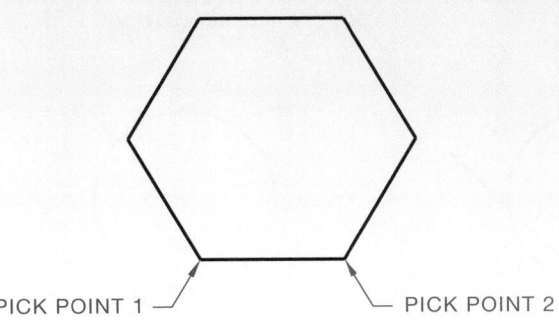

FIGURE 6.80 ■ Drawing a polygon automatically by picking the end points of a side.

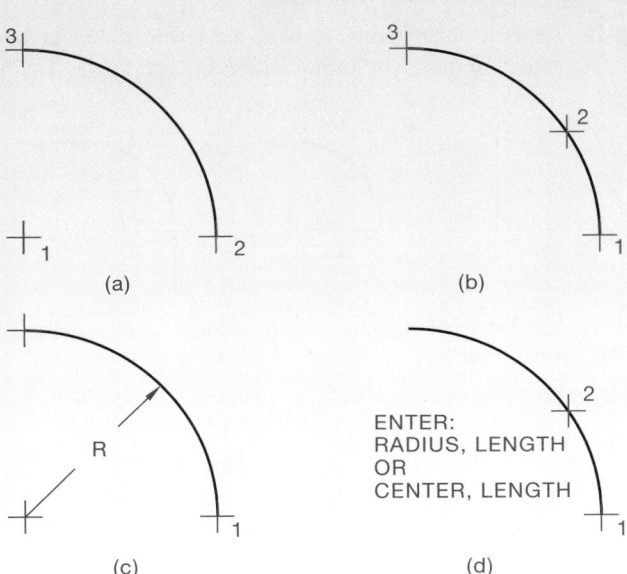

FIGURE 6.78 ■ Four ways to draw an arc.

(Continued)

CADD APPLICATIONS *(continued)*

computer then connects these two points with the straight lines needed to form the rectangle. Figure 6.81 illustrates this process. Rectangles and boxes can be drawn using the LINE command, but it takes longer.

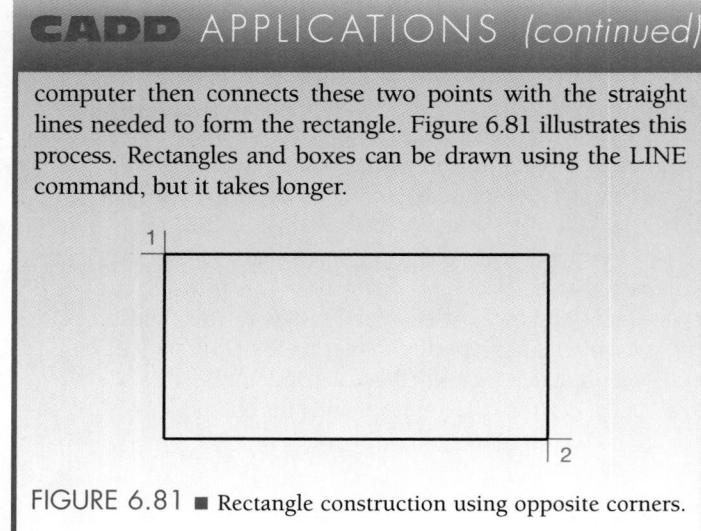

FIGURE 6.81 ■ Rectangle construction using opposite corners.

CONSTRUCTING CIRCLES AND TANGENCIES

To be tangent to a circle or arc, a line must touch the circle or arc at only one point, and a line drawn from the center of the circle or arc must be perpendicular to the tangent line at the point of tangency. (See Figure 6.82.) Lines drawn tangent to circles or arcs are common in engineering drafting. It is important that tangent lines be drawn with a smooth transition at the point of tangency. Figure 6.83 shows some common tangency examples.

Drawing Arcs and Tangencies

When drafting tangent features it is usually best to lightly draw the centerlines, then draw the arcs or circles, and finally draw the tan-

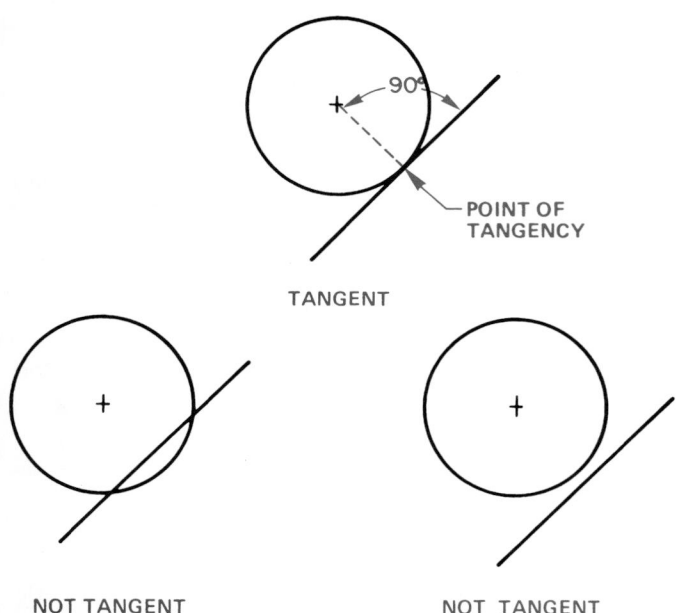

FIGURE 6.82 ■ Tangency.

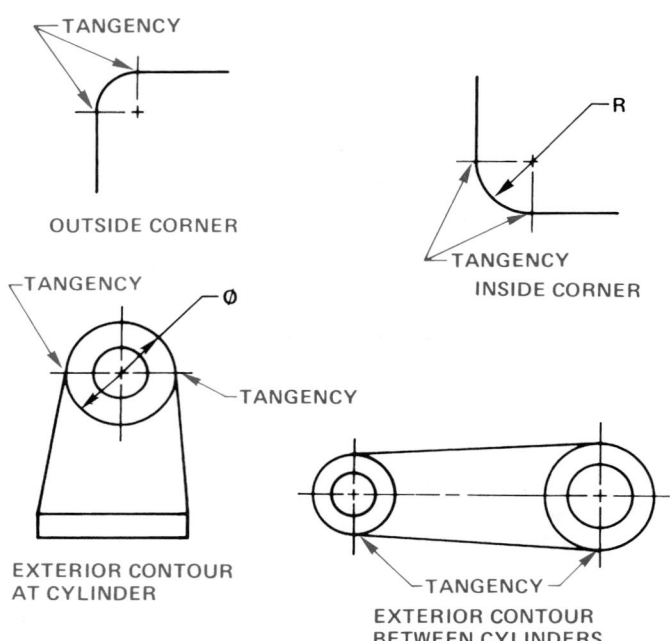

FIGURE 6.83 ■ Common tangencies.

gent lines. It is easier to draw a line tangent to a circle or arc than it is to draw a circle or arc tangent to a line. (See Figure 6.84.) When possible, use a template to draw the arcs and circles. Keep in mind that template circles are calibrated in diameters while arcs have radius dimensions. So, for example, a .50-in. radius requires using a 1.00-in. diameter circle. The combination of templates and a drafting machine provides the easiest and fastest results. Use a compass only if necessary. Figure 6.85 shows the relationship between a well-drawn and a poorly drawn tangency.

Draw an arc tangent to a given acute or obtuse angle in much the same way as previously shown for lines and circles. Given the acute and obtuse angles in Figure 6.86, draw an arc with a 12 mm radius tangent to the sides of each angle.

Step 1. Select a circle template with a 24 mm diameter circle. Move the template into position at each angle, leaving a

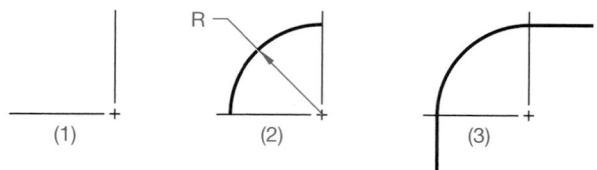

FIGURE 6.84 ■ Drawing lines tangent to an arc and an arc at a 90° corner.

FIGURE 6.85 ■ Good and poor tangency examples.

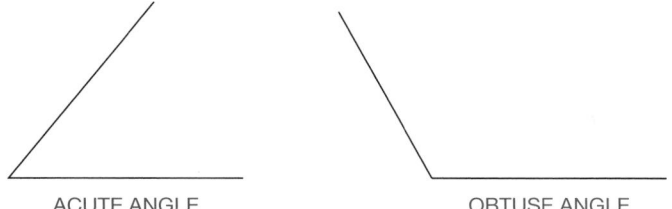

FIGURE 6.86 ■ Construct an arc tangent to given angles.

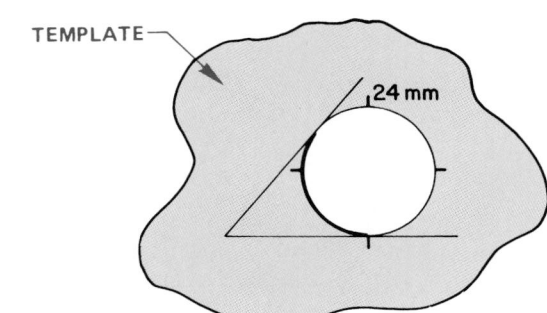

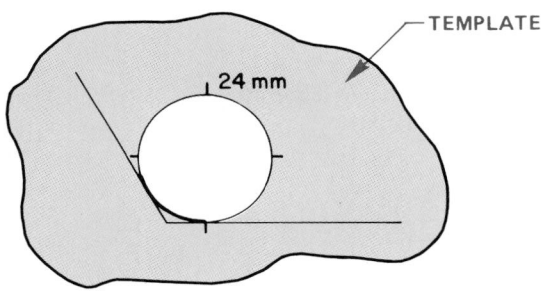

FIGURE 6.87 ■ Step 1, draw a tangent arc with a template.

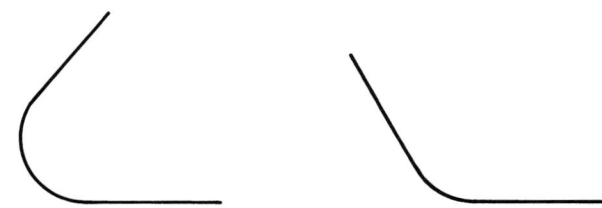

FIGURE 6.88 ■ Step 2, complete tangent arcs.

small space for the lead or pen point thickness, and draw the arc as shown in Figure 6.87.

Step 2. Remove the template and draw the sides of each angle to the points of tangency of the arcs. (See Figure 6.88.)

If the arc center is required for dimensioning purposes, use the following procedure:

Step 1. Establish the arc centers by lightly drawing lines parallel to the given sides at a distance from the sides equal to the radius of the arc. In this example the radius is 12 mm. (See Figure 6.89.)

Step 2. Set the compass at the given radius, 12 mm, and draw two arcs tangent to the given sides. Connect the sides to the points of the tangency of the arcs as shown in Figure 6.90.

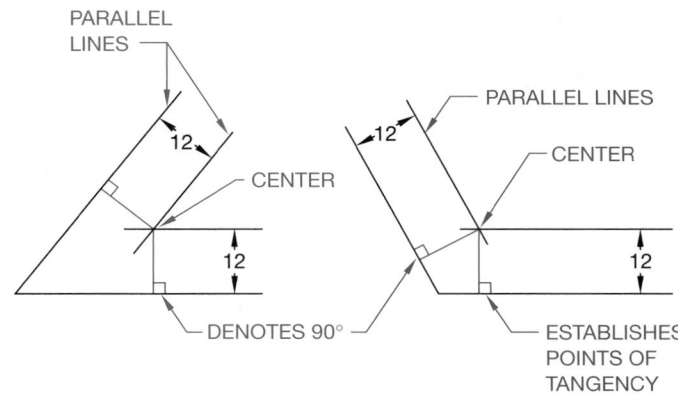

FIGURE 6.89 ■ Step 1, drawing tangent arcs with a compass. This procedure also works for arcs at 90° corners.

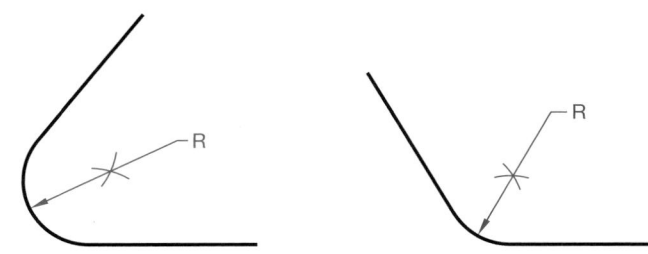

FIGURE 6.90 ■ Step 2, complete tangent arcs.

Arc Tangent to a Given Line and a Circle

The important point to remember is that the distance from the point of tangency to the center of the tangent arc is equal to the radius of the tangent arc. Given the machine part in Figure 6.91, draw an arc tangent to the cylinder and base with a 24 mm radius.

Step 1. Draw a line parallel to the base at a distance equal to the radius of the arc, 24 mm. Then draw an arc that intersects the line drawn with a radius equal to the given arc plus the radius of the cylinder (24 + 11 = 35 mm). (See Figure 6.92.)

Step 2. Use a compass or a template to draw the required radius from the center point established in step 1. (See Figure 6.93.)

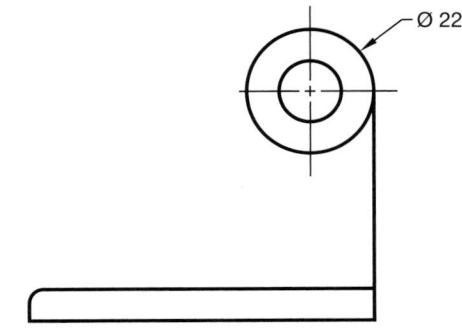

FIGURE 6.91 ■ Construct an arc tangent to a given line and circle.

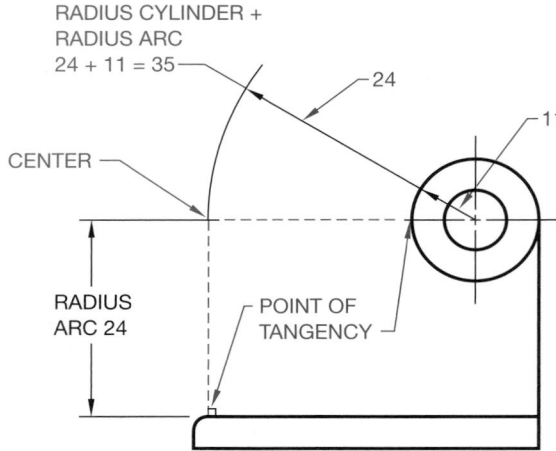

FIGURE 6.92 ■ Step 1.

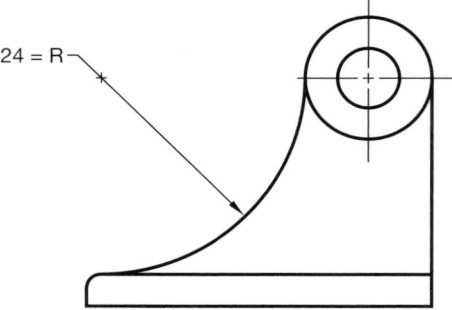

FIGURE 6.93 ■ Step 2.

The previous example demonstrated an internal arc tangent to a circle or arc and a line. A similar procedure is used for an external arc tangent to a given circle and line as shown in Figure 6.94. In this case the radius of the cylinder (circle) is subtracted from the arc radius to get the center point.

The following examples, Figures 6.95 through 6.97, are situations that commonly occur on machine parts where an arc may be tangent to given cylindrical or arc shapes. Keep in mind that the methods used to draw the tangents are the same as just described. The key is that the center of the required arc is always placed at a distance equal to its radius from the points of tangencies. To achieve this, you will have to either add or subtract the given radius and the required radius, depending on the situation. Use a template to draw circles and arcs when possible.

Ogee Curve

An S curve, commonly called an *ogee* curve, occurs in situations where a smooth contour is needed between two offset features. The following steps show how to draw an S curve with equal radii.

Step 1. Given the offset points A and B in Figure 6.98, draw a line between points A and B and bisect that line to find point C.

Step 2. Draw the perpendicular bisector of lines AC and CB. From point A, draw a line perpendicular to the lower line

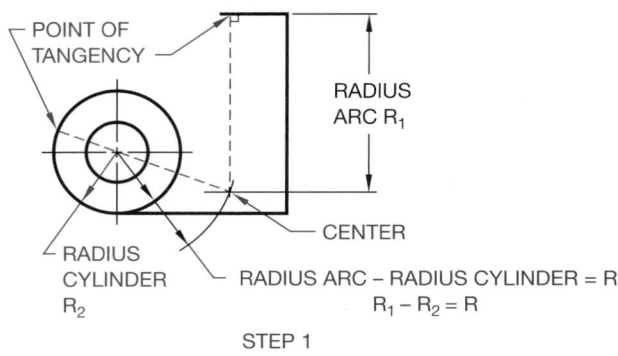

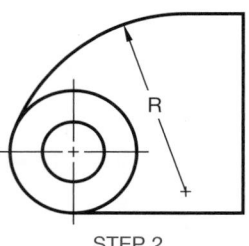

FIGURE 6.94 ■ Construct an external arc tangent to a line and cylinder.

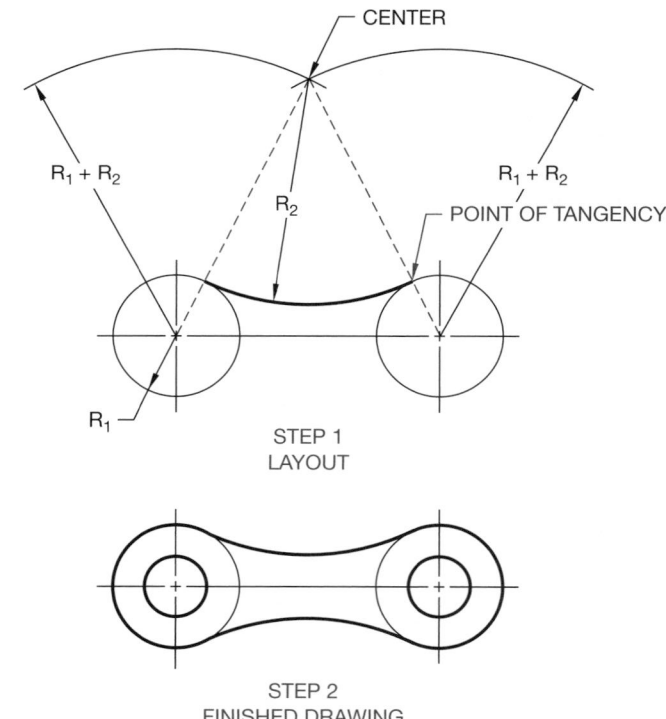

FIGURE 6.95 ■ Chain link.

that intersects the AC bisector at X. From point B draw a line perpendicular to the upper line that intersects the CB bisector at Y. (See Figure 6.99.)

Step 3. With X and Y as the centers, draw a radius from A to C and a radius from B to C as shown in Figure 6.100.

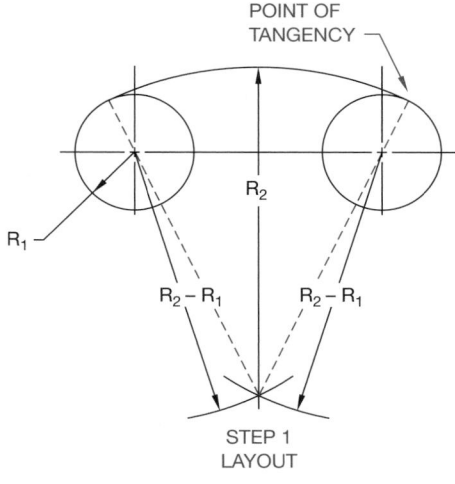

POINT OF
TANGENCY

R_1

R_2

$R_2 - R_1$ $R_2 - R_1$

STEP 1
LAYOUT

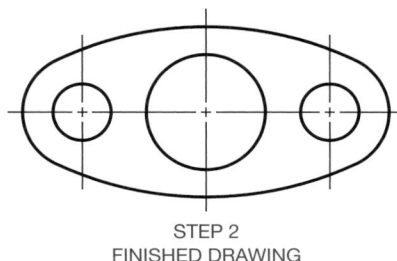

STEP 2
FINISHED DRAWING

FIGURE 6.96 ■ Gasket.

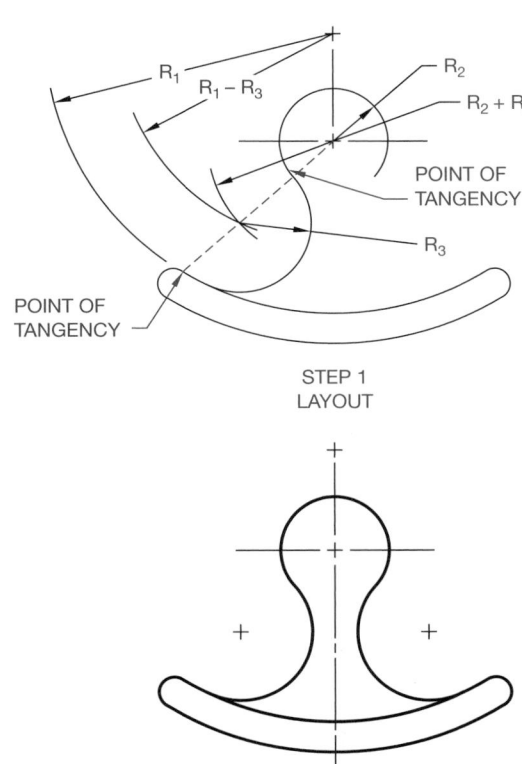

R_1

$R_1 - R_3$

R_2

$R_2 + R_3$

POINT OF
TANGENCY

R_3

POINT OF
TANGENCY

STEP 1
LAYOUT

STEP 2
FINISHED DRAWING

FIGURE 6.97 ■ Hammer head.

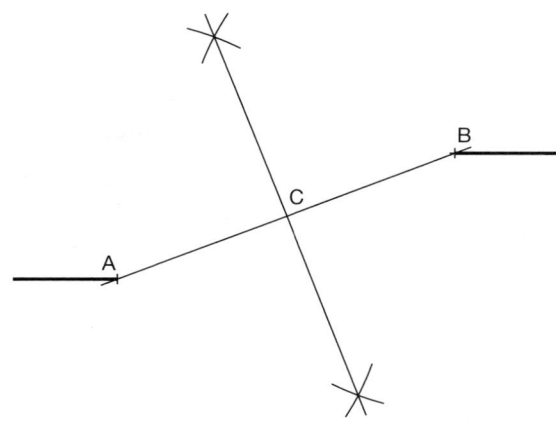

B

C

A

FIGURE 6.98 ■ Step 1, construct an S curve.

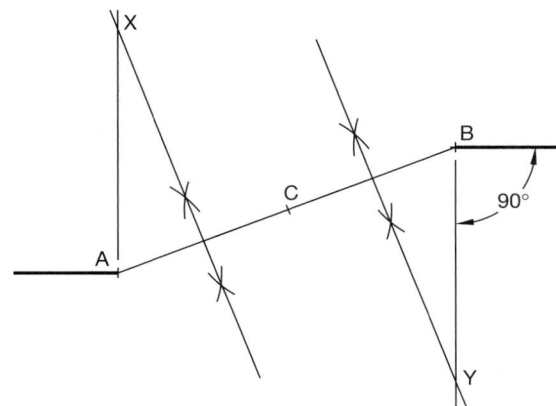

X

B

C

90°

A

Y

FIGURE 6.99 ■ Step 2, construct an S curve.

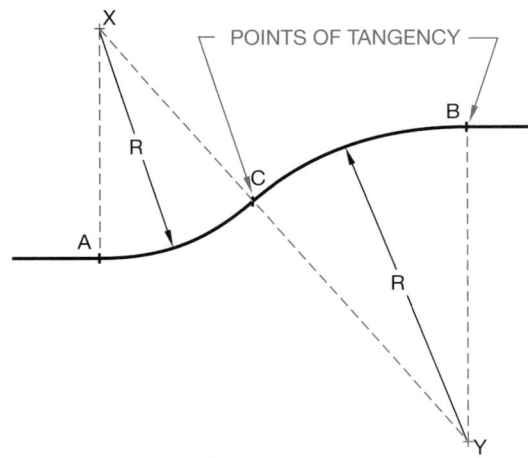

X

POINTS OF TANGENCY

B

R

C

A

R

Y

FIGURE 6.100 ■ Step 3, construct an S curve.

Step 4. If the S curve line, AB, represents the centerline of the part, then develop the width of the part parallel to the centerline using concentric arcs to complete the drawing. (See Figure 6.101.)

When an S curve has unequal radii, the procedure is similar to the previous example, as shown in Figure 6.102.

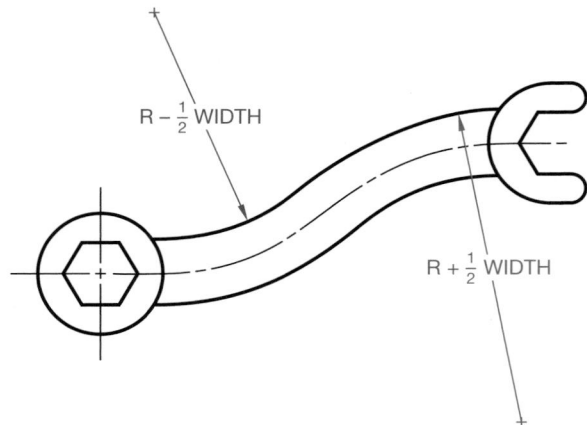

FIGURE 6.101 ■ Step 4, a wrench with a complete S curve.

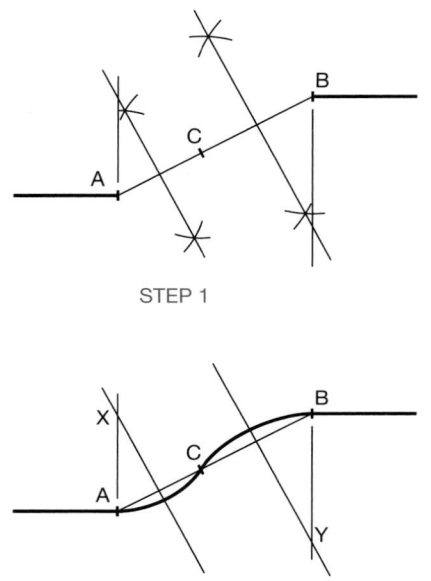

STEP 1

STEP 2

FIGURE 6.102 ■ Constructing an S curve with unequal radii.

CONSTRUCTING AN ELLIPSE

When a circle is viewed at an angle, an ellipse is observed. The angular relationship of an ellipse to a circle is shown in Figure 6.103. If a surface with a through hole is inclined 45°, the representation is a 45° ellipse, as shown in Figure 6.104. An ellipse has a major diameter and a minor diameter. In Figure 6.104 the minor diameter is established by projecting the top and bottom edge from the slanted surface. The major diameter, in this case, is equal to the diameter of the circle.

If possible, always use a template to draw elliptical shapes. Ellipse templates are available from 10° to 85°. The ellipse shown in Figure 6.104 could have been drawn by projecting the centerlines and then lining up the center marks of a 45° ellipse template. (See Figure 6.105.)

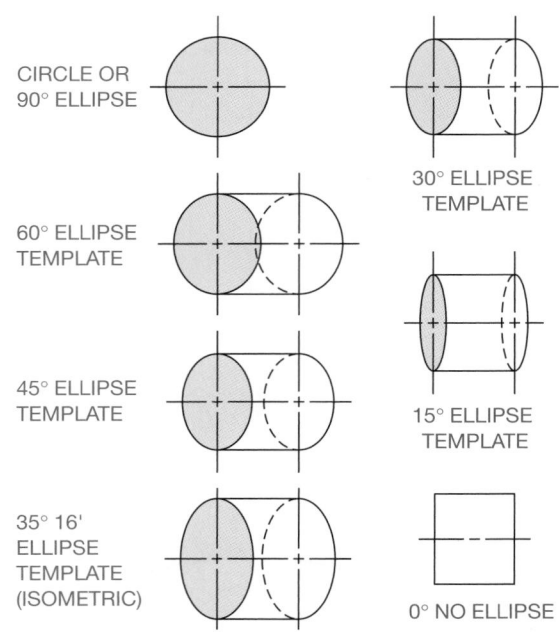

FIGURE 6.103 ■ Ellipses are established by their relationship to a circle turned at various angles.

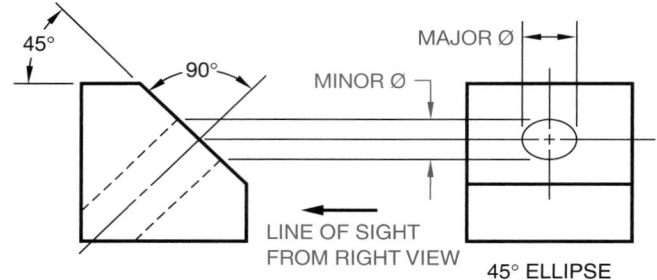

FIGURE 6.104 ■ Elliptical view.

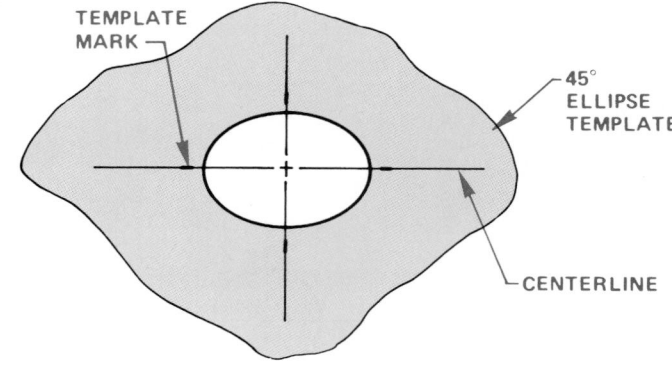

FIGURE 6.105 ■ Drawing an ellipse with a template.

Approximating the Ellipse

If the elliptical shape does not exactly fit the available template sizes, use one that is close. How close depends on your company standards. Most companies would rather have you draw a close representation than take the time to lay out an ellipse geometrically.

CADD APPLICATIONS

Drawing objects with tangencies using manual drafting techniques is often tricky. The feature must touch at exactly the point of tangency. While the methods discussed in this chapter allow you to do this with consistent success, the use of CADD systems makes this task automatic. For example, AutoCAD has a TTR (Tangent, Tangent, Radius) option in the CIRCLE command. This allows you to automatically draw a circle tangent to lines, circles, or arcs simply by picking the objects and specifying the desired radius of the circle as shown in Figure 6.106.

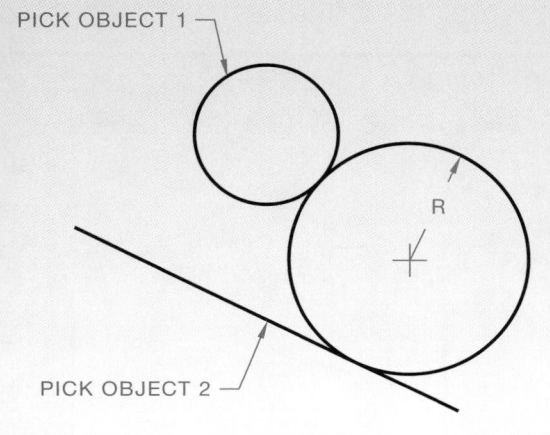

PICK OBJECT 1

R

PICK OBJECT 2

FIGURE 6.106 ■ Automatically drawing tangencies with CADD.

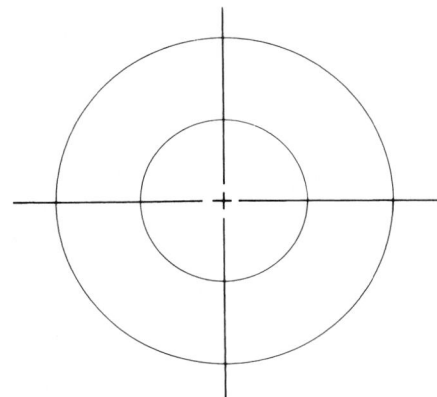

FIGURE 6.107 ■ Step 1, constructing an ellipse, concentric circle method.

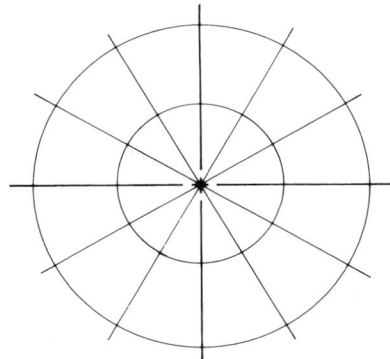

FIGURE 6.108 ■ Step 2, concentric circle method.

Using the Concentric Circle Construction Method

When templates are not available or are not close enough for the ellipse that must be drawn, several methods can be used to draw elliptical shapes. One of the most practical is the concentric circle method described in the following steps:

Step 1. Use construction lines to draw two concentric circles, one with a diameter equal to the minor ellipse diameter and the other with a diameter equal to the major ellipse diameter. (See Figure 6.107.)

Step 2. Use the drafting machine or triangles to divide the circles into at least 12 equal parts; 30° and 60° each way from horizontal. More parts give better accuracy, but also consume more time. (See Figure 6.108.)

Step 3. From the points of intersection of each line at the circumference of the circles, draw horizontal lines from the minor diameter and vertical intersecting lines from the major diameter, essentially creating a series of right triangles, although the intersection points only are needed. (See Figure 6.109.)

Step 4. Using an irregular curve carefully connect the points created in step 3 to complete the ellipse as shown in Figure 6.109.

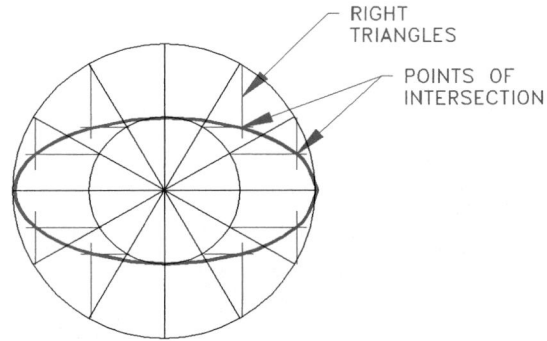

RIGHT TRIANGLES

POINTS OF INTERSECTION

FIGURE 6.109 ■ Steps 3 and 4, complete ellipse.

WEB SITE RESEARCH

The following Web site can provide you with additional information:

www.industrialpress.com/handbook/trig.html—Industrial Press. On-line trigonometry tables.

PROFESSIONAL PERSPECTIVE

Geometric constructions are found in nearly every engineering drafting assignment. Some geometric constructions are as simple as a radius corner while others are very complex. As a professional drafter, you should be able to quickly identify the type of geometric construction involved in the problem and solve it with one of the techniques you have learned. When you solve these problems, accuracy and carefully drawn construction lines are critical. Use the best tools available. For example, a professional drafter hardly ever draws an ellipse using construction methods; he or she uses the large variety of ellipse templates available.

Always use a template if you can, although, if all else fails, you do know how to construct the ellipse. Even though the geometric construction methods are presented for the manual drafter, the principles are the same for the CADD drafter. You should always refer to these principles and techniques to ensure that your CADD application has performed the task correctly. You will find with CADD that some of the very time-consuming constructions, such as dividing a space into equal parts or drawing a pentagon, are not only extremely fast but also are almost perfectly accurate.

MATH APPLICATION

INTERIOR ANGLES OF REGULAR POLYGONS

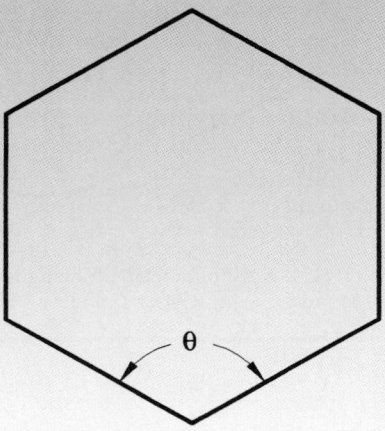

FIGURE 6.110 ■ Regular hexagon.

CHAPTER 6

Geometric Construction Test

DIRECTIONS

Answer the questions with short complete statements or drawings as needed.

QUESTIONS

1. What is known to be true about the opposite angles created by intersecting lines?
2. Define parallel lines.
3. Define perpendicular.
4. How many degrees are there in a circle?
5. How are angles sized?
6. Name four basic types of angles.
7. Define triangle.
8. Show how an angle is named.
9. Name and show an example of three types of triangles.
10. Show two methods of naming triangles.
11. Name three types of quadrilaterals.
12. What name is given to the distance from the center to the circumference of a circle?
13. What is the name of the distance that goes across a circle through its center?

14. What is a circle called that touches the flats of a polygon?
15. What is a circle called that touches the corners of a polygon?
16. Show and label an example of concentric and eccentric circles.
17. Show an example of a line tangent to a circle and two circles tangent to each other.
18. When should templates be used to make circles and arcs?
19. When should templates be used to make elliptical shapes?
20. Which of the following methods could be used to most easily produce circles and arcs: compass, templates, computer-aided drafting.

CHAPTER 6 — Geometric Construction Problems

DIRECTIONS

1. Problems may be completed using manual or computer-aided drafting depending on your course guidelines.
2. Use very lightly drawn lines (construction lines) for all preliminary work and darken in only the completed product. *Do not* erase the construction lines when you complete the problems. This will help your instructor observe your drawing technique.
3. Geometric construction problems are presented as written instructions, drawings, or a combination of both. If the problems are presented as drawings without dimensions, transfer the drawing from the text, using dividers and scales, to your drawing sheet. Draw dimensioned problems full scale (1:1) unless otherwise specified. Draw all object, hidden, and centerlines. Do not draw dimensions.

PROBLEM 6-1   Draw two tangent circles with their centers on a horizontal line. Circle 1 has a 64 mm diameter and circle 2 has a 50 mm diameter.

PROBLEM 6-2   Make a perpendicular bisector of a horizontal line that is 79 mm long.

PROBLEM 6-3   Make an angle 48° with one side vertical. Bisect the angle.

PROBLEM 6-4   Divide a 96 mm line into 7 equal spaces.

PROBLEM 6-5   Draw two parallel horizontal lines each 50 mm long with one 44 mm above the other. Divide the space between the lines into 8 equal spaces.

PROBLEM 6-6   Transfer triangle a,b,c to a new position with side c 45° from horizontal.

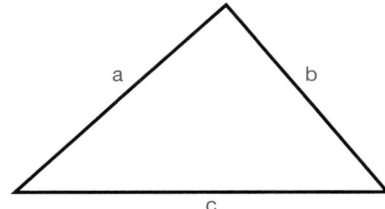

PROBLEM 6-7   Make a right triangle with one side 35 mm long and the other side 50 mm long.

PROBLEM 6-8   Make a right triangle with the hypotenuse 75 mm long and one side 50 mm long.

PROBLEM 6-9   Make an equilateral triangle with 65 mm sides.

PROBLEM 6-10   Make a square with a distance of 50 mm across the flats.

PROBLEM 6-11   Draw a hexagon with a distance of 2.5 in. across the flats.

PROBLEM 6-12   Draw a hexagon with a distance of 2.5 in. across the corners.

PROBLEM 6-13   Draw an octagon with a distance of 2.5 in. across the flats.

PROBLEM 6-14   Draw a rectangle 50 mm × 75 mm with 12 mm radius tangent corners.

PROBLEM 6-15   Draw two separate angles, one a 30° acute angle and the other a 120° obtuse angle. Then draw a .5-in. radius arc tangent to the sides of each angle.

PROBLEM 6-16   Given the following incomplete part, draw an inside arc with a 2.5 in. radius tangent to ⌀A and line B.

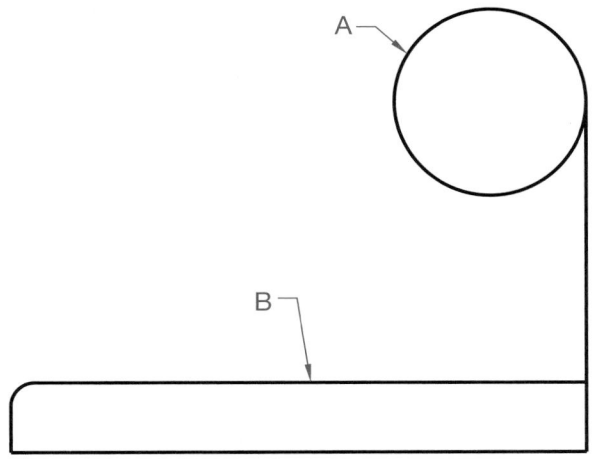

PROBLEM 6-17   Draw two circles with 25 mm diameters and 38 mm between centers on horizontal. On the upper side between the two circles draw an inside radius 44 mm. On the lower side draw an outside radius 76 mm.

PROBLEM 6-18 Draw an S curve with equal radii between points A and B given below:

PROBLEM 6-19 Draw two lines, one on each side of the ogee curve drawn in problem 4hr. –18 and .5 in. away.

PROBLEM 6-20 Make an ellipse with a 38 mm minor diameter and a 50 mm major diameter.

PROBLEM 6-21 Corner arcs (in.)
Part Name: Plate Spacer
Material: .250-thick Aluminum

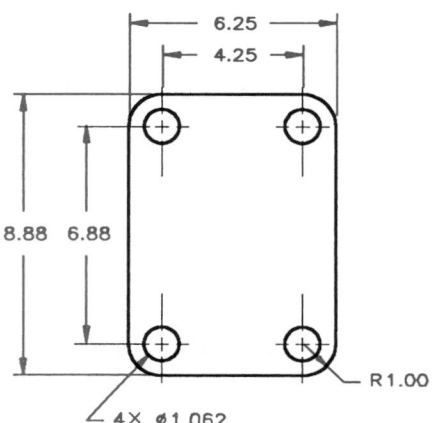

PROBLEM 6-22 Circles (in.)
Part Name: Flange
Material: .50-thick Cast Iron (CI)

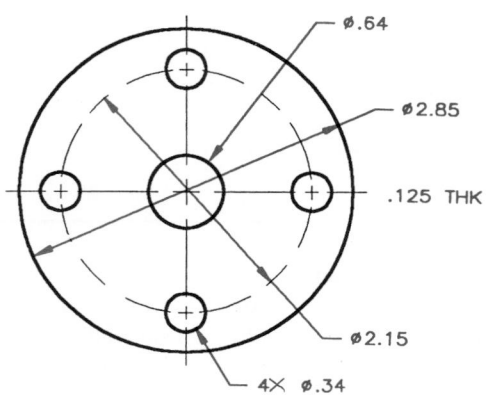

PROBLEM 6-23 Hexagon (in.)
Part Name: Sleeve
Material: Bronze
Draw the hexagon with circles view only.

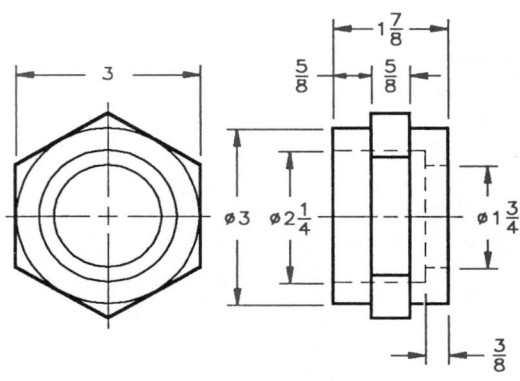

Note: This engineer's sketch shows dimensions to hidden features, which is not a standard practice.

PROBLEM 6-24 Tangencies (metric)
Part Name: Coupler
Material: SAE 1040

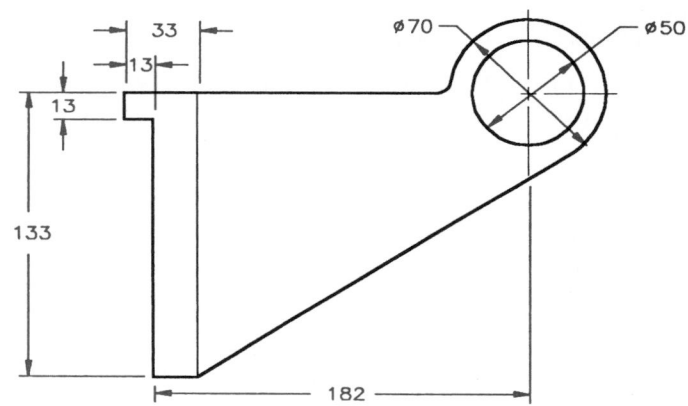

PROBLEM 6-25 Tangencies (in.)
Part Name: Hanger
Material: Mild Steel (MS)

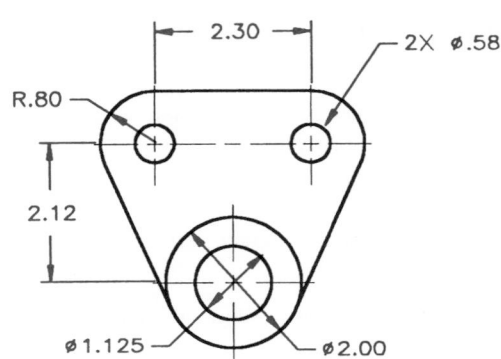

PROBLEM 6-26 Tangencies (in.)
Part Name: Bracket
Material: Aluminum

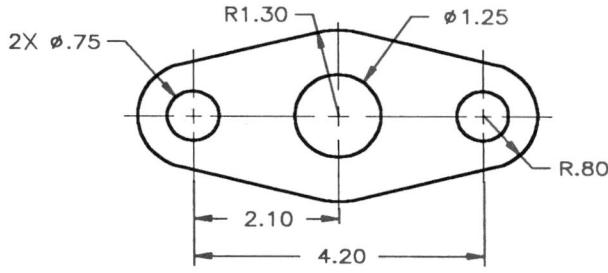

PROBLEM 6-28 Circle and arc object lines and centerlines (metric)
Part Name: Gasket
Material: .06-thick Cork

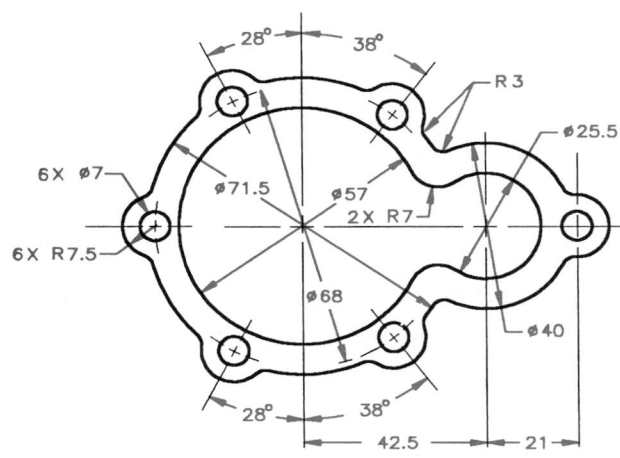

PROBLEM 6-27 Arcs and tangencies (in.)
Part Name: Gusset
Material: .250-thick Aluminum

Specific instructions: sizes are given as limit dimensions. Limit dimensioning is discussed in Chapter 8. For now, when a dimension is shown as .531/.468, for example, you produce the drawing using an even numeral between the two. In this case you draw using a dimension of .500. *Courtesy TEMCO.*

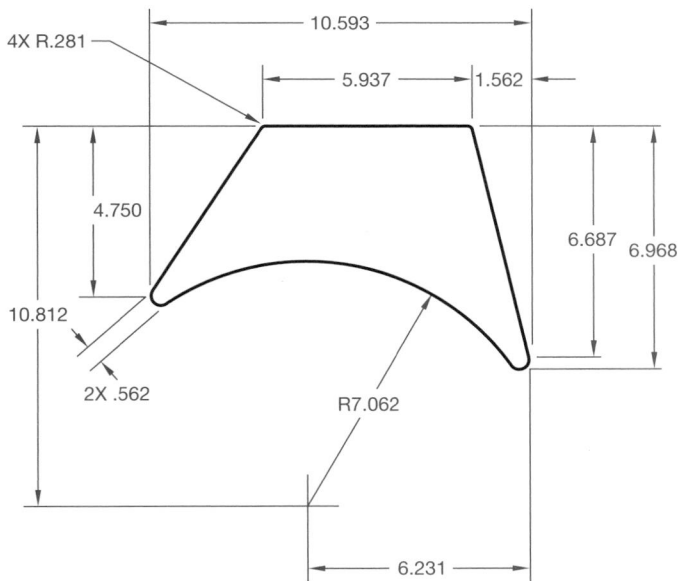

PROBLEM 6-29 Arcs and angles (in.)
Part Name: Support Brace
Material: .125-thick Copper

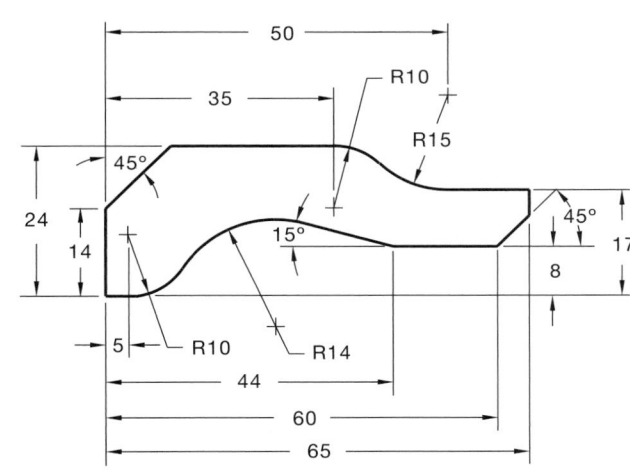

PROBLEM 6–30 Arcs and circles (in.)
Part Name: Gasket
Material: .0625-thick Bronze

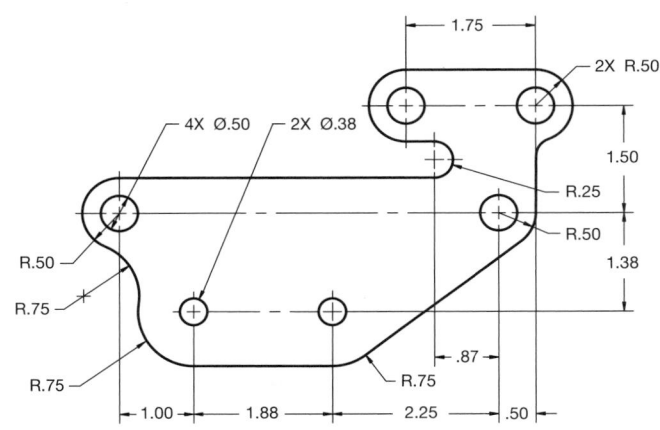

PROBLEM 6–31   Ogee curve and arcs (in.)

Part Name: Wrench

Material: .25-thick Cast Iron

Fillets and rounds R.125

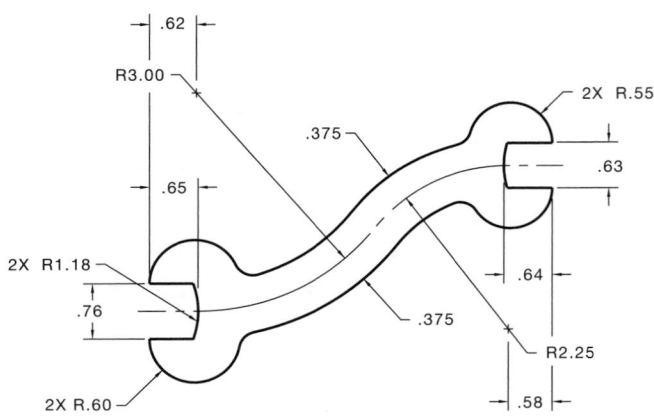

PROBLEM 6–32   Ogee curve, hexagon, and arcs (in.)

Part Name: Wrench

Material: .25-thick Cast Iron

Fillets and rounds R.125

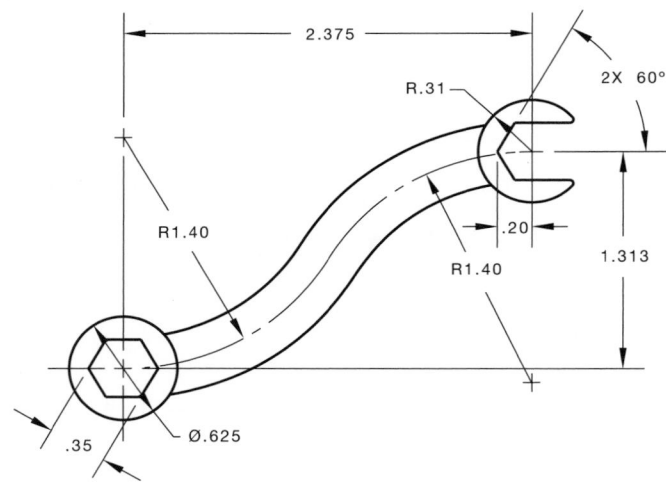

PROBLEM 6-33   Arc tangencies (in.)

Part Name: Hammer Head

Material: Cast Iron

Establish unknown dimensions to your own specifications.

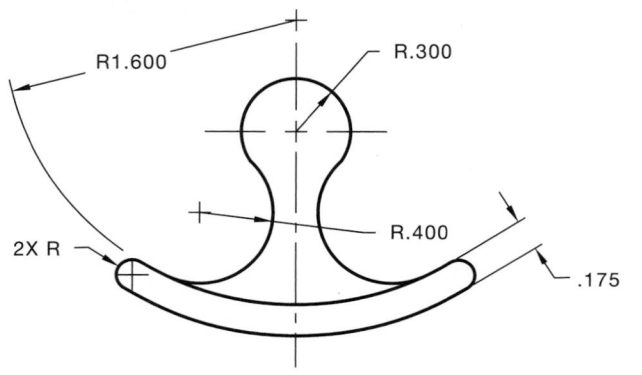

PROBLEM 6-34   Design problem

Part Name: Sailboat

Design the sailboat to your own specifications using the techniques discussed in this chapter.

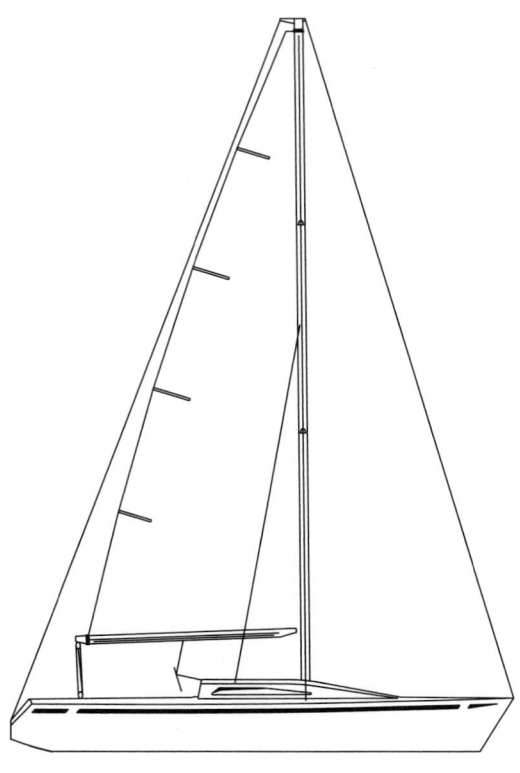

MATH PROBLEMS

For Problems 6-35 through 6-40, find the interior angles of these types of regular polygons:

PROBLEM 6-35   Octagon.

PROBLEM 6-36   Equilateral triangle.

PROBLEM 6-37   Pentagon (5 sides).

PROBLEM 6-38   Decagon (10 sides).

PROBLEM 6-39   Square.

PROBLEM 6-40   How many sides does a regular polygon have if its interior angle is 140°?

CHAPTER 7

Multiviews

OBJECTIVES

After completing this chapter, you will:

- Prepare single- and multiview drawings.
- Select appropriate views for presentation.
- Draw view enlargements.

- Establish runouts.
- Explain the difference between first- and third-angle projection.
- Prepare formal multiview drawings from an engineer's sketch and actual industrial layouts.

THE ENGINEERING DESIGN APPLICATION

The engineer has just handed you a sketch of a new part design. (See Figure 7.1.) The engineer explains that a multiview drawing is needed on the shop floor as soon as possible so a prototype can be manufactured and tested. As the engineering drafter, it is your responsibility to create a drawing that shows all necessary manufacturing data. Based on drafting guidelines, you first decide which view should be the front view. Having established the front view, you must now determine what other views are required in order to depict all of the features of the part. Using visualization techniques based on the glass box helps you to decide which views are needed. (See Figure 7.2.) Unfolding the box puts all of the views in their proper positions, so sketch the layout and complete the drawing. (See Figure 7.3.)

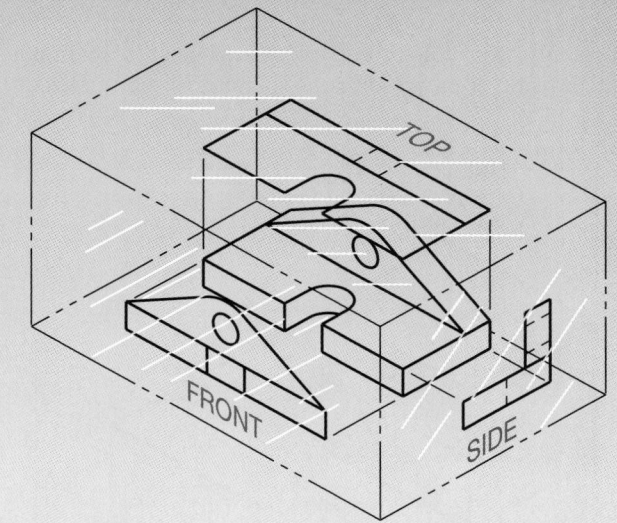

FIGURE 7.2 ■ Using the glass box principle to visualize the needed views.

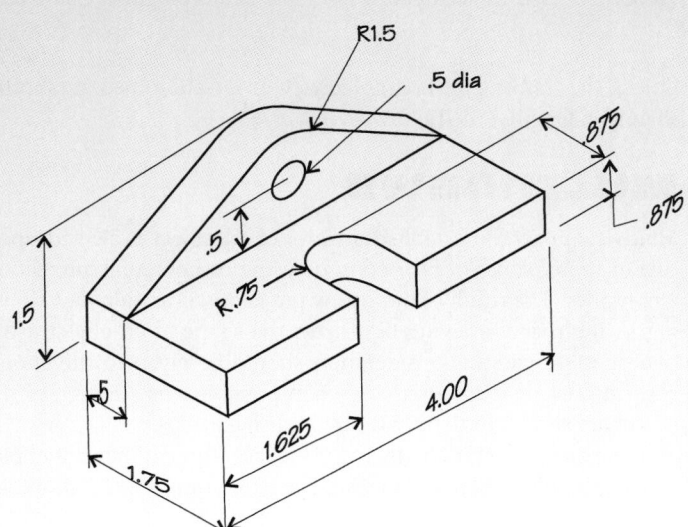

FIGURE 7.1 ■ Engineer's rough sketch.

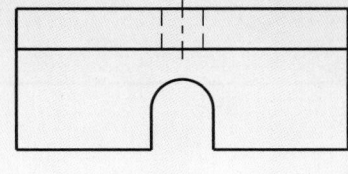
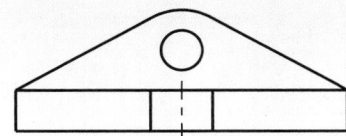
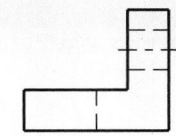

FIGURE 7.3 ■ A multiview drawing of the part without dimensions.

ASME/ANSI This chapter is developed in accordance with the ASME/ANSI standard for multiview presentation, titled *Multi and Sectional View Drawings,* ANSI Y14.3, published by The American Society of Mechanical Engineers (ASME). This standard is available from the American National Standards Institute, 1430 Broadway, New York, NY 10018, or The Society of Mechanical Engineers, 345 E 47th Street, New York, NY 10017. The content of this discussion provides an in-depth analysis of the techniques and methods of multiview presentation.

Orthographic projection is any projection of the features of an object onto an imaginary plane called a plane of projection. The projection of the features of the object is made by lines of sight that are perpendicular to the plane of projection. When a surface of the object is parallel to the plane of projection, the surface appears in its true size and shape on the plane of projection. In Figure 7.4, the plane of projection is parallel to the surface of the object. The line of sight (projection from the object) is perpendicular to the plane of projection. Notice also that the object appears three-dimensional (width, height, and depth) while the view on the plane of projection has only two dimensions (width and height). In situations where the plane of projection is not parallel to the surface of the

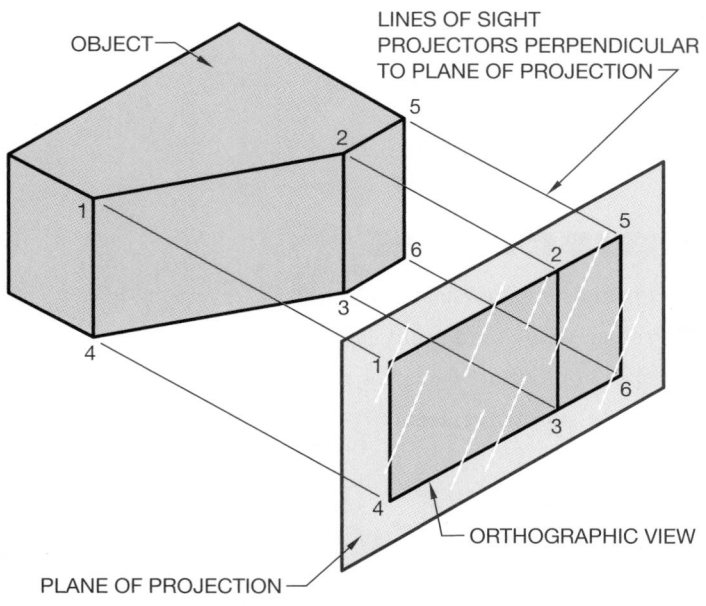

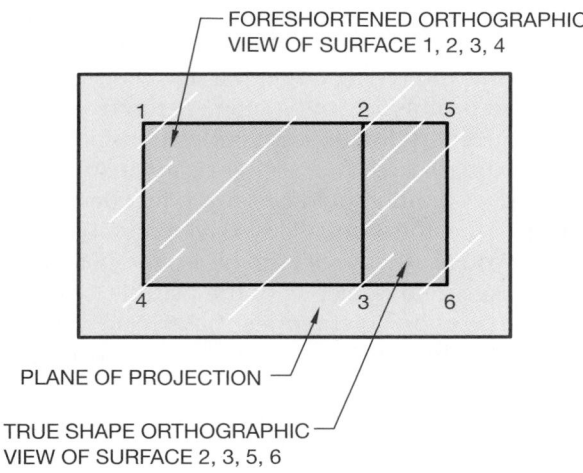

FIGURE 7.5 ■ Projection of a foreshortened orthographic surface.

object, the resulting orthographic view is foreshortened, or shorter than true length. (See Figure 7.5.)

MULTIVIEWS

Multiview projection establishes views of an object projected upon two or more planes of projection by using orthographic projection techniques. The result of multiview projection is a multiview drawing. A multiview drawing represents the shape of an object using two or more views. Consideration should be given to the choice and number of views used so, when possible, the surfaces of the object are shown in their true size and shape.

It is easier for an individual to visualize a three-dimensional picture of an object than it is to visualize a two-dimensional drawing. In mechanical drafting, however, the common practice is to prepare completely dimensioned detail drawings using two-dimensional views, known as *multiviews.* Figure 7.6 shows an object represented by a three-dimensional drawing, also called a *pictorial,* and three

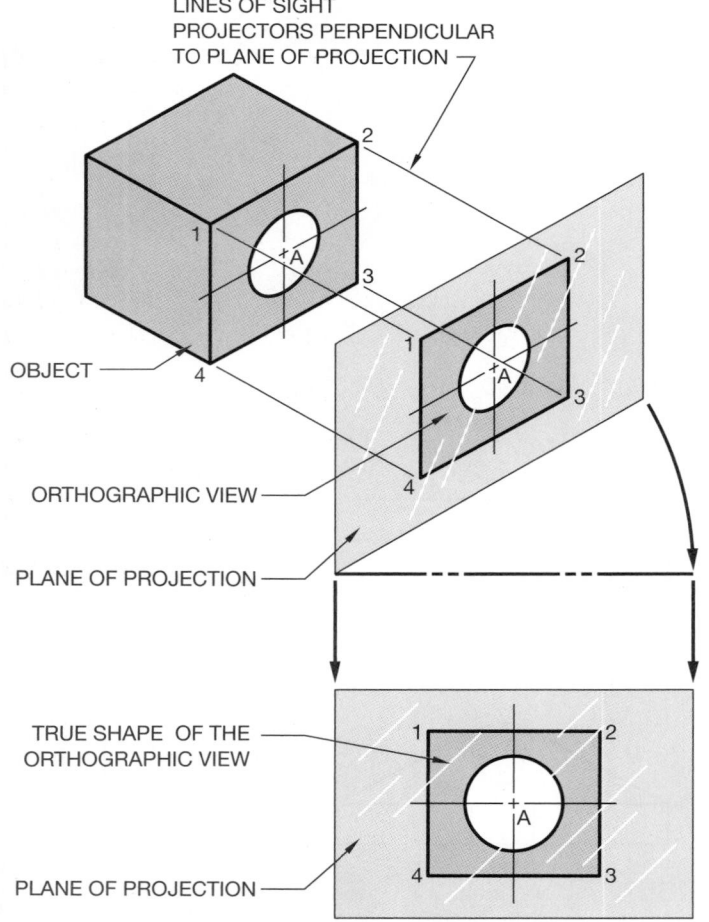

FIGURE 7.4 ■ Orthographic projection to form orthographic view.

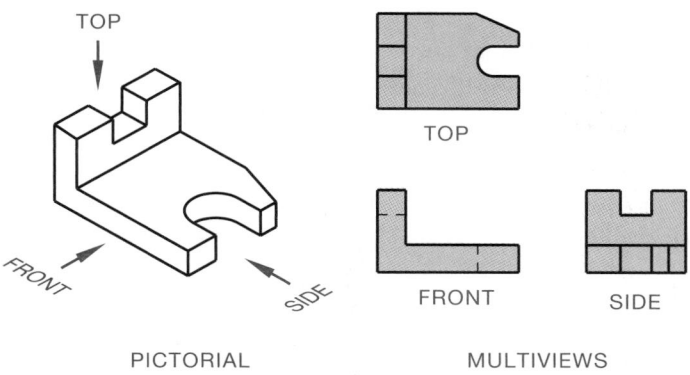

FIGURE 7.6 ■ Pictorial versus multiview.

two-dimensional views, or multiviews, also known as orthographic projection. The multiview method of drafting represents the *shape* description of the object.

Glass Box

If the object in Figure 7.6 is placed in a glass box so the sides of the glass box are parallel to the major surfaces of the object, you can project those surfaces onto the sides of the glass box and create multiviews. Imagine the sides of the glass box are the planes of projection that were previously discussed. (See Figure 7.7.) If you look at all sides of the glass box, you have six total views: FRONT, TOP, RIGHT-SIDE, LEFT-SIDE, BOTTOM, and REAR. Now unfold the glass box as if the corners were hinged about the front (except the rear view) as demonstrated in Figure 7.8. These hinge lines are commonly called *fold lines*.

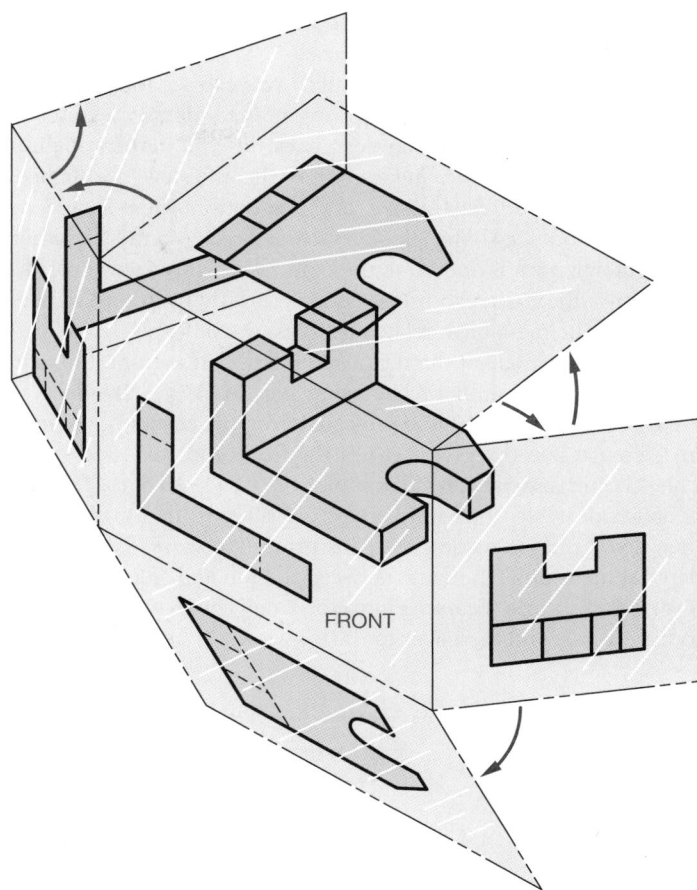

FIGURE 7.8 ■ Unfolding the glass box at hinge lines, also called fold lines.

Completely unfold the glass box onto a flat surface and you have the six views of an object represented in multiview. Figure 7.9 shows the glass box unfolded. Notice the views are labeled FRONT, TOP, RIGHT, LEFT, REAR, and BOTTOM. This is the arrangement that the views are always found when using multiviews in third-angle projection. *Third-angle projection* is the principal multiview

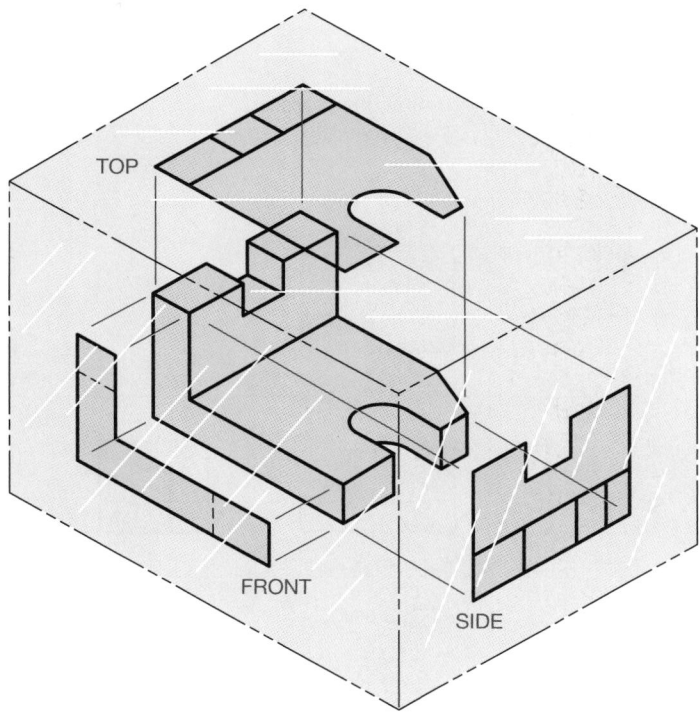

FIGURE 7.7 ■ The glass box principle.

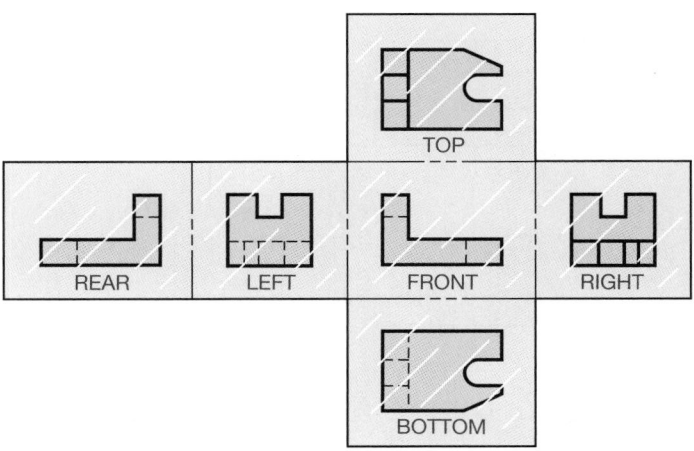

FIGURE 7.9 ■ Glass box unfolded.

projection used by United States industries. *First-angle projection* is commonly used in European countries.

Look at Figure 7.9 in more detail so you can see the items that are common between views. Knowing how to identify features of the object that a common between views aids you in the visualization of multiviews. Notice in Figure 7.10 that the views are aligned. The top view is directly above, and the bottom view is directly below, the front view. The left-side view is directly to the left, while the right-side view is directly to the right, of the front view. This format allows the drafter to project points directly from one view to the next to help establish related features on each view.

Now, look closely at the relationship among the front, top, and right-side views. A similar relationship exists with the left-side view. Figure 7.11 shows a 45° line projected from the corner of the fold, or reference line (hinge), between the front, top, and side views. This 45° line is used as an aid in projecting views. All of the features established on the top view can be projected to the 45° line and then down onto the side view. This projection works because the depth dimension is the same between the top and side views. The reverse is also true. Features from the side view may be projected to the 45° line and then over to the top view.

The same concept of projection achieved in Figure 7.11 using the 45° line also works by using a compass at the intersection of the horizontal and vertical fold lines. The compass establishes the common relationship between the top and side views as shown in Figure 7.12. Another method commonly used to transfer the size of features from one view to the next is to use dividers to transfer distances from the fold line of the top view to the fold line of the side view. The relationships between the fold line and these two views are the same, as shown in Figure 7.13.

The front view is usually the most important view and the one from which the other views are established. There is always one dimension common between adjacent views. For example, the width is common between the front and top views and the height between the front and side views. Knowing this allows you to relate information from one view to another. Take one more look at the relationship among the six views shown in Figure 7.14.

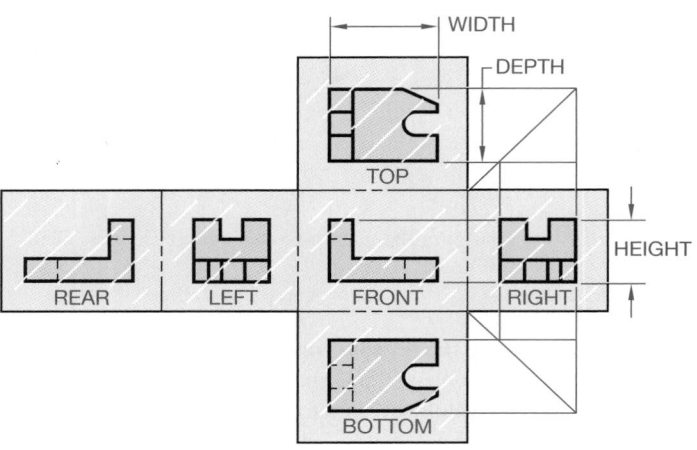

FIGURE 7.10 ■ View alignment.

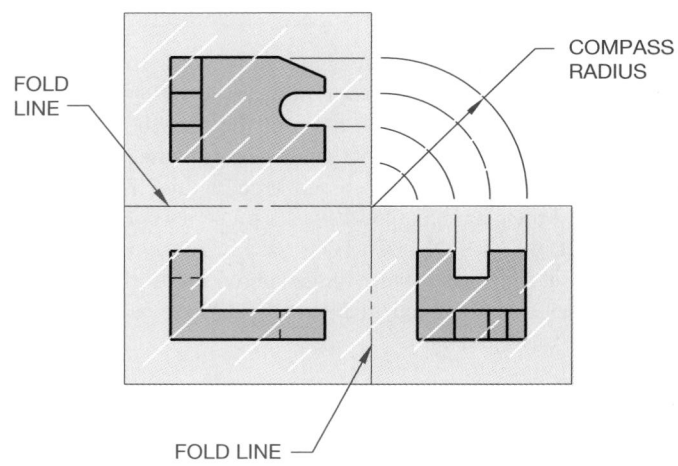

FIGURE 7.12 ■ Projection with a compass.

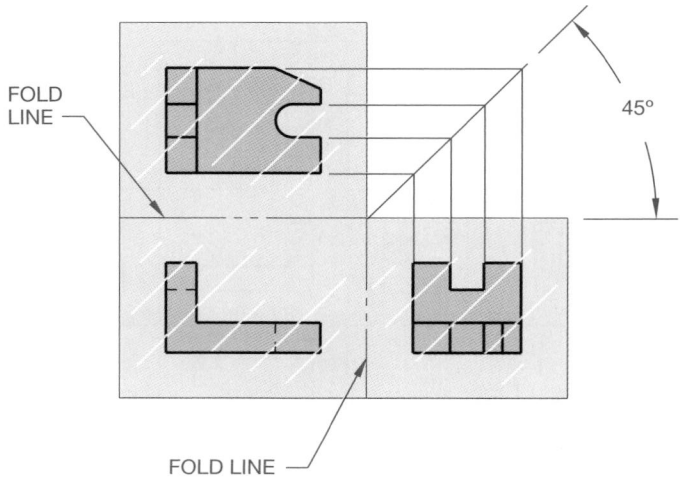

FIGURE 7.11 ■ 45° projection line.

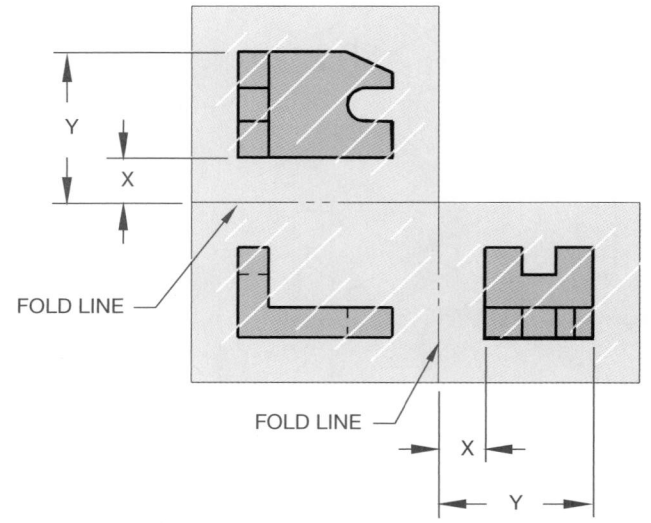

FIGURE 7.13 ■ Using dividers to transfer view projections.

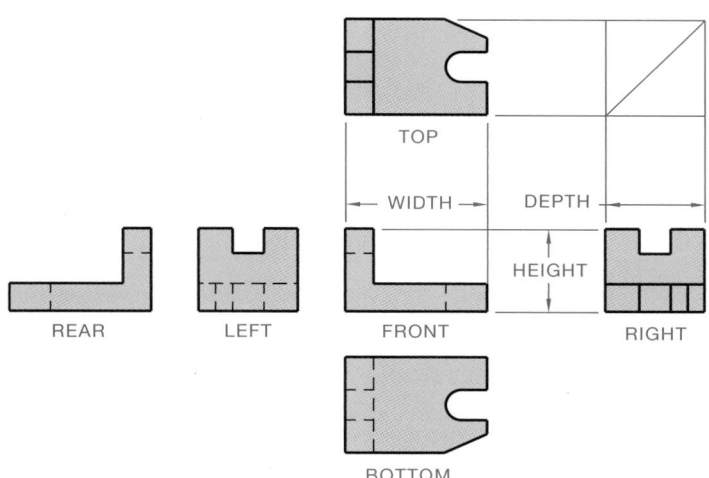

FIGURE 7.14 ■ Multiview orientation.

VIEW SELECTION

Although there are six primary views that you can select to completely describe an object, it is seldom necessary to use all six. As a drafter, you must decide how many views are needed to properly represent the object. If you draw too many views you are wasting time, which costs your employer money. If you draw too few views, you may not have completely described the object. The manufacturing department then has to waste time trying to determine the complete description of the object, which again costs your employer money.

Selecting the Front View

Usually, you should select the front view first. The front view is generally the most important view and, as you learned from the glass box description, the front view is the origin of all other views. There is no exact way for everyone to always select the same front view, but there are some guidelines to follow. The front view should:

■ Represent the most natural position of use.

■ Provide the best shape description or most characteristic contours.

■ Have the longest dimension.

■ Have the fewest hidden features.

■ Be the most stable and natural position.

Take a look at the pictorial drawing in Figure 7.15. Notice the front view selection. This front view selection violates the best shape description and fewest hidden features guidelines. However, the selection of any other view as the front would violate other rules, so in this case there is possibly no absolutely correct answer. Given the pictorial drawings in Figure 7.16, identify the view that you believe would be the best front view for each.

Other View Selection

Use the same rules when selecting other views needed as you do when selecting the front view:

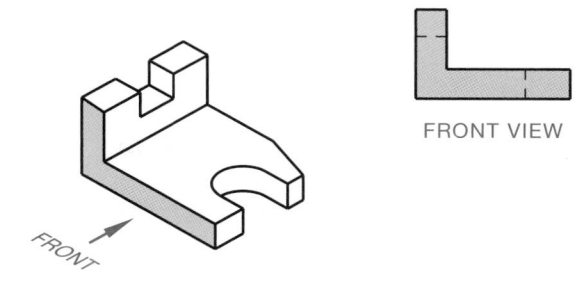

FIGURE 7.15 ■ Front view selection.

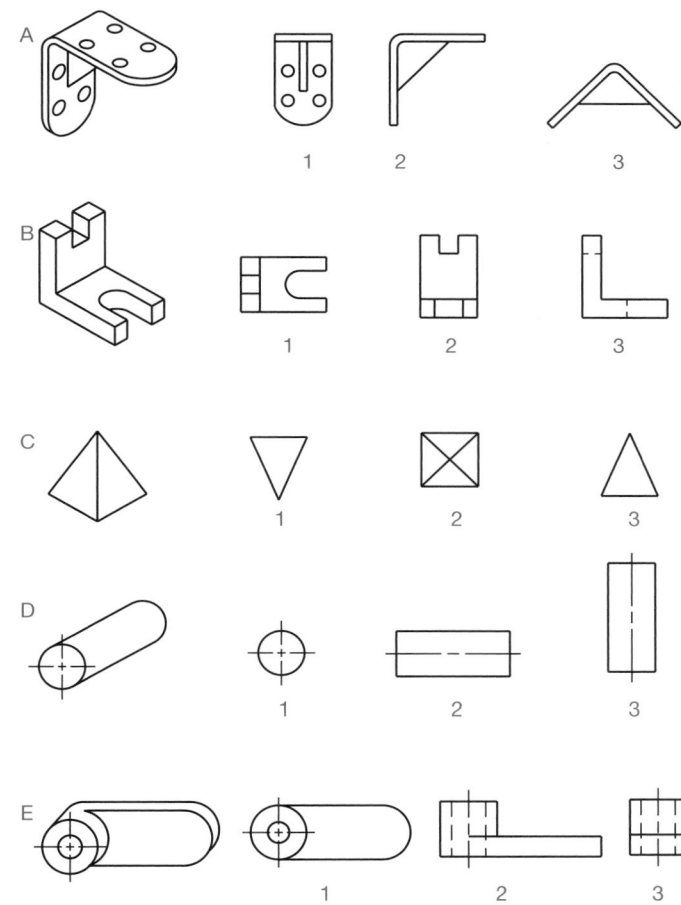

FIGURE 7.16 ■ Select the best front views that correspond to the pictorial drawings at the left. You may make a first and second choice, if you wish.

A. 2,1. B. 3,2. C. 3. D. 2,1. E. 2,1.

■ Most contours.

■ Longest side.

■ Least hidden features.

■ Best balance or position.

Given the six views of the object in Figure 7.17, which views would you select to completely describe the object? If your selection is the front, top, and right side, then you are correct. Now, take a closer look. Figure 7.18 shows the selected three views. The front view shows the best position and the longest side, the top view

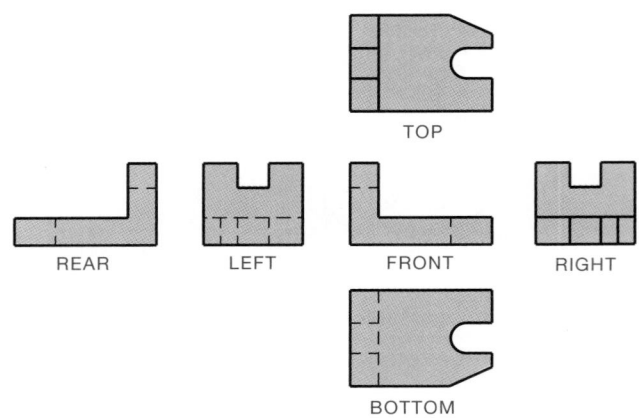

FIGURE 7.17 ■ Select the necessary views to describe the object.

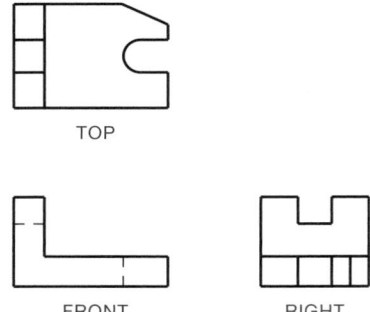

FIGURE 7.18 ■ The selected views.

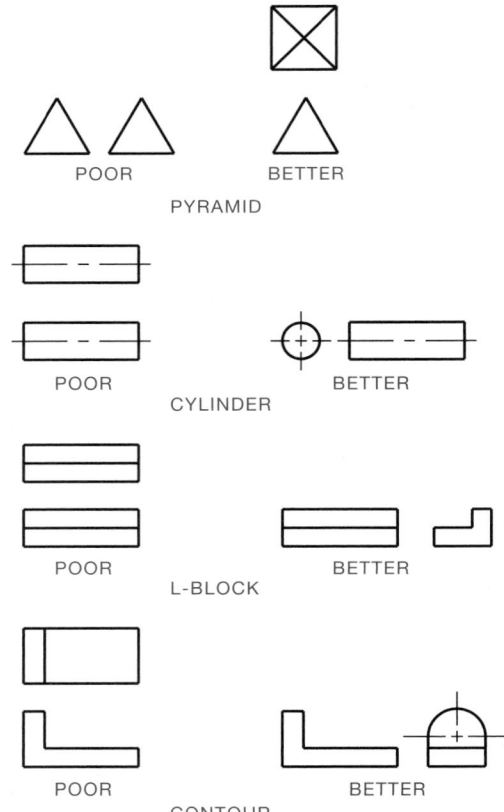

FIGURE 7.19 ■ Selecting two views.

clearly represents the angle and the arc, and the right-side view shows the notch. Any of the other views have many more hidden features. You should always avoid hidden features if you can.

It is not always necessary to select three views. Some objects can be completely described with two views or even one view. When selecting fewer than three views to describe an object, you must be careful which view other than the front view you select. (See Figure 7.19.)

One-view drawings are also often practical. When an object has shape and a uniform thickness, more than one view is unnecessary. Figure 7.20 shows a gasket drawing where the thickness of the part is identified in the materials specifications of the title block. The types of parts that may fit into this category include gaskets, washers, spacers, and similar features.

Although two-view drawings are generally considered the industry's minimum recommended views for a part, other objects that are clearly identified by dimensional shape may be drawn with one view, as in Figure 7.21.

In Figure 7.21, the shape of the pin is clearly identified by the 25 mm diameter. In this case, the second view would be a circle and would not necessarily add any more valuable information to the drawing. The primary question to keep in mind is, can the part be easily manufactured, from the drawing without confusion? If there is any doubt, then the adjacent view should probably be drawn. As an entry-level drafter, you should ask your drafting supervisor to clarify the company policy regarding the number of views to be drawn.

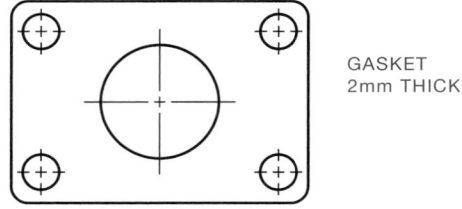

GASKET
2mm THICK

FIGURE 7.20 ■ One-view drawing.

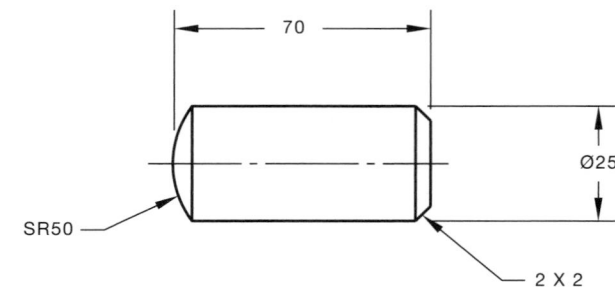

FIGURE 7.21 ■ One-view drawing.

Contour Visualization

Some views do not clearly identify the shape of certain contours. You must then draw the adjacent views to visualize the contour. (See Figure 7.22.)

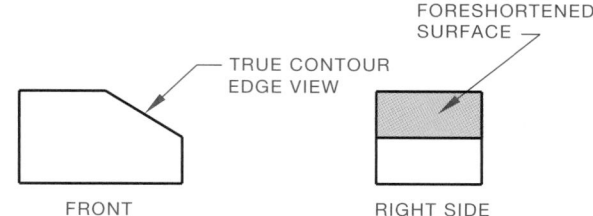

FIGURE 7.22 ■ Contour representation.

The edge view of a surface shows that surface as a line. The true contour of the slanted surface is seen as an edge in the front view of Figure 7.22 and describes the surface as an angle, while the surface is foreshortened, slanting away from your line of sight, in the right-side view. In Figure 7.23, select the front view that properly describes the given right-side view. All three front views in Figure 7.23 could be correct. The side view does not help the shape description of the front view contour. (See Figure 7.24.)

Cylindrical shapes appear round in one view and rectangular in another view, as seen in Figure 7.25. Both views in Figure 7.25 may be necessary as one shows the diameter shape and the other shows the length. The ability to visualize from one view to the next is a critical skill for a drafter. You may have to train yourself to look at two-dimensional objects and picture three-dimensional shapes. You can also use some of the techniques discussed here to visualize features from one view to another.

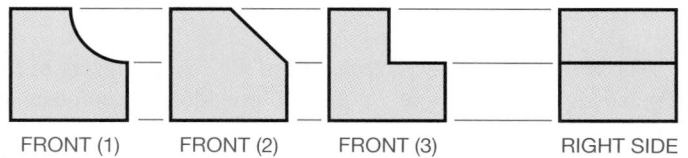

FIGURE 7.23 ■ Select the front view that properly goes with the given right-side view.

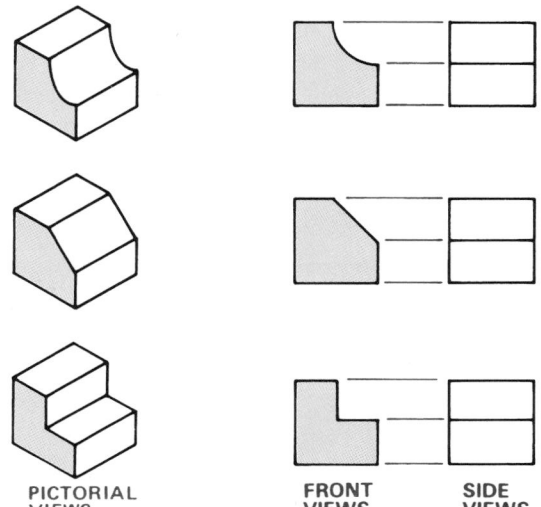

FIGURE 7.24 ■ The importance of a view that clearly shows the contour of a surface.

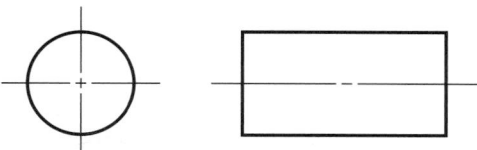

FIGURE 7.25 ■ Cylindrical shape representation.

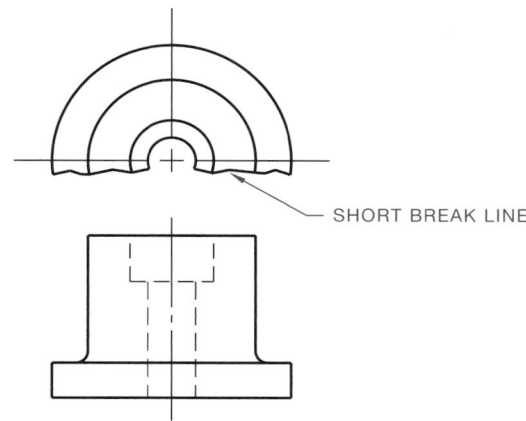

FIGURE 7.26 ■ Partial view.

Partial Views

When symmetrical objects are drawn in limited space or when there is a desire to save valuable drafting time, partial views may be used. The top view in Figure 7.26 is a partial view. Notice how the short break line is used to show that a portion of the view is omitted. Caution should be exercised when using partial views, as confusion could result in some situations. If the partial view reduces clarity, then draw the entire view. When using CADD, it is easier to draw the full view.

View Enlargement

When part of a view has detail that cannot be clearly dimensioned due to the drawing scale or complexity, a view enlargement may be used. To establish a view enlargement, a thick phantom line circle is placed around the area to be enlarged. This circle is broken at a convenient location and an identification letter is centered in the break. The height of this letter is generally .18 to .25 in. (6 mm) so it stands out from other lettering. Arrowheads are then placed on the line next to the identification letter, as shown in Figure 7.27a. An enlarged view of the detailed area is then shown in any convenient location in the field of the drawing. Common enlargement scales are 2:1, 4:1, and 10:1. The view identification and scale is then placed below the enlarged view, such as DETAIL A. The lettering height is placed as a title, which is .18 to .25 in. (6 mm). The word DETAIL can be the same height or less than the identification letter depending on company preference. (See Figure 7.27b.)

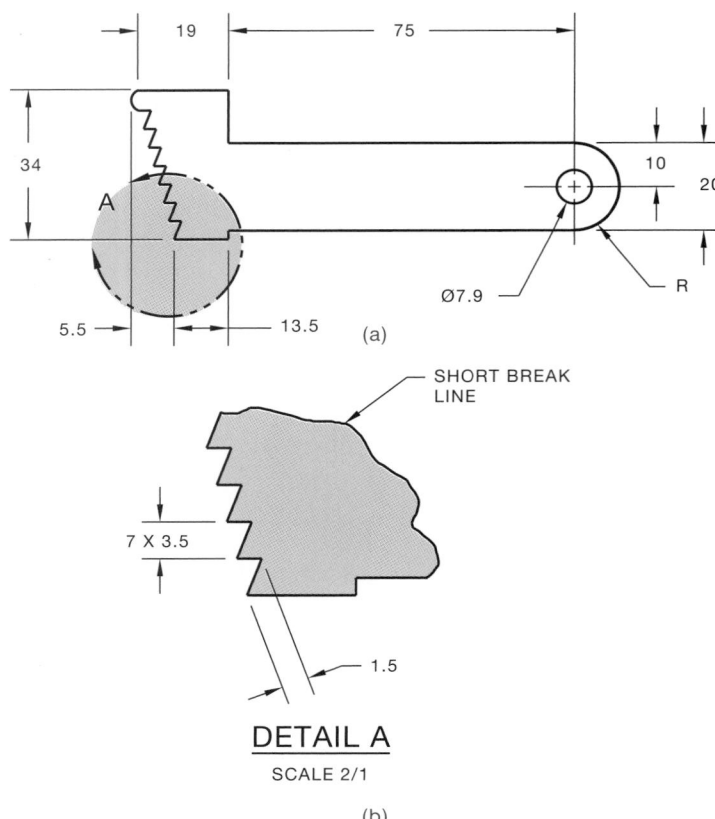

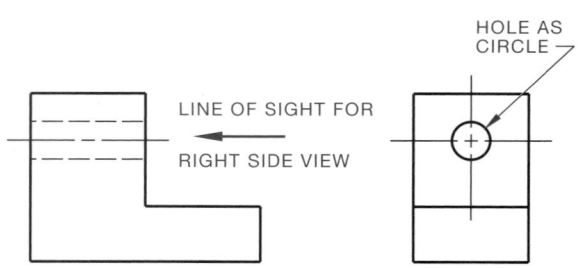

FIGURE 7.27 ■ View enlargement example.

PROJECTION OF CIRCLES AND ARCS

Circles on Inclined Planes

When the line of sight in multiviews is perpendicular to a circular feature, such as a hole, the feature appears round, as shown in Figure 7.28. When a circle is projected onto an inclined surface, its view is elliptical in shape, as shown in Figure 7.29. The ellipse shown in the top and right-side views of Figure 7.29 was established by projecting the major diameter from the top to the side view and the minor diameter to both views from the front view, as shown in Figure 7.30. The major diameter in this example is the hole diameter.

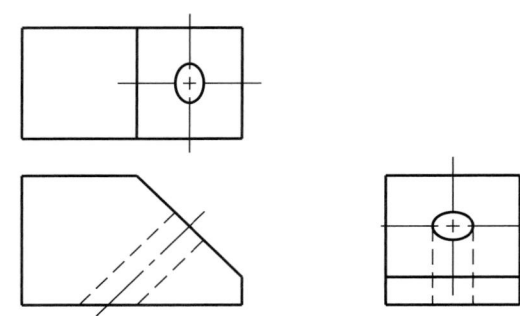

FIGURE 7.29 ■ Hole represented on an inclined surface as an ellipse.

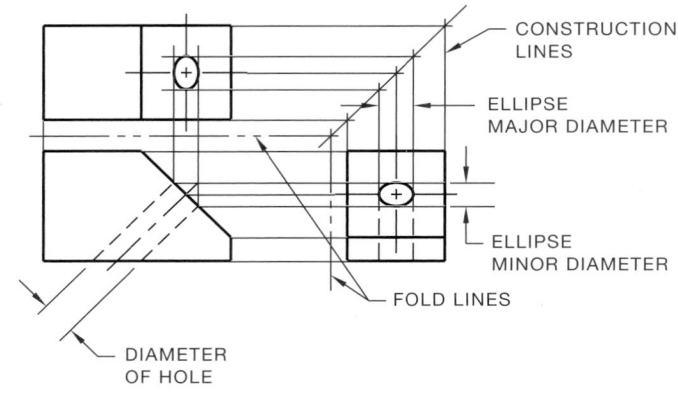

FIGURE 7.30 ■ Establish an ellipse in the inclined surface.

The rectangular areas projected from the two diameters in the top and right-side views of Figure 7.30 provide the parameters of the ellipse to be drawn in these views. The easiest method of drawing the ellipse is to find an ellipse template that has a shape that fits within the area.

Ellipse templates, as discussed in Chapters 1 and 3, designate elliptical shapes by the angle of the circle in relationship to the line of sight. If your slanted surface is 30°, then a 30° ellipse template would be used to draw the elliptical shape. Keep in mind that the other view would require a 60° ellipse. When the angle is 45°, then both views would be the same. The ellipse template selected should have a shape that fits into the rectangle constructed, as shown in Figure 7.31. If you cannot find a representative ellipse that fits exactly, then use the closest size possible, or verify the drawing pro-

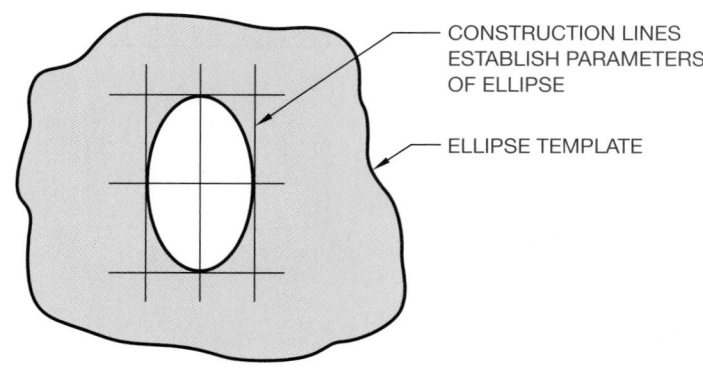

FIGURE 7.28 ■ View of a hole projected as a circle and its hidden view through the part.

FIGURE 7.31 ■ Ellipse template.

cedure with your supervisor. Establishing an elliptical shape with geometric construction is much more time-consuming but may be necessary when templates are not available. The elliptical shapes are a representation and may not have to be perfect, depending on company policy. Also, notice that the hidden lines for the hole has been omitted in the top view of Figure 7.30. These lines would require a great amount of extra work without adding significantly to the function of the drawing. Such a practice of omitting lines for an accurate representation may be used with caution. Remember that the drawing must easily communicate the manufacture of the part.

Arcs on Inclined Planes

When a curved surface from an inclined plane must be drawn in multiview, a series of points on the curve establishes the contour. Begin by selecting a series of points on the curved contour as shown in the right-side view of Figure 7.32. Project these points from the right side to the inclined front view. From the point of intersection on the inclined surface in the front view, project lines to the top view. Then project corresponding points from the right side to the 45° projection line and onto the top view. Corresponding lines create a pattern of points, as shown in the top view of Figure 7.32. You may also use a compass or dividers to transfer measurements from the fold lines, as previously discussed. After the series of points is located in relationship to the front and top views, connect the points with an irregular curve. See the completed curve in Figure 7.33.

Fillets and Rounds

Fillets are slightly rounded inside curves at corners, generally used to ease the machining of inside corners or to allow patterns to

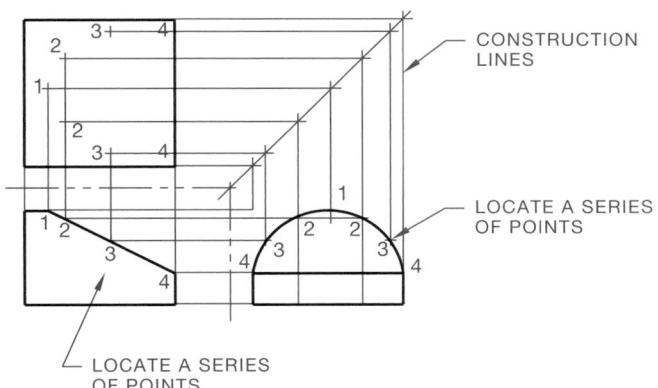

FIGURE 7.32 ■ Locating an inclined curve in multiview.

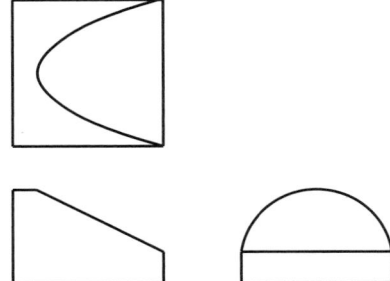

FIGURE 7.33 ■ Completed curve.

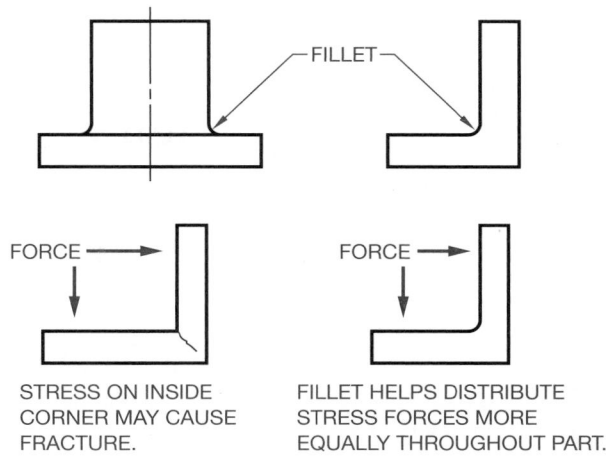

FIGURE 7.34 ■ Fillets.

release more easily from casting and forgings. Fillets may also be designed into a part to allow additional material on inside corners for stress relief. (See Figure 7.34.)

Other than the concern of stress factors on parts, certain casting methods require that inside corners have fillets. The size of the fillet often depends on the precision of the casting method. For example, very precise casting methods may have smaller fillets than green-sand casting, where the exactness of the pattern requires large inside corners.

Rounds are rounded outside corners that are used to relieve sharp exterior edges. Rounds are also necessary in the casting and forging process for the same reasons as fillets. Figure 7.35 shows rounds represented in views.

A machined edge causes sharp corners, which may be desired in some situations. However, if these sharp corners are to be rounded, the extent of roundness depends on the function of the part. When a sharp corner has only a slight relief, it is referred to as a break corner, as shown in Figure 7.36.

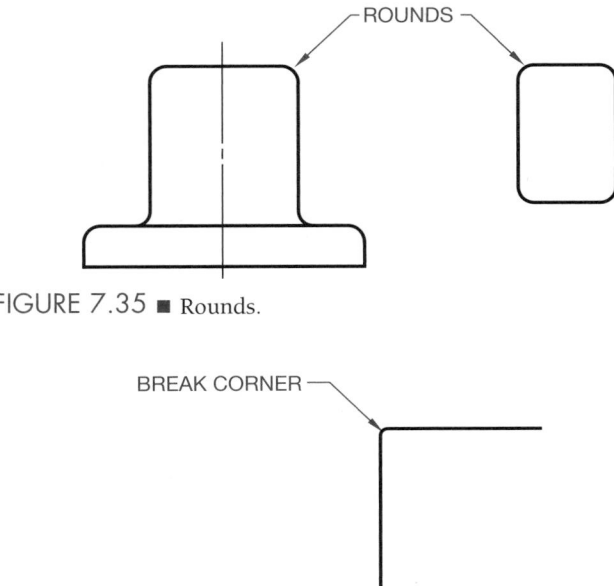

FIGURE 7.35 ■ Rounds.

FIGURE 7.36 ■ Break corner.

FILLETS AND ROUNDS

CADD programs allow you to draw a fillet or round between any two lines or a line and an arc simply by picking the two lines and specifying the fillet radius. The AutoCAD command for this operation is called FILLET, but it draws any radius exterior or interior corner. If the lines intersect, they are automatically trimmed and the fillet is drawn. If the lines do not meet, they are automatically extended and the fillet is drawn. If all of the lines of an object are connected as one object (referred to as a polyline in AutoCAD), picking anywhere on the line automatically fillets all corners. Figure 7.37 shows some before and after examples of using the FILLET command.

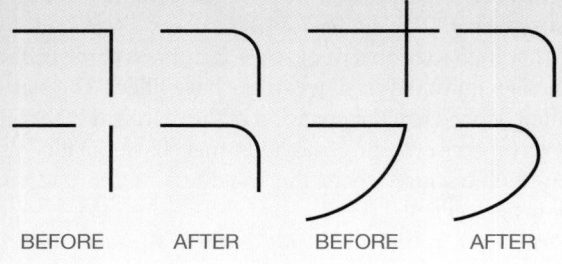

BEFORE   AFTER   BEFORE   AFTER

FIGURE 7.37 ■ Automatically drawing fillets and rounds using CADD.

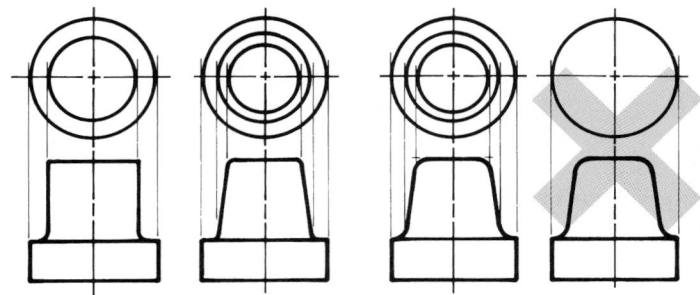

FIGURE 7.39 ■ Rounded curves and cylindrical shapes in multiview.

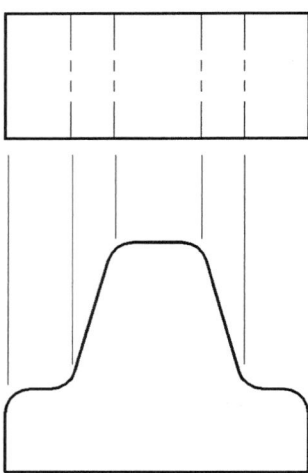

FIGURE 7.40 ■ Contour in multiview.

Rounded Corners in Multiview

An outside or inside slightly rounded corner of an object is represented in multiview as a contour only. The extent of the round or fillet is not projected into the view as shown in Figure 7.38. Cylindrical shapes may be represented with a front and top view, where the front identifies the height and the top shows the diameter. Figure 7.39 shows how these cylindrical shapes should be represented in multiview. Figure 7.40 shows the representation of the contour of an object as typically displayed in multiview using phantom lines. This is done to clearly accent the rounded feature.

Runouts

The intersections of features with circular objects are projected in multiview to the extent where one shape runs into the other. The characteristics of the intersecting features are known as *runouts*. The runout of features intersecting cylindrical shapes is projected from the point of tangency of the intersecting feature, as shown in Figure 7.41. Notice also that the shape of the runout varies when drawn at the cylinder depending on the shape of the intersecting feature. Rectangular-shaped features have a fillet at the runout, while curved (elliptical or round) features contour toward the centerline at the runout. Runouts may also exist when a feature such as a web intersects another feature, as shown in Figure 7.42.

LINE PRECEDENCE

When drawing multiviews, it is common for one type of line to fall in line with a different line type. As a drafter, you need to decide which line to draw. This is known as *line precedence*. You draw the line that is most important based on these rules:

■ Object lines take precedence over hidden lines and centerlines.

■ Hidden lines take precedence over centerlines.

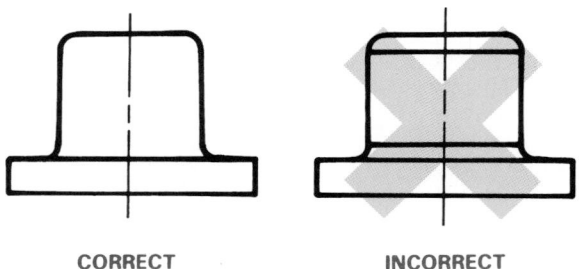

CORRECT   INCORRECT

FIGURE 7.38 ■ Rounds and fillets in multiview.

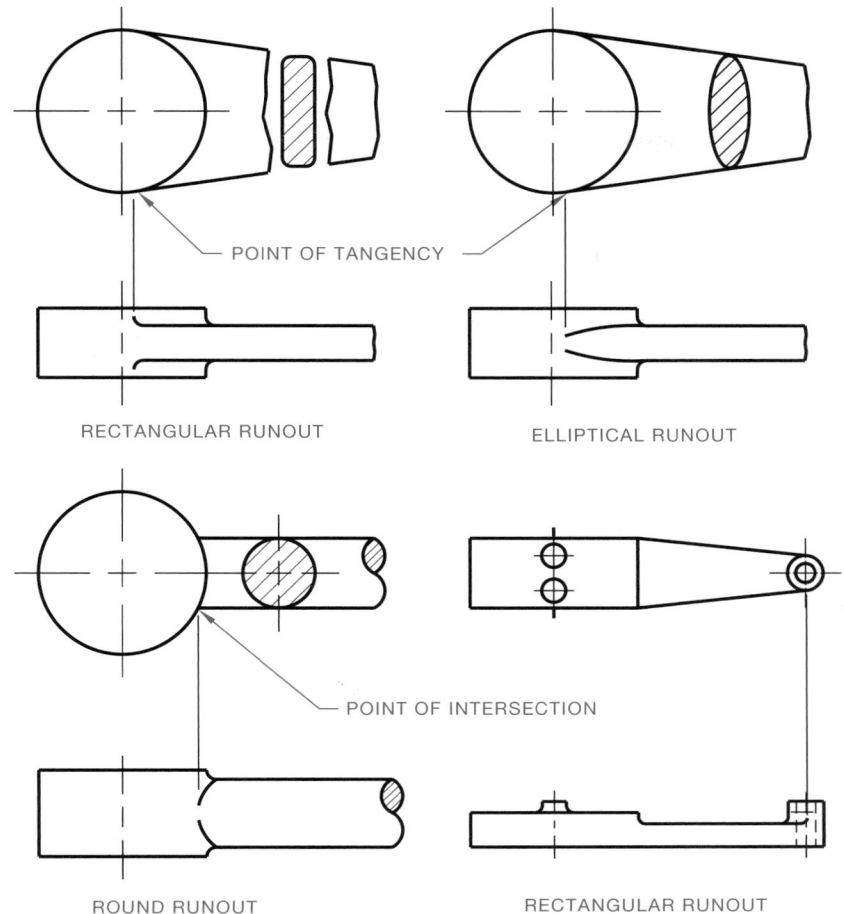

POINT OF TANGENCY

RECTANGULAR RUNOUT

ELLIPTICAL RUNOUT

POINT OF INTERSECTION

ROUND RUNOUT

RECTANGULAR RUNOUT

FIGURE 7.41 ■ Runouts.

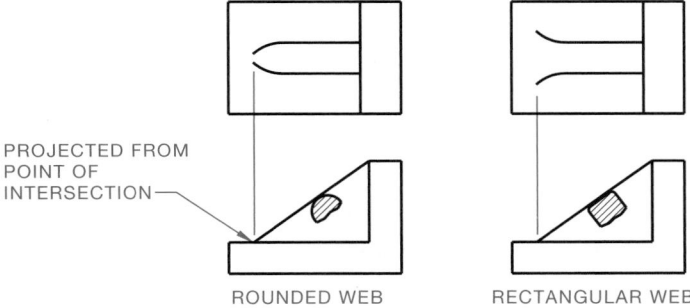

PROJECTED FROM POINT OF INTERSECTION

ROUNDED WEB

RECTANGULAR WEB

FIGURE 7.42 ■ Other types of runouts.

■ In sectioning (covered in Chapter 14), cutting plane lines take precedence over centerlines.

When an object line is drawn over a centerline, the ends or tails of the centerline may be drawn slightly beyond the outside of the view. Figure 7.43 shows examples of line precedence.

THIRD-ANGLE PROJECTION

The method of multiview projection described in this chapter is also known as *third-angle projection*. This is the method of view arrangement that is commonly used in the United States. In the previous discussion on multiview projection, the object was placed in a glass box so the sides of the glass box were parallel to the major surfaces of the object. Next, the object surfaces were projected onto the adjacent surfaces of the glass box. This achieved the same effect as if the viewer's line of sight were perpendicular to the surface of the box and looking directly at the object, as shown in Figure 7.44. With the multiview concept in mind, assume an area of space is divided into four quadrants, as shown in Figure 7.45.

If the object were placed in any of these quadrants, the surfaces of the object would be projected onto the adjacent planes. When placed in the first quadrant, the method of projection is known as first-angle projection. Projections in the other quadrants are termed second-, third-, and fourth-angle projections. Second- and fourth-angle projections are not used, though first- and third-angle projections are very common.

Third-angle projection, as commonly used in the United States, is achieved when you take the glass box from Figure 7.44 and place it in quadrant three from Figure 7.45. Figure 7.46 shows the relationship of the glass box to the projection planes in the third-angle projection. In this quadrant, the projection plane is between the viewer's line of sight and the object. When the glass box in the third-angle projection quadrant is unfolded, the

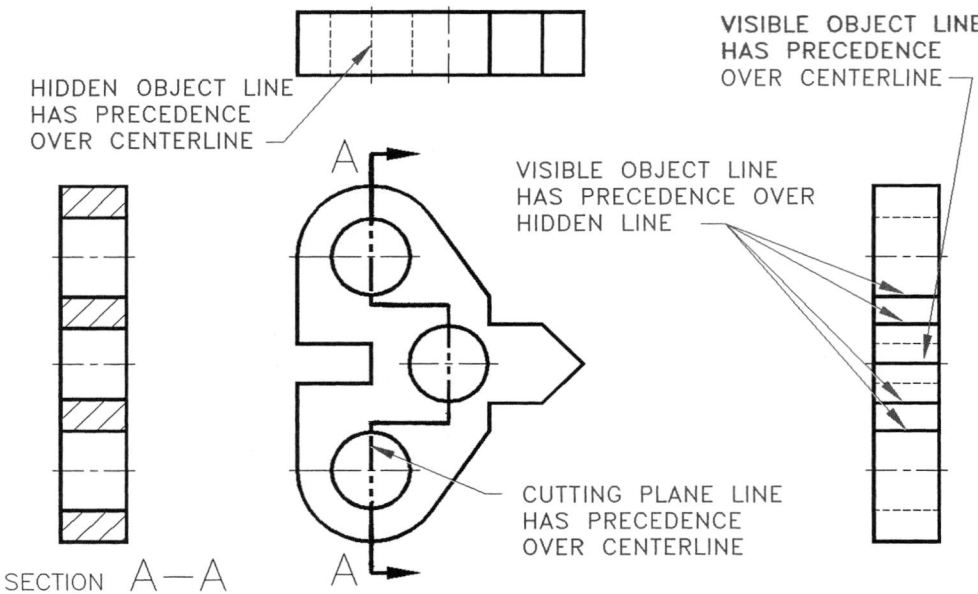

HIDDEN OBJECT LINE HAS PRECEDENCE OVER CENTERLINE

VISIBLE OBJECT LINE HAS PRECEDENCE OVER CENTERLINE

VISIBLE OBJECT LINE HAS PRECEDENCE OVER HIDDEN LINE

A

CUTTING PLANE LINE HAS PRECEDENCE OVER CENTERLINE

SECTION A—A

A

FIGURE 7.43 ■ Line precedence.

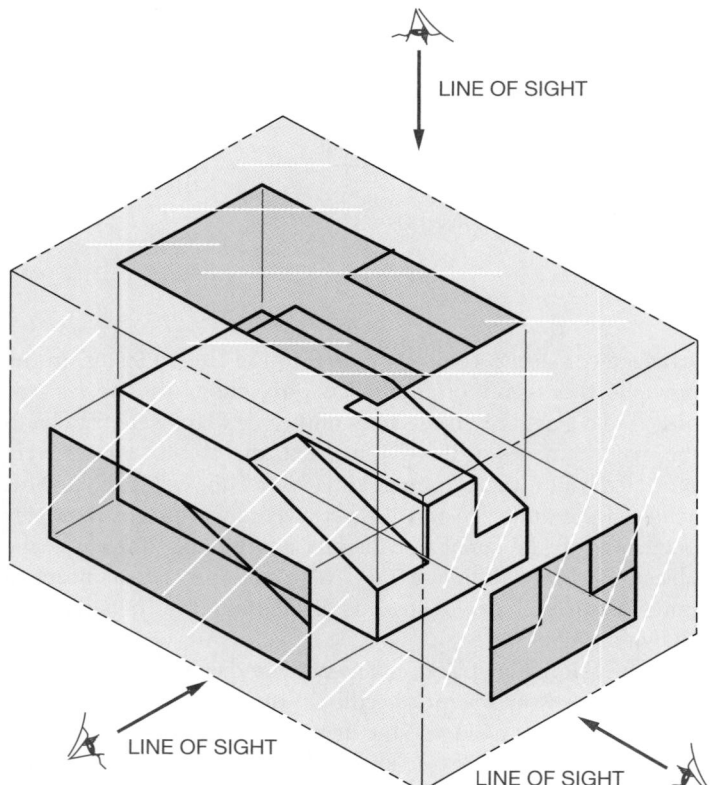

LINE OF SIGHT

LINE OF SIGHT

LINE OF SIGHT

FIGURE 7.44 ■ Glass box in third-angle projection.

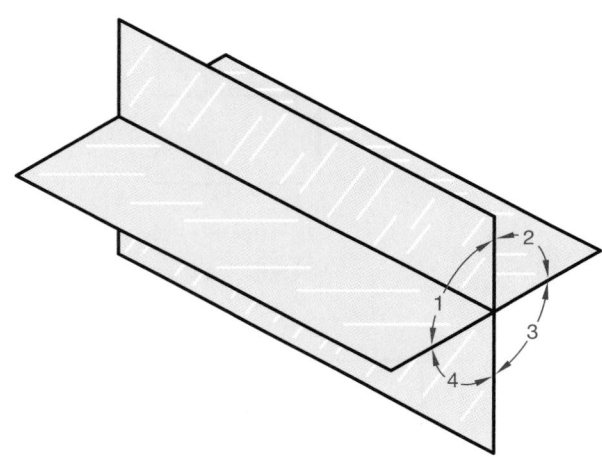

FIGURE 7.45 ■ Quadrants of spatial visualization.

FIRST-ANGLE PROJECTION

First-angle projection is commonly used in Europe and other countries of the world. This method of projection places the glass box in the first quadrant. Views are established by projecting surfaces of the object onto the surface of the glass box. In this projection arrangement, however, the object is between the viewer's line of sight and the projection plane, as you can see in Figure 7.49. When the glass box in the first-angle projection quadrant is unfolded, the result is the multiview arrangement shown in Figure 7.50.

A first-angle projection drawing may be accompanied by a symbol on or adjacent to the drawing title block. The standard first-angle projection symbol is shown in Figure 7.51. Figure 7.52 shows a comparison of the same object in first- and third-angle projections.

result is the multiview arrangement previously discussed and shown in Figure 7.47.

A third-angle projection drawing may be accompanied by a symbol on or next to the drawing title block. The standard third-angle projection symbol is shown in Figure 7.48.

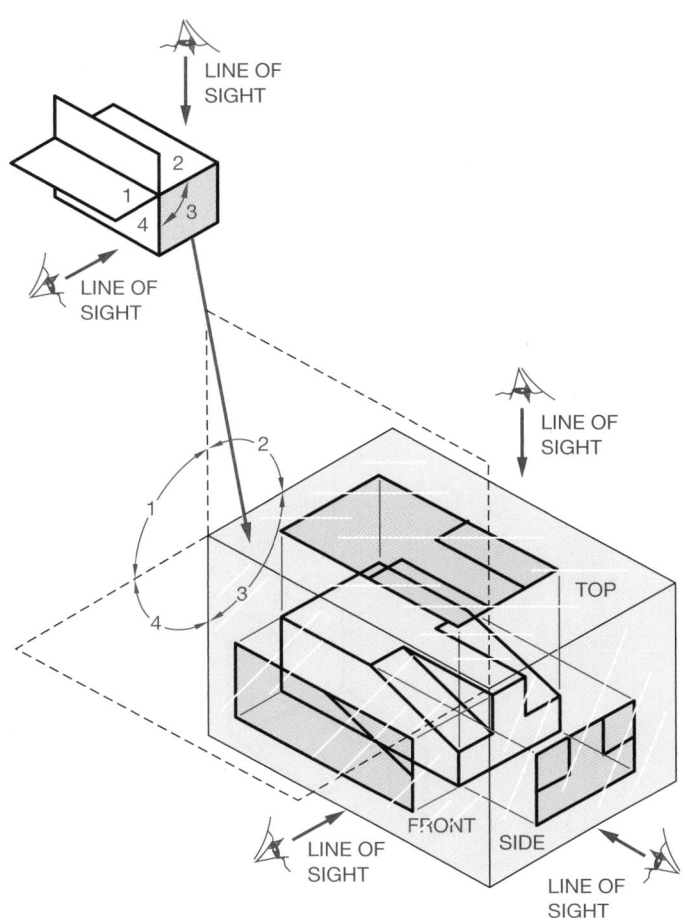

FIGURE 7.46 ■ Glass box placed in the third-quadrant for third-angle projection.

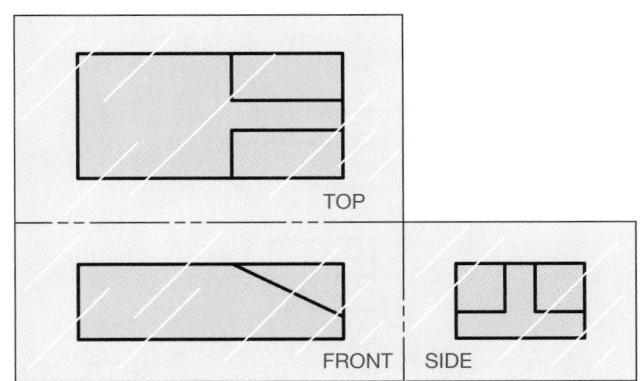

FIGURE 7.47 ■ Third-angle projection.

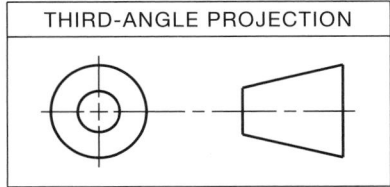

FIGURE 7.48 ■ Third-angle projection symbol.

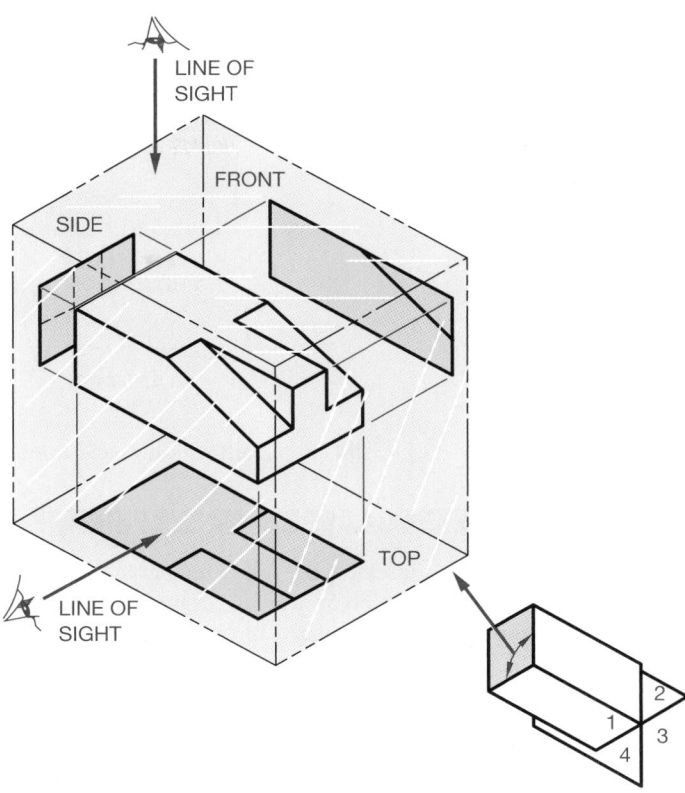

FIGURE 7.49 ■ Glass box in first-angle projection.

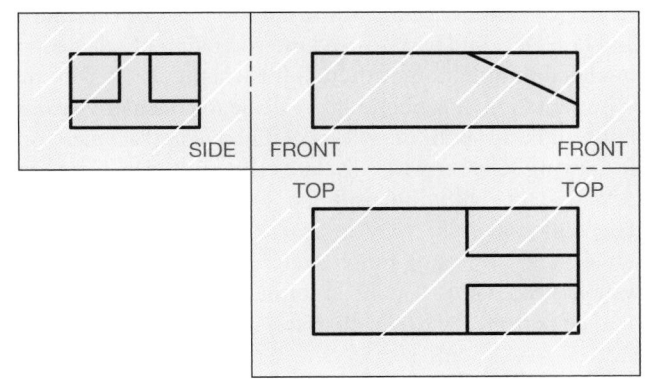

FIGURE 7.50 ■ First-angle projection.

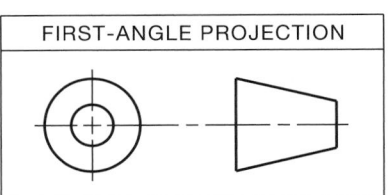

FIGURE 7.51 ■ First-angle projection symbol.

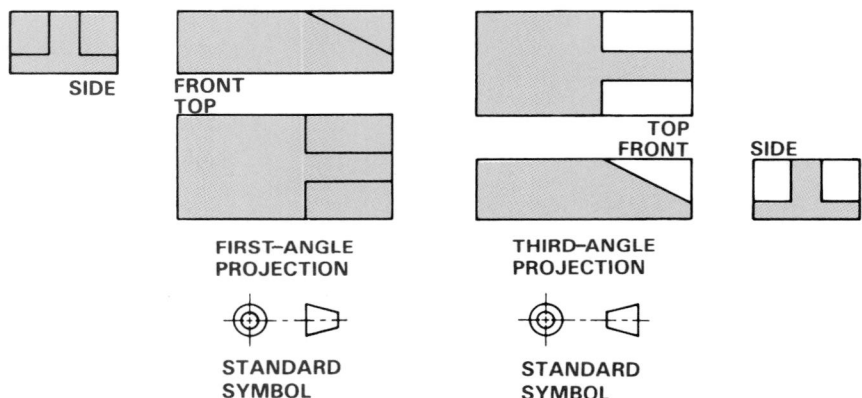

FIGURE 7.52 ■ First-angle and third-angle projection.

RECOMMENDED REVIEW

It is recommended that you review the following parts of Chapter 2 before you begin working on multiview drawings. This will refresh your memory about how related lines are properly drawn:

■ Object Lines.
■ Viewing- and Cutting-Planes.
■ Hidden Lines.
■ Break Lines.
■ Centerlines.
■ Phantom Lines.

LAYOUT

Many factors influence the drawing layout. Your prime goal should be a clear and easy-to-read interpretation of selected views with related information. Although this chapter deals with multiview presentation, it is not totally realistic to consider view layout without thinking about the effects of dimension placement on the total drawing. Chapter 12 correlates the multiview drawings of *shape description* with dimensioning, known as *size description*.

The initial steps in view layout should be performed using rough sketches. By using rough sketches, you can analyze which views you need before you begin formal drafting. Sketches do not have to be perfect. Try to sketch as fast as you can to save time.

Sketching the Layout

Consider the engineering sketch in Figure 7.53 as you evaluate the proper view layout.

Step 1. Select the front view using the rules discussed in this chapter. Sketch the front view that you have picked. Try to keep your sketch proportional to the actual object, as in Figure 7.54. However, keep in mind that a sketch does not have to be perfect. It should be done quickly to save time, while helping you lay out the drawing.

Step 2. Select the other views needed to completely describe the shape of the V-BLOCK MOUNT, as shown in Figure 7.55.

The front, top, and left-side views clearly define the shape of the V-BLOCK MOUNT. Now lay out the formal drawing using the

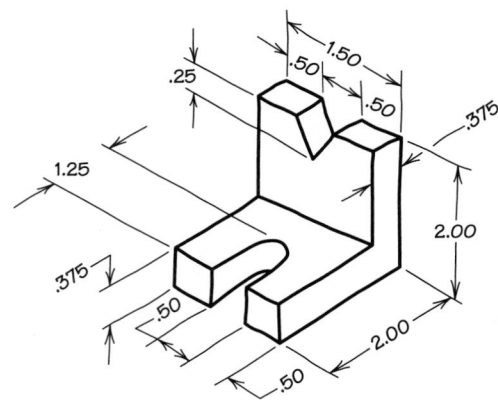

FIGURE 7.53 ■ Engineering sketch.

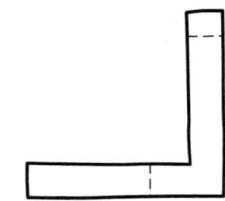

FIGURE 7.54 ■ Sketch the front view.

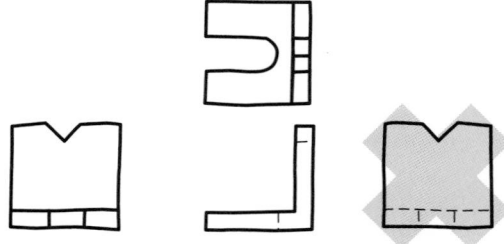

FIGURE 7.55 ■ Sketch the required views.

sketch as a guide. Several factors must be considered before you begin:

1. Size of drawing sheet.

2. Scale of the drawing.

3. Number and size of views.

4. Amount of blank space required for future revisions.

5. Dimensions and notes (not drawn at this time).

Drawing the Layout

Step 1. Use an A3-size (297 × 420 mm) or B-size (11 × 17 in.) drawing sheet. The recommended working area is shown in Figure 7.56. The amount of blank area on a drawing depends on company standards. Some companies want the drawing to be easy to read with no crowding; others may want as much information as possible on a sheet. Generally, .50 in. (12.7 mm) should be the minimum space between the drawing and border line. More space is preferred. An area for future revision generally (but not always) should be left between the title block and upper right corner. An area for general notes should be available

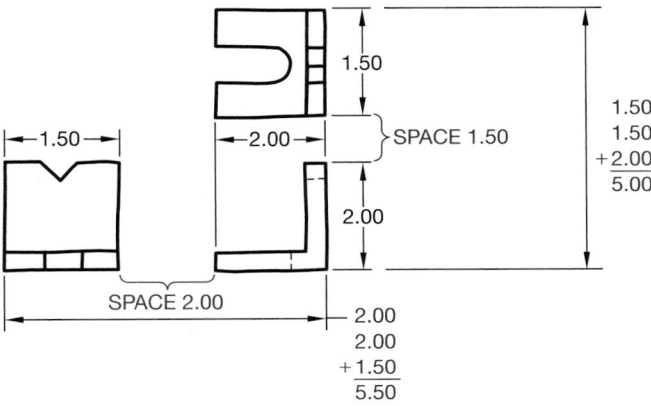

FIGURE 7.56 ■ Rough sketch with overall dimensions and selected space between views.

to the left of the title block for ASME/ANSI standard layout or in the upper left corner for Military standard layout. The remaining area is the space available for drawing. On Figure 7.56, this area is about 10 × 6 in. (250 × 15 mm).

Step 2. After determining the approximate working area, use your rough sketch as a guide to establish the actual size of the drawing by adding the overall dimensions and the space between views. (See Figure 7.56.) The amount of space selected will not crowd the views. The amount to select is an arbitrary decision and one where the drafter must use good judgment. Keep in mind that, for now, you will not consider dimensions when evaluating space requirements. The effect of dimensions on drawing layouts is discussed in Chapter 14.

Step 3. Now, in each direction subtract the total drawing size from the total space available and divide by two. This gives you the boundaries of the drawing area. Use construction lines to block out the total drawing areas that you have selected. (See Figure 7.57.)

Calculations:

Total space height	6.00
Total drawing height	−5.00
	1.00 ÷ 2 = .50
Total space width	10.00
Total drawing width	−5.50
	4.50 ÷ 2 = 2.25

Step 4. Within the area established, use construction lines to block out the views that you selected in the rough sketch. Use multiview projection as discussed in this chapter, beginning with the front view. (See Figure 7.58.)

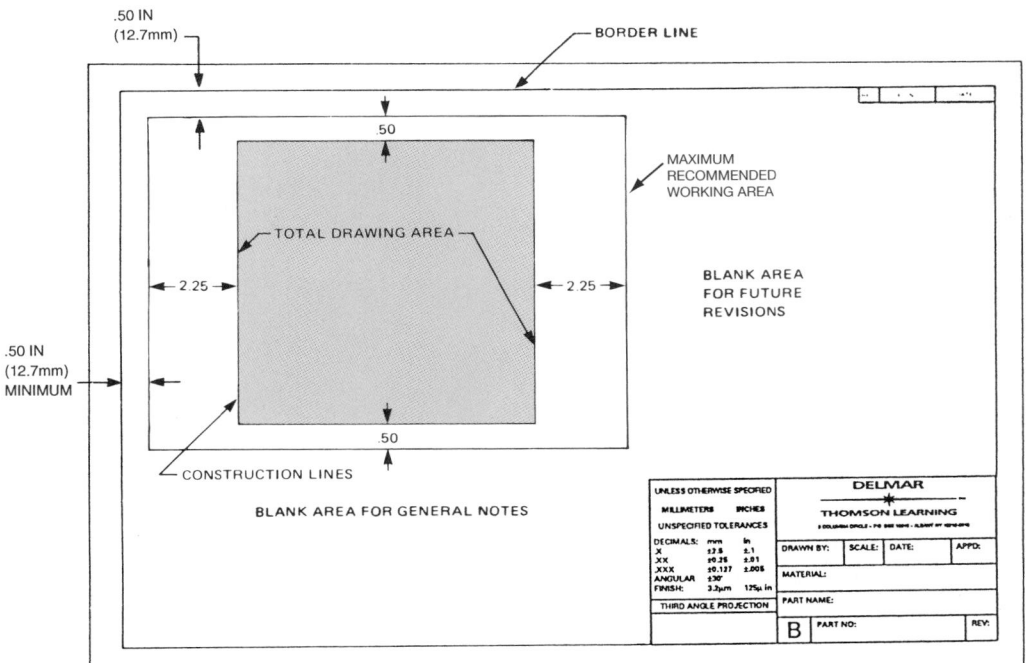

FIGURE 7.57 ■ The recommended working area and layout of the total drawing area.

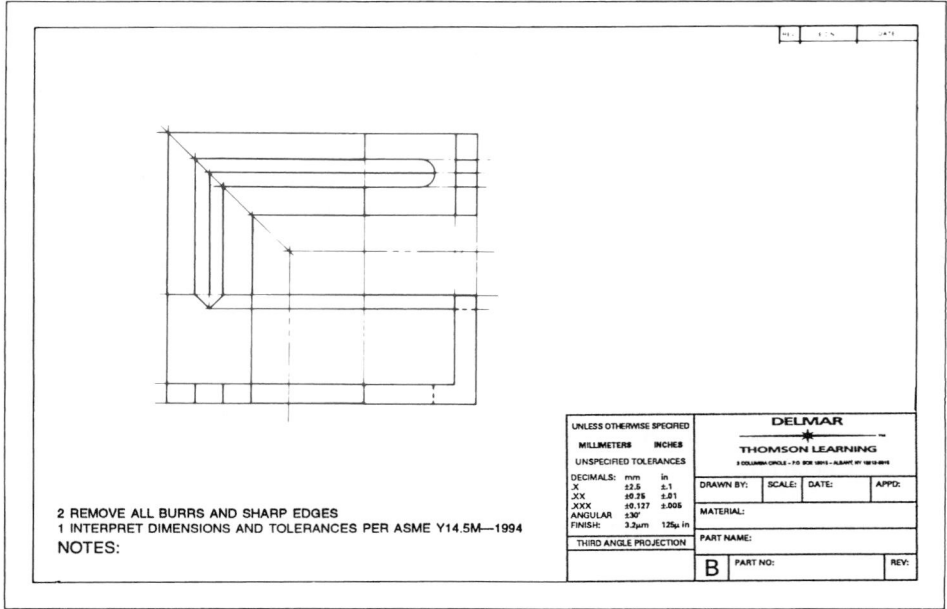

FIGURE 7.58 ■ Lay out the views using construction lines.

The construction lines, if properly drawn, will not have to be erased. Remember that construction lines are drawn very lightly with a 6H, 4H, or light-blue lead. In any case, the construction lines should not reproduce in the diazo or photocopy process, if properly drawn.

Step 5. Complete your drawing by using proper techniques to draw the finish lines. To help keep the final drawing neat, remember to:
a. Work from top to bottom.
b. Work from left to right if right-handed or from right to left if left-handed.

c. Draw thin lines first.
d. Draw circles and arcs next.
e. Draw object lines.
f. Do all lettering and place a clean paper under your hand while lettering.
g. Avoid moving equipment or your hands over completed areas.
h. Keep your equipment and hands clean.
i. Cover large completed areas with blank paper to help avoid smudging.

Figure 7.59 shows the completed drawing.

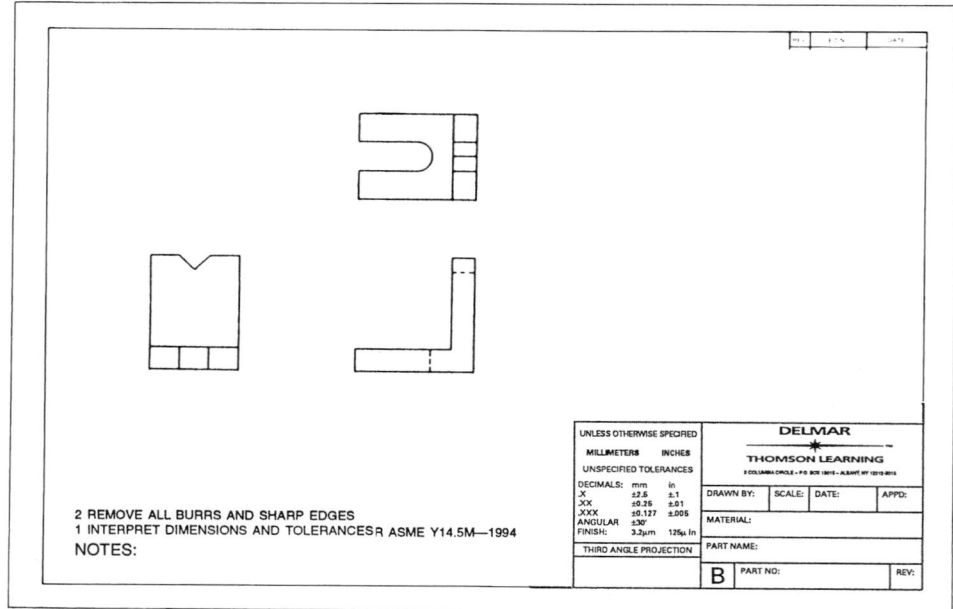

FIGURE 7.59 ■ Completed multiview drawing without dimensions.

When setting up a drawing for completion in a CADD system, many of the same considerations are made. A sketch is used to determine the total drawing area just as with manual drafting. The sheet size can then be determined and established with a Startup Wizard that guides you through the setup process and helps you establish the paper size, or you can use a command such as LIMITS in AutoCAD. The LIMITS command asks you to specify the length and height of the drawing area by establishing the lower left corner and the upper right corner. One nice thing about using CADD is that if you make a mistake in calculating the right sheet size, you can change any time with the LIMITS command.

Another option is to use a template. A *template* in CADD is a file that contains standard settings that are applied to the new drawing. These settings can contain predefined drawing layouts, borders, title blocks, and other common drafting components, features, and standards. Templates are normally available in all of the standard inch and metric sheet sizes. Figure 7.60 shows an ANSI A title block template from AutoCAD.

When using CADD, the views are generally drawn full size and then scaled as needed to fit into a paper layout before plotting or printing.

An earlier discussion explained and showed you how to set up multiviews by projecting features from one view to another. The lines that you use to project between views are construction lines. They are used for layout purposes only and may either be erased or left on the drawing if they are very light and do not reproduce. When using a CADD system, construction lines may also be used. Construction lines in a CADD system are generally placed on their own layer and in their own color. After they have been used to serve their purpose, such as establish views, their layer may be frozen or turned off. The

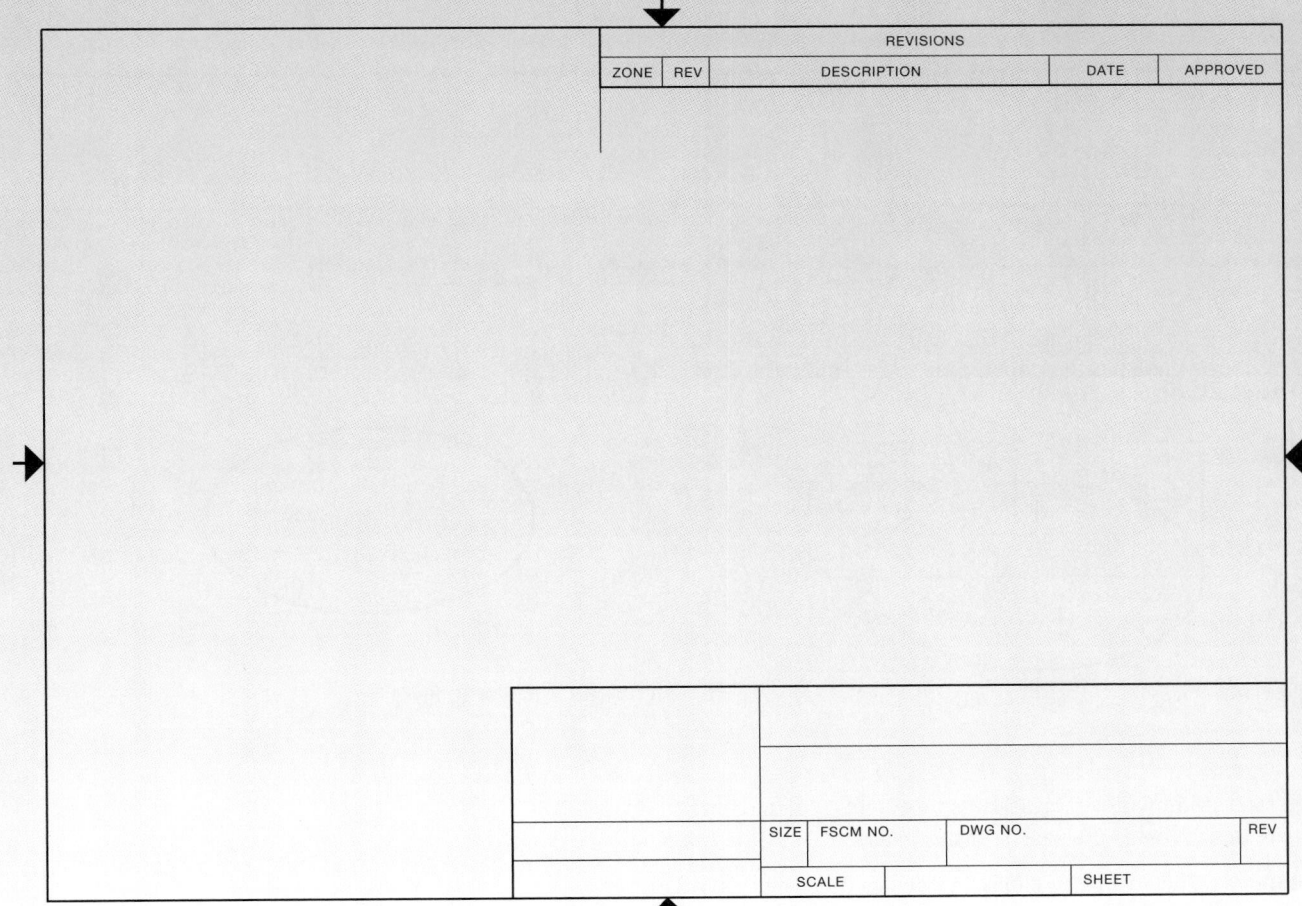

FIGURE 7.60 ■ A CADD ANSI A title block template.

(Continued)

CADD APPLICATIONS

construction lines are only used for layout and construction and are not part of the drawing file when they are turned off. The AutoCAD program has construction lines that are drawn using the XLINE command. The XLINE command creates construction lines of infinite length.

They may be drawn horizontally, vertically at an angle, or offset a given distance from another line or object. The XLINE may also be used to bisect an angle. Figure 7.61 shows how construction lines may be used in CADD to project features between views.

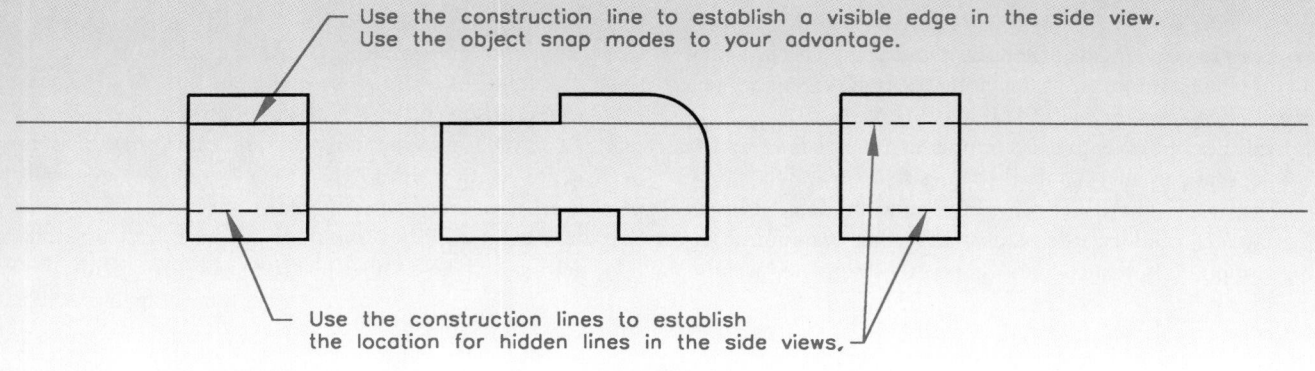

Use the construction line to establish a visible edge in the side view. Use the object snap modes to your advantage.

Use the construction lines to establish the location for hidden lines in the side views.

FIGURE 7.61 ■ CADD drawing with construction lines created using the XLINE command.

MATH APPLICATION

WEIGHT OF AN ELLIPTICAL PLATE

Problem: A large steel plate in the shape of an ellipse is to be moved. The plate has the dimensions shown in Figure 7.63.

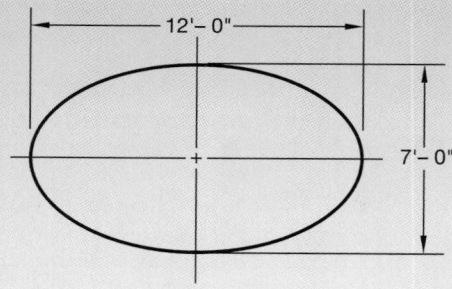

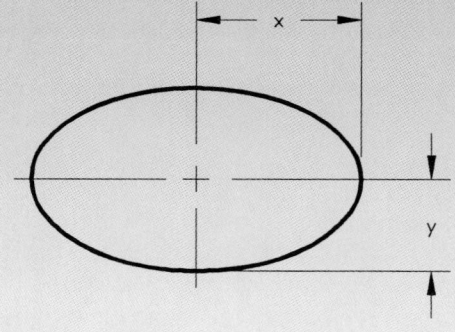

FIGURE 7.64 ■ Ellipse.

The steel is known by its thickness to weigh 19 pounds per square foot. What is the weight of this plate?

Solution: The area of an ellipse, as well as other figures, is found in the Math Instruction Appendix in the Online Companion. It is given by the formula, Area = $\pi x y$, as defined by Figure 7.64.

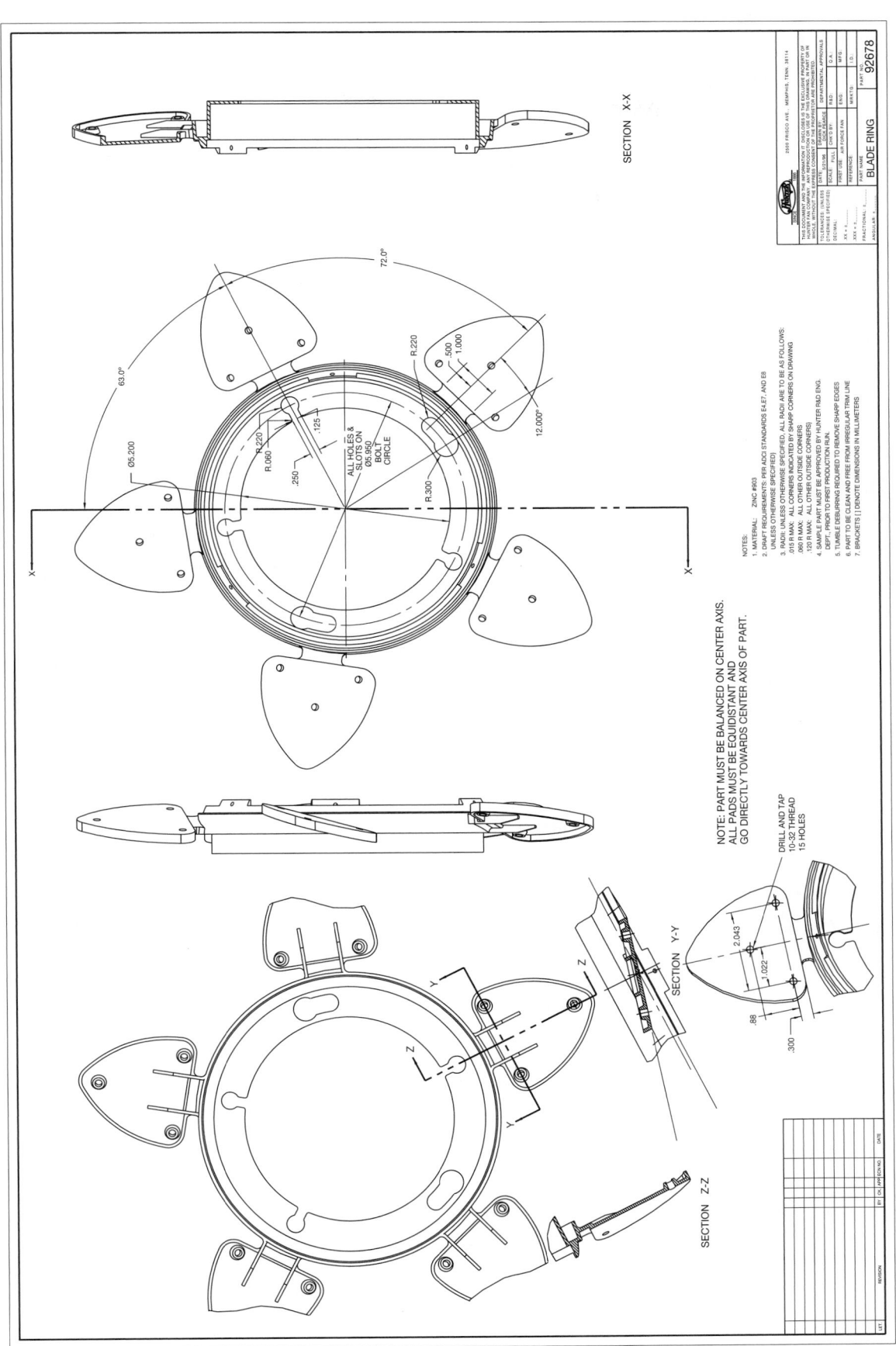

FIGURE 7.62 ■ A complex part displayed on the drawing using several carefully selected views. Courtesy of Hunter Fan Company.

PROFESSIONAL PERSPECTIVE

If you follow the view selection guidelines discussed in this chapter, you should be able to handle any drafting project. Establishing and laying out the necessary views of a part can be one of the most challenging aspects of drafting. The views that you select and the way you lay out the drawing can make the difference between having a drawing that is easy to read and understand or a drawing that is confusing and difficult to read. The practice of view projection that you learned in this chapter provides you with the foundation for effectively selecting and laying out multiviews. Look at the drawing in Figure 7.62, page 189. This is a complex part that is laid out with a number of multiviews that help the reader see every surface of the part. Spend as much time as you can to look at drawings that have been created by professional drafters in an effort to become familiar with practices that are used in industry.

CHAPTER 7

Multiviews Test

DIRECTIONS

Answer the questions with short, complete statements or drawings as needed.

QUESTIONS

1. Why is it important to prepare a sketch before you begin the actual drawing?
2. Explain in brief statements five basic guidelines to follow when selecting the front view of an object for multiview presentation.
3. List two reasons why views are aligned on a drawing.
4. What methods are used to space views on a drawing?
5. Show an example of third-angle projection.
6. Show an example of first-angle projection.
7. Given the pictorial drawings below, draw the six basic views of each using multiview projection. Show all construction lines. Use your dividers to transfer dimensions from the pictorials. Make your drawings 2× as large as the given.

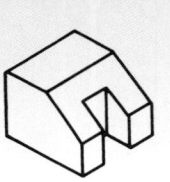 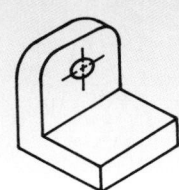 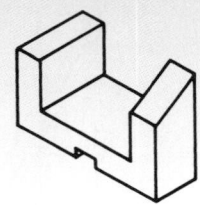

8. Given the objects below, complete the unfinished views using proper runouts. Transfer the drawings to your answer sheet. Make your drawings 2× as large as the given.

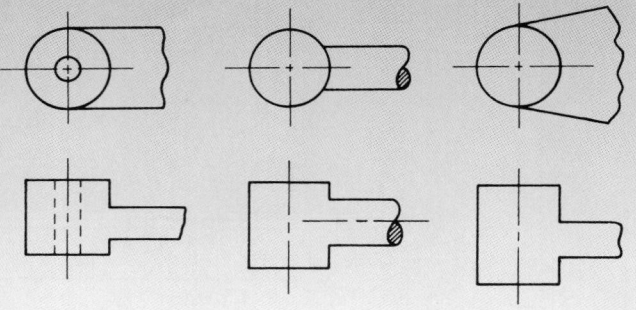

CHAPTER 7

Multiviews Problems

DIRECTIONS

You are given an object with its surfaces projected to the planes of the glass box. This can help you visualize as you draw the same views on the folded out planes of the glass box at the left. Measure the given objects and transfer the measurements to your manual or CADD drawing. You can also photocopy the textbook pages and draw the views on the copies. Confirm the preferred method with your instructor.

PART 1

PROBLEM 7-1

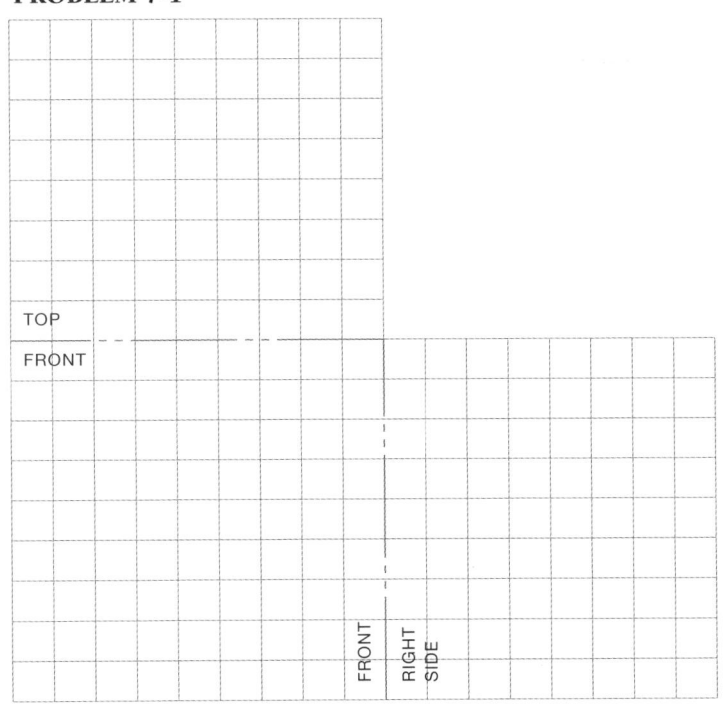

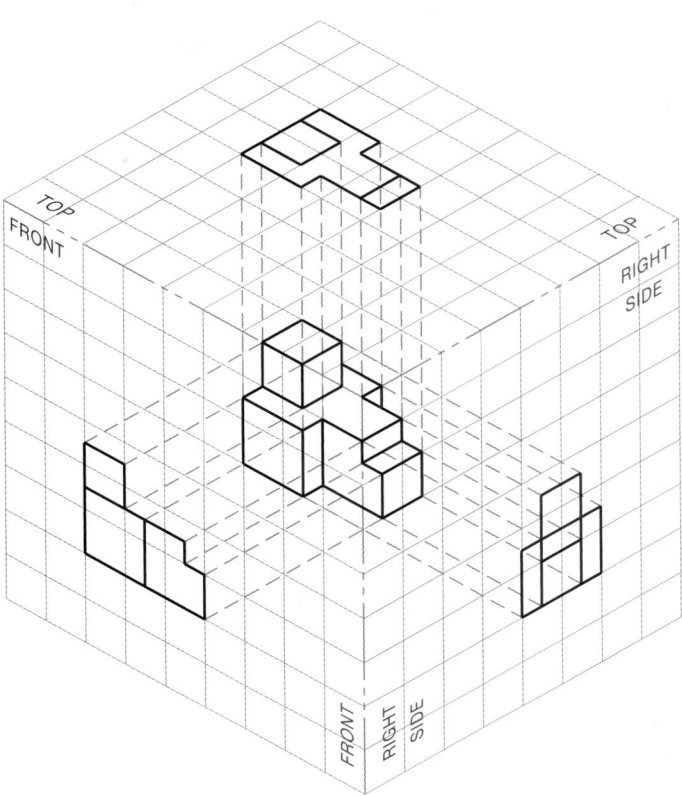

PROBLEM 7-2

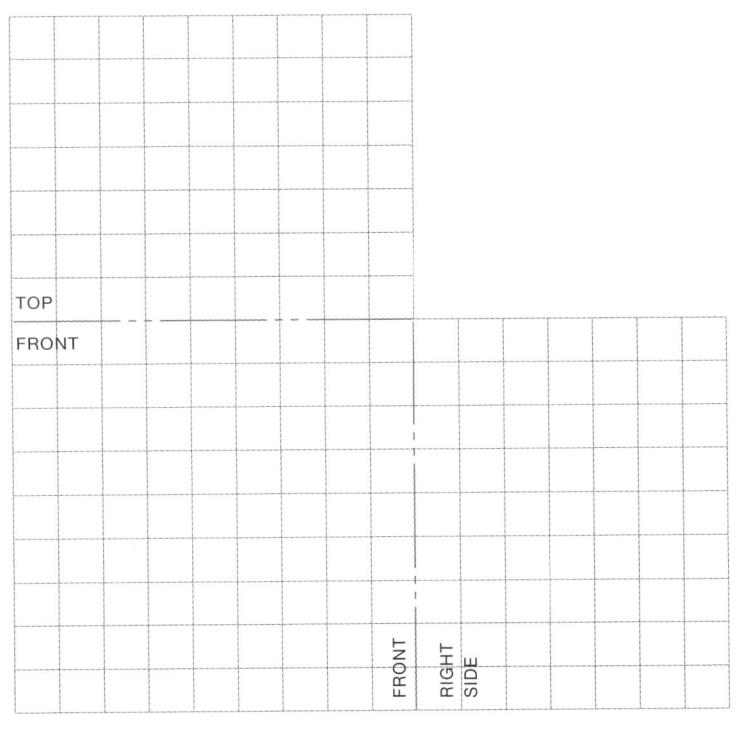

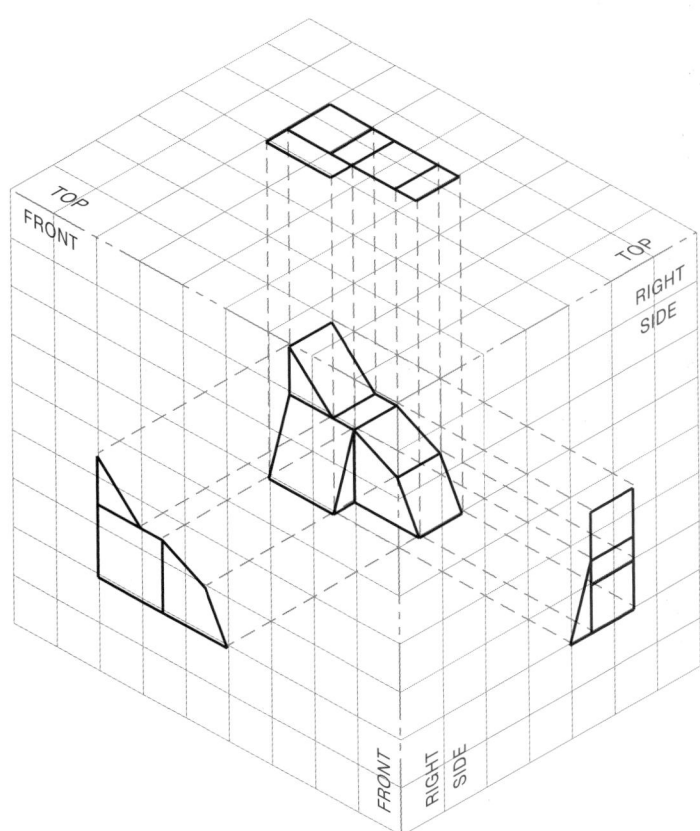

PROBLEM 7-3

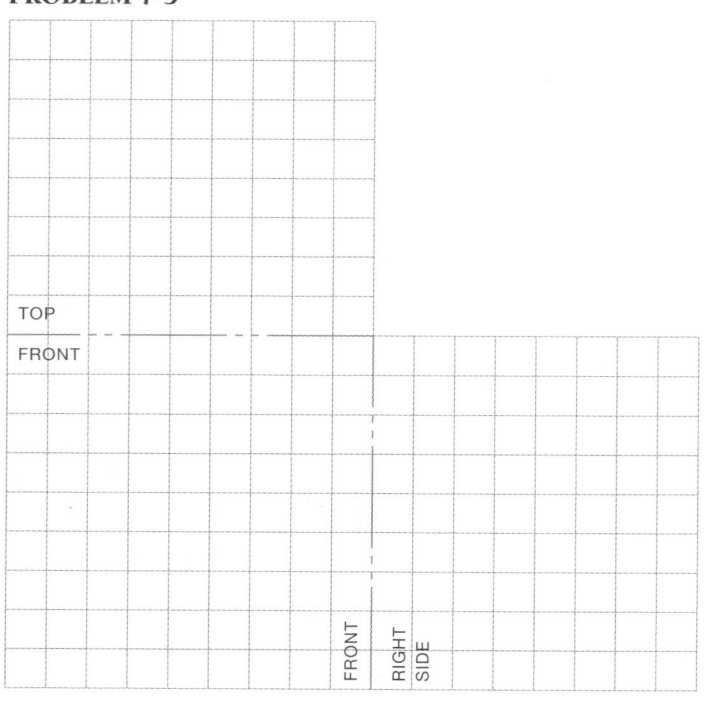

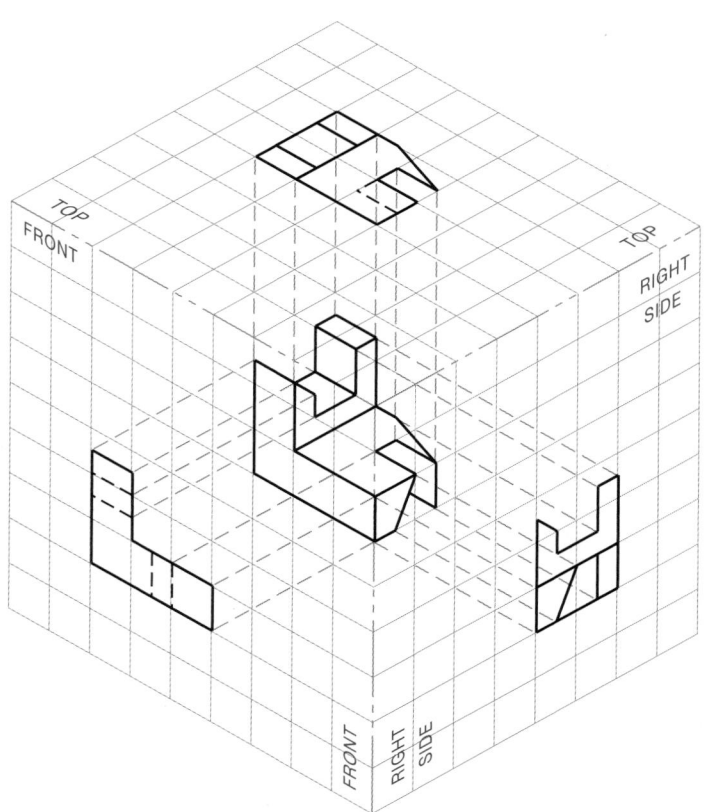

PROBLEM 7-4

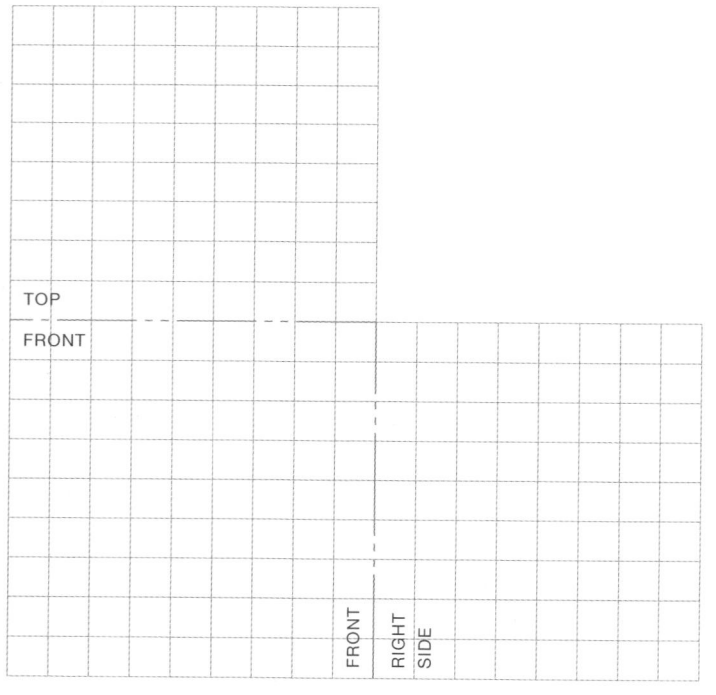

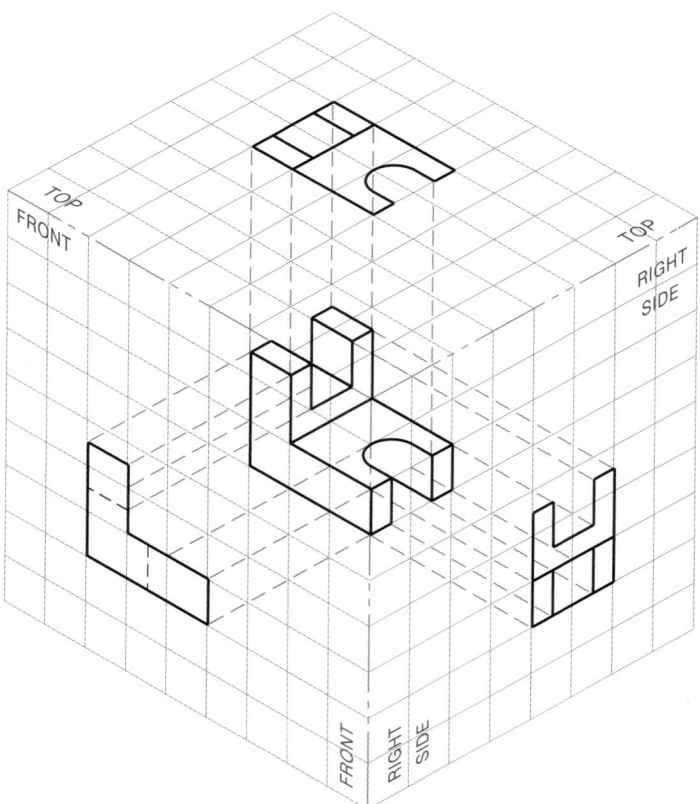

PROBLEM 7-5

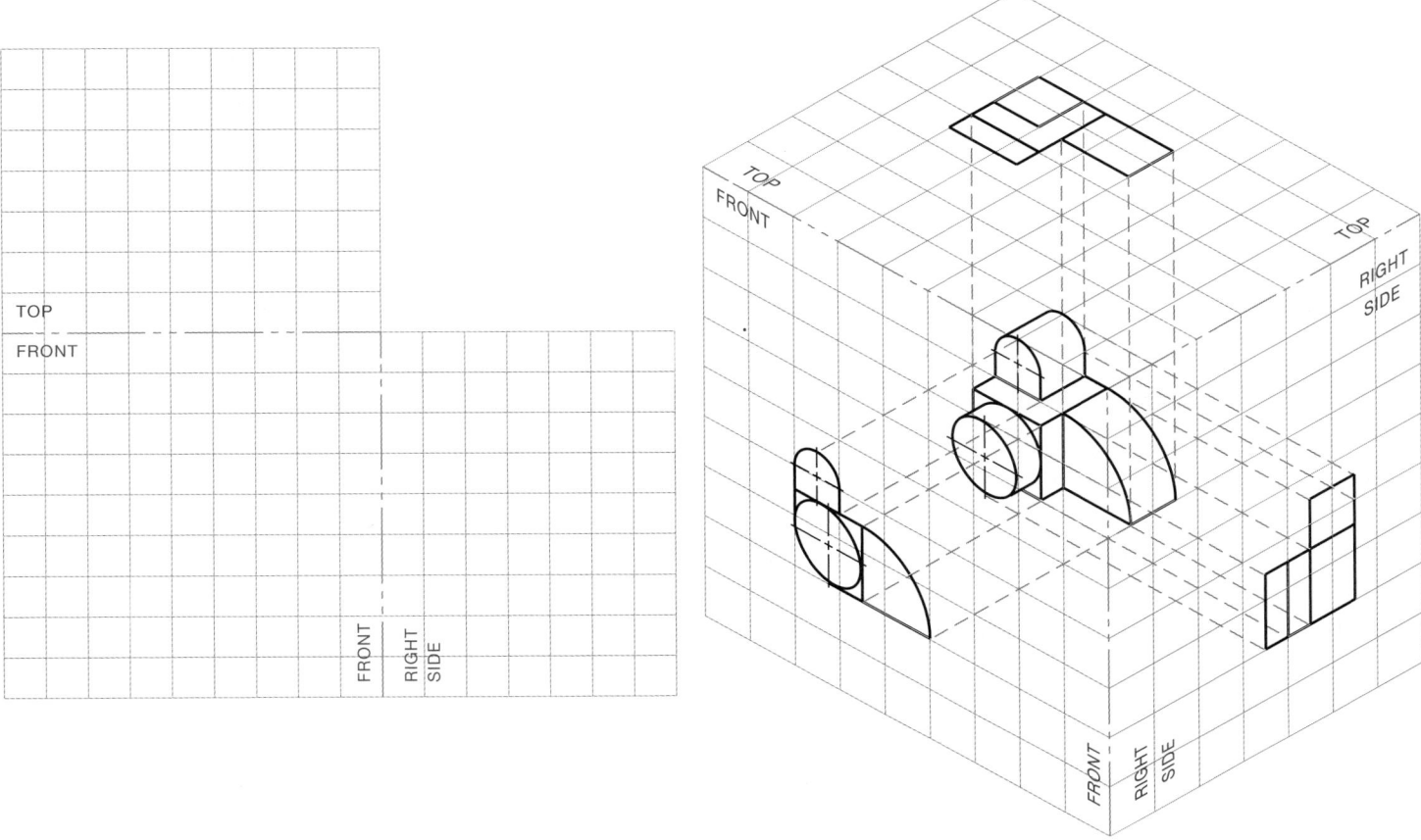

PART 2

DIRECTIONS

The following problems provide you with views that contain missing lines or missing views. Draw the missing lines or missing views as appropriate. Some problems provide you with a pictorial view to aid in visualization. You do not need to draw the pictorial view.

Measure the given views and transfer the measurements to your manual or CADD drawing. You can also photocopy the textbook pages and draw the views on the copies. Confirm the preferred method with your instructor.

PROBLEM 7-6

PROBLEM 7-8

PROBLEM 7-7

PROBLEM 7-9

PROBLEM 7-10

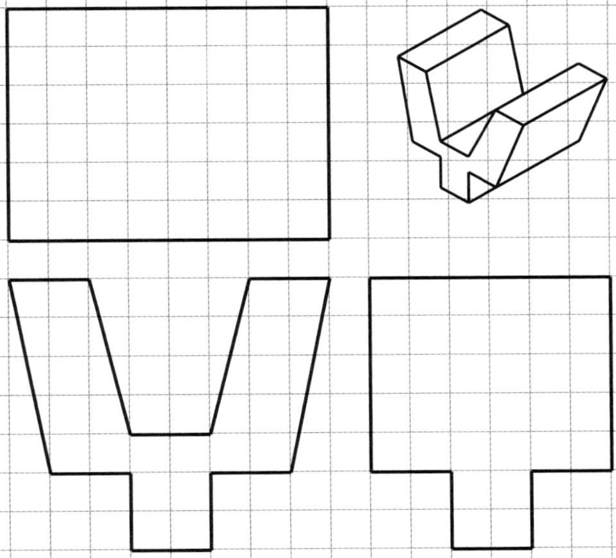

PROBLEM 7-12

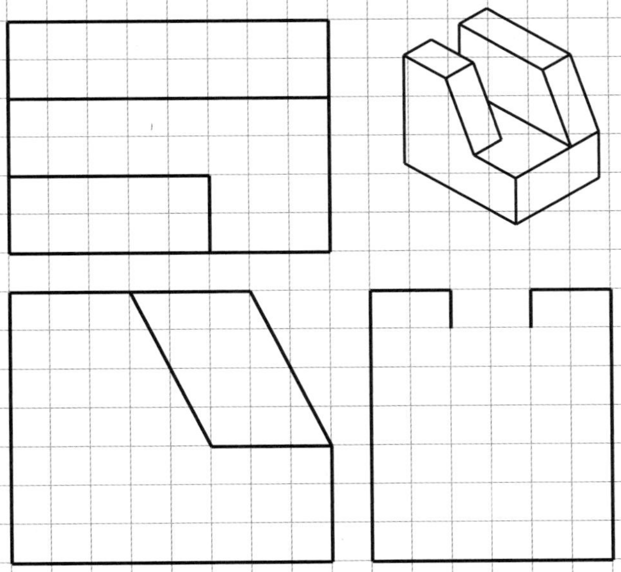

PROBLEM 7-11

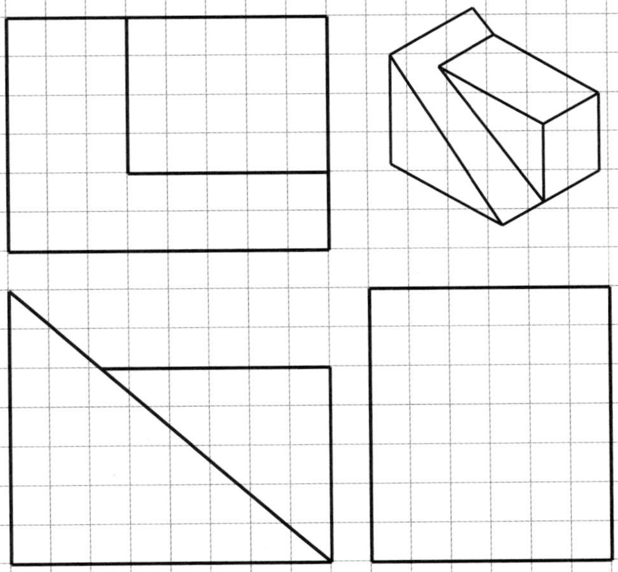

PROBLEM 7-13

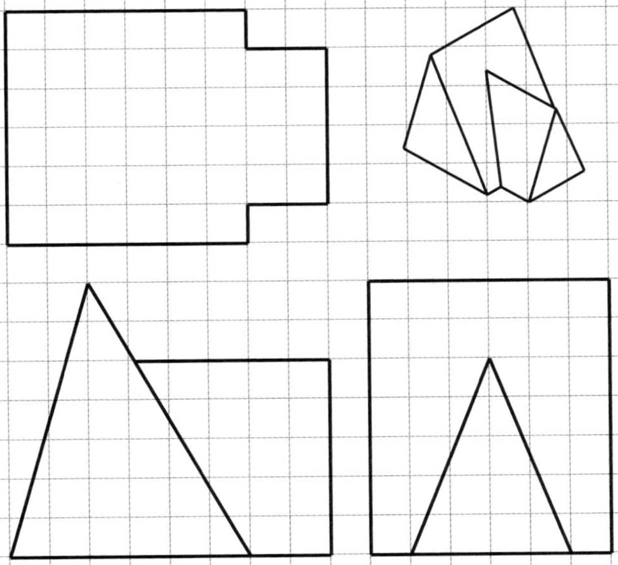

PROBLEM 7-14

PROBLEM 7-16

PROBLEM 7-15

PROBLEM 7-17

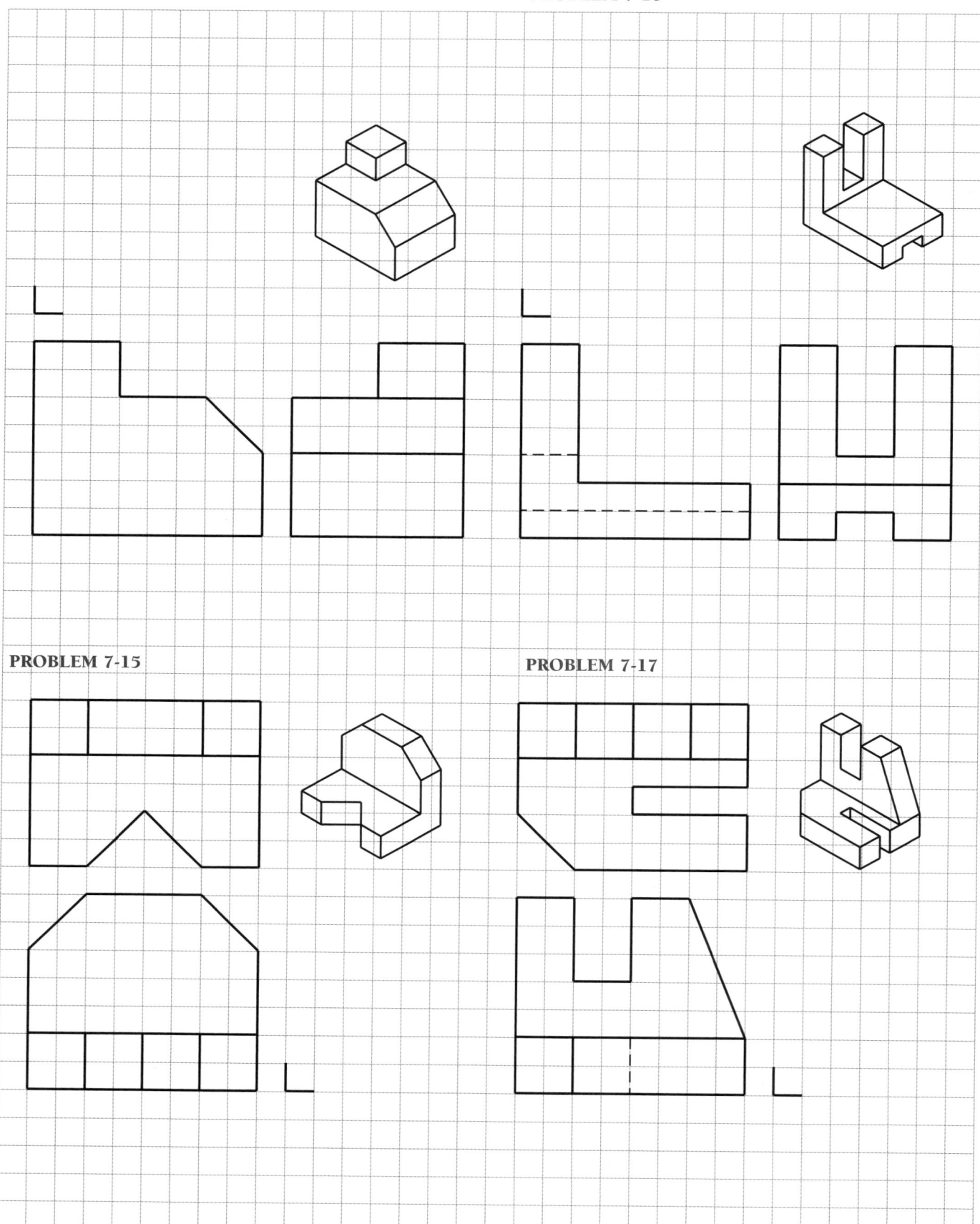

PROBLEM 7-18

PROBLEM 7-20

PROBLEM 7-19

PROBLEM 7-21

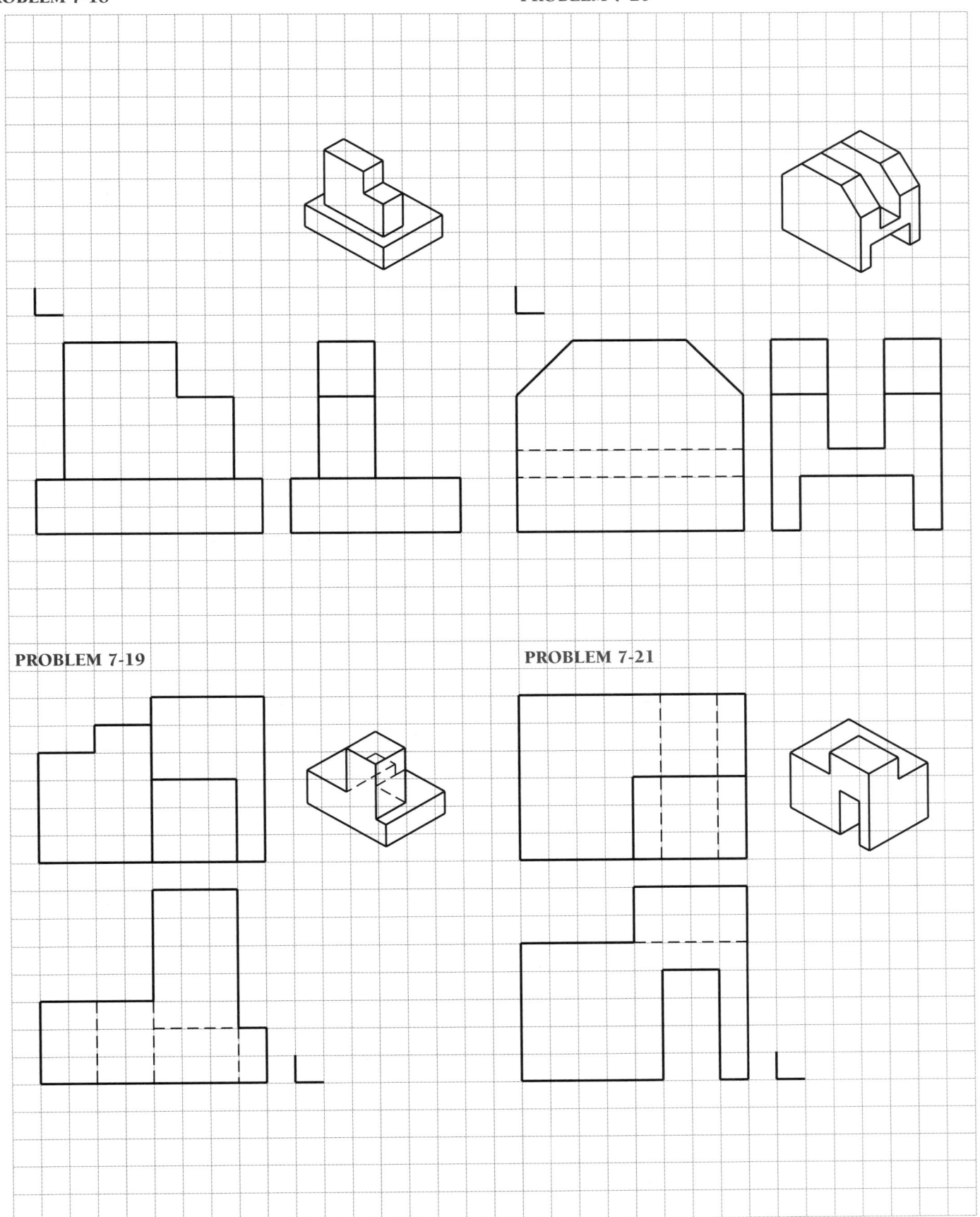

PROBLEM 7-22

PROBLEM 7-25

PROBLEM 7-23

PROBLEM 7-26

PROBLEM 7-24

PROBLEM 7-27

PROBLEM 7-28

PROBLEM 7-31

PROBLEM 7-29

PROBLEM 7-32

PROBLEM 7-30

PROBLEM 7-33

PROBLEM 7-34

PROBLEM 7-37

PROBLEM 7-35

PROBLEM 7-38

PROBLEM 7-36

PROBLEM 7-39

PROBLEM 7-40

PROBLEM 7-43

PROBLEM 7-41

PROBLEM 7-44

PROBLEM 7-42

PROBLEM 7-45

PART 3

DIRECTIONS

These problems provide you with views that contain missing lines or missing views. Draw the missing lines or missing views as appropriate. Pictorial views are provided to aid in visualization. You do not need to draw the pictorial view. Measure the given views and transfer the measurements to your manual or CADD drawing. Set up your drawings with a properly sized border and title block. Properly complete the information in the title block.

PROBLEM 7-46 Pocket Block

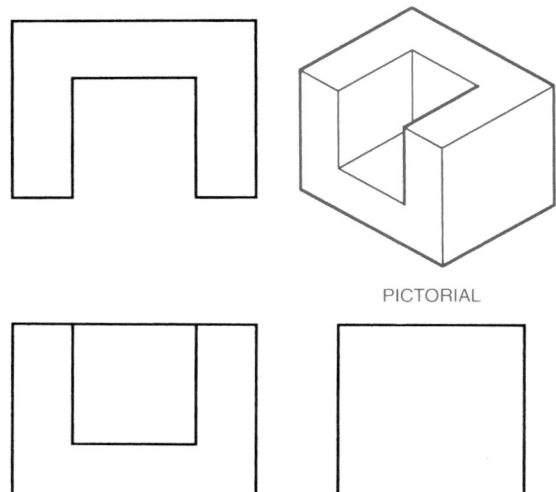

PICTORIAL

PROBLEM 7-47 Angle Gage

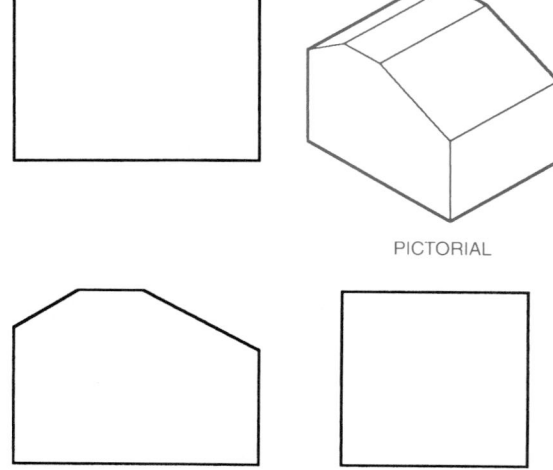

PICTORIAL

PROBLEM 7-48 Base

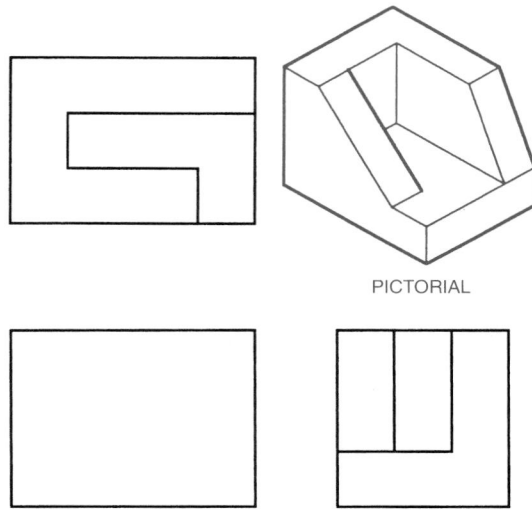

PICTORIAL

PROBLEM 7-49 Corner Block

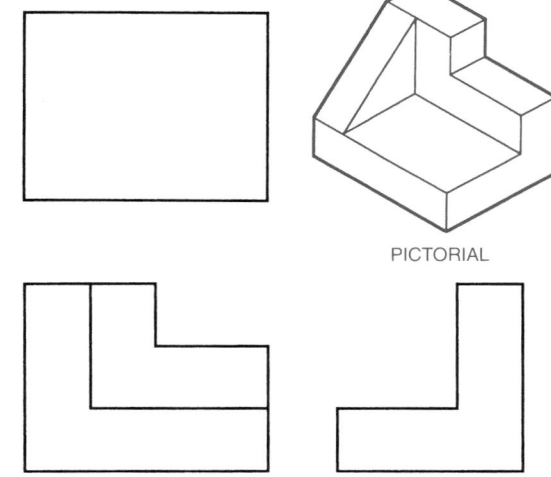

PICTORIAL

PROBLEM 7-50 Cylinder Block

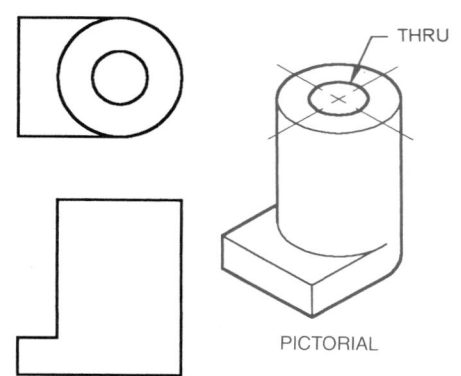

THRU

PICTORIAL

PROBLEM 7-51   Shaft Block

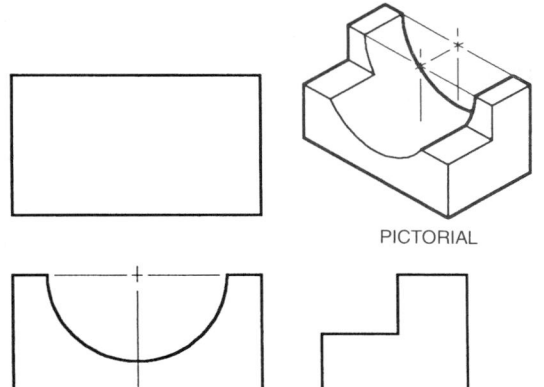

PICTORIAL

PROBLEM 7-52   Gib

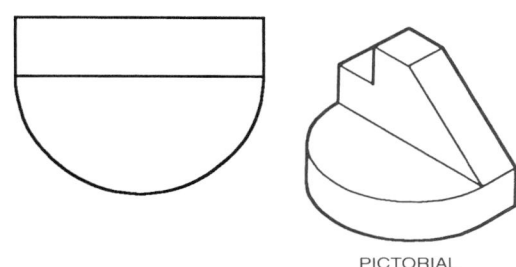

PICTORIAL

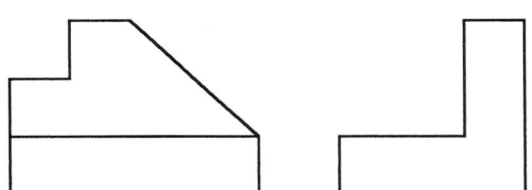

PROBLEM 7-53   Eccentric

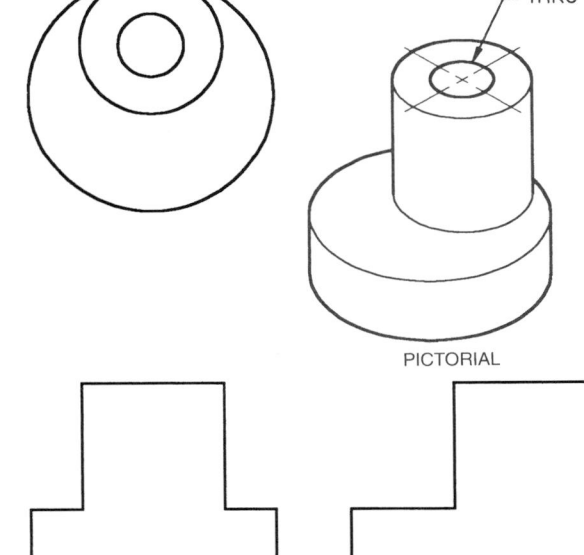

THRU

PICTORIAL

PROBLEM 7-54   Guide Block

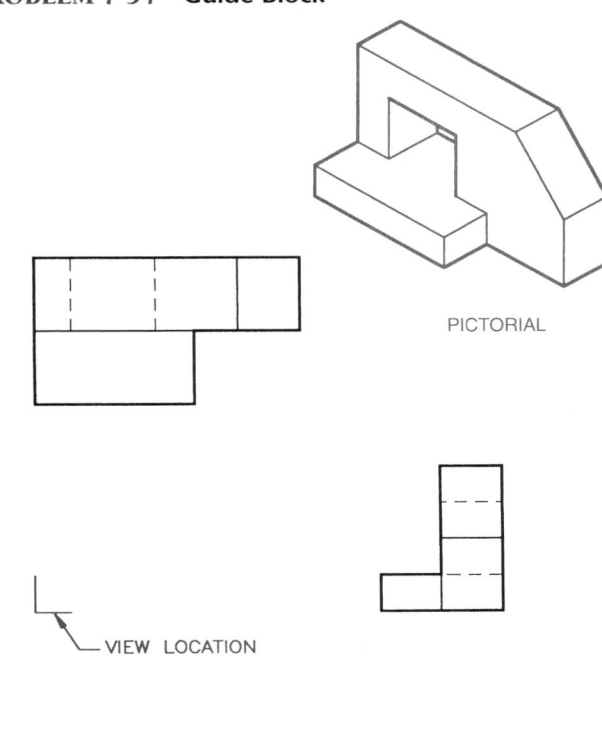

PICTORIAL

VIEW LOCATION

PROBLEM 7-55   Key Slide

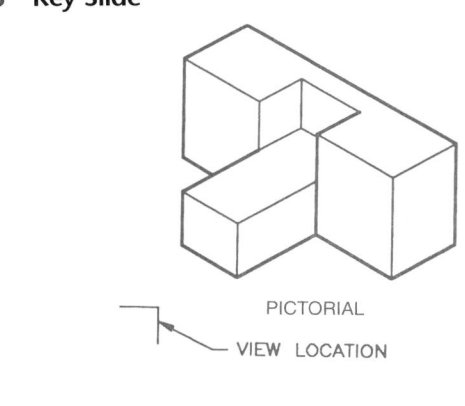

PICTORIAL

VIEW LOCATION

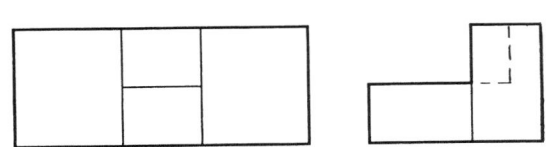

PROBLEM 7-56   Angle Bracket

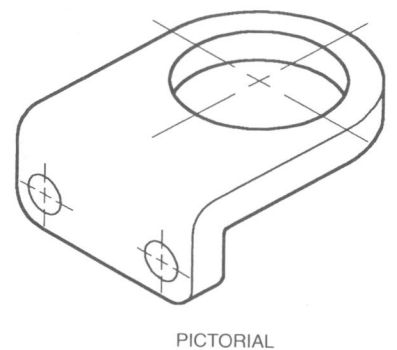

PICTORIAL

VIEW LOCATION

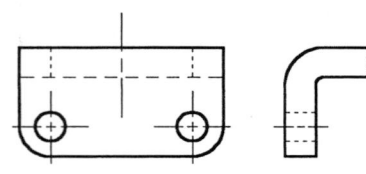

PROBLEM 7-57   Clevis

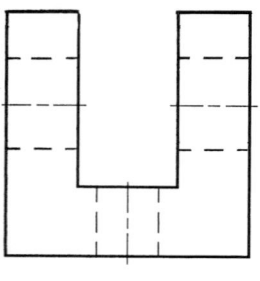

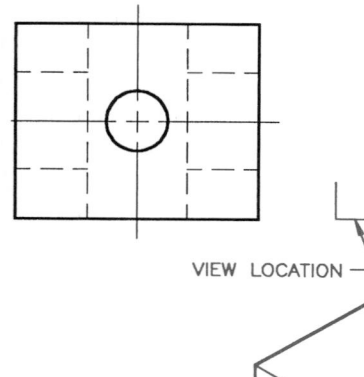

VIEW LOCATION

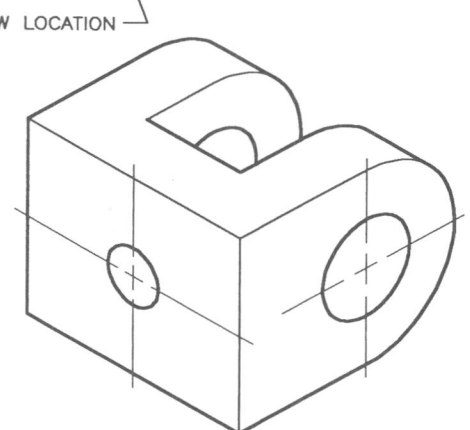

PART 4

DIRECTIONS

These problems provide you with pictorial views that contain dimensions. A suggested view layout is provided for your reference. Draw the views. The pictorial views are provided to aid in visualization. You do not need to draw the pictorial view. Use the given dimensions to create your manual or CADD drawing. Set up your drawings with a properly sized border and title block. Properly complete the information in the title block. Do not draw the dimensions.

PROBLEM 7-58   V-block (metric)

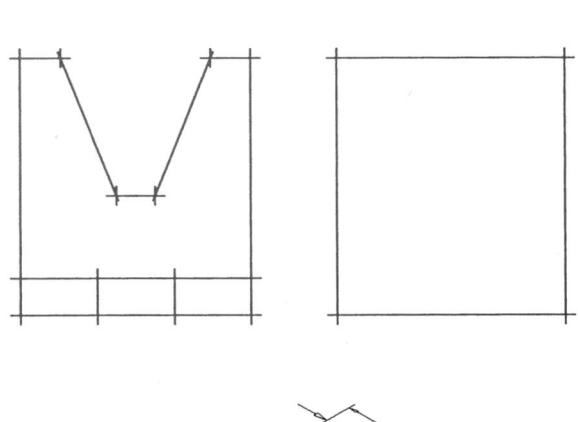

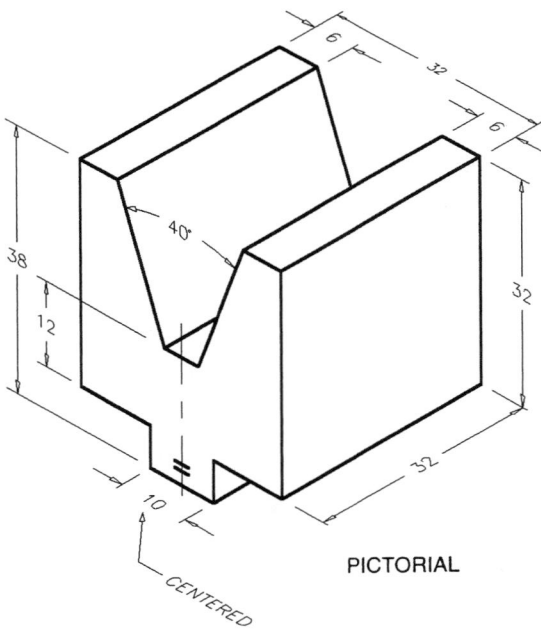

PICTORIAL

PROBLEM 7-59   Guide Base (in.)

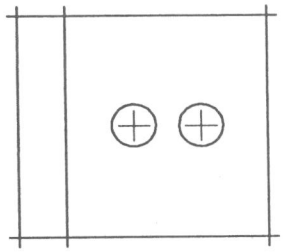

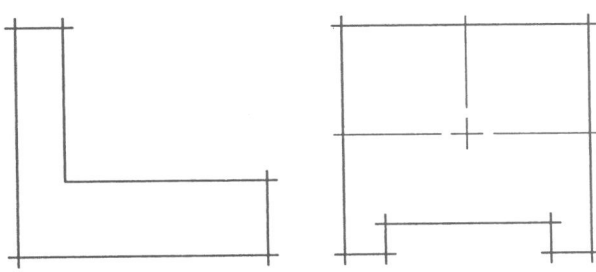

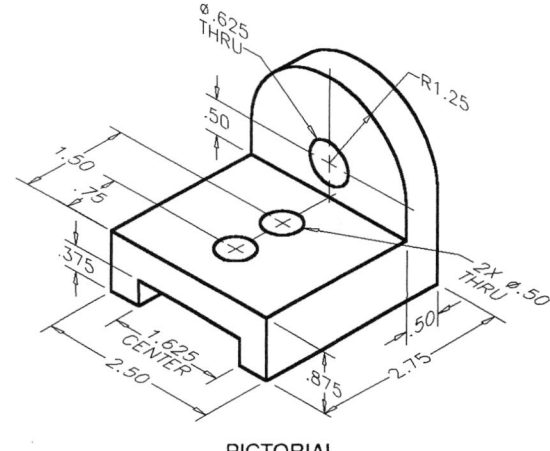

PICTORIAL

PART 5

DIRECTIONS

These problems provide you with a pictorial view or multiview lay-out that contains dimensions. Use the given information to select and draw the necessary multiviews using manual drafting or CADD. Do not draw the pictorial view. Set up your drawings with a properly sized border and title block. Properly complete the infor-mation in the title block. Do not draw the dimensions.

PROBLEM 7-60   Cylindrical object (in.)
Part Name: Sleeve Bearing
Material: Phosphor Bronze
Problem based on original art courtesy Production Plastics.

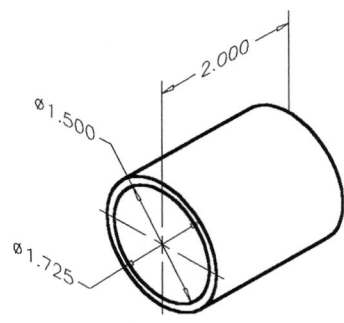

PROBLEM 7-61   One view (in.)
Part Name: Pin
Material: SAE1035

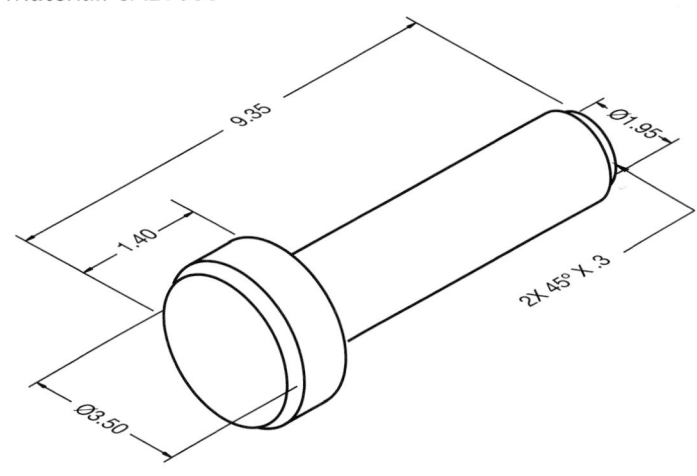

PROBLEM 7-62   Arc and hole (in.)
Part Name: Door Lock
Material: Mild Steel

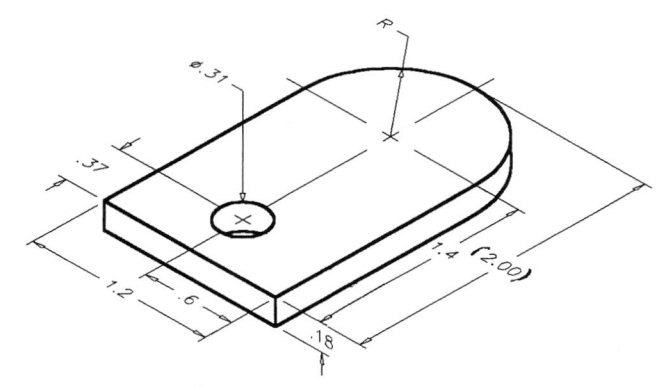

PROBLEM 7-63   Two views (in.)
Part Name: Washer
Material: SAE 1060

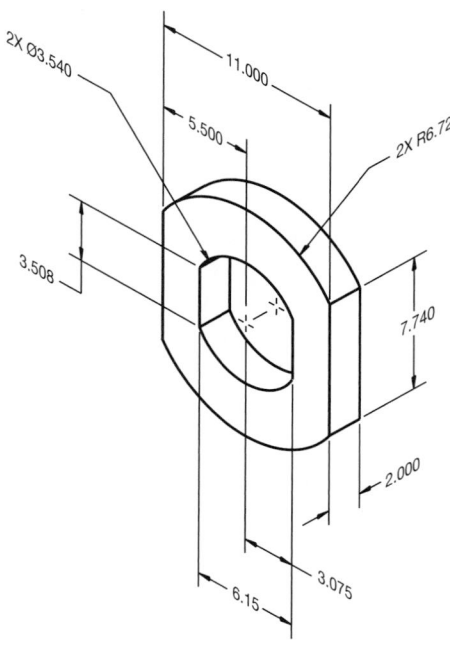

PROBLEM 7-64   Planes and slot (in.)
Part Name: Step Block
Material: Mild Steel

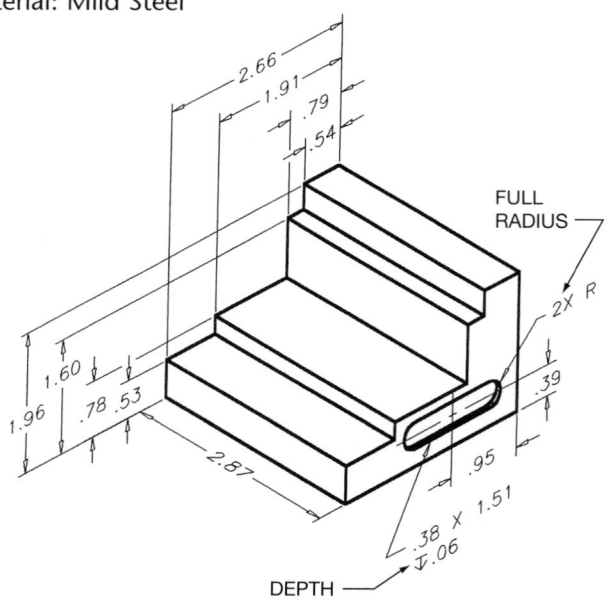

PROBLEM 7-65   Cylinders and circles (in.)
Part Name: Roll End Bearing
Material: Phosphor Bronze
Problem based on original art courtesy Production Plastics.

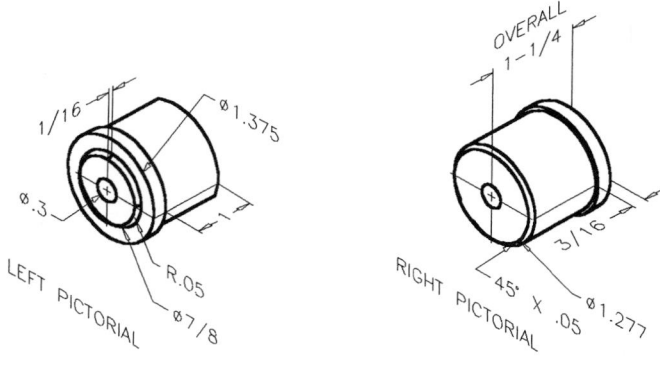

PROBLEM 7-66   Cylinders and circles (in.)
Part Name: Roll End Bearing
Material: Phosphor Bronze

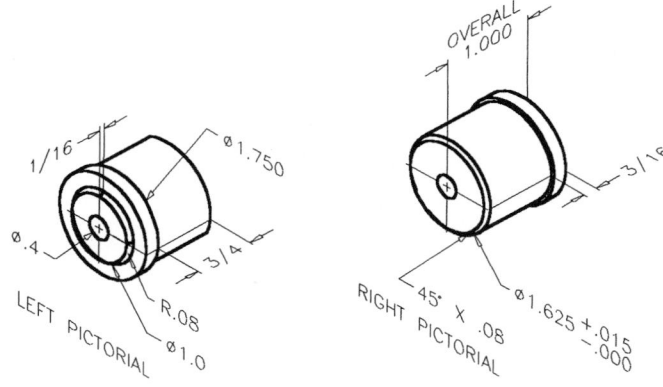

PROBLEM 7-67   Two views with holes (in.)
Part Name: Pivot Bracket
Material: SAE 3135

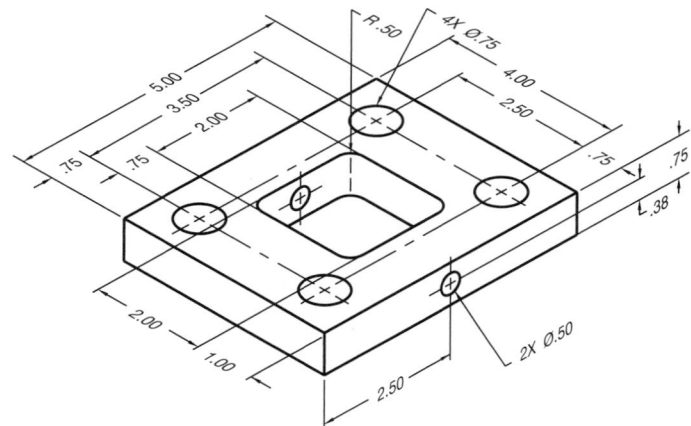

PROBLEM 7-68 Three views with holes and slot (in.)
Part Name: Sliding Bracket
Material: Phosphor Bronze

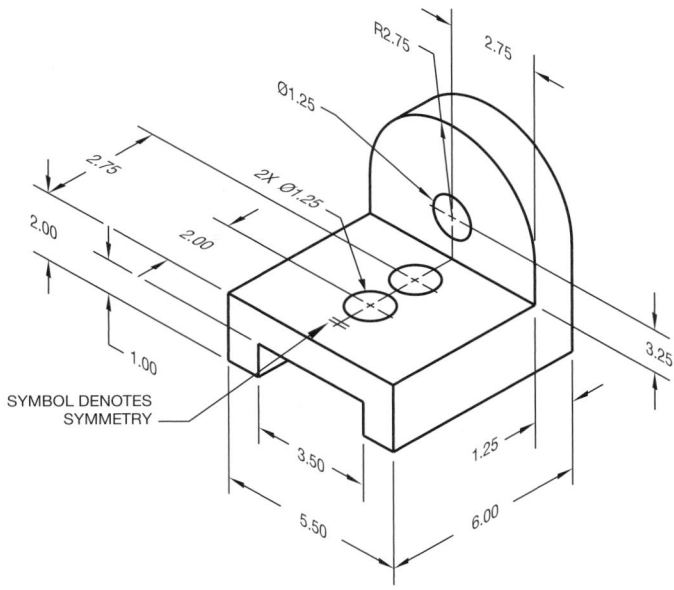

PROBLEM 7-70 View enlargement (in.)
Part Name: Drill Gauge
Material: 16-GA Mild Steel

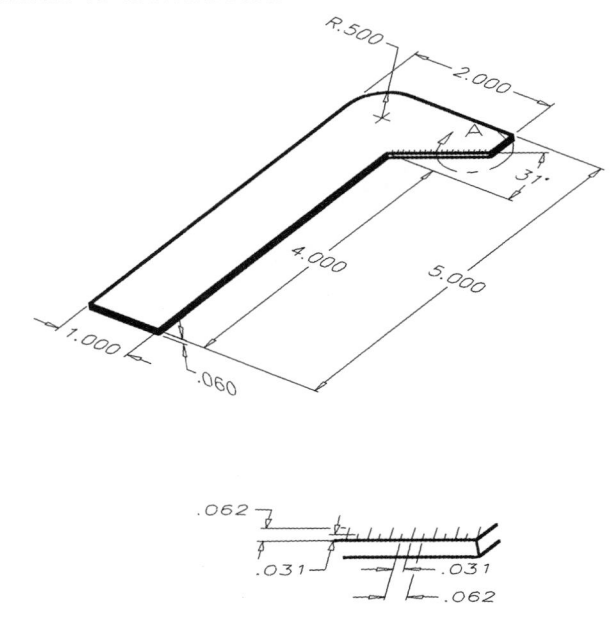

DETAIL A
SCALE 4:1

PROBLEM 7-69 Circle arcs and planes (in.)
Part Name: V-block Clamp
Material: SAE 1080

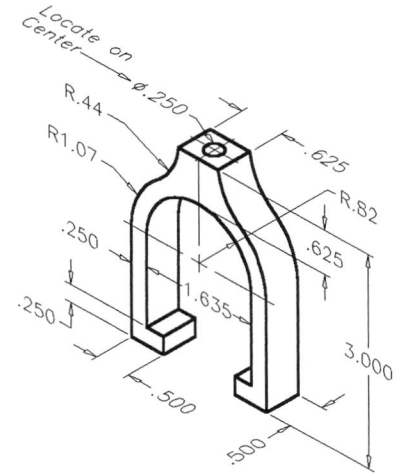

PROBLEM 7-71 Two views with angled surface, counter-
sinks (in.)
Part Name: Angle Bracket
Material: SAE 1145

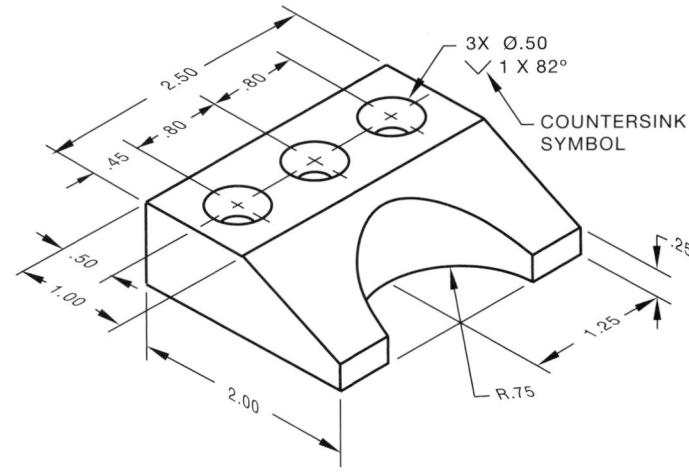

PROBLEM 7-72 **Multiple features (in.)**
Part Name: Lock Ring
Material: SAE 1020

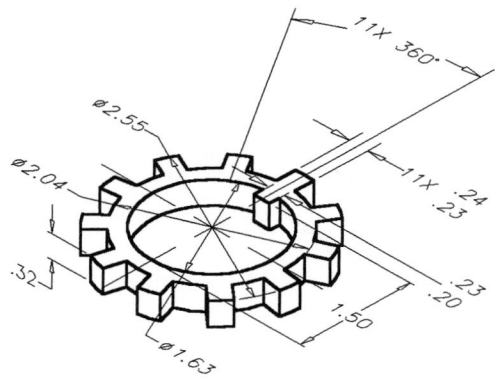

PROBLEM 7-73 **Angles and holes (in.)**
Part Name: Bracket
Material: SAE 1020

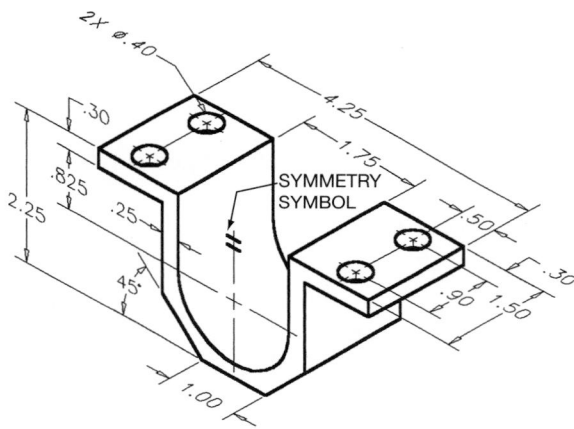

PROBLEM 7-74 **Multiviews (metric)**
Part Name: Guide Rail
Material: SAE 4310

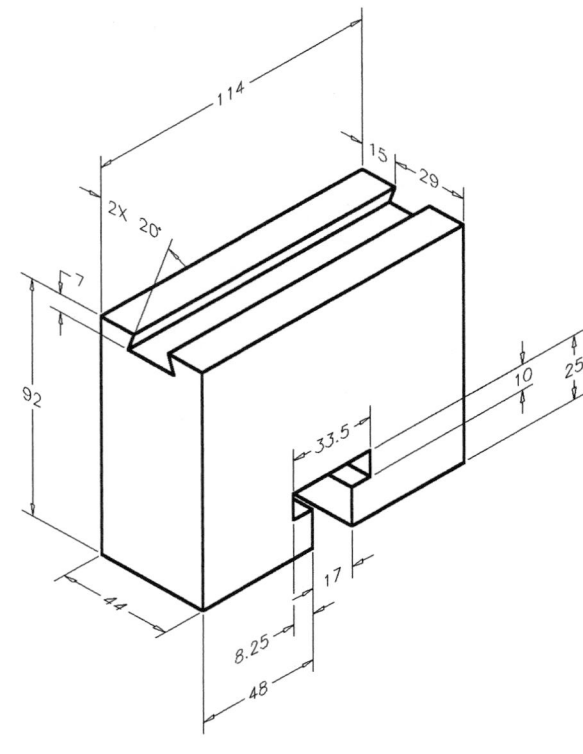

PROBLEM 7-75 **Multiviews (in.)**
Part Name: Support Bracket
Material: Plastic

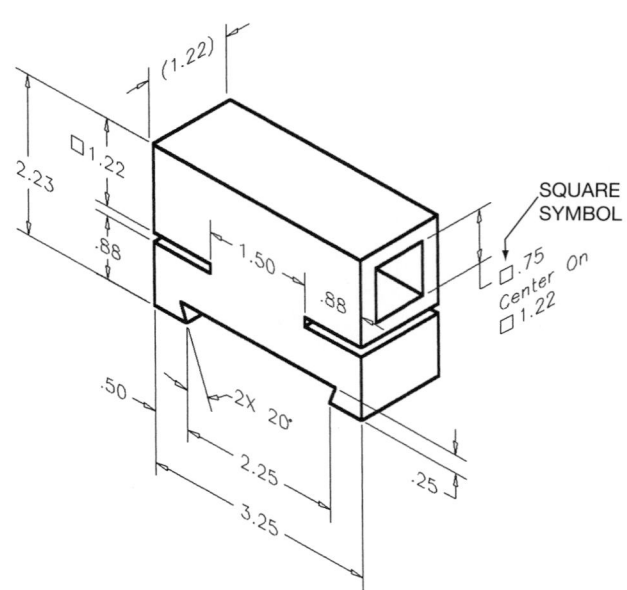

PROBLEM 7-76 **Multiviews (in.)**
Part Name: V-block
Material: A-Steel

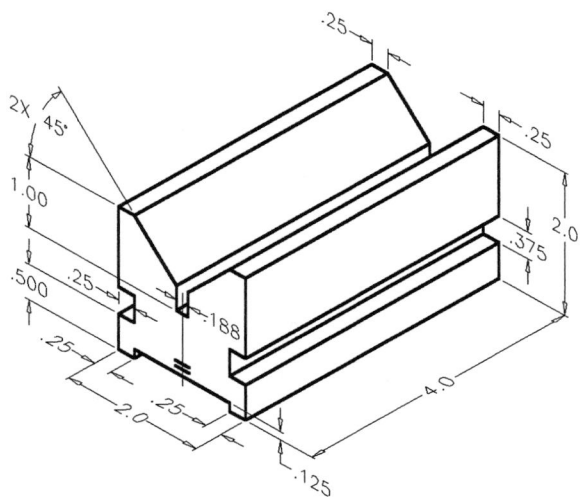

PROBLEM 7-77 **Multiviews, angles, and holes (metric)**
Part Name: Angle Bracket
Material: Mild Steel
Center 2X Ø15 holes and provide location dimension.

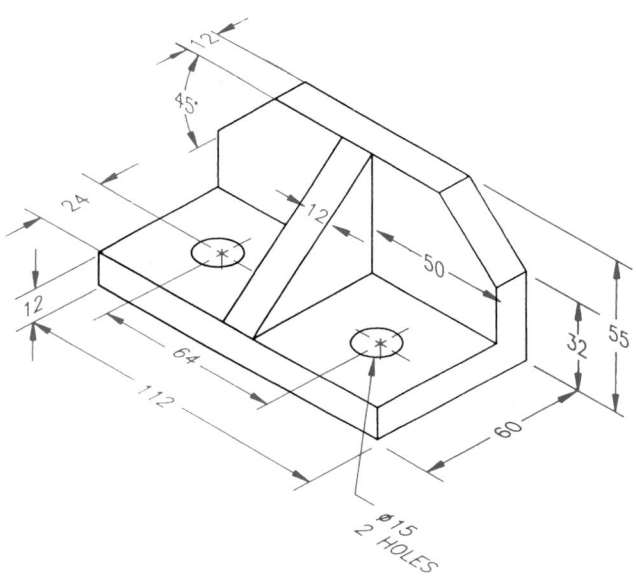

PROBLEM 7-78 **Multiviews (in.)**
Part Name: Support
Material: Mild Steel

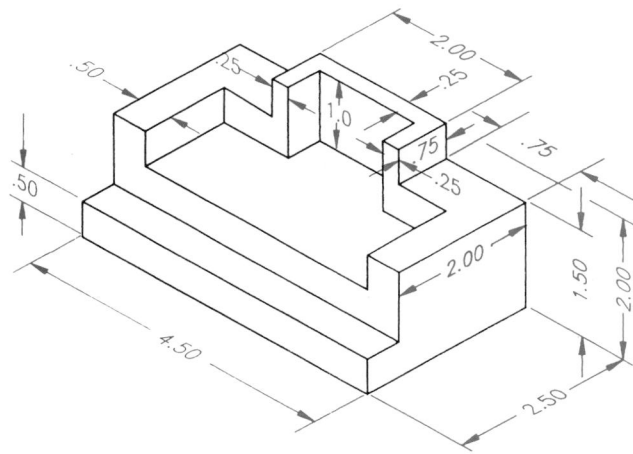

PROBLEM 7-79 **Multiviews, circles, and arcs (in.)**
Part Name: Chain Link
Material: SAE 4320

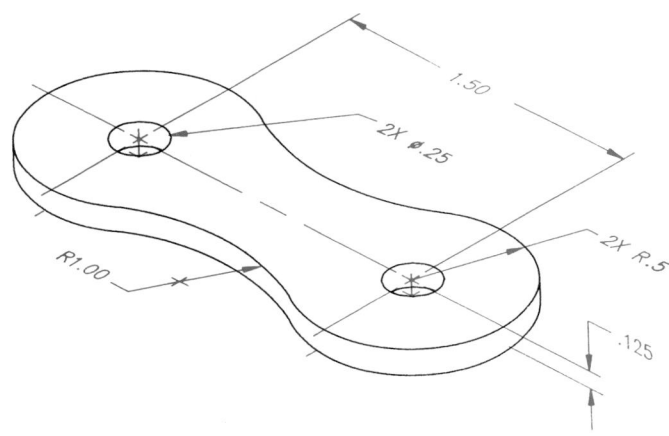

PROBLEM 7-80 **Multiviews, circles, and arcs (in.)**
Part Name: Pivot Bracket
Material: Cold-Rolled Steel

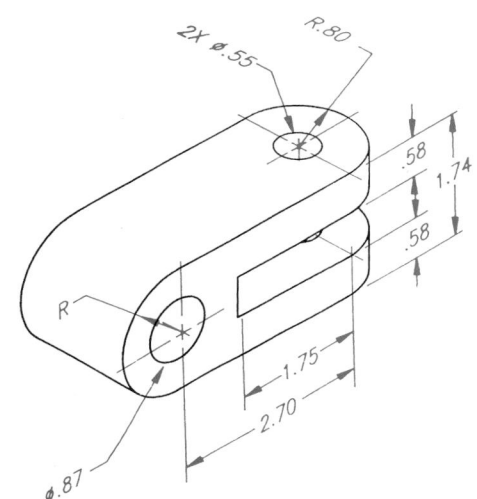

PROBLEM 7-81 **Multiview arcs and circles (metric)**
Part Name: Hinge Bracket
Material: Cast Aluminum

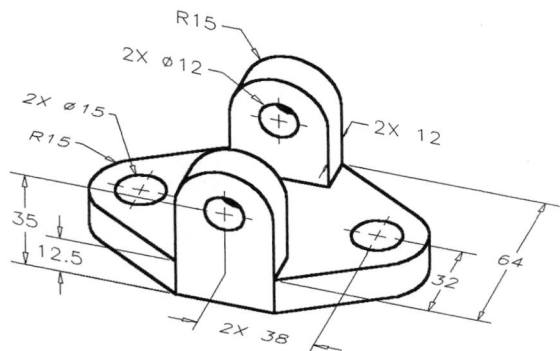

PROBLEM 7-82 **Multiview circles and arcs (in.)**
Part Name: Bearing Support
Material: SAE 1040

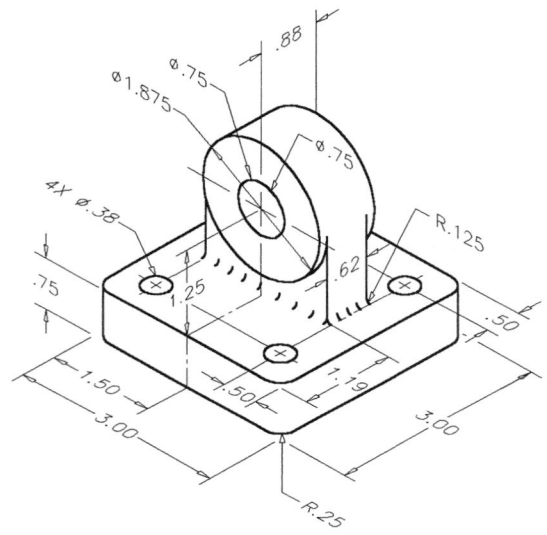

PROBLEM 7-83 **Multiviews (metric)**
Part Name: Gate Latch Base
Material: Aluminum

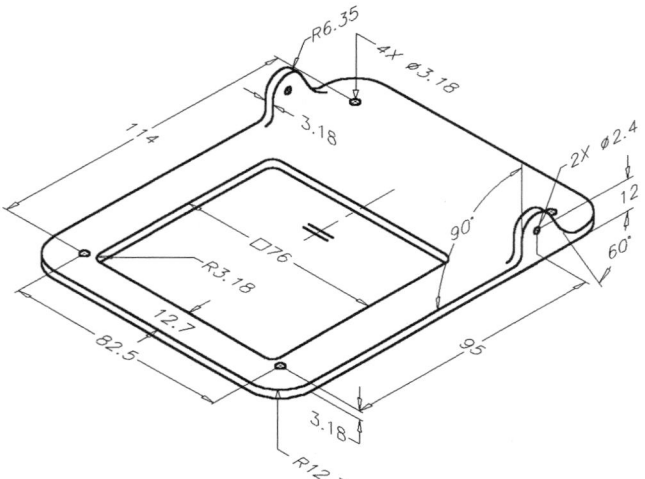

PROBLEM 7-84 **Multiviews (in.)**
Part Name: Mounting Bracket
Material: SAE 1020
All fillets R.13

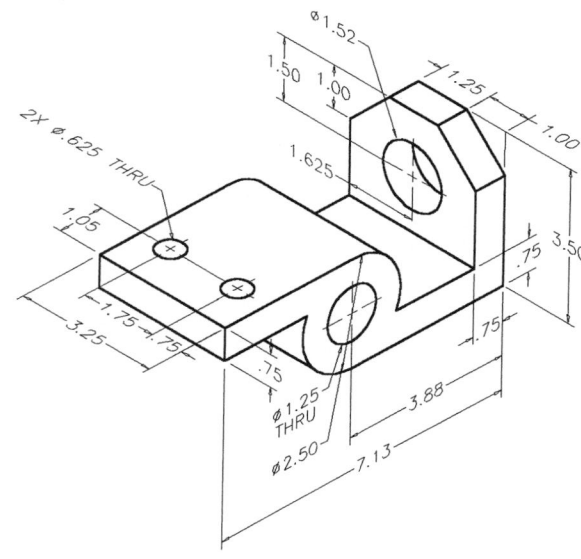

PROBLEM 7-85 **Multiviews (in.)**
Part Name: Support Base
Material: SAE 1040

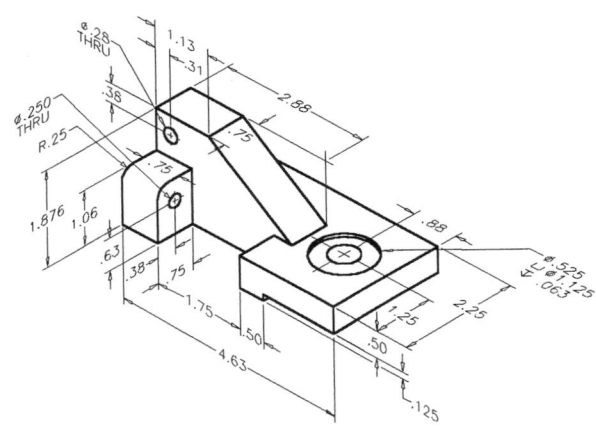

MATH PROBLEMS
Find the areas of these figures:

PROBLEM 7-86

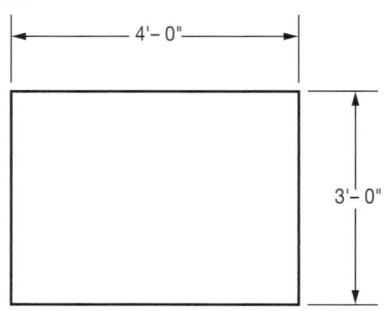

PROBLEM 7-87

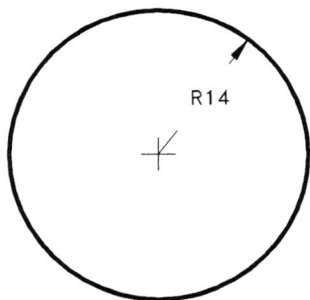

PROBLEM 7-88

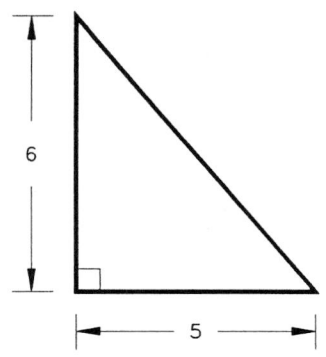

PROBLEM 7-89

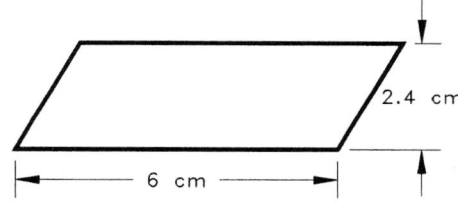

PROBLEM 7-90

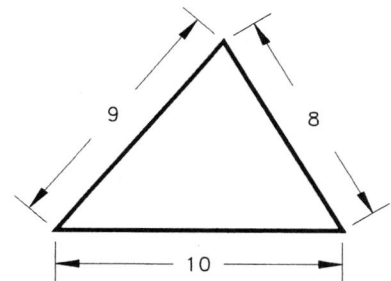

PROBLEM 7-91

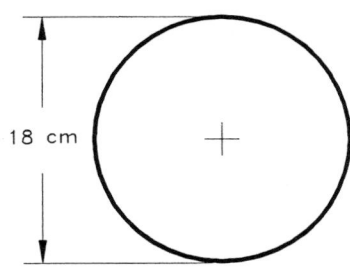

PROBLEM 7-92 What is the weight of a steel plate in the shape of a circle with a diameter of 10'? The steel weighs 14 lb per ft^2.

Find the volume of:

PROBLEM 7-93 A spherical holding tank 15' in diameter.

PROBLEM 7-94 A cylinder with a diameter of 10' and a height of 4'.

PROBLEM 7-95 A 7'- high cone having a base with a radius of 3'.

Auxiliary Views

OBJECTIVES

After completing this chapter, you will:

- Describe the purpose of an auxiliary view.
- Explain how an auxiliary view is projected.
- Discuss and draw viewing-plane lines related to auxiliary views.
- Draw primary and secondary auxiliary views along with the related multiviews from given engineering problems.

THE ENGINEERING **DESIGN** APPLICATION

You have been asked to create a multiview detail drawing for a guide bracket, and the engineer has provided a sketch of the part. (See Figure 8.1.) As you study the sketch, you discover that one of the surfaces on the bracket is not parallel to any of the six principal viewing planes. You determine that your multiview drawing will need an auxiliary view in order to show the angled face in its true size and shape. Through sketching your layout ideas, you conclude that a complete auxiliary view would not add clarity to the drawing and decide instead on a partial auxiliary view. The final layout is shown in Figure 8.2.

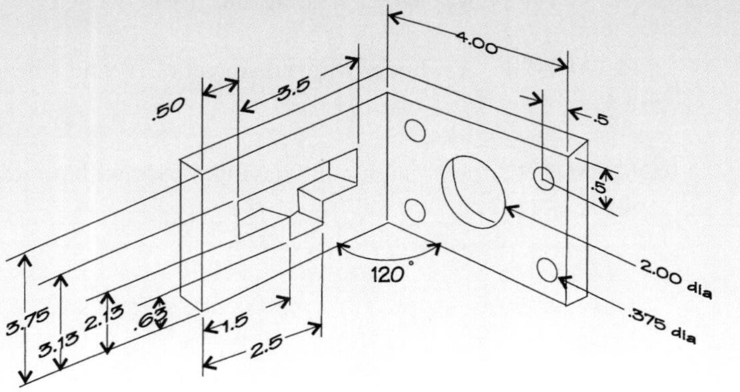

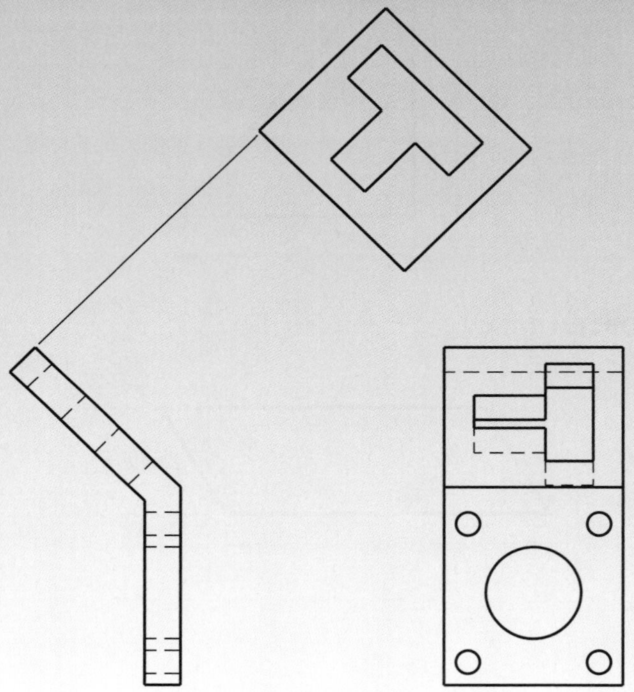

FIGURE 8.2 ■ Final layout with partial auxiliary view (dimensions not included).

FIGURE 8.1 ■ Engineer's rough sketch.

AUXILIARY VIEWS

ANSI/ASME The standard for auxiliary view presentation is found in ANSI Y14.3 titled *Multi and Sectional View Drawings,* American National Standards Institute (ANSI), published by the American Society of Mechanical Engineers (ASME).

Auxiliary views are used to show the true size and shape of a surface that is not parallel to any of the six principal views. When a surface feature is not perpendicular to the line of sight, the feature is said to be foreshortened, or shorter than true length. These foreshortened views do not give a clear or accurate representation of the feature. It is not proper to place dimensions on foreshortened views of objects. Figure 8.3 shows three views of an object with an inclined foreshortened surface.

An auxiliary view allows you to look directly at the inclined surface in Figure 8.3 so you can view the surface and locate the hole in its true size and shape. An auxiliary view is projected from the inclined surface in the view where that surface appears as a line. The projection is at a 90° angle. (See Figure 8.4.) The height dimension, *H,* is taken from the view that shows the height in its true length.

Notice in Figure 8.4 that the auxiliary view shows only the true size and shape of the inclined surface. This is known as a *partial aux-*

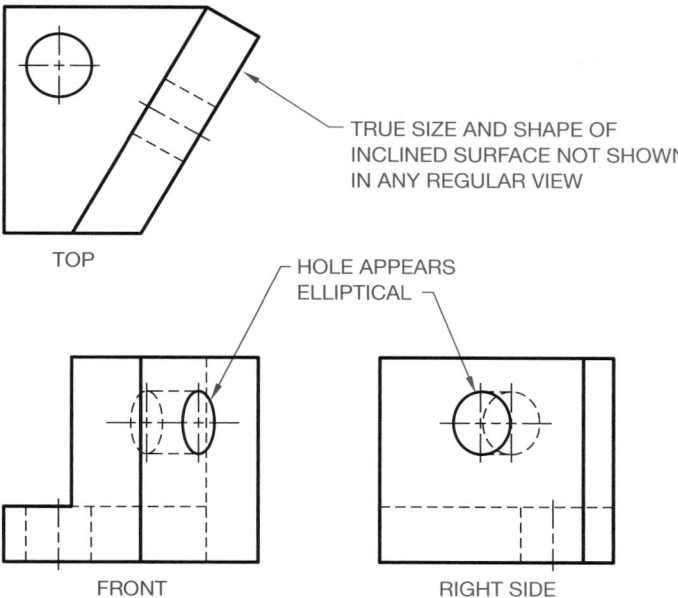

FIGURE 8.3 ■ Foreshortened surface auxiliary view needed.

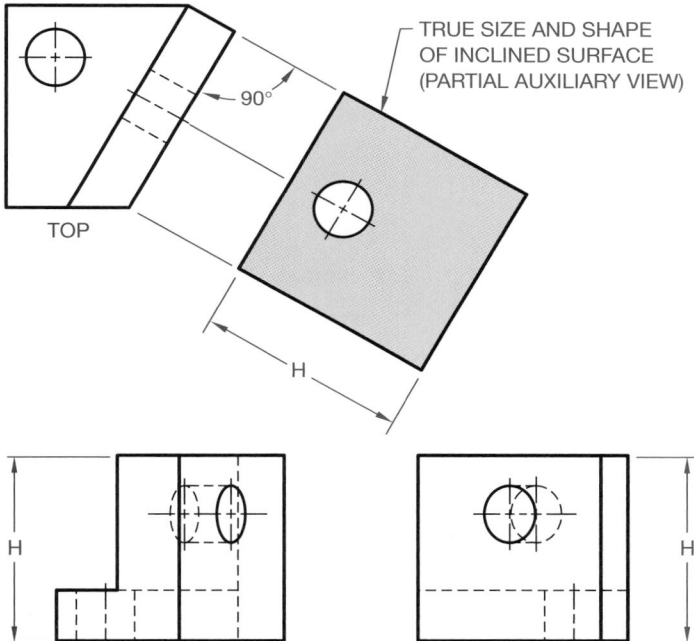

FIGURE 8.4 ■ Partial auxiliary view.

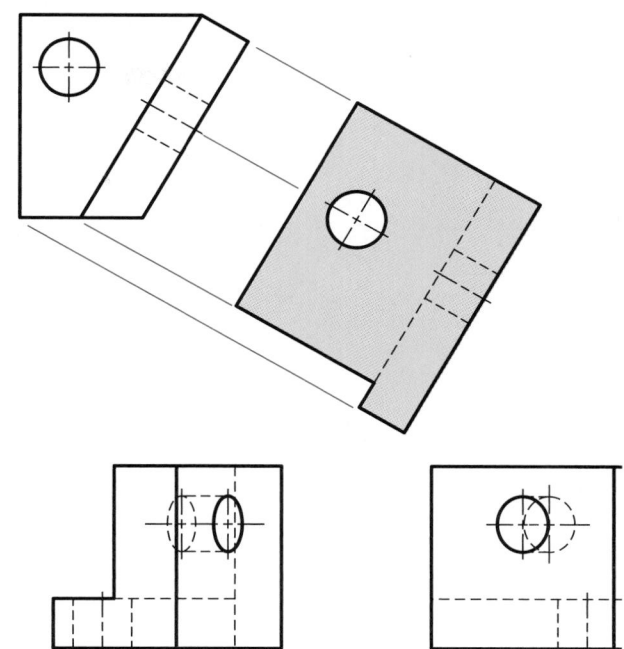

FIGURE 8.5 ■ Complete auxiliary view.

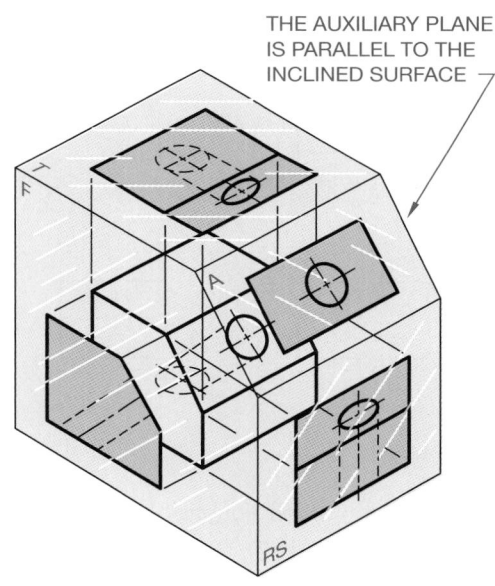

FIGURE 8.6 ■ The object in a glass box.

iliary view. A full auxiliary view, Figure 8.5, also shows all the other features of the object projected onto the auxiliary plane. Normally the information needed from the auxiliary view is the inclined surface only, and the other areas do not usually add clarity to the view. However, each object must be considered separately. Usually, you do not need to draw those areas that do not add clarity to the drawing.

Look at the glass box principle that was discussed earlier as it applies to auxiliary views, as shown in Figure 8.6. Figure 8.7 shows the glass box unfolded. Notice that the fold line between the front view and the auxiliary view is parallel to the edge view of the slanted surface.

The auxiliary view is projected from the edge view of the inclined surface, which establishes true length. The true width of the inclined surface may be transferred from the fold lines of the top or side views as shown in Figure 8.8. Projection lines are drawn as very light construction lines and should not reproduce. Hidden lines are generally not shown on auxiliary views unless the use of hidden lines helps clarify certain features. The auxiliary view may be projected directly from the inclined surface, as in Figure 8.4. When this is done, a centerline or projection line may continue between the views to indicate alignment and view relationship. This centerline or projection line is often used as an extension line for dimensioning purposes. If view alignment is

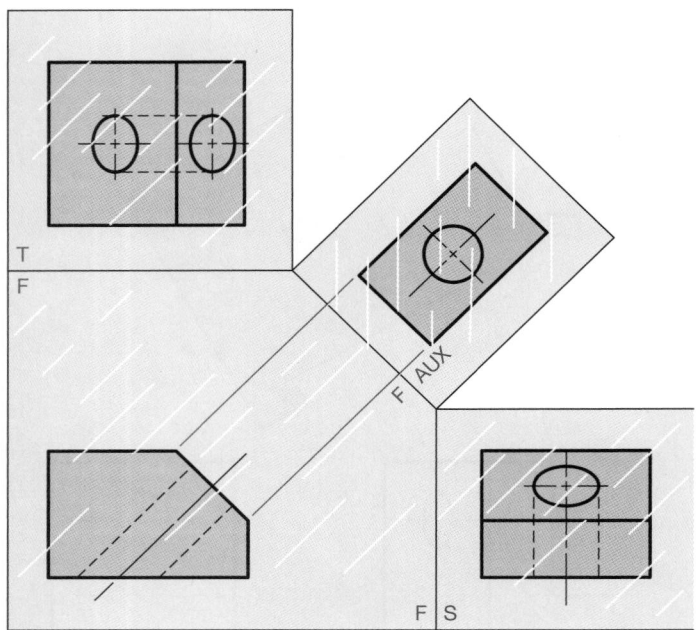

FIGURE 8.7 ■ The glass box unfolded.

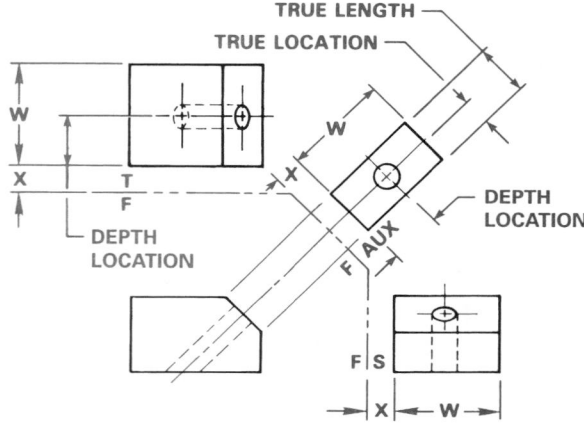

FIGURE 8.8 ■ Establishing auxiliary view with fold lines.

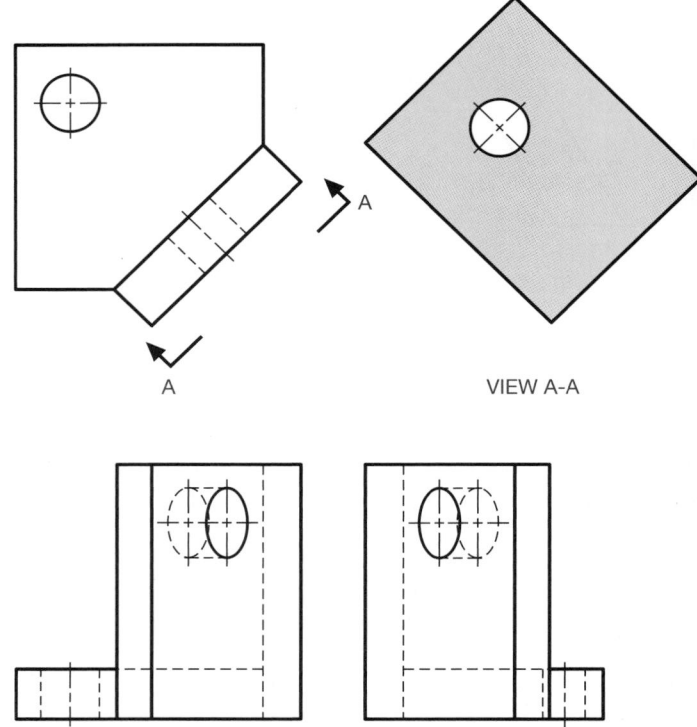

FIGURE 8.9 ■ Establishing auxiliary view with viewing plane.

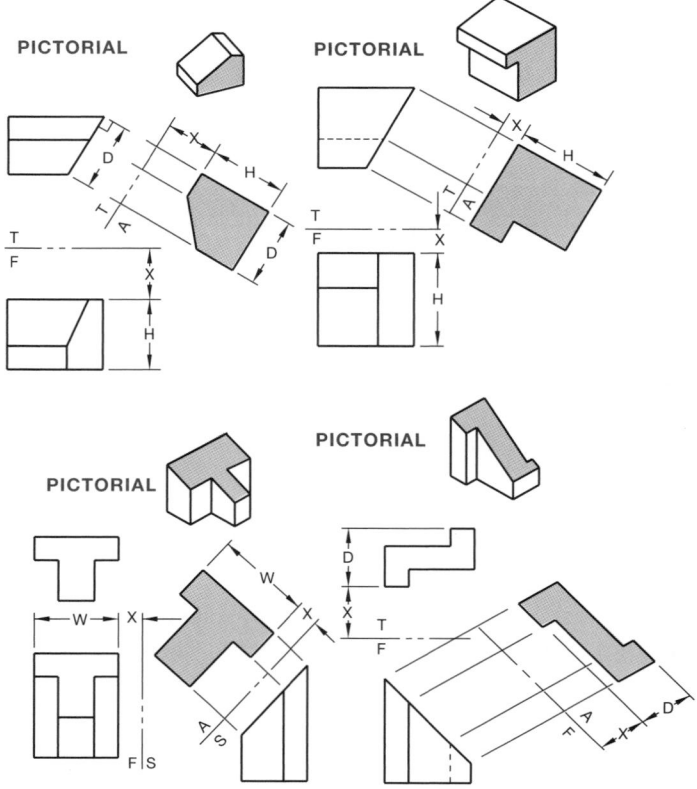

FIGURE 8.10 ■ Partial auxiliary view examples.

clearly obvious, the projection line may be omitted. When it is not possible to align the auxiliary view directly from the inclined surface, then a viewing-plane line may be used and the view placed in a convenient location on the drawing (Figure 8.9).

The viewing-plane line is labeled with letters so the view can be clearly identified. This is especially necessary when several viewing-plane lines are used to label different auxiliary views. Place views in the same relationship as the viewing-plane lines indicate. Do not rotate the auxiliary but leave it in the same relationship as if it were projected from the slanted surface. Where multiple views are used, orient views from left to right and from top to bottom. The views are labeled to correlate with their viewing plane, such as VIEW A-A. The word VIEW and A-A are title height of ³⁄₁₆" or ¼" (6 mm), or the word VIEW can be smaller than the A-A depending on company preference.

Figure 8.10 shows some other examples of partial auxiliary views in use. It is important to visualize the relationship of the slanted surfaces, edge view, and auxiliary view. If you have trouble with visualization, it is possible to establish the auxiliary view through the mechanics of view projection. Use the following steps:

Step 1. Number each corner of the inclined view so the numbers coincide from one view to the other, as shown in Figure 8.11. Carefully project one point at a time from view to view. Some points have two numbers depending on the view.

Step 2. With the corresponding points numbered in each view, draw an auxiliary fold line parallel to the edge view of the slanted surface. The auxiliary fold line may be any convenient distance from the edge view. Project the points on the edge view perpendicular (90°) across the auxiliary fold line. (See Figure 8.12.)

Step 3. Establish the distance to each point from the adjacent view and transfer the distance to the auxiliary view as shown in Figure 8.13a. Sometimes it is helpful to sketch a small pictorial to assist in visualization. (See Figure 8.13b.)

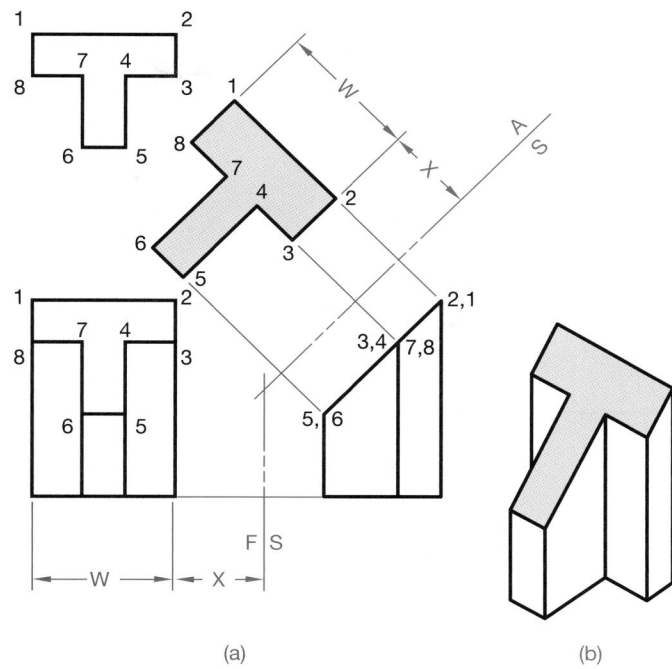

(a)   (b)

FIGURE 8.13 ■ **(a)** Step 3, layout auxiliary view. **(b)** Pictorial to help visualize object.

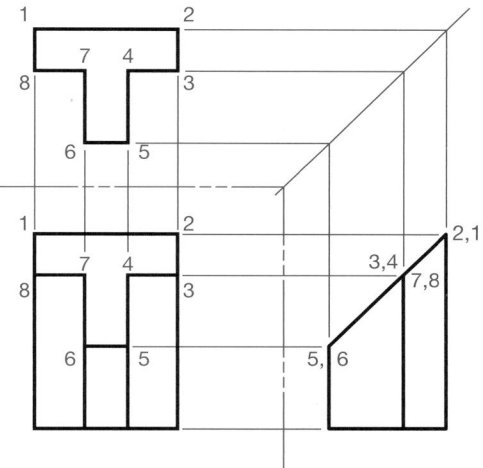

FIGURE 8.11 ■ Step 1, multiview layout.

PLOTTING CURVES IN AUXILIARY VIEWS

Curves are plotted in auxiliary views in the same manner as those shapes described previously. When corners exist, they are used to lay out the extent of the auxiliary surface. An auxiliary surface with irregular curved contours requires that the curve be divided into elements so the element points can be transferred from one view to the next, as shown in Figure 8.14.

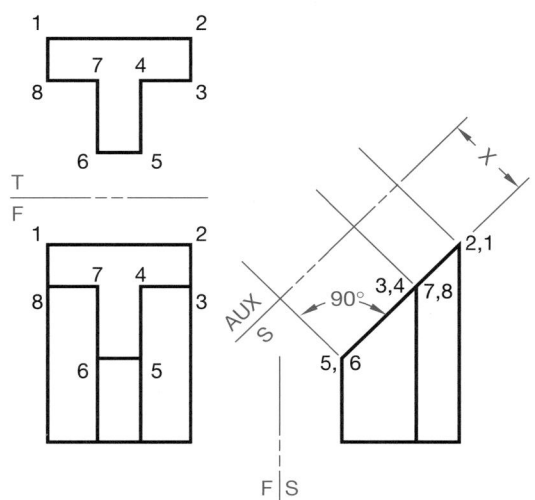

FIGURE 8.12 ■ Step 2, establish the auxiliary fold line.

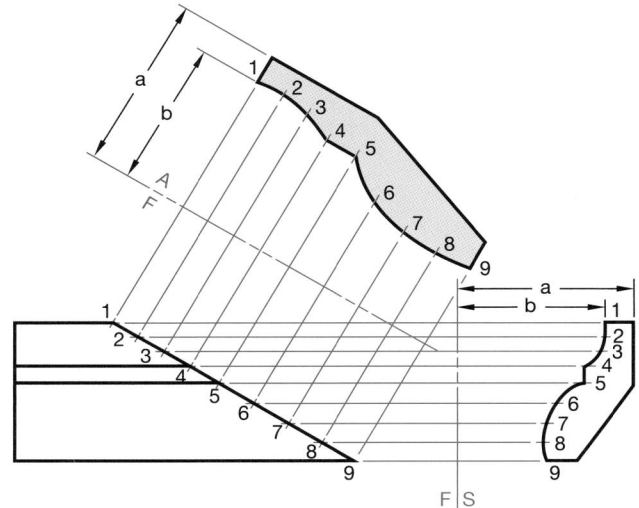

FIGURE 8.14 ■ Plotting curves in auxiliary view.

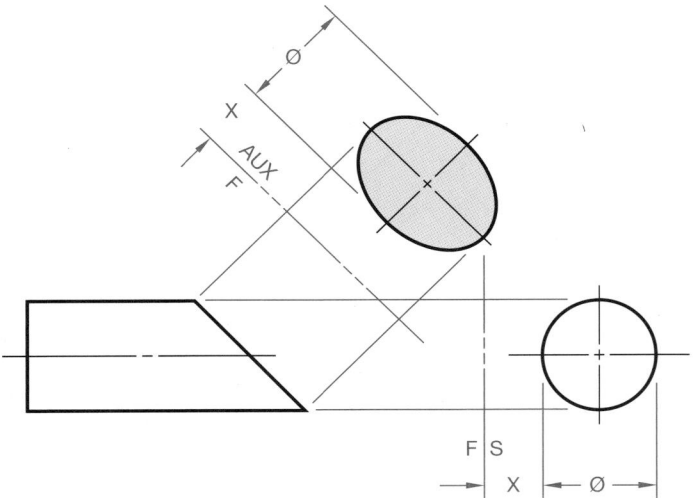

FIGURE 8.15 ■ Elliptical auxiliary view.

The contour of elliptical shapes may be plotted, or if the shape coincides with an available elliptical template, a template should always be used to save time. (See Figure 8.15.)

ENLARGEMENTS

In some situations you may need to enlarge an auxiliary view so a small detail can be shown more clearly. Figure 8.16 shows an object with a foreshortened surface. The two principal views are clearly dimensioned at a 1:1 scale. The 1:1 scale is too small to clarify the shape and size of the slot through the part. A viewing-plane line is placed to show the relationship of the auxiliary view. The auxiliary view can then be drawn in any convenient location at any desired scale; in this case a 2:1 scale is used.

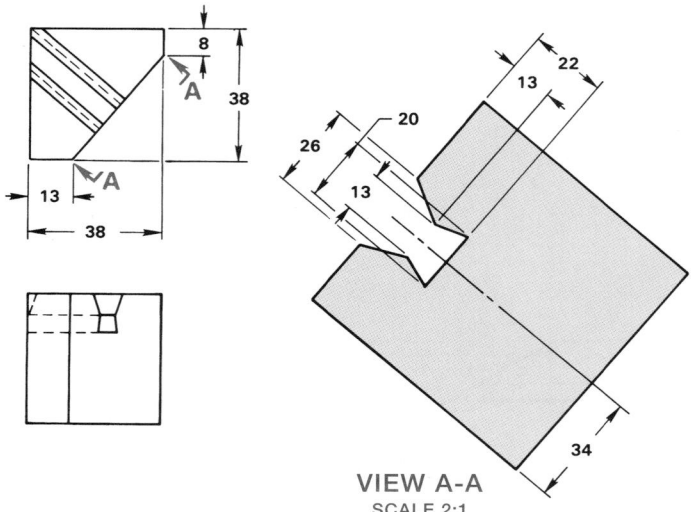

VIEW A-A
SCALE 2:1

FIGURE 8.16 ■ Auxiliary view enlargement.

SECONDARY AUXILIARY VIEWS

There are some situations when a feature of an object is in an oblique position in relationship to the principal planes of projection. These inclined, or slanting, surfaces do not provide an edge view in any of the six possible multiviews. The inclined surface in Figure 8.17 is foreshortened in each view, and an edge view also does not exist. From the discussion on primary auxiliary views you realize that projection must be from an edge view to establish the auxiliary.

In order to obtain the true size and shape of the inclined surface in Figure 8.17, a secondary auxiliary view is needed. The following steps may be used to prepare a secondary auxiliary.

Step 1. Only two principal views are necessary to work from, as the third view does not add additional information. Establish an element in one view that is true in length, as shown in Figure 8.18. Label the corners of the inclined surface and draw a fold line perpendicular to one of the true-length elements.

Step 2. The purpose of this step is to establish a primary auxiliary view that displays the slanted surface as an edge. Project the slanted surface onto the primary auxiliary plane, as shown in Figure 8.19. Doing so results in the inclined surface appearing as an edge view or line.

Step 3. Now, with an edge view established, the next step is the same as the normal auxiliary view procedure. Draw a fold line parallel to the edge view. Project points from the edge view perpendicular to the secondary fold line to

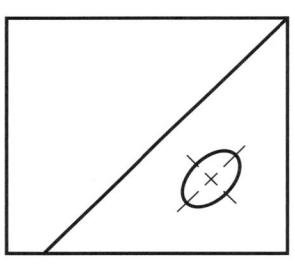

HOLE PERPENDICULAR TO SLANTED SURFACE

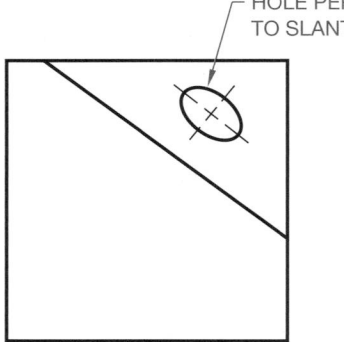

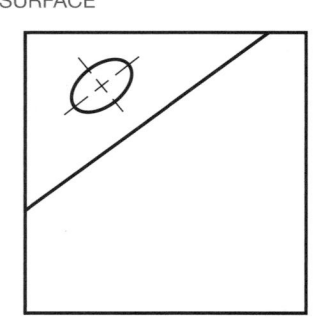

FIGURE 8.17 ■ Oblique surface. There is no edge view in the principal views.

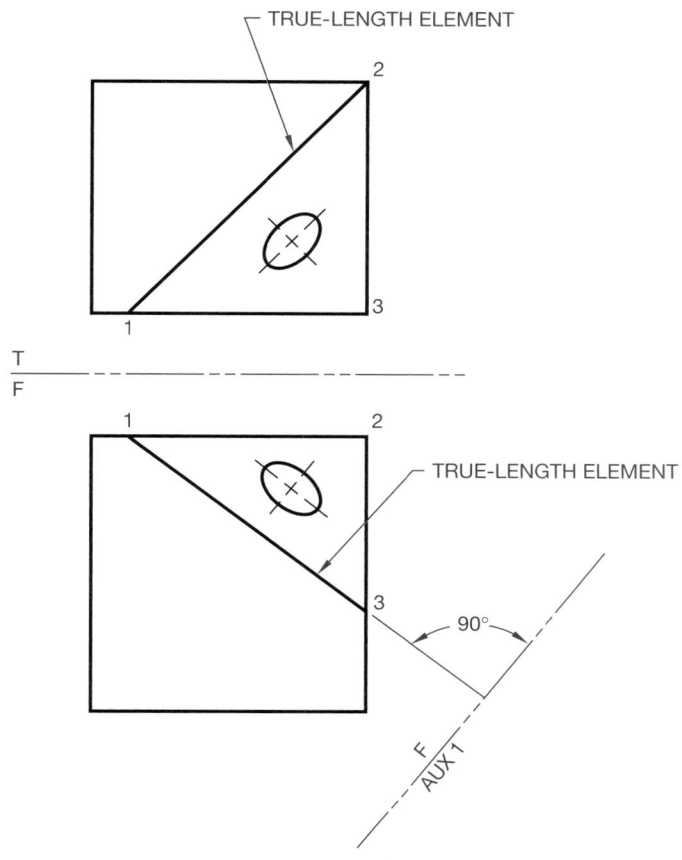

FIGURE 8.18 ■ Step 1, draw a fold line perpendicular to a true-length element.

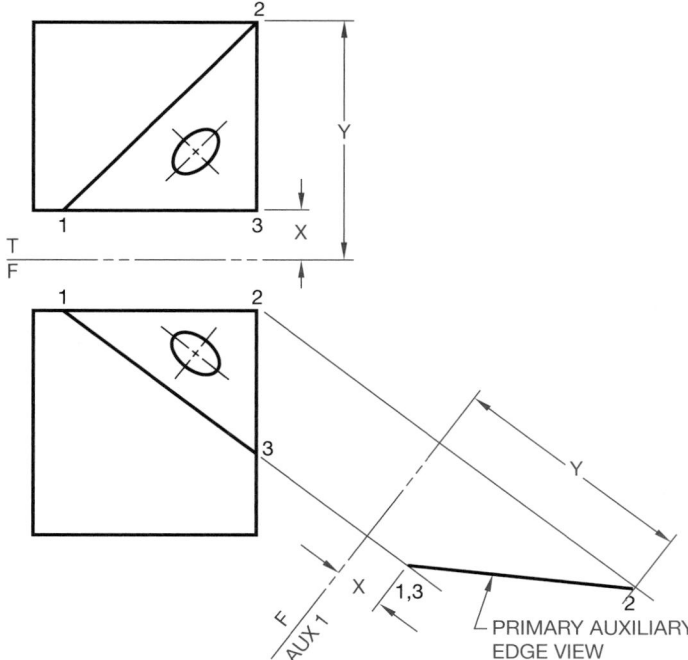

FIGURE 8.19 ■ Step 2, primary auxiliary edge view of oblique surface.

establish points for the secondary auxiliary, as shown in Figure 8.20, page 218.

In Figure 8.20, page 218, the primary auxiliary view established all corners of the inclined surface in a line, or edge view. This edge view is necessary so the perpendicular line of projection (sight) for the secondary auxiliary assists in establishing the true size and shape of the surface. In many situations both the primary and secondary auxiliary views are used to establish the relationship between features of the object. (See Figure 8.21.) The primary auxiliary view shows the relationship of the inclined feature to the balance of the part, and the secondary auxiliary view shows the true size and shape of the inclined features plus the true location of the holes.

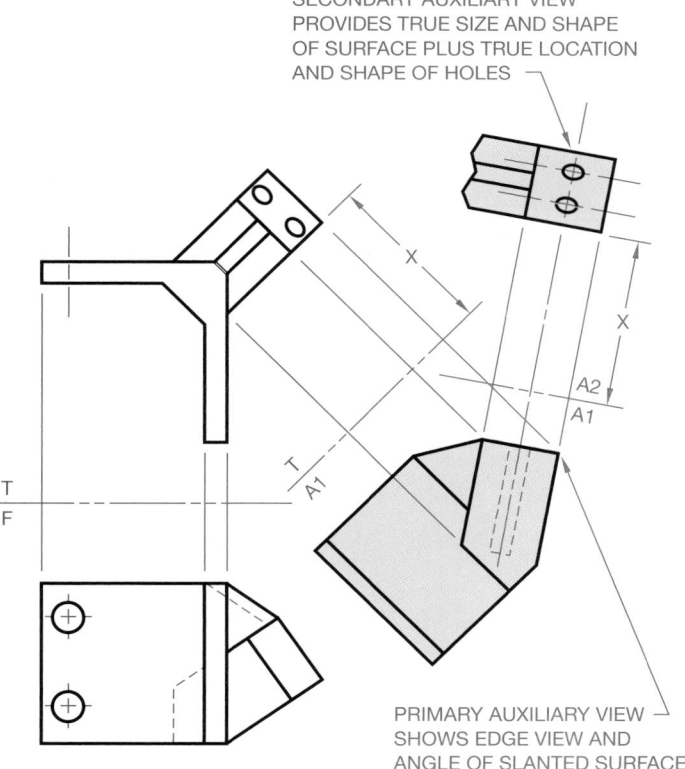

FIGURE 8.21 ■ The primary and secondary auxiliary views.

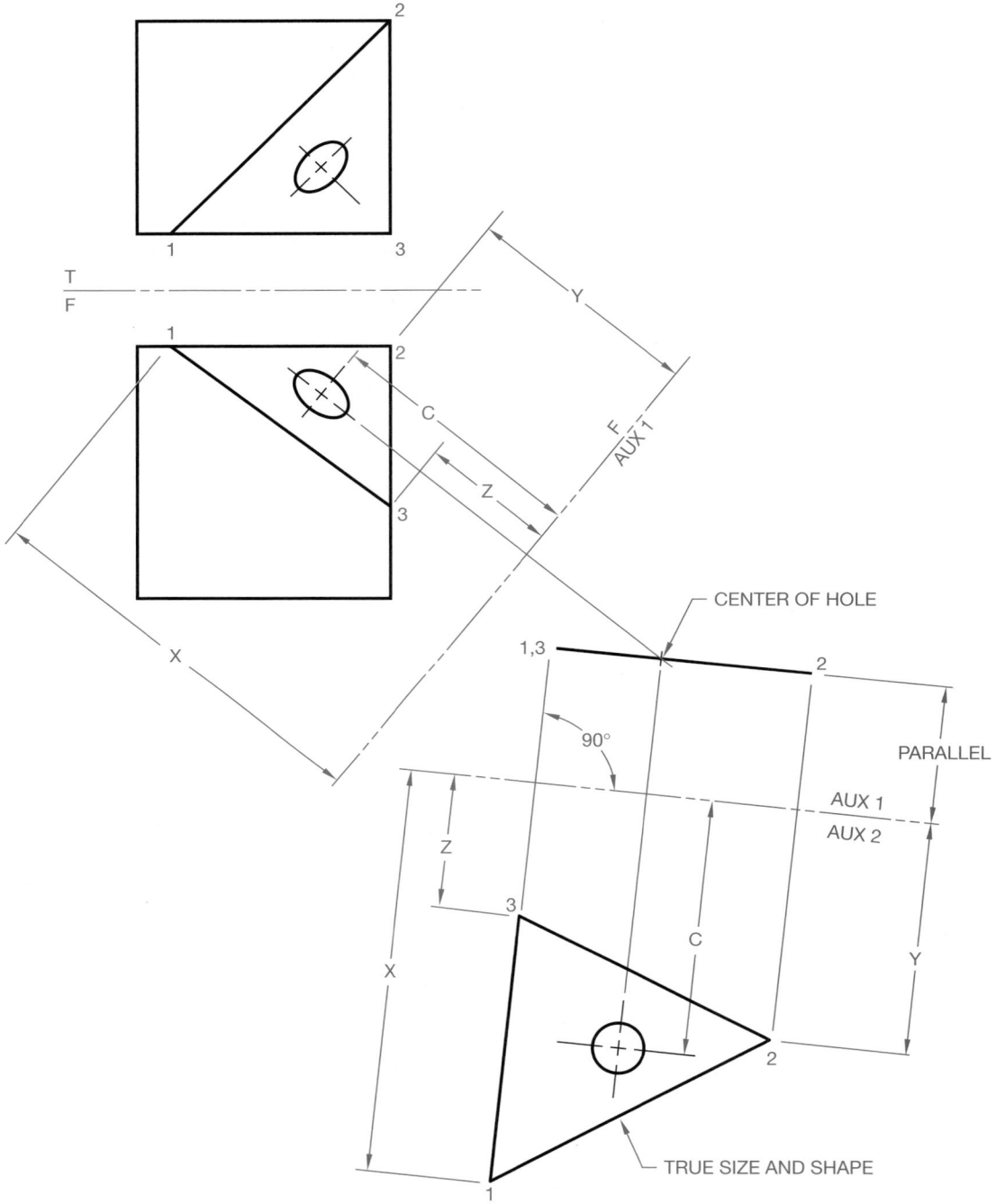

FIGURE 8.20 ■ Step 3, secondary auxiliary projected from the edge view to get the true size and shape of the oblique surface.

MULTIVIEW AND AUXILIARY VIEW DRAWING

So far in your study you have learned that multiview and auxiliary view drawings are the orthographic projection of two or more planes at right angles to each other. One important aspect of this technology is accuracy. As a manual drafter, the accuracy depends on your sharp pencil and a keen eye for technique. Even solutions in the best manual drafting have a degree of error due to line widths or the slight tilt of the pencil in relation to the paper. Multiview and auxiliary view drawings produced with CADD are, on the other hand, nearly perfect. The reason for this perfection has to do with the capabilities of the hardware and software, and the fact that the computer is actually displaying the mathematical counterpart of the graphic analysis. In other words, the views are a representation of the mathematical coordinates of the problem being presented. Therefore, the drawing accuracy is only controlled by the accuracy potential of the computer. Figure 8.22 shows a multiview drawing on a CADD monitor.

Most CADD systems provide the drafter with "drawing aids" such as a screen grid similar to the light-blue lines preprinted on vellum to assist the manual drafter. The screen grid is displayed as dots. These grid dot patterns may be displayed in any convenient increment to assist in accurately placing the views. The drafter may also establish designated increments for the screen cursor crosshairs to snap. This allows the drafter to accurately determine distances on the drawing without guesswork. Another asset of the CADD system is the display of drawing distance and coordinates. For example, when you draw a line on the screen, the computer displays the exact line length. Drawing multiviews with these kinds of drawing aids makes the job easy and accurate. When drawing auxiliary views, the CADD drawing aids are flexible and may be rotated at the angle of the auxiliary view to assist in accurately laying out the view as shown in Figure 8.23.

A CADD system also allows you to work in layers and colors for a clearer picture of the problem and solution. For example, you can set a layer for projection lines shown in any color, such as red. This helps you visualize the entire drawing and allows you to accurately project a feature from one view to the next. The layers may be turned on or off at your convenience. When you complete the drawing and you are ready to make a plot, all you have to do is turn off the layer with projection lines and the drawing is ready to plot.

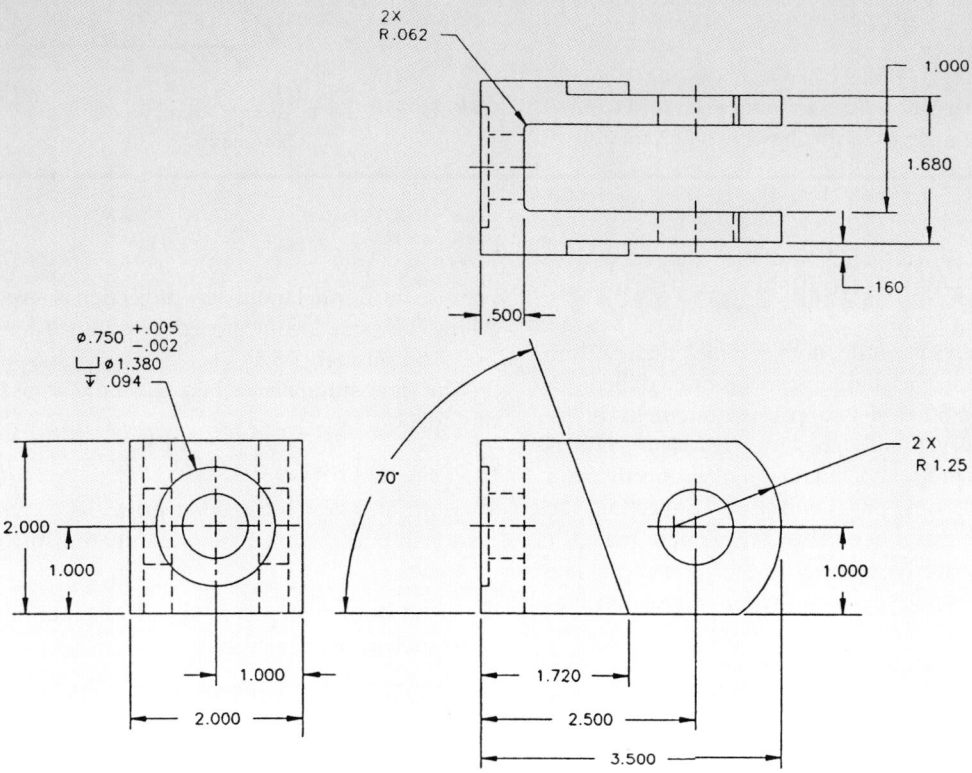

FIGURE 8.22 ■ Using CADD to draw multiviews.

(Continued)

CADD APPLICATIONS *(continued)*

FIGURE 8.23 ■ Using the rotated grid points in CADD to help draw the auxiliary view.

The lines that you use to project between views are construction lines. The construction lines are only used for layout and construction and are not part of the drawing when they are turned off. The AutoCAD program has construction lines that are drawn using the XLINE command. The XLINE command creates construction lines of infinite length. They may be drawn horizontally, vertically, at an angle, or offset a given distance from another line or object. The XLINE may also be used to bisect an angle. When laying out auxiliary views, the XLINE command Angle option may be used. This allows you to project from an inclined surface to establish the correct true size and shape, and location of the auxiliary view as shown in Figure 8.24.

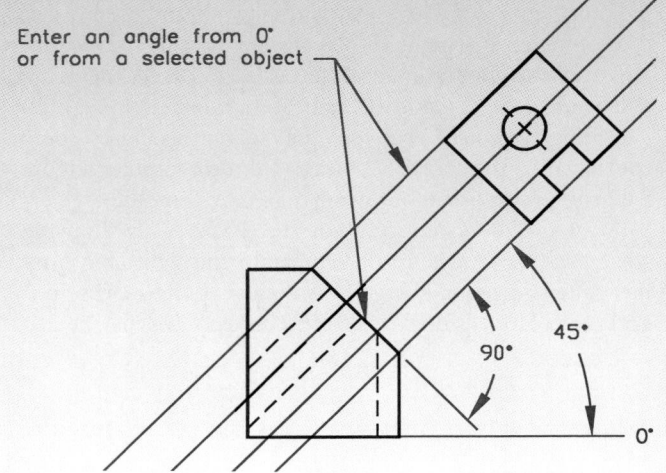

FIGURE 8.24 ■ Laying out an auxiliary view using the XLINE command.

AUXILIARY VIEW LAYOUT

Working with auxiliary views is a little more complex than dealing with the normal multiviews. The auxiliary view is often difficult to place in relationship to the other views. Space requirements many times cause the auxiliary view to interfere with other views. Whenever possible, it is best to project the auxiliary view directly from the inclined surface with one projection line connecting the inclined surface to the auxiliary view. This arrangement makes it easier for the reader to correctly interpret the relationship of the views. This method is shown in the layout steps provided in Figures 8.11, 8.12, and 8.13.

When drawing space is limited, it is possible to use the viewing-plane method to display the auxiliary view. This technique is shown in Figure 8.10. The advantage of the viewing-plane method is that it allows you to place the auxiliary view in any convenient place on the drawing. Multiple auxiliary views should be placed in a group arranged from right to left and from top to bottom on the drawing. One unorthodox method that is often used by entry-level drafters is to place viewing planes all over the drawing and shift views from their normal position. This practice should be avoided, because the normal multiview projection is always best. You should follow the same layout procedure outlined for multiviews in Figures 7.53 through 7.57.

The steps summarized here work the same for manual drafting or CADD:

■ Select and sketch the front view.

■ Select and sketch the other principal views in proper relationship to the front view, eliminating any unnecessary views.

■ An additional step is to sketch the auxiliary view or views in their proper positions.

■ Establish the working area and the sheet size to be used.

■ Determine the total drawing area.

■ Lay out the views using construction lines.

■ Complete the formal drawing.

Instruction on the development of views often focuses on the use of three views, such as a front, top, and side view, or a front, top, and auxiliary view. While this view orientation is common, the actual view selection totally depends on the part you are drawing. Many parts can require a number of views that can include several multiviews and an auxiliary view. The combination of views can even include a sectional view. Section views are covered in detail in Chapter 14. Spend some time looking at as many real-world drawings that you can find. You may discover that some actual industry drawing could be done in a different manner, and that you might have ideas for improving the way drawings are done. This is your challenge as you begin to create your own drawings. The actual industry drawing shown in Figure 8.25, page 222, shows a front view, top view, right-side view, a view created with a viewing plane (VIEW A-A), a sectional view (SECTION A-A), and an auxiliary view projected from the sectional view. This example demonstrates the complexity of some drawings.

PROFESSIONAL PERSPECTIVE

For many people, one of the most difficult aspects of engineering drafting is the need to visualize two-dimensional views. This "spacial" ability is natural for some, while others must carefully analyze every part of a multiview and auxiliary view drawing in order to fully visualize the product. Ideally, you should be able to look at the views of an object and readily formulate a pictorial representation in your mind. The principal reason that multiviews are aligned is to assist in the visualization and interpretation process. The front view is the most important view, and that is why it contains the most significant features. As you look at the front view, you should be able to gain a lot of information about the object, then visually project from the front view to other views to gain an understanding of the entire product.

It is up to the drafter to select the best views to completely describe the object. Too few views make the object difficult or impossible to interpret, while too many views make the drawing too complex and waste valuable drafting time. If you follow the guidelines set up in this chapter, you should be able to successfully complete the task. Remember, always begin with a sketch. With the sketch you can quickly determine view arrangement and spacing. Without the preliminary sketch, you might use a lot of drafting time and end up discovering that the layout does not work as you expected.

PROJECTED AREA

Problem: Suppose the steel plate in the shape of an ellipse with an area of 65.98 ft² (from the Math Application in Chapter 5) is set at a 60° angle on the ground. With the sun overhead, what is the area of the *shadow* of the plate on the ground? (See Figure 8.26.)

Solution: The shadow's area is called the *projected area* in math. It is found by multiplying the true area by the *cosine* of the inclined angle. Using a calculator, cos 60° = .5, so the area of the shadow is 65.98 × .5 = 32.99 or 33 ft². The principle of multiplying true area by cosine to obtain projected area applies to all 2-D shapes. Math Applications in later chapters explore right triangles and trigonometry functions (like cosine) further.

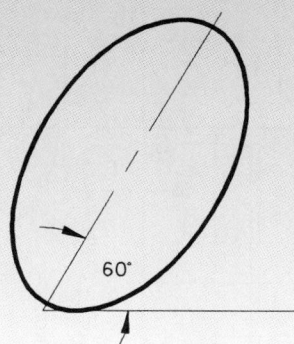

FIGURE 8.26 ■ Tilted ellipse.

MATH APPLICATION

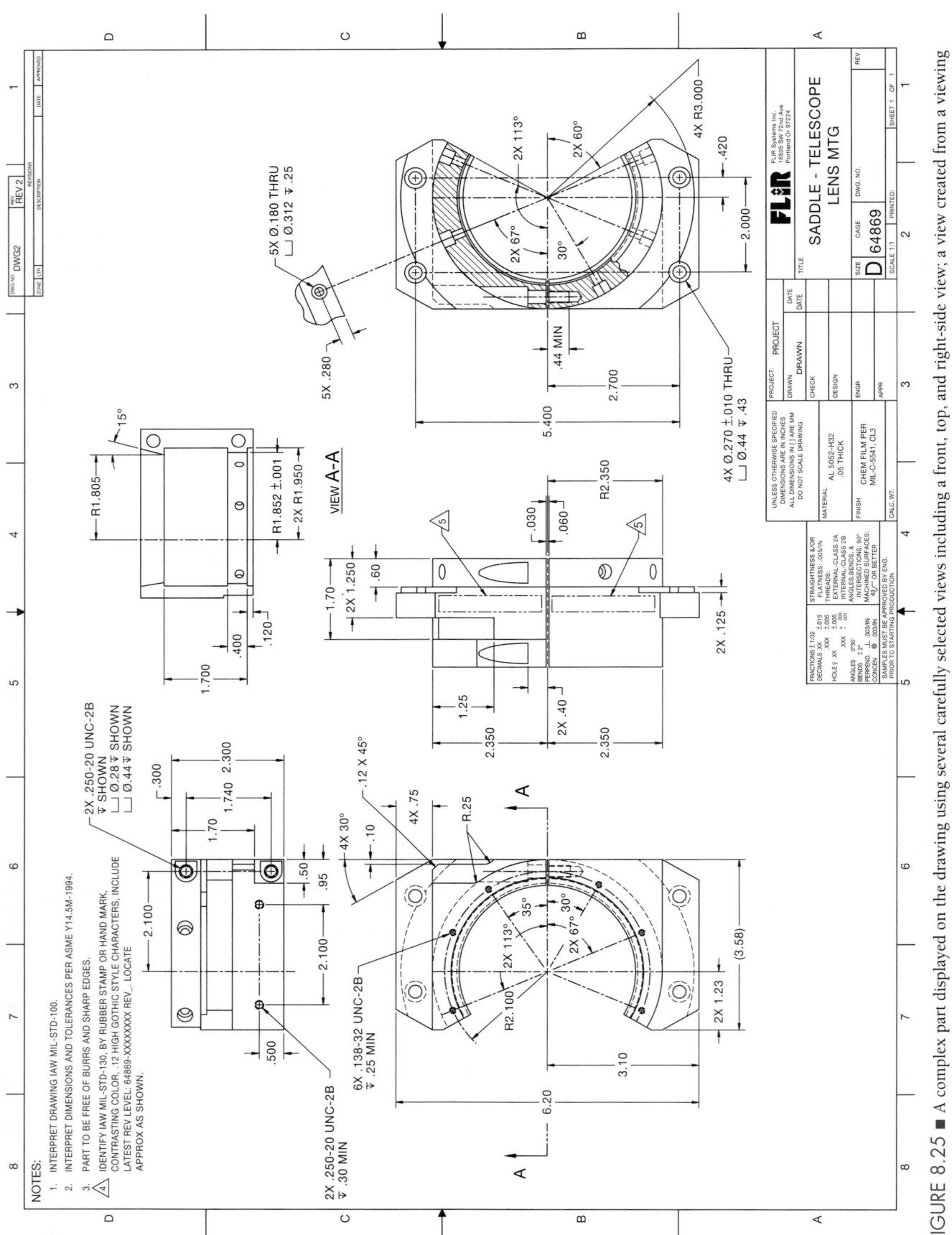

FIGURE 8.25 ■ A complex part displayed on the drawing using several carefully selected views including a front, top, and right-side view; a view created from a viewing plane; a sectional view; and an auxiliary view. *Courtesy Flir Systems.*

CHAPTER 8

Auxiliary Views Test

DIRECTIONS

Answer the questions with short, complete statements or drawings as needed.

1. Describe the purpose of auxiliary views.
2. When projecting an auxiliary view from the edge view of the surface, what is the angle of projection in relationship to the edge view?
3. Discuss how viewing-plane lines may be used to establish an auxiliary view.
4. When would it be necessary to enlarge an auxiliary view?
5. Describe the procedure for drafting an enlarged auxiliary view.
6. Describe a situation when a secondary auxiliary view would be necessary.

7. Identify the ANSI document that governs auxiliary views.
8. What method is used to show the alignment between the auxiliary view and the direct projection from the inclined surface?
9. If the axis of a hole is perpendicular to an inclined surface, what is the shape of the hole in an auxiliary view of the inclined surface?
10. If the axis of a hole is at 45° to an inclined surface, what is the shape of the hole in an auxiliary view of the inclined surface?

CHAPTER 8

Auxiliary Views Problems

DIRECTIONS

You are given an object with its surfaces projected to the planes of the glass box. This can help you visualize as you draw the same views on the folded out planes of the glass box at the left. Measure the given objects and transfer the measurements to your manual or CADD drawing. You can also photocopy the textbook pages and draw the views on the copies. Confirm the preferred method with your instructor.

PART 1

PROBLEM 8-1

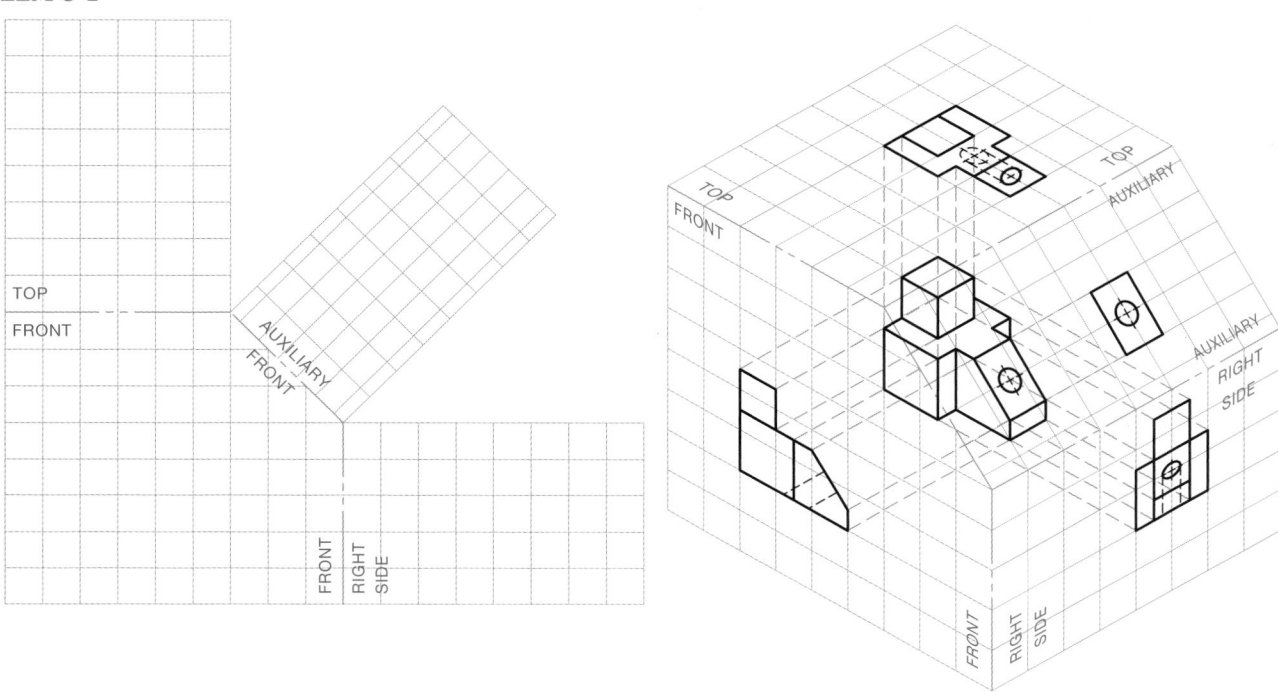

PROBLEM 8-2

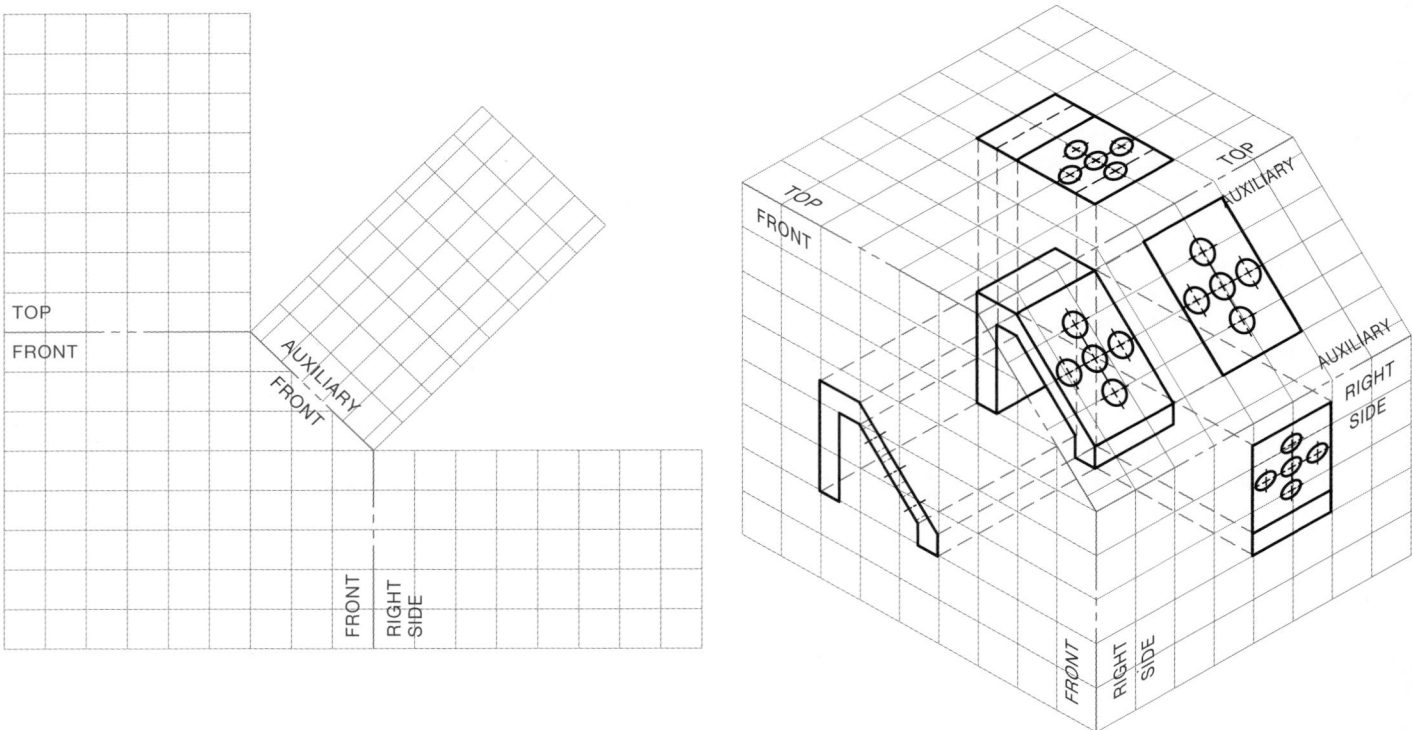

PART 2

DIRECTIONS

These problems provide you with views that contain missing auxiliary views. Draw the missing auxiliary views as appropriate. These problems provide you with a pictorial view to aid in visualization. You do not need to draw the pictorial view. Measure the given views and transfer the measurements to your manual or CADD drawing. You can also photocopy the textbook pages and draw the views on the copies. Confirm the preferred method with your instructor.

PROBLEM 8-3   **PROBLEM 8-5**

PROBLEM 8-4   **PROBLEM 8-6**

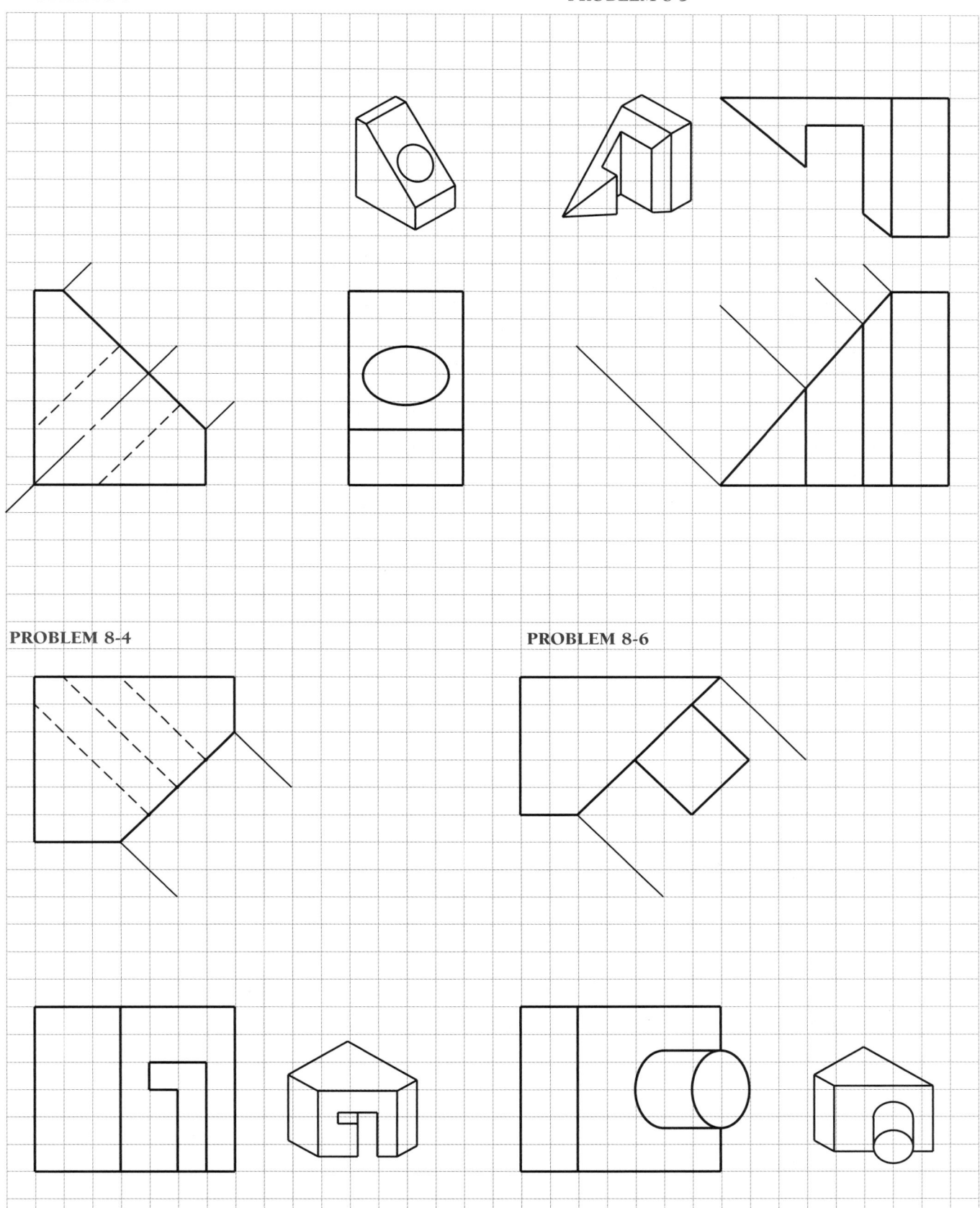

PART 3

DIRECTIONS

These problems provide you with views that contain missing auxiliary views. Draw the missing auxiliary views as appropriate. The recommended auxiliary view projection and a started view or view location is given for your reference. Measure the given views and transfer the measurements to your manual or CADD drawing. You can also photocopy the textbook pages and draw the views on the copies. Confirm the preferred method with your instructor.

PROBLEM 8-7

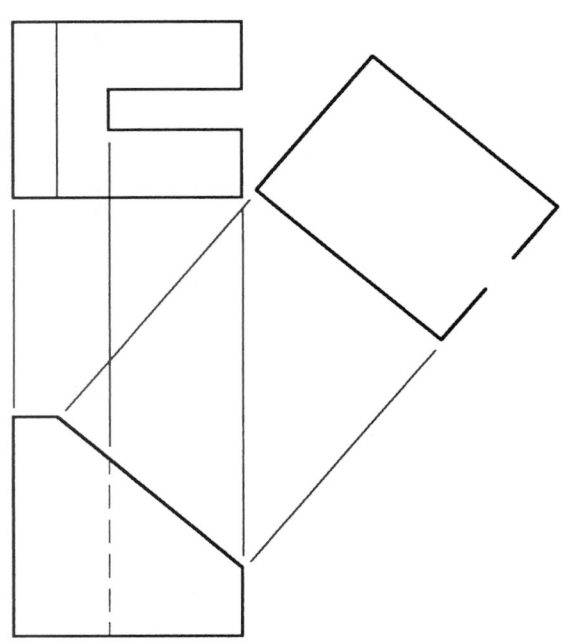

PROBLEM 8-9

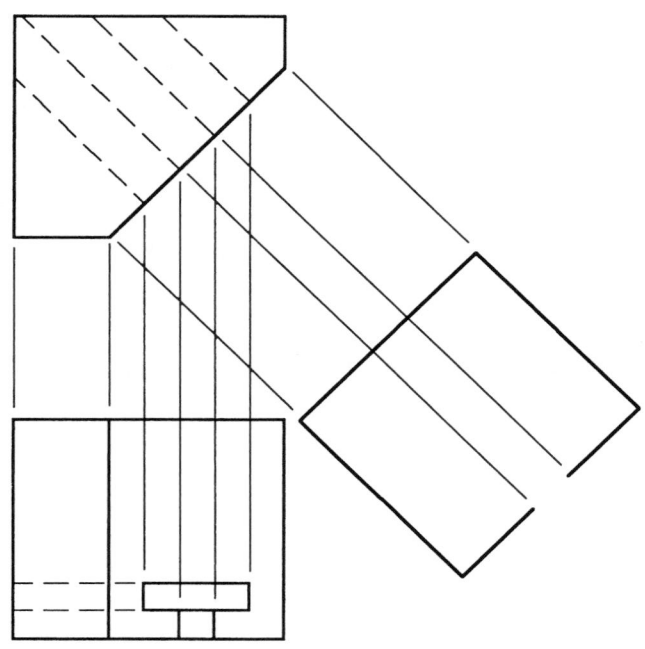

PROBLEM 8-8

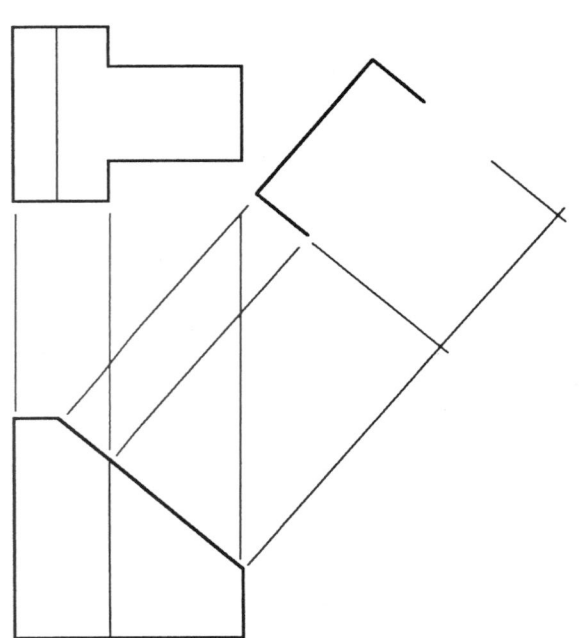

PROBLEM 8-10

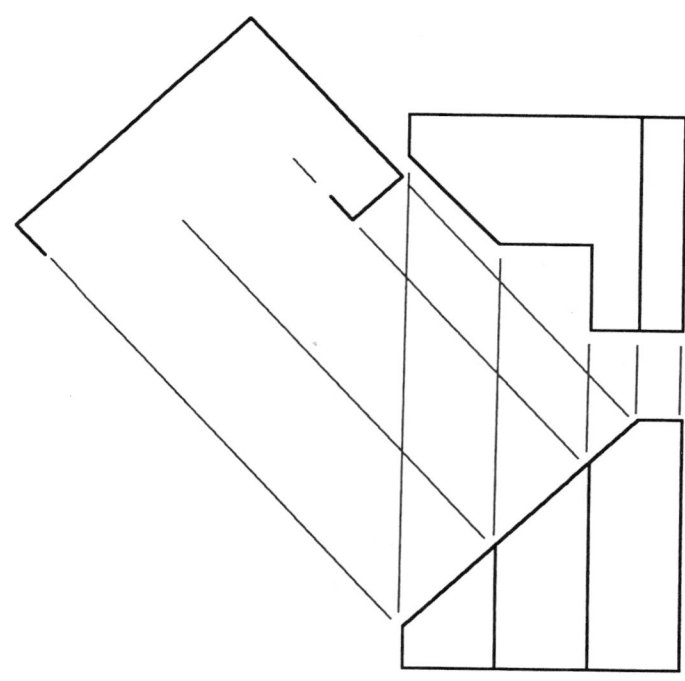

PROBLEM 8-11

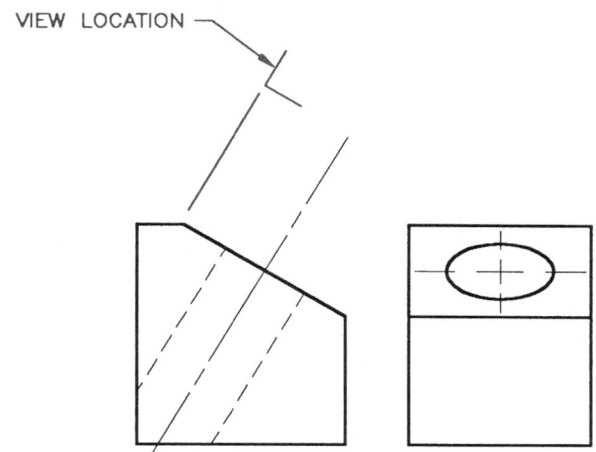

VIEW LOCATION

PROBLEM 8-13

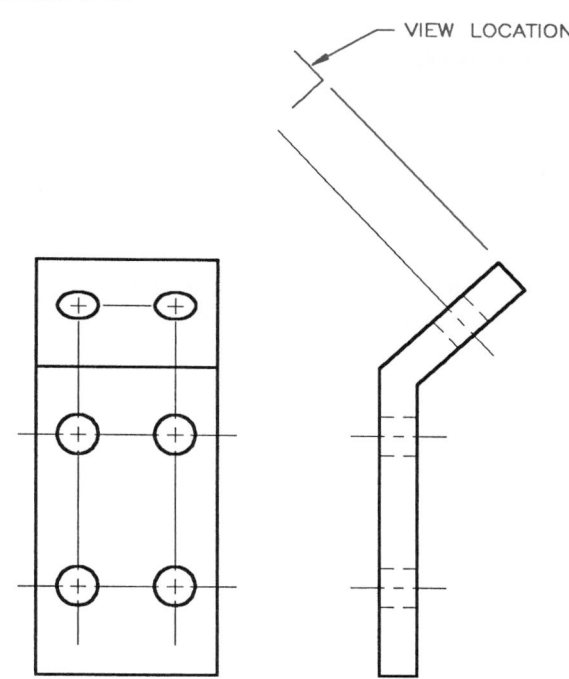

VIEW LOCATION

PROBLEM 8-12

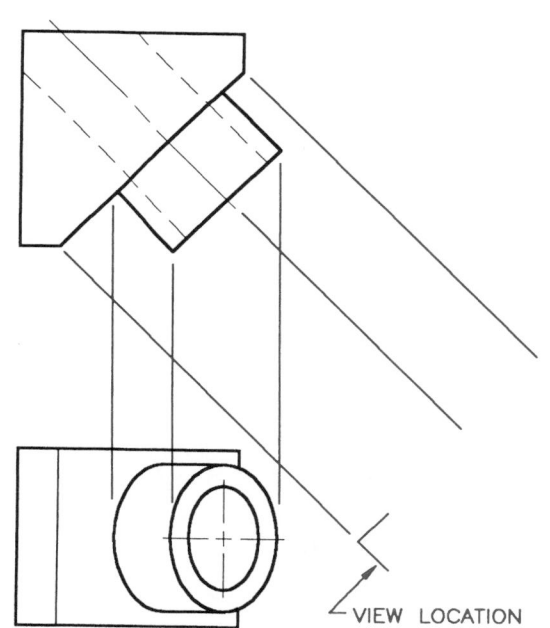

VIEW LOCATION

PROBLEM 8-14

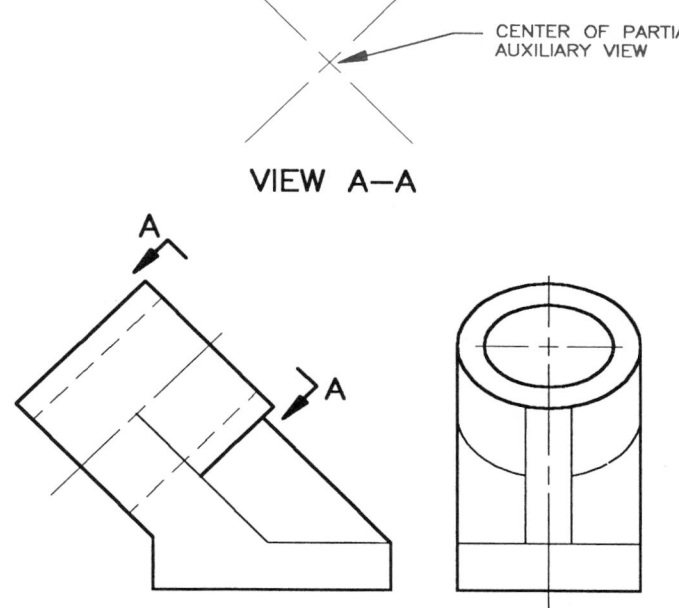

CENTER OF PARTIAL
AUXILIARY VIEW

VIEW A—A

PART 4

DIRECTIONS

These problems provide you with dimensioned views and a proposed auxiliary view location, or pictorial views that contain dimensions and a recommended view layout. Draw the views. The pictorial views are provided to aid in visualization and to display the dimensions. You do not need to draw the pictorial view. Use the given dimensions to create your manual or CADD drawing. Set up your drawings with a properly sized border and title block. Properly complete the information in the title block. Do not draw the dimensions.

PROBLEM 8-15 Primary auxilary view (in.)
Title: Angle V-block
Material: SAE 4320

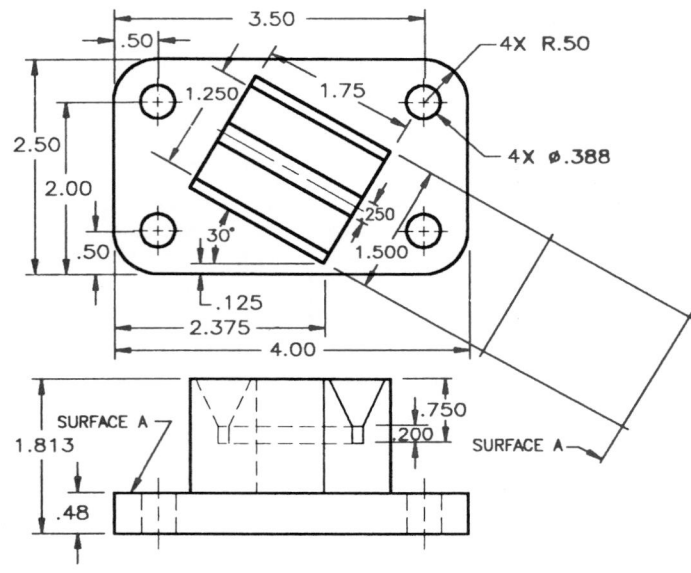

PROBLEM 8-16 Primary auxiliary view (in.)
Title: Cylinder Support
Material: Cast Iron

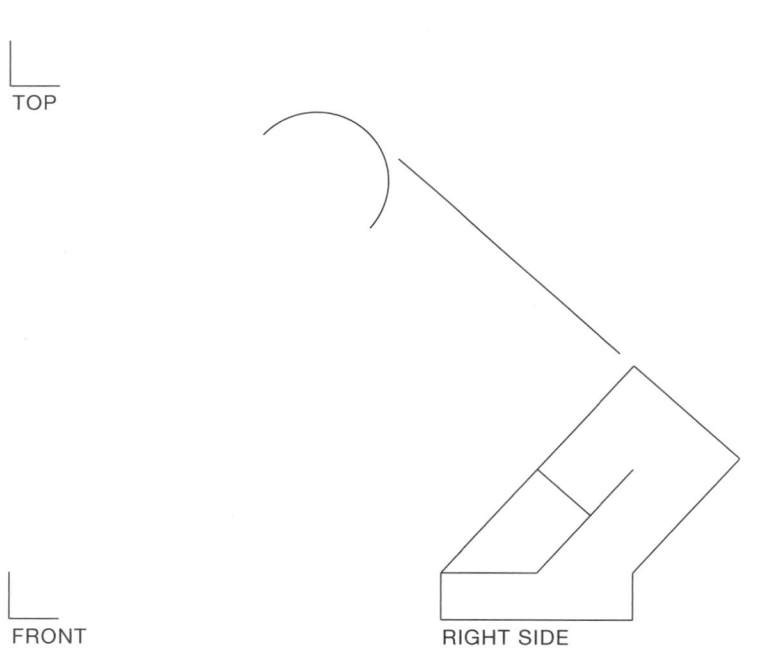

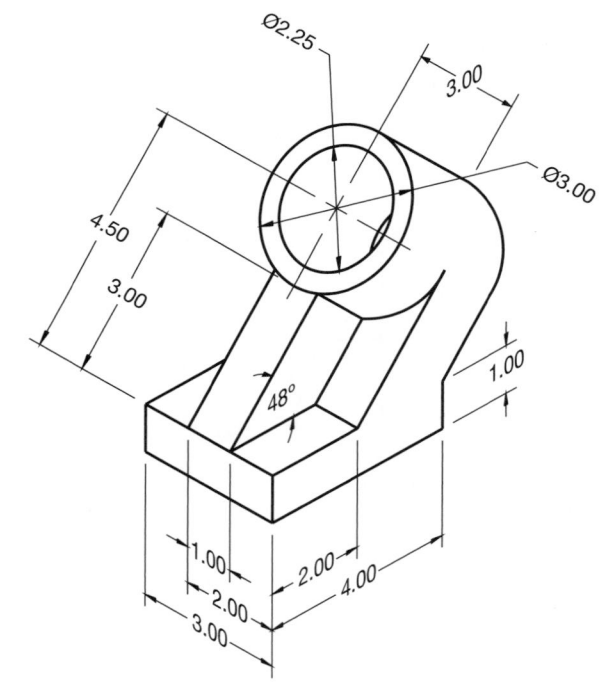

PROBLEM 8-17   Primary auxiliary view (in.)
Title: 135 Bracket
Material: Aluminum

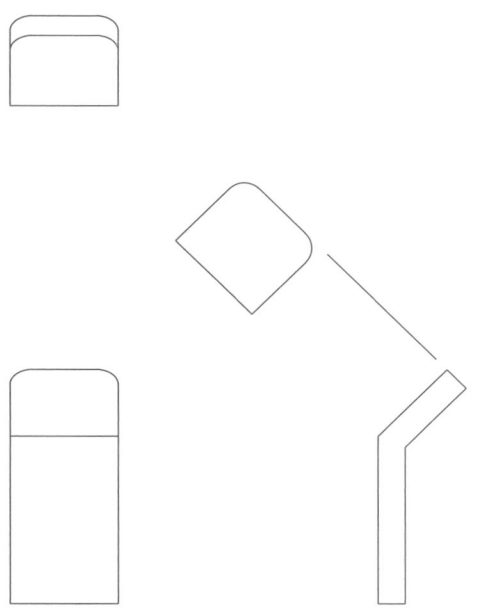

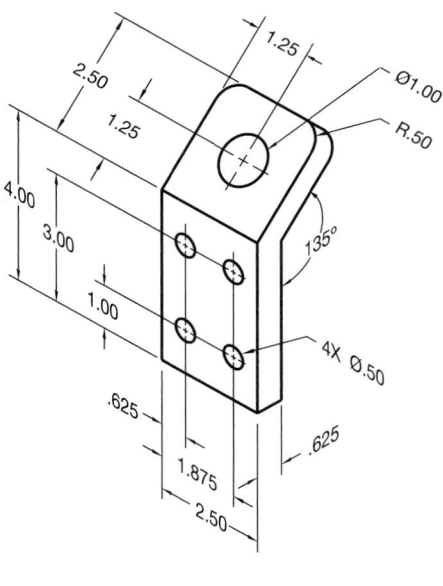

SUGGESTED VIEW LAYOUT

PROBLEM 8-18   Primary auxiliary view (in.)
Title: 135-1 Bracket
Material: Aluminum

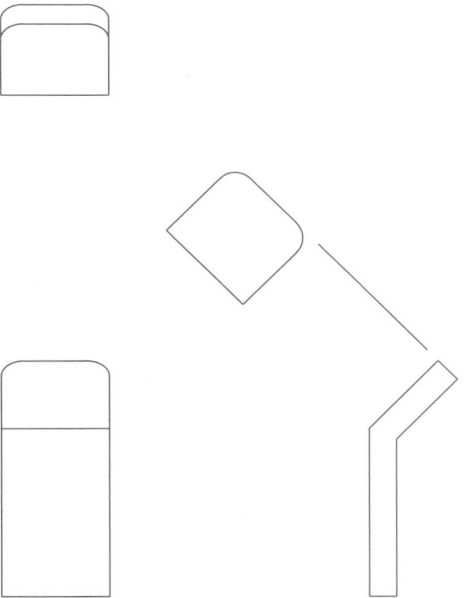

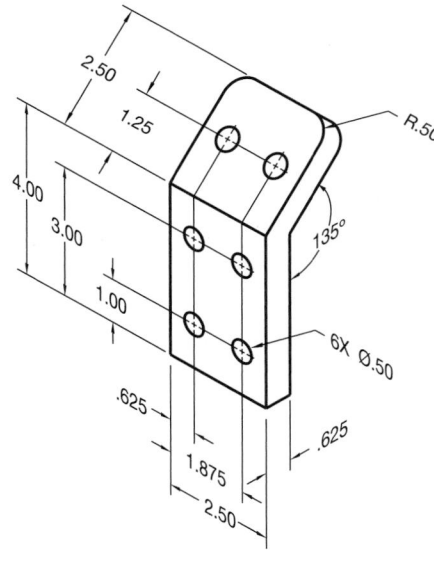

SUGGESTED VIEW LAYOUT

PROBLEM 8-19 Primary auxiliary view (in.)
Title: Support Base
Material: Cast Aluminum

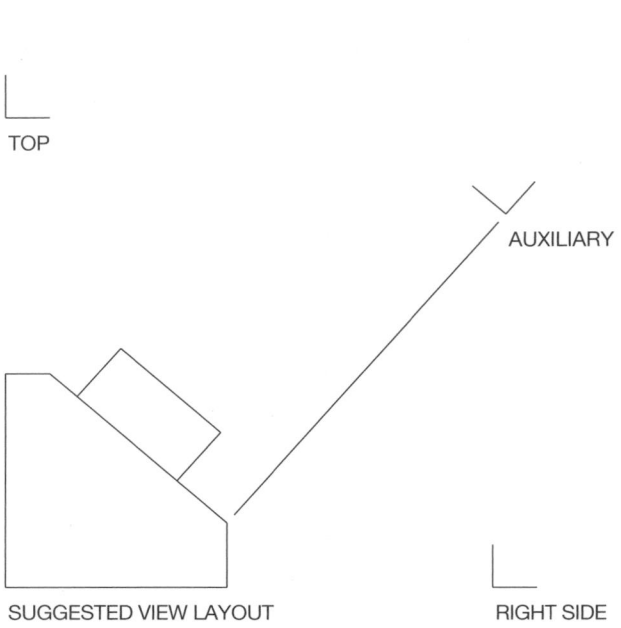

TOP

AUXILIARY

SUGGESTED VIEW LAYOUT

RIGHT SIDE

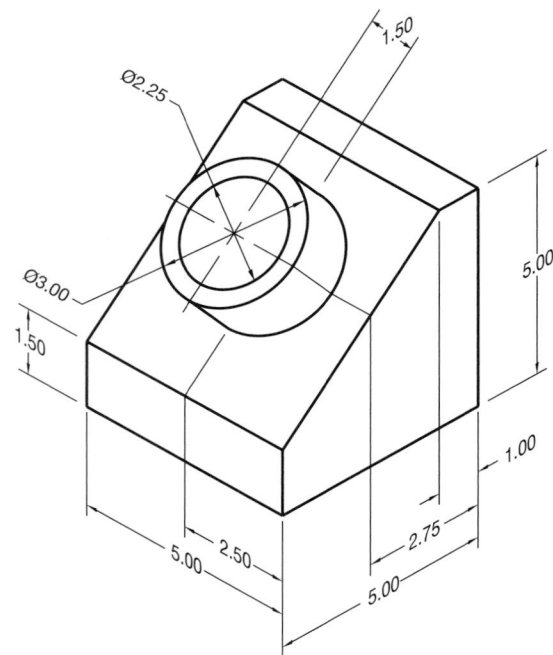

PROBLEM 8-20 Primary auxiliary view (in.)
Title: Shaft Support
Material: SAE 1020

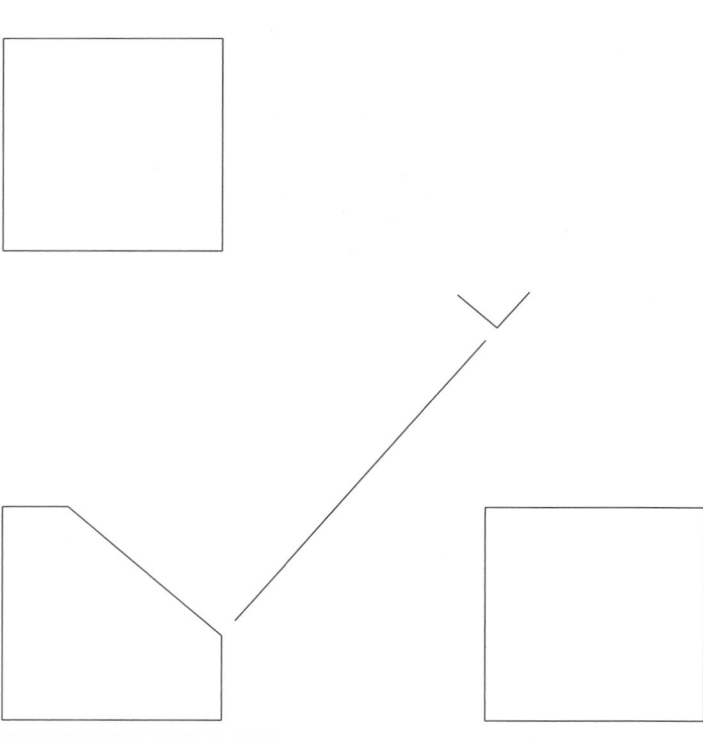

SUGGESTED VIEW LAYOUT

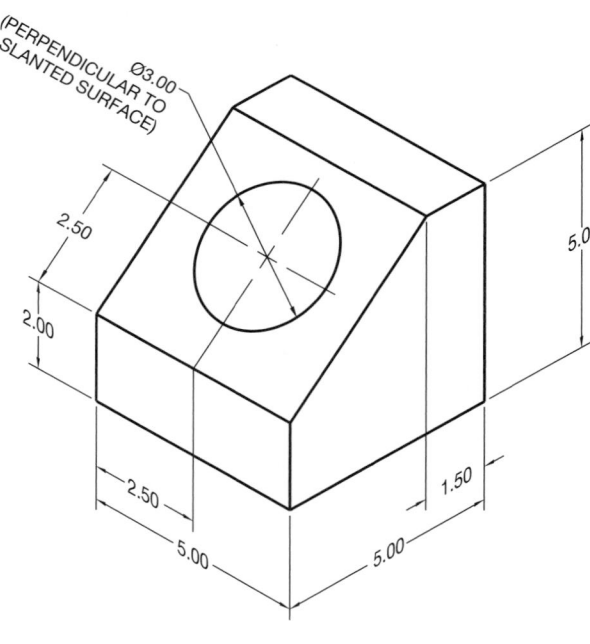

PROBLEM 8-21   Primary auxiliary view (in.)
Title: Spacer
Material: Cast Iron

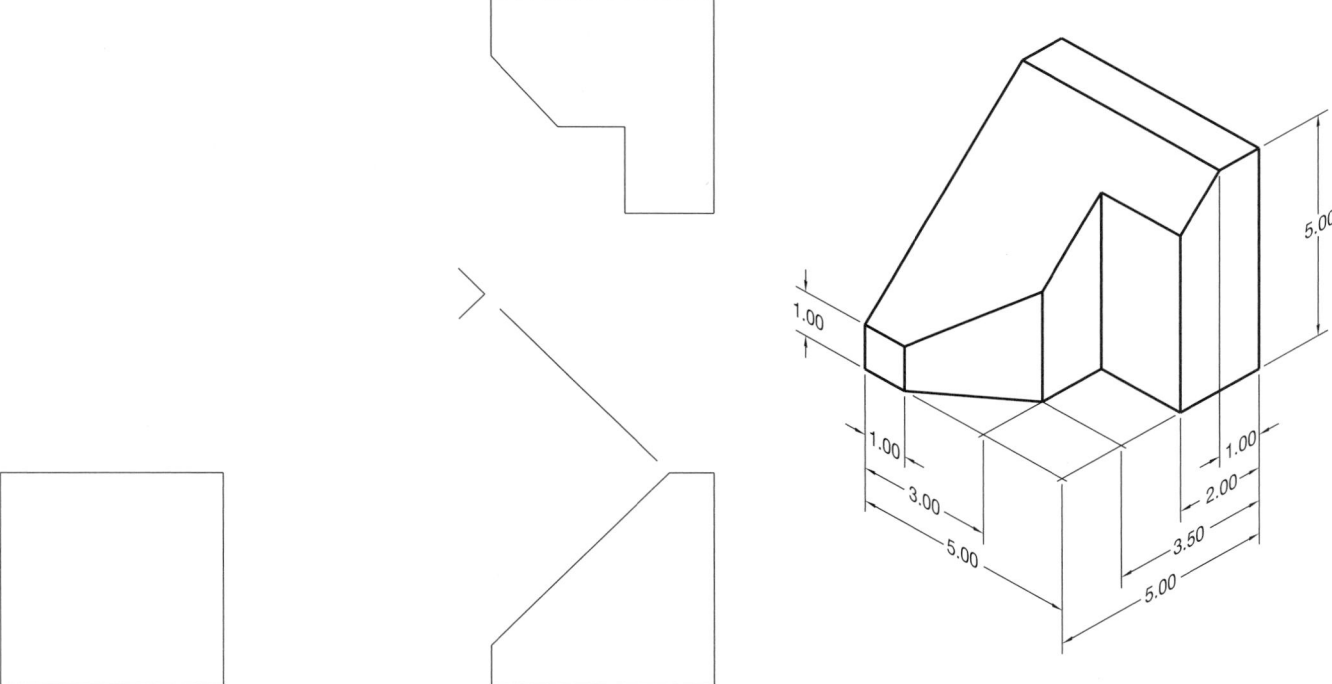

SUGGESTED VIEW LAYOUT

PROBLEM 8-22   Primary auxiliary view (in.)
Title: T-Block
Material: SAE 4320

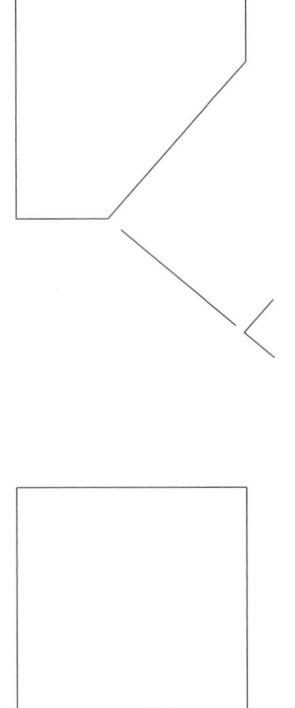

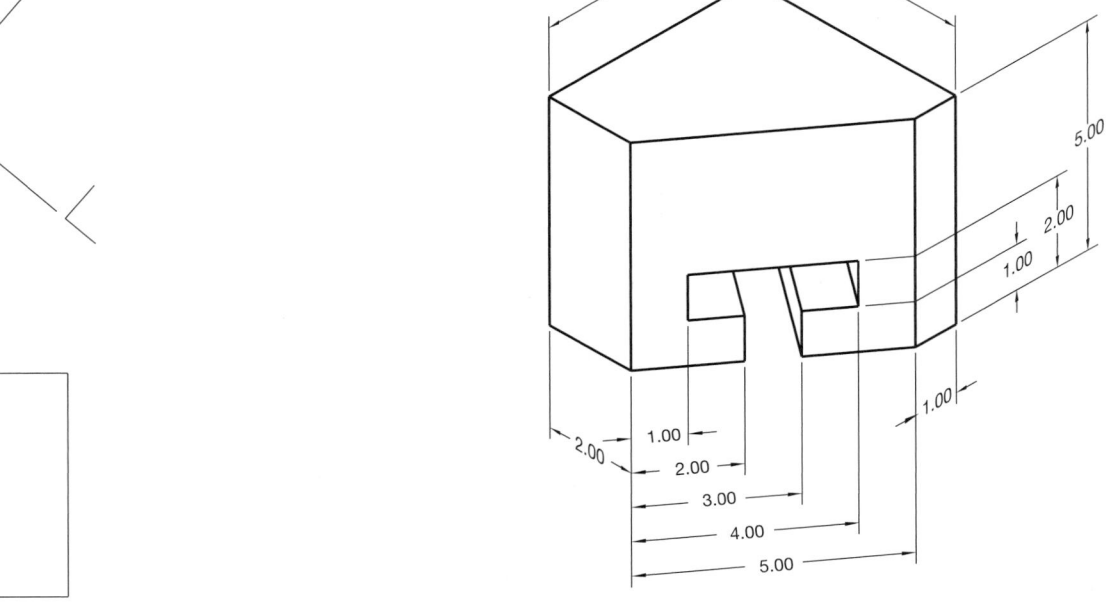

SUGGESTED VIEW LAYOUT

PROBLEM 8-23   Primary auxiliary view (in.)

Title: T-Wedge

Material: Mild Steel

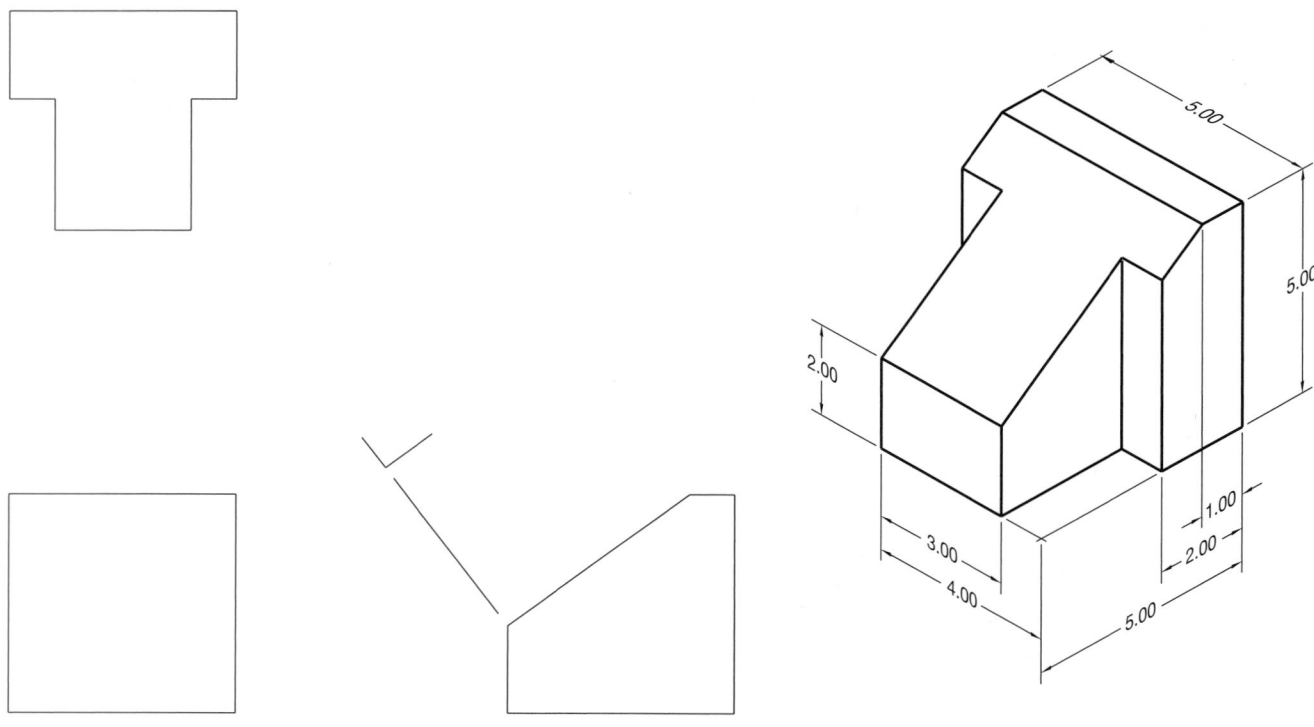

SUGGESTED VIEW LAYOUT

PROBLEM 8-24   Primary auxiliary view (in.)

Title: Brace

Material: Cast Iron

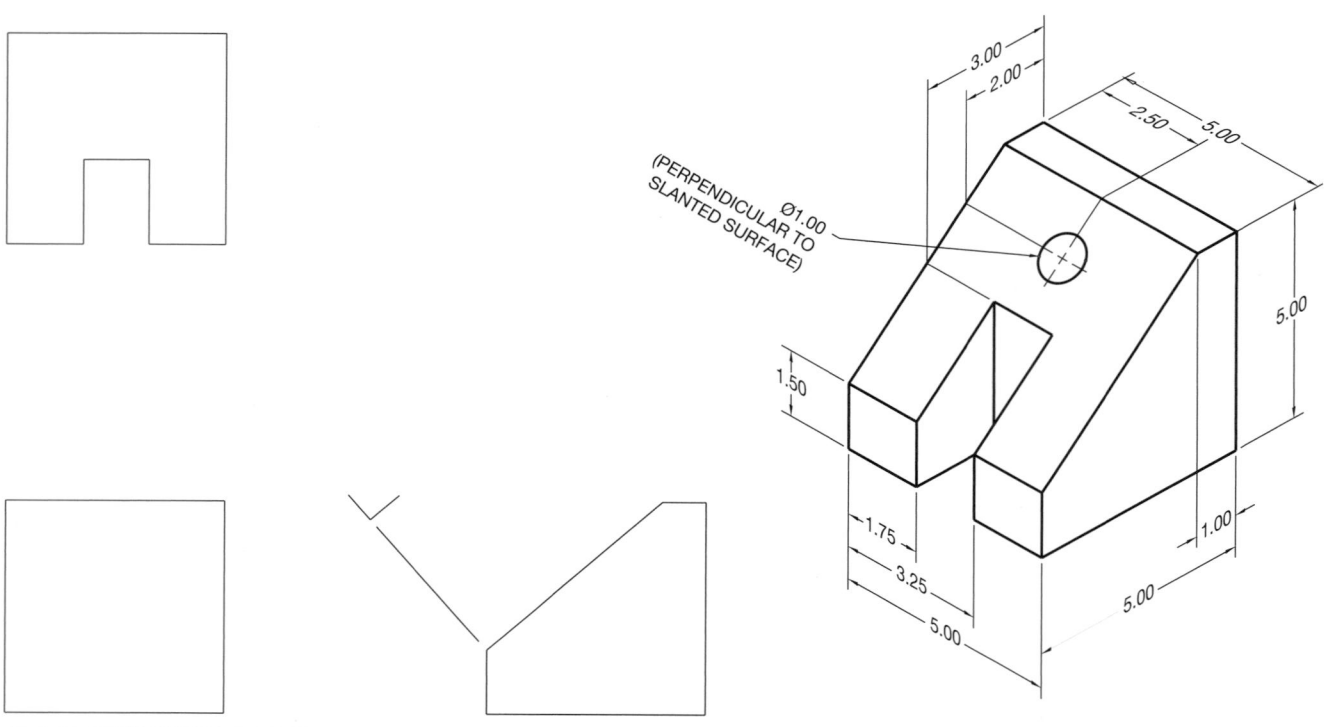

SUGGESTED VIEW LAYOUT

PART 5
DIRECTIONS

These problems provide you with a pictorial view with dimensions. Use the given information to select and draw the necessary multi-views and auxiliary view or views using manual drafting or CADD. Do not draw the pictorial view. Set up your drawings with a properly sized border and title block. Properly complete the information in the title block. Do not draw the dimensions.

PROBLEM 8-25   Primary auxiliary view (in.)
Part Name: Angle Base
Material: Mild Steel

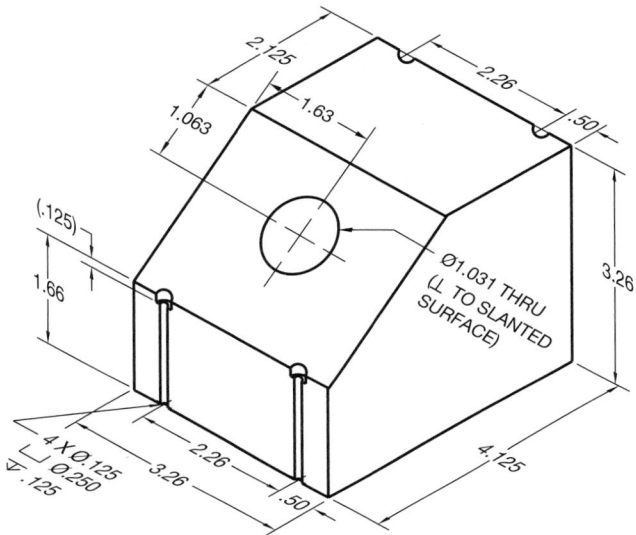

PROBLEM 8-26   Primary auxiliary view (metric)
Part Name: Belt Guide
Material: Aluminum

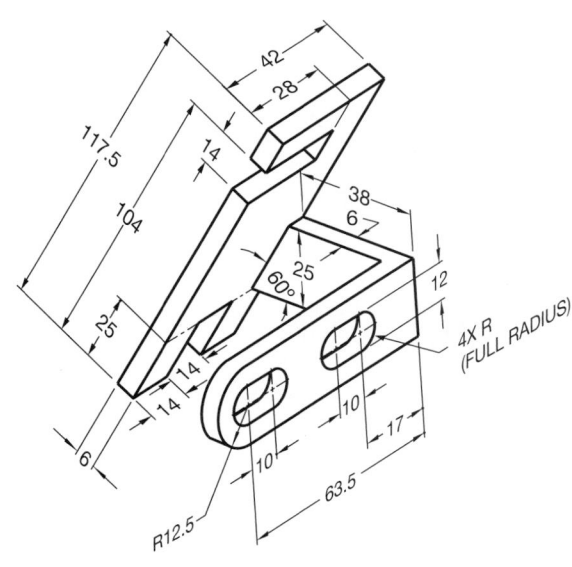

PROBLEM 8-27   Primary auxiliary view (in.)
Part Name: Angle V-Block
Material: Mild Steel
BASE: .50 THICK

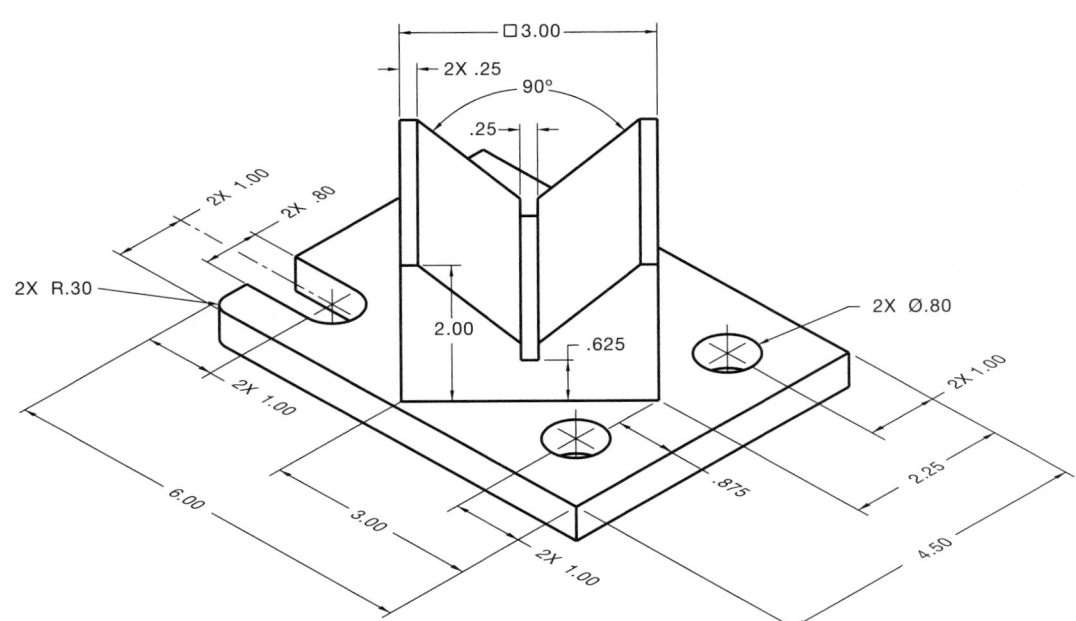

PROBLEM 8-28 Primary auxiliary views (metric)
Part Name: Pivot Link
Material: Aluminum Number 195

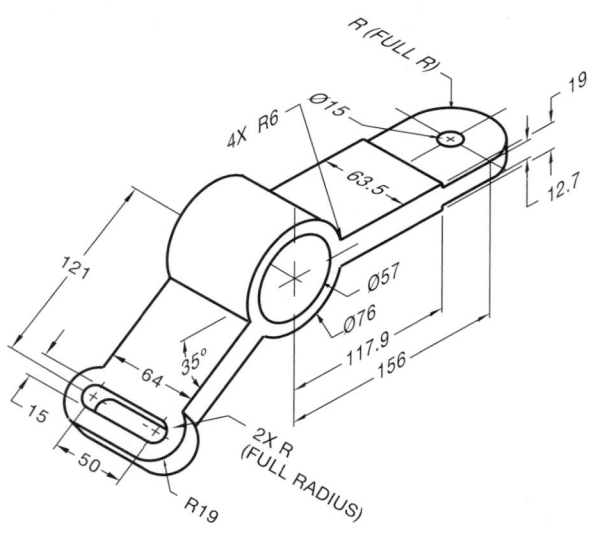

PROBLEM 8-30 Design change
Use Problem 8-29 to change the 6.00 length to 3.25.

PROBLEM 8-31 Secondary auxiliary view (metric)
Part Name: Skewed Face Plate
Material: SAE 1040

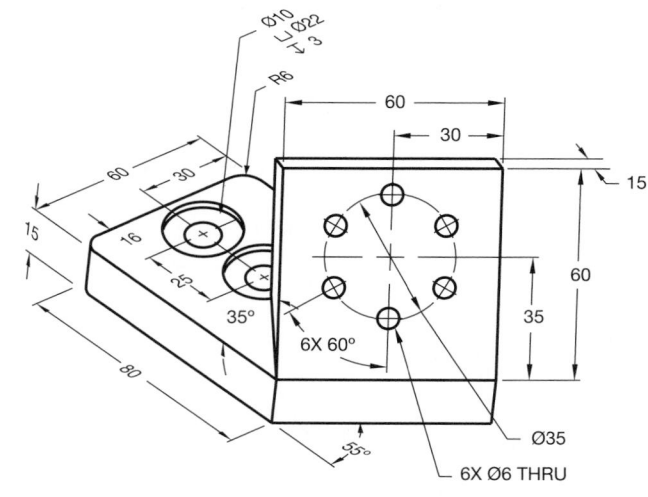

PROBLEM 8-29 Primary auxiliary views (in.)
Part Name: Mounting Bracket
Material: Mild Steel

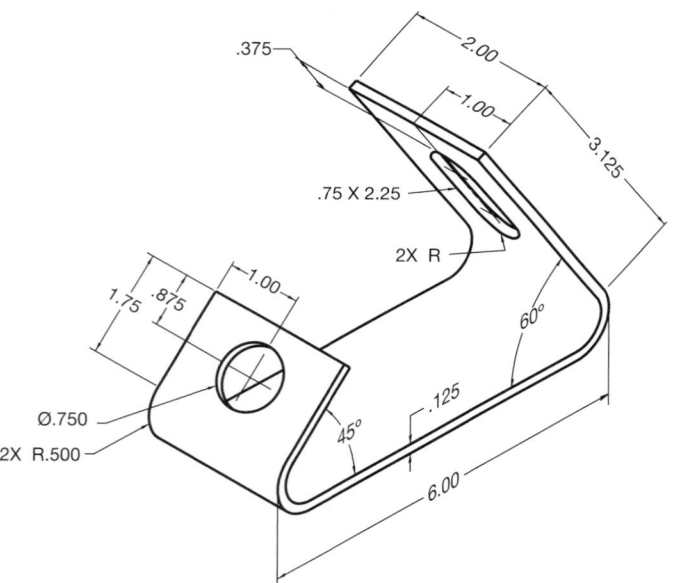

PROBLEM 8-32 Secondary auxiliary views (in.)
Part Name: Angle Bracket
Material: Cast Iron

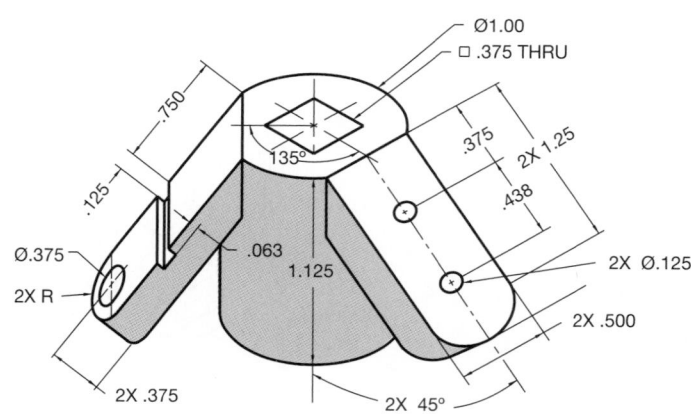

PROBLEM 8-33 **Primary and secondary auxiliary views (metric)**

Part Name: Chassis Switch Plate

Material: 4-mm-thick .416 Stainless Steel

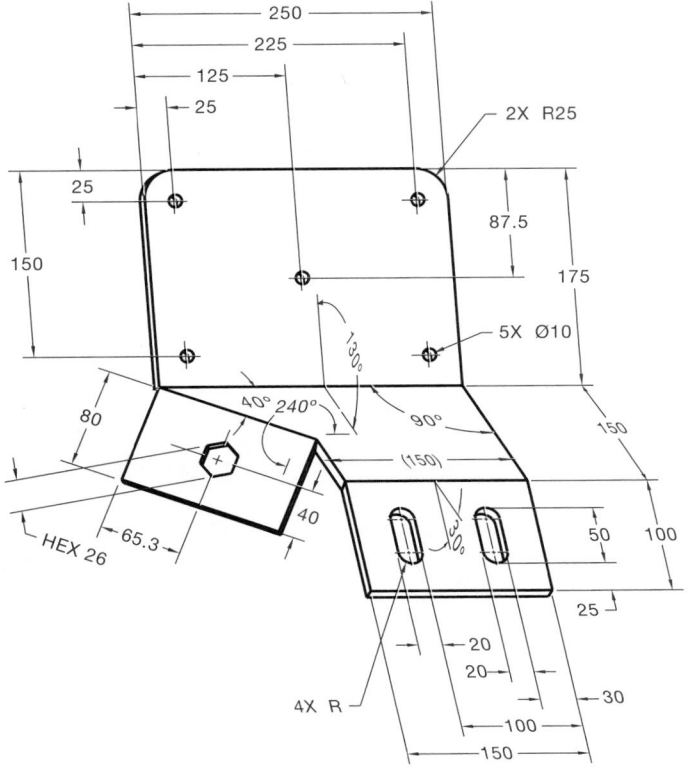

MATH PROBLEMS

A solar panel is in the shape of a 4' × 8' rectangle. With the sun overhead, what is the area of the shadow it casts on the ground if it is inclined at the following angles?

Number as:

PROBLEM 8-34 10°

PROBLEM 8-35 20°

PROBLEM 8-36 30°

PROBLEM 8-37 80°

PROBLEM 8-38 90°

PROBLEM 8-39 What should be the area of a solar panel inclined at 45° if the projected area from an overhead sun is to be 100 ft²?

PROBLEM 8-40 A circular sign with an area of 77.8 ft² casts a shadow on the ground. The area of the shadow is 26.6 ft². If the sun is overhead, what is the angle that the sign is inclined from the ground?

PROBLEM 8-41 A long post is sticking up from the ground at an angle of 35°. The sun is overhead and casting a shadow 98.3 feet long onto the ground. How long is the post?

PROBLEM 8-42 A square sign measuring 3 meters on an edge is inclined at an angle of 25° to the ground. What is its projected area onto the ground?

PROBLEM 8-43 A solar panel in the shape of a 4' × 8' rectangle casts a shadow with an area of 16 ft² when the sun is overhead. To what angle is it inclined?

CHAPTER

Descriptive Geometry I

OBJECTIVES

After completing this chapter, you will be able to:

■ Project a point onto various projection planes when given two adjacent views of the point.

■ Project a line onto various projection planes; project a line onto a projection plane that will show the line in *true length*; and project the true-length line onto a projection plane where it will *appear* as a point when given two adjacent views of the line.

■ Project a plane onto various projection planes; project a plane onto a projection plane that will show the plane as an *edge view*; and project the edge view of the plane onto a projection plane that will show the plane in *true shape* when given two adjacent views of the plane.

THE ENGINEERING DESIGN APPLICATION

Descriptive geometry principles are valuable for determining true shapes of planes, angles between two lines, two planes, or a line and a plane, and for locating the intersection between two planes, a cone and a plane, or two cylinders. Problems are solved *graphically* by projecting points onto selected adjacent projection planes in an imaginary projection system.

In this chapter, an imaginary projection plane system will be used based upon the conventional third-angle projection system used in positioning standard multiview drawing in CADD drawings. Most of the problems can be solved using just two basic concepts. The first concept is correctly projecting a *point* onto an imaginary projection plane, and the second is knowing when and how to create a *fold line* (either *parallel* or *perpendicular* to a line) to get the desired results.

In addition to the practical applications of descriptive geometry, these concepts can provide valuable training in visualizing how lines and planes are orientated in space. For those who cannot visualize well, the system of solving descriptive geometry problems used in this chapter works well by memorizing and/or following the step-by-step cookbook approach shown for each example.

There are usually several acceptable ways to solve descriptive geometry problems. An attempt has been made in this chapter to include at least one, and in some cases two, understandable solutions for each problem.

Whether solving descriptive geometry problems by using a straightedge and a triangle on paper or by using CADD, the most important first step in solving the problem is to *define the problem*. The quickest and usually the easiest way to do this is to develop a freehand sketch of the given problem, being sure to include all of the known information.

Secondly, sketch the correct process of your solution, but do not be too concerned with accuracy at this point. When you believe you have sketched a correct process, then draw out the solution, as accurately as possible, using a straightedge and a triangle, or with CADD where, if the process is correct, the solution will be perfect. The CADD system also allows you to work in layers and colors for a clearer picture of the problem and solutions, although if you are using a straightedge and triangle, you can obtain the same effect if different colored pencil leads are used for different lines when solving the problems.

BASIC CONCEPTS

The Descriptive Geometry Projection Layout Format

The way descriptive geometry problems will be laid out in this chapter is based upon the third-angle projection system used on all regular drawings. An example of the imaginary glass box system drawn about an object is shown in Figure 9.1. The unfolded mul-

tiprojection system is shown in a flat layout projection of that same object in Figure 9.2 where all the projected images of the six standard sides of the object are shown projected onto the six standard views.

In Figure 9.3, the image of a *real point* located in the space of the imaginary glass box is shown projected onto the three visible sides of the glass box. The real point floating within the glass box is represented with a star and identified with any uppercase letter. Uppercase letter *A* is used in this example, but any letter can be used. The

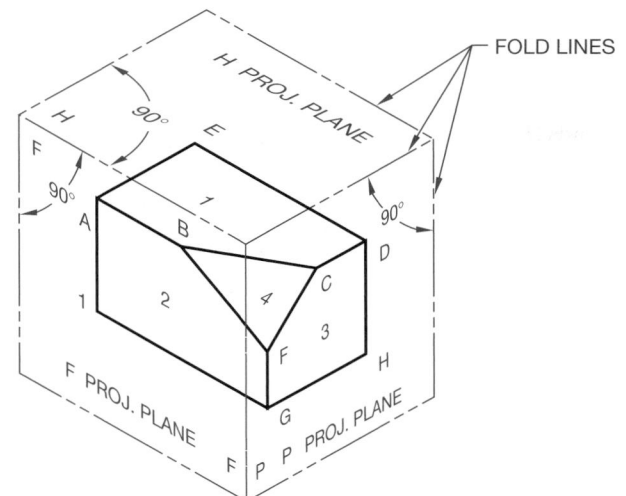

FIGURE 9.1 ■ 3-D box layout projection.

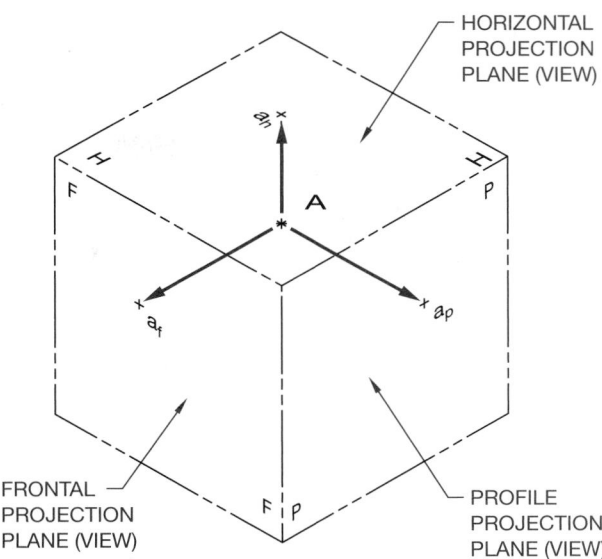

FIGURE 9.3 ■ Image of real-point projection.

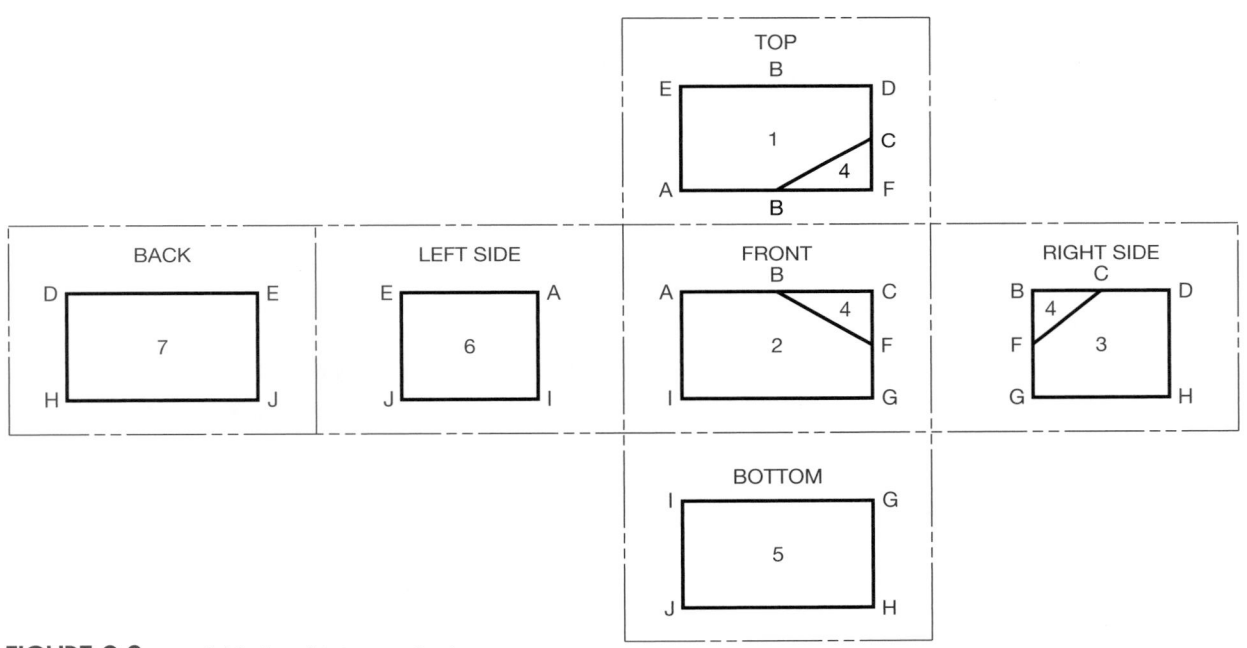

FIGURE 9.2 ■ Unfolded multiview projection system.

real point is shown projected with *projectors* radiating *perpendicular* (90°) from the real point onto the three sides of the imaginary glass box. The real point will not be used to solve any of the problems. Where the real point is projected onto the frontal projection plane of the imaginary glass box, a *cross* is located. This cross represents the image of the real point projected onto the frontal plane and is identified as a_f. The lowercase letter *a* corresponds to the real point, and the lowercase subscript letter *f* corresponds to the frontal projection plane. On the profile projection plane, the cross representing the projected point is identified as a_p, and on the horizontal projection plane, the cross representing the projected point is identified as a_h.

Thick phantom lines called *fold lines* represent the edges of the imaginary glass box. Two uppercase letters corresponding to the two adjacent (next to) projection planes identify each fold line. In this example, *F/H* identifies the fold line between the frontal and horizontal projection planes, *F/P* identifies the fold line between the frontal and profile projection planes, and *H/P* identifies the fold line between the horizontal and profile projection planes.

Figure 9.4 shows the images of the same point *A* shown in Figure 9.3, but in a multiview flat layout system. The point is identified in the same manner as in Figure 9.3. The projection planes are now called *views* just as they are in the standard third-angle projection system used in the other CADD classes. Each adjacent view

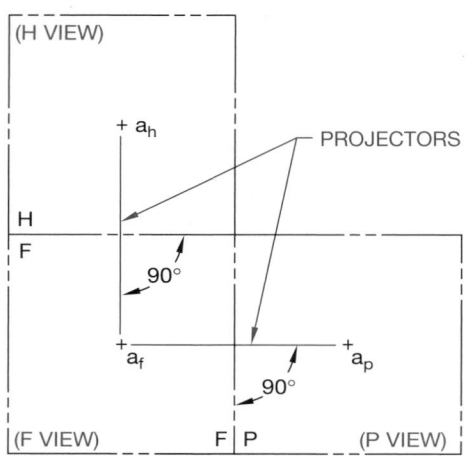

FIGURE 9.4 ■ Flat layout of one-point projection.

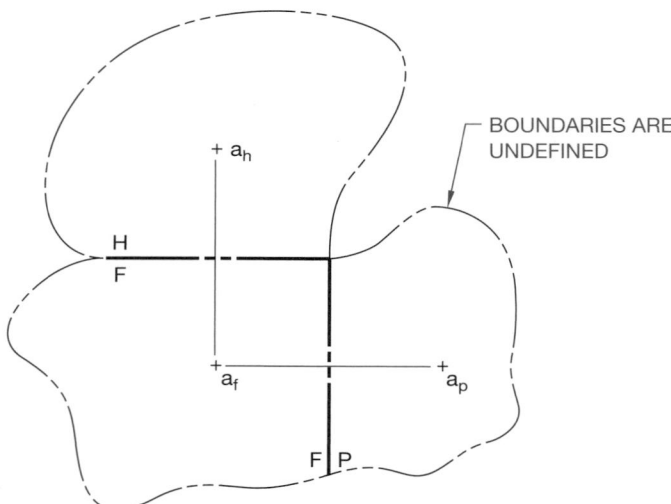

FIGURE 9.5 ■ Flat layout with unidentified boundaries omitted.

is shown separated by the same fold line as seen in Figure 9.3 with the same identification uppercase letters. The other fold lines shown that do not have uppercase identification letters merely represent the other imaginary boundaries of each view. These fold lines are not identified because there was no adjacent view given in Figure 9.3. The lines of projection (projectors) from point *A* are also included. Notice that the projectors cross each fold line at a 90°, or right, angle. This will be important to remember when you begin projecting points.

Figure 9.5 shows the transition from the complete multiview flat layout system to using the minimal information layout system that will be used in this chapter. This is exactly the same multiview flat layout system as shown in Figure 9.4, but the unidentified fold lines are now omitted, because they represented arbitrary boundaries that are not used.

Rule 1

All given problems in descriptive geometry must include *two adjacent views* of the given points, lines, and/or planes. Two adjacent views are needed to solve the problems because one view contains only two of the three dimensions needed to solve descriptive geometry problems. In Figure 9.6, notice that the front view includes the up, down, right, and left dimensions only, while the adjacent top and profile views contain the additional forward and back (depth) dimensions needed to complete a three-dimensional system.

Rule 2

When a projector crosses over a fold line, it must always cross *perpendicular* to that fold line. This represents a change of 90° in direction from one view to another as shown in Figure 9.7.

PROJECTION OF A POINT (Basic Concept 1)

Knowing how to project an image of a real point in the flat projection system is fundamental to solving most of the problems in applied descriptive geometry. This concept is simple but must be understood completely before attempting to solve the descriptive geometry problems.

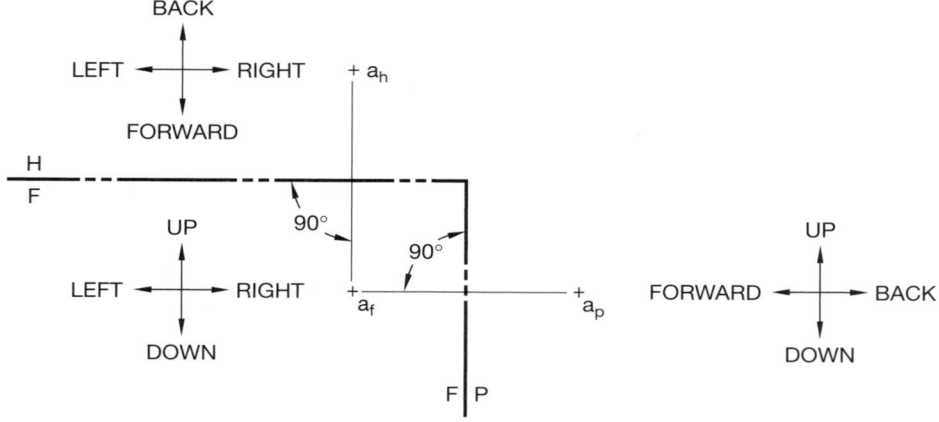

FIGURE 9.6 ■ Example showing each view contains only four different directions.

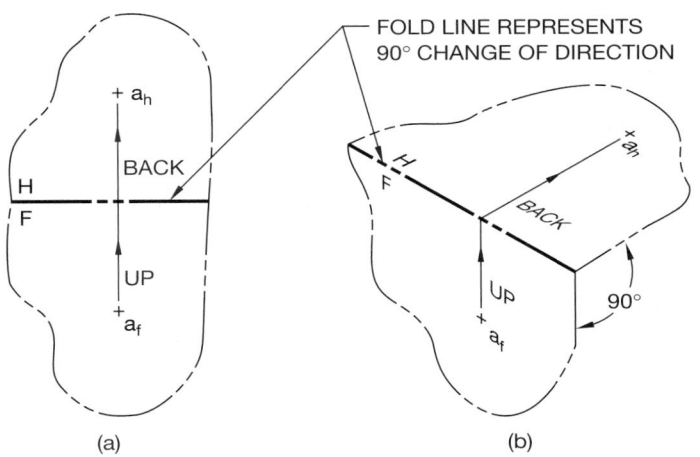

(a)   (b)

FIGURE 9.7 ■ **(a)** Showing fold line changing directions in flat layout. **(b)** Showing fold line changing directions in 3-D layout.

Problem

Project a given point correctly onto a new view. Things that must be given to solve this problem:

1. Two adjacent views of the point.

2. A fold line located correctly between the two given adjacent views.

3. Another fold line that represents one edge of the new view onto which the point is to be projected.

As shown in Figure 9.8, point *A* is located in the front view with a cross and is identified by the letters a_f. In the horizontal view, point *A* is also shown with a cross, but is identified with the letters a_h. These two views are adjacent as indicated by the fold line *F/H* located between them. The combined dimensions shown in both views supply the needed three dimensions to solve the problem. When a projector is drawn connecting both views of point *A*, it crosses the fold line *F/H* at a 90° angle. This verifies that point *A* is correctly located in both adjacent views. If a projector connected to both views of point *A* (a_f to a_h) did not cross the fold line at 90°, the fold line could be located at a wrong angle between the points, as shown in Figure 9.9a. Or the problem could be that the drawing really shows one view of two different points, as shown in Figure 9.9b. If either condition exists, not enough information has been

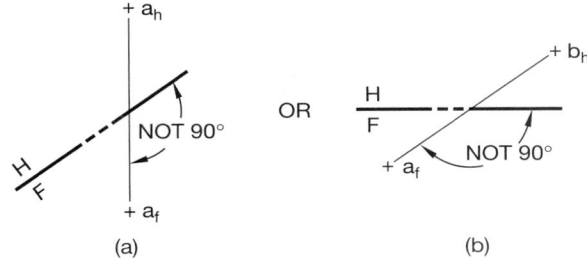

(a)   (b)

FIGURE 9.9 ■ **(a)** Fold line placed incorrectly. **(b)** Two different points assumed to be the same.

given to solve the problem. Finally, another fold line *F/P* representing one known edge of the *P* view is given. (See Figure 9.8 again.) This represents the edge of the new view (view *P*) onto which point *A* is to be projected.

Solving the Problem

Given the problem shown in Figure 9.10:

Step 1: Draw a projector (projector 1) from point a_f to point a_h. (See Figure 9.11.) Projector 1 must cross the fold line *F/H* at a 90° angle making it perpendicular to the fold line *F/H*.

Step 2: Draw a new projector (projector 2) from point a_f perpendicular to the fold line *F/P* and extend projector 2 into the *P* view at any given distance at this time. Point *A* will lie somewhere along projector 2 in the *P* view. (See Figure 9.12.) Next, draw in a little angle bracket close to where projector 2 and the fold line *F/P* intersect just to reinforce the idea that a projector must always be drawn at a 90° angle to a fold line and that it was done in this case.

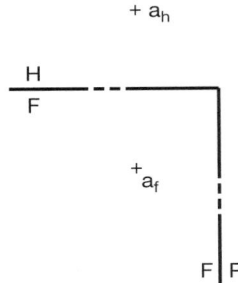

FIGURE 9.10 ■ A given problem.

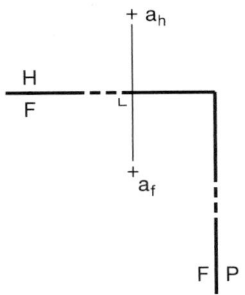

FIGURE 9.8 ■ Correctly given problem to project a point.

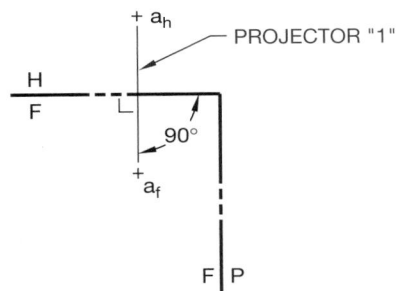

FIGURE 9.11 ■ Drawing the first projector.

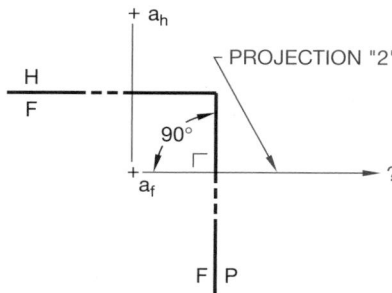

FIGURE 9.12 ■ Drawing the second projector.

Step 3: Measure the perpendicular distance from point a_h to the fold line H/F in the H view. Identify this distance with a bracket and the number 1. Transfer this distance 1 along projector 2, measuring from the fold line F/P. Identify this distance with a similar bracket and the same number, 1. Do this just to reinforce the idea that this distance 1 must be the same distance as the other distance 1 and that it was correctly identified. Make a cross representing point A on projector 2 at this distance. Point A now has been correctly projected onto the P view. Identify the point (cross) as a_p. (See Figure 9.13.)

Note: This problem represents only *one point*, point A, projected onto three different views, *not three points* projected onto three different views.

To project a point into an auxiliary view, the same process is used. The only difference is that the fold line representing the edge of the given auxiliary view is at a different angle than that in the standard views. (See Figure 9.14.)

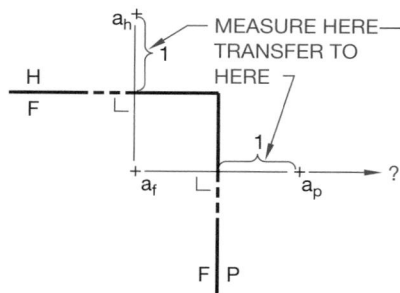

FIGURE 9.13 ■ Transferring dimension 1.

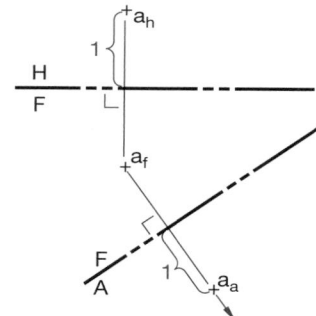

FIGURE 9.14 ■ Locating a point in an auxiliary view.

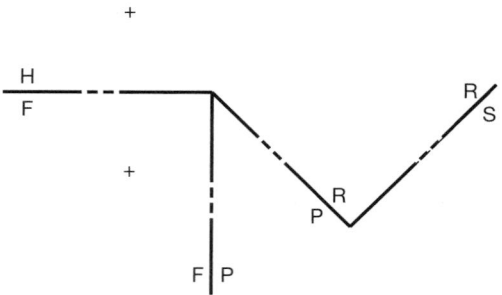

FIGURE 9.15 ■ A given multiview problem.

To project a point through successive (more than one) auxiliary views, you will need to repeat the three-step process for each given auxiliary view. The only "tricky" part of this process will be to understand *where* you get the dimension from to locate the point in each additional auxiliary view. The problem given in Figure 9.15 is similar to the problem given in Figure 9.10, but two additional fold lines, P/R and R/S, have been added. This problem requires you to find point A in the auxiliary view R and in the auxiliary view S.

Step 1: *Project point A onto the P view.* To solve this problem, you must first project point A onto the P view just as was done for the problem in Figure 9.10.

Step 2: *Project point A onto the R view.* To project point A onto the R view, imagine that only views F, P, and the edge (fold line P/R) of view R exist as shown in Figure 9.16. Now project point A onto the R view using the same three steps as used before.

Step 2a: Draw a third projector, projector 3, from point a_p *perpendicular* to the fold line P/R and extend projector 3 into the R view at any given distance at this time. Point A will lie somewhere along projector 3 in the R view. (See Figure 9.16 again.)

Next, draw in a little angle bracket close to where projector 3 and the fold line P/R intersect, again just to reinforce the idea that a projector must always be drawn at a 90° angle to the fold line and that it was done in this case.

Step 2b: Locate the correct placement of point A in the R view. When determining the correct distance for point A in the R view, you need to be careful where to find this correct

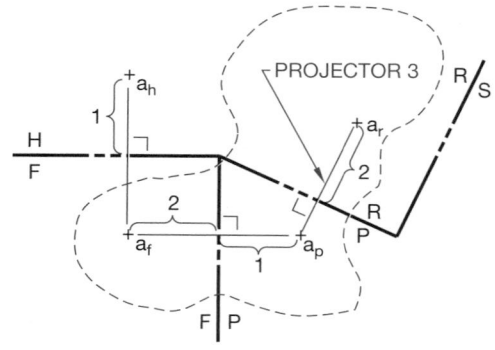

FIGURE 9.16 ■ Transferring distances 1 and 2 to the correct views.

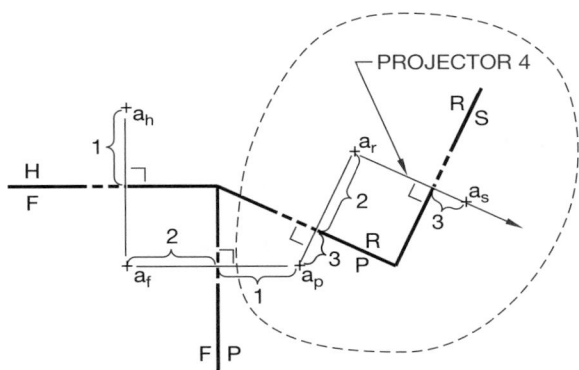

FIGURE 9.17 ■ Transferring distance 3 to the S view.

distance. Notice it will be from the *F/P* fold line to point *a_f* *skipping* the *P* view. Identify this distance with a bracket and the number 2.

Transfer this distance 2 measuring from the fold line *P/R* along projector 3. Identify this distance with a bracket and the same number 2 just to confirm that the correct distance was used.

Identify the point *A* with a cross and the letters *a_r*.

Step 3: *Project point A onto the S view.* To project point *A* onto the *S* view, imagine that only views *P*, *R*, and the edge (fold line *R/S*) of view *S* exist as shown in Figure 9.17. Notice that each part of the problem is being solved using groups of three elements, the *two known views and the one unknown view*. Repeat the process to project point *A* onto the *S* view the same way as was done for the *R* view.

Step 3a: Draw a fourth projector, projector 4, from point *a_r* *perpendicular* to the fold line *R/S* and extend projector 4 into the *R* view at any given distance at this time. Point *A* will lie somewhere along projector 4 in the *S* view. (See Figure 9.17 again).

Next, draw in a little angle bracket close to where projector 4 and the fold line *R/S* intersect, again just to reinforce the idea that a projector must always be drawn at a 90° angle to the fold line and that it was done in this case.

Step 3b: Locate the correct placement of point *A* in the *S* view. When determining the correct distance for point *A* in the *S* view, you need to be careful where to find this correct distance. Notice it will be from the *P/R* fold line to point *a_p* *skipping* the *R* view. Identify this distance with a bracket and the number 3.

Transfer this distance 3 measuring from the fold line *R/S* along projector 4. Identify this distance with a bracket and the same number 3 just to confirm that the correct distance was used.

Identify the point *A* with a cross and the letters *a_s*.

LINES

■ **Straight line:** The shortest distance between two points. (See Figure 9.18.)

FIGURE 9.18 ■ Two points representing a straight line.

■ **True-length line:** The actual measured length of a line. (See Figure 9.19.)

■ **Foreshortened line:** A line that *appears* shorter than it really is. (See Figure 9.20.)

■ **Parallel lines:** Lines that are equal distance from each other throughout their lengths. (See Figure 9.21.) Parallel lines will appear parallel in all views with only *two exceptions*. You will not be able to see that the lines are parallel when one line is located behind the other; they will appear as *one line*. And lines will not appear parallel where both lines are viewed at the ends. Here they will appear as *points*. (See Figure 9.22.)

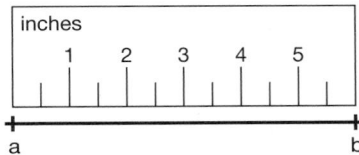

FIGURE 9.19 ■ True length represented by true measurement.

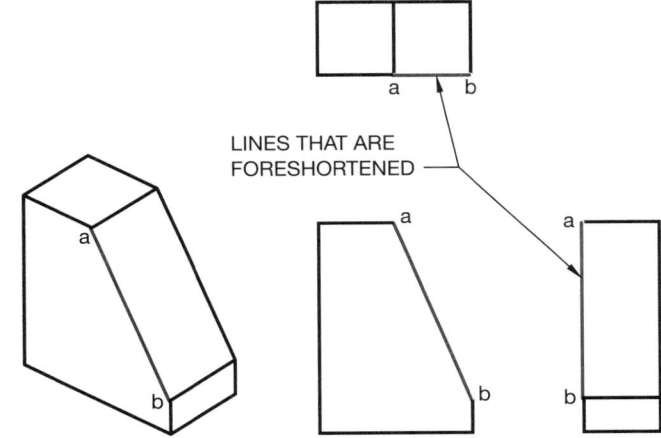

FIGURE 9.20 ■ A foreshortened line represented by line *A-B*.

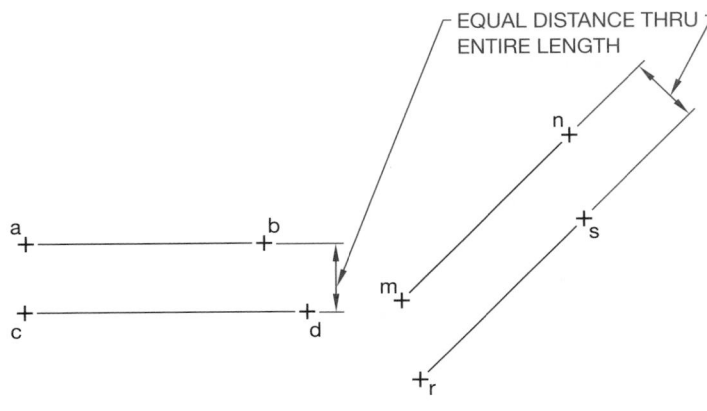

FIGURE 9.21 ■ Parallel lines.

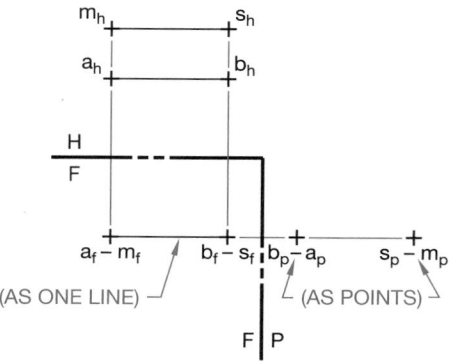

FIGURE 9.22 ■ Parallel lines viewed as one line in the first view and as points in the second view.

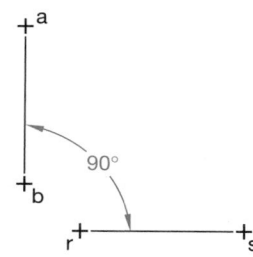

FIGURE 9.23 ■ Perpendicular lines.

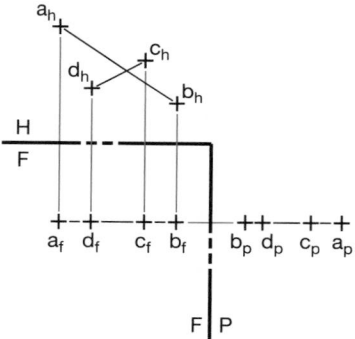

FIGURE 9.24 ■ Intersecting lines.

■ **Perpendicular lines:** Lines that are positioned at a 90° angle to each other. (See Figure 9.23.)

■ **Intersecting lines:** Lines that actually intersect each other. They will appear to intersect in all views except where they appear as one line. (See Figure 9.24.)

■ **Skew lines:** Lines that are *nonparallel* and *nonintersecting*. (See Figure 9.25.)

PROJECTION OF A LINE (Basic Concept 2)

Projecting a line in the flat projection systems is done in exactly the same way as projecting two points in space. Remember, a straight line is defined as the shortest distance between *two points*.

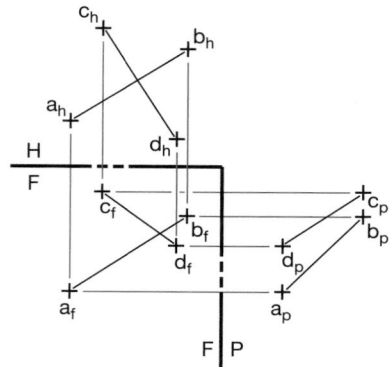

FIGURE 9.25 ■ Skew lines.

Problem

Project line *A-B* correctly onto the *P* view. (See Figure 9.26.) Things that must be given to solve this problem:

1. Two adjacent views of the line.

2. A fold line located correctly between the two given adjacent views.

3. Another fold line that represents one edge of the new view onto which the line is to be projected.

Step 1: Draw a projector (projector 1) from point a_f to point a_h and another projector (projector 2) from point b_f to point b_h. (See Figure 9.27.) Projectors 1 and 2 must cross the fold line *F/H* at a 90° angle making them perpendicular to the fold line *F/H*. Next, draw little angle brackets close to where projectors 1 and 2 and the fold line *F/H* intersect just to reinforce the idea that a projector must always be drawn at a 90° angle to a fold line and that it was done in this case. (See Figure 9.27 again.)

Step 2: Draw a new projector (projector 3) from point a_f *perpendicular* to the fold line *F/P* and extend projector 3 into the *P* view at any given distance at this time. Point *A* will lie somewhere along projector 3 in the *P* view. (See Figure 9.28.)

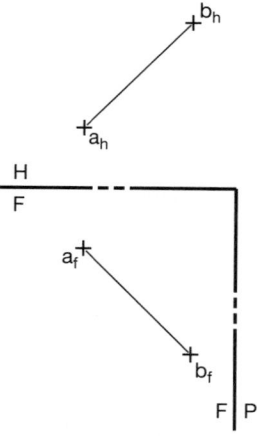

FIGURE 9.26 ■ Setup to project a line in space.

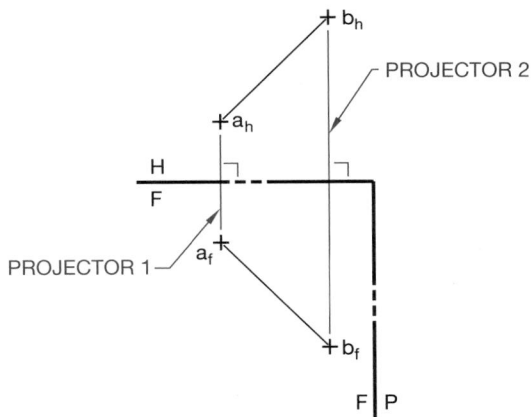

FIGURE 9.27 ■ Confirming views are correct by connecting projectors.

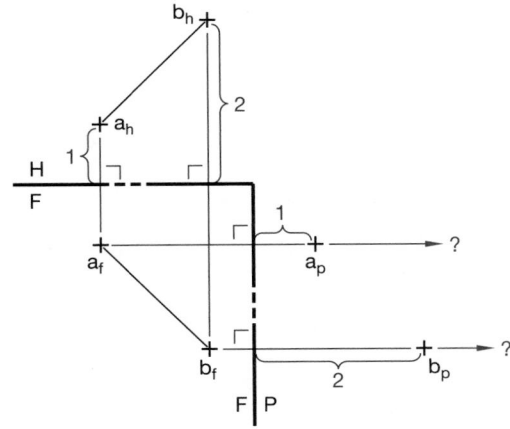

FIGURE 9.29 ■ Transferring distances 1 and 2 correctly.

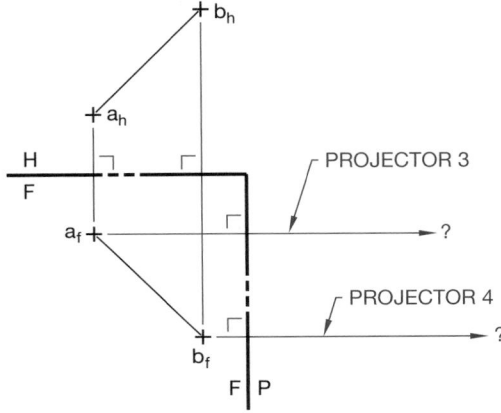

FIGURE 9.28 ■ Locating projectors correctly into new view.

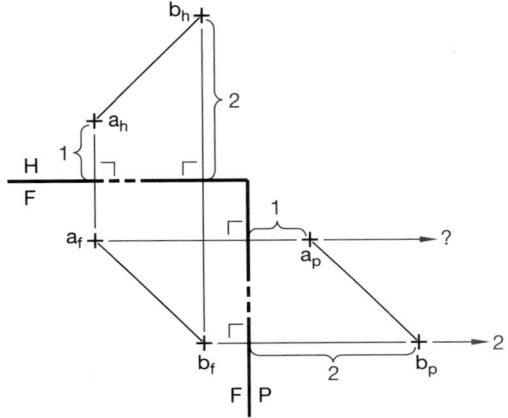

FIGURE 9.30 ■ Showing problem solved.

Draw another new projector (projector 4) from point b_f perpendicular to the fold line *F/P* and extend projector 4 into the *P* view at any given distance at this time. Point *B* will lie somewhere along projector 4 in the *P* view. (See Figure 9.28 again.)

Next, draw little angle brackets close to where projectors 3 and 4 and the fold line *F/P* intersect just to reinforce the idea that projectors must always be drawn at a 90° angle to a fold line and that it was done in this case. (See Figure 9.28 again.)

Step 3: Measure the perpendicular distance from point a_h to the fold line *H/F* in the *H* view. Identify this distance with a bracket and the number 1. Transfer this distance 1 along projector 3 measuring *from* the fold line *F/P.* Identify this distance with a similar bracket and the same number 1. Do this just to reinforce the idea that this distance 1 must be the same distance as the other distance 1 and that it was correctly identified. Make a cross representing point *A* on projector 3 at this distance. Point *A* now has been correctly projected onto the *P* view. Identify the point (cross) as $a_p.$ (See Figure 9.29.)

Next, measure the perpendicular distance from point b_h to the fold line *H/F* in the *H* view. Identify this distance with a bracket and the number 2. Transfer this distance 2 along projector 4 measuring *from* the fold line *F/P.* Identify this distance with a similar bracket and the same number 2. Do this just to reinforce the idea that this distance 2 must be the same distance as the other distance 2 and that it was correctly identified. Make a cross representing point *B* on projector 4 at this distance. Point *B* now has been correctly projected onto the *P* view. Identify the point (cross) as $b_p.$ (See Figure 9.29 again.)

Connect the new points a_p and b_p with a line. This represents the line a_p-$b_p.$ (See Figure 9.30.)

PROJECTION OF A LINE TO FIND ITS TRUE LENGTH (Basic Concept 3)

Knowing how to project a line in space has little value in itself. But knowing how to project the line in a certain manner will enable you to find the line in *true length* and where the true-length line will appear as a point. These are valuable concepts to know when trying to determine the true distance between lines or the true distance between a line and a plane; when finding the *true shape* of a plane or the true angle between planes; and when dealing with other advanced concepts.

There are two basic ways to find the true length of a line that appears foreshortened in the two given views of a problem. The first is the *fold-line method,* and the second is the *revolution method.* Both methods are presented because each method of finding the true length of a line has certain advantages over the other when used in solving certain advanced descriptive geometry problems, although either method, fold-line or revolution, can be used equally as well to find the true length of a line.

Fold-line Method of Finding True Length of a Line

The fold-line method used to find the true length of a line represents the concept of the viewer moving around the line in such a way that both ends of the line are the same distance from the viewer. (See Figure 9.31.)

Problem

Find the true length of line *C-D* by the fold-line method. (See Figure 9.32.) Things that must be given to solve this problem:

1. Two adjacent views of the line.

2. A fold line located correctly between the two given adjacent views.

Step 1: *Draw a new fold line parallel to the given line C-D in either view.* A new fold line *F/R* as shown in Figure 9.33a or a new fold line *H/A* as shown in Figure 9.33b can be

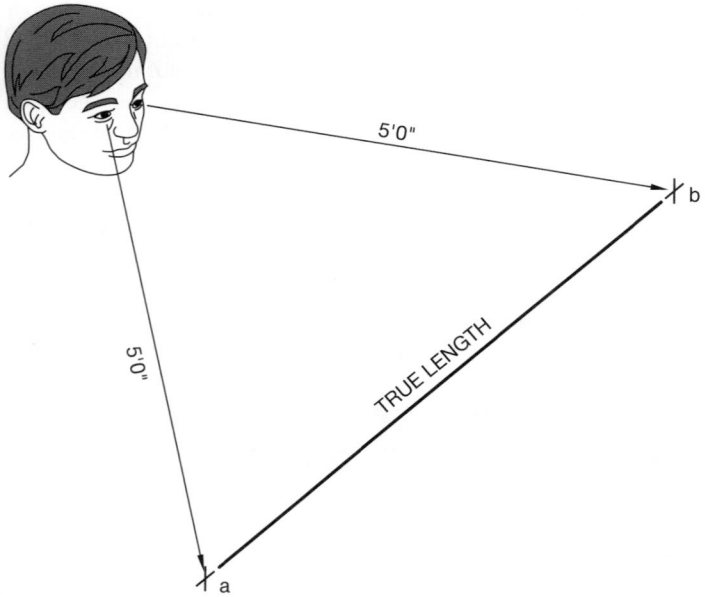

FIGURE 9.31 ■ A line viewed in true length.

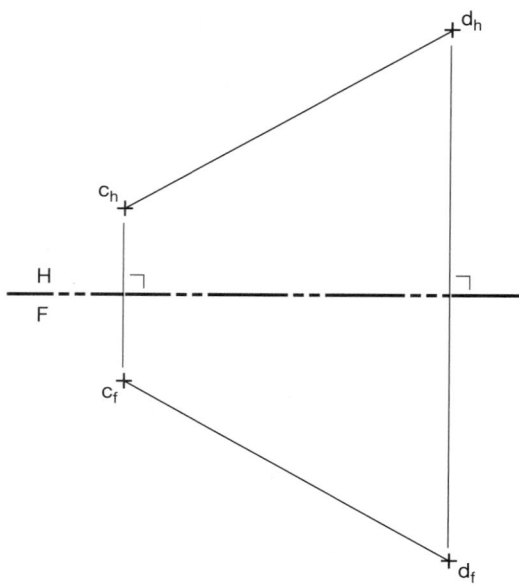

FIGURE 9.32 ■ Setup to find a line in true length by the fold-line method.

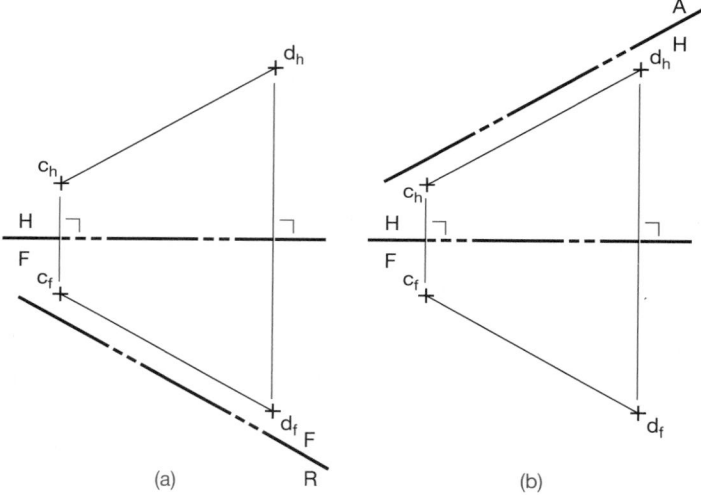

FIGURE 9.33 ■ **(a)** Possible placement of a new fold line next to the front view. **(b)** Possible placement of a new fold line next to the top view.

drawn parallel to line c_f-d_f or line c_h-d_h on either side. However, drawing the new fold line *F/R* or *H/A* between line c_f-d_f or line c_h-d_h and the existing fold line *H/F* would make solving the problem much more complicated and confusing, so this is *not* recommended. (See Figure 9.34a and b.)

Also, the distance the fold line *F/R* or *H/A* is positioned *parallel* from line c_f-d_f or line c_h-d_h is not critical and will not change the solution. (See Figure 9.35a.) The fold line *F/R* or *H/A* could actually be drawn on top of line c_f-d_f or line c_h-d_h, but again it is *not* recommended. (See Figure 9.35b.)

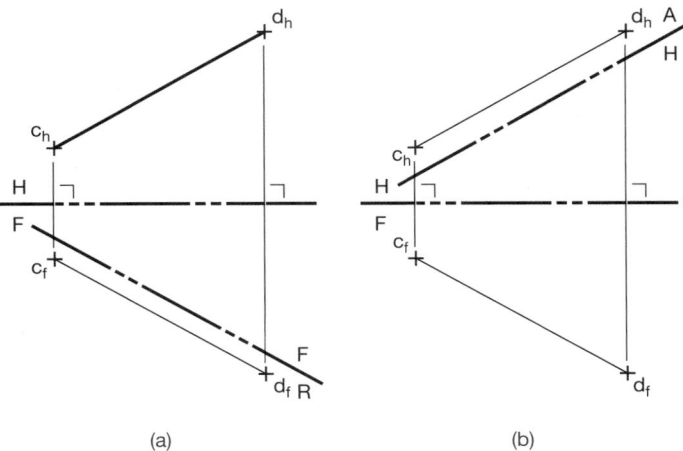

FIGURE 9.34 ■ (a) New fold-line placement next to front view that is *not* recommended. (b) New fold-line placement next to top view that is *not* recommended.

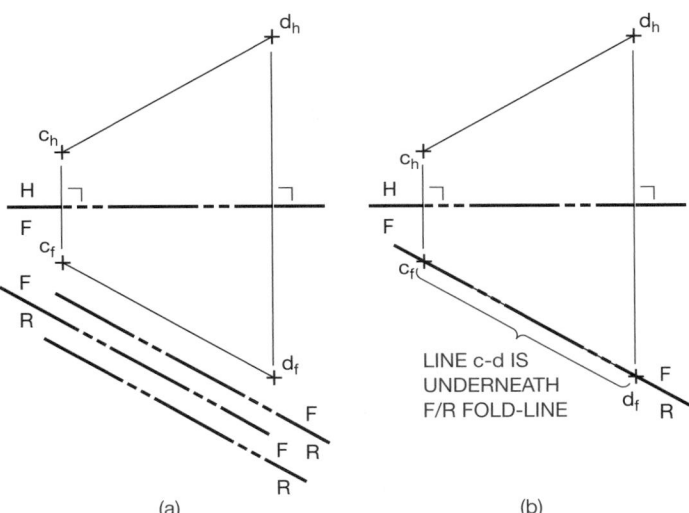

FIGURE 9.35 ■ (a) Other good possible locations for the new fold line *F/R*. (b) Fold line could work on top of the given line, but this is *not* recommended.

Step 2: *Project the line onto the new view.* Project the line c_f-d_f onto view R or project the line c_h-d_h onto view A just the same way as was previously explained in projecting a line onto a new view. The line c_r-d_r or c_a-d_a is the line C-D shown in true length. (See Figure 9.36a and b.)

Hint: Finding the true length of line C-D in both the A and the R views could be used to double-check that you have found the correct true length of line C-D. Line c_r-d_r must be the *same* length as line c_a-d_a. If this is not the case, one line or both lines are projected incorrectly, and you have not really found the true length of line C-D.

The trick to solving this problem by the fold-line method is to remember to create a new fold line *parallel* to the given line. Then project the line (end points) just as you learned previously when projecting points or a line onto a new view.

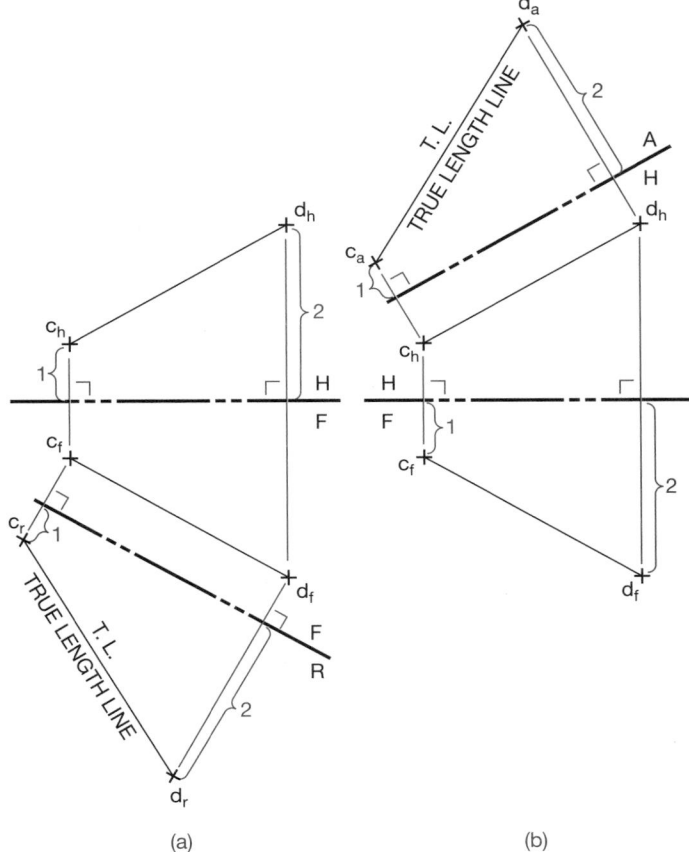

FIGURE 9.36 ■ (a) Creating true length from the front view. (b) Creating true length from the top view.

Revolution Method of Finding True Length of a Line

The revolution method of finding the true length of a line represents the concept of moving or *revolving the line* in space in such a way that both ends of the line are the same distance from the viewer. (See Figure 9.37.) An advantage of using this method over the fold-line method is that you can find the true length of a line in the two given views, thus eliminating the need to create a new view. This will also give more accuracy in the solution when using the board method.

Problem

Find the true length of line C-D by the revolution method. (See Figure 9.38.) Things that must be given to solve this problem:

1. Two adjacent views of the line C-D.

2. The fold line F/H located correctly between the two given adjacent views.

Step 1: *Move point d_h around point c_h or point c_h around point d_h in the H view to find the true length in the F view.* To rotate point d_h around point c_h graphically, place a compass needle end on point c_h and the compass lead point on point d_h. Draw an arc from point d_h until it is the

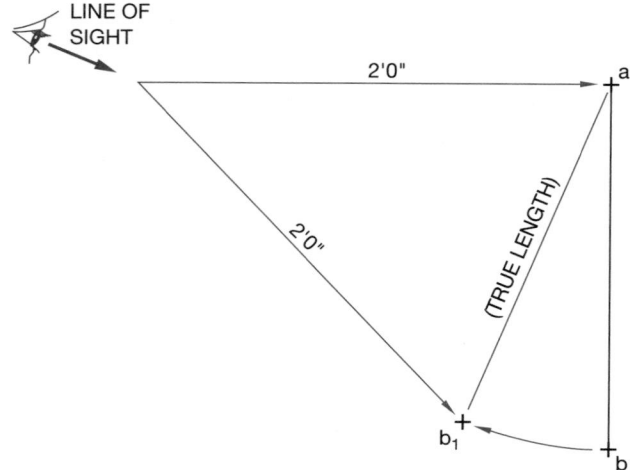

FIGURE 9.37 ■ Revolved line to see in true length.

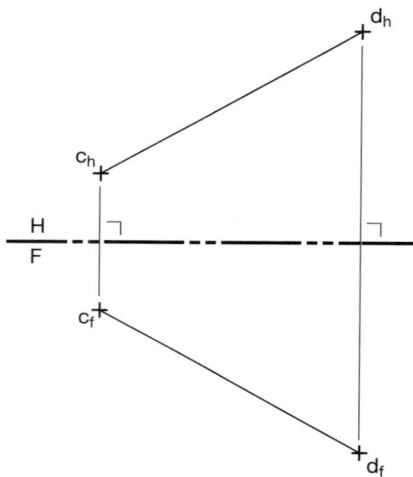

FIGURE 9.38 ■ Setup problem to find true length.

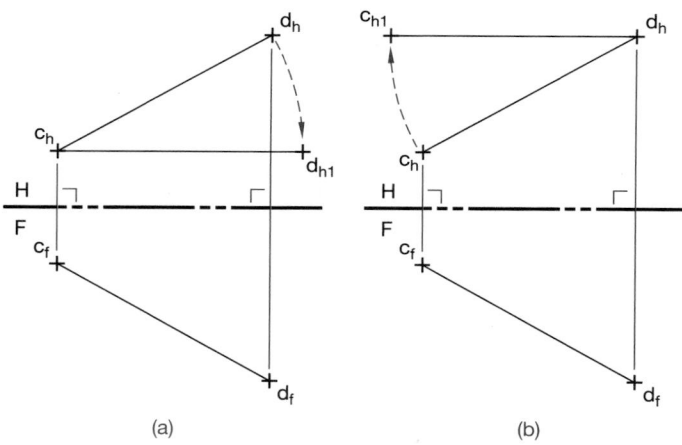

FIGURE 9.39 ■ **(a)** Rotating around point *C* with a compass. **(b)** Rotating around point *D* with a compass.

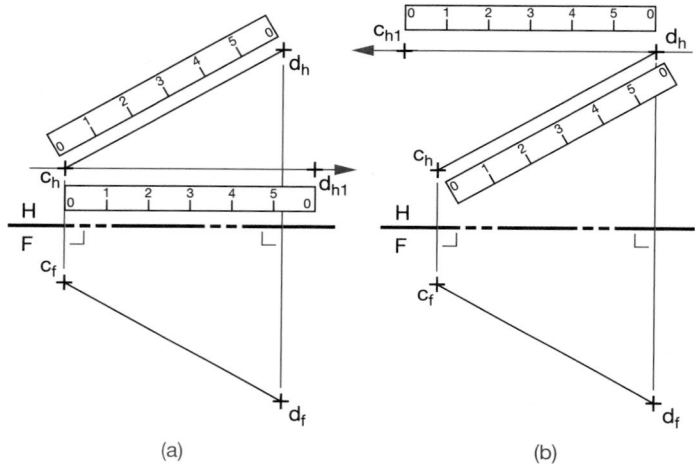

FIGURE 9.40 ■ **(a)** Rotating without using a compass around point *C*, step 1. **(b)** Rotating without using a compass around point *D*, step 1.

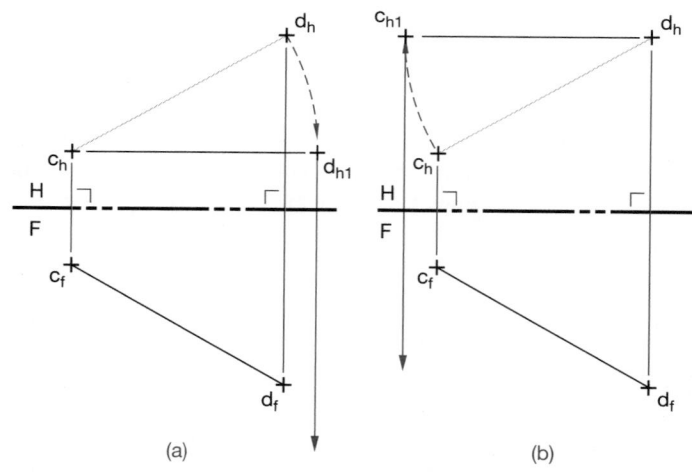

FIGURE 9.41 ■ **(a)** Rotating without using a compass around point *C*, step 2. **(b)** Rotating without using a compass around point *D*, step 2.

same distance from the fold line *H/F* as point c_h. Make a cross at this place and label the cross as d_{h1}. Then connect point c_h to point d_{h1} with a straight line. (See Figure 9.39a.) To rotate point d_h around point c_h, use the same process, but put the compass point on point c_h and proceed just as in the previous example. (See Figure 9.39b.)

If you do not have a compass, you can rotate point d_h by the following way. (See Figure 9.40.) Draw a light construction line parallel to fold line *H/F* through point c_h. Measure the length of line c_h-d_h and transfer this length from point c_h along the construction line. As in the previous example, identify the other end of this length with a cross and as d_{h1}. Then connect point c_h to point d_{h1} with a straight line.

Step 2: *Project point d_{h1} or point c_{h1} onto the F view.* Draw a projector from point d_{h1} *perpendicular* to and on through the *H/F* fold line and beyond point d_f in the *F* view. (See Figure 9.41a.) To project point c_{h1}, see Figure 9.41b.

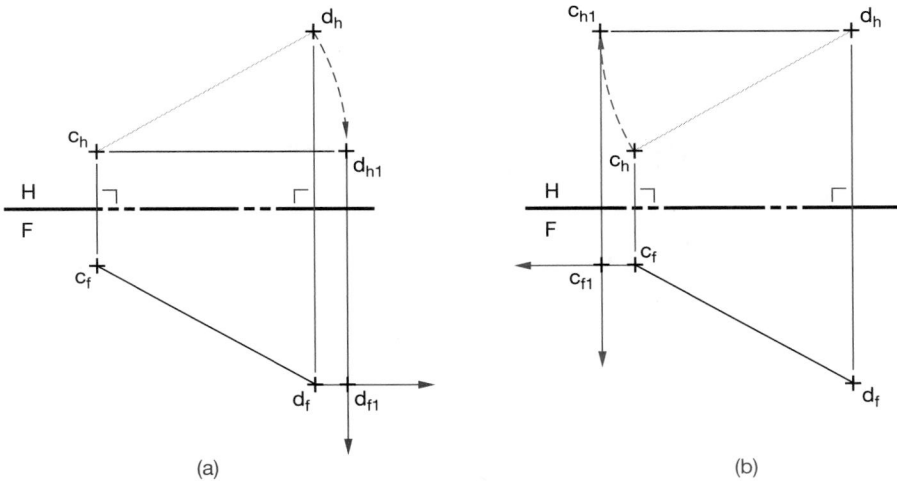

FIGURE 9.42 ■ (a) Rotating without using a compass around point C, step 3.
(b) Rotating without using a compass around point D, step 3.

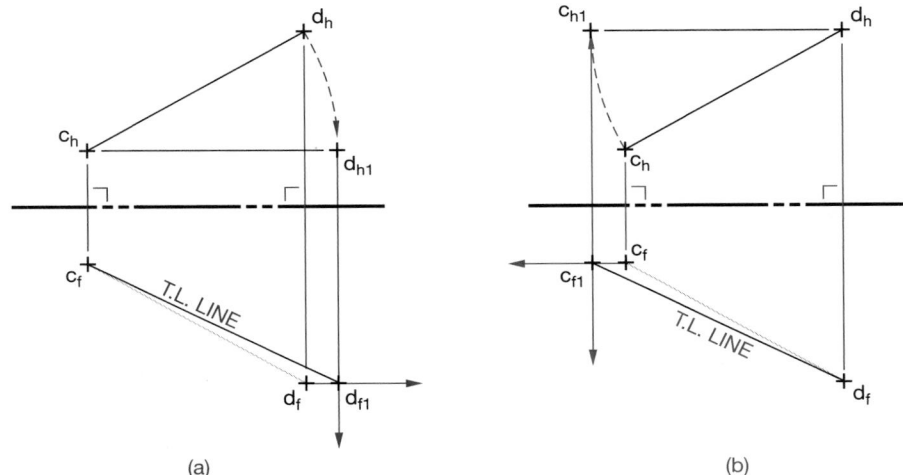

FIGURE 9.43 ■ (a) Rotating without using a compass around point C, step 4.
(b) Rotating without using a compass around point D, step 4.

Step 3: *Project point d_f across the projector of d_{h1} or project point c_f across the projector of c_{h1}.* Draw a projector from point d_f parallel to fold line H/F crossing the d_{h1} projector in the F view. Make a cross at this intersection point and label the cross d_{f1}. (See Figure 9.42a.) If projecting point c_f, see Figure 9.42b.

Step 4: *Create the true length of line C-D.* Connect point c_f to point d_{h1} with a straight line. Line c_f-d_{h1} is the true length of line C-D. Identify the true-length line c_f-d_{h1} with the initials T.L. (See Figure 9.43a.) If point c_h was rotated, connect point d_f to point c_{f1} and identify this true-length line d_f-c_{f1} with the initials T.L. This is also the true length of line C-D. (See Figure 9.43b.)

Note: The true length of the line could just as well have been found in the H view if the revolution process was started in the F view.

PROJECTION OF A TRUE-LENGTH LINE TO FIND WHERE IT APPEARS AS A POINT (Basic Concept 4)

Knowing how to find where a true-length line appears as a point is necessary to being able to find where an oblique plane appears as an edge view. When a true-length line is viewed at the end, it will appear as a *point*. (See Figure 9.44.)

Problem

Find where true-length line A-B appears as a point. (See Figure 9.45.) Things that must be given or must be found first to solve this problem:

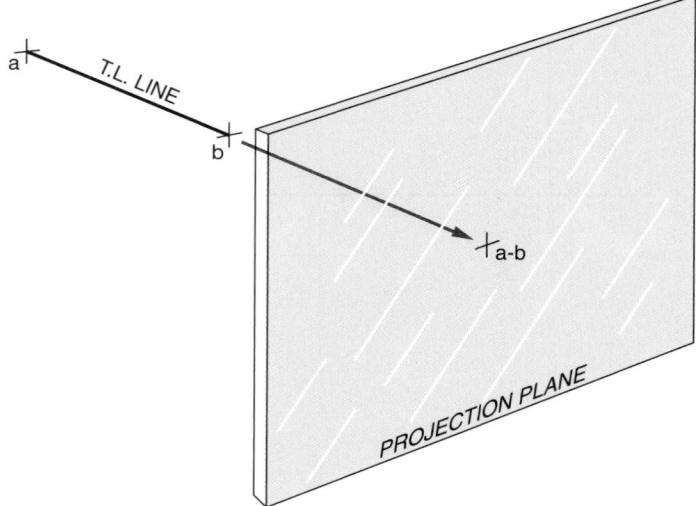

FIGURE 9.44 ■ Viewing a true-length line where it appears as a point.

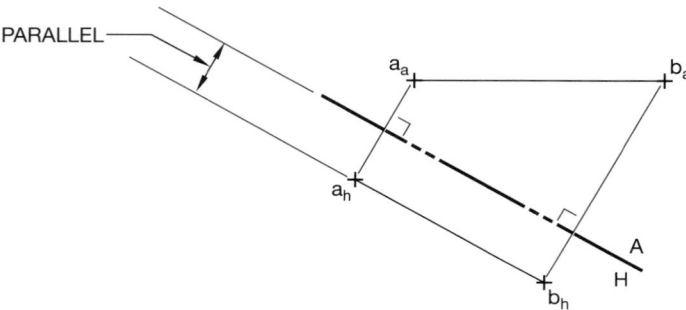

FIGURE 9.45 ■ Problem setup to find true-length line as a point.

1. Two adjacent views of line *A-B* where line *A-B* is in true length in at least in one of the views. *Hint:* To verify if line a_a-b_a is in true length in the *A* view, line a_h-b_h in the *H* view must be *parallel* to the fold line *H/A*. (See Figure 9.45.)

2. The fold line *H/A* must be located correctly between the two given adjacent views.

Step 1: *Draw a new fold line perpendicular to line A-B.* Extend the true-length line a_a-b_a with a light construction line. At a short convenient distance, draw a new fold line *A/R* perpendicular to this light construction line. Next, draw in a little angle bracket close to where the light construction line and the fold line *A/R* intersect just to reinforce the idea that the new fold line must always be drawn at a 90° angle to the true-length line. (See Figure 9.46.) This *A/R* fold line can be drawn at any perpendicular distance from the true-length line a_a-b_a. It could be drawn through point b_a or through the true-length line a_a-b_a itself, but this is not recommended because it is a very confusing way to solve the problem.

Step 2: *Project line A-B onto the R view.* Project line a_a-b_a onto the *R* view just the same way it was explained earlier in pro-

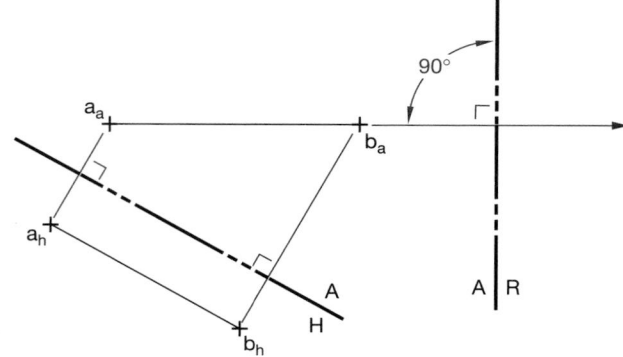

FIGURE 9.46 ■ Locating new fold line.

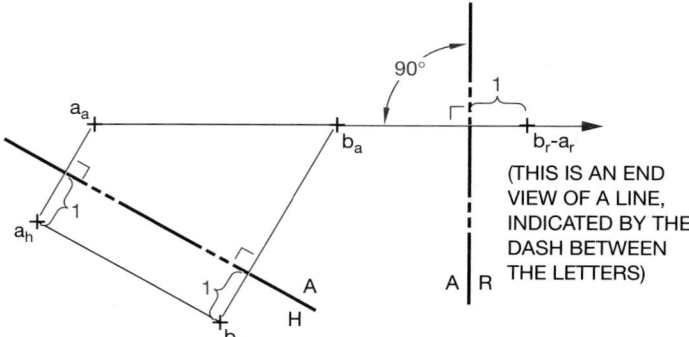

(THIS IS AN END VIEW OF A LINE, INDICATED BY THE DASH BETWEEN THE LETTERS)

FIGURE 9.47 ■ Projecting line *A-B* onto the *R* view.

jecting a line in space. Include all the distance brackets, projectors, etc. as before. When you do, you will notice that point a_a will project on top of point b_a in the *R* view creating what looks like one point. Identify this one point as a_r-b_r so as to remember that this is a line appearing as a point. The dash is included to further clarify that this is the end of the line *A-B*, not just a point. (See Figure 9.47.)

PLANES

Without question, problems involving planes and plane surfaces are some of the most common found in industry. The basic principles involving planes are applicable in most industrial fields. A *plane* is a flat surface that is not curved or warped, and in which a straight line will lay. Figure 9.48 shows four different ways a plane may be represented in graphic form: by three points not in a straight line, by one straight line and one point not on the line, by two intersecting straight lines, and by two parallel straight lines.

Note: Planes may be any shape (rectangular, circular, irregular, etc.), but their true shapes are found in exactly the same way by using at least three points on the plane that are not in a straight line to define the plane.

Inclined Plane

When a plane appears as a line (edge) in at least one of the three standard views of third-angle projection and foreshortened in the

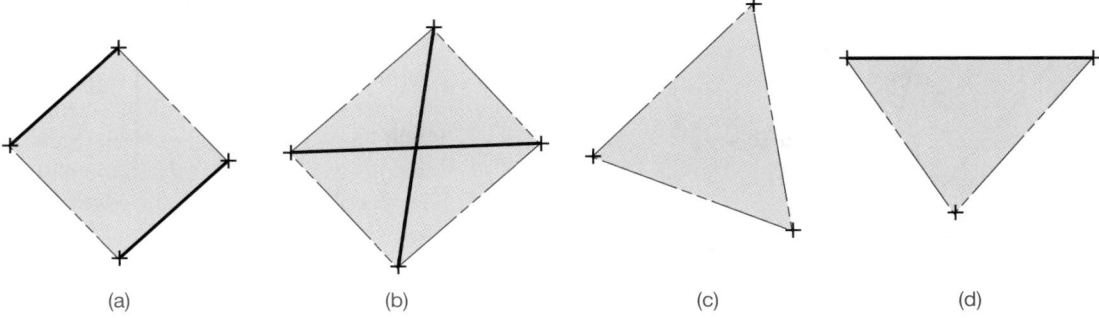

FIGURE 9.48 ■ Different ways of representing planes: **(a)** three points not in a straight line; **(b)** one straight line and one point not on the line; **(c)** two intersecting straight lines; **(d)** two parallel straight lines.

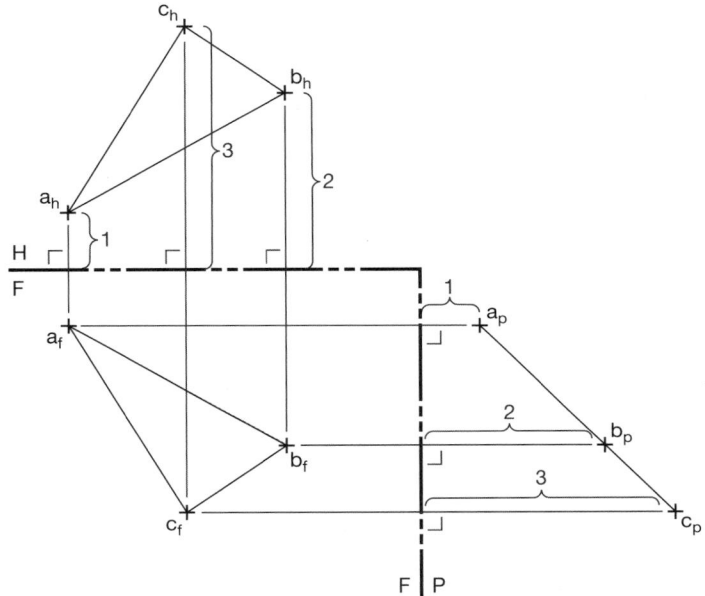

FIGURE 9.49 ■ An inclined plane.

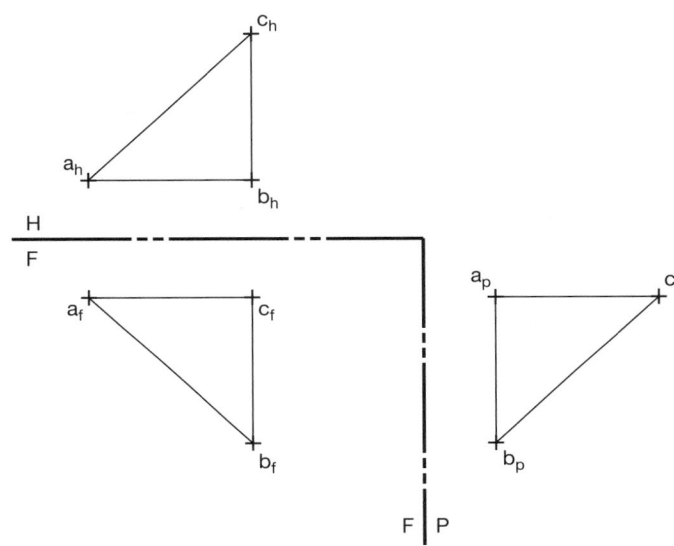

FIGURE 9.50 ■ An oblique plane.

other views, it is considered an inclined plane. The true shape of the inclined plane can be found in a first auxiliary view. (See Figure 9.49.)

Oblique Plane

When a plane does not appear as a line (edge) in any of the standard views of third-angle projection, it is considered an oblique plane, and the true shape of the plane can be found only in a secondary auxiliary view (See Figure 9.50.)

True Shape of a Plane

The true shape is the actual or real *size* and *shape* of the plane. All measurements and angles can be measured on this view of the plane. In Figure 9.51, plane *A-B-C* is parallel to the horizontal view and will appear in true size and shape on this view.

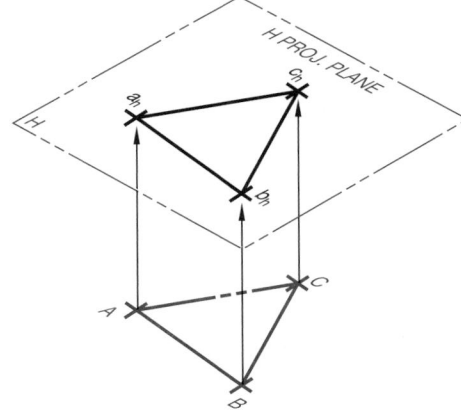

FIGURE 9.51 ■ The true shape of a horizontal plane.

PROJECTION OF A PLANE

Knowing how to project a plane in space has little value in itself other than it is a part of the process in finding the true shape of a plane. Using the concept that a plane may represented by three

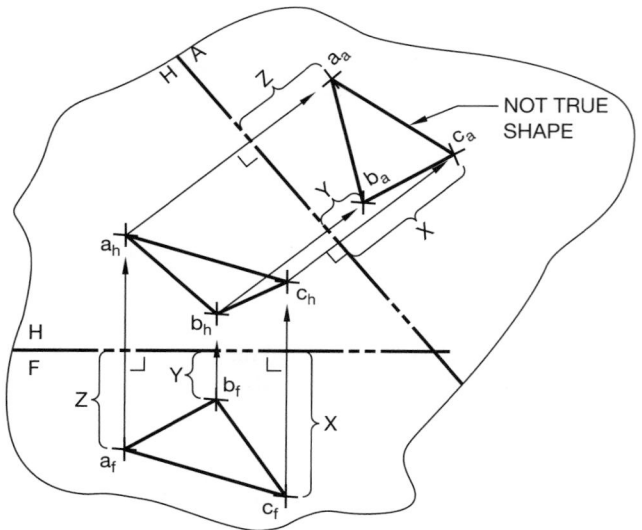

FIGURE 9.52 ■ Projecting a plane onto an auxiliary view.

points not in a straight line that are connected with straight lines, project the three points (plane) onto a new view in exactly the same manner as one point was projected, but do it three times. The projected plane may not necessarily be in true shape. (See Figure 9.52.)

Finding the True Shape of an Inclined *Plane*

An inclined plane appears as an edge view (line) in at least one of the standard views of third-angle projection. If in a given problem one of the two given views of the plane appears as a line, it can be considered an inclined plane. (See Figure 9.53.) The true shape of the inclined plane can be found by either the fold-line or the revolution method.

Problem

Find the true shape of the inclined plane *A-B-C* by the fold-line method. (See Figure 9.53 again.) Things that must be given to solve this problem:

1. Two adjacent views of the plane where one view of the plane appears as a line.

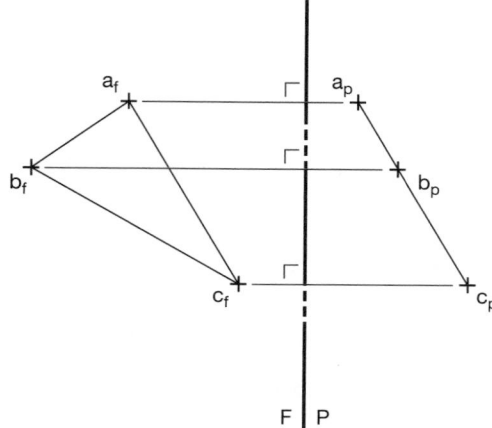

FIGURE 9.53 ■ Inclined plane setup.

2. A fold line located correctly between the two adjacent views.

Step 1: *Draw a new fold line parallel to the edge view of plane A-B-C.* Just as when finding the true length of a line, the new fold line can be placed on either side of the edge view at any space from or directly on top of the edge view as long as it is placed *parallel* to the edge view. (See Figure 9.54.) However, placing the new fold line *P/R* about where it is placed in Figure 9.55 works well.

Step 2: *Project the edge view of plane A-B-C onto the new view.* Project the edge view of plane a_p-b_p-c_p (a line with three points) onto view *R* just the same way as projecting a line with two points onto view *R*. Plane a_r-b_r-c_r is the true shape of the inclined plane. Put the initials T.S. somewhere on the true shape of the plane. (See Figure 9.56.)

Problem

Find the true shape of the inclined plane *A-B-C* by the revolution method. (See Figure 9.57.) Things that must be given to solve this problem:

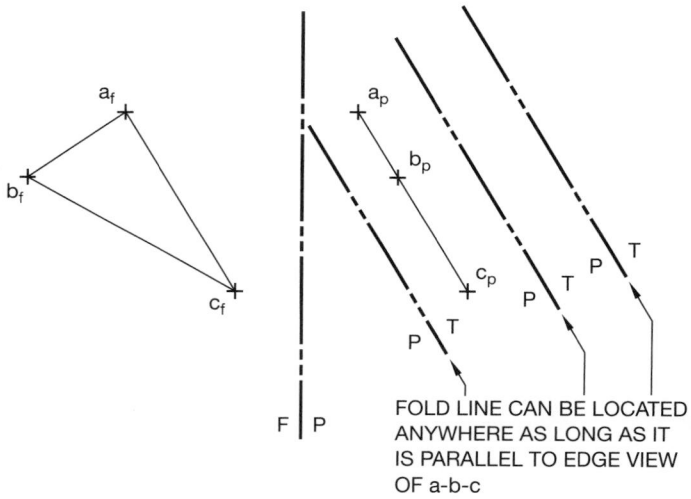

FIGURE 9.54 ■ Possible locations of a new fold line.

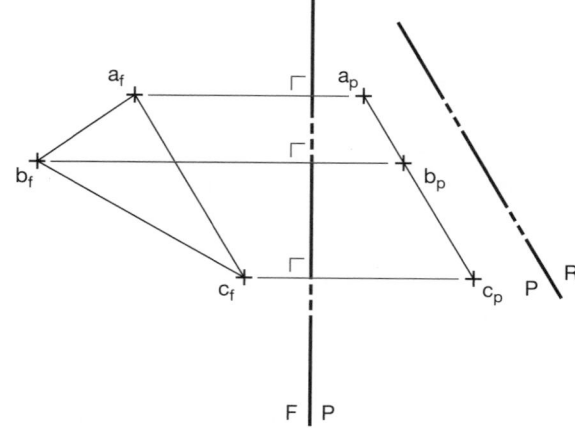

FIGURE 9.55 ■ Locating new fold line for this problem.

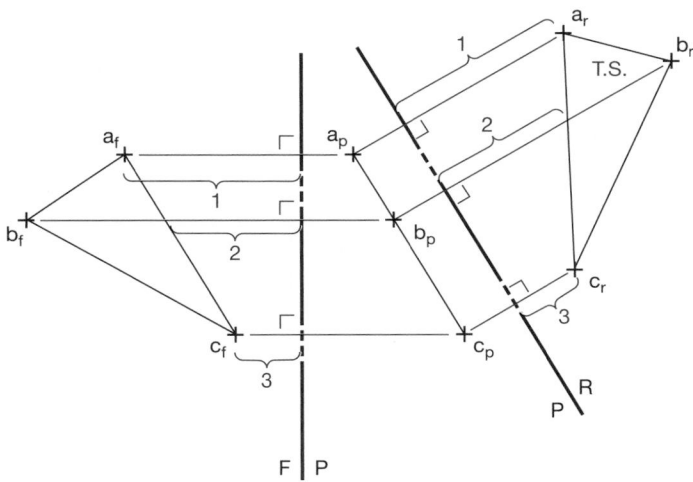

FIGURE 9.56 ■ Creating true shape of plane *A-B-C*.

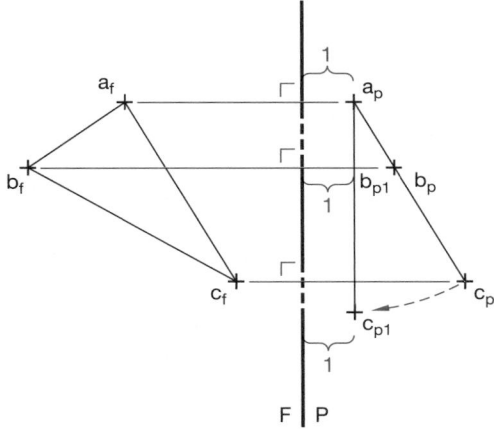

FIGURE 9.58 ■ Step 1 to find true shape of an inclined plane, using a compass.

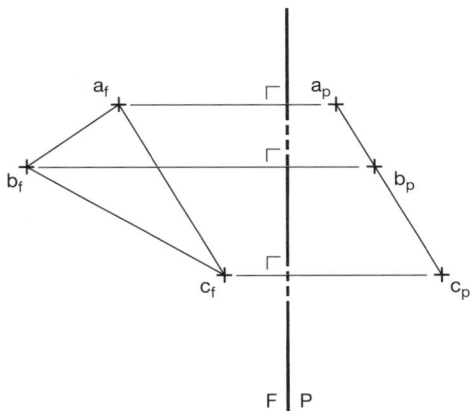

FIGURE 9.57 ■ Problem setup.

1. Two adjacent views of the plane where one view of the plane appears as a line.

2. A fold line located correctly between the two adjacent views.

Step 1: *Revolve the edge of the plane where all the points of the plane A-B-C are the same distance from the existing fold line.* There are several ways to attack this problem. For this example, points b_p and c_p will be revolved around point a_p where they will be the same distance from the fold line *F/P*. Or, in other words, the edge view of plane *A-B-C* will be revolved *parallel* to the existing *F/P* fold line.

To rotate points b_p and c_p around point a_p graphically, place a compass needle end on point a_p and the compass lead point on point b_p. Draw an arc from point b_p until it is the same distance from the fold line *F/P* as point a_p. Make a cross at this place and label the cross as b_{p1}. Again, place a compass needle end on point a_p and the compass lead point on point c_p. Draw an arc from point c_p until it is the same distance from the fold line *F/P* as point a_p. Make a cross at this place and label the cross as c_{p1}. Then connect point a_p to points b_{p1} and c_{p1} with a straight line. (See Figure 9.58.)

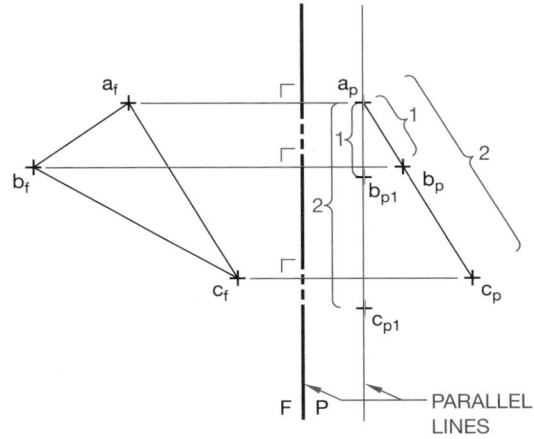

FIGURE 9.59 ■ Step 1 to find true shape of an inclined plane, without a compass.

If you do not have a compass, you can rotate the points b_p and c_p by the following way. (See Figure 9.59.) Draw a light construction line *parallel* to fold line *F/P* through point a_p. Measure the length of line a_p-b_p and transfer this length from point a_p along the light construction line. Identify the other end of this length with a cross and as b_{p1}. Then measure the length of line a_p-c_p and transfer this length from point a_p along the light construction line. Identify the other end of this length with a cross and as c_{p1}. Connect points a_p, b_{p1}, and c_{p1} with a straight line.

Step 2: *Project points b_{p1} and c_{p1} onto the F view.* Draw a projector from point b_{p1} *perpendicular* to and on through the *F/P* fold line and beyond point b_f in the *F* view. Then draw a projector from point c_{p1} *perpendicular* to and on through the *F/P* fold line and beyond point c_f in the *F* view. (See Figure 9.60.)

Step 3: *Project point b_f across the projector of b_{p1} and project point c_f across the projector of c_{p1}.* Draw a projector from point b_f *parallel* to fold line *F/P* crossing the b_{p1} projector in the *F* view. Make a cross at this intersection point and label the cross b_{f1}. Next draw a projector from point c_f *parallel* to

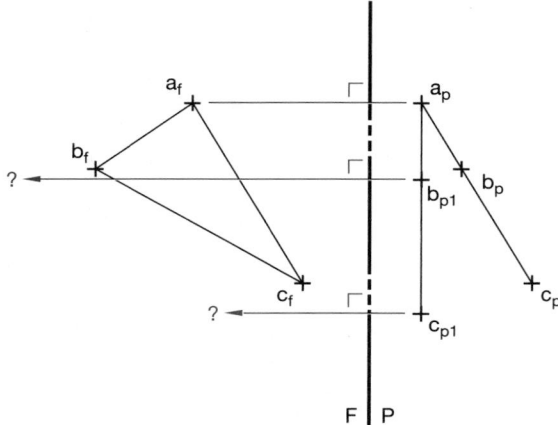

FIGURE 9.60 ■ Step 2 to find true shape of an inclined plane.

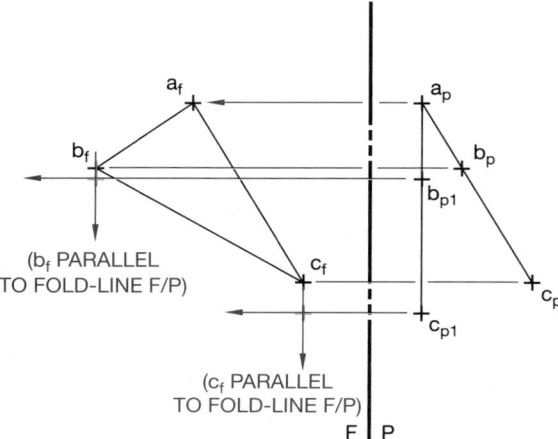

FIGURE 9.61 ■ Step 3 to find true shape of an inclined plane.

fold line *F/P* crossing the c_{p1} projector in the *F* view. Make a cross at this intersection point and label the cross c_{f1}. (See Figure 9.61.)

Step 4: *Create the true shape of plane A-B-C.* Connect point a_f to points b_{f1} and c_{f1} with straight lines. Label this plane with the initials T.S. to identify this plane as the true shape. (See Figure 9.62.)

Hint: The size of the true shape of a plane can never appear smaller than any other view of the plane. It can appear the same size, in which case the other view is also a true shape. If when you think you have found the true shape of the plane, it appears smaller in some other view, that indicates you have made some error and you have not really found the true shape.

Finding the True Shape of an Oblique *Plane*

An oblique plane will not appear as an edge view (line) in either of the two given views of a problem, nor will it appear as an edge view (line) in any of the standard views of third-angle projection. To find the true shape of an oblique plane, the plane must first be found in a view where it appears as an edge view (line).

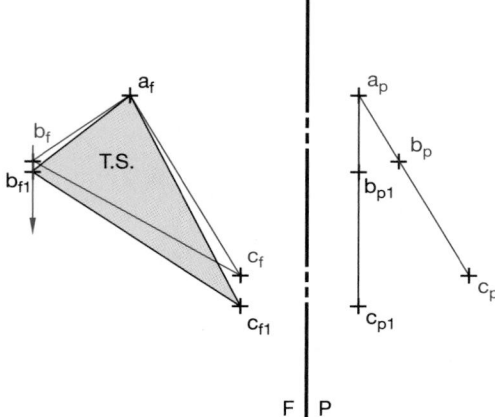

FIGURE 9.62 ■ Step 4 to find true shape of an inclined plane.

Problem

Find the true shape of the oblique plane *A-B-C*. (See Figure 9.63.) Things that must be given to solve this problem:

1. Two adjacent views of the plane where neither view of the plane appears as a line.

2. A fold line located correctly between the two adjacent views.

Step 1: *Find a view where the oblique plane A-B-C appears as an edge.* To find the true shape of an oblique plane, you must first find an edge view of the oblique plane. The concept that will be used in this explanation is based upon the idea that when you find a view that shows the end view of any true-length line in the plane (a point), you will also see the oblique plane as an edge view (line). There are at least two good ways to do this.

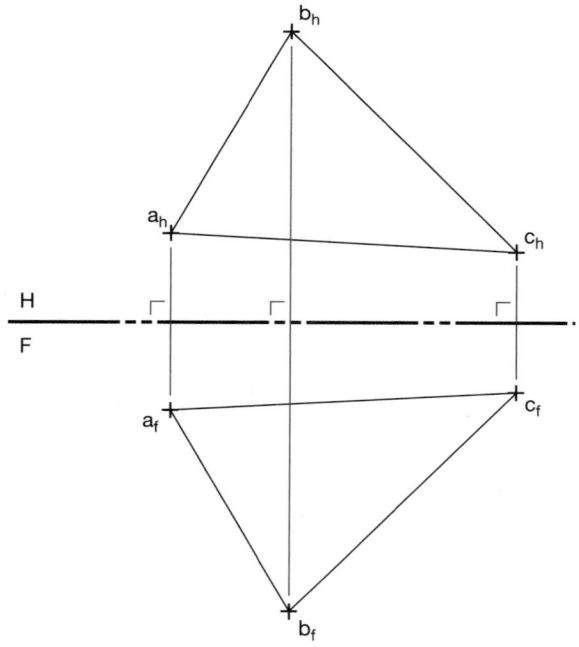

FIGURE 9.63 ■ Problem setup for an oblique plane.

First Method:

Step 1: *Find one of the edges of the oblique plane A-B-C in true length.* (See Figure 9.64.)

(a) Draw a new fold line *parallel* to any side of the oblique plane in either view. In this case, side b_f-c_f was chosen at random. The new fold line *F/X* was drawn parallel to side b_f-c_f.

(b) Project points a_f, b_f, and c_f onto view X just as individual points have been projected in the past. Make sure to include all measurement brackets and perpendicular bracket reminders. Identify each of the new point locations with a cross and the correct letters and subscript letters.

(c) Connect each of the points a_x, b_x, and c_x with straight lines to form another view of the oblique plane A-B-C where the side b_x-c_x is in true length. (See Figure 9.64 again.)

Step 1b: *Find the true-length line b_x-c_x in a view where it appears as a point.* (See Figure 9.65.)

(a) Draw a new fold line *X/Y* perpendicular to the true-length line b_x-c_x of the oblique plane a_x-b_x-c_x.

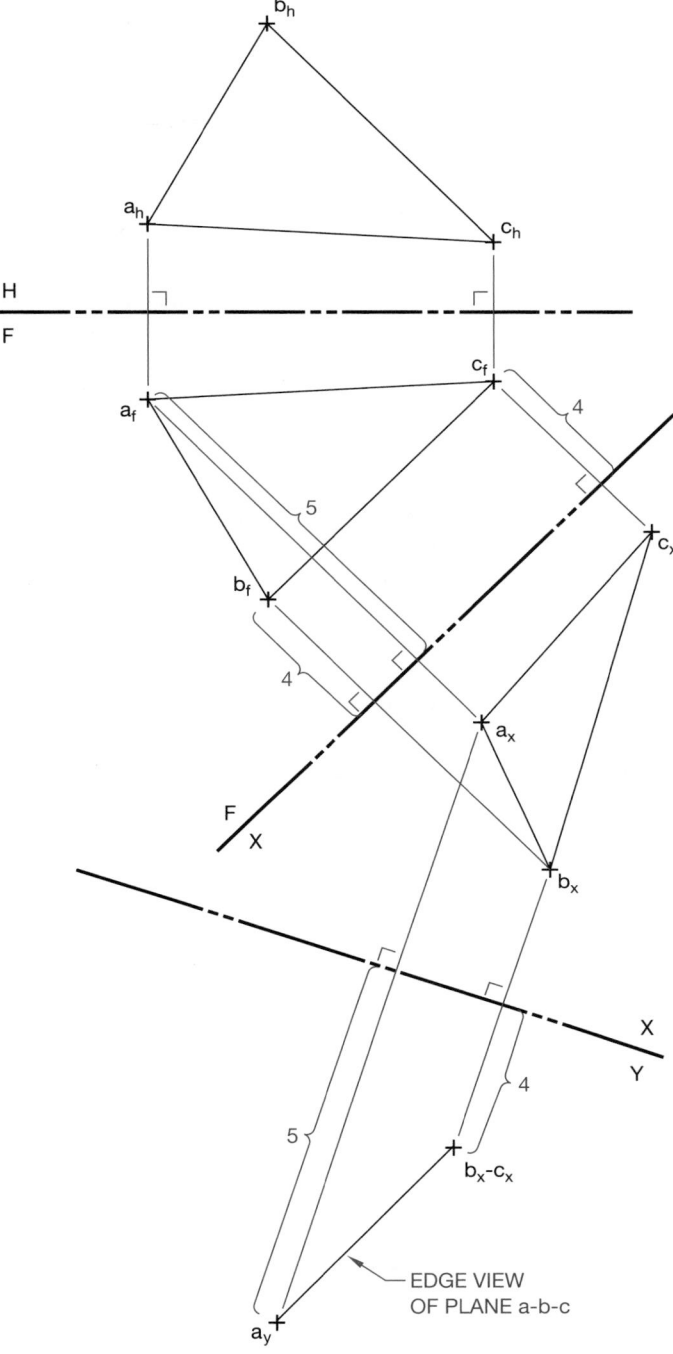

FIGURE 9.65 ■ Step 1b to find true shape of an oblique plane.

(b) Project a_x, b_x, and c_x onto view Y just as individual points have been projected in the past. Make sure to include all measurement brackets and perpendicular bracket reminders. Identify each of the new point locations with a cross and the correct letters and subscript letters. Notice that when points c_x and b_x are projected onto the Y view, they are superimposed on top of each other creating what looks like one point, b_y-c_y.

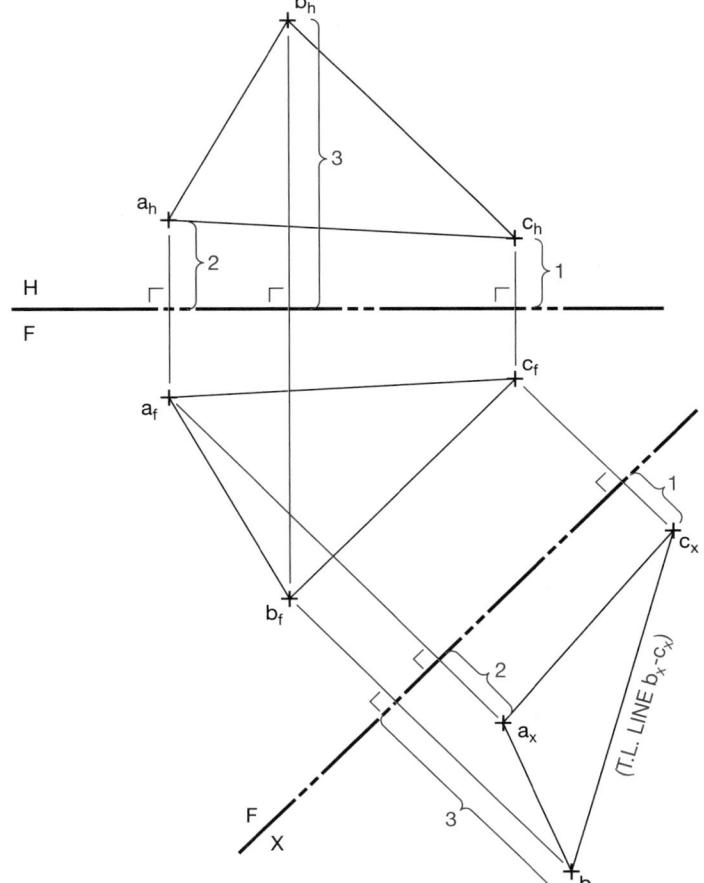

FIGURE 9.64 ■ Step 1a to find true shape of an oblique plane.

(c) Connect point a_y to b_y-c_y with a straight line to form the edge view of the oblique plane *A-B-C*. (See Figure 9.65 again.)

Second Method:

Step 1a: *Create a true-length line in one of the original views of the oblique plane. (See Figure 9.66.)*

(a) Draw a new line on either of the two existing views of the oblique plane *A-B-C* parallel to the fold line *H/F*. In this case, a new line c_f-x_f was drawn parallel to the fold line *H/F* in the *F* view using the existing point c_f as one end and making only one new point. The new line could have been drawn anywhere on the plane as long as it was drawn parallel to the fold line.

(b) Project the new point x_f onto the *H* view. Draw a projector from point x_f perpendicular to the fold line *F/H* and on through until it crosses the a_h-b_h line. Identify this point of intersection in the *H* view with a cross and the letters x_h.

(c) Connect point c_h to x_h to create the true-length line c_h-x_h. (See Figure 9.66)

Step 1b: *Find the true-length line c_h-x_h in a view where it appears as a point. (See Figure 9.67.)*

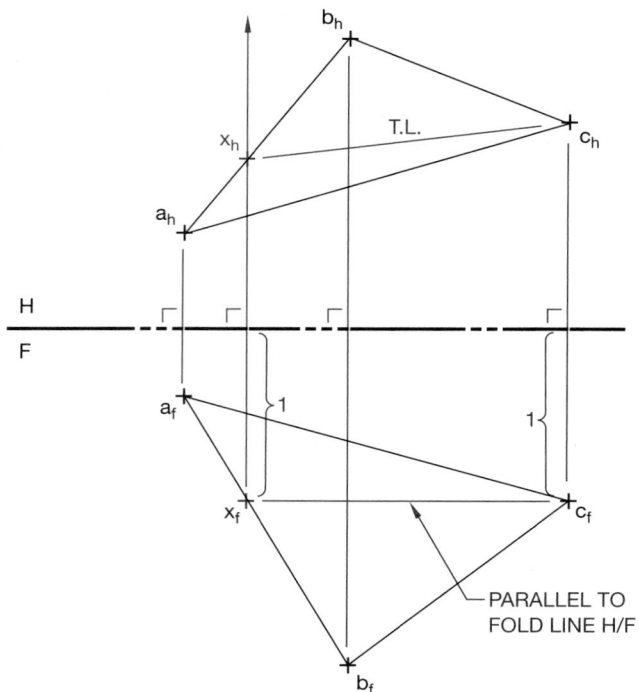

FIGURE 9.66 ■ Another method of finding edge view of an oblique plane, step 1a.

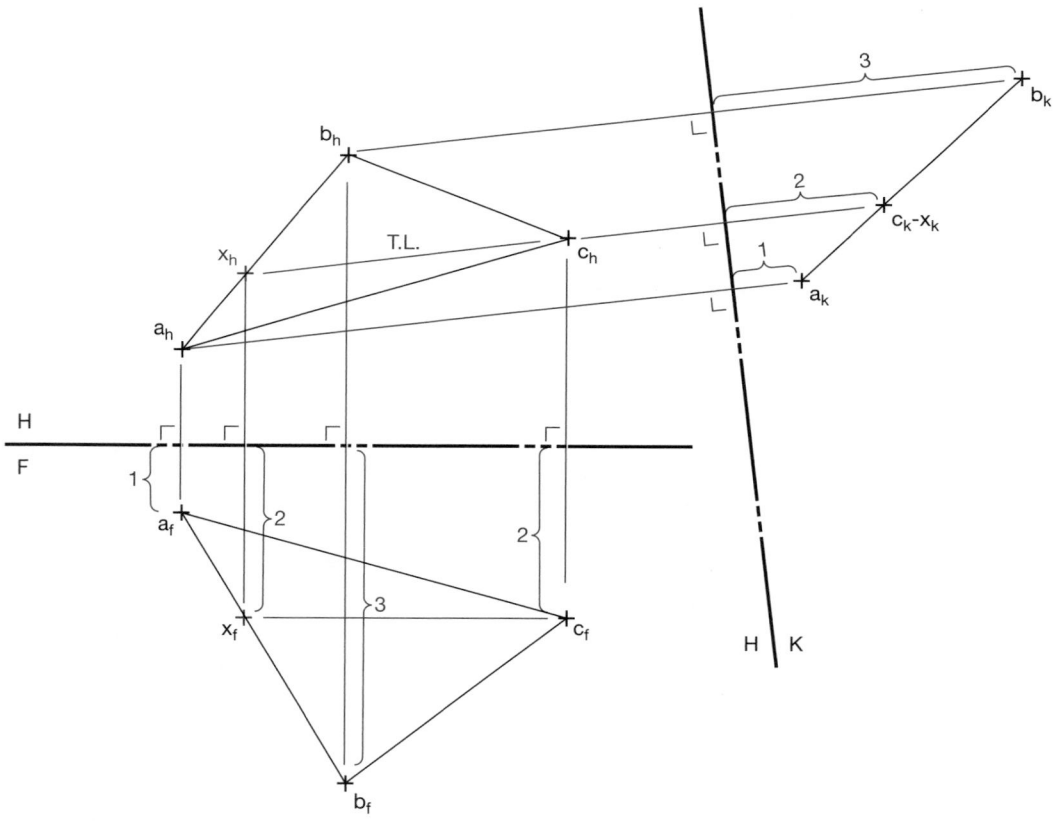

FIGURE 9.67 ■ Step 1b of finding edge view with the second method.

(a) Draw a new fold line *H/K perpendicular* to the true-length line c_h-x_h of the oblique plane a_h-b_h-c_h.

(b) Project a_h, b_h, c_h, and x_h onto view *K* just as individual points have been projected in the past. Make sure to include all measurement brackets and perpendicular bracket reminders. Identify each of the new point locations with a cross and the correct letters and subscript letters. Notice that when points c_h and x_h are projected onto the *K* view, they are superimposed on top of each other creating what looks like one point, c_k-x_k.

(c) Connect points a_k and b_k to c_k-x_k with a straight line to form the edge view of the oblique plane *A-B-C*. (See Figure 9.67 again.)

Hint: Look for a true-length line already in the problem, and you will be able to skip step 1 in either method. (See Figure 9.68.)

Step 2: To finish solving the problem, use the same method as that used in finding the true shape of an inclined plane as described earlier in this chapter. You can use either the fold-line or the revolution method to solve the problem. (See Figures 9.69 and 9.70 for a quick review.)

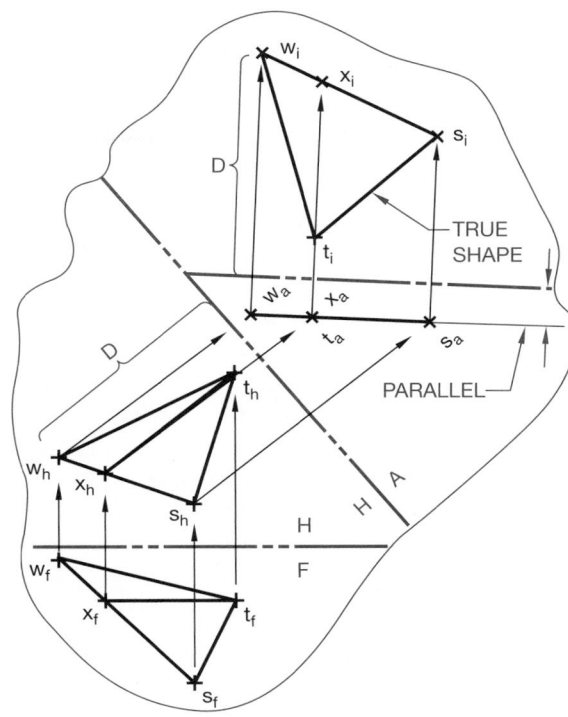

FIGURE 9.69 ■ Finding the true shape of a plane by the fold-line method.

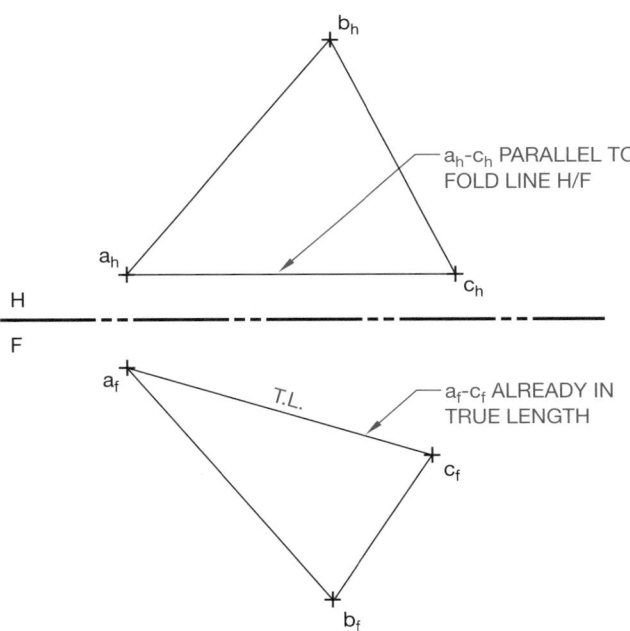

FIGURE 9.68 ■ An oblique plane with an existing true-length line a_f-c_f.

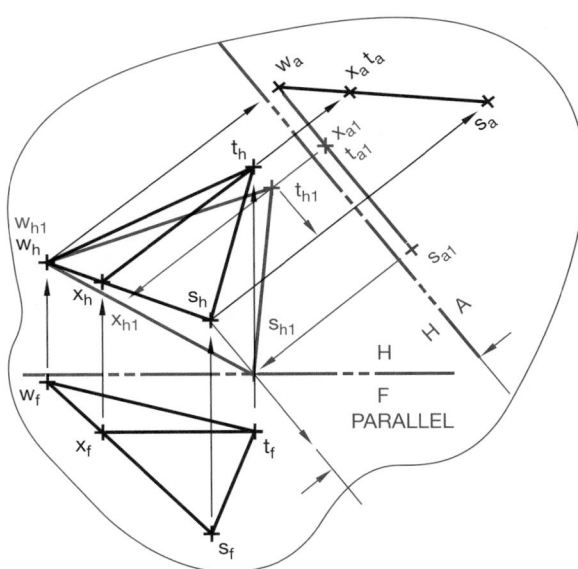

FIGURE 9.70 ■ Finding the true shape of a plane by the revolution method.

PROFESSIONAL PERSPECTIVE

Finding the correct solutions to descriptive geometry problems is mainly dependent on three things: *defining your problem,* using a problem-solving *method,* and *accuracy* in constructing the solution. There are different techniques that can be used to solve each problem graphically, and the challenge is to select the best method. Extreme accuracy is also required when drawing and constructing the various lines on the board.

For people who tend to visualize objects easily, descriptive geometry problems are not too difficult solve. Others who do not visualize easily will need to follow the step-by-step cookbook procedure explained in this chapter and memorize specific procedures to obtain the correct answers. Either way, solving descriptive geometry problems can and should be fun and intriguing.

MATH APPLICATION

It is important to be able to convert from one type of scale measurement to another when setting up descriptive geometry problems in order to have accuracy in your answers. Otherwise, the process for solution may be correct, but the answer will not be. Some examples of conversions follow.

1 in. = 2.54 cm   1 ft. = 30.48 cm
1 cm = .394 in. = ⅜ + in.   1 m = 100 cm = 39.37 in.

You may also need to calculate areas of planes in descriptive geometry problems. The area formulas are included for review. (See Figures 9.71 through 9.76.)

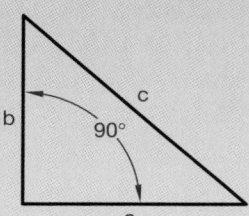

FIGURE 9.71 ■ Area of a right triangle = *ab*/2.

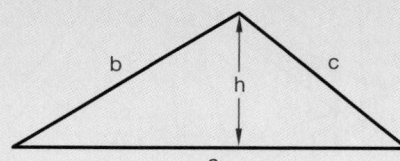

FIGURE 9.72 ■ Area of a general triangle = *ah*/2.

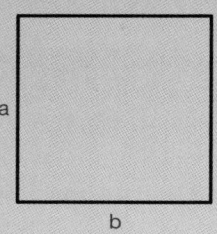

FIGURE 9.73 ■ Area of a square = a^2.

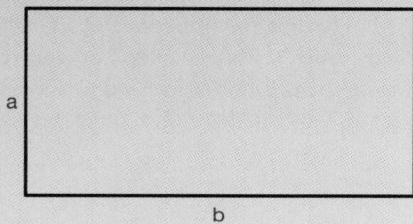

FIGURE 9.74 ■ Area of a rectangular feature = *ab*.

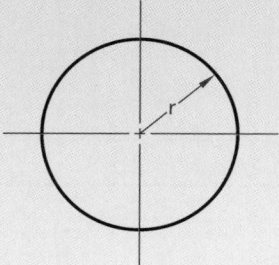

FIGURE 9.75 ■ Area of a circle = πr^2.

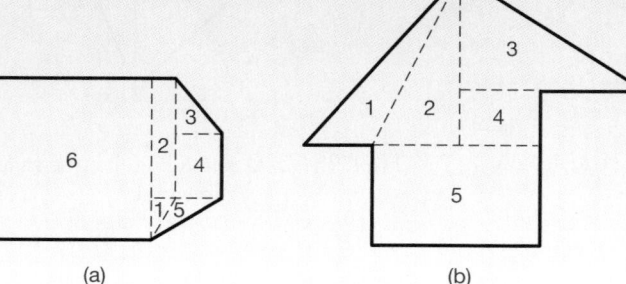

(a)   (b)

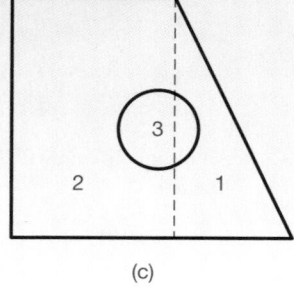

(c)

FIGURE 9.76 ■ Calculating areas of odd-shaped planes: (**a**) area separated into a rectangle and plane; (**b**) area separated into a triangle and a rectangle; (**c**) area separated into a rectangle and triangle minus area of a circle.

CHAPTER 9

Descriptive Geometry I Test

DIRECTIONS

Answer the following questions with short, complete statements and/or create thumbnail sketches to explain your answer.

1. Show by thumbnail sketches examples of the concepts:
   a. Parallelism
   b. Perpendicularity
   c. Fold line
   d. True-length line
   e. Plane
2. A point is always projected at what angle to a fold line?
3. What is the difference between a projection plane and a view?
4. Crossing over or projecting over a fold line represents a _____ degree change of direction.
5. What does the end view of a point look like?
6. In geometry problems, two _____ views must be given to solve the problem.
7. What three things must be given to project a point correctly in space?
8. If a point is identified as a_h, the subscript h refers to what?
9. If a point is identified as b_{f1}, the 1 in the subscript refers to what?
10. A projector must always project to a fold line at what angle?
11. A straight line in space can be described as what?
12. What does the end view of a true-length line appear as?
13. What are skew lines?
14. What are intersecting lines?
15. What three things must be given to project a line in space?
16. When finding the true length of a line by the fold-line method, which moves, the viewer or the given line?

17. When finding the true length of a line by the revolution method, which moves, the viewer or the given line?
18. To find the true length of a line by the fold-line method, a new fold line must be created and placed in what position to the given line?
19. How does the distance that you place a new fold line parallel to a given line affect finding the true length of that line?
20. Explain how you can rotate a line in space graphically when doing board solutions without the use of a compass.
21. List four ways to describe a flat plane.
22. Any surface of a plane seen in true size can also be seen in _____.
23. An edge view of a plane is represented by _____.
24. A plane is a _____ surface.
25. An inclined plane appears as a _____ in at least one of the three standard views of projection and _____ in the other views.
26. An oblique plane does not appear as _____ in any of the three standard views of projection.
27. When finding the true shape of an inclined plane, one of the first steps is to create a new fold line and place it _____ to one of the edges of the given plane.
28. When trying to find the true shape of a plane, one of the first steps is to determine if the plane is inclined or _____.
29. There is/are _____ method(s) used to solve descriptive geometry problems.
30. If descriptive geometry problems are hard for you to solve, you may have to use _____ to solve descriptive geometry problems.

CHAPTER 9

Descriptive Geometry I Problems

DIRECTIONS

Carefully study the problems before you begin working. Solve Problems 9-1 through 9-25 on graph paper or CAD, showing and identifying all projectors, brackets with distances used, and point identifications in the required view(s). If using graph paper, use a separate sheet of graph paper for each problem and the graph lines included with each problem for proper setup. Use a scale of at least three squares to one to obtain accuracy in the problem solutions.

PROBLEM 9-1 Point projection. Locate point *A* in the *P* view.

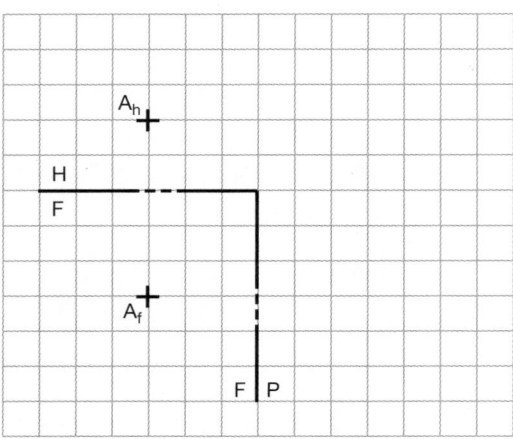

PROBLEM 9-2 Point projection. Locate point *R* in the *A* view.

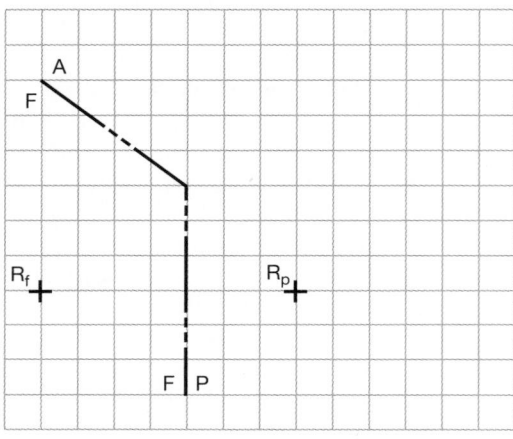

PROBLEM 9-3 Point projection. Locate point *T* in the *A* view.

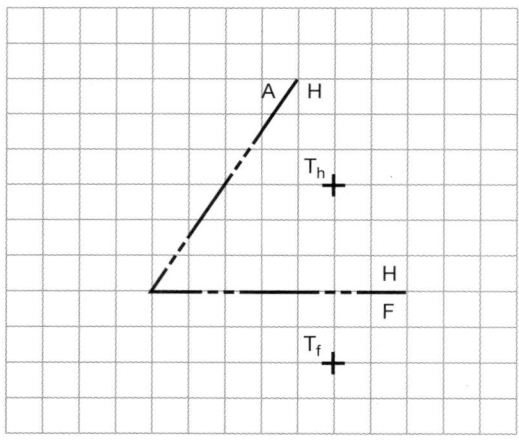

PROBLEM 9-4 Point projection. Locate point *F* in views *A*, *G*, and *S*.

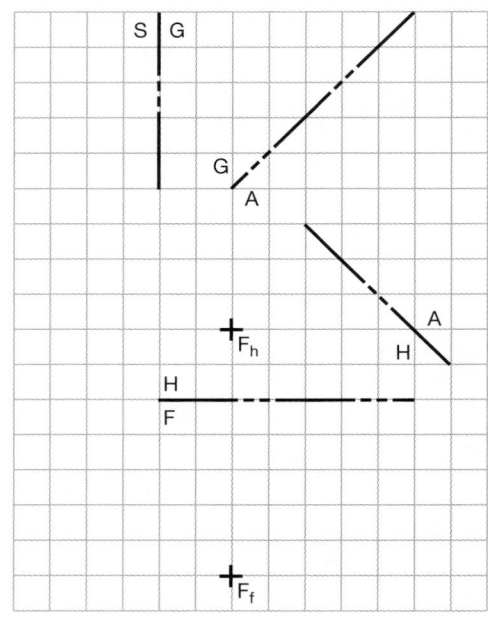

PROBLEM 9-5 Point projection. Locate point *W* in views *B*, *C*, *F*, *G*, *R*, and *S*.

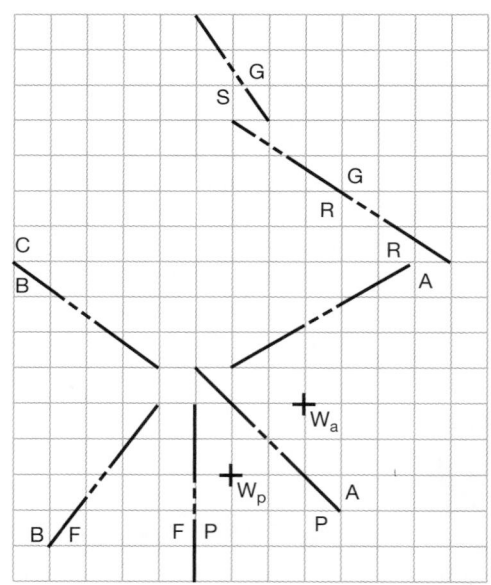

PROBLEM 9-6   Line Projection. Locate line *A-B* in the *H* view.

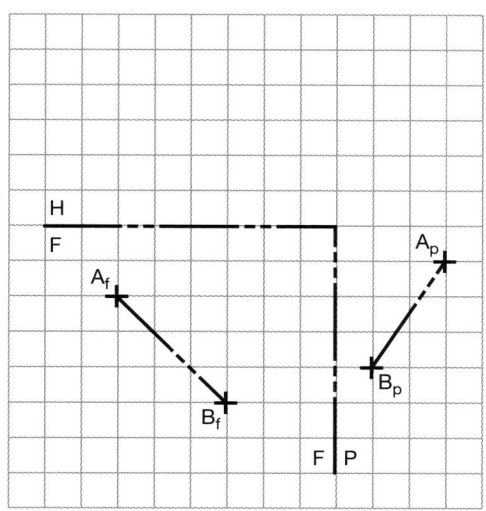

PROBLEM 9-7   Line projection. Locate line *S-T* in the *H* view.

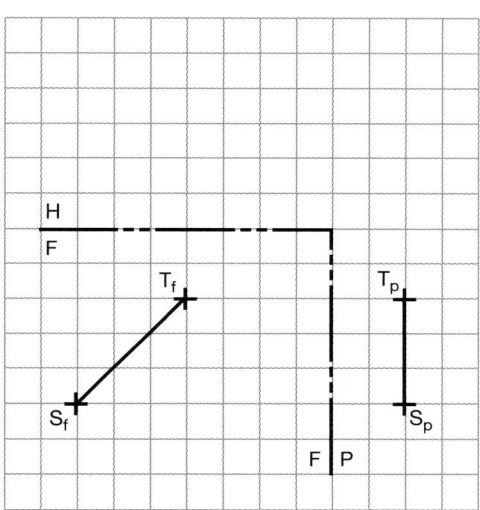

PROBLEM 9-8   Line projection. Locate line *M-N* in the *R* view.

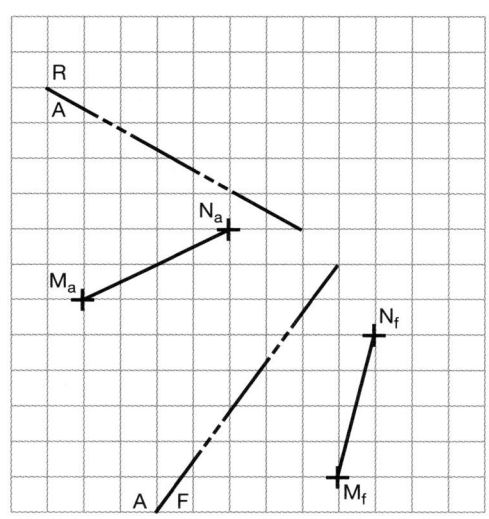

PROBLEM 9-9   Line projection. Locate line *D-E* in the *P* view.

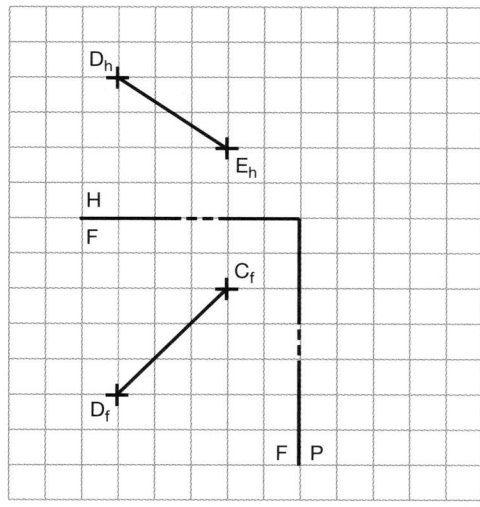

PROBLEM 9-10   True length. Find the true length of line *J-K* by the (a) fold-line method and (b) revolution method.

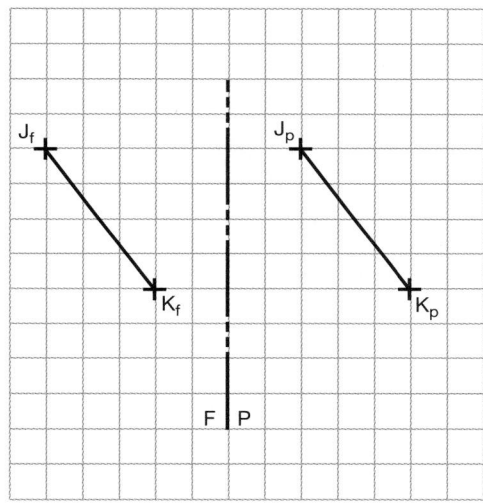

PROBLEM 9-11   True length. Find the true length of line *G-H* by (a) fold-line method and (b) revolution method.

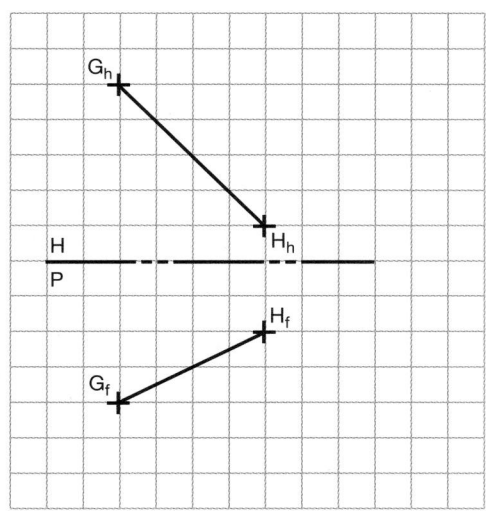

PROBLEM 9-12 True length. Find the true length of line *R-S* by the (a) fold-line method and (b) revolution method.

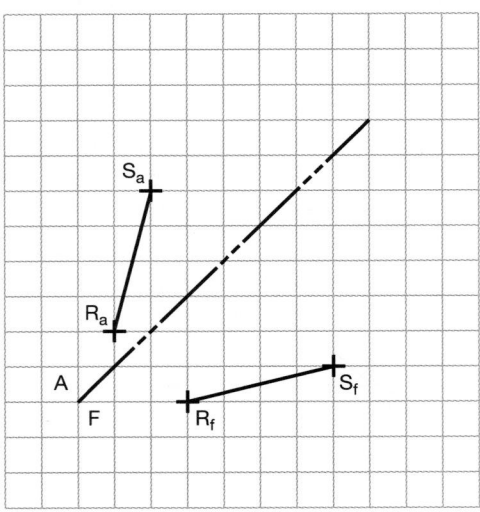

PROBLEM 9-13 True length. Find the true length of line *C-D* by the (a) fold-line method and (b) revolution method.

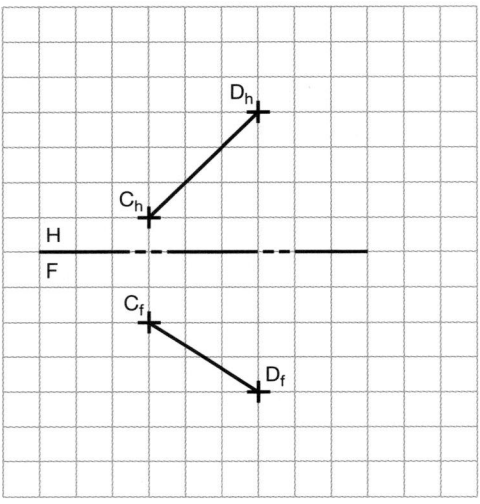

PROBLEM 9-14 End view. Locate line *A-B* in a view where it appears as a point.

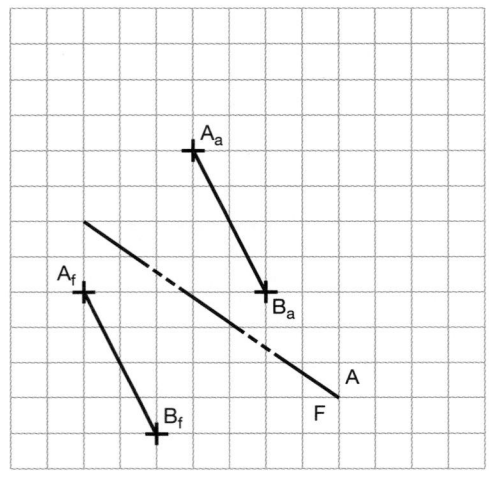

PROBLEM 9-15 End view. Locate line *G-H* in a view where it appears as a point.

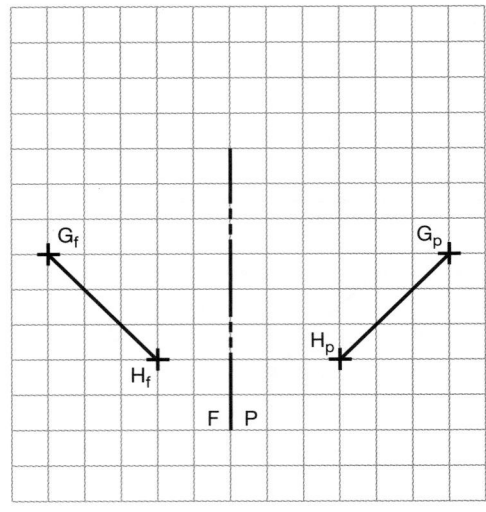

PROBLEM 9-16 End view. Locate line *M-N* in a view where it appears as a point.

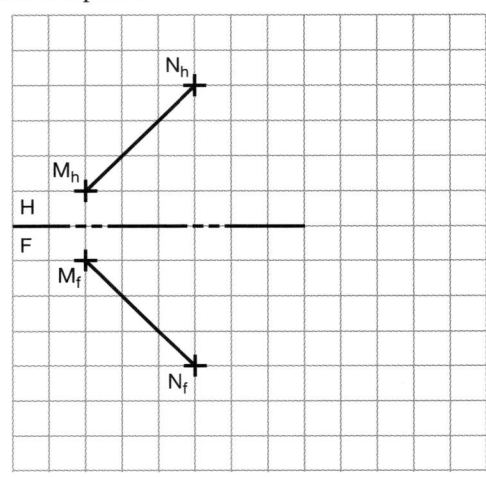

PROBLEM 9-17 End view. Locate lines *M-N* and *A-B* in a view where they appear as points.

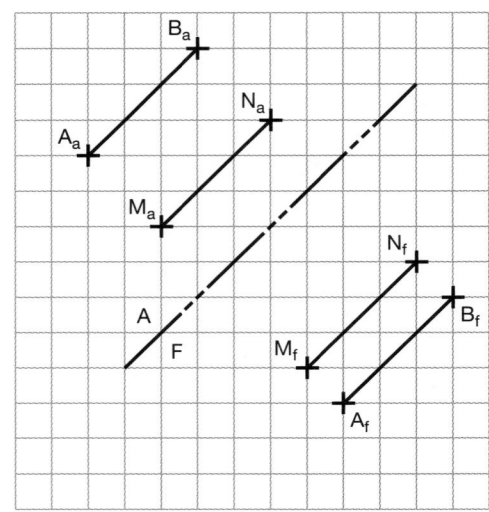

PROBLEM 9-18 Projection of a plane. Complete the missing *P* view of plane *A-B-C*.

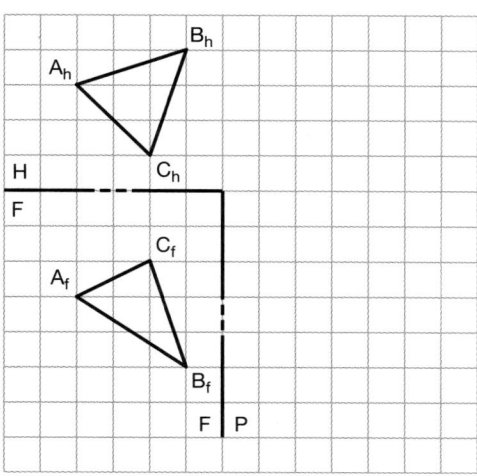

PROBLEM 9-19 Projection of a plane. Complete the missing *F* view of plane *R-S-T-U*.

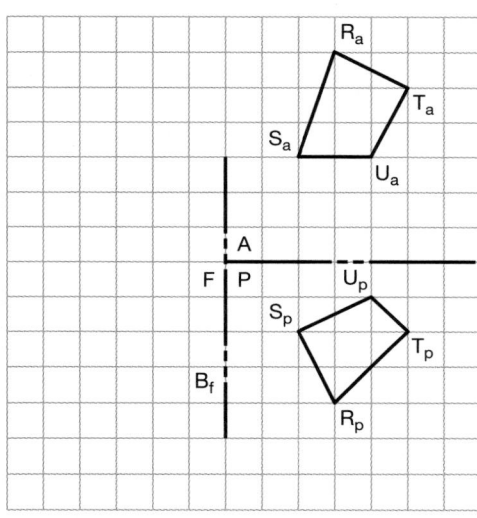

PROBLEM 9-20 Projection of a plane. Complete the missing *R* view of plane *X-Y-Z*.

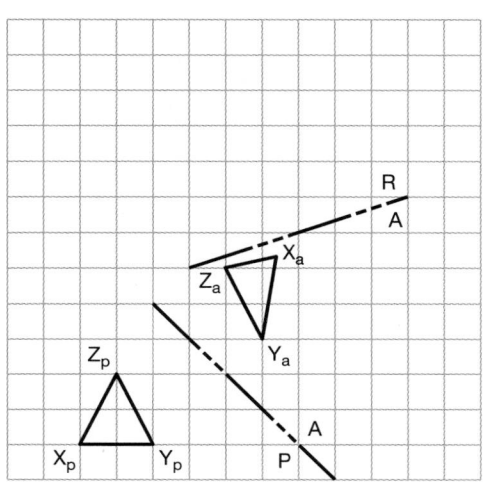

PROBLEM 9-21 Projection of a plane. Complete the missing *F* view of plane *M-N-O-P-R-S*.

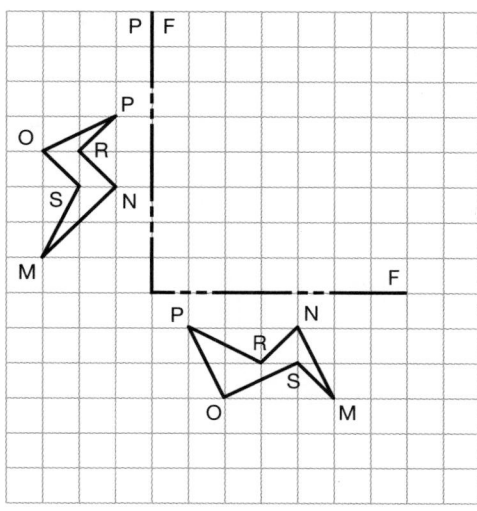

PROBLEM 9-22 Edge view of a plane. Locate plane *A-B-C* in a view where it appears as a line.

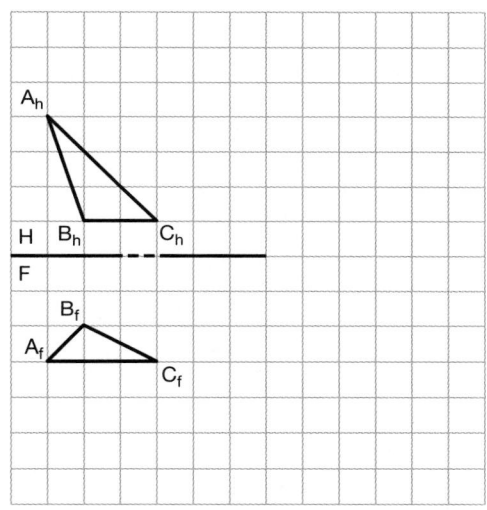

PROBLEM 9-23 Edge view of a plane. Locate plane *W-X-Y-Z* in a view where it appears as a line.

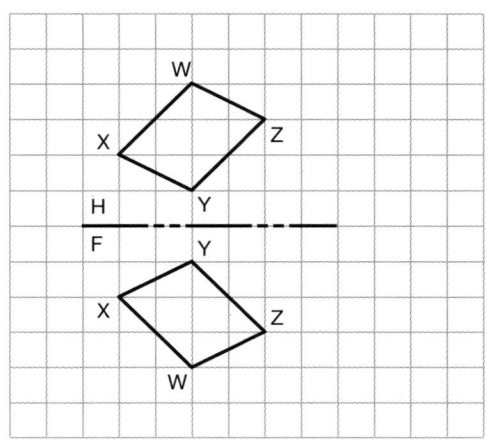

PROBLEM 9-24 True shape of a plane. Locate plane *R-S-T* in a view where you can see it in true shape.

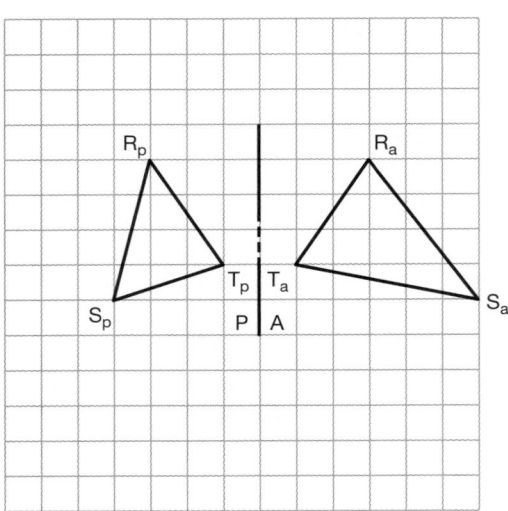

PROBLEM 9-25 True shape of a plane. Locate plane *A-B-C-D-E-F* in a view where you can see it in true shape.

Note: Opposite sides of this figure are parallel.

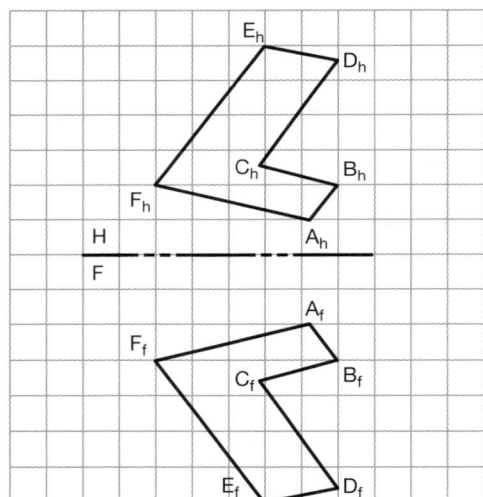

PROBLEM 9-26 Find the true shape of surface *A* by the fold-line or revolution method.

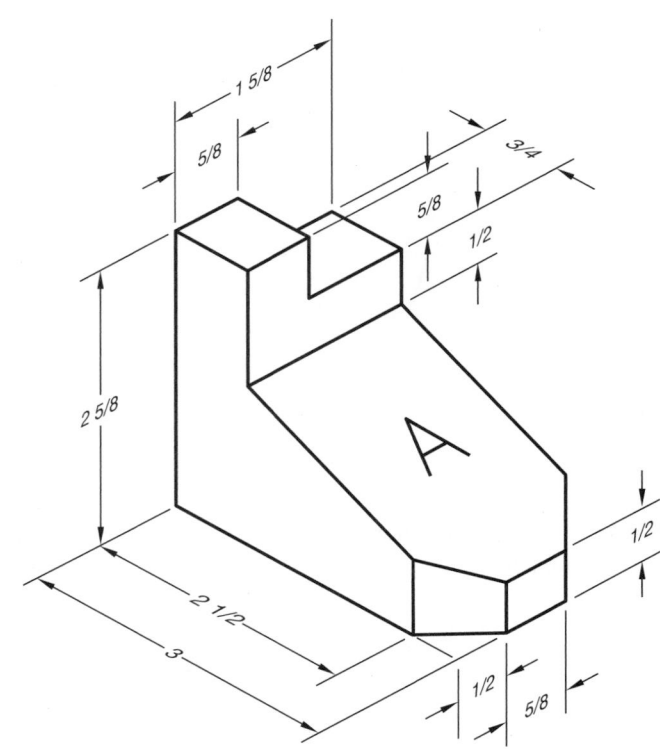

PROBLEM 9-27 Find the true shape of surface *B* by the fold-line or revolution method.

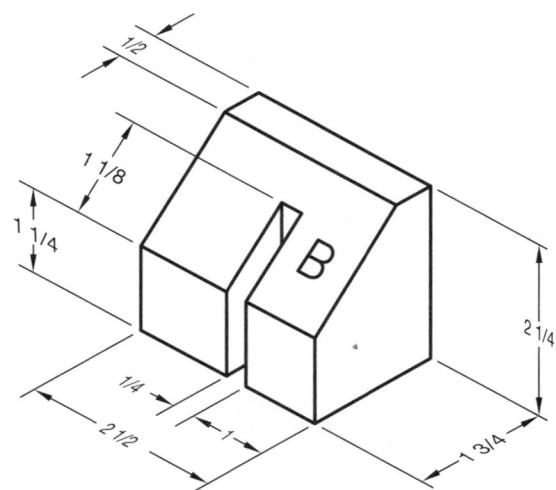

DIRECTIONS

For Problems 9-26 through 9-32, you will need to determine and draw the required multiviews used to set up the problems to establish the true shapes of the surfaces. Do *not* include the given dimensions on your drawing solution: they are only included to help you to create the needed views for each problem.

PROBLEM 9-28 Find the true shapes of surfaces *A* and *B* by the fold-line method.

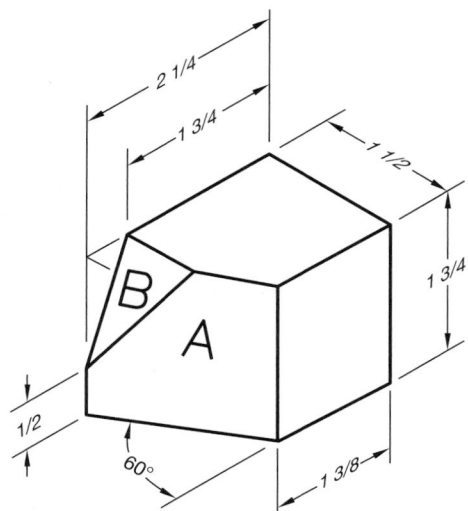

PROBLEM 9-29 Find the true shape of surface *D*.

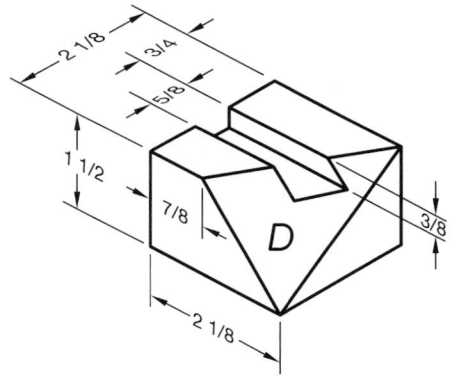

PROBLEM 9-30 True shape of a plane. Find the true shapes of the surfaces *A*, *B*, *C*, *D*, *E*, and *F* of the concrete pier block.

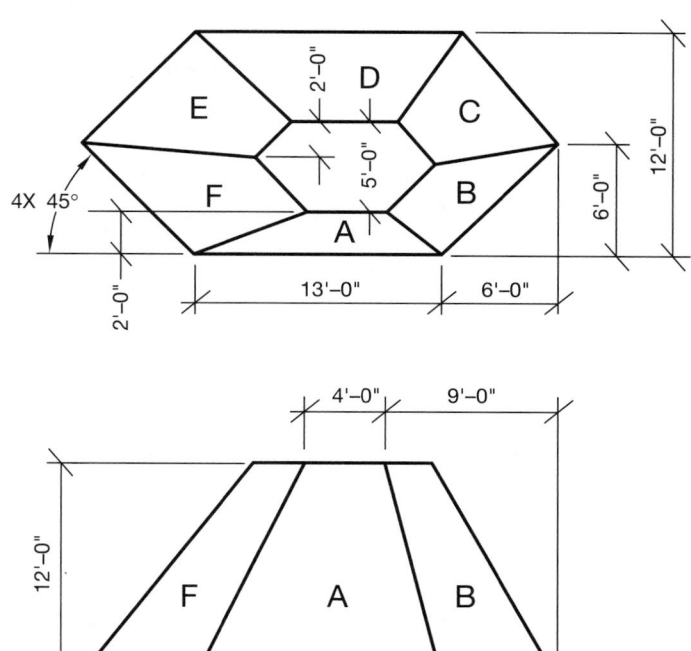

PROBLEM 9-31 From the given engineer's sketch, draw the top, front, and appropriate projection planes of the sloped surface. Use the fold-line method to create the true shape of the sloped surface.

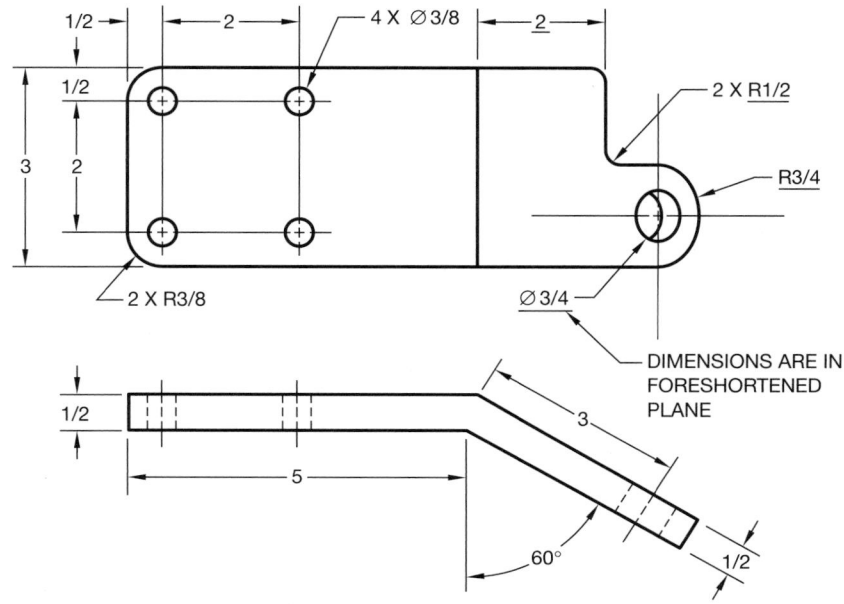

PROBLEM 9-32 From the given engineer's sketch, draw the top, front, and appropriate projection plane of the sloped surface. Use the fold-line method to create the true shape of the sloped surface.

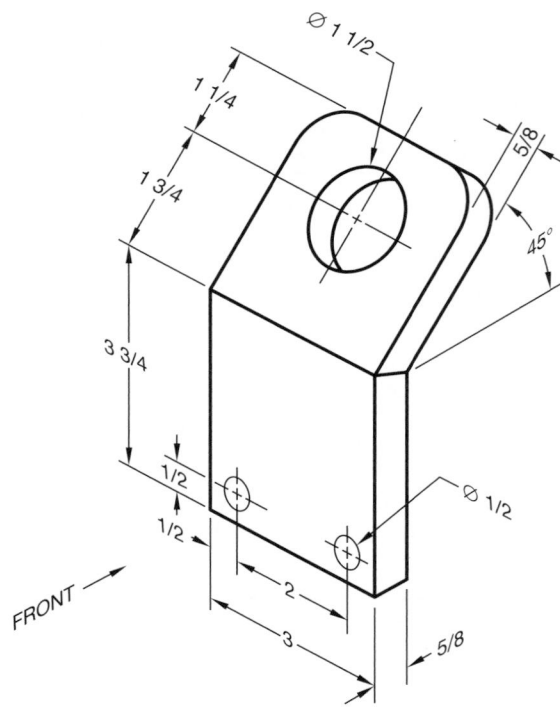

MATH PROBLEMS

Convert the following inch and feet dimensions to metric dimensions.

PROBLEM 9-33 3 in.

PROBLEM 9-34 100 ft.

PROBLEM 9-35 5"-6'

Convert the following metric dimensions to inch dimensions.

PROBLEM 9-36 50 cm

PROBLEM 9-37 2.54 cm

PROBLEM 9-38 150 mm

Convert the following given measurements to the required measurements.

PROBLEM 9-39 32 feet equal how many meters?

PROBLEM 9-40 104 inches equal how many millimeters?

PROBLEM 9-41 .3125 decimal inch equals how many mm?

PROBLEM 9-42 66 inches equal how many mm?

PROBLEM 9-43 5 miles equal how many km?

PROBLEM 9-44 Determine the area of the feature shown in the following drawing.

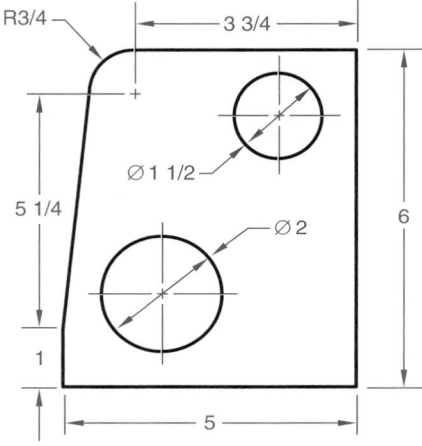

OBJECTIVES

After completing this chapter, you will:

■ Determine bearing, slope, and percent of grade.

■ Establish the true angle between a line and a plane.

■ Find the true distance between a plane and a point.

■ Find the true distance between two skew lines.

■ Find the piercing point of a line and a plane.

■ Determine the visibility of lines.

■ Find the true angle between intersecting planes.

■ Graphically solve vector force problems.

THE ENGINEERING DESIGN APPLICATION

A sample descriptive geometry problem solved by the design process follows.

1. *Identify and define the problem.* Given an isometric and partial detail drawing of a building with a chip blower and its discharge pipe location, determine the location and the shape of the roof opening shown in Figure 10.1. Also, the angle between the pipe and the roof must be determined to complete the design of the pipe support bracket. Finally, determine the slope of the discharge pipe. The partial plane view shows the location of a chip blower in a 40' wide × 100' long metal building. The building has 16' high walls and a 4/12 slope on the roof. The blower point 1 is 4' above the floor and the discharge point 2 is 28' above the floor.

2. *Research and generate solutions.* To see how much can be understood from the problem statement, take it word by word. "Determine the location of the roof opening" means there will be a pipe intersecting or going through the roof. This is the concept of a line intersecting a plane. "Determine the shape of the roof opening" explains itself. The true shape of the opening is asked for here. You are finding the true shape of a plane to determine the angle between the pipe and the roof. The concept of the angle between a line and a plane is used to determine the slope of the discharge pipe. Therefore, the problem requires you to find:

   a. The intersection of a line and a plane.
   b. The true shape of a plane.
   c. The angle between a line and a plane.
   d. The slope of a line.

3. *Evaluate the possible solutions.* Draw the top, front, and side projection planes or, as in architectural drawing, the top, front elevation, and side elevation of the given problem. This will help in the evaluation if there is doubt about which part of the information is useful. Only one solution for each part has been determined, so this step is limited in its usefulness.

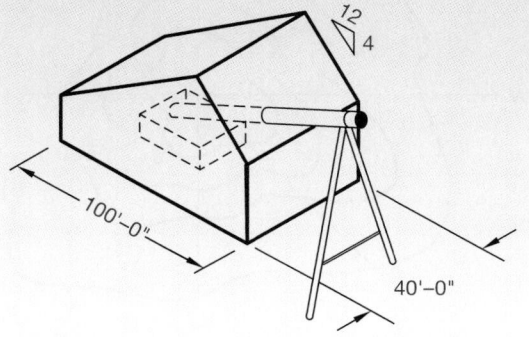

FIGURE 10.1 ■ Problem layout with partial detail.

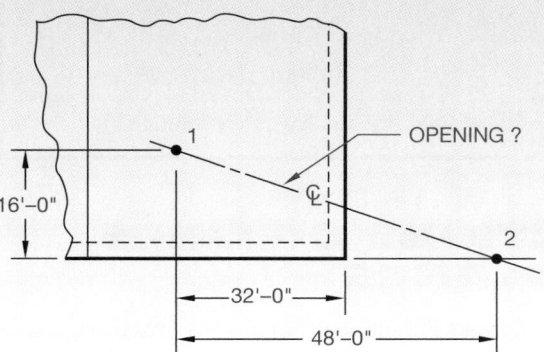

(Continued)

THE ENGINEERING **DESIGN** APPLICATION (continued)

4. *Use the best solution.* Decide which views or projection planes will give the necessary information. The front, top, and side views do not give the true lengths, shapes, or angles needed. Auxiliary views must be found and constructed, as in Figure 10.2. The roof plane must be seen as a true shape in order to see the true shape of the hole in it. You must see the edge of the roof plane and the true length of the pipe in the same view to determine where they intersect and the angle between them. To find the slope of the discharge pipe, you must see the true length of the centerline of the pipe in an elevation view.

5. *Evaluate the best solution.* Check the answers by other methods if possible. Does everything in the answer make sense?

6. *Finalize the solution.* Present the solution to your employer or the instructor with the completed notations.

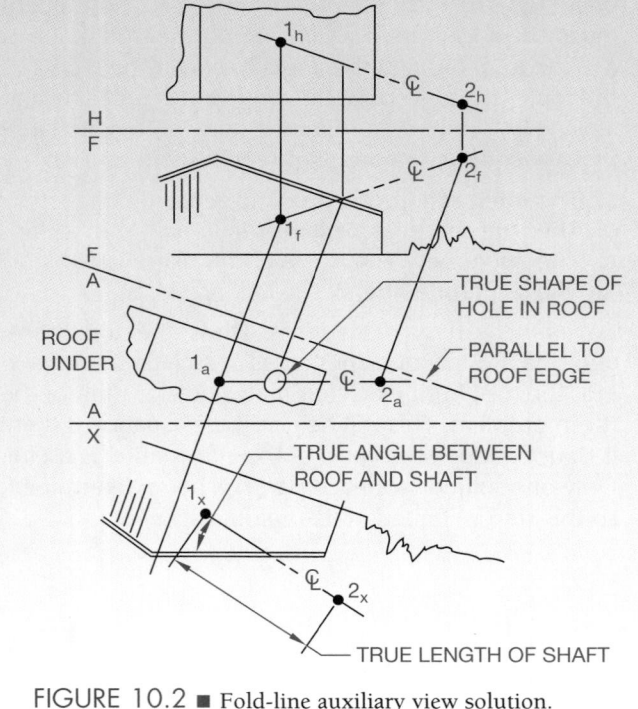

FIGURE 10.2 ■ Fold-line auxiliary view solution.

DIRECTIONS OF LINES AND PLANES

The procedure to locate lines and planes in space was discussed in Chapter 9. To solve many engineering problems, the direction of lines and planes must also be known. The direction of lines and planes is identified in space by a variety of ways, depending on their uses.

Horizontal Directions

Bearings are generally used for map directions and can only be determined in a *horizontal view* or a top view. Knowing the bearing angle of a line or plane is especially useful in mining and civil engineering. The direction of north is always understood to be toward the top of a drawing layout unless otherwise indicated by a north-pointing arrow pointing in another direction. South is always opposite north, while west is to the left and east is to the right forming the word *WE* as shown in Figure 10.3. A bearing angle is always 90° or less and will be identified either from the north or south. Lines pointing directly north are identified as due north. Also, lines pointing directly east, west, or south are identified as due east, due west, and due south, respectively. The bearings of several angles are shown in Figure 10.4. Notice that the bearing is first identified by

FIGURE 10.3 ■ Compass directions.

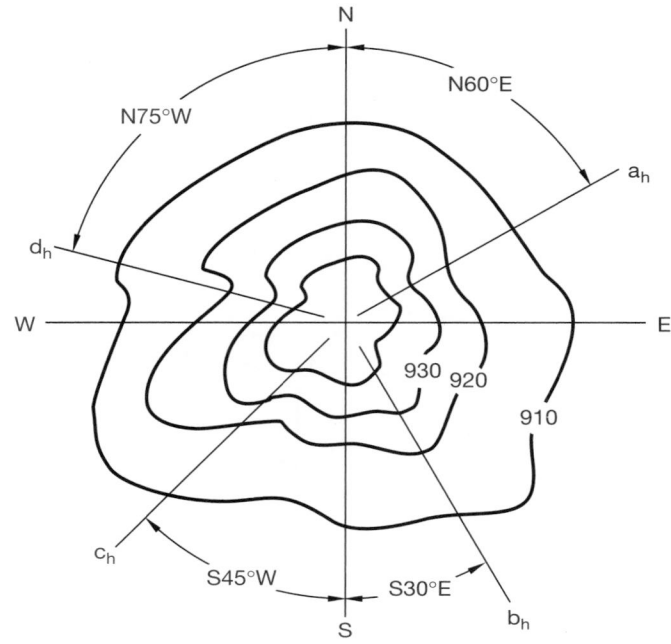

FIGURE 10.4 ■ Bearing angles of lines on a contour map.

a north or south direction indicator. The acute angle formed from the vertical direction and east or west follows next. Lastly, the corresponding side of the acute angle direction east or west is noted. Examples include: N 60° E, N 75° W, S 30° E, and S 45° W.

Bearing Is Found in the Horizontal View

The bearing of a line can only be found in the horizontal view. The line does *not* have to be in true length to represent the bearing.

There is really no such thing as the bearing of a plane; but in mining applications, it is common practice to locate a *stratum* by using the bearing of a horizontal line in the plane of a stratum. Stratums in their simplest form are horizontal layers of rock. Within a specific area, it is reasonable to assume that a stratum is uniform in thickness and that it lies between two parallel planes. For the discussion in this text, these conditions will be assumed. The bearing of a horizontal line in a stratum is called the *strike*. To find the strike of plane M-N-O in Figure 10.5, you can use the same technique we used in Chapter 9 by locating a horizontal line in the front view and finding that same line in the horizontal view where it appears as true length.

Step 1. Draw horizontal line m_f-s_f in the F view.

Step 2. Find line M-S in the H view. (Line m_h-s_h being shown in true length is immaterial, because this is where the

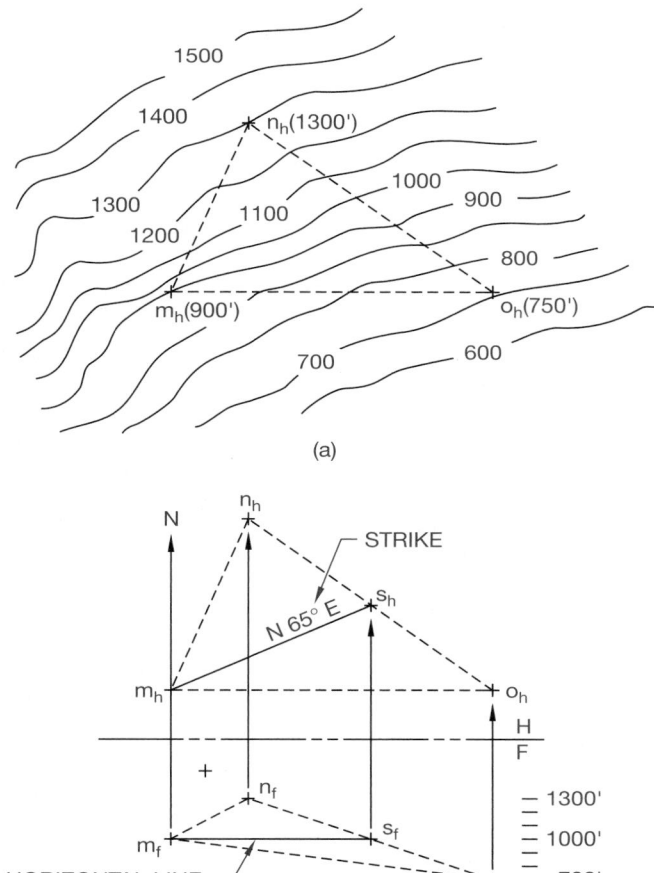

(a)

(b)

FIGURE 10.5 ■ Stratums: (a) horizontal view of a stratum; (b) finding the strike of the stratum.

bearing of the line is shown whether it is in true length or not.)

Step 3. Measure the angle of line m_h-s_h from the north or south direction. This bearing N 65° E of line M-N as seen in the horizontal view is the strike of the plane.

This discussion of the application of descriptive geometry to mining problems is brief for the purposes of this chapter. It will be necessary to study in more detail the concepts of contour lines, saddles, ridges, ravines, stratas, veins, faults, outcrops, cuts, and fills to adequately utilize descriptive geometry for civil engineering applications. Many of the calculations to determine cross sections for highways, and the designs and drawings from many civil engineering projects, are currently being made using computers.

Vertical Directions

The vertical direction of a line can only be found when the following two conditions exist: *the line must be in an elevation view* and *the line must appear in true length*. The vertical direction of a line is identified by the acute angle formed between the vertical line and the horizon (ground). This angle will vary from true horizontal (0°) to vertical (+90°). If the angle is between the horizon and +90°, or the line goes upward from its origin, the angle is called *inclination* and is assigned a positive value. If the angle is between the horizon and –90°, or the line goes downward from its origin, it is called *declination* and is assigned a negative value. (See Figure 10.6.) The terminology used to identify the inclination of a line is specified differently for various engineering fields. Each of the terms that follow are those ways that express the relationship of a line to a horizontal plane. In all cases, the same two conditions must exist to be able to identify the true direction of a vertical line. The line must be in an elevation plane where it appears in true length.

Rules for Finding the Slope of a Line

Always start from the horizontal view when solving problems that require you to find the slope of a line. For example, when using the fold-line technique, place the fold line parallel to the line where it exits on the horizontal view. When using the revolution method, revolve the line where it exists in the horizontal view parallel to the existing fold line. Either method will result in showing the line in

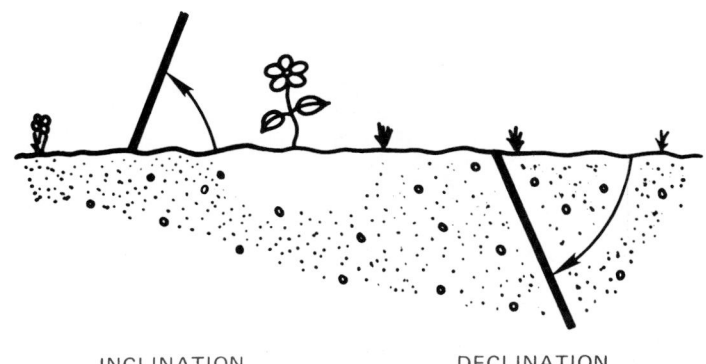

INCLINATION   DECLINATION

FIGURE 10.6 ■ Inclination and declination.

true length in an elevation view. This is the only place you can measure the true slope of a line.

The *slope angle* or *slope* of a line is the angle in degrees that the line makes with the horizontal (ground or level) plane. The slope angle can be expressed by the following formula:

Slope angle (s) is the angle where tangent (arctan s) $= \dfrac{Rise}{Run}$

Again, the true slope of a line can only be seen in an elevation view where the line appears in true length. (See Figure 10.7.)

The *grade* of a line is expressed in percent form and is found by using the following formula:

$$Percent\ grade = \frac{Rise}{Run} \times 100$$

This is illustrated in Figure 10.8a. Observe also that the grade is the tangent of the slope angle multiplied by 100. From trigonometry, we know that the tangent of an angle is defined as dividing the length of the opposite side of a triangle of the included angle by the adjacent side of that angle as seen in Figure 10.8b. One of the most common ways of measuring run and rise is to use an engineer's scale, which has divisions in multiples of ten, or set the computer scale in CAD to the appropriate scale. Percent of grade is simply another way to express slope; therefore, the conditions for finding the percent of grade of a line are the same as for finding the true

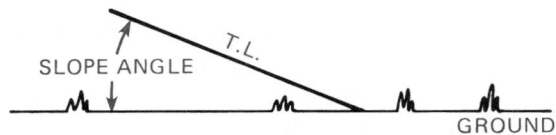

FIGURE 10.7 ■ Slope angle.

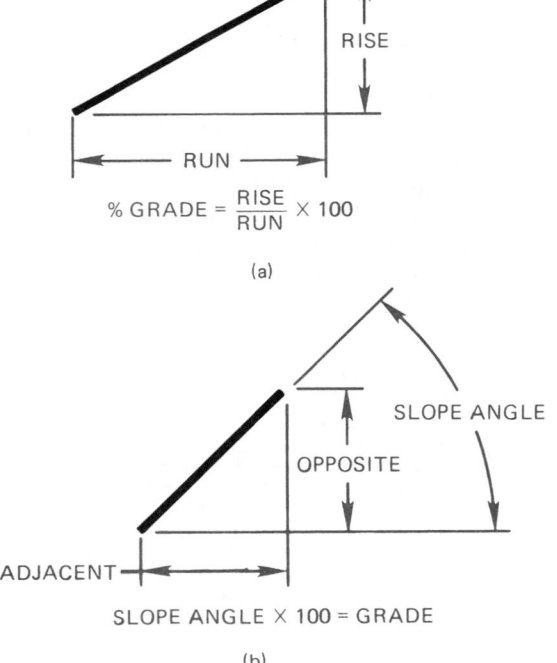

(a)

(b)

FIGURE 10.8 ■ Calculating the grade of a line: (a) formula one; (b) formula two.

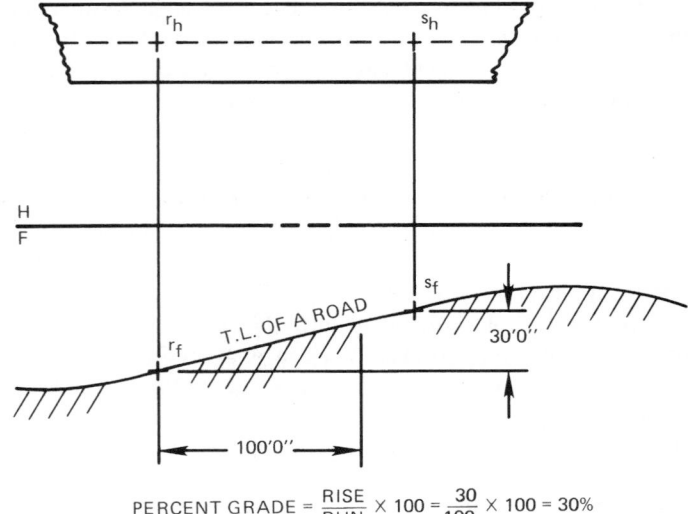

PERCENT GRADE $= \dfrac{RISE}{RUN} \times 100 = \dfrac{30}{100} \times 100 = 30\%$

FIGURE 10.9 ■ Grade of a road.

slope of a line. (See Figure 10.9.) The term *grade* is usually used when referring to a line that has an angle with less than a 45° inclination. Grade is usually associated with civil engineering but not always. Slope is usually applied to a line that has an angle with greater than 45° inclination, such as the slope of a roof on a house. Slope is used more with architectural engineering but not always.

In architectural engineering, the slope may also be referred to as the *pitch*. When describing the angle the roof plane has with the horizon, it is expressed as the ratio of the vertical rise to 12' of span. The graphical method of showing pitch on a drawing is illustrated in Figure 10.10.

In structural engineering, slope is usually referred to as the *bevel* of a line. Figure 10.11 shows the bevel of several beams. In civil engineering, the terms *grade* and *slope* are frequently used as shown in Figure 10.12a. In addition, *batter* is applied when dealing with the slope of concrete footings as shown in Figure 10.12b. *Dip* is used when finding the slope of a stratum (plane). The dip is the slope of the strike. Figure 10.13 shows how to find the dip of a stratum.

In review, *slope, grade, pitch, bevel, batter,* and *dip* are all representations of the angle between a true-length line and the horizon. Finding this angle graphically is done with the same procedure in each situation.

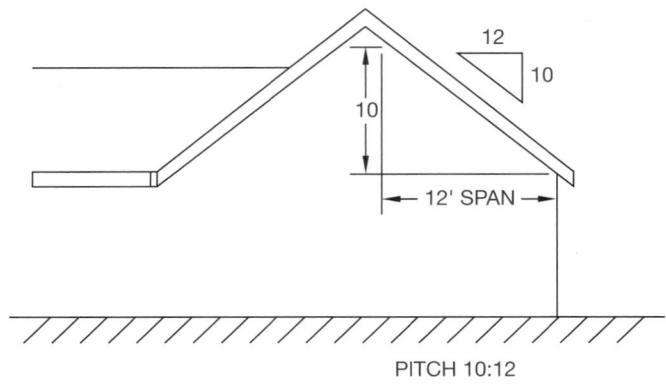

PITCH 10:12

FIGURE 10.10 ■ Pitch of a roof.

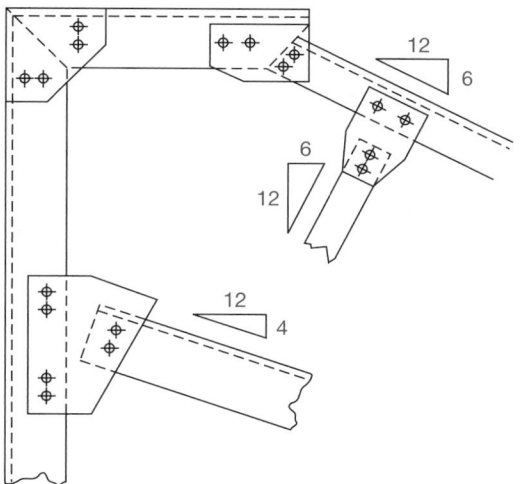

FIGURE 10.11 ■ Bevel of a beam.

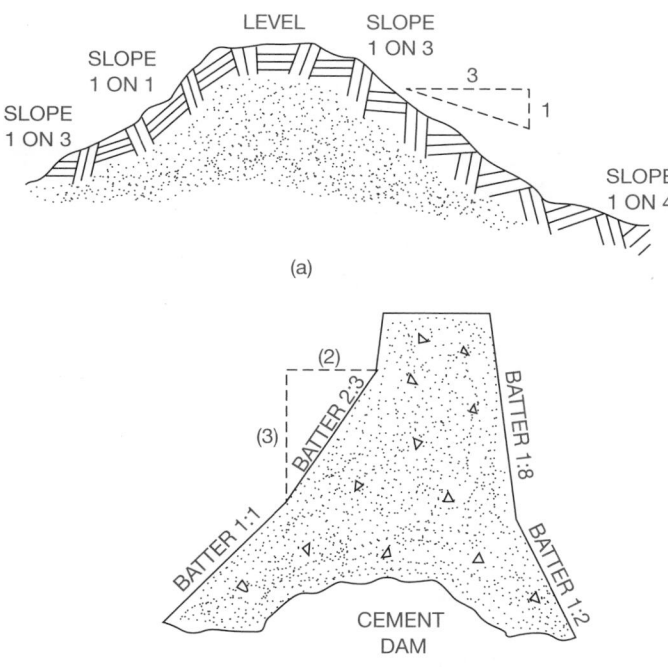

FIGURE 10.12 ■ Terms in civil engineering: (a) slopes of earth dam; (b) batter of concrete dam.

TRUE DISTANCE BETWEEN A POINT AND A LINE

To find the true distance between a point and a line, you must create a view that will show the line and the point appearing as *two points*. This occurs when you are viewing the end of the line and the point on one view. As with most problems, there are several

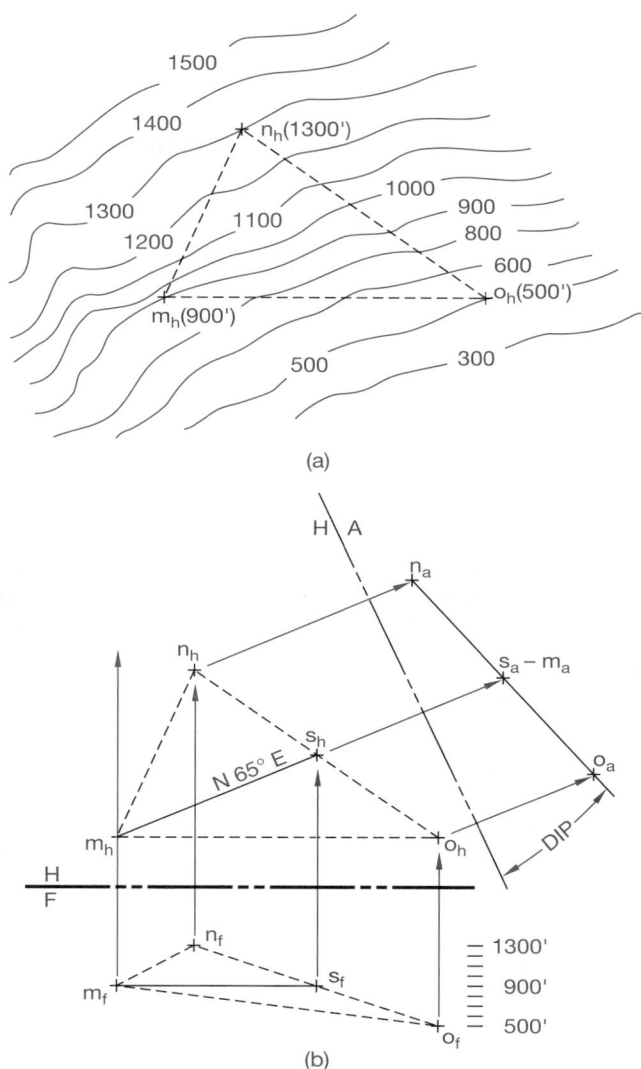

FIGURE 10.13 ■ Fold-line method of finding the dip of a stratum.

ways to solve this problem. Figure 10.14 shows how to find this distance by the fold-line method.

Finding the Distance between Two Points with the Fold-Line Method

The first step in the solution of this type of problem is always to place a fold line parallel to the given line. The line will be in true length when it is projected onto the new view. The second step is to place another fold line perpendicular to the true-length line. The line will then appear as a point when it is projected onto the next new view. After projecting the point onto the two new views, it is just a matter of measuring the distance between two points.

Step 1. Find the true length of line R-S and the location of point X on the A view.

Step 2. Find the end view of the true-length line R-S and the location of point X on the I view. The distance measured between point X and the end view of line R-S on the I view is the true distance between point X and line R-S.

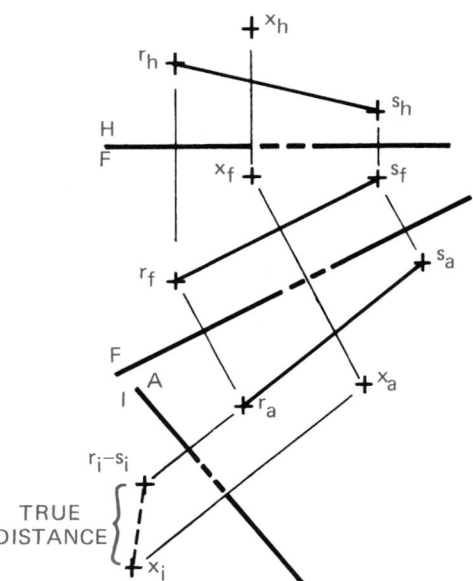

FIGURE 10.14 ■ True distance between a line and a point.

TRUE DISTANCE BETWEEN TWO SKEW LINES

An example of two *skew lines* (two nonparallel, nonintersecting lines) is shown in Figure 10.15. The true distance between two skew lines is the perpendicular between them. To be able to determine the true distance between the two skew lines, you must create a view where one line appears as a point. (*Note:* While the one line appears as a point, the other line will not necessarily appear in true length.) The perpendicular distance between the line appearing as a point and the other line is the true distance between both lines. (See Figure 10.16.)

Step 1. Find the true length of one of the lines. In this example, line R-S was selected to find the true length. The true

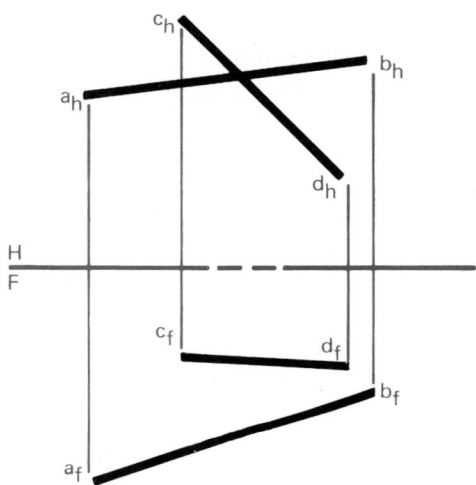

FIGURE 10.15 ■ Skew lines.

length of line R-S and the location of line M-N has been found on the A view. (See Figure 10.16a.)

Step 2. Create the view where line R-S appears as a point. Line R-S appears as a point in the I view. Line M-N is also carried onto this view. (See Figure 10.16b.)

Step 3. On the I view, the true distance can be measured between line R-S and line M-N by constructing a perpendicular line from line R-S (appearing as a point) and line M-N. (See Figure 10.16c.)

Sometimes a problem may require you to adjust the distance between two skew lines or make sure there is enough clearance between the two lines. An application for determining the true distance or clearance determination might be for the location of a valve in a piping system or the clearance between power lines and a nearby metal structure in an electrical power substation. When you have found the true distance between the two skew lines as in step 3 in the previous problem, it is just a matter of adjusting the point (line R-S) in or out along the perpendicular distance from line M-N, as shown in Figure 10.16d. You then can relocate the new position of line R-S in all the previous views. If line R-S cannot be moved in the original problem, you will need to solve the problem by starting with line M-N rather than line R-S as in this example.

INTERSECTION OF A LINE AND A PLANE

The location and anchoring of guy wires; the location of holes for shafts, cables, pipes, and wires; and the angles that are created by these features are important considerations for the designer in structural and mining problems.

Finding the Piercing Point of a Line and a Plane

To find the point of intersection, *the piercing point,* of a line and a plane, you must create a view where the *plane appears as an edge.* You then can project this apparent point of intersection back into the original views.

Use the following steps if you have a problem that requires finding the intersection of the roof and the circular shaft as shown in Figure 10.17a (piercing point problem.)

Step 1. Find a view that shows the roof plane as an edge. Figure 10.17b shows that the P view contains the roof plane as an edge. When the centerline of the shaft is projected onto the P view, the apparent point of intersection, w_p, of the centerline of the shaft and the roof plane can be seen.

Step 2. Project this apparent point of intersection, w_p, back onto the previous views. As Figure 10.17c shows, a projector is projected from the A view back to the F view intersecting the centerline of the shaft. This is the actual point of intersection, point w_f of the roof plane shown on the F view. You can also show the proper location of point W in the H view by the normal procedure of locating a point in space.

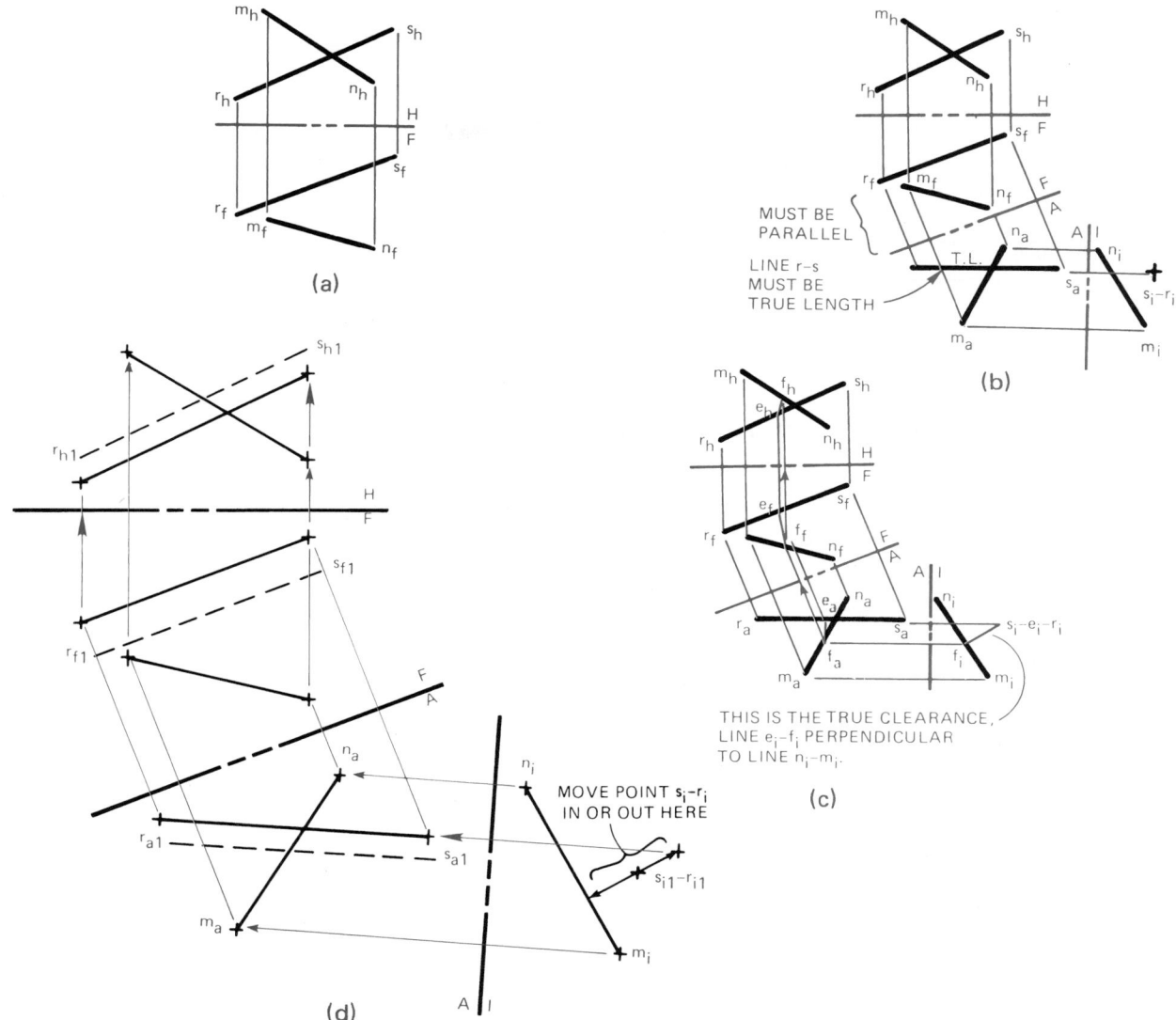

FIGURE 10.16 ■ Finding the true distance between skew lines: (a) step 1; (b) step 2; (c) step 3; (d) adjusting the true distance.

VISIBILITY

When two lines are shown in space, sometimes it is difficult to know which line is on top of which or if they intersect. Problems dealing with the intersection of planes require you to know which line is in front of the other in order to correctly solve and complete the problem as you will see later in this chapter. For example, look at Figure 10.18a. Looking at the F view, you cannot tell if lines A-B and R-S intersect or if line A-B is in front of or behind line R-S. There is a simple way to determine which situation exists.

Step 1. On the F view shown in Figure 10.18b, project a projector from the apparent point of intersection of lines A-B and R-S perpendicular to fold line F/H. Continue projecting until it crosses one or both of the lines on the H view.

Step 2. If the projector passes through the apparent point of intersection on the H view, these would be intersecting lines and this would be the real point of intersection. But

this is *not* the case in our example. The projector crosses line R-S in the H view first and verifies that line R-S is closest to us in the F view.

Step 3. We can apply the same procedure when determining the visibility of lines R-S and A-B on the H view. Figure 10.18c shows a projector being projected from the apparent point of intersection on the H view perpendicular to the H/F fold line. The projector is then continued on until it crosses one or both of lines A-B and R-S in the F view. The projector crosses line A-B in the F view first and verifies that line A-B is closest to us in the H view.

INTERSECTION OF PLANES

Structural skins, houses, sheet metal forms, and many other things have intersecting surfaces. Determining the lines of intersection is

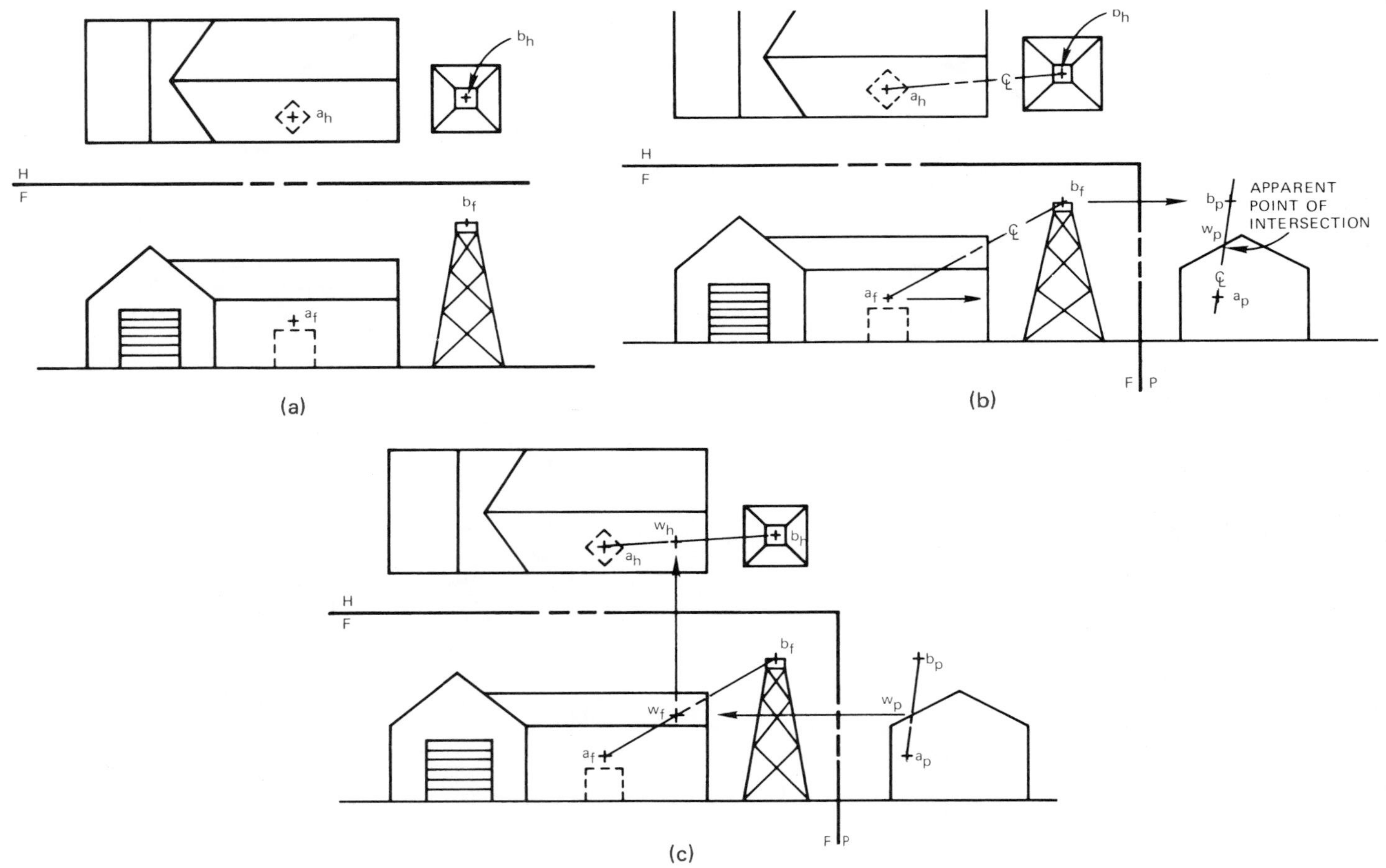

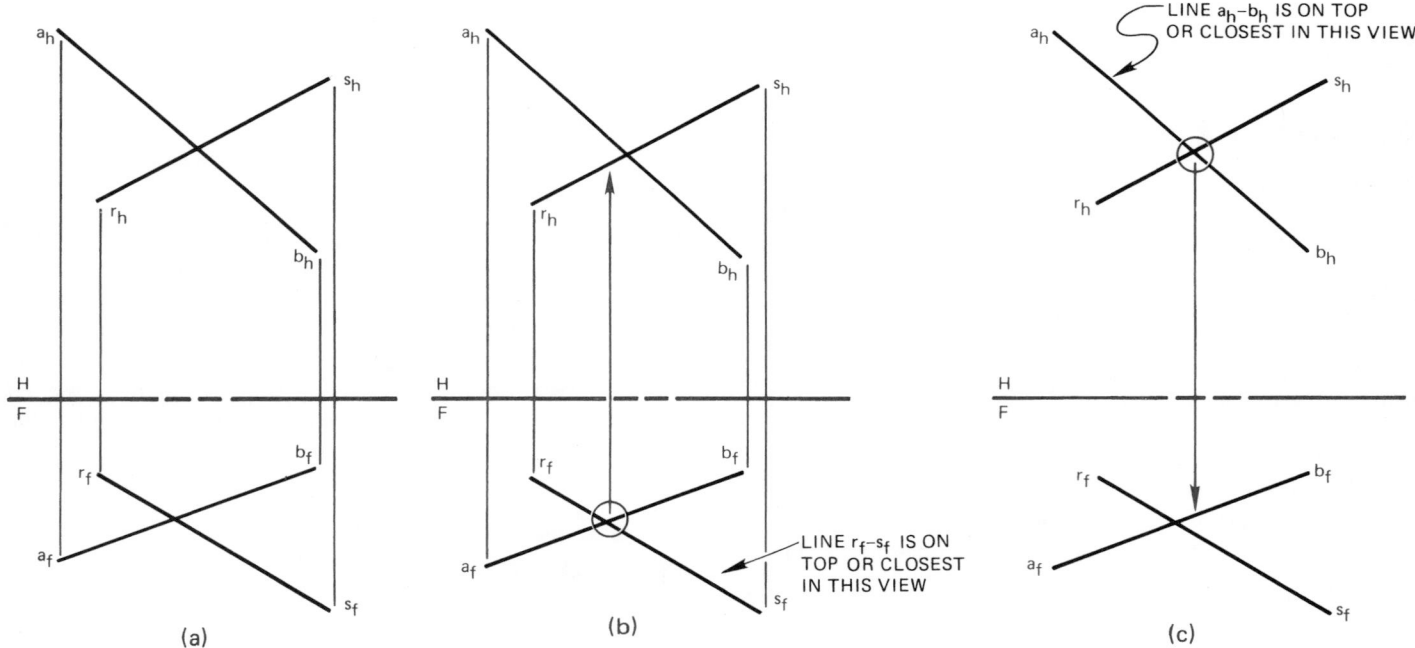

FIGURE 10.17 ■ Piercing point problem: (a) problem setup; (b) step 1, connecting two points and finding the apparent point of intersection; (c) step 2, finding the actual piercing point.

FIGURE 10.18 ■ Visibility: (a) two lines in space; (b) finding visibility of lines in the F view; (c) finding visibility of lines in the H view.

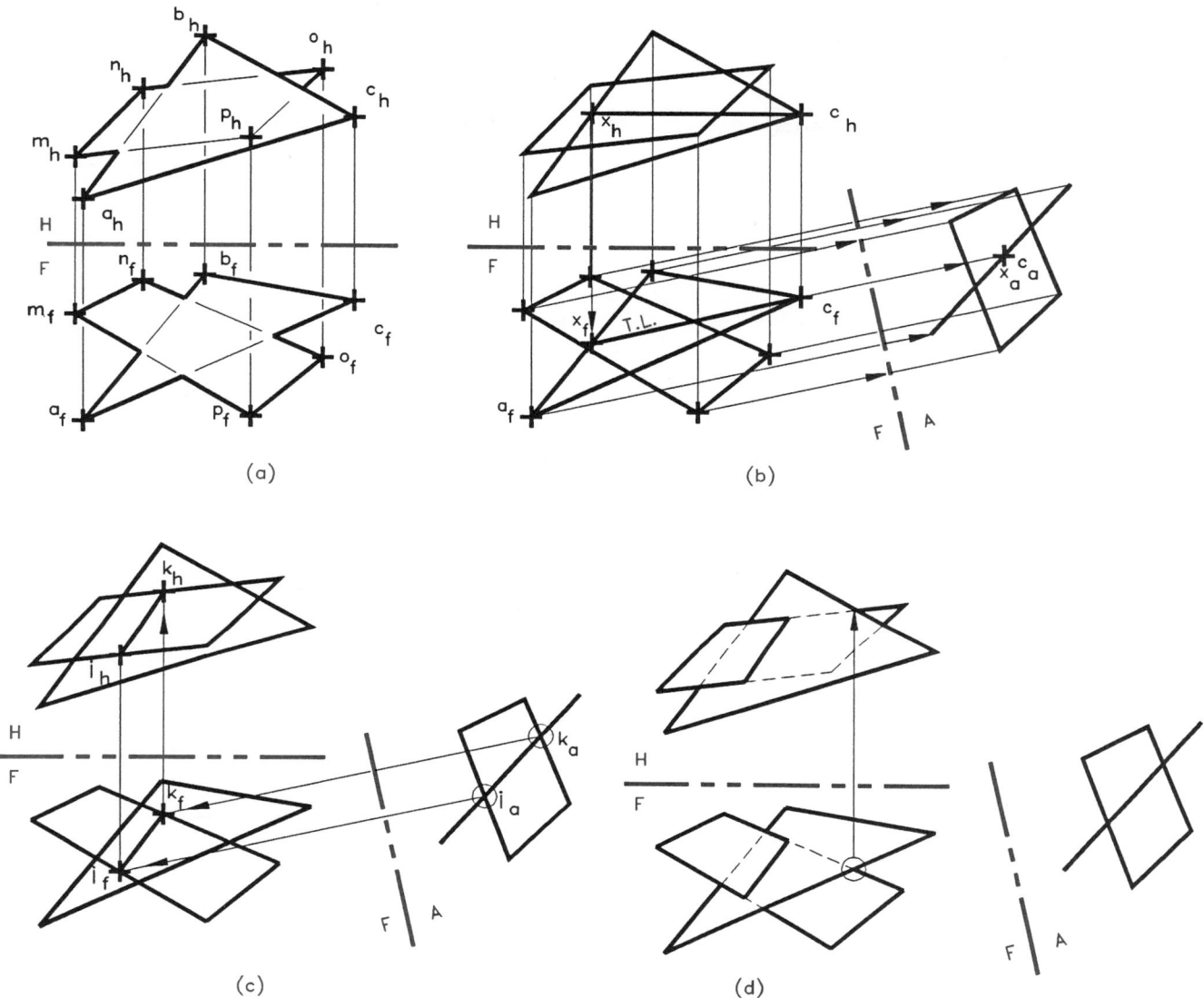

FIGURE 10.19 ■ Two planes that appear to intersect: (**a**) problem setup; (**b**) step 1, finding the edge view of a plane; (**c**) step 2, finding the line of intersection of two planes; (**d**) step 3, determining visibility and completing the visible and hidden lines of two planes.

essential in completing each view of a drawing and eventually determining the shape and size of each surface for the construction of each part in the fabrication shop. If two planes are not parallel, they will intersect forming a straight line that will be common to both planes. Only two points that lay on both planes will need to be found to fix the position of this line of intersection that is common to both planes. To find the line of intersection of two planes, a view must be created that shows both planes where one appears as an edge view. Figure 10.19a shows two views in which the two planes appear to intersect.

Step 1. Create a projection plane where one of the planes appears as an edge view. Figure 10.19b shows finding plane A-B-C as an edge view in the A view. You already know how to do this from Chapter 9.

Step 2. Figure 10.19c shows the common points J and K of intersection of the two planes. These points can now be projected back onto the F and H views and completed as the line of intersection.

Step 3. Identify the visibility of the intersecting lines of each plane on the F and H views by the process previously explained. You will need to do this in order to show which lines are visible and which lines are hidden on the two planes. Figure 10.19d shows this process completed.

DIHEDRAL ANGLES

Working with structural skins or various sheet metal projects requires knowing the exact angle between those surfaces or planes. The true angle between two planes, or *dihedral angle,* is found in a view in which both of the intersecting planes appear as an edge. (See Figures 10.20 and 10.21.) Figure 10.22a shows the top and front views of a sheet metal funnel on the H and F views.

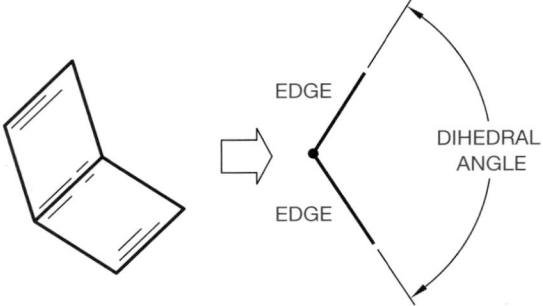

FIGURE 10.20 ■ Pictorial representation of a dihedral angle.

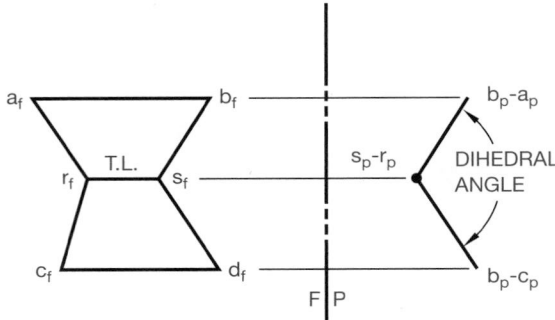

FIGURE 10.21 ■ Descriptive geometry flat layout showing the dihedral angle of two planes.

(a)

(b)

(c)

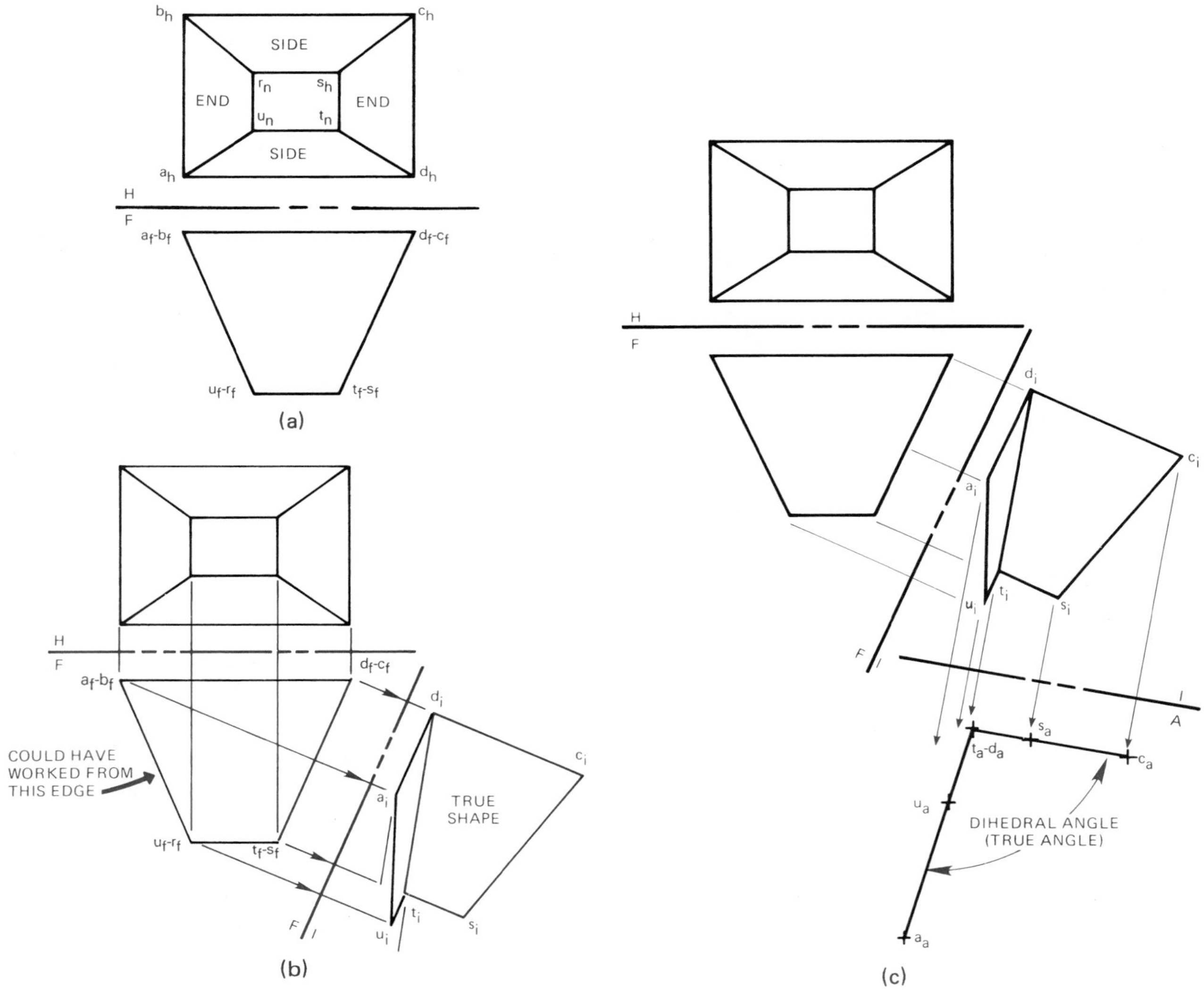

FIGURE 10.22 ■ Dihedral angle problem: **(a)** two views of a sheet metal funnel; **(b)** step 1, finding the true length of the line of intersection; **(c)** step 2, finding the end view of a true-length line and creating the dihedral angle of two planes.

The problem here is to determine the dihedral angle between the front side and the right side of the funnel.

Steps in Finding the True Dihedral Angle

To find the true dihedral angle between two planes, you need to find the end view of the line that is created by the intersection of the two planes. That first step is to place a fold line parallel to this line of intersection. Then, place a fold line perpendicular to this true-length line.

Step 1. Create a view in which you show the true length of the line of intersection of the two planes you are trying to determine the dihedral angle. In Figure 10.22b, the true length of line D-T, the line of intersection, has been found in the I view. You already know how to find the true length of a line from Chapter 9. Then, project only the points that identify the rest of the two intersecting planes onto the I view as shown.

Step 2. Create a view that shows the intersecting line D-T appearing as a point. In Figure 10.22c, line D-T has been found on the A view. It appears as a point. Notice also that when you project all the other points of the two planes onto the A view, they will form two intersecting straight lines. The angle formed between these two planes is the true of dihedral angle. This angle can be measured with a protractor or calculated on the computer.

AN INTRODUCTION TO VECTOR GEOMETRY

Vector analysis is a branch of mathematics that includes the manipulation of vectors. Vector geometry is included in this chapter because vectors are a vital part of many of the engineering sciences. Graphic solutions are usually sufficiently accurate for engineering analysis of forces and other directional quantities in structural design, mechanics, and other physical sciences. Vectors can be manipulated in some very complex ways, but the discussion here is limited to vector addition and its practical purposes. Since vectors are segments of straight lines, the principles of descriptive geometry are directly applicable to the graphic solution of problems involving vectors.

A *vector quantity* is one that has *magnitude* and *direction*, such as displacement, acceleration, momentum, force, velocity, and torque. The speed of a space shuttle, for example, describes only the magnitude of its velocity, which is a physical or *scalar* quality. The velocity of that space shuttle, on the other hand, is a vector quality that describes not only its speed but its direction of motion in space. A vector quantity is usually represented by a straight line with an attached arrowhead pointed in the same direction in which the quantity is acting. The length of the vector is drawn proportionally to the magnitude of that quantity. An example of a *vector diagram* representation of a force of 10 lb directed N 45° E is shown in Figure 10.23. To solve problems using vector geometry, it is essential to know the difference between a vector diagram and a *space diagram*. Figure 10.24 shows the space diagram of the 10-lb vector. Notice that the space diagram is *not* drawn to scale. It shows

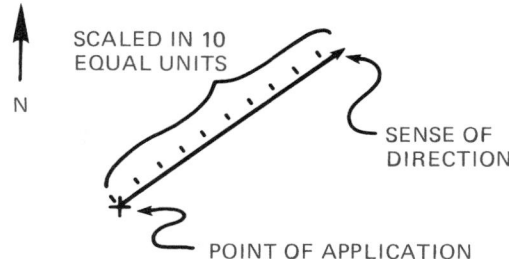

FIGURE 10.23 ■ Vector representation/vector diagram.

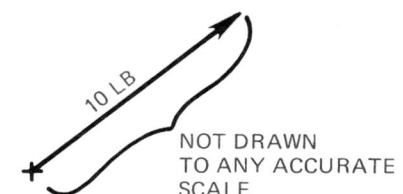

FIGURE 10.24 ■ Space diagram.

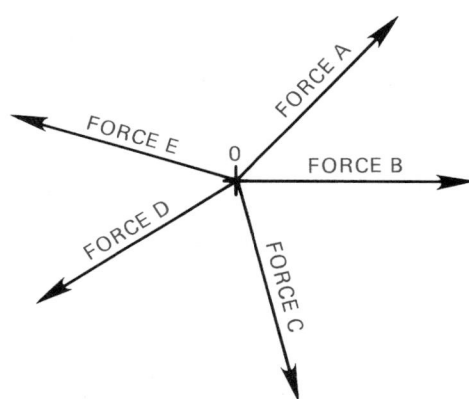

FIGURE 10.25 ■ Space diagram of concurrent forces.

sense and the accurate direction. When two or more vectors act on an object, it is called a vector or *force system*. The discussion in this chapter will be limited to forces that are *concurrent*, because they are the most common. In concurrent forces, vectors pass through a common point, as seen in Figure 10.25. Forces with a line of action lying in one plane are called *coplanar* forces. Figure 10.26 shows five coplanar vectors pictorially and with their corresponding vector diagram. When those forces lie in more than one plane but still meet at one point, they are identified as *noncoplanar*, as shown pictorially and in their corresponding vector diagram in Figure 10.27.

Addition of Concurrent Coplanar Vectors

To determine the net effect of a vector system, vector quantities must be combined. The graphic addition of concurrent coplanar vectors simply requires you to draw a vector diagram locating the vectors *head-to-tail* in their correct direction. The sum, or the space that is left open, is called the *resultant*. The resultant has the same effect as the total of the individual vectors added together.

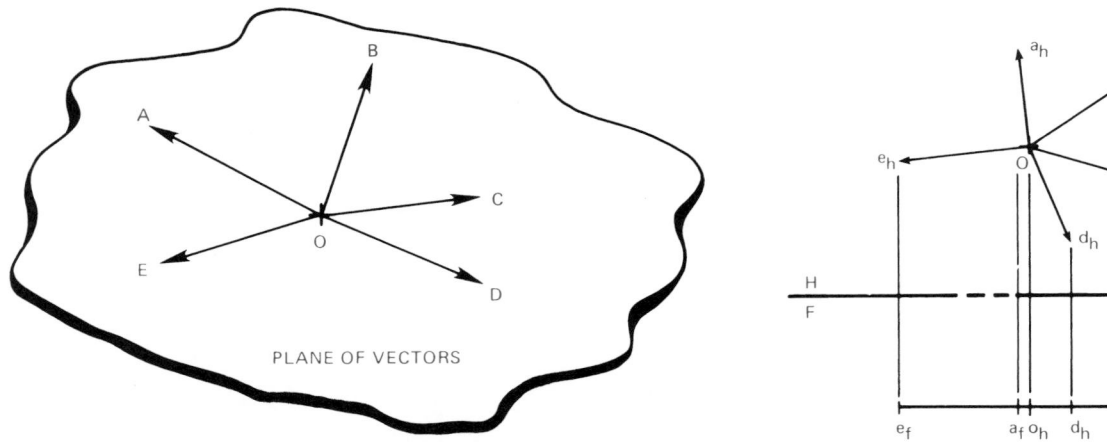

FIGURE 10.26 ■ Concurrent coplanar vector system.

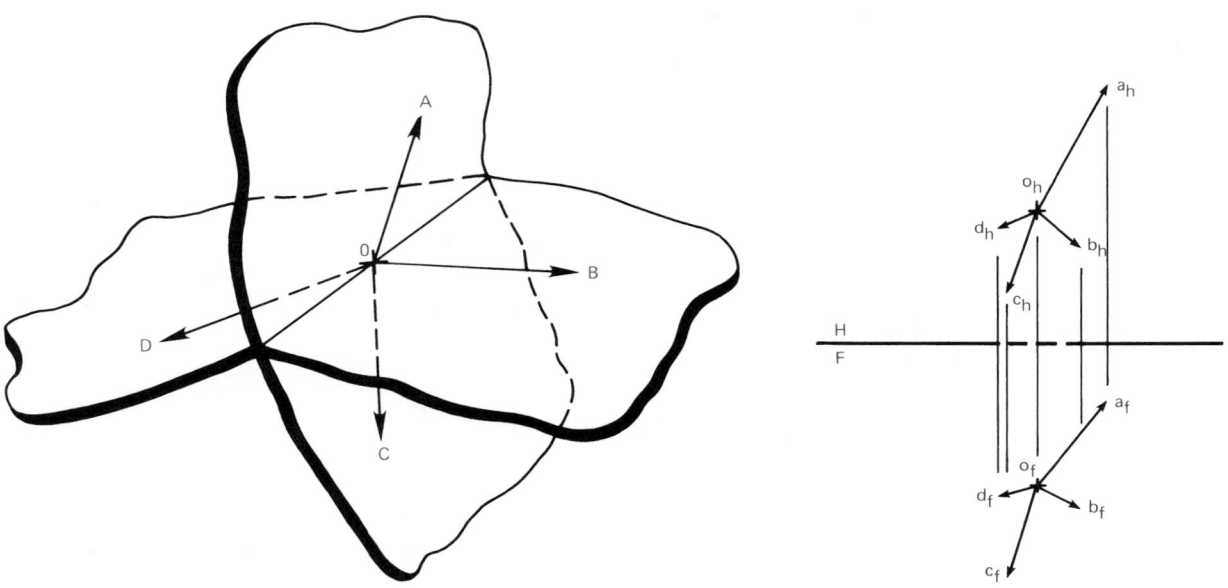

FIGURE 10.27 ■ Concurrent noncoplanar vector system.

In Figure 10.28a, the two vectors M and N shown in the space diagram are to be added. The magnitude of vector M is 40 lb and vector N is 60 lb. Remember that the space diagram shows the *correct direction,* so in the vector diagram, Figure 10.28b, the direction of the vectors is duplicated and the vector length is scaled to the same units. After placing each vector head-to-tail, in any order, the space left is connected with a straight line. This line is called the resultant and the direction arrow is in the same direction as the added vectors. Measuring the length of the resultant to scale gives a length of 80 lb. If the direction of the resultant is reversed as shown in Figure 10.28c, it is called the *equilibrant.* An equilibrant is a force that will balance all other forces through a point. An equilibrant is the same as the resultant of zero. If you have just two vectors to be added, the *parallelogram method* can be used. This method requires more construction, in that the sides of the parallelogram are drawn parallel to the given vectors. Then the diagonal of the parallelogram is drawn, which is the

resultant of the two vectors, as shown in Figure 10.28d. When you have three or more vectors, similar methods can be used. In Figure 10.29a, the vectors are drawn to scale, using the space diagram. In Figure 10.29b, the correct direction is obtained as previously explained. The closing side of the vector polygon is the *resultant.* As before, if the equilibrant is required, the direction of the arrowhead of the resultant is reversed, as shown in Figure 10.29c.

Resolution of Concurrent Coplanar Vectors

Two or more vectors can be added to form a single resultant vector. By reversing that process, a single vector can be resolved into two or more component vectors. In the case of coplanar vectors, only two can be unknown in magnitude or direction. Most commonly, a vector must be resolved into two other vectors having specified directions. In Figure 10.30a, a given vector T is to be resolved into

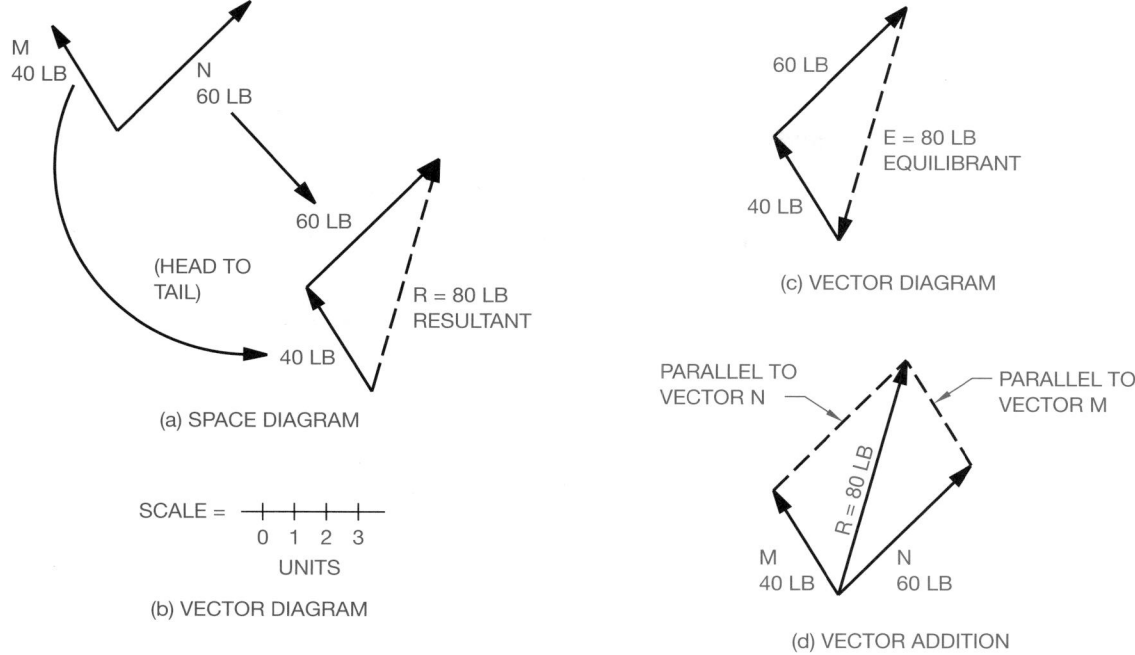

FIGURE 10.28 ■ Addition of concurrent coplanar vectors: (**a**) space diagram; (**b**) vector diagram showing the resultant; (**c**) vector diagram showing the equilibrant; (**d**) vector addition by parallelogram method.

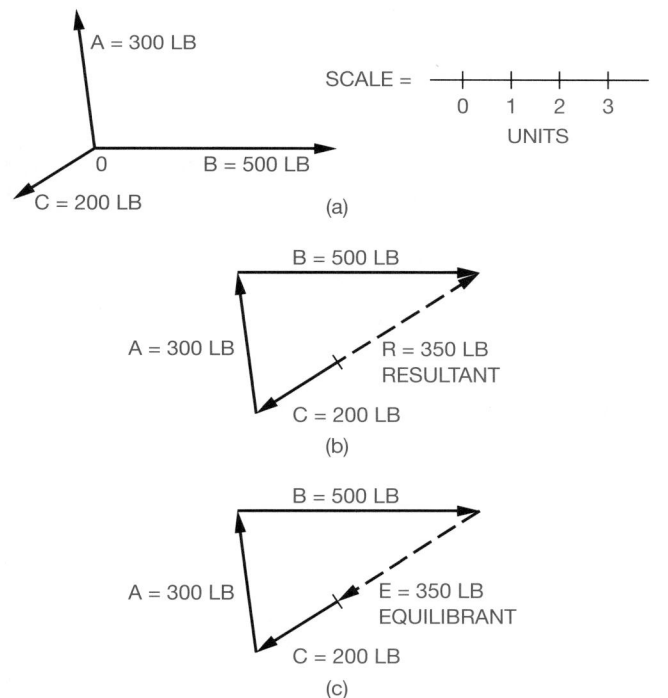

FIGURE 10.29 ■ Addition of concurrent coplanar vectors: (**a**) space diagram; (**b**) vector diagram showing the resultant; (**c**) vector diagram showing the equilibrant.

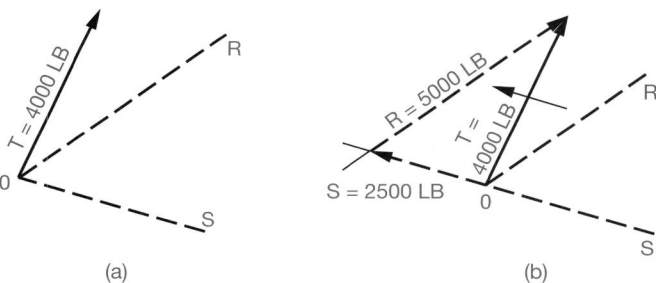

FIGURE 10.30 ■ Resolution of a single vector into two components: (**a**) space diagram; (**b**) vector diagram.

two component vectors having direction of OR and OS. The magnitude of each will need to be found. By using the parallelogram method for the addition of two vectors, the solution is obtained in reverse. Vector R is moved parallel until it crosses the arrow end of the given vector. (See Figure 10.30b.) The S vector is extended

through the tail end of the given vector. Where the two moved vectors cross will determine the lengths of each. The direction of each of the two vectors will be opposite that of the given vector.

Addition of Concurrent Noncoplanar Vectors

The addition of noncoplanar vectors is the same as for coplanar vectors, except that three dimensions are used now. Remember that noncoplanar means *in more than one plane,* so the principles of descriptive geometry will be needed to solve these problems. Figure 10.31a shows the space diagram on the F and H views for the given vector quantities. To solve this problem, you must create a projection plane that shows the true length of the resultant force.

Step 1. Set up the space diagram on a frontal view and the adjacent horizontal view as shown in Figure 10.31a.

Step 2. Find the resultant vector on the H and F views just as you would for coplanar vectors with one major exception.

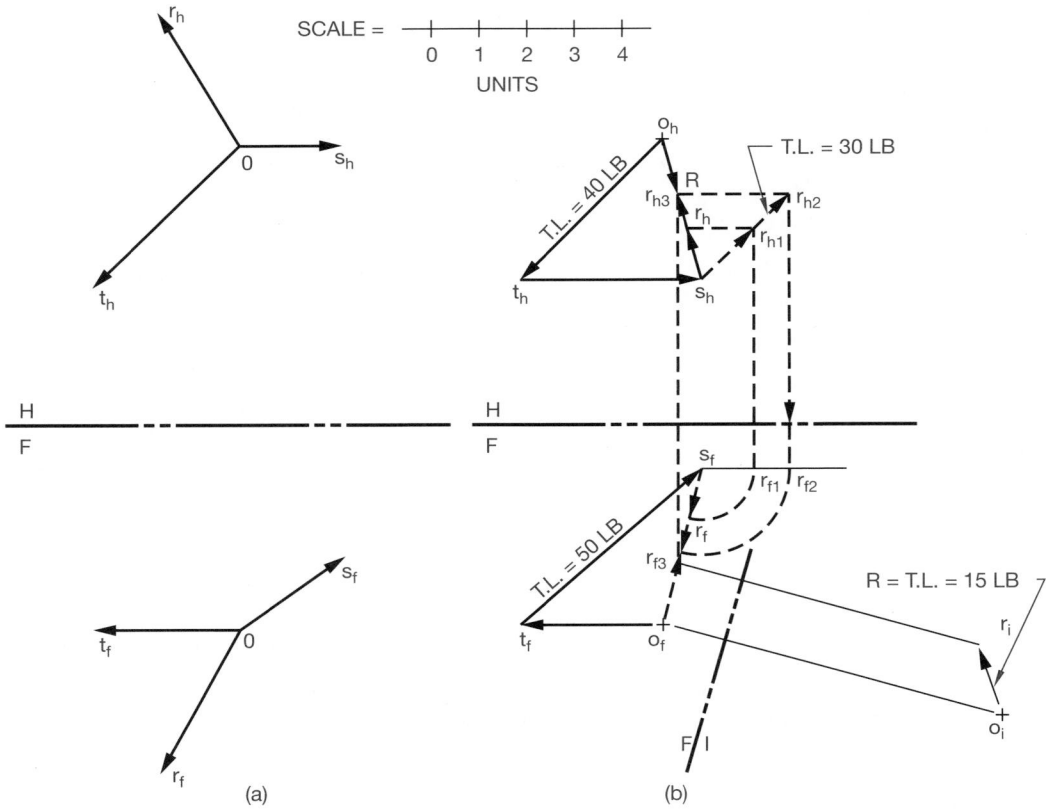

FIGURE 10.31 ■ Addition of concurrent noncoplanar vectors: (**a**) space diagram; (**b**) vector diagram.

You will need to find the true lengths of those vectors, which are *not* shown in true length on either the H view or the F view. To do this you can use either the fold-line or the revolution method. Then project the true lengths back onto the H and F views correctly. In Figure 10.31b, the S vector is already shown in true length on the F view. Therefore, you can scale the S vector on the F view. The T vector is shown already in true length on the H projection plane. Therefore, you can scale the T vector on the H view. However, the R vector is not shown in true length on either the H view or the F view. To find the true length of the R vector, use the same procedure as you did when you had to find the slope of a line when given only the bearing of that line as explained in Chapter 9. In this case, line s_h-r_h was established with an arbitrary length after the vector diagram was completed and placed in the correct direction on the H view. Then, the corresponding vector diagram was completed in the F view. Vector R was placed in the correct direction as shown in the space diagram on the F view with the corresponding length as projected from the H view. The true length of line s_f-r_f was found in the H view. Notice that the line needed to be lengthened to equal the scaled 30 lb. Line s_h-r_{h2} now represents the true length of vector R on the H view. This length was then projected into the vector diagrams as shown in Figure 10.31b.

Step 3. Find the true length of the resultant force, the space left in the vector diagrams. The resultant force is *not* shown in true length on either the H or F views, so the fold-line method was used this time to find the true length of the resultant force on the I view. The resultant force is measured at 15 lb as shown on the I view. (See Figure 10.31b.)

FORCES IN EQUILIBRIUM

Equilibrium is a condition in which the series of forces acting upon a body or structure equals zero and it will remain at rest. Graphically, the vector polygon closes, with vectors in a continuous head-to-tail arrangement. (See Figure 10.32.) When a body is

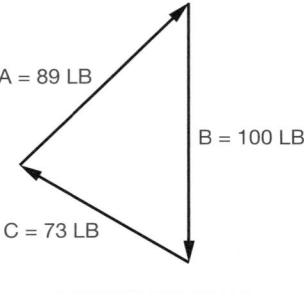

VECTOR DIAGRAM

FIGURE 10.32 ■ Vector diagram showing forces in equilibrium.

known to be in equilibrium under the action of certain known and unknown forces, these graphic conditions can be used to evaluate the unknown forces. However, the number of unknown forces is strictly limited. Coplanar vector systems can have up to three unknown forces. Vector diagrams can more easily be constructed if you isolate the structure or part to which the known forces are applied. This isolation simply means free from all adjacent bodies and is referred to as *free-body*. Construction of a free-body diagram should be the first step in the analysis of every equilibrium problem.

Finding the Equilibrium of Two Unknowns in a Coplanar System

The following steps may be used to find the equilibrium of two unknowns in a coplanar system when you are given a weight that is supported by two ropes as shown in Figure 10.33a.

Step 1. Imagine the ropes are cut, making the knot an isolated free-body, as shown in Figure 10.33b. Draw a free-body diagram with the tension forces in the free-body diagram being represented with force vectors pointing away from the knot.

Step 2. Make a vector diagram. Start with the known vector force of 100 lb by drawing it to scale with the proper direction. The direction of forces A and B are known but not their magnitudes. It was previously explained that for forces to be in equilibrium, the vector diagram must close. As shown in Figure 10.33c, the vectors A and B are drawn parallel to their corresponding vectors in the free-body diagram shown in Figure 10.33b. When the vectors are drawn properly, point L is established in the vector diagram.

Step 3. Scale the magnitudes of vectors A and B. In this example, both vector A and vector B measure to be 60 lb.

A similar type of problem occurs when two directions are unknown. In this situation, you are given the space diagram in Figure 10.34a which shows a weight of 10 lb supported by two cables that pass over pulleys, which, in turn, support two other weights of 6 and 8 lb.

Step 1. Draw the space diagram. The forces acting in the cables are known, but the direction of the vectors is not. As shown in Figure 10.34b, start with the vector with the force and direction known. The vector that represents 10 lb is drawn in the down position. As shown, draw the other two vector forces at any two different directions as shown in Figure 10.34b. The squiggly lines indicate that the real directions are unknown.

Step 2. Draw the free-body diagram. This is solved by using the method of creating a triangle when three sides of the triangle are known. This concept was covered in Chapter 6 on geometric constructions. Start with the known force of 10 lb. Draw it in its correct direction and magnitude. Next, with a radius equal to 6 lb, an arc is drawn with A as its center. (See Figure 10.34c.) With a radius equal to 8 lb, another arc is drawn with B as its center. These arcs intersect at C, completing the triangle.

Step 3. Determine the directions of the 6-lb and 8-lb vectors. The directions are indicated by the angles shown and are determined by placing the arrowheads in a head-to-tail manner in the diagram. (See Figure 10.34c.)

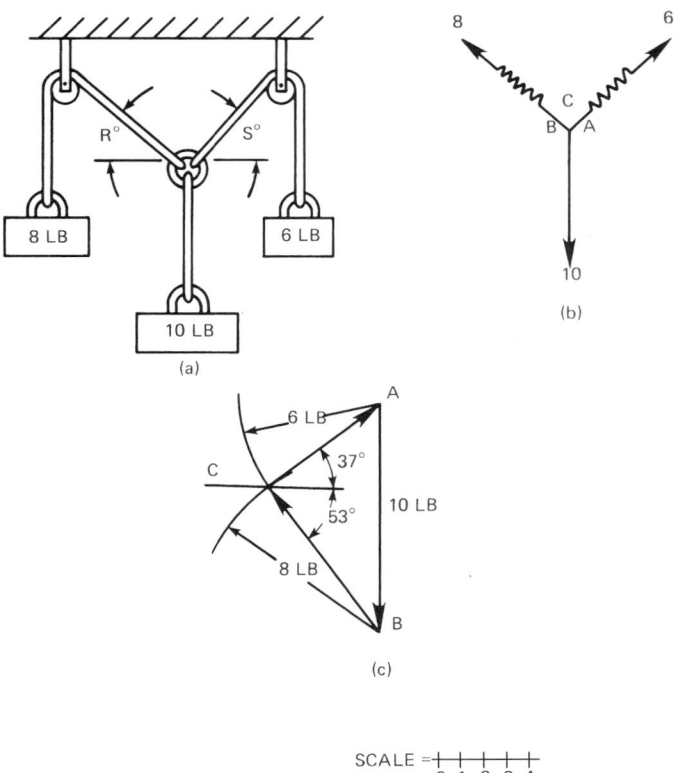

(a)

(b)

(c)

SCALE = 0 1 2 3 4

FIGURE 10.34 ■ Forces in equilibrium with two directions unknown: **(a)** space diagram; **(b)** free-body diagram; **(c)** vector diagram.

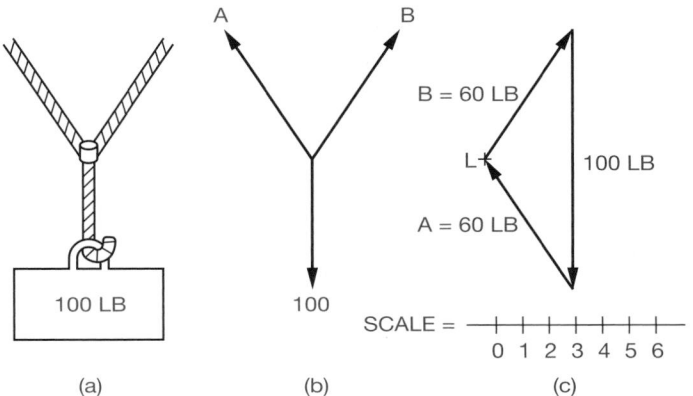

(a)   (b)   (c)

B = 60 LB

L

100 LB

A = 60 LB

SCALE = 0 1 2 3 4 5 6

FIGURE 10.33 ■ Forces in equilibrium with two magnitudes unknown: **(a)** space diagram; **(b)** free-body diagram; **(c)** vector diagram.

In this chapter, many new concepts have been introduced. But as you may have noticed, the techniques used to solve each of the unknown angles and locations of the intersections are the same as those used in Chapter 9 with, for the most part, only the terminology changed. Again, accuracy in the construction and drawing of lines is essential unless you use CAD. In most cases, the problem situations are more complex; so it is essential to understand how to break each one down into various simple elements. To solve vector problems, the correct terminology must be learned for the different types of vectors and diagrams. This will be especially true if there has been no previous experience with the related physics. Problem solving, while difficult for many, can be made easier if a logical system, such as the one used at the beginning of this chapter, is used. This system is essentially the one most engineers or anyone can best use to solve any type of problem.

CHAPTER
10
Descriptive Geometry II Problems

DIRECTIONS

Carefully study the problems before you begin working. Solve Problems 10-1 through 10-19 on graph paper or with CAD, showing and identifying all projectors, brackets with distances used, and point identifications in the required view(s). If using graph paper, use a separate sheet of graph paper for each problem and the graph lines included with each problem for proper setup. Use a scale of at least three squares to one to obtain accuracy in the problem solutions.

PROBLEM 10-1 Given the adjacent horizontal and front views showing the cable R-S, find its bearing, slope, and grade.

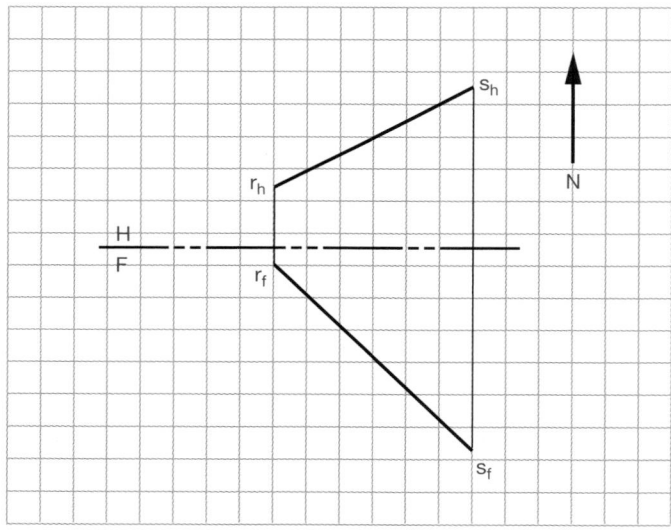

PROBLEM 10-2   True length and slope. A crane is mounted on the bow of a boat as shown. Find the true length and the slope of the cable A-B and the support structure B-C.

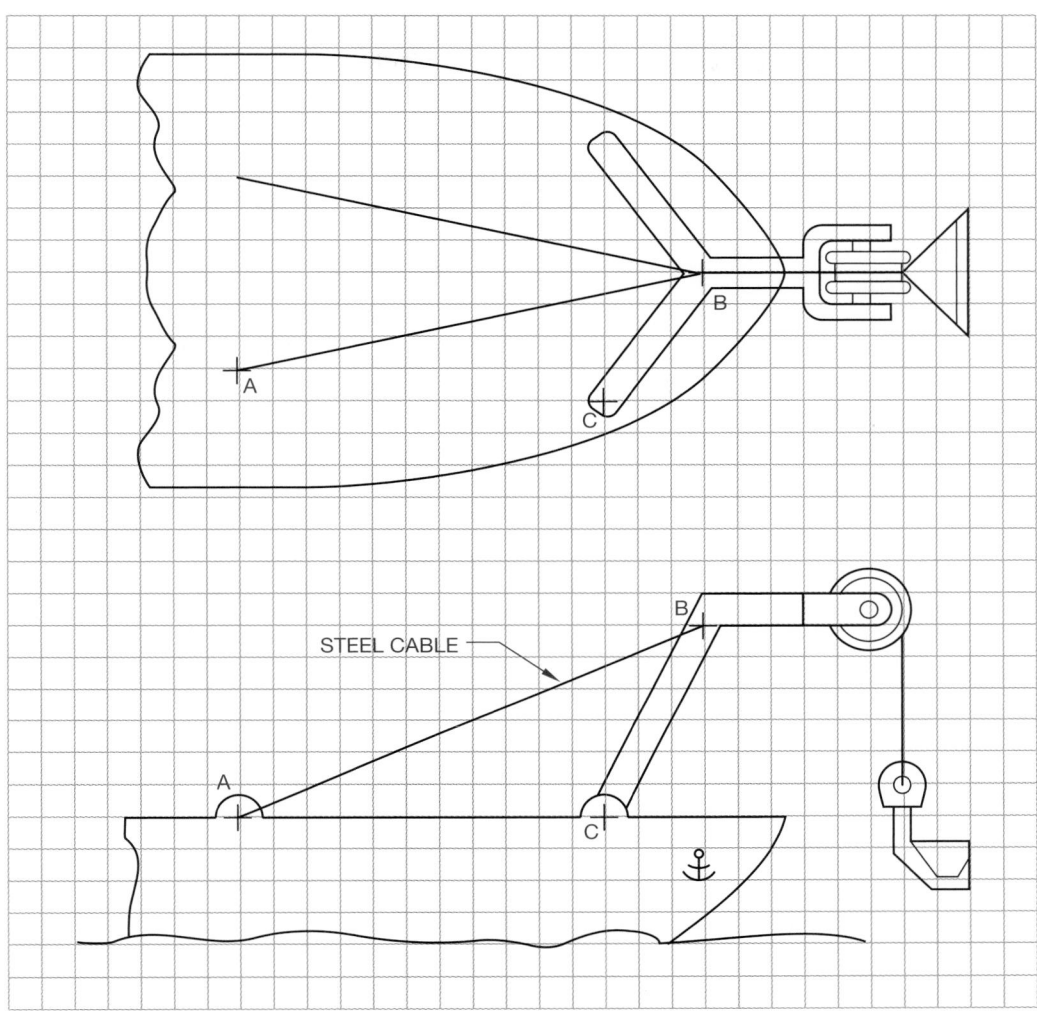

PROBLEM 10-3   Bearing, slope, and true length. From the given data, draw separate adjacent horizontal and vertical views that contain the following lines:

Line	Bearing	Slope	True Length
B-C	N 22° W	+30°	3 in.
R-S	S 75° E	−10°	4 in.
M-N	N 60° E	+15°	5 in.

PROBLEM 10-4   Slope and grade. In this problem, two sections of pipe that are connected together are given. Using only the centerlines of each pipe, draw the necessary views to answer the following questions. Leave all construction lines on the drawing. Identify on the drawing where the answers appear:

a. Identify and calculate the true length of pipe A-B.
b. Identify and calculate the slope in degrees of pipe B-C.
c. Identify and calculate the grade in percent of pipe A-B.
d. Identify and calculate the true angle between the pipes.

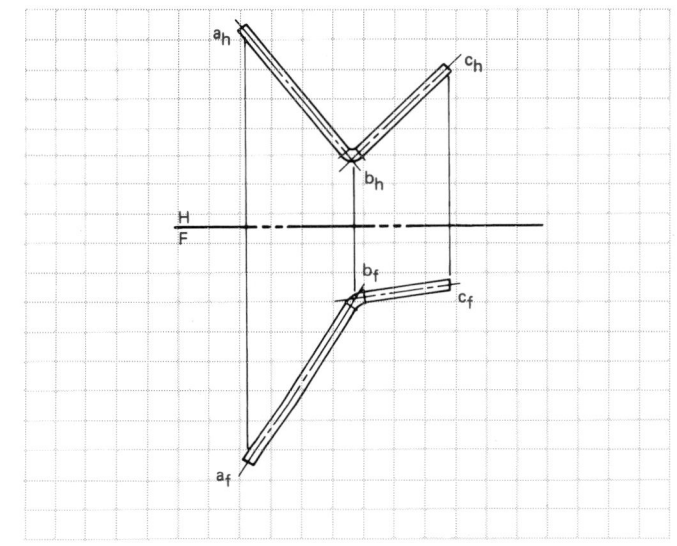

PROBLEM 10-5 True length, true angle. A designer's sketch, on the right, shows the adjacent front and horizontal views of the roof plane A-B-C-D and the location where the antenna is to be mounted on the roof plane. The top end of the antenna is to project 8' above the ridge of the house.

a. Identify and calculate the true length of the antenna in feet and inches.

b. Find the place where the bottom of the antenna touches the roof in the front view.

c. Create a view that will show the true angle made by the roof plane and then calculate the true angle in degrees.

d. Show in the horizontal and front views three equally spaced guy wires anchored *on the roof plane*. Two guy wires are attached 5' from the base of the antenna along the roof plane, and the third is attached at the point shown in the horizontal view. All three guy wires are anchored to the antenna at a distance of 6' above the ridge.

e. Identify and calculate the total length of guy wire used to stabilize the antenna to the nearest foot.

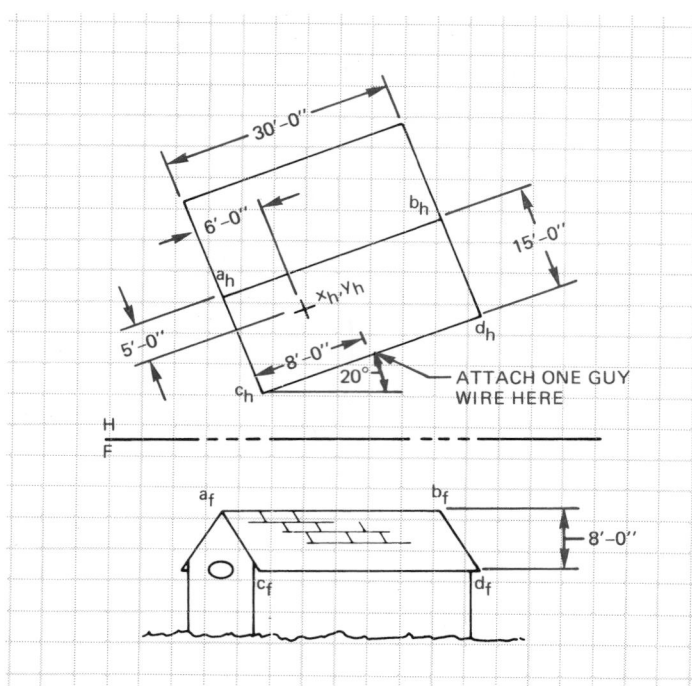

PROBLEM 10-6 True distance between point and a plane. Find the shortest distance between cable R-T and point E.

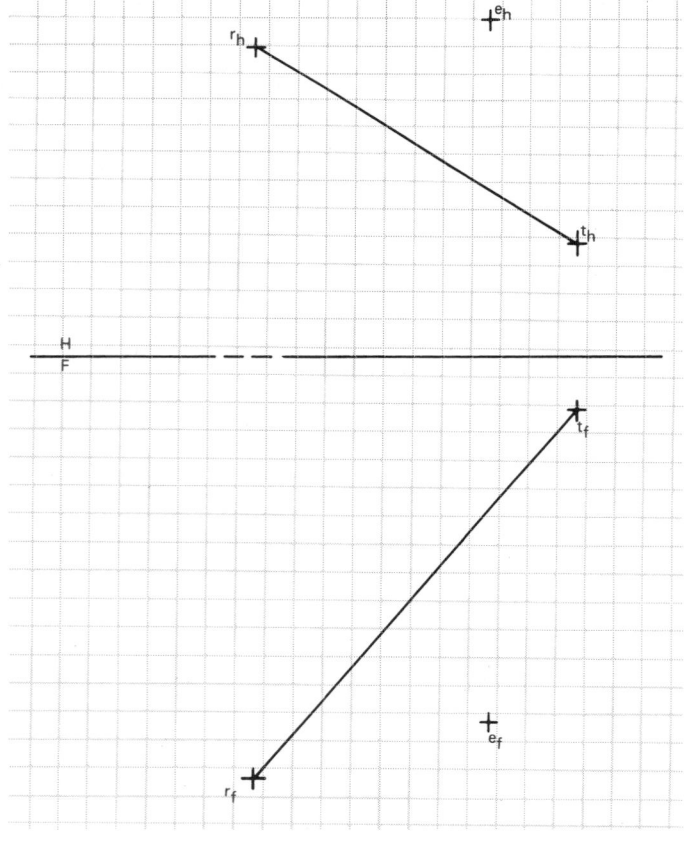

PROBLEM 10-7 Skew lines. It is important to keep the clearance between the two cable segments A-B and M-N of a diameter of ½" each at a minimum of one inch and a maximum of two inches. Create a view where you can verify how far apart they are in the setup. If they do not fit this criteria, readjust cable A-B in all views so it will be met. (Drawing scale: Each square = ¼")

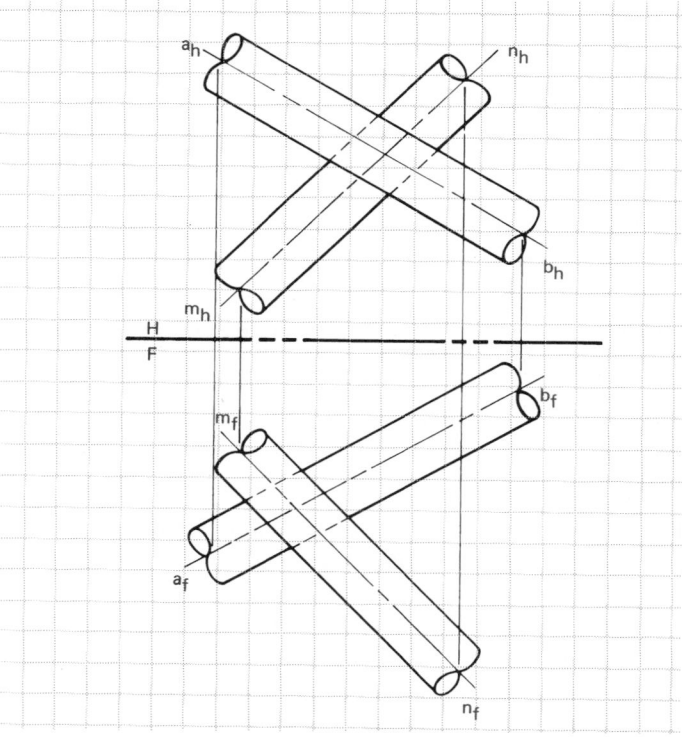

PROBLEM 10-8 True distance between parallel lines. Two parallel cable segments M-N and R-S are shown in adjacent horizontal and front views. Create a view that will show the true distance between the cable segments.

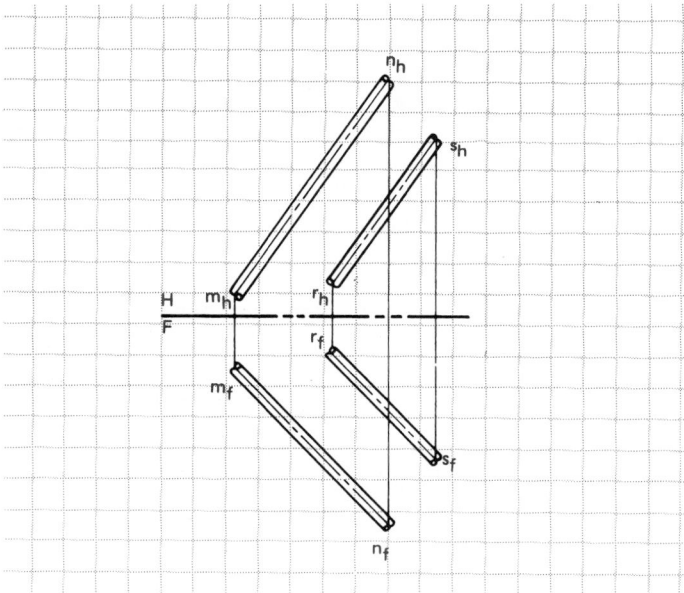

PROBLEM 10-10 Piercing point. A vein of ore has been determined by plane D-H-G. A mine tunnel, R-S, is being dug toward this vein. Create a view that will show the true distance of how much further the tunnel will need to be extended from point R to reach the vein. Calculate the distance to the nearest foot. (Drawing scale: Four squares = 200')

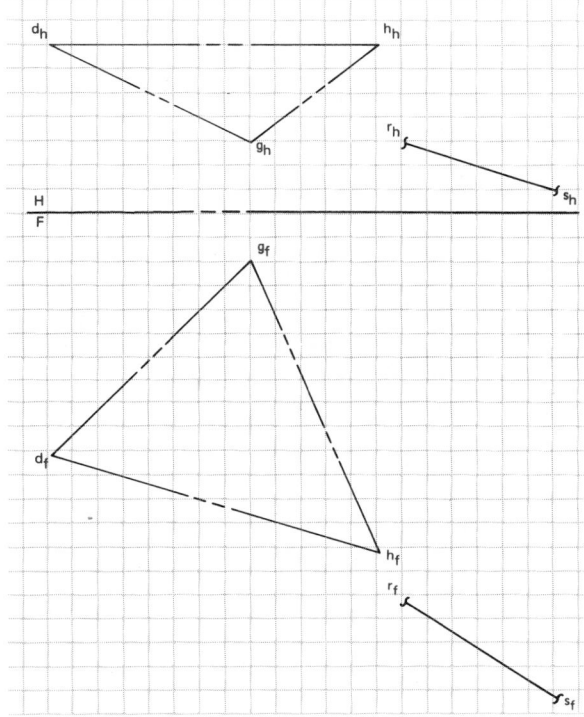

PROBLEM 10-9 True angles. Two parallel pipe corner segments, K-P and R-S, are shown in adjacent horizontal and front views. Create a view that will show the true angle of the pipe segments.

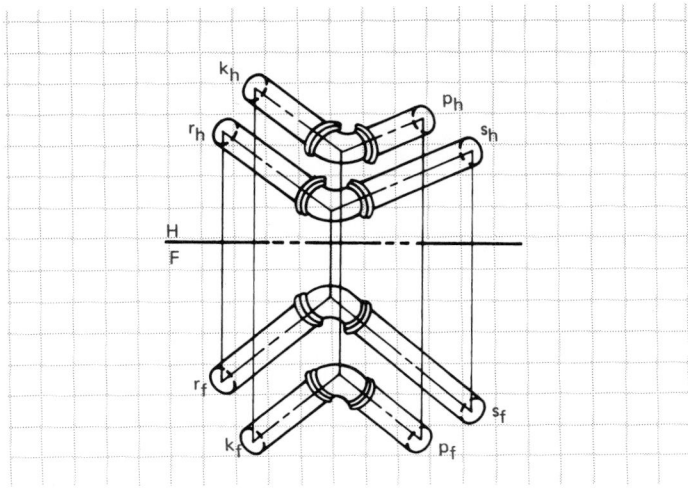

PROBLEM 10-11 Visibility. The adjacent front and horizontal projection planes of three cable segments A-B, M-N, and R-S are shown. Determine which cable segments are in front of which in each projection plane. Then, correctly complete the cable segment drawings by adding the missing object lines in each projection plane. (Drawing scale: One square= ¼")

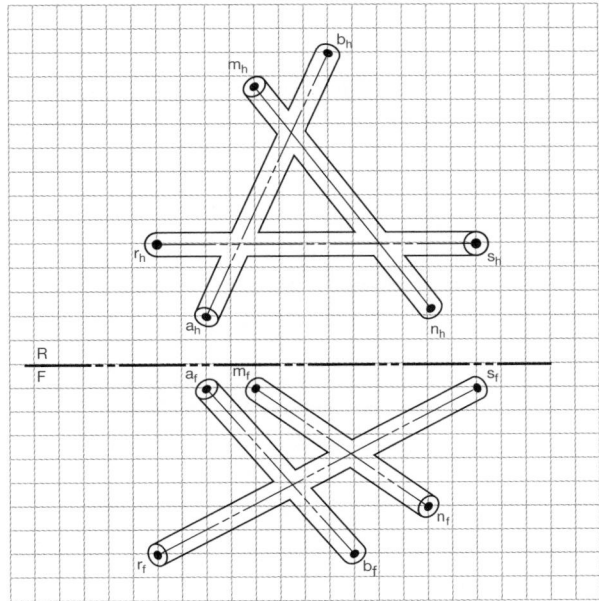

PROBLEM 10-12 True shapes and angles between planes. A designer's sketch of a roof plan of a residential house with specifications for size of the roof and chimney flashing detail is given. It is necessary to determine some information about the roof area and the amount of roof flashing needed to keep the building job on schedule. Use the sketch and detail information to develop the necessary projection planes that will supply the information needed. You need to:

a. Create the views that will show how many squares of roofing will be needed to cover the entire roof area. (100 ft² of roof surface = 1 square.) Calculate to the nearest square how much roofing will be needed to cover the roof, with any extra to be left for repairs.

b. Calculate how much 18" wide flashing will be needed to flash all the ridges, valleys, and hips to the nearest foot.

c. Create the views that will show all the pitches of the roof. Calculate each of the pitches in the correct pitch ratio.

d. Create the view that will show the ridge angle needed to bend the chimney flashing correctly. Calculate the ridge angle to the nearest degree.

HEIGHT ABOVE GROUND OF:
RIDGE = 24'–0"
EAVES = 17'–0"

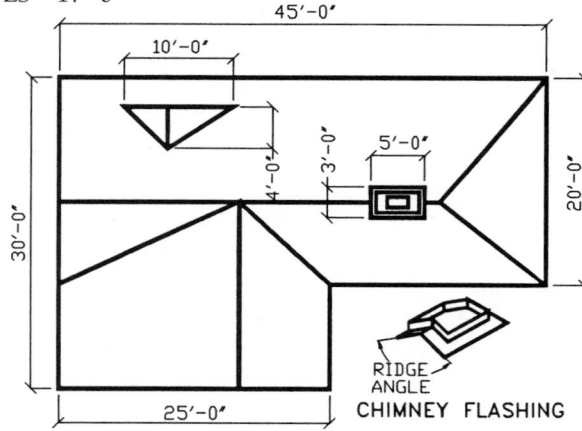

CHIMNEY FLASHING

PROBLEM 10-13 Dihedral angle. Given are two adjacent front and horizontal views of a laundry chute. Find the true angle between side A and side B. (Drawing scale: One square = ¼")

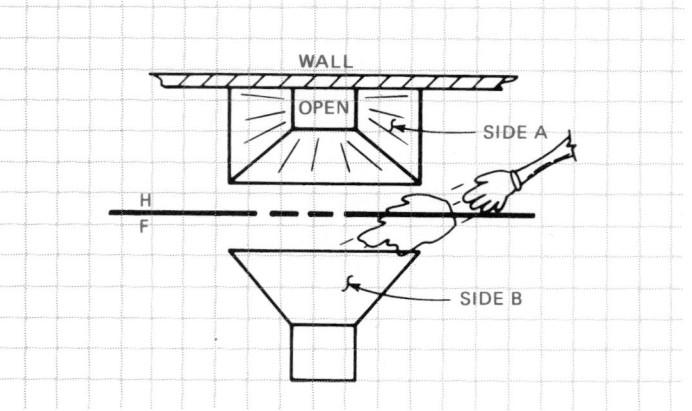

PROBLEM 10-14 True shapes and dihedral angle. This sketch of a sand collection hopper was given to you by an engineer. He asked you to do the following:

a. Create a view that will show the true shape of plane A including the hinged door.

b. Create a view that will show how far the top of the door, which is hinged at the bottom, will be from the bottom of the hopper when the door is hanging down in the open position. Show the measurement of this distance in feet and inches.

c. Create a view that will show the true angle between plane A and plane B. Show the measurement of this angle to the nearest degree.

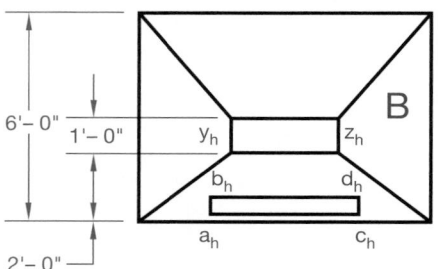

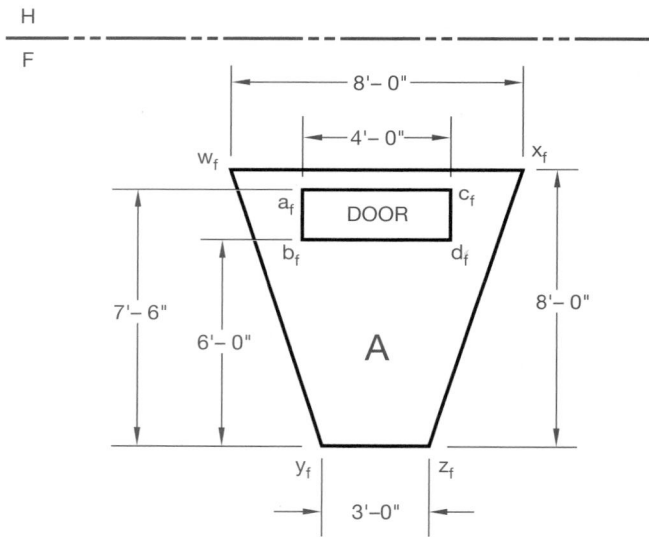

PROBLEM 10-15   Vectors. Draw the correct vector diagram to determine the resultant force of the concurrent coplanar forces below. Measure the resultant force to the nearest 10 lb.

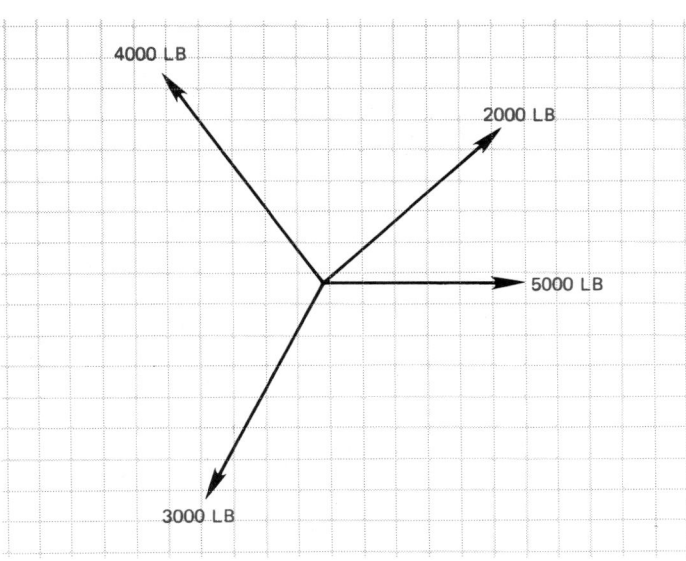

PROBLEM 10-16   Vectors. Given two adjacent front and horizontal views showing three concurrent noncoplanar forces, find the resultant force of vectors A-B, A-C, and A-D to the nearest pound.

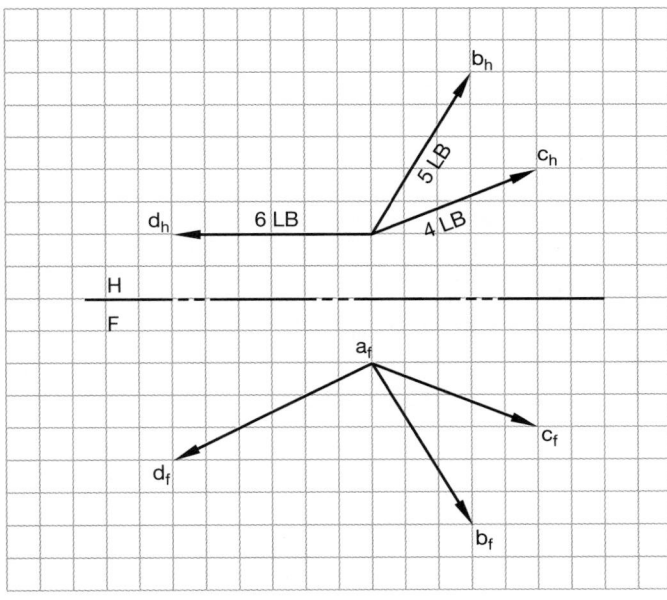

PROBLEM 10-17   Vectors. Given the front view showing the space diagram of three forces, find the tensions in members A-B and A-C acted upon by the given force.

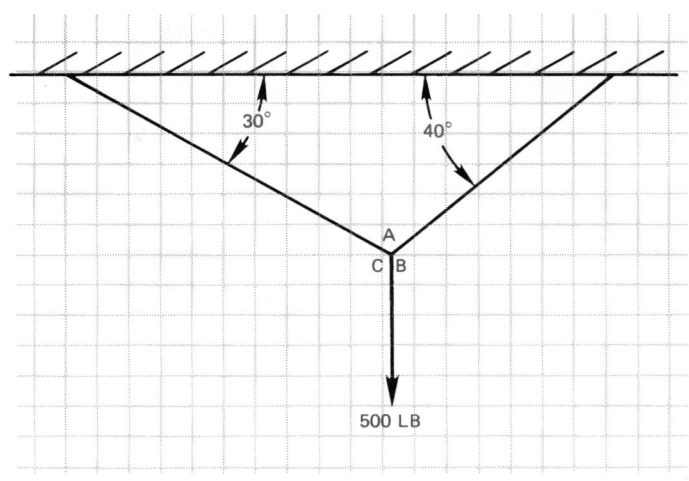

PROBLEM 10-18   Vectors. Given the front view showing a space diagram of three forces in equilibrium, find angles A and B to the nearest degree.

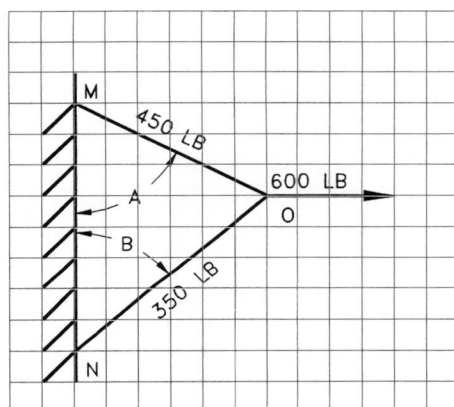

PROBLEM 10-19   Given the following information on sloping lines, determine their slope, slope angle, and percent of grade.

1. Line A-B: run = 50 feet; rise = 34 feet
2. Line C-D: run = 100 feet; rise = 25 feet
3. Line D-E: run = 148 yards; rise = 100 yards
4. Line F-G: run = 65 inches; rise = 0 inches
5. Line H-J: run = 100 meters; rise = 75 meters

CHAPTER

Manufacturing Materials and Processes

OBJECTIVES

After completing this chapter, you will:

■ Define and describe various manufacturing materials, material terminology, numbering systems, and material treatment.

■ Discuss casting processes and terminology.

■ Explain the forging process and terminology.

■ Describe manufacturing processes.

■ Define and draw the representation of various machined features.

■ Explain tool design and drafting practices.

■ Draw a basic machine tool.

■ Discuss the statistical process quality control assurance system.

■ Evaluate the results of an engineering and manufacturing problem.

■ Explain the use of computer-aided manufacturing (CAM) in today's industry.

■ Discuss robotics in industry.

■ Identify a variety of manufacturing processes used to create plastic products.

THE ENGINEERING DESIGN APPLICATION

Your company produces die cast aluminum parts. A customer has explained that any delay in parts shipment will require them to shut down a production line. It would be very costly. It is critical to your customer that certain features on their castings fall within specification limits, and the engineering department has decided to develop an early warning system in an attempt to prevent production delays.

You are asked to develop new quality control charts with two levels of control. The maximum allowable deviation is indicated by the upper and lower specification limits as specified by the dimensional tolerances on the part drawing. A mean value is established between the nominal dimension and dimensional limits, and these are set up as upper and lower control limits. The chart instructs the inspector to immediately notify the supervisor if the feature falls outside of control limits and to halt production if it is outside of specification limits. This provides a means of addressing problems before they can interfere with production, shifting the emphasis from revision to prevention. A sample chart with two levels of control is shown in Figure 11.1.

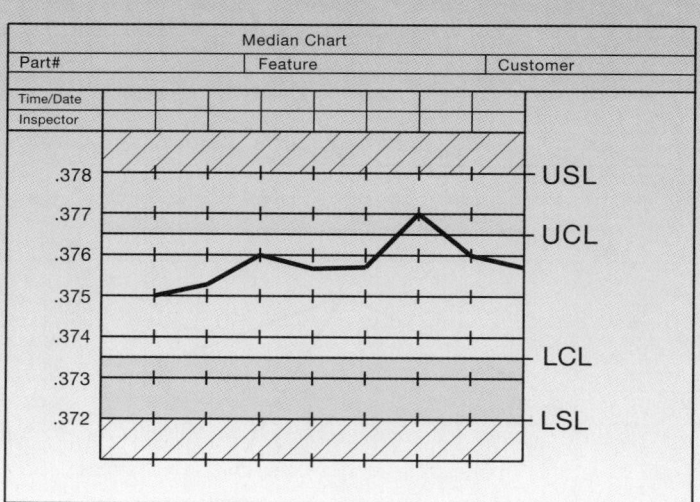

FIGURE 11.1 ■ Sample quality control chart with both control and specifications limits.

INTRODUCTION

A number of factors influence product manufacturing. Beginning with research and development (R & D), a product should be designed to meet a market demand, have good quality, and be economically produced. The sequence of product development begins with an idea and results in a marketable commodity, as shown in Figure 11.2.

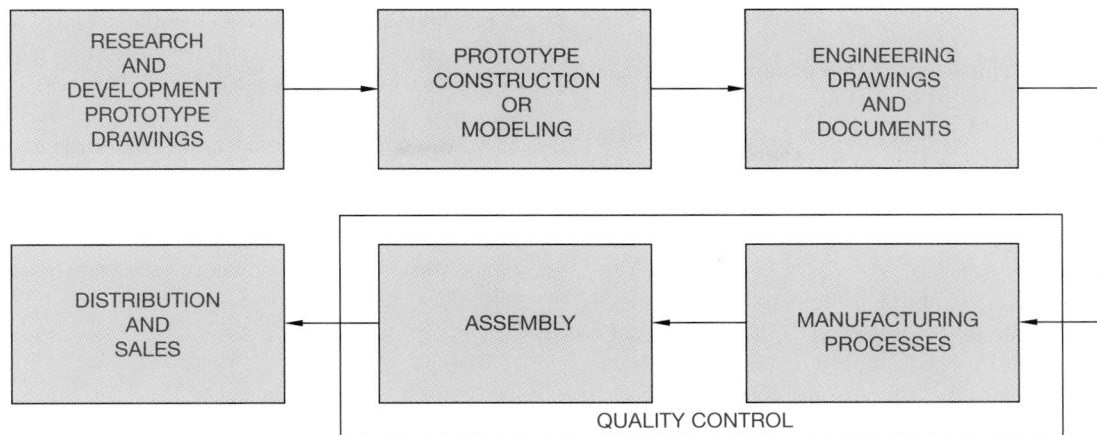

FIGURE 11.2 ■ Sequence of product development.

MANUFACTURING MATERIALS

There is a wide variety of materials available for product manufacturing that fall into three general categories: metal, plastic, and inorganic materials. Metals are classified as *ferrous, nonferrous,* and *alloys.* Ferrous metals contain iron, such as cast iron and steel. Nonferrous metals do not have iron content; for example, copper and aluminum. Alloys are a mixture of two or more metals.

Plastics or *Polymers* have two types of structure: thermoplastic and thermoset. *Thermoplastic* material may be heated and formed by pressure and, upon reheating, the shape can be changed. *Thermoset* plastics are formed into a permanent shape by heat and pressure and may not be altered by heating after curing. Plastics are molded into shape and only require machining for tight tolerance situations or when holes or other features are required that would be impractical to produce in a mold. It is common practice to machine some plastics for parts, such as gears and pinions.

Inorganic materials include carbon, ceramics, and composites. Carbon and graphite are classified together and have properties that allow molding by pressure. These materials have low *tensile strength* (ability to be stretched) and high compressive strength with increasing strength at increased temperatures. Ceramics are clay, glass, refractory, and inorganic cements. Ceramics are very hard, brittle materials that are resistant to heat, chemicals, and corrosion. Clay and glass materials have an amorphus structure, while *refractories* must be bonded together by applying temperatures. Due to their great heat resistance, refractories are used for high-temperature applications, such as furnace liners. *Composites* are two or more materials that are bonded together by adhesion. *Adhesion* is a force that holds together the molecules of unlike substances when the surfaces come in contact. These materials generally require carbide cutting tools or special methods for machining.

Ferrous Metals

The two main types of ferrous metals are cast iron and steel. These are the metals that contain iron. There are many classes of cast iron and steel that are created for specific types of application based on their composition.

Cast Iron

There are several classes of cast iron, including gray, white, chilled, alloy, malleable, and nodular cast iron. Cast iron is primarily an alloy of iron and 1.7% to 4.5% of carbon, with varying amounts of silicon, manganese, phosphorus, and sulfur.

Gray Cast Iron

Gray iron is a popular casting material for automotive cylinder blocks, machine tools, agricultural implements, and cast iron pipe. Gray cast iron is easily cast and machined. It contains 1.7% to 4.5% carbon and 1% to 3% silicon.

> **ANSI/ASTM** The American National Standards Institute (ANSI) and *American Society for Testing Materials* (ASTM) specifications A48–76 group gray cast iron into two classes: easy to manufacture (20A, 20B, 20C, 25A, 25B, 25C, 30A, 30B, 30C, 35A, 35B, 35C); and more difficult to manufacture (40B, 40C, 45B, 45C, 50B, 50C, 60B, 60C). The prefix denotes the minimum tensile strength in thousand pounds per square inch.

White Cast Iron

White cast iron is extremely hard, brittle, and has almost no ductility. *Ductility* is the ability to be stretched, drawn, or hammered thin without breaking. Caution should be exercised when using this material, because thin sections and sharp corners may be weak and the material is less resistant to impact uses. This cast iron is suited for products with more compressive strength requirements than gray cast iron (compare over 200,000 pounds per square inch [psi] with 65,000 to 160,000 psi). White cast iron is used where high wear resistance is required.

Chilled Cast Iron

When gray iron castings are chilled rapidly, an outer surface of white cast iron results. This material has the internal characteristics of gray cast iron and the surface advantage of white cast iron.

Alloy Cast Iron

Elements such as nickel, chromium, molybdenum, copper, or manganese may be alloyed with cast iron to increase the properties of strength, wear resistance, corrosion resistance, or heat resistance. Alloy iron castings are commonly used for such items as pistons, crankcases, brake drums, and crushing machinery.

Malleable Cast Iron

The term *malleable* means the ability to be hammered or pressed into shape without breaking. Malleable cast iron is produced by heat-treating white cast iron. The result is a stronger, more ductile, and shock-resistant cast iron that is easily machinable.

> **ANSI/ASTM** The specifications for malleable cast iron are found in ANSI/ASTM A47–77.

Nodular Cast Iron

Special processing procedures, along with the addition of magnesium or cerium bearing alloys, result in a cast iron with spherical-shaped graphite rather than flakes, as in gray cast iron. The results are iron castings with greater strength and ductility. Nodular cast iron may be chilled to form a wear-resistant surface ideal for use in crankshafts, anvils, wrenches, or heavy-use levers.

Steel

Steel is an alloy of iron containing 0.8% to 1.5% carbon. Steel is a readily available material that may be worked in either a heated or cooled state. The properties of steel can be changed by altering the carbon content and heat treating. *Mild steel* (MS) is low in carbon (less than 0.3%), and is commonly used for forged and machined parts, but may not be hardened. *Medium carbon steel* (0.3% to 0.6% carbon) is harder than mild steel yet remains easy to forge and machine. *High carbon steel* (0.6% to 1.50% carbon) may be hardened by heat treating, but is difficult to forge, machine, or weld.

Hot-rolled steel (HRS) characterizes steel that is formed into shape by pressure between rollers or by forging when in a red-hot state. When in this hot condition, the steel is easier to form than when it is cold. An added advantage of hot forming is a consistency in the grain structure of the steel, which results in a stronger, more ductile metal. The surface of hot-rolled steel is rough, with a blue-black oxide buildup. The term *cold-rolled steel* (CRS) implies the additional forming of steel after initial hot rolling. The cold-rolling process is used to clean up hot-formed steel, provide a smooth, clean surface, ensure dimensional accuracy, and increase the tensile strength of the finished product.

Steel alloys are used to increase such properties as hardness, strength, corrosion resistance, heat resistance, and wear resistance. Chromium steel is the basis for stainless steel and is used where corrosion and wear resistance is required. Manganese alloyed with steel is a purifying element that adds strength for parts that must be shock and wear resistant. Molybdenum is added to steel when the product must retain strength and wear resistance at high temperatures. When tungsten is added to steel, the result is a material that is very hard and ideal for use in cutting tools. Tool steels are high in carbon and/or alloy content so that the steel will hold an edge when cutting other materials. When the cutting tool requires deep cutting at high speed, then the alloy and hardness characteristics are improved for a classification known as *high-speed steel*. Vanadium alloy is used when a tough, strong, nonbrittle material is required.

Steel castings are used for machine parts where the use requires heavy loads and the ability to withstand shock. These castings are generally stronger and tougher than cast iron. Steel castings have uses as turbine wheels, forging presses, gears, machinery parts, and railroad car frames.

Stainless Steel

Stainless steels are high-alloy chromium steels that have excellent corrosion resistance. In general, stainless steels contain at least 10.5% chromium, with some classifications of stainless steel having between 4% and 30% chromium. In addition to corrosion resistance properties, stainless steel can have oxidation and heat resistance, and can have very high strength. Stainless steel is commonly used for restaurant and hospital equipment and for architectural and marine applications. The high strength-to-weight ratio also makes stainless steel a good material for some aircraft applications.

Stainless steel is known for its natural luster and shine that makes it look like chrome. It is also possible to add color to stainless steel, which is used for architectural products such as roofing, hardware, furniture, kitchen, and bathroom fixtures.

The American Iron and Steel Institute (AISI) identifies stainless steels with a system of 200, 300, or 400 series numbers. The 200 series stainless steels contain chromium, nickel, and manganese; the 300 series steels have chromium and nickel; and the 400 series steels are straight-chromium stainless steels.

Steel Numbering Systems

AISI/SAE The American Iron and Steel Institute (AISI) and the Society of Automotive Engineers (SAE) provide similar steel numbering systems. Steels are identified by four numbers, except for some chromium steels, which have five numbers. For a steel with the identification SAE 1020, the first two numbers (10) identify the type of steel and the last two numbers (20) specify the approximate amount of carbon in hundredths of a percent (0.20% carbon). The letter *L* or *B* may be placed between the first and second pair of numbers. When this is done, the *L* means that lead is added to improve machinability, and the *B* identifies a boron steel. The prefix *E* means that the steel is made using the electric furnace method. The prefix *H* indicates that the steel is produced to hardenability limits. Steel that is degassed and deoxidized before solidification is referred to as *killed steel* and is used for forging, heat treating, and difficult stampings. Steel that is cast with little or no degasification is known as *rimmed steel,* and has applications where sheets, strips, rods, and wires with excellent surface finish or drawing requirements are needed. General applications of SAE steels are shown in Appendix B, Table 25. For a more in-depth analysis of steel and other metals, refer to the *Machinery's Handbook.** Additional information about numbering systems is provided later in this chapter.

*Erick Oberg, Franklin D. Jones, and Holbrook L. Horton, *Machinery's Handbook,* 25th ed. (New York: Industrial Press Inc., 1996).

Hardening of Steel

The properties of steel may be altered by *heat treating*. Heat treating is a process of heating and cooling steel using specific, controlled conditions and techniques. When steel is initially formed and allowed to cool naturally, it is fairly soft. *Normalizing* is a process of heating the steel to a specified temperature and then allowing the material to cool slowly by air, which brings the steel to a normal state. In order to harden the steel, the metal is first heated to a specified temperature which, varies with different steels. Next the steel is quenched, which means to cool suddenly by plunging into water, oil, or other liquid. Steel may also be *case hardened* using a process known as *carburization*. Case hardening refers to the hardening of the surface layer of the metal. Carburization is a process where carbon is introduced into the metal by heating to a specified temperature range while in contact with a solid, liquid, or gas material consisting of carbon. This process is often followed by quenching to enhance the hardening process. *Tempering* is a process of reheating a normalized or hardened steel through a controlled process of heating the metal to a specified temperature, followed by cooling at a predetermined rate to achieve certain hardening characteristics. For example, the tip of a tool may be hardened while the balance of the tool remains unchanged.

Under certain heating and cooling conditions and techniques, steel may also be softened using a process known as *annealing*.

Hardness Testing

There are several methods of checking material hardness. The techniques have common characteristics based on the depth of penetration of a measuring device or other mechanical systems that evaluate hardness. The Brinell and Rockwell hardness tests are popular. The Brinell test is performed by placing a known load, using a ball of a specified diameter, in contact with the material surface. The diameter of the resulting impression in the material is measured and the Brinell Hardness Number (BHN) is then calculated. The Rockwell hardness test is performed using a machine that measures hardness by determining the depth of penetration of a spherical-shaped device under controlled conditions. There are several Rockwell hardness scales depending on the type of material, the type of penetrator, and the load applied to the device. A general or specific note on a drawing that requires a hardness specification may read: CASE HARDEN 58 PER ROCKWELL "C" SCALE. For additional information, refer to the *Machinery's Handbook*.

Nonferrous Metals

Metals that do not contain iron have properties that are better suited for certain applications where steel may not be appropriate.

Aluminum

Aluminum is corrosion resistant, lightweight, easily cast, conductive of heat and electricity, may be easily *extruded*, and is very malleable. Extruding is shaping the metal by forcing it through a die. Pure aluminum is seldom used, but alloying it with other elements provides materials that have an extensive variety of applications. Some aluminum alloys lose strength at temperatures generally above 121° C (250° F), while they gain strength at cold temperatures. There are a variety of aluminum alloy numerical designations used. A two- or three-digit number is used, and the first digit indicates the alloy type, as follows: 1 = 99% pure, 2 = copper, 3 = manganese, 4 = silicon, 5 = magnesium, 6 = magnesium and silicon, 7 = zinc, 8 = other. For designations above 99%, the last two digits of the code are the amount over 99%. For example, 1030 means 99.30% aluminum. The second digit is any number between 0 and 9 where 0 means no control of specific impurities and numbers 1 through 9 identify control of individual impurities.

Copper Alloys

Copper is easily rolled and drawn into wire, has excellent corrosion resistance, is a great electrical conductor, and has better ductility than any metal except for silver and gold. Copper is alloyed with many different metals for specific advantages, which include improved hardness, casting ability, machinability, corrosion resistance, elastic properties, and lower cost.

Brass

Brass is a widely used alloy of copper and zinc. Its properties include corrosion resistance, strength, and ductility. For most commercial applications, brass has about a 90% copper and 10% zinc content. Brass may be manufactured by any number of processes including casting, forging, stamping, or drawing. Its uses include valves, plumbing pipe and fittings, and radiator cores. Brass with greater zinc content may be used for applications requiring greater ductility, such as cartridge cases, sheet metal, or tubing.

Bronze

Bronze is an alloy of copper and tin. Tin in small quantities adds hardness and increases wear resistance. Tin content in coins and medallions, for example, ranges from 4% to 8%. Increasing amounts of tin also improves the hardness and wear resistance of the material, but causes brittleness. Phosphorus added to bronze (phosphor bronze) increases its casting ability and aids in the production of more solid castings, which is important for thin shapes. Other materials, such as lead, aluminum, iron, and nickel, may be added to copper for specific applications.

Precious and Other Specialty Metals

Precious metals include gold, silver, and platinum. These metals are valuable because they are rare, costly to produce, and have specific properties that influence use in certain applications.

Gold

Gold for coins and jewelry is commonly hardened by adding copper. Gold coins, for example, are 90% gold and 10% copper. The term *carat* is used to refer to the purity of gold, where $\frac{1}{24}$ gold is one carat. Therefore, $24 \times \frac{1}{24}$ or 24 carats is pure gold. Fourteen-carat gold, for example, is $\frac{14}{24}$ gold and $\frac{10}{24}$ copper. Gold is extremely malleable, corrosion resistant, and is the best conductor of electricity. In addition to use in jewelry and coins, gold is used as a conductor in some electronic circuitry applications.

Gold is also used in applications where resistance to chemical corrosion is required.

Silver

Silver is alloyed with 8% to 10% copper for use in jewelry and coins. Sterling silver for use in such items as eating utensils and other household items is $^{925}/_{1000}$ silver. Silver is easy to shape, cast, or form, and the finished product may be polished to a high-luster finish. Silver has uses similar to gold because of corrosion resistance and ability to conduct electricity.

Platinum

Platinum is rarer and more expensive than gold. Industrial uses include applications where corrosion resistance and a high melting point are required. Platinum is used in catalytic converters because it has the unique ability to react with and reduce carbon monoxide and other harmful exhaust emissions in automobiles. The high melting point of platinum makes it desirable in certain aerospace applications.

Columbium

Columbium is used in nuclear reactors because it has a very high melting point, 2403°C (4380°F), and is resistant to radiation.

Titanium

Titanium has many uses in the aerospace and jet aircraft industries because it has the strength of steel, the approximate weight of aluminum, and is resistant to corrosion and temperatures up to 427°C (800°F).

Tungsten

Tungsten has been used extensively as the filament in light bulbs because of its ability to be drawn into very fine wire and its high melting point. Tungsten, carbon, and cobalt are formed together under heat and pressure to create tungsten carbide, the hardest manmade material. Tungsten carbide is used to make cutting tools for any type of manufacturing application. Tungsten carbide saw blade inserts are used in saws for carpentry so the cutting edge will last longer. Such blades make a finer and faster cut than plain steel saw blades.

A Unified Numbering System for Metals and Alloys

Many numbering systems have been developed for the identification of metals. The organizations that developed these numbering systems include the American Iron and Steel Institute (AISI), Society of Automotive Engineers (SAE), American Society for Testing Materials (ASTM), American National Standards Institute (ANSI), Steel Founders Society of America, American Society of Mechanical Engineers (ASME), American Welding Society (AWS), Aluminum Association, and Copper Development Association, as well as military specifications by the U.S. Department of Defense and federal specifications by the General Accounting Office.

A combined numbering system created by the ASTM and the SAE was established in an effort to coordinate all the different numbering systems into one system. This system avoids the possibility that the same number might be used for two different metals. This combined system is the Unified Numbering System (UNS). The UNS is an identification numbering system for commercial metals and alloys; it does not provide metal and alloy specifications. The UNS system is divided into the following categories.

UNS Series	Metal
Nonferrous Metals and Alloys	
A00001 to A99999	Aluminum and aluminum alloys
C00001 to C99999	Copper and copper alloys
E00001 to E99999	Rare-earth and rare-earth-like metals and alloys
L00001 to L99999	Low-melting metals and alloys
M00001 to M99999	Miscellaneous metals and alloys
P00001 to P99999	Precious metals and alloys
R00001 to R99999	Reactive and refractory metals and alloys
Z00001 to Z99999	Zinc and zinc alloys
Ferrous Metals and Alloys	
D00001 to D99999	Specified mechanical properties steels
F00001 to F99999	Cast irons
G00001 to G99999	AISI and SAE carbon and alloy steels, excluding tool steels
H00001 to H99999	AISI H-classification steels
J00001 to J99999	Cast steels, excluding tool steels
K00001 to K99999	Miscellaneous steels and ferrous alloys
S00001 to S99999	Stainless steels
T00001 to T99999	Tool steels

The prefix letters of the UNS system often match the type of metal being identified. For example, A for aluminum, C for copper, and T for tool steels. Elements of the UNS numbers typically match numbers provided by other systems; for example, SAE1030 is G10300 in the UNS system.

PLASTICS

A general definition of *plastic* is any complex, organic, polymerized compound capable of being formed into a desired shape by molding, casting, or spinning. Plastic retains its shape under ordinary conditions of temperature. *Polymerization* is a process of joining two or more molecules to form a more complex molecule with physical properties that are different from the original molecules. The terms *plastic* and *polymer* are often used to mean the same thing. Plastics can range in any state from liquid to solid. The main elements of plastic are generally common petroleum products, crude oil, and natural gas.

There are many types of plastics that are available for use in the design and manufacture of a product. These plastics generally fall into two main categories, as you learned earlier; these are thermoplastics and thermosets. The *thermoplastics* can be heated and formed by pressure, and the shape can change when reheated. *Thermosets* are formed into shape by heat and pressure and cannot be changed into a different shape after curing. Most plastic products are made with thermoplastics, because they are easy to make into shapes by heating, forming, and cooling. Thermoset plastics are the choice when the end product is used in an application where heat exists, such as the distributor cap and other plastic parts found on or near the engine of your car. In addition to thermoplastics and thermosets, there are elastomers. *Elastomers* are polymer-based materials that have elastic qualities not found in the two types of plastics previously defined. Elastomers are generally able to be stretched at least equal to their original length and return to their original length after stretching.

Thermoplastics

Although there are thousands of different thermoplastic combinations, the following gives you some of the more commonly used alternatives. You may recognize some of them by their acronyms, such as PVC.

Acetal—Acetal is a rigid thermoplastic that has good corrosion resistance and machinability. While it will burn, it is good in applications where friction, fatigue, toughness, and tensile strength are factors. Some applications include gears, bushings, bearings, and products that come in contact with chemicals or petroleum.

Acrylic—Acrylics are used when a transparent plastic is needed in clear or a variety of colors. Acrylics are commonly used in products that you see through or through which light passes because, in addition to transparency, they are scratch and abrasion resistant, and hold up well in the weather. Examples of use include windows, light fixtures, and lenses.

Acrylic-styrene-acrylonitrile (ASA)—ASA has very good weatherability for use as siding, pools and spas, exterior car and marine parts, outdoor furniture, and garden equipment.

Acrylonitrile-butadiene-styrene (ABS)—ABS is one of the most commonly used plastics because of its excellent impact strength, reasonable cost, and ease of processing. ABS also has good dimensional stability, temperature resistance above 212°F (100°C), and chemical and electrical resistance. ABS is commonly used in products that you see daily, such as electronics enclosures, knobs, handles, and appliance parts.

Cellulose—There are five versions of cellulose, they are nitrate, acetate, butyrate, propionate, and ethyl cellulose.

■ *Nitrate* is tough, but very flammable, explosive, and difficult to process. Nitrate is commonly used to make products such as photo film, combs, brushes, and buttons.

■ *Acetate* is not explosive, but it is slightly flammable, is not solvent resistant, and becomes brittle with age. Acetate has the advantage of being transparent, can be made in bright colors, is tough, and is easy to process. It is often used for transparency film, magnetic tape, knobs, and sunglass frames.

■ *Butyrate* is similar to acetate, but is used in applications where moisture resistance is needed. Butyrate is used for exterior light fixtures, handles, film, and outside products.

■ *Propionate* has good resistance to weathering; is tough and impact resistant; and has reduced brittleness with age. Propionate is used for flashlights, automotive parts, small electronics cases, and pens.

■ *Ethyl cellulose* has very high shock resistance and durability at low temperatures, but has poor weatherability.

Fluoroplastics—There are four types of fluoroplastics that have similar characteristics. These plastics are very resistant to chemicals, friction, and moisture. They have excellent dimensional stability for use as wire coating and insulation, nonstick surfaces, chemical containers, O-rings, and tubing.

Ionomers—Ionomers are very tough; resistant to abrasion, stress, cold, and electricity; and very transparent. They are commonly used for cold food containers, other packaging, and film.

Liquid crystal polymers—This plastic can be made very thin with excellent temperature, chemical, and electrical resistance. Uses include cookware and electrical products.

Methyl pentenes—Methyl pentenes have excellent heat and electrical resistance, and are very transparent. These plastics are used for items such as medical containers and products, cooking and cosmetic containers that require transparency, and pipe and tubing.

Polyallomers—This plastic is rigid with very high impact and stress fracture resistance at temperatures between –40° to 210°F (–40° to 99°C). Polyallomers are used when constant bending is required in the function of the material design.

Polyamide (nylon)—Commonly called nylon, this plastic is tough; abrasion, heat, and friction resistant; and strong. Nylon is corrosion resistant to most chemicals, but is not as dimensionally stable as other plastics. Nylon is used for combs, brushes, tubing, gears, cams, and stocks.

Polyarylate—This material is impact, weather, electrical, and extremely fire resistant. Typical uses include electrical insulators, cookware, and other options where heat is an issue.

Polycarbonates—Excellent heat resistance, impact strength, dimensional stability, and transparency are positive characteristics of this plastic. Additionally, this material does not stain or corrode, but has moderate chemical resistance. Common uses include food containers, power tool housings, outside light fixtures, and appliance and cooking parts.

Polyetheretherketone (PEEK)—PEEK has excellent heat, fire, abrasion, and fatigue resistance qualities. Typical uses include high electrical components, aircraft parts, engine parts, and medical products.

Polyethylene—This plastic has excellent chemical resistance and has properties that make it good to use for slippery or nonstick surfaces. Polyethylene is a common plastic that is used for chemical, petroleum, and food containers; plastic bags; pipe fittings; and wire insulation.

Polyimides—This plastic has excellent impact strength, wear resistance, and very high heat resistance, but it is difficult to produce. Products made from this plastic include bearings, bushings, gears, piston rings, and valves.

Polyphenylene oxide (PPO)—PPO has an extensive range of temperature use from –275° to 375°F (–170° to 191°C), and is fire retardant and chemical resistant. Applications include containers that require super-heated steam, pipe and fittings, and electrical insulators.

Polyphenylene sulfide (PPS)—PPS has the same characteristics as PPO, but it is easier to manufacture.

Polypropylene—This is an inexpensive plastic to produce and has many desirable properties including heat, chemical, scratch, and moisture resistance. It is also resistant to continuous bending applications. Products include appliance parts, hinges, cabinets, and storage containers.

Polystyrene—This plastic is inexpensive and easy to manufacture, has excellent transparency, and is very rigid. However, it can be brittle and has poor impact, weather, and chemical resistance. Products include model kits, plastic glass, lenses, eating utensils, and containers.

Polysulfones—This material resists electricity and some chemicals, but can be damaged by certain hydrocarbons. While somewhat difficult to manufacture, it does have good structural applications at high temperatures, and can be made in several colors. Common applications are hot water products, pump impellers, and engine parts.

Polyvinylchloride (PVC)—PVC is one of the most common products found for use as plastic pipes and vinyl house siding because of its ability to resist chemicals and the weather.

Thermoplastic polyesters—There are two types of this plastic that exhibit strength and good electrical, stress, and chemical resistance. Common uses include electrical insulators, packaging, automobile parts, and cooking and chemical use products.

Thermoplastic rubbers (TPR)—This resilient material has uses where tough, chemical-resistant plastic is needed. Uses include tires, toys, gaskets, and sports products.

Thermosets

Thermoset plastics make up only about 15% of the plastics used because they are more expensive to produce; they are generally more brittle than thermoplastics; and once they are molded into shape, they cannot be remelted. However, their use is important in products that require a rigid and harder plastic than thermoplastic materials, and in applications where heat could melt thermoplastics. The following provides information about common thermoset plastics.

Alkyds—These plastics can be used in molding processes, but they are generally used as paint bases.

Melamine formaldehyde—This is a rigid thermoset plastic that is easily molded, economical, nontoxic, tough, and abrasion and temperature resistant. Common uses include electrical devices, surface laminates, plastic dishes, cookware, and containers.

Phenolics—The use of this material dates back to the late 1800s. This plastic is hard and rigid, has good compression strength, is tough, and does not absorb moisture; but it is brittle. Phenolic plastics are commonly used for the manufacture of electrical switches and insulators, electronics circuit boards, distributor caps, and binding material and adhesive.

Unsaturated polyesters—It is common to use this plastic for reinforced composites, also known as reinforced thermoset plastics (RTP). Typical uses are for boat and recreational vehicle construction, automobiles, fishing rods, tanks, and other structural products.

Urea formaldehyde—These plastics are used for many of the same applications as the previously described plastics, but they do not hold up to sunlight exposure. Common uses are for construction adhesives; as well as limited internal plastic electrical devices are made.

Elastomers

The two main types of plastics have been introduced as thermoplastics and thermosets. However, there are elastomers, which are types of plastics that are elastic, much like rubber. Elastomers can be referred to as synthetic rubbers. Synthetic rubbers produce almost twice as many products as natural rubber. *Natural rubber* is a material that starts as the sap from some trees. Natural rubber and many synthetic rubbers are processed by combining with adhesives and using a process called vulcanization. *Vulcanization* is basically the heating of the material in a steel mold that forms the desired shape. The following provides you with information about the most commonly used elastomers.

Butyl rubber—This material has a low air penetration ability and very good resistance to ozone and aging, but has poor petroleum resistance. Common uses include tire tubes and puncture-proof tire liners.

Chloroprene rubber (Neoprene)—The trade name Neoprene was the first commercial synthetic rubber. This material has better weather, sunlight, and petroleum resistance than natural rubber. It is also very flame resistant, but does not resist electricity. Common uses include automotive hoses and other products where heat is found, gaskets, seals, and conveyor belts.

Chlorosulfonated polyethylene (CSM)—CSM has excellent chemical, weather, heat, electrical, and abrasion resistance. It is typically used in chemical tank liners and electrical resistors.

Epichlorohydrin rubber (ECO)—This material has great petroleum resistance at very low temperatures. For this reason, it is used in cold weather applications such as snow handling equipment and vehicles.

Ethylene proplene rubber (EPM) and ethylene proplene diene monomer (EPDM)—This is a family of materials that have excellent weather, electrical, aging, and good heat resistance. These materials are used for weather stripping, wire insulation, conveyor belts, and many outdoor products.

Fluoroelastomers (FPM)—FPM materials have excellent chemical and solvent resistance up to 400°F (204°C). FPM is expensive to produce, so it is used only when its positive characteristics are needed.

Nitrile rubber—This material resists swelling when immersed in petroleum. Nitrile rubber is used for any application involving fuels and hydraulic fluid, such as hoses, gaskets, O-rings, and shoe soles.

Polyacrylic rubber (ABR)—This material is able to resist hot oils and solvents. ABR is commonly used in situations such as transmission seals where it is submerged in oil.

Polybutadiene—This material has qualities that are similar to those of natural rubber. It is commonly mixed with other rubbers to improve tear resistance.

Polyisoprene—This material was developed during World War II to help with a shortage of natural rubber, and it has the same chemical structure as natural rubber. However, this synthetic is more expensive to produce.

Polysulfide rubber—The advantage of this rubber is that it is petroleum, solvent, gas, moisture, weather, and age resistant. The disadvantage is that it is low in tensile strength, resilience, and tear resistance. Applications include caulking and putty, sealants, and castings.

Polyurethane—This material has the ability to act like rubber or hard plastic. Due to the combined rubber and hard characteristics, products include rollers, wear pads, furniture, and springs. In addition to these products, polyurethene is used to make foam insulation and floor coverings.

Silicones—This material has a wide range of makeup from liquid to solid. Liquid and semiliquid forms are used for lubricants. Harder forms are used where nonstick surfaces are required.

Styrene butadiene rubber (SBR)—SBR is very economical to produce and is heavily used for tires, hoses, belts, and mounts.

Thermoplastic elastomers (TPE)—The other elastomers typically require the fairly expensive vulcanization process to produce. TPE, however, is able to be processed with injection molding just like other thermoplastics, and any scraps can be reused. There are different TPEs, and in general, they are less flexible than other types of rubber.

MANUFACTURING PROCESSES

Casting, forging, and machining processes are used extensively in the manufacturing industry. It is a good idea for the entry-level drafter or pre-engineer to be generally familiar with types of casting, forging, and machining processes, and to know how to prepare related drawings. Any number of process methods may be used by industry. For this reason, it is best for the beginning drafting technician to remain flexible and adapt to the standards and techniques used by the specific company. As the drafting technician gains knowledge of company products, processes, and design goals, he or she may begin to produce designs. It is common for a drafter to become a designer after three years of practical experience.

Castings

Castings are the end result of a process called founding. *Founding,* or *casting,* as the process is commonly called, is the pouring of molten metal into a hollow or wax-filled mold. The mold is made in the shape of the desired casting. There are several casting methods used in industry. The results of some of the processes are castings that are made to very close tolerances and with smooth finished surfaces. In the simplest terms, castings are made in three separate steps:

1. A pattern that is the same shape as the desired finished product is constructed.

2. Using the pattern as a guide, a mold is made by packing sand or other material around the pattern.

3. When the pattern is removed from the mold, molten metal is poured into the hollow cavity. After the molten metal solidifies, the surrounding material is removed and the casting is ready for cleanup or machining operations.

Sand Casting

Sand casting is the most commonly used method of making castings. There are two general types of sand castings: green sand and dry sand molding. *Green sand* is a specially refined sand that is mixed with specific moisture, clay, and resin, which work as binding agents during the molding and pouring procedures. New sand is light brown in color; the term "green sand" refers to the moisture content. In the *dry sand* molding process, the sand does not have any moisture content. The sand is bonded together with specially formulated resins. The end result of the green sand or the dry sand molds is the same.

Sand castings are made by pounding or pressing the sand around a split pattern. The first or lower half of the pattern is placed upside down on a molding board, then sand is pounded or compressed around the pattern in a box called a *drag.* The drag is then turned over, and the second or upper half of the pattern is formed when another box, called a *cope,* is packed with sand and joined to the drag. A fine powder is used as a parting agent between the cope and drag at the parting line. The parting line is the separating joint between the two parts of the pattern or mold. The entire box, made up of the cope and drag, is referred to as a *flask.* (See Figure 11.3.)

Before the molten metal can be poured into the cavity, a passageway for the metal must be made. The passageway is called a *runner* and *sprue.* The location and design of the sprue and runner are important to allow for a rapid and continuous flow of metal. Additionally, vent holes are established to allow for gases, impurities, and metal to escape from the cavity. Finally, a riser (or group of risers) is used, depending on the size of the casting, to allow for the excess metal to evacuate from the mold and, more importantly, to help reduce shrinking and incomplete filling of the casting. (See Figure 11.4.) After the casting has solidified and cooled, the filled risers, vent holes, and runners are removed.

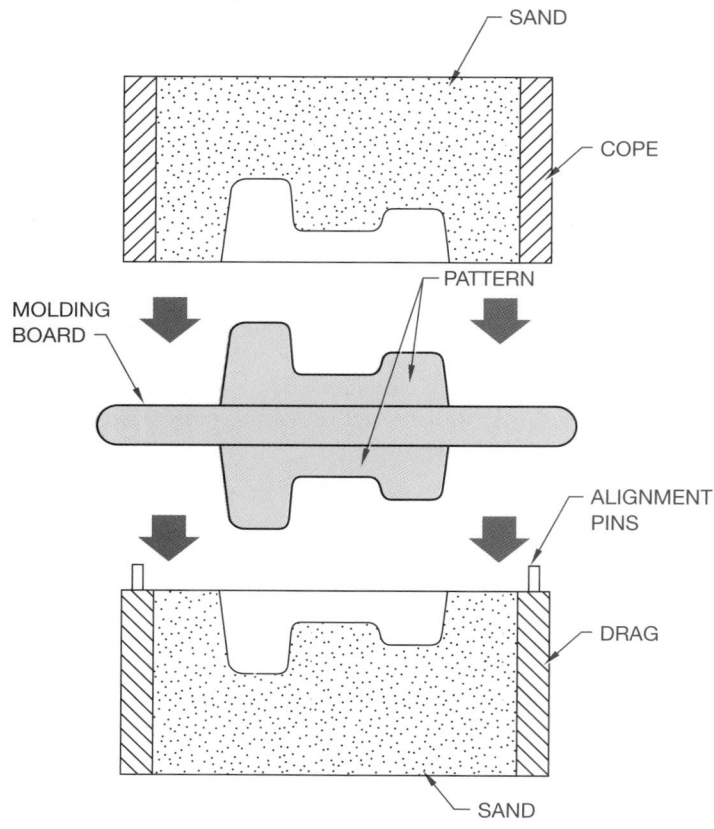

FIGURE 11.3 ■ Components of sand casting process.

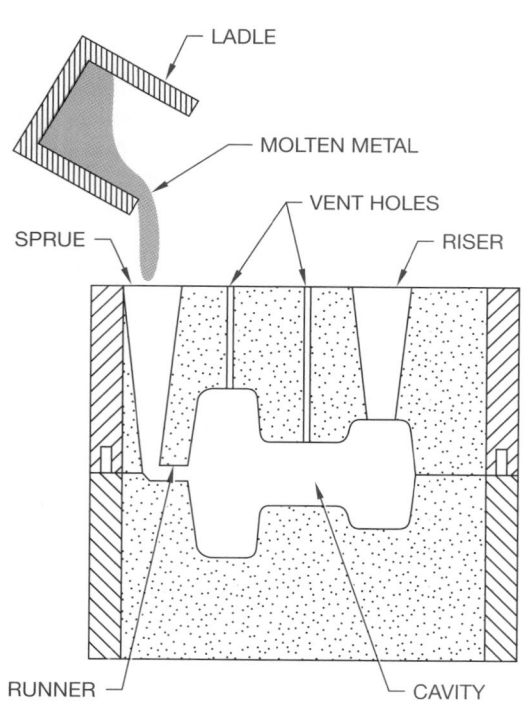

FIGURE 11.4 ■ Pouring molten metal into a sand casting mold.

Cores

In many situations a hole or cavity is desired in the casting to help reduce the amount of material removal later or to establish a wall thickness. When this is necessary, a core is used. Cores are made from either clean sand mixed with binders, such as resin, and baked in an oven for hardening, or ceramic products when a more refined surface finish is required. When the pattern is made, a place for positioning the core in the mold is established; this is referred to as the core print. After the mold is made, the core is then placed in position in the mold at the core print. The molten metal, when poured into the mold, flows around the core. After the metal has cooled, the casting is removed from the flask and the core is cleaned out, usually by shaking or tumbling. Figure 11.5 shows cores in place. Cored features help reduce casting weight and save on machining costs. Cores used in sand casting should generally be more than one inch (25.4 mm) in cross section. Cores used in precision casting methods may have much closer tolerances and fine detail. Certain considerations must be taken for supporting very large or long cores when placed in the mold. Usually in sand casting, cores require extra support when they are three times longer than the cross-sectional dimension. Depending on the casting method, the material, the machining required, and the quality, the core holes should be a specified dimension smaller than the desired end product if the hole is to be machined to its final dimension. Cores in sand castings should be between .125 to .5 in. (3.2 to 12.7 mm) smaller than the finished size.

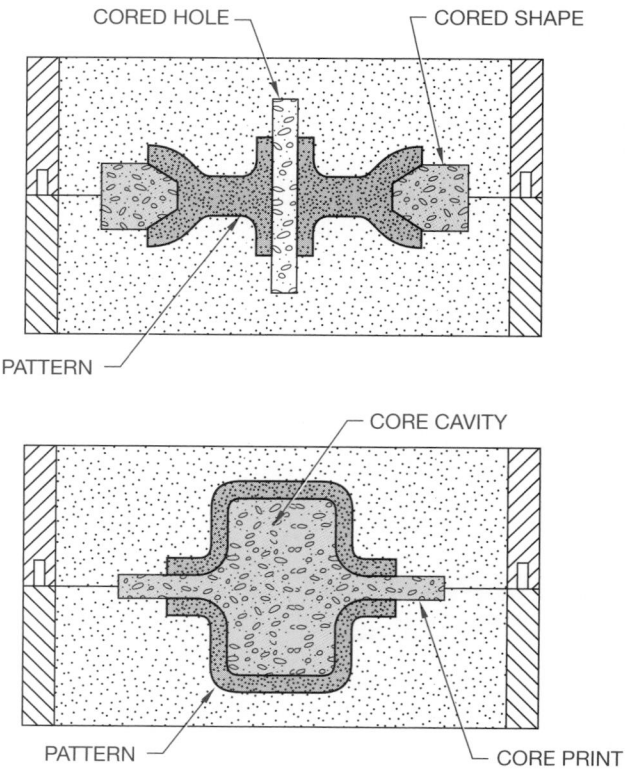

FIGURE 11.5 ■ Cores in place.

Centrifugal Casting

Objects with circular or cylindrical shapes lend themselves to *centrifugal casting*. In this casting process, a mold is revolved very rapidly while molten metal is poured into the cavity. The molten metal is forced outward into the mold cavity by centrifugal forces. No cores are needed because the fast revolution holds the metal against the surface of the mold. This casting method is especially useful for casting cylindrical shapes such as tubing, pipes, or wheels. (See Figure 11.6.)

Die Casting

Some nonferrous metal castings are made using the *die casting* process. Zinc alloy metals are the most common, although brass, bronze, aluminum, and other nonferrous products are also made using this process. Die casting is the injection of molten metal into a steel or cast iron die under high pressure. The advantage of die casting over other methods, such as sand casting, is that castings can be produced quickly and economically on automated production equipment. When multiple dies are used, a number of parts can be cast in one operation. Another advantage of die casting is

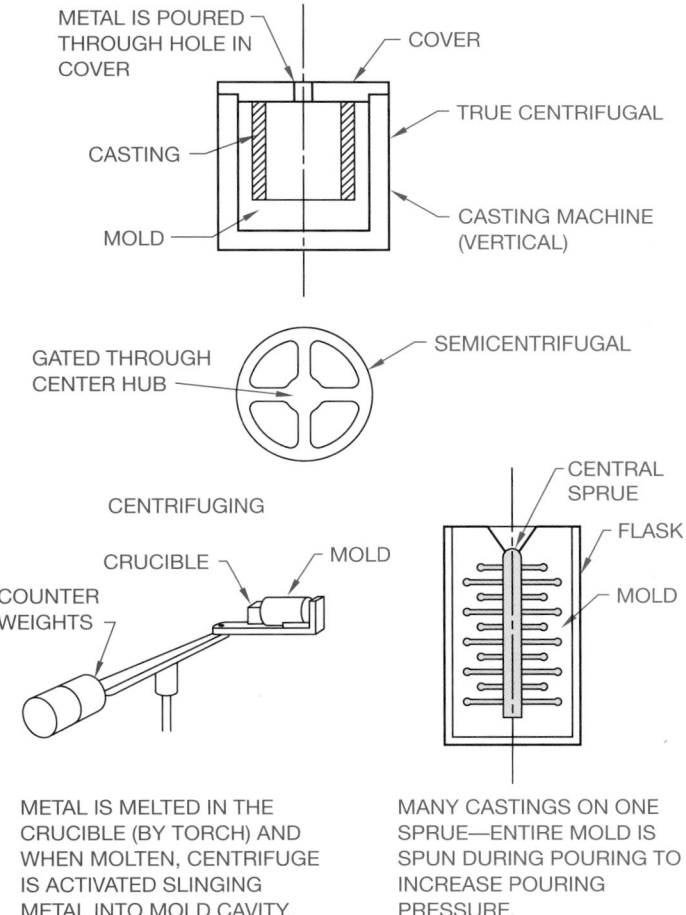

FIGURE 11.6 ■ Centrifugal casting.

that high-quality precision parts can be cast with fine detail and a very smooth finish.

Permanent Casting

Permanent casting refers to a process in which the mold can be used many times. This type of casting is similar to sand casting in that molten metal is poured into a mold. It is also similar to die casting because the mold is made of cast iron or steel. The result of permanent casting is a product that has better finished qualities than can be gained by sand casting.

Investment Casting

Investment casting is one of the oldest casting methods. It was originally used in France for the production of ornamental figures. The process used today is a result of the *cire perdue*, or *lost-wax*, casting technique that was originally used. The reason that investment casting is called lost-wax casting is that the pattern is made of wax. This wax pattern allows for the development of very close tolerances, fine detail, and precision castings. The wax pattern is coated with a ceramic paste. The shell is allowed to dry and then is baked in an oven to allow the wax to melt and flow out; thus "lost" as the name implies. The empty ceramic mold has a cavity that is the same shape as the precision wax pattern. This cavity is then filled with molten metal. When the metal solidifies, the shell is removed and the casting is complete. Generally very little cleanup or finishing is required on investment castings. (See Figure 11.7.)

Forgings

Forging is a process of shaping malleable metals by hammering or pressing between dies that duplicate the desired shape. The forging process is shown in Figure 11.8. Forging may be accomplished on hot or cold materials. Cold forging is possible on certain materials or material thicknesses where hole punching or bending is the required result. Some soft, nonferrous materials may be forged into shape while cold. Ferrous materials such as iron and steel must be heated to a temperature that results in an orange-red or yellow color. This color is usually achieved between 982° to 1066°C. Forging is used for a large variety of products and purposes. The advantage of forging over casting or machining operations is that the material is not only shaped into the desired form, but in the process it retains its original grain structure. Forged metal is generally stronger and more ductile than cast metal, and exhibits a greater resistance to fatigue and shock than machined parts. Notice in Figure 11.9 that the grain structure of the forged material remains parallel to the contour of the part, while the machined part cuts through the cross section of the material grain.

Hand Forging

Hand forging is an ancient method of forming metals into desired shapes. The method of heating metal to a red color and then beating it into shape is called *smithing* or, more commonly, *blacksmithing*. Blacksmithing is used only in industry for finish work,

HOW IT WORKS

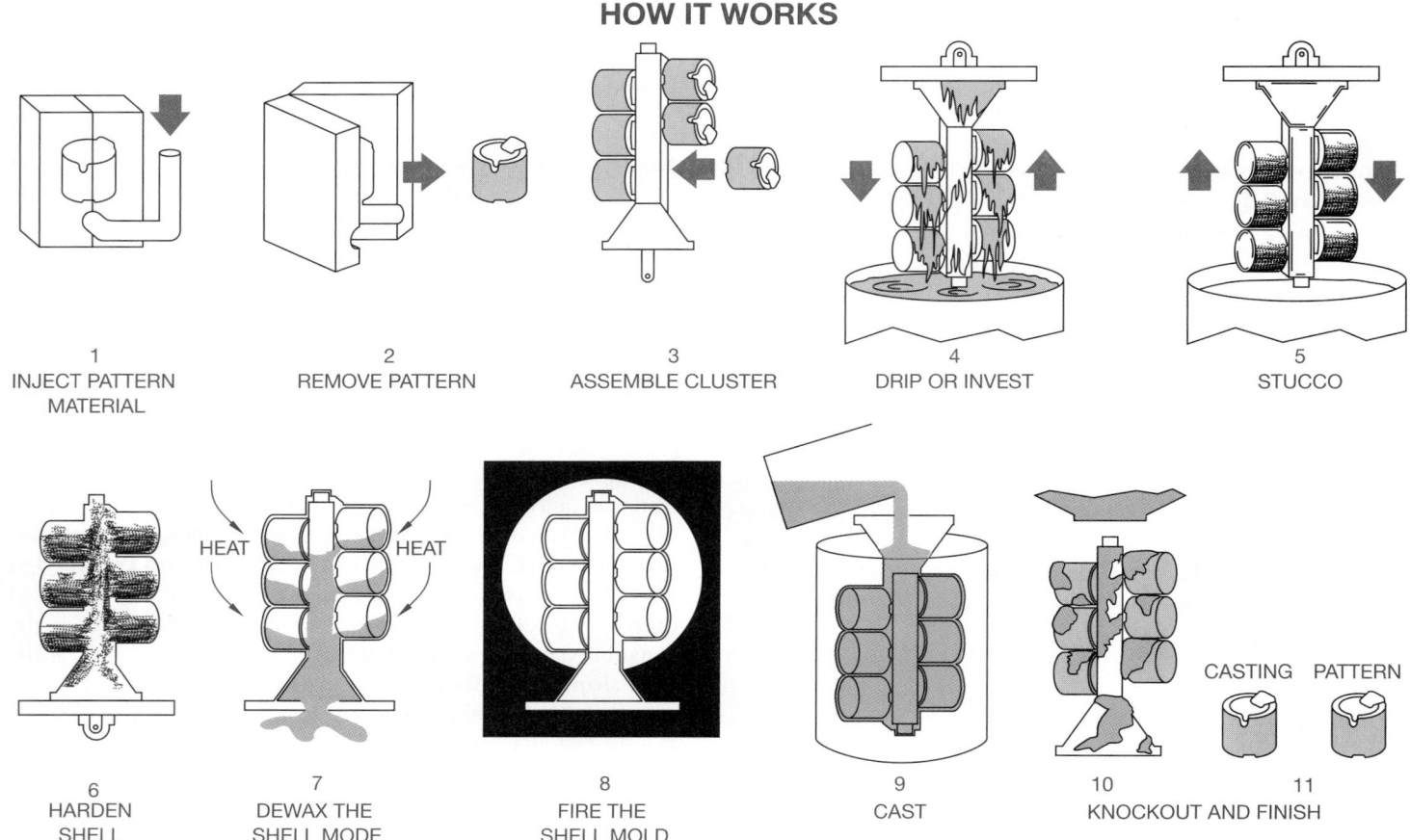

1	2	3	4	5
INJECT PATTERN MATERIAL	REMOVE PATTERN	ASSEMBLE CLUSTER	DRIP OR INVEST	STUCCO

6	7	8	9	10	11
HARDEN SHELL	DEWAX THE SHELL MODE	FIRE THE SHELL MOLD	CAST	KNOCKOUT AND FINISH	

FIGURE 11.7 ■ Investment casting. *Courtesy Precision Castparts Corporation.*

but is still used for horseshoeing and the manufacture of specialty ornamental products.

Machine Forging

Types of machine forging include upset, swaging, bending, punching, cutting, and welding. *Upset* forging is a process of forming metal by pressing along the longitudinal dimension to decrease the length while increasing the width. For example, bar stock is upset forged by pressing dies together from the ends of the stock to establish the desired shape. *Swaging* is the forming of metal by using concave tools or dies that result in a reduction in material thickness. *Bending* is accomplished by forming metal between dies, changing it from flat stock to a desired contour. Bending sheet metal is a cold forging process in which the metal is bent in a machine called a break. *Punching* and *cutting* are performed when the die penetrates the material to create a hole of any desired shape and depth or to remove material by cutting away. In forge *welding*, metals are joined together under extreme pressure. Material that is welded in this manner is very strong. The resulting weld takes on the same characteristics as the metal before joining.

Mass production forging methods allow for the rapid production of the high-quality products shown in Figure 11.10. In machine forging, the dies are arranged in sequence so that the finished forging is done in a series of steps. Complete shaping may take place after the material has been moved through several stages. Additional advantages of machine forging include:

1. The part is formed uniformly throughout the length and width.

2. The greater the pressure exerted on the material, the greater the improvement of the metallic properties.

3. Fine grain structure is maintained to help increase the part's resistance to shock.

4. A group of dies may be placed in the same press.

Metal Stamping

Stamping is a process that produces sheet metal parts by the quick downward stroke of a ram die that is in the desired shape. The machine is called a *punch press*. The punch press can "punch" holes of different sizes and shapes, cut metal, or form a variety of shapes. Automobile parts such as fenders and other body panels are often produced using stamping. If the punch press is used to create holes, then the ram has a die in the shape of the desired hole or holes that pushes through the sheet metal. The punch press can also produce a detailed shape by pressing sheet metal between a die set, where the shape of the desired part is created between the die on the bed of the machine and the matching die on the ram. Stamping is gen-

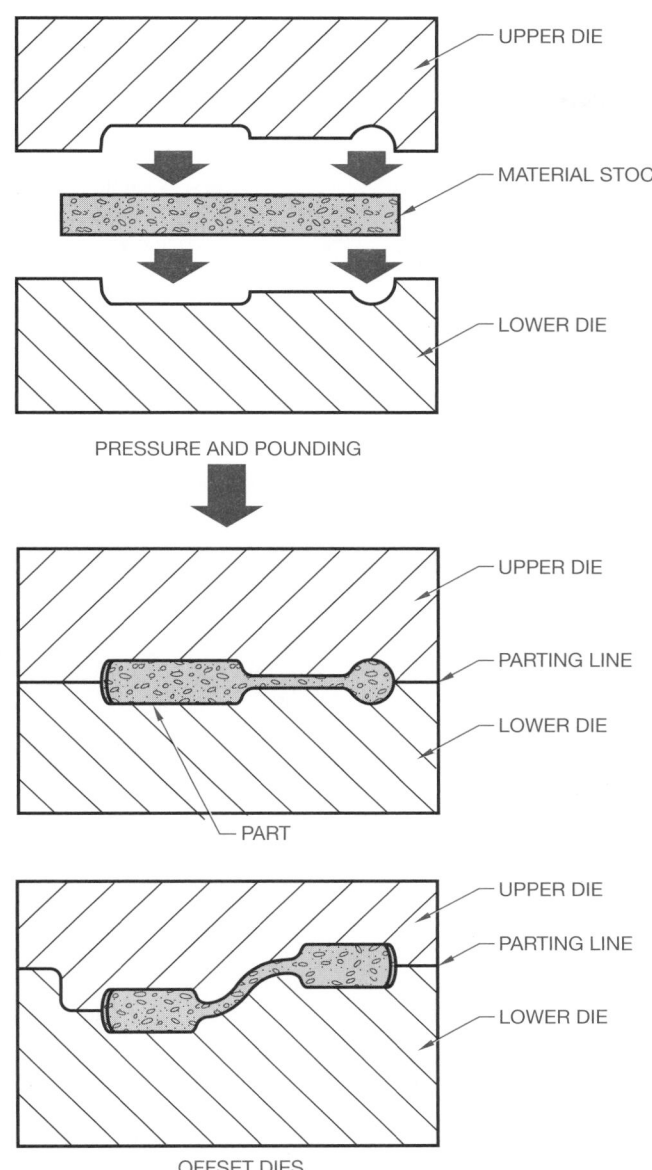

FIGURE 11.8 ■ The forging process.

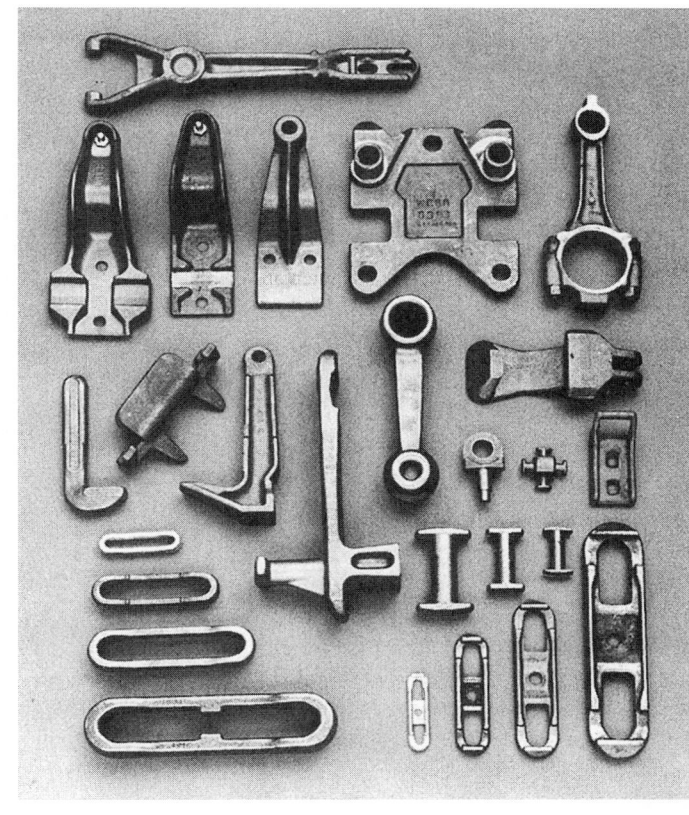

FIGURE 11.10 ■ Forged products. *Courtesy Jarvis B. Webb Co.*

erally done on cold sheet metal, as compared with forging that is done on hot metal that is often much thicker. The stamping process is useful in producing a large number of parts. The process can often fabricate thousands of parts per hour. This is a very important consideration for mass production.

Powder Metallurgy

The powder metallurgy process takes metal-alloyed powders and feeds them into a die where they are compacted under

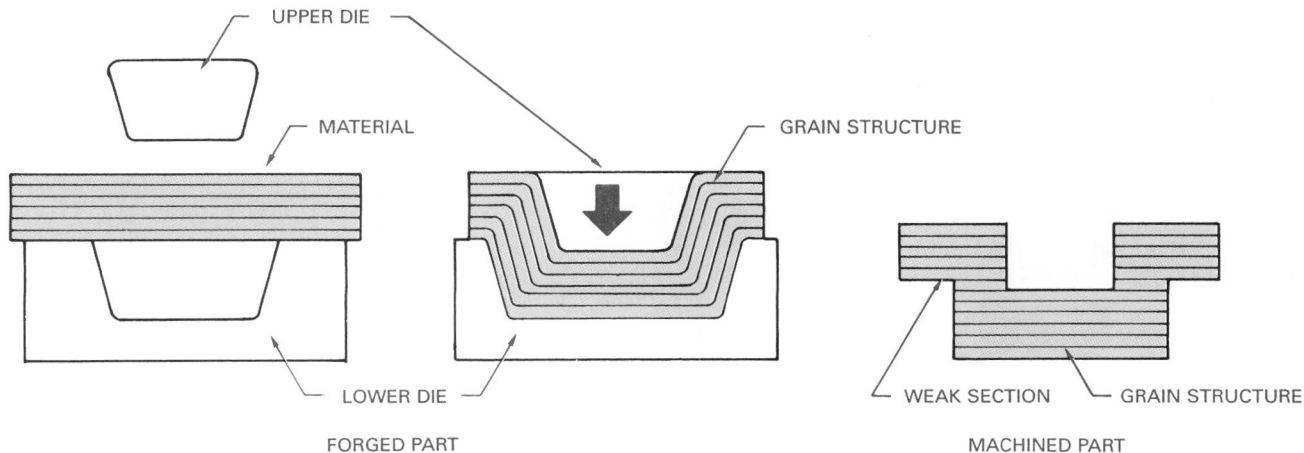

FIGURE 11.9 ■ Forging compared to machining.

CADD APPLICATIONS

COMPUTER NUMERICAL CONTROL (CNC) MACHINE TOOLS

Computer aided design and drafting (CADD) has a direct link to computer-aided manufacturing (CAM) in the form of *computer numerical control* (CNC) machine tools. A flowchart for the CNC process is shown in Figure 11.11. Figure 11.12 shows a CNC machine. In most cases, the drawing is generated in a computer and this information is sent directly to the machine tool for production. In some applications, information in the computer is transferred to a punched numerical control tape. The numerical control tape may be used as a backup system to the CNC, or the tape may be used as a copy for the numerical control program. Nearly all types of machine tools have been designed to operate from numerical control tape. However, this method is becoming obsolete.

Many CNC systems offer a microcomputer base that incorporates a monitor display and full alphanumeric keyboard, making programming and on-line editing easy and fast. For example, data input for certain types of machining may result in the programming of one of several identical features while the other features are oriented and programmed

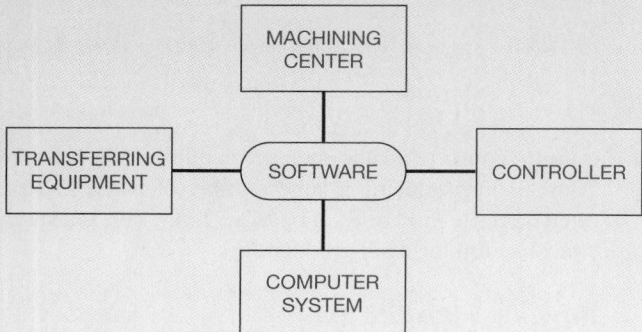

FIGURE 11.11 ■ The computer numerical control (CNC) process.

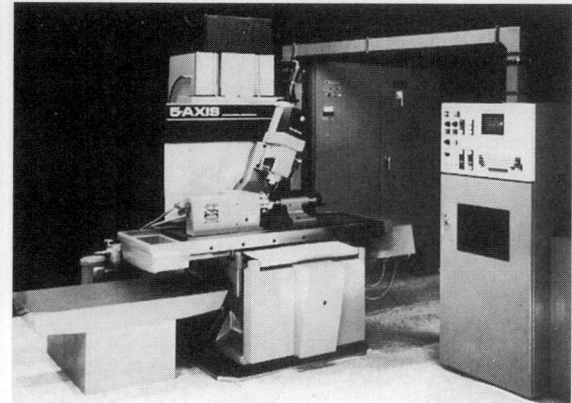

FIGURE 11.12 ■ A CNC machine. *Courtesy Boston Digital Corporation.*

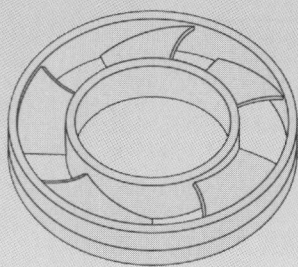

FIGURE 11.13 ■ CNC programming of a part with five equally spaced blades. This figure demonstrates the CNC programming of one of the five equally spaced blades and the automatic orientation and programming of the other four blades. *Courtesy Boston Digital Corporation.*

automatically, such as the five equally spaced blades of the centrifugal fan shown in Figure 11.13. Among the advantages of CNC machining are increased productivity, reduction of production costs, and manufacturing versatility.

The drafter has a special challenge when preparing drawings for CNC machining. The drawing method must coordinate with a system of controlling a machine tool by instructions in the form of numbers. The types of dimensioning systems that relate directly to CNC applications are tabular, arrowless, datum, and related coordinate systems. The main emphasis is to coordinate the dimensioning system with the movement of the machine tools. This can be best accomplished with datum dimensioning systems in which each dimension originates from a common beginning point. Programming the CNC system requires that cutter compensation be provided for contouring. This task is automatically computer-calculated in some CNC machines, as shown in Figure 11.14. Definitions and examples of the dimensioning systems are discussed in Chapter 12, Dimensioning and Tolerancing.

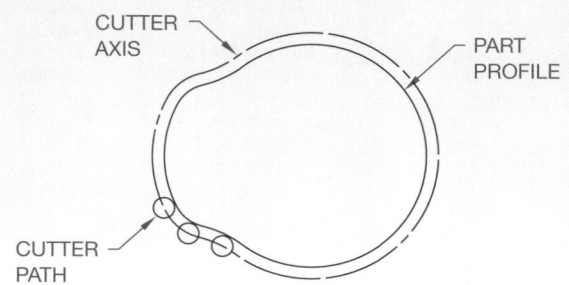

FIGURE 11.14 ■ Automatic cutter compensation for profile machining. *Courtesy Boston Digital Corporation.*

pressure to form the desired shape. The compacted metal is then removed from the die and heated at temperatures below the melting point of the metal. This heating process is referred to as *sintering,* which forms a bond between the metal powder particles. This is the conventional powder metallurgy process that is called P/M.

Powder metallurgy processes can be a cost-effective alternative to casting, forging, and stamping. Powder metallurgy manufacturing can produce quality, precision parts at the rate of thousands per hour.

Metal injection molding (MIM) is another powder metallurgy process that can produce very complex parts. This process injects a mixture of powder metal and a binder into a mold under pressure. The molded product is then sintered to create properties in the metal particles that are close to a casting.

Powder forging (P/F) is another powder metallurgy process that places the formed metal particles in a closed die where pressure and heat are applied. This is similar to the forging process. This process produces precision products that have good impact resistance and fatigue strength.

MACHINE PROCESSES

The concepts covered in this chapter serve as a basis for many of the dimensioning practices presented in Chapter 12. A general understanding of machining processes and the drawing representations of these processes is a necessary prerequisite to dimensioning practices. Problem applications are provided in the following chapters because a knowledge of dimensioning practice is important and necessary to be able to complete manufacturing drawings.

Machine Tools

Drilling Machine

The *drilling machine,* often referred to as a *drill press* (Figure 11.15), is commonly used to machine-drill holes. Drilling machines are also used to perform other operations, such as reaming, boring, countersinking, counterboring, and tapping. During the drilling procedure, the material is held on a table while the drill or other tool is held in a revolving spindle over the material. When drilling begins, a power or hand-feed mechanism is used to bring the rotating drill in contact with the material. Mass production drilling machines are designed with multiple spindles. Automatic drilling procedures are available on turret drills. These turret drills allow for automatic tool selection and spindle speed. Several operations can be performed in one setup—for example, drilling a hole to a given depth and tapping the hole with a specified thread.

Grinding Machine

A *grinding machine* uses a rotating abrasive wheel, rather than a cutting tool, for the purpose of removing material. (See Figure 11.16.) The grinding process is generally used when a smooth,

FIGURE 11.15 ■ Drilling machine. *Courtesy Delta International Machining Corporation.*

accurate surface finish is required. Extremely smooth surface finishes can be achieved by honing or lapping. *Honing* is a fine abrasive process often used to establish a smooth finish inside cylinders. *Lapping* is the process of creating a very smooth surface finish using a soft metal impregnated with fine abrasives, or fine abrasives mixed in a coolant that floods over the part during the lapping process.

Lathe

One of the earliest machine tools, the *lathe* (Figure 11.17), is used to cut material by turning cylindrically shaped objects. The material to be turned is held between two rigid supports, called *centers,* or in a holding device called a *chuck* or *collet,* as shown in Figure 11.18. The material is rotated on a spindle while a cutting tool is brought into contact with the material. The cutting tool is supported by a tool holder on a carriage that slides along a bed as the lathe operation continues. The turret is used in mass production manufacturing where one machine setup must perform several operations. A turret lathe is designed to carry several cutting tools in place of the lathe tailstock or on the lathe carriage. The operation of the turret provides the operator with an automatic selection of cutting tools at preestablished fabrication stages. Figure 11.19 shows an example of eight turret stations and the tooling used.

FIGURE 11.16 ■ Grinding machine. *Courtesy Litton Industrial Automation.*

FIGURE 11.17 ■ Lathe. *Courtesy Hardinge Brothers, Inc.*

Milling Machine

The *milling machine* (Figure 11.20) is one of the most versatile machine tools. The milling machine uses a rotary cutting tool to remove material from the work. The two general types of milling machines are *horizontal* and *vertical* mills. The difference is in the position of the cutting tool, which may be mounted on either a horizontal or vertical spindle. In the operation, the work is fas-

tened to a table that is mechanically fed into the cutting tool, as shown in Figure 11.21. There are a large variety of milling cutters available that influence the flexibility of operations and shapes that can be performed using the milling machine. Figure 11.22 shows a few of the milling cutters available. Figure 11.23 shows a series of milling cutters grouped together to perform a milling operation. End milling cutters, as shown in Figure 11.24, are

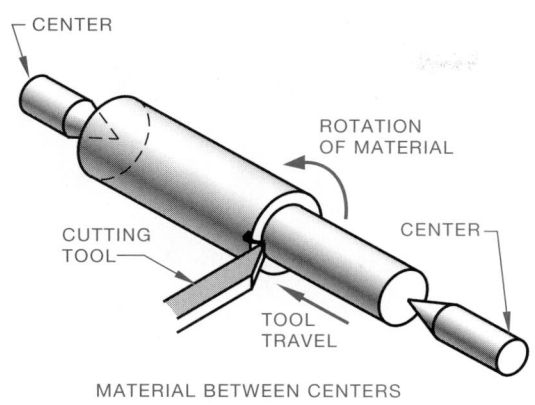

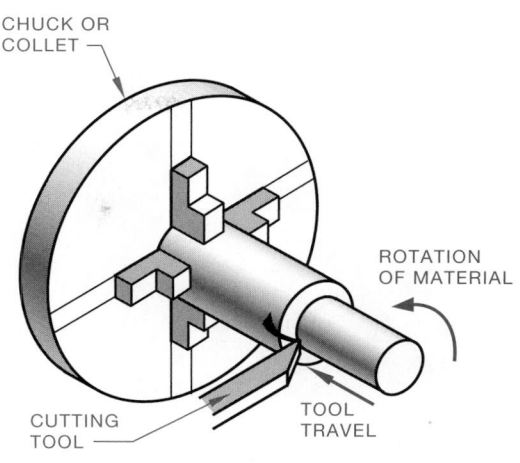

FIGURE 11.18 ■ Holding material in a lathe.

STATION		TOOLING	
---	---	---	
	Model	Description
1A		Drill with Bushing
1B		Center Drill with Bushing
1C	T20 -5/8	Adjustable Revolving Stock Stop
2A		Drill with Bushing
2B		Boring Bar with Bushing
2C		Threading Tool with Bushing
3A		Grooving Tool
3C		Insert Turning Tool
4A		Center Drill with Bushing
4B		Flat Bottom Drill with Bushing
4C		Insert Turning Tool
5A		Drill with Bushing
5B		Step Drill with Bushing
5C		Insert Turning Tool
6A	T8 -5/8	Knurling Tool
6B		Drill with Bushing
6C		Insert Turning Tool
7A		Grooving Tool
7B		Drill with Bushing
7C		Insert Threading Tool
8A	TT -5/8	"Collet Type" Releasing Tap Holder
		Tap Collet
		Tap
8C	TE -5/8	Tool Holder Extension
	T19 -5/8	Floating Reamer Holder
		Reamer with Bushing

FIGURE 11.19 ■ A turret with eight tooling stations and the tools used at each station. *Courtesy Toyoda Machinery USA, Inc.*

designed to cut on the end and the sides of the cutting tool. Milling machines that are commonly used in high-production manufacturing often have two or more cutting heads that are available to perform multiple operations. The machine tables of standard horizontal or vertical milling machines move from left to right (x-axis), forward and backward (y-axis), and up and down (z-axis), as shown in Figure 11.25.

The Universal Milling Machine

Another type of milling machine, known as the *universal milling machine,* has table action that includes x-, y-, and z-axis movement plus angular rotation. The universal milling machine looks much the same as other milling machines, but has the advantage of additional angular table movement as shown in Figure 11.26. This

FIGURE 11.20 ■ Close-up of a horizontal milling cutter.

FIGURE 11.21 ■ Material removal with a vertical machine cutter. *Courtesy The Cleveland Twist Drill Company.*

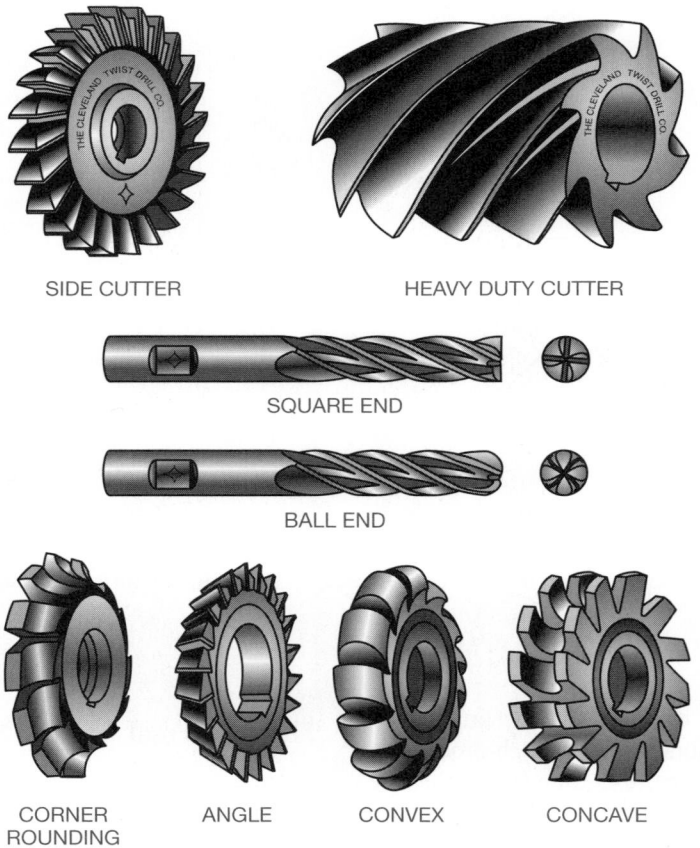

SIDE CUTTER

HEAVY DUTY CUTTER

SQUARE END

BALL END

CORNER
ROUNDING

ANGLE

CONVEX

CONCAVE

FIGURE 11.22 ■ Milling cutters. *Courtesy The Cleveland Twist Drill Company.*

FIGURE 11.23 ■ Grouping horizontal milling cutters for a specific machining operation. *Courtesy The Cleveland Twist Drill Company.*

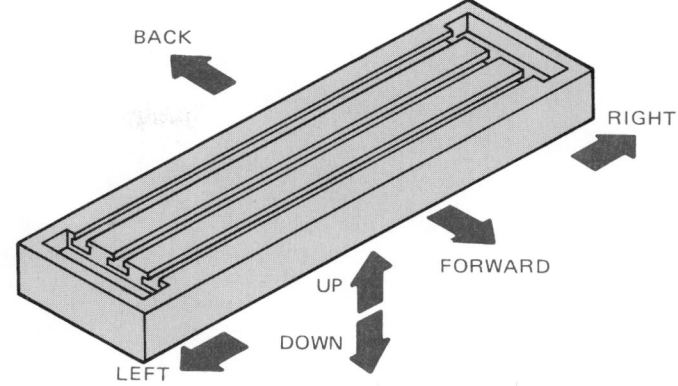

TABLE MOVEMENTS OF THE PLAIN MILLING MACHINE

FIGURE 11.25 ■ Table movements on a standard milling machine.

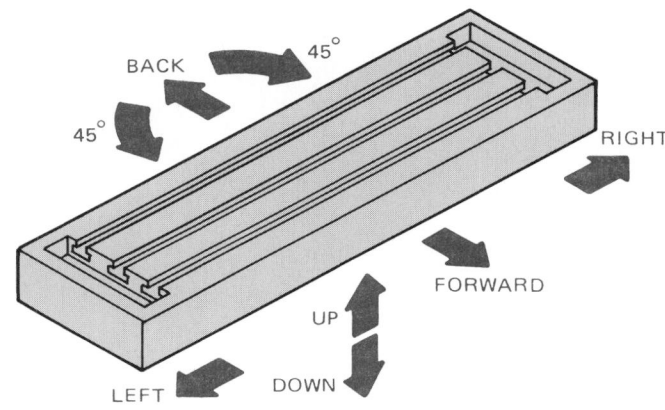

TABLE MOVEMENTS OF THE UNIVERSAL MILLING MACHINE

FIGURE 11.26 ■ Table movements on a universal milling machine.

Saw Machines

Saw machines may be used as cutoff tools to establish the length of material for further machining, or saw cutters can be used to perform certain machining operations such as cutting a narrow slot (*kerf*).

Two types of machines that function as cutoff units only are the *power hacksaw* and the *band saw*. These saws are used to cut a wide variety of materials. The hacksaw, as shown in Figure 11.27, operates using a back-and-forth motion. The fixed blade in the power hacksaw cuts material on the forward motion. The metal cutting band saw is available in a vertical or horizontal design, as shown in Figure 11.28. This type of cutoff saw has a continuous band that runs either vertically or horizontally around turning wheels. Vertical band saws may also be used to cut out irregular shapes.

Saw machines are also made with circular abrasive or metal cutting wheels. The *abrasive saw* may be used for high-speed cutting where a narrow saw kerf is desirable or when very hard materials must be cut. One advantage of the abrasive saw is its ability to cut a variety of materials—from soft aluminum to case-hardened steels.

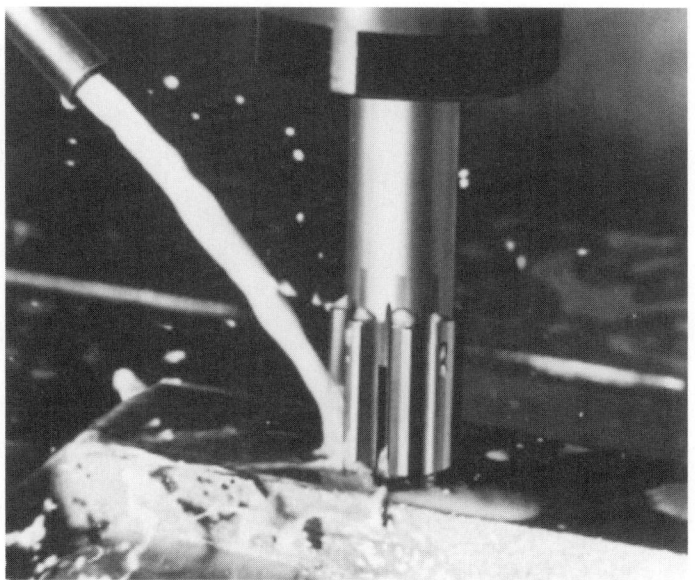

FIGURE 11.24 ■ End mill operation. *Courtesy The Cleveland Twist Drill Company.*

additional table movement allows the universal milling machine to produce machined features, such as spirals, not possible on conventional machines.

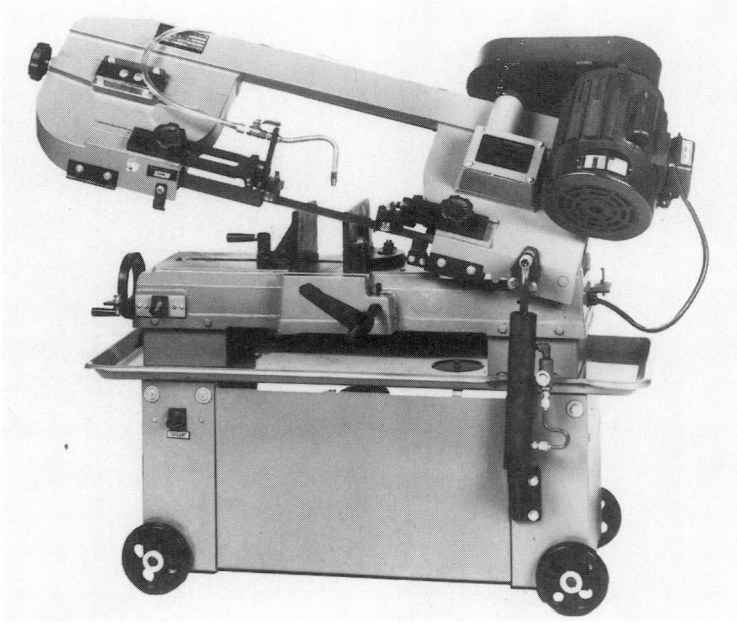

FIGURE 11.27 ■ Power hacksaw. *Courtesy JET Equipment and Tool.*

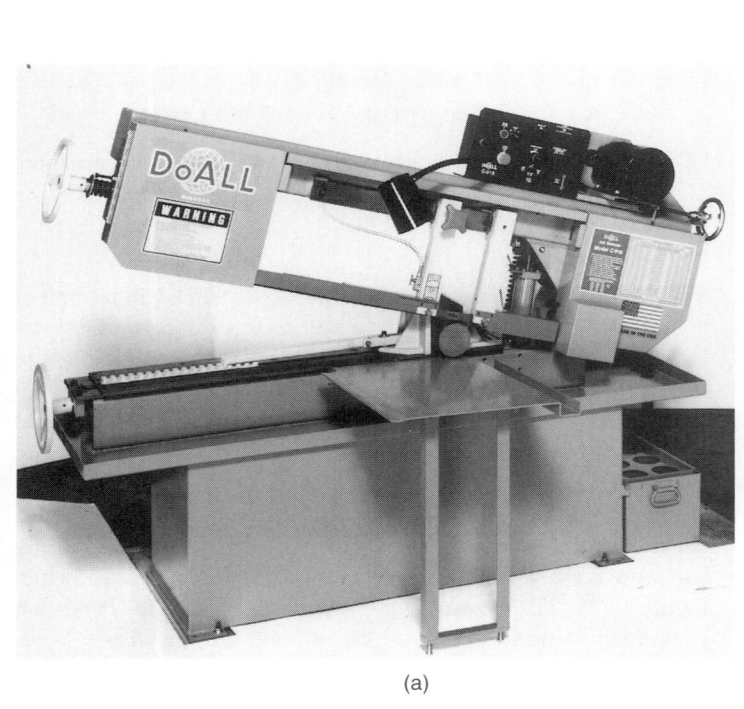

(a)

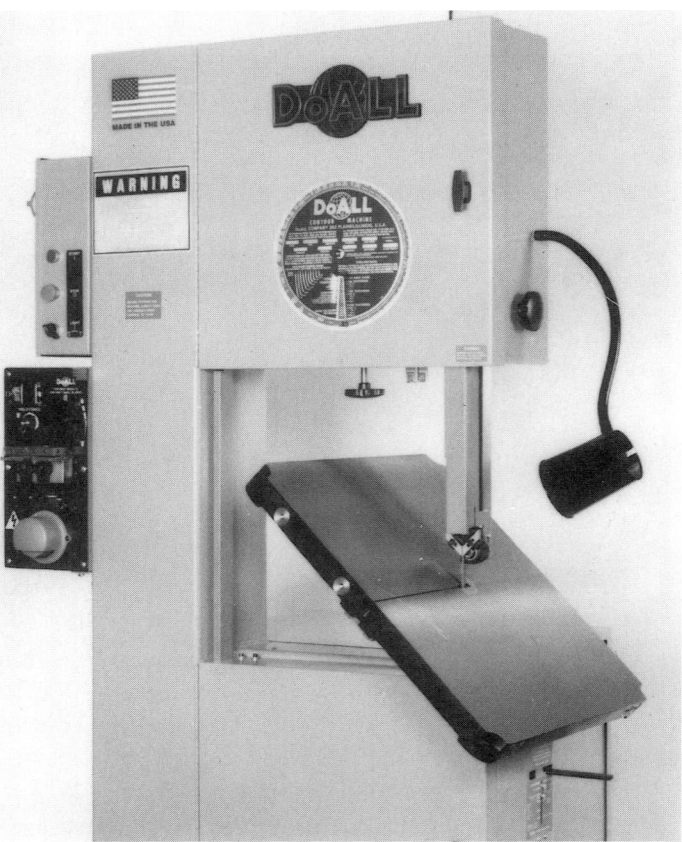

(b)

FIGURE 11.28 ■ **(a)** Horizontal band saw; **(b)** vertical band saw. *Courtesy DoAll Company.*

FIGURE 11.29 ■ Circular blade saw. *Courtesy The Cleveland Twist Drill Company.*

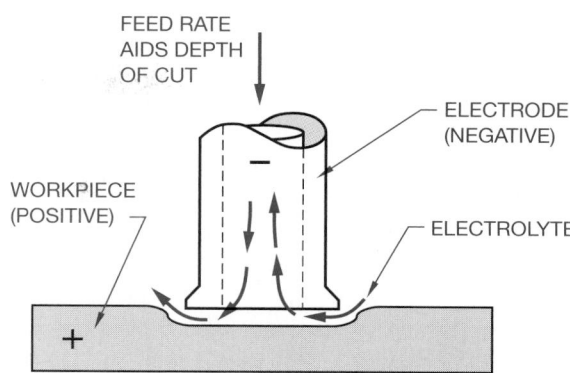

FIGURE 11.30 ■ Electrochemical machining (ECM).

(Cutting a variety of metals on the band or the power hacksaw requires blade and speed changes.) A disadvantage of the abrasive saw is the expense of abrasive discs. Many companies use this saw only when versatility is needed. The abrasive saw is usually found in the grinding room where abrasive particles can be contained, but may also be used in the shop for general purpose cutting. Metal cutting saws with teeth, also known as *cold* saws, are used for precision cutoff operations, cutting saw kerfs, slitting metal, and other manufacturing uses. Figure 11.29 shows a circular saw blade.

Water jet cutting is used on composite materials and thin metal with a computer-controlled 55,000 pounds per square inch (psi) water jet. Cuts are made with holding tolerances of .0008 in. (0.020 mm) and without generating heat.

Shaper

The *shaper* is used primarily for production of horizontal, vertical, or angular flat surfaces. Shapers are generally becoming out of date and are rapidly being replaced by milling machines. A big problem with the shaper in mass production industry is that it is very slow and cuts only in one direction. One of the main advantages of the shaper is its ability to cut irregular shapes that cannot be conveniently reproduced on a milling machine or other machine tools. However, other more advanced multiaxis machine tools are now available that quickly and accurately cut irregular contours.

Chemical Machining

Chemical machining uses chemicals to remove material accurately. The chemicals are placed on the material to be removed while other areas are protected. The amount of time the chemical remains on the surface determines the extent of material removal. This process, also known as *chemical milling,* is generally used in situations where conventional machining operations are difficult. A similar method, referred to as *chemical blanking,* is used on thin material to remove unwanted thickness in certain areas while maintaining "foil" thin material at the machined area. Material may be machined to within .00008 in. (0.002 mm) using this technique.

Electrochemical Machining

Electrochemical machining (ECM) is a process in which a direct current is passed through an electrolyte solution between an electrode and the workpiece. Chemical reaction, caused by the current in the electrolyte, dissolves the metal, as shown in Figure 11.30.

Electrodischarge Machining

In *electrodischarge machining* (EDM), the material to be machined and an electrode are submerged in a dielectric fluid that is a nonconductor, forming a barrier between the part and the electrode. A very small gap of about .001 inch is maintained between the electrode and the material. An arc occurs when the voltage across the gap causes the dielectric to break down. These arcs occurs about 25,000 times per second, removing material with each arc. The compatibility of the material and the electrode is important for proper material removal. The advantages of EDM over conventional machining methods include its success in machining intricate parts and shapes that otherwise cannot be econmomically machined, and its use on materials that are difficult or impossible to work with, such as stainless steel, hardened steels, carbides, and titamium.

Electron Beam (EB) Cutting and Machining

In this type of chemical machining, an electron beam generated by a heated tungsten filament is used to cut or "machine" very accurate features into a part. this process may be used to machine holes as small as .0002 in. (0.005 mm) or contour irregular shapes with tolerances of .0005 in. (0.013 mm). *Electron beam cutting* techniques are versatile and may be used to cut or machine any metal or nonmetal.

Ultrasonic Machining

Ultrasonic machining, also known as impact grinding, is a process in which a high-frequency mechanical vibration is maintained in a tool designed to a specific shape. The tool and material to be machined are suspended in an abrasive fluid. The combination of the vibration and the abrasive causes the material removal.

Laser Machining

The laser is a device that amplifies focused light waves and concentrates them in a narrow, very intense beam. The term LASER comes from the first letters of the words "*Light Amplification by Stimulated Emission of Radiation.*" Using this process, materials are cut or machined by instant temperatures up to 75,000°F (41,649°C). *Laser machining* may be used on any type of material and produces smooth surfaces without burrs or rough edges.

Machined Features and Drawing Representations

The following discussion provides a brief definition of the common manufacturing-related terms. The figures that accompany each definition show an example of the tool, a pictorial of the feature, and the drawing representation. The terms are organzied in categories of related features, rather than alphabetical order.

Drill

A *drill* is used to machine new holes or enlarge existing holes in material. The drilled hole may go through the part, in which case the note THRU can be added to the diameter dimension. When the views of the hole clearly show that the hole goes through the part, then the note THRU many be omitted. When the hole does not go through, the depth must be specified. This is referred to as a *blind hole.* The drill depth is th total usable depth to where the drill point begins to taper. A drill is a conical-shaped tool with cutting edges, normally used in a drill press. The drawing representation of a drill point is a 120° total angle. (See Figure 11.31.)

Ream

The tool is called a *reamer.* The reamer is used to enlarge or finish a hole that has been drilled, bored, or cored. A cored hole is cast in place, as previously discussed. A reamer removes only a small amount of material: for example, .005 to 0.16 in. depending on the size of a hole. The intent of a reamed hole is to provide a smooth surface finish and a closer tolerance than is available with the existing hole. A reamer is a conical-shaped tool with cutting edges similar to a drill; however, a reamer will not create a hole as with a drill. Reamers may be used on a drill press, lathe, or mill. (See Figure 11.32.)

Bore

Boring is the process of enlarging an existing hole. The purpose may be to make a drilled or cored hole in a cylinder or part concentric with or perpendicular to other features of the part. A boring tool is used on machines such as a lathe, milling machine, or vertical bore mill for removing internal material. (See Figure 11.33.)

Counterbore

The *counterbore* is used to enlarge the end(s) of a machined hole to a specified diameter and depth. The machined hole is made first, and then the counterbore is aligned during the

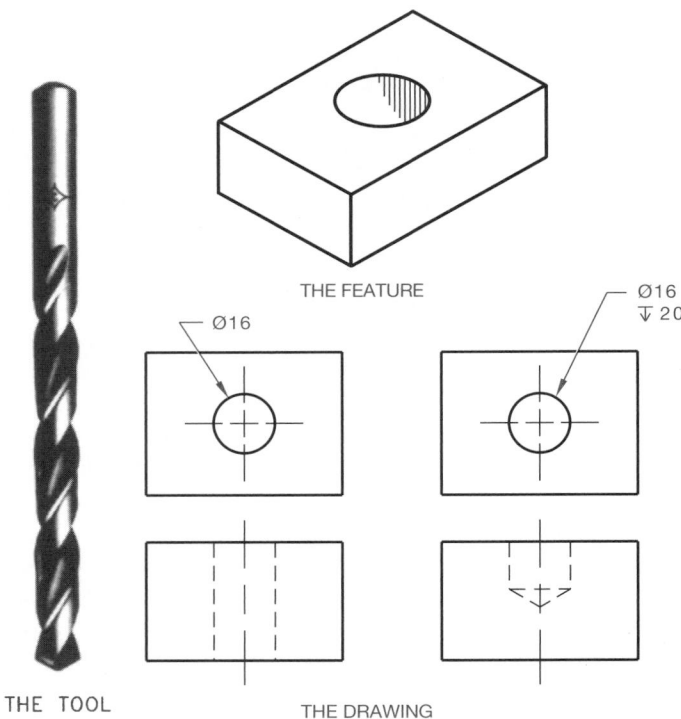

FIGURE 11.31 ■ Drill. *Tool photo courtesy The Cleveland Twist Drill Company.*

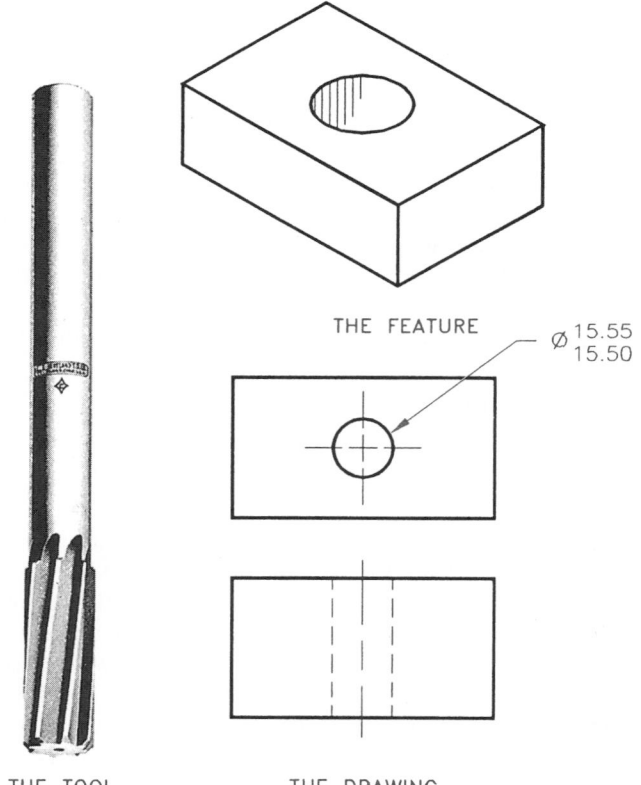

FIGURE 11.32 ■ Reamer. *Tool photo courtesy The Cleveland Twist Drill Company.*

machining process by means of a pilot shaft at the end of the tool. Counterbores are usually made to recess the head of a fastener below the surface of the object. You should be sure that the diameter and depth of the counterbore are adequate to accommodate the fastener head and fastening tools. (See Figure 11.34.)

Countersink

A *countersink* is a conical feature in the end of a machined hole. Countersinks are used to recess the conically shaped head of a fastener, such as a flathead machine screw. The drafter should specify the countersink note so the fastener head is recessed slightly below the surface. The total countersink angle should match the desired

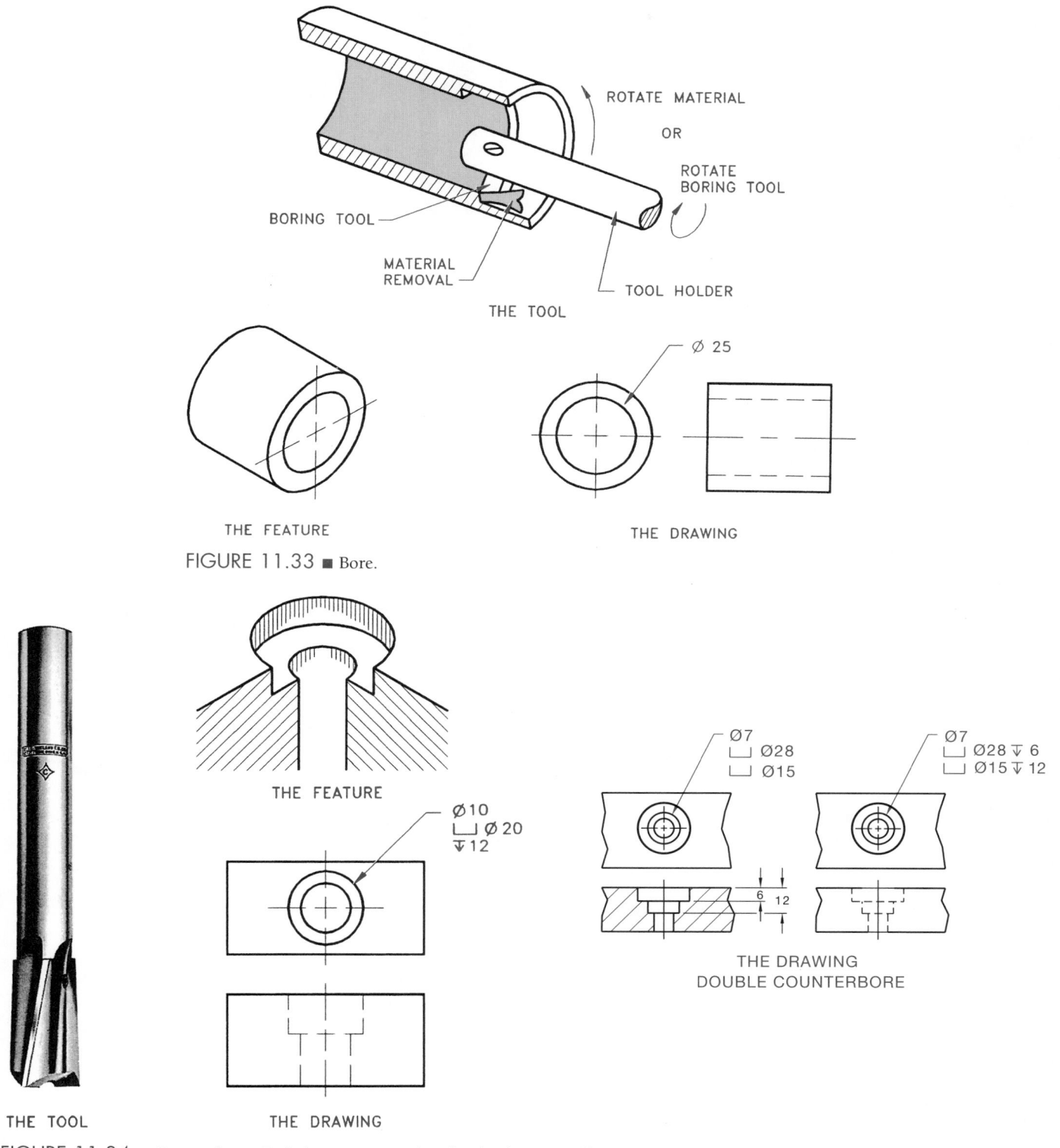

FIGURE 11.33 ■ Bore.

FIGURE 11.34 ■ Counterbore. *Tool photo courtesy The Cleveland Twist Drill Company.*

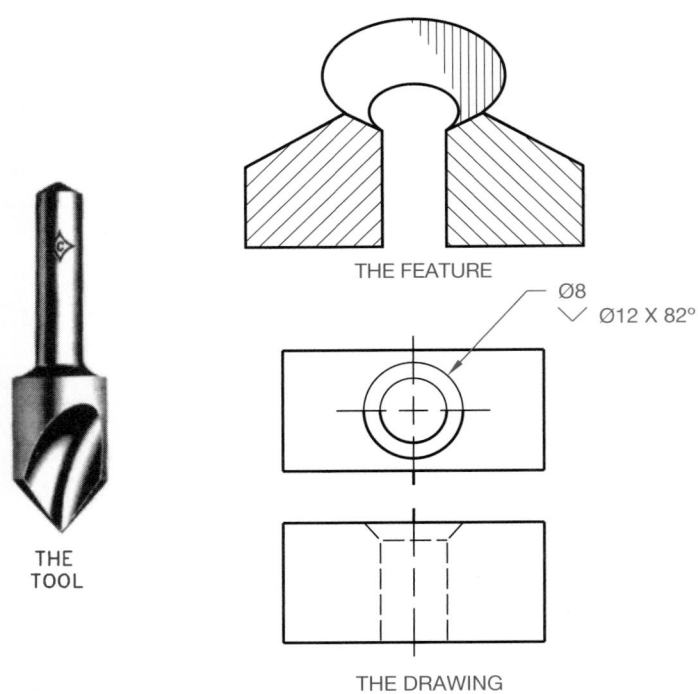

FIGURE 11.35 ■ Countersink. *Tool photo courtesy The Cleveland Twist Drill Company.*

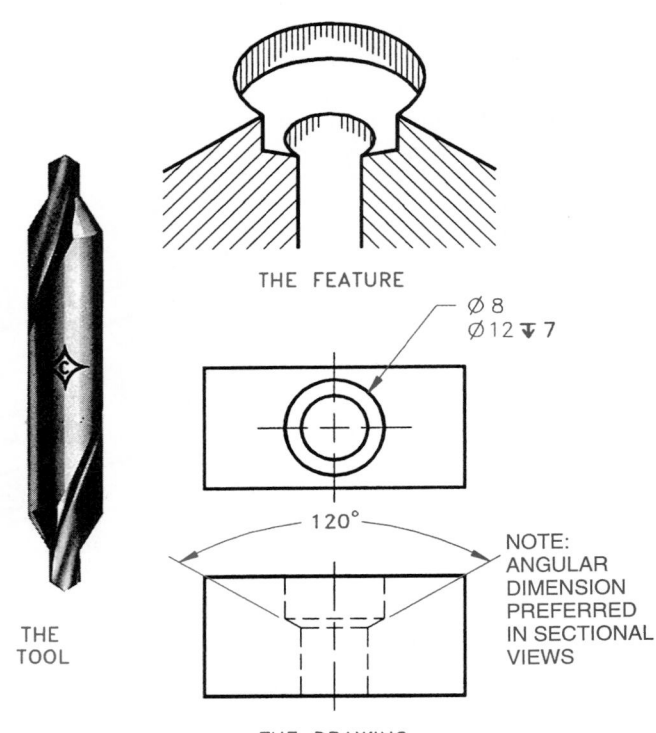

FIGURE 11.36 ■ Counterdrill. *Tool photo courtesy The Cleveland Twist Drill Company.*

screw head. This angle is generally 80° to 82° or 99° to 101°. (See Figure 11.35.)

Counterdrill

A *counterdrill* is a combination of two drilled features. The first machined feature may go through the part, while the second feature is drilled, to a given depth, into one end of the first. The result is a machined hole that looks similar to a countersink-counterbore combination. The angle at the bottom of the counterdrill is a total of 120°, as shown in Figure 11.36.

Spotface

A *spotface* is a machined, round surface on a casting, forging, or machined part on which a bolt head or washer can be seated. Spotfaces are similar in characteristics to counterbores, except that a spotface is generally only about 2 mm or less in depth. Rather than a depth specification, the dimension from the spotface surface to the opposite side of the part may be given. This is also true for counterbores; however, the depth dimension is commonly provided in the note. When no spotface depth is given, the machinist will spotface to a depth that establishes a smooth cylindrical surface. (See Figure 11.37.)

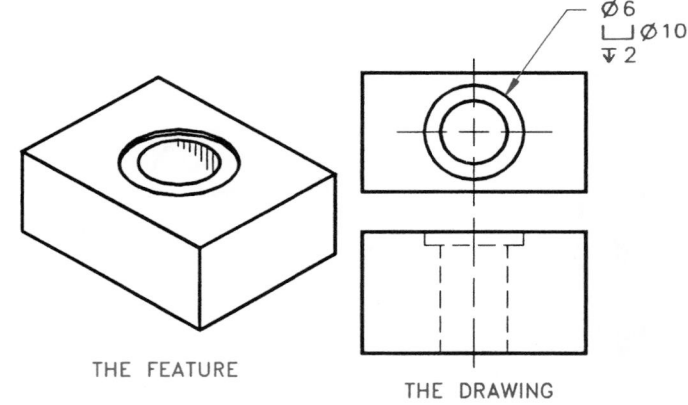

FIGURE 11.37 ■ Spotface.

Boss

A *boss* is a circular pad on forgings or castings that projects out from the body of the part. While more closely related to castings and forgings, the surface of the boss is often machined smooth for a bolt head or washer surface to seat on. Also, the boss commonly has a hole machined through it to accommodate the fastener's shank. (See Figure 11.38.)

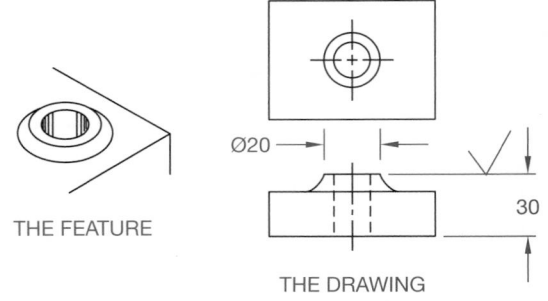

FIGURE 11.38 ■ Boss.

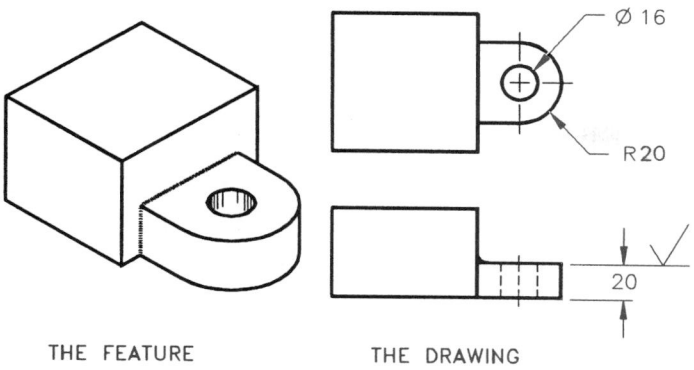

FIGURE 11.39 ■ Lug.

Lug

Generally cast or forged into place, a *lug* is a feature projecting out from the body of a part, usually rectangular in cross section. Lugs are used as mounting brackets or function as holding devices for machining operations. Lugs are commonly machined with a drilled hole and a spotface to accommodate a bolt or other fastener. (See Figure 11.39.)

Pad

A *pad* is a slightly raised surface projecting out from the body of a part. The pad surface can be any size or shape. The pad may be cast, forged, or machined into place. The surface is often machined to accommodate the mounting of an adjacent part. A boss is a type of pad, although the boss is always cylindrical in shape. (See Figure 11.40.)

Chamfer

A *chamfer* is the cutting away of the sharp external or internal corner of an edge. Chamfers may be used as a slight angle to relieve a sharp edge, or to assist the entry of a pin or thread into the mating feature. (See Figure 11.41.) Verify alternate methods of dimensioning chamfers in Chapter 12.

Fillet

A *fillet* is a small radius formed between the inside angle of two surfaces. Fillets are often used to help reduce stress and strengthen an

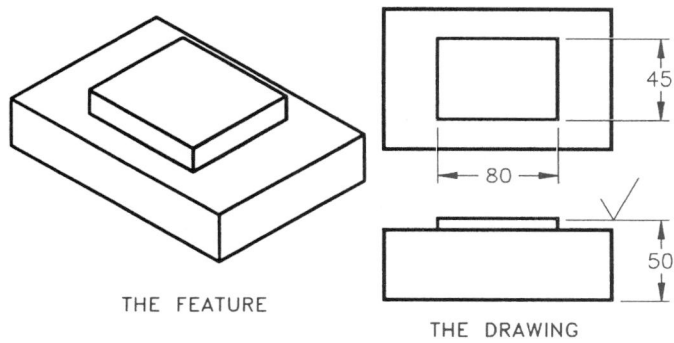

FIGURE 11.40 ■ Pad.

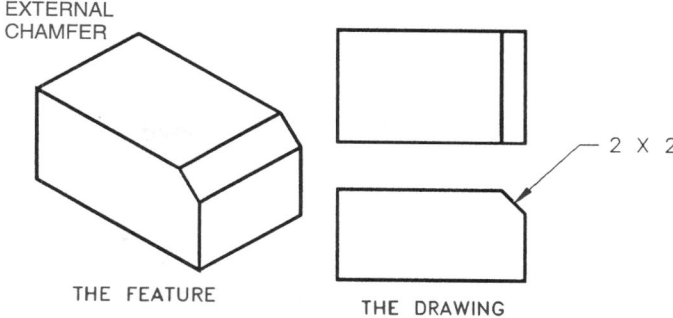

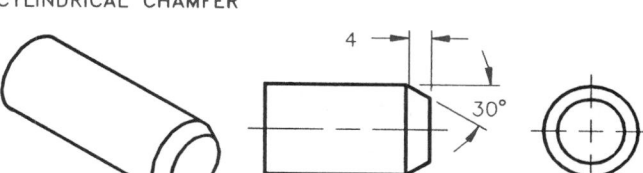

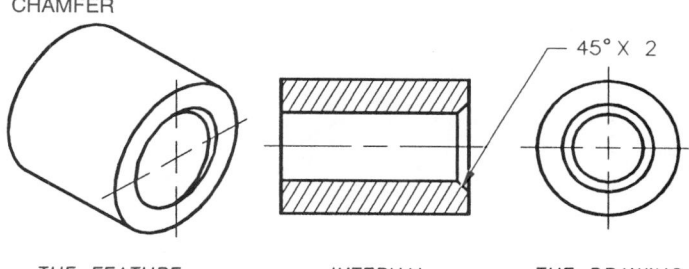

FIGURE 11.41 ■ Chamfers.

inside corner. Fillets are common on the inside corners of castings and forgings to strengthen corners. Fillets are also used to help a casting or forging release a mold or die. Fillets are arcs given as radius dimensions. The fillet size depends on the function of the part and the manufacturing process used to make the fillet. (See Figure 11.42.)

Round

A *round* is a small-radius outside corner formed between two surfaces. Rounds are used to refine sharp corners, as shown in Figure 11.43. In some situations where a sharp corner must be relieved and a round is not required, a slight corner relief may be used, referred to as a *break corner.* The note BREAK CORNER may be used on the drawing. Another option is to provide a note that specifies REMOVE ALL BURRS AND SHARP EDGES. *Burrs* are machining fragments that are often left on a part after machining.

Dovetail

A *dovetail* is a slot with angled sides that may be machined at any depth and width. Dovetails are commonly used as a sliding mechanism between two mating parts. (See Figure 11.44.)

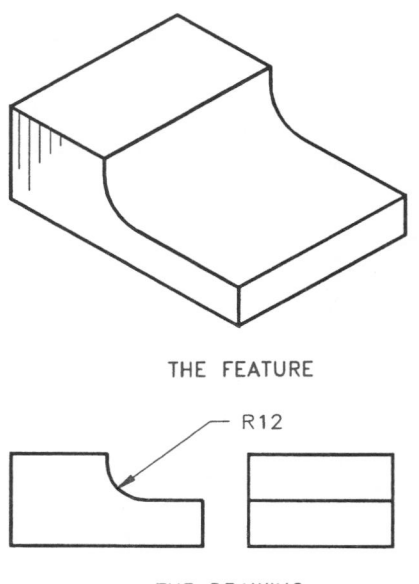

THE FEATURE

R12

THE DRAWING

FIGURE 11.42 ■ Fillet.

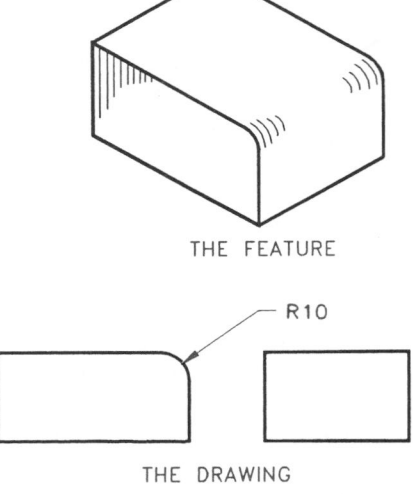

THE FEATURE

R10

THE DRAWING

FIGURE 11.43 ■ Round.

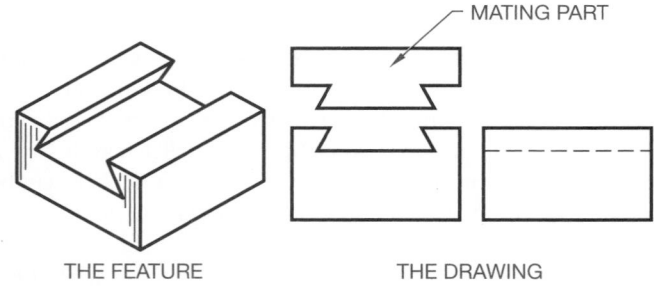

MATING PART

THE FEATURE   THE DRAWING

FIGURE 11.44 ■ Dovetail.

Kerf

A *kerf* is a narrow slot formed by removing material while sawing or using some other machining operation. (See Figure 11.45.)

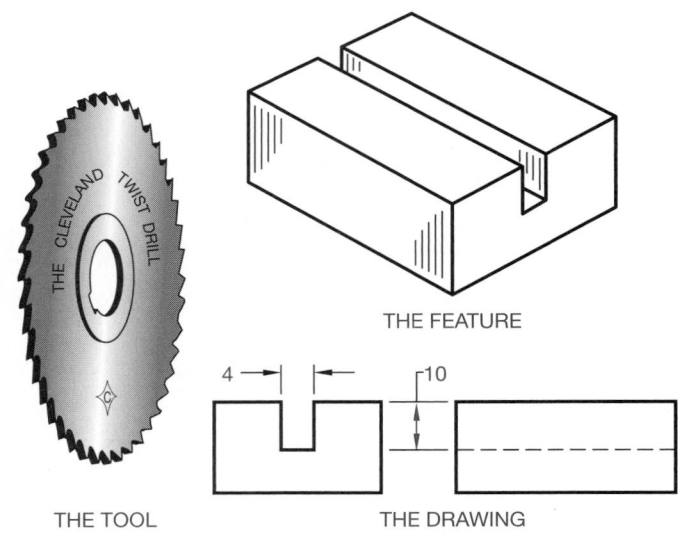

THE FEATURE

4   10

THE TOOL   THE DRAWING

FIGURE 11.45 ■ Kerf. *Tool photo courtesy The Cleveland Twist Drill Company.*

Key, Keyseat, Keyway

A *key* is a machine part that is used as a positive connection for transmitting torque between a shaft and a hub, pulley, or wheel. The key is placed in position in a *keyseat,* which is a groove or channel cut in a shaft. The shaft and key are then inserted into a hub, wheel, or pulley where the key mates with a groove, called a *keyway.* There are several different types of keys. The key size is often determined by the shaft size. (See Figure 11.46.) Types of keys and key sizes are discussed in Chapter 13, Fasteners and Springs.

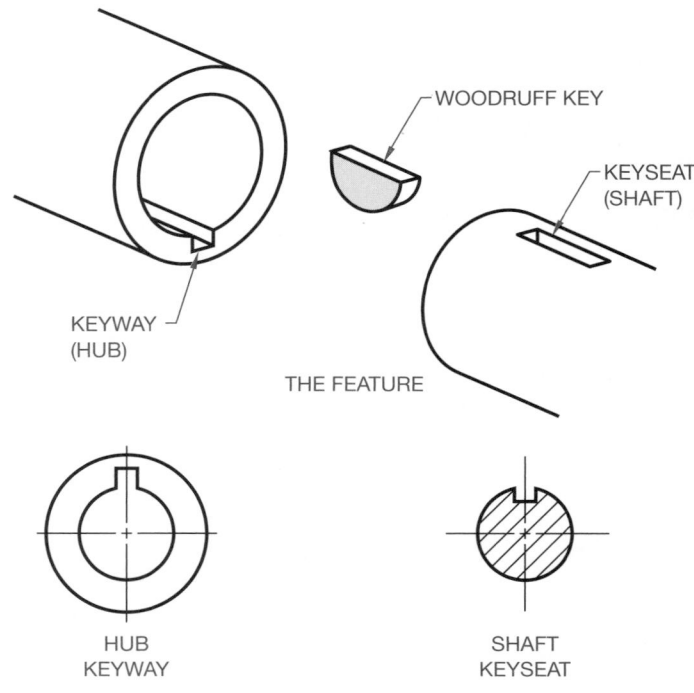

WOODRUFF KEY

KEYSEAT (SHAFT)

KEYWAY (HUB)

THE FEATURE

HUB KEYWAY   SHAFT KEYSEAT

THE DRAWING

FIGURE 11.46 ■ Key, keyseat, keyway.

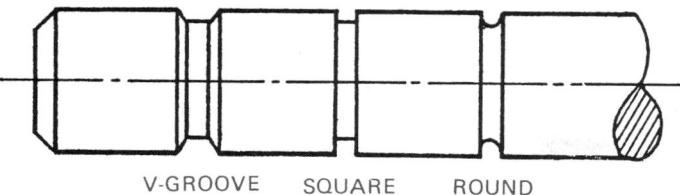

FIGURE 11.47 ■ Neck.

Neck

A *neck* is the result of a machining operation that establishes a narrow groove on a cylindrical part or object. There are several different types of neck grooves, as shown in Figure 11.47. Dimensioning necks is clearly explained in Chapter 12.

Spline

A *spline* is a gearlike, serrated surface on a shaft and in a mating hub. Splines are used to transmit torque and allow for lateral sliding or movement between two shafts or mating parts. A spline can be used to take the place of a key when more torque strength is required or when the parts must have lateral movement. (See Figure 11.48.)

Threads

There are many different forms of threads available that are used as fasteners to hold parts together, to adjust parts in alignment with each other, or to transmit power. Threads that are used as fasteners are commonly referred to as *screw threads*. *External threads* are thread forms on an external feature, such as a bolt or shaft. The machine tool used to make external threads is commonly called a *die*. Threads may be machined on a lathe using a thread-cutting tool. (See Figure 11.49.) *Internal threads* are threaded features on the inside of a hole. The machine tool that is commonly used to cut internal threads is called a *tap*. (See Figure 11.50.)

T-slot

A *T-slot* is a slot of any dimension that is cut to resemble a "T." The T-slot may be used as a sliding mechanism between two mating parts. (See Figure 11.51.)

Knurl

Knurling is a cold forming process used to uniformly roughen a cylindrical or flat surface with a diamond or straight pattern.

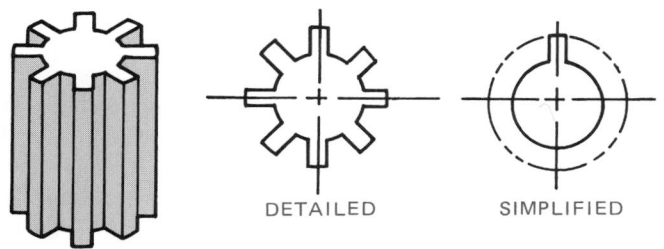

THE FEATURE

THE DRAWING

FIGURE 11.48 ■ Spline.

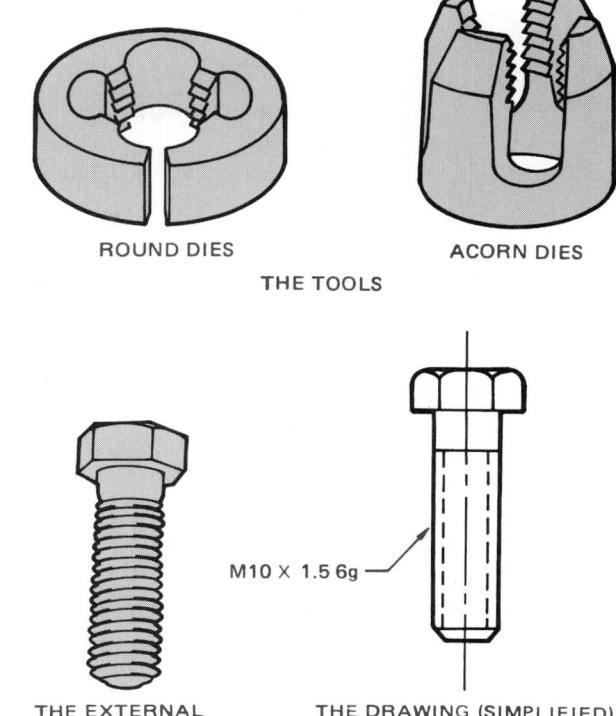

ROUND DIES   ACORN DIES

THE TOOLS

M10 X 1.5 6g

THE EXTERNAL THREAD   THE DRAWING (SIMPLIFIED)

FIGURE 11.49 ■ External thread.

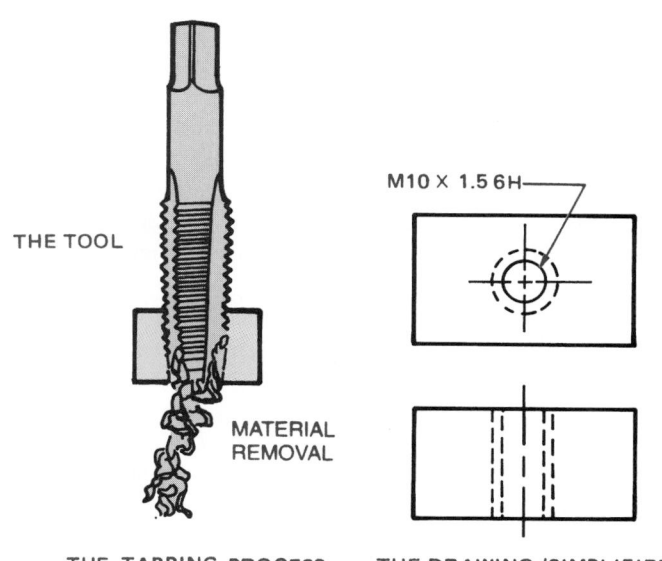

THE TOOL

MATERIAL REMOVAL

M10 X 1.5 6H

THE TAPPING PROCESS   THE DRAWING (SIMPLIFIED)

FIGURE 11.50 ■ Internal thread.

Knurls are often used on handles or other gripping surfaces. Knurls may also be used to establish an interference (press) fit between two mating parts. The actual knurl texture is not displayed on the drawing. (See Figure 11.52.)

Surface Texture

Surface texture, or *surface finish*, is the intended condition of the material surface after manufacturing processes have been

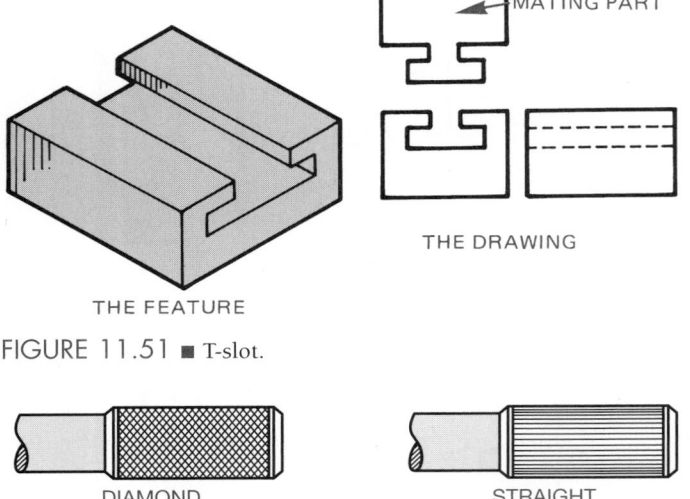

FIGURE 11.51 ■ T-slot.

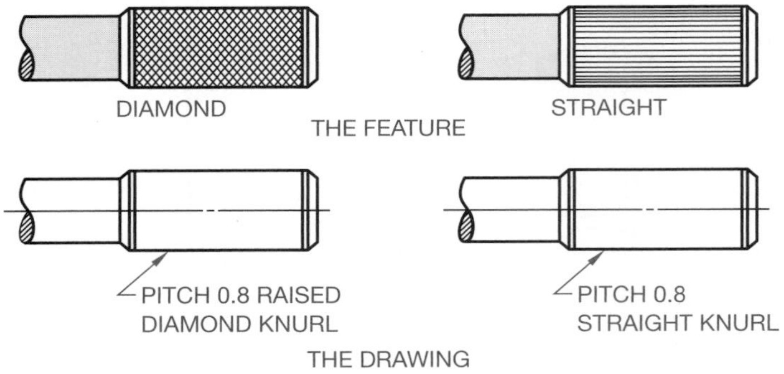

FIGURE 11.52 ■ Knurl.

implemented. Surface texture includes such characteristics as roughness, waviness, lay, and flaws. Surface roughness is one of the most common characteristics of surface finish, which consists of the finer irregularities of the surface texture due, in part, to the manufacturing process. The surface roughness is measured in micrometers (μm) or microinches (μin.) Micro (μ) means millionth. The roughness averages for different manufacturing processes are shown in Appendix B, Table 26. Surface finish is considered a drafting specification and therefore is discussed in detail in Chapter 12.

Design and Drafting of Machined Features

Drafters should gain a working knowledge of the machining processes and capabilities of their companies. Drawings should be prepared that allow machining within the capabilities of the machinery available. If the local machinery will not produce parts that have certain functional requirements, then the machining may need to be performed elsewhere, or the company may have to purchase the necessary equipment. However, the first consideration should be looking for the least-expensive method to get the desired result. Avoid overmachining. Machining processes are expensive, so drawing requirements that are not necessary for the function of the part should be avoided. For example, surface finishes become more expensive to machine as the roughness height decreases. So if a surface roughness of 125 microinches is adequate, then do not use a 32-microinch specification just because you like smooth surfaces. For example, note the difference between 63- and 32-microinch finishes. A 63-microinch finish is a good machine finish that may be performed using sharp tools at high speeds with extra

fine feeds and cuts. The 32-microinch callout, on the other hand, requires extremely fine feeds and cuts on a lathe or milling machine and in many cases requires grinding. The 32-microinch finish is more expensive to perform.

In a manufacturing environment where cost and competition are critical considerations, a great deal of thought must be given to meeting the functional and appearance requirements of a product for the least possible cost. It generally does not take very long for an entry-level drafter to pick up these design considerations by communicating with the engineering and manufacturing departments. Many drafters become designers, checkers, or engineers within a company by learning the product and being able to implement designs based on the company's manufacturing capabilities.

Manufacturing Plastic Products

The previous discussions gave you some general information about the different types of plastics and synthetic rubbers, and many of the products that are commonly made from these materials. The traditional plastic product manufacturing processes include injection molding, extrusion, blow molding, compression molding, transfer molding, and thermoforming. These processes are explained in the following.

The Injection Molding Process

Injection molding is the most commonly used process for creating thermoplastic products. The process involves injecting molten plastic material into a mold that is in the form of the desired part or product. The mold is in two parts that are pressed together during the molding process. The mold is then allowed to cool so the plastic can solidify. When the plastic has cooled and solidified, the press is opened and the part is removed from the mold. The injection molding machine has a hopper where either powder or granular material is placed. The material is then heated to a melting temperature. The molten plastic is then fed into the mold by an injection nozzle or a screw injection system. The injection system is similar to a large hypodermic needle and plunger that pushes the molten plastic material into the mold. The most commonly used injection system is the screw machine. This machine has a screw design that transports the material to the injection nozzle. While the material moves toward the mold, the screw also mixes the plastic to a uniform consistency. This mixing process is an important advantage over the plunger system. Mixing is especially important when color is added or when using recycled material. The process of creating a product and the parts of the screw injection and injection nozzle systems are shown in Figure 11.53.

The Extrusion Process

The extrusion process is used to make continuous shapes such as moldings, tubing, bars, angle, hose, weather stripping, films, and any product that has a constant shape. This process creates the desired continuous shape by forcing molten plastic through a metal die. The extrusion process typically uses the same type of injection nozzle or a screw injection system that is used in the injection molding process. The contour of the die establishes the shape of the extruded plastic. Figure 11.54 shows the extrusion process in action.

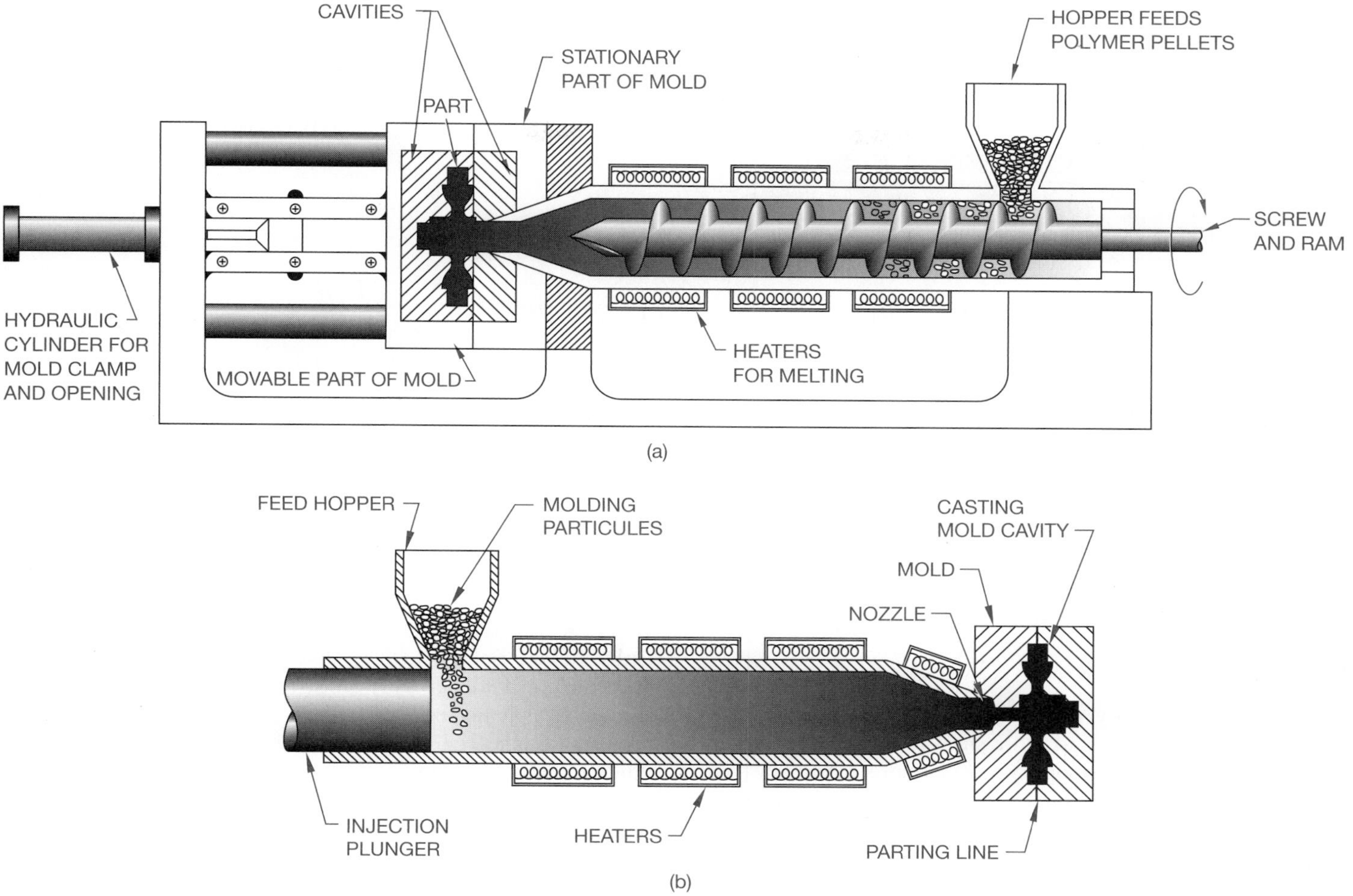

CAVITIES

PART

STATIONARY
PART OF MOLD

HOPPER FEEDS
POLYMER PELLETS

SCREW
AND RAM

HYDRAULIC
CYLINDER FOR
MOLD CLAMP
AND OPENING

MOVABLE PART OF MOLD

HEATERS
FOR MELTING

(a)

FEED HOPPER

MOLDING
PARTICULES

CASTING
MOLD CAVITY

MOLD

NOZZLE

INJECTION
PLUNGER

HEATERS

PARTING LINE

(b)

FIGURE 11.53 ■ **(a)** The process of creating a product and the parts of the screw injection system. **(b)** The process of creating a product and the parts of the nozzle injection system.

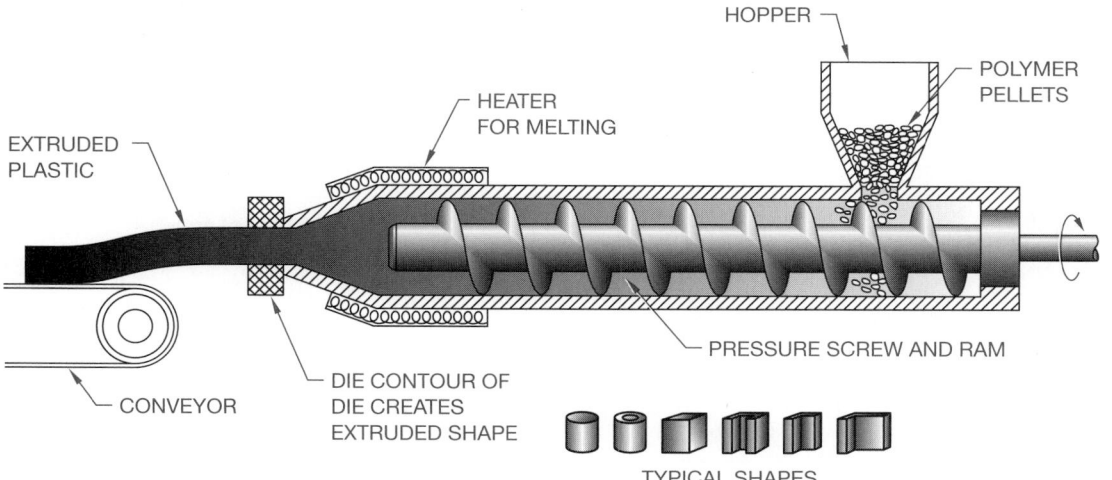

EXTRUDED
PLASTIC

HEATER
FOR MELTING

HOPPER

POLYMER
PELLETS

CONVEYOR

DIE CONTOUR OF
DIE CREATES
EXTRUDED SHAPE

PRESSURE SCREW AND RAM

TYPICAL SHAPES

FIGURE 11.54 ■ The extrusion porcess in action.

The Blow Molding Process

The blow molding process is commonly used to produce hollow products such as bottles, containers, receptacles, and boxes. This process works by blowing hot polymer against the internal surfaces of a hollow mold. The molten plastic enters around a tube that also forces air inside the material, which forces it against the interior surface of the mold. The polymer expands to a uniform thickness against the mold. The mold is formed in two halves, so when the

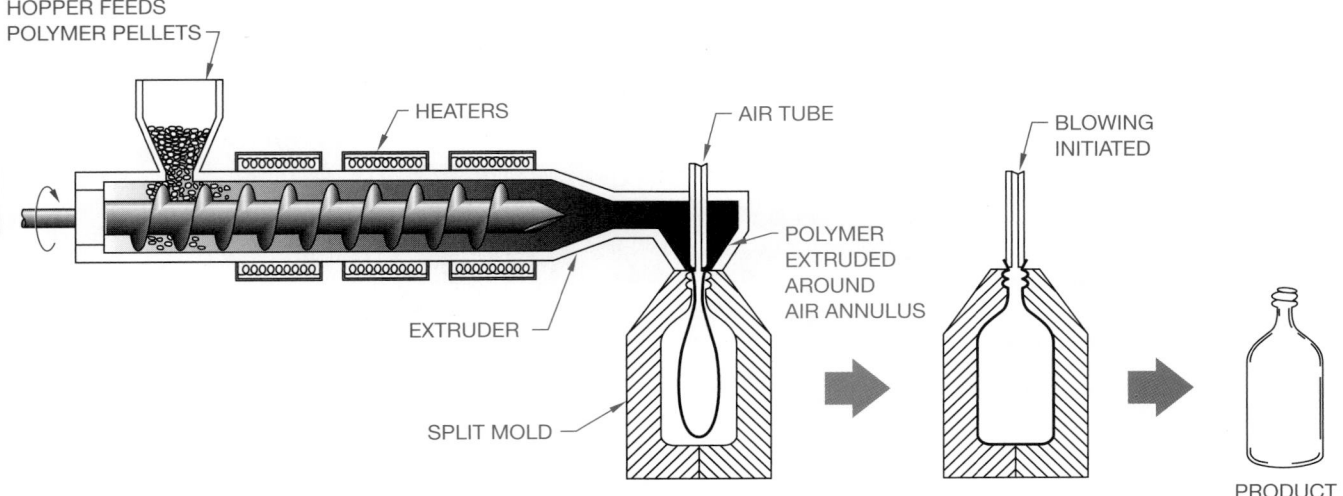

FIGURE 11.55 ■ The blow molding process.

plastic cools, the mold is split to remove the product. The blow molding process is shown in Figure 11.55.

The Calendering Process

The calendering process is generally used to create products such as vinyl flooring, gaskets, and other sheet products. This process fabricates sheet or film thermoplastic or thermoset plastics by passing the material through a series of heated rollers. The space between the sets of rollers gets progressively smaller until the distance between the last set of rollers establishes the desired material thickness. Figure 11.56 illustrates the calendering process used to produce sheet plastic products.

The Rotational Molding Process

This process is typically used to produce large containers such as tanks, hollow objects such as floats, and other similar types of large hollow products. This process works by placing a specific amount of polymer pellets into a metal mold. The mold is then heated as it is rotated. This forces the molten material to form a thin coating against the sides of the mold. When the mold is cooled, the product is removed. The rotational molding process is shown in Figure 11.57.

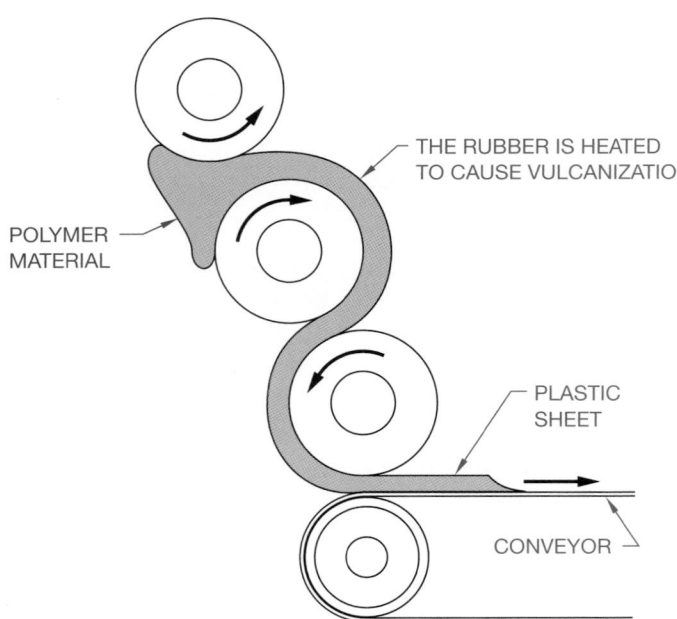

FIGURE 11.56 ■ The calendering process used to produce sheet plastic products.

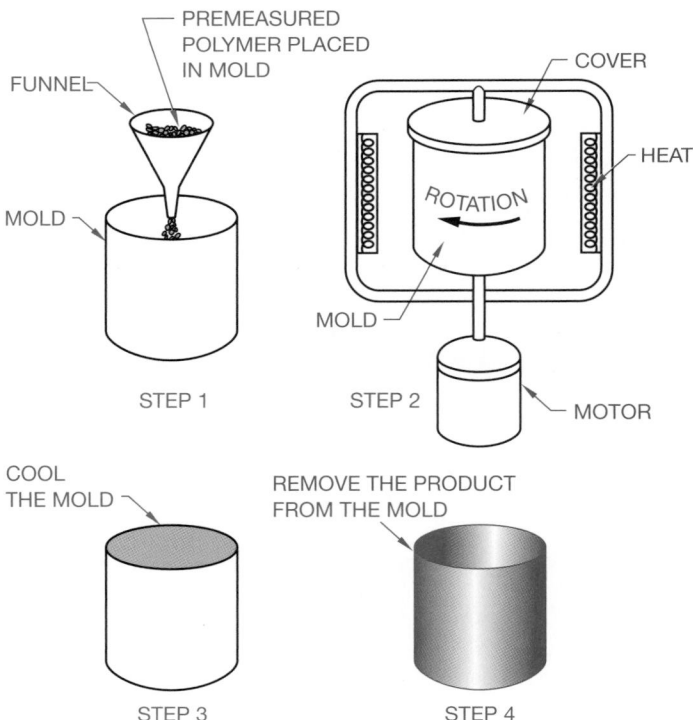

FIGURE 11.57 ■ The rotational molding process.

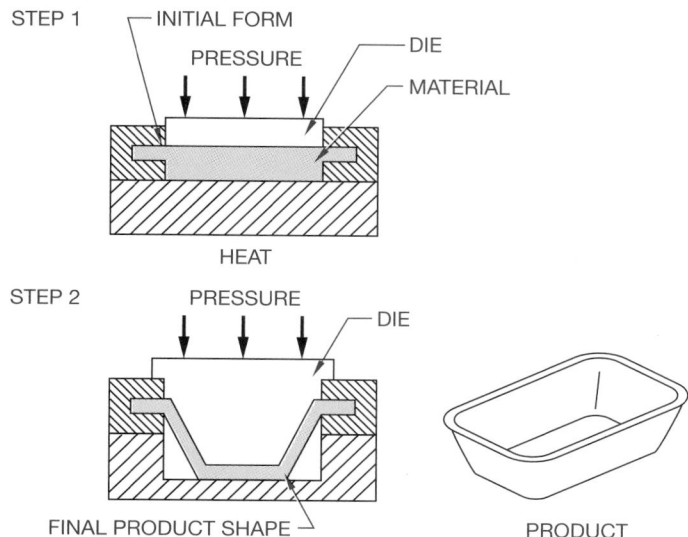

FIGURE 11.58 ■ The solid phase forming process.

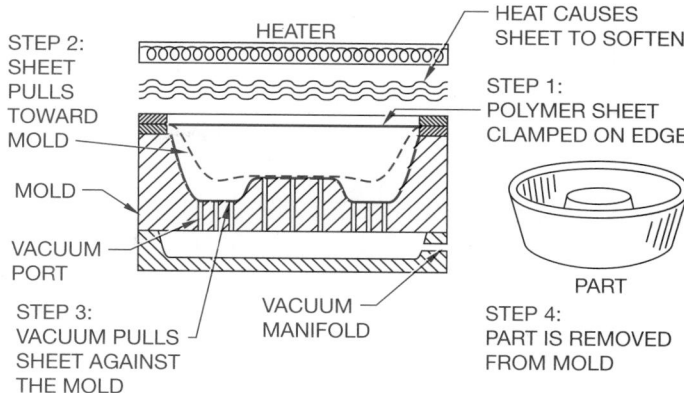

FIGURE 11.59 ■ Thermoforming of plastic. Vacuum pressure is commonly used to suck the hot material down against the mold.

The Solid Phase Forming Process

This process can be used to make a variety of objects including containers, electrical housings, automotive parts, and anything that has detailed shapes. Products can be stronger using this process, because the polymer is heated but not melted. The solid phase forming process works by placing material into an initial hot die where it takes the preliminary shape. As the material cools, a die that matches the shape of the desired product forms the final shape. This process is similar to metal forging. Figure 11.58 shows the solid phase forming process.

Thermoforming of Plastic

This process is similar to solid phase forming, but can be done without a die. This process is used to make all types of thin-walled plastic shapes such as containers, guards, fenders, and other similar products. The process works by taking a sheet of material and heating it until it softens and sinks down by its own weight into a mold that conforms to the desired final shape. Vacuum pressure is commonly used to suck the hot material down against the mold as shown in Figure 11.59.

Free-form Fabrication of Plastic

The free-form fabrication process uses a computer model that is traced in thin cross sections to control a laser that deposits layers of liquid resin or molten particles of plastic material to form the desired shape. Using the laser to fuse several thin coatings of powder polymer to form the desired shape is also an option for this process. This process is often referred to by the trade name Stereolithography. Figure 11.60 shows the free-form process.

Thermoset Plastic Fabrication Processes

The manufacture of thermoset products can be more difficult than thermoplastic fabrication, because thermosets cannot be remelted once they have been melted and formed the first time. This char-

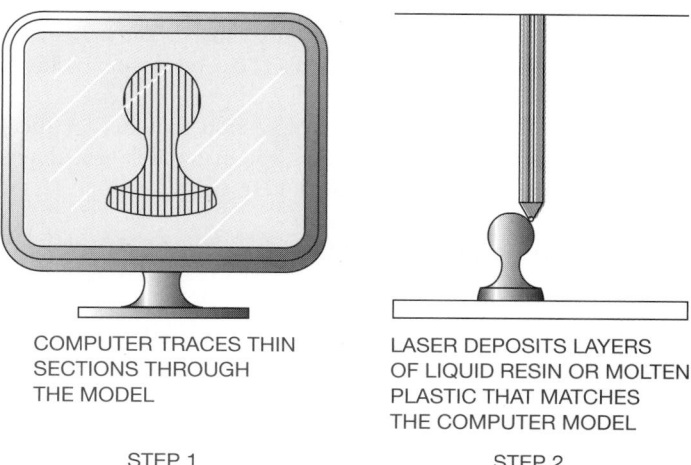

FIGURE 11.60 ■ The free-form fabrication process.

acteristic makes it necessary to keep the manufacturing equipment operational for as long of a period as possible. When process equipment is shut down, it must be thoroughly cleaned before the thermoset material solidifies. The injection molding process that was discussed earlier can be used, but extreme care must be taken to ensure that the equipment is kept clean. For this reason, other methods are commonly used for manufacturing thermoset products. Thermoset plastic production generally takes longer than thermoplastic production because thermoset plastics require a longer cure time. The most common production practices are casting, compression molding, foam molding, reaction injection molding, transfer molding, sintering, and vulcanization.

Casting Thermoset Plastics

The method used to cast plastics is very similar to the permanent casting of metals that was explained earlier in this chapter. When casting plastic, the molten polymer or resin is poured into a metal flask with a mold that forms the desired shape of the product. The plastic product is removed when it has solidified. The casting process is shown in Figure 11.61.

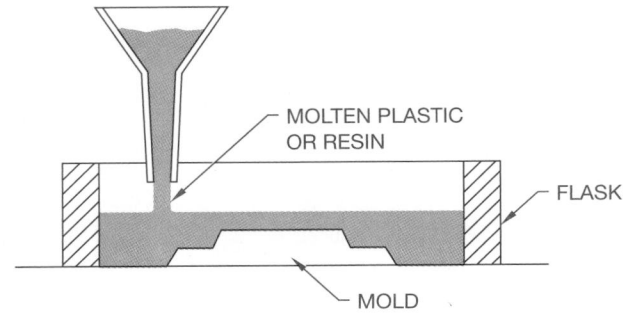

FIGURE 11.61 ■ Casting thermoplastics.

Compression Molding and Transfer Molding

These are common fabrication processes for thermosets. These processes use a specific amount of material that is heated and placed in a closed mold where additional heat and pressure are applied until the material takes the desired shape. The material is then cured and removed from the mold. The compression molding process is shown in Figure 11.62.

Transfer molding is similar to compression molding for thermoset plastic products. In this process, the material is heated and then forced under pressure into the mold.

Foam Molding and Reaction Injection Molding

Foam molding is similar to casting, but this process uses a foam material that expands during the cure to fill the desired mold. The foam molding process can be used to make products of any desired shape, or sheets of foam products. The foam molding process is used to make a duck decoy in Figure 11.63.

The reaction injection molding process is similar to the foam process, only because the material that is used expands to fill the mold. This process is often used to fabricate large parts such as automobile dashboards and fenders. Polymer chemicals are mixed together under pressure and then poured into the mold where they react and expand to fill the mold.

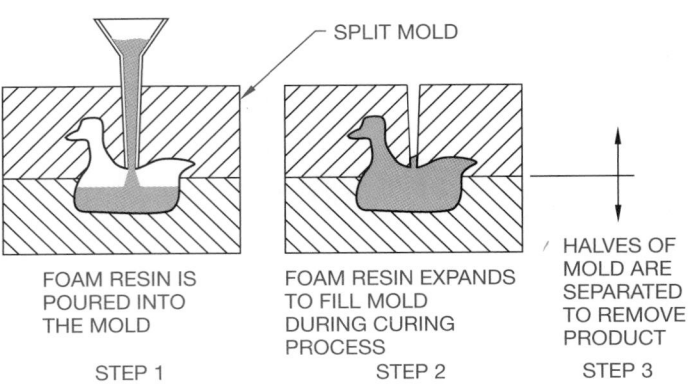

FIGURE 11.63 ■ The foam molding process.

The Sintering Process

Sintering is a process that takes powdered particles of material and produces detailed products under heat, pressure, and chemical reaction. The powder particles do not melt, but they do join together into a dense and solid structure. This process is used when creating products for high-temperature applications.

The Vulcanization Process

Vulcanization is used to make rubber products such as tires and generally other circular and cylindrical shapes. This process works by wrapping measured layers of the polymer around a steel roll in the form of the desired product. The steel roll and material is placed inside an enclosure where steam is introduced to create heat and pressure. This process forces the molecules of the material to form a solid rubber coating on the roll. The vulcanization process is shown in Figure 11.64.

Manufacturing Composites

Composites are also referred to as *reinforced plastics*. Earlier in this chapter, there was a discussion about composite materials. These materials combine polymers with reinforcing material such as glass,

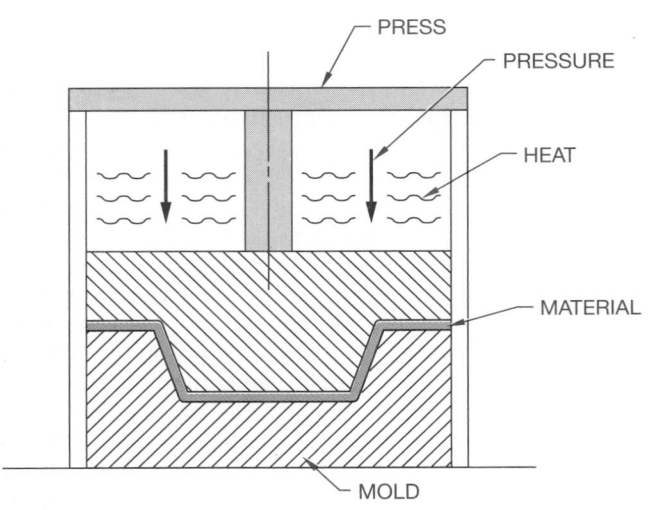

FIGURE 11.62 ■ The compression molding process.

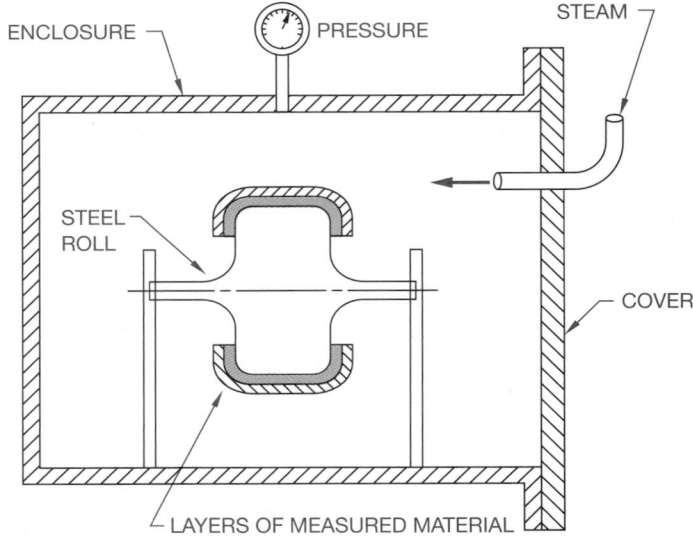

FIGURE 11.64 ■ The vulcanization process.

graphite, thermoplastic fibers, cotton, paper, and metal. The result is very strong material for use in products such as boats, planes, automobile parts, fishing rods, electrical devices, corrosion-resistant containers, and structural members. Composites are less expensive to produce and can be stronger and are lighter in weight than many metals including aluminum, steel, and titanium. The basics of producing composite products are the layering of polymer and reinforcing material in alternate coats or in combination.

The *layering process* combines alternating layers of polymer resin with reinforcing material such as glass. The number of layers determines the desired thickness. During the curing process, the resin saturates the reinforcing material creating a unified composite. Rolling or spraying over the reinforcing material can apply layering in the resin. A similar method uses a spray of resin combined with pieces of reinforcing material. This method is called *chopped fiber spraying*.

Another process uses machines to wind resin-saturated reinforcement fibers around a shaft. That process is referred to as *filament winding*.

A process called *compression molding* is similar to the compression molding process that was explained earlier for thermoset plastic products. The difference is that mixing the polymer with reinforcing fibers creates composites. A similar process takes *continuous reinforcing* strands through a resin bath and then through a forming die. The die forms the desired shape. This method is often used to create continuous shapes of uniform cross section.

A process called *resin transfer molding* is used to make quality composite products with a smooth surface on both sides. This method places reinforcing material into a mold and then pumps resin into the mold.

A process referred to as *vacuum bag forming* uses vacuum pressure to force a thin layer of sheet-reinforced polymer around a mold. Figure 11.65 shows a variety of processes used to make composite products.

TOOL DESIGN

In most production machining operations, special tools are required to either hold the workpiece or guide the machine tool. Tool design involves knowledge of kinematics (study of mechanisms), machining operations, machine tool function, material handling, and material characteristics. *Tool design* is also known as *jig and fixture design*. In mass production industries, jigs and fixtures are essential to ensure that each part is produced quickly and accurately within the dimensional specifications. These tools are used to hold the workpiece so that machining operations are performed in the required positions. Application examples are shown in Figure 11.66. Jigs are either fixed or moving devices that are used to hold the workpiece in position and guide the cutting tool. Fixtures do not guide the cutting tool, but are used in a fixed position to hold the workpiece. Fixtures are often used in the inspection of parts to ensure that the part is held in the same position each time a dimensional or other type of inspection is made.

Jig and fixture drawings are prepared as an assembly drawing where all of the components of the tool are shown as if they were assembled and ready for use, as shown in Figure 11.67. Components of a jig or fixture often include such items as fast-acting clamps, spring-loaded positioners, clamp straps, quick-release locating pins, handles, knobs, and screw clamps, as shown in Figure 11.68, page 320. Normally the part or workpiece is drawn in position using phantom lines, in a color such as red, or a combination of phantom lines and color.

*The Tool Design Process**

Hammers, pliers, screwdrivers, wrenches, and other related items are tools, but they are not the kind of tools that are created by *tool designers*. Tools, as referred to in this discussion, are specially designed and built manufacturing aids, normally in production, that are used to assist operators in the manufacture of specific parts. There are many different kinds of tools that fit this definition. These include machining fixtures, welding fixtures, drill fixtures, drill jigs, inspection fixtures, progressive dies, injection molds, and many others. Tools can be very small and simple, to very large and complicated. Some tools are mechanisms that may resemble machinery, but there is a clear difference between tools and machines. Tools are dedicated to a specific part, family of parts, or product, while machines are for general usage across many parts and products. This differentiation is made more for accounting and tax purposes than it is for technical reasons. Machines are capital investments, while tools are expense items related to a particular part or product. In many cases, the tool may be of such complexity that from a technical perspective, it is a machine, but from an accounting perspective, it is a tool. The following are some very simplified definitions of the typical types of tools that are used in manufacturing.

■ *Drill jigs* are used in drilling operations using hand drills, drill presses, or radial drills, rather than milling machines. The drill jig registers the part relative to the critical datums. *Datums* are important points, axes, surfaces, or planes from which features are established and dimensioned. Datums are discussed in detail in Chapters 12 and 15.

JIG AND FIXTURE DESIGN

Computer-aided design and drafting (CADD) jig and fixture design programs make it possible for the designer or engineer to electronically construct the tooling for a specific application directly on the computer terminal. As with other CADD programs, tooling component libraries, consisting of fully detailed and accurate jig and fixture components, are available. The speed of tooling design sharply increases as compared to manual methods. Items such as clamps, locator pins, rests, or fixture bases can be retrieved from the library and inserted into your design quickly by using the specific functions provided by the software.

CADD APPLICATIONS

*Courtesy of Martin Soll.

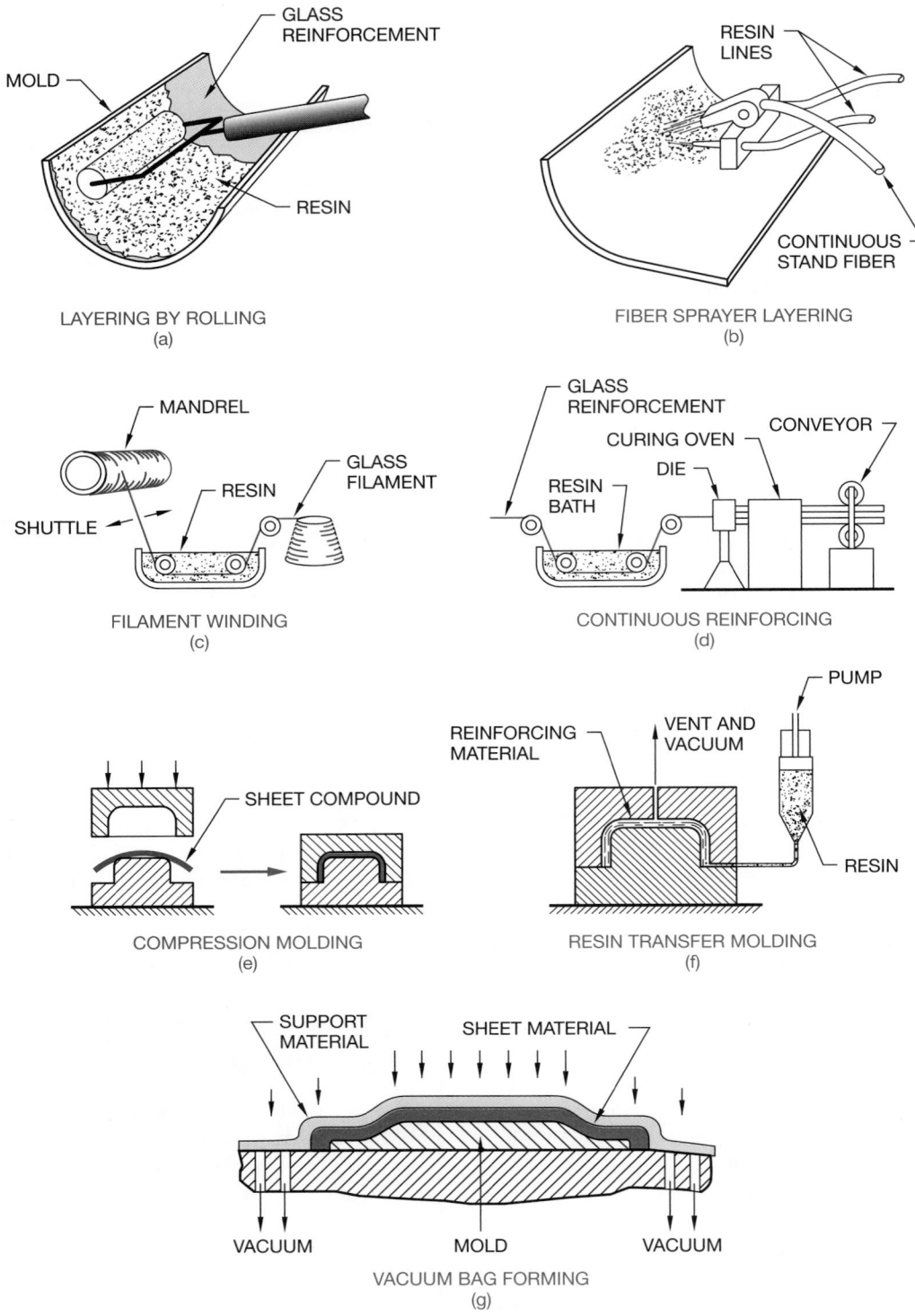

FIGURE 11.65 ■ Common processes used to make composite products. (**a**) The layering process using rolling. (**b**) The layering process using spraying. (**c**) Filament winding. (**d**) The continuous reinforcing strands process. (**e**) Compression molding. (**f**) Resin transfer molding. (**g**) Vacuum bag forming.

The operator then drills through a drill bushing or bushings in the jig, to locate the hole or holes in the part.

■ *Drill fixtures* are sometimes referred to as *holding fixtures*. Drill fixtures are used for drilling operations using milling machines, either manual or computer numerical control (CNC). The drill fixture registers the part relative to the critical datums, but does not have drill bushings. The machine axes are used to locate the hole or holes. Drill

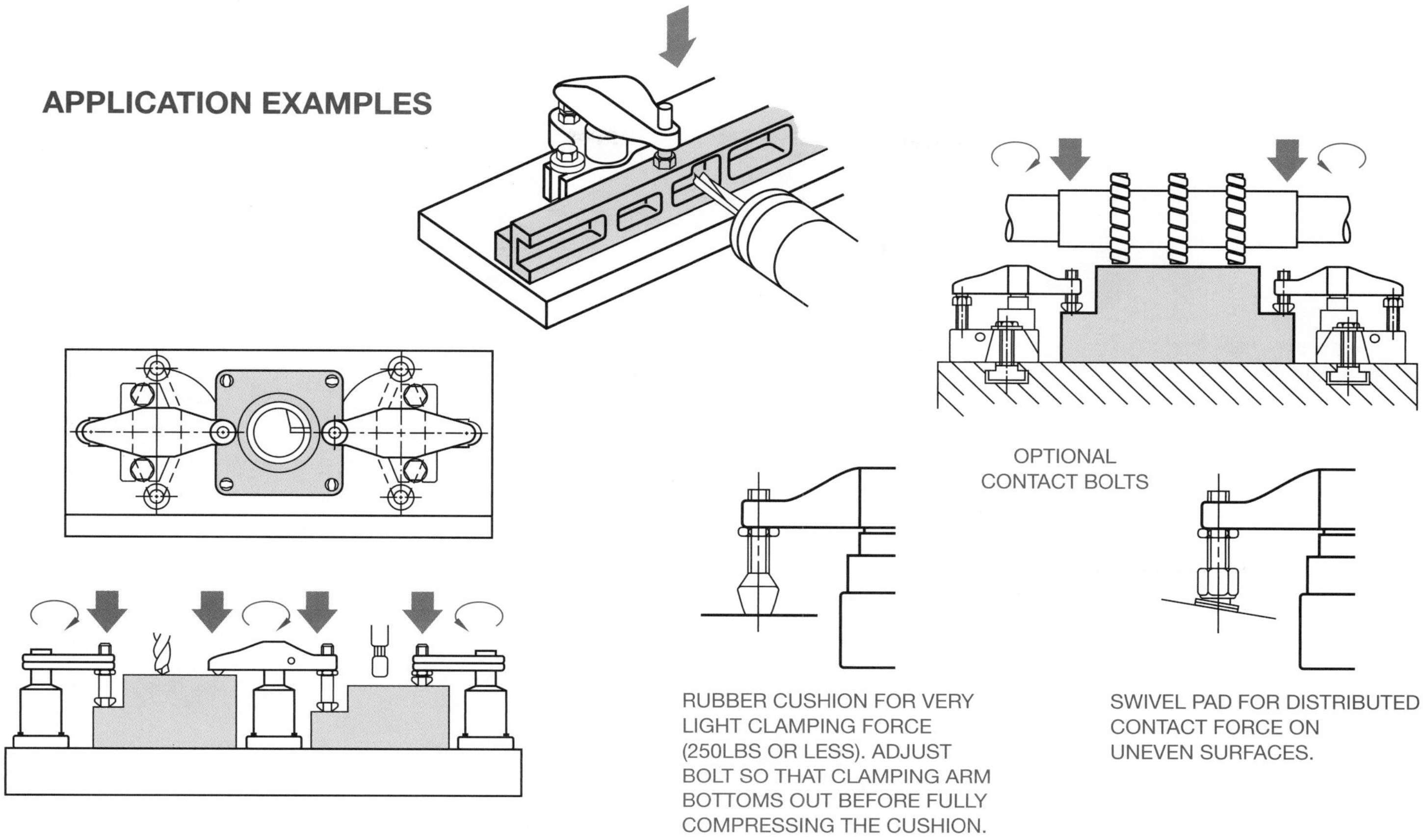

APPLICATION EXAMPLES

RUBBER CUSHION FOR VERY LIGHT CLAMPING FORCE (250LBS OR LESS). ADJUST BOLT SO THAT CLAMPING ARM BOTTOMS OUT BEFORE FULLY COMPRESSING THE CUSHION.

OPTIONAL CONTACT BOLTS

SWIVEL PAD FOR DISTRIBUTED CONTACT FORCE ON UNEVEN SURFACES.

FIGURE 11.66 ■ Fixture application examples. *Courtesy Carr Lane Manufacturing Company.*

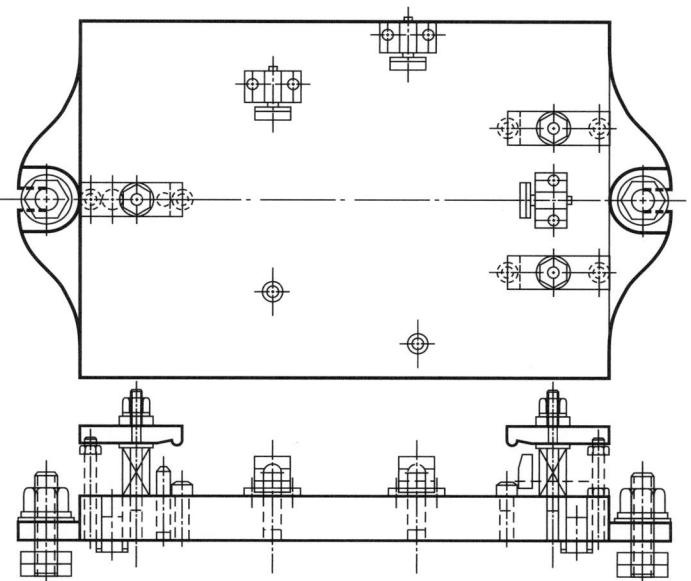

FIGURE 11.67 ■ Fixture assembly drawing. *Courtesy Carr Lane Manufacturing Company.*

fixtures must resist the force of the drilling operation. Typically, drilling forces are relatively low, and only in one direction.

■ *Machining fixtures* are used for machining operations using milling machines that are either manual or CNC. The machining fixture registers the part relative to the critical datums. The machine axes are used to locate the feature or features to be machined. Machining fixtures must resist the force of the machining operation; typically, machining forces are relatively high, and can be in any direction.

■ *Welding fixtures* are more accurately termed *welding jigs,* but the names have become synonymous in the industry. Welding fixtures are used to hold two or more pieces in the proper position and orientation so that the pieces can be welded together.

■ *Inspection fixtures* are very common in the casting industry. They are used to hold a part, registering it on the critical datums, while an inspector checks critical feature sizes and/or locations.

■ *Progressive dies* are the tooling used in punch press operations. Continuous stock of relatively thin material is fed into the punch press from coils. As the punch press cycles, the material is cut to appropriate shape, holes punched, or other operations performed to make a finished part. Cookie cutters, scissors, and paper hole punches are comparable to progressive dies.

Figure 11.69 shows a very simple drill jig tool. This drill jig is designed and manufactured to assist a shop employee in performing a specific task on a production part. The part is very simple, and

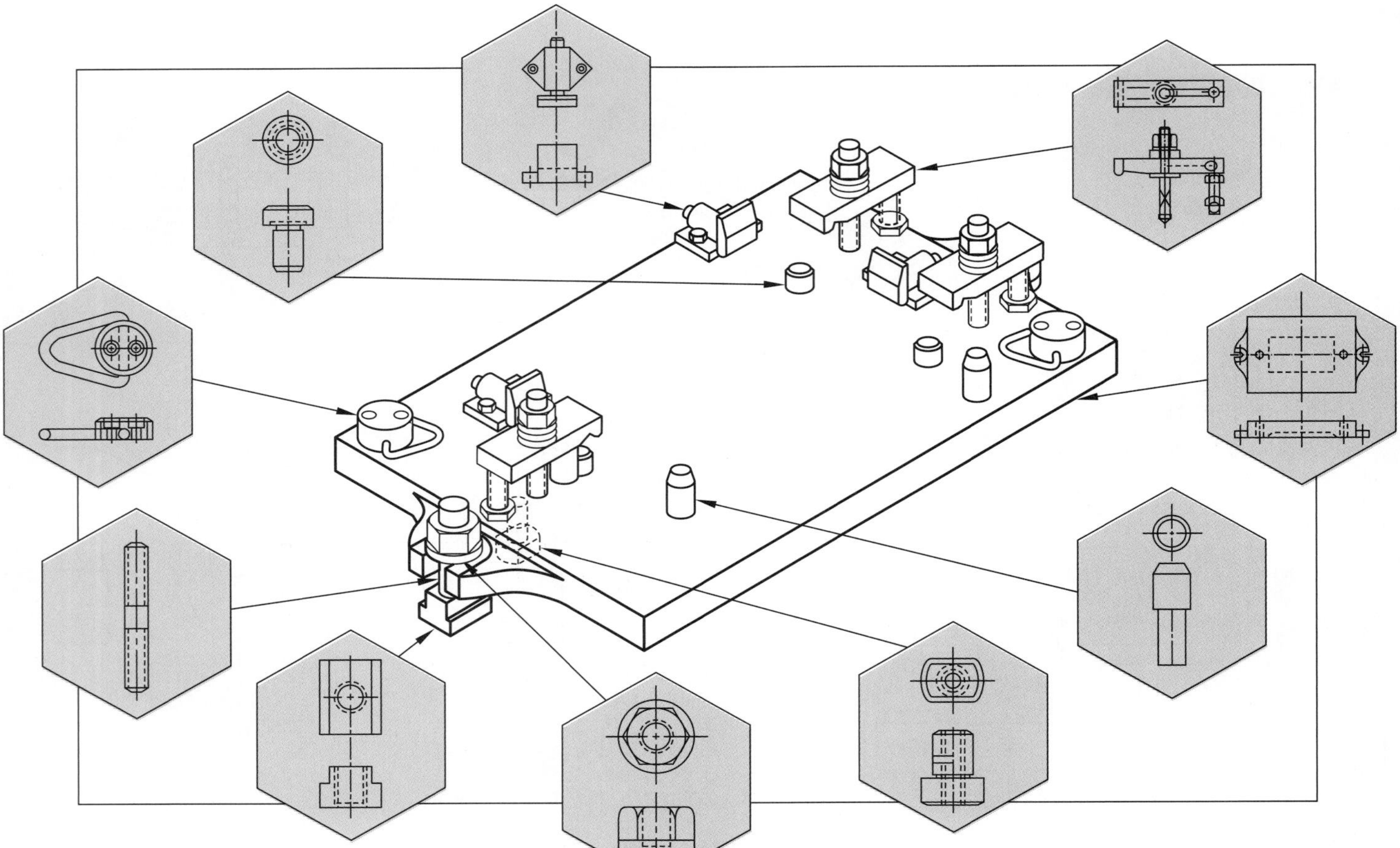

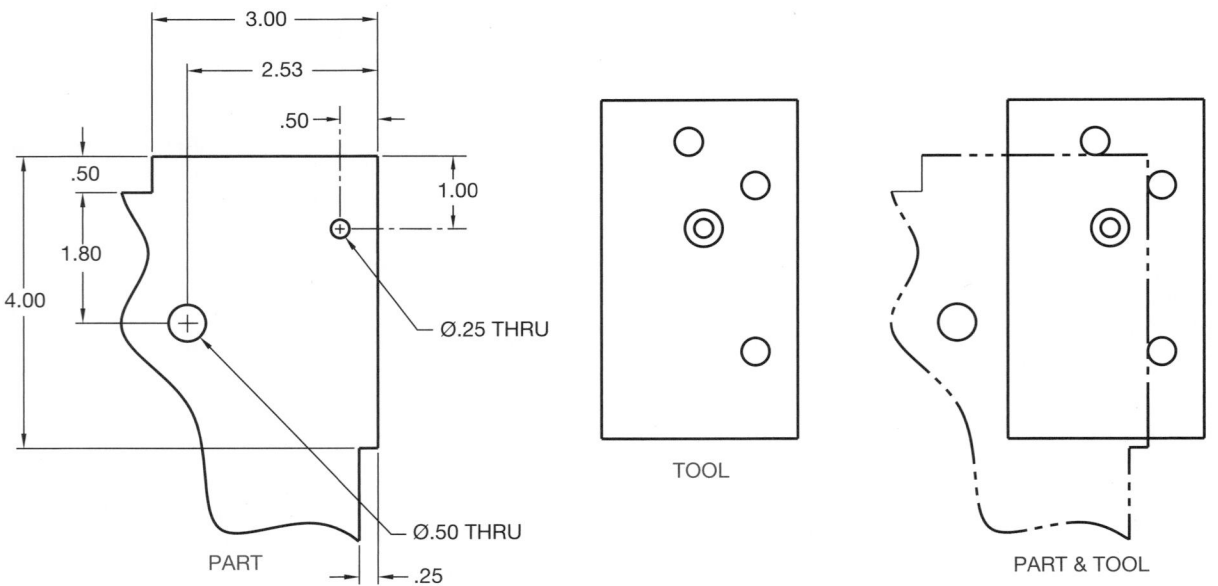

FIGURE 11.68 ■ Fixture components. *Courtesy Carr Lane Manufacturing Company.*

FIGURE 11.69 ■ A simple example of a PART and a TOOL. The task of the employee is to drill a .25-in. diameter hole in a specific location. It is very efficient for the employee to use the TOOL to locate the hole. He or she simply places the TOOL on top of the PART, holds it with one hand, and drills the hole through the PART, using the hole in the TOOL to locate the hole in the PART.

the employee could measure and drill the hole without any tool; but for production work, the tool makes the location of the hole much faster, and the accuracy of the hole location is less dependent on the employee's abilities. This type of drill jig is referred to as a *pickoff jig; pickoff* because it sits on the part, rather than the part being put into the fixture, and *jig* because it actually locates a feature, which is the .25-in. diameter hole in this example.

Now that you have an idea of what makes up a tool, the following gives you the three basic elements of tool design:

1. Tool design is the art of visualizing how shop personnel will accomplish a specific task.
2. Conceptualizing hardware to assit in the accomplishment of that task.
3. Creating drawings so the hardware can be manufactured.

Visualizing and *conceptualizing* are not to be confused with inventing. Tool designers are not inventors. In fact, a good tool designer may not consider him or herself to be creative. Tool designers use existing products and items whenever possible. Many times, an existing design can be modified to satisfy a new requirement, and good tool designers often save their employer time and money by using existing designs. A tool designer *must* be familiar with shop practices, and *must* be able to visualize shop personnel accomplishing specific tasks within the shop environment. This visualization leads to a concept of tools that the shop personnel may use to assist in the accomplishment of the task. The tool designer *must* also be very good at print reading.

The part print contains all the information the tool designer needs. The tool designer must be able to find the important information. Looking back to the example in Figure 11.69, you are shown only a small corner of the part. Even this small corner has much more information than is required for this drill jig. Print reading is the first step in the tool design process. The information necessary to design the drill jig must be found on the part print. Some dimensions are directly related to the drill jig, some are incidental to the tool designer, and some are totally unrelated to this drill jig. Looking at Figure 11.69, it is relatively easy to *sort out the applicable information.* Dimensions defining the feature size and location are directly related to the required tool. Dimensions that define the size and locations of other features that may impact the tool are of incidental interest. Dimensions that define the size and locations of features that do not impact the design of the tool are not relevant to the tool designer. More complicated part prints may require in-depth study to find the relevant information.

Tool designers often receive their assignments from a manufacturing engineer (ME). The ME defines the specific task, and how it is to be accomplished. Consider the example in Figure 11.69 again. The tool designer would receive a tooling design request (TDR) from the ME. The TDR would have an engineering drawing of the part attached, and might contain the following information:

PART NAME: ----------------- PART
PART NO: ---------------------- XXXX-XX
OPERATION: ------------------- DRILL .25" DIA HOLE
MACHINE: -------------------- USE HAND DRILL
FIXTURE: --------------------- PICKOFF JIG LOCATING
                                THE HOLE .50" AND 1.00"
                                FROM THE EDGES OF THE PLATE

The decision to use a pickoff jig and hand drill rather than a radial drill or even a CNC machine would be made prior to the tool designer's involvement in the project. If the ME had determined that the best way to accomplish this task was to put this part on a CNC machine to drill the hole, then the TDR would be written accordingly. The tool designer would design a drill fixture that would be mounted on the CNC machine table. The fixture would hold and clamp the PART. The PART would be located with accurate registration to the two critical edges that define the location of the .25-in. diameter hole.

The Tool Designer's Tools

A tool designer uses manual drafting or CAD practices to design a fixture. He or she must capture the concept on paper or computer screen in order for the fixture to be built. The tool designer must also be familiar with standard tooling components that are available from numerous manufacturers. Various components, such as rest pads, clamps, pins, and drill bushings to name a few, are available in a wide variety of sizes and shapes. Remember, tool designers do not create something that already exists, but use standard components whenever possible.

Tools, in general, must possess some of the following certain qualities:

■ Reliability.

■ Repeatability.

■ Ease of use.

■ Ease of manufacture.

■ Ease of maintenance and repair.

Figure 11.70 shows the design that would go to the tool room for a machinist to use to build the pickoff drill jig previously discussed and shown in Figure 11.69. The drawing shown is a combination assembly drawing and detail drawing. More complex fixtures require the fixture assembly to be drawn on one sheet, and the component details to be drawn on separate sheets. Detail and assembly drawings are covered in Chapter 18, Working Drawings.

Notice that the PART in Figure 11.69 is dimensioned using conventional dimension and extension lines, and the DRILL JIG drawing in Figure 11.70 is dimensioned using arrowless dimensioning.

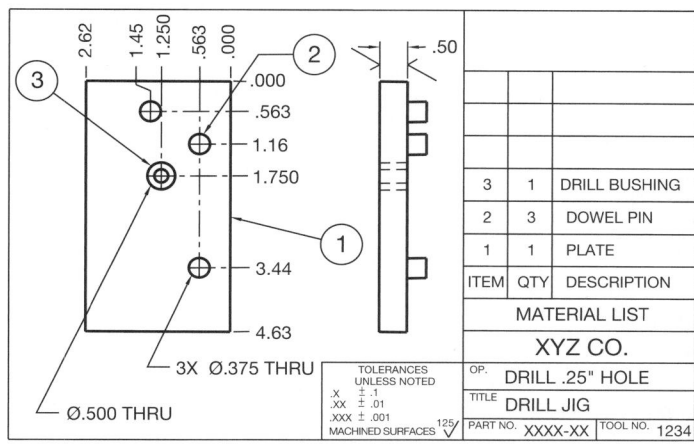

FIGURE 11.70 ■ A simplified example of a tool design drawing.

Either type of dimensioning practice can be used on both drawings. Chapter 12, Dimensioning and Tolerancing, completely explains each type of dimensioning practice.

The PLATE (Figure 11.70, Item 1) is the only piece that must be made in the tool room. The *tool room* is the shop where tools are manufactured. The DOWEL PIN (Figure 11.70, item 2) and the DRILL BUSHING (Figure 11.70, item 3) are purchased components. The following evaluates if this jig would meet the list of quality requirements for tools:

- *Reliability*—Yes, this jig is simple, and very little could break or cause problems.

- *Repeatability*—Yes, again the simplicity makes repeatability inherent. The drill jig registers directly on the datum surfaces from which the hole location is defined.

- *Ease of use*—Yes, the jig offers the operator a simple means to locate the hole.

- *Ease of manufacture*—Yes, with only one manufactured part, and four purchased parts, it will be very easy to make.

- *Ease of maintenance and repair*—Yes, should something break or wear out, the component parts can be easily replaced.

In summary, you have learned about tools, tool design, and tool designers. Tools are shop aids, sometimes simple, and sometimes very complex. Tool design is the process of turning a concept of a tool into drawings so that the fixture can be manufactured. Tool designers are the people who imagine the concepts and turn them into drawings.

COMPUTER-INTEGRATED MANUFACTURING (CIM)

Completely automated manufacturing systems combine computer-aided design and drafting (CADD), computer-aided engineering (CAE), and computer-aided manufacturing (CAM) into a controlled system known as *computer-integrated manufacturing* (CIM). In Figure 11.71, CIM brings together all the technologies in a management system, coordinating CADD, CAM, CNC, robotics, and material handling from the beginning of the design process through the packaging and shipment of the product. The computer system is used to control and monitor all the elements of the manufacturing system. Before a more complete discussion of the CIM systems, it is a good idea to define some of the individual elements.

Computer-Aided Design and Drafting (CADD)

Engineers and designers use the computer as a flexible design tool. Designs can be created graphically using 3-D models. The effects of changes can be quickly seen and analyzed. This allows the engineer to perform experimentation, perform stress analysis, and make calculations right on the computer. In this manner, engineers can be more creative and improve the quality of the product at less cost. Computer-aided drafting is a partner to the design process. Accurate quality 2-D drawings are created from the designs. Drafting with the computer has increased productivity from two to twenty times over manual techniques depending on the project and skill of the drafter.

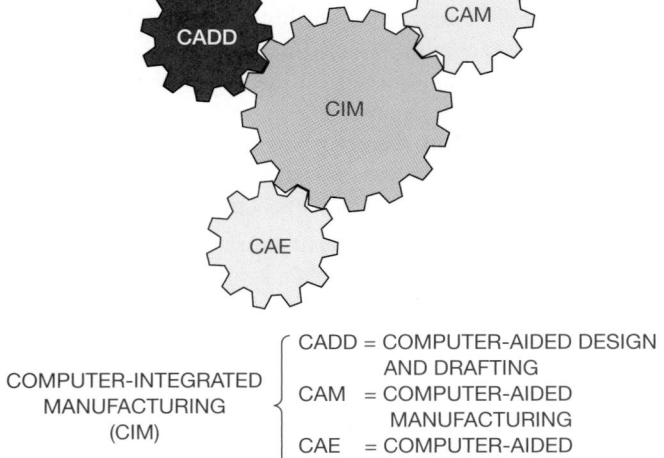

COMPUTER-INTEGRATED MANUFACTURING (CIM)

CADD = COMPUTER-AIDED DESIGN AND DRAFTING
CAM = COMPUTER-AIDED MANUFACTURING
CAE = COMPUTER-AIDED ENGINEERING

FIGURE 11.71 ■ Computer-integrated manufacturing (CIM) is the bringing together of the technologies in a management system.

The computer has also revolutionized the storage of drawings. There is no longer a need for rooms full of drawing file cabinets.

Computer-Aided Engineering (CAE)

Three-dimensional models of the product are used for *finite element analysis*. This is done in a program in which the computer breaks the model up into finite elements, which are small rectangular or triangular shapes. Then the computer is able to analyze each element and determine how it will act under given conditions. It also evaluates how each element acts with the other elements and with the entire model. CAE also allows the engineer to simulate function and motion of the product without the need to build a real prototype. In this manner, the product can be tested to see if it works as it should. This process has also been referred to as *predictive engineering,* in which a computer software prototype, rather than a physical prototype, is made to test the function and performance of the product. In this manner, design changes may be made right on the computer screen. Key dimensions are placed on the computer model, and by changing a dimension, you can automatically change the design or you can change the values of variables in mathematical engineering equations to automatically alter the design.

Computer Numerical Control (CNC)

Computer numerical control is the use of a computer to write, store, and edit numerical control programs to operate a machine tool. Numerical control (NC) is a method of controlling a machine tool using coded computer language.

Computer-Aided Manufacturing (CAM)

CAM is a concept that surrounds any use of computers to aid in any manufacturing process. CAD and CAM work best when product programming is automatically performed from the model geometry created during the CAD process. Complex 3-D geometry can be quickly and easily programmed for machining.

Computer-Aided Quality Control (CAQC)

Information on the manufacturing process and quality control is collected by automatic means while parts are being manufactured. This information is fed back into the system and compared to the design specifications or model tolerances. With this type of monitoring of the mass production process, it can be ensured that the highest product quality is maintained.

Robotics

According to the Society of Manufacturing Engineers (SME) Robot Institute of America, a *robot* is a reprogrammable multifunctional manipulator designed to move material, parts, tools, or specialized devices through variable programmed motions for the performance of a variety of tasks. *Reprogrammable* means the robot's operating program may be changed to alter the motion of the arm or tooling. *Multifunctional* means the robot is able to perform a variety of operations based on the program and tooling it uses. A typical robot is shown in Figure 11.72.

To make a complete *manufacturing cell*, the robot is added to all of the other elements of the manufacturing process. The other elements of the cell include the computer controller, robotic program, tooling, associated machine tools, and material handling equipment.

Closed-loop (servo) and *open-loop (nonservo)* are the two types of control used to position the robot tooling. The robot is constantly

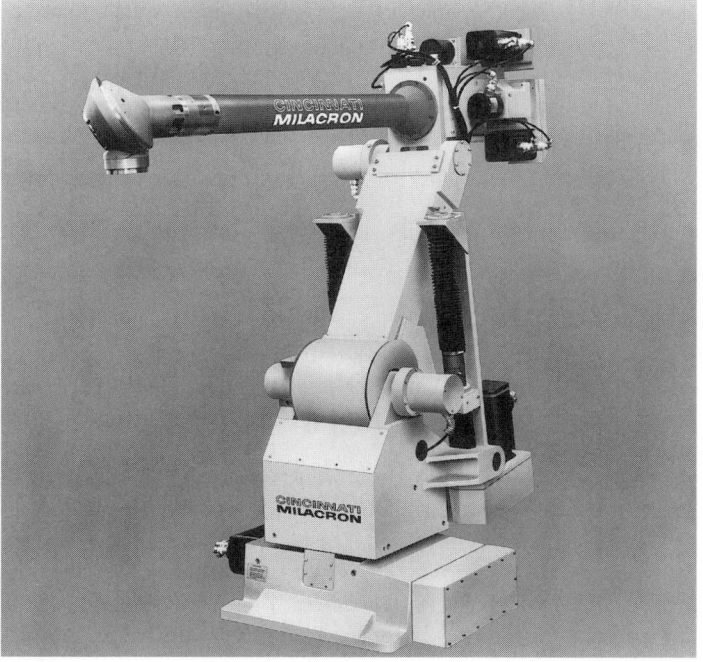

FIGURE 11.72 ■ A typical robot. *Courtesy Cincinnati Milacron Marketing Company.*

monitored by position sensors in the closed-loop system. The movement of the robot arm must always conform to the desired path and speed. Open-loop robotic systems do not constantly monitor the position of the tool while the robot arm is moving. The control happens at the end of travel where limit controls position the accuracy of the tool at the desired place.

The Human Factor

The properly operating elements of the CIM system provide the best manufacturing automation available. What has been described in this discussion might lead you to believe that the system can be set up, turned on, and the people can go home. Not so. An important part of CIM is the human element. People are needed to handle the many situations that happen in the manufacturing cell. The operators constantly observe the machine operation to ensure that material is feeding properly and the quality is maintained. In addition to monitoring the system, people load raw material into the system, change machine tools when needed, maintain and repair the system as needed, and handle the completed work when the product is finished.

INTEGRATION OF COMPUTER-AIDED DESIGN AND COMPUTER-AIDED MANUFACTURING (CAD/CAM)

Computer-aided design (CAD) and computer-aided manufacturing (CAM) can be set up to create a direct link between the design and manufacture of a product. The CAD program is used to create the product geometry. This can be in the form of 2-D multiview drawings as discussed in Chapter 7, Multiviews, of this text, or as 3-D models as explained in Chapter 20, Solid Modeling, Animation, and Virtual Reality. The drawing geometry is then used in the CAM program to generate instructions for the computer numerical control (CNC) machine tools employing stamping, cutting, burning, bending, and other types of operations discussed throughout this chapter. This is commonly referred to as *CAD/CAM integration*. The CAD and CAM operations can be performed on the same computer, or the CAD work can be done at one location and the CAM program can be created at another location. If both the design and manufacturing are done within the company at one location, then the computers can be linked through the local area network. If the design is done at one location and the manufacturing done at another, then the computers can be linked through the Internet, or files can be transferred on disk or CD.

CAD/CAM is commonly used in modern manufacturing because it increases productivity over conventional manufacturing methods. The CAD geometry or model is created during the design and drafting process and is then used directly in the CAM process for the development of the CNC programming. The coordination can also continue into computerized quality control.

The CAD/CAM integration process allows the CAM program to import data from the CAD software. The CAM program then uses a series of commands to instruct CNC machine tools by setting up tool paths. The tool path includes the selection of specific tools to accomplish the desired operation. This can also include specifying tool feed rates and speeds, selecting tool paths and cutting methods, activating tool jigs and fixtures, and selecting coolants for material removal. Some CAM programs automatically calculate the tool offset based on the drawing geometry. An example of tool offset is shown in Figure 11.73. CAM programs such as SurfCAM and MasterCAM directly integrate the CAD drawing geometry from programs such as AutoCAD as a reference. The CAM programmer then just establishes the desired tool and tool path. The final CNC program is generated when the postprocessor is run. A *postprocessor* is an integral piece of software that converts a generic, CAM system tool path into usable CNC machine code (G-code).

The *CNC program* is a sequential list of machining operations in the form of a code that is used to machine the part as needed. Typically known as G & M code, these codes invoke preparatory functions and control tool and machine movement, spindle speed and direction, and other operations such as clamping, part manipulation, and on-off switching. The machine programmer is trained to select the proper machine tools. A separate tool is used for each operation. These tools can include milling, drilling, turning, threading, and grinding. The CNC programmer orders the machining sequence, selects the tool for the specific operation, and determines the tool feed rate and cutting speed depending on the material.

CAM software programs such as SurfCAM and MasterCAM increase productivity by assisting the CNC programmer in creating the needed CNC code. CAD drawings from programs such as AutoCAD or solid models from software such as Solid Works can be transferred directly to the CAM program. The following sequence of activities is commonly used to prepare the CAD/CAM integration:

1. Create the part drawing or model using programs such as AutoCAD, Autodesk Inventor, or Solid Works. However, programmers commonly originate the model in the CAD/CAM software, such as MasterCAM or SurfCAM.

2. Open the CAD file in the CAM program.

3. Run the CAM software program such as SurfCAM or MasterCAM to establish the following:
   - Choose the machine or machines needed to manufacture the part.
   - Select the required tooling.
   - Determine the machining sequence.
   - Calculate the machine tool feed rates and speeds based on the type of material.
   - Verify the CNC program using the software's simulator.
   - Create the CNC code.

4. Prove-out the program on the CNC machine tool.

5. Run the program to manufacture the desired number of parts.

STATISTICAL PROCESS CONTROL (SPC)

A system of quality improvement is helpful to anyone who turns out a product or is engaged in a service and who wishes to improve the quality of work and at the same time increase the output, all with less labor and at reduced cost. Competition is here to stay, regardless of the nature of the business. Improved quality means less waste and less rework, resulting in increased profits and an improved market position. Customarily, many managers see quality as a drag on profits. Quality is often placed after cost and delivery, because some managers believe that high quality can be achieved only through costly and slow inspection processes. Many managers are now seeing the influence of quality on sales because high quality has become an important criterion in their customers' purchase decisions. In addition, poor quality is expensive. It is estimated that between 15% and 40% of the American manufacturer's product cost is a result of unacceptable output. Regardless of the goods or service produced, it is always less costly to do it right the first time. Improved quality improves productivity, increases sales, reduces cost, and improves profitability. The net result is continued business success.

Traditionally, a type of quality control/detection system has been used in most organizations in the United States. This system comprises customer demand for a product, which is then manufactured in a process made up of a series of steps or procedures. Input to the process includes machines, materials, workforce, methods, and environment, as shown in Figure 11.74. Once the product or service is produced, it goes on to an inspection operation where decisions are made to ship, scrap, rework, or otherwise correct any defects when discovered (if discovered). In actuality, if nonconforming products are being produced, then some are being shipped. Even the best inspection process screens out only a portion of the defective goods. Problems inherent in this system are that it does

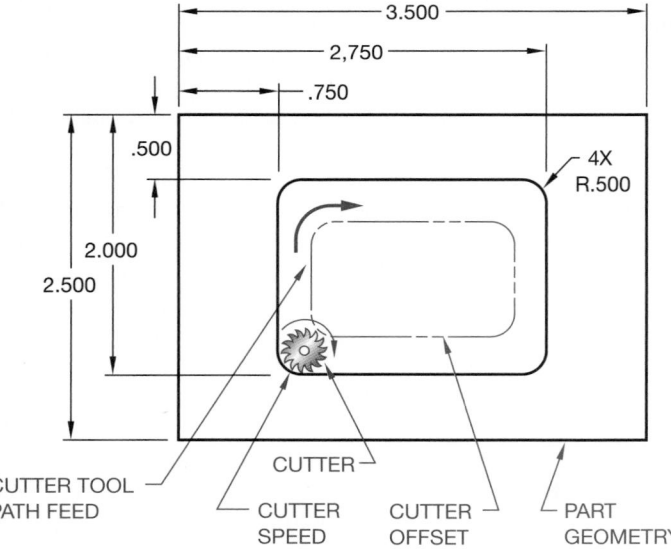

FIGURE 11.73 ■ The machine tool cutter, tool path, cutter speed, and cutter offset.

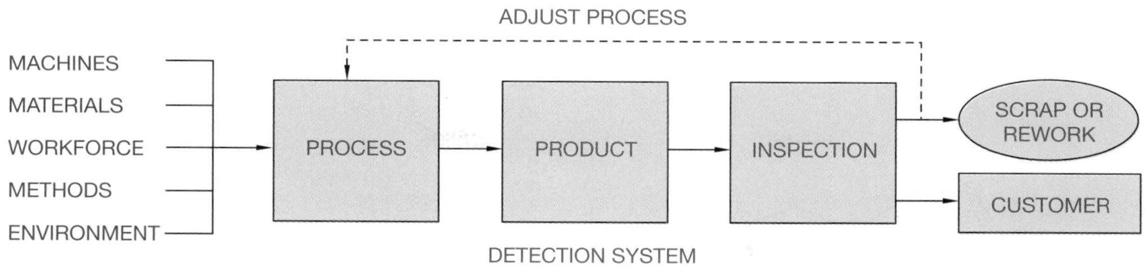

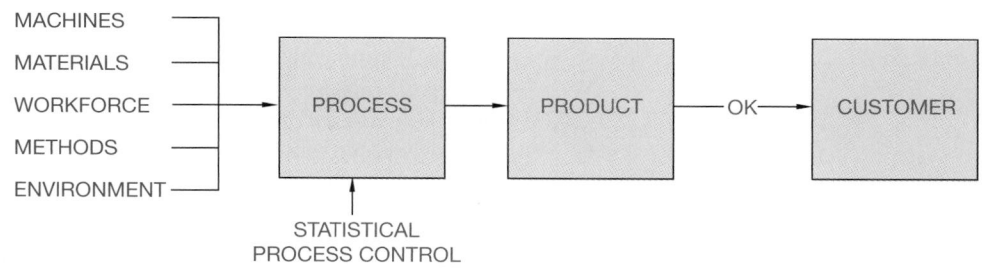

FIGURE 11.74 ■ Quality control/detection system.

FIGURE 11.75 ■ Quality control/prevention system.

not work very well and it is costly. American businesses have become accustomed to accepting these limitations as the "costs of doing business."

The most effective way to improve quality is to alter the production process, the system, rather than the inspection process. This entails a major shift in the entire organization from the detection system to a prevention mode of operation. In this system, the elements (inputs, process, product or service, customer) remain the same, but the inspection method is significantly altered or eliminated. A primary difference between the two systems is that in the prevention system, statistical techniques and problem-solving tools are used to monitor, evaluate, and provide guidance for adjusting the process to improve quality. *Statistical process control* (SPC) is a method of monitoring a process quantitatively and using statistical signals to either leave the process alone, or change it. (See Figure 11.75.) It involves several fundamental elements:

1. The process, product, or service must be measured. It can be measured using either variables (a value that varies), or attributes (a property or characteristic) from the data collected. The data should be collected as close to the process as possible. If you are collecting data on a particular dimension of a manufactured part, it should be collected by the machinist who is responsible for holding that dimension.

2. The data can be analyzed using control charting techniques. Control charting techniques use the natural variation of a process, determining how much the process can be expected to vary if the process is operationally consistent. The control charts are used to evaluate whether the process is operating as designed, or if something has changed.

3. Action is taken based on signals from the control chart. If the chart indicates that the process is *in control* (operating consistently), then the process is left alone at this point. On the other hand, if the process is found to be *out of control* (changing more than its normal variability allows), then action is taken to bring it back into control. It is also important to determine how well the process meets specifications and how well it accomplishes the task. If a process is not in control, then its ability to meet specifications is constantly changing. Process capability cannot be evaluated unless the process is in control. Process improvement generally involves changes to the process system that will improve quality or productivity or both. Unless the process is consistent over time, any actions to improve it may be ineffective.

Manufacturing quality control often uses computerized monitoring of dimensional inspections. When this is done, a chart is developed that shows feature dimensions obtained at inspection intervals. The chart shows the expected limits of sample averages as two dashed parallel horizontal lines, as shown in Figure 11.76. It is important not to confuse control limits with tolerances—they are not related to each other. The control limits come from the manufacturing process as it is operating. For example, if you set out to make a part $1.000 \pm .005$ in., and you start the run and periodically take five samples and plot the averages (x) of the samples, the sample averages will vary less than the individual parts. The control limits represent the expected variation of the sample averages if the process is stable. If the process shifts or a problem occurs, the control limits signal that change. Notice in Figure 11.76 that the x values represent the average of each five samples; $\bar{x}$ is the average of averages over a period of sample taking. The *upper control limit* (UCL) and the *lower control limit* (LCL) represent the expected

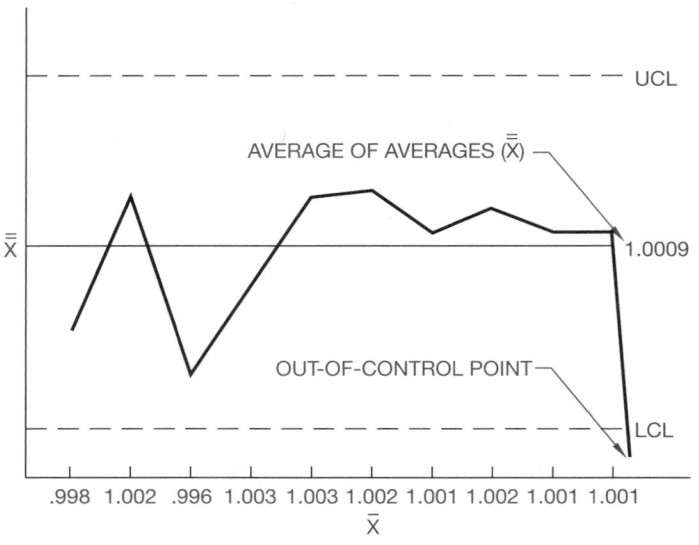

FIGURE 11.76 ■ Quality control chart.

variation of the sample averages. A sample average may be "out of control," yet remain within tolerance. During this part of the monitoring process, the out-of-control point represents an individual situation that may not be a problem; however, if samples continue to be measured out of control limits, then the process is out of control (no longer predictable) and therefore may be producing parts outside the specification. Action must then be taken to bring the process back into statistical control, or 100% inspection must be resumed. The SPC process only works when a minimum of twenty-five sample means are in control. When this process is used in manufacturing, part dimensions remain within tolerance limits and parts are guaranteed to have the quality designed.

WEB SITE RESEARCH

The following Web sites can assist you in doing additional research on subjects such as manufacturing materials, manufacturing practices, tooling, and related areas:

www.allamericanproducts.com—All American Products: tooling components.

www.ab.com—Allen-Bradley: industrial electrical controls.

www.aluminum.org—Aluminum Association.

www.asme.org—American Society of Mechanical Engineers.

www.aise.org—Association of Iron and Steel Engineers.

www.building.org—Plastic materials directory.

www.carrlane.com—Carr Lane Manufacturing: tooling components.

www.copper.org—Copper Development Association.

www.destaco.com—De-Sta-Co Industries: toggle clamps.

www.emtec.org—Edison Materials Technology Center: manufacturing materials research.

www.idesinc.com—IDES Inc. plastic materials information.

www.industrialpress.com—*Machinery's Handbook:* manufacturing materials and processes.

www.jergensinc.com—Jergens Inc.: tooling components.

www.mitutoyo.com—Mitutoyo: precision manufacturing devices.

www.mjvail.com—M. J. Vail Co., Inc.: wholesale distributor.

www.properties.copper.org—Standard properties for copper and copper alloys.

www.reidtool.com—Reid Tool Supply Co.: tooling components.

www.stratasys.com—Stratasys Inc.: rapid prototyping.

www.wolverinetool.com—Wolverine Tool Co.: heavy-duty toggle clamps.

www.worldsteel.org—International Iron and Steel Institute.

PROFESSIONAL PERSPECTIVE

Mechanical drafting and manufacturing are very closely allied. The mechanical drafter should have a general knowledge of manufacturing methods, including machining processes. It is common for the drafter to consult with engineers and machinists regarding the best methods to implement a drawing from the design to the manufacturing processes. Part of design problem solving is to create a design that is not only functional, but one that can also be manufactured using available technology and at a cost that justifies the end product. The solutions to these types of concerns depend on how familiar the drafter and designer are with the manufacturing capabilities of the company. For example, if the designer over-tolerances a part or feature, the result could be a rejection of the part by manufacturing or an expensive machining operation. The drafter should also know something about how machining processes operate so that drawing specifications do not call out something that is not feasible to manufacture. The drafter should be familiar with the processes that are used in machine operations so the notes that are placed on a drawing conform to the proper machining techniques. Notes for machining processes are given on a drawing in the same manner they are performed in the shop. For example, the note for a counterbore is given as the diameter and depth of the hole (first process) and then the diameter and depth of the counterbore (second process). Also, the drafter should know the specification given will, in fact, yield the desired result. One interesting aspect of the design drafter's job is the opportunity to communicate with people in manufacturing and come up with a design that enables the product to be easily manufactured.

FINDING MEANS FOR SPC

Here is a set of data consisting of weekly samplings of a 2-in. bolt making machine. The process is considered "out of control" if sample averages exceed a UCL of 2.020 in. or are less than an LCL of 1.980 in. In preparation of a chart, find $\bar{x}$ for each week, and find $\bar{\bar{x}}$ for the set of 4 weeks. Was the process "out of control" during any week?

Week 1	Week 2	Week 3	Week 4
1.970	2.022	1.980	2.003
2.013	2.014	2.001	1.989
2.051	2.011	1.898	1.993
1.993	2.001	1.888	2.019
1.992	2.009	1.979	2.003

Solution: In the world of math statistics, a bar over a letter denotes an average (mean). The formula for finding the average of x values is usually written as $\bar{x} = \Sigma x/n$ where x is an individual data value, and n is the total number of data values. The Σ is the capital Greek letter sigma, which stands for repeated addition. The formula is the mathematical way of saying to add up all the data and then divide by the total number of data. Applying this idea to each week separately:

$$\bar{x} \text{ for Week 1} = (1.970 + 2.013 + 2.051 + 1.993 + 1.992) \div 5$$
$$= (10.019) \div 5 = 2.0038 \text{ or } \textbf{2.004}$$

Similarly,

$$\bar{x} \text{ for Week 2} = 2.0114 \text{ or } \textbf{2.011}$$
$$\bar{x} \text{ for Week 3} = 1.9492 \text{ or } \textbf{1.949}$$
$$\bar{x} \text{ for Week 4} = 2.0014 \text{ or } \textbf{2.001}$$

$\bar{\bar{x}}$ is the mean *of these means*. It is found by the same method:

$$\bar{\bar{x}} = (2.0038 + 2.0114 + 1.9492 + 2.0014) \div 4$$
$$= 1.99145 \text{ or } \textbf{1.991}$$

The mean for Week 3 is lower than the LCL, so the process was "out of control" then, and corrective measures should have been taken at the end of that week.

MATH APPLICATION

CHAPTER 11

Manufacturing Materials and Processes Test

DIRECTIONS

Answer the questions with short, complete statements or drawings as needed.

PART 1 • MANUFACTURING MATERIALS

1. Define *ferrous metals*.
2. Identify the two main types of ferrous metals.
3. Define *nonferrous metals*.
4. What is another name for plastics?
5. Define *thermoplastic*.
6. Define *thermoset*.
7. Define *tensile strength*.
8. Why are refractories used for such applications as furnace liners?
9. What are composites?
10. Describe the characteristics of gray cast iron.
11. Given the gray cast iron material specification 30A, what does the prefix 30 denote?
12. Define *ductility*.
13. What are the properties of white cast iron?
14. Which cast iron has the internal characteristics of gray cast iron and the exterior properties of white cast iron?
15. Define *malleable*.
16. Identify at least two uses for nodular cast iron.
17. How is it possible to alter the properties of steel?
18. Name a steel that is low in carbon and is commonly used for forged and machined parts.
19. Describe high carbon steel.
20. Describe the difference between hot- and cold-rolled steel.
21. Describe the properties of the following steel alloying elements: manganese, molybdenum, and tungsten.
22. Describe stainless steel.
23. Give the typical contents of stainless steel.
24. Identify the contents of the 200, 300, and 400 series stainless steels as identified by the American Iron and Steel Institute.
25. Identify at least four common uses for stainless steel.
26. Given the steel identification number SAE 1020, describe the components: SAE, 10, and 20.
27. Identify the steel recommended for the following general applications: agricultural steel, bolts and screws, car and truck gears, transmission shafts.
28. Define *heat treating*.

29. Define *normalizing*.
30. Define *case hardening*.
31. Define *carburization*.
32. Define *tempering*.
33. How is the Rockwell hardness test performed?
34. Describe the properties of aluminum.
35. Define *extruded*.
36. Identify the alloying elements in brass.
37. Identify the alloying elements in bronze.
38. What is the advantage of adding phosphorus to bronze?
39. Describe at least one industrial use for gold.
40. Identify the metal that has the weight advantage of aluminum and the strength of steel.
41. What are the elements of and process for making tungsten carbide?

In Questions 42 through 56, identify the type of metal or metals that are part of the given Unified Numbering System (UNS) series:

42. A00001 to A99999
43. C00001 to C99999
44. E00001 to E99999
45. L00001 to L99999
46. M00001 to M99999
47. P00001 to P99999
48. R00001 to R99999
48. Z00001 to Z99999
49. D00001 to D99999
50. F00001 to F99999
51. G00001 to G99999
52. H00001 to H99999
53. J00001 to J99999
54. K00001 to K99999
55. S00001 to S99999
56. T00001 to T99999
57. Give the general definition of *plastic*.
58. Name the process of joining two or more molecules to form a more complex molecule with physical properties that are different from the original molecules.
59. Identify the term that refers to polymer-based materials that have elastic qualities not found in thermoplastics and thermosets.
60. Give the name and basic characteristics of ABS.
61. Give the name and basic characteristics of ASA.
62. Give the name of the thermoplastic that has the versions nitrate, acetate, butyrate, propionate, and ethyl cellulose.
63. Name the type of thermoplastics that is very resistant to chemicals, friction, and moisture. These materials have excellent dimensional stability for use as wire coating and insulation, nonstick surfaces, chemical containers, O-rings, and tubing.
64. Identify the common name for the polyamide thermoplastic.
65. Name a thermoplastic that is rigid with very high impact and stress fracture resistance at temperatures between −40° and 210°F (−40° to 99°C), and is used when constant bending is required in the function of the material design.

66. Give the name of an inexpensive plastic to produce that has many desirable properties including heat, chemical, scratch, and moisture resistance. It is also resistant to continuous bending applications. Products include appliance parts, hinges, cabinets, and storage containers.
67. Name the plastic that is inexpensive and easy to manufacture, has excellent transparency, and is very rigid. However, it can be brittle and has poor impact, weather, and chemical resistance. Products include model kits, plastic glass, lenses, eating utensils, and containers.
68. Name the plastic that is commonly called PVC.
69. Identify at least two applications for PVC and its characteristics for these applications.
70. Briefly discuss the major differences between thermoplastics and thermosets.
71. Name the plastic material that dates back to the late 1800s. This plastic is hard and rigid, has good compression strength, is tough, and does not absorb moisture, but it is brittle. These plastics are commonly used for the manufacture of electrical switches and insulators, electronics circuit boards, distributor caps, and binding material and adhesive.
72. Name the material that is commonly used as a plastic for reinforced composites, also known as reinforced thermoset plastics (RTP). Typical uses are for boat and recreational vehicle construction, automobiles, fishing rods, tanks, and other structural products.
73. Identify the type of plastics that are also referred to as synthetic rubber.
74. Give the term that refers to the heating of the material in a steel mold that forms the desired shape.
75. Name the material that has the trade name Neoprene and was the first commercial synthetic rubber. This material has better weather, sunlight, and petroleum resistance than natural rubber. It is also very flame resistant, but does not resist electricity. Common uses include automotive hoses and other products where heat is found, gaskets, seals, and conveyor belts.
76. Name the material that has a wide range of makeup from liquid to solid. Liquid and semiliquid forms are used for lubricants. Harder forms are used where nonstick surfaces are required.

PART 2 • MANUFACTURING PROCESSES

77. What is another name for casting?
78. Define *casting*.
79. Name the most commonly used method of making castings.
80. Discuss the function of cores in the casting process.
81. List at least two advantages of using cores.
82. Describe centrifugal casting.
83. List at least two advantages of the die casting process.
84. Describe the permanent casting process.
85. Name the casting technique that is referred to as "lost wax."
86. Why is shrinkage allowance required in casting design?
87. What is the estimated shrinkage for most irons?

88. Define *draft*.
89. In casting design, is draft added to the minimum or maximum design sizes of the part?
90. Identify at least two reasons why fillets are used in casting design.
91. List at least three considerations that influence the amount of extra material that must be left on a casting for machining allowance.
92. What is the recommended standard finish allowance for iron or steel?
93. Describe hot spots in casting.
94. Define *forging*.
95. What is the grain structure advantage of using the forging process to manufacture a part rather than making a machined part?
96. Describe upset forging.
97. Describe swaging.
98. Describe bending.
99. Describe punching.
100. List at least two advantages of machine forging.
101. Briefly describe the stamping process.
102. Give the name of the machine that performs the stamping operation.
103. Name the process that takes metal-alloyed powders and feeds them into a die where they are compacted under pressure to form the desired shape. The compacted metal is then removed from the die and heated at temperatures below the melting point of the metal.
104. What is the process called that takes the compacted metal that has been removed from the die in the previous question and then heats it at temperatures below the melting point of the metal?
105. Briefly describe the metal injection molding (MIM) process.
106. Name the manufacturing process that is the most commonly used process for creating thermoplastic products. The process involves injecting molten plastic material into a mold that is in the form of the desired part or product. The mold is in two parts that are pressed together during the molding process. The mold is then allowed to cool so the plastic can solidify. When the plastic has cooled and solidified, the press is opened and the part is removed from the mold.
107. Name the manufacturing process that is used to make continuous shapes such as moldings, tubing, bars, angle, hose, weather stripping, films, and any product that has a constant shape. This process creates the desired continuous shape by forcing molten plastic through a metal die.
108. Name the manufacturing process that is commonly used to produce hollow products such as bottles, containers, receptacles, and boxes. This process works by blowing hot polymer against the internal surfaces of a hollow mold.
109. Give the name of the manufacturing process that is generally used to create products such as vinyl flooring, gaskets, and other sheet products. This process fabricates sheet or film thermoplastic or thermoset plastics by passing the material through a series of heated rollers.

110. Name the process that works by placing a specific amount of polymer pellets into a metal mold. The mold is then heated as it is rotated. This forces the molten material to form a thin coating against the sides of the mold. When the mold is cooled, the product is removed.
111. Give the name of the process that works by placing material into an initial hot die where it takes the preliminary shape. As the material cools, a die that matches to the shape of the desired product forms the final shape.
112. Name the process that works by taking a sheet of material and heating it until it softens and sinks down by its own weight into a mold that conforms to the desired final shape. Vacuum pressure is commonly used to suck the hot material down against the mold.
113. Give the name of the process that uses a computer model that is traced in thin cross sections to control a laser that deposits layers of liquid resin or molten particles of plastic material to form the desired shape.
114. Name the common fabrication process for thermosets that uses a specific amount of material that is heated and placed in a closed mold where additional heat and pressure are applied until the material takes the desired shape. The material is then cured and removed from the mold.
115. Give the name of the process that is similar to casting, but this process uses a foam material that expands during the cure to fill the desired mold.
116. Give the name of the material that is also commonly referred to as reinforced plastic.
117. Briefly describe the layering process that is used for making reinforced plastic.
118. Give the name of a process where resin is used to make quality composite products with a smooth surface on both sides. This method places reinforcing material into a mold and then pumps resin into the mold.
119. Name the process that uses vacuum pressure to force a thin layer of sheet-reinforced polymer around a mold.
120. Give the name of the process that uses machines to wind resin-saturated reinforcement fibers around a shaft.

PART 3 • MACHINE PROCESSES

121. List at least two reasons why the mechanical drafter should be familiar with machining processes.
122. List four types of machining operations that can be performed on a drilling machine.
123. Identify one of the primary functions of the grinding machine.
124. Describe the main function of a lathe.
125. Describe a feature of milling machines that influences the ability of operations and shapes that can be performed.
126. Identify two types of saw tools that can be used for cutoff and machining operations.
127. Describe chemical machining.
128. Describe electrochemical machining (ECM).
129. Describe electrodischarge machining (EDM).
130. Describe electron beam (EB) machining.
131. What is another name for ultrasonic machining?

132. Describe the basic function of a laser device.
133. Laser machining may be used on what materials?
134. What does the abbreviation *CNC* mean?
135. Briefly explain the CNC process. Describe computer-integrated manufacturing.
136. What is the purpose of boring?
137. Identify at least one function of a counterbore.
138. Describe drill depth.
139. Describe and give one application of a knurl.
140. Describe the primary function of a key.
141. Define *keyseat*.
142. What is the function of a reamer?
143. What action will the machinist take when no spotface depth is given?
144. Define *surface texture*.
145. In what units is surface roughness height measured?
146. Discuss the results of designing a part with specifications that require overmachining.
147. Describe the difference between jigs and fixtures.
148. Normally a jig or fixture is drawn as an assembly of the unit ready for use, and the workpiece or part to be held is drawn in position. How is the workpiece drawn in relationship to the jig or fixture?
149. Describe the use of drill jigs.
150. Drill fixtures are sometimes referred to as what?
151. Describe the use of drill fixtures.
152. Explain how machining fixtures work.
153. Describe the function of a welding fixture.
154. Briefly explain the use of inspection fixtures.
155. Define and describe the use of progressive dies.
156. What is a pickoff jig?
157. Why is it important for a tool designer to be a good print reader?
158. Identify at least four qualities that tools must possess.
159. Name the quality control/detection system that uses statistical techniques and problem solving to monitor, evaluate, and provide guidance for adjusting the process to improve quality.
160. Describe the basic function of control charting in quality control.

CHAPTER 11
Manufacturing Materials and Processes Problems

This chapter is intended as a reference for manufacturing materials and processes. The concepts discussed serve as a basis for further study in the following chapters. A thorough understanding of dimensioning practices is necessary before complete manufactured products can be drawn. Problem assignments ranging from basic to complex manufacturing drawings are assigned in the following chapters.

PROBLEMS 11-1 THROUGH 11-14   Make a two-view drawing of each of the following machined features using manual or computer-aided drafting as required by your instructor. Prepare the drawings on 8-1/2" × 11" vellum using manual drafting or CADD unless otherwise specified by your instructor. Dimensioning may be omitted; however, all line representations should be properly drawn using correct techniques. Machined features to be drawn:

PROBLEM 11-1.   Counterbore

PROBLEM 11-2.   Chamfer

PROBLEM 11-3.   Countersink

PROBLEM 11-4.   Counterdrill

PROBLEM 11-5.   Drill (not through the material)

PROBLEM 11-6.   Fillet

PROBLEM 11-7.   Round

PROBLEM 11-8.   Spotface

PROBLEM 11-9.   Dovetail

PROBLEM 11-10.   Kerf

PROBLEM 11-11.   Keyseat

PROBLEM 11-12.   Keyway

PROBLEM 11-13.   T-slot

PROBLEM 11-14.   Knurl

PROBLEMS 11-15 THROUGH 11-48   The following topics require research and/or industrial visitations. It is recommended that you research current professional magazines, visit local industries, or interview professionals in the related field. Your reports should emphasize the following:

■ Product

■ The process

■ Special manufacturing considerations

■ The link between manufacturing and engineering

■ Current technological advances

Select one or more of the topics listed below or as assigned by your instructor and write a 500-word report for each.

PROBLEM 11-15.   Casting

PROBLEM 11-16.   Forging

PROBLEM 11-17. Conventional machine shop

PROBLEM 11-18. Computer numerical control (CNC) machining

PROBLEM 11-19. Surface roughness

PROBLEM 11-20. Tool design

PROBLEM 11-21. Chemical machining

PROBLEM 11-22. Electrochemical machining (ECM)

PROBLEM 11-23. Electrodischarge machining (EDM)

PROBLEM 11-24. Electron beam (EB) machining

PROBLEM 11-25. Ultrasonic machining

PROBLEM 11-26. Laser machining

PROBLEM 11-27. Statistical process control (SPC)

PROBLEM 11-28. Thermoplastics

PROBLEM 11-29. Thermosets

PROBLEM 11-30. Elastomers

PROBLEM 11-31. Metal stamping

PROBLEM 11-32. Powder metallurgy

PROBLEM 11-33. Manufacturing thermoplastic products

PROBLEM 11-34. Manufacturing thermoset plastic products

PROBLEM 11-35. Manufacturing composites

PROBLEM 11-36. Cast iron

PROBLEM 11-37. Steel

PROBLEM 11-38. Aluminum

PROBLEM 11-39. Copper alloys

PROBLEM 11-40. Precious and other specialty metals

PROBLEM 11-41. Computer-aided manufacturing (CAM)

PROBLEM 11-42. Computer-aided engineering (CAE)

PROBLEM 11-43. Computer-integrated manufacturing (CIM)

PROBLEM 11-44. Robotics

PROBLEM 11-45. Computer-aided design

PROBLEM 11-46. Computer-aided drafting

PROBLEM 11-47. Manufacturing cell

PROBLEM 11-48. CAD/CAM integration

PROBLEM 11-49. Look at the PART in Figure 11.69, page 320, and answer the following:

a. Which dimensions are critical to this drill jig?

b. Which dimensions are of incidental interest to the tool designer?

c. Which dimensions are not relevant to this drill jig?

PROBLEM 11-50 Make a photocopy of the drawing below and then sketch a drill jig on the drawing to satisfy the following TDR. You can also redraw the drawing using manual drafting or CADD and then draw the drill jig, if preferred by your course objectives:

PART NAME: ------------------ PART
PART NO: ---------------------- XXXX-XX
OPERATION: ------------------ DRILL .50" DIA HOLE
MACHINE: -------------------- USE HAND DRILL
FIXTURE: --------------------- PICKOFF JIG LOCATING
THE HOLE 1.80" AND 2.53"
FROM THE EDGES OF THE PLATE

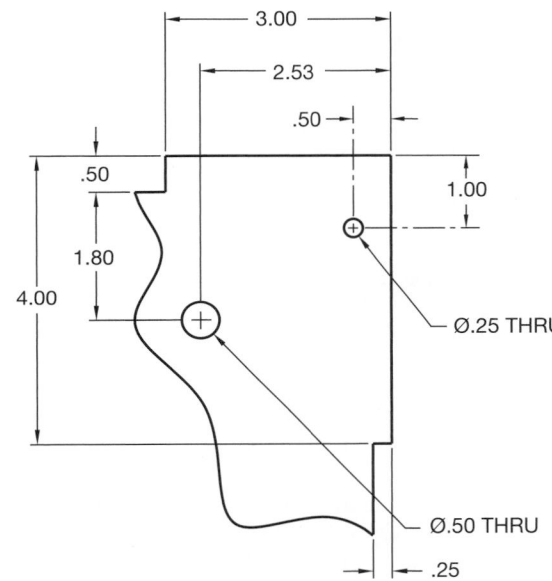

MATH PROBLEMS

Here is a set of six monthly samples from a process. The UCL is 5.05 and the LCL is 4.95.

Jan	Feb	Mar	Apr	May	Jun
5.06	5.03	5.04	5.04	4.95	4.95
4.94	5.04	5.04	5.07	4.95	4.97
4.99	5.05	5.05	5.05	4.93	4.98
4.99	4.99	5.06	5.02	4.92	5.01
5.00	4.98	5.06	5.00	4.95	5.00

PROBLEM 11-51. Find $\bar{x}$ for each month.

PROBLEM 11-52. Find $\bar{\bar{x}}$ for the six month period.

PROBLEM 11-53. Conclude whether the process was "out of control" for any of the months.

Dimensioning and Tolerancing

OBJECTIVES

After completing this chapter, you will:

■ Identify and use common dimensioning systems.

■ Explain and apply dimensioning standards based on ASME Y14.5M.

■ Apply proper specific notes for manufacturing features.

■ Place proper general notes and delta notes on a drawing.

■ Interpret and use correct tolerancing techniques.

■ Prepare completely dimensioned multiview drawings from engineering sketches and industrial drawings.

■ Apply draft angles as needed to a drawing.

■ Dimension CAD/CAM machine tool drawings.

■ Prepare casting and forging drawings.

■ Provide surface finish symbols on drawings.

■ Solve tolerance problems including limits and fits.

■ Use an engineering problem as the basis for your layout techniques.

■ Describe the purpose of ISO 9000 Quality Systems Standard and related standards.

THE ENGINEERING DESIGN APPLICATION

As mentioned later in this chapter, a complete detail drawing is made up of multiviews and dimensions. The layout techniques of a detail drawing must include an analysis of how the views and dimensions go together to create the finished drawing. The most effective way to form preliminary layout ideas is through the use of rough sketches as described in Chapter 3.

Consider the engineering sketch in Figure 12.1 as you evaluate how to prepare a complete detail drawing. The procedure is the same with CADD or manual drafting.

Step 1. Select and make a rough sketch of the proper multiviews. Leave plenty of space between views so dimensions may be added. The engineer's sketch is not always accurate. It is the drafter's responsibility to convert the engineering ideas from the rough stage to a formal drawing. The information may be correct; however, the organization of information and proper layout are the drafter's duties. (See Figure 12.2.)

Step 2. Place dimensions and notes on your multiview sketch. (See Figure 12.3.) Keep the following dimensioning rules in mind as you select and place the dimensions:

■ Begin with smallest dimension closest to the object and place dimensions that progressively increase in size further away from the object.

■ Do not crowd dimensions.

■ Dimension between views where possible.

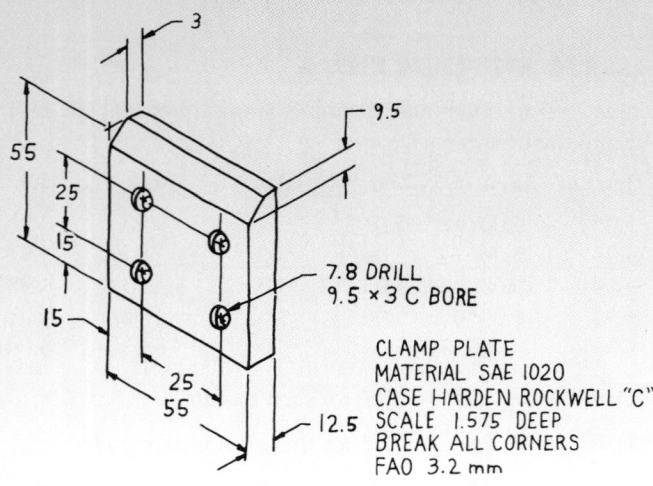

FIGURE 12.1 ■ Engineering sketch.

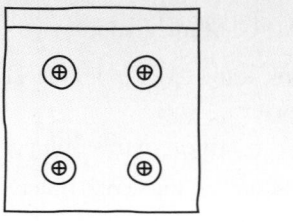

FIGURE 12.2 ■ Rough sketch of selected multiviews.

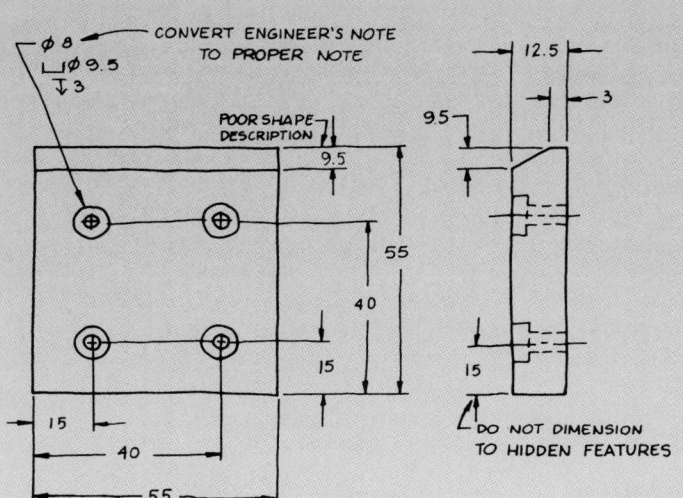

FIGURE 12.3 ■ Rough sketch with dimensions and notes placed. In this rough sketch, as in the real world, the dimensions and features are often out of proportion.

■ Dimension to views that show the best shape of features.

■ Group dimensions when possible.

■ Do not dimension size or location dimensions to hidden features.

■ Stagger adjacent dimension numerals.

■ Use leaders to properly label specific notes.

■ Convert all information to proper drafting standards.

Evaluate these basic rules as you decide where dimensions should be placed.

Step 3. Determine the scale to use based on the amount of fine detail to be shown and the total size of the object.

The clamp plate can easily be drawn full scale.

Step 4. Determine the sheet size based on the drawing scale to be used, the total drawing area (including views and dimensions), and the amount of clear area needed for general notes and future revisions. Refer to Chapter 6.

Look at Figure 12.3 as you follow the calculations to determine the length and height of the drawing area needed for the clamp plate.

Length of drawing area (in inches)	
Front view width	2.00
Space between views:	
Front view to .50 dimension	.75
.50 to 1.00 dimension	.50
1.00 to 2.00 dimension	.50
2.00 to .375 dimension	.50
.375 dimension to right-side view	.75
Right-side view depth	+.50
Drawing area total length	5.50

The space from the view to the first dimension and the spaces between dimensions are drafting decisions. The first space is often, but not always, larger than distances to additional dimension lines. Each dimension line after the first is equally spaced so that dimensions are not crowded.

Height of drawing area (in inches)	
Height of views	2.00
Front view to .50 dimension	.75
.50 to 1.00 dimension	.50
1.00 to 2.00 dimension	.50
Right-side view to .125 dimension	.75
.125 to .50 dimension	+.50
There should be enough space for the CBORE note.	
Drawing area total height	5.00

The total length and height of the drawing area is 5.50 x 5.00. Keep in mind that these calculations are approximate. There should be adequate space on your drawing for some flexibility. Select a B (A3 metric) size drafting sheet for this drawing. Figure 12.4 shows the 5.50 x 5.00 drawing area blocked out on a B-size sheet. Before you proceed, be sure that your equipment, table, and hands are clean.

Step 5. Use construction lines to draw the selected multiviews.

Step 6. Use construction lines to place hidden, extension, center, dimension, and leader lines, using selected spacing and placement from rough sketch. Also, use guidelines to prepare for the placement of lettering. This drafter chose to change the hole locations to datum dimensioning. This method is defined and advantages given later in this chapter. (See Figure 12.5.)

Step 7. Complete the detail drawing using proper drafting technique in the following sequence:

1. Darken all horizontal thin lines from top to bottom and all vertical thin lines from left to right if right-handed or right to left if left-handed.

2. Draw all circles, arcs, and object lines.

3. Draw all straight object lines in the same order as thin lines.

4. Do all lettering from the top to the bottom of the sheet. Letter all specific and general notes and fill in the title block. To help keep your drawing clean, place a blank sheet of paper under your hand while lettering.

5. Draw all arrowheads.

(Continued)

THE ENGINEERING **DESIGN** APPLICATION *(continued)*

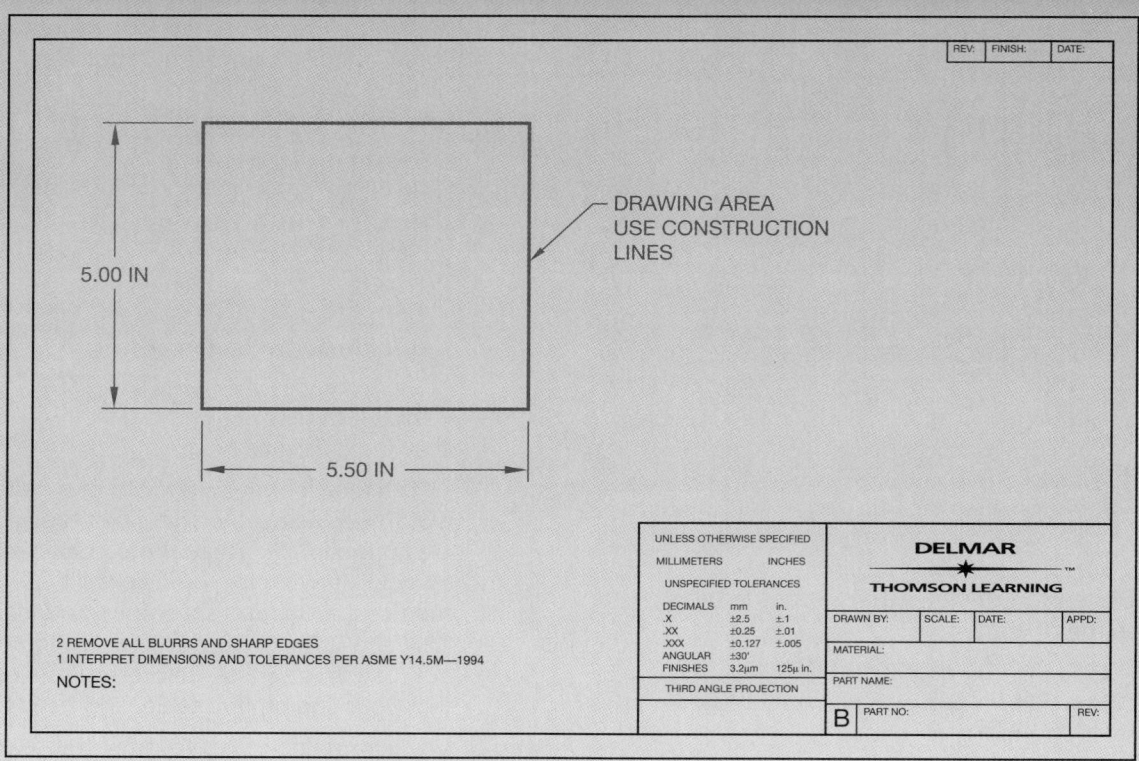

FIGURE 12.4 ■ Establish the approximate drawing area.

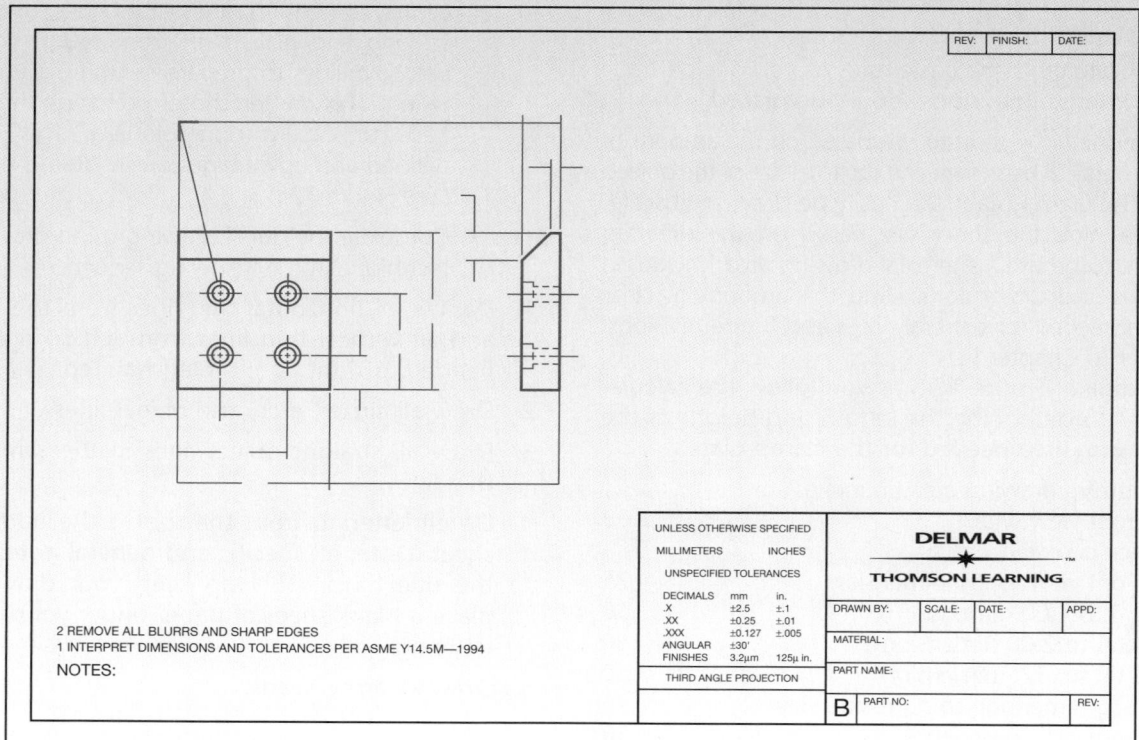

FIGURE 12.5 ■ Lay out the multiview's object lines, hidden lines, centerlines, extension and dimension lines, leader lines; establish guidelines for lettering.

6. Check your line quality by placing the original over a light table. If light passes through the lines or lettering, the work is not dark enough. Another way to check your work is to run a test-strip diazo copy. If the quality of the print is not adequate, you may need to work more on the drawing. A print will not improve the quality of a drawing.

Figure 12.6 shows the completed detail drawing of the clamp plate.

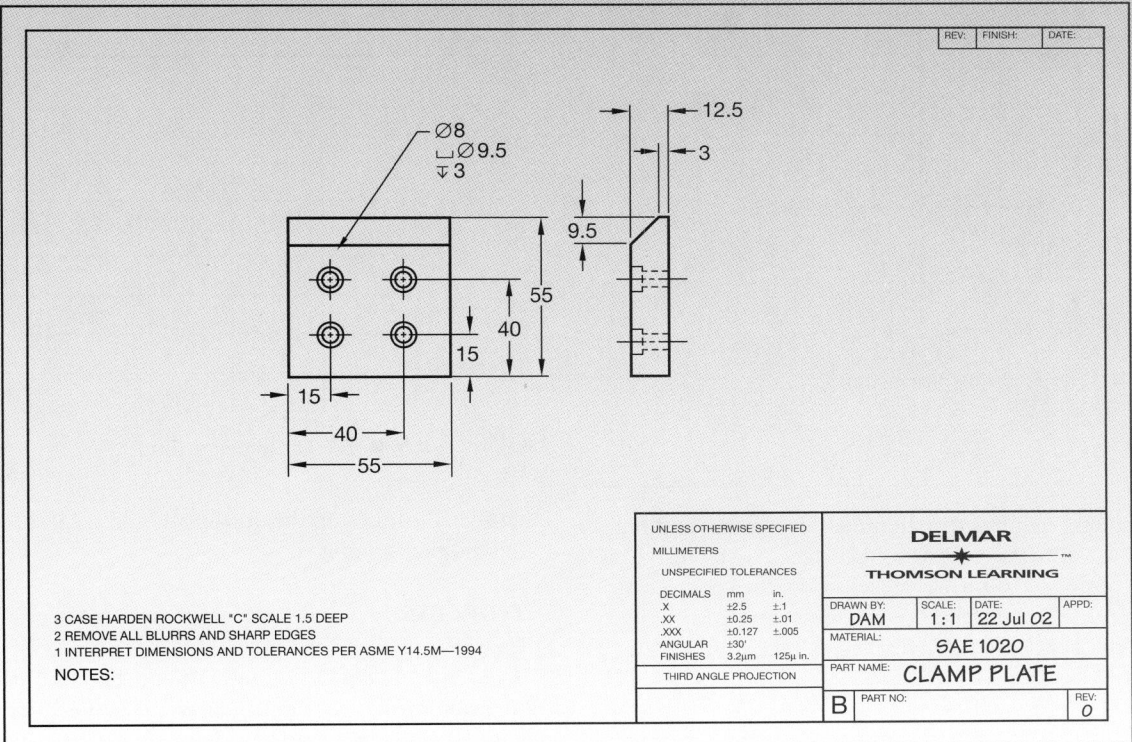

FIGURE 12.6 ■ Complete the drawing and add all lettering.

ANSI/ASME The standard adopted by the American National Standards Institute (ANSI) and published by the American Society of Mechanical Engineers (ASME) is titled *Dimensioning and Tolerancing*, ASME Y14.5M. This standard is available through the ASME at 345 East 47th Street, New York, NY 10017. The standard that controls general dimensional tolerances found in the title block or in general notes is ASME Y14.1M, *Metric Drawing Sheet Size and Format*.

A complete detail drawing includes multiviews and dimensions, which provide both shape and size description. There are two classifications of dimensions: size and location. *Size dimensions* are placed directly on a feature to identify a specific size or may be connected to a feature in the form of a note. The relationship of features of an object is defined with *location dimensions*.

Notes are a type of dimension that generally identify the size of a feature or features with more than a numerical specification. For example, a note for a counterbore will give size and identification of the machine process used in manufacturing. There are basically two types of notes: local (or specific) notes and general notes. *Local* notes are connected to specific features on the views of the draw-

ing. *General* notes are placed separate from the views and relate to the entire drawing.

It is important for the drafter to effectively combine shape and size descriptions so the drawing is easy to read and understand. There are many techniques that will help implement this goal. The drafter should carefully evaluate the dimensioning rules while preparing detail drawings and should keep in mind at all times never to crowd information on a drawing. It is better to use larger paper than to crowd the drawing, unless otherwise indicated by your company.

DIMENSIONING SYSTEMS
Unidirectional

Unidirectional dimensioning is commonly used in mechanical drafting. It requires that all numerals, figures, and notes be lettered horizontally and be read from the bottom of the drawing sheet. Figure 12.7 shows unidirectional dimensioning in use. The dimension in parentheses (80) is a reference dimension, which is discussed later in this chapter.

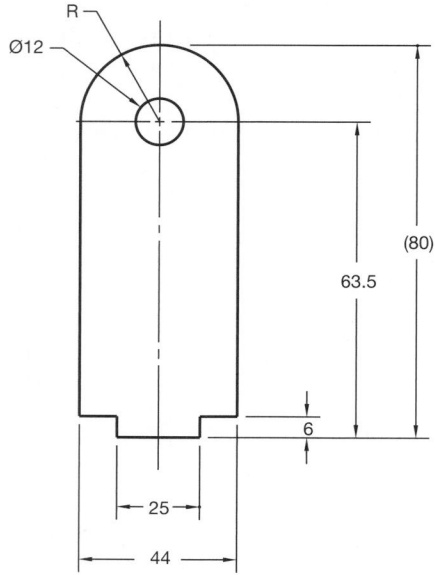

FIGURE 12.7 ■ Unidirectional dimensioning.

Aligned

Aligned dimensioning requires that all numerals, figures, and notes be aligned with the dimension lines so that they may be read from the bottom (for horizontal dimensions) and from the right side (for vertical dimensions). This method of dimensioning is commonly used in architectural and structural drafting. (See Figure 12.8.)

Tabular

Tabular dimensioning is a system in which size and location dimensions from datums or coordinates (x, y, z axes) are given in a table

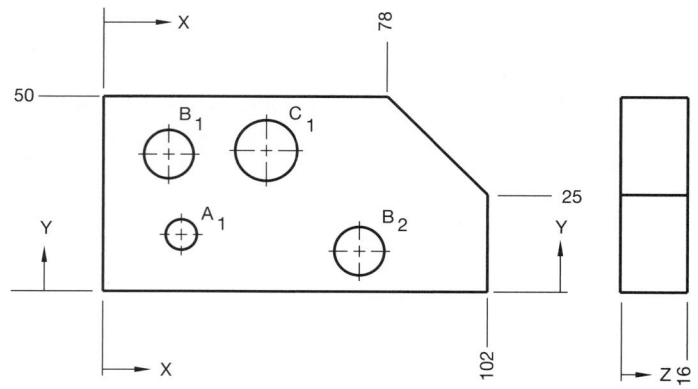

HOLE SYMBOL	HOLE DIA	LOCATION		DEPTH Z
		X	Y	
A₁	6	15	14	THRU
B₁	9	12	38	9
B₂	9	57	7	12
C₁	12	43	38	THRU

FIGURE 12.9 ■ Tabular dimensioning.

identifying features on the drawing. Figure 12.9 shows a method of tabular dimensioning.

Arrowless

Also known as dimensioning without dimension lines, *arrowless* dimensioning is similar to tabular dimensioning in that features are identified with letters and keyed to a table. Location dimensions are established with extension lines as coordinates from determined datums. (See Figure 12.10.)

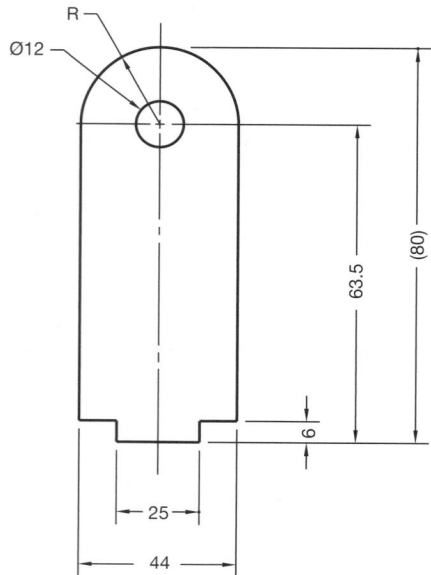

FIGURE 12.8 ■ Aligned dimensioning.

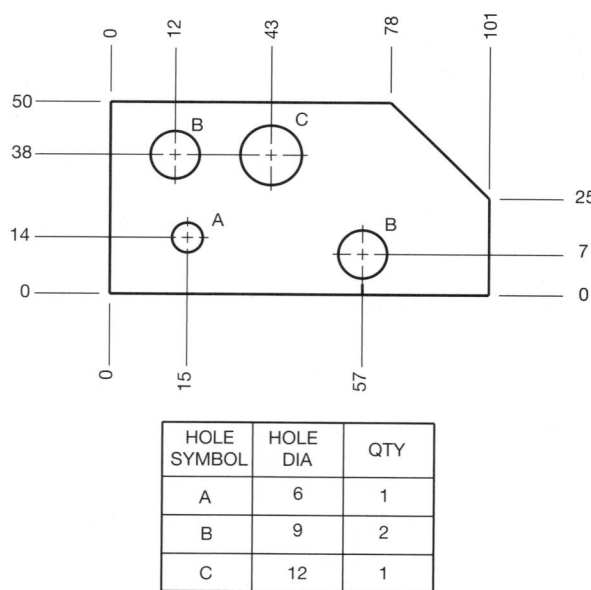

HOLE SYMBOL	HOLE DIA	QTY
A	6	1
B	9	2
C	12	1

FIGURE 12.10 ■ Arrowless dimensioning.

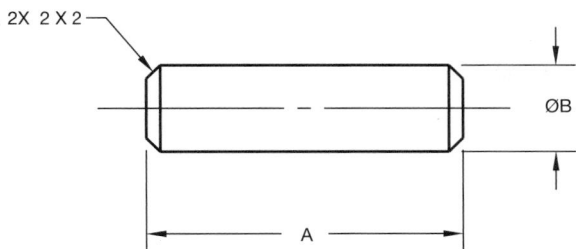

LENGTH A	B=20.3	B=38.1	B=50.8	B=57.2
	PART NO.	PART NO.	PART NO.	PART NO.
76	DP20.3-76.2	DP38.1-76.2	DP50.8-76.2	DP57.2-76.2
101	DP20.3-101.6	DP38.1-101.6	DP50.8-101.6	DP57.2-101.6
127	DP20.3-127	DP38.1-127	DP50.8-127	DP57.2-127
152	DP20.3-152.4	DP38.1-152.4	DP50.8-152.4	DP57.2-152.4

FIGURE 12.11 ■ Chart drawing.

Chart Drawing

Chart drawings are used when a particular part or assembly has one or more dimensions that change depending on the specific application. For example, the diameter of a part may remain constant with several alternate lengths required for different purposes.

The variable dimension is usually labeled on the drawing with a letter in the place of the dimension. The letter is then placed in a chart where the changing values are identified. Figure 12.11 is a chart drawing that shows two dimensions that have alternate sizes. The view drawn represents a typical part, and the dimensions are labeled A and B. The correlated chart identifies the various (A) lengths available at given (B) diameters. The chart in this example also shows purchase part numbers for each specific item. This method of dimensioning is commonly used in vendor or specification catalogues for alternate part identification.

DIMENSIONING RULES

ANSI/ASME The ANSI/ASME standard document for dimensioning is titled *Dimensioning and Tolerancing* ASME Y14.5M. This standard is available from the American National Standards Institute, 1430 Broadway, New York, NY 10018. This standard is all metric as denoted by the *M* following the number. The following fundamental rules for dimensioning are adapted from ASME Y14.5M.

1. Each dimension shall have a tolerance, except for those dimensions specifically identified as reference, maximum, minimum, or stock. The tolerance may be applied directly to the dimension or indicated by a general note located in the title block of the drawing.

2. Dimensioning and tolerancing shall be complete to the extent that there is full understanding of the characteristics of each features. Neither measuring the drawing nor assumption of a distance or size is permitted, except draw-

ings such as loft, printed wiring, templates, and master layouts prepared on stable material, provided the necessary control dimensions are given.

3. Each necessary dimension of an end product shall be shown. No more dimensions than those necessary for complete definition shall be given. The use of reference dimensions on a drawing should be minimized.

4. Dimensions shall be selected and arranged to suit the function and mating relationship of a part and shall not be subject to more than one interpretation.

5. The drawing should define a part without specifying manufacturing methods. For example, give only the diameter symbol for a hole without a process note such as drill, ream, or punch. However, in those cases in which manufacturing, processing, quality assurance, or environmental information is essential to the definition of engineering requirements, it shall be specified on the drawing or in document references on the drawing.

6. It is permissible to identify as nonmandatory certain processing dimensions that provide for finish allowance, shrink allowance, and other requirements, provided the final dimensions are given on the drawing. Nonmandatory processing dimensions shall be identified by an appropriate note, such as NONMANDATORY (MFG DATA.)

7. Dimensions should be arranged to provide required information for optimum readability. Dimensions should be shown in true profile views and refer to visible outlines.

8. Wires, cables, sheets, rods, and other materials manufactured to gage or code numbers shall be specified by linear dimensions indicating the diameter or thickness. Gage or code numbers may be shown in parentheses following the dimension.

9. A 90° angle is implied where centerlines and lines depicting features are shown on a drawing at right angles and no angle is specified. The tolerance for these 90° angles is the same as the general angular tolerance specified in the title block or general notes.

10. A 90° basic angle applies where centerlines of features in a pattern or surfaces shown at right angles on the drawing are located or defined by basic dimensions and no angle is specified.

11. Unless otherwise specified, all dimensions are applicable to 20° C (68° F). Compensation may be made for measurements made at other temperatures.

12. All dimensions and tolerances apply in a *free state condition* except nonrigid parts. Free state condition describes distortion of a part after removal of forces applied during manufacturing. *Nonrigid* parts are those that may have dimensional fluctuation due to thin wall characteristics.

13. Unless otherwise specified, all geometric tolerances apply for full depth, length, and width of the feature.

14. Dimensions apply on the drawing where specified.

Dimensioning Definitions

Actual Size

The part size as measured after production is known as actual size, also known as produced size.

Allowance

The tightest possible fit between two mating parts. MMC external feature – MMC internal feature = allowance.

Basic Dimension

A basic dimension is considered to be a theoretically exact size, location, profile, or orientation of a feature or point. The basic dimension provides a basis for the application of tolerance from other dimensions or notes. Basic dimensions are drawn with a rectangle around the numerical value. For example: $\boxed{.625}$ or $\boxed{30°}$. This is discussed more in Chapter 13.

Bilateral Tolerance

A bilateral tolerance is allowed to vary in two directions from the specified dimension.

Inch examples are $.250^{+.002}_{-.005}$ and $.500 \pm .005$.

Metric examples are $12^{+0.1}_{-0.2}$ and 12 ± 0.2.

Datum

A datum is considered to be a theoretically exact surface, plane, axis, center plane, or point from which dimensions for related features are established.

Datum Feature

The datum feature is the actual feature of the part that is used to establish a datum.

Dimension

A dimension is a numerical value used on a drawing to describe size, shape, location, geometric characteristic, or surface texture.

Feature

A feature is any physical portion of an object, such as a surface or hole.

Geometric Tolerance

The general term applied to the category of tolerances used to control form, profile, orientation, location, and runout. See Chapter 13.

Least Material Condition (LMC)

Least material condition is the opposite of maximum material condition. The LMC is the lower limit for an external feature and the upper limit for an internal feature.

Limits of Dimension

The limits of a dimension are the largest and smallest possible boundary to which a feature may be made as related to the tolerance of the dimension. Consider the following inch dimension and tolerance: $.750 \pm .005$. The limits of this dimension are calculated as follows: $.750 + .005 = .755$ upper limit and $.750 - .005 = .745$ lower limit. For the metric dimension 19.00 ± 0.15, $19.00 + 0.15 = 19.15$ is the upper limit, and $19.00 - 0.15 = 18.85$ is the lower limit.

Maximum Material Condition (MMC)

The maximum material condition, given the limits of the dimension, is the situation where a feature contains the most material possible. MMC is the largest limit for an external feature and the smallest limit for an internal feature.

Nominal Size

A dimension used for general identification such as stock size or thread diameter.

Reference Dimension

A dimension usually without tolerance, used for information purposes only. A reference is a repeat of a given dimension or established from other values shown on the drawing. It does not govern production or inspection. Reference dimensions are enclosed in parentheses () on the drawing. See Figure 12.7 and Figure 12.8.

Specified Dimension

The specified dimension is the part of the dimension from which the limits are calculated. For example, the specified dimension of $.625 \pm .001$ in. is $.625$. In the following metric example, 15 ± 0.1 mm, 15 is the specified dimension.

Tolerance

The tolerance of a dimension is the total permissible variation in size or location. Tolerance is the difference of the lower limit from the upper limit. For example, the limits of $.005 \pm .005$ in. are $.505$ and $.495$ making the tolerance equal to $.010$ in. The tolerance for the following metric example, 12 ± 0.1 mm, is 0.2 mm.

Unilateral Tolerance

A unilateral tolerance is a tolerance that has a variation in only one direction from the specified dimension, as in $.875^{+.000}_{-.002}$ in. or $22^{\ 0}_{-0.2}$, $22^{+0.2}_{\ 0}$ mm.

Dimensioning Units

The metric *International System of Units* (SI) is featured predominately in this text because SI units (millimeters) supersede United States (U.S.) customary units specified on engineering drawings.

Metric units expressed in millimeters or U.S. customary units expressed in decimal inches are considered the standard units of linear measurement on engineering documents and drawings. The selection of millimeters or inches depends on the needs of the individual company. When all dimensions are either in millimeters or inches, the general note, UNLESS OTHERWISE SPECIFIED, ALL DIMENSIONS ARE IN MILLIMETERS (or INCHES), should be lettered on the drawing. Inch dimensions should be followed by *IN.* on predominantly millimeter drawings, and *mm* should follow millimeters on predominantly inch-dimensioned drawings.

While **not** an ANSI standard, some companies use a method of placing inch and metric equivalents together on each dimension. This is known as *dual dimensioning*. The inch dimension may be followed by millimeters in brackets or separated by a slash; for example, 1.00 [25.4] or 1.00/25.4. A general note should accompany the drawing to identify this practice: DIMENSIONS IN [ ] ARE MILLIMETERS (or INCHES), or INCH/MILLIMETERS or MILLIMETERS/INCHES.

Decimal Points

Dimension numerals that contain decimal points should be allowed adequate space at the decimal so there is no crowding with the numerals. A minimum of two-thirds the height of the letters is recommended. The decimal should be clear and bold and in line with the bottom of the numerals; for example, 1.750.

A specified dimension in inches is expressed to the same number of decimal places as its tolerance, and zeros are added to the right of the decimal point if needed; for example, .500±.002.

Where millimeter dimensions are less than one, a zero shall precede the decimal, as in 0.8. Decimal inch dimensions do not have a zero before the decimal, as in .75. Proper spacing of numerals with a decimal is shown in Figure 12.12.

Neither the decimal point nor a zero is shown when the metric dimension is a whole number; for example, 24. Also, when the metric dimension is greater than a whole number by a fraction of a millimeter, the last digit to the right of the decimal point is not followed by a zero, as in 24.5. This is true unless tolerance numerals are involved. (Refer to Figure 12.16). Both the plus and minus values of a metric or inch tolerance have the same number of decimal places and zeros are added to fill in where needed.

Fractions

Fractions are used on engineering drawings, but they are not as common as decimal inches or millimeters. Fraction dimensions generally mean a larger tolerance than decimal numerals. When fractions are used on a drawing, the fraction numerals should be the same size as other numerals on the drawing. The fraction bar should be drawn in line with the direction that the dimension reads. For unidirectional dimensioning, the fraction bars are all horizontal. For aligned dimensioning, the fraction bars are hori-

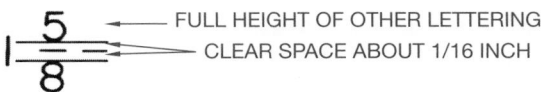

FOR TITLE BLOCKS, MATERIAL STOCK SIZES, AND GENERAL NOTES

FIGURE 12.13 ■ Numerals in fractions.

zontal for dimensions that read from the bottom of the sheet and vertical for dimensions that read from the right. The fraction numerals should not be allowed to touch the fraction bar. A space of 1.5 mm or .06 in. is recommended between the bar and the fraction numerals. In a few situations—for example, when a fraction is part of a general note, material specification, or title—the fraction bar may be placed diagonally, as shown in Figure 12.13. This is also a common practice using CADD, although fractions can be placed as previously described.

Arrowheads

Arrowheads are used to terminate dimension lines and leaders. Properly drawn arrowheads should be drawn three times as long as they are high. All arrowheads on a drawing should be the same size. Do not use small arrowheads in small spaces. Limited-space dimensioning practice is covered in this chapter (see Figure 12.16).

Some companies require that arrowheads be drawn with an arrow template, while others accept properly drawn freehand arrowheads. (See Figure 12.14.) Individual company preference dictates if arrowheads are filled in solid or left open as shown. CADD users often prefer the open style, which helps increase regeneration and plotting speed. However, others prefer the appearance of the filled-in arrowhead. Reduced regeneration and plotting speed is less of a concern with current technology.

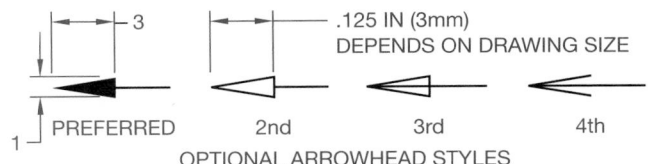

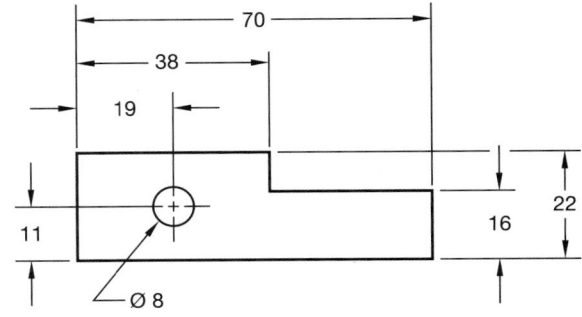

FIGURE 12.12 ■ Spacing of numerals with a decimal.

FIGURE 12.14 ■ Arrowheads.

DIMENSIONING FUNDAMENTALS

Dimension Line Spacing

Dimension lines should be placed at a uniform distance from the object, and all succeeding dimension lines should be equally spaced. Figure 12.15 shows the minimum acceptable distances for spacing dimension lines. In actual practice the minimum distance is too crowded. Judgment should be used based on space available and information presented. Never crowd dimensions, if at all possible. Always place the smallest dimensions closest to the object and progressively larger dimensions outward from the object. Group dimensions and dimension between views when possible.

Relationship of Dimension Lines to Numerals

Dimension numerals are centered on the dimension line. Numerals are commonly all the same height and are lettered horizontally (unidirectional). A space equal to at least half the height of lettering should be provided between numerals in a tolerance. The numeral, dimension line, and arrowheads should be placed between extension lines when space allows. When space is limited, other options should be used. Figure 12.16 shows several dimensioning options. Evaluate each of the examples carefully as you dimension your own drawing assignments. Figure 12.17 shows some correct and incorrect dimensioning practices. Keep in mind that some computer-aided drafting programs do not necessarily acknowledge all of the rules or accepted examples. Some flexibility on your part may be needed to become accustomed to some of the potential differences.

Chain Dimensioning

Chain dimensioning, also known as point-to-point dimensioning, is a method of dimensioning from one feature to the next. Each

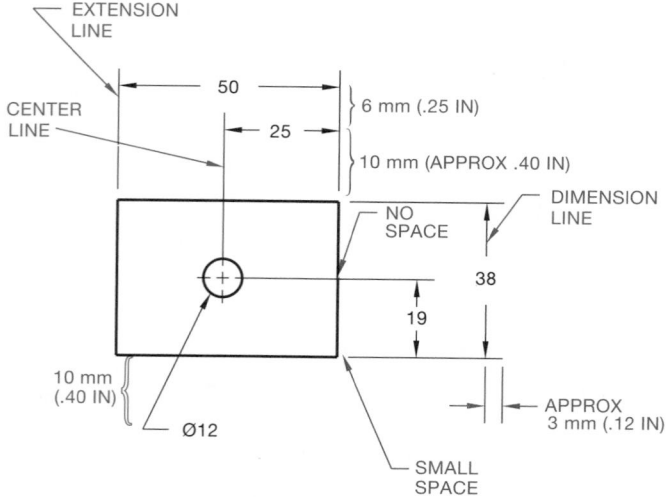

FIGURE 12.15 ■ Minimum dimension line spacing.

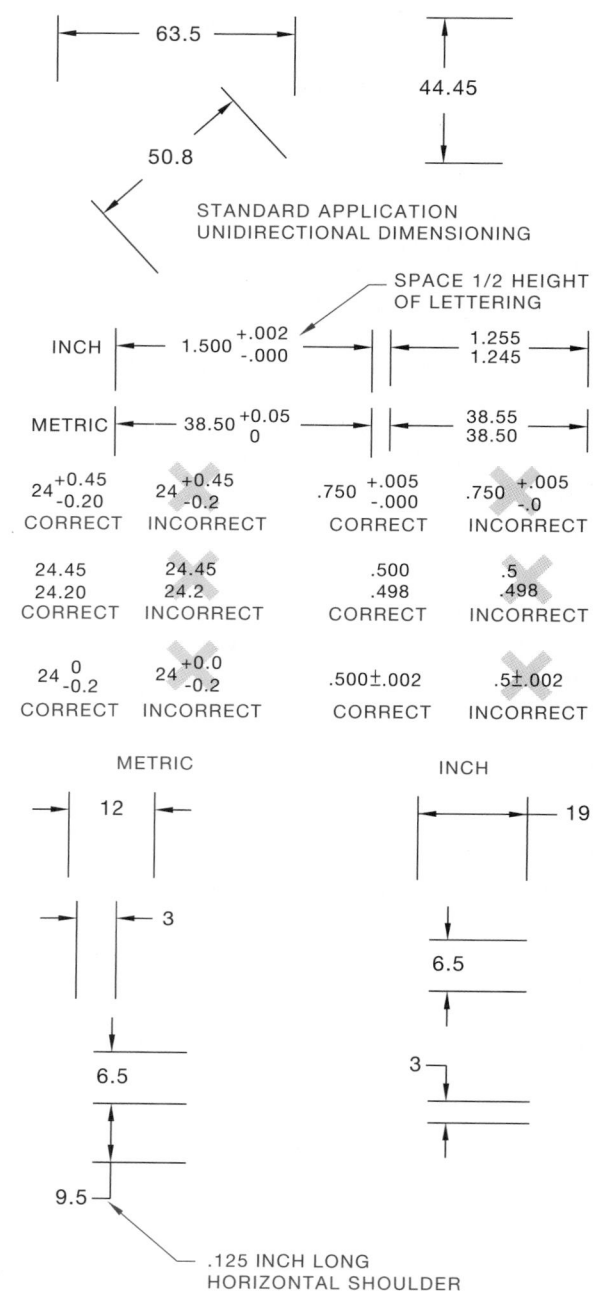

FIGURE 12.16 ■ Dimensioning applications to limited spaces.

dimension is dependent on the previous dimension or dimensions. This is a common practice, although caution should be used as the tolerance of each dimension builds on the next, which is known as *tolerance buildup,* or *stacking.* Figure 12.18 also shows the common mechanical drafting practice of providing an overall dimension while leaving one of the intermediate dimensions blank. The overall dimension is often a critical dimension that should stand independent in relationship to the other dimensions. Also, if all dimensions are given, then the actual size may not equal the given overall dimension due to tolerance buildup. An example of tolerance buildup is when three chain dimensions

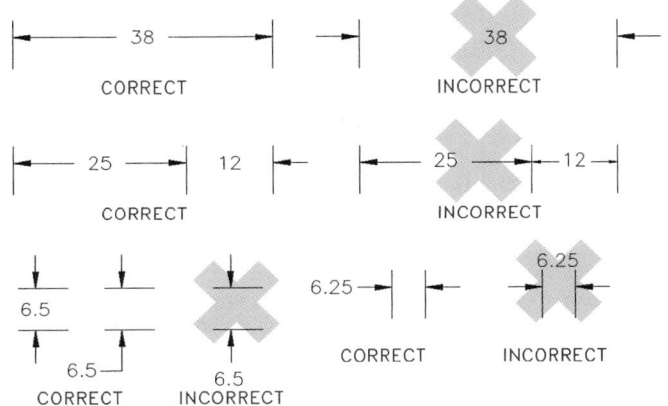

FIGURE 12.17 ■ Correct and incorrect dimensioning practices.

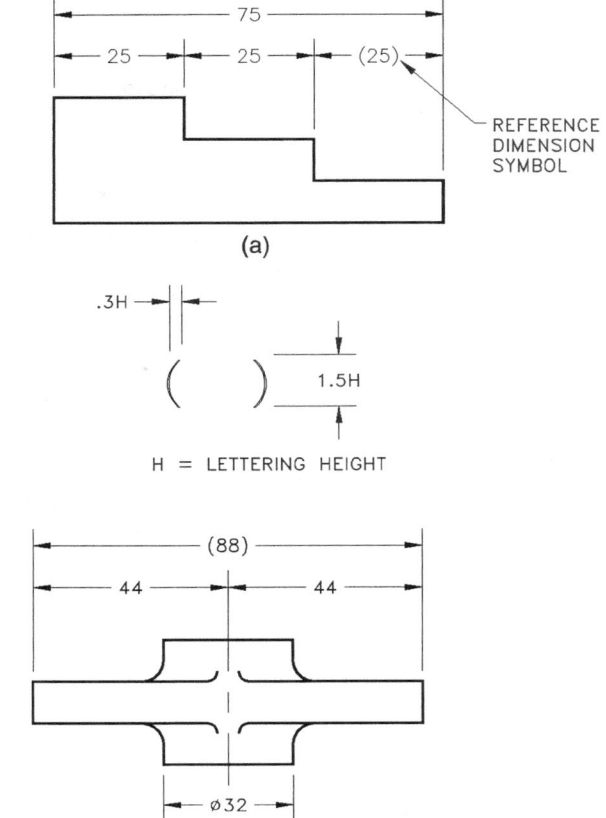

FIGURE 12.19 ■ Reference dimension examples, and reference dimension symbol.

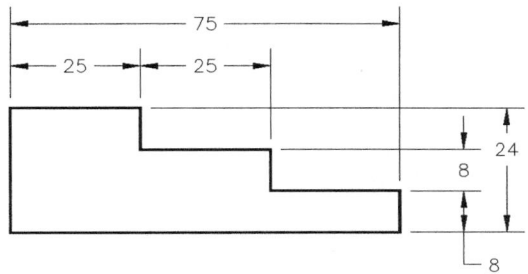

FIGURE 12.18 ■ Chain dimensioning.

have individual tolerances of ±.15 and each feature is manufactured at or toward the +.15 limit: the potential tolerance buildup is three times +.15 for a total of .45. The overall dimension would have to carry a tolerance of ±0.45 to accommodate this buildup. If the overall dimension is critical, such an amount may not be possible. Thus, either one intermediate dimension should be omitted or the overall dimension omitted. The exception to this rule is when a dimension is given only as reference. A reference dimension is enclosed in parentheses, as in Figure 12.19a. Figure 12.19b shows the overall dimension of an object as a reference. Chain dimensioning is commonly used in architectural drafting.

Datum Dimensioning (Stagger Adjacent Dimensions)

Datum dimensioning is a common method of dimensioning machine parts whereby each feature dimension originates from a common surface, axis, or center plane. (See Figure 12.20.) Each dimension in datum dimensioning is independent so there is no tolerance buildup. Figure 12.21 shows how dimensions can be symmetrical about a center plane used as a datum. Also notice in Figure 12.21 how dimension numerals are staggered rather than being stacked directly above one another. Always **stagger adjacent dimensions** when possible. Doing so helps clarity and reduces crowding. Figure 12.21 also shows use of the symmetrical symbol.

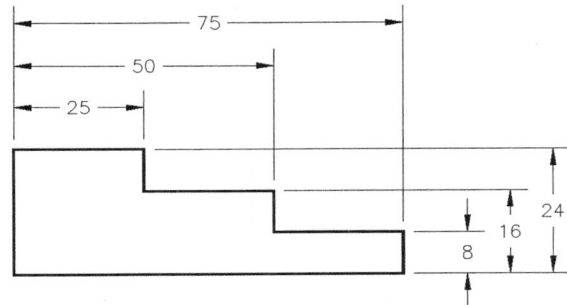

FIGURE 12.20 ■ Datum dimensioning from a common surface.

Both halves of the object are the same. Part of the right side is removed, and a short break line is used to save space. Use this practice only if necessary, because it can cause confusion if improperly read.

Dimension cylindrical shapes in the view where the cylinders appear rectangular. The diameters are identified by the diameter symbol, and the circular view may be omitted. (See Figure 12.22.) Square features may be dimensioned in a similar manner using the square symbol shown in Figure 12.23.

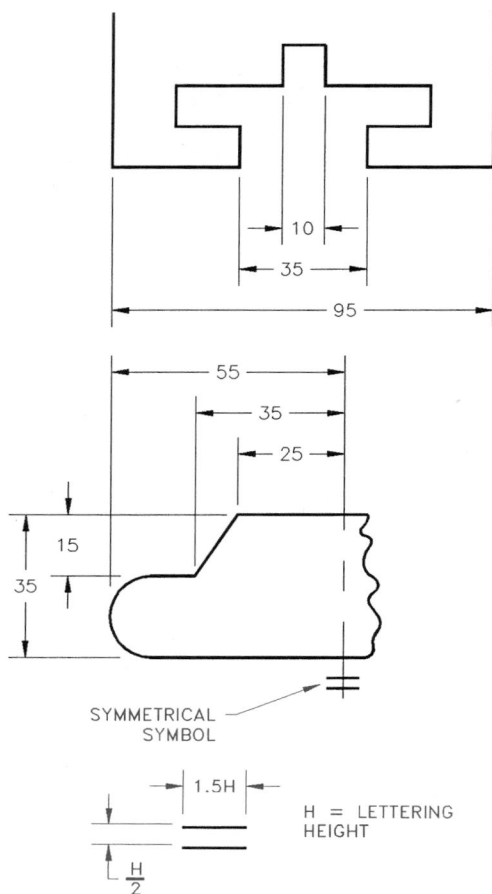

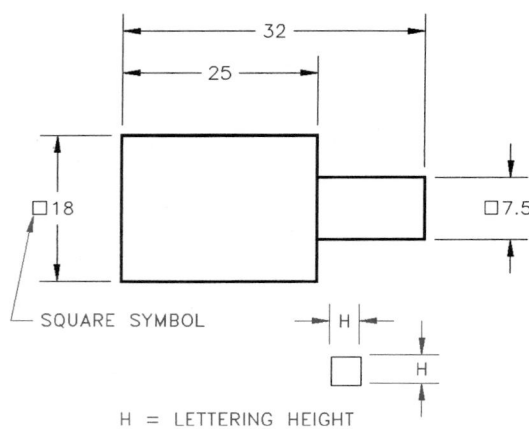

FIGURE 12.23 ■ Dimensioning square features, and square symbol.

FIGURE 12.21 ■ Datum dimensioning from center plane datums and symmetrical symbol.

PREFERRED DIMENSIONING PRACTICES

The drawings in Figure 12.24 show both correct and incorrect dimensioning techniques. These are mostly suggested options for correct and incorrect dimensioning. Good judgment should be used with each case.

Always try, when possible, to:

■ Avoid crossing extension lines, but **do not** break them when they do cross.

■ **Never** cross extension lines over dimension lines, or break the extension line over the dimension line, unless there is **absolutely no other solution**.

■ Break extension lines when they cross **over or near** an arrowhead.

■ **Avoid** dimensioning over or through the object.

■ **Avoid** dimensioning to hidden features.

■ **Avoid** unnecessary long extension lines.

■ **Avoid** using any line of the object as an extension line.

■ Dimension **between views** when possible.

■ **Group** adjacent dimensions.

■ Dimension to views that provide the **best shape description**.

Dimensioning Angles

Angular surfaces may be dimensioned as coordinates, as angles in degrees, or as a flat taper. (See Figure 12.25.) Angles are calibrated in degrees, symbol °. (There are 360° in a circle. Each degree contains 60 minutes, symbol '. Each minute has 60 seconds, symbol ". 1° = 60'; 1' = 60".) Notice in the angular method in Figure 12.25, the dimension line for the 45° angle is drawn as an arc. The radius of this arc is centered at the vertex of the angle.

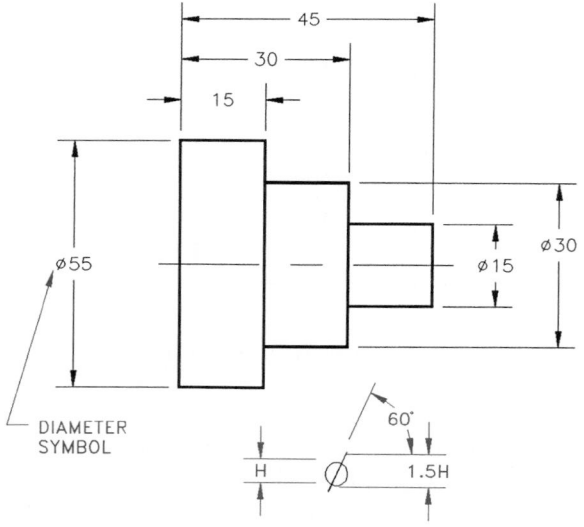

FIGURE 12.22 ■ Dimensioning cylindrical shapes, and diameter symbol. Note: Some CADD programs do not automatically provide the properly sized diameter symbol, as shown in this example.

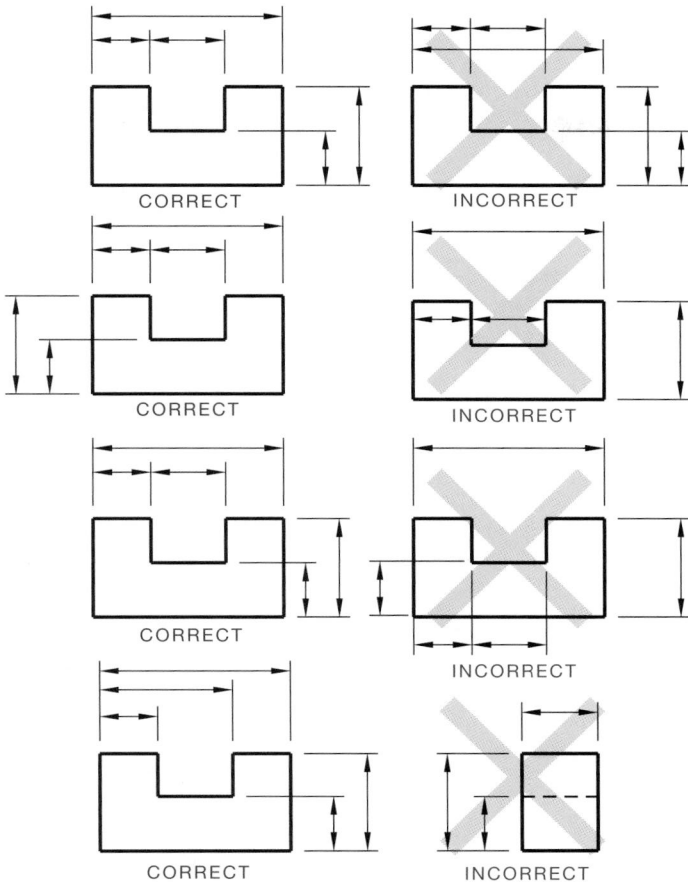

FIGURE 12.24 ■ Correct and incorrect dimensioning examples.

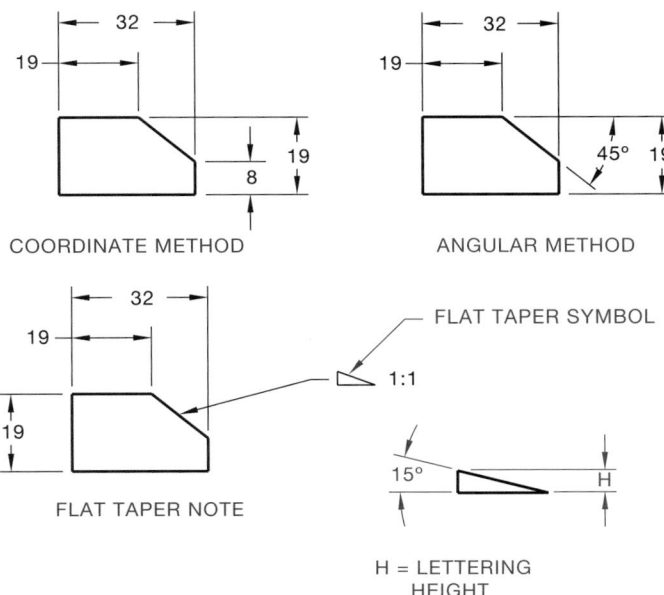

FIGURE 12.25 ■ Dimensioning angular surfaces and flat taper symbol.

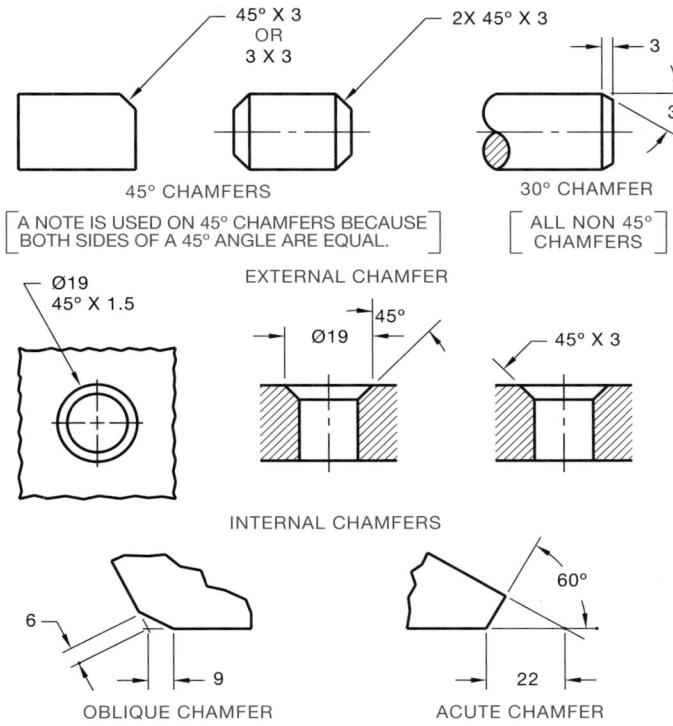

FIGURE 12.26 ■ Dimensioning chamfers.

Dimensioning Chamfers

A *Chamfer* is a slight surface angle used to relieve a sharp corner. Chamfers of 45° are dimensioned with a note, while other chamfers require an angle and size dimension, as seen in Figure 12.26. A note is used on 45° chamfers because both sides of a 45° angle are equal.

Dimensioning Conical Shapes

Conical shapes should be dimensioned when possible in the view where the cone appears as a triangle, as in Figure 12.27. A conical

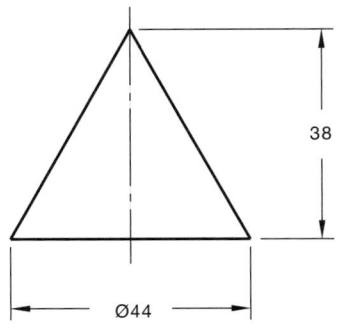

FIGURE 12.27 ■ Dimensioning conical shapes.

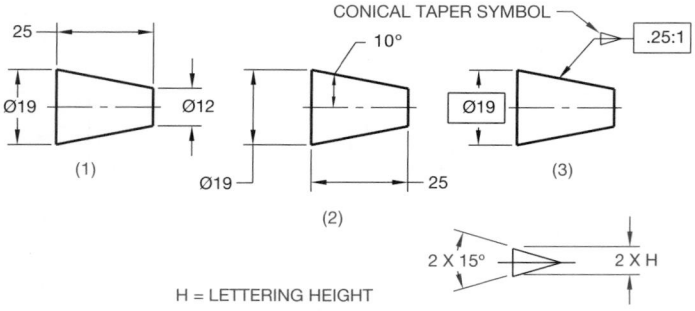

FIGURE 12.28 ■ Dimensioning conical tapers, and the conical taper symbol.

taper may be treated in one of three possible ways, as shown in Figure 12.28.

Dimensioning Hexagons and Other Polygons

Dimension hexagons and other polygons across the flats in the views where the true shape is shown. Provide a length dimension in the adjacent view, as shown in Figure 12.29.

Dimensioning Arcs

Arcs are dimensioned with leaders and radius dimensions in the views where they are shown as arcs. The leader may extend from the center to the arc or may point to the arc, as shown in Figures 12.30 and 12.31. The letter *R* precedes all radius dimensions. Depending on the situation, arcs may be dimensioned with or without their centers located. Figure 12.31 shows a very large arc with the center moved closer to the object. To save space, a break line is used in the leader and the shortened locating dimension. The length of an arc may also be dimensioned one of three ways, as shown in Figure 12.32.

In a situation where an arc lies on an inclined plane and the true representation is not shown, the note TRUE R may be used to spec-

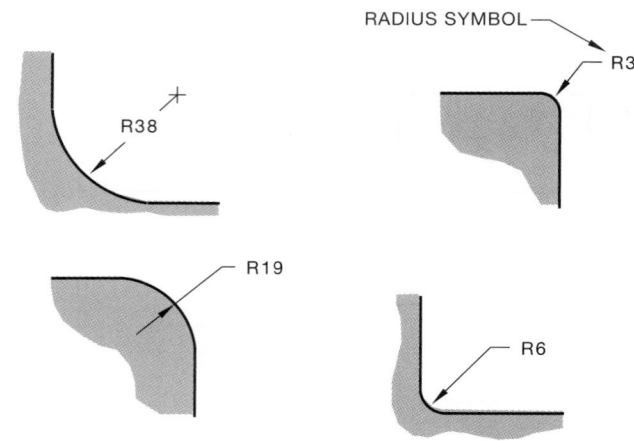

FIGURE 12.30 ■ Dimensioning arcs—no centers located, and the radius symbol.

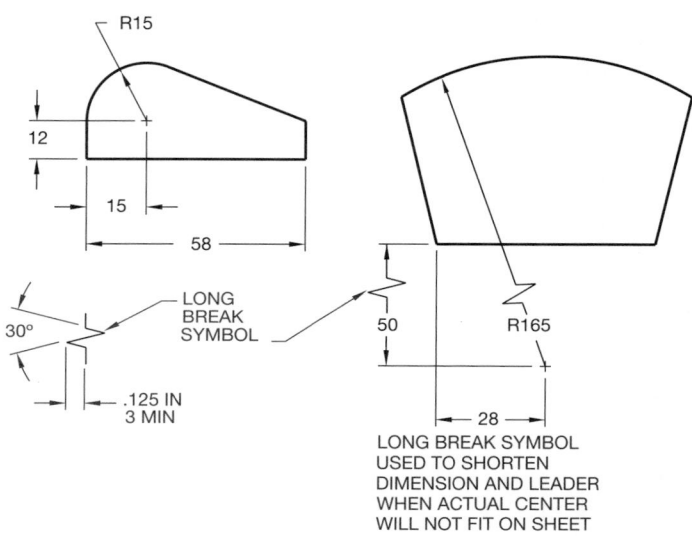

FIGURE 12.31 ■ Dimensioning arcs—with centers located, and the long break symbol.

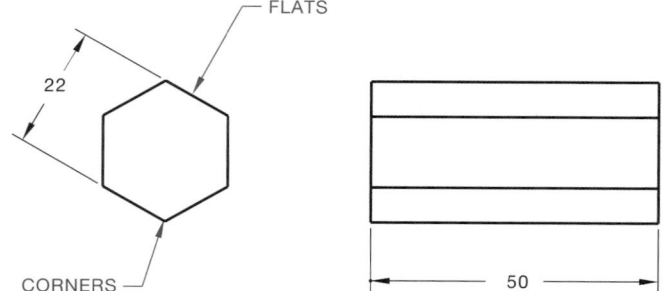

FIGURE 12.29 ■ Dimensioning hexagons.

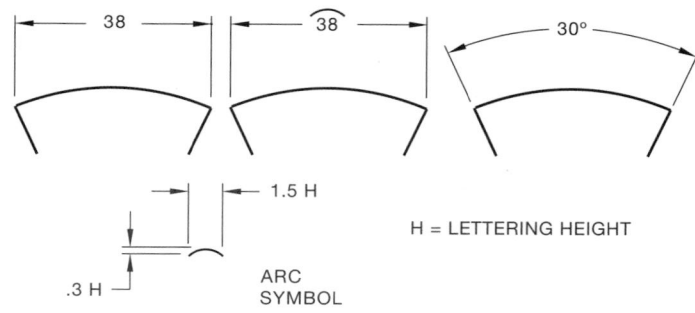

FIGURE 12.32 ■ Dimensioning arc length and the arc symbol.

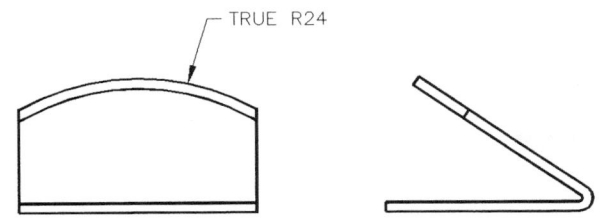

FIGURE 12.33 ■ Dimensioning a true radius in an inclined plane.

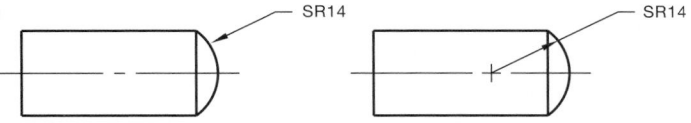

FIGURE 12.34 ■ Dimensioning a spherical radius.

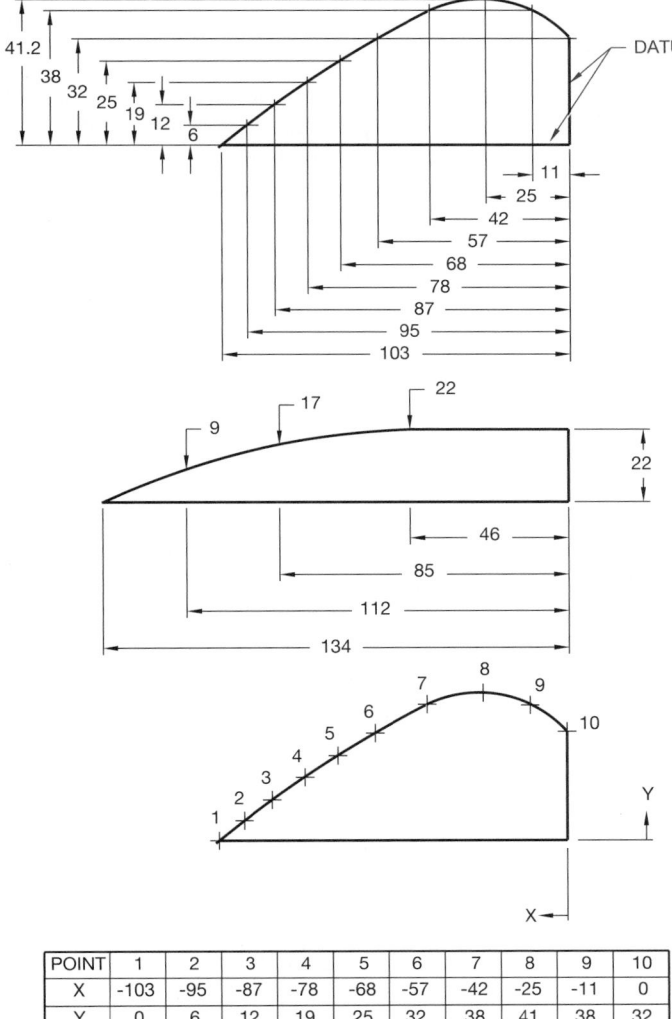

POINT	1	2	3	4	5	6	7	8	9	10
X	-103	-95	-87	-78	-68	-57	-42	-25	-11	0
Y	0	6	12	19	25	32	38	41	38	32

FIGURE 12.35 ■ Dimensioning contours not defined as arcs.

ify the actual radius. However, dimensioning the arc in an auxiliary view of the inclined surface would be better if possible. (See Figure 12.33).

The symbol *CR* refers to controlled radius. *Controlled radius* means that the limits of the radius tolerance zone must be tangent to the adjacent surfaces, and there can be no reversals in the contour. The *CR* control is more restrictive than use of the *R* radius symbol where reversals in the contour of the radius are permitted.

A spherical radius may be dimensioned with the abbreviation *SR* preceding the numerical value, as shown in Figure 12.34.

Dimensioning Contours Not Defined as Arcs

Coordinates or points along the contour are located from common surfaces or datums as shown in Figure 12.35. Figure 12.36 shows a curved contour dimensioned using oblique extension lines. While this technique may be used, it is not as common as the previous method.

Locating a Point Established by Extension Lines

If the sides of the object in Figure 12.37 were extended beyond the bend, they would meet at the intersection of the extension lines. This imaginary point is where the dimension often originates in this type of situation.

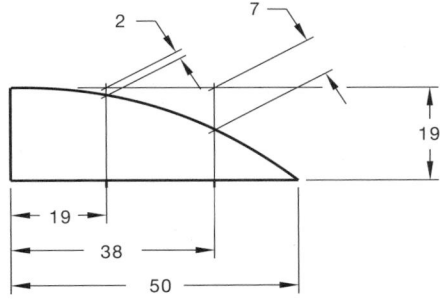

FIGURE 12.36 ■ Dimensioning a curved contour using oblique extension lines.

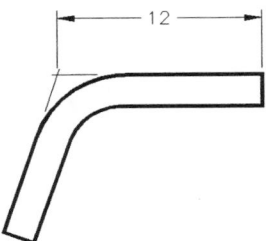

FIGURE 12.37 ■ Locating a point established by extension lines.

DIMENSIONING

The proper placement and use of dimensions is one of the most difficult aspects of drafting. It requires thought and planning. The drafter must first determine the type of dimension that suits the application, then place it on the drawing. The dimensioning options for the CADD drafter are many, and the proper usage of these options should be learned early and practiced whenever possible. A variety of dimensioning possibilities may be located on the CADD menu under the heading Dimension Parameters, Automatic Dimensioning, or simply, Dimensions. The menu may show each of the possibilities, or they may be listed on the screen after selecting the DIMENSION or DIM command.

Two basic types of dimensions are horizontal and vertical. These can be labeled on the menu as HORIZ and VERT, or DMH (dimension horizontal) and DMV (dimension vertical). AutoCAD, for example, has the DIMLINEAR (DIMLIN) command that allows you to automatically place any straight dimension whether it is horizontal, vertical, or at an angle. All you have to do is pick the extension line origins and the dimension line location. To place a horizontal dimension, first select the DIMLIN command. You may then be prompted to select the first extension line origin by placing the screen cursor so that the crosshair is on the first corner, as shown in step 1 of Figure 12.38. Pick this point and then locate the crosshairs on the opposite end of the feature and pick that point, as shown in step 2 of Figure 12.38. You have now established the length of the feature. The dimension is calculated by the computer. Next, you need to pick the distance away from the feature where the dimension line should

be located. After this step is done, the dimension is automatically drawn on the screen. (See Figure 12.38.)

Another way to place dimensions on a drawing is to pick the extension line origins on the object. The CADD dimensioning then automatically calculates the extension line offset and places the extension lines, dimension lines, text, and arrowheads. This can be done in either a datum or chain dimensioning format, as shown in Figure 12.39. When you are dimensioning circles, the dimension line and text numeral are automatically placed inside the circle by picking the circle, or you have the option of dimensioning with a leader. Figure 12.40 illustrates some of the options. Dimensioning arcs is also easy. All you have to do is pick the arc to be dimensioned and enter the dimension text; the leader with text numeral is placed on the drawing automatically, as shown in Figure 12.41.

Some CADD programs also allow you to automatically dimension a group of objects just by selecting them. All you have to do is select a group of objects followed by picking the dimension line location, and the CADD system automatically draws the extension lines, dimension lines, and arrowheads, and places the dimension text. The AutoCAD command that performs this operation is QDIM.

CADD systems allow you to draw dimensions for any desired application, such as mechanical, architectural, structural, or civil. The standard that you use to set up the way dimensions appear on your drawing is called a *dimension style*. There is a wide variety of variables that you can set when creating a dimension style. For example, if you want to

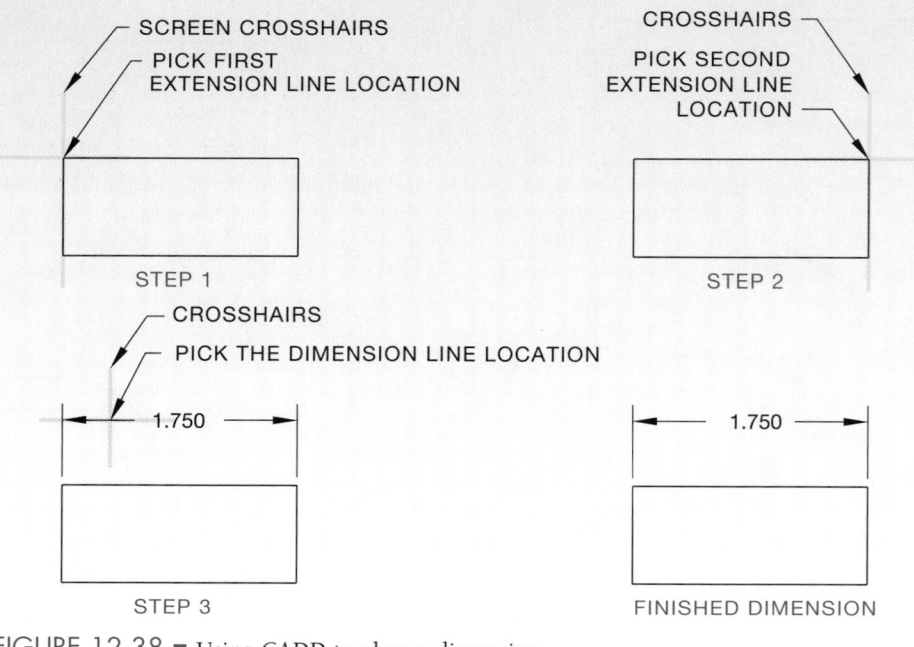

FIGURE 12.38 ■ Using CADD to place a dimension.

create a dimension style for mechanical drafting, you might want to set the following variables:

■ Dimension text placed in a space provided in the dimension line as in Figure 12.39.

■ Dimension text height of .125 in. (3 mm).

■ Dimension text using a ROMANS font.

■ .125 in (3 mm) long arrowheads.

■ Three place decimals dimension values, such as 2.625.

■ The gap from the object to the start of the extention line, such as .062 in. (1.5 mm).

■ The distance the extension line extends beyond the last dimension line, such as .125 in. (3 mm).

■ Tolerance specifications, if any.

■ Plus other available options depending on the way you want the dimensions to look.

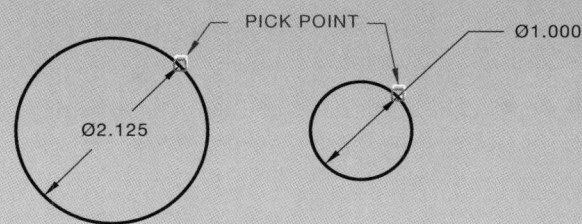

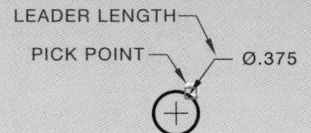

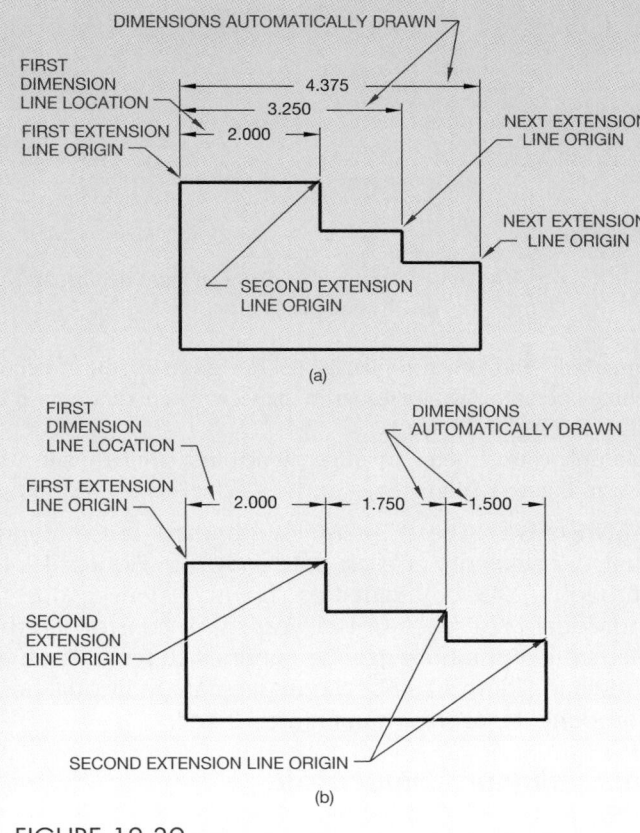

FIGURE 12.39 ■ Locating dimensions at extension line origins for **(a)** datums and **(b)** chain dimensioning.

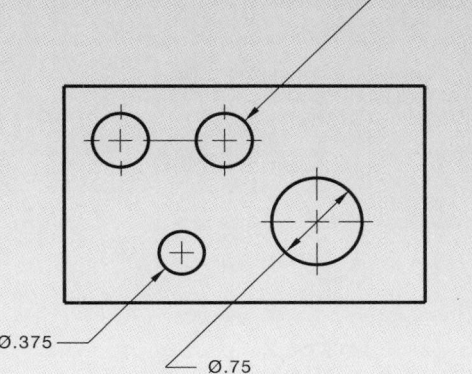

FIGURE 12.40 ■ Dimensioning circles with CADD.

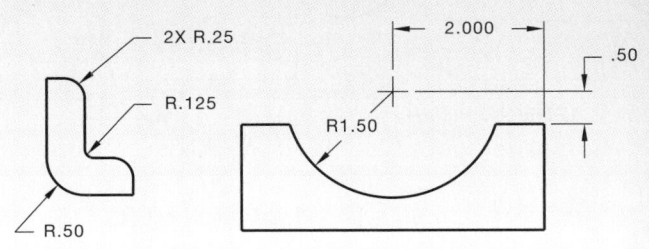

FIGURE 12.41 ■ Dimensioning arcs with CADD.

NOTES FOR SIZE FEATURES

Holes

Hole sizes are dimensioned with leaders to the view where they appear as circles, or dimensioned in a sectional view. When leaders are used to establish notes for holes, the shoulder should be centered on the beginning or the end of the note.

When a leader begins at the left side of a note, it should originate at the beginning of the note. When a leader begins at the right side of a note, it should originate at the end of the note. (See Figure 12.42.)

Figure 12.43 shows the leader should touch the side of the circle and, if it were to continue, would intersect the center of the circle. Leaders may be drawn at any angle, preferably between 15° and 75° from horizontal. Do not draw horizontal or vertical leaders.

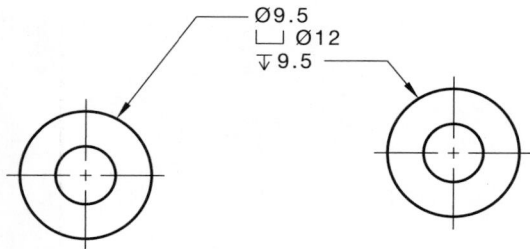

FIGURE 12.42 ■ Leader orientation to the note. Center leader shoulder at beginning or end of note.

CORRECT   INCORRECT

FIGURE 12.43 ■ Leader orientation to the circle. Leader arrow should *point* to center.

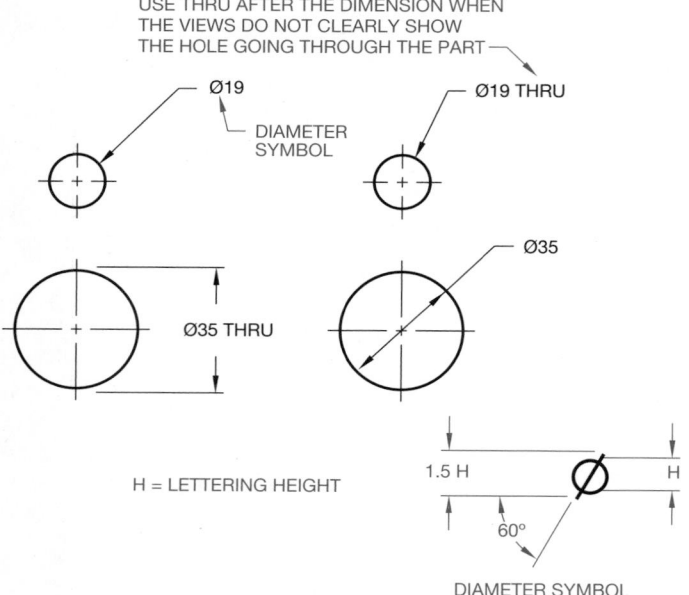

USE THRU AFTER THE DIMENSION WHEN THE VIEWS DO NOT CLEARLY SHOW THE HOLE GOING THROUGH THE PART

Ø19

DIAMETER SYMBOL

Ø19 THRU

Ø35 THRU

Ø35

H = LETTERING HEIGHT

1.5 H   H

60°

DIAMETER SYMBOL

FIGURE 12.44 ■ Dimensioning hole diameters, and the diameter symbol.

A hole through a part may be noted if not obvious. The diameter symbol precedes the diameter numeral, as dimensioned in Figure 12.44. If a hole does not go through the part, the depth must be noted in the circular view or in section, as shown in Figure 12.45.

Counterbore

A counterbore is often used to machine a diameter below the surface of a part so a bolt head or other fastener may be recessed.

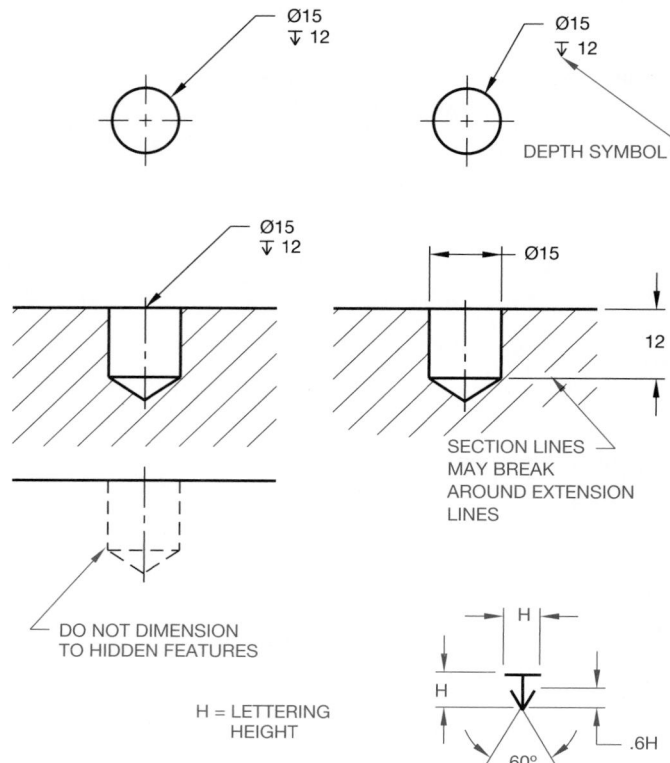

Ø15 ▼ 12

Ø15 ▼ 12

DEPTH SYMBOL

Ø15 ▼ 12

Ø15

12

SECTION LINES MAY BREAK AROUND EXTENSION LINES

DO NOT DIMENSION TO HIDDEN FEATURES

H = LETTERING HEIGHT

H

H

.6H

60°

FIGURE 12.45 ■ Dimensioning hole diameters and depths, and the depth symbol.

Counterbore and other similar notes are given in the order of machine operations with a leader in the view where they appear as circles. (See Figure 12.46a.)

Multiple counterbores are dimensioned in a similar manner as shown in Figure 12.46d.

ASME/ANSI The ASME/ANSI standard recommends that the elements of each note shown in Figures 12.46 through 12.49 be aligned as shown. However, due to individual preference or drawing space constraints, the elements of the note may be confined to fewer lines, as shown in Figure 12.46b. **Never** separate individual note components, as shown in Figure 12.46c.

Countersink or Counterdrill

A countersink, or counterdrill, is also often used to recess the head of a fastener below the surface of a part. (See Figure 12.47.)

Spotface

A spotface is used to provide a flat bearing surface for a washer face or bolthead. (See Figure 12.48.) Follow the counterbore guidelines when lettering the spotface note.

Multiple Features

When a part has more than one feature of the same size, they may be dimensioned with a note that specifies the number of like fea-

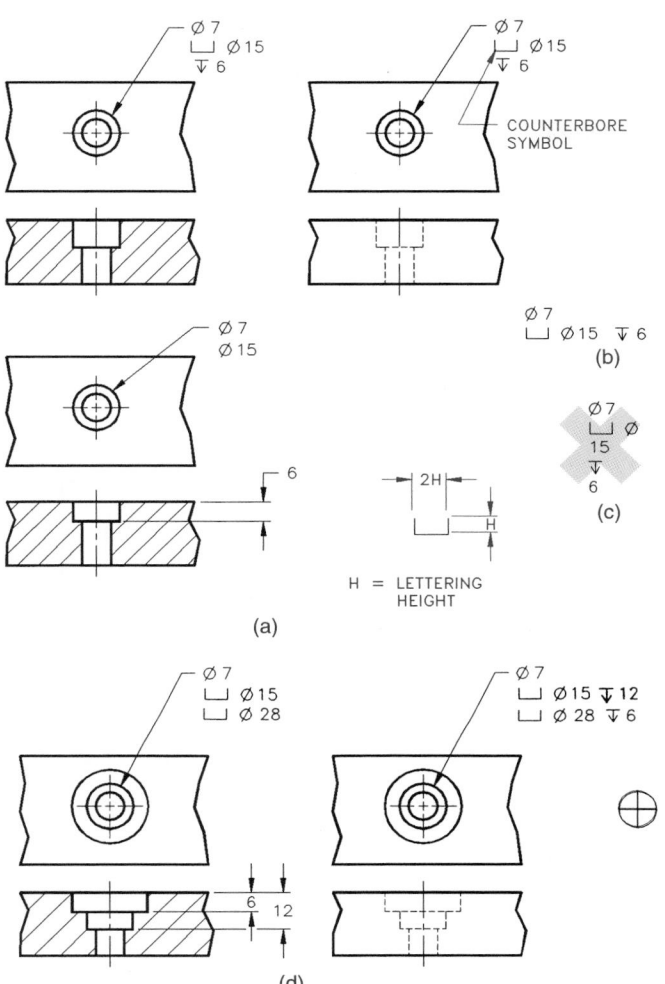

(a)

(b)

(c)

(d)

FIGURE 12.46 ▪ (a) Counterbore note; (b) alternate note with elements of note grouped on one line; (c) never split individual note elements. The proper counterbore/spotface symbol is also displayed.

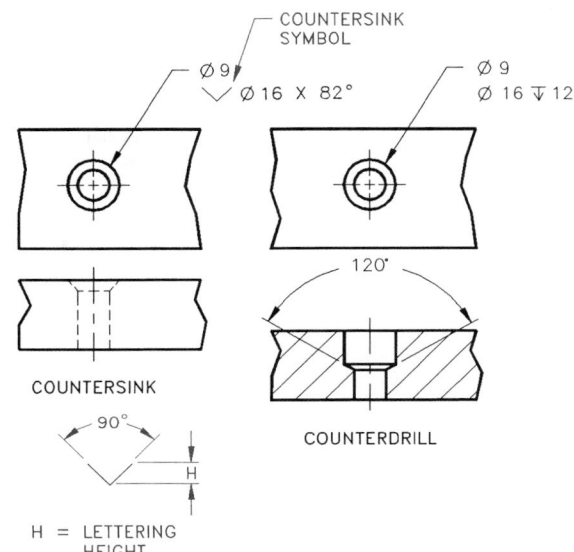

FIGURE 12.47 ▪ Countersink, counterdrill notes, and dimensions. The properly drawn countersink symbol is also shown.

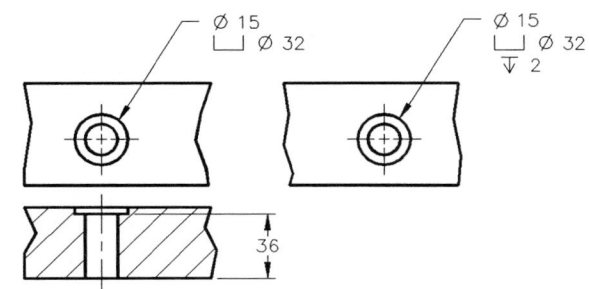

FIGURE 12.48 ▪ Spotface note and dimensions.

tures, as shown in Figure 12.49. If a part contains several features all the same size, the methods shown in Figure 12.50 may be used.

Slots

Slotted holes may be dimensioned in one of three ways, as shown in Figure 12.51. These methods are used only when the ends are fully rounded and tangent to the sides. When the ends of a slot or external feature have a radius greater than the width of the feature, then the size of the radius must be given, as shown in Figure 12.52.

Keyseats

Keyseats are dimensioned in the view that clearly shows their shape by width, depth, length, and location, as in Figure 12.53.

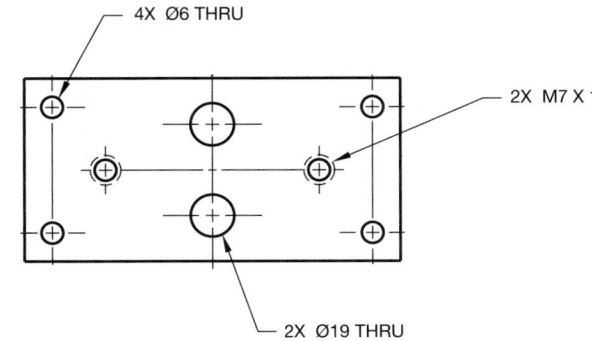

FIGURE 12.49 ▪ Dimension notes for multiple features.

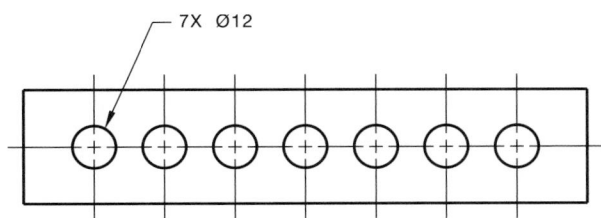

FIGURE 12.50 ▪ Dimension notes for multiple features all of common size.

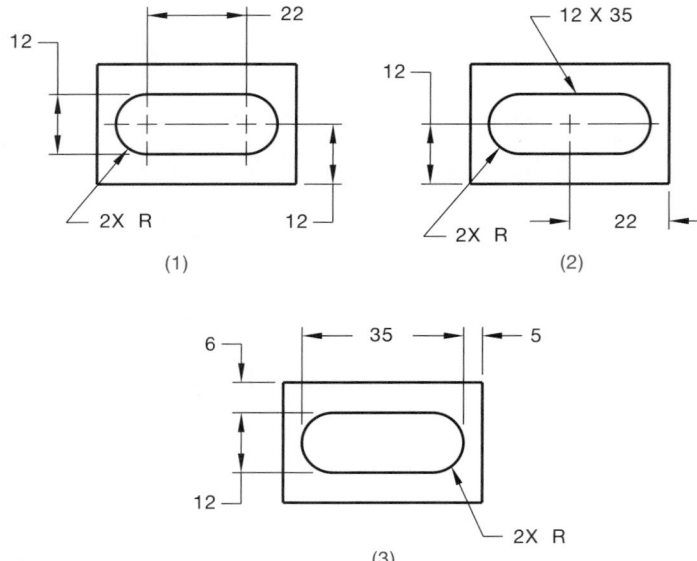

(1)

(2)

(3)

FIGURE 12.51 ■ Dimensioning slotted holes with full radius ends.

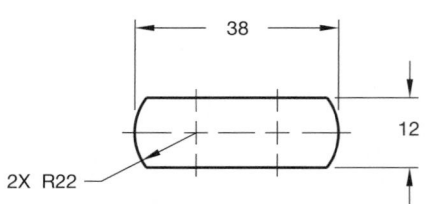

FIGURE 12.52 ■ Dimensioning slot or external feature with end radius larger than feature width.

Knurls

Knurls are dimensioned with notes and leaders that point to the knurl in the rectangular view, as shown in Figure 12.54a. ASME/ANSI **does not** recommend showing a knurl representation on the view as in Figure 12.54b, although some companies prefer this practice.

Necks and Grooves

Necks and grooves may be dimensioned as shown in Figure 12.55.

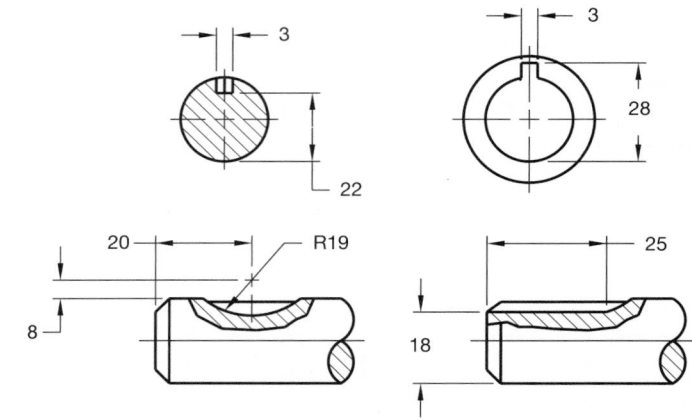

FIGURE 12.53 ■ Dimensioning keyseats.

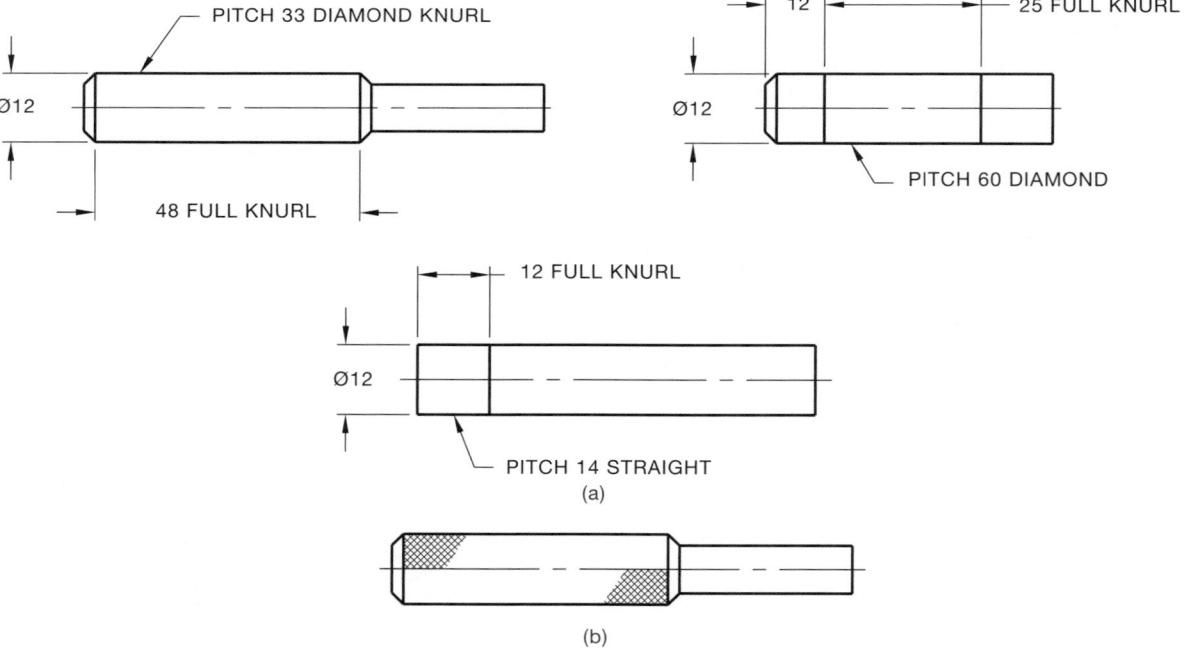

(a)

(b)

FIGURE 12.54 ■ (a) Dimensioning knurls. (b) Optional knurl representation.

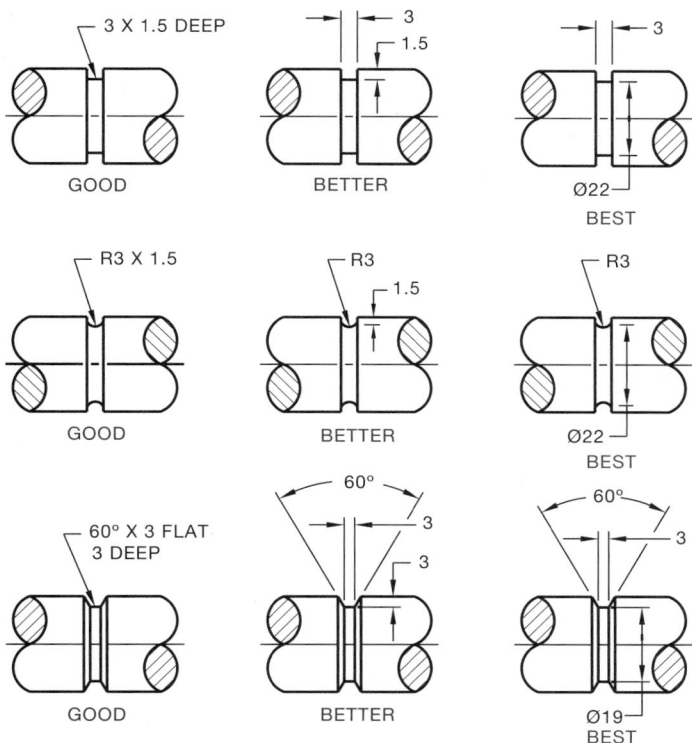

FIGURE 12.55 ■ Dimensions for necks and grooves.

LOCATION DIMENSIONS

In general, dimensions identify either a size or a location. The previous dimensioning discussions focused on size dimensions. Location dimensions to cylindrical features, such as holes, are given to the center of the feature in the view where they appear as a circle, or in a sectional view. Rectangular shapes may be located to their sides, and symmetrical features may be located to their centerline or center plane. Some location dimensions also control size.

Locating Holes

Locate a hole center in the view where the hole appears as a circle, as shown in Figure 12.56.

Rectangular Coordinates

Linear dimensions are used to locate features from planes or centerlines, as shown in Figure 12.57.

Polar Coordinates

Angular dimensions locate features from planes or centerlines, as shown in Figure 12.58.

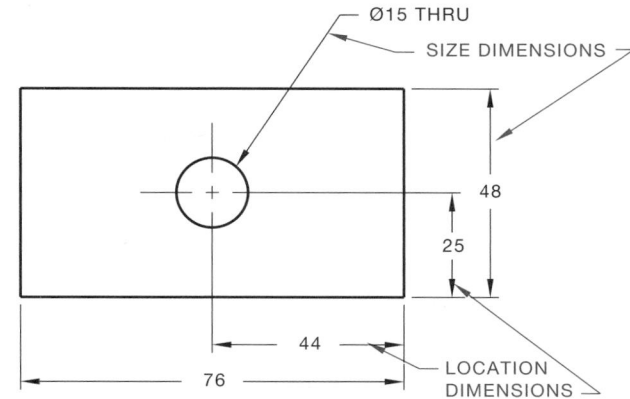

FIGURE 12.56 ■ Size and location dimensions.

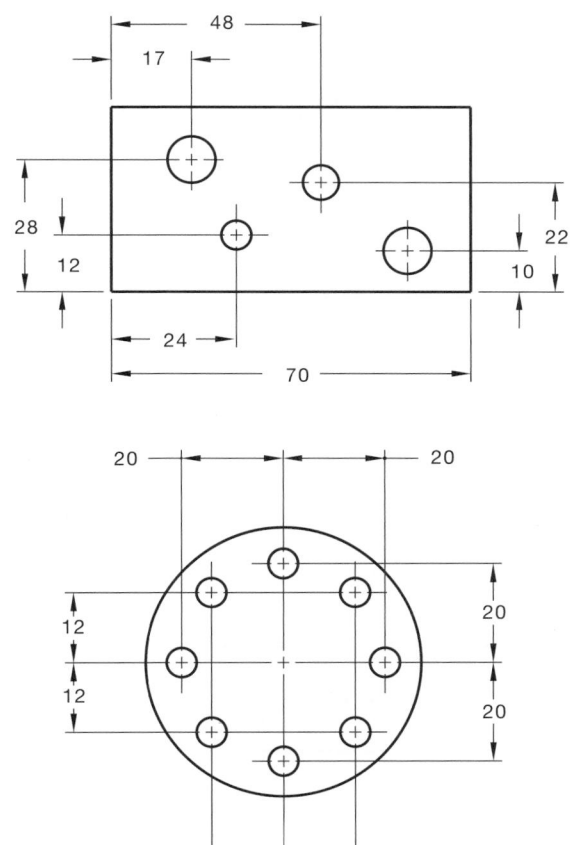

FIGURE 12.57 ■ Rectangular coordinate location dimensions.

Repetitive Features

Repetitive features may be located by noting the number of times a dimension is repeated, and giving one typical dimension and the total length as reference. This method is acceptable for chain dimensioning. (See Figure 12.59.) Locating multiple tabs is shown in Figure 12.60. This method also works for slots. When repetitive

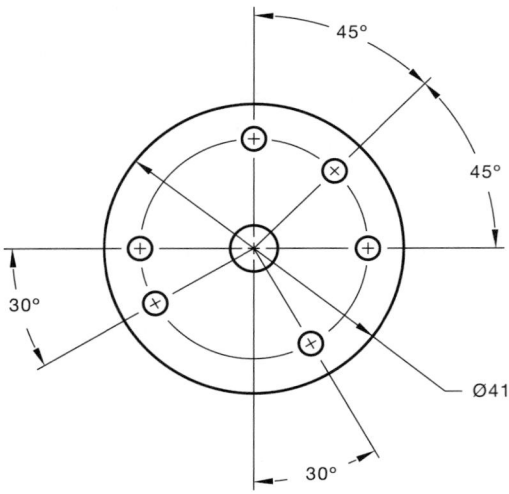

FIGURE 12.58 ■ Polar coordinate dimensions. If all features are equally spaced, a note such as 6X 60° may be used in one location.

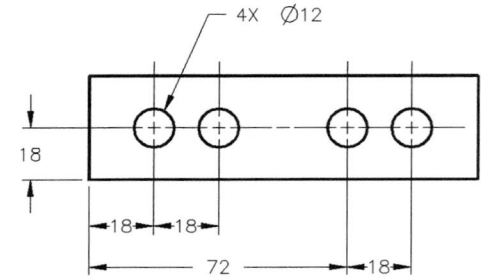

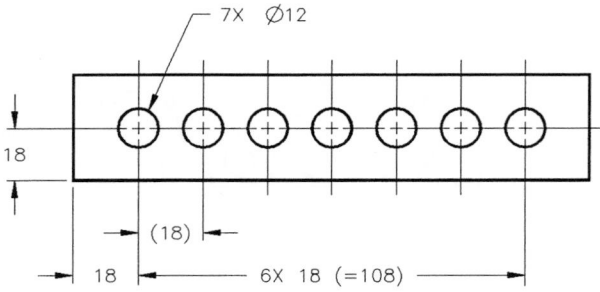

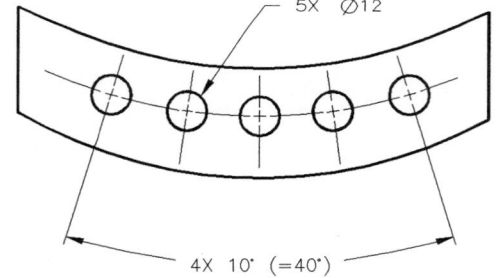

FIGURE 12.59 ■ Dimensioning repetitive features.

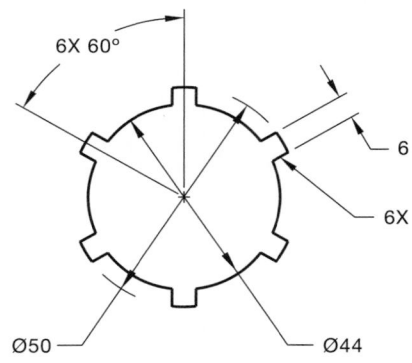

FIGURE 12.60 ■ Locating multiple tabs.

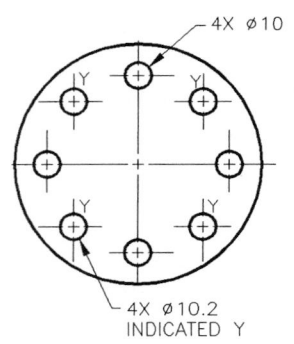

FIGURE 12.61 ■ Dimensioning similar sized multiple features.

features on an object are nearly the same size, they may be shown with an identification letter, such as *Y*. (See Figure 12.61.)

DIMENSION ORIGIN

When the dimension between two features must clearly denote from which feature the dimension originates, the dimension origin symbol may be used. This method of dimensioning means that the origin feature must be established first and the related feature may then be dimensioned from the origin. (See Figure 12.62.)

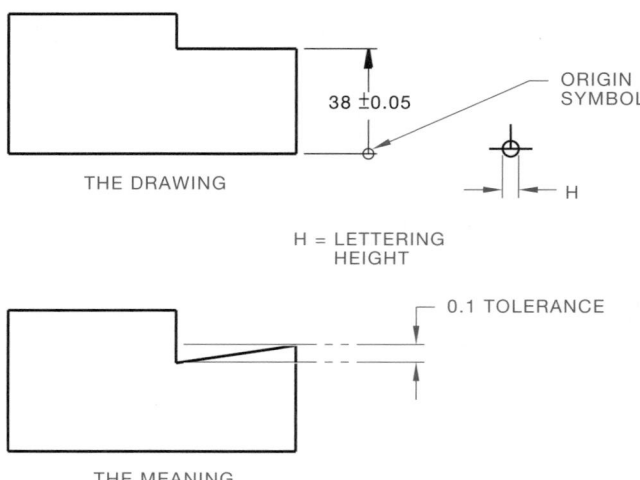

FIGURE 12.62 ■ Dimension origin, and the dimension origin symbol.

DIMENSIONING AUXILIARY VIEWS

Dimensions should be placed on views that provide the best size and shape description of an object. In many instances the surfaces of a part are foreshortened and require auxiliary views to completely describe true size, shape, and the location of features. When foreshortened views occur, dimensions should be placed on the auxiliary view for clarity. In unidirectional dimensioning, the dimension numerals are placed horizontally so they read from the bottom of the sheet. When aligned dimensioning is used, the dimension numerals are placed in alignment with the dimension lines. (See Figure 12.63.)

GENERAL NOTES

ASME/ANSI General notes are located next to the title block for ASME/ANSI drawings. The exact location of the general notes depends on specific company standards. A common location for general notes is the lower left corner of the drawing, usually .5 in. each way from the border line. Military standards specify general notes be located in the upper left corner of the drawing.

The notes that were discussed previously are classified as specific notes because they referred to specific features of an object. *General notes,* on the other hand, relate to the entire drawing. Each drawing contains a certain number of general notes either in or near the drawing title block. (See Figure 12.64.) General notes are concerned with such items as the following:

■ Material specifications.
■ Dimensions: inches or millimeters.
■ General tolerances.
■ Confidential note, copyrights, or patents.
■ Drawn by.
■ Scale.
■ Date.

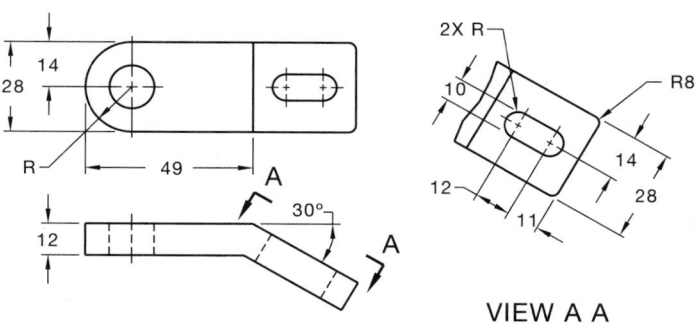

UNIDIRECTIONAL DIMENSIONING
ON AN AUXILIARY VIEW

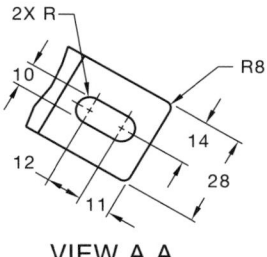

VIEW A A

ALIGNED DIMENSIONING
ON AN AUXILIARY VIEW

FIGURE 12.63 ■ Dimensioning auxiliary views.

■ Part name.
■ Drawing size.
■ Part number.
■ Number of revisions.
■ First-angle or third-angle projection symbol.
■ ANSI/ASME, MIL (Military), or other standard reference.
■ General machining, finish, or paint specifications.
■ Identification of features that are the same throughout the drawing.

UNLESS OTHERWISE SPECIFIED	DR	DATE			C/C
DIMENSIONS IN INCHES	CHKD	DATE		**Althin**	
	APVD	DATE		Medical, Inc. Portland, OR	
DECIMALS: ANGLES: X +/− .020 XX +/− .010 +/−.5° XXX +/− .005	APVD	DATE	TITLE		
	MATERIAL				
REMOVE ALL BURRS AND BREAK SHARP EDGES .02 MAX					
SURFACE ROUGHNESS 63/	FINISH		DWG SIZE	SCALE	SHEET OF
DIMENSIONS BEFORE PLATING OR COATING			C	DWG NO.	REV
3rd ANGLE PROJECTION	DO NOT SCALE DRAWING				

FIGURE 12.64 ■ General title block information. *Courtesy Althin Medical, Inc.*

USING LAYERS FOR DIMENSIONING

A time-tested method of creating several drawings containing different information on the same outline is known as *overlay drafting*. A base drawing is generated, then an overlay is placed over it and a specific type of information is drawn on the overlay. For example, an architectural floor plan is drawn as the base, and then overlays are drawn for the plumbing plan, electrical plan, and so on. The base can be combined with any individual overlay or combinations thereof to create a print of just the information required. Also, several drafters can be given copies of the base and each can create an overlay containing specific information. This enables several drafters to essentially work on the same drawing at the same time; a great time-saver. The final drawing is then composed of several layers or levels.

The concept of layering is a prime component of CADD systems, some having the ability to display over 200 layers. This not only allows different drafters to work on the same drawing at the same time, but also addresses a shortcoming of the computer itself: the speed at which a drawing can be redrawn on the screen. The less information that is on the screen at one time, the less time it takes to be redrawn, and the less clutter there is to get in the way of the drafter. Figure 12.65 shows a simple example of the layering concept.

Several commands exist to allow the operator to work with the different layers. The LAYER command itself lets the operator assign a layer number to a drawing. RELAYER gives the drafter the chance to change the numbers of the drawing layers, and DISPLAY LAYER, or LAYER CONTROL, provides for the display of any combination of layers.

Color is a valuable tool when working with layers, and more companies are realizing this fact. Many color systems enable the user to assign any color desired to any layer. Some systems have a limited number of colors, while others possess the ability to create thousands of hues. Color gives even more meaning to the layering concept, because several layers displayed on the screen at the same time can be distinguished from each other. The colors on the screen can be transferred to the plotter by use of the PEN command. This command allows specific colors and widths of pens in the pen plotter to be selected.

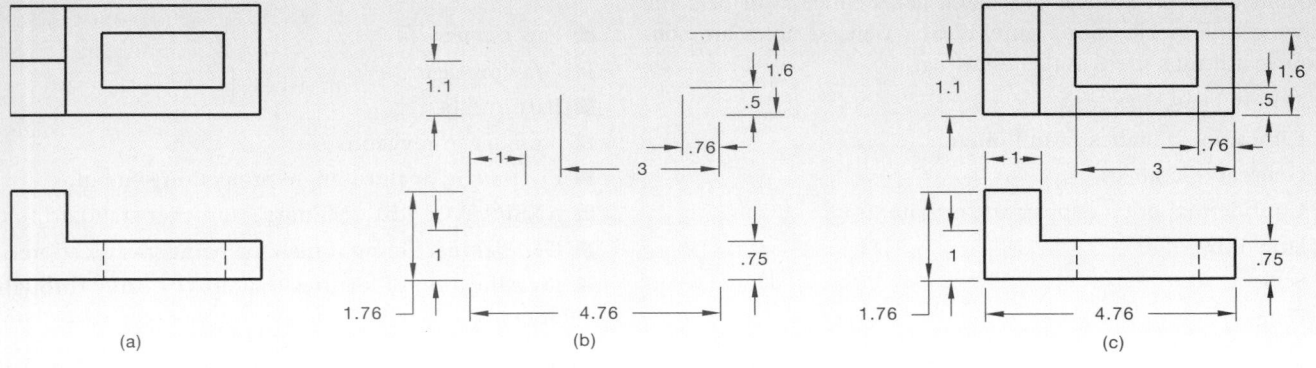

(a) (b) (c)

FIGURE 12.65 ■ Layering enables the drafter to focus on one aspect of the drawing at a time.

Figure 12.66 shows some general notes that could commonly be included on ASME/ANSI standard format drawings. Other common locations for general notes are the lower right corner of the drawing directly above the title block or just to the left of the title block.

Notice in Figure 12.66 that the word NOTES is lettered first, followed by the first, second, and additional notes. Depending on company standards, the word NOTES is lettered .18 to .25 in. high, but some companies prefer to omit the word NOTES as an unnecessary preface to obvious general notes. The space between notes is from one-half to full height of the lettering, and the notes are often lettered .125 in. in height. The first note, INTERPRET DIMENSIONS AND TOLERANCES PER ASME Y14.5M—1994, should be included on all new drawings. Additional notes depend on the information required to support the drawing.

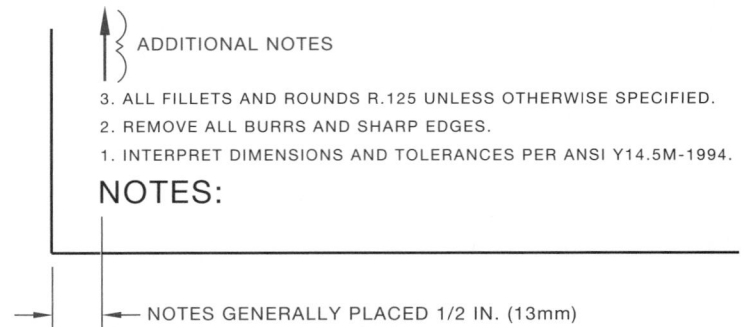

ADDITIONAL NOTES

3. ALL FILLETS AND ROUNDS R.125 UNLESS OTHERWISE SPECIFIED.

2. REMOVE ALL BURRS AND SHARP EDGES.

1. INTERPRET DIMENSIONS AND TOLERANCES PER ANSI Y14.5M-1994.

NOTES:

NOTES GENERALLY PLACED 1/2 IN. (13mm)
EACH WAY FROM THE CORNER

FIGURE 12.66 ■ General notes located in the lower left corner of the sheet.

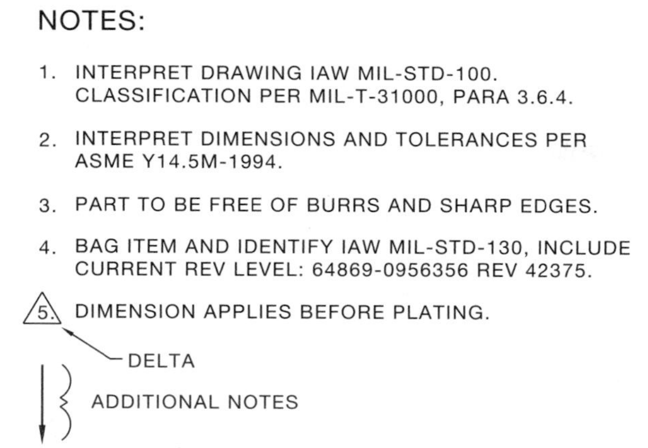

NOTES:

1. INTERPRET DRAWING IAW MIL-STD-100. CLASSIFICATION PER MIL-T-31000, PARA 3.6.4.

2. INTERPRET DIMENSIONS AND TOLERANCES PER ASME Y14.5M-1994.

3. PART TO BE FREE OF BURRS AND SHARP EDGES.

4. BAG ITEM AND IDENTIFY IAW MIL-STD-130, INCLUDE CURRENT REV LEVEL: 64869-0956356 REV 42375.

$\triangle$5. DIMENSION APPLIES BEFORE PLATING.

— DELTA

ADDITIONAL NOTES

FIGURE 12.67 ■ Notes can be conveniently placed to read from the first note downward with CADD. This makes it easy to continue from one note to the next. Editing notes is easy with CADD, because all you have to do is move the group of notes up to place additional notes when needed.

The technique of placing general notes that read from the bottom up as in Figure 12.66 has been done for the ease of manual drafting. The idea here is that additional notes can be added easily without the need to erase everything and start over. With the use of CADD, the need to place notes in this manner is up to the preference of the company or school. Placing notes with CADD is actually easier when the notes are entered from top to bottom. If additional notes are needed at a later time, the CADD drafter can easily move the existing notes up and add the new notes below. Figure 12.67 shows general notes placed with CADD so they read from top to bottom. These notes are placed in the lower left corner of the drawing. When notes are placed to conform to Military standards, then the notes are always placed in the upper left corner and they read from top to bottom.

Delta Notes

The term *delta* refers to a triangle placed on the drawing for reference. The triangle is commonly placed next to a dimension such as 2.625 $\triangle$5, or other location where it applies to a feature or item. This is used to refer the reader to a general note that relates to this item. This method is often used when it applies to a specific feature, but placing the note directly on the drawing is very difficult because the note is too long, or the note applies to several dimensions or features. Look at the note $\triangle$5 in Figure 12.67. This note relates to the same delta item located somewhere on the drawing. Some companies use symbols other than a triangle. Hexagons and circles can also be used, but the triangle is common.

TOLERANCING
Definitions

As previously mentioned, a tolerance is the total permissible variation in a size or location dimension.

A *specified dimension* is that part of the dimension from which the limits are calculated. For example, 15.8 is the specified dimension of 15.8±0.2.

A *bilateral tolerance* is allowed to vary in two directions from the specified dimension, as in $6.5^{+0.1}_{-0.3}$ or 6.5 ± 0.2.

A *unilateral tolerance* varies in only one direction from the specified dimension: $22^{\;0}_{-0.2}$ or $19.5^{+0.5}_{\;0}$.

Limits are the largest and smallest possible sizes a feature may be as related to the tolerance of the dimension.

Example 1: 19.0 ± 0.1
Upper limit: $19.0 + 0.1 = 19.1$
Lower limit: $19.0 - 0.1 = 18.9$

Example 2: $9.5^{\;0}_{-0.5}$
Upper limit: $9.5 + 0 = 9.5$
Lower limit: $9.5 - 0.5 = 9.0$

The tolerance, being the total permissible variation in the dimension, is easily calculated by subtracting the lower limit from the upper limit.

Example 3: 22.0 ± 0.1
Upper limit: 22.1
Lower limit: −21.9
Tolerance: 0.2

Example 4: $31.75^{+0.10}_{\;0}$
Upper limit: 31.85
Lower limit: −31.75
Tolerance: 0.10

All dimensions on a drawing have a tolerance except reference, maximum, minimum, or stock size dimensions. Dimensions on a drawing may read as in Figure 12.68 with general tolerances specified in the title block of the drawing. General tolerance specifications, as given in a typical industry title block, are shown in Figure 12.69 where x refers to one-place decimal dimensions, xx is for two-place decimals, and xxx refers to the tolerance for three-place decimal dimensions. Using this title block as an example, the tolerances of the dimensions in Figure 12.68 are as follows (given in inches):

2.500 has .xxx ± .005 applied; 2.500 ± .005, tolerance equals .010.

2.50 has .xx ± .010 applied; 2.50 ± .010, tolerance equals .020.
2.5 has .x ± .020 applied; 2.5 ± .020, tolerance equals .040.
30° has Angles ± 0.5° applied; 30° ± 0.5°, tolerance equals 1°.

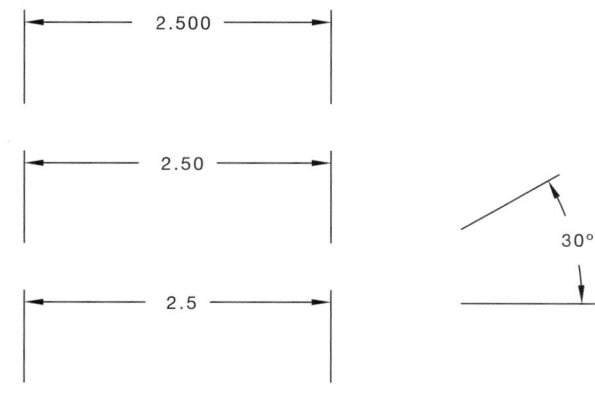

FIGURE 12.68 ■ Typical drawing dimensions.

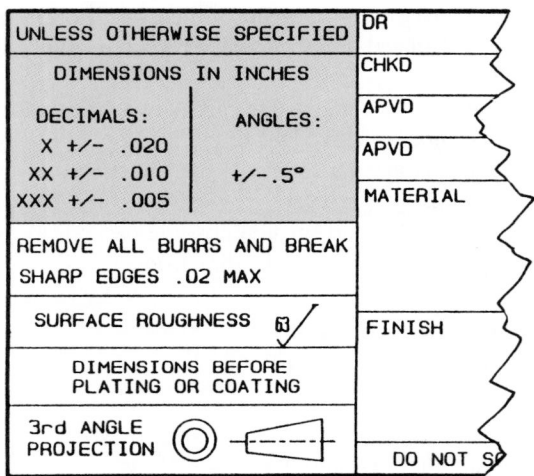

FIGURE 12.69 ■ General tolerances from a company title block. *Courtesy Althin Medical, Inc.*

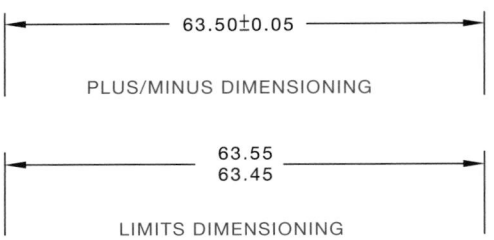

FIGURE 12.70 ■ Specific tolerance dimensions.

Dimensions that require tolerances different from the general tolerances given in the title block must be specified in the dimension. These are referred to as *specific tolerance* dimensions and may be represented with *plus/minus* or *limits* dimensioning as in Figure 12.70.

Statistical Tolerancing

Statistical tolerancing is the assigning of tolerances to related dimensions in an assembly based on the requirements of statistical process control (SPC). SPC is discussed in Chapters 11 and 15. Statistical tolerancing is displayed in dimensioning as shown in Figure 12.71a. When the feature could be manufactured using SPC or conventional means, it is necessary to show both the statistical tolerance and the conventional tolerance as in Figure 12.71b. The appropriate general note should also accompany the drawing as shown in Figure 12.71.

Maximum and Least Material Conditions

Maximum material condition, abbreviated MMC, is the condition of a part or feature when it contains the most amount of material. The key is *most material*. The MMC of an external feature is the upper limit. (See Figure 12.72.) The MMC of an internal feature is the lower limit. (See Figure 12.73.)

The *least material condition* (LMC) is the opposite of MMC. LMC is the least amount of material possible in the size of a feature.

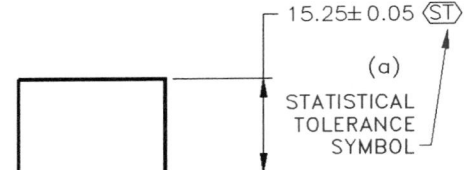

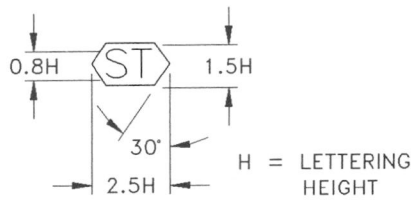

FEATURES IDENTIFIED AS STATISTICALLY TOLERANCED SHALL BE PRODUCED WITH STATISTICAL PROCESS CONTROLS.

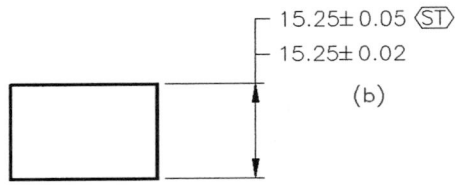

FEATURES IDENTIFIED AS STATISTICALLY TOLERANCED SHALL BE PRODUCED WITH STATISTICAL PROCESS CONTROLS, OR TO THE MORE RESTRICTIVE ARITHMETIC LIMITS.

FIGURE 12.71 ■ Statistical tolerancing application and notes, and the statistical tolerancing symbol.

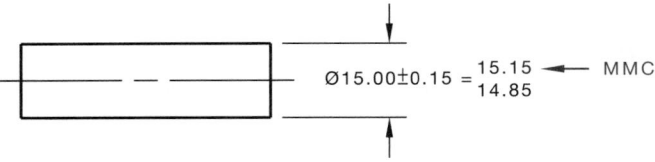

FIGURE 12.72 ■ Maximum material condition (MMC) of an external feature. Plus/minus dimension shown.

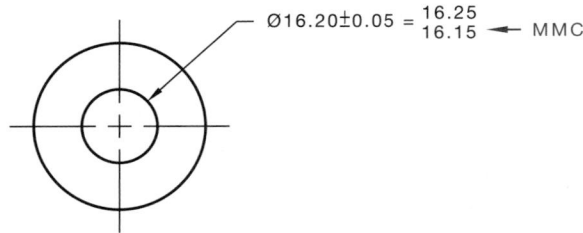

FIGURE 12.73 ■ Maximum material condition (MMC) of an internal feature. Plus/minus dimension shown.

The LMC of an external feature is its lower limit. The LMC of an internal feature is its upper limit.

Clearance Fit

A *Clearance fit* is a condition when, due to the limits of dimensions, there is always a clearance between mating parts. The features in

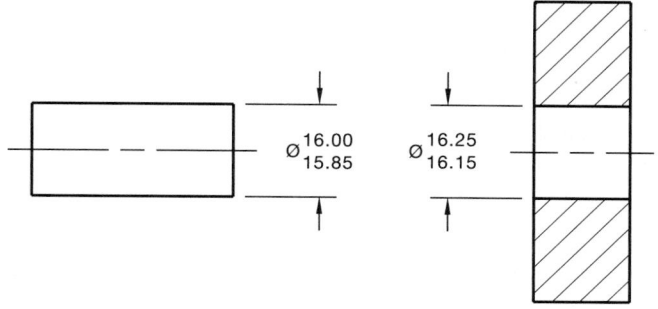

FIGURE 12.74 ■ A clearance fit between two mating parts. Limits dimensions shown.

Figure 12.74 have a clearance fit. Notice that the largest limit of the shaft is smaller than the smallest hole limit.

Allowance

The *allowance* of a clearance fit between mating parts is the tightest possible fit between the parts. The allowance is calculated with the formula:

$$\begin{array}{l} \text{MMC Internal Feature} \\ \underline{-\text{MMC External Feature}} \\ \text{Allowance} \end{array}$$

The allowance of the parts in Figure 12.74 is:

MMC Internal Feature	16.15
−MMC External Feature	−16.00
Allowance	0.15

Interference Fit

An *interference fit,* also known as force or shrink fit, is the condition that exists when, due to the limits of the dimensions, mating parts must be pressed together. Interference fits are used, for example, when a bushing must be pressed onto a housing or when a pin is pressed into a hole. (See Figure 12.75.)

Types of Fits

Selection of Fits

In selecting the limits of size for any application, the type of fit is determined first based on the use or service required from the equipment being designed; then the limits of size of the mating

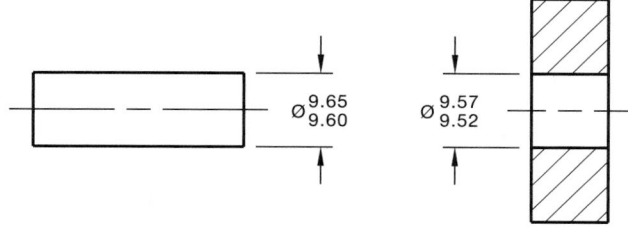

FIGURE 12.75 ■ An interference fit between two mating parts. Limits dimensions shown.

parts are established to ensure that the desired fit will be produced. The number of standard fits described here cover most applications.

Designation of Standard ANSI Fits

Standard fits are designated by means of the following symbols, which facilitate reference to classes of fit for educational purposes. The symbols are not intended to be shown on manufacturing drawings; instead, sizes should be specified on drawings. The letter symbols used are as follows:

RC	Running, or Sliding, Clearance Fit
LC	Locational Clearance Fit
LT	Transition Clearance, or Interference, Fit
LN	Locational Interference Fit
FN	Force, or Shink, Fit

These letter symbols are used in conjunction with numbers representing the class of fit: thus FN 4 represents a class 4 force fit.

Description of Standard ANSI Fits

The classes of fits are arranged in three general groups known as running and sliding fits, locational fits, and force fits. Standard Fit Tables are given in Appendix B.

Running and Sliding Fits (RFC)

Running and sliding fits are intended to provide a similar running performance with suitable lubrication allowance, throughout their range of sizes. The clearances for the first two classes, used chiefly as sliding fits, increase more slowly with the diameter than do the clearances for the other classes, so that accurate location is maintained even at the expense of free relative motion. These fits may be described as follows:

■ RC1—Close sliding fits are intended for the accurate location of parts that must assemble without perceptible play.

■ RC2—Sliding fits are intended for accurate location, but with greater maximum clearance than class RC1. Parts made to this fit move and turn easily but are not intended to run freely, and in the larger sizes may seize with small temperature changes.

■ RC3—Precision running fits are about the closest fits that can be expected to run freely, and are intended for precision work at slow speeds and light journal pressures, but are not suitable where appreciable temperature differences are likely to be encountered.

■ RC4—Close running fits are intended chiefly for running fits on accurate machinery with moderate surface speeds and journal pressures, where accurate location and minimum play is desired.

■ RC5 and RC6—Medium running fits are intended for high running speeds, or heavy journal pressures, or both.

■ RC7—Free running fits are intended for use where accuracy is not essential, or where large temperature variations are likely to be encountered, or under both these conditions.

■ RC8 and RC9—Loose running fits are intended for use where wide commercial tolerances may be necessary, together with an allowance, on the external member.

Locational Fits (LC, LT, and LN)

Locational fits are fits intended to determine only the location of the mating parts; they may provide rigid or accurate location, as with interference fits, or provide some freedom of location, as with clearance fits. Accordingly, they are divided into three groups: clearance fits (LC), transition fits (LT), and interference fits (LN). These fits are described as follows:

■ LC—Locational clearance fits are intended for parts that are normally stationary, but which can be freely assembled or disassembled. They range from snug fits for parts requiring accuracy of location, through medium clearance fits for parts such as spigots, to looser fastener fits where freedom of assembly is of prime importance.

■ LT—Locational transition fits are a compromise between clearance and interference fits. They are for applications where accuracy of location is important but either a small amount of clearance or interference is permissible.

■ LN—Locational interference fits are used where accuracy of location is of prime importance, and for parts requiring rigidity and alignment with no special requirements for bore pressure. Such fits are not intended for parts designed to transmit frictional loads from one part to another by virtue of the tightness of fit. Such conditions are covered by force fits.

Force Fits (FN)

Force, or shrink, fits constitute a special type of interference fit normally characterized by maintenance of constant bore pressures throughout its range of sizes. The interference, therefore, varies almost directly with diameter, and the difference between its minimum and maximum values is small so as to maintain the resulting pressures within reasonable limits. These fits are described as follows:

■ FN1—Light drive fits are those requiring light assembly pressures and producing more or less permanent assemblies. They are suitable for thin sections or long fits, or in external cast iron members.

■ FN2—Medium drive fits are suitable for ordinary steel parts, or for shrink fits on light sections. They are about the tightest fits that can be used with high-grade cast iron external members.

■ FN3—Heavy drive fits are suitable for heavy steel parts or for shrink fits in medium sections.

■ FN4 and FN5—Force fits are suitable for parts that can be highly stressed, or for shrink fits where the heavy pressing forces required are impractical.

Establishing Dimensions for Standard ANSI Fits

The fit used in a specific situation is determined by the operating requirements of the machine. When the type of fit has been established, the engineering drafter refers to tables that show the standard hole and shaft tolerances for the specified fit. One source of these tables is the *Machinery's Handbook*. Tolerances are based on

NOMINAL SIZE RANGE IN INCHES	RC4 STANDARD TOLERANCE LIMITS HOLE	SHAFT
0–.12	+.0006	–.0003
	0	–.0007
.12–.24	+.0007	–.0004
	0	–.0009
.24–.40	+.0009	–.0005
	0	–.0011
.40–.71	+.0010	–.0006
	0	–.0013
.71–1.19	+.0012	–.0008
	0	–.0016
1.19–1.97	+.0016	–.0010
	0	–.0020

FIGURE 12.76 ■ Standard RC4 fits for nominal sizes ranging from 0 to 1.97 inches. Standard fit tables are given in Appendix B, Table 28.

the type of fit and nominal size ranges, such as 0–.12, .12–.24, .24–.40, .40–.71, .71–1.19, and 1.19–1.97 in. So, if you have a 1-in. nominal shaft diameter and an RC4 fit, refer to Figure 12.76 to determine the shaft and hole limits. The hole and shaft limits for a 1-in. nominal diameter are:

Upper hole limit = 1.000 + .0012 = 1.0012

Lower hole limit = 1.000 + 0 = 1.000

Upper shaft limit = 1.000 – .0008 = .9992

Lower shaft limit = 1.000 – .0016 = .9984

You then dimension the hole as ∅1.0012–1.0000, and the shaft as ∅.9992–.9984.

Standard ANSI/ISO Metric Limits and Fits

The standard for the control of metric limits and fits is governed by the document ANSI B4.2, *Preferred Metric Limits and Fits*. The system is based on symbols and numbers that relate to the internal or external application and the type of fit. The specifications and terminology for fits are slightly different from the ANSI standard fits previously described. The metric limits and fits are divided into three general categories: clearance fits, transition fits, and interference fits. *Clearance fits* are generally the same as the running and sliding fits explained earlier. With clearance fits, a clearance always occurs between the mating parts under all tolerance conditions. With *transition fits*, a clearance or interference may result due to the range of limits of the mating parts. When *interference fits* are specified, a press or force situation exists under all tolerance conditions. Refer to Figure 12.77 for the ISO symbol and descriptions of the different types of metric fits.

The metric limits and fits may be designated in a dimension one of three ways. The method used depends on individual company standards and the extent of use of the ISO system. When most companies begin using this system, the tolerance limits are calculated and shown on the drawing followed by the tolerance symbol in

| TYPE OF FIT | ISO SYMBOL | | DESCRIPTION OF FIT |
	HOLE	SHAFT	
CLEARANCE FIT	H11/c11	C11/h11	*Loose running*
	H9/d9	D9/h9	*Free running*
	H8/f7	F8/h7	*Close running*
	H7/g6	G7/h6	*Sliding*
	H7/h6	H7/h6	*Locational clearance*
TRANSITION FIT	H7/k6	K7/h6	*Locational transition*
	H7/n6	N7/h6	*Locational transition*
INTERFERENCE FIT	H7/p6[1]	P7/h6	*Locational interference*
	H7/s6	S7/h6	*Medium drive*
	H7/u6	U7/h6	*Force*

FIGURE 12.77 ■ Description of metric fits. Standard fit tables are given in Appendix B, Table 28.

parentheses; for example, 25.000–24.979 (25 h7). The symbol in parentheses represents the basic size, 25, and the shaft tolerance code, h7. The term *basic size* is the dimension from which the limits are calculated just like the term *specified dimension* that was previously introduced. When companies become accustomed to using the system, they represent dimensions with the code followed by the limits in parentheses, as follows: 25 h7 (25.000–24.979). Finally, when a company has used the system long enough for interpreters to understand the designations, the code is placed alone on the drawing, like this: 25 h7. When it is necessary to determine the dimension limits from code dimensions, use the charts in ANSI B4.2 or the *Machinery's Handbook*. For example, if you want to determine the limits of the mating parts with a basic size of 30 and a close running fit, refer to the chart shown in Figure 12.78. The hole limits for the 30 mm basic size are ∅30.033–30.000, and the shaft dimension limits are ∅29.980–29.959.

| BASIC SIZE | CLOSE RUNNING FIT | |
	HOLE (h8)	SHAFT (f7)
20	20.033	19.993
	20.000	19.959
25	25.033	24.980
	25.000	24.959
30	30.033	29.980
	30.000	29.959
40	40.039	39.975
	40.000	39.950
50	50.039	49.975
	50.000	49.950

FIGURE 12.78 ■ Tolerances of close running fits for basic sizes ranging from 20 to 50 mm. Standard fit tables are given in Appendix B, Table 28.

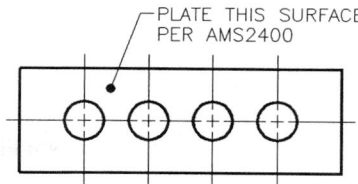

FIGURE 12.79 ■ A dot replaces the arrowhead on a leader connecting a specific note to a surface.

DIMENSIONS APPLIED TO PLATINGS AND COATINGS

When platings such as chromium, copper, and brass, or coatings such as galvanizing, polyurethane, and silicone, are applied to a part or feature, the specified dimensions should be defined in relationship to the coating or plating process. A general note that indicates that the dimensions apply before or after plating or coating is commonly used and specifies the desired variables; for example, DIMENSIONAL LIMITS APPLY BEFORE (AFTER) PLATING (COATING). A leader connecting a specific note to a surface may also be used. Notice the dot replaces the arrowhead when the leader points to the surface in Figure 12.79.

MAXIMUM AND MINIMUM DIMENSIONS

In some situations, a dimension with an unspecified tolerance may require that the general tolerance be applied in one direction only from the specified dimension. For example, when a 12 mm radius shall not exceed 12 mm, the dimension reads R12 MAX. Therefore, when it is desirable to establish a maximum or minimum dimension, the abbreviations MAX or MIN are applied to the dimension. A dimension with a specified tolerance reads as previously discussed:

for example, $R12_{-0.05}^{0}$, or $R12_{0}^{+0.05}$.

CASTING DRAWING AND DESIGN

The end result of a casting drawing is the fabrication of a pattern. The preparation of casting drawings depends on the casting process used, the material to be cast, and the design or shape of the part. The drafter makes the drawing of the part the same as the desired end result after the part has been cast. The drafter may need to take certain casting characteristics into consideration, and the patternmaker needs to adjust the size and shape of the pattern to take into account characteristics that the drafter does not intentionally apply on the drawing.

Shrinkage Allowance

When metals are heated and then cooled, they shrink until the final temperature is reached. The amount of shrinkage depends on the material used. The shrinkage for most iron is about .125 in. per ft., .250 in. per ft. for steel, .125 to .156 in. per ft. for aluminum, .22 in. per ft. for brass, and .156 in. per ft. for bronze. Values for shrinkage allowance are approximate since the exact allowance depends on the size and shape of the casting and the contraction of the casting during cooling. The drafter normally does not need to take shrinkage into consideration because the patternmaker uses shrink rules that use expanded scales to take into account the shrinkage of various materials.

Draft

Draft is the taper allowance on all vertical surfaces of a pattern, which is necessary to facilitate the removal of the pattern from the mold. Draft is not necessary on horizontal surfaces because the pattern easily separates from these surfaces without sticking. Draft angles begin at the parting line and taper away from the molding material. (See Figure 12.80.) The draft is added to the minimum design sizes of the product. Draft varies with different materials, size and shape of the part, and casting methods. Little if any draft is necessary in investment casting. The factors that influence the amount of draft are the height of vertical surfaces, the quality of the pattern, and the ease with which the pattern must be drawn from the mold. A typical draft angle for cast iron and steel is .125 in. per ft. Whether the drafter takes draft into consideration on a drawing depends on company standards. Some companies leave draft angles to the patternmaker, while others require drafters to place draft angles on the drawing.

Fillets and Rounds in Casting

One of the purposes of fillets and rounds on a pattern is the same as that of draft angles: to allow the pattern to eject freely from the mold. Also, the use of fillets on inside corners helps reduce the tendency of cracks to develop during shrinkage. (See Figure 12.81.) The radius for fillets and rounds depends on the material to be cast, the casting method, and the thickness of the part. The recommended radii for fillets and rounds used in sand casting is determined by part thickness as shown in Figure 12.82.

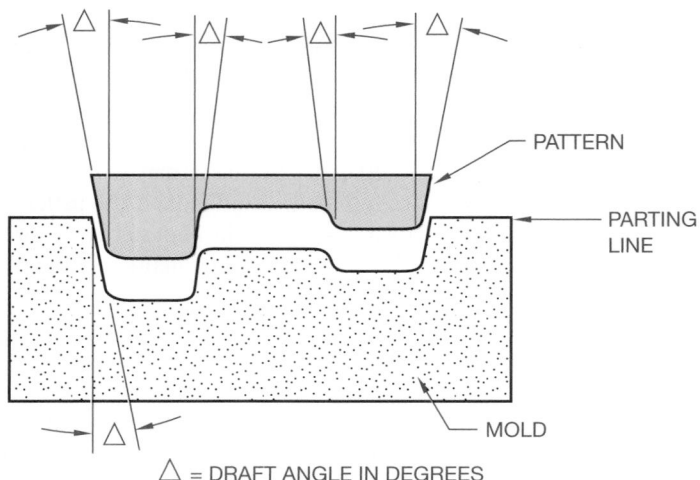

△ = DRAFT ANGLE IN DEGREES

FIGURE 12.80 ■ Draft angles for castings.

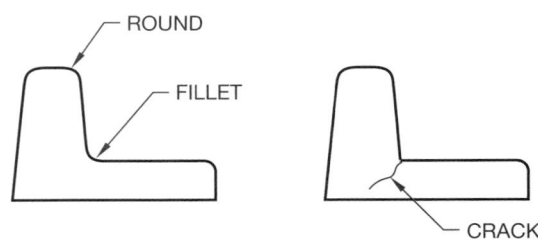

FIGURE 12.81 ■ Fillets and rounds for castings.

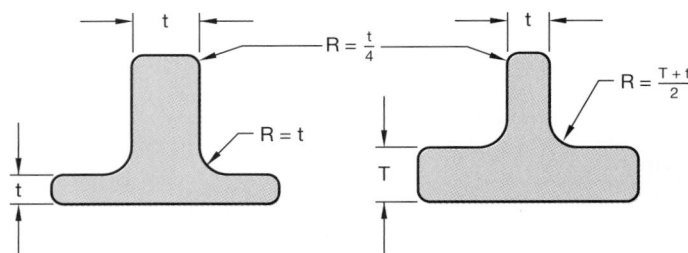

FIGURE 12.82 ■ Recommended fillet and round radii for sand castings.

MACHINING ALLOWANCE

Extra material must be left on the casting for any surface that will be machined. As with other casting design characteristics, the machining allowance depends on the casting process, material, size, and shape of the casting, and the machining process to be used for finishing. The standard finish allowance for iron and steel is .125 in., and for nonferrous metals such as brass, bronze, and aluminum, .062 in. In some situations the finish allowance may be as much as .5 or .75 in. for castings that are very large or have a tendency to warp.

Other machining allowances may be the addition of lugs, hubs, or bosses on castings that are otherwise hard to hold. The drafter can add these items to the drawing or the patternmaker may add them to the pattern. These features may not be added for product function, but they serve as aids for chucking or clamping the casting in a machine. (See Figure 12.83.)

There are several methods that can be used to prepare drawings for casting and machining operations. The method used depends on company standards. A commonly used technique is to prepare two drawings, one a casting drawing and the other a machining drawing. The casting drawing, as shown in Figure 12.84, shows the part as a casting. Only dimensions necessary to make the casting are shown in this drawing. The other drawing is of the same part, but this time only machining information and dimensions are given. The actual casting goes to the machine shop along with the machining drawing so the features can be machined as specified. The machining drawing is shown in Figure 12.85.

Another method of preparing casting and machining drawings is to show both casting and machining information together on one drawing. This technique requires the patternmaker to add machining allowances. The drawing may have draft angles specified in the form of a note. The patternmaker must add the draft angles to the finished sizes given. With draft angles and finish allowances omit-

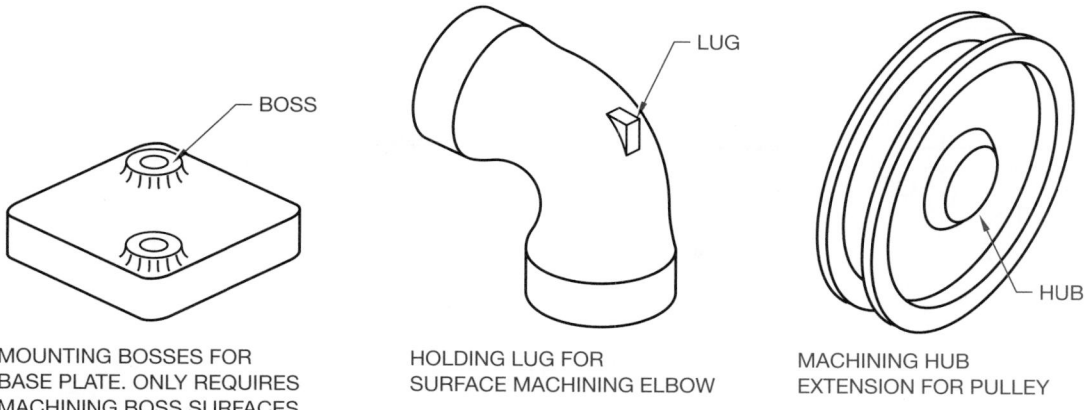

BOSS

MOUNTING BOSSES FOR
BASE PLATE. ONLY REQUIRES
MACHINING BOSS SURFACES.

LUG

HOLDING LUG FOR
SURFACE MACHINING ELBOW

HUB

MACHINING HUB
EXTENSION FOR PULLEY

FIGURE 12.83 ■ Cast features added for machining.

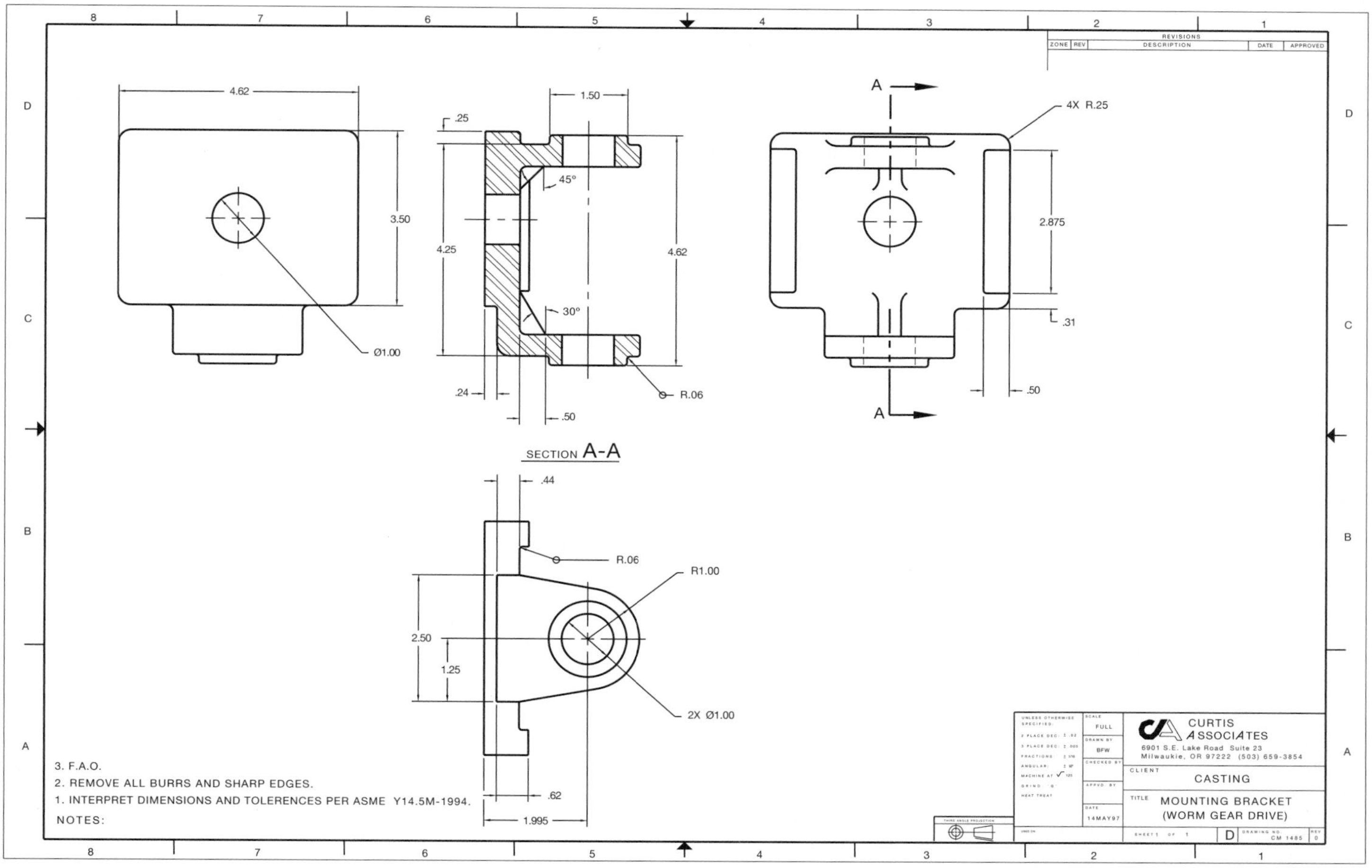

FIGURE 12.84 ■ Casting drawing. *Courtesy Curtis Associates.*

ted, the designer or drafter may need to consult with the pattern-maker to ensure that the casting is properly made. A combination casting/machining drawing is shown in Figure 12.86.

Another technique is to draw the part as a machining drawing and then use phantom lines to show the extra material for machining allowance and draft angles, as shown in Figure 12.87, page 364.

Forging, Design, and Drawing

Draft for forgings serves much the same purpose as draft for castings. The draft associated with forging is found in the dies. The sides of the dies must be angled to facilitate the release of the metal during the forging process. If the vertical sides of the dies do not

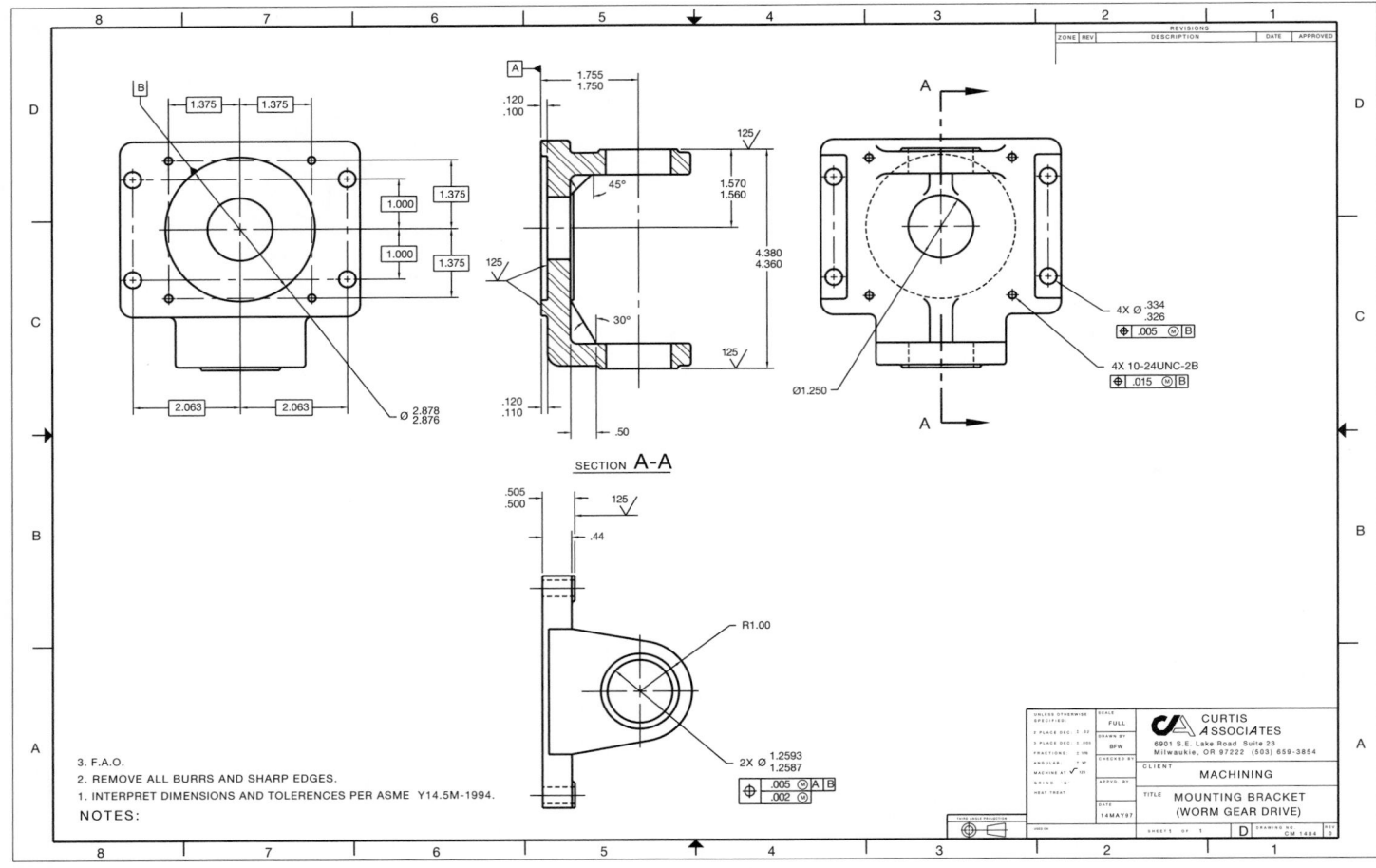

FIGURE 12.85 ■ Machining drawing. *Courtesy Curtis Associates.*

have draft angle, then the metal would become stuck in the die. Internal and external draft angles may be specified differently because the internal drafts in some materials may have to be greater to help reduce the tendency of the part to stick in the die. While draft angles may change slightly with different materials, the common exterior draft angle recommended is 7°. The internal draft angles for most soft materials is also 7°, but the recommended interior draft for iron and steel is 10°.

The application of fillets and rounds to forging dies is to improve the ejection of the metal from the die. Another reason, similar to that for casting, is increased inside corner strength. One factor that applies to forgings as different from castings is that fillets and rounds that are too small in forging dies may substantially reduce the life of the dies. Recommended fillet and round radius dimensions are shown in Figure 12.88.

Forging Drawings

A number of methods may be used in the preparation of forging drawings. One technique used in forging drawings that is clearly different from the preparation of casting drawings is the addition of draft angles. Casting drawings usually do not show draft angles; forging detail drawings usually do show draft.

Before a forging can be made, the dimensions of the stock material to be used for the forging must be determined. Some compa-

nies leave this information to the forging shop to determine; other companies have their engineering department make these calculations. After the stock size is determined, a drawing showing size and shape of the stock material is prepared. The blank material is dimensioned, and the outline of the end product is drawn inside the stock view using phantom lines. (See Figure 12.89.)

Forgings are made with extra material added to surfaces that must be machined. Forging detail drawings may be made to show the desired end product with the outline of the forging shown in phantom lines at areas that require machining. (See Figure 12.90, page 365.) Notice the double line around the perimeter showing draft angle. Another option used by some companies is to make two separate drawings, one a forging drawing and the other a machining drawing. The forging drawing shows all of the views, dimensions, and specifications that relate only to the production of the forging, as shown in Figure 12.91, page 365. The machining drawing gives views, dimensions, and specifications related to the machining processes, as shown in Figure 12.92, page 366.

Drawings for Plastic Part Manufacturing

As discussed in Chapter 11, there is a wide variety of plastic materials and manufacturing processes for creating plastic parts. Much like castings and forgings of metal parts, plastic manufacturing often requires draft angles to be applied to the design. This depends

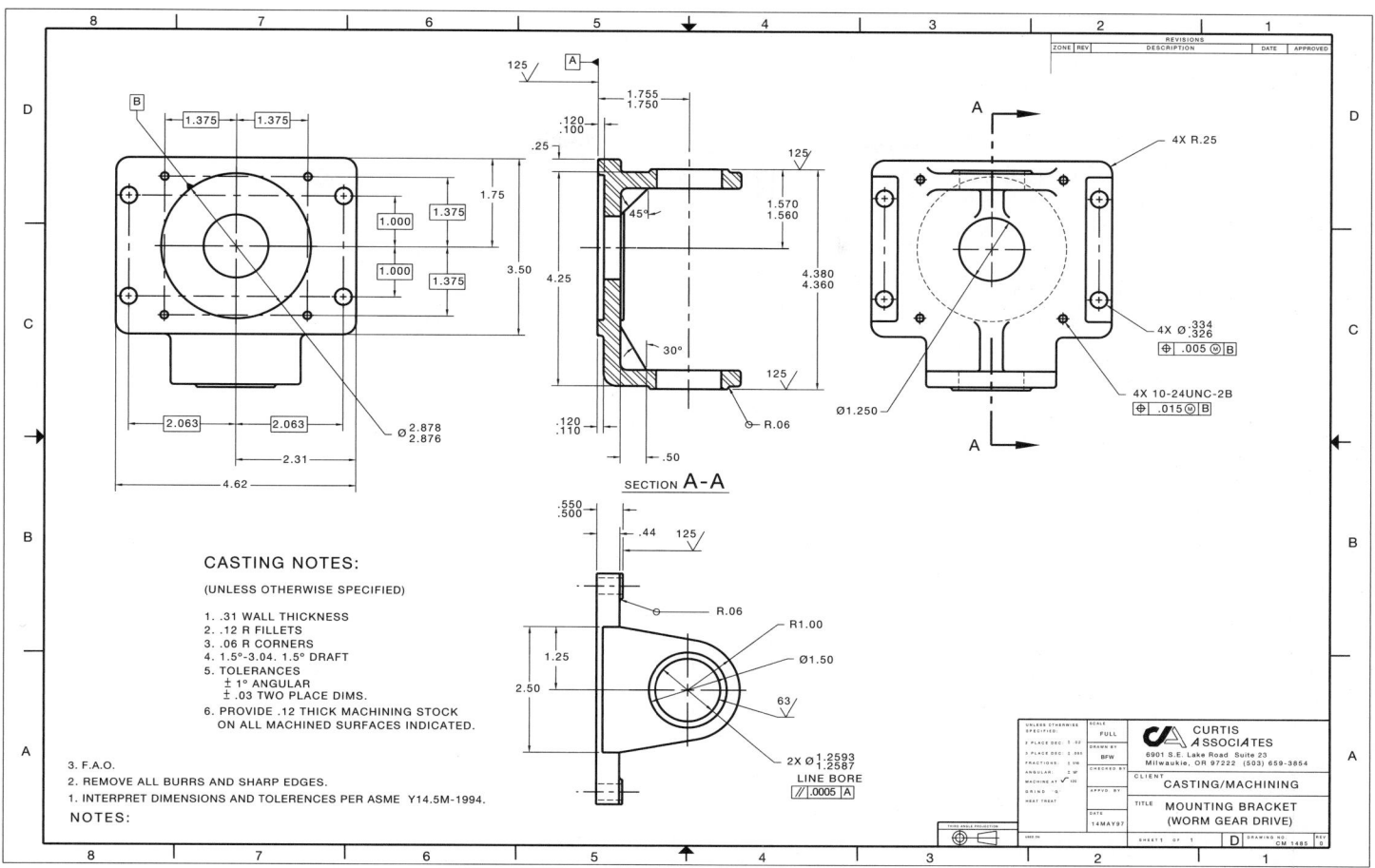

FIGURE 12.86 ■ Drawing with casting and machining information. *Courtesy Curtis Associates.*

on the type of plastic and the manufacturing method used. Draft angles allow the finished plastic part to be ejected or removed from the mold without difficulty. Also associated with plastic parts, as with metal castings and forgings, is a parting line. The *parting line* is the location on the part where it has been separated from the mold. When labeled on the drawing, the abbreviation PL is used. Figure 12.93, page 366, shows some examples of parting lines.

Draft angles can be specified in a note, such as .010 MAX DRAFT ANGLE. In this case, the patternmaker uses this amount of draft as a guide and produces a pattern that has draft angles that are less than .010 on each side where needed for the manufacturing process. These draft angles are generally established within the specified part dimensions rather than added on to the part dimensions. A draft angle tolerance can also be specified as a zone. This can be shown on the drawing as in Figure 12.94, page 366, or it can be specified as a tolerance in a general note or in the title block. A general note might read: ALL DRAFT ANGLES .010, or ALL DRAFT ANGLES 6°. The engineer or the mold maker determines the amount of draft angle.

Another method of specifying draft is the plus draft and minus draft methods. This is abbreviated as +DFT or –DFT and is placed with the feature dimensions on the part. In the +DFT application, the draft is added to the dimension for external dimensions and removed from internal dimensions as shown in Figure 12.95a, page 366. In the –DFT method, the draft is removed from the external dimension and added to the internal dimension as shown in Figure

12.95b, page 366. Both +DFT and –DFT can be combined on a drawing as shown in Figure 12.95c, page 366.

MACHINED SURFACES

Surface Finish Definitions

Surface Finish

Surface finish refers to the roughness, waviness, lay, and flaws of a surface. Surface finish is the specified smoothness required on the finished surface of a part that is obtained by machining, grinding, honing, or lapping. The drawing symbol associated with surface finish is shown in Figure 12.96, page 366.

Surface Roughness

Surface roughness refers to fine irregularities in the surface finish and is a result of the manufacturing process used. Roughness height is measured in micrometers, µm (millionths of a meter), or in microinches, µin (millionths of an inch).

Surface Waviness

Surface waviness is the often widely spaced condition of surface texture usually caused by such factors as machine chatter, vibrations, work deflection, warpage, or heat treatment. Waviness is rated in millimeters or inches.

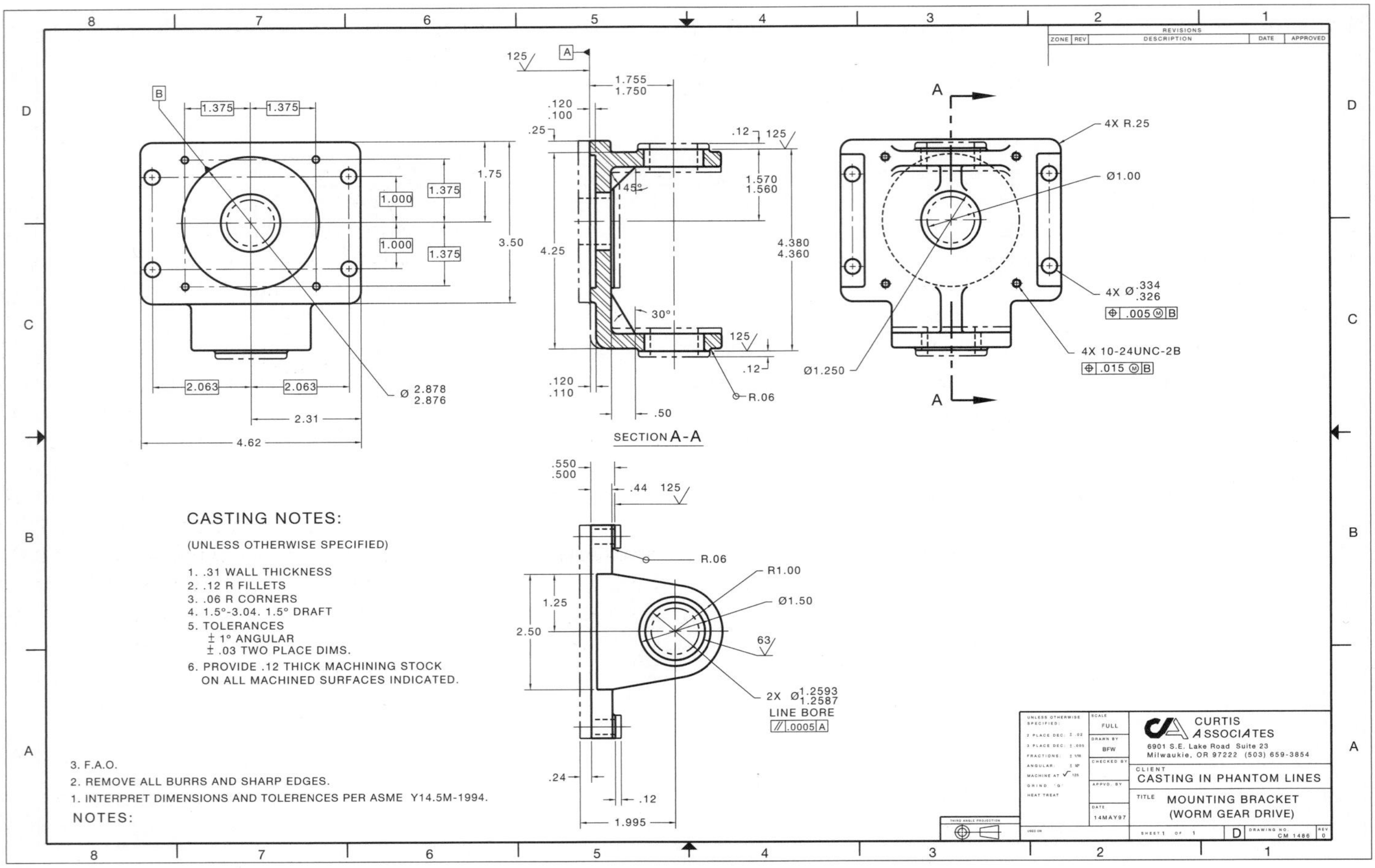

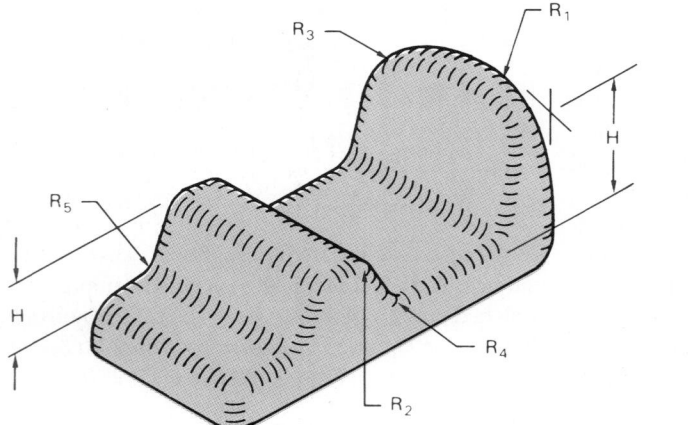

FIGURE 12.87 ■ Phantom lines used to show machining allowances. *Courtesy Curtis Associates.*

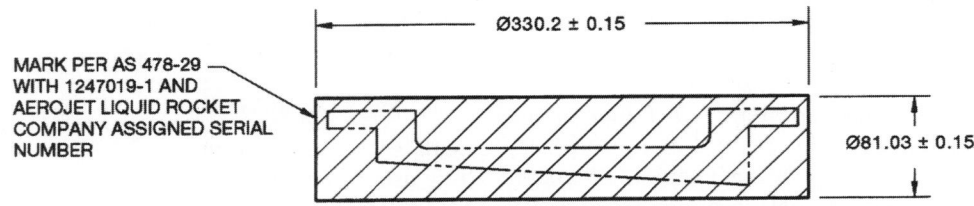

FIGURE 12.88 ■ Recommended fillets and rounds for forgings.

DIMENSIONS IN MILLIMETERS					
H	R₁	R₂	R₃	R₄	R₅
6	1.5	1.5	4.5	3	3
13	1.5	1.5	4.5	3	3
25	3	3	9	6	9
50	4.5	6	13	13	15
75	6	7.5	16	16	25
100	7.5	10.5	25	25	35
125	9	13	23	32	44
150	10.5	15	32	38	50

FIGURE 12.89 ■ Blank material for forging process. *Courtesy Aerojet Propulsion Division.*

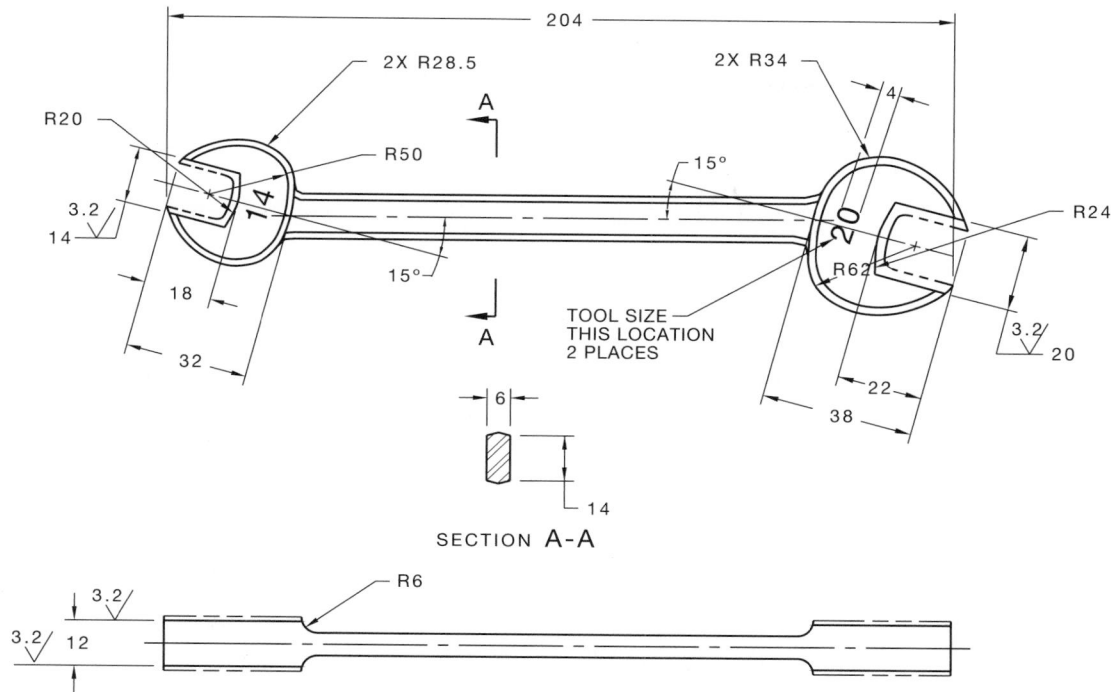

FIGURE 12.90 ■ Phantom lines used to show machining allowance on a forging drawing.

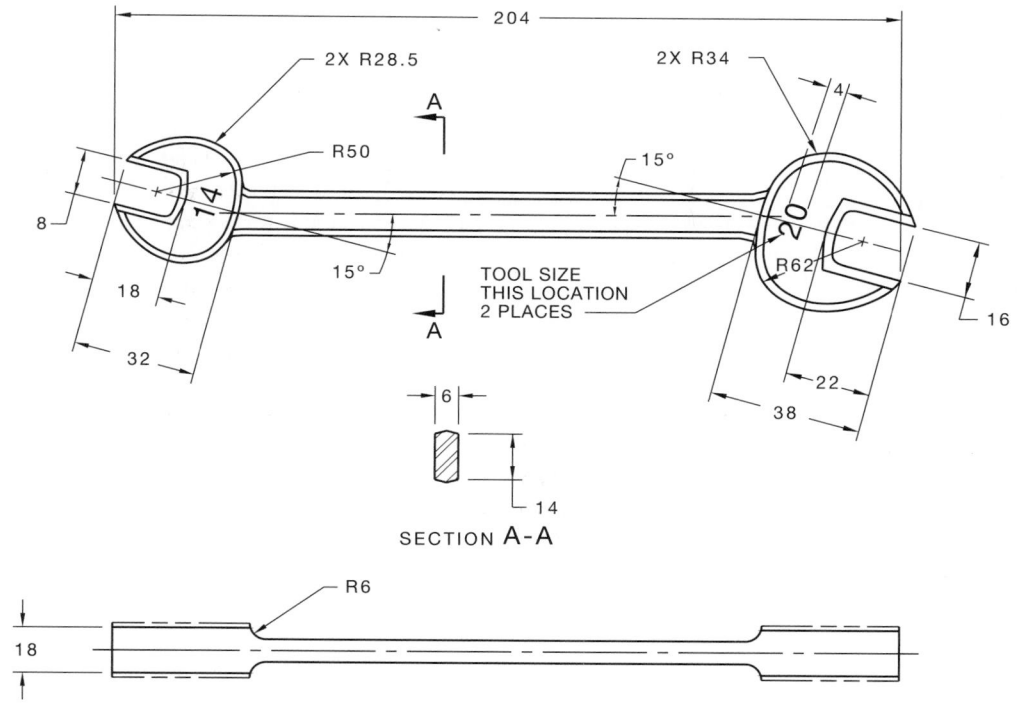

FIGURE 12.91 ■ Forging drawing with forging dimensions only.

Lay

Lay is the term used to describe the direction or configuration of the predominant surface pattern. The lay symbol is used if considered essential to a particular surface finish. The characteristic lay symbol may be attached to the surface finish symbol, as shown in Figure 12.97.

Surface Finish Symbol

Some of the surfaces of an object may be machined to certain specifications. When this is done, a surface finish symbol is placed on the view where the surface or surfaces appear as lines (edge view). (See Figure 12.98.) The finish symbol on a machine drawing alerts

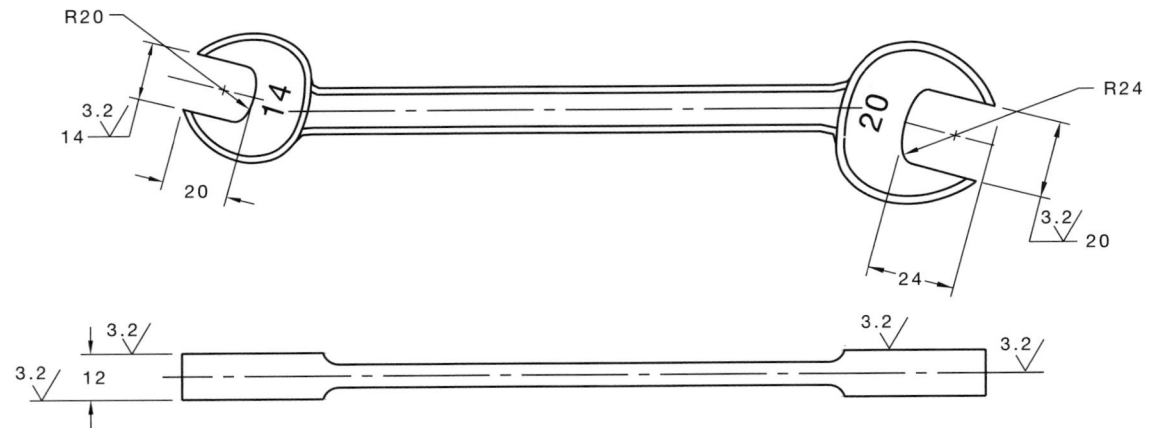

FIGURE 12.92 ■ Machining drawing with machining dimensions only.

FIGURE 12.93 ■ Parting lines labeled on a drawing.

FIGURE 12.96 ■ Surface finish symbol.

the machinist that the surface must be machined to the given specification. The finish symbol also tells the pattern or die maker that extra material is required in a casting or forging.

The surface finish symbol is properly drawn using a thin line as detailed in Figure 12.99. The numerals or letters associated with the surface finish symbol should be the same height as the lettering used on the drawing dimensions and notes.

Often only the surface roughness height is used with the surface finish symbol, which means 3.2 micrometers. When other characteristics of a surface texture are specified, they are shown in the format represented in Figure 12.100. For example, roughness width cutoff is a numerical value that establishes the maximum width of surface irregularities to be included in the roughness height mea-

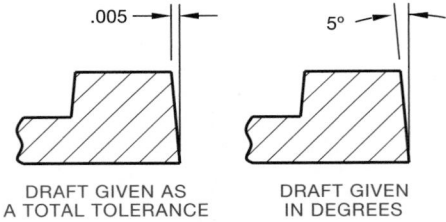

FIGURE 12.94 ■ Draft angle tolerance shown on a drawing.

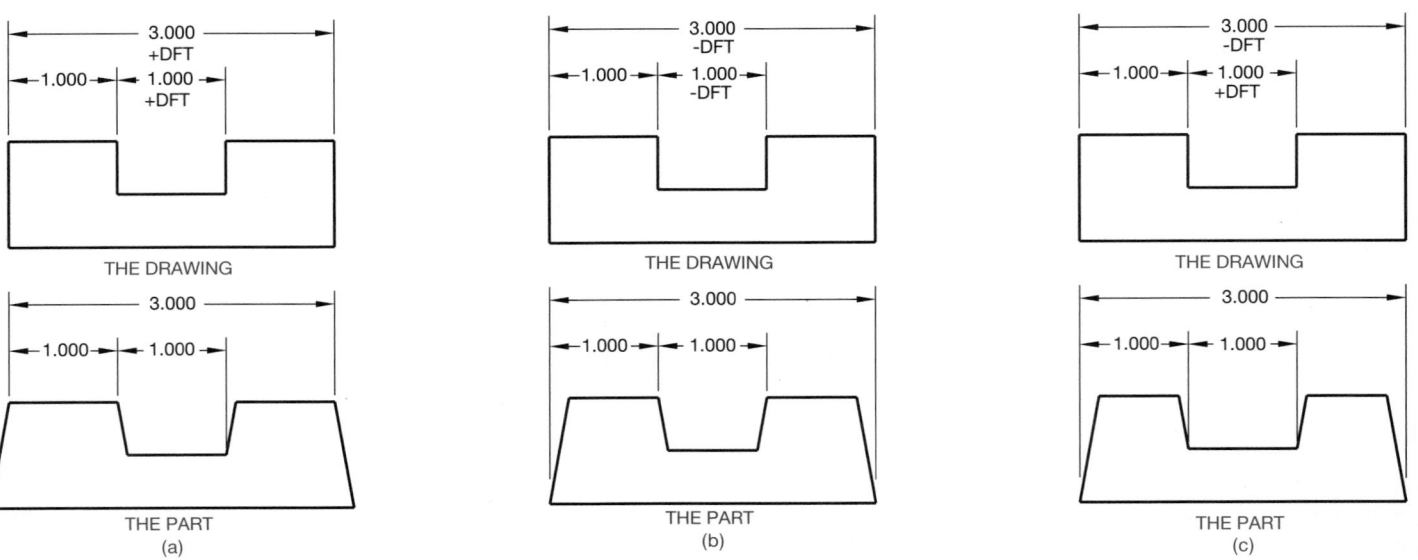

FIGURE 12.95 ■ Draft can be specified on a drawing with (**a**) the plus draft method, (**b**) the minus draft method, or (**c**) a combination of both.

ILLUSTRATION   DEFINITION   SYMBOL

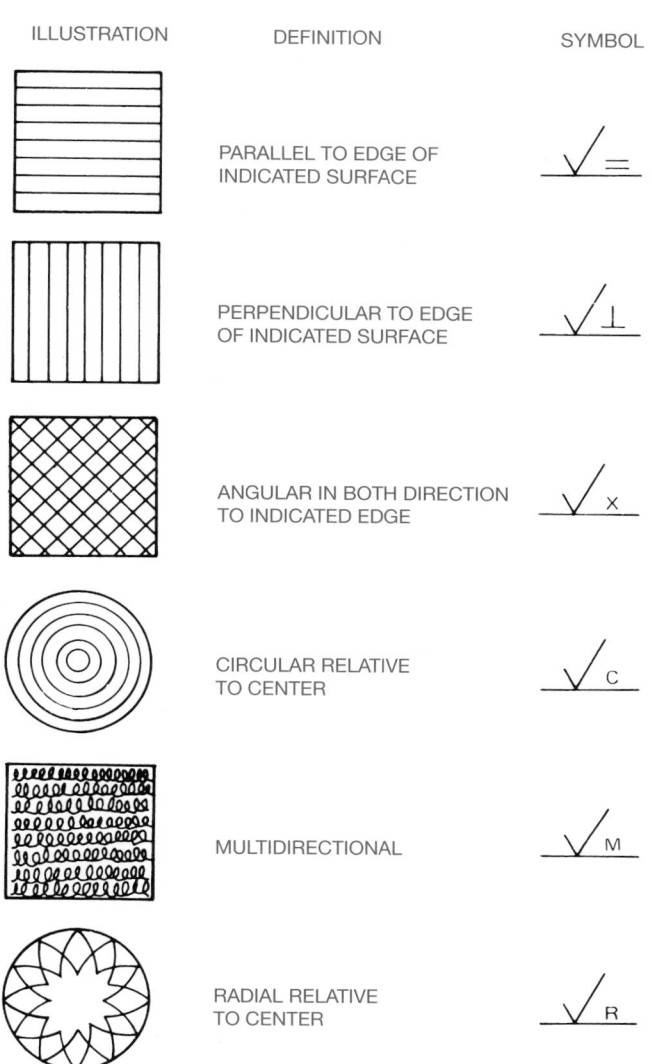

PARALLEL TO EDGE OF
INDICATED SURFACE

PERPENDICULAR TO EDGE
OF INDICATED SURFACE

ANGULAR IN BOTH DIRECTION
TO INDICATED EDGE

CIRCULAR RELATIVE
TO CENTER

MULTIDIRECTIONAL

RADIAL RELATIVE
TO CENTER

FIGURE 12.97 ■ Characteristic lay added to the surface finish symbol.

surement. Standard roughness width cutoff values for inch specifications are .003, .010, .030, .100, .300, and 1.000; .030 is implied
when no specification is given.

Figure 12.101 is a magnified pictorial representation of the characteristics of a surface finish symbol. Figure 12.102 shows some
common roughness height values in micrometers and microinches,
a description of the resulting surface, and the process by which the
surface may be produced. When a maximum and minimum limit
is specified, the average roughness height must lie within the two
limits.

When a standard or general surface finish is specified in the
drawing title block or in a general note, then a surface finish symbol without roughness height specified is used on all surfaces that
are the same as the general specification. When a part is completely
finished to a given specification, then the general note FINISH ALL
OVER, or abbreviation FAO, or FAO 125 μIN, may be used. The
placement of surface finish symbols on a drawing can be accomplished a number of ways, as shown in Figure 12.103. Additional
elements may be applied to the surface finish symbol, as shown in
Figure 12.104.

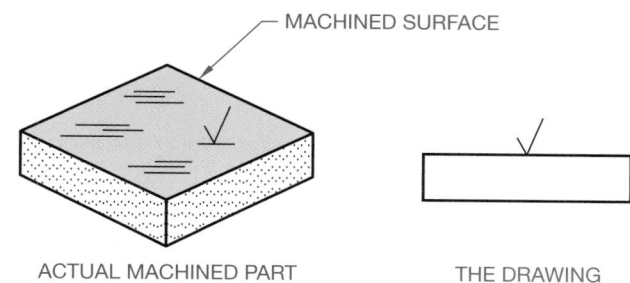

MACHINED SURFACE

ACTUAL MACHINED PART   THE DRAWING

FIGURE 12.98 ■ Standard surface finish symbol placed on the
edge view. The symbol should always be placed
horizontally when unidirectional dimensioning
is used.

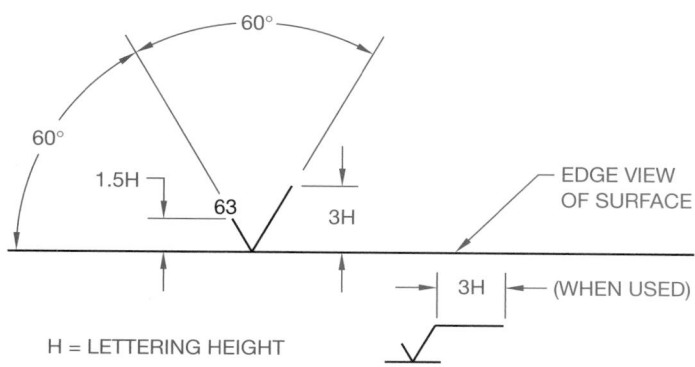

FIGURE 12.99 ■ Properly drawn surface finish symbol.

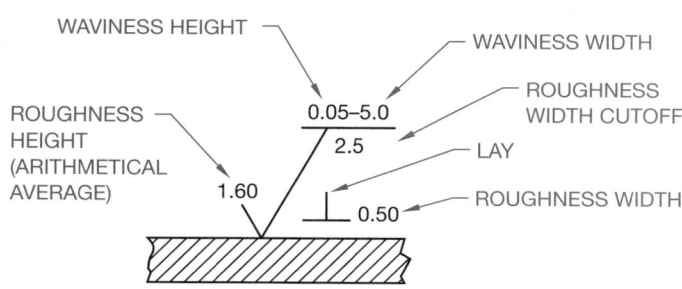

FIGURE 12.100 ■ Elements of a complete surface finish symbol.

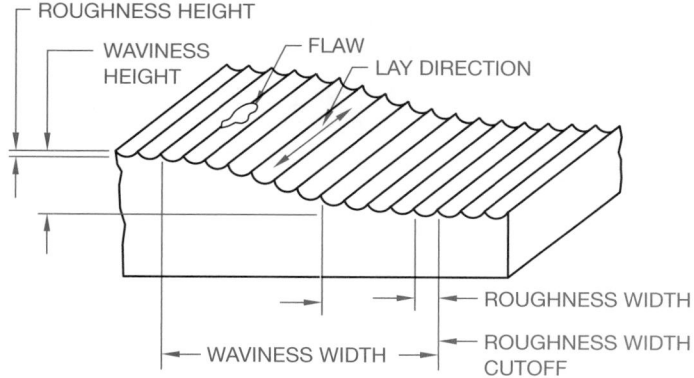

FIGURE 12.101 ■ Surface finish characteristics magnified.

MICROMETERS	ROUGHNESS HEIGHT RATING MICRO INCHES	SURFACE DESCRIPTION	PROCESS
25	1000	VERY ROUGH	SAW AND TORCH CUTTING, FORGING, OR SAND CASTING.
12.5	500	ROUGH MACHINING	HEAVY CUTS AND COARSE FEEDS IN TURNING, MILLING, AND BORING.
6.3	250	COARSE	VERY COARSE SURFACE GRIND, RAPID FEEDS IN TURNING, PLANING, MILLING, BORING, AND FILING.
3.2	125	MEDIUM	MACHINING OPERATIONS WITH SHARP TOOLS, HIGH SPEEDS, FINE FEEDS, AND LIGHT CUTS.
1.6	63	GOOD MACHINE FINISH	SHARP TOOLS, HIGH SPEEDS, EXTRA-FINE FEEDS AND CUTS.
0.80 / 0.40	32 / 16	HIGH-GRADE MACHINE FINISH	EXTREMELY FINE FEEDS AND CUTS ON LATHE, MILL, AND SHAPERS REQUIRED. EASILY PRODUCED BY CENTERLESS, CYLINDRICAL, AND SURFACE GRINDING.
0.20	8	VERY FINE MACHINE FINISH	FINE HONING AND LAPPING OF SURFACE.
0.050 0.100	2 – 4	EXTREMELY SMOOTH MACHINE FINISH	EXTRA-FINE HONING AND LAPPING OF SURFACE. MIRROR FINISH.
0.025	1	SUPER FINISH	DIAMOND ABRASIVES.

FIGURE 12.102 ■ Common roughness height values with a surface description and associated machining process.

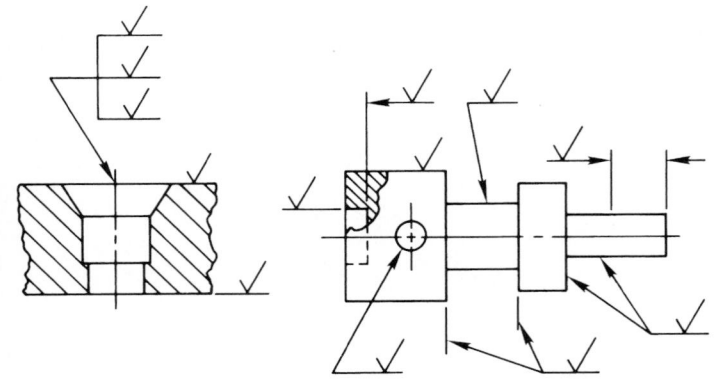

FIGURE 12.103 ■ Proper placement of surface finish symbols.

 STANDARD SURFACE FINISH SYMBOL WITH ROUGHNESS HEIGHT ONLY SPECIFIED.

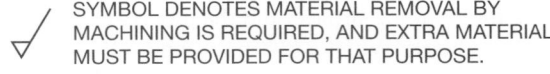 SYMBOL DENOTES MATERIAL REMOVAL BY MACHINING IS REQUIRED, AND EXTRA MATERIAL MUST BE PROVIDED FOR THAT PURPOSE.

2.5 THE NUMBER TO THE LEFT OF THE SYMBOL MAY BE USED TO SPECIFY THE AMOUNT OF STOCK TO BE REMOVED BY MACHINING. GIVEN IN MILLIMETERS OR IN INCHES.

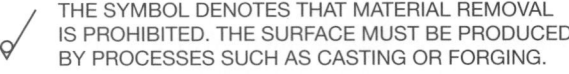 THE SYMBOL DENOTES THAT MATERIAL REMOVAL IS PROHIBITED. THE SURFACE MUST BE PRODUCED BY PROCESSES SUCH AS CASTING OR FORGING.

FIGURE 12.104 ■ Material removal elements added to the surface finish symbol.

DIMENSIONING SOFTWARE PACKAGES

Most of the major CADD software packages provide for a variety of dimensioning applications depending on the user's need. For example, you may use unidirectional or aligned dimensioning, and datum or chain dimensioning based on your company's standards and procedures. You can also alter the way dimensions are presented by changing any one or more of these variables:

■ Space between dimension lines.

■ Style and height of text.

■ Size and shape of arrowheads.

■ Extension line length beyond the last dimension line.

■ Space between the object and the start of the extension line.

■ Location of the text in relation to the dimension line.

Dimensioning *default* values are normally set to ASME standards, but the user has the flexibility to change these values for specific applications. A default value refers to a standard preset value. Figure 12.105 shows some of the flexibility available in changing predefined default values.

In addition to the major CADD software packages, there are hundreds of third-party software packages. The term *third-party* refers to programs that support or may be used in conjunction with the main package. In third-party software for dimensioning, the principal emphasis seems to be on datum, tabular, arrowless, and geometric tolerancing packages. These are the types of software that help increase speed and productivity and simplify the dimensioning process. For example, some of the programs have automatic dimensioning, which allows the drafter to pick all of the items for dimensioning and, when finished, press the ENTER key or select AUTOMATIC, depending on the program, and watch everything be dimensioned. Some packages also automatically generate a tabular chart for hole reference. A drawing showing arrowless datum dimensioning is in Figure 12.106.

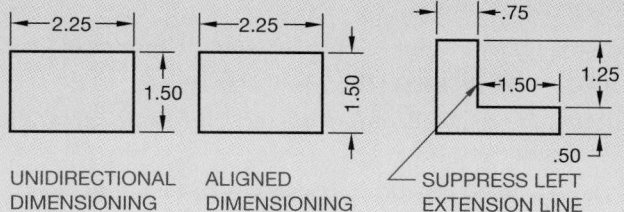

UNIDIRECTIONAL DIMENSIONING

ALIGNED DIMENSIONING

SUPPRESS LEFT EXTENSION LINE

SUPPRESS EXTENSION LINE, UNIDIRECTIONAL DIMENSIONS, TEXT OUTSIDE OF EXTENSION LINES.

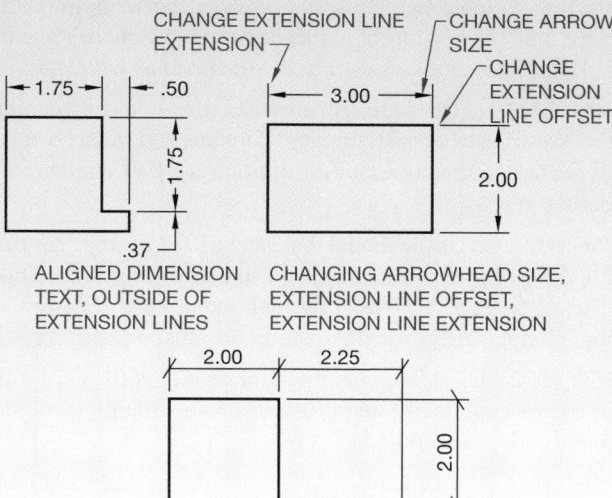

CHANGE EXTENSION LINE EXTENSION

CHANGE ARROW SIZE

CHANGE EXTENSION LINE OFFSET

ALIGNED DIMENSION TEXT, OUTSIDE OF EXTENSION LINES

CHANGING ARROWHEAD SIZE, EXTENSION LINE OFFSET, EXTENSION LINE EXTENSION

ARCHITECTURAL APPLICATION OF DIMENSION TEXT ABOVE DIMENSION LINE, TICK MARKS

FIGURE 12.105 ■ Changing dimension default values for specific applications.

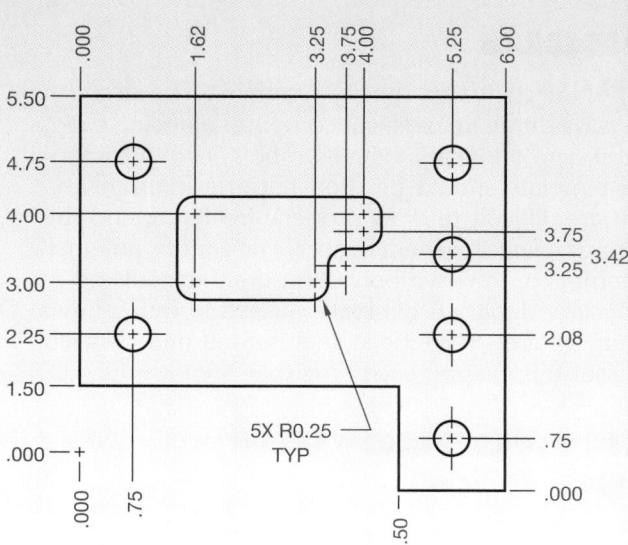

5X R0.25 TYP

FIGURE 12.106 ■ Datum dimensioning with a third-party software package, Auto-DATUM. *Courtesy CADMASTER, Inc.*

DESIGN AND DRAFTING OF MACHINED FEATURES

Drafters should gain a working knowledge of machining processes and the machining capabilities of the company for which they work. Drawings should be prepared that allow machining within the capabilities of the machinery available. The first consideration should be the least-expensive method to get the desired result. Avoid overmachining. Machining processes are expensive, so do not call for requirements on a drawing that are not necessary for the function of the part. For example, surface finishes become more expensive to machine as the roughness height decreases. So, if a surface roughness of 125 microinches is adequate, then do not use a 32-microinch specification just because you like smooth surfaces. Another example is demonstrated by the difference between the 63- and 32-microinch finish. A 63-microinch finish is a good machine finish that may be performed using sharp tools at high speeds with extra-fine feeds and cuts. The 32-microinch callout, on the other hand, requires extremely fine feeds and cuts on a lathe or milling machine and in many cases requires grinding. The 32 finish is more expensive to perform. In a manufacturing environment where cost and competition are critical considerations, a great deal of thought must be given to meeting the functional and appearance requirements of a product at the least possible cost. It generally does not take very long for an entry-level drafter to pick up these design considerations by communicating with engineering and manufacturing department personnel. Many drafters become designers, checkers, or engineers within a company by learning the product and being able to implement designs based on manufacturing capabilities.

SYMBOLS

ASME/ANSI standards specify symbols to be used on drawings. With the influence of computer-aided drafting, the use of symbols is easy. When symbols are drawn manually, a template should be used because many of the symbols are difficult to draw properly freehand, and the template drawing is generally neater and more uniform. Any attempt to draw symbols using drafting tools other than specially designed templates takes too much time. Individual symbols have been represented and detailed throughout this chapter each time they are introduced.

AN INTRODUCTION TO ISO 9000

The ISO 9000 Quality Systems Standard was established to encourage the development of standards, testing, quality control, and the certification of companies, organizations, and institutions where these practices are implemented. *ISO* stands for the International Organization for Standardization. The ISO is an international organization that is made up of nearly 100 countries. The United States is a member country that is represented by the American National Standards Institute (ANSI). ISO 9000 can certify companies, organizations, and institutions when their engineering, drafting standards, manufacturing, and quality control meet the requirements of a model quality management system established by the ISO 9000 organization. A certification is obtained by passing an inspection by an independent representative of the ISO. ISO 9000 certification is also referred to as registration. There are a number of reasons why a company may want to become ISO 9000 certified. The reasons can include:

- In order to do business with customers that require ISO 9000 certification. This includes agencies of the U.S. government.
- To compete in a competitive market that requires strict attention to quality control.
- Save cost and improve profits.
- Improve and maintain customer confidence.
- A desire to improve product quality and place emphasis on customer relations.
- To manufacture and sell products in the European Union markets where this certification is required.
- To require that subcontractors meet the same expectations for quality control.
- The ISO 9000 certification also satisfies requirements established by other local and national organizations.
- Improve company standards.

The ISO 9000 Quality Systems Standard is made up of a series of five international standards that provide leadership in the development and completion of a successful quality management system. These five standards are briefly described as follows:

ISO 9000-1 This standard provides direction and definitions that describe what each standard contains and assists companies in the selection and use of the appropriate ISO standard for the desired results.

ISO 9001 This is the model that can be used by any organization for designing, documenting, and implementing ISO standards. The model takes a product through the process of design, drafting, manufacturing, quality control, installation, and service.

ISO 9002 This standard is the same as ISO 9001, except that it does not contain the requirement of documenting the design and development process.

ISO 9003 This standard is for companies or organizations that only need to demonstrate through inspection and testing methods that they are providing the desired product or service.

ISO 9004-1 This is a set of guidelines that can be used to assist organizations in the development and implementation of a quality management system.

In addition to the ISO 9000 series are the QS 9000 and AS 9000 standards. The QS 9000 is the Automotive Requirements, and the AS 9000 is the Aerospace Standard. These standards contain all of ISO 9001 plus requirements beyond ISO 9001. These standards were developed by the U.S. automotive and aerospace industries for specific applications and needs that customize the standard for their industries.

PROFESSIONAL PERSPECTIVE

The proper placement and use of dimensions is one of the most difficult aspects of drafting. It requires careful thought and planning. First, you must determine the type of dimension that fits the application, then place it on the drawing. While the CADD system makes the actual placement of dimensions quick and easy, it does not make the planning process any easier. Making preliminary sketches is very important before beginning a drawing. The preliminary sketch allows you to select and place the views and then place the dimensions. When you were drawing only the views the space requirements were not as critical, but a drawing without dimensions is unrealistic.

The problem with dimensions is that they take up a lot of space. One of the big issues an entry-level drafter faces is determining the space requirements for dimensions. In many cases these requirements are underestimated. As a rule of thumb, if you think the drawing is crowded on a particular size sheet, then play it safe and use the next larger size. Most companies want the drawing information spread out and easy to read, but be careful, because some companies want the drawing crowded with as much information as you can get on a sheet. This text advocates a clean and easy-to-read drawing that is not crowded. Out on the job, however, you must do what is required by your employer. If you are drawing an uncrowded drawing, you should consider leaving about one-quarter of the drawing space clear of view and dimensional information. Usually this space is above and to the left of the title block. This space provides adequate room for general notes and engineering changes.

Follow the dimensioning rules, guidelines, and examples discussed in this chapter and use proper dimensioning standards. Try as hard as you can to avoid breaking dimensioning standards. Following are some of the pitfalls to watch out for as a beginning drafter:

■ Do not crowd dimensions. Keep your dimension line spacing equal and far enough apart to clearly separate the dimension numerals.

■ Do not dimension to hidden features. This also means do not dimension to the centerline of a hole in the hidden view. Always dimension to the view where the feature appears as an object line.

■ Dimension to the view that shows the most characteristic shape of an object or feature. For example, dimension shapes where you see the contour, dimension holes to the circular view, and dimension cylindrical shapes in the rectangular view.

■ Do not stack adjacent dimension numerals; stagger them so they are easier to read.

■ Group dimensions as much as you can. It is better to keep dimensions concentrated to one location or one side of a feature rather than spread around the drawing. This makes the drawing easier to read.

■ If you are using manual drafting, use a template to draw symbols. If you use computer-aided drafting, make a standard symbol library or use software with symbols available. Symbols speed up the drafting process, and they clearly identify the feature. For example, if you draw a single view of a cylinder, the diameter ⌀ dimension is identified in the rectangular view and drawing the circular view may not be necessary.

■ Look carefully at the figures in this text and use them as examples as you prepare your drawings.

Try to put yourself in the place of the person who has to read and interpret your drawing. Make the drawing as easy to understand as possible. Keep the drawing as uncluttered and as simple as possible, yet still complete. Figure 12.107 shows an industry drawing of a complex part. Notice how the drafter carefully selected views and dimension placement to make the drawing as easy to read as possible.

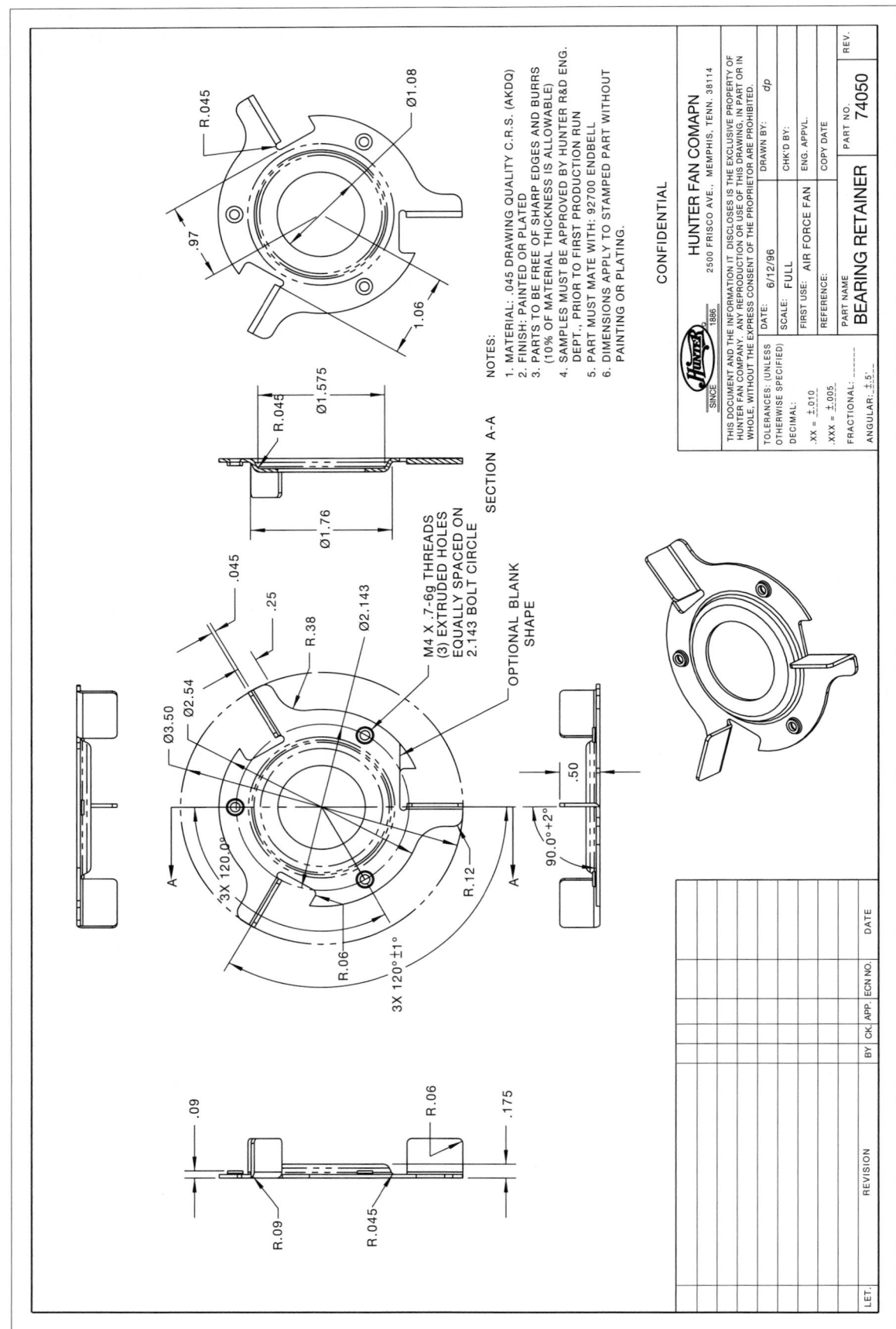

FIGURE 12.107 ■ A complex drawing with dimensions placed in the proper locations to aid in reading the drawing. Notice the pictorial drawing in the bottom center. This is shown to help you visualize the part. *Courtesy Hunter Fan Company, Memphis, TN.*

DIMENSIONING FOR CADD/CAM

The implementation of computer-aided design and drafting (CADD) and computer-aided manufacturing (CAM) in industry is best accomplished when common control of the computer exists between engineering and manufacturing. The success of this automation is, in part, relative to the standardization of operating and documentation procedures. CADD can be accomplished through the same coordinate dimensioning systems previously described in this chapter. Standard dimensioning systems are used to establish a geometric model of the part, which in turn is displayed at a CADD workstation. The data retrieved from this model is the mathematical description of the part to be produced. The drafter must dimension the part completely and accurately so that each contour or geometric shape of the part is continuous. The dimensioning systems that locate features or points on a feature in relation to x, y, and z axes derived from a common origin are most effective, such as datum, tabular, arrowless, and polar coordinate dimensioning. The x, y, and z axes originate from three mutually perpendicular planes which are generally the geometric counterpart of the sides of the part when the surfaces are at right angles as seen in Figure 12.108. If the part is cylindrical, two of the planes intersect at right angles to establish the axis of the cylinder and the third is perpendicular to the intersecting planes as shown in Figure 12.109. The x, y, and z coordinates that are used to establish features on a drawing are converted to x, y, and z axes that correspond to the linear and rotary motions that occur in CAM.

The position of the part in relation to mathematical quadrants determines whether the x, y, and z values are positive or negative. The preferred position is the mathematical quadrant that allows for programming positive commands for the machine tool. Notice in Figure 12.110 that the positive x and y values occur in quadrant 1.

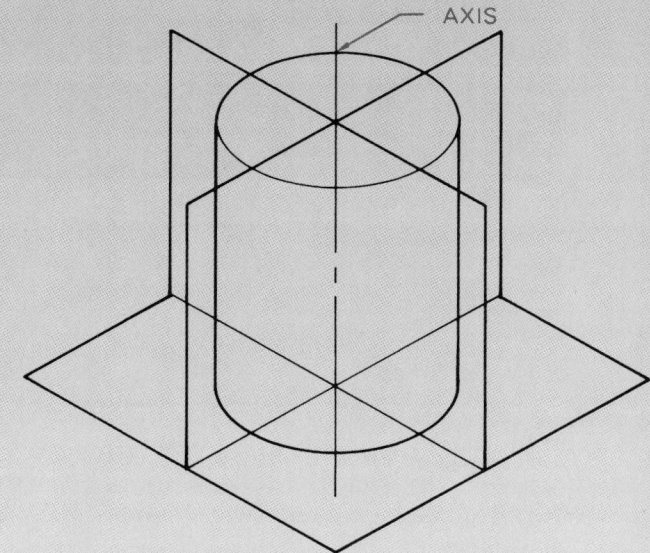

FIGURE 12.109 ■ The planes related to the axis of a cylindrical feature.

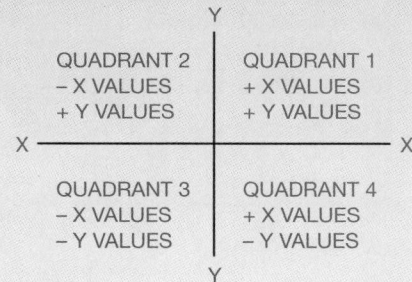

FIGURE 12.110 ■ Determination of X and Y values in related quadrants.

In CAD/CAM and computer-integrated manufacturing (CIM) programs, the drawing is made on the computer screen and sent directly to the computer numerical control machine tool without generating a hard copy of the drawing. In this situation it becomes important for the drafter to understand the machine tool operation. Figure 12.111a shows a drawing created for CAD/CAM, and Figure 12.111b shows the same drawing as represented on the computer screen prior to generating the machine tool program. Notice, the dimensions are not displayed on the drawing in Figure 12.111b. The data used to input the information from the original design drawing is used, by the computer, to establish the machine tool paths. The CAD/CAM program also allows the operator to show the tool path, as shown in Figure 12.112. The tool path display allows the operator to determine if the machine tool will perform the assigned machining operation.

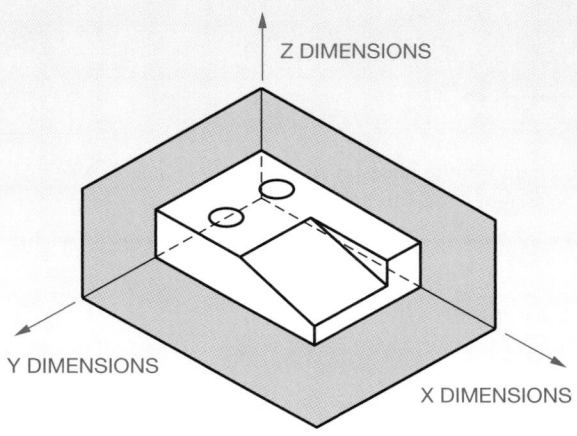

FIGURE 12.108 ■ Features of a part dimensioned in relation to X, Y, and Z axes.

(Continued)

CADD APPLICATIONS *(continued)*

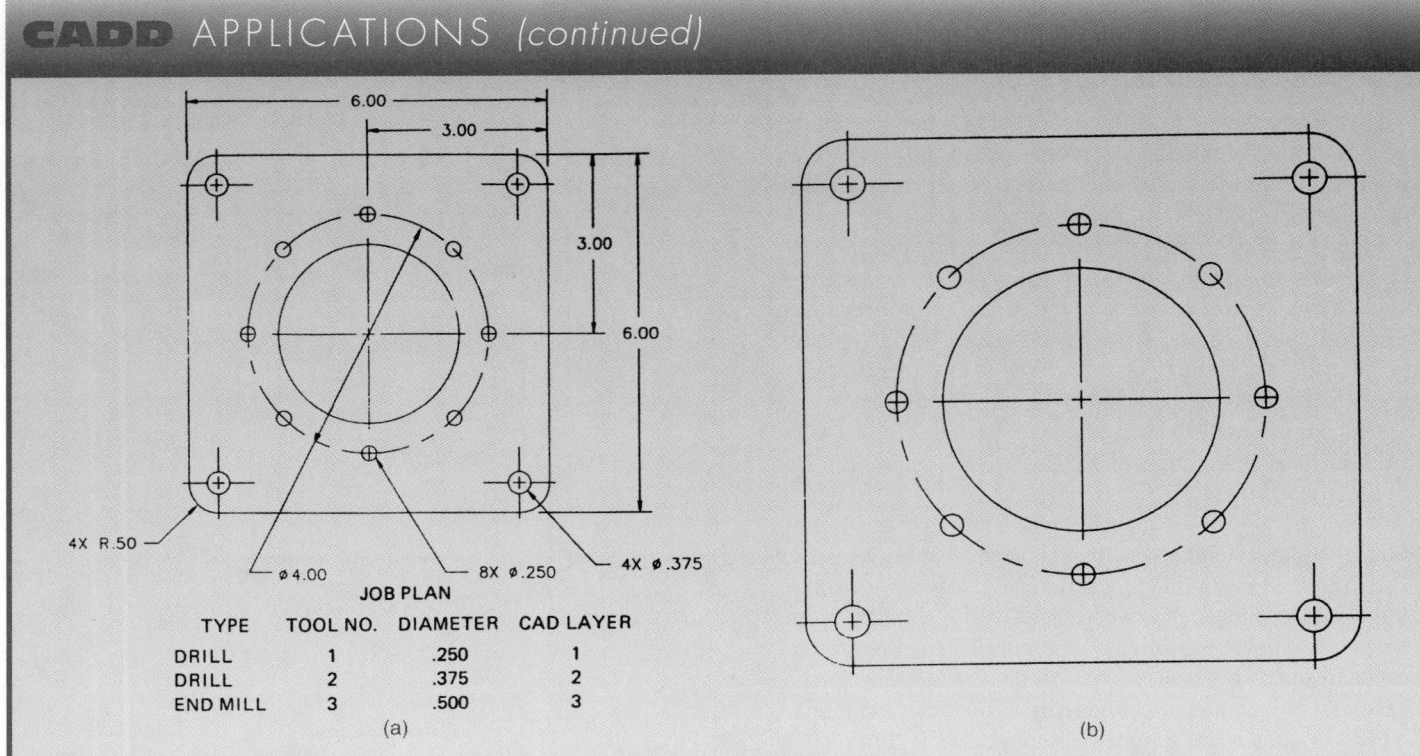

JOB PLAN			
TYPE	TOOL NO.	DIAMETER	CAD LAYER
DRILL	1	.250	1
DRILL	2	.375	2
END MILL	3	.500	3

(a)

(b)

FIGURE 12.111 ■ **(a)** A drawing created for CAD/CAM. **(b)** Drawing from (a) represented on the computer screen prior to generating the machine tool program.

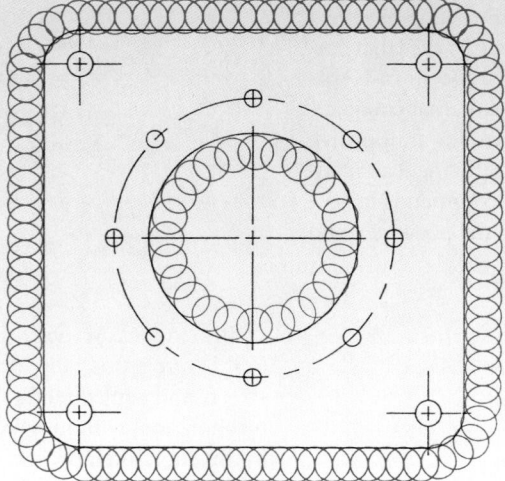

FIGURE 12.112 ■ The tool path display, shown in color, allows the operator to determine if the machine tool will perform the assigned machining operations.

RECOMMENDED REVIEW

It is recommended that you review the following parts of Chapter 2 before you begin working on fully dimensioned multiview drawings. This will refresh your memory about how related lines are properly drawn.

- Object Lines.
- Viewing Planes.
- Extension Lines.
- Hidden Lines.
- Break Lines.
- Dimension Lines.
- Centerlines.
- Phantom Lines.
- Leader Lines.

Also review Chapters 7 and 8 covering multiview and auxiliary view drawings.

Web Site Research

The following Web sites can provide you additional information for research or further study into topics covered in this chapter:

http://www.asme.org/asme/8.html—Find information and publications related to the American Society of Mechanical Engineers, including ASME Y14.5M, *Dimensioning and Tolerancing.*

http://www.ansi.org—The American National Standards Institute. Information about national and international drafting standards, including ASME Y14.5M, *Dimensioning and Tolerancing.*

http://www.adda.org—American Design Drafting Association, *Drafting Reference Guide.*

http://www.industrialpress.com—Information about the *Machinery's Handbook.* This is a valuable resource for manufacturing standards, sizes, tolerances, fits, materials, and anything else you can think of for design and drafting.

http://www.industrialpress.com/handbook/trig.html—On-line trigonometry tables.

http://www.industrialpress.com/handbook/prime.html—Prime numbers and factoring information.

FINDING DIAGONALS

Suppose you need the distance from one point on a drawing to another, such as between points A and B in Figure 12.113.

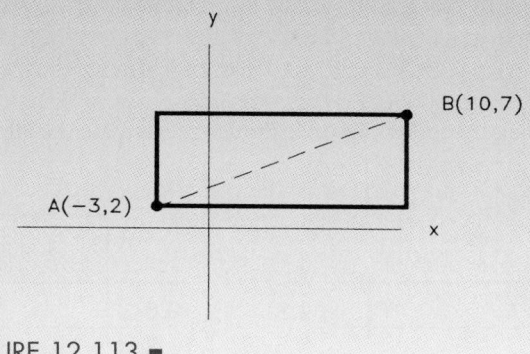

FIGURE 12.113 ■

Find the distance from C to D in Figure 12.114.

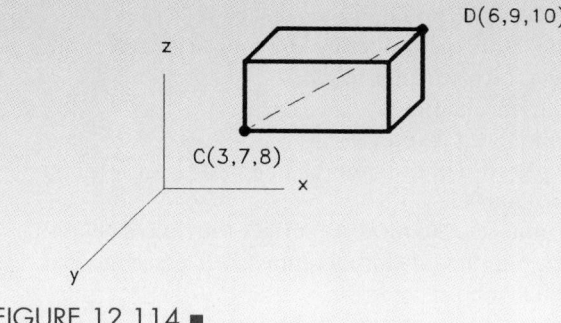

FIGURE 12.114 ■

MATH APPLICATION

CHAPTER 12
Dimensioning and Tolerancing Test

DIRECTIONS

Answer the questions with short complete statements or drawings as needed.

PART I • GENERAL

1. Name two classifications of dimensions.
2. Identify and describe two types of notes.
3. Define *unidirectional dimensioning.*
4. Specify the ANSI document that governs the standard for dimensioning and tolerancing.

5. Define the following terms; examples may be used if appropriate: *actual size; bilateral tolerance; dimension; feature; limits of dimension; specified dimension; tolerance; unilateral tolerance.*

6. What are the recommended standard units of linear measurements on engineering documents?

7. When all dimensions are metric, what is the general note that should accompany the drawing?

8. Define *dual dimensioning,* and show an example of a dual dimensioning general note.

9. How should the decimal in numerals be treated?

10. Should all of the dimension arrowheads on a drawing be the same size?

11. What is the recommended length-to-height ratio of arrowheads?

12. Discuss proper dimension line spacing.

13. Identify a possible disadvantage of chain dimensioning.

14. Describe datum dimensioning.

15. How are notes for holes dimensioned on an engineering document?

16. Describe how holes are located.

17. Define *maximum material condition* (MMC).

18. Define *least material condition* (LMC).

19. Describe a clearance fit.

20. Describe running and sliding fits (RC), locational fits (LC, LT, LN), and force fits (FN).

21. List three factors that influence sheet size selection.

22. Identify two factors that influence drawing scale selection.

23. Define *casting.* (See Chapter 9.)

24. Define *core.* (See Chapter 9.)

25. Define *forging.* (See Chapter 9.)

26. Explain the purpose of draft angle on a casting or forging.

27. Define *surface finish.*

28. Identify the units used to measure surface roughness height.

29. Show three examples of the recommended placement of surface finish symbols.

30. Describe the surface condition and process used to establish the following surface roughness heights given in micrometers: 12.5, 6.3, 3.2, 1.6, 0.80, 0.20, 0.050.

31. When using manual drafting, explain why you should place general notes that read from the bottom up when placing ASME/ANSI general notes in the lower left corner of the drawing.

32. When using CADD, explain why it may be preferred to place general notes that read from the top down when placing ASME/ANSI general notes in the lower left corner of the drawing.

33. Where are the general notes placed when using Military standards?

34. Describe a delta note and when it is used.

35. Define *parting line.*

36. What does it mean when a note such as .010 MAX DRAFT ANGLE is applied to the drawing for a plastic part?

37. Explain what +DFT means when applied to a dimension.

38. What does −DFT mean when applied to a dimension?

39. What does ISO stand for?

40. Briefly explain the purpose of the ISO 9000 Quality Systems Standard.

41. Give the name of the organization that represents the United States in the ISO 9000.

42. Give at least five reasons why an organization might want to have an ISO 9000 registration.

43. Explain the purpose of the ISO 9000-1.

44. Briefly describe the function of the ISO 9001.

45. Describe the ISO 9002 standard.

46. What is the purpose of the ISO 9003 standard?

47. Explain the purpose of the ISO 9004-1 standard.

48. Give the name of the ISO 9001 standard that has been specifically related to the automotive industry.

49. Name the ISO 9001 standard that has been specifically related to the aerospace industry.

PART 2 · GENERAL TOLERANCING

50. Identify the tolerance and limits of each of the following dimensions:
Metric:　25.5 ± 0.1;　19 ± 0.25.　Inch:　.375 ± .003; 1.6250 ± .0005.

51. Given the following CAD drawing, calculate the allowance:

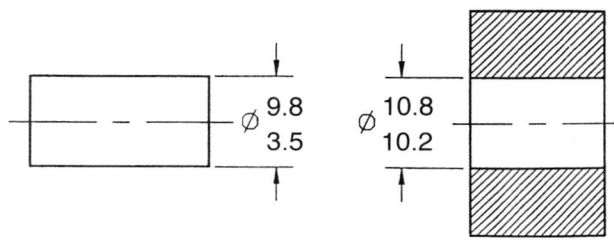

52. From the following list of given conditions, calculate the limits of the shaft and the limits of the hole. (Review the allowance calculation formula.)
   a. Metric dimensions.
   b. A clearance fit.
   c. Allowance = 0.05.
   d. Specified dimension of shaft = 12.
   e. Shaft tolerance = 0.26 BILATERAL. (Remember the tolerance is the total permissible variation.)
   f. Hole tolerance = 0.18.

53. Name five items that a CADD system needs in order to place a dimension.

54. Define *layering.*

55. List at least two reasons why CADD layering can assist the dimensioning process.

CHAPTER 12

Dimensioning and Tolerancing Problems

DIRECTIONS

1. From the selected sketch, determine which view should be the front view. Then determine which other views, if any, you need to draw to fully display the part in a multiview drawing.

2. Make a multiview sketch of the selected problem as close to correct proportions as possible. Be sure to indicate where you intend to place the dimension lines, extension lines, arrowheads, and hidden features, to help you determine the spacing for your final drawing.

3. Using the sketch you have just developed as a guide, make an original multiview drawing on an adequate size drawing sheet and at an appropriate scale. Include all dimensions needed using unidirectional dimensioning. Use computer-aided drafting if specified by your course outline. Include the following general notes at the lower left corner, .5 in. each direction from the borderline using ⅛-in.-high lettering. Letter the word NOTES with ³⁄₁₆-in.-high lettering.

   1. INTERPRET DIMENSIONS AND TOLERANCES PER ASME Y14.5M—1994.
   2. REMOVE ALL BURRS AND SHARP EDGES.

Notes

Additional notes may be required depending on the specifications of each individual assignment. Remember that each problem assignment is given as an engineer's layout to help simulate actual drafting conditions.

Over 135 additional problems are available by completing the multiview and auxiliary view drawings that you created in Chapters 7 and 8. If you are using manual drafting, continue where you left off by adding the given dimensions and notes. If you are using CADD, open the previous file and complete the given dimensions and notes. Add the general notes and follow the directions given in the preceding instructions.

Dimensions and views on engineers' layouts may not be placed in accordance with acceptable standards. You need to carefully review the chapter material when preparing the layout sketch. In some problems, the engineer's layout may "assume" certain information, such as the symmetry of a part or the alignment of holes. You need to place enough dimensions or draw lines between features to "fully" dimension the part. Use a mechanical drafting title block with the following unspecified tolerances and third-angle projection symbol unless otherwise specified by your instructor. Use manual drafting or CADD as appropriate with your course objectives.

UNSPECIFIED TOLERANCES:

DECIMALS	mm	IN.
X	± 2.5	± .1
XX	± 0.25	± .01
XXX	± 0.127	± .005
ANGULAR ± 30'		
FINISH	3.2μm	125μIN.

PROBLEM 12-1 Basic practice (metric)
Part Name: Step Block
Material: SAE 1020

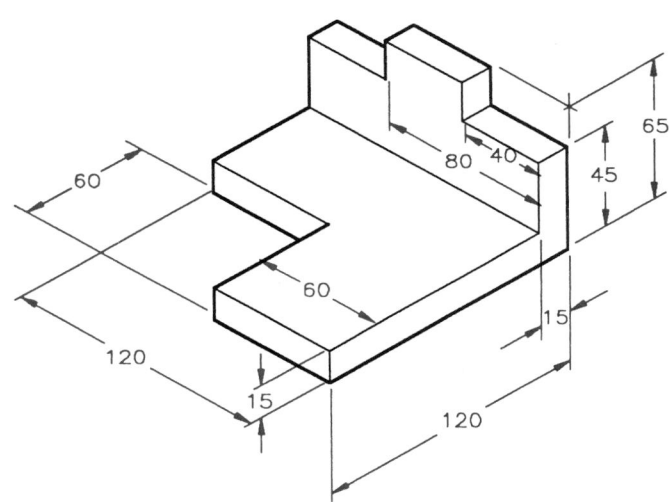

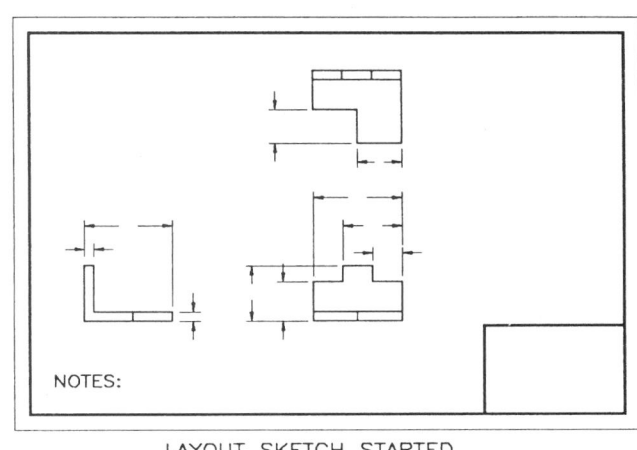

LAYOUT SKETCH STARTED

PROBLEM 12-2   Basic practice (metric)
Part Name: Machine Tool Wedge Plate
Material: SAE 4320

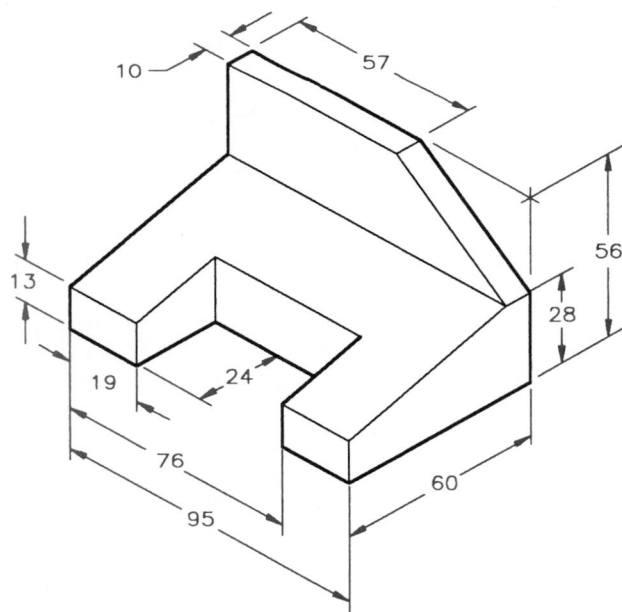

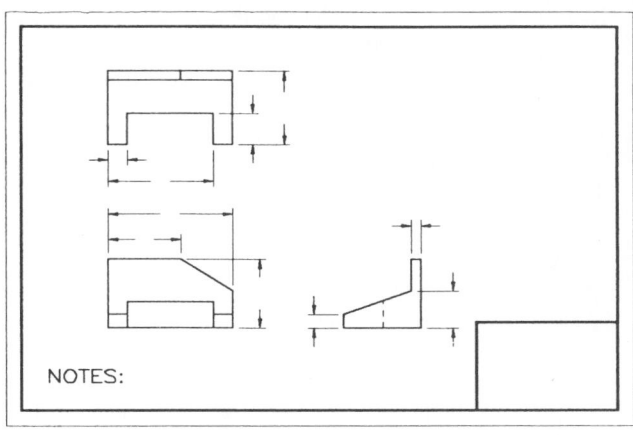

NOTES:

LAYOUT SKETCH STARTED

PROBLEM 12-3   Dimensioning basic practice (in.)
Part Name: V-guide
Material: SAE 4320
Fractions: ± ⅟₃₂

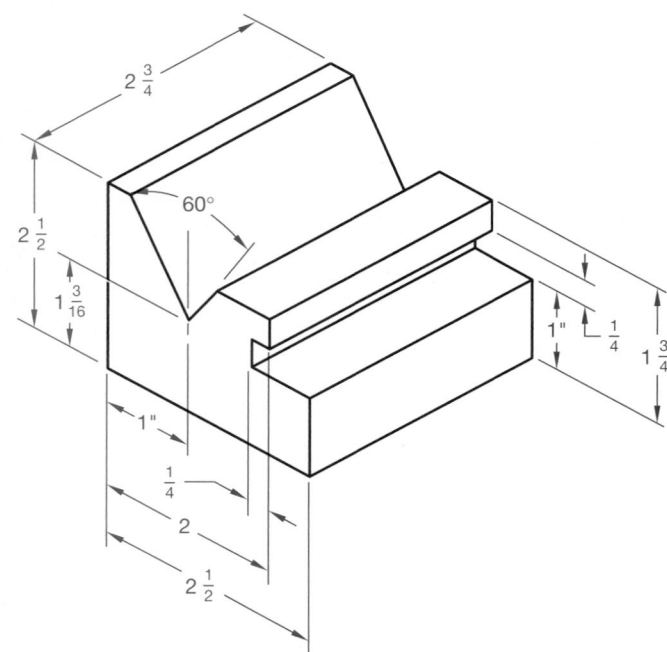

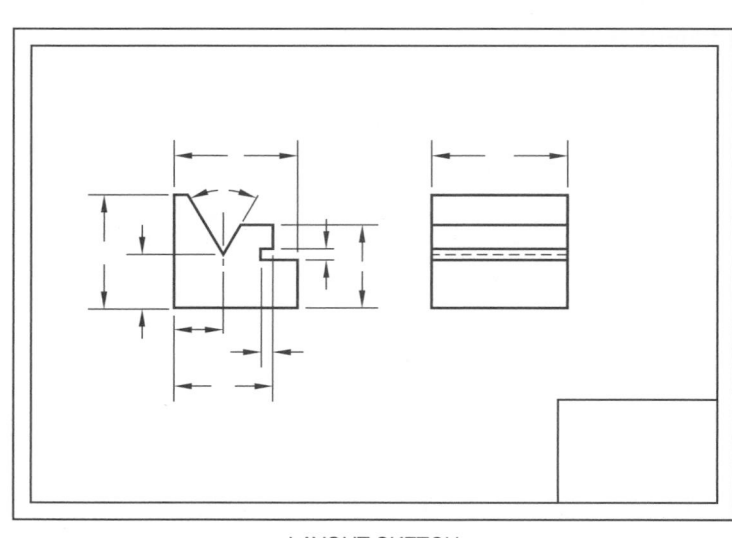

LAYOUT SKETCH
OPTIONAL SINGLE VIEW WITH LENGTH
GIVEN IN GENERAL NOTE OR TITLE BLOCK

PROBLEM 12-4   Circles, arcs, and counterbores (in.)
Part Name: Rest Pad
Material: SAE 1040
Fillets: R.125

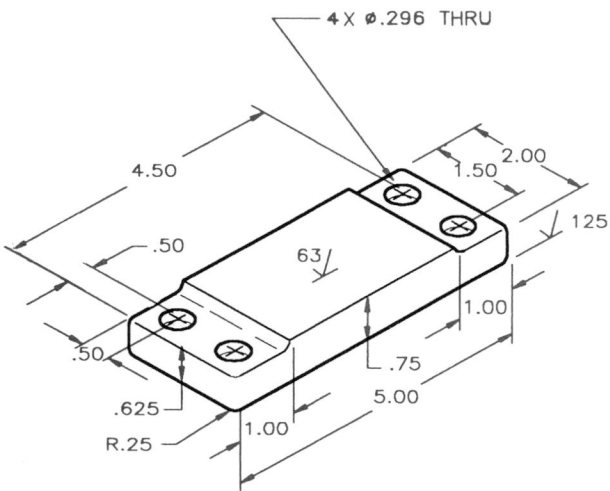

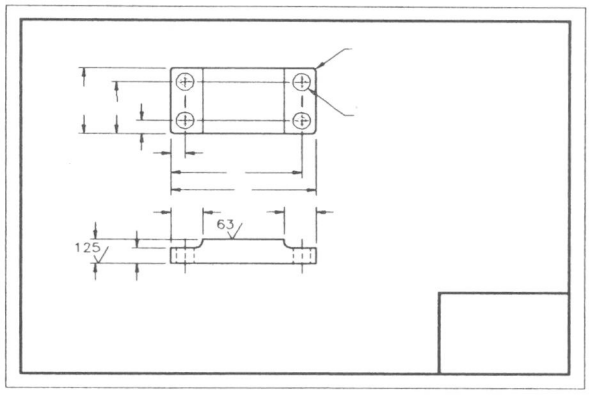

LAYOUT SKETCH

PROBLEM 12-5   Limited space (metric)
Part Name: Angle Bracket
Material: Mild Steel (MS)

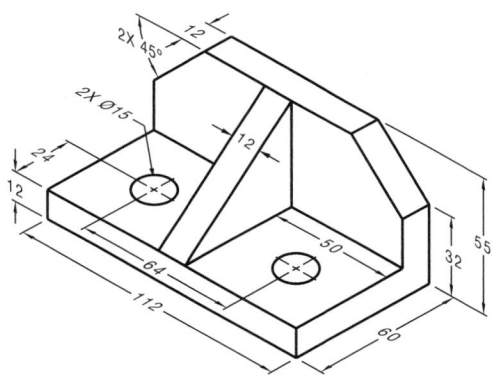

PROBLEM 12-7   Limited spaces (metric)
Part Name: Selector Slide Kicker
Material: Al 1510

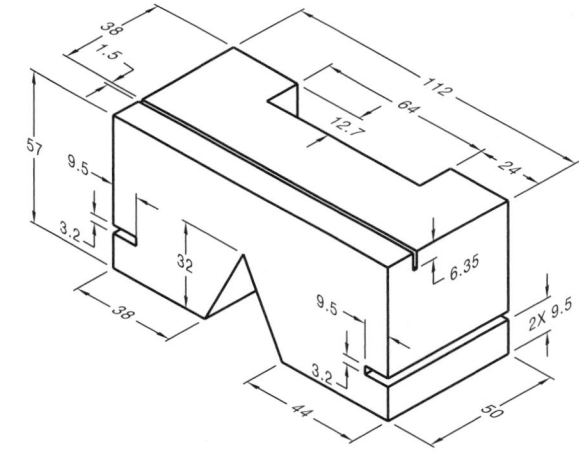

PROBLEM 12-6   Dimensioning contours and limited spaces (in.)
Part Name: Support
Material: Aluminum

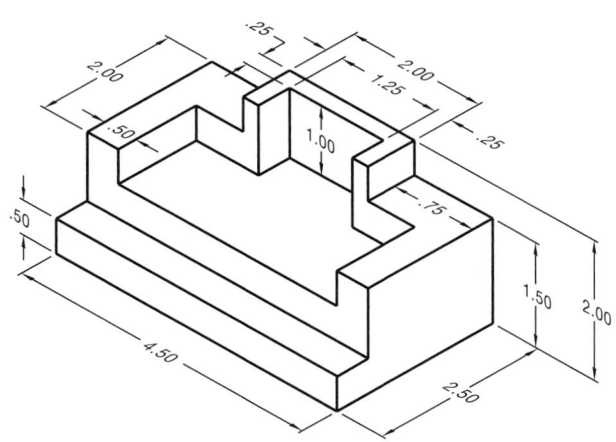

PROBLEM 12-8   Circles and arcs (in.)
Part Name: Chain Link
Material: SAE 4320

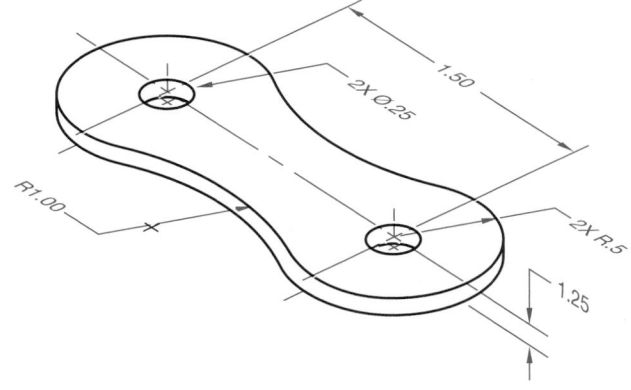

PROBLEM 12-9   Holes and limited space (in.)
Part Name: Pivot Bracket
Material: SAE 1040

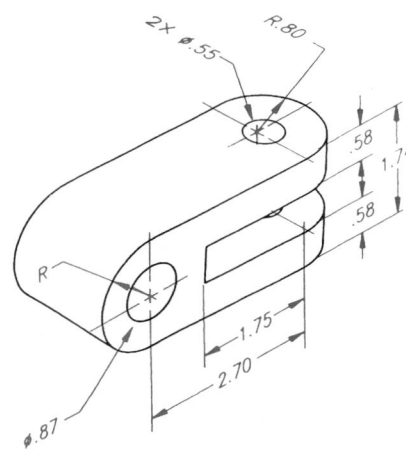

PROBLEM 12-10   Holes, angles, and arcs (in.)
Part Name: Journal Bracket
Material: Cast Iron (CI)

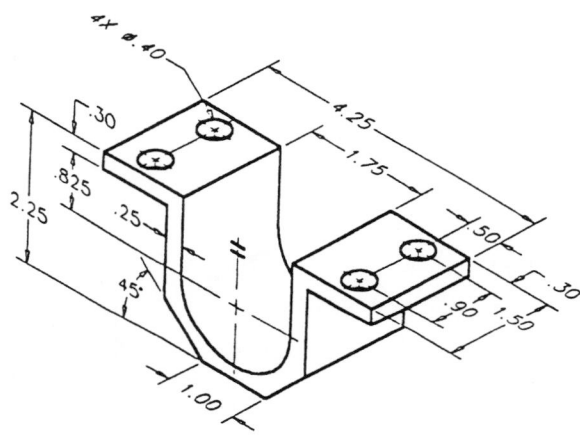

PROBLEM 12-11   Dimensioning multiple features (in.)
Part Name: Lock Ring
Material: SAE 1020

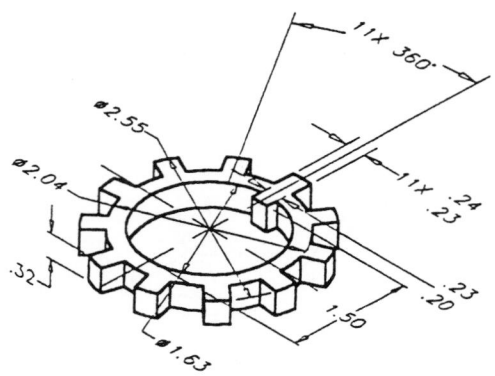

PROBLEM 12-12   Single view (in.)
Part Name: Idler Gear Shaft
Material: MIL-S-7720

Problem based on original art courtesy Aerojet TechSystems Co.

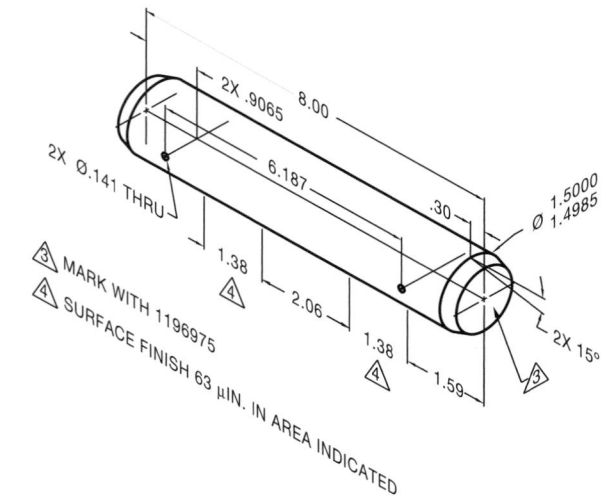

PROBLEM 12-13   Circles and arcs (in.)
Part Name: Bearing Support
Material: SAE 1040
Fillets: R.6

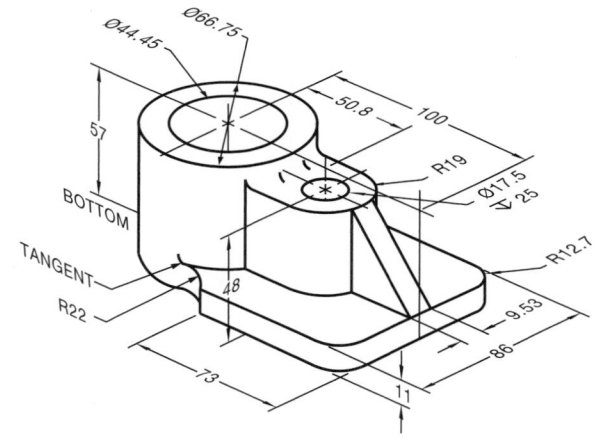

PROBLEM 12-14   Circles and arcs (metric)
Part Name: Hinge Bracket
Material: Cast Aluminum

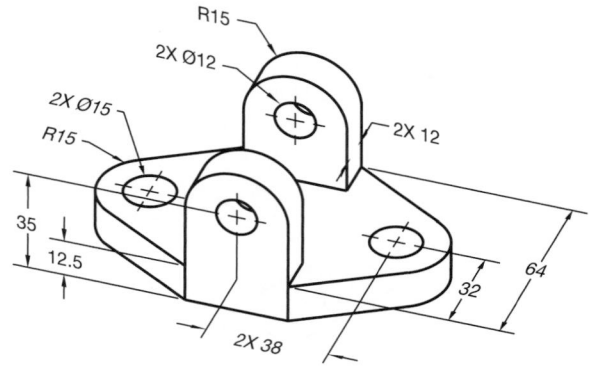

PROBLEM 12-15   Machine features (in.)
Part Name: Spacer
Material: SAE 1030

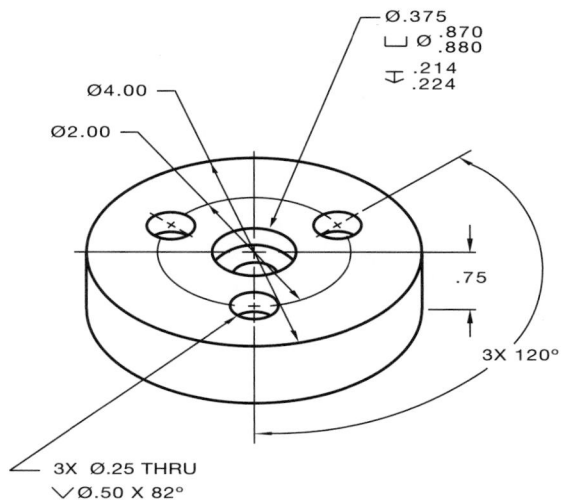

PROBLEM 12-16 Polar coordinate dimensioning (in.)
Part Name: Spacer
Material: Plastic

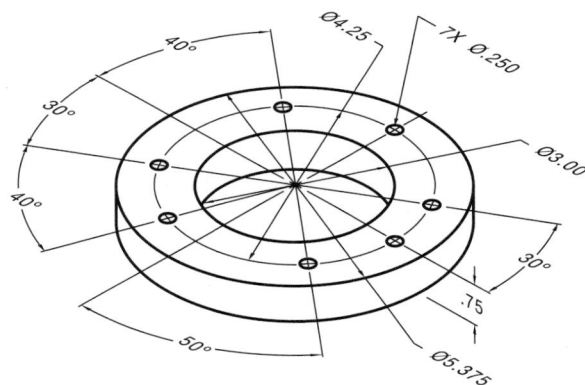

PROBLEM 12-17   Repetitive features (in.)
Part Name: Slot Plate
Material: Aluminum

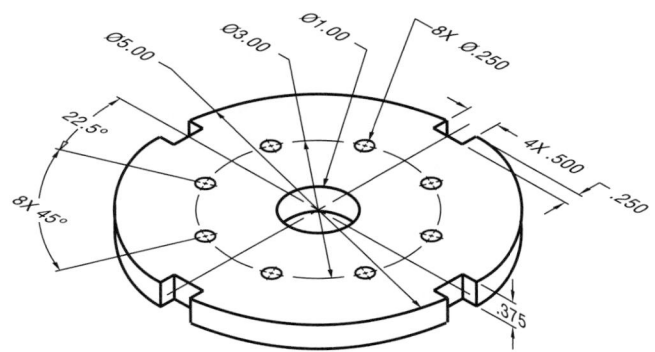

PROBLEM 12-18   Tabular dimensioning (metric)
Part Name: Mounting Base
Material: Stainless Steel

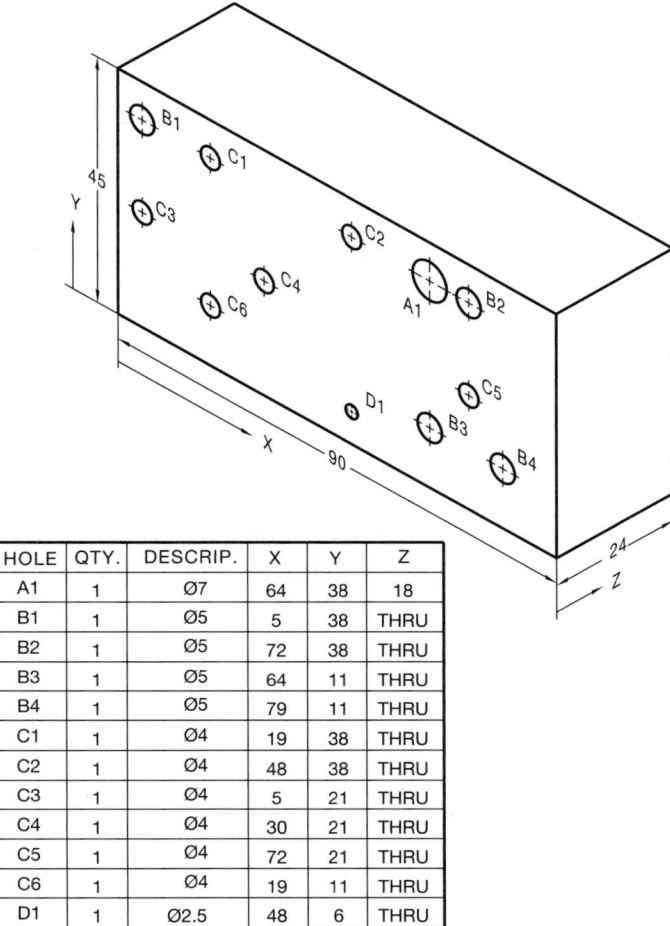

HOLE	QTY.	DESCRIP.	X	Y	Z
A1	1	Ø7	64	38	18
B1	1	Ø5	5	38	THRU
B2	1	Ø5	72	38	THRU
B3	1	Ø5	64	11	THRU
B4	1	Ø5	79	11	THRU
C1	1	Ø4	19	38	THRU
C2	1	Ø4	48	38	THRU
C3	1	Ø4	5	21	THRU
C4	1	Ø4	30	21	THRU
C5	1	Ø4	72	21	THRU
C6	1	Ø4	19	11	THRU
D1	1	Ø2.5	48	6	THRU

PROBLEM 12-19   Dimensioning circles, arcs, and slots (in.)
Part Name: Top Pipe Support Bracket
Material: SAE 1020
Fillets: R.12

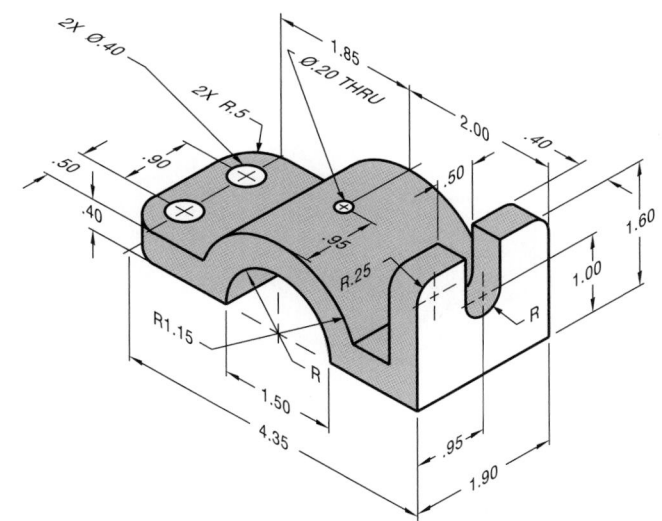

PROBLEM 12-20   Chain dimensioning (metric)

Part Name: Control Housing Cover

Material: Cast Iron

Note: Do not draw a sectional view. Sections are covered in Chapter 14. Consider a bottom view to show wall thickness.

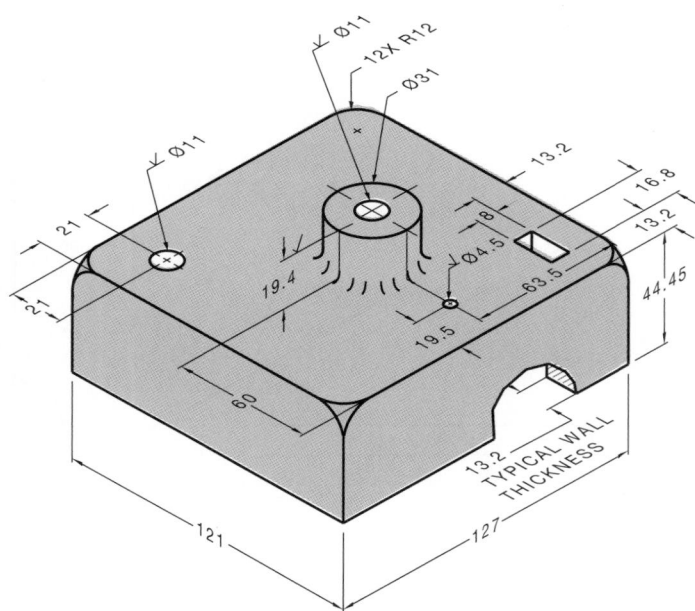

PROBLEM 12-22   Dimensioning auxiliary views (in.)

Part Name: Connector

Material: SAE 4320

Note: Finish all over 63 µin.

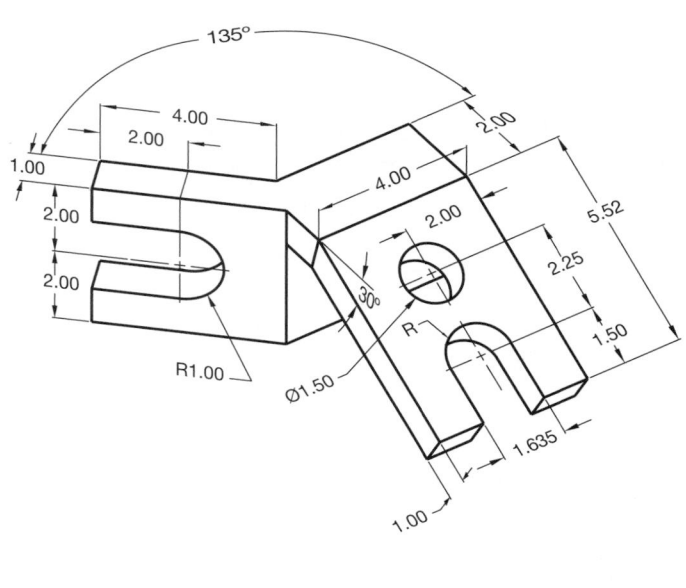

PROBLEM 12-21   Dimensioning compound circles and arcs (in.)

Part Name: Multiple Shaft Support

Material: SAE 1030

Fillets and rounds: R.08

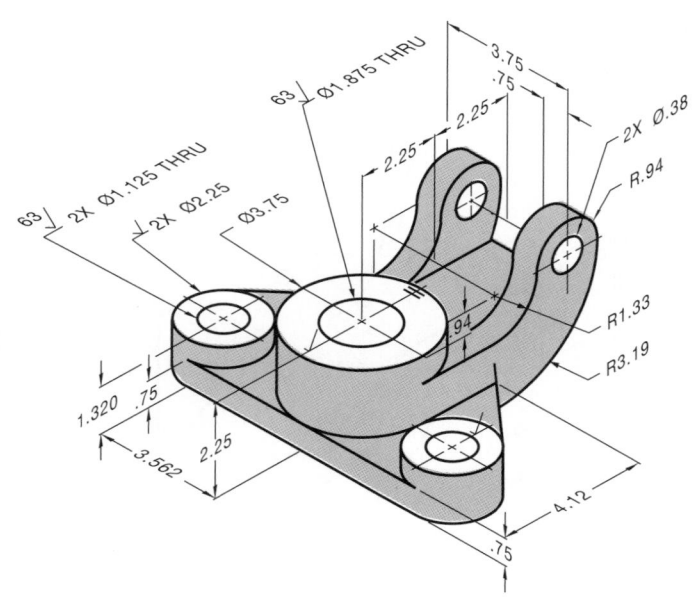

PROBLEM 12-23   Dimensioning small spaces (in.)

Part Name: Guide Bracket

Material: SAE 4020

Fillets and rounds: R.125

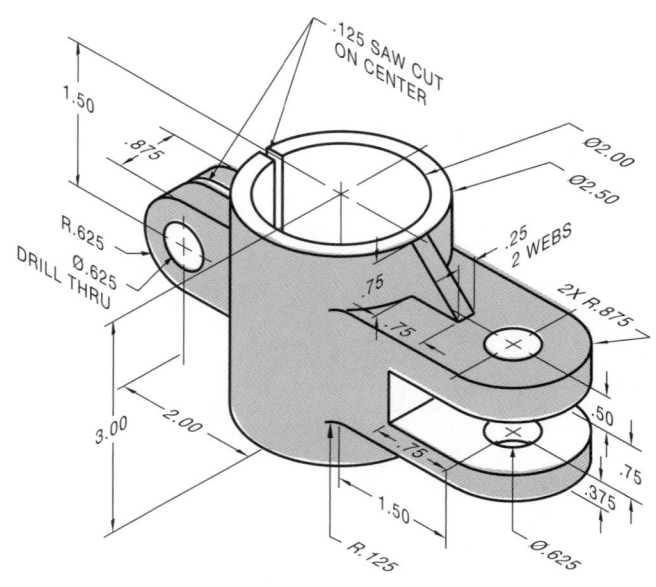

PROBLEM 12-24 Chart drawing (in.)

Part Name: Tank Bracket

Material: 11 GA A570-30

Problem based on original art courtesy TEMCO.

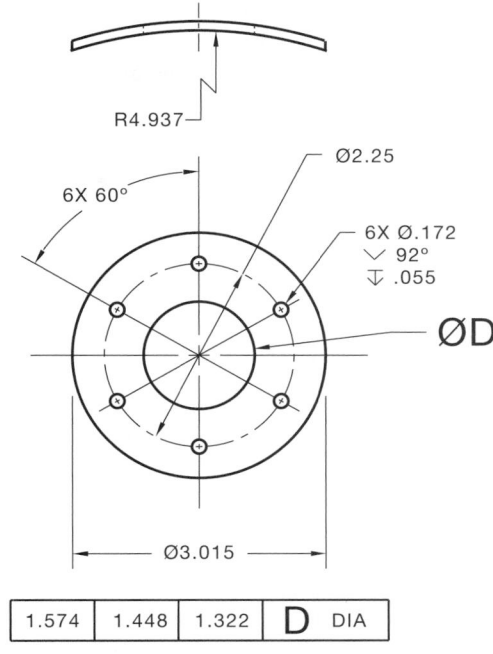

1.574	1.448	1.322	D	DIA

PROBLEM 12-25 Arrowless dimensioning (in.)

Part Name: Metering Box Gasket

Material: Neoprene

Notes: Dimensions have been corrected to allow for steel rule blade.

SPECIFIC INSTRUCTIONS:

Convert the engineering sketch to arrowless tabular dimensioning. Set up a table that identifies hole diameters, arc radii, and location dimensions from x and y coordinates. Dimensions above Y are + dimensions and below Y are – dimensions. Dimensions to the right of X are + dimensions and to the left of X are – dimensions.

Problem based on original art courtesy Vellumoid Inc.

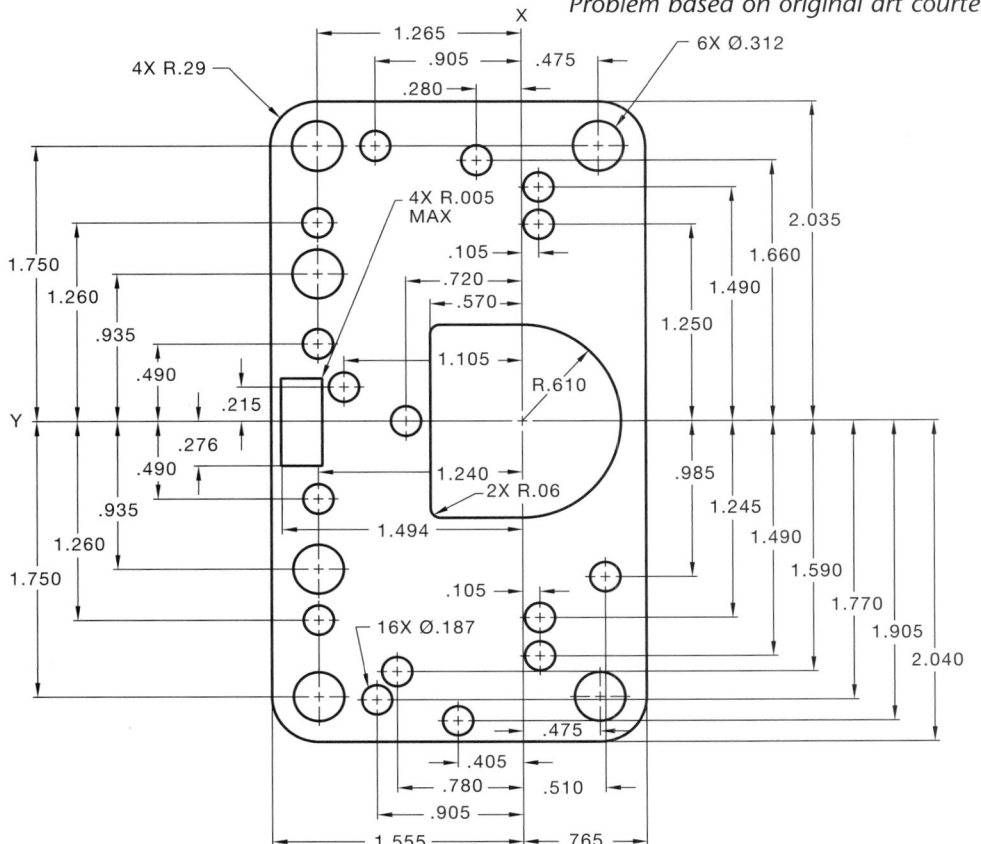

PROBLEM 12-26   Casting drawing (metric)

Part Name: Slider

Material: ASTM 60 CI

Hardness: Brinell 180-220

SPECIFIC INSTRUCTIONS:

Draw part as finished. Include all machining and casting dimensions on one detail drawing. The patternmaker will apply unspecified allowances during production.

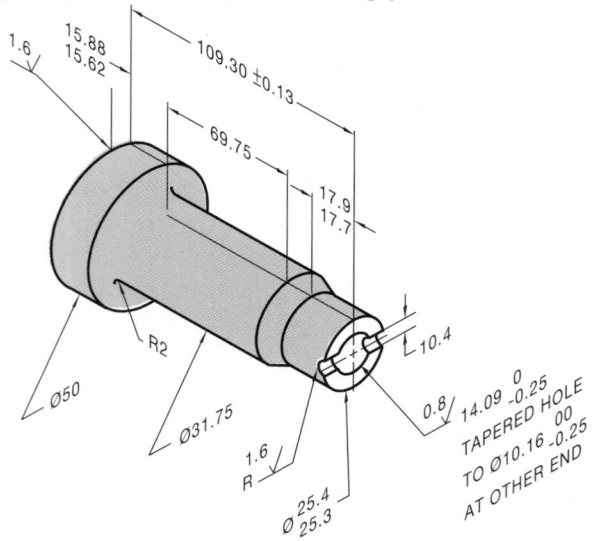

PROBLEM 12-27   Forging and machining drawings (metric)

Part Name: Pump Pivot Support

Material: HRMS (Hot-rolled Mild Steel)

SPECIFIC INSTRUCTIONS:

Prepare two drawings: one showing only forging-related views, dimensions, and notes; and the other showing only machining-related views, dimensions, and notes. Provide the draft angles recommended for steel (refer to chapter). Add 3 mm to forging where finish surface identified. Do not draw thread symbol or thread note (M12 × 1.75).

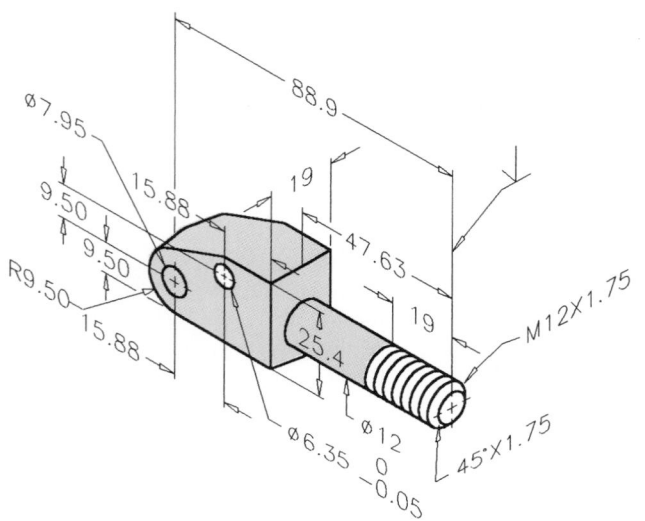

PROBLEM 12-28   Angles, holes, and arcs (in.)

Part Name: Mounting Bracket

Material: SAE 1020

Fillets: R.13

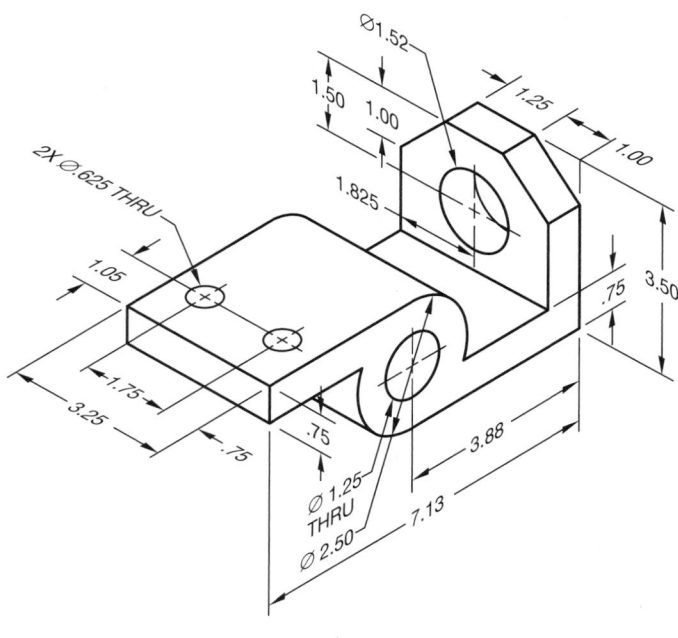

PROBLEM 12-29   Contours, machine features, and limited space (in.)

Part Name: Support Base

Material: SAE 1040

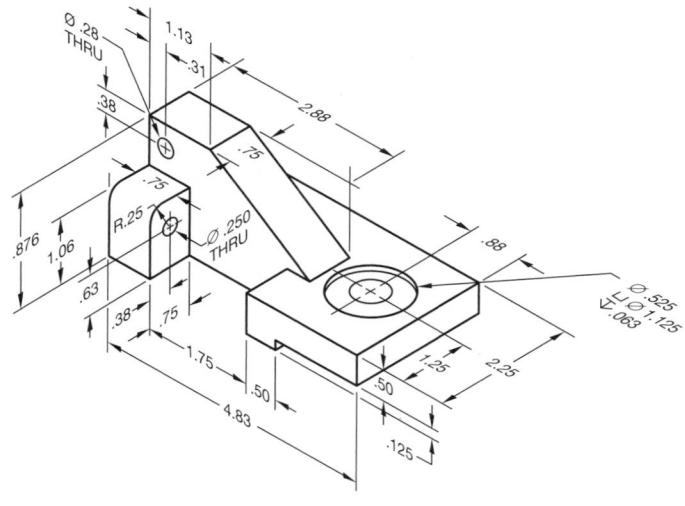

PROBLEM 12-30   CAD/CAM (in.)

Part Name:Spacer
Material: SAE 1030

JOB PLAN			
Tool Type	Tool No.	Diameter	CAD Layer
Drill	1	.250	1
Drill	2	.375	2
End Mill	3	.500	3

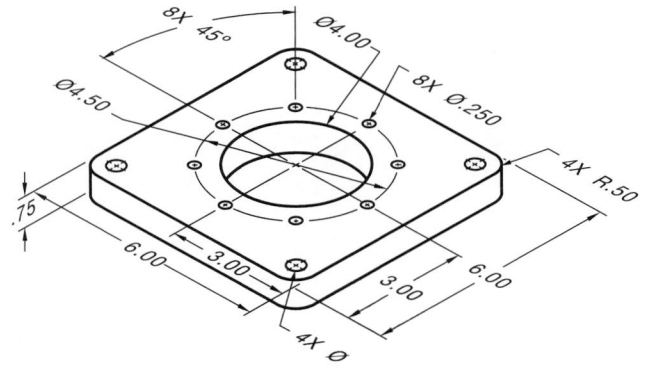

PROBLEM 12-31   CAD/CAM (in.)

Part Name: Cover Plate
Material: .500-thick Aluminum

JOB PLAN				
Tool Type	Tool No.	Diameter	Length	CAD Layer
End Mill	1	.750	1.000	1
End Mill	2	.500	1.000	2
End Mill	3	.250	1.000	3
Spot Drill	4	.125	1.000	4
Drill	5	.500	1.000	5
Drill	6	1.000	1.000	6
Clamps				7

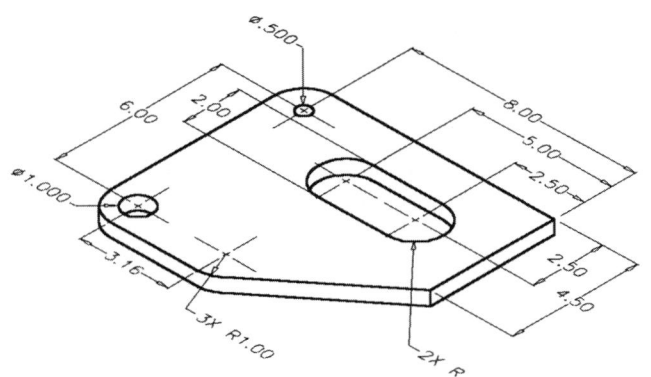

MATH PROBLEMS

Find the distance between these points on a 2-D drawing:

PROBLEM 12-32   (5, 7) and (8, 9)

PROBLEM 12-33   (3, –4) and (7, 10)

PROBLEM 12-34   (–2, –5) and (8, 3)

PROBLEM 12-35   (6, 4) and (6, 12)

Find the distance between these points on a 3-D drawing:

PROBLEM 12-36   (0, 0, 0) and (5, 7, 9)

PROBLEM 12-37   (1, 2, 3) and (12, 18, 20)

PROBLEM 12-38   (–1, 0, 6) and (1, 3, 9)

PROBLEM 12-39   (6, 12, 3) and (–2, 12, 8)

PROBLEM 12-40   What is the straight-line distance (to the nearest $\frac{1}{16}$") from one corner of a piece of 4' by 8' plywood to the opposite corner?

PROBLEM 12-41   What is the straight-line distance (to the nearest $\frac{1}{4}$") from one corner at the floor of a room to the opposite corner at the ceiling if the dimensions of the room are 20' high by 30' wide by 50' long?

Fasteners and Springs

OBJECTIVES

After completing this chapter, you will:

■ Draw screw thread representations and provide correct thread notes.

■ Prepare drawings for fastening devices.

■ Draw completely dimensioned spring representations.

■ Prepare formal drawings from engineers' sketches and actual industrial layouts.

THE ENGINEERING DESIGN APPLICATION

You are asked to make a drawing from an engineer's notes for a fastening device that is needed for one of the products the engineer is designing. The engineer's notes look like Figure 13.1. You may be thinking:

■ A 10-gage shaft threaded with 32—UNF—2A. Maybe you can use a 10—32UNF—2A threaded rod and cut it to .375 lengths.

■ Then you need to chamfer both ends with .015 × 35° and provide a slot in one end that is .030 wide and .047 deep.

10 - 32 UNF - 2A THREADED
MATERIAL .375 LG.
W/ .015 LATERAL X 35°
CHAMFERS ON BOTH ENDS
PROVIDE A .030 WIDE X
.047 DEEP SLOT ON
ONE END.

FIGURE 13.1 ■ Engineer's notes.

So, you go to work preparing a drawing like the one shown in Figure 13.2. The point here is that the engineer provided a lot of information in the thread note. In fact, standard screws and other fasteners may be completely described without a drawing. A written specification can be used to completely describe an object. For example, ½–13UNC–2 × 1.5 LG FILLISTER HEAD MACHINE SCREW.

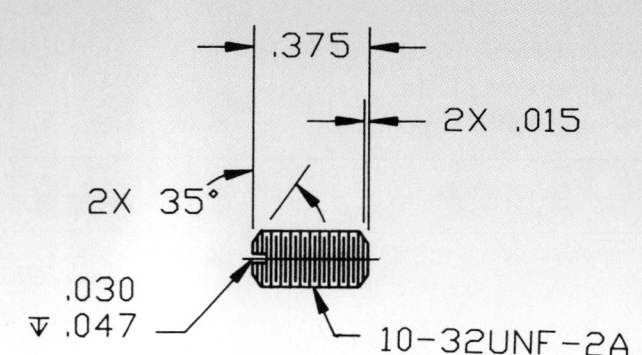

FIGURE 13.2 ■ CADD drawing from engineer's notes.

ANSI This chapter introduces you to the methods of specifying and drafting fasteners and springs. Fasteners include screw threads, keys, pins, rivets, and weldments. There are two types of springs, helical and flat. The American National Standards Institute documents that govern the standards for fasteners and springs are *Screw Thread Representation*, ANSI Y14.6; *Screw Thread Representation (Metric Supplement)*, ANSI Y14.6aM; *Symbols for Welding and Nondestructive Testing Including Brazing*, ANSI/AWS A2.4; and *Mechanical Spring Representation*, ANSI Y14.13M.

SCREW THREAD FASTENERS

The standardization of screw threads was achieved among the United States, United Kingdom, and Canada in 1949. A need for interchangeability of screw thread fasteners was the purpose of this standardization and resulted in the Unified Thread Series. The Unified Thread Series is now the American standard for screw threads. Prior to 1949 the United States standard was the American National screw threads. The unification standard occurred as a result of combining some of the characteristics of the American

National screw threads with the United Kingdom's long accepted Whitworth screw threads. Screw thread systems were revised again in 1974 for metric application. The modifications were minor and based primarily on metric translation. In order to emphasize that the Unified screw threads evolve from inch calibrations, the term Unified Inch Screw Threads is used while the term Unified Screw Threads Metric Translation is used for the metric conversion.

Screw threads are a helix or conical spiral formed on the external surface of a shaft or internal surface of a cylindrical hole, as shown in Figures 13.3 and 13.4. Screw threads are used for an unlimited number of services, such as for holding parts together as fasteners, for leveling and adjusting objects, and for transmitting power from one object or feature to another.

Screw Thread Terminology

Refer to Figure 13.3 and Figure 13.4 as a reference for the following definitions related to external and internal threads.

Axis

The thread axis is the centerline of the cylindrical thread shape.

Body

That portion of a screw shaft that is left unthreaded.

Chamfer

An angular relief at the last thread to help allow the thread to more easily engage with a mating part.

Classes of Threads

A designation of the amount of tolerance and allowance specified for a thread.

Crest

The top of external and the bottom of internal threads.

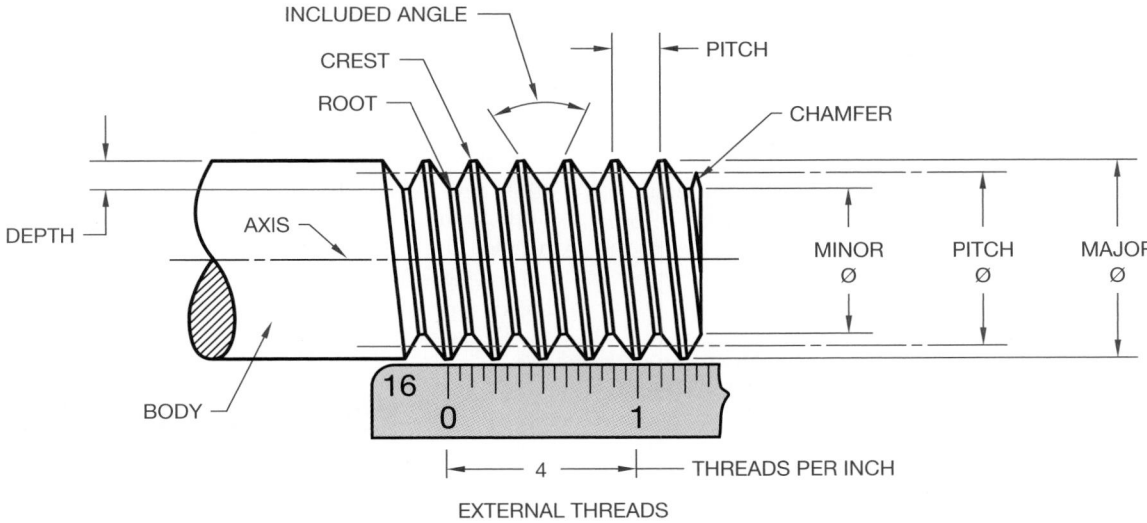

FIGURE 13.3 ■ External screw thread components.

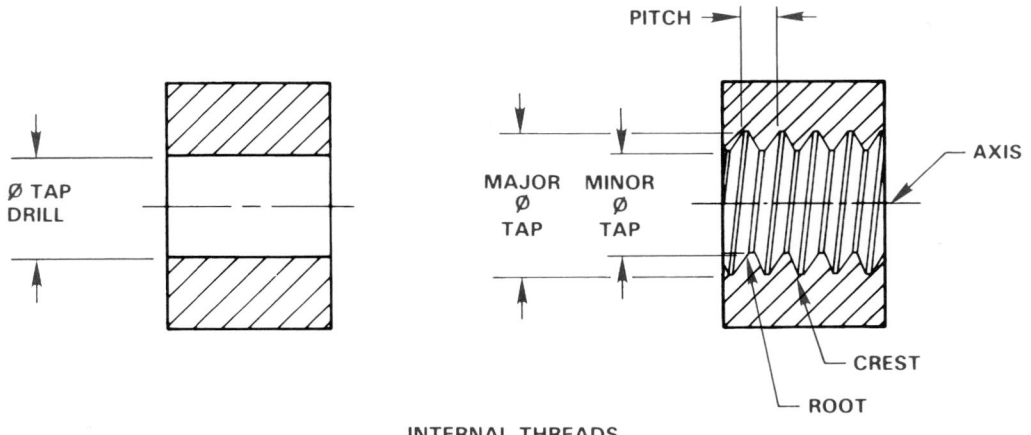

FIGURE 13.4 ■ Internal screw thread components.

Depth of Thread

Depth is the distance between the crest and the root of a thread, measured perpendicular to the axis.

Die

A machine tool used for cutting external threads.

Fit

Identifies a range of thread tightness or looseness.

Included Angle

The angle between the flanks (sides) of the thread.

Lead

The lateral distance a thread travels during one complete rotation.

Left-hand Thread

A thread that engages with a mating thread by rotating counterclockwise, or with a turn to the left when viewed toward the mating thread.

Major Diameter

The distance on an external thread from crest to crest through the axis. For an internal thread the major diameter is measured from root to root across the axis.

Minor Diameter

The dimension from root to root through the axis on an external thread and measured across the crests through the center for an internal thread.

Pitch

The distance measured parallel to the axis from a point on one thread to the corresponding point on the adjacent thread.

Pitch Diameter

A diameter measured from a point halfway between the major and minor diameter through the axis to a corresponding point on the opposite side.

Right-hand Thread

A thread that engages with a mating thread by rotating clockwise, or with a turn to the right when viewed toward the mating thread.

Root

The bottom of external and the top of internal threads.

Tap

A tap is the machine tool used to form an interior thread. Tapping is the process of making an internal thread.

Tap Drill

A tap drill is used to make a hole in material before tapping.

Thread

The part of a screw thread represented by one pitch.

Thread Form

The design of a thread determined by its profile.

Thread Series

Groups of common major diameter and pitch characteristics determined by the number of threads per inch.

Threads per Inch

The number of threads measured in one inch. The reciprocal of the pitch in inches.

THREAD-CUTTING TOOLS

A tap is a machine tool used to form an internal thread as shown in Figure 13.5. A die is a machine tool used to form external threads. (See Figure 13.6.)

A tap set is made up of a taper tap, a plug tap, and a bottoming tap as shown in Figure 13.7. The taper tap is generally used for starting a thread. The threads are tapered to within ten threads from the end. The tap is tapered so the tool more evenly distributes the cutting edges through the depth of the hole. The plug tap has the threads tapered to within five threads from the end. The plug tap

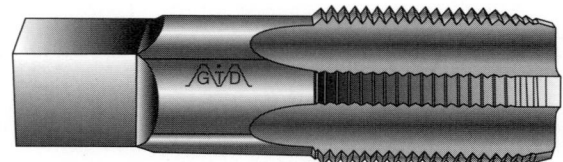

FIGURE 13.5 ■ Tap. *Courtesy Greenfield Tap & Die, Division of TRW, Inc.*

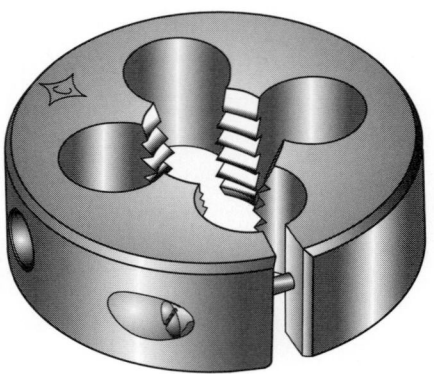

FIGURE 13.6 ■ Die. *Courtesy The Cleveland Twist Drill, an Acme-Cleveland Company.*

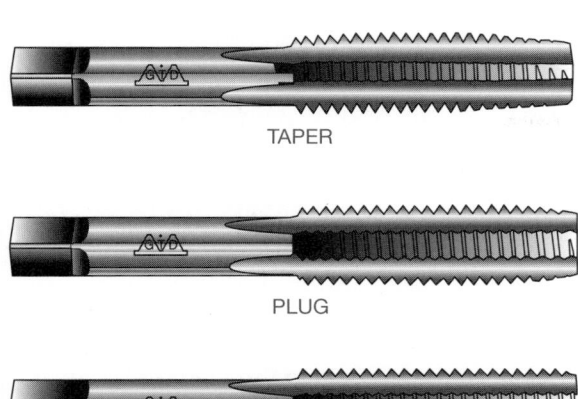

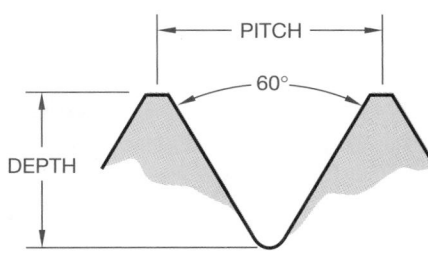

FIGURE 13.9 ■ Unified thread form.

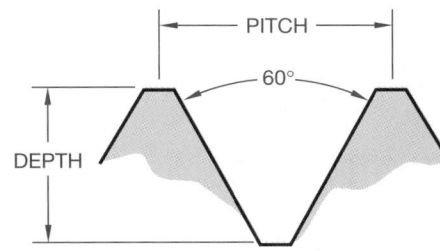

FIGURE 13.10 ■ American National thread form.

FIGURE 13.7 ■ Tap set includes taper, plug, and bottoming taps.
Courtesy Greenfield Tap & Die, Division of TRW, Inc.

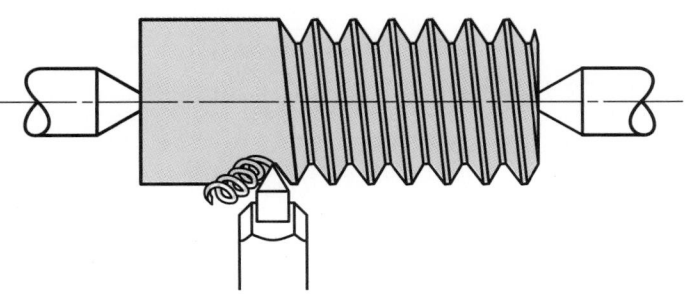

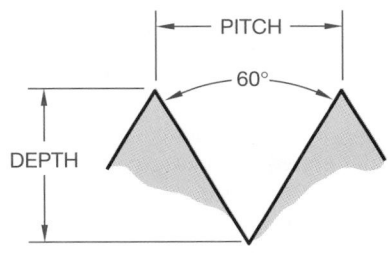

FIGURE 13.11 ■ Sharp-V thread form.

FIGURE 13.8 ■ Thread cutting on a lathe.

can be used to completely thread through material or thread a *blind hole* (a hole that does not go through the material) if full threads are not required all the way to the bottom. The bottoming tap is used when threads are needed to the bottom of a blind hole.

The die is a machine tool used to cut external threads. Thread-cutting dies are available for standard thread sizes and designations.

External and internal threads may also be cut on a lathe. A lathe is a machine that holds a piece of material between two centers or in a chucking device. The material is rotated as a cutting tool removes material while traversing along a carriage which slides along a bed. Figure 13.8 shows how a cutting tool can make an external thread.

THREAD FORMS

Unified threads are the most common threads used on threaded fasteners. Figure 13.9 shows the profile of a Unified thread.

American National threads, shown in profile in Figure 13.10, are similar to the Unified thread but have a flat root. Still in use today, the American National thread has generally replaced the sharp-V thread form.

The *sharp-V thread,* although not commonly used, is a thread that will fit and seal tightly. It is difficult to manufacture since the sharp crests and roots of the threads are easily damaged. (See Figure 13.11.) The sharp-V thread was the original United States standard thread form.

Metric thread forms vary slightly from one European country to the next. The International Organization for Standardization (ISO) was established to standardize metric screw threads. The ISO thread specifications are similar to the Unified thread form. (See Figure 13.12.)

Whitworth threads are the original British standard thread forms developed in 1841. These threads have been referred to as parallel screw threads. The Whitworth thread forms are primarily being used for replacement parts. (See Figure 13.13.)

Square thread forms, shown in Figure 13.14, have a longer pitch than Unified threads. Square threads were developed as threads that would effectively transmit power; however, they are difficult to

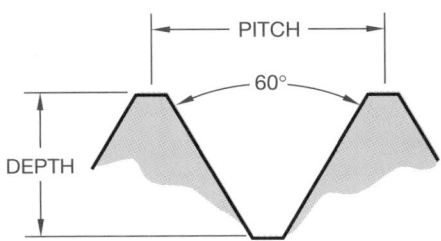

FIGURE 13.12 ■ Metric thread form.

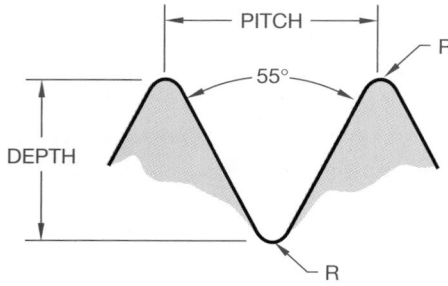

FIGURE 13.13 ■ Whitworth thread form.

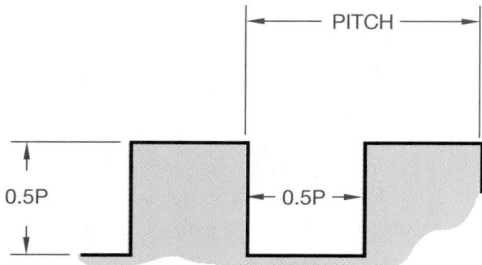

FIGURE 13.14 ■ Square thread form.

manufacture because of their perpendicular sides. These are modified square threads with 10° sides. The square thread is generally replaced by Acme threads.

Acme thread forms are commonly used when rapid traversing movement is a design requirement. Acme threads are popular on such designs as screw jacks, vice screws, and other equipment and machinery that requires rapid screw action. A profile of the Acme thread form is shown in Figure 13.15.

Buttress threads are designed for applications where high stress occurs in one direction along the thread axis. The thread flank or side that distributes the thrust or force is within 7° of perpendicularity to the axis. This helps reduce the radial component of the thrust. The buttress thread is commonly used in situations where tubular features are screwed together and lateral forces are exerted in one direction. (See Figure 13.16.)

Dardelet thread forms are primarily used in situations where a self-locking thread is required. These threads resist vibrations and remain tight without auxiliary locking devices. (See Figure 13.17.)

Rolled thread forms are used for screw shells of electric sockets and lamp bases. (See Figure 13.18.)

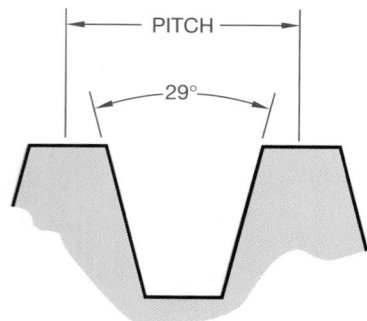

FIGURE 13.15 ■ Acme thread form.

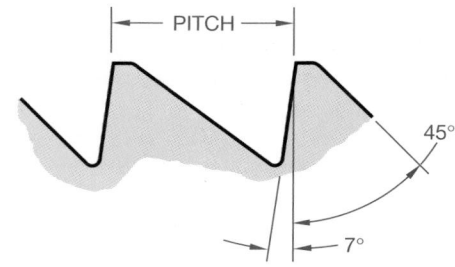

FIGURE 13.16 ■ Buttress thread form.

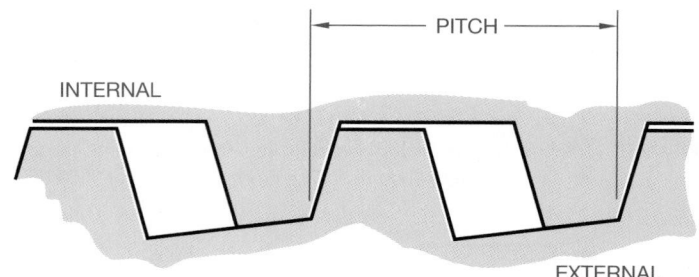

FIGURE 13.17 ■ Dardelet self-locking thread form.

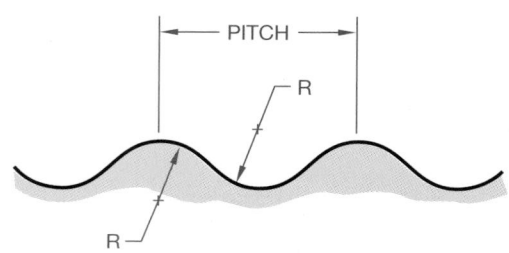

FIGURE 13.18 ■ Rolled thread form.

American National Standard taper pipe threads are the standard threads used on pipes and pipe fittings. These threads are designed to provide pressure-tight joints or not, depending on the intended function and materials used. American pipe threads are measured by the *nominal pipe* size, which is the inside pipe diameter. For example, a ½-in. pipe size has an outside pipe diameter of .840 in. (See Figure 13.19.) Pipe thread design considerations and drafting practices are discussed later in this text.

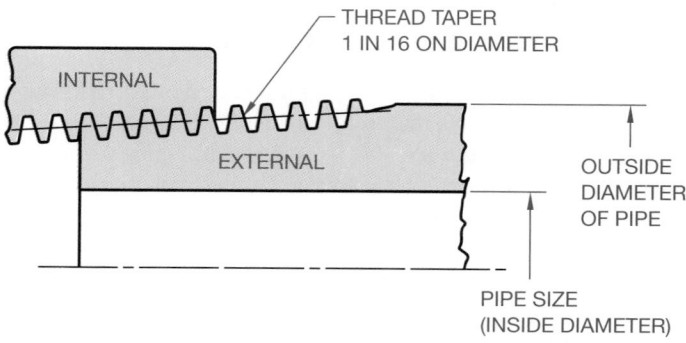

FIGURE 13.19 ■ American National Standard taper pipe thread form.

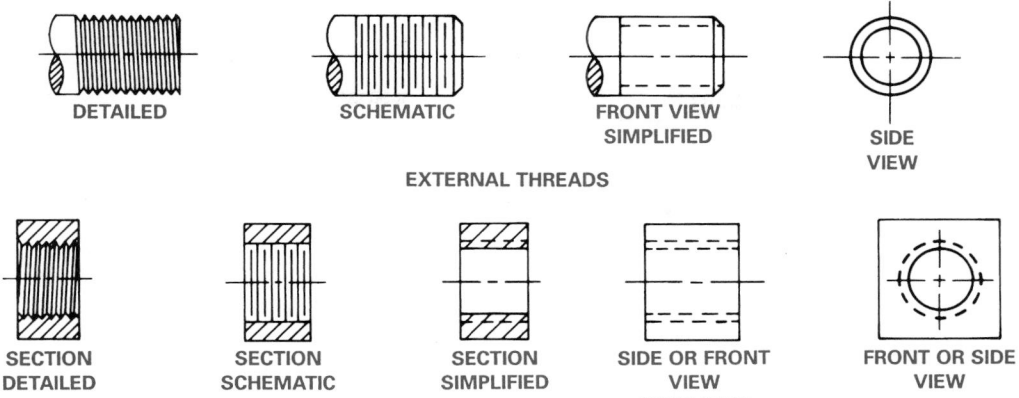

FIGURE 13.20 ■ Thread representations.

THREAD REPRESENTATIONS

There are three methods of thread representation in use: detailed, schematic, and simplified, as shown in Figure 13.20. The *detailed* representation may be used in special situations that require a pictorial display of threads such as in a sales catalog or a display drawing. Detailed thread representations are not commonly used on most manufacturing drawings since they are much too time-consuming to draw. *Schematic* representations are also not commonly used in industry. Although they do not take the time of detailed symbols, they do require extra time to draw. Some companies, however, may continue to use the schematic thread representation.

The actual use and purpose of the drawing helps determine which thread symbol to use. It may even be possible to mix representations on a particular drawing if clarity is improved. The simplified representation is the most common method of drawing thread symbols. *Simplified* representations clearly describe threads and they are easy and quick to draw. They are also very versatile as they can be used in all situations, while the other representations cannot be used in all situations. Figure 13.21 shows simplified threads in different applications.

Also, notice how the use of a thread chamfer slightly changes the appearance of the thread. Chamfers are commonly applied to the first thread to help start a thread in its mating part.

When an internal screw thread does not go through the part, it is common to drill deeper than the depth of the required thread when possible. This process saves time and reduces the chance of breaking a tap. The thread may go to the bottom of a hole, but to produce it requires an extra process using a bottoming tap. Figure 13.22 shows a simplified representation of a thread that does not go through. The bolt should be shorter than the depth of thread so the bolt does not hit bottom. Notice in Figure 13.22 the hidden lines representing the major and minor thread diameters are spaced far enough apart to be clearly separate. This spacing is important because on some threads the difference between the major and minor diameters is very small and, if drawn as they actually appear,

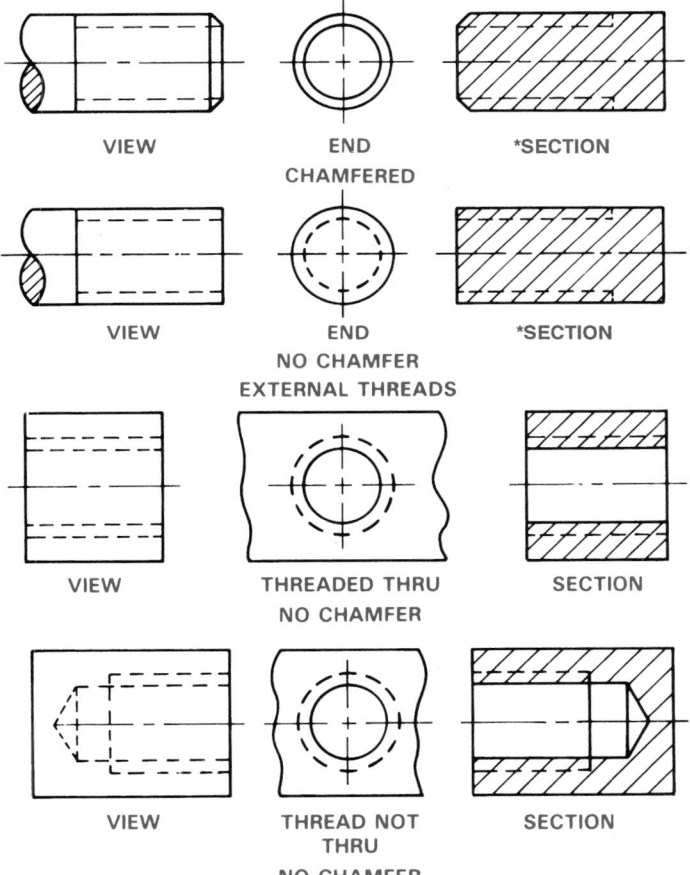

FIGURE 13.21 ■ Simplified thread representations. *Threaded shafts are not sectioned unless there is a need to expose an internal feature.

the lines would run together. The hidden line dashes are also drawn staggered for clarity. Figure 13.23 shows a bolt fastener as it would appear drawn in assembly with two parts using simplified thread representation.

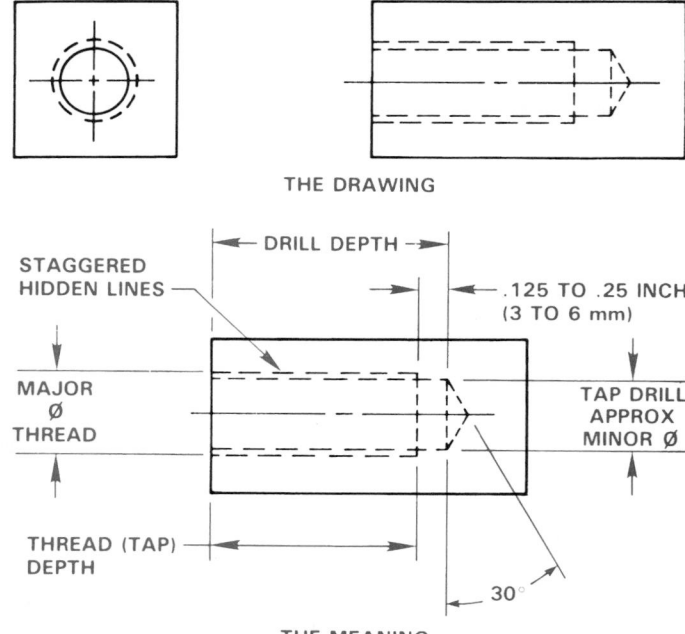

FIGURE 13.22 ■ The simplified internal thread that does not go through the part.

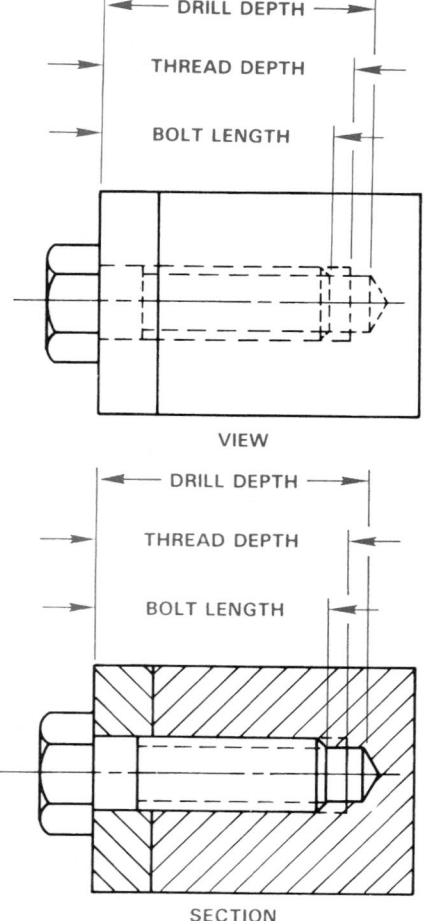

FIGURE 13.23 ■ The simplified external thread in assembly.

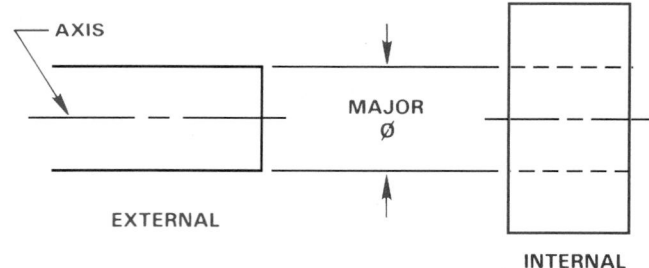

FIGURE 13.24 ■ Step 1, the simplified major thread diameter.

Drawing Simplified Threads

Simplified representations are the easiest thread symbols to draw and are the most commonly used in industry. The following steps show how to draw simplified threads.

Step 1. Draw the major thread diameter as an object line for external threads and hidden line for internal threads as in Figure 13.24.

Step 2. Draw the minor thread diameter, which is about equal to the tap drill size (found in a tap drill chart). If the minor diameter and major diameter are too close together, then slightly exaggerate the space. The minor diameter is a hidden line for the external thread and a hidden line staggered with the major diameter lines for the internal thread. The minor diameter is an object line for the internal thread in section. (See Figure 13.25.)

The simplified thread representation can be drawn with the same steps when CADD is used. For example, when using AutoCAD, the object lines are drawn with a continuous line type and the hidden lines are drawn with a hidden line type. The line weights can be set to display the proper line contrast between object lines and hidden lines. The lines can also be put on their own layers as needed. Simplified thread symbols can also be inserted as blocks in AutoCAD and scaled as needed upon insertion.

Drawing Schematic Threads

Schematic thread representations are drawn to approximate the appearance of threads by spacing lines equal to the pitch

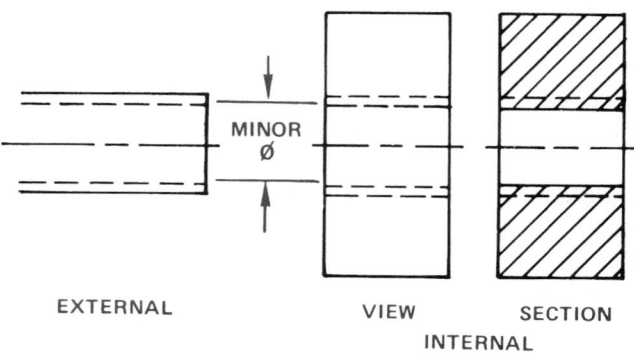

FIGURE 13.25 ■ Step 2, the simplified minor diameter.

of the thread. When there are too many threads per inch to draw easily, then the distance can be exaggerated for clarity. The following steps show how to draw schematic thread representations.

Step 1. Draw the major diameter of the thread. Schematic symbols can only be drawn in section for internal threads. Then lay out the number of threads per inch (if convenient and space is available) using a thin line at each space. Figure 13.26 uses eight threads per inch.

Step 2. Draw a thick line equal in length to the minor diameter between each pair of thin lines drawn in step 1. (See Figure 13.27.)

The schematic thread representation can be drawn with the same steps when using CADD. For example, when using AutoCAD, all of the lines are drawn with a continuous line type. The line weights can be set to display the proper line contrast. The shorter lines representing the minor diameter are drawn the same thickness as the outside object lines, and the longer lines representing the major diameter are drawn as thin lines that are the same thickness as other thin lines in your drawing. The lines can be put on their own layers as needed. The lines representing the schematic threads can also be copied as needed with a command such as ARRAY that makes multiple copies of an object. Schematic thread symbols can be inserted as blocks in AutoCAD and scaled as needed upon insertion.

Drawing Detailed Threads

Detailed thread representations are the most difficult and time-consuming thread symbols to draw. They may be necessary for some applications as they most closely approximate the actual thread. Detailed external and sectioned threads may be drawn but detailed internal threads may not be drawn in multiview. Detailed internal threads may be drawn only in section. The following steps show how to draw detailed thread representations for external and internal threads in section.

Step 1. Use construction lines to lightly draw the major and minor diameters of the thread as shown in Figure 13.28.

Step 2. Divide one edge of the thread into equal parts; in this case, eight threads per inch so the pitch is .125 in., as shown in Figure 13.29. If the pitch is too small (much less than .125 in.), then exaggerate the distance. Remember, these symbols are representations and while they should have an appearance close to the actual thread, they do not have to be exact if the result is too difficult and time-consuming to draw.

Step 3. Stagger the opposite side one-half pitch and draw parallel thin lines equal to the spaces established in step 2. (See Figure 13.30.)

Step 4. Draw Vs at 60° to form the root and crest of each thread. (See Figure 13.31.)

Step 5. Complete the detailed thread representation by connecting the roots of opposite threads by drawing parallel lines as shown in Figure 13.32. Accuracy is very important if detailed threads are to turn out satisfactorily.

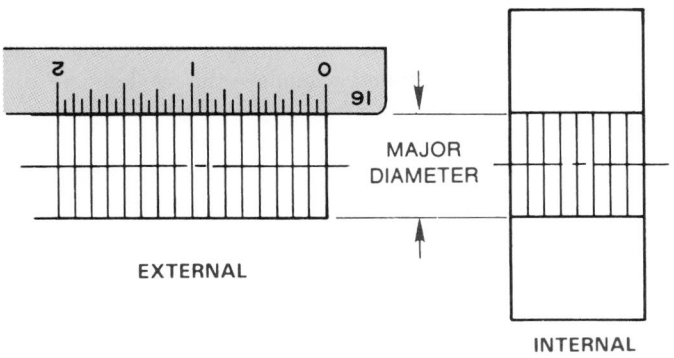

FIGURE 13.26 ■ Step 1, the schematic thread.

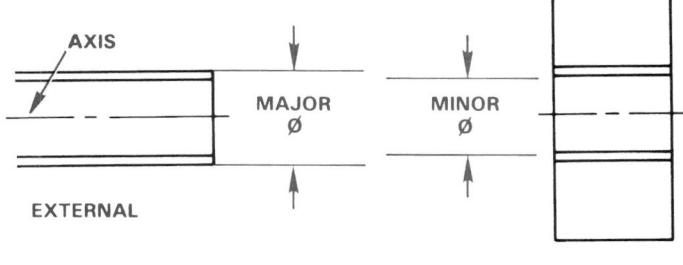

FIGURE 13.28 ■ Step 1, the detailed thread.

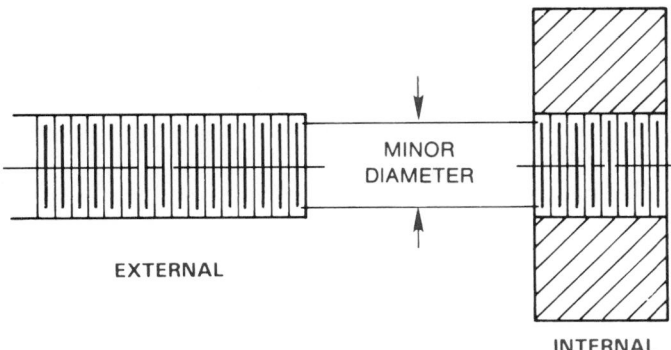

FIGURE 13.27 ■ Step 2, the schematic thread.

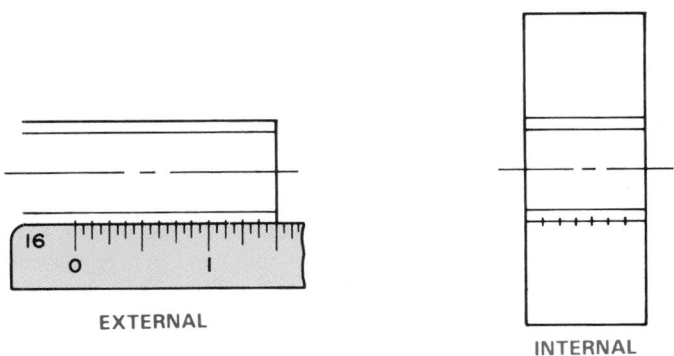

FIGURE 13.29 ■ Step 2, the detailed thread.

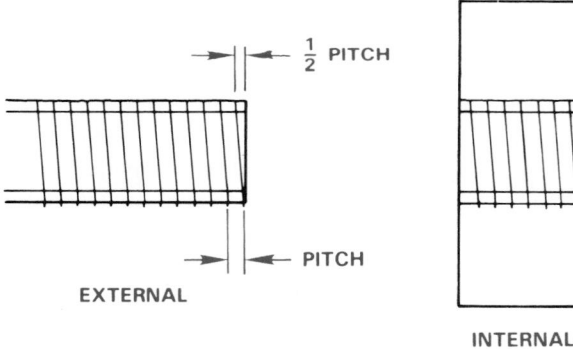

FIGURE 13.30 ■ Step 3, the detailed thread.

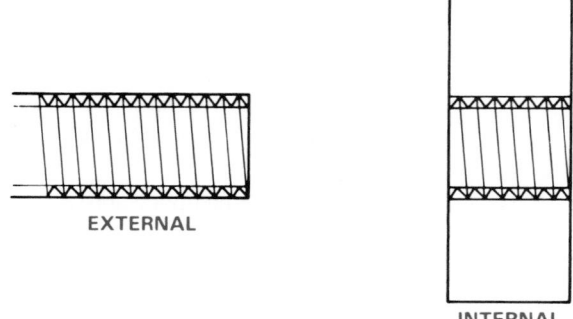

FIGURE 13.31 ■ Step 4, the detailed thread.

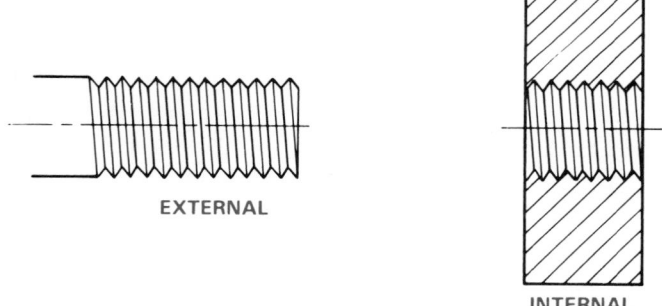

FIGURE 13.32 ■ Step 5, the complete detailed thread representation.

The detailed thread representation can be drawn with the same steps when CADD is used. For example, when using AutoCAD, the object lines are drawn with a continuous line type. You can draw one V thread form and then use the ARRAY command to conveniently copy the desired number of threads onto one side of the fastener. Then use the MIRROR command to copy the threads to the other side, and the MOVE command to move them over one-half pitch. Then draw one set of major diameter and minor diameter lines followed by using the ARRAY command to duplicate them across the entire thread. Use the TRIM command as needed to finish the thread detail. The thread representation can be put on its own layer as needed. Detailed thread symbols can also be created as a block and then scaled as needed upon insertion.

Detailed thread representations may be used to draw any thread form using the same steps previously shown. The difference occurs

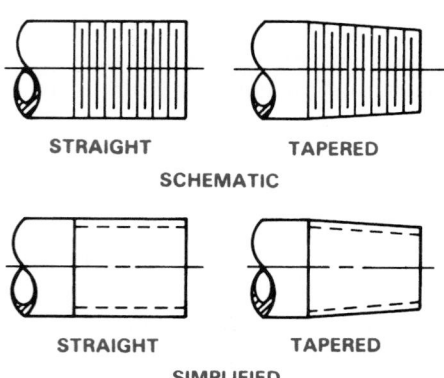

FIGURE 13.33 ■ Straight and tapered pipe thread representations.

in drawing the profile of the thread. For example, Vs are used to draw Unified, sharp-V, American Standard, and metric threads. The profile changes for other threads. The detailed thread drawings for American National Standard taper pipe threads are the same as for Unified threads except that the major and minor diameters taper at a rate of .0625 in. per inch. Figure 13.33 shows that pipe threads may be drawn tapered or straight depending on company preference. The thread note clearly defines the type of thread. Additional information about pipe thread design and drafting is provided later in this chapter.

THREAD NOTES

Simplified, schematic, and detailed thread representations clearly show where threads are displayed on a drawing. However, the representations alone do not give the full information about the thread. As the term *representation* implies, the symbols are not meant to be exact but they are meant to describe the location of a thread when used. The information that clearly and completely identifies the thread being used is the *thread note*. The thread note must always be in the same order and be accurate; otherwise the thread may be manufactured incorrectly.

Metric Threads

ISO The metric thread notes shown below are the recommended standard as specified by the International Organization for Standardization (ISO). The note components are described as follows:

M 10 X 1.5—6H
(A) (B) (C)   (D) (E) (F)

(A) M is the symbol for ISO metric threads.

(B) The nominal major diameter in millimeters, followed by the symbol X, meaning *by*.

(C) The thread pitch in millimeters, followed by a dash (—).

(D) The number may be a 3, 4, 5, 6, 7, 8, or 9, which identifies the grade of tolerance from fine to coarse. The larger the number, the larger the tolerance. Grades 3 through 5 are fine, and 7 through 9 are coarse. Grade 3 is very fine and grade 9 is very coarse. Grade 6 is the most commonly used and is the medium tolerance metric thread. The grade 6 metric thread is comparable to the class 2 Unified screw thread. A letter placed after the number gives the thread tolerance class of the internal or external thread. Internal threads are designated by uppercase letters such as *G* or *H*, where *G* means a tight allowance and *H* identifies an internal thread with no allowance. The term *allowance* refers to the tightness of fit between the mating parts. External threads are defined with lowercase letters such as *e, g,* or *h*. For external threads, *e* denotes a large allowance, *g* is a tight allowance, and *h* establishes no allowance. Grades and tolerances below 5 are intended for tight fits with mating parts; those above 7 are a free class of fit intended for quick and easy assembly. When the grade and allowance are the same for both the major diameter and the pitch diameter of the metric thread, then the designation is given as shown, 6H. In some situations where precise tolerances and allowances are critical between the major and pitch diameters, separate specifications could be used, for example, 4H 5H, or 4g 5g, where the first group (4g) refers to the grade and allowance of the pitch diameter, and the second group (5g) refers to the grade and tolerance of the major diameter. A fit between a pair of threads is indicated in the same thread note by specifying the internal thread followed by the external thread specification separated by a slash, for example, 6H/6g.

(E) A blank space at (E) denotes a right-hand thread, a thread that engages when turned to the right. A right-hand thread is assumed unless an LH is lettered in this space. LH, which describes a left-hand thread, must be specified for a thread that engages when rotated to the left.

(F) The depth of internal threads or the length of external threads in millimeters is provided at the end of the note. When the thread goes through the part, this space is left blank, although some companies prefer to letter the description THRU.

Unified and American National Threads

The thread note shall always be drawn in the order shown. The components of the note are described as follows:

½—13 UNC—2A
(A) (B) (C)   (D)(E)(F)(G)   (H)

(A) The major diameter of the thread in inches followed by a dash (—). The major diameter is generally given as a fractional value.

(B) Number of threads per inch.

(C) Series of threads are classified by the number of threads per inch as applied to specific diameters and thread forms, such as coarse or fine threads. UNC (in the example) means Unified National Coarse. Others include UNF for Unified National Fine, UNEF for Unified National Extra Fine, or UNS for Unified National Special. The UNEF and UNS thread designations are for special combinations of diameter, pitch, and length of engagement. American National screw threads are identified with UN for external and inter-

nal threads, or UNR, a thread designed to improve fatigue strength of external threads only. The series designation is followed by a dash (—).

(D) Class of fit is the amount of tolerance. 1 means a large tolerance, 2 is a general-purpose moderate tolerance, and 3 is for applications requiring a close tolerance.

(E) A means an external thread (shown in the example) while B means an internal thread. (B replaces A in this location.) The A or B may be omitted if the thread is clearly external or internal, as shown on the drawing.

(F) A blank space at (F) means a right-hand thread. A right-hand thread is assumed. LH in this space identifies a left-hand thread.

(G) A blank space at (G) identifies a thread with a single lead, that is, a thread that engages one pitch when rotated 360°. If a double or triple lead is required, then the word DOUBLE or TRIPLE must be lettered here.

(H) This location is for internal thread depth or external thread length in inches. When the drawing clearly shows that the thread goes through, this space is left blank. If clarification is needed, then the word THRU may be lettered here.

Other Thread Forms

Other thread forms, such as Acme, are noted on a drawing using the same format. For example, ⅜—8 ACME—2G describes an Acme thread with a ⅜-in. major diameter, 8 threads per inch, and a general purpose (G) class 2 thread fit. For a complete analysis of threads and thread forms, refer to the *Machinery's Handbook* published by Industrial Press, Inc.

American National Standard taper pipe threads are noted in the same manner with the letters NPT (National pipe thread) used to designate the thread form. A typical note may read ¾—14 NPT. Additional information about pipe thread design is provided later in this chapter.

Thread Notes on a Drawing

The thread note is usually applied to a drawing with a leader in the view where the thread appears as a circle for internal threads as shown in Figure 13.34. External threads may be dimensioned with a leader as shown in Figure 13.35, with the thread length given as a dimension or at the end of the note. An internal thread that does not go through the part may be dimensioned as in Figure 13.36. Some companies may require the drafter to indicate the complete process required to machine a thread including noting the tap drill size, tap drill depth if not through, the thread note, and thread depth if not through. (See Figure 13.37.) A thread chamfer may also be specified in the note as shown in Figure 13.38.

Many companies require only the thread note and depth. The complete process is determined in manufacturing.

MEASURING SCREW THREADS

When measuring features from prototypes or existing parts, the screw thread size can be determined on a fastener or threaded part

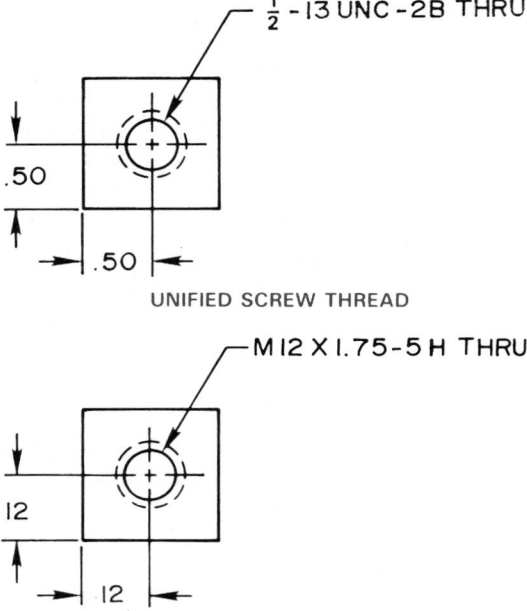

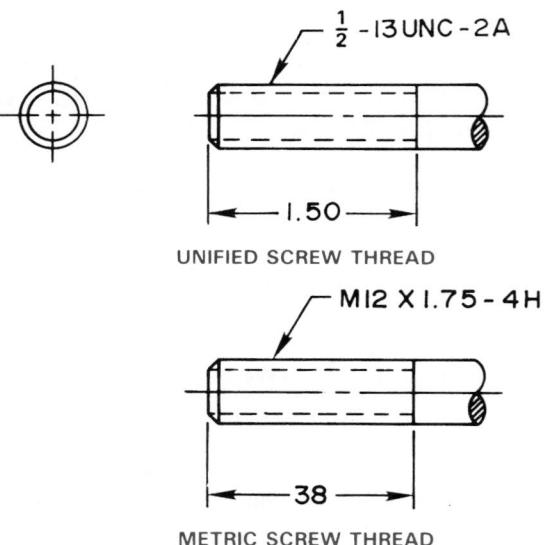

FIGURE 13.34 ■ Drawing and noting internal screw threads (simplified representation).

FIGURE 13.35 ■ Drawing and noting external screw threads.

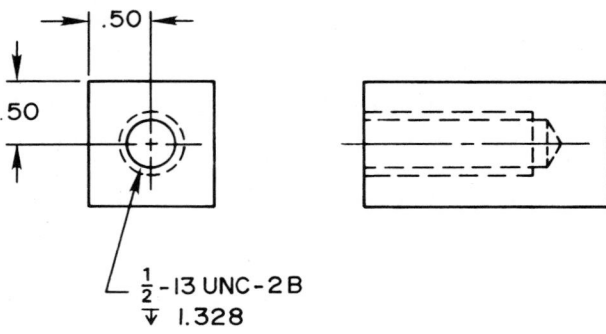

UNIFIED SCREW THREAD

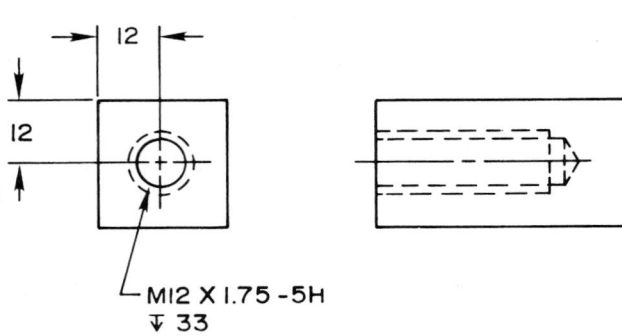

METRIC SCREW THREAD

FIGURE 13.36 ■ Drawing and noting internal screw threads with a given depth.

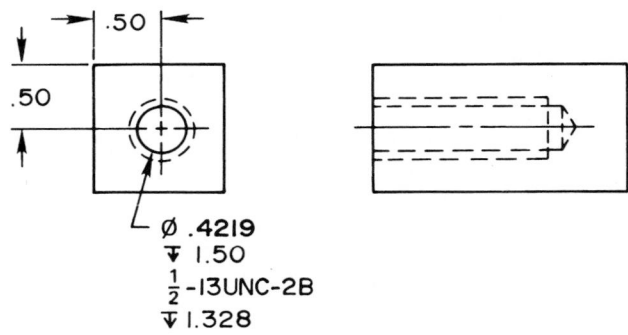

FIGURE 13.37 ■ Showing tap drill depth and thread depth.

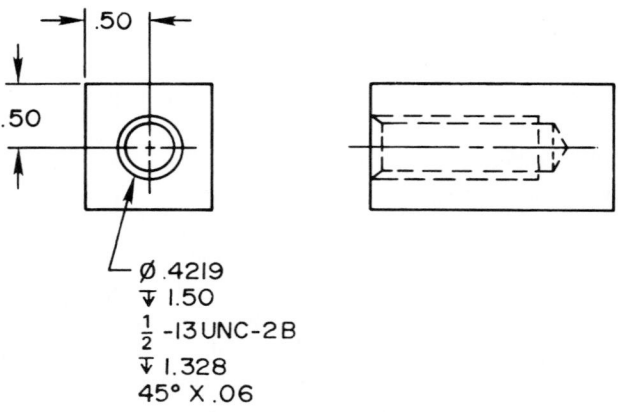

FIGURE 13.38 ■ Showing tap drill and thread depth with a chamfer.

by measurement. Measure the major diameter with a vernier caliper or micrometer. Determine the number of threads per inch when a rule or scale is the only available tool by counting the number of threads between inch graduations. The quickest and easiest way to determine the thread specification is with a screw pitch gage, which is a set of thin leaves with teeth on the edge of each leaf that correspond to standard thread sections. Each leaf is stamped to show the number of threads per inch. Therefore, if the major diameter measures .625 in. and the number of teeth per inch is 18, then by looking at a thread variation chart you find that you have a ⅝—18 UNF thread.

DRAWING FASTENERS

The use of simplified thread representations in manual drafting is mostly due to ease and time savings, while schematic thread representations are used to add a little realism to drawings. Detailed thread representations, on the other hand, are usually avoided unless a specific reason requires this time-consuming drawing method. With CADD the drafter has any type of thread representation available in an instant. With this flexibility, CADD drafters often select the detailed thread because it makes a drawing look more realistic and artistic. However, choosing simplified or schematic thread representations still makes sense from an economic standpoint, because screen regeneration and plotting time are less. CADD fastener software allows you to select the following variables:

■ Type of screw, such as machine, cap, or self-tapping.

■ Type of nut, if an internal threaded feature.

■ Type of head, if any.

■ Type of thread representation: simplified, schematic, or detailed.

■ Threading of holes or studs is also an option.

After selecting these variables, you give the thread specification. The software uses parametrics based on these variables to draw and label the exact thread required in a fraction of the time it would take to manually draw the same fastener. A sample CADD fastener overlay template is shown in Figure 13.39.

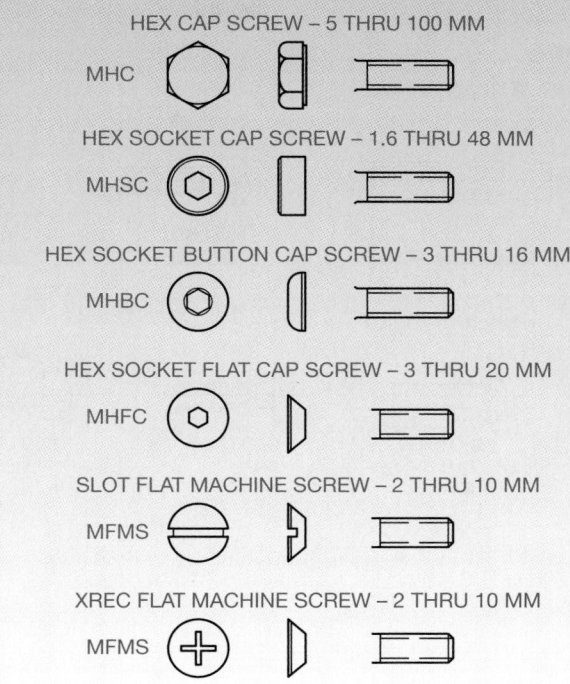

FIGURE 13.39 ■ A CADD fastener symbol library. *Courtesy CAD Tech Corporation.*

CADD APPLICATIONS

THREADED FASTENERS

Bolts and Nuts

A *bolt* is a threaded fastener with a head on one end and is designed to hold two or more parts together with a nut or threaded feature. The *nut* is tightened upon the bolt or the bolt head may be tightened into a threaded feature. Bolts can be tightened or released by torque applied to the head or to the nut. Bolts are identified by a thread note, length, and head type; for example, ⅜—11 UNC—2 × 1½ LONG HEXAGON HEAD BOLT. Figure 13.40 shows various types of bolt heads. Figure 13.41 shows common types of nuts. Nuts are classified by thread specifications and type. Nuts are available with a flat base or a washer face.

Machine Screws

Machine screws are a thread fastener used for general assembly of machine parts. Machine screws are available in coarse (UNC) and fine (UNF) threads, in diameters ranging from .060 in. to .5 in., and in lengths from ⅛ in. to 3 in. Machine screws are specified by thread, length, and head type. Machine screws have no chamfer. There are several types of heads available for machine design flexibility. (See Figure 13.42.)

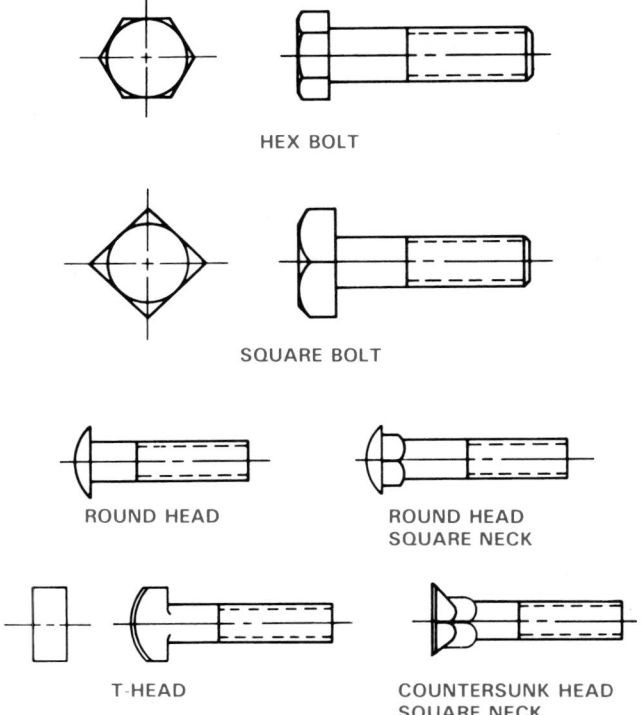

FIGURE 13.40 ■ Bolt head types.

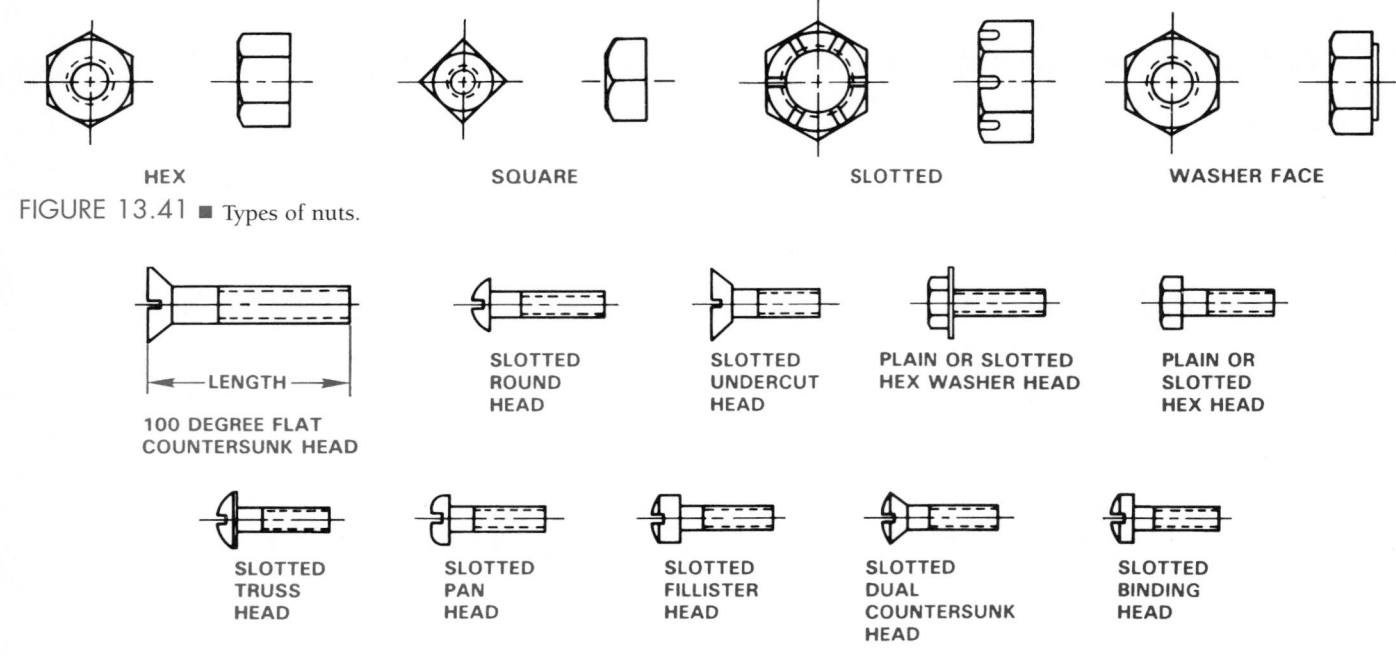

FIGURE 13.41 ■ Types of nuts.

FIGURE 13.42 ■ Types of machine screw heads.

Cap Screws

Cap screws are fine-finished machine screws that are generally used without a nut. Mating parts are fastened where one feature is threaded. Cap screws have a variety of head types and range in diameter from .060 in. to 4 in., with a large range of lengths. Lengths vary with diameter, for example, lengths increase in $\frac{1}{16}$ in. increments for diameters up to 1 in. For diameters larger than 1 in., lengths increase in increments of $\frac{1}{8}$ in. or $\frac{1}{4}$ in.; the other extreme is a 2-in. increment for lengths over 10 in. Cap screws have a chamfer to the depth of the first thread. Standard cap screw head types are shown in Figure 13.43.

Set Screws

Set screws are used to help prevent rotary motion and to transmit power between two parts such as a pulley and shaft. Purchased with or without a head, set screws are ordered by specifying thread, length, head or headless, and type of point. Headless set screws are available in slotted and with hex or spline sockets. The shape of a set screw head is usually square. Standard square-head set screws have cup points, although other points are available. Figure 13.44 shows optional types of set screw point styles.

Thread Design Guidelines*

The ASME Y14 series of standards includes the ASME Y14.6 and Y14.6aM, *Screw Thread Representation* (inch and metric series).

Courtesy of Dick Button, Standards Coordinator
Fisher Controls International, Inc.,
Marshalltown, IA.

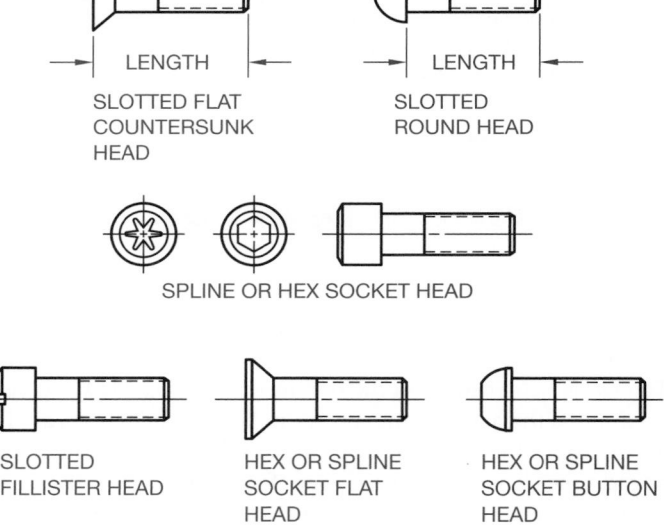

FIGURE 13.43 ■ Cap screw head styles.

While these standards provide the guidelines for representing threads on the drawing, they do not include design criteria.

The focus of this discussion is to provide design guidelines for threaded features that better accommodate manufacturing practices and tooling. Standard thread forming tools generally have lead in chamfers that produce two or more incomplete threads on the leading edge of the tool as shown in Figure 13.45. These incomplete threads are called *runout*. Designing features that do not provide enough allowance for runout or sufficient room for the tool affect tool life and reduce the probability of a good thread.

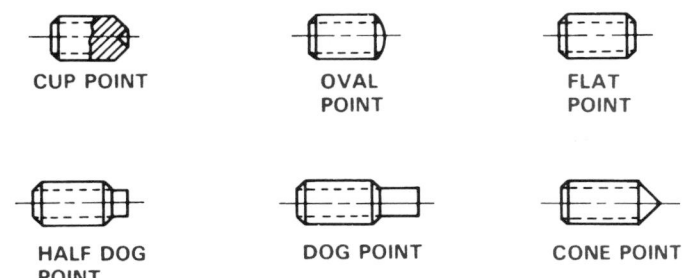

FIGURE 13.44 ■ Set screw point styles.

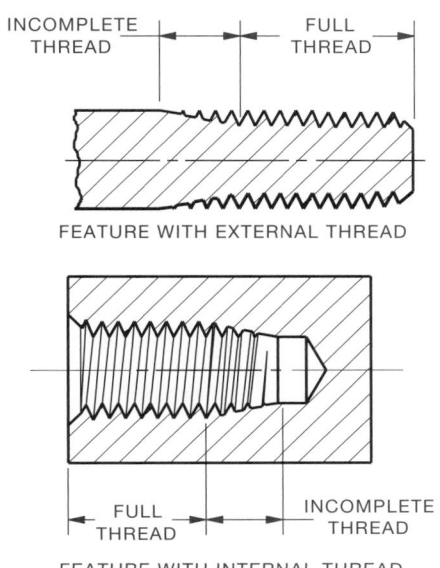

FIGURE 13.45 ■ Standard thread forming tools generally have lead in chamfers that produce two or more incomplete threads on the leading edge of the tool.

Designing Threaded Holes

Holes threaded to receive a fastener are produced as either a through hole or a blind hole. Through holes are preferred over blind holes from a manufacturing standpoint. This eliminates consideration of incomplete threads, facilitates chip disposal, and allows use of most effective production methods. Unless the thickness of the material to be threaded is considerably greater than the required thread length, or unless design requirements prevent it, the pilot hole (tap drill) for the thread should be drilled through the section. For example, if the total material thickness is equal to or less than the thread depth plus 2 times the allowance A shown in Figure 13.46, then make the feature a through hole.

A *blind hole* is a hole that does not go through. If a blind hole is required, the part design should allow the pilot hole to be considerably deeper than the required thread length to maximize machining procedures. There must be an allowance for the chamfer (tap lead) that is greater than the typical lead for the style of threading tool expected to be used. The most common method of producing internal threads is with taps. ASME/ANSI B94.9 covers designs of ten different types of taps, furnished in taper, plug, or bottoming chamfers. This was introduced earlier in this chapter.

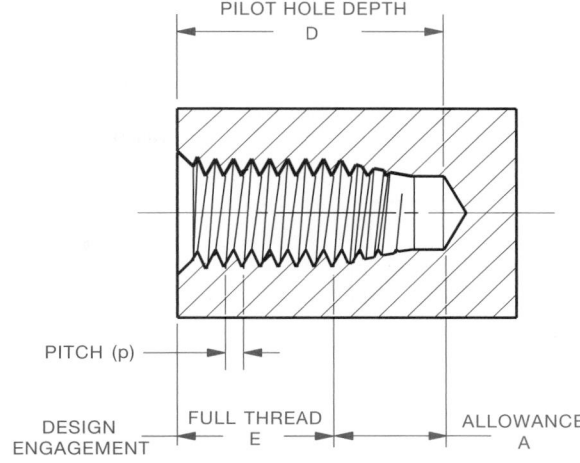

$$D = E + A$$
$$A = 5p + \text{ADDITIONAL}$$

ADDITIONAL TO BE ESTABLISHED LOCALLY BASED ON EXPERIENCE WITH PARTICULAR MANUFACTURING ENVIRONMENT. CONSIDER 1 TO 2 EXTRA PITCHES.

FIGURE 13.46 ■ Specifications for a threading for a blind hole.

Taper taps generally have lead in chamfers equal to or greater than 10 pitches. They are typically used with through holes in high-speed or high-quantity production applications.

Bottoming taps have lead in chamfers equal to 2 pitches. They are used for special applications where through holes are not acceptable and the space is limited for pilot hole depth allowance. These should be avoided if at all possible by considering a different design approach. It is difficult obtain a good start for the tapping process because there is such a short chamfer. Also, depending on the type of material being machined, bottoming taps may not produce "clean" threads.

Plug taps have lead in chamfers of 5 to 7 pitches and are the most commonly used production tool for thread forming. Typical is the cutting tap, but gaining popularity is the cold forming tap. Because cold forming does not remove material and cold work hardens some materials, this method, when done correctly, usually produces a stronger thread.

When determining pilot hole depth, this formula should be a minimum:

Pilot hole depth = Required thread depth + [5p]

(p = pitch, distance between two thread crests, as shown in Figures 13.3 and 13.4).

Additional allowance for chips from the machining is beneficial. According to ASME/ANSI Y14.6, the dimension specified on the drawing for thread depth is to be the minimum measured distance to the last full thread, which means crowding the allowance for chips, tool chamfer, and general machining tolerances could jeopardize the quality of the thread.

A thread chamfer (or countersink) in the pilot hole before tapping helps prevent burrs from the tapping process. A reasonable countersink is a 90° inclusive, 0.01 to 0.02 in. (0.2 to 0.5 mm) larger than the major diameter of the thread.

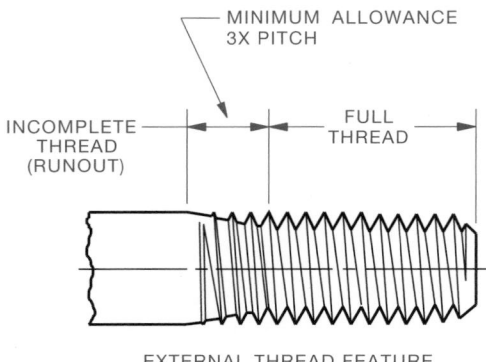

EXTERNAL THREAD FEATURE

FIGURE 13.47 ■ Tool runout is the incomplete threads that a tool can go beyond the required full thread length.

Designing External Threads

Externally threaded features are used with a variety of applications that require considerations for manufacture as well as design function. This discussion deals with the standard Unified or metric (M) profile V thread. These are used primarily for fastening and joining.

Just as with the internal thread production, there must be allowance for incomplete threads due to tool runout. Generally an allowance of three pitch lengths minimum should be used for design of tool runout. *Tool runout* is the distance a tool may go beyond the required full thread length as shown in Figure 13.47. Whenever design permits, this allowance should be increased. If design requirements do not permit the full allowance for standard tools, other possibilities must be investigated with your manufacturing source. For long-term cost-effectiveness of a product, it is better to design around standard tooling.

There are three physical possibilities to consider when designing external threaded features:

1. Apply the thread to a straight rod sized to the major diameter of the thread. This includes continuously threaded studs and rods as shown in Figure 13.48a.

2. Apply the thread to a feature diameter that is reduced in size from its adjacent feature, creating a shoulder as shown in Figure 13.48b.

3. Apply the thread to a feature diameter that is larger in size than the adjacent feature. (See Figure 13.48c.)

External threads should have a lead-in chamfer for tooling. The applied chamfer diameter should not exceed the minor diameter of the thread as indicated in ASME B1.1, *Unified Inch Screw Threads,* or ASME B1.1M, *Metric Screw Thread—M Profile.* If a relief is used to provide room for tool runout, the applied relief diameter should not exceed the minor diameter of the thread as shown in Figure 13.49. Chamfer and relief calculations and tool runout lengths for drawing application can be made available to designers and drafters by preparing a chart of allowance values to be used for commonly used thread sizes as displayed in Table 13-1. Each design and manufacturing group must determine an acceptable allowance for their environment and product. The design that is applied to large assemblies may not be applicable to small precision products.

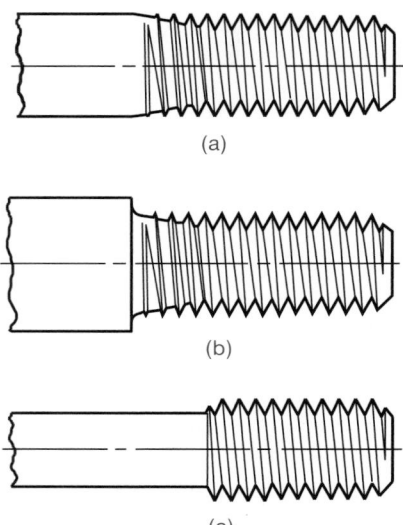

FIGURE 13.48 ■ There are three physical properties to consider when designing external threads: (**a**) Apply the thread to a straight rod sized to the major diameter of the thread. (**b**) Apply the thread to a feature diameter that is reduced in size from its adjacent feature, creating a shoulder. (**c**) Apply the thread to a feature diameter that is larger in size than the adjacent feature.

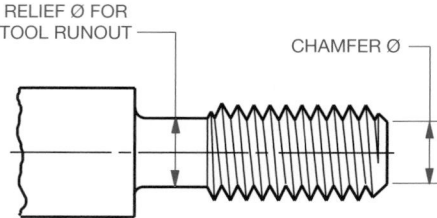

FIGURE 13.49 ■ If a relief is used to provide room for tool runout, the applied relief diameter should not exceed the minor diameter of the threads.

When designing a threaded joint with a shoulder, there must be the provision of either a counterbore on the internal threaded feature, as shown in Figure 13.50a, or a relief on the external threaded feature, as shown in Figure 13.50b, to accommodate the tool runout.

Designs using threads with greater than normal length of engagement require special designations and gaging tolerances to properly verify the length during the manufacturing process. Typical applications are an internal thread in soft material such as plastic or aluminum, where longer engagement is required to increase sheer strength, or a continuously threaded rod used in an assembly for an adjusting mechanism. In either case, binding of the assembly may occur if the special length of thread is not properly noted. For example, if a standard thread gage is used to verify a long external threaded feature, the gage typically covers only 15 pitches. However, if the thread is assembled into a feature that covers greater than 15 pitches, the gage area will not have verified the amount of thread relationship equal to that of the assembly. Nor-

TABLE 13-1 EXAMPLE OF CHART FOR RELIEF OR CHAMFER ALLOWANCE FOR INCH THREADS (Use with Figure 13.49)

Threads per Inch	For Relief Ø or Chamfer Ø, Subtract from Major Ø
20	.08
18	.08
16	.09
14	.10
13	.11
12	.12
11	.13
10	.14
8	.17

The allowances for this table were created by determining the difference between the maximum major and minor Ø, then adding .02.

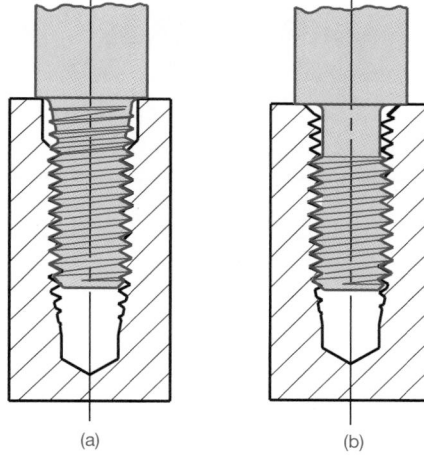

FIGURE 13.50 ■ When designing a threaded joint with a shoulder, provide (**a**) a counterbore on the internal threaded feature or (**b**) a relief on the external threaded feature.

mal length of engagement and gaging lengths are defined in ASME B1.1 and B1.13M, section 5, *Allowances and Tolerances.* Normal length of engagement and full thread length of a threaded feature are not necessarily the same.

Designing Pipe Threads

Pipe threads appear to be such a simple concept, and yet, depending on the desired function, they can consume many hours of engineering time attempting to define them adequately and manufacturing time trying to get the expected results.

There are three major functions of pipe threads:

1. An assembly that creates a pressure-tight joint, either with a sealer or without a sealer.
2. An assembly with a free/loose fitting mechanical joint that is not pressure-tight.
3. An assembly with a rigid mechanical joint that is not pressure-tight.

The assembly with a rigid mechanical joint that is not pressure-tight is used for structural purposes such as railings or racks.

The assembly with a free/loose fitting mechanical joint that is not pressure-tight can be used for fixture joints, mechanical joints with locknuts, and hose couplings that have a gasket to seal the joint.

The assembly that creates a pressure-tight joint, either with a sealer or without a sealer for pressure-tight joints, can be obtained by these two separate pipe thread types:

■ Taper pipe threads for general use, NPT.

■ Dryseal pipe threads, NPTF.

The dryseal pipe thread form is based on the NPT thread, but with some modifications and greater accuracy of manufacture, it can make a pressure-tight joint without a sealer. General use NPT threads must have a sealing compound applied to get a pressure-tight joint.

> ANSI/ASME inch series NPT has wide use throughout the world due to the U.S. influences and increase of piping products and construction. There is also the DIN (German) metric pipe thread and the ISO pipe thread that is based on the British Whitworth pipe thread. The ISO pipe thread uses a designation that is similar to the NPT designation and a few of the sizes appear to assemble with the NPT thread, so be aware of this when dealing with products or manufacturing from other countries.

Taper pipe threads, NPT, are the focus of this discussion. ANSI/ASME B1.20.1, *Pipe Threads, General Purpose (Inch),* is the standard for manufacture and gaging of NPT threads. ANSI/ASME Y14.6, *Screw Thread Representation,* is the standard for presentation and designation on drawings. Figure 13.51 shows recommended methods of representing NPT threads on detail drawings. An introduction to this was provided earlier in this chapter.

The drawing angle for representing the taper of a pipe thread should be 1½° to 2° from the axis of the thread, or a 3½° included

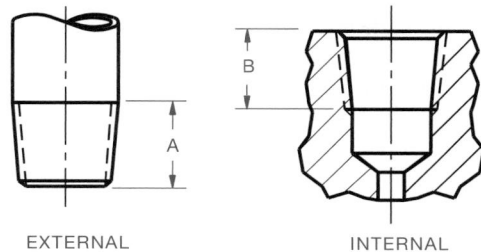

EXTERNAL   INTERNAL

FIGURE 13.51 ■ Recommended methods of representing NPT threads on detail drawings. The dimensions A and B are for reference to Table 13-2 and are not shown on the drawing.

TABLE 13-2		
NPT Size	A	B
1/16 – 27	.38	.31
1/8 – 27	.38	.31
1/4 – 18	.58	.50
3/8 – 18	.61	.50
1/2 – 14	.78	.62
3/4 – 14	.78	.62
1 – 11 1/2	.97	.75
1 1/4 – 11 1/2	1.00	.75
1 1/2 – 11 1/2	1.03	.78
2 – 11 1/2	1.06	.78

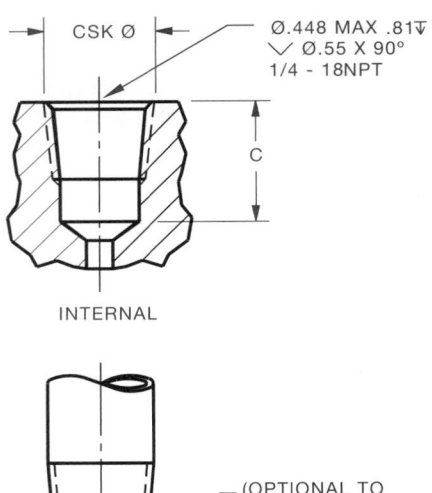

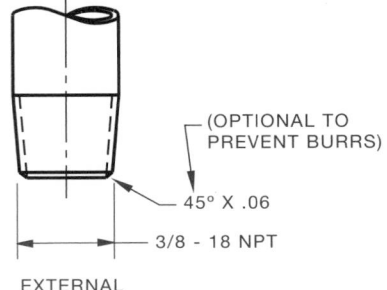

FIGURE 13.52 ■ Methods for showing NPT thread specifications on a drawing.

angle. The values A and B in Table 13-2 refer to Figure 13.51, and provide a suggested distance on the drawing to represent where the thread ends, or the runout of imperfect threads. These dimensions are for drawing representation purposes only and must not be used for manufacture. *Do not* specify a dimension on the drawing.

When defining the criteria on the drawing for internal threads, two major factors for producing a good thread are the drill diameter and the depth of the pilot hole. There are many published drill guides for pipe thread tapping and most list a single drill size. The *Machinery's Handbook* is a good reference. However, various materials require adjustments to the pilot hole size due to differing machining characteristics. A better approach is to specify the pilot hole size on the drawing as a maximum size, then let manufacturing put in a pilot hole size that suits the process. When specifying the depth of the pilot hole, provide clearance for the tap lead that permits cutting into the hole far enough to produce thread for full gaging. Most standard pipe thread taps require at least a 7 pitch lead.

Specify the requirements on the drawing according to the examples in Figure 13.52. The dimensions given in Table 13-3 provide a guideline for values that can be used in dimensioning the pipe thread. These values have been derived from calculations and experience over many years of manufacturing NPT threads on products. These figures and values should not be taken as absolute for every application and process. The production of NPT threads has always been a bit of an art form, and it takes cooperation between engineering and manufacturing at each factory to develop the best criteria.

External NPT threads applied to a feature that has a shoulder similar to the example in Figure 13.52 must have room for the tooling. The part should be designed to provide as much room as possible. Dimension D from Figure 13.53 must be greater than dimension A in Figure 13.51. See the values from Table 13-2 for dimension A.

In the process of manufacturing taper pipe threads, several considerations for producing a good joint are:

■ Pilot holes that are round.

■ Minimal waviness on produced thread, both internal and external (watch tool alignment).

■ Full crests for at least the first three threads, internal.

TABLE 13-3 MAXIMUM HOLE SIZES FOR NPT THREADS

NPT Thread		Produced Hole	Countersink Ø	Dimension C
Size	Pitch	MAX Ø	(Tol ± 0.02)	MIN
1/16	27	.253	.34	.56
1/8	27	.345	.44	.56
1/4	18	.448	.56	.81
3/8	18	.583	.69	.81
1/2	14	.721	.86	1.06
3/4	14	.931	1.06	1.06
1	11 1/2	1.168	1.34	1.25
1 1/4	11 1/2	1.513	1.69	1.31
1 1/2	11 1/2	1.752	1.91	1.31
2	11 1/2	2.226	2.38	1.31

■ L1 gaging that goes deep instead of shallow.

■ Proper installation torque. Too much torque can be damaging.

Lag Screws and Wood Screws

Lag screws are designed to attach metal to wood or wood to wood. Before assembly with a lag screw, a pilot hole is cut into the wood. The threads of the lag screw then form their own mating thread in the wood. Lag screws are sized by diameter and length. Wood screws are similar in function to lag screws and are available in a wide variety of sizes, head styles, and materials.

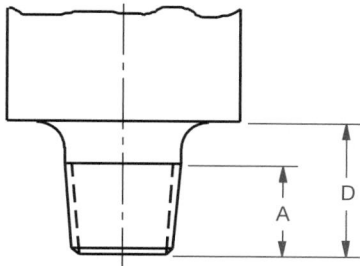

FIGURE 13.53 ■ Applying an NPT thread to a feature that has a shoulder. Dimension D must be greater than dimension A, which is found in Table 13-2. These dimensions are for reference only and are not displayed on the drawing.

Self-tapping Screws

Self-tapping screws are designed for use in situations where the mating thread is created by the fastener. These screws are used to hold two or more mating parts when one of the parts becomes a fastening device. A clearance fit is required through the first series of features or parts while the last feature receives a pilot hole similar to a tap drill for unified threads. The self-tapping screw then forms its own threads by cutting or displacing material as it enters the pilot hole. There are several different types of self-tapping screws with head variations similar to cap screws. The specific function of the screw is important as these screws may be designed for applications ranging from sheet metal to hard metal fastening.

Thread Inserts

Screw thread inserts are helically formed coils of diamond-shaped wire made of stainless steel or phosphor bronze. The inserts are used by being screwed into a threaded hole to form a mating internal thread for a threaded fastener. Inserts are used to repair worn or damaged internal threads and to provide a strong thread surface in soft materials. Some screw thread inserts are designed to provide a secure mating of fasteners in situations where vibration or movement could cause parts to loosen. Figure 13.54 shows the relationship among the fastener, thread insert, and tapped hole.

Self-clinching Fasteners

A *self-clinching fastener* is any device, usually threaded, that displaces the material around a mounting hole when pressed into a properly sized drilled or punched hole. This pressing or squeezing process causes the displaced sheet material to cold flow into a specially designed annular recess in the shank or pilot of the fastener. A serrated clinching ring, knurl, ribs, or hex head prevents the fastener from rotating in the metal when tightening torque is applied to the mating screw or nut. When properly installed, self-clinching fasteners become a permanent and integral part of the panel, chassis, bracket, printed circuit board, or other item in which they are installed. They meet high performance standards and can allow for easier disassembly of components for repair or service.

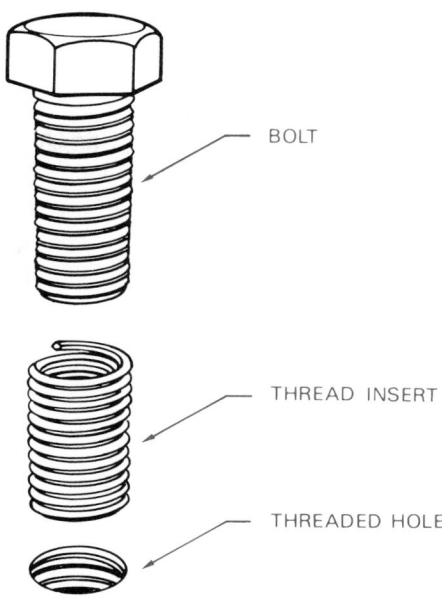

FIGURE 13.54 ■ Thread insert.

Self-clinching fasteners generally take less space and require fewer assembly operations than caged and other types of locking nuts. They also have greater reusability and more holding power than sheet metal screws. They are used mainly where good pullout and torque loads are required in sheet metal that is too thin to provide secure fastening by any other method.

Self-clinching fasteners traditionally fall into the categories of nuts, spacers and standoffs, and studs.

Self-clinching Nuts

These types feature thread strengths greater than those of mild screws and are commonly used wherever strong internal threads are needed for component attachment or fabrication assembly. During installation, a clinching ring locks the displaced metal behind the fastener's tapered shank, resulting in a high push-out resistance. High torque-out resistance is achieved when the knurled platform is embedded in the sheet metal. The clinching action of self-clinching nuts occurs on the fastener side of the thin sheet, with the reverse side remaining flush. A self-clinching nut is shown in Figure 13.55a.

Self-clinching Spacers and Standoffs

Self-clinching spacers and standoffs are used where it is necessary to space or stack components away from a panel. Thru-threaded or blind types are generally standard, but variations have been developed to meet emerging applications, primarily in the electronics industry. These types include standoffs with concealed heads, others that allow boards to snap into place for easier assembly and removal, and those designed specifically for use in printed circuit boards. Figure 13.55b shows an example of a self-clinching standoff.

PEM® SELF-CLINCHING NUTS

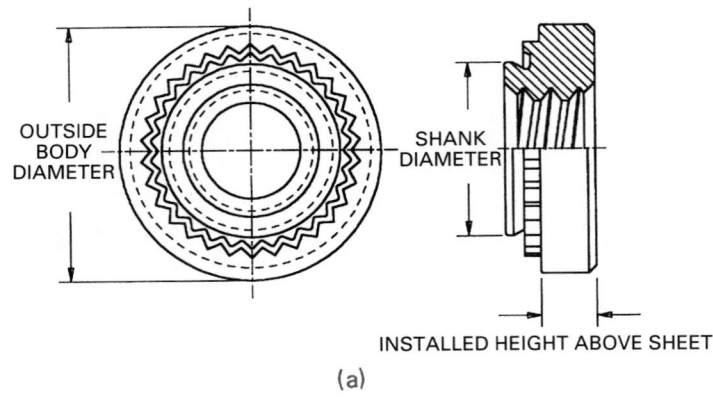

(a)

PEM® SELF-CLINCHING STANDOFFS

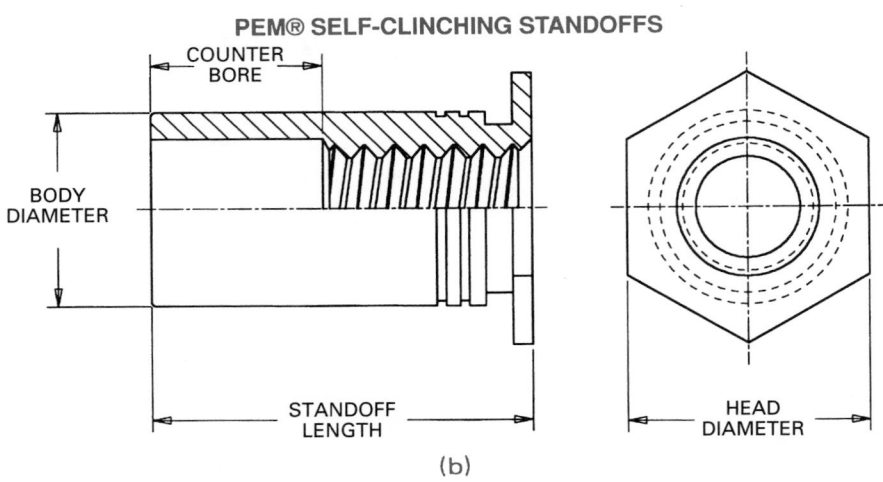

(b)

PEM® SELF-CLINCHING STUDS

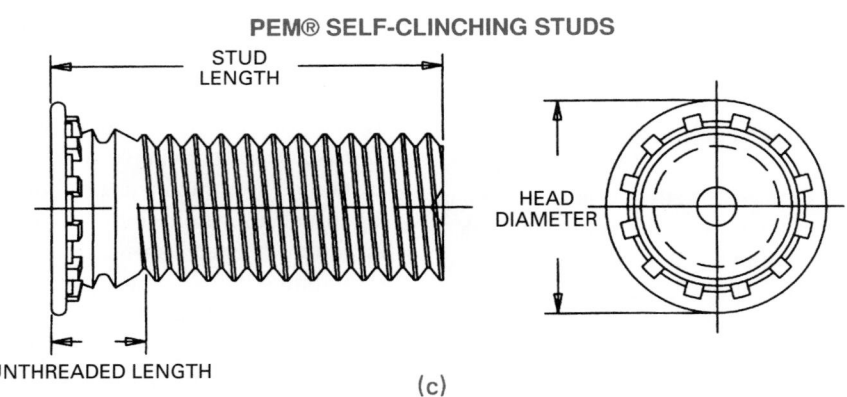

(c)

FIGURE 13.55 ■ Self-clinching fasteners: (**a**) a self-clinching nut; (**b**) a self-clinching standoff; (**c**) a self-clinching stud. *Courtesy of Penn Engineering & Manufacturing Corp. and Hammer Inc. Advertising & Public Relations.*

Self-clinching Studs

Self-clinching studs are externally threaded self-clinching fasteners that are used where the attachment must be positioned before being fastened. Flush-head studs are normally specified, but variations are available for desired high torque, thin sheet metal, or electrical applications. Figure 13.55c shows a self-clinching stud.

How to Draw Various Types of Screw Heads

As you have found from the previous discussion, screw fasteners are classified in part by head type, and there is a large variety of head styles. A valuable drafting reference is the *Machinery's Handbook*, which clearly lists specifications for all common head types.

Hexagon head fasteners are generally drawn with the hexagon positioned across the corners vertically in the front view. When projected across the flats, the hex head looks like a square head. The following steps show how to draw a hex head bolt. Hex nuts are drawn in the same manner.

Step 1. Use construction lines to draw the end view of the hexagon with the distance across the flats. A ¾-in. nominal thread measures 1⅛ in. across the flats. Position the bolt head so a front view projection is across the corners as shown in Figure 13.56.

Step 2. Block out the major diameter and length, and bolt head height. The same ¾-in. hex bolt has a head height, H, of ¹¹⁄₃₂ in. (See Figure 13.57.)

Step 3. Project the hexagon corners to the front view. Then establish radii *B*, centers with 60° angles as shown in Figure 13.58. Draw the radii.

Step 4. Draw a 30° chamfer tangent on each side to the small radius arcs. Use object lines to complete both views and draw the thread representation. (See Figure 13.59.)

Square-head bolts are drawn in the same manner as hex-head bolts.

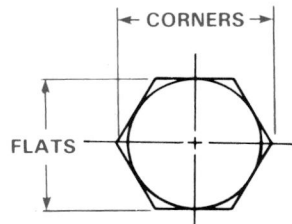

FIGURE 13.56 ■ Step 1, the flats and corners of a hexagon.

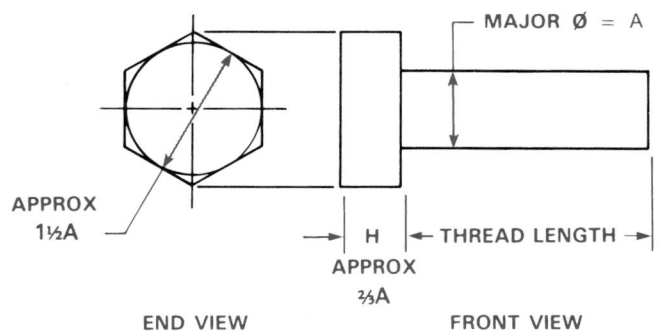

FIGURE 13.57 ■ Step 2, layout with construction lines.

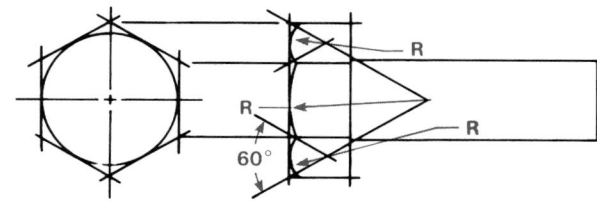

FIGURE 13.58 ■ Step 3, hex head layout.

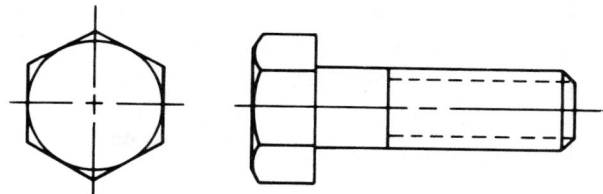

FIGURE 13.59 ■ Step 4, completed drawing.

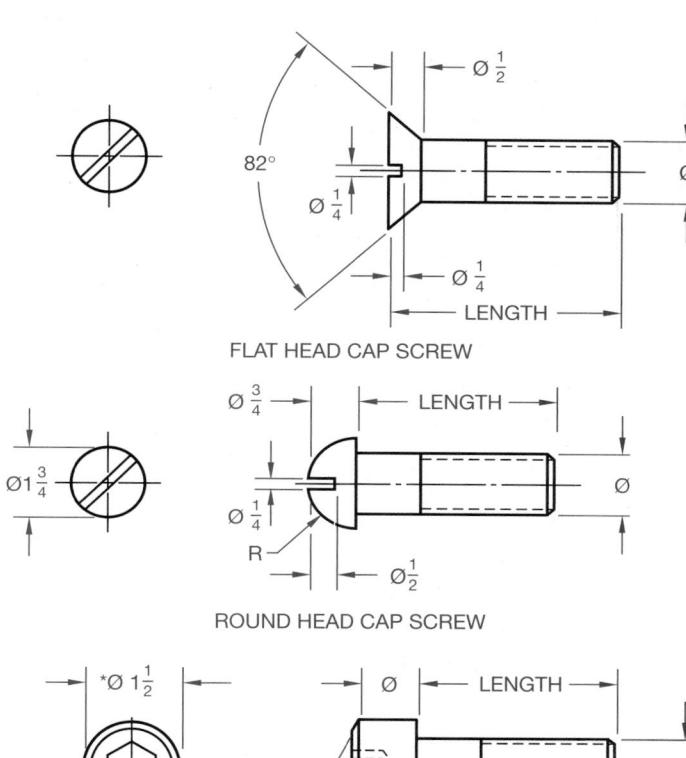

FLAT HEAD CAP SCREW

ROUND HEAD CAP SCREW

HEX SOCKET HEAD CAP SCREW

FIGURE 13.60 ■ Layout specifications for common cap screws. *For metric screw sizes, the ratio of screw diameter to head diameter is 1.5 for 12 mm and above, but for sizes 3 mm to 10 mm it is 1.5 times plus 1 mm. Consult the *Machinery's Handbook* for sizes 2 mm and smaller.

A few common cap screw heads are shown with approximate layout dimensions in Figure 13.60.

Using Templates to Draw Fasteners

When bolt heads, screw heads, or nuts are drawn only occasionally, then it may be satisfactory to use the layout techniques. However, whether you draw bolt or screw heads a few times or many times each day, it is always better to use templates. There is a large variety of templates designed to allow you to quickly and easily draw bolt or screw heads and detailed screw threads.

Using CADD to Draw Fasteners

A CADD system can be used to very quickly and accurately draw bolt heads, screw heads, and nuts. The basic commands that are commonly used include LINE, ARC, CIRCLE, and POLYGON. The fastener heads are commonly constructed from a series of lines, arcs, circles, and polygons. The most common polygon is the hexagon which creates the "hex" head bolt or screw, and the nut. The different types of fasteners can be created as blocks and then inserted and scaled into the drawing as needed. This practice saves a great deal of time. Special applications software is also available for inserting fasteners into the drawing. This was discussed earlier in this chapter.

Nuts

A *nut* is used as a fastening device in combination with a bolt to hold two or more pieces of material together. The nut thread must match the bolt thread for acceptable mating. Figure 13.61 shows the nut and bolt relationship. The hole in the parts must be drilled larger than the bolt for clearance.

There are a variety of nuts in hexagon or square shapes. Nuts are also designed slotted to allow them to be secured with a pin or key. Acorn nuts are capped for appearance. Self-locking nuts are available with neoprene gaskets that help keep the nut tight when movement or vibration is a factor. Figure 13.62 shows some common nuts.

WASHERS

Washers are flat, disk-shaped objects with a center hole to allow a fastener to pass through. Washers are made of metal, plastic, or other materials for use under a nut or bolt head, or at other machinery wear points, to serve as a cushion or a bearing surface, to prevent leakage, or to relieve friction or locking device. (See Figure 13.63.) Washer thickness varies from .016 in. to .633 in.

DOWEL PINS

Dowel pins used in machine fabrication are metal cylindrical fasteners that retain parts in a fixed position or keep parts aligned. Generally, depending on the function of the parts, one or two dowel pins are sufficient for holding adjacent parts. Dowel pins must generally be pressed into a hole with an interference tolerance of between .0002 in. to .001 in. depending on the material and the function of the parts. Figure 13.64 shows the section of two adjacent parts and a dowel pin. Figure 13.65 is a chart drawing of some standard dowel pins.

TAPER AND OTHER PINS

For applications that require perfect alignment of accurately constructed parts, *tapered dowel pins* may be better than straight dowel pins. Taper pins are also used for parts that have to be taken apart frequently or where removal of straight dowel pins may cause excess hole wear. Figure 13.66 shows an example of a taper pin assembly. Taper pins, shown in Figure 13.67, range in diameter, *D*, from 7/0, which is .0625 in., to .875 in., and lengths, *L*, vary from .375 in. to 8 in.

Other types of pins serve functions similar to taper pins, such as holding parts together, aligning parts, locking parts, and transmitting power from one feature to another. Other common pins are shown in Figure 13.68.

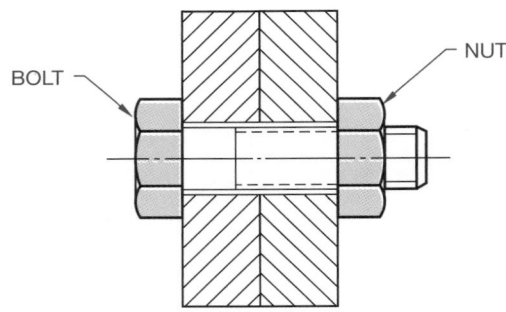

FIGURE 13.61 ■ Nut and bold relationship known as a floating fastener.

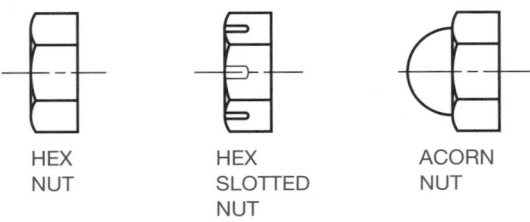

FIGURE 13.62 ■ Common nuts.

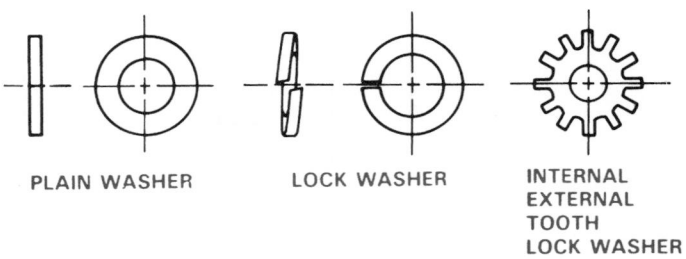

FIGURE 13.63 ■ Types of washers.

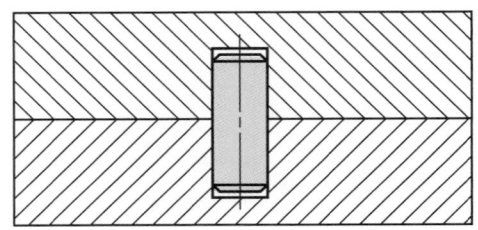

FIGURE 13.64 ■ Dowel pin in place, sectional view.

DOWEL PINS
HARDENED-PRESS FIT

1/16 TO 3/8 DIAMETERS
3/16 TO 1-1/2 LENGTHS

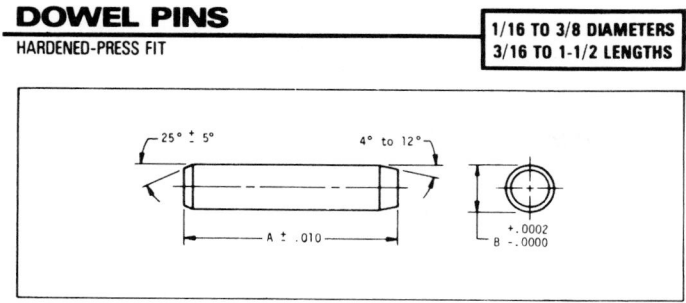

Material: 416 Stainless Steel (Clear Passivate) Hardened to: Rockwell C 36-40
Conforms to specification MS-16555
For 303 stainless pins, please see Cat. No. EPS-A1 and EPS-B1

	1/16 DIA. PIN B = .0626		3/32 DIA. PIN B = .0938		1/8 DIA. PIN B = .1251		5/32 DIA. PIN B = .1563	
A	Cat. No.	Price	Cat. No.	Price	Cat. No.	Price	Cat. No.	Price
3/16	EPS-D1-1	$.09	*EPS-D2-1	$.14	-		-	
1/4	EPS-D1-2	.09	*EPS-D2-2	.15	*EPS-D3-2	$.14	-	
5/16	EPS-D1-3	.10	EPS-D2-3	.11	*EPS-D3-3	.15	-	
3/8	EPS-D1-4	.10	EPS-D2-4	.11	EPS-D3-4	.13	*EPS-D4-4	$.17
7/16	EPS-D1-5	.10	EPS-D2-5	.11	EPS-D3-5	.13	*EPS-D4-5	.17
1/2	EPS-D1-6	.11	EPS-D2-6	.11	EPS-D3-6	.13	EPS-D4-6	.17

FIGURE 13.65 ■ Dowel pin chart drawing. *Courtesy NORDEX, Inc.*

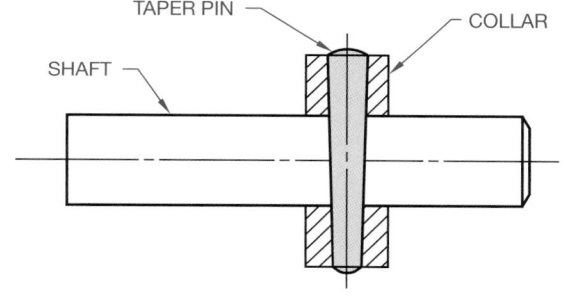

FIGURE 13.66 ■ Taper pin in assembly, sectional view.

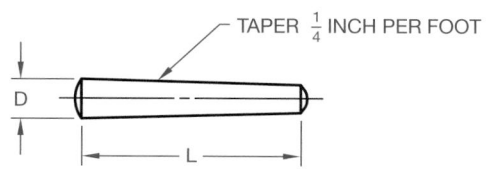

FIGURE 13.67 ■ Taper pin.

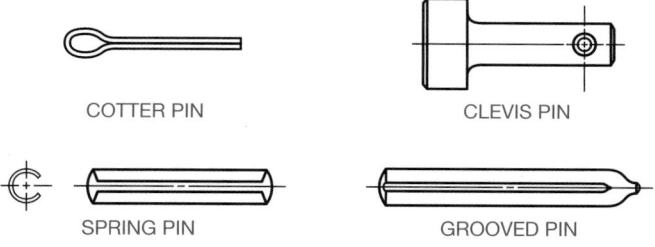

FIGURE 13.68 ■ Other common pins.

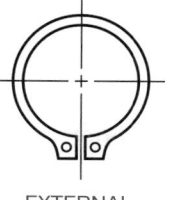
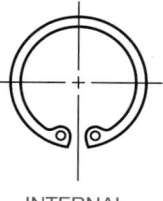

FIGURE 13.69 ■ Retaining rings.

EXTERNAL   INTERNAL

RETAINING RINGS

Internal and external retaining rings are available as fasteners to provide a stop or shoulder for holding bearings or other parts on a shaft. They are also used internally to hold a cylindrical feature in a housing. Common retaining rings require a groove in the shaft or housing for mounting with a special plier tool. Also available are self-locking retaining rings for certain applications. (See Figure 13.69.)

KEYS, KEYWAYS, AND KEYSEATS

Standards for keys were established to control the relationship among key sizes, shaft sizes, and tolerances for key applications. A *key* is an important machine element, which is employed to provide a positive connection for transmitting torque between a shaft and hub, pulley, or wheels. The key is placed in position in a *keyseat*, which is a groove or channel cut in a shaft. The shaft and key are then inserted into the hub, wheel, or pulley, where the key mates with a groove called a *keyway*. Figure 13.70 shows the relationship among the key, keyseat, keyway, shaft, and hub.

Standard key sizes are determined by shaft diameter. For example, shaft diameters ranging from ⅞ in. to 1¼ in. would require a ¼-in. nominal width key. Keyseat depth dimensions are established in relationship to the shaft diameter. Figure 13.71 shows the standard dimensions for the related features. For a shaft with a ¾-in. diameter, the recommended shaft dimension, *S,* would be .676 in. and the hub dimension, *T,* would be .806 in. when using a rectangular key. Types of keys are shown in Figure 13.72.

RIVETS

A *rivet* is a metal pin with a head used to fasten two or more materials together. The rivet is placed through holes in mating parts and the end without a head extends through the parts to be headed-over (formed into a head) by hammering, pressing, or forging. The end with the head is held in place with a solid steel bar known as a dolly while a head is formed on the other end. Rivets are classified by body diameter, length, and head type. (See Figure 13.73.)

SPRINGS

A *spring* is a mechanical device, often in the form of a helical coil, that yields by expansion or contraction due to pressure, force, or stress applied. Springs are made to return to their normal form when the force or stress is removed. Springs are designed to store energy for the purpose of pushing or pulling machine parts by

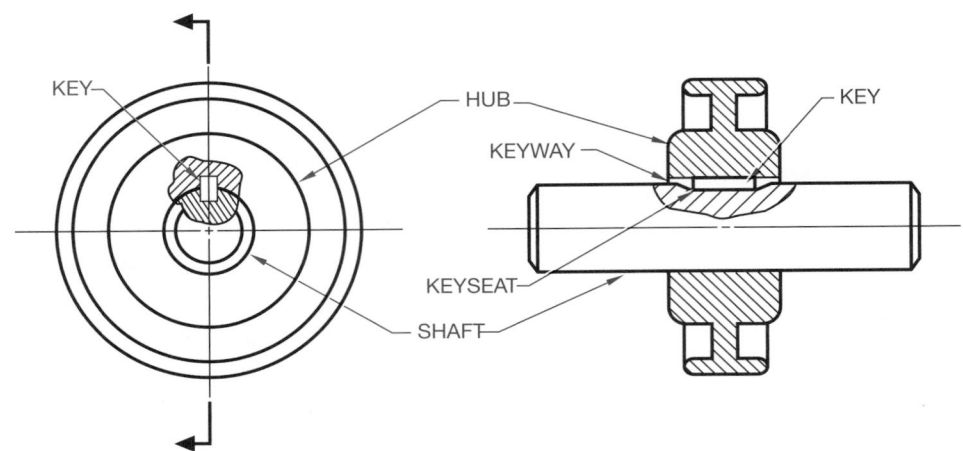

FIGURE 13.70 ■ Relationship between the key, keyseat, keyway, shaft, and hub.

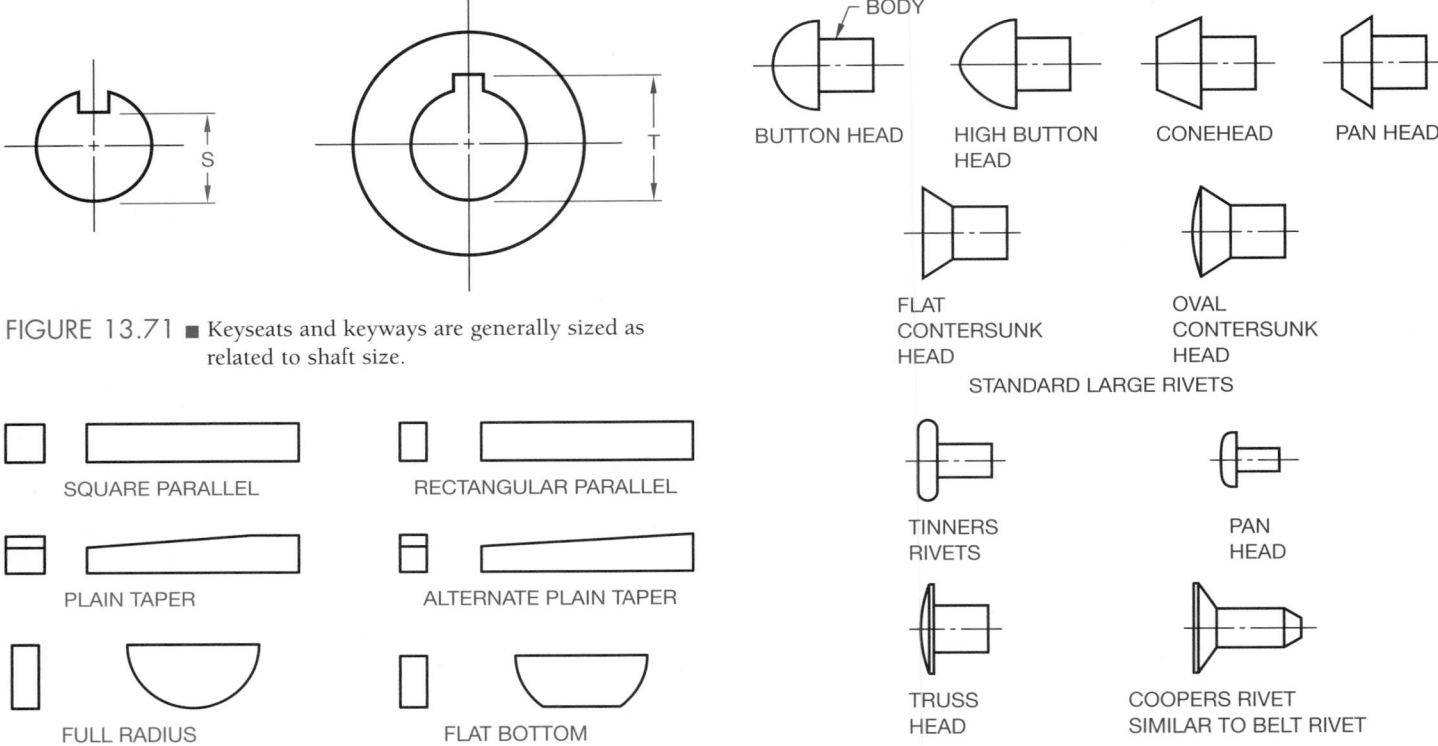

FIGURE 13.71 ■ Keyseats and keyways are generally sized as related to shaft size.

SQUARE PARALLEL

RECTANGULAR PARALLEL

PLAIN TAPER

ALTERNATE PLAIN TAPER

FULL RADIUS WOODRUFF

FLAT BOTTOM WOODRUFF

FIGURE 13.72 ■ Types of keys.

BODY

BUTTON HEAD

HIGH BUTTON HEAD

CONEHEAD

PAN HEAD

FLAT CONTERSUNK HEAD

OVAL CONTERSUNK HEAD

STANDARD LARGE RIVETS

TINNERS RIVETS

PAN HEAD

TRUSS HEAD

COOPERS RIVET SIMILAR TO BELT RIVET

STANDARD SMALL RIVETS

FIGURE 13.73 ■ Common rivets.

reflex action into certain desired positions. Improved spring technology provides springs with the ability to function for a long time under high stresses. The effective use of springs in machine design depends on five basic criteria including material, application, functional stresses, use, and tolerances.

Continued research and development of spring materials have helped to improve spring technology. The spring materials most commonly used include high-carbon spring steels, alloy spring steels, stainless spring steels, music wire, oil-tempered steel, copper-based alloys, and nickel-based alloys. Spring materials, depending on use, may have to withstand high operating temperatures and high stresses under repeated loading.

Spring design criteria are generally based on material gage, kind of material, spring index, direction of the helix, type of ends, and function. Spring wire gages are available from several different sources ranging in diameter from number 7/0 (.490 in.) to number 80 (.013 in.). The most commonly used spring gages range from 4/0 to 40. There are a variety of spring materials available in round or square stock for use depending on spring function and design stresses.

The spring index is a ratio of the average coil diameter to the wire diameter. The index is a factor in determining spring stress, deflection, and the evaluation of the number of coils needed and the spring diameter. Recommended index ratios range between 7 and 9 although other ratios commonly used range from 4 to 16.

COMPRESSION
SPRING

EXTENSION
SPRING

FIGURE 13.74 ■ Compression and extension spring.

The direction of the helix is a design factor when springs must operate in conjunction with threads or with one spring inside of another. In such situations the helix of one feature should be in the opposite direction of the helix for the other feature.

Compression springs are available with ground or unground ends. Unground, or rough, ends are less expensive than ground ends. If the spring is required to rest flat on its end, then ground ends should be used. Spring function depends on one of two basic factors, compression or extension. Compression springs release their energy and return to their normal form when compressed. Extension springs release their energy and return to the normal form when extended. (See Figure 13.74.)

Spring Terminology

The springs shown in Figure 13.75 show common characteristics.

Ends

Compression springs have four general types of ends: open or closed ground ends and open or closed unground ends, shown in Figure 13.76. Extension springs have a large variety of optional ends, a few of which are shown in Figure 13.77.

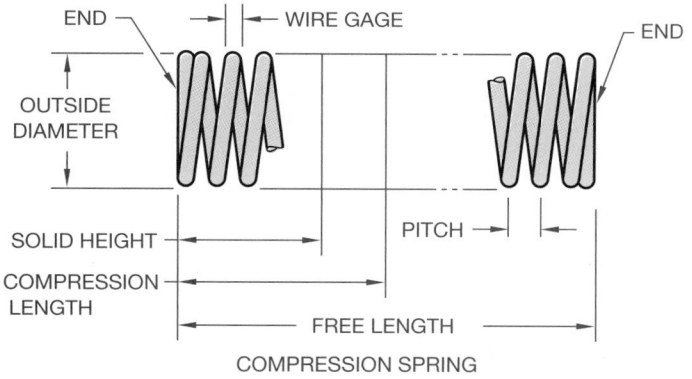

COMPRESSION SPRING

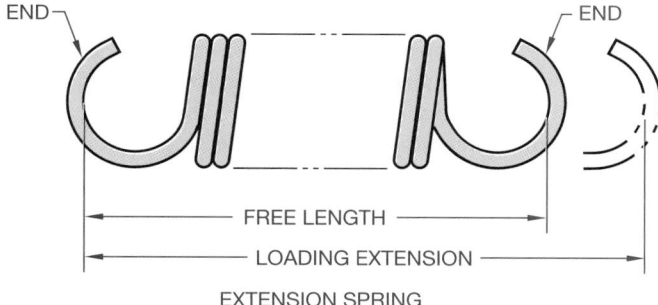

EXTENSION SPRING

FIGURE 13.75 ■ Spring characteristics.

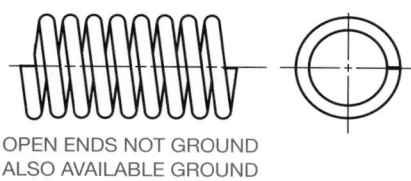

OPEN ENDS NOT GROUND
ALSO AVAILABLE GROUND
RIGHT-HAND HELIX

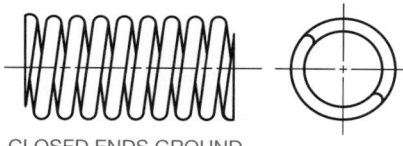

CLOSED ENDS GROUND
ALSO AVAILABLE NOT GROUND
LEFT-HAND HELIX

FIGURE 13.76 ■ Helix direction and compression spring end types.

IN LINE MACHINE LOOP
AND HOOK ALSO AVAILABLE
AT RIGHT ANGLES

FULL LOOP ON SIDE WITH
SMALL EYE ON CENTER
ALSO AVAILABLE WITH
FULL LOOP CENTERED

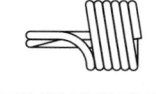

DOUBLE TWISTED
FULL LOOP

SMALL
OFFSET HOOK

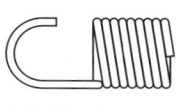

LONG ROUNDED
END

CONED END WITH
SHORT SWIVEL EYE

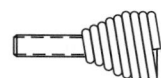

CONED END WITH
SWIVEL BOLT

MANY OTHER COMBINATIONS ARE AVAILABLE

FIGURE 13.77 ■ Extension spring end types.

Helix Direction

The helix direction may be specified as right-hand or left-hand. (See Figure 13.76.)

Free Length

The length of the spring when there is no pressure or stress to affect compression or extension.

Compression Length

The compression length is the maximum recommended design length for the spring when compressed.

Solid Height

The solid height is the maximum compression possible. The design function of the spring should not allow the spring to reach solid height when in operation unless this factor is a function of the machinery.

Loading Extension

The extended distance to which an extension spring is designed to operate.

Pitch

The pitch is one complete helical revolution, or the distance from a point on one coil to the same corresponding point on the next coil.

Torsion Springs

Torsion springs are designed to transmit energy by a turning or twisting action. *Torsion* is defined as a twisting action that tends to turn one part or end around a longitudinal axis while the other part or end remains fixed. Torsion springs are often designed as antibacklash devices or as self-closing or self-reversing units. (See Figure 13.78.)

Flat Springs

Flat springs are arched or bent flat-metal shapes designed so when placed in machinery they cause tension on adjacent parts. The tension may be used to level parts, provide a cushion, or position the relative movement of one part to another. One of the most common examples of flat springs is leaf springs on an automobile.

Spring Representations

There are three types of spring representations, detailed, schematic, and simplified, as seen in Figure 13.79. Detailed spring drawings are used in situations that require a realistic representation, such as vendors, catalogs, assembly instructions, or detailed assemblies. Schematic spring representations are commonly used on drawings. The single-line schematic symbols are easy to draw and clearly represent springs without taking the additional time required to draw a detailed spring. The use of simplified spring drawings is limited to situations where the clear resemblance of a spring is not necessary. While very easy to draw, the simplified spring symbol must be

Material: 302 Stainless Steel (Spring Tamper) Passivated

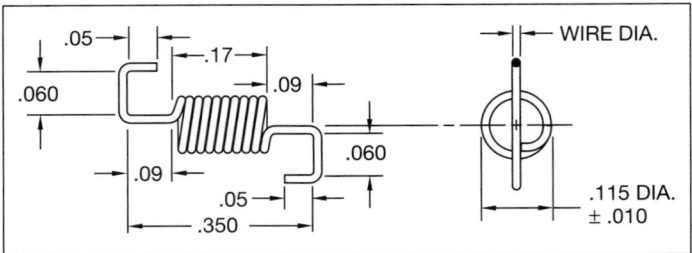

FIGURE 13.78 ■ Torsion spring, also called antibacklash spring. *Courtesy NORDEX, Inc.*

DETAILED

SCHEMATIC

SIMPLIFIED

FIGURE 13.79 ■ Spring representations.

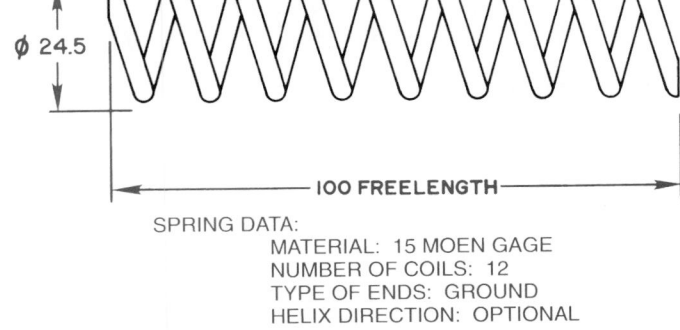

SPRING DATA:
MATERIAL: 15 MOEN GAGE
NUMBER OF COILS: 12
TYPE OF ENDS: GROUND
HELIX DIRECTION: OPTIONAL

FIGURE 13.80 ■ Detailed spring drawing with spring data chart.

accompanied by clearly written spring specifications. The simplified spring representation is not very useful in assembly drawings or other situations that require a visual comparison of features.

Spring Specifications

No matter which representation is used, several important specifications must accompany the spring symbol. Spring information is generally lettered in the form of a specific or general note.

Spring specifications include outside or inside diameter, wire gage, kind of material, type of ends, surface finish, free and compressed length, and number of coils. Other information, when required, may include spring design criteria and heat treatment specifications. The information is often provided on a drawing as shown in Figure 13.80. The material note is usually found in the title block.

Drawing Detailed Spring Representations

Detailed spring drawing requires a great degree of accuracy. First determine the outside diameter, number of coils, wire diameter, and free length or compressed length.

The following steps show how to draw a detailed spring representation with the following specifications:

Material: 2.5 mm diameter high-carbon spring steel

Outside Diameter: 16 mm

Free Length: 50 mm

Number of Coils: 6

Step 1. Using construction lines, draw a rectangle equal to the outside diameter (16 mm) wide and the free length (50 mm) long. (See Figure 13.81.)

Step 2. Along one inside edge of the length of the rectangle, lay out seven equally spaced full circles with a 2.5 mm diameter (wire size). The layout may be done by dividing the length into six equal spaces. (See Figure 13.82.) On the other inside edge, lay out a half circle at each end. Begin-

FIGURE 13.81 ■ Step 1, preliminary spring layout.

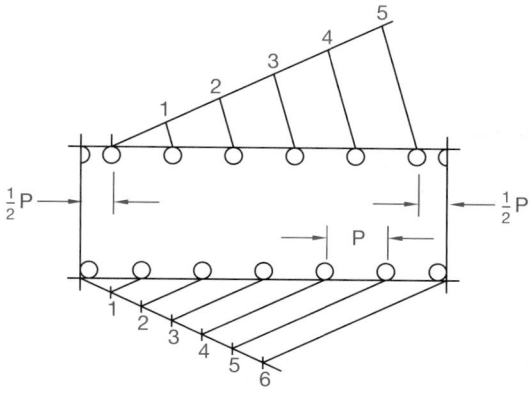

FIGURE 13.82 ■ Step 2, spacing the coils.

FIGURE 13.83 ■ Step 3, connect the coils to complete the detailed spring representation.

FIGURE 13.84 ■ Detailed spring representation in section.

ning at a distance of ½ P away from one end, draw the first of six full (2.5 mm) circles with equal spaces between them.

Step 3. Connect the circles drawn in step 2 to make the coils. Draw lines from a point of tangency on one circle to a corresponding point on a circle on the other side. Draw the last element on each side down the edge of the rectangle for ground ends. To draw unground ends, make the last element terminate at the axis of the spring. (See Figure 13.83.)

For detailed coils in a longitudinal sectional view, leave the circles from step 2 and draw that part of the spring that appears as if the front half were removed as shown in Figure 13.84. Then fill in or section-line the circles.

Drawing Schematic Spring Representations

Schematic spring symbols are much easier to draw than detailed representations while clearly resembling a spring. The following steps show how to draw a schematic spring symbol for the same spring previously drawn.

Step 1. Use construction lines to draw a rectangle equal in size to the outside diameter by the free length as seen in Figure 13.81.

Step 2. Establish six equal spaces at P distance along one edge of the rectangle. Along the opposite edge begin and end with

DRAWING SPRINGS

Parametric capabilities are the key to drafting and design flexibility with CADD. *Parametric* means that you have the ability to change a value or values in a situation and automatically have the entire design changed to fit the circumstances. Many CADD packages have this feature available, and CADD programs for spring design are no exception. For instance, if you are designing a compression spring, all you have to do is make the selection from a menu and then enter the following information as prompted by the computer:

■ Wire diameter.

■ Number of coils.

■ Outside diameter.

■ Free length.

■ Type of ends.

When you enter this data, the computer calculates the rest of the information, such as free length and compression length, and then asks if you want an external or sectional view. After all the information is complete, the SPRING program automatically draws and dimensions the spring on the screen, followed by prompts for you to rotate or move the spring to the desired location.

a space equal to ½ P. Establish five equal spaces between the ½ P ends. (See Figure 13.85.)

Step 3. Beginning on one side, draw the elements of each spring coil as shown in Figure 13.86.

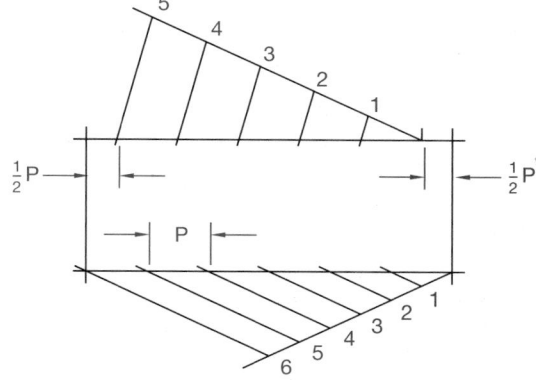

FIGURE 13.85 ■ Step 2, spacing the coils.

FIGURE 13.86 ■ Step 3, complete the schematic representation.

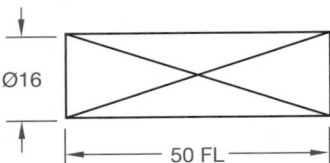

FIGURE 13.87 ■ Simplified spring representation.

Drawing a Simplified Spring Representation

Simplified spring representations are drawn by making a rectangle equal in size to the major diameter and free length and then drawing diagonal lines as shown in Figure 13.87.

RECOMMENDED REVIEW

It is recommended that you review the following parts of Chapter 2 before you begin working on fully dimensioned fastener and spring drawings. This will refresh your memory about how related lines are properly drawn.

- Object Lines.
- Viewing Planes.
- Extension Lines.
- Hidden Lines.
- Break Lines.
- Dimension Lines.
- Centerlines.
- Phantom Lines.
- Leader Lines.

Also review Chapters 7 and 8 covering multiview and auxiliary view drawings, and Chapter 12 covering dimensioning practices.

WEB SITE RESEARCH

The following Web sites can provide you with additional information for research or further study into topics covered in this chapter:

http://www.asme.org/asme/8.html—Find information and publications related to the American Society of Mechanical Engineers.

http://www.ansi.org—The American National Standards Institute. Information about national and international drafting standards.

http://www.adda.org—American Design Drafting Association, *Drafting Reference Guide.*

http://www.industrialpress.com—Information about the *Machinery's Handbook.* This is a valuable resource for all types of fasteners and fastener specifications.

http://www.cadtechcorp.com—CADD fasteners software applications.

PROFESSIONAL PERSPECTIVE

There are three basic methods for drawing screw thread and spring representations: using construction techniques, using a template, or using CADD. The manual construction techniques should be avoided in most cases because they are very time-consuming. However, in some special applications these methods may be used effectively. The manual drafter should use a template when at all possible to draw screw threads, fastener heads, and spring representations. For the CADD user, there are many software programs available that allow drawing of screw threads in simplified, schematic, or detailed representations. A complete variety of head types can be drawn in addition to springs in simplified, schematic, or detailed representations. This type of CADD software makes an otherwise complex task very simple and saves a lot of drafting time. The entry-level

drafter often spends unnecessary time trying to make a screw thread representation look exactly like the real thing. It is important to have an accurate drawing, but with regard to screw threads, the thread note tells the reader the exact thread specifications. This is why threads are shown on a drawing as a representation rather than a duplication of the real thing. The simplified representation is the most commonly used technique, because it is fast and easy to draw.

Your goal as a professional drafter is to make the drawing communicate so that there is no question about what is intended. In simple terms, make the drawing clear and accurate using the easiest method, and make sure the thread note and specifications are complete and correct as shown in the real world example found in Figure 13.88.

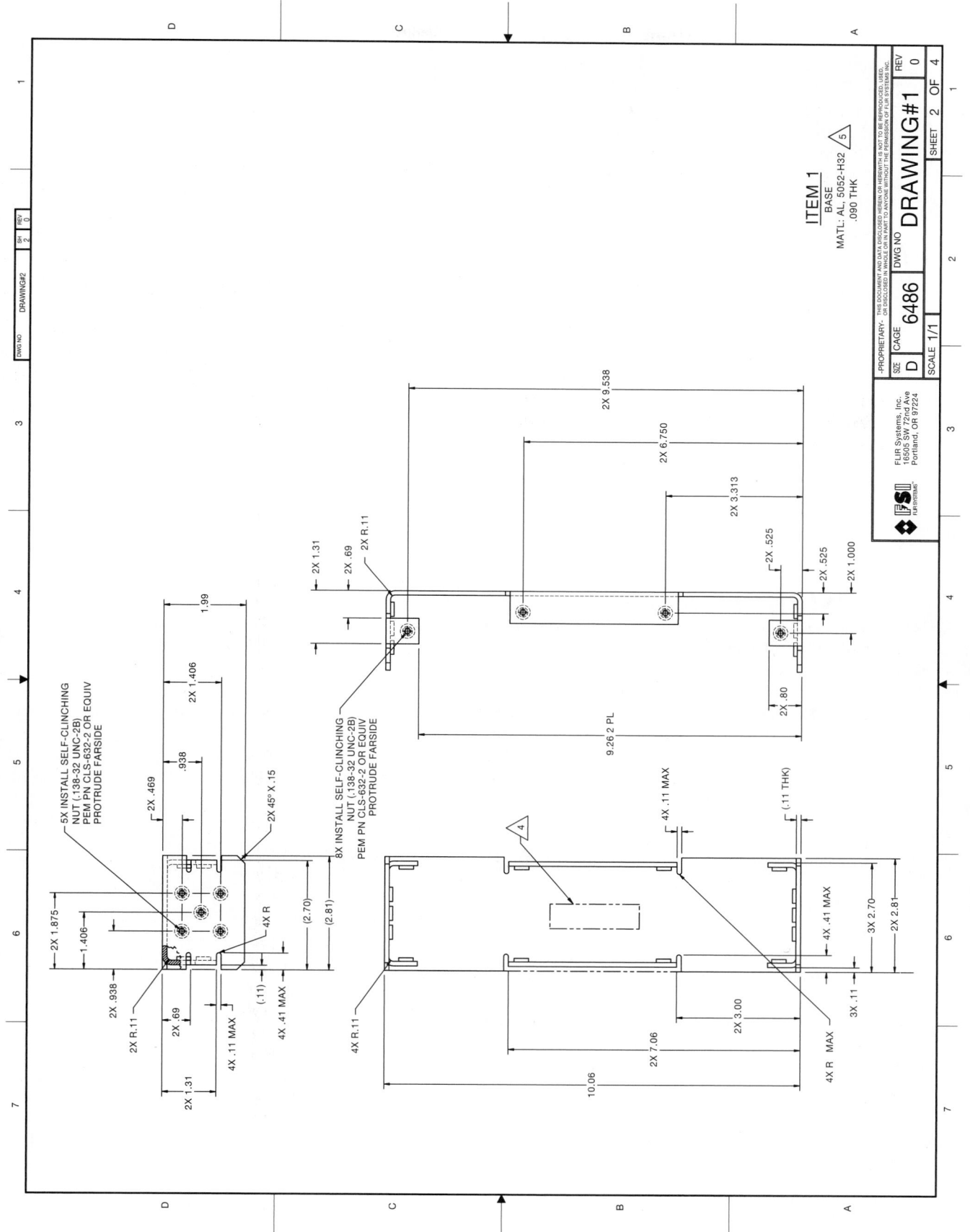

FIGURE 13.88 ■ An actual industry drawing with fasteners displayed and specified. *Courtesy Flir Systems, Inc.*

MATH APPLICATION

RATIO AND PROPORTION OF A TAPER PIN

PROBLEM: The taper of a pin is .25:12. What will the taper be for the 3-in. long pin shown in Figure 13.89?

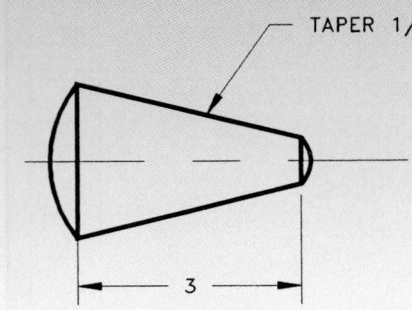

TAPER 1/4 IN. PER FOOT

3

FIGURE 13.89 ■ Taper pin.

CHAPTER 13

Fasteners and Springs Test

DIRECTIONS

Answer the questions with short complete statements or drawings as needed.

QUESTIONS

Define the following screw thread terms:

1. Axis
2. Classes of threads
3. Fit
4. Die
5. Major diameter
6. Minor diameter
7. Pitch
8. Tap
9. Tap drill
10. Thread
11. Threads per inch
12. Name and describe three types of taps.
13. Name the three methods of thread representation.
14. What thread representation is most commonly used and why?
15. Why is it common, when possible, to drill deeper than the intended depth of the tap?
16. Why should an internal tap be deeper than the fastener thread?
17. Identify the components of the following metric thread note.

   M 2.5 × 0.45—6g

18. Identify the components of the following Unified screw thread note.

   ⅞—14 UNF—2 B

19. When is the abbreviation LH used at the end of a thread note?
20. What assumption is made about the thread if LH is not placed at the end of the thread note?
21. Identify two functions of thread inserts.
22. What are set screws used for?
23. What type of information is required to completely identify a bolt or screw?
24. Identify two applications for taper pins.
25. Identify two applications for washers.
26. What are retaining rings used for?
27. What is the principal application for keys?
28. How are standard key sizes determined?
29. Define *keyseat*.
30. What are rivets used for, and how are they applied?
31. Standard thread forming tools generally have lead in chamfers that produce how many incomplete threads on the leading edge of the tool?
32. Define *runout*.
33. Give two problems that can occur when designing features that do not provide enough allowance for runout or sufficient room for the tool.

34. Define *blind hole.*
35. Give three reasons why through threaded holes are preferred over blind threaded holes from a manufacturing standpoint.
36. Taper taps generally have how many lead in chamfers?
37. Bottoming taps generally have how many lead in chamfers?
38. Plug taps generally have how many lead in chamfers?
39. Name the type of tap that is the most commonly used production tool for thread forming.
40. Give the minimum formula that you would use when determining pilot hole depth.
41. According to ASME/ANSI Y14.6, what is the general guideline for the dimension specified on a drawing for thread depth in a blind hole?
42. What does a thread chamfer or countersink in the pilot hole help prevent?

43. Give the specifications for a reasonable chamfer or countersink on the pilot hole.
44. Give the general rule that can be used when designing the minimum tool runout for external threads.
45. External threads should have a lead in chamfer for tooling. Give the general guideline for the applied chamfer diameter for this application.
46. If a relief is used on external threads, the applied relief diameter should not exceed what specification?
47. Identify the two optional provisions that must be used when designing a threaded joint with a shoulder.
48. List the three major functions of pipe threads.
49. Give the drawing angle that is recommended for representing a taper pipe thread.
50. Define *self-clinching fastener.*

CHAPTER 13 *Fasteners and Springs Problems*

DIRECTIONS

1. From the selected problems, determine which views and dimensions should be used to completely detail the part. Use simplified representation unless otherwise specified in the instructions or by your instructor.
2. Make a multiview sketch, to proper proportions, including dimensions and notes.
3. Using the sketch as a guide, draw an original multiview drawing on an adequately sized drawing sheet. Add all necessary dimensions and notes using unidirectional dimensioning. Use manual or computer-aided drafting as required by your course guidelines.
4. Include the following general notes at the lower left corner of the sheet .5 in. each way from the corner border lines:

   1. INTERPRET DIMENSIONS AND TOLERANCES PER ASME Y14.5M—1994.
   2. REMOVE ALL BURRS AND SHARP EDGES.

NOTES:

Additional general notes may be required depending on the specifications of each individual assignment. The following should be part of your title block unless otherwise specified by your instructor:

UNSPECIFIED TOLERANCES:

DECIMALS	mm	IN.
X	± 2.5	± .1
XX	± 0.25	± .01
XXX	± 0.127	± .005
ANGULAR ± 30'		
FINISH	3.2μm	125μIN.

PROBLEM 13-1 **(in.)**
Part Name: Full Dog Point Gib Screw
Material: 10-32 UNF-2A × .75 Long

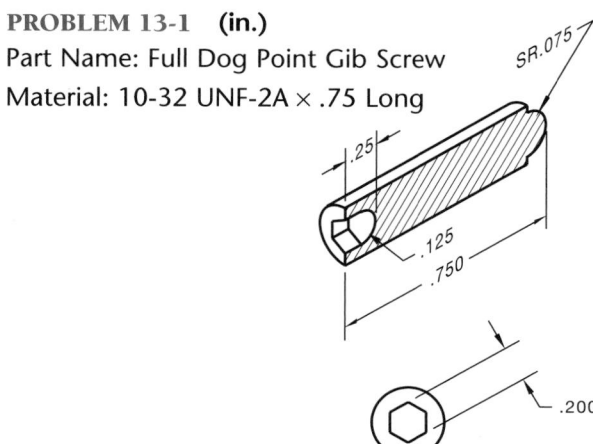

SR.075

.25
.125
.750
.200

SIDE VIEW

PROBLEM 13-2 **(in.)**
Part Name: Thumb Screw
Material: SAE 1315 Steel

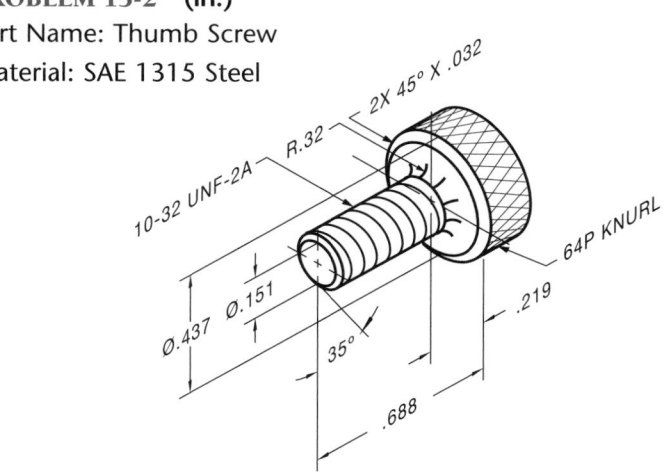

2X 45° X .032
R.32
10-32 UNF-2A
64P KNURL
Ø.437
Ø.151
35°
.219
.688

PROBLEM 13-3 (in.)
Part Name: H-Step Threading Screw
Material: SAE 3130

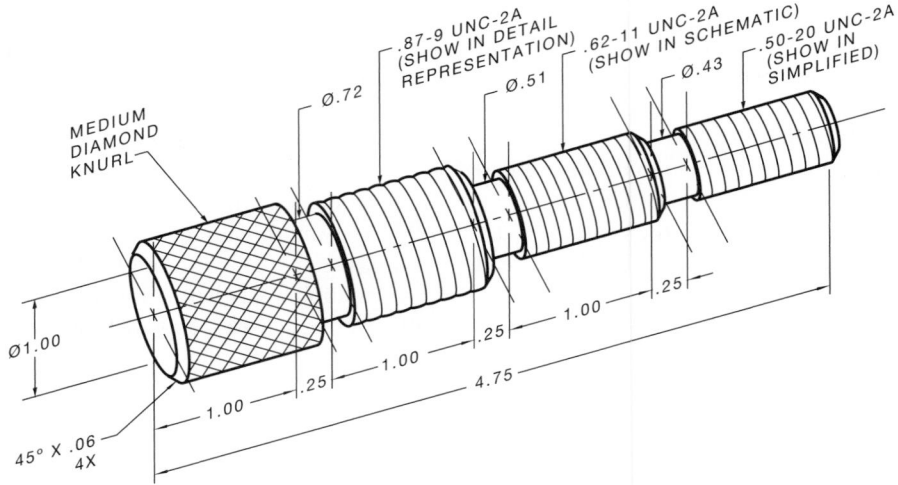

PROBLEM 13-4 (metric)
Part Name: Knurled Hex Soc Hd Step Screw
Material: SAE 1040
Case Harden: 1.6-mm deep per Rockwell C Scale
Finish: 2 μm; Black Oxide

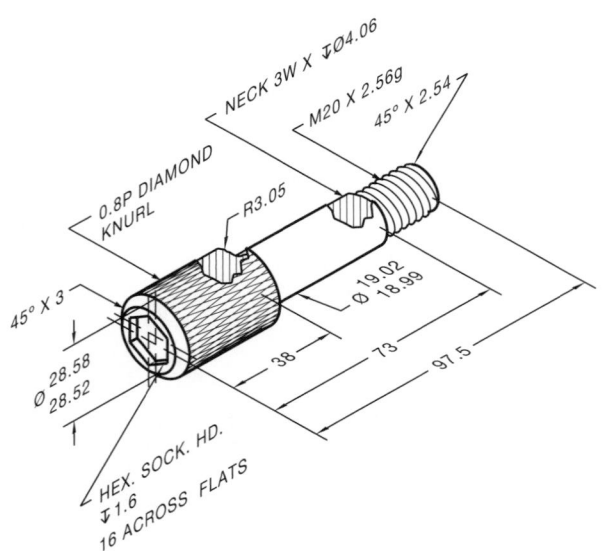

PROBLEM 13-5 (in.)
Part Name: Machine Screw
Material: Stainless Steel
Finish All Over: 2 μm

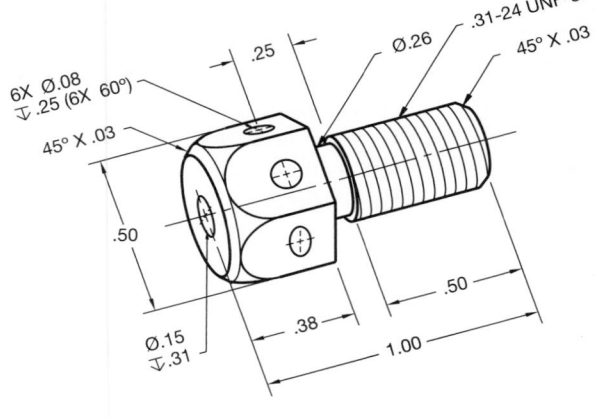

PROBLEM 13-6   (metric)
Part Name: Lathe Dog
Material: Cast Iron

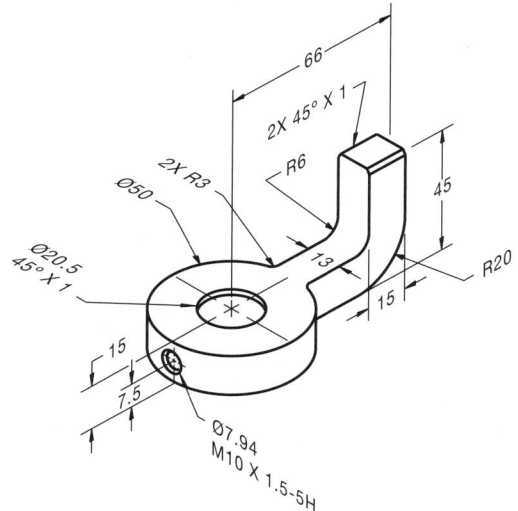

PROBLEM 13-9   (in.)
Part Name: Shoulder Screw
Material: SAE 4320 Steel

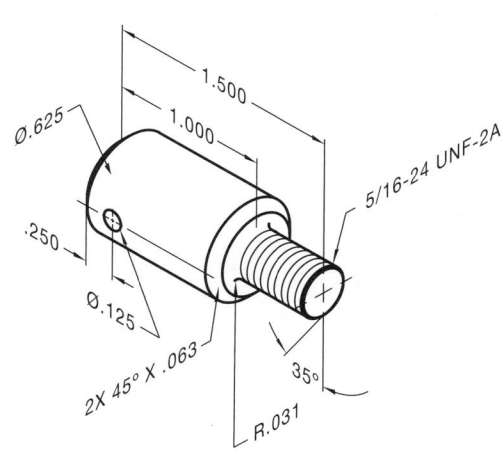

PROBLEM 13-7   (in.)
Part Name: Threaded Step Shaft
Material: SAE 1030

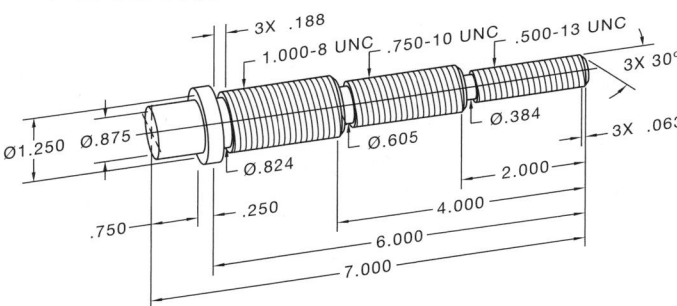

PROBLEM 13-10   (in.)
Part Name: Stop Screw
Material: SAE 4320
Hex Depth: .175
Note: Medium Diamond Knurl at head.

PROBLEM 13-8   (in.)
Part Name: Washer Face Nut
Material: SAE 1330 Steel

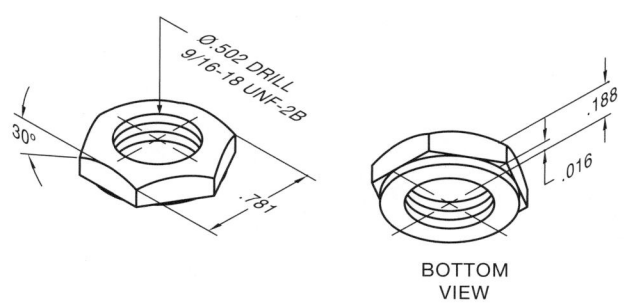

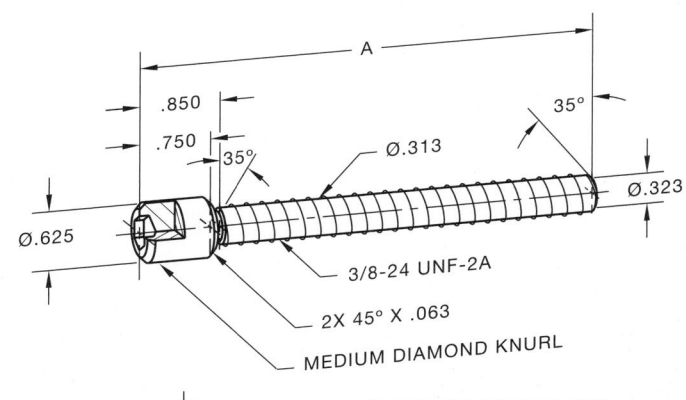

CHART A	
PART #	LENGTH
1DT-1011	4.438
1DT-1012	3.688

PROBLEM 13-11 (metric)
Part Name: Vice Base
Material: Cast Iron

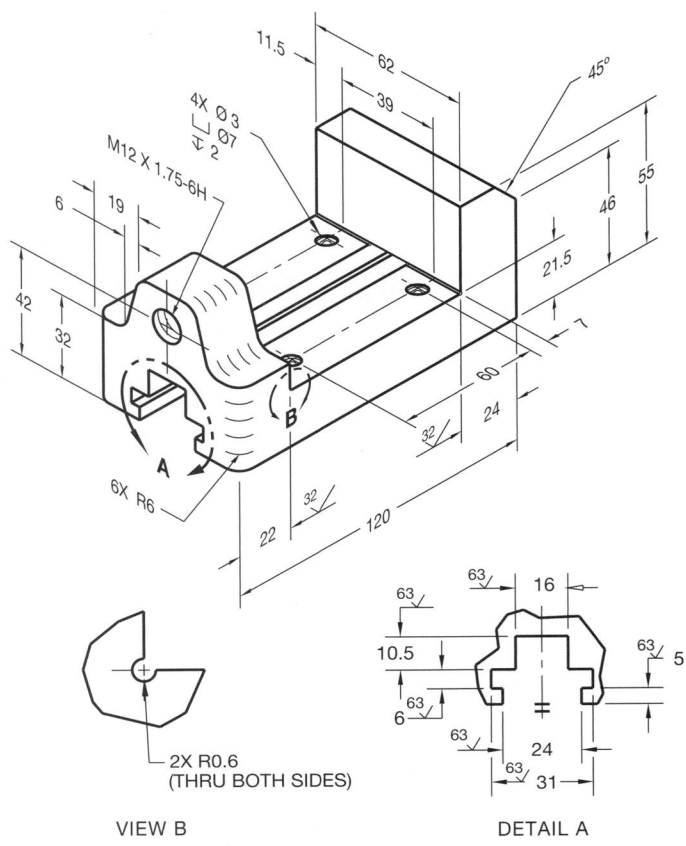

VIEW B

DETAIL A

PROBLEM 13-12 (metric)
Part Name: Compression Spring (use schematic representation)

Material: 2.5 mm Steel Spring Wire

Ends: Plain Ground

Outside Diameter: 25

Free Length: 75

Number of Coils: 16

Finish: Chrome Plate

PROBLEM 13-13 (metric)
Part Name: Flat Spring
Material: 3.5 mm Spring Steel
Finish: Black Oxide
Heat Treat: 1-mm deep Rockwell C Scale

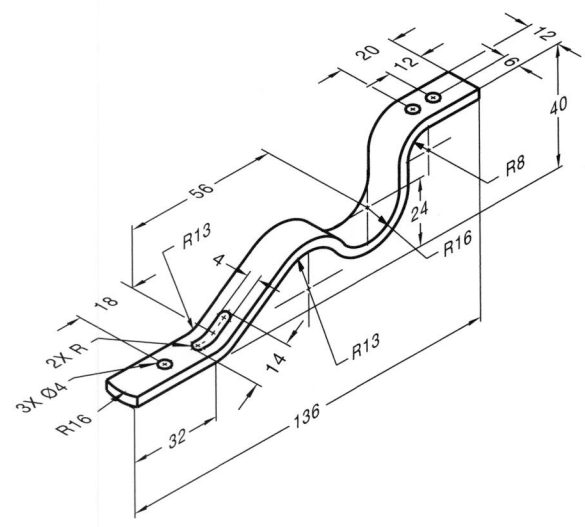

PROBLEM 13-14 (in.)
Part Name: Retaining Ring
Material: Stainless Steel

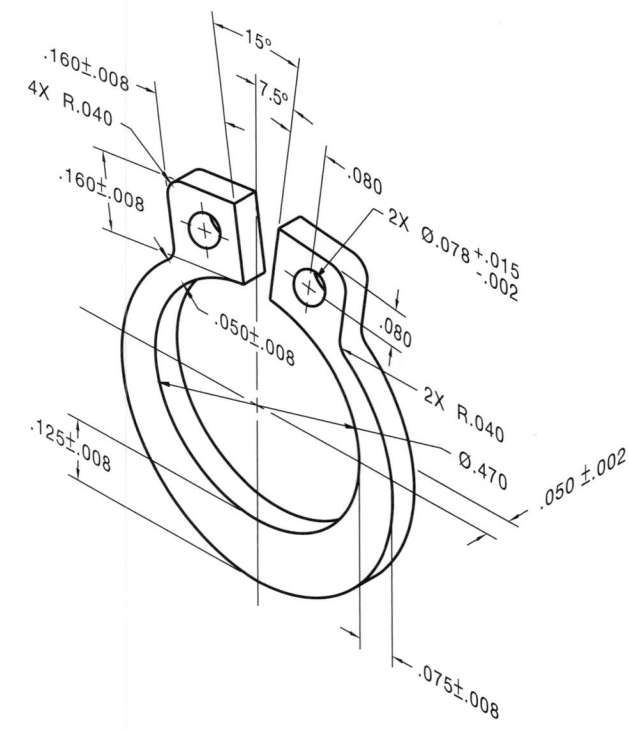

PROBLEM 13-15 (in.)
Part Name: Half Coupling
Material: Ø1.250 C1215 Steel
Problem based on original art courtesy TEMCO.

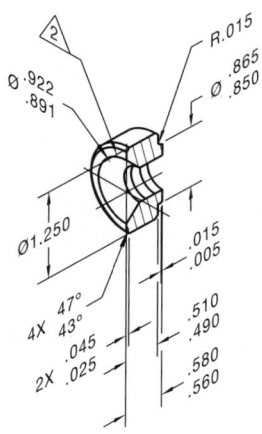

HALF OF PART SHOWN FOR CLARITY.

OUTSIDE SURFACE OF COUPLING MUST BE
FREE OF OXIDIZATION AND INK.

.500-14 NPTF: L-1 GAGE, PLUS 1/MINUS 1
TURNS FROM NOMINAL. TAP FROM THIS END.

PROBLEM 13-16 (in.)
Part Name: Collar
Material: SAE 1020

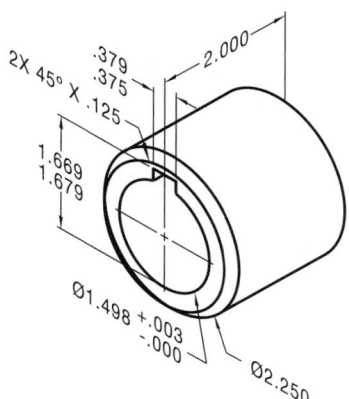

PROBLEM 13-17 (in.)
Part Name: Bearing Nut
Material: SAE 1040

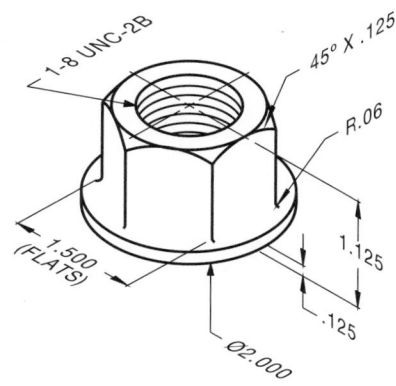

PROBLEM 13-18 (in.)
Part Name: Adjustment Screw
Material: SAE 2010 Steel

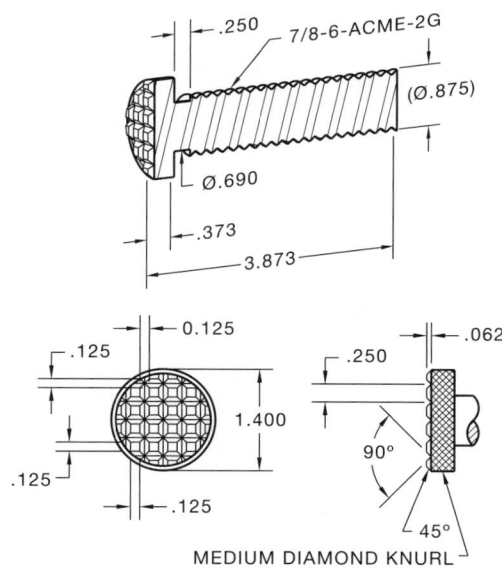

MEDIUM DIAMOND KNURL

PROBLEM 13-19 (in.)
Part Name: Screw Shaft
Material: ¾ Hex × 4⅛ Stock Mild Steel

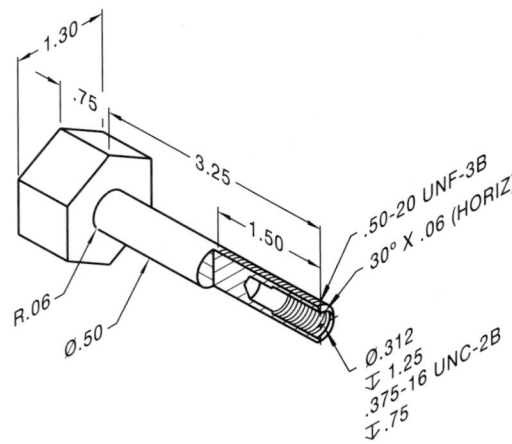

PROBLEM 13-20 (in.)
Part Name: Packing Nut
Material: Bronze
A: Spanner slots .250 wide × .063 deep.

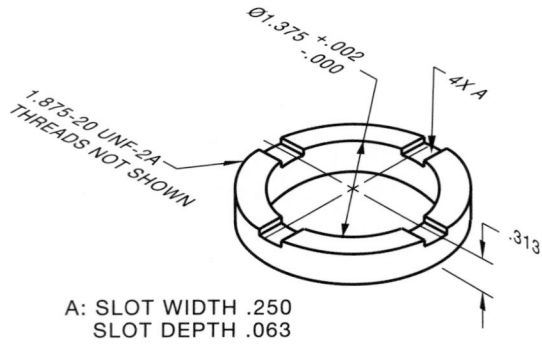

A: SLOT WIDTH .250
SLOT DEPTH .063

PROBLEM 13-21 (in.)
Part Name: Pump Pivot Support
Material: Cold-rolled Mild Steel

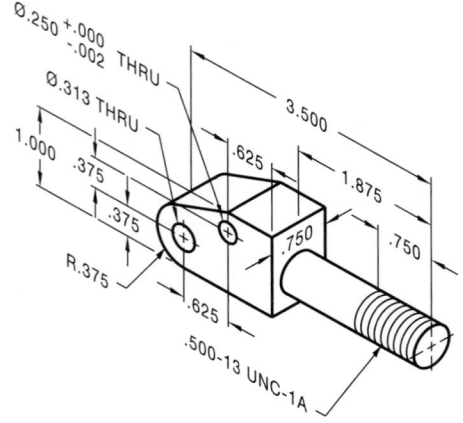

PROBLEM 13-22 (in.)
Part Name: Set Screw
Material: Steel
SPECIFIC INSTRUCTIONS:
Prepare a detailed drawing from the written instructions below.

10-32 UNF-2A THREADED
MATERIAL .375 LG.
W/ .015 LATERAL X 35°
CHAMFERS ON BOTH ENDS
PROVIDE A .030 WIDE X
.047 DEEP SLOT ON
ONE END.

MATH PROBLEMS

PROBLEM 13-23 The taper of a pin is .5:12. What would be the taper for a 5-in. long pin?

PROBLEM 13-24 A lathe operator turns out 9 brass bushings in 2 hours 15 minutes. At the same rate, how many bushings can be turned in 8 hours?

PROBLEM 13-25 Seventeen drills cost $27.50. What would seven drills cost?

PROBLEM 13-26 The scale on a landscape map is 1":50'. If two points on the map are 3.5" apart, what is the actual distance between the two points?

PROBLEM 13-27 A crew of twelve people assembles 45 units a day. If production is to be increased to 80 units per day, how many people will be needed?

PROBLEM 13-28 If 5 dozen shop cloths are enough for six mechanics, how many cloths will be needed for eleven mechanics?

PROBLEM 13-29 A shop manual chain hoist requires a 22-lb pull to lift one ton (2000 lb). How much pull would it take to lift a 1500-lb load?

Sections, Revolutions, and Conventional Breaks

OBJECTIVES

After completing this chapter, you will:

■ Draw proper cutting-plane line representations.

■ Draw sectional views, including full, half, aligned, broken-out, auxiliary, revolved, and removed sections.

■ Identify features that should remain unsectioned in a sectional view.

■ Prepare drawings with conventional revolutions and conventional breaks.

■ Modify the standard sectioning techniques as applied to specific situations.

■ Make sectional drawings from given engineers' sketches and actual industrial layouts.

THE ENGINEERING DESIGN APPLICATION

You are working with an engineer on a special project and she gives you a rough sketch of a part from which you are to make a formal drawing. At first glance, you think it is an easy part to draw, but with further study you realize that it is more difficult than it looks. The rough sketch, shown in Figure 14.1, has a lot of hidden features. So your first thought is, "What do I use for the front view?" You decide that a front exterior view would show only the diameter and length. So instead of drawing an outside view, you decide to use a full section to expose all of the interior features and, at the same time, give the overall length and diameter. "This is great!" you say. Then you realize there are 6 holes spaced 60° apart that remain as hidden features in the left-side view. This is undesirable, because you do not want to section all of the holes. So you decide to use a broken-out section to expose only two holes. This allows you to dimension the 6×60° and the $6 \times \emptyset 6^{+0.2}_{0}$ holes all at the same time. Now with all this thinking and planning out of the way, it only takes you three hours to completely draw and dimension the part, and you are ready to give the formal drawing, shown in Figure 14.2, to the checker.

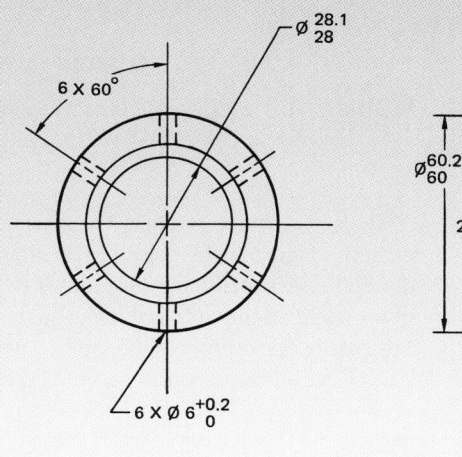

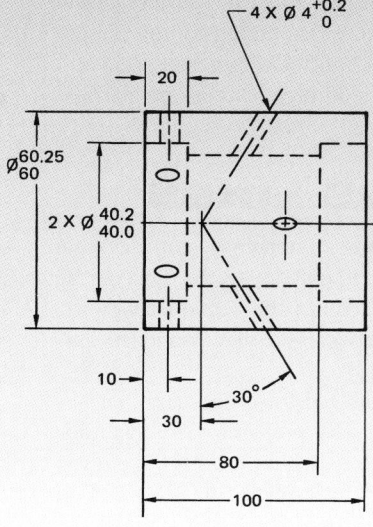

FIGURE 14.1 ■ Engineer's layout drawing.

(Continued)

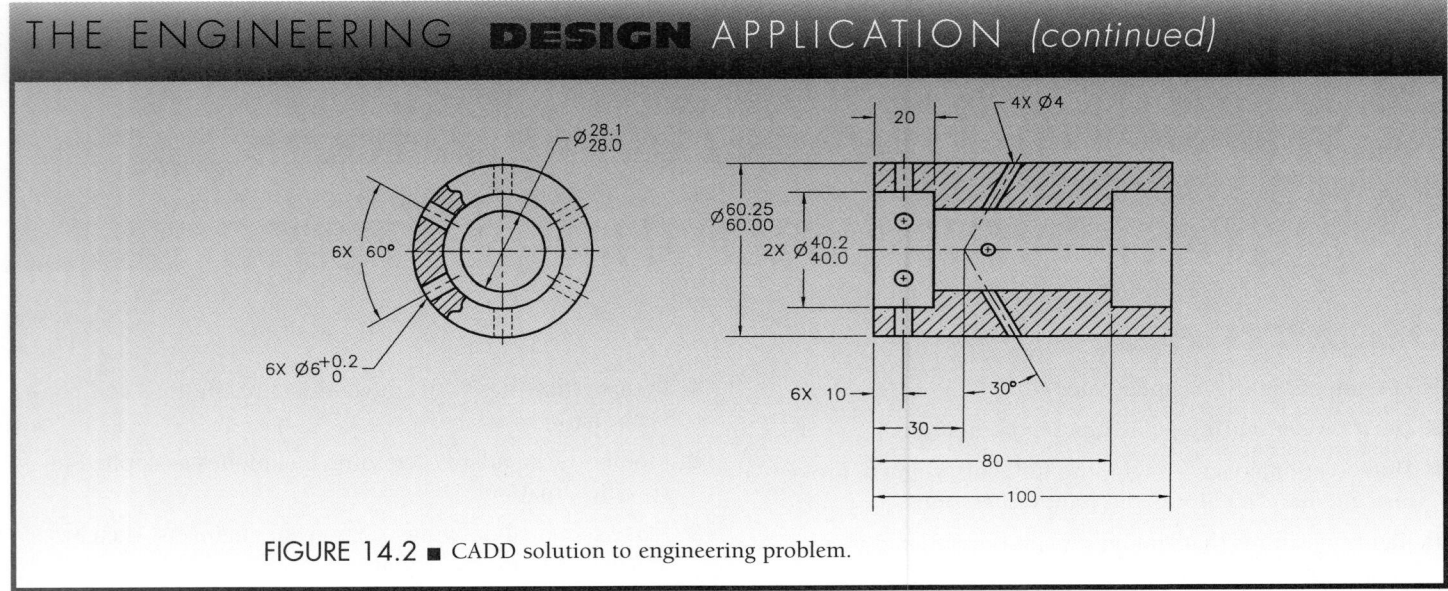

THE ENGINEERING DESIGN APPLICATION *(continued)*

FIGURE 14.2 ■ CADD solution to engineering problem.

ANSI The American National Standards Institute document that governs sectioning techniques is titled *Multi and Sectional View Drawings,* ANSI Y14.3. The engineering standard *Line Conventions and Lettering,* ASME Y14.2M, covers the principles of drawing recommended section lines. The content of this chapter is based upon the ANSI standard and provides an in-depth analysis of the techniques and methods of sectional view presentation.

SECTIONING

Sections, or *sectional views,* are used to describe the interior portions of an object that are otherwise difficult to visualize. Interior features that are described using hidden lines may not appear as clear as if they were exposed for viewing as visible features. It is also a poor practice to dimension to hidden features, but the sectional view allows you to expose the hidden features for dimensioning. Figure 14.3 shows an object in conventional multiview representation and using a sectional view. Notice how the hidden features are clarified in the sectional view.

CUTTING-PLANE LINES

The sectional view is created by placing an imaginary cutting plane through the object as if you were to cut away the area to be exposed. The adjacent view then becomes the sectional view by removing the portion of the object between the viewer and the cutting plane. (See Figure 14.4).

The sectional view should be projected from the view that has the cutting plane as you would normally project a view in multiview. The *cutting-plane line* is a thick line that represents the cutting plane as shown in Figure 14.4. The cutting-plane line is capped on the ends with arrowheads that show the direction of sight of the sectional view. When the extent of the cutting plane is obvious, only the ends of the cutting-plane line may be used as shown in Figure 14.5. Such treatment of the cutting plane also helps keep the view clear of excess lines.

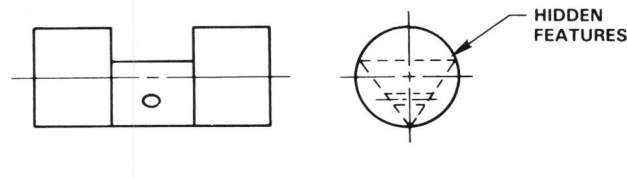

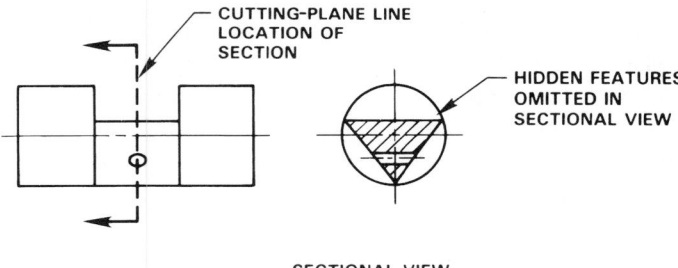

FIGURE 14.3 ■ Conventional multiview compared to a sectional view.

If lack of space restricts the normal placement of a sectional view, the view may be placed in an alternate location. When this is done, the sectional view should not be rotated but should remain in the same orientation as if it were a direct projection from the cutting plane. The cutting planes and related sectional views should be labeled with letters beginning with A, as shown in Figure 14.6.

The cutting-plane line may be completely omitted when the location of the cutting plane is clearly obvious. (See Figure 14.7.) When in doubt, use the cutting-plane line.

SECTION LINES

Section lines are thin lines used in the view of the section to show where the cutting-plane line has cut through material. (See Figure 14.8.) Also refer to Chapter 3. Section lines are usually drawn

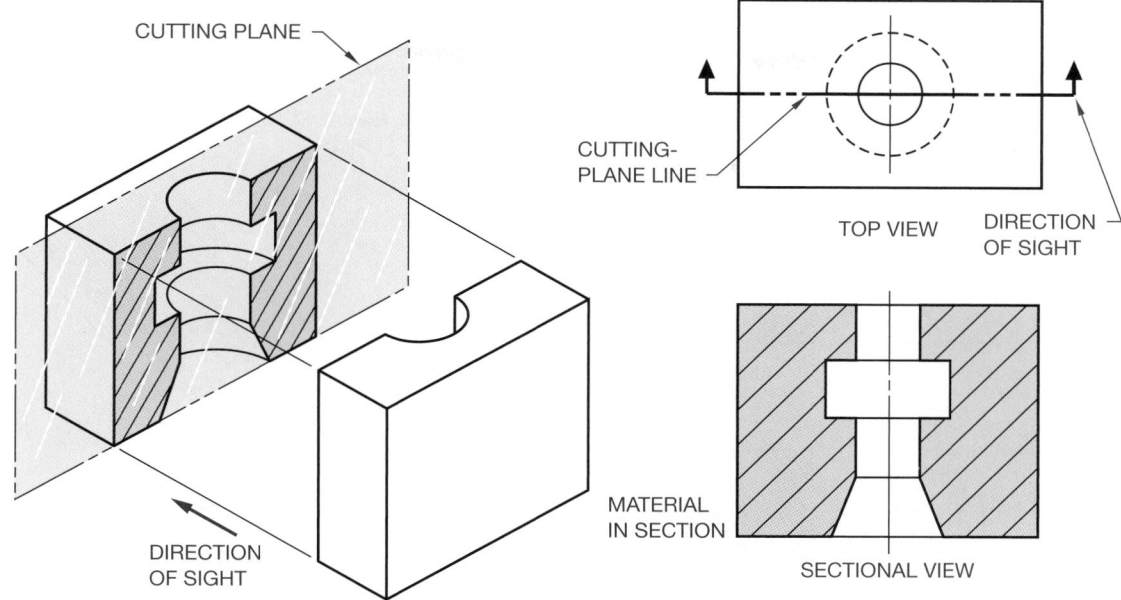

FIGURE 14.4 ■ Cutting-plane line.

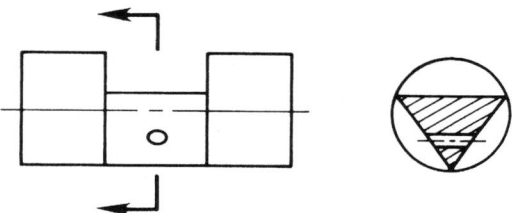

FIGURE 14.5 ■ Simplified cutting-plane line.

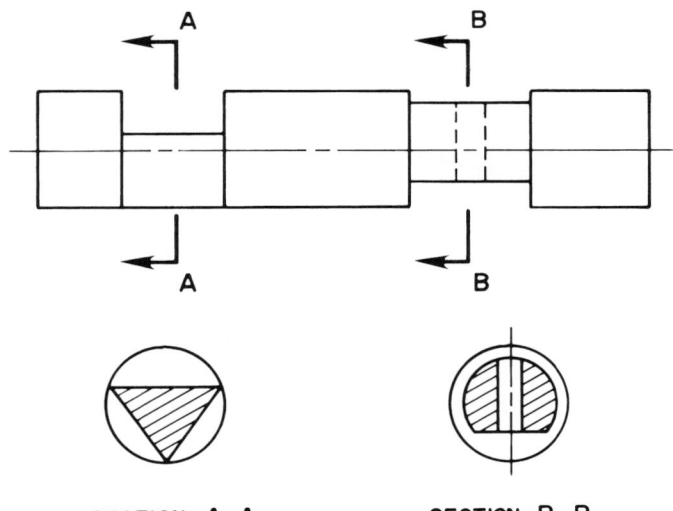

SECTION A-A   SECTION B-B

FIGURE 14.6 ■ Labeled cutting-plane lines and related sectional views.

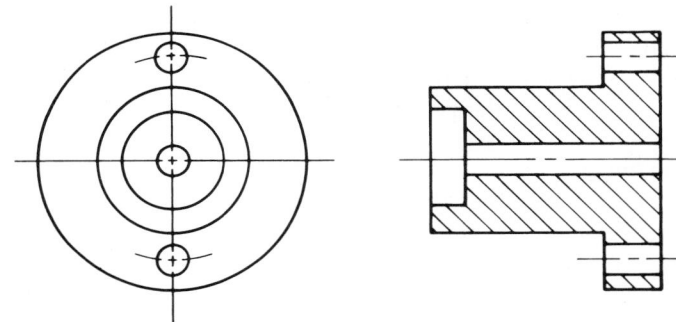

FIGURE 14.7 ■ An obvious cutting-plane line may be omitted.

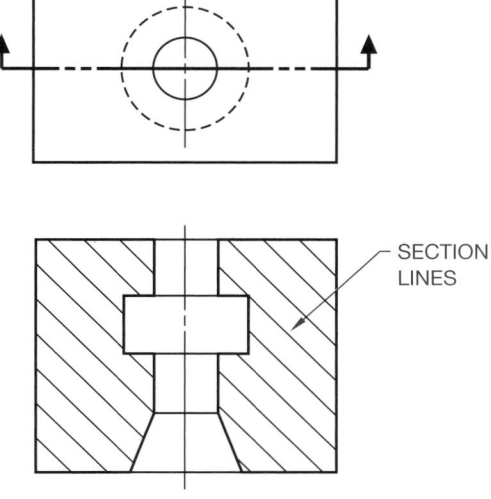

FIGURE 14.8 ■ Section lines.

equally spaced at 45° but may not be parallel or perpendicular to any line of the object. Any convenient angle may be used to avoid placing section lines parallel or perpendicular to other lines of the object. Angles of 30° and 60° are common. Section lines that are more than 75° or less than 15° from horizontal should be avoided.

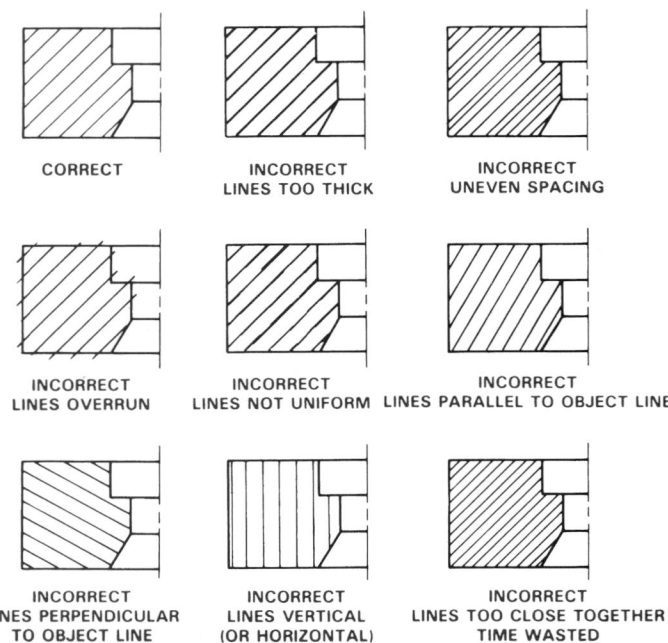

CORRECT | INCORRECT LINES TOO THICK | INCORRECT UNEVEN SPACING

INCORRECT LINES OVERRUN | INCORRECT LINES NOT UNIFORM | INCORRECT LINES PARALLEL TO OBJECT LINE

INCORRECT LINES PERPENDICULAR TO OBJECT LINE | INCORRECT LINES VERTICAL (OR HORIZONTAL) | INCORRECT LINES TOO CLOSE TOGETHER TIME WASTED

FIGURE 14.9 ■ Common section line errors.

Section lines must never be drawn either horizontally or vertically. Figure 14.9 shows some common errors in drawing section lines. Section lines should be drawn in opposite directions on adjacent parts, and when several parts are adjacent, any suitable angle may be used to make the parts appear clearly separate. When a very large area requires section lining, you may elect to use outline section lining.

Equally spaced section lines specify either a general material designation or cast iron. This method of drawing section lines is quick and easy with the actual material identification located in the drawing title block. The other option is to use coded section lining, which is more time-consuming to draw than general section lining. Coded section lining may be used effectively when a section is taken through an assembly of adjacent parts of different materials as seen in Figure 14.10.

General section lines are evenly spaced. The amount of space between lines depends on the size of the part. Very large parts have larger spacing than very small parts. Use your own judgment. The

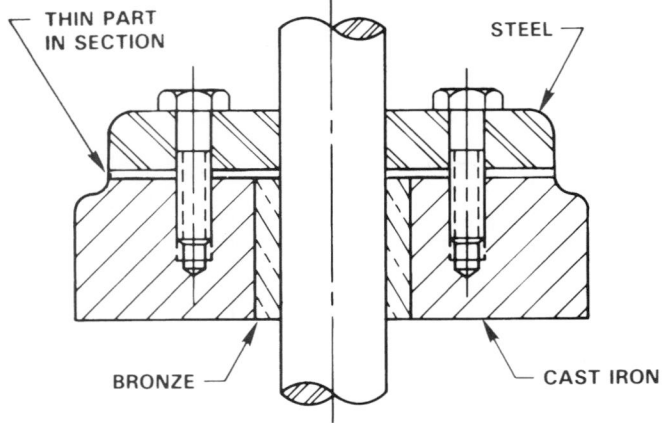

THIN PART IN SECTION | STEEL

BRONZE | CAST IRON

FIGURE 14.10 ■ Assembly section, coded section lines.

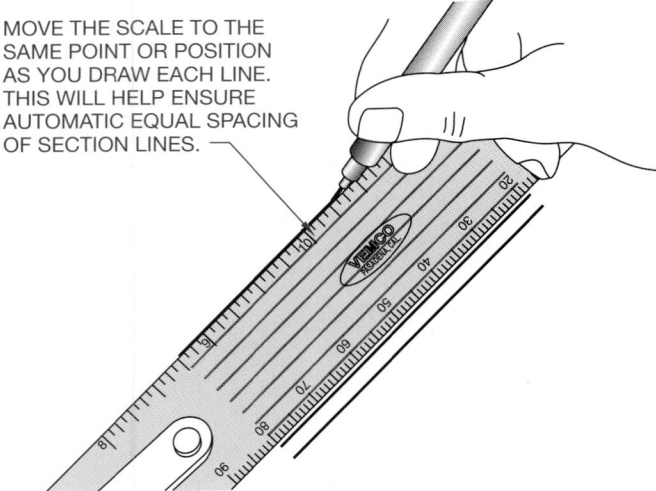

MOVE THE SCALE TO THE SAME POINT OR POSITION AS YOU DRAW EACH LINE. THIS WILL HELP ENSURE AUTOMATIC EQUAL SPACING OF SECTION LINES.

FIGURE 14.11 ■ Equally spacing section lines with the drafting machine scale.

key is to clearly represent section lines without an unnecessary expenditure of time. There are a number of ways to space section lines equally without measuring each space. One good way is to calibrate the space using the same point or mark on the drafting machine scale as you draw each line. This method works with clear plastic scales as shown in Figure 14.11. Another technique is to use the Ames Lettering Guide for drawing equally spaced section lines. Refer to Chapter 3 to review the use of this tool.

FULL SECTIONS

A *full section* is drawn when the cutting plane extends completely through the object, usually along a center plane as shown in Figure 14.12. The object shown in Figure 14.12 could have used two full sections to further clarify hidden features. In such a case, the cutting planes and related views are labeled. (See Figure 14.13.)

HALF SECTIONS

A *half section* may be used when a symmetrical object requires sectioning. The cutting-plane line of a half section actually removes one quarter of the object. The advantage of a half section is that the sectional view shows half of the object in section and the other half of the object as it normally appears. Thus the name half section. (See Figure 14.14.) Notice that a centerline is used in the sectional view to separate the sectioned portion from the unsectioned portion. Hidden lines are generally omitted from sectional views unless their use improves clarity.

OFFSET SECTIONS

Staggered interior features of an object may be sectioned by allowing the cutting-plane line to *offset* through the features as shown in Figure 14.15. Notice in Figure 14.15 that there is no line in the sectional view indicating a change in direction of the cutting-plane line. Normally the cutting-plane line in an offset section extends completely through the object to clearly display the location of the section.

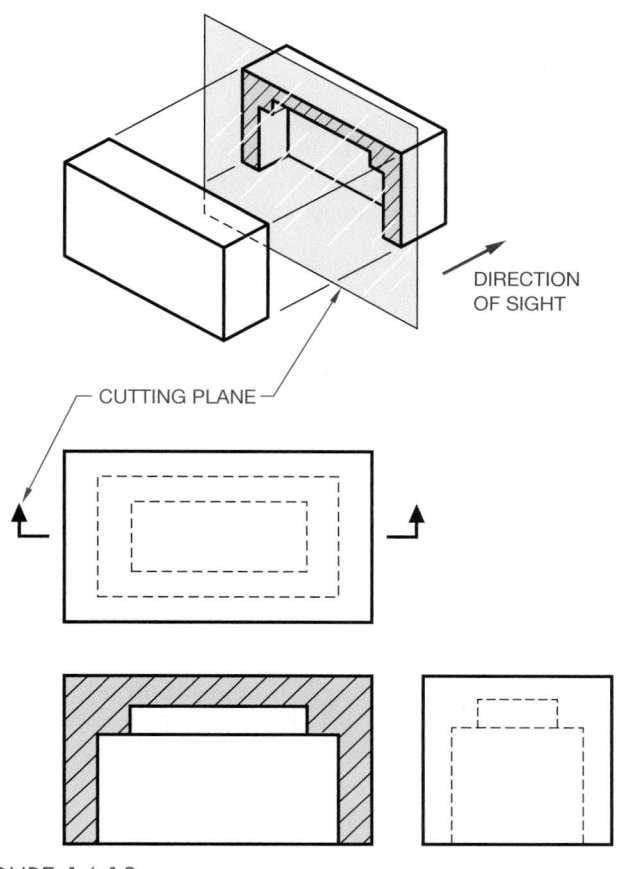

FIGURE 14.12 ■ Full section.

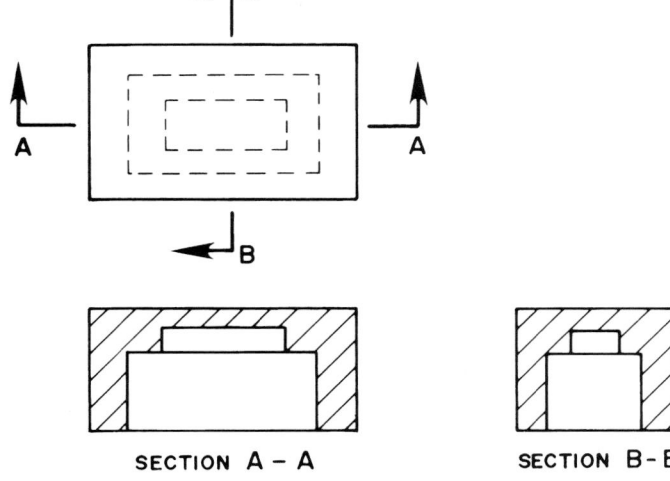

SECTION A – A   SECTION B-B

FIGURE 14.13 ■ Full sections.

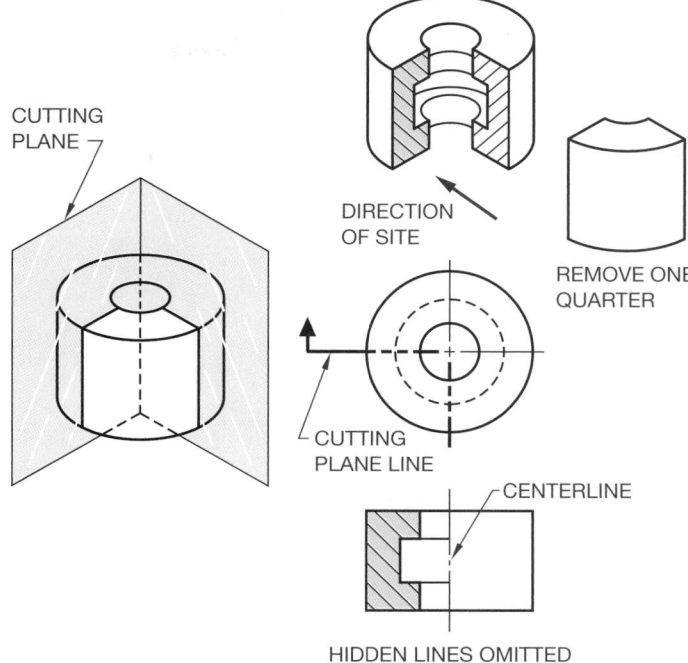

FIGURE 14.14 ■ Half section.

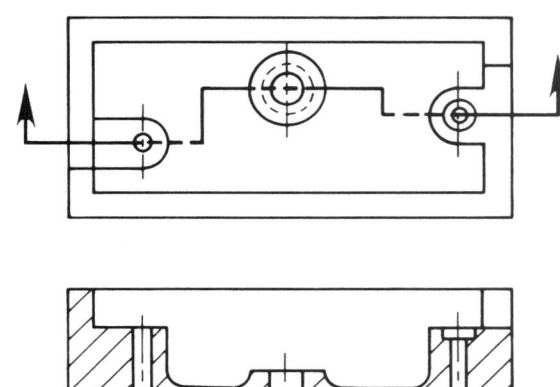

FIGURE 14.15 ■ Offset section.

ALIGNED SECTIONS

Similar to the offset section, the *aligned section* cutting-plane line also staggers to pass through offset features of an object. Normally the change in direction of the cutting-plane line is less than 90° in an aligned section. When this section is taken, the sectional view is drawn as if the cutting plane is rotated to a plane perpendicular to the line of sight as shown in Figure 14.16.

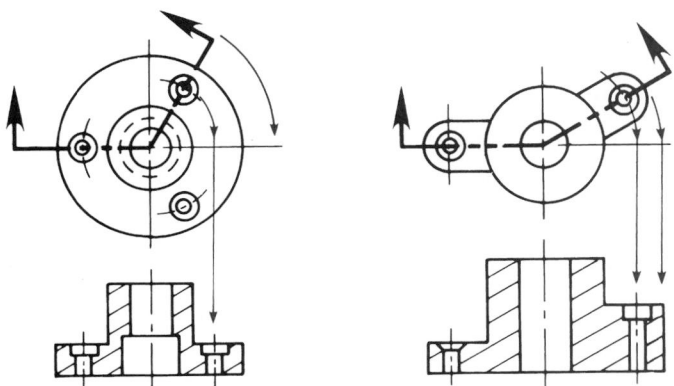

FIGURE 14.16 ■ Aligned section.

UNSECTIONED FEATURES

Specific features of an object are commonly left unsectioned in a sectional view if the cutting-plane line passes through the feature and parallel to it. The types of features that are left unsectioned for clarity are bolts, nuts, rivets, screws, shafts, ribs, webs, spokes, bearings, gear teeth, pins, and keys. (See Figure 14.17.) When the cutting-plane line passes through the previously described features perpendicular to their axes, then section lines are shown as seen in Figure 14.18.

While it is most common to draw the outline of the previously discussed features and display them without section lines, a less used method called alternate section lines may be used. The normally unsectioned feature is drawn using hidden lines and every other section line is drawn through the feature, as shown in Figure 14.19.

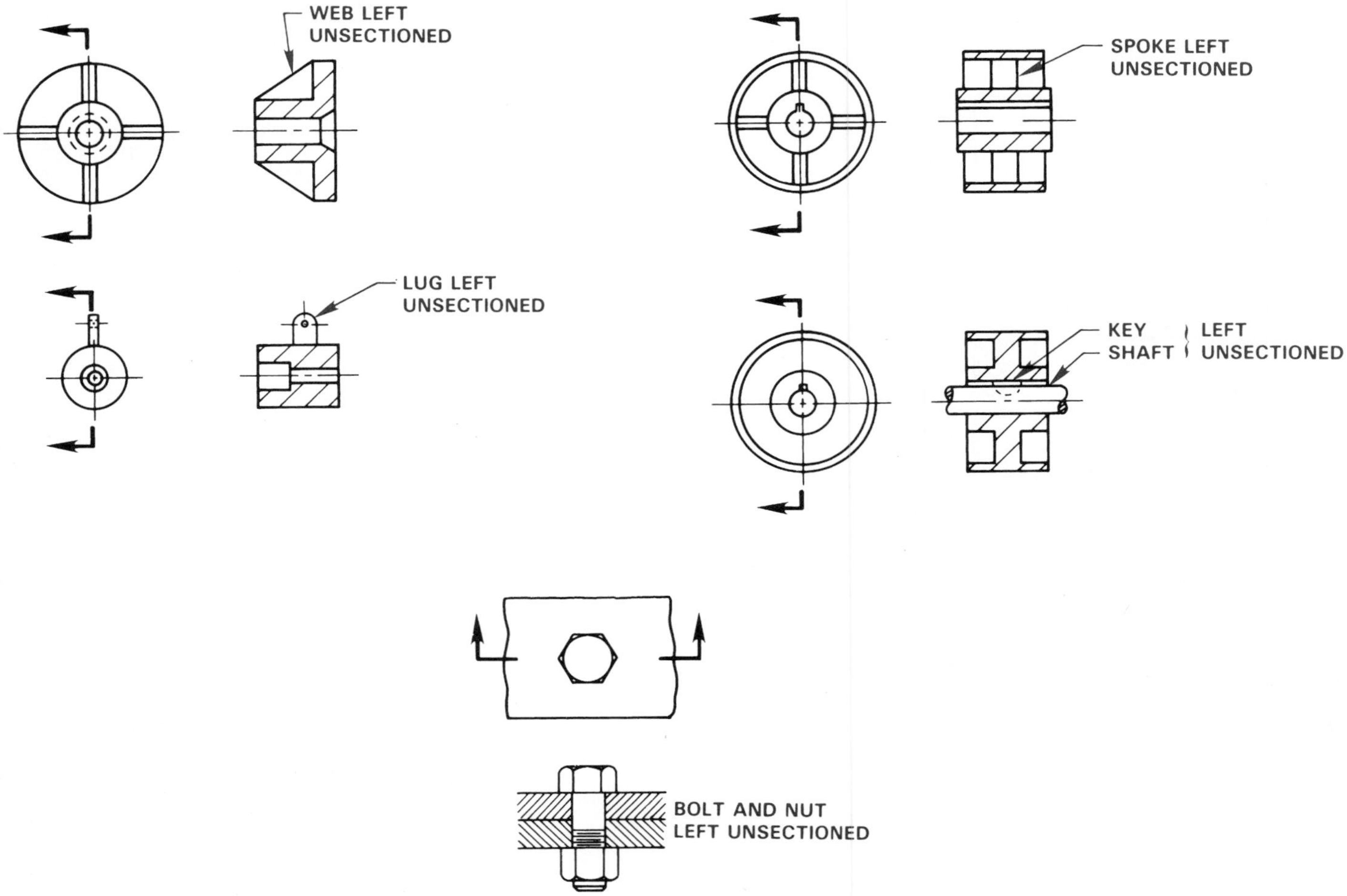

WEB LEFT UNSECTIONED

SPOKE LEFT UNSECTIONED

LUG LEFT UNSECTIONED

KEY ⎱ LEFT
SHAFT ⎰ UNSECTIONED

BOLT AND NUT LEFT UNSECTIONED

FIGURE 14.17 ■ Certain features are not sectioned when a cutting-plane line passes parallel to their axes.

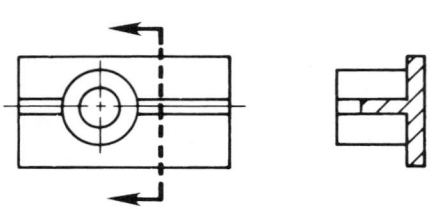

FIGURE 14.18 ■ Cutting-plane perpendicular to normally unsectioned features.

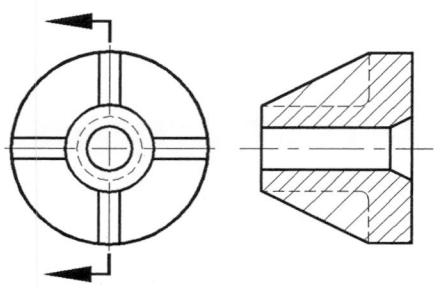

FIGURE 14.19 ■ Alternate section line method.

CADD APPLICATIONS

SECTIONING

Everyone should have the chance to spend what often seems like hours drawing section lines in a complex sectional view to really appreciate the speed and accuracy of drawing section lines with a computer. One important aspect of drawing section lines manually is getting the lines uniform and equally spaced. This task is automatic with CADD. Additionally, most CADD drafting packages have more section line symbols available than you would ever have the need to use. CADD section line commands such as HATCH or BHATCH are used to select one of a variety of patterns, and you have the opportunity to change the section line scale if you wish. Figure 14.20 shows three different section line scale factors available.

When using CADD to section-line an area, you need to define the boundaries of the area. In other words, tell the computer where the section lines are to be placed so you do not end up with lines in an unwanted area. If you want to section-line inside a circle or square, all you need to do is pick the circle or square. (See Figure 14.21.)

Some CADD programs allow you to draw section lines in an enclosed area simply by picking a point within the area. In AutoCAD, for example, the enclosed area is called a boundary. You get an error message, "BOUNDARY AREA NOT CLOSED," if there is a gap in the boundary. A common boundary gap might be a corner where lines do not meet. You may have to "ZOOM" in on the corner to actually see a very small gap. If this happens, correct the problem and try to section the area again as shown in Figure 14.22.

The computer makes drawing section lines for any drafting field quick and easy. There are standard material section lines for the mechanical drafter, and brick and other patterns for architectural, structural, civil, and other engineering drafting disciplines. A large variety of section line symbols and graphic patterns is available in most CADD packages.

SCALE = 1   SCALE = 2   SCALE = 3

FIGURE 14.20 ■ Three different section line scale factors for CADD applications.

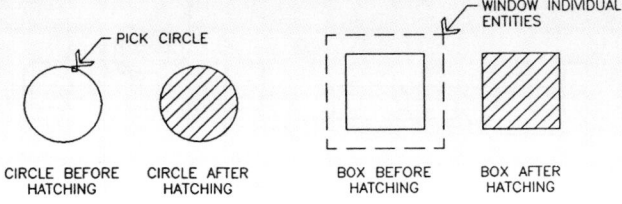

CIRCLE BEFORE HATCHING   CIRCLE AFTER HATCHING   BOX BEFORE HATCHING   BOX AFTER HATCHING

PICK CIRCLE

WINDOW INDIVIDUAL ENTITIES

FIGURE 14.21 ■ Selecting objects for sectioning with CADD.

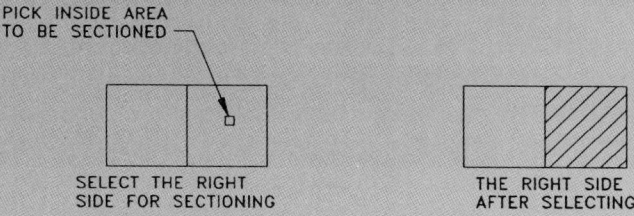

PICK INSIDE AREA TO BE SECTIONED

SELECT THE RIGHT SIDE FOR SECTIONING

THE RIGHT SIDE AFTER SELECTING

FIGURE 14.22 ■ Recommended drawing sequence for section lining adjacent areas with CADD.

When using a program such as AutoCAD, for example, the section lines are drawn using the BHATCH command. This command opens the Boundary Hatch dialog box, where any enclosed area can be sectioned automatically with any one of many different patterns. The patterns are called Hatch patterns. AutoCAD has many predefined Hatch patterns. These patterns can be selected by name by accessing the "Pattern:" drop-down list in the Boundary Hatch dialog box. You can also pick the ellipsis (. . .) button next to the "Pattern:" drop-down list to visually select the desired pattern in the Hatch Pattern palette shown in Figure 14.23.

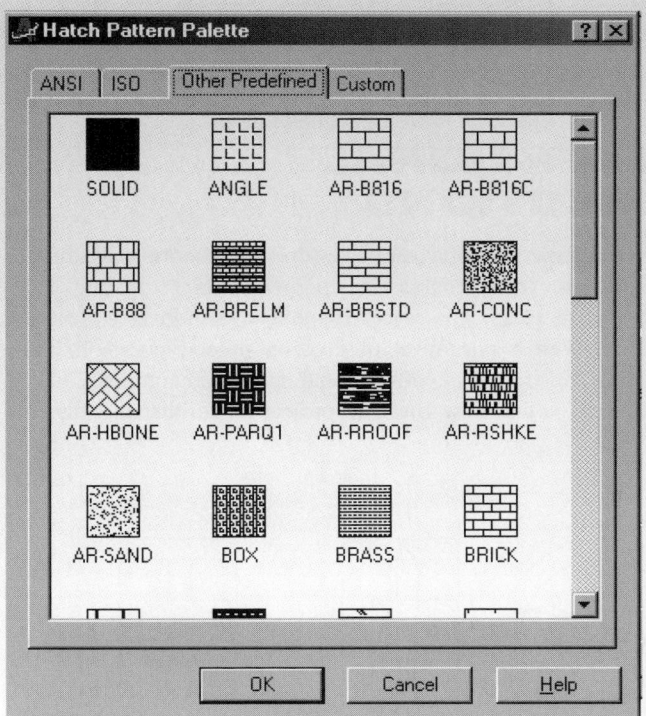

FIGURE 14.23 ■ AutoCAD Hatch Pattern palette where many different section line symbols and patterns are available for use in your drawings.

INTERSECTIONS IN SECTION

When a section is drawn through a small intersecting shape, the true projection is ignored. The detail is too complex to represent. (See Figure 14.24a and b.) However, larger intersecting features are drawn as their true representation. (See Figure 14.24c and d.) The professional decision is up to you.

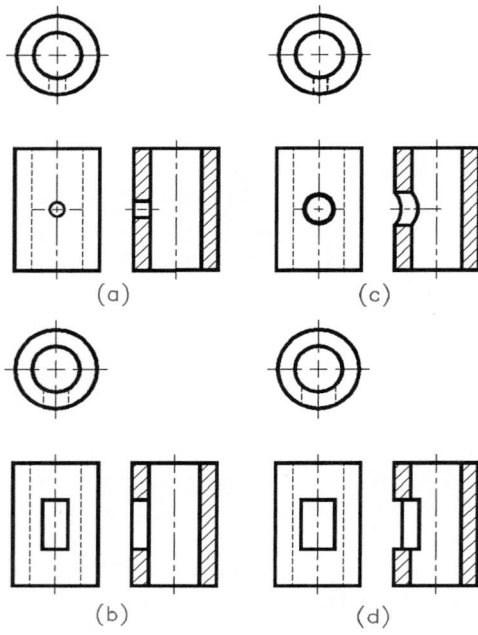

FIGURE 14.24 ■ Intersections in section.

CONVENTIONAL REVOLUTIONS

When the true projection of a feature results in foreshortening, the feature should be revolved onto a plane perpendicular to the line of sight as in Figure 14.25. The revolved spoke shown gives a clear representation with a minimum of drafting time. Figure 14.26 shows another illustration of conventional revolution compared to true projection. Notice how the true projection results in a dis-

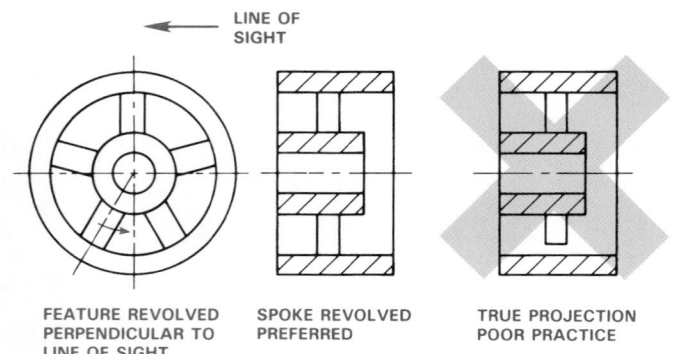

FEATURE REVOLVED
PERPENDICULAR TO
LINE OF SIGHT

SPOKE REVOLVED
PREFERRED

TRUE PROJECTION
POOR PRACTICE

FIGURE 14.25 ■ Conventional revolution.

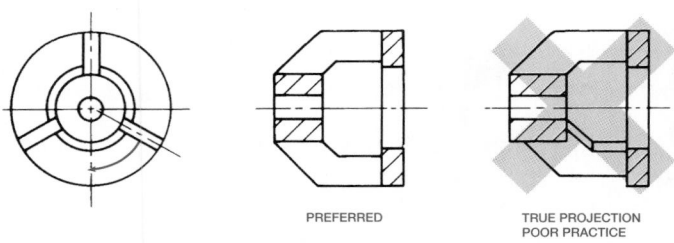

PREFERRED

TRUE PROJECTION
POOR PRACTICE

FIGURE 14.26 ■ Conventional revolution in section.

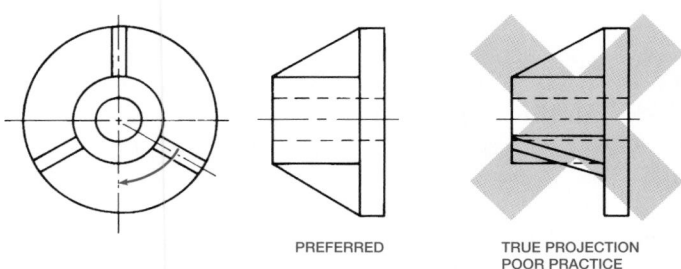

PREFERRED

TRUE PROJECTION
POOR PRACTICE

FIGURE 14.27 ■ Conventional revolution in multiview.

torted and foreshortened representation of the spoke. The revolved spoke in the preferred view is clear and easy to draw. The practice illustrated here also applies to features of unsectioned objects in multiview as shown in Figure 14.27.

BROKEN-OUT SECTIONS

Often a small portion of a part may be broken away to expose and clarify an interior feature. This technique is called a *broken-out section*. There is no cutting-plane line used, as you can see in Figure 14.28. A short break line is generally used with a broken-out section.

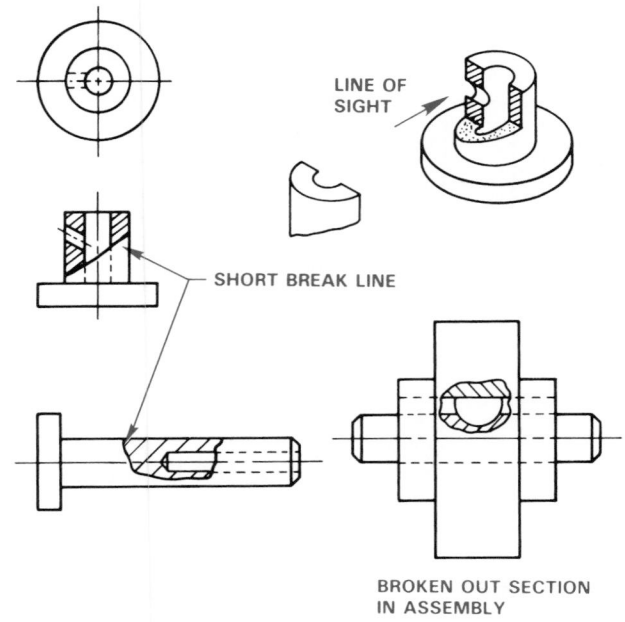

LINE OF
SIGHT

SHORT BREAK LINE

BROKEN OUT SECTION
IN ASSEMBLY

FIGURE 14.28 ■ Broken-out sections.

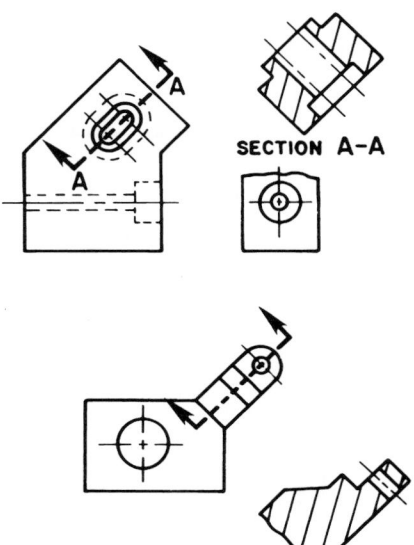

FIGURE 14.29 ■ Auxiliary sections.

AUXILIARY SECTIONS

A section that appears in an auxiliary view is known as an *auxiliary section*. Auxiliary sections are generally projected directly from the view of the cutting plane. If these sections must be moved to other locations on the drawing sheet, they should remain in the same relationship (not rotated) as if taken directly from the view of the cutting plane. (See Figure 14.29.)

CONVENTIONAL BREAKS

When a long object of constant shape throughout its length requires shortening, *conventional breaks* may be used. These breaks may be used effectively to save time, paper, or space or to increase the scale of an otherwise very long part. Figure 14.30 shows typical conventional breaks. Used on metal shapes, the short break line is drawn thick, freehand, and slightly irregular. Notice that the actual length can be given with a long break line used as a dimension line. For wood shapes, the short break line is drawn freehand as a thick, very irregular line.

The break line for solid round shapes may be drawn freehand or with an irregular curve or template. The shape widths should be approximately ⅓ radius and should be symmetrical about the horizontal centerline and the vertical guidelines as shown.

The break lines for tubular round shapes may be drawn freehand or with an irregular curve or template. The total shape widths should be approximately ½ radius and should be symmetrical about the horizontal centerline and vertical guidelines as shown.

REVOLVED SECTIONS

When a feature has a constant shape throughout the length that cannot be shown in an external view, a *revolved section* may be used. The desired section is revolved 90° onto a plane perpendicular to the line of sight as shown in Figure 14.31. Revolved sections may be represented on a drawing one of two ways as shown in Figure

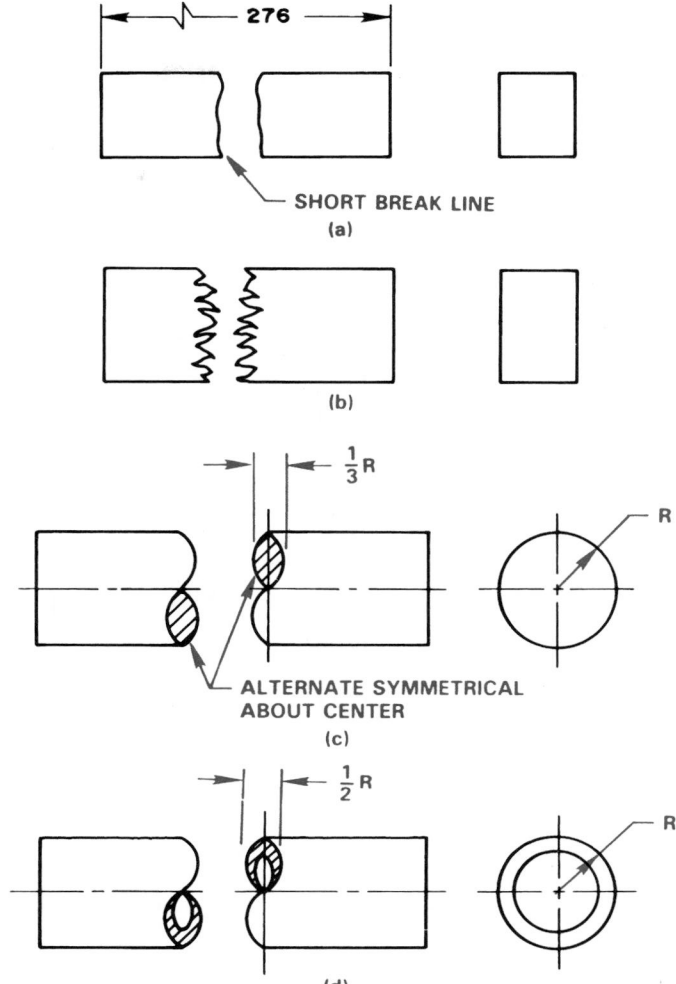

FIGURE 14.30 ■ Conventional breaks for various shapes are (a) metal shapes, (b) wood shapes, (c) cylindrical solid shapes, (d) cylindrical tubular shapes.

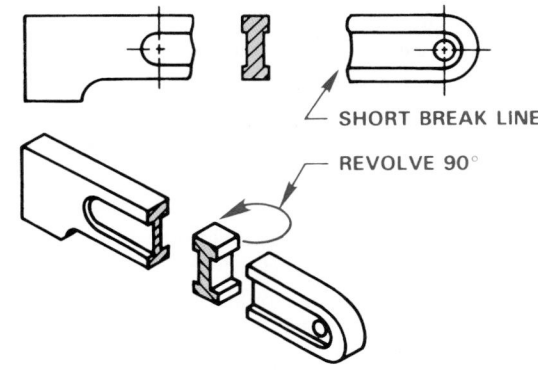

FIGURE 14.31 ■ Revolved section.

14.32. In Figure 14.32a the revolved section is drawn on the part. The revolved section also may be broken away as seen in Figure 14.32b. The surrounding space may be used for dimensions as shown in Figure 14.33.

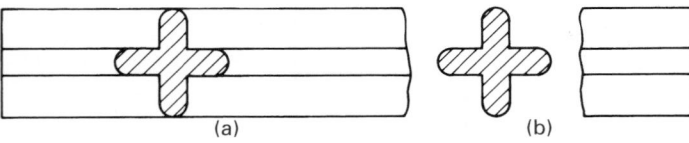

FIGURE 14.32 ■ **(a)** Revolved section not broken away; **(b)** revolved section broken away.

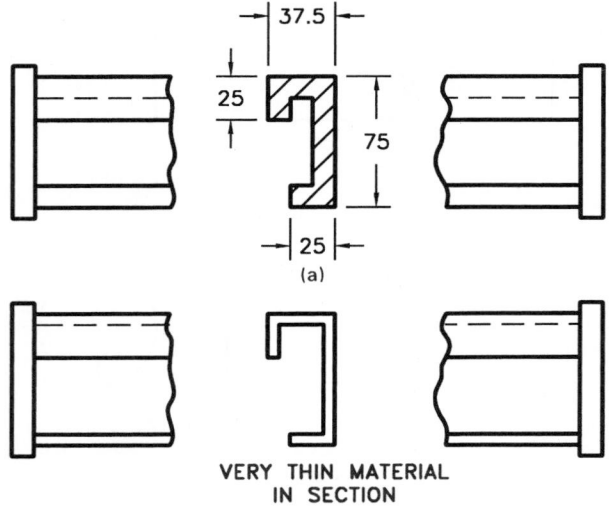

FIGURE 14.33 ■ **(a)** Dimensioning a broken-away revolved section; **(b)** a revolved section through thin material. Section lines are omitted when thin material less than 4 mm thick is sectioned.

Notice in Figure 14.33 that very thin parts, less than 4 mm thick, in sections may be left unsectioned. This practice is also common when sectioning a gasket or similar feature.

REMOVED SECTIONS

Removed sections are similar to revolved sections except they are removed from the view. A cutting-plane line is placed through the object where the section is taken. Removed sections are not generally placed in direct alignment with the cutting-plane line but are placed in a surrounding area as shown in Figure 14.34.

Removed sections may be preferred when a great deal of detail makes it difficult to effectively use a revolved section. An additional advantage of the removed section is that it may be drawn to a larger scale so close detail may be more clearly identified as shown in Figure 14.35. The sectional view should be labeled, as shown, with SECTION A-A and the revised scale which in this example is 2:1. The predominant scale of the principal views is shown in the title block.

Multiple removed sections are generally arranged on the sheet in alphabetical order from left to right and top to bottom. (See Figure 14.36.) The cutting planes and related sections are labeled alphabetically excluding the letters *I*, *O*, and *Q*, as they may be mistaken for numbers. When the entire alphabet has been used,

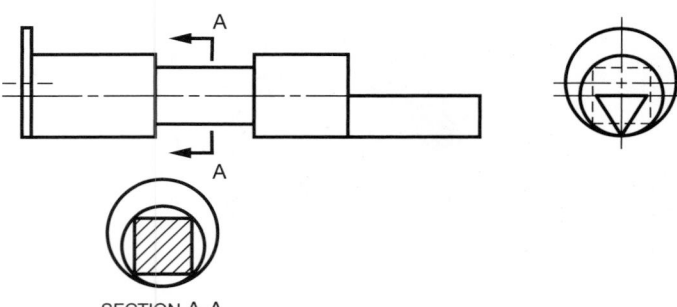

SECTION A-A

FIGURE 14.34 ■ Removed section.

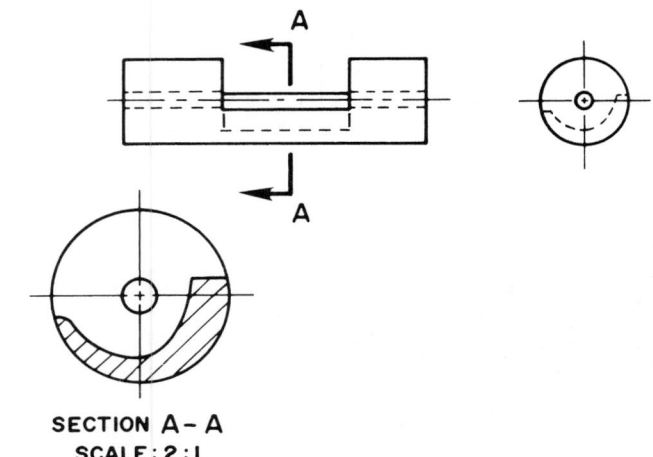

SECTION A- A
SCALE: 2:1

FIGURE 14.35 ■ Enlarged removed section.

label your sections with double letters beginning with *AA*, *BB*, and so on.

Another method of drawing a removed section is to extend a centerline adjacent to a symmetrical feature and revolve the section of the centerline as shown in Figure 14.37. The removed section may be drawn at the same scale or enlarged as necessary to clarify detail.

RECOMMENDED REVIEW

It is recommended that you review the following parts of Chapter 3 before you begin working on fully dimensioned sectioning drawings. This will refresh your memory about how related lines are properly drawn.

■ Object Lines.

■ Viewing Planes.

■ Extension Lines.

■ Hidden Lines.

■ Break Lines.

■ Dimension Lines.

■ Centerlines.

■ Phantom Lines.

■ Leader Lines.

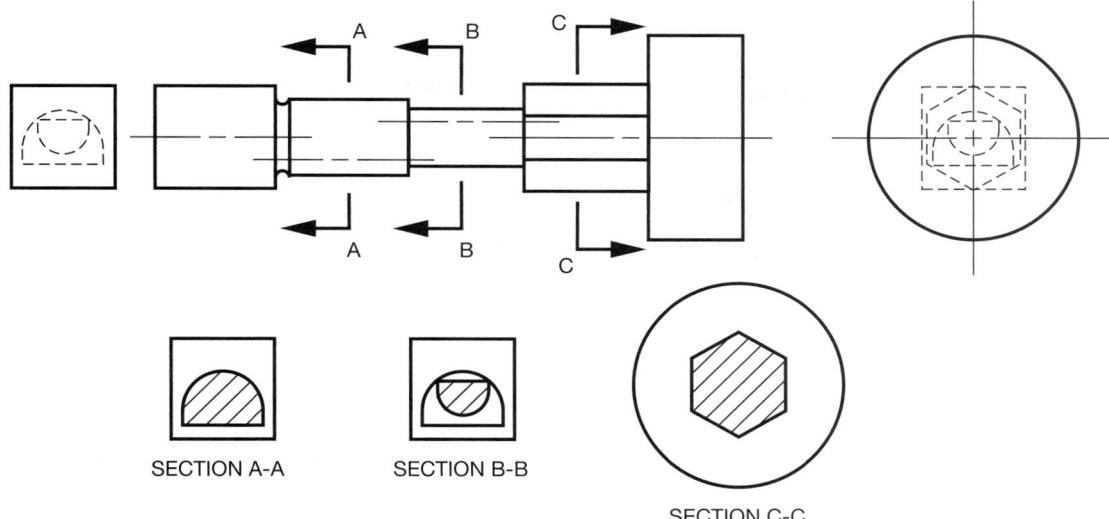

FIGURE 14.36 ■ Multiple removed sections. Hidden lines in the profile views help illustrate the importance of sectional views.

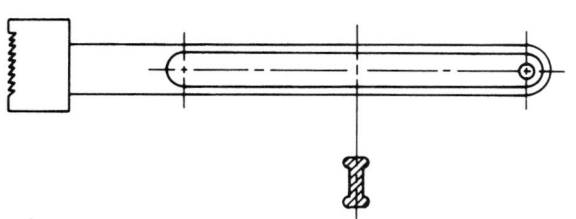

FIGURE 14.37 ■ Alternate removed section method.

Study these topics more carefully as related directly to sectioning practices.

■ Cutting Planes.

■ Section Lines.

■ Coded Section Lines.

Also review Chapters 7 and 8 covering multiview and auxiliary view drawings and Chapter 10 for recommended dimensioning practices.

PROFESSIONAL PERSPECTIVE

Your role as an engineering drafter is communication. You are the link between the engineer and manufacturing, fabrication, or construction. Up to this point you have learned a lot about how to perform this communication task, but the hard part comes when you have to apply what you have learned. Your goal is to make every drawing clear, complete, and easy to interpret. There are a lot of factors to consider, including:

■ Selection of the front view and related views.

■ Deciding if sections are to be used.

■ Placement of dimensions in a logical format so that the size description and located features are complete.

This is a difficult job, but you are motivated to do it. And just when you have mastered how to select and dimension multiviews, you are faced with preparing sectional drawings, which means a dozen main options and endless applications. (But don't worry because you have made it this far!) Here are some key points to consider:

■ You need to completely show and describe the outside of the part first.

■ If there are a lot of hidden lines for internal features, you know immediately that a section is needed, because you cannot dimension to hidden lines.

■ Analyze the hidden lines and decide on the type of section that will work best.

■ Review each type of sectioning technique until you find one that looks best for the application.

■ Be sure you clearly label the cutting-plane lines and related multiple sections so they correlate.

■ Half sections are good for showing both the outside and inside of the object at the same time, but be cautious about their use, because sometimes they confuse the reader.

You're not alone in this new venture. The problem assignments in this chapter recommend a specific sectioning technique during this learning process.

Figure 14.38 shows a carefully created drawing of a complex part. Along with front, left-side and rear views, there is a full section and auxiliary section.

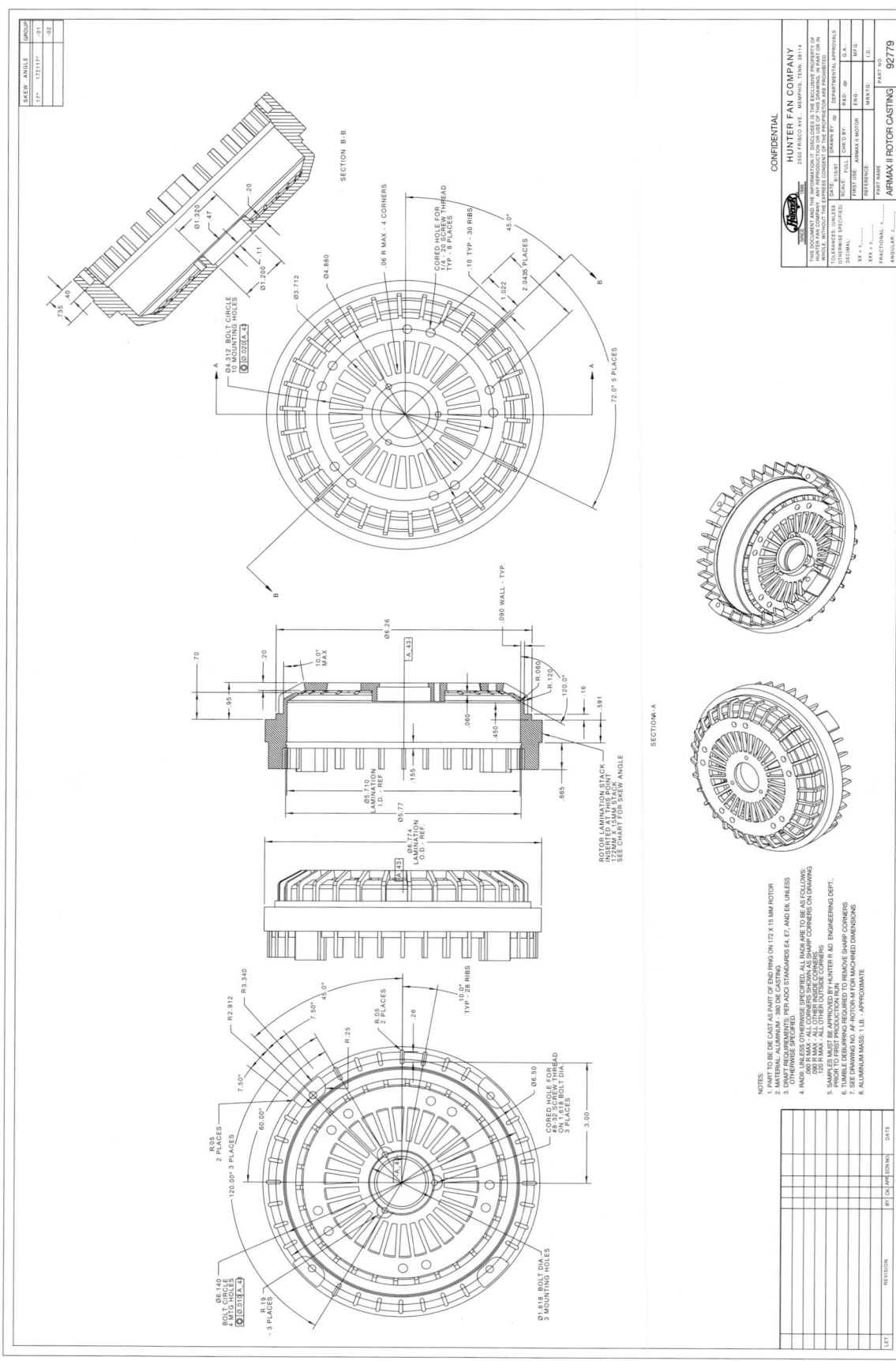

FIGURE 14.38 ■ An actual industry drawing of a complex part, using front, left-side, and rear views along with a full and auxiliary section. *Compliments of Hunter Fan Company, Memphis, TN.*

DISTANCE BETWEEN HOLES ON A BOLT CIRCLE

MATH APPLICATION

PROBLEM: Find the center-to-center distance between adjacent holes on the bolt circle shown in Figure 14.39.

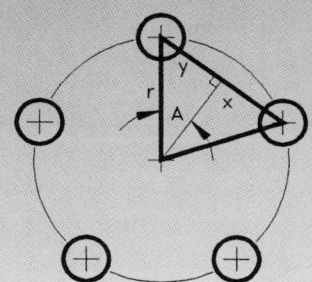

FIGURE 14.40 ■ Bolt circle with a right triangle drawn.

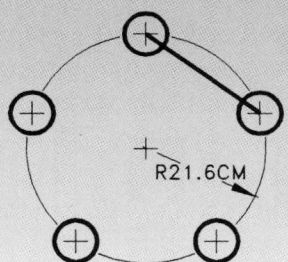

FIGURE 14.39 ■ Bolt circle.

SOLUTION: Problems like this involve finding the side of a right triangle. The first step is to construct a right triangle onto the drawing, as shown in Figure 14.40. From the definition of *sine*. (See Math Instruction Appendix in the Online Companion.)

CHAPTER 14

Sectioning Test

DIRECTIONS
Answer the questions with short, complete statements or drawings as needed on an A-size drawing sheet.

QUESTIONS

1. Describe the potential advantage of a sectional view over a multiview.
2. Explain and use a sketch to show the relationship of the cutting-plane line arrowheads to the sectional view.
3. How much of the object does a full section actually remove?
4. At what angles may section lines be drawn?
5. What are coded section lines?
6. Describe the purpose and advantage of a half section.
7. What type of line is used in a half-sectional view to separate the sectioned area from the nonsectioned area?
8. List four items that would remain unsectioned in a sectional view when the cutting-plane line cuts through parallel to the axes of the items.
9. Describe the error found in drawings (a), (b), (c), (d), and (e) on page 434.
10. Describe a situation when a cutting-plane line may be omitted in sectioning drawings.
11. Discuss the advantage of using conventional revolutions.
12. Discuss the cutting-plane line associated with a broken-out section.
13. What type of break line is generally associated with a broken-out section?
14. In what situations are conventional breaks commonly used?
15. Discuss the difference between a revolved and removed section. An example may be used.
16. Show an example of conventional breaks for the following shapes: (a) Solid cylindrical shape Ø.5 in.; (b) Tube Ø.5 in. with .06 in. wall thickness; (c) Solid steel □ .5 in.

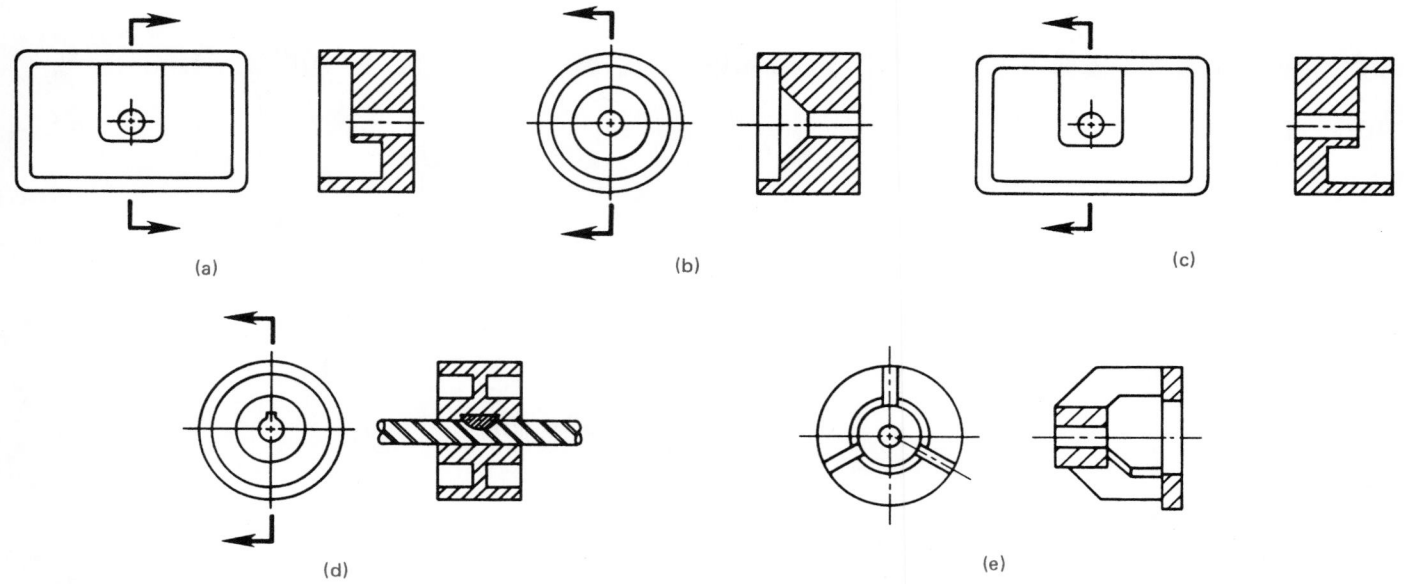

(a) (b) (c)

(d) (e)

14 *Sectioning Problems*

DIRECTIONS

PART 1

The following problems give you a finished view with a proposed cutting-plane line. Complete the other view by drawing the section that properly correlates with the finished view and given cutting-plane line. A small pictorial view is given to assist you in visualization. Do not draw the pictorial view. Measure the given views and transfer the measurements to your manual or CADD drawing. You can also photocopy the textbook page and draw the completed views on the copy. Confirm the preferred method with your instructor.

PROBLEM 14-1

PROBLEM 14-2

FULL SECTION

ALIGNED SECTION

PROBLEM 14-3

PROBLEM 14-4

2X Ø.50
⌴ .75

HALF SECTION

OFFSET SECTION

PART 2

1. From the selected engineer's sketch or layout, determine the views and sections that are needed.
2. Make a sketch of the selected views and sections as close to correct proportions as possible. Do not spend a lot of time, as the sketch is only a guide. Indicate where the cutting-plane lines and dimensions are to be placed.
3. Using the sketch you have developed as a guide, draw an original sectioned multiview drawing on an adequately sized drawing sheet. Select a scale that properly details the part on the selected sheet size. Use unidirectional dimensioning. Use manual or computer-aided drafting as required by your course guidelines.
4. Include the following general notes at the lower left corner of the sheet .5 in. each way from the corner border lines:

   1. INTERPRET DIMENSIONS AND TOLERANCES PER ASME Y14.5M—1994.
   2. REMOVE ALL BURRS AND SHARP EDGES.

NOTES:

Additional notes may be required depending on the specifications of each individual assignment. A tolerance block is recommended as shown in problems for Chapter 12 unless otherwise specified.

PROBLEM 14-5   Full Section (in.)

Part Name: Fitting

Material: Bronze

Finish All Over: 63 μin.

Note: Refer to Chapter 10 for proper dimensioning practices. Engineering sketches may not display correct practices. This problem may require a front view, side view showing the hexagon, and a full section to expose the interior features for dimensioning.

UNSPECIFIED TOLERANCES:

DECIMALS	mm	IN.
X	± 2.5	± .1
XX	± 0.25	± .01
XXX	± 0.127	± .005
ANGULAR ± 30'		
FINISH	3.2μm	125μIN.

5. The engineering layouts may not be dimensioned properly. Verify the correct practice before placing dimensions; for example, the diameter symbol should precede the diameter dimension, and leaders should not cross over dimension lines. Check other line and dimensioning techniques for proper standards. Actual industrial drawings are provided as advanced problems throughout.

PROBLEM 14-6   Full section and view enlargement (in.)

Part Name: Spring

Material: SAE 1060

Problem based on original art courtesy Stanley Hydraulic Tools, a Division of the Stanley Works.

Heat Treat:

1. Austenitize at 1475° F.
2. Direct quench in agitated oil.
3. Temper to R_c C 44–46.

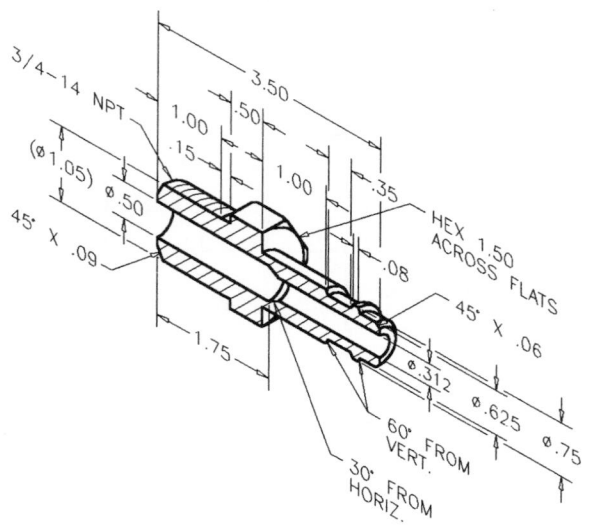

PROBLEM 14-7 Full section (metric)
Part Name: Hydraulic Valve Cylinder
Material: Phosphor Bronze
All Fillets and Rounds: R.1

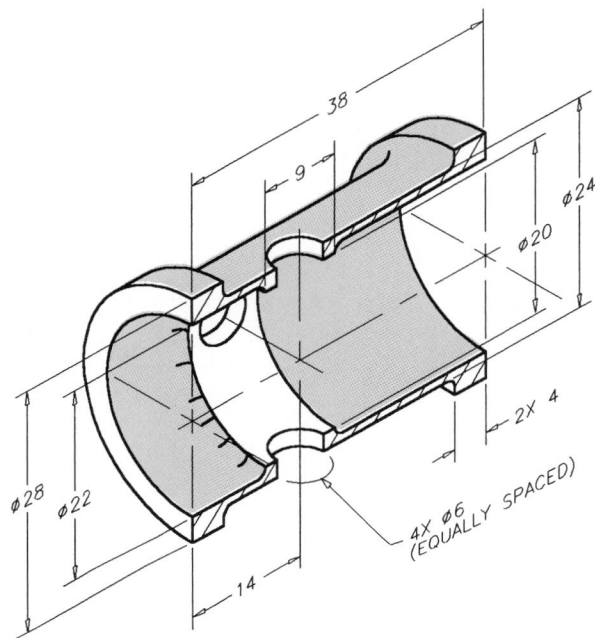

PROBLEM 14-8 Full section (in.)
Part Name: Machine Plate
Material: 6160 T6 Steel

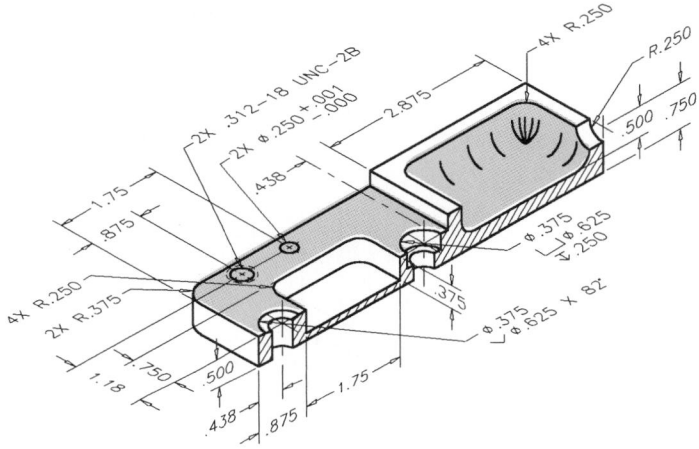

PROBLEM 14-9 Full section (in.)
Part Name: Face Plate
Material: 6160 T6

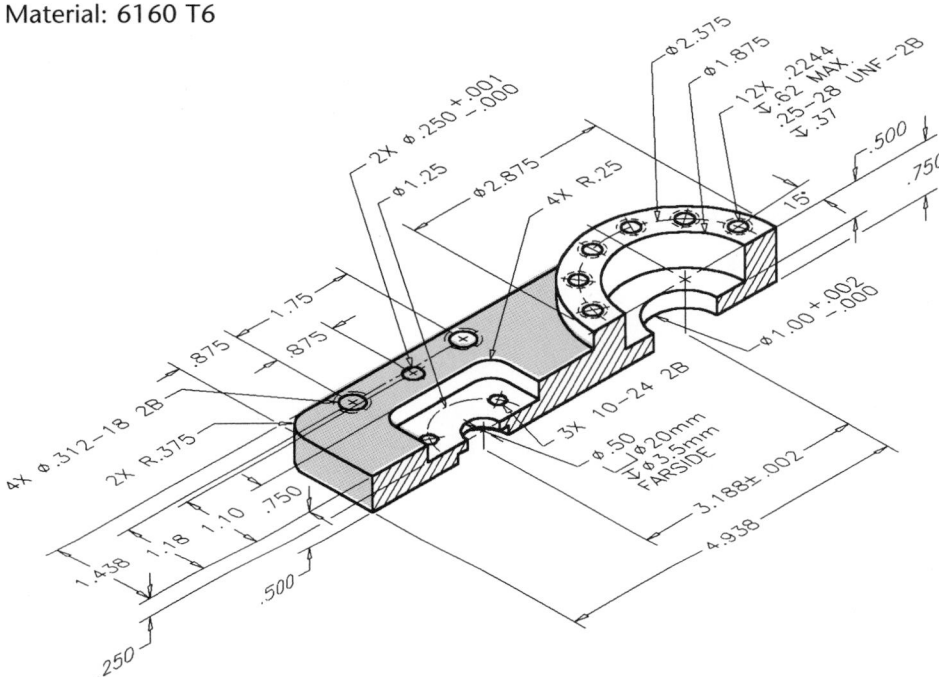

PROBLEM 14-10   **Full section, dimensioning cylindrical shapes (in.)**
Part Name: Plug
Material: Phosphor Bronze

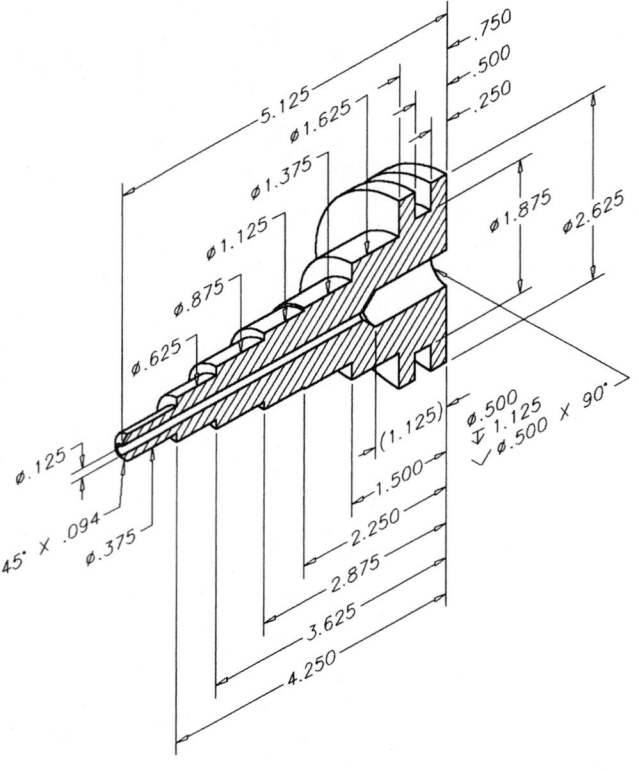

PROBLEM 14-12   **Full and broken-out section (metric)**
Part Name: Hydraulic Valve Cylinder
Material: Phosphor Bronze

Note: Proposed sections and section lines are not given in engineer's layout.

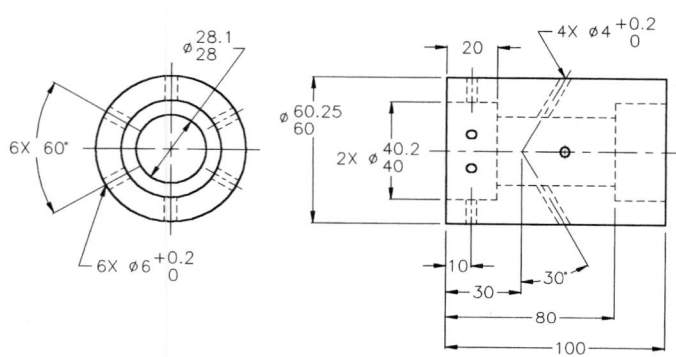

PROBLEM 14-13   **Full section (in.)**
Part Name: Hanger
Material: SAE 1030
Fillets and Rounds: R.062

PROBLEM 14-11   **Full section or half section (in.)**
Part Name: Hub
Material: Cast Iron
SPECIFIC INSTRUCTIONS:

Convert the broken-out section in the given drawing to a full section.

Note: Type of section used affects view selection.

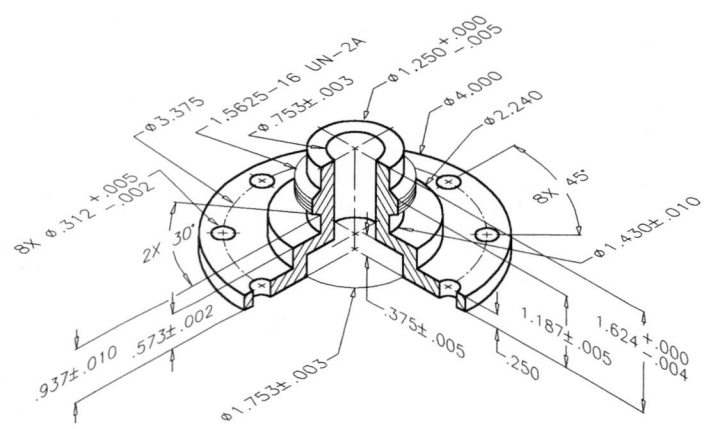

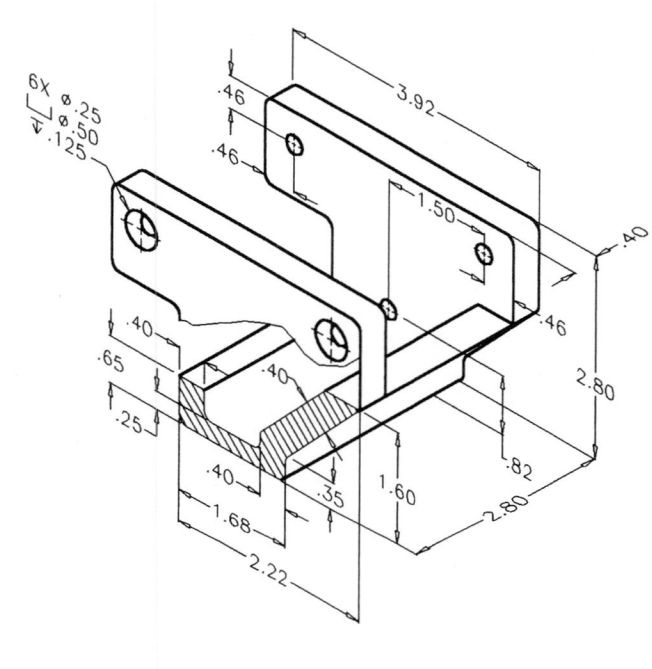

PROBLEM 14-14   Advanced full section from engineer's sketch (metric)

Part Name: Bearing Housing

Material: SAE 1015

Courtesy Aerojet TechSystems Co.

Note: Use dimensioning symbols based on the ASME/ANSI standards discussed in Chapter 10.

Note the following abbreviations:

CBORE—counterbore

SF—spotface

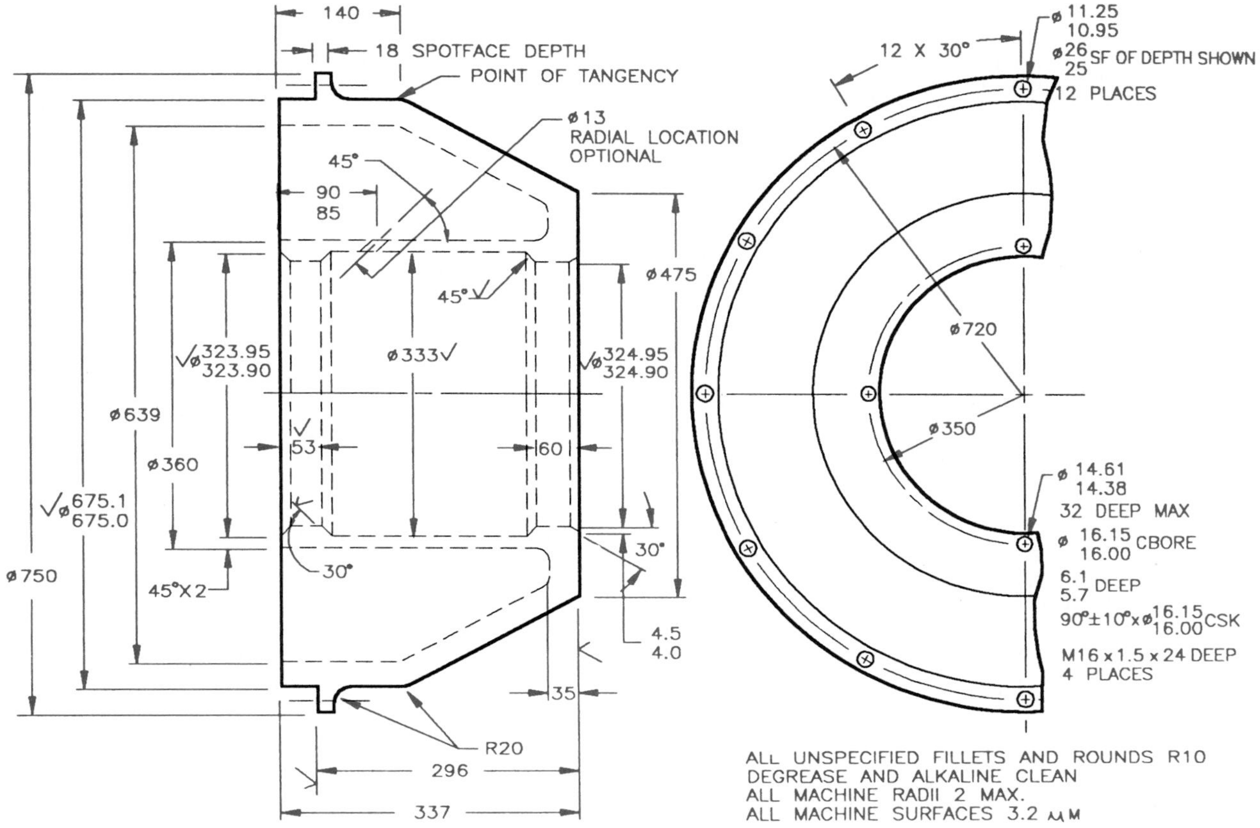

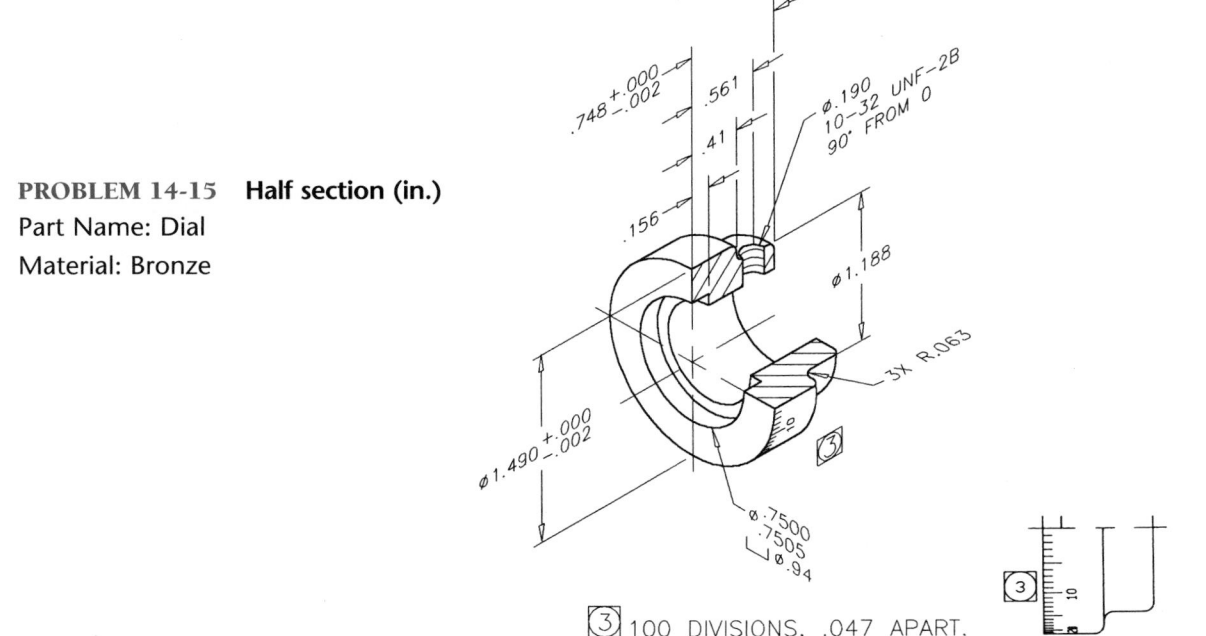

PROBLEM 14-15   Half section (in.)

Part Name: Dial

Material: Bronze

PROBLEM 14-16   Half section and auxiliary view (metric)

Part Name: Nozzle Base

Material: Titanium, ASTM-B367 Grade C3

Notes:

Clean casting by mechanical blasting prior to alpha case removal. Chemically remove alpha case prior to inspection of castings using procedure approved by procurement activity. Cast surfaces shall be visually inspected and be free from cracks, tears, laps, shrinkage, and porosity.

Radiographic inspect castings per MIL-STD-271 acceptance criteria for porosity and inclusions per ASTM-E-446 and for cavity shrinkage per ASTM-E-192 (plates for .75 wall thickness). Severity levels of casting defects shall be no greater than tabulated here for the indicated casting areas.

Note: Provide adequate views to avoid crossing extension lines over dimension lines.

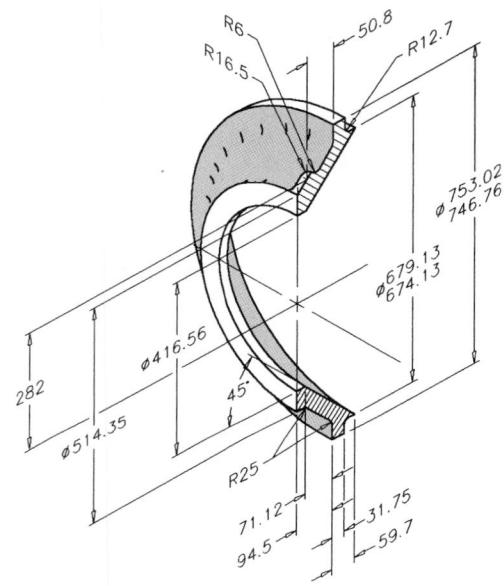

PROBLEM 14-17   Half section and broken-out section (in.)

Part Name: Bench Block

Material: AISI 1018

Case Harden: .020 Deep 59-60 Rockwell C Scale or AISI 4140

Oil Quench 40-45C.

REAM TOLERANCE: $+.0002$ / $-.0000$

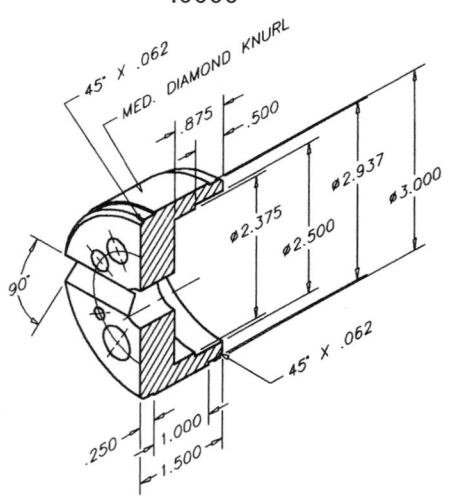

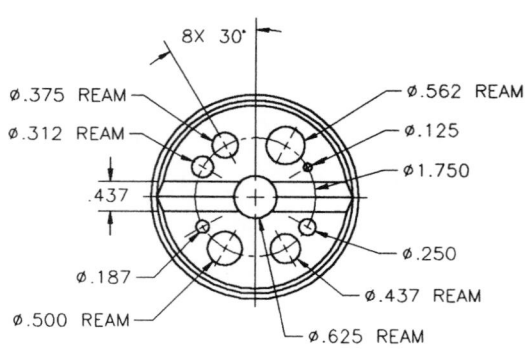

FRONT VIEW

PROBLEM 14-18   Half section (metric)

Part Name: Idler Pulley

Material: SAE 4310

Fillets and Rounds: R3.2 mm

Note: Many diameter dimensions should be placed on left- and right-side views.

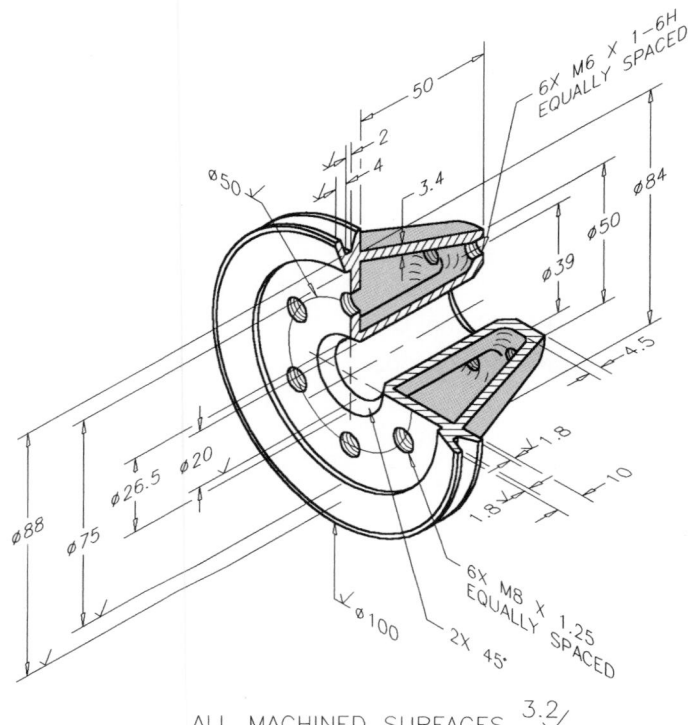

ALL MACHINED SURFACES 3.2

PROBLEM 14-19   Offset section (in.)
Part Name: Die Casting
Material: SAE 6150

SPECIFIC INSTRUCTIONS:
Convert all dimensions to ANSI Y14.5M standards as shown and discussed in this text.

Courtesy of Kris Altmiller.

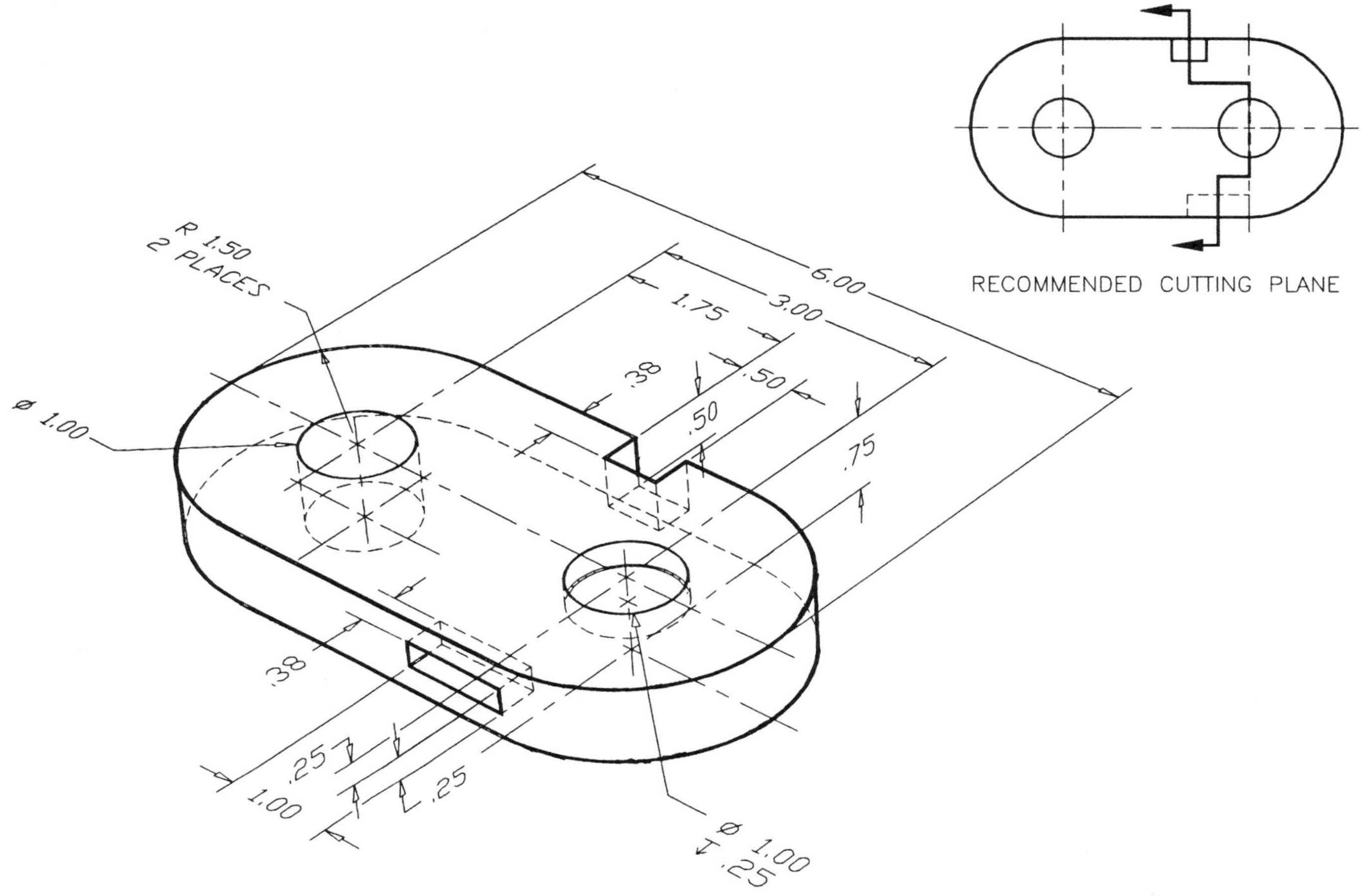

RECOMMENDED CUTTING PLANE

PROBLEM 14-20   Offset section (in.)
Part Name: Drill Plate
Material: SAE 1020
Case Harden: 55 Rockwell C Scale
Fillets and Rounds: R. 12
FAO: 63 μin.

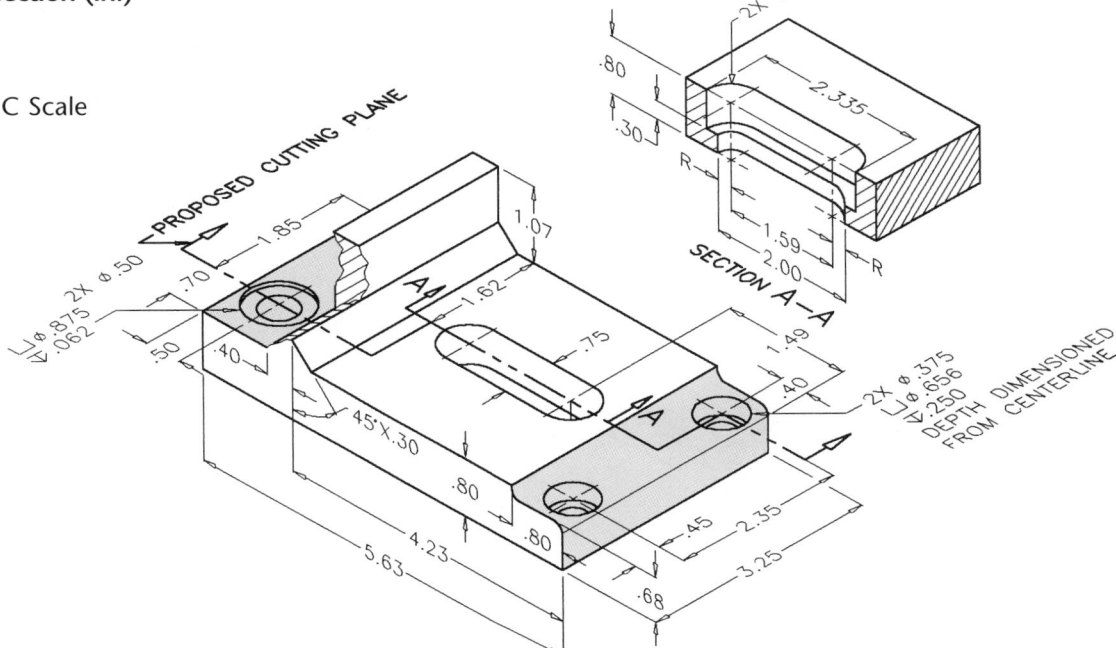

PROBLEM 14-21   Aligned section (in.)
Part Name: Hub
Material: SAE 3145
Fillets and Rounds: R. 125

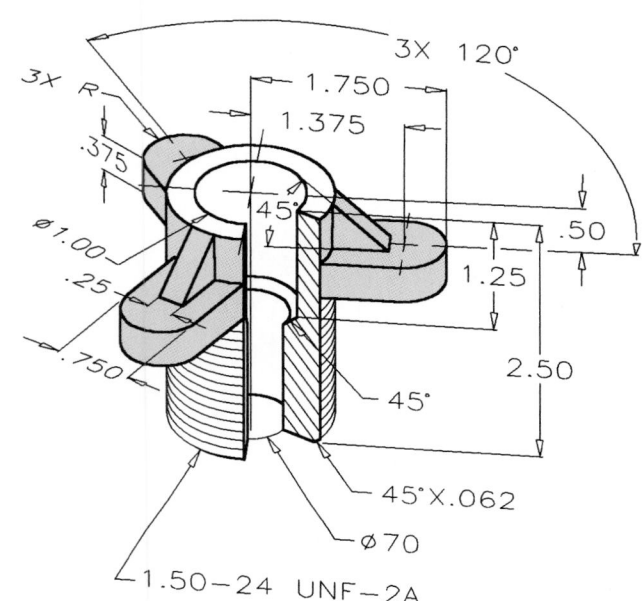

PROBLEM 14-22   Broken-out section (in.)
Part Name: Taper Shaft
Material: SAE 4320
FAO: 16 μin.

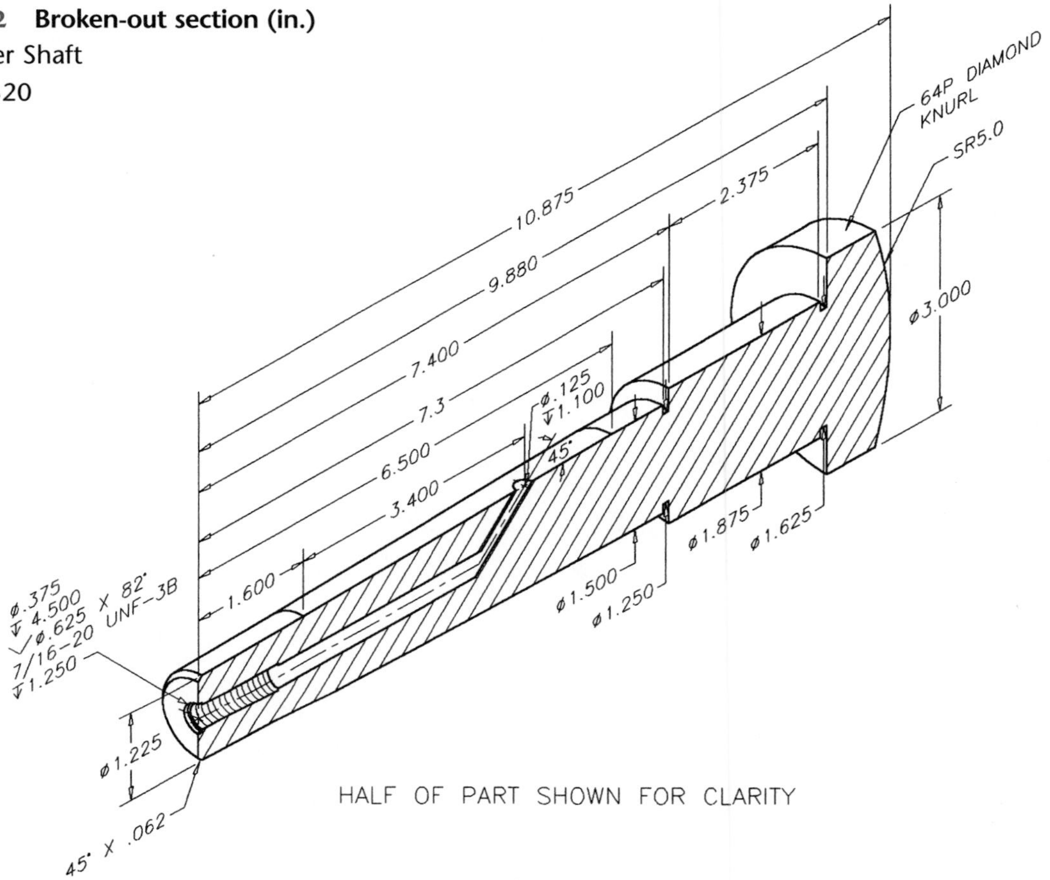

HALF OF PART SHOWN FOR CLARITY

PROBLEM 14-23   Auxiliary section (metric)

Part Name: Gear Base

Material: SAE 2340

Finish All Over: 0.8 μm

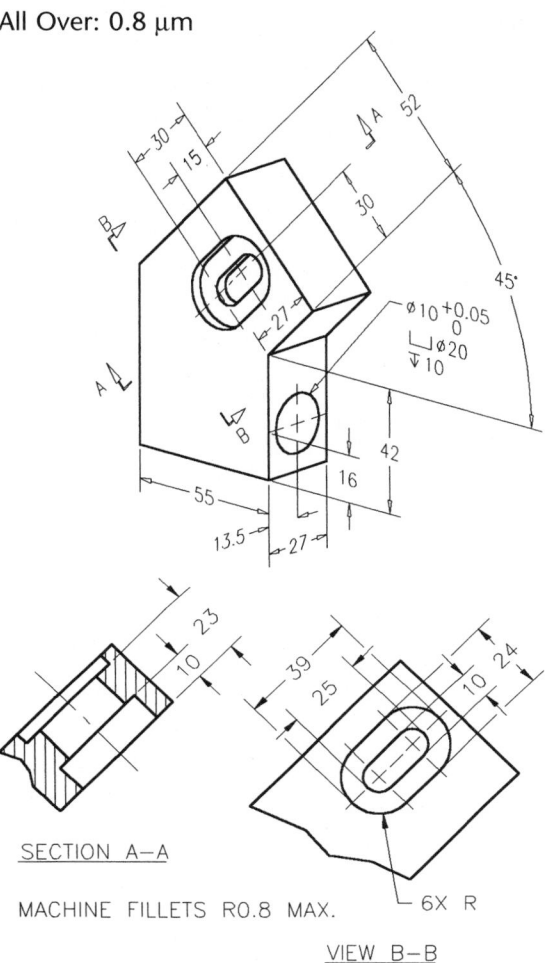

SECTION A–A

MACHINE FILLETS R0.8 MAX.

VIEW B–B

PROBLEM 14-24   Broken-out section, view enlargement (in.)

Part Name: Clamp Cap

Material: Cast Aluminum

Fillets and Rounds: R.06

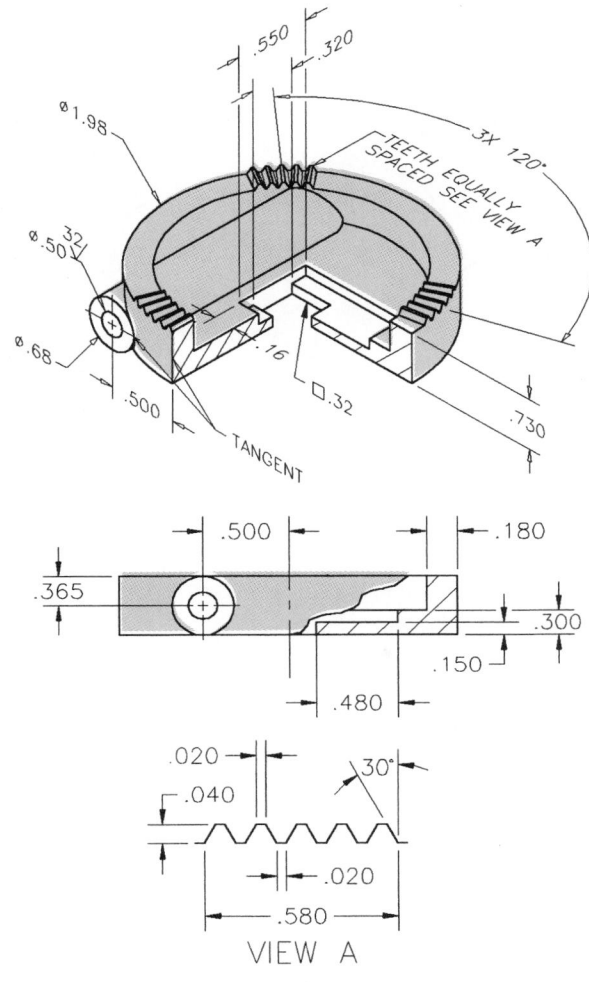

VIEW A

PROBLEM 14-25   Broken-out section (metric)

Part Name: 50mm 45° Elbow

Material: Cast Iron

Fillets and Rounds: R4 mm

Consider bottom view or auxiliary view for hole pattern dimensions.

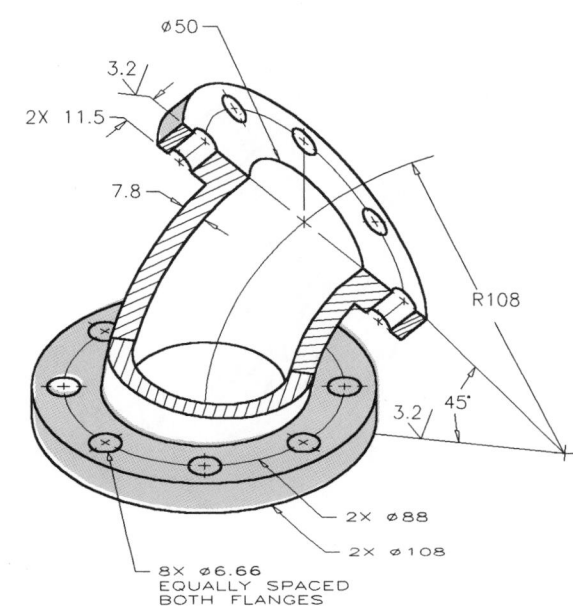

PROBLEM 14-26   Revolved section (in.)

Part Name: End Loading Arm

Material: SAE 2310

Fillets and Rounds: R.25

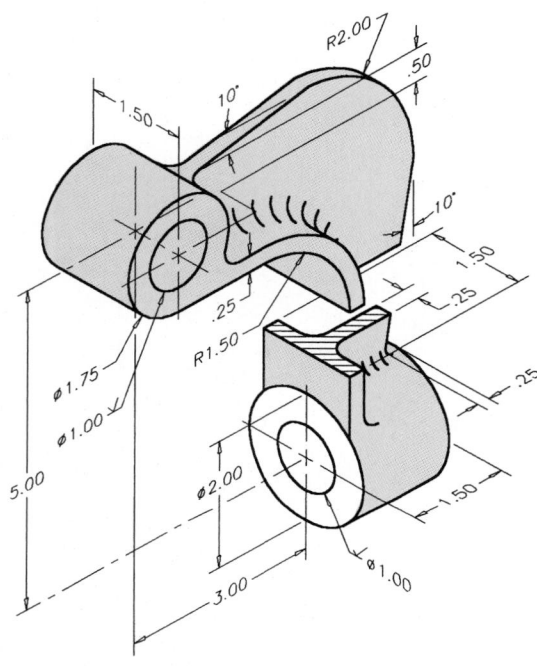

PROBLEM 14-27   Revolved section (in.)

Part Name: Offset Handwheel

Material: Bronze

All Fillets and Rounds: R.12

Finishes: 125 μin.

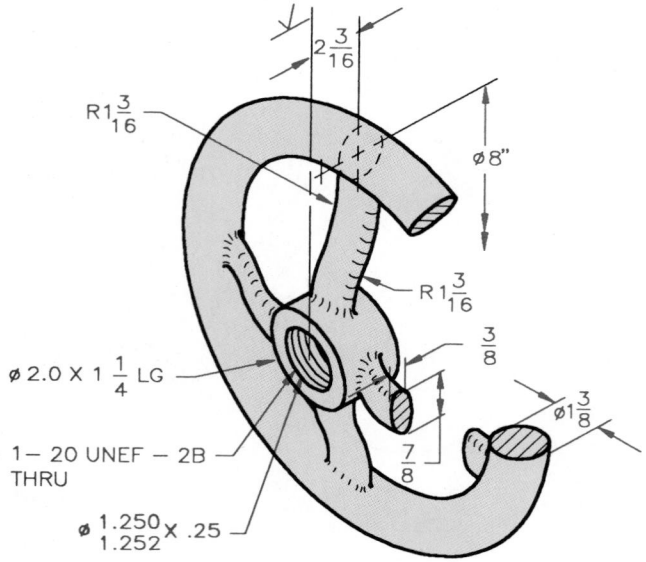

PROBLEM 14-28   Broken-out, revolved section, view enlargement (in.)

Part Name: Pipe Wrench Handle

Material: SAE 5120

Fillets and Rounds: R.06

SPECIFIC INSTRUCTIONS: Engineering sketch is given in fractional inches, convert all size dimensions to two-place decimals and location dimensions to three-place decimals for final drawing.

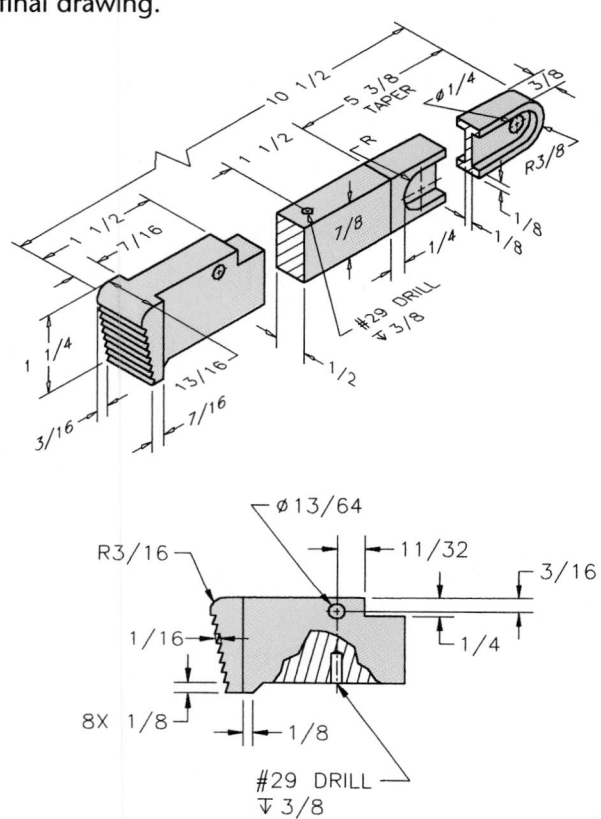

PROBLEM 14-29   Removed sections (metric)

Part Name: Valve Stem

Material: Phosphor Bronze

Finish All Over: 1 μm

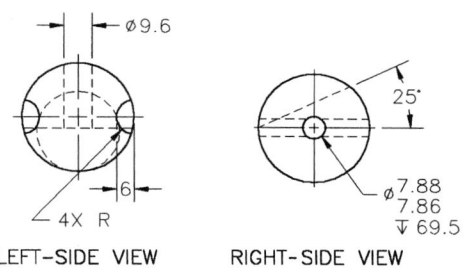

LEFT-SIDE VIEW   RIGHT-SIDE VIEW

NOTE: Some hidden lines omitted for clarity.

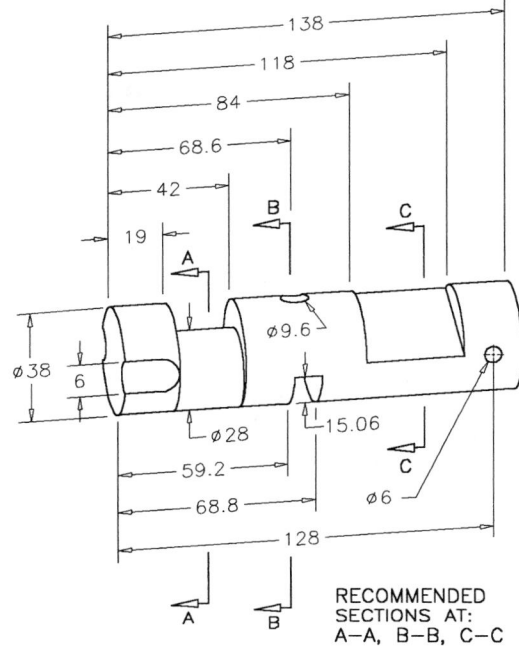

RECOMMENDED
SECTIONS AT:
A–A, B–B, C–C

PROBLEM 14-30   Offset section (in.)

Part Name: Die Plate Casting

Material: SAE 3120

Courtesy of Kris Altmiller.

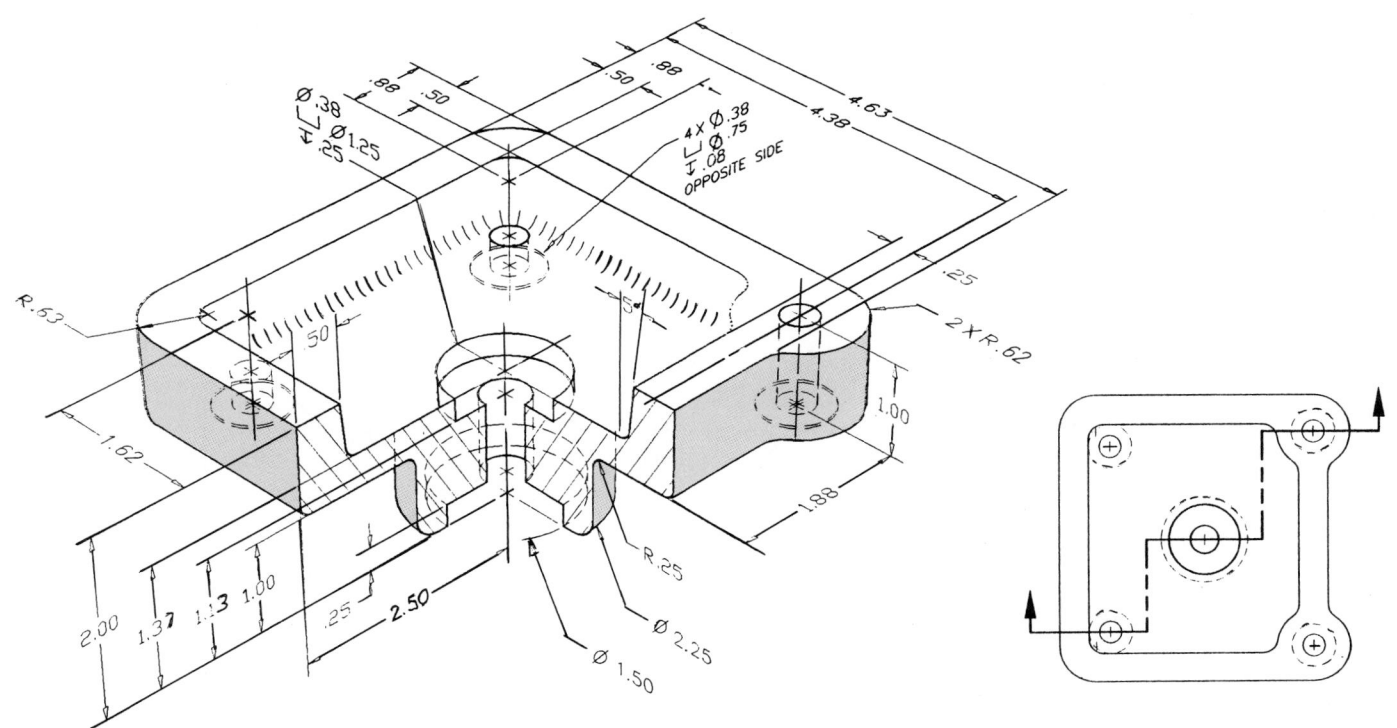

RECOMMENDED CUTTING PLANE

ALL FILLETS AND ROUNDS R.25 UNLESS OTHERWISE SPECIFIED

PROBLEM 14-31   Broken-out section (in.)
Part Name: Slide Bar Connector
Material: SAE 4120
SPECIFIC INSTRUCTIONS: Convert point-to-point dimensioning to datum dimensioning.

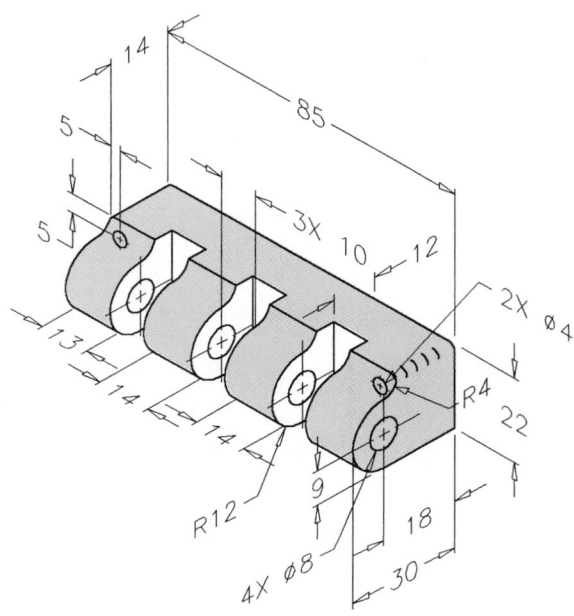

PROBLEM 14-33   Conventional break (in.)
Part Name: Leg
Material: Ø2.00 Schedule 40 A120
Problem based on original art courtesy TEMCO.

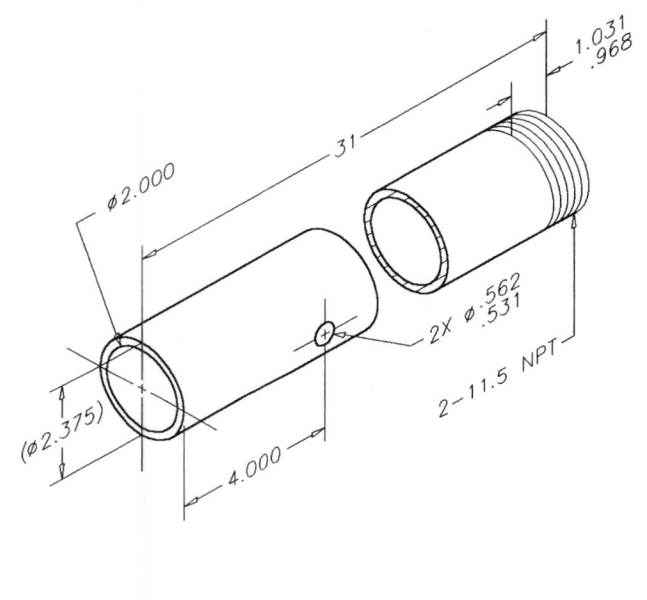

PROBLEM 14-32   Broken-out section (in.)
Part Name: Drain Tube
Material: Ø1.00 × .065
Problem based on original art courtesy TEMCO.

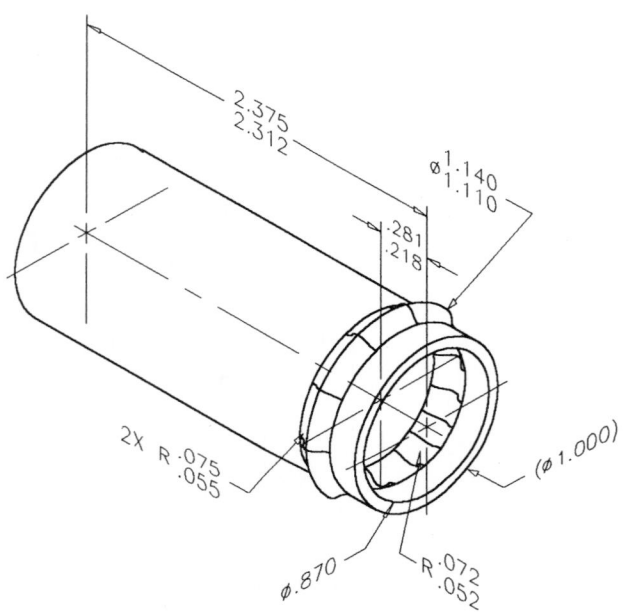

PROBLEM 14-34   Offset section (in.)
Part Name: Cover
Material: CI

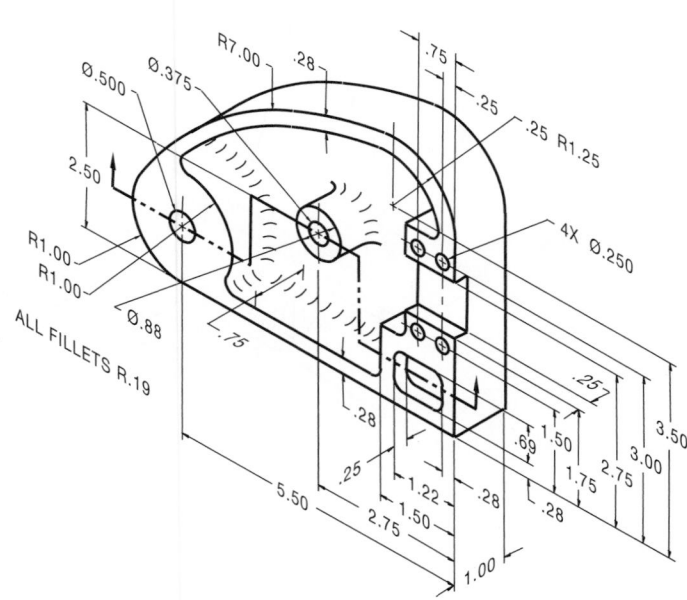

PROBLEM 14-35   Revolved section, conventional revolution (in.)

Part Name: Crank Arm

Material: Cast Steel 80,000

All Fillets and Rounds: R.12

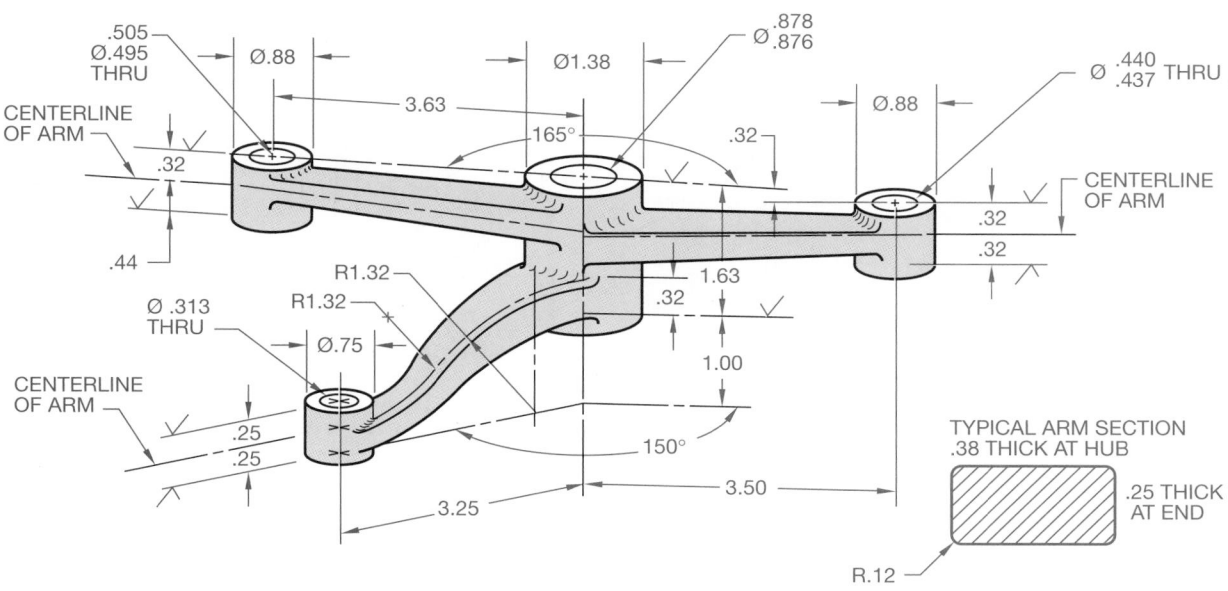

PROBLEM 14-36   You select the sectioning technique (in.)

Part Name: Base Support

Material: Mild Steel

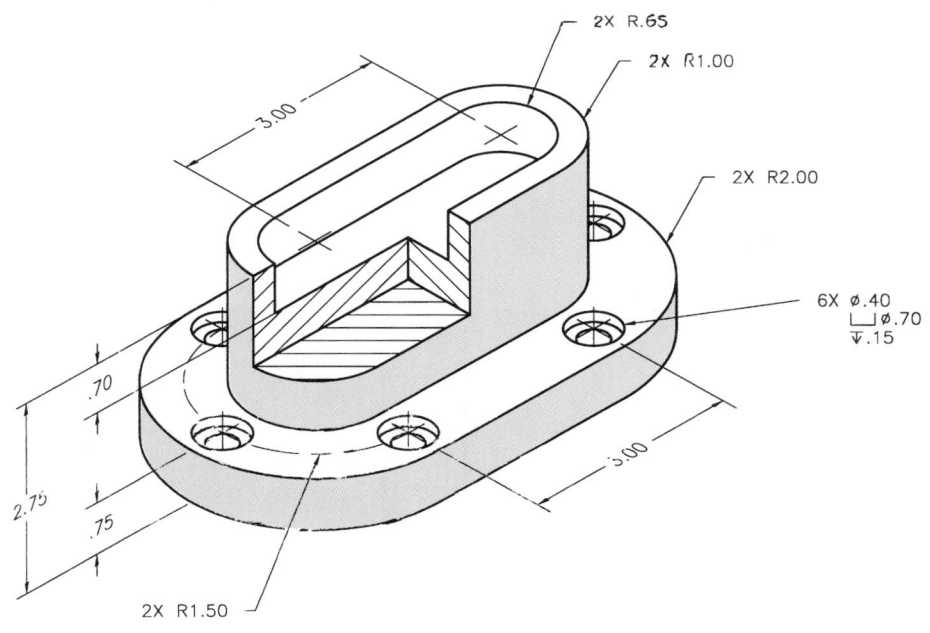

PROBLEM 14-37   You select the sectioning technique (in.)
Part Name: Mounting Base
Material: Cast Aluminum

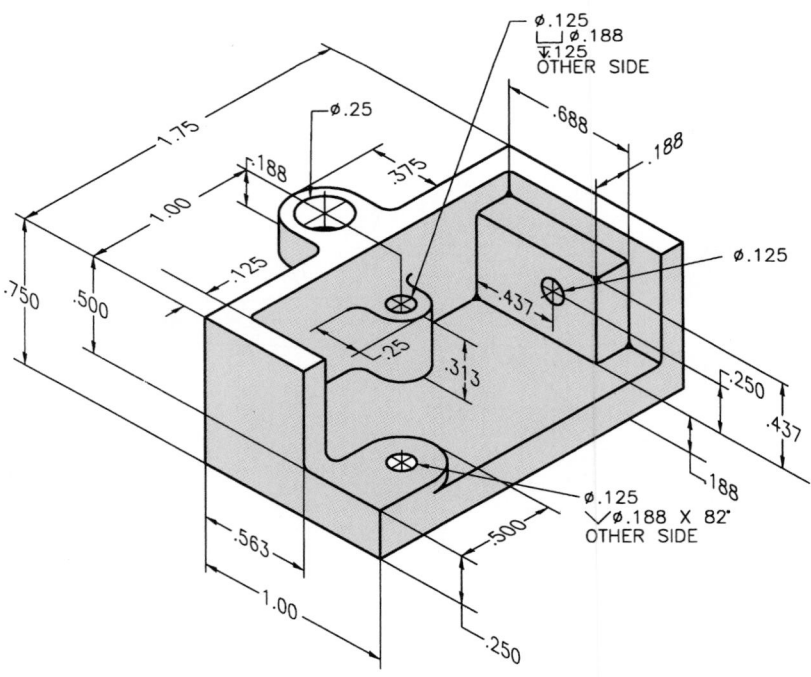

PART 3 DIRECTIONS

The following problems are actual industry drawings that require redrawing using CADD. A common responsibility of an entry-level engineering drafter is to convert existing drawings to CADD. This is a good way to become proficient and gain product knowledge. Use the same instructions given in Part II, and convert all conventions and dimensions to the proper ASME standards as you redraw the given projects.

PROBLEM 14-38   Full section from actual industry drawing (in.)
Part Name: Bolt Guard
Material: Aluminum

Courtesy Stanley Hydraulic Tools, a Division of the Stanley Works.

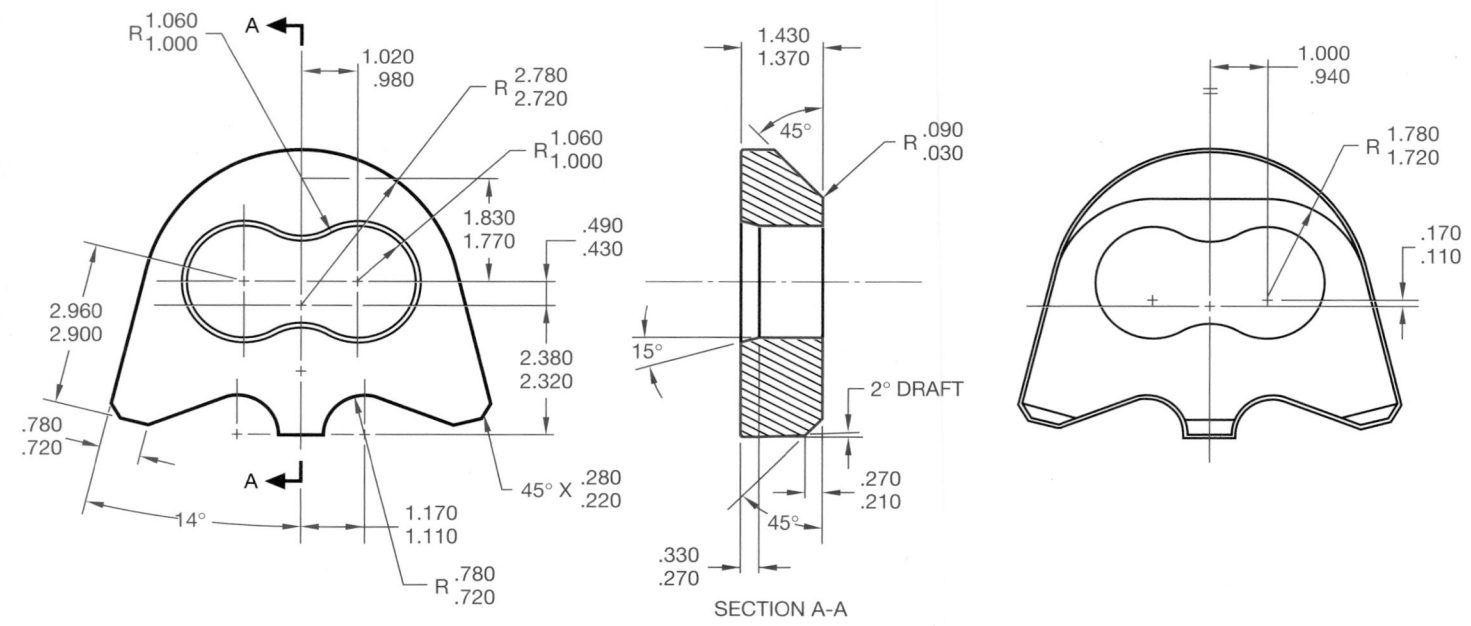

SECTION A-A

PROBLEM 14-39   Advanced full section and partial auxiliary section from actual industry drawing (in.)
Part Name: Accumulator Plug

Material: Phosphor Bronze
Courtesy Stanley Hydraulic Tools, a Division of the Stanley Works.

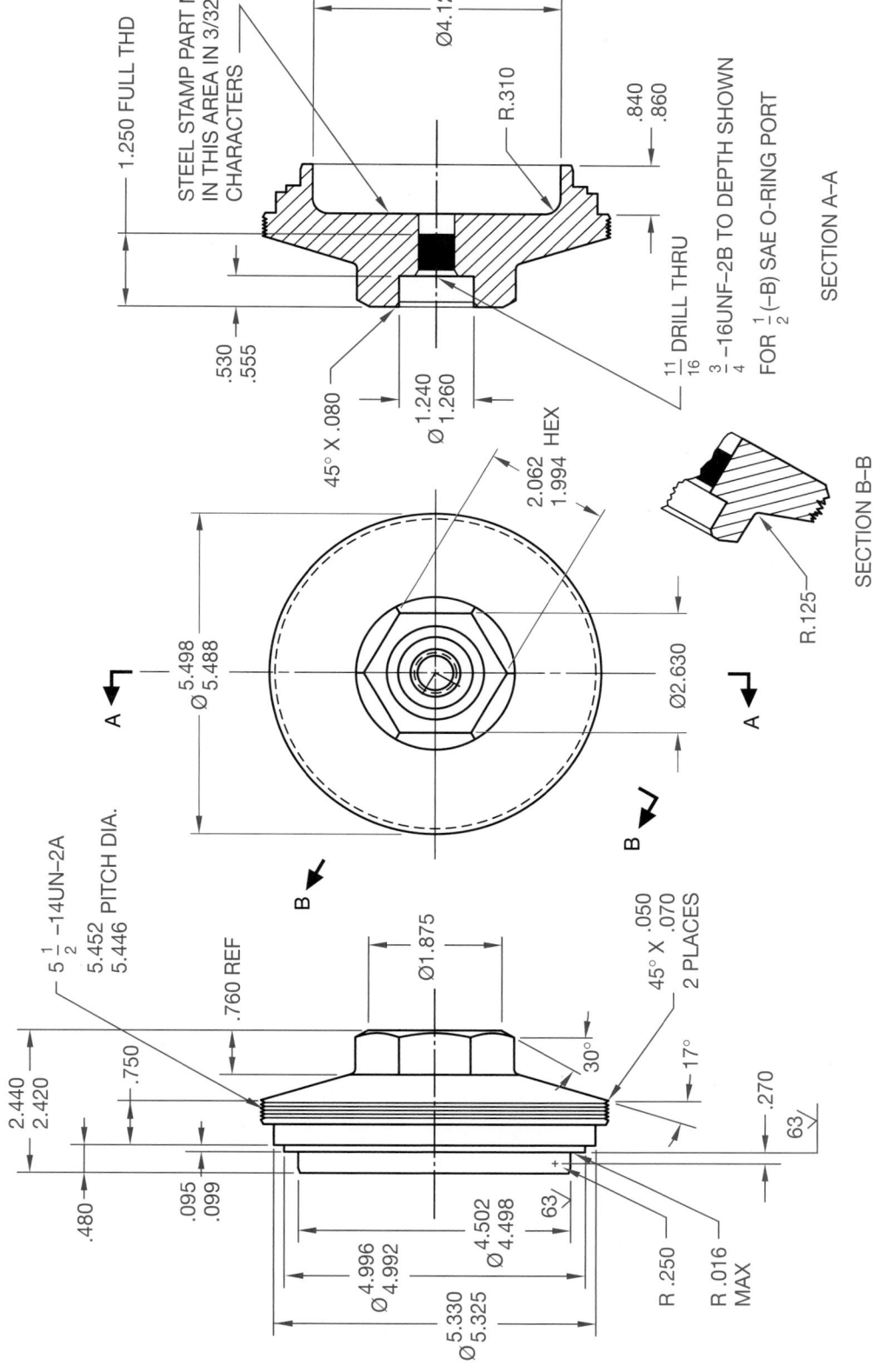

PROBLEM 14-40 **Advanced full section and view enlargement from actual industry drawing (in.)**

Part Name: Swivel

Material: SAE 5150

Courtesy Stanley Hydraulic Tools, a Division of the Stanley Works.

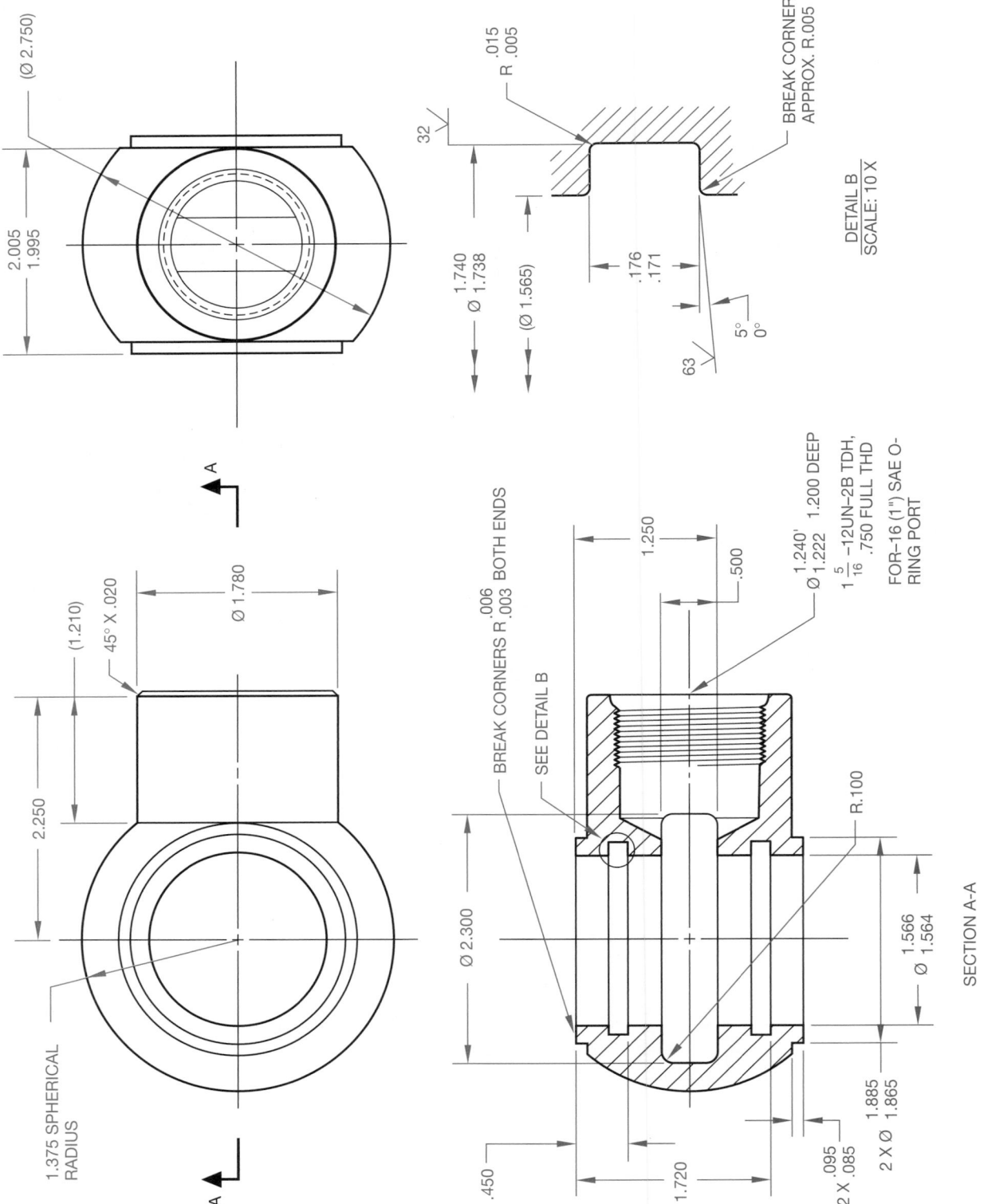

PROBLEM 14-41 Advanced full section, auxiliary section, and view enlargement from actual industry drawing (in.)

Part Name: Bulk Head

Material: Cast Iron

Courtesy Stanley Hydraulic Tools, a Division of the Stanley Works.

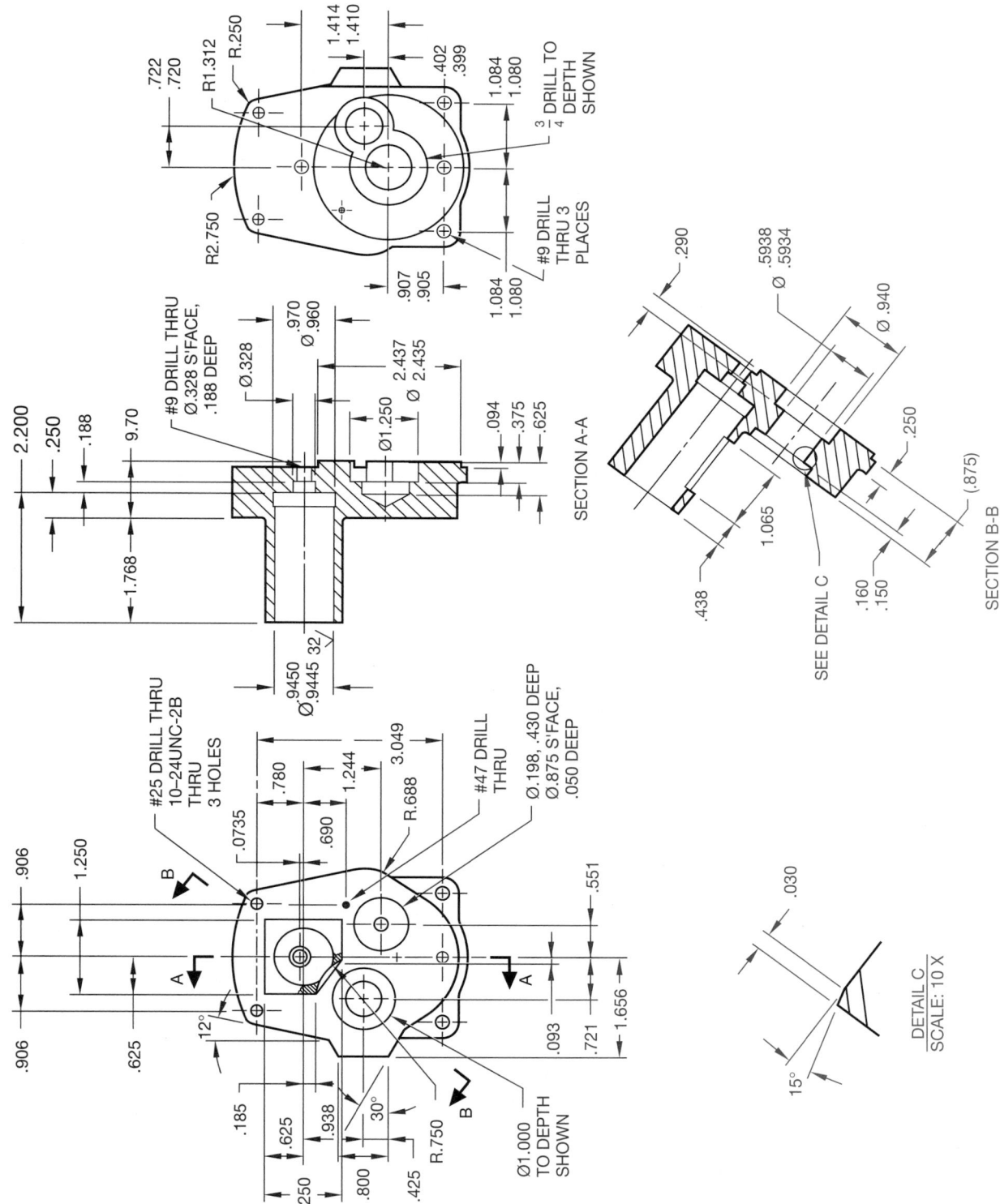

PROBLEM 14-42 Advanced removed sections and view enlargements from actual industry drawing (in.)

Part Name: Swivel Stem

Material: SAE 4340

Courtesy Stanley Hydraulic Tools, a Division of the Stanley Works.

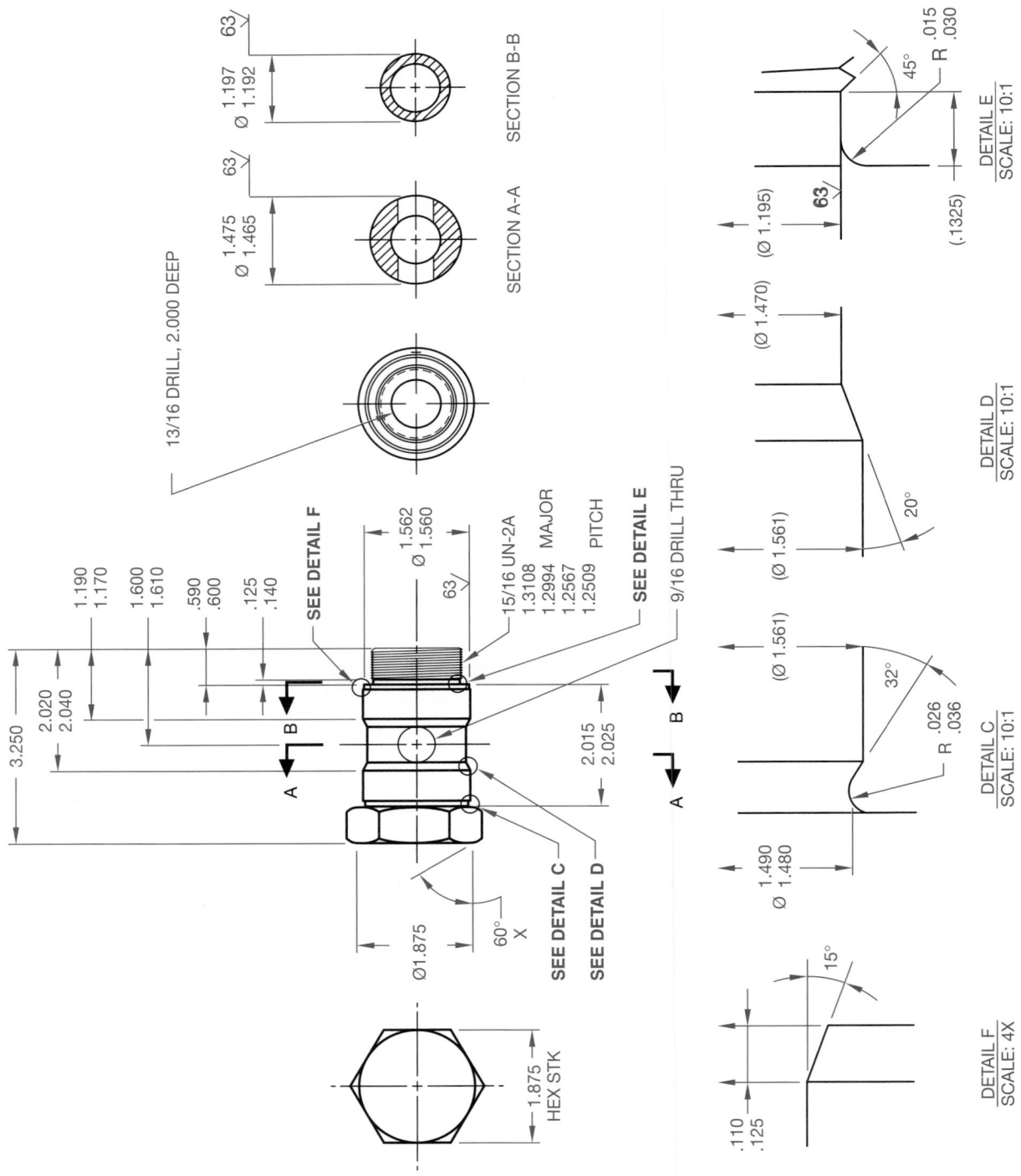

PROBLEM 14-43   Advanced half and full sections from actual industry drawing (in.)

Part Name: Diffuser Casting

Material: Titanium

SPECIFIC INSTRUCTIONS:

Convert all dimensions to ASME Y14.5M standards as shown and discussed in this text. Display the casting mate-rial using phantom lines as shown in the engineer's layout. Use ANSI standard cutting-plane line.

ALTERNATE INSTRUCTIONS:

Omit phantom line technique for showing casting and make two drawings: one a casting drawing with only casting information, and the other a machining drawing with only machining dimensions and information. Refer to Chapter 11 for a review of this practice.

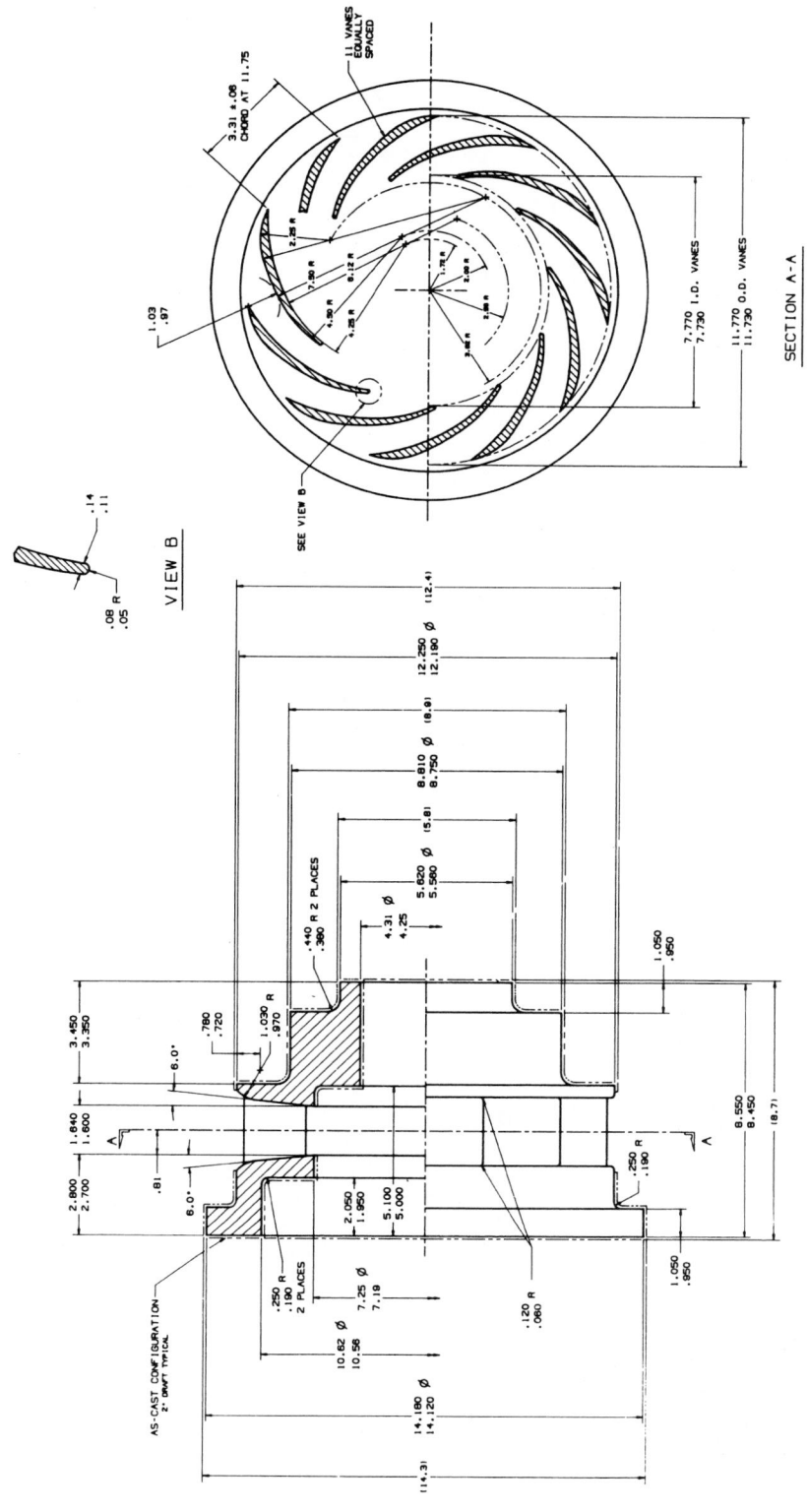

PROBLEM 14-44   **Advanced aligned section from actual industry drawing (in.)**

Part Name: Crankshaft Adapter

Material: CI

Courtesy American Hoist and Derrick Company.

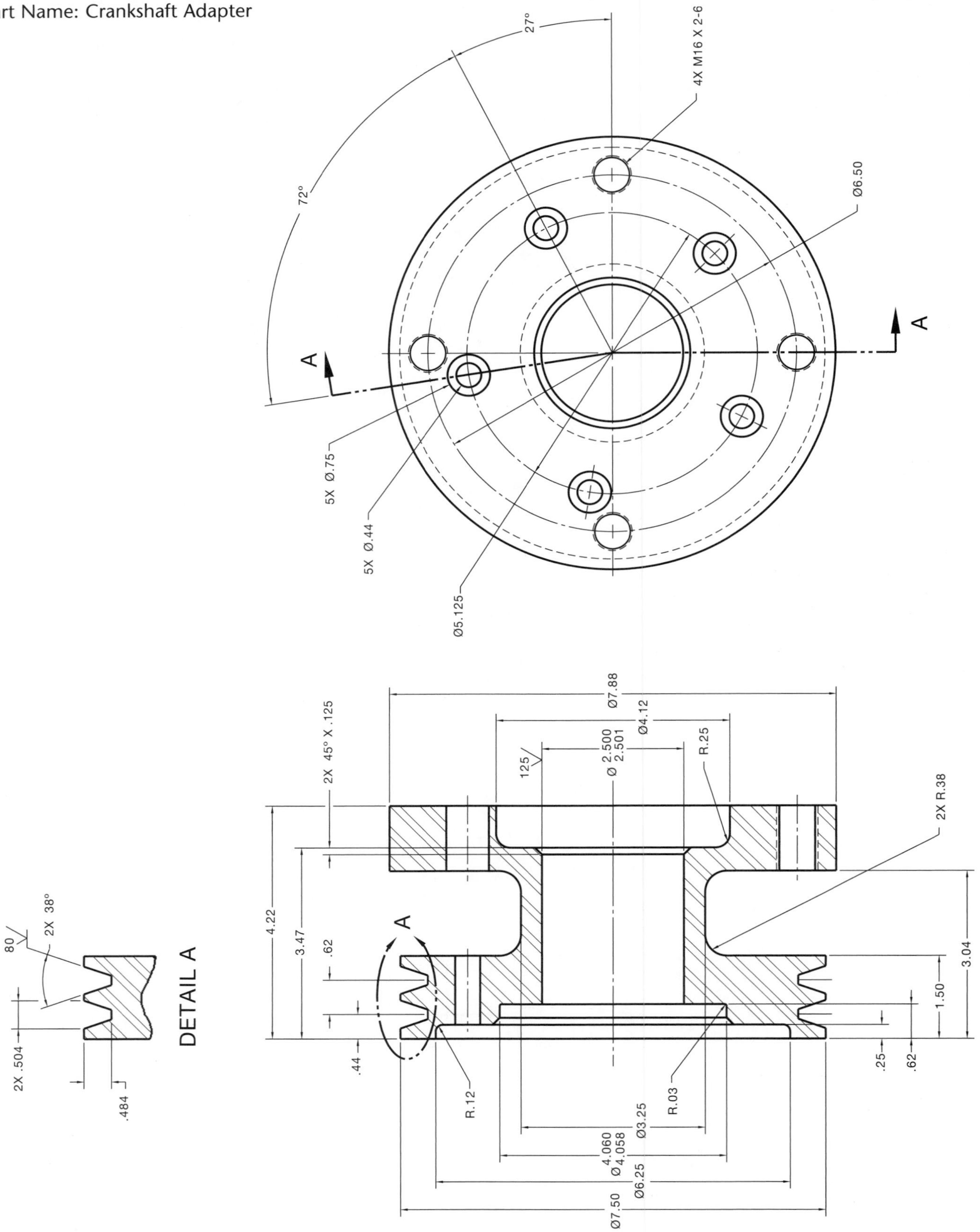

27°

72°

4X M16 X 2-6

Ø6.50

A

A

5X Ø.75

5X Ø.44

Ø5.125

DETAIL A

80

2X 38°

2X .504

.484

Ø7.88

Ø4.12

2X 45° X .125

125

Ø 2.500
Ø 2.501

R.25

2X R.38

4.22

3.47

.62

.44

R.12

Ø 4.060
Ø 4.058

Ø3.25

R.03

Ø6.25

Ø7.50

.25

.62

1.50

3.04

SECTION A-A

**PROBLEM 14-45 Advanced aligned sections from actual
    industry drawing (metric)**

Part Name: Steering Pump Adapter

Material: HC-80

Courtesy Hyster Company.

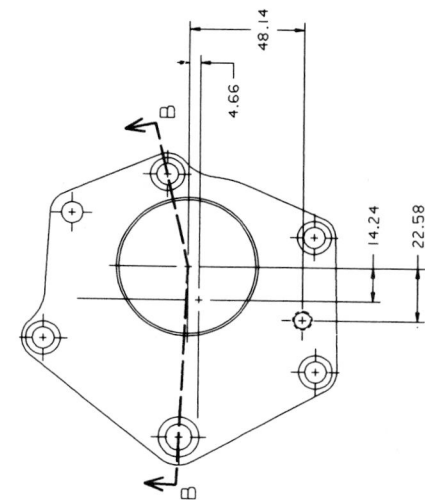

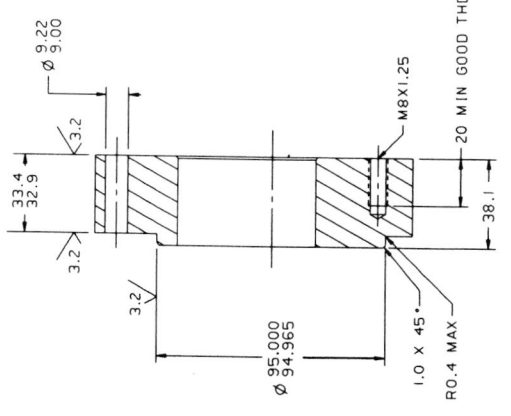

SECTION A - A

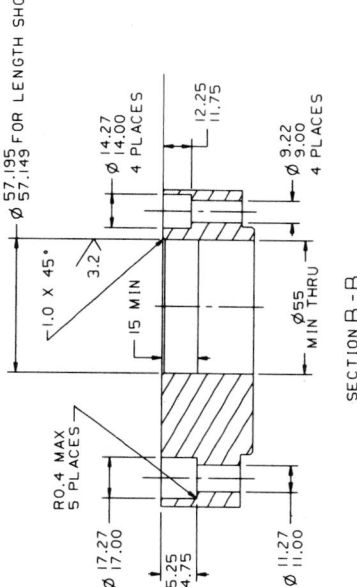

SECTION B - B

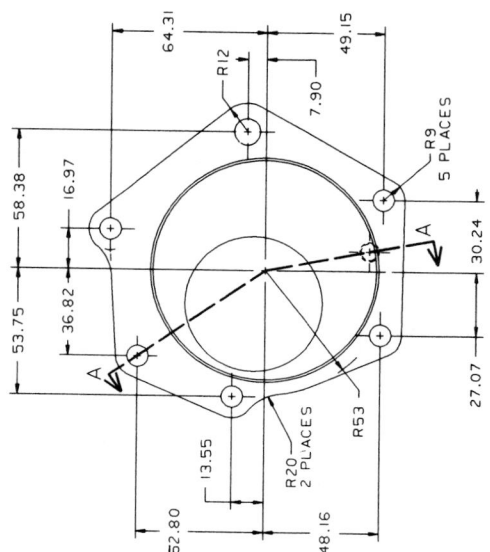

PROBLEM 14-46 **Advanced offset section, enlarged views, from actual industry drawing (metric)**

Part Name: Pad-Monotrol

Material: Hi-mount Pro-fax 8623

SPECIAL INSTRUCTIONS:

Avoid leaders crossing dimension lines.

Courtesy Hyster Company.

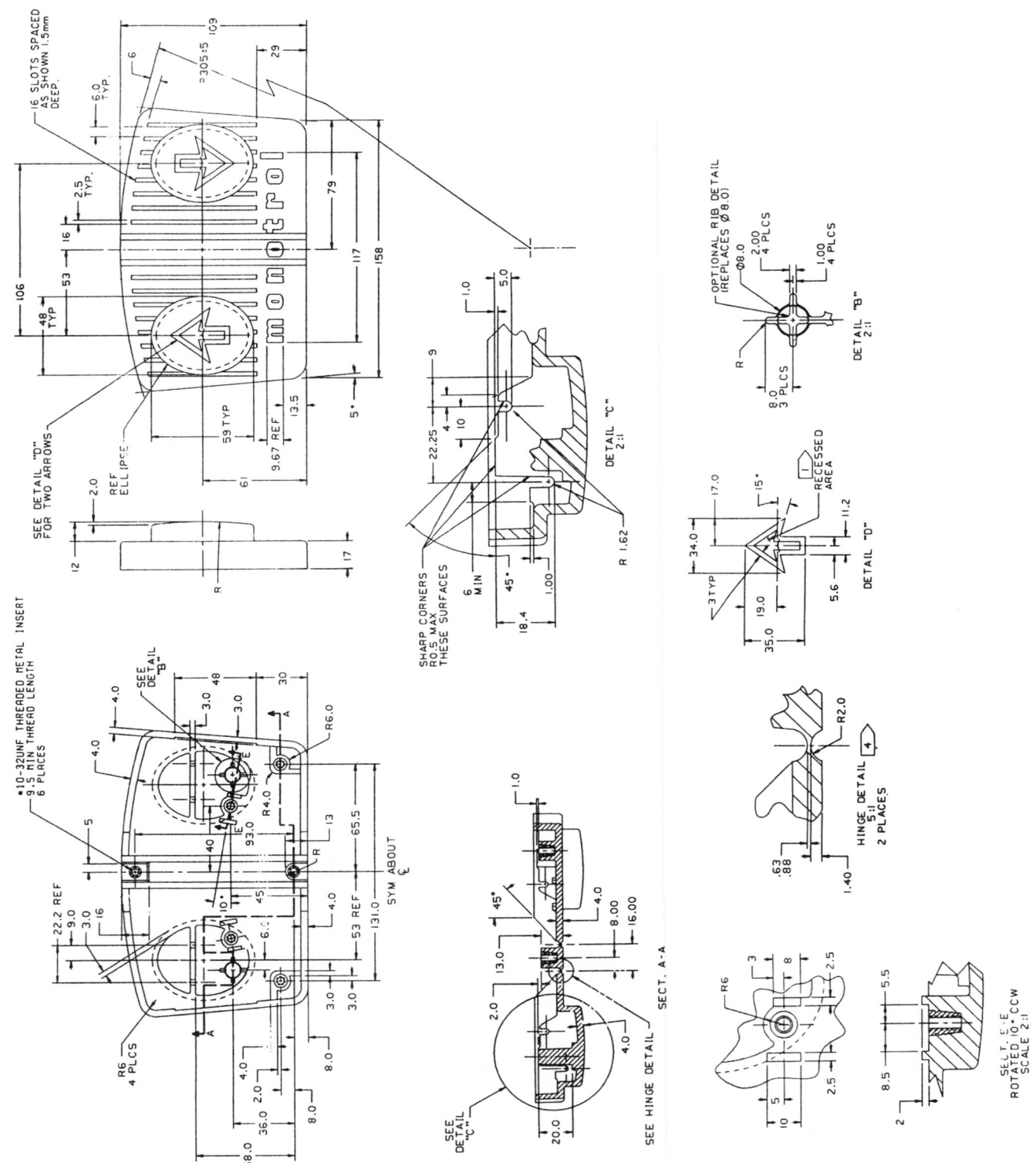

MATH PROBLEMS

PROBLEM 14-47   A bolt circle with radius 14 in. has six equally spaced holes. Find the center-to-center distance between the adjacent holes.

PROBLEM 14-48   A bolt circle with radius 16 in. has eight equally spaced holes. Find the center-to-center distance between the adjacent holes.

PROBLEM 14-49   Find side y of the right triangle shown below.

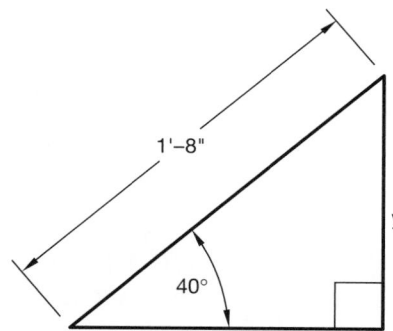

PROBLEM 14-50   Find side x of the right triangle shown below. (Hint: Use cosine.)

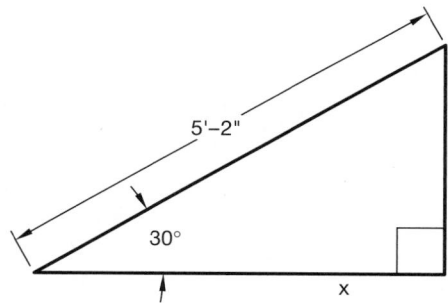

PROBLEM 14-51   Find angle A of the right triangle shown below.

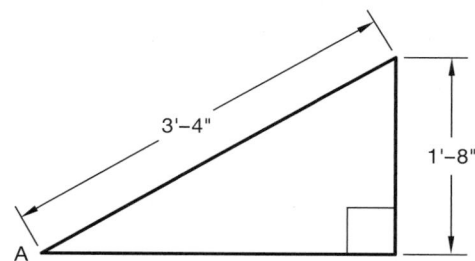

PROBLEM 14-52   Find angle Ø of the right triangle shown below.

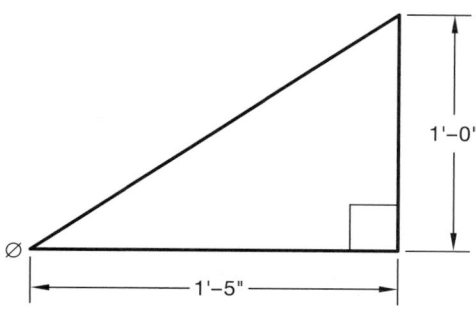

PROBLEM 14-53   Find side y of the right triangle shown below.

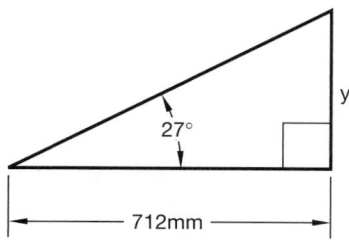

PROBLEM 14-54   Find side r of the right triangle shown below.

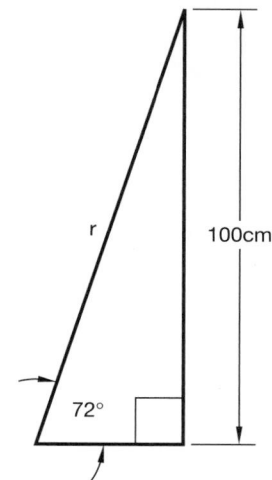

Geometric Tolerancing

OBJECTIVES

After completing this chapter, you will:

- Label datum features on a drawing.

- Establish basic dimensions where appropriate.

- Place proper feature control frames on drawings establishing form, profile, orientation, runout, and location geometric tolerances.

- Use and interpret material condition symbols.

- Determine the virtual condition of features.

- Use geometric tolerancing to completely dimension objects from an engineer's sketch and actual industrial layouts.

- Solve an engineering problem from engineer's notes.

THE ENGINEERING DESIGN APPLICATION

There have been some problems in manufacturing one of the parts due to the extreme tolerance variations possible. The engineer requests that you revise the drawing shown in Figure 15.1 and gives you these written instructions:

- Establish datum A with three equally spaced datum target points at each end of the ⌀28.1–28.0 cylinder.

- Establish datum B at the left end surface.

- Make the bottom surfaces of the 2X ⌀40.2–40.0 features perpendicular to datum A by 0.06.

- Provide a cylindricity tolerance of 0.3 to the outside of the part.

- Make the 2X ⌀40.2–40.0 features concentric to datum A by 0.1.

- Locate the 6X ⌀6 + 0.2 holes with reference to datum A at MMC and datum B with a position tolerance of 0.05 at MMC.

- Locate 4X ⌀4 holes with reference to datum A at MMC and datum B with a position tolerance of 0.04 at MMC.

This is an easy job because you did the original drawing on CADD and you have a custom geometric tolerancing package. You go back to the workstation and within one hour you have the check plot shown in Figure 15.2 ready for evaluation.

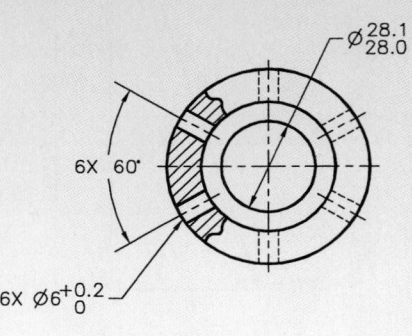

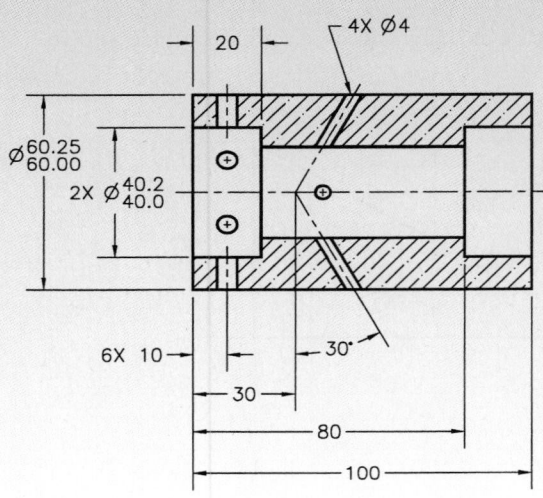

FIGURE 15.1 ■ The original drawing to be revised with geometric tolerancing added from engineer's notes.

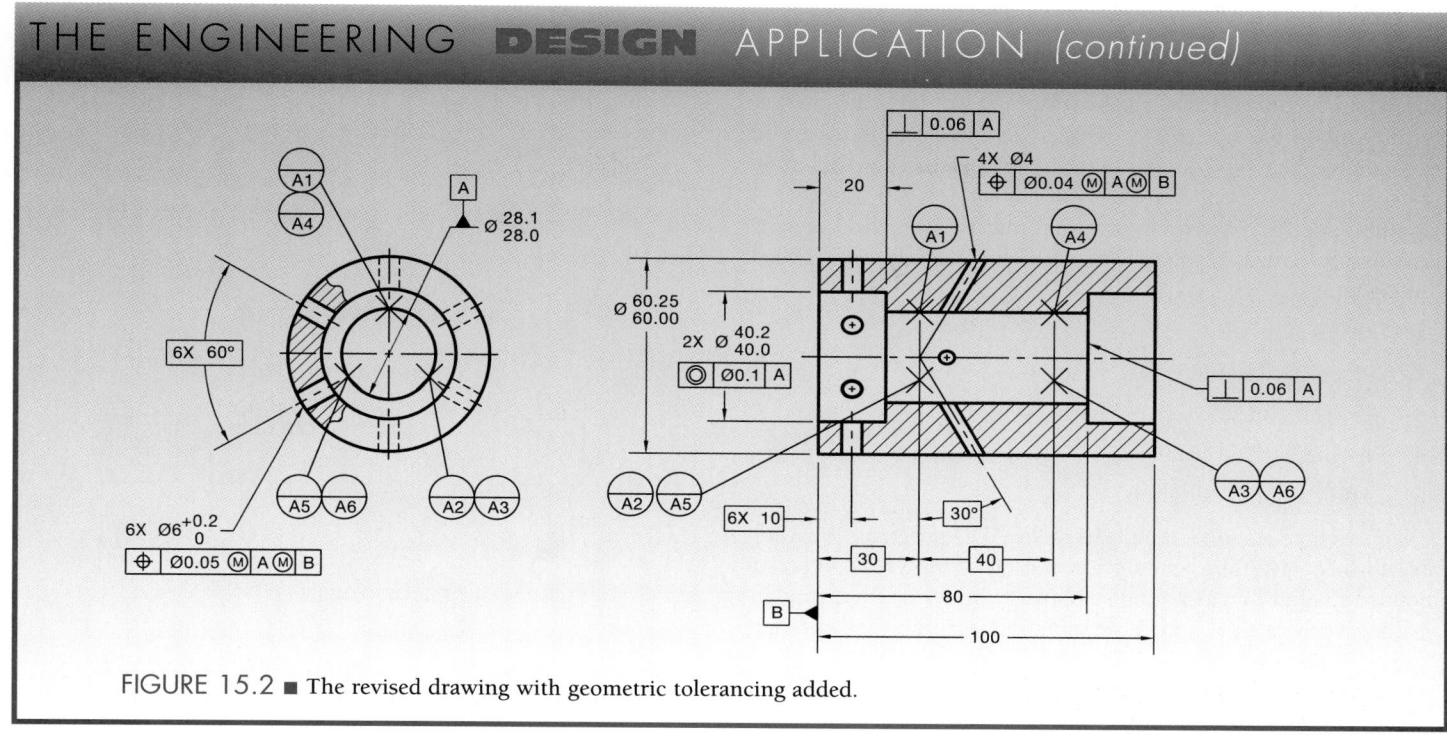

FIGURE 15.2 ■ ■ The revised drawing with geometric tolerancing added.

Geometric tolerancing (GT) is the dimensioning and tolerancing of individual features of a part where the permissible variations relate to characteristics of form, profile, orientation, runout, or the relationship between features. This subject is commonly referred to as geometric dimensioning and tolerancing (GD & T).

ANSI The standards for geometric dimensioning and tolerancing (GD & T) are governed by the American National Standards document ASME Y14.5M—1994 titled *Dimensioning and Tolerancing,* published by the American Society of Mechanical Engineers.

GENERAL TOLERANCING

Tolerancing was introduced in Chapter 12 with definitions and examples of how dimensions are presented on a drawing. Review Chapter 12 to establish a solid understanding of dimensioning and tolerancing terminology.

The term *general tolerancing* as used here implies dimensioning without the use of geometric tolerancing. General tolerancing applies a degree of form and location control by increasing or decreasing the tolerance. The *limits of size* of a feature control the amount of variation in size and geometric form. This is the boundary between maximum material condition (MMC) and least material condition (LMC). MMC and LMC are defined later in this chapter. The form of the feature may vary between the upper limit and lower limit of a size dimension. This is known as *extreme form variation.* (See Figure 15.3.) The addition of GT is necessary to control form but will still permit a relatively large size tolerance.

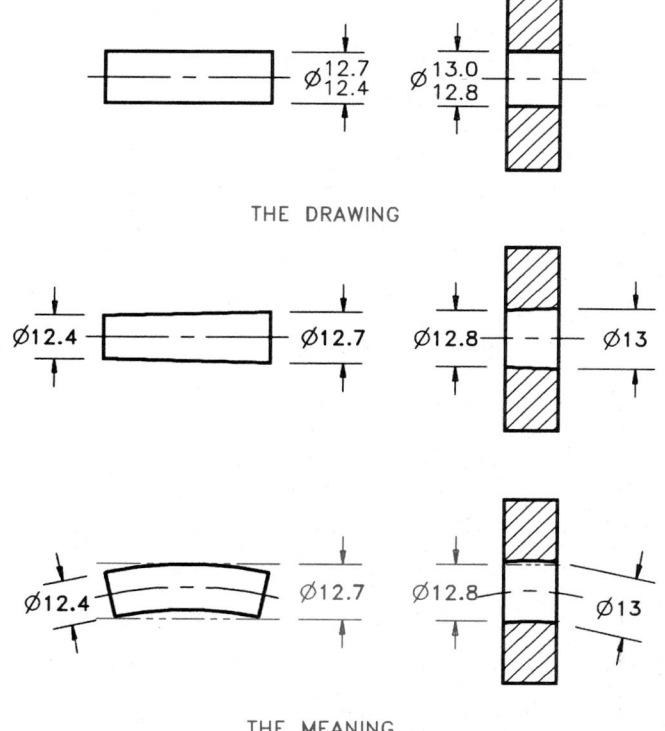

FIGURE 15.3 ■ Extreme form variations of given tolerances.

SYMBOLOGY

Symbols on drawings represent specific information that would otherwise be difficult and time-consuming to duplicate in note form. Symbols must be clearly drawn to required size and shape so they communicate the desired meaning uniformly. The exact drafting specifications are given when each symbol is first introduced. Use these specifications to prepare your own CADD blocks or symbols library. Geometric tolerancing symbols are divided into the following types:

1. Datum symbols.

2. Geometric characteristic symbols.

3. Material condition (modifying) symbols.

4. Feature control frame.

5. Supplementary symbols.

A variety of professional templates and CADD programs are available to help save time when preparing GT symbols. It is recommended that all dimensioning symbols be drawn with a template or CADD program for good representation, accuracy, and speed.

DATUM FEATURE SYMBOLS

Datums are considered to be theoretically perfect surfaces, planes, points, or axes that are established from the true geometric counterpart of the datum feature. The datum feature, as shown in Figure 15.4, is the actual feature of the part that is used to establish the datum. Datums are used to originate size and location dimensions. Without the use of GT, datums are often implied. When dimensions originate at a common surface, for example, that surface is assumed to be the datum. The only problem with this method is that the manufacturing operation may not read the datum in the same way as the engineering department, or the implied datum may be ignored all together. With the use of defined datums, each part is made with dimensions that begin from the same origin; there is no variation.

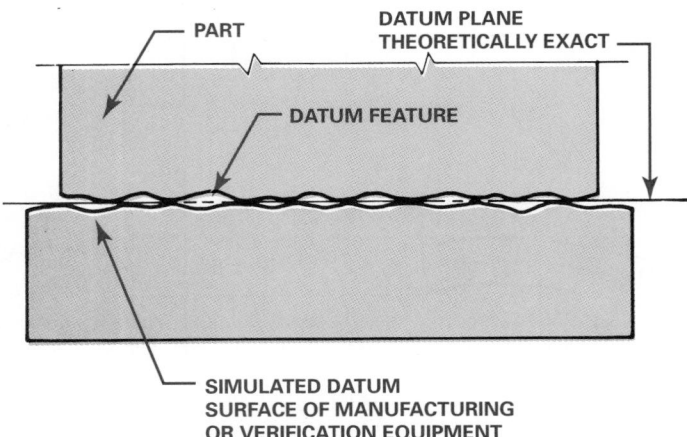

FIGURE 15.4 ■ Magnified representation of a datum feature.

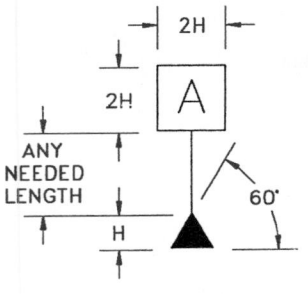

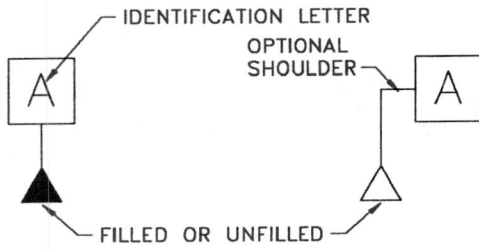

FIGURE 15.5 ■ Datum feature symbol.

Datum feature symbols are commonly drawn using thin lines with the symbol size related to the size of the lettering on a drawing, as shown in Figure 15.5. The triangular base may be filled or unfilled but should be consistently applied in a drawing. However, the filled triangle helps the symbol show up better on a drawing.

Datum feature symbols are placed in the view that shows the edge of a surface or are attached to a diameter or symmetrical dimension when associated with a centerline or center plane. (See Figure 15.6.)

DATUM REFERENCE FRAME

Datum referencing is used to relate features of a part to an appropriate datum or datum reference frame. A datum indicates the origin of a dimensional relationship between a toleranced feature and a designated feature or features on a part. The datum identification symbol on a drawing relates to an actual feature on the part known as the *datum feature*. The part is then placed on the surface of an inspection table or manufacturing verification equipment. This inspection table or equipment is referred to as the *simulated datum*. The *datum* is the theoretically exact plane, axis, or point that is established by the true geometric counterpart of the specified datum feature. Measurements cannot be made from the true geometric counterpart, because it is theoretical or assumed to exist. The machine tables, surface plates, or inspection tables are of such high quality that they are used to simulate the datums from which measurements are taken and dimensions are verified. In this way, each dimension always originates from the same reliable location. Dimensions are never taken or verified from one surface of the part to another.

Sufficient datum features, those most important to the design of a part, are chosen to position the part in relation to a set of three

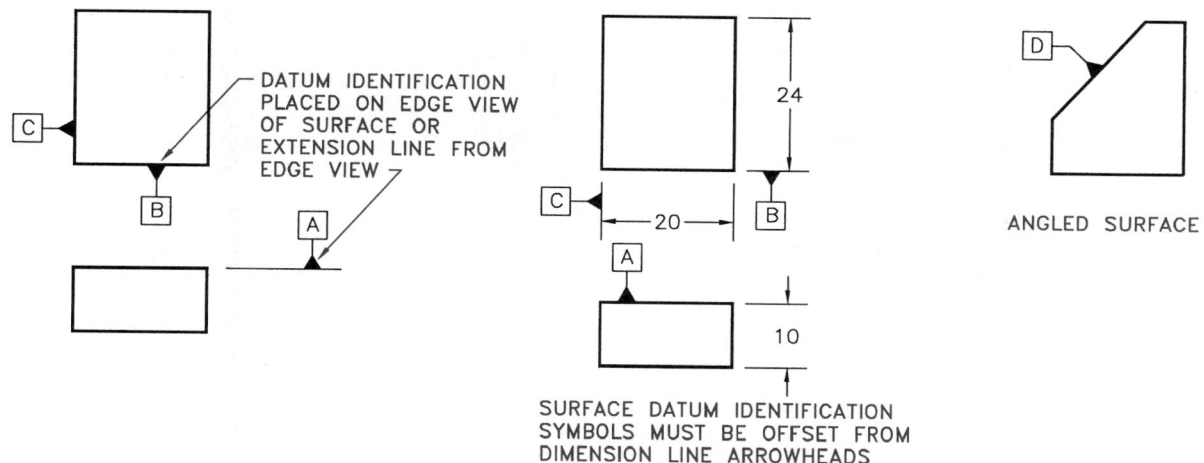

ANGLED SURFACE

SURFACE DATUM IDENTIFICATION
SYMBOLS MUST BE OFFSET FROM
DIMENSION LINE ARROWHEADS

SURFACE DATUMS

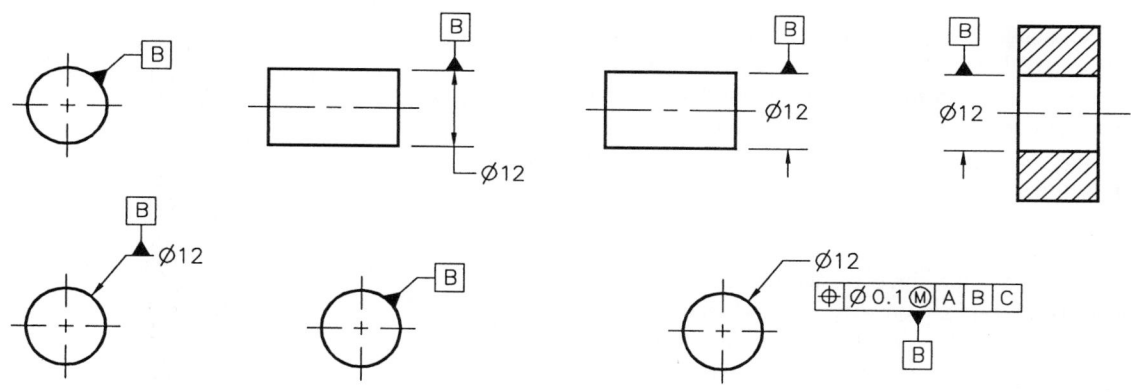

AXIS DATUMS

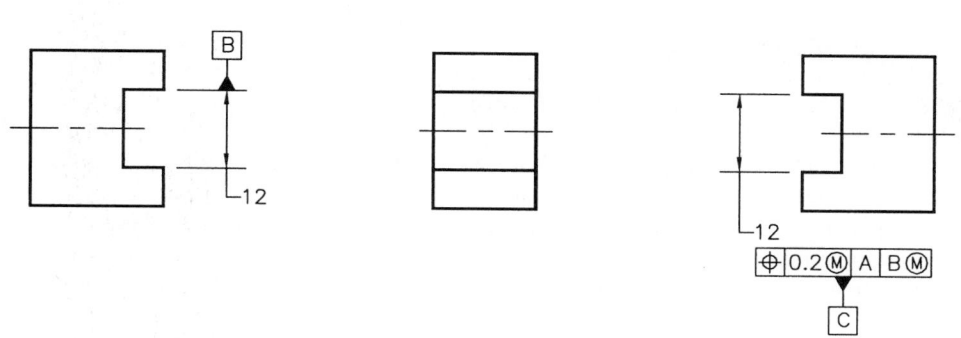

AXIS AND CENTER PLANE DATUM IDENTIFICATION
SYMBOLS MUST ALIGN WITH OR REPLACE THE
DIMENSION LINE ARROWHEAD OR BE PLACED ON
THE FEATURE, LEADER SHOULDER, OR FEATURE
CONTROL FRAME.

CENTER PLANE DATUMS

FIGURE 15.6 ■ Datum identification.

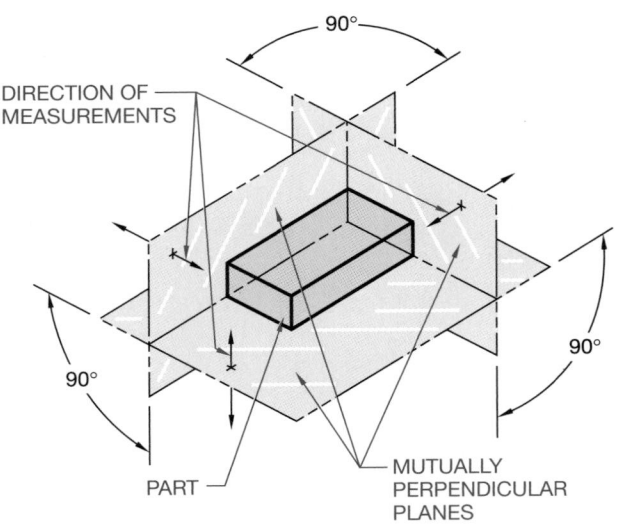

FIGURE 15.7 ■ Datum reference frame.

mutually perpendicular planes, jointly called a *datum reference frame*. This reference frame exists in theory only and not on the part. Therefore, it is necessary to establish a method for simulating the theoretical reference frame from the actual features of the part. This simulation is accomplished by positioning the part on appropriate datum features to adequately relate the part to the reference frame and to restrict the motion of the part in relation to it. (See Figure 15.7.) These planes are simulated in a mutually perpendicular relationship to provide direction as well as the origin for related dimensions and measurements. Thus when a part is positioned on the datum reference frame, dimensions related to the datum reference frame by a feature control frame or note are thereby mutually perpendicular. This theoretical reference frame constitutes the three-plane dimensioning system used for datum referencing.

DATUM FEATURES

A datum feature is selected on the basis of its geometric relationship to the toleranced feature and the requirements of the design. To ensure proper part interface and assembly, corresponding features of mating parts are also selected as a datum feature where practical. Datum features must be readily recognizable on the part. Therefore, in the case of symmetrical parts or parts with identical features, physical identification of the datum feature on the part may be necessary. A datum feature should be accessible on the part and be of sufficient size to permit processing operations.

The three reference planes are mutually perpendicular unless otherwise specified. Planes in the datum reference frame are placed in order of importance. For example, the plane that originates most size and location dimensions, or has the most functional importance, is the *primary*, or first, datum plane. The next datum is the *secondary*, or second, datum plane and the third element of the frame is called the *tertiary*, or third, datum plane. The desired order of precedence is indicated by entering the appropriate datum reference letters from left to right in the feature control frame. For instructional purposes it may be convenient to label datums as A,

B, and C, although in industry other letters are also used to identify datums such as D, E, and F, or X, Y, and Z. The letters that should be avoided are O, Q, and I. Figure 15.8a shows a part and the planes that are chosen as datum features. Notice in Figure 15.8b how the order of precedence of datum features relates the part to the datum reference frame. The datum features are identified as surfaces A, B, and C. These surfaces are most important to the design and function of the part. Surfaces A, B, and C are the primary, secondary, and tertiary datum features respectively since they appear in that order in the feature control frame.

Multiple datum frames may be established for some parts depending on the complexity and function of the part. The relationship between datum frames is often controlled by a representative angle. (See Figure 15.9.)

Partial Surface Datums

In some situations it may be more realistic to apply a datum feature symbol to a portion of a surface rather than the entire surface. For example, when a long part has related features located in one or more concentrated places, then the datum may be located next to the important features. When this is done, a chain line is used to identify the extent of the datum feature. The location and length of the chain line must be dimensioned. (See Figure 15.10.)

Coplanar Surface Datums

Coplanar surfaces are two or more surfaces that are in the same plane. The relationship of coplanar surface datums may characterize the surfaces as one plane or datum in correlated geometric tolerance callouts as shown in Figure 15.11.

DATUM TARGET SYMBOLS

In many situations it is not possible to establish an entire surface or surfaces as datums, or it may not be practical to coordinate a partial datum surface with the part features. When this happens due to the size or shape of the part, then datum targets may be used to establish datum planes. Datum targets are especially useful on parts with surface or contour irregularities such as sand castings or forgings, on some sheet metal parts subject to bowing or warpage, or on weldments where heat can induce warpage. Datum targets are designated points, lines, or surface areas that are used to establish the datum reference frame. The primary datum is established by three points or contact locations. The three locations are placed on the part so they form a stable, well-spaced pattern and are not in a line. The secondary datum is created by two points or locations. The tertiary datum is established by one point or location. Remember the purpose of the datum frame is to prepare a stable position for the part dimensions to be established in relationship to corresponding datums. For this reason, the three points on the primary datum are used to provide stability similar to a three-legged stool. The two points on the secondary datum provide the needed amount of stability when the object is placed against the secondary datum. Finally,

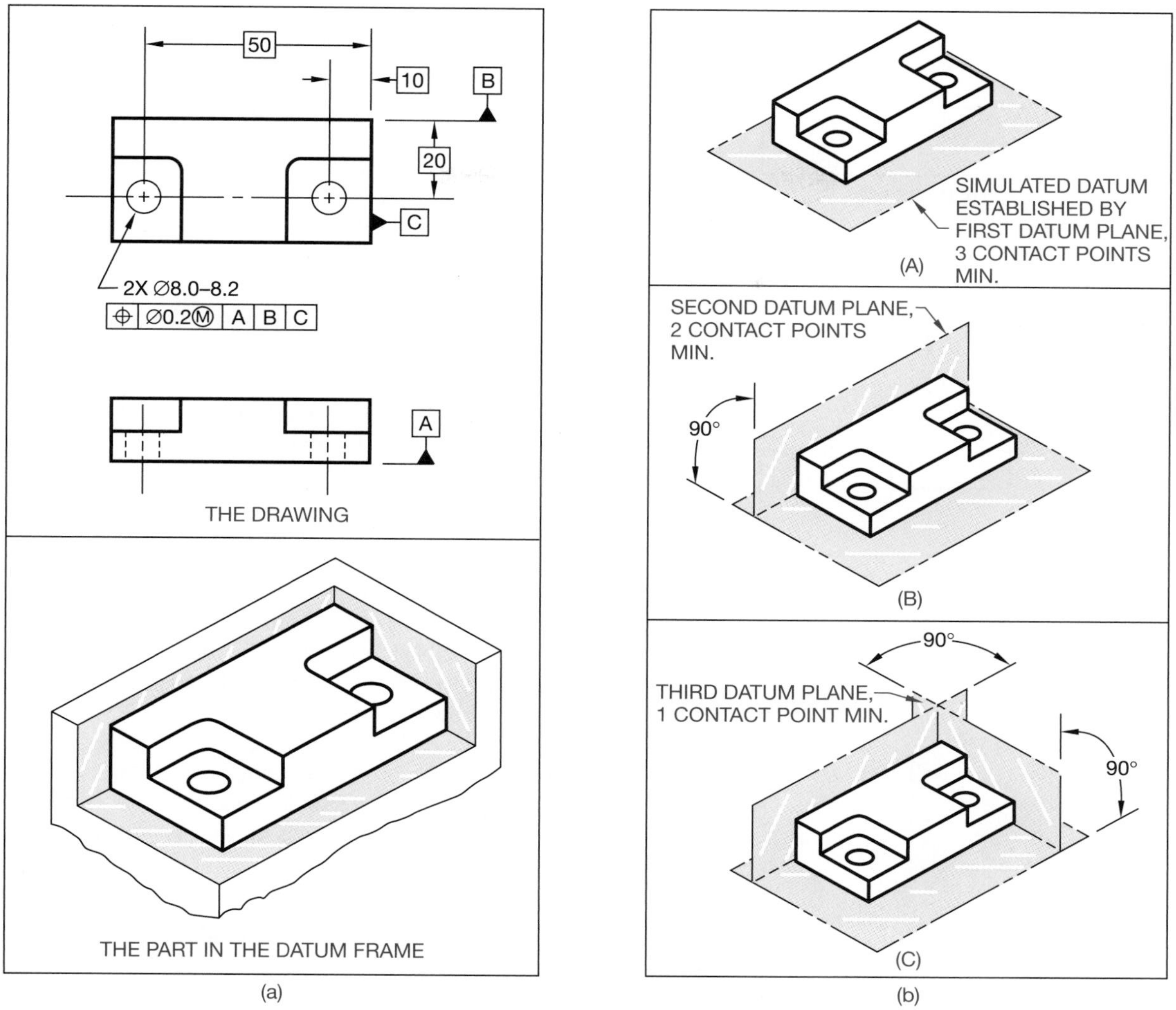

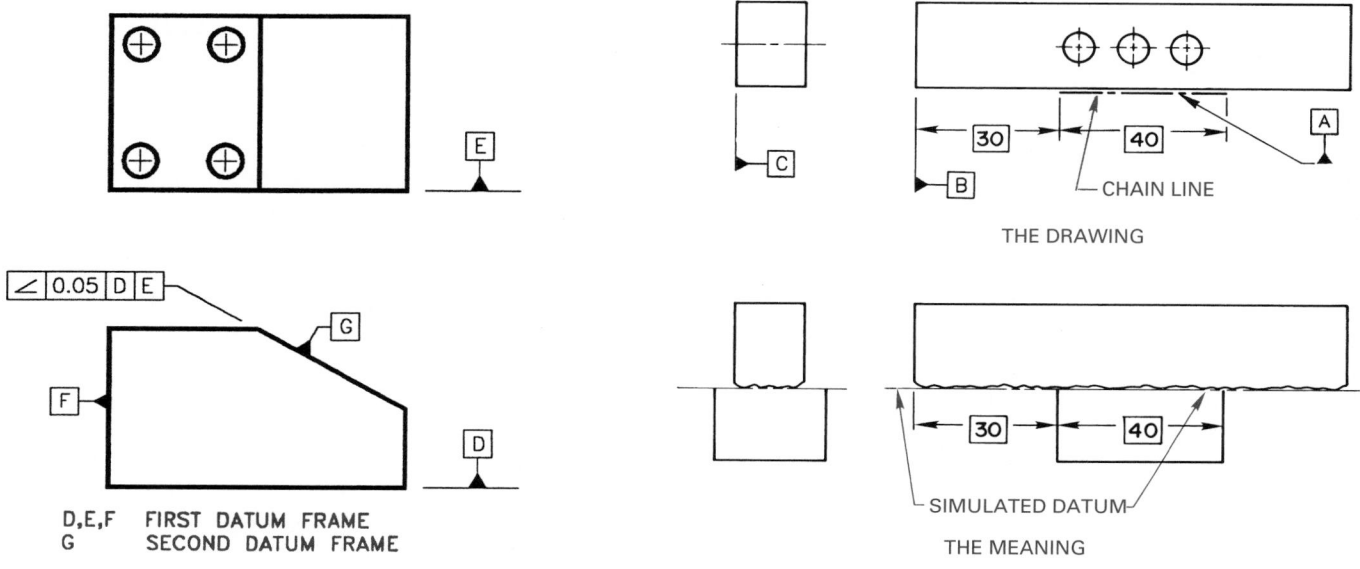

FIGURE 15.8 ■ (a) Part in which datum features are plane surfaces; (b) sequence of datum features relates part to datum reference frame.

FIGURE 15.9 ■ Multiple datum frames.

FIGURE 15.10 ■ Partial surface datums.

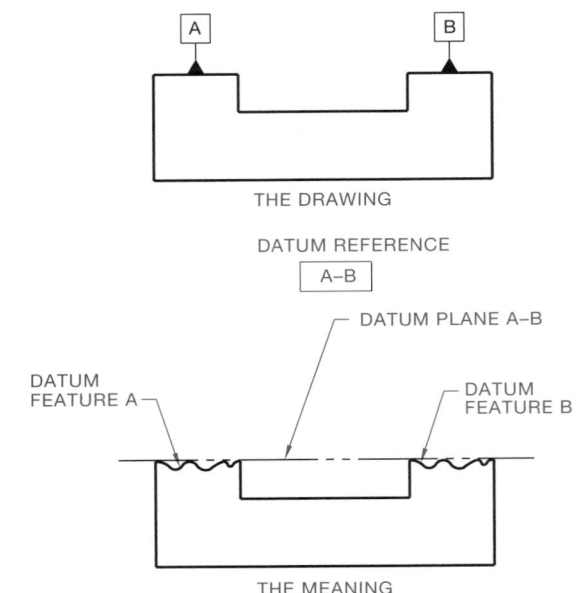

FIGURE 15.11 ■ Coplanar surface datums.

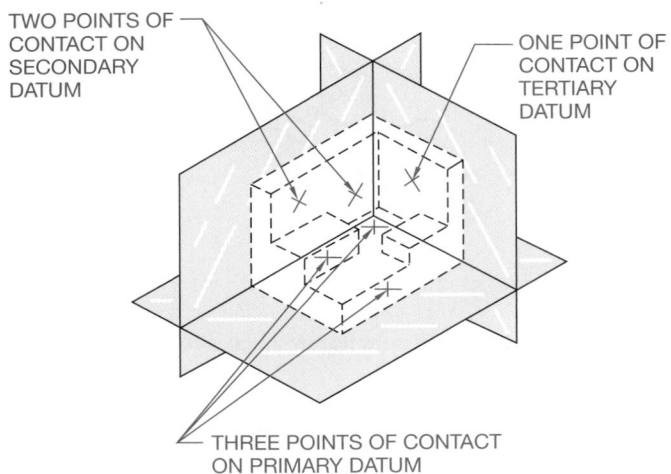

FIGURE 15.12 ■ Datum frame established by datum target symbols.

the tertiary datum requires only one point of contact to complete the stability between the three datum frame elements. (See Figure 15.12.)

The datum target symbol is drawn as a circle using thin lines and is connected with a leader that points to a target point, line, or surface area. The datum target symbol is divided into two halves by a horizontal line. The top half of the symbol is reserved for identification of the datum target area size when used. The bottom half of the symbol is used for datum target identification. For example, if there are three datum target points on a datum, the first point may be labeled A1, the second A2, and the third A3. The datum target point is located on the surface or edge view from adjacent datums with basic dimensions. (See Figure 15.13.) Chain dimensioning is commonly used to locate datum target points and target areas. The location dimensions must originate

from datums. Datum target areas are located to their centers. (See Figure 15.14.) When the surface of a part has an irregular contour or different levels, the contact points, lines, or areas may lie on the same plane and different lengths of locating pins may be used. The pins used to make point contact are usually rounded. The pins for target area contact have a flat surface equal in diameter to the specified target area.

DATUM AXIS

The datum frame established by a cylindrical object is the base of the part. The two theoretical planes, represented in Figure 15.15 by the X and Y center planes, cross to establish the datum axis. The actual secondary datum is the cylindrical surface. The center planes located at X and Y are used to indicate the direction of dimensions that originate from the datum axis. (See Figure 15.16.)

Datum target points, lines, or surface areas may also be used to establish a datum axis. A primary datum axis may be established by two sets of three equally spaced targets: a set at one end of the cylinder and the other set near the other end as shown in Figure 15.17, page 467. When two cylindrical features of different diameter are used to establish a datum axis, then the datum target points are identified in correspondence to the adjacent cylindrical datum feature as shown in Figure 15.18, page 467. Cylindrical datum target areas and circular datum target lines may also be used to establish the datum axis of cylindrically shaped parts as shown in Figure 15.19, page 467. A secondary datum axis is established by placing three equally spaced targets on the cylindrical surface. (See Figure 15.20, page 468.)

FEATURE CONTROL FRAME

The feature control frame is used to relate a geometric tolerance to a part feature. The elements in a feature control frame must always be in the same order. The most basic format is when a geometric characteristic and related tolerance are applied to an individual feature as shown in Figure 15.21, page 468. The next expanded format is when a geometric characteristic, tolerance zone descriptor, and material condition symbol are used in the feature control frame. (See Figure 15.22, page 468.) One, two, or three datum references may be included in a feature control frame as shown in Figure 15.23, page 468. The material condition symbol that applies to the feature tolerance is always placed after the tolerance. When a material condition symbol is applied to a datum reference, it is placed after the datum reference and in the same compartment as shown in Figure 15.24, page 468.

The feature control frame is drawn with thin lines and connected to the feature with a leader or extension line. (See Figure 15.25, page 468.) The feature control frame may be combined with a datum feature symbol when the feature controlled by the geometric tolerance also serves as a datum. The datum feature symbol and the datum reference in the feature control frame are considered separately. When a datum feature symbol and a feature control frame are combined, the feature control frame should be shown first. The datum feature symbol is centered on the base of the feature control frame. (See Figure 15.26, page 468.)

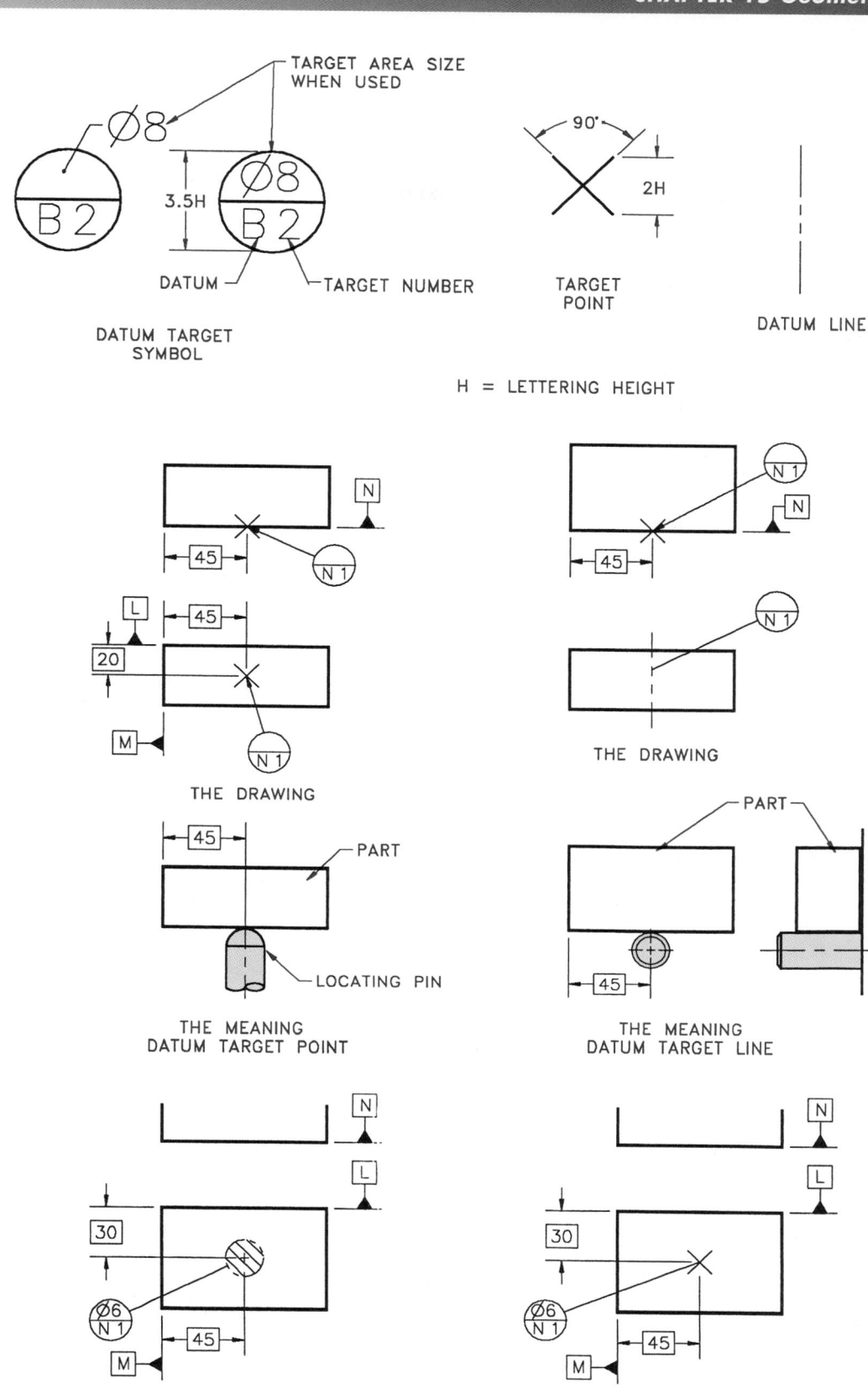

FIGURE 15.13 ■ Datum target symbol, target point, datum line, and target area.

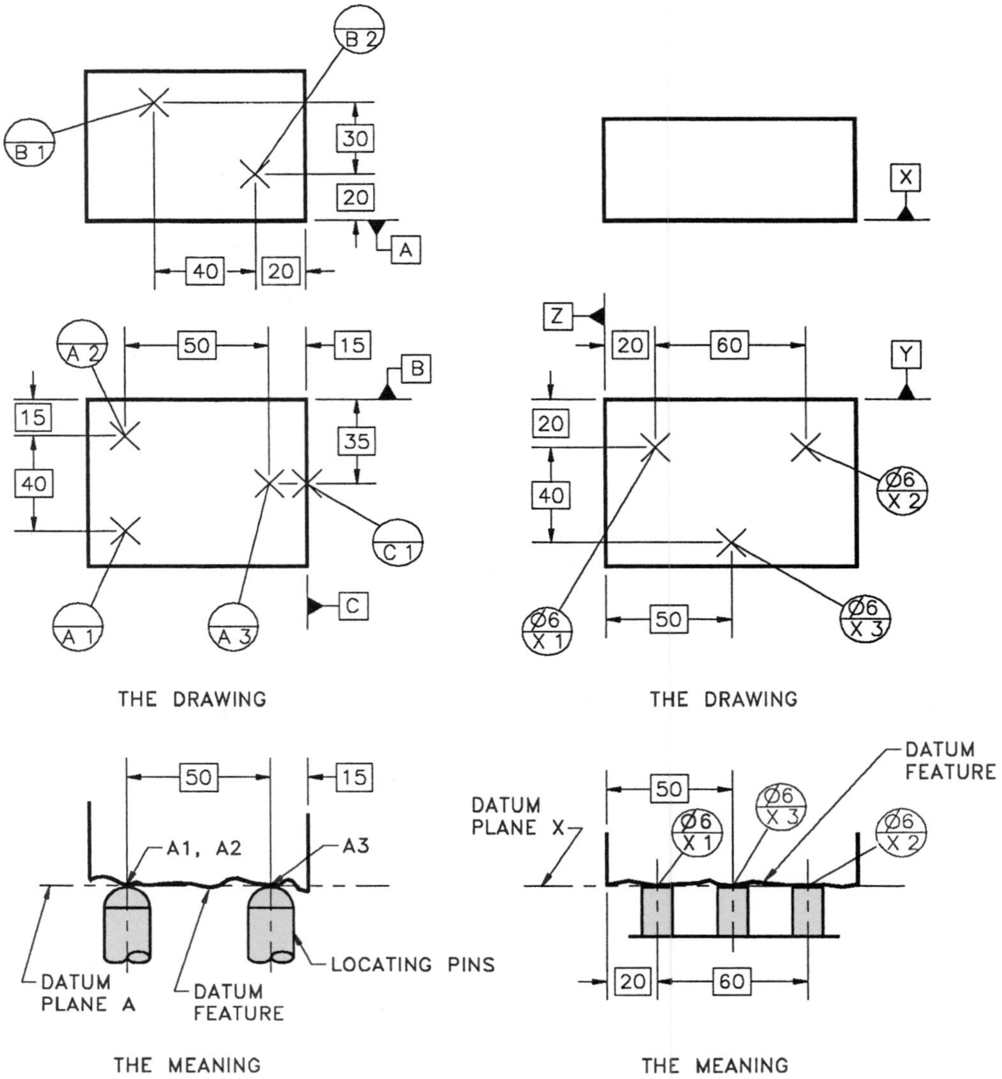

THE DRAWING

THE DRAWING

THE MEANING

THE MEANING

FIGURE 15.14 ■ Locating datum targets.

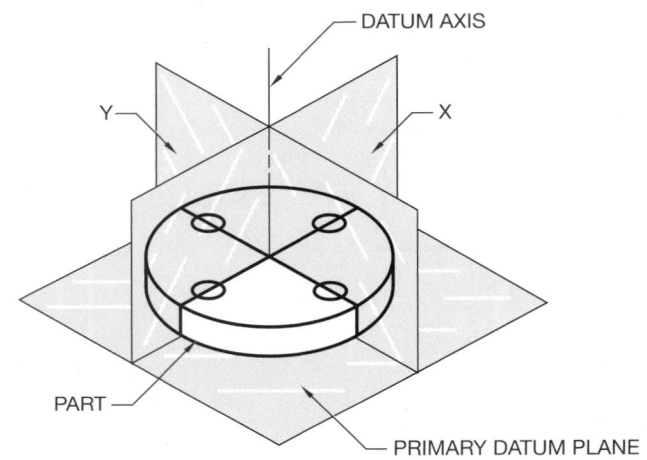

FIGURE 15.15 ■ Datum frame for cylindrical object.

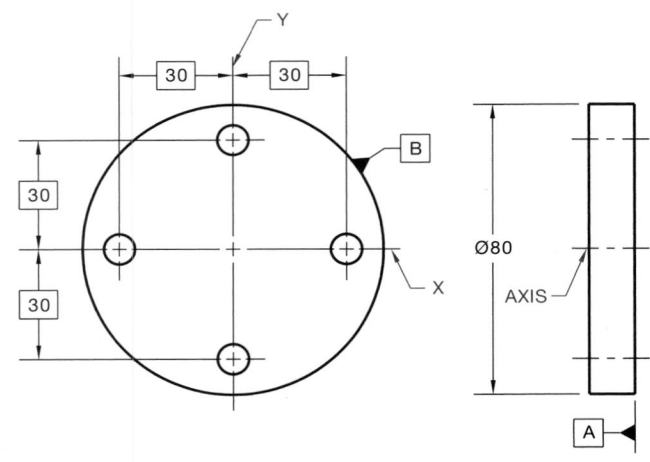

FIGURE 15.16 ■ Datum axis.

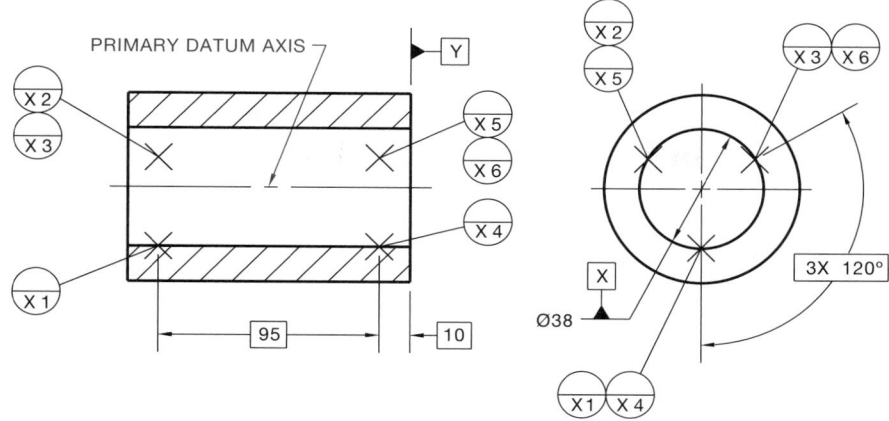

FIGURE 15.17 ■ Establishing a primary datum axis with two sets of three equally spaced targets.

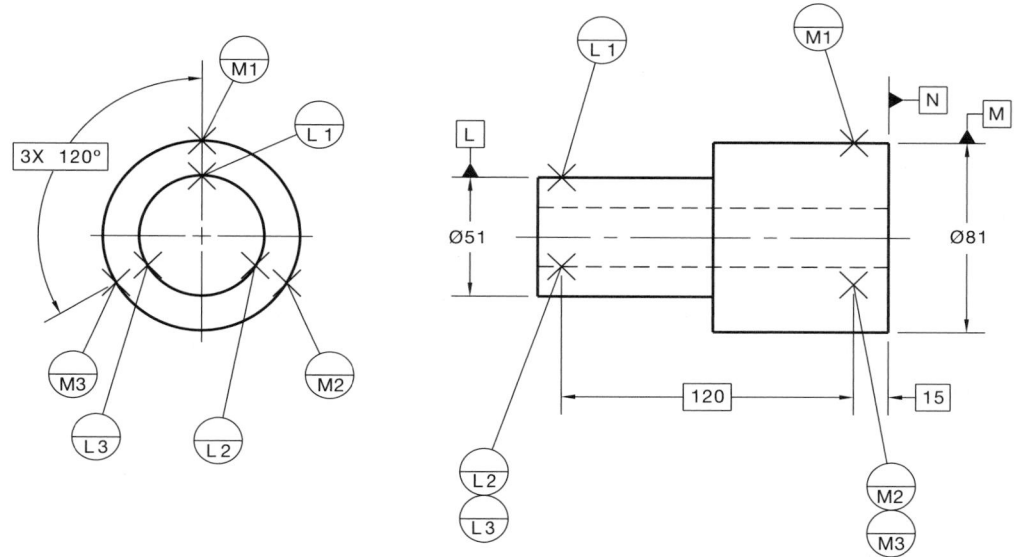

FIGURE 15.18 ■ Datum targets identified on adjacent cylindrical features.

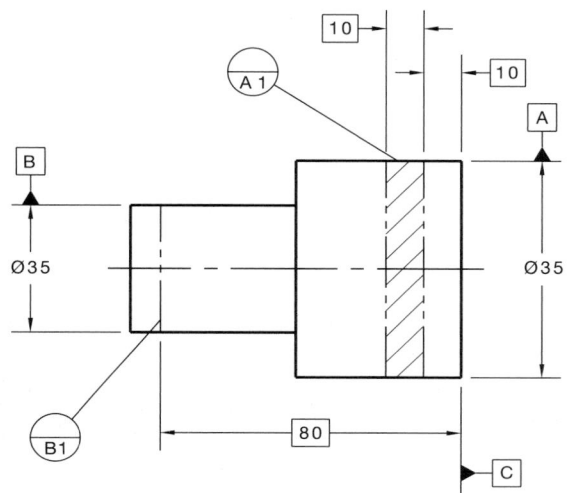

FIGURE 15.19 ■ Cylindrical datum target areas and circular datum target lines.

BASIC DIMENSIONS

A basic dimension is defined as any size or location dimension that is used to identify the theoretically exact size, profile, orientation, or location of a feature or datum target. Basic dimensions are the basis from which permissible variations are established by tolerances on other dimensions in notes or in feature control frames. A basic dimension is described on a drawing by placing a thin-line box around the dimension. Metric and inch basic dimension numerals follow the same rules applied to other dimension numerals. Refer to Chapter 12. Metric basic dimensions contain only the number of decimal places needed to display the intended control. For inch basic dimensions, the basic dimension value is expressed in the same number of decimal places as the related tolerance. (See Figure 15.27.)

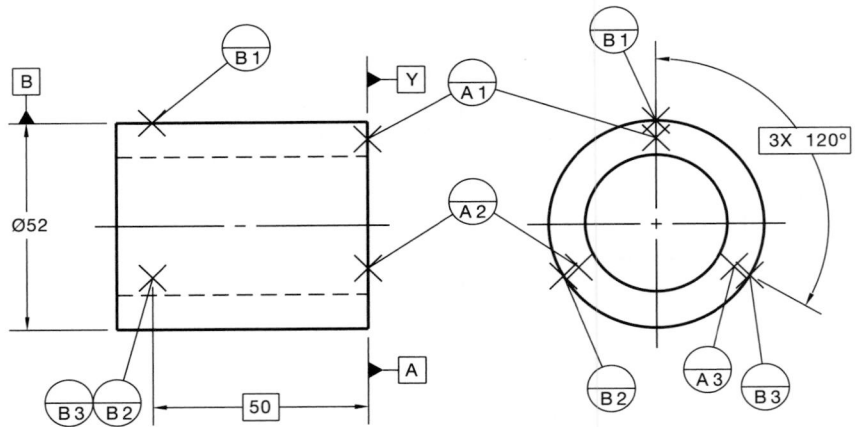

FIGURE 15.20 ■ Establishing a secondary datum on a cylindrical object with three equally spaced targets.

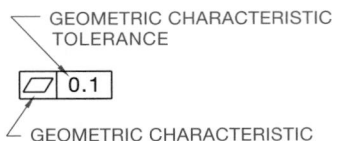

FIGURE 15.21 ■ Feature control frame with geometric characteristics and related tolerance.

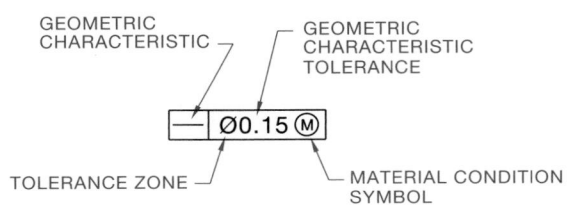

FIGURE 15.22 ■ Feature control frame with geometric characteristic, tolerance zone descriptor, and material condition symbol.

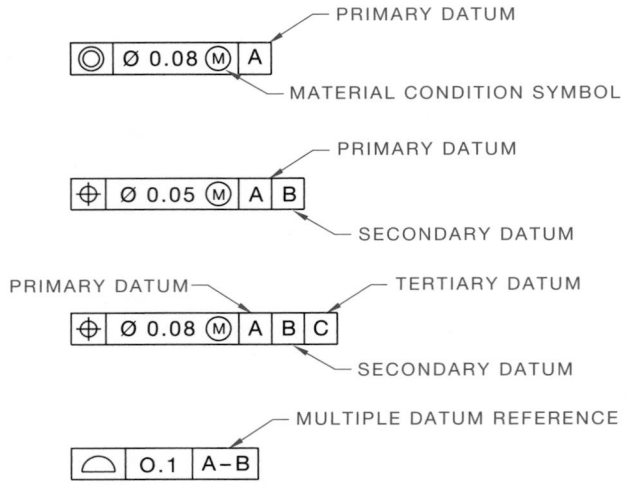

FIGURE 15.23 ■ Applying one, two, or three datum references to the feature control frame.

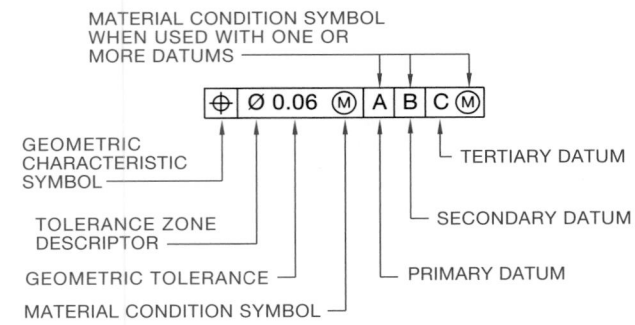

FIGURE 15.24 ■ Elements of a feature control frame.

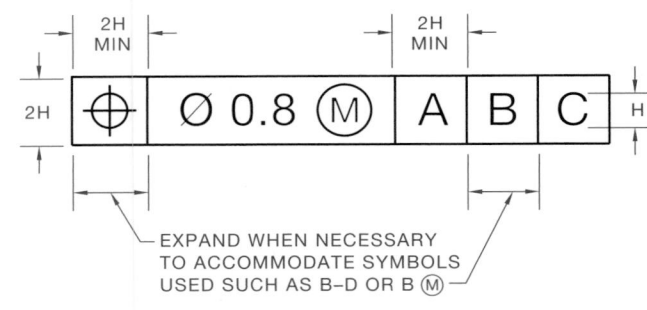

H = LETTERING HEIGHT

FIGURE 15.25 ■ Detailed feature control frame and application.

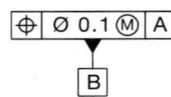

FIGURE 15.26 ■ Combination datum feature symbol and feature control frame.

GEOMETRIC TOLERANCES

Geometric tolerances are divided into five types: form, profile, orientation, runout, and location. These tolerances are subdivided into thirteen characteristics plus modifying terms, all of which are discussed in this section.

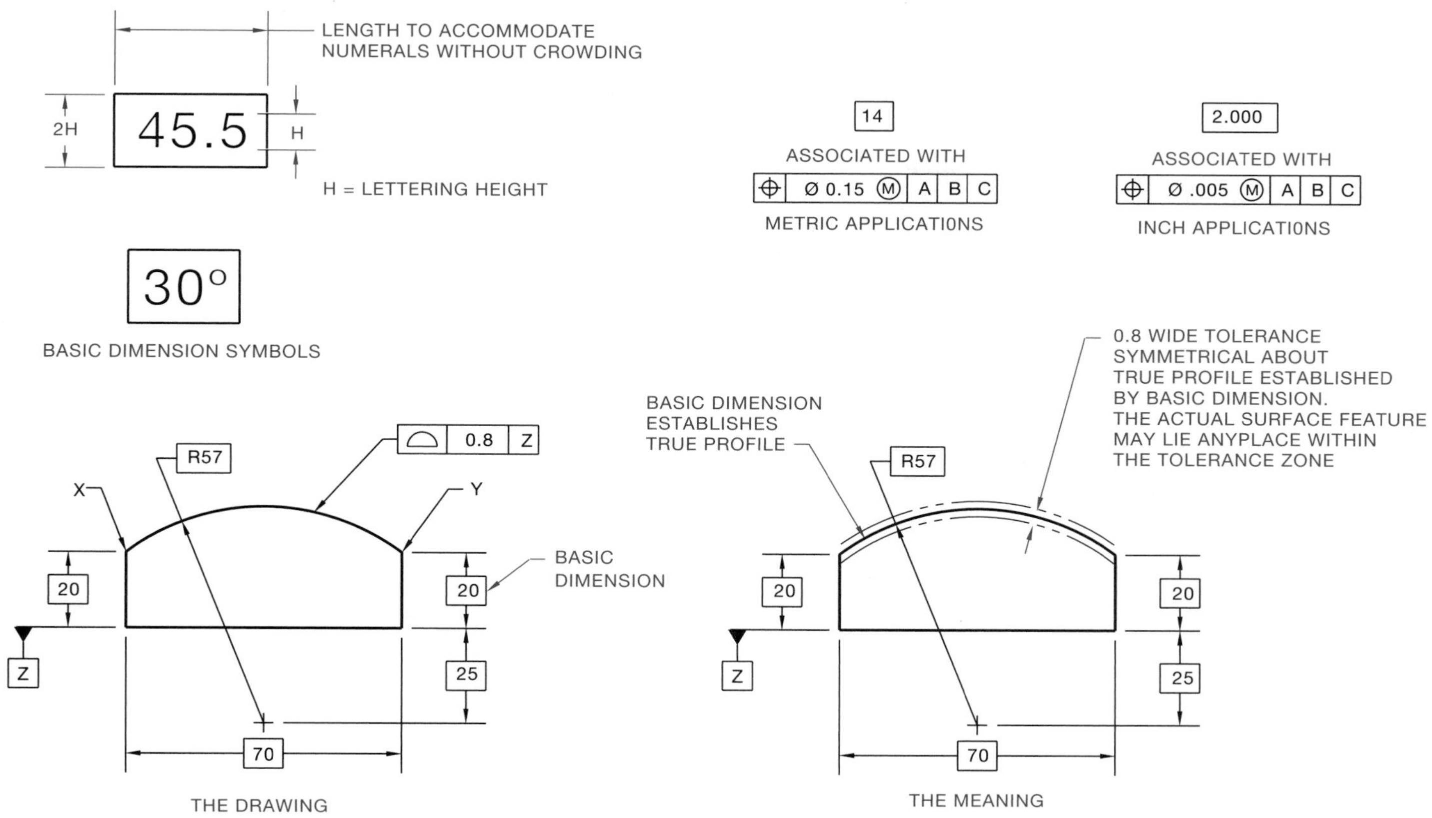

LENGTH TO ACCOMMODATE
NUMERALS WITHOUT CROWDING

2H 45.5 H

H = LETTERING HEIGHT

30°

BASIC DIMENSION SYMBOLS

14

ASSOCIATED WITH

⊕ | Ø 0.15 Ⓜ | A | B | C

METRIC APPLICATIONS

2.000

ASSOCIATED WITH

⊕ | Ø .005 Ⓜ | A | B | C

INCH APPLICATIONS

THE DRAWING

R57

X

⌒ | 0.8 | Z

Y

20

Z

20

25

BASIC
DIMENSION

70

THE DRAWING

BASIC DIMENSION
ESTABLISHES
TRUE PROFILE

0.8 WIDE TOLERANCE
SYMMETRICAL ABOUT
TRUE PROFILE ESTABLISHED
BY BASIC DIMENSION.
THE ACTUAL SURFACE FEATURE
MAY LIE ANYPLACE WITHIN
THE TOLERANCE ZONE

R57

20

Z

20

25

70

THE MEANING

FIGURE 15.27 ■ Basic dimensions.

Form Tolerance

A tolerance of form is commonly applied to individual features or
elements of single features and is not related to datums. The
amount of given form variation must fall within the specified size
tolerance zone.

Straightness

The straightness symbol is detailed in Figure 15.28. Perfect
straightness exists when a surface element or the axis of a part is a
straight line. A straightness tolerance allows for a specified amount
of variation from a straight line. The straightness feature control
frame may be attached to the surface of the object with a leader or
combined with the diameter dimension of the part. When
straightness is connected to the surface of the feature with a leader,
surface straightness is implied as shown in Figure 15.29. When the
straightness feature control frame is combined with the diameter
dimension of the part, then *axis straightness* is specified. (See Fig-

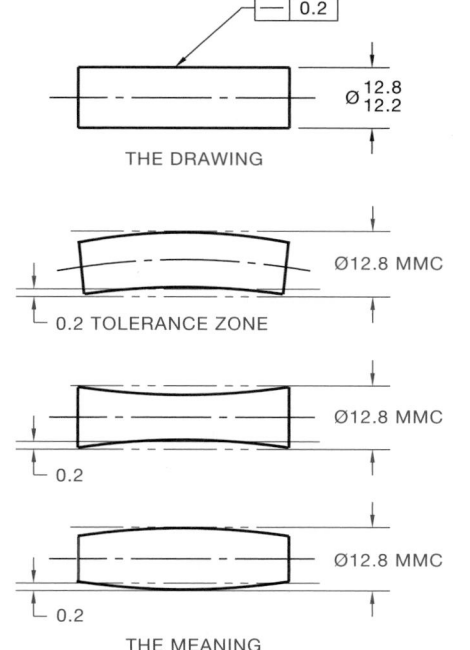

— | 0.2

Ø 12.8 / 12.2

THE DRAWING

Ø12.8 MMC

0.2 TOLERANCE ZONE

Ø12.8 MMC

0.2

Ø12.8 MMC

0.2

THE MEANING

FIGURE 15.29 ■ Surface straightness.

ure 15.30.) The tolerance zone shape is cylindrical for this application. While a straightness tolerance is common on cylindrical
parts, straightness may also be applied to the surface of noncylin-
drical parts. When specifying center plane feature straightness, a
center plane is implied and the diameter tolerance zone descriptor

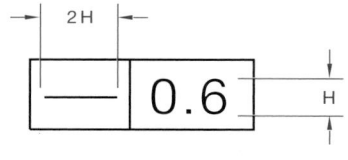

2H

0.6 H

H = LETTERING HEIGHT

FIGURE 15.28 ■ Straightness feature control frame.

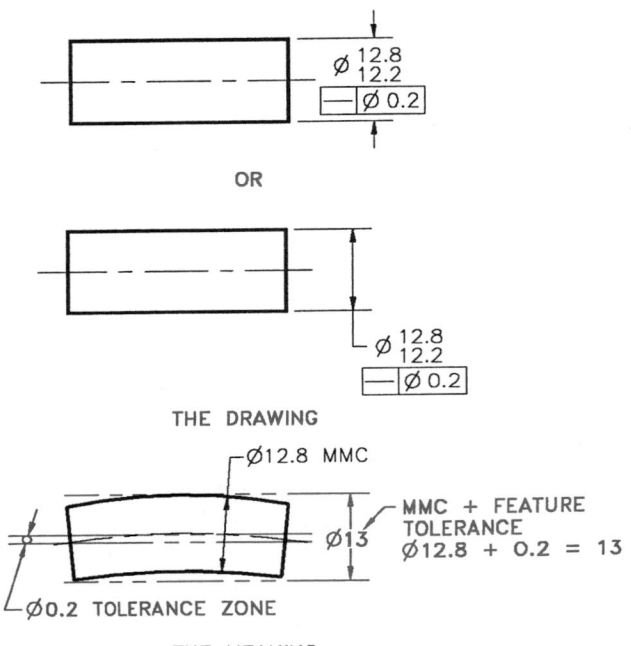

FIGURE 15.30 ■ Axis straightness.

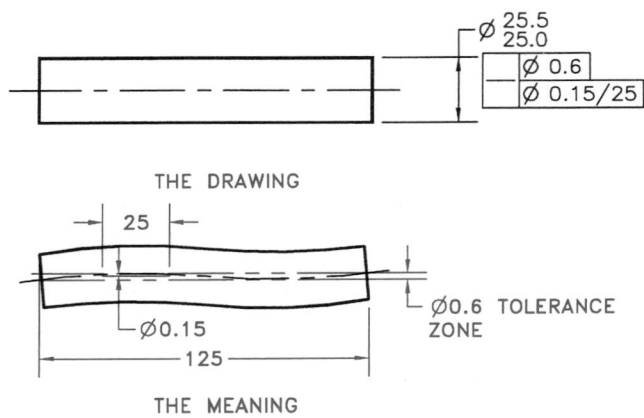

FIGURE 15.31 ■ Unit straightness.

is omitted. The straightness geometric tolerance must always be less than the size tolerance, thereby refining the size tolerance.

Unit straightness is a situation where a specified tolerance is given per unit of length, for example 25 mm units, and a greater amount of tolerance is provided over the total length of the part. (See Figure 15.31.) This is intended to control excessive waviness in a long part.

Flatness

The flatness tolerance symbol is detailed in relationship to the feature control frame in Figure 15.32. A surface is considered *flat* when all of the surface elements lie in one plane. A flatness geometric tolerance callout allows for a specified amount of surface variation from a flat plane. The flatness tolerance zone establishes two parallel planes. The actual surface of the object may not extend beyond the boundary of the size tolerance zone and, when associ-

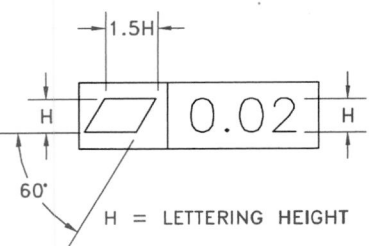

FIGURE 15.32 ■ Flatness feature control frame.

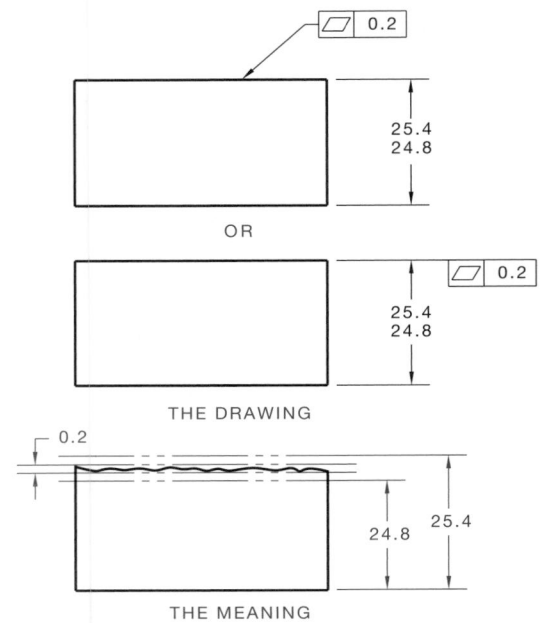

FIGURE 15.33 ■ Flatness representation.

ated with the size dimension, the flatness tolerance must be smaller than the size tolerance. The flatness feature control flame may be connected to the edge view of the surface with a leader or with an extension line. (See Figure 15.33.)

Unit flatness may be specified when it is desirable to control the flatness of surface units. The unit size may be 25 × 25 mm or 1.00 × 1.00 in. This is used to prevent abrupt surface variation. A unit flatness callout may be presented with or without a separate tolerance for the total area. Unit flatness, just as unit straightness, often has a total tolerance in order to avoid a situation where the unit tolerance gets out of control. The size of the unit area is given after the unit tolerance specification. The feature control frame is doubled in height and the total area flatness is given first as shown in Figure 15.34.

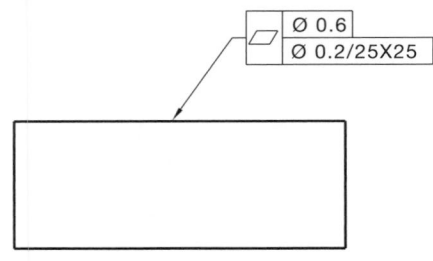

FIGURE 15.34 ■ Unit flatness.

H = LETTERING HEIGHT

FIGURE 15.35 ■ Circularity feature control frame.

Circularity

The circularity geometric characteristic symbol is detailed in a feature control frame in Figure 15.35. Circularity may be applied to cylindrical, conical, or spherical shapes. Circularity exists when all of the elements of a circle are the same distance from the center. *Circularity* is a cross-sectional evaluation of the feature to determine if the circular surface lies between a tolerance zone that is made up of two concentric circles. The cross-sectional tolerance zone is established perpendicular to the axis of the feature. The term *circular* or *line element* is used to refer to a cross-sectional or single-line tolerance zone as opposed to a blanket or entire surface tolerance zone. The circularity geometric tolerance zone is a radius dimension. The circularity feature control frame may be connected to the part with a leader in the circular or rectangular view as shown in Figure 15.36. The circularity tolerance must always be less than the size tolerance except for parts subject to free state variations. *Free state variation* is a term used to describe distortion of a part after removal of forces applied during manufacturing, such as a rubber gasket or sheet metal. The part may have to meet the tolerance specifications while in free state, or it may be necessary to hold features in a simulated mating part to verify dimensions. The free state symbol shown in Figure 15.37 is placed in the feature control frame after the geometric tolerance and material condition symbol, if any.

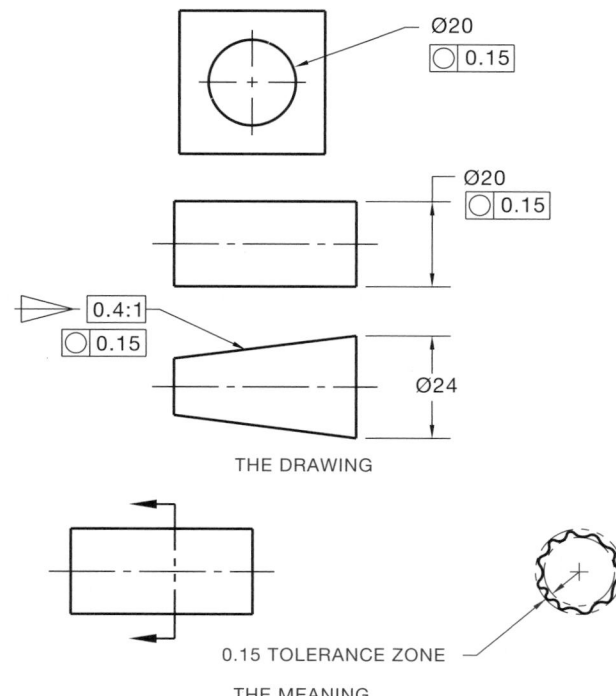

THE DRAWING

THE MEANING

FIGURE 15.36 ■ Circularity representation.

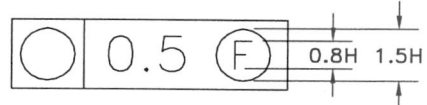

H = LETTERING HEIGHT

FIGURE 15.37 ■ Using the free state symbol.

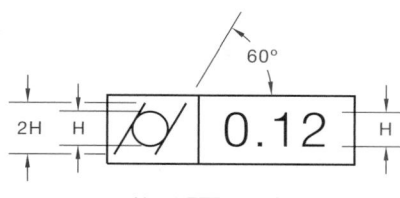

H = LETTERING HEIGHT

FIGURE 15.38 ■ Cylindricity feature control frame.

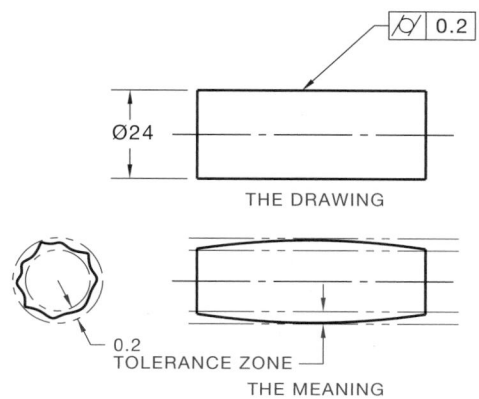

THE DRAWING

0.2 TOLERANCE ZONE

THE MEANING

FIGURE 15.39 ■ Cylindricity representation.

Cylindricity

The cylindricity geometric characteristic symbol is detailed in Figure 15.38. *Cylindricity* is similar to circularity in that both have a radius tolerance zone. The difference is that circularity is a cross-sectional tolerance which results in a feature that must lie between two concentric circles, while cylindricity is a blanket tolerance that results in a feature lying between two concentric cylinders. The cylindricity feature control frame may be connected with a leader to either the circular or rectangular view. (See Figure 15.39.) The cylindricity tolerance must be less than the size tolerance.

Profile Tolerance

Profile is used to control form or a combination of size, form, and orientation. Profile callouts are commonly used on arcs, curves, or irregular shaped features but can also be applied to plane surfaces. The shape and size of the profile may be defined with basic dimensions or tolerance dimensions. When tolerance dimensions are used to establish profile, the profile tolerance zone must be within the size tolerance. The profile feature control frame is connected to the longitudinal view with a leader line. The two types of profile are profile of a line and profile of a surface. (See Figure 15.40.)

The *profile of a line* is a cross-sectional or single-line tolerance that extends through the specified feature. The profile of a line may

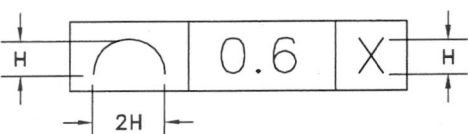

PROFILE OF A LINE

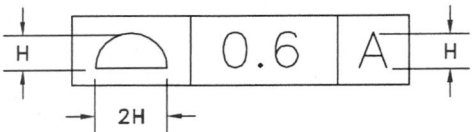

PROFILE OF A SURFACE

FIGURE 15.40 ■ Profile feature control frames.

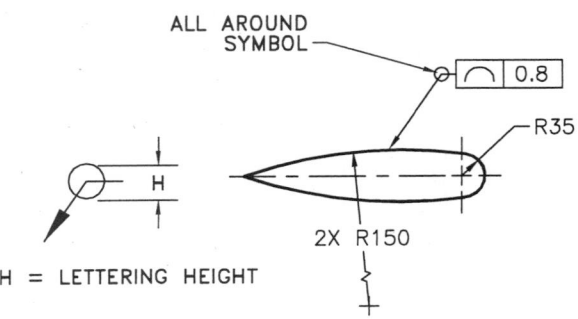

THE DRAWING

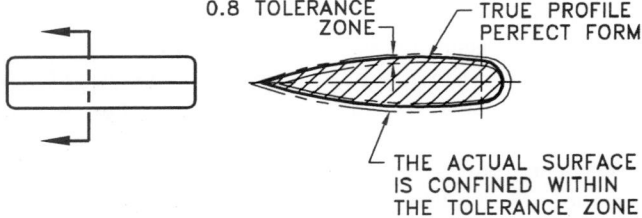

THE MEANING

FIGURE 15.41 ■ Profile of a line all around, and the all around symbol.

be controlled in relation to datums or without datums. The profile of a line may be all around the part by using the all around symbol on the leader, as seen in Figure 15.41, or between two specified points on the part by using the between symbol and identifying the points, as shown in Figure 15.42.

The *profile of a surface* is a blanket tolerance zone that affects all the surface elements of a feature equally. Profile of a surface is generally referenced to one or more datums so proper orientation of the profile boundary can be maintained. Profile of a surface may be applied all around a part as in Figure 15.43 or between two specified points as shown in Figure 15.44.

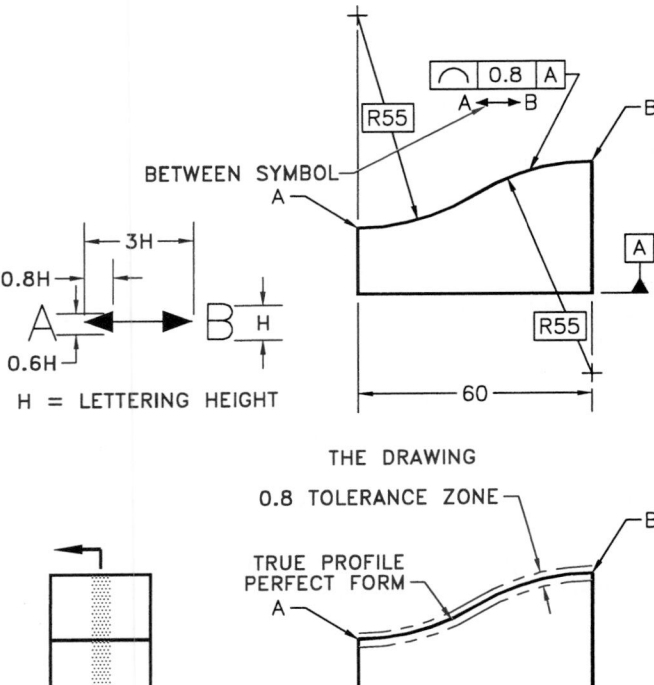

FIGURE 15.42 ■ Profile of a line between two given points, and the between symbol.

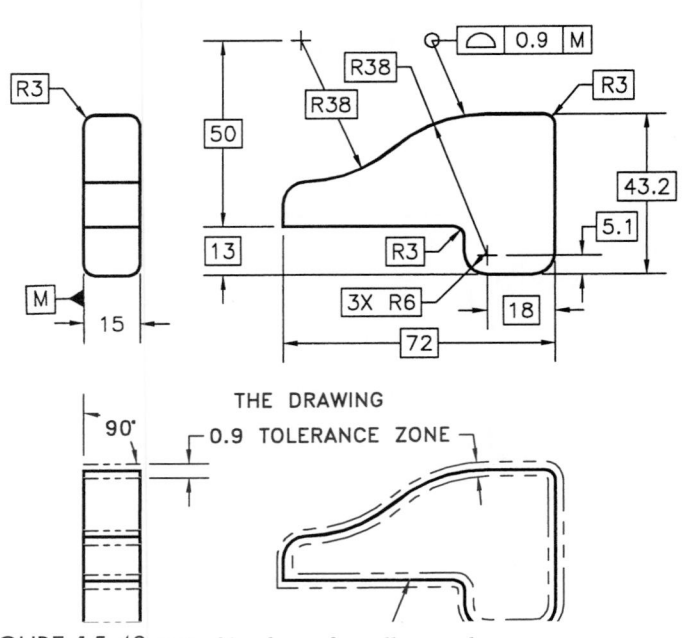

FIGURE 15.43 ■ Profile of a surface all around.

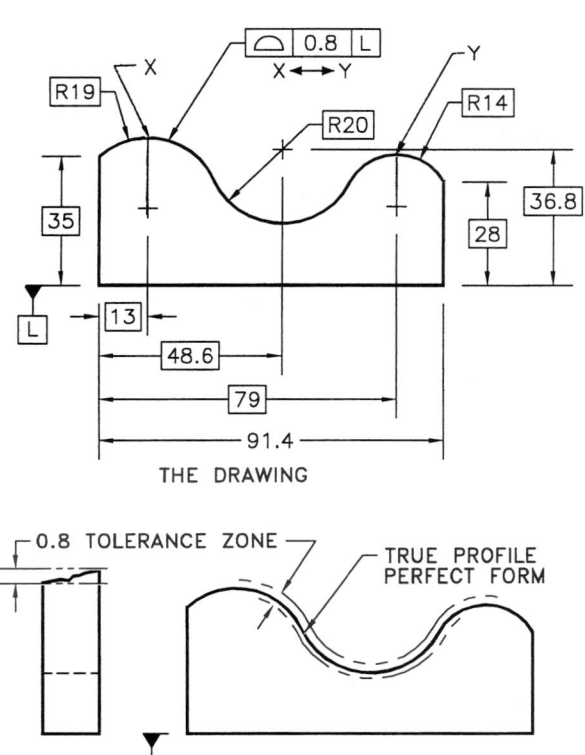

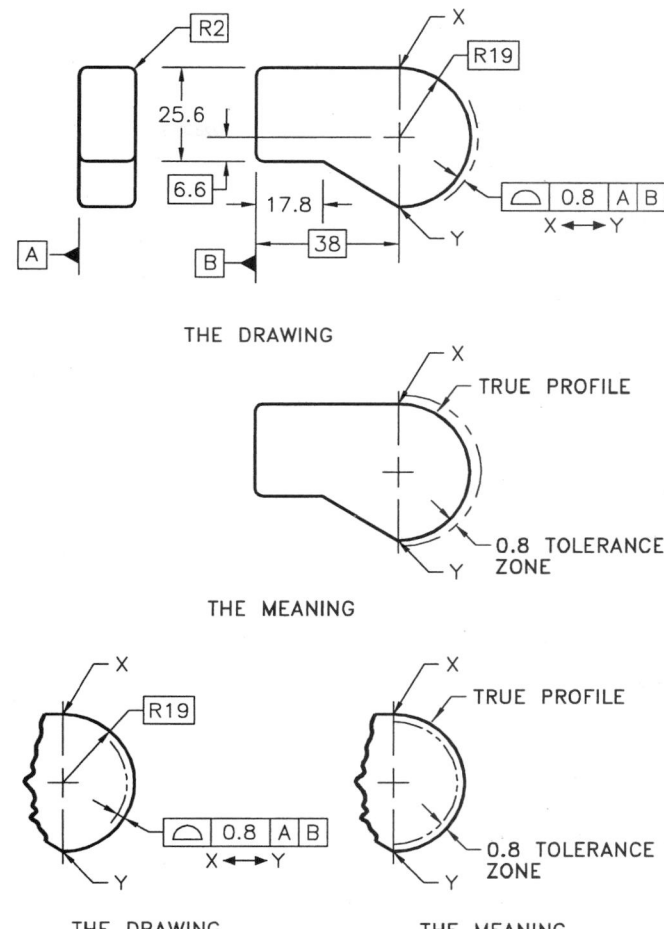

FIGURE 15.44 ■ Profile of a surface between two given points.

The profile of a line or the profile of a surface is a bilateral tolerance zone when the leader line points to the surface of the part without any additional symbology as was shown in Figures 15.40 through 15.44. A *bilateral profile tolerance* means that the tolerance zone is split equally on each side of the specified perfect form. Either type of profile callout may also have a unilateral tolerance zone specified. A unilateral tolerance zone places the entire zone on only one side of the true profile or perfect form. This is accomplished on a drawing by placing a phantom line parallel to the surface where the leader arrowhead touches the part. The phantom line is placed any clear distance from the part surface and is either inside or outside depending on the specified direction of the unilateral tolerance from the true profile. The actual feature is confined between the true profile and the given tolerance reference. (See Figure 15.45.)

The profile of coplanar surfaces may also be specified by placing a phantom line between the surfaces in the view where they appear as edges. The number of coplanar surfaces is identified below the feature control frame with a note, such as 2 SURFACES. The profile feature control frame is then connected to the phantom line with a leader. (See Figure 15.46.) Profile may also be used to control the angle of an inclined surface in relationship to a datum. (See Figure 15.47.)

Orientation Tolerance

Orientation tolerances refer to a specific group of geometric characteristics that establish a relationship between the features of an

FIGURE 15.45 ■ A unilateral profile representation.

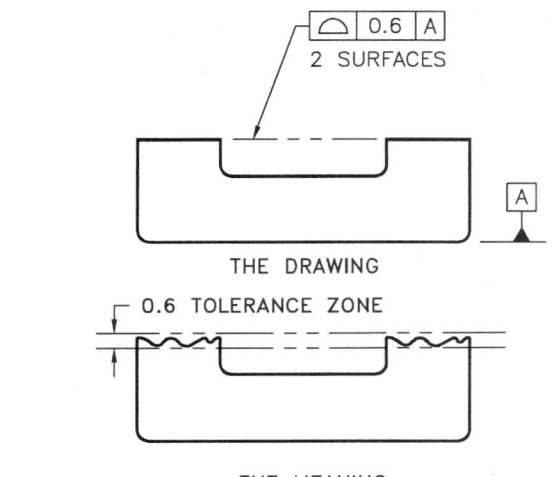

FIGURE 15.46 ■ Profile of coplanar surfaces.

object. The tolerances that orient one feature to another are parallelism, perpendicularity, angularity and, in some applications, profile. Orientation tolerances require that one or more datums be used to establish the relationship between features. Parallelism, perpendicularity, and angularity also control flatness or cylindricity

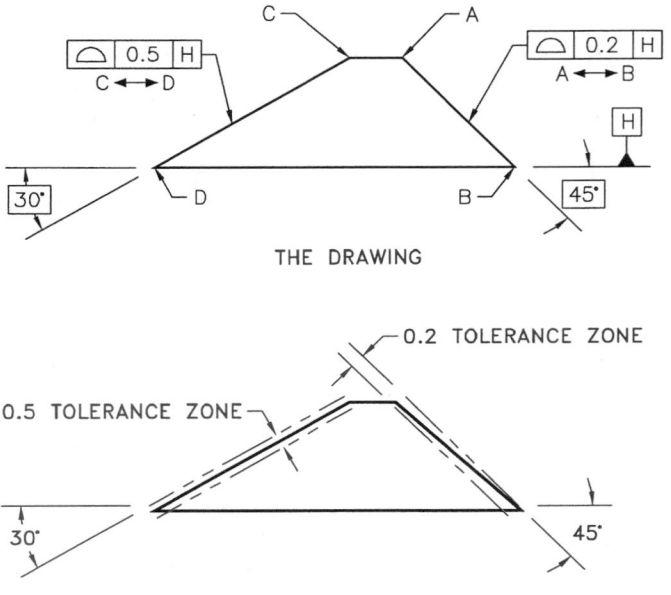

FIGURE 15.47 ■ Profile of an inclined surface.

depending on the shape of the feature being controlled. The geometric tolerance must fall within the size tolerance of the part.

Parallelism

The parallelism geometric characteristic symbol is detailed in Figure 15.48. A *parallelism* tolerance zone requires that the actual feature is between two parallel planes or lines that are parallel to a datum. The parallelism feature control frame may be attached to the feature with a leader or on an extension line of the surface. (See Figure 15.49.) Unless otherwise specified, a parallelism tolerance zone is a total tolerance that covers the entire surface. If a single-line element is to be specified rather than the surface, then the note EACH ELEMENT, or EACH RADIAL ELEMENT for an arc shaped surface, must be added below the feature control frame. (See Figure 15.50.) This technique also applies to perpendicularity and angularity.

In certain instances parallelism may also be applied to a cylindrical feature. The tolerance zone is a cylindrical shape that is parallel to a datum axis reference. The feature control frame that relates to an axis specification is attached to the diameter dimension, and a diameter zone descriptor should precede the geometric tolerance. (See Figure 15.51.)

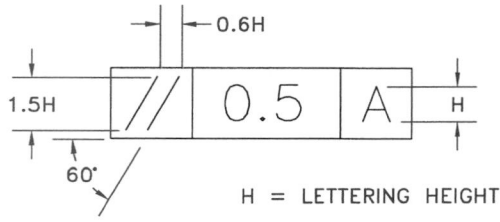

FIGURE 15.48 ■ Parallelism feature control frame.

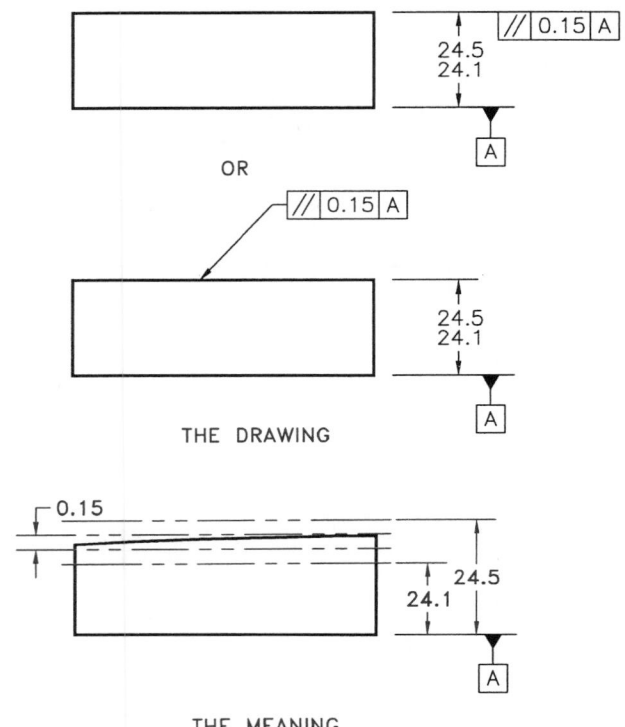

FIGURE 15.49 ■ Parallelism representation.

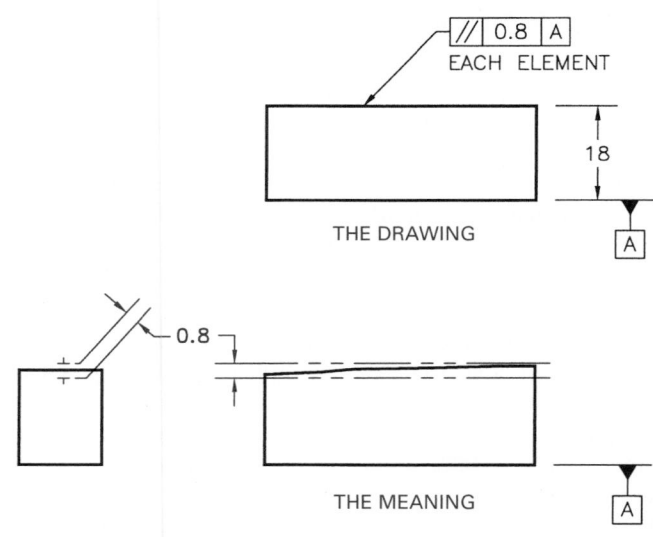

FIGURE 15.50 ■ Single-line element parallelism.

Perpendicularity

The perpendicularity geometric characteristic symbol is detailed in Figure 15.52. A *perpendicularity* geometric tolerance requires that a given feature be located between two parallel planes or lines, or within a cylindrical tolerance zone that is a basic 90° to a datum. The perpendicularity feature control frame may be connected to the feature surface with a leader or an extension line, or attached to the diameter dimension for axis perpendicularity. (See Figure 15.53.)

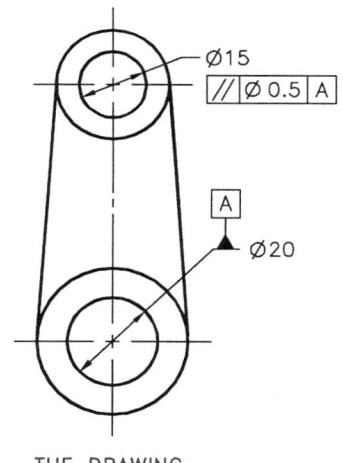

THE DRAWING

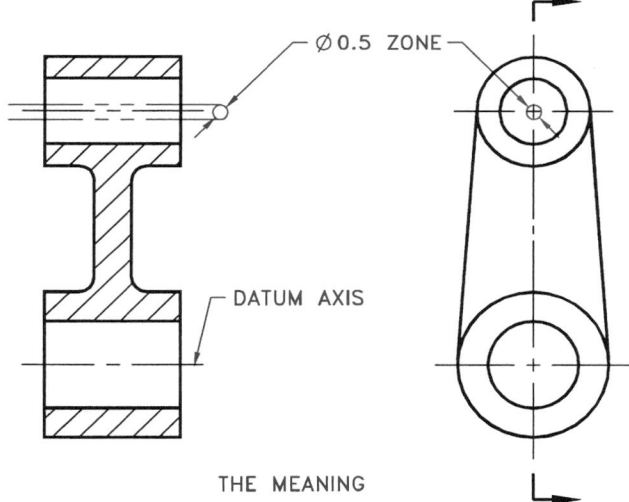

THE MEANING

FIGURE 15.51 ■ Axis parallelism.

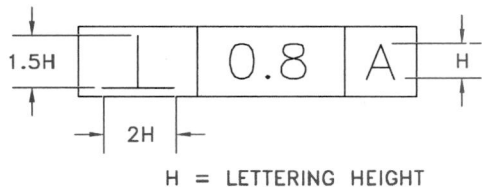

H = LETTERING HEIGHT

FIGURE 15.52 ■ Perpendicularity feature control frame.

Angularity

The angularity geometric characteristic symbol is represented in Figure 15.54. An *angularity* tolerance zone places a given feature between two parallel planes that are at a specified basic angular dimension from a datum. The basic angle from the datum may be any amount except 90°. The angularity feature control frame may be connected to the surface with a leader or from an extension line, or attached to the diameter dimension for axis angularity. (See Figure 15.55.)

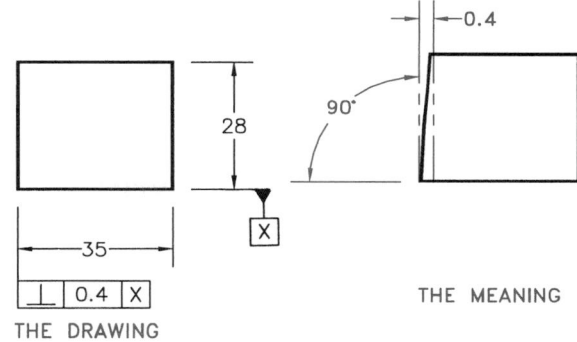

THE DRAWING

THE MEANING

SURFACE PERPENDICULARITY

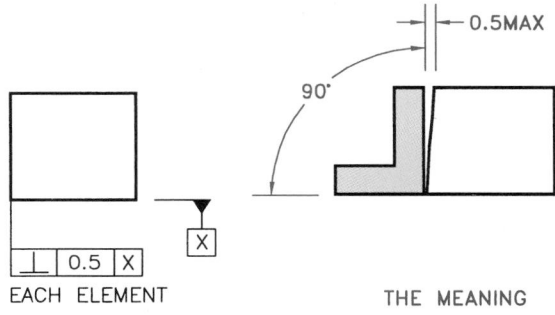

EACH ELEMENT

THE DRAWING

THE MEANING

LINE ELEMENT PERPENDICULARITY

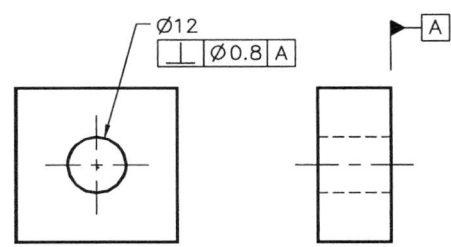

THE DRAWING

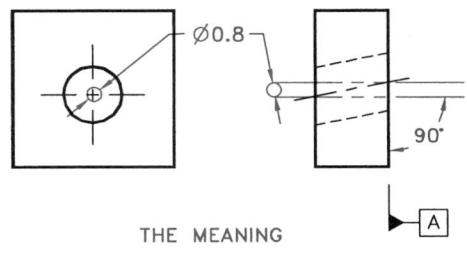

THE MEANING

AXIS PERPENDICULARITY

FIGURE 15.53 ■ Perpendicularity representations.

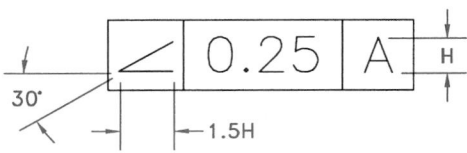

H = LETTERING HEIGHT

FIGURE 15.54 ■ Angularity feature control frame.

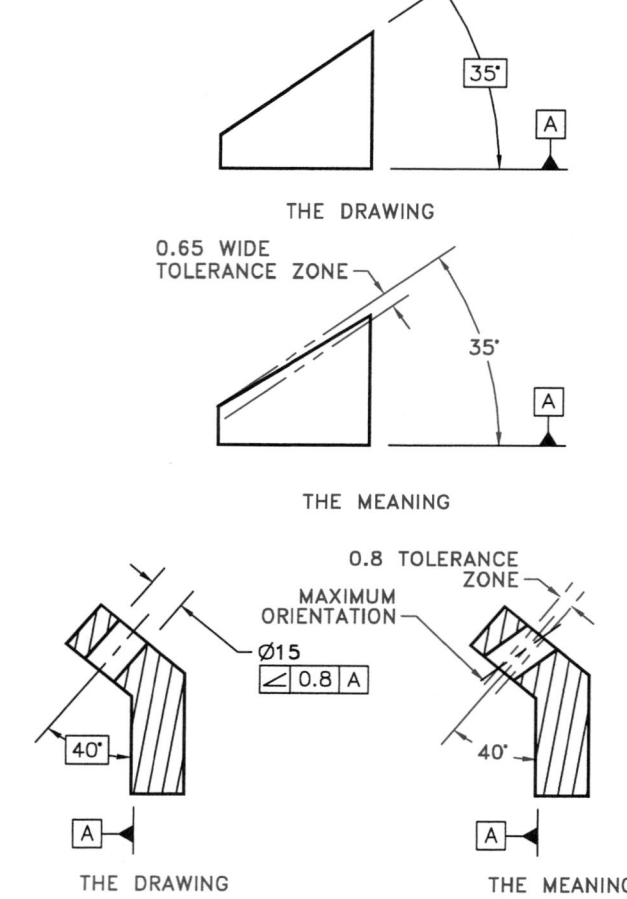

FIGURE 15.55 ■ Angularity representations.

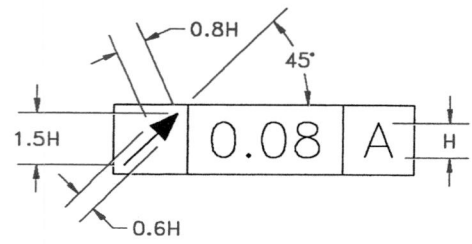

H = LETTERING HEIGHT

CIRCULAR RUNOUT

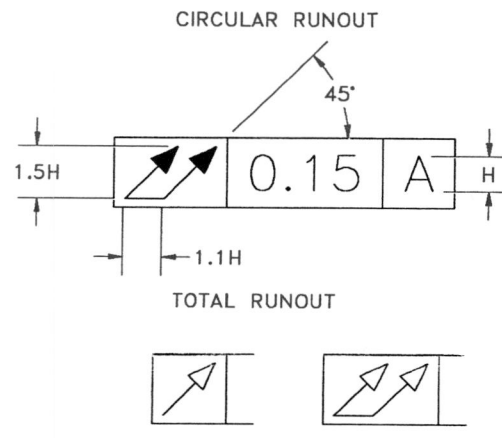

TOTAL RUNOUT

OPEN ARROWS OPTIONAL

FIGURE 15.56 ■ Runout feature control frames.

Runout Tolerance

Runout is used to control the relationship of radial features to a datum axis and features that are 90° to a datum axis. These types of features include cylindrical, tapered, and curved shapes as well as plane surfaces that are at right angles to a datum axis. A runout tolerance is determined when a dial indicator is placed on the surface to be inspected and the part is rotated 360°. The *full indicator movement* (FIM) on the dial indicator must not exceed the amount of specified tolerance. The runout feature control frame is attached to the required surface with a leader. The runout tolerance applies to the length of the intended surface or until there is a break, or change in shape or diameter of the surface. There are two types of runout, circular and total. (See Figure 15.56.)

Circular Runout

Circular runout provides control of single circular elements of a surface. Circular runout controls circularity and the relationship of the common axes (coaxial) of parts. This tolerance, established by the FIM of a dial indicator, is placed at one location on the feature as the part is rotated 360°. (See Figure 15.57.)

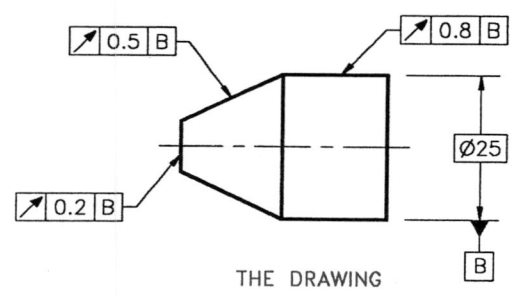

THE DRAWING

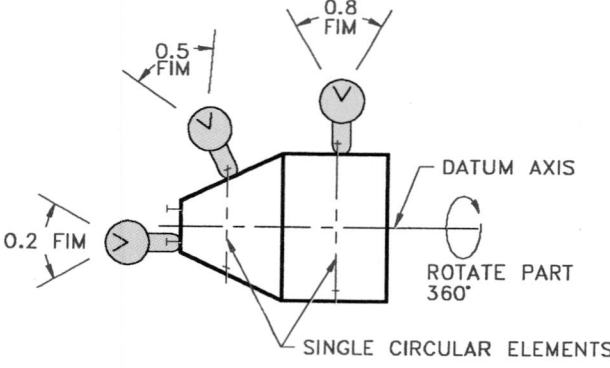

THE MEANING

FIGURE 15.57 ■ Circular runout.

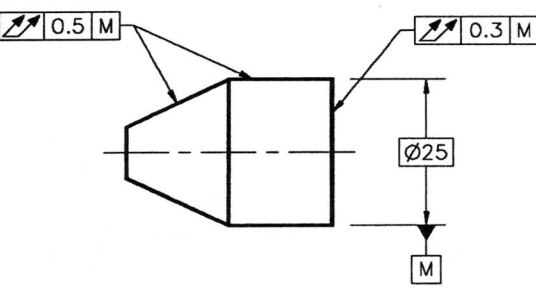

THE DRAWING

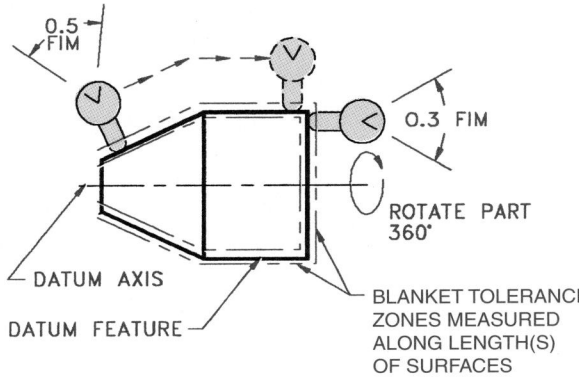

THE MEANING

FIGURE 15.58 ∎ Total runout.

Total Runout

Total runout provides composite control of all surface elements. The tolerance is applied simultaneously to all circular and profile measuring positions as the part is rotated 360°. (See Figure 15.58.) Where applied to surfaces constructed around a datum axis, total runout is used to control cumulative variations of circularity, straightness, coaxiality, angularity, taper, and profile of a surface. Where applied to surfaces constructed at right angles to a datum axis, total runout controls cumulative variations of perpendicularity to detect wobble and flatness, and to detect concavity or convexity.

A portion of a surface may have a specified runout tolerance if it is not desired to control the entire surface. This is done by placing a chain line in the linear view adjacent to the desired location. The chain line is located with basic dimensions as shown in Figure 15.59. When a part has compound datum features, that is, where

more than one datum feature is used to establish a common datum, then the combined datum features are shown separated by a dash in the feature control frame. The datums are of equal importance. (See Figure 15.59.)

MATERIAL CONDITION SYMBOLS

Material condition symbols are used in conjunction with the feature tolerance or datum reference in the feature control frame. The *material condition symbols* are required to establish the relationship between the size or location of the feature and the geometric tolerance. The use of different material condition symbols alters. the effect of this relationship. The material condition modifying elements are maximum material condition, MMC; regardless of feature size, RFS; and least material condition, LMC. There is no symbol for RFS because it is assumed for all geometric tolerances and datum references unless another material condition symbol is specified. The standard material condition symbols are shown in Figure 15.60.

Perfect Form Envelope

The form of a feature is controlled by the size tolerance limits. The envelope, or boundary, of these size limits is established at MMC. Remember from the discussion in Chapter 12 that MMC is the largest limit for an external feature and the smallest limit for an internal feature. The key is *most material*. The true geometric form of the feature is at MMC. This is known as the *perfect form boundary*. This is also referred to as the perfect form envelope. If the part feature is produced at MMC, it is considered to be at perfect form. When it is desired to permit a surface or surfaces of a feature to exceed the boundary of perfect form at MMC, a note such as PERFECT FORM AT MMC NOT REQUIRED is specified, exempting

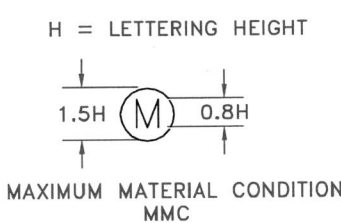

H = LETTERING HEIGHT

MAXIMUM MATERIAL CONDITION
MMC

NO SYMBOL.
RFS IS IMPLIED UNLESS
OTHERWISE SPECIFIED

REGARDLESS OF FEATURE SIZE
RFS

LEAST MATERIAL CONDITION
LMC

FIGURE 15.60 ∎ Material condition symbols.

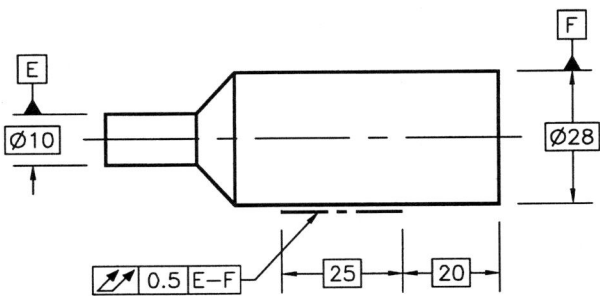

FIGURE 15.59 ∎ Partial surface runout.

GEOMETRIC TOLERANCING

The implementation of geometric tolerancing and dimensioning into a mechanical drafting CADD program is quite practical. The GT symbology makes this application a bonus to the mechanical drafting system.

Some dimensioning and tolerancing guidelines for use in conjunction with CADD/CAM are outlined, in part, from ASME Y14.5M as follows:

1. Major features of the part should be used to establish the basic coordinate system but are not necessarily defined as datums.

2. Subcoordinated systems that are related to the major coordinates are used to locate and orient features on a part.

3. Define part features in relation to three mutually perpendicular reference planes and along features that are parallel to the motion of CAM equipment.

4. Establish datums related to the function of the part and relate datum features in order of precedence as a basis for CAM usage.

5. Completely and accurately dimension geometric shapes. Regular geometric shapes may be defined by mathematical formulas, although a profile feature that is defined with mathematical formulas should not have coordinate dimensions unless required for inspection or reference.

6. Coordinate or tabular dimensions should be used to identify approximate dimensions on an arbitrary profile.

7. Use the same type of coordinate dimensioning system on the entire drawing.

8. Continuity of profile is necessary for CADD. Clearly define contour changes at the change or point of tangency. Define at least four points along an irregular profile.

9. Circular hole patterns may be defined with polar coordinate dimensioning.

10. When possible, dimension angles in degrees and decimal parts of degrees; for example, 45° 30′ = 45.5°.

11. Base dimensions at the mean of a tolerance because the numerical control (NC) programmer will normally split a tolerance and work to the mean. Establish dimensions without limits that conform to the NC machine capabilities and part function where possible. Bilateral profile tolerances are also recommended for the same reason.

12. Geometric tolerancing is necessary to control specific geometric form and location, but is not required if general tolerancing provides sufficient form or location control.

The standard ASME Y14.5.1, *Mathematical Definition of Y14.5,* is the mathematical model of all specifications found in ASME Y14.5M. This allows for convenient digitizing of a part so all gaging is in a CADD system.

---

the pertinent size dimension. If a feature is produced at LMC, the opposite of MMC, the form is allowed to vary within the geometric tolerance zone or to the extent of the MMC envelope.

Regardless of Feature Size

Regardless of feature size (RFS) means that the geometric tolerance applies at any produced size. The tolerance remains as the specified value regardless of the actual size of the feature. Regardless of feature size is implied (assumed) for all geometric characteristics and related datums unless otherwise specified.

Effect of RFS on Surface Straightness

RFS is implied for the straightness geometric characteristic. When a surface straightness tolerance is used by connecting a leader from the feature control frame to the surface of the part, then the geometric tolerance remains the same regardless of the feature size, and

the actual size may not exceed the perfect form envelope at MMC. An acceptable part may be produced between the given size tolerance, and any straightness irregularity may not be greater than the specified geometric tolerance. Perfect form is required at MMC. (See Figure 15.61.)

Effect of RFS on Axis Straightness

When the straightness tolerance is applied to the axis of the feature by a relationship with the diameter dimension, RFS is implied unless otherwise specified, and the actual feature size plus the geometric tolerance may exceed the MMC perfect form envelope. (See Figure 15.62.)

Datum Feature RFS

Datum features that are influenced by size variations, such as diameters and widths, are also subject to variations in form. RFS

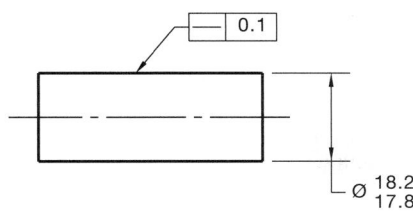

THE DRAWING

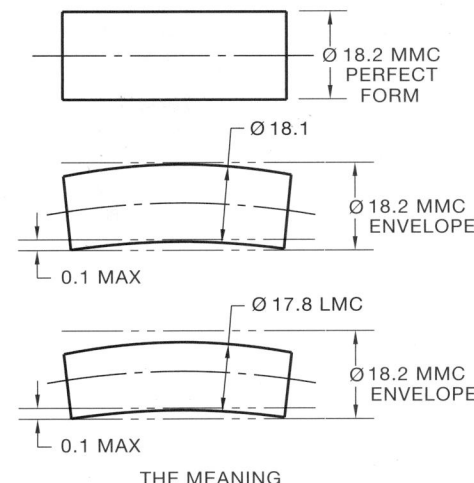

THE MEANING

FIGURE 15.61 ■ Effect of regardless of feature size, RFS, on surface straightness. Perfect form is required at MMC.

ACTUAL SIZE	GEOMETRIC TOLERANCE
18.2 MMC	0–PERFECT FORM
18.1	0.1
18.0	0.1
17.9	0.1
17.8 LMC	0.1

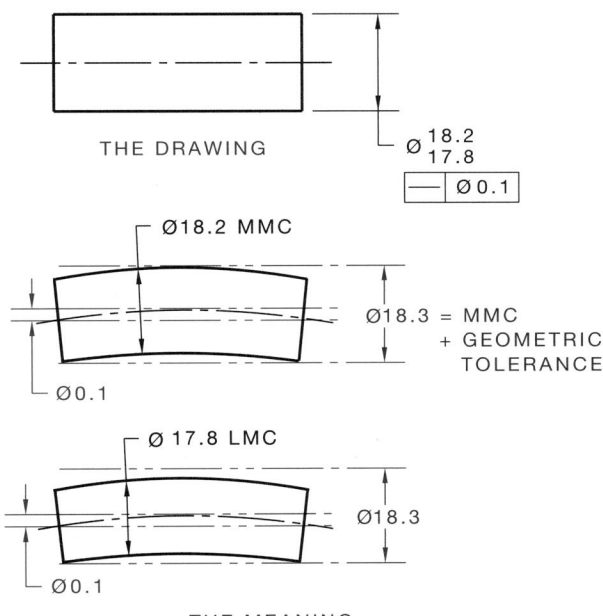

ACTUAL SIZE	GEOMETRIC TOLERANCE
18.2 MMC	0.1
18.1	0.1
18.0	0.1
17.9	0.1
17.8 LMC	0.1

FIGURE 15.62 ■ Effect of RFS on axis straightness.

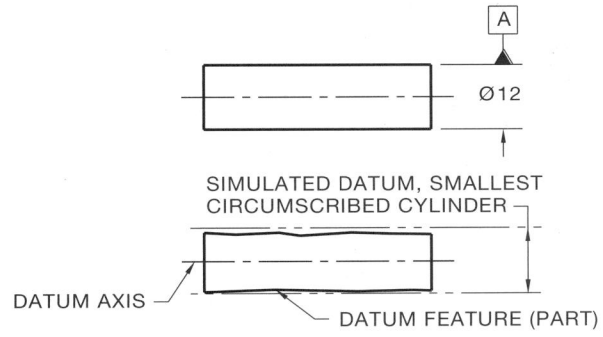

PRIMARY EXTERNAL DATUM AXIS – RFS

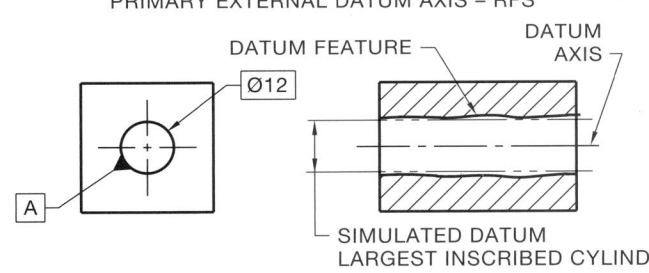

PRIMARY INTERNAL DATUM AXIS – RFS

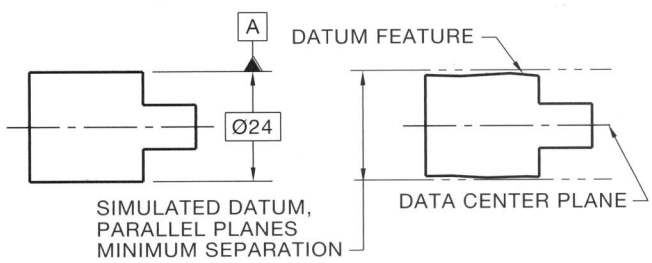

PRIMARY EXTERNAL DATUM CENTER PLANE – RFS

FIGURE 15.63 ■ Effect of RFS on the primary datum feature with axis and center plane datums.

is implied unless otherwise specified. When a datum feature has a size dimension and a geometric form tolerance, the size of the simulated datum is the MMC size limit. This rule applies except for axis straightness where the envelope is allowed to exceed MMC. Figure 15.63 shows the effect of RFS on the primary datum feature with axis and center plane datums. When the

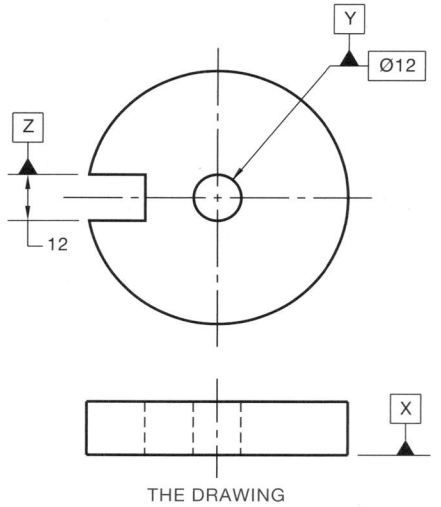

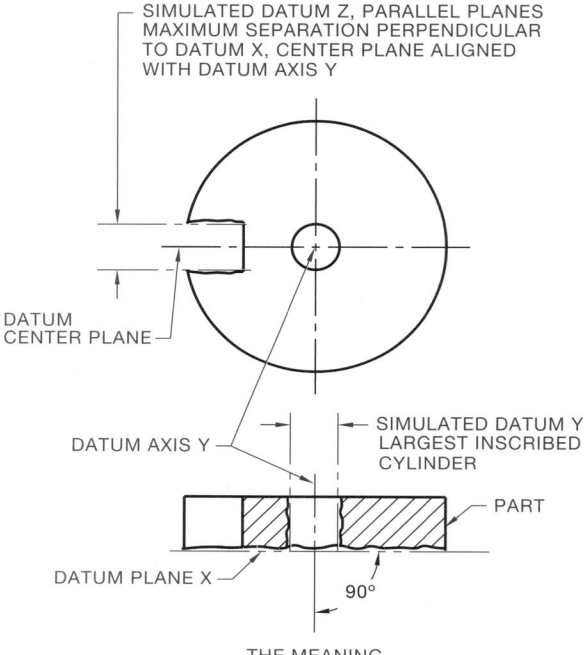

FIGURE 15.64 ■ The secondary or tertiary datum relationship to the primary datum.

datum features are secondary or tertiary, then the axis or center plane shall also have an angular relationship to the primary datum. (See Figure 15.64.)

Maximum Material Condition

The use of MMC in conjunction with the geometric tolerance in a feature control frame means that the given tolerance applies at the MMC produced size. As the feature dimension departs from MMC the geometric tolerance is allowed to increase equal to the change.

The maximum amount of change is at the LMC produced size. MMC must be specified for any geometric characteristic where its application is desired.

Effect of MMC on Form Tolerance

When MMC is specified in conjunction with an axis straightness callout, the MMC feature size envelope is exceeded by the given geometric tolerance. The given geometric tolerance is held at the MMC produced size, then as the actual produced size departs from MMC the geometric tolerance is allowed to increase equal to the change from MMC to a maximum amount of departure at LMC. (See Figure 15.65.) The same situation may also be demonstrated with a perpendicularity tolerance where a datum axis is perpendicular to a given datum as in Figure 15.66.

Least Material Condition

The application of LMC is not as prevalent as MMC. LMC may be used when it is desirable to control minimum wall thick-

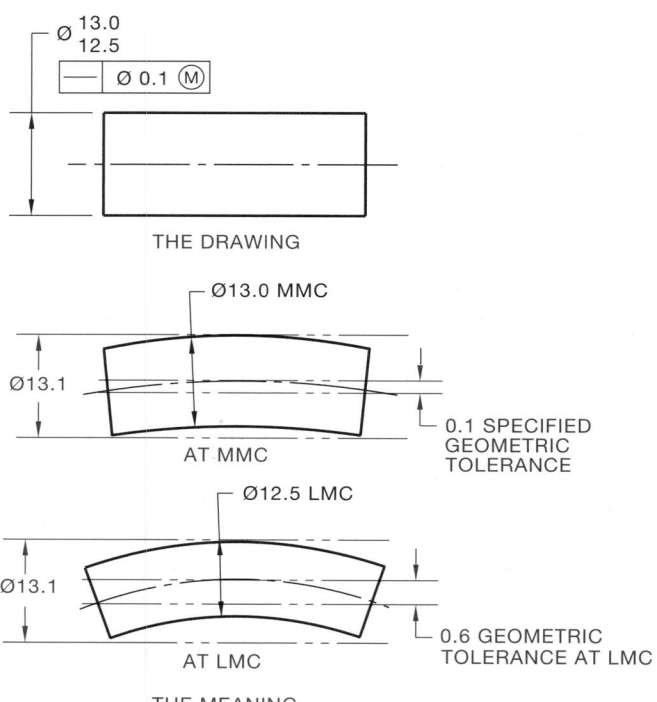

ACTUAL SIZE	GEOMETRIC TOLERANCE
13.0 MMC	0.1
12.9	0.2
12.8	0.3
12.7	0.4
12.6	0.5
12.5 LMC	0.6

FIGURE 15.65 ■ Effect of maximum material condition, MMC, on form tolerance, axis straightness.

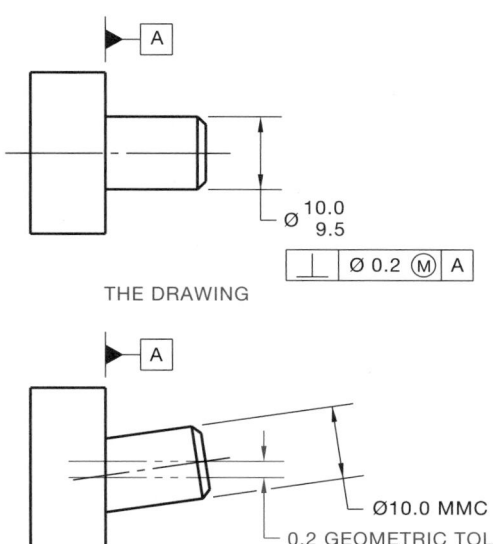

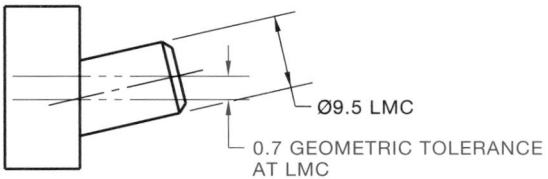

THE MEANING

ACTUAL SIZE	GEOMETRIC TOLERANCE
10.0 MMC	0.2
9.9	0.3
9.8	0.4
9.7	0.5
9.6	0.6
9.5 LMC	0.7

FIGURE 15.66 ■ Effect of MMC on form tolerance, axis perpendicularity.

ness. This is discussed with positional geometric tolerance at LMC later. When an LMC material condition symbol is used in conjunction with a geometric tolerance, the specified tolerance applies at the LMC produced size. GDT controls are refinement of size controls. As the actual produced size deviates from LMC toward MMC, the tolerance is allowed an increase equal to the amount of change from LMC. (See Figure 15.67.)

LOCATION TOLERANCE

Location tolerances include concentricity, symmetry, and position. *True position* is the theoretically exact location of the axis or center plane of a feature. The true position is located using basic dimensions. The locational tolerance specifies the amount that the axis of the feature is allowed to deviate from true position. All geometric tolerances imply RFS. MMC or LMC must be specified if desired. Reference to true position dimensions must be provided as basic

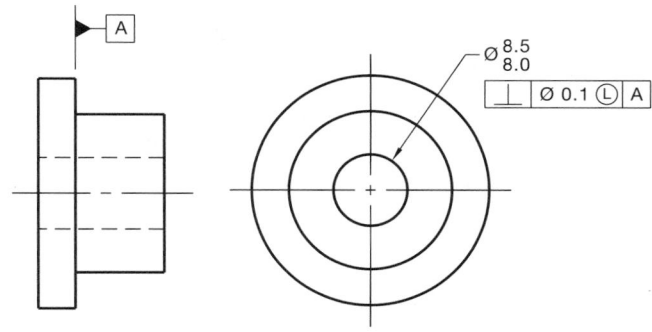

ACTUAL SIZE	GEOMETRIC TOLERANCE
8.0 MMC	0.6
8.1	0.5
8.2	0.4
8.3	0.3
8.4	0.2
8.5 LMC	0.1

FIGURE 15.67 ■ Application of least material condition, LMC.

dimensions at each required location on the drawing or a general note may be used to specify that UNTOLERANCED DIMENSIONS LOCATING TRUE POSITION ARE BASIC. Basic dimensions are theoretically perfect, so that datum or chain dimensioning may be used equally to locate true position because there is no tolerance buildup. The location feature control frame is generally added to the note or dimension of the related feature.

Concentricity

The concentricity geometric characteristic symbol is detailed in Figure 15.68. *Concentricity* is the relationship of the axes of cylindrical shapes. Perfect concentricity exists when the axes of two or more cylindrical features are in perfect alignment. (See Figure 15.69.) A concentricity geometric tolerance allows for a

H = LETTERING HEIGHT

FIGURE 15.68 ■ Concentricity feature control frame.

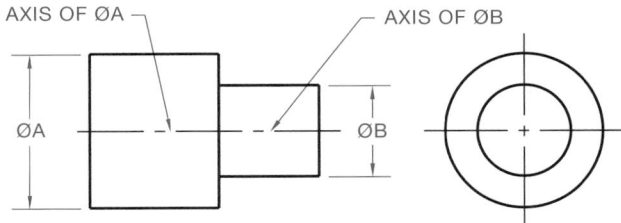

FIGURE 15.69 ■ Perfect concentricity when both axes coincide.

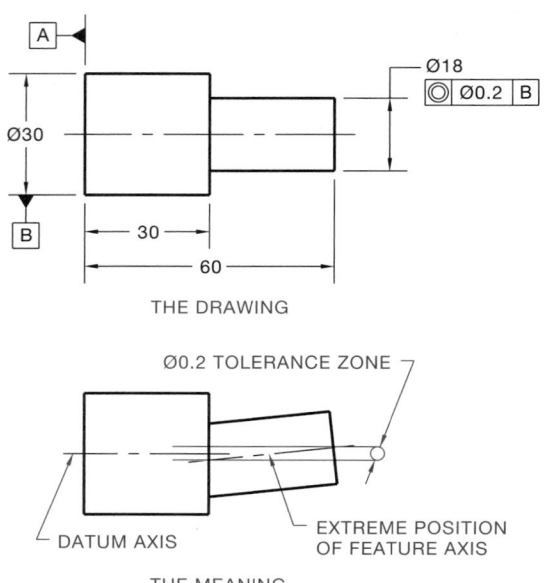

FIGURE 15.70 ■ Concentricity represented.

specified amount of deviation of the axes of concentric cylinders as shown in Figure 15.70. The geometric tolerance and related datums imply RFS. It is difficult to control the axis relationship specified by concentricity, so runout is often used. Where the balance of a shaft is critical, concentricity may be used.

Position Tolerance of Symmetrical Features

Position tolerancing may be used to locate the center plane of one or more features relative to a datum center plane. Figure 15.71

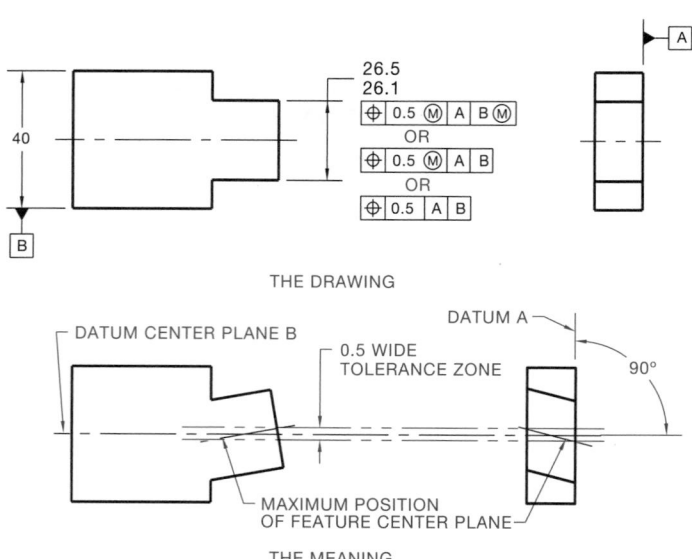

FIGURE 15.71 ■ Position geometric characteristic specifying symmetry.

shows optional feature control frames where the position geometric tolerance and related datum reference are controlled on an RFS or MMC basis. When the position tolerance is at MMC, the related datum feature may be RFS, MMC, or LMC depending on design requirements. If the position tolerance is held RFS, then the related datum reference is RFS. RFS is implied unless otherwise specified.

Symmetry

Symmetry is the center plane relationship between two or more features. Perfect symmetry exists when the center plane of two or more features is in alignment. The geometric characteristic symbol used to specify symmetry is detailed in the feature control frame shown in Figure 15.72. The *symmetry* geometric tolerance is a zone in which the median points of opposite symmetrical surfaces align with the datum center plane. The symmetry geometric tolerance and related datum reference are applied only on an RFS basis. (See Figure 15.73.) A zero position tolerance at MMC is used when it is required to control the symmetrical relationship of features within their size limits. An explanation of zero position tolerance at MMC follows.

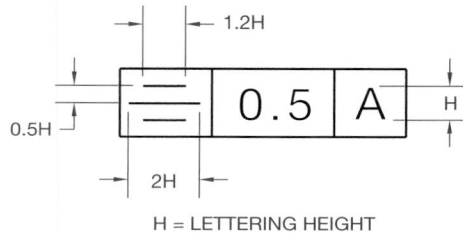

FIGURE 15.72 ■ The symmetry symbol detailed in a feature control frame.

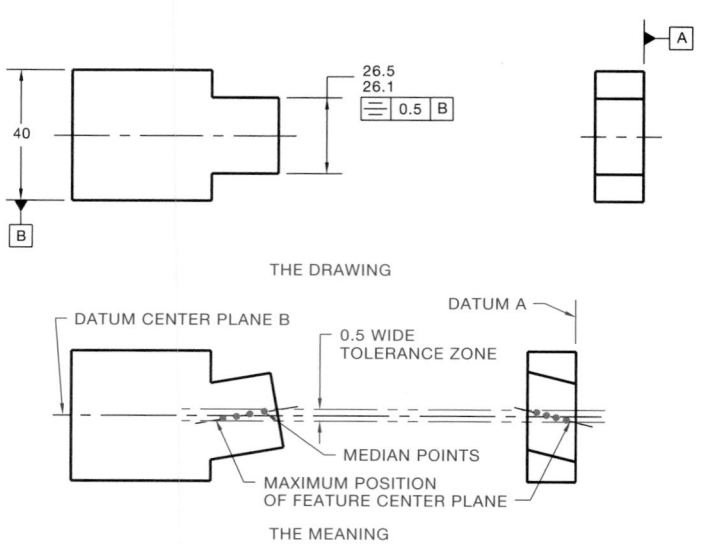

FIGURE 15.73 ■ Symmetry application.

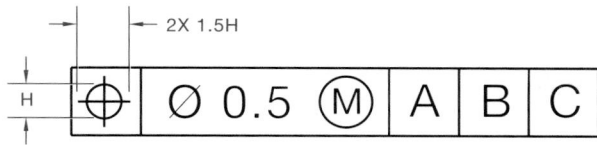

FIGURE 15.74 ■ Position feature control frame.

Position

The position geometric characteristic symbol, also used for symmetry, is detailed in Figure 15.74. RFS is implied for the position geometric tolerance and related datum reference unless otherwise specified. MMC or LMC is applied to the position tolerance and datum reference as needed for design. The maximum material condition application is common, although some cases require the use of regardless of feature size or least material condition. The feature control frame is applied to the note or size dimension of the feature. A *positional* tolerance defines a zone within which the center, axis, or center plane of a feature or size is permitted to vary from the true, theoretically exact, position. Basic dimensions establish the true position from specified datum features and between interrelated features. A *positional* tolerance is indicated by the position symbol, a tolerance, and appropriate datum references placed in a feature control frame. The location of each feature is given by basic dimensions. Dimensions locating true position must be excluded from the drawing's general tolerances by applying the basic dimension symbol to each basic dimension or by specifying on the drawing or drawing reference the general note UNTOLERANCED DIMENSIONS LOCATING TRUE POSITION ARE BASIC.

Positional Tolerance at Maximum Material Condition (MMC)

Positional tolerance at MMC means that the given tolerance is held at the MMC produced size. Then, as the feature size departs from MMC, the position tolerance increases equal to the amount of change from MMC to the maximum change at LMC. Positional tolerance at MMC may be defined by the feature axis or surface. The datum references commonly establish true position perpendicular to the primary datum and the coordinate location dimensions to the secondary and tertiary datums. The position tolerance zone is a cylinder equal in diameter to the given position tolerance. The cylindrical tolerance zone extends through the thickness of the part unless otherwise specified. The actual centerline of the feature may be anywhere within the cylindrical tolerance zone. (See Figure 15.75.) Another explanation of position may be related to the surface of a hole. The hole surface cannot be inside a cylindrical tolerance zone established by the MMC diameter of the hole less the position tolerance. (See Figure 15.76.)

Zero Positional Tolerance at MMC

When the positional tolerance is associated with the MMC symbol, the tolerance is allowed to exceed the specified amount when the

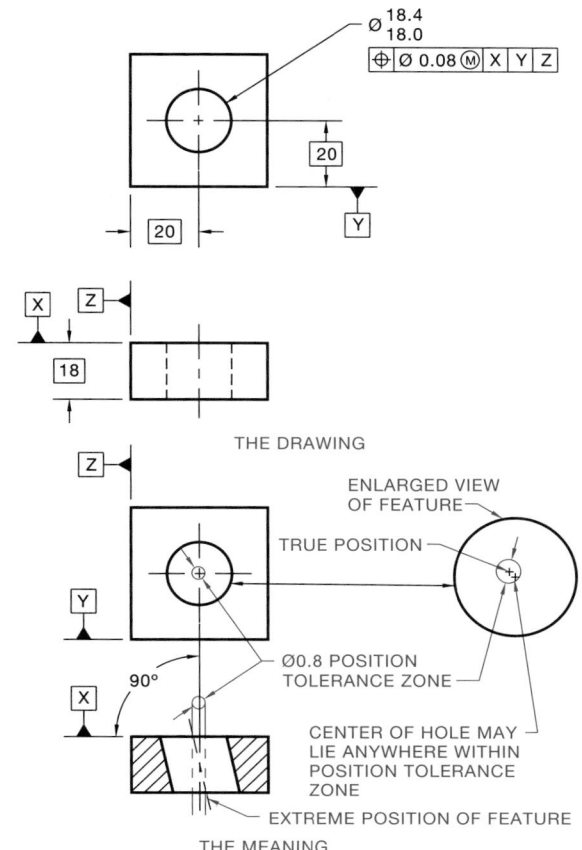

ACTUAL SIZE	POSITIONAL TOLERANCE
18.0 MMC	0.08
18.1	0.18
18.2	0.28
18.3	0.38
18.4 LMC	0.48

FIGURE 15.75 ■ Positional tolerance at maximum material condition, MMC.

actual feature size departs from MMC. When the actual sizes of features are manufactured very close to MMC, it is critical that the feature axis or surface not exceed the boundaries discussed previously. When parts are rejected due to this situation, it is possible to increase the acceptability of mating parts by reducing the MMC size of the feature to minimum allowance with the mating part (virtual condition) and providing a zero positional tolerance at MMC. The positional tolerance is dependent on the feature size. When zero positional tolerance is used, no positional tolerance is allowed when the part is produced at MMC. True position is required at MMC. As the actual size departs from MMC, the positional tolerance increases equal to the amount of change to the maximum tolerance at LMC. (See Figure 15.77.)

Positional Tolerance at RFS

RFS may be applied to the positional tolerance when it is desirable to maintain the given tolerance at any produced size. This application results in close positional control. (See Figure 15.78.) No additional tolerance is permitted based on size of feature.

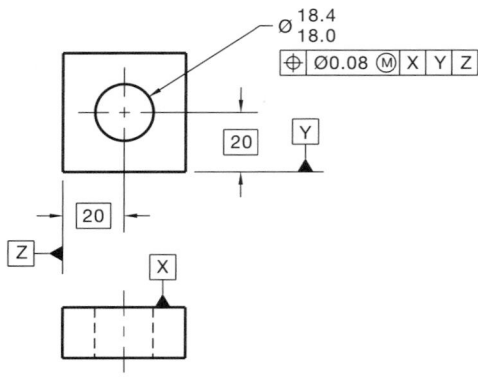

THE DRAWING

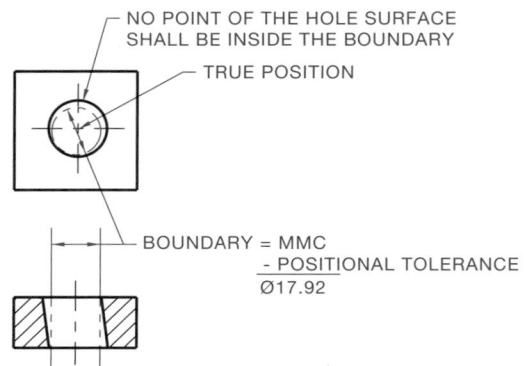

FIGURE 15.76 ■ Position Boundary = MMC – Position Tolerance

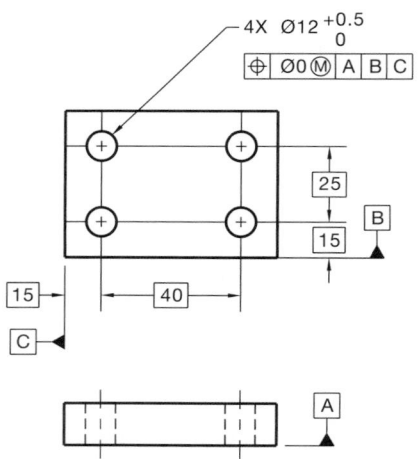

ACTUAL SIZE	POSITIONAL TOLERANCE
12.0 MMC ZERO ALLOWANCE 12mm FASTENER AT MMC	0
12.1	0.1
12.2	0.2
12.3	0.3
12.4	0.4
12.5	0.5

FIGURE 15.77 ■ Zero positional tolerances at MMC.

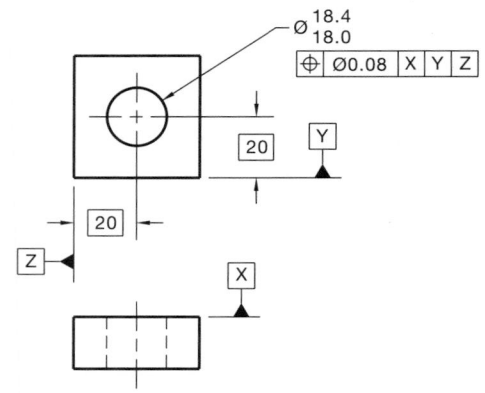

ACTUAL SIZE	POSITIONAL TOLERANCE
18.0 MMC	0.08
18.1	0.08
18.2	0.08
18.3	0.08
18.4 LMC	0.08

FIGURE 15.78 ■ Positional tolerance at RFS.

Positional Tolerance at LMC

Positional tolerance at LMC is used to control the relationship of the feature surface and the true position at largest hole size. The function of the LMC specification is generally for the control of minimum-edge distances. When LMC is used, the given positional tolerance is held at the LMC produced size where perfect form is required. As the actual size departs from LMC toward MMC, the positional tolerance zone is allowed to increase equal to the amount of change. The maximum positional tolerance is at the MMC produced size. (See Figure 15.79.)

DATUM PRECEDENCE AND MATERIAL CONDITION

The effect of material condition on the datum and related feature may be altered by changing the datum precedence and the applied material condition symbol. The datum precedence is established by the order of placement in the feature control frame. The first datum listed is the primary datum; subsequent datums are secondary and tertiary. Figure 15.80 shows the effect of altering datum precedence and material condition.

POSITION OF MULTIPLE FEATURES

The location of multiple features is handled in a manner similar to the location of a single feature. The true positions of the features are located with basic dimensions using rectangular or polar coordinates. The features are then identified by quantity, size, and position. (See Figure 15.81.) When two or more separate patterns are referenced to the same datums and with the same datum precedence, then the patterns are functionally the same. Verification of location and size is performed together. If

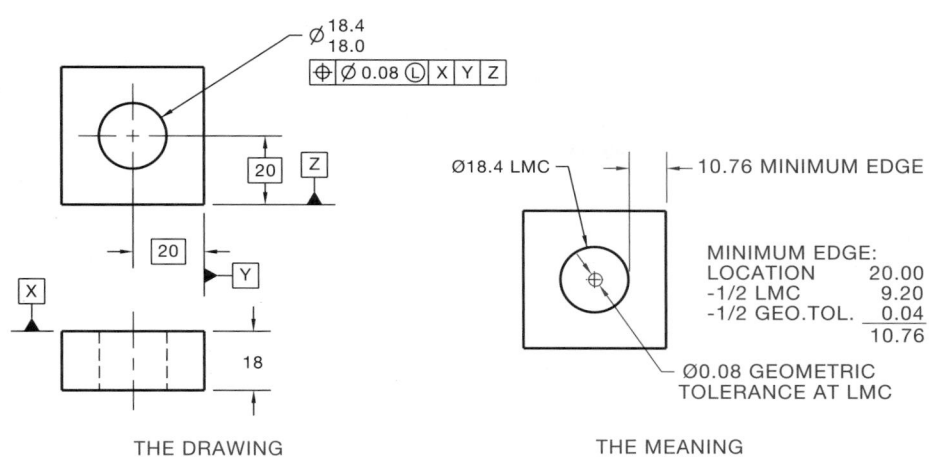

ACTUAL SIZE	POSITIONAL TOLERANCE
18.0 MMC	0.48
18.1	0.38
18.2	0.28
18.3	0.18
18.4 LMC	0.08

FIGURE 15.79 ■ Positional tolerance at LMC.

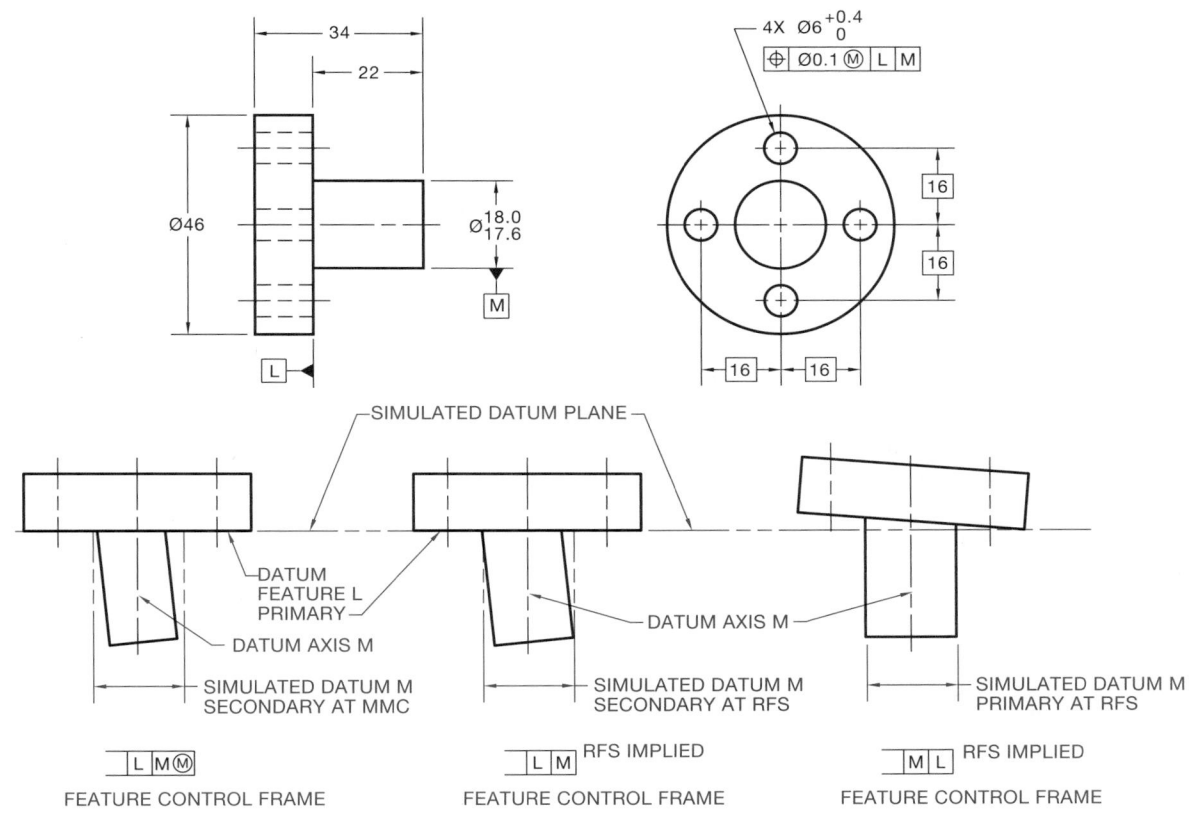

FIGURE 15.80 ■ The effect of altering datum precedence and material condition.

this situation occurs and the interrelationship between patterns is not desired, then the specific note SEP REQT, meaning separate requirement, shall be placed beneath each affected feature control frame.

COMPOSITE POSITIONAL TOLERANCING

In some situations it may be permissible to allow the location of individual features in a pattern to differ from the tolerance related to the items as a group. When this is done, the group of features is given a positional tolerance that is greater in diameter than the zone specified for the individual features. The tolerance zone of the individual features must fall within the group zone. The positional tolerance of the individual elements controls the perpendicularity of the features. An individual tolerance zone may extend partly beyond the group zone only if the feature axis does not fall outside the confines of both zones. When such is the case, the feature control frame is expanded in height and divided into two parts. The upper part, known as the *pattern locating control,*

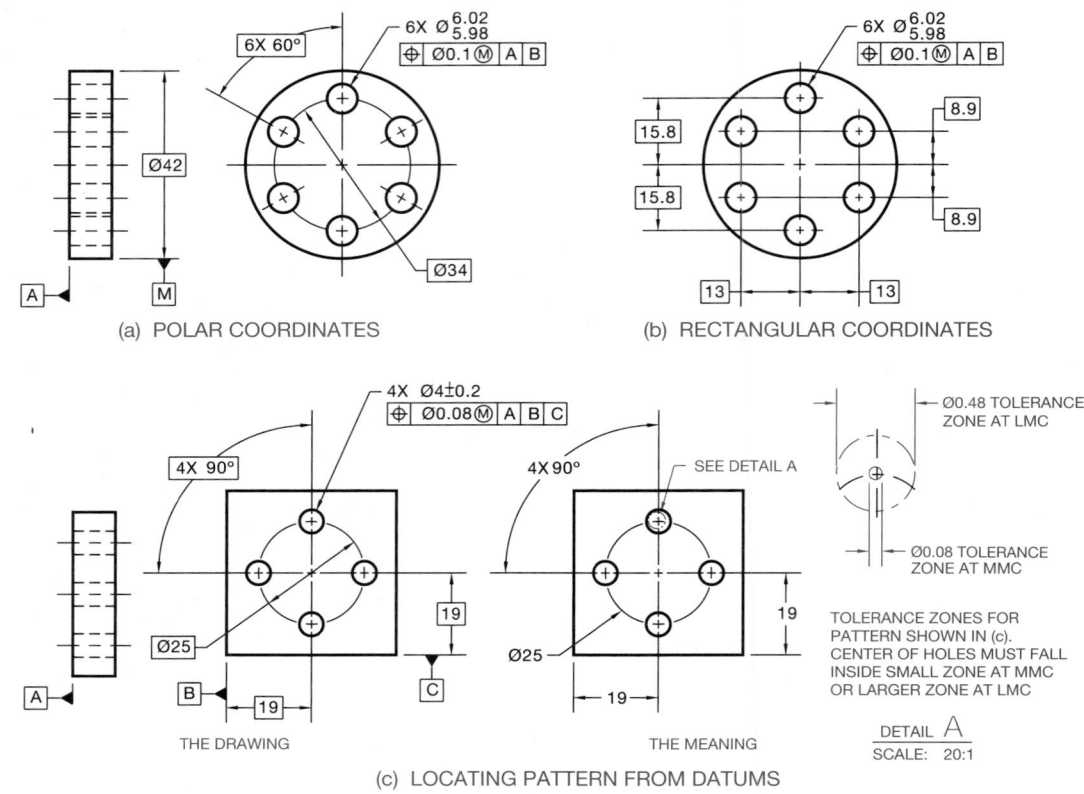

(a) POLAR COORDINATES

(b) RECTANGULAR COORDINATES

(c) LOCATING PATTERN FROM DATUMS

FIGURE 15.81 ■ Position of multiple features.

specifies the larger positional tolerance for the pattern of features as a group. The lower entry, commonly called the *feature relating control*, specifies the smaller positional tolerance for the individual features within the pattern. The pattern locating control is located first with basic dimensions. The feature relating control is established at the actual position of the feature center. Only the primary datum is represented in the feature relating control. (See Figure 15.82.)

TWO SINGLE-SEGMENT FEATURE CONTROL FRAMES

The composite position tolerance is specified by a feature control frame doubled in height with one position symbol shown in the first compartment and only a single datum reference for orientation given for the feature relating control. The *two single-segment feature control frame* is similar, except there are two position symbols, each displayed in a separate compartment, and a two-datum reference in the lower feature control frame. Look at Figure 15.83. The top feature control frame is the pattern locating control and works as previously discussed. The lower feature control frame is the feature relating control in which two datums control the orientation and alignment with the pattern locating control. This type of position tolerance provides a tighter relationship of the holes within the pattern than the composite position tolerance.

POSITIONAL TOLERANCE OF TABS

Positional tolerancing of tabs may be accomplished by identifying related datums, dimensioning the relationship between the tabs, and providing the number of units followed by the size and feature control frame. (See Figure 15.84.)

POSITIONAL TOLERANCE OF SLOTTED HOLES

The positional tolerance of slotted holes may be accomplished by locating the slot centers with basic dimensions and providing a feature control frame to both the length and width of the slotted holes. (See Figure 15.85.) The word BOUNDARY is placed below each feature control frame. This means that each slotted feature is controlled by a theoretical boundary of identical shape that is located at true position. The size of each slot

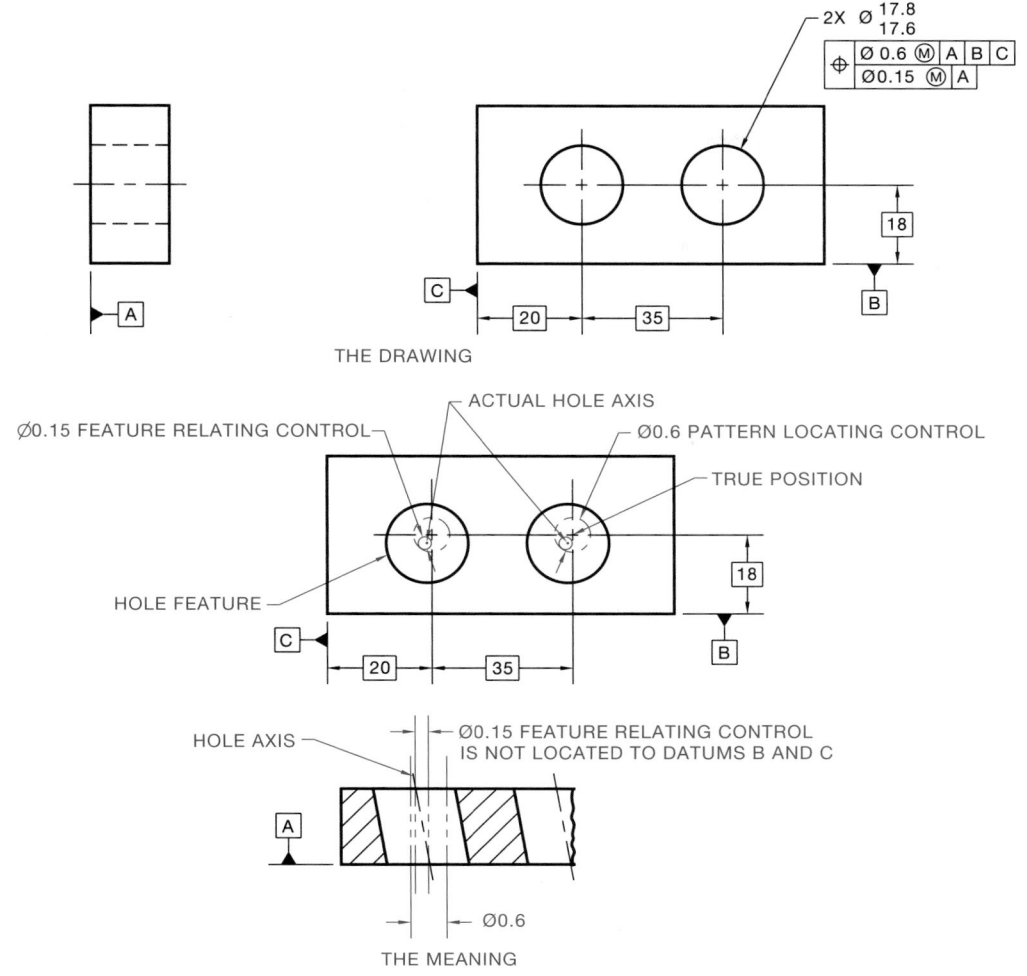

FIGURE 15.82 ■ Composite positional tolerance.

must remain within the size limits and no portion of the surface may enter the theoretical boundary, which is the size of the boundary calculated by MMC (each dimension, length and width) - Position Tolerance. Normally there is a greater position tolerance for the length than the width. When the position tolerance is the same for the length and the width, the feature control frame is separated from the size dimensions and connected to one of the slots with a leader.

POSITIONAL TOLERANCE OF THREADED FEATURES AND GEARS

The orientation or positional tolerance of screw threads applies to the datum axis. The datum feature is established by a pitch diameter cylinder. If a different datum feature is required, the note should be lettered below the feature control frame, for example, MINOR DIA or MAJOR DIA. Similarly, the intended datum feature for gears

or splines should be identified below the feature control frame. The options include MAJOR DIA, PITCH DIA, or MINOR DIA.

POSITIONAL TOLERANCE OF COUNTERBORED HOLES

Counterbored holes are a type of coaxial feature. *Coaxial* means that one or more features have the same axis. There are three different ways to provide positional tolerancing to counterbored holes as shown in Figure 15.86.

POSITIONAL TOLERANCING OF COAXIAL HOLES

Coaxial holes that are aligned through different features may be controlled with positional tolerancing using a feature relating zone

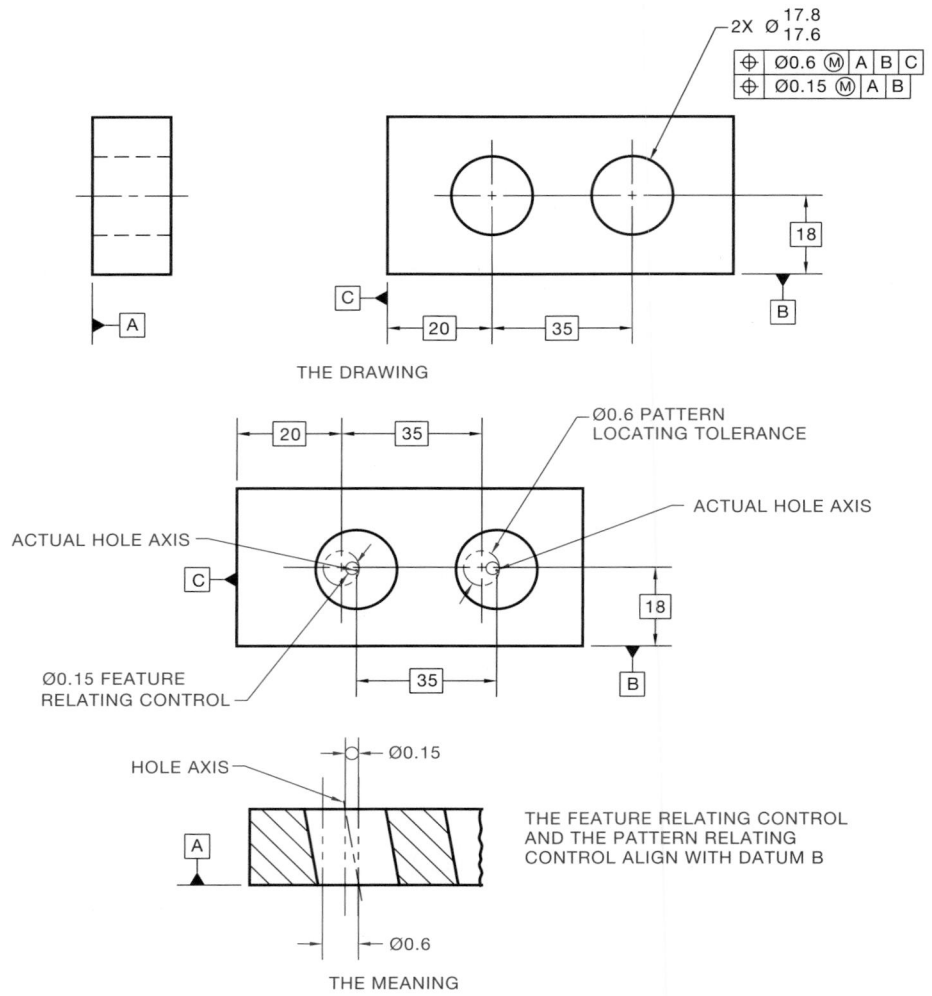

FIGURE 15.83 ■ Two single-segment feature control frame.

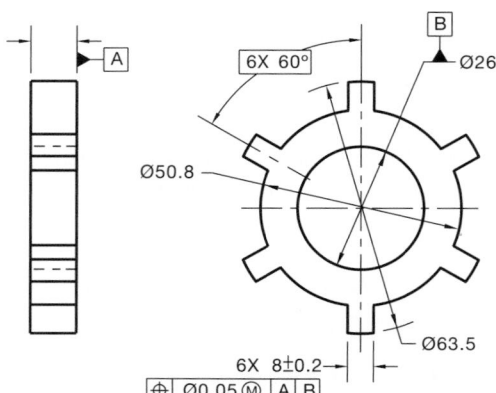

FIGURE 15.84 ■ Positional tolerance of tabs, similar practice for slots.

for each hole and a pattern locating zone for the holes as a group. Some possible options are shown in Figure 15.87.

POSITIONAL TOLERANCING OF NONPARALLEL HOLES

Positional tolerancing may be applied to holes that are not parallel to each other. The axis may not be perpendicular to a surface, as shown in Figure 15.88. The feature is located with a base longitudinal and angular dimension.

PROJECTED TOLERANCE ZONE

The standard application of a positional tolerance implies that the cylindrical zone extends through the thickness of the part or feature. In some situations where there is the possibility of interference with mating parts, the tolerance zone could be extended or

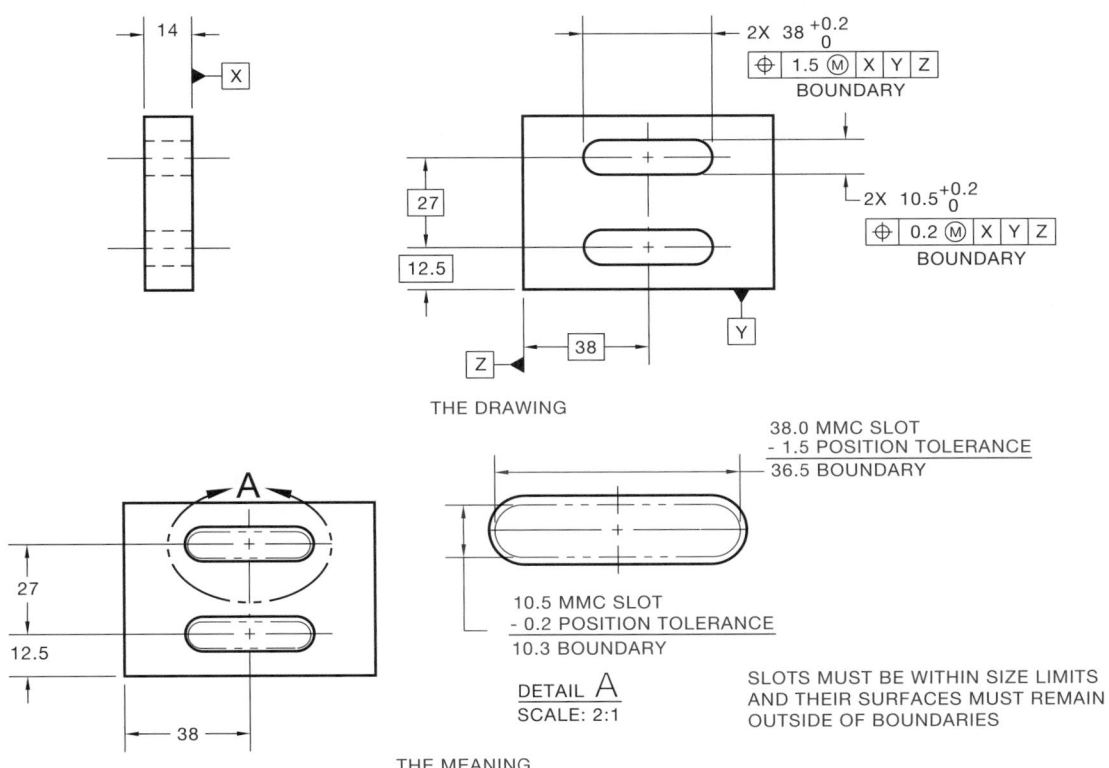

FIGURE 15.85 ■ Positional tolerance for slotted holes.

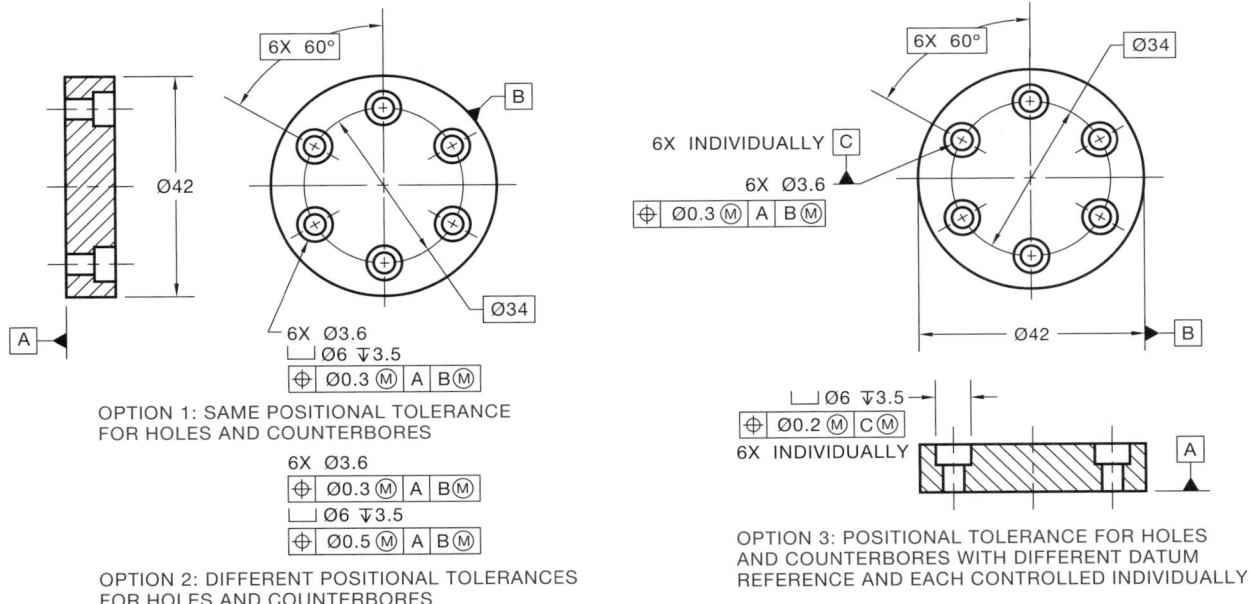

FIGURE 15.86 ■ Positional tolerance of counterbored holes.

projected away from the primary datum controlling the axis of the related feature. The amount of projection is equal to the thickness of the mating part. This works because the axis of the threaded or press fit feature is limited by the exact location and angle of the thread or hole within which it is assembled. This type of application is especially useful when the mating features are screws, pins,

or studs. These types of conditions are referrred to as *fixed fasteners* because the fastener is fixed in the mating part and there is no clearance allowance. The projected tolerance zone may be handled in one of two ways using the projected tolerance zone symbol. (See Figure 15.89.) When calculating the positional tolerance zones that are applied to the parts of a fixed fastener, the following

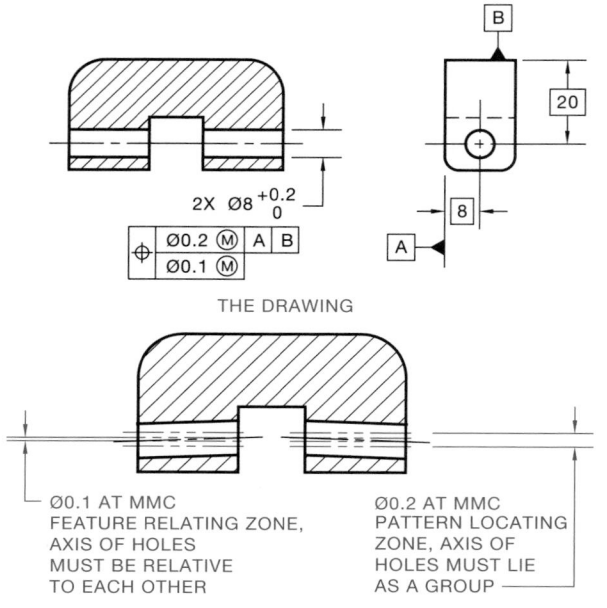

Ø0.1 AT MMC
FEATURE RELATING ZONE,
AXIS OF HOLES
MUST BE RELATIVE
TO EACH OTHER

Ø0.2 AT MMC
PATTERN LOCATING
ZONE, AXIS OF
HOLES MUST LIE
AS A GROUP

THE MEANING
SAME SIZE COAXIAL HOLES

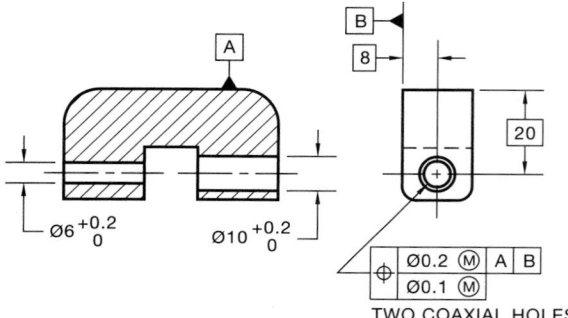

COAXIAL HOLES OF DIFFERENT SIZE

FIGURE 15.87 ■ Positional tolerance for coaxial holes.

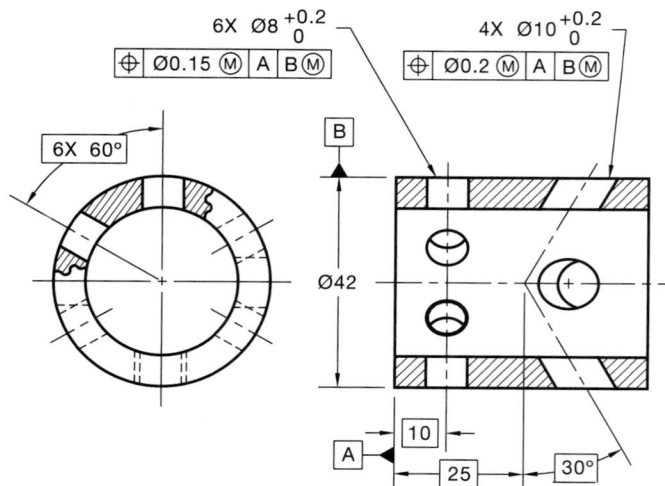

FIGURE 15.88 ■ Positional tolerancing of nonparallel holes.

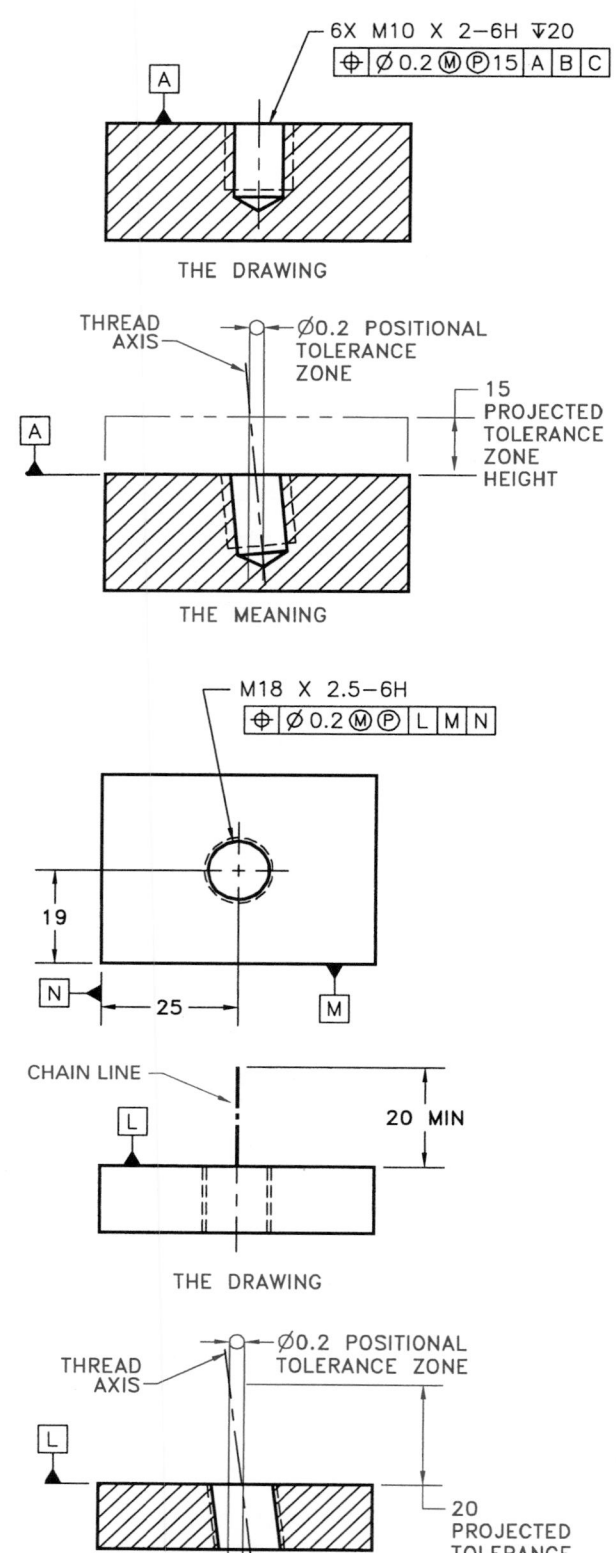

FIGURE 15.89 ■ Projected tolerance zone.

GEOMETRIC TOLERANCING SYMBOLS

Make a sketch of the desired symbol and decide how it will be placed on a drawing. Name the symbol; this is also the CADD file name. Select a point on each symbol that is a convenient insertion point when placing the symbol on a drawing. Figure 15.90 shows the insertion point determined for a feature control frame.

After you have drawn a specific symbol, store it as a symbol by using commands such as SYMBOL, BLOCK, or WBLOCK. When you have created a group of symbols, it is time to organize the symbols into a symbol library in a folder named GD & T SYMBOLS.

Symbols can be selected for use by entering a command such as INSERT. Some systems automatically display the symbol at the crosshairs location on the screen. Then as the point device is moved, the symbol also moves on the screen. When the symbol is positioned in the desired location, a button is pushed that places the symbol. Some symbols may require informational prompts such as ZONE DESCRIPTOR, GEOMETRIC TOLERANCE,

MATERIAL CONDITION, and/or DATUM REFERENCE after placement.

If you choose not to design your own custom geometric tolerancing symbols, there are predesigned GD & T symbol libraries available for most major software packages. Refer to the third-party applications catalog with your software to find any software available.

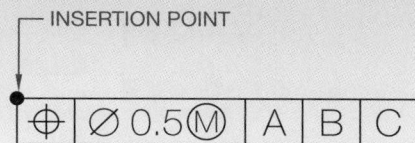

FIGURE 15.90 ■ The insertion point for geometric tolerancing symbols may be located in any convenient position. This example shows the upper left corner as a convenient insertion point.

---

formula may be used: MMC Hole − MMC Fastener (nominal thread size) ÷ 2 = Positional Tolerance Zone of Each Part. In some situations it may be desirable to provide more tolerance to one part than another, for example, 60% to the threaded part and 40% to the unthreaded part.

When parts are assembled with fasteners, such as bolts and nuts or rivets, and where all the parts have clearance holes to accommodate the fasteners, the application is referred to as a *floating fastener.* Floating fasteners require that the fastening device be secured on each side of the part, such as with a bolt and nut. With a fixed fastener, one of the parts is a fastening device. Greater tolerance flexibility with floating fasteners is due to the fastener clearance at each part. When calculating the positional tolerance zone for floating fastener parts, the following formula may be used: MMC Hole − MMC Fastener (nominal thread size) = Positional Tolerance Zone of Each Part.

GEOMETRIC DIMENSIONING AND TOLERANCING WITH AUTOCAD

AutoCAD provides the ability to add GD & T symbols to your drawings. The feature control frame and related GD & T symbols may be created using the TOLERANCE and LEADER commands.

These commands may be used to access the Geometric Tolerance dialog box which contains the feature control frame symbols. (See Figure 15.91.)

The Geometric Tolerance dialog box is divided into compartments that relate to the compartments found in the feature control frame. These compartments allow you to create the desired feature control frame by entering geometric characteristic symbols, diameter symbols, geometric tolerance values, material condition symbols, and datum references as needed. Pick the Sym box to open the Symbol dialog box shown in Figure 15.92. Pick the desired geometric characteristic symbol from the Symbol dialog box. The selected symbol is then displayed in the Geometric Tolerance dialog box. (See Figure 15.91.) If a diameter symbol is desired, pick the first box in the Tolerance compartment of the Geometric Tolerance dialog box. (See Figure 15.91.) Now type the desired geometric tolerance in the Tolerance text box as shown in Figure 15.91. Pick the third box in the Tolerance area if a material condition symbol is desired. This displays the Material Condition dialog box shown in Figure 15.93. Select the desired symbol to have it displayed in the Geometric Tolerance dialog box. (See Figure 15.91.) If datum references are applied to the feature control frame, type the desired datum reference letters in the Datum 1, Datum 2, and Datum 3 text boxes. Pick the box to the right of each datum reference text box if you want to open the Material Condition text box again to add desired material condition symbols with the datum reference.

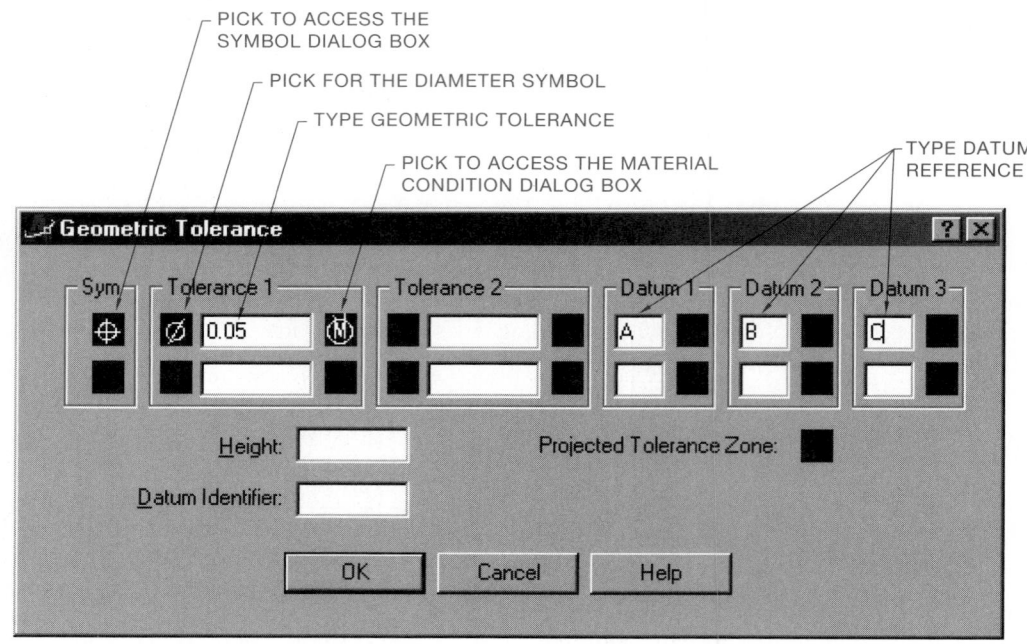

PICK TO ACCESS THE
SYMBOL DIALOG BOX

PICK FOR THE DIAMETER SYMBOL

TYPE GEOMETRIC TOLERANCE

PICK TO ACCESS THE MATERIAL
CONDITION DIALOG BOX

TYPE DATUM
REFERENCE

FIGURE 15.91 ■ The AutoCAD Geometric Tolerance dialog box.

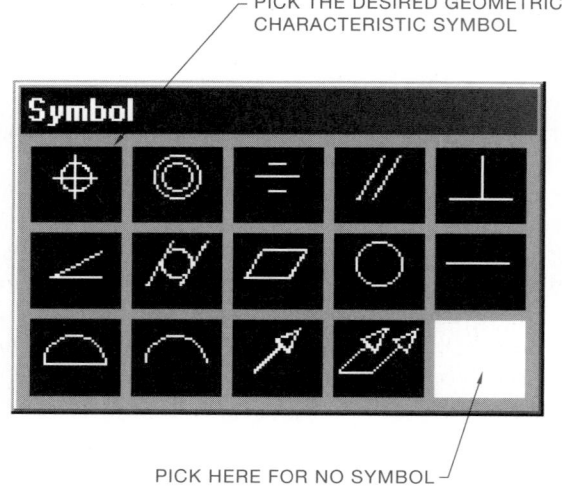

PICK THE DESIRED GEOMETRIC
CHARACTERISTIC SYMBOL

PICK HERE FOR NO SYMBOL

FIGURE 15.92 ■ The AutoCAD Symbol dialog box. Pick the
desired symbol or pick the blank box for no
symbol.

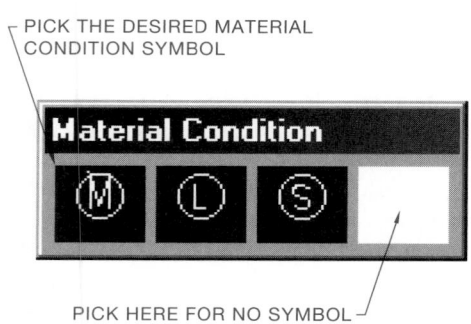

PICK THE DESIRED MATERIAL
CONDITION SYMBOL

PICK HERE FOR NO SYMBOL

FIGURE 15.93 ■ The AutoCAD Material Condition dialog box.
Pick the desired symbol or pick the blank box
for no symbol.

Double feature control frames may also be created for applications such as unit straightness, unit flatness, composite profile tolerance, composite positional tolerance, or coaxial positional tolerance. The projected tolerance zone symbol can also be added to the feature control frame by picking the Projected Tolerance Zone box shown in Figure 15.91. Add the desired height of the projected tolerance zone by typing the value in the Height text box.

VIRTUAL CONDITION

The tolerances of a feature that relate to size, form, orientation, and location, including the possible application of MMC or RFS, are determined by the function of the part. Consideration must be given to the collective effect of these factors in determining the clearance between mating parts and in establishing gage feature sizes. The boundary created by the combined effects of size, MMC, and the geometric tolerance is known as the *virtual condition*. Virtual condition is the sole condition where a feature size may be outside of MMC. Controlling the clearance between mating parts is critical to the design process. When features are dimensioned using a combination of size and geometric tolerances, the resulting effects of the specifications should be considered to ensure that parts will always fit together.

When a positional tolerance is applied to an internal feature, the Virtual Condition = MMC Hole – Positional Tolerance. This calculation determines the maximum feature size that should be allowed to fit within the hole. (See Figure 15.94.)

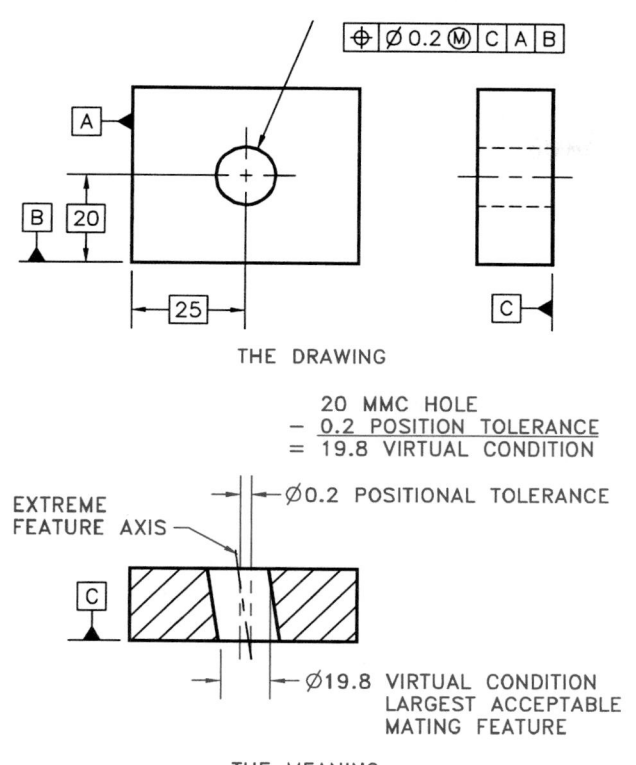

FIGURE 15.94 ■ Virtual Condition of Hole = MMC − Positional Tolerance

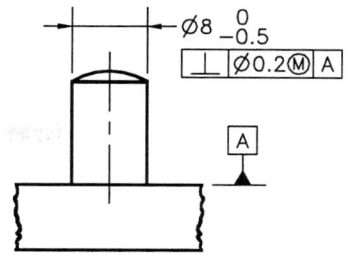

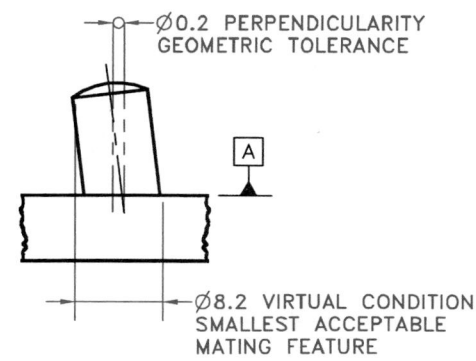

FIGURE 15.95 ■ Virtual Condition Perpendicular Pin = MMC Feature + Geometric Tolerance

When perpendicularity is applied to an external diameter, such as a pin, the Virtual Condition = MMC Feature + Perpendicularity Geometric Tolerance. The virtual condition determines the smallest acceptable mating feature that fits over the given part while maintaining a positive connection between the surfaces at datum A as shown in Figure 15.95.

STATISTICAL TOLERANCING WITH GEOMETRIC CONTROLS

Methods of tolerancing for statistical process control (SPC) were introduced in Chapter 12. Statistical tolerances may also be used with geometric controls by placing the statistical tolerancing symbol in the feature control frame as shown in Figure 15.96.

COMBINATION CONTROLS

In some situations, compatible geometric characteristics may be combined in one feature control frame or separate frames associated with the same surface. This is normally done when the combined effect of two different geometric characteristics and tolerance zones is desired. The profile tolerance may be used to illustrate the combination of geometric characteristics.

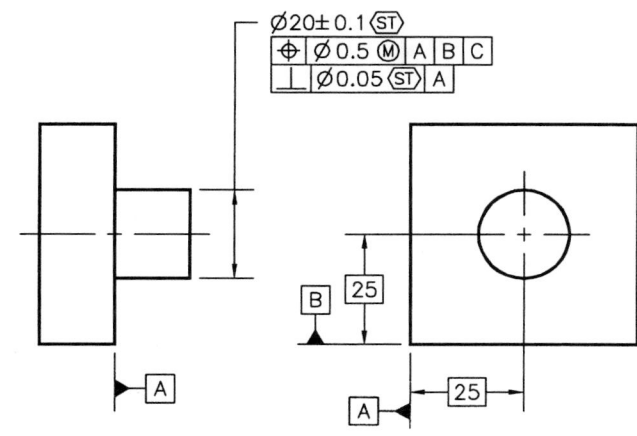

FIGURE 15.96 ■ Statistical tolerancing with geometric controls.

Profile and parallelism may be combined to control the profile of a surface plus the parallelism of each element to a datum. Profile and runout may be combined to control the line elements within the profile specification and circular elements within the runout tolerance as shown in Figure 15.97. The combined control of parallelism and perpendicularity may be represented on the same feature as shown in Figure 15.98. Other combined geometric characteristics may be used when

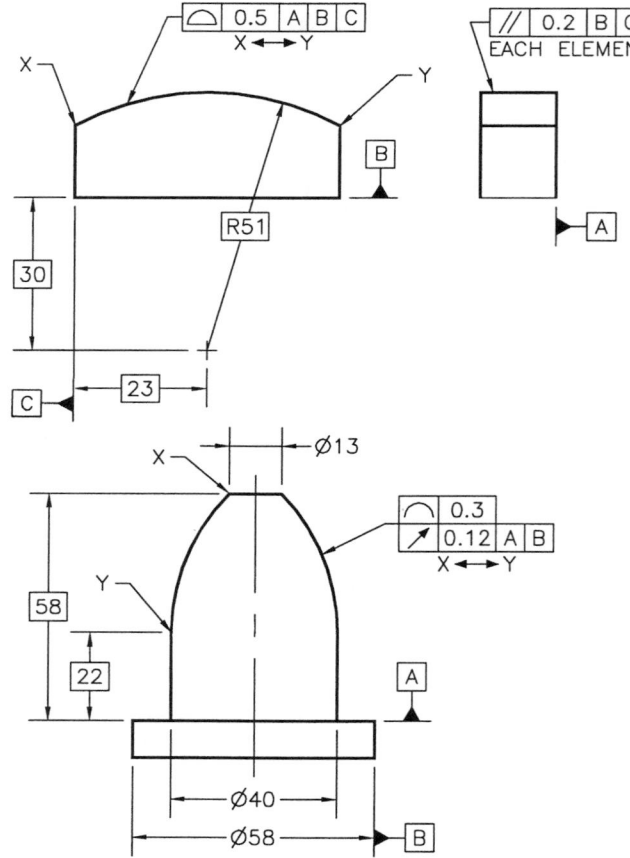

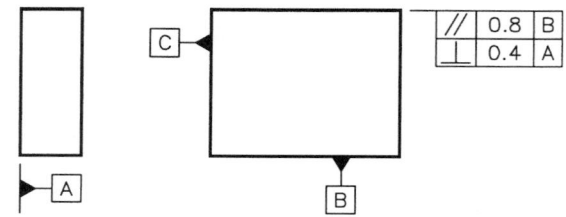

FIGURE 15.97 ■ Combination controls (profile and runout).

FIGURE 15.98 ■ Combination control (parallelism and perpendicularity).

the callouts are compatible and the design function of the part requires such specific controls.

WEB SITE RESEARCH

The following Web sites can provide you additional information for research or further study into topics covered in this chapter:

http://www.asme.org/asme/8.html—Find information and publications related to the American Society of Mechanical Engineers, including ASME Y14.5M, *Dimensioning and Tolerancing.*

http://www.ansi.org—The American National Standards Institute. Information about national and international drafting standards, including ASME Y14.5M, *Dimensioning and Tolerancing.*

http://www.adda.org—American Design Drafting Association, *Drafting Reference Guide.*

http://www.industrialpress.com—Information about the *Machinery's Handbook.* This is a valuable resource for manufacturing standards, sizes, tolerances, fits, materials, and anything else you can think of for design and drafting.

http://www.industrialpress.com/handbook/trig.html—On-line trigonometry tables.

http://www.industrialpress.com/handbook/prime.html—Prime numbers and factoring information.

http://www.goodheartwillcox.com/pinfo/—*Geometric Dimensioning and Tolerancing Basic Fundamentals,* by David A. Madsen. Comprehensive basic text workbook covering geometric dimensioning and tolerancing. Complete with tests, print reading exercises, and drafting problems.

PROFESSIONAL PERSPECTIVE

Written By Michael A. Courtier, Supervisor/Instructor
CAD/CAM/ CAE Employee Training and Development
Boeing Aerospace

In working with engineers and drafters for years on the subject of geometric tolerancing, I find that questions keep popping up in several key areas—datums; limits of size; geometric tolerances as a refinement of size; the tolerance zones (2-D or 3-D, cylindrical or total wide, etc.); and the apparent overlap of controls (circularity, cylindricity, circular runout, total runout, position). It is in these "weak" spots that you should dig a little deeper and fill in your knowledge.

New and old users alike tend to miss the key rules of thumb in selecting *datums,* and most textbooks on the subject do not address this area.

■ Make sure you select datums that are real, physical surfaces of the part. I have seen many poor drawings with datums on centerlines, at the centers of spheres, or even at the center of gravity.

■ Be sure to select large enough datum surfaces. It is hard to chuck up on a cylinder $\frac{1}{10}$" long, or get three noncolinear points of contact on a plane of similar dimensions.

■ Choose datums that are accessible. If manufacturing or quality assurance has to build a special jig to reach the datum, extra costs are incurred for the ultimate part.

■ Given the above constraints, select functional surfaces and corresponding features of mating parts where possible.

Remember, datums are the *basis* for subsequent dimensions and tolerances, so they should be selected carefully. You might also be well advised to assign a little form, orientation, or runout control to the datum surface to ensure an accurate basis.

In the area of *limits of size,* there are three possible interpretations of a size dimension. Every class I have taught has been divided on what the proper extent of a size dimension's control should be—the overall size (or geometric form), the size at various cross-sectional checks, or both. The correct answer, per definition in the standard, is *BOTH!*

The words *refinement of size* are a source of general confusion to new users of geometric tolerancing. Let us say a part is dimensioned on the drawing as .500 ± .005" thick, with parallelism control to .003" on the upper surface. If a production part has a convex upper surface with low points at .504, then the high point can only be .505, *not* .507. In other words, only .001" of the .003" of parallelism control can be used since parallelism is *a refinement of size within the overall size limits.*

Another one of the often misunderstood aspects of geometric tolerancing is the *tolerance zone.* Many engineers, technicians, drafters, machinists, and inspectors misinterpret the *shape* of the tolerance zone. I recommend that all new users make a chart showing the tolerance zones for each type of tolerance and a sampling of its various applications. This chart should include—at a minimum—the 2-D zones between two lines or concentric circles (e.g., straightness, circularity) versus the 3-D zones between parallel planes (e.g., flatness, perpendicularity); the cylindrical versus the total wide zones (e.g., straightness to an axis versus surface straightness, or position of a hole versus position of a slot); the cross-sectional zones versus the total surface zones (e.g., circularity versus cylindricity); and many, many others. If you can properly visualize the tolerance zone, you are well on your way to the proper interpretation of the geometric tolerance.

One more thing to watch out for is the *apparent overlap of controls.* While it may appear that there is overlap, there are subtle yet distinct differences between all the geometric tolerance controls. Consider cylindrical shapes, for instance. You can use a wide variety of controls for them, including straightness, circularity, cylindricity, concentricity, circular runout, total runout, and position, to name a few. Remember that the first three are form controls, so no datum relationship is controlled. Straightness can control the axis or the surface merely by placing the feature control frame in a different location on the drawing. Straightness is a longitudinal control, whereas circularity is a circular element control (both are 2-D); cylindricity is a 3-D control of the surface relative to itself (no datum). The last four controls are with respect to a datum (or datums), but they have their differences also. Concentricity and the runouts are always RFS, whereas positional tolerancing can be LMC or MMC and get bonus tolerances. Runout is a rotational consideration, while positional tolerancing does not imply rotation.

In summary, ASME Y14.5M is a very powerful tool for engineering drawings. Digging deep into the subtleties will separate the amateur from the professional, who knows it will save time and money in every discipline from design to manufacturing to quality assurance testing. Used properly, the engineering design team can convey more information about the overall design to downstream areas or subcontracting concerns that may have no knowledge of the final assembly. The proper assessment of the symbology requires no interpreter for the reader who is well versed in this international sign language called *geometric tolerancing.* Figure 15.99, page 496, is an example of an actual industry drawing. In general, the part is fairly basic, but the geometric tolerancing is very specific and leaves no doubt about the interpretation.

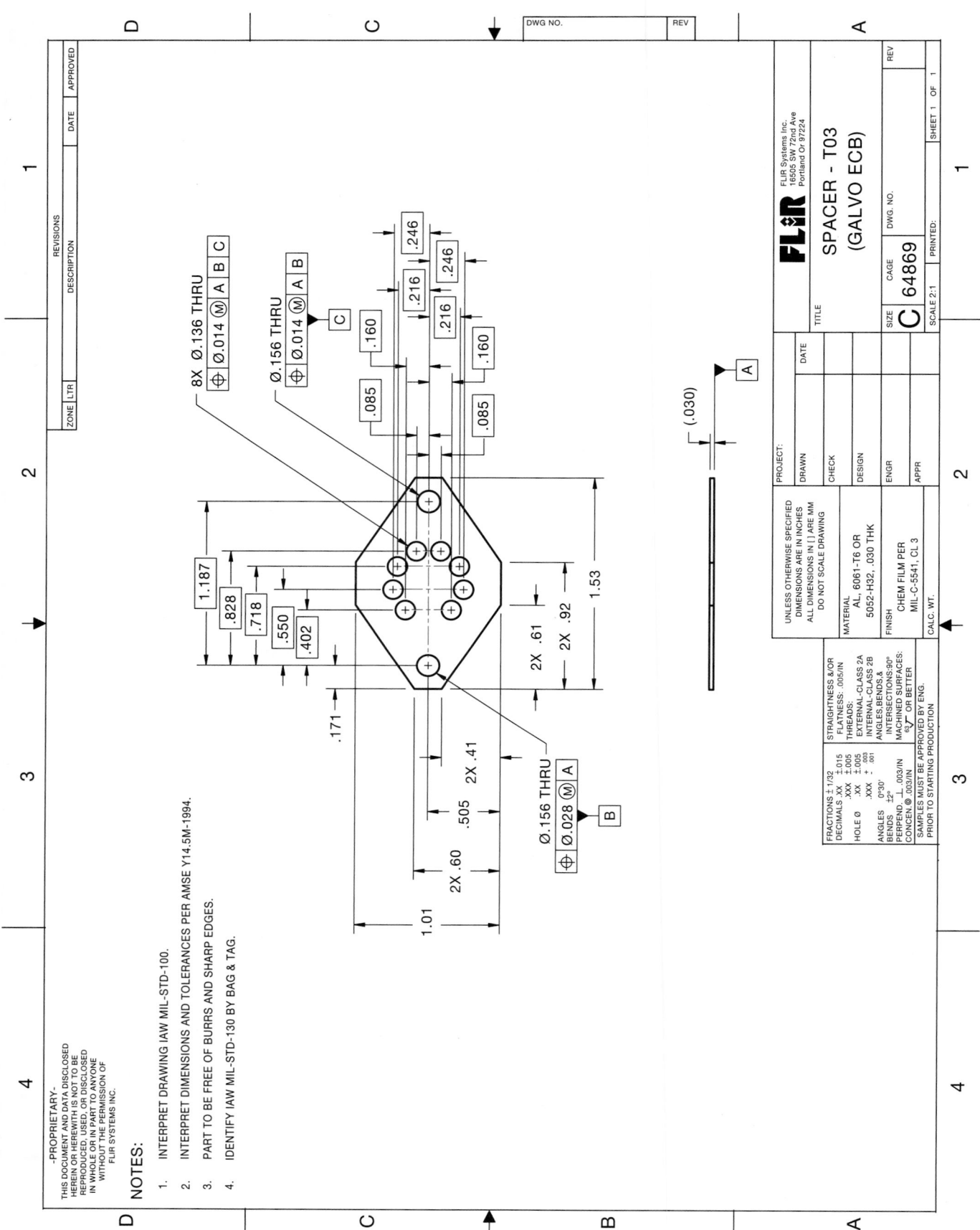

FIGURE 15.99 ■ An actual industry drawing of a part with geometric tolerancing applied. *Courtesy Flir Systems, Inc.*

ROUNDING NUMBERS

Calculators and computers usually indicate more precision in their answers than is warranted by the original data. In addition to the information on rounding in the Math Instruction Appendix found in the Online Companion, here is the mathematical convention for rounding calculations:

When Adding or Subtracting:
1. Round each figure to the coarsest (least "precise") piece of data.
2. Do the arithmetic.

For example, to add this group 4.39 + 7.9 + 6.42, first round each number to the nearest tenth (because of the 7.9) giving 4.4 + 7.9 + 6.4. Then, add for the answer of **18.7**.

When Multiplying or Dividing:
1. Do the arithmetic.

2. Round the answer to the least number of significant digits in the original data.

Digits are *significant* when they give us information other than holding a decimal point in place; 13.09 has four significant digits, but 13,000,000 has only two.

For example, to calculate 13.78 × 6.1 × 4.385, first multiply, giving 368.59433. Because the 6.1 has only two significant digits, we round the answer to **370**, which has only two significant figures (the zero of 370 being merely a place-keeper).

When working with complicated formulas or when running computer programs, do not round off until the final answer is obtained. Constants in formulas are not considered data. When using a geometrical formula with π in it, it is best to push the calculator's π button instead of manually entering 3.14.

MATH APPLICATION

CHAPTER 15

Geometric Tolerancing Test

DIRECTIONS

Answer the questions with short complete statements or drawings as needed.

PART 1 • QUESTIONS

1. Give a brief definition of geometric tolerancing (GT).
2. Give an example of how general tolerancing has some control on form.
3. Define *datum*.
4. Describe or show an example of how a datum feature symbol is shown on a plane surface and a centerline axis.
5. What is another name for the three-plane dimensioning system?
6. Name the elements of the three-plane dimensioning system.
7. Discuss the order of precedence of datum features.
8. When is it practical to use a partial surface datum?
9. Describe how coplanar surface datums are drawn.
10. Show an example of how datum target points are located and identified on a primary datum surface.
11. Show an example of how a datum target line may be shown on a drawing.
12. Show an example of how datum target areas may be shown on a drawing.
13. What is a feature control frame used for?
14. Show and label the complete order of elements in a feature control frame.

15. Define *basic* and show an example of basic dimension.
16. List the geometric characteristics for form, profile, orientation, location, and runout.
17. Show examples of how the following geometric characteristics are represented on a drawing: straightness, flatness, circularity, cylindricity, profile of a surface, parallelism, perpendicularity, angularity, and runout.
18. Which geometric tolerance is more confining, circularity or cylindricity?
19. Show an example of how unit straightness may be shown on a drawing.
20. Which profile tolerance is most confining?
21. Must the parallelism geometric tolerance zone be within the related size tolerance?
22. What does the note EACH ELEMENT denote when placed below a feature control frame?
23. Is a basic angle required to establish an angular surface for an angularity geometric tolerance?
24. Name the runout tolerance that is most confining.
25. How are the specified limits of total runout determined?

26. Describe or show an example of how a surface portion can have a specified runout tolerance.
27. Define *perfect form envelope*.
28. Clearly define regardless of feature size (RFS).
29. When is RFS implied?
30. Clearly define maximum material condition (MMC) as related to the effect of MMC on the geometric tolerance.
31. Clearly define least material condition (LMC) as related to the effect of LMC on the geometric tolerance.
32. True or False: position tolerances must have a correlated material condition symbol (MMC or LMC) applied to the tolerance and related datums.
33. Define the concentricity geometric characteristic.
34. Define the symmetry geometric characteristic.
35. Describe the position tolerance zone and how it is located.
36. Define *true position*.
37. Describe a projected tolerance zone.
38. Show an example of a combination control.

PART 2 • CALCULATIONS

1. Given:
   a. Shaft ∅24.00/23.92.
   b. Straightness geometric tolerance 0.02.

   What is the geometric tolerance at the actual sizes specified below for the type of straightness and material condition shown?

Actual Size	Surface Straightness		Axis Straightness	
	RFS		RFS	MMC
24.00				
23.99				
23.98				
23.96				
23.94				
23.92				

2. Given:
   a. Positional tolerance ∅0.02 at true position in reference to datums L, M, N.
   b. Hole size ∅8.50/8.40.

   What is the positional tolerance using different material condition symbols at the actual sizes shown in the table?

Actual Sizes	Material Condition Applied to Tolerance		
	MMC	RFS	LMC
8.50			
8.49			
8.48			
8.46			
8.44			
8.42			
8.40			

3. If the positional tolerance of the hole in problem 2 above is zero at MMC, then what would the positional tolerance be at the actual produced sizes given below?

Actual Sizes	MMC
8.50	
8.48	
8.46	
8.44	
8.42	
8.40	

4. What is the virtual condition of a ∅12.2/12.0 hole that is located with a positional tolerance of ∅0.05 at MMC?
5. What is the virtual condition of a ∅6.0/5.9 pin established with perpendicularity to a datum A by ∅0.02 at MMC?
6. Calculate the positional tolerance for the location of holes with the following specifications:
   a. Floating fastener.
   b. Fastener: M20 × 2.5.
   c. Hole through two parts: ∅21.2/20.8.

   Positional tolerance for holes in part 1 equals _____, part 2 equals _____.
7. Calculate the positional tolerance for the location of holes with the following specifications:
   a. Fixed fastener.
   b. Part 1 hole: ∅9.0/8.6.
   c. Part 2 hole: M8 × 1.25.
   d. Equal positional tolerance for each part.

   Positional tolerance for holes in part 1 equals _____, part 2 equals _____.

CHAPTER 15

Geometric Tolerancing Problems

DIRECTIONS

1. From the selected problems, determine which views and dimensions should be used to completely detail the part. Use simplified representation unless otherwise specified in the instructions or by your instructor.

2. Make a multiview sketch, to proper proportions, including dimensions and notes.

3. Using the sketch as a guide, draw an original multiview drawing on an adequately sized drawing sheet. Add all necessary dimensions and notes using unidirectional dimensioning. Use manual or computer-aided drafting as required by your course guidelines.

4. Include the following general notes at the lower left corner of the sheet .5 in. each way from the corner border lines:

1. INTERPRET DIMENSIONS AND TOLERANCES PER ASME Y14.5M—1994.
2. REMOVE ALL BURRS AND SHARP EDGES.

NOTES:

Additional general notes may be required depending on the specifications of each individual assignment. The following should be part of your title block unless otherwise specified by your instructor:

UNSPECIFIED TOLERANCES:

DECIMALS	mm	IN.
X	± 2.5	± .1
XX	± 0.25	± .01
XXX	± 0.127	± .005
ANGULAR ± 30'		
FINISH	3.2μm	125μIN.

PROBLEM 15-1   **Geometric tolerancing (metric)**

Part Name: Flow Pin

Material: Bronze

Finish: Finish All Over 0.20 μm.

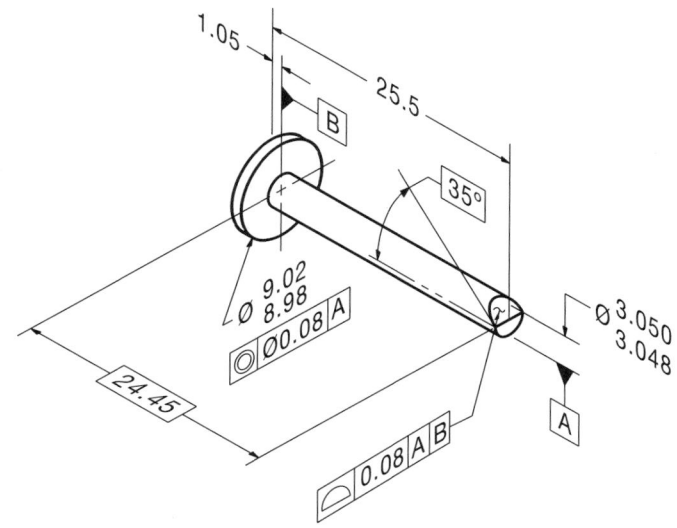

PROBLEM 15-2   **Geometric tolerancing (metric)**

Part Name: LN2 Test Pump Lock Nut

Material: AMS 5732.

Additional General Notes:

1. △3 : Mark per AS478 Class D with 1193125 and applicable dash number.

2. Finish All Over 1.6 μm.

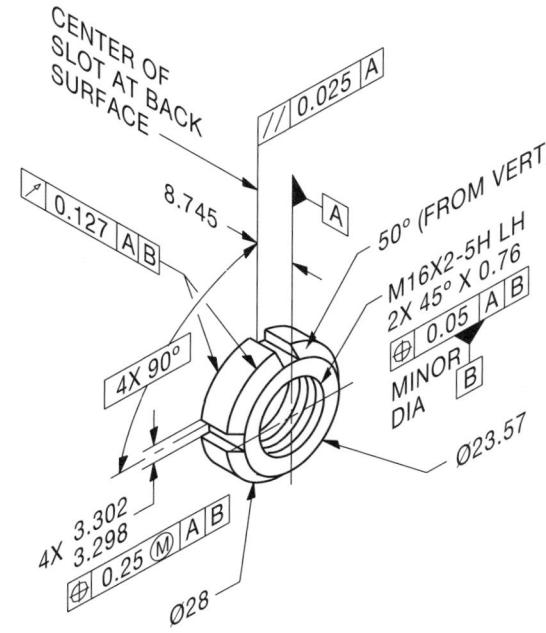

PROBLEM 15-3 (in.)
Part Name: Half Coupling

Material: ∅1.250 6061-T6 Aluminum

SPECIFIC INSTRUCTIONS:
Provide MMC material condition after position tolerance except for RFS at threads.
Problem based on original art courtesy TEMCO.

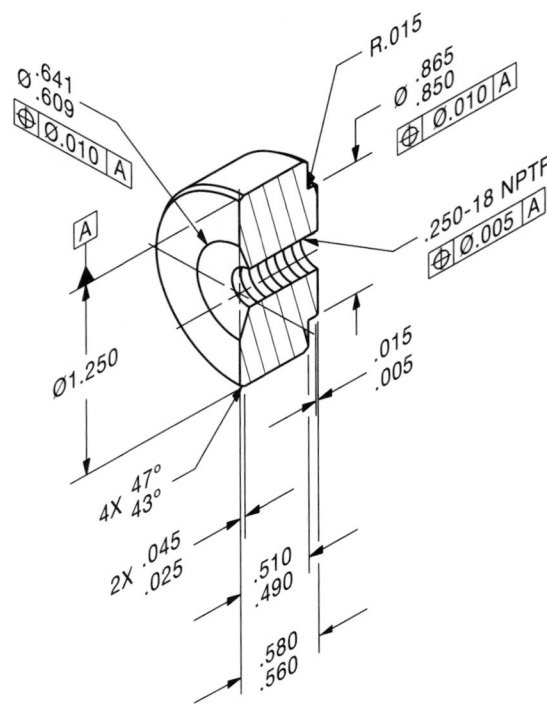

PROBLEM 15-5 (in.)
Part Name: Half Coupling

Material: ∅1.625 6061-T6511

Problem based on original art courtesy TEMCO.

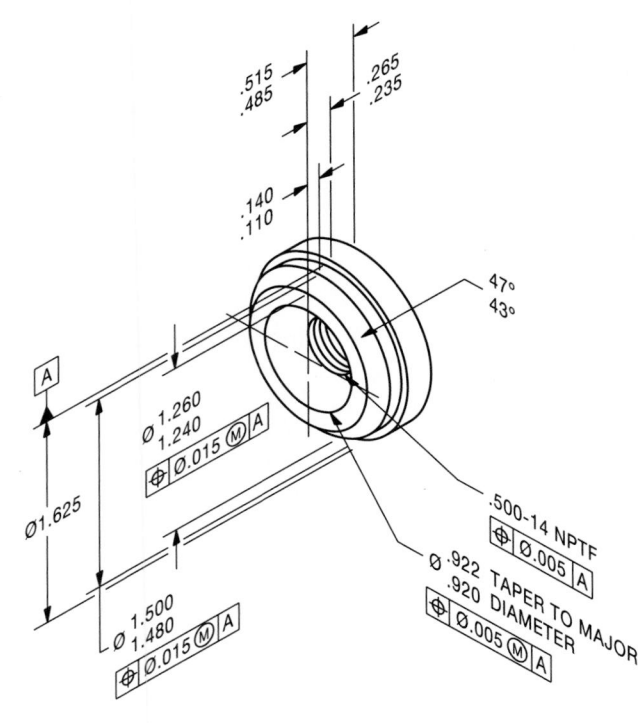

PROBLEM 15-4 (in.)
Part Name: Coupling

Material: AISI 1010, Killed

SPECIFIC INSTRUCTIONS:
Provide MMC material condition after position tolerance except for RFS at threads.
Problem based on original art courtesy TEMCO.

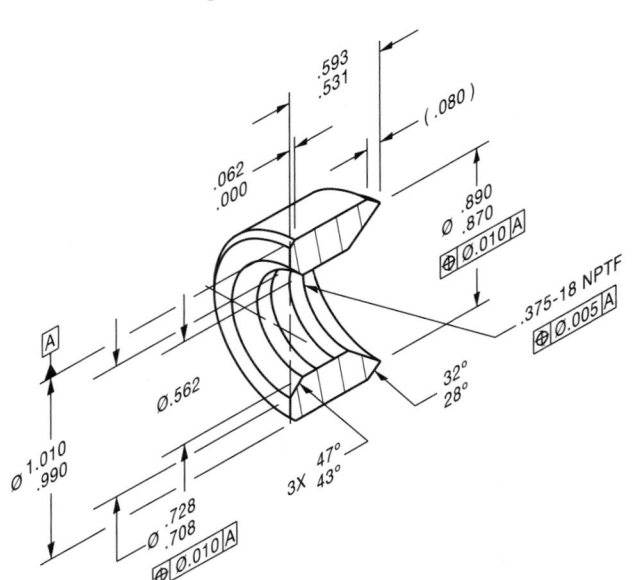

PROBLEM 15-6 (metric)
Part Name: Spline Plate

Material: SAE 3135

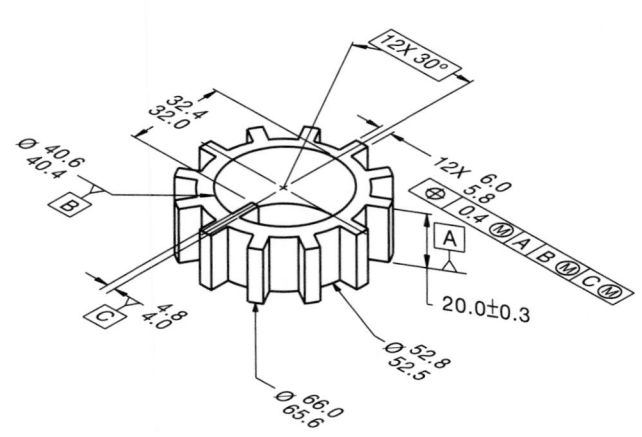

PROBLEM 15-7 **(in.)**
Part Name: Nut
Material: No. 10 Bronze

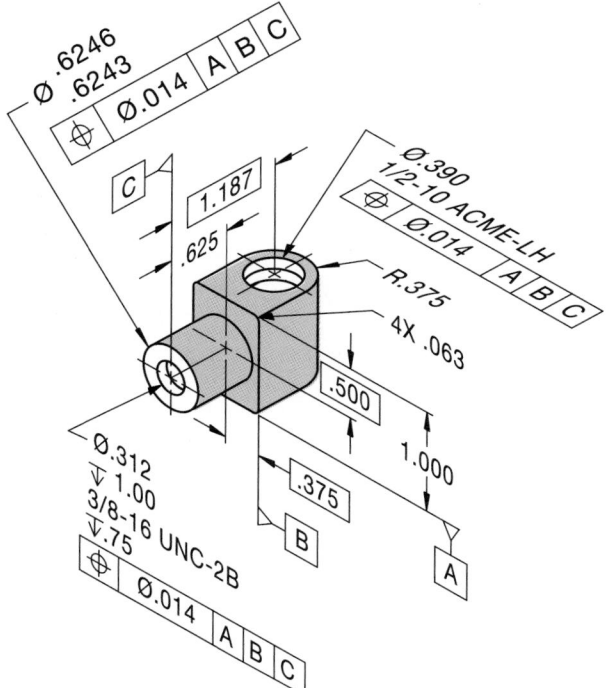

PROBLEM 15-9 **(in.)**
Part Name: Thrust Washer
Material: SAE 5150

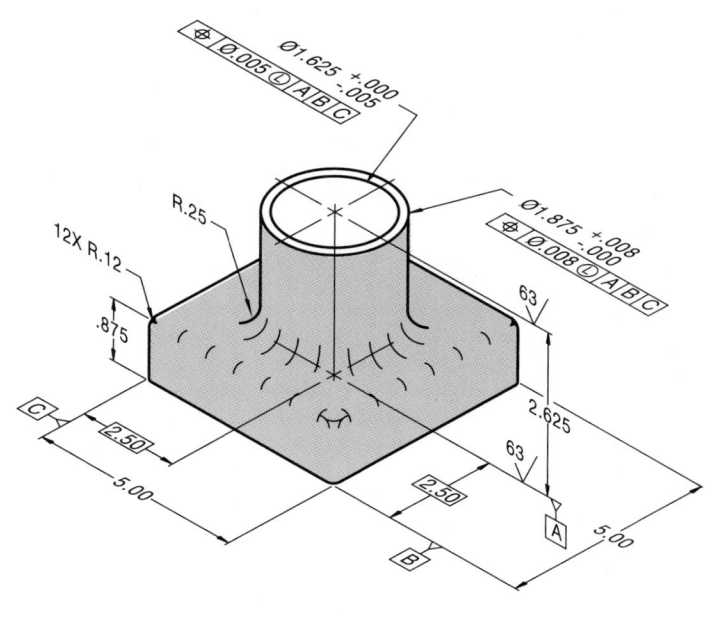

PROBLEM 15-8 **(metric)**
Part Name: Coupling Bracket
Material: SAE 4310 Steel

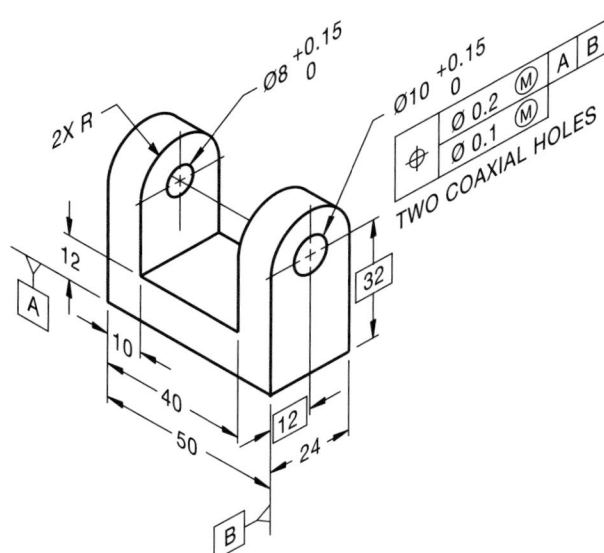

PROBLEM 15-10 **(metric)**
Part Name: Spacer
Material: SAE 4310

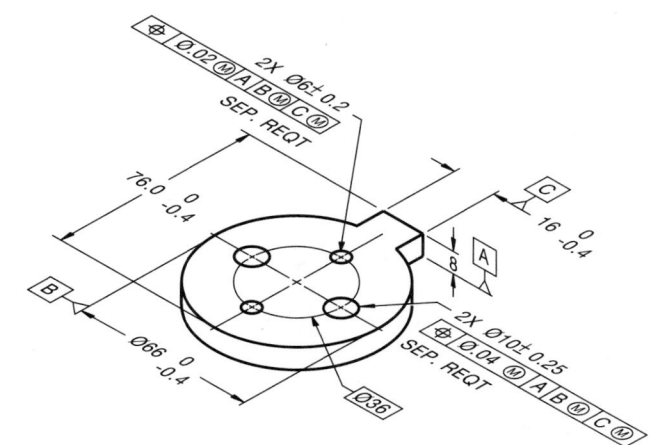

PROBLEM 15-11 (metric)
Part Name: Bearing Support
Material: SAE 1040

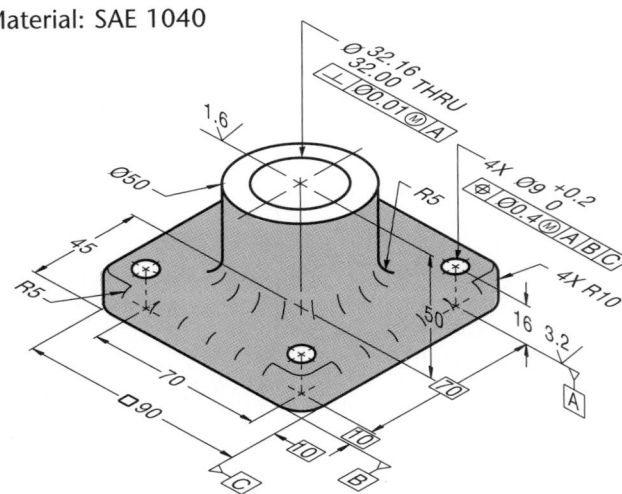

PROBLEM 15-12 (metric)
Part Name: Lock Nut
Material: SAE 3130

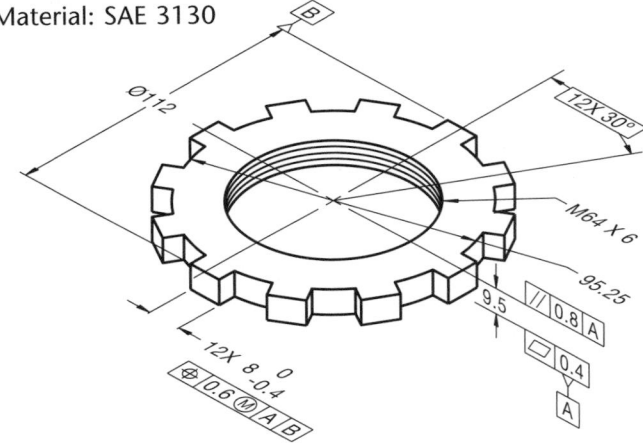

PROBLEM 15-13 (in.)
Part Name: Cover Plate
Material: Phosphor Bronze

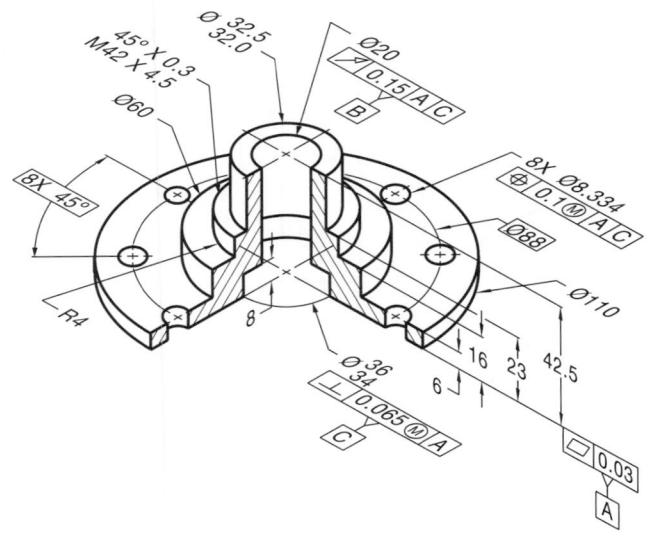

PROBLEM 15-14 (in.)
Part Name: Angle Support Mounting
Material: SAE 3110

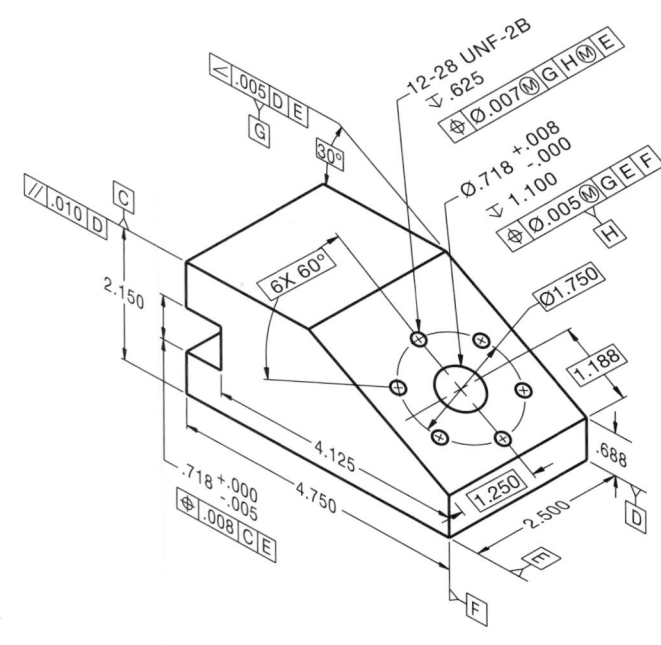

PROBLEM 15-15 (metric)
Part Name: Hub
Material: SAE 3310

PROBLEM 15-16   Geometric tolerancing (metric)
Part Name: Fixture MIBRDA—1265
Material: SAE 4320
Harden: Brinell 200–240
Additional General Notes:

1. Finish All Over 0.80 μm.

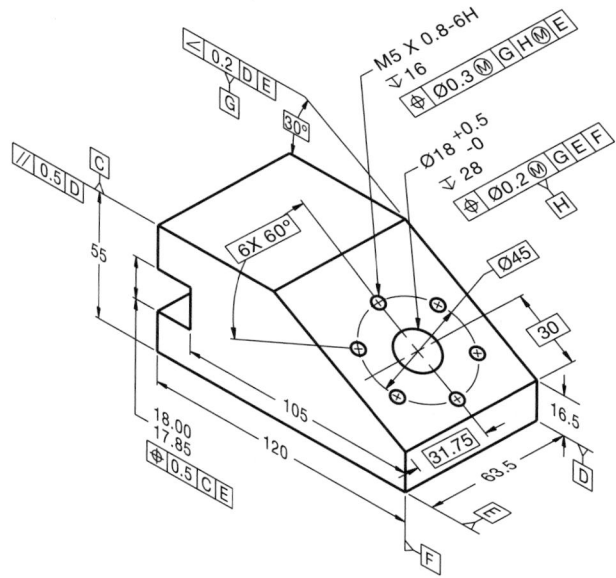

PROBLEM 15-17   Geometric tolerancing (metric)
Part Name: Mounting Bracket
Material: Stainless Steel
Additional General Notes:

1. All Fillets and Rounds R24.
2. Finish All Over 1.6 μm.

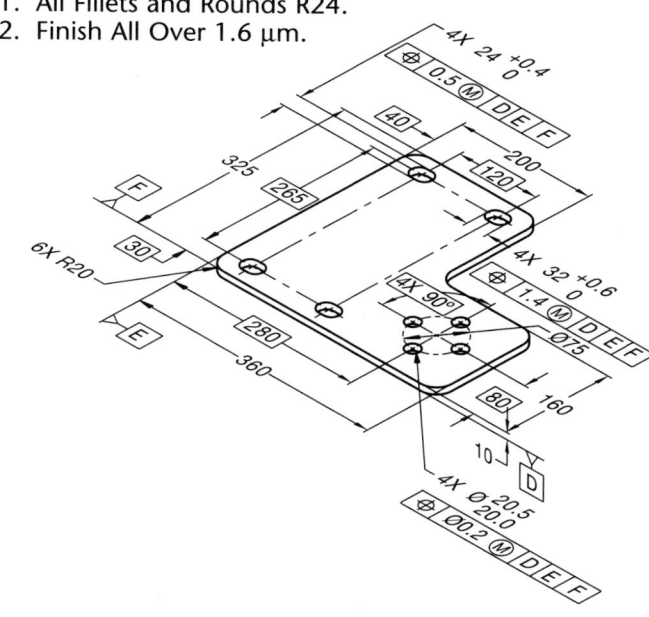

PROBLEM 15-18   Geometric tolerancing (metric)
Part Name: Oscillator Housing
Material: Phosphor Bronze
Additional General Notes:

1. Finish All Over 0.80 μm.

SPECIFIC INSTRUCTIONS:

Use the following engineer's notes to complete the geometric tolerancing of the Oscillator Housing:

■ Establish datum A with three equally spaced datum target points at each end of the Ø28.1–28.0 cylinder.

■ Establish datum B at the left end surface.

■ Make the bottom surfaces of the 2X Ø40.2–40.0 features perpendicular to datum A by 0.06.

■ Provide a cylindricity tolerance of 0.3 to the outside of the part.

■ Make the 2X Ø40.2–40.0 features concentric to datum A by 0.1.

■ Locate the 6X Ø6 holes with reference to datum A at MMC and datum B with a position tolerance of 0.05 at MMC.

■ Locate the 4X Ø4 holes with reference to datum A at MMC and datum B with a position tolerance of 0.04 at MMC.

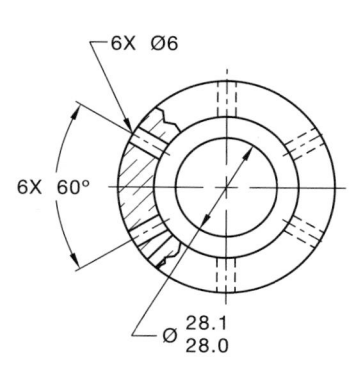

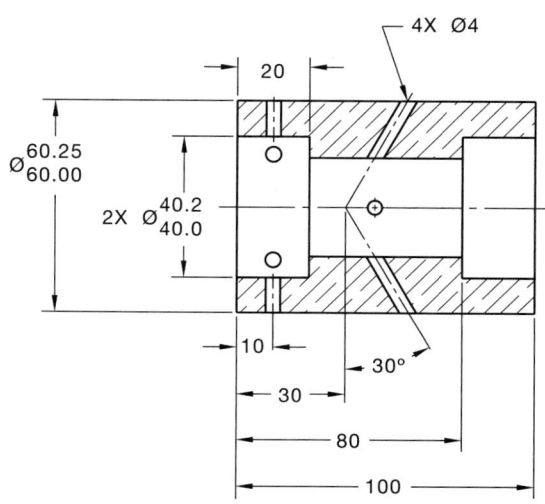

PROBLEM 15-19   (in.)
Part Name: Hub
Material: SAE 4320

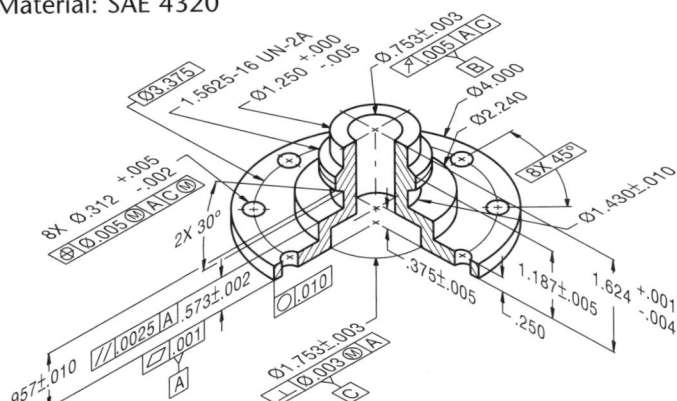

PROBLEM 15-20   (in.)
Part Name: Lock Spacer
Material: SAE 1030

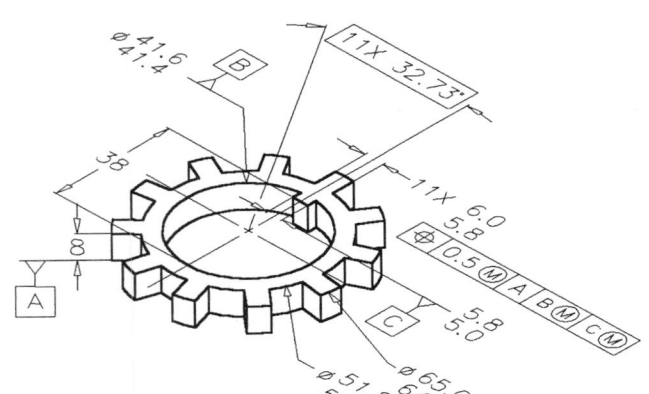

PROBLEM 15-21   (metric)
Part Name: Mounting Plate
Material: SAE 4140

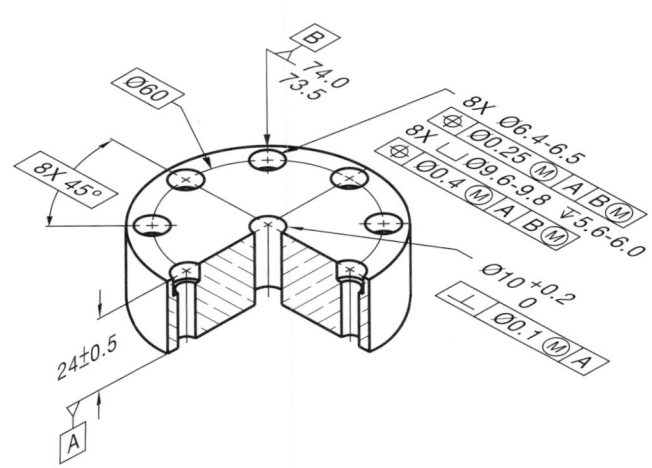

PROBLEM 15-22   Geometric tolerancing (metric)
Part Name: Side Panel Mounting Plate
Material: SAE 30308

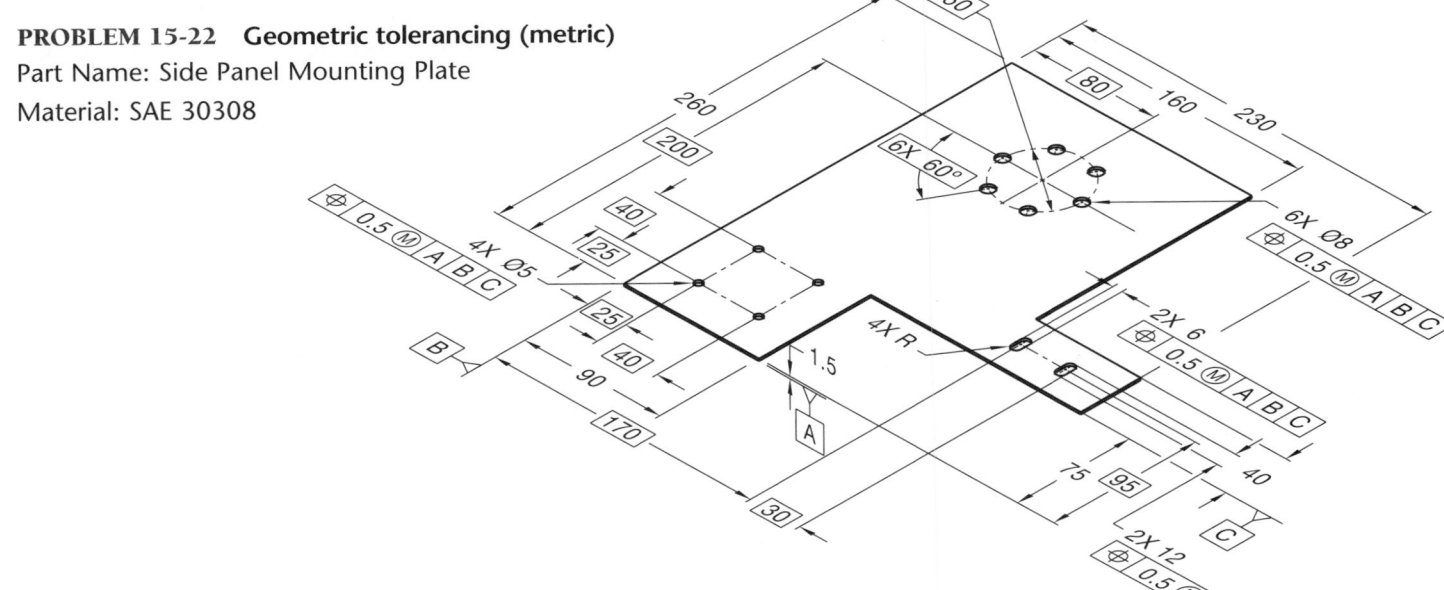

PROBLEM 15-23   Advanced geometric tolerancing from actual industry drawing (metric)

Part Name: Pinion Gear Shaft

Material: CRES 15-5PH ASTM A564

Additional General Notes:

1. Finish All Over 1.6 μm.

2. Heat Treat Per Mil-H-6875 to H1100 Condition.

3. Penetrant Inspect Finished Part Per Mil-Std-271, Group III. No evidence of Linear Indications Permitted.

4. Part to be Clean and Free of Foreign Debris.

Problem based on original art courtesy Aerojet TechSystems Co.

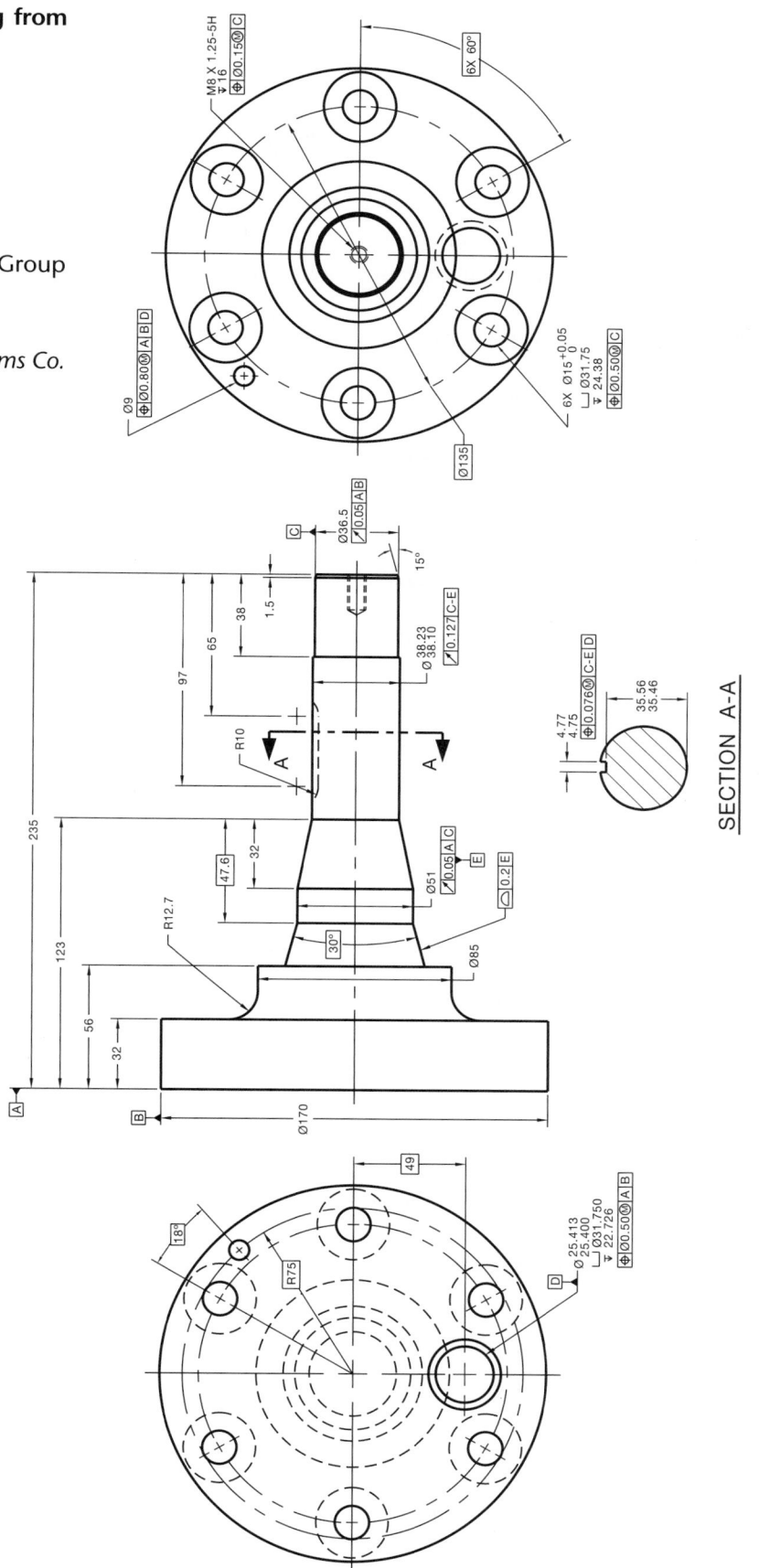

SECTION A-A

PART 2 • DESIGN PROBLEM

DIRECTIONS

Given the engineering layout below, design and detail the plate, angle, and yoke per ASME Y14.5M standards. Use arrowless dimensioning unless otherwise specified by your instructor. Use the following information and specifications:

■ Not all features are shown in all views.

■ Some features that are hidden from view are not drawn with hidden lines.

■ Item 4 (Pin) is a purchase part. Do not detail this part.

PROBLEM 15-24   *Problem courtesy of Martin Soll.*

■ Pin specifications: ∅2.0000±.0001

Straightness within .0002 over full length.

Clearance fit in Angle = .005/.007

Clearance fit in Yoke = .003/.005

■ Dowell pin specifications: ∅.5000±.0005

Interference fit in Plate = .001/.003

Clearance fit in Angle and Yoke = .002/.005

■ Be sure the pin will slide smoothly through the three holes, and the 3.000+.020/−.000 spacing is maintained.

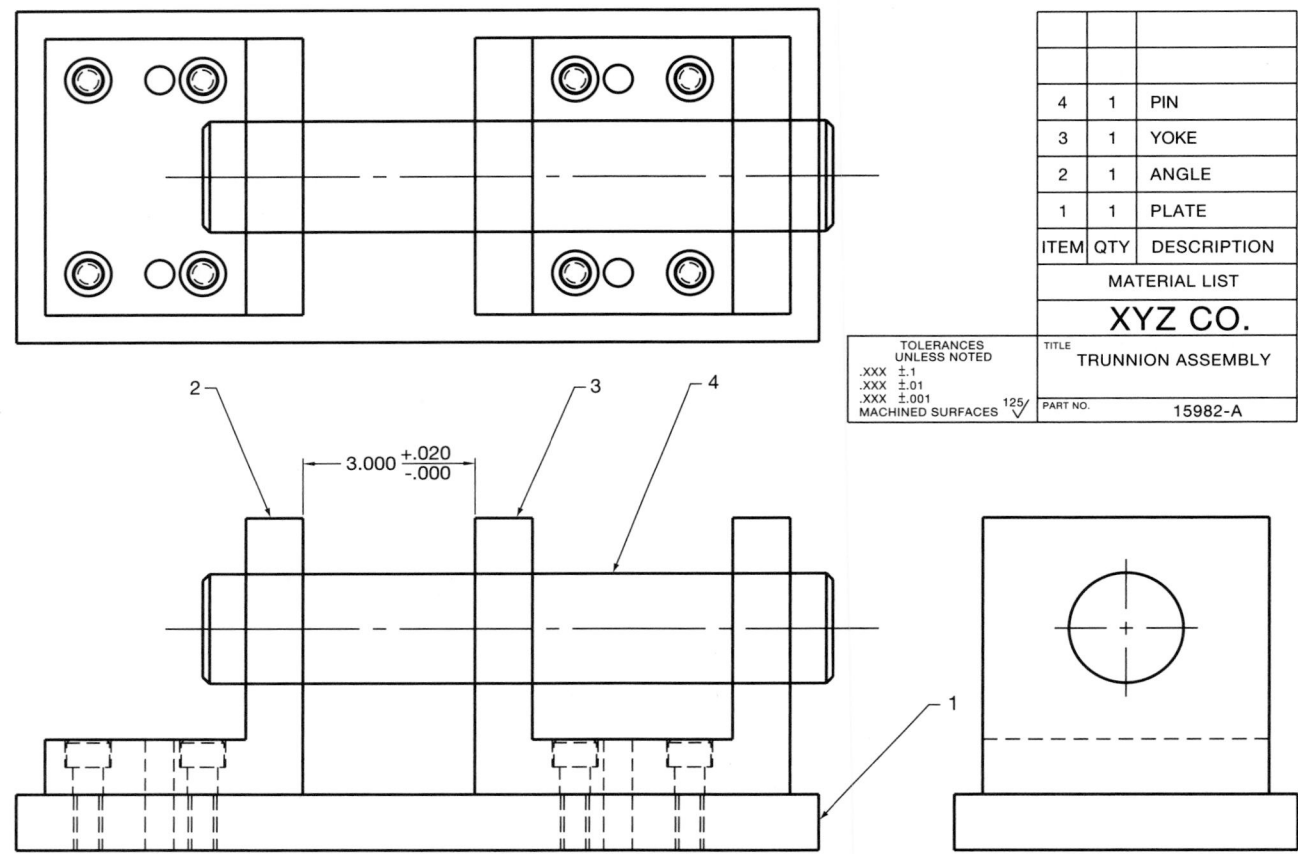

4	1	PIN
3	1	YOKE
2	1	ANGLE
1	1	PLATE
ITEM	QTY	DESCRIPTION
MATERIAL LIST		

XYZ CO.

TITLE: TRUNNION ASSEMBLY

PART NO.: 15982-A

TOLERANCES UNLESS NOTED
.XXX ±.1
.XXX ±.01
.XXX ±.001
MACHINED SURFACES 125/∀

MATH PROBLEMS

Round to the nearest tenth.

PROBLEM 15-25   4.849

PROBLEM 15-26   3.650

PROBLEM 15-27   .275

PROBLEM 15-28   5.249

Round to the nearest hundredth.

PROBLEM 15-29   4.849

PROBLEM 15-30   7.0574

PROBLEM 15-31   .27499

Perform the following calculations, rounding appropriately.

PROBLEM 15-32 $3.7 + 4.19 + 8.00004$

PROBLEM 15-33 $7.8 \times 6.3 \times 5.29$

PROBLEM 15-34   Find C using the formula $C = \pi D$ for D = 6542 cm.

Mechanisms: Linkages, Cams, Gears, and Bearings

OBJECTIVES

After completing this chapter, you will:

- Draw linkage diagrams.
- Create cam displacement diagrams.
- Design cam profile drawings from previously drawn cam displacement diagrams.
- Make detail gear drawings using simplified representations and gear data charts.

- Establish unknown data for gear trains.
- Calculate bearing information from specifications.
- Design a complete gear reducer from engineering data and sketches.

THE ENGINEERING DESIGN APPLICATION

Your latest assignment is to produce a drawing of an in-line follower plate cam profile for the manufacturing department. The engineer has provided the specifications in a cam displacement diagram. (See Figure 16.1.) Your manufacturing department is using a sophisticated computer-aided manufacturing (CAM) system and needs a CADD drawing from which the required tooling paths will be established.

Using a CADD system, the process is fairly simple and completely accurate. Using appropriate tech-

niques, you find and draw the base circle and the prime circle. Next, you create an array of lines extending outward from the center of the circle at 30° increments. (See Figure 16.2.) Finally, by utilizing offsets of the base circle as indicated by the engineer's diagram, you find all the specified control points. Using a spline curve that intersects each of these points, you create a profile drawing to generate tooling paths for the CAM system. (See Figure 16.3.)

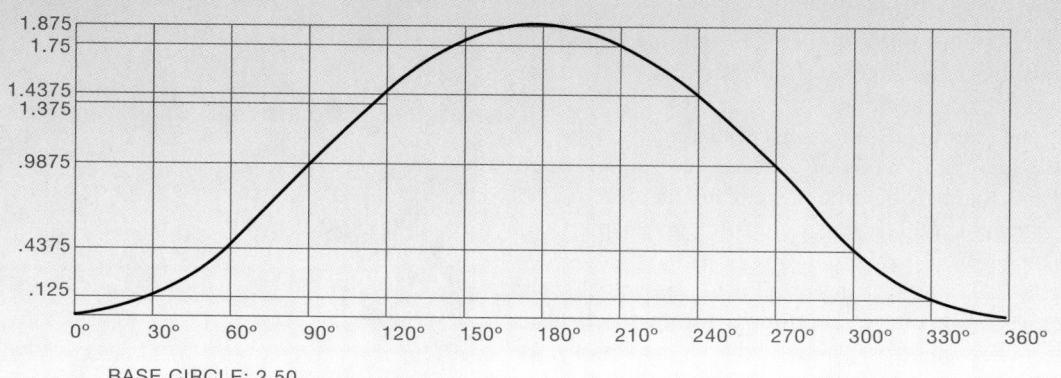

BASE CIRCLE: 2.50
PRIME CIRCLE: 2.50+.50=3.00

FIGURE 16.1 ■ Engineer's cam displacement diagram.

(Continued)

THE ENGINEERING **DESIGN** APPLICATION *(continued)*

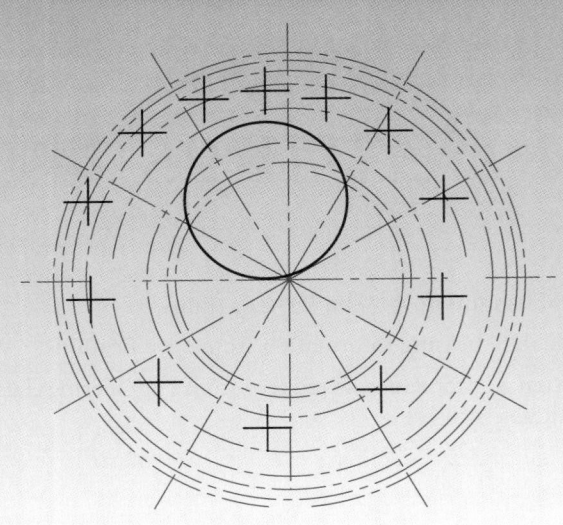

FIGURE 16.2 ■ Construction of a plate cam profile for computer-aided manufacturing (CAM).

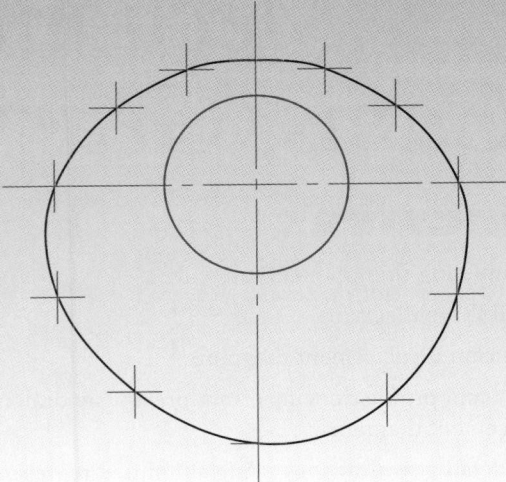

FIGURE 16.3 ■ The plate cam profile used to generate the CAM tool path.

MECHANISM DESIGN

Computer-aided design and drafting is a natural in mechanism design and drafting. Mechanism design drawings are created using symbols and techniques that are easily adapted to CADD systems. You will see later that linkage mechanisms, for example, are often designed by displaying the mechanism in several different positions. Using CADD to do this allows you to place each different position of movement on a separate layer and in a different color. Any one or more layers may be turned on or off at the designer's convenience to evaluate the function of the mechanism. In addition, the CADD system is much faster and more accurate than manual design and drafting methods.

There are computer programs available to make the design and drafting of mechanisms easy. For example, there are programs that may be used to simulate the movement of linkage mechanisms. The illustrations in this chapter were created using a CADD system.

Using CADD in linkage design allows you to quickly and accurately establish a series of positions for the mechanism. After the first position is determined, you may easily use commands such as COPY and ROTATE to place the linkages in alternate positions. You can even set up a continuous slide show with each position of the mechanism as a screen display in the slide show. The term *slide show* refers to screen displays that are sequenced in a specific order and for a given period of time. If you allow the slide show to run fairly fast,

the viewer gets a good understanding and display of the mechanism in action. This type of a CADD display can be established in 2-D, or 3-D as shown in Figure 16.4.

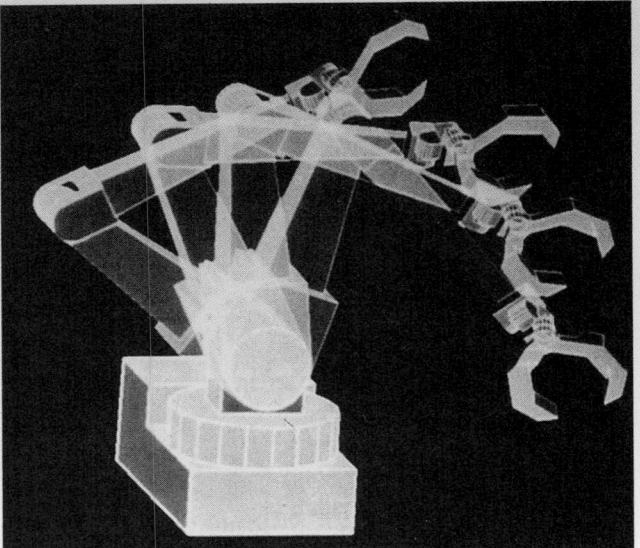

FIGURE 16.4 ■ Linkage mechanism designed with 3-D CADD showing the movement at several positions. *Courtesy Spectragraphics.*

CADD APPLICATIONS

MECHANISMS

A *mechanism* is an arrangement of parts in a mechanical device or machine. This chapter deals with the design and drafting of elements of a mechanism, including linkages, cams, gears, and bearings. The study of mechanisms is part of the physical sciences known as *mechanics*. Mechanics includes statics and dynamics. *Statics* is the study of physics dealing with nonmoving objects acting as weight. *Dynamics* is the branch of physics that studies the motion of objects and the effects of the forces that cause motion. Dynamics is divided into two categories, kinetics and kinematics, *Kinetics* is an element of physics that deals with the effects of forces that cause motion in mechanisms. The linkages, cams, and gears discussed in this chapter relate to the branch of physics known as kinematics. *Kinematics* is the study of mechanisms without reference to the forces that cause the movement.

Mechanisms in Our Daily Lives

Mechanisms play an important role in our daily activities. Every modern convenience, from the toaster used to make part of your breakfast to the automobile that you drive to work, is made up of one or more mechanisms. The car, for example, is a complex combination of mechanisms that includes all of the types of mechanisms to be discussed in this chapter.

LINKAGES

The elements of any mechanism are referred to as *links*. Links or linkages may be defined as any rigid element of the mechanism. In actual practice all of the mechanism components are links, including levers, bars, sliders, cams, or gears. The frame of the device is even considered a fixed link and is an important part of the mechanism. The first part of this chapter separates the types of linkages that deal primarily with motion caused by levers, rockers, cranks, and sliders.

LINKAGE SYMBOLS

One of the nice things about designing linkage mechanisms is that the drawings are in the form of schematic or single-line representations. After the complete design is created using these schematic drawings, the designer may go to work creating the actual components in relation to the schematic design. Schematic drawings have only a few basic components as shown in Figure 16.5. The symbols may be drawn proportional to the examples. The actual scale depends on the size of the drawing. Typical sizes are shown for most applications in this chapter. You can see here how easy it would be to develop a CADD symbols library for use when drawing these mechanisms.

TYPES OF LINKAGES

There are many different combinations of linkage mechanisms. These devices are broken into a few basic elements. The illustrations in this discussion show you the linkage mechanism in a simple pictorial drawing of the actual device and in the schematic representation.

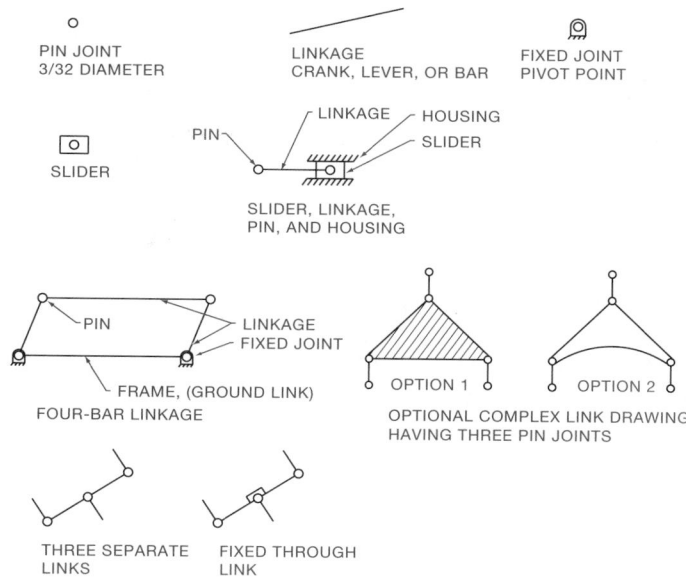

FIGURE 16.5 ■ Linkage diagram symbols.

Crank Mechanism

A *crank* is a link that makes a complete revolution around a fixed point. When working with the crank, keep in mind that the crank link is a fixed distance equal to the radius of the movement as shown in Figure 16.6.

Lever, or Rocker, Mechanism

A *lever*, or *rocker*, is a link that moves back and forth or oscillates through a given angle as illustrated in Figure 16.7.

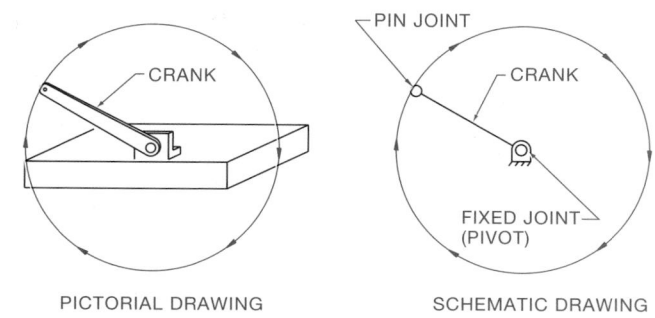

FIGURE 16.6 ■ Crank mechanism.

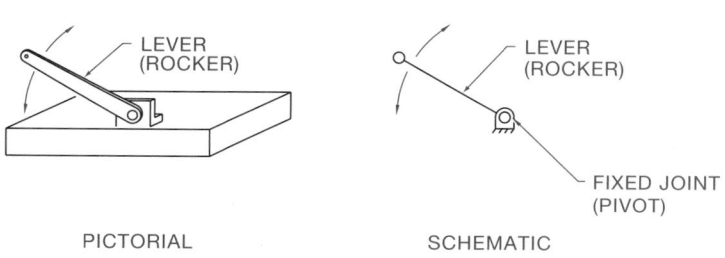

FIGURE 16.7 ■ Lever or rocker mechanism.

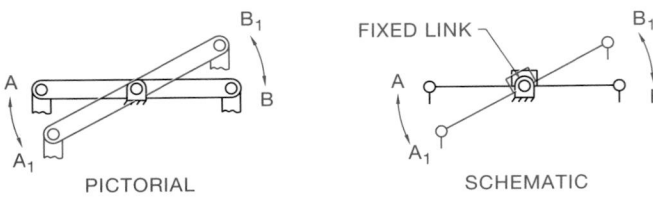

FIGURE 16.8 ■ Rocker arm and fixed link symbol.

Rocker Arm Mechanism

A *rocker arm* is different from the rocker previously discussed because it has a pivot point near the center and oscillates through a given angle as shown in Figure 16.8. Notice the symbol for a fixed link. This symbol is used to indicate that the link between points A and B remains rigid.

Bell Crank Mechanism

Another form of a rocker arm is the *bell crank*. The bell crank is a more complex form of the linkage mechanism. The bell crank drawing shown in Figure 16.9 is a three-joint mechanism where the distances between points A, B, and C are fixed. Other bell crank designs may contain more than three pivot points depending on the design requirements. Notice that the schematic representation in Figure 16.9 shows two alternatives. Consult with your instructor or employer on the technique to use.

Four-bar Linkage

The most commonly used linkage mechanism is called a four-bar linkage. There are many alternate designs of the four-bar linkage, but the basic form has four links, one of which is the ground link or machine frame. One of the rotating links is called the driver or crank and the other is called the follower or rocker. The link connected between the crank and rocker is called the connecting rod or coupler. The two pivoted links both rotate through 360°, or one rotates while the other oscillates, or they both oscillate, depending on the lengths and arrangement of the links. (See Figure 16.10.)

Normally when determining the function of a four-bar linkage, the designer must show the extreme right and extreme left positions. If the crank rotates and the rocker oscillates, then the angle of oscillation may be established with this technique. For example, determine the angle of oscillation for rocker link CD in the four-bar linkage shown in Figure 16.11. If you are using manual drafting, it is best to use a sharp pencil and make careful measurements. Color pencils are often helpful in showing the different positions. If you

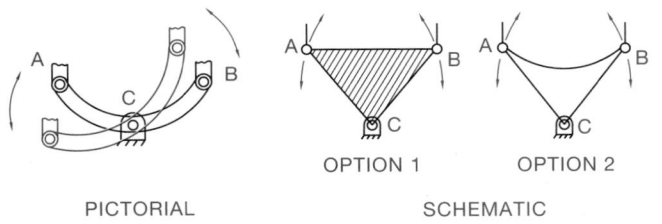

FIGURE 16.9 ■ Bell crank mechanism.

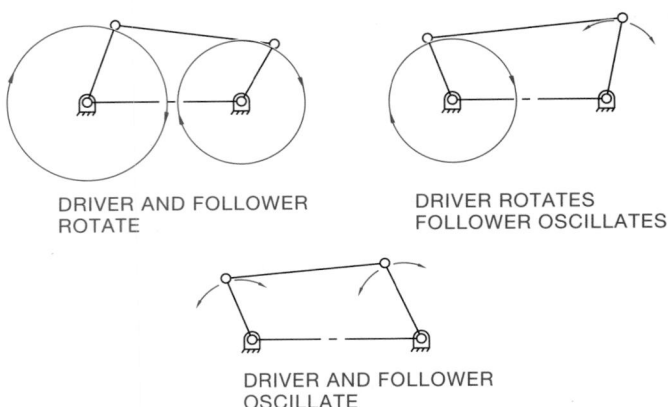

FIGURE 16.10 ■ Four-bar linkage movements.

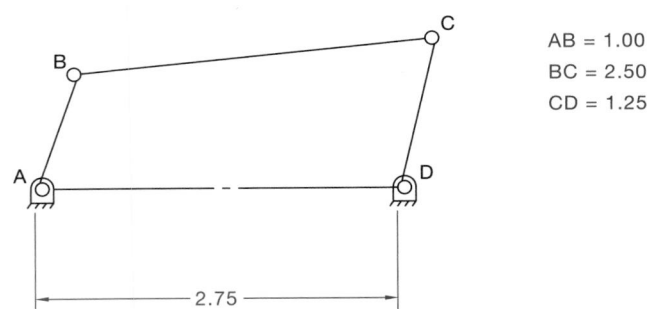

AB = 1.00
BC = 2.50
CD = 1.25

FIGURE 16.11 ■ Four-bar linkage example.

are using CADD to draw the diagrams, then the different positions may be shown on different layers and in different colors for a clear analysis of the operation. Follow these steps:

Step 1. Draw a circle through point B with the center at A and a
and circle through C with the center at D. Determine the
Step 2. extreme right position of link CD by adding the length of links AB and BC (1.00 + 2.50 = 3.50). This puts AB and BC in a straight line as shown in Figure 16.12.

Step 3. Establish the extreme left position of link CD by subtracting the length of link AB from BC (2.50 − 1.00 = 1.50). This places link AB and BC in a straight line to the left as shown in Figure 16.13.

Step 4. Determine the angle between the extreme right and extreme left positions of CD as shown in Figure 16.14.

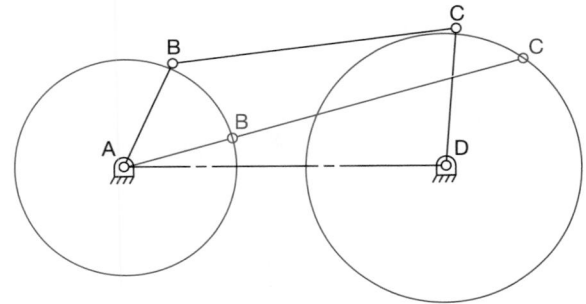

FIGURE 16.12 ■ Steps 1 and 2, four-bar linkage solution.

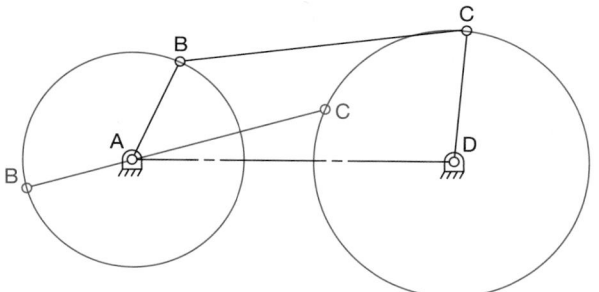

FIGURE 16.13 ■ Step 3, four-bar linkage solution, extreme left position.

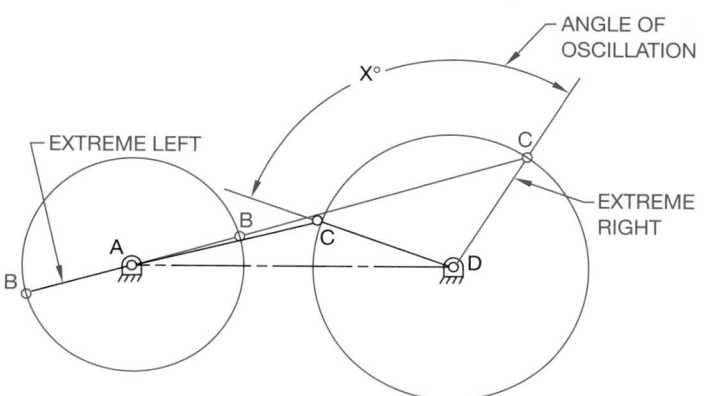

FIGURE 16.14 ■ Step 4, four-bar linkage solution, extreme right position.

Slider Crank Mechanism

A *slider crank* is a linkage mechanism that is commonly used in machines such as engine pistons, pumps, and clamping devices where a straight line motion is required. Figure 16.15 shows the type of slider design used to move a piston back and forth in a straight line. Notice that the distance the slider travels from the extreme left to the extreme right position is called the *stroke*. When working with the design of an engine, it is important to establish the piston stroke and diameter. The distance the piston travels is the stroke. A combination of the stroke and piston diameter determines the piston displacement. Therefore, a four-cylinder 2000 cc engine has a displacement of 500 cc per piston cylinder. You can graphically show the stroke of a piston if you have an engine with a crank length of 1.25 in. and a connecting rod length of 3.25 in. Determine the length of the piston stroke as follows:

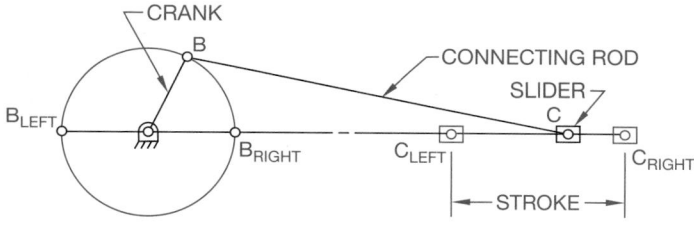

FIGURE 16.15 ■ Slider mechanism.

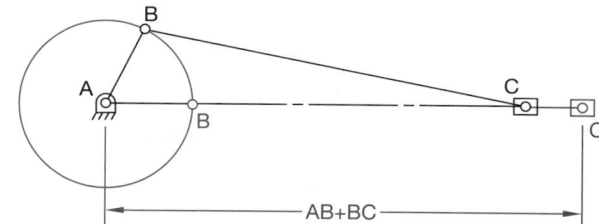

FIGURE 16.16 ■ Steps 1 and 2, slider mechanism, extreme right position.

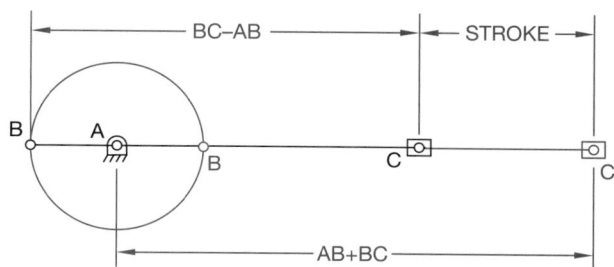

FIGURE 16.17 ■ Step 3, slider mechanism, extreme left position.

Step 1. Draw an arc with point A as its center and AB = 1.25 in.
and as the radius. Show the connecting rod attached to the
Step 2. piston as link BC. Add AB to BC (1.25 + 3.25 = 4.50). Measure this distance from point A along the centerline through A and C. This establishes the extreme right position as shown in Figure 16.16.

Step 3. Subtract AB from BC (3.25 − 1.25 = 2.00). Measure from point A along the centerline through A and C. This establishes the extreme left position of point C. Measure the distance between the extreme left and right positions to determine the stroke as shown in Figure 16.17.

Combination Four-bar Linkage and Slider Mechanism

The design of linkage mechanisms is only limited by the imagination of the designer. There may be any variety of combinations. In the previous examples, the extreme right and left positions were established to determine the function of the mechanism. However, in many situations the designer must establish many positions to completely analyze the movement. One such example is shown in Figure 16.18. This mechanism is a combination four-bar linkage and slider. The objective is to determine the path of point P as the link AB rotates 360°. In order to effectively solve a problem of this type, it is necessary to plot the path of point P by moving link AB a minimum of 30° increments. This is best accomplished if you use a different color or line type for each position. It can become confusing; therefore, careful labeling of each position is important.

Step 1. Draw a circle through point B with A as the center. Point B must remain on this circle through each movement. Draw an arc through point C with D as the center. Point C must remain on this arc through each movement of the mechanism.

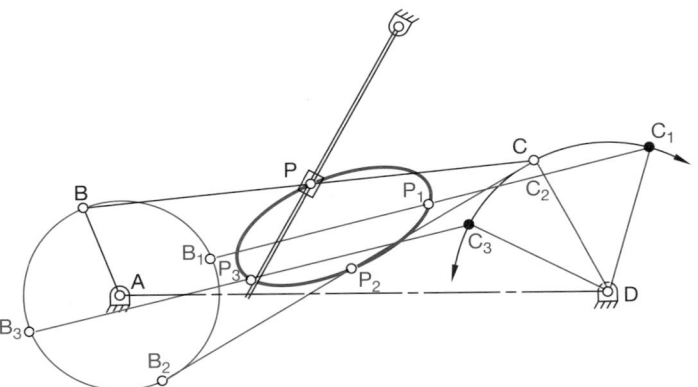

FIGURE 16.18 ■ Solution to four-bar linkage/slider mechanism example.

Step 2. While 30° increments are recommended, this example uses 90° increments for convenience and clarity. Move point B clockwise to B_1. (See Figure 16.18.) From B_1 draw an arc with a radius equal to BC until it intersects arc CD, establishing point C_1. Draw link B_1C_1. Measure distance BP and transfer it to link B_1C_1 thus establishing point P_1.

Step 3. Follow the same procedure for positions B_2 and B_3. This determines the path of coupler point P at positions P_2 and P_3.

CADD APPLICATIONS

CAMS AND GEARS

Designing cams and gears is easy with CADD if a parametric program is used. In a system of this type all you have to do is change the variables and automatically create a new cam or gear. For cam design the variables are:

■ Type of cam motion.

■ Follower displacement.

■ The specific rise, dwell, and fall configuration.

■ Prime circle and base circle diameters.

■ In-line or offset follower.

■ Hub diameter, hub projection, face width, shaft, and keyway specifications.

In gear design the variables are:

■ Type of gear.

■ Pitch diameter, base circle diameter, pressure angle.

■ Diametral pitch.

■ Hub diameter, hub projection, face width, shaft, and keyway specifications.

The program automatically calculates the rest of the data and draws the cam profile or a detail drawing of the gear in simplified or detailed representation.

CAMS

A *cam* is a rotating mechanism that is used to convert rotary motion into a corresponding straight motion. The timing involved in the rotary motion is often the main design element of the cam. For example, a cam may be designed to make a follower rise a given amount in a given degree of rotation, then remain constant for an additional period of rotation, and finally fall back to the beginning in the last degree of rotation. The total movement of the cam follower happens in one 360° rotation of the cam. This movement is referred to as the displacement. Cams are generally in the shape of irregular plates, grooved plates, or grooved cylinders. The basic components of the cam mechanism are shown in Figure 16.19.

Cam Types

There are basically three different types of cams: the plate cam, face cam, and drum cam. (See Figure 16.20.) The plate cam is the most commonly used type of cam.

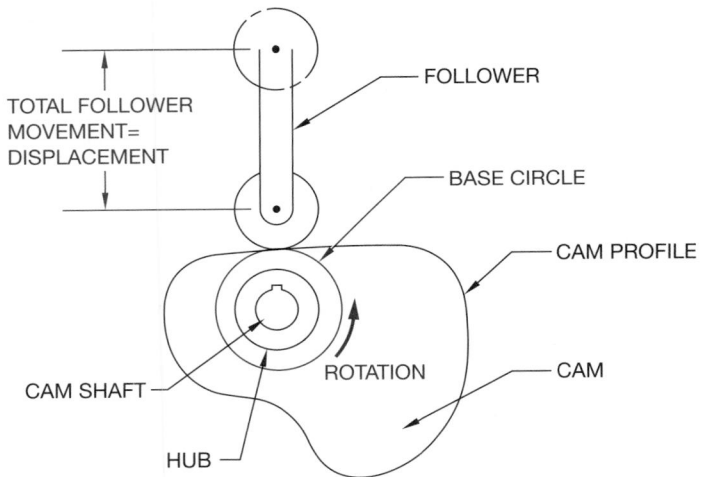

FIGURE 16.19 ■ Elements of a cam mechanism.

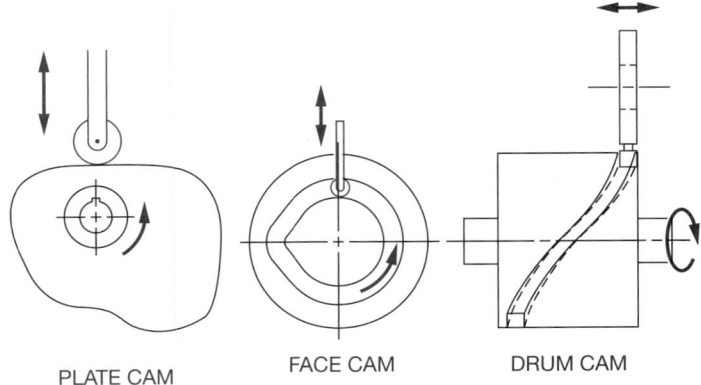

PLATE CAM   FACE CAM   DRUM CAM

FIGURE 16.20 ■ Types of cam mechanisms.

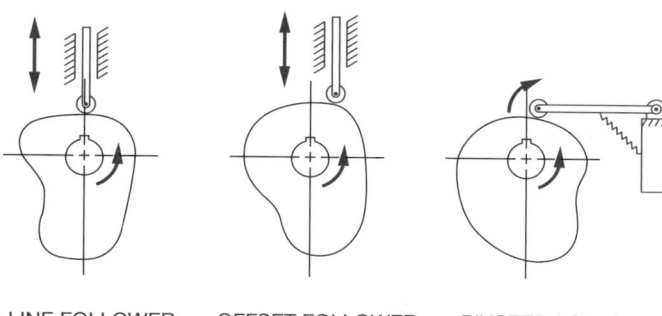

IN-LINE FOLLOWER   OFFSET FOLLOWER   PIVOTED FOLLOWER

FIGURE 16.21 ■ Types of cam roller followers.

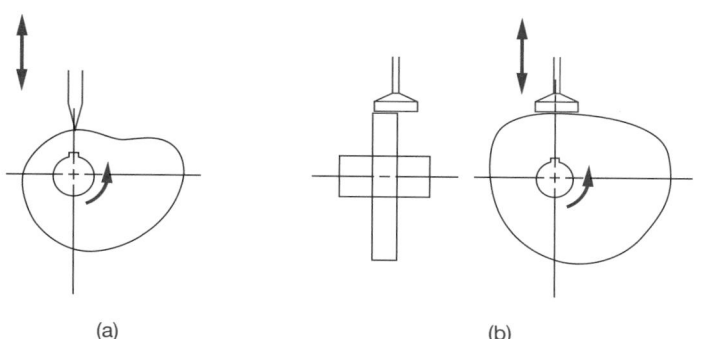

(a)   (b)

FIGURE 16.22 ■ (a) Knife-edged cam follower. (b) Flat-faced cam follower.

Cam Followers

There are several types of cam followers. The type used depends on the application. The most common type of follower is the roller follower. The roller follower works well at high speeds, reduces friction and heat, and keeps wear to a minimum. The arrangement of the follower in relation to the cam shaft may differ depending on the application. The roller followers shown in Figure 16.21 include: the in-line follower where the axis of the follower is in line with the cam shaft; the offset roller follower; and the pivoted follower. The pivoted follower requires spring tension to keep the follower in contact with the cam profile.

Another type of cam follower is the knife-edged follower shown in Figure 16.22a. This follower is used for only low-speed and low-force applications. The knife-edged follower has a low resistance to wear, but is very responsive and may be effectively used in situations that require abrupt changes in the cam profile.

The flat-faced follower shown in Figure 16.22b is used in situations where the cam profile has a steep rise or fall. Designers often offset the axis of the follower. This practice causes the follower to rotate while in operation. This rotating action allows the follower surface to wear evenly and last longer.

CAM DISPLACEMENT DIAGRAMS

Cams are generally designed to achieve some type or sequence of a timing cycle in the movement of the follower. There are several predetermined types of motion from which cams are designed. These

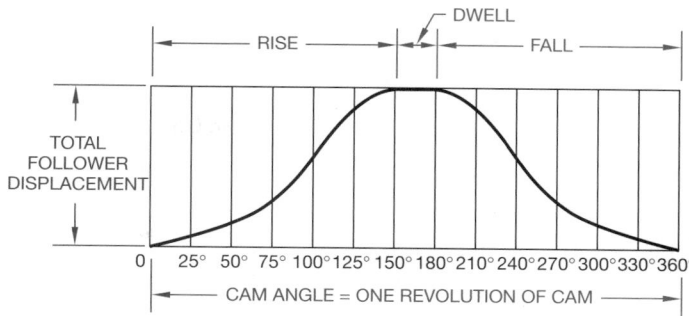

FIGURE 16.23 ■ Cam displacement diagram.

forms of motion may be used alone, in combination, or custom designed to suit specific applications. The following discussion shows you how to set up a cam displacement diagram given a specific type of cam motion. The cam displacement diagram is similar to a graph representing the cam profile in a flat pattern of one complete 360° revolution of the cam. The terms associated with the displacement diagram include *cycle, period, rise, fall, dwell,* and *displacement*. A complete cam cycle has taken place when the cam rotates 360°. A period of the cam cycle is a segment of follower operation such as rise, dwell, or fall. Rise exists when the cam is rotating and the follower is moving upward. Fall is when the follower is moving downward. Dwell exists when the follower is constant, not moving either up or down. A dwell is shown in the displacement diagram as a horizontal line for a given increment of degrees. When developing the cam displacement diagram, the height of the diagram is drawn to scale and is equal to the total follower displacement. (See Figure 16.23.) The horizontal scale is equal to one cam revolution or 360°. The horizontal scale may be drawn without scale. Some engineering drafters prefer to make this scale equal to the circumference of the base circle, or any convenient length. The base circle is an imaginary circle with its center at the center of the cam shaft and its radius tangent to the cam follower at zero position. The horizontal scale is then divided into increments of degrees. Each rise and fall is divided into six increments. So, if the follower rises 150°, then each increment is 150°/6 = 25°. If the cam falls between 180° and 360°, this represents a total fall of 180°. The fall increments are 180°/6 = 30° as shown in Figure 16.23.

Simple Harmonic Motion

Simple harmonic motion may be used for high-speed applications if the rise and fall are equal at 180°. Moderate speeds are recommended if the rise and fall are unequal or if there is a dwell in the cycle. This application causes the follower to jump if the speeds are too high.

Draw a cam displacement diagram using simple harmonic motion when the total displacement is 2.00 in. and the cam follower rises the length of the total displacement in 180° and falls back to 0° in 180°. Use the following procedure to set up the displacement diagram:

Step 1. Draw a rectangle equal in height (vertical scale) to the total displacement of 2.00 in. and equal in length (horizontal scale) to 360°. The horizontal scale should have 6°–30°

increments for the rise from 0° to 180° and 6°–30° increments for the fall from 180° to 360°. The horizontal scale may be any convenient length. Draw a thin vertical line from each horizontal increment as shown in Figure 16.24.

Step 2. Draw a half circle at one end of the displacement diagram equal in diameter to the rise of the cam. Divide the half circle into six equal parts as shown in Figure 16.25.

Step 3. The cam follower begins its rise at 0°. The rise continues by projecting point 1 on the half circle over to the first (30°) increment on the horizontal scale. Continue this process for points 2, 3, 4, 5, and 6 on the half circle, each intersecting the next increment on the horizontal scale.

Step 4. Notice the pattern of points created in the preceding step. Use your irregular curve to carefully connect these points. If you are using CADD, use your curve-fitting command to draw the cam profile.

Step 5. Develop the fall profile by projecting the points from the half circle in the reverse order discussed in step 3. (See Figure 16.26.)

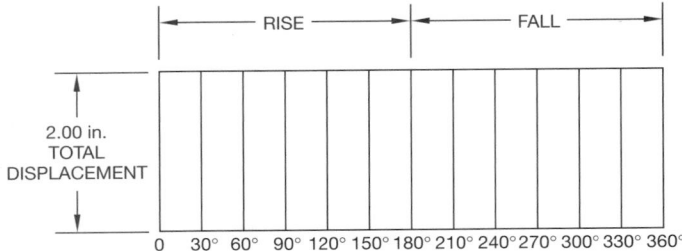

FIGURE 16.24 ■ Step 1, layout for simple harmonic motion cam displacement diagram.

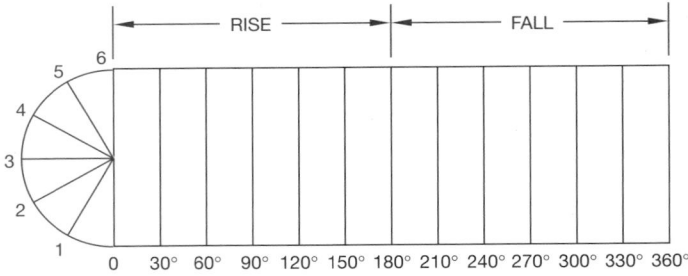

FIGURE 16.25 ■ Step 2, layout for simple harmonic motion cam displacement diagram.

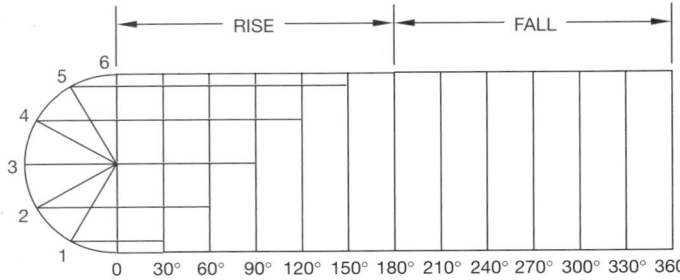

FIGURE 16.26 ■ Steps 3, 4, and 5, layout for simple harmonic motion cam displacement diagram.

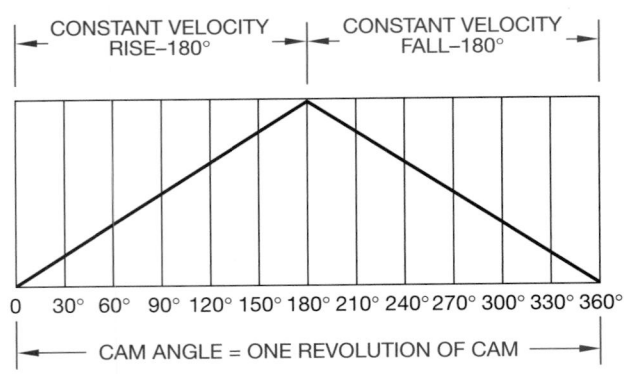

FIGURE 16.27 ■ Cam displacement diagram for constant velocity motion.

Constant Velocity Motion

Constant velocity motion is also known as straight-line motion. This is used for the feed control of some machine tools, when it is required for the follower to rise and fall at a uniform rate. Constant velocity motion is only used at slow speeds because of the abrupt change at the beginning and end of the motion period. The displacement diagram is easy to draw. All you have to do is draw a straight line from the beginning of the rise or fall to the end as shown in Figure 16.27.

Modified Constant Velocity Motion

Modified constant velocity motion was designed to help reduce the abrupt change at the beginning and end of the motion period. This type of motion may be adjusted to accomplish specific results by altering the degree of modification. This is done by placing a curve at the beginning and end of the rise and fall. The radius of this arc depends on the amount of smoothing required, but the radius normally ranges from one-third to full displacement. If the motion were modified to one-third the displacement, then the cam displacement diagram is drawn as shown in Figure 16.28.

Uniform Accelerated Motion

Uniform accelerated motion is designed to reduce the abrupt change at the beginning and end of a period. It is recommended for

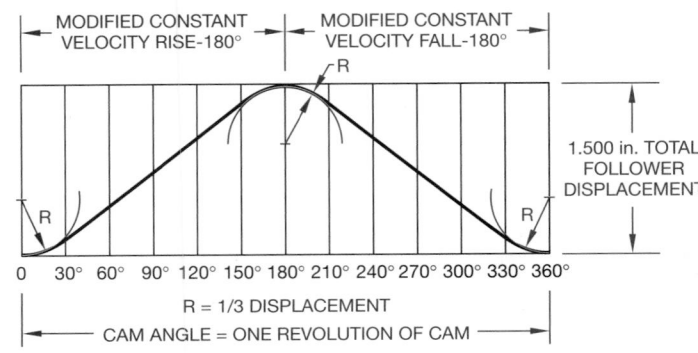

FIGURE 16.28 ■ Cam displacement diagram for modified constant velocity motion.

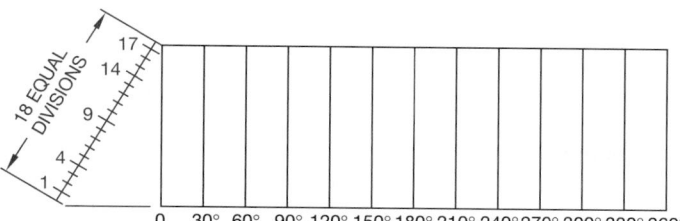

FIGURE 16.29 ■ Steps 1 and 2, layout for uniform accelerated motion cam displacement diagram.

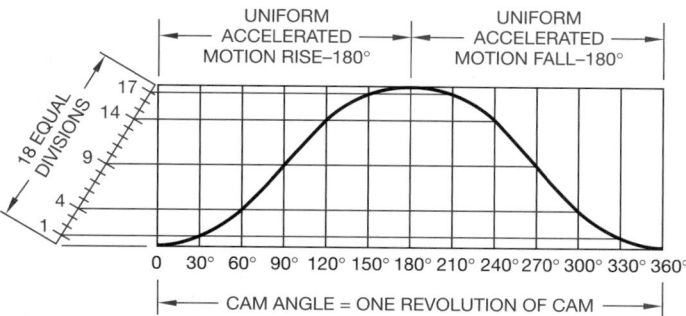

FIGURE 16.30 ■ Step 3, layout for uniform accelerated motion cam displacement diagram.

moderate speeds, especially when associated with a dwell. The advantage of this motion is its use when constant acceleration for the first half of the rise and constant deceleration for the second half of the rise are required.

Use the following technique to draw a uniform accelerated motion displacement diagram where the follower rises a total of 2.00 in. in 180° and falls back to 0° in 180°:

Step 1. and Step 2.   Set up the displacement diagram with the height equal to 2.00 in. total rise and the horizontal scale divided into 30° increments. Keep in mind that the horizontal scale is divided into 30° increments if the rise and fall is in 180° each. If the rise, for example, were 120°, then the increments would be 120°/6 = 20° each. Set up a scale with eighteen equal divisions at one end of the displacement diagram and mark off the first, fourth, ninth, fourteenth and seventeenth divisions as shown in Figure 16.29.

Step 3.   Establish the rise by projecting from the first division on the scale to the 30° increment on the diagram, then continue with the fourth division to 60°, and so on until each division is used. Continue this same procedure in reverse order to establish the profile of the fall. Connect all of the points to complete the displacement diagram as shown in Figure 16.30.

Cycloidal Motion

Cycloidal motion is the most popular cam profile development for smooth-running cams at high speeds. The term *cycloidal* comes from the word *cycloid*. A cycloid is a curved line generated by a point on the circumference of a circle as the circle rolls along a

straight line. The cycloidal cam motion is developed in this same manner, and the result is the smoothest possible cam profile. Cycloidal motion is a little more complex to set up than other types of motion. Use the following procedure to develop a cam displacement diagram for cycloidal motion with a total rise of 2.500 in. in 180°:

Step 1.   Begin the displacement diagram with a total rise of 2.500 in. in 180°. Only half of the diagram is shown for this example.

Step 2.   Draw a circle tangent to and centered on the total displacement at one end of the diagram. This circle must have a circumference equal to the displacement. Calculate the diameter using the formula $D = C/\pi$. In this cased $D = 2.500/3.1414 = .7958$. Round off to .80 for manual drafting or use as is for a CADD application. Then, beginning at the point where the circle is tangent to the diagram, divide the circle into six equal parts and number them as shown in Figure 16.31.

Step 3.   Draw a vertical line through the center of the circle equal in length to the displacement. Divide this line into six equal parts as shown in Figure 16.32.

Step 4.   Use a radius equal to the radius of the circle to draw arcs from the divisions on the vertical line as shown in Figure 16.33. These arcs should intersect the dashed lines drawn from points 2, 3, 5, and 6 on the circle. Where the dashed lines and the arcs intersect, draw horizontal lines into the

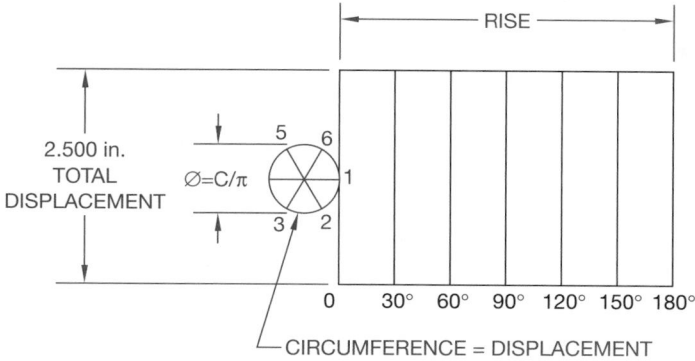

FIGURE 16.31 ■ Steps 1 and 2, layout for cycloidal motion cam displacement diagram.

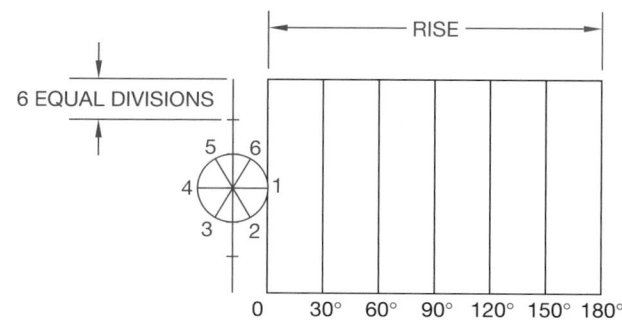

FIGURE 16.32 ■ Step 3, layout for cycloidal motion cam displacement diagram.

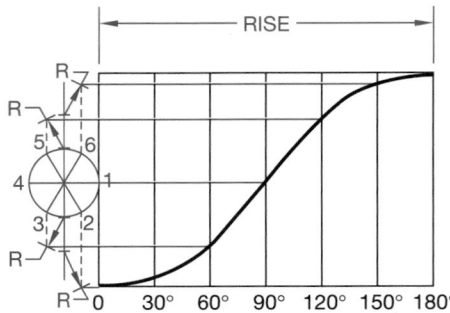

FIGURE 16.33 ■ Step 4, layout for cycloidal motion cam displacement diagram.

displacement diagram, intersecting the appropriate increment from the horizontal scale. Connect the points of intersection with a smooth curve as shown in Figure 16.34.

Developing a Cam Displacement Diagram with Different Cam Motions

Most cam profiles are not as simple as the preceding examples. Many designs require more than one type of cam motion and may also incorporate dwell. Construct a cam displacement diagram from the following information:

■ Total displacement equals 2.00 in.

■ Rise 1.00 in. simple harmonic motion in 120°.

■ Dwell for 30°.

■ Rise 1.00 in. modified constant velocity motion in 90°.

■ Fall 2.00 in. uniform accelerated motion in 90°.

■ Dwell for 30° through the balance of the cycle.

Look at Figure 16.34 as you review the method of development for each type of cam motion displayed.

CONSTRUCTION OF AN IN-LINE FOLLOWER PLATE CAM PROFILE

Each of the cam displacement diagrams may be used to construct the related cam profiles. The *cam profile* is the actual contour of the cam. In operation, the cam follower is stationary and the cam rotates on the cam shaft. In cam profile construction, however, the cam is drawn in one position and the cam follower is moved to a series of positions around the cam in relationship to the cam displacement diagram. The following technique is used to draw the cam profile for the cam displacement diagram given in Figure 16.35 and for a cam with a 2.50 in. base circle, a .75 in. cam follower, and counterclockwise rotation:

Step 1.   Refer to Figure 16.35. Use construction lines for all preliminary work. Draw the cam follower in place near the top of the sheet with phantom lines. Draw the base circle (∅2.50 in.). The base circle is tangent to the follower at the 0° position. Draw the prime circle 2.50 + .75 = ∅3.25 in. The *prime circle* passes through the center of the follower at the 0° position. The cam displacement diagram is placed on Figure 16.35 for easy reference. In actual practice, you may have the displacement diagram next to the cam profile drawing or on a CADD layer to be inserted on the screen for reference.

Step 2.   Begin working in a direction opposite from the rotation of the cam. Since this cam rotates counterclockwise, work clockwise. Starting at 0° draw an angle equal to each of the horizontal scale angles on the displacement diagram. In this case the angles are 30° increments. This is not true with all cam displacement diagrams. Some have varying increments.

Step 3.   Notice the measurements labeled A through L on the displacement diagram in Figure 16.35. Begin by transferring the distance A along the 30° element (in the profile construction drawing) by measuring from the prime circle

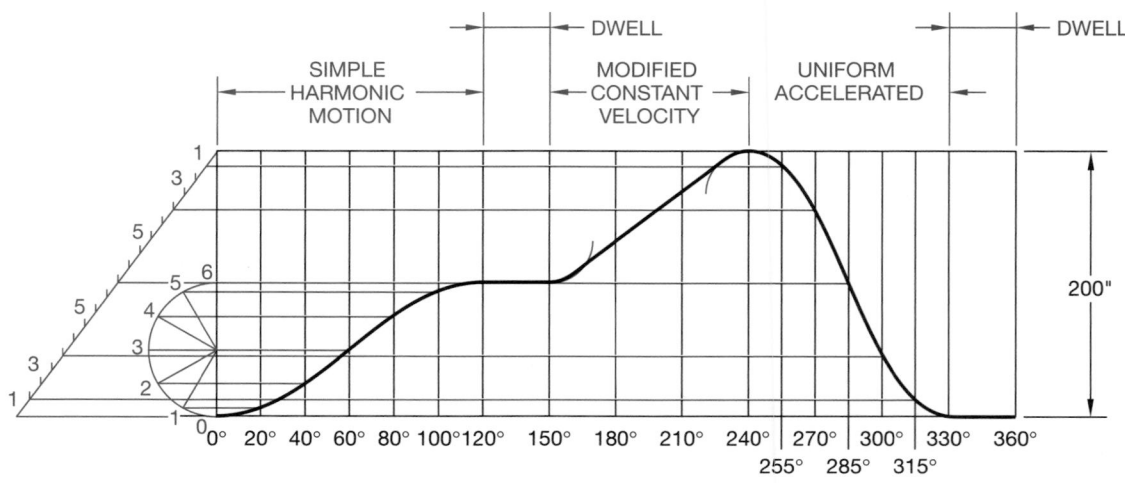

FIGURE 16.34 ■ The development of a cam displacement diagram with different cam motions.

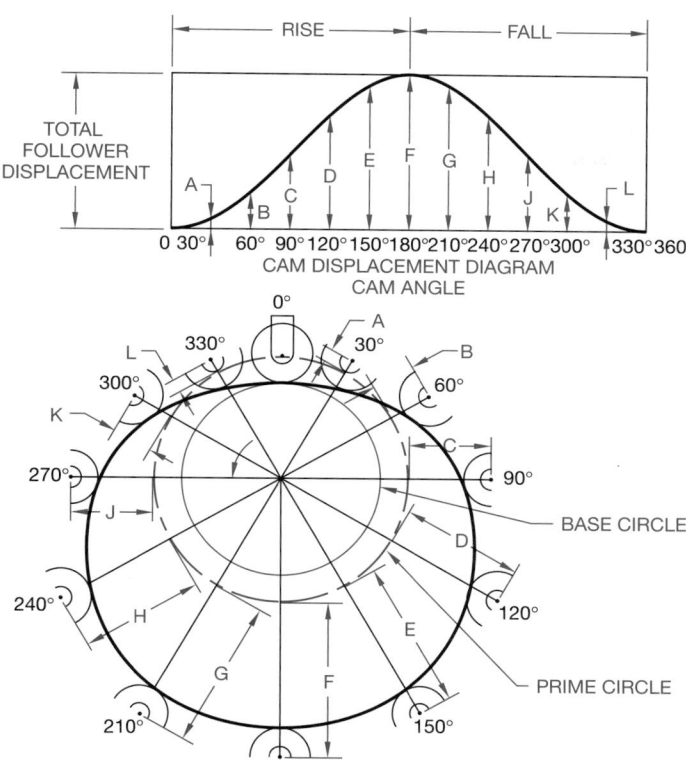

FIGURE 16.35 ■ Construction of an in-line follower plate cam profile.

along this line. This establishes the center of the cam follower at this position. Do the same with each of the measurements B through L located on each corresponding increment.

Step 4. Lightly draw the cam followers in position with centers located at each of the points found in step 3. (See Figure 16.35.)

Step 5. Draw the cam profile by connecting a smooth curve tangent to the cam followers at each position. This is one place where CADD drafting plays an important role regarding accuracy and speed.

Preparing the Formal Plate Cam Drawing

You are ready to prepare the formal plate cam drawing after you have constructed the plate cam profile using the previously described techniques. The previous drawing may be used because all lines were drawn as construction lines or on a CADD layer that may be turned off when the formal drawing is complete. The information needed on the plate cam drawing includes:

■ Cam profile.

■ Hub dimensions, including cam shaft, outside diameter, width, keyway dimensions.

■ Roller follower placed in one convenient location, such as 60°, using phantom lines.

■ The drawing is set up as a chart drawing where A° equals the angle of the follower at each position, and R equals the radius from the center of the cam shaft to the center of the follower at each position.

■ A chart giving the values of the angles A and the radii R at each follower position.

■ Side view showing the cam plate thickness and set screw location with thread specification, if used.

■ Tolerances, unless otherwise specified.

Establish all measurements for dimensions A and R at each of the follower positions. This may be done graphically by measuring from the profile construction, or mathematically using trigonometry. If a CADD system is used, the measurements may be taken directly from the layout. Figure 16.36 shows a formal plate cam drawing.

CONSTRUCTION OF AN OFFSET FOLLOWER PLATE CAM PROFILE

When an offset cam follower is used, the method of construction is a little more complex than the technique used for the in-line follower. For this example, the follower is offset .75 in. as shown m Figure 16.37. Prepare the cam profile drawing as follows:

Step 1. Refer to Figure 16.37. Use construction lines or a construction CADD layer for all preliminary work. Draw the cam follower in place near the top of the sheet with phantom lines. Draw the base circle (∅2.50 in.). The base circle is tangent to the follower at 0° position. Draw the prime circle (2.50 + .75 = ∅3.25 in.). The prime circle passes through the center of the follower at 0° position. Draw the offset circle (∅1.50 in.). The offset circle is drawn with a radius equal to the follower offset distance. The cam displacement diagram is placed on Figure 16.37 for easy reference. In actual practice, you may have the displacement diagram next to the cam profile drawing or on a CADD layer to be inserted on the screen for reference.

Step 2. Begin working in a direction opposite from the rotation of the cam. Since this cam rotates counterclockwise, work clockwise. Starting at 0° on the offset circle, draw an angle equal to each of the horizontal scale angles on the displacement diagram. In this case the angles are 30° increments. This is not true with all cam displacement diagrams. Some have varying increments.

Step 3. Notice where each of the angle increments drawn in step 2 intersects the offset circle. At each of these points, draw another line tangent to the offset circle. Make these lines long enough to pass beyond the reference circle.

Step 4. Notice the measurements labeled A through L on the displacement diagram in Figure 16.37. Begin by transferring the distance A along the line tangent to the 30° element by measuring from the prime circle along this line. This establishes the center of the cam follower at this position.

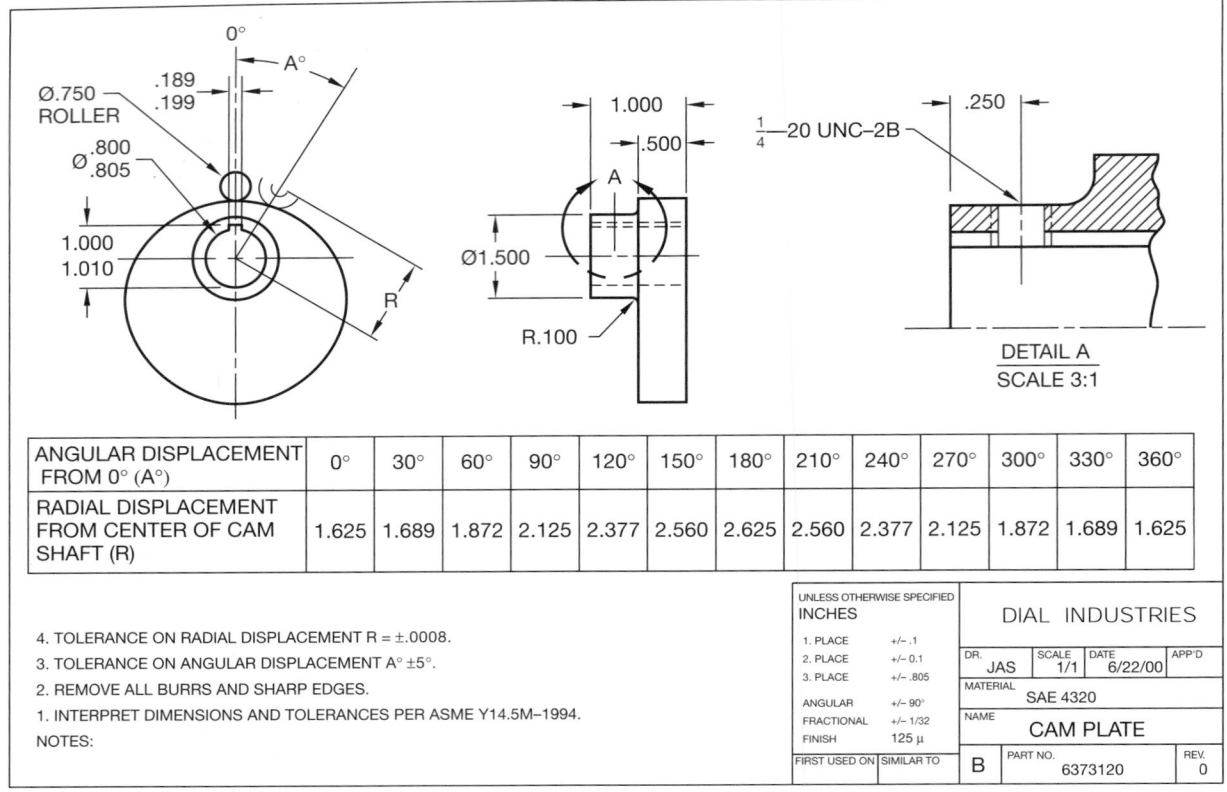

ANGULAR DISPLACEMENT FROM 0° (A°)	0°	30°	60°	90°	120°	150°	180°	210°	240°	270°	300°	330°	360°
RADIAL DISPLACEMENT FROM CENTER OF CAM SHAFT (R)	1.625	1.689	1.872	2.125	2.377	2.560	2.625	2.560	2.377	2.125	1.872	1.689	1.625

4. TOLERANCE ON RADIAL DISPLACEMENT R = ±.0008.

3. TOLERANCE ON ANGULAR DISPLACEMENT A° ±5°.

2. REMOVE ALL BURRS AND SHARP EDGES.

1. INTERPRET DIMENSIONS AND TOLERANCES PER ASME Y14.5M–1994.

NOTES:

FIGURE 16.36 ■ Formal plate cam drawing. *Courtesy Dial Industries.*

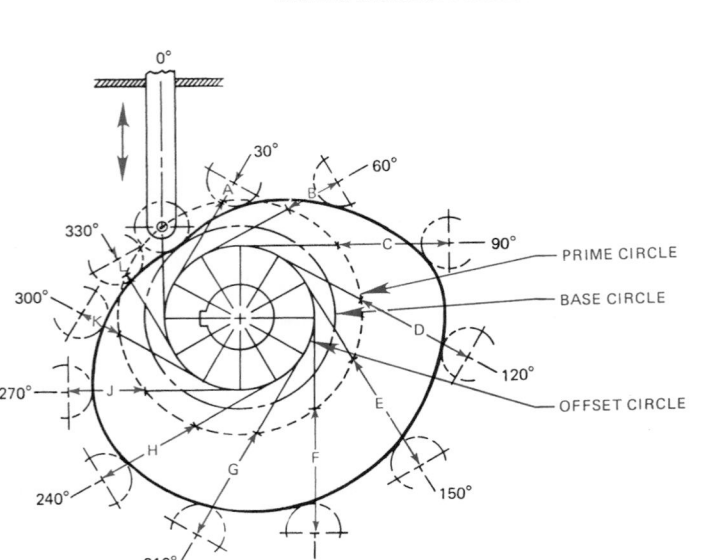

CAM DISPLACEMENT DIAGRAM

FIGURE 16.37 ■ Construction of an offset follower plate cam profile.

Do the same with each of the measurements B through L located on the tangent line from each corresponding increment.

Step 5. Lightly draw the cam followers in position with centers located at each of the points found in step 4. (Refer to Figure 16.37.)

Step 6. Draw the cam profile by connecting a smooth curve tangent to the cam followers at each position. This is one place where CADD drafting plays an important role regarding accuracy and speed.

DRUM CAM DRAWING

Drum cams are used when it is necessary for the follower to move in a path parallel to the axis of the cam. The drum cam is a cylinder with a groove machined in the surface the shape of the cam profile. The cam follower moves along the path of the groove as the drum is rotated. The displacement diagram for a drum cam is actually the pattern of the drum surface as if it were rolled out flat. The height of the displacement diagram is equal to the height of the drum. The length of the displacement diagram is equal to the circumference of the drum. Refer to the drum cam drawing in Figure 16.38 as you follow the construction steps:

Step 1. Draw the top view showing the diameter of the drum, the cam shaft and keyway, and the roller follower in place at 0°. Draw the front view as shown in Figure 16.38. Draw

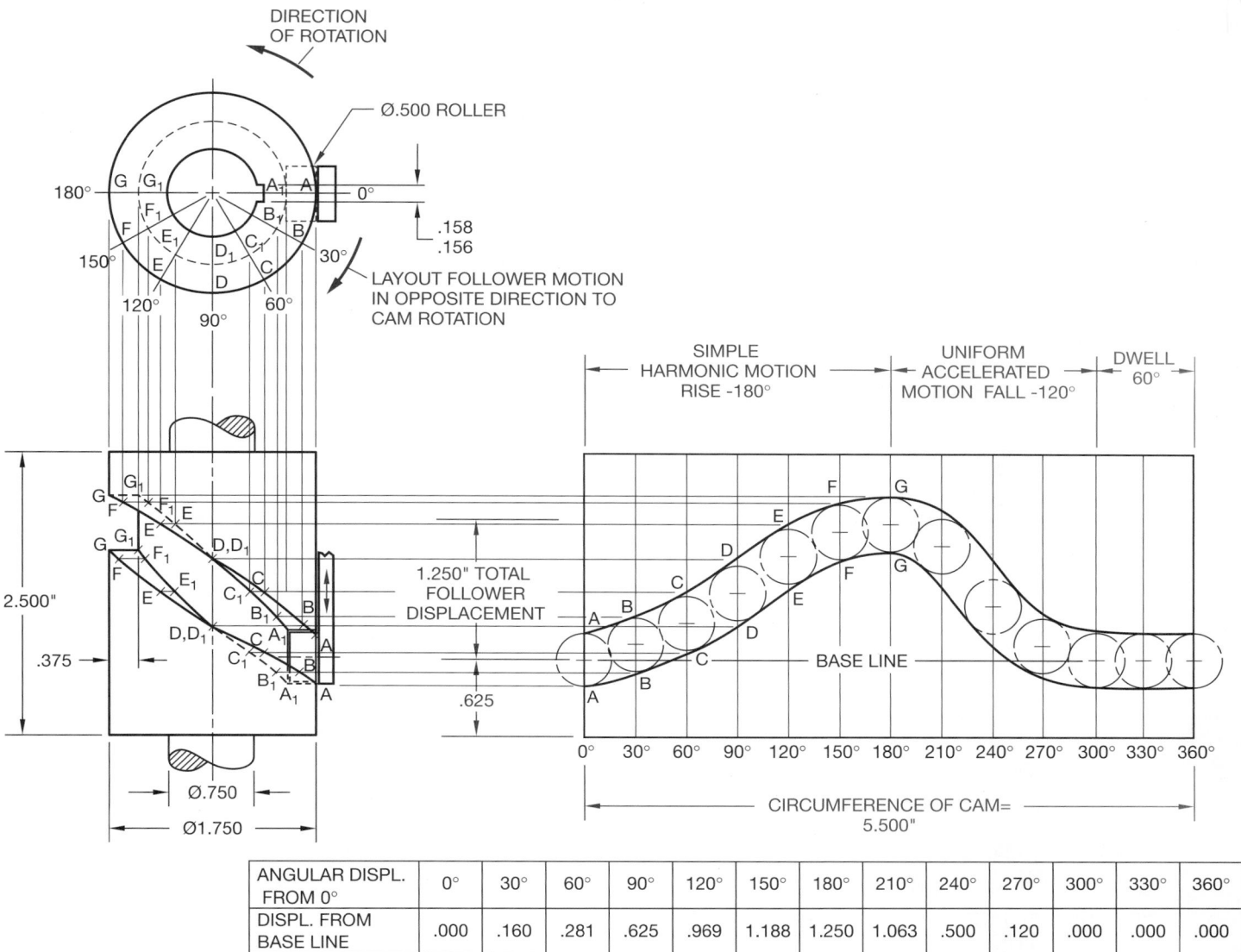

ANGULAR DISPL. FROM 0°	0°	30°	60°	90°	120°	150°	180°	210°	240°	270°	300°	330°	360°
DISPL. FROM BASE LINE	.000	.160	.281	.625	.969	1.188	1.250	1.063	.500	.120	.000	.000	.000

2. TOLERANCE ON ANGULAR DISPLACEMENT ± .5°.
1. TOLERANCE ON DISPLACEMENT FROM BASELINE ± .0008.
NOTES:

FIGURE 16.38 ■ Construction of a drum cam drawing.

the outline of the cam displacement diagram equal to the height and circumference of the drum.

Step 2. Draw the roller follower on the displacement diagram at each angular interval.

Step 3. Draw curves tangent to the top and bottom of each roller follower position. These curves represent the development of the groove in the surface of the cam.

Step 4. In the top view, draw radial lines from the cam shaft center equal to the angle increments shown on the displacement diagram. Be sure to lay out the angles in a direction opposite the cam rotation. The points where these lines intersect the depth and the outside circumference of the groove are labeled A, A₁, B, B₁, respectively. Notice that the

same corresponding points are labeled on the displacement diagram.

Step 5. From points A, A₁ on the displacement diagram, project horizontally until each point intersects a vertical line from the same corresponding point in the top view. This establishes the points in the front view along the outer and inner edges of the groove, both top and bottom. Continue this process for each pair of points on the drum cam displacement diagram.

GEARS

Gears are toothed wheels used to transmit motion and power from one shaft to another. Gears are rugged and durable and can transmit power with up to 98% efficiency with long service life. Gear design involves a combination of material, strength, and wear characteristics. Most gears are made of cast iron or steel, but plastic and brass and bronze alloys are also used for some applications. Gear selection and design are often done through vendors' catalogs or the use of standard formulas. A gear train exists when two or more gears are in combination for the purpose of transmitting power. Generally, two gears in mesh are used to increase or reduce speed, or change the direction of motion from one shaft to another. When two gears are in mesh, the larger is called the *gear* and the smaller is called the *pinion*. (See Figure 16.39.)

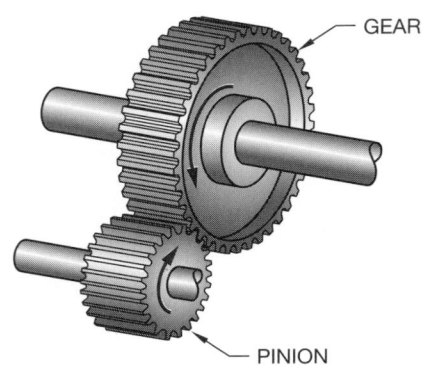

FIGURE 16.39 ■ The gear and pinion.

AGMA/ANSI Gear selection generally follows the guidelines of the American Gear Manufacturers Association (AGMA) or the American National Standards Institute (ANSI).

It is important for engineering drafters to fully understand gear terminology and formulas. However, many drafters will not draw gears, because gears are commonly supplied as purchase parts. When this happens, the drafter may be required to make gear selections for specific applications or draw gears on assembly drawings. Details of gears are often drawn using simplified techniques as described in this chapter. In some situations, the gears are drawn as they actually exist for display on assembly drawings or in catalogs. When this is necessary, CADD can make the job easy and increase productivity. If manual drafting is used, a gear tooth template is needed. Gear data is readily available in the *Machinery's Handbook*.

GEAR STRUCTURE

Gears are made in a variety of structures depending on the design requirements, but there is some basic terminology that can typically be associated with gear structure. The elements of the gear structure shown in Figure 16.40 include the outside diameter, face, hub, bore, and keyway.

Types of Hubs

The hubs are the lugs or shoulders projecting from one or both faces of some gears. These may be referred to as A, B, or C hubs. An A hub is also called a flush hub because there is no projection from the gear face. B hubs have a projection on one side of the gear, while C hubs have projections on both sides of the gear face. (See Figure 16.41.)

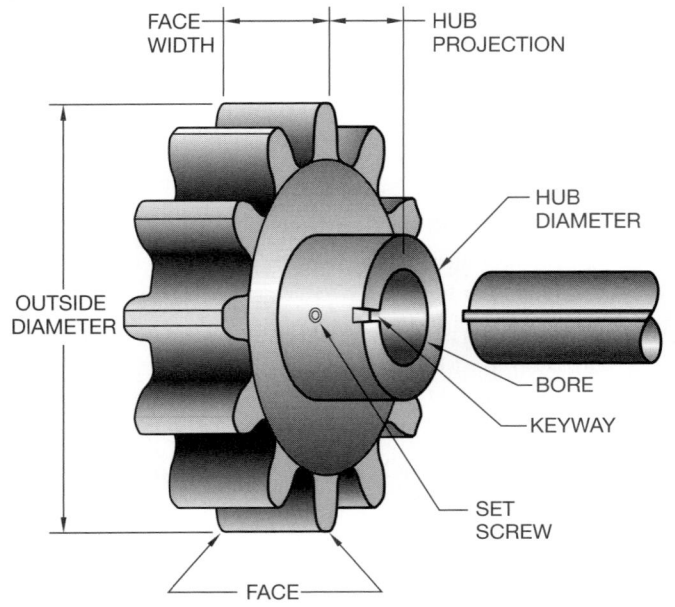

FIGURE 16.40 ■ Elements of the gear structure.

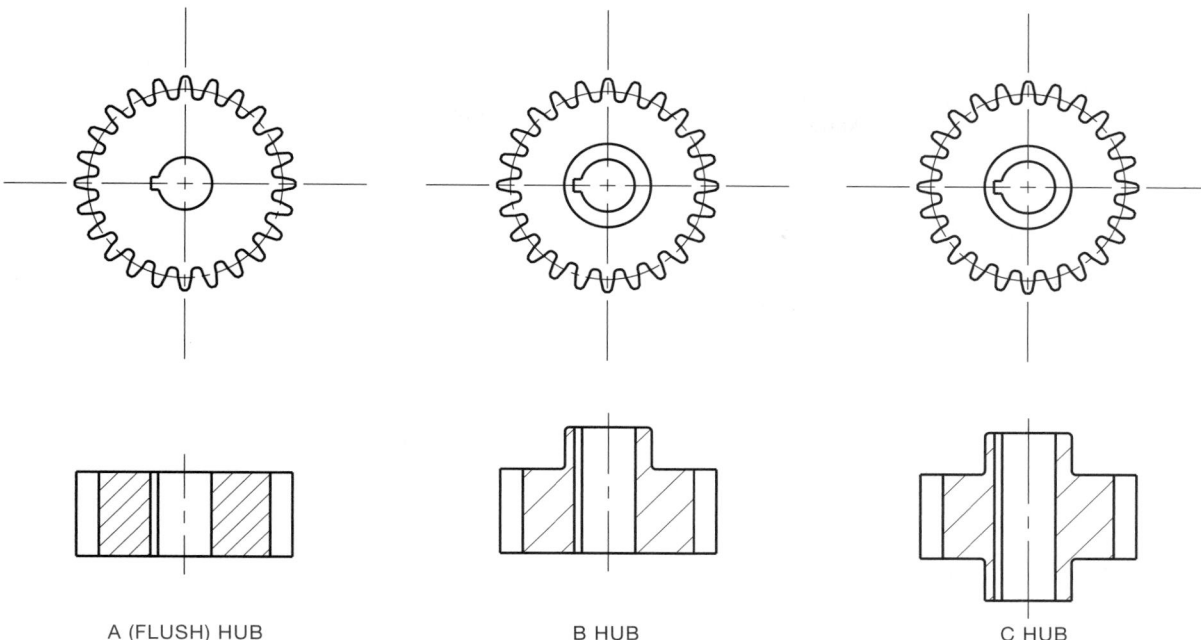

A (FLUSH) HUB   B HUB   C HUB

FIGURE 16.41 ■ Types of gear hubs.

Keyways, Keys, and Set Screws

Gears are usually held on the shaft with a key, keyway, and set screw. Refer to Chapter 11 for information on the manufacture of keyways, Chapter 12 for proper dimensioning, and Chapter 13 for screw thread specifications. One or more set screws are usually used to keep the key secure in the keyway. (See Figure 16.42.)

SPLINES

Splines are teeth cut in a shaft and a gear or pulley bore and are used to prevent the gear or pulley from spinning on the shaft. Splines are often used when it is necessary for the gear or pulley to easily slide on the shaft. Splines may also be nonsliding and in all cases are stronger than keyways and keys. The standardization of splines is established by the Society of Automotive Engineers (SAE) so that any two parts with the same spline specifications should fit together. The following is an example of an SAE spline specification:

SAE 2 1/2–10   B SPLINE
 (A)   (B)   (C) (D)

The note components are described below:
   (A) Society of Automotive Engineers.
   (B) The outside diameter of the spline.
   (C) The number of teeth.
   (D) A = this is a fixed nonsliding spline.
       B = this spline slides under no-load conditions.
       C = this spline slides under load conditions.

Involute Spline

Another spline standard is the *involute spline*. The teeth on the involute spline are similar to the curved teeth found on spur gears. The spline teeth generally have a shorter whole depth than standard spur gears, and the pressure angle is normally 30°.

GEAR TYPES

The most common and simplest form of gear is the spur gear. Bevel gears and worm gears are also used. Gear types are designed based on one or more of the following elements:

■ The relationship of the shafts: either parallel, intersecting, nonintersecting shafts, or rack and pinion.

■ Manufacturing cost.

■ Ease of maintenance in service.

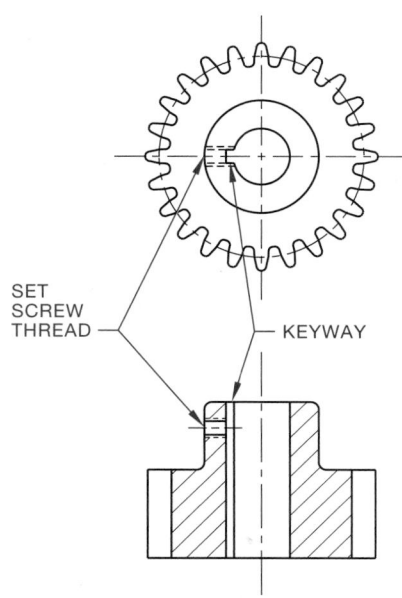

SET
SCREW
THREAD   KEYWAY

FIGURE 16.42 ■ Gear, keyways, and set screw.

■ Smooth and quiet operation.

■ Load-carrying ability.

■ Speed reduction capabilities.

■ Space requirements.

Parallel Shafting Gears

Many different types of mating gears are designed with parallel shafts. These include spur and helical gears.

Spur Gears

There are two basic types of spur gears: external and internal spur gears. (See Figure 16.43.) When two or more spur gears are cut on a single shaft, they are referred to as cluster gears. External spur gears are designed with the teeth of the gear on the outside of a cylinder. External spur gears are the most common type of gear used in manufacturing. Internal spur gears have the teeth on the inside of the cylindrical gear. The advantages of spur gears over other types are their low manufacturing cost, simple design, and ease of maintenance. The disadvantages include less load capacity and higher noise levels than other types.

Helical Gears

Helical gears have their teeth cut at an angle that allows more than one tooth to be in contact. (See Figure 16.44.) Helical gears carry more load than equivalent-sized spur gears and operate more quietly and smoothly. The disadvantage of helical gears is that they develop *end thrust*. End thrust is a lateral force exerted on the end of the gear shaft. Thrust bearings are required to reduce the effect of this end thrust. Double helical gears are designed to eliminate the end thrust and provide long life under heavy loads. However, they are more difficult and costly to manufacture. The herringbone gear shown in Figure 16.45 is a double helical gear without space between the two opposing sets of teeth.

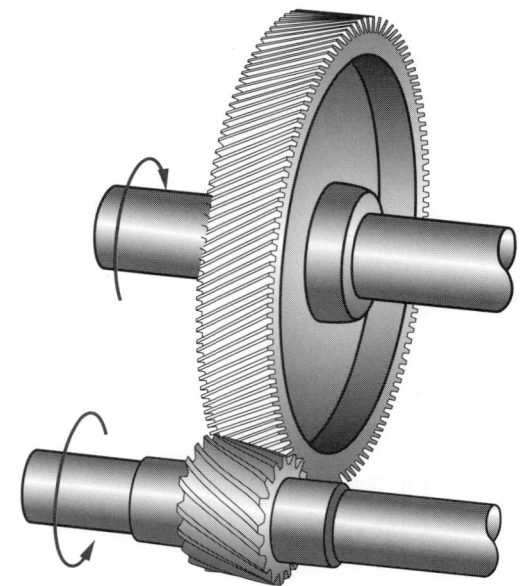

FIGURE 16.44 ■ Helical gear.

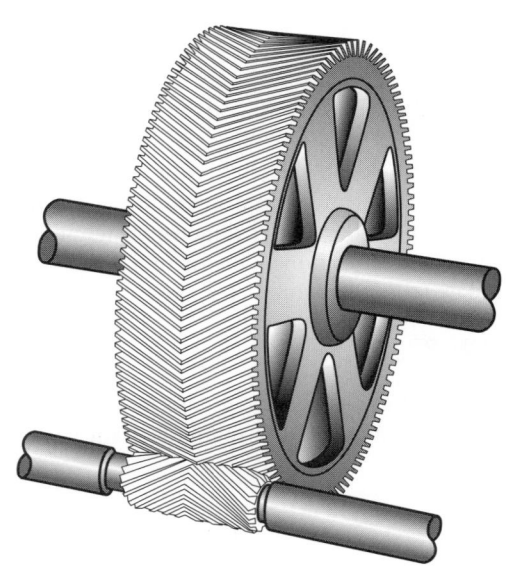

FIGURE 16.45 ■ Herringbone gear.

Intersecting Shafting Gears

Intersecting shafting gears allow for the change in direction of motion from the gear to the pinion. Different types of intersecting shafting gears include bevel and face gears.

Bevel Gears

Bevel gears are conical in shape, allowing the shafts of the gear and pinion to intersect at 90° or any desired angle. The teeth on the bevel gear have the same shape as the teeth on spur gears except they taper toward the apex of the cone. Bevel gears provide for a speed change between the gear and pinion. (See Figure 16.46.) Miter gears are the same as bevel gears except both the gear and pinion are the same size and are used when shafts must intersect at

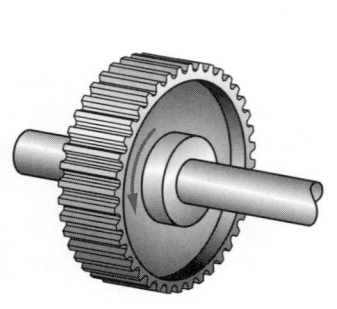

EXTERNAL SPUR
GEAR

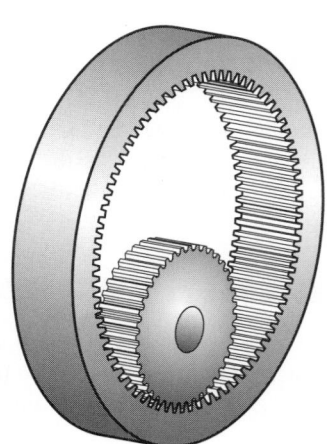

INTERNAL SPUR
GEAR

FIGURE 16.43 ■ Spur gears.

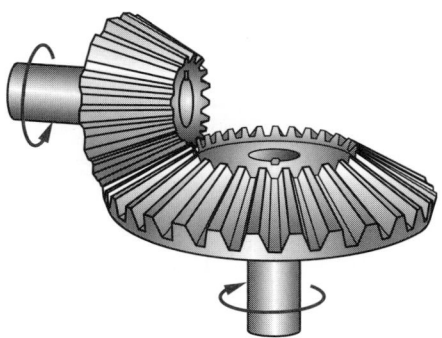

FIGURE 16.46 ■ Bevel gears.

90° without speed reduction. Spiral bevel gears have the teeth cut at an angle, which provides the same advantages as helical gears over spur gears.

Face Gears

The face gear is a combination of bevel gear and spur pinion, or bevel gear and helical pinion. This combination is used when the mounting accuracy is not as critical as with bevel gears. The load-carrying capabilities of face gears is not as good as that of bevel gears.

Nonintersecting Shafting Gears

Gears with shafts that are at right angles but not intersecting are referred to as nonintersecting shafts. Gears that fall into this category are crossed helical, hypoid, and worm gears.

Crossed Helical Gears

Also known as right angle helical or spiral gears, crossed helical gears provide for nonintersecting right angle shafts with low load-carrying capabilities. (See Figure 16.47.)

FIGURE 16.47 ■ Crossed helical gears. *Courtesy Browning Mfg. Division of Emerson Electric Co.*

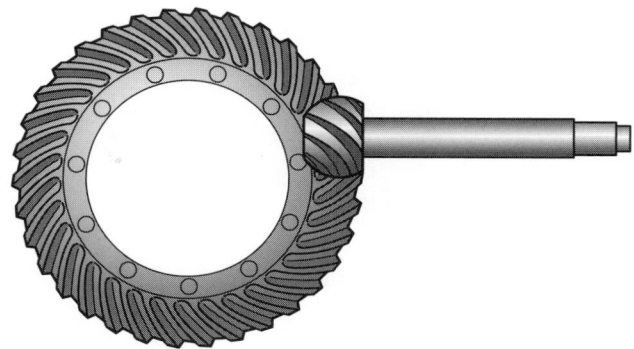

FIGURE 16.48 ■ Hypoid gears.

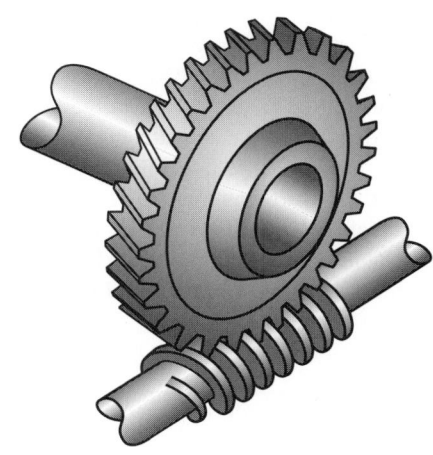

FIGURE 16.49 ■ Worm and worm gear.

Hypoid Gears

Hypoid gears have the same design as bevel gears except the gear shaft axes are offset and do not intersect. (See Figure 16.48.) The gear and pinion are often designed with bearings mounted on both sides for improved rigidity over standard bevel gears. Hypoid gears are very smooth, strong, and quiet in operation.

Worm Gears

A worm and worm gear are shown in Figure 16.49. This type of gear commonly used when a large speed reduction is required in a small space. The worm may be driven in either direction. When the gear is not in operation, the worm automatically locks in place. This is a particular advantage when it is important for the gears to have no movement of free travel when the equipment is shut off.

Rack and Pinion

A rack and pinion is a spur pinion operating on a flat straight bar rack. (See Figure 16.50.) The rack and pinion is used to convert rotary motion into straight-line motion.

SPUR GEAR DESIGN

Spur gear teeth are straight and parallel to the gear shaft axis. The tooth profile is designed to transmit power at a constant rate, and

FIGURE 16.50 ■ Rack and pinion.

with a minimum of vibration and noise. To achieve these require-ments, an *involute curve* is used to establish the gear tooth profile. An involute curve is a spiral curve generated by a point on a chord as it unwinds from the circle. The contour of a gear tooth, based on the involute curve, is determined by a base circle, the diameter of which is controlled by a pressure angle. The *pressure angle* is the direction of push transmitted from a tooth on one gear to a tooth on the mating gear or pinion. (See Figure 16.51.) Two standard pressure angles, 14.5° and 20°, are used in spur gear design. The most commonly used pressure angle is 20° because it provides a stronger tooth for quieter running and heavier load-carrying char-acteristics. One of the basic rules of spur gear design is to have no fewer than 13 teeth on the running gear and 26 teeth on the mat-ing gear.

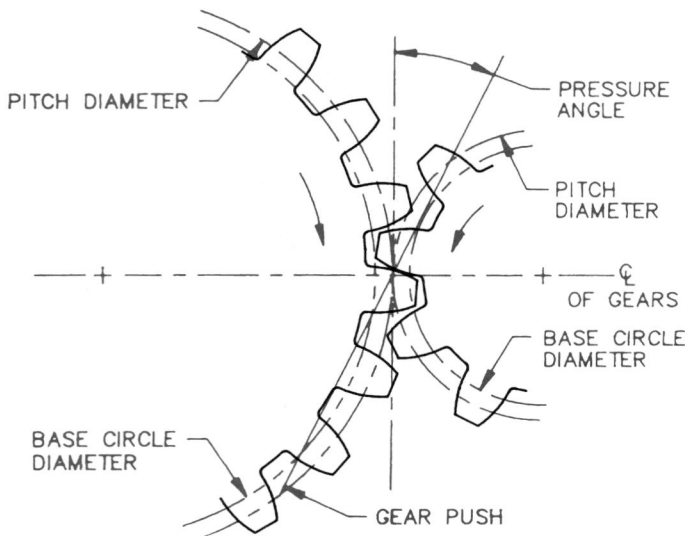

FIGURE 16.51 ■ The spur gear pressure angle and related terminology.

Standard terminology and formulas control the drawing require-ments for spur gear design and specifications. Figure 16.52 shows a pictorial representation of the spur gear teeth with the compo-nents labeled. As an engineering drafter, it is important that you

Term	Description	Formula
Pitch Diameter (D)	The diameter of an imaginary pitch circle on which a gear tooth is designed. Pitch circles of two spur gears are tangent.	$D = N/P$
Diametral Pitch (P)	A ratio equal to the number of teeth on a gear per inch of pitch diameter.	$P = N/D$
Number of Teeth (N)	Number of teeth on a gear.	$N = D \times P$
Circular Pitch (p)	The distance from a point on one tooth to the corresponding point on the adjacent tooth, measured on the pitch circle.	$p = 3.1416 \times D/N$ $p = 3.1416/P$
Center Distance (C)	The distance between the axis of two mating gears	$C =$ sum of pitch DIA/2
Addendum (a)	The radial distance from the pitch circle to the top of the tooth.	$a = 1/P$
Dedendum (b)	The radial distance from the pitch circle to the bottom of the tooth. (This formula is for 20° teeth only.)	$b = 1.250/P$
Whole Depth (h_t)	The full height of the tooth. It is equal to the sum of the addendum and the dedendum.	$h_t = a + b$ $h_t = 2.250/P$
Working Depth (h_k)	The distance that a tooth occupies in the mating space. A distance equal to two times the addendum.	$h_k = 2a$ $h_k = 2.000/P$
Clearance (c)	The radial distance between the top of a tooth and the bottom of the mating tooth space. It is also the difference between the addendum and dedendum.	$c = b - a$ $c = .250/P$
Outside Diameter (D_o)	The overall diameter of the gear. It is equal to the pitch diameter plus two dedendums.	$D_o = D + 2b$
Root Diameter (D_r)	The diameter of a circle coinciding with the bottom of the tooth spaces.	$D_r = D - 2b$
Circular Thickness (t)	The length of an arc between the two sides of a gear tooth measured on the pitch circle.	$t = 1.5708/P$
Chordal Thickness (t_c)	The straight line thickness of a gear tooth measured on the pitch circle.	$t_c = D \sin (90°/N)$
Chordal Addendum (a_c)	The height from the top of the tooth to the line of the chordal thickness.	$a_c = a + t^2/4D$
Pressure Angle (∅)	The angle of direction of pressure between contacting teeth. It determines the size of the base circle and the shape of the involute teeth.	
Base Circle Diameter (D_B)	The diameter of a circle from which the involute tooth form is generated.	$D_B = D \cos ∅$

FIGURE 16.52 ■ Gear terminology and formulas.

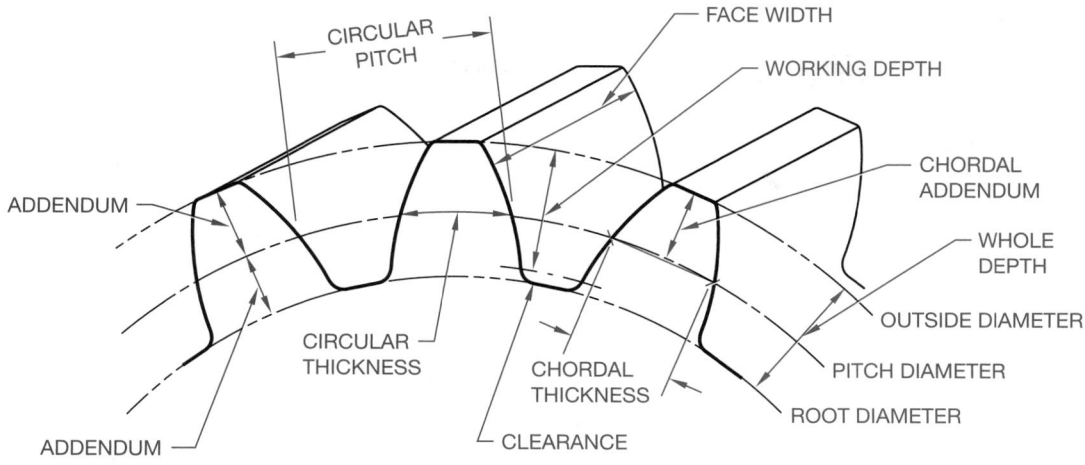

FIGURE 16.52 (continued)

become familiar with the terminology and associated mathematical formulas used to calculate values.

Diametral Pitch

The diametral pitch actually refers to the tooth size and has become the standard for tooth size specifications. As you look at Figure 16.53, notice how the tooth size increases as the diametral pitch decreases. One of the most elementary rules of gear tooth design is that mating teeth must have the same diametral pitch.

GEAR ACCURACY
AGMA

A system has been established by the American Gear Manufacturers Association (AGMA) for the classification of gears based on the accuracy of the maximum tooth-to-tooth and total composite tolerances allowed. This number is called the AGMA *quality number.* The AGMA quality numbers and corresponding maximum tolerances are established by diametral pitch and pitch diameter. The AGMA quality numbers are listed in *AGMA Gear Handbook, 2000-A88.* The higher the AGMA quality number, the tighter the toler-

ance. For example, an AGMA Q6 allows approximately .004 total composite error, Q10 is less than .001, and Q13 is less than .0004. The *AGMA Gear Handbook* also displays a list of gear applications and the quality number suggested for each application.

DRAWING SPECIFICATIONS AND TOLERANCES

The final step in designing a gear is the presentation of a drawing that displays the dimensioned gear using multiviews and gear data charts. While companies do not always provide the complete gear data charts on their drawings, errors can happen when complete data is not provided. Gear data formulas and sample gear data charts are shown with examples in this chapter. However, the AGMA recommended information for spur and helical gears is also in Appendix I.

DESIGNING AND DRAWING SPUR GEARS

ANSI The drafting standard that governs gear drawings is the American National Standards document ANSI Y14.7.1, *Gear Drawing Standards—Part I.*

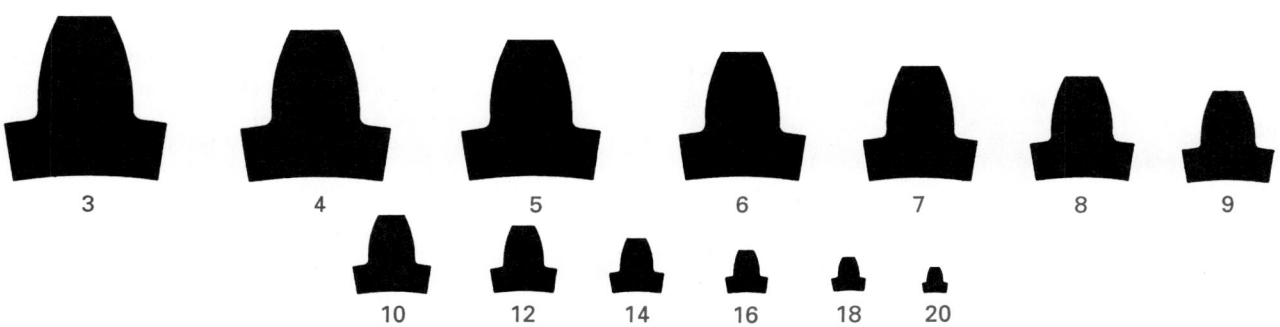

FIGURE 16.53 ■ Diametral pitch.

GEAR TEETH

There are custom CADD programs that allow the gear teeth to be drawn automatically. To do this, the computer prompts the drafter for variables such as pitch diameter, diametral pitch, pressure angle, and number of teeth. The program then automatically notifies the drafter whether the information is accurate, provides all additional data, and creates a detail drawing of the gear. CADD was used to draw the teeth in the gear displayed in Figure 16.54. The CADD drafter may easily draw gears displayed in detailed representation, or use the simplified technique to save regeneration and plotting time.

Some software programs do more than assist in the design process. For example, objects such as gear teeth can be subjected to simulated tests and stress analysis on the computer screen as shown in Figure 16.55.

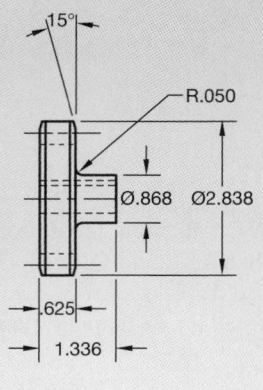
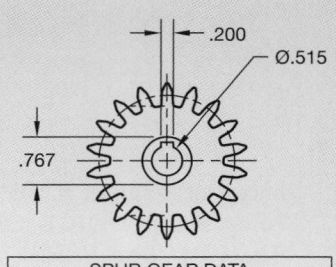

SPUR GEAR DATA	
DIAMETRAL PITCH	8
NUMBER OF TEETH	18
PRESSURE ANGLE	20°
PITCH DIAMETER	2.250
BASE CIRCLE DIAMETER	2.114
CIRCULAR PITCH	.393
CIRCULAR THICKNESS	.196
ROOT DIAMETER	1.9375

5. PROFILE TOLERANCE .003
4. PITCH TOLERANCE .003
3. ALL TOOTH ELEMENT SPECIFICATIONS ARE FROM DATUM A
2. INTERPRET GEAR DATA PER ANSI Y14.7.1.
1. INTERPRET DIMENSIONS AND TOLERANCES PER ASME Y14.5M.
NOTES:

FIGURE 16.54 ■ Gear detailed drawing using CADD to automatically draw teeth and gear data chart from given specifications.

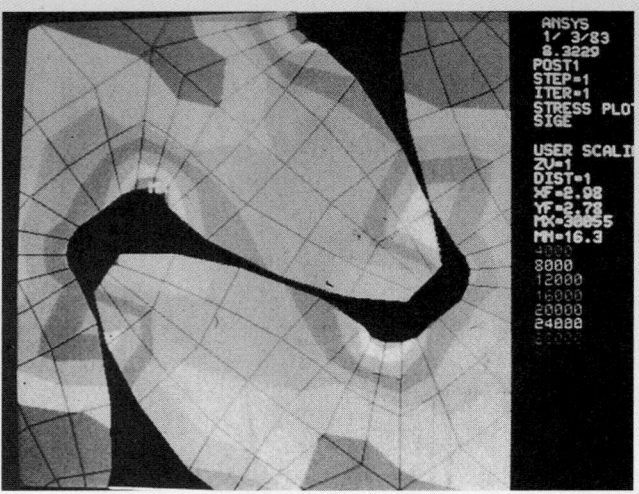

FIGURE 16.55 ■ Objects such as these gear teeth can be subjected to simulated tests and stress analysis on the computer screen. *Courtesy Swanson Analysis Systems, Inc.*

Because gear teeth are complex and time-consuming to draw, simplified representations are used to make the practice easier. (See Figure 16.56.) The simplified method shows the outside diameter and the root diameters as phantom lines and the pitch diameter as a centerline in the circular view. In addition, a side view is often required to show width dimensions and related features. (See Figure 16.56a.) If the gear construction has webs, spokes, or other items that require further clarification, then a full section is normally used. (See Figure 16.56b.) Notice in the cross section that the gear tooth is left unsectioned and the pitch diameter is shown as a centerline.

When cluster gears are drawn, the circular view may show both sets of gear tooth representations in simplified form, or two circular views may be drawn. (See Figure 16.57.) When cluster gears are more complex than those shown here, multiple views and removed sections may be required.

One or more teeth may be drawn for specific applications. For example, when a tooth must be in alignment with another feature of the gear, the tooth may be drawn as shown in Figure 16.58.

Gear drawings typically have a chart that shows the manufacturing information associated with the teeth, and with related part detail dimensions placed on the specific views.

DESIGNING SPUR GEAR TRAINS

A gear train is an arrangement of two or more gears connecting driving and driven parts of a machine. Gear reducers and transmissions are examples of gear trains. The function of a gear train is to:

■ Transmit motion between shafts.

■ Decrease or increase the speed between shafts.

■ Change the direction of motion.

It is important for you to understand the relationship between two mating gears in order to design gear trains. When gears are designed, the end result is often a specific *gear ratio*. Any two

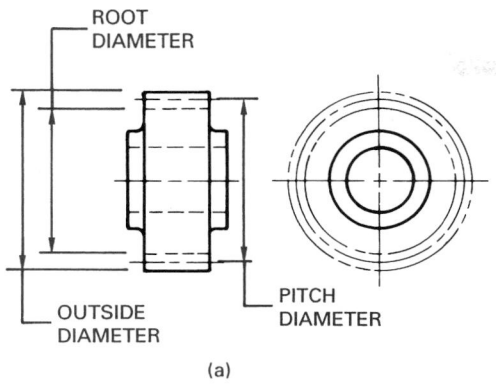

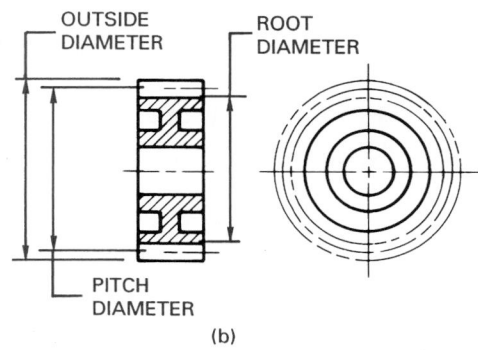

FIGURE 16.56 ■ Typical spur gear drawings using simplified gear teeth representation.

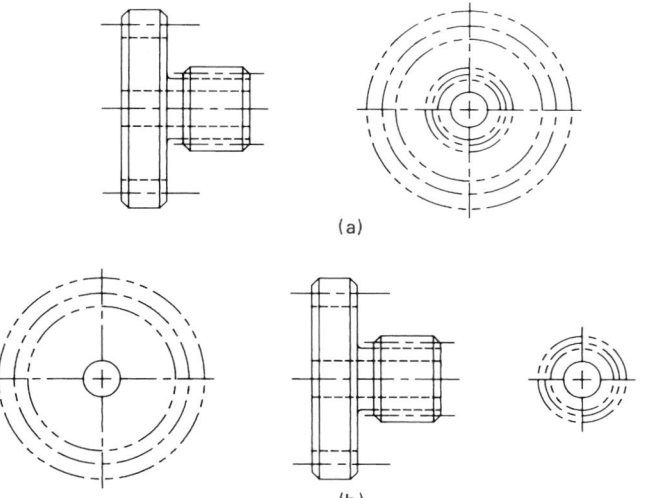

FIGURE 16.57 ■ Cluster gear drawings.

gears in mesh have a gear ratio. The gear ratio is expressed as a proportion, such as 2:1 or 4:1, between two similar values. The gear ratio between two gears is the relationship between the following characteristics:

■ Number of teeth.

■ Pitch diameters.

■ Revolutions per minute (rpm).

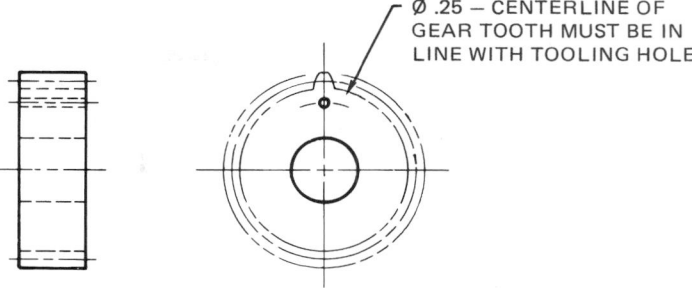

FIGURE 16.58 ■ Showing the relationship of one gear tooth to another feature on the gear.

If you have gear A (pinion) mating with gear B, as shown in Figure 16.59, the gear ratio is calculated by dividing like values of the smaller gear into the larger gear as follows:

$$\frac{\text{Number Teeth}_{\text{Gear B}}}{\text{Number Teeth}_{\text{Gear A}}} = \text{Gear Ratio}$$

$$\frac{\text{Pitch Diameter}_{\text{Gear B}}}{\text{Pitch Diameter}_{\text{Gear A}}} = \text{Gear Ratio}$$

$$\frac{\text{rpm}_{\text{Gear A}}}{\text{rpm}_{\text{Gear B}}} = \text{GearRatio}$$

Now, calculate the gear ratio for the two mating gears in Figure 16.59 if gear A has 18 teeth, 6 in. pitch diameter, and operates at 1200 rpm, and gear B has 54 teeth, 18 in. pitch diameter, and operates at 400 rpm:

Number of Teeth	=	54/18	= 3:1
Pitch Diameter	=	18/6	= 3:1
rpm	=	1200/400	= 3:1

You can solve for unknown values in the gear train if you know the gear ratio you want to achieve, the number of teeth and pitch diameter of one gear, and the input speed. For example, gear A has 18 teeth, a pitch diameter of 6 in., and an input speed of 1200 rpm, and the ratio between gear A and gear B is 3:1. In order to keep this information well organized, it is recommended that you set up a chart similar to the one shown in Figure 16.59. The unknown values are shown in color for your reference. Determine the number of teeth for gear B as follows:

$$\text{Teeth}_{\text{Gear A}} \times \text{Gear Ratio} = 18 \times 3 = \text{Teeth}_{\text{Gear B}} = 54$$

Or, if you know that gear B has 54 teeth and the gear ratio is 3:1, then:

$$\frac{\text{Teeth}_{\text{Gear B}}}{\text{Gear Ratio}} = \frac{54}{3} = \text{Teeth}_{\text{Gear A}} = 18$$

Determine the rpm of gear B:

$$\frac{\text{rpm}_{\text{Gear A}}}{\text{Gear Ratio}} = \frac{1200}{3} = \text{rpm}_{\text{Gear B}} = 400$$

Determine the pitch diameter of gear B:

$$\text{Pitch Diameter}_{\text{Gear A}} \times \text{Gear Ratio} = 6 \text{ in.} \times 3 = \text{Pitch Diameter}$$

$$\text{Gear B} = 18 \text{ in.}$$

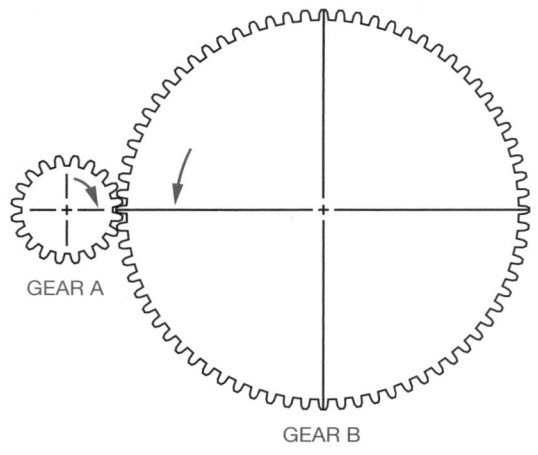

GEAR A

GEAR B

	(D) PITCH DIAMETER	(N) NUMBER OF TEETH	(P) DIAMETRAL PITCH	RPM	DIRECTION
GEAR A	6	18	3	1200	C.WISE
GEAR B	18	54	3	400	C.C.WISE

FIGURE 16.59 ■ Calculating gear data.

In some situations it is necessary for you to refer to the formulas given in Figure 16.52 to determine some unknown values. In this case it is necessary to calculate the diametral pitch using the formula P = N/D, where P = diametral pitch, N = number of teeth, and D = pitch diameter:

$$P_{Gear\ A} = \frac{N}{D} = \frac{18}{6} = 3$$

Because the diametral pitch is the tooth size and the teeth for mating gears must be the same size, the diametral pitch of gear B is also 3.

Keep in mind that the preceding example presents only one set of design criteria. Other situations may be different. Always solve for unknown values based on information that you have and work with the standard formulas presented in this chapter. Following are some important points to keep in mind as you work with the design of gear trains:

■ The rpm of the larger gear is always slower than the rpm of the smaller gear.

■ Mating gears always turn in opposite directions.

■ Gears on the same shaft (cluster gears) always turn in the same direction and at the same speed (rpm).

■ Mating gears have the same size teeth (diametral pitch).

■ The gear ratio between mating gears is a ratio between the number of teeth, the pitch diameters, and the rpms.

■ The distance between the shafts of mating gears is equal to $1/2D_{Gear\ A} + 1/2D_{Gear\ B}$. This distance is 3 + 9 = 12 in. between shafts in Figure 16.59.

DESIGNING AND DRAWING THE RACK AND PINION

A rack is a straight bar with spur teeth used to conver rotary motion to reciprocating motion. Figure 16.60 shows a spur gear pinion mating with a rack. Notice the circular dimensions of the pinion become linear dimensions on the rack (Linear Pitch = Circular Pitch).

When preparing a detailed drawing of the rack, the front view is normally shown with the profile of the first and last tooth drawn. Phantom lines are drawn to represent the top and root and a centerline is used for the pitch line. (See Figure 16.61.) Related tooth and length dimensions are also placed on the front view, and a depth dimension is placed on the side view. A chart is then placed in the field of the drawing to identify specific gear-cutting information.

DESIGNING AND DRAWING BEVEL GEARS

Bevel gears are usually designed to transmit power between intersecting shafts at 90°, although they may be designed for any angle. Some gear design terminology and formulas relate specifically to the construction of bevel gears. These formulas and a drawing of a bevel gear and pinion are shown in Figure 16.62. Most of the gear terms discussed for spur gears apply to bevel gears. In many cases information must be calculated using the formulas provided in the spur gear discussion shown in Figure 16.52.

The drawing for a bevel gear is similar to the drawing for a spur gear, but in many cases only one view is needed. This view is often a full section. Another view may be necessary to show the dimensions for a keyway. As with other gear drawings, a chart is used to specify gear-cutting data. (See Figure 16.63.) Notice in this example that only a partial side view is used to specify the bore and keyway.

DESIGNING AND DRAWING WORM GEARS

Worm gears are used to transmit power between nonintersecting shafts, and they are used for large speed reductions in a small space

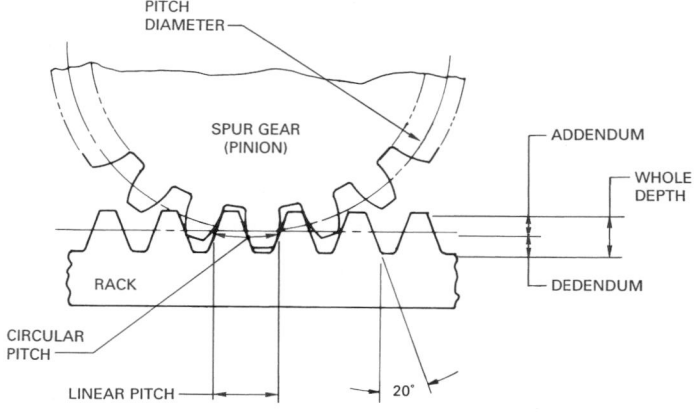

FIGURE 16.60 ■ Rack and pinion terminology.

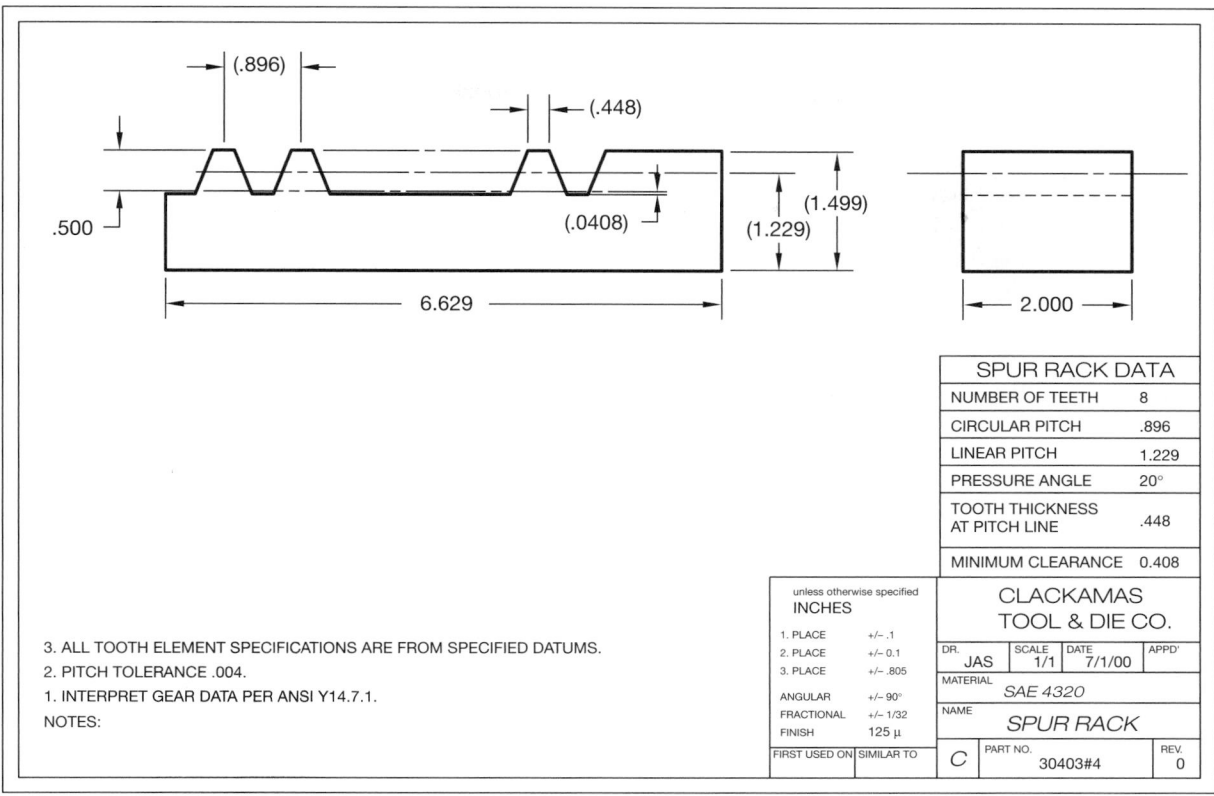

SPUR RACK DATA	
NUMBER OF TEETH	8
CIRCULAR PITCH	.896
LINEAR PITCH	1.229
PRESSURE ANGLE	20°
TOOTH THICKNESS AT PITCH LINE	.448
MINIMUM CLEARANCE	0.408

3. ALL TOOTH ELEMENT SPECIFICATIONS ARE FROM SPECIFIED DATUMS.
2. PITCH TOLERANCE .004.
1. INTERPRET GEAR DATA PER ANSI Y14.7.1.
NOTES:

unless otherwise specified INCHES	
1. PLACE	+/– .1
2. PLACE	+/– 0.1
3. PLACE	+/– .805
ANGULAR	+/– 90°
FRACTIONAL	+/– 1/32
FINISH	125 µ

CLACKAMAS TOOL & DIE CO.

DR. JAS	SCALE 1/1	DATE 7/1/00	APPD'

MATERIAL   *SAE 4320*

NAME   *SPUR RACK*

| FIRST USED ON | SIMILAR TO | C | PART NO. 30403#4 | REV. 0 |

FIGURE 16.61 ■ A detailed drawing of a rack.

as compared to other types of gears. Worm gears are also strong, move in either direction, and lock in place when the machine is not in operation. A single lead worm advances one pitch with every revolution. A double lead worm advances two pitches with each revolution. As with bevel gears, the worm gear and worm have specific terminology and formulas that apply to their design. The representative worm gear and worm technology and design formulas are shown in Figure 16.64.

When drawing the worm gear, the same techniques are used as discussed earlier. Figure 16.65, page 532 shows a worm gear drawing using a half-section method. A side view is provided to dimension the bore and keyway, and a chart is given for gear-cutting data.

A detailed drawing of the worm is prepared using two views. The front view shows the first gear tooth on each end with phantom lines between for a simplified representation. The side view is the same as a spur gear drawing with the keyway specifications. The gear-cutting chart is then placed on the field of the drawing or over the title block. Figure 16.66, page 532 shows the worm drawing using CADD to provide a detailed representation of the worm teeth in the front view rather than phantom lines as in the simplified representation.

PLASTIC GEARS

The following is taken in part from *Plastic Gearing* by William McKinlay and Samuel D. Pierson, published by ABA/PGT Inc., Manchester, CT. Gears can be molded of many engineering plastics in various grades and in filled varieties. Filled plastics are those in

which a material has been added to improve the mechanical properties. The additives normally used in gear plastics are glass, polytetrafluoroethylene (PTFE), silicones, and molybdenum disulphide. Glass fiber reinforcement can double the tensile strength and reduce the thermal expansion. Carbon fiber is often used to increase strength. Silicones, PTFE, and molybdenum disulphide are used to act as built-in lubricants and provide increased wear resistance. Plastic gears are designed in the same manner as gears made from other materials. However, the physical characteristics of plastics make it necessary to follow gear design practices more closely than when designing gears that are machined from metals.

Advantages of Molded Plastic Gears

Gears molded of plastic are replacing stamped and cut metal gears in a wide range of mechanisms. Designers are turning to molded plastic gears for one or more of the following reasons:

■ Reduced cost.

■ Increased efficiency.

■ Self-lubrication.

■ Increased tooth strength with nonstandard pressure angles.

■ Reduced weight.

■ Corrosion resistance.

■ Less noise.

■ Available in colors.

Term	Description	Formula
Pitch Diameter (D)	The diameter of the base of the pitch cone.	$D = N/P$
Pitch Cone	In Figure 16.64, the pitch cone is identified as XYZ.	$\tan \varnothing_a = D_a/D_b$
Pitch Angle (∅)	The angle between an element of a pitch cone and its axis. Pitch angles of mating gears depend on their relative diameters (gear a and gear b).	$\tan \varnothing_b = D_a/D_b$
Cone Distance (A)	Slant height of pitch cone.	$A = D/2 \sin \varnothing$
Addendum Angle (δ)	The angle subtended by the addendum. It is the same for mating gears.	$\tan \delta = a/A$
Dedendum Angle (Ω)	The angle subtended by the dedendum. It is the same for mating gears.	$\tan \Omega = b/A$
Face Angle (∅o)	The angle between the top of the teeth and the gear axis.	$\varnothing_o = \varnothing - 8$
Root Angle (∅r)	The angle between the bottom of the tooth space and the gear axis.	$\varnothing_r = \varnothing - \Omega$
Outside Diameter (Do)	The diameter of the outside circle of gear.	$D_o = D + 2a \, (\cos\varnothing)$
Crown Height (X)	The distance between the cone apex and the outer tip of the gear teeth.	$X = .5(D_o)/\tan\varnothing_o$
Crown Backing (Y)	The distance from the rear of the hub to the outer tip of the gear tooth, measured parallel to the axis of the gear.	
Face Width	A distance that should not exceed one-third of the cone distance (A).	

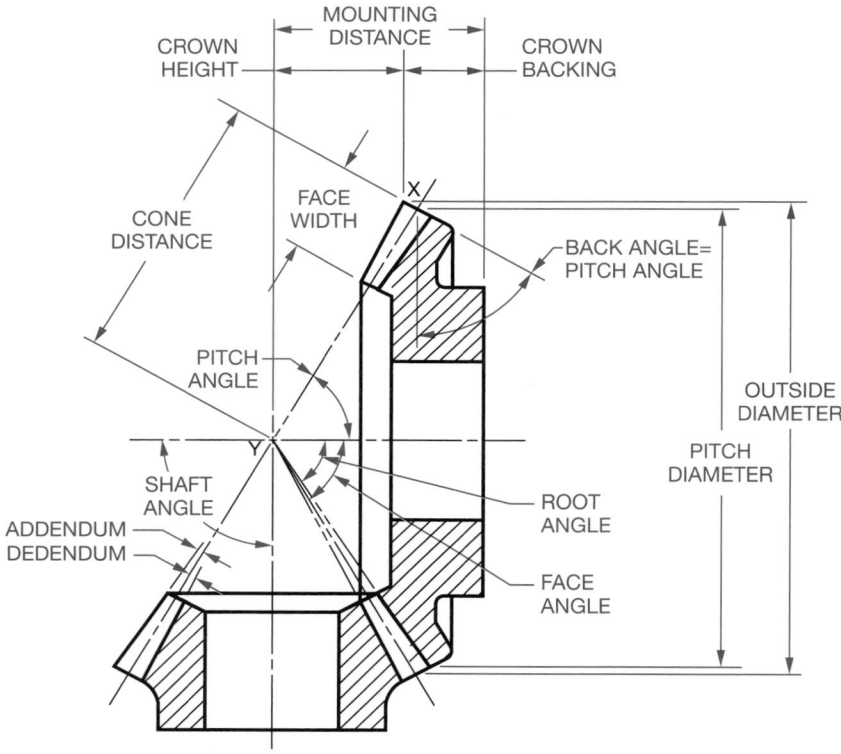

FIGURE 16.62 ■ Bevel gear terminology and formulas.

Disadvantages of Molded Plastic Gears

Plastic gearing has the following limitations when compared with metal gearing:

■ Lower strength.

■ Greater thermal expansion and contraction.

■ Limited heat resistance.

■ Size change with moisture absorption.

Accuracy of Molded Plastic Gears

Today's technology permits a very high degree of precision. In general, tooth-to-tooth composite tolerances can be economically held to .0005 or less for fine pitch gears. Total composite tolerance varies depending on configuration, evenness of product cross section, and the selection of the molding material.

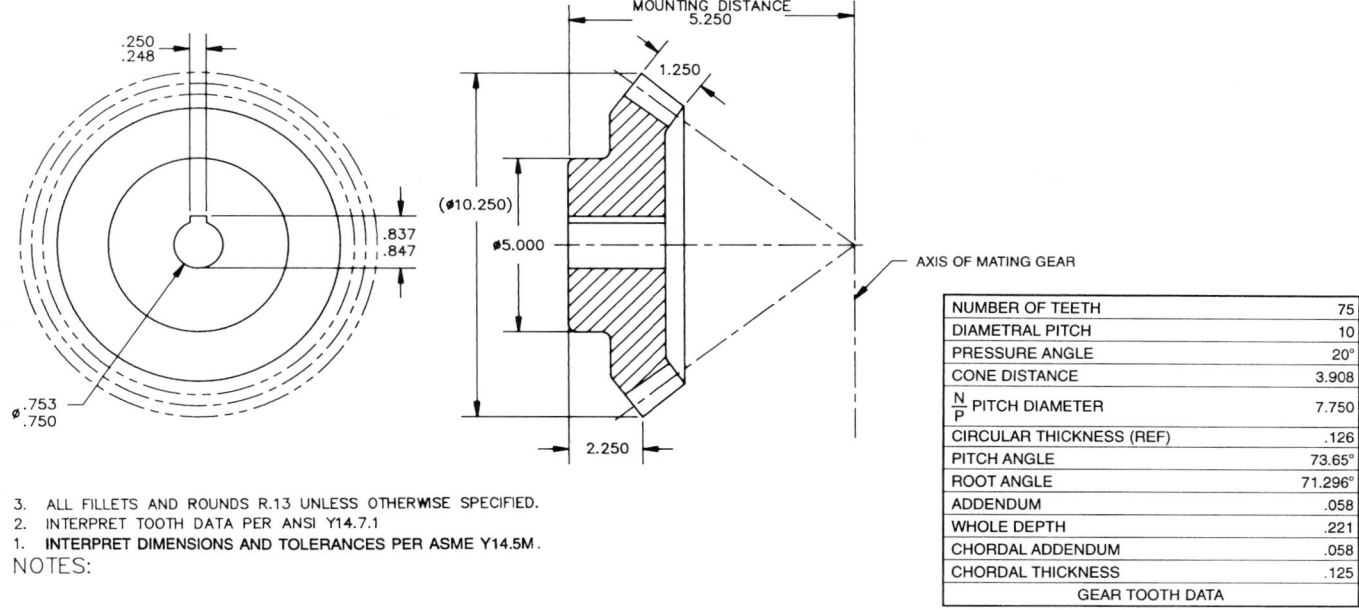

GEAR TOOTH DATA	
NUMBER OF TEETH	75
DIAMETRAL PITCH	10
PRESSURE ANGLE	20°
CONE DISTANCE	3.908
$\frac{N}{P}$ PITCH DIAMETER	7.750
CIRCULAR THICKNESS (REF)	.126
PITCH ANGLE	73.65°
ROOT ANGLE	71.296°
ADDENDUM	.058
WHOLE DEPTH	.221
CHORDAL ADDENDUM	.058
CHORDAL THICKNESS	.125

3. ALL FILLETS AND ROUNDS R.13 UNLESS OTHERWISE SPECIFIED.
2. INTERPRET TOOTH DATA PER ANSI Y14.7.1
1. **INTERPRET DIMENSIONS AND TOLERANCES PER ASME Y14.5M.**
NOTES:

FIGURE 16.63 ■ A detailed drawing of a bevel gear.

Term	Description	Formula
Pitch Diameter (worm) (D_w)		$D_w = 2C - D_g$
Pitch Diameter (gear) (D_g)		$D_g = 2C - D_w$
Pitch (P)	The distance from one tooth to the corresponding point on the next tooth measured parallel to the worm axis. It is equal to the circular pitch on the worm gear.	$P = L/T$
Lead (L)	The distance the thread advances axially in one revolution of the worm.	$L = D_g/R$ $L = P \times T$
Threads (T)	Number of threads or starts on worm.	$T = L/P$
Gear Teeth (N)	Number of teeth on worm gear.	$N = \pi D_g/P$
Ratio (R)	Divide number of gear teeth by the number of worm threads.	$R = N/T$
Addendum (a)	For single and double threads.	$a = .318P$
Whole Depth (WD)	For single and double threads.	$WD = .686P$

NOTE: REFER TO THE FORMULAS ON PAGE 524 FOR ADDITIONAL CALCULATIONS.

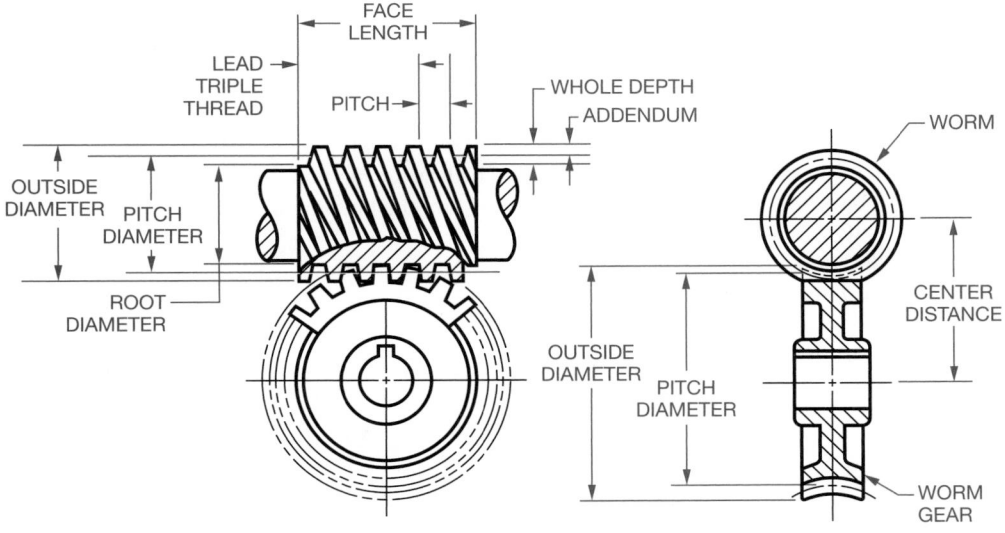

FIGURE 16.64 ■ Worm and worm gear terminology and formulas.

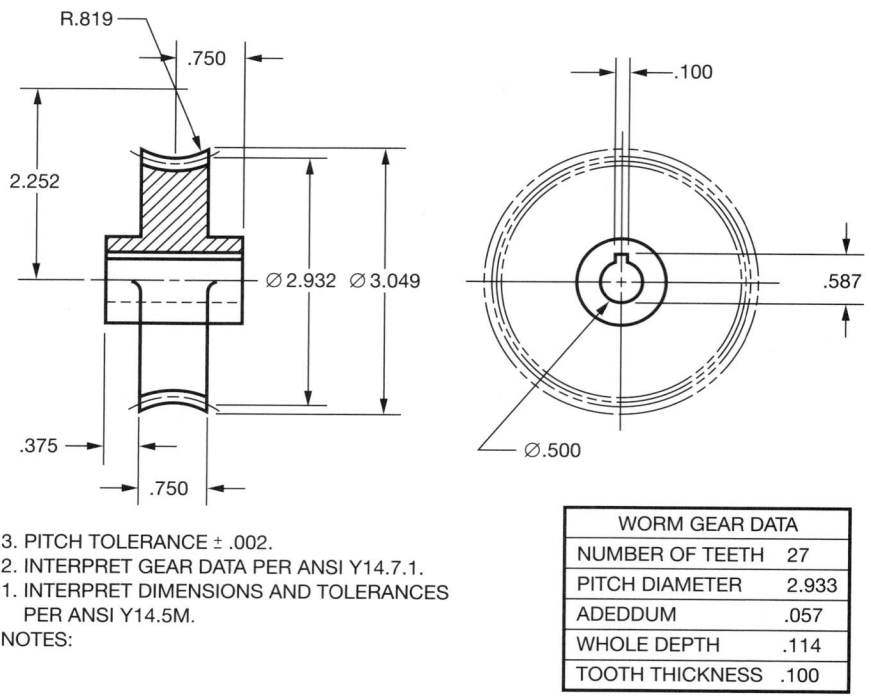

WORM GEAR DATA	
NUMBER OF TEETH	27
PITCH DIAMETER	2.933
ADEDDUM	.057
WHOLE DEPTH	.114
TOOTH THICKNESS	.100

3. PITCH TOLERANCE ± .002.
2. INTERPRET GEAR DATA PER ANSI Y14.7.1.
1. INTERPRET DIMENSIONS AND TOLERANCES
   PER ANSI Y14.5M.
NOTES:

FIGURE 16.65 ■ A detailed drawing of a worm gear.

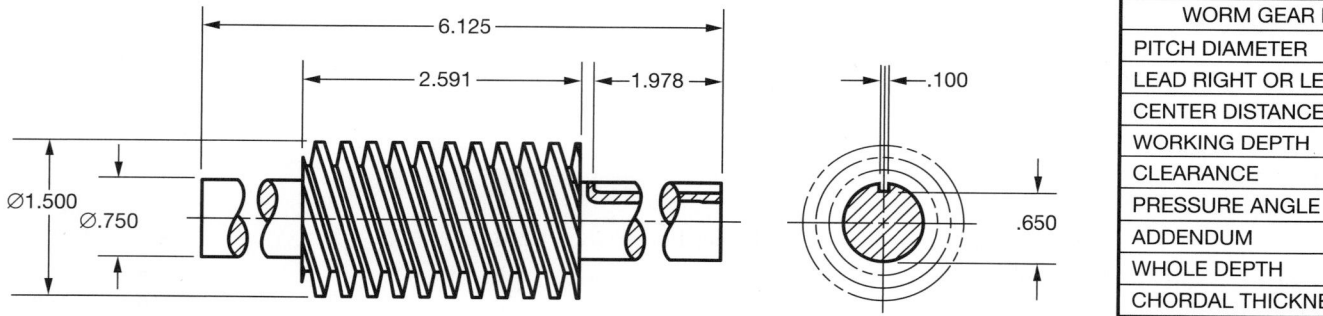

WORM GEAR DATA	
PITCH DIAMETER	1.251
LEAD RIGHT OR LEFT	RIGHT
CENTER DISTANCE	1.858
WORKING DEPTH	.218
CLEARANCE	.0312
PRESSURE ANGLE	20°
ADDENDUM	.125
WHOLE DEPTH	.25
CHORDAL THICKNESS	.125

2. PITCH TOLERANCE ± .002.
1. INTERPRET GEAR DATA PER ANSI Y14.7.1.
NOTES:

FIGURE 16.66 ■ A detailed drawing of a worm.

BEARINGS

Bearings are mechanical devices used to reduce friction between two surfaces. They are divided into two large groups known as plain and rolling element bearings. Bearings are designed to accommodate either rotational or linear motion. Rotational bearings are used for radial loads, and linear bearings are designed for thrust loads. Radial loads are loads that are distributed around the shaft. Thrust loads are lateral. Thrust loads apply force to the end of the shaft. Figure 16.67 shows the relationship between rotational and linear motion.

Plain Bearings

Plain bearings are often referred to as sleeve, journal bearings, or bushings. Their operation is based on a sliding action between mating parts. A clearance fit between the inside diameter of the

bearing and the shaft is critical to ensure proper operation. Refer to fits between mating parts in Chapter 12 for more information. The bearing has an interference fit between the outside of the bearing and the housing or mounting device as shown in Figure 16.68.

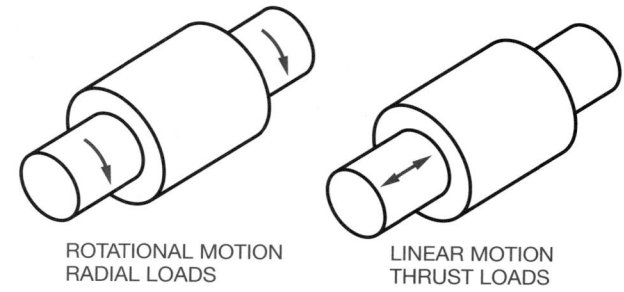

ROTATIONAL MOTION
RADIAL LOADS

LINEAR MOTION
THRUST LOADS

FIGURE 16.67 ■ Radial and thrust loads.

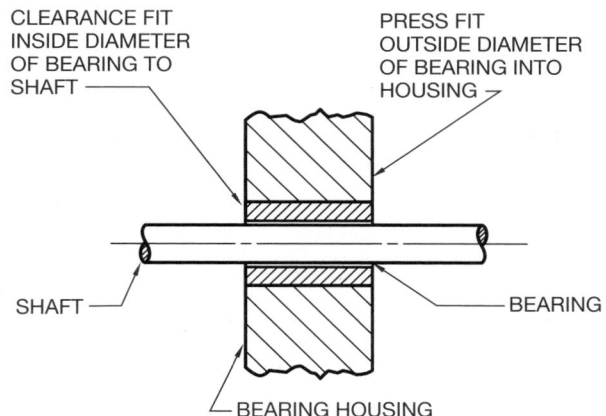

FIGURE 16.68 ■ Plain bearing terminology and fits.

The material from which plain bearings are made is important. Most plain bearings are made from bronze or phosphor bronze. Bronze bearings are normally lubricated, while phosphor bronze bearings are commonly impregnated with oil and require no additional lubrication. Phosphor bronze is an excellent choice when antifriction qualities are important and where resistance to wear and scuffing are needed.

Rolling Element Bearings

Ball and roller bearings are the two classes of rolling element bearings. Ball bearings are the most commonly used rolling element bearings. In most cases, ball bearings have higher speed and lower load capabilities than roller bearings. Even so, ball bearings are manufactured for most uses. Ball bearings are constructed with two grooved rings, a set of balls placed radially around the rings, and a separator that keeps the balls spaced apart and aligned as shown in Figure 16.69.

Single-row ball bearings are designed primarily for radial loads, but they can accept some thrust loads. Double-row ball bearings may be used where shaft alignment is important. Angular contact ball bearings support a heavy thrust load and a moderate radial load. Thrust bearings are designed for use in thrust load situations only. When both thrust and radial loads are necessary, both radial

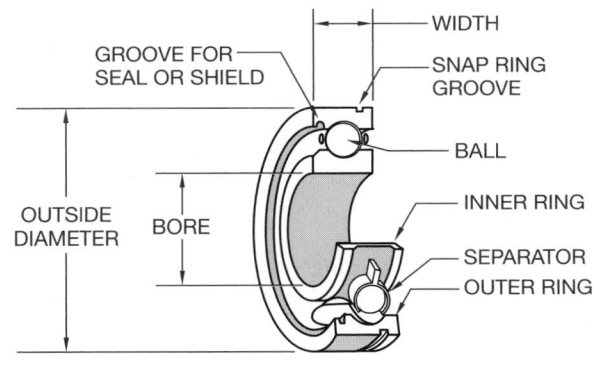

FIGURE 16.69 ■ Ball bearing components.

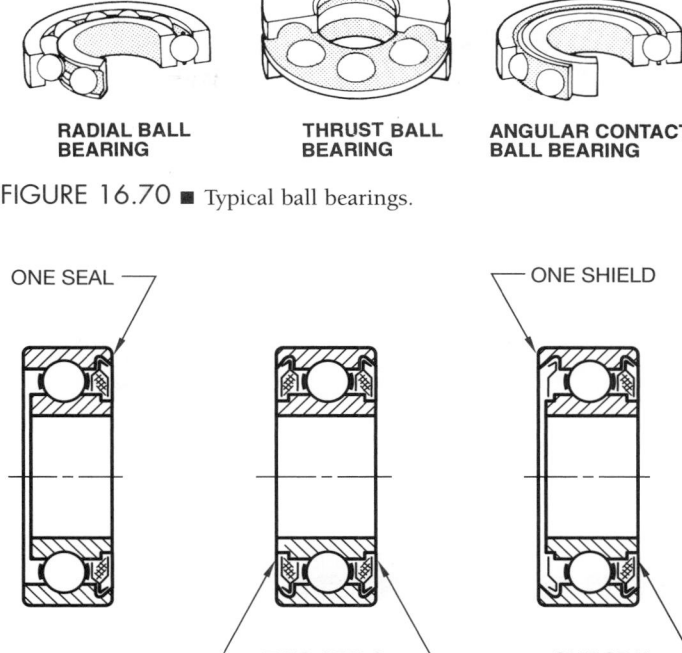

FIGURE 16.70 ■ Typical ball bearings.

FIGURE 16.71 ■ Bearing seals and shields.

and thrust ball bearings are used together. Some typical ball bearings are shown in Figure 16.70.

Ball bearings are available with shields and seals. A shield is a metal plate on one or both sides of the bearing. The shields act to keep the bearing clean and retain the lubricant. A sealed bearing has seals made of rubber, felt, or plastic placed on the outer and inner rings of the bearing. The sealed bearings are filled with special lubricant by the manufacturer. They require little or no maintenance in service. Figure 16.71 shows the shields and seals used on ball bearings.

Roller bearings are more effective than ball bearings for heavy loads. Cylindrical roller bearings have a high radial capacity and assist in shaft alignment. Needle roller bearings have small rollers and are designed for the highest load-carrying capacity of all rolling element bearings with shaft sizes under 10 in. Tapered roller bearings are used in gear reducers, steering mechanisms, and machine tool spindles. Spherical roller bearings offer the best combination of high load capacity, tolerance to shock, and alignment, and are used on conveyors, transmissions, and heavy machinery. Some common roller bearings are displayed in Figure 16.72.

DRAWING BEARING SYMBOLS

Only engineering drafters who work for a bearing manufacturer normally make detailed drawings of roller or ball bearings. In most other industries, bearings are not drawn because they are purchase parts. When bearings are drawn as a representation on assembly drawings or product catalogs, they are displayed using symbols.

CYLINDRICAL ROLLER
BEARING

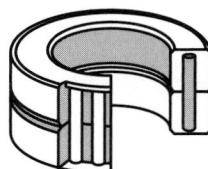

NEEDLE ROLLER
BEARING

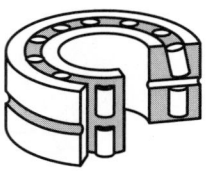

SPHERICAL ROLLER
BEARING

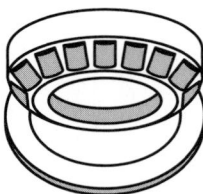

TAPERED ROLLER
BEARING

FIGURE 16.72 ■ Typical roller bearings.

BEARING CODES

Bearing manufacturers use similar coding systems for the identification and ordering of different bearing products. The bearing codes generally contain the following type of information:

- Material.
- Bearing type.
- Bore size.

CADD APPLICATIONS

BEARING SYMBOLS

When CADD is used, the bearing symbols may be drawn and saved in a symbols library for immediate use at any time. This use of CADD helps increase productivity and accuracy over other drafting techniques. (See Figure 16.73.)

BALL BEARINGS			ROLLER BEARINGS			
RADIAL	ANGULAR CONTACT	THRUST	CYLINDRICAL	SPHERICAL	TAPERED	NEEDLE

FIGURE 16.73 ■ Bearing symbols drawn using CADD.

- Lubricant.
- Type of seals or shields.

A sample bearing numbering system is shown in Figure 16.74.

BEARING SELECTION

A variety of bearing types are available from manufacturers. Bearing design may differ depending on the use requirements. For example, a supplier may have light, medium, and heavy bearings available. Bearings may have specially designed outer and inner rings. Bearings are available open (without seals or shields), or with one or two shields or seals. Light bearings are generally designed to accommodate a wide range of applications involving light to medium loads combined with relatively high speeds. Medium bearings have heavier construction than light bearings and provide a greater radial and thrust capacity. They are also able to withstand greater shock than light bearings. Heavy bearings are often designed for special service where extra heavy shock loads are required. Bearings may also be designed to accommodate radial loads, thrust loads, or a combination of loading requirements.

Bearing Bore, Outside Diameters, and Width

Bearings are dimensioned in relation to the bore diameter, outside diameter, and width. These dimensions are shown in Figure 16.75. After the loading requirements have been established, the bearing is selected in relationship to the shaft size. For example, if an approximate ∅1.5 inch shaft size is required for a medium service bearing, then a vendor's catalog chart similar to the one shown in Figure 16.76 is used to select the bearing.

Referring to the chart shown in Figure 16.76, notice that the first column is the vendor's bearing number, followed by the bore size (B). To select a bearing for the approximate 1.5 inch shaft, go to the chart and pick the bore diameter of 1.5748, which is close to 1.5. This is the 308K bearing. The tolerance for this bore is specified in the chart as 1.5748 + .0000 and − .0005. Therefore, the limits dimension of the bore in this example is 1.5748 − 1.5743. The outside diameter is 3.5433 + .0000 − .0006 (3.5433 − 3.5427). The width of this bearing is .906 + .000 − .005 (.906 − .901). The fillet radius is the maximum shaft or housing fillet radius in which the bearing corners will clear. The fillet radius for the 308K bearing is R.059. Notice that the dimensions are also given in millimeters.

Shaft and Housing Fits

Shaft and housing fits are important, because tight fits may cause failure of the balls, rollers, lubricant, or overheating. Loose fits can cause slippage of the bearing in the housing, resulting in overheating, vibration, or excessive wear.

Shaft Fits

In general, for precision bearings, it is recommended that the shaft diameter and tolerance be the same as the bearing bore diameter and tolerance. The shaft diameter used with the 308K bearing is dimensioned ∅1.5748 − 1.5743.

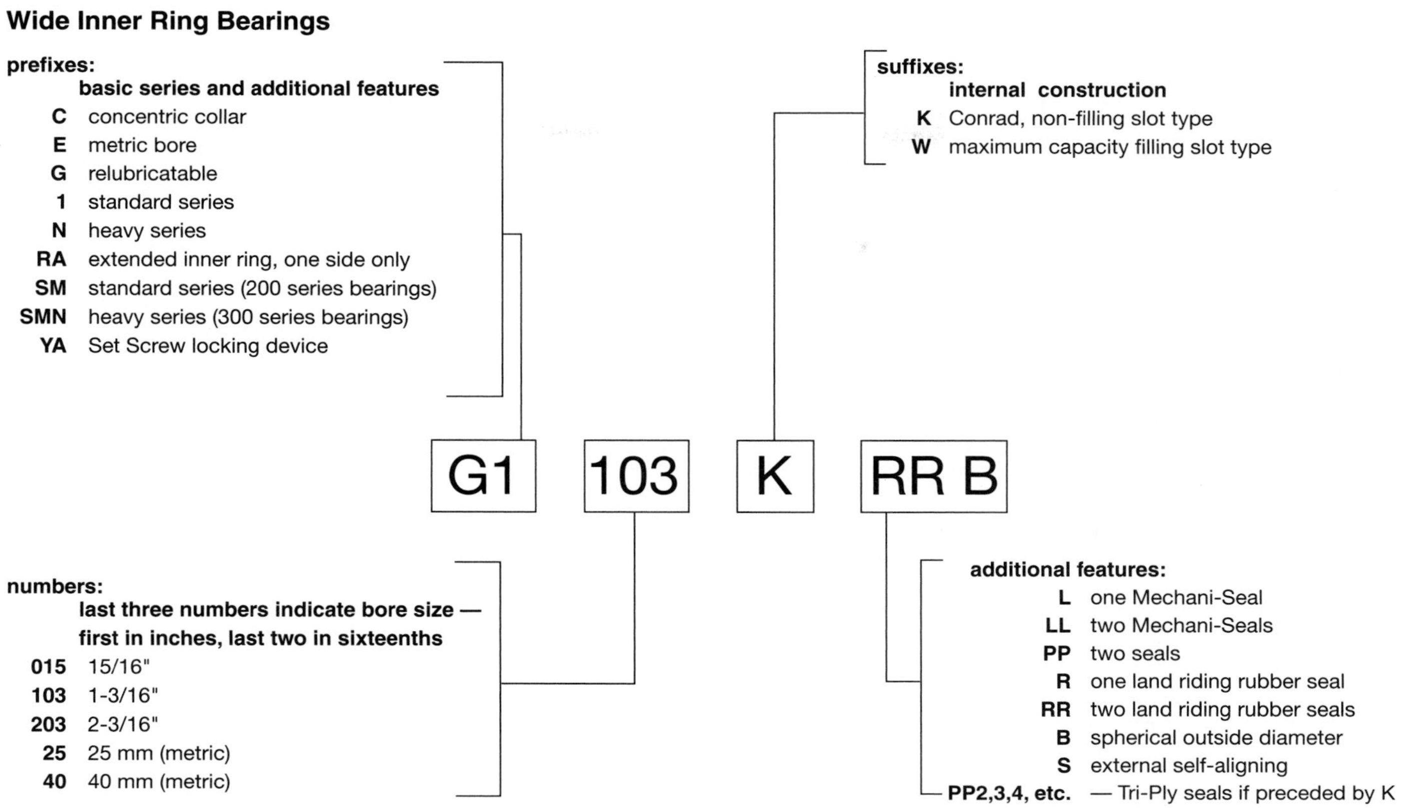

Wide Inner Ring Bearings

prefixes:

basic series and additional features

C	concentric collar
E	metric bore
G	relubricatable
1	standard series
N	heavy series
RA	extended inner ring, one side only
SM	standard series (200 series bearings)
SMN	heavy series (300 series bearings)
YA	Set Screw locking device

suffixes:

internal construction

K	Conrad, non-filling slot type
W	maximum capacity filling slot type

$$\boxed{G1} \quad \boxed{103} \quad \boxed{K} \quad \boxed{RR\ B}$$

numbers:

last three numbers indicate bore size —
first in inches, last two in sixteenths

015	15/16"
103	1-3/16"
203	2-3/16"
25	25 mm (metric)
40	40 mm (metric)

additional features:

L	one Mechani-Seal
LL	two Mechani-Seals
PP	two seals
R	one land riding rubber seal
RR	two land riding rubber seals
B	spherical outside diameter
S	external self-aligning
PP2,3,4, etc.	— Tri-Ply seals if preceded by K

FIGURE 16.74 ■ A sample bearing numbering system. *Courtesy The Torrington Company.*

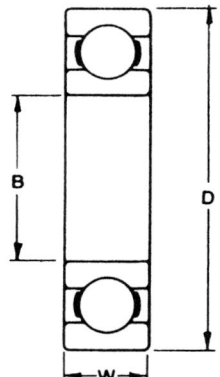

FIGURE 16.75 ■ Bearing dimensions. *Courtesy The Torrington Company.*

Housing Fits

In most applications with rotating shafts, the outer ring is stationary and should be mounted with a push of the hand or light tapping. In general, the minimum housing diameter is .0001 larger than the maximum bearing outside diameter and the maximum housing diameter is .0003 larger than the minimum housing diameter. With this in mind, the housing diameter for the 308K bearing is 3.5433 + .0001 = 3.5434 and 3.5434 + .0003 = 3.5437. The housing diameter limits are 3.5437 – 3.5434.

The Shaft Shoulder and Housing Shoulder Dimensions

Next, you should size the shaft shoulder and housing shoulder diameters. The shaft shoulder and housing shoulder diameter dimensions are represented in Figure 16.77 as S and H. The shoulders should be large enough to rest flat on the face of the bearing and small enough to allow bearing removal. Refer to the chart in Figure 16.78 to determine the shaft shoulder and housing shoulder diameters for the 308K bearing selected in the preceding discussion. Find the basic bearing number 308 and determine the limits of the shaft shoulder and the housing shoulder. The shaft shoulder diameter is 2.00 – 1.93 and the housing shoulder diameter is 3.19 – 3.06. Now you are ready to detail the bearing location on the housing drawing and on the shaft drawing. A partial detailed drawing of the shaft and housing for the 308K bearing is shown in Figure 16.79.

Surface Finish of Shaft and Housing

The recommended surface finish for precision bearing applications is 32 microinches (0.80 micrometer) for the shaft finish on shafts under 2 inches in diameter. For shafts over 2 inches in diameter, a 63 microinch (1.6 micrometer) finish is suggested. The housing diameter may have a 125 microinch (3.2 micrometer) finish for all applications.

DIMENSIONS — TOLERANCES

Bearing Number	Bore B		tolerance + .0000" + .000 mm to minus		Outside Diameter D		tolerance + .0000" + .000 mm to minus		Width W +.000", −.005" + .00 mm, − 13 mm		Fillet Radius[1]		WL		Static Load Rating C$_o$		Extended Dynamic Load Rating C$_E$	
	in.	mm	in.	mm	in.	mm	in.	mm	in.	mm	.in	mm	lbs.	kg	lbs.	N	lbs.	N
300K	.3937	10	.0003	.008	1.3780	35	.0005	.013	.433	11	.024	.6	.12	.054	850	3750	2000	9000
301K	.4724	12	.0003	.008	1.4567	37	.0005	.013	.472	12	.039	1.0	.14	.064	850	3750	2080	9150
302K	.5906	15	.0003	.008	1.6535	42	.0005	.013	.512	13	.039	1.0	.18	.082	1270	5600	2900	13200
303K	.6693	17	.0003	.008	1.8504	47	.0005	.013	.551	14	.039	1.0	.24	.109	1460	6550	3350	15000
304K	.7874	20	.0004	.010	2.0472	52	.0005	.013	.591	15	.039	1.0	.31	.141	1760	7800	4000	17600
305K	.9843	25	.0004	.010	2.4409	62	.0005	.013	.669	17	.039	1.0	.52	.236	2750	12200	5850	26000
306K	1.1811	30	.0004	.010	2.8346	72	.0005	.013	.748	19	.039	1.0	.78	.354	3550	15600	7500	33500
307K	1.3780	35	.0005	.013	3.1496	80	.0005	.013	.827	21	.059	1.5	1.04	.472	4500	20000	9150	40500
308K	1.5748	40	.0005	.013	3.5433	90	.0006	.015	.906	23	.059	1.5	1.42	.644	5600	24500	11000	49000
309K	1.7717	45	.0005	.013	3.9370	100	.0006	.015	.984	25	.059	1.5	1.90	.862	6700	3000	13200	58500
310K	1.9685	50	.0005	.013	4.3307	110	.0006	.015	1.063	27	.079	2.0	2.48	1.125	8000	35500	15300	68000
311K	2.1654	55	.0006	.015	4.7244	120	.0006	.015	1.142	29	.079	2.0	3.14	1.424	9500	41500	18000	80000
312K	2.3622	60	.0006	.015	5.1181	130	.0008	.020	1.220	31	.079	2.0	3.89	1.765	10800	48000	20400	90000

[1]Maximum shaft or housing fillet radius which bearing corners will clear.

FIGURE 16.76 ■ Bearing selection chart. *Courtesy The Torrington Company.*

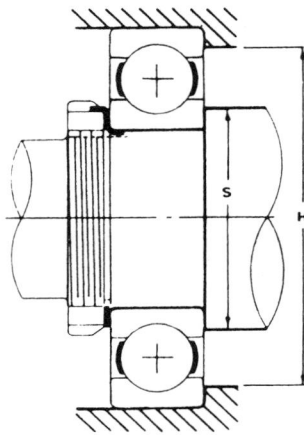

FIGURE 16.77 ■ Shaft shoulder and housing shoulder dimensions. *Courtesy The Torrington Company.*

Bearing Lubrication

It is necessary to maintain a film of lubrication between the bearing surfaces. The factors to consider when selecting lubrication requirements include the:

■ Type of operation, such as continuous or intermittent.

■ Service speed in rpm (revolutions per minute).

■ Bearing load, such as light, medium, or heavy.

Bearings may also be overlubricated, which may cause increased operating temperatures and early failure. Selection of the proper lubrication for the application should be determined by the manufacturer's recommendations. The ability of the lubricant is due, in part, to *viscosity*. Viscosity is the internal friction of a fluid, which makes it resist a tendency to flow. Fluids with low viscosity flow more freely than those with high viscosity. The chart in

Extra-Light • 9100 Series									Light • 200, 7200WN Series									Medium • 300, 7300WN Series								
Basic Bearing Number	Shoulder Diameters								Basic Bearing Number	Shoulder Diameters								Basic Bearing Number	Shoulder Diameters							
	shaft, S				housing, H					shaft, S				housing, H					shaft, S				housing, H			
	max.		min.		max.		min.			max.		min.		max.		min.			max.		min.		max.		min.	
	in.	mm	in.	mm	in.	mm	in.	mm		in.	mm	in.	mm	in.	mm	in.	mm		in.	mm	in.	mm	in.	mm	in.	mm
9100	.52	13.2	.47	11.9	.95	24.1	.91	23.1	200	.56	14.2	.50	12.7	.98	24.9	.97	24.6	300	.59	15.0	.50	12.7	1.18	30.0	1.15	29.2
9101	.71	18.0	.55	14.0	1.02	25.9	.97	24.6	201	.64	16.3	.58	14.7	1.06	26.9	1.05	26.7	301	.69	17.5	.63	16.0	1.22	31.0	1.21	30.7
9102	.75	19.0	.67	17.0	1.18	30.0	1.13	28.7	202	.75	19.0	.69	17.5	1.18	30.0	1.15	29.2	302	.81	20.6	.75	19.0	1.42	36.1	1.40	35.6
9103	.81	20.6	.75	19.0	1.30	33.0	1.25	31.8	203	.84	21.3	.77	19.6	1.34	34.0	1.31	33.3	303	.91	23.1	.83	21.1	1.61	40.9	1.60	40.6
9104	.98	24.9	.89	22.6	1.46	37.1	1.41	35.8	204	1.00	25.4	.94	23.9	1.61	40.9	1.58	40.1	304	1.06	26.9	.94	23.9	1.77	45.0	1.75	44.4
9105	1.18	30.0	1.08	27.4	1.65	41.9	1.60	40.6	205	1.22	31.0	1.14	29.0	1.81	46.0	1.78	45.2	305	1.31	33.3	1.14	29.0	2.17	55.1	2.09	53.1
9106	1.38	35.1	1.34	34.0	1.93	49.0	1.88	47.8	206	1.47	37.3	1.34	34.0	2.21	56.1	2.16	54.9	306	1.56	39.6	1.34	34.0	2.56	65.0	2.44	62.0
9107	1.63	41.4	1.53	38.9	2.21	56.1	2.15	54.6	207	1.72	43.7	1.53	38.9	2.56	65.0	2.47	62.7	307	1.78	45.2	1.69	42.9	2.80	71.1	2.72	69.1
9108	1.81	46.0	1.73	43.9	2.44	62.0	2.39	60.7	208	1.94	49.3	1.73	43.9	2.87	72.9	2.78	70.6	308	2.00	50.8	1.93	49.0	3.19	81.0	3.06	77.7
9109	2.03	51.6	1.94	49.3	2.72	69.1	2.67	67.8	209	2.13	54.1	1.94	49.3	3.07	78.0	2.97	75.4	309	2.28	57.9	2.13	54.1	3.58	90.9	3.41	86.6
9110	2.22	56.4	2.13	54.1	2.91	73.9	2.86	72.6	210	2.34	59.4	2.13	54.1	3.27	83.1	3.17	80.5	310	2.50	63.5	2.36	59.9	3.94	100.1	3.75	95.2
9111	2.48	63.0	2.33	59.2	3.27	83.1	3.22	81.8	211	2.54	64.5	2.41	61.2	3.68	93.5	3.56	90.4	311	2.75	69.8	2.56	65.0	4.33	110.0	4.13	104.9
9112	2.67	67.8	2.53	64.3	3.47	88.1	3.42	86.9	212	2.81	71.4	2.67	67.8	3.98	101.1	3.87	98.3	312	2.94	74.7	2.84	72.1	4.65	118.1	4.44	112.8

FIGURE 16.78 ■ Shaft shoulder and housing shoulder dimension selection chart. *Courtesy The Torrington Company.*

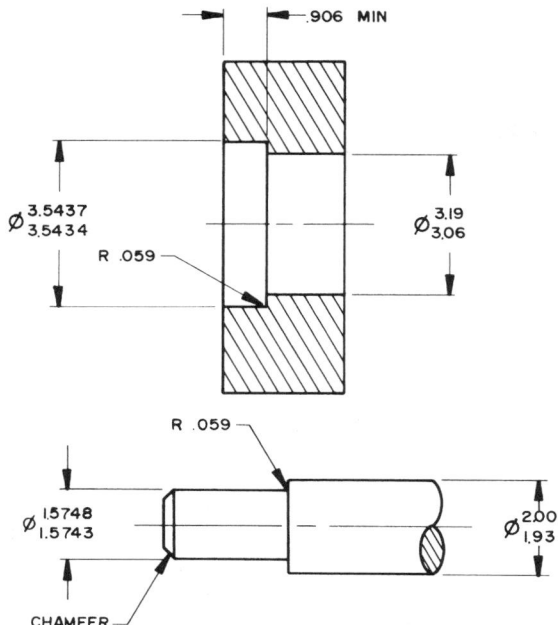

FIGURE 16.79 ■ A partial detail drawing of the shaft and housing for the 308K bearing.

Figure 16.80 shows the selection of oil viscosity based on temperature ranges and speed factors.

Oil Grooving of Bearings

In situations where bearings or bushings do not receive proper lubrication, it may be necessary to provide grooves for the proper flow of lubrication to the bearing surface. The bearing grooves help provide the proper lubricant between the bearing surfaces, and maintain adequate cooling. There are several methods of designing paths for the lubrication to the bearing surfaces as shown in Figure 16.81.

Sealing Methods

Machine designs normally include means for stopping leakage and keeping out dirt and other contaminants when lubricants are involved in the machine operation. This is accomplished using static or dynamic sealing devices. Static sealing refers to stationary devices that are held in place and stop leakage by applied pressure. Static seals such as gaskets do not come in contact with the moving parts of the mechanism. Dynamic seals are those that contact the moving parts of the machinery, such as packings.

Gaskets are made from materials that prevent leakage or access of dust contaminants into the machine cavity. Silicone rubber gasket materials are used in applications such as water pumps, engine filter housings, and oil pans. Gasket tapes, ropes, and strips provide good cushioning properties for dampening vibration, and the adhesive sticks well to most materials. Non-stick gasket materials such as paper, cork, and rubber are avail-

Oil Viscosities and Temperature Ranges for Ball Bearing Lubrication

Maximum Temperature Range Degrees F	Optimum Temperature Range Degrees F	Speed Factor, S_i (inner race bore diameter (inches) × RPM)	
		Under 1000	Over 1000
		Viscosity	
−40 to +100	−40 to −10	80 to 90 SSU (at 100 deg. F)	70 to 80 SSU (at 100 deg. F)
−10 to +100	−10 to +30	100 to 115 SSU (at 100 deg. F)	80 to 100 SSU (at 100 deg. F)
+30 to +150	+30 to +150	SAE 20	SAE 10
+30 to +200	+150 to +200	SAE 40	SAE 30
+50 to +300	+200 to +300	SAE 70	SAE 60

FIGURE 16.80 ■ Selection of oil viscosity based on temperature ranges and speed factors.

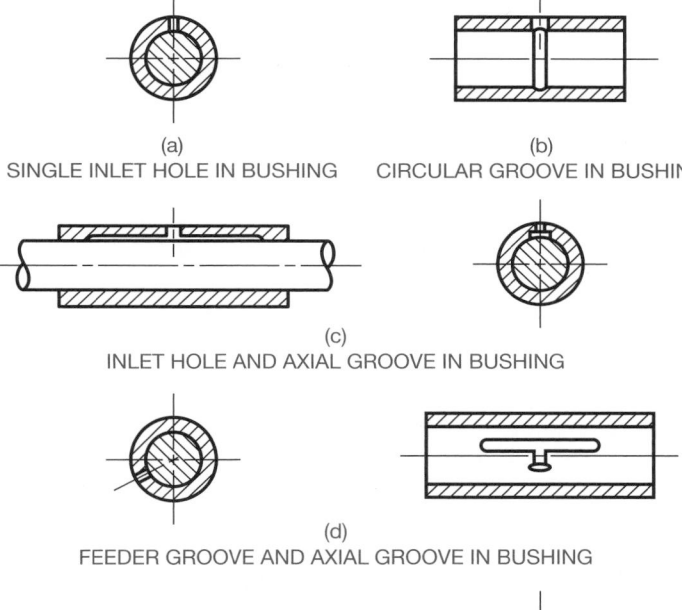

(a) SINGLE INLET HOLE IN BUSHING
(b) CIRCULAR GROOVE IN BUSHING
(c) INLET HOLE AND AXIAL GROOVE IN BUSHING
(d) FEEDER GROOVE AND AXIAL GROOVE IN BUSHING
(e) FEEDER GROOVE AND STRAIGHT AXIAL GROOVE IN THE SHAFT

FIGURE 16.81 ■ Methods of designing paths for lubrication to bearing surfaces.

able for certain applications. Figure 16.82 shows a typical gasket mounting.

Dynamic seals include packings and seals that fit tightly between the bearing or seal seat and the shaft. The pressure applied by the seal seat or the pressure of the fluid causes the sealing effect. Molded lip packings that provide sealing as a result of the pressure generated by the machine fluid are available.

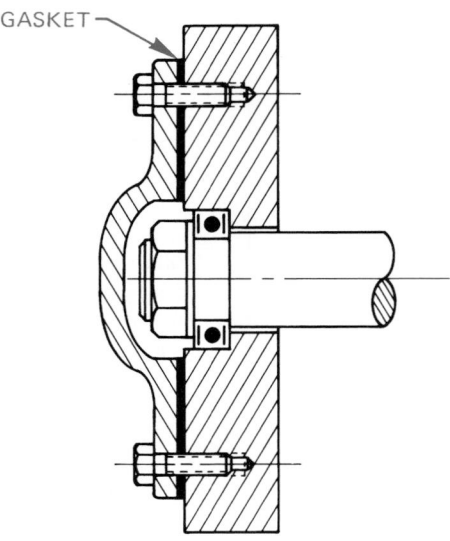

FIGURE 16.82 ■ Typical gasket mounting.

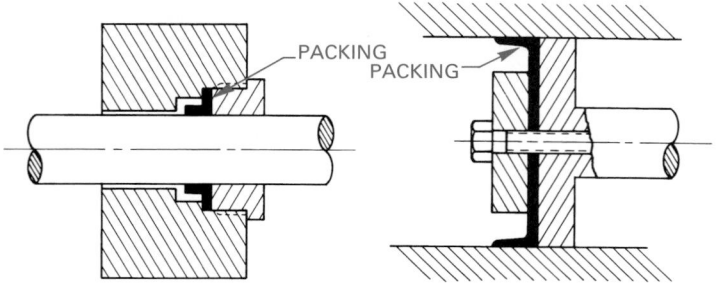

FIGURE 16.83 ■ Molded lip packings.

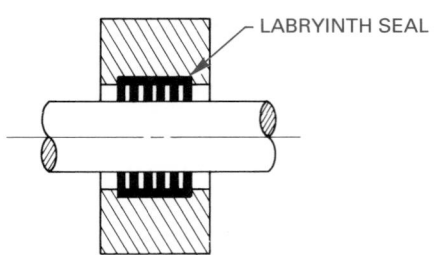

FIGURE 16.84 ■ Labyrinth seal.

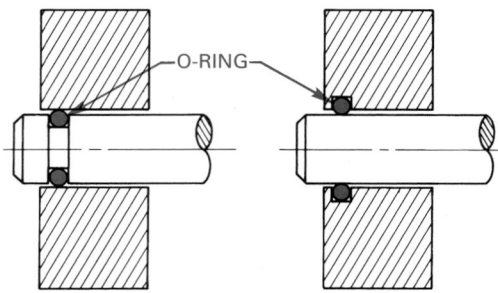

FIGURE 16.85 ■ O-ring seals.

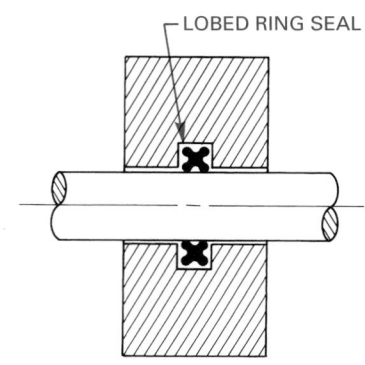

FIGURE 16.86 ■ Lobed ring seal.

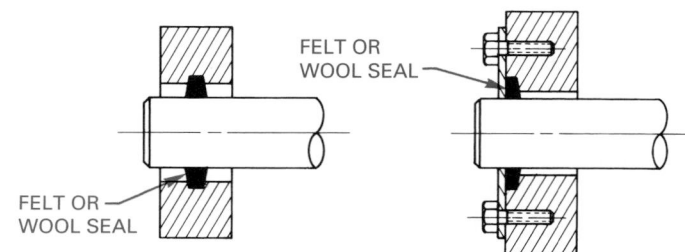

FIGURE 16.87 ■ Felt and wool seals.

Figure 16.83 shows examples of molded lip packings. Molded ring seals are placed in a groove and provide a positive seal between the shaft and bearing or bushing. Types of molded ring seals include labyrinth, O-ring, lobed ring, and others. *Labyrinth,* which means maze, refers to a seal that is made of a series of spaced strips that are connected to the seal seat, making it difficult for the lubrication to pass. Labyrinth seals are used in heavy machinery where some leakage is permissible. (See Figure 16.84.) The O-ring seal is the most commonly used seal because of its low cost, ease of application, and flexibility. The O-ring may be used for most situations involving rotating or oscillating motion. The O-ring is placed in a groove that is machined in either the shaft or the housing as shown in Figure 16.85. The lobed ring has rounded lobes that provide additional sealing forces over the standard O-ring seal. A typical lobed ring seal is shown in Figure 16.86.

Felt and wool seals are used where economical cost, lubricant absorption, filtration, low friction, and a polishing action are required. However, the ability to completely seal the machinery is not as positive as with the seals described earlier. (See Figure 16.87.)

Bearing Mountings

There are a number of methods used for holding the bearing in place. Common techniques include a nut and lock washer, a nut and lock nut, or a retaining ring. Other methods may be designed to fit the specific application or requirements, such as a shoulder plate. Figure 16.88 shows some examples of mountings.

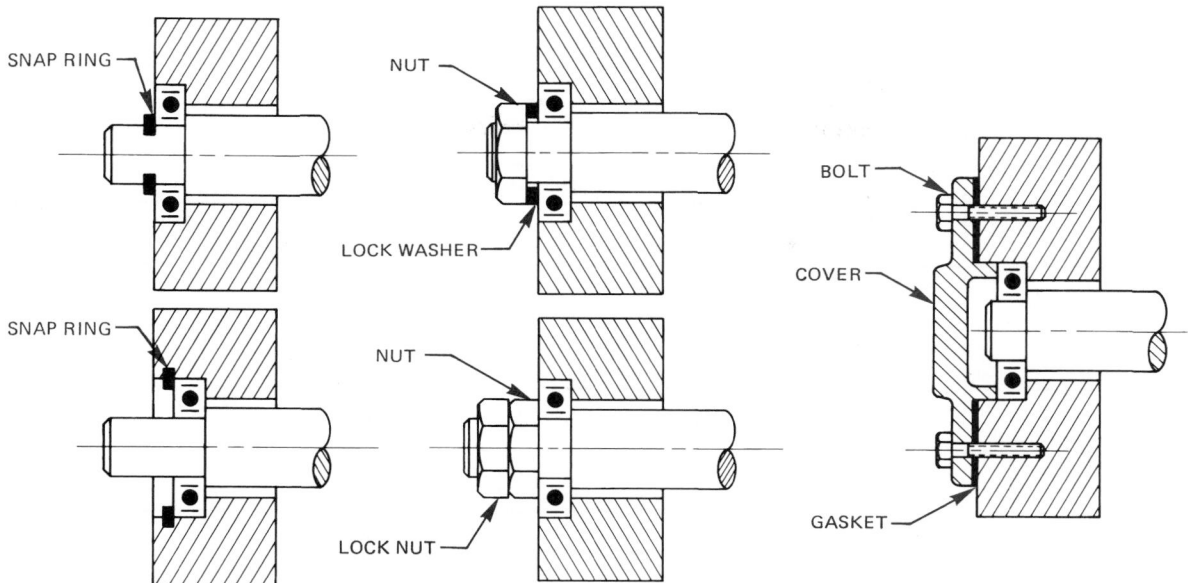

FIGURE 16.88 ■ Typical bearing mountings.

GEAR AND BEARING ASSEMBLIES

Gear and bearing assemblies show the parts of the complete mechanism as they appear assembled. (See Figure 16.89.) When drawing assemblies, you need to use as few views as possible, but enough to adequately display how all the parts fit together. In some situations all that is needed is a full sectional view that displays all of the internal components. An exterior view such as a front or top view plus a section sometimes works. Dimensions are normally omitted from the assembly unless the dimensions are needed for assembly purposes. For example, when a specific dimension regarding the relationship of one part to another is required to properly assemble the parts, each part is identified with a number in a circle. This circle is referred to as a balloon. The balloons are connected to the part being identified with a leader. The balloons are about one-half inch in diameter. The identification number is quarter-inch-high lettering or text. The balloon numbers correlate with a parts list. The parts list is normally placed on the drawing above or adjacent to the title block, or on a separate sheet as shown in Figure 16.90. Assembly includes torque data and lubricant information.

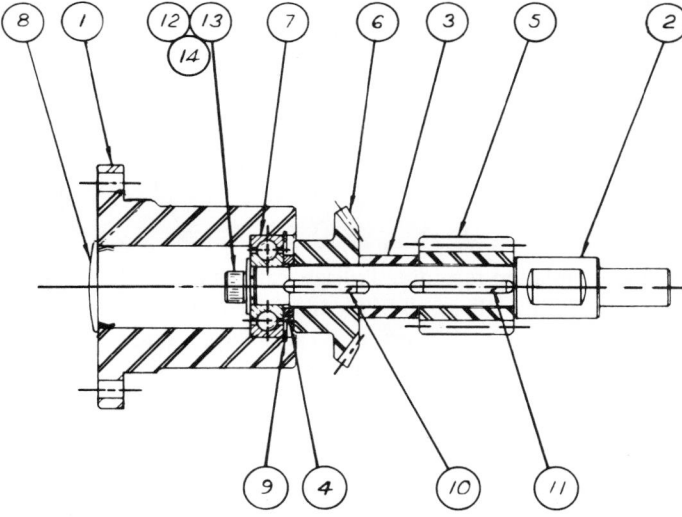

FIGURE 16.89 ■ Assembly drawing. *Courtesy Curtis Associates.*

ASSEMBLY <u>Cross Shaft Assembly</u>   USED ON _____

NUMBER OF UNITS _____   DATE _____

HAVE	NEED	P/Ø NO. W/Ø NO.	DET. NO.	PART NO.	DWG.	QTY.	PART NAME	DESCRIPTION	VENDOR
			1		B	1	Bearing Retainer	Ø 3" C.D. Bar	
			2		B	1	Cross Shaft	Ø 3/4" C.D. Bar	
			3		A	1	Spacer	Ø 3/4 O.D. × 11GA Wall Tube × .738 Thick	
			4		A	1	Spacer	Ø 3/4 O.D. × 11GA Wall Tube × .125 Thick	
			5		-	1	Steel Worm 12 D.P. Single Thread	Boston Gear #H 1056 R.H.	
			6		-	1	Bevel Gear 20° P.A. 2" P.D.	Boston Gear #HL 149 Y-G	
			7		-	1	Ball Bearing .4724 Ø Bore	T.R.W. or Equivalent #MRC 201-S22	
			8		-	1	End Plug		
			9		-	1	Snap Ring	Waldes-Truarc #N 5000-125	
			10		-	1	Key Stock	1/8 Sq. × 3/4 Lg.	
			11		-	1	Key Stock	1/8 Sq. × 1 Lg.	
			12		-	1	Socket Head Cap Screw	1/4 UNC × 3/4 Lg.	
			13		-	1	Lockwasher	1/4 Nominal	
			14		-	1	Flat Washer	1/4 Nominal	

FIGURE 16.90 ■ Parts list. *Courtesy Curtis Associates.*

PROFESSIONAL PERSPECTIVE

In every situation regarding linkage, cam, and gear design, there is a need to investigate all of the manufacturing alternatives and provide a solution the customer can afford. Competition is so fierce in the manufacturing industry that you should evaluate each design to find the best way to produce the product. The best way to understand this concept as an entry-level engineering drafter is to talk to experienced designers, engineers, and machinists. Go to the shop to see how things are done and determine the drawing requirements directly from the people who know. According to one design engineer, "If you don't know how it is going to be made, you're not a good drafter. Know the manufacturing capabilities of each piece of equipment." In addition to your drafting courses, it is a good idea to take some manufacturing technology classes. Math is also an important part of your program. Drafters in this field use a lot of geometry and trigonometry. After you have a strong educational background, the experienced engineer says, "Keep an open mind and look at all the alternatives."

For example, you work as an engineering drafter for a foundry. For years the flasks have been handled either by hand or with a lift truck. Both of these methods are time-consuming and dangerous, and the company has some contracts that require founding some castings that are too heavy to handle. So, the engineering department plans the design for a hydraulic flask handler. Your team is responsible for the handling mechanism that must be designed with this criteria:

■ Use a hydraulic piston with a 6-in. stroke.

■ Handle up to 44-in. wide flasks.

■ Take up no more than 28 in. in overall height.

Your team begins the problem-solving process and comes up with these steps to solve the problem:

■ Develop a single-line schematic kinematic diagram representing the movement of the mechanism based on the design information.

■ Identify and locate the available materials needed to build the flask handler.

The preliminary design is shown in Figure 16.91. All your team needs to do now is ask the other design teams for input and, after revisions are made, prepare a complete set of working drawings for fabrication.

PROFESSIONAL PERSPECTIVE (continued)

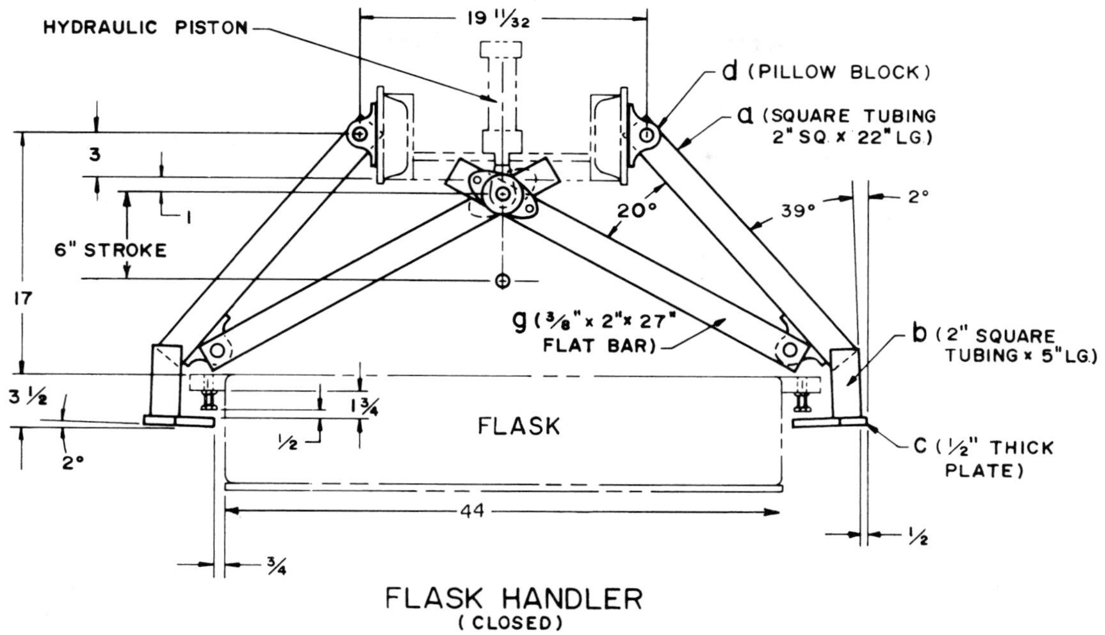

FIGURE 16.91 ■ A preliminary design is created.

WEB SITE RESEARCH

The following Web sites can assist you in doing additional research on subjects such as standards, gears, cams, linkages, bearings, and related areas:

www.asme.org—American Society of Mechanical Engineers for ANSI Y14.7.1.
www.industrialpress.com—*Machinery's Handbook*. Gear, cam, linkage, and bearing related information and specifications.
www.gearmfg.com—American Gear Manufacturers Association.
www.abapgt.com—Plastic gearing technology.
www.torrington.com—Bearings.
www.thomasregional.com—Gear, cam, and bearing products.
www.source4industries.com—Bearings.

MATH APPLICATIONS

LENGTH OF A CONNECTING ROD

Problem: Find the length of the connecting rod for the slider mechanism of Figure 16.92.

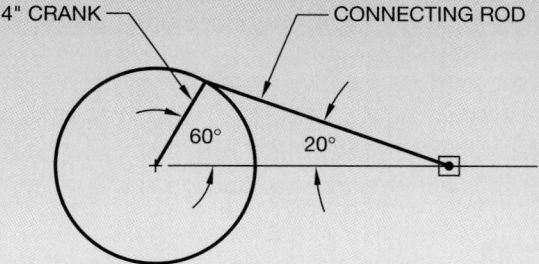

FIGURE 16.92 ■ Slider mechanism.

Solution: Using the Law of Sines (from the Math Instruction Appendix found in the Online Companion),

$$\frac{a}{\sin A} = \frac{b}{\sin B}$$

$$\frac{14''}{\sin 20^\circ} = \frac{b}{\sin 60^\circ}$$

Solving this proportional equation for b gives b = 35.4".

CHAPTER 16

Mechanisms Test

DIRECTIONS

Answer the questions with short complete statements or drawings as needed.

QUESTIONS

1. Draw a four-bar linkage and label the four links.
2. List at least two uses for a slider crank mechanism.
3. Explain why schematic symbols are used for linkage drawings.
4. The total movement or displacement of a slider mechanism is called _____.
5. Describe the function of a cam mechanism.
6. List three types of cam mechanisms.
7. Explain the relationship of the motion of the cam follower to the cam shaft for the three types of cams described in question 6.
8. Name the most commonly used cam follower.
9. Describe the purpose of the cam displacement diagram and how it relates to the cam profile construction.
10. List at least four types of cam motion.
11. What type of cam motion is best for smooth-running high-speed applications?
12. Which type of cam motion is suited for only very slow-speed applications?
13. Explain why constant velocity motion is modified with a curve at the bottom and top of the period.
14. The length of the displacement diagram for a drum cam is equal to _____.
15. Define *gear.*
16. List at least two functions of a gear.
17. Identify the parts of the spur gear shown in the drawing to the right.
18. List three types of parallel shafting gears.
19. Describe the function of a bevel gear.
20. Describe the function of a worm gear and worm.

21. Given a spur gear with 54 teeth, diametral pitch of 6, and a pitch diameter of 9.000 in., calculate the following:
    a. Addendum
    b. Dedendum
    c. Circular pitch
    d. Outside diameter
    e. Root diameter
    f. Circular thickness
    g. Chordal thickness
    h. Base circle diameter
22. What is the gear ratio if the gear in question 21 mates with a pinion having 18 teeth?
23. Explain the difference between plain and roller element bearings.
24. What is the best material for use in the manufacture of plain bearings?
25. Describe the difference between shielded and sealed bearings.

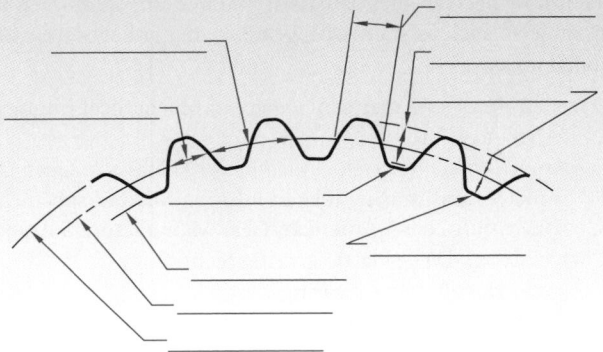

CHAPTER 16

Mechanisms Problems

DIRECTIONS

Please read problems carefully before you begin working. Complete each problem on an appropriately sized sheet. Precision work is important for accurate solutions.

LINKAGE PROBLEMS

PROBLEM 16-1 On the right is a pictorial drawing and a linkage schematic of a vise grip in the closed position. Reproduce the schematic exactly as shown and also show the open position in a second color.

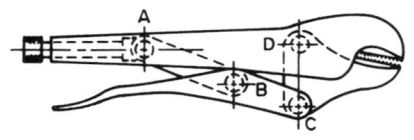

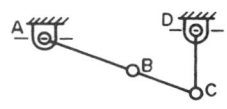

PROBLEM 16-2 Following is a pictorial drawing and a linkage schematic of a toggle clamp in the closed position. Reproduce the schematic exactly as shown and also show the open position in a second color.

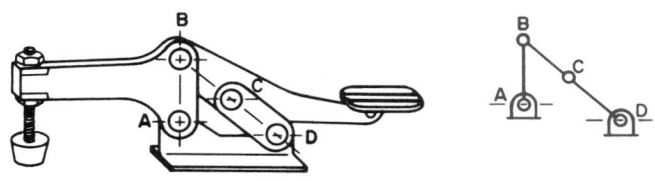

PROBLEM 16-3 Determine the extreme right and left positions of link CD in the figure shown below. Determine and label the angle through which CD oscillates.

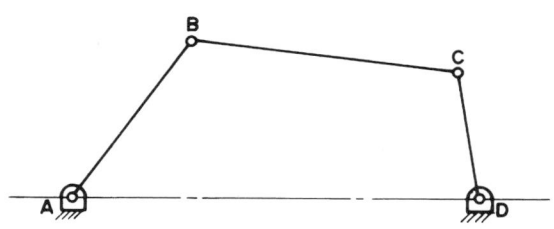

PROBLEM 16-4 Draw the mechanism below in the position shown. Using different colors, draw the mechanism in the extreme right and left positions. Dimension the stroke.

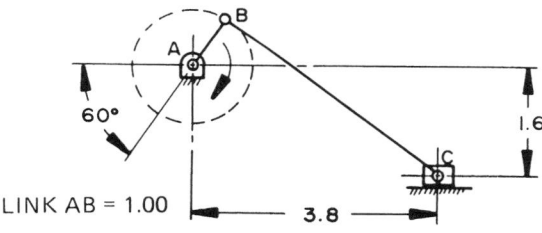

LINK AB = 1.00

PROBLEM 16-5 Draw the combination bellcrank slider mechanism in the position shown below. Determine the stroke of the slider if A moves to position A¹. Note: Position of features in sketch may be out of proportion.

AB = 1.500"
BC = 2.125"
CD = 3.375"

ANGLE ABC = 80°

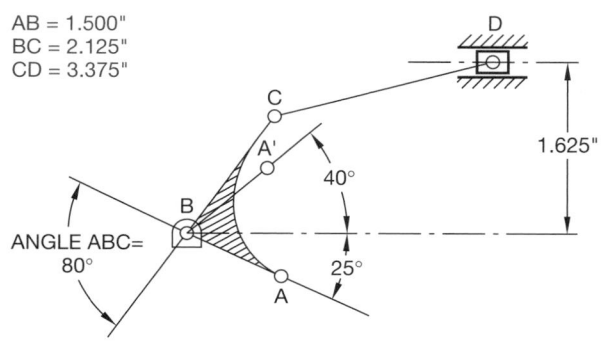

PROBLEM 16-6 The following drawing is a linkage schematic of an oscillating law sprinkler. The spray tube, shown in section, is part of link CD. Link AB moves through 360°, while points A and D are stationary. Determine and dimension the angle of oscillation through which the spray moves.

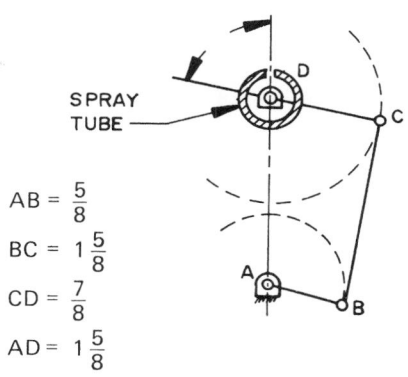

$$AB = \frac{5}{8}$$

$$BC = 1\frac{5}{8}$$

$$CD = \frac{7}{8}$$

$$AD = 1\frac{5}{8}$$

PROBLEM 16-7 Given the windshield wiper mechanism shown below, determine and dimension the angle of oscillation of the wiper blades. The electric motor rotates link ED continuously through 360°. ABC is one link with a 90° angle at B. A tension spring is located in the center of link BG.

AB = FG = ED = 1.25 in.   BC = .875 in.
AF = BG = 16 in.   CD = 7.625 in.

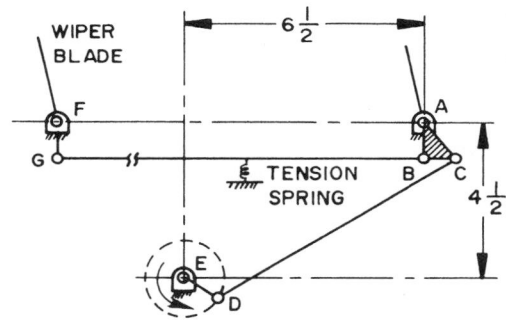

PROBLEM 16-8 Draw the mechanism shown below. Using different colors, show the path of point D in a total of five equally spaced positions, including the extreme right and left positions.

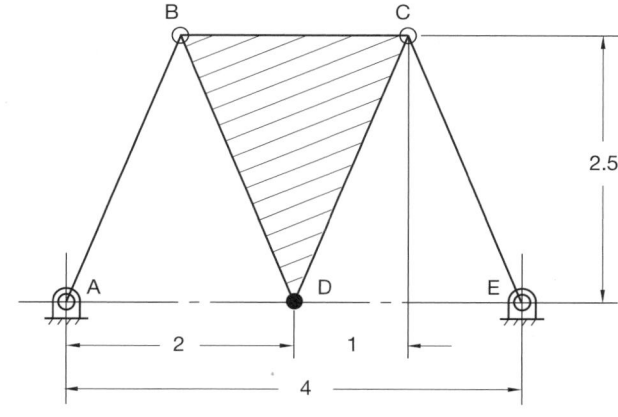

PROBLEM 16-9 Draw the mechanism in the position shown below. Draw the path of point P as linkage AB moves every 30° through a total of 360°. Use a different color and/or line type for each position.

Dimension the angle through which CD oscillates.
BCP is a through link.
Point P slides on EF.
EF = 6.5 in.

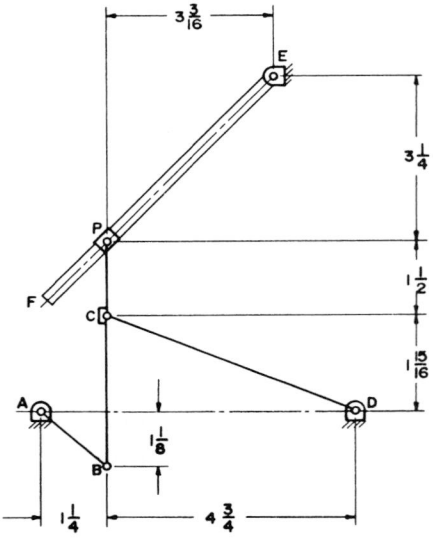

PROBLEM 16-10 Given the mechanism shown below, rotate link BC at 60° intervals clockwise through 360°. Plot and draw the paths of points D, E, and F. Dimension the angle of oscillation of link AF.

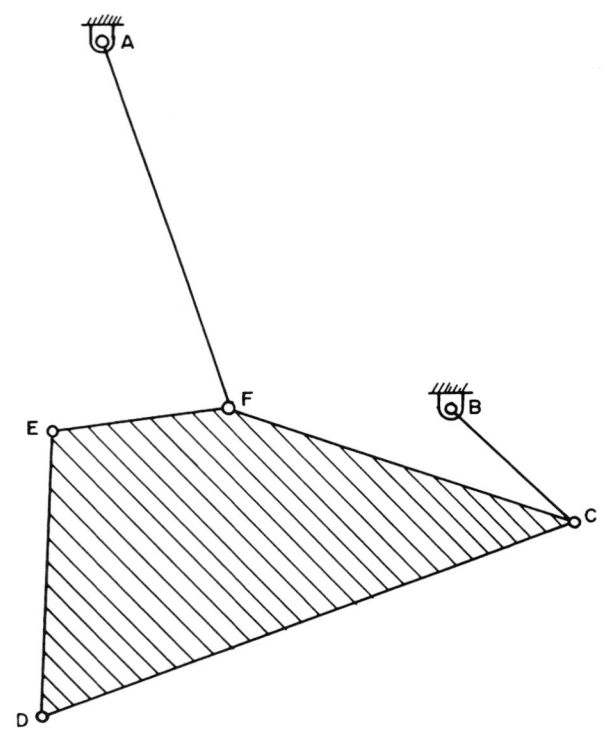

PROBLEM 16-11 Given the assembly drawing of the foundry flask handler shown below, draw a mechanism schematic showing the two extreme positions of movement. The handler is operated by a hydraulic piston with a 6-in. stroke.

a is welded to b.
b is welded to c.

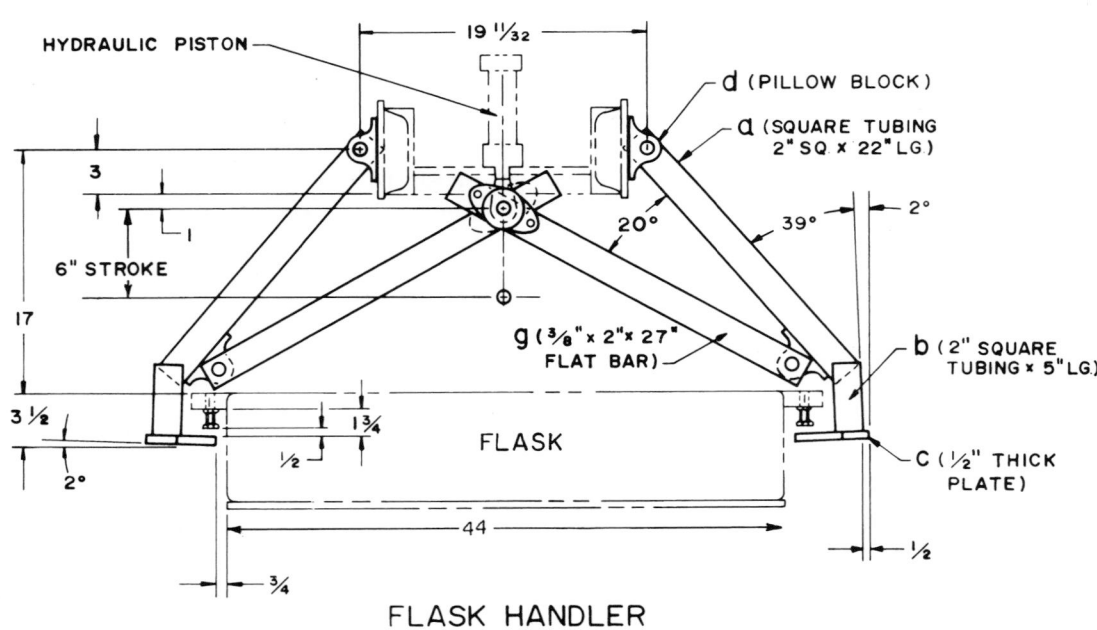

FLASK HANDLER
(CLOSED)

PROBLEM 16-12 Given the pivot hoist shown below, draw and scale exactly as shown. Rotate ADE clockwise so link AE is horizontal. Determine and dimension the extended length of the spring between C and E. Determine and dimension the angle between AE and CE when E is in the new position.

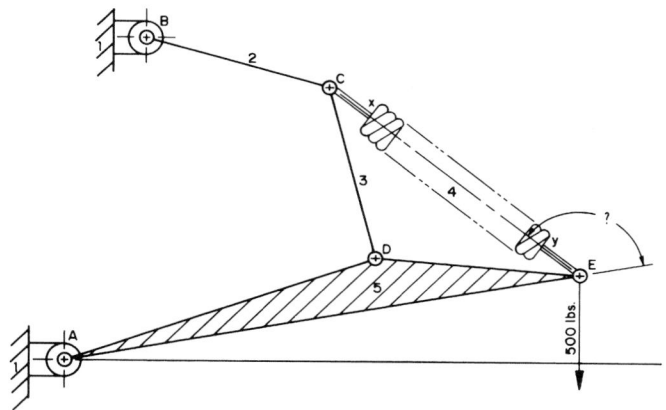

PROBLEM 16-13 Given the figure shown below, determine and dimension the stroke of point E moving in a straight line as link AB rotates 360°. Also dimension the angle of oscillation of link CD.

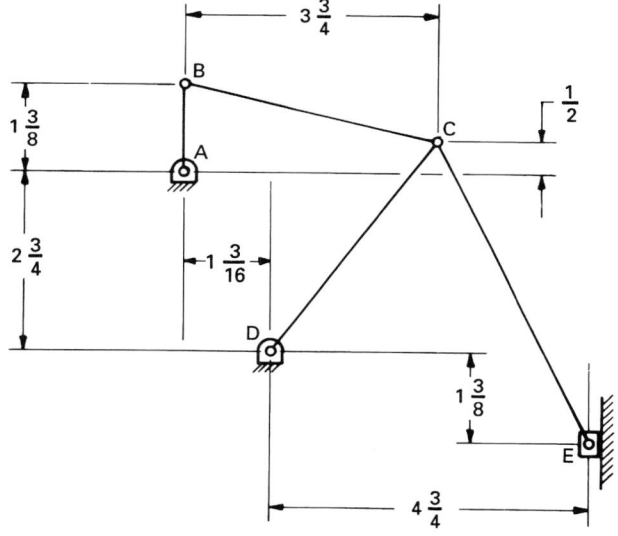

PROBLEM 16-14 Find two examples of linkage mechanisms at home or school. Explain in a short complete statement the function of each mechanism. Using schematic representations, show and dimension the extreme positions of each mechanism.

CAM DISPLACEMENT DIAGRAMS

PROBLEM 16-15 Construct a cam displacement diagram for a cam follower that rises in simple harmonic motion a total of 2.00 in. in 150°, dwells for 30°, falls 2.00 in. simple harmonic motion in 120°, and dwells for 60°. Draw the horizontal scale 6.00 in.

PROBLEM 16-16 Construct a cam displacement diagram for a cam follower that rises in uniform accelerated motion a total of 2.00 in. in 180°, dwells for 30°, falls 2.00 in. uniform accelerated motion in 120°, and dwells for 30°. Draw the horizontal scale 6.00 in.

PROBLEM 16-17 Construct a cam displacement diagram for a cam follower that rises in modified constant velocity for 3.00 in. in 180°, falls 3.00 in. modified constant velocity motion in 120°, and dwells for 60°. Use a modified constant velocity motion designed with one-third of the displacement.

PROBLEM 16-18 Construct a cam displacement diagram for a cam follower that rises 2.000 in. cycloidal motion in 120°, dwells for 60°, and falls 2.000 in. in cycloidal motion in 180°.

PROBLEM 16-19 Construct a cam displacement diagram for a cam follower that rises in simple harmonic motion a total of 1.250 in. in 90°, dwells for 60°, rises .750 in. in 45° simple harmonic motion, falls 2.00 in. with cycloidal motion in 120°. Draw the horizontal scale 12 in.

PROBLEM 16-20 Construct a cam displacement diagram for a cam follower that rises in modified constant velocity motion (modified to one-third the displacement) for 3.000 in. in 180°, dwells for 30°, and falls 3.000 in. simple harmonic motion in 30°, and dwells to the end of the cycle. Draw the horizontal scale 12 in.

PROBLEM 16-21 Construct a cam displacement diagram for a cam follower that rises 3.500 in. in 90° cycloidal motion, dwells for 45°, falls 2.500 in. cycloidal motion in 135°, dwells for 30°, falls 1.000 in. simple harmonic motion in 30°, and dwells to the end of the cycle. Draw the horizontal scale 12 in.

PROBLEM 16-22 Construct a cam displacement diagram for a cam follower that rises in cycloidal motion for 3.000 in. in 90°, dwells for 30°, falls 1.000 in. in simple harmonic motion in 90°, dwells for 30°, and falls the remaining 2.000 in. in uniform accelerated motion in 120°. Draw the horizontal scale 12 in.

PROBLEM 16-23 Construct a cam displacement diagram for a cam follower that rises in cycloidal motion a total of 2.1875 in. in 150°, dwells for 30°, falls back to the original level in simple harmonic motion in 150°, then dwells through the remainder of the cycle. Use a 12-in. horizontal scale.

PROBLEM 16-24 Construct a cam displacement diagram for a cam follower that raises a .375 in. diameter in-line roller follower 1.500 in. in uniform accelerated motion in 150°, dwells 45°, falls with modified constant velocity (one-third displacement) in 120°, and dwells the remainder of the cycle.

PROBLEM 16-25 Construct a cam displacement diagram for a cam follower that rises 3.500 in. 90° cycloidal motion, dwells for 30°, falls 2.250 in. cycloidal motion in 150°, falls 1.250 in. simple harmonic motion in 60°, and dwells for 30°. Draw the horizontal scale equal in circumference to a 3.000-in. diameter circle.

CAM PROFILE DRAWINGS

PROBLEM 16-26   Use the displacement diagram constructed in Problem 16-15 and the information given in the illustration below to lay out the plate cam profile drawing. The cam rotates counterclockwise. Make a two-view detailed drawing of the cam and dimension it as shown in this chapter.

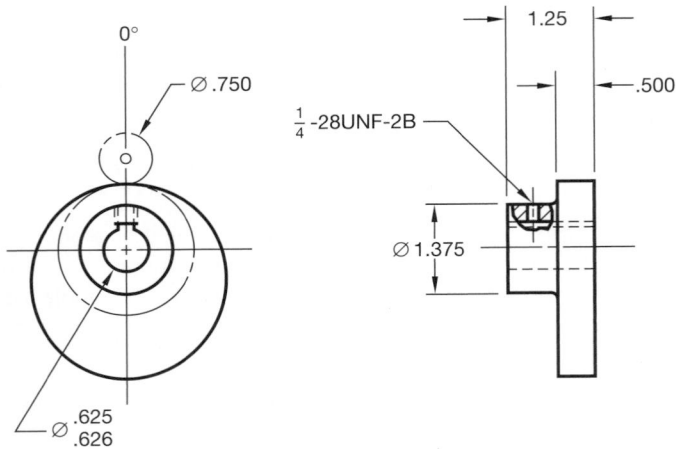

IN-LINE ROLLER FOLLOWER	= ⌀.750
BASE CIRCLE	= ⌀2.000
KEY SIZE USED	= 1/4 × 1/4 SQ. KEY
PLATE THICKNESS	= .500
HUB THICKNESS	= .750
HUB DIAMETER	= ⌀1.375
SHAFT DIAMETER	= ⌀.625/.626

PROBLEM 16-27   Use the displacement diagram constructed in Problem 16-18 and the information given in the illustration for Problem 16-26 to lay out the plate cam profile. The cam is rotating clockwise. Make a two-view detailed drawing of the cam, properly toleranced and dimensioned.

PROBLEM 16-28   Make a two-view detailed drawing of a plate cam using the displacement diagram from Problem 16-24. Completely dimension the drawing using the following information:

The cam rotates counterclockwise.
In-line roller follower = ⌀.375   Hub projection = .500
Base circle = ⌀2.750   Plate thickness = .375
Shaft = ⌀1.000   All dimensions in inches.
Hub diameter = ⌀1.250

PROBLEM 16-29   Make a two-view detailed drawing of a plate cam using the displacement diagram from Problem 16-20. Completely dimension the drawing using the following information:

The cam rotates counterclockwise.
In-line roller follower = ⌀.750   Hub projection = .500
Base circle = ⌀2.000   Plate thickness = .500
Shaft = ⌀.625   All dimensions in inches.
Hub diameter = ⌀1.250

PROBLEM 16-30   Make a two-view detailed drawing of a plate cam using the displacement diagram from Problem 16-18. Completely dimension the drawing using the information shown below:

The cam rotates clockwise.

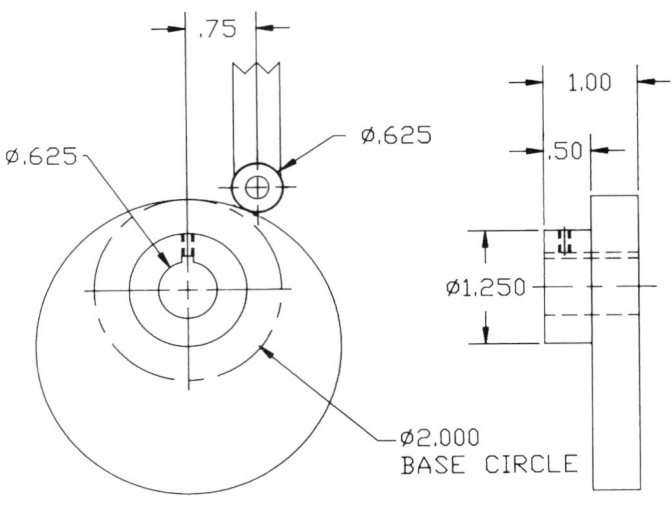

PROBLEM 16-31   Use the displacement diagram constructed in Problem 16-25 and the following illustration to lay out the profile of the groove in the drum cam. The cam rotates clockwise. Make a two-view detailed drawing of the drum cam, with tolerances and dimensions as discussed and shown in this chapter.

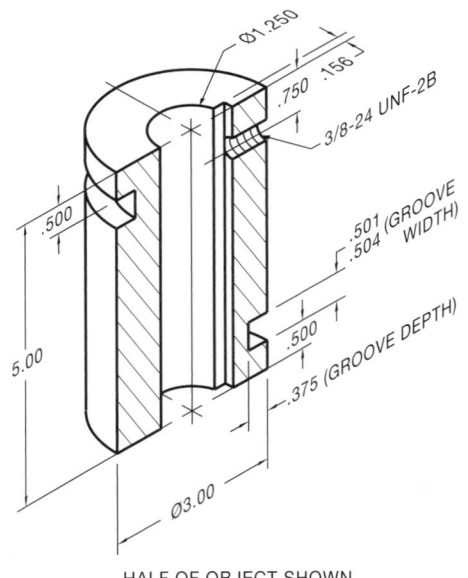

HALF OF OBJECT SHOWN

GEAR PROBLEMS

PROBLEM 16-32 Given the gear train and chart shown, calculate or determine the missing information to complete the chart.

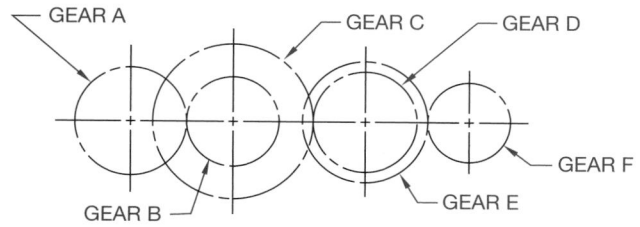

Gear	Diametral Pitch (P)	Number of teeth (N)	Pitch Diameter (D)	RPM	Direction	Center Distance	Gear Ratio
A	4		7.5"	240	Clockwise		
B		18					
C			10.0"	400			
D	5	40					
E	7						
F		14		1500			

PROBLEM 16-33 Given the ten-gear power transmission and chart shown, calculate or determine the missing information to complete the chart.

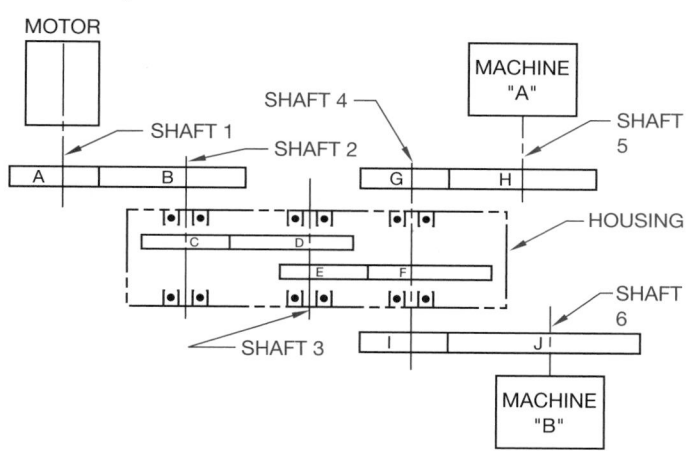

					Gear Data Block	
Gear	Pitch Diameter	No. of Teeth	Diametral Pitch	RPM	Ctr. Distance Between Mating Gears	Direction
A	3.00			3600	4.00	Counter-clockwise
B			5			
C		48				
D			12	1080		
E	4.00	40				
F		100				
G	5.00				6.00	
H			6			
I			4			
J		40		108		

PROBLEM 16-34 Given the following information, use ANSI standards to make a detailed drawing of the spur gear shown at right:

20 teeth	Face width = 2.500 in.
Diametral pitch = 5	Shaft diameter = Ø1.125
20° pressure angle	Keyway for a .25 in. square key

Place the centerline of the keyway in line with a radial line through the center of one tooth (one tooth profile needed to show alignment). Include the necessary spur gear data in a chart placed over the title block. Use the formulas given in this chapter to solve for unknown values.

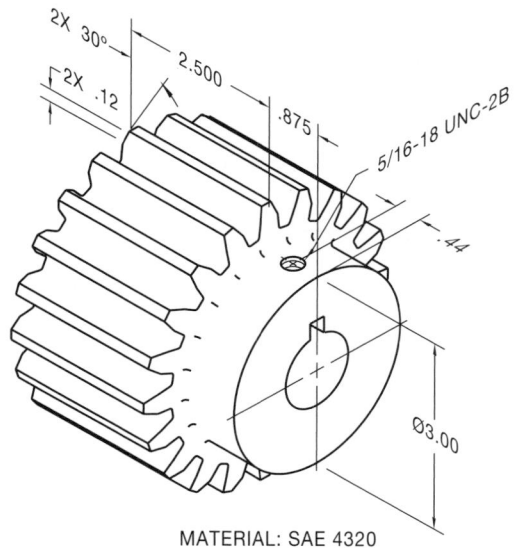

MATERIAL: SAE 4320

PROBLEM 16-35   Use ANSI standards to make a detailed drawing of a rack that mates with the spur gear in Problem 16-34. The overall length is 24 in.

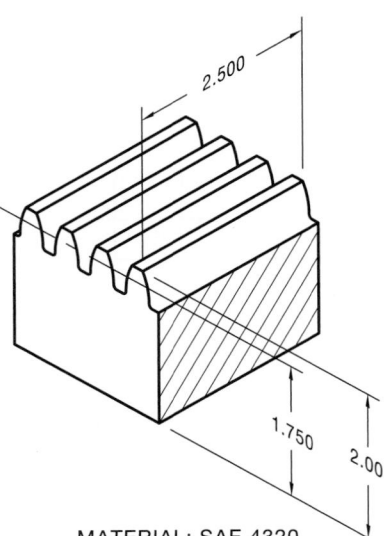

MATERIAL: SAE 4320

PROBLEM 16-36   Use ASME standards to make a detailed drawing of a straight bevel gear given the following information and the illustration shown below:

Pitch diameter = Ø8.000 in.   Circular thickness = 4.0939
Pressure angle = 20°   Pitch angle = 65°
32 teeth   Root angle = 62.15°
Diametral pitch = 4   Addendum = .3022
Face width = 1.400   Whole depth = .5493
Shaft diameter = Ø1.125 in.   Cordal addendum = .0496
Use a .250 square key.   Cordal thickness = .7841
Core distance = 4.401

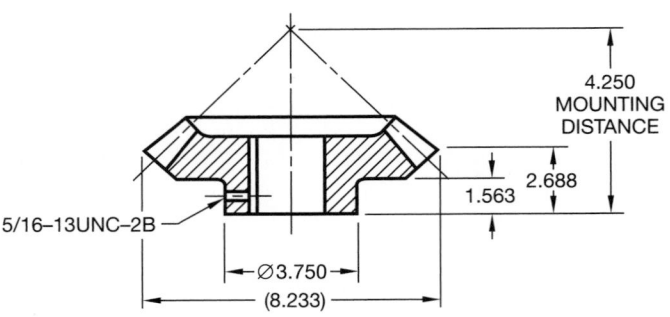

PROBLEM 16-37   Prepare detail drawing of the pinion and gear from the following gear data:

Dimensions for 20° straight bevel gear 90° shaft angle

	Pinion	Gear
Number of teeth	22	75
Diametral pitch	10	10
Face width	1.25	1.25
Pressure angle	20°	20°
Shaft angle	90°	90°
Working depth	0.200	0.200
Whole depth	0.221	0.221
Pitch diameter	2.200	7.500
Pitch angle	16.348°	73.652°
Cone distance	3.908	3.908
Circular pitch	0.314	0.314
Addendum	0.142	0.058
Dedendum	0.077	0.161
Clearance	0.021	0.021
Dedendum angle	1.126°	2.356°
Face angle of blank	18.704°	74.778°
Root angle	15.222°	71.296°
Outside diameter	2.473	7.533
Pitch apex to crown	3.710	1.044
Circular thickness	0.188	0.126
Backlash	0.002	0.002
Chordal thickness	0.186	0.125
Chordal addendum	0.146	0.058
Tooth angle	107.149 min	107.149 min
Limit point width	0.046	0.046
Tool advance	0.002	0.002

BEARING PROBLEMS

PROBLEM 16-38   Use the charts shown in this chapter to establish the following medium-service bearing dimensions for an approximate Ø1.25-in. shaft.

Bearing catalog number _____
Bore _____
Outside diameter _____
Width _____
Fillet radius _____
Shaft shoulder diameter _____
Housing shoulder diameter _____
Shaft diameter _____
Housing diameter _____

PROBLEM 16-39 Use the charts shown in this chapter to establish the following medium-service bearing dimensions for an approximate ∅3.5-in. shaft.

Bearing catalog number _____
Bore _____
Outside diameter _____
Width _____
Fillet radius _____
Shaft shoulder diameter _____
Housing shoulder diameter _____
Shaft diameter _____
Housing diameter _____

PROBLEM 16-40 Use the charts shown in this chapter to establish the following medium-service bearing dimensions for an approximate ∅20-mm shaft.

Bearing catalog number _____
Bore _____
Outside diameter _____
Width _____
Fillet radius _____
Shaft shoulder diameter _____
Housing shoulder diameter _____
Shaft diameter _____
Housing diameter _____

PROBLEM 16-41 Use the charts shown in this chapter to establish the following medium-service bearing dimensions for an approximate ∅60-mm shaft.

Bearing catalog number _____
Bore _____
Outside diameter _____
Width _____
Fillet radius _____
Shaft shoulder diameter _____
Housing shoulder diameter _____
Shaft diameter _____
Housing diameter _____

LINKAGE DESIGN PROBLEMS

PROBLEM 16-42 Given the following drawing as an example, design a backhoe that will dig a 20-ft. deep trench. Draw an assembly drawing with dimensions specified between linkages. Show the backhoe in the fully closed position, half extended position, fully extended horizontal, and fully extended at the maximum trench depth.

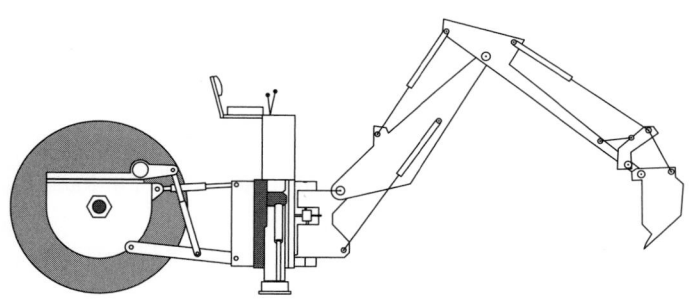

PROBLEM 16-43 Given the following drawing as an example, design a similar material handling lift that will lift a maximum of 12 ft. vertically and 16 ft. horizontally. The telescoping actuator can operate a maximum 4-ft. extension.

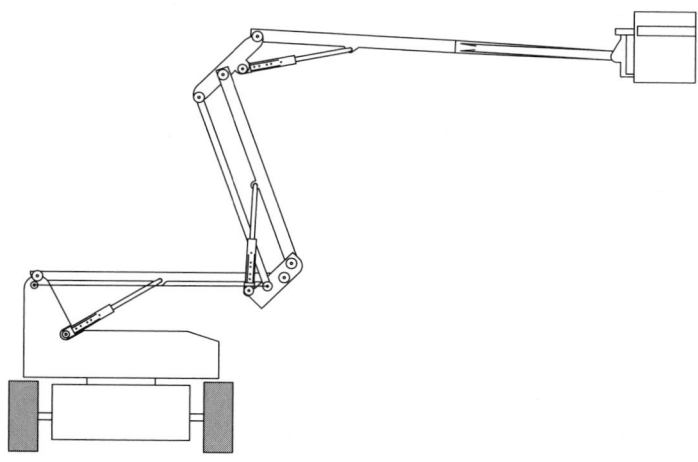

GEAR DESIGN, BEARING SELECTION, SHAFT DESIGN PROBLEM

PROBLEM 16-44 Design a two-speed gear reducer that will operate eight to ten hours per day and receive moderate shock while in operation. Use the following information:

■ A 5-HP 1750 electric motor supplies the input power.
■ There are six gears arranged approximately as shown on page 550.
■ Gear C-D is a cluster gear sliding on the countershaft.
■ The output speed is 625 rpm when gear C is engaged with gear E.
■ The output speed is 437.5 rpm when gear D is engaged with gear F.
■ Gear A has 32 teeth, gear B has 64 teeth, gear C has 25 teeth, and gear D has 24 teeth.

Use the gear and bearing information from this chapter to design the gear reducer and do the following:

1. Determine the diametral pitches for all six gears.

2. Determine the number of teeth for gears E and F.

3. Use tolerances, surface finishes, and fit as discussed in this chapter.

4. Use manufactures' catalogs shown in this chapter or supplied by your instructor to select standard parts.

5. Make a detailed drawing of the cluster gear, including spur gear data charts for both gears on one sheet.

6. Make detailed drawings of the three shafts—input, output, and countershaft—each on a separate sheet. The shafts are approximately ∅1.250 in. Design the shafts based on bearing specifications and fits. Design the keyseats based on

the shaft diameter given and as specified in the *Machinery's Handbook* or other source.

7. Make an assembly drawing of the entire product. Prepare a complete parts list placed over the title block. All parts should be ballooned on the drawing to coordinate with the parts list. The dimensions given in the illustration below are for reference only. Detailed dimensions are not required on the assembly drawing.

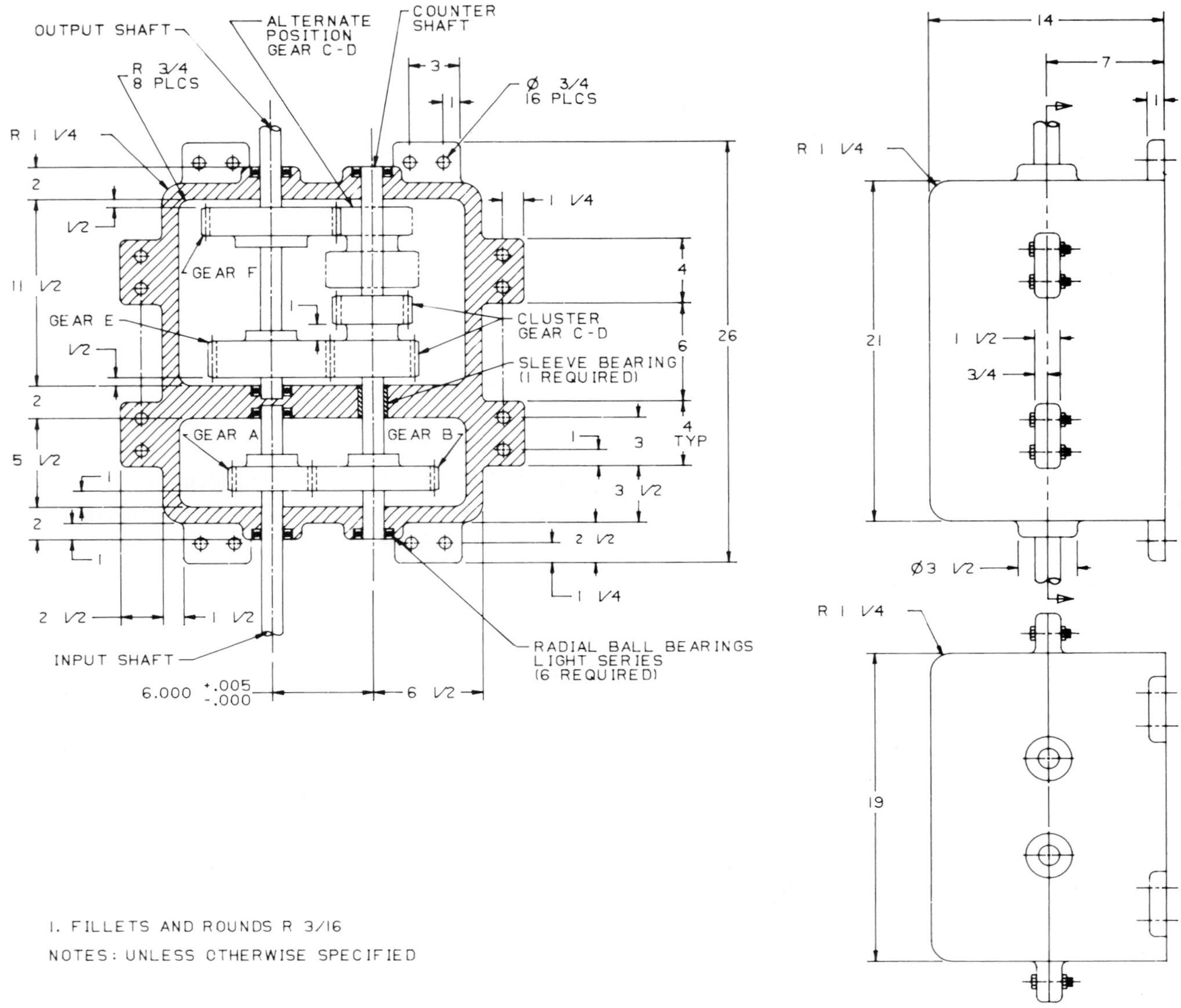

MATH PROBLEMS

For the oblique triangle shown below (which is not drawn to scale):

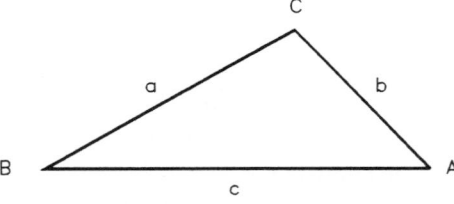

PROBLEM 16-45 Find side b given a = 125, A = 54.7°, B = 65.2°.

PROBLEM 16-46 Find side c given b = 321, A = 75.3°, C = 38.5°. (Hint: A + B + C = 180°.)

PROBLEM 16-47 Find angle C given b = 50.4, c = 33.3, B = 118.5°.

PROBLEM 16-48 Find angle A given b = 51.5, a = 62.5, B = 40.7°, and given that angle A is greater than 90°.

PROBLEM 16-49 Find side b given a = 320, c = 475, A = 35.3°, and given that angle C is less than 90°. (Hint: Find angle C first.)

PROBLEM 16-50 Find side a given b = 50.4, c = 33.3, B = 118.5°.

CHAPTER

Belt and Chain Drives

OBJECTIVES

After completing this chapter, you will:

■ Design belt drive systems using vendor specifications.

■ Design chain drive systems from vendor specifications.

■ Draw sprocket and chain designs from given engineering layouts.

THE ENGINEERING DESIGN APPLICATION

The problem is to design a two-belt sheave for Class D conventional industrial belts. A *sheave* is a grooved wheel in a pulley block. The ID should be 5.00 and the OD 6.375. Provide 1.126 between belt centers. Keep the weight down as much as possible; try for .375 web thickness. Design for a ∅.875 shaft. Refer to the *Machinery's Handbook* for the bore diameter and the recommended keyway for a parallel rectangular key. Your research concludes that you need the following specifications:

■ 38° V in sheave for belts.

■ ∅.874–.875 bore.

■ .203 keyway × 1.032 (measured from bottom of bore).

■ You decide on a ∅2.00 hub for good material strength.

The problem solution is shown in Figure 17.1.

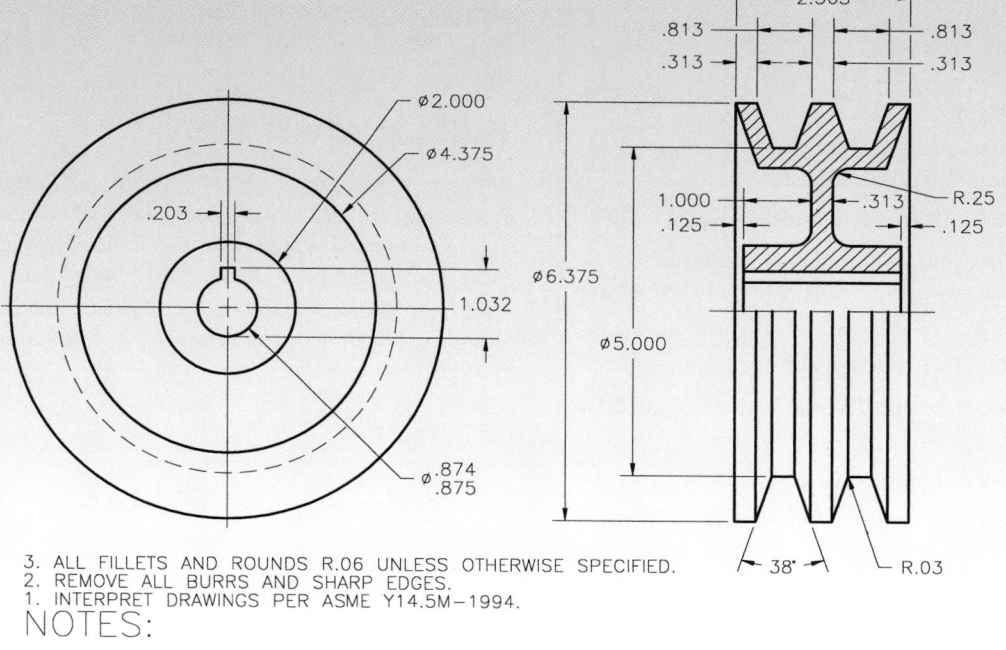

NOTES:
1. INTERPRET DRAWINGS PER ASME Y14.5M–1994.
2. REMOVE ALL BURRS AND SHARP EDGES.
3. ALL FILLETS AND ROUNDS R.06 UNLESS OTHERWISE SPECIFIED.

FIGURE 17.1 ■ Solution to engineering problem.

The function of belt, chain, and gear drives is to transmit power between rotating shafts. It is important to evaluate the advantages and disadvantages of each type of mechanism for the proper design and implementation of the final product. Gears were discussed in Chapter 16 because they fall into a class of linkage mechanism where two or more elements of the linkage system are in direct contact with each other. Gears are also the most rugged and durable of the mechanisms, transmitting motion with little or no slippage. They can satisfy high power demands at efficiencies of up to 98%.

ADVANTAGES OF GEAR DRIVES

Gear drives are commonly used in some applications rather than belt or chain drives for the following reasons:

■ They are more rugged and durable than most belt and chain drives.

■ They are more efficient than belt or chain drives.

■ They are more practical when space limitations require the shortest distance between centers.

■ They have great maximum speed ratios.

■ They work better when high horsepower and load capabilities are important.

■ They are preferable when nonparallel shafting is the design requirement.

■ They may be used where timing and synchronization are required. Chain drives may work effectively in these applications, but belt drives are not recommended because of slippage.

ADVANTAGES OF BELT DRIVES

Belt drives are often the choice of the designer when some of the following characteristics are important:

■ They are up to 95% efficient, especially at high speeds.

■ They are designed to slip when an overload occurs.

■ They resist abrasion.

■ They require no lubrication, because there is no metal-to-metal contact, except at shaft bearings.

■ They are normally smooth running and less noisy.

■ Belt drives, especially those using flat belts, may be used more effectively when very long shaft center distances are required.

■ They can operate effectively at high speed ranges.

■ They have flexible shaft center distances, where gear drives are restricted.

■ They are less expensive than gear or chain drives.

■ They are easy to assemble and install and have flexible tolerances.

■ They absorb shock well.

■ They are easy and inexpensive to maintain.

ADVANTAGES OF CHAIN DRIVES

Chain drive mechanisms have certain design advantages over both gear and belt drives:

■ They have flexible shaft center distances, where gear drives are restricted.

■ They are less expensive than gear drives, in most cases.

■ They have simpler installation and assembly tolerances than gear drives.

■ They provide better shock-absorbing qualities than gears.

■ They have no slippage as compared to belt drives, resulting in more efficient operation.

■ They have lower loads on shaft bearings because tension is not required as with belt drives.

■ They are easy to install.

■ They are not affected by sun, heat, or oil and do not deteriorate with age as do belts.

■ They are more effective at lower speeds than belts.

■ They require little adjustment, while belt drives require frequent adjustment.

BELTS AND BELT DRIVES

Belts are used to transmit power from one shaft to another smoothly, quietly, and inexpensively. Belts are made of continuous construction from materials such as rubberized cord, leather strands, fabric, or, more commonly, reinforced nylon, rayon, steel, or glass fiber. Belts operate on shaft-mounted pulleys and sheaves. *Pulleys* are wheels constructed with a groove in the circumference to match the shape of the belt and transmit power to the belt. *Sheaves* are the same as pulleys with the differentiation being that sheaves are generally the drive and pulleys are the driven. Sheaves and pulleys both guide the belt in its operation, so the terms are normally used interchangeably.

BELT TYPES

Different types of belts are designed for individual applications. The type of belt you are probably most familiar with is the V belt, which is used for most automobile and household applications. Other belts, such as flat belts, are used for long center distances and high-speed applications. Positive drive belts fit notched pulleys and are designed for more positive power transmission than other belt designs.

V Belts

V belts are, as just mentioned, the most commonly used belts. These belts have a wide range of applications and operating conditions. They operate efficiently at speeds between 1500 and 6500 fpm (feet per minute) and a temperature range of −30° F to 185° F.

SAE V-belt manufacturers code the belts according to shape to help ensure interchangeability between suppliers. V belts fall into three main categories: automotive, agricultural, and industrial. Automotive belts are manufactured in accordance with six Society of Automotive Engineers (SAE) approved sectional shapes. Agricultural V belts are similar in shape to automotive V belts. They are identified with a letter *H.* Industrial V belts fall into the categories of conventional, narrow, light-duty, and double V cross section.

A comparison of the standard V-belt cross sections is shown in Figure 17.2.

The load-carrying capability of a V belt is provided by reinforcing cords in the belt. The reinforcement is made of rayon, nylon, steel, or glass fibers. Figure 17.3 shows the construction of a V belt with steel reinforcing cables. The reinforcing cords are usually embedded in a soft rubber material called a cushion section. The cushion section is surrounded by a hard rubber or other material and the internal structure is covered with an abrasion-resistant material.

The *pitch line* is the only portion of the belt that does not change length as the belt bends around the pulley. The pitch line is used to determine the pitch diameter of the pulley. The *pitch diameter* is the effective diameter of the pulley, which is used to establish the speed ratio. Most pulleys are made of cast iron or steel. The construction and terminology of a pulley and mating belt are shown in Figure 17.4.

Flat Belts

Flat belts are typically used where high-speed applications are more important than power transmission, and when long center distances are necessary. Flat belts are less efficient at moderate speeds, because they tend to slip under load. Flat belts are also

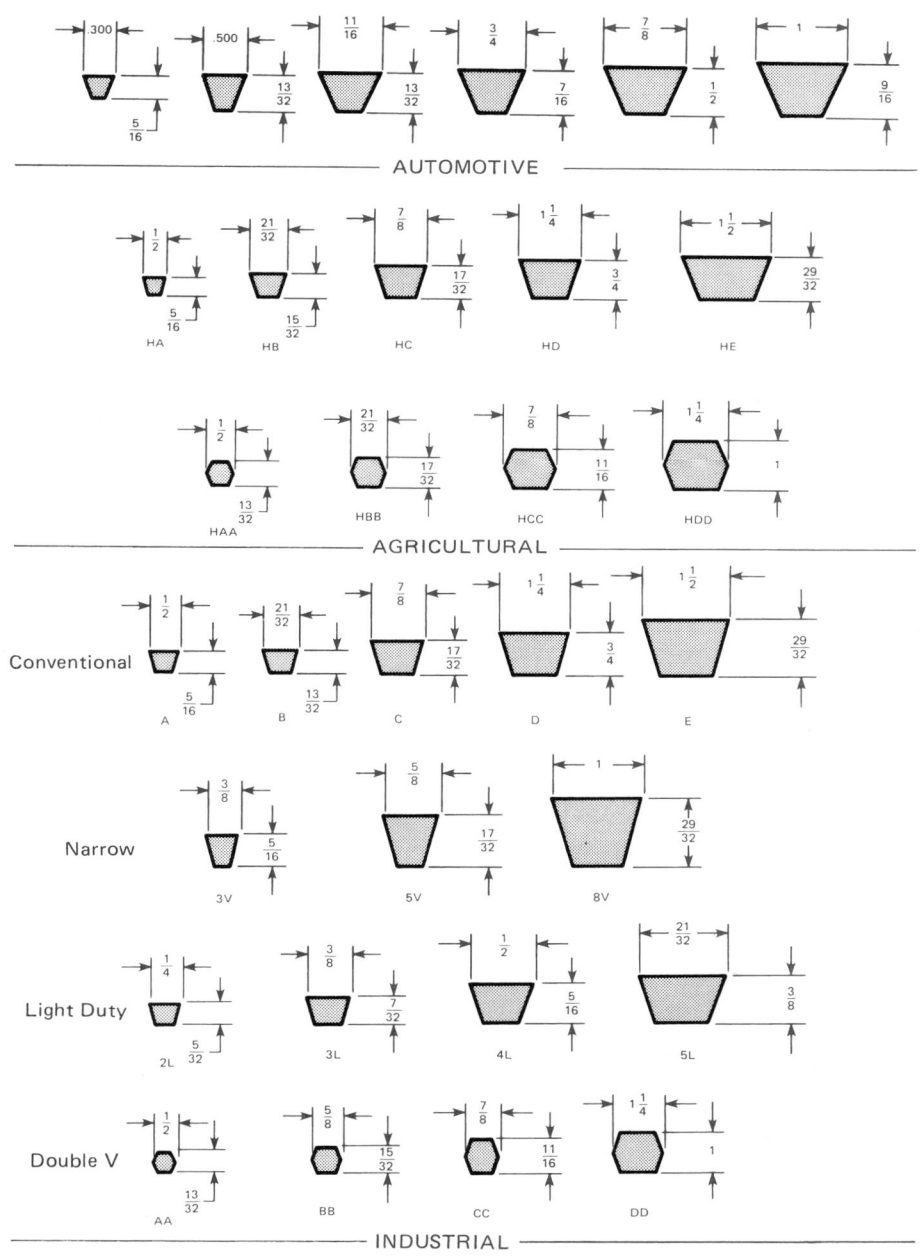

FIGURE 17.2 ■ Standard V-belt types.

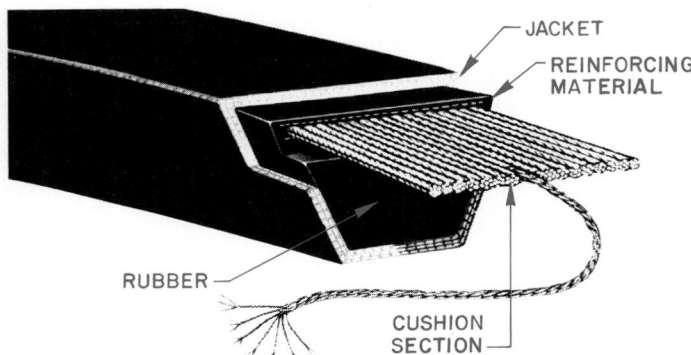

FIGURE 17.3 ■ V belt with steel reinforcing cables. *Courtesy Browning Mfg. Emerson Power Transmission.*

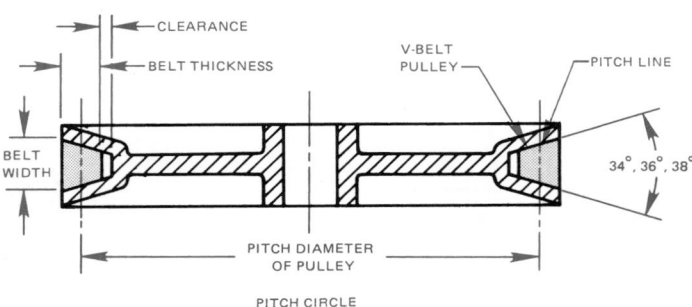

FIGURE 17.4 ■ V belt and pulley (sheave) cross section.

used where drives with nonparallel shafts are required, because the belt may be twisted to accommodate the relationship of the shafts.

Positive Drive Belts

Positive drive belts have a notched underside that contacts a pulley with the same design on the circumference. These belts have some of the same advantages as chain and gear drives due to their positive contact between the belt and the pulley. These belts are more suitable for operations requiring high efficiency, timing, or constant velocity. They may also be used to decrease the pulley size and give the same operating performance of a larger sized V-belt pulley. A sample positive drive belt is shown in Figure 17.5.

FIGURE 17.5 ■ Positive drive belt. *Courtesy Browning Mfg. Emerson Power Transmission.*

TYPICAL BELT DRIVE ARRANGEMENTS

Belt drives are normally arranged with or without an idler. The *idler* is often used to help maintain constant tension on the belt. (See Figure 17.6.)

When belts are in operation, they stretch. Machines that are operated by belt drives must have an idler, an adjustable motor base, or both to provide for belt adjustment. Maintenance of the tension between the shafts is important in controlling the efficiency of the belt drive. Figure 17.7 shows an assembly drawing of an adjustable motor base. There are many different types of adjustable motor bases, including the sliding and tilting types.

In some cases adjustable motor bases are not practical. When this occurs, idler pulleys are used to keep belts tight. A V-belt and flat-belt idler are displayed in Figure 17.8. When idler pulleys are used, their size and location in relation to the belt drive are important. When referring to the placement of the idler pulley, the terms *slack side* and *tight side* are used. When the belt is in operation, the slack side is the belt on the top of the drive and the tight side is normally on the bottom of the drive when the driver and driven shafts are in horizontal alignment as shown in Figure 17.9. There are several typical design positions for the location of the idler pulley in relation to the driver pulley. An

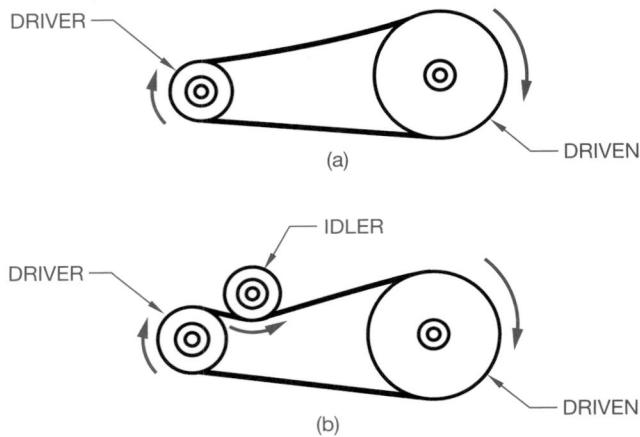

FIGURE 17.6 ■ **(a)** V-belt drive arrangement. **(b)** Belt drive with an idler.

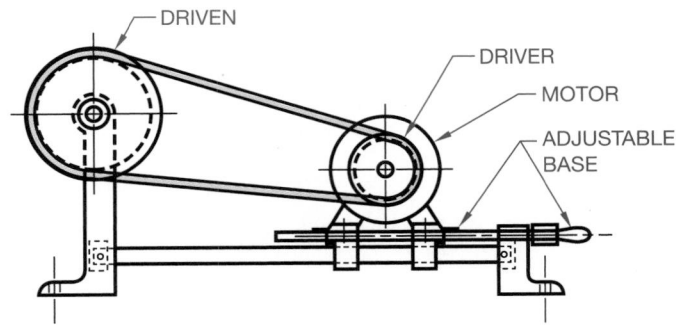

FIGURE 17.7 ■ V-belt drive arrangement with adjustable motor base.

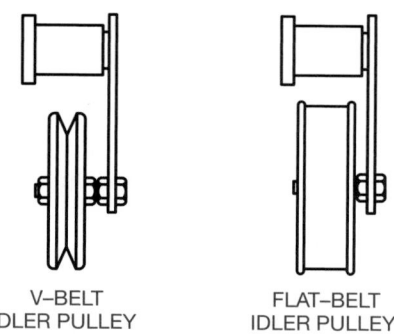

FIGURE 17.8 ■ Idler pulleys. *Courtesy Lovejoy, Inc., Downers Grove, IL.*

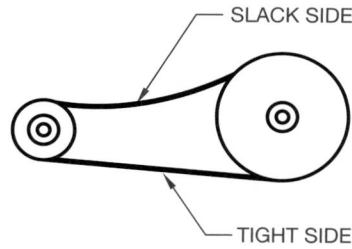

FIGURE 17.9 ■ Belt drive slack side and tight side.

inside idler pulley is placed near the driver on the slack side of the belt. It should be the same diameter or larger than the driver. An inside idler pulley may also be placed on the tight side near the driven pulley. An outside idler pulley is placed near the driver on the outside of the belt on the slack side. The best design for this application is to have the idler pulley slightly larger in diameter than the driver pulley. When an outside idler pulley is placed on the tight side, it is normally placed near the driven pulley. (See Figure 17.10.)

BELT DRIVE SELECTION

Belts are selected for a specific application based on the function of the machinery. The factors influencing the selection and design of belt drives include:

■ Horsepower of driving shaft.

■ Speed of driving shaft.

■ Service conditions.

■ Type of machine.

■ Center distance between shafts.

■ Belt length.

■ Number of belts required.

Designing V-belt Drives

Belt manufacturers provide formulas and catalogs of the design and selection of belt drives based on the criteria described previously. The following discussion is based on engineering data provided in the Browning Power Transmission catalog, produced by the Browning Manufacturing Division of Emerson Power Transmission Company. In order to select a drive using standard drive selection tables, this sample problem is used, courtesy of Browning Manufacturing Division of Emerson Power Transmission Company.

A drive is required for a 20-HP (horsepower) motor driving a fan 16 hours per day. The motor speed is 3600 rpm and the shaft size is 1.625 in. The fan speed is approximately 2250 rpm and the fan shaft is 1.438 in. The center distance is 38 in. minimum and 41 in. maximum.

Step 1. Determine the design horsepower. The design HP of a machine takes into account the type of service and the efficiency of a particular machine. Design horsepower is calculated using the formula:

$$\text{Design HP} = \text{Rated HP} \times \text{Service Factor}$$

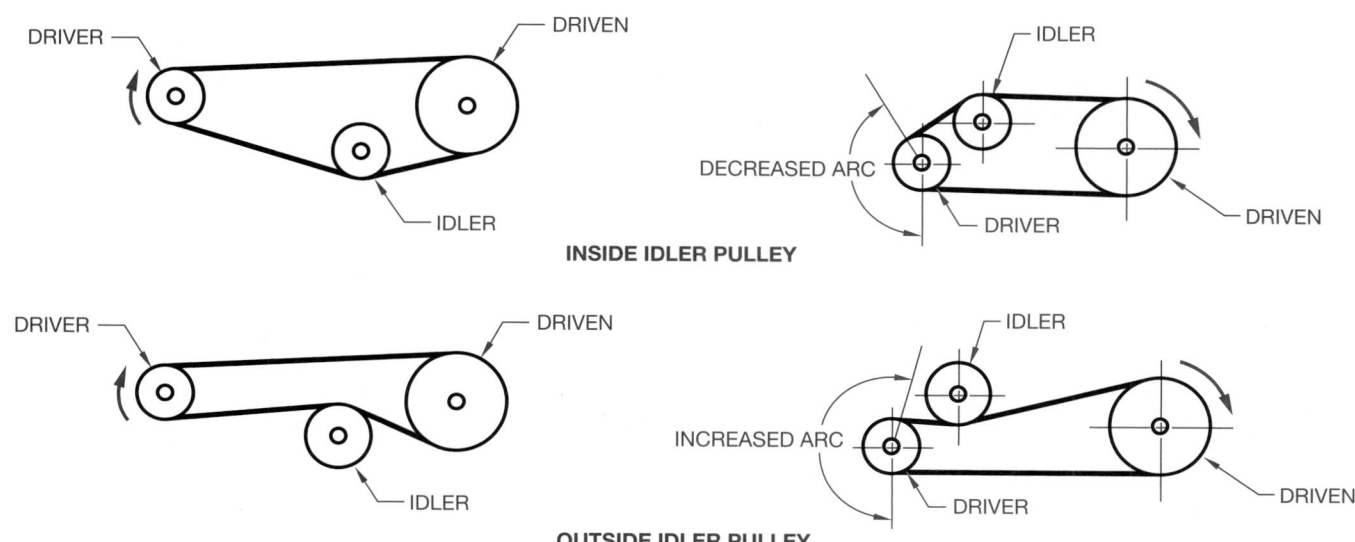

FIGURE 17.10 ■ Idler pulley arrangements.

	TYPES OF DRIVING UNITS		
	AC Motors; Normal Torque, Squirrel Cage, Synchronous and Split Phase, DC Motors; Shunt Wound. Multiple Cylinder Internal Combustion Engines.		
TYPES OF DRIVEN MACHINES	Intermittent Service (3-5 Hours Daily or Seasonal)	Normal Service (8-10 Hours Daily)	Continuous Service (16-24 Hours Daily)
Agitators for Liquids Blowers and Exhausters Centrifugal Pumps and Compressors Fans up to 10 HP Light Duty Conveyors	1.0	1.1	1.2
Belt Conveyors For Sand, Grain, etc. Dough Mixers Fans Over 10 HP Generators Line Shafts Laundry Machinery Machine Tools Punches-Presses-Shears Printing Machinery Positive Displacement Rotary Pumps Revolving and Vibrating Screens Speed Reducers, All Types	1.1	1.2	1.3
Brick Machinery Bucket Elevators Exciters Piston Compressors Conveyors (Drag-Pan-Screw) Hammer Mills Paper Mill Beaters Piston Pumps Positive Displacement Blowers Pulverizers Saw Mill and Woodworking Machinery Textile Machinery	1.2	1.3	1.4
Crushers (Gyratory-Jaw-Roll) Mills (Ball-Rod-Tube) Hoists Rubber Calendars-Extruders-Mills	1.3	1.4	1.5

A minimum Service Factor of 2.0 is suggested for equipment subject to choking.*

> ***CAUTION—Drives requiring high Overload Services Factors, such as crushing machinery, certain reciprocating compressors, etc. subjected to heavy shock load without suitable fly wheels, may need heavy duty web type sheaves rather than standard arm type. For any such application, consult the Browning Engineers.**

FIGURE 17.11 ■ Service factors for belt drives. *Courtesy Browning Mfg. Emerson Power Transmission.*

The rated horsepower is the horsepower specified on the driver motor. For example, the label on the motor may read 2 HP. The service factor is determined by the type of machinery, the number of hours of daily operation, and the type of driven unit. Service factors may be selected from the chart shown in Figure 17.11. Based on the example problem, a fan operating 16 hours per day has a service factor of 1.3. Calculate the design HP:

$$20 \text{ HP} \times 1.3 = 26 \text{ Design HP}$$

Step 2. Determine the driven speed. To establish the required driven speed, refer to Figure 17.12. Find the "Nominal Driven Speeds and Horsepower per Belt" column. Under the column labeled "3500 RPM Motor," find the "Nominal Driven RPM" column. Select the RPMs equal to or less than the given fan speed of 2250 rpm. Fifteen combinations that provide 2244 rpm for the driven pulley are highlighted in Figure 17.12.

Step 3. Determine the available selections. Refer again to Figure 17.12 and look under the "HP per Belt" column to determine which sheave combination results in the smallest sheave diameters, the least number of belts, and the

lightest belt selection for the sample problem. To begin this process, assume your selection will be found in the "Gripnotch" column under HP per Belt. If you use a "Super" belt, another selection would be made. Read down the "Gripnotch" column until you find the largest HP per belt (12.94) without exceeding the Design HP (26) when multiplied by 1, 2, or 3 belts. Your goal is to use the least number of belts. Therefore, the best solution in this case is 2 belts × 12.94 = 25.88. 25.88 is less than the 26-Design HP. These items must be equal to or less than the given data. Of the fifteen selections from step 2, the proposed selections are highlighted in Figure 17.12. They are:

> Pitch Diameter (PD) Driver = 5.4 in.
> PD Driven = 8.4 in.
> Belt Section = B
> Number of Belts = 1 minimum and 3 maximum (exact number to be determined later)
> HP per Belt = 12.94

Step 4. Determine the belt part number for the approximate center distance and the F factor for the belt. The selected center distance, as stated in the problem, must be between 38 and 41 in. The F factor is a correction factor for the loss of arc of contact. The standard arc contact for a belt to sheave is 180°. When the belt does not contact 180°, there is loss or gain or arc contact. When there is loss or gain or arc contact, a correction factor must be used to establish the corrected horsepower and number of belts required. Look at Figure 17.13 which is an extension of the chart in Figure 17.12. Find the center distance (CD) in the horizontal column that is closest to the required center distance. If you cannot find an acceptable CD between the given requirements, find a pair of CDs as highlighted in Figure 17.13. In this case, you have located:

> Belt B90 with a CD of 35.0 in. and F of 0.99
> Belt B103 with a CD of 41.5 and F of 1.02

To establish the exact belt for this application, you must *interpolate*. Interpolate means to make an approximation between existing known values. Interpolate to find the correct belt for the application using this method:

> For every inch of belt length difference between belts, there is about ½ in. of center distance change. The belt numbers represent a relationship. For example, the B90 belt is 13 in. shorter than the B103 belt (B103 – B90 = 13). For this example, if you subtract 3 in. from a B103 belt, you get a B100 belt. If there is ½ in. of CD change for every inch of belt length, then 3 in. of length change equals 1.5 in. of CD change. Therefore, the B100 belt has a CD of 40 in.

Now determine the F factor of the B100 belt by comparing the values of the known belts:

B90	CD = 35 in.	F = .99
B100	CD = 40 in.	F = ?
B103	CD = 41.5 in.	F = 1.02

3500 and 1750 RPM Motors

Ratio	Driver P.D.	Driven P.D.	Belt Section	Grooves Min.	Grooves Max.	3500 RPM Nominal Driven RPM	3500 Super	3500 Gripnotch	1750 RPM Nominal Driven RPM	1750 Super	1750 Gripnotch	Belt No.	C.D.	F.	Belt No.	C.D.	F.	Belt No.	C.D.	F.
1.54	10.4	16.0	B	1	3	—	—	—	1136	14.97	17.10	B67	13.4	0.87	B74	16.9	0.90	B81	20.5	0.94
1.54	13.0	20.0	C	1	12	—	—	—	1136	27.81	33.05	C85	17.7	0.85	C100	25.3	0.89	C112	31.3	0.91
1.55	3.8	5.9	B	1	3	2258	4.33	8.03	1129	3.32	5.33	B28	7.2	0.73	B45	15.7	0.84	B56	21.3	0.89
1.55	4.0	6.2	A	1	6	2258	5.19	6.35	1129	3.31	4.17	A26	5.5	0.76	A36	10.6	0.84	A46	15.6	0.89
1.55	4.0	6.2	B	1	6	2258	5.00	8.71	1129	3.75	5.76	B28	6.8	0.71	B45	15.3	0.84	B56	20.9	0.87
1.55	9.6	14.9	5V	2	10	—	—	—	1129	27.32	32.36	5V670	14.0	0.85	5V800	20.6	0.89	5V950	28.1	0.93
1.56	3.2	5.0	A	1	6	2244	3.62	4.35	1122	2.32	2.92	A24	6.1	0.78	A35	11.7	0.84	A45	16.7	0.88
1.56	3.2	5.0	B	1	2	2244	2.18	5.87	1122	1.99	4.00	B28	8.4	0.73	B45	16.9	0.84	B56	22.4	0.89
1.56	3.6	5.6	A	1	6	2244	4.43	5.38	1122	2.82	3.56	A24	5.3	0.76	A35	10.9	0.84	A45	15.9	0.88
1.56	3.6	5.6	B	1	6	2244	3.63	7.33	1122	2.88	4.89	B28	7.6	0.73	B45	16.1	0.84	B56	21.6	0.89
1.56	3.9	6.1	B	1	3	2244	4.67	8.38	1122	3.54	5.55	B28	7.0	0.71	B45	15.4	0.84	B56	21.0	0.87
1.56	4.1	6.4	B	1	3	2244	5.33	9.05	1122	3.97	5.97	B28	6.6	0.71	B45	15.1	0.84	B56	20.6	0.87
1.56	4.1	6.4	3V	1	4	2244	7.72	8.76	1122	4.47	4.90	3V280	5.6	0.79	3V335	8.4	0.84	3V400	11.7	0.89
1.56	4.5	7.0	A	1	3	2244	6.07	7.50	1122	3.90	4.92	A29	6.0	0.78	A39	11.0	0.84	A49	16.1	0.90
1.56	4.5	7.0	5V	2	5	2244	—	17.45	1122	—	10.28	5V500	15.9	0.82	5V630	22.4	0.86	5V800	30.9	0.90
1.56	4.8	7.5	A	1	1	2244	6.56	8.15	1122	4.26	5.36	A31	6.3	0.79	A40	10.9	0.85	A49	15.4	0.90
1.56	5.4	8.4	B	1	3	2244	9.06	12.94	1122	6.66	8.65	B34	6.9	0.74	B47	13.5	0.84	B58	19.0	0.88
1.56	5.4	8.4	5V	2	5	2244	—	24.53	1122	—	14.42	5V500	14.1	0.82	5V630	20.6	0.86	5V800	29.1	0.90
1.56	5.5	8.6	A	1	3	2244	7.57	9.53	1122	5.05	6.36	A36	7.4	0.81	A45	12.0	0.87	A54	16.5	0.92
1.56	6.4	10.0	A	1	3	2244	8.58	11.03	1122	6.02	7.58	A42	8.6	0.84	A50	12.6	0.89	A58	16.7	0.93
1.57	2.8	4.4	A	1	2	2229	2.76	3.26	1115	1.80	2.28	A24	6.9	0.78	A35	12.5	0.84	A45	17.5	0.90
1.57	3.0	4.7	A	1	3	2229	3.19	3.81	1115	2.06	2.60	A24	6.5	0.78	A35	12.1	0.84	A45	17.1	0.90
1.57	3.5	5.5	A	1	3	2229	4.23	5.13	1115	2.69	3.40	A24	5.5	0.76	A35	11.0	0.84	A45	16.0	0.88
1.57	3.7	5.8	A	1	3	2229	4.62	5.63	1115	2.94	3.71	A24	5.1	0.75	A35	10.6	0.84	A45	15.7	0.88
1.57	4.2	6.6	A	1	6	2229	5.56	6.83	1115	3.55	4.48	A27	5.5	0.76	A37	10.6	0.84	A47	15.6	0.89
1.58	4.8	7.6	A	1	6	2215	6.56	8.16	1108	4.26	5.36	A31	6.3	0.79	A40	10.8	0.85	A49	15.3	0.90
1.58	5.0	7.9	B	1	1	2215	8.04	11.85	1108	5.86	7.86	B32	6.6	0.73	B46	13.7	0.84	B57	19.2	0.87
1.58	5.2	8.2	A	1	10	2215	7.16	8.97	1108	4.71	5.94	A34	7.0	0.80	A43	11.5	0.86	A52	16.1	0.92
1.58	5.7	9.0	A	1	3	2215	7.83	9.90	1108	5.27	6.64	A37	7.4	0.82	A46	12.0	0.88	A55	16.5	0.93
1.58	6.0	9.5	A	1	1	2215	8.18	10.41	1108	5.60	7.05	A39	7.8	0.82	A47	11.8	0.88	A55	15.9	0.92
1.58	6.6	10.4	B	1	3	2215	11.51	15.74	1108	8.97	10.95	B43	8.8	0.80	B53	13.9	0.86	B63	19.0	0.89
1.58	6.7	10.6	A	1	1	2215	8.85	11.46	1108	6.33	7.98	A44	8.9	0.85	A52	12.9	0.89	A60	16.9	0.94
1.58	6.9	10.9	B	1	1	2215	11.95	16.30	1108	9.52	11.50	B45	9.2	0.81	B55	14.3	0.86	B65	19.3	0.90
1.58	7.2	11.4	C	2	6	—	—	—	1108	12.92	19.41	C51	12.2	0.76	C75	24.2	0.84	C96	34.8	0.89
1.58	7.4	11.7	5V	2	8	—	—	—	1108	18.97	23.39	5V500	9.8	0.79	5V630	16.4	0.85	5V800	24.9	0.90

FIGURE 17.12 ■ Motor speed and drive selection table. *Courtesy Browning Mfg. Emerson Power Transmission.*

3500 and 1750 RPM Motors

Nominal Center Distance (C.D.) and Arc-Length Factor (F) using Browning Belts | Sheave Combination

Belt No.	C.D.	F.	Belt No.	C.D.	F.	Belt No.	C.D.	F.	Belt No.	C.D.	F.	Belt No.	C.D.	F.	Belt No.	C.D.	F.	Driver P.D.	Driven P.D.	Ratio
5V1120	36.7	0.95	5V1320	46.7	0.98	5V1600	60.7	1.03	5V1900	75.7	1.06	5V2240	92.7	1.08	5V3550	158.2	1.17	9.6	14.9	1.55
A55	21.7	0.95	A65	26.7	0.98	A75	31.7	1.01	A85	36.7	1.05	A95	41.7	1.06	A180	84.2	1.19	3.2	5.0	1.56
B67	27.9	0.93	B78	33.4	0.97	B88	38.4	1.00	B99	43.9	1.02	B136	62.5	1.09	B360	173.7	1.31	3.2	5.0	1.56
A55	20.9	0.95	A65	25.9	0.98	A75	30.9	1.01	A85	35.9	1.04	A95	40.9	1.06	A180	83.4	1.19	3.6	5.6	1.56
B67	27.2	0.93	B78	32.7	0.97	B88	37.7	0.99	B99	43.2	1.02	B136	61.7	1.09	B360	172.9	1.31	3.6	5.6	1.56
B67	26.5	0.93	B78	32.0	0.97	B88	37.0	0.99	B99	42.5	1.01	B136	61.0	1.09	B360	172.3	1.31	3.9	6.1	1.56
B67	26.1	0.93	B78	31.6	0.97	B88	36.6	0.99	B99	42.1	1.01	B136	60.6	1.09	B360	171.9	1.31	4.1	6.4	1.56
3V475	15.5	0.92	3V560	19.7	0.95	3V670	25.2	1.00	3V800	31.7	1.03	3V950	39.2	1.07	3V1400	61.7	1.15	4.1	6.4	1.56
A59	21.1	0.94	A69	26.1	0.99	A79	31.1	1.02	A89	36.1	1.04	A100	41.6	1.07	A180	81.6	1.19	4.5	7.0	1.56
5V1000	40.9	0.95	5V1250	53.5	1.00	5V1600	71.0	1.04	5V2000	91.0	1.08	5V2500	116.0	1.11	5V3550	168.5	1.17	4.5	7.0	1.56
A58	19.9	0.94	A67	24.5	0.97	A76	29.0	1.01	A85	33.5	1.04	A94	38.0	1.05	A180	81.0	1.19	4.8	7.5	1.56
B69	24.5	0.92	B80	30.0	0.97	B90	35.0	0.99	B103	41.5	1.02	B144	62.0	1.11	B360	169.3	1.31	5.4	8.4	1.56
5V1000	39.1	0.95	5V1250	51.6	0.99	5V1600	69.1	1.04	5V2000	89.2	1.08	5V2550	114.2	1.11	5V3550	166.7	1.17	5.4	8.4	1.56
A63	21.0	0.96	A72	25.5	0.98	A81	30.0	1.01	A90	34.5	1.05	A100	39.5	1.07	A180	79.6	1.19	5.5	8.6	1.56
A66	20.7	0.97	A74	24.7	0.99	A82	28.7	1.01	A90	32.7	1.03	A98	36.7	1.07	A180	77.7	1.19	6.4	10.0	1.56
A55	22.5	0.95	A65	27.5	0.98	A75	32.5	1.02	A85	37.5	1.05	A95	42.5	1.06	A180	85.0	1.19	2.8	4.4	1.57
A55	22.1	0.95	A65	27.1	0.98	A75	32.1	1.01	A85	37.1	1.05	A95	42.1	1.06	A180	84.6	1.19	3.0	4.7	1.57
A55	21.1	0.95	A65	26.1	0.98	A75	31.1	1.01	A85	36.1	1.04	A95	41.1	1.06	A180	83.6	1.19	3.5	5.5	1.57
A55	20.7	0.93	A65	25.7	0.98	A75	30.7	1.01	A85	35.7	1.04	A95	40.7	1.05	A180	83.2	1.19	3.7	5.8	1.57
A57	20.6	0.94	A67	25.6	0.99	A77	30.6	1.01	A87	36.6	1.04	A97	40.6	1.07	A180	82.2	1.19	4.2	6.6	1.57
A58	19.9	0.94	A67	24.4	0.97	A76	28.9	1.01	A85	33.4	1.04	A94	37.9	1.05	A180	80.9	1.19	4.8	7.6	1.58
B68	24.7	0.92	B79	30.2	0.97	B89	35.2	0.99	B100	40.7	1.02	B140	60.7	1.10	B360	170.0	1.31	5.0	7.9	1.58
A61	20.6	0.95	A70	25.1	0.98	A79	29.6	1.00	A88	34.1	1.04	A97	38.6	1.07	A180	80.1	1.19	5.2	8.2	1.58
A64	21.0	0.96	A73	25.5	0.98	A82	30.1	1.01	A91	34.6	1.05	A103	40.6	1.07	A180	79.1	1.19	5.7	9.0	1.58
A63	19.9	0.96	A71	23.9	0.98	A79	27.9	1.00	A87	31.9	1.02	A95	35.9	1.05	A180	78.5	1.19	6.0	9.5	1.58
B73	24.0	0.93	B82	29.0	0.94	B92	33.5	0.97	B105	40.0	1.03	B144	59.5	1.10	B360	166.8	1.31	6.6	10.4	1.58
A68	21.0	0.97	A76	25.0	0.99	A84	29.0	1.01	A92	33.0	1.03	A103	38.5	1.05	A180	77.0	1.18	6.7	10.6	1.58
B75	24.3	0.94	B84	28.8	0.96	B94	33.9	0.98	B112	42.9	1.04	B154	63.9	1.11	B360	166.2	1.31	6.9	10.9	1.58
C112	42.8	0.94	C136	54.8	0.98	C162	67.8	1.02	C210	91.8	1.08	C270	120.8	1.14	C420	195.8	1.24	7.2	11.4	1.58
5V1000	34.9	0.93	5V1250	47.4	0.99	5V1600	65.0	1.03	5V2000	85.0	1.07	5V2500	110.0	1.11	5V3550	162.5	1.17	7.4	11.7	1.58

FIGURE 17.13 ■ Nominal center distances, number of belts, arc length factor chart. *Courtesy Browning Mfg. Emerson Power Transmission.*

Establish the difference in F factor between the B90 and B103 belts.

$$1.02 - .99 = .03 \text{ (difference in F factor)}$$

Establish the difference in CD between the B90 and B103 belts:

$$41.5 - 35 = 6.5 \text{ in. (difference in CD)}$$

Now divide the difference in F factor by the difference in CD to determine the F factor per inch of CD:

$$.03 \text{ divided by } 6.5 = .0046$$
(F factor per inch of CD)

The B100 belt has a 40-in. CD, which is 1.5 in. less than the B103 belt; so, the B100 belt has an F factor of 1.5 × .0046 less than the B103 belt:

$$1.5 \times .0046 = .0069$$
F 1.02 (B103 belt) − .0069 = 1.01
(rounded from 1.031)
F factor for the B100 belt.

Therefore, the B100 belt has an F factor of 1.01.

Step 5. Determine the corrected horsepower and number of belts for the drive. The corrected horsepower is an adjustment of the design horsepower in relation to the F factor. It is calculated with this formula:

$$\text{F Factor} \times \text{HP per Belt} = \text{Corrected Horsepower}$$

The corrected horsepower for the sample problem is:

$$1.01 \text{ (F)} \times 12.94 \text{ (HP per belt)} = 13.06$$
Corrected HP per Belt

Now determine the number of belts required for the drive using this formula:

Design HP/Corrected HP = Number of Belts

If the number calculated in this formula is one or less, then one belt is required; or two belts for two or less; three belts for three or less; and so on. The number of belts for the sample problem is:

$$26 \text{ (Design HP)}/13.06 \text{ (Corrected HP)} = 1.99$$
= 2 Belts Required

In conclusion, the design solution to the sample problem is has follows:

26 Design HP
12.94 Design HP per Belt
13.06 Corrected HP per Belt
B100 Belt Number
40 in. Center Distance
2 Belts Required
⌀5.4 in. Driver Pitch Diameter (PD)
⌀8.4 in. Driven PD
B Belt Section

Determine the Belt Drive Ratio

The belt drive ratio is the relationship between the drive and driven pulleys. The ratio may be determined by dividing the rpm of the driver by the rpm of the driven pulley. If you use the sample problem presented earlier, for example, the motor rpm from Figure 17.12 is 3500 rpm, and the nominal driven rpm from the same chart is 2244 rpm. Therefore, the Ratio = 3500 rpm/2244 rpm = 1.5597 = 1.56. Now, refer back to Figure 17.12 and notice the 1.56 highlighted in the Ratio column at the far left. This 1.56 ratio coincides with the other items selected for the application.

Determine the Belt Velocity

Belt velocity is the speed a belt travels in feet per minutes (fpm) for a given application. In some cases, the belt velocity may be important. In many applications, though, it is not a factor. Belt velocity becomes important at very high speeds. For example, cast iron sheaves should not travel at speeds greater than 6500 fpm. Another factor that influences design at high speeds is vibration, which may require balancing of the sheaves. Belt velocity is calculated using this formula:

$$\text{Belt Velocity fpm} = \text{Pitch Diameter (PD)}$$
Sheave × 0.2618 × Sheave Speed rpm
(0.2618 is a constant to be used for all applications)

The belt velocity for the driver sheave of the sample problem is:

$$⌀5.4 \text{ (PD Sheave)} \times 0.2618 \times 3500 \text{ rpm} = 4948 \text{ fpm}$$

CHAIN DRIVES

Chain drives, like gear and belt drives, are used to transmit power from one shaft to another. As discussed earlier, chain drives have certain advantages over gear and belt drives. In general, chain drives are less efficient and durable than gear drives, but they offer greater power capacity, more position power transmission, durability, and longer service life than belts, especially at high temperatures.

In chain drive applications, toothed wheels called *sprockets* mate with a chain to transmit power from one shaft to another. (See Figure 17.14.)

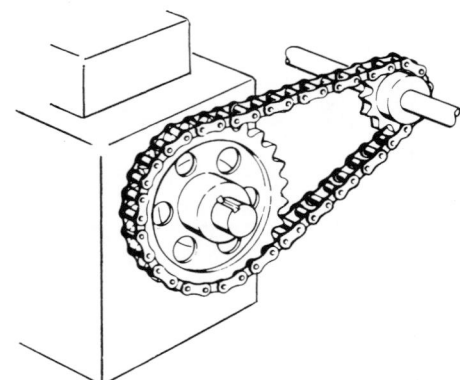

FIGURE 17.14 ■ Chain drive sprockets and chain. *Courtesy Morse Industrial.*

CHAIN DRIVE SPROCKETS

Sprockets are designed for various applications. Depending on the shaft mounting requirements, sprockets are available with a hub on one or both sides or without hubs. Sprockets are designed with a solid web in most applications, although weight is reduced by designing recessed webs or spokes.

Sprockets are machined from cast iron, steel, aluminum alloy, powdered metal, or plastic. Split sprockets are used when mounting between bearings requires easy installation. Some applications have sprockets mounted to a removable steel or cast iron hub when frequent replacement is required.

CHAIN CLASSIFICATION AND TYPES

Chains are classified in relation to the accuracy of construction between the sprocket and the chain links. There are three broad categories: precision, nonprecision, and light-duty chains. Precision chains are designed for smooth, free-running operation at high speed and high power. Common precision chains include the roller, offset sidebar, silent inverted tooth, and double pitch chains. Nonprecision chains do not have as high a degree of precision between the sprocket and chain links. These lower cost chains are generally used for low-speed applications on machinery rated below 50 HP. Nonprecision chains include detachable, pintle, and welded chains. Light-duty chains are designed for application in low-power situations such as equipment control mechanisms for computers and typewriters or appliance controls.

PRECISION CHAINS
Roller Chain

Roller chain is the most commonly used power transmission chain. Roller chain is rated up to 630 HP for single chain drives referred to as single-strand chain. Roller chains may be added in multiple units, known as multiple-strand chains, to substantially increase the horsepower under which they may operate. Roller chains provide efficient, quiet operation, and should be lubricated for the best service life. Figure 17.15 shows the parts of a typical roller chain.

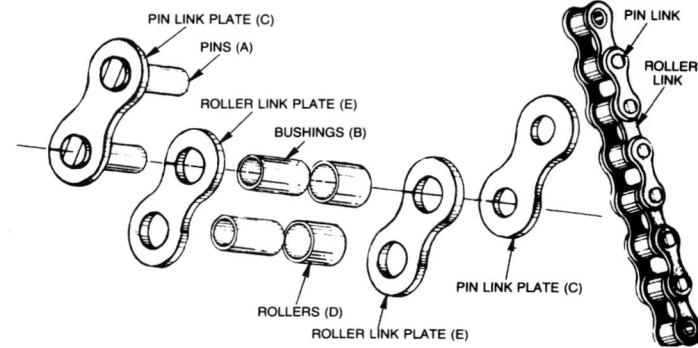

FIGURE 17.15 ■ Roller chain assembly. *Courtesy Rainbow Industrial Products Corporation.*

SPROCKETS

The same simplified representation used in gear CADD drawings may also be used for sprocket CADD drawings, or custom software is available to automatically generate detailed drawings from given data (see Figure 17.16).

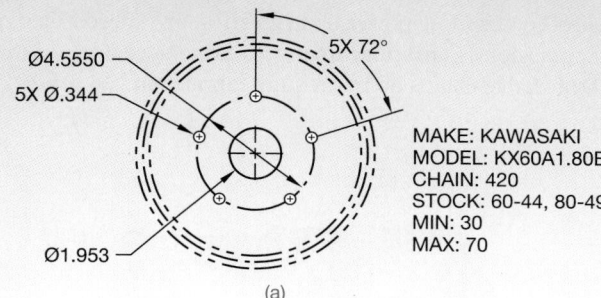

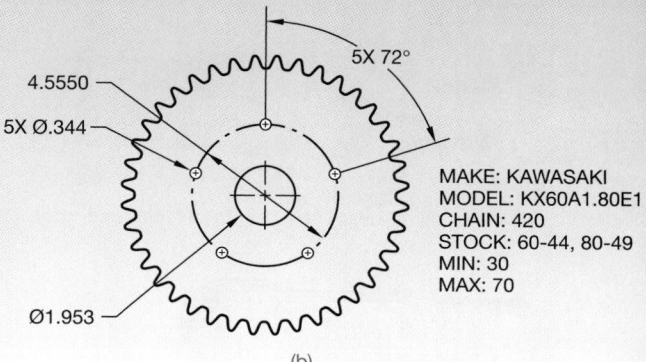

FIGURE 17.16 ■ Sprocket drawing: (**a**) simplified representation. (**b**) detailed representation.

FIGURE 17.17 ■ Single-strand and multiple-strand roller chain. *Courtesy Morse Industrial.*

The *chain pitch* is important when designing a chain drive. The chain pitch is the distance from the center of one pin to the center of the other pin in one link. Figure 17.17 shows a single-strand chain with the pitch dimensioned and a multiple-strand chain. Roller chains are available with pitch lengths ranging from ¼ to 3 in.

Double Pitch Roller Chain

Double pitch roller chain is designed mostly for situations with long center distances such as conveyors. While the efficiency of the double pitch chain is as good as the roller chain, it is intended for lighter duty operation than regular pitch roller chain. (See Figure 17.18.)

Offset Sidebar Roller Chain

The offset sidebar, shown in Figure 17.19, is the least expensive precision chain. It is designed to carry heavier loads than non-

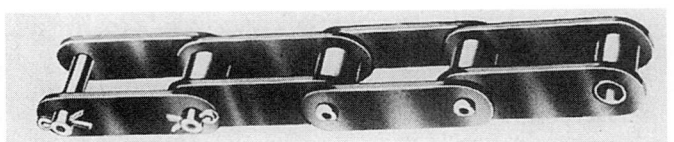

FIGURE 17.18 ■ Double pitch roller chain. *Courtesy Morse Industrial.*

FIGURE 17.19 ■ Offset sidebar roller chain.

FIGURE 17.20 ■ Inverted tooth silent chain. *Courtesy Morse Industrial.*

precision chains. Offset sidebar chains are designed to handle loads up to 425 HP and speeds up to 36 feet per second. One of the advantages of the offset sidebar chain over the other chain types is its open construction, which allows it to withstand dirt and contaminants that might cause other precision chains to bind and wear out rapidly. For this reason and because it is rugged and durable, offset sidebar chains are often used to drive construction machinery.

Inverted Tooth Silent Chain

The most expensive precision chain to manufacture is the inverted tooth silent chain shown in Figure 17.20. This chain is used where high speed and smooth, quiet operation are required in rigorous applications. Lubrication is required to keep these chains running trouble free. Applications include machine tools, pumps, and power drive units.

NONPRECISION CHAINS

Detachable Chain

The detachable chain shown in Figure 17.21 is the lightest, simplest, and least expensive of all chains. The detachable chain is capable of transmitting power up to 25 HP at low speeds. One common application is farm machinery requiring speeds of up to 350 fpm. Detachable chains do not require lubrication.

FIGURE 17.21 ■ Detachable chains.

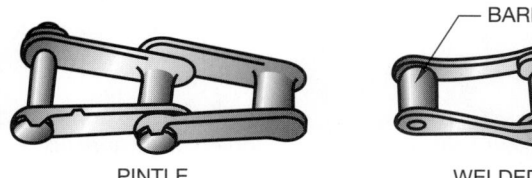

FIGURE 17.22 ■ Pintle and welded steel chains.

Pintle and Welded Steel Chains

The pintle and welded steel chains shown in Figure 17.22 are a combination design between the detachable and the offset sidebar roller chain. They are used for applications similar to the detachable chain where more rigorous service is required for up to 40 HP and 425 fpm. Lubrication is not required, so they may also be used where lubrication is not effective, such as dusty or wet conditions.

LIGHT-DUTY CHAIN
Bead Chain

Bead chains, Figure 17.23, are commonly used for control mechanisms for light-duty, low-power applications. Bead chains are rated from 15 to 75 pounds and are available in standard bead diameters of ³⁄₃₂, ⅛, ³⁄₁₆, and ¼ in.

CHAIN DRIVE ARRANGEMENTS

Chain drives have some of the same characteristics as belt drives in the following ways:

■ Chains elongate when in operation, thus requiring shaft adjustment or link removal to maintain proper tightness.

■ Chains have a tight side and a slack side when in operation.

The arrangement and direction of rotation of the driver sprocket and the driven sprocket are important to the proper function of the chain drive mechanism. For chain drives with long center distances, the slack side should be on the bottom, because upper side slack could cause the chain on the top meet the bottom as the chain elongates. Figure 17.24 shows the recommended and not recommended arrangements.

On short center drive arrangements, as shown in Figure 17.25, the slack side should be on the bottom because slack on the top could cause the chain to jump out of the sprocket. The relationship of the shaft centers is also important. The shaft centers should be

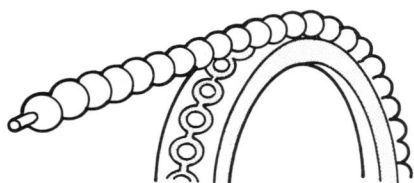

FIGURE 17.23 ■ Bead chain.

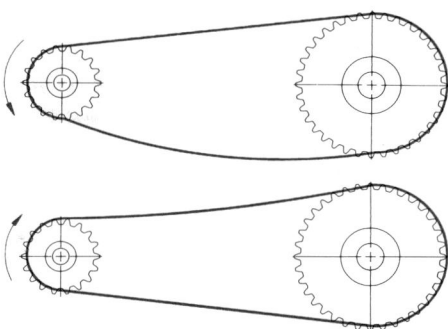

FIGURE 17.24 ■ Chain drives with long center distances should have slack on bottom side.

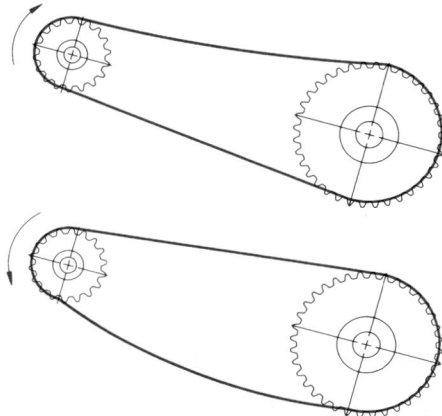

FIGURE 17.25 ■ Chain drives with short center distances should have slack on the lower side.

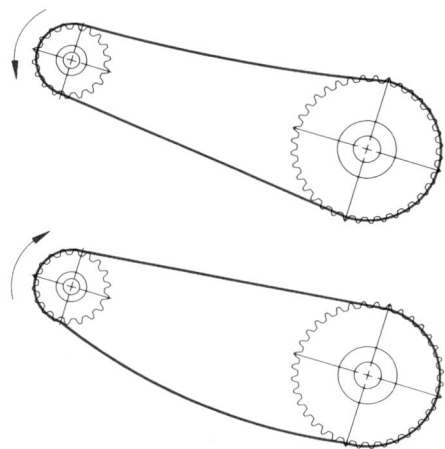

FIGURE 17.26 ■ Chain drive shaft centers should be on a horizontal line or on a line inclined not more than 45°, in which case the slack may be on the upper or lower side.

designed in a horizontal position or inclined no more than 45°. When the shafts are inclined up to 45°, the slack side may be on either the top or bottom. (See Figure 17.26.)

Shaft arrangements between 45° and vertical should be avoided. However, when they are required due to the machine design, the chain should be kept tighter than normal applications. This type of

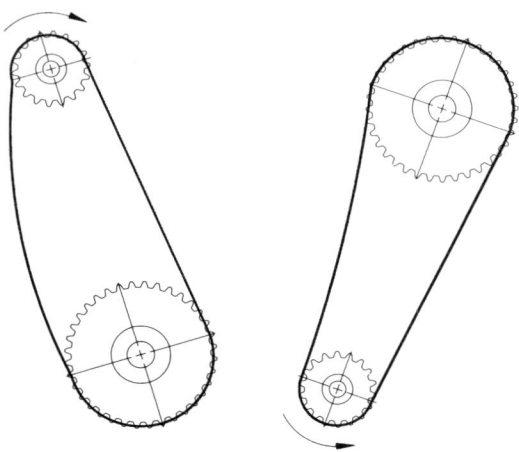

FIGURE 17.27 ■ When the shaft centers are aligned more than 45° from horizontal, the slack side should be in the side closest to horizontal.

arrangement requires that the chain be tightened frequently. Properly designed situations with the shafts inclined greater than 45° provide for the slack to be on the side closest to horizontal. (See Figure 17.27.)

ROLLER CHAIN DRIVE SELECTION

Designing roller chain drives includes selecting the sprocket sizes, the chain pitch, the chain length, the distance between shaft centers, and the lubrication requirements. Factors that contribute to selecting the proper chain drive variables are rpm of the sprockets, the horsepower (HP) of the motor, and the load conditions.

Chain manufacturers provide formulas and catalogs for the design and selection of chain drives. The following discussion is based on engineering data provided in the Morse Power Transmission Products catalog, provided by the Morse Industrial Products Division, Emerson Power Transmission Corporation.

A chain drive is required for a 2-HP electric motor operating laundry machinery. The drive sprocket turns at 1800 rpm, which produces 900 rpm at the driven sprocket. The approximate center distance is 12 in.

Step 1. Determine the design horsepower. The design horsepower (HP) of a machine takes into account the type of service and the efficiency of the machine. A service factor is applied to the horsepower rating of the drive motor to determine the design horsepower of a chain drive. The service factors are established by the power source, the nature of the load, and the strain or shock on the drive. Three basic operating characteristics are used to establish the service factors. (See Figure 17.28.)

Now, determine the service factor for the given problem. You have a 2-HP electric motor operating laundry machinery. Laundry machinery operates under moderate shock. Refer to the table in Figure 17.29 to determine the

Smooth: Running load is fairly uniform. Starting and peak loads may be somewhat greater than running load, but occur infrequently.

Moderate Shock: Running load is variable. Starting and peak loads are considerably greater than running load and occur frequently.

Heavy Shock: Starting loads are extremely heavy. Peak loads and overloads occur continuously and are of maximum fluctuation.

FIGURE 17.28 ■ Chain drive operating classifications. *Courtesy Morse Industrial.*

service factor. An electric motor operating under moderate shock has a service factor of 1.3.

Establish the design horsepower using the formula:

Motor HP × Service Factor = Design HP

For the sample problem, the design HP is:

2 HP × 1.3 = 2.6 Design HP

Step 2. Determine the number of teeth of the drive sprocket. Now that you know the design HP (2.6 HP) and the rpm of the drive sprocket (1800 rpm), you can determine the number of teeth of the drive sprocket. The drive sprocket is normally the "Small" sprocket as shown in the left column of the chart in Figure 17.30. Look at Figure 17.30 and find the rpm—Small Sprocket highlighted along the top horizontal column. Move down the 1800 rpm column until you find a design HP equal to or greater than the design HP of 2.6 for the sample problem. Find 2.93 highlighted in Figure 17.30. Now move horizontally to the left from 2.93 to find the number of teeth of the small sprocket in the left column. Notice that 40 teeth is highlighted in Figure 17.30.

Step 3. Determine the chain pitch. The *chain pitch* is the center distance between pins of one chain link. (See Figure 17.31.) The number of teeth of the small sprocket is taken from the horsepower rating table for a chain with a specific pitch. If you look back at Figure 17.30, you can see the pitch identified in the heading as "No. 25–¼" pitch." Chains are manufactured with standard size pitches available in increments from ¼ to 3 in. Chains are also classified by number; for example, a ¼-in. pitch is No. 25, a ⅜-in. pitch is No. 35, a ½-in. pitch is No. 40, and a ⅝-in. pitch is No. 50.

Step 4. Determine the ratio and the number of teeth for the driven sprocket. The chain drive ratio is the relationship between the drive and driven sprockets. The ratio may be determined by dividing the rpm of the driver by the rpm of the driven sprocket. If you use the sample problem, the driver is 1800 rpm and the driven is 900 rpm. The ratio is 1800 rpm/900 rpm = 2:1. Now, multiply the number of teeth of the small sprocket by the ratio to get the number of teeth for the large sprocket: 40 (teeth small sprocket) × 2 (ratio) = 80 teeth large sprocket.

service factors

Service Factors are selected below for various applications after first determining the prime mover or power source type.

service factor table

Prime Mover	TYPE
Internal Combustion Engine with Hydraulic Coupling or Torque Converter Electric Motor Turbine Hydraulic Motor	A
Internal Combustion Engine with Mechanical Drive	B

APPLICATION	Type of Prime Mover A	B
AGITATORS (paddle or propeller) Pure liquid Liquids—variable density	1.1 1.2	1.3 1.4
BAKER MACHINERY Dough Mixer	1.2	—
BLOWERS	See Fans	
BREWING & DISTILLING EQUIPMENT Bottling Machinery Brew Kettles, cookers, mash tubs Scale Hopper—Frequent starts	1.0 1.0 1.2	— — —
BRICK & CLAY EQUIPMENT Auger machines, cutting table Brick machines, dry press, & granulator Mixer, pug mill, & rolls	1.3 1.4 1.4	1.5 1.6 1.6
CENTRIFUGES	1.4	1.6
COMPRESSORS Centrifugal & rotary (lobe) Reciprocating 1 or 2 cyl. 3 or more	1.1 1.6 1.3	1.3 1.8 1.5
CONSTRUCTION EQUIPMENT OR OFF-HIGHWAY VEHICLES Drive duty, power take-off, accessory drives	Consult Morse	
CONVEYOR Apron, bucket, pan & elevator Belt (ore, coal, sand, salt) Belt—light package, oven Screw & flight (heavy duty)	1.4 1.2 1.0 1.6	1.6 1.4 1.2 1.8
CRANES & HOISTS Main hoist—medium duty Main hoist—heavy duty, skip hoist	1.2 1.4	1.4 1.6

APPLICATION	Type of Prime Mover A	B
CRUSHING MACHINERY Ball mills, crushing rolls, Jaw crushers	1.6	1.8
DREDGES Conveyors, cable reels Jigs & screens Cutter head drives Dredge pumps	1.4 1.6 Consult Morse See Pumps	1.6 1.8
FANS & BLOWERS Centrifugal, propeller, vane Positive blowers (lobe)	1.3 1.5	1.5 1.7
GRAIN MILL MACHINERY Sifters, purifiers, separators Grinders and hammer mills Roller mills	1.1 1.2 1.3	1.3 1.4 1.5
GENERATORS & EXCITERS	1.2	1.4
MACHINE TOOLS Grinders, lathes, drill press Boring mills, milling machines	1.0 1.1	— —
MARINE DRIVES	Consult Morse	
MILLS Rotary type: Ball, Pebble, Rod, Tube, Roller Dryers, Kilns, & tumbling barrels Metal type: Draw bench carriage & main drive Forming Machines	1.5 1.6 1.5 Consult Morse	1.7 1.8 —
MIXERS Concrete Liquid & Semi-liquid	1.6 1.1	1.8 1.3
OIL INDUSTRY MACHINERY Compounding Units Pipe line pumps Slush pumps Draw works Chillers, Paraffin filter presses, Kilns	1.1 1.4 1.5 1.8 1.5	1.3 1.6 1.7 2.0 1.7

APPLICATION	Type of Prime Mover A	B
PAPER INDUSTRY MACHINERY Agitators, bleachers Barker—mechanical Beater, Yankee Dryer Calendars, Dryer & Paper Machines Chippers & winder drums	1.1 1.6 1.3 1.2 1.5	1.3 1.8 1.5 1.4 1.7
PRINTING MACHINERY Embossing & flat bed presses, folders Paper cutter, rotary press & linotype machine Magazine & newspaper presses	1.2 1.1 1.5	— — —
PUMPS Centrifugal, gear, lobe & vane Dredge Pipe line Reciprocating 3 or more cyl. 1 or 2 cyl.	1.2 1.6 1.4 1.3 1.6	1.4 1.8 1.6 1.5 1.8
RUBBER & PLASTICS INDUSTRY EQUIPMENT Calendars, rolls, tubers Tire-building and Banbury Mills Mixers and sheeters Extruders	1.5 1.6 1.5	1.7 1.8 1.7
SCREENS Conical & revolving Rotary, gravel, stone & vibrating	1.2 1.5	1.4 1.7
STOKERS	1.1	—
TEST STANDS & DYNAMOMETERS	Consult Morse	
TEXTILE INDUSTRY Spinning frames, twisters, wrappers & reels Batchers, calendars & looms	1.0 1.1	— —

FIGURE 17.29 ■ Chain drive service factors. *Courtesy Morse Industrial.*

Step 5. Determine the center distance in chain pitches. As a general rule, the preferred center distance between shafts is between 30 and 50 chain pitches. The maximum recommended spacing between centers is 80 pitches. This is to help ensure that there is clearance between the two sprockets and to allow for a minimum of 120° of chain contact around the small sprocket. (See Figure 17.32.) The center distance between shafts should be no less than the difference between sprocket diameters for ratios greater than 3:1. To determine the center distance in chain pitches, divide the center distance by the chain pitch. For the sample problem, this is:

$$\frac{12 \text{ in (Center Distance)}}{0.25 \text{ in. (Pitch)}} = 48 \text{ Chain Pitches}$$

Step 6. Determine the chain length. First, determine the chain length in pitches using the formula:

$$2C + \frac{S}{2} + \frac{K}{C}, \text{ where:}$$

C = Center distances between sprockets in chain pitches.
S = Number of teeth in small sprocket plus number of teeth in large sprocket.
K = Number of teeth in large sprocket minus number of teeth in small sprocket equals D. Look at the table in Figure 17.33 to find the value K corresponding to D.

For the same problem:

C = 48 (from step 5)
S = 40 + 80 = 120
K = 80 − 40 = 40
D = 40

No. 25—¼" pitch   standard single strand roller chain

No. of Teeth Small Spkt.	Revolutions per Minute—Small Sprocket																								
	50	100	300	500	700	900	1200	1500	1800	2100	2500	3000	3500	4000	4500	5000	5500	6000	6500	7000	7500	8000	8500	9000	10000
12	0.03	0.06	0.16	0.25	0.34	0.43	0.55	0.68	0.80	0.92	1.07	1.26	1.45	1.57	1.32	1.12	0.97	0.86	0.76	0.68	0.61	0.56	0.51	0.47	0.40
13	0.04	0.06	0.17	0.27	0.37	0.47	0.60	0.74	0.87	1.00	1.17	1.38	1.58	1.77	1.49	1.27	1.10	0.96	0.86	0.77	0.69	0.63	0.57	0.53	0.45
14	0.04	0.07	0.19	0.30	0.40	0.50	0.65	0.80	0.94	1.08	1.27	1.49	1.71	1.93	1.66	1.42	1.23	1.08	0.96	0.86	0.77	0.70	0.64	0.59	0.50
15	0.04	0.07	0.20	0.32	0.43	0.54	0.70	0.86	1.01	1.17	1.36	1.61	1.85	2.08	1.84	1.57	1.36	1.20	1.06	0.95	0.86	0.78	0.71	0.65	0.56
16	0.04	0.08	0.22	0.34	0.47	0.58	0.76	0.92	1.09	1.25	1.46	1.72	1.98	2.23	2.03	1.73	1.50	1.32	1.17	1.05	0.94	0.86	0.78	0.72	0.61
17	0.05	0.08	0.23	0.37	0.50	0.62	0.81	0.99	1.16	1.33	1.56	1.84	2.11	2.38	2.22	1.90	1.64	1.44	1.28	1.14	1.03	0.94	0.86	0.79	0.67
18	0.05	0.09	0.25	0.39	0.53	0.66	0.86	1.05	1.24	1.42	1.66	1.96	2.25	2.53	2.42	2.07	1.79	1.57	1.39	1.25	1.12	1.02	0.93	0.86	0.73
19	0.05	0.09	0.26	0.41	0.56	0.70	0.91	1.11	1.31	1.50	1.76	2.07	2.38	2.69	2.62	2.24	1.94	1.70	1.51	1.35	1.22	1.11	1.01	0.93	0.79
20	0.06	0.10	0.28	0.44	0.59	0.74	0.96	1.17	1.38	1.59	1.86	2.19	2.52	2.84	2.83	2.42	2.10	1.84	1.63	1.46	1.32	1.20	1.09	1.00	0.86
21	0.06	0.11	0.29	0.46	0.62	0.78	1.01	1.24	1.46	1.68	1.96	2.31	2.66	2.99	3.05	2.60	2.26	1.98	1.76	1.57	1.42	1.29	1.17	1.08	0.92
22	0.06	0.11	0.31	0.48	0.66	0.82	1.07	1.30	1.53	1.76	2.06	2.43	2.79	3.15	3.27	2.79	2.42	2.12	1.88	1.69	1.52	1.38	1.26	1.16	0.99
23	0.06	0.12	0.32	0.51	0.69	0.86	1.12	1.37	1.61	1.85	2.16	2.55	2.93	3.30	3.50	2.98	2.59	2.27	2.01	1.80	1.62	1.47	1.35	1.24	1.06
24	0.07	0.13	0.34	0.53	0.72	0.90	1.17	1.43	1.69	1.94	2.27	2.67	3.07	3.46	3.73	3.18	2.76	2.42	2.15	1.92	1.73	1.57	1.44	1.32	1.12
25	0.07	0.13	0.35	0.56	0.75	0.94	1.22	1.50	1.76	2.02	2.37	2.79	3.21	3.61	3.96	3.38	2.93	2.57	2.28	2.04	1.84	1.67	1.53	1.40	1.20
26	0.07	0.14	0.37	0.58	0.79	0.98	1.28	1.56	1.84	2.11	2.47	2.91	3.34	3.77	4.19	3.59	3.11	2.73	2.42	2.17	1.95	1.77	1.62	1.49	1.27
28	0.08	0.15	0.40	0.63	0.85	1.07	1.38	1.69	1.99	2.29	2.68	3.15	3.62	4.09	4.54	4.01	3.47	3.05	2.70	2.42	2.18	1.98	1.81	1.66	1.42
30	0.08	0.16	0.43	0.68	0.92	1.15	1.49	1.82	2.15	2.46	2.88	3.40	3.90	4.40	4.89	4.45	3.85	3.38	3.00	2.68	2.42	2.20	2.01	1.84	1.57
32	0.09	0.17	0.46	0.73	0.98	1.23	1.60	1.95	2.30	2.64	3.09	3.64	4.18	4.72	5.25	4.90	4.25	3.73	3.30	2.96	2.67	2.42	2.21	2.03	1.73
35	0.10	0.19	0.51	0.80	1.08	1.36	1.76	2.15	2.53	2.91	3.41	4.01	4.61	5.20	5.78	5.60	4.86	4.26	3.78	3.38	3.05	2.77	2.53	2.32	1.98
40	0.12	0.22	0.58	0.92	1.25	1.57	2.03	2.48	2.93	3.36	3.93	4.64	5.32	6.00	6.68	6.85	5.93	5.21	4.62	4.13	3.73	3.38	3.09	2.83	2.42
45	0.13	0.25	0.66	1.05	1.42	1.78	2.31	2.82	3.32	3.82	4.47	5.26	6.05	6.82	7.58	8.17	7.08	6.21	5.51	4.93	4.45	4.04	3.69	3.38	2.89
	TYPE A				**TYPE B**								**TYPE C**												

TYPE A: Manual or Drip Lubrication (500 fpm max.)
TYPE B: Bath or Disc Lubrication (3500 fpm max.)
TYPE C: Oil Stream Lubrication (Up to max. speed shown).
The limiting RPM for each lubrication type is read from the column to the left of the boundary line shown.
The ratings on this page are in accordance with the standards of the American Chain Association, Copyright 1974.

For Multi. Strand Chain use

No. of Strands	Strand Factor
2	1.7
3	2.5
4	3.3

FIGURE 17.30 ■ Horsepower ratings for single-strand roller chain. *Courtesy Morse Industrial.*

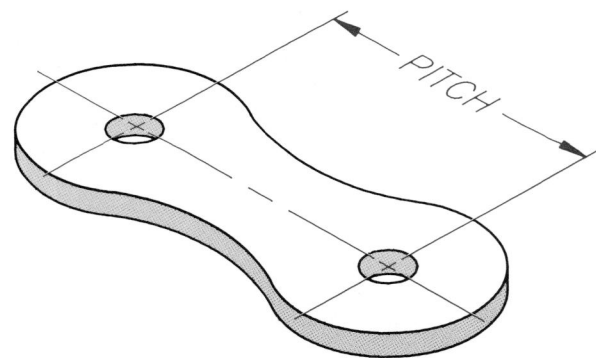

FIGURE 17.31 ■ Chain pitch.

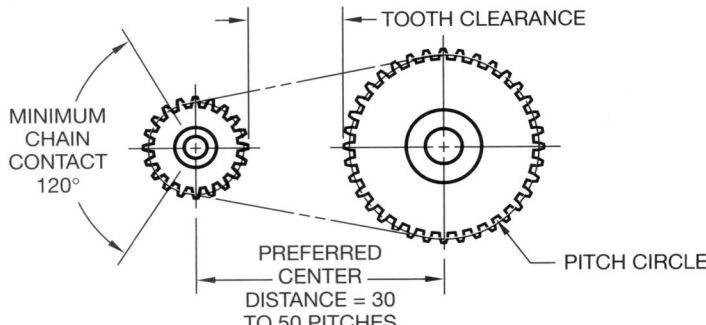

FIGURE 17.32 ■ The center distance between shafts and a 120° minimum recommended chain contact.

Look at Figure 17.33 to find K = 40.53.

$$2C + \frac{S}{2} + \frac{K}{C} = 2(48) + \frac{120}{2} + \frac{40.53}{48} =$$
$$96 + 60 + .84 = 156.84$$

The chain length in pitches must be the next higher whole number, because you cannot have a partial chain pitch. So, 156.84 = 157 chain length in pitches. Whenever possible, use an even number of chain pitches. Using an odd number of chain pitches requires an offset link, which is not preferred. For this problem, therefore, use 158 chain length in pitches rather than 157. Now, multiply the chain length in pitches by the chain pitch to find the chain length in inches:

$$158 \times .25 = 39.5 \text{ inches of chain length}$$

Step 7. Determine chain lubrication. Correct lubrication is important in maintaining long life for chain drives. Lubrication methods are governed by the speed and function of the machine. Periodic manual lubrication works well for slow-speed chain drives. To do this, use a brush to apply

D	K	D	K	D	K	D	K
32	25.94	63	100.54	94	223.82	125	395.79
33	27.58	64	103.75	95	228.61	126	402.14
34	29.28	65	107.02	96	233.44	127	408.55
35	31.03	66	110.34	97	238.33	128	415.01
36	32.83	67	113.71	98	243.27	129	421.52
37	34.68	68	117.13	99	248.26	130	428.08
38	36.58	69	120.60	100	253.30	131	434.69
39	38.53	70	124.12	101	258.39	132	441.36
40	40.53	71	127.69	102	263.54	133	448.07
41	42.58	72	131.31	103	268.73	134	454.83
42	44.68	73	134.99	104	273.97	135	461.64
43	46.84	74	138.71	105	279.27	136	468.51
44	49.04	75	142.48	106	284.67	137	475.42
45	51.29	76	146.31	107	290.01	138	482.39
46	53.60	77	150.18	108	295.45	139	489.41
47	55.95	78	154.11	109	300.95	140	496.47
48	58.36	79	158.09	110	306.50	141	503.59
49	60.82	80	162.11	111	312.09	142	510.76
50	63.33	81	166.19	112	317.74	143	517.98
51	65.88	82	170.32	113	323.44	144	525.25
52	68.49	83	174.50	114	329.19	145	532.57
53	71.15	84	178.73	115	334.99	146	539.94
54	73.86	85	183.01	116	340.84	147	547.36
55	76.62	86	187.34	117	346.75	148	554.83
56	79.44	87	191.73	118	352.70	149	562.36
57	82.30	88	196.16	119	358.70	150	569.93
58	85.21	89	200.64	120	364.76	151	577.56
59	88.17	90	205.18	121	370.86	152	585.23
60	91.19	91	209.76	122	377.02	153	592.96
61	94.25	92	214.40	123	383.22	154	600.73
62	97.37	93	219.08	124	389.48	155	608.56

FIGURE 17.33 ■ Sprocket teeth factors "K." *Courtesy Morse Industrial.*

a medium-consistency mineral oil while a machine is stopped. For moderate-speed chain drives, a drip system is often designed to keep the chain lubricated. For high-speed chain drives, an oil bath or an oil spray is often provided. To determine the type of oil application recommended for the sample problem, refer back to the chart in Figure 17.30. Notice the notes Type A, Type B, and Type

PROFESSIONAL PERSPECTIVE

This chapter covers the use of vendors' catalogs to find specific belt and chain drives for a given application. As a drafter in any engineering field, you will find that using vendors' catalogs and specifications is a very important part of your job. Most companies have a complete library of catalogs of purchase parts. Purchase parts, often referred to as standard parts, are products that can be purchased already made. Common purchase parts are bolts, nuts, belts, chains, and sprockets. You should become familiar with the types of purchase parts your company uses and how to find them in the vendors' catalogs. Also determine the suppliers that have the best prices, highest quality, and fastest availability.

C at the bottom of the chart. These refer to the recommended lubrication as follows:

Type A = Manual or drip lubrication.
Type B = Bath or disc lubrication.
Type C = Oil stream lubrication.

Notice the HP rating of 2.93 for the sample problem falls in the Type B lubrication category.

WEB SITE RESEARCH

The following Web sites can assist you in doing additional research on subjects such as standards, gears, cams, linkages, bearings, and related areas:

www.sae.org—Society of Automotive Engineers (SAE).
www.industrialpress.com—*Machinery's Handbook.* Gear, cam, linkage, and bearing related information and specifications.
www.phoenix-mfg.com—Phoenix Manufacturing. Pulleys and idlers.
www.dxpe.com—Industrial V belts.
www.rainbow-ram.com—Rainbow Industrial Products. Roller chain products.
www.igus.com—IGUS Corporation. Bearings and chains.

MATH APPLICATION

BELT LENGTH

Problem: Find the length of a driving belt running around two pulleys of radii 14" and 8" if the distance between the centers of the two pulleys is 25", as shown in Figure 17.34.

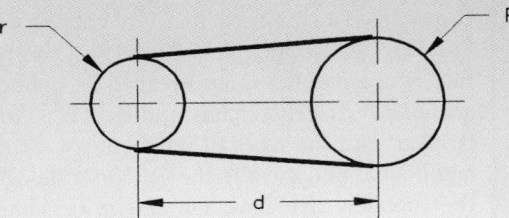

FIGURE 17.34 ■ Drive belt around pulleys.

Solution: The formula for belt length between two pulleys is

$$L = 2(d)(\cos\theta) + \pi(R + r) + 2\theta(R - r)$$
$$\text{where } \theta = \text{Inv sin}\left(\frac{R - r}{d}\right)$$

"Inv sin" stands for "inverse sine" and means "angle whose sine is." For example, Inv sin .7 is equal to 44.4° (or .775 radians) because sin 44.4° = .7. It is important to note that this formula requires angle θ to be in *radian* measure. The first step in this rather complicated formula to find L is to determine θ. It is easiest to put your calculator in the "radian mode" to work this problem:

$$\theta = \text{Inv sin}\left(\frac{14 - 8}{25}\right) = \text{Inv sin }.24 = .2424 \text{ radians}$$

Then substituting into the formula for L:

$$L = 2(25)(\cos.2424) + \pi(14 + 8) + 2(.2424)(14 - 8)$$
$$L = 2(25)(.9708) + \pi(22) + 2(.2424)(6)$$
$$L = 48.54 + 69.12 + 2.91 = \mathbf{120.57"}$$

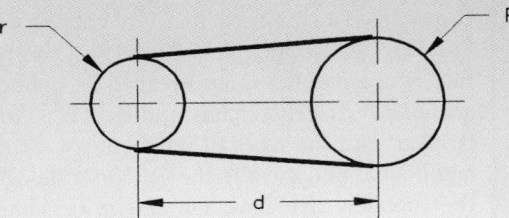

CHAPTER 17

Belt and Chain Drives Test

DIRECTIONS

Answer the following questions with short complete statements or drawings as needed on an A-size drawing sheet.

QUESTIONS

1. List at least five advantages of gear drives.
2. List at least five advantages of belt drives.
3. What would be three disadvantages of belt drives?
4. List at least five advantages of chain drives.
5. Most V belts function most effectively at what speed and temperature?
6. Are the terms *pulley* and *sheave* used interchangeably?
7. Define *pitch line*.
8. What is the function of an idler in belt drive design?
9. Name the three main categories of V belts.
10. Why are adjustable motor bases used in belt drives?
11. Describe the typical use for flat belts.
12. Describe positive drive belts.
13. List at least five factors that are involved in the selection of a correct belt drive.
14. What is the ratio of the driver pulley to the driven pulley if the driver operates at 2400 rpm and the driven operates at 600 rpm?
15. What is the belt velocity for the driver described in question 14 if the pitch diameter is 4.5 in.?
16. What is the most commonly used power chain in industry?
17. Define *sprocket*.
18. The preferred position of chain drive shafts is in a horizontal line or on a line inclined not more than how many degrees?
19. List at least four of the items that must be determined when designing roller chain drives.
20. If a small sprocket has 12 teeth and travels at 1600 rpm, and the large sprocket travels at 400 rpm, how many teeth does the large sprocket have?

21. What is the center distance between shafts in chain pitches if the center distance measures 16 in. and a No. 25 chain is used?

22. Why is correct lubrication of chain drives important?

23. Identify at least three methods of lubricating chain drives.

24. When establishing chain length in chain pitches, why is it recommended that the chain length in chain pitches be rounded up to an even number of chain pitches?

CHAPTER 17

Belt and Chain Drives Problems

DIRECTIONS

Solve Problems 17-1 through 17-8 using the information and tables given in this chapter and in Figure 17.35. All motors are AC motors with normal torque, squirrel cage, synchronous, and split phase, or shunt wound DC motors, or multiple cylinder internal combustion engines. Include the following information with each solution:

- Service factor.
- Design horsepower.

- Driven speed.
- Pitch diameter driver and driven.
- Belt section.
- Number of belts.
- Belt number and center distance.
- Corrected horsepower.
- Belt drive ratio.
- Belt velocity of driver sheave.

PART 1

PROBLEM 17-1 A 3-HP, 1750-rpm motor is to operate a furnace blower having a shaft speed of approximately 1115 rpm under normal service. The center distance between the motor and blower shafts is about 16 in.

PROBLEM 17-2 A 3-HP, 1750-rpm motor is used to operate a drill press speed reducer under intermittent service. The spindle speed is about 1136 rpm. The center distance between the motor and spindle shafts is about 20.5 in.

PROBLEM 17-3 A 1½-HP, 1750-rpm electric motor is used to operate a woodworking band saw with the blade turning at 1144 rpm, intermittent service. The center-to-center distance is about 16 in.

PROBLEM 17-4 A 2-HP electric motor with a shaft speed of 1750 rpm operates a printing machine at normal service. The

shaft on the printing machine is to operate at 1167 rpm. The center-to-center distance is about 18 in.

PROBLEM 17-5 A 2-HP electric motor with a shaft speed of 1750 rpm operates a punch machine at continuous service. The shaft on the punch machine is to operate at 1108 rpm. The center-to-center distance is about 17 in.

PROBLEM 17-6 A 1.5-HP motor with a shaft speed of 1750 rpm operates a compressor at normal service. The shaft on the compressor is to operate at 1167 rpm. The center-to-center distance is about 18 in.

PROBLEM 17-7 A 2-HP electric motor with a shaft speed of 1750 rpm operates a printing machine at normal service. The shaft on the printing machine is to operate at 1115 rpm. The center-to-center distance is about 17.5 in.

1.50	12.0	18.0	A	1	10	—	—	—	1167	10.57	13.63	A77	15.3	0.96	A81	17.3	0.98	A85	19.4	0.99
1.50	12.0	18.0	C	2	12	—	—	—	1167	25.92	31.20	C75	15.1	0.82	C96	25.7	0.88	C112	33.8	0.92
1.50	13.0	19.5	A	1	3	—	—	—	1167	11.07	14.40	A83	16.3	0.98	A86	17.8	0.99	A89	19.4	0.99
1.51	3.7	5.6	A	1	3	2318	4.61	5.62	1159	2.94	3.71	A24	5.3	0.76	A35	10.8	0.84	A45	15.8	0.88
1.51	3.9	5.9	B	1	3	2318	4.64	8.35	1159	3.53	5.53	B28	7.1	0.73	B45	15.7	0.84	B56	21.2	0.89
1.53	3.4	5.2	B	1	6	2288	2.90	6.59	1144	2.43	4.44	B28	8.1	0.73	B45	16.6	0.84	B56	22.1	0.89
1.53	3.6	5.5	A	1	3	2288	4.42	5.37	1144	2.81	3.55	A24	5.4	0.76	A35	11.0	0.84	A45	16.0	0.88
1.53	3.6	5.5	3V	1	4	2288	6.28	7.30	1144	3.63	4.07	3V250	5.3	0.78	3V300	7.8	0.83	3V355	10.6	0.87
1.53	3.8	5.8	A	1	6	2288	4.81	5.87	1144	3.06	3.86	A24	5.0	0.76	A35	10.6	0.84	A45	15.6	0.88
1.53	3.8	5.8	B	1	6	2288	4.31	8.01	1144	3.31	5.32	B28	7.3	0.73	B45	15.8	0.84	B56	21.3	0.89
1.54	2.4	3.7	A	1	1	2273	1.85	2.12	1136	1.27	1.61	A24	7.8	0.79	A35	13.3	0.86	A45	18.3	0.90
1.54	2.6	4.0	A	1	2	2273	2.31	2.70	1136	1.54	1.94	A24	7.4	0.79	A35	12.9	0.86	A45	18.0	0.90
1.54	3.5	5.4	A	1	3	2273	4.22	5.12	1136	2.69	3.40	A24	5.6	0.76	A35	11.1	0.84	A45	16.1	0.88
1.54	3.7	5.7	A	1	3	2273	4.62	5.62	1136	2.94	3.71	A24	5.2	0.76	A35	10.7	0.84	A45	15.7	0.88
1.54	3.9	6.0	B	1	3	2273	4.66	8.36	1136	3.53	5.54	B28	7.0	0.73	B45	15.6	0.84	B56	21.1	0.89

FIGURE 17.35 ■ Nominal center distances, number of belts, arc length factor chart.

PROBLEM 17-8 Given the following layout, prepare a detailed drawing of the pulley on appropriately sized vellum or using the CADD system, unless otherwise specified by your instructor.

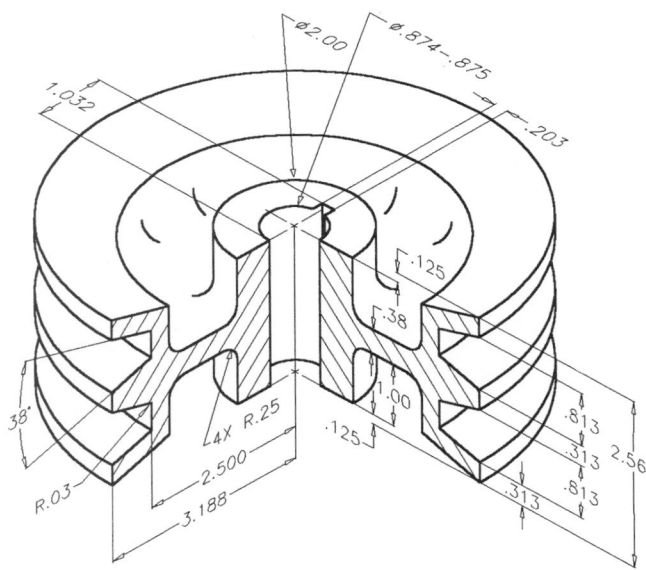

DIRECTIONS

Solve Problems 17-9 through 17-15 using the information and tables given in this chapter. Include the following information with each solution:

■ Service factor.

■ Design horsepower.

■ Number of teeth drive sprocket and driven sprocket.

■ Chain pitch.

■ Ratio.

■ Center distance in chain pitches.

■ Chain length.

■ Lubrication.

PROBLEM 17-9 A chain drive is required for a 1.5-HP electric motor operating a paper machine. The drive sprocket turns 1500 rpm and the driven sprocket turns 600 rpm. The shaft center distance is about 19 in.

PROBLEM 17-10 A chain drive is required for a ⅒-HP electric motor operating a speed reducer for a belt light package conveyor. The drive sprocket turns 50 rpm and the driven sprocket turns 17 rpm. The shaft center distance is about 17 in.

PROBLEM 17-11 A chain drive is required for a 1-HP electric motor operating a centrifuge. The drive sprocket turns 10,000 rpm and the driven sprocket turns 4000 rpm. The shaft center distance is about 18 in.

PROBLEM 17-12 A chain drive is required for a 2-HP electric motor operating a centrifugal pump. The drive sprocket turns 5500 rpm and the driven sprocket turns 1550 rpm. The shaft center distance is about 14 in.

PROBLEM 17-13 A chain drive is required for a 5-HP electric motor operating a flour grinder. The drive sprocket turns 3500 rpm and the driven sprocket turns 1500 rpm. The shaft center distance is about 18 in.

PROBLEM 17-14 A chain drive is required for a 2.5-HP internal combustion tractor engine operating a generator. The drive sprocket turns 3000 rpm and the driven sprocket turns 2000 rpm. The shaft center distance is about 20 in.

PROBLEM 17-15 A chain drive is required for a 2-HP electric motor operating a pure liquid agitator. The drive sprocket turns 2100 rpm and the driven sprocket turns 700 rpm. The shaft center distance is about 20 in.

PART 2

DIRECTIONS

Given the following engineering layouts, prepare detailed drawings of the sprockets. These problems may be done using manual drafting; however, the problems may be completed much faster if computer-aided drafting is available. *Problems courtesy of Production Plastics, Inc.*

Note: Dimensions given on engineering problems may not meet ASME standards. Convert all dimensions to comply with ASME Y14.5M—1994 as specified in Chapters 10 and 13 of this text.

UNSPECIFIED TOLERANCES:

DECIMALS	mm	IN.
X	± 2.5	± .1
XX	± 0.25	± .01
XXX	± 0.127	± .005
ANGULAR ± 30'		
FINISH	3.2μm	125μIN.

PROBLEM 17-16 **Chain saver**

CHAIN PITCH : 2.609
BARREL DIA. : .88

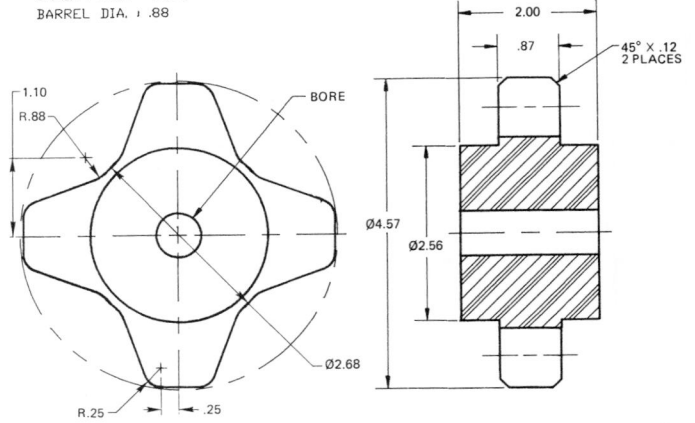

PROBLEM 17-17   FWB SD 18 sprocket

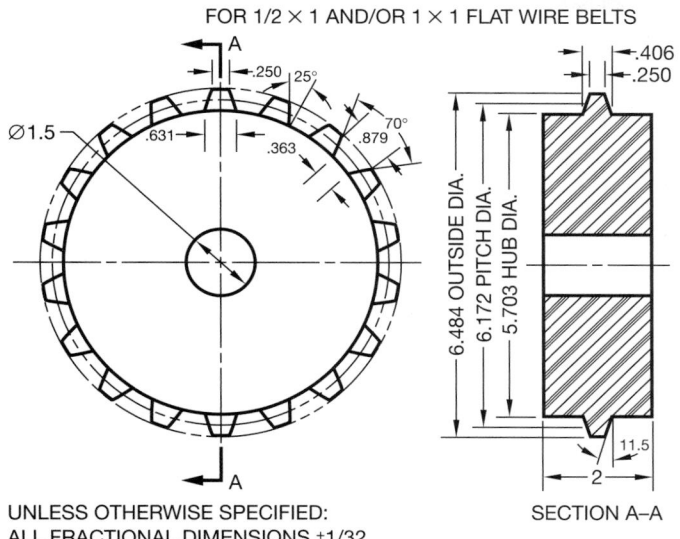

UNLESS OTHERWISE SPECIFIED:
ALL FRACTIONAL DIMENSIONS ±1/32
ALL TWO-PLACE DECIMALS ±.010
ALL THREE-PLACE DECIMALS ±.005

PROBLEM 17-18   Chain saver sprocket

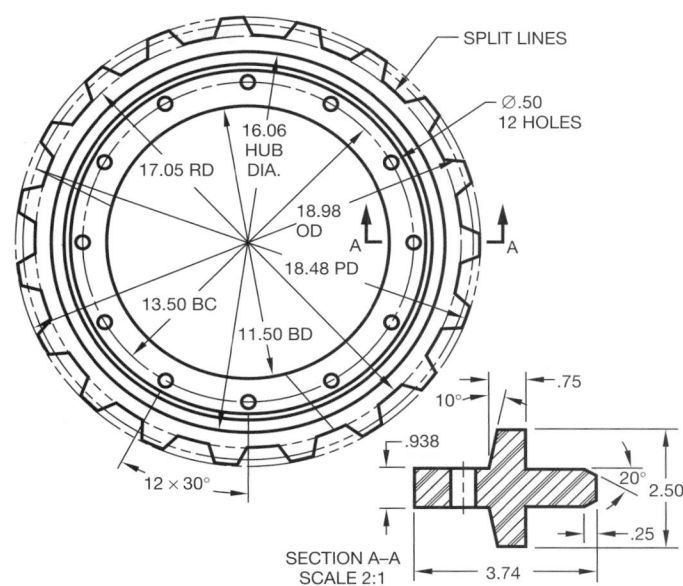

2. SPROCKET CONSISTS OF (2) 6 TOOTH SEGMENTS, AND (1) 7-TOOTH SEGMENT.
1. TOLERANCES TO BE +.003/- .000 ON BD, ALL OTHERS ±.005.
NOTES:

PART 3

DIRECTIONS

Draw the following chain chart drawings. These are excellent computer-aided drafting problems, if available. *Courtesy HHK Chain Corporation.*

PROBLEM 17-19   Heavy series roller chain

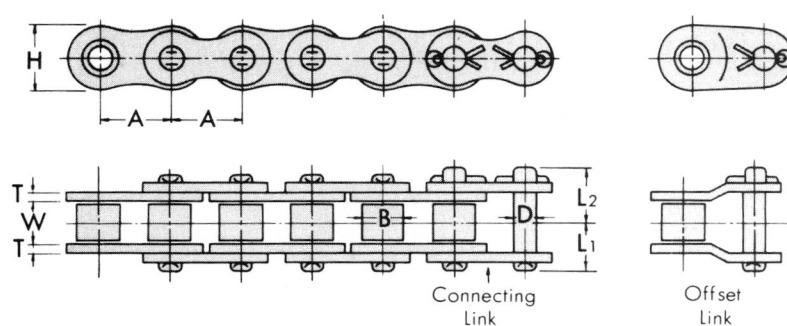

Connecting Link

Offset Link

● **HEAVY SERIES ROLLER CHAIN**   DIMENSIONS—INCHES

HKK Chain No.	Pitch	Roller		Pin	Plate		Overall Dimension		Average Ultimate Strength (lbs.)	Average Weight Per Foot (lbs.)
		Dia.	Width	Dia.	Height	Thickness				
	A	B	W	D	H	T	L₁	L₂		
HKK 60H	3/4	0.469	1/2	0.234	0.689	0.125	0.565	0.644	11,300	1.17
HKK 80H	1	0.625	5/8	0.312	0.921	0.156	0.699	0.817	20,000	2.05
HKK 100H	1-1/4	0.750	3/4	0.375	1.154	0.187	0.831	0.969	28,900	3.01

These chains are constructed with riveted pins. On request, cotter chain is available for all sizes.
Multiple strand roller chains are available.

PROBLEM 17-20   Conveyor-type standard roller chain

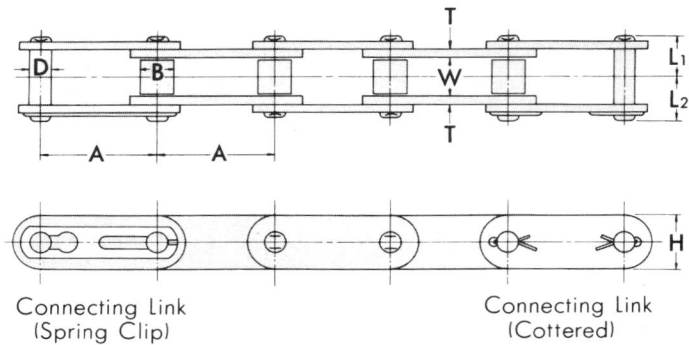

Connecting Link
(Spring Clip)

Connecting Link
(Cottered)

Standard Roller Type

DIMENSIONS—INCHES

HKK Chain No. (ANSI No.)	Pitch	Roller		Pin	Plate		Overall Dimension		Average Ultimate Strength (lbs.)	Average Weight Per Foot (lbs.)	
		Dia.	Width	Dia.	Height	Thickness					
	A	B	W	D	H	T	L₁	L₂		Riveted	Cottered
STANDARD ROLLER TYPE											
C 2040	1	0.312	5/16	0.156	0.450	0.060	0.323	0.378	4,300	0.32	—
C 2050	1-1/4	0.400	3/8	0.200	0.591	0.080	0.400	0.467	7,200	0.55	—

PROBLEM 17-21   Offset sidebar chain

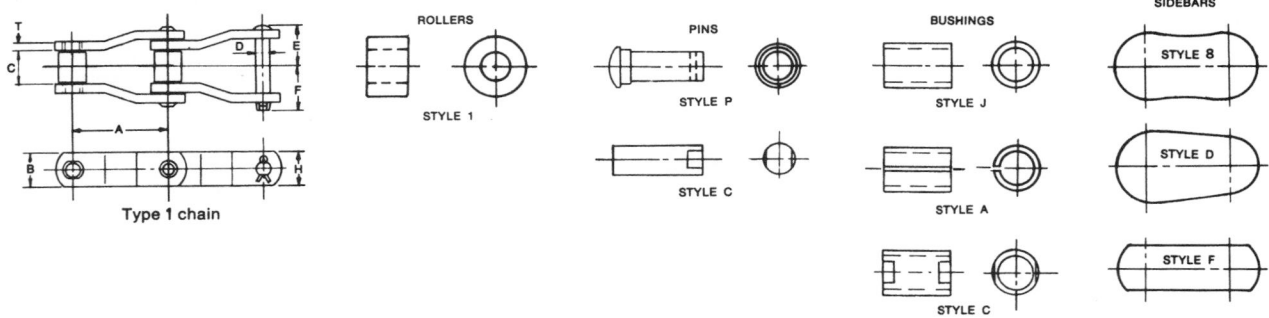

Type 1 chain

(Units in Inches or Pounds)

Chain number	Chain type	Ave. pitch	Roller			Pin				Sidebar			Bush	Average ultimate strength	Average working load	No. of links/ approx. 10 ft.	Weight /ft.
			dia.	width	Style	dia.	to rivet	to end	Style	height	thick.	Style	Style				
		A	B	C		D	E	F		H	T						
ED 501	1	1.500	.813	.906	1	.313	.9	1.0	C	1.000	.156	F	C	10.000	1.350	80	2.6
ED 1534	1	1.500	.875	1.000	1	.438	1.0	1.2	"	1.375	.188	8	A	34.000	2.100	80	4.2
ED 620	1	1.654	.875	1.000	1	.375	.9	1.0	"	1.125	.125	A	C	8.000	1.350	73	2.9
ED 622	1	1.654	.875	1.000	1	.438	1.1	1.2	"	1.125	1.88	"	A	20.000	2.100	73	3.1

PROBLEM 17-22 Sticker chain for the forest industries

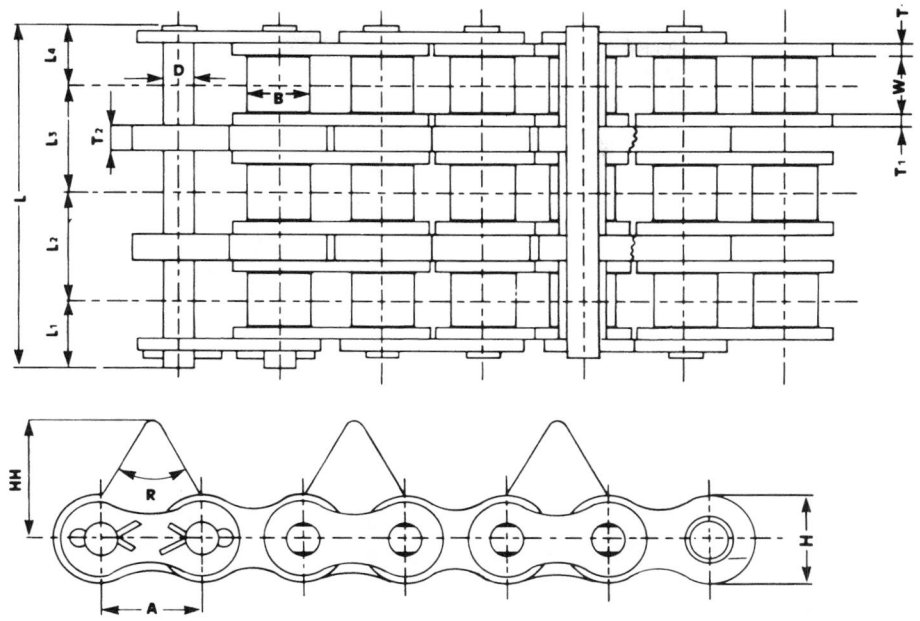

HKK No.	Pitch	Roller		Pin	Plate		Overall Dimension					Sticker Height	Sticker Thickness	Sticker Angle	Average Ultimate Strength (LBS.)	Average Weight per ft. (LBS.)
		Dia.	Width	Dia.	Height	Thickness										
	A	B	W	D	H	T₁	L	L₁	L₂	L₃	L₄	HH	T₂	R		
108110	1.000	0.625	0.625	0.312	0.914	0.125	3.678	0.745	1.153	1.153	0.627	1.250	0.125	60°	53,000	5.54
108139	1.000	0.625	0.625	0.312	0.914	0.125	3.678	0.745	1.153	1.153	0.627	1.250	0.250	60°	53,100	5.54
110131	1.250	0.750	0.750	0.375	1.154	1.585	4.503	0.913	1.409	1.409	0.772	1.535	0.312	80°	78,600	8.05

MATH PROBLEMS

Find the belt length for these pulley radii and center-to-center distances:

PROBLEM 17-23 r = 8", R = 10", d = 27"

PROBLEM 17-24 r = 12", R = 12", d = 100"

PROBLEM 17-25 r = 5", R = 30", d = 40"

PROBLEM 17-26 r = 7", R = 15", d = 30"

PROBLEM 17-27 r = 4", R = 8", d= 35"

PROBLEM 17-28 r = 101.6 mm, R = 203.2 mm, d = 889 mm

Working Drawings

OBJECTIVES

After completing this chapter, you will:

- Draw complete sets of working drawings, including details, assemblies, and parts lists.
- Prepare written specifications of purchase parts for the parts list.
- Properly group information on the assembly drawing with identification numbering systems.
- Explain the engineering change process and prepare engineering changes.

THE ENGINEERING DESIGN APPLICATION

The engineering department has been asked to develop a new product line consisting of an adjustment knob replacement kit. As the engineering drafter, you have been supplied with appropriate engineering sketches and asked to develop a complete drawing package.

In the new product development process, several departments within your company will need specific information. The manufacturing department will need the dimensional data as well as material and finish specifications. Since one of the items in the kit is a purchase part, the purchasing department must be provided with the necessary information. Packaging information will be required to get the product ready for shipping, and the sales department will need some presentation drawings to show potential clients. Additionally, customer assembly drawings are to be included in an instruction sheet to be supplied with the kit.

It is up to you to develop a complete set of working drawings. Figures 18.1 through 18.6 show all of the drawings and information used to complete this project.

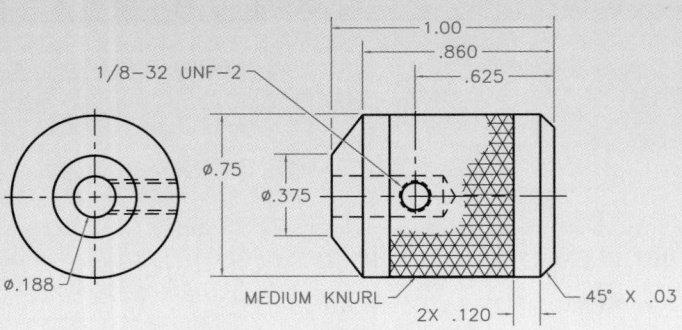

FIGURE 18.1 ■ Detail drawings provide all of the necessary manufacturing data.

B	ADDED KNURL	3/7	RJ
A	Ø.375 WAS .438	2/9	RJ
REV	DESCRIPTION	DATE	BY

FIGURE 18.2 ■ Engineering changes must be recorded.

2	1	RAK–002	1/8–32UNF SET SCREW	ACME PART #15–125–32A
1	1	RAK–001	ADJUSTMENT KNOB	A360
ITEM	QTY	PART #	DESCRIPTION	MATERIAL

FIGURE 18.3 ■ Purchasing information is included in the bill of materials or parts list.

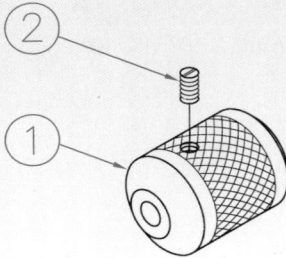

FIGURE 18.4 ■ Isometric assembly drawings show preassembly data and packaging information.

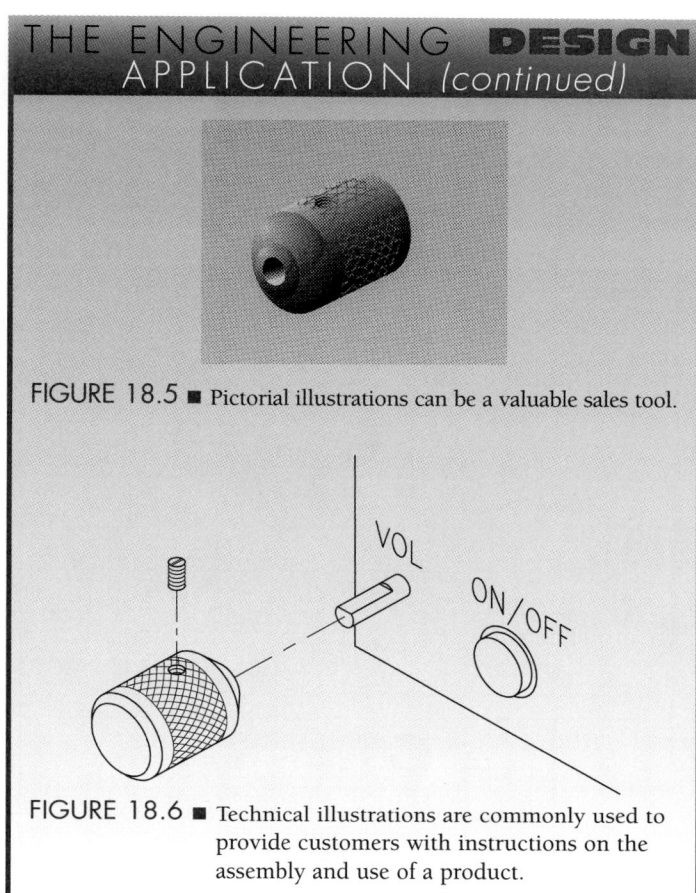

THE ENGINEERING **DESIGN**
APPLICATION *(continued)*

FIGURE 18.5 ■ Pictorial illustrations can be a valuable sales tool.

FIGURE 18.6 ■ Technical illustrations are commonly used to provide customers with instructions on the assembly and use of a product.

Most of the drawings that have been shown as examples or assigned as problems in this text are called *detail drawings*. Detail drawings are the kind of drawings that most entry-level mechanical drafters prepare. When a product is designed and drawings are made for manufacturing, each part of the product must have a drawing. These drawings of individual parts are referred to as detail drawings. Component parts are assembled to create a final product, and the drawing that shows how the parts go together is called the *assembly drawing*. Associated with the assembly drawing and coordinated to the detail drawings is the parts list. When the detail drawings, assembly drawing, and parts list are combined, they are referred to as a complete set of *working drawings*. Working drawings, then, are a set of drawings that supply all the information necessary to manufacture any given product. A set of working drawings includes all the information and instructions needed for the purchase or construction of parts and the assembly of those parts into a product.

DETAIL DRAWINGS

Detail drawings, used by workers in manufacturing, are drawings of each part contained in the assembly of a product. The only parts that may not have to be drawn are standard parts. *Standard parts* are items that can be purchased from an outside supplier more economically than they can be manufactured. Examples of standard, or purchased parts are common bolts, screws, pins, keys, and any other product that can be purchased from a vendor. Standard parts

1/2 - 13UNC-2 X 1.5 LG, SOCKET HEAD CAP SCREW

FIGURE 18.7 ■ Written description of standard or purchase part.

do not have to be drawn because a written description clearly identifies the part as shown in Figure 18.7. Detail drawings contain some or all of the following items:

1. Necessary multiviews.
2. Dimensional information.
3. Identity of the part, project name, and part number.
4. General notes and specific manufacturing information.
5. The material of which the part is made.
6. The assembly that the part fits (could be keyed to the part number).
7. Number of parts required per assembly.
8. Name(s) of person(s) who worked on or with the drawing.
9. Engineering changes and related information.

In general, detail drawings have information that is classified into three groups:

1. Shape description, which shows or describes the shape of the part.
2. Size description, which shows the size and location of features on the part.
3. Specifications regarding items such as material, finish, and heat treatment.

Figure 18.8 shows an example of a detail drawing.

Monodetail and Multidetail Drawings

Detail drawings may be prepared with one part per sheet, referred to as a *monodetail* drawing, as in Figure 18.8, or with several parts grouped on one sheet, which is called a *multidetail* drawing. The method of presentation depends on the choice of the individual company. The drawing assignments in this text have been done with one detail drawing per sheet, which is common industry practice. Some companies, however, draw many details per sheet. The advantage of monodetail drawings is that each part stands alone so the drawing of the part can be distributed to manufacturing without several other parts attached. Drawing sheet sizes vary depending on the part size, scale used, and information presented. This procedure requires that drawings be filed with numbers that allow the parts to be located in relation to the assembly. The advantage of multidetail drawings per sheet is one of economics. The drafter may be able to draw several parts on one sheet depending on the size of the parts, the scale used, and the information associated with the parts. If there are six parts on one sheet, then the number of title blocks that must be completed is reduced by five. The company may use one standard sheet size and encourage drafters to place as many parts as possible on one sheet. When this practice is used, there may be a group of sheets with parts detailed for one assembly. The sheet numbers correlate the sheets to the assembly, and each part is keyed to the assembly. The sheets are given page numbers identifying the page number and total number of pages in the set. For example, if there are three pages in a set, then the first page is

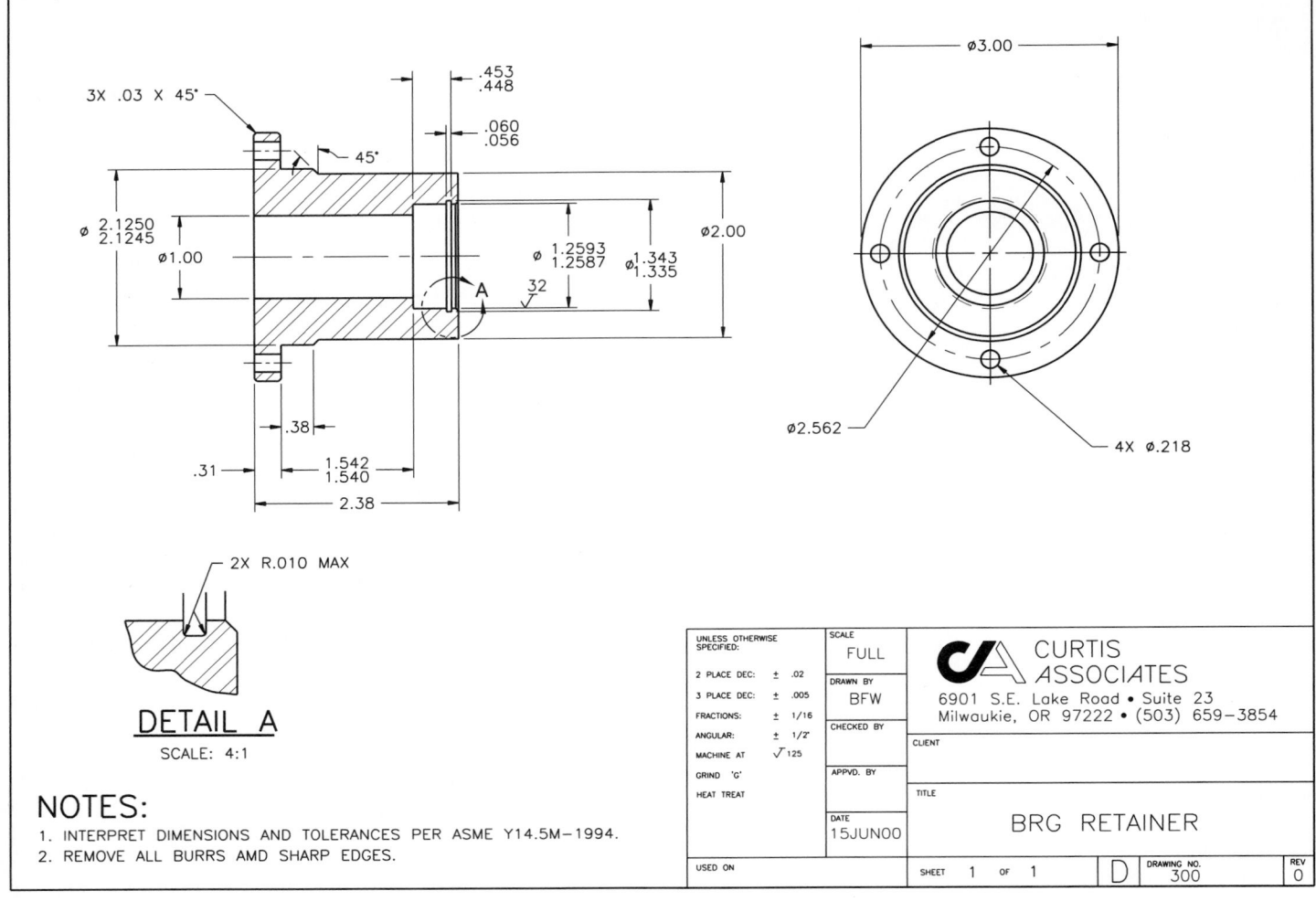

FIGURE 18.8 ■ Monodetail drawing (one part per sheet). *Courtesy Curtis Associates.*

identified as 1 of 3, the second as 2 of 3, and the third as 3 of 3. Figure 18.9 shows an example of a *multidetail* drawing with several detail drawings on one sheet. Some companies may use both methods at different times depending on the purpose of the drawings and the type of product. For example, it is more common for the parts of a weldment to be drawn grouped on sheets as opposed to one per sheet because the parts may be fabricated at one location in the shop.

Sheet Layout

The way in which a detail drawing is laid out normally depends on company practice. Some companies want the drawing to be crowded on as small of a sheet as possible. However, a drawing is easy to read when it is set up on a sheet that provides enough room for views and dimensions without crowding. For example, Figure 18.8 displays two views of a part with dimensions spaced out far enough to make the drawing clear and easy to read. Notice that there is enough clear space available on the drawing to do future revisions without difficulty. An area needs to be kept free of the drawing to provide for the general notes. This area is often in the lower left corner, or above the title block when using ASME stan-

dards, or in the upper left corner when using Military standards. An area should also be left clear of drawing content to provide for engineering change documentation. Preparing engineering changes is covered later in this chapter. The standard location for engineering change information is the upper right corner of the drawing. It is normally recommended that the space between the revision block and the title block be left clear. Some companies place their revision column above or to the left of the title block, although the ASME standard places this in the upper right corner.

Advanced sheet layout planning is critical when doing manual drafting, because poor space planning can cause a need for redrawing when engineering changes are done. Proper planning for sheet layout is also important when using CADD, but the sheet size can be quickly and easily changed at any time if needed to provide additional space.

Steps in Making a Detail Drawing

Review Chapter 7, covering the layout of multiviews, Chapter 8 for auxiliary views, and Chapter 12, which explains in detail how to lay out a dimensioned drawing. Review Chapter 13 for fasteners, and Chapter 14 for sectioning techniques.

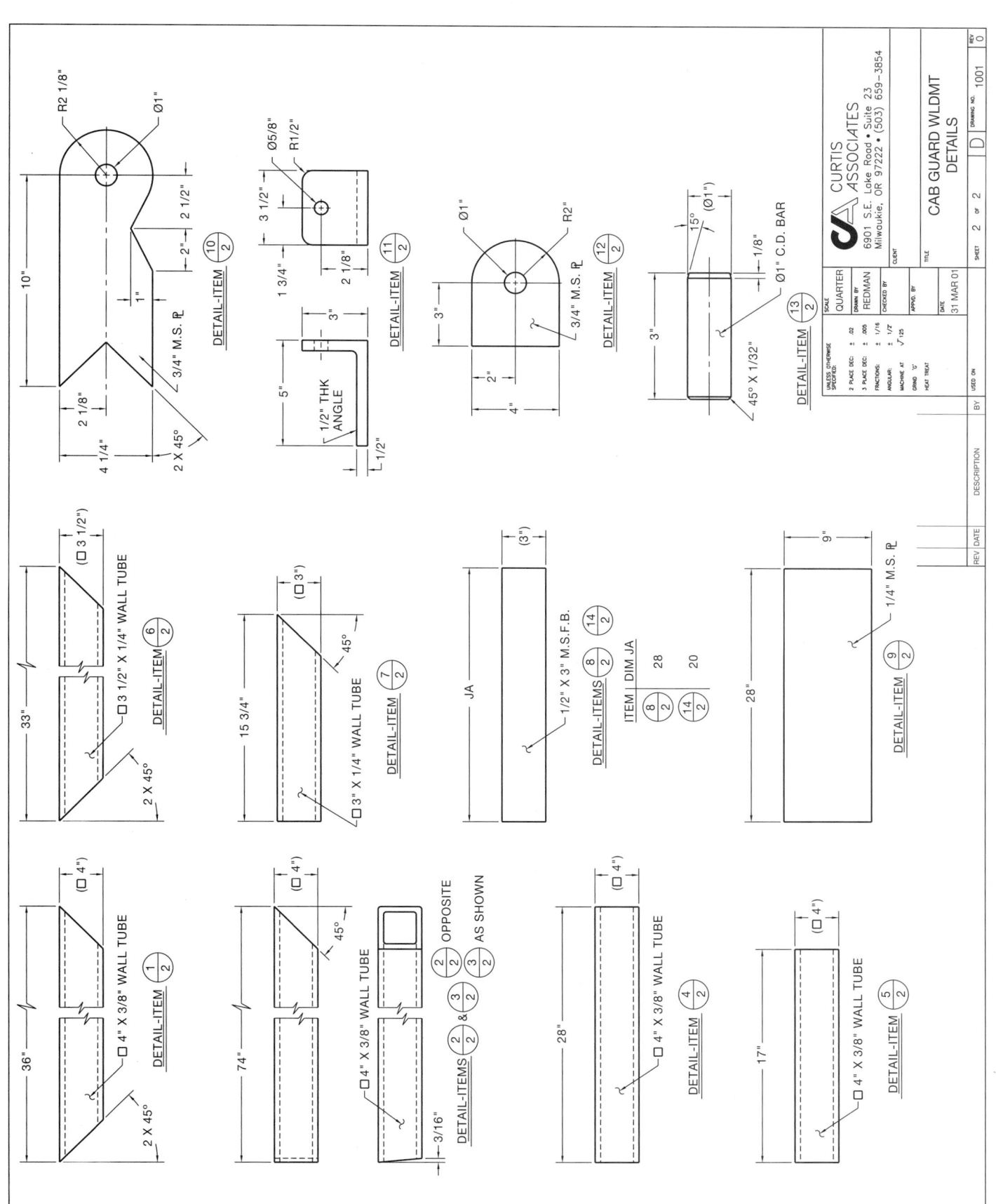

FIGURE 18.9 ■ Multidetail drawing (several detail drawings on one sheet). *Courtesy Curtis Associates.*

Detail Drawing Manufacturing Information

Detail drawings may be drawn to suit the needs of the manufacturing processes. A detail drawing may have all of the information necessary to completely manufacture the part—for example, casting and machining information on one drawing. In some situations a completely dimensioned machining drawing may be sent to the pattern or die maker. The pattern or die is then made to allow for extra material where machined surfaces are specified. When company standards require, two detail drawings may be prepared for each part. One detail gives views and dimensions that are necessary only for the casting or forging process. Another detail is drawn that does not give the previous casting or forging dimensions but provides only the dimensions needed to perform the machining operations on the part. Examples of these drawings are given in Chapter 12.

ASSEMBLY DRAWINGS

Most products are composed of several parts. A drawing showing how all of the parts fit together is called an *assembly drawings*. Assembly drawings may differ in the amount of information provided, and this decision often depends on the nature or complexity of the product. Assembly drawings are generally multiview drawings. The drafter's goal in the preparation of assembly drawings is to use as few views as possible to completely describe how each part goes together. In many cases a single front view does the job. (See Figure 18.10.) Full sections are commonly associated with assembly drawings because the full section may expose the assembly of most or all of the internal features, as shown in Figure 18.11. If one section or view is not enough to show how the parts fit together, then a number of views or sec-

VersaCad Sample Drawing

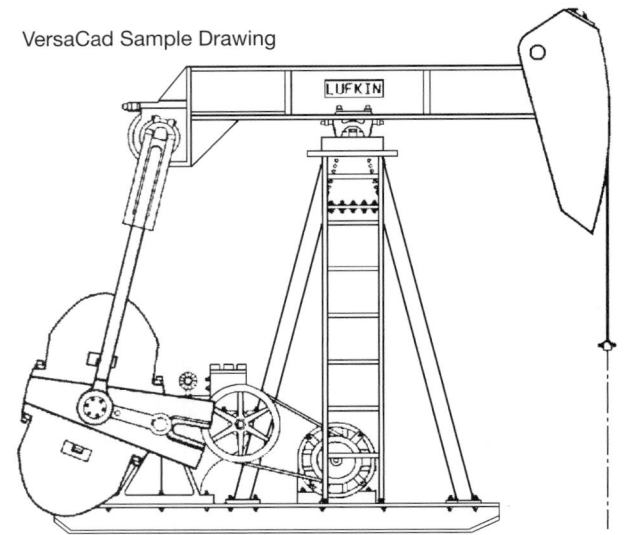

FIGURE 18.10 ■ Assembly drawing with single front view. *Courtesy T&W Systems, Inc.*

tions may be necessary. In some situations a front view or group of views with broken-out sections is the best method of showing the external features while exposing some of the internal features. (See Figure 18.12.) The drafter must make the assembly drawing clear enough for the assembly department to put the product together. Other elements of assembly drawings that make them different from detail drawings is that they usually contain few or no hidden lines or dimensions. Hidden lines should be avoided on assembly drawings unless absolutely necessary for clarity. The common practice is to draw an exterior view to clarify outside features and a sectional view to expose

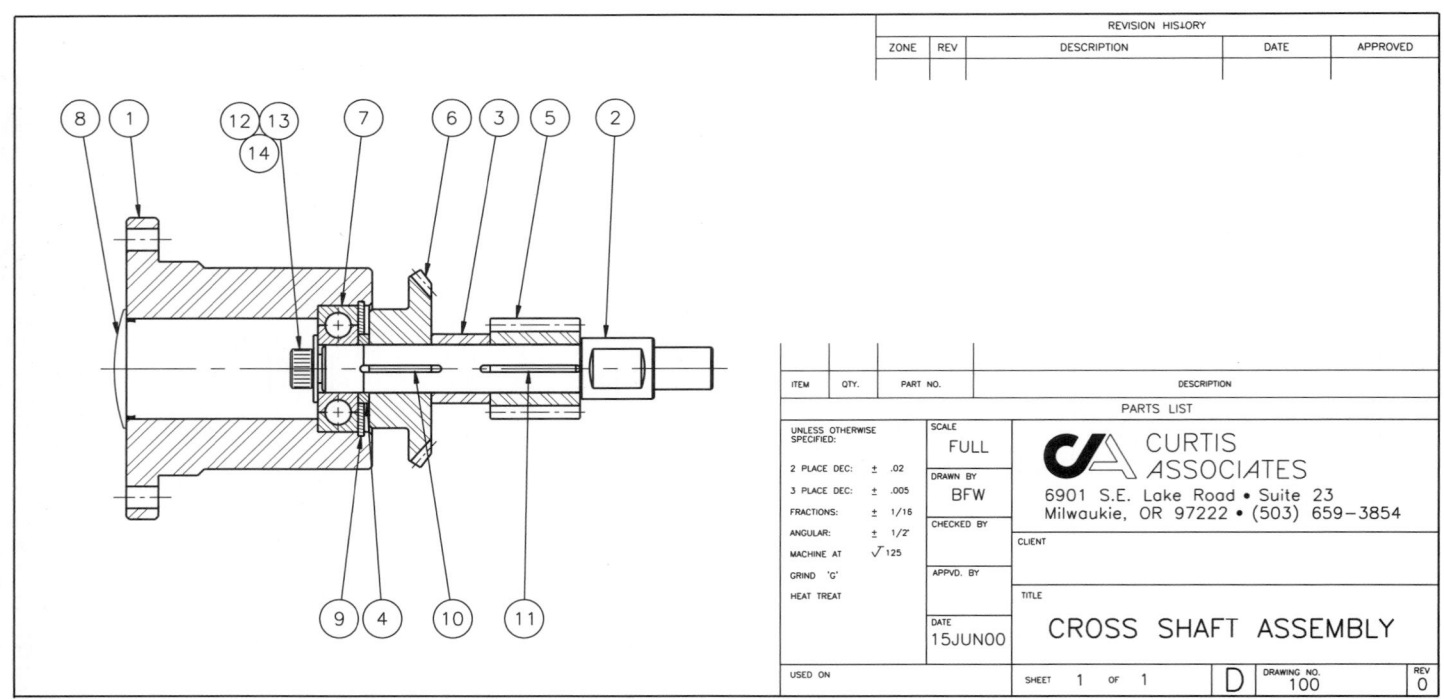

FIGURE 18.11 ■ Assembly in full section. *Courtesy Curtis Associates.*

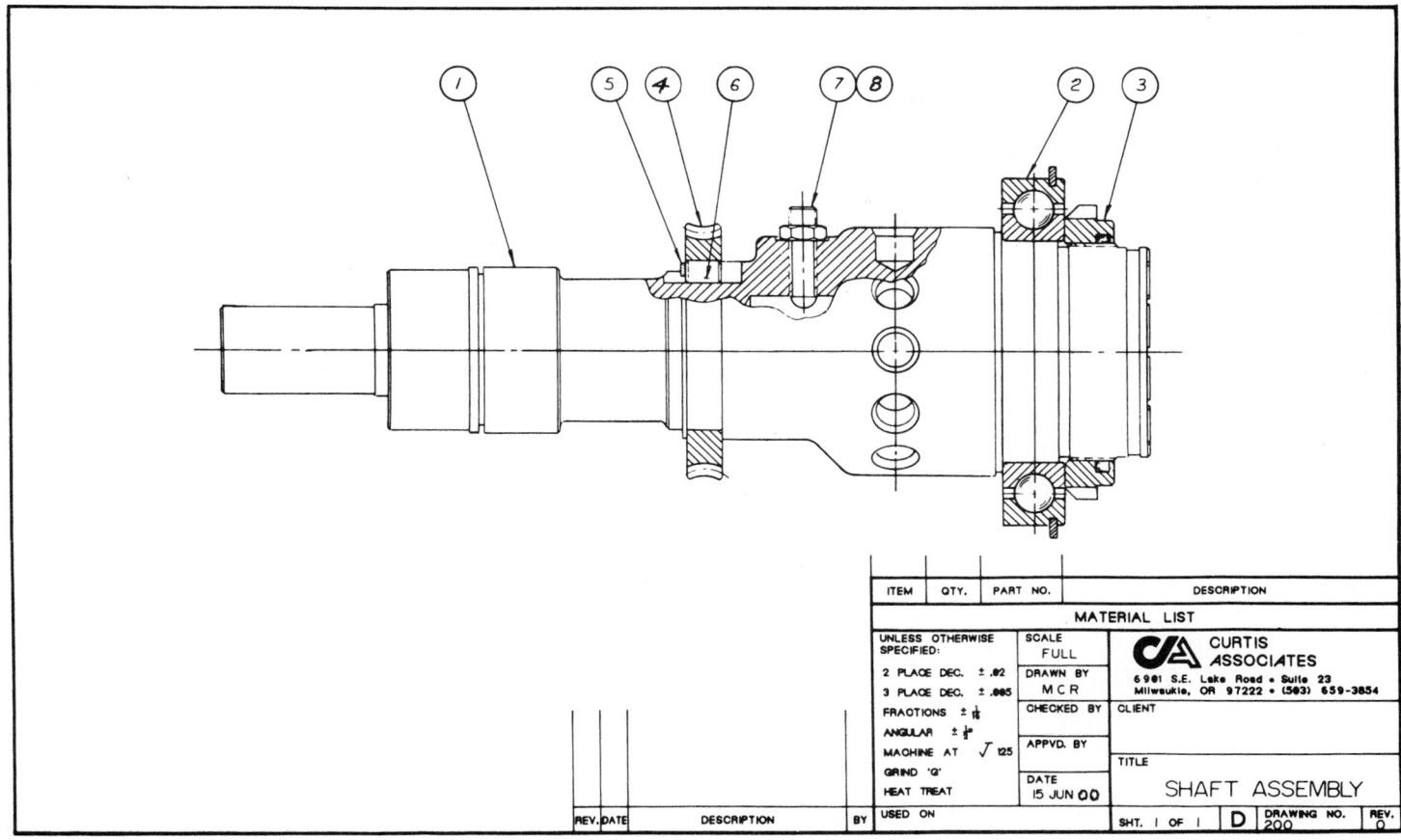

FIGURE 18.12 ■ Assembly with broken-out sections. *Courtesy Curtis Associates.*

interior features. Dimensions serve no purpose on an assembly drawing unless the dimensions are used to show the assembly relationship of one part to another. These *assembly dimensions* are only necessary when a certain distance between parts must exist before proper assembly can take place. Machining processes and other specifications are generally not given on an assembly drawing unless a machining operation must take place after two or more parts are assembled. Other assembly notes may include bolt tightening specifications, assembly welds, or cleaning, painting, or decal placement that must take place after assembly. Figure 18.13 shows a process note applied to an assembly drawing. Some company standards or drawing presentations prefer that assemblies be drawn with sectioned parts shown without section lines, although this is not a common practice. Typically, sectioned assemblies do not show section lines on objects such as fasteners, pins, keys, and shafts. Figure 18.14 shows a full-section assembly with parts left unsectioned.

Assembly drawings may contain some or all of the following information:

1. One or more views. Auxiliary views can be used as needed.
2. Sections necessary to show internal features, function, and assembly.
3. Any enlarged views necesssary to show adequate detail.
4. Arrangement of parts.
5. Overall size and specific dimensions necessary for assembly.

6. Manufacturing processes necessary for or during assembly.
7. Reference or item numbers that key the assembly to a parts list and to the details.
8. Parts list or bill of materials.

TYPES OF ASSEMBLY DRAWINGS

There are several different types of assembly drawings used in industry:

1. Layout, or design, assembly.
2. General assembly.
3. Working-drawing or detail assembly.
4. Erection assembly.
5. Subassembly.
6. Pictorial assembly.

Layout Assembly

Engineers and designers may prepare a design layout in the form of a sketch or as an informal drawing. These engineering design drawings are used to establish the relationship of parts in a product assembly. From the layout the engineer prepares sketches or informal detail drawings for prototype construction. This research

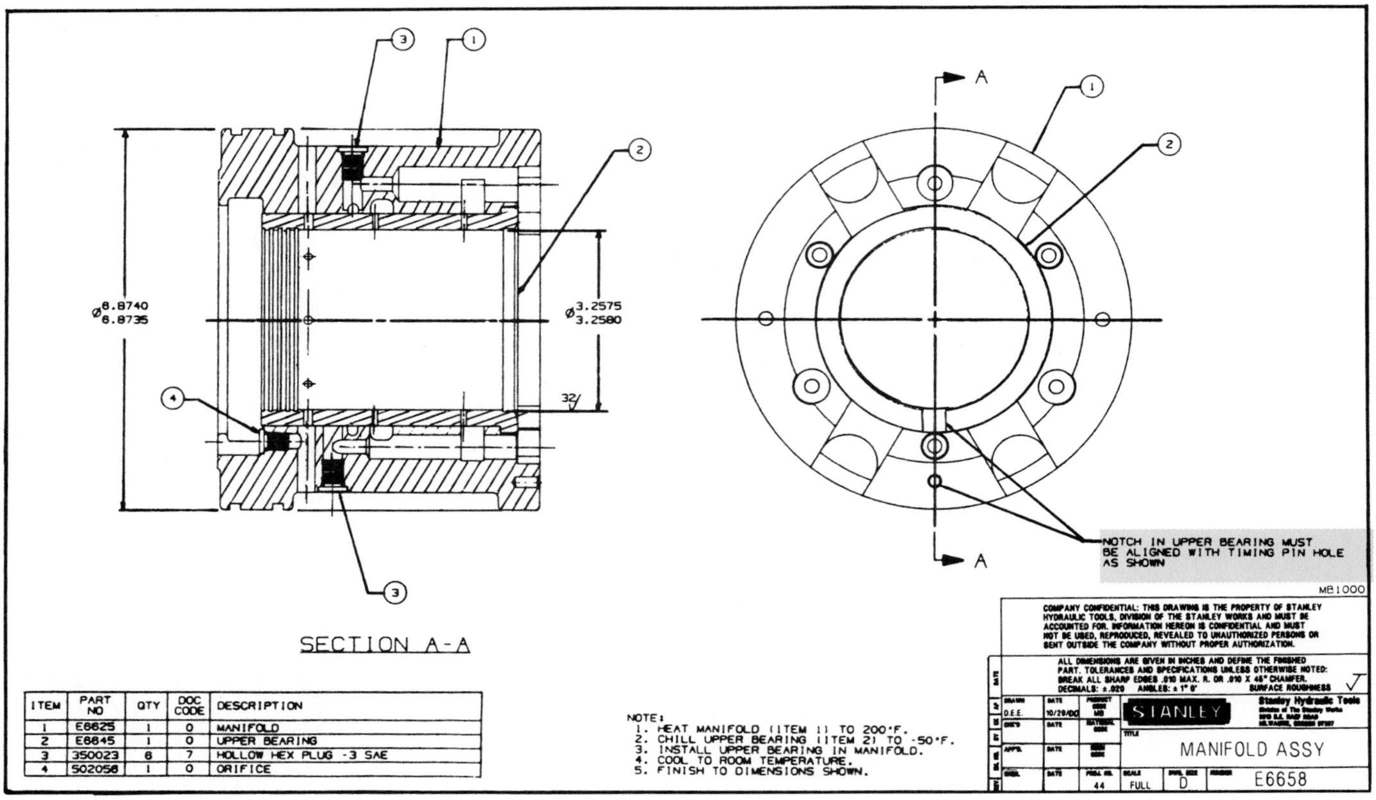
SECTION A-A
Ø6.8740
6.8735
Ø3.2575
3.2580
32
A
A
NOTCH IN UPPER BEARING MUST BE ALIGNED WITH TIMING PIN HOLE AS SHOWN
MB1000
NOTE:
1. HEAT MANIFOLD (ITEM 1) TO 200°F.
2. CHILL UPPER BEARING (ITEM 2) TO -50°F.
3. INSTALL UPPER BEARING IN MANIFOLD.
4. COOL TO ROOM TEMPERATURE.
5. FINISH TO DIMENSIONS SHOWN.
ITEM | PART NO | QTY | DOC CODE | DESCRIPTION
1 | E6625 | 1 | 0 | MANIFOLD
2 | E6645 | 1 | 0 | UPPER BEARING
3 | 350023 | 6 | 7 | HOLLOW HEX PLUG -3 SAE
4 | 502056 | 1 | 0 | ORIFICE
STANLEY
MANIFOLD ASSY
E6658
D.E.E.
10/29/00
FULL
44
D

FIGURE 18.13 ■ Process note on assembly drawing. *Courtesy Stanley Hydraulic Tools Division of The Stanley Works.*

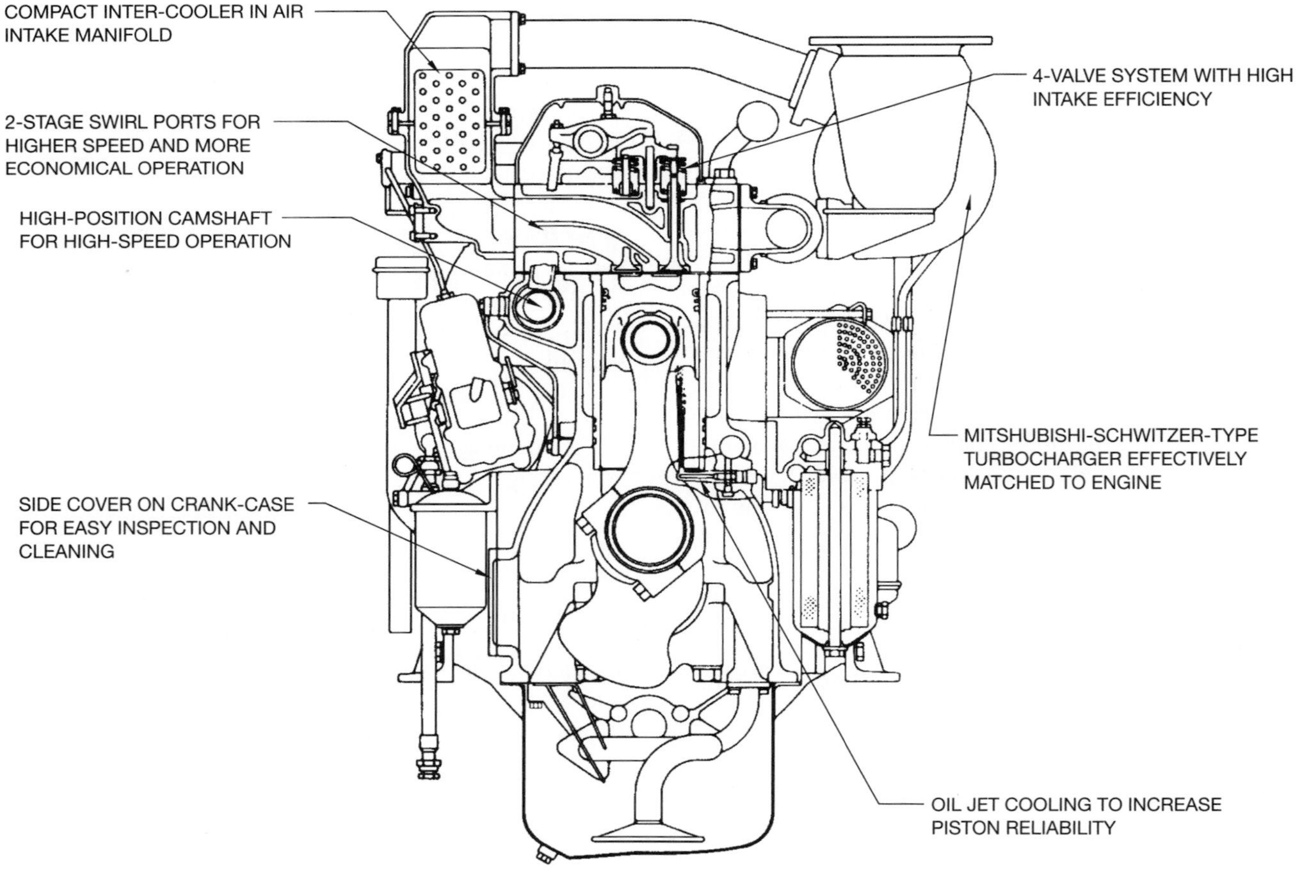
COMPACT INTER-COOLER IN AIR INTAKE MANIFOLD
2-STAGE SWIRL PORTS FOR HIGHER SPEED AND MORE ECONOMICAL OPERATION
HIGH-POSITION CAMSHAFT FOR HIGH-SPEED OPERATION
SIDE COVER ON CRANK-CASE FOR EASY INSPECTION AND CLEANING
4-VALVE SYSTEM WITH HIGH INTAKE EFFICIENCY
MITSHUBISHI-SCHWITZER-TYPE TURBOCHARGER EFFECTIVELY MATCHED TO ENGINE
OIL JET COOLING TO INCREASE PISTON RELIABILITY

FIGURE 18.14 ■ Full-section assembly with section lines omitted. *Courtesy Mitsubishi Heavy Industries America, Inc.*

product in the development stage. The limits of operation are shown in phantom lines.

General Assembly

General assemblies are the most common types of assemblies that are used in a complete set of working drawings. A set of working drawings contains three parts: detail drawings, an assembly drawing, and a parts list.

Working-drawing, or Detail, Assembly

When a drawing is created where details of parts are combined on the same sheet with an assembly of those parts, a detail assembly is the result. While this practice is not as common as general assemblies, it is a practice at some companies. The use of working-drawing assemblies may be a company standard, or this technique may be used in a specific situation even when it is not considered a normal procedure at a particular company.

The detail assembly may be used when the end result dictates that the details and assembly be combined on as few sheets as possible. An example may be a product with few parts that is produced only once for a specific purpose. (See Figure 18.16.)

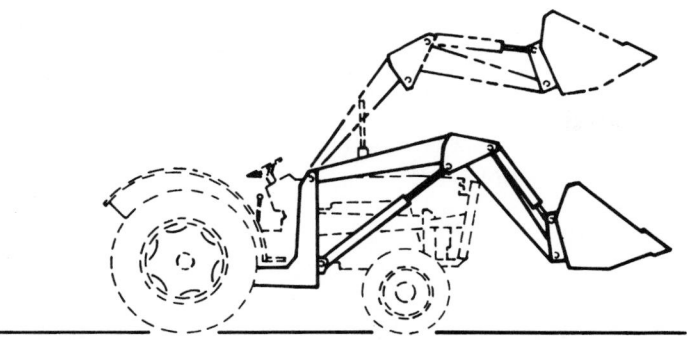

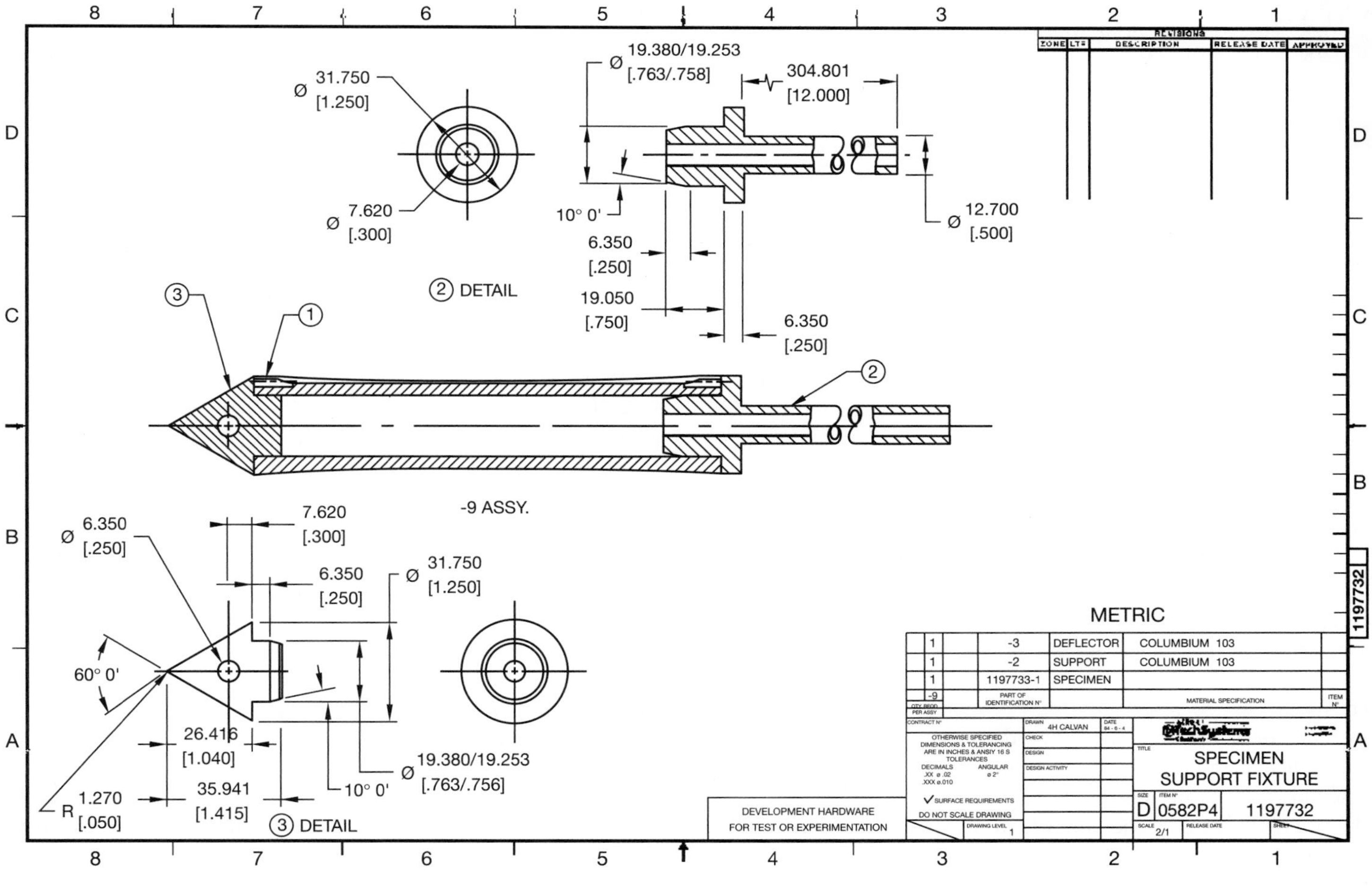

FIGURE 18.15 ■ Layout assembly. The limits of operation are shown in phantom lines.

and development (R & D) is the first step in the process of taking a design from an idea to a manufactured product. Layout, or design, assemblies may take any form depending on the drafting ability of the engineer, the time frame for product implementation, the complexity of the product, or company procedures. In many companies the engineers work with drafters who help prepare formal drawings from engineering sketches or informal drawings. The R & D department is one of the most exciting places for a drafter to be. Figure 18.15 shows a simple layout assembly of a

FIGURE 18.16 ■ Working-drawing, or detail, assembly. *Courtesy Aerojet Propulsion Division.*

WORKING DRAWINGS

Using CADD to prepare working drawings may increase your productivity while you are drawing details. Some drafters indicate a 1:1 productivity ratio while others boast up to a 10:1 ratio for CADD versus manual drafting. One engineer explained that it normally takes a drafter 100 hours of training and daily use of the CADD system before he or she is on a 1:1 level of productivity with the manual drafter. After the initial 100 hours, the productivity of the CADD drafter increases. This all depends on some contributing factors such as:

■ The complexity of the part.
■ The experience of the drafter with the CADD system.
■ The ability of the individual or group to customize the CADD system for specific applications.
■ The number of common symbols, notes, or details that can be used on a variety of drawings.

After all of the detail drawings are made, the drafter can easily and quickly prepare the assembly drawing. It used to take the manual drafter hours or even days to complete the assembly drawing, depending on the complexity of the product. With CADD even the most complex drawings can be done in a short period of time. The assembly should be drawn after the detail drawings of the individual parts are complete because the drafter can use the details to complete the assembly. To do this, follow these helpful techniques.

■ Do all detail drawings first.
■ Have the views on a layer separate from the dimensions.
■ Make the principal view of each detail drawing a BLOCK or SYMBOL that can be sent between drawings.
■ Start the assembly drawing by bringing the main detail view into the layout and position it where the other detail views can be conveniently added.
■ Bring each detail view into position to begin building the assembly. Use a command such as SCALE to increase or reduce the size of each detail view to fit the scale of the assembly.
■ After all of the detail views are assembled, you can use the editing functions to erase, trim, or otherwise clean up the assembly.
■ Add balloons, notes, and dimensions (if any).
■ Add the parts list. It is best to have a standard parts list format that can be called up and added to any drawing. Then all you have to do is add the text and the drawing is done.

This method saves a lot of time because you do not have to start over again drawing each part in the assembly.

Erection Assembly

Erection assemblies usually differ from general assemblies in that dimensions and fabrication specifications are commonly included. Typically associated with products that are made of structural steel, or cabinetry, election assemblies are used for both fabrication and assembly. Figure 18.17 shows an erection assembly with multiviews, fabrication dimensions, and an isometric drawing that also helps display how the parts fit together.

Subassembly

The complete assembly of a product may be made up of several component assemblies. These individual unit assemblies are called *subassemblies*. A complete set of working drawings may have several subassemblies, each with its own detail drawings, and the general assembly. The general assembly of an automobile, for example, includes the subassemblies of the drive components, the engine components, and the steering column, just to name a few. A subassembly, such as an engine, may have other subassemblies, such as the carburetor or the generator. Figure 18.18 shows a subassembly with a parts list.

Pictorial Assembly

Pictorial assemblies are used to display a pictorial rather than multiview representation of the product, which may be used in other types of assembly drawings. Pictorial assemblies may be made from photographs or artistic renderings. The pictorial assembly may be as simple as the isometric drawing in Figure 18.17, which was used to more clearly assist workers in the assembly of the product. Pictorial assemblies are commonly used in product catalogs or brochures. The purpose of these pictorial representations may be for sales promotion, customer self-assembly, or maintenance procedures. (See Figure 18.19.) Pictorial assemblies may also take the form of exploded technical illustrations, also commonly known as illustrated parts breakdowns. These exploded multiview or isometric pictorials are used in vendors' catalogs and instruction manuals, for maintenance or assembly. (See Figure 18.20.)

IDENTIFICATION NUMBERS

Identification or item numbers are used to key the parts from the assembly drawing to the parts list. Identification numbers are generally placed in balloons. *Balloons* are circles that are connected to the

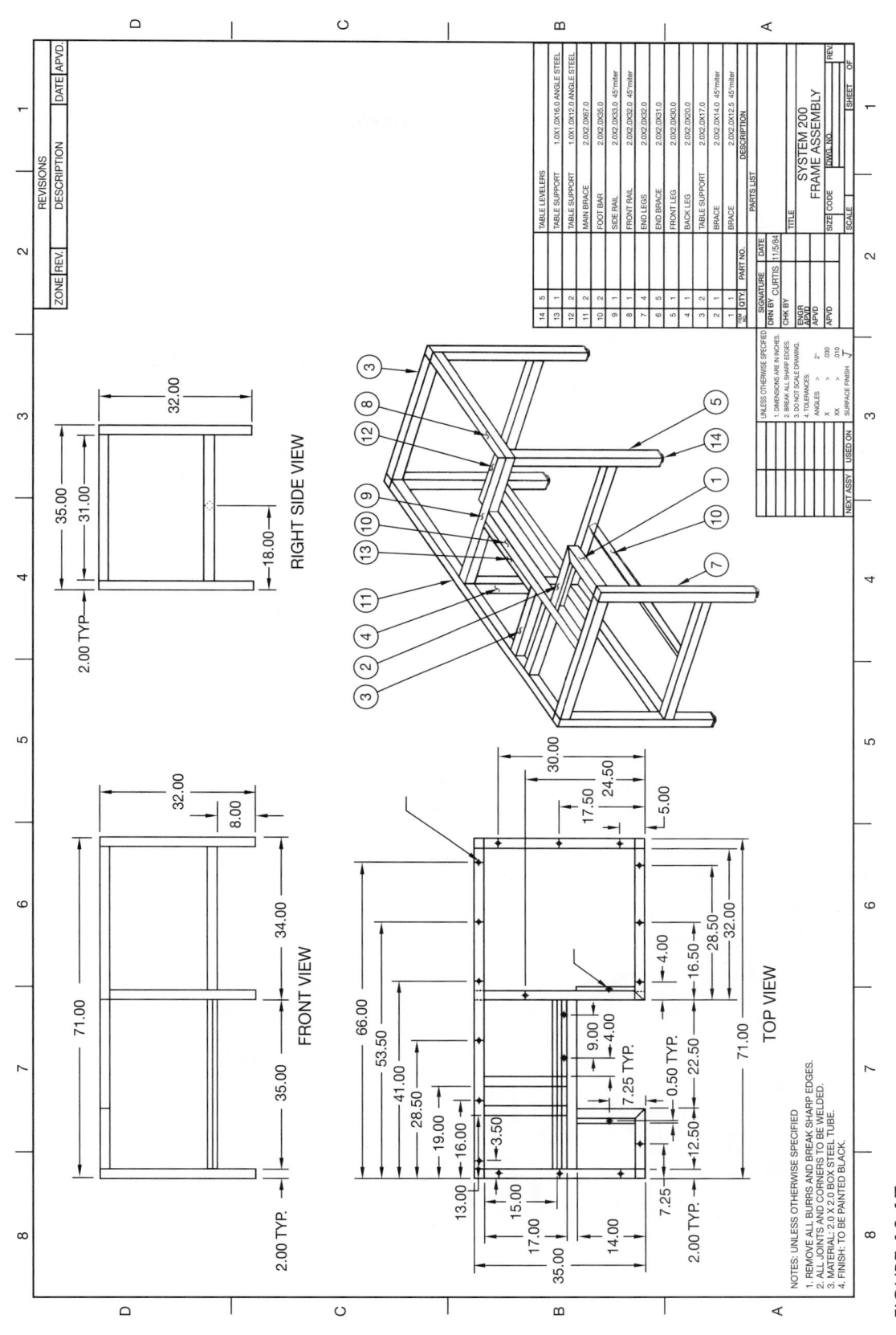

FIGURE 18.17 ■ *Erection assembly. Courtesy EFT Systems.*

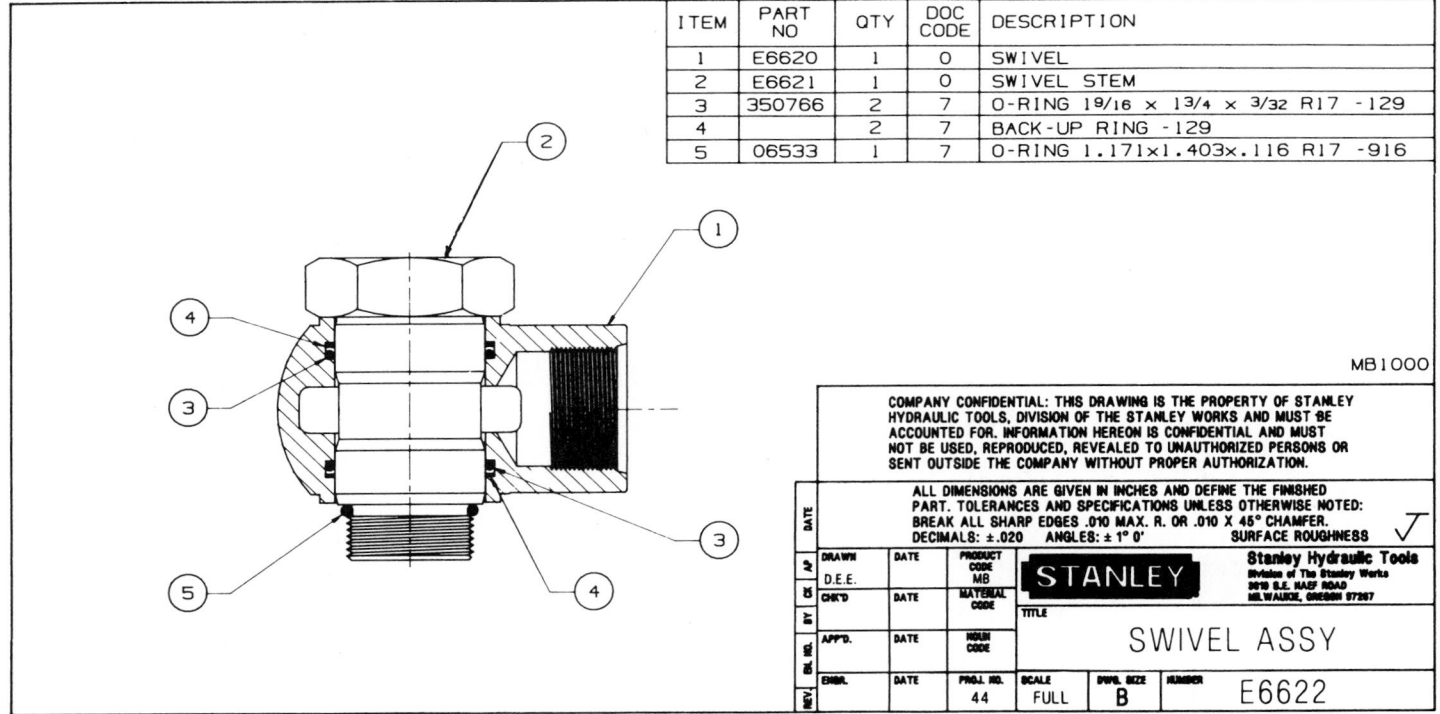

ITEM	PART NO	QTY	DOC CODE	DESCRIPTION
1	E6620	1	0	SWIVEL
2	E6621	1	0	SWIVEL STEM
3	350766	2	7	O-RING 19/16 × 13/4 × 3/32 R17 -129
4		2	7	BACK-UP RING -129
5	06533	1	7	O-RING 1.171×1.403×.116 R17 -916

MB1000

COMPANY CONFIDENTIAL: THIS DRAWING IS THE PROPERTY OF STANLEY HYDRAULIC TOOLS, DIVISION OF THE STANLEY WORKS AND MUST BE ACCOUNTED FOR. INFORMATION HEREON IS CONFIDENTIAL AND MUST NOT BE USED, REPRODUCED, REVEALED TO UNAUTHORIZED PERSONS OR SENT OUTSIDE THE COMPANY WITHOUT PROPER AUTHORIZATION.

ALL DIMENSIONS ARE GIVEN IN INCHES AND DEFINE THE FINISHED PART. TOLERANCES AND SPECIFICATIONS UNLESS OTHERWISE NOTED: BREAK ALL SHARP EDGES .010 MAX. R. OR .010 X 45° CHAMFER. DECIMALS: ±.020   ANGLES: ± 1° 0'   SURFACE ROUGHNESS

STANLEY

Stanley Hydraulic Tools
Division of The Stanley Works
3810 S.E. NAEF ROAD
MILWAUKIE, OREGON 97267

TITLE

SWIVEL ASSY

DRAWN D.E.E.	DATE	PRODUCT CODE MB
CHK'D	DATE	MATERIAL CODE
APP'D.	DATE	NOUN CODE
ENGR.	DATE	PROJ. NO. 44	SCALE FULL	DWG. SIZE B	NUMBER E6622

FIGURE 18.18 ■ *Subassembly with parts list. Courtesy Stanley Hydraulic Tools Division of The Stanley Works.*

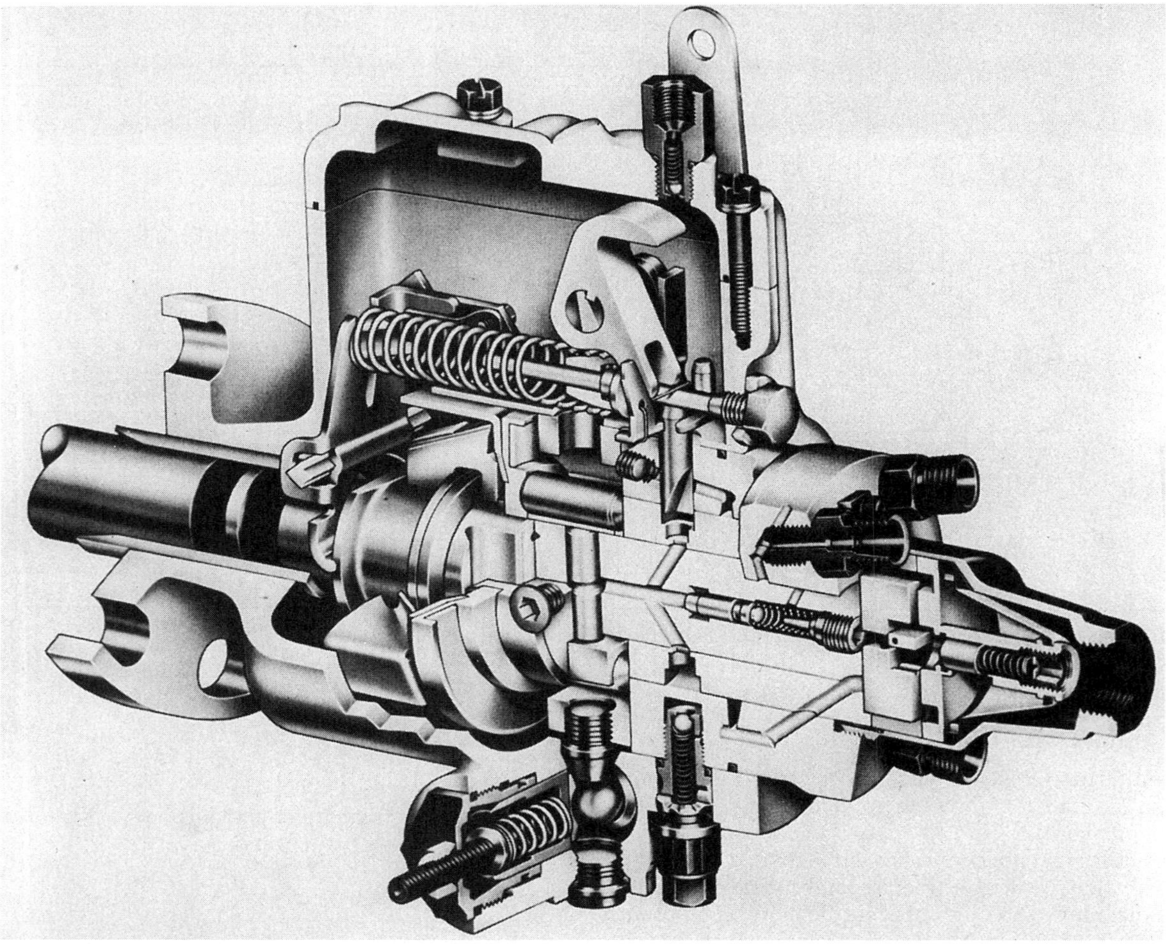

FIGURE 18.19 ■ *Pictorial assembly. Courtesy Stanadyne Diesel Systems.*

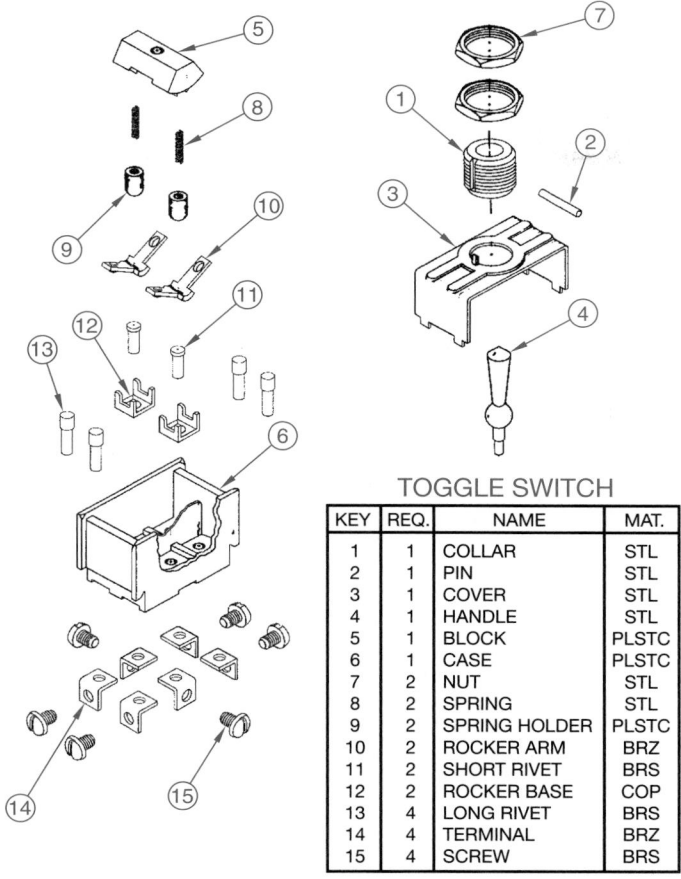

FIGURE 18.20 ■ Exploded isometric assembly.

TOGGLE SWITCH

KEY	REQ.	NAME	MAT.
1	1	COLLAR	STL
2	1	PIN	STL
3	1	COVER	STL
4	1	HANDLE	STL
5	1	BLOCK	PLSTC
6	1	CASE	PLSTC
7	2	NUT	STL
8	2	SPRING	STL
9	2	SPRING HOLDER	PLSTC
10	2	ROCKER ARM	BRZ
11	2	SHORT RIVET	BRS
12	2	ROCKER BASE	COP
13	4	LONG RIVET	BRS
14	4	TERMINAL	BRZ
15	4	SCREW	BRS

related part with a leader line. Several of the assembly drawings in this chapter show examples of identification numbers and balloons. Numbers in balloons are common, although some companies prefer to use identification letters. Balloons are drawn between .375 and 1 in. in diameter depending on the size of the drawing and the amount of information that must be placed in the balloon. All balloons on the drawing are the same size. The leaders that connect the balloons to the parts are thin lines that may be presented in any one of several formats depending on company standards. Figure 18.21 shows the common methods that are used to connect balloon leaders. Notice that the leaders may terminate with arrowheads or dots. Whichever method of connecting balloons is used, the same technique should be used throughout the entire drawing.

The item numbers in balloons should be grouped so they are in an easy-to-read pattern. This is referred to as *information grouping.* Good information grouping occurs when balloons are in alignment in a pattern that is easy to follow, as opposed to scattered about the drawing. (See Figure 18.22.) In some situations when particular part groups are so closely related that individual identification is difficult, the identification balloons may be grouped next to each other. For example, a cap screw, lock washer, and flat washer may require that the balloons be placed in a cluster or side by side, as shown with items 12, 13, and 14 in Figure 18.11.

In some cases the balloons key the detail drawings to the assembly and parts list, and also key the assembly drawing and parts list to the page on which the detail drawing is found. (See Figure 18.23.)

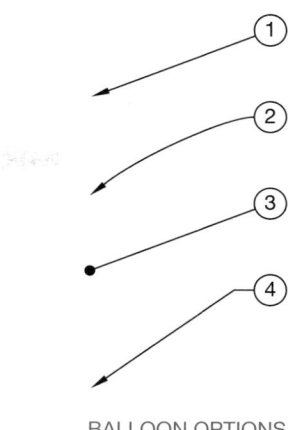

BALLOON OPTIONS

FIGURE 18.21 ■ Balloons and styles of leaders.

Figure 18.24 is an assembly drawing and parts list that is located on page 1 of a two-page set of working drawings. Notice how the balloons key the parts from the assembly and parts list to the detail drawings found on page 2. The page 2 details are located in Figure 18.9 (see page 575).

PARTS LISTS

The parts list is usually combined with the assembly drawing yet remains one of the individual components of a complete set of working drawings. The information that is associated with the parts list generally includes:

1. Item number—from balloons.

2. Quantity—the number of that particular part needed for this assembly.

3. Part or drawing number, which is a reference back to the detail drawing.

4. Description, which is usually a part name or complete description of a purchase part or stock specification including sizes or dimensions.

5. Material identification—the material that the part is made of.

6. Information about vendors for purchase parts.

Parts lists may or may not contain all of the previous information, depending on company standards. The elements listed 1 through 4 are the most common items. When all six elements are provided, the parts list may also be called a *list of materials,* or *bill of materials.*

When drawn on the assembly drawing, the parts list may be located above the title block, in the upper right or left corner, or in a convenient location on the drawing field. The location depends on company standards, although the position over the title block is most common. The information on the parts list is usually presented with the first item number followed by consecutive item numbers. When the parts list is so extensive that the columns fill the page, a new group of columns is added next to the first. If additional parts are added to the assembly, space on the parts list is available. This is the reason that parts list data is provided from the bottom of the sheet upward or the top of the sheet downward.

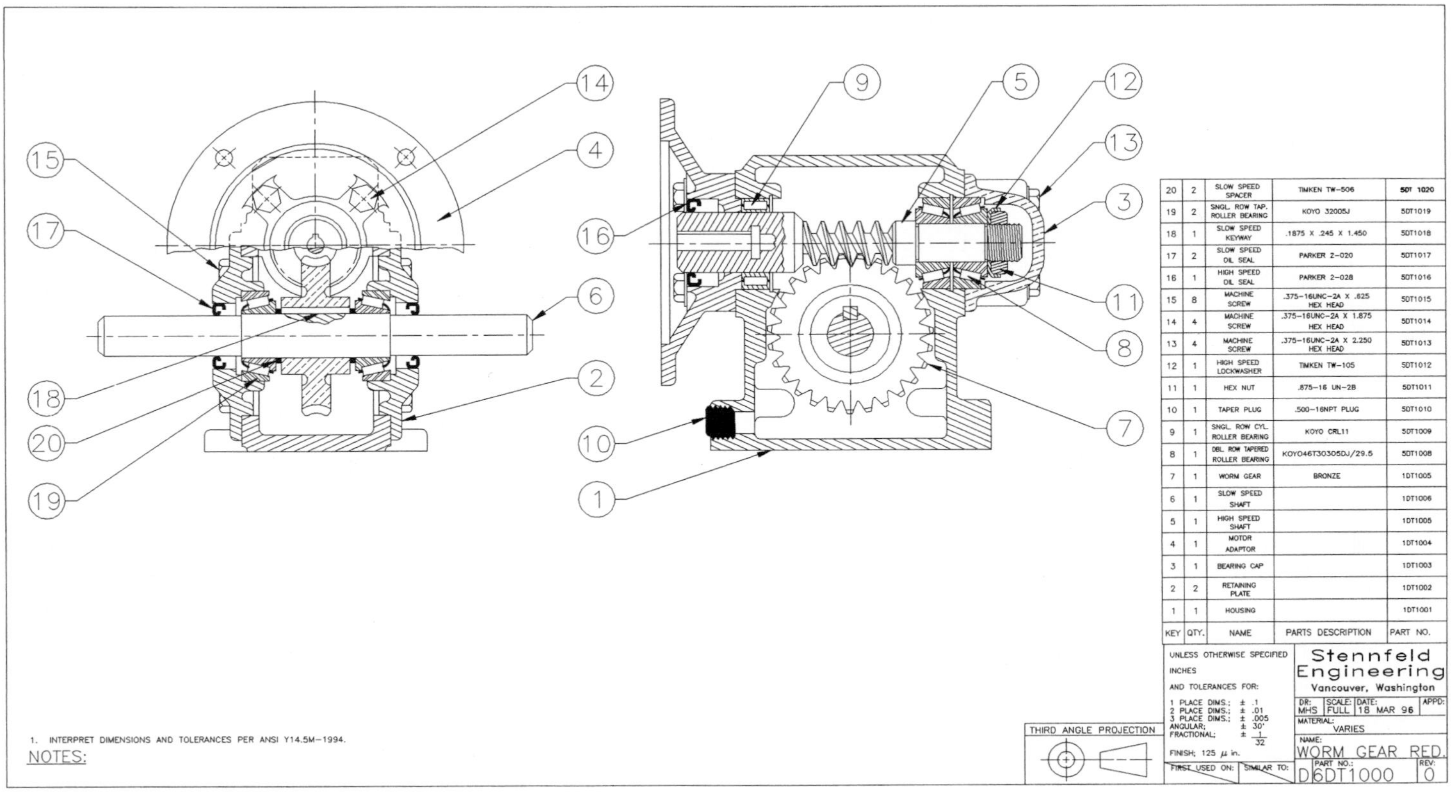

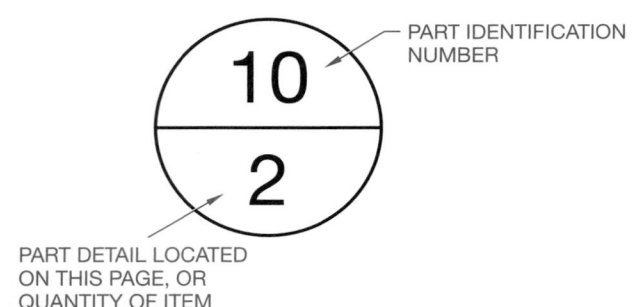

KEY	QTY.	NAME	PARTS DESCRIPTION	PART NO.
20	2	SLOW SPEED SPACER	TIMKEN TW-506	5DT 1020
19	2	SNGL. ROW TAP. ROLLER BEARING	KOYO 32005J	5DT1019
18	2	SLOW SPEED KEYWAY	.1875 X .245 X 1.450	5DT1018
17	2	SLOW SPEED OIL SEAL	PARKER 2-020	5DT1017
16	1	HIGH SPEED OIL SEAL	PARKER 2-028	5DT1016
15	8	MACHINE SCREW	.375-16UNC-2A X .825 HEX HEAD	5DT1015
14	4	MACHINE SCREW	.375-16UNC-2A X 1.875 HEX HEAD	5DT1014
13	4	MACHINE SCREW	.375-16UNC-2A X 2.250 HEX HEAD	5DT1013
12	1	HIGH SPEED LOCKWASHER	TIMKEN TW-105	5DT1012
11	1	HEX NUT	.875-16 UN-2B	5DT1011
10	1	TAPER PLUG	.500-16NPT PLUG	5DT1010
9	1	SNGL. ROW CYL. ROLLER BEARING	KOYO CRL11	5DT1009
8	1	DBL. ROW TAPERED ROLLER BEARING	KOYO46T30305DJ/29.5	5DT1008
7	1	WORM GEAR	BRONZE	1DT1005
6	1	SLOW SPEED SHAFT		1DT1006
5	1	HIGH SPEED SHAFT		1DT1005
4	1	MOTOR ADAPTOR		1DT1004
3	1	BEARING CAP		1DT1003
2	2	RETAINING PLATE		1DT1002
1	1	HOUSING		1DT1001

UNLESS OTHERWISE SPECIFIED
INCHES
AND TOLERANCES FOR:
1 PLACE DIMS.: ± .1
2 PLACE DIMS.: ± .01
3 PLACE DIMS.: ± .005
ANGULAR: ± 30'
FRACTIONAL: ± 1/32
FINISH: 125 μ in.

THIRD ANGLE PROJECTION

Stennfeld Engineering
Vancouver, Washington

DR: MHS | SCALE: FULL | DATE: 18 MAR 98 | APPD:
MATERIAL: VARIES
NAME: WORM GEAR RED.
PART NO.: D6DT1000 | REV: 0

1. INTERPRET DIMENSIONS AND TOLERANCES PER ANSI Y14.5M-1994.
NOTES:

FIGURE 18.22 ■ Assembly drawing and parts list. *Courtesy Mark Stennfeld.*

PART IDENTIFICATION NUMBER

10

2

PART DETAIL LOCATED ON THIS PAGE, OR QUANTITY OF ITEM

FIGURE 18.23 ■ Balloon with page identification.

Some companies prepare parts lists on a computer so information can be retrieved and edited easily. When this is done, the parts list may be plotted or typed directly on the drawing original or on an adhesive Mylar® sheet that is added to the drawing. If the entire drawing is prepared on a CADD system, then the parts list can easily be plotted on the original.

The parts list is not always placed on the assembly drawing. Some companies prefer to prepare parts lists on separate sheets, which allows for convenient filing. This method also allows for the parts list to be computer-generated or typed separately from the drawings. Separate parts lists are usually prepared on a computer so information may be edited conveniently. Another option is to have parts lists typed on a standard parts list form (see Figure 18.25).

PURCHASE PARTS

Purchase or standard parts, as previously mentioned, are parts that are manufactured and available for purchase from an outside vendor. It is generally more economical for a company to buy such items than to make them. Parts that are available from suppliers can be found in the *Machinery's Handbook, Fastener's Handbook,* or vendors' catalogs. Purchase parts do not require a detail drawing, because a written description completely describes the part. For this reason, the purchase parts found in a given assembly must be described clearly and completely in the parts list. Some companies have a purchase parts book or computer directory that is used to record all purchase parts used in their product line. The standard parts book gives a reference number for each part, which is placed on the parts list for convenient identification.

ENGINEERING CHANGES

ASME The standards controlling the use of engineering changes are specified in ASME Y14.35M.

Engineering change documents are used to initiate and record changes to products in the manufacturing industry. Changes to engineering drawings can be requested from any branch of a company that deals directly with production and distribution of the product. For example, the engineering department may implement product changes as research and development results show a need for upgrading. Manufacturing requests change as problems arise in product fabrication or assembly. The sales staff may also initiate change proposals that stem from customer complaints. Engineer-

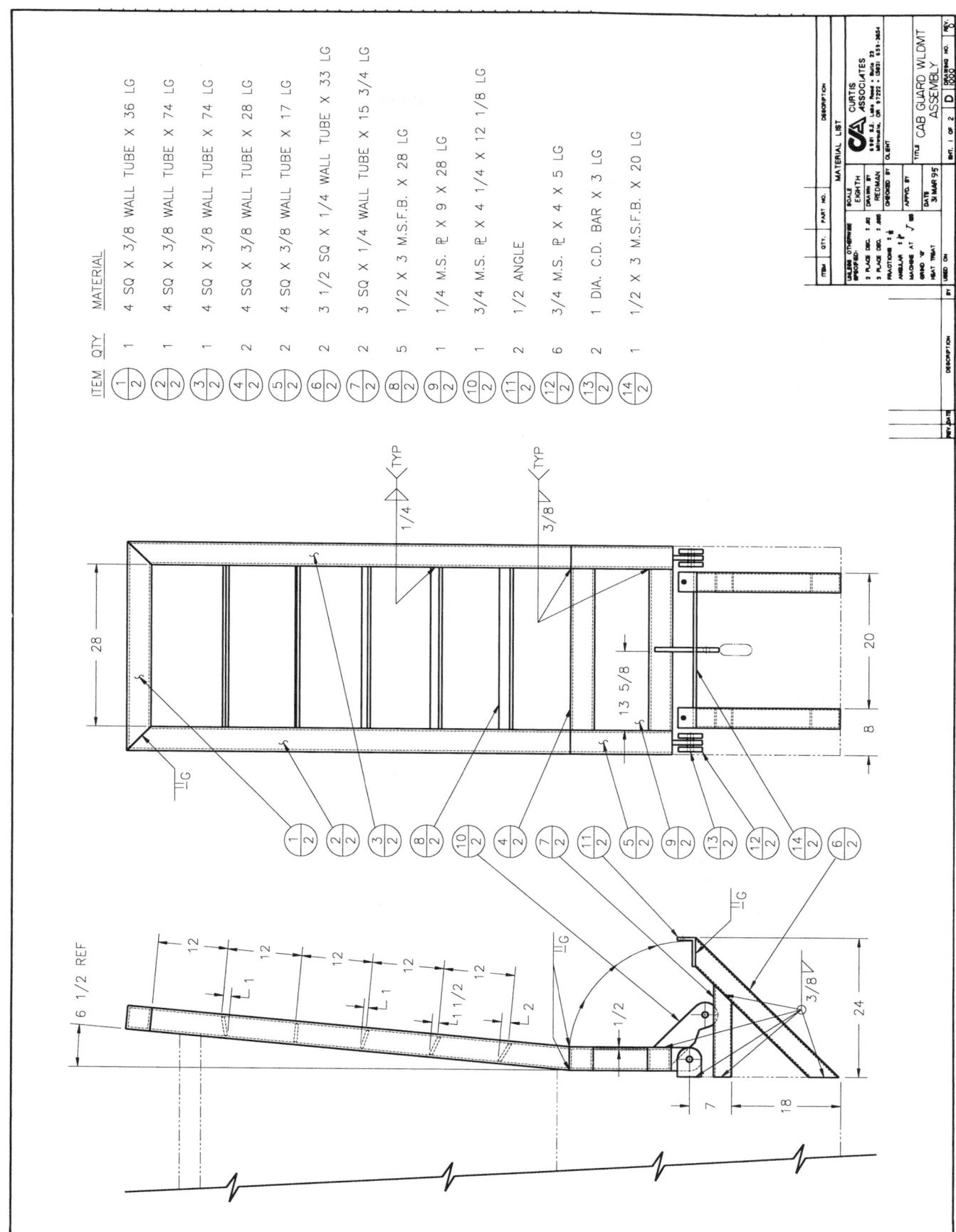

FIGURE 18.24 ■ Assembly and parts list with page identification balloons. *Courtesy Curtis Associates.*

ASSEMBLY Cross Shaft Assembly   USED ON _____

NUMBER OF UNITS _____ DATE _____

HAVE	NEED	P/Ø NO. W/Ø NO.	DET. NO.	PART NO.	DWG.	QTY.	PART NAME	DESCRIPTION	VENDOR
			1		B	1	Bearing Retainer	Ø 3" C.D. Bar	
			2		B	1	Cross Shaft	Ø 3/4" C.D. Bar	
			3		A	1	Spacer	Ø 3/4 O.D × 11GA Wall Tube × .738 Thick	
			4		A	1	Spacer	Ø 3/4 O.D × 11GA Wall Tube × .125 Thick	

FIGURE 18.25 ■ Parts list separate from assembly drawing. *Courtesy Curtis Associates.*

ing changes are treated in the same way as original drawings; they are initiated, approved, sent to the drafting department, and filed when completed. It is a good idea to leave the area below the revision block clear for future changes.

Engineering Change Request (ECR)

Before a drawing can be changed by the drafting department, an engineering change request (ECR) is needed. This ECR is the document used to initiate a change in a part or assembly. The ECR is usually attached to a print of the part affected. The print and the ECR show by sketches and written descriptions what changes will be made. The ECR also contains a number that becomes the reference record of the change to be made. Figure 18.26 shows a sample ECR form. Generally when a drafter receives an ECR, the following procedure is used:

1. Get the drawing original, sepia, or computer file.

2. Make the requested change using the same media as that used for the original.

3. Fill out the engineering documents using freehand lettering or a word processor.

4. Distribute a changed copy of the drawing and the completed engineering change forms for release approval.

5. Refile the drawing original.

Engineering Change Notice (ECN)

Records of changes are kept so reference can be made between the existing and proposed product. To make sure that records of these changes are kept, special notations are made on the drawing, and engineering change records are kept. These records are commonly known as engineering change notices (ECN) or engineering change orders (ECO).

When a change is to be made, an engineering change request initiates the change to the drafting department. The drafter then alters the original drawing or computer file to reflect the change request. When the drawing of a part is changed, a revision letter is placed next to the change. For example, the first change is lettered *A*, the second change is *B*, and so on. The letters *I, O, S, X,* and *Z* are not used. When all of the available letters *A* through *Y* have been used, double letters such as *AA* and *AB*, or *BA* and *BB* are used next. A circle drawn around the revision number helps the identification stand out clearly from the other drawing numerals. Figure 18.27 shows a part as it exists and also after a change has been made.

When the drafter chooses to make the change by not altering the drawing of the part but only changing the dimension, that dimension is labeled not-to-scale. The method of making the change will be the decision of the drafting department based on the extent of the change and the time required to make it. The not-to-scale symbol may be used to save time. Figure 18.28 shows a change made to a part with the new dimension identified as not-to-scale with a thick straight line placed under the new dimension. With CADD, it is generally easy and preferred to make the change to scale and avoid using the not-to-scale method.

After the part has been changed on the drawing and the proper revision letter is placed next to the change, the drafter records the change in the ECN column of the drawing. The location of the ECN column varies with different companies. This ECN identification may be found next to the title block or in a corner of the drawing.

ASME The ASME document, *Decimal Inch Drawing Sheet Sizes and Format,* ASME Y14.1, recommends the ECN column be placed in the upper right corner of a drawing.

The drafter records the revision number (number of times revised), ECN number (usually given on the ECR), and the date of the change. Some companies also have a column for a brief description of the change (as recommended by ANSI) and an approval. Figure 18.29 shows an expanded and condensed ECN column format. Notice that changes are added in alphabetic order from top to bottom.

The condensed ECN column in Figure 18.29b shows that the drawing has been changed twice. The first change was initiated by ECN number 2604 on September 10, 1994, and the second by ECN 2785 on November 18, 1995. Some companies may also identify the number of times a drawing has been changed by providing the letter of the current change in the title block, as shown in Figure 18.30. The drafter alters this letter to reflect each change.

When the drafter has made all of the drawing changes as specified on the ECR, then an engineering change notice (ECN) is filled out. The ECN completes the process and is filed for future reference. Usually the ECN completely describes the part as it existed before the change and the change that was made as presented on the ECR.

Aerojet Liquid Rocket Company
ENGINEERING CHANGE REQUEST & ANALYSIS

DATE	ECRANG	PAGE

ORIGINATOR NAME		DEPT	EXT	DATE	DOCUMENT NEED DATE	PROPOSED EFFECTIVITY

PART DOCUMENT NO	CURRENT REV TR	PART DOCUMENT NAME

USED ON NEXT ASSY NO	PROGRAMS AFFECTED	PROGRAMS AFFECTED

	YES	NO
QTSS ITEM		
REQUAL REQD		
REVISE QTSS FORM		

DESCRIPTION OF CHANGE

JUSTIFICATION OF CHANGE

PROJECT ENGINEER SIGNATURE	DEPT	EXT	DATE	CCB REP SIGNATURE	DATE

DESIGN ENGINEERING TECHNICAL EVALUATION

DESIGN ENGINEER SIGNATURE	DEPT	EXT	DATE	CAUSING CONT OR WORK ORDER NO	

ANY AFFECT ON	YES	NO		YES	NO		YES	NO
1 PERFORMANCE			5 WEIGHT			9 OPERATIONAL COMPUTER PROGRAMS		
2 INTERCHANGEABILITY			6 COST SCHEDULE			10 RETROFIT		
3 RELIABILITY			7 OTHER END ITEMS			11 END ITEM IDENT		
4 INTERFACE			8 SAFETY EMI			12 VENDOR CHANGE CRITICAL ITEMS ONLY		

CCB DECISION	SIGNATURES	DEPT	CON CUR	DIS SENT
	QUALITY ASSURANCE			
	MANUFACTURING			
	ENGINEERING			
	PRODUCT SUPPORT			
	MATERIAL			
	TEST OPERATION			

CCB CHAIRMAN SIGNATURE	DATE	CLASS I ☐ CLASS II ☐
CUSTOMER SIGNATURE	DATE	EFFECTIVITY

FIGURE 18.26 ■ Sample engineering change request (ECR) form. *Courtesy Aerojet Propulsion Division.*

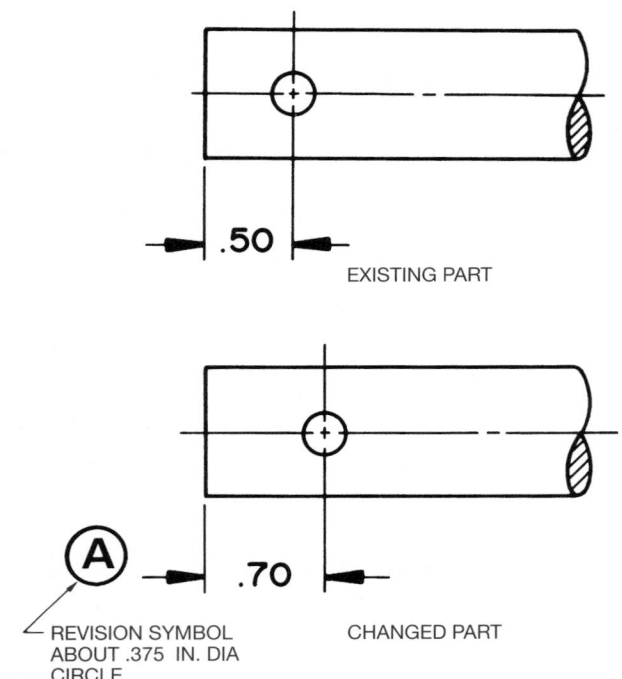

FIGURE 18.27 ■ An existing part and the same part after a change has been made.

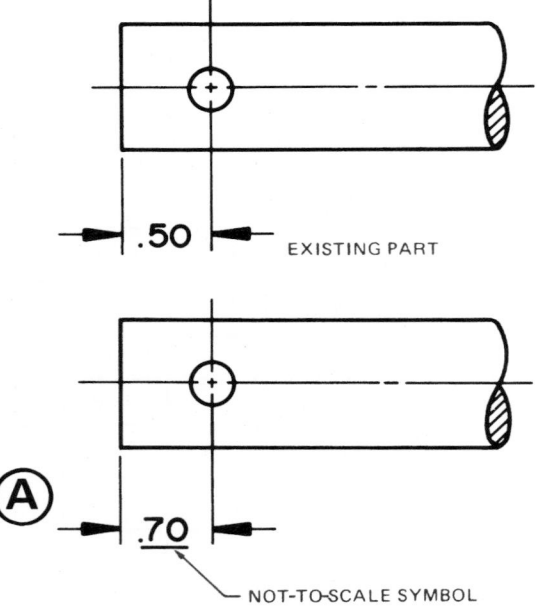

FIGURE 18.28 ■ An existing part and the same part changed using the not-to-scale symbol.

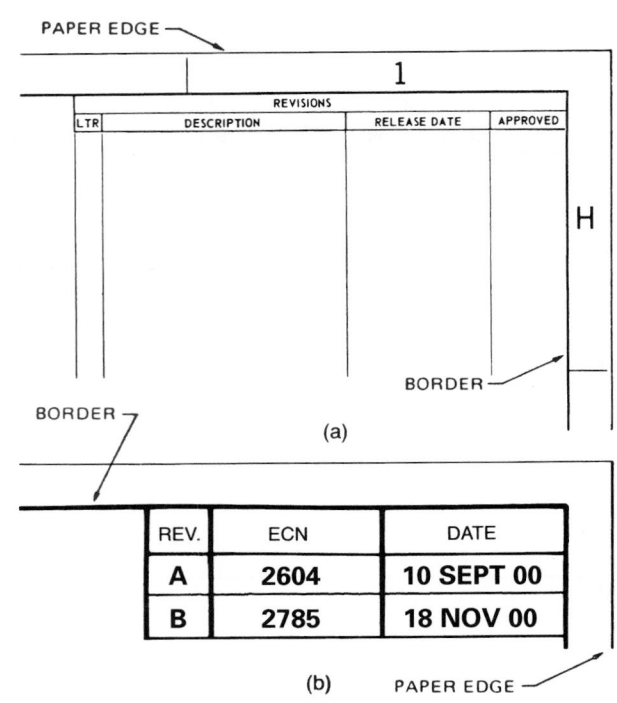

FIGURE 18.29 ■ (a) Expanded revision columns. Courtesy Aerojet Propulsion Division. (b) Condensed revision columns.

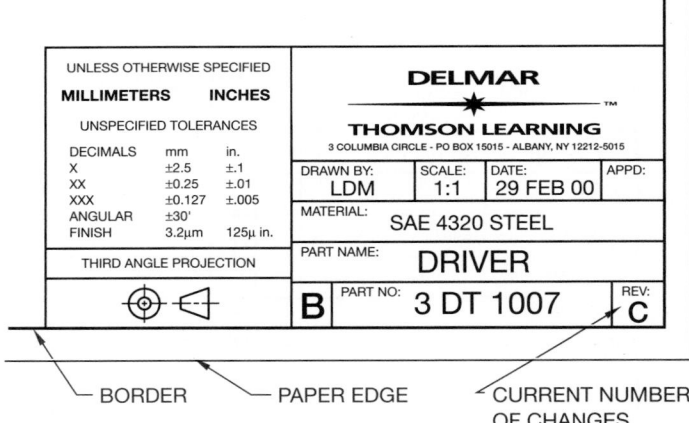

FIGURE 18.30 ■ Title block displaying the current number of times the drawing has been changed.

Figure 18.31 shows a typical ECN form. With a changed drawing and a filed ECN, anyone can verify what the part was before the change and the reason for the change. The ECN number is a reference for reviewing the change. This process allows the history of a product to be maintained and permit manufacture of a product that has not been produced in some time (replacement parts for repair).

The general engineering change elements, terminology, and techniques are consistent among companies, although the actual format of engineering change documents may be considerably different. Some formats are very simple while others are much more detailed. One of the first tasks of an entry-level drafter is to become familiar with the specific method of preparing engineering changes.

DRAWING FROM A PROTOTYPE

Occasionally the engineering drafter has to take an existing product or structure and draw a set of working drawings. This may be a prototype. A *prototype* is a model or original design that has not been released for production. The prototype is often used for test-

			DATE	DOCUMENT CHANGE NOTICE	DWG LVL	DWG FORM	DOCUMENT NUMBER	REV LTB

☐ ACDN ☐ DCN

AEROJET
@TechSystems
COMPANY
SACRAMENTO CALIFORNIA
FSCM NO 05824

ECRA NO   RELEASE DATE

DOCUMENT TITLE

PREPARED BY

SHEET OF

APPROVALS

SH	ZONE	ITEM	DESIGN	DESIGN ACTIVITY	CHECK	STRESS	WT
						MAT'L	CMO

FIGURE 18.31 ■ Sample engineering change notice. *Courtesy Aerojet Propulsion Division.*

ing and performance evaluations. In most cases there are prototype drawings from which the prototype was built, but during the modification process changes may have been made. Your job, then, is to complete the production drawings, which are established by measuring the prototype, redrawing the prototype drawings with changes, or a combination of these methods. These revised drawings are often referred to as *as-built drawings*. In this type of situation, you need to be well versed in the use of measuring equipment such as calipers, micrometers, and surface gages. You need to consult with the engineer, the testing department, and the shop personnel who made any modifications to the prototype. With this research you can fully understand how the new drawings should be created. You need to be careful to make the drawings accurate and take into account the manufacturing capabilities of your company. Then the set of working drawings that you complete is checked by the checker and the engineer before the product is released for production. Once the product is released for production, changes are often costly because tooling and patterns may have to be altered. These types of changes are usually submitted as engineering change requests (ECR). They must be approved by the engineering and manufacturing departments.

ANALYSIS OF A SET OF WORKING DRAWINGS

The Design Sketches

The design of a product generally begins in the research and development (R & D) department of a company. The engineer or designer may prepare engineering sketches or layout drawings. Figure 18.32 shows the engineer's sketches for a screwdriver to be manufactured by the company. Notice in this example that the engineer's sketch may be rough. It is your responsibility, as a professional drafter, to prepare the final drawings in accordance with ASME/ANSI, MIL, or the appropriate standards adopted by the company.

The Detail Drawings

The engineer's sketches go to the drafting department for the preparation of formal drawings. Depending on the complexity of the

PROFESSIONAL PERSPECTIVE

Many entry-level drafters are often involved in preparing detail drawings or making changes to detail drawings. When a drafter has gained valuable experience with drafting practices, standards, and company products, there is usually an opportunity to advance to a design drafter position. Drafting jobs at these levels can be exciting, as the drafters work closely with engineers to create new and updated product designs. An individual drafter in a small company may be teamed with an engineer to implement designs. In larger companies, a team of drafters, designers, and engineers may work together to design new products. These are the types of situations where a drafter may have the opportunity to prepare some or all of the drawings for a complete set of working drawings. Generally in these research and development departments of companies the preliminary product drawings are used to build prototypes of the designs for testing. After sufficient tests have been performed on the product, the drawings are revised and released for production. The new product will now become reality.

project, the engineer and drafter may work together, or there may be a team of engineers and drafters working on various components of the project. The drafter may first complete the detail drawings as shown in Figure 18.33 and Figure 18.34.

The Assembly Drawing

As discussed earlier in this chapter, the assembly drawing is used to show the assembly department how the parts fit together and if there are any special instructions required for the assembly process. When CAD is used, the drafter often freezes the dimensions and isolates the individual parts, which are brought together in a separate drawing. Using CAD allows the drafter to scale the parts into one common size and then move them together in the assembled position. Assembly notes and dimensions, if any, are

Ø.15 DRILL X 1.25
.047
.62
.093 ON ASSEMBLY
.25
.37
.75
.56
5.00
4.00
.75
.250
Ø.50
Ø 1.00
Ø.890
.25
R .40
6X R.230

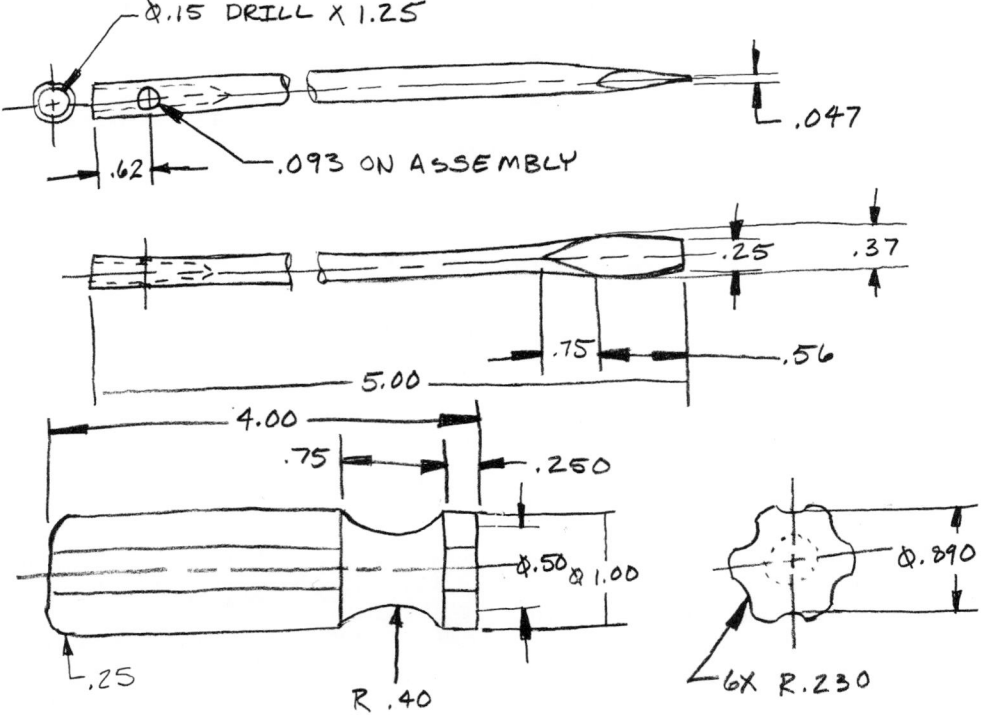

FIGURE 18.32 ■ Very rough engineer's sketch. Engineering sketches may not be as good as the one shown here. The engineering drafter must convert the sketch to proper ASME, MIL, or ISO standards as appropriate.

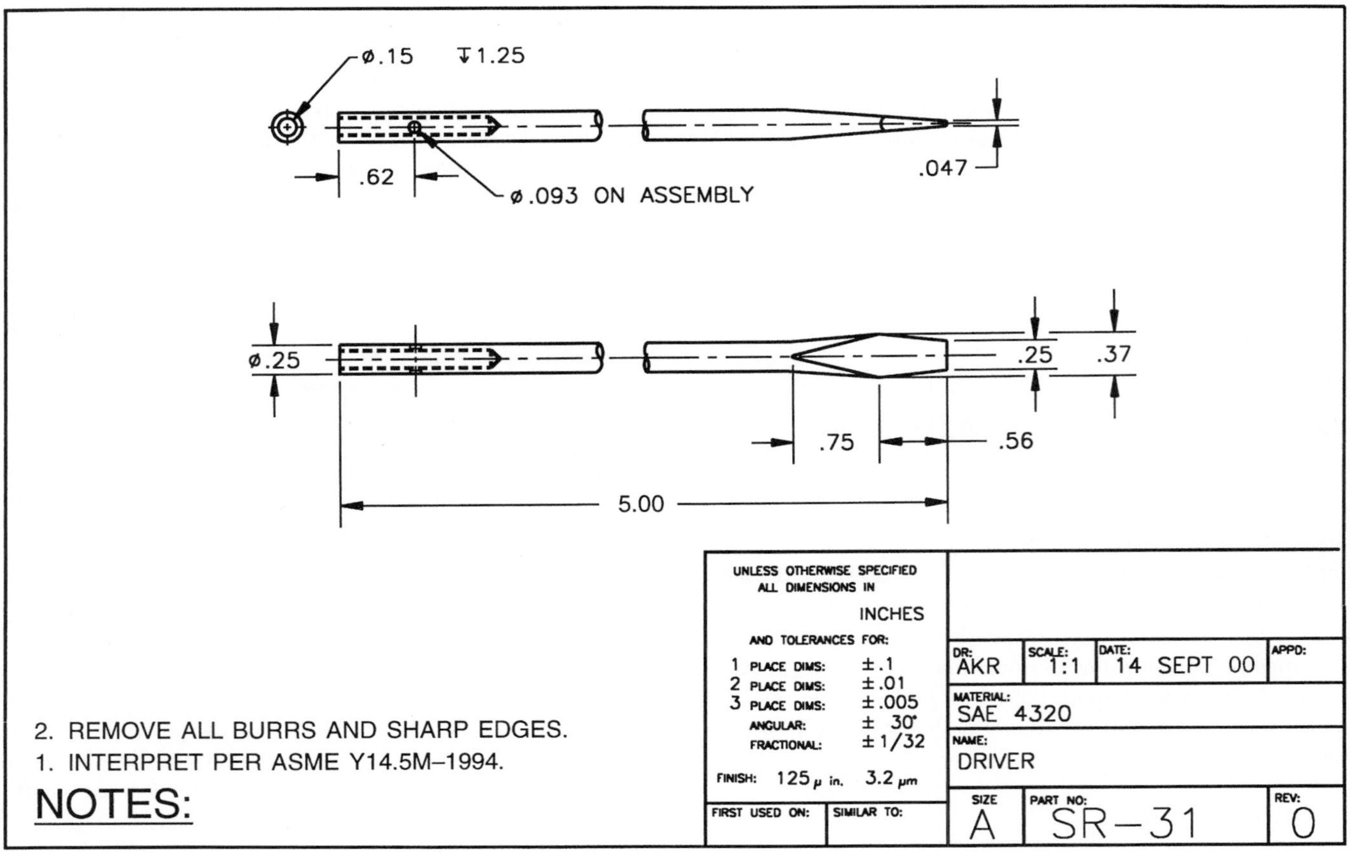

Ø.15 ↧1.25
.62
Ø.093 ON ASSEMBLY
.047
Ø.25
.25
.37
.75
.56
5.00
UNLESS OTHERWISE SPECIFIED ALL DIMENSIONS IN
INCHES
AND TOLERANCES FOR:
1 PLACE DIMS: ±.1
2 PLACE DIMS: ±.01
3 PLACE DIMS: ±.005
ANGULAR: ± 30'
FRACTIONAL: ±1/32
FINISH: 125 μ in. 3.2 μm
FIRST USED ON:
SIMILAR TO:
DR: AKR
SCALE: 1:1
DATE: 14 SEPT 00
APPD:
MATERIAL: SAE 4320
NAME: DRIVER
SIZE A
PART NO: SR–31
REV: 0
2. REMOVE ALL BURRS AND SHARP EDGES.
1. INTERPRET PER ASME Y14.5M–1994.
NOTES:

FIGURE 18.33 ■ The DRIVER detail drawing before an ECR is issued. Notice the "0" in the lower right corner. This indicates an original unchanged drawing.

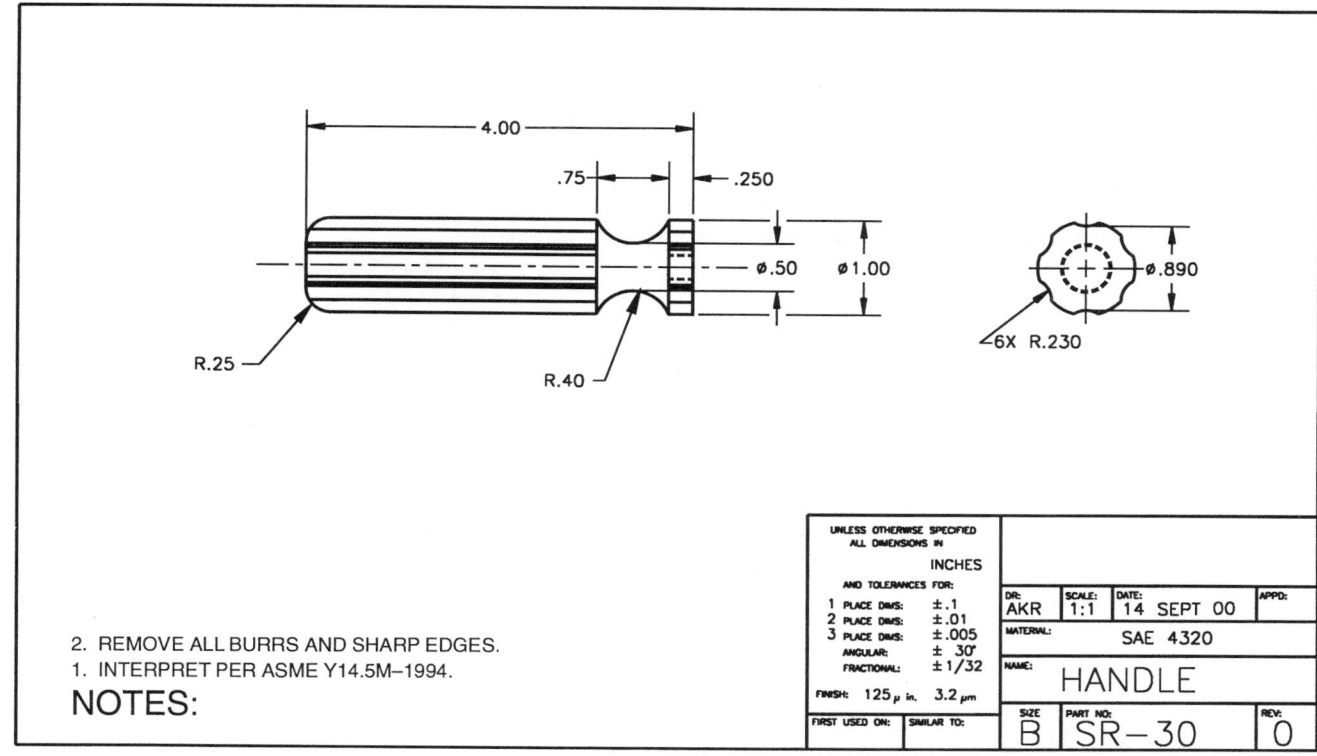

FIGURE 18.34 ■ Detail drawing of the HANDLE.

added. Each part receives a balloon and a correlating parts list is created as shown for the screwdriver design in Figure 18.35. When the complete set of working drawings has been drawn and approved, the product is released to production.

The Product Catalog

The power of 3-D CAD applications makes it possible for companies to display their products using pictorial assemblies or exploded pictorial assemblies. Figure 18.36 shows a pictorial assembly of the screwdriver that was designed in the previous discussion.

Engineering Changes

When a product is in production, a change can be requested from any department including manufacturing, engineering, or sales. These changes may be based on new ways to manufacture, redesign, customer feedback, or a variety of issues. Before a drawing can be changed by the drafting department, it usually requires an engineering change request (ECR). The ECR is also referred to as an engineering change order (ECO). The ECR is the document that is used to initiate a change in a part or assembly. The ECR is often attached to a redlined print of the part or drawing affected. The print and the ECR normally contain a change number that is used later by the drafter to properly document the change. Every company has an ECR form that looks different. The ECR form shown in Figure 18.37 displays a change requested for the DRIVER shown in Figure 18.33.

Companies use various methods to change drawings depending on the complexity of the change, the company standards, and the

time allowed. If dimensions are changed and the feature is not drawn to its proper scale, the not-to-scale symbol should be used. Changes made to CAD drawings are usually easier to accurately scale than those created using manual drawing. When CAD is used, the company often desires to have the changed drawing accurately reflect the product.

After a change is made on the drawing, a balloon is placed next to the drawing revision and a letter identifying the change is placed in the balloon. Balloons are usually ⅜ to ½ inch (8 to 12 mm) in diameter depending on the drawing size. Some companies use a square, a hexagon, or another symbol for change balloons. The letter in the balloon designates the change. An *A* is used for the first change, a *B* for the second change, and so on. Some companies use R1, R2, and R3. However, letters are common to differentiate from balloon numbers on assembly drawings. Next, the drafter records the change in the revision block of the drawing. The revision block is usually in the upper right corner in accordance with ANSI standards. The current change is also recorded in the title block as "A," "B," or "C" as appropriate. An original drawing generally has a "O" in the title block that indicates no changes have been made. Figure 18.38 shows the DRIVER in Figure 18.33 changed in accordance with the ECR in Figure 18.37.

The next procedure in the change process is for the drafter to fill out the engineering change notice (ECN). The change documents are a formal part of the drafting process and must be professionally prepared, recorded, and filed for future reference. The ECN should be hand-lettered, typed, or electronically prepared if the document is available in a word processor. The ECN form used in companies and the process for filling out the forms vary. The ECN form shown in Figure 18.39, page 594, has been

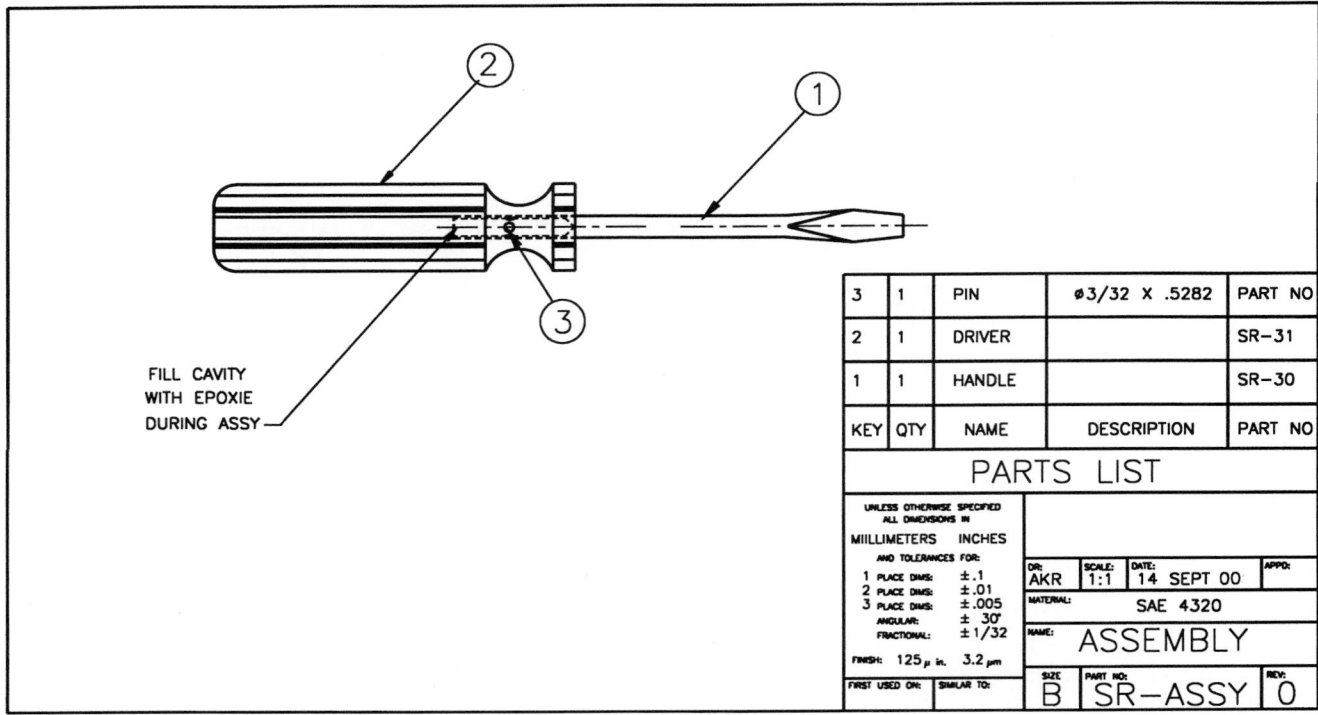

3	1	PIN	⌀3/32 X .5282	PART NO
2	1	DRIVER		SR-31
1	1	HANDLE		SR-30
KEY	QTY	NAME	DESCRIPTION	PART NO

PARTS LIST

UNLESS OTHERWISE SPECIFIED ALL DIMENSIONS IN MILLIMETERS INCHES AND TOLERANCES FOR: 1 PLACE DIMS: ±.1 2 PLACE DIMS: ±.01 3 PLACE DIMS: ±.005 ANGULAR: ± 30' FRACTIONAL: ± 1/32 FINISH: 125 μ in. 3.2 μm			

DR: AKR	SCALE: 1:1	DATE: 14 SEPT 00	APPD:
MATERIAL:		SAE 4320	
NAME:		ASSEMBLY	
FIRST USED ON:	SIMILAR TO:	SIZE B	PART NO: SR-ASSY	REV: 0

FILL CAVITY WITH EPOXIE DURING ASSY

FIGURE 18.35 ■ Assembly drawing.

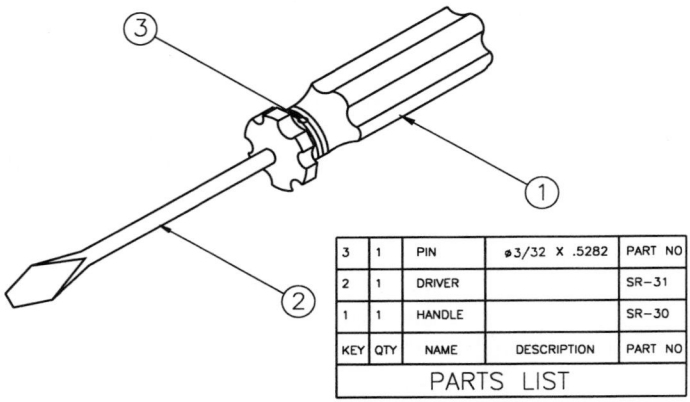

3	1	PIN	⌀3/32 X .5282	PART NO
2	1	DRIVER		SR-31
1	1	HANDLE		SR-30
KEY	QTY	NAME	DESCRIPTION	PART NO

PARTS LIST

FIGURE 18.36 ■ Pictorial assembly drawing for the company catalog.

created from a combination of formats found in industry. The large numbers on the ECN refer to the changes requested for the DRIVER in Figure 18.33 and the ECR in Figure 18.37. The following steps are used when filling out this ECN form.

1. Put the ECN number in at the top right side of page, on this example. (Get this number from the engineering change request.) In this case, it is 1171.

2. List quantity of the part affected. (On this drawing, a detail-only part can be affected.) *Note:* Quantity = 1.

3. List drawing size (A, B, C, D, etc.) and part number. This is an A size, and the part number is SR-31.

4. List revision. (In this case, it is the first one; if it were a new drawing, N or O would be put here.) Put revision letter (A, B, C, D, etc.) here. Some companies use numbers.

5. List title for drawing or part here (DRIVER).

6. List changes you made. Describe the changes made in short terms.

7. List what is to be done with parts already made under the old design. Use a symbol (A, U, T, S) to show this. The definitions of these letters are conveniently located on the ECN form.

8. List your reasons for the change in brief but complete statements so in the future someone can find out why the change was made.

9. To help the machinist, list if casting or forgings will be affected.

10. You will need to initial your name or that of the engineer you are working for, so if questions arise about the changes, those asking will know who to query.

11. Your supervisor should approve the changes you have made.

12. Pages are listed as 1/1 or 1/3, if three pages are used to complete one change. If there are three pages, the pages would be numbered 1/3, 2/3, 3/3, or 1 of 3, 2 of 3, 3 of 3.

13. The date you made the change must be recorded.

The part or product is released again for continued manufacturing after the engineering change documents have been properly prepared and approved.

WEB SITE RESEARCH

The following Web sites can assist you in doing additional research on subjects such as manufacturing materials, manufacturing practices, tooling, and related areas:

www.asme.org—American Society of Mechanical Engineers.
www.industrialpress.com—*Machinery's Handbook.* Manufacturing materials and processes.

Engineering Change Request

Request to be completed
Through ECN# | 1171

Engineering:

Engineer

Description of Request:

Refer to part number SR–31 (Driver)

Change 5" length to 6"

Change Hole location dimension from .62 to .69

Date: 20 JUL 01

Manufacturing:

Date:

Sales:

Reason for Request:

Longer tool dimension will make part more stable when locked in handle.

Hole location was design error.

Date:

| Castings & Forgings Affected? (yes or no) NO | Disposition of Production Stock: | ☐ | Use in production | ☐ |
| Approved: | Date: 20 JUL 01 | Drawing by: Student | Scrap ☒ | Transfer to service stock | ☐ |

FIGURE 18.37 ■ The engineering change request (ECR).

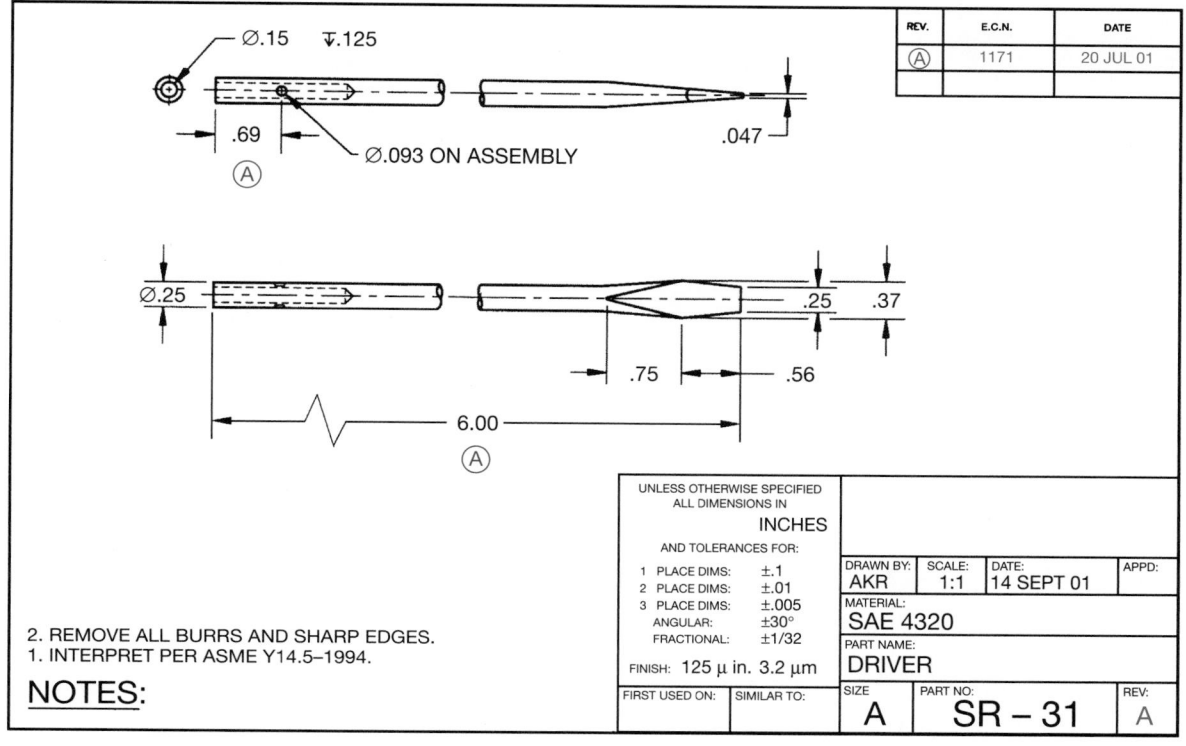

FIGURE 18.38 ■ Changes to the DRIVER based on the ECR shown in Figure 18.37. Notice the balloons next to each change, the identification of the change in the upper right corner, and the current change "A" in the lower right corner.

Engineering Change Request

ECN NO. 1171　①

Disposition of production in stock:
A = Alter or rework U = Use in production
T = Transfer to service stock S = Scrap

Qty.		Drawing Size Part No.	R/N	Description	Change	Other Usage in Production	D/S
01	1	A SR–31	A	DRIVER	6" was 5"		S
02	②	③	④	⑤	.69 Hole Location Dimension		⑦
03					was .62		
04						⑥	
05							
06							
07							
08							
09							
10							
11							
12							
13							
14							
15							
16							
17							
18							

Reason: ⑧
Line 1–This dimension needed to be loner to help make the part more stable when locked in handle.
Line 2–3 this was a design error.

Castings & forgings affected? ☐ Yes ☒ No ⑨	Design engineer: 24 NOV 01 ⑩	Supervisor approval: 25 NOV 01 ⑪	Release date: 25 NOV 01 ⑬	Page ⑫ 1/1

FIGURE 18.39 ■ The completed engineering change notice (ECN) form. The circled numbers refer to the steps for completing the form given in the text. The ECN form must be professionally typed or hand-lettered.

DISTANCE BETWEEN TANGENT CIRCLES

MATH APPLICATION

Problem: It is desired to find dimension z, the distance between the centers of the two smaller circles of the illustration shown in Figure 18.40.

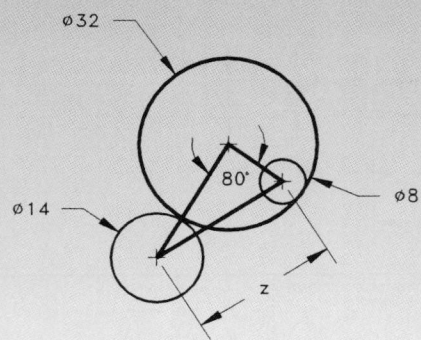

FIGURE 18.40 ■ Tangent circles.

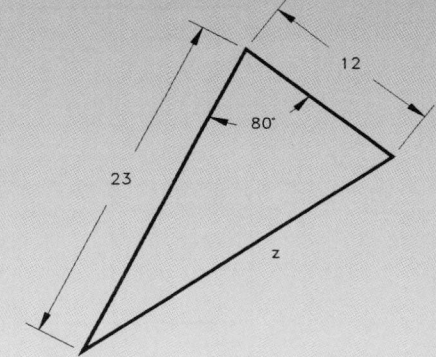

FIGURE 18.41 ■ Oblique triangles for the tangent circles.

Solution: It is necessary to visualize the triangle connecting the centers of the circles and make use of the known radii to calculate two of the sides. The side of the triangle connecting the 32" and 14" diameter circle centers must be the sum of their radii, 16 + 7 = 23", and the side connecting the 32" and 8" diameter circle centers must be the difference of their radii, 16 − 4 = 12". (See Figure 18.41.)

CHAPTER 18

Working Drawings Test

Answer the questions with short complete statements or drawings as needed.

QUESTIONS

1. Name three elements that make up a complete set of working drawings.
2. Define *detail drawing*.
3. Identify six items of information that may be found on a detail drawing.
4. List the three general groups of information that are usually found on detail drawings.
5. Describe one advantage of having one detail per sheet.
6. Describe one advantage of drawing several details per sheet.
7. List five elements to consider when selecting paper size for detail drawings.
8. Define *assembly drawing*.
9. List six items of information that may be found on an assembly drawing.
10. List the six types of assembly drawings.
11. Define *balloons*.
12. What is used to connect balloons to their related parts?
13. Describe proper information grouping.
14. List five items that are normally found on the parts list.
15. Where is the parts list normally located?
16. When are parts lists prepared on separate sheets?
17. What is the advantage of computerized parts lists?
18. What are purchase parts?
19. Provide an example of a purchase part.
20. What does the abbreviation ECR denote?
21. What does the abbreviation ECN denote?
22. How is a revision identified on a drawing?
23. How are revision numbers or letters placed on a drawing so they clearly stand out from other drawing numerals or letters?
24. Give an example of how a dimension is shown not to scale.

CHAPTER 18

Working Drawings Problems

DIRECTIONS

1. From the selected engineering sketches or layouts, prepare a complete set of working drawings, including details assembly and parts list. Determine which views and dimensions should be used to completely detail each part. Also, determine the views, parts list, dimensions, and notes, if any, for the assembly drawing. Use ANSI/ASME standards. Use manual drafting or CADD as required by your course guidelines and objectives.
2. The complete set of working drawings should be prepared with one detail drawing per sheet using multiview projection and with the assembly drawing and parts list on one sheet unless otherwise specified. All purchase (standard) parts will be completely identified in the parts list. Using the sketches as a guide, draw original multiview drawings on adequately sized sheets. Add all necessary dimensions and notes using unidirectional dimensioning. Use computer-aided drafting if required by your course guidelines. Many of the problems

are designed to be manufactured as projects in the manufacturing (machine) technology lab.
3. Some problems in this chapter contain errors, missing information, or slight inaccuracies. This is intentional and is meant to encourage you to apply appropriate problem-solving methods, engineering, and drafting standards in order to solve the problems. This is meant to force you to think about each part and how parts fit together in the assembly. As in "real-world" projects, the engineering problem should be considered as a basis for your preliminary layouts. Always question inaccuracies in project designs and consult with the proper standards and other sources. In some cases, an error might be the source of engineering changes provided by the instructor; however, this is determined by your specific course objectives. Other situations may require that corrections be made during the development of the original design drawings. This is not intended as a source

(Continued)

DIRECTIONS (Continued)

of frustration, but is considered to be part of the engineering drafter's daily responsibility in project development.

4. Include the following general notes at the lower left corner of the sheet .5 in. each way from the corner border lines for each detail drawing: (*Note:* Number 2 does not apply to the assembly drawings.)

Notes:

1. INTERPRET DIMENSIONS AND TOLERANCES PER ASME Y14.5M—1994.
2. REMOVE ALL BURRS AND SHARP EDGES.

Additional general notes may be required depending on the specifications of each individual assignment. The following should be part of your title block unless otherwise specified by your instructor:

UNSPECIFIED TOLERANCES:

DECIMALS	mm	IN.
X	± 2.5	± .1
XX	± 0.25	± .01
XXX	± 0.127	± .005
ANGULAR ± 30'		
FINISH	3.2μm	125μIN.

PROBLEM 18-1 Working-drawing assembly (metric)

Assembly Name: Plumb Bob

SPECIFIC INSTRUCTIONS:

Prepare a working-drawing assembly that has a detail drawing of each part, an assembly drawing, and a parts list on one sheet.

PARTS LIST:

ITEM	QTY	NAME	MATERIAL
1	1	PLUMB BOB	BRONZE
2	1	CAP	BRONZE

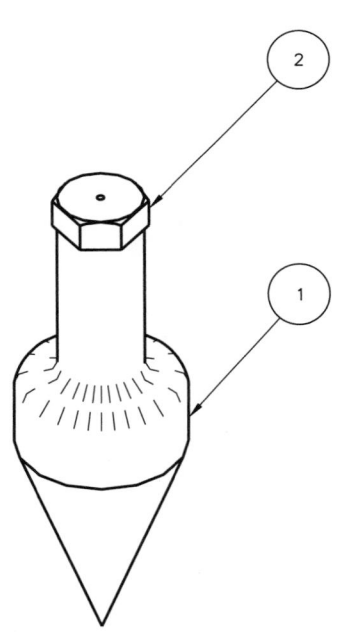

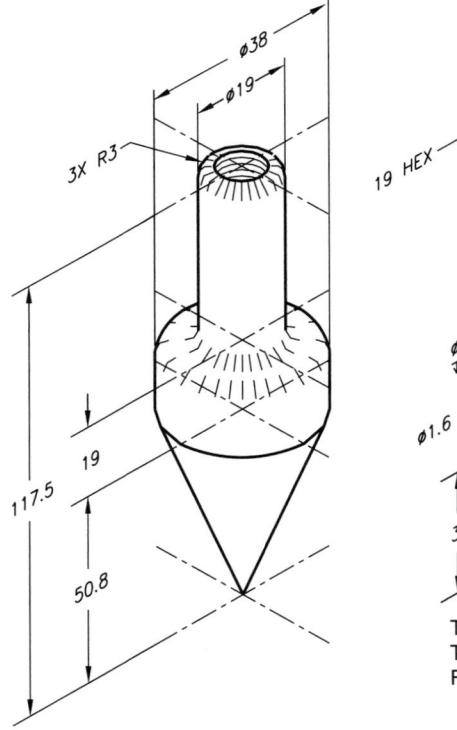

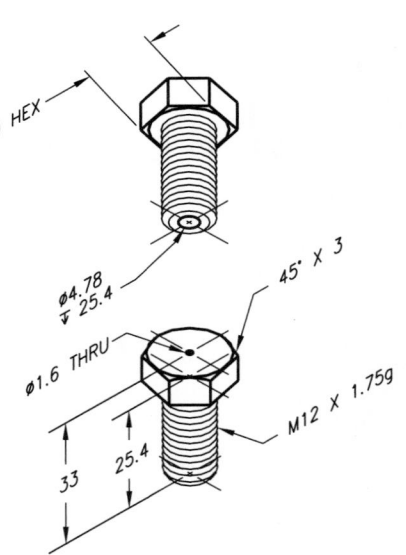

TWO PICTORIAL VIEWS OF
THE CAP ARE PROVIDED
FOR CLARITY

PROBLEM 18-2   Working drawing (in.)

Assembly Name: Hammer

SPECIFIC INSTRUCTIONS:

Prepare a detail drawing for the hammer head and two optional hammer handles on one sheet. Make the assembly drawing and parts list on another sheet.

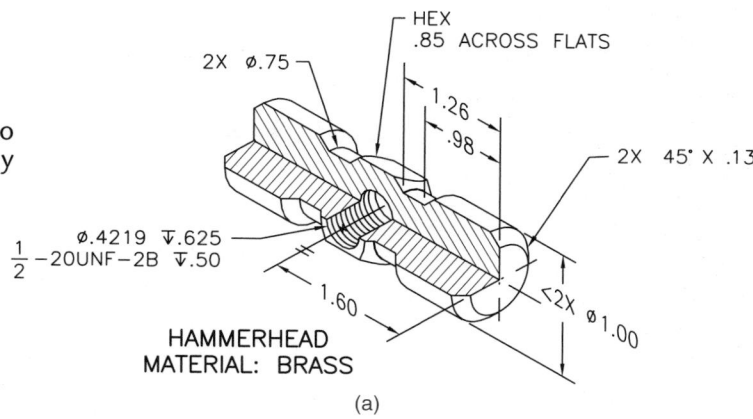

HAMMERHEAD
MATERIAL: BRASS

(a)

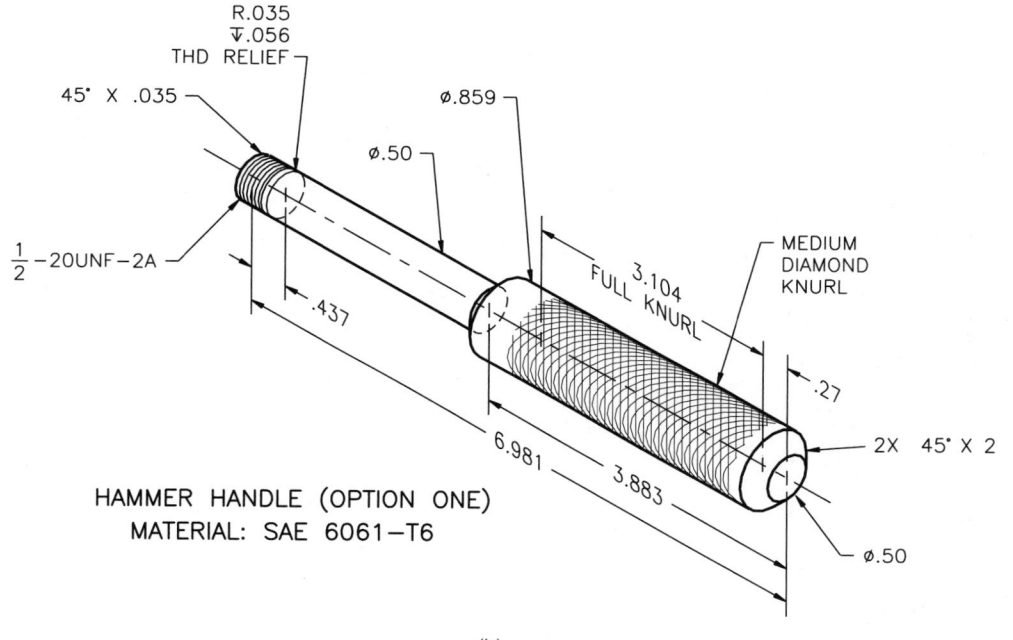

HAMMER HANDLE (OPTION ONE)
MATERIAL: SAE 6061–T6

(b)

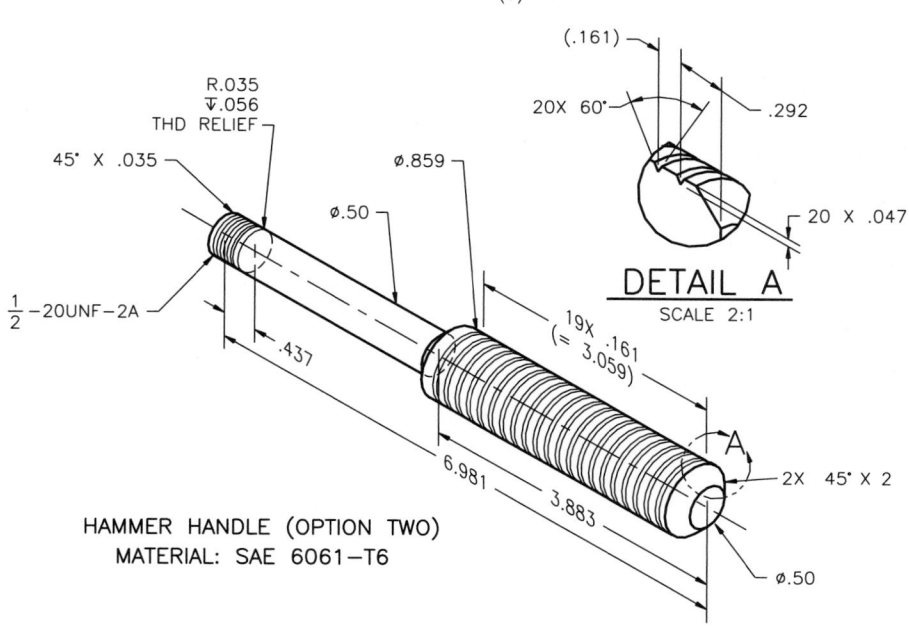

HAMMER HANDLE (OPTION TWO)
MATERIAL: SAE 6061–T6

(c)

PROBLEM 18-3   **Working drawing (in.)**

Assembly Name: Key Holder

SPECIFIC INSTRUCTIONS:

Prepare a detail drawing for each part, an assembly drawing, and a parts list on one sheet, unless otherwise specified
by your instructor.

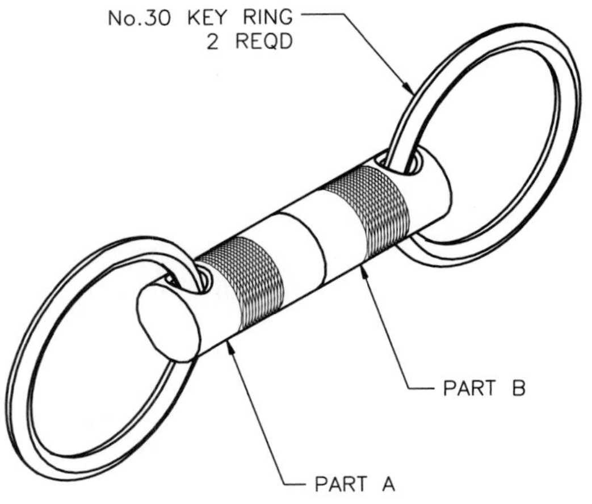

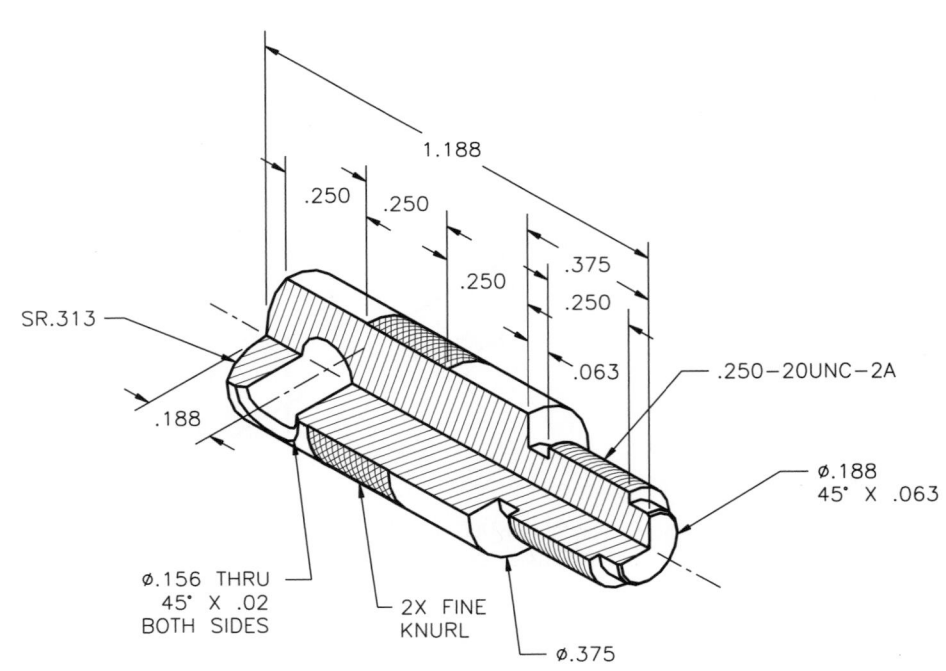

No.30 KEY RING
2 REQD

PART B

PART A

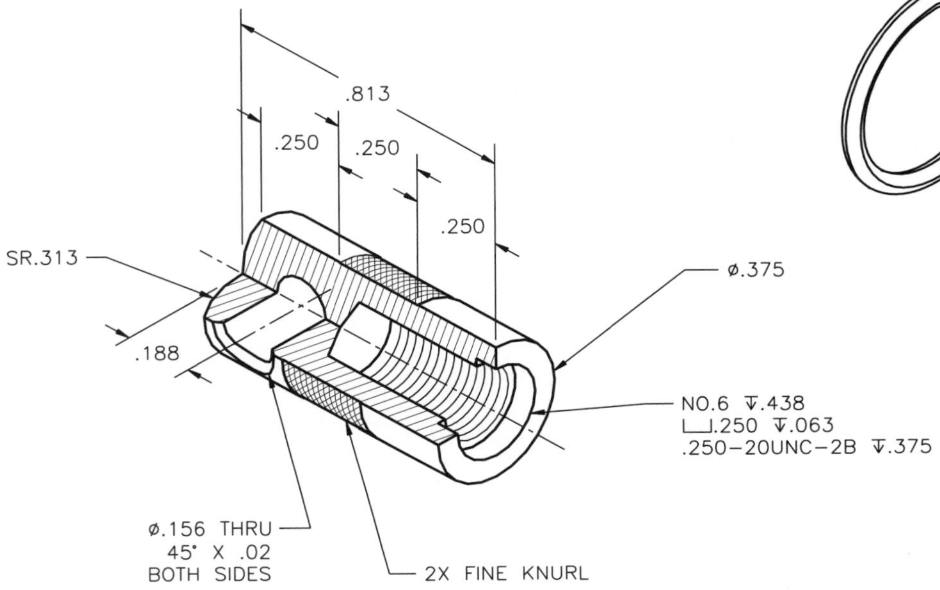

.813

.250   .250

.250

SR.313

.188

Ø.375

NO.6 ▼.438
⌴.250 ▼.063
.250−20UNC−2B ▼.375

Ø.156 THRU
45° X .02
BOTH SIDES

2X FINE KNURL

1.188

.250   .250

.250

.375

.250

SR.313

.188

.063

.250−20UNC−2A

Ø.188
45° X .063

Ø.156 THRU
45° X .02
BOTH SIDES

2X FINE
KNURL

Ø.375

PROBLEM 18-4 **Working drawing (metric)**

Assembly Name: C-clamp

SPECIFIC INSTRUCTIONS:

Prepare a complete set of working drawings with all of the detail drawings on one sheet and the assembly drawings and parts list on another sheet. Use multiview projection for view layout.

PARTS LIST:

ITEM	QTY	NAME	MATERIAL
1	1	PIN	CRMS
2	1	BODY	SAE 4320
3	1	SCREW	SAE 4320
4	1	SWIVEL	SAE 1020

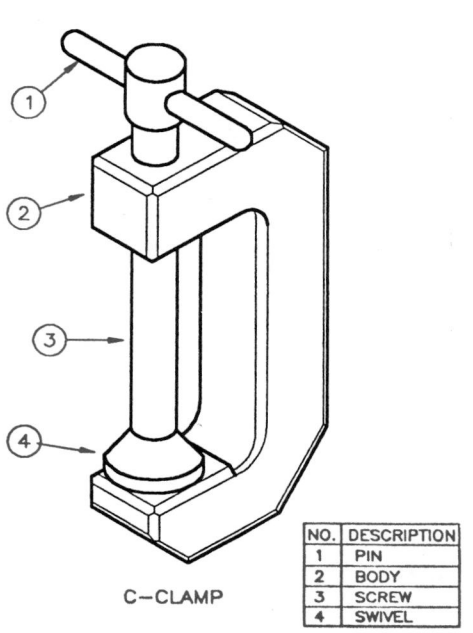

C—CLAMP

NO.	DESCRIPTION
1	PIN
2	BODY
3	SCREW
4	SWIVEL

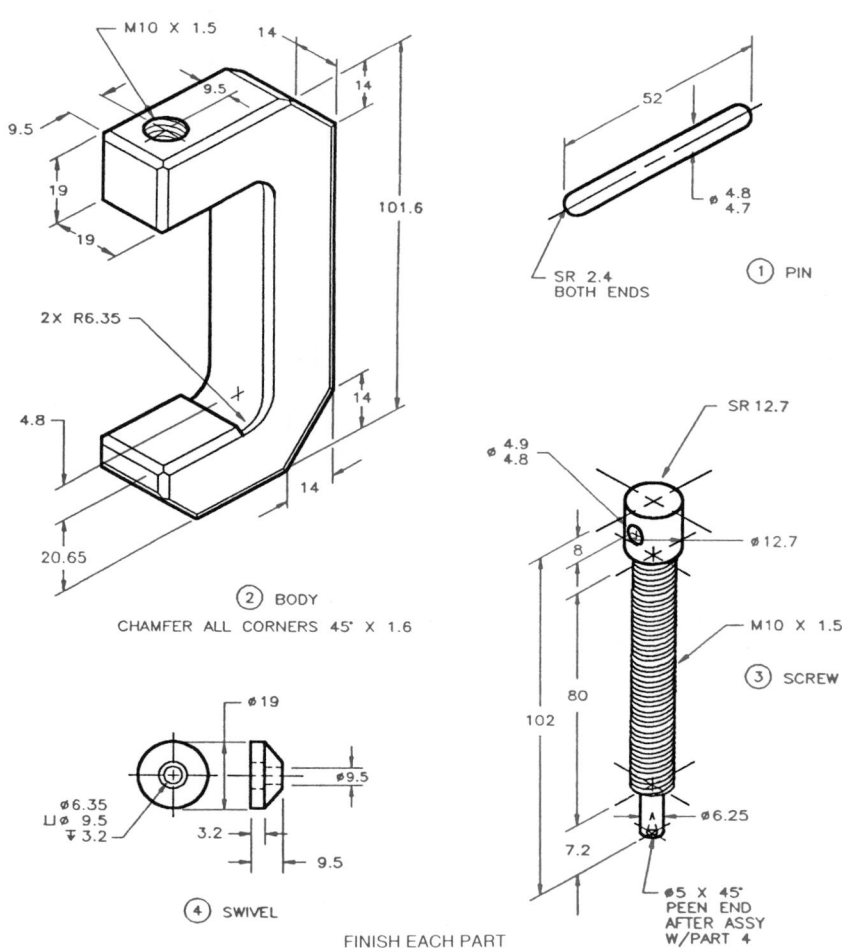

② BODY
CHAMFER ALL CORNERS 45° X 1.6

① PIN

③ SCREW

④ SWIVEL

FINISH EACH PART
ALL OVER 0.8 μM

PROBLEM 18-5 **Working drawing (in.)**

Assembly Name: Mill Work Stop

SPECIFIC INSTRUCTIONS:

Prepare a complete set of working drawings with one detail drawing per sheet and the assembly drawing with parts list on one sheet. Notice: part number 10, TEE STUD, is cut from standard 3/8-16UNC ALL THREAD. Parts 12 and 13, BLANK

KNOBS, are purchased as knobs without the threads machined. You will draw the knobs as a "representation" of the purchase part and give the thread note as identified in the parts list. The actual dimensions of the knob to be purchased are not important; however, accurate specifications for the threads to be machined into the purchase part are important.

NOTE: Some design is required throughout the completion of the project.

PARTS LIST:

ITEM	QTY	NAME	DESCRIPTION	MATERIAL
1	1	VERTICAL MEMBER	$3/4 \times 2 \times 4$	SAE 6061
2	1	BASE MEMBER	$1 \times 2 \times 3.5$	SAE 6061
3	2	RIB	1/4 PLATE	SAE 6061
4	1	ARM	$3/8 \times 1 \times 5\text{-}1/2$	SAE 1018
5	1	CLAMP	$1/2 \times 1 \times 1\text{-}5/16$	SAE 1018
6	1	TEE PLATE DRILL AND TAP AT CENTER FOR 3/8-16UNC-2 THREAD THRU	$5/16 \times 1 \times 1\text{-}1/4$	SAE 1018
7	1	STOP ROD	$\varnothing 5/16 \times 6$	SAE 303 SS
8	1	WING NUT	$1/4\text{-}28 \times 1/2$	
9	1	BUSHING	$\varnothing 1/2 \times .31$	SAE 1018
10	1	TEE STUD	$3/8\text{-}16 \times 2$ LG ALL THREAD	
11	1	CARRIAGE BOLT	5/16-18UNC-2	
12	2	BLANK KNOB	VLIER HK-3 (DRILL AND TAP FOR 3/8-16UNC-2B)	
13	1	BLANK KNOB	VLIER HK-3 (DRILL AND TAP FOR 3/8-16UNC-2B)	

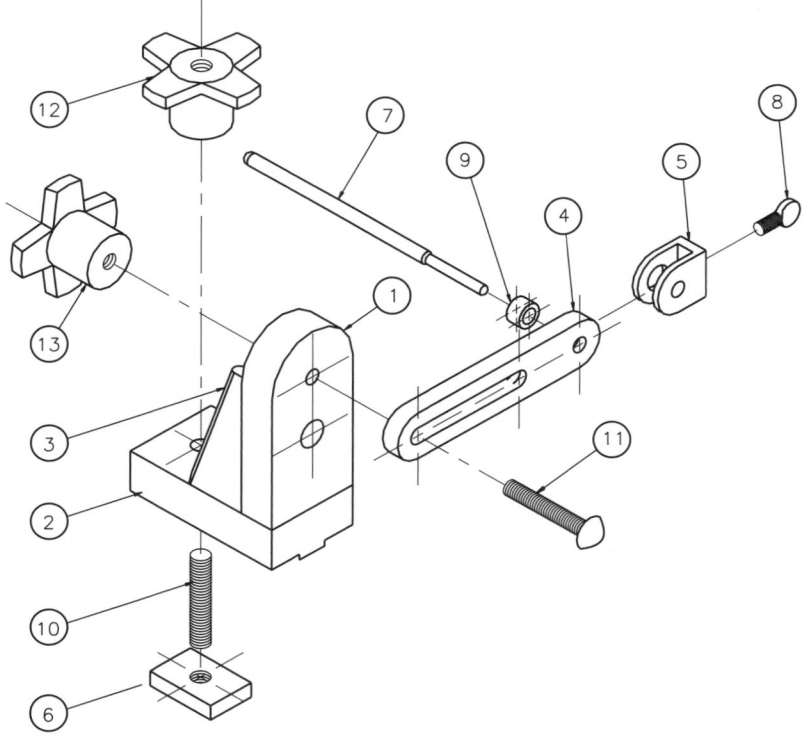

PROBLEM 18-5 (continued)

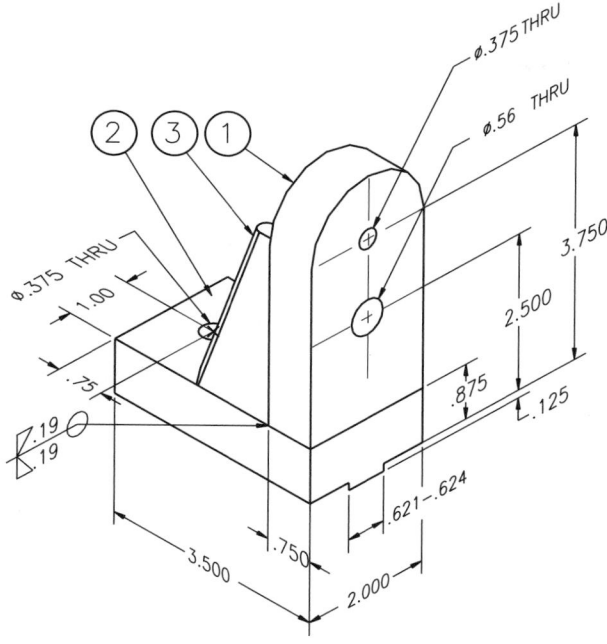

DEBUR ALL EDGES .020

BODY ASSEMBLY WELDMENT
FINISH: COLOR ANODIZED

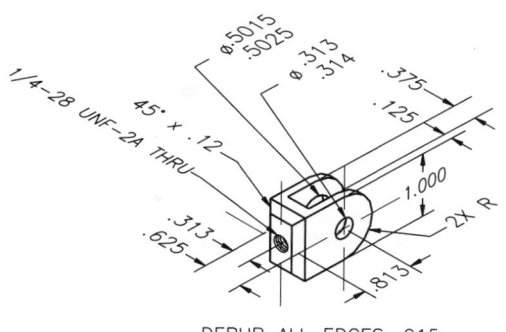

DEBUR ALL EDGES .015
EXCEPT ø.313–.314

⑤ CLAMP
FINISH: C HARDEN .005/.010
BLACK OXIDE

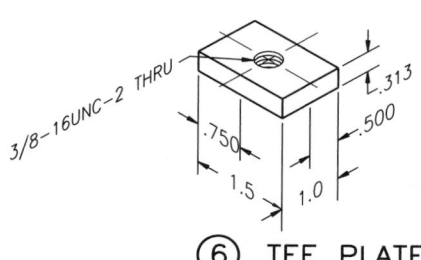

⑥ TEE PLATE

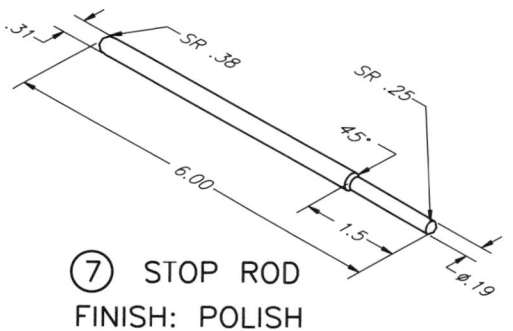

⑦ STOP ROD
FINISH: POLISH

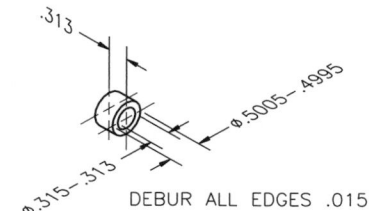

DEBUR ALL EDGES .015

⑨ BUSHING
FINISH: C HARDEN .005/.010
BLACK OXIDE

③ RIB

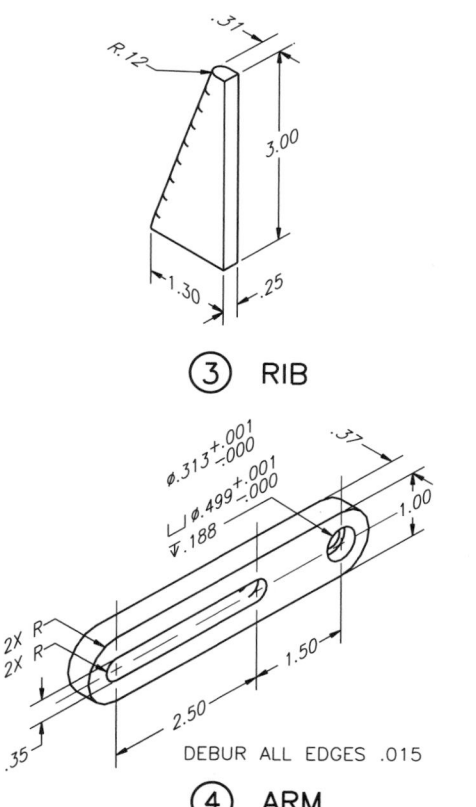

DEBUR ALL EDGES .015

④ ARM
FINISH: C HARDEN .005/.010
BLACK OXIDE

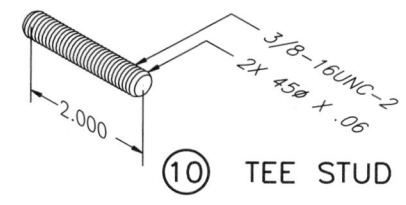

⑩ TEE STUD

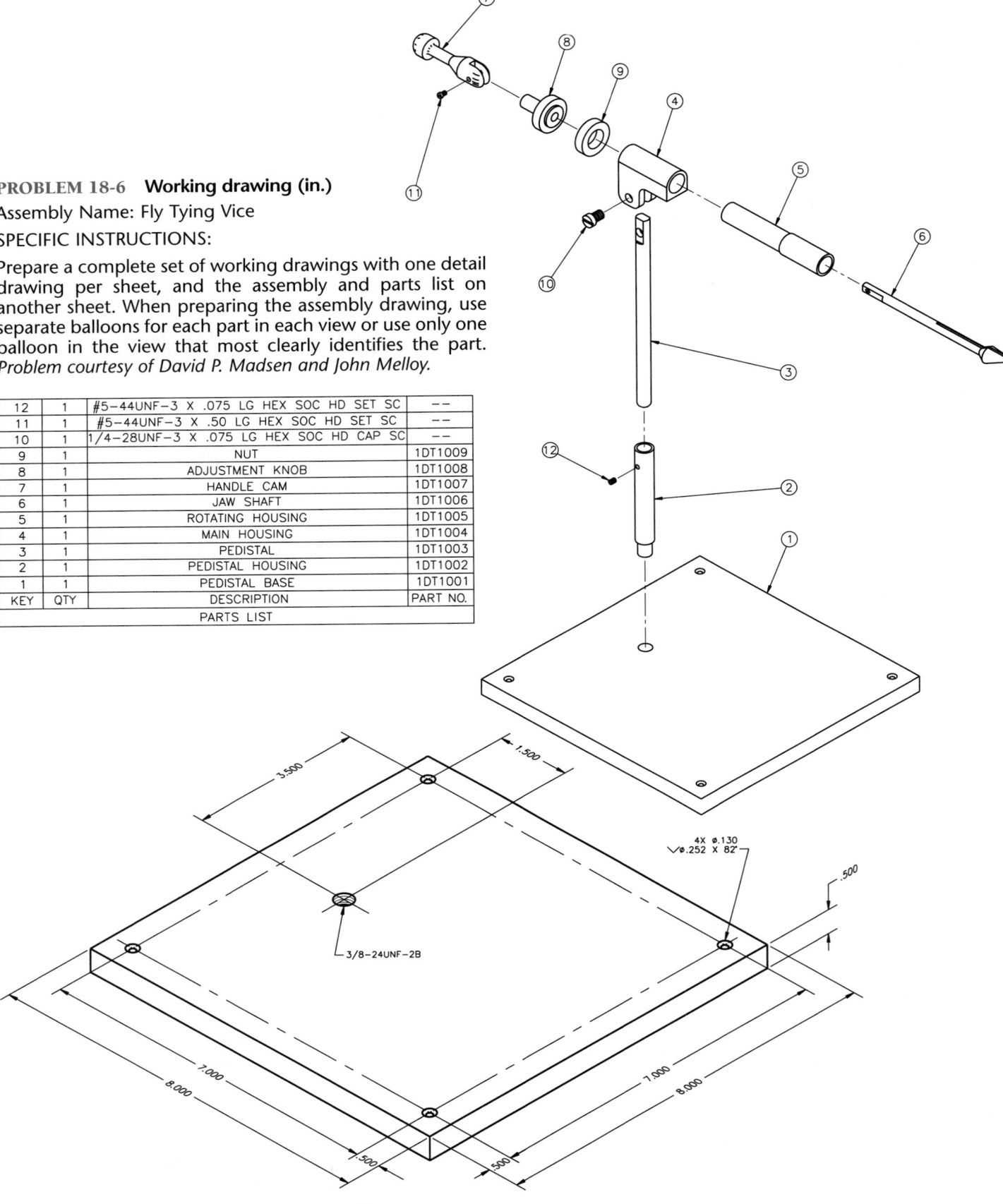

PROBLEM 18-6 **Working drawing (in.)**

Assembly Name: Fly Tying Vice

SPECIFIC INSTRUCTIONS:

Prepare a complete set of working drawings with one detail drawing per sheet, and the assembly and parts list on another sheet. When preparing the assembly drawing, use separate balloons for each part in each view or use only one balloon in the view that most clearly identifies the part. *Problem courtesy of David P. Madsen and John Melloy.*

12	1	#5−44UNF−3 X .075 LG HEX SOC HD SET SC	−−
11	1	#5−44UNF−3 X .50 LG HEX SOC HD SET SC	−−
10	1	1/4−28UNF−3 X .075 LG HEX SOC HD CAP SC	−−
9	1	NUT	1DT1009
8	1	ADJUSTMENT KNOB	1DT1008
7	1	HANDLE CAM	1DT1007
6	1	JAW SHAFT	1DT1006
5	1	ROTATING HOUSING	1DT1005
4	1	MAIN HOUSING	1DT1004
3	1	PEDISTAL	1DT1003
2	1	PEDISTAL HOUSING	1DT1002
1	1	PEDISTAL BASE	1DT1001
KEY	QTY	DESCRIPTION	PART NO.
		PARTS LIST	

④ **PEDISTAL BASE**

PROBLEM 18-6 (continued)

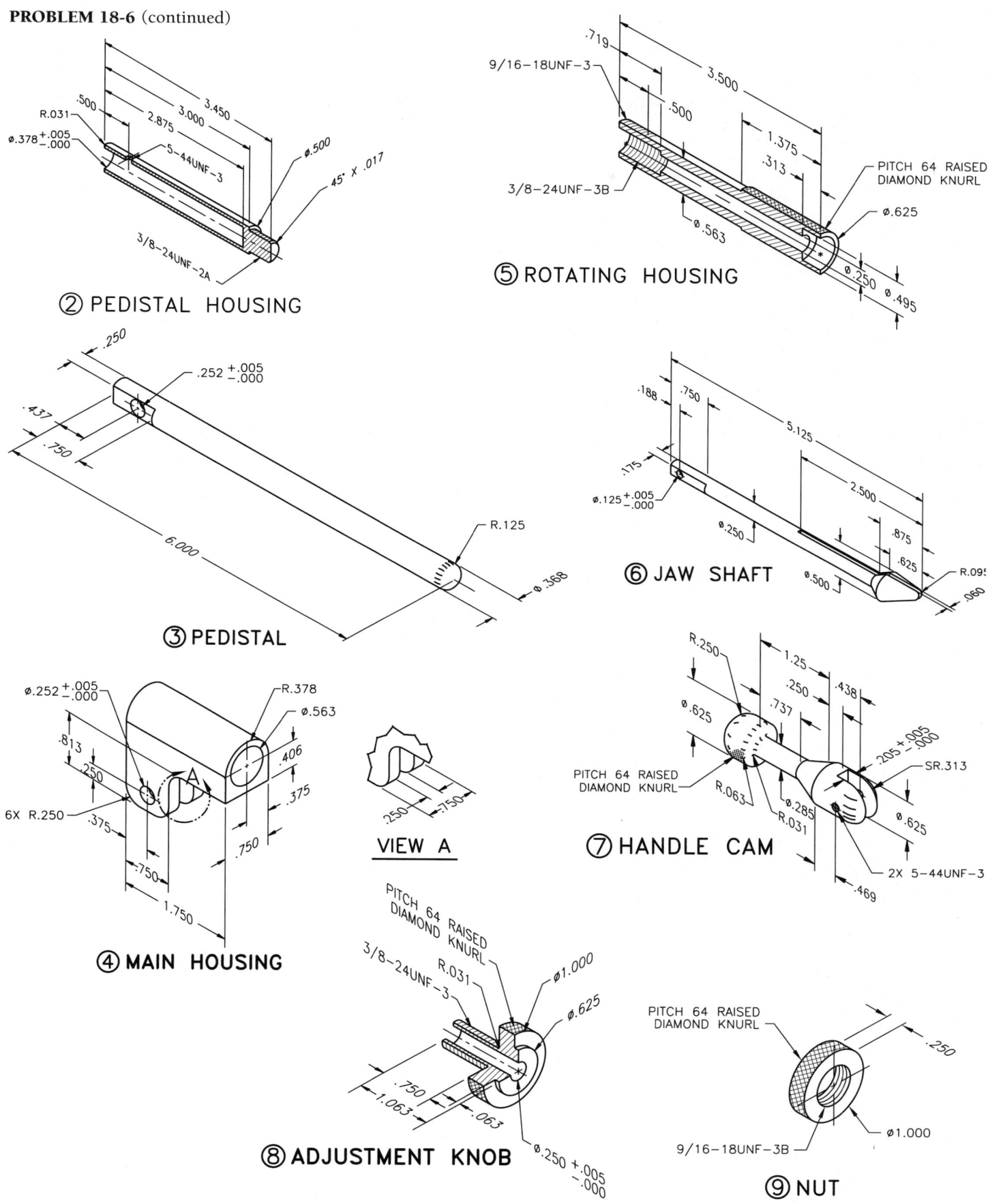

② PEDISTAL HOUSING

③ PEDISTAL

④ MAIN HOUSING

VIEW A

⑤ ROTATING HOUSING

⑥ JAW SHAFT

⑦ HANDLE CAM

⑧ ADJUSTMENT KNOB

⑨ NUT

PROBLEM 18-7   Working drawing (in.)

Assembly Name: Cork Screw

SPECIFIC INSTRUCTIONS:

Prepare a complete set of working drawings with one detail drawing per sheet, and the assembly and parts list on another sheet. When preparing the assembly drawing, use separate balloons for each part in each view or use only one balloon in the view that most clearly identifies the part. Establish tolerances between mating parts. Determine the dimensions for the Pin, Part 6 and create needed detail or purchase part specifications. *Problem courtesy of Kasey Cassell.*

PARTS LIST:

ITEM	QTY	NAME	DESCRIPTION
1	1	BODY	Stainless Steel
2	1	HANDLE	Stainless Steel
3	2	ARM	Stainless Steel
4	2	UPPER PIN	Stainless Steel
5	2	LOWER PIN	Stainless Steel
6	1	PIN	Stainless Steel
7	1	CORK SCREW	Stainless Steel

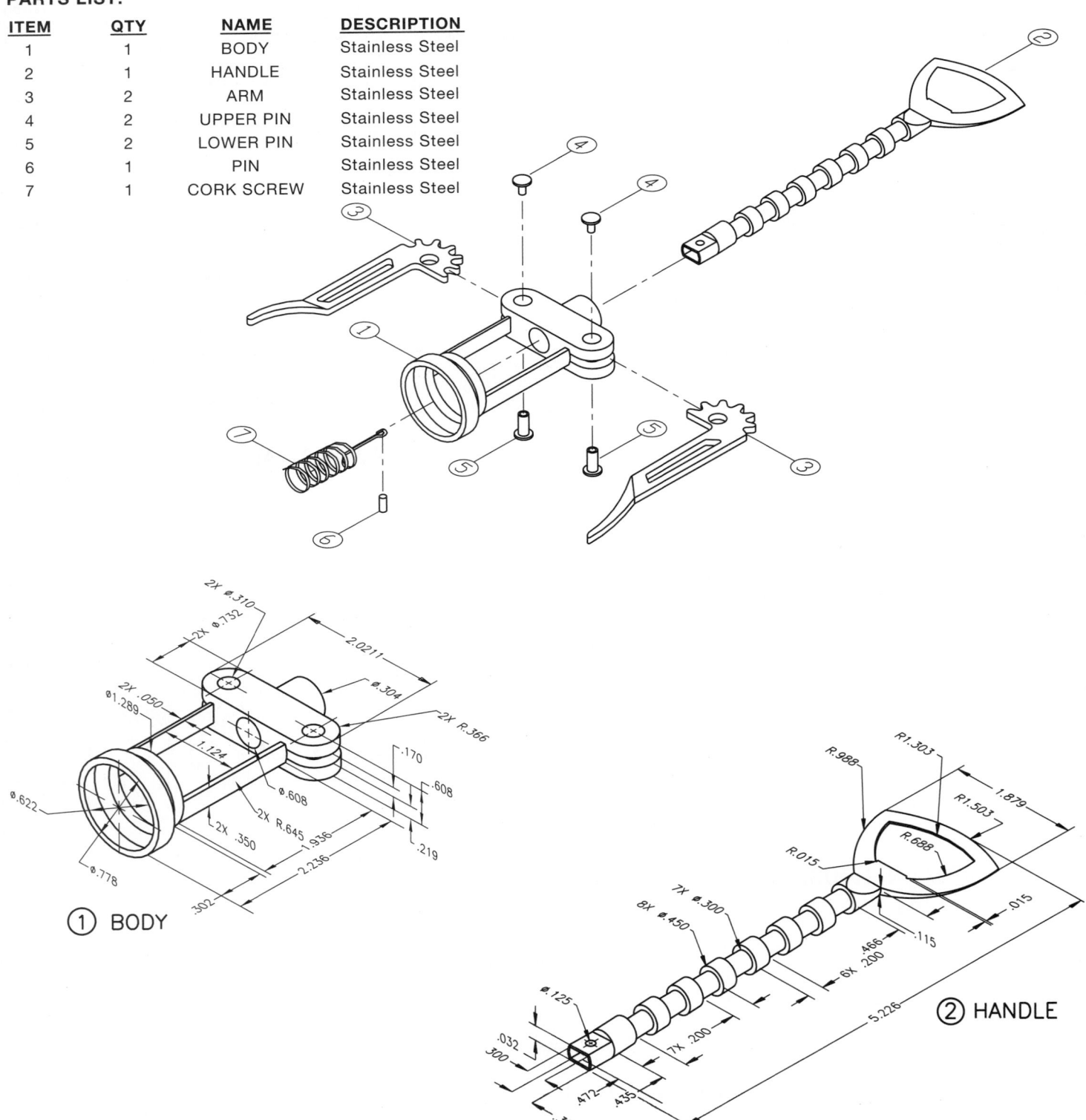

PROBLEM 18-7 (continued)

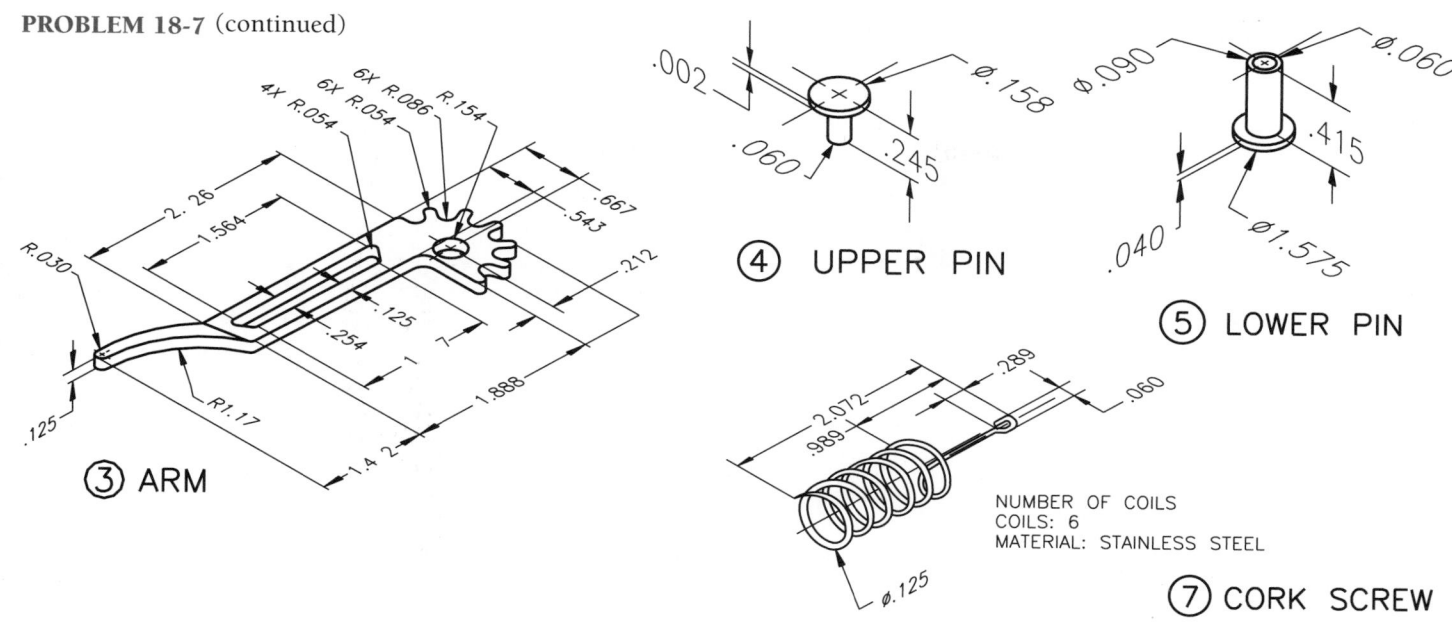

③ ARM

④ UPPER PIN

⑤ LOWER PIN

NUMBER OF COILS
COILS: 6
MATERIAL: STAINLESS STEEL

⑦ CORK SCREW

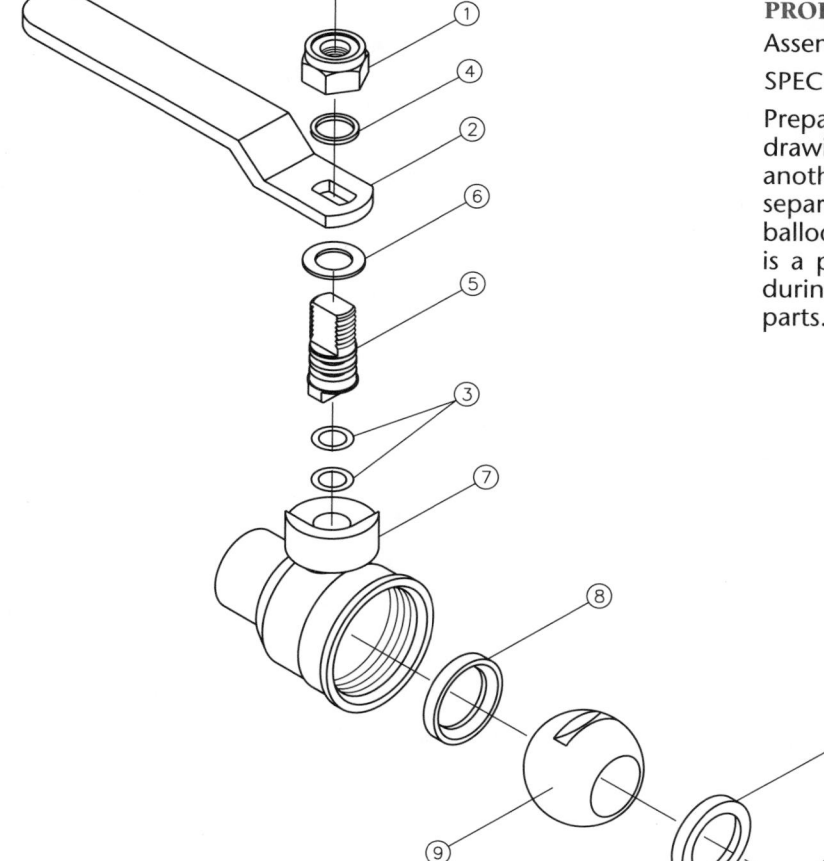

PROBLEM 18-8 **Working drawing (in.)**

Assembly Name: Ball Valve

SPECIFIC INSTRUCTIONS:

Prepare a complete set of working drawings with one detail drawing per sheet, and the assembly and parts list on another sheet. When preparing the assembly drawing, use separate balloons for each part in each view or use only one balloon in the view that most clearly identifies the part. This is a prototype project. Errors are likely. Verify dimensions during assembly. Establish tolerances between mating parts. *Problem courtesy of Chris Hurford.*

10	1	END CAP	BRASS	1DT1010
9	1	BALL	STAINLESS STEEL	1DT1009
8	2	WASHER	PLASTIC	1DT1008
7	1	VALVE BODY	BRASS	1DT1007
6	1	WASHER	PLASTIC	1DT1006
5	1	STEM	STAINLESS STEEL	1DT1005
4	1	WASHER	PLASTIC	1DT1004
3	2	O-RING	PARKER #2-008	5DT1003
2	1	HANDLE	HANDLE	1DT1002
1	1	NUT	5/16-24UNF-2	5DT1001
KEY	QTY	NAME	DESCRIPTION	PART NO.
PARTS LIST				

PROBLEM 18-8 (continued)

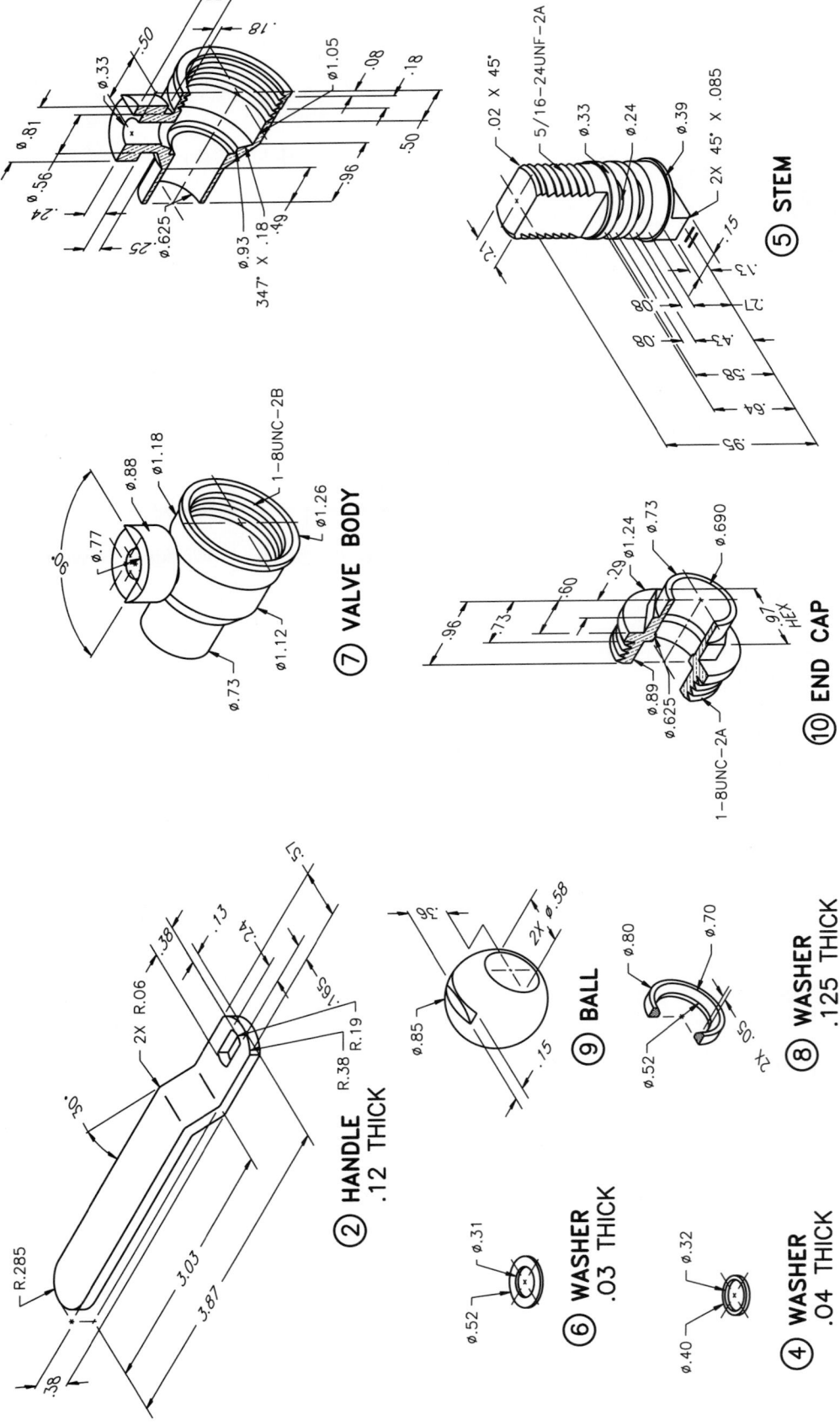

⑤ STEM

⑦ VALVE BODY

⑩ END CAP

② HANDLE .12 THICK

⑨ BALL

⑧ WASHER .125 THICK

⑥ WASHER .03 THICK

④ WASHER .04 THICK

PROBLEM 18-9 Working drawing (metric)

Assembly Name: Tool Holder

SPECIFIC INSTRUCTIONS:

Prepare a complete set of working drawings with one detail drawing per sheet, and the assembly and parts list on another sheet. When preparing the assembly drawing, use separate balloons for each part in each view, or use only one balloon in the view that most clearly identifies the part.

PARTS LIST:

ITEM	QTY	NAME	MATERIAL
1	1	PARTING TOOL, 3/32 in. × 1/2 in. PURCHASE PART	TOOL STEEL
2	1	TOOL HOLDER BODY	06 STEEL
3	1	ADJUSTMENT SCREW	SAE 1035 STEEL
4	1	SHIM	SAE 4320 STEEL
5	1	KNURL NUT	SAE 3130 STEEL
6	1	WASHER	SAE 1060 STEEL
7	1	STUD	SAE 1035 STEEL
8	1	M 10 × 1.5 HEX NUT PURCHASE PART	

4 SHIM
FAO 3.2μM

3 ADJUSTMENT SCREW
FAO 3.2μM

7 STUD
FAO 3.2μM

5 KNURL NUT
FAO 3.2μM

6 WASHER
FAO 3.2μM

2 TOOL HOLDER BODY
FAO 3.2μM UNLESS OTHERWISE SPECIFIED

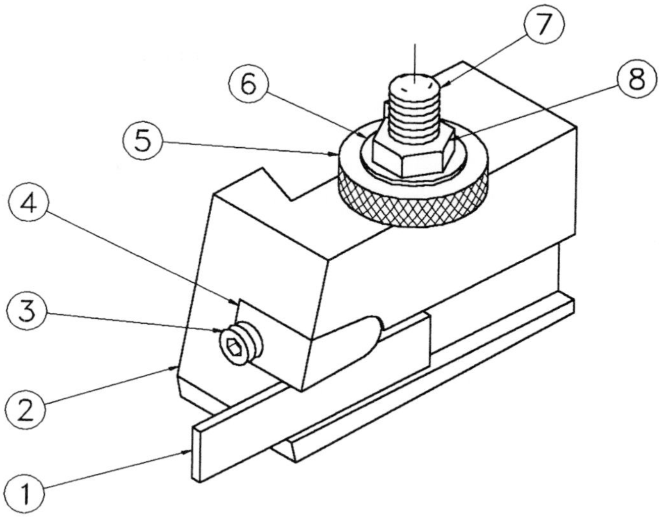

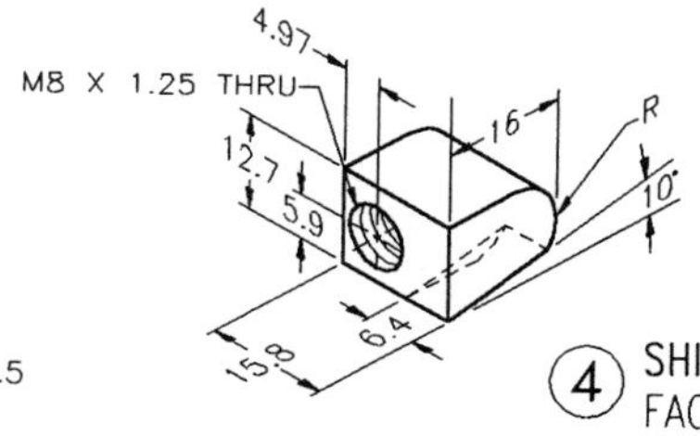

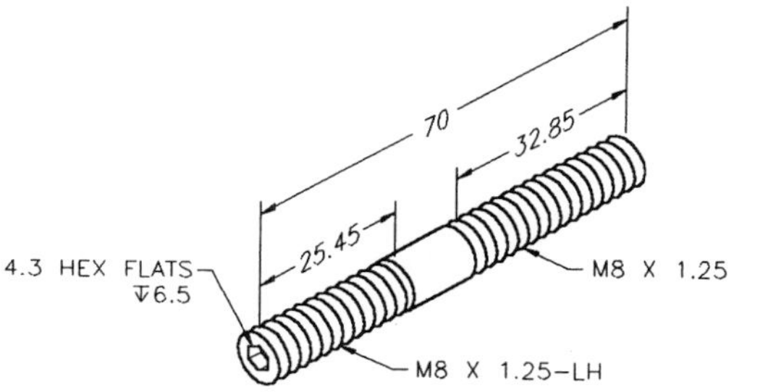

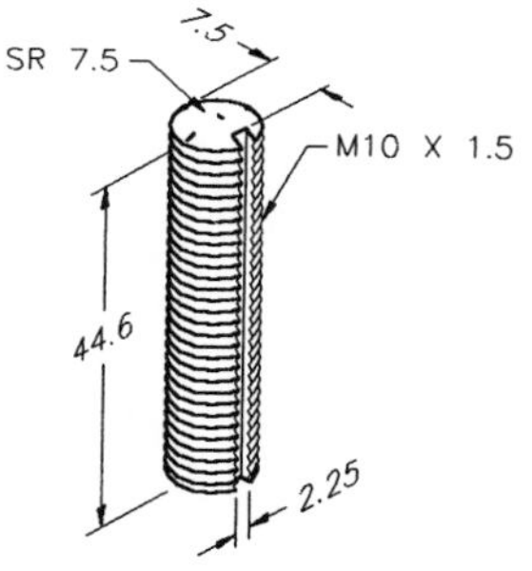

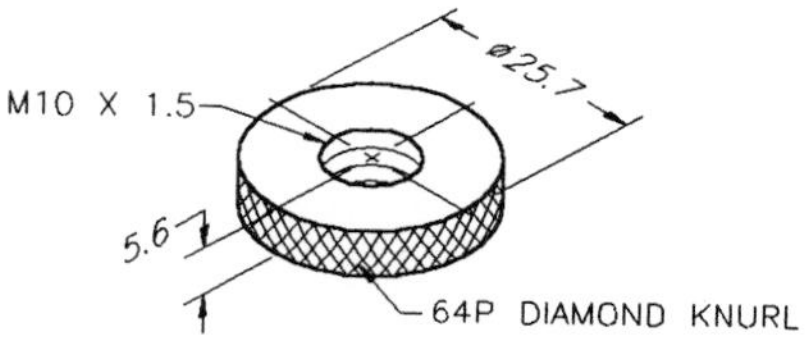

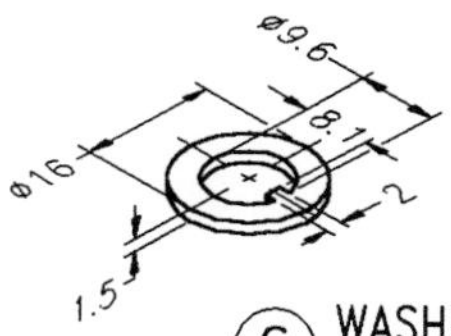

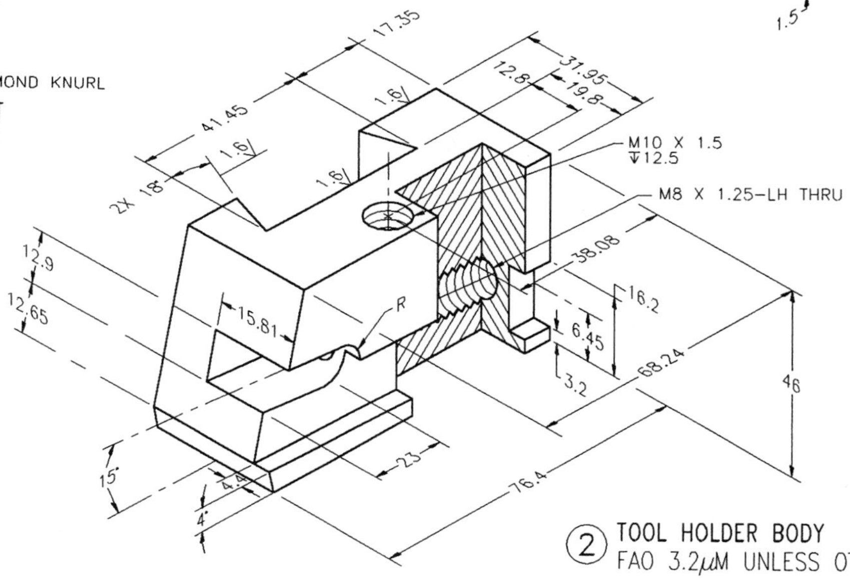

PROBLEM 18-10 **Working drawing (in.)**

Assembly Name: Adjustable Attachment

SPECIFIC INSTRUCTIONS:

Prepare a complete set of working drawings with detail drawings of individual parts combined on one or more sheets depending on the size of sheet selected. The assembly drawing and parts list will be combined on one sheet.

PARTS LIST:

ITEM	QTY	NAME	DESCRIPTION	MATERIAL
1	1	CAP SCREW	.25-20UNC-2 × .75 HEX SOC HEAD	STL
2	1	FRAME		SAE 4340
3	1	ADJUSTING SCREW		SAE1045
4	1	SET SCREW	8-32UNC-2 HEX SOC FLAT POINT	STL
5	1	TAPER PIN	0 × .625	STL
6	1	KNURL KNOB		SAE 1024
7	1	CLAMP SCREW		SAE 2330
8	1	PILOT SCREW		SAE 2330
9	1	ADJUSTING NUT		SAE 3130

ADJUSTABLE ATTACHMENT

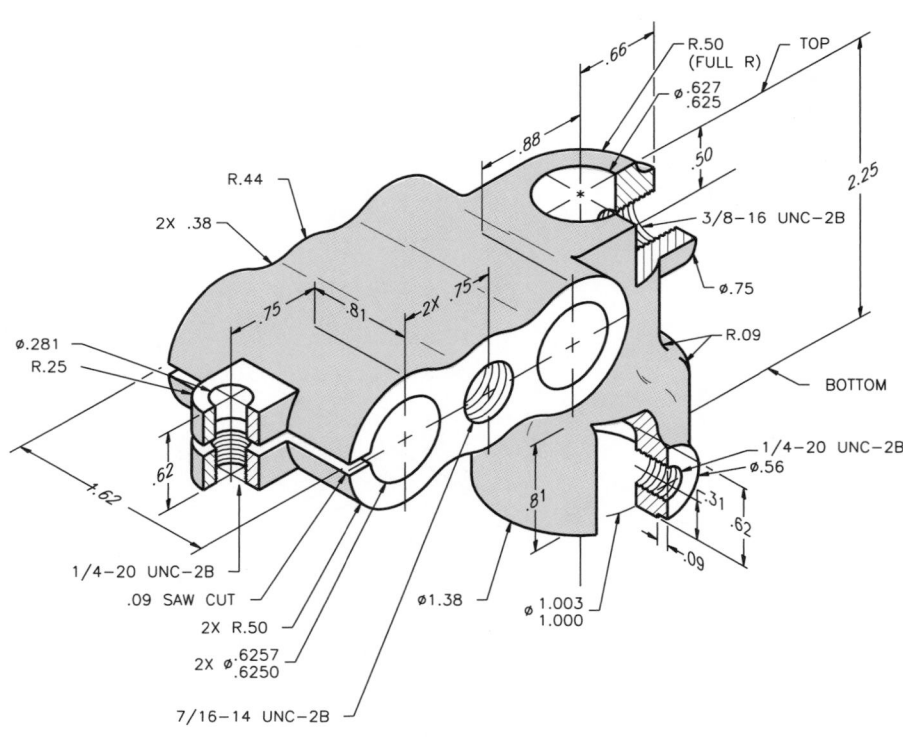

② **FRAME**

SAE 4340
REMOVE ALL BURRS & SHARP EDGES
FAO
FILLETS AND ROUNDS R.125
1 REQ'D

PROBLEM 18-10 (continued)

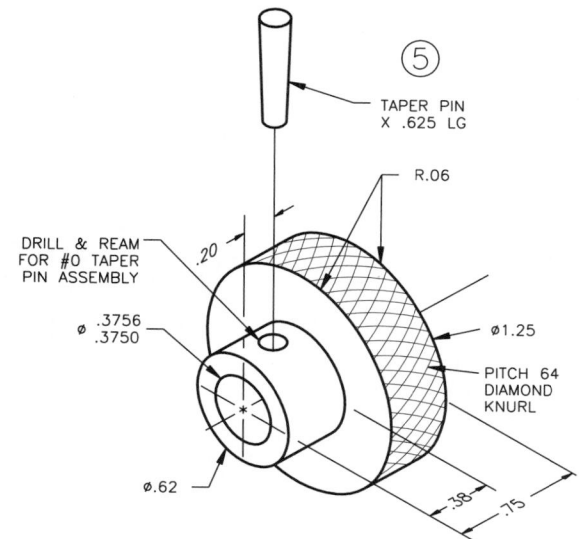

TAPER PIN
X .625 LG

R.06

DRILL & REAM
FOR #0 TAPER
PIN ASSEMBLY

.20

ø .3756
.3750

ø1.25

PITCH 64
DIAMOND
KNURL

ø.62

.38

.75

⑥ **KNURL KNOB**
SAE 1024
REMOVE ALL BURRS & SHARP EDGES
FAO
1 REQD

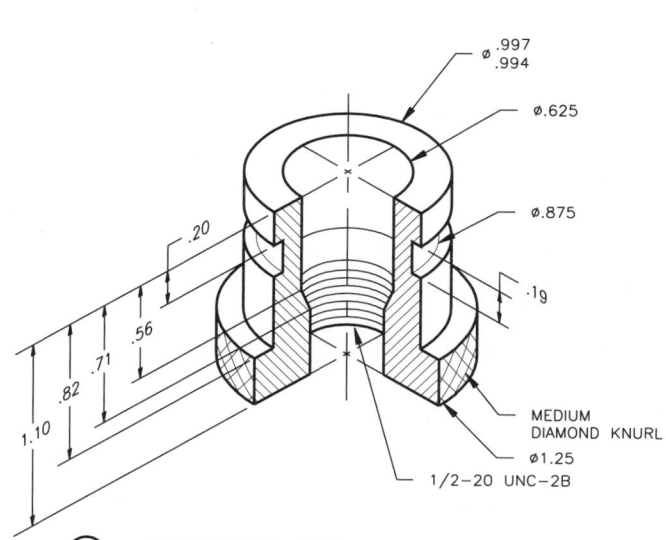

ø .997
.994

ø.625

ø.875

.20

.56

.71

.82

.19

1.10

MEDIUM
DIAMOND KNURL

ø1.25

1/2-20 UNC-2B

⑨ **ADJUSTING NUT**
SAE 3130
REMOVE ALL BURRS & SHARP EDGES
1 REQD

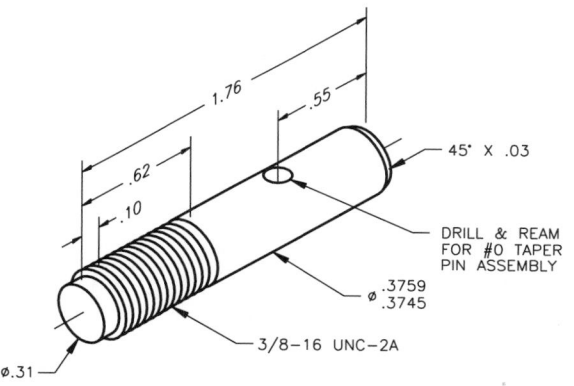

1.76

.55

.62

.10

45° X .03

DRILL & REAM
FOR #0 TAPER
PIN ASSEMBLY

ø .3759
.3745

3/8-16 UNC-2A

ø.31

⑦ **CLAMP SCREW**
SAE 2330
REMOVE ALL BURRS & SHARP EDGES
1 REQD

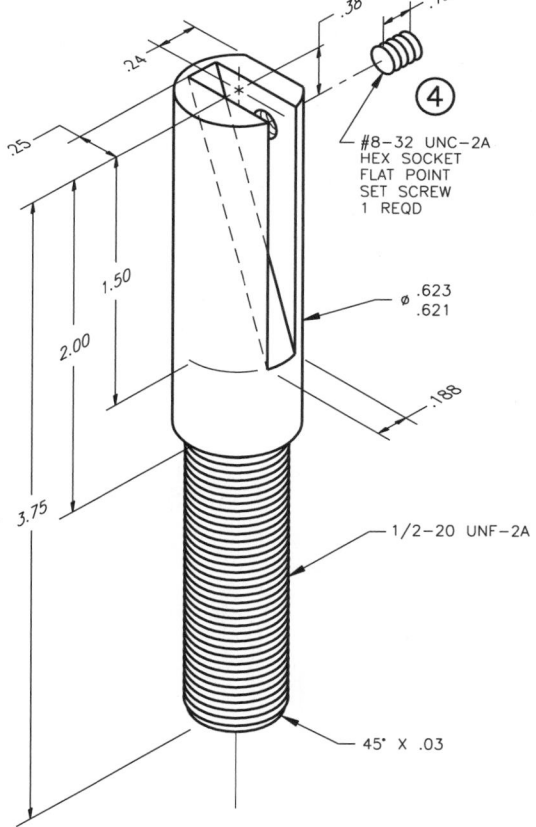

.38

.18

.24

.25

④

#8-32 UNC-2A
HEX SOCKET
FLAT POINT
SET SCREW
1 REQD

1.50

2.00

ø .623
.621

.188

3.75

1/2-20 UNF-2A

45° X .03

③ **ADJUSTING SCREW**
SAE 1045
REMOVE ALL BURRS & SHARP EDGES
FAO
1 REQD

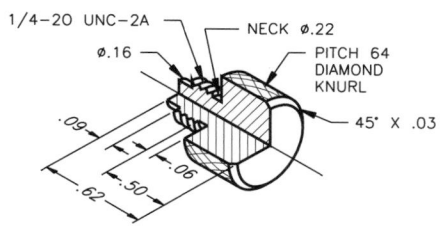

1/4-20 UNC-2A

NECK ø.22

ø.16

PITCH 64
DIAMOND
KNURL

.09

45° X .03

.06

.50

.62

⑧ **PILOT SCREW**
SAE 2330
REMOVE ALL BURRS & SHARP EDGES
FAO
1 REQD

PROBLEM 18-11   **Working drawing (in.)**

Assembly Name: Precision Vice

PARTS LIST:

ITEM	QTY	NAME	DESCRIPTION	MATERIAL
1	1	BASE		SAE 1040
2	1	BALL WASHER		SAE 1040
3	1	NUT		SAE 1040
4	1	CAP SCREW	⁵⁄₁₆-24NF × 1½ HEX SOCKET HEAD	STL
5	1	JAW		SAE 1040
6	1	JAW INSERT		SAE 4330
7	2	MACH SCREW	¼-28NF × ½ FLAT HD	STL
8	2	CAP SCREW	¼-28NF × ¾ HEX SOCKET HEAD	STL
9	1	JAW INSERT		SAE 4330

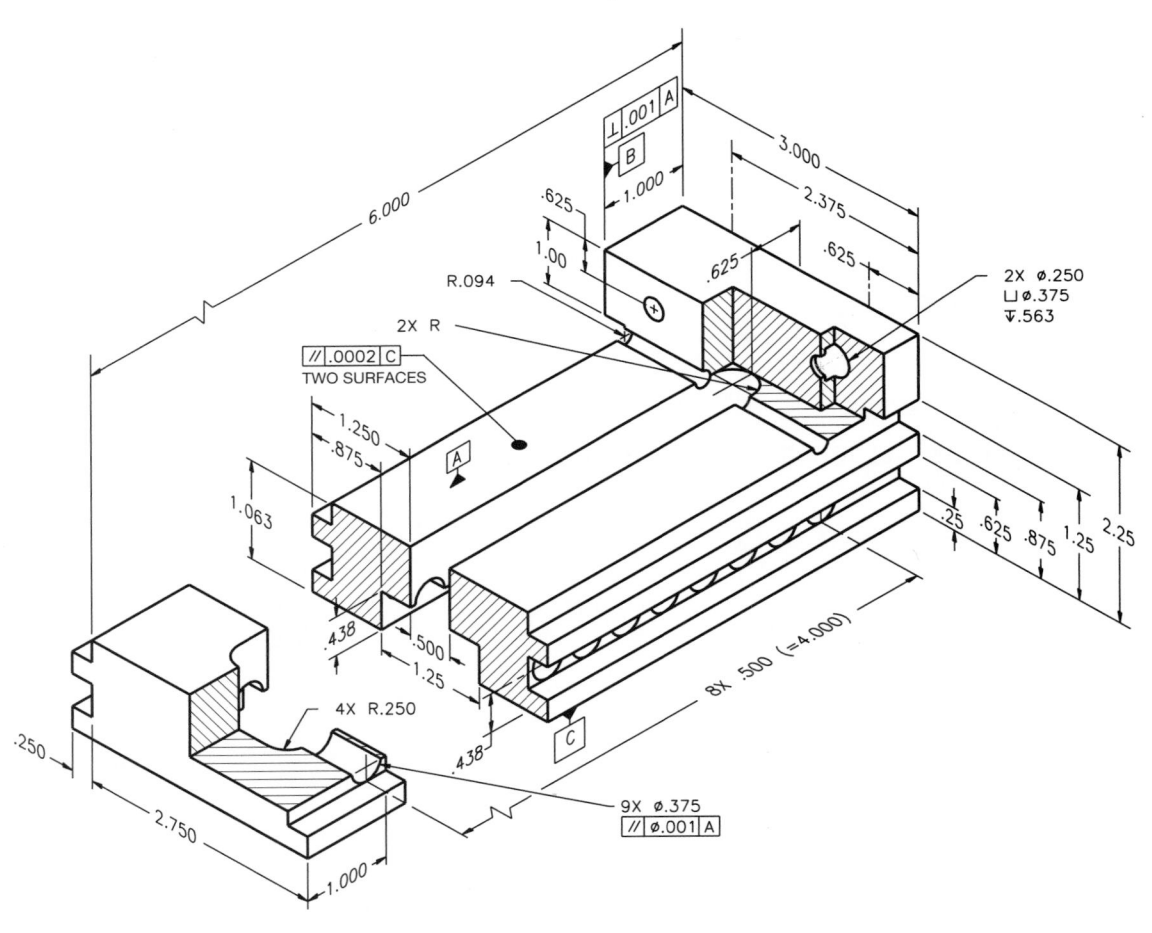

① **BASE**
FAO 32μIN

PROBLEM 18-11 (continued)

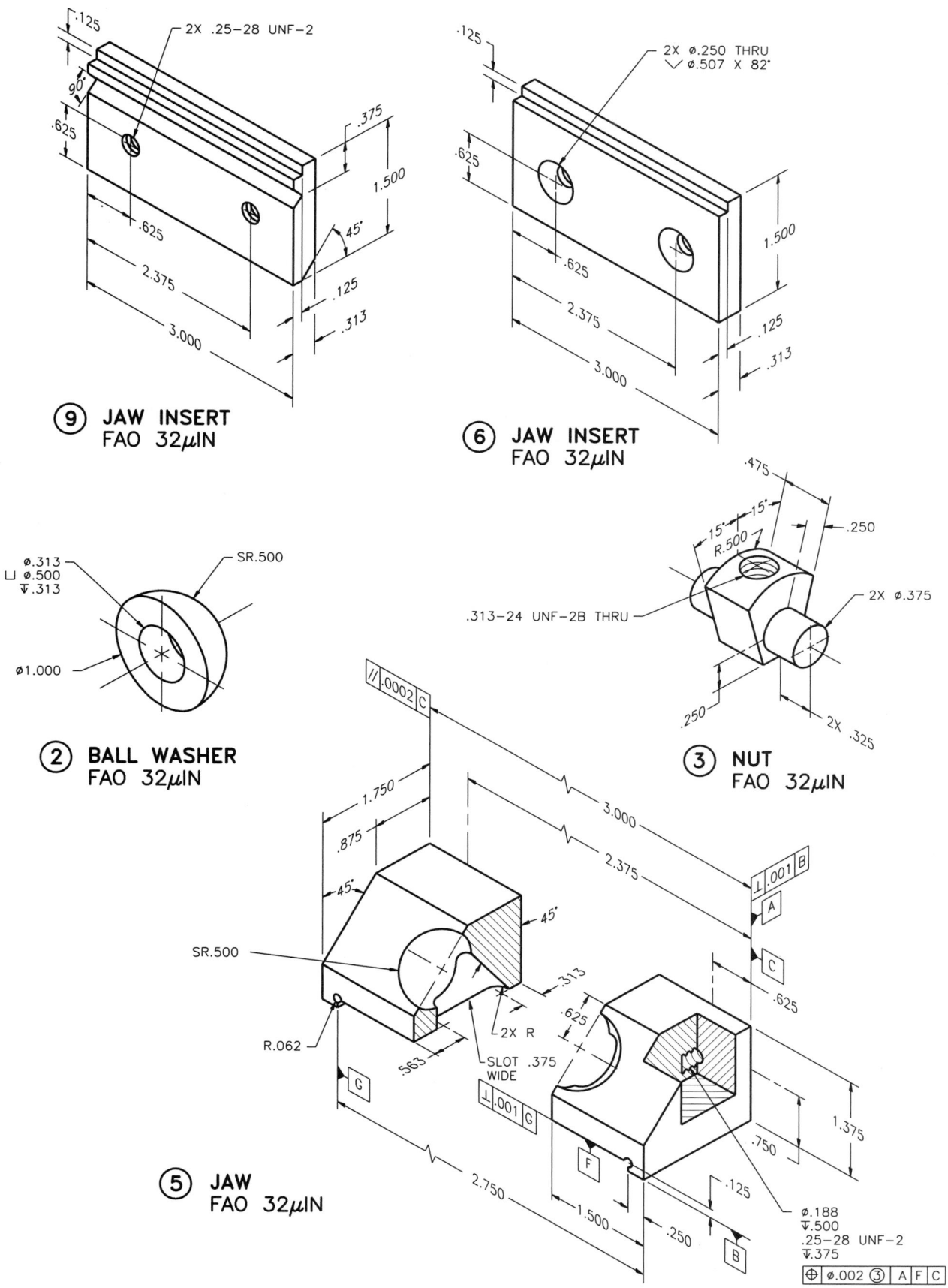

.125
2X .25−28 UNF−2
90°
.625
.375
1.500
.625
45°
2.375
.125
.313
3.000

⑨ JAW INSERT
FAO 32μIN

.125
2X ⌀.250 THRU
⌵ ⌀.507 X 82°
.625
1.500
.625
2.375
.125
.313
3.000

⑥ JAW INSERT
FAO 32μIN

⌀.313
⊔ ⌀.500
⌵.313
SR.500
⌀1.000

② BALL WASHER
FAO 32μIN

.475
15° 15°
R.500
.250
.313−24 UNF−2B THRU
2X ⌀.375
.250
2X .325

③ NUT
FAO 32μIN

∥ .0002 C
1.750
.875
45°
SR.500
45°
3.000
2.375
⊥ .001 B
A
C
.625
R.062
.313
2X R
.625
SLOT .375
WIDE
⊥ .001 G
G
.563
F
1.375
.750
2.750
1.500
.250
.125
⌀.188
⌵.500
.25−28 UNF−2
⌵.375
⊕ ⌀.002 ③ A F C
B

⑤ JAW
FAO 32μIN

PROBLEM 18-12   Working drawing (in.)

Assembly Name: Machine Vice

SPECIFIC INSTRUCTIONS:

When preparing the assembly drawings, use separate balloons for each part in each view, or use only one balloon in the view that most clearly identifies the part.

PARTS LIST:

ITEM	QTY	NAME	DESCRIPTION	MATERIAL
1	1	SCREW		SAE 4320
2	1	HANDLE		MS
3	2	CAP		MS
4	2	MACHINE SCREW	(10).190-32UNF-2 × 6 SLOT FIL HD	STL
5	1	SET SCREW	1/4-20UNC-2 × .250 FULL DOG POINT	STL
6	1	MOVABLE JAW		SAE 1020
7	1	MOVABLE JAW PLATE		SAE 4320
8	2	MACHINE SCREW	(10).190-32UNF-2 × .875 SLOT FIL HD	STL
9	1	FIXED JAW PLATE		SAE 4320
10	1	BODY		SAE 4320
11	1	GUIDE		SAE 1020
12	2	MACHINE SCREW	1/4-20UNC-2 × .500 SLOT FIL HD	STL

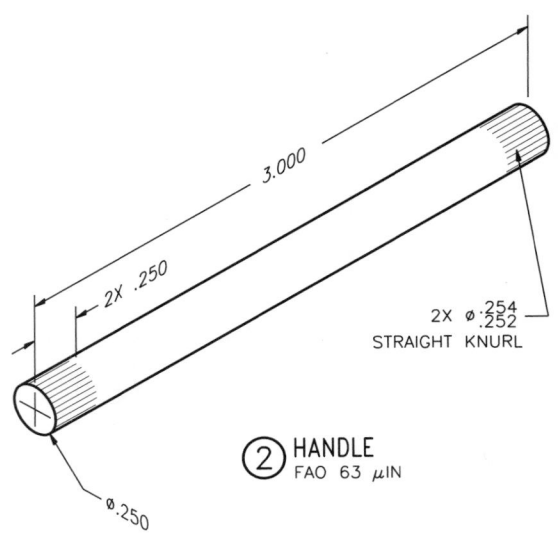

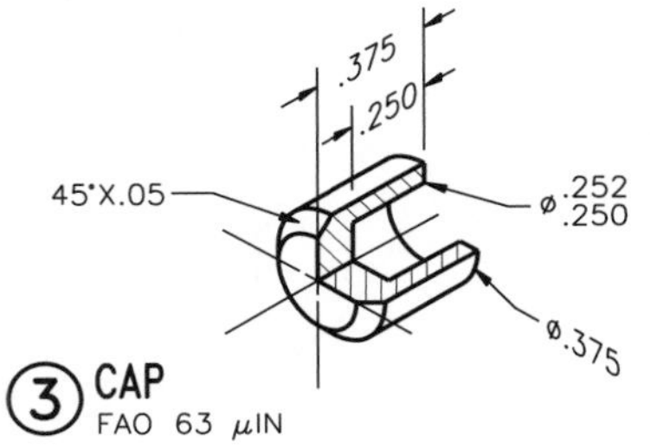

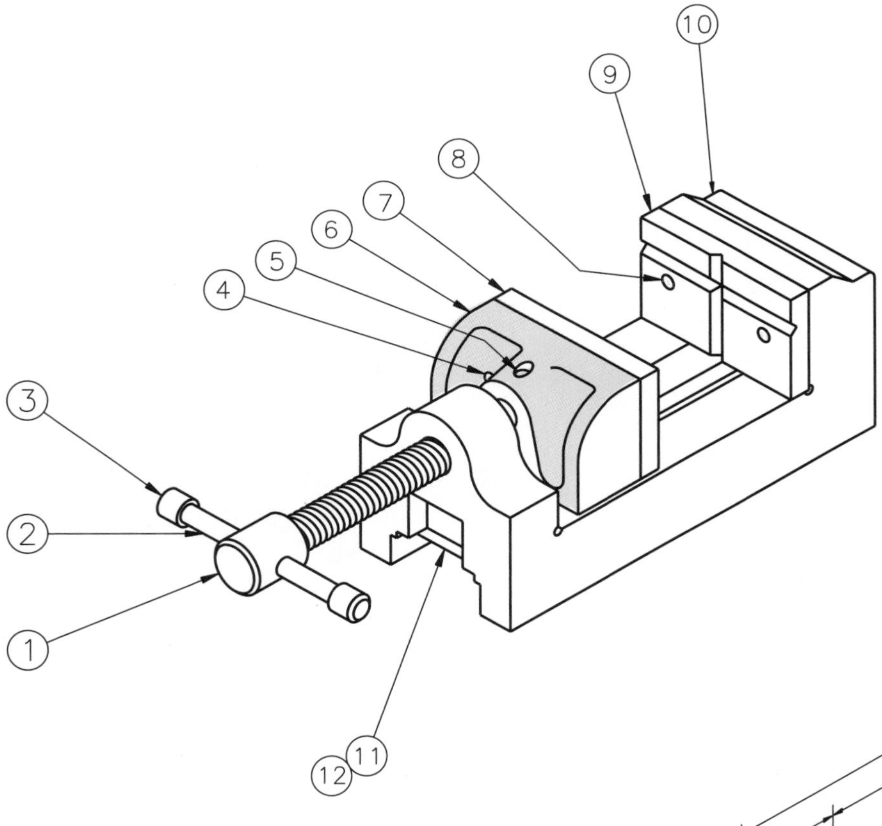

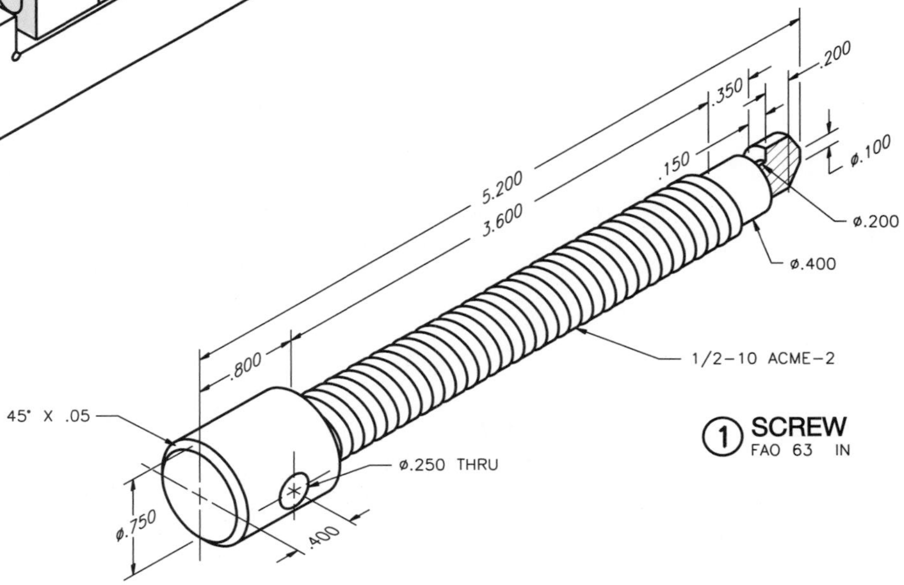

PROBLEM 18-12 (continued)

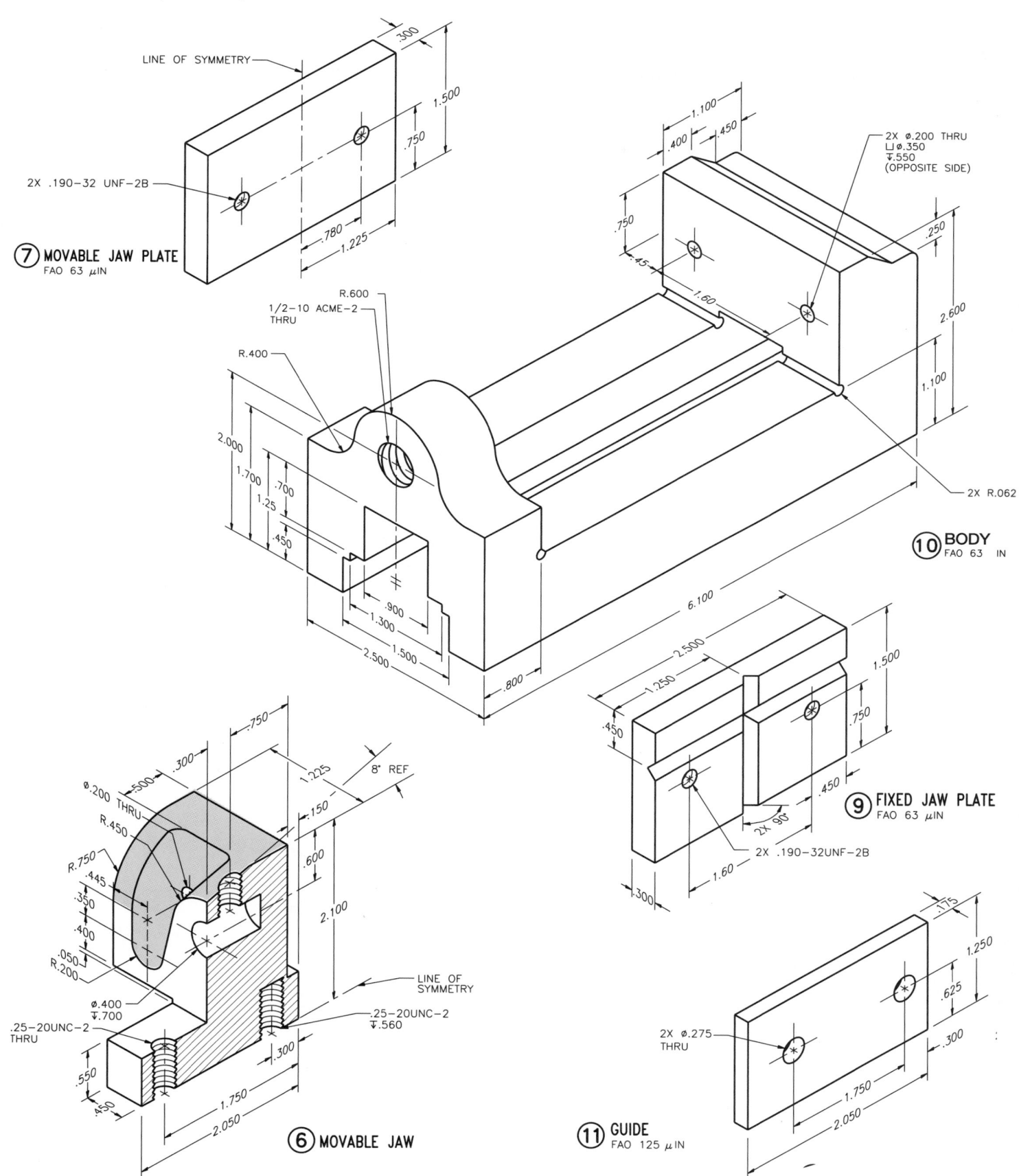

LINE OF SYMMETRY

.300
1.500
.750
2X .190−32 UNF−2B
.780
1.225

⑦ **MOVABLE JAW PLATE**
FAO 63 μIN

1.100
.400
.450
2X ∅.200 THRU
⊔∅.350
▼.550
(OPPOSITE SIDE)
.750
45°
.250
1.60
2.600
1.100
2X R.062

⑩ **BODY**
FAO 63 IN

R.600
1/2−10 ACME−2
THRU
R.400
2.000
1.700
.700
1.25
.450
.900
1.300
2.500
1.500
.800
6.100

.750
.300
.500
∅.200 THRU
R.450
R.750
.445
.350
.400
.050
R.200
∅.400
▼.700
.25−20UNC−2
THRU
1.225
8° REF
.150
.600
2.100
LINE OF SYMMETRY
.25−20UNC−2
▼.560
.550
.450
.300
1.750
2.050

⑥ **MOVABLE JAW**

2.500
1.250
.450
1.500
.750
.450
2X 90°
1.60
.300
2X .190−32UNF−2B

⑨ **FIXED JAW PLATE**
FAO 63 μIN

.175
1.250
.625
.300
2X ∅.275
THRU
1.750
2.050

⑪ **GUIDE**
FAO 125 μIN

PROBLEM 18-13   **Working drawing (in.)**

Assembly Name: Arbor Press

PARTS LIST:

ITEM	QTY	NAME	DESCRIPTION	MATERIAL	ITEM	QTY	NAME	DESCRIPTION	MATERIAL
1	1	BASE		SAE 1020	9	1	COVER PLATE		SAE 1020
2	1	TABLE PIN		SAE 1020	10	4	CAP SCREWS	8-32UNC-2 × .50 HEX SOC	STL
3	1	TABLE		SAE 1020					
4	2	BALL END		SAE 1020	11	1	RACK		SAE 4320
5	1	SLEEVE		SAE 1020	12	1	SCREW		SAE 1040
6	1	HANDLE		SAE 1020	13	1	COLUMN		SAE 1020
7	1	GEAR		SAE 4320	14	1	MACHINE SCREW	3/8-16UNC-2 × 1.00 HEX HEAD	STL
8	1	RACK PAD		SAE 4320					

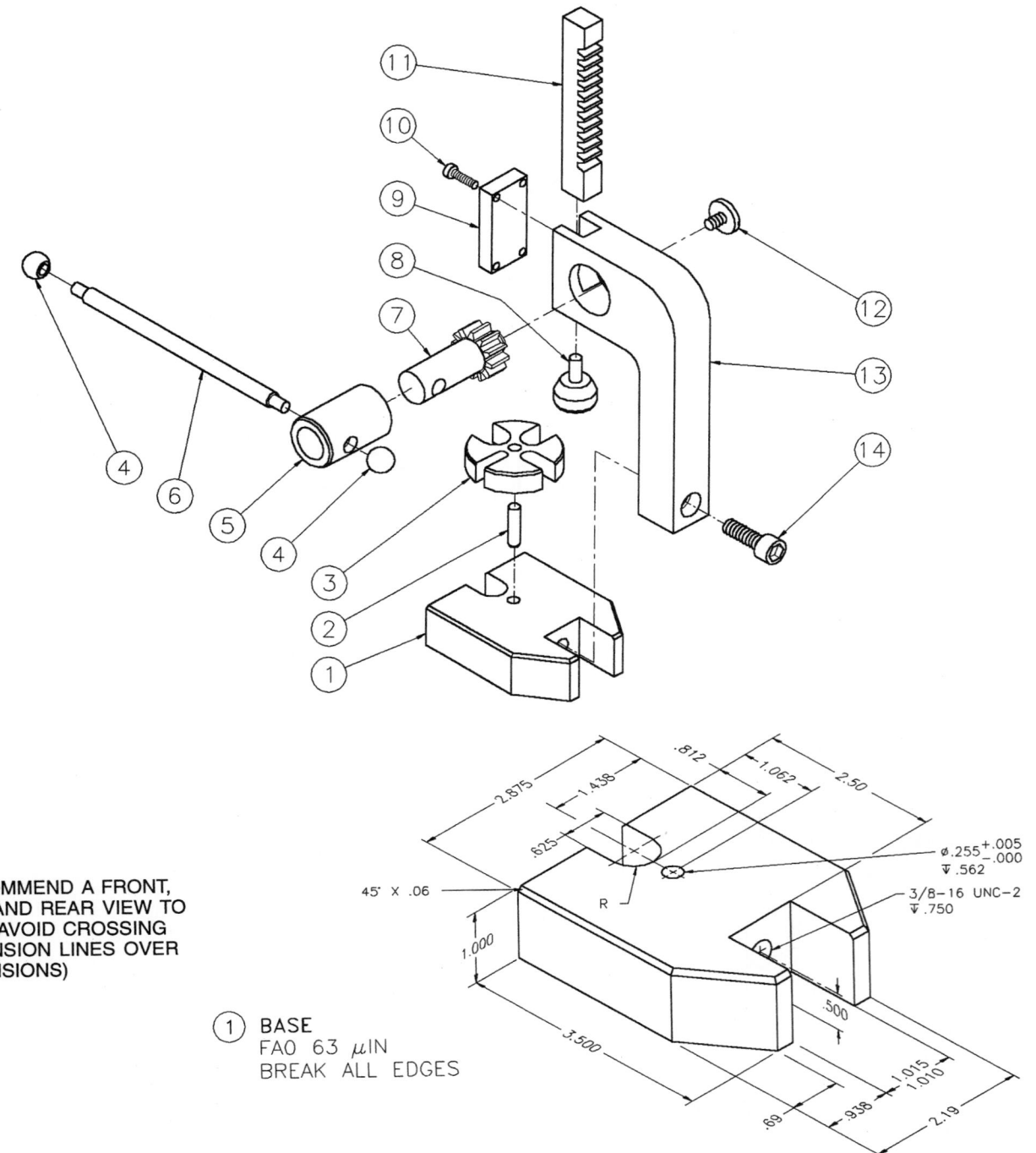

(RECOMMEND A FRONT,
SIDE AND REAR VIEW TO
HELP AVOID CROSSING
EXTENSION LINES OVER
DIMENSIONS)

1 BASE
FAO 63 μIN
BREAK ALL EDGES

PROBLEM 18-13 (continued)

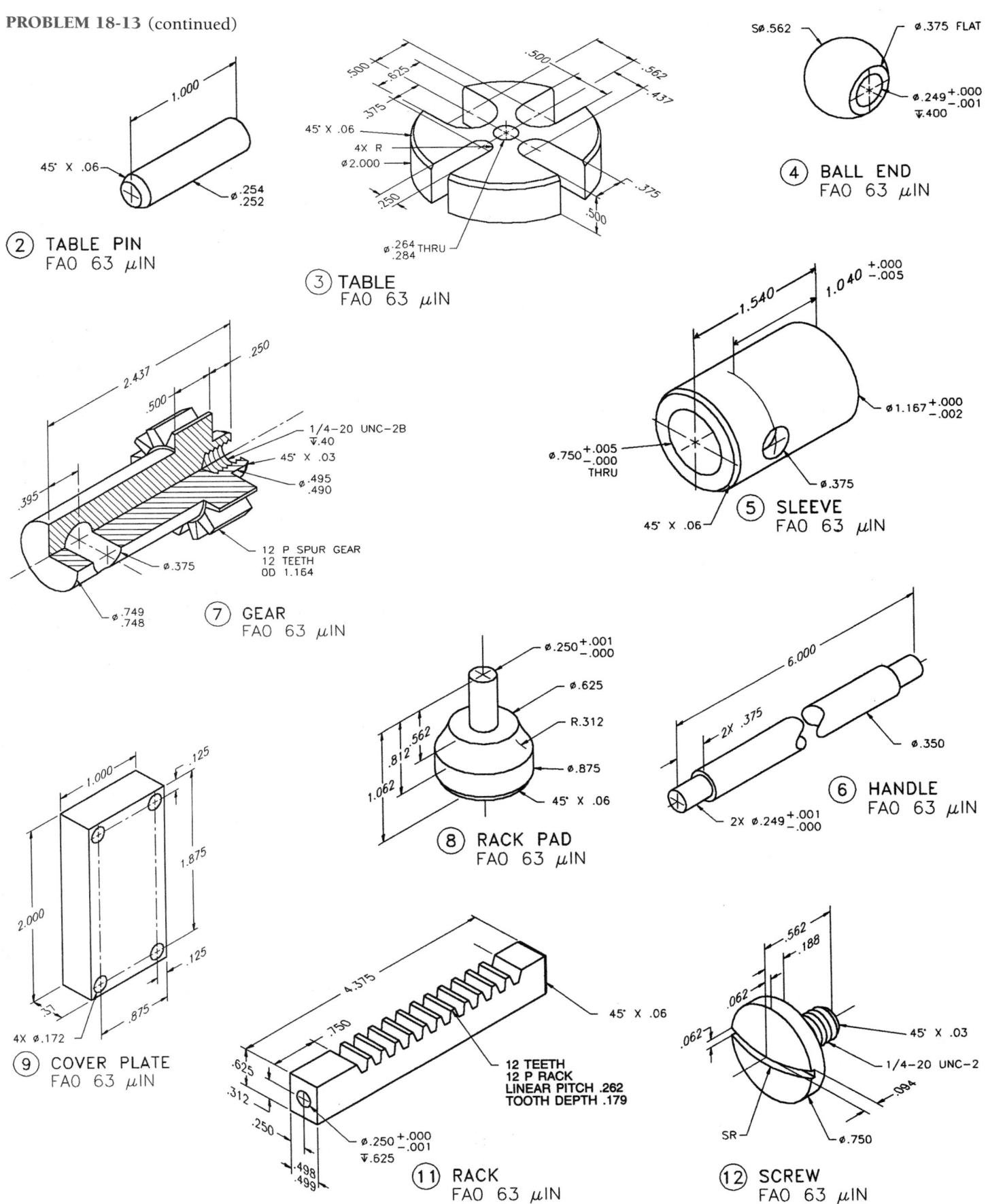

② TABLE PIN
FAO 63 μIN

③ TABLE
FAO 63 μIN

④ BALL END
FAO 63 μIN

⑤ SLEEVE
FAO 63 μIN

⑦ GEAR
FAO 63 μIN

⑧ RACK PAD
FAO 63 μIN

⑥ HANDLE
FAO 63 μIN

⑨ COVER PLATE
FAO 63 μIN

⑪ RACK
FAO 63 μIN

⑫ SCREW
FAO 63 μIN

PROBLEM 18-13 (continued)

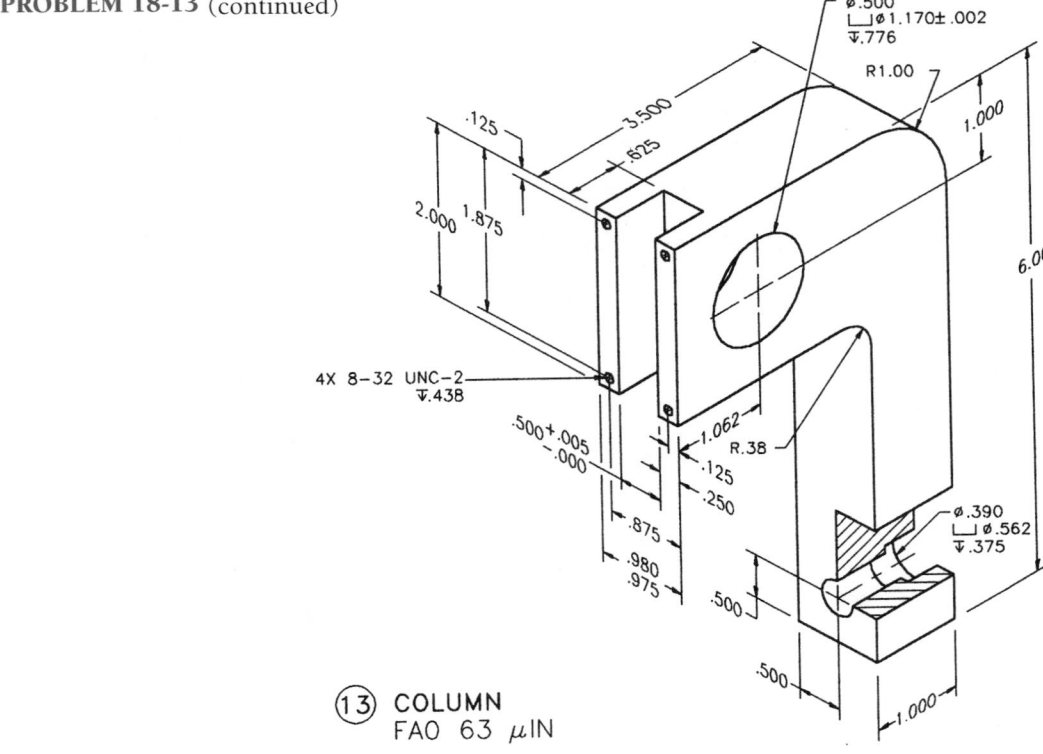

⑬ COLUMN
FAO 63 μIN

PROBLEM 18-14   Working drawing (in.)

Assembly Name: Pen Light

SPECIFIC INSTRUCTIONS:

Prepare a complete set of working drawings with one detail drawing per sheet, and the assembly and parts list on another sheet. When preparing the assembly drawing, use separate balloons for each part in each view or use only one balloon in the view that most clearly identifies the part. This is an engineering prototype project. Errors are likely. Verify dimensions during assembly. Establish tolerances between mating parts. *Problem courtesy of Kim P. Lockwood.*

KEY	QTY	NAME	DESCRIPTION	PART NO.
14	1	FIBER OPTIC	FIBER-OPTIC 3-1/2" × .12"	5DT1014
13	1	FIBER OPTIC HOLDER	(PVF) POLYVINYLIDENE FLOURIDE	1DT1013
12	1	LAMP LENS	POLYCARBONATE	1DT1012
11	1	LAMP	HIGH INTENSITY XENON	5DT1011
10	1	LAMP MODULE	VACUUM METALLIZED	1DT1010
9	1	BULB CASING	POLYCARBONATE	1DT1009
8	1	CATALYST	DE OXO TYPE D	5DT1008
7	1	PLUG	PHOSPHOR BRONZE	1DT1007
6	1	WASHER	PHOSPHOR BRONZE	1DT1006
5	1	SPRING	PHOSPHOR BRONZE	1DT1005
4	1	PRONG	PHOSPHOR BRONZE	1DT1004
3	1	PARKER O-RING	PARKER #2-014	5DT1003
2	1	BODY	POLYCARBONATE	1DT1002
1	1	MAGNET	MAGNET	5DT1001
KEY	**QTY**	**NAME**	**DESCRIPTION**	**PART NO.**
PARTS LIST				

PROBLEM 18-14 (continued)

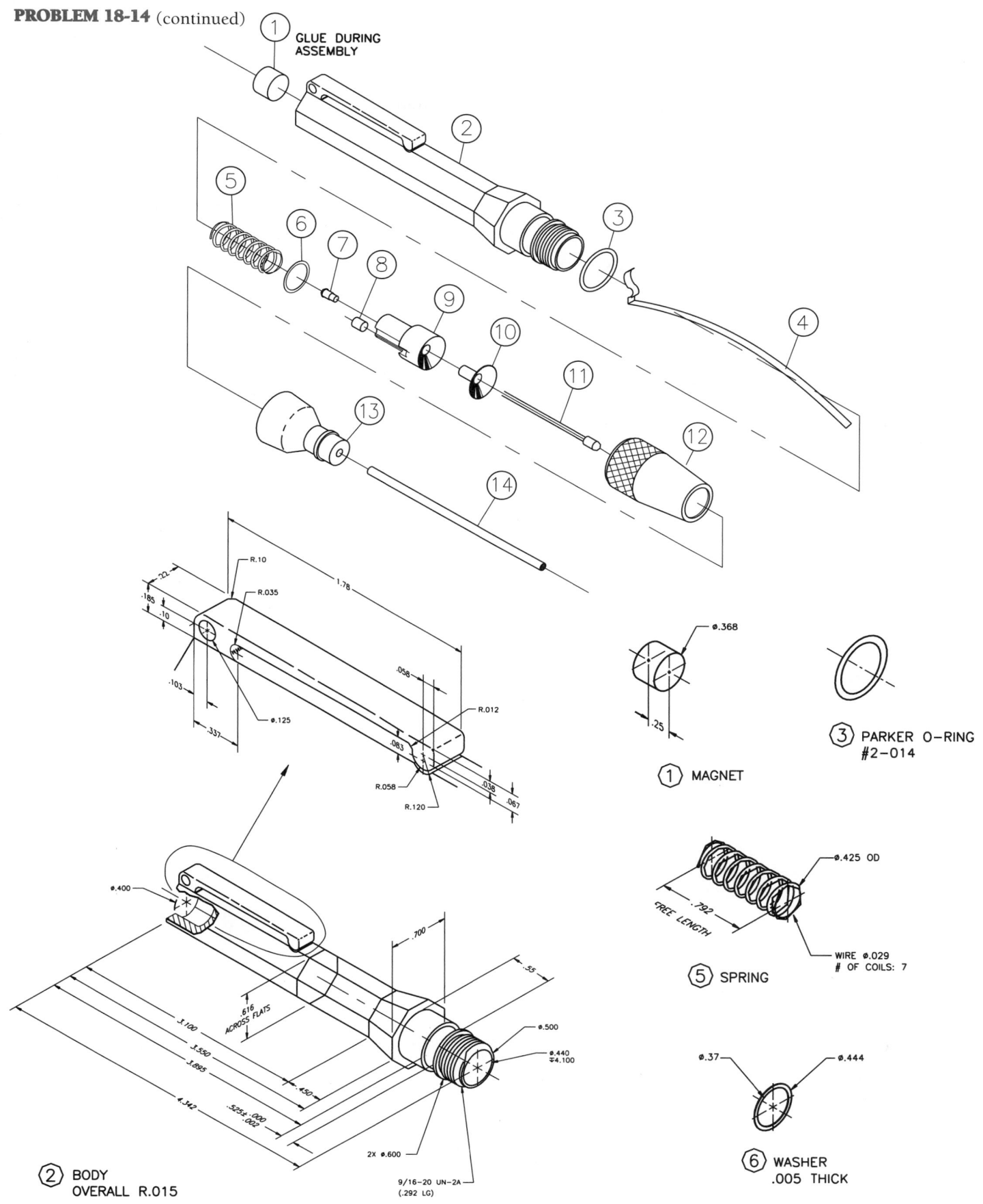

GLUE DURING ASSEMBLY

① MAGNET

③ PARKER O-RING #2-014

⑤ SPRING

⑥ WASHER .005 THICK

② BODY OVERALL R.015

PROBLEM 18-14 (continued)

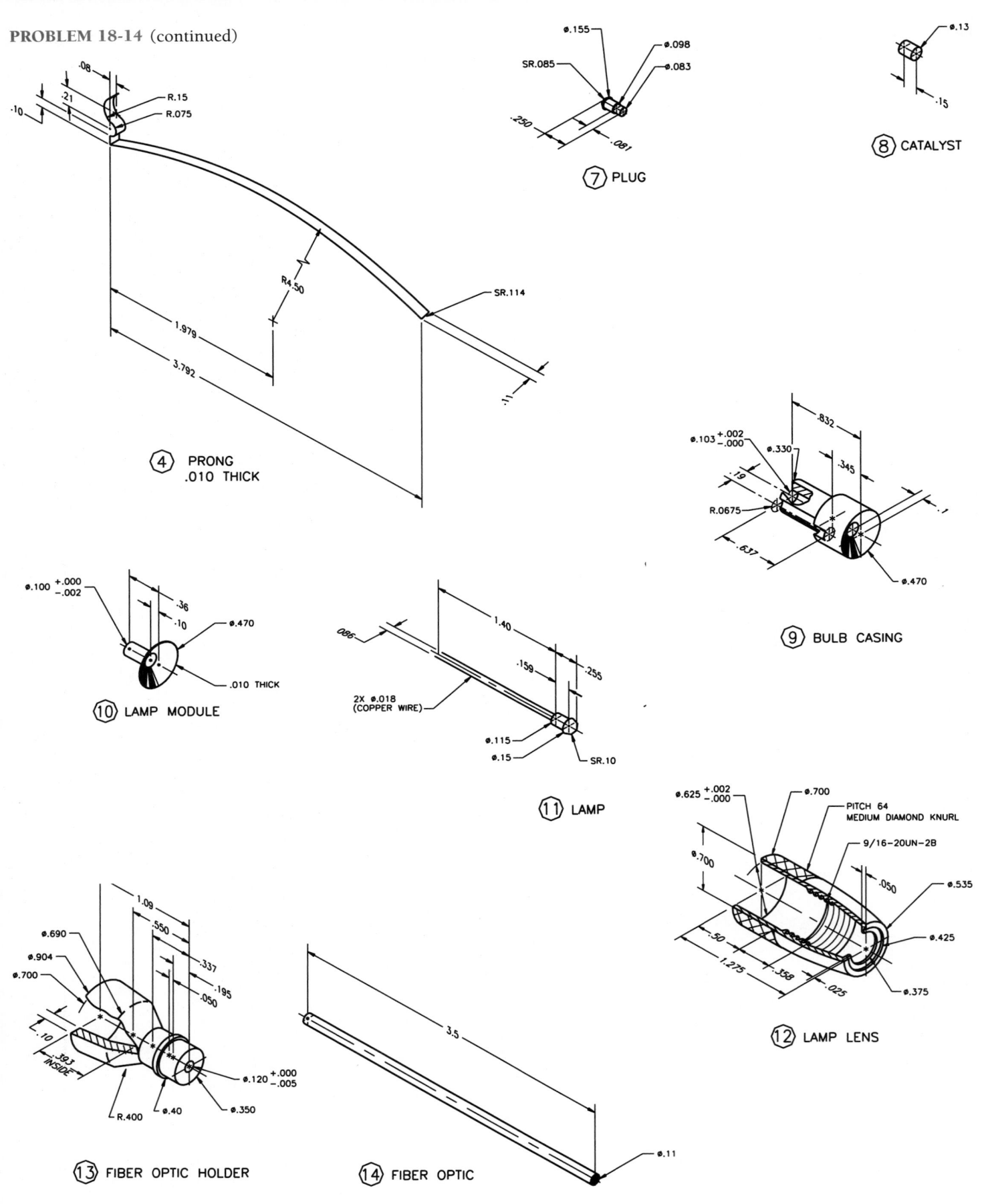

.08

.21

.10

R.15
R.075

R4.50

1.979

3.792

SR.114

.11

④ PRONG
.010 THICK

⑦ PLUG
.155
SR.085
ø.098
ø.083
.250
.081

⑧ CATALYST
ø.13
.15

⑨ BULB CASING
.832
ø.103 +.002 −.000
ø.330
.345
.19
R.0675
.637
ø.470

⑩ LAMP MODULE
ø.100 +.000 −.002
.36
.10
ø.470
.010 THICK

⑪ LAMP
.086
1.40
.159
.255
2X ø.018
(COPPER WIRE)
ø.115
ø.15
SR.10

⑫ LAMP LENS
ø.625 +.002 −.000
ø.700
PITCH 64
MEDIUM DIAMOND KNURL
9/16−20UN−2B
.700
.050
ø.535
.700
ø.425
.50
1.275
.358
.025
ø.375

⑬ FIBER OPTIC HOLDER
1.09
.550
.337
ø.690
.195
ø.904
.050
ø.700
.10
.393
INSIDE
ø.120 +.000 −.005
R.400
ø.40
ø.350

⑭ FIBER OPTIC
3.5
ø.11

PROBLEM 18-15 **Working drawing (in.)**

Assembly Name: Fluorescent Light Fixture

SPECIFIC INSTRUCTIONS:

Prepare a complete set of working drawings with one detail drawing per sheet, and the assembly and parts list on another sheet. When preparing the assembly drawing, use separate balloons for each part in each view or use only one balloon in the view that most clearly identifies the part. This is an engineering prototype project. Errors are likely. Verify dimensions during assembly. Establish tolerances between mating parts. There are several purchase parts in this assembly. Manufacturer research is needed to locate available products. *Problem courtesy of John Melloy.*

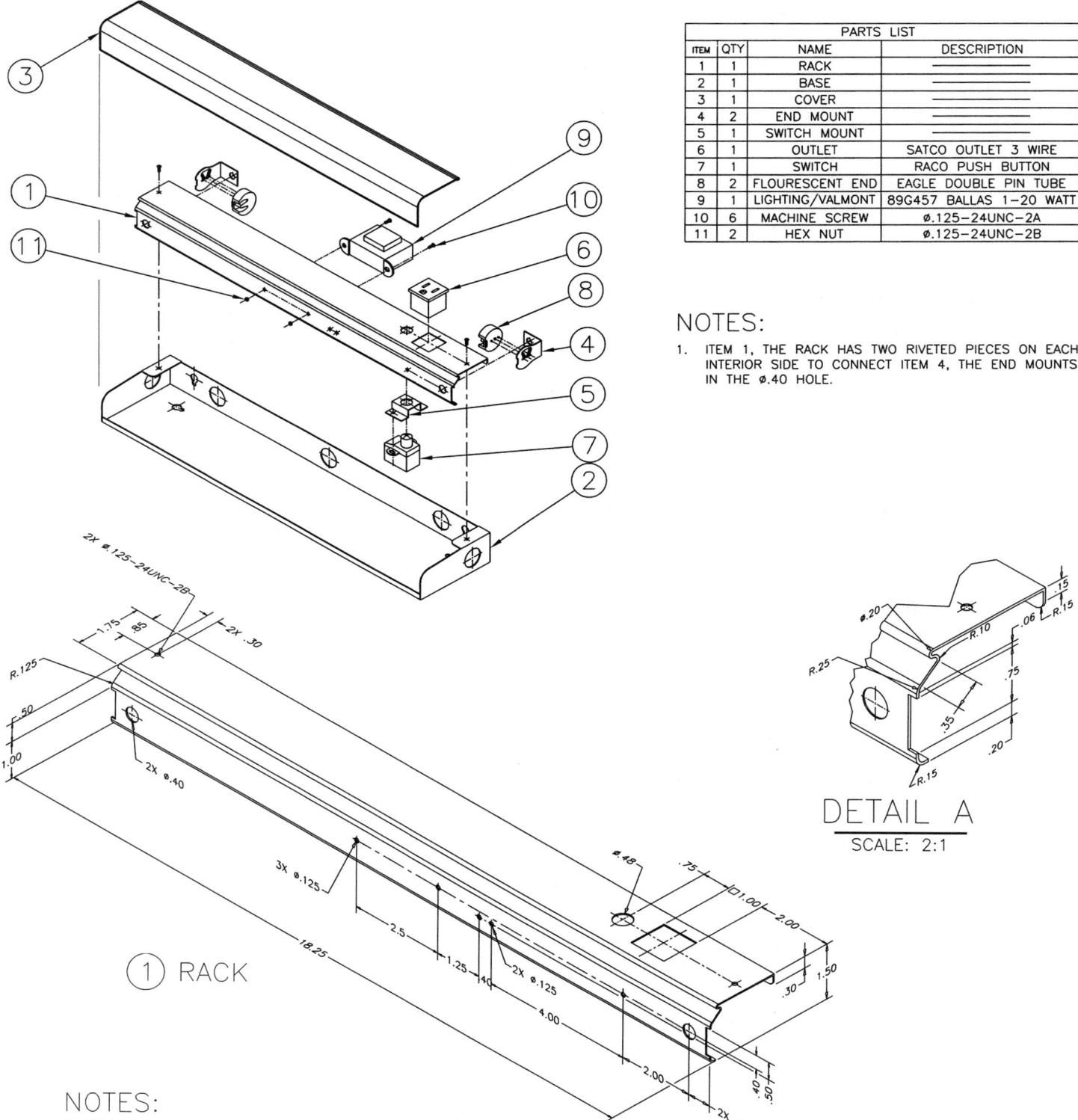

PARTS LIST			
ITEM	QTY	NAME	DESCRIPTION
1	1	RACK	——————
2	1	BASE	——————
3	1	COVER	——————
4	2	END MOUNT	——————
5	1	SWITCH MOUNT	——————
6	1	OUTLET	SATCO OUTLET 3 WIRE
7	1	SWITCH	RACO PUSH BUTTON
8	2	FLOURESCENT END	EAGLE DOUBLE PIN TUBE
9	1	LIGHTING/VALMONT	89G457 BALLAS 1—20 WATT
10	6	MACHINE SCREW	Ø.125—24UNC—2A
11	2	HEX NUT	Ø.125—24UNC—2B

NOTES:

1. ITEM 1, THE RACK HAS TWO RIVETED PIECES ON EACH INTERIOR SIDE TO CONNECT ITEM 4, THE END MOUNTS IN THE Ø.40 HOLE.

DETAIL A
SCALE: 2:1

① RACK

NOTES:
1. THE 2X Ø.40 HOLES ARE RIVETED ON INTERIOR SIDE.

PROBLEM 18-15 (continued)

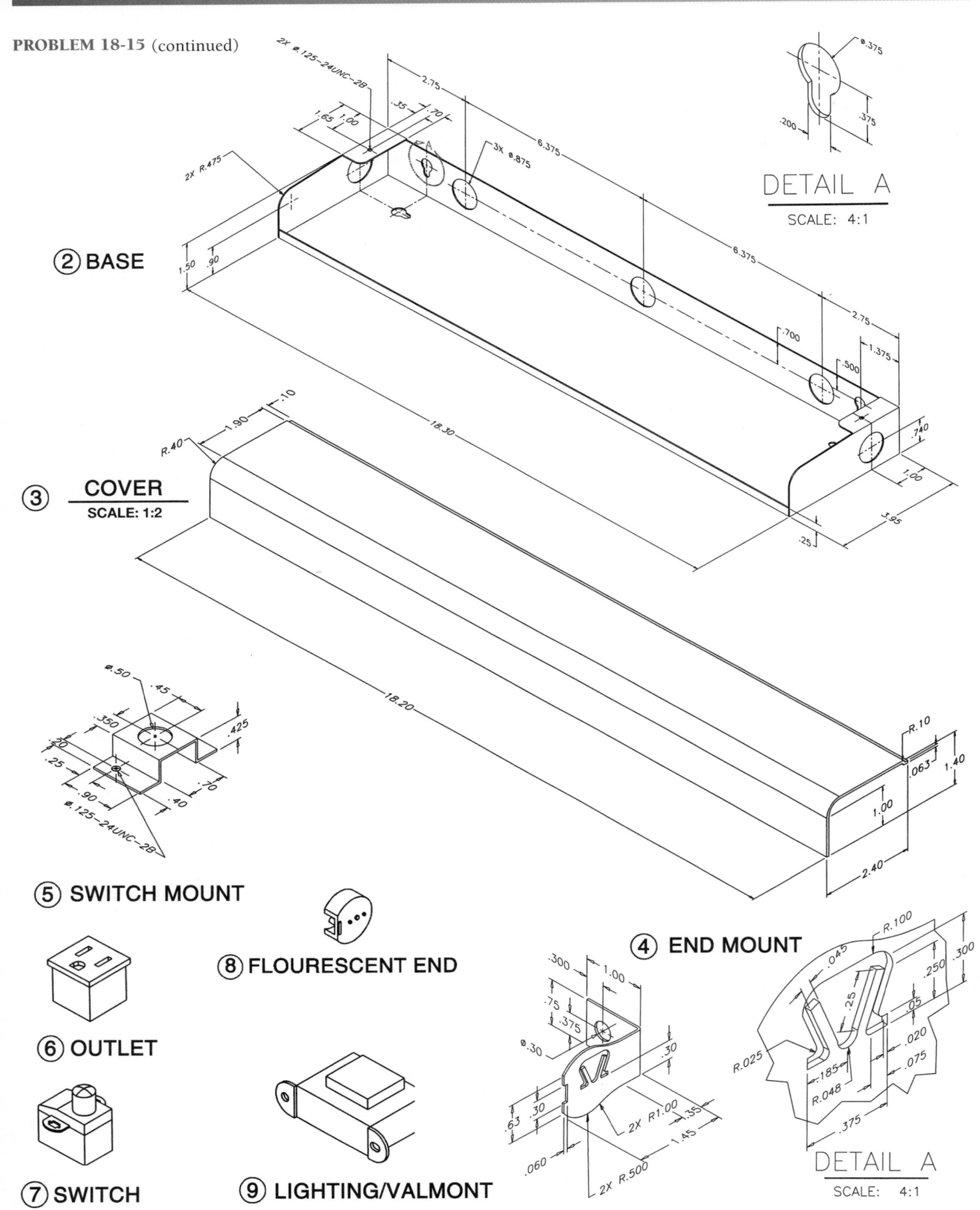

DETAIL A

SCALE: 4:1

② BASE

③ COVER

SCALE: 1:2

⑤ SWITCH MOUNT

⑧ FLOURESCENT END

④ END MOUNT

⑥ OUTLET

⑦ SWITCH

⑨ LIGHTING/VALMONT

DETAIL A

SCALE: 4:1

PROBLEM 18-16 **Working drawing (in.)**

Assembly Name: Oil Pump

SPECIFIC INSTRUCTIONS:

Prepare a complete set of working drawings with one detail drawing per sheet, and the assembly and parts list on another sheet. When preparing the assembly drawing, use separate balloons for each part in each view or use only one balloon in the view that most clearly identifies the part. This is an engineering prototype project. Errors are likely. Verify dimensions during assembly. Establish tolerances between mating parts. There are several purchase parts in this assembly. Manufacturer research is needed to locate available products. *Problem courtesy of Richard Hertel.*

17	1/4 - 20 X 2 1/2" HEX HEAD BOLT	4
16	DRIVE GEAR (FEED)	1
15	IDLER GEAR (FEED)	1
14	1/4 - 20 X 2" HEX HEAD BOLT	2
13	CHECK BALL	1
12	PSI CONTROL SPRING	1
11	END CAP	2
10	PLUNGER SPRING	1
9	PLUNGER VALVE	1
8	OIL PUMP BODY	1
7	IDLER GEAR (RETURN)	1
6	DRIVE GEAR (RETURN)	1
5	OIL SEAL	1
4	DRIVE SHAFT	1
3	END COVER	1
2	1/4 - 20 X 3/4" BUTTON HEAD BOLT	2
1	PUMP COVER	1
ITEM NO.	NOMENCLATURE OR DESCRIPTION	QTY REQD
	PARTS LIST	

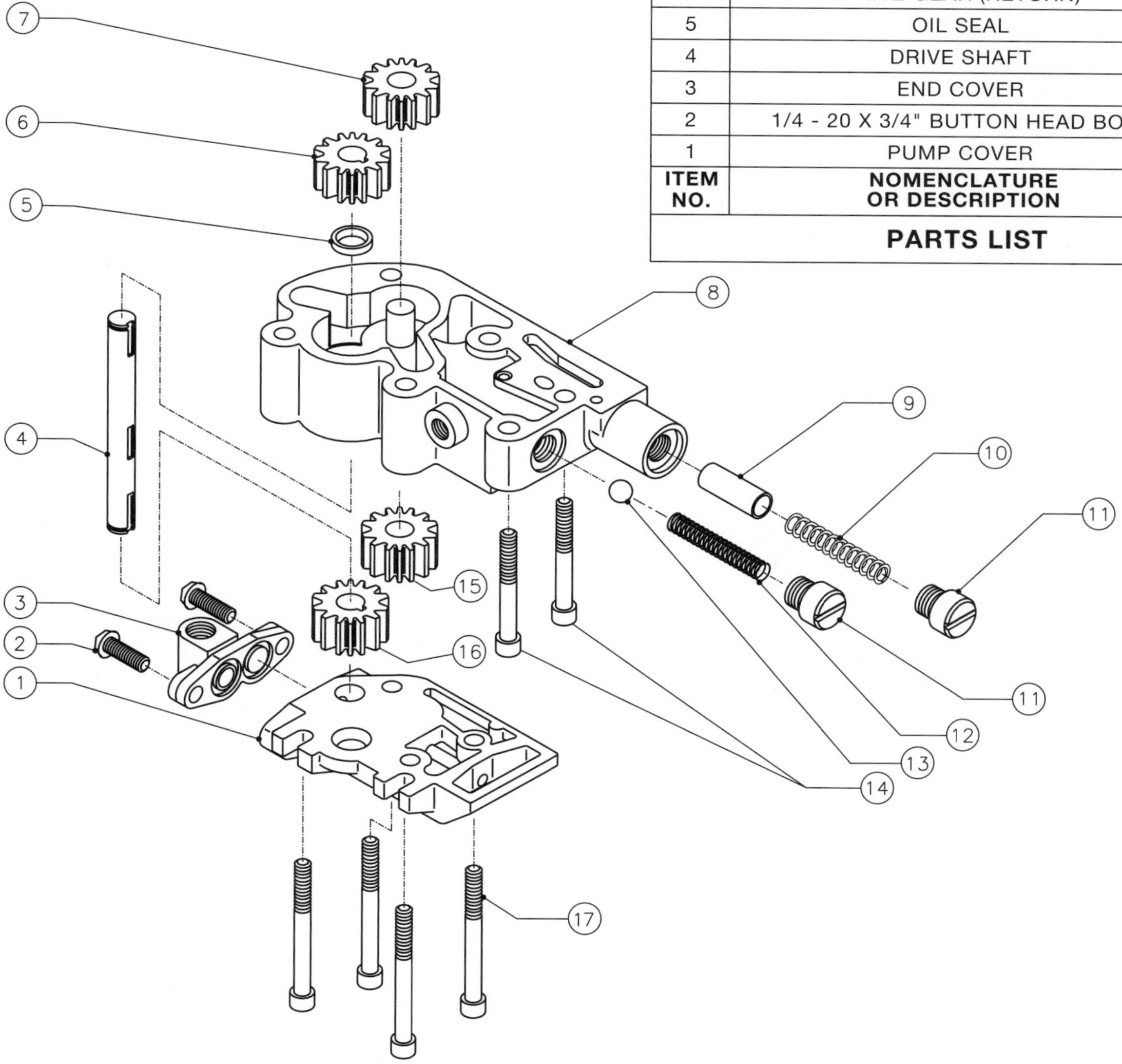

PROBLEM 18-16 (continued)

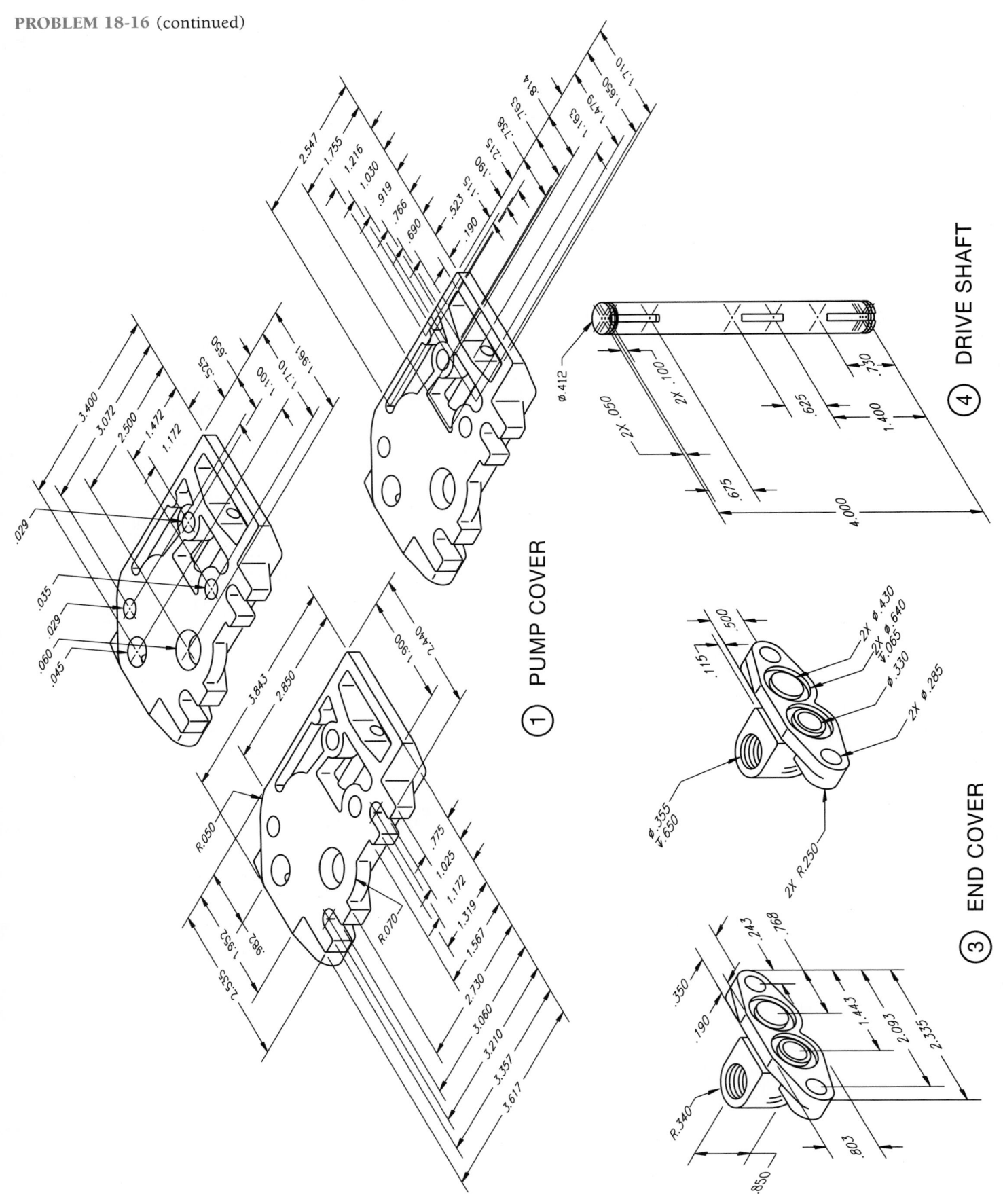

① PUMP COVER

③ END COVER

④ DRIVE SHAFT

PROBLEM 18-16 (continued)

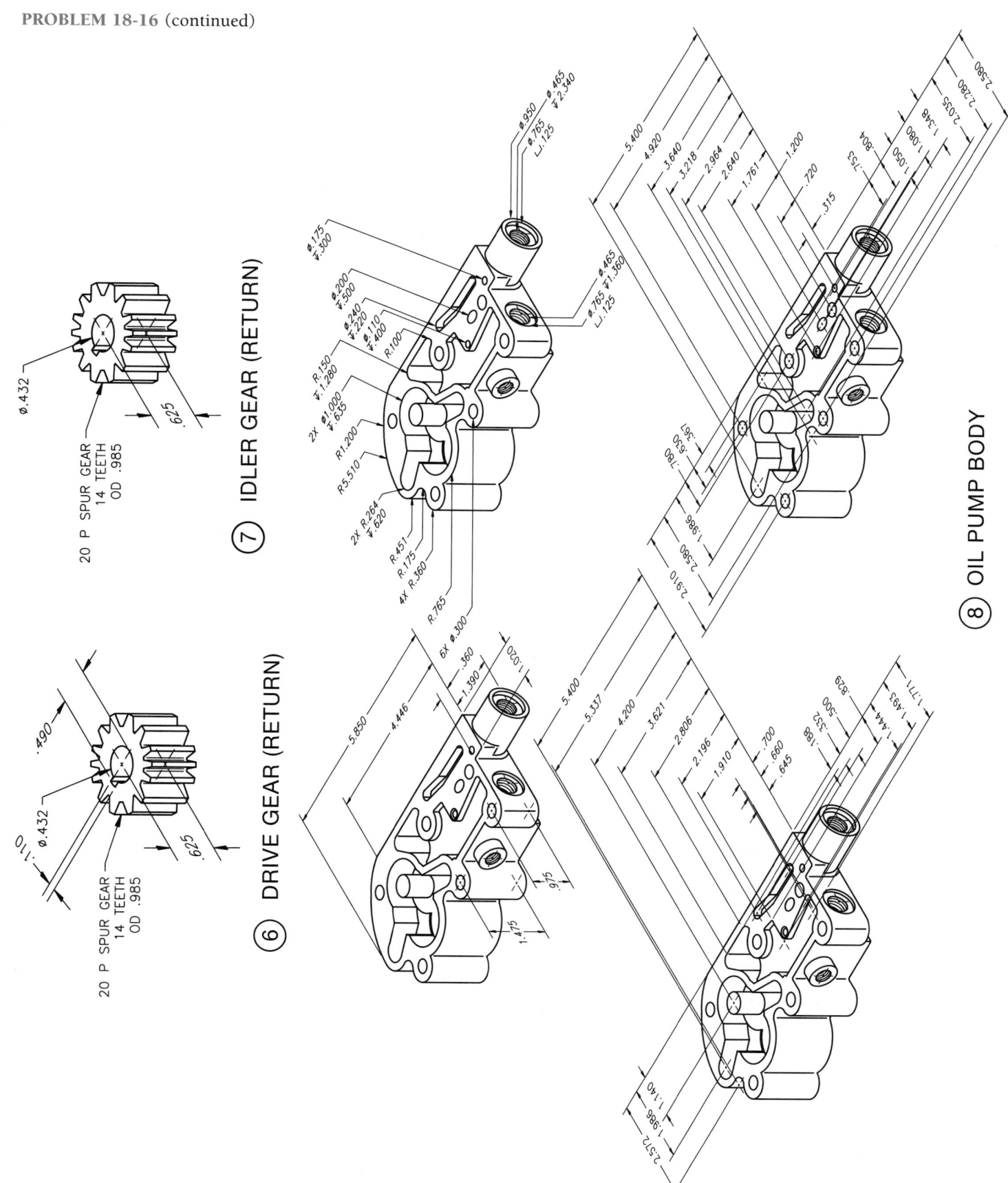

PROBLEM 18-16 (continued)

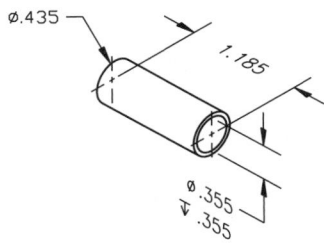

Ø.435
1.185
Ø.355
∇ .355

⑨ PLUNGER VALVE

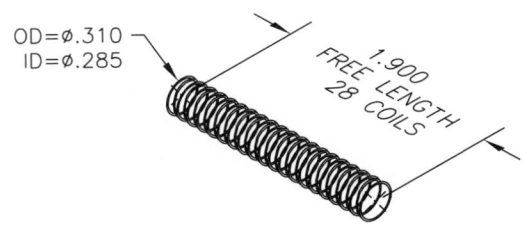

OD=Ø.310
ID=Ø.285
1:900
FREE LENGTH
28 COILS

⑫ PSI CONTROL SPRING

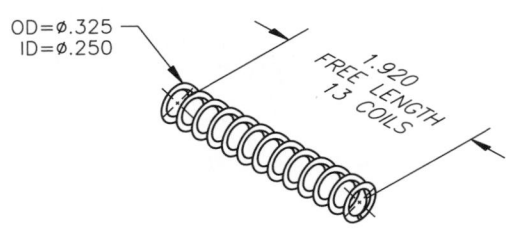

OD=Ø.325
ID=Ø.250
1:920
FREE LENGTH
13 COILS

⑩ PLUNGER SPRING

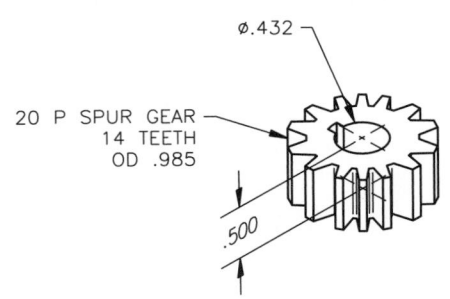

Ø.432
20 P SPUR GEAR
14 TEETH
OD .985
.500

⑮ IDLER GEAR (FEED)

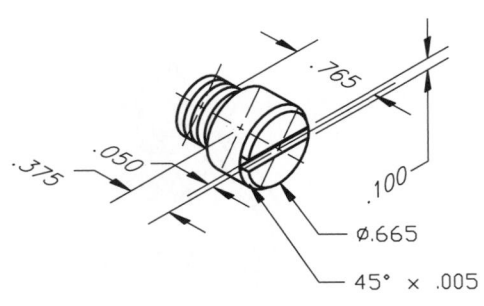

.765
.050
.375
.100
Ø.665
45° × .005

⑪ END CAP

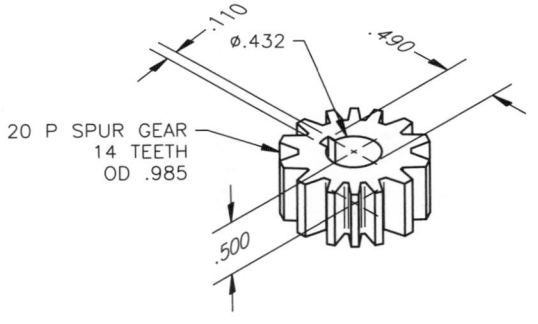

.110
Ø.432
.490
20 P SPUR GEAR
14 TEETH
OD .985
.500

⑯ DRIVE GEAR (FEED)

PROBLEM 18-17 **Working drawing (in.)**

Assembly Name: Landing Gear Retract Assembly

SPECIFIC INSTRUCTIONS:

Prepare a complete set of working drawings with one detail drawing per sheet, and the assembly and parts list on another sheet. When preparing the assembly drawing, use separate balloons for each part in each view or use only one balloon in the view that most clearly identifies the part. This is an engineering prototype project. Errors are likely. Verify dimensions during assembly. Establish tolerances between mating parts. There are several purchase parts in this assembly. Manufacturer research is needed to locate available products. *Problem courtesy of Mike McIntyre.*

PARTS LIST:

ITEM	QTY	NAME	DESCRIPTION	PART NO.	ITEM	QTY	NAME	DESCRIPTION	PART NO.
1	1	AIR CYLINDER	CLIPPERD	CL0002	8	1	CENTER BLOCK	ALUMINUM	LGR005
2	2	E-CLIP	⅛"	EC0018	9	1	45° BEVEL GEAR	MODIFIED	BG0002
3	1	BACKPLATE	ALUMINUM	LGR001	10	1	SPACER	ALUMINUM	LGR007
4	1	¼" × 2" BOLT	2" HARD BOLT	BT0002	11	1	SIDE FRAME (RT)	ALUMINUM	LGR002
5	1	SIDE FRAME (LF)	ALUMINUM	LGR003	12	1	WASHER	BRASS	WA0014
6	11	CAP SCREW	6-32UNF −2 × ¼"	CS0012	13	1	PIVOT PIN	HARD STEEL	LGR008
7	1	SLIDE PIN	STEEL	LGR006	14	1	45° BEVEL GEAR	BROWNING	BG0001

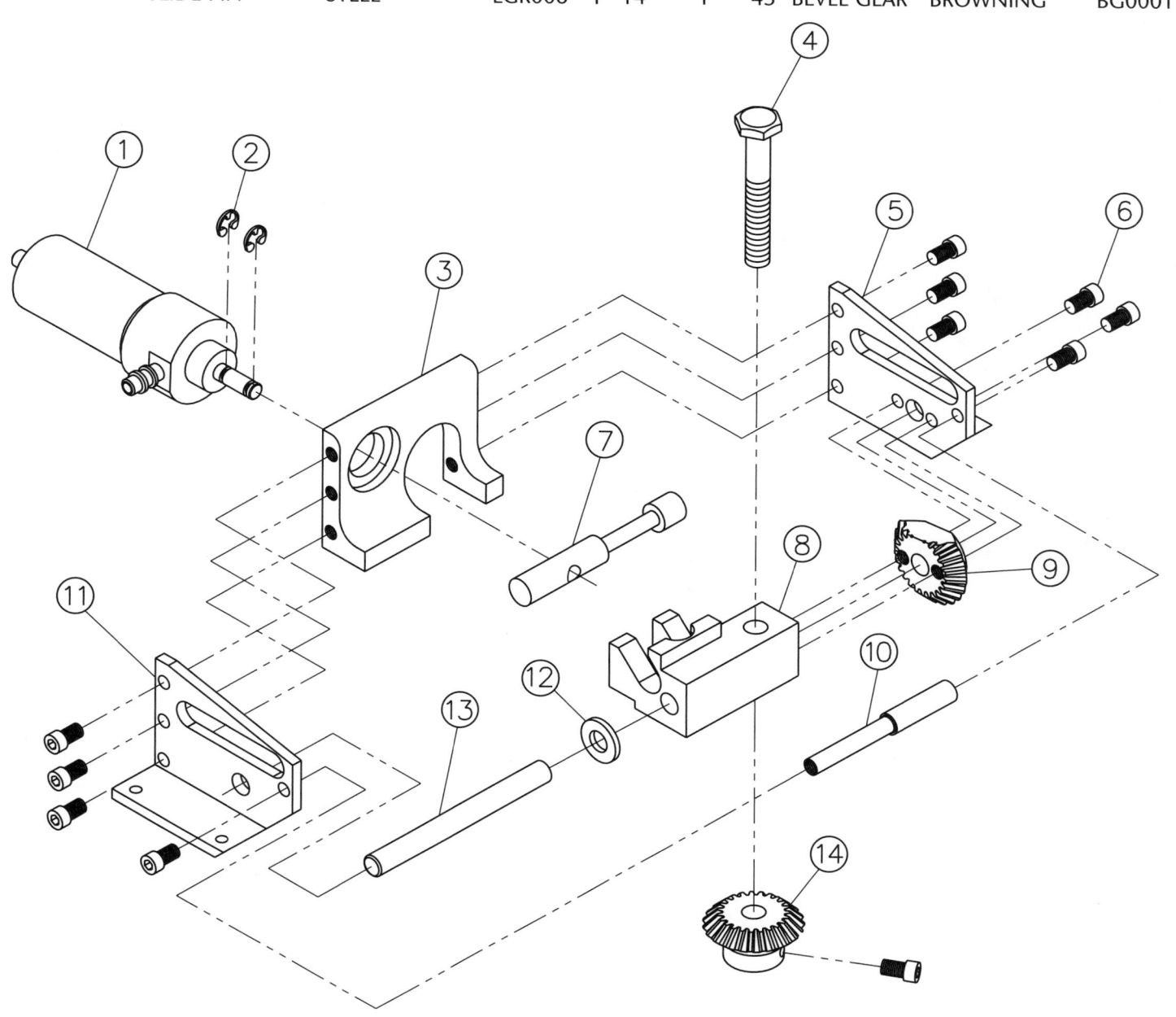

PROBLEM 18-17 (continued)

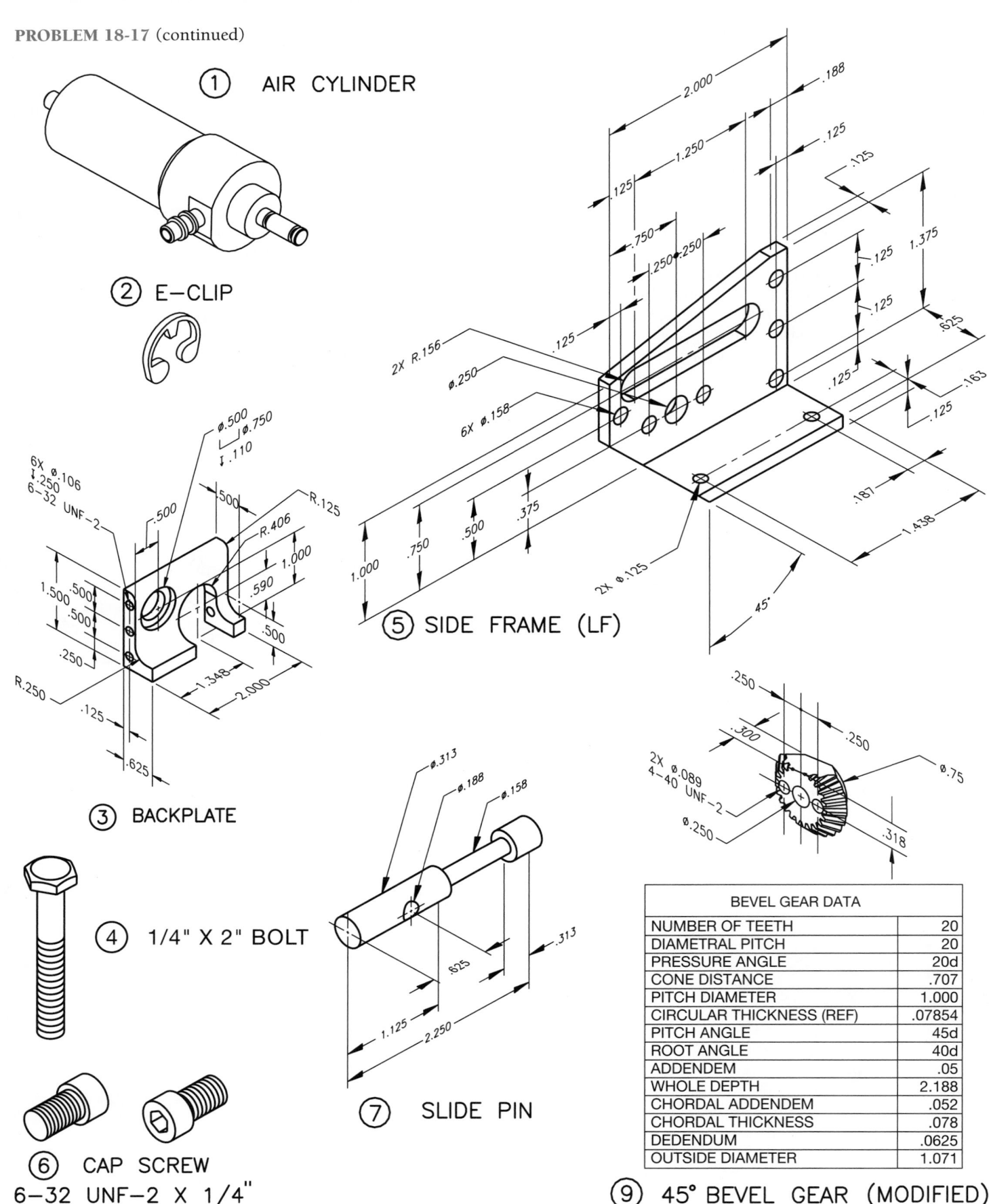

① AIR CYLINDER

② E–CLIP

③ BACKPLATE

④ 1/4" X 2" BOLT

⑥ CAP SCREW
6–32 UNF–2 X 1/4"

⑦ SLIDE PIN

⑤ SIDE FRAME (LF)

BEVEL GEAR DATA	
NUMBER OF TEETH	20
DIAMETRAL PITCH	20
PRESSURE ANGLE	20d
CONE DISTANCE	.707
PITCH DIAMETER	1.000
CIRCULAR THICKNESS (REF)	.07854
PITCH ANGLE	45d
ROOT ANGLE	40d
ADDENDEM	.05
WHOLE DEPTH	2.188
CHORDAL ADDENDEM	.052
CHORDAL THICKNESS	.078
DEDENDUM	.0625
OUTSIDE DIAMETER	1.071

⑨ 45° BEVEL GEAR (MODIFIED)

PROBLEM 18-17 (continued)

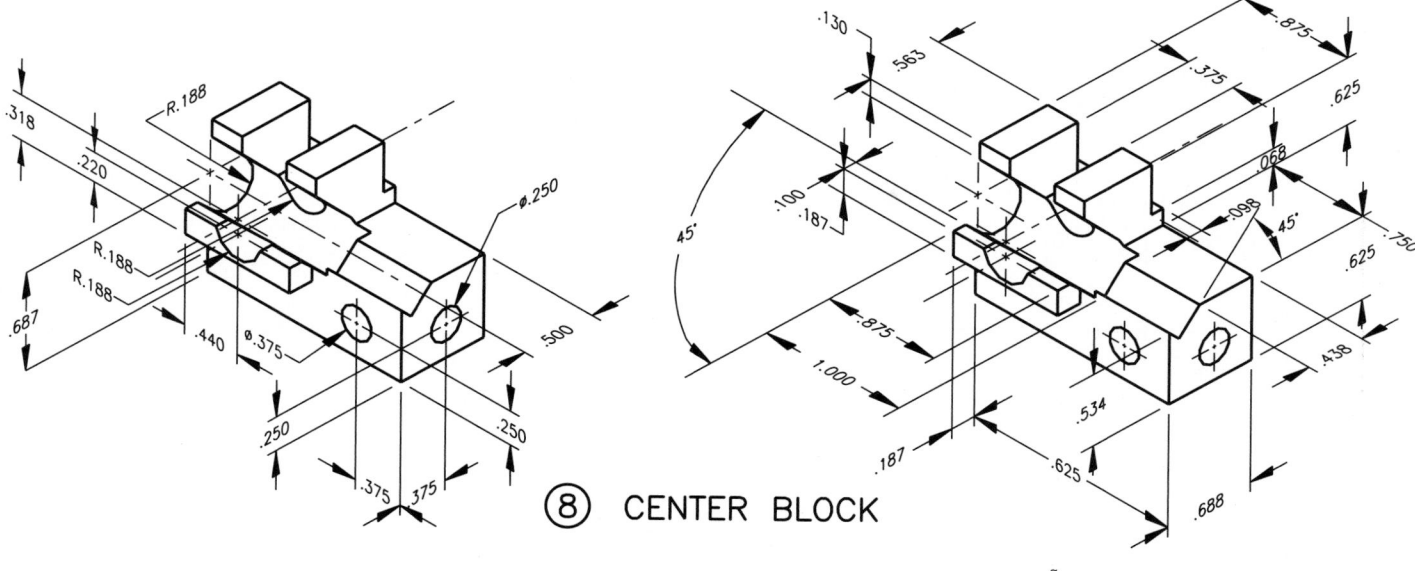

⑧ CENTER BLOCK

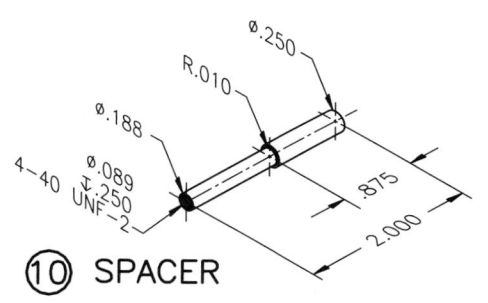

⑩ SPACER

⑫ WASHER
1/4" BRASS

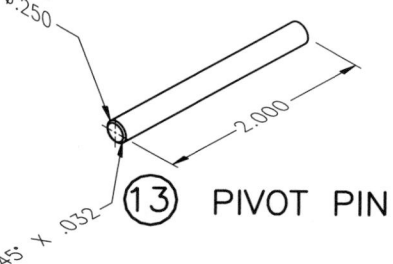

⑬ PIVOT PIN

BEVEL GEAR DATA	
NUMBER OF TEETH	20
DIAMETRAL PITCH	20
PRESSURE ANGLE	20d
CONE DISTANCE	.707
PITCH DIAMETER	1.000
CIRCULAR THICKNESS (REF)	.07854
PITCH ANGLE	45d
ROOT ANGLE	40d
ADDENDEM	.05
WHOLE DEPTH	2.188
CHORDAL ADDENDEM	.052
CHORDAL THICKNESS	.078
DEDENDUM	.0625
OUTSIDE DIAMETER	1.071

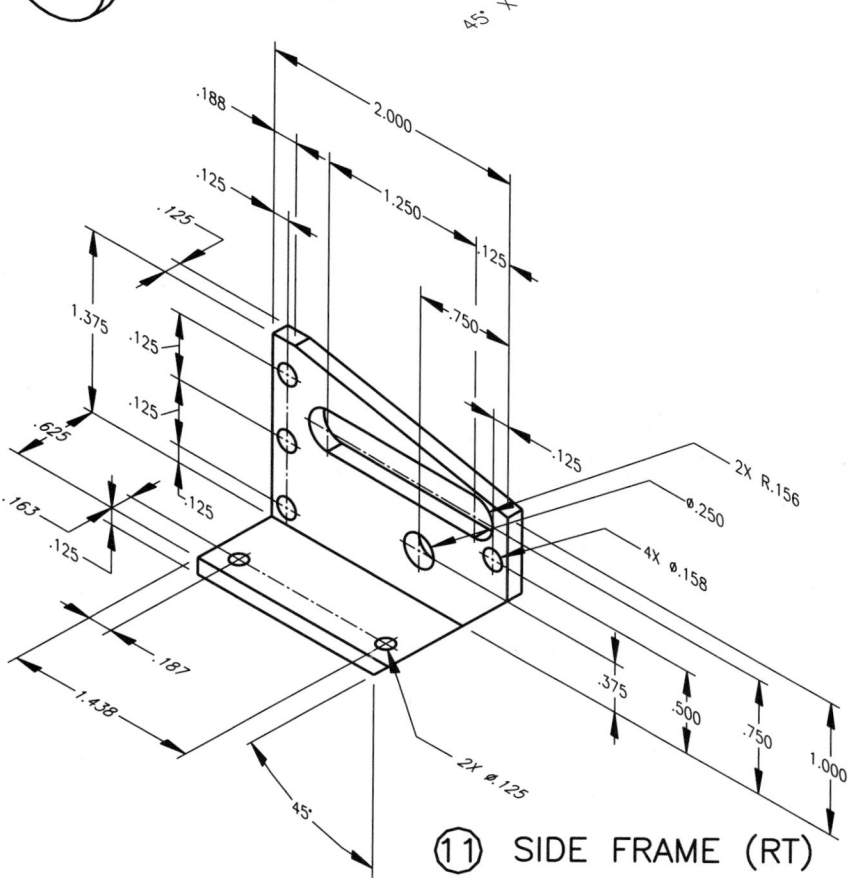

⑪ SIDE FRAME (RT)

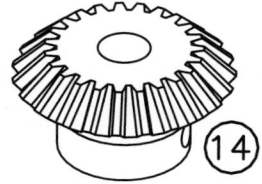

⑭ 45° BEVEL GEAR

PROBLEM 18-18 Working drawing (in.)
Assembly Name: Hydraulic Jack

KEY	QTY	NAME	DESCRIPTION	PART NO.
31	5	PLUG	⅛-27 NPT	5DT1031
30	1	HANDLE	NEEDLE VALVE	1DT1030
29	1	VALVE	NEEDLE	1DT1029
28	1	NUT	NEEDLE VALVE	1DT1028
27	1	SPRING	∅.343 × 1.250 CLOSED ENDS, 9 COILS	5DT1027
26	1	BALL	∅.3125	5DT1026
25	1	SPRING	∅.281 × 1.000 CLOSED ENDS, 6 COILS	5DT1025
24	1	BALL	∅.250	5DT1024
23	1	PIN	PIVOT	1DT1023
22	1	GUIDE	BRONZE	1DT1022
21	1	NUT	.437-20 UNF-2B	5DT1021
20	1	WASHER	PISTON	1DT1020
19	1	CUP	∅1.500 LEATHER	5DT1019
KEY	**QTY**	**NAME**	**DESCRIPTION**	**PART NO.**
			PARTS LIST	

KEY	QTY	NAME	DESCRIPTION	PART NO.
18	1	NUT	.437-14 UNC-2B	5DT1018
17	1	NUT	10-32 UNF-2B	5DT1017
16	1	WASHER	ANS TYPE B PLAIN NO. 10 N SERIES	5DT1016
15	1	CUP SEAL	∅.445 NEOPRENE	5DT1015
14	1	SUPPORT	PUMP PIVOT	1DT1014
13	1	PIN	STOP	1DT1013
12	1	PIN	DRIVE	1DT1012
11	1	PLUNGER	PUMP	1DT1011
10	1	HANDLE	PUMP	1DT1010
9	1	SOCKET	PUMP HANDLE	1DT1009
8	1	NUT	PACKING	1DT1008
7	1	PACKING	∅1.500 LEATHER	5DT1007
6	1	CAP	TOP	1DT1006
5	1	TUBE	RESERVOIR	1DT1005
4	1	SCREW	JACK	1DT1004
3	1	PISTON	JACK	1DT1003
2	1	TUBE	CYLINDER	1DT1002
1	1	BASE	JACK	1DT1001
KEY	**QTY**	**NAME**	**DESCRIPTION**	**PART NO.**
			PARTS LIST	

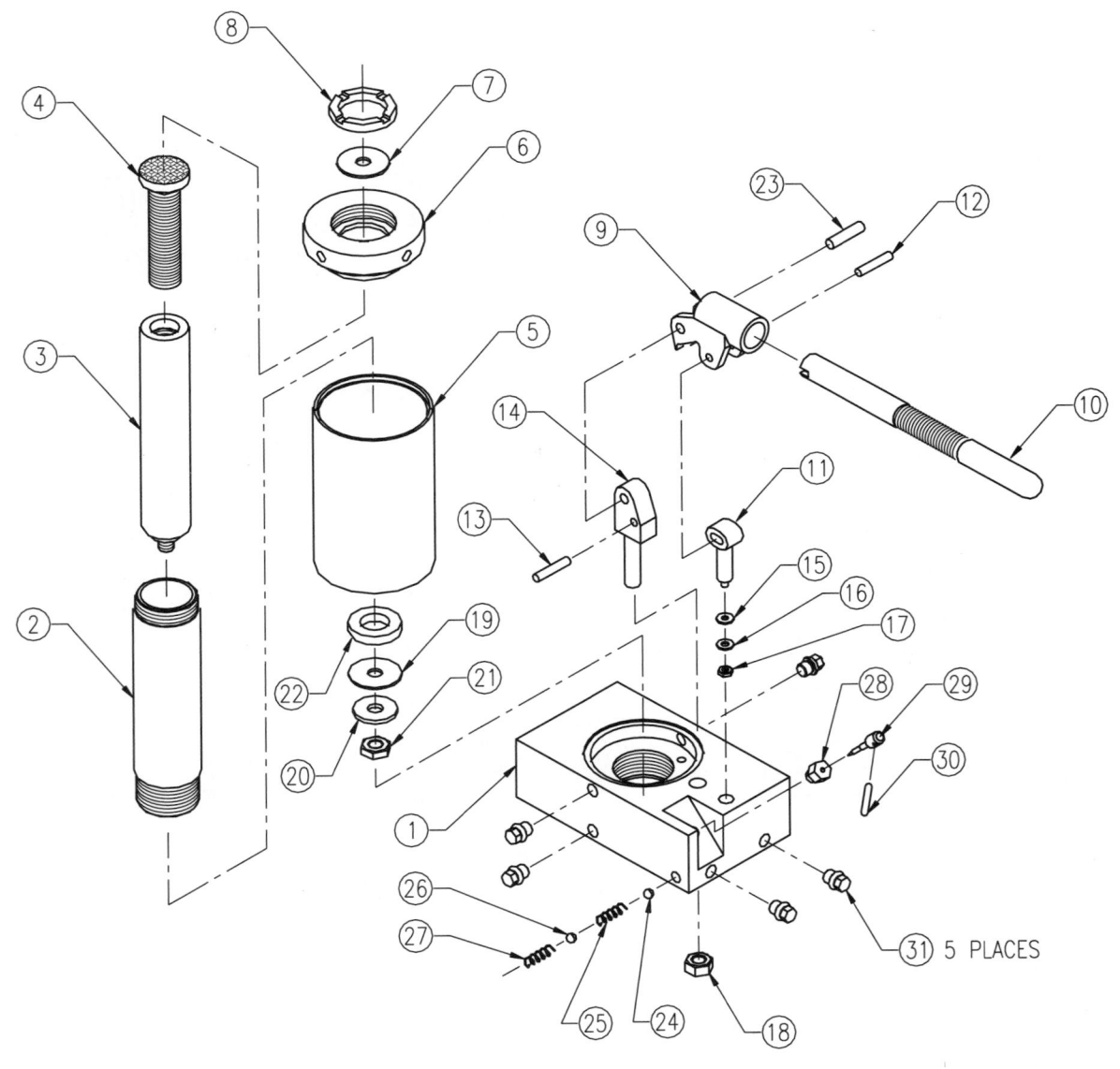

31 5 PLACES

PROBLEM 18-18 (continued)

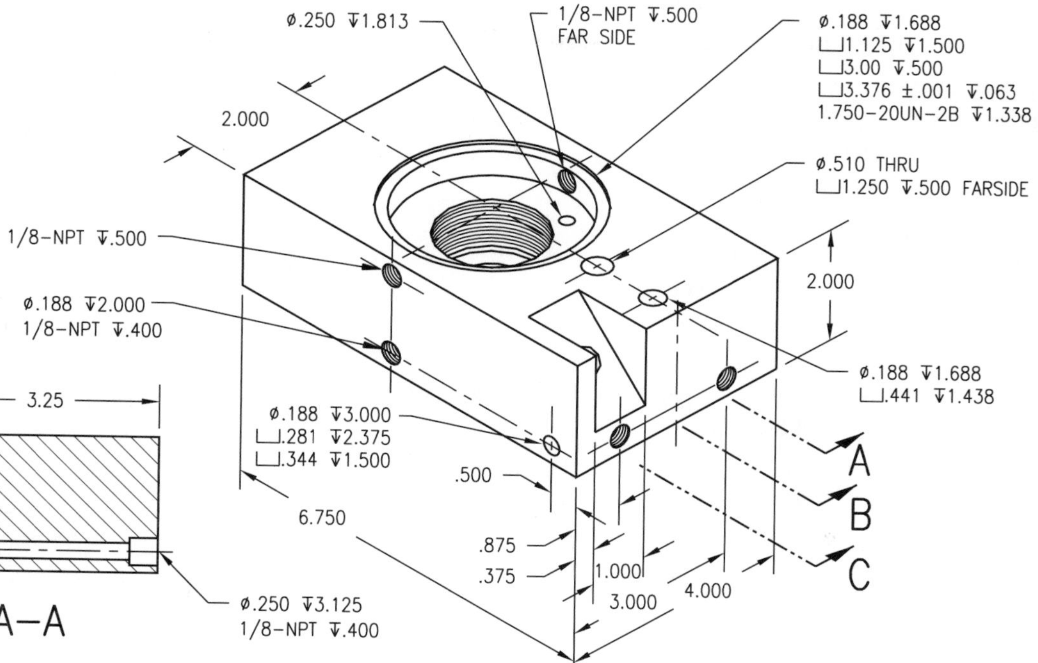

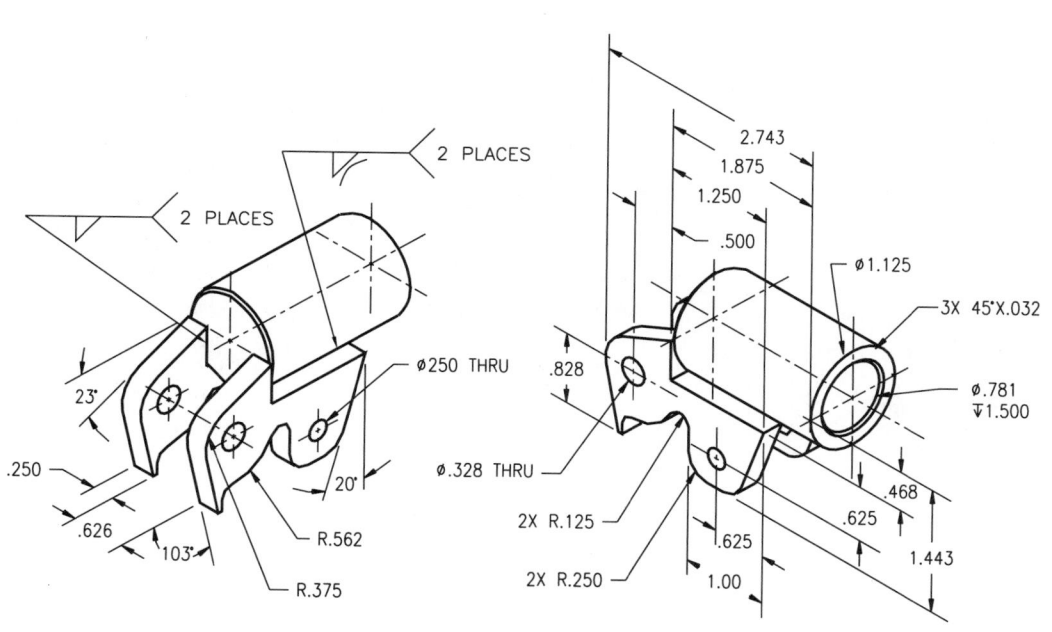

Ø.250 ⬇1.813

1/8-NPT ⬇.500
FAR SIDE

Ø.188 ⬇1.688
⌴1.125 ⬇1.500
⌴3.00 ⬇.500
⌴3.376 ±.001 ⬇.063
1.750-20UN-2B ⬇1.338

2.000

Ø.510 THRU
⌴1.250 ⬇.500 FARSIDE

1/8-NPT ⬇.500

2.000

Ø.188 ⬇2.000
1/8-NPT ⬇.400

Ø.188 ⬇1.688
⌴.441 ⬇1.438

Ø.188 ⬇3.000
⌴.281 ⬇2.375
⌴.344 ⬇1.500

.500

A
B
C

6.750

.875
.375
1.000
3.000
4.000

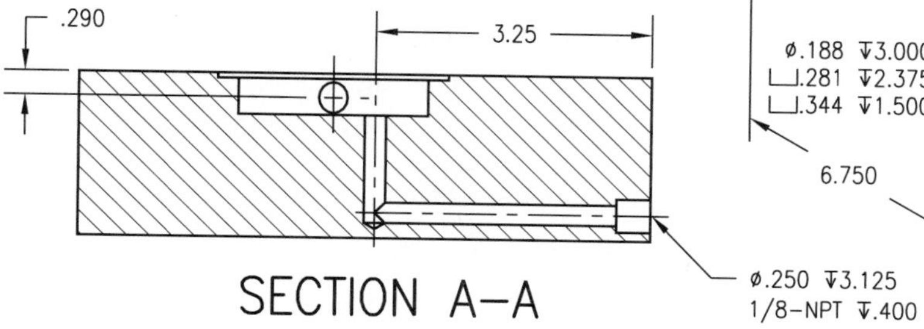

.290

3.25

SECTION A–A

Ø.250 ⬇3.125
1/8-NPT ⬇.400

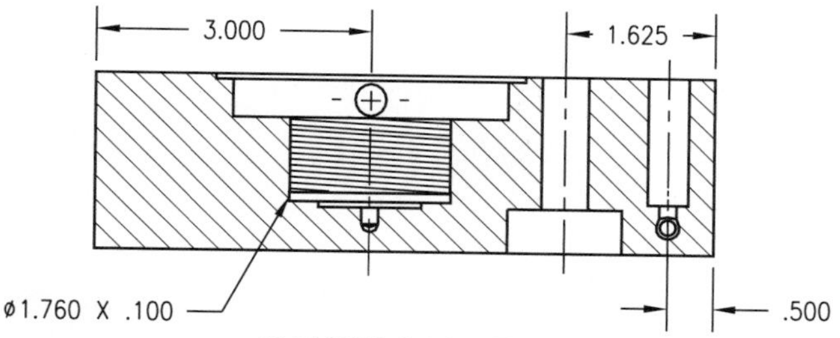

3.000

1.625

Ø1.760 X .100

.500

SECTION B–B

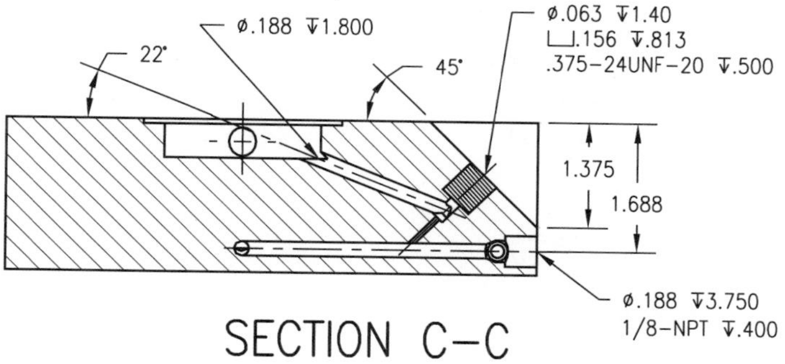

22°

Ø.188 ⬇1.800

45°

Ø.063 ⬇1.40
⌴.156 ⬇.813
.375-24UNF-20 ⬇.500

1.375

1.688

Ø.188 ⬇3.750
1/8-NPT ⬇.400

SECTION C–C

2 PLACES

2 PLACES

23°

Ø250 THRU

.250

.626

103°

20°

R.562

R.375

2.743

1.875

1.250

.500

Ø1.125

3X 45°X.032

.828

Ø.781
⬇1.500

Ø.328 THRU

2X R.125

2X R.250

.468

.625

.625

1.00

1.443

PROBLEM 18-19 **Working drawing (in.)**

Assembly Name: Worm Gear Reducer

Advanced project: design changes may be required.

PARTS LIST:

ITEM	QTY	NAME	DESCRIPTION
1	1	HOUSING	
2	2	RETAINING PLATE	
3	1	BEARING CAP	
4	1	MOTOR ADAPTER	
5	1	HIGH SPEED SHAFT	
6	1	SLOW SPEED SHAFT	
7	1	WORM GEAR	BRONZE
8	1	DBL. ROW TAPERED ROLLER BEARING	KOYO46T30305DJ/29.5
9	1	SNGL. ROLE CYL. ROLLER BEARING	KOYO CRL11
10	1	TAPER PLUG	500-16NPT PLUG
11	1	HEX NUT	.875-16 UN-28

ITEM	QTY	NAME	DESCRIPTION
12	4	HIGH SPEED LOCKWASHER	TIMKEN TW-105
13	4	MACHINE SCREW	.375-16UNC-2A X 1.813 HEX HEAD
14	4	MACHINE SCREW	.375-16UNC-2A X 1.625 HEX HEAD
15	8	MACHINE SCREW	.375-16UNC-2A X .625 HEX HEAD
16	1	HIGH SPEED OIL SEAL	PARKER 2-028
17	2	SLOW SPEED OIL SEAL	PARKER 2-020
18	1	SLOW SPEED KEYWAY	.1875 X .245 X 1.450
19	1	SNGL. ROW TAP ROLLER BEARING	KOYO 32005J
20	2	SLOW SPEED SPACER	TIMKEN TW-506

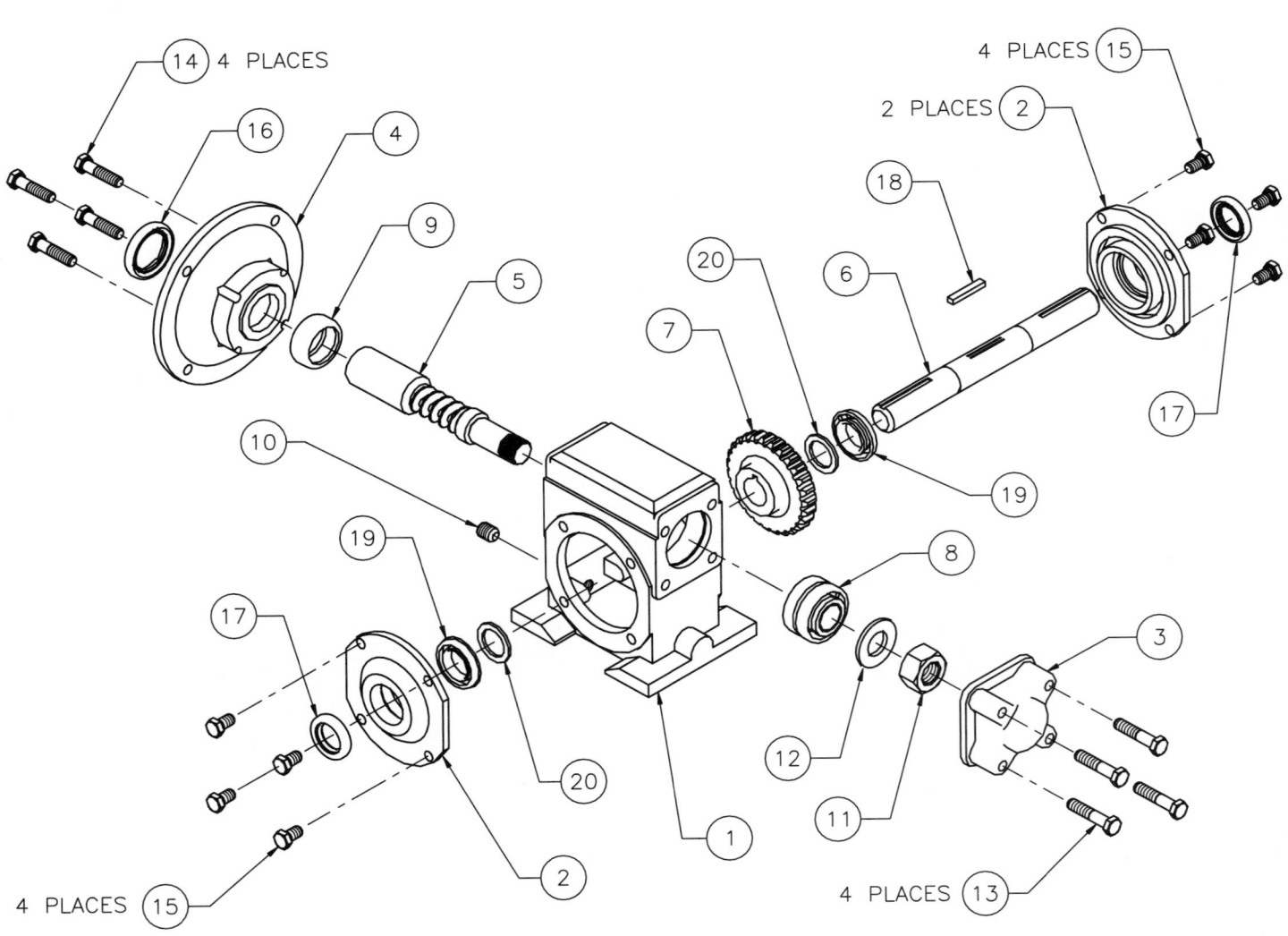

PROBLEM 18-19 (continued)

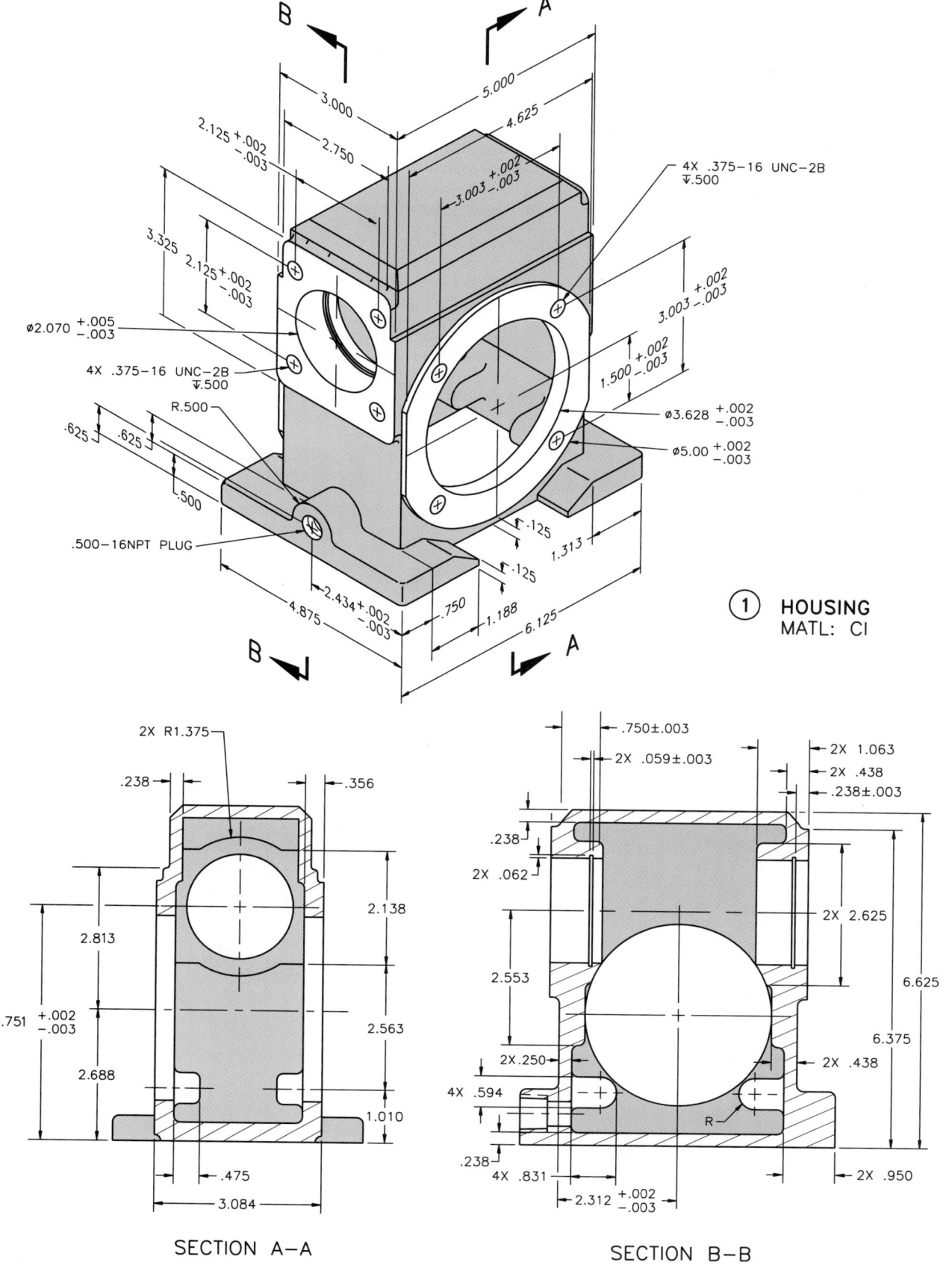

① **HOUSING**
MATL: CI

SECTION A–A

SECTION B–B

PROBLEM 18-19 (continued)

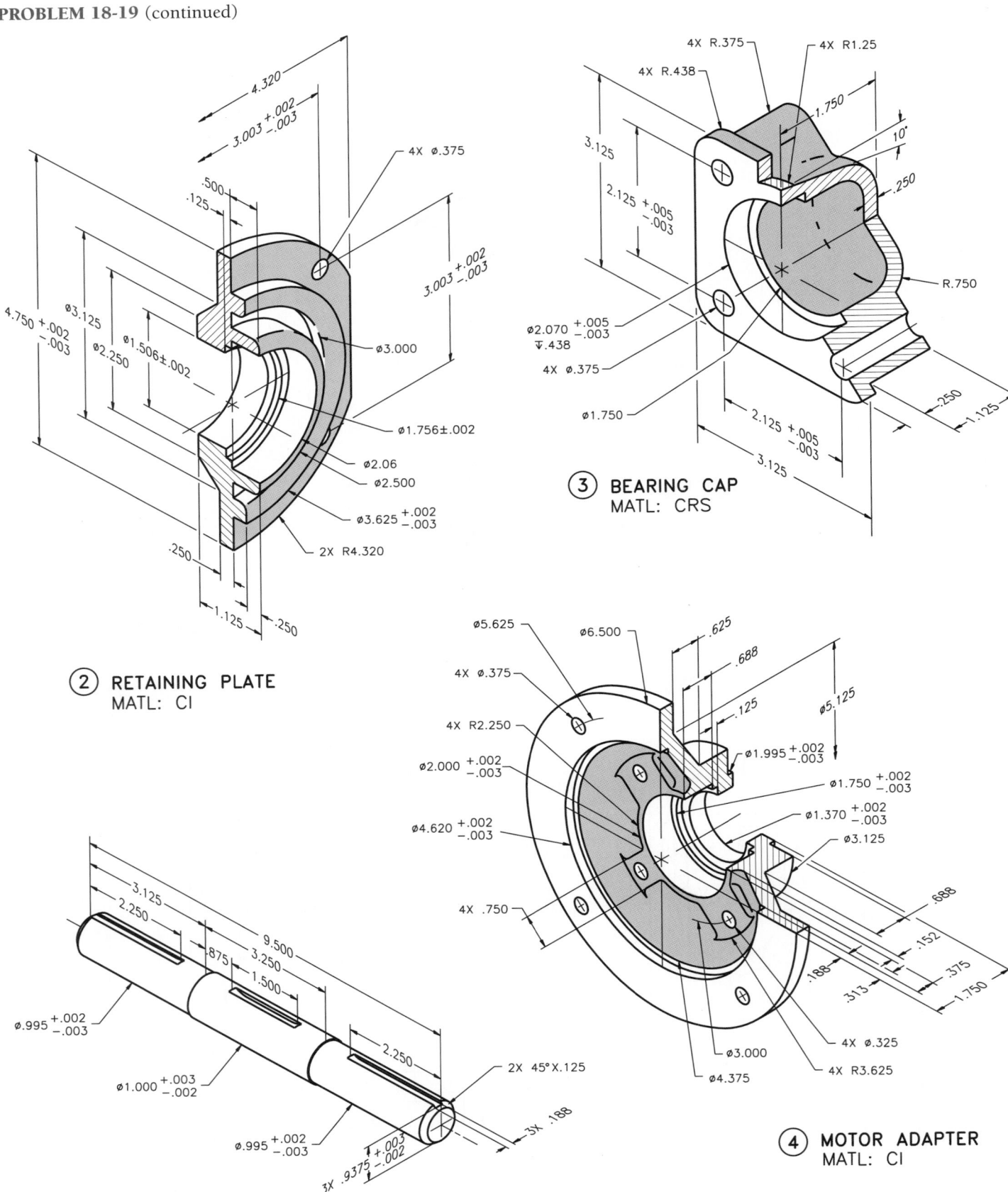

② **RETAINING PLATE**
MATL: CI

③ **BEARING CAP**
MATL: CRS

④ **MOTOR ADAPTER**
MATL: CI

⑥ **SLOW SPEED SHAFT**
MATL: SAE 4320

PROBLEM 18-19 (continued)

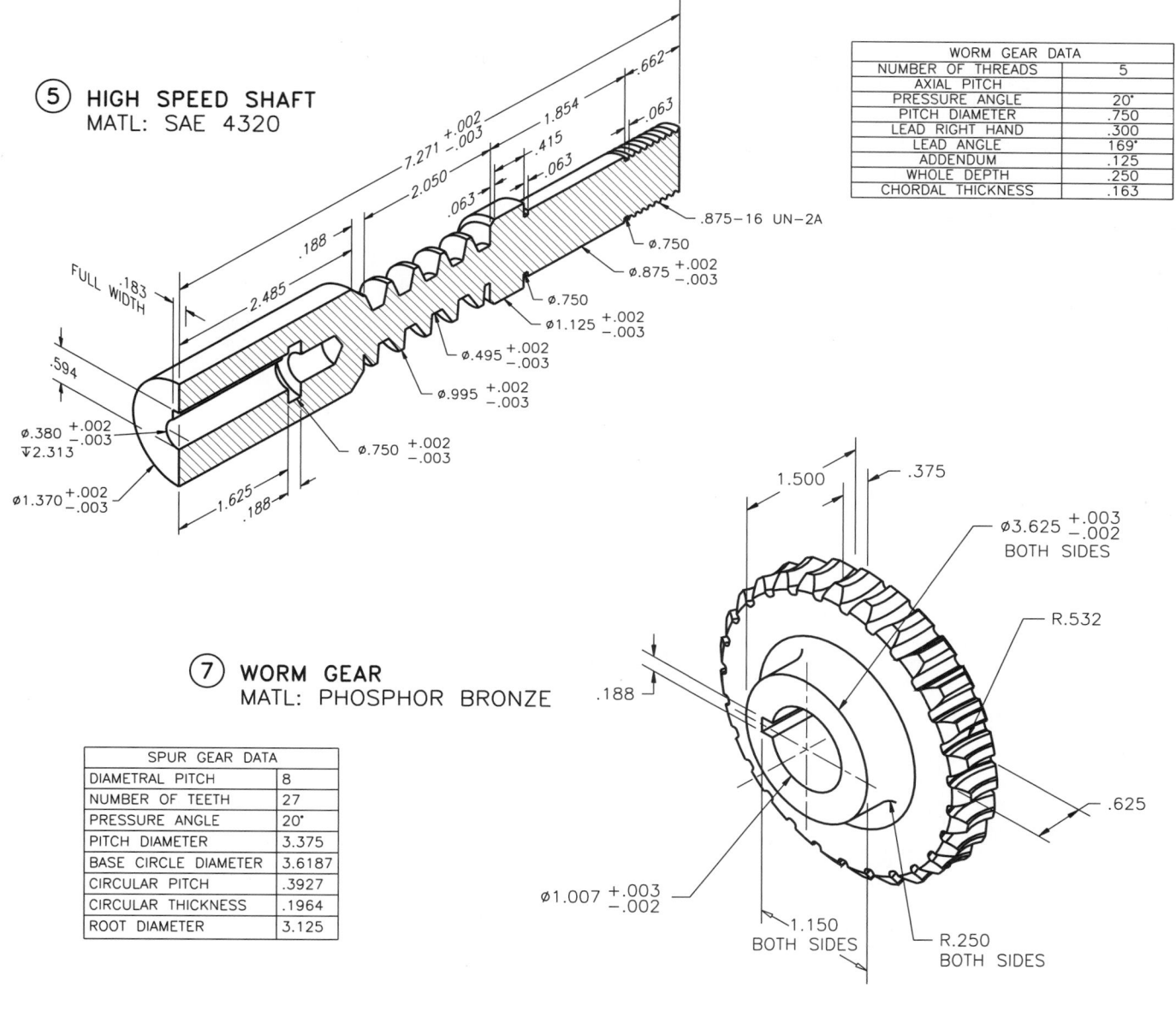

⑤ **HIGH SPEED SHAFT**
MATL: SAE 4320

WORM GEAR DATA	
NUMBER OF THREADS	5
AXIAL PITCH	
PRESSURE ANGLE	20°
PITCH DIAMETER	.750
LEAD RIGHT HAND	.300
LEAD ANGLE	169°
ADDENDUM	.125
WHOLE DEPTH	.250
CHORDAL THICKNESS	.163

⑦ **WORM GEAR**
MATL: PHOSPHOR BRONZE

SPUR GEAR DATA	
DIAMETRAL PITCH	8
NUMBER OF TEETH	27
PRESSURE ANGLE	20°
PITCH DIAMETER	3.375
BASE CIRCLE DIAMETER	3.6187
CIRCULAR PITCH	.3927
CIRCULAR THICKNESS	.1964
ROOT DIAMETER	3.125

PROBLEM 18-20 Working drawing (in.)

Assembly Name: Table Vice

SPECIFIC INSTRUCTIONS:

Prepare a complete set of working drawings with one detail drawing per sheet, and the assembly and parts list on another sheet. When preparing the assembly drawing, use separate balloons for each part in each view or use only one balloon in the view that most clearly identifies the part. This is an engineering prototype project. Errors are likely. Verify dimensions during assembly. Establish tolerances between mating parts. *Problem courtesy of Jack Pitcher.*

(Continued)

PROBLEM 18-20 (continued)

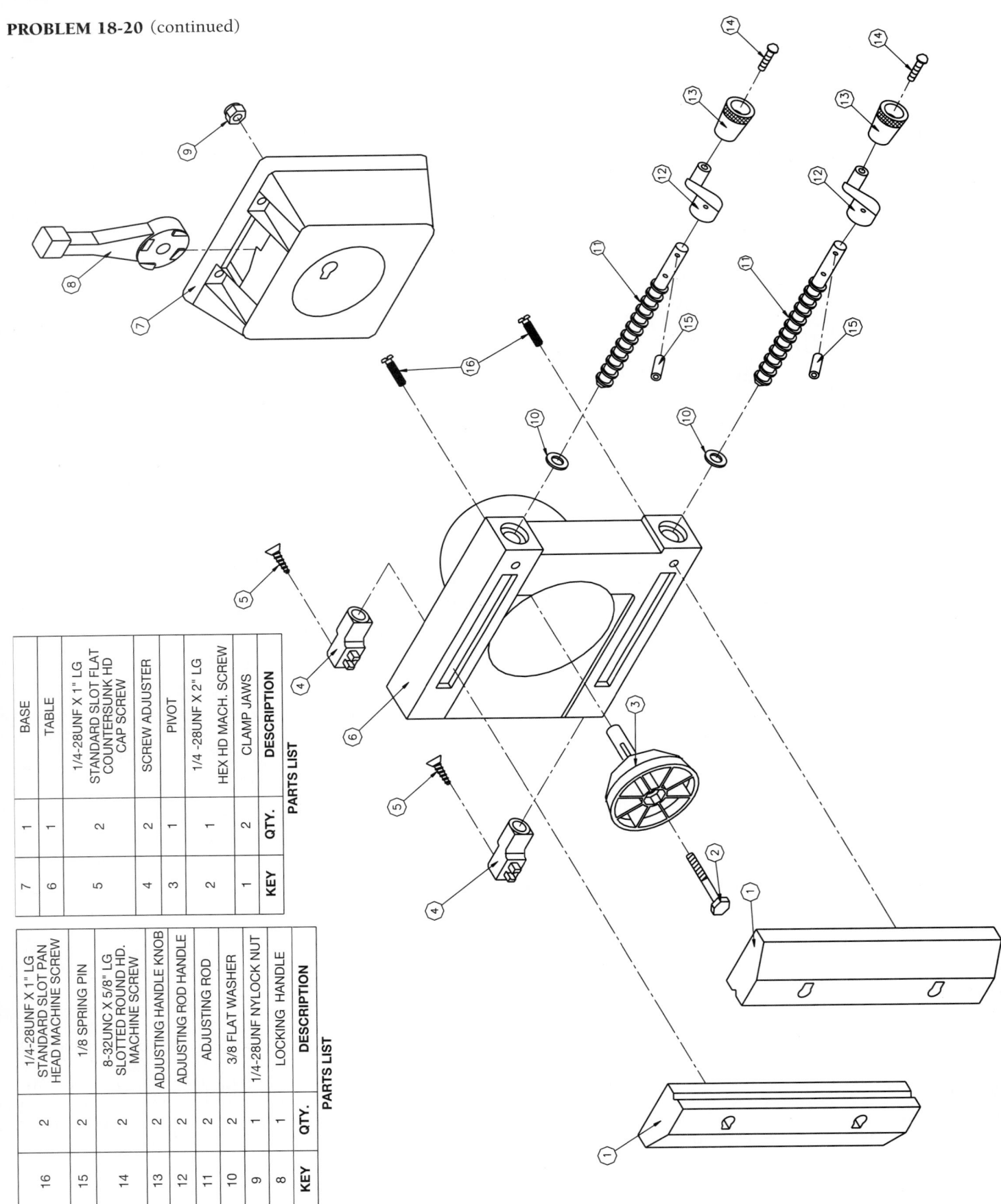

KEY	QTY.	DESCRIPTION
7	1	BASE
6	1	TABLE
5	2	1/4-28UNF X 1" LG STANDARD SLOT FLAT COUNTERSUNK HD CAP SCREW
4	2	SCREW ADJUSTER
3	1	PIVOT
2	1	1/4-28UNF X 2" LG HEX HD MACH. SCREW
1	2	CLAMP JAWS

PARTS LIST

KEY	QTY.	DESCRIPTION
16	2	1/4-28UNF X 1" LG STANDARD SLOT PAN HEAD MACHINE SCREW
15	2	1/8 SPRING PIN
14	2	8-32UNC X 5/8" LG SLOTTED ROUND HD. MACHINE SCREW
13	2	ADJUSTING HANDLE KNOB
12	2	ADJUSTING ROD HANDLE
11	2	ADJUSTING ROD
10	2	3/8 FLAT WASHER
9	1	1/4-28UNF NYLOCK NUT
8	1	LOCKING HANDLE

PARTS LIST

PROBLEM 18-20 (continued)

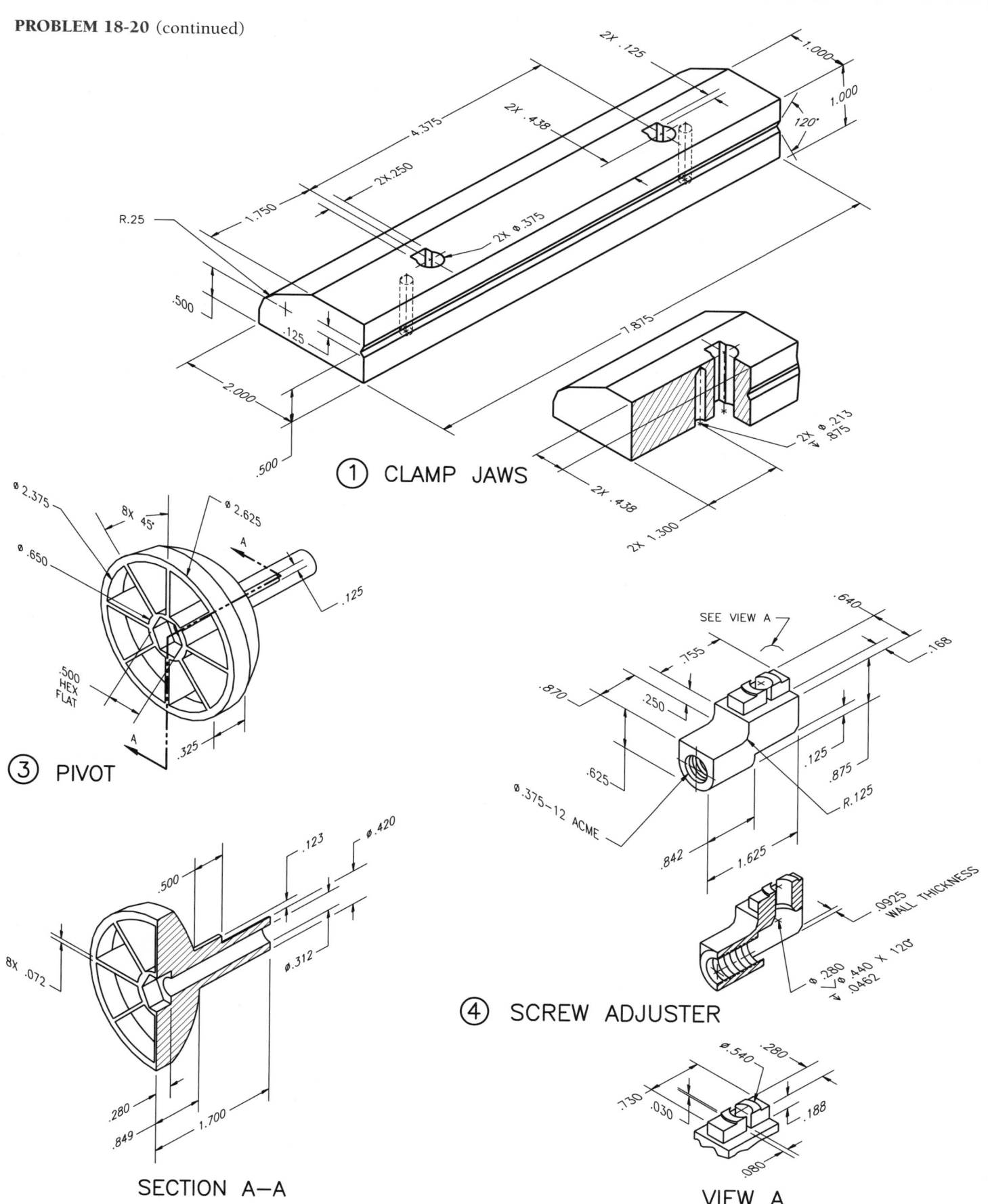

① CLAMP JAWS

③ PIVOT

④ SCREW ADJUSTER

SECTION A–A

VIEW A

PROBLEM 18-20 (continued)

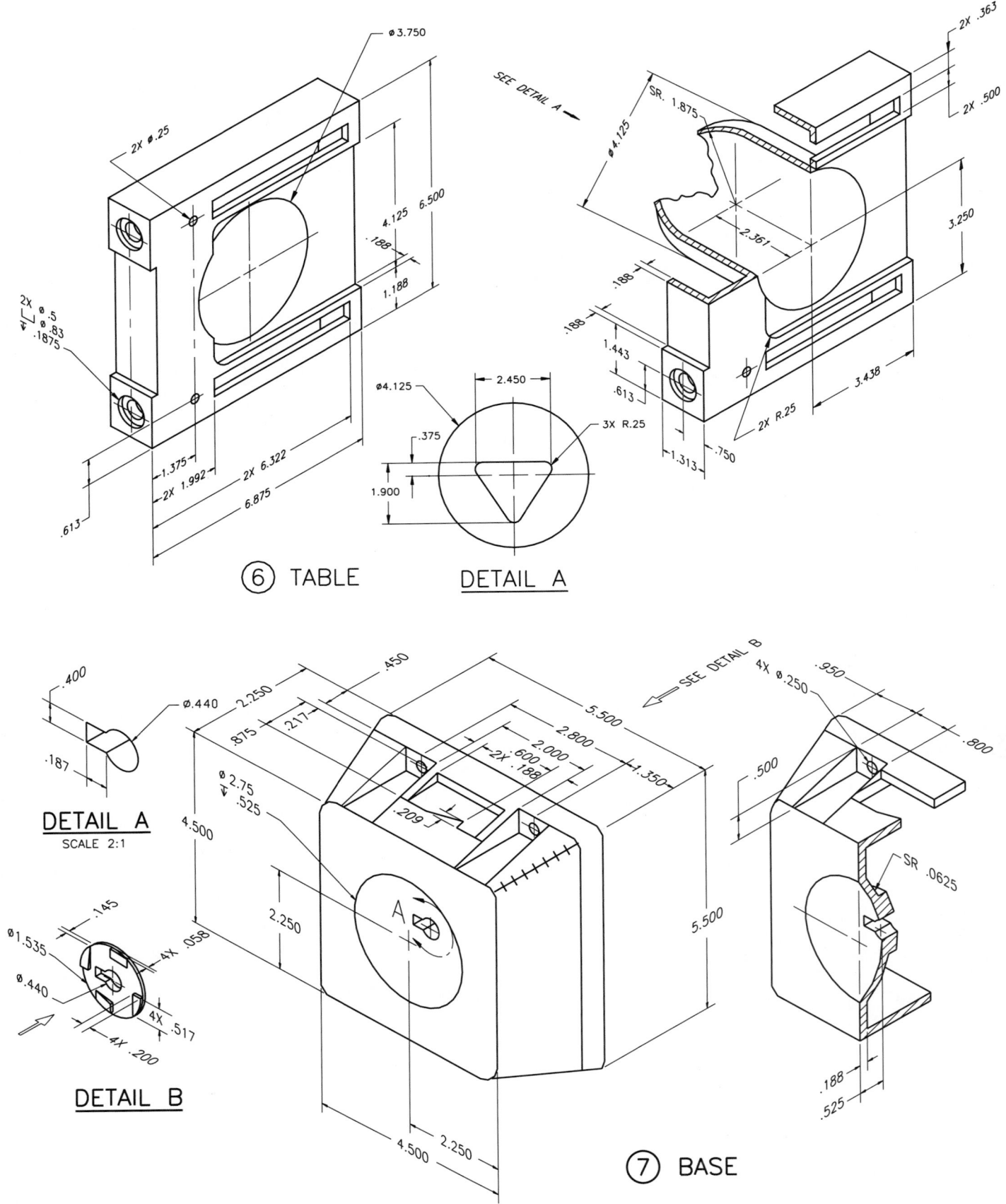

⑥ TABLE

DETAIL A

DETAIL A
SCALE 2:1

DETAIL B

⑦ BASE

PROBLEM 18-20 (continued)

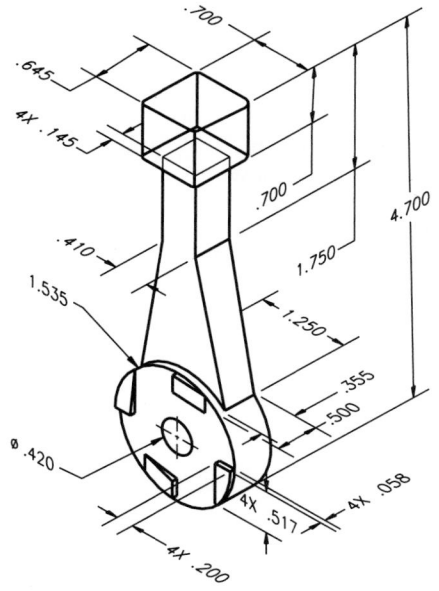

⑧ LOCKING HANDLE

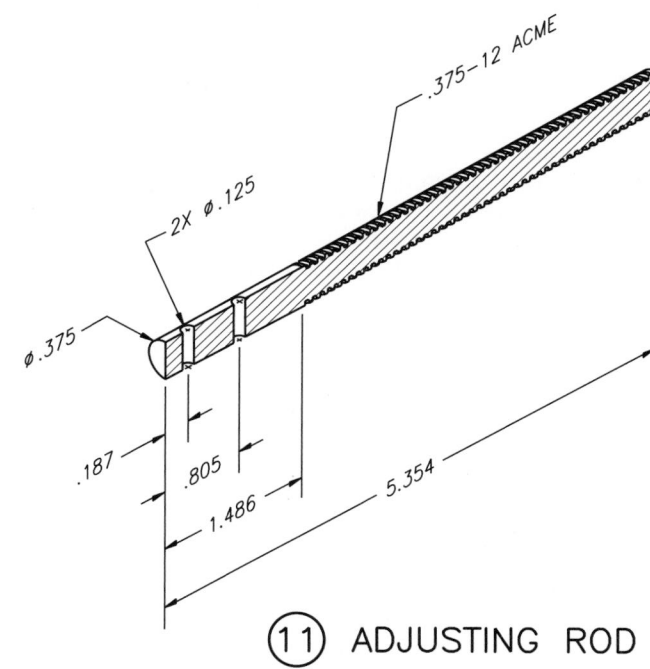

⑪ ADJUSTING ROD

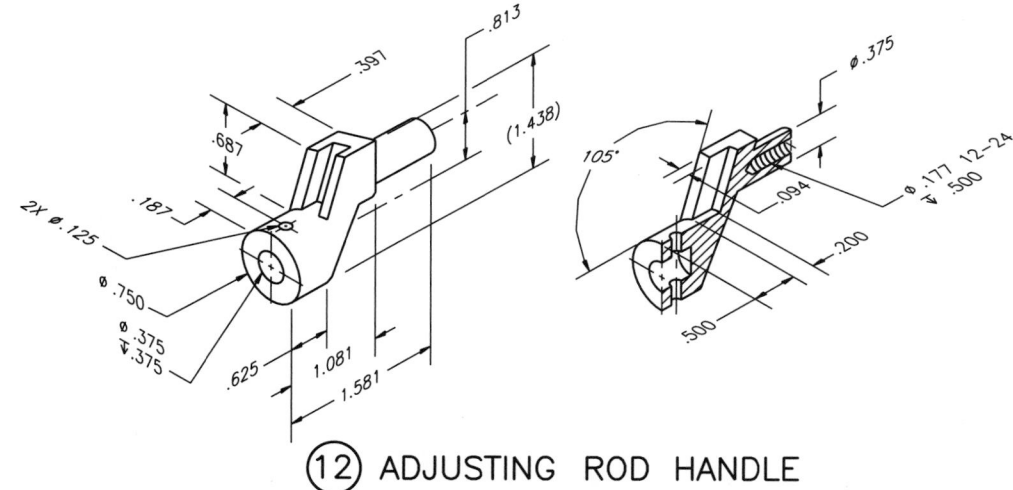

⑫ ADJUSTING ROD HANDLE

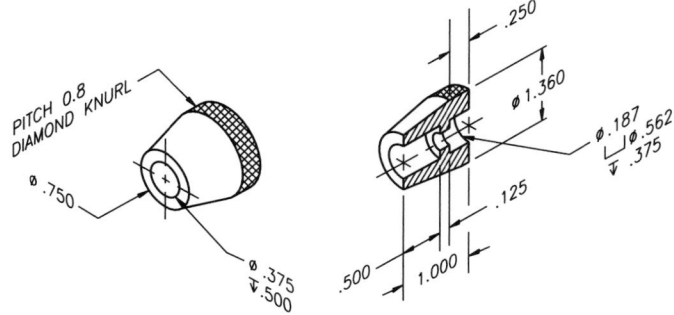

⑬ ADJUSTING HANDLE KNOB

ADVANCED TEAM DESIGN PROBLEM

PROBLEM 18-21   Working drawing (in.)

Assembly Name: Motor

SPECIFIC INSTRUCTIONS:

This is a project that is suited to a team approach. It is recommended that different parts of the project be assigned to each team member. Team members must coordinate mating parts during the development of the working drawings. This is an engineering prototype project. Errors are likely. Verify dimensions during assembly. Establish tolerances between mating parts. If errors are encountered, the team members should use the engineering design process to resolve conflicts and establish design alternatives.

Prepare a complete set of working drawings with one detail drawing per sheet, and the subassembly, assembly, and parts lists on separate sheets. When preparing the subassembly and assembly drawings, use separate balloons for each part in each view or use only one balloon in the view that most clearly identifies the part. *Problem courtesy of Timothy Y. Taylor and Matthew Bell.*

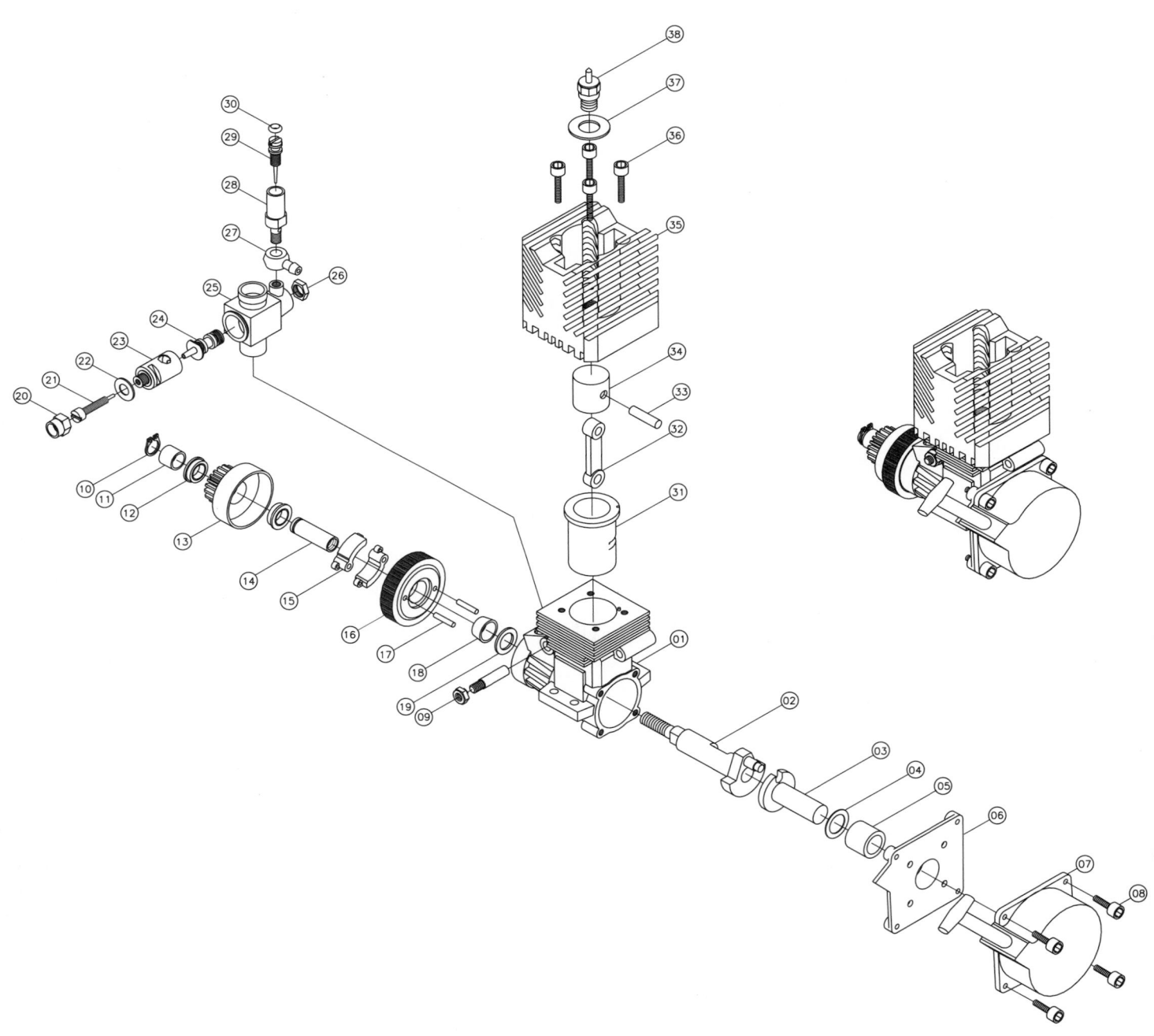

PROBLEM 18-21 (continued)

		O.S. MAX GAS MOTOR		
ITEM NO.	PART OR IDENTIFYING NO.	NOMENCLATURE OR DESCRIPTION OF MAIN ASSEMBLY	QTY REQD
01	MAIN HOUSING	REFER TO FOLLOWING PAGES	1
02	CRANK SHAFT	REFER TO FOLLOWING PAGES	1
03	PULLEY SHAFT	REFER TO FOLLOWING PAGES	1
04	WASHER	I.D.Ø.316 O.D.Ø.475 × .015THK	1
05	SLEEVE	REFER TO FOLLOWING PAGES	1
06	FACE PLATE	REFER TO FOLLOWING PAGES	1
07	STARTER HOUSING	DESIGN OR FIND ALTERNATE PURCHASE	1
08	CAP SCREW	3-48UNC-2 × .250"	4
09	SCREW LOCK	REFER TO FOLLOWING PAGES	1
10	HORSESHOE CLIP	REFER TO FOLLOWING PAGES	1
11	CLUTCH SLEEVE	REFER TO FOLLOWING PAGES	1
12	CLUTCH BEARING	ESTABLISH PURCHASE PART	2
13	CLUTCH GEAR	REFER TO FOLLOWING PAGES	1
14	CLUTCH SLEEVE	REFER TO FOLLOWING PAGES	1
15	BRAKE PAD	REFER TO FOLLOWING PAGES	2
16	CLUTCH HOUSING	REFER TO FOLLOWING PAGES	1
17	CLUTCH PIN	Ø.075 × .375	2
18	BEVELED SLEEVE	REFER TO FOLLOWING PAGES	1
19	WASHER	I.D. Ø.025 × .O.D. Ø.042 × .015THK	1
20	HEX CAP	REFER TO FOLLOWING PAGES	1
21	AIR ADJUSTER	REFER TO FOLLOWING PAGES	1
22	WASHER	REFER TO FOLLOWING PAGES	1
23	MIXTURE CONTROL	REFER TO FOLLOWING PAGES	1
24	GAS REGULATOR	REFER TO FOLLOWING PAGES	1
25	CARBURETOR HOUSING	REFER TO FOLLOWING PAGES	1
26	HEX NUT	1/4-16UNC-2	1
27	FUEL INTAKE	REFER TO FOLLOWING PAGES	1
28	FUEL ADJUSTER	REFER TO FOLLOWING PAGES	1
29	FUEL INTAKE ADJUSTER PIN	REFER TO FOLLOWING PAGES	1
30	RUBBER SEAL	ESTABLISH PURCHASE PART	1
31	PISTON SHAFT	REFER TO FOLLOWING PAGES	1
32	CRANK ARM	REFER TO FOLLOWING PAGES	1
33	PISTON PIN	REFER TO FOLLOWING PAGES	1
34	PISTON	REFER TO FOLLOWING PAGES	1
35	HEAT SINK	REFER TO FOLLOWING PAGES	1
36	HEX SCREW	3-48UNC-2 × .375"	1
37	WASHER	ESTABLISH PURCHASE PART	1
38	GLOW PLUG	ESTABLISH PURCHASE PART	1

PROBLEM 18-21 (continued)

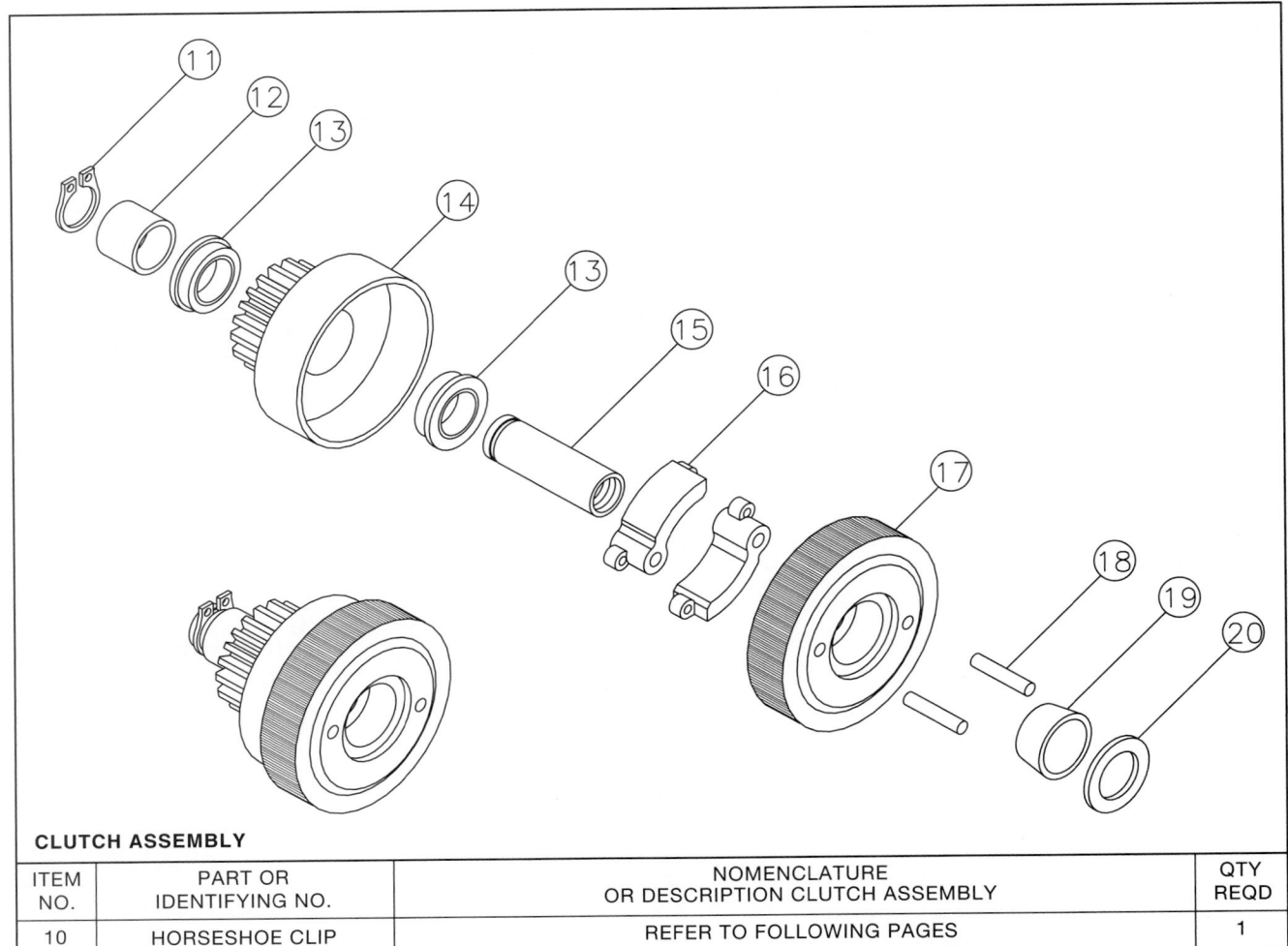

CLUTCH ASSEMBLY

ITEM NO.	PART OR IDENTIFYING NO.	NOMENCLATURE OR DESCRIPTION CLUTCH ASSEMBLY	QTY REQD
10	HORSESHOE CLIP	REFER TO FOLLOWING PAGES	1
11	CLUTCH SLEEVE	REFER TO FOLLOWING PAGES	1
12	CLUTCH BEARING	ESTABLISH PURCHASE PART	2
13	CLUTCH GEAR	REFER TO FOLLOWING PAGES	1
14	CLUTCH SLEEVE	REFER TO FOLLOWING PAGES	1
15	BRAKE PAD	REFER TO FOLLOWING PAGES	2
16	CLUTCH HOUSING	REFER TO FOLLOWING PAGES	1
17	CLUTCH PIN	Ø.075 × .375	2
18	BEVELED SLEEVE	REFER TO FOLLOWING PAGES	1
19	WASHER	I.D. Ø.025 × O.D. Ø.042 × .015THK	1

PROBLEM 18-21 (continued)

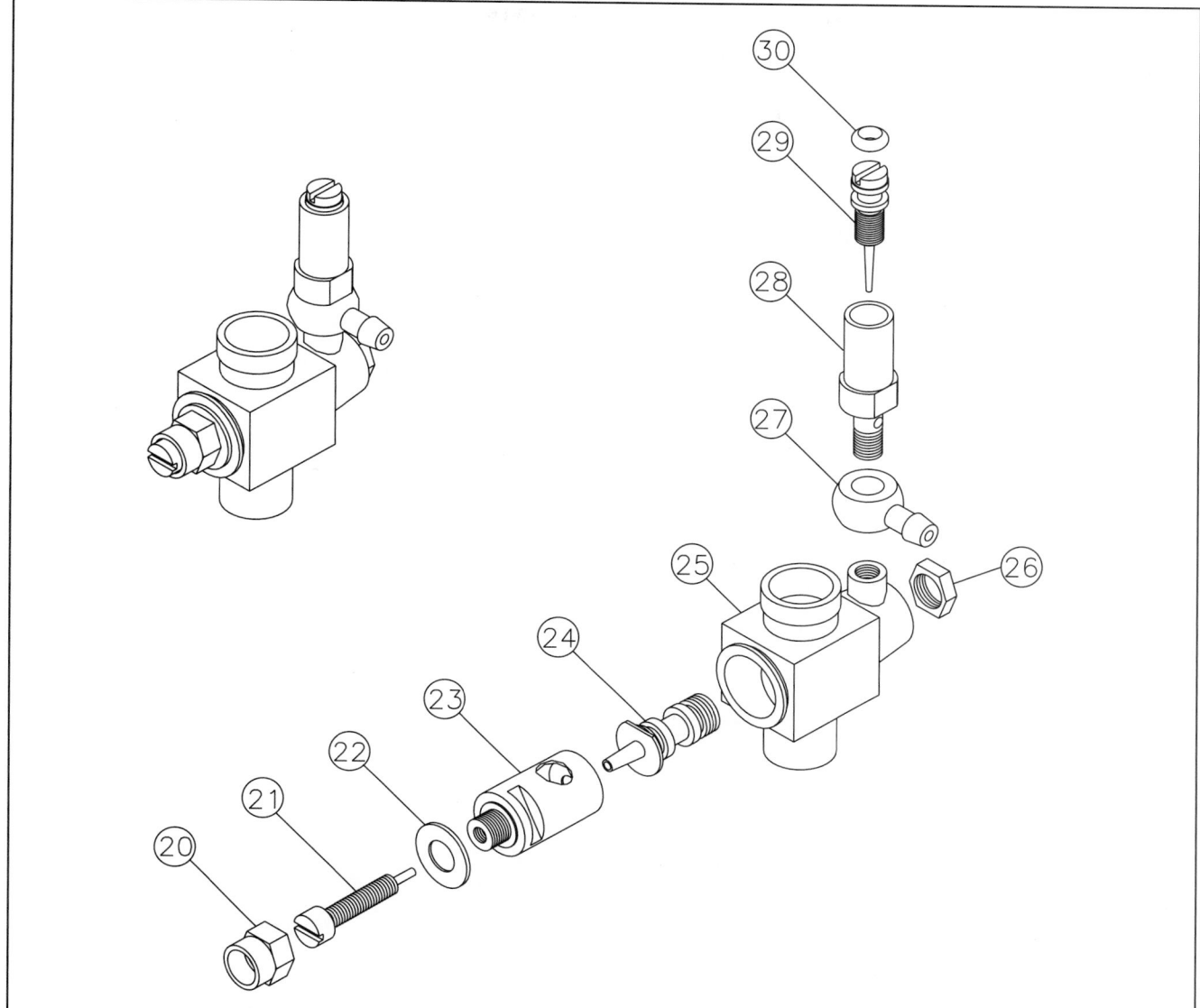

CARBURETOR SUBASSEMBLY

ITEM NO.	PART OR IDENTIFYING NO.	NOMENCLATURE OR DESCRIPTION	QTY REQD
20	HEX CAP	REFER TO FOLLOWING PAGES	1
21	AIR ADJUSTER	REFER TO FOLLOWING PAGES	1
22	WASHER	REFER TO FOLLOWING PAGES	1
23	MIXTURE CONTROL	REFER TO FOLLOWING PAGES	1
24	GAS REGULATOR	REFER TO FOLLOWING PAGES	1
25	CARBURETOR HOUSING	REFER TO FOLLOWING PAGES	1
26	HEX NUT	1/4-16UNC-2	1
27	FUEL INTAKE	REFER TO FOLLOWING PAGES	1
28	FUEL ADJUSTER	REFER TO FOLLOWING PAGES	1
29	FUEL INTAKE ADJUSTER PIN	REFER TO FOLLOWING PAGES	1
30	RUBBER SEAL	ESTABLISH PURCHASE PART	1

PROBLEM 18-21 (continued)

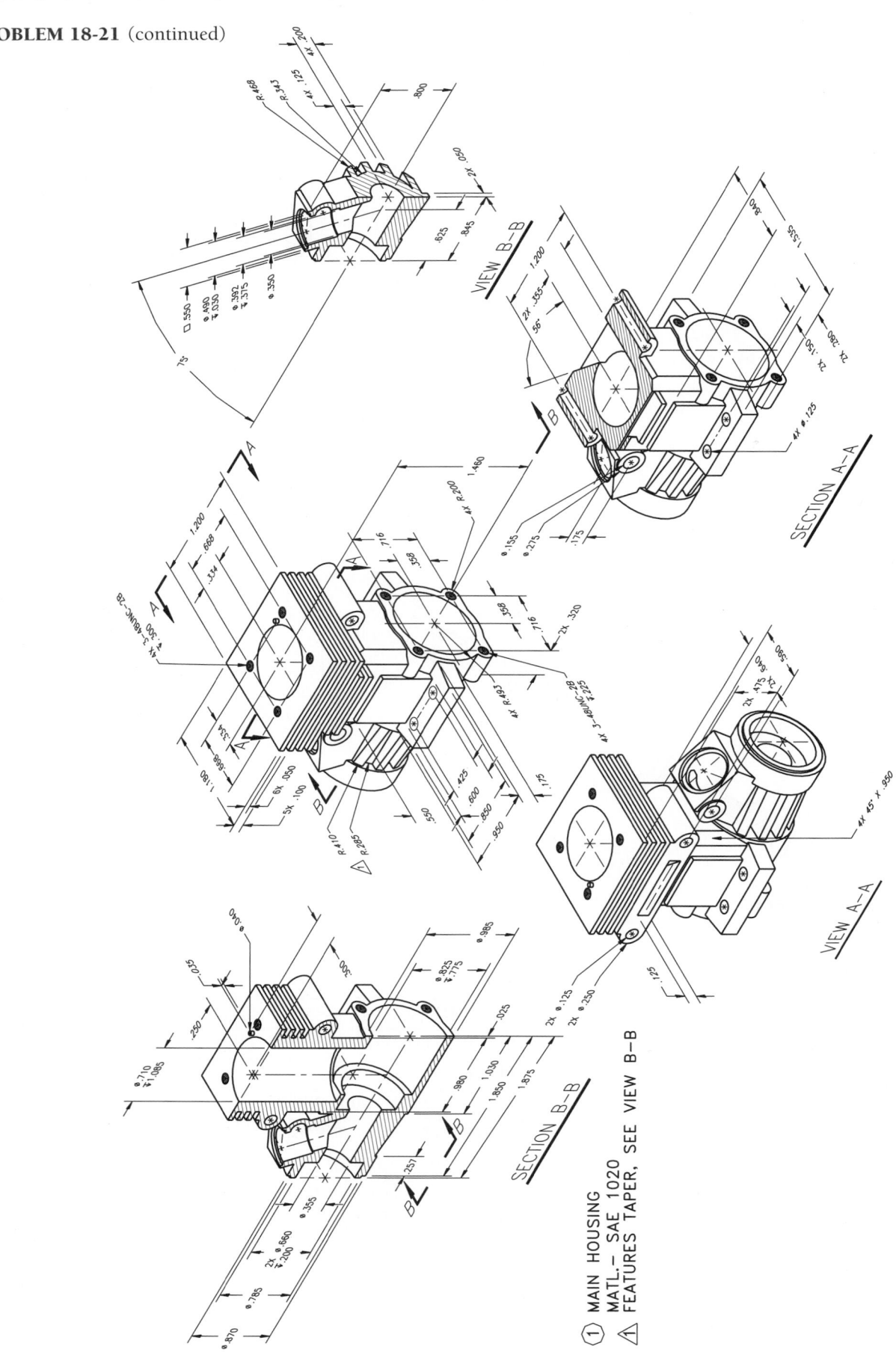

VIEW B–B

SECTION A–A

VIEW A–A

SECTION B–B

1 MAIN HOUSING
MATL. – SAE 1020
⚠ FEATURES TAPER, SEE VIEW B–B

PROBLEM 18-21 (continued)

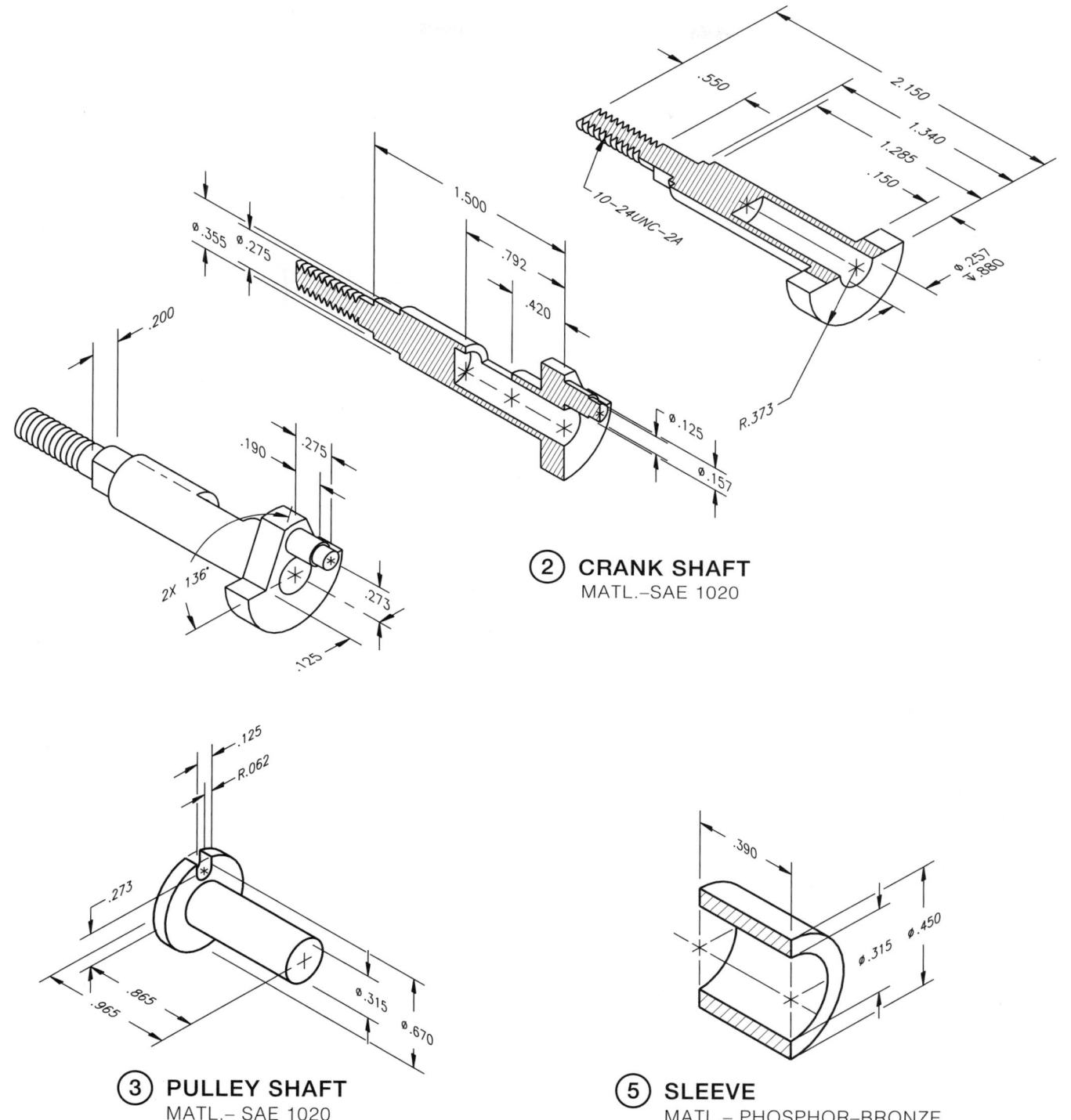

.550   2.150
1.340
1.285
.150

10-24UNC-2A

Ø.257
▼.880

R.373

Ø .355   Ø .275

1.500
.792
.420

Ø.125
Ø.157

.200

.190   .275

2X 136°

.273
.125

(2) **CRANK SHAFT**
MATL.–SAE 1020

.125
R.062

.273

.865
.965

Ø.315
Ø.670

(3) **PULLEY SHAFT**
MATL.– SAE 1020

.390

Ø.315   Ø.450

(5) **SLEEVE**
MATL.– PHOSPHOR–BRONZE

PROBLEM 18-21 (continued)

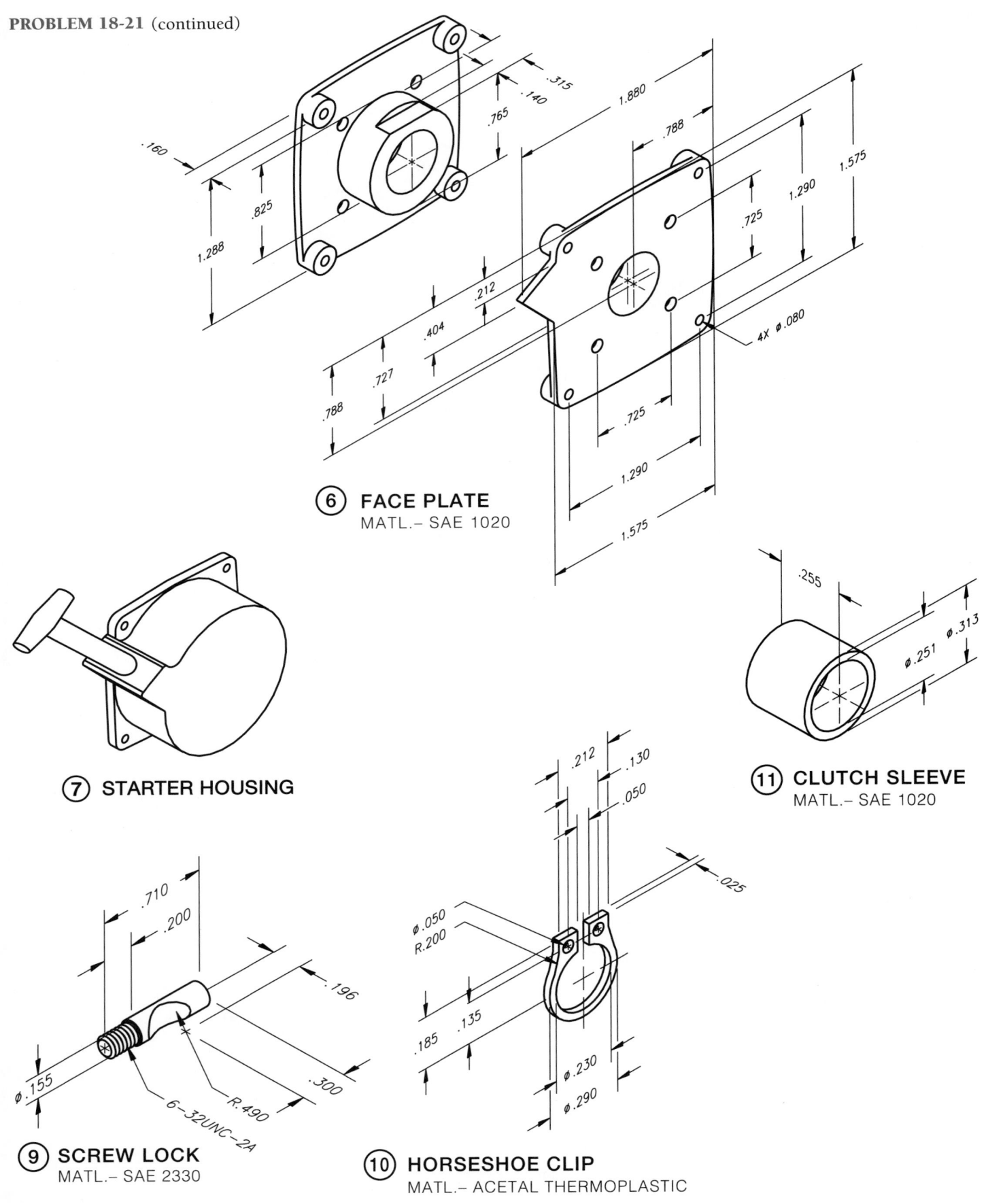

⑥ **FACE PLATE**
MATL.– SAE 1020

⑦ **STARTER HOUSING**

⑪ **CLUTCH SLEEVE**
MATL.– SAE 1020

⑨ **SCREW LOCK**
MATL.– SAE 2330

6–32UNC–2A

⑩ **HORSESHOE CLIP**
MATL.– ACETAL THERMOPLASTIC

PROBLEM 18-21 (continued)

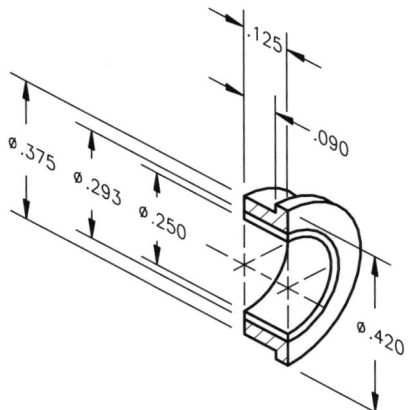

Ø.375
Ø.293
Ø.250
.125
.090
Ø.420

(12) **CLUTCH BEARING**
MATL.– BRONZE

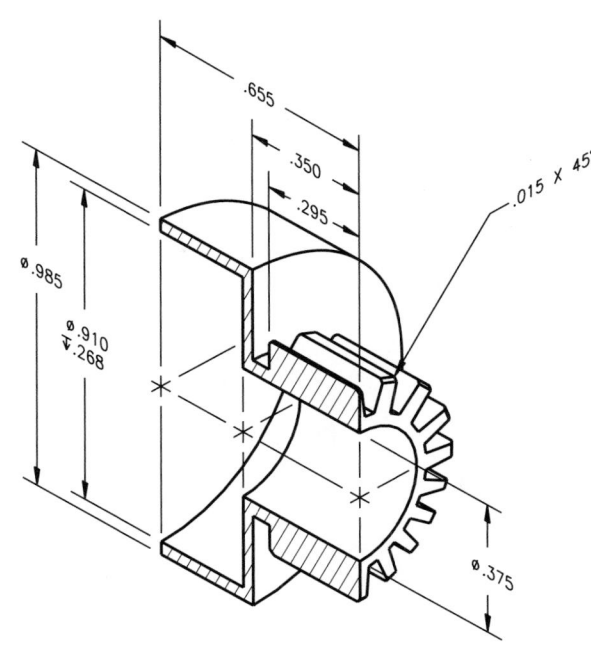

.655
.350
.295
.015 X 45°
Ø.985
Ø.910
▼.268
Ø.375

(13) **CLUTCH GEAR**
MATL.– SAE 1020

SPUR GEAR DATA	
DIAMETRAL PITCH	29
NUMBER OF TEETH	18
PRESSURE ANGLE	20°
BASE CIRCLE DIAMETER	.583
CIRCULAR PITCH	.108
CIRCULAR THICKNESS	.054
ROOT DIAMETER	.5344

.762
.197
.075
.045
Ø.250
Ø.235
Ø.095 HEX
Ø.200–20UNC–2

(14) **CLUTCH SLEEVE**
MATL.– SAE 1060

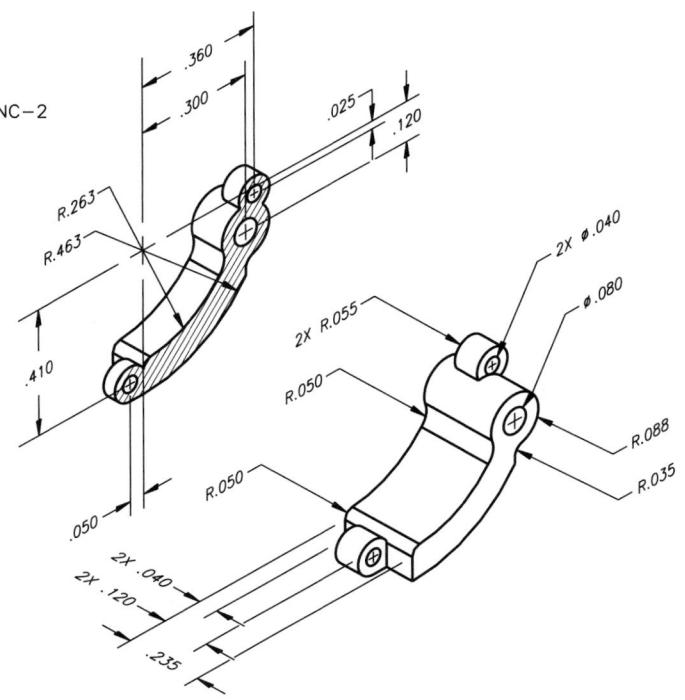

.360
.300
.025
.120
R.263
R.463
2X Ø.040
Ø.080
2X R.055
R.050
R.088
R.035
.410
R.050
R.050
.050
2X .040
2X .120
.235

(15) **BRAKE PAD**
MATL.– ACETAL THERMOPLASTIC

PROBLEM 18-21 (continued)

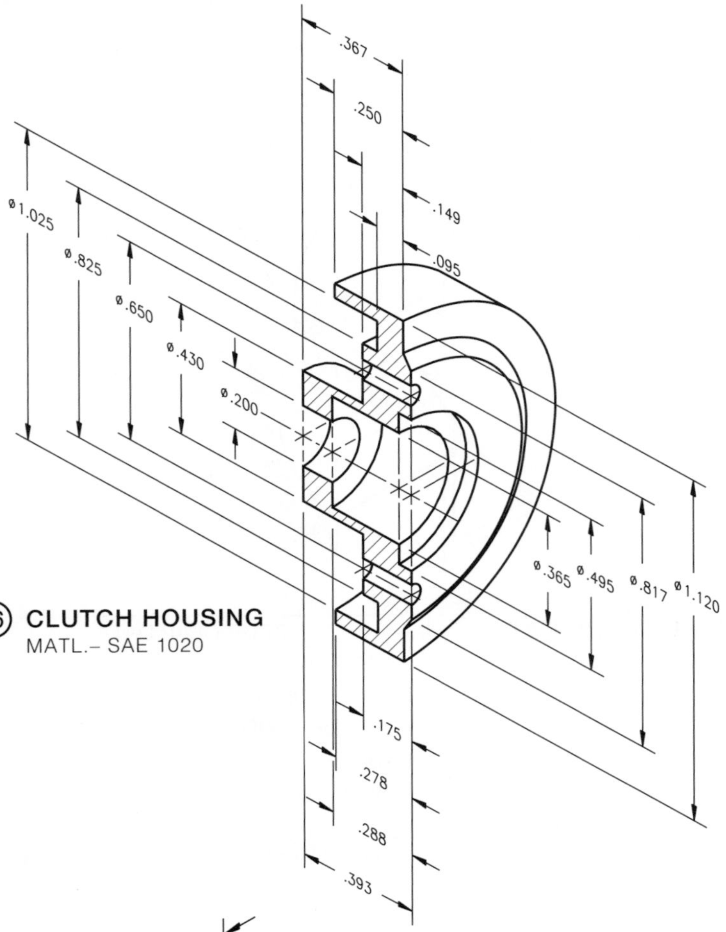

.367
.250
.149
.095

ø1.025
ø.825
ø.650
ø.430
ø.200

ø.365 ø.495 ø.817 ø1.120

(16) **CLUTCH HOUSING**
MATL.– SAE 1020

.175
.278
.288
.393

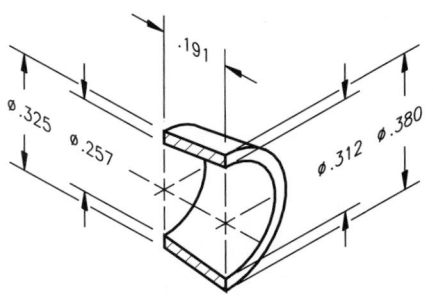

.191
ø.325
ø.257
ø.312 ø.380

(18) **BEVELED SLEEVE**
MATL.– PHOSPHOR–BRONZE

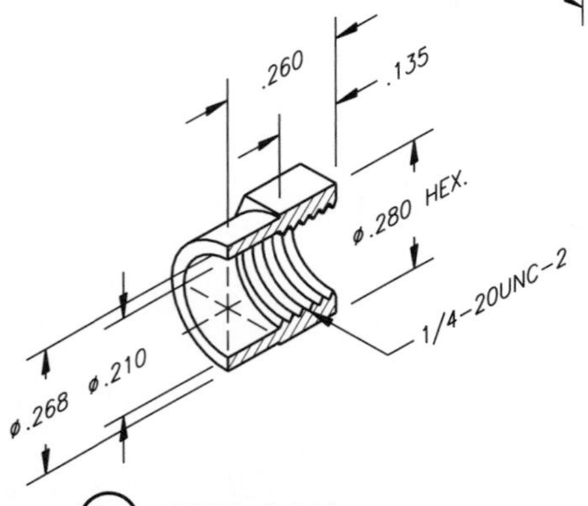

.260
.135
ø.280 HEX.
1/4–20UNC–2

ø.268 ø.210

(20) **HEX CAP**
MATL.– SAE 1020

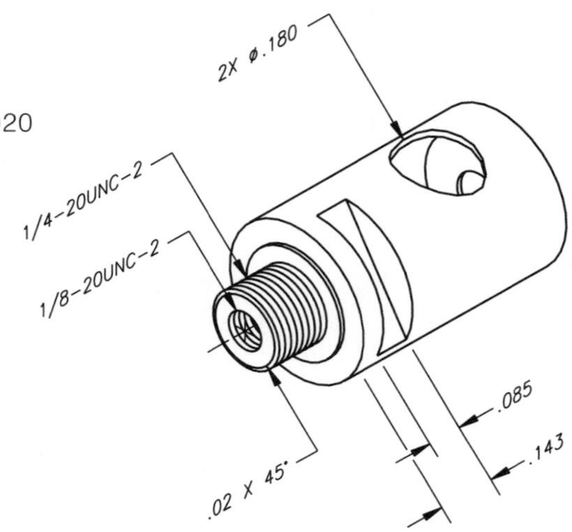

2X ø.180
1/4–20UNC–2
1/8–20UNC–2
.085
.143
.02 X 45°

.739
.602
.592
.212
ø.417
ø.384
.128
ø.275
.060
.018
.084
.592

(23) **MIXTURE CONTROL**
MATL.– SAE 1020

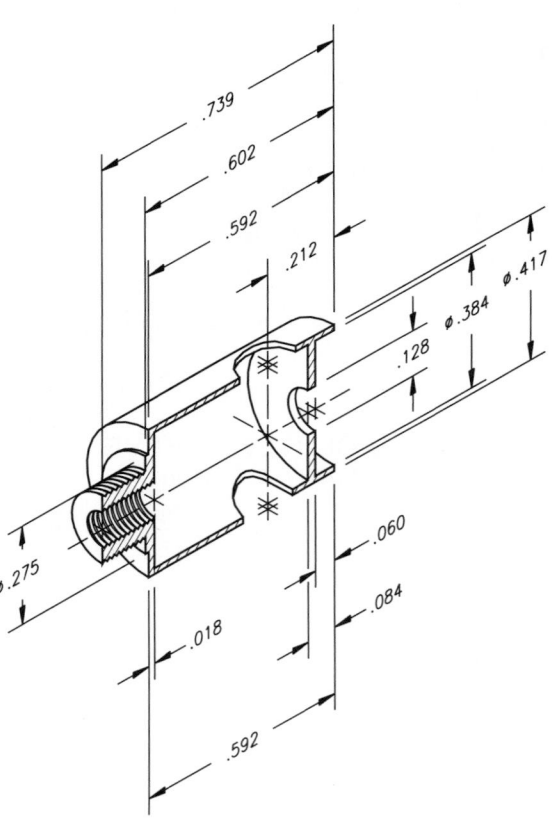

PROBLEM 18-21 (continued)

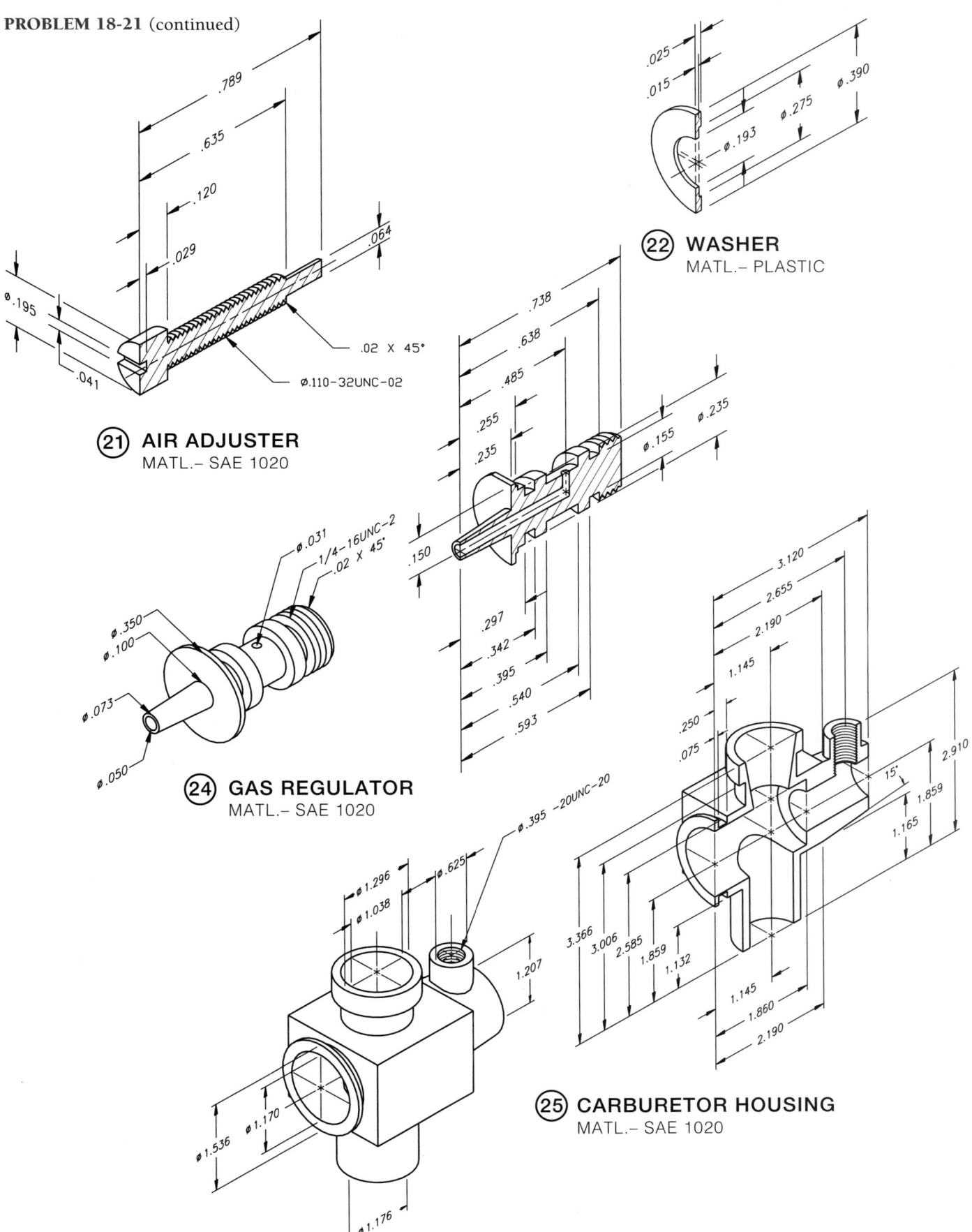

.789
.635
.120
.029
.064
Ø.195
.041
.02 X 45°
Ø.110–32UNC–02

21 **AIR ADJUSTER**
MATL.– SAE 1020

22 **WASHER**
MATL.– PLASTIC

.025
.015
Ø.193
Ø.275
Ø.390

.738
.638
.485
.255
.235
Ø.155
Ø.235
.150
.297
.342
.395
.540
.593

Ø.031
1/4–16UNC–2
.02 X 45°
Ø.350
Ø.100
Ø.073
Ø.050

24 **GAS REGULATOR**
MATL.– SAE 1020

Ø.395 –20UNC–20
Ø1.296
Ø1.038
Ø.625
1.207
Ø1.536
Ø1.170
Ø1.176

3.120
2.655
2.190
1.145
.250
.075
15°
2.910
1.859
1.165
3.366
3.006
2.585
1.859
1.132
1.145
1.860
2.190

25 **CARBURETOR HOUSING**
MATL.– SAE 1020

PROBLEM 18-21 (continued)

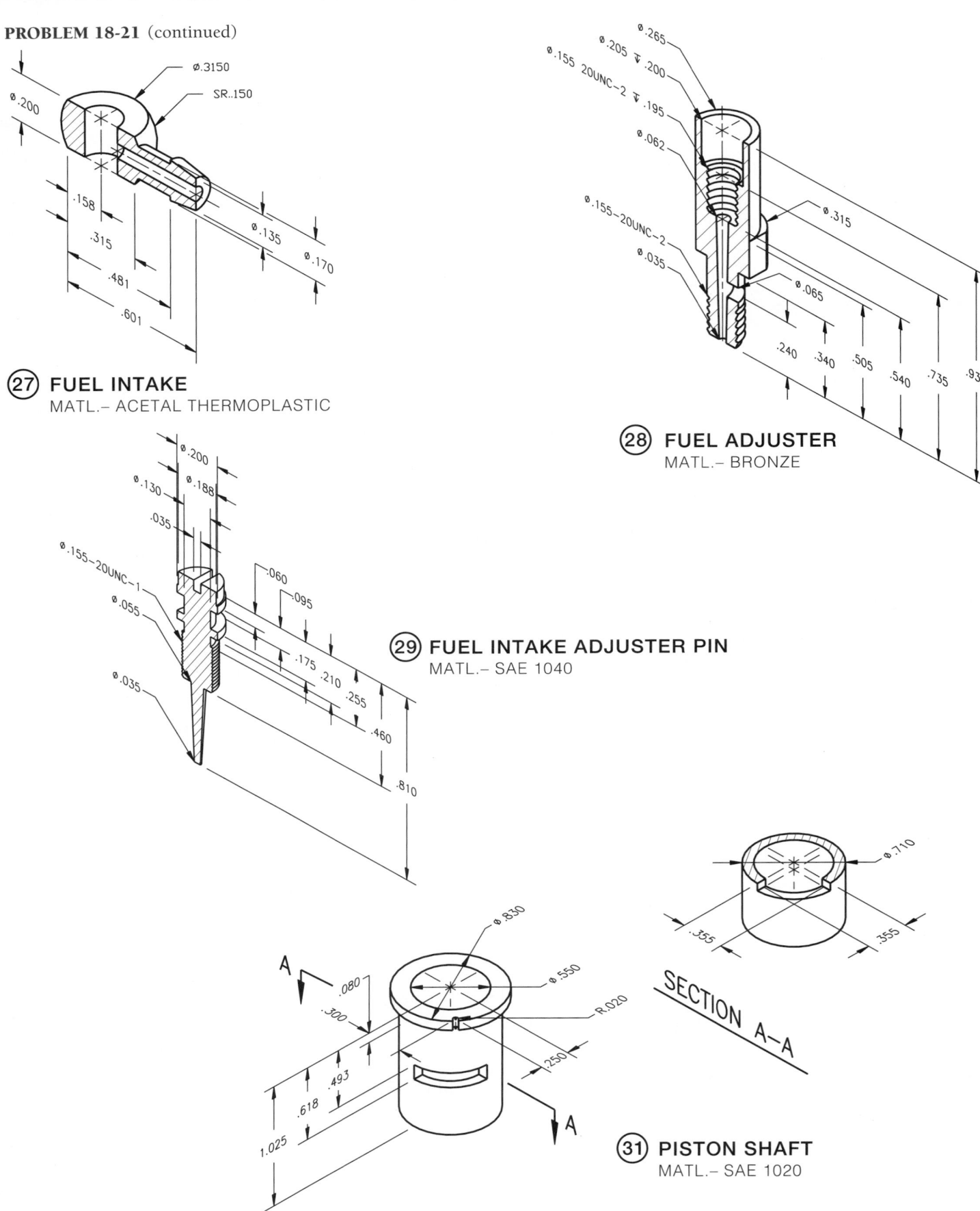

Ø.3150
SR..150
Ø.200
.158
.315
.481
.601
Ø.135
Ø.170

(27) **FUEL INTAKE**
MATL.– ACETAL THERMOPLASTIC

Ø.265
Ø.205 .200
Ø.155 20UNC–2 .195
Ø.062
Ø.155–20UNC–2
Ø.035
Ø.315
Ø.065
.240 .340 .505 .540 .735 .935

(28) **FUEL ADJUSTER**
MATL.– BRONZE

Ø.200
Ø.130 Ø.188
.035
Ø.155–20UNC–1
Ø.055
Ø.035
.060
.095
.175 .210 .255
.460
.810

(29) **FUEL INTAKE ADJUSTER PIN**
MATL.– SAE 1040

Ø.830
Ø.550
R.020
.080
.300
.250
.493
.618
1.025

A
A

Ø.710
.355 .355

SECTION A–A

(31) **PISTON SHAFT**
MATL.– SAE 1020

PROBLEM 18-21 (continued)

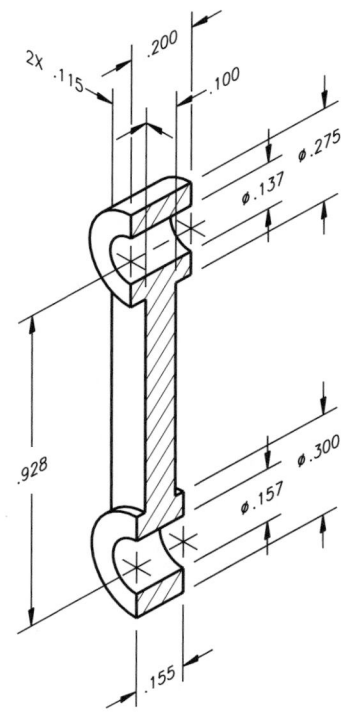

.200
2X .115
.100
ø.137 .275
.928
ø.300
ø.157
.155

③② **CRANK ARM**
MATL.– SAE 1020

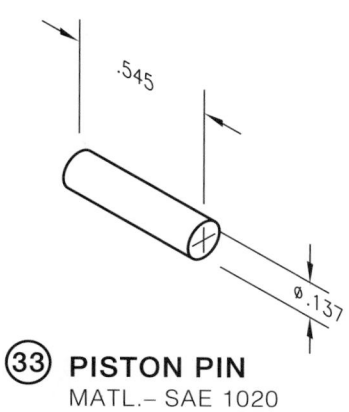

.545
ø.137

㉝ **PISTON PIN**
MATL.– SAE 1020

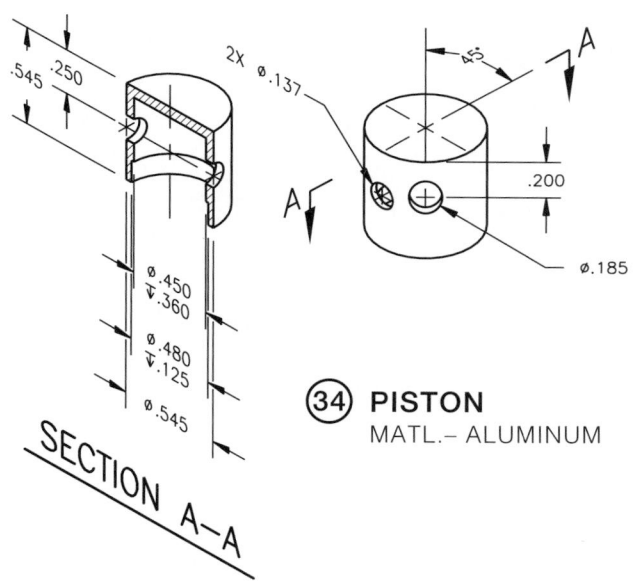

.545 .250
2X ø.137
45°
.200
ø.185
ø.450
.360
ø.480
.125
ø.545

SECTION A–A

㉞ **PISTON**
MATL.– ALUMINUM

PROBLEM 18-21 (continued)

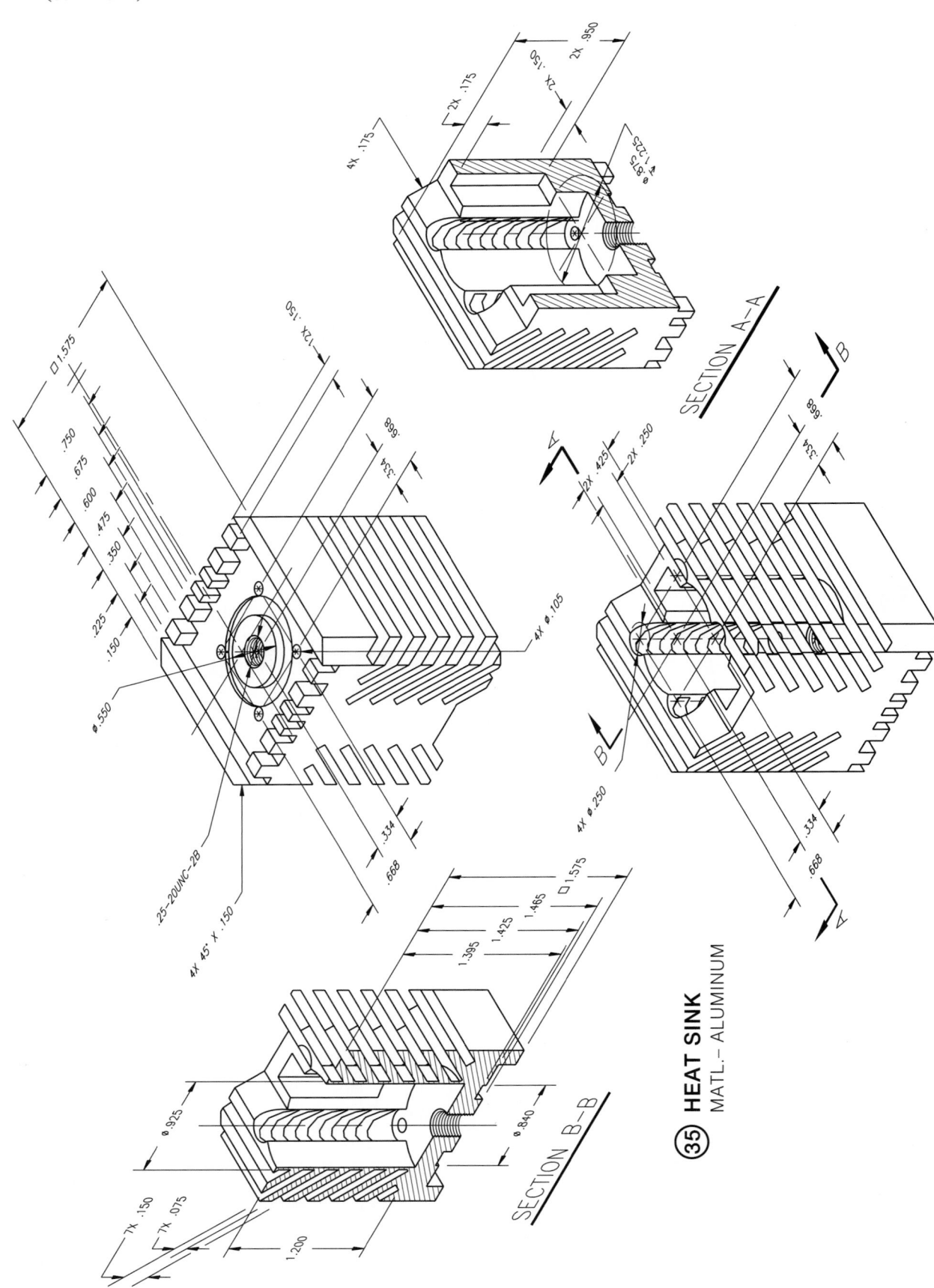

SECTION A–A

SECTION B–B

SECTION B–B

(35) **HEAT SINK**
MATL.– ALUMINUM

PROBLEM 18-22   Engineering Changes

Your instructor will use your complete set of working drawings to prepare a series of engineering changes.

MATH PROBLEMS

PROBLEM 18-23   Find dimension L in the following figure.

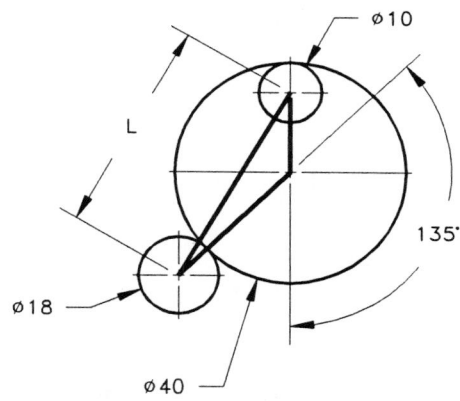

PROBLEM 18-24   Find angle θ in the following drawing.

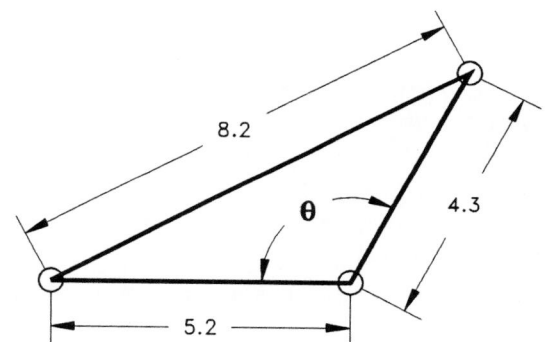

Refer to the following drawing of a general oblique triangle (not drawn to scale).

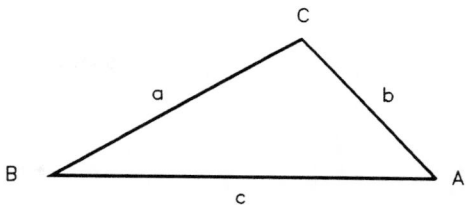

PROBLEM 18-25   Find angle A if a = 15.4, b = 13.1, and c = 12.6.

PROBLEM 18-26   Find side c if C = 61.5°, a = 7.56, and b = 8.94.

PROBLEM 18-27   Find angle B if A = 37.6°, b = 7.63, and c = 8.72. (Hint: Find side a first.)

PROBLEM 18-28   Find angle C if a = 10.1, b = 15.3, and c = 12.7

Pictorial Drawings and Technical Illustrations

OBJECTIVES

After completing this chapter, you will:

- Draw three-dimensional objects using 3-D coordinates.
- Construct objects using isometric, dimetric, or trimetric methods.
- Construct objects using oblique drawing methods.
- Draw objects using one-, two-, or three-point perspective.
- Apply a variety of shading techniques to pictorial drawings.
- Given an orthographic engineering sketch of a part or assembly, draw it in pictorial form using proper line contrasts and shading techniques.

THE ENGINEERING DESIGN APPLICATION

The design and manufacturing department has proposed a new part using 2-D orthographic sketches. The sketches are extremely crude, and it is your task to construct an isometric drawing that can be used for visualization purposes and to construct a prototype. Your first task is to create a readable 2-D sketch from the sketch provided. Check with the designers to verify dimensions and sizes. The engineering sketch shown in Figure 19.1a is used to draw an isometric view of the object.

After the 2-D sketch is completed and checked by the designers, begin construction of the isometric view. Use the following steps to construct the part.

1. Choose the view of the object that best shows most of the features of the part. Orient this view facing to the left or right.
2. Use the centerline layout method to locate the axis lines of the circular features.
3. Lay out additional thicknesses and features using the coordinate, or box-in, method.
4. Use proper ellipses to draw circles and arcs.
5. If inking, use different pen widths for line contrast.
6. Apply shading as required.

The completed object is shown in Figure 19.1b.

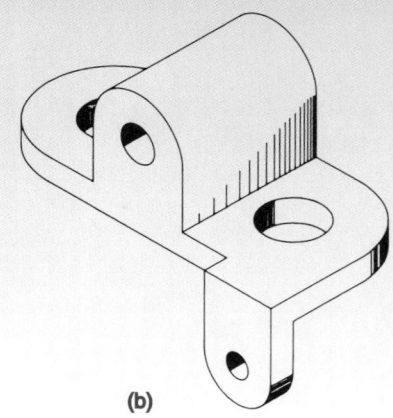

FIGURE 19.1 ■ Using an engineering sketch to create an isometric drawing.

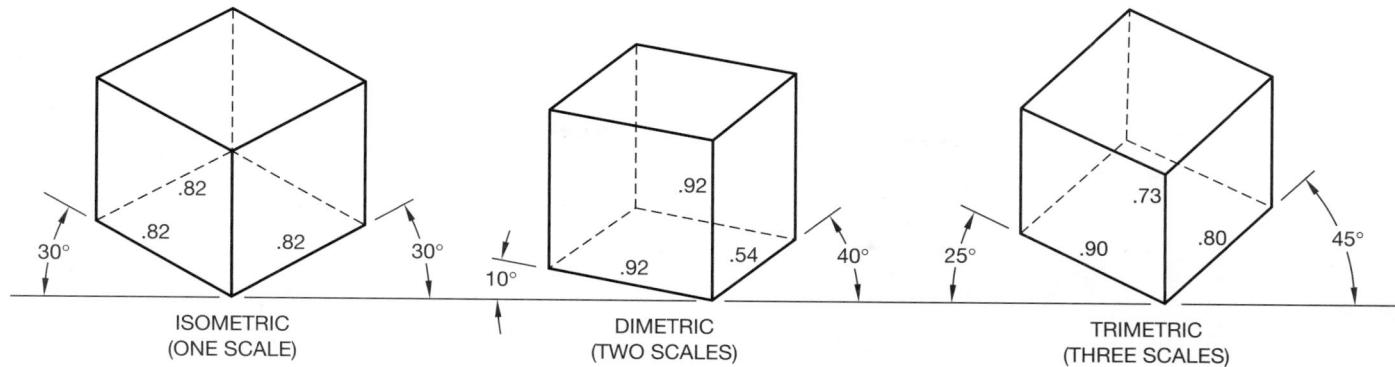

ISOMETRIC
(ONE SCALE)

DIMETRIC
(TWO SCALES)

TRIMETRIC
(THREE SCALES)

FIGURE 19.2 ■ Three types of axonometric projections.

PICTORIAL DRAWINGS

Most products are made from orthographic drawings that allow you to view an object with the line of sight perpendicular to the surface you are looking at. The one major shortcoming of this form of drawing is the lack of depth. Certain situations demand a single view of the object that provides a more realistic representation. This realistic single view is achieved with pictorial drawing.

The most common forms of pictorial drawing used in engineering drawing are *isometric* and *oblique*. These two basic forms of pictorial drawing are easy to master, if you can visualize objects in orthographic projection and three dimensions.

Isometric drawing belongs to a family of pictorial representation known as *axonometric* projection. Two other similar forms of drawing occupy this group. *Dimetric* projection involves the use of two different scales. Isometric uses single scale for all axis, the simplest of the three. *Trimetric* projection is the most involved of the three and uses three different scales for measurement. Figure 19.2 illustrates the differences in scale between isometric and a common dimetric and trimetric drawing.

The terms drawing and projection should be clarified. A *projection* is an exact representation of an object projected onto a plane from a specific position. The observer's line of sight to the various points on the object passes through a projection plane. The representation on a drawing sheet of the points of the object on the projection plane becomes the *drawing*. Exact projections of objects are time-consuming to make and often involve the use of odd angles and scales. Therefore most drafters and illustrators work with axonometric drawing techniques rather than true projection techniques. Creating an axonometric drawing involves the use of approximate scales and angles that are close enough to the projection scales and angles to be acceptable.

The most realistic type of pictorial illustration is perspective drawing. The use of vanishing points in the projection of these drawings gives them the depth and distortion that you see with your eyes. You will examine each of these types of pictorial drawings in this chapter and discuss step-by-step construction methods for them. In addition you will see how pictorial drawings can be drawn with a computer drafting system.

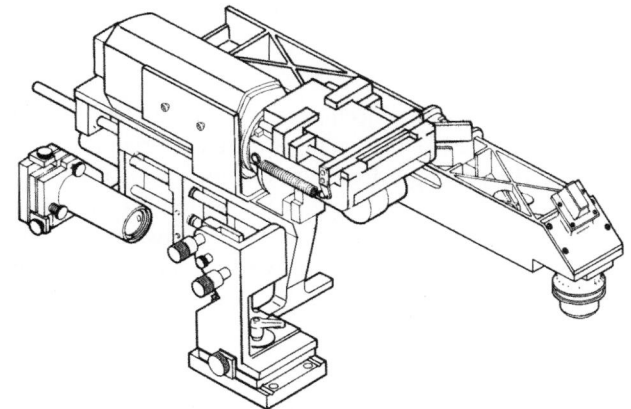

FIGURE 19.3 ■ Pictorial drawing. *Courtesy Industrial Illustrators, Inc.*

Technical Illustration

Pictorial drawing is a term that is often used interchangeably with technical illustration. But pictorial drawing includes only line drawings done in one of several three-dimensional methods, whereas technical illustration involves the use of a variety of artistic and graphic arts skills and a wide range of media in addition to pictorial drawing techniques. Figure 19.3 is an example of a pictorial drawing, and Figure 19.4 shows a technical illustration. Pictorial drawings are most often the basis for technical illustrations.

Uses of Pictorial Drawings

Pictorial drawings are excellent aids in the design process, for they allow designers and engineers to view the objects at various stages in their development. Pictorial drawings are used in instruction manuals, parts catalogs, advertising literature, technical reports and presentations, and as aids in the assembly and construction of products.

ISOMETRIC PROJECTIONS AND DRAWINGS

The word *isometric* means equal (iso) measure (metric). The three principal planes and edges make equal angles with the plane of

FIGURE 19.4 ■ Cutaway technical illustration. *Courtesy Industrial Illustrators, Inc.*

projection. An isometric projection is achieved by first revolving the object, in this case a 1-in. cube, 45° in a multiview drawing, Figure 19.5a, then tilting it forward until the diagonal line AE is perpendicular to the projection plane, as seen in the side view of Figure 19.5b. This creates an angle of 35° 16' between the vertical axis AD and the plane of projection. When viewed in the isometric or front view, this axis appears vertical. The remaining two principal axes, AB and AC, are at 30° to a horizontal line.

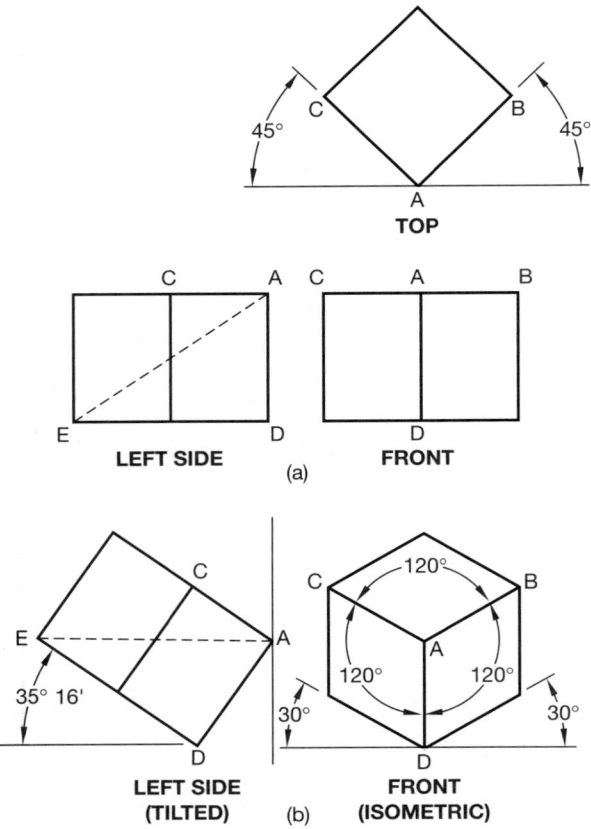

FIGURE 19.5 ■ Construction of an isometric projection.

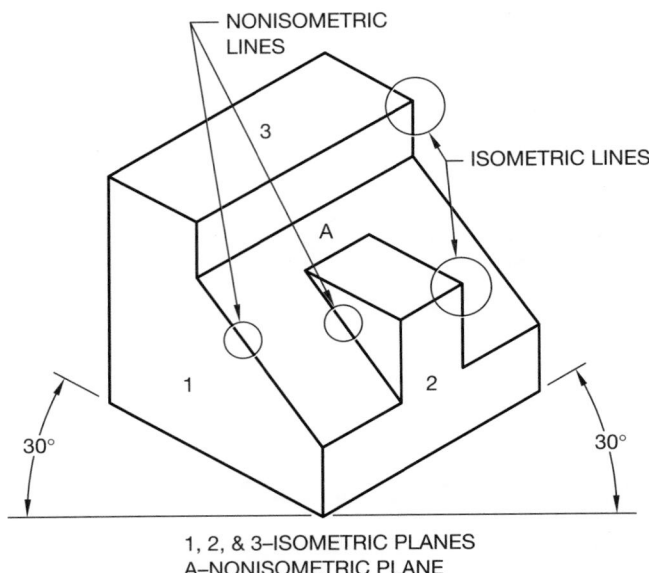

1, 2, & 3–ISOMETRIC PLANES
A–NONISOMETRIC PLANE

FIGURE 19.6 ■ Isometric and nonisometric planes.

The three principal axes are called isometric lines, and any line parallel to them is also an isometric line. These lines can all be measured. Any lines not parallel to these three axes are nonisometric lines and cannot be measured. The angles between each of these three isometric axes are 120°. The three planes between the isometric axes, and any plane parallel to them, are called isometric planes. (See Figure 19.6.)

Isometric Scale

The *isometric projection* is a true representation of an object rotated and tilted in the manner just described. An isometric projection must be drawn using an isometric scale. An isometric scale is created by first laying a regular scale at 45° and projecting the increments of that scale vertically down to a blank scale drawn at an angle of 30°. The resulting isometric scale is seen in Figure 19.7. A 1-in. measurement on the regular scale now measures .816 in. on the isometric scale.

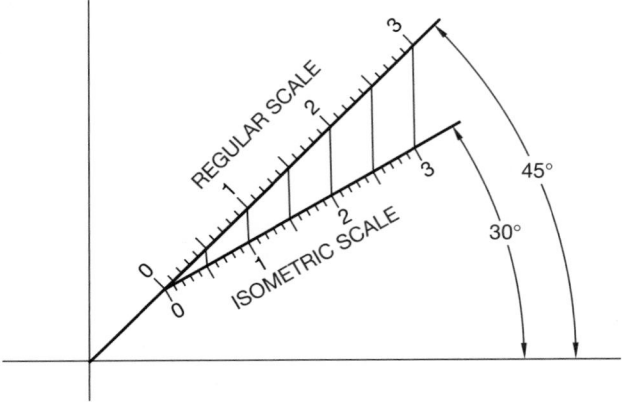

FIGURE 19.7 ■ Projection of regular scale to isometric scale.

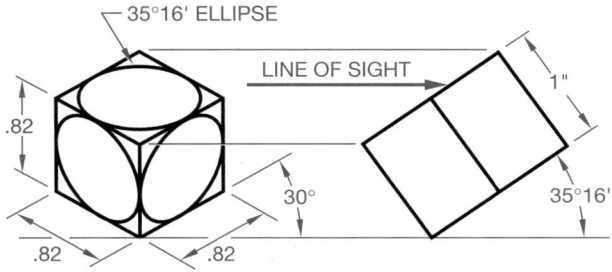

ISOMETRIC DRAWING
BASED ON TRUE MEASUREMENT IN ISOMETRIC VIEW
(PREFERRED METHOD)

ISOMETRIC PROJECTION
BASED ON TRUE MEASUREMENT IN ORTHOGRAPHIC
(SELDOM USED)

FIGURE 19.8 ■ The differences between isometric drawing and isometric projection.

An *isometric drawing* is done using a regular scale. This is most common in industry because it does not involve the creation of special scales. The only difference between an isometric drawing and an isometric projection is the size. The drawing appears slightly larger than the projection. Figure 19.8 illustrates the differences between the drawing and the projection.

TYPES OF ISOMETRIC DRAWINGS

Isometric drawing is a form of pictorial drawing in which the receding axes are drawn at 30° from the horizontal, as shown in Figure 19.5. There are three basic forms of isometric drawing: these are known as regular, reverse, and long-axis isometric.

Regular Isometric

The top of an object can be seen in the regular isometric form of drawing. An example is shown in Figure 19.9a. This is the most common form of isometric drawing, and when using it, the illustrator can choose to view the object from either side.

Reverse Isometric

The only difference between reverse and regular isometric is that you can view the bottom of the part instead of the top. The 30° axis lines are drawn downward from the horizontal line instead of upward. Figure 19.9b shows an example of reverse isometric.

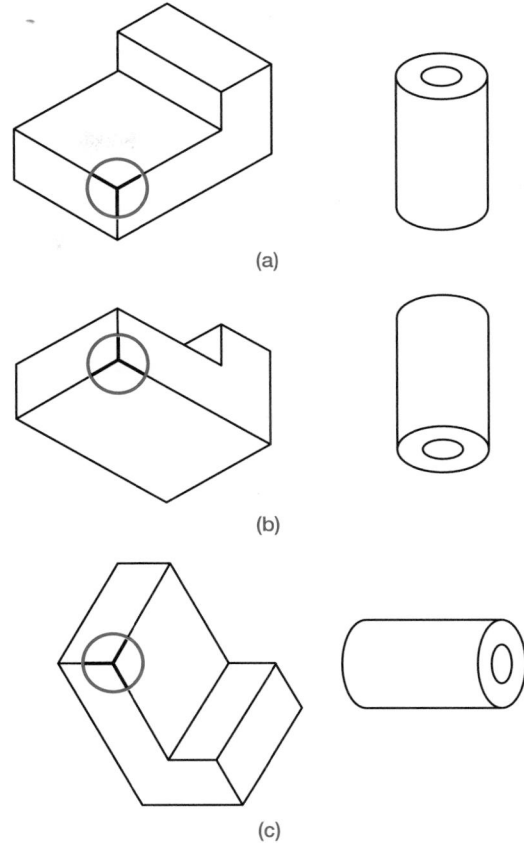

(a)

(b)

(c)

FIGURE 19.9 ■ Isometric axis variations: (a) regular; (b) reverse; (c) long-axis.

Long-axis Isometric

The long-axis isometric drawing is normally used for objects that are long, such as shafts. Figure 19.9c shows an example of the long-axis form.

The designer or illustrator should choose the view that will give the most realistic presentation of the object. For example, if the object is normally seen from below, then the reverse isometric would be the proper form to use.

ISOMETRIC CONSTRUCTION TECHNIQUES

Just as objects differ in their geometric makeup, so do the construction methods used to draw the object. Different techniques exist to assist the drafter in constructing the various shapes.

Box, or Coordinate, Method

The most common form of isometric construction is the box, or coordinate, method and is used on objects that have angular or radial features. The orthographic views of the object to be drawn are shown in Figure 19.10a.

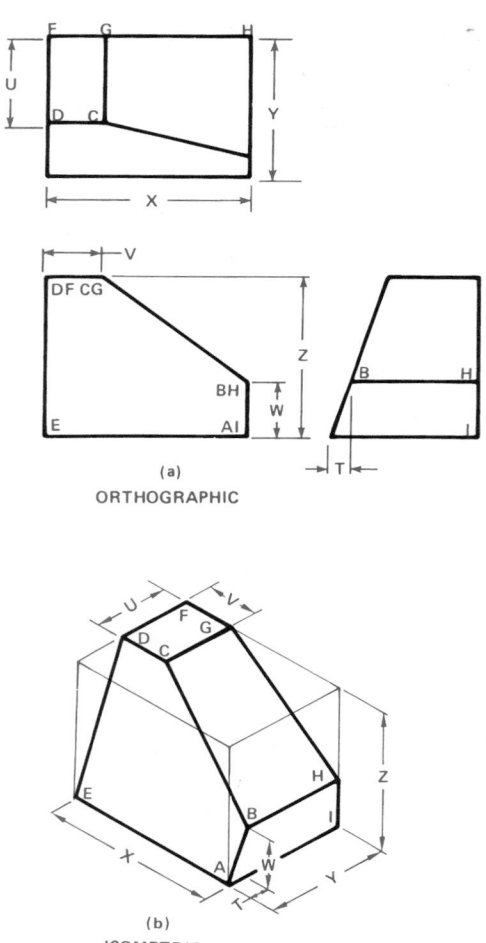

(a)
ORTHOGRAPHIC

(b)
ISOMETRIC

FIGURE 19.10 ■ Box method of isometric construction: (a) orthographic view: (b) isometric view.

First an isometric box the size of the overall dimensions of the
object (X, Y, and Z) must be drawn. (See Figure 19.10b.) Then the
measurements of the features of the object can be transferred to
the isometric box. To locate points D and C, just measure dimension U from points F and G. Point E is located at the lower left
corner of the box. Draw a line from E to D. Next locate point H
by measuring up distance W from the bottom right corner of the
box. Draw a light construction line at a 30° angle toward the front
vertical axis. Now draw a line from A parallel to line ED until it
intersects the line from H. This intersection will be point B. The
location of point B can also be found by measuring the horizontal dimension T from point A, then the vertical dimension W.

It may be necessary to draw construction lines on the orthographic views and transfer measurements directly from these views
to the isometric view. This method may work well with irregularly
shaped objects such as the one shown in Figure 19.11.

Centerline Layout Method

The centerline method begins with the skeleton of the object, the
centerlines. This technique is used on objects with many circles and
arcs. The use of this method is seen in Figure 19.12a through d. The
center points of all of the circles and arcs should be located first.

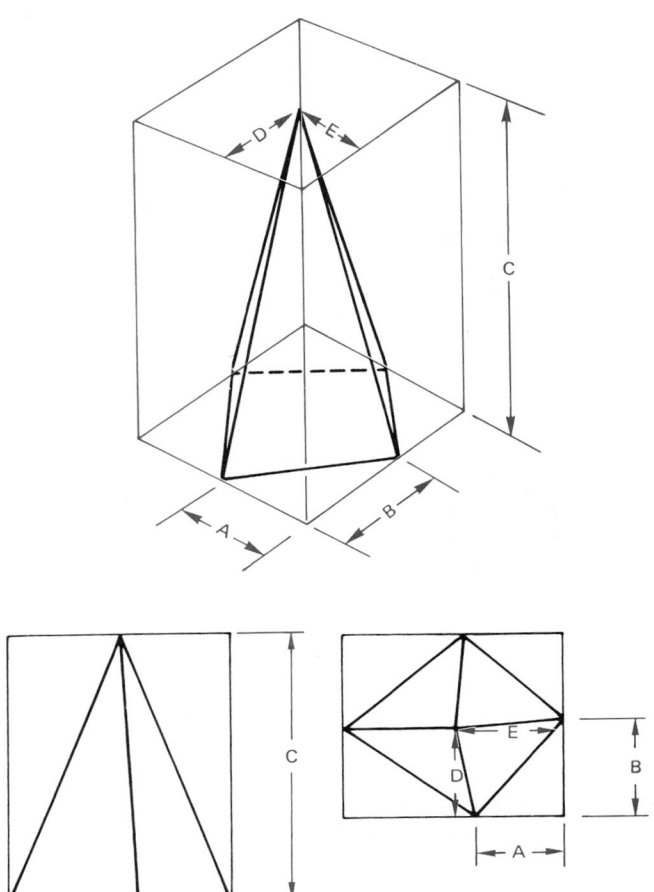

FIGURE 19.11 ■ Isometric box method for an irregular object.

Begin with a good reference point from which you can work, such
as the bottom of the object in Figure 19.12a. Center points A and B
are first established, then vertical axis lines are drawn up from these.
Now the vertical locations of center points C, D, and E can be measured as seen in Figure 19.12b. Then determine the various sizes of
the ellipses and draw them at their proper locations as shown in Figure 19.12c. The finished drawing is shown in Figure 19.12d.

Circle and Arc Construction

After the center points of circles and arcs are located, you need to
select the correct size of circle from an isometric ellipse template.
The isometric ellipse template is only useful for drawing ellipses on
isometric surfaces. Angled ellipse templates are necessary for new
isometric surfaces. Each ellipse on the template has several tick
marks around it. A description of these tick marks is shown in Figure 19.13. An isometric ellipse is measured on the marks at 30°
angles from the large (horizontal) diameter of the ellipse. When
aligning an isometric ellipse on the centerlines of your layout,
always align the tick marks on the *minor* diameter of the ellipse
with the *axis* of the feature, as shown in Figure 19.14. When you
do this, two other sets of tick marks on the template will align with
the ellipse centerlines.

Arc locations are found in the same way as circles, and the use
of the isometric ellipse template is also the same, except just a

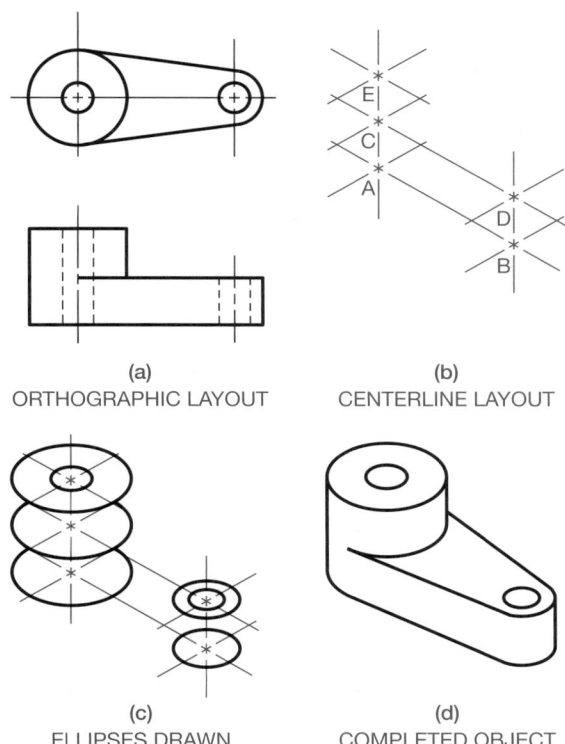

(a)
ORTHOGRAPHIC LAYOUT

(b)
CENTERLINE LAYOUT

(c)
ELLIPSES DRAWN

(d)
COMPLETED OBJECT

FIGURE 19.12 ■ Isometric centerline layout method: (a) orthographic layout; (b) centerline layout; (c) ellipses drawn; (d) completed object.

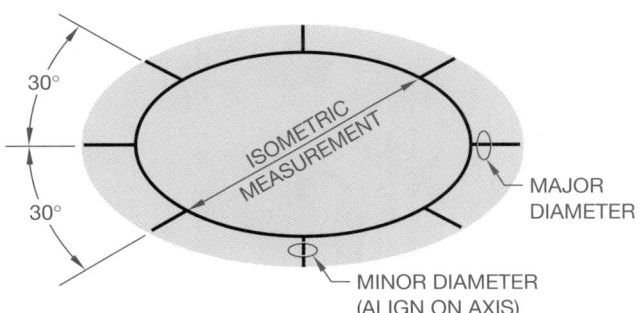

FIGURE 19.13 ■ Meaning of isometric ellipse template markings.

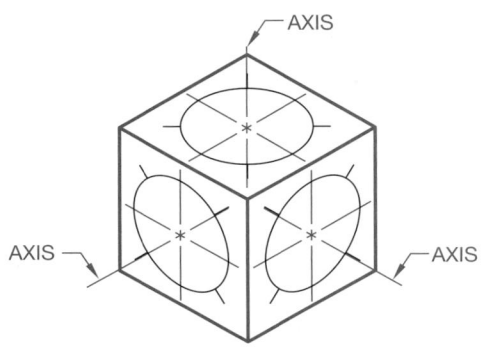

FIGURE 19.14 ■ Align minor diameter of isometric ellipse template on axis of hole.

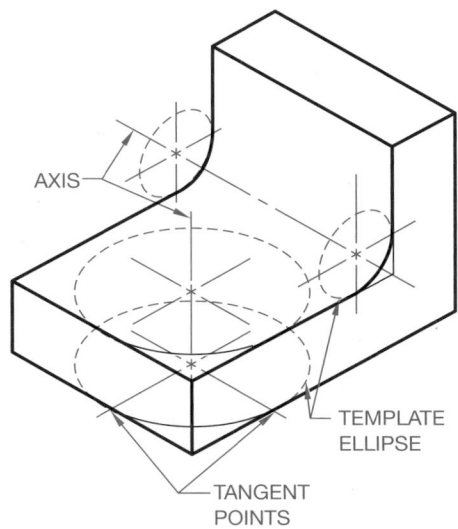

FIGURE 19.15 ■ Isometric arc layout using ellipse template.

portion of the circle is drawn. First, find the center point of the arc, then the tangent points. (See Figure 19.15.) Draw the axis line of the arc lightly and again align the minor diameter tick marks with the axis line. Now draw only the portion of the ellipse that is needed for the arc.

Constructing isometric arcs is easy with CADD software such as AutoCAD. Simply use the isocircle option of the ELLIPSE command. The isometric crosshairs can be toggled while the ellipse is attached to the crosshairs, and you can see the ellipse in the left, top, and right-side orientations.

Fillets and Rounds

The key to drawing isometric circles and arcs is getting the ellipse aligned properly. Study the object in Figure 19.16. Determine the axis of the circular feature, then align the minor diameter tick marks of the template with the axis of the fillet or round.

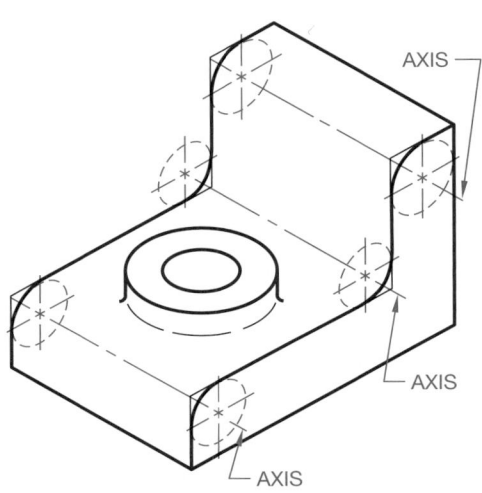

FIGURE 19.16 ■ Isometric fillet and round layout.

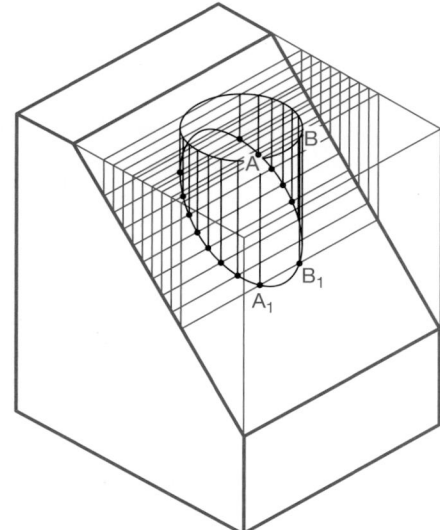

FIGURE 19.17 ■ Construction of ellipse on isometric oblique plane.

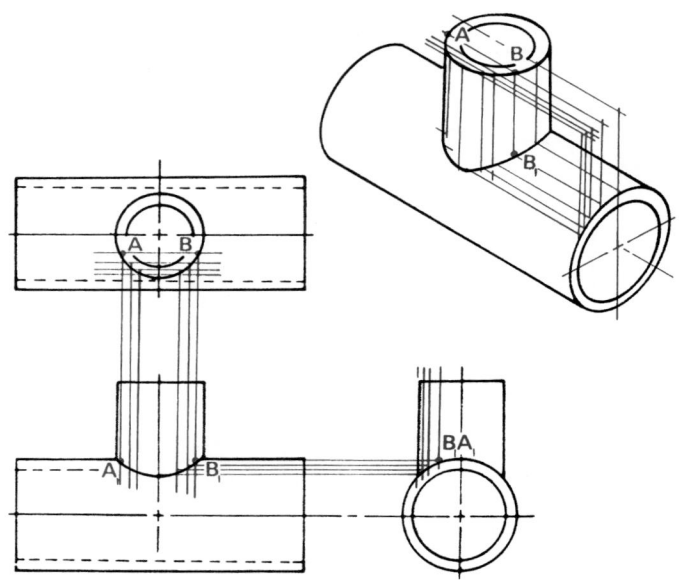

FIGURE 19.18 ■ Isometric construction of two intersecting cylinders.

Intersections

A common type of intersection is that of a cylindrical hole passing through an oblique plane. First draw an isometric ellipse at the top of the boxed surface of the object. (See Figure 19.17.) Then draw a series of construction lines parallel to one side of the box that pass through the ellipse. Project each of these lines down the sides of the box and across the oblique surface. Each line forms a trapezoid or parallelogram depending on the shape of the object. The points at which each line intersects the ellipse (A and B, for example) should be projected straight down to the opposite side of the parallelogram. These new points, A_1 and B_1, are now points on the perimeter of the hole in the oblique surface.

Another common form of intersection occurs when two cylinders meet. This construction is similar to the one just described. In Figure 19.18, the center point location of the branch cylinder is determined, and the ellipse is drawn. Next draw as many construction lines passing through the ellipse as needed to produce a smooth curve on the intersection. Project these lines to the end of the main cylinder, down to the edge, then back along the cylinder. Now project points on the ellipse down to the corresponding construction line on the cylinder, A to A_1 and B to B_1, for example. Then connect the points using an irregular curve or polyline in AutoCAD to produce the intersection.

Sections

Full and half sections are common in technical illustration and should be drawn along the isometric axes. The section lines in full sections should all be drawn in the same direction, while those on a half section should appear in opposite directions. See the section lines illustrated in Figure 19.19a and b. The section lines in offset

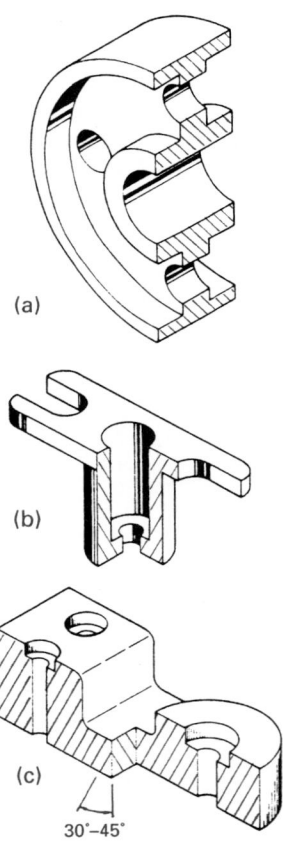

FIGURE 19.19 ■ Isometric sections: (a) full; (b) half; (c) offset.

sections should change directions with each jog in the part, as seen in Figure 19.19c.

There is no preferred way to draw an isometric section. One technique that you might try is to imagine that the cutting-plane

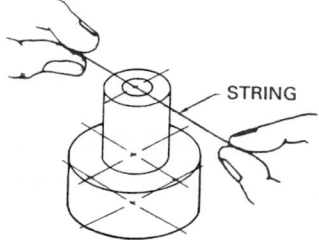

(a) STRING HELD OVER OBJECT ALONG CENTERLINE

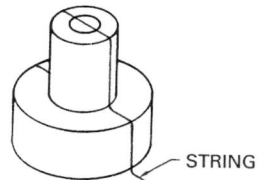

(b) STRING FALLS ALONG CENTERLINES

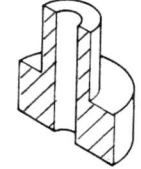

(c) COMPLETED FULL SECTION

FIGURE 19.20 ■ Visualize a string dropped along the cutting plane to create an isometric full section.

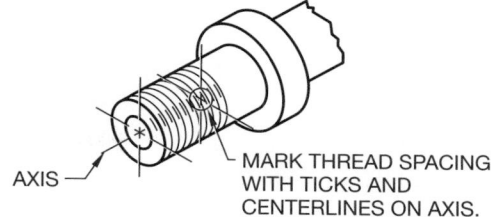

FIGURE 19.21 ■ Isometric thread representation.

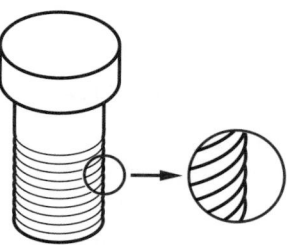

FIGURE 19.22 ■ Detailed isometric thread representation.

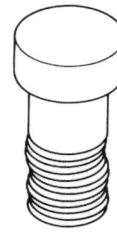

FIGURE 19.23 ■ Off-center ellipses produce "wandering" threads.

line is a long string, and you are stretching it out just above the axis along which you plan to cut. If you let the string fall, it would come to rest along the axis where the cut is to be made, as illustrated in Figure 19.20.

Threads

ASME/ANSI Screw threads can be drawn in isometric by first measuring equal spaces along the shaft or hole to be threaded. Then, using the same size ellipse as the diameter of the shaft or hole, draw a series of parallel ellipses. (See Figure 19.21.) These ellipses represent the crests of the threads. This method achieves a simple isometric representation of threads.

A more detailed thread appearance can be achieved by giving each ellipse rounded ends instead of butting the ellipse into the straight sides of the shaft. (See Figure 19.22.) When using this technique, begin drawing the threads from the head of the shaft. With manual drawing, be sure to lay out guidelines for the shaft diameter so the ends of the ellipse do not fall short of the edge of the shaft or go beyond it, as seen in Figure 19.23. This can create a phenomenon known as wandering threads.

Drawing threads, like those shown in Figure 19.22, is quick using software such as AutoCAD. Draw the uppermost thread using an arc or ellipse and edit as needed by trimming or break-

ing the ellipse so it fits properly. Then use a command like ARRAY, and create a rectangular array of the required number of threads.

Spheres

A sphere drawn isometrically is nothing more than a true circle. Figure 19.24 illustrates the construction of a sphere using three isometric ellipses drawn to represent the perpendicular axes of the circle. If you need to draw a sphere, remember to

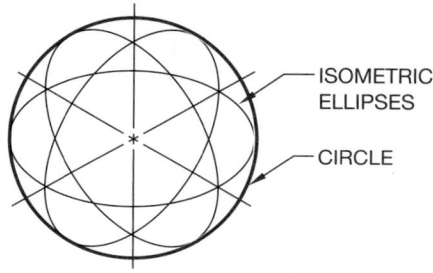

FIGURE 19.24 ■ Isometric sphere construction.

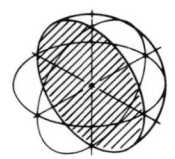
(a) HALF SPHERE

(b) THREE-QUARTER SPHERE

x = FLAT SIDE DIAMETER

(c) SPHERE WITH FLAT SIDE

FIGURE 19.25 ■ Portions of isometric spheres: (a) half sphere; (b) three-quarter sphere; (c) sphere with flat side.

choose a circle that is 1¼ times larger than the actual sphere, because isometric drawings are 1¼ times larger than the actual representation.

Should you need to draw spheres that have been cut in some manner, refer to Figure 19.25. This figure illustrates a half sphere, three-quarter sphere, and a sphere with a flat side.

Isometric Dimensioning

It is not common for isometric drawings to be dimensioned; however, some isometric piping drawings rely heavily on dimensioning to get their message across. One technique that can be used when dimensioning is to letter vertical strokes parallel to extension lines, as shown in Figure 19.26. This gives the appearance that the dimension is lying in the plane of the extension lines. A second simple technique is to draw the heel of the arrowhead parallel with the extension line. This further emphasizes the plane that the

dimension is lying in. (See Figure 19.26a.) Examples of unidirectional, aligned, and one-plane (horizontal) isometric dimensioning are shown in Figure 19.26b.

DIMETRIC PICTORIAL REPRESENTATION

Dimetric projection is similar to isometric projection, but instead of all three axes forming equal angles with the plane of projection, only two axes form equal angles. These can be greater than 90° and less than 180°, but cannot have an angle of 120° because that is the isometric angle. The third axis may have an angle less or greater than the two equal axes, depending on the angles chosen. The two equal angles are said to be foreshortened because they are not measured at full scale. They are foreshortened equally. Since the third axis is projected at a different angle, it is foreshortened at a different scale. Full-size and foreshortened approximate scales are used on the dimetric axes. Some common approximate dimetric angles and scales are shown in Figure 19.27.

TRIMETRIC PICTORIAL REPRESENTATION

The term trimetric refers to three measurements. It is a type of pictorial drawing in which none of the three principal axes makes an equal angle with the plane of projection. Since all three angles are unequal, the scales used to measure on the three axes are also unequal. Trimetric projection provides an infinite number of projections.

Trimetric projection is similar to dimetric projection. Figure 19.28 shows some common trimetric angles for the width and depth axes, plus the scales to use on each of the three axes. The size of angle ellipse to use on each principal plane is also indicated.

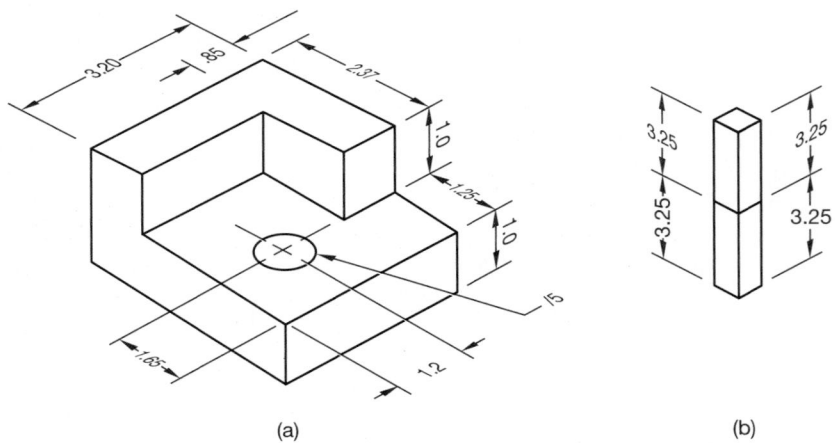

(a)   (b)

FIGURE 19.26 ■ Isometric dimensioning: (a) dimensions parallel to extension lines; (b) various styles of dimensioning.

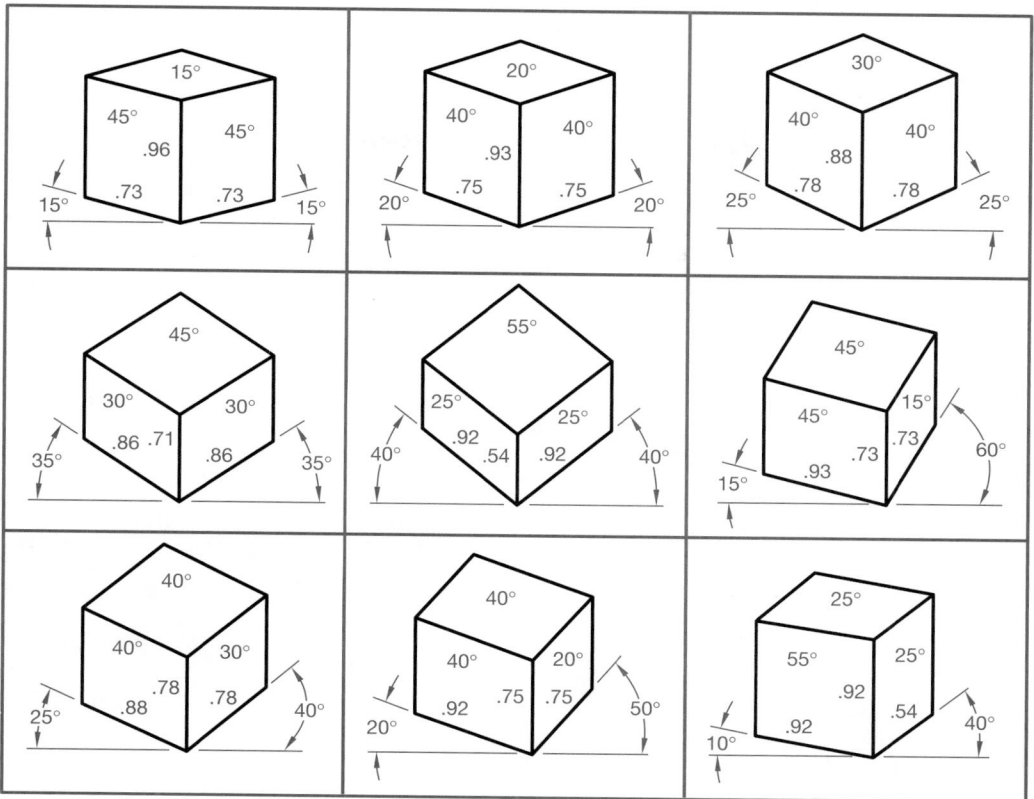

FIGURE 19.27 ▪ Common approximate dimetric angles, scales, and ellipse angles.

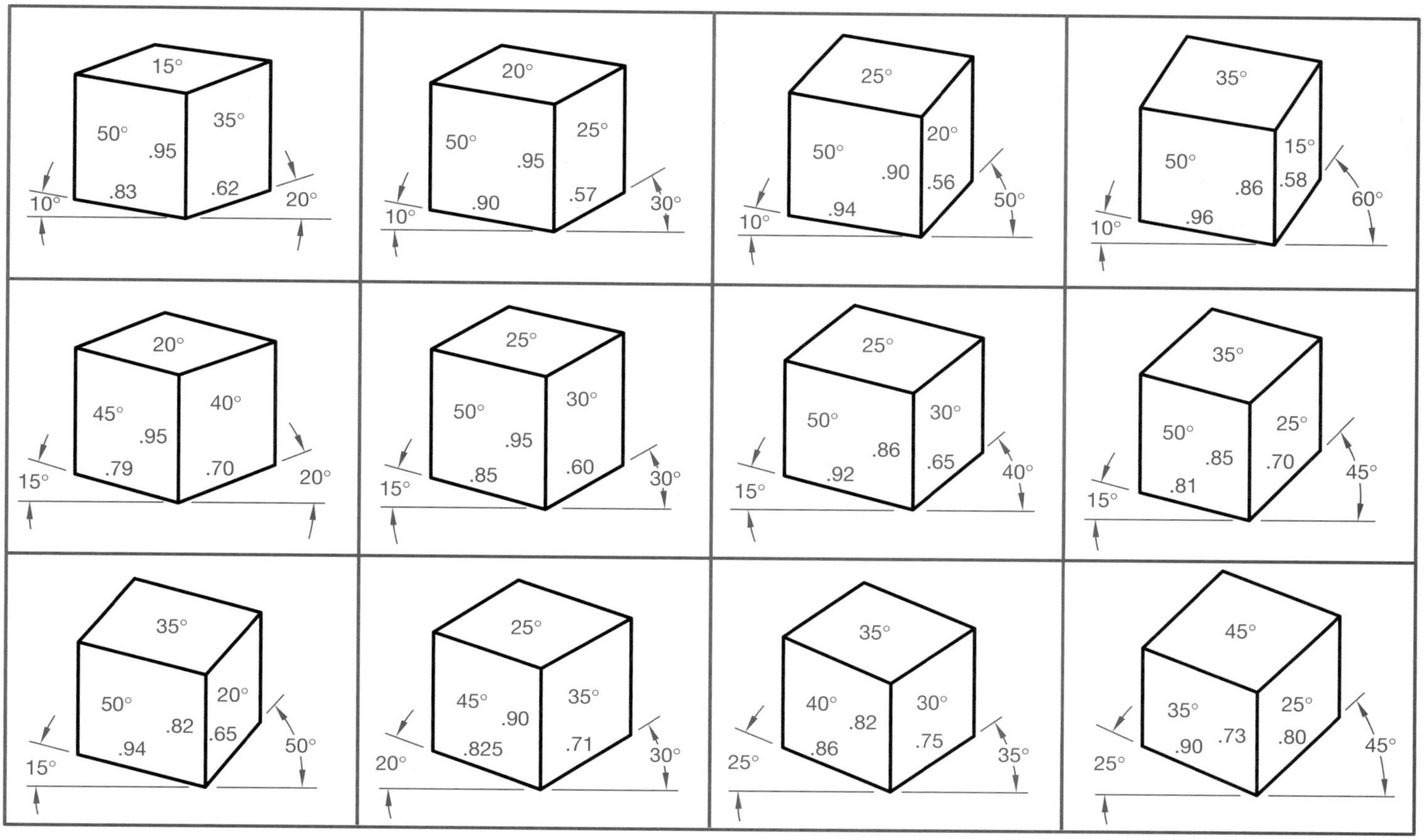

FIGURE 19.28 ▪ Common approximate trimetric angles, scales, and ellipse angles.

3-D CAPABILITIES

Computer-generated pictorial line drawings have given drafters, designers, and engineers the ability to draw an object once and then create as many different displays of that object as they can think of. Three-dimensional wire-form capabilities, seen in Figure 19.29, are available with most commercial CADD systems. They allow the operator to view the object as if it were constructed of wire. All edges can be seen at the same time. This is somewhat limiting because complex parts soon become a maze of lines. The solid form, shown in Figure 19.30, is a step closer to reality because it shows only the surfaces that would actually be seen.

The greatest realism in pictorial presentation can be achieved by CADD systems that possess color solids-modeling capabilities. These systems allow the operator to draw the object, shade or color its different parts or surfaces, and rotate it in any direction desired. The operator can also cut into the object at any location to view internal features and then rotate it to achieve the best view. This screen display can then be sent to a color hardcopy unit, a pen plotter, an electrostatic plotter, a 35-mm slide production unit, or videotape.

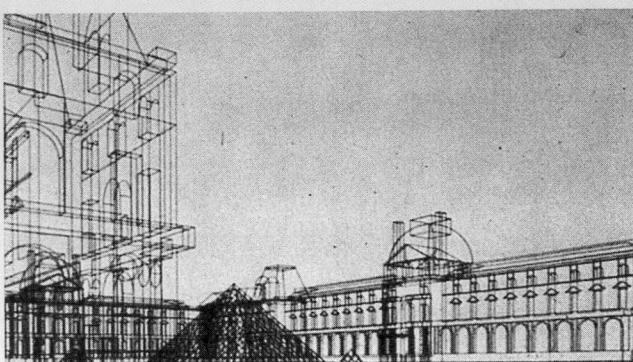

FIGURE 19.29 ■ The Louvre art museum in Paris, France, drawn as a wire form. *Courtesy Computervision Corporation.*

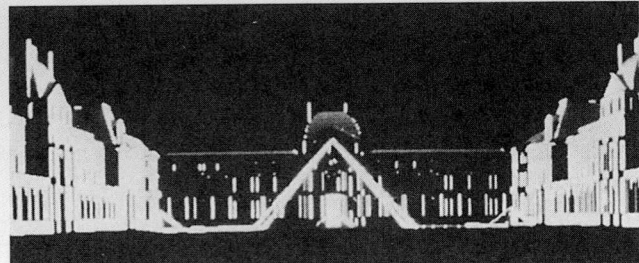

FIGURE 19.30 ■ Another view of the Louvre, drawn as a solid model. *Courtesy Computervision Corporation.*

EXPLODED PICTORIAL DRAWING

Complicated parts and mechanisms are often illustrated as an exploded assembly in order to show the relationship of the parts in the most realistic manner. (See Figure 19.31a.)

An exploded assembly is a collection of parts, each drawn in the same axonometric method. Any of the pictorial drawing methods mentioned in this chapter can be used to create exploded assemblies. The most important aspect of this type of drawing is to select the viewing direction that illustrates as much of the assembly as possible.

Centerlines are used in exploded views to represent lines of explosion. These lines aid the eye in following a part to its position in the assembly. Centerlines should avoid crossing other centerlines and parts. Centerlines should always be drawn parallel to the axis lines of the drawing regardless of the drawing method selected. The neatest presentation is achieved by leaving gaps where the centerline intersects the part to which it applies and the feature it is mating with. This is shown in Figure 19.31b.

When drawing an exploded assembly with CADD, it is not necessary to draw all the parts in their final positions. Draw each part as needed, placing construction lines for the centerlines, or "lines of explosion." When all parts are completed, move them around as needed to construct the final layout.

OBLIQUE DRAWING

Oblique drawing is a form of pictorial drawing in which the plane of projection is parallel to the front surface of the object. The lines of sight are at an angle to the plane of projection and are parallel to each other. This allows the viewer to see three faces of the object. The front face, and any surface parallel to it, is shown in true shape and size, while the other two faces are distorted in relation to the angle and scale used. Oblique drawing is useful if one face of an object needs to be shown without distortion. Choose the orthographic face with the most curves/circles because these are more difficult to draw on other surfaces due to distortion.

There are three methods of oblique drawing: cavalier, cabinet, and general.

Cavalier Oblique

The *cavalier* projection is one in which the receding lines are drawn true size, or full scale. This form of oblique drawing is usually drawn at an angle to a horizontal of 45°, which approximates a viewing angle of 45°. Because of the scale used on the receding axis, it is not a good idea to draw long objects with the long axis perpendicular to the front face, or projection plane. Objects that have a depth that is smaller than the width can be drawn in the cavalier form without too much distortion. Figure 19.32 shows an object drawn in cavalier oblique.

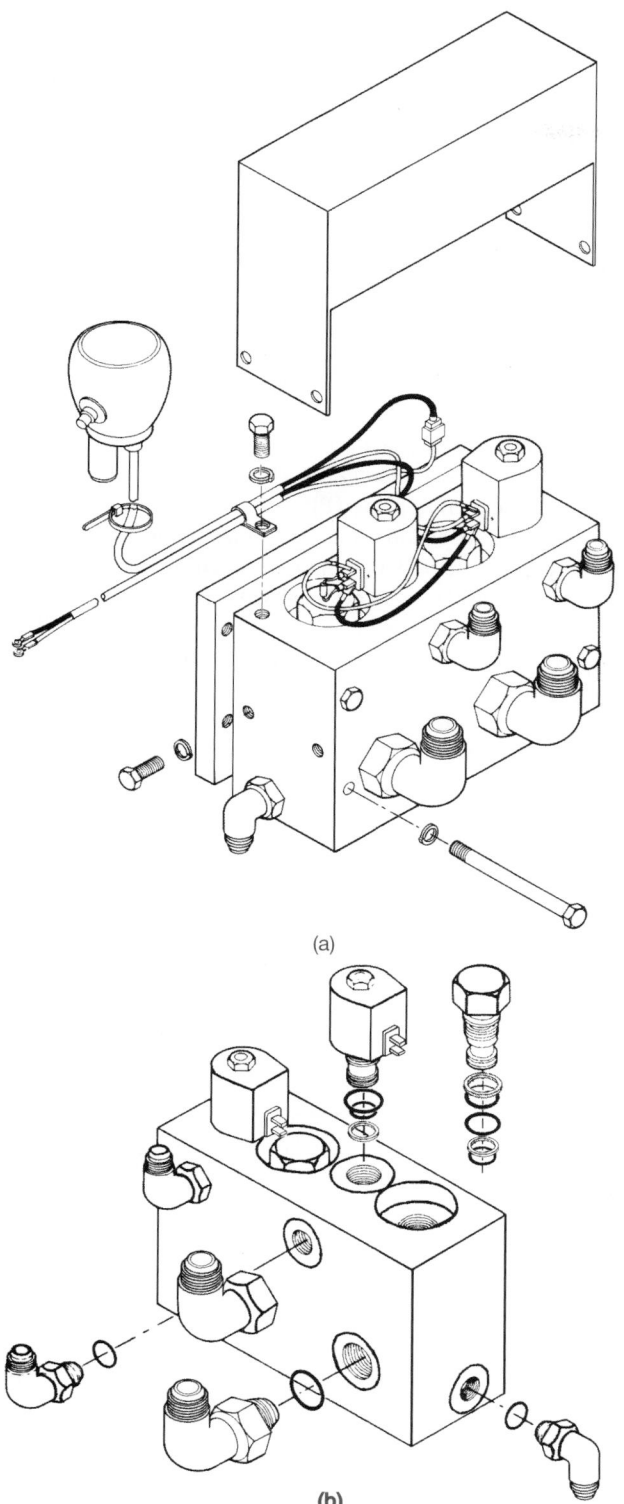

(a)

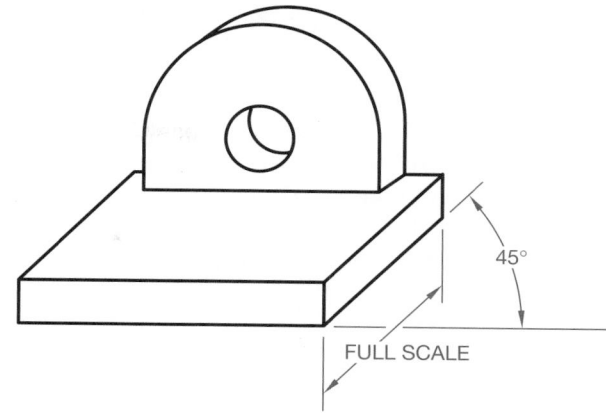

FIGURE 19.32 ■ Cavalier oblique.

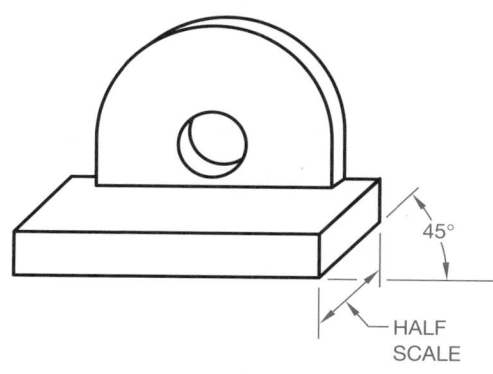

FIGURE 19.33 ■ Cabinet oblique.

(b)

FIGURE 19.31 ■ Exploded assemblies. *Courtesy Industrial Illustrators, Inc.*

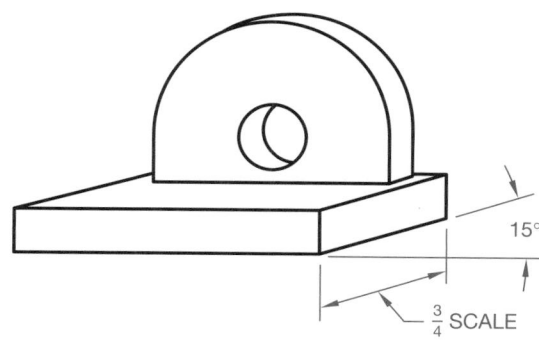

FIGURE 19.34 ■ General oblique.

the appearance of too much distortion. Cabinet makers often manually drew their cabinet designs using this type of oblique drawing. A cabinet drawing is shown in Figure 19.33.

General Oblique

The *general* oblique drawing is normally drawn at an angle other than 45°, and the scale on the receding axis is also different from those used in cavalier and cabinet. The most common angles for a general oblique drawing are 30°, 45°, and 60°, but any angle can be used. Any scale from half to full size can be used. A general oblique drawing is shown in Figure 19.34.

Cabinet Oblique

A *cabinet* oblique drawing is also drawn with a receding angle of 45°, but the scale along the receding axis is half size. Objects having a greater depth than width can be drawn in this form without

PERSPECTIVE DRAWING

Perspective drawing is the most realistic form of pictorial illustration. It reflects the phenomenon of objects appearing smaller the farther away they are until they vanish at a point on the horizon. Lines that are drawn parallel to the principal orthographic planes appear to converge on a vanishing point in perspective drawing.

Computer-aided drafting enables drafters and designers to create pictorial and perspective drawings without much knowledge of the construction techniques involved in either kind of drawing. Architects and drafters have long been using perspective grids to aid in the creation of perspective drawings. But the knowledge of how perspective drawings are made is valuable even for those who use drafting aids and computer systems.

The three types of perspective drawing techniques take their names from the number of vanishing points used in each. *One-point*, or *parallel*, perspective has one vanishing point and is used most often when drawing interiors of rooms. *Two-point* or *angular*, perspective is the most popular and is used to illustrate exteriors of houses, small buildings, civil engineering projects, and, occasionally, machine parts. The third type, *three-point* perspective, has three vanishing points and is used to illustrate objects having great vertical measurements, such as tall buildings. It is a lengthy process to draw in three-point perspective; therefore, it is not used as often as two-point perspective.

The following discussions cover the construction of perspective drawings using manual drawing methods or CADD 2-D drawing methods. Keep in mind that once an object is drawn in 3-D using some CADD systems, it can then be displayed in a perspective format.

General Concepts

Two principal components of a perspective drawing are the eye of the person viewing the object, and the location of the person in relation to the object. The eye level of the observer is the *horizon line* (HL). This line is established in the elevation (front) view. (See Figure 19.35.) The position of the observer in relation to the object is the *station point* (SP) and is established in relation to the plan (top) view. The location of the station point determines how close the observer is to the object, and the angle at which the observer is viewing the object. The *ground line* (GL) is the line on which the object rests. The *picture plane* (PP), or plane of projection, is the surface (drawing sheet) on which the object is projected. The picture plane can be situated anywhere between the observer and the object. The picture plane can also be located beyond the object. The observer's lines of sight, or visual rays, determine what will show on the picture plane and where. Finally, the three vanishing points discussed in the following sections are referred to as vanishing point right (VPR), vanishing point left (VPL), and vanishing point vertical (VPV).

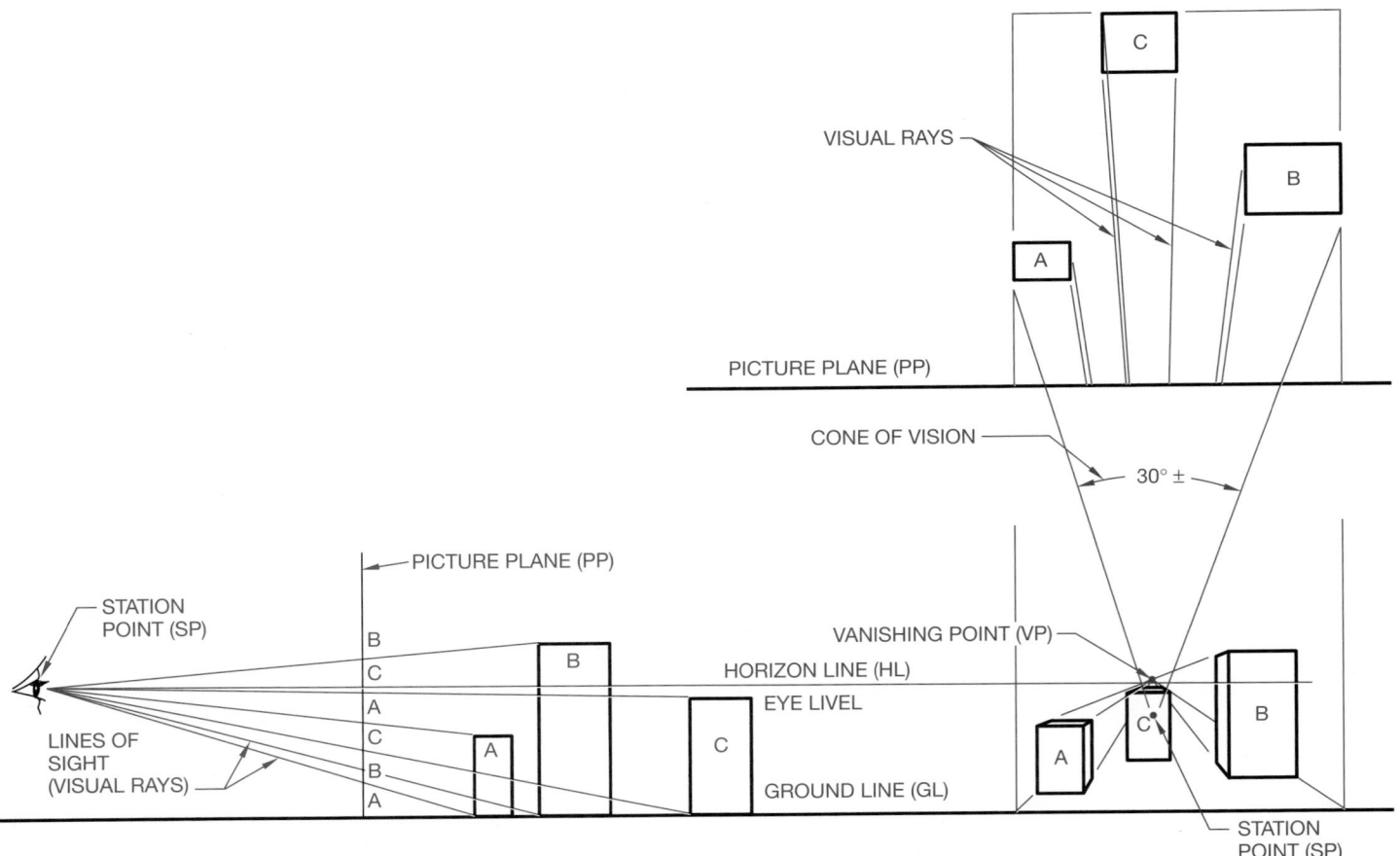

FIGURE 19.35 ■ Principal components of perspective drawings.

ONE-POINT PERSPECTIVE

This form of perspective, also known as parallel perspective, has only one vanishing point. The plan view is oriented so that the front surface of the object is parallel to the picture plane. The elevation view is situated below and to the right or left of the plan and rests on the ground line. The following steps will enable you to construct a one-point perspective. Refer to Figure 19.36 as you read the instructions.

Step 1. Locate the station point between the picture plane and the elevation view of the object. The station point can be anywhere, depending on the part of the object you wish to view. The visual rays from the station point to the extreme corners of the object should form an included angle of approximately 30° to provide the most realistic perspective.

Step 2. Determine the eye level of the observer in the elevation view. If you wish to look over the object to see the top surface, the horizon line should be drawn above the elevation view. The horizon line can be drawn at any eye level.

Step 3. The vanishing point is located on the horizon line directly in line with the station point.

Step 4. Project all points in the plan view that touch the picture plane to the corresponding points in the elevation view. These lines are true scale.

Step 5. Draw visual rays from the station point to the rear corners of the object in the plan view.

Step 6. The points where the visual rays intersect the picture plane are projected vertically down to the perspective view. When drawing complex objects, work with only a portion of these points at one time to avoid confusion.

Step 7. Project points A, B, and C toward the vanishing point to intersect the corresponding projectors from the picture plane. The horizontal line B'C' can then be drawn. Remember, any line that is parallel to the PP in the plan view must be parallel to the GL in the perspective view.

Step 8. Project the height of the object from the elevation view to points D and E on the two *true-height lines* (THL). True-height lines are projected from points that touch the PP.

Step 9. Project the height at points D and E back toward the vanishing point to intersect the projectors from the rear portion of the object in the plan view.

Step 10. Complete the object by connecting the ends of the sloping portion and projecting the base of the sloped feature toward the vanishing point.

TWO-POINT PERSPECTIVE

The two-point perspective method is also termed angular perspective because it is turned so that two of its principal planes are at an angle to the picture plane. This is the most popular form of perspective drawing. Its two vanishing points allow parallel lines on the two principal planes to be projected in two directions, thus giving another dimension to the depth of the perspective.

We will use the same object drawn previously in the discussion of one-point perspective. The object has been positioned at an angle in the plan view, with one corner touching the PP. The elevation view, SP, HL, and GL have all been established. Refer to Figure 19.37 as you read the instructions.

Step 1. Draw a line from the station point, parallel to each side of the object in the plan view, to intersect the PP.

Step 2. Project the points on the PP down to the HL. This establishes vanishing point right (VPR) and vanishing point left (VPL) on the horizon line.

Step 3. Project point A on the PP down to the GL. This becomes the true-height line AB in the perspective view.

Step 4. Begin blocking in the object by projecting points A and B to the two vanishing points. This establishes the two angular, or perspective, sides of the object.

Step 5. Project visual rays from the SP to the extreme corners of the object in the plan view. Remember that this cone of vision formed by the visual rays should be approximately 30°.

Step 6. Project the intersection of these visual rays with the PP down to intersect with the projectors from points A and B to the two vanishing points. This blocks in the two sides of the object.

Step 7. Draw lines from the SP to points C and D. Where these lines intersect the PP, project vertical lines down into the perspective view. The height of the object at points A and A' can now be projected toward the VPL to intersect with the projectors from points C and D.

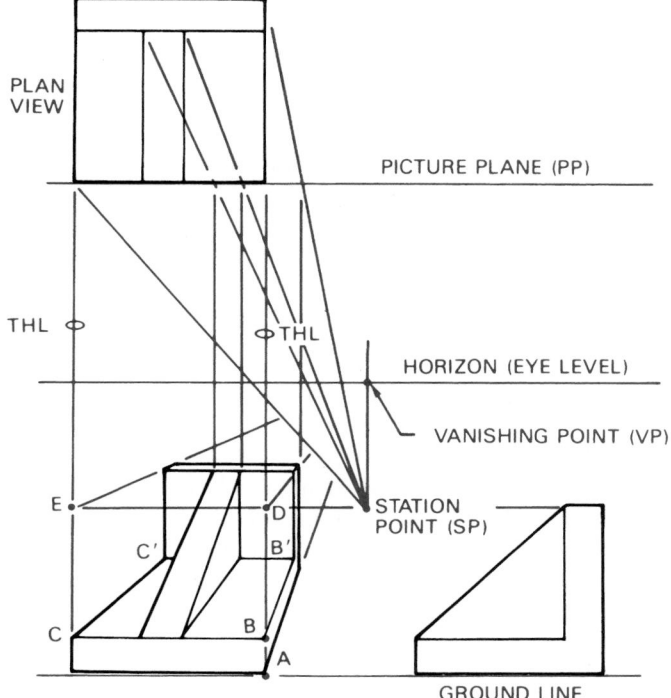

FIGURE 19.36 ■ Constructing a one-point perspective drawing.

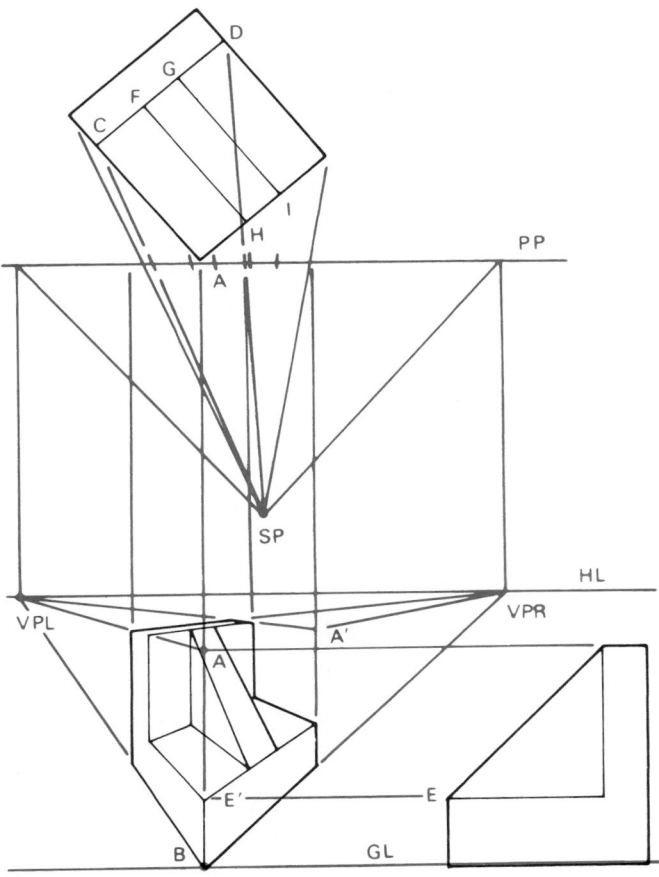

FIGURE 19.37 ■ Constructing a two-point perspective drawing.

Step 8. The height at E can be projected to E' on the THL. Project E' toward both vanishing points to create the basic shape of the part.

Step 9. Project points F, G, H, and I to the station point. Where these lines intersect the PP, drop vertical projectors down to the perspective view. These lines will intersect corresponding points on the perspective drawing. Connect the points as shown in Figure 19.37 to complete the perspective view.

THREE-POINT PERSPECTIVE

Three vanishing points are used in three-point perspective. These drawings require more time to construct than do two-point perspectives, and often occupy a considerable area on the drawing sheet. Three-point perspective drawings are used when certain effects are needed for visual stimulation. You will examine the method of constructing a three-point perspective that requires the least amount of drawing space. Refer to Figure 19.38 as you read the instructions.

Step 1. Draw an equilateral triangle to occupy as much of the sheet as possible. Label the three corners as vanishing points VPR, VPL, and VPV.

Step 2. Construct perpendicular bisectors of each of the three sides. Their intersection at the center of the triangle should be labeled SP (station point). (See Figure 19.38a.)

Step 3. Draw a horizontal line through the SP. Measure and mark the length of the object to the left of the SP. Measure and mark the width of the object to the right of the SP. This will allow you to draw an object in perspective as if you had turned the plan view 45° to the pictureplane line. (See Figure 19.38b.)

Step 4. Draw a line parallel to line VPV, VPR. This becomes the height measuring line. Measure the height of the object from the SP down to the left along this line, as shown in Figure 19.38b. Note that numbers have been placed at intervals along the measuring lines. These indicate measurements of features on the object.

Step 5. Project lines from points 1 and 2 to the left of the SP to VPR. Next, draw a line from point 1 at the right of the SP to VPL. Draw lines from points 1 and 2 on the height measuring line VPR. (See Figure 19.38c.) You have now blocked in the overall measurements of the object. The following points have been located: lower front corner (A), upper right corner (B), upper far corner (C) and upper left corner (D). Note in Figure 19.38c that the station point (SP) is the upper front corner of a box that encloses the object.

Step 6. Project points B and D to VPV, then project A to VPR and VPL. The intersections of these lines are points E and F, the lower left and right corners respectively. Point G is the height of the front corner of the object as shown in Figure 19.38c. Keep in mind that any surface of the object that is parallel to one of the principal planes of the orthographic view must project to one of the vanishing points as you draw the remainder of the object's features. The completed object is shown in Figure 19.38d.

CIRCLES AND CURVES IN PERSPECTIVE

In most cases, circles in perspective will appear as ellipses. But if a surface of an object is parallel to the picture plane, any circle on that surface will appear as a circle. (See Figure 19.39.) Circles located in planes that are at an angle to the picture plane will appear as ellipses and can be drawn by a method of intersecting lines projected from the elevation and plan view, which is referred to as the coordinate method.

The object having the circles should first be drawn in plan and elevation, and all of the necessary lines and points for your perspective drawing should be determined and placed on the drawing. (See Figure 19.40.) Next divide the circle in the elevation view into a convenient number of pie-shaped sections. Project the intersections of these section lines with the circle to the top and side of the object. The points along the side of the object can then be projected onto the perspective view. Now transfer the distances formed by the intersection of the top of the object and the lines

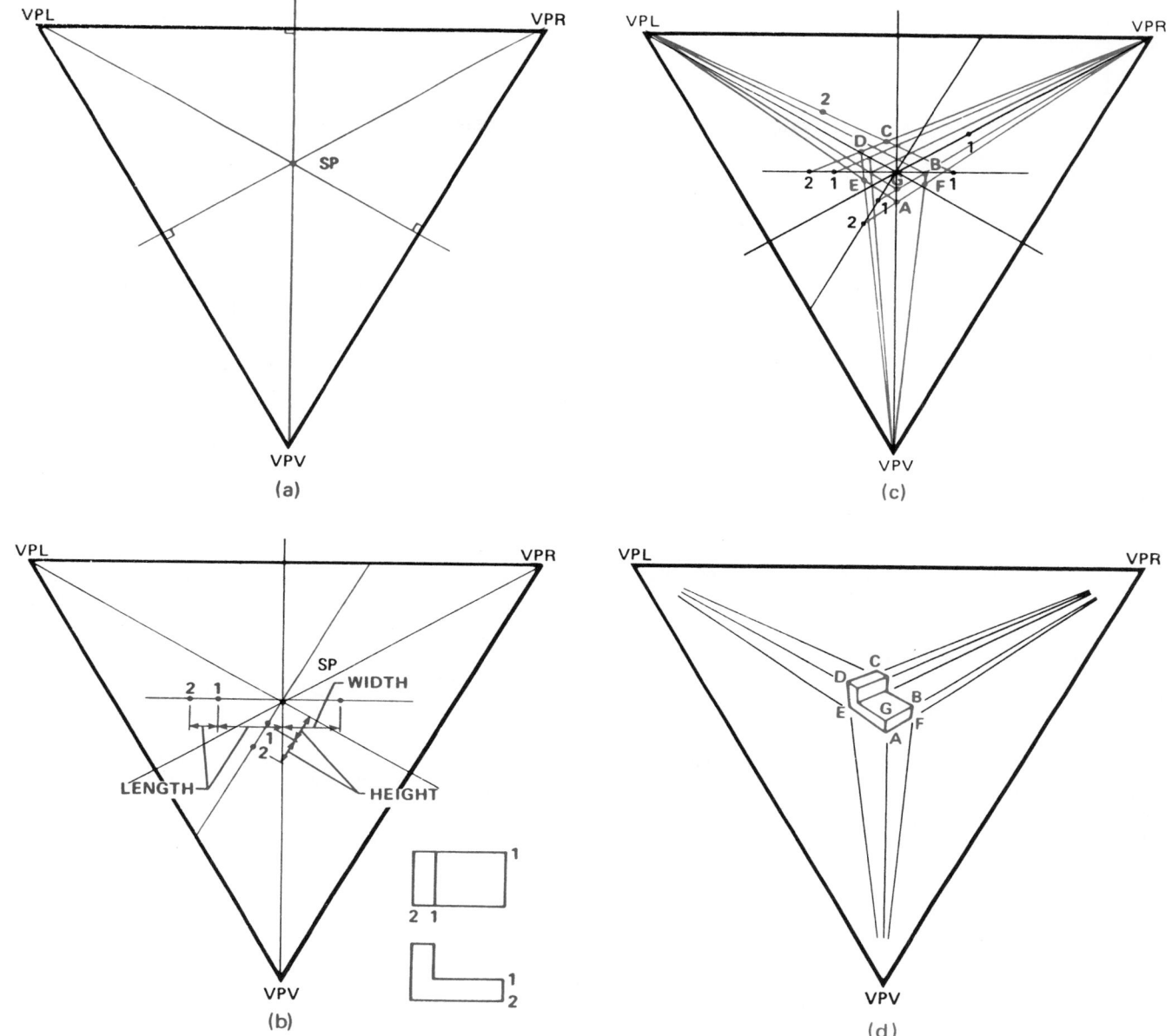

FIGURE 19.38 ■ Constructing a three-point perspective drawing: (a) establishing vanishing points and the station point; (b) establishing the length, width, and height of the object; (c) establishing the outline of the object; (d) the completed object.

projected from the pie-shaped sections in the elevation view to the plan view. Next draw visual rays from the SP to the points in the plan view. Where the visual rays intersect the picture-plane line, project the points straight to the perspective view. These projectors will intersect the ones drawn from the elevation view. Connect these intersection points with an irregular curve or an ellipse template.

The same method can be used to draw irregular curves. Establish a grid, or coordinates, on the curve in the elevation view and place the same divisions on a plan view. Project these two sets of coordinates to the perspective view and then use an irregular curve to connect the points. (See Figure 19.41.)

BASIC SHADING TECHNIQUES
Line Shading

Most pictorial drawings are created to illustrate shape description and to portray the relationship of parts in an assembly clearly. Highly artistic renderings are not normal for most industrial purposes; therefore, any shading techniques used should be simple while conveying the desired effect.

The objects shown in Figure 19.42 illustrate the most basic form of shading, called line-contrast shading. Vertical lines opposite the

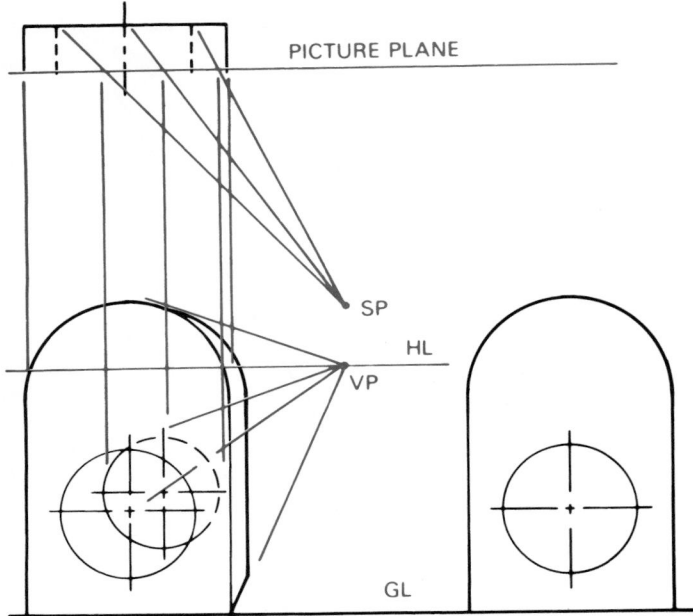

FIGURE 19.39 ■ Circles in surfaces parallel to the picture plane.

light source and bottom edges of the part are drawn with the thick lines. Some illustrators will outline the entire object in a heavy line to make it stand out as shown in Figure 17.42b.

Straight-line shading is a series of thin straight lines that can be varied to achieve any desired shading. They can be used on flat and curved surfaces. Note in Figure 19.43 the two ways that straight lines can be used on curved surfaces.

Additional emphasis can be given to curved surfaces with the use of block shading. Figure 19.44 shows how block shading on one or both sides of a curved surface can produce a highlight effect. One to three lines of shading are normally used to achieve the block effect. The total amount of block shading should be approximately one-third the width of the object.

Stipple shading is just dots. (See Figure 19.45.) The closer they are, the darker the shading. Although stippling takes longer than line shading, the results can be quite pleasing.

Fillets and Rounds

Fillets and rounds can be depicted in three ways, which are shown in Figure 19.46. Example (a) uses three lines to indicate the two outside edges (tangent points) of the radius and the centerline of the radius. Note the occasional gaps in the outside line.

With the technique in example (b), the same size ellipse is moved along the axis of the radius and repeated at regular intervals. It is important to draw construction lines first for the outside edges of the fillet or round. This ensures that all of the ellipses are aligned. Draw these ellipses with a thin pen or pencil point. Using a CADD system, you can copy or array the arc or ellipse and do not need to draw the construction lines.

The example at (c) is the simplest and requires the least amount of time to draw. A broken line is drawn along the axis of the radius to indicate the curved surface.

LAYOUT TECHNIQUES

The best way to begin any pictorial drawing, especially if you are working from an orthographic drawing, is to make a quick free-hand sketch. This will enable you to see the part in a 3-D form before you try to create a drawing. Always know what the final drawing is going to look like before you begin.

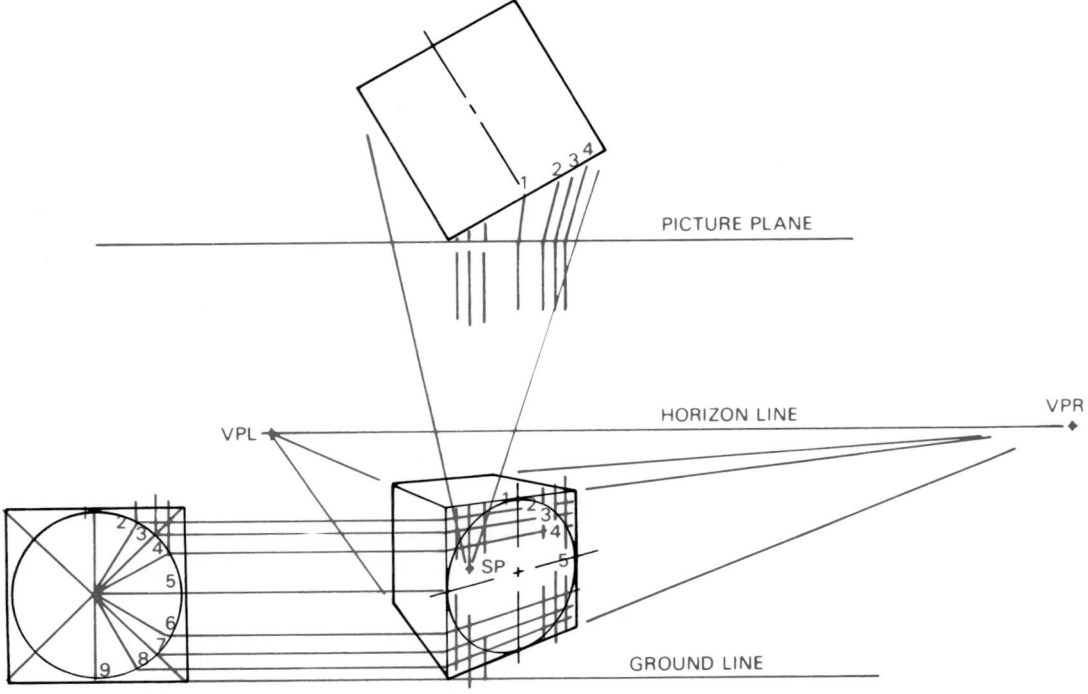

FIGURE 19.40 ■ Circles plotted in perspective by the coordinate method.

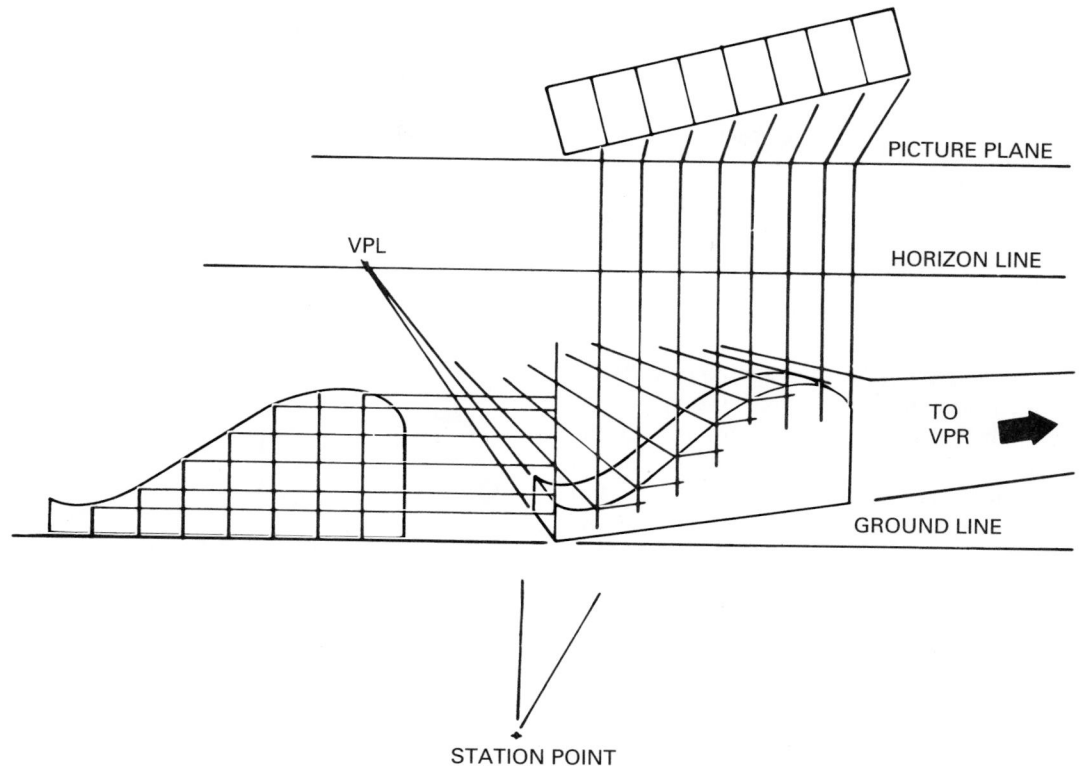

FIGURE 19.41 ■ Irregular curve plotted in perspective by the coordinate method.

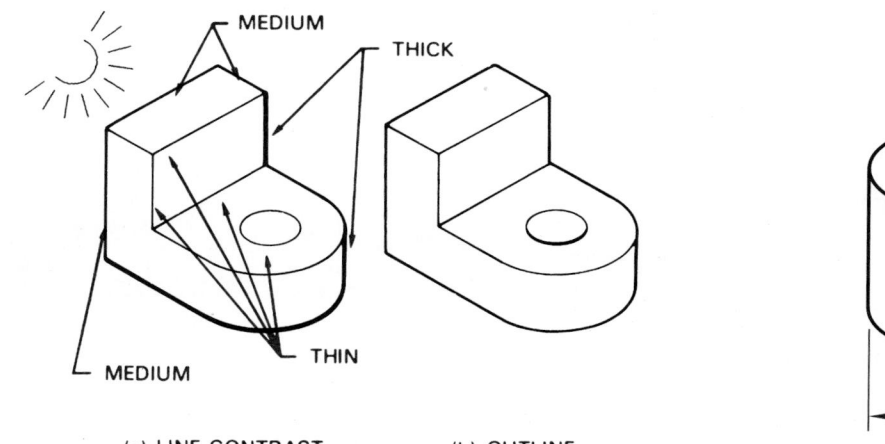

(a) LINE CONTRAST (b) OUTLINE

FIGURE 19.42 ■ Line-contrast shading.

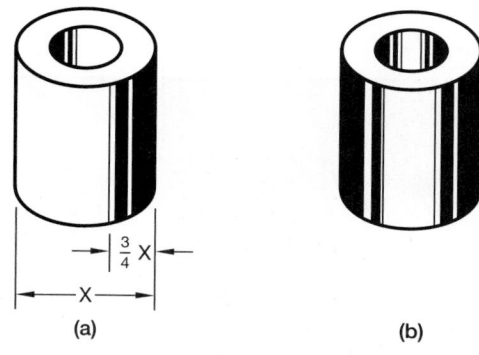

(a) (b)

FIGURE 19.44 ■ Block shading.

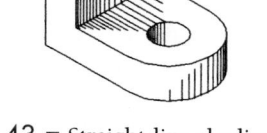

FIGURE 19.43 ■ Straight-line shading.

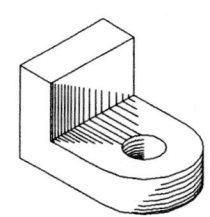
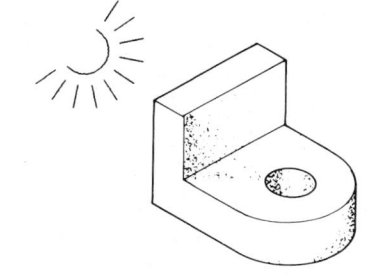

FIGURE 19.45 ■ Stipple shading.

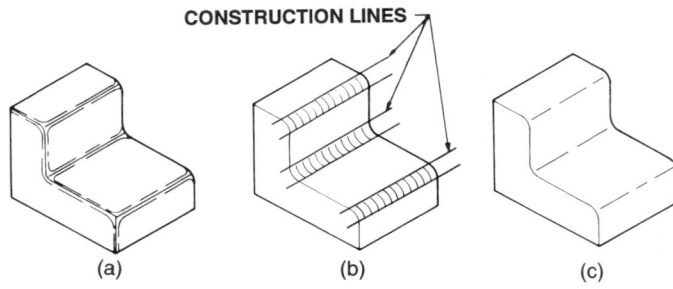

CONSTRUCTION LINES

(a)　　(b)　　(c)

FIGURE 19.46 ■ Depicting fillets and rounds.

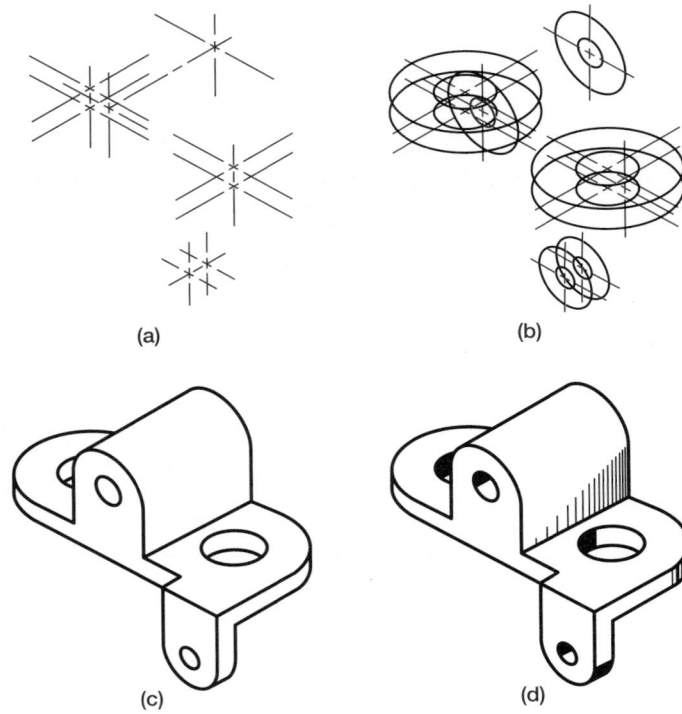

(a)　　(b)

(c)　　(d)

FIGURE 19.47 ■ (a) Lay out centerlines and axis lines of circular features; (b) draw circular features using ellipses; (c) join ellipses with straight lines; (d) complete the drawing with shading.

Objects that have boxy or angular shapes should be drawn by first laying out the overall sizes of the part. Box in the shape. Then begin to cut away from the box using measurements from the part or drawing. That is called the coordinate, or "box-in," method. Use construction lines to assist in the layout, be it a manual or a CADD drawing. If the object has circular features, use the following layout techniques.

Step 1. Lay out the centerlines and axis lines of these features first. (See Figure 19.47a.) They act as the skeleton of the part.

Step 2. Draw the circular features using ellipses. (See Figure 19.47b.)

Step 3. Join ellipses with straight lines. (See Figure 19.47c.)

Step 4. Add additional features such as shading, notes, and item tags. If you are drawing manually, trace the object lines in pencil or ink and add shading as required. Any notes or item tags that are needed can be added last. If you are working with a CADD drawing, you can edit the drawing so only the object lines show. If you have created a special

"construction" layer for your CADD drawing, turn this layer off when the drawing is completed. The finished drawing is shown in Figure 19.47d.

MATH APPLICATION

An isometric drawing is constructed using the actual measurements of the part. But because isometric drawing does not take into account "foreshortening," the object appears larger in the drawing than it is in reality. To compensate for this, isometric projection can be used to show the object in its foreshortened appearance. If a part measured one inch on a side, a true isometric projection would create an isometric line that measured .816 in.

In order to construct an isometric view of an object that is a true isometric projection, it is necessary to multiply the actual measurements by .816 to obtain the drawing measurements. The following measurements are converted as follows:

$$1.45 \text{ in.} \times .816 = \textbf{1.183 in.}$$
$$3.67 \text{ in.} \times .816 = \textbf{2.995 in.}$$
$$.89 \text{ in.} \times .816 = \textbf{.726 in.}$$

PROFESSIONAL PERSPECTIVE

Pictorial drawing requires a good ability to visualize objects in three-dimensional form. Companies that produce parts catalogs, instruction manuals, and presentation drawings require the services of a technical illustrator or someone skilled in pictorial drawing. A part of your professional portfolio should be 8" × 10" photo reductions or laser prints of your best pictorial drawings. Reductions of large drawings can look good, and small mistakes tend to fade away when reduced.

The field of technical illustration is much more limited than engineering and drafting but is wide open to the freelancer or student looking for quick jobs. Always keep examples of a variety of pictorial drawing types in your portfolio. Your portfolio should be a three-ring binder with tabbed dividers for the different types of designs and drawings you have created. Manila pocket folders that fit into the binder are excellent for holding folded blueprints and reduced pictorial drawings. Remember that often a small freelance job can lead to bigger jobs and even to owning your own company.

CHAPTER 19

Pictorial Drawings and Technical Illustrations Test

MULTIPLE CHOICE

Select the best choice a, b, c, or d that answers or completes each question or statement.

1. Isometric drawing belongs to a family of pictorial representation known as:
   a. Orthometric   c. Diametric
   b. Axonometric   d. Trimetric

2. What is the angle in degrees of a true isometric ellipse?
   a. 34° 16'   c. 35° 15'
   b. 36° 15'   d. 35° 16'

3. What is the angle between each of the three isometric axes?
   a. 130°   c. 120°
   b. 125°   d. 140°

4. Which type of isometric drawing is best for representing cylindrical parts?
   a. Long-axis   c. Regular
   b. Short-axis   d. Reverse

5. Which layout method is best for objects with many circles and arcs?
   a. Box-in   c. Diametric
   b. Long-axis   d. Center line

6. The diameter of an isometric circle is measured:
   a. Along its major diameter
   b. Along its minor diameter
   c. 35° 16' from horizontal diameter
   d. 30° from horizontal diameter

7. What is the most realistic form of pictorial drawing?
   a. Perspective   c. Isometric
   b. Dimetric   d. Trimetric

8. Which of the following is *not* a form of oblique drawing?
   a. Cabinet   c. Long-axis
   b. General   d. Cavalier

9. Where are the vanishing points located in perspective drawing?
   a. Ground line   c. Horizon line
   b. Picture plane   d. Eye level

10. What is the best angle for the cone of vision in perspective drawing?
    a. 45°   c. 40°
    b. 30°   d. 35°

11. Points of the object in the plan view of a perspective drawing that touch the _____ line are the only lines that are true scale in the perspective view.
    a. Ground   c. Picture plane
    b. Horizon   d. Eye level

12. The vanishing point is located directly above or below the _____ in one-point perspective drawing.
    a. Ground line   c. Picture plane
    b. Horizon line   d. Station point

13. Which of the following is *not* a commonly used form of shading?
    a. Stipple   c. Fillet
    b. Block   d. Line contrast

14. Block shading normally covers _____ of the object surface.
    a. ½   c. ⅛
    b. ¼   d. ⅓

CHAPTER 19

Pictorial Drawings and Technical Illustrations Problems

DIRECTIONS

Choose the best axis to show as many features of the object as possible in your axonometric or oblique drawing.

For axonometric problems—problems numbered 19-1 through 19-10—draw isometric, dimetric, or trimetric as assigned. For oblique problems—problems numbered 19-11 through 19-17—draw cavalier, cabinet, or general

oblique as assigned. Remember that circular features are best shown in the front plane of the oblique view.

For perspective problems—problems numbered 19-18 through 19-25—make a one-, two-, or three-point perspective drawing as assigned except for problem 19-18, which should be done as a one-point perspective view. All objects may be turned at any angle on the picture-plane line for viewing from the station point except problem 19-18, which should be drawn in the direction indicated.

1. Make a freehand sketch of the object to assist in visualization and layout of axonometric and oblique problems.
2. For axonometric and oblique problems, select a scale to fit the drawing comfortably on an A- or B-size drawing sheet. Use a C- or D-size drawing sheet for drawing an initial layout of the perspective problems on sketch or butcher paper.

3. Dimension axonometric or oblique problems only if assigned by your instructor. Do not place dimensions on a perspective view.
4. Perspective objects without dimensions can be measured directly and scaled up as indicated or assigned.
5. Trace the perspective view in pencil or ink on vellum or polyester film.
6. Make your drawings using a CADD system if appropriate with course guidelines.

PROBLEM 19-1   Axonometric projection

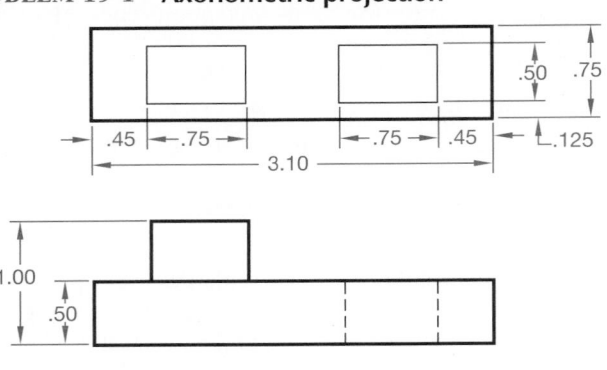

PROBLEM 19-2   Axonometric projection

PROBLEM 19-3   Axonometric projection

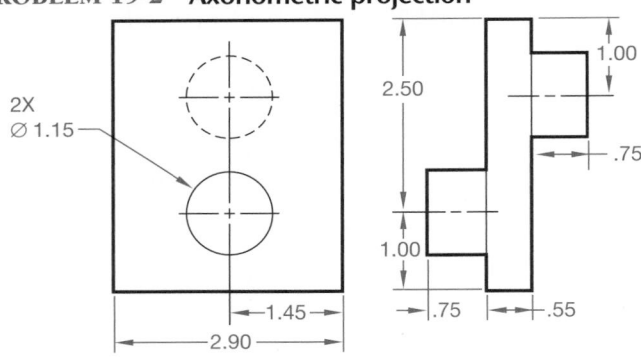

PROBLEM 19-4   Axonometric projection (metric)

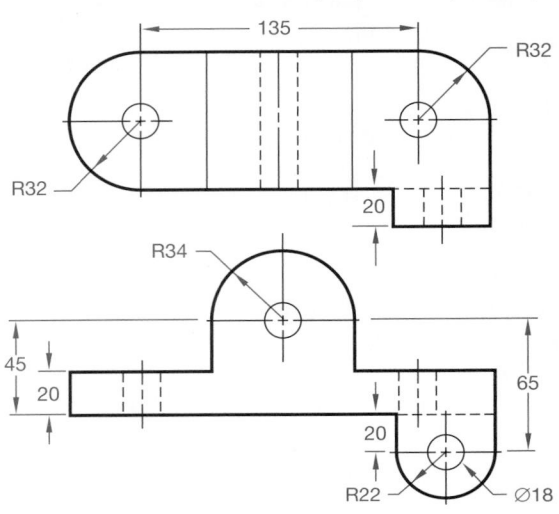

PROBLEM 19-5   Axonometric projection

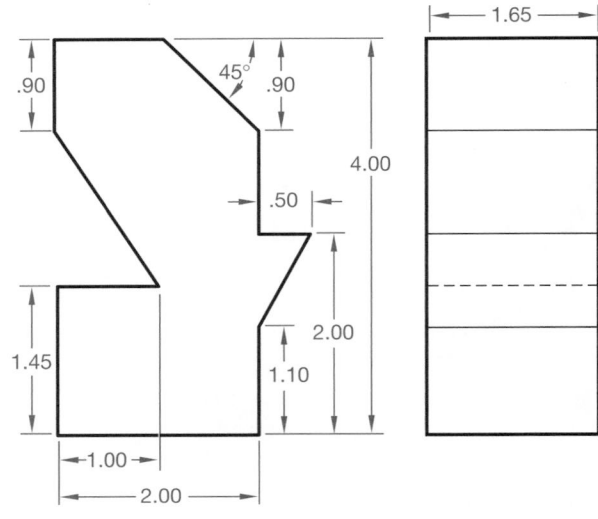

PROBLEM 19-6 **Axonometric projection (metric)**

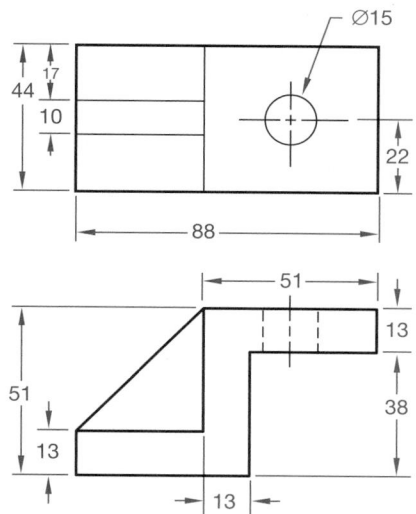

PROBLEM 19-7 **Axonometric projection**

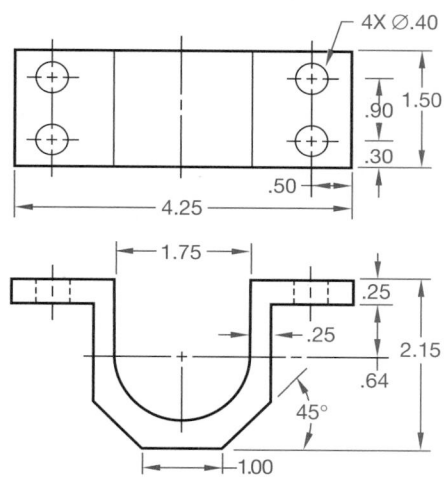

PROBLEM 19-8 **Axonometric projection (metric)**

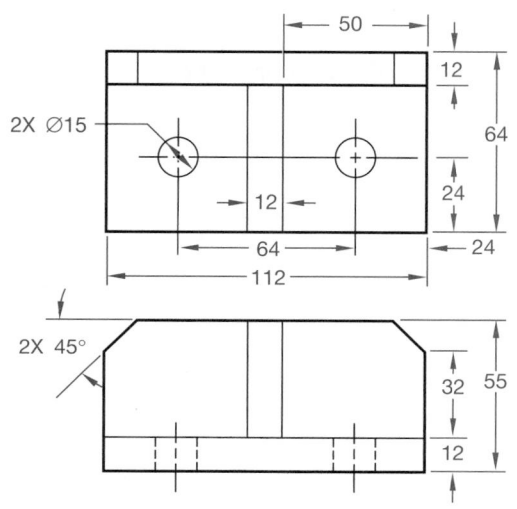

PROBLEM 19-9 **Axonometric projection (metric)**

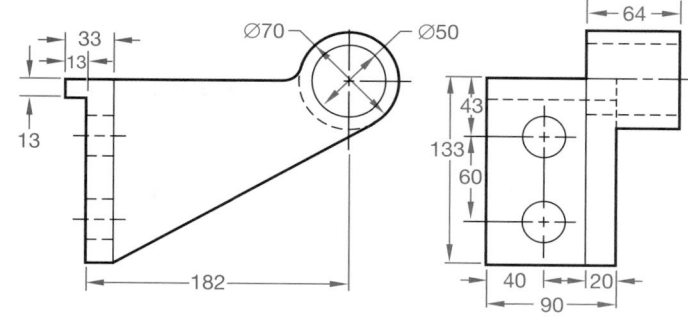

PROBLEM 19-10 **Axonometric projection**

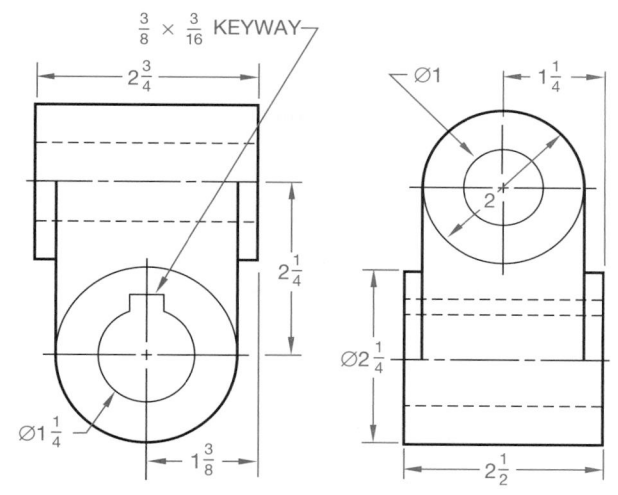

PROBLEM 19-11 **Oblique projection**

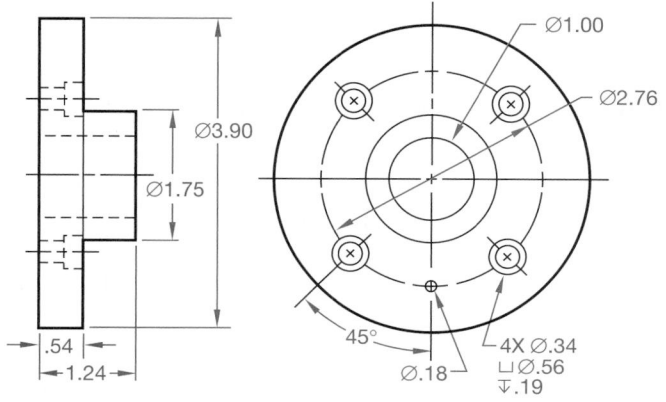

PROBLEM 19-12 Oblique projection

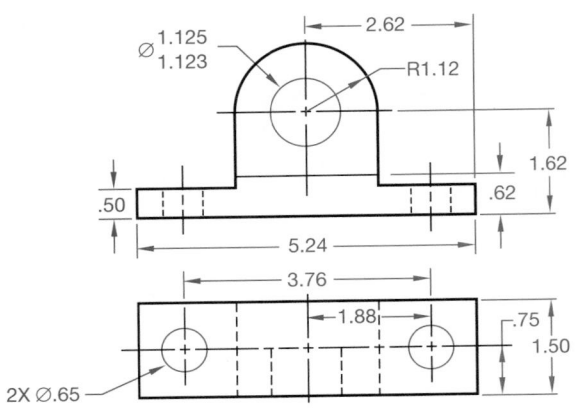

PROBLEM 19-15 Oblique projection

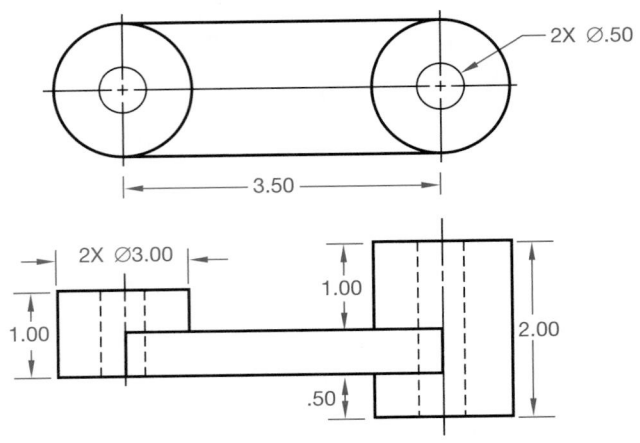

PROBLEM 19-13 Oblique projection

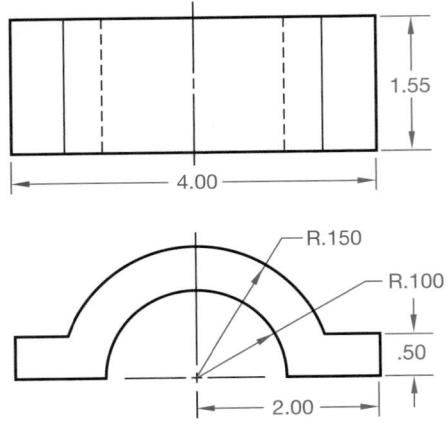

PROBLEM 19-16 Oblique projection

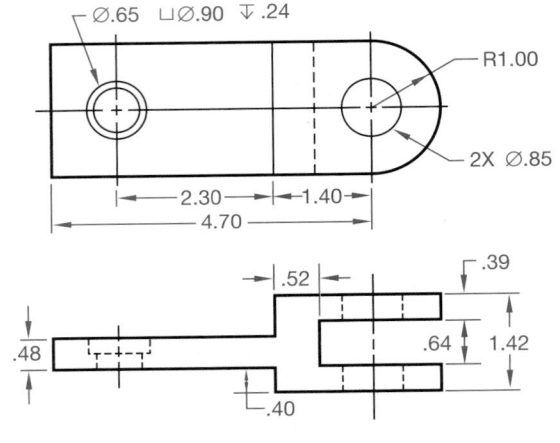

PROBLEM 19-14 Oblique projection (metric)

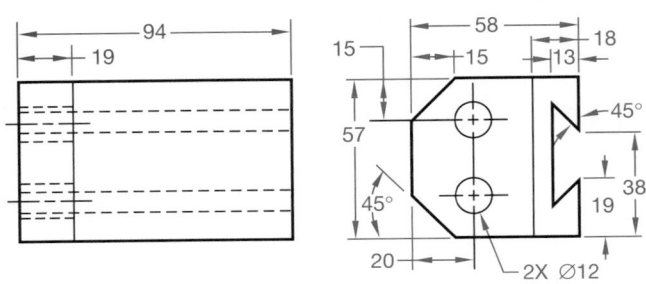

PROBLEM 19-17 Oblique projection

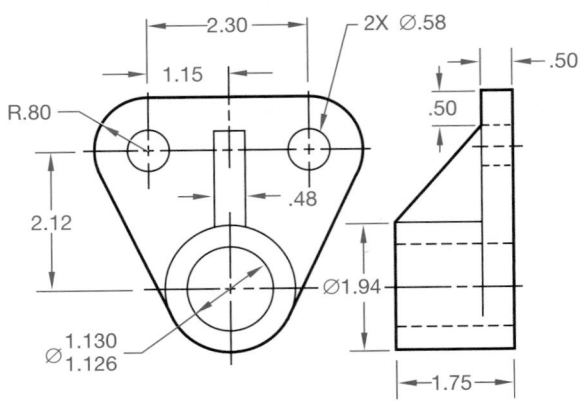

PROBLEM 19-18　Perspective

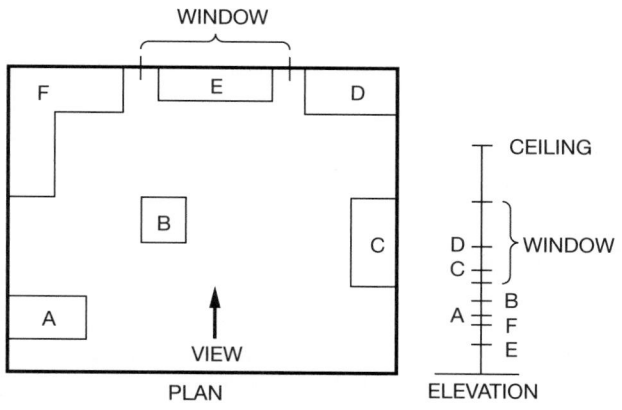

PROBLEM 19-19　Perspective

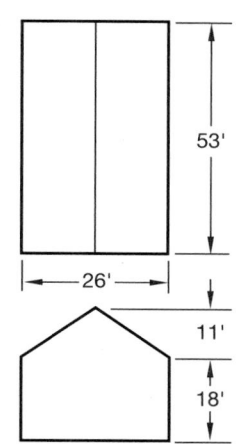

PROBLEM 19-20　Perspective

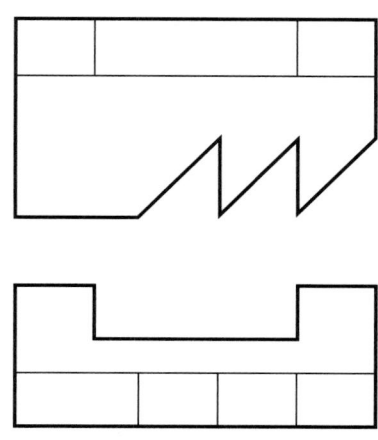

MEASURE AND INCREASE 2X

PROBLEM 19-21　Perspective

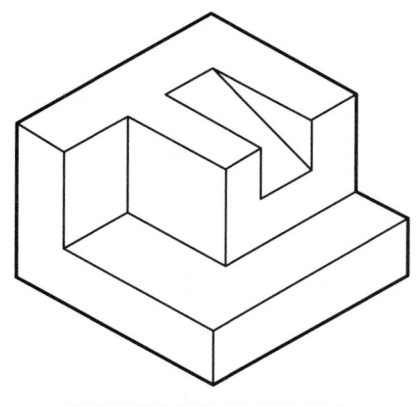

MEASURE AND INCREASE 2X

PROBLEM 19-22　Perspective

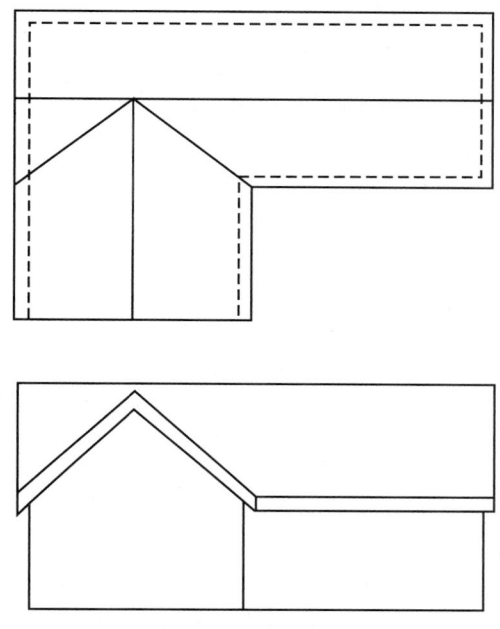

MEASURE AND INCREASE 2X

PROBLEM 19-23　Perspective

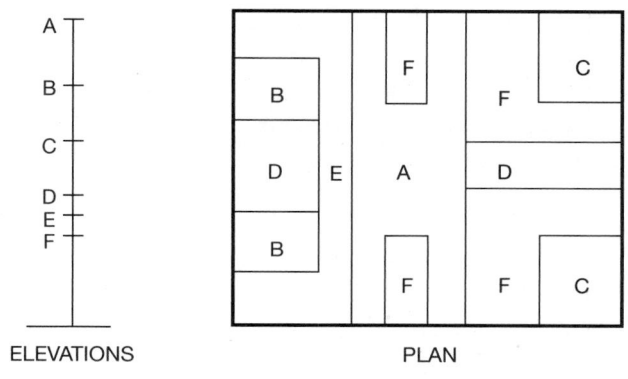

MEASURE AND INCREASE 3X

PROBLEM 19-24 Perspective

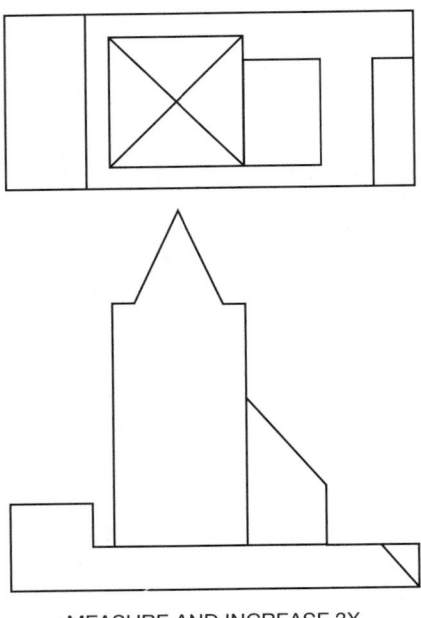

MEASURE AND INCREASE 3X

PROBLEM 19-25 Perspective

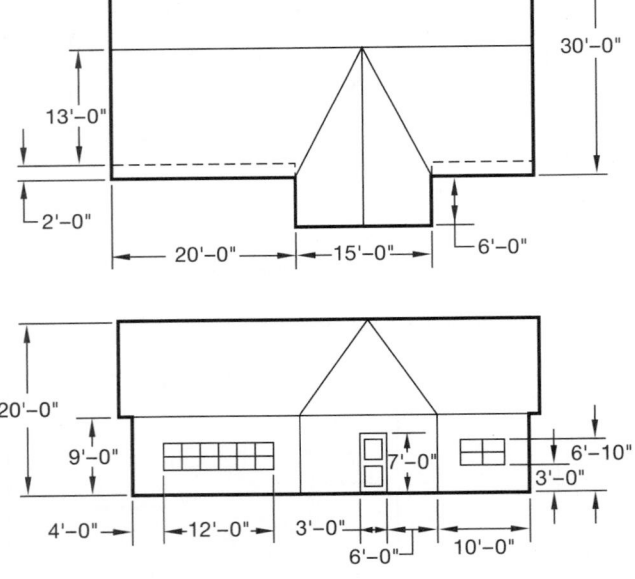

MATH PROBLEMS

PROBLEM 19-26 Provide the measurements required to construct isometric, dimetric, and trimetric projections using the following actual dimensions of the parts. The dimetric and trimetric angles are given.

DIMENSION		ISOMETRIC	DIMETRIC	TRIMETRIC
Left axis	1.55	_____	(10°)_____	(25°)_____
Right axis	2.95	_____	(40°)_____	(45°)_____
Vertical axis	2.07	_____	_____	_____

PROBLEM 19-27 Provide the dimetric projection measurements for an object with the following actual dimensions. The left and right axis angles are given.

	DIMENSION	DRAWING MEASUREMENT
Left axis 25°	2.87	_____
Right axis 40°	4.46	_____
Vertical	3.22	_____

PROBLEM 19-28 Provide the trimetric projection measurements for an object with the following actual dimensions. The left and right axis angles are given.

	DIMENSION	DRAWING MEASUREMENT
Left axis 10°	4.95	_____
Right axis 60°	3.18	_____
Vertical	2.86	_____

PROBLEM 19-29 If a ¾ scale is used to draw the receding axis of a general oblique object, how long would you draw the following dimensions on the receding axis?

3.23 _____

6.54 _____

2.48 _____

4.29 _____

PROBLEM 19-30 If a cylindrical object has a diameter of 3.75 in., what is the approximate width of the block shading that should be applied to it?

Solid Modeling, Animation, and Virtual Reality

OBJECTIVES

After completing this chapter, you will:

- Explain the difference between wireframe, surface, and solid modeling.
- Discuss the difference between and importance of the solid modeling kernels.
- Discuss the importance of parametric modeling within the engineering design process.
- Understand how to design with parametric modelers.
- Explain how to work with parts, assemblies, and features.
- Understand the significance of built-in features.
- Understand how to edit solid models based on history.
- Understand the importance of animation within the design process.
- Understand the importance of virtual reality within the design process.

THE ENGINEERING DESIGN APPLICATION

Today's business world is extremely competitive in nature. Cross-functional teams must work to produce high-quality products that are brought to the marketplace at the correct time. If a product is introduced to the market too late, the competition may have already captured the majority of the customers, leaving only a minority behind for the latecomer. The latecomer in this case has a major hurdle to overcome, that of winning customers over from the competition. Research clearly shows that once customers have selected a product, they will be reluctant to change. Therefore, it is critical to gain early entry into the marketplace and to capture a customer base. Furthermore, early entry into the marketplace means an extended sales life for the product. Simply put getting to the marketplace first is a major strategy for success.

Design teams are one business strategy that is being used by business and industry to accelerate the design process and thus get product to the market. Traditionally the design process happened in a linear fashion. For example, the design engineer would complete a design and then pass it to manufacturing. When manufacturing received the design, they would often have to make changes, changes that required time. The design process would get longer and longer as manufacturing and the engineer passed the design back and forth requesting changes.

Traditionally these changes were very difficult to make. Some of this can be contributed to the fact that communication methods were not very flexible. Drawing documentation was done by hand, and often nonengineers had a difficult time understanding the drawings. Furthermore, changes required arduous erasing and sometimes redrawing of prints that were stamped with approvals

and passed back to manufacturing. Lack of flexibility was improved with the introduction of AutoCAD, which allowed the draftsperson to make changes relatively quickly and then print out a new drawing.

Today teams consisting of individuals from marketing, design, engineering, and manufacturing work together using integrated CAD solid modeling software. This software allows for changes to be made in a more fluid manner. Changes can be made by marketing and design while the engineers' requirements are protected through constraints. Furthermore, parts within the solid modeling environment can be related to one another, so that if one part changes, the associated parts will also update.

This interoperability is critical in today's competitive marketplace. Again, the winners in today's competitive marketplace are those companies that can bring innovative and high-value products and services to the customer ahead of their competition. Reaching the marketplace with innovation before the competition has several critical benefits, including:

- The product's sales life is lengthened,
- The early-to-market company gains a pricing advantage.
- The early-to-market company can start product development later and therefore can use more up-to-date technology to develop that product.
- The early-to-market company has a marketplace advantage by gaining customers who potentially become loyal customers.

As an engineering drafter it will be part of your job to make things happen and make them happen quickly so that your company can gain these advantages and win in today's marketplace.

TYPES OF 3D CAD MODELS

Multiview drawings are very useful for documenting engineering design requirements. However, the person viewing the document must be able to interpret the drawing in order to visualize your design. This may be difficult for individuals who have not been trained to interpret multiview projections. Often a communication gap is created between those who can interpret multiview projections and those who cannot. A 3-D representation of your design is a good tool to help you bridge this communication gap. Within the CAD environment there are three different 3-D modeling techniques that can help you with this visualization. These are wireframe, surface, and solid modeling techniques. It is important that you know the difference between these models and that you know their limitations.

Wireframe Modeling

Wireframe modelers create basic 3-D CAD models. The CAD file contains information about the objects' vertices and edges. A wireframe model appears as though it were constructed from wire coat hangers, thus the name wireframe (see Figure 20.1). Although these models provide an outline of the 3-D object, they can be difficult to interpret. This can be overcome on some models by hiding the lines that fall behind object features (see Figure 20.2), but sometimes even a hidden line representation can be difficult to visualize.

Wireframe models have both advantages and disadvantages. Wireframe CAD files are small in size and therefore regenerate quickly on the computer screen. Their size is limited because only information about the objects' vertices and edges are stored in the CAD file. A wireframe model does not contain information about the objects' surfaces or the objects' volume. Complex surfaces, which are found on many of today's products, cannot be represented. Furthermore, wireframe models are not very useful for doing mass calculations or generating machining code.

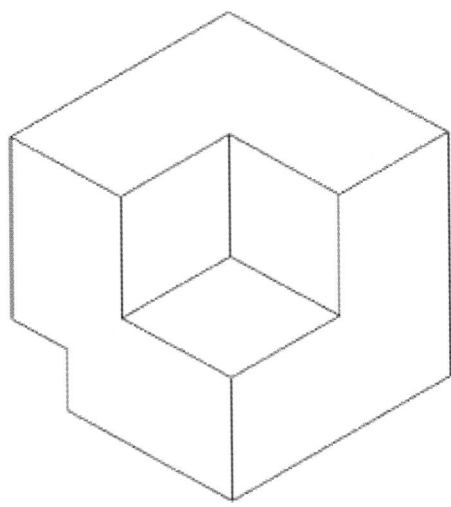

FIGURE 20.2 ■ Wireframe model with hidden lines improves visibility of object.

Surface Modeling

Surface CAD models also contain information on the objects' vertices and edges. In addition, information about the objects' surfaces is included. When rendered, surface models provide a realistic appearance of the design. These modelers allow designers to create complex curves and forms and have given rise to more organic and ergonomic product shapes. Therefore, surface modelers are used by industrial designers who are primarily concerned with the external product appearance. Figure 20.3 provides an example of these complex organic forms that are common on the many home appliances.

Surface models also have inherent advantages and disadvantages. Surface models can be rendered providing a visual means of communicating your design intent to those who have a hard time visualizing a design from a multiview drawing. Surface models are extremely useful for creating organic shapes adding to the desirability of the product's overall appearance. Despite these advantages, surface models are limited in mechanical design because they do not contain volumetric information. Similar to wireframe models, surface models are limited when it comes to doing mass calculations or generating machining code.

Solid Modeling

As you progress from wireframe to surface and then to *solid* models, additional information is included in the CAD database. Additional information allows you to perform supplementary operations and analysis on those models. Solid models include information that allows designers to perform checks for interference, mass calculations, simulations, and computer numeric control (CNC) code generation.

Although solid modelers contain the information to do these calculations, not all solid models are created equal. Depending on the method of creation, a solid may have more or less functionality. Solid models range from the very basic type that use Boolean operations for creating solid models to hybrid modeler applications that

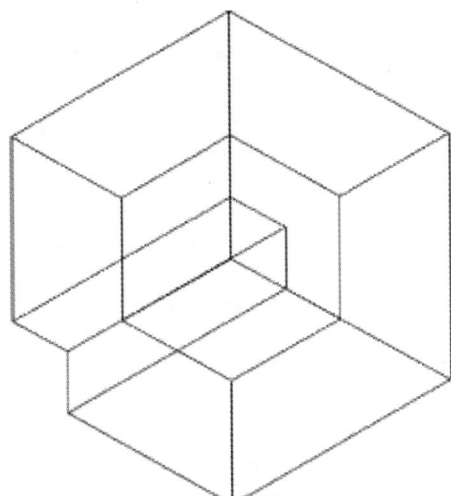

FIGURE 20.1 ■ A wireframe model that represents edges and contours of an object.

FIGURE 20.3 ■ Product design with complex forms and shapes. *Courtesy Unigraphics Solutions Inc.*

Characteristic	Basic Boelon Modeler	Parametric Modeler	Hybrid Modeler
Analysis	No	Yes–add on software	Yes
Animation	No	Yes–add on software	Yes
Feature based	No	Yes	Yes
History	No	Yes	Yes
Intelligence	No	Yes	Yes
Surface modeling	No	Limited	Yes
Updateable	No	Yes	Yes
Volumetric Information	Yes	Yes	Yes

FIGURE 20.4 ■ Characteristics of Boolean, parametric, and hybrid solid modelers.

incorporate wireframe, surface, and solid models into one software package. The table in Figure 20.4 provides an illustration of the range of solid modelers and the characteristics included in each.

Basic Solids (Dumb Solids)

Solid models that are created using *basic* Boolean processes are often referred to as "dumb solids." Dumb solids contain volumetric information, but they do not include intelligence about dimensions, parameters, or features. Boolean operations are common in most solid modelers and allow for various solid shapes to be joined, subtracted, or intersected. Figure 20.5 provides an illustration of these three basic operations. Although these operations are used in most solid modelers, basic Boolean solid modelers do not store a history of how the solid model was created, and therefore the database contains little intelligence or history about the model. Those who have worked with dumb solids quickly realize the disadvantage of being unable to update and change a solid. For example, once a hole has been created within an object (this is often done with a subtractive Boolean operation), it can be eliminated only

with a repair or join operation. If the hole cannot be repaired, the designer will be required to recreate the model starting from the beginning.

This is very problematic in today's business environments where engineering drafters are required to design products quickly and in cooperation with cross-functional team members. The rapid pace of product development and team member input requires that the drafter remain extremely flexible with the design. Dumb solids do not allow for this flexibility. Stated simply, engineering drafters who are working with dumb solids are at a disadvantage because they are unable to respond to the rapid pace of today's product development environment.

Parametric Solids (Intelligent Solids)

Parametric solids, or "intelligent solids," have the advantage of being able to be updated and changed. Parametric solid modeling software stores information in the database on the history of how the part was created. This history includes information on sequence (which object was created first, second, etc.), features (e.g., holes, chamfers, etc.), dimensions, and design parameters (e.g., relationships between holes). All of this information is considered to be model "intelligence," which is used to help control and maintain your design. Furthermore, the engineering drafter can incorporate design intent by placing constraints or requirements on related features. Constraints are parameters that are required for your design. You build these constraints into your model, and the CAD software protects them and makes sure they are not violated.

Hybrid Modelers

Hybrid modelers have been developed to incorporate the functionality of all the 3-D modelers discussed previously. In the work world, industrial designers, design engineers, and manufacturing engineers have to work together to successfully bring a product to

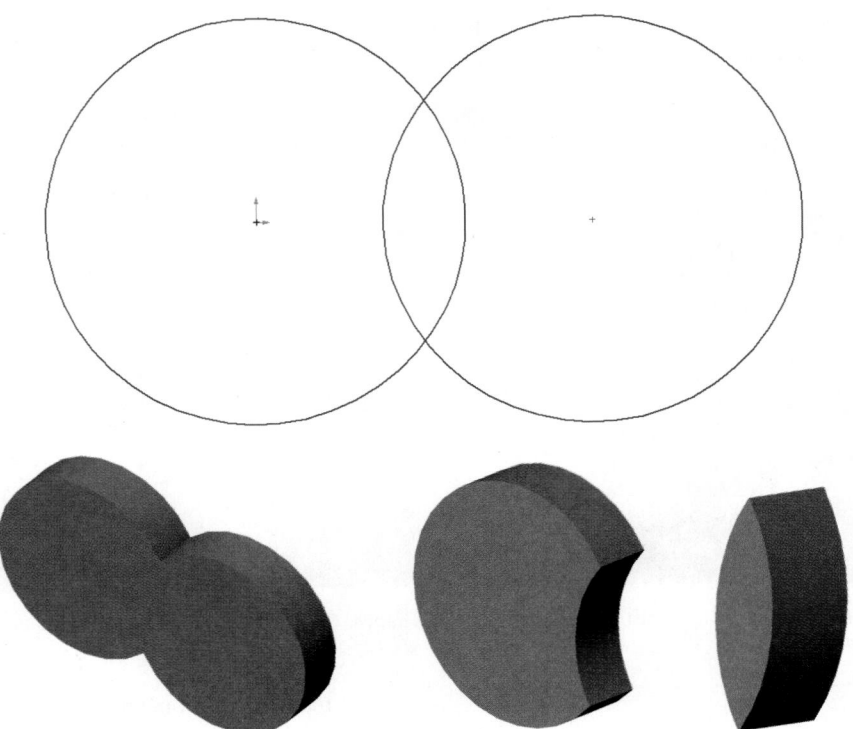

FIGURE 20.5 ■ Join, subtraction, and intersection of Boolean operations.

market. The hybrid modeler allows these product development professionals to work with wireframe, surface, or solid models in a step-by-step approach to bringing a design to completion.

For example, the design engineer can quickly take a surface model created by an industrial designer and translate it into a solid model using a surface-to-solid converter. The surface model allows the industrial designer to create organic freeform shapes, while the solid model allows the engineer to run various analyses, interference checks, and simulations.

MODELING KERNEL

The *modeling kernel* is the underlying software code that allows the solid modeler to perform operations such as creating basic geometric shapes. Unless you will be developing your own solid modeling software, you do not need to understand the inner workings of the modeling kernel. However, you should be aware of which kernel your solid modeler uses, so that you are aware of its limitations, because its confines will be the limitations for your CAD solid model software.

ACIS and Parasolids are two kernels that are used by a number of popular solid modelers. Companies such as AutoCAD purchase a license to use a modeling kernel as the foundation for building their software. The functionality within the kernel is extremely specialized for CAD/solid modeling, allowing the developer to quickly build on applications without reinventing the basic code. Both the ACIS and Parasolid kernels have similar functionality, and the code for both is constantly updated to include additional functionality. If you are considering the purchase of a software package, make sure you know the limitation of the kernel. For example, some kernels do not have the ability to generate lofts, a complex extrusion using

two or more objects. Without this functionality you may be unable to complete your model.

The modeling kernel has a significant impact on interoperability, or how well a solid model will translate from one software package (e.g., SolidWorks) to another (e.g., ProEngineer). Although ongoing efforts focus on interoperability between solid modelers, data are often lost in translation. Frequently in the translation "intelligent solids" are converted to "dumb solids," and all the information regarding dimensions, parameters, features, and overall development history is lost. Although it may be possible to open these "dumb solid" files in another software package, the engineering drafter cannot update the solids based on the parametric or historical information. This forces the drafter to either recreate the model from scratch or use the same software that was originally used to create the model. It is for this reason that many designers use more than one solid modeler.

PARAMETRIC MODELING AND THE DESIGN PROCESS

In today's competitive marketplace, the engineering drafter needs to understand how the design fits within the function of the product development cycle. As presented in Chapter 5, to remain competitive, a systems approach to engineering design must be adopted. *Parametric* solid modeling has become a very important innovation that is helping to meet the demands of this new systems approach.

Traditionally engineering, manufacturing, and marketing have operated as independent business functions. This independent nature created an environment where engineers designed a product with very little input from the other functional areas of the company. Furthermore, if employees within the other functional areas of the company wanted to make changes to the product, they had to go through a lengthy change order process. This design cycle process left much to be desired as manufacturing had to solve problems created by designs that were difficult to manufacture, and marketing/sales had to promote and sell products that often were not valued by the customer. Today companies that operate in this fashion have a difficult time staying in business. If a product development company cannot produce products of high quality that meet their customers' needs, their competitors will drive them out of the marketplace.

Furthermore, companies are turning out new products at a pace not seen prior to this time in history. Five years ago computer companies produced a new computer every 12 months; today they produce a new model every few months. Design engineers must respond to the demands of their cross-functional team and the pace of the marketplace by making changes to their designs while controlling for critical functionality. Parametric modeling provides an extremely flexible and descriptive CAD platform on which to work to control designs while allowing for flexibility.

Designing with Parametric Modelers

Designing with parametric modelers requires design engineers to protect their critical design parameters while allowing for change. The idea is to build protection into your parametric models so that your design intent is protected. You build this protection by adding parametric constraints to your model. These constraints can take the form of dimensions or geometric relationships.

Whenever you add constraints to a model through parametric dimensions or relationships, it is important that the information is accurate and that it indeed reflects your design. You should take an active approach to designing with parametric relationships through preplanning. This planning process should include a detailed evaluation of the part being drawn and its relationship to other features or parts in its assembly. You may want to ask the following questions related to your part: What features must remain constant? What dimensions are critical and cannot change? How does this part relate to other assembly parts? Which features will need to change over time?

After determining the required constraints, you can begin to create your parametric model. Most parametric modelers follow a similar creation process. Once you learn the process in one software, such as SolidWorks, you can easily follow the logic in another, such as ProEngineer. Parametric modeling begins with a sketch or profile of the object you are creating. When creating this sketch, you must think in terms of objects, which are created through extrusion, sweeps, or revolutions, because it is through these processes that your sketch will become a 3-D feature. A feature is the basic building block for the parametric model. Features have both geometric constraints and dimensional constraints that define them. Furthermore, these constraints are stored and can be used to change the model in the future. For example, Figure 20.6 shows how a basic sketch can be turned into three different features using an extrude, sweep, and revolve process. Features such as those shown in Figure 20.6 are the basic building blocks of parametric design.

When you create a sketch, you will add design constraints that reflect your design intent. Design intent is created and maintained using both dimension and physical conditions or constraints. Constraints are sketch parameters that will be held constant within the model and, therefore, cannot be violated. Figure 20.7 lists the geometric constraints (relationships) that can be applied in SolidWorks. In this table, each relationship (constraint) is listed along with the objects that must be selected and the result of the operation.

It is important to know that most solid modelers will automatically apply geometric constraints to your sketch. Solid modeling software has been programmed to guess with a predetermined approximation about sketch geometry and apply "best fit" constraints. For example, if you sketch two lines that are approximately parallel to each other, the software may automatically apply a parallel constraint. If this does not match your design intent, you will have to remove that constraint, which is holding those lines parallel.

As mentioned earlier, constraints hold design elements constant. If these constraints are violated or contradictions occur, the solid modeling software will warn you that the sketch is overconstrained. An overconstrained sketch is one that has too many things controlling its geometry. This overconstrained condition prevents the solid modeling software from understanding the sketch, which means that the software is confused by the contradiction. Therefore, it is critical that the overconstrained condition be resolved before the sketch is extruded, swept, or revolved into a solid feature.

Figure 20.8 provides an example of an overconstrained sketch. The sketch is overconstrained because the radius dimension and the width dimension contradict each other. This contradiction is actually set up because both of the vertical lines in the sketch have geometric constraints that hold them as vertical. In this example, if either of the dimensional constraints were to be changed, the solid modeling software would be presented with a contradiction and would not be able to solve the sketch.

A sketch can also be underconstrained. An underconstrained sketch has sketch elements that have ambiguity or are free to move and change. This underconstrained condition may be desirable as you begin your design, allowing more flexibility. However, as you perfect your design and more fully understand the design intent, it is important to protect that intent through constraints. Figure 20.9 shows a sketch that is underconstrained. Although the length of the sketch has been dimensioned, the height remains free to change. By adding a height dimension, this sketch becomes fully constrained (see Figure 20.10).

Figure 20.11, page 684, shows a basic sketch in which relationships are added, moving the sketch from an underconstrained to a fully constrained and then to an overconstrained condition. Solid modelers usually provide a visual indication of the constrained condition. For example, within SolidWorks the color indicates the overconstrained, underconstrained, or fully constrained nature of each line. A blue line indicates a underconstrained condition, black indicates a fully constrained condition, and red indicates an overconstrained sketch.

When applying design constraints add only those that are required to protect your design intent. Adding unnecessary constraints only applies restrictions that are not needed and that may limit future changes.

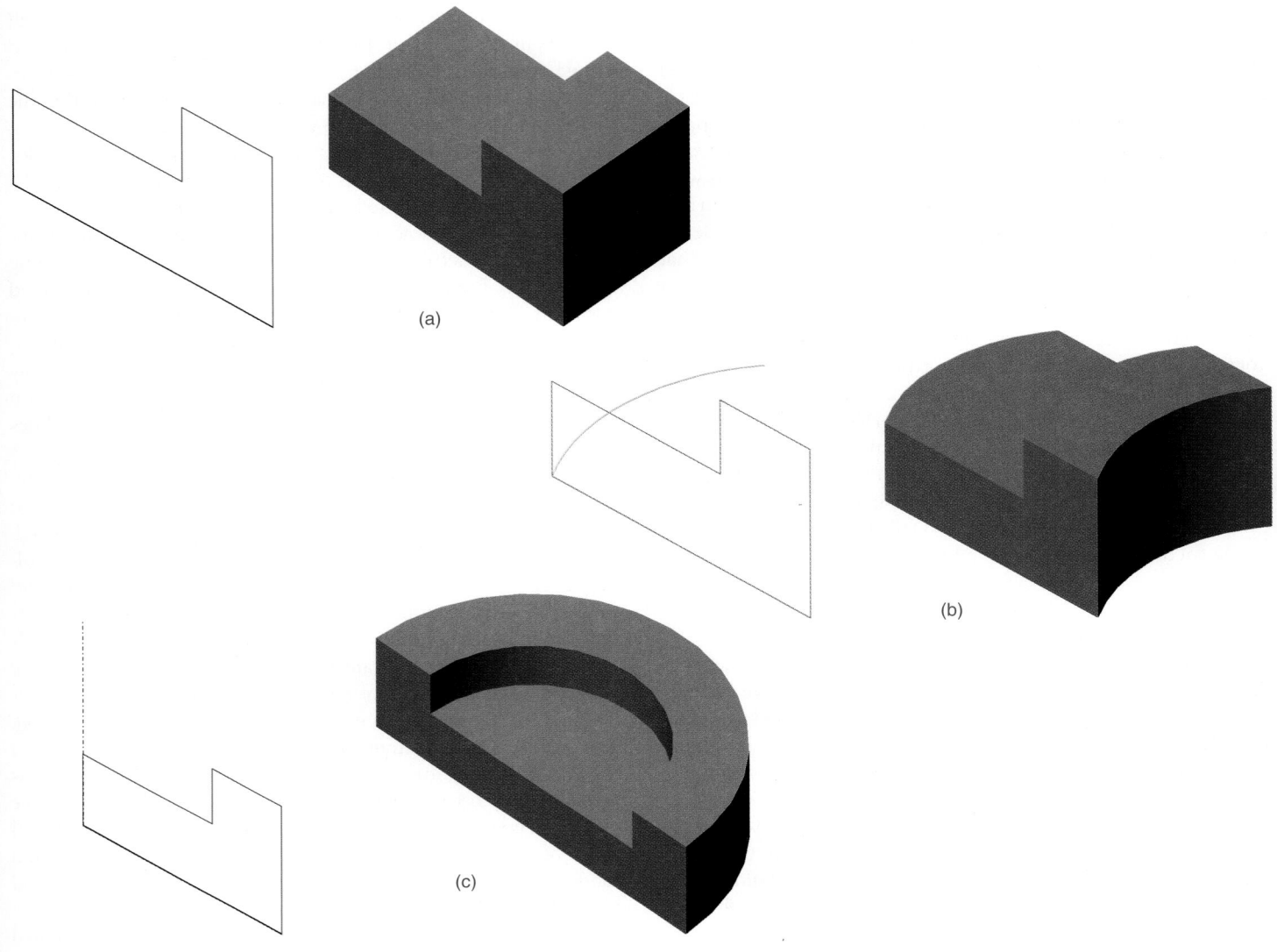

FIGURE 20.6 ■ Extrude, sweep, and revolve features.

WORKING WITH PARTS, ASSEMBLIES, AND FEATURES

Engineered products typically consist of individual parts that are fitted together to create the final assembly. As a design drafter you will develop many solid parts that will come together in an assembly to create the finished product. Figure 20.12 illustrates how parts make up subassemblies and then full assemblies. Solid modelers are organized around this part and assembly concept. When you begin work on your design you will work on an individual part. Each part you create will be created in a separate CAD file and then combined into an assembly file. Selecting a part with which to begin will depend on the nature of the design. You may choose to work on the base part of the assembly or on a part that dictates the nature of other parts in the assembly.

You begin a part by selecting and creating a base feature. The base feature is the first feature you draw. This feature should be selected because it provides a good "base" for the other features on the part. For example, the simple bracket in Figure 20.13 has a natural base and orientation. The part naturally rests on the large flat support where it is bolted into place, which makes for an appropriate base feature.

Creating the sketch for this base feature can be done in three different ways. Figure 20.14, page 686, provides an illustration of each sketch variation that can be used to create the base of the simple bracket shown in Figure 20.13. When selecting a sketch orientation, your design intent should be considered. For example, if the length and width dimensions may change in the future, it will be advantageous to include those in your sketch. The first sketch in Figure 20.14 includes both length and width dimensions in the sketch. Because they were created within the sketch, these dimensions will be available for change in the future.

Again, when you draw a sketch, solid modelers will apply sketch constraints automatically based on predetermined parameters. For example, if a line is sketched and it is horizontal, the solid modeler will place a horizontal constraint on that line. In Figure 20.15, page 686, the base of the bracket has been sketched and the CAD oper-

Geometric relation	Selection	Result
Coincident	A point and a line, arc, or ellipse	The point lies on the line, arc, or ellipse.
Collinear	Two or more lines	The items lie on the same infinite line.
Concentric	Two or more arcs, or a point and an arc.	The arcs share the same centerpoint.
Coradial	Two or more arcs.	The items share the same centerpoint and radius
Equal	Two or more lines, or two or more arcs.	The line lengths or radii remain equal.
Fix	Any item.	The item's size and location are fixed. However, the end points of a fixed line are free to move along the infinite line that underlies it. Also, the endpoints of an arc or elliptical segment are free to move along the underlying full circle or ellipse.
Horizontal or Vertical	One or more lines or two or more points.	The lines become horizontal or vertical (as defined by the current sketch space). Points are aligned horizontally or vertically.
Merge points	Two sketch points or endpoints.	The two points are merged into a single point.
Intersection	Two lines and one point.	The point remains at the intersection of the lines.
Midpoint	A point and a line.	The point remains at the midpoint of the line.
Parallel	Two or more lines.	The items are parallel to each other.
Perpendicular	Two lines.	The two items are perpendicular to each other.
Pierce	A sketch point and an axis, edge, line, or spline.	The sketch point is coincident to where the axis, edge, or curve pierces the sketch plane.
Symmetric	A centerline and two points, lines, arcs, or ellipses.	The items remain equidistant from the centerline, on a line perpendicular to the centerline.
Tangent	An arc, ellipse, or spline, and a line, or arc.	The two items remain tangent.

FIGURE 20.7 ■ Geometric constraints that can be applied in SolidWorks. *Courtesy SolidWorks Corporation.*

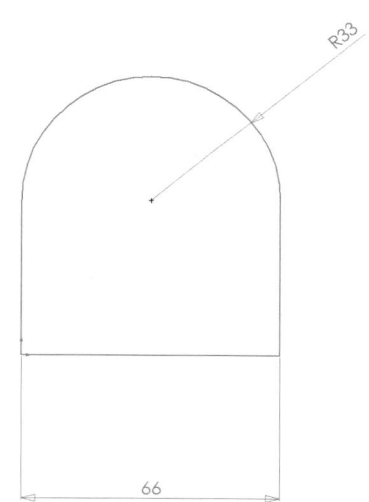

FIGURE 20.8 ■ An example of an overconstrained sketch.

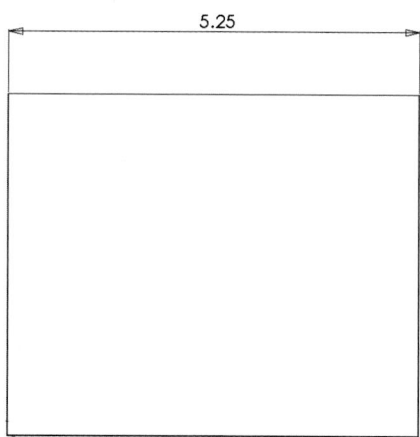

FIGURE 20.9 ■ An example of an underconstrained sketch.

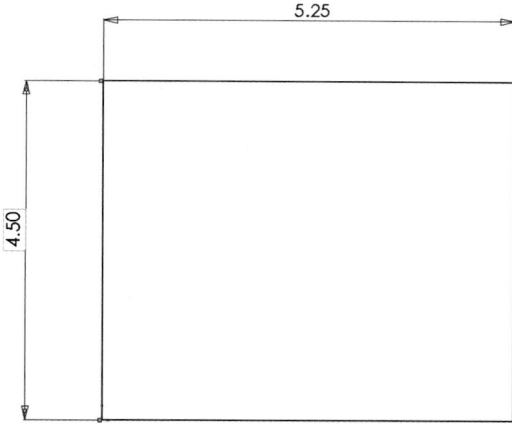

FIGURE 20.10 ■ An example of a fully constrained sketch.

ator has not added constraints. However, in actuality four constraints exist, two vertical and two horizontal constraints. It is important to identify the constraints that your software adds and decide if they are appropriate for your design. If they are not appropriate, remove them from the sketch before they cause problems.

After the sketch has been appropriately constrained, it can be extruded, swept, or revolved to the desired depth. Figure 20.16 shows the extrusion of the base feature to a depth of 1/2 inch. All additional features that are added will become a part of the base feature. To create another feature, you must tell the solid modeling software which surface the new feature is to start from. For example, to add the boss (raised feature) to the base feature, a plane of reference must be selected on which to create a new sketch. A plane of reference is the surface on which you will create the next sketch. When a face is selected in SolidWorks, it changes color to indicate that selection. Figure 20.17 shows how the front face has been selected (indicated in green) as the reference plane for creating the next sketch. Figure 20.18, page 687, shows how that sketch is extruded to create the desired design. The hole features were created through the same sequence of sketch plane selection, sketch, and extrusion.

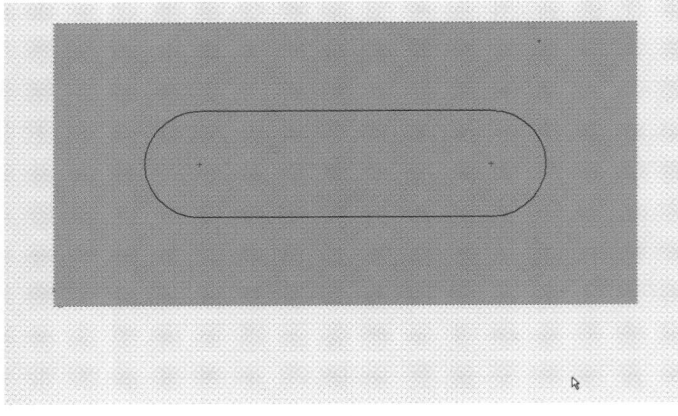

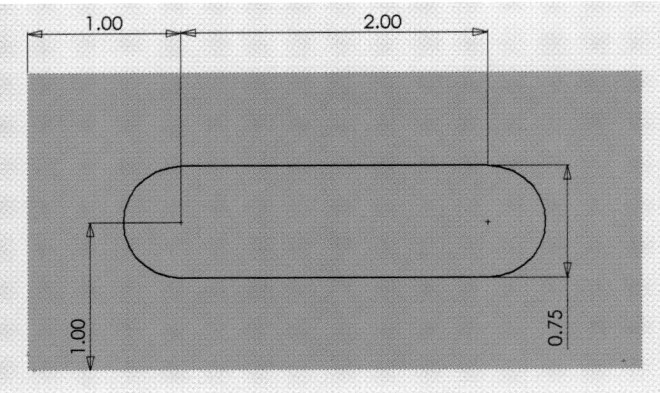

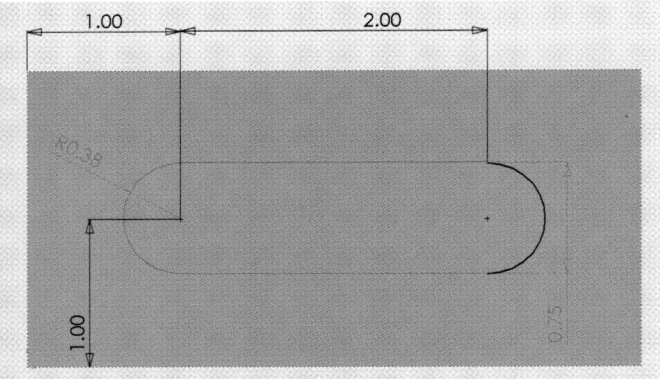

FIGURE 20.11 ■ Sketch in various constrained conditions—underconstrained, fully constrained, and overconstrained.

BUILT-IN FEATURES

In engineering design there are many features that are common between parts. These features usually coincide with a machining process that is used to create the geometric shape. For example, many mechanical parts require that edges be rounded (rounds) and corners be filleted. The cellular phone in Figure 20.19 is a good example. Notice how all the parts edges and corners are rounded as an important part of the design. These features are often required by manufacturing because sharp edges and corners are difficult to machine into an injection mold tool, and, furthermore, sharp corners may cause the part to stick in that production tool.

These round and fillet features are used frequently and are added to most parts; therefore, automating their use can save a designer a considerable amount of time. Most solid modeling software include these as built-in or automated features. Parametric modelers build these features automatically and maintain their attributes so they can easily be edited. Parametric modeling software also has a built-in standard hole feature. Counterbore and countersinks are two common hole types that are used to receive bolt heads, nuts, and screws. These holes can be added quickly using the hole feature within the solid modeler. Figure 20.20 shows the hole feature dialog box for creating the hole feature within SolidWorks.

EDITING FEATURE HISTORY

One of the advantages of historical-based modeling is that you can modify the model after it has been created. First, you identify which feature of the part you want to edit. Most parametric modelers allow you to identify the feature in two ways: first, by selecting the feature itself, and second, by identifying the feature in a design tree. A design tree is a historical reference of how your part was created. In SolidWorks the design tree places the base feature at the top of the tree and any additional features below in a tree-type arrangement. Figure 20.21 shows the design tree layout for a solid model of a wrench that was created in SolidWorks.

Once the feature has been identified in the design tree, it can be opened for editing. This includes opening any of the original sketches that were drawn to create the feature. For example, you can open the original sketch of the wrench shown in Figure 20.22, page 688, and edit the length/width or original shape of the handle.

ANIMATION

Animation programs like 3-D Studio Viz from Autodesk let you render your solid models into 3-D motion simulations. Computer-based animations are made by recording a series of still images in various positions of incremental movement that, when played back, no longer appear individually as static images but appear to be unbroken motion. This is similar to a "flipbook" where the views on each page change slightly, and as you riffle through the pages the appearance of motion is created. Animations are used in many ways in the design engineering process including, to name just a few, to fly through the design site, verify movements of parts in an assembly, and provide illustrations of assembly for training production workers.

As the animator it is important that you plan your animation. In a similar process to video or movie production, you should consider the overall length of your animation. The length will determine how many frames you will need to produce and give you an idea of how long it will take your computer to render the animation. Keep in mind the basic principle of frames per second (the number of images needed to produce one second of film). Production film runs at 24 frames per second; therefore, a film producer must make 24 images for each second of animation that he or she wishes to create.

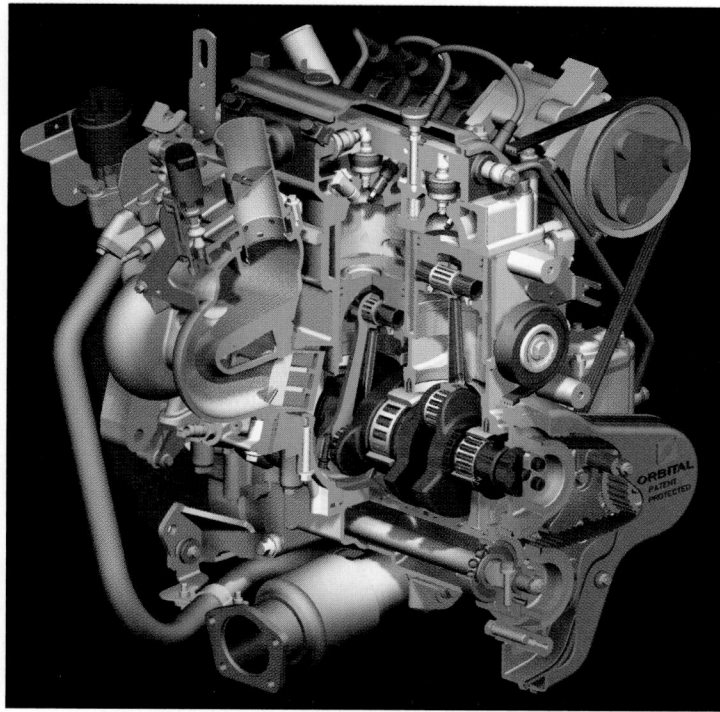

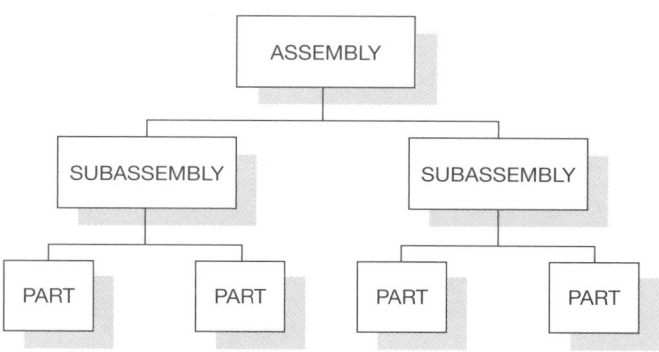

FIGURE 20.12 ■ Orbital engine assembly consists of parts and subassemblies. *Courtesy Unigraphics Solutions Inc.*

In planning to build your animation, think in terms of individual actions. For example, if you decide that the action you need to animate will take 5 seconds to complete, your computer will have to generate 120 frames of animation (multiply 5 seconds of movement by a frame speed per second of 24). Even on a high speed computer rendering these frames can be computationally intense. The time per frame will vary depending on the complexity of the objects, saturation of color, and the resolution of the images. Typically, for each frame that takes 2 to 3 minutes to render, 4 to 6 hours of rendering time are added for just the 5-second animation.

Storyboarding

It is always a good idea to do some preproduction work before you record your animation. *Storyboarding* is a process by which you sketch out the key events of the animation. These sketches will help ensure that the key scenes will be included to complete your story or demonstration. The process of storyboarding is used by

FIGURE 20.13 ■ An example of a simple bracket designed using feature based modeling.

video producers to preplan their production before costly studio editing time is incurred (professional video editing suites can run upward of $300 an hour). As discussed earlier, rendering can take days to complete even on a high speed computer. If scenes are left out of your animation, the animation will have to be redone, causing you to lose significant time and money. Renderings, like video productions, are different from live action film productions where improvising takes place. Improvising does not occur while rendering animations, and therefore they must be precisely preplanned.

When storyboarding your animation, keep your focus on your audience. This focus should include the overall length of the animation, key points that must be demonstrated, and how these key points are to be best illustrated. Storyboarding is a simple process that can be done on note cards or plain paper. Include sketches of the key scenes that show how these key events should be illustrated, and the time allotted for each.

Inverse Kinematics

Inverse kinematics (IK) is a method used for manipulating how solid objects move in an assembly. IK joins solid objects together using natural links or joints such as that illustrated in the sequence of frames of the universal joint shown in Figure 20.23, page 688. For example, IK relationships may lock the rotation of an object around one particular axis. Adding this type of information allows the solid assembly to move as the finished product would move. IK is used extensively to animate both human joint and mechanical joint movements.

Building and simulating an IK model involves a number of steps. These steps include:

■ Building a solid model of each jointed component

■ Linking the solid model together by defining the joints

■ Defining the joint behavior at each point (i.e., direction of rotation)

■ Animating the IK assembly, which can be accomplished through an animation sequence

Animation can be applied to anything; it can be used in conjunction with any product to aid its explanation. For example, a robotic arm could be more easily understood if all the parts were moving. An animation that showed the parts of an assembly actually coming together in sequence would be an impressive way to

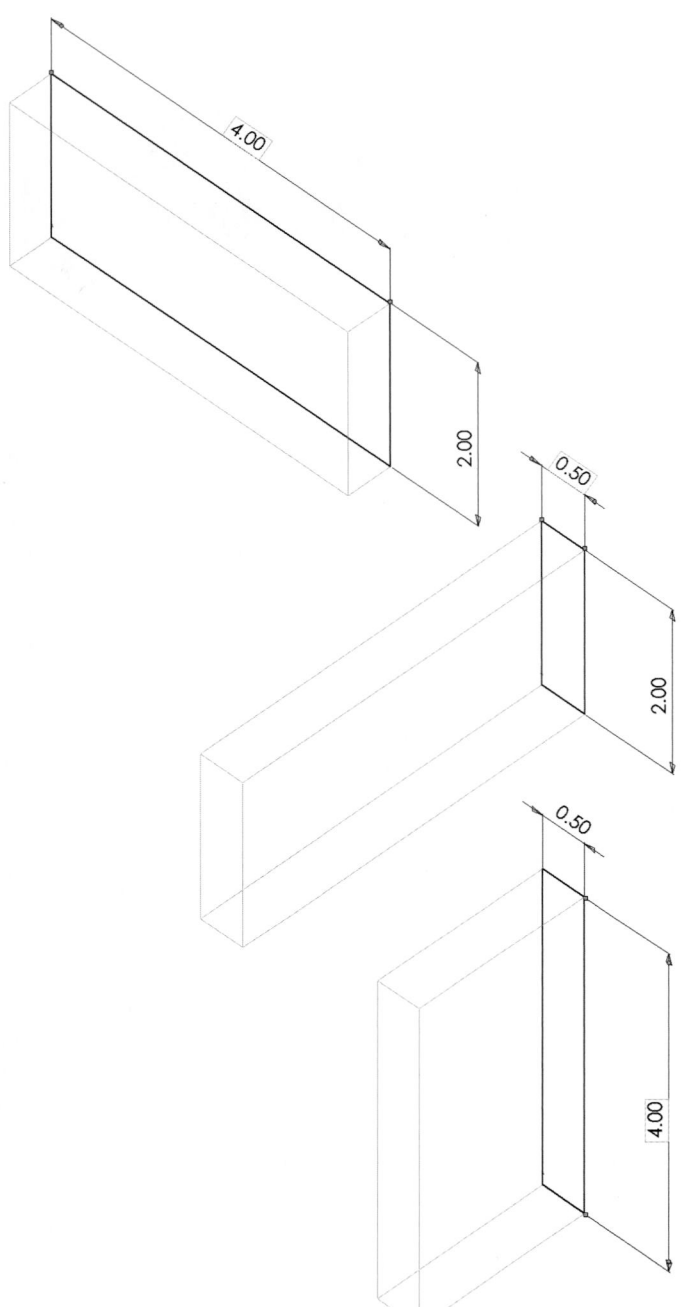

FIGURE 20.14 ■ Three different ways to create the base feature of the simple bracket shown in Figure 20.13.

train assembly workers. Animation can be used purely to illustrate a product in an environment, making for an unforgettable presentation that retains people's attention. The options for animation are limited only by creativity and imagination.

Exporting Your Animations

Most rendering software allows the user to preview the animation sequences before rendering is executed. This feature is a good way to verify that your animation will meet your expectations. When finished, you will select a rendering output file format and instruct

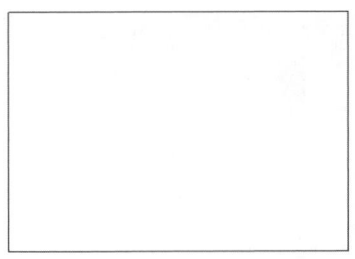

FIGURE 20.15 ■ This object was created with no constraints added; however, four automatic constraints were added to the sketch, including two vertical and two horizontal constraints.

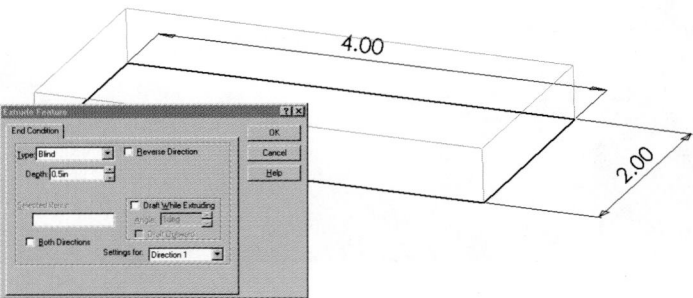

FIGURE 20.16 ■ An example of an object being extruded to ½ inch in SolidWorks. *Courtesy SolidWorks Corporation.*

FIGURE 20.17 ■ An example of how a new sketch plane is selected on SolidWorks. *Courtesy SolidWorks Corporation.*

your software to render your animation to a file. Animation software will render to a number of different file formats that will allow for convenient playback. Common files formats are AVI, Quicktime, and MPEG.

VIRTUAL REALITY

Virtual reality (VR) is a term used to describe a system that enables one or more users to move and react in a computer-simulated environment. Users manipulate virtual objects using various types of devices as though they were real objects. This simulated world gives participants a feeling of being immersed in a real world.

Virtual reality requires special interface devices that transmit the sights, sounds, and sensations of the simulated world. In return, these devices record the speech and movement of the user and transmit them back to the simulation software program.

To interact visually with the simulated world, the user wears a head-mounted display (HMD), which directs computer images at

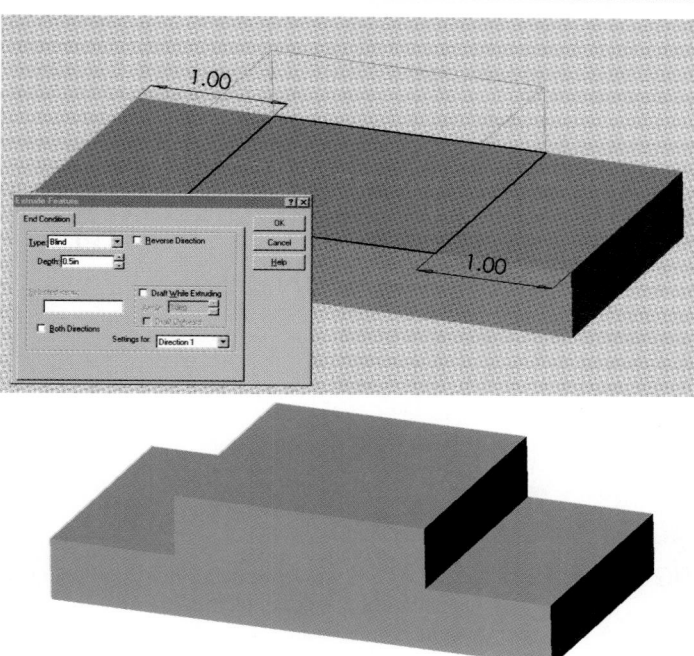

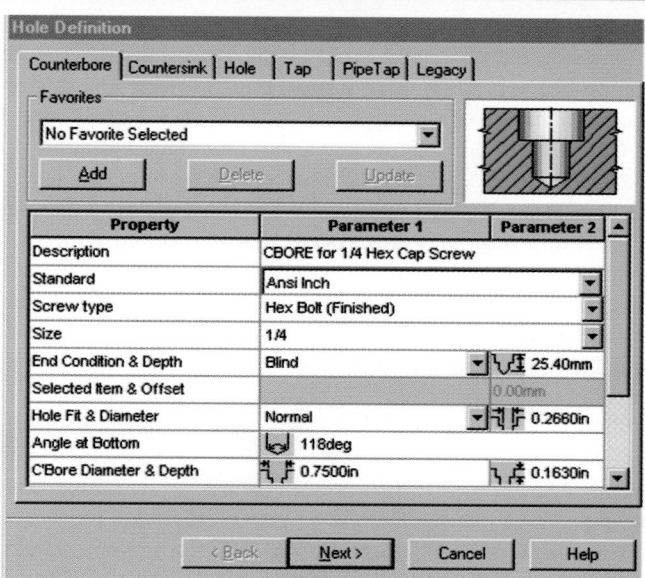

FIGURE 20.18 ■ An example of an extruded boss feature.

FIGURE 20.20 ■ Hole feature dialog box in SolidWorks. *Courtesy SolidWorks Corporation.*

FIGURE 20.19 ■ Plastic cellular telephone parts with visible rounds and fillets.

each eye (see Figure 20.24). The HMD tracks the user's head including the direction in which the user is looking. Using this information, the HMD receives updated images from the computer system, which is continually recalculating the virtual world based on the user's head motions. The computer generates new views at a rate of 10 times a second, which prevents the user's view from appearing halting and jerky and from lagging behind the user's movements. The HMD can also deliver sounds to the user via earphones. The tracking feature of the HMD can also be used to update the audio signal to simulate surround sound effects.

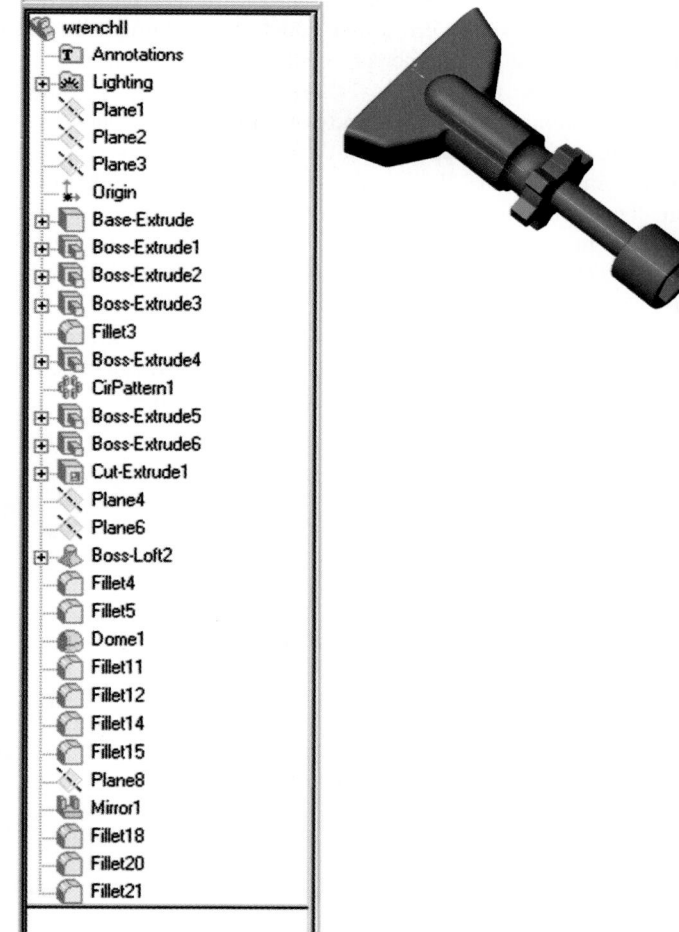

FIGURE 20.21 ■ An example of a design history tree within Solid-Works. *Courtesy SolidWorks Corporation.*

One of the problems of HMDs is the uncomfortable nature of wearing heavy headmounted display gear. To overcome this problem, two alternative concepts have been introduced, the BOOM and the CAVE. The BOOM (Binocular Omni-Orientation Monitor) from Fakespace is a head-coupled stereoscopic display device (see Figure 20.25). The heavy display, instead of being worn by the user, is attached to a multilink arm system that is counterbalanced. The user can guide the counterbalanced display while looking into it like binoculars. The system is guided by a tracking system that is attached to the counterbalanced arms. The CAVE (Cave Automatic Virtual Environment) projects stereo images on the walls and floor of a room (see Figure 20.26). CAVE was developed at the University of Illinois at Chicago to allow users to wear lightweight stereo glasses and to walk around freely inside the virtual environment. Several persons can participate within the CAVE environment as the tracking system follows the lead viewer's position and movements.

The most challenging physical sensation to simulate in a virtual world is the sense of touch. A *haptic interface* is a device that relays the sense of touch and other physical sensations. The user's hand and finger movements can be tracked, allowing the user to reach into the virtual world and handle objects. Haptic interfaces of this type hold great potential for design engineers, allowing various team members to manipulate a product design in a virtual environment in a natural way.

Although the user can handle an object, it is difficult to generate sensations associated with human touch, for example, the sensations that are felt when a person touches a soft surface, picks up a heavy object, or runs a finger across a bumpy surface. To simulate these sensations, very accurate and fast computer-controlled motors generate force by pushing against the user. These haptic devices are synchronized with HMD sight and sound, and the motors must be small enough to be worn without interfering with the user's natural movement. A simple haptic device is the desktop stylus (see Figure 20.27). This device can apply a small force, through a mechanical linkage, to a stylus held in the user's hand. When the stylus encounters a virtual object, the user is provided feedback that simulates the interaction. In addition, if the stylus is dragged across a textured surface, it responds with the proper vibration.

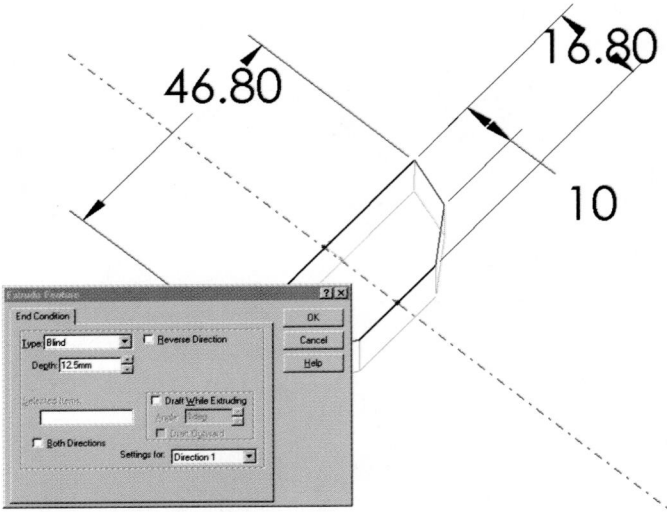

FIGURE 20.22 ■ An example of how a part can be edited using the design history tree. In this example the designer can edit any of the dimensions embedded in the history of the tool handle.

In the future, engineers may use VR to increase productivity in a variety of areas, including virtual mock-up, assembly, and design reviews. These applications may include the realistic-simulation of human factors, such as snap-fits, key component function, and experiencing of virtual forms. Virtual assemblies may include fit evaluation, maintenance path planning, manufacturability analysis, and assembly training.

An emerging area in the world of virtual reality is Web-enabled *virtual reality modeling language* (VRML). VRML is a formatting language that is used to publish virtual 3-D settings, called "worlds," on the World Wide Web. Once the developer has placed the world on the Internet, the user can view it using a Web browser plug-in. This web-browser plug-in contains controls that allow the user to move around in the virtual world, as the user would like to experience it. Currently VRML is a standard authoring language that provides authoring tools for the creation of 3-D worlds with integrated hyper-

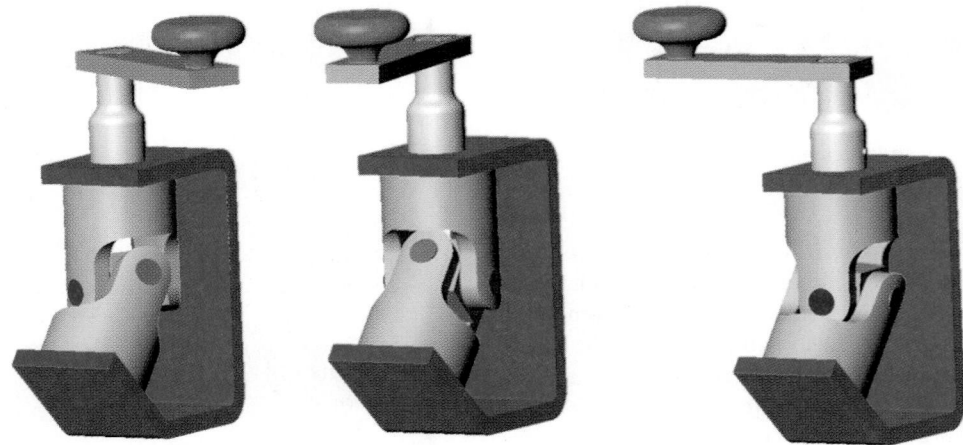

FIGURE 20.23 ■ An example of inverse kinematics (IK) is used to animate a simple part. This simple universal joint has been constrained around its axis. When the crank handle is turned, the constrained assembly rotates appropriately.

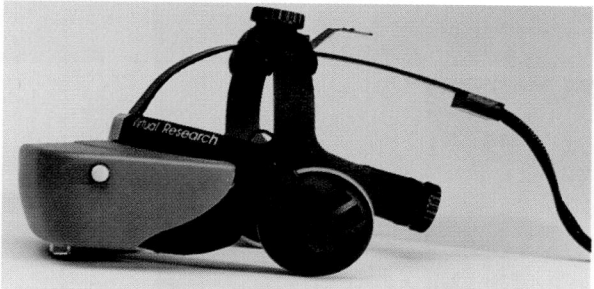

FIGURE 20.24 ■ Virtual reality (VR) glasses are a head-mounted display that allows the wearer to view 3-D images. *Courtesy Virtual Research.*

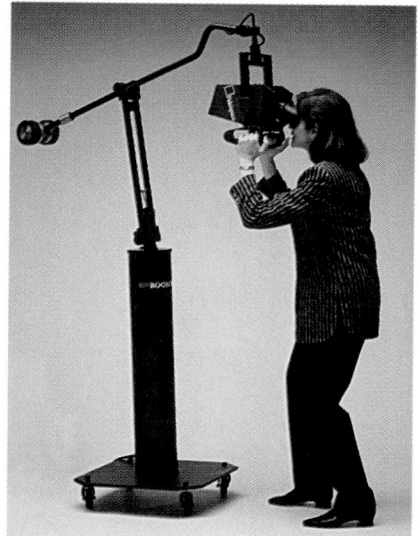

FIGURE 20.25 ■ Binocular Omni-Orientation Monitor (BOOM) allows viewers to view 3-D images without wearing a head-mounted display. *Courtesy Fakespace Labs, Inc.*

FIGURE 20.26 ■ Cave Automatic Virtual Environment (CAVE) allows viewers to view 3-D images projected on the walls of a room. *Courtesy Fakespaces Systems, Inc.*

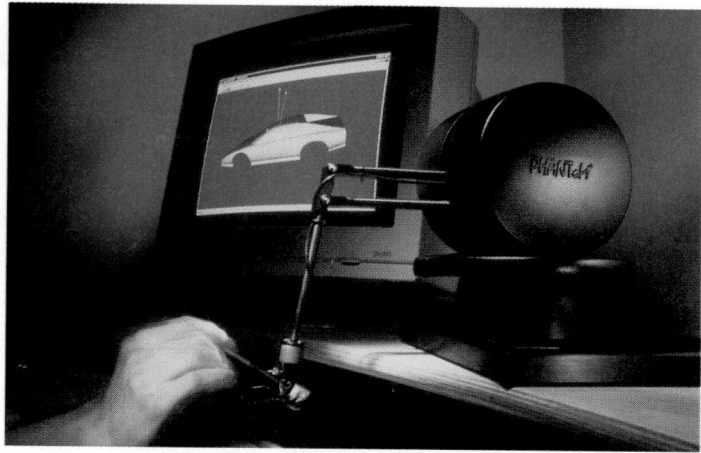

FIGURE 20.27 ■ Phantom haptic device provides haptic feedback to the user. *Courtesy SensAble Technologies.*

links. The current version of VRML is viewed using a basic computer monitor and, therefore, is not fully immersive. However, the future of VRML should incorporate the use of HMDs and haptic devices, making for more truly immersive environments.

PROFESSIONAL PERSPECTIVE

CAD solid modeling has significantly changed the process by which products at Waterpik® Technologies, Inc., are developed. Waterpik Technologies, Inc. uses ProEngineer to develop its products. This high-end solid modeler has allowed the product development team to reduce the product lifecycle (the time it takes a product to get to the market) and has allowed them to remain competitive in today's ultra-aggressive marketplace.

Waterpik Technologies began using ProEngineer only 10 years ago. Within this short period of time, the number and diversity of products that have been introduced has changed significantly. Waterpik will introduce 30 new products in 2001 as opposed to only four 10 years ago. Furthermore, these 30 products are being developed using a smaller number of employees than that used to develop the four in previous years. Simply stated, a smaller number of people have been able to produce a significantly larger number of products. These gains in product development have been made possible because of the advent of the CAD solid model.

CAD solid models have provided an important tool for improving communications. At Waterpik the CAD solid model has been used for demonstrating future products. Before solid models, industrial engineers relied on basic multiview drawings or hand-made models. Multiview drawings are difficult for the untrained person to read and interpret. Therefore, a communication gap was created between those who could not read the drawings (marketing, sales) and those who could (engineering, manufacturing). The CAD solid model has given marketing, sales, and engineering a way to visualize the design

(Continued)

before any physical models are created. This improvement in communications, has shortened the time needed to make a product and has improved the product's design.

Maybe even more significant is the impact that solid modeling has had on the appearance of the products at Waterpik. "Solid modeling has significantly changed the complexity of our products. Before we had to design things that were easy to machine. This meant products with straight lines or minimal curves. Today our products don't have a straight line on them" (Tim Hanson, Waterpik Technologies). These changes can be seen by comparing Waterpik's designs from before and after the introduction of solid modeling. The toothbrush shown in Figure 20.28 was produced in 1989, before solid modeling. Notice that the product's shape is based on simple forms and straight lines. In contrast, the toothbrush shown in Figure 20.29 was produced in 1998, using solid modeling. This product has very few simple forms and is comprised of a number of complex surfaces. Furthermore, the Flosser shown in Figure 20.30 has an extremely complex form that incorporates a variety of shapes and complex curves. These complex shapes and forms were simply not possible before CAD solid modeling was introduced.

All of these advances in speed, communication, and product design occurred at Waterpik because of solid modeling. "Without the solid-modeling tool we would not be able to produce the number of high-quality products that we produce today" (Tim Hanson, Waterpik Technologies).

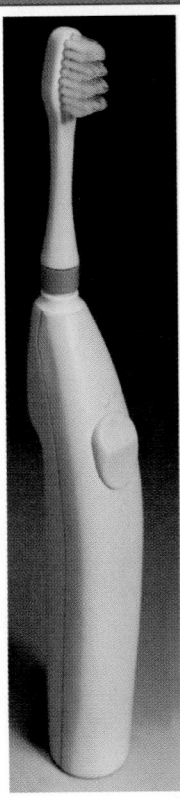

FIGURE 20.29 ■ An example of a toothbrush design after the introduction of solid modeling. *Courtesy Waterpik Technologies, Inc.*

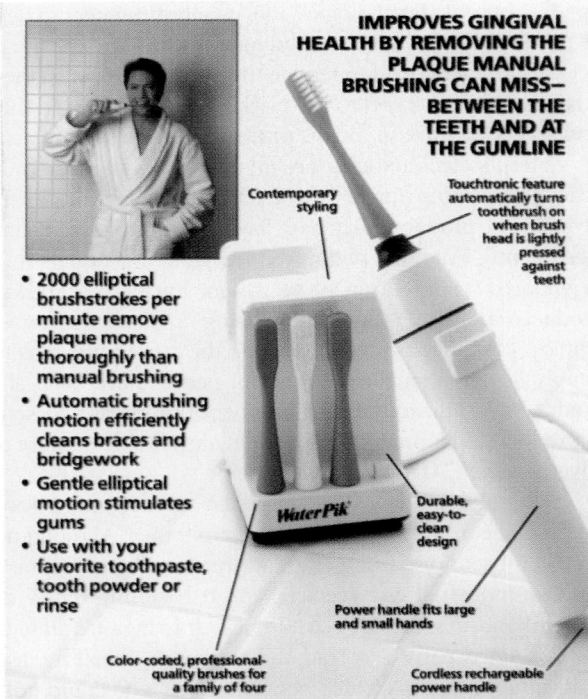

IMPROVES GINGIVAL HEALTH BY REMOVING THE PLAQUE MANUAL BRUSHING CAN MISS— BETWEEN THE TEETH AND AT THE GUMLINE

Contemporary styling

Touchtronic feature automatically turns toothbrush on when brush head is lightly pressed against teeth

- 2000 elliptical brushstrokes per minute remove plaque more thoroughly than manual brushing
- Automatic brushing motion efficiently cleans braces and bridgework
- Gentle elliptical motion stimulates gums
- Use with your favorite toothpaste, tooth powder or rinse

WaterPik

Durable, easy-to-clean design

Power handle fits large and small hands

Color-coded, professional-quality brushes for a family of four

Cordless rechargeable power handle

FIGURE 20.28 ■ An example of a toothbrush design before the introduction of solid modeling. *Courtesy Waterpik Technologies, Inc.*

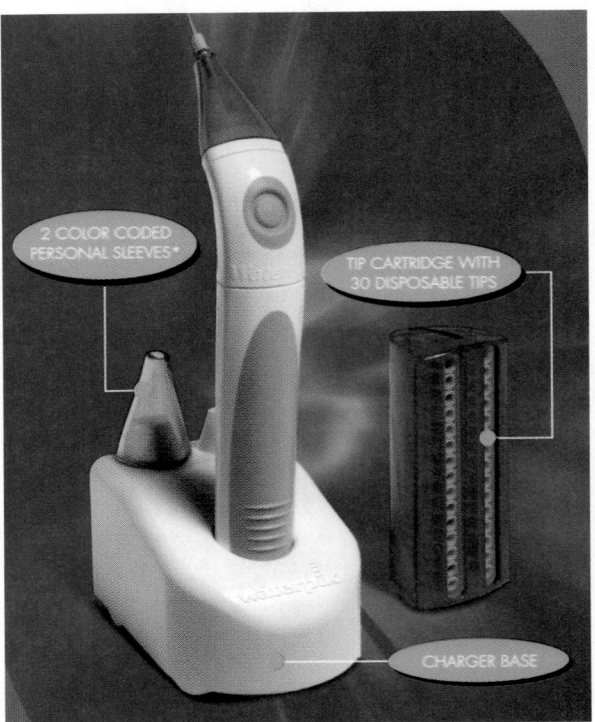

2 COLOR CODED PERSONAL SLEEVES*

TIP CARTRIDGE WITH 30 DISPOSABLE TIPS

CHARGER BASE

FIGURE 20.30 ■ An example of a freeform design of the Teledyne Flosser. *Courtesy Waterpik Technologies, Inc.*

CHAPTER 20

Solid Modeling, Animation, and Virtual Reality Problems

This chapter is intended as a reference for solid modeling, animation, and virtual reality as used within the engineering design process. As technologies surrounding these concepts become more advanced, their use within the engineering design process changes. Those working in the engineering design process must continually learn and make decisions about whether new technologies will improve their job performance. The following problems have been designed to help you explore these concepts further and collect additional resources on these topics.

PROBLEM 20-1 This assignment requires that you visit your library or online resource to locate articles related to the engineering design process. Select one topic from the following list. Research professional and trade journals to find a related article. Write a critique of that article. Your article critique should include an introduction, body, and critical review of the article. The introduction should include an overview of the article. The body should contain the main points of the article. The critique should provide a critical review of the article. In order to complete the critique section, you may want to ask yourself the following questions: For whom is the book or article written? Are the author's viewpoints consistent with other authorities? Are the author's viewpoints supported with valid reasons? Does the author adequately cover the subject? Where do you agree or disagree with the author? Why? How might the article be strengthened? What implications do you find for your field of work?

■ Solid modeling
■ Animation
■ Virtual reality
■ Parametric modeling

PROBLEM 20-2 This assignment requires that you visit the World Wide Web to find additional information on the use of solid modeling, animation, or virtual reality. Select one of the following topics and locate two companies that provide related services. Develop an oral presentation on each company's services. Include examples of the technologies each company uses to provide these services.

■ Product development
■ Aerospace engineering
■ Automotive engineering
■ Heavy equipment engineering
■ Production/manufacturing

PROBLEM 20-3 This assignment requires that you conduct research on solid modeling applications. Visit two solid modeling software companies via the World Wide Web. Develop a report about those companies. Provide information about the following items related to each company.

■ Functionality of software (e.g., analysis tools, surface modeling)
■ Major customers/users of software
■ Kernel that is used within the software
■ How each company makes revisions or updates to their software
■ Other pertinent information

CHAPTER 20

Solid Modeling, Animation, and Virtual Reality Test

DIRECTIONS

Answer the questions with short, complete statements.

PART 1 • 3-D CAD MODELS

1. List both the advantages and disadvantages for each modeling type (wireframe, surface, and solid models).
2. Define wireframe modeling.
3. Define surface modeling.
4. Define solid modeling.
5. Why would you want to use a solid 3-D CAD model of your design?
6. List three areas where CAD models can be used within the product development process.

PART 2 • SOLID MODELING

7. Define solid modeling.
8. Explain what is meant by an "intelligent" solid.
9. What are the advantages of using solid models?

PART 3 • PARAMETRIC MODELING

10. Define parametric modeling.
11. Define overconstrained.
12. Define underconstrained.
13. Explain the advantages of parametric modeling over solid modeling

PART 4 • THE PARAMETRIC MODELING PROCESS

14. Explain how the parametric modeling supports the systems approach to engineering design.
15. Describe the basic steps in the parametric modeling process
16. Define parts.
17. Define assembly.
18. Define feature.

PART 5 • PARAMETRIC BUILT-IN FEATURES

19. What are built-in features?
20. What are the advantages of using built-in features?
21. List four-built in features.

PART 6 • PARAMETRIC HISTORY TREES

22. Define history trees.
23. What are the advantages to having a history tree?
24. List two editing functions that can be done using the history tree.

PART 7 • THE ANIMATION PROCESS

25. Define animation.
26. Define storyboarding.
27. Define inverse kinematics.
28. What can be done with inverse kinematics?
29. Explain the importance of preplanning your animations.

PART 8 • VIRTUAL REALITY

30. Define virtual reality.
31. What is the purpose of the HMD?
32. What role does virtual reality play in the engineering design process?
33. Define haptic.
34. What does VRML stand for, and what role may VRML play in the future of virtual reality?

CHAPTER 21

Welding Processes and Representations

OBJECTIVES

After completing this chapter, you will:

■ Identify welding processes.

■ Draw welding representations and provide proper welding symbols and notes.

■ Draw weldments from engineering sketches and actual industrial layouts.

THE ENGINEERING DESIGN APPLICATION

The engineer has just handed you a sketch showing the details for a couple of weldments. The welding details are not shown as standard welding symbols but rather as written instructions, and it is up to you to design the appropriate weld symbols for the weldment drawing.

The first weldment sketch shows an assembly of two aluminum bars connected in a lap joint; the instructions are as follows: *Use ³/₈" plug weld, ⁷/₁₆" deep with a 45° included angle. Space the welds at 3" on center, for the length of the material.*

Since aluminum welding requires special considerations, you examine the thickness of the material and decide on a *gas tungsten arc welding* process. Using this information, you design the weld symbol shown in Figure 21.1.

The second weldment sketch shows two low carbon steel plates joined in a T-joint. The instructions are: *Use ¼" intermittent fillet weld on both sides. The welds should be 2" long with a pitch of 10" and staggered. Field weld at installation site.* The low carbon steel material and T-joint suggest a *shielded metal arc welding* process, and you design an appropriate symbol (Figure 21.2).

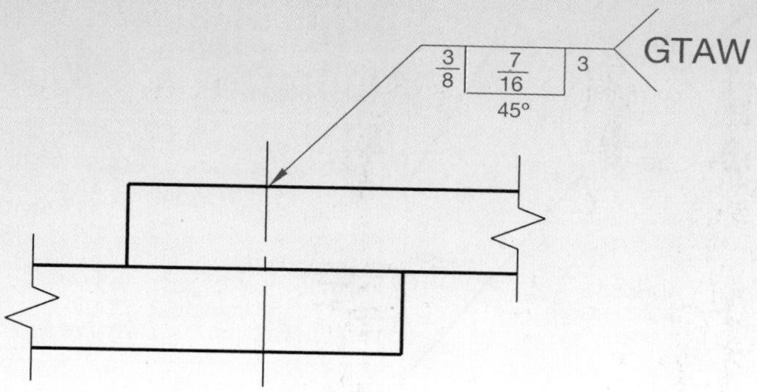

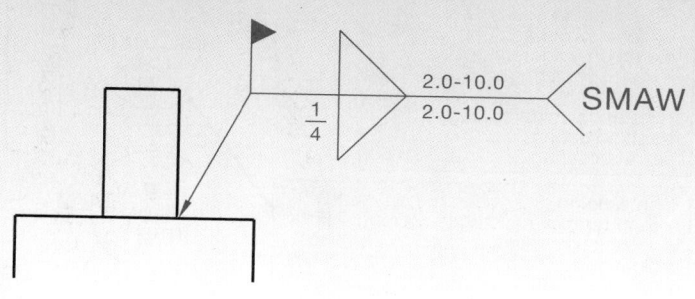

FIGURE 21.1 ■ Plug weld symbol.

FIGURE 21.2 ■ Fillet weld symbol.

Welding is a process of joining two or more pieces of like metals by heating the material to a temperature high enough to cause softening or melting. The location of the weld is where the materials actually combine the grain structure from one piece to the other. The parts that are welded become one, and the properly welded joint is as strong or stronger than the original material. Welding may be performed with or without pressure applied to the materials. Some materials may actually be welded together by pressure alone. Most welding operations, however, are performed by filling a heated joint between pieces with molten metal.

Welding is actually a method of fastening adjacent parts. Welding was not discussed with fasteners in Chapter 13 because the weld is a more permanent fastening application than screw threads or pins, for example. Welding is a common fastening method used in many manufacturing applications and industries from automobile to aircraft manufacturing, and from computers to ship building. Some of the advantages of welding over other fastening methods include better strength, better weight distribution and reduction, a possible decrease in the size of castings or forgings needed in an assembly, and a potential savings of time and manufacturing costs.

WELDING PROCESSES

There are a large number of welding processes available for use in industry, as shown in Figure 21.3. The most common welding processes include oxygen gas welding, shielded metal arc welding, gas tungsten arc welding, and gas metal arc welding.

Oxygen Gas Welding

Oxygen gas welding, commonly known as *oxyfuel* or *oxyacetylene welding*, may also be performed with such fuels as natural gas, propane, or propylene. Oxyfuel welding is most typically used to fabricate thin materials, such as sheet metal and thin-wall pipe or tubing. Oxyfuel processes are also used for repair work and metal cutting. One advantage of oxyfuel welding is that the equipment and operating costs are less than with other methods. But other welding methods have advanced over the oxyfuel process because they are faster, cleaner, and cause less material

distortion. Common oxyfuel welding and cutting equipment is shown in Figure 21.4.

Also associated with oxyfuel applications are soldering, brazing, and braze welding. These methods are more of a bonding process than welding, as the base material remains solid while a filler metal is melted into a joint. Soldering and brazing differ in application temperature. *Soldering* is done below 450° C and brazing above 450° C. Like alloys may be used depending on their melting temperatures. The filler generally associated with soldering is solder. *Solder* is an alloy of tin and lead. The filler metal associated with brazing is an alloy of copper and zinc. *Brazing* is a process of joining two very closely fitting metals by heating the pieces, causing the filler metal to be drawn into the joint by capillary action. *Braze welding* is more of a joint filling process that does not rely on capillary action. Another process that uses an oxyfuel mixture is *flame cutting*. This process uses a high-temperature gas flame to preheat the metal to a kindling temperature, at which time a stream of pure oxygen is injected to cause the cutting action.

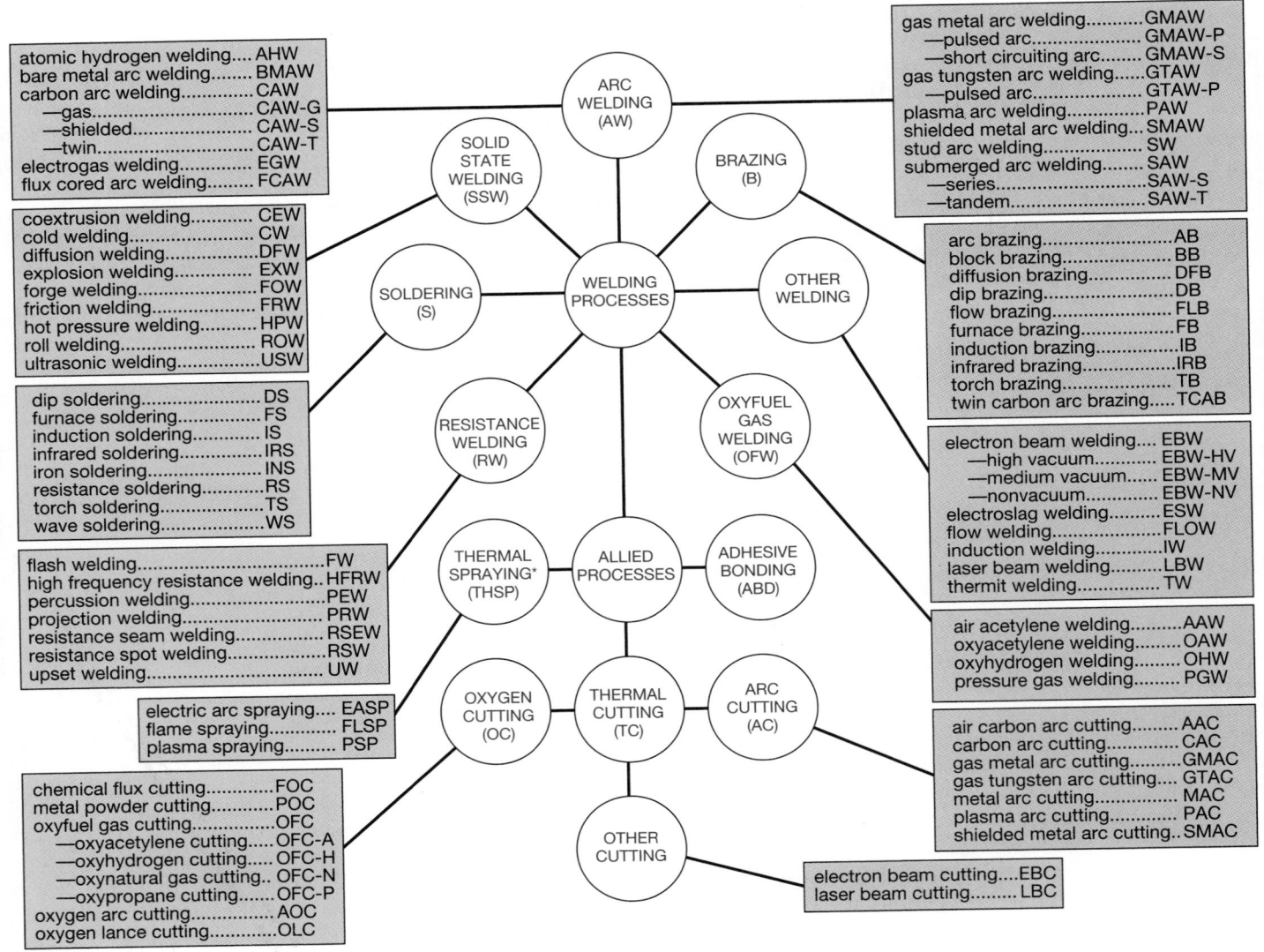

FIGURE 21.3 ■ Master chart of welding and allied processes. *Courtesy American Welding Society.*

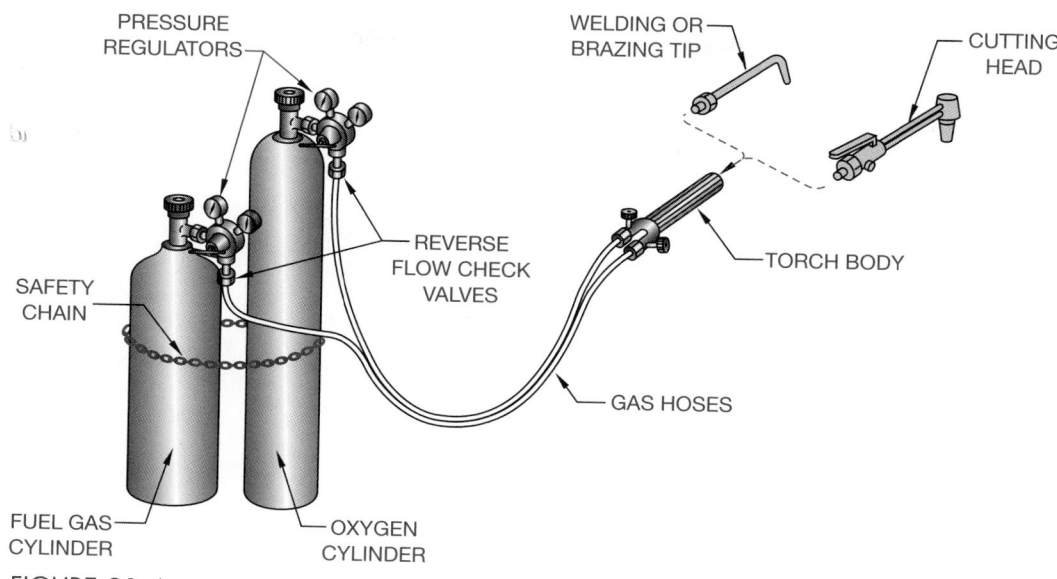

FIGURE 21.4 ■ Oxyfuel welding and cutting equipment.

Shielded Metal Arc Welding

Shielded metal arc or *stick electrode* welding is the most traditionally used welding method. High-quality welds on a variety of metals and thicknesses can be made rapidly with excellent uniformity. This method uses a flux-covered metal electrode to carry an electrical current forming an arc that melts the work and the electrode. The molten metal from the electrode mixes with the melting base material, forming the weld. Shielded metal arc welding is popular because of low-cost equipment and supplies, flexibility, portability, and versatility. Figure 21.5 shows a shielded metal arc welding setup.

Gas Tungsten Arc Welding

The *gas tungsten arc welding* process is sometimes referred to as *TIG (tungsten inert gas)* welding, or as *Heliarc®*, which is a trademark of the Union Carbide Corporation. Gas tungsten arc welding can be performed on a wider variety of materials than shielded metal arc welding, and produces clean, high-quality welds. This welding process is useful for certain materials and applications. Gas tungsten arc welding is generally limited to thin materials, high-integrity joints, or small parts, because of its slow welding speed and high cost of equipment and materials. (See Figure 21.6.)

Gas Metal Arc Welding

Another welding process that is extremely fast, economical, and produces a very clean weld is *gas metal arc welding*. This process may be used to weld thin material or heavy plate. It was originally used previously for welding aluminum using a metal inert gas shield, a process which became referred to as MIG. The present application employs a current-carrying wire that is fed into a joint between pieces to form the weld. This welding process is used in

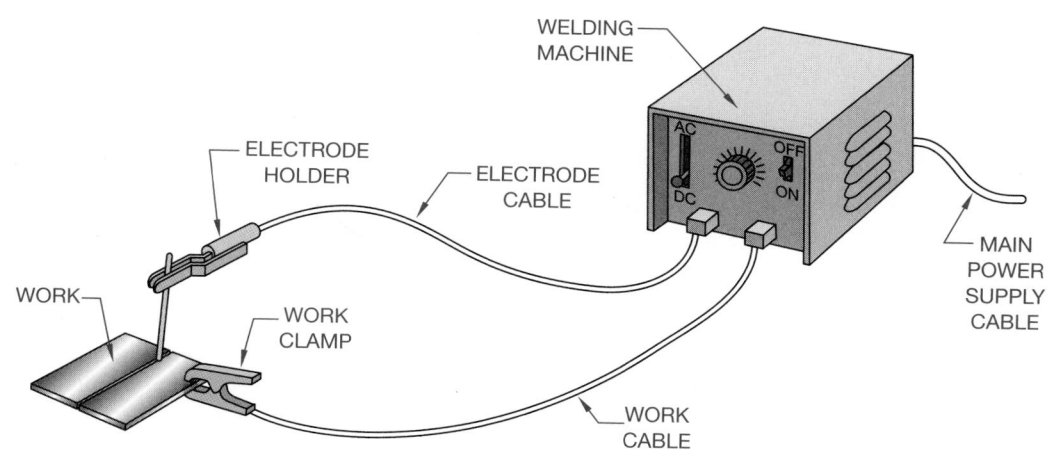

FIGURE 21.5 ■ Shielded metal arc welding equipment.

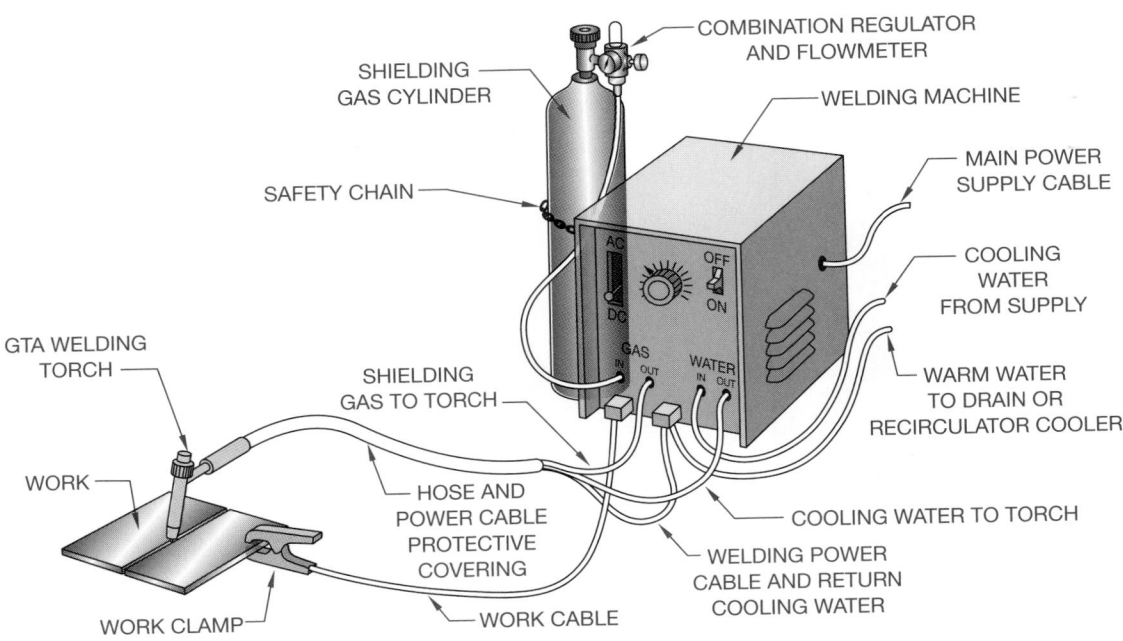

FIGURE 21.6 ■ Gas tungsten arc welding equipment.

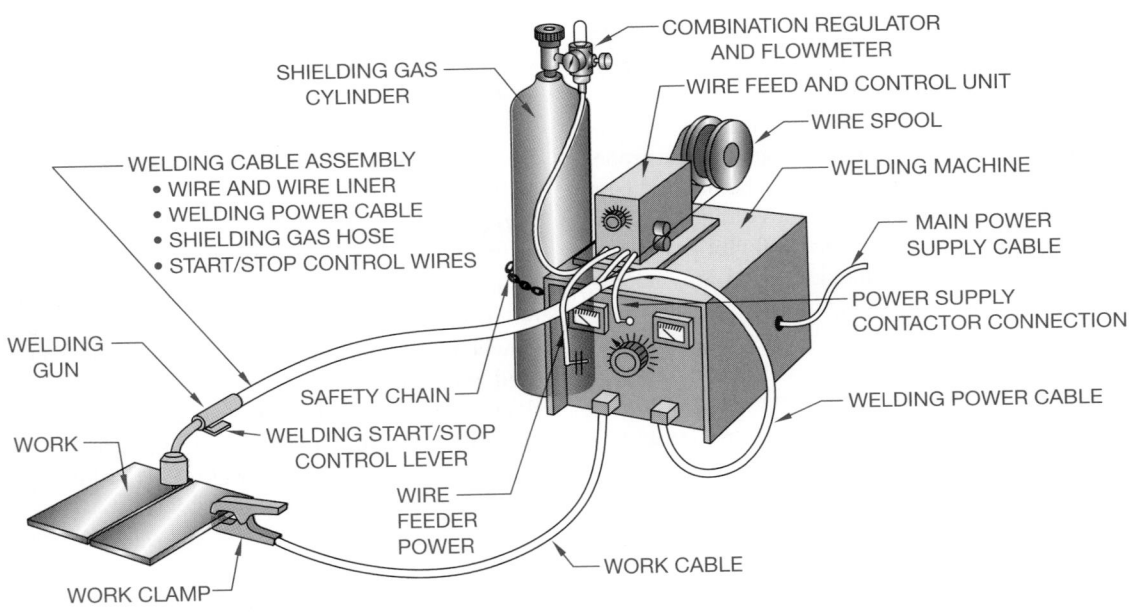

FIGURE 21.7 ■ Gas metal arc welding equipment.

industry with automatic or robotic welding machines to produce rapidly made, high-quality welds in any welding position. While the capital expense of the equipment remains high, the cost is declining due to its popularity. Figure 21.7 shows gas metal arc welding equipment.

ELEMENTS OF WELDING DRAWINGS

Welding drawings are made up of several parts to be welded together. These drawings are usually called *weldments,* or *weld-ing assemblies* or *subassemblies.* The welding assembly will typically show the parts together in multiview with all of the fabrication dimensions, types of joints, and weld symbols. Welding symbols identify the type of weld, the location of the weld, the welding process, the size and length of the weld, and other weld information. The welding assembly has a list of materials that generally provides a key to the assembly, the number of each part, part size, and material. Figure 21.8 shows a welding subassembly. When additional clarity of component parts must be given, then detailed drawings of each part are prepared, as shown in Figure 21.9.

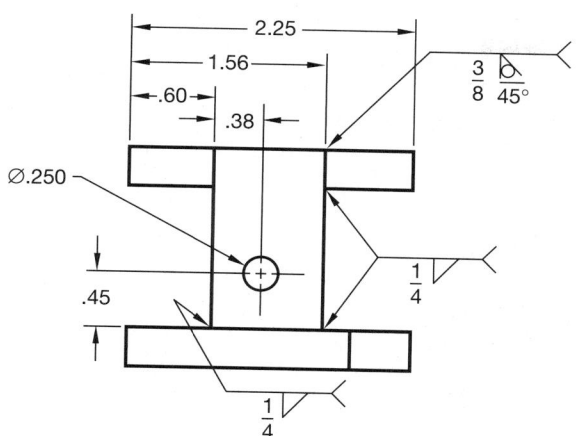

FIGURE 21.8 ■ Welded subassembly.

Welding Symbols

AWS The welding symbol represents complete information about the weld. The welding symbols discussed here are in accordance with the American Welding Society document AWS A2.4.

There are a few basic components of a welding symbol, beginning with the reference line, tail, and leader which are drawn as thin lines, as shown in Figure 21.10. The reference line is usually the first welding symbol element to be located on the drawing. The reference line leader may be drawn using the same rules associated with leaders for notes. The leader may be drawn at any angle, although angles less than 15° or greater than 75° should be avoided. Also, leaders for notes typically run from the shoulder directly to the feature. This practice should be used for welding leaders, although sometimes welding leaders bend to point into difficult-to-reach places, or with

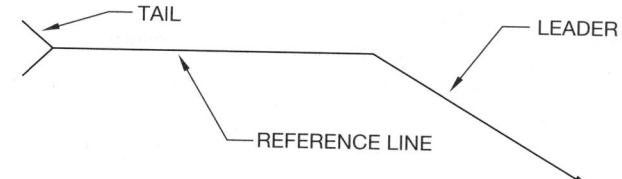

FIGURE 21.10 ■ Welding symbols: reference line, tail, and leader.

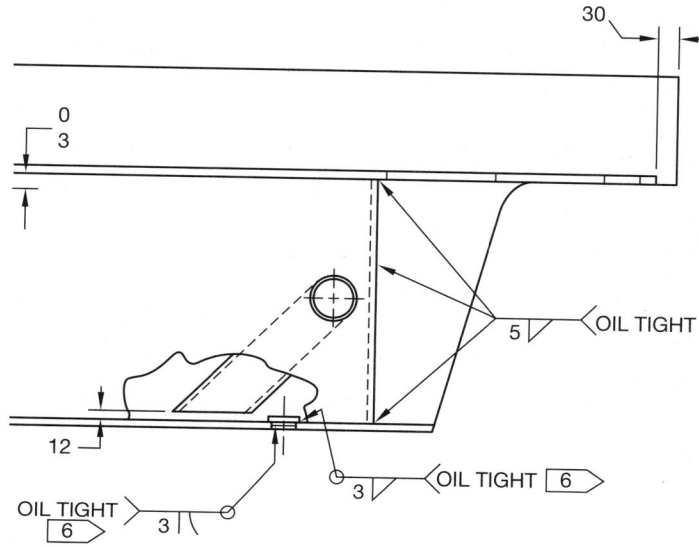

FIGURE 21.11 ■ Welding symbol leader use. *Courtesy Hyster Company.*

more than one leader extending from the same reference line, as shown in Figure 21.11.

After the reference line has been established, additional information is placed on the reference line to continue the weld specification. Figure 21.12 shows the standard location of welding symbol elements as related to the reference line, tail, and leader.

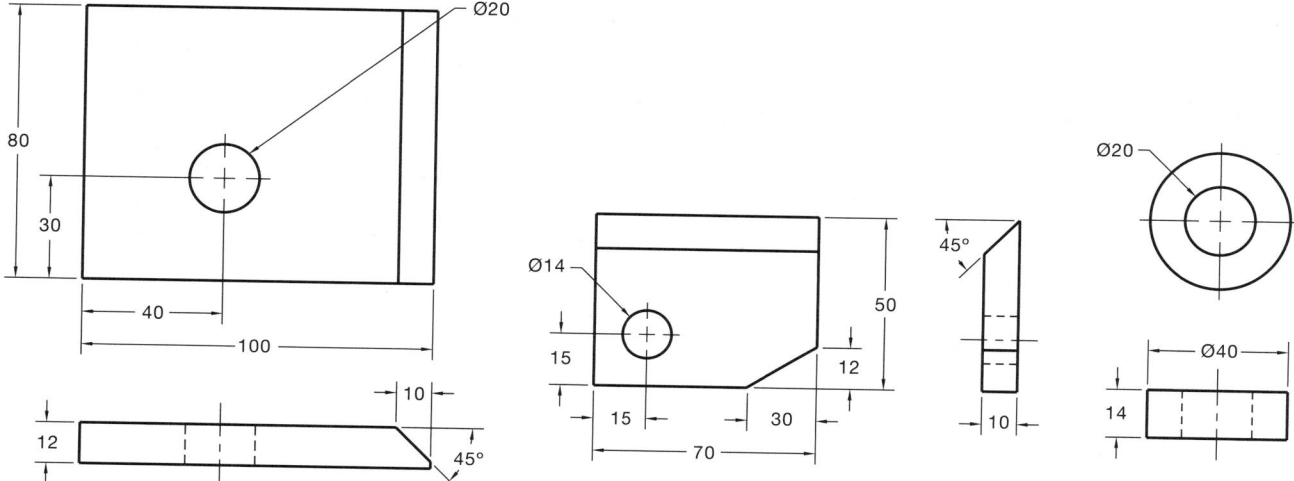

FIGURE 21.9 ■ Drawings of each part of the welded subassembly.

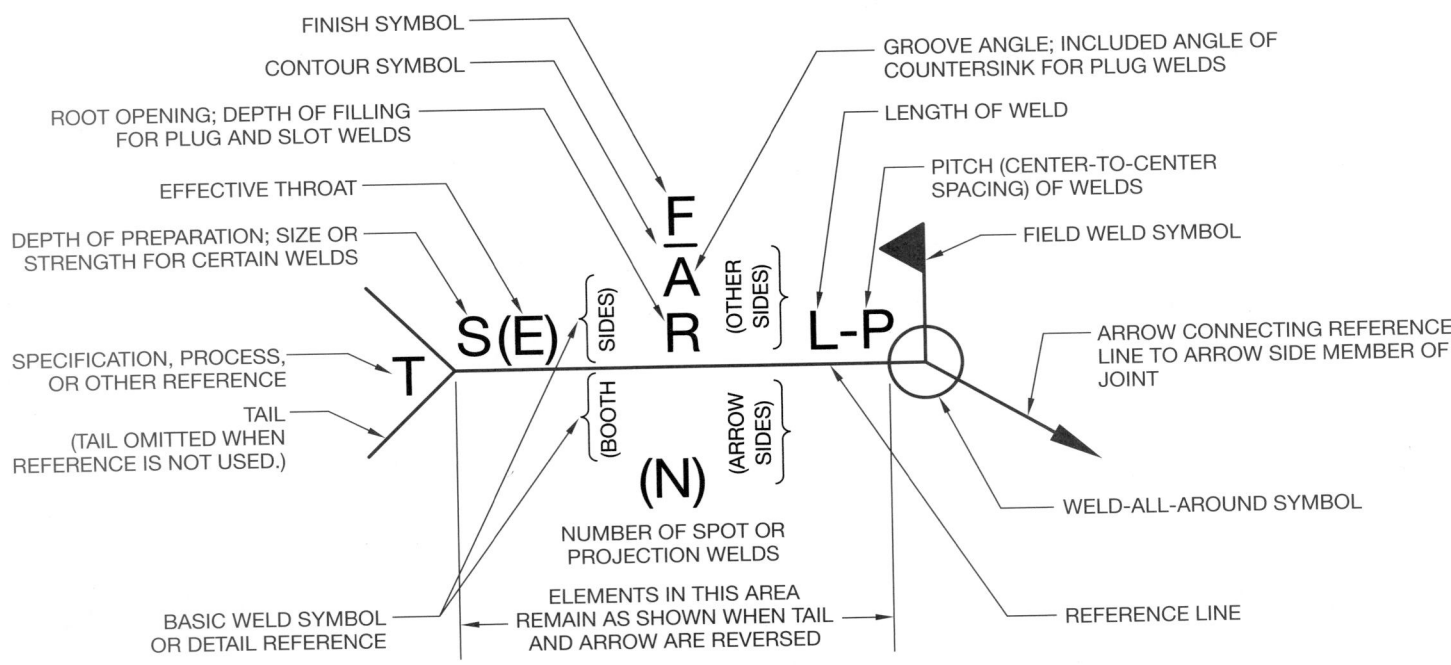

FIGURE 21.12 ■ Standard location of elements of a welding symbol. *Courtesy American Welding Society.*

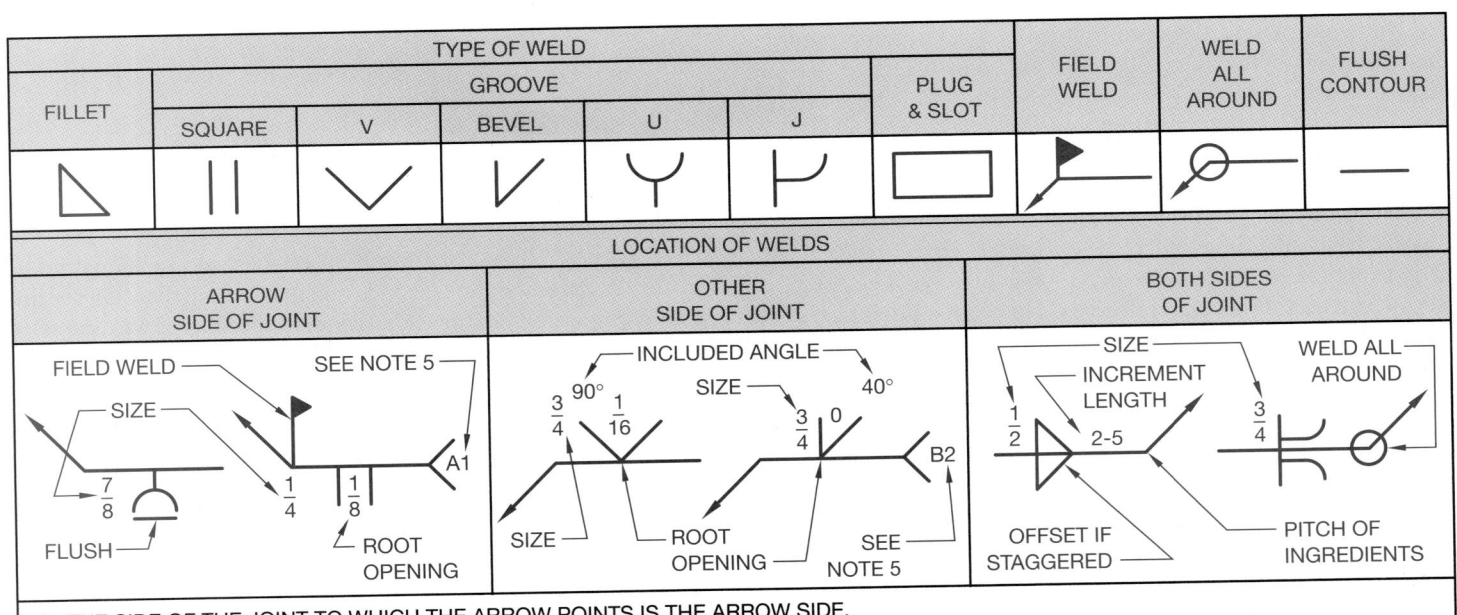

1. THE SIDE OF THE JOINT TO WHICH THE ARROW POINTS IS THE ARROW SIDE.
2. BOTH-SIDES WELDS OF SAME TYPE ARE OF SAME SIZE UNLESS OTHERWISE SHOWN.
3. SYMBOLS APPLY BETWEEN ABRUPT CHANGES IN DIRECTION OF JOINT OR AS DIMENSIONED (EXCEPT WHERE ALL AROUND SYMBOL IS USED).
4. ALL WELDS ARE CONTINUOUS AND OF USER'S STANDARD PROPORTIONS, UNLESS OTHERWISE SHOWN.
5. TAIL OF ARROW USED FOR SPECIFICATION REFERENCE. (TAIL MAY BE OMITTED WHEN REFERENCE NOT USED.)
6. DIMENSIONS OF WELD SIZES, INCREMENT LENGTHS, AND SPACING IN INCHES.

FIGURE 21.13 ■ Standard welding symbols. *Courtesy American Welding Society.*

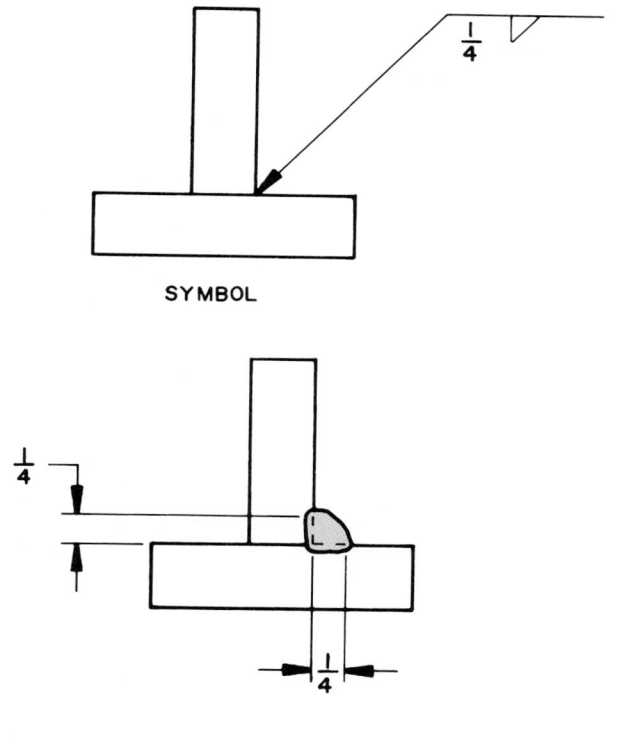

FIGURE 21.14 ■ Fillet weld with equal legs.

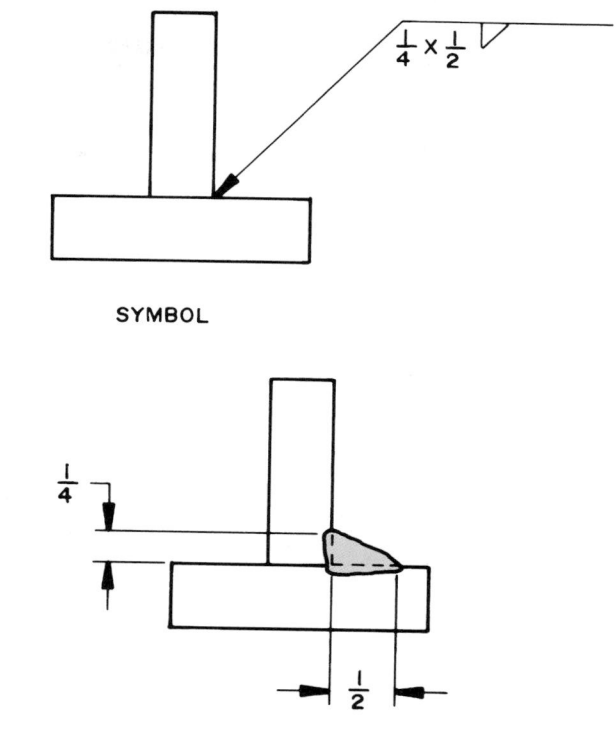

FIGURE 21.15 ■ Fillet weld with different leg lengths.

Types of Welds

The next information that is applied to the reference line is the type of weld. The type of weld is associated with the weld shape and/or the type of groove to which the weld is applied. Figure 21.13 shows the information that is associated with the types of welds.

Fillet Weld

A *fillet weld* is formed in the internal corner of the angle formed by two pieces of metal. The size of the fillet weld is shown on the same side of the reference line as the weld symbol and to the left of the symbol. When both legs of the fillet weld are the same, the size is given once, as shown in Figure 21.14. When the leg lengths are different in size, the vertical dimension is followed by the horizontal dimension. (See Figure 21.15.)

Square Groove Weld

A *square groove weld* is applied to a butt joint between two pieces of metal. The two pieces of metal are spaced apart a given distance, known as the *root opening*. If the root opening distance is a standard in the company, then this dimension is assumed. If the root opening is not standard, then the specified dimension is given inside of the square groove symbol, as shown in Figure 21.16.

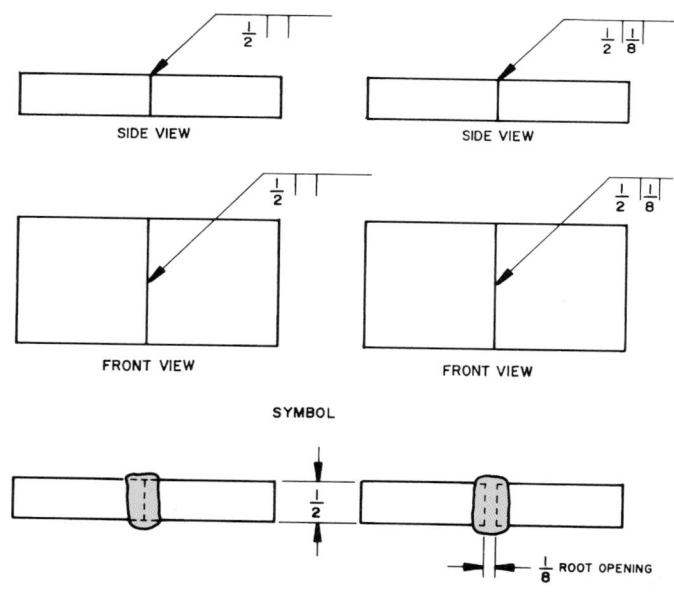

FIGURE 21.16 ■ Square groove weld showing the root opening.

V Groove Weld

A *V groove weld* is formed between two adjacent parts when the side of each part is beveled to form a groove between the parts in the shape of a V. The included angle of the V may be given with or without a root opening, as shown in Figure 21.17.

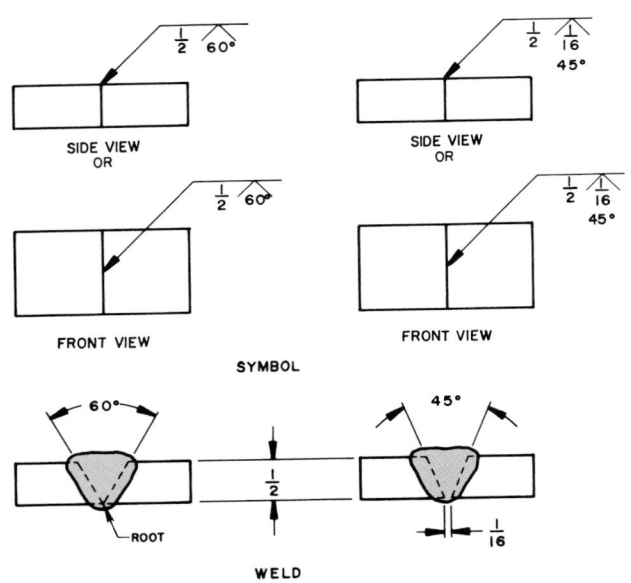

FIGURE 21.17 ■ V groove weld.

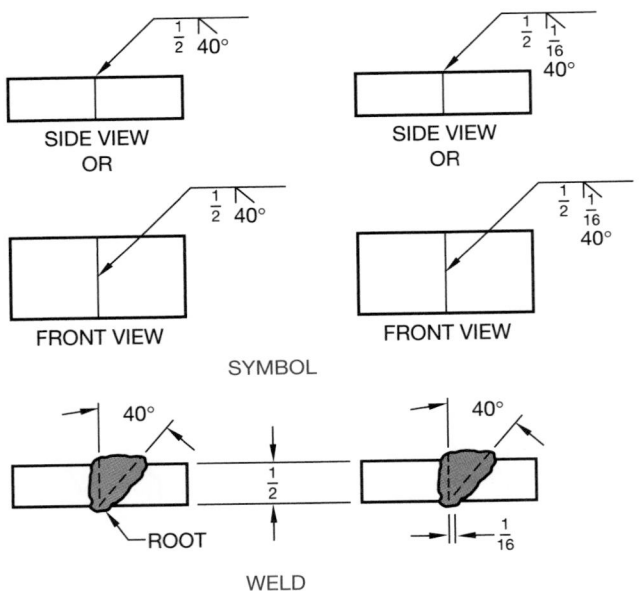

FIGURE 21.18 ■ Bevel groove weld.

Bevel Groove Weld

The *bevel groove weld* is created when one piece is square and the other piece has a beveled surface. The bevel weld may be given with a bevel angle and a root opening, as shown in Figure 21.18.

U Groove Weld

A *U groove weld* is created when the groove between two parts is in the form of a U. The angle formed by the sides of the U shape, the root, and the weld size are generally given. (See Figure 21.19.)

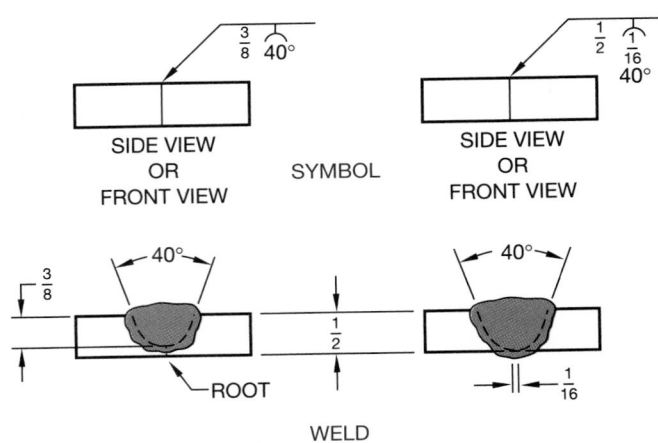

FIGURE 21.19 ■ U groove weld.

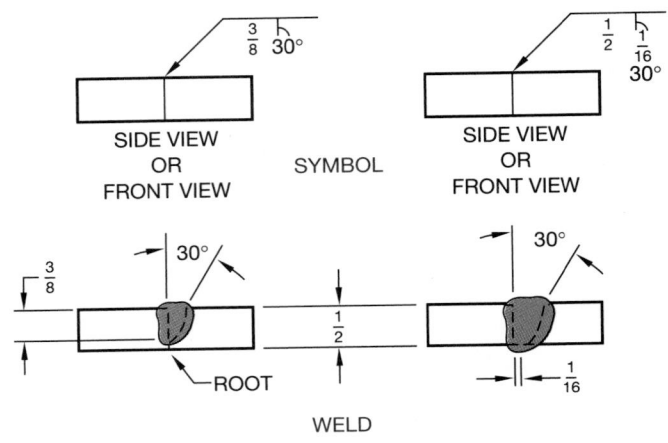

FIGURE 21.20 ■ J groove weld.

J Groove Weld

The *J groove weld* is necessary when one piece is a square cut and the other piece is in a J-shaped groove. The included angle, the root opening, and the weld size are given, as shown in Figure 21.20.

Plug Weld

A *plug weld* is made in a hole in one piece of metal that is lapped over another piece of metal. These welds are specified by giving the weld size, angle, depth, and pitch. (See Figure 21.21a.) The same type of weld may be applied to a slot. This is referred to as a slot weld as shown in Figure 21.21b.

Field Weld

A *field weld* is a weld that is performed in the field as opposed to in a fabrication shop. The reason for this application may be that the individual components may be easier to transport disassembled, or the mounting procedure may require job site installation. The field weld symbol is a flag attached to the reference line at the leader intersection, as shown in Figures 21.12 and 21.13.

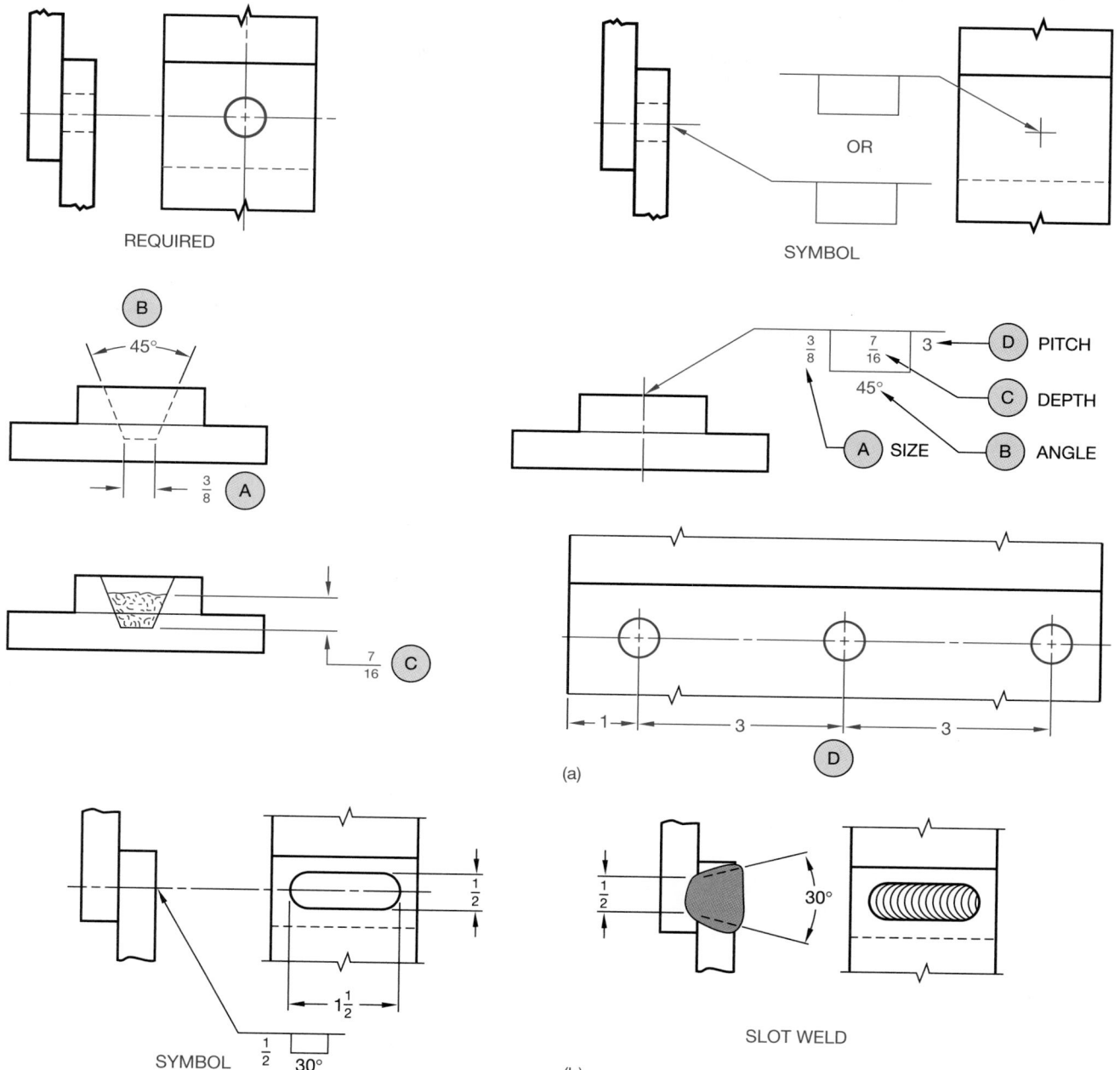

FIGURE 21.21 ▪ (a) Plug welds. (b) Slot weld.

Weld-all-around

When a welded connection must be performed all around a feature, the *weld-all-around* symbol is attached to the reference line at the junction of the leader. (See Figure 21.22.)

Flush Contour Weld

Generally the surface contour of a weld is raised above the surface face. If this is undesirable, a flush surface symbol must be applied to the weld symbol. When the *flush contour weld* symbol is applied without any further consideration, then the welder must perform this effect without any finishing. The other option is to specify a flush finish using another process. The letter designating the other process is placed above the flush contour symbol for another side application, or below the flush contour symbol for an arrow-side application. The options include: C = chipping, G = grinding, M = machining, R = rolling, or H = hammering. (See Figure 21.23.)

Weld Length and Increment

When a weld is not continuous along the length of a part, then the weld length should be given. In some situations, the weld along the length of a feature is given in lengths spaced a given distance apart.

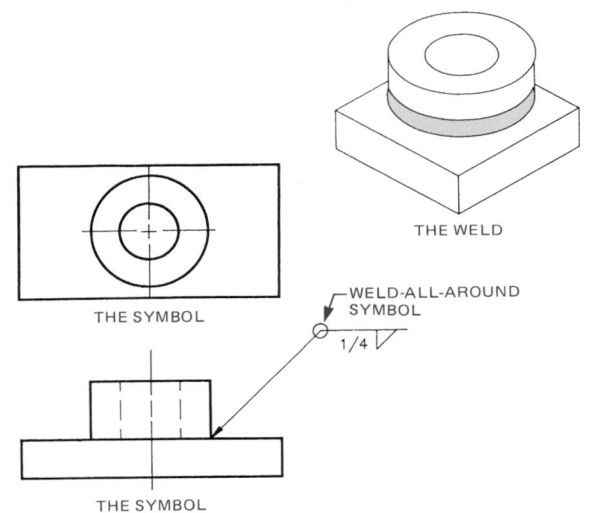

FIGURE 21.22 ■ Weld-all-around.

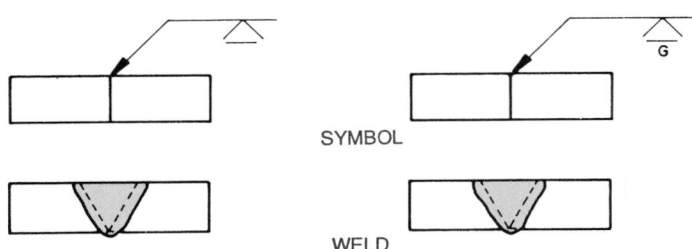

FIGURE 21.23 ■ Flush contour weld.

The distance from one point on a weld length to the same corresponding point on the next weld is called the *pitch;* generally from center to center of welds. The weld length and increment are shown to the right of the weld symbol, as shown in Figure 21.24.

Spot Weld

Spot welding is a process of resistance welding where the base materials are clamped between two electrodes and a momentary electric current produces the heat for welding at the contact spot. Spot welding is generally associated with welding sheet metal lap seams. The size of the spot weld is given as a diameter to the left of the symbol. The center-to-center pitch is given to the right of the symbol. (See Figure 21.25.) The strength of the spot welds may be given as minimum shear strength in pounds per spot to the left of the symbol, as shown in Figure 21.26a. When a specific number of spot welds is required in a seam or joint, then the quantity is placed above or below the symbol in parentheses, as shown in Figure 21.26b.

Seam Weld

A *seam weld,* another type of resistance weld, is a continuous weld made between or upon overlapping members. The continuous weld may consist of a single weld bead or a series of overlapping spot welds. The dimensions of seam welds, shown on the same side

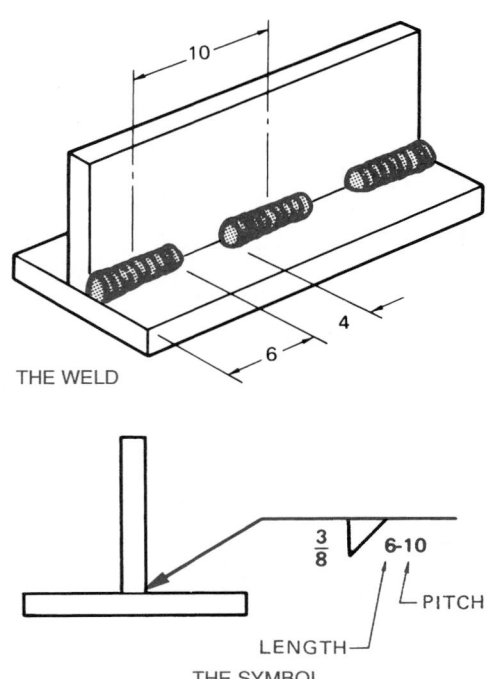

FIGURE 21.24 ■ Intermittent fillet weld.

of the reference line as the weld symbol, relate to either size or strength. The weld size width is shown to the left of the symbol in fractional or decimal inches, or millimeters. (See Figure 21.27a.) The weld length, when specified, is provided to the right of the symbol, as shown in Figure 21.27b. The strength of a seam weld is expressed as minimum acceptable shear strength in pounds per linear inch, placed to the left of the weld symbol, as shown in Figure 21.28.

Weld Symbol Leader Arrow Related to Weld Location

Welding symbols are applied to the joint as the basic reference. All joints have an *arrow side* and an *other side.* When fillet and groove welds are used, the welding symbol leader arrows connect the symbol reference line to one side of the joint known as the arrow side. The side opposite the location of the arrow is called the other side. If the weld is to be deposited on the arrow side of the joint, the proper weld symbol is placed *below* the reference line, as shown in Figure 21.29a. If the weld is to be deposited on the side of the joint opposite the arrow, then the weld symbol is placed *above* the reference line. (See Figure 21.29b.) When welds are to be deposited on both sides of the joint, then the same weld symbol is shown *above and below* the reference line, as shown in Figure 21.29c and d.

For plug, spot, seam, or resistance welding symbols, the leader arrow connects the welding symbol reference line to the outer surface of one of the members of the joint at the center line of the desired weld. The member that the arrow points to is considered the arrow side member. The member opposite of the arrow is considered the other side member.

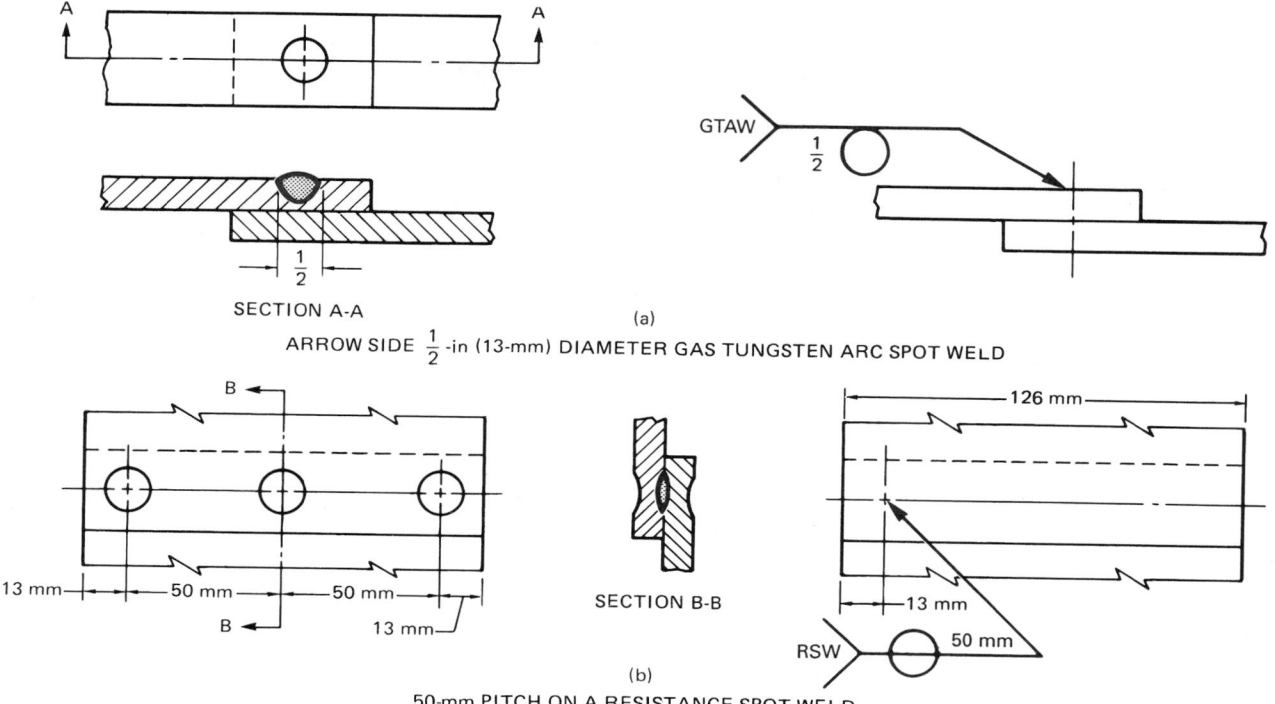

SECTION A-A

(a)

ARROW SIDE $\frac{1}{2}$-in (13-mm) DIAMETER GAS TUNGSTEN ARC SPOT WELD

SECTION B-B

(b)

50-mm PITCH ON A RESISTANCE SPOT WELD

FIGURE 21.25 ■ Designating spot welds.

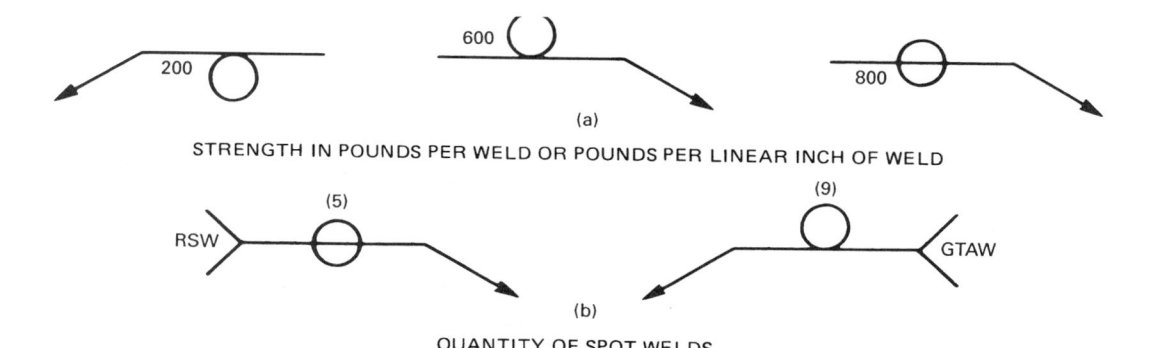

(a)

STRENGTH IN POUNDS PER WELD OR POUNDS PER LINEAR INCH OF WELD

(b)

QUANTITY OF SPOT WELDS

FIGURE 21.26 ■ Designating strength and number of spot welds.

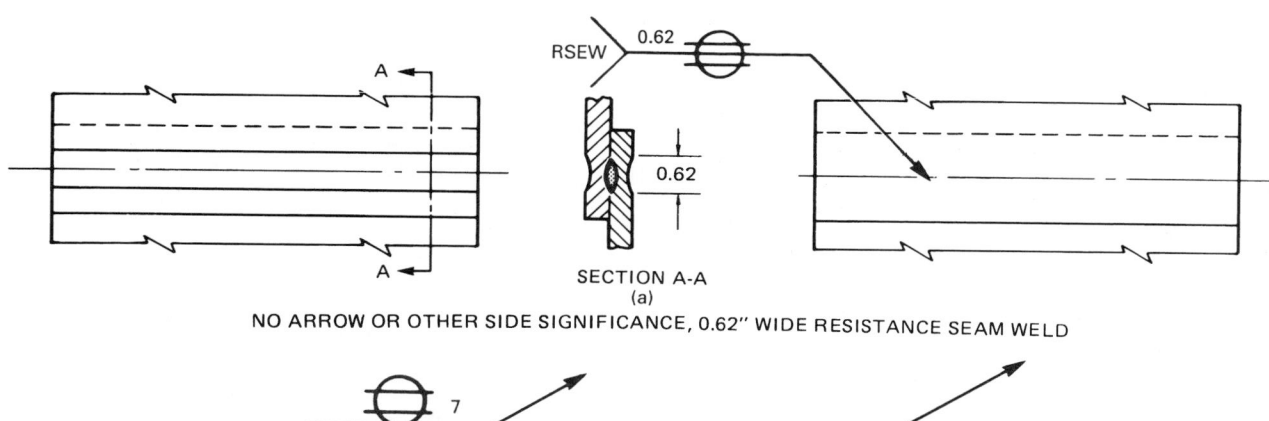

SECTION A-A

(a)

NO ARROW OR OTHER SIDE SIGNIFICANCE, 0.62″ WIDE RESISTANCE SEAM WELD

(b)

LENGTH OF SEAM WELD

FIGURE 21.27 ■ Indicating the length of a seam weld.

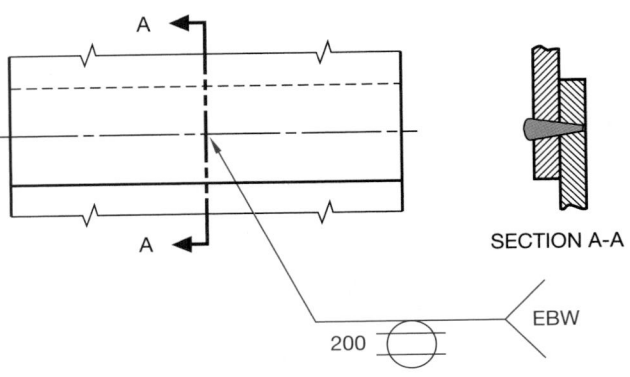

FIGURE 21.28 ■ Indicating the strength of a seam weld.

Additional Weld Characteristics

Weld Penetration

Unless otherwise specified, a weld penetrates through the thickness of the parts at the joint. The size of the groove weld remains to the left of the weld symbol. Figure 21.30a shows the size of grooved welds with partial penetration. Notice in Figure 21.30b that a weld with partial penetration may specify the depth of the groove followed by the depth of weld penetration in parentheses, with both items placed to the left of the weld symbol.

When single-groove and symmetrical double-groove welds penetrate completely through the parts being joined, the weld size may be omitted, as shown in Figure 21.31. The depth of penetration of flare-formed groove welds is assumed to extend to the tangent points of the members, as shown in Figure 21.32.

Flange Welds

Flange welds are used on light-gauge metal joints where the edges to be joined are flanged or flared. Dimensions of flange welds are placed to the left of the weld symbol. Further, the radius and height of the weld above the point of tangency are indicated by showing both the radius and the height separated by a plus (+) symbol. The size of the flange weld is then placed outward of the flange dimensions. (See Figure 21.33.)

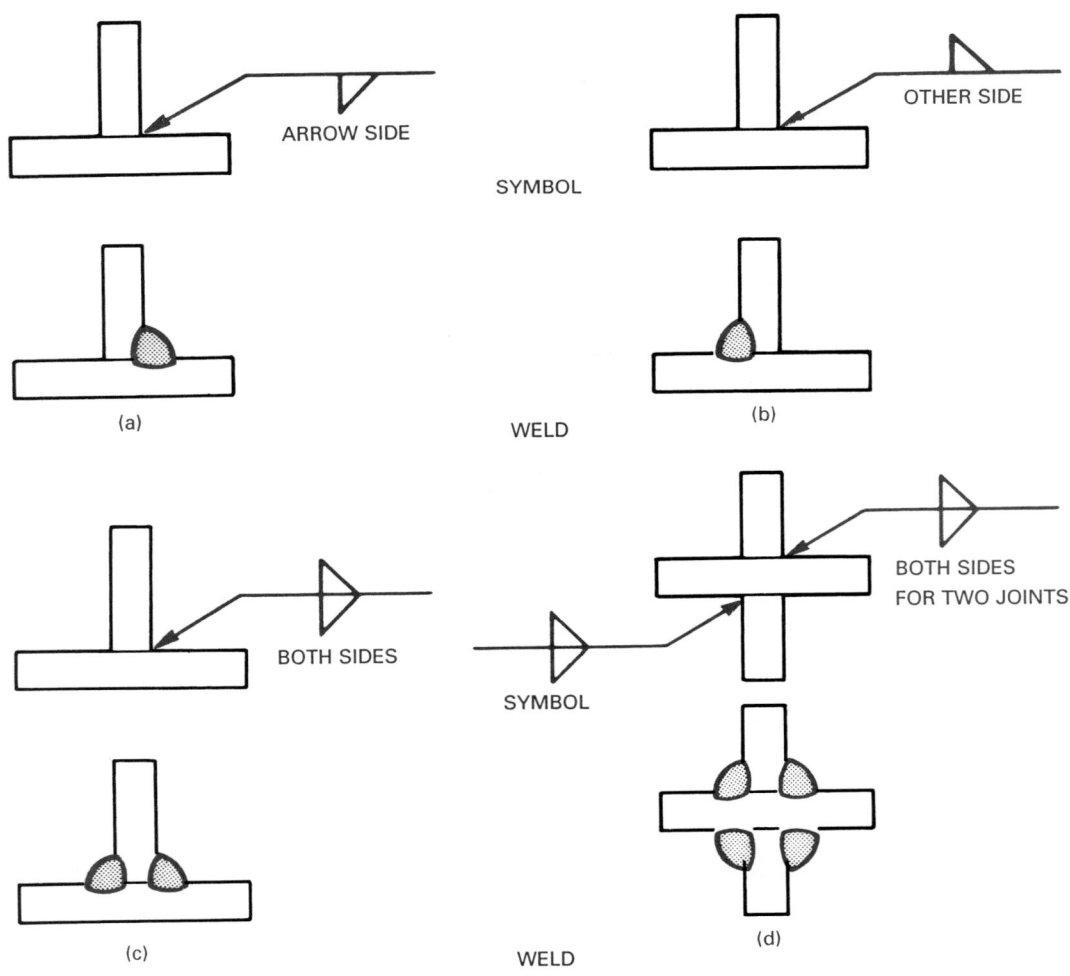

FIGURE 21.29 ■ Designating weld locations. *Courtesy American Welding Society.*

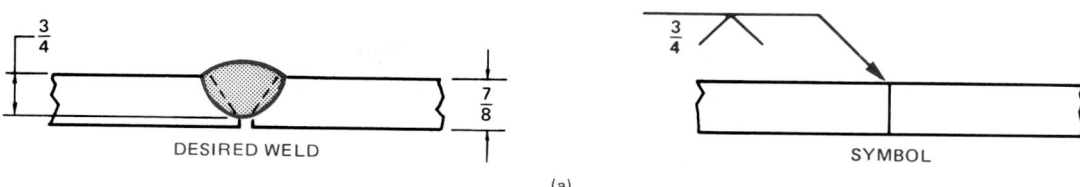

DESIGNATING THE SIZE OF GROOVED WELDS WITH PARTIAL PENETRATION

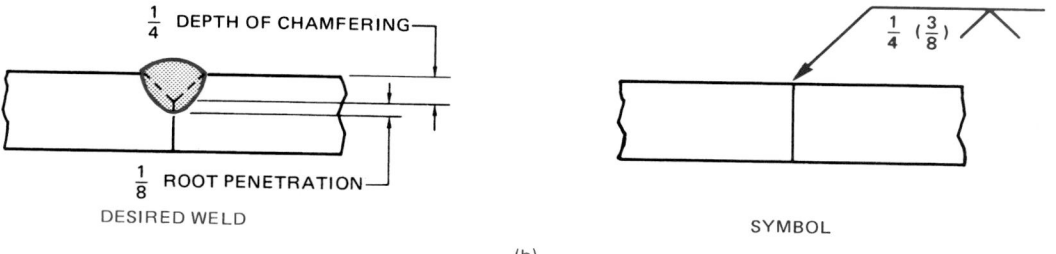

SHOWING SIZE AND ROOT PENETRATION OF GROOVED WELDS

FIGURE 21.30 ■ (a) Designating the size of grooved welds with partial penetration. (b) Showing size and penetration of grooved welds. *Courtesy American Welding Society.*

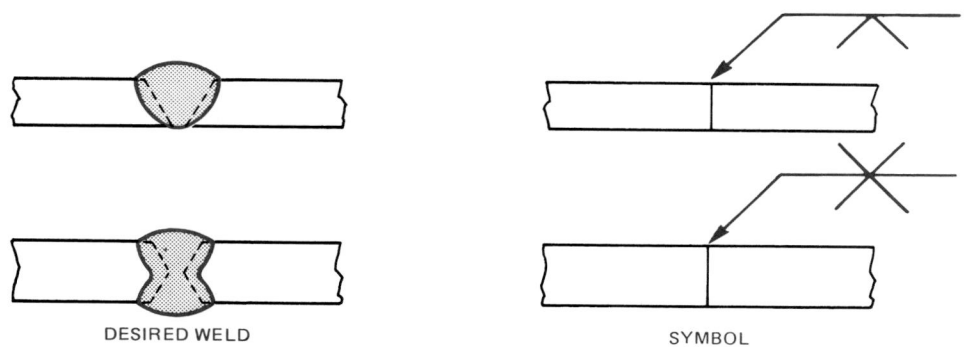

FIGURE 21.31 ■ Designating single- and double-groove welds with complete penetration. *Courtesy American Welding Society.*

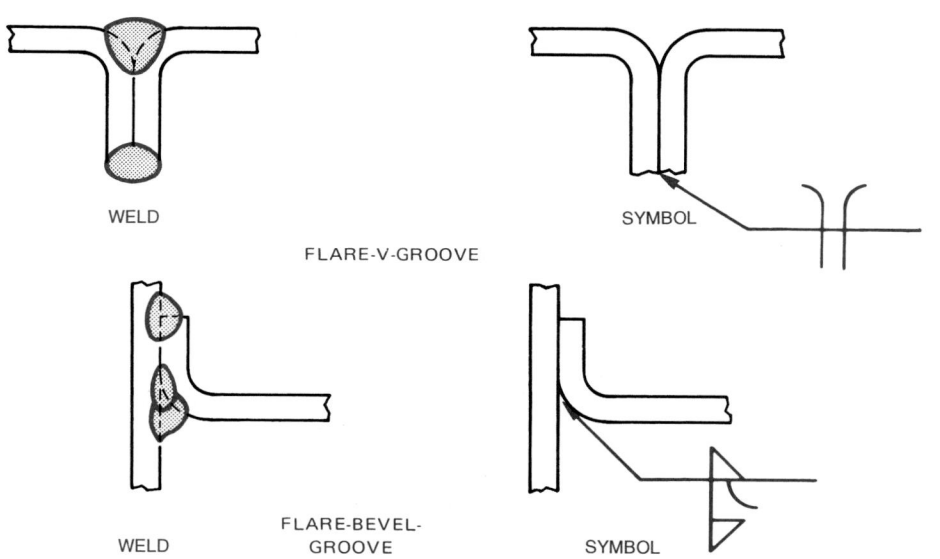

FIGURE 21.32 ■ Designating flare-V and flare-bevel-groove welds. *Courtesy American Welding Society.*

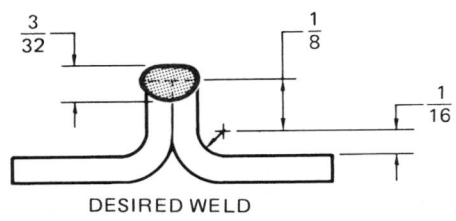

DESIRED WELD

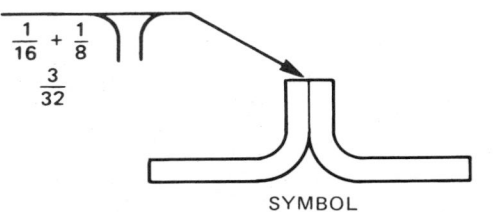

SYMBOL

FIGURE 21.33 ■ Applying dimensions to flange welds. *Courtesy American Welding Society.*

Welding Process Designation

The tail is added to the welding symbol when it is necessary to designate the welding specification, procedures, or other supplementary information needed to fabricate the weld. (See Figure 21.34.)

Weld Joints

The types of weld joints are often closely associated with the types of weld grooves already discussed. The weld grooves may be applied to any of the typical joint types. The weld joints used in most weldments are the butt, lap, tee, outside corner, and edge joints shown in Figure 21.35.

WELDING TESTS

There are two types of welding tests—destructive and nondestructive tests.

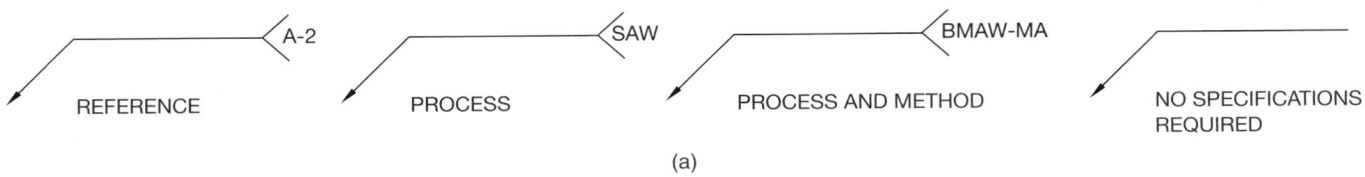

(a)

	Welding Process	Letter Designation
Brazing	Torch brazing	TB
	Induction brazing	IB
	Resistance brazing	RB
Flow Welding	Flow welding	FLOW
Induction Welding	Induction welding	IW
Arc Welding	Bare metal arc welding	BMAW
	Submerged arc welding	SAW
	Shielded metal arc welding	SMAW
	Carbon arc welding	CAW
	Oxyhydrogen welding	OHW
Gas Welding	Oxyacetylene welding	OAW

The following suffixes may be added if desired to indicate the method of applying the above processes:

	Automatic welding	-AU
	Machine welding	-ME
	Manual welding	-MA
	Semiautomatic welding	-SA

(b)

FIGURE 21.34 ■ (a) Locations for weld specifications, processes, and other references on weld symbols. (b) Designation of welding processes.

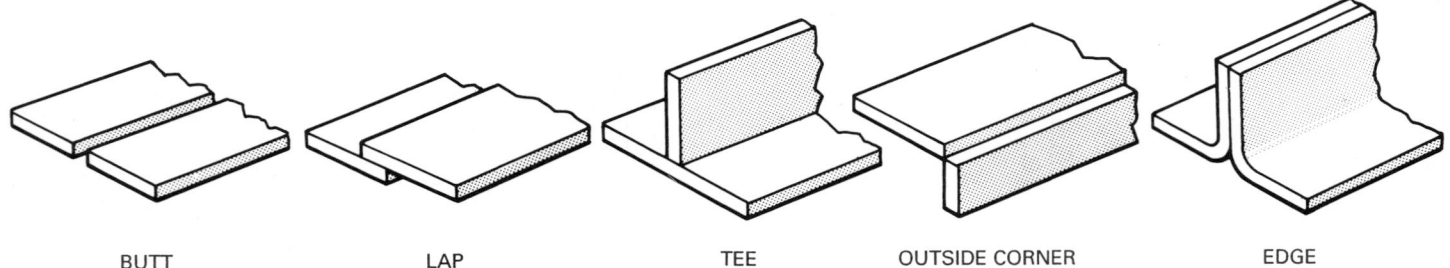

BUTT   LAP   TEE   OUTSIDE CORNER   EDGE

FIGURE 21.35 ■ Types of joints.

WELDING SYMBOLS AND APPLICATIONS

Computer-aided design and drafting has entered every engineering drafting field including those using welding processes. Most software packages use a main menu that remains the same for every application. Industry-specific menu overlays are then used to customize the main template for individual applications. The advantage of this is that the drafter becomes familiar with the commands on the main

menu, because it is used all of the time. Then as specific needs arise, such as drawing welding symbols, customized symbol libraries are placed over the main menu in locations provided for replaceable tablet menus. This kind of flexibility allows increased efficiency and makes the CADD system as flexible and powerful as possible. A custom tablet overlay featuring a welding symbol library menu is shown in Figure 21.36.

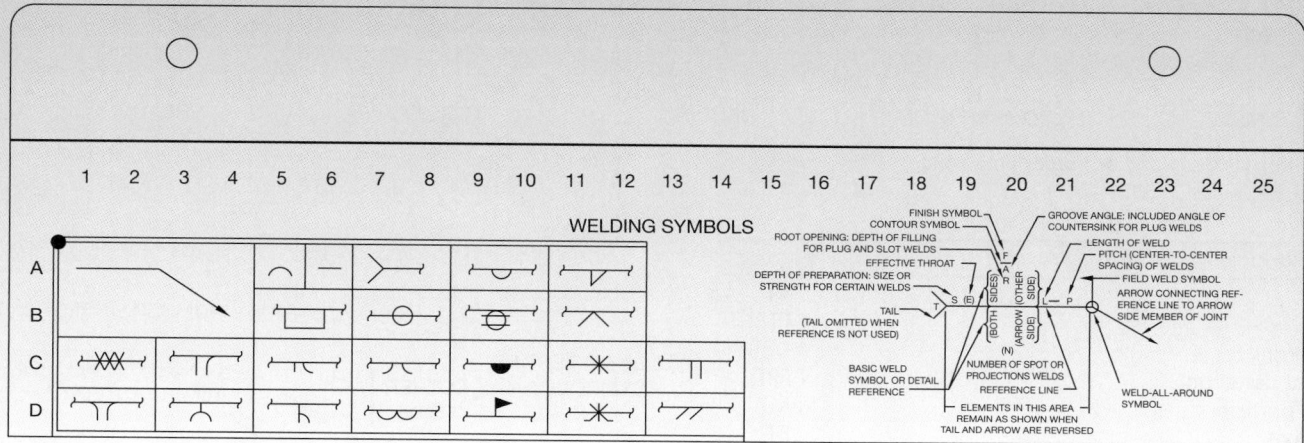

FIGURE 21.36 ■ A CADD welding symbols library template overlay. This template overlay also has welding standards applications reproduced for your reference. *Courtesy Drafting Technology Services, Inc., Bartlesville, OK.*

Destructive Tests (DT)

Destructive tests (DT) use the application of a specific force on the weld until the weld fails. These types of tests may include the analysis of tensile, compression, bending, torsion, or shear strength. Figure 21.37 shows the relationship of forces that may be applied to a weld. The *continuity* of a weld is when the desired characteristics of the weld exist throughout the weld length. *Discontinuity* or lack of continuity exists when a change in the shape or structure exists. The types of problems that alter the desired weld characteristic may include cracks, bumps, seams or laps, or changes in density. The intent of destructive testing is to determine how much of a discontinuity can exist in a weld before the weld is considered to be flawed. Parts may be periodically selected for destructive testing. The weld that is tested is unfit for any further use.

Nondestructive Tests (NDS)

Nondestructive tests (NDS) are tests for potential defects in welds that are performed without destroying or damaging the weld or the part. The types of nondestructive tests and the corresponding symbol for each test are shown in Figure 21.38. The testing symbols are

used in conjunction with the weld symbol to identify the area to be tested and the type of test to be used. (See Figure 21.39.)

The location of the testing symbol above, below, or placed in a break on the reference line has the same reference to the weld joint as the weld symbol application. Test symbols below the reference line mean arrow side tests. Symbols above the reference line are for other side tests, and a test symbol placed in a break on the line indicates no preference of side to be tested. Test symbols placed on both sides of the reference line require the weld to be tested on both sides of the joint. (See Figure 21.40.)

Two or more different tests may be required on the same section or length of weld. Methods of combining welding test symbols to indicate more than one test procedure are shown in Figure 21.41. The length of the weld to be tested may be shown to the right of the test symbol, or may be provided as a dimension line giving the extent of the test length, as shown in Figure 21.42. The number of tests to be made may be identified in parentheses below the test symbol for arrow side tests, or above the symbol for other side tests, as shown in Figure 21.43. The welding symbols and nondestructive testing symbols can be combined as shown in Figure 21.44. The combination symbol is appropriate to help the welder and inspector identify welds that require special attention. When a radiograph test is needed, a special symbol and the angle of radiation may be specified, as shown in Figure 21.45.

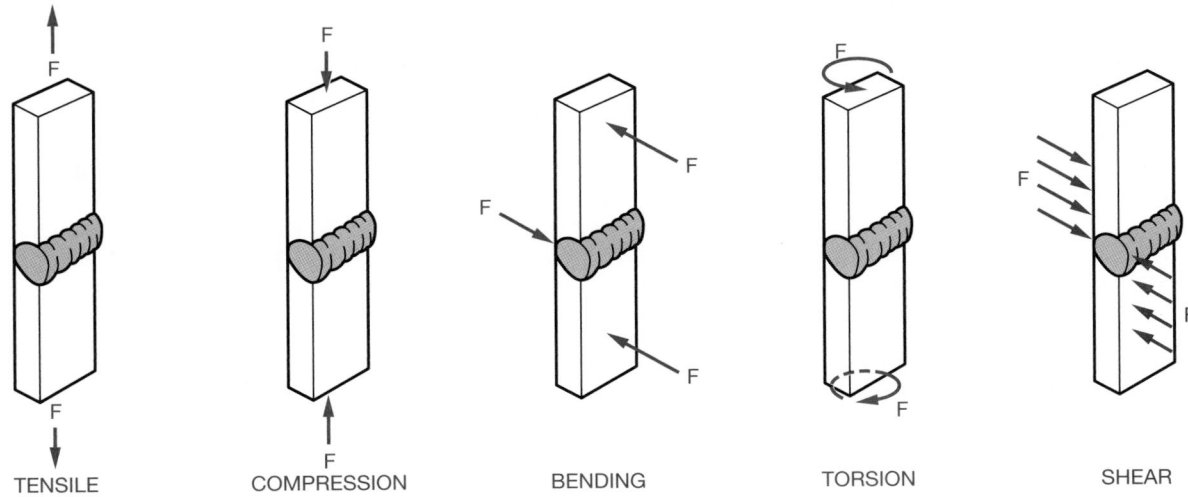

FIGURE 21.37 ■ Forces on a weld.

Type of Nondestructive Test	Symbol
Visual	VT
Penetrant	PT
Dye penetrant	DPT
Fluorescent penetrant	FPT
Magnetic particle	MT
Eddy current	ET
Ultrasonic	UT
Acoustic emission	AET
Leak	LT
Proof	PRT
Radiographic	RT
Neutron radiographic	NRT

FIGURE 21.38 ■ Standard nondestructive testing symbols.

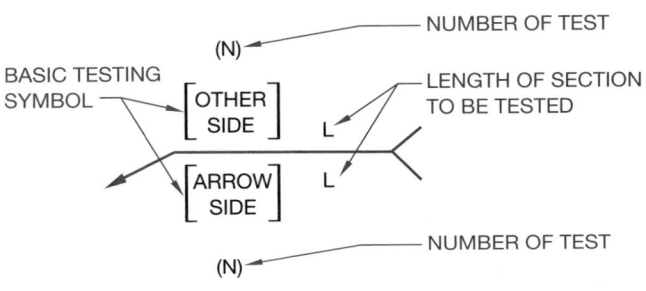

FIGURE 21.39 ■ Basic nondestructive testing symbol.

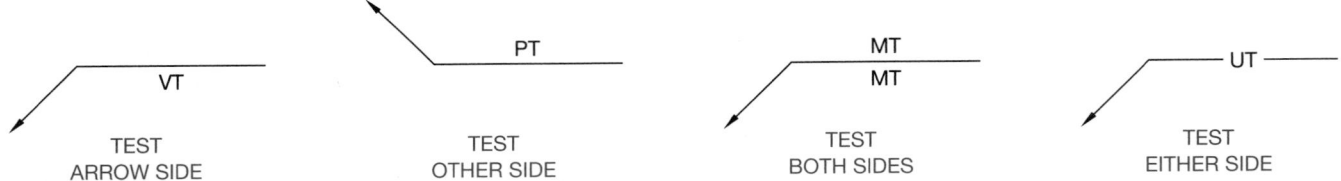

FIGURE 21.40 ■ Testing symbols used to indicate what side is to be tested.

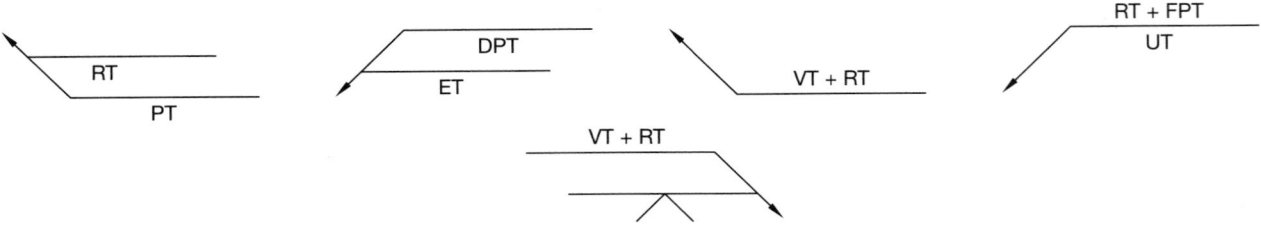

FIGURE 21.41 ■ Methods of combining testing symbols.

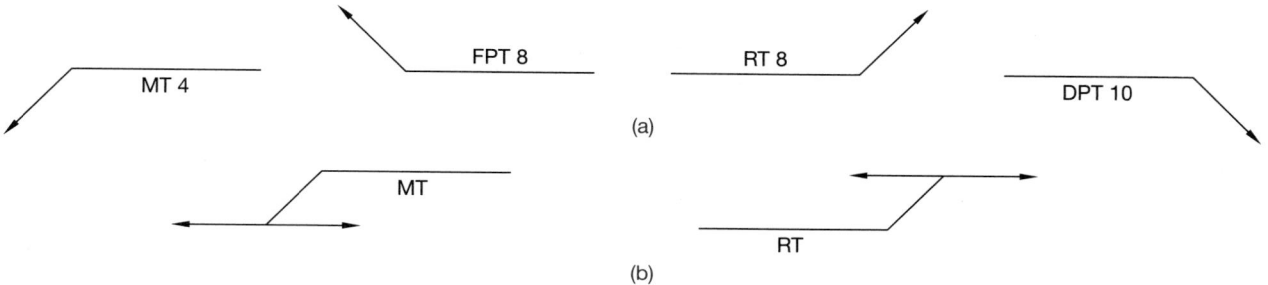

(a)

(b)

FIGURE 21.42 ■ ■ Two methods of designating the length of weld to be tested.

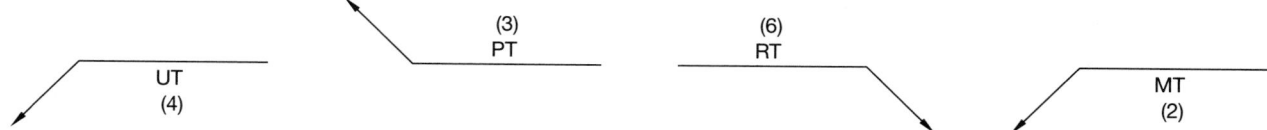

FIGURE 21.43 ■ ■ Method of specifying number of tests to be made.

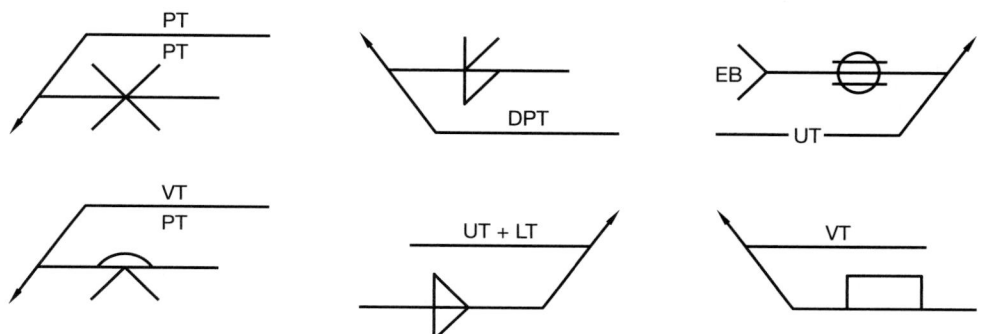

FIGURE 21.44 ■ ■ Combination welding and nondestructive testing symbols.

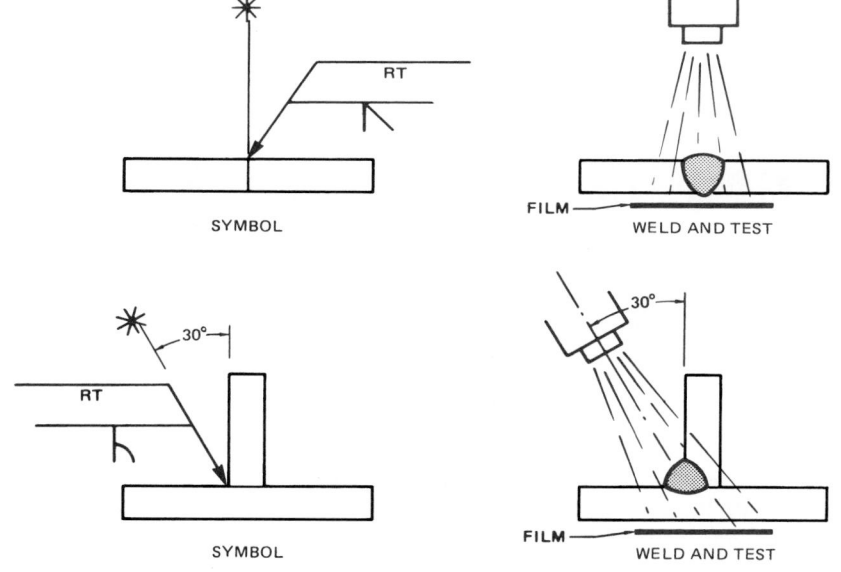

FIGURE 21.45 ■ ■ Combination symbol for welding and radiation location for testing.

WELDING SPECIFICATIONS

A *welding specification* is a detailed statement of the legal requirements for a specific classification or type of product. Products manufactured to code or specification requirements commonly must be inspected and tested to ensure compliance.

There are a number of agencies and organizations that publish welding codes and specifications. The application of a particular code or specification to a weldment can be the result of one or more of the following requirements:

■ Local, state, or federal government regulations.

■ Bonding or insurance company requirements.

■ Customer requirements.

■ Standard industrial practice.

Commonly used codes include:

■ No. 1104, American Petroleum Institute (API). Used for pipeline specifications.

■ Section IX, American Society of Mechanical Engineers (ASME). Used to specify welds for pressure vessels.

■ D1.1, American Welding Society (AWS). Welding specifications for bridges and buildings.

■ AASHT, American Association of State Highway and Transportation Officials.

■ AIAA, Aerospace Industries Association of America.

■ AISC, American Institute of Steel Construction.

■ ANSI, American National Standards Institute.

■ AREA, American Railway Engineering Association.

■ AWWA, American Water Works Association.

■ AAR, Association of American Railroads.

■ MILSTD, Military Standards, Department of Defense.

■ SAE, Society of Automotive Engineers.

Note in the problem assignments that the welding specifications and notes are provided under SPECIFICATIONS and in the general notes.

PREQUALIFIED WELDED JOINTS

AISC/AWS The American Institute of Steel Construction (AISC) and the Structural Welding Code of the American Welding Society (AWS) exempt from tests and qualification most of the common welded joints used in steel construction. These specific common welded joints are referred to as *prequalified*. Work on prequalified welded joints must be in accordance with the Structural Welding Code. Generally, fillet welds are considered prequalified if they conform to the requirements of the AISC and the AWS code.

WELD DESIGN

In most cases, the weld should be about as wide as the thickness of the metal to be welded. Undersized or oversized welds can result in joint failure. Undersized welds may not have enough area to hold the parts together under loading conditions. Oversized welds could result in a joint that is too stiff. The lack of flexibility in the oversized weld could cause the metal near the joint to become overstressed and result in failure adjacent to the weld joint. Welds between materials of different thickness should provide for more weld next to the thickest piece. If this design precaution is not taken, the weld may result in good bonding to the thin material and poor bonding to the thick material because the welding heat is more concentrated at the thinner material. Another option would be to reduce the thickness of the thicker material, or to build up the thickness of the thinner material at the weld joint.

Welding design is generally an engineering decision. When weldments are used in the design, the drafter must be able to recognize what the weld means and establish the best location on the drawing for the weld symbol. As entry-level drafters become familiar with company products and the welds used in them, under certain conditions they may be given more design-related tasks.

PROFESSIONAL PERSPECTIVE

Welding processes are used in many manufacturing situations from heavy equipment manufacturing to electronic chassis fabrication to steel building construction. While the specific applications in these fields are different, the display of welding symbols is similar. If there is a strong chance that you will be employed as an engineering drafter in a field that performs a lot of welding operations, then it is a good idea for you to take a class in welding technology. Several classes may even be necessary for you to gain a good understanding of the welding methods and materials. The welding drafting standards and techniques discussed in this chapter are in accordance with the American Welding Society and the American Institute of Steel Construction. If you work in either mechanical engineering or steel construction and fabrication, you should become familiar with the welding applications demonstrated and discussed in these standards.

As an entry-level drafter your drafting assignments will probably be engineer's sketches, or prints marked for revision. As you gain some experience, the engineer may explain verbally or in writing what needs to be done. As you gain experience, you will start to design with welding symbols based on the methods used in the past, material thickness, and welding processes. Assume you are working as an engineering drafter with a mechanical engineer and one of the parts for the project you are working on is a weldment. You have a good idea about what to do, but you ask the engineer for input. The engineer says, "Use a fillet weld-all-around on the diameter 41 side and a 4.5 mm deep bevel weld on the diameter 44 side, but be sure the bevel weld is ground smooth so it doesn't interfere with the mating part." You say, "Okay, now should the process be shielded metal arc welding?" The engineer indicates with a nod, "Yes, you have it!" Now you go back to work and come up with the drawing shown in Figure 21.46.

PROFESSIONAL
PERSPECTIVE *(continued)*

FIGURE 21.46 ■ A welding drawing from engineering information.

MATH APPLICATIONS

PERCENTAGE OF WELDED PARTS

Problem: Out of a lot of 1600 welded parts, 2 percent were rejected. How many were accepted?

Solution: This is a problem in percentages. (See Math Instruction Appendix in the Online Companion.) Following percentage problem-solving terminology, the *base* is 1600 and the *rate* is 2 percent. The number of rejected parts is the *part*. To find the part, multiply: $1600 \times .02$ and get 32. Then the number of accepted parts must be $1600 - 32$, or **1568**.

The math problems for this chapter involve finding the base and rate as well as the part.

CHAPTER
21
Welding Processes and Representations Test

DIRECTIONS
Answer the questions with short complete statements or drawings as needed.

QUESTIONS

1–10. Given the following welding symbols with letters (a, b, c, etc.) pointing to various components, identify the components related to each letter.

1.

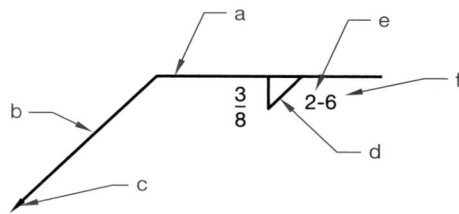

2.

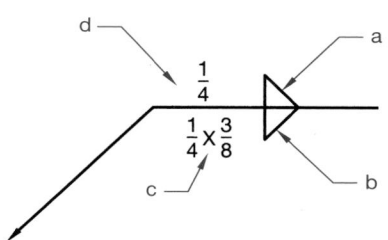

3.

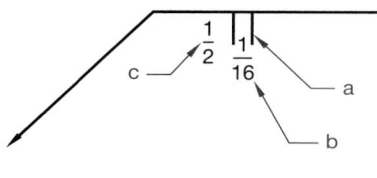

4.

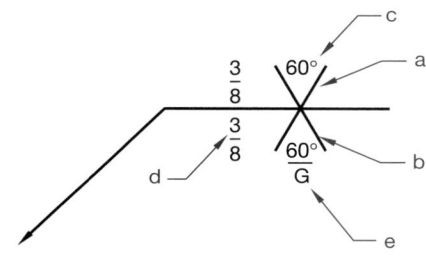

5.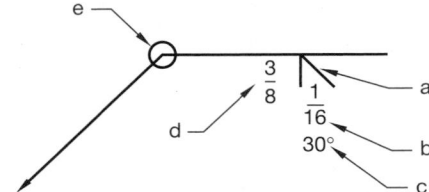

6.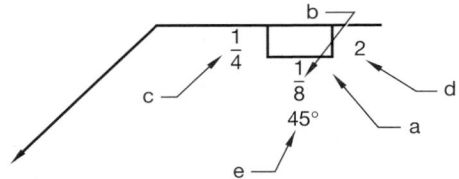

7.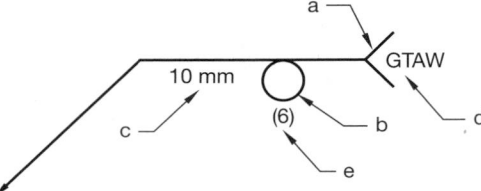

8.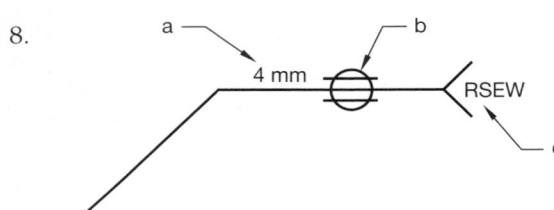

9.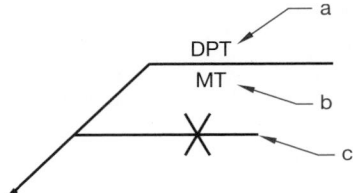

10.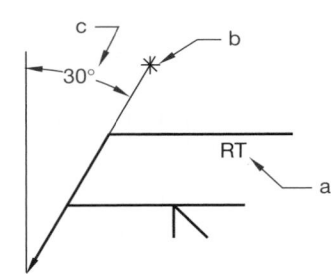

11. Define *field weld* and show an example of the field weld symbol.
12. List five types of weld joints.
13. Name four categories of the most common welding processes.
14. Describe what is meant by arrow side, other side, and both sides.
15. When is it possible to omit the weld size?
16. What are prequalified welded joints?

CHAPTER

21 *Welding Processes and Representations Problems*

DIRECTIONS

1. Given the engineer's sketches and layouts, draw the required number of views and proper welding symbols for each problem. Completely dimension unless otherwise specified.

2. For Problems 21-1 through 21-7: Given the drawings showing actual welds or drawings and specifications, draw the necessary views and proper welding symbols.

PROBLEM 21-1   Fillet (inches)

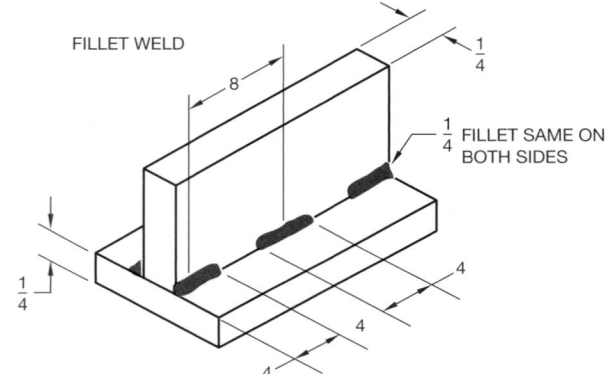

PROBLEM 21-2   Fillet and bevel groove (inches)

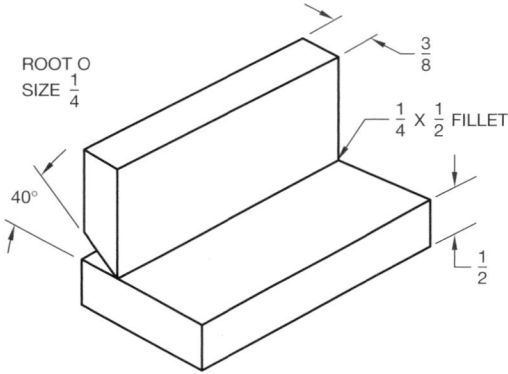

PROBLEM 21-3   V groove (inches)

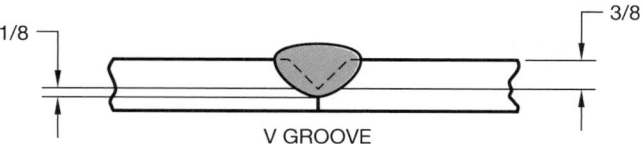

V GROOVE

PROBLEM 21-4   Double V groove (inches)

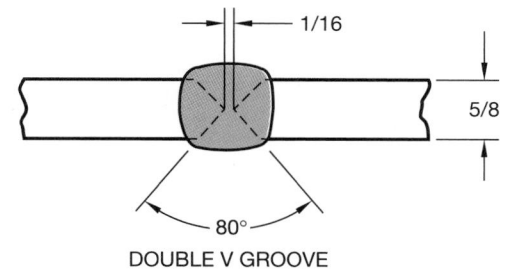

80°

DOUBLE V GROOVE

PROBLEM 21-5   Flange (inches)

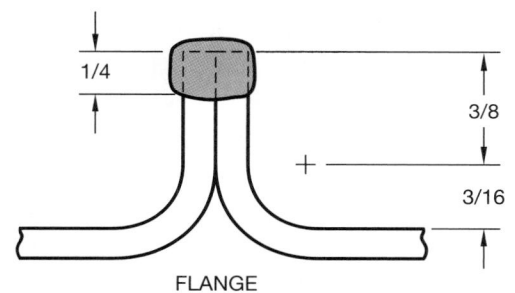

FLANGE

PROBLEM 21-6   J groove weld with radiation test (inches)

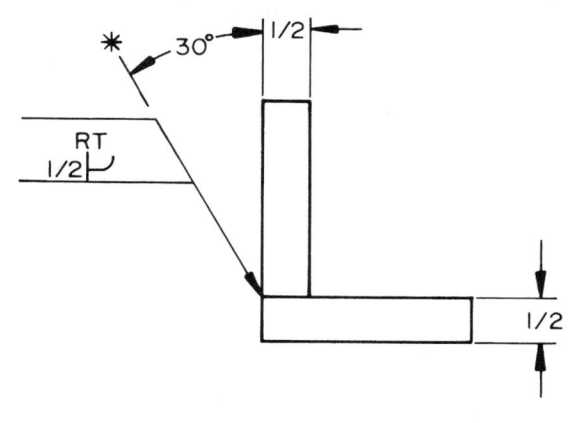

PROBLEM 21-7   Resistance spot weld (metric)

SPOT WELDS (RESISTANCE SPOT WELDS)
(METRIC)

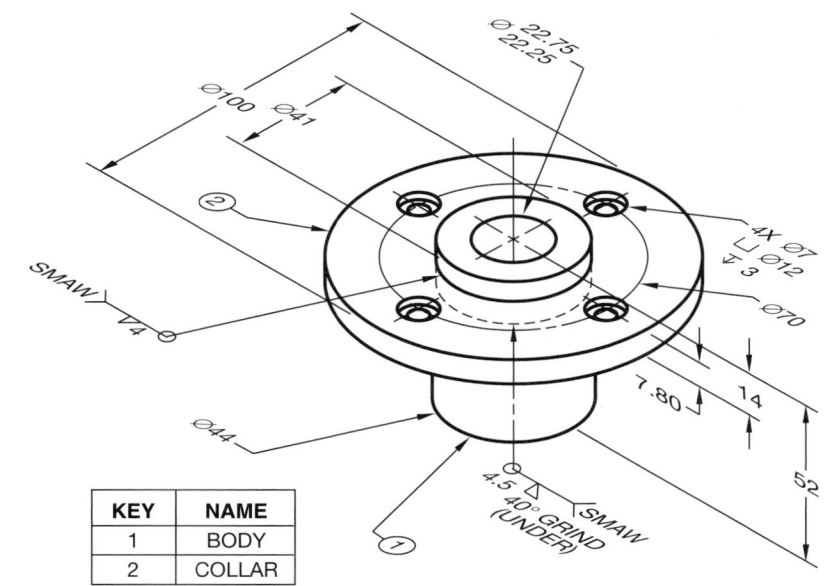

PROBLEM 21-8   (metric)
Part Name: Spring Housing Weldment Material: MS

KEY	NAME
1	BODY
2	COLLAR

PROBLEM 21-9 (inches)

Drawing Name: Column Base Plate Detail Material: MS

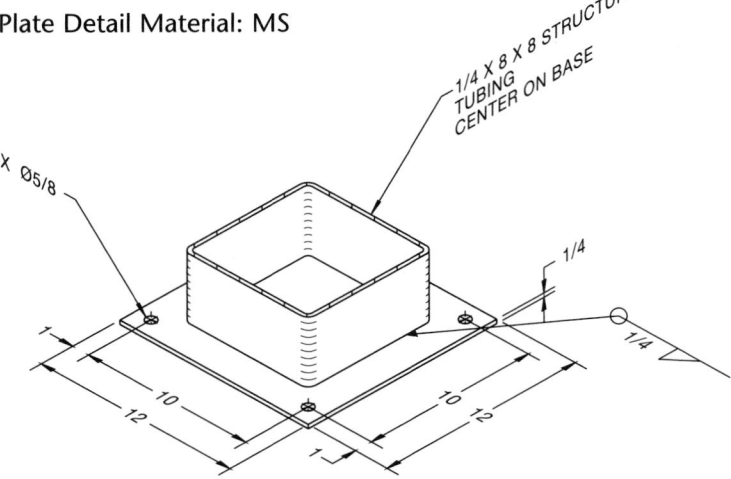

PROBLEM 21-10 (inches)

Drawing Name: Column Base Detail Material: MS

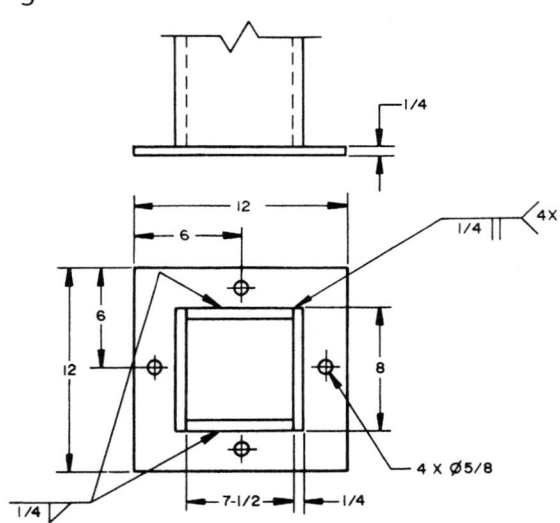

PROBLEM 21-11 (inches)

Drawing Name: Column Intersection Detail Material: MS

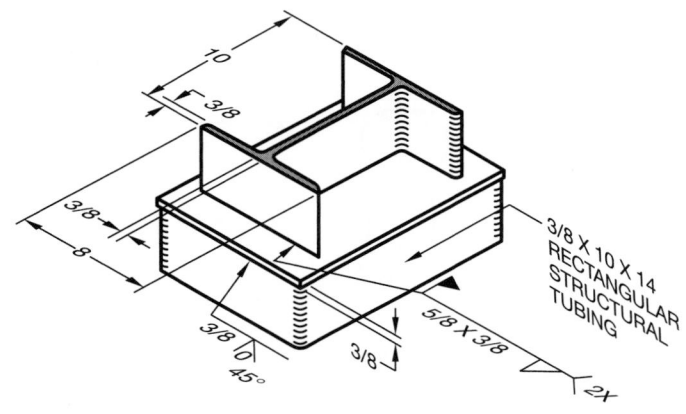

PROBLEM 21-12 (inches)

Drawing Name: Framed Beam Connection

Material: Steel

Dimensions of Members: W10 × 39: depth = 9-7/8", width = 8", web thickness = 5/16", flange thickness = 1/2".

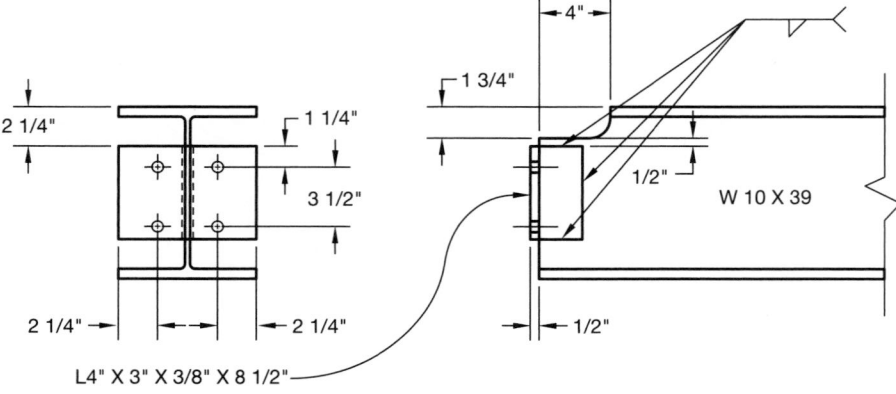

PROBLEM 21-13 (inches)

Drawing Name: Framed Beam Connection

Material: Steel

Dimensions of Members: W12 × 40: depth = 12", width = 8", web thickness = 3/16", flange thickness = 1/2".

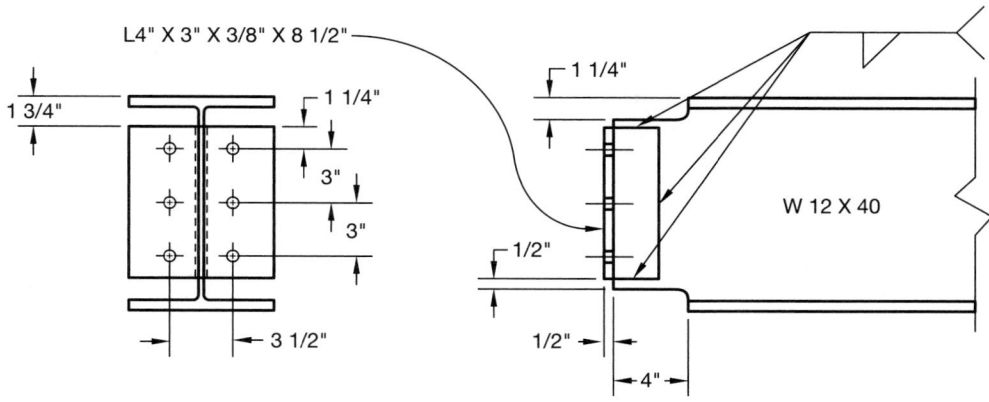

PROBLEM 21-14 (inches)

Drawing Name: Motor Support

Material: See Materials List

Dimensions of Members: C4 × 5.4: depth = 4", width = 1–⅜", web thickness = 3/16", flange thickness = 5/16". *Courtesy Production Plastics.*

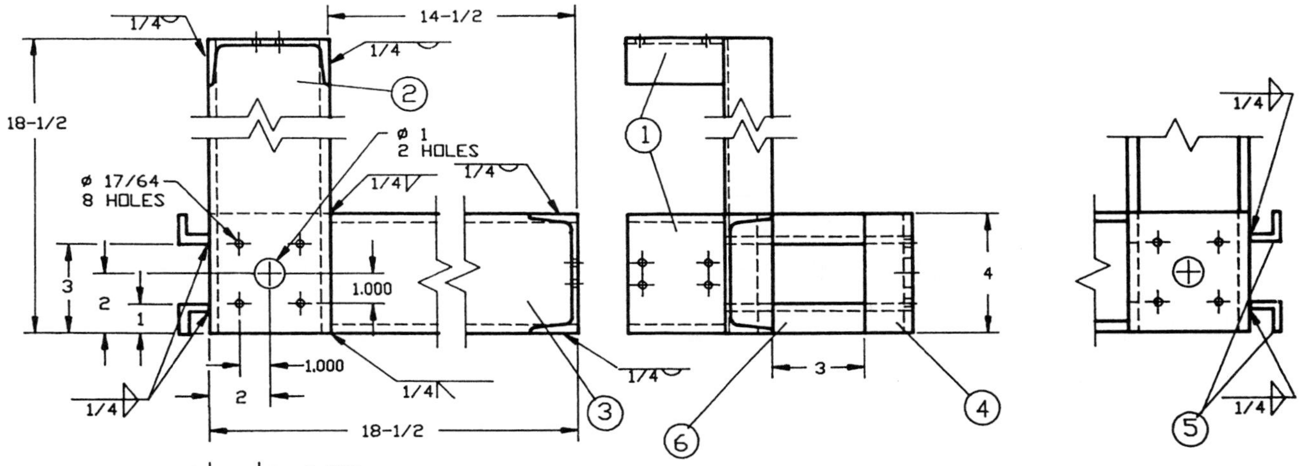

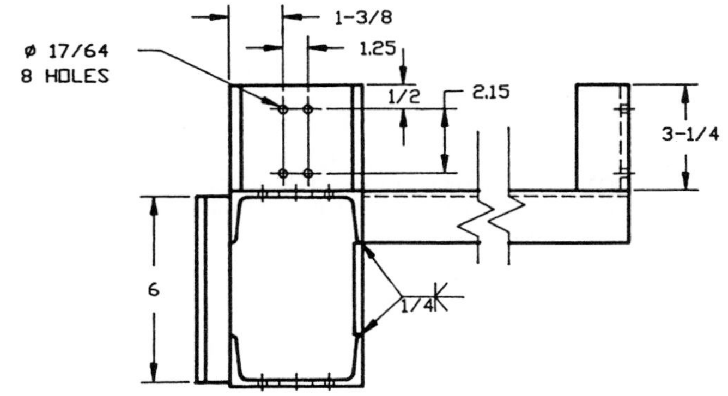

NO.	DESCRIPTION	REQD.
6	1 X 1/4 BAR – 3 LG	2
5	1 X 1 X 1/4 ANGLE – 6 LG	2
4	C4 X 5.4 – 4 LG	1
3	C4 X 5.4 – 14-1/4 LG	2
2	C4 X 5.4 – 18-1/2 LG	1
1	C4 X 5.4 – 3-1/4 LG	2

PROBLEM 21-15   (metric)

Drawing Name: Mounting Bracket

Material: See Parts List

Problem based on original art courtesy TEMCO.

2	WELD NUT	OHIO/ SF 3306 (3/8-16)
1	ANGLE	M821W200L2000 (2X2X3/16 A36)
-01	NAME OF PART	P/N DESCRIPTION MATERIAL

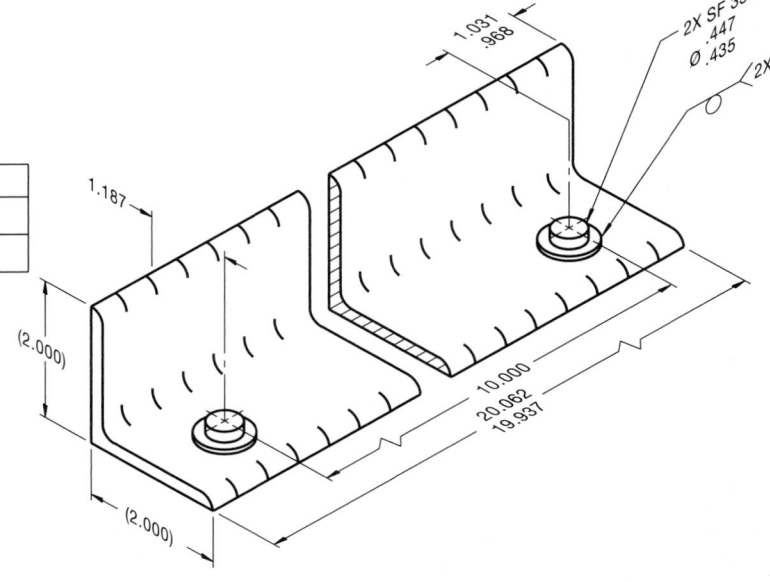

PROBLEM 21-16   (metric)

Drawing Name: Hydraulic Tank Weldment

SPECIFIC INSTRUCTIONS:

Drawing not to scale; make your drawing proportional to the problem and reproduce on C-size vellum. Some dimensions given for reference. Include all general notes as shown. *Courtesy HYSTER Company.*

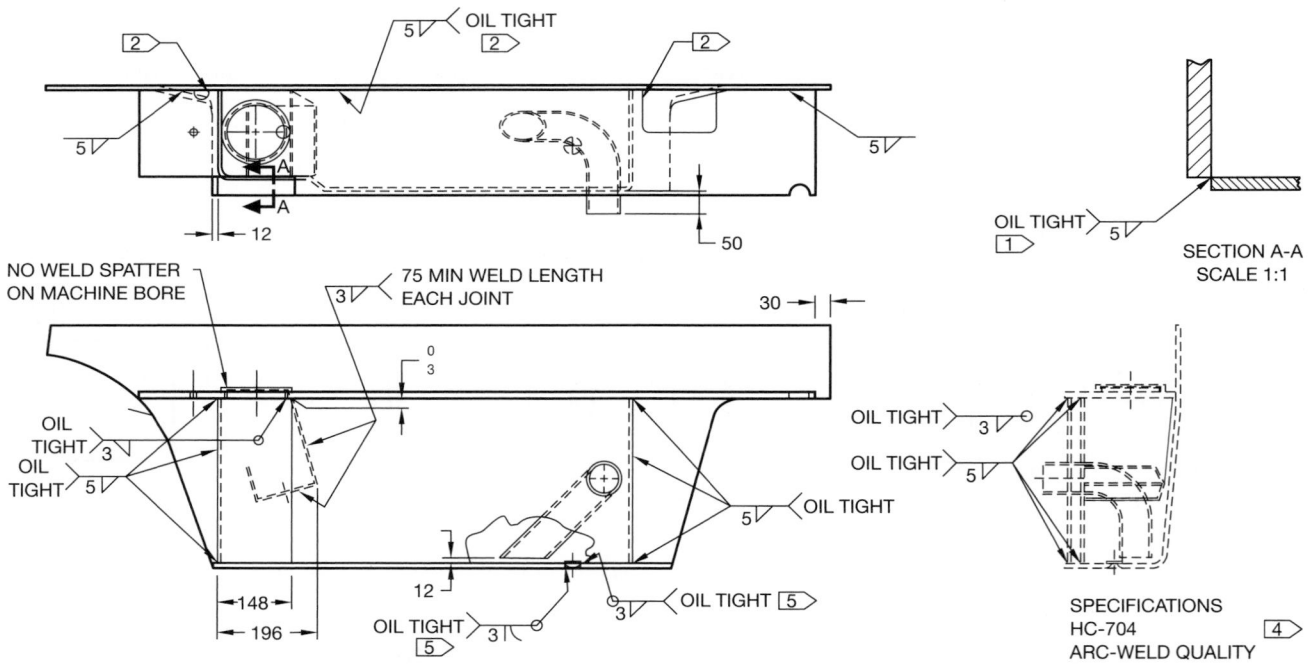

1 ▷ BOTH ENDS OF WELD MUST WRAP AROUND AND TIE IN WITH WELD BETWEEN ITEMS 2 & 3 ON BOTTOM SIDE OF ITEM 2.

2 ▷ BOTH ENDS OF WELD MUST WRAP AROUND AND TIE IN WITH WELDS BETWEEN ITEMS 2 & 3 AND ITEMS 1 & 3 ON BOTTOM SIDE OF ITEM 2 IN AREAS SHOWN.

3. INSIDE OF TANK MUST BE CLEAN AND FREE OF RUST, MILL SCALE, AND FOREIGN MATERIAL. INLET AND OUTLET HOLES TO BE PLUGGED. PRECAUTIONS SHOULD BE TAKEN DURING STORAGE TO MAINTAIN RUST-FREE CONDITION, SUCH AS USE OF A RUST INHIBITOR DURING THE TANK WASH PROCESS OR A SUBSEQUENT OIL COATING.

4. TANK MUST NOT LEAK WHEN AN INTERNAL PRESSURE OF 41 kPa IS APPLIED.

5 ▷ WELD CHOICE IS MFG OPTION. NO WELD SPLATTER IN THREADED HOLE.

PROBLEM 21-17 (metric)
DRAWING NAME: Hydraulic Tank Weldment
SPECIFIC INSTRUCTIONS:

Drawing not to scale; make your drawing proportional to the problem and reproduce on B-size-vellum. Some dimensions given for reference. Include all general notes as shown.

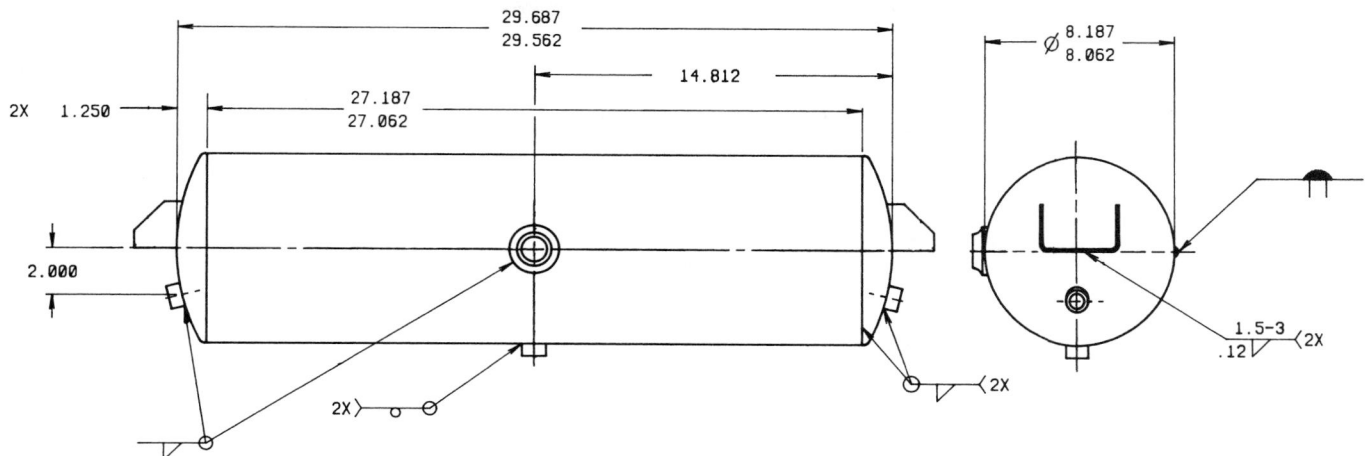

PROBLEM 21-18 (metric)
Drawing Name: LPG Bracket Weldment
SPECIFIC INSTRUCTIONS:

Drawing not to scale; make your drawing proportional to the problem and reproduce on C-size vellum. Some dimensions given for reference. Include all general notes as shown. *Courtesy HYSTER Company.*

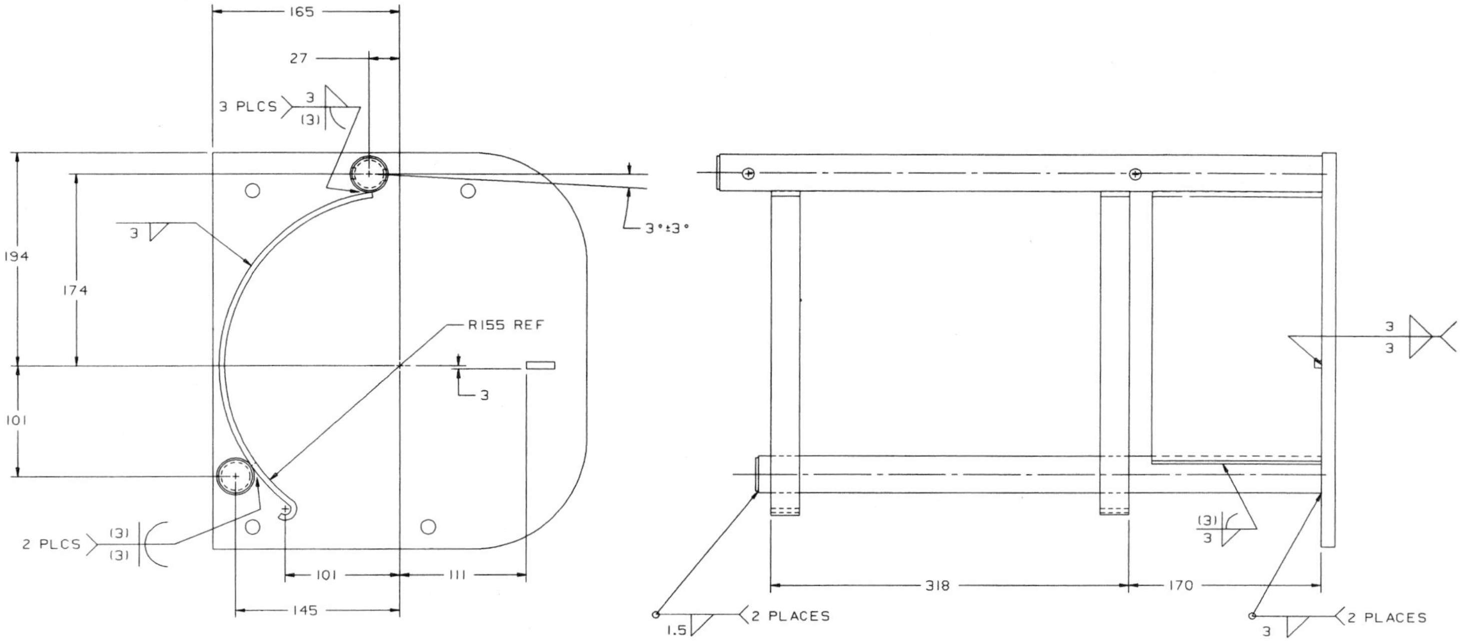

ITEM	QTY	DESCRIPTION
1	2	PLATE
2	1	PIPE
3	2	PLATE
4	1	PIPE
5	1	PLATE
6	1	PLATE
7	1	PLATE

SPECIFICATIONS
HC-704 ARC WELD QUALITY

PROBLEM 21-19 (inches)

DRAWING NAME: Cab Guard Weldment

SPECIFIC INSTRUCTIONS:

Drawing not to scale; make your drawing proportional to the problem. Some dimensions given for reference. Include all general notes as shown. *Problem based on original art courtesy Curtis Associates.*

ITEM	QTY	MATERIAL
1/2	1	4 SQ X 3/8 WALL TUBE X 36 LG
2/2	1	4 SQ X 3/8 WALL TUBE X 74 LG
3/2	1	4 SQ X 3/8 WALL TUBE X 74 LG
4/2	2	4 SQ X 3/8 WALL TUBE X 28 LG
5/2	2	4 SQ X 3/8 WALL TUBE X 17 LG
6/2	2	3 1/2 SQ X 1/4 WALL TUBE X 33 LG
7/2	2	3 SQ X 1/4 WALL TUBE X 15 3/4 LG
8/2	5	1/2 X 3 M.S.F.B. X 28 LG
9/2	1	1/4 M.S. ℞ X 9 X 28 LG
10/2	1	3/4 M.S. ℞ X 4 1/4 X 12 1/8 LG
11/2	2	1/2 ANGLE
12/2	6	3/4 M.S. ℞ X 4 X 5 LG
13/2	2	1 DIA. C.D. BAR X 3 LG
14/2	1	1/2 X 3 M.S.F.B. X 20 LG

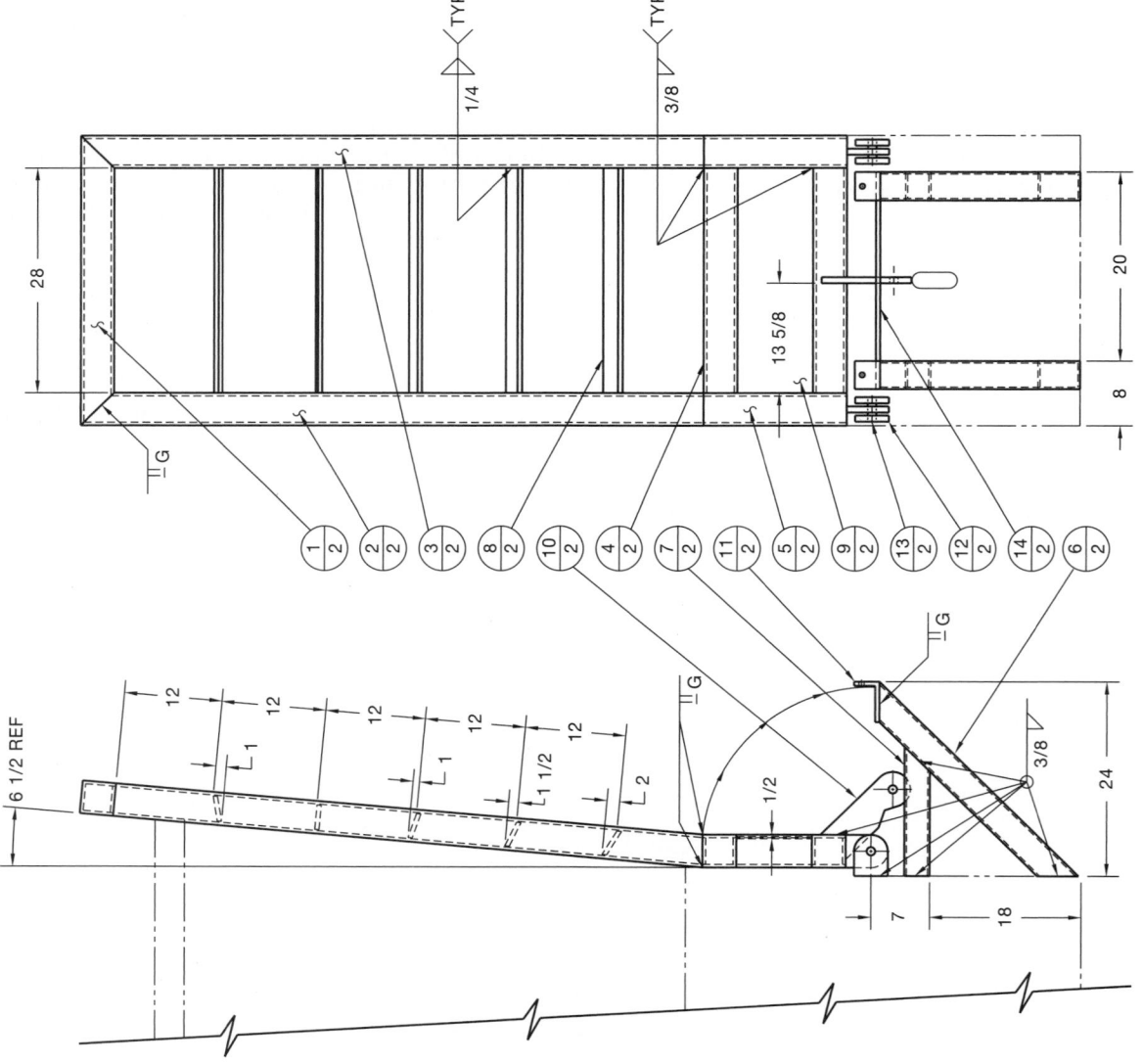

PROBLEM 21-19 (continued)

Refer to the Cab Guard Weldment Details below for part dimensions. *Problem based on original art courtesy Curtis Associates.*

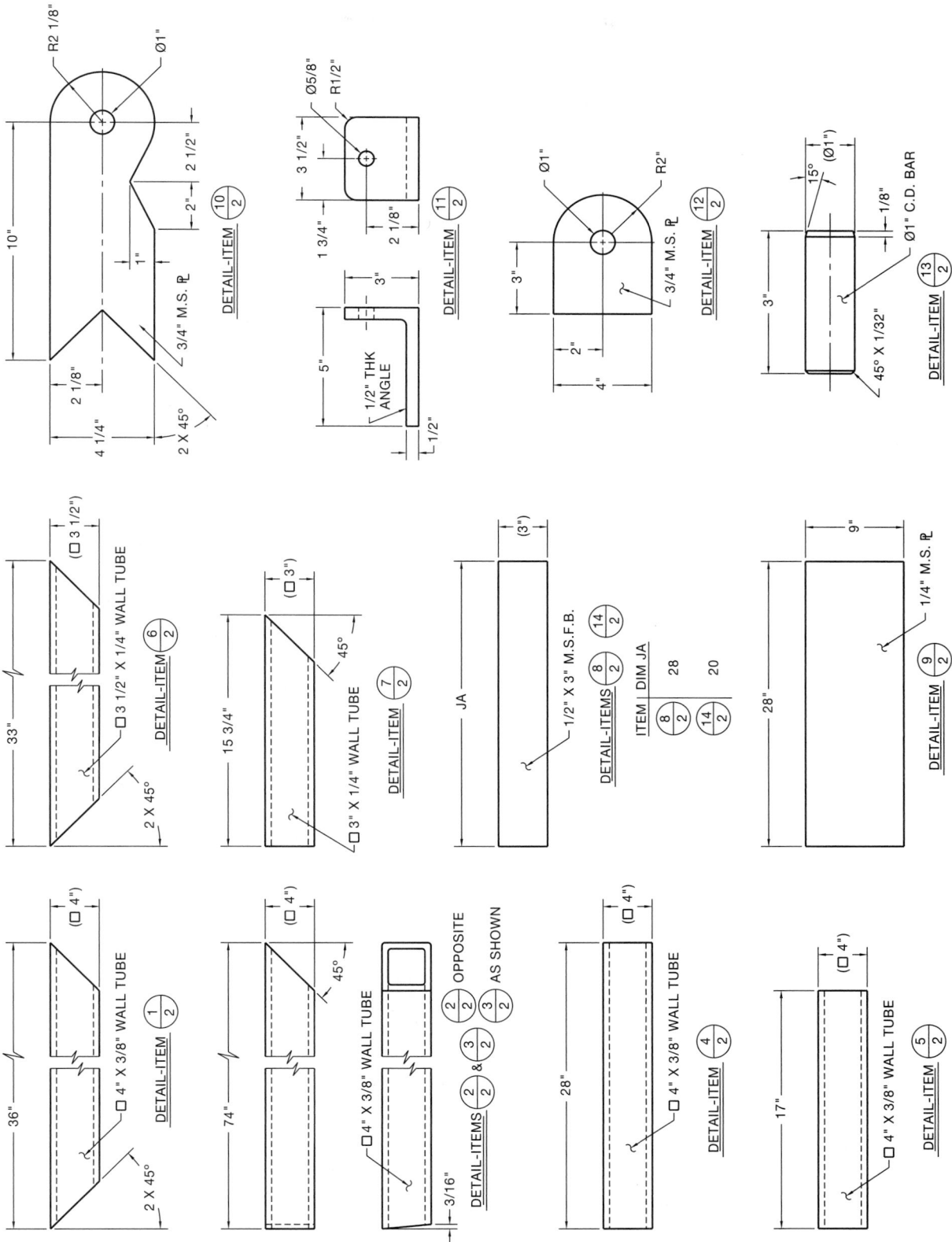

PROBLEM 21-20 (inches)

Drawing Name: Beam Column Connector Detail

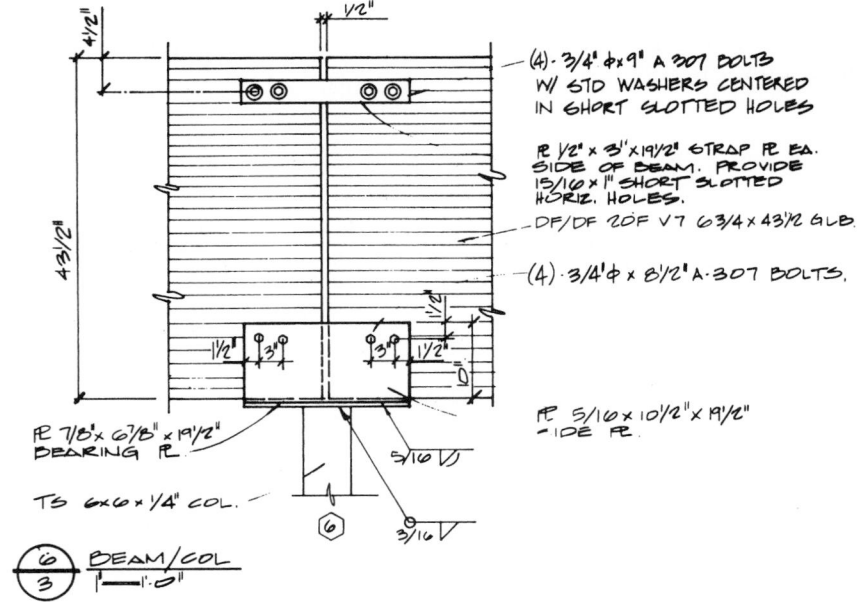

PROBLEM 21-21 (inches)

Drawing Name: Beam Wall Connector Detail

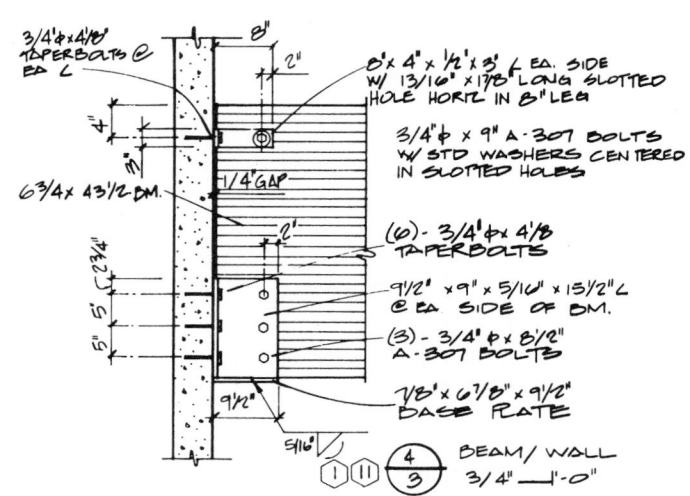

PROBLEM 21-22 (inches)

Drawing Name: Beam-to-Beam Connector Detail

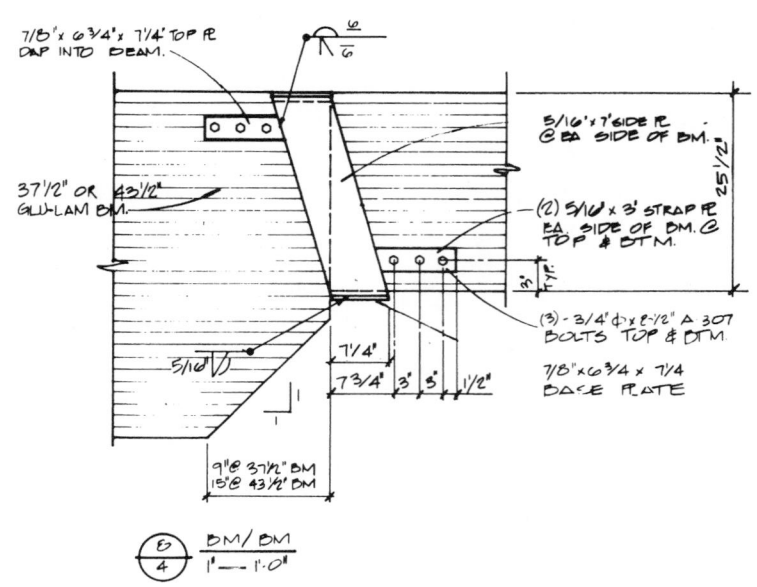

PROBLEM 21-23   (inches)

Drawing Name: Mast Brace Weldment
Courtesy American Hoist and Derrick Company.

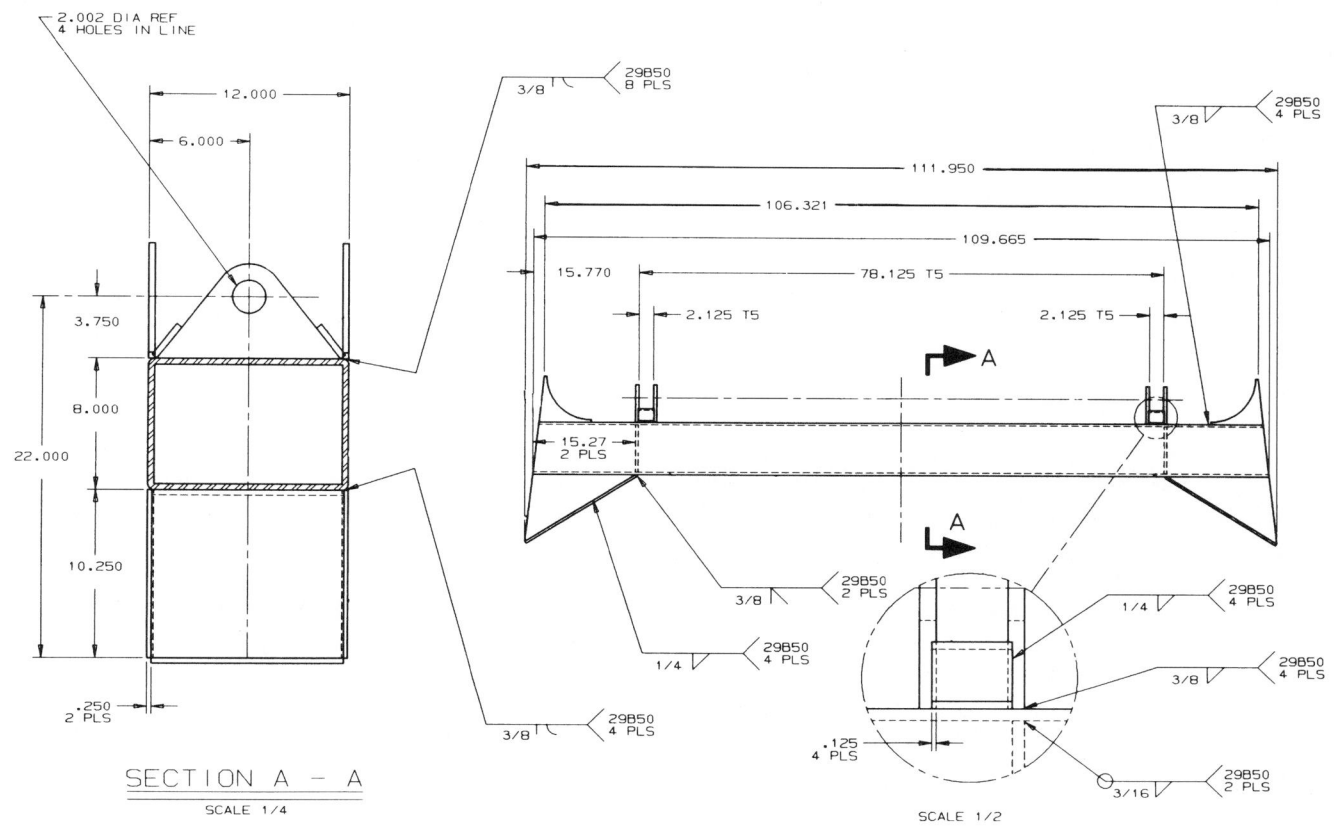

SECTION A – A
SCALE 1/4

SCALE 1/2

MATH PROBLEMS

PROBLEM 21-24   Twenty-seven percent of the shop output is done on automatic welding machines. If the total output has a value of $57,750.00, what is the value of the work done on the automatic machines?

PROBLEM 21-25   If ⅞ of the total weight of rod used on a job is 67 lb, what is 37½ percent of the total?

PROBLEM 21-26   A battery-operated tool has a useful life of 110 hours. If the battery is used 50 hours, what percent of its useful life is left?

PROBLEM 21-27   What percent is wasted when 2.4 of every 120 sheets of metal are spoiled?

PROBLEM 21-28   The repairs on an electric motor cost 24.5 percent of the original cost of $750.00 What amount was charged for repairs?

PROBLEM 21-29   How much should an article sell for to make a 35 percent profit on a cost of $125.00?

PROBLEM 21-30   A circuit requires 90 volts at a certain point. The true voltage is 72. What is the percent error?

PROBLEM 21-31   Which is the better buy for a piece of computer software?

a. A sale price of 20 percent off the list price of $560

b. A sale price of $440

c. A "half-off sale" on the regular price of $900

PROBLEM 21-32   What is the weight of the welding material used on a job if 8 percent of the total weight of 12,550 lb is welding material?

PROBLEM 21-33   A gas metal arc welding machine costing $4,100 last year costs $4,600 this year. What percent of last year's cost is this year's cost?

Industrial Process Piping

OBJECTIVES

After completing this chapter, you will:

■ Describe the different kinds of pipe and their uses.

■ Define the methods of pipe connection and their applications.

■ Identify pipe fittings and valves.

■ Draw dimensioned, single-line and double-line piping drawings using pipe fittings and valves.

■ Construct piping isometric and spool drawings using piping plans and elevations.

■ Given an engineering sketch of a piping arrangement, construct an isometric sketch, and calculate linear dimensions and straight lengths of pipe in angular pipe runs.

THE ENGINEERING DESIGN APPLICATION

This problem requires that the student use an engineering sketch (see Figure 22.1) to create an isometric freehand sketch, then calculate the lengths of straight runs of pipe shown as "1" and "2" in the sketch. If you are not familiar with isometric drawing, see Chapter 19. The following instructions should be used in this problem.

Step 1. Sketch the pipe run in isometric format.

Step 2. Calculate the lengths of the hypotenuse of triangles A and B. (See Figure 22.1.)

Step 3. Use the Pythagorean theorem or trigonometric formulae. Each method is shown in the solution.

Step 4. Calculate overall lengths of straight runs of pipe (1 and 2) including fittings.

Step 5. Add the fitting lengths and subtract from the overall length of the pipe run. Use Appendix A to find pipe fitting and flange dimensions. All fittings and flanges are 150# rating.

(Continued)

The term *piping* can refer to any kind of pipe used in a wide range of applications. The word *plumbing* refers to the small diameter pipes within our houses that carry water, gas, and wastes. Pipes of this kind can be copper, steel, cast iron, and plastic. Large underground pipes that transport water and gas to our homes from public utilities, and collect the waste water that is taken to treatment plants, are known as *civil piping* because of their municipal nature. Steel, cast iron, and concrete are the principal materials used for this piping.

Industrial plants involve a process in which raw materials are converted to finished products. Water, air, and steam are used in many industries for processing the raw materials. Piping used to transport fluids between storage tanks and processing equipment is called *process piping*. (See Figure 22.2.)

Large diameter pipes called *pipelines* carry crude oil, water, petroleum products, gases, coal slurries, and a variety of liquids hundreds of miles. This type of pipe is known as *transportation piping,* as shown in Figure 22.3.

WHERE IS INDUSTRIAL PIPING USED?

A web of piping lies beneath the ground in your neighborhood. Miles of pipes for water, sewers, storm drains, natural gas, and electrical power provide vital services. (See Figure 22.4.) Oil and gas pipelines provide the compounds that enable us to drive to work and school daily. The petrochemical plants that process the crude oil into a tremendous variety of products employ thousands of miles of piping. (See Figure 22.5.)

This paper was once a tree that was chopped into chips, cooked into a pulpy stew, and pumped via pipes through a series of processes in a paper mill. The food you consume is processed with the aid of pipes filled with water, chemicals, and liquid mixtures. The beverages you drink were moved in pipes through the various stages of production. The electric wires that provide power are resting inside pipes (electrical conduit). The electric power inside the wires is produced at steam and nuclear power plants containing intricate webs of piping. (See Figure 22.6.) Even the air you are breathing, if you are in a school or office, was probably brought to you through pipe-shaped ducts.

When you think about all the uses for pipe, and realize how seldom it is seen, it becomes obvious why piping drafting and design has often taken a back seat to other visible manifestations of drafting.

WHAT IS PIPE DRAFTING?

Pipe drafting is a specialized field that calls upon the drafter's skill of visualization and the ability to see pipe and fittings in several planes (depth) in an orthographic view. Pipe fittings must often be turned and rotated at odd angles. Visualization can be confusing for beginning pipe drafters and engineers, therefore ample study

THE ENGINEERING DESIGN APPLICATION (continued)

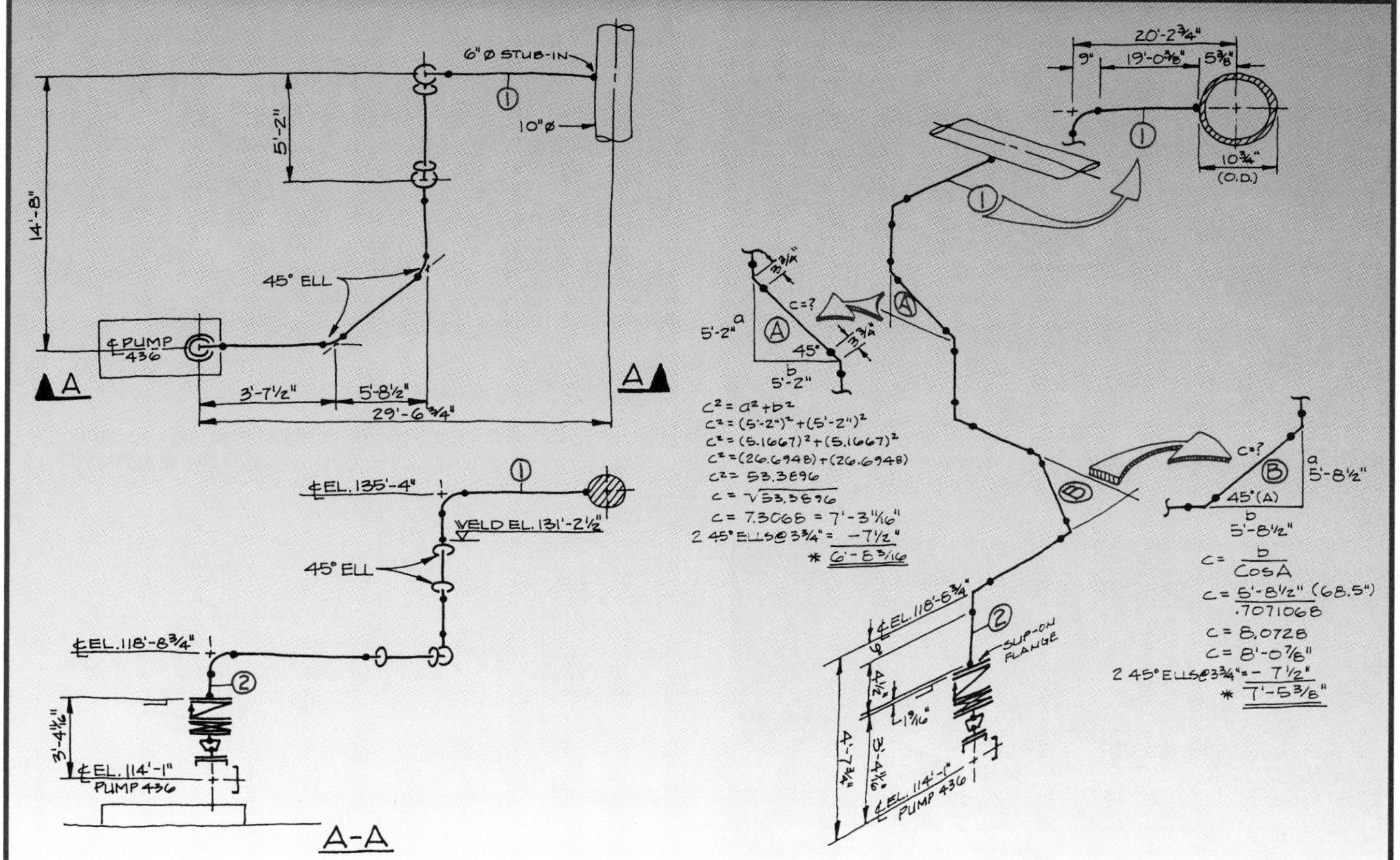

FIGURE 22.1 ■ This engineering sketch is used to construct an isometric sketch and to calculate the lengths of straight pipe in the angular offsets.

FIGURE 22.2 ■ Process piping transports fluids between storage tanks and processing equipment at this oil refinery. *Courtesy Amoco Corporation.*

FIGURE 22.3 ■ This pipeline is a form of transportation piping. *Courtesy Amoco Corporation.*

FIGURE 22.4 ■ Underground piping supplies natural gas for homes and industries. *Courtesy Pacific Gas and Electric.*

FIGURE 22.6 ■ Piping is an integral part of steam generation power plants. *Courtesy Washington Public Power Supply System.*

FIGURE 22.5 ■ Petroleum refineries employ miles of interwoven piping. *Courtesy Amoco Corporation.*

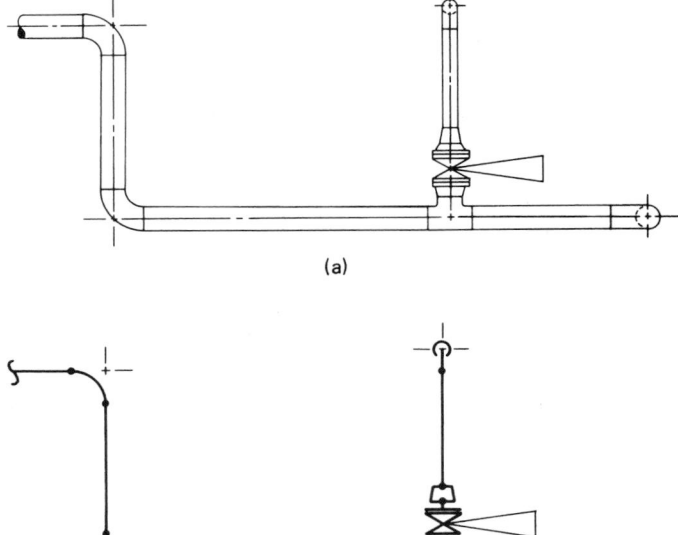

(a)

(b)

FIGURE 22.7 ■ Pipe can be drawn in (a) double-line or (b) single-line representation.

should be given to pictorial (isometric and 3-D) drawing techniques. Pipe can be drawn in two forms, double line and single line, as shown in Figure 22.7. The single-line method usually gives beginners the most visualization problems. This method is discussed later in this chapter.

The principal area of learning, other than refining the ability to visualize, is the realm of pipe fittings and joining methods. A major portion of this chapter will be spent on these areas. Those who may have visualization problems should spend additional time studying and drawing pipe fittings.

Pipe drafting involves the creation of a variety of drawing types from maps (site plans) to mechanical drawings (piping details). One of the easiest types of drawings to construct is the *flow diagram* as shown in Figure 22.8. This is a schematic, nonscale diagram that illustrates the layout and composition of a system using symbols. The *piping drawing* is the most complex. It is a scale drawing that provides plan, elevation, and section views. All equipment, fittings, dimensions, and notes are shown on this type of drawing. Examples of plans and sections are shown in Figure 22.9a and Figure 22.9b. The drafter must use flow diagrams, structural, mechanical, and instrumentation drawings, and vendor catalogs to construct these drawings.

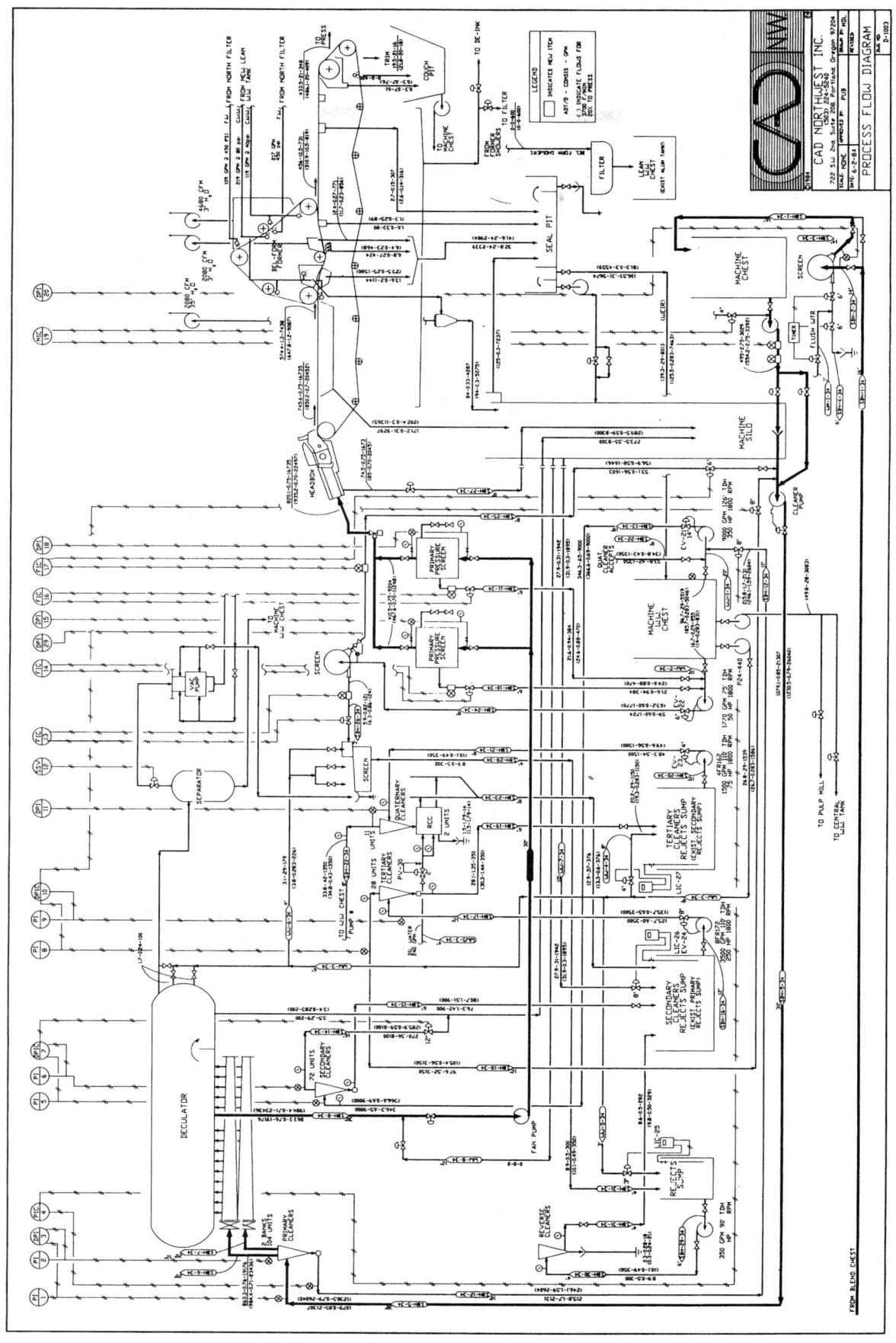

FIGURE 22.8 ■ This piping flow diagram is a nonscale view of a piping system. *Courtesy Autodesk, Inc.*

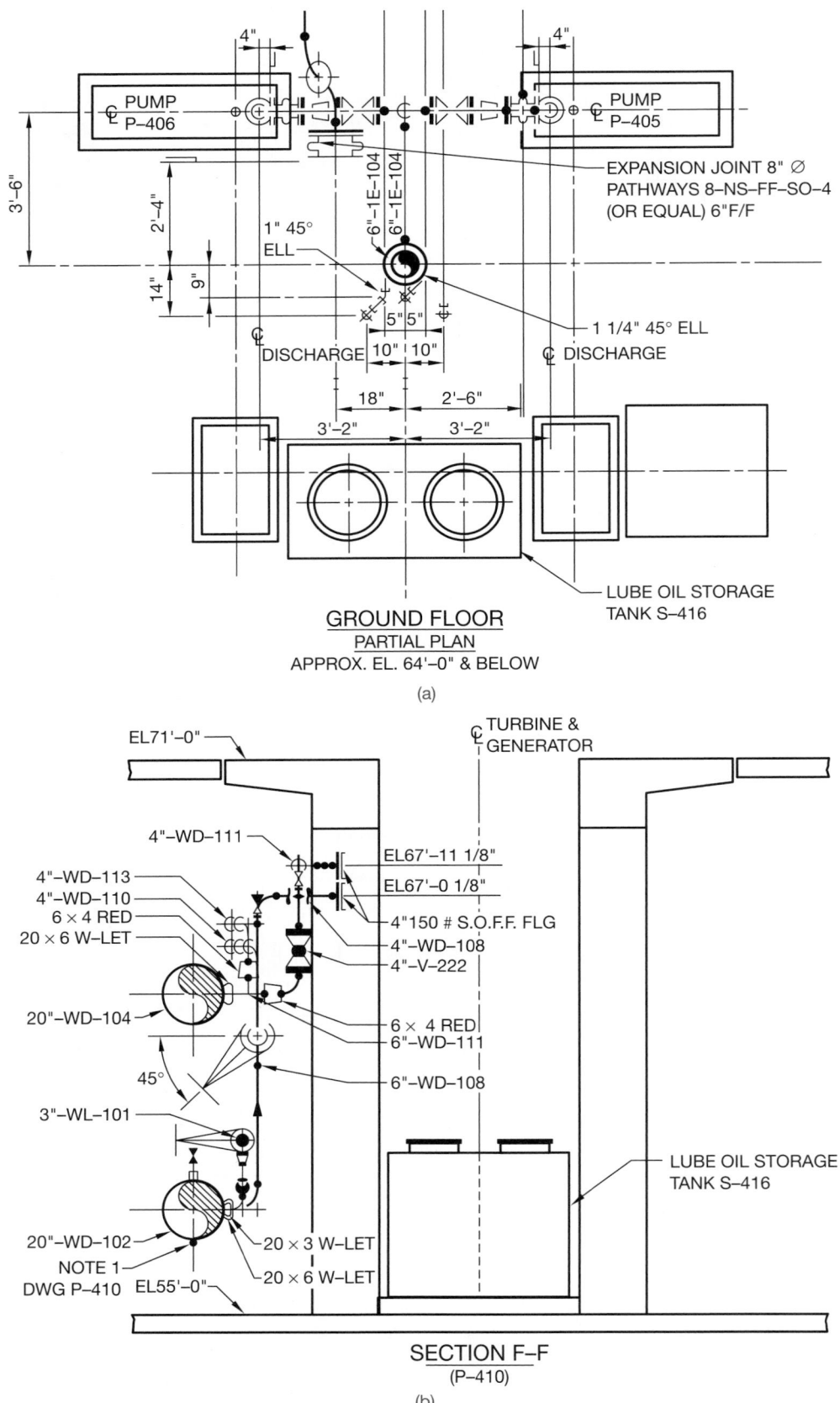

FIGURE 22.9 ■ (a) The piping plan; (b) section containing detailed piping information. *Courtesy Schuchart and Associates.*

The piping *isometric* is a pictorial drawing that illustrates a pipe run in three-dimensional form. (See Figure 22.10.) Information from this drawing is obtained from the piping drawings. The drafter's ability to view piping runs in three planes is important when working with piping isometrics. Some pipe drawings are done in isometric form because lines can be measured and fits checked. Isometrics are used extensively in chemical process industries.

Subassemblies of pipe and fittings are constructed by pipe fitters and welders who use the piping *spool* drawing. This drawing is usually drawn orthographically and shows all of the pipe and fittings needed to assemble a segment of piping. Spools are often non-scale drawings, but show all dimensions needed for assembly. A typical spool drawing is shown in Figure 22.11.

TYPES OF PIPE

When an engineer or designer decides to use a specific type of pipe, that decision is based on a number of considerations. Principal among those are temperature, pressure, and corrosion. Safety and cost are also important factors. And finally, any decision must comply with the project specifications and local and national codes.

ANSI The American National Standards Institute (ANSI) specifications that apply to pipes and fittings are listed in Figure 22.12 should you wish to research certain dimensions or properties.

Cast Iron

The most common types of pipe used for commercial and industrial applications are steel and cast iron.

AWWA Cast iron is specified by the American Water Works Association (AWWA) for underground water lines. Cast iron pipe is corrosion resistant and has a heavy wall construction. It is good for use under pavement because its long life obviates the need to dig up the pavement due to frequent leaks. Cast iron is used extensively for water, gas, and sewage piping.

Steel

Carbon steel pipe is the preferred above-ground pipe used in industry today. It is strong, relatively durable, and it can be welded and machined. It is also not as expensive as other materials. In conditions of high temperature it tends to lose strength, so stainless steel and other alloys that withstand high temperatures are then used.

Steel pipe is manufactured by several processes. *Forged* pipe is formed to a special *outside diameter* or O.D. (usually one inch greater than the finished pipe), then bored to the required *inside diameter* (I.D.). Seamless pipe is created by piercing a solid billet and rolling the resulting cylinder to the required diameter. *Welded* pipe is formed from plate steel. The edges are welded together in a *lap* weld or a *butt* weld. (See Figure 22.13, page 730.)

Copper and Copper Alloys

Copper pipe and copper alloys can be manufactured by the hot piercing and rolling process or by an extrusion method. Copper

PIPING

Most piping drawings involve the use of symbols. Symbols need to be drawn only once, saved in a symbol library, and displayed on a menu attached to a digitizer or menu tablet. When a symbol is needed, it is simply picked from the menu, placed on the drawing, and scaled or rotated as needed. If your CADD system does not have a piping symbol library yet, it will save you time in the future if you begin to create one now.

Software companies create standard and specialty symbols that are assembled in libraries. The tablet menu for the CADD package has a "user" area that was designed for symbol overlays. A productive part of the piping drafter's job is the use and placement of these standard symbols and shapes on new drawings.

Many of the symbol programs are "intelligent." That is, they contain pieces of information called *attributes* or *tags* that give meaning to the symbol. For example, a valve symbol may contain hidden attributes pertaining to its diameter, pressure rating, material, operating mechanism, specific type of styling, weight, price, and manufacturer. When the valve symbol is used on a drawing, all of the attributes become a part of the drawing's database.

Drafters using CADD systems should keep an easily accessible, up-to-date record of all piping details. They may be needed again on the same project or on another one. Time savings are involved by just placing an entire detail on a drawing instead of constructing it again. An individual detail may not be exactly the same as a previous one, but even revising an existing detail is much quicker than redrawing it.

CADD APPLICATIONS

pipe is corrosion resistant and has good heat transfer properties, but has a low melting point and is expensive. It is good for instrument lines and food processing, although stainless steel is being used more frequently in these applications. Copper pipe is also used in residential water lines.

Copper tubing is softer and more pliable than rigid copper and brass pipe, and is used for steam, air, and oil piping. It is also used extensively for *steam tracing,* the small diameter pipe that is attached to larger pipes to protect them from freezing or to keep the fluids inside the pipe warm. The outside diameter of copper tubing is the same as its *nominal pipe size* (NPS). For example, a ³/₄" tube has an outside diameter of ³/₄".

Plastic

Thermoplastic pipe was first developed in Germany in 1921, but was not used to any great extent in the United States until 1951. This first type of plastic pipe was *polyvinyl chloride* (PVC). PVC piping is now used for acids, salt solutions, alcohols, crude oil, and a variety of highly corrosive chemicals.

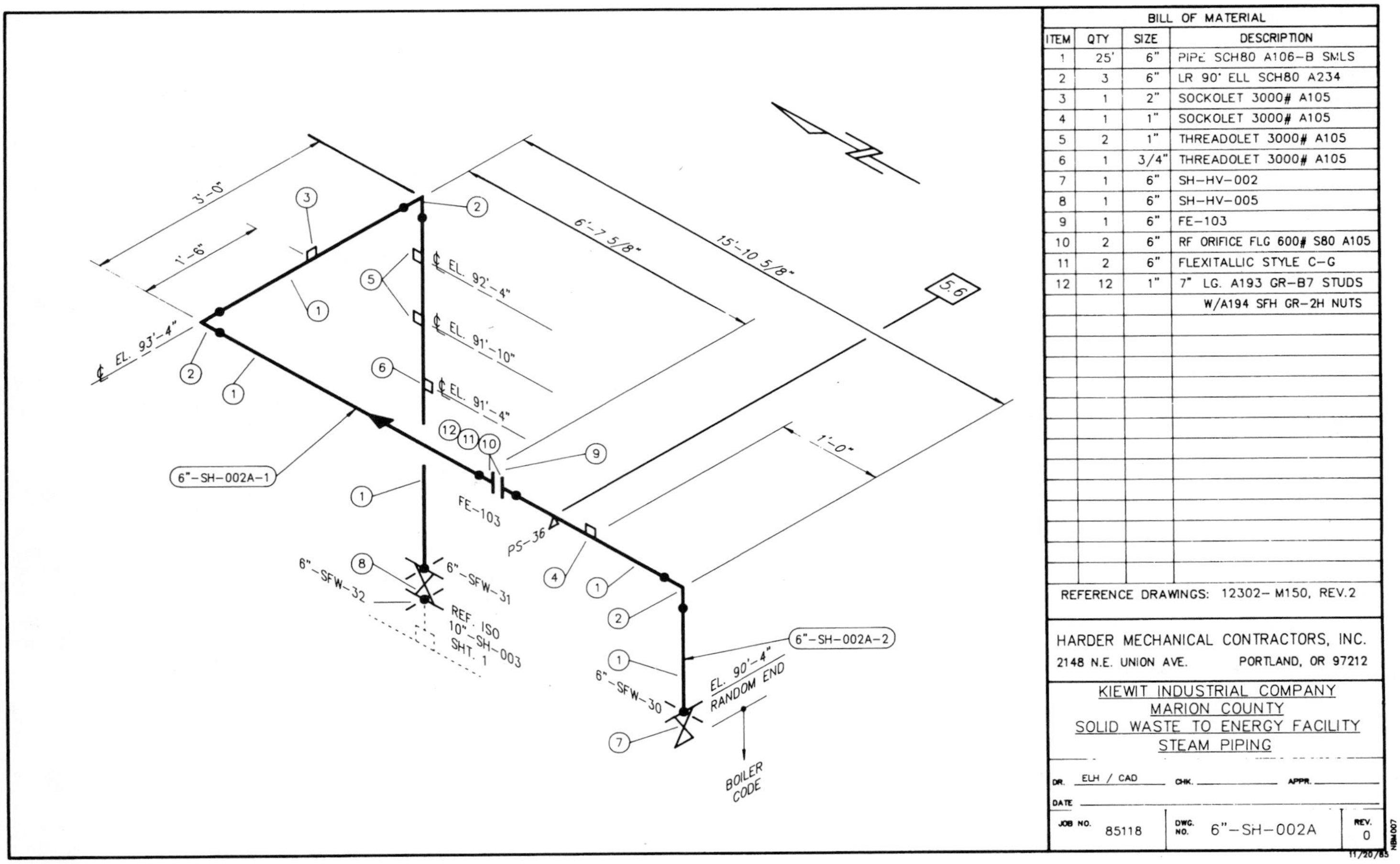

BILL OF MATERIAL			
ITEM	QTY	SIZE	DESCRIPTION
1	25'	6"	PIPE SCH80 A106-B SMLS
2	3	6"	LR 90° ELL SCH80 A234
3	1	2"	SOCKOLET 3000# A105
4	1	1"	SOCKOLET 3000# A105
5	2	1"	THREADOLET 3000# A105
6	1	3/4"	THREADOLET 3000# A105
7	1	6"	SH-HV-002
8	1	6"	SH-HV-005
9	1	6"	FE-103
10	2	6"	RF ORIFICE FLG 600# S80 A105
11	2	6"	FLEXITALLIC STYLE C-G
12	12	1"	7" LG. A193 GR-B7 STUDS
			W/A194 SFH GR-2H NUTS

REFERENCE DRAWINGS: 12302- M150, REV.2

HARDER MECHANICAL CONTRACTORS, INC.
2148 N.E. UNION AVE.   PORTLAND, OR 97212

KIEWIT INDUSTRIAL COMPANY
MARION COUNTY
SOLID WASTE TO ENERGY FACILITY
STEAM PIPING

DR. ELH / CAD	CHK.	APPR.

DATE

JOB NO. 85118	DWG. NO. 6"-SH-002A	REV. 0

FIGURE 22.10 ■ The piping isometric shows a pictorial view of a single run of pipe. *Courtesy Harder Mechanical Contractors, Inc.*

ITEM	QTY.	SIZE	DESCRIPTION		UNIT LIST PRICE	DISCOUNT	TOTAL PRICE
1	8'-10 7/16"	3"	PE/PE PIPE TY316L SCH. 10S				
2	2'-7 1/2"	3"	PE/PE PIPE TY316L SCH. 10S				
3	1	3"	LR 90° ELL TY316L SCH. 10S				
4	2	3"	SK-1-P FACE RING TY316L				
5	2	3"	SK-39-P BACKING FLG.				
A	4	3"	FLAME CUT PIPE P.E.				
B	3	3"	BUTT WELD				

HARDER MECHANICAL CONTRACTORS, INC.
2148 N.E. UNION AVENUE   PORTLAND, OR. 97212

PROJECT
TRI-CITY SERVICE DISTRICT
CLACKAMAS COUNTY, OREGON
SEWAGE TREATMENT PLANT

JOB 84011	SURF PREP	WELDING		TOTAL MATERIAL PRICE	
CHK	SPEC NO. 15001-19	STRS RELF		TOTAL LABOR PRICE	
APPR	WEIGHT	X-RAY		TOTAL MAT'L HANDLING	
LINE # AREA 540		TEST			
SHEET ___ OF ___	MARK 3"-DG-101-10	REV		TOTAL PRICE	

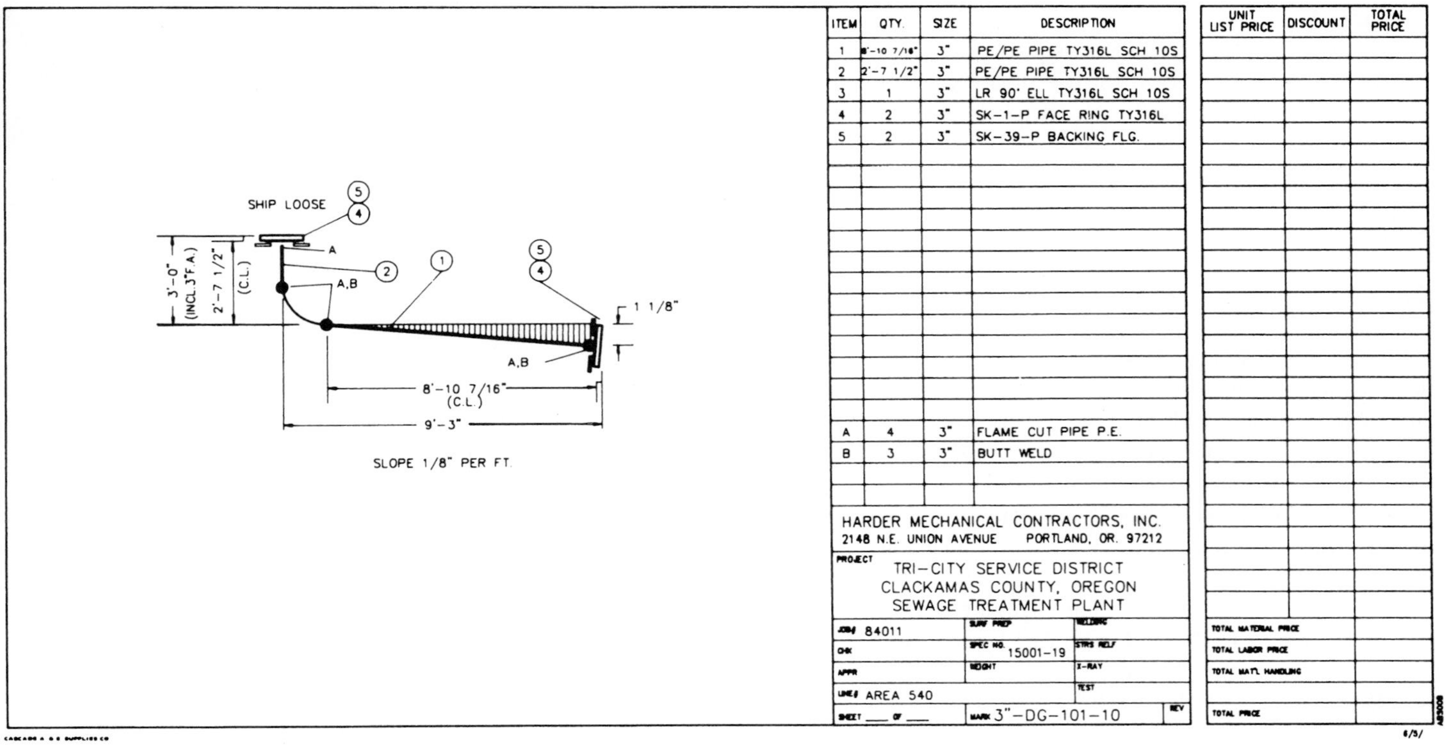

FIGURE 22.11 ■ A piping spool drawing is a subassembly of pipe and fittings. *Courtesy Harder Mechanical Contractors, Inc.*

SPECIFICATION NUMBER	DESCRIPTION
Graphic Standards	
ASA Z32.2.3–1949	Graphical Symbols for Pipe Fittings, Valves, and Piping
ANSI Y32.4–1977	Graphic Symbols for Plumbing Fixtures for Diagrams Used in Architecture and Building Construction
ASA Y32.11–1961	Graphical Symbols for Process Flow Diagrams
Pipe, Steel	
ANSI/ASTM A430–79	Austenitic Steel Forged and Bored Pipe for High-Temperature Service
ANSI/ASTM A524–80	Seamless Carbon Steel Pipe for Process Piping
ANSI B36.19–1976	Stainless Steel Pipe
ANSI B36–10–1979	Welded and Seamless Wrought Steel Pipe
ANSI/ASTM A135–79	Electric-Resistance-Welded Steel Pipe
Pipe and Fittings, Plastic	
ANSI/ASTM D1527–77	Acrylonitrile-Butadiene-Styrene (ABS) Plastic Pipe, Schedules 40 and 80
ANSI/ASTM F443–77	Bell-End Chlorinated Poly(Vinyl Chloride) (CPVC) Pipe
ANSI/ASTM D2446–73	Cellulose Acetate Butyrate (CAB) Plastic Pipe (SDR-PR) and Tubing
ANSI/ASTM D1503–73	Cellulose Acetate Butyrate (CAB) Plastic Pipe, Schedule 40
ANSI/ASTM F441–77	Chlorinated Poly(Vinyl Chloride) (CPVC) Plastic Pipe, Schedule 40, 80, and 120
ANSI/ASTM D2104–74	Polyethylene (PE) Plastic Pipe, Schedule 40
ASTM D2469–76	Socket-Type Acrylonitrile-Butadiene-Styrene (ABS) Plastic Pipe Fittings, Schedule 80
ANSI/ASTM D2466–78	Socket-Type Poly(Vinyl Chloride) (CPVC) Plastic Pipe Fittings, Schedule 40
Pipe, Miscellaneous	
ANSI/ASTM C700–78a	Extra Strength and Standard Strength Clay Pipe and Perforated Clay Pipe
ANSI/ASTM B43–80	Seamless Red Brass Pipe, Standard Sizes
ANSI/ASTM C599–70(1977)	Process Glass Pipe and Fittings
ANSI A 40.5–1943	Threaded Cast-Iron Pipe for Drainage, Vent, and Waste Services
ANSI/AWWA C151/A21.51–(1981)	Ductile-Iron Pipe, Centrifugally Cast, in Metal Molds or Sand-Lined Molds for Water or Other Liquids
Fittings and Flanges	
ANSI B16.26–1975	Cast Copper Alloy Fittings for Flared Copper Tubes
ANSI B16.1–1975	Cast Iron Pipe Flanges and Flanged Fittings, Class 25, 125, 250, and 800
ANSI B16.4–1977	Cast-Iron Threaded Fittings, Class 125 and 250
ANSI B16.9–1978	Factory-Made Wrought Steel Buttwelding Fittings
ANSI B16.42–1979	Fittings, Class 150 and 300, Ductile Iron Pipe Flanges and Flanged Fittings
ANSI/AWWA C207–78	Flanges for Water Works Service, 4 in. through 144 in. Steel
ANSI B16.11–1980	Forged Steel Fittings, Socket-Welding and Threaded
ANSI/AWWA C606–81	Joints, Grooved and Shouldered Type
ANSI B16.3–1977	Malleable-Iron Threaded Fittings, Class 150 and 300
ANSI B16.21–1978	Nonmetallic Flat Gaskets for Pipe Flanges
ANSI B16.5–1981	Pipe Flanges and Flanged Fittings, Steel Nickel Alloy and Other Special Alloys
ANSI B16.39–1977	Pipe Unions (Class 150, 250, and 300), Malleable Iron Threaded
ANSI B16.20–1973	Ring-Joint Gaskets and Grooves for Steel Pipe Flanges
ANSI B16.36–1975	Steel Orifice Flanges, Class 300, 600, 900, 1500, and 2500 (includes supplement ANSI B16.36a–1979)
ANSI B16.22–1980	Wrought Copper and Copper Alloy Solder-Joint Pressure Fittings
ANSI B16.28–1978	Wrought Steel Buttwelding Short Radius Elbows and Returns
Valves	
ANSI B16.34–1981	Valves, Flanged and Butt-Welding End-Steel, Nickel Alloy, and Other Special Alloys
ANSI B16.10–1973	Face-to-Face and End-to-End Dimensions of Ferrous Valves
ANSI/FCI 74-1–1979	Valves Spring Loaded Lift Disc Check
ANSI/AWWA C509–80	Valves, 3 through 12 NPS, for Water and Sewage Systems, Resilient-Seated Gate
ANSI/AWWA C500–80	Valves 3 through 48 NPS, for Water and Sewage Systems, Gate
Pipe Hangers	
ANSI/MSS SP–58–1979	Pipe Hangers and Supports—Materials, Design, and Manufacture

FIGURE 22.12 ■ ANSI specifications for pipe, fittings, valves, and pipe hangers.

Polyethylene pipe (PE) was introduced into the United States in 1947. Plastic piping can handle temperatures up to 150° F and is suitable for water and vent piping of corrosive and acidic gases. PE piping is used as conduit for electrical and phone lines, water lines, farm sprinkler systems, salt water disposal, and chemical waste lines. Other forms of plastic piping include *acrylonitrile-butadiene-styrene* (ABS), which is popular for sewage piping, *cellulose-acetate-butyrate* (CAB), and *fiberglass-reinforced pipe* (FRP).

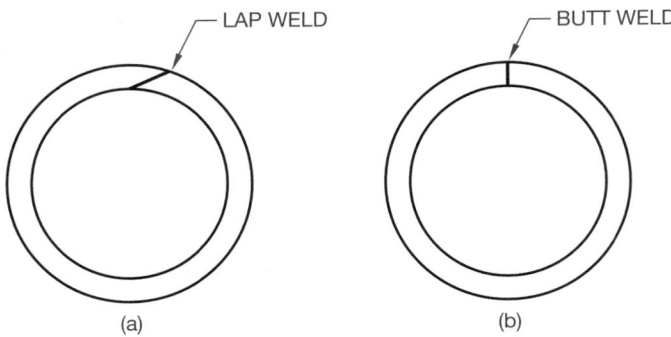

FIGURE 22.13 ■ (a) Lap-welded steel pipe; (b) butt-welded steel pipe.

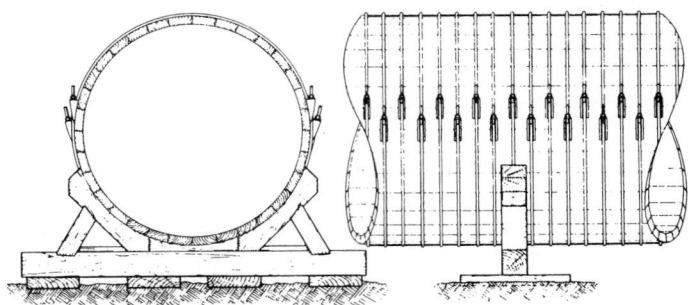

FIGURE 22.14 ■ Front and side views of a wood stave pipe resting in its foundation cradle. *Courtesy National Tank and Pipe Co.*

Clay

Clay pipe is formed under high pressure from fire clays or shales, or a combination of the two. It is dried, then fired at 2100° F. Under these extreme temperatures, the clay particles actually fuse or "weld" together to form a solid, durable pipe. It is one of the most corrosion-proof pipes available for use in sanitary and industrial sewers, and can carry any known chemical waste except hydrofluoric acid.

Glass

The chemical resistance, transparency, and cleanliness of glass makes it popular for applications in the chemical, food and beverage, and pharmaceutical industries. It can withstand temperatures up to 450° F.

Wood

Continuous stave wood pipe is used in the Pacific Northwest because of the abundance of redwood and Douglas fir. Tongue and groove staves are milled to exact radii for the specific diameter of pipe. These staves are fitted together, then strapped with wire, flat steel bands, or steel rod threaded on the ends and tightened with bolts. (See Figure 22.14.) The wood used for this pipe is not resistant, especially when water under constant pressure is flowing through it. Wood stave pipe is used almost exclusively for transporting water, and is available in sizes from 10 in. to 16 ft. in diameter.

Steel Tubing

Tubing is small-diameter pipe and is often flexible, thus eliminating the need for many common fittings. Tubing is specified by its outside diameter and wall thickness. Its uses include external heating applications, boilers, superheaters, and hydraulic lines in the automotive and aircraft industries.

Pipe Sizes and Wall Thickness

Pipe is available in sizes from $\frac{1}{8}$" to 44" in diameter. Sizes greater than this can be ordered but are not normally stocked by suppliers. The typical range of commonly stocked pipe is from $\frac{1}{2}$" to 24". These are the sizes most often used in process piping. Sizes from $\frac{1}{8}$" to $\frac{1}{2}$" are used for instrument lines and service piping.

Pipe is specified by its *nominal pipe size* (NPS). The NPS for pipe $\frac{1}{8}$" to 12" is the inside diameter (I.D.), and pipe 14" and above uses the outside diameter (O.D.) as the NPS. (See Figure 22.15.) If you wish to determine the actual inside diameter of a pipe, you should double the wall thickness and subtract that number from the outside diameter.

ANSI The wall thickness of pipe varies in relation to the size and weight of the pipe. The ANSI specifications for wall thickness (B36.10M) classify pipe as *schedule* (SCH) numbers. These numbers range from 10 to 160, the higher number representing the thicker wall. The ANSI schedule numbers incorporate the classifications of the ASTM and ASME, which use the designations of standard (STD), extra strong (XS), and double extra strong (XXS). These three classifications are drawn from manufacturers' dimensions. The STD wall thickness compares to SCH 40, XS to SCH 80, and XXS has no comparable schedule number (it is a thicker wall than SCH 160).

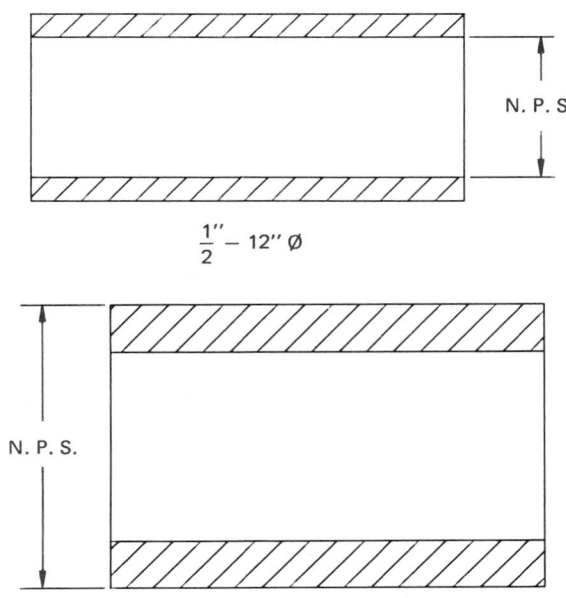

FIGURE 22.15 ■ How to measure nominal pipe size (NPS).

PIPE CONNECTION METHODS

Butt-welded Connection

Butt welding is the most common method of joining pipes used in industry today. It is used on steel pipe 2" in diameter and more to create permanent systems.

Butt-welded pipe and fittings provide a uniform wall thickness throughout the system. The smooth inside surface creates gradual direction changes, thus generating little turbulence. One circumferential weld is required to join two pieces of pipe (the number of welds may vary depending on pipe NPS). Figure 22.16 shows a cross section of a butt weld. The weld is strong, leak-proof, and relatively maintenance free. Pipe joined in this manner is self-contained, withstands high temperature and pressure, is easy to insulate, and requires less space for construction and hanging than do other methods.

Pipe and fittings joined by butt welding are prepared with a *beveled end* (BE). This type of end preparation provides space for the welding operation. (See Figure 22.17.)

Socket-welded Connection

The *socket-welded* connection shown in Figure 22.18 is used on pipe 2 in. and smaller. It forms a reliable leak-proof connection. The pipe has *plain end* (PE) preparation and slips into the fitting. One exterior weld is required, thus no weld material protrudes into the pipe. Since the pipe is slipped inside the fitting, the connection is self-aligning. Socket welding small diameter pipe is less expensive than other welded systems.

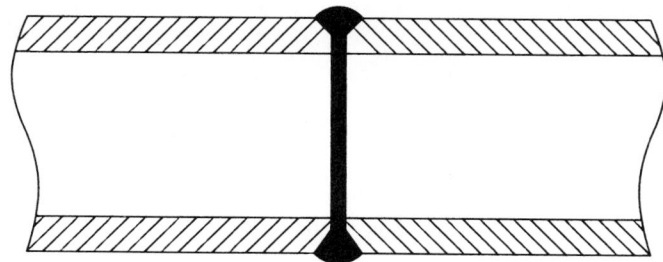

FIGURE 22.16 ■ Cross section of a butt-welded pipe connection.

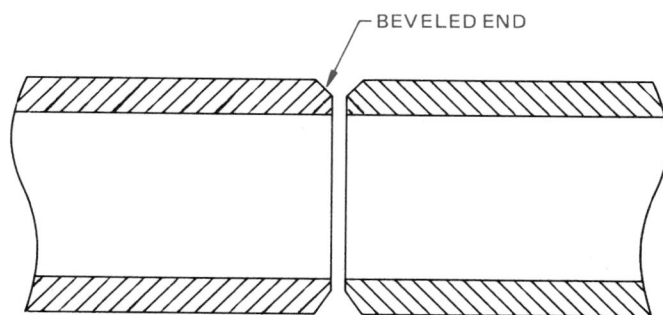

FIGURE 22.17 ■ Beveled-end pipe is used for butt-welded pipe connections.

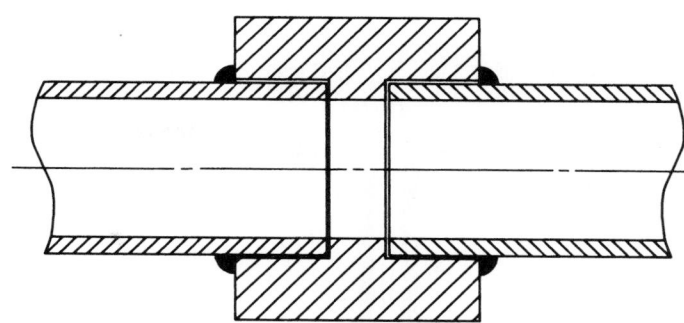

FIGURE 22.18 ■ The socket-welded connection forms a leak-proof joint.

Screwed Connection

Screwed connections are used on steel, malleable iron, cast iron, and cast brass pipe less than 2½" in diameter. It is the least leak-proof of the pipe joining methods, and is used where the temperature and pressure are low. The end preparation for screwed pipe is termed *threaded and coupled* (T&C) because a coupling is usually supplied with a straight length of pipe. American Standard pipe threads are the most common. The tightest fit is achieved by using fittings with straight threads and pipe with tapered threads. Pipe sealing compounds and teflon tape aid in producing a tighter fit. A typical coupling and pipe screwed connection is shown in Figure 22.19.

Flanged Connection

The *flanged* method of connecting pipes utilizes a fitting called a flange. The flange has an outside diameter greater than the pipe, and contains several bolt holes. Flanges are welded to the ends of pipe and then sections of pipe can be bolted together, as shown in Figure 22.20. Most flanges are forged steel, cast steel, or iron. Flanged pipe is easily assembled and disassembled, but is considerably heavier than butt-welded pipe. It occupies more space and is also more expensive to support or hang.

Two steel flanges bolted together do not form a tight fit. There must be some sort of sealing material placed between the two flange faces. This sealer is called a *gasket*. It can be soft or semimetallic material. When the bolts are tightened, the gasket is squeezed until it is pressed into the machining grooves of the flange faces.

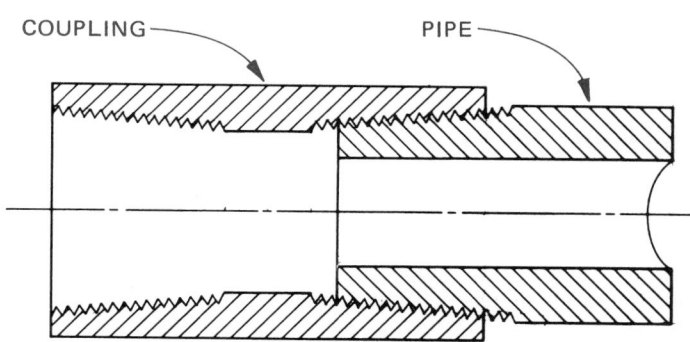

FIGURE 22.19 ■ Cross section of a screwed connection.

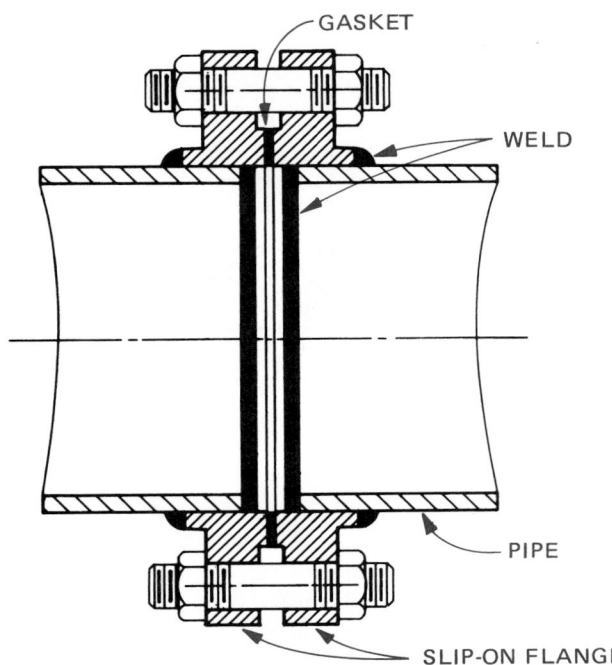

FIGURE 22.20 ■ Cross section of a typical flanged connection.

Soft gaskets can be cork, rubber, asbestos, or a combination of materials. Semimetallic gaskets contain both metal and soft material. The soft material provides resilience and gives a tight seal while the metal helps to retain the gasket in place against high pressure and temperature. Gaskets are chosen from a wide range of materials that resist deterioration caused by high temperatures, and that are not chemically affected by the fluids in the pipe.

Soldered Connection

Copper and brass water tubes are most often joined by *soldered* connections. Domestic water systems where temperatures and pressures are low are the most common use of this method. Both rigid and soft pipe can be soldered.

Bell (Hub) and Spigot Connection

Underground sewage, water, and gas lines are common applications of the *bell and spigot* connection method. (See Figure 22.21.) Cast iron pipe is used where pressures and loads are low. There are many variations of the bell and spigot method, but they all utilize some sort of sealer. The most common is lead and oakum (a fibrous sealer), although cement is used in certain situations.

Mechanical Unit

The *mechanical joint* is a modification of the bell and spigot connection in which flanges and bolts are used with gaskets, packing rings, or grooved pipe ends providing a seal. Some mechanical joints allow for angular deflection and lateral expansion of the pipe. This type of connection is used in low pressure applications, and where vibration may be excessive. Figure 22.22 illustrates a mechanical joint.

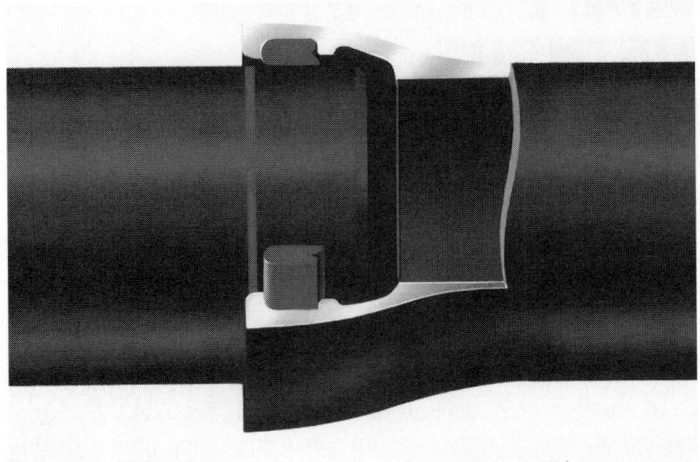

FIGURE 22.21 ■ Bell and spigot connection. *Courtesy Griffin Pipe Products Co.*

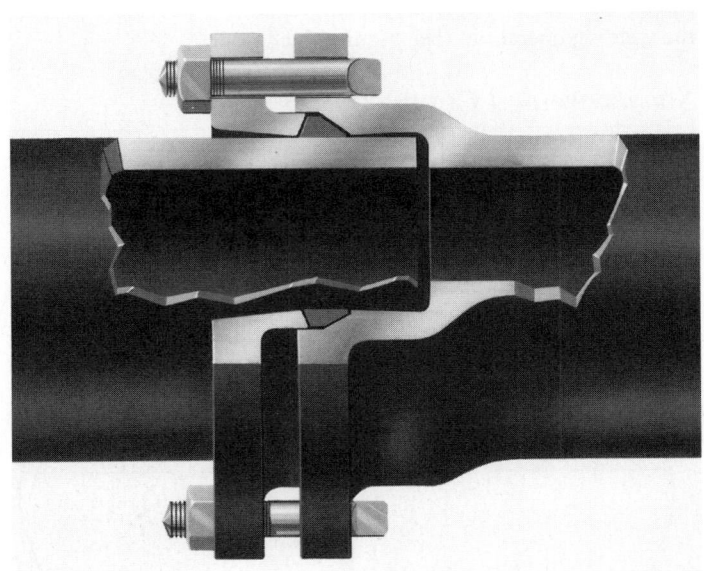

FIGURE 22.22 ■ Typical mechanical joint connection. *Courtesy Griffin Pipe Products Co.*

Solvent Welding

This type of pipe connection is also known as *cementing* or *gluing*. It is used on plastic pipe. The solvent is applied to the pipe end, which is then inserted in the "socket" opening of a fitting and twisted. The solvent creates a bonding that acts much the same as a weld. Plastic pipe can also be joined by *hot gas welding* in which a torch is used to melt the pipe and fitting together.

Flaring

Soft copper pipe and tubing can be joined by a method called *flaring*. A special tool is clamped on the pipe and the flaring tip is inserted into the end of the pipe. As the tip is rotated, it is forced into the pipe, spreading the end of the pipe open, or "flaring" it. The end of the pipe resembles an old blunderbuss rifle. A fitting called a *swage* is then used to connect the flared end to pipe or a fitting.

FITTINGS

Pipe fittings enable pipe to change direction and size, and provide for branches and connections. Each type of pipe and connection method discussed previously uses the same type of fittings. But special fittings may be required by the nature of the connection method used. The following is a general discussion of common fittings used in industry today.

Welded Fittings

Seamless forged steel fittings are prepared with a beveled end to accommodate butt welding. A welding ring (see Figure 22.23a) is placed between pipe and fitting to aid in alignment, provide even spacing, and to prevent weld material from falling into the pipe (see Figure 22.23b). Since fittings have the same schedule numbers and wall thickness as pipe, a smooth inside surface is produced at the joints when fittings and pipe of the same schedule number are welded.

LONG NUBS   SHORT NUBS

(a)

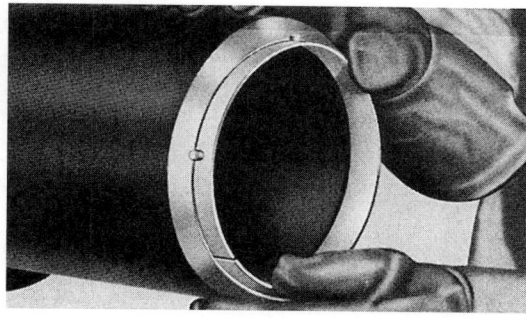

(1)

This photograph shows the ease with which a welder can fit a backing ring into the beveled end of pipe. This is an NPS 6″ schedule 80 pipe. Tack welds are not necessary. The spacer nubs melt and become part of the weld metal.

(2)

The adjoining pipe is slipped onto the backing ring until the edge of the bevel touches the spacer nubs. Short pieces of pipe will be self supported by the close tolerances of the ring diameter.

(3)

Note the close fit between the ring and the pipe . . . the minimum gap at the split. The carefully aligned spacer nubs center the backing ring perfectly across the welding groove and provide exact, even spacing between the pipe ends. The metal arrow locates the position of the spacer nubs in X-ray photographs.

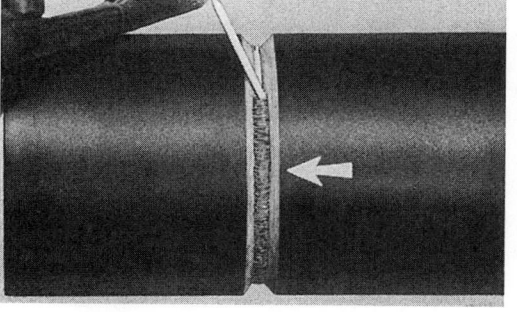

(4)

The weld is started at the bottom of the pipe. As this weld is being made in a fixed horizontal position, the split in the ring is on the bottom of the weld to prevent icicle formation inside the pipe. The ring is designed to allow a gap of only 1/16″ at the split, minimizing the operation of filling the gap.

(b)

FIGURE 22.23 ■ **(a)** Common welding rings with long nubs and short nubs; **(b)** pipe-joining process using a welding ring. *Courtesy ITT Grinnell Corp.*

The most familiar fittings are *elbows*. These and other welded fittings are seen in Figure 22.24. Standard shapes are the 90° and 45° elbows. The 90° *reducing elbow* reduces the pipe size in addition to changing direction. A reversal in direction can be achieved by using a 180° elbow. Standard elbows can also be cut to any angle required.

A *mitered* elbow is produced by cutting and welding straight pipe. The mitered elbow can be composed of one, two, three, or more welded joints. (See Figure 22.25.) It is not used often since it produces considerably more turbulence than standard elbows. An installation of the mitered elbow is seen in Figure 22.26.

90° ELBOW 90° REDUCING ELBOW 45° ELBOW

STRAIGHT TEE REDUCING TEE CROSS

CONCENTRIC REDUCER ECCENTRIC REDUCER CAP

STRAIGHT LATERAL REDUCING LATERAL

FIGURE 22.24 ■ Welded seamless fittings. *Courtesy ITT Grinnell Corp.*

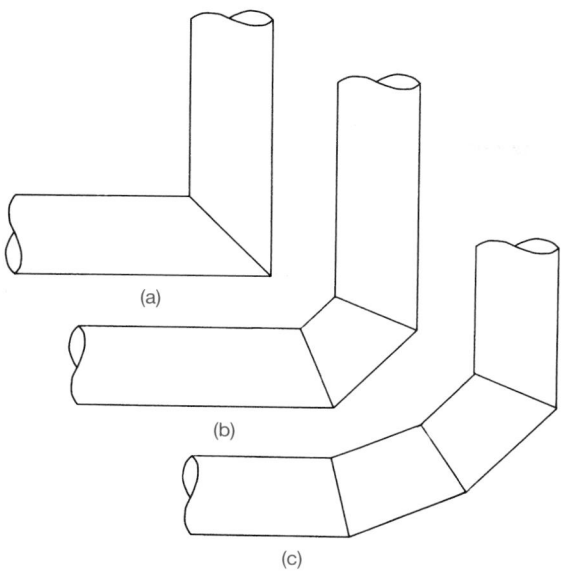

FIGURE 22.25 ∎ The mitered elbow is composed of several cut pipe segments: (a) two-piece miter; (b) three-piece miter; and (c) four-piece miter.

FIGURE 22.26 ∎ Mitered elbows in a vacuum pump installation at a paper mill. *Courtesy Ingersoll-Rand.*

THREE-DIMENSIONAL SYSTEM MODELS

In addition to increasing productivity in generating drawings and materials lists, some CADD systems enable the drafter or designer to create a 3-D, full-color model of the system. This model can be shown in wire or solid form, colored and shaded in any fashion, and rotated to suit the needs of the user. (See Figure 22.27.) Hard copy prints of any aspect of the piping system can be generated for use in the design or construction of the system.

A CADD-generated model can be used to create a scale model and/or orthographic and pictorial drawings for the project. This is a major step in the effort to generate a single data base for an entire construction project that is accessible to all engineers, technicians, drafters, and clients. Once all the design information is in the system, any type of data, in the form of plotted drawings, reports, tables, material take-off, estimates, bills, and a variety of screen displays can be requested. As changes are made to the design, it is reflected in all displays, hard copy prints, and plotted drawings.

Future product and industrial piping design and construction may be handled completely by a computer system. This central system containing an extensive database, including building or site sizes and construction codes, may be able to automatically route and lay out pipe, valves, fittings, and equipment. The more routine tasks will be performed by the computer, and the overall design of the system will be the

FIGURE 22.27 ∎ This 3-D model of a piping system has been shaded and rendered to achieve a realistic presentation. *Courtesy International Software Systems, Inc.*

function of the engineer, technician, or design drafter. Therefore, a knowledge of CADD and a continuing interest in computer and software advances will enable the piping engineer or drafter/designer to learn and change as our technology advances.

CADD APPLICATIONS

The diameter of the pipe can be changed by *reducers*. The *concentric reducer* tapers the pipe equally about the axis centerline of the pipe. The *eccentric reducer* has a flat side that allows one side of the pipe—typically the top or bottom—to remain level.

The *straight tee* is a standard fitting that creates a branch or inlet. The straight tee has the same diameter on all three openings. The *reducing tee* has a smaller opening on the outlet side of the fitting. A branch or inlet can also be created with a *lateral*. This fitting provides a 45° branch of the *run* pipe. The run is the straight portion of the fitting. The lateral is available in both straight and reducing forms. The *true* Y is similar to the lateral, but produces branches that are at a 90° angle, thus forming a "Y" with the run pipe. A branching fitting similar to the tee is the *cross*. It has two branches opposite each other. It can be either straight or reducing. It is employed in special situations and tight spaces, but is seldom used because of its expense.

There are several small fittings, less expensive than regular fittings, that are used to create branch connections in new installations or on pipe that is already assembled. They are commonly called *weldolets*®, but are manufactured to accept welded, screwed, socket-welded, and brazed pipe. These little fittings can be welded to an elbow (*elbolet*®) at a 45° angle to the run pipe (*latrolet*®), or at a 90° angle to the run pipe (*weldolet*®). (See Figure 22.28.)

Dimensions for seamless welded fittings can be found in Appendix F.

Screwed Fittings

All of the welded fittings are available in threaded form. In addition to these, there are several special fittings that are used with screwed pipe. Appendix G provides dimensions for galvanized malleable iron fittings.

■ *Union*—Composed of two threaded sleeves and a threaded union ring, this fitting provides a connection point in a straight run of pipe. It enables pipe to be broken apart without tearing down the entire run of pipe. Figure 22.29 illustrates screwed fittings.

■ *Coupling*—Threaded at both ends (TBE) with internal threads, it is used to attach two lengths of pipe together.

■ *Half coupling*—Threaded at one end (TOE), this fitting is often welded to pipes and used for instrument connections.

■ *Street elbow*—This 90° elbow is threaded at one end with internal threads and the other end with external threads. It can be attached directly to a fitting, thus eliminating the need of a short piece of threaded pipe (*nipple*).

■ *Bushing*—A reducing fitting used to connect small pipe to larger fittings.

■ *Plug*—A fitting with external threads on one end that is used to seal the screwed end of a fitting.

FLANGES

The component that creates a bolted connection point in welded pipe is the *flange*. It is a circular piece of steel containing a center bore that matches the pipe I.D. to which it is attached. It has several bolt holes evenly spaced around the center bore. Flanges are used most extensively on welded pipe, but are also available for most other types of pipe. Figure 22.30 shows several types of flanges. Flange dimensions are given in Appendix F.

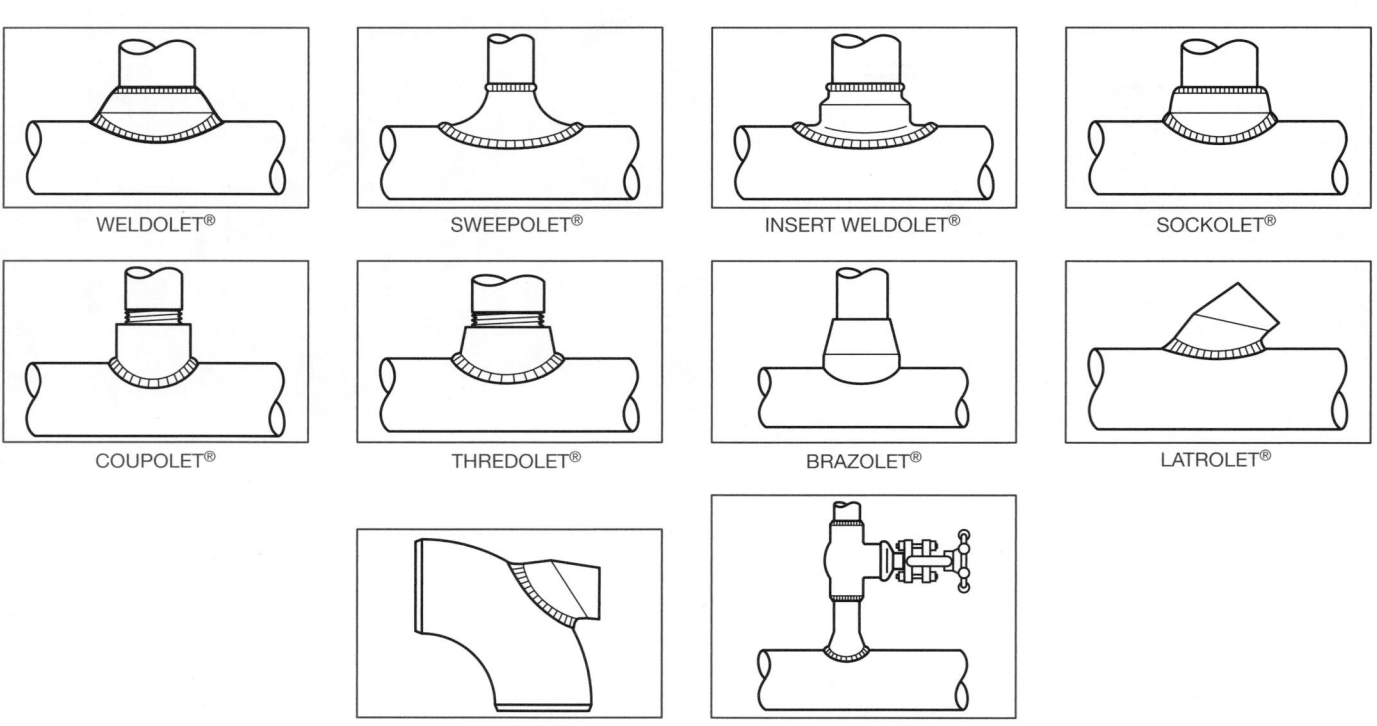

WELDOLET®   SWEEPOLET®   INSERT WELDOLET®   SOCKOLET®

COUPOLET®   THREDOLET®   BRAZOLET®   LATROLET®

ELBOLET®   NIPOLET®

FIGURE 22.28 ■ Branch connections can create inlets or outlets of varying sizes and angles to the main run of pipe and at less expense than regular fittings. *Courtesy Bonney Forge Division, Gulf and Western.*

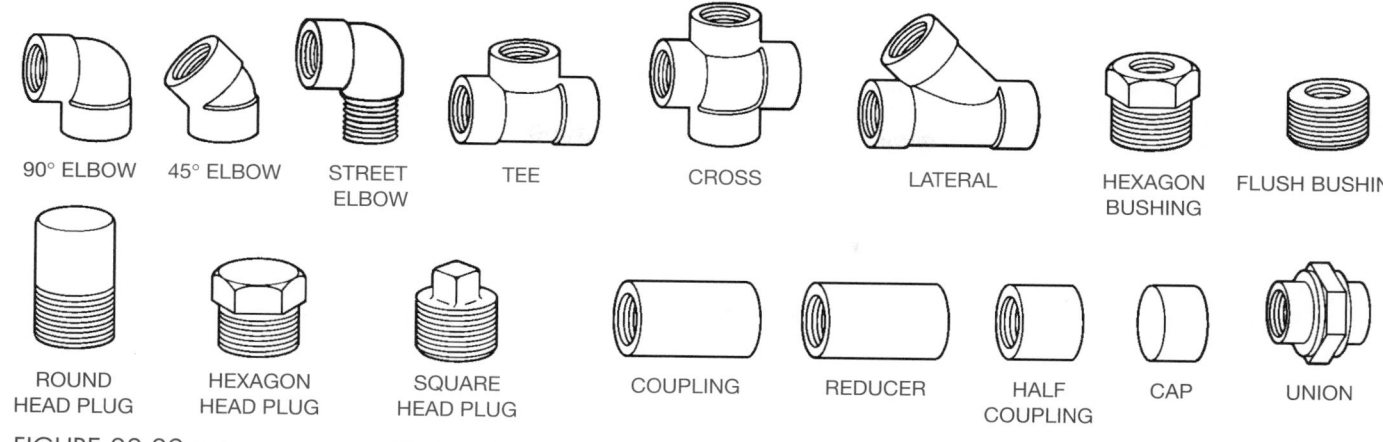

FIGURE 22.29 ■ Common screwed fittings. *Courtesy Alaskan Copper Works.*

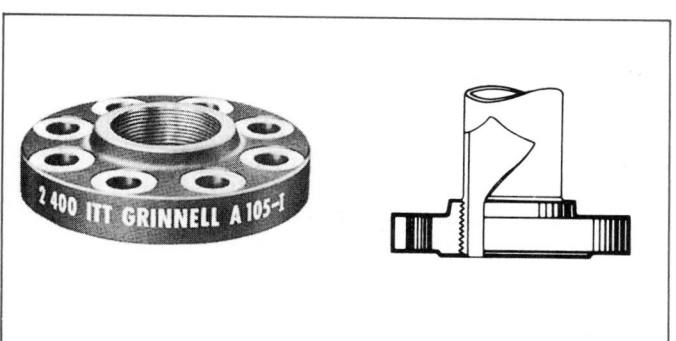

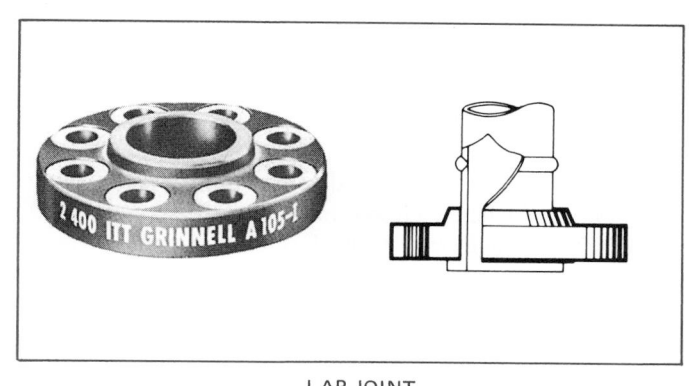

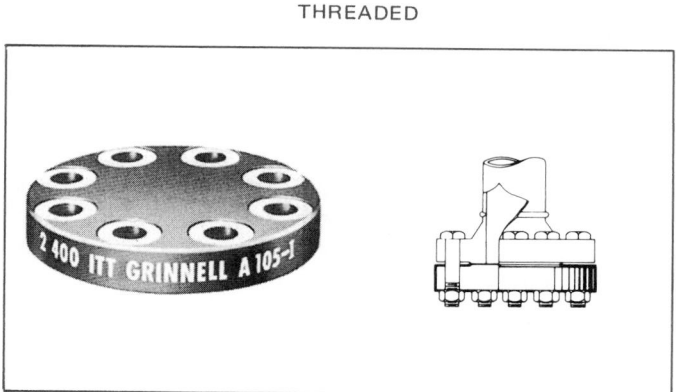

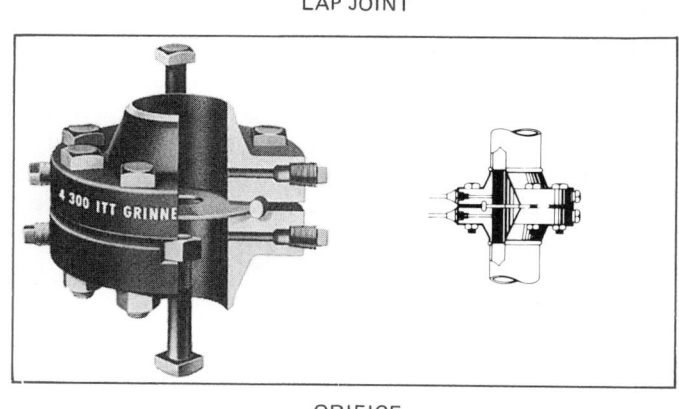

FIGURE 22.30 ■ Common flanges used on butt-welded steel pipe. *Courtesy ITT Grinnell Corp.*

Slip-on Flange

The *slip-on* flange can be used only on straight pipe because it is bored to slip over the end of the pipe. Two welds are required to attach this flange to pipe. (See Figure 22.30.) There are two types of slip-on flanges.

ASME When the Type 1 slip-on flange is used, the pipe is set back from the face of the flange. It has two-thirds the strength of a weld-neck flange and is limited to 300 lb service by the ASME Pressure Piping Code. The Type 2 slip-on flange allows the pipe to be welded flush with the face of the flange, which is then machined flat. This type of flange is used on lines having 400 lb pressure and above.

Weld-neck Flange

The *weld-neck* flange is forged steel and prepared with a beveled end for butt welding to pipe or fittings. It is always used when a flange must be attached directly to a fitting.

Blind Flange

A pipe can be temporarily sealed with a *blind flange.* It is basically a steel plate with bolt holes in it.

Stub-end or Lap-joint Flange

The *stub-end* or *lap-joint* flange is composed of two parts, the stub end and the flange ring. This flange ring can be carbon steel if expensive pipe such as stainless steel is used. Only the stub end need be stainless.

Reducing and Expander Flange

A change in line size can be achieved with a *reducing flange,* but it should not be used where increased turbulence would be undesirable. A reduction in line size can also be created with an *expander flange.* This fitting is a flange/reducer combination and can be used in place of a weld-neck flange and a reducer.

Orifice Flange

The flow rate inside a pipe can be measured using *orifice flanges* and an *orifice plate.* The two orifice flanges are drilled and tapped to accommodate tubing and a pressure gauge used to measure the flow rate. The orifice plate is a flat disc with a small hole drilled in its center. The two flanges are welded to the pipe, the orifice plate and gaskets are placed between them, and the flanges are bolted together. As fluid flows through the hole in the orifice, a pressure differential is created on either side of the plate. This differential can be read on the pressure gauge.

Flange Faces

The *facing* of a flange is the type of machining that is done to the contact surface of the flange. A common flange used in industry is the *raised face* (RF) flange. The face of this flange protrudes a certain distance beyond the flange itself. The *male and female* facings interlock with each other. A recessed face on the female flange accepts both the gasket and the raised face of the male flange. The *tongue-and-groove* facing is interlocking. The male tongue-and-groove flange is ¼" longer than the female flange to accommodate the height of the tongue The tongue fits into the groove creating a tighter lock and seal. Figure 22.31 shows cross sections of the flange faces mentioned here.

The *ring-type joint* (RTJ) flange connection requires two flanges that both have a machined groove in the face. This groove is cut to accept an oval or octagonal ring. The grooves on both flanges fit over the ring which creates a seal. Examples of the ring joint flange are seen in the cutaway photo of the orifice flange in Figure 22.30, and the section view in Figure 22.31.

VALVES

Valves are the components in the piping system that control and regulate fluids. Valves not only provide on/off service in a pipe, but are also used to regulate the flow of fluid in the pipe, maintain a constant pressure, prevent dangerous pressure buildup, and prevent backflow in the pipe. A variety of valves are used to perform these tasks.

On/off

The *gate* valve is used exclusively to provide on/off service in a pipe. It is the most common valve used in industry today. (See Figure 22.32.) A gate or disc is moved up and down inside the valve manually or automatically. The design of the gate valve is such that fluid can flow through it with a minimum of friction and pressure loss. The valve *seat* does not interfere with the straight line flow of the fluid. The seat is the material with which the gate makes contact to create a seal. The gate valve is designed specifically for on or off service and infrequent operation. It is unsuitable for throttling or regulating flow. A partially open disc (gate) can cause erosion and wear of the downstream side of the seat and the disc itself.

The sealing mechanism inside a *ball* valve is just that—a ball. The ball has a hole through it that matches the I.D. of the pipe. (See Figure 22.33.) It is a quick-opening valve, requiring only a one-quarter turn, and is used extensively because of its tight seat. Plastics, nylon, and synthetic rubber are used for the seat material and enable this valve to achieve a tight seal. The ball valve is popular because it has a low profile and low torque requirement for operation. It is also easy and inexpensive to maintain and repair. It is not used in large diameter lines (greater than 12") because the pressure in the line makes it difficult to open and close.

The *plug* valve, also known as *cock* valve, is similar in design to the ball valve and requires only a one-quarter turn to open and close. The opening through the plug can be either rectangular or round. (See Figure 22.34.) When open, there is a pressure drop through the valve, but a high flow efficiency due to the contours of the valve. It has low throttling ability and is best used for on/off service. It can achieve a tighter shutoff than a gate valve, but is normally used on smaller diameter lines.

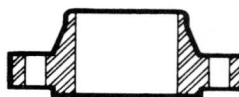

The Raised Face is the most common facing employed with steel flanges; it is ¹⁄₁₆" high for Class 150 and Class 300 flanges and ¼" high for all other pressure classes. The facing is machine-tool finished with spiral or concentric grooves (approximately ¹⁄₆₄" deep on approximately ¹⁄₃₂" centers) to bite into and hold the gasket. Because both flanges of a pair are identical, no stocking or assembly problems are involved in its use. Raised face flanges generally are installed with soft flat ring composition gaskets. The width of the gasket is usually less than the width of the raised face. Faces for use with metal gaskets preferably are smooth finished.

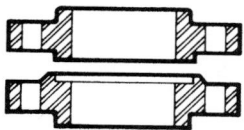

Male-and-Female Facings are standardized in both large and small types. The female face is ³⁄₁₆" deep and the male face ¼" high and both are usually smooth finished since the outer diameter of the female face acts to locate and retain the gasket. The width of the large male and female gasket contact surface, like the raised face, is excessive for use with metal gaskets. The small male and female overcomes this but provides too narrow a gasket surface for screwed flanges assembled with standard weight pipe.

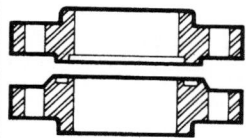

Tongue-and-Groove Facings are also standardized in both large and small types. They differ from male-and-female in that the inside diameters of tongue and groove do not extend to the flange bore, thus retaining the gasket on both its inner and outer diameter; this removes the gasket from corrosive or erosive contact with the line fluid. The small tongue-and-groove construction provides the minimum area of flat gasket it is advisable to use, thus resulting in the minimum bolting load for compressing the gasket and the highest joint efficiency possible with flat gaskets.

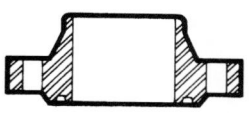

Ring Joint Facing is the most expensive standard facing but also the most efficient, partly because the internal pressure acts on the ring to increase the sealing force. Both flanges of a pair are alike, thus reducing the stocking and assembling problem found with both male-and-female and tongue-and-groove joints. Because the surfaces the gasket contacts are below the flange face, the ring joint facing is least likely of all facings to be damaged in handling or erecting. The flat bottom groove is standard.

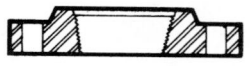

Flat Faces are a variant of raised faces, sometimes formed by machining off the ¹⁄₆₄" raised face of Class 150 and Class 300 flanges. Their chief use is for mating with Class 125 and Class 250 cast iron valves and fittings. A flat-faced steel flange permits employing a gasket whose outer diameter equals that of the flange or is tangent to the bolt holes. In this manner the danger of cracking the cast iron flange when the bolts are tightened is avoided.

FIGURE 22.31 ■ Cross section views of common flange faces. *Courtesy ITT Grinnell Corp.*

Regulating

The most common type of regulating valve is the *globe* valve. It is normally used in pipe up to 3" in diameter, but can be used in lines with diameters up to 12". Fluid flowing through the globe valve travels in an "S" pattern, which enables the valve to maintain a

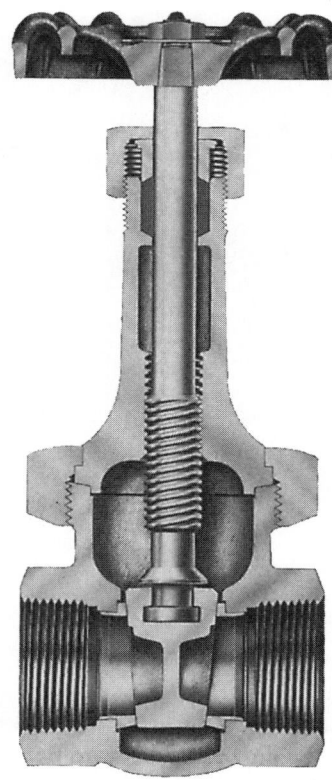

FIGURE 22.32 ■ The gate valve provides on/off service. *Courtesy Crane Co.*

FIGURE 22.33 ■ The ball valve is quick-opening and is used in pipe less than 12" in diameter. *Courtesy The Wm. Powell Co.*

close control on the flow and to achieve a tight, positive shutoff. This flow pattern is seen in Figure 22.35. The internal design of the globe valve creates a high flow resistance, leading to a significant pressure drop through the valve. Body pockets in the valve are not drained when the flow is stopped. Situations that require frequent valve operation and maintenance are suited for globe valves because the discs and seats are easily replaced.

A special type of globe valve that creates a 90° direction change in the pipe is an *angle* valve. It is similar in design to a globe valve (see Figure 22.36), and is used in place of a globe valve and a 90° elbow to save money. In high-stress situations, the angle valve is not used.

Low-pressure situations may warrant the use of the *butterfly* valve. The disc is mounted on a stem that is turned a one-quarter

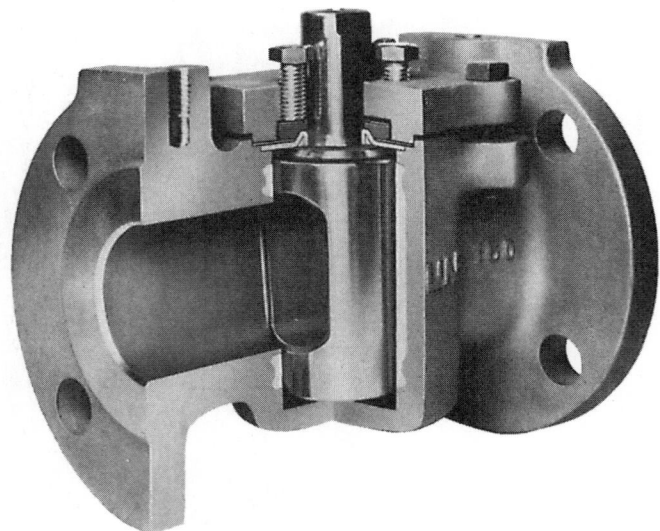

FIGURE 22.34 ■ The plug valve achieves tighter shutoff than a gate valve but is used on smaller diameter lines. *Courtesy Xomox Corp.*

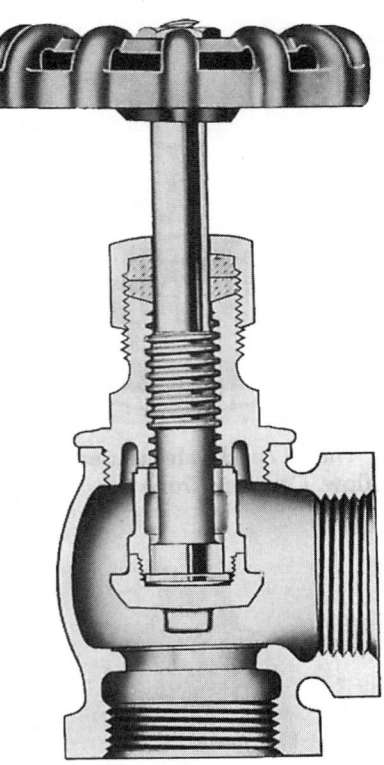

FIGURE 22.36 ■ The angle valve replaces the globe valve and an elbow. *Courtesy Crane Co.*

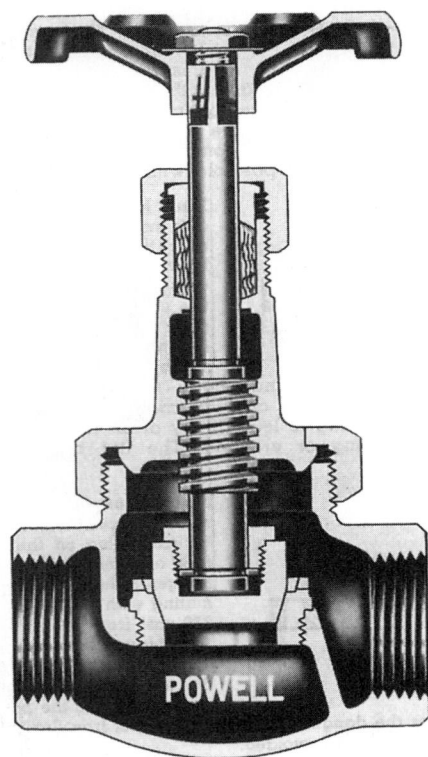

FIGURE 22.35 ■ The globe valve achieves close flow control but creates high flow resistance. *Courtesy The Wm. Powell Co.*

turn to open and close. The entire disc moves inside the valve. (See Figure 22.37.) It is a simple operating mechanism that is excellent for regulating flow. The design creates a minimum pressure drop through the valve. It is light and inexpensive, and it requires a small installation space and is easy to maintain. All-plastic butterfly valves are available.

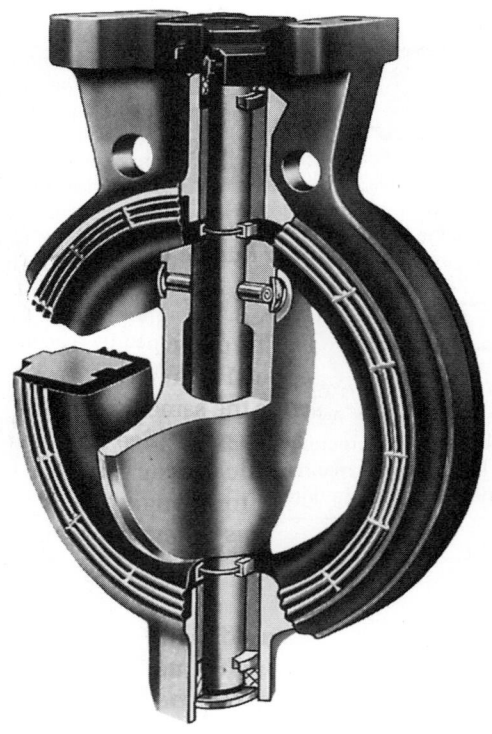

FIGURE 22.37 ■ The butterfly valve is excellent for regulating flow and creates a minimum pressure drop. *Courtesy Crane Co.*

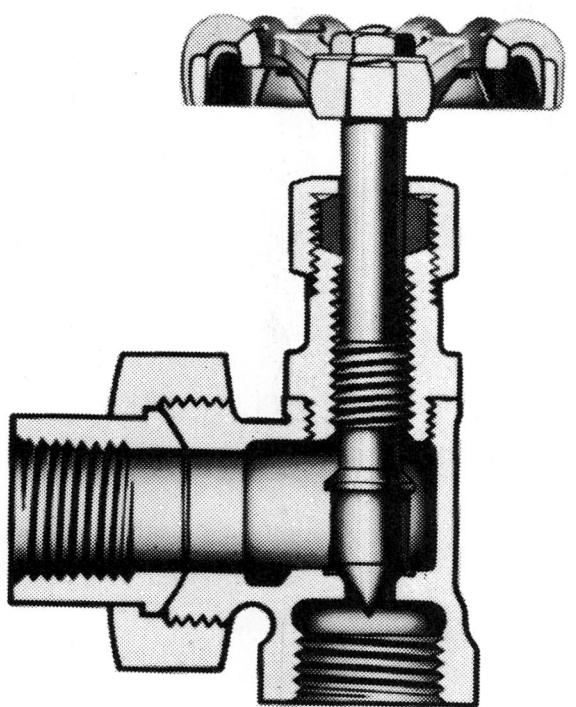

FIGURE 22.38 ■ The needle valve is good for accurate throttling on instrument and meter lines. *Courtesy Crane Co.*

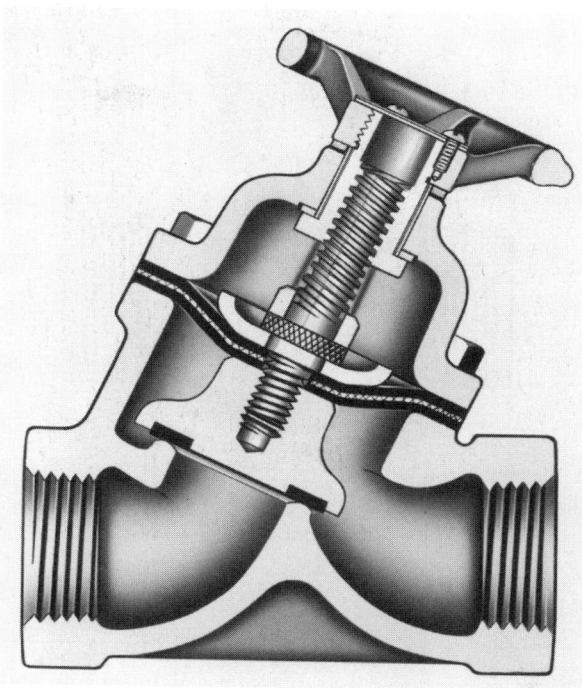

FIGURE 22.39 ■ The diaphragm valve protects the valve mechanism from the fluid. *Courtesy Crane Co.*

The *needle* valve is so named because the end of the stem is needle shaped. (See Figure 22.38.) The stem has fine threads enabling this valve to be adjusted exactly to achieve accurate throttling. The flow through the needle valve changes direction much like the flow through a globe valve. The needle valve is used for high-temperature and high-pressure service in instrument, gauge, and meter lines.

Some types of service require that the working parts of the valve be sealed from the fluid stream. The protection of the valve parts is important when working with fluids that are corrosive, viscous, fibrous, or contain suspended solids and sludges. The protection of the *fluid* is most important when dealing with food and beverages. The *diaphragm* valve suits this need. In place of a disc or other type of sealing mechanism, a diaphragm of rubber, neoprene, butyl, silicone, or other flexible material is used. The diaphragm is pushed down by the stem and creates a seal against a seat on the bottom of the valve. (See Figure 22.39.) The diaphragm also serves to protect the working parts of the valve. It has a smooth, streamlined flow, is easy to maintain, achieves positive flow control, and is leak-tight. It is suited for both on/off and regulating services up to 400° F.

Checking Backflow

Certain situations require that fluids be prevented from flowing backward in the pipe should a power failure or pump breakdown occur. *Check* valves prevent backflow, closing when the fluid stops flowing.

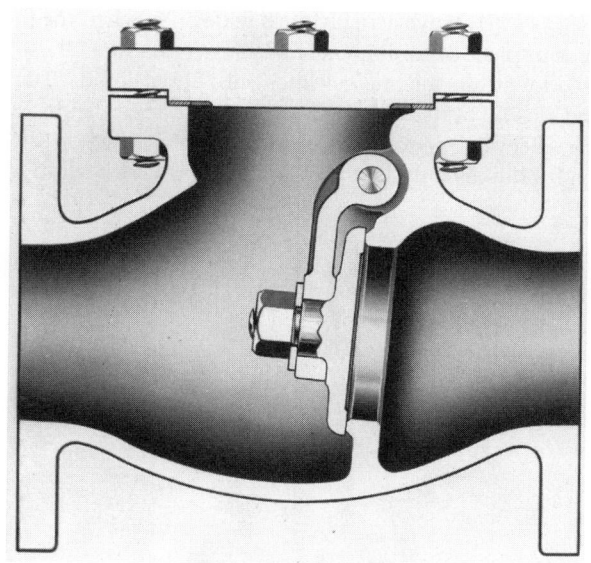

FIGURE 22.40 ■ The disc inside the swing check valve operates by gravity to check backflow. *Courtesy Crane Co.*

The *swing check* valve as shown in Figure 22.40 is similar in construction to the gate valve and is used with it. The disc inside the swing check valve operates by gravity or the weight of the disc. It is best used with low-velocity liquids. The *lift check* valve (see Figure 22.41) is similar to the globe valve and is used with it. It operates by gravity and is available in either horizontal or vertical models.

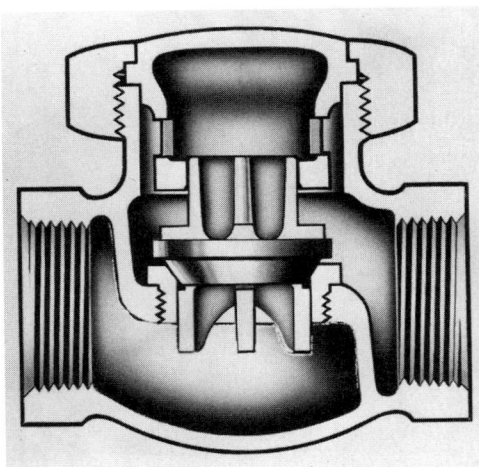

FIGURE 22.41 ■ The lift check valve is used with the globe valve to check backflow. *Courtesy Crane Co.*

Safety

The high temperatures and pressures within many industrial processes are potentially dangerous and must be controlled and regulated to prevent serious accidents. The responsibility for keeping pressures at or below a given point is the task of *safety* and *relief* valves.

The *pop safety* valve actually pops wide open when the pressure in a pipe or piece of equipment reaches a set pressure. These valves are used for steam, air, and gas lines only, never liquids. The opening and closing of this valve is instantaneous. (See Figure 22.42.)

The relief valve performs the same function as the pop safety, but is used for liquids only. (See Figure 22.43.) It is also set to open at

(a)

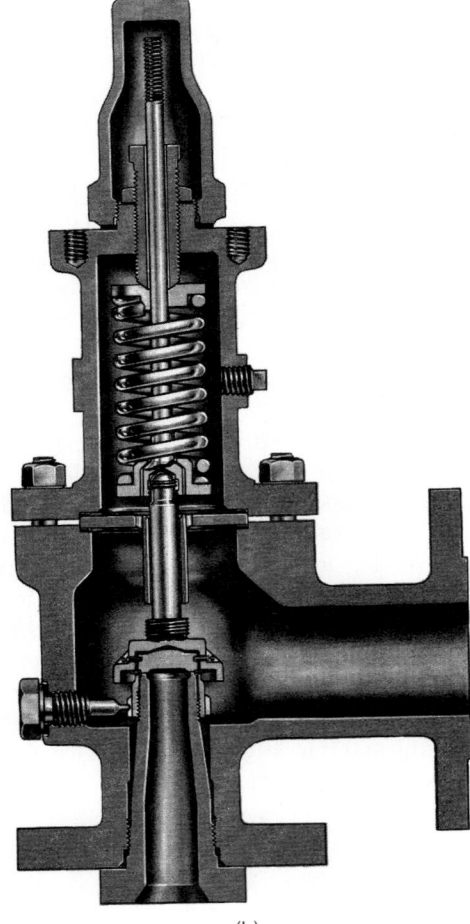

(b)

FIGURE 22.42 ■ The pop safety valve provides momentary release of pressure. *Courtesy Kunkle Valve Co., Inc.*

FIGURE 22.43 ■ (a) Relief valve exterior; (b) cutaway view; provides slow release of fluids and pressure. *Courtesy Kunkle Valve Co., Inc.*

(a)

(b)

FIGURE 22.44 ■ (a) Cage guided control valve exterior view; (b) cutaway view of same valve. *Courtesy Dezurik Valve Co.*

a specific pressure, but it opens and closes slowly in response to changing pressure. This type of valve is found on most home hot water heaters.

Control

The complex processes used in industry today demand instantaneous control and adjustment of flow, pressure, and temperature. That is where *control* valves enter in. Most any type of valve can be a control valve. The example in Figure 22.44 illustrates one of the types used. The distinguishing component on the control valve is the controller or *actuator*. This is the mechanism that operates the valve. It can be operated by an electric motor, air cylinder, or hydraulic cylinder.

Pressure Regulating

Processes relying on steam require a constant flow at a steady rate. This task falls on *pressure regulators*. This mechanism reduces the incoming pressure of the steam to the required service pressure. The regulator maintains the pressure at the specified level and provides a uniform flow. (See Figure 22.45.)

FIGURE 22.45 ■ Pressure regulator. *Courtesy Spence Engineering.*

PIPE DRAFTING

Pipe drafting positions are found with consulting engineering companies, large industrial construction firms that maintain engineering offices, and manufacturers of equipment that utilize piping. The entry-level drafter in these companies may begin by drawing revisions (markups) to existing drawings. These may be flow diagrams (nonscale schematic drawings) or piping drawings. As experience is gained, the drafter is assigned more complex drawings that must be constructed from engineering sketches.

Engineers, designers, and technicians design the piping system, select and size the equipment and pipe, and give this information to the drafter. The drafter's job is to construct the required type of drawing or model, often using vendor catalogs, brochures, and charts, or other drawings containing reference information. Once a working knowledge of the subject is gained, the drafter may be given the responsibility of designing and laying out pipe runs. At this point the drafter begins to utilize his or her knowledge of pipe fittings and reference information.

Plans and Sections (Elevations)

The most common and the most complex type of piping drawings are plans and sections that show building outlines, equipment, pipe, and pipe supports. Examples of a plan and several sections taken from that plan are shown in Figure 22.46. Study these drawings carefully. Look at the plan and sections together to try to determine where specific pipe runs are located. A little time spent now will aid you in working with this type of drawing in the future.

Equipment shown on piping plan and section drawings is seldom drawn in detail, but instead an outline is used to illustrate clearances and access. Tanks, vessels, heat exchangers, pumps, columns, compressors, boilers, dryers, and reactors are just a few examples of some of the equipment with which a piping drafter may work.

In addition to major pieces of equipment, the drafter may be required to indicate pipe supports and hangers. Note in Figure 22.47 (see page 749) the extensive use of pipe hangers in the form of pipe clamps and iron rods. Pipe supports must also come from the bottom up, as seen in Figure 22.48 (see page 750). These are vital to a piping system because they provide stability and anchor points for the pipe. A few common pipe supports and hangers and their application in a sample piping system are shown in Figure 22.49 (see page 750). If the hanger or support is complex or of a special design, it is often illustrated on a "detail" drawing, such as that in Figure 22.50 (see page 751). When standard pipe clamps, anchors, hangers, and supports are used, they may be indicated by a symbol or note such as "P.S."

Single-line Drawings

A time-saving method of drawing pipe is the single-line method. The centerline of the pipe is used to represent the pipe. It may be used to better illustrate new pipe that is added to an existing installation, or to show small diameter pipe. Standard pipe symbols are used in single-line drawings, and the pipe and fittings are often drawn as heavy lines to provide contrast with other lines on the drawing.

Double-line Drawings

The double-line method is the easiest to interpret, for it looks like the actual pipe. But it takes longer to display on a monitor and plot with a pen plotter. This method is used extensively by equipment manufacturers to illustrate standard installation and in presentation drawings that are seldom revised. (See Figure 22.51, page 752.) The double-line method is also used to portray large diameter pipe, to show existing pipe at an industrial facility, or for small views that may be more easily visualized as double-line drawings (see Figure 22.52, page 753). When new or small pipe is shown in the single-line method on the same drawing, the contrast between the two styles aids in interpretation. The single-line method is seldom used for large diameter pipe because clearances, interferences, spacing, and distances are not as easily seen. The double-line method is preferred where these factors are critical.

When laying out either type of piping drawing, it is best to begin with background information such as building outlines or structural steel column reference points. From these points the equipment locations can be determined. Pumps are often represented by centerlines. Other large equipment must be drawn. Pipe centerline locations must then be determined. CADD drafters can use pipe centerline location lines of a different color layer or as tick marks on the screen, then insert fittings and draw pipe as they go.

Fitting and Valve Symbols

A major part of the piping drafter's job is drawing fittings and valves. Appendix H provides commonly used valve symbols. Manual drafting piping templates are available in a variety of sizes. Some companies may even have their own special templates made for them.

Dimensions and Notes

Piping drawings are normally dimensioned in a style similar to architectural drawings, but some companies may use the mechanical style of broken dimension lines. The former type of dimension lines are unbroken, and values are written above horizontal lines and to the left of and reading up on vertical dimension lines. Values are given in feet and inches with a dash between the two. Refer to the plan in Figure 22.46a for examples of piping plan dimensions.

Piping elevations or sections are not normally dimensioned like plan views. Instead of linear dimensions, elevations are used. Occasionally elevations for the bottom of the pipe (B.O.P.) are required. The sections shown in Figure 22.46b–e, pages 746–749, illustrate the

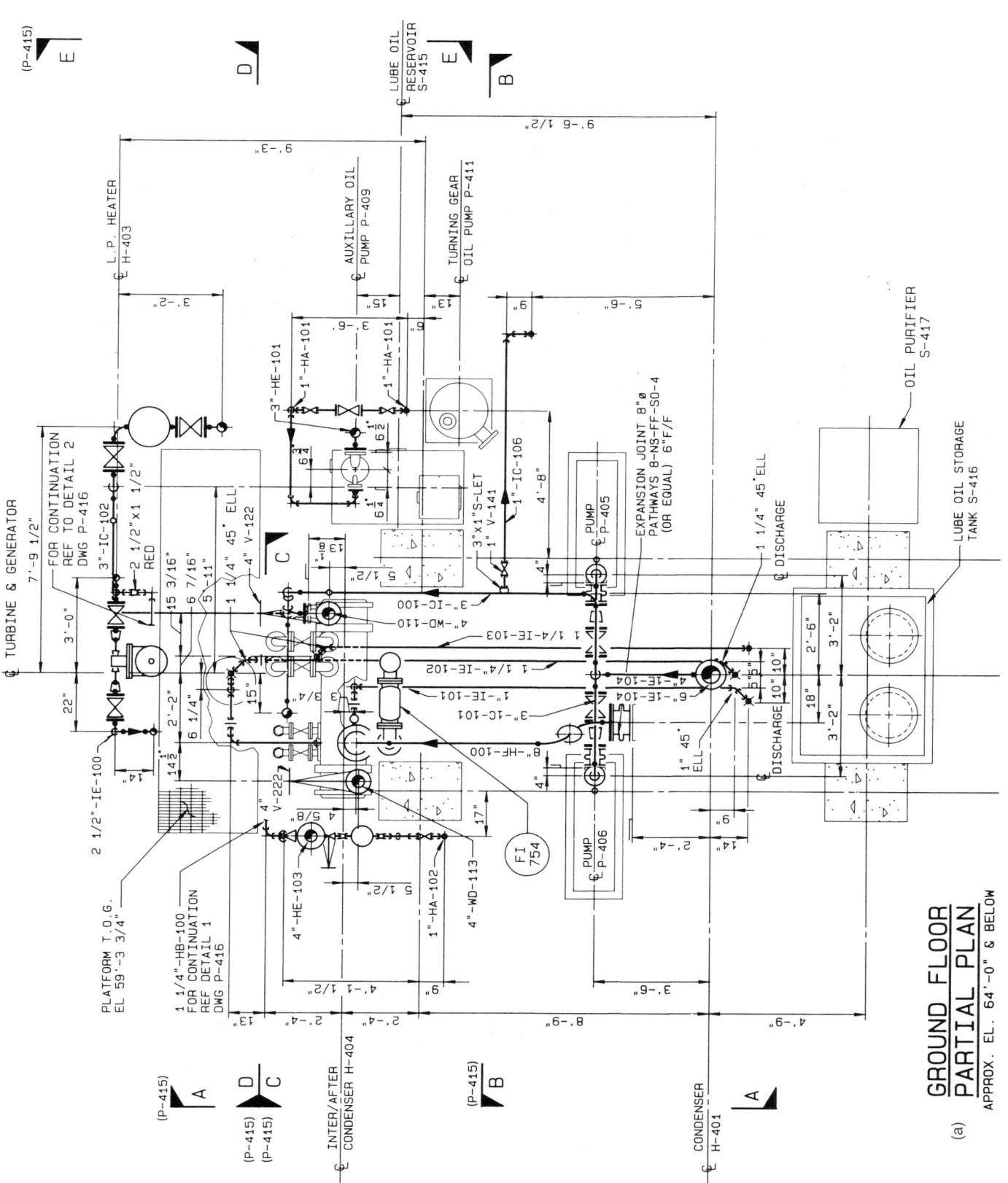

(a)

GROUND FLOOR PARTIAL PLAN

APPROX. EL. 64'-0" & BELOW

FIGURE 22.46 ■ (a) Piping plan (partial) of a turbine generator installation; (b–e) piping sections are taken from plan in (a). *Courtesy Schuchart and Associates.*

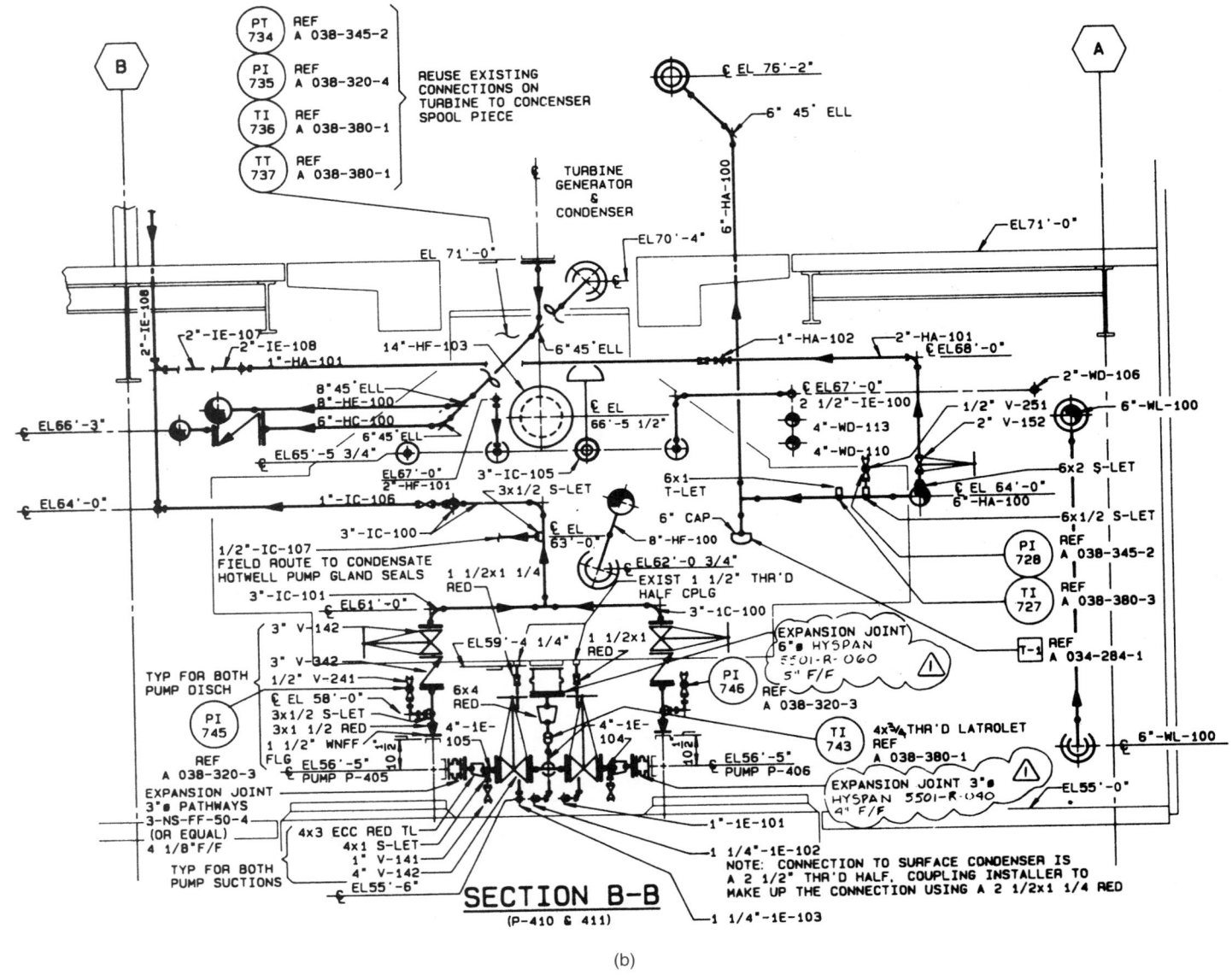

SECTION B-B
(P-410 & 411)

(b)

FIGURE 22.46 ■ (continued)

use of elevations to call out the centerlines of pipes and the tops of concrete (T.O.C.), or steel, for example.

Lay out dimensions before any other written information is added to the drawing. Indicated pipe lengths and dimensions to pipe direction changes should be given from reference points such as equipment and structural steel centerlines. Location dimensions of fittings, flanges, and valves (if needed) should be given to center points or flange faces. Pipe fittings and valves that are attached without pipe between them are often left undimensioned. This is termed "fitting-to-fitting," and their location is simply the result of the combined fitting lengths, and will not vary.

Dimension lines are important and should not be broken if passing through other lines. Pipes, equipment, or extension lines that must pass through dimension lines should be broken and the dimension line should remain intact.

Once dimension lines have been placed on the drawing, local notes and callouts can be given. These, too, should be placed close to where they apply without crowding other entities or creating a confused layout. Try to group local notes together when possible. This enables those who must interpret the drawing to find similar information without too much searching.

Pipe specification numbers can be written near the pipe and connected to it with a leader line, or the numbers can reside inside a symbol called a *line balloon*. The line balloon is either rounded or squared on the ends and is 1/4" wide and approximately 1" long. The preferred location for the line balloon is inside the pipe, but it may be placed outside the pipe and used with a leader line if space is unavailable. Some examples of pipe specification numbers are shown in Figure 22.53, page 753.

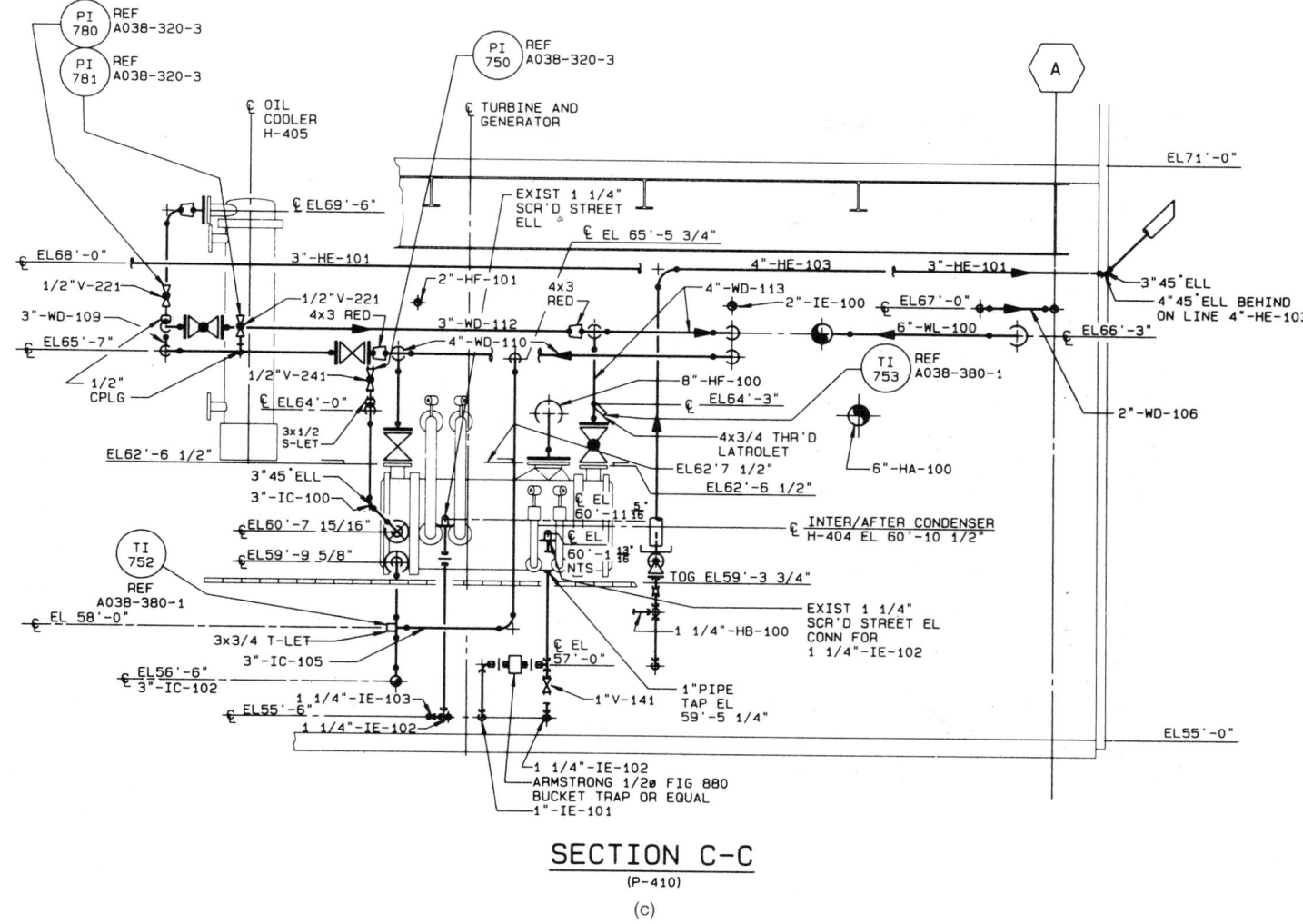

SECTION C-C
(P-410)

(c)

FIGURE 22.46 ■ (continued)

When checking your drawing, be sure that all of the following items have been given:

1. Pipe length dimensions.
2. Dimension locations of all pipe direction changes.
3. Locations of all fittings and valves if required.
4. Elevation location of all pipe direction changes in section views.
5. Size and type of valves.
6. Size and type of fittings (if not readily identified).
7. Pipe diameter, contents, and identification number.
8. Pipe flow arrows.
9. Equipment names and numbers.

PIPING DETAILS

A *piping detail* drawing can be made of any connection, application, or installation that cannot be readily distinguished from the other piping drawings. Details can often quickly clarify a point that would have taken valuable time to determine. Common piping detail drawings are used for the following:

■ Special pipe connections or fittings, or special valve arrangements.

■ Small diameter pipe and fitting assemblies.

■ Special pipe support arrangements.

■ Tank attachment details.

■ Minor structural alterations (concrete, wood, and steel).

■ Operating and installation procedures.

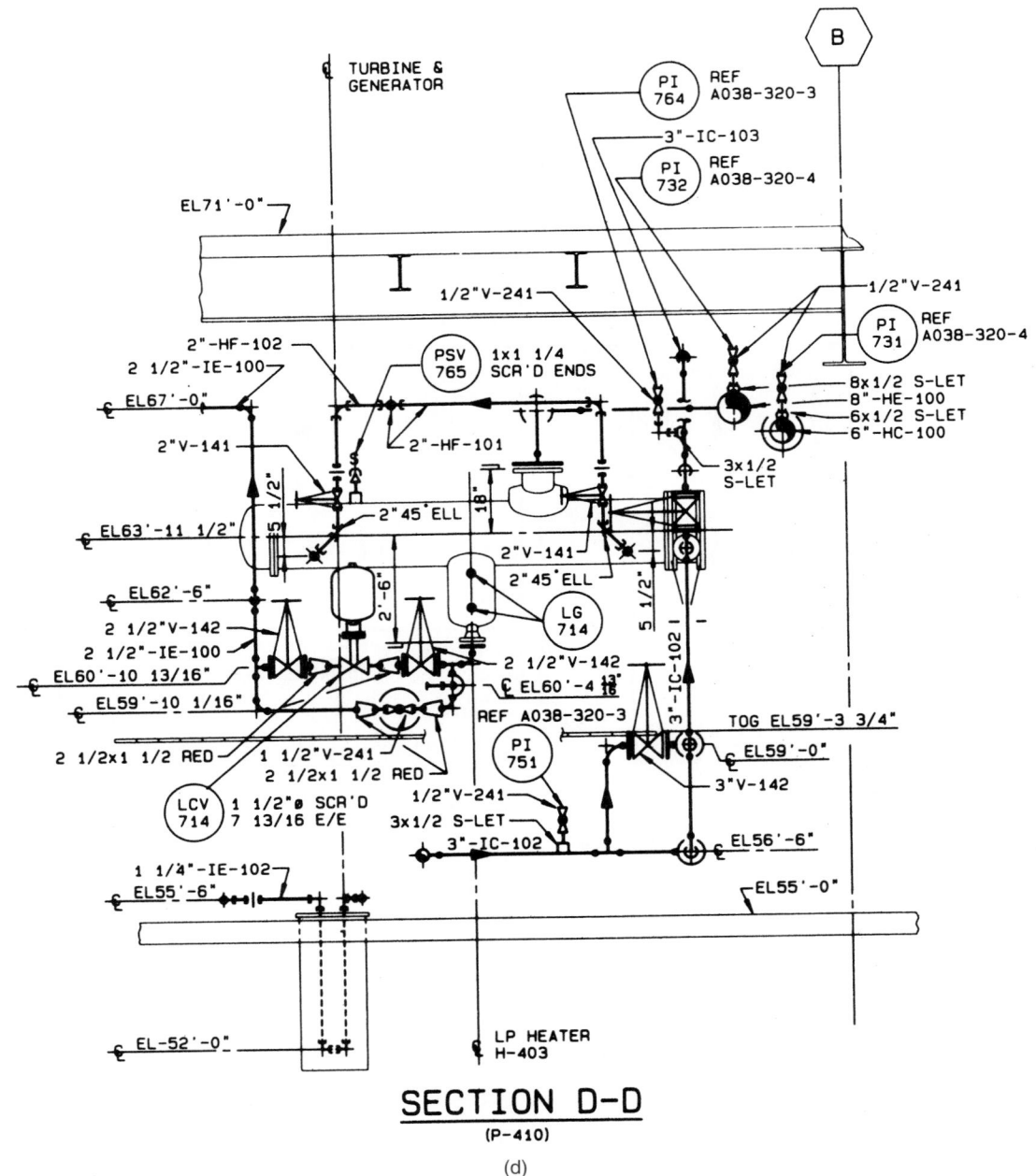

FIGURE 22.46 ■ (continued)

The engineers and drafters on the job must determine when piping details are needed and if they are to be drawn orthographically or pictorially. Some common piping details are shown in Figure 22.54, page 754.

Piping Isometrics and Spools

A *piping isometric* drawing is a pictorial view of a piping system complete with fittings, valves, dimensions, notes, and possibly instrumentation. This single view of the piping system eliminates the need for additional views. It is often done after the plans and elevations are completed.

The advantages of a piping isometric are many. A single run of pipe from beginning to end can be shown in one view and is more easily visualized than corresponding orthographic views. Many companies draw piping isometrics to scale, which makes layout and visualization more accurate. Some firms (especially mechanical contractors) do *not* draw piping isometrics to scale. If an isometric is not drawn to scale, it eliminates the need for large sheets of paper and is most often drawn on B-size media. Dimensions are given on the isometric view, and straight lengths of pipe are drawn

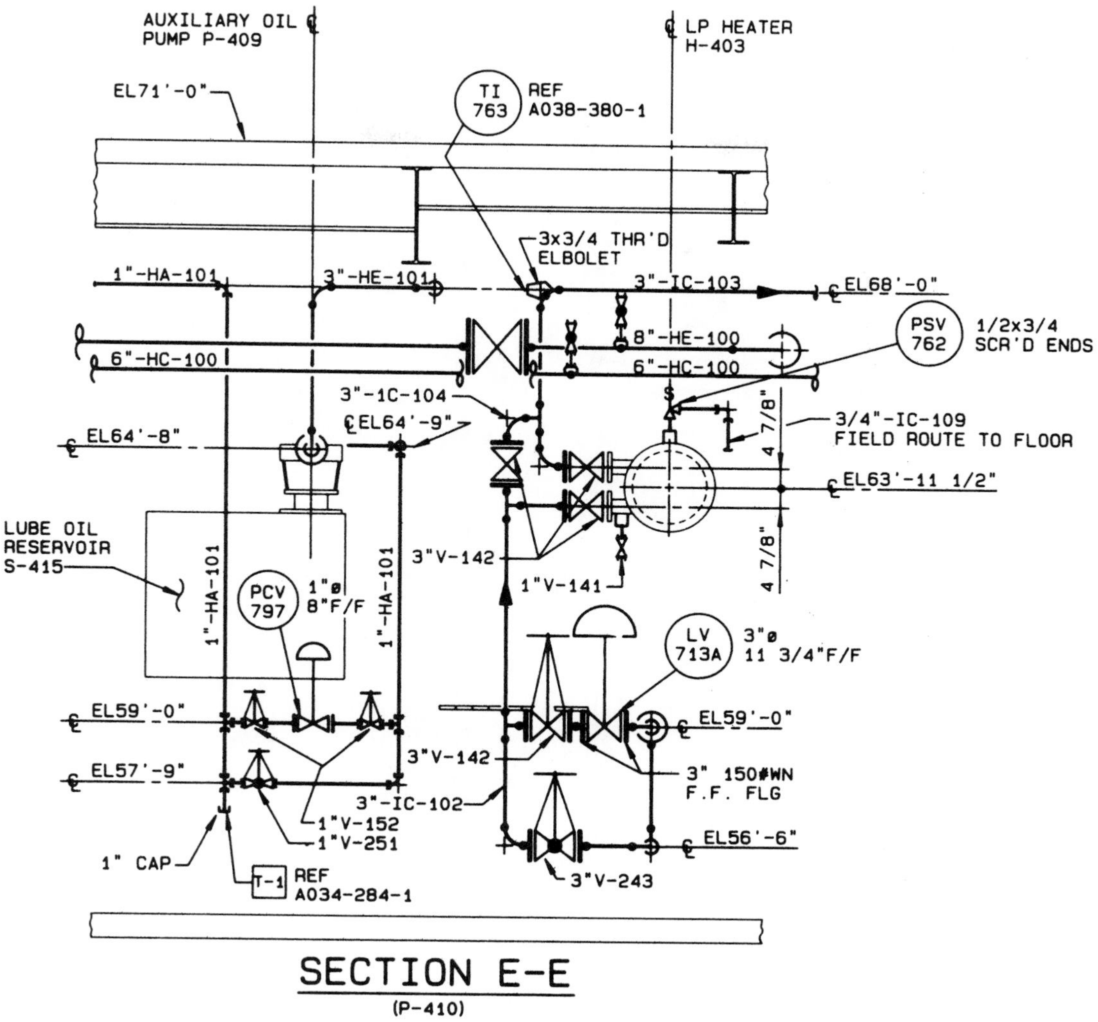

AUXILIARY OIL
PUMP P-409

LP HEATER
H-403

EL71'-0"

TI 763 REF A038-380-1

3x3/4 THR'D
ELBOLET

1"-HA-101

3"-HE-101

3"-IC-103

EL68'-0"

8"-HE-100

PSV 762   1/2x3/4
SCR'D ENDS

6"-HC-100

6"-HC-100

3"-IC-104

3/4"-IC-109
FIELD ROUTE TO FLOOR

EL64'-8"

EL64'-9"

4 7/8"

EL63'-11 1/2"

LUBE OIL
RESERVOIR
S-415

1"-HA-101

3"-V-142

1"-HA-101

1"V-141

4 7/8"

PCV 797   1"ø
8"F/F

LV 713A   3"ø
11 3/4"F/F

EL59'-0"

EL59'-0"

3"-V-142

3" 150#WN
F.F. FLG

EL57'-9"

3"-IC-102

1"V-152
1"V-251

1" CAP

T-1 REF
A034-284-1

3"V-243

EL56'-6"

SECTION E-E
(P-410)

(e)

FIGURE 22.46 ■ (continued)

FIGURE 22.47 ■ Pipe hangers are used to suspend pipe and conduit from the ceiling in this paper mill expansion project. *Courtesy Publisher's Paper Co.*

FIGURE 22.48 ■ Pipe supports anchored to the ground are used on this air compressor piping at a chemical processing plant. *Courtesy Ingersoll-Rand.*

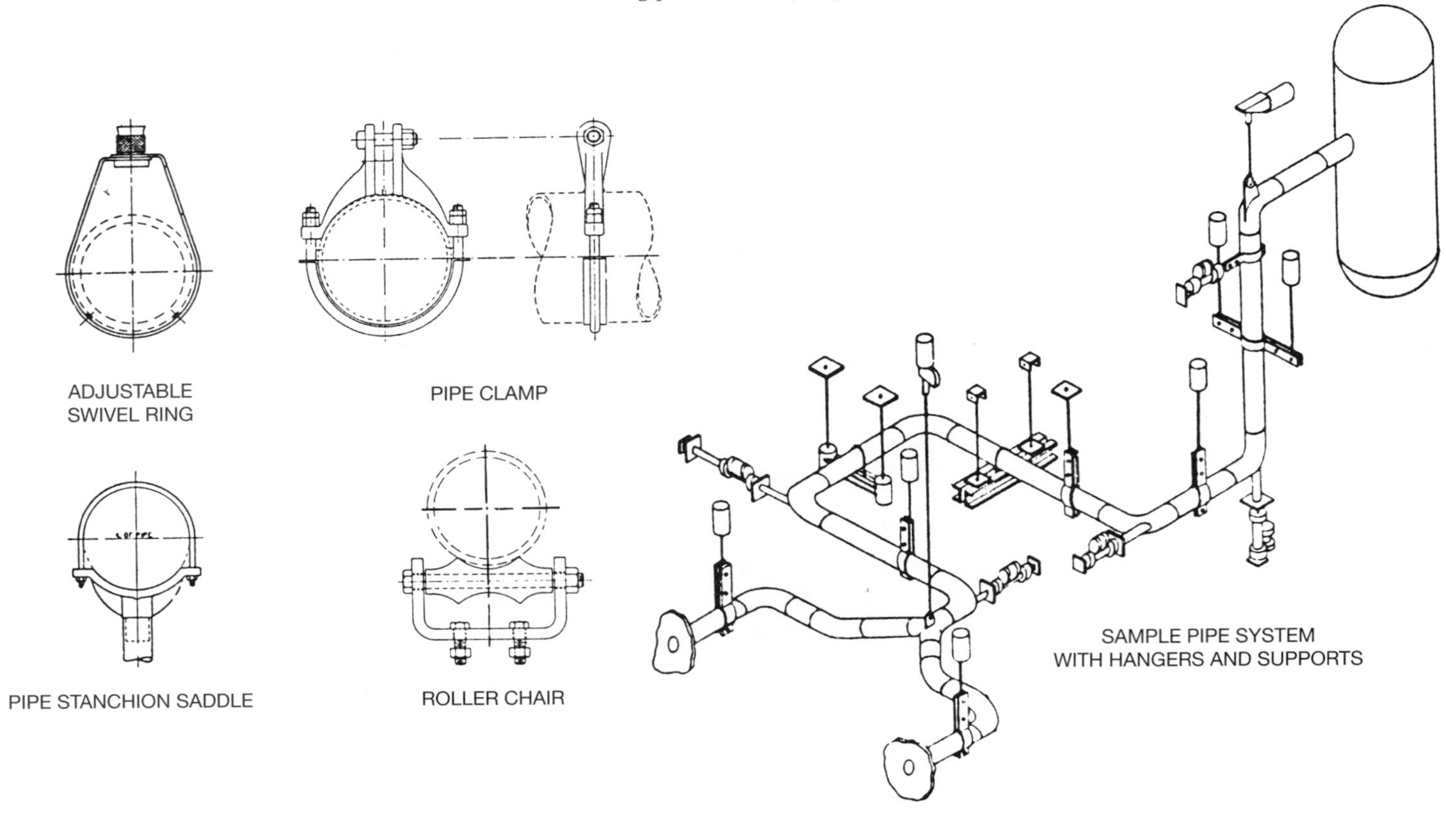

ADJUSTABLE
SWIVEL RING

PIPE CLAMP

PIPE STANCHION SADDLE

ROLLER CHAIR

SAMPLE PIPE SYSTEM
WITH HANGERS AND SUPPORTS

FIGURE 22.49 ■ Examples of common pipe hangers and supports. *Courtesy ITT Grinnell Corp.*

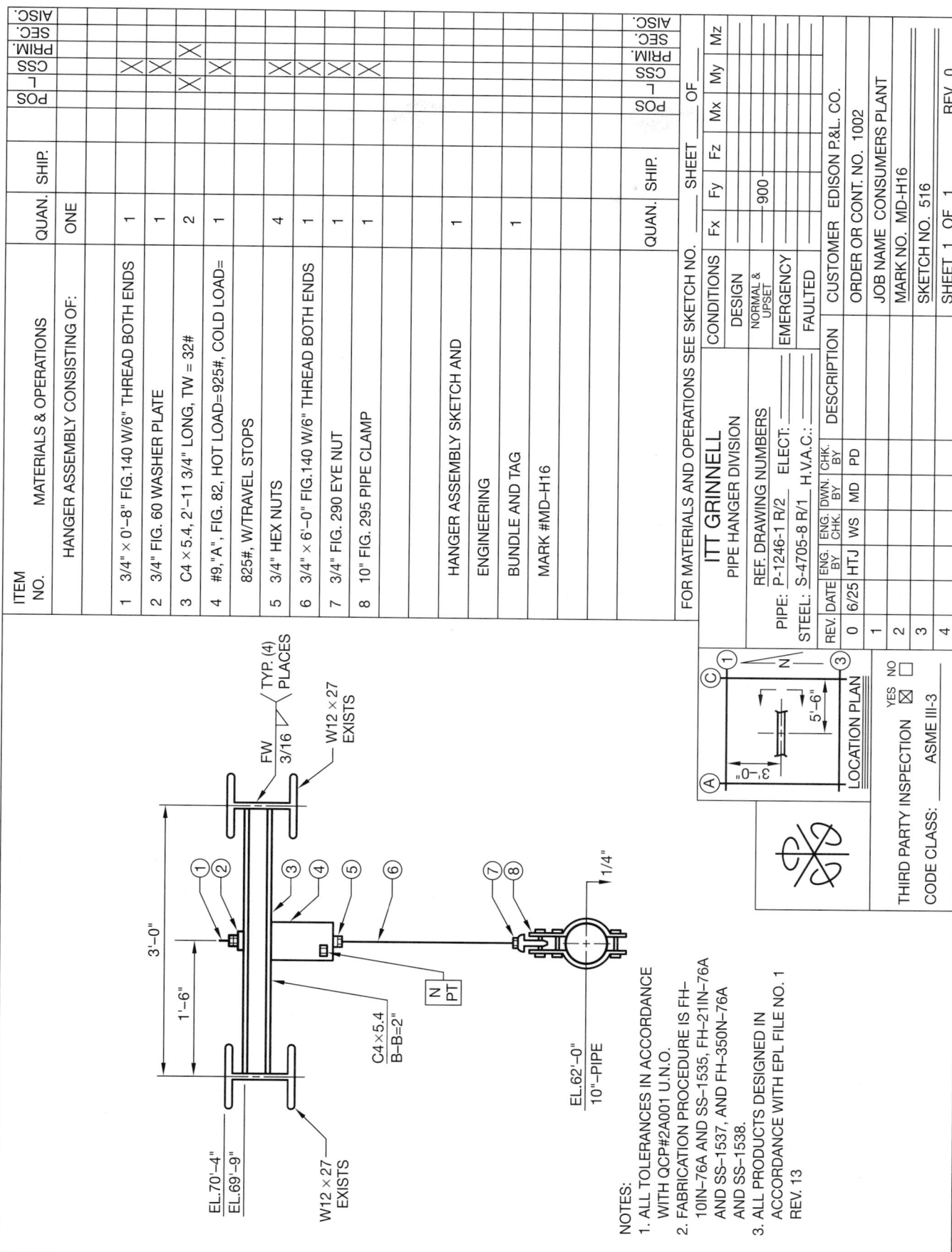

ITEM NO.	MATERIALS & OPERATIONS	QUAN. SHIP.	POS	L	CSS	PRIM.	SEC.	AISC.
	HANGER ASSEMBLY CONSISTING OF:	ONE						
1	3/4" × 0'-8" FIG.140 W/6" THREAD BOTH ENDS	1			X			
2	3/4" FIG. 60 WASHER PLATE	1			X			
3	C4 × 5.4, 2'-11 3/4" LONG, TW = 32#	2		X				
4	#9, "A", FIG. 82, HOT LOAD=925#, COLD LOAD= 825#, W/TRAVEL STOPS	1		X	X			
5	3/4" HEX NUTS	4			X			
6	3/4" × 6'-0" FIG.140 W/6" THREAD BOTH ENDS	1			X			
7	3/4" FIG. 290 EYE NUT	1			X			
8	10" FIG. 295 PIPE CLAMP	1			X			
	HANGER ASSEMBLY SKETCH AND							
	ENGINEERING	1						
	BUNDLE AND TAG	1						
	MARK #MD-H16							
		QUAN. SHIP.	POS	L	CSS	PRIM.	SEC.	AISC.

FOR MATERIALS AND OPERATIONS SEE SKETCH NO. _____ SHEET _____ OF _____

ITT GRINNELL PIPE HANGER DIVISION

CONDITIONS	Fx	Fy	Fz	Mx	My	Mz
DESIGN						
NORMAL & UPSET		-900-				
EMERGENCY						
FAULTED						

REF. DRAWING NUMBERS
PIPE: P-1246-1 R/2 ELECT: _____
STEEL: S-4705-8 R/1 H.V.A.C.: _____

REV.	DATE	ENG. BY	ENG. CHK. BY	DWN. BY	CHK. BY	DESCRIPTION
0	6/25	HTJ	WS	MD	PD	
1						
2						
3						
4						

CUSTOMER EDISON P.&L. CO.
ORDER OR CONT. NO. 1002
JOB NAME CONSUMERS PLANT
MARK NO. MD-H16
SKETCH NO. 516
SHEET 1 OF 1 REV. 0

LOCATION PLAN

5'-6"
3'-0"
N

THIRD PARTY INSPECTION YES ☒ NO ☐
CODE CLASS: _____ ASME III-3

FW 3/16 ◁ TYP. (4) PLACES
W12 × 27 EXISTS

EL.70'-4"
EL.69'-9"
W12 × 27 EXISTS

3'-0"
1'-6"

①② ③④ ⑤ ⑥ ⑦⑧

C4 × 5.4
B–B=2"

N
PT

EL.62'-0"
10"–PIPE

1/4"

NOTES:
1. ALL TOLERANCES IN ACCORDANCE WITH QCP#2A001 U.N.O.
2. FABRICATION PROCEDURE IS FH–10IN–76A AND SS–1535, FH–21IN–76A AND SS–1537, AND FH–350N–76A AND SS–1538.
3. ALL PRODUCTS DESIGNED IN ACCORDANCE WITH EPL FILE NO. 1 REV. 13

FIGURE 22.50 ■ Detail drawing of pipe hanger installation. *Courtesy ITT Grinnell Corp.*

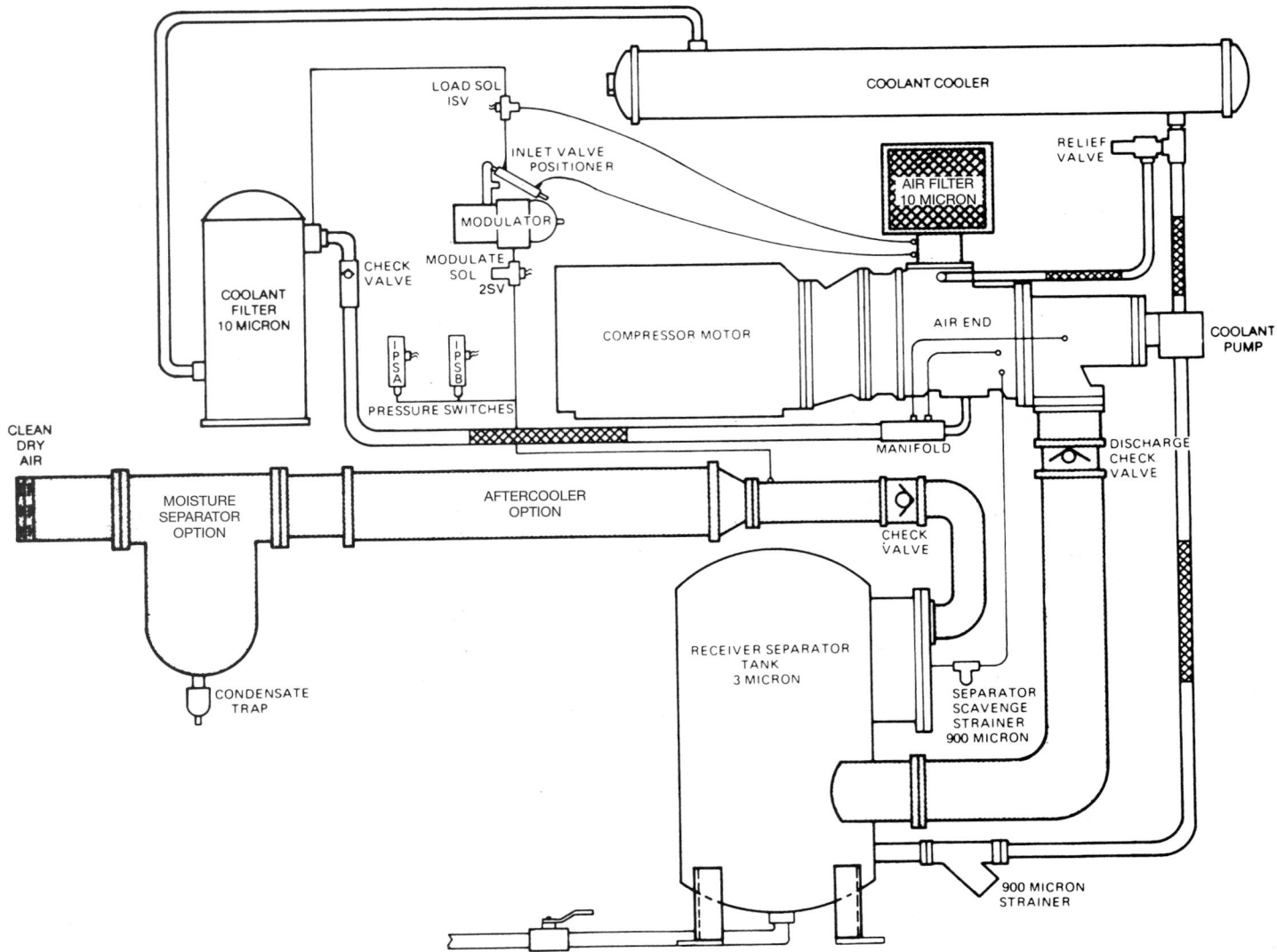

FIGURE 22.51 ■ ■ Manufacturer's double-line piping drawing of a compressed air system. *Courtesy Ingersoll-Rand.*

proportionally to indicate long and short pieces. Note this in the isometric shown in Figure 22.55.

It is easier to check the drawing for interferences and clearances if the isometric is drawn to scale. A bill of materials is often included on the piping isometric, which enables the pipe fitter, checker, and purchasing agent to cross-check the drawing with a list of the materials required.

Some companies may illustrate a system or portions of a system in a pictorial manner for purposes of orientation, training, assembly, and installation. It is not often practical for large and complex systems to be shown in isometric form if manual drafting is used. But the development of more powerful CADD systems is enabling companies to create complete pictorially modeled systems.

Piping systems are constructed by pipe fitters who must initially weld or thread pipe fittings and pipe together to create pipe

spools. (See Figure 22.56, page 756.) The pipe fitters work from assembly instructions called *spool drawings* or spool sheets. The pipe assemblies (spools) are then transported to the job site and installed. A pile of pipe spools awaiting installation is shown in Figure 22.57, page 756.

Pipe spools are drawn in isometric or orthographic style on B-size paper and may or may not be drawn to scale. (See Figure 22.11, page 728.) Spool sheets are assembly drawings that contain complete dimensions and a bill of materials (B.O.M.) that indicates the exact size and specifications for each fitting. Extra pipe is usually added to pipe lengths to compensate for errors. If the drawing is orthogonal, it only shows the views required to present all of the lengths of straight pipe and fittings. Valves are never shown on spool drawings because they are installed at the job site when the spools are erected.

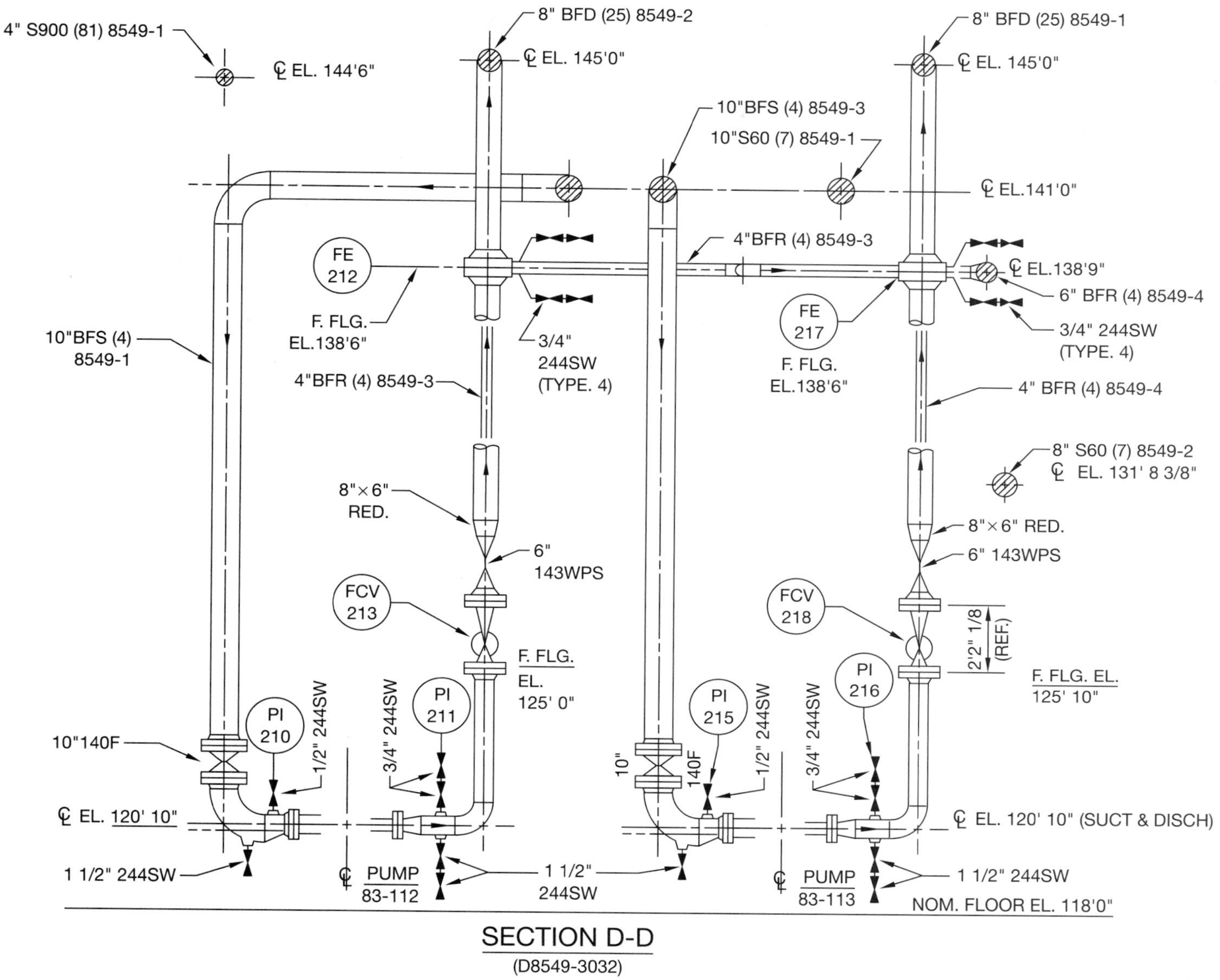

4" S900 (81) 8549-1 ⌀ EL. 144'6"

8" BFD (25) 8549-2   ⌀ EL. 145'0"

8" BFD (25) 8549-1   ⌀ EL. 145'0"

10"BFS (4) 8549-3
10"S60 (7) 8549-1   ⌀ EL.141'0"

FE 212

4"BFR (4) 8549-3   ⌀ EL.138'9"
6" BFR (4) 8549-4

F. FLG. EL.138'6"

3/4" 244SW (TYPE. 4)

10"BFS (4) 8549-1

3/4" 244SW (TYPE. 4)

FE 217

F. FLG. EL.138'6"

4"BFR (4) 8549-3

4" BFR (4) 8549-4

8"×6" RED.

8" S60 (7) 8549-2
⌀ EL. 131' 8 3/8"

6" 143WPS

8"×6" RED.

6" 143WPS

FCV 213

FCV 218

PI 210

PI 211

PI 215

PI 216

2'2" 1/8 (REF.)

F. FLG. EL. 125' 10"

F. FLG. EL. 125' 0"

10"140F

1/2" 244SW

3/4" 244SW

10" 140F

1/2" 244SW

3/4" 244SW

⌀ EL. 120' 10"

⌀ EL. 120' 10" (SUCT & DISCH)

1 1/2" 244SW

PUMP 83-112

1 1/2" 244SW

PUMP 83-113

1 1/2" 244SW

NOM. FLOOR EL. 118'0"

SECTION D-D
(D8549-3032)

FIGURE 22.52 ■ Sections from a typical double-line piping drawing. *Courtesy Schuchart and Associates.*

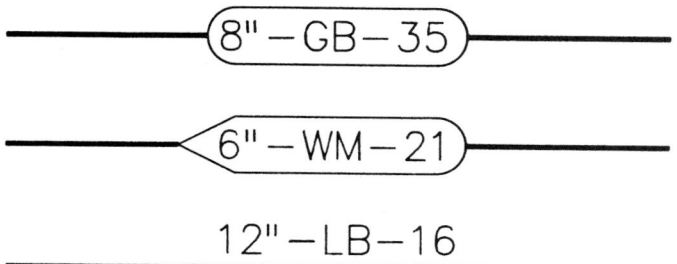

8" – GB – 35

6" – WM – 21

12" – LB – 16

FIGURE 22.53 ■ Methods of indicating pipe line specifications.

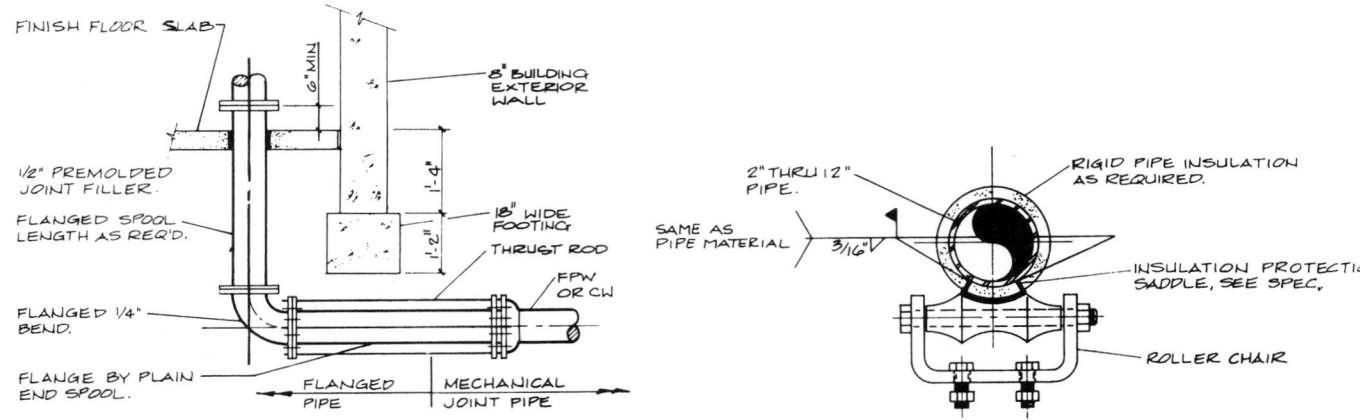

FINISH FLOOR SLAB

1/2" PREMOLDED JOINT FILLER.

FLANGED SPOOL LENGTH AS REQ'D.

FLANGED 1/4" BEND.

FLANGE BY PLAIN END SPOOL.

6" MIN

8" BUILDING EXTERIOR WALL

18" WIDE FOOTING

THRUST ROD

FPW OR CW

1'-4"

1'-2"

FLANGED PIPE | MECHANICAL JOINT PIPE

NOTE:
1. SEE MECHANICAL DRAWINGS FOR LOCATIONS AND SIZES.
2. SEE FOUNDATION DRAWINGS FOR ANY DIFFERENCES IN FOOTING DEPTHS.

⑧ THRUST TIE

(a)

2" THRU 12" PIPE

SAME AS PIPE MATERIAL

3/16"

RIGID PIPE INSULATION AS REQUIRED.

INSULATION PROTECTION SADDLE, SEE SPEC.

ROLLER CHAIR

⑤ ROLLERCHAIR PIPE SUPPORT

(c)

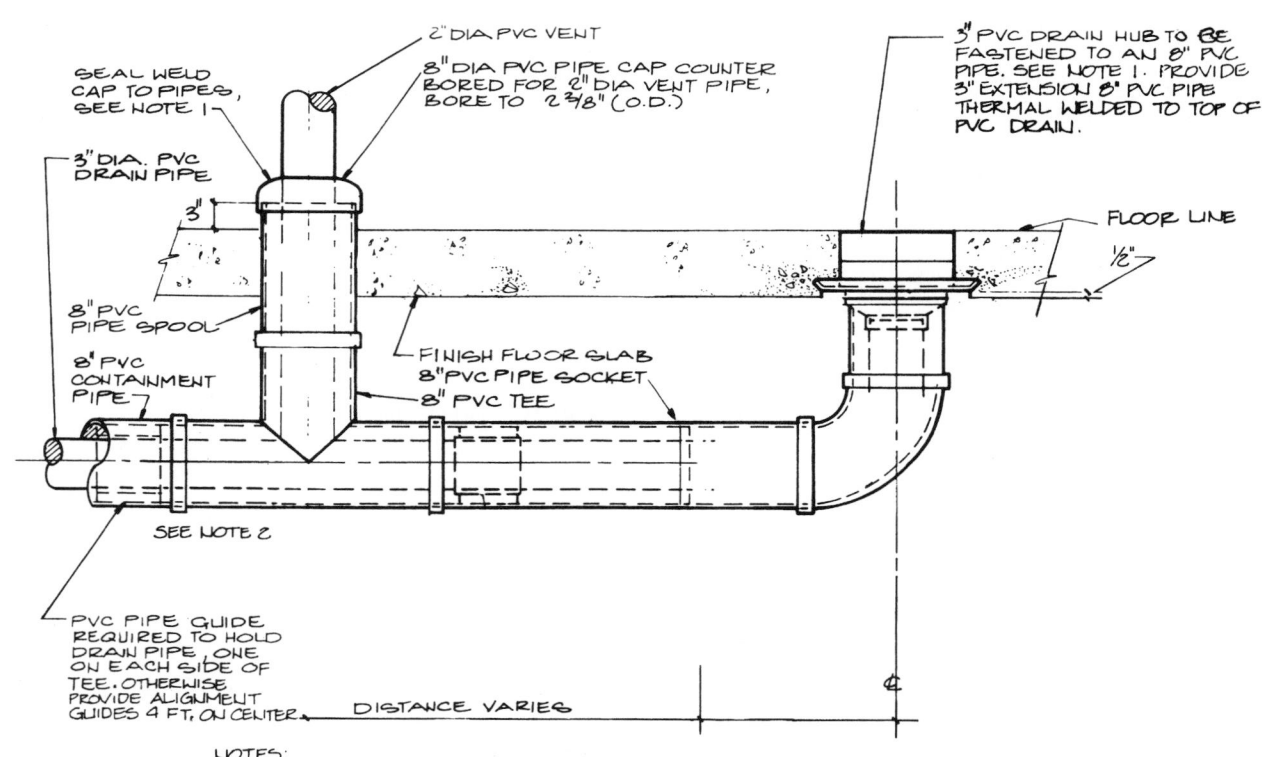

SEAL WELD CAP TO PIPES, SEE NOTE 1.

3" DIA. PVC DRAIN PIPE

8" PVC PIPE SPOOL

8" PVC CONTAINMENT PIPE

SEE NOTE 2

PVC PIPE GUIDE REQUIRED TO HOLD DRAIN PIPE, ONE ON EACH SIDE OF TEE. OTHERWISE PROVIDE ALIGNMENT GUIDES 4 FT. ON CENTER.

2" DIA PVC VENT

8" DIA PVC PIPE CAP COUNTER BORED FOR 2" DIA VENT PIPE, BORE TO 2 3/8" (O.D.)

3"

FINISH FLOOR SLAB
8" PVC PIPE SOCKET
8" PVC TEE

3" PVC DRAIN HUB TO BE FASTENED TO AN 8" PVC PIPE. SEE NOTE 1. PROVIDE 3" EXTENSION 8" PVC PIPE THERMAL WELDED TO TOP OF PVC DRAIN.

FLOOR LINE

1/2"

DISTANCE VARIES

NOTES:
1. THERMAL WELD ALL SEAMS WHERE POSSIBLE.
2. TERMINATE 8" CONTAINMENT PIPE 3" INTO TRENCH OR SUMP.
3. SLOPE PIPE AND CONTAINMENT PIPE 1/8"/FT. TOWARD TRENCH OR SUMP.
4. PROVIDE RUNNING TRAP ON 3" PIPE INTO TRENCH OR SUMP.

FIGURE 22.54 ■ (a and b) Piping details of buried pipe; (c) pipe hanger. *Courtesy CH₂M Hill—I.D.C.*

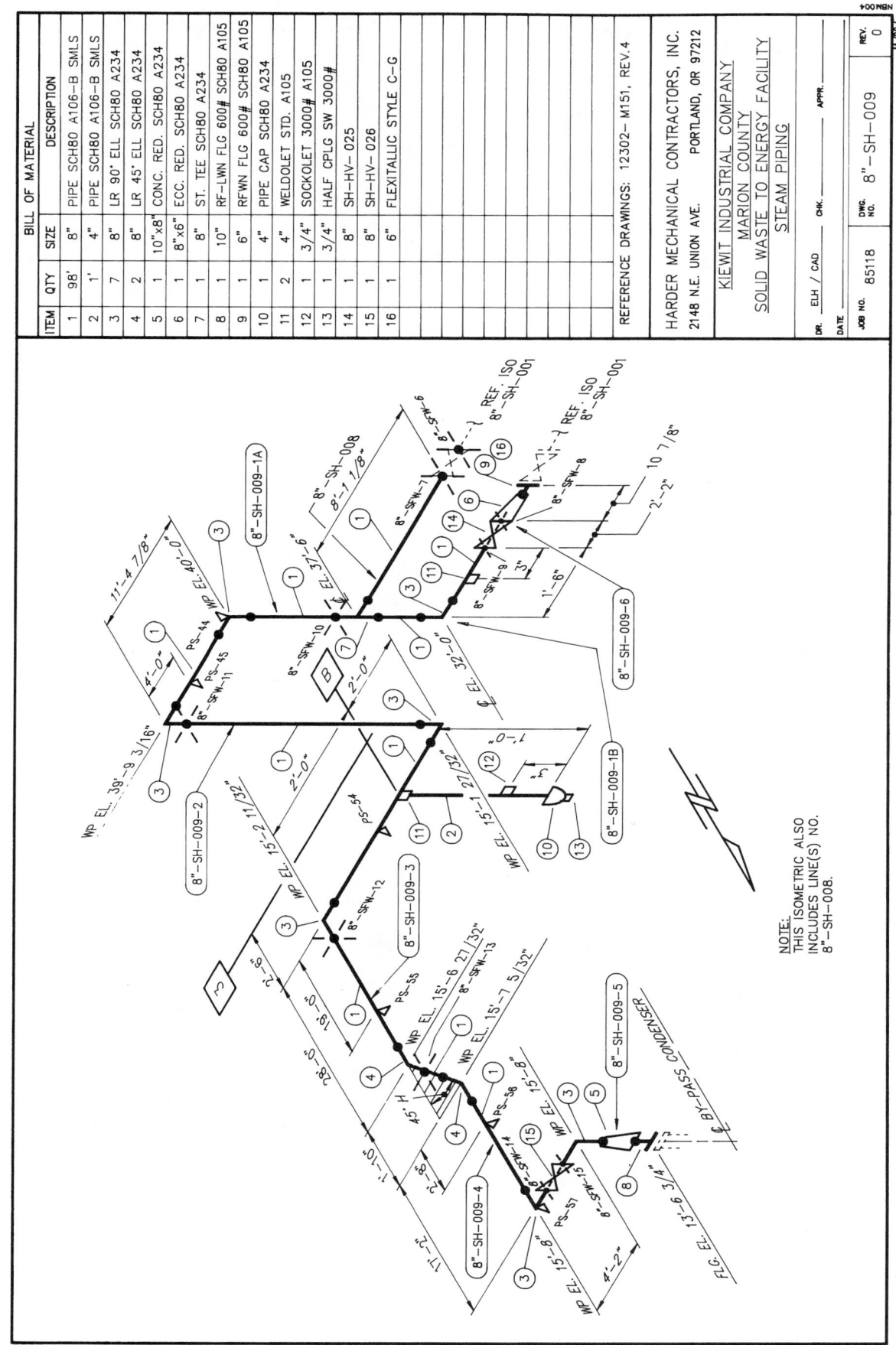

FIGURE 22.55 ■ (a) Piping isometric with nonscale proportional dimensioning. *Courtesy Harder Mechanical Contractors, Inc.*

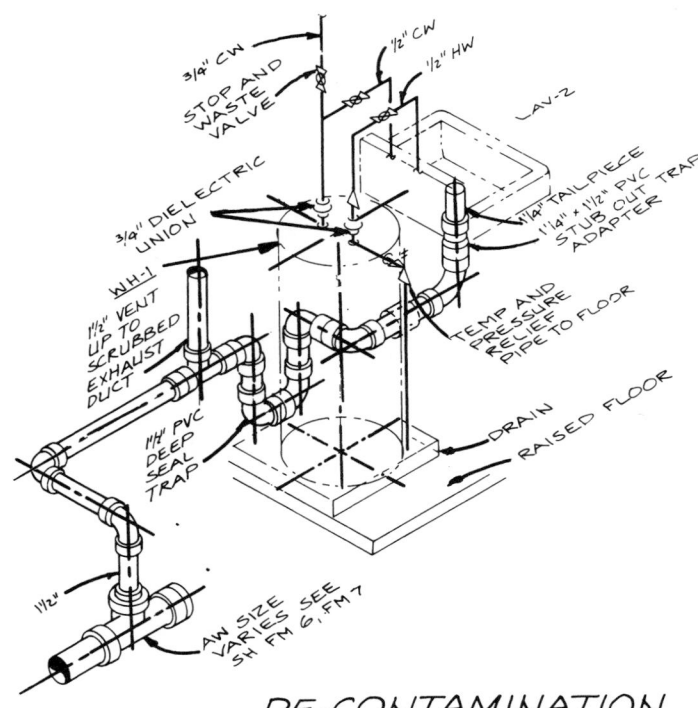

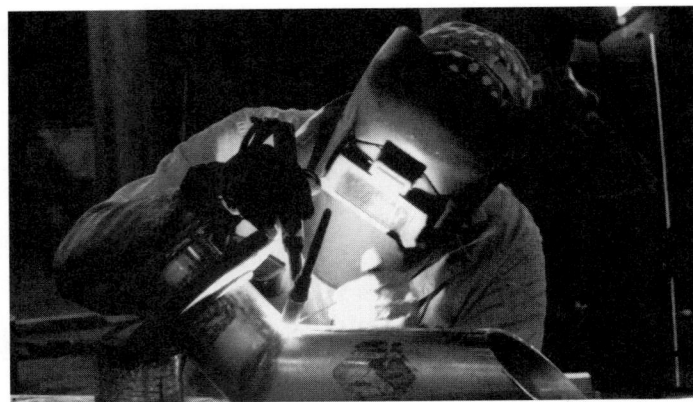

FIGURE 22.56 ■ Pipe fitter assembling a piping spool. *Courtesy Alaskan Copper Works.*

DE-CONTAMINATION
WASH SINK PIPING
DETAIL ③

FIGURE 22.55 ■ **(b)** Nonscale double-line piping isometric. *Courtesy CH₂M Hill—I.D.C.*

FIGURE 22.57 ■ Assembled pipe spools awaiting installation. *Courtesy Alaskan Copper Works.*

CADD APPLICATIONS

3-D DIGITIZING

Three-dimensional digitizers allow you to pick points on an object in order to input 3-D model data into CADD packages such as AutoCAD and CADKEY. This form of data input is called *reverse engineering*, because the real part is used to construct a model. (See Figure 22.58.)

The 3-D digitizer has an articulated arm that can be moved and rotated in order to pick any point on an existing object or model. The data is transmitted as 3-D points to the CADD software through a cable attached to a serial port on the computer. This process is useful for collecting CADD data of an existing object in order to create a computer model. The computer model can then be revised in order to construct a new part.

This CADD input is also useful for digitizing an existing scale model of an object, building, or industrial site. The digitized data from the model can then be used to create construction drawings and 3-D computer models and renderings.

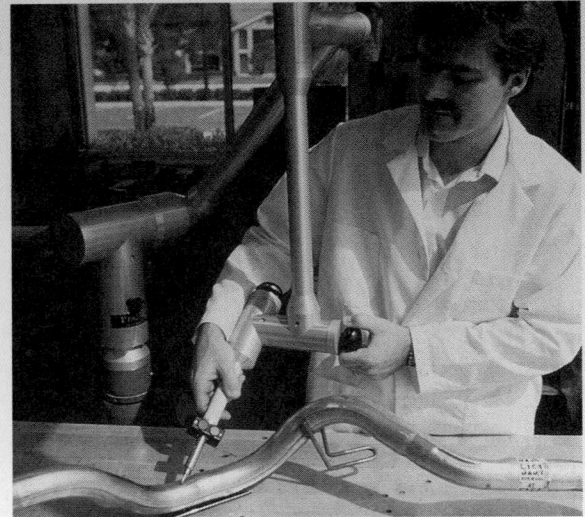

FIGURE 22.58 ■ The dimensions of an existing pipe can be input to the CADD software using a 3-D digitizer. *Courtesy FARO Technologies, Inc.*

Models

The construction of the model (see Figure 22.59) may take months, even years, making it impossible to ship to the job site during most of the construction phases. Photos of the model are used at the job site for checking purposes. Drawings of the system must still be generated for use in construction.

One of the greatest advantages is the ability to see readily clearances and interferences. (See Figure 22.60.) These can be quickly transferred to the drawings, thus eliminating many errors that might occur if only drawings were used. Some companies may use a three-dimensional digitizing system to locate

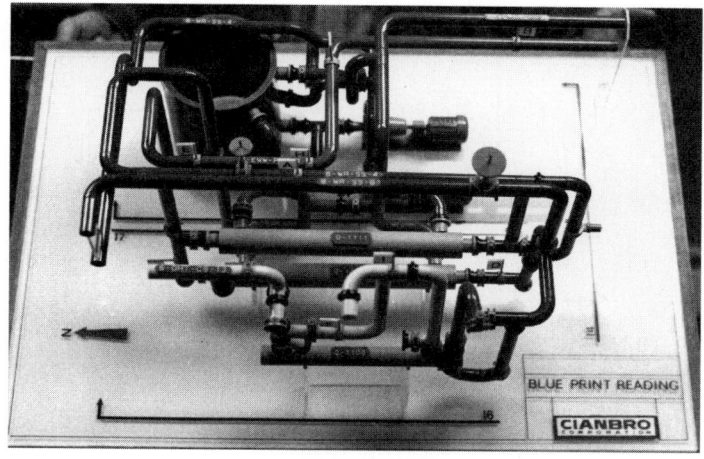

FIGURE 22.61 ■ Piping model kits are good visualization tools for the classroom. *Courtesy Engineering Model Associates, Inc.*

FIGURE 22.59 ■ Complex piping models can require months to build. *Courtesy Engineering Model Associates, Inc.*

FIGURE 22.60 ■ Designers can readily check for interferences on a piping model. *Courtesy Engineering Model Associates, Inc.*

pipe, equipment, and structural steel on the model. This information is then fed into the CADD system and used to generate drawings.

Equipment on the model is often shown in the form of block shapes, details seldom being required. Equipment is created from standard, inexpensive materials such as rigid foam, plastic, and wood. Piping and structural steel components are available from model supply vendors.

Upon completion, the model is shipped to the job site where it is used in the final stages of the construction project. It is also used to check the design drawings of the system. Operators, maintenance technicians, and engineers can be trained on the workings of the system using the model. Revisions, alterations, and design changes can be accomplished efficiently using the model to check for clearances, and the addition of new equipment and piping locations.

Piping model kits are available for use in piping and blueprint reading classes. (See Figure 22.61.) These kits come with all the parts necessary to construct a small piping and equipment arrangement. These models aid in visualization and enable the student to better construct piping drawings.

LAYOUT TECHNIQUES

When constructing scaled piping drawings, use the following layout procedures:

Step 1. Locate all building outlines, concrete foundations, structural steel columns, and equipment. Draw the centerlines of the pipe that connects the equipment. (See Figure 22.62a.)

Step 2. Insert fittings and valves in the pipe centerlines and draw pipe, as shown in Figure 22.62b.

Step 3. Place dimensions and elevations on the drawing, remembering to keep dimensions close to where they apply. Locate pipe specification symbols and text in the pipe runs. Add all notes and textual information required. (See Figure 22.62c.)

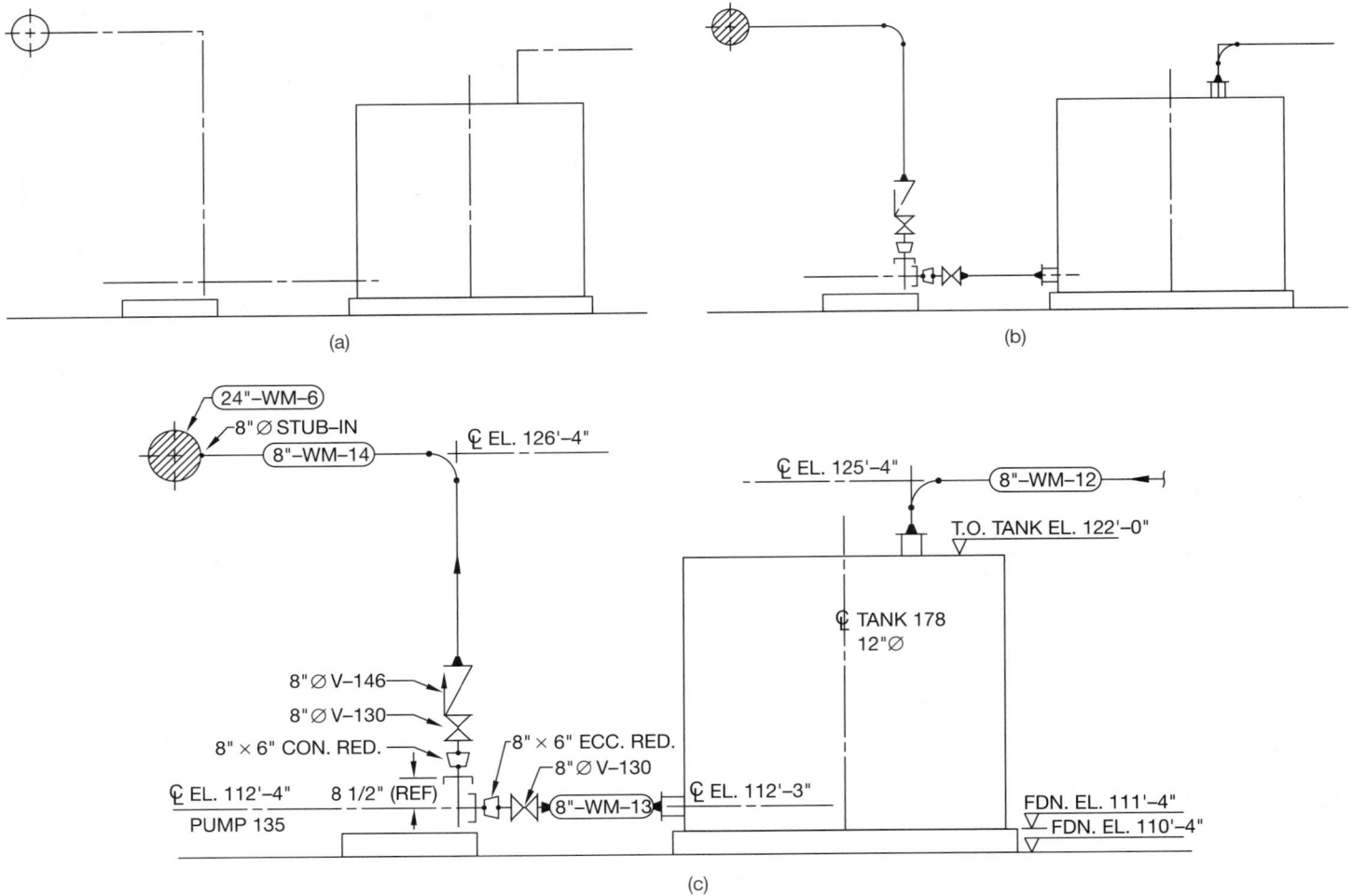

FIGURE 22.62 ■ **(a)** Equipment outlines and pipe centerlines are located; **(b)** pipe, fittings, and valves are added; **(c)** dimensions, elevations, and text are added to complete the drawing.

PROFESSIONAL PERSPECTIVE

The field of process piping design and drafting can be an excellent opportunity for a person to gain entry into engineering. Many consulting engineering companies will pay for the education of employees who wish to upgrade their skills or work toward an engineering degree. This is often a good method of becoming a designer, technician, or engineer, because you can get valuable job training while working toward a degree. Such job training is not possible if you are going to school full time.

A good piping drafter, designer, or engineer is one who is aware of the actual job site requirements and problems. These "field" situations are often considerably different from the layout that is designed in the office. Therefore, if you are planning to work in the industrial piping profession, do your best to work on projects in which you can gain field experience. Field work is especially important when adding new equipment and pipe in an existing facility. There are many stories of inexperienced piping designers who have created a design in the office, and then gone to the field to find a new 4" pipe routed directly through an existing 12" pipe, exactly according to the plan!

MATH APPLICATION

Industrial pipe design often requires the use of fittings such as 45° elbows to route pipe past obstructions. Exact dimensions of all lengths of pipe and fittings must be calculated before the pipe can be drawn and assembled. When 45° elbows are used, the piping designer must use the Pythagorean theorem to solve for one side of the triangle.

The designer has laid out a run of pipe shown in Figure 22.63. In order to find the true length, or *travel*, of the angled run of pipe, a 45° triangle is applied to the pipe. The length of one side of the triangle is determined by finding the difference between the two elevations given, 2'–9". The two adjacent sides each have a length of 2'–9". Solve for the angled side of the triangle, or *hypotenuse*. (See Figure 22.64.)

Therefore, the square root of the hypotenuse equals the travel of the angled run of pipe.

This problem is solved as follows:

$$a^2 + b^2 = c^2$$
$$(2'-9'')^2 + (2'-9'')^2 = c^2$$
$$(33)^2 + (33)^2 = c^2$$
$$1089 + 1089 = c^2$$
$$2178 = c^2$$
$$\sqrt{2178} = c$$
$$c = 46.669''$$
$$c = 3'-10\ 11/16''$$

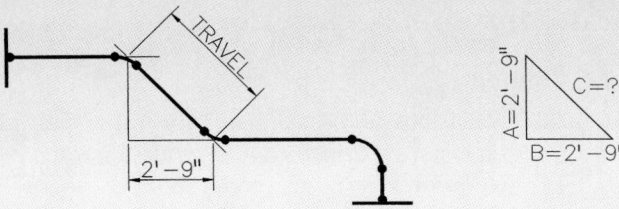

FIGURE 22.63 ■ Find the true length, or travel, of the angled run of pipe.

The Pythagorean theorem is stated as follows: The sum of the squares of the two sides is equal to the square of the hypotenuse; or:

$$a^2 + b^2 = c^2$$

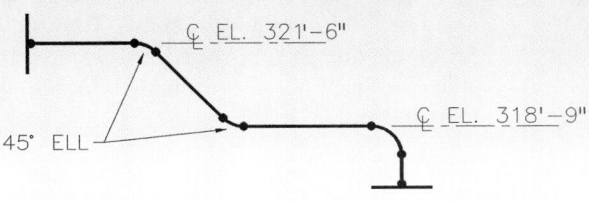

FIGURE 22.64 ■ Solve for the travel (angled) side of the triangle.

Industrial Process Piping Test

DIRECTIONS

Answer the following questions with short, complete statements.

1. What is the difference between plumbing and piping?
2. In what two forms is piping drawn?
3. What is one of the major uses of cast iron pipe?
4. What type of steel pipe withstands high temperatures?
5. What is steam tracing?
6. What type of pipe is one of the most corrosion proof?
7. What type of pipe is best used for hydraulic lines in automobiles and aircraft?
8. Define *NPS*.
9. What is the meaning of "pipe schedule number"?
10. What designations do the ASTM and ASME use to classify pipe strength?
11. How does butt welding differ from socket welding?
12. What provides the seal in a flanged connection?
13. What type of pipe is commonly used with a bell and spigot connection?
14. What type of pipe is solvent welding used on?
15. Describe the functions of the following pipe fittings: 180° elbow, reducing tee, eccentric reducer, lateral, elbolet, union, and bushing.
16. When is a weld-neck flange used?
17. What is placed between two orifice flanges, and what is its function?
18. Name two types of on/off valves.
19. What is one of the major drawbacks of the globe valve?

20. What type of regulating valve is operated with a one-quarter turn?
21. What type of check valve is used with a globe valve?
22. What type of valve is used to relieve pressure in pipes carrying liquids?
23. Why are detailed representations of equipment not necessary on piping plans?
24. When would the double-line method of pipe drawing be used?

25. How are pipe length dimensions handled in piping sections?
26. What is meant by "fitting-to-fitting"?
27. When would a piping detail drawing be needed?
28. Why are piping isometric drawings often not drawn to scale?
29. What is the relationship between piping isometrics and piping spool drawings?
30. What are the benefits of piping models?
31. What is in store for the future of CADD piping?

CHAPTER 22

Industrial Process Piping Problems

DIRECTIONS

1. Please read problems carefully before you begin working. Your instructor will assign one or more of the following problems. Complete each problem on an appropriately sized drawing sheet or use the size indicated in the specific instructions.

2. Refer to Appendices F and G for fitting and valve dimensions. Consult with your instructor to determine if specific vendor catalogs should be used to obtain dimensions.

3. Construct your drawings using a CADD system, if indicated in course guidelines.

PROBLEM 22-1 Draw the following fittings in double-line representation. Use 8" diameter NPS and draw at a scale of ½" = 1'–0".

∎ 90° elbow

∎ 45° elbow

∎ Straight tee

∎ Concentric reducer

∎ Eccentric reducer

Draw a front view and each of the four orthographic views (top, bottom, left, and right sides).

PROBLEM 22-2 Draw the fittings listed in Problem 22-1 in single-line representation. Use the same NPS and the same scale.

PROBLEM 22-3 Draw each of the five pipe assemblies shown on the right (a–e) in double-line form. The view given is the front view. Draw each of the other four orthographic views (top, bottom, left, and right sides). Use the following information when doing this problem:

∎ Draw at a scale of ³⁄₈" = 1'–0".

∎ Use pipe diameters indicated.

∎ Draw the fittings to scale but straight lengths of pipe can be drawn proportional to the sketch.

∎ Pipe that appears to be going away from the viewer should be drawn "fitting-to-fitting" if indicated with "F-F" on the sketch. This means that there is no straight pipe in the run that is going away from the viewer. Runs that do contain pipe are not shown with "F-F."

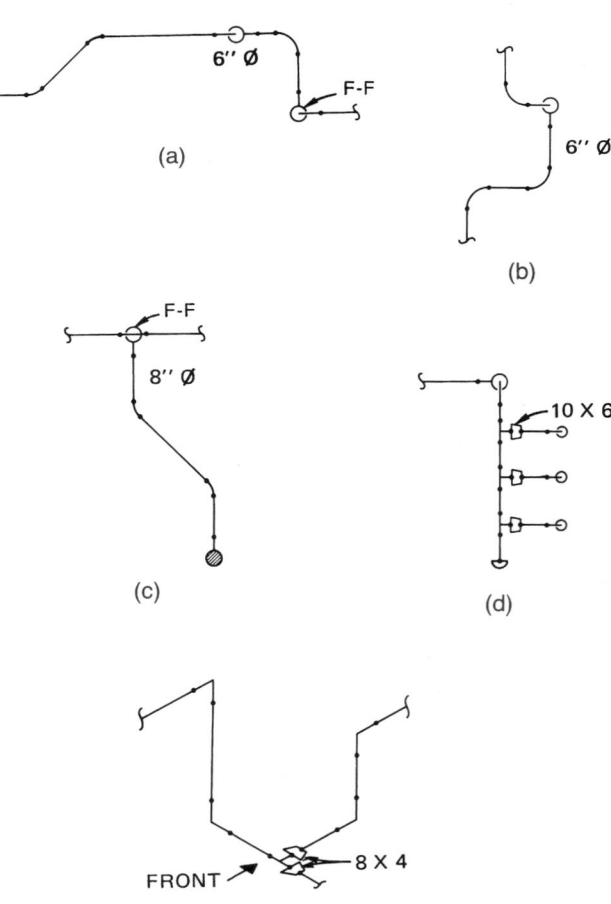

PROBLEM 22-4 Draw the assemblies shown in Problem 22-3 in single-line form. Follow the instructions given in Problem 22-3.

PROBLEM 22-5 Redraw the Ground Floor Partial Plan shown in Figure 22.46a, page 745, on C-size media. Use the following information:

■ Draw at a scale of $1/2$" = 1'–0".

■ Pipe more than 3" in diameter should be drawn in double-line form.

■ Use line balloons to indicate pipe specifications.

■ Lettering should be $1/8$" high.

Use Figure 22.46b–e, pages 746–749, as references for this drawing.

PROBLEM 22-6 Redraw section E-E (Figure 22.46e) in double-line form at a scale of $1/2$" = 1'–0". Use standard dimensions for fittings and valves given in the appendices.

PROBLEM 22-7 Draw a piping detail of the suction piping of pump P-405 in section B-B, Figure 22.46.

PROBLEM 22-8 Redraw section B-B (Figure 22.46b) in single-line form at a scale of $1/2$" = 1'–0".

PROBLEM 22-9 Draw the sections shown in Figure 22.46 in single-line form. Use a scale of $3/8$" = 1'–0" on C-size media.

PROBLEM 22-10 Redraw section E-E, Figure 22.46e, at a scale of $1/2$" = 1'–0". Pipe that is greater than 3" in diameter, draw as a single line. Pipe less than 3" diameter, draw as double line.

PROBLEM 22-11 Draw isometric views of the five pipe assemblies shown in Problem 22-3. Draw in single-line form.

PROBLEM 22-12 Draw a piping isometric of lines 1"–HA–101 in section E-E (Figure 22.46e). Use B-size media.

PROBLEM 22-13 Draw a plan and elevation of 8"–SH–009 (Figure 22.56a) in single-line form.

PROBLEM 22-14 Draw a plan and elevation of 8"–SH–009 (Figure 22.56a) in double-line form.

PROBLEM 22-15 Draw the spools required for the isometric 4"–SH–004 shown below. Draw one per sheet of B-size media. Include a bill of materials for each spool. The field weld connection point between the two largest spools is indicated by an "X" just above El. 36'–6". *Courtesy Harder Mechanical Contractors, Inc.*

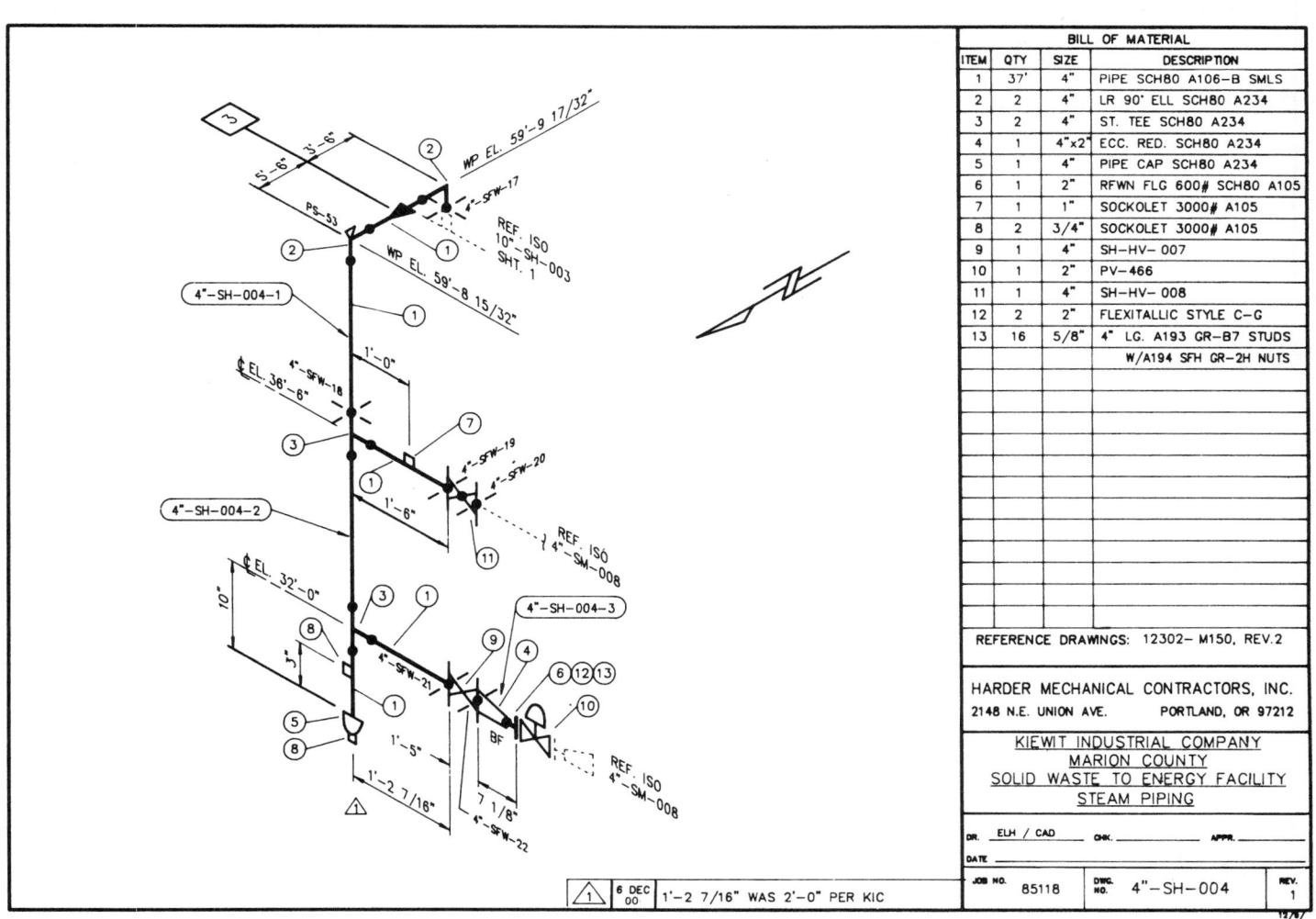

BILL OF MATERIAL			
ITEM	QTY	SIZE	DESCRIPTION
1	37'	4"	PIPE SCH80 A106-B SMLS
2	2	4"	LR 90° ELL SCH80 A234
3	2	4"	ST. TEE SCH80 A234
4	1	4"x2"	ECC. RED. SCH80 A234
5	1	4"	PIPE CAP SCH80 A234
6	1	2"	RFWN FLG 600# SCH80 A105
7	1	1"	SOCKOLET 3000# A105
8	2	3/4"	SOCKOLET 3000# A105
9	1	4"	SH–HV– 007
10	1	2"	PV–466
11	1	4"	SH–HV– 008
12	2	2"	FLEXITALLIC STYLE C–G
13	16	5/8"	4" LG. A193 GR–B7 STUDS
			W/A194 SFH GR–2H NUTS

REFERENCE DRAWINGS: 12302– M150, REV.2

HARDER MECHANICAL CONTRACTORS, INC.
2148 N.E. UNION AVE.   PORTLAND, OR 97212

KIEWIT INDUSTRIAL COMPANY
MARION COUNTY
SOLID WASTE TO ENERGY FACILITY
STEAM PIPING

DR. ELH / CAD   CHK. _____   APPR. _____
DATE

| JOB NO. 85118 | DWG. NO. 4"–SH–004 | REV. 1 |

1 / 6 DEC 00 / 1'–2 7/16" WAS 2'–0" PER KIC

PROBLEM 22-16   Draw a plan and elevation of 14"–SL–011 in both single- and double-line forms. Place both drawings on C-size media. Choose a scale that allows both drawings to fit uncrowded on the sheet. *Courtesy Harder Mechanical Contractors, Inc.*

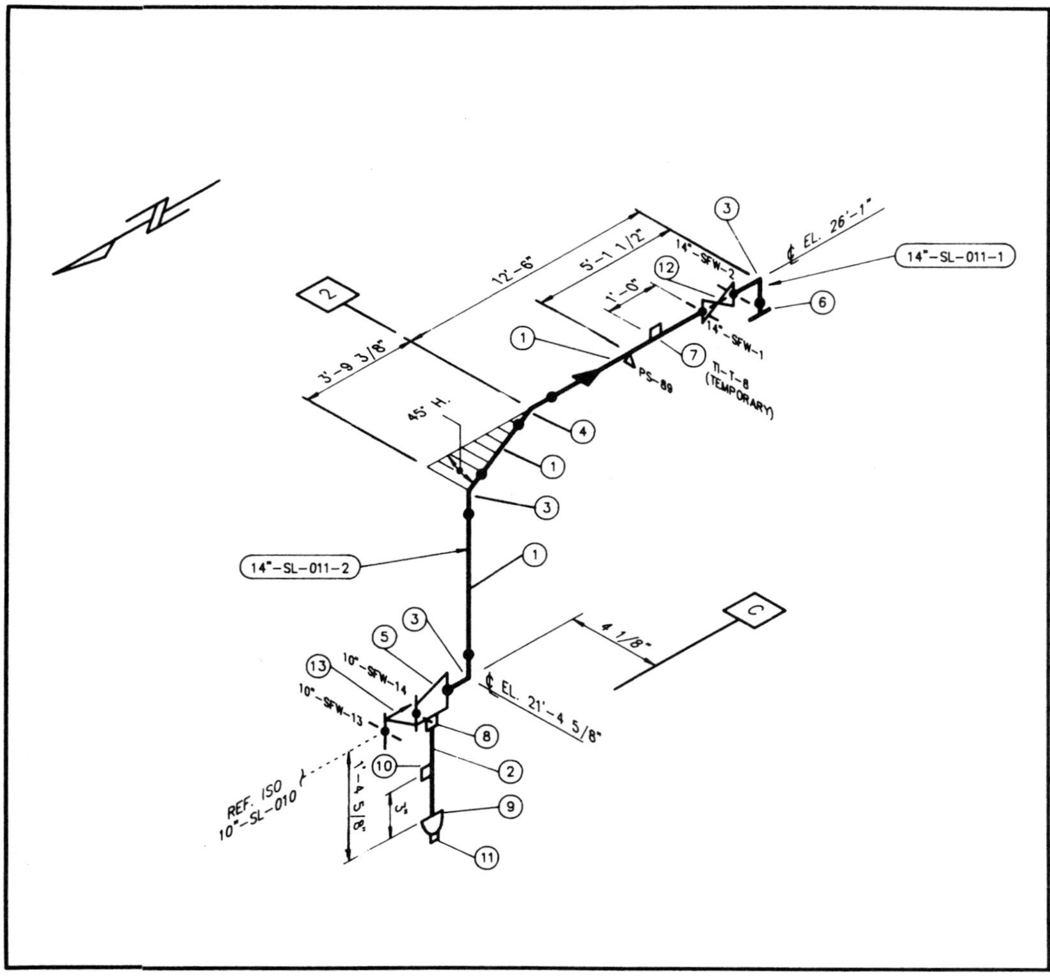

BILL OF MATERIAL			
ITEM	QTY	SIZE	DESCRIPTION
1	12'	14"	PIPE STD A53–B SMLS
2	1'	4"	PIPE STD A53–B SMLS
3	3	14"	LR 90° ELL STD A234
4	1	14"	LR 45° ELL STD A234
5	1	14"x10"	ECC. RED. STD A234
6	1	14"	RFWN FLG. 150# STD. A105
7	1	1"	THREADOLET 3000# A105
8	1	4"	WELDOLET STD A105
9	1	4"	PIPE CAP STD A234
10	2	3/4"	SOCKOLET 3000# A105
11	1	3/4"	HALF CPLG SW 3000#
12	1	14"	SL–HV–011
13	1	10"	XV–612

REFERENCE DRAWINGS: 12302– M151, REV.4

HARDER MECHANICAL CONTRACTORS, INC.
2148 N.E. UNION AVE.   PORTLAND, OR 97212

KIEWIT INDUSTRIAL COMPANY
MARION COUNTY
SOLID WASTE TO ENERGY FACILITY
STEAM PIPING

DR. ELH / CAD   CHK. _____   APPR. _____
DATE _____
JOB NO. 85118 | DWG. NO. 14"–SL–011 | REV. 0

PROBLEM 22-17   Draw the spools required for line 6"–SH–002A (Figure 22.10, page 728) on B-size media.

PROBLEM 22-18   Draw the spool for pipe 3"–1C–102 in section E-E (Figure 22.46e, page 749) from the 3" V–43 valve at elevation 56'–6" to the 3" V–142 valve at approximately the 64' elevation mark. Draw *no* valves in the spool. Include a bill of materials.

PROBLEM 22-19   Draw a plan and elevation view of the double-line piping isometric drawing shown on the right. Use threaded fittings. Have the instructor assign pipe size and dimensions to the problem. Draw on B-size media. *Courtesy Armstrong Machine Works.*

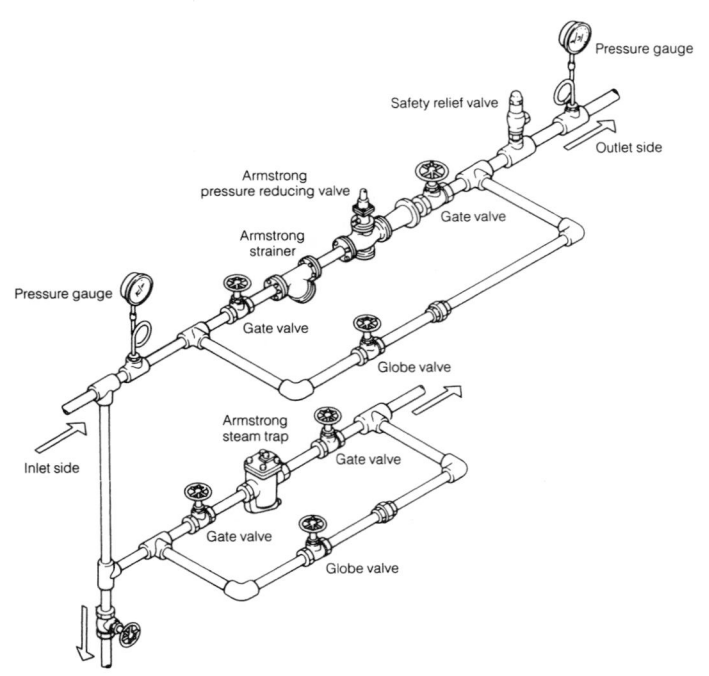

PROBLEM 22-20 Draw the piping assembly shown. The assembly is shown in isometric format, and the north arrow is pointing to the lower right of the drawing. Draw the assembly in isometric form, and orient the drawing with the north arrow pointing to the upper right of the drawing. Completely dimension the drawing as shown.

■ Use a B-size sheet of vellum or drafting film.

■ Draw using a scale of $^3/_8$" = 1'–0" for valves and fittings and no scale for pipe lengths.

■ Include the list of materials on your drawing.

Courtesy Willamette Industries, Inc.

PROBLEM 22-21 Draw a plan view of the piping assembly shown in Problem 22-20. Use break lines on the long piece of pipe, so it fits on the drawing sheet. Draw two elevation views of the piping assembly and place them on the same sheet as the plan view. See your instructor for the specific views to draw.

MATH PROBLEMS

Refer to Appendix F for fitting, flange, and valve dimensions to solve the following problems.

PROBLEM 22-22 The drawing shown below is an assembly of pipe, fittings, flanges, and valves. Only one dimension is given for the short, straight length of pipe. This dimension is the length of pipe between two welds. You are to provide all of the missing dimensions. Gaskets are inserted between flanges and valves. All gaskets are ¹/₁₆" thick. Use Appendix F to find dimensions for valves and fittings.

A = _____
B = _____
C = _____

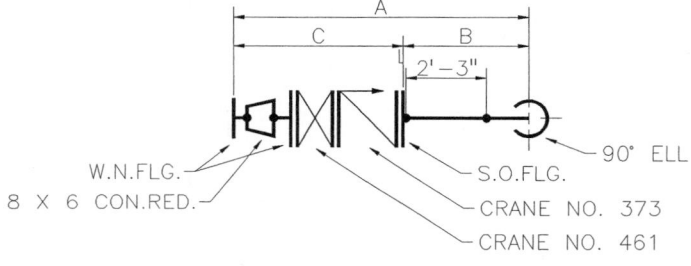

- Use a C-size sheet of vellum or drafting film.
- Draw using a scale of ³/₈" = 1'–0".
- Show the single-line pipe and fittings as thick lines.

PROBLEM 22-23 You are given a run of pipe shown below. Using the dimensions shown, provide the missing dimensions. Write your answers in the spaces provided.

A = _____
B = _____

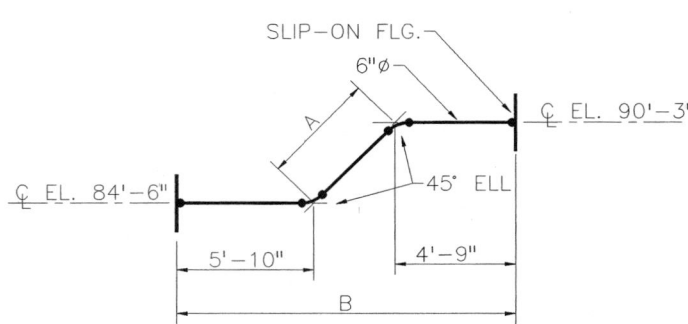

PROBLEM 22-24 Calculate the travel of the angled pipe shown in the figure below. Write your answer in the space provided.

A = _____

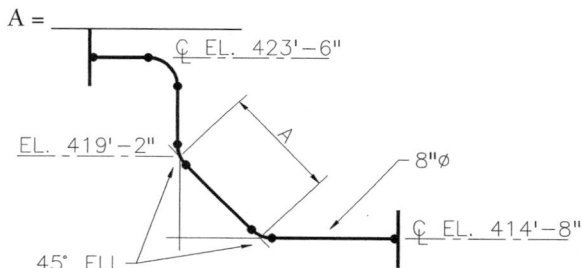

Structural Drafting

OBJECTIVES

After completing this chapter, you will:

■ Identify, describe, and draw various components of the following commercial construction methods: concrete, concrete block, wood, heavy timber, laminated beam, and steel.

■ Prepare a complete set of structural drawings.

■ Draw a site plan and grading plan.

■ Draw commercial structural drawings from engineering sketches.

THE ENGINEERING **DESIGN** APPLICATION

In many situations the drafter prepares formal drawings from engineering calculations and sketches. Registered structural engineers, architects, or designers with experience and training may prepare structural calculations and designs. In many situations the entry-level engineering drafter may not fully understand the calculations, but can interpret the results of the calculations because the engineer or designer normally highlights the solution to be placed on the drawing by clearly underlining or placing a box around it, as shown in Figure 23.1. When engineering calculations result in a design sketch, the drafter's job is to convert the sketch to a formal drawing, as shown in Figure 23.2b.

The quality and completeness of the engineering sketch often depends on the experience of the drafter. If the drafter is right out of school, the engineer may have to take a little more time providing detailed information until the drafter gains experience. Once the drafter has had some experience, the engineering sketches may be less complete. For example, if the engineering sketch in Figure 23.2a had omitted the note: 3/16" PLY SHIM EA SIDE GLUE AND NAIL W/4-ROWS 8–10d COMMON, an experienced drafter would have realized that shims are required when two beams of different thicknesses are joined. In some situations, an experienced drafter can work directly from the engineering calculations without sketches. In all situations it is very important for the drafter to maintain a current set of vendor's catalogs. These catalogs give the detailed information necessary for reproducing and labeling the drawings. The kinds of

FIGURE 23.1 ■ Engineering calculations typically contain a problem to be solved, mathematical solutions, and specifications to be placed on the drawing. The drawing note or information is placed in a box or otherwise highlighted.

information found in a vendor's catalog that will enable the drafter to complete the drawing in Figure 23.2b are the dimensions and specifications for the MST 27 straps, CC 5-1/4 COLUMN CAP, and L50 or A35 framing anchor.
(Continued)

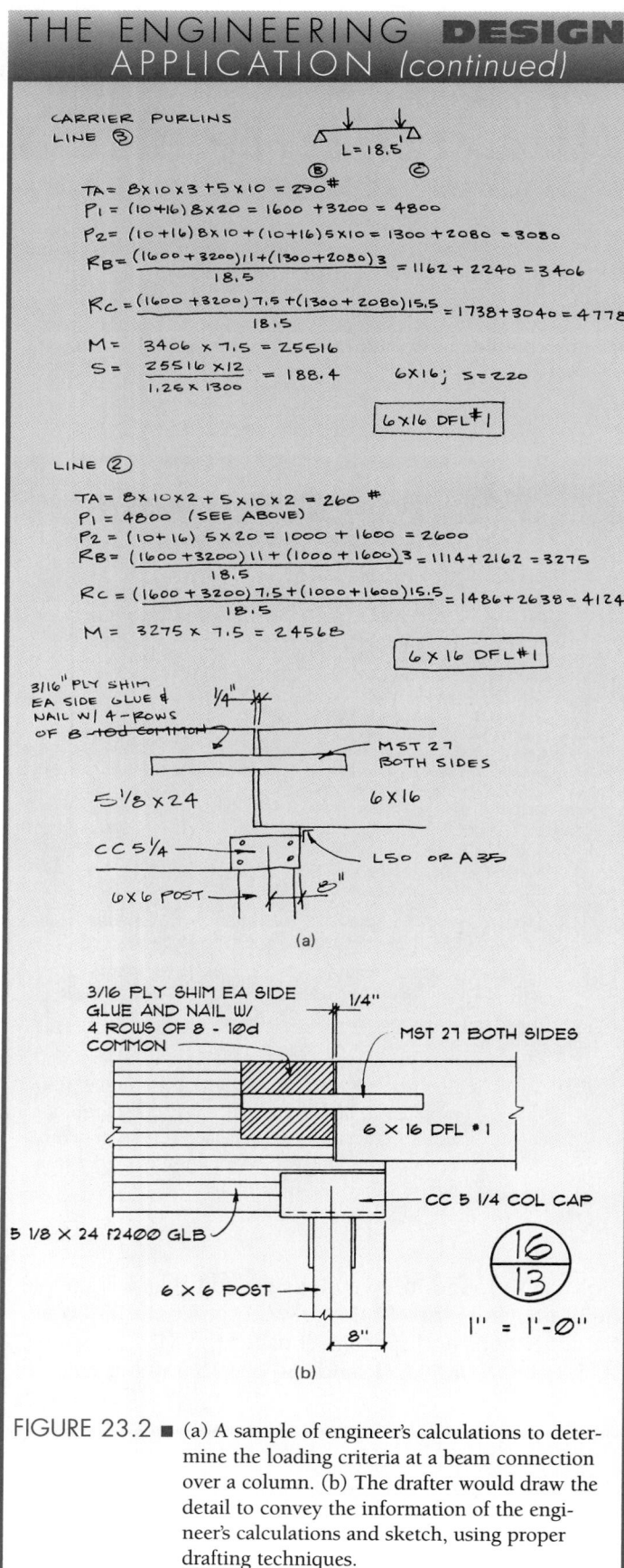

THE ENGINEERING DESIGN APPLICATION (continued)

FIGURE 23.2 ■ (a) A sample of engineer's calculations to determine the loading criteria at a beam connection over a column. (b) The drafter would draw the detail to convey the information of the engineer's calculations and sketch, using proper drafting techniques.

A structural engineer works with architects and building designers to engineer the structural components of a building. Structural engineering is generally associated with commercial steel and concrete buildings, and in some situations, it is also used for the structural design of residential buildings. The structural engineer works with civil engineers to design bridges and other structures related to road and highway construction. Mechanical engineers can also be involved in the structural project by designing machinery supports and foundations. There is a wide variety of projects where structural drafting may be involved. Structural drafting techniques are generally the same as mechanical drafting, although a combination of mechanical and architectural methods are used.

Structural engineering drawings and detail drawings show in a condensed form the final results of designing. Drawings, general notes, schedules, and specifications serve as instructions to the contractor. The drawings must be complete and have sufficient detail so no misinterpretation can be made. Structural drawings are usually independent of architectural drawings and other drawings, such as plumbing or heating, ventilating, and air conditioning (HVAC), or, when necessary, structural drawings are clearly cross-referenced to architectural drawings.

LINE WORK

The lines used in structural drafting are generally the same as those used in mechanical drafting. There are a few exceptions. For example, the object line may be drawn thicker than normal when a shape or feature requires extra emphasis. Object lines may also be thinner than the standard when used on a small-scale drawing. (See Figure 23.3.) Dimension lines may be drawn with a space provided for the numeral as in mechanical drafting, or the numeral may be placed above the dimension line as in architectural drafting. The dimension line may be capped on the end with arrowheads, slashes, or dots. (See Figure 23.4.) Cutting-plane lines and symbols are provided one of several ways, as shown in Figure 23.5.

LETTERING

The quality of lettering in structural drafting is equally as important as it is in other drafting fields. There is a great deal of lettering on structural drawings. A well-developed, legible style of lettering makes the job easier and adds professional quality. Lettering is similar to the Gothic style used in mechanical drafting; however, the structural drafter has more freedom of style, but usually less freedom than is allowed in architectural drafting. The best structural lettering is simple, easy to read, and quick to produce. Structural drafters often prefer slanted letters. Lettering height on drawings is typically $1/8''$ to $5/32''$ (3–4 mm) for all lettering except titles, which are $3/16''$ to $1/4''$ (5–6 mm) in height. If drawings are to be microfilmed, $5/32''$ lettering is used. Figure 23.6 shows lettering used in structural drafting. Other lettering rules are presented in Chapter 2.

COORDINATION OF WORKING DRAWINGS

The structural drawings are a portion of a complete set of working drawings. A set of drawings for a commercial project can have over

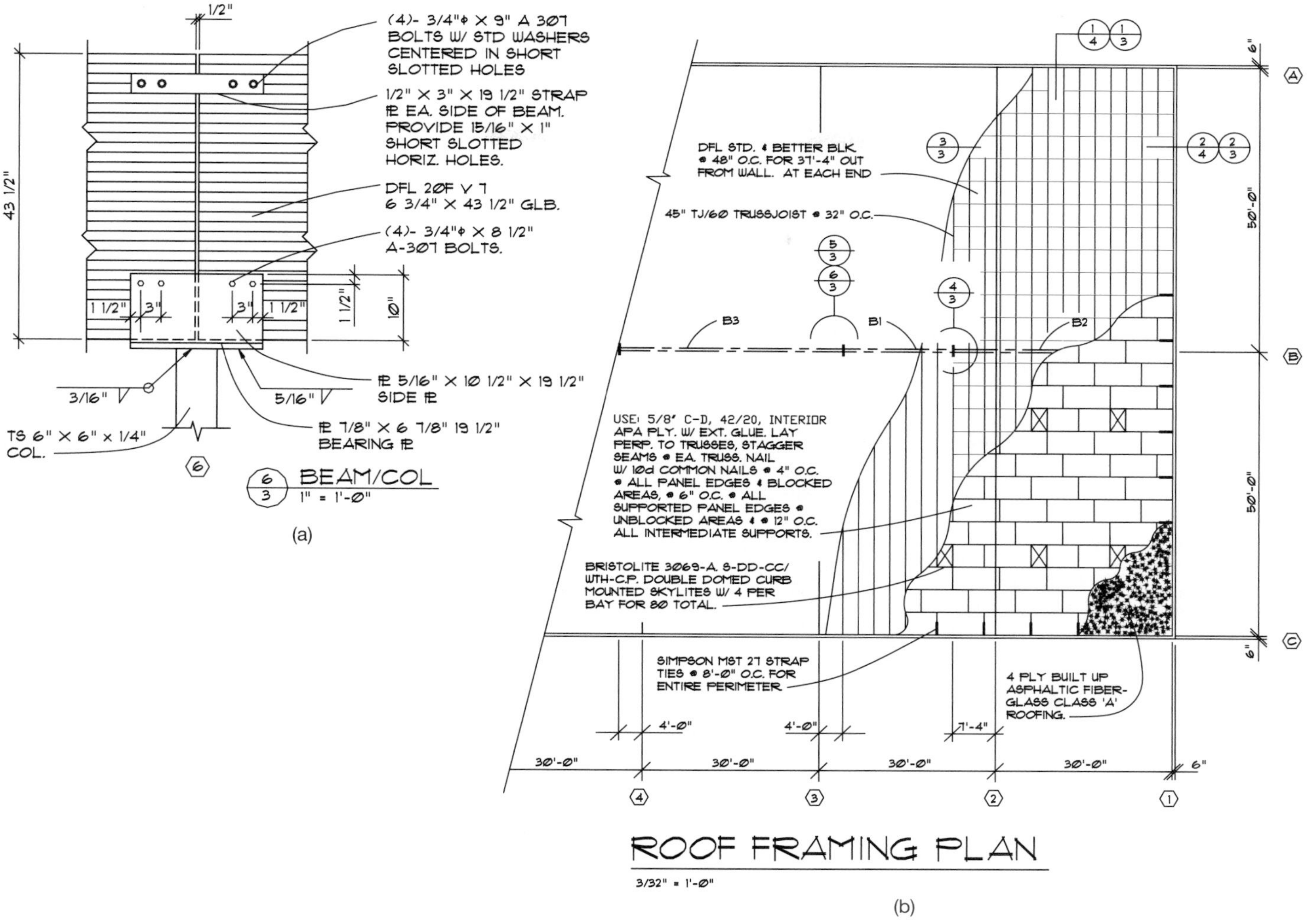

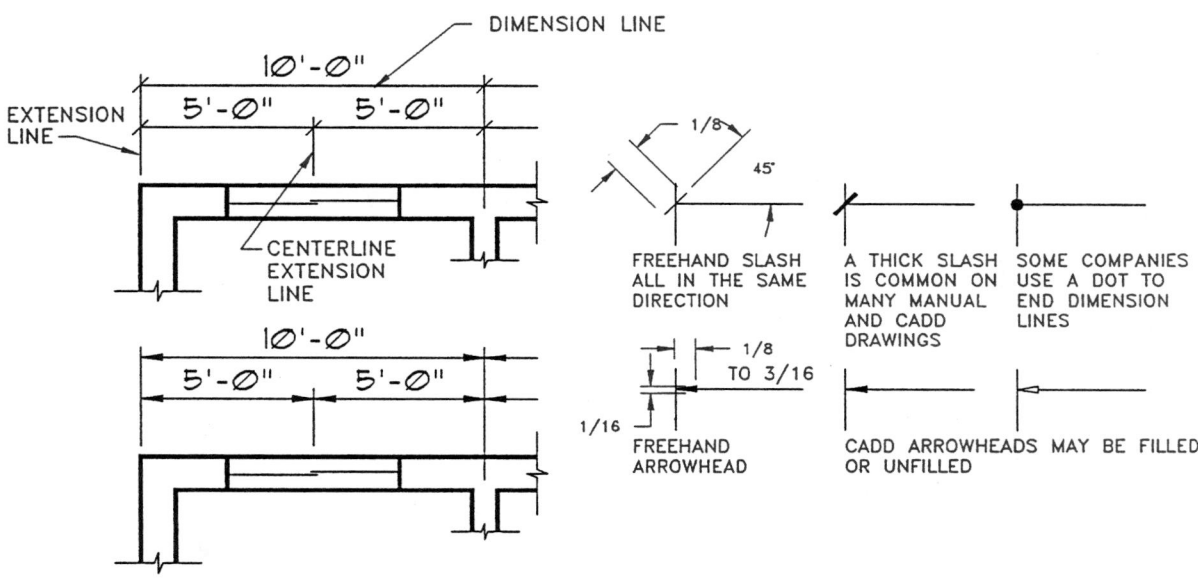

FIGURE 23.3

(a) Thick object lines for emphasis used on a beam/column detail; **(b)** thin object lines on small-scale drawing used on a panelized roof framing system.

FIGURE 23.4

■ Dimension line examples.

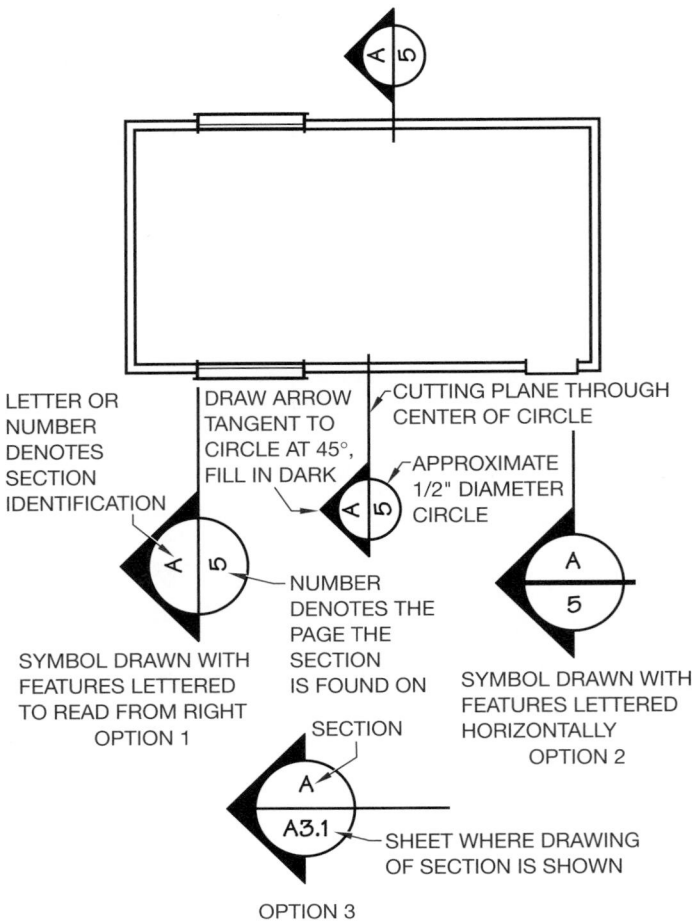

LETTER OR NUMBER DENOTES SECTION IDENTIFICATION

DRAW ARROW TANGENT TO CIRCLE AT 45°, FILL IN DARK

CUTTING PLANE THROUGH CENTER OF CIRCLE

APPROXIMATE 1/2" DIAMETER CIRCLE

NUMBER DENOTES THE PAGE THE SECTION IS FOUND ON

SYMBOL DRAWN WITH FEATURES LETTERED TO READ FROM RIGHT
OPTION 1

SYMBOL DRAWN WITH FEATURES LETTERED HORIZONTALLY
OPTION 2

SECTION

SHEET WHERE DRAWING OF SECTION IS SHOWN

OPTION 3

FIGURE 23.5 ■ Cutting-plane line symbols.

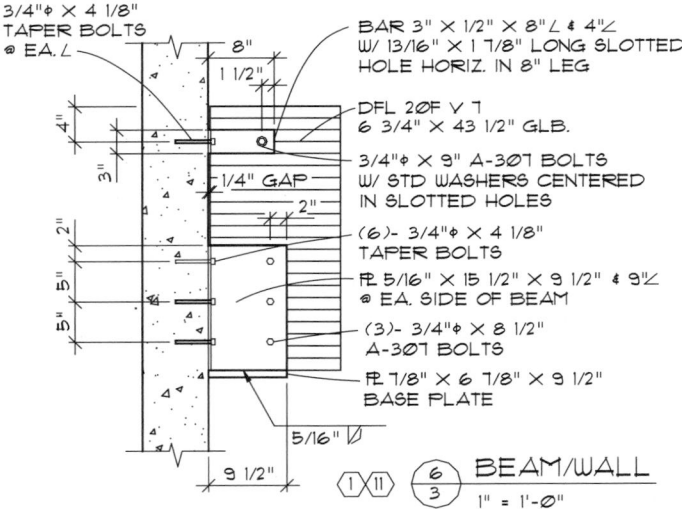

FIGURE 23.6 ■ Common lettering used in structural drafting used on a beam construction detail.

100 pages. Due to the complex nature of the drawings and the number of disciplines involved, the set of drawings is generally divided into several major groups. The major groups are generally architectural, structural, mechanical, plumbing, electrical, and fix-

ture drawings. The architect normally prepares the architectural drawings. The architect then coordinates with consulting engineering firms that prepare the drawings for their specific disciplines. The structural engineering firm generally prepares the related structural drawings.

Numbering the Pages

A page numbering system needs to be used because of the large number of pages in a set of commercial drawings. When a set of drawings has a number of pages that are easy to manage, it is common to see pages numbered in consecutive order such as 1, 2, 3, 4. If a sheet has sub pages, then the sheet number might be followed by letters alphabetically, such as 4A, 4B, 4C, 4D. The *AIA Handbook of Professional Practice* recommends a decimal page numbering system. This numbering system is established from the following sample general categories:

Architectural
T1   Title sheet, site demolition and survey
A1   Site plan and details
A2   Grading plan and details
A3   First floor plan and details
A4   Second floor plan, sections and details
A5   Enlarged plans, interior elevations and details
A6   Exterior elevations and details
A7   Building and wall sections and details
A8   Roof plan and details
A9   Details
A10   Reflected ceiling plan and details

Civil
U1   Site utilities
U2   Erosion control plans and details
U3   Public utility plan
U4   Utility details

Landscape
L1   Irrigation system plan and legend
L2   Irrigation system details and notes
L3   Planting plan, details and notes

Structural
S1   Foundation plan
S2   Foundation details
S3   Second floor framing plan
S4   Roof framing plan
S5   Details

Mechanical
M1   First floor plumbing plan, legends
M2   Second floor plumbing plan, notes
M3   Details and schedules

M4   First floor HVAC plans and legends

M5   Second floor HVAC plan

M6   Roof mounted HVAC equipment plan

M7   HVAC details and schedules

Electrical

E1   Notes, legend, riser

E2   First floor lighting plan

E3   Second floor lighting plan

E4   First floor power plan

E5   Second floor power plan

E6   Roof mounted equipment power plan

E7   Details and schedules

E8   First floor communications plan and legend

E9   Second floor communications plan

Fixtures

F1   Fixture plan and schedules

F2   Details

The general groups and elements within the groups can differ depending on the company practice and the building being designed. Each category within a general group can have additional pages. These pages are numbered with the sequential decimals of .1, .2, .3. So, the architectural drawings might have pages such as:

A1.1, A1.2, A1.3

A2.1, A2.2, A2.3, A2.4

The structural drawings can have a series of sheets that are numbered such as:

S1.1, S1.2

S2.1, S2.2, S2.3

S3.1, S3.2, S3.3, S3.4, S3.5

The decimal sheet numbering system can broken down even further by adding .01, .02, .03 to the existing numbers as needed. For example, additional pages in the series of S3.1 are numbered S3.1.01, S3.1.02, S3.1.03.

Coordinating the Details and Sections

Now that a page numbering system has been established, you need to coordinate the elements of drawings between pages. Figure 23.5 shows some examples of how cutting-plane line symbols are correlated to the drawing. Construction details and sections are commonly labeled in a similar manner. The details might be numbered in consecutive order and correlated to the page where the detail is found as shown in Figure 23.7. Sections are commonly labeled with letters in alphabetical order, but some companies use numbers. Details are generally labeled with numbers.

Laying Out Details and Sections

Detail and section drawing can be placed with the drawing where they relate if space is available. For example, the foundation details and sections can be on the same sheet as the foundation if they can be placed there clearly and while maintaining a drawing that is easy to read and well organized. If there is not enough space on the same sheet, then the details and sections can be placed on other sheets. Details should be grouped together and organized from left to right and top to bottom in an aligned, orderly manner. If the details are numbered, they should be organized in numerical order. The sections should also be grouped together and organized from left to right and top to bottom. If the sections are labeled with letters, they should be placed in alphabetical order.

STRUCTURAL DRAFTING RELATED TO CONSTRUCTION SYSTEMS

Different types of construction methods relate directly to the materials to be used, the area of the country where the construction will take place, the type of structure to be built, and even the office practices of the architect or engineer. The structural drafter should have a knowledge of construction materials and techniques. This chapter provides an introduction to construction techniques and materials. Additional resources should be used for reference, as each construction method discussed has volumes of both general and vendor information available. Another valuable way to learn about construction is to visit job sites to talk to builders and see firsthand how things are done.

CONCRETE CONSTRUCTION

Concrete is a mixture of Portland cement, sand, gravel, and water. This mixture is poured into *forms* that are built of wood or other materials to contain the mix in the desired shape until it is hard. Concrete is a fundamental material used for building *foundations*. Concrete is also used in commercial applications for wall and floor systems. Residential buildings use concrete foundations with or without steel reinforcing, while commercial buildings usually use steel-reinforced concrete, depending on the structural requirements.

Concrete can be either poured in place at the job site, formed at the job site and lifted into place, or formed off-site and delivered ready to be erected into place. Concrete alone has excellent compression qualities. Steel added to concrete improves the tension properties of the material. Concrete poured around steel bars placed in the forms is known as *reinforced concrete*. The steel bars are referred to as *rebar*. Steel is the best choice for reinforcing concrete, as its coefficient of thermal expansion is almost the same as cured concrete. The resulting structure has concrete to resist the compressive stress, and steel to resist the tensile stress which is caused by the loads acting on the structure.

Steel reinforcing is available in a number of shapes and configurations. The most common steel reinforcing is *plain* and *deformed*, in round or square bars. Plain reinforcing bars are smooth surface steel. Deformed reinforcing bars have raised ridges to hold better in concrete. (See Figure 23.8.)

Deformed steel bars have surface projections that increase the adhesion between the concrete and steel. (See Figure 23.8.) Steel

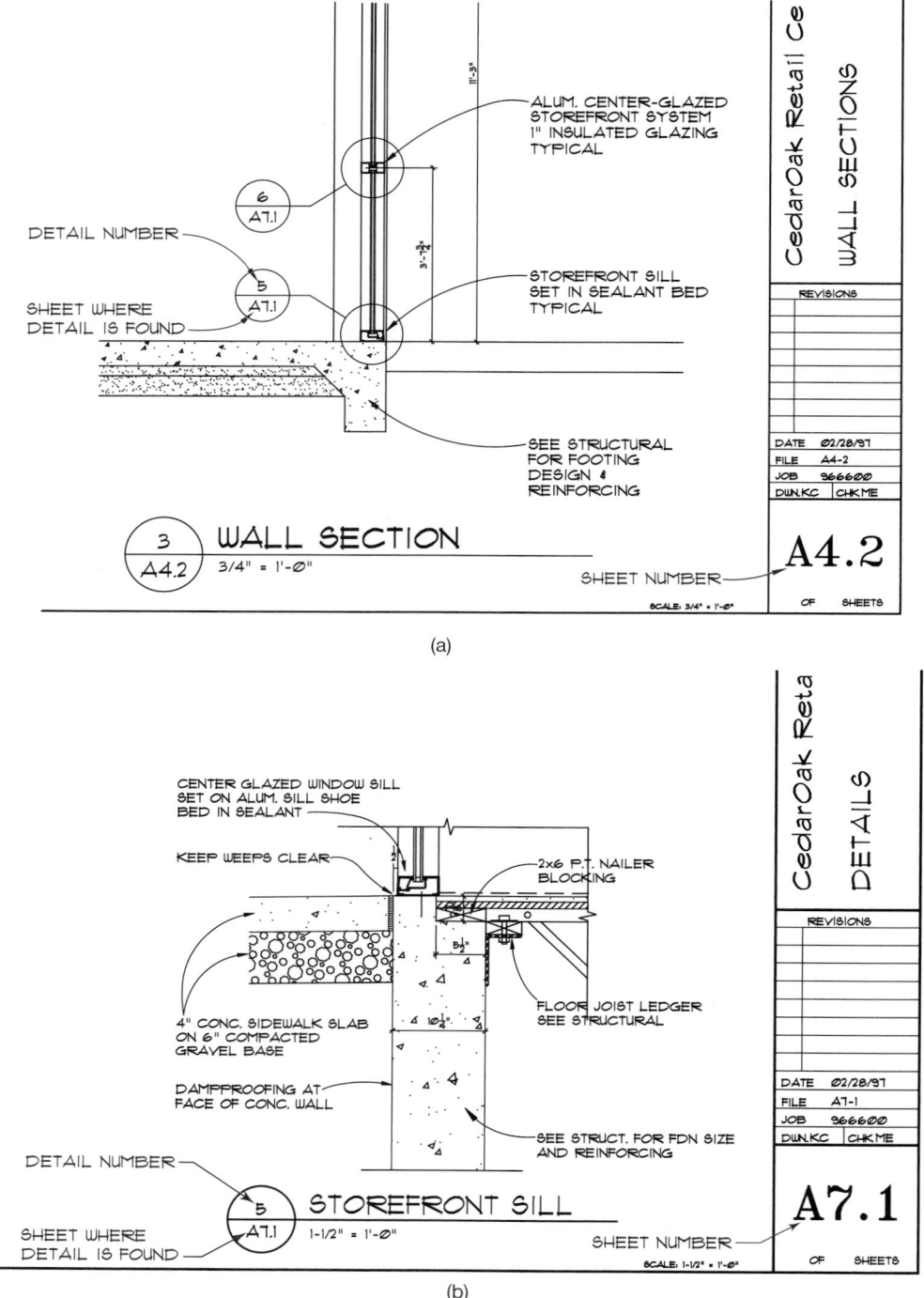

FIGURE 23.7 ■ Coordinating the details and sections to the sheet number. (a) Detail 5 is found on sheet number A7.1. (b) Sheet number A7.1 displays detail number 5. *Courtesy Ankrom Moisan Architects.*

reinforcing bars are sized by number, starting at no. 2, which is ⅛" in diameter, and increasing in size at approximately ⅛" intervals. No. 4 rebar is ½" in diameter. Another kind of steel reinforcing is wire fabric, which is available as welded wire or expanded metal. Welded wire fabric is commonly used to reinforce concrete slabs. Welded wire fabric is sized by the gage and spacing of the wires. For example, the callout WWF 6 × 6 10/10 means welded wire

fabric (WWF), 6 in on center (6 × 6), #10 gage wires (10/10). (See Figure 23.9.)

Poured-in-place Concrete

Commercial and residential applications for concrete poured in place are similar except that the size of the *casting* (resulting con-

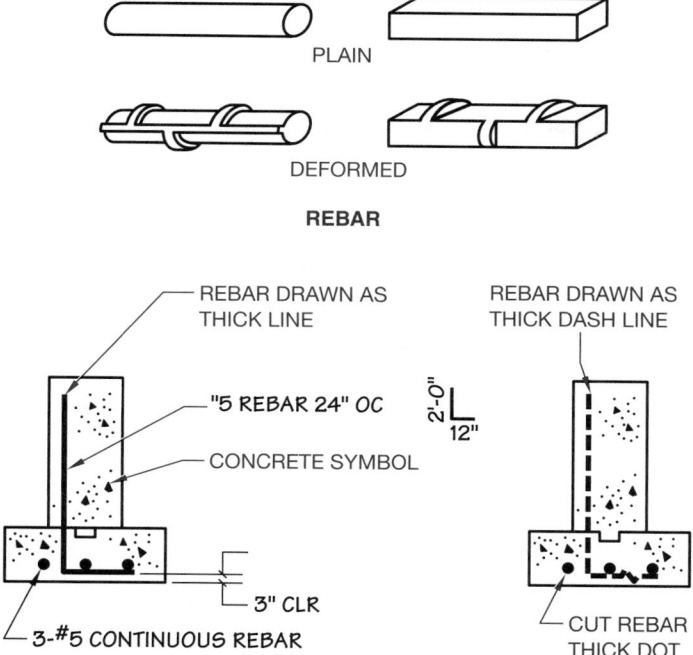

FIGURE 23.8 ■ Representation of reinforcing bars, *rebar*.

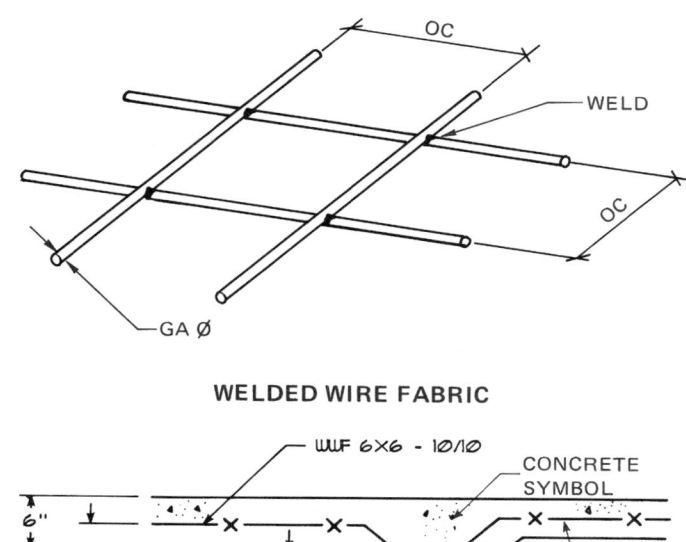

FIGURE 23.9 ■ Welded wire fabric representation.

crete structure) and the amount of reinforcing are generally more extensive in commercial construction. In addition to the foundation and on-grade (ground) floor systems, concrete is often used for walls, columns, and floors above ground. Lateral soil pressure acting on concrete structures tends to bend the wall inward, thus placing the soil side of the wall in compression and the interior side of the wall in tension. Steel reinforcing is used to increase the concrete's ability to withstand this *tensile stress,* as shown in Figure 23.10. Steel-reinforced walls and columns are constructed by set-

ting steel reinforcing in place and then surrounding it by wooden forms to contain the concrete. Once the concrete has been poured and allowed to set (harden), the forms are removed. As a drafter you will be required to draw details showing the sizes of the feature to be constructed and steel placement within the structure. This typically consists of drawing the vertical steel and the horizontal ties. Ties are wrapped around vertical steel in a column or placed horizontally in a wall or slab to help keep the structure from separating, and they keep the rebar in place while the concrete is being poured into the forms. Figure 23.11 shows two examples of column reinforcing. The drawings required to detail the construction of a rectangular concrete column are shown in Figure 23.12.

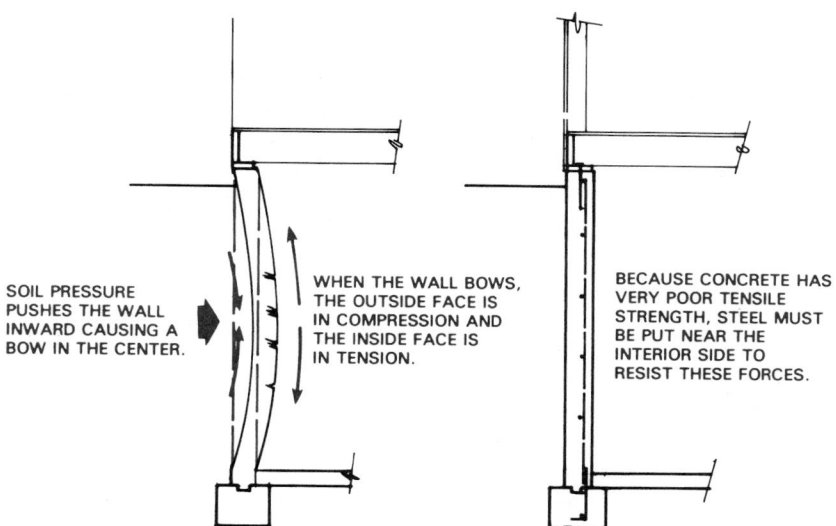

FIGURE 23.10 ■ Stresses created in a concrete wall from horizontal forces. The wall serves as a beam spanning between each floor, and the soil is the supported load.

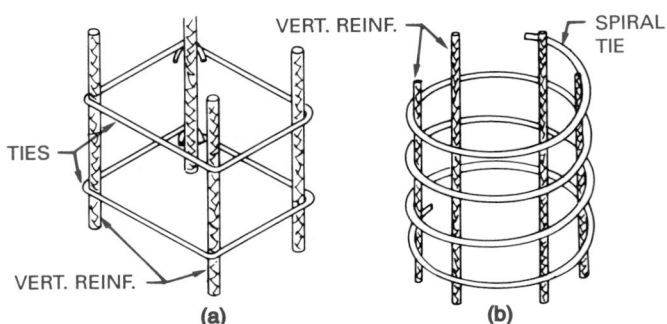

FIGURE 23.11 ■ Examples of column reinforcing: (a) square column ties; (b) round column spiral tie.

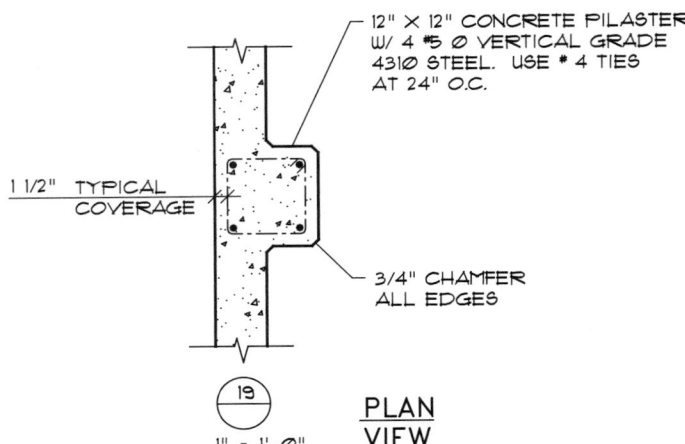

FIGURE 23.12 ■ Typical reinforcing for a rectangular concrete column.

Concrete is also used on commercial projects to build aboveground floor systems. The floor slab may be either supported by a steel deck or self-supported. The steel deck system is typically used on structures constructed with a steel frame. (See Figure 23.13.) Two of the most common poured-in-place concrete floor systems are the *ribbed* and *waffle* floor methods, as shown in Figure 23.14. The ribbed system is used in many office buildings. The ribs serve as floor joists to support the slab, but are actually part of the slab. Spacing of the ribs varies depending on the *span* and the amount and size of reinforcing. The span is the horizontal distance between two supporting members. The waffle system is used to provide added support for the floor slab and is typically used in the floor systems of parking garages.

Precast Concrete

Precast concrete construction consists of forming the concrete component off-site at a fabrication plant and transporting it to the construction site. Figure 23.15 shows a precast beam being lifted into place. Drawings for precast components must show how precast members are to be constructed and methods of transporting and lifting the member into place. Precast members often have an exposed metal flange so the member can be connected to other parts of the structure. Common details used for wall connections are shown in Figure 23.16.

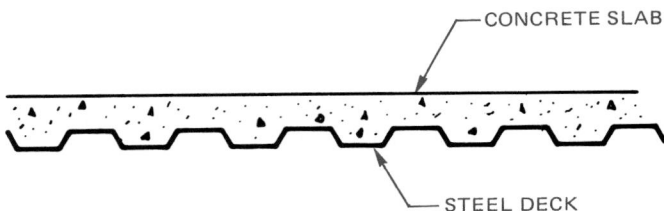

FIGURE 23.13 ■ Representation of a steel deck system.

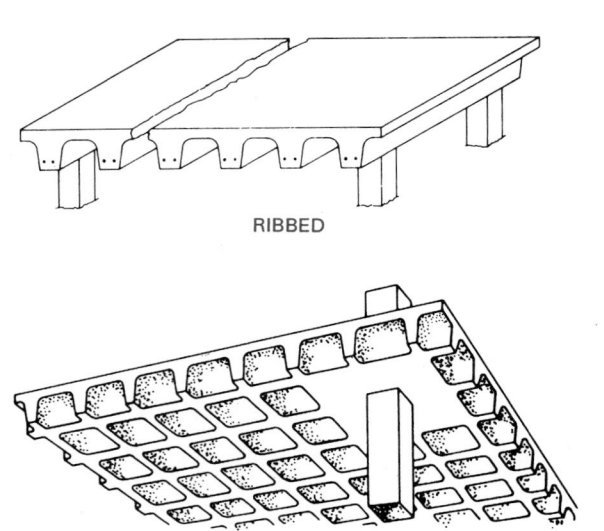

FIGURE 23.14 ■ Two common concrete floor systems.

FIGURE 23.15 ■ Precast concrete beams and panels are often formed off-site, delivered to the job site, and then set into place with a crane.

Many concrete structures are precast and *prestressed*. Concrete is prestressed by placing steel cables, wires, or bars held in tension between the concrete forms while the concrete is poured around them. Once the concrete has hardened and the forms are removed, the cables act like big springs. As the cables attempt to regain their original shape, compression pressure is created within the concrete. The compressive stresses built into the concrete member helps prevent cracking and deflection. Prestressed concrete members are

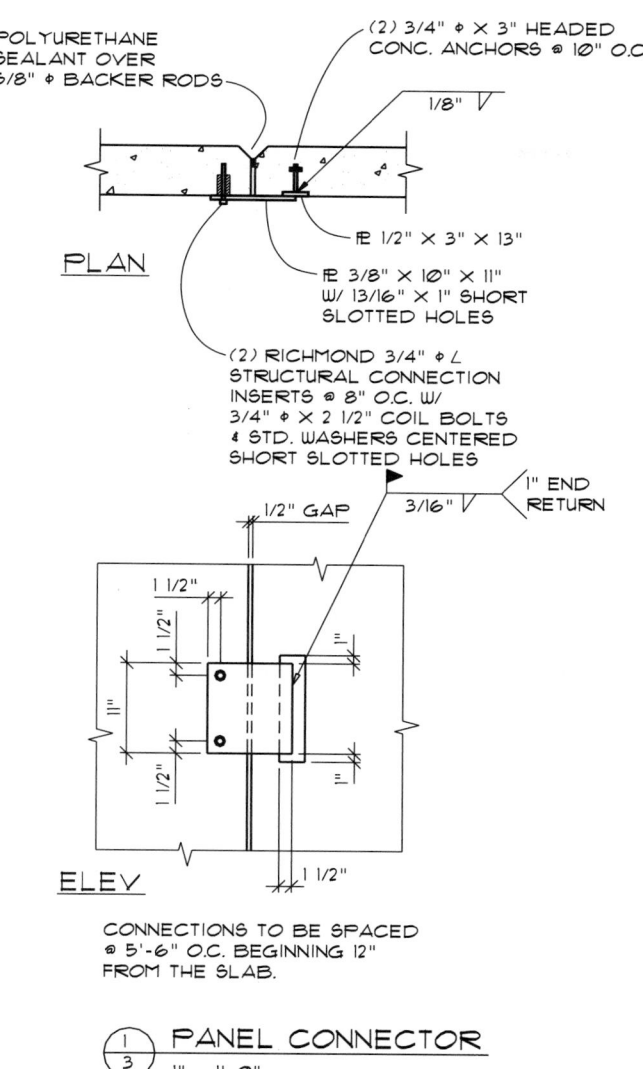

PLAN

POLYURETHANE SEALANT OVER 5/8" ⌀ BACKER RODS

(2) 3/4" ⌀ X 3" HEADED CONC. ANCHORS @ 10" O.C.

1/8"

℄ 1/2" X 3" X 13"

℄ 3/8" X 10" X 11" W/ 13/16" X 1" SHORT SLOTTED HOLES

(2) RICHMOND 3/4" ⌀ ∠ STRUCTURAL CONNECTION INSERTS @ 8" O.C. W/ 3/4" ⌀ X 2 1/2" COIL BOLTS & STD. WASHERS CENTERED SHORT SLOTTED HOLES

1" END RETURN

1/2" GAP

3/16"

1 1/2"

1 1/2"

1"

1 1/2"

ELEV

1 1/2"

CONNECTIONS TO BE SPACED @ 5'-6" O.C. BEGINNING 12" FROM THE SLAB.

PANEL CONNECTOR

1/3   1" = 1'-0"

FIGURE 23.16 ■ Typical panel connection detail.

generally reduced in size in comparison with the same design features of a standard precast concrete member. Prestressed concrete components are commonly used for the structural beams of buildings and bridges. Common prestressed concrete shapes are shown in Figure 23.17.

Tilt-up Precast Concrete

Tilt-up construction is a precast concrete method using formed wall panels that are lifted or "tilted" into place. Panels may be formed and poured either at or off the job site. Forms for a wall are constructed in a horizontal position and the required steel placed in the form. Concrete is then poured around the steel and allowed to harden. The panel is lifted into place once it has reached its desired hardening and design strength. When using this type of construction, the drafter usually draws a plan view to specify the panel locations, as shown in Figure 23.18. The location and size of steel placement and openings are also important, as shown in the panel elevation in Figure 23.19.

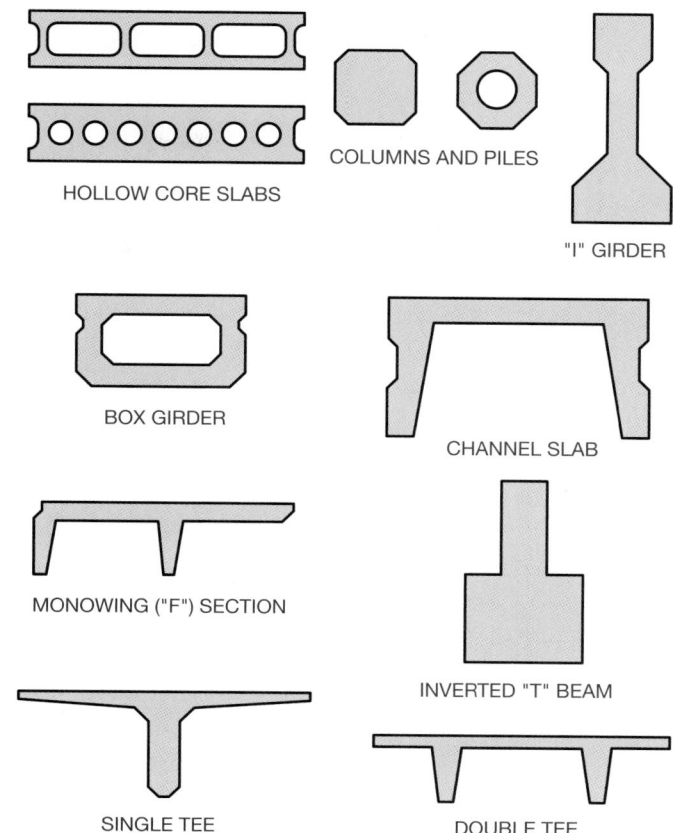

HOLLOW CORE SLABS

COLUMNS AND PILES

"I" GIRDER

BOX GIRDER

CHANNEL SLAB

MONOWING ("F") SECTION

INVERTED "T" BEAM

SINGLE TEE

DOUBLE TEE

FIGURE 23.17 ■ Common prestressed concrete shapes.

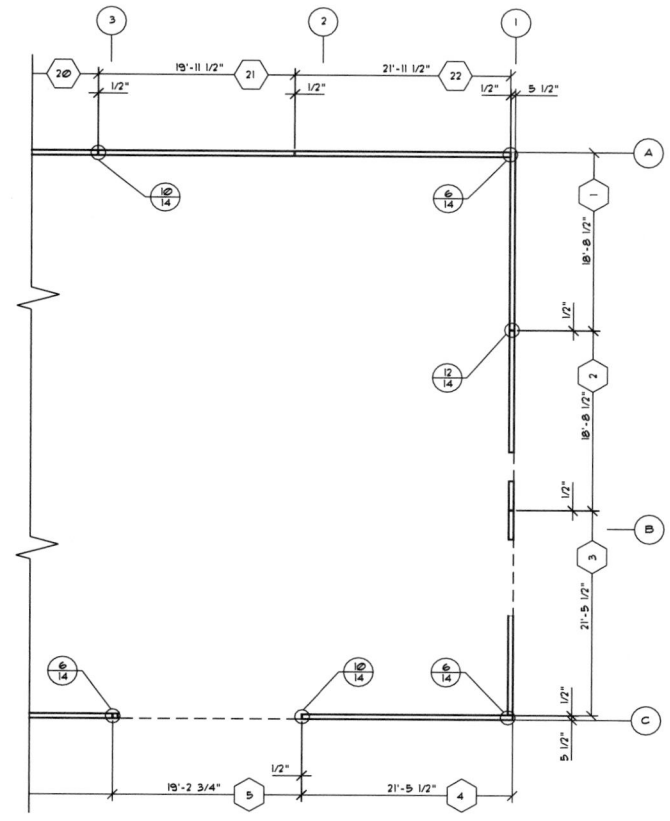

FIGURE 23.18 ■ Tilt-up panel plan.

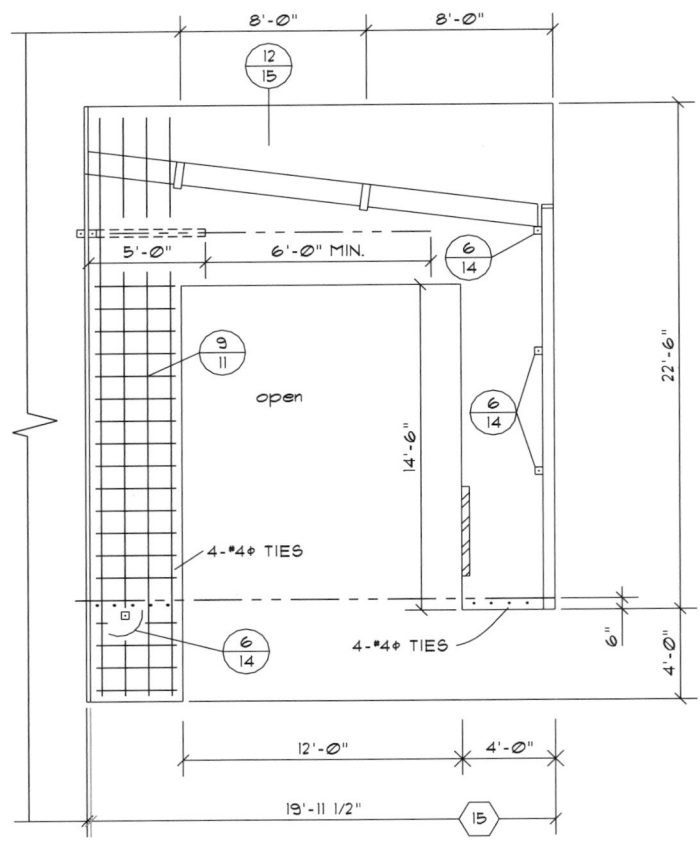

FIGURE 23.19 ■ Steel and opening locations specified in precast panel elevation.

Figure 23.20 shows a precast concrete panel drawing. A precast concrete beam drawing is displayed in Figure 23.21, and a precast concrete slab drawing is shown in Figure 23.22. Construction details are commonly used in prestressed concrete construction. Figure 23.23, page 778, shows a precast concrete panel installed on a poured concrete wall and footing.

Standard Structural Callouts for Concrete Reinforcing

When specifying anchor bolts (AB) on a drawing, give the quantity, diameter, type, length, spacing on center (OC), and projection of the thread out of the concrete. The quantity can be displayed followed by a dash or in parentheses. The w/ is the abbreviation for *with*, and can also be abbreviated w/. For example:

(12)-3/4" ∅ × 12" STD AB 24" OC W/3" PROJ.

When specifying rebar on a drawing, give the quantity (if required), bar size, length (if required), spacing in inches on center, horizontal or vertical, and bend information (if required). The (if required) note means that this specific information is given only if needed. These specifications are not needed if general space requirements control the application. Deformed steel rebar is assumed unless otherwise specified. For example:

#4 @ 12" O.C.
(25)#8 @ 24" O.C.HORIZ AND 16" O.C.VERT
#6 @ 24" O.C.EA WAY (or EW)

#5 @ 16" O.C.× 12" O.C.

Dimensions of bends are to be provided in feet and inches without the (') and (") marks given, for example:

#5 AT 16" OC VERT W/90° × 12 BEND

or the bend diagram may be drawn, as shown in Figure 23.24, page 778.

When specifying welded wire on a drawing, give the designation WWF, the wire spacing in inches OC, and the wire size. For example:

WWF 6 × 6-6/6

Reinforcing steel rebar lap splices may be shown on the drawing by giving the length of the lap and a lap location dimension. For example:

#5 REBAR 16" O.C.W/24" MIN SPLICE AT FOOTING

Welded wire fabric lap splices need not be shown as a specific note on the drawing; however, a general note should clarify the amount of allowable splice overlap, in inches, at the cross wires.

Clear distances should be given from the surface of the concrete to the rebar. This dimension is assumed to be to the edge of the rebar or clarified by the abbreviation CLR, as in 3" CLR. If this dimension is designated to the centerline of the rebar, then OC must be specified.

When structural members are embedded in concrete, the size of holes for rebar to pass through should be specified with the structural member callout. For example:

#5 AT 16" OC W/13/16 ∅ HOLES

Recommended hole sizes for rebar passing through steel or timber are as follows:

Bar #	Hole ∅	Bar #	Hole ∅
4	11/16	8	1 1/4
5	13/16	9	1 3/8
6	1	10	1 9/16
7	1 1/8		

When rebar must be driven through timber, a tighter hole tolerance is recommended. For example:

#8 REBAR × 6'–0" at 12" OC W/1-1/8" ∅ HOLES
AT TIMBER

This application for given rebar is as follows:

Bar #	Hole ∅	Bar #	Hole ∅
4	9/16	8	1 1/8
5	1 1/16	9	1 1/4
6	7/8	10	1 7/16
7	1		

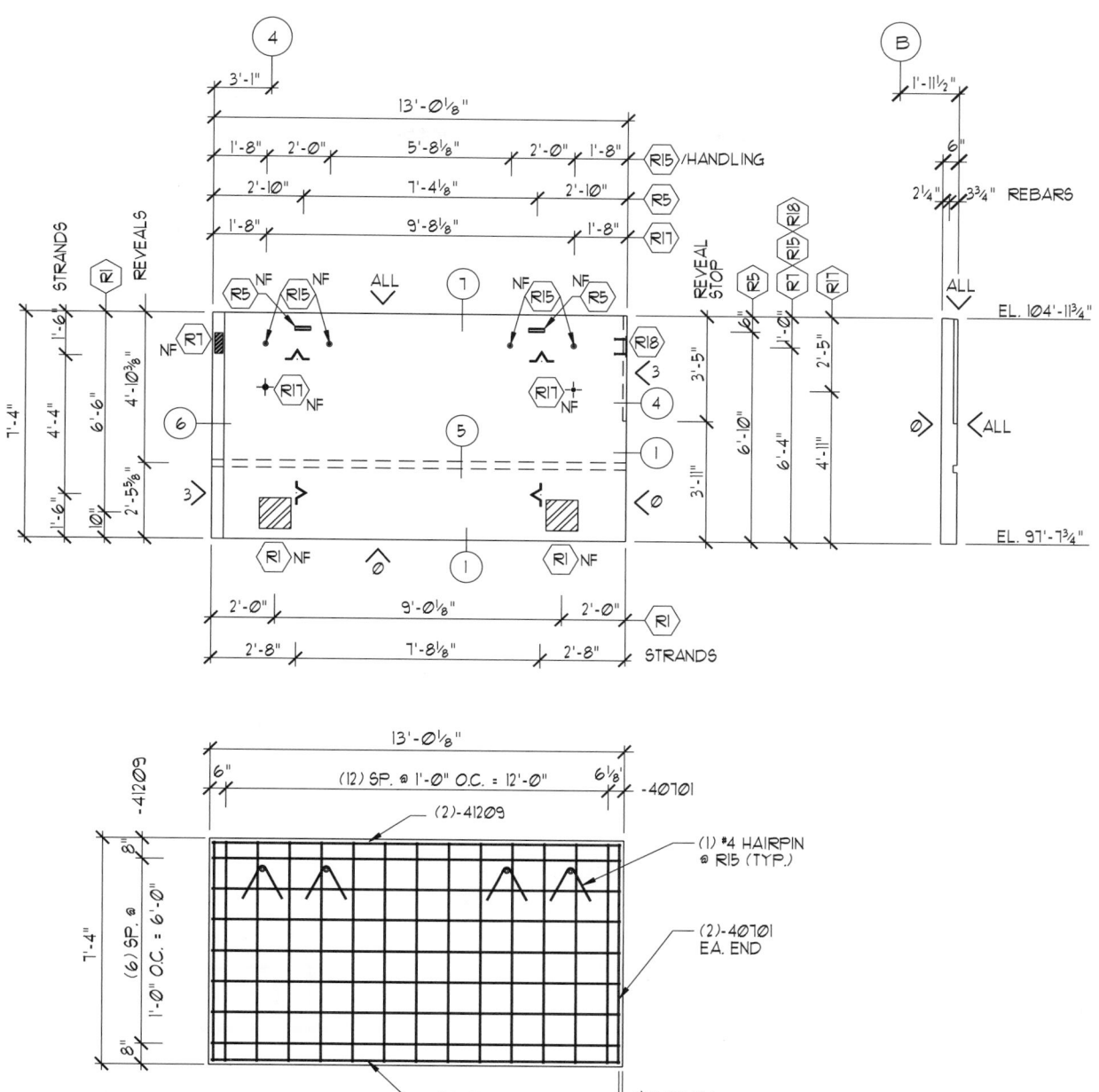

FIGURE 23.20 ■ Precast concrete panel drawing. *Courtesy of Morse Bros. Inc.*

Slab thickness, concrete wall, or concrete beam and column cross-sectional dimensions should be given in inches or feet and inches. A complete slab-on-grade callout may read as follows:

4" CONC SLAB ON FIRM UNDISTURBED SOIL
OR 4" SAND FILL

or

6" THICK 3000 PSI CONC SLAB WITH WWF 6 × 6—
W2.9 × W2.9 3" CLR AT 4" MIN 3/4" MINUS FILL COM-
PACTED TO 5% OF SOIL DENSITY

Concrete footing thickness should be given in inches or feet and inches. The footing width may be given followed by the thickness all in one note.

Concrete Structural Engineering Drawings

Structural engineering drawings show general information that is required for sales, marketing, engineering, or erection purposes. Concrete structures are drawn in plan (top) view, elevation (side) view, or sectional views at a scale that is dependent on the size of the structure, the amount of detail to be shown, or the size of paper used. Dimensions and notes are used to completely describe the construction characteristics. Concrete material symbols are used where appropriate. Rebar is shown as a very thick line or a thick dashed line in longitudinal view, or as a round dot where the bar appears cut.

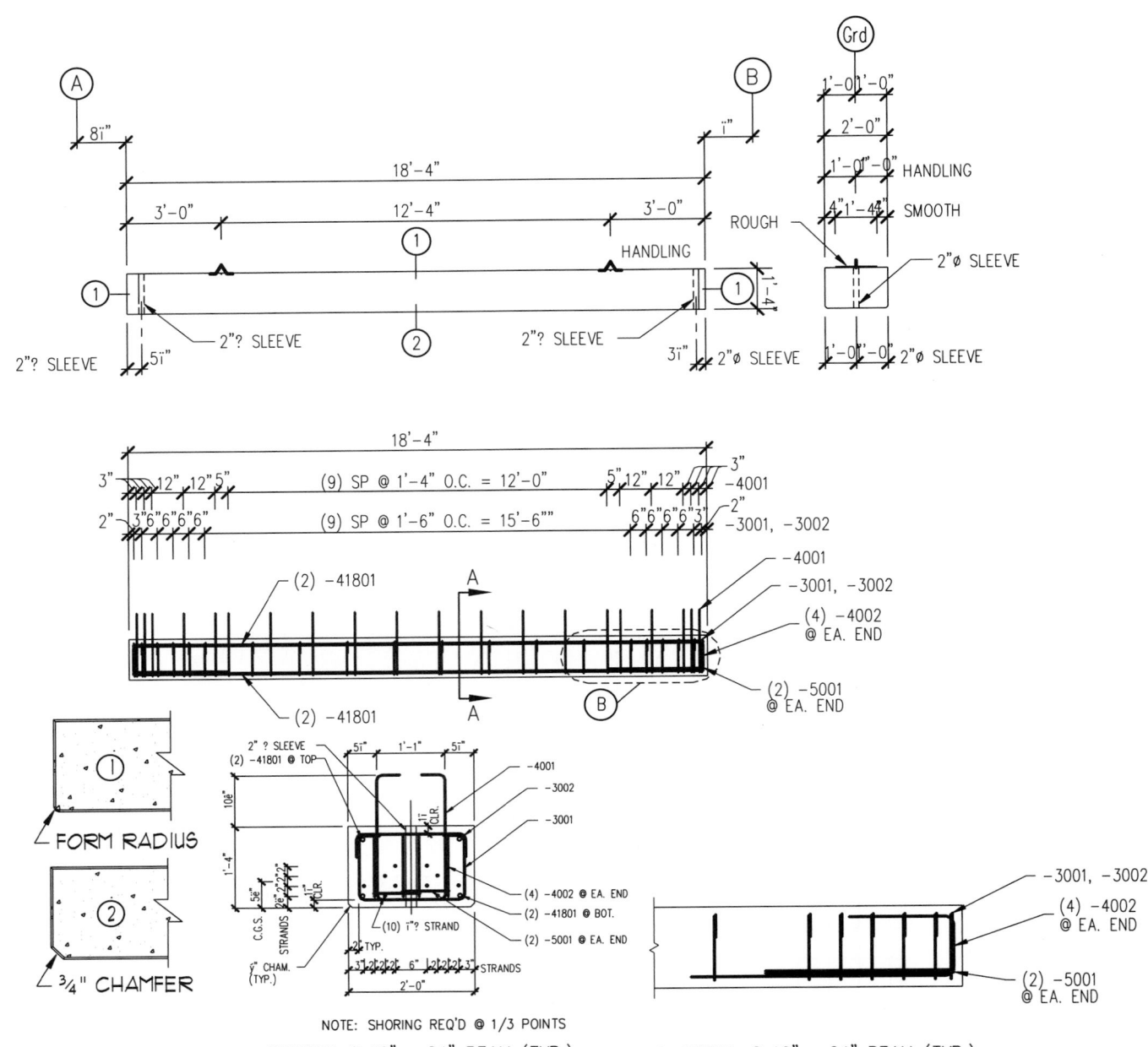

FIGURE 23.21 ■ Precast concrete beam drawing. *Courtesy of Morse Bros. Inc.*

Figure 23.25 illustrates a concrete foundation and related concrete and concrete block details as a typical example of the type of concrete drawings created by the architectural structural drafter.

CONCRETE BLOCK CONSTRUCTION

Concrete block construction is often used for residential foundations, but is also used in some above-ground construction. In commercial applications, concrete blocks are used to form the wall systems for many types of buildings. Concrete blocks provide a durable construction material and are relatively inexpensive to install and main-

tain. In residential and light commercial applications, foam-filled blocks provide excellent insulating characteristics and are often used in desert climates. Blocks are commonly manufactured in nominal size *modules* of $8 \times 8 \times 16$ in., $4 \times 8 \times 16$ in., or $6 \times 8 \times 16$ in. The actual size of the block is smaller than the nominal size so that *mortar* (grout) joints can be included in the final size. Although the building designer determines the size of the structure, it is important that the drafter be aware of the modular principles of concrete block construction. Wall lengths, opening locations, and wall and opening heights must be based on the modular size of the block being used. Failure to maintain the modular layout can result in a tremendous increase in labor costs to cut and lay the blocks.

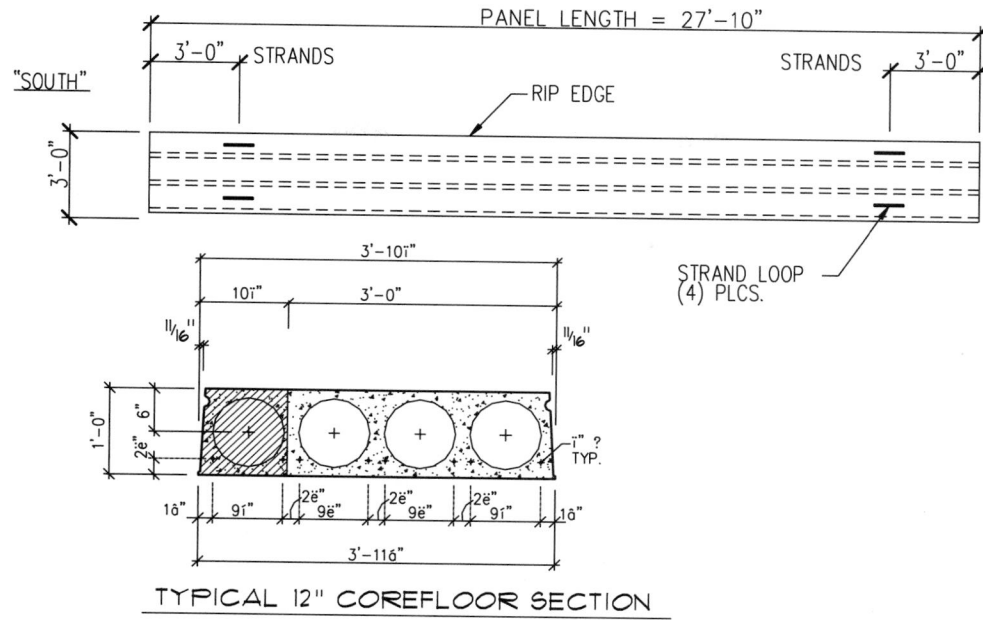

FIGURE 23.22 ■ Precast concrete slab drawing. *Courtesy of Morse Bros. Inc.*

The drafter's responsibility when working with concrete block structures is also to detail steel reinforcing patterns. Concrete blocks are often reinforced with a wire mesh at every other course of blocks. Where the risk of seismic activity must be considered, concrete block structures are often required to have reinforcing steel placed within the wall to help tie the blocks together. The steel is placed in a block that has a channel or cell running through it. This cell is then filled with grout or concrete to form a *bond beam* within the wall. The bond beam solidifies and ties the block structure together. A typical bond beam concrete block structure is detailed in Figure 23.26, page 780. Figure 23.27 (see page 781) shows typical concrete block sizes and reinforcing methods. Openings for windows and doors require steel reinforcing for both poured-in-place and concrete block construction. (See Figure 23.28, page 782.)

When the concrete blocks are required to support a load from a beam, a *pilaster* is often placed in the wall to help transfer the beam loads down the wall to the footing. Pilasters are also used to provide vertical support to the wall when the wall is required to span long distances. Examples of pilasters are shown in Figure 23.29 (see page 782). A structural detail is shown in Figure 23.30 (see page 783).

WOOD CONSTRUCTION

Wood frame construction is typical in residential construction, and is also used in some commercial construction applications, especially for multifamily dwellings and office buildings. Other commercial uses include partition framing, upper-level floor framing and roof framing. Residential and commercial wood wall construction is essentially the same, with the main difference often being the type of covering used. In commercial applications, wood walls may require special finishes to meet fire protection requirements, as shown in Figure 23.31, page 783.

Joists, trusses, and panelized systems are the most commonly used wood roof framing systems. These systems allow for the members to be placed 24 or 32 in. OC. Figure 23.3b (see page 767) shows a truss roof system drawing. Panelized roof systems use beams placed 20 to 30 ft. apart with smaller beams called *purlins* placed between the main beams on 8 ft. centers. Joists that are 2 or 3 in. wide are placed between the purlins at 24 in. OC. The roof is then covered with plywood sheathing. Figure 23.32, page 783, shows an example of how a panelized roof system is constructed. A roof framing plan for a panelized roof system is shown in Figures 23.3b, page 767, and 23.33, page 784.

When wood frame construction joins concrete or concrete block construction, a rigid connection between the two systems is required. Several methods of connecting wood to concrete block are shown in Figure 23.34 (see page 785).

Heavy Timber Construction

Large wood members are sometimes used for the structural frame of a building. This method of construction is used for appearance, structural purpose, and when material is available to suit the application. Heavy timbers have structural advantages for short spans and excellent fire retardant qualities. In a fire, heavy wood members will char on the exposed surfaces while maintaining structural integrity long after a steel beam of equal size has failed. A heavy timber roof framing plan is shown in Figure 23.33.

Laminated Beam Construction

It is difficult and often impossible to produce large-size wood timbers in long lengths, especially when lumber is currently sawn from smaller and smaller logs. When long-span wood beams are

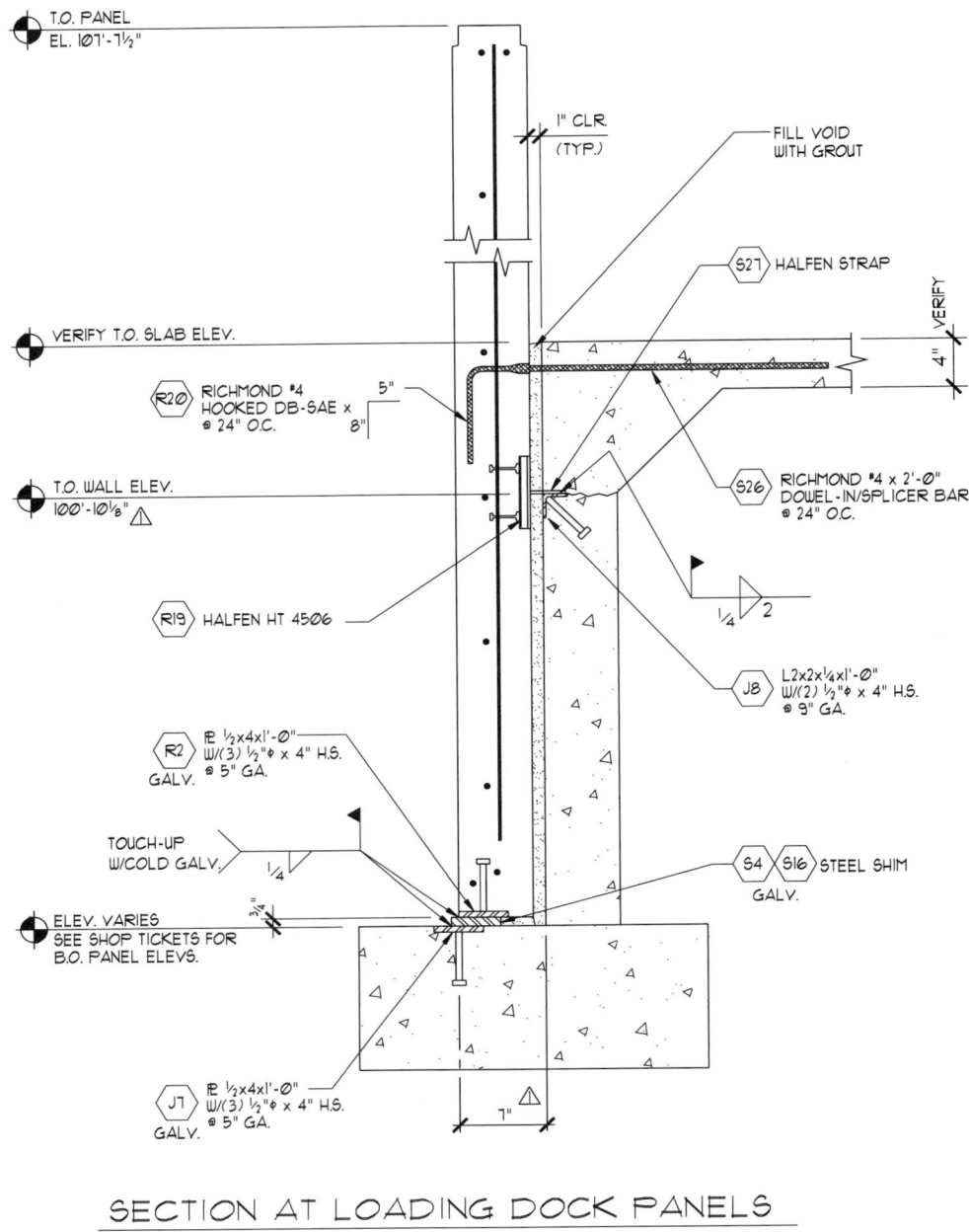

T.O. PANEL
EL. 107'-7½"

1" CLR.
(TYP.)

FILL VOID
WITH GROUT

S27 HALFEN STRAP

VERIFY T.O. SLAB ELEV.

4" VERIFY

R20 RICHMOND #4
HOOKED DB-SAE x
@ 24" O.C.

5"

8"

S26 RICHMOND #4 x 2'-0"
DOWEL-IN/SPLICER BAR
@ 24" O.C.

T.O. WALL ELEV.
100'-10⅛"

R19 HALFEN HT 4506

¼ 2

J8 L2x2x¼x1'-0"
W/(2) ½"φ x 4" H.S.
@ 9" GA.

R2 ℞ ½x4x1'-0"
W/(3) ½"φ x 4" H.S.
@ 5" GA.
GALV.

TOUCH-UP
W/COLD GALV.

¼

3¾"

ELEV. VARIES
SEE SHOP TICKETS FOR
B.O. PANEL ELEVS.

S4 S16 STEEL SHIM
GALV.

J7 ℞ ½x4x1'-0"
W/(3) ½"φ x 4" H.S.
@ 5" GA.
GALV.

7"

SECTION AT LOADING DOCK PANELS

1" = 1'-0"

REF. DBA SK-N

FIGURE 23.23 ■ Precast concrete panel, and concrete wall, footing, and slab detail drawing.
Courtesy of Morse Bros. Inc.

#4 @ 16" O.C.

3'-6"

1'-6"

FIGURE 23.24 ■ Rebar bend diagram.

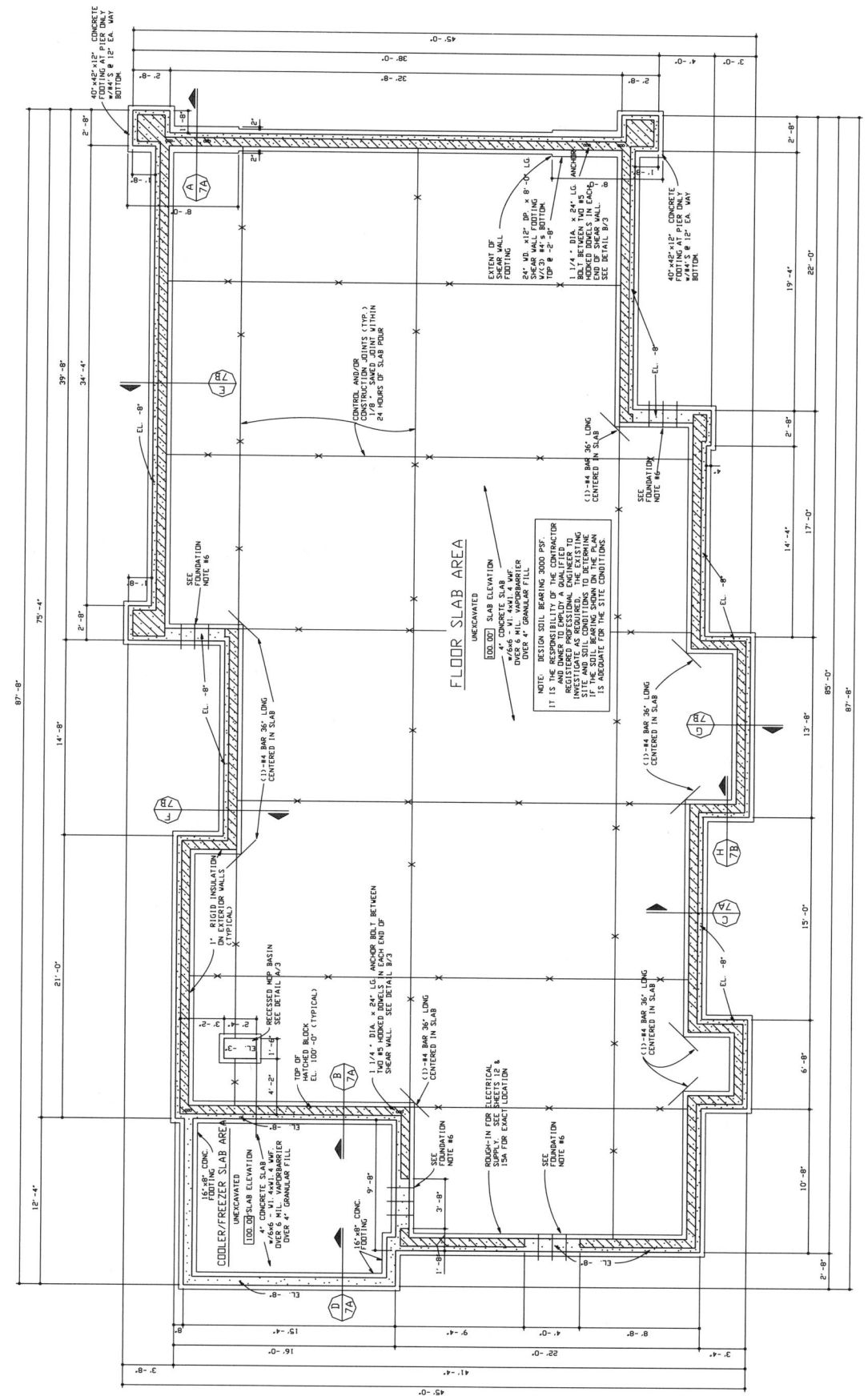

FIGURE 23.25 ■ Concrete foundation and related concrete and concrete block details.

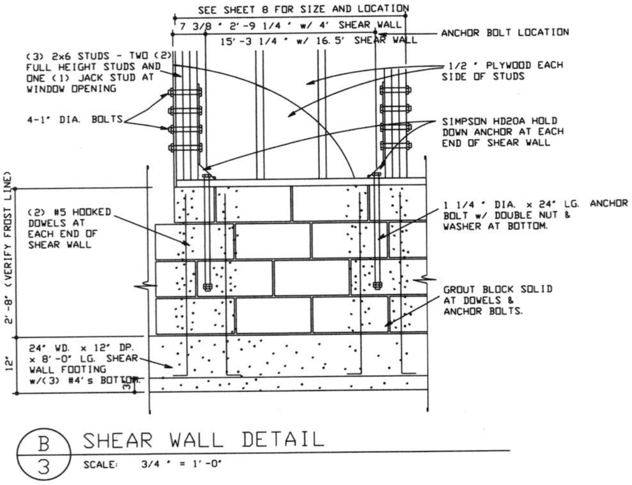

SEE SHEET 8 FOR SIZE AND LOCATION
7 3/8 " 2'-9 1/4 " w/ 4' SHEAR WALL
15'-3 1/4 " w/ 16.5' SHEAR WALL
ANCHOR BOLT LOCATION
(3) 2x6 STUDS - TWO (2) FULL HEIGHT STUDS AND ONE (1) JACK STUD AT WINDOW OPENING
1/2 " PLYWOOD EACH SIDE OF STUDS
4-1" DIA. BOLTS
SIMPSON HD20A HOLD DOWN ANCHOR AT EACH END OF SHEAR WALL
(2) #5 HOOKED DOWELS AT EACH END OF SHEAR WALL
1 1/4 " DIA. x 24" LG. ANCHOR BOLT w/ DOUBLE NUT & WASHER AT BOTTOM.
GROUT BLOCK SOLID AT DOWELS & ANCHOR BOLTS.
24" WD. x 12" DP. x 8'-0" LG. SHEAR WALL FOOTING w/(3) #4's BOTTOM
2'-8" (VERIFY FROST LINE)
12"
B
3
SHEAR WALL DETAIL
SCALE: 3/4 " = 1'-0"

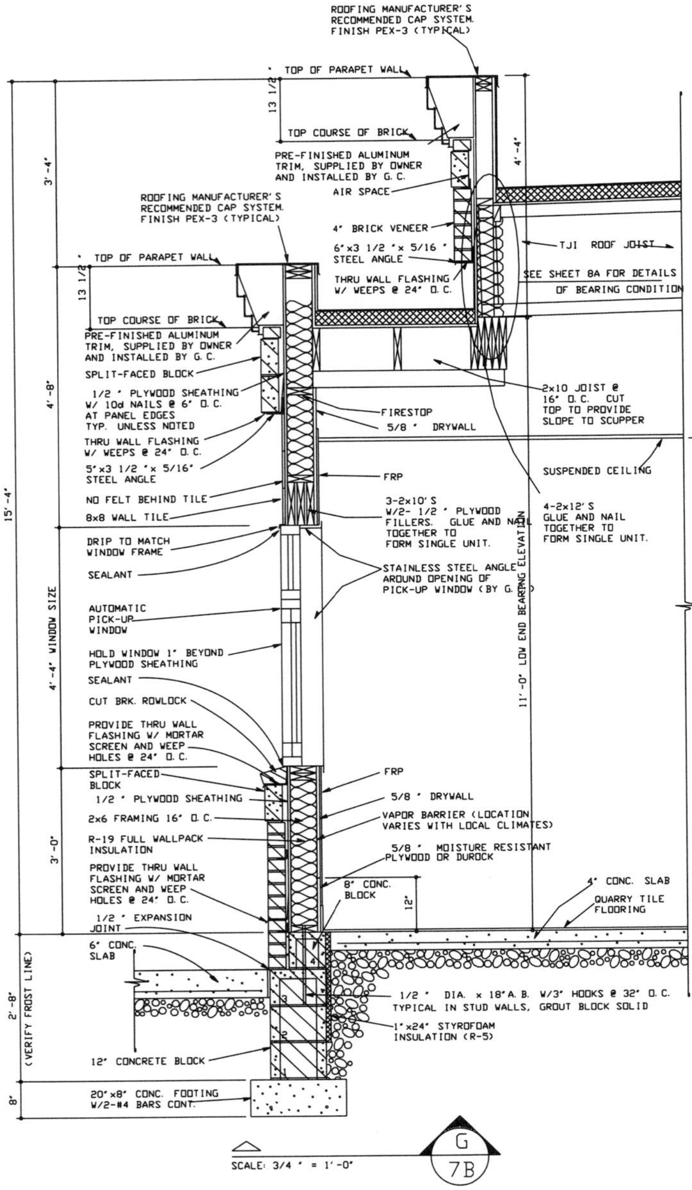

ROOFING MANUFACTURER'S RECOMMENDED CAP SYSTEM. FINISH PEX-3 (TYPICAL)
TOP OF PARAPET WALL
13 1/2
TOP COURSE OF BRICK
PRE-FINISHED ALUMINUM TRIM, SUPPLIED BY OWNER AND INSTALLED BY G.C.
AIR SPACE
4'-4"
ROOFING MANUFACTURER'S RECOMMENDED CAP SYSTEM. FINISH PEX-3 (TYPICAL)
4" BRICK VENEER
6"x3 1/2 "x 5/16 " STEEL ANGLE
TJI ROOF JOIST
THRU WALL FLASHING W/ WEEPS @ 24" O.C.
SEE SHEET 8A FOR DETAILS OF BEARING CONDITION
3'-4"
TOP OF PARAPET WALL
13 1/2
TOP COURSE OF BRICK
PRE-FINISHED ALUMINUM TRIM, SUPPLIED BY OWNER AND INSTALLED BY G.C.
SPLIT-FACED BLOCK
1/2 " PLYWOOD SHEATHING W/ 10d NAILS @ 6" O.C. AT PANEL EDGES TYP. UNLESS NOTED
FIRESTOP
5/8 " DRYWALL
2x10 JOIST @ 16" O.C. CUT TOP TO PROVIDE SLOPE TO SCUPPER
THRU WALL FLASHING W/ WEEPS @ 24" O.C.
5"x3 1/2 "x 5/16" STEEL ANGLE
4'-8"
FRP
SUSPENDED CEILING
NO FELT BEHIND TILE
8x8 WALL TILE
3-2x10'S W/2- 1/2 " PLYWOOD FILLERS. GLUE AND NAIL TOGETHER TO FORM SINGLE UNIT.
4-2x12'S GLUE AND NAIL TOGETHER TO FORM SINGLE UNIT.
15'-4"
DRIP TO MATCH WINDOW FRAME
SEALANT
STAINLESS STEEL ANGLE AROUND OPENING OF PICK-UP WINDOW (BY G.C.)
AUTOMATIC PICK-UP WINDOW
4'-4" WINDOW SIZE
11'-0" LOW END BEARING ELEVATION
8'-10"
HOLD WINDOW 1" BEYOND PLYWOOD SHEATHING
SEALANT
CUT BRK. ROWLOCK
PROVIDE THRU WALL FLASHING W/ MORTAR SCREEN AND WEEP HOLES @ 24" O.C.
SPLIT-FACED BLOCK
FRP
1/2 " PLYWOOD SHEATHING
5/8 " DRYWALL
2x6 FRAMING 16" O.C.
VAPOR BARRIER (LOCATION VARIES WITH LOCAL CLIMATES)
R-19 FULL WALLPACK INSULATION
5/8 " MOISTURE RESISTANT PLYWOOD OR DUROCK
3'-0"
PROVIDE THRU WALL FLASHING W/ MORTAR SCREEN AND WEEP HOLES @ 24" O.C.
8" CONC. BLOCK
12"
4" CONC. SLAB
QUARRY TILE FLOORING
1/2 " EXPANSION JOINT
6" CONC. SLAB
1/2 " DIA. x 18" A.B. W/3" HOOKS @ 32" O.C. TYPICAL IN STUD WALLS, GROUT BLOCK SOLID
1"x24" STYROFOAM INSULATION (R-5)
2'-8" (VERIFY FROST LINE)
12" CONCRETE BLOCK
20"x8" CONC. FOOTING W/2-#4 BARS CONT.
8"
SCALE: 3/4 " = 1'-0"
G
7B

FOUNDATION NOTES

1. ALL FOOTINGS ARE 20"x8" w/2-#4 BARS, CONTINUOUS, UNLESS NOTED OTHERWISE.
2. VERIFY DEPTH OF FOOTING BELOW GRADE WITH LOCAL CODE.
3. CONCRETE BLOCK COURSING STARTS AT TOP OF FOOTING WITH CRS #1
4. SEE ELECTRICAL SHEETS FOR LOCATIONS OF CONDUIT RUNS.
5. SEE PLUMBING PLAN FOR LOCATION OF PIPING RUNS.
6. AT DOOR OPENINGS:
   - PROVIDE 3-#4 BARS AT 12" ON CENTER x 2'-0" LONG, W/8" PROJECTION AS INDICATED ON FOUNDATION PLAN
   - POUR SLAB THROUGH OPENING.
   - PROVIDE CONTROL JOINT CENTERED ON WALL.
7. ALL FOOTINGS ARE DESIGNED FOR 3000 P.S.F. SOIL BEARING PRESSURE. VERIFY SOIL BEARING CAPACITY BEFORE PLACING FOOTINGS.
8. TYPICAL IN STUD WALLS, GROUT BLOCK SOLID

FIGURE 23.25 ■ (continued)

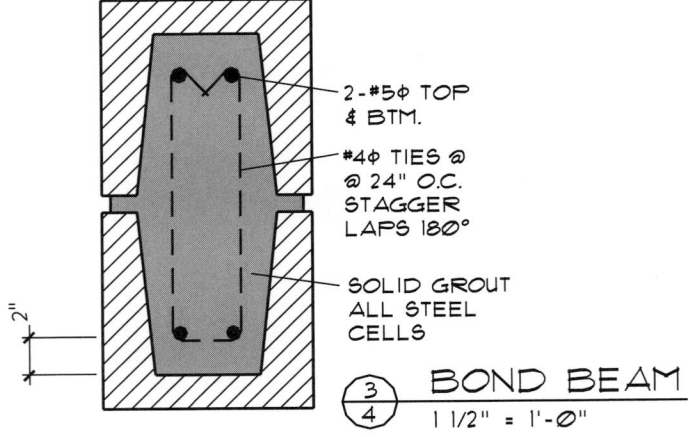

2-#5φ TOP & BTM.
#4φ TIES @ @ 24" O.C. STAGGER LAPS 180°
SOLID GROUT ALL STEEL CELLS
2"
3
4
BOND BEAM
1 1/2" = 1'-0"

FIGURE 23.26 ■ Concrete block bond beam reinforcing detail.

required, the solution is *laminated* (lam) beams, as shown in Figure 23.35, page 786. Notice the beams labeled GLB 1, GLB 2, GLB 3, and GLB 4 in Figure 23.33. These are *glu-lam beams* (GLB). The intermediate members labeled P1 and P2 are purlins, as previously discussed. Laminated beams are manufactured from smaller, equally sized members glued together to form a larger beam. Laminated beams are used to support heavy loads or accommodate long spans and are also used when the wood appearance is important. The common types of laminated beams are the *single span straight, Tudor arch,* and *three-hinged arch* beams. (See Figure 23.36, page 786.)

The single span beam is commonly referred to as glu-lam and noted on a drawing with the abbreviation GLB for glu-lam beam. A glu-lam beam can be used to replace a much larger wood timber due to increased structural qualities. Figure 23.37, page 786, compares the strength of common wood members with laminated beams. In some applications, laminated beams have a *camber* or

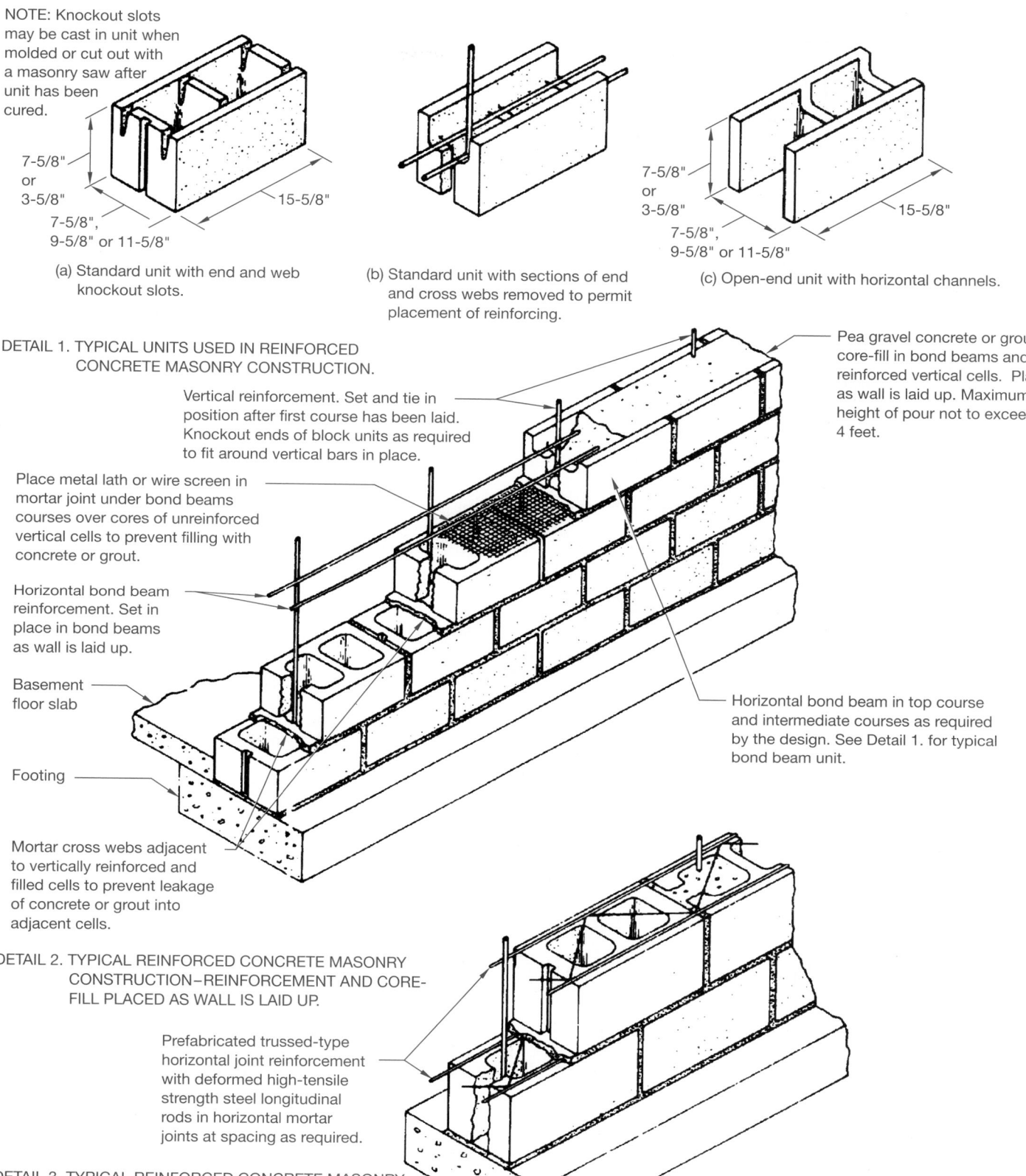

NOTE: Knockout slots may be cast in unit when molded or cut out with a masonry saw after unit has been cured.

7-5/8" or 3-5/8"

7-5/8", 9-5/8" or 11-5/8"

15-5/8"

(a) Standard unit with end and web knockout slots.

(b) Standard unit with sections of end and cross webs removed to permit placement of reinforcing.

7-5/8" or 3-5/8"

7-5/8", 9-5/8" or 11-5/8"

15-5/8"

(c) Open-end unit with horizontal channels.

DETAIL 1. TYPICAL UNITS USED IN REINFORCED CONCRETE MASONRY CONSTRUCTION.

Vertical reinforcement. Set and tie in position after first course has been laid. Knockout ends of block units as required to fit around vertical bars in place.

Place metal lath or wire screen in mortar joint under bond beams courses over cores of unreinforced vertical cells to prevent filling with concrete or grout.

Horizontal bond beam reinforcement. Set in place in bond beams as wall is laid up.

Basement floor slab

Footing

Pea gravel concrete or grout core-fill in bond beams and reinforced vertical cells. Place as wall is laid up. Maximum height of pour not to exceed 4 feet.

Horizontal bond beam in top course and intermediate courses as required by the design. See Detail 1. for typical bond beam unit.

Mortar cross webs adjacent to vertically reinforced and filled cells to prevent leakage of concrete or grout into adjacent cells.

DETAIL 2. TYPICAL REINFORCED CONCRETE MASONRY CONSTRUCTION–REINFORCEMENT AND CORE-FILL PLACED AS WALL IS LAID UP.

Prefabricated trussed-type horizontal joint reinforcement with deformed high-tensile strength steel longitudinal rods in horizontal mortar joints at spacing as required.

DETAIL 3. TYPICAL REINFORCED CONCRETE MASONRY CONSTRUCTION USING HORIZONTAL JOINT REINFORCEMENT IN LIEU OF BOND BEAMS TO PROVIDE LATERAL REINFORCEMENT.

FIGURE 23.27 ▪ Suggested construction details for reinforced concrete masonry foundation walls. *Courtesy National Concrete Masonry Association.*

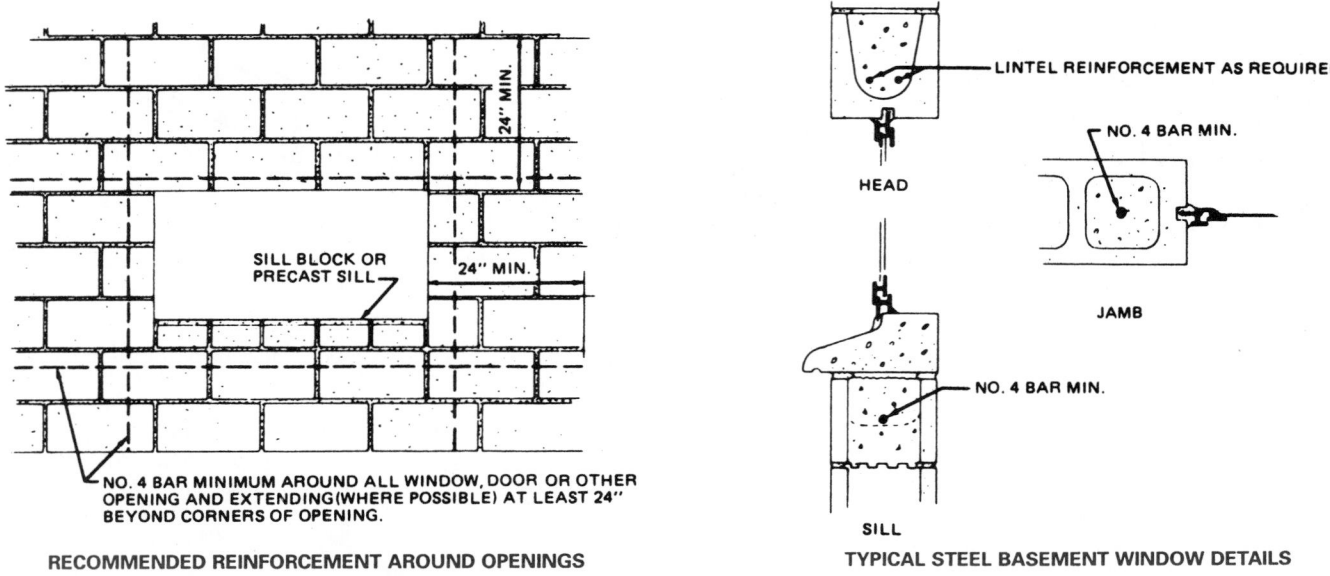

FIGURE 23.28 ■ Typical reinforcing and window detail at opening in reinforced concrete masonry. *Courtesy National Concrete Masonry Association.*

RECOMMENDED REINFORCEMENT AROUND OPENINGS

SILL BLOCK OR PRECAST SILL

24" MIN.

24" MIN.

NO. 4 BAR MINIMUM AROUND ALL WINDOW, DOOR OR OTHER OPENING AND EXTENDING (WHERE POSSIBLE) AT LEAST 24" BEYOND CORNERS OF OPENING.

TYPICAL STEEL BASEMENT WINDOW DETAILS

LINTEL REINFORCEMENT AS REQUIRED

HEAD

NO. 4 BAR MIN.

JAMB

NO. 4 BAR MIN.

SILL

8 × 8 × 16

Alternate courses

HOLLOW UNIT

4 × 8 × 16 solid units laid flat

4 × 8 × 16 solid units set on end

Alternate courses

SOLID UNIT

Special pilaster unit

Hollow cores filled with concrete or mortar

Alternate courses

FILLED CELL HOLLOW UNIT

12 × 8 × 16

Alternate courses

HOLLOW UNIT

4 × 8 × 16 solid units set on end

Alternate courses

4 × 8 × 16 solid units laid flat

SOLID UNIT

Special pilaster unit with hollow core reinforced

Wood sash jamb unit

REINFORCED FILLED CELL HOLLOW UNIT

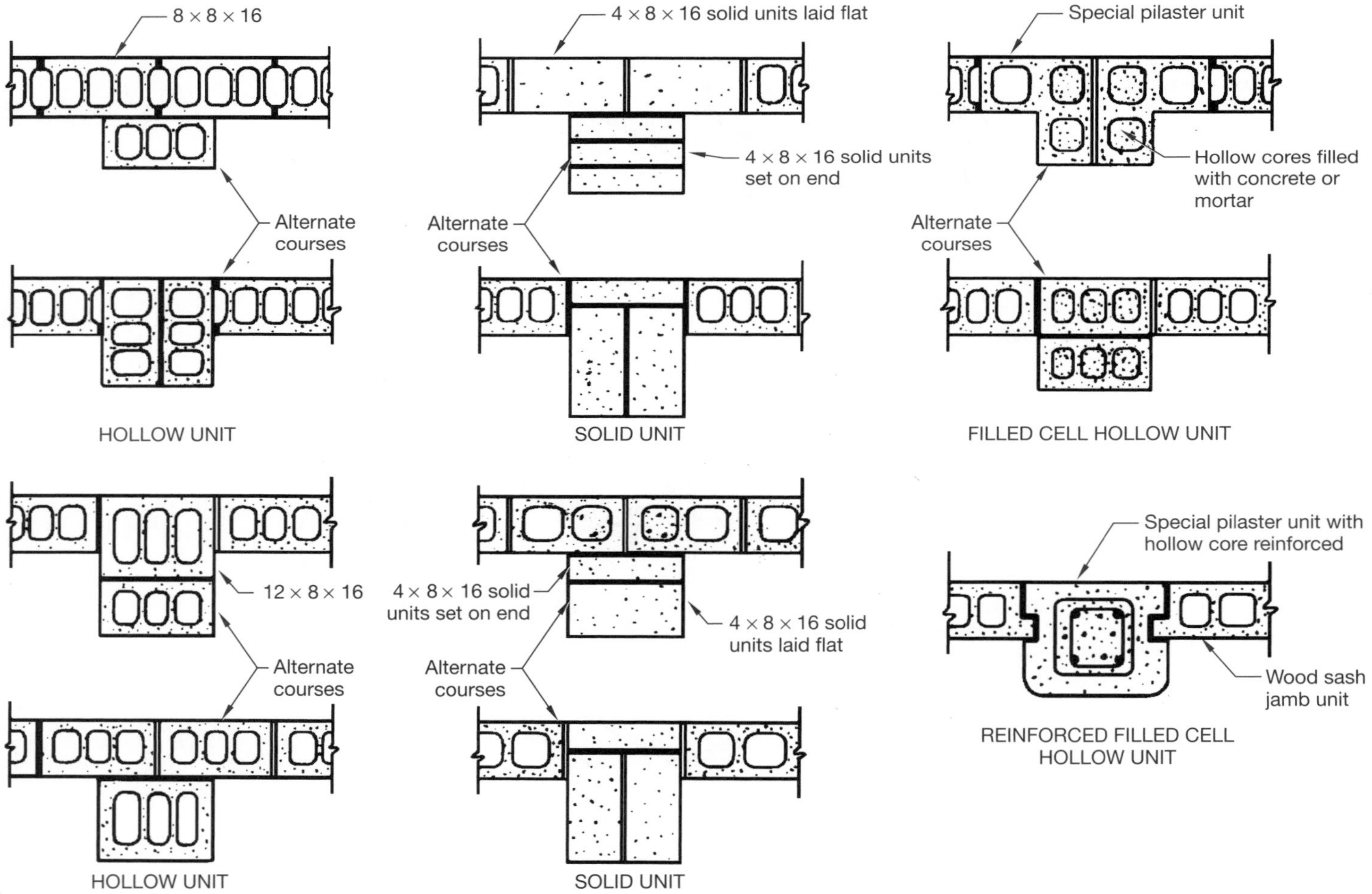

FIGURE 23.29 ■ Typical concrete masonry pilaster designs. *Courtesy National Concrete Masonry Association.*

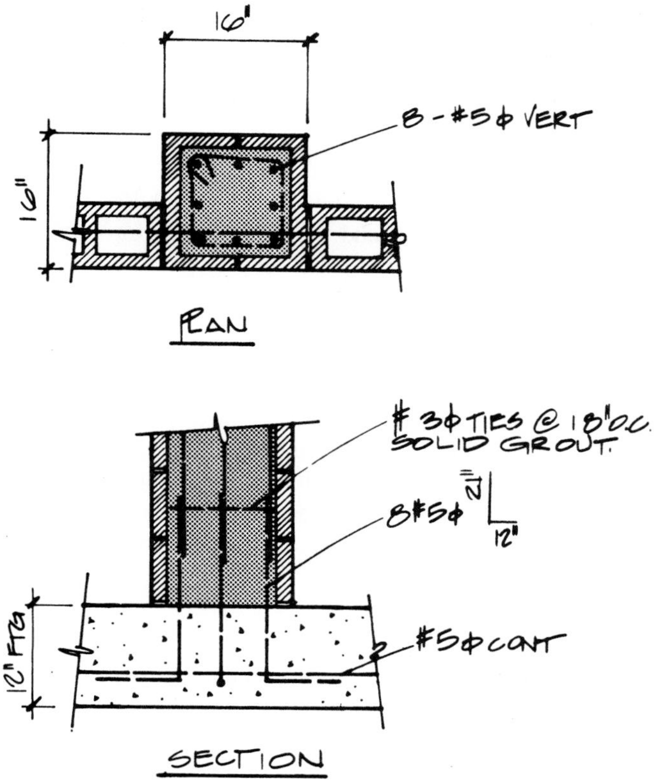

16"

16"

B - #5 φ VERT

<u>PLAN</u>

#3 φ TIES @ 18" O.C.
SOLID GROUT.

8 #5 φ
12"

12" FTG

#5 φ CONT

<u>SECTION</u>

FIGURE 23.30 ■ Reinforced pilaster detail.

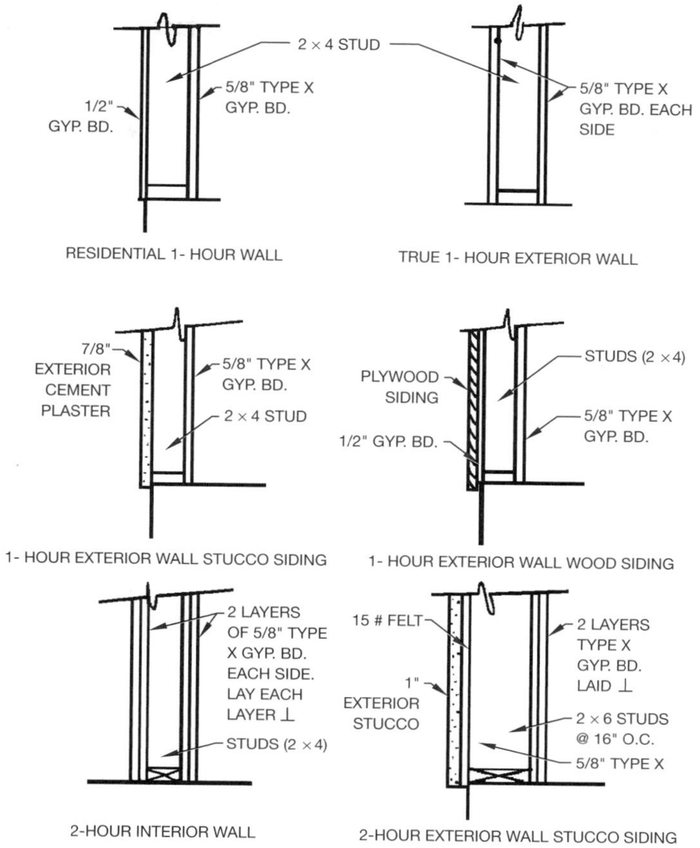

1/2" GYP. BD.

2 × 4 STUD

5/8" TYPE X GYP. BD.

5/8" TYPE X GYP. BD. EACH SIDE

RESIDENTIAL 1- HOUR WALL

TRUE 1- HOUR EXTERIOR WALL

7/8" EXTERIOR CEMENT PLASTER

5/8" TYPE X GYP. BD.

2 × 4 STUD

PLYWOOD SIDING

STUDS (2 ×4)

1/2" GYP. BD.

5/8" TYPE X GYP. BD.

1- HOUR EXTERIOR WALL STUCCO SIDING

1- HOUR EXTERIOR WALL WOOD SIDING

2 LAYERS OF 5/8" TYPE X GYP. BD. EACH SIDE. LAY EACH LAYER ⊥

STUDS (2 ×4)

15 # FELT

1" EXTERIOR STUCCO

2 LAYERS TYPE X GYP. BD. LAID ⊥

2 × 6 STUDS @ 16" O.C.

5/8" TYPE X

2-HOUR INTERIOR WALL

2-HOUR EXTERIOR WALL STUCCO SIDING

FIGURE 23.31 ■ Separation walls often require special treatment to achieve the needed fire rating for certain types of construction.

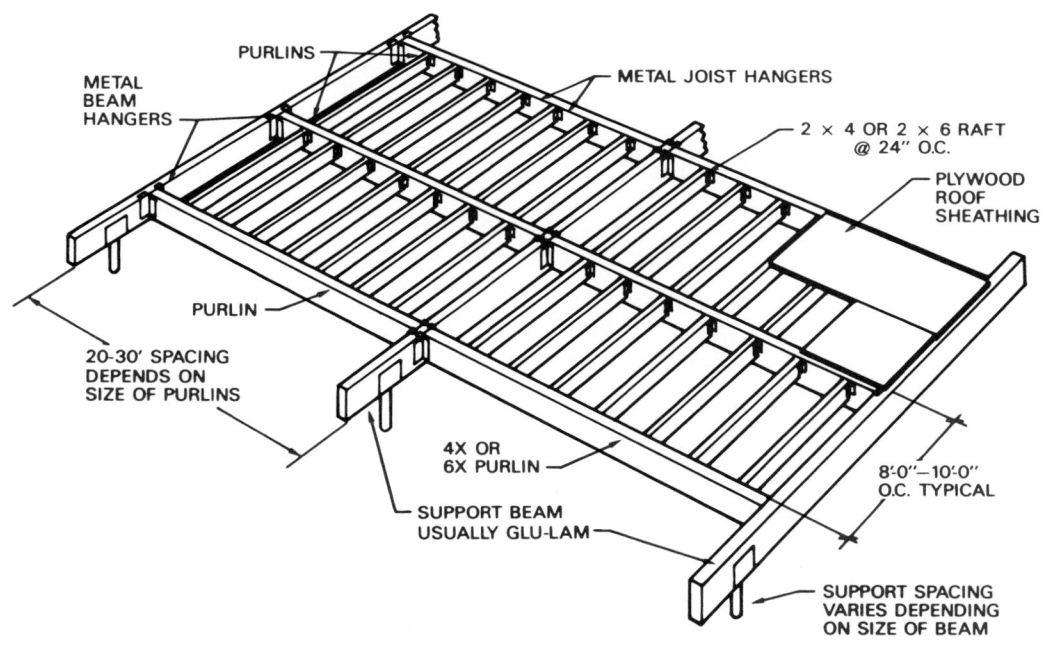

METAL BEAM HANGERS

PURLINS

METAL JOIST HANGERS

2 × 4 OR 2 × 6 RAFT @ 24" O.C.

PLYWOOD ROOF SHEATHING

PURLIN

20-30' SPACING DEPENDS ON SIZE OF PURLINS

4X OR 6X PURLIN

SUPPORT BEAM USUALLY GLU-LAM

8'-0"—10'-0" O.C. TYPICAL

SUPPORT SPACING VARIES DEPENDING ON SIZE OF BEAM

FIGURE 23.32 ■ A panelized roof system is often used to provide roofing for large area with limited supports.

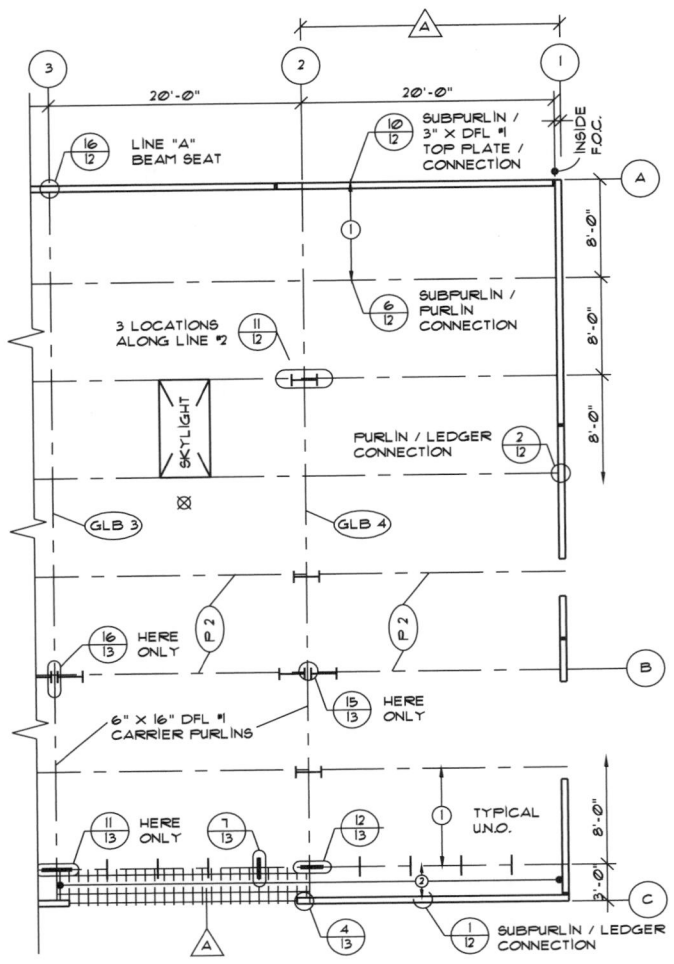

FIGURE 23.33 ■ A panelized roof and heavy timber framing plan. *Courtesy Structureform Masters, Inc.*

curve built into the beam. The camber is designed into the beam to help resist the loads to be carried.

The Tudor and three-hinged arch members are a post-and-beam system combined into one member. These beams are specified on plans in a method similar to other beams. The drafter's major responsibility when working with either heavy timber construction or laminated beams is in the drawing of connection details. Beam to beam, beam to column, and column to support are among the most common details drawn by drafters. Common types of manufactured connectors used to connect timbers are shown in Figure 23.38. The drafter is required to draw the fabrication details for a connector when the size of the beam does not match existing connectors, as shown in Figure 23.3a (see page 767).

Wood members can be attached to concrete in several ways. Two of the most common methods are by the use of a pocket or seat, as shown in Figure 23.39, or a metal connector.

Engineered Wood Products

Engineered wood products are a variety of products that have been designed to replace conventional lumber and provide many advan-

tages as well as reduce the industry dependence on natural lumber. Common types of engineered wood products include I-joists, laminated veneer lumber (LVL), and related products such as rim board.

I-Joists

I-joists are generally made of softwood veneers such as fir or pine that are bonded together or solid wood to make the top and bottom flanges. The web or core is then made of composite wood. The name comes from the I shape that is shown in Figure 23.40, page 788.

I-joists provide stronger, more stable performance; are lighter weight, easier to handle, and faster to install; and make more efficient use of valuable wood resources than conventional lumber. They also provide longer lengths and prestamped knockouts for wire or plumbing runs. I-joists are drawn just like other joists in the plan view using solid lines or centerlines. The shape of the I-joist showing the top and bottom flanges and the core is displayed when drawing in sectional view.

Laminated Veneer Lumber

Laminated veneer lumber (LVL) is an engineered structural member that is manufactured by bonding wood veneers with an exterior adhesive. LVL has almost no shrinking, checking, twisting, or splitting; it is excellent for floor and roof framing supports or as headers for doors, windows, and garage doors and columns. LVL has no camber, so LVL products provide flatter, quieter floors. This lumber is available in lengths that are longer than conventional lumber, so builders can avoid on-site splicing and multiple nailing, and save on labor costs. Common sizes range from 1-3/4" to 7" wide, with depths from 5 1/2" to 18" and lengths of up to 66'. Figure 23.41, page 788, shows an example of LVL sizes. Figure 23.42, page 788, shows a comparison of physical properties of glu-laminated lumber, laminated veneer lumber, and solid lumber. You can see the physical advantage of the LVL product. The physical properties vary depending on the type of material and the manufacturer. LVL members are drawn like other beams, using a thick solid line or a thick centerline symbol.

Rim Board

As part of a complete engineered wood construction system, rim board is used to cap the ends of the I-joist construction just like the rim joist used in conventional framing systems. The advantages of engineered rim joists is that they exactly match I-joist depths, require no special backing for siding products, and are quick and easy to install. Rim boards have almost no shrinking, checking, twisting, or splitting. Rim boards cannot be used as headers or beams.

Plywood Lumber Beams

Plywood lumber beams are also an engineered wood product that are made to precise specifications in size and nailing requirements for specific applications. Plywood lumber beams are also known as box beams. You can see their boxlike construction in Figure 23.43, page 788.

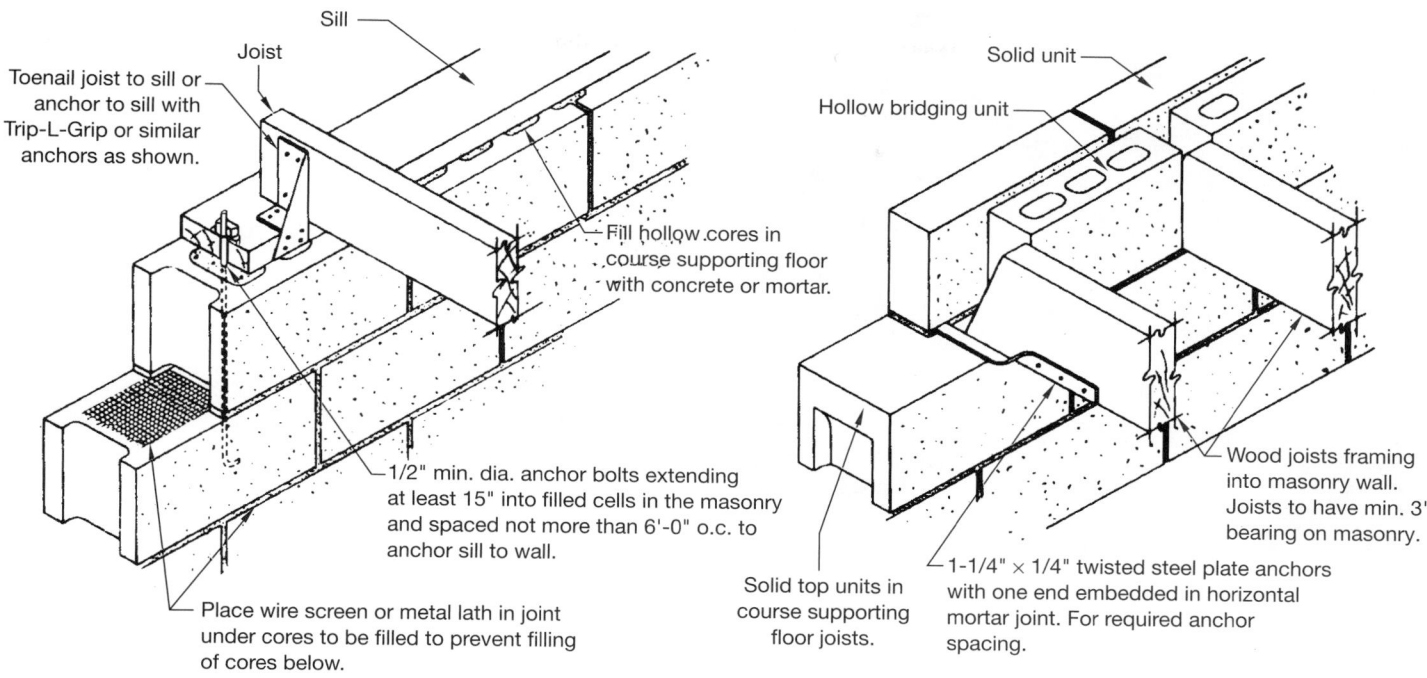

Toenail joist to sill or anchor to sill with Trip-L-Grip or similar anchors as shown.

Sill

Joist

Fill hollow cores in course supporting floor with concrete or mortar.

1/2" min. dia. anchor bolts extending at least 15" into filled cells in the masonry and spaced not more than 6'-0" o.c. to anchor sill to wall.

Place wire screen or metal lath in joint under cores to be filled to prevent filling of cores below.

ANCHORAGE OF WOOD JOISTS TO FOUNDATION IN WOOD FRAME CONSTRUCTION

Solid unit

Hollow bridging unit

Wood joists framing into masonry wall. Joists to have min. 3" bearing on masonry.

1-1/4" × 1/4" twisted steel plate anchors with one end embedded in horizontal mortar joint. For required anchor spacing.

Solid top units in course supporting floor joists.

ANCHORAGE OF WOOD JOISTS TO FOUNDATION IN MASONRY CONSTRUCTION

SUGGESTED METHODS OF ANCHORING WOOD JOISTS BEARING ON CONCRETE MASONRY FOUNDATION WALLS.

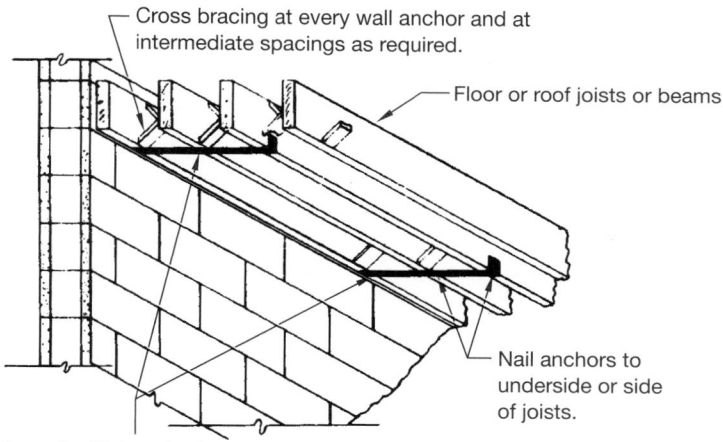

Cross bracing at every wall anchor and at intermediate spacings as required.

Floor or roof joists or beams

Nail anchors to underside or side of joists.

Wall anchors at required intervals. Anchors should have split end embedded in mortar joint or end bent down into block core and core filled with mortar. Length of anchor should be sufficient to engage at least three joists.

TYPICAL DETAILS FOR ANCHORAGE OF CONCRETE MASONRY WALLS TO PARALLEL WOOD JOISTS OR BEAMS.

FIGURE 23.34 ■ Metal angles and straps are typically used to ensure a rigid connection between the wall and floor or roof system. *Courtesy National Concrete Masonry Association.*

FIGURE 23.35 ■ Timbers are often used in commercial construction because of their beauty and structural qualities. The structural system of St. Philip's Episcopal Church is formed by glue-laminated Southern pine arches and beams under a sweeping canopy of Southern pine roof decking. *Courtesy Southern Forest Products Association.*

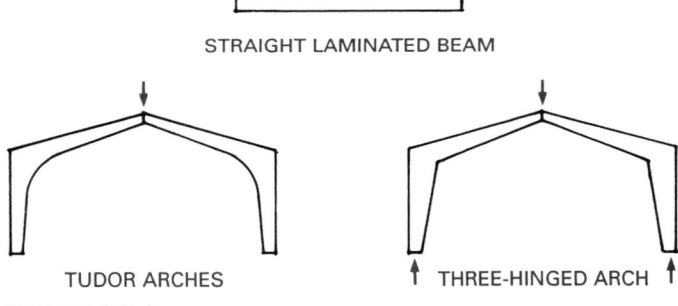

FIGURE 23.36 ■ Common laminated beam shapes.

Stressed Skin Panels

Stressed skin panels use solid lumber or engineered lumber stringers and headers with a plywood skin on the top and bottom as shown in Figure 23.44. These panels can be used for walls, floors, and roof systems. They allow the builder to quickly erect a building because they cover a large area. They are commonly used in remote areas where they can be shipped in and quickly installed.

Standard Wood Structural Callouts

When specifying the callouts for structural wood sizes on a drawing, give the *nominal* (rough, before planing) cross-sectional dimensions for sawn lumber and timbers. For example:

$$4 \times 12, 6 \times 14$$

The *net* (actual) cross-sectional dimension is given for glu-lams. The net dimensions of a glu-lam are followed by the abbreviation GLB and any other specification. For example:

$$6\text{-}1/8 \times 14 \text{ GLB } f2400$$
$$(f2400 = \text{actual units of stress in fiber bending})$$

If the net cross-sectional dimensions for lumber are required, they may be given in parentheses after the nominal dimensions. If lengths (LG) are required for wood members, they should be given in feet and inches. For example:

$$6\text{-}3/4 \times 18 \text{ GLB} \times 24'\text{-}0'' \text{ LG}$$

Glu-lam beams can also be designated with a manufacturing number such as:

$$6\text{-}3/4 \times 18 \text{ 22FV4}$$

The $6\text{-}3/4 \times 18$ gives the width and depth as usual. The 22FV4 designation represents the following values:

22F This is the fiber bending, f or F_b as given in previous examples. Different wood species have different fiber bending values.

V4 This is a designation that varies depending on the wood species and the application for the beam. This specification

SPECIS AND COMMERCIAL GRADE	EXTREME FIBER BENDING F_b	HORIZONTAL SHEAR F_v	COMPRESSION PERPENDICULAR TO GRAIN $F_c \perp$	MODULUS OF ELASTICITY E
DFL #1	1350	85	385	1,600,600
22FV4				
DF/DF	2200	165	385	1,700,000
Hem-Fir #1	1050	70	245	1,300,000
22F E2				
HF/HF	2200	155	245	1,400,000
SPF	900	65	265	1,300,000
22F-E-1				
SP/SP	2200	200	385	1,400,00

FIGURE 23.37 ■ Comparative values of common framing lumber with laminated beams of equal material. Values based on the *Uniform Building Code.*

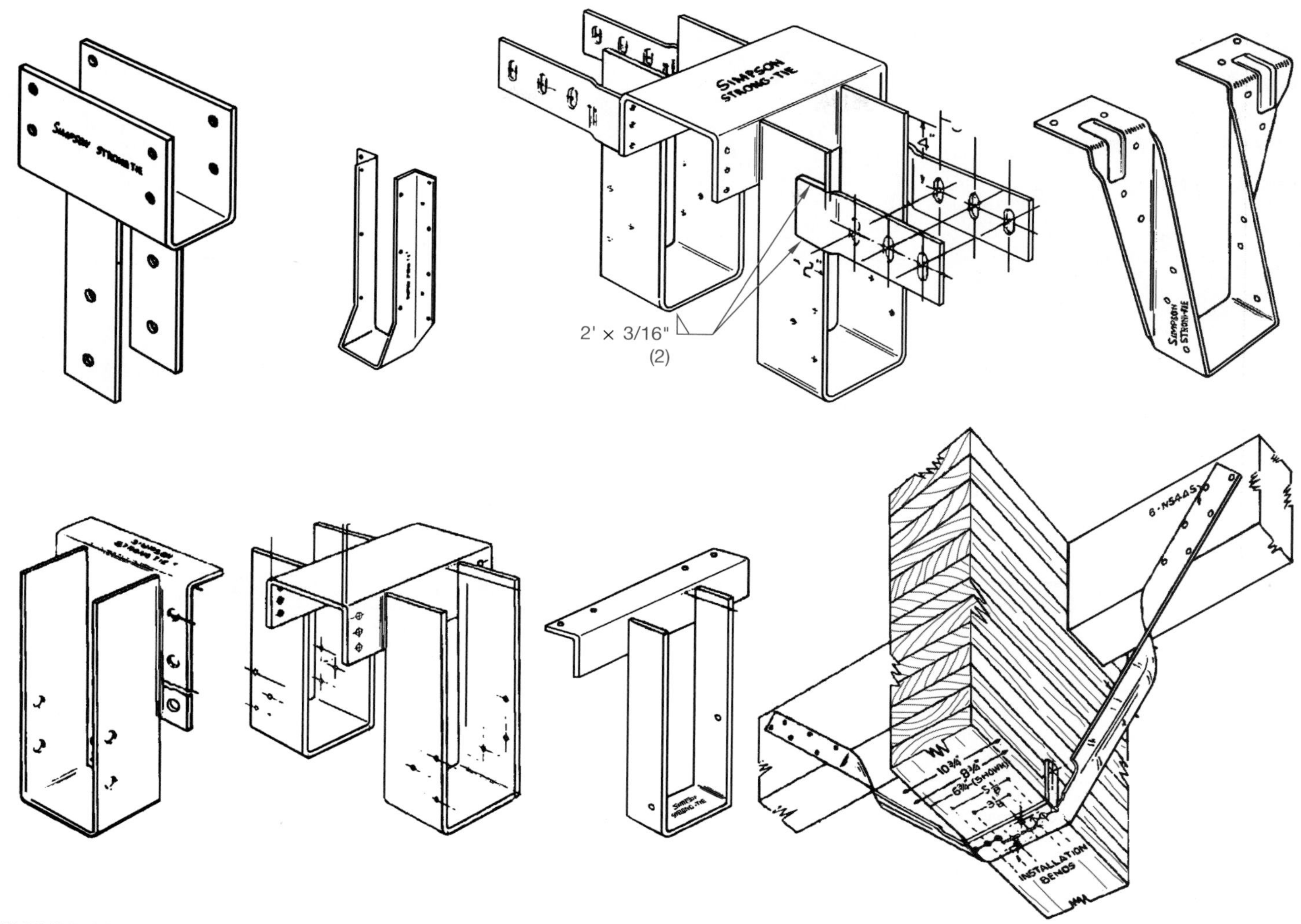

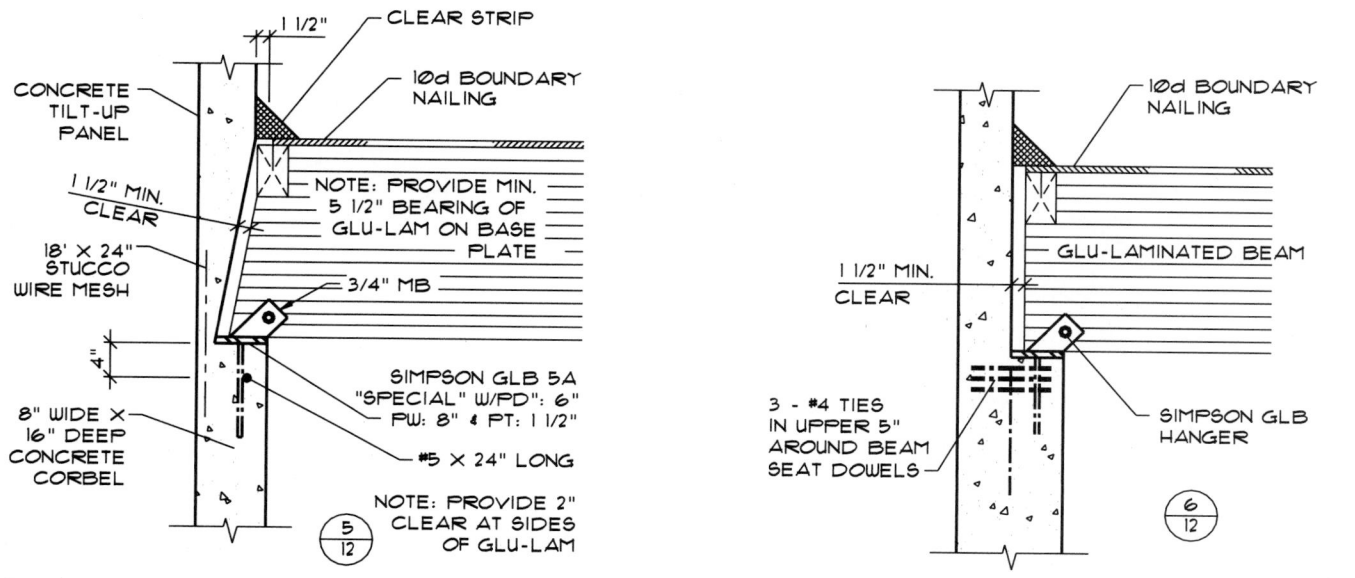

FIGURE 23.38 ■ Common manufactured metal beam connectors. *Courtesy Simpson Strong-Tie Company, Inc.*

FIGURE 23.39 ■ Wood beams are often designed to rest on a ledge or pocket. *Courtesy Structureform Masters, Inc.*

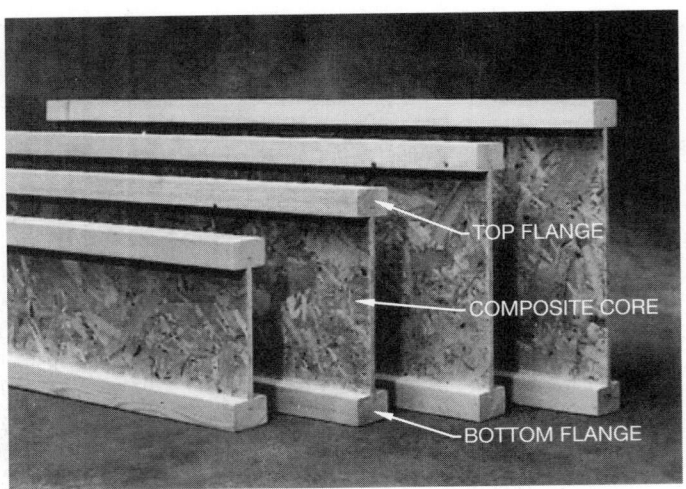

FIGURE 23.40 ■ ■ The I-joist has top and bottom flanges connected by a composite wood core. *Courtesy Louisiana-Pacific.*

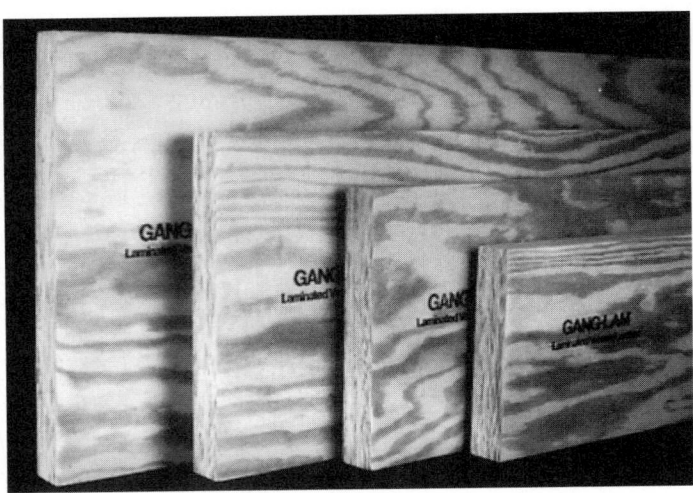

FIGURE 23.41 ■ Laminated veneer lumber (LVL) is manufactured by bonding layers of wood veneer with an exterior adhesive. *Courtesy Louisiana-Pacific.*

Property	LVL [a]	Glued Laminated [b]	Solid Lumber [c]
Extreme fiber in bending (F_b) psi	3000	1600–2400	760–2710
Horizontal shear (H_v) psi	290	140–650	95–120
Compression perpendicular to grain $(F_{c\perp})$ psi	3180	375–650	565–660
Compression parallel with grain (F_c) psi	—	1350–2300	625–1100
Tension parallel with grain (F_t) psi	2300	1050–1450	450–1200
Modulus of elasticity (E) million psi	2.0	1.2–1.8	1.3–1.9

[a] Limited to one size member
[b] Data from one manufacturer
[c] Data for one species of lumber

FIGURE 23.42 ■ Properties of laminated veneer lumber, glu-laminated members, and solid lumber.

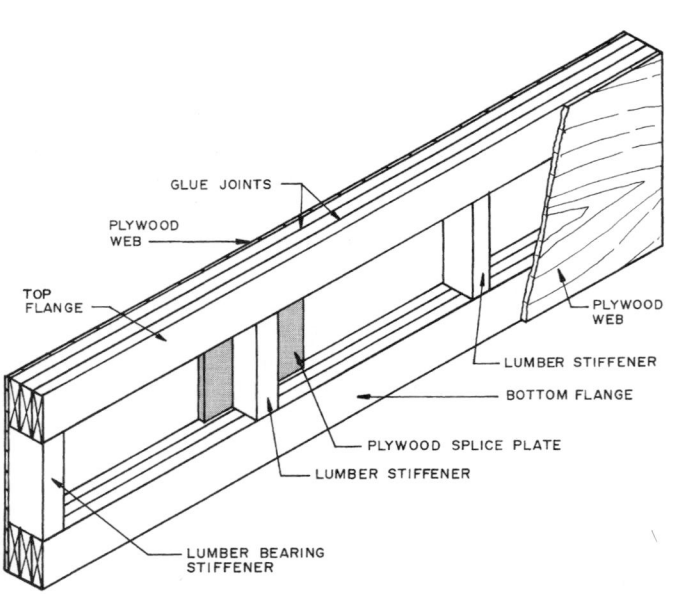

FIGURE 23.43 ■ A typical plywood-lumber beam (box beam) with plywood webs and solid wood flanges and stiffeners.

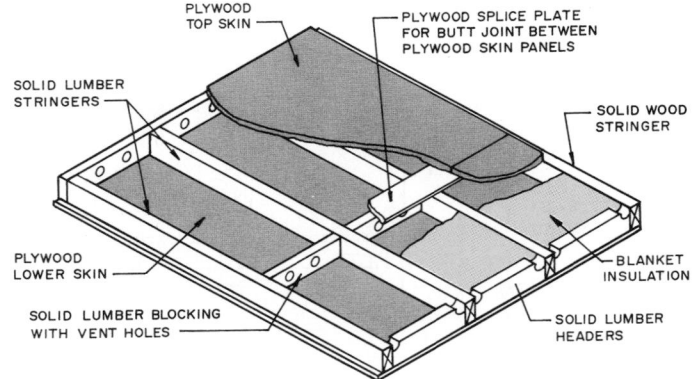

FIGURE 23.44 ■ A stress skin panel.

might be V1, V2, V3, or E1, E2, E3, for example. The options include the species of laminations outside and inside of the beam. This also relates to the application for the beam, such as down loading, up loading, and cantilever. *Down loading* refers to a beam that is designed to support weight from above. *Up loading* refers to a beam that can accept forces that push up on the beam, such as wind loads or the forces of weight applied to an adjacent cantilever. Some beams are designed to accept both down loading and up loading. *Cantilever* beams are designed to project beyond their supporting wall or column. Refer to building codes for glu-lam beams or manufacturers' catalogs for specific information about these glu-lam beam designations.

The specifications for laminated veneer lumber (LVL) vary slightly between manufacturers, but the elements of the note are similar. For example, an LVL beam or header might be specified like this:

1 3/4" × 14" BCI VERSALAM — D

The 1 3/4" × 14" is the width and depth, and the other elements are described as follows:

BCI   This is the manufacturer, in this case Boise Cascade Corporation.

VERSALAM   This is a specific LVL product of the Boise Cascade Corporation called Versa-Lam®.

D   This identifies the wood species, Douglas fir in this case.

The specifications for I-joists can also vary between manufacturers, but there are similar components, such as:

14" BCI 90XL @16" OC.

The following identifies each of the elements of this I-joist note:

14"   This is the depth of the I-joist.

BCI   This is the manufacturer, in this case Boise Cascade Corporation.

90XL   This identifies the strength characteristics based on a specific BCI I-joist.

@16" OC.   This is the spacing on center of the I-joists.

When wood is used for floor or ceiling joists, rafters or trusses, or stud walls, the member size should be followed by the OC spacing in inches up to 24" OC and feet and inches for over 24" OC. For example, abbreviations are shown in parentheses:

2 × 8 CEILING JOISTS (CJ) 16" OC,
4 × 14 @ 4'–0" OC
2 × 6 STUDS 24" OC,
2 × 12 FLOOR JOISTS (FJ) 12" OC

All lumber and timber on a drawing should be dimensioned to the centerlines unless dimensioning to the face of the member is otherwise required by the engineer or architect. The exception to this rule is dimensioning to the top of a beam or other structural timber.

When specifying plywood on a drawing, give the thickness, group, face veneer grades, identification index (if required), and glue type for each different panel used. Also include nailing, blocking, and edge spacing requirements if specific applications are required. Dimension to the face of the panel when surface location dimensions are required. The following are plywood callout examples:

1. 1/2" CDX 32/16 PLYWOOD SHEATHING.

2. 1/2" CDX SHTG W/10d NAILS 3" OC @ SEAMS AND 8" OC FIELD. (Note: *SEAMS* = edges or splices; *FIELD* = within the sheet at supports.)

3. 1/2" GROUP 1, CC, 24/0 EXTERIOR APA PLYWOOD W/8d NAILS 4" OC @ EDGES AND @ 6" OC FIELD.

4. 5/8" GROUP 2, UNDERLAYMENT CC-PTS EXTERIOR APA PLYWOOD W/8d RINGSHANK NAILS @ 3" OC EDGES AND 6" OC FIELD, BLOCK ALL PANEL EDGES PERPENDICULAR TO SUPPORTS. (Note: *CC* = CC grade outside veneer, *PTS* = plugged and touch sanded.)

When lumber decking is used, the lumber size and specification is followed by nailing information, if required. For example:

3 × 8 T&G RANDOM CONTROLLED DECKING W/20d TOE-NAIL EACH SUPPORT AND 30d RINGSHANK FACE NAIL EACH SUPPORT.

Random controlled means various lengths, usually 4'–0" modules placed so there are not two adjacent splices in the same support. *T&G* refers to tongue and groove.

STEEL CONSTRUCTION

Steel construction can be divided into three categories: steel studs, prefabricated steel structures, and steel-framed structures.

Steel Studs

Prefabricated steel studs are used in many types of commercial structures. Steel studs offer lightweight, noncombustible, corrosion-resistant framing for interior partitions and load-bearing exterior walls up to four stories high. Steel members are available for use as studs or joists. Members are designed for rapid assembly and are predrilled for electrical and plumbing conduits. The standard 24" spacing reduces the number of studs required by about one-third when compared with wood studs spaced 16" OC. Steel stud widths range from 3-5/8" to 10", but can be manufactured in any width. The material used to make studs ranges from 12- to 20-gage steel depending on the design loads to be supported. Steel studs are mounted in a channel track at the top and bottom of the wall or partition. This channel serves to tie the studs together at the ends. Horizontal bridging is often placed through the predrilled holes in the studs and then welded to the studs to serve as fire blocking within walls and as mid supports. The components of steel stud framing are shown in Figure 23.45.

Prefabricated Steel Structures

Prefabricated or metal buildings, as they are often called, have become a common type of construction for commercial and agricultural structures in many parts of the country. Drafters who are involved in the preparation of drawings for premanufactured structures may work in a structural engineering office or for a building manufacturer. Standardized premanufactured steel buildings are sold as modular units with given spans, wall heights, and lengths in 12' or 20' increments. Most manufacturers provide a wide variety of design options for custom applications that may be required by the client. One advantage of these structures is faster erection time as compared with other construction methods.

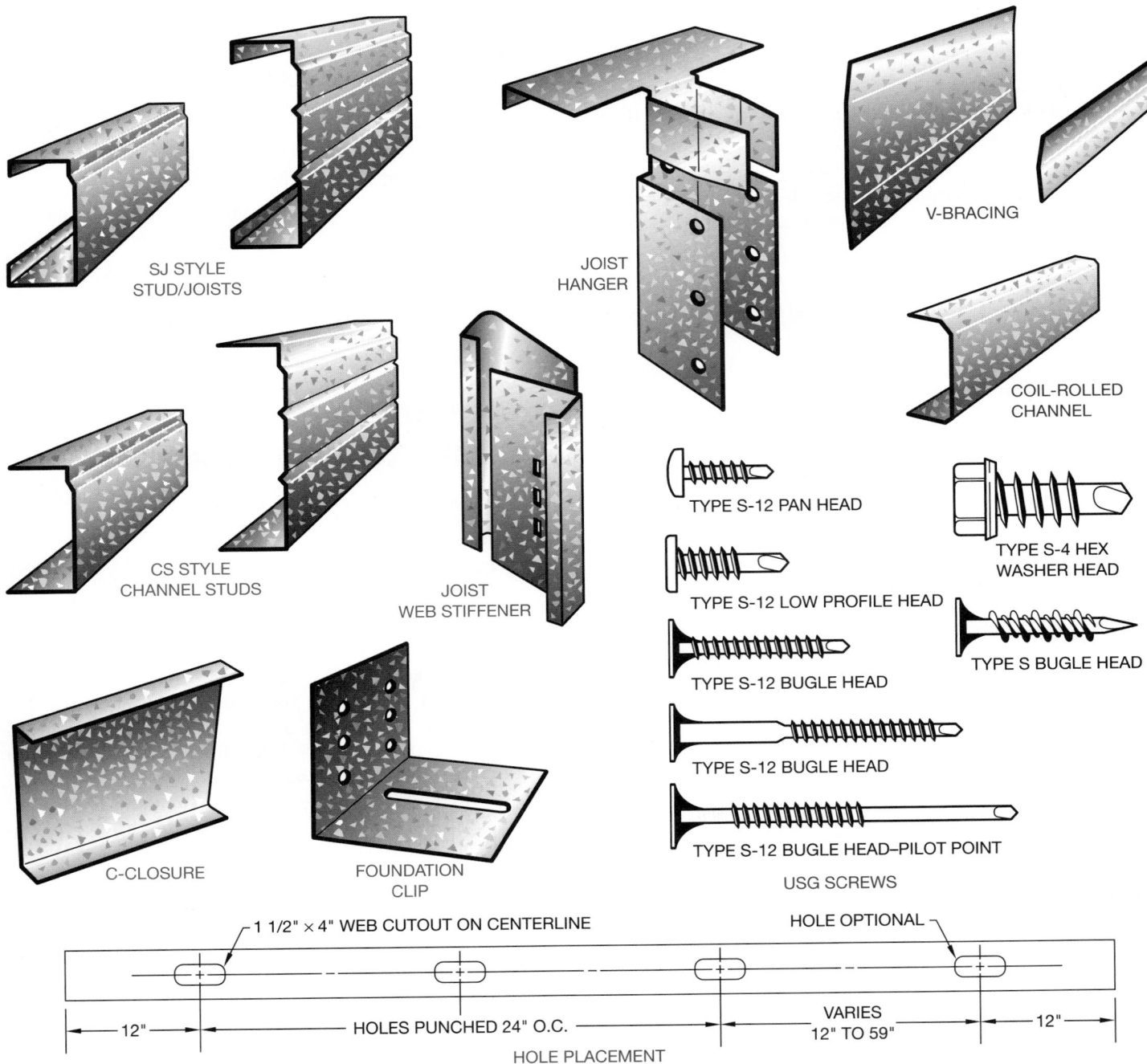

FIGURE 23.45 ■ Common components of steel stud construction. *Courtesy United States Gypsum Company.*

The Structural System

The structural system is made up of the frame that supports the walls and roof. There are several different types of structural systems commonly used, as shown in Figure 23.46. The wall system is horizontal *girts* attached to the vertical structure and metal wall sheets attached to the girts. The roof system is horizontal purlins attached to the structure and metal sheets attached to the purlins. (See Figure 23.47.) Steel wall and roof sheets are available from many vendors in a variety of patterns and may be purchased plain, galvanized, or prepainted. A sample pattern design is shown in Figure 23.48.

Steel-framed Structures

Steel-framed buildings require structural engineering and shop drawings similar to those used for concrete structures. As a drafter in an engineering or architectural firm, you will most likely be drafting engineering drawings similar to the one shown in Figure 23.49.

AISC Drafters in this field must become familiar with the *Manual of Steel Construction* published by the American Institute of Steel Construction, Inc. (AISC). In addition to

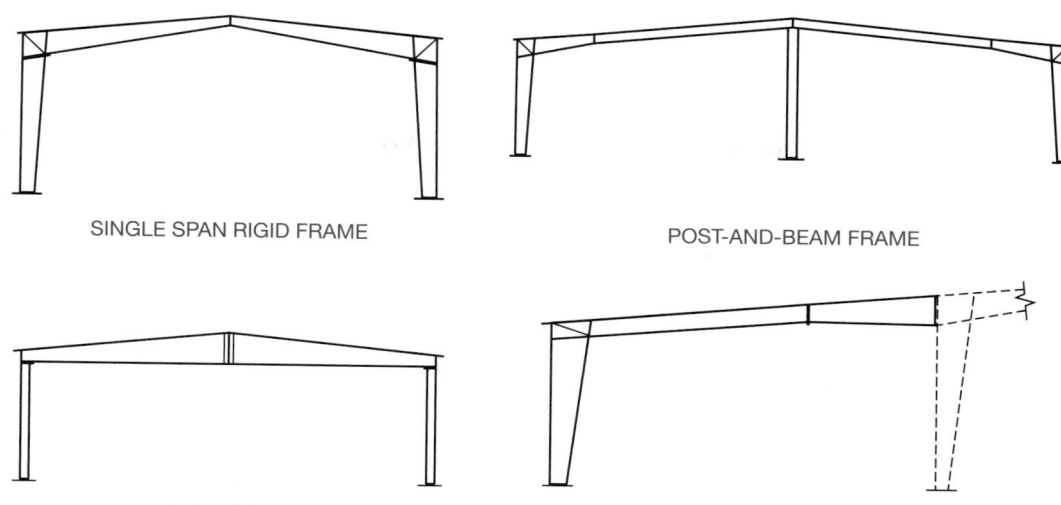

SINGLE SPAN RIGID FRAME

POST-AND-BEAM FRAME

TAPERED BEAM

LEAN-TO FRAME

FIGURE 23.46 ■ Common prefabricated structural systems.

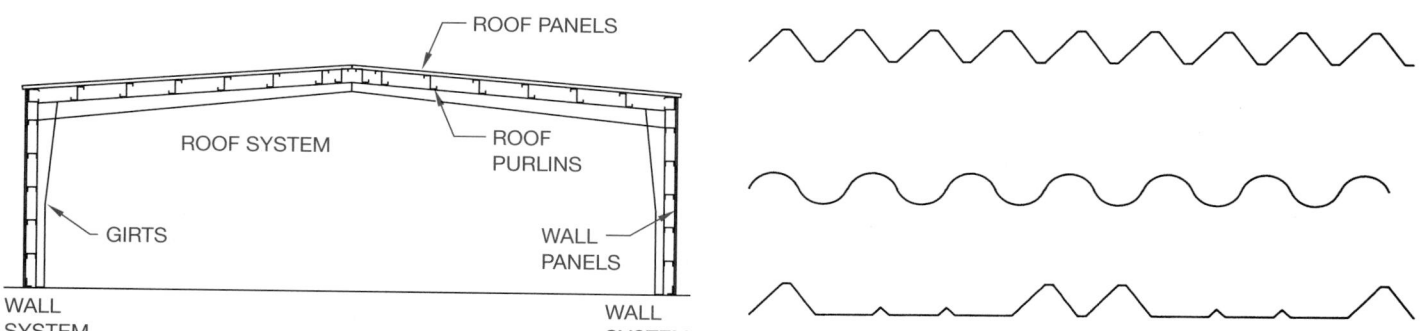

FIGURE 23.47 ■ Components of the prefabricated structural system.

FIGURE 23.48 ■ Typical cross sections of sheet metal siding and roofing material; many pattern shapes and finish colors are available.

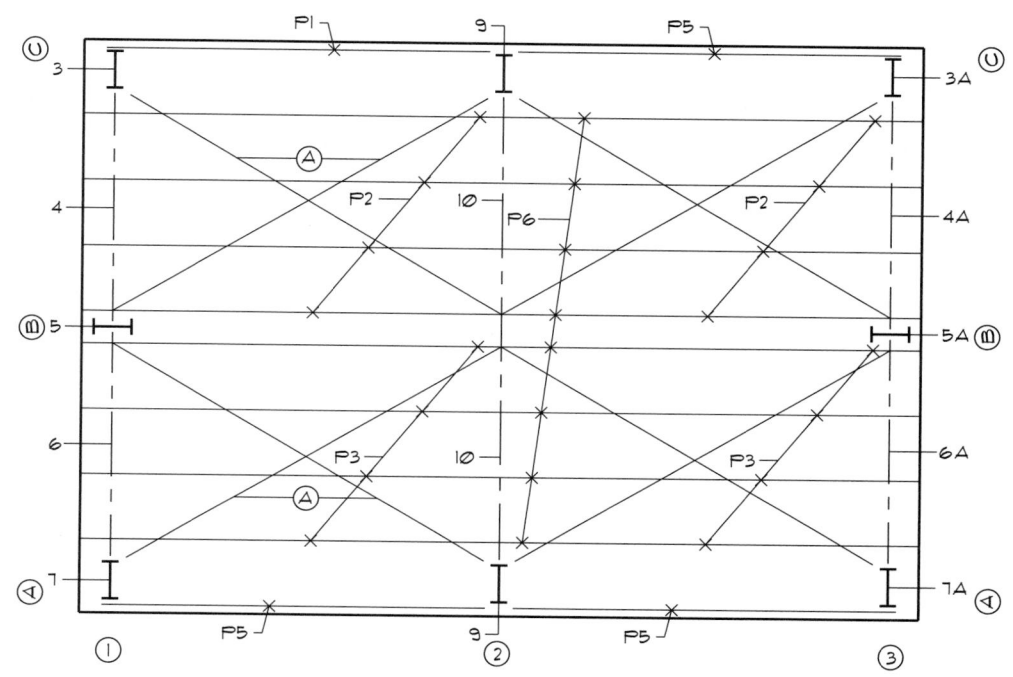

FIGURE 23.49 ■ Structural steel engineering drawing. *Courtesy Pacific Building Systems.*

the local and national building codes, this manual is one of your prime references, as it helps you determine dimensions and properties of common steel shapes. Another manual that provides information on dimensions for detailing and properties for design work related to steel structural materials is *Structural Steel Shapes,* published by U.S. Steel Corporation.

Common Structural Steel Materials

Structural steels are commonly identified as plates, bars, or shape configurations. *Plates* are flat pieces of steel of various thickness used at the intersection of different members and for the fabrication of custom connectors. Figure 23.50 shows an example of a steel connector that uses top, side, and bottom plates. Plates are typically specified on a drawing by giving the thickness, width, and length in that order. The symbol ℙ is often used to specify plate material. For example:

ℙ 1/4 × 6 × 10 (with or without inch marks)

Bars are the smallest of structural steel products and are manufactured in round, square, rectangular, flat, or hexagonal cross sections. Bars are often used as supports or braces for other steel parts or connectors. Flat bars are usually specified on a drawing by giving the width, thickness, and length in that order. For example:

BAR 3 × 1/2 × 1'–6"

Structural steel is also available in several different manufactured shapes, as shown in Figure 23.51. When specifying a steel shape on a drawing, the shape identification letter is followed by the member's depth, the "by" sign (×), and the weight in number of pounds per linear foot. For example:

W 12 × 22 or C 6 × 10.5

as shown in Figure 23.52. In the AISC *Manual of Steel Construction,* specific information regarding dimensions for detailing and dimensioning is clearly provided along with typical connection details.

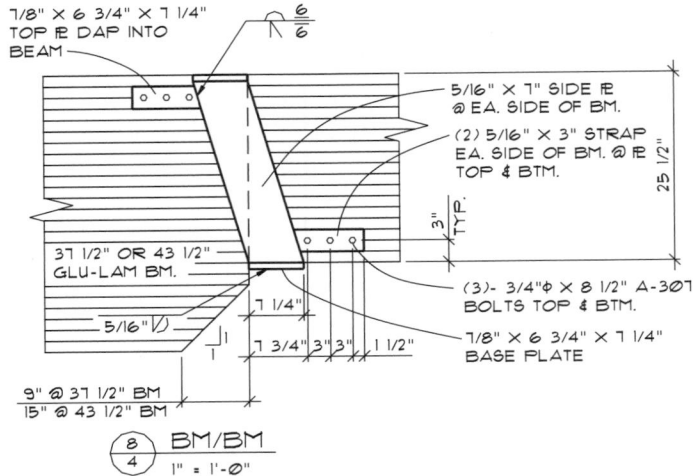

FIGURE 23.50 ■ Steel plates used to fabricate a beam connector.

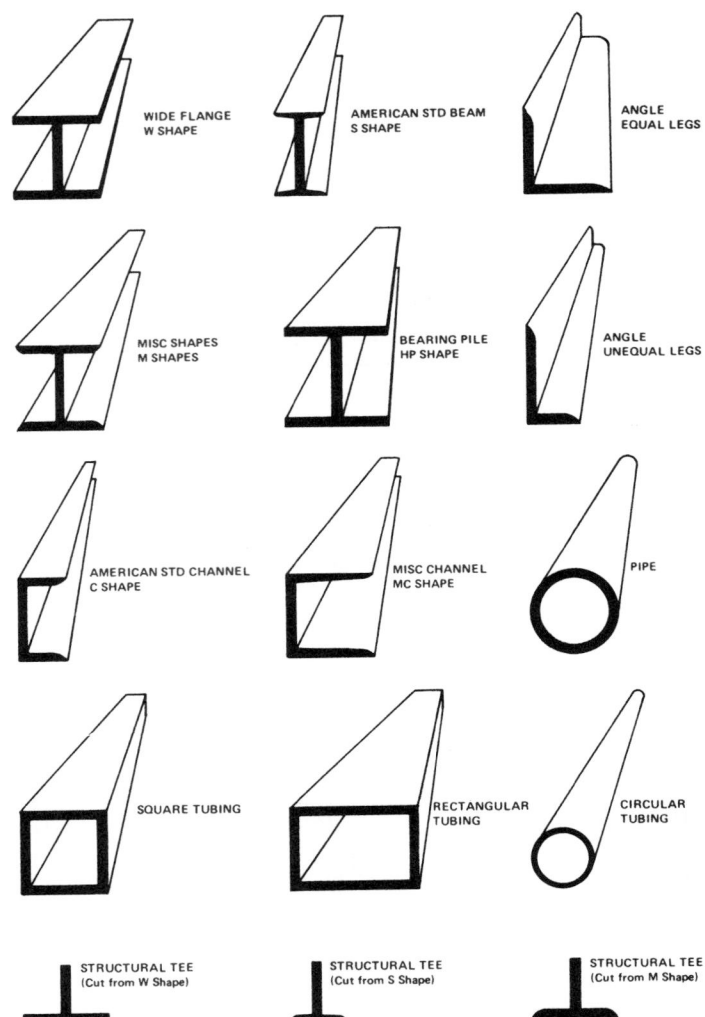

FIGURE 23.51 ■ Standard structural steel shapes.

The representative pages for the W 12 × 22 wide flange and the C 6 × 10.5 channel from Figure 23.52 are shown in Figure 23.53. The W, S, and M shapes all have an I-shaped cross section and are often referred to as "I" beams. The three differ in the width of their flanges. In addition to varied flange widths, the S shape flanges are tapered, making them stronger than equivalent sized "W" beams and suitable for train rail or monorail beams. The W shape is commonly used for columns. All can be used for horizontal or vertical members.

Angles are structural steel components that have an L shape. The legs of the angle may be either equal or unequal in length but are usually equal in thickness. Channels have a squared C cross-sectional area and are designated with the letter C or MC. Structural tees are produced from W, S, and M steel shapes. Common designations include WT, ST, and MT.

Structural tubing is manufactured in square, rectangular, and round cross-sectional configurations. These members are used as columns to support loads from other members. Tubes are also commonly used for beams and truss members. Tubes are specified by the size of the outer wall, followed by the thickness of the wall.

WIDE FLANGE SHAPES

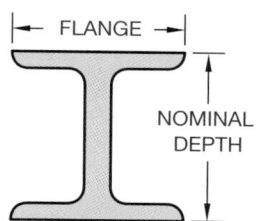

FLANGE

NOMINAL
DEPTH

THE DESIGNATION FOR A WIDE FLANGE LOOKS LIKE THIS:

W 12 × 22

WHERE: **W** IS THE SHAPE.
12 IS THE NOMINAL DEPTH.
22 IS THE NUMBER OF POUNDS
PER LINEAL FOOT.

CHANNELS

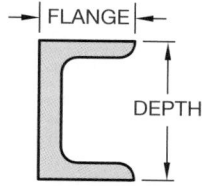

FLANGE

DEPTH

THE DESIGNATION FOR A CHANNEL SHAPE LOOKS LIKE THIS:

C 6 × 10.5

WHERE: **C** IS THE SHAPE.
6 IS THE DEPTH.
10.5 IS THE NUMBER OF POUNDS
PER LINEAL FOOT.

FIGURE 23.52 ■ Dimensional elements of the wide flange and channel shapes.

Steel pipe is also commonly used for columns and bracing. Available steel pipe are standard, extra strong, and double-extra strong. The wall thickness increases with each type.

A variety of templates are available for structural drafting to assist in drawing steel shapes. Many CADD programs are available also to increase structural drafting productivity.

Structural Steel Callouts

Many structural steel materials are specified by shape designation, flange width, and weight in pounds per linear foot. For example:

W 24 × 120

The structural steel shapes that fall into this category are as follows:

W—wide flange shapes
S—American standard beams
M—miscellaneous beam and column shapes
C—American standard channels
MC—miscellaneous channel shapes
WT—structural tees cut from W shapes
ST—structural tees cut from S shapes
MT—structural tees cut from M shapes
T—structural tees
Z—zee shapes
HP—steel H piling

Structural materials that are specified by shape designation, type, diameter or outside dimension, and wall thickness are in the following table. (Note: If length dimensions are required, they are given at the end of the callout in feet and inches except for plates, which should be given only in inches.)

Designation	Shape	Example	Meaning
PL	plate	PL 1/2 × 6 × 8	THK × WIDTH × LENGTH
BAR	square bar	BAR 1-1/4□*	1-1/4" WIDTH × THICKNESS
	round bar	BAR 1-1/4∅	1-1/4" DIAMETER
	flat bar	BAR 2 × 3/8	WIDTH × THICKNESS
PIPE	pipe	PIPE 3∅STD	OD (OUTSIDE DIA) SPEC
TS	structural tubing		
	square	TS 4 × 4 × .250	OUTWIDTH × OUTHEIGHT × WALL THICKNESS
	rectangular	TS 6 × 3 × .375	SAME
	round	TS 4 OD × .188	OD × WALL THICKNESS
L	angle		
	unequal leg	L3 × 2 × 1/4	LEG × LEG × THICKNESS
	equal leg	L3 × 3 × 1/2	LEG × LEG × THICKNESS

* Some companies use ⌀ for the square symbol.

When plate material is to be bent, the minimum bend radius should be given with the plate callout, and the length of the bend legs are dimensioned on the drawing. For example:

3/8 × 10 W/MIN BEND R 5/8"

Location dimensions for structural steel components should be as shown in Figure 23.54. Drafters in architectural or structural engineering firms typically draw structural drawings as shown in Figure 23.55.

Shop Drawings

Shop drawings are used to break each individual component of a structural engineering drawing down into fabrication parts. Shop drawings are also referred to as fabrication drawings, and are generally drawn by the drafter in the fabrication company. Many large companies do both structural and shop drawings. Depending on the complexity of the structure, the structural and shop drawings may be combined on the same drawing. An example of a shop drawing is shown in Figure 23.56, page 796.

COMMON CONNECTION METHODS

Bolting

ASTM Bolts are used for many connections in lumber and steel construction. A bolt specification includes the diameter, length, and strength of the bolt. Washers or plates are specified so the bolt head will not pull through the hole made for the bolt. Bolt strength is classified in accordance with the American Society for Testing Materials (ASTM) specifications. Refer to Chapter 13, Fasteners and Springs, for more information.

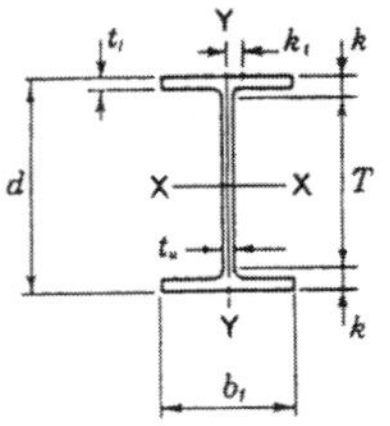

W SHAPES
Dimensions

Designation	Area A	Depth d		Web Thickness t_w		Web $\frac{t_w}{2}$	Flange Width bf		Flange Thickness tf		Distance T	Distance k	Distance k_1
	in.²	in.		in.		in.	in.		in.		in.	in.	in.
W 12 × 336	98.8	16.82	16 7/8	1.775	1 3/4	7/8	13.385	13 3/8	2.955	2 15/16	9 1/2	3 11/16	1 1/2
× 305	89.6	16.32	16 3/8	1.625	1 5/8	13/16	13.235	13 1/4	2.705	2 11/16	9 1/2	3 7/16	1 7/16
× 279	81.9	15.85	15 7/8	1.530	1 1/2	3/4	13.140	13 1/8	2.470	2 1/2	9 1/2	3 3/16	1 3/8
× 252	74.1	15.41	15 3/8	1.395	1 3/8	11/16	13.005	13	2.250	2 1/4	9 1/2	2 15/16	1 15/16
× 230	67.7	15.05	15	1.285	1 5/16	11/16	12.895	12 7/8	2.070	2 1/16	9 1/2	2 3/4	1 1/4
× 210	61.8	14.71	14 3/4	1.180	1 3/16	5/8	12.790	12 3/4	1.900	1 7/8	9 1/2	2 5/8	1 1/4
× 190	55.8	14.38	14 3/8	1.060	1 1/16	9/16	12.670	12 5/8	1.735	1 3/4	9 1/2	2 7/16	1 3/16
× 170	50.0	14.03	14	0.960	15/16	1/2	12.570	12 5/8	1.560	1 9/16	9 1/2	2 1/4	1 1/8
× 152	44.7	13.71	13 3/4	0.870	7/8	7/16	12.480	12 1/2	1.400	1 3/8	9 1/2	2 1/8	1 1/16
× 136	39.9	13.41	13 3/8	0.790	13/16	7/16	12.400	12 3/8	1.250	1 1/4	9 1/2	1 15/16	1
× 120	35.3	13.12	13 1/8	0.710	11/16	3/8	12.320	12 3/8	1.105	1 1/8	9 1/2	1 13/16	1
× 106	31.2	12.89	12 7/8	0.610	5/8	5/16	12.220	12 1/4	0.990	1	9 1/2	1 11/16	15/16
× 96	28.2	12.71	12 3/4	0.550	9/16	5/16	12.160	12 1/8	0.900	7/8	9 1/2	1 5/8	7/8
× 87	25.6	12.53	12 1/2	0.515	1/2	1/4	12.125	12 1/8	0.810	13/16	9 1/2	1 1/2	7/8
× 79	23.2	12.38	12 3/8	0.470	1/2	1/4	12.080	12 1/8	0.735	3/4	9 1/2	1 7/16	7/8
× 72	21.1	12.25	12 1/4	0.430	7/16	1/4	12.040	12	0.670	11/16	9 1/2	1 3/8	7/8
× 65	19.1	12.12	12 1/8	0.390	3/8	3/16	12.000	12	0.605	5/8	9 1/2	1 5/16	13/16
W 12 × 58	17.0	12.19	12 1/4	0.360	3/8	3/16	10.010	10	0.640	5/8	9 1/2	1 3/8	13/16
× 53	15.6	12.06	12	0.345	3/8	3/16	9.995	10	0.575	9/16	9 1/2	1 1/4	13/16
W 12 × 50	14.7	12.19	12 1/4	0.370	3/8	3/16	8.080	8 1/8	0.640	5/8	9 1/2	1 3/8	13/16
× 45	13.2	12.06	12	0.335	5/16	3/16	8.045	8	0.575	9/16	9 1/2	1 1/4	13/16
× 40	11.8	11.94	12	0.295	5/16	3/16	8.005	8	0.515	1/2	9 1/2	1 1/4	3/4
W 12 × 35	10.3	12.50	12 1/2	0.300	5/16	3/16	6.560	6 1/2	0.520	1/2	10 1/2	1	9/16
× 30	8.79	12.34	12 3/8	0.260	1/4	1/8	6.520	6 1/2	0.440	7/16	10 1/2	15/16	1/2
× 26	7.65	12.22	12 1/4	0.230	1/4	1/8	6.490	6 1/2	0.380	3/8	10 1/2	7/8	1/2
W 12 × 22	6.48	12.31	12 1/4	0.260	1/4	1/8	4.030	4	0.425	7/16	10 1/2	7/8	1/2
× 19	5.57	12.16	12 1/8	0.235	1/4	1/8	4.005	4	0.350	3/8	10 1/2	13/16	1/2
× 16	4.71	11.99	12	0.220	1/4	1/8	3.990	4	0.265	1/4	10 1/2	3/4	1/2
× 14	4.16	11.91	11 7/8	0.200	3/16	1/8	3.970	4	0.225	1/4	10 1/2	11/16	1/2

AMERICAN INSTITUTE OF STEEL CONSTRUCTION

(a)

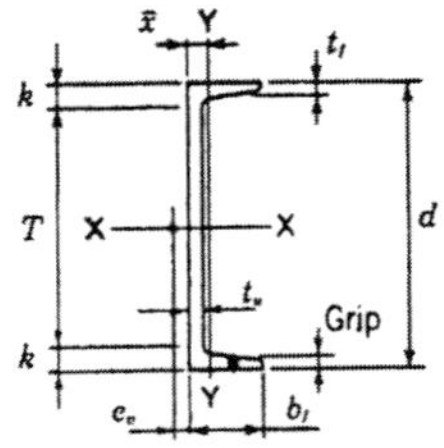

CHANNELS
AMERICAN STANDARD
Dimensions

Designation	Area A	Depth d	Web Thickness t_w		Web $\frac{t_w}{2}$	Width bf		Flange Average Thickness tf		Distance T	Distance k	Grip	Max. Flge. Fastener
	in.²	in.	in.		in.	in.		in.		in.	in.	in.	in.
C 15 × 50	14.7	15.00	0.716	11/16	3/8	3.716	3 3/4	0.650	5/8	12 1/8	1 7/16	5/8	1
× 40	11.8	15.00	0.520	1/2	1/4	3.520	3 1/2	0.650	5/8	12 1/8	1 7/16	5/8	1
× 33.9	9.96	15.00	0.400	3/8	3/16	3.400	3 3/8	0.650	5/8	12 1/8	1 7/16	5/8	1
C 12 × 30	8.82	12.00	0.510	1/2	1/4	3.170	3 1/8	0.501	1/2	9 3/4	1 1/8	1/2	7/8
× 25	7.35	12.00	0.387	3/8	3/16	3.047	3	0.501	1/2	9 3/4	1 1/8	1/2	7/8
× 20.7	6.09	12.00	0.282	5/16	1/8	2.942	3	0.501	1/2	9 3/4	1 1/8	1/2	7/8
C 10 × 30	8.82	10.00	0.673	11/16	5/16	3.033	3	0.436	7/16	8	1	7/16	3/4
× 25	7.35	10.00	0.526	1/2	1/4	2.886	2 7/8	0.436	7/16	8	1	7/16	3/4
× 20	5.88	10.00	0.379	3/8	3/16	2.739	2 3/4	0.436	7/16	8	1	7/16	3/4
× 15.3	4.49	10.00	0.240	1/4	1/8	2.600	2 5/8	0.436	7/16	8	1	7/16	3/4
C 9 × 20	5.88	9.00	0.448	7/16	1/4	2.648	2 5/8	0.413	7/16	7 1/8	15/16	7/16	3/4
× 15	4.41	9.00	0.285	5/16	1/8	2.485	2 1/2	0.413	7/16	7 1/8	15/16	7/16	3/4
× 13.4	3.94	9.00	0.233	1/4	1/8	2.433	2 3/8	0.413	7/16	7 1/8	15/16	7/16	3/4
C 8 × 18.75	5.51	8.00	0.487	1/2	1/4	2.527	2 1/2	0.390	3/8	6 1/8	15/16	3/8	3/4
× 13.75	4.04	8.00	0.303	5/16	1/8	2.343	2 3/8	0.390	3/8	6 1/8	15/16	3/8	3/4
× 11.5	3.38	8.00	0.220	1/4	1/8	2.260	2 1/4	0.390	3/8	6 1/8	15/16	3/8	3/4
C 7 × 14.75	4.33	7.00	0.419	7/16	3/16	2.299	2 1/4	0.366	3/8	5 1/4	7/8	3/8	5/8
× 12.25	3.60	7.00	0.314	5/16	3/16	2.194	2 1/4	0.366	3/8	5 1/4	7/8	3/8	5/8
× 9.8	2.87	7.00	0.210	3/16	1/8	2.090	2 1/8	0.366	3/8	5 1/4	7/8	3/8	5/8
C 6 × 13	3.83	6.00	0.437	7/16	3/16	2.157	2 1/8	0.343	5/16	4 3/8	13/16	5/16	5/8
× 10.5	3.09	6.00	0.314	5/16	3/16	2.034	2	0.343	5/16	4 3/8	13/16	3/8	5/8
× 8.2	2.40	6.00	0.200	3/16	1/8	1.920	1 7/8	0.343	5/16	4 3/8	13/16	5/16	5/8
C 5 × 9	2.64	5.00	0.325	5/16	3/16	1.885	1 7/8	0.320	5/16	3 1/2	3/4	5/16	5/8
× 6.7	1.97	5.00	0.190	3/16	1/8	1.750	1 3/4	0.320	5/16	3 1/2	3/4	—	—
C 4 × 7.25	2.13	4.00	0.321	5/16	3/16	1.721	1 3/4	0.296	5/16	2 5/8	11/16	5/16	5/8
× 5.4	1.59	4.00	0.184	3/16	1/16	1.584	1 5/8	0.296	5/16	2 5/8	11/16	—	—
C 3 × 6	1.76	3.00	0.356	3/8	3/16	1.596	1 5/8	0.273	1/4	1 5/8	11/16	—	—
× 5	1.47	3.00	0.258	1/4	1/8	1.498	1 1/2	0.273	1/4	1 5/8	11/16	—	—
× 4.1	1.21	3.00	0.170	3/16	1/16	1.410	1 3/8	0.273	1/4	1 5/8	11/16	—	—

AMERICAN INSTITUTE OF STEEL CONSTRUCTION

(b)

FIGURE 23.53 ■ Dimensional information for (a) W12×22, and (b) C6×10.5. *Courtesy American Institute for Steel Construction, Manual of Steel Construction.* Additional samples of structural steel dimensions may be found in Appendix E.

Standard Structural Bolt Callouts

When specifying bolts on a drawing, give the quantity, diameter, bolt type if special, length in inches, and ASTM specification. If special washer and nut requirements are present, this should also be specified in the bolt callout. Hexagon head bolts and hexagon nuts are assumed unless either is specified differently in the callout. Examples of bolt callouts are:

2-3/4"∅ BOLTS ASTM A503
4-1/2"∅ × 10" BOLTS
2-5/8"∅ × 6" CARRIAGE BOLTS
6-3/4"∅ BOLTS W/MALLEABLE IRON WASHERS AND HEAVY HEX NUTS
4-1/2"∅ GALVANIZED BOLTS

Give the hole diameter when holes for standard bolts must be specified on the drawing. In general, holes should be 1/16" larger in diameter than the specified bolt for standard steel-to-steel, wood-to-wood, or wood-to-steel construction up to 1" in diameter, and 1/8" larger for hole sizes over 1" in diameter, than the specified bolt for standard steel-to-concrete or wood-to-concrete applications unless otherwise specified by the engineer. For example:

2-3/4"∅ BOLTS FIELD DRILL 13/16"∅ HOLES.

The following are recommended standard hole diameters in inches for given bolt sizes:

Bolt ∅	Standard Hole ∅	Concrete Hole ∅	Oversize Hole ∅
1/2	9/16	5/8	11/16
5/8	11/16	3/4	13/16
3/4	13/16	7/8	15/16
7/8	15/16	1	1 1/16
1	1 1/16	1 1/8	1 1/4
1 1/8	1 1/4	1 1/4	1 7/16
1 1/4	1 3/8	1 3/8	1 9/16
1 3/8	1 1/2	1 1/2	1 11/16
1 1/2	1 5/8	1 5/8	1 13/16

WIDE FLANGES AND OTHER I SHAPES — TO CENTERLINE IN BOTH DIRECTIONS

CHANNELS — CENTERLINE ALONG X–X AXIS AND BACK FACE OF WEB ALONG Y–Y AXIS

Note: An exception for wide flanges and channels is when specifying top of beam.

ANGLES — OUTSIDE FACE OF LEGS

PLATES — TO THE CENTERLINE IN BOTH DIRECTIONS WHEN DIMENSIONING IN THE PLAN VIEW OF THE PLATE AND TO THE FACE OF THE PLATE WHEN DIMENSIONING IN SECTION OR PROFILE.

PLAN   PROFILE

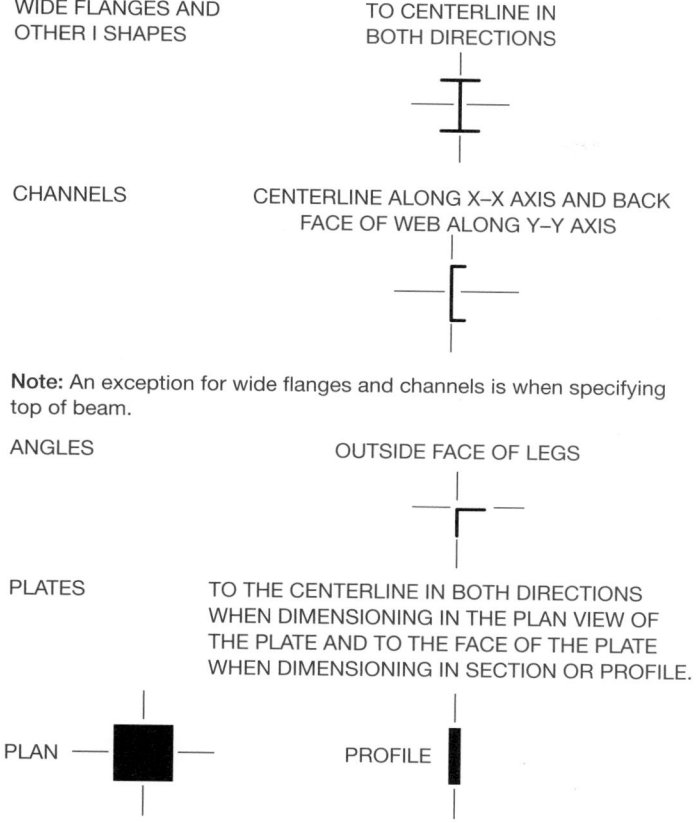

FIGURE 23.54 ■ Location dimensions for structural components.

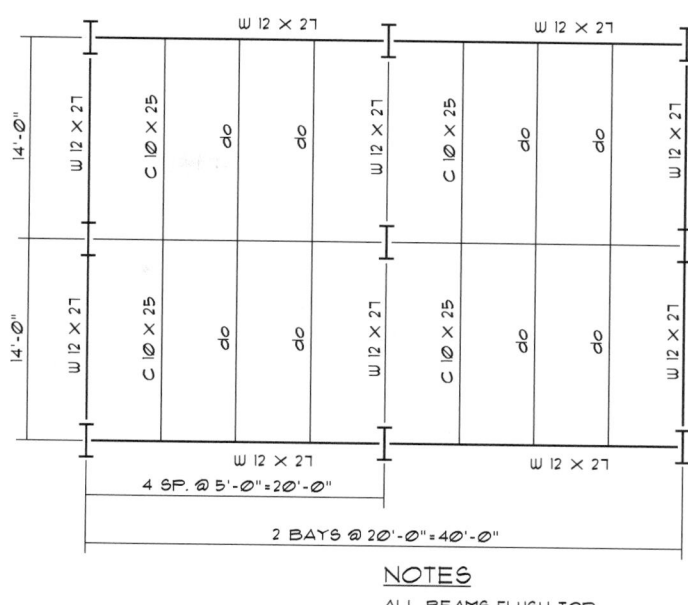

NOTES
ALL BEAMS FLUSH TOP
TOP OF STEEL ELEVATION 113'-0"

FIGURE 23.55 ■ Structural steel engineering drawing.

When specifying lag bolts on a drawing, the lead or tap hole diameter should be given with the bolt specification. For example:

2-3/4"∅ × 8 LAG BOLTS W/7/16"∅ LEAD HOLES.

The following are tap hole diameters in inches for lag bolts used in Douglas fir, larch, or Southern pine:

Bolt ∅	Lead Hole ∅
3/8	1/4
1/2	5/16
5/8	7/16
3/4	1/2
7/8	5/8
1	3/4

Bolts on a drawing are located to their centerlines. When counterbores are required, provide the specification to the location of the counterbore. For example:

∅2-3/4" BOLT ∅3-1/4" CBORE × 7/8" DEEP

ASME Y14.5M symbols for counterbore and depth may also be used (see Chapter 12).

Nails

Nails used for the fabrication of wood-to-wood members are sized by the term *penny* and denoted by the letter *d*. Penny is weight classification for nails—the number of pounds per 1000 nails. So, one thousand 16d nails weigh 16 pounds. Nails are also sized by diameter when over 60d. Standard nail sizes (2d to 20d) and nail types are shown in Figure 23.57. When specifying nails, the penny weight should be given plus the quantity and spacing if required. The specification for special nails should be given when required. Nailing callout examples include:

4–16d NAILS
5–20d GALV NAILS EA SIDE
10d NAILS 4" OC AT SEAMS AND 12" OC IN FIELD
30d NAILS ALTERNATELY STAGGERED 12" OC TOP AND BOTTOM BOTH SIDES
8d RINGSHANK NAILS 6" OC

If pilot holes are required for nailing, the diameter of the pilot hole should be given after the nail callout. For example:

4–20d NAILS W/5/32"∅ HOLES

Verify pilot hole diamaters with manufacturers' recommendations for types of woods and applications.

Welding

Welds are classified according to the type of joint on which they are used. The four common welds used in construction are the fillet, back, plug or slot, and groove welds. A welding symbol is used to designate the type and dimensional specifications of the weld. Review Chapter 12 for an in-depth coverage of welding symbols before doing the drawings in this chapter.

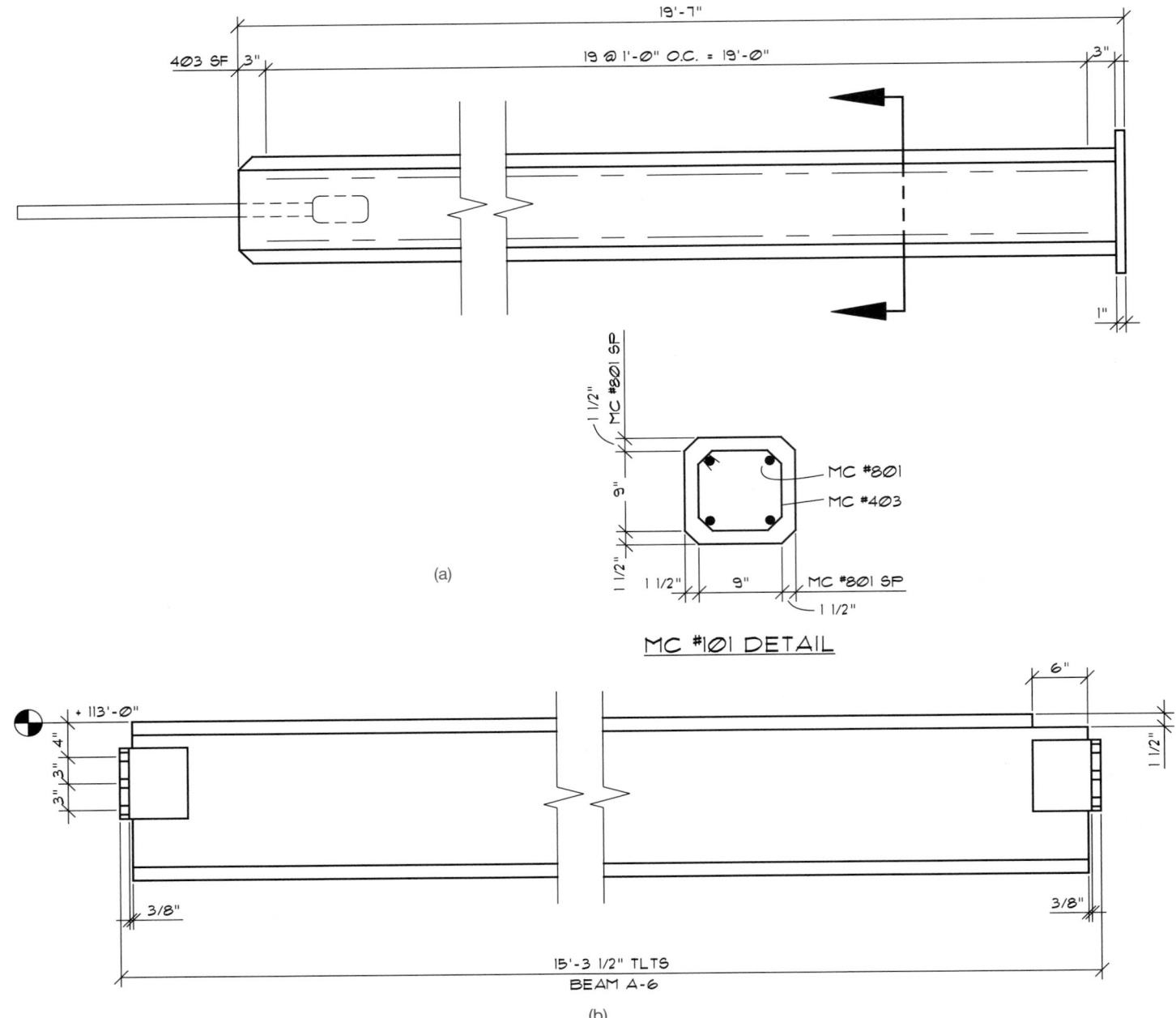

FIGURE 23.56 ■ (a) Precast concrete fabrication detail; (b) structural steel fabrication drawing used to show how individual components are to be made.

Fabrication Methods

There are an unlimited number of fabrication methods. As a drafter you must become familiar with the fabrications used. Many techniques are typical while others require special design. The drafter either draws from engineering sketches or refers to previously drawn examples. Once a certain amount of experience has been gained, the drafter is able to establish drawings from either written or verbal instructions, or from a particular given situation.

AISC The AISC *Manual of Steel Construction* provides a number of common connection details. Several fabrica-

tion details given without specifications or dimensions are shown in Appendix E.

COMPONENTS IN A SET OF STRUCTURAL DRAWINGS

A complete set of structural drawings may have a number of sheets with specific elements of the building provided in a general format or with specific construction elements shown in cross sections or clear details. While not all complete sets of structural drawings

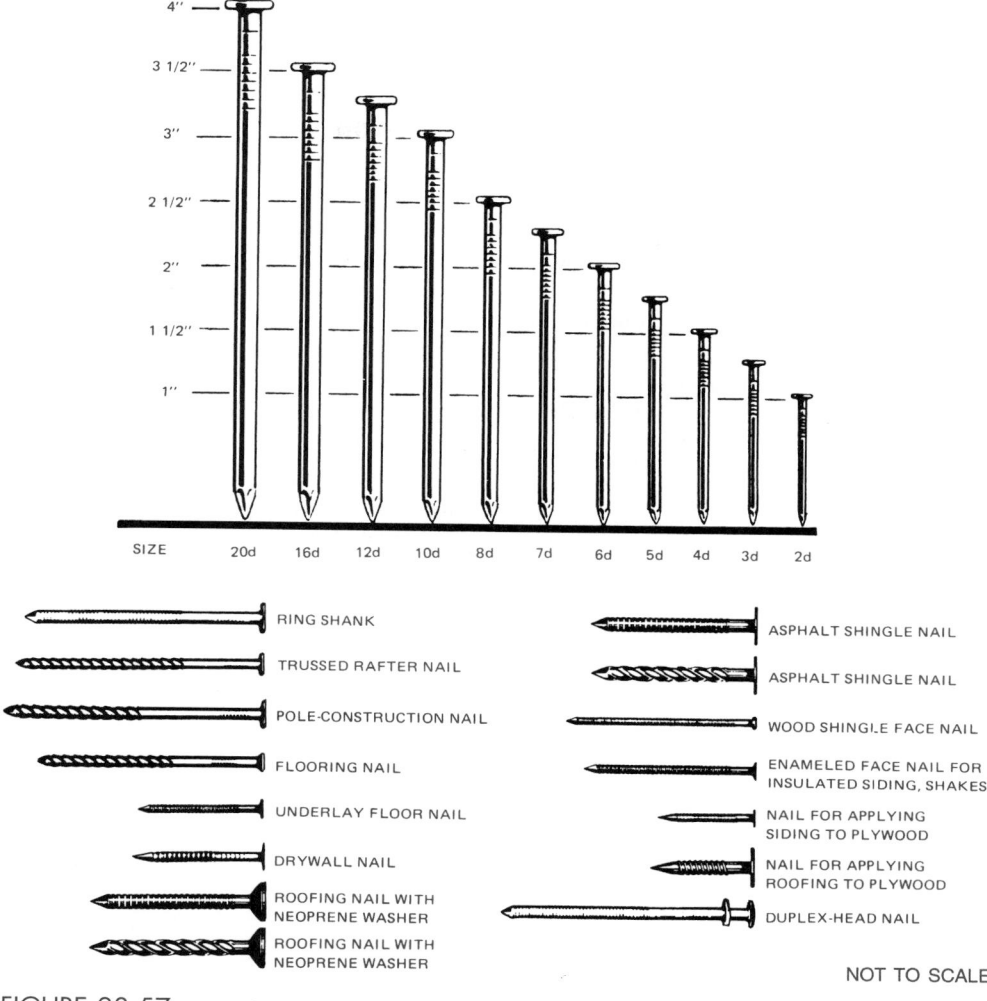

FIGURE 23.57 ■ Standard nail sizes and types.

STRUCTURAL DRAFTING

The procedure for preparing structural drawings works especially well with CADD. When a floor plan is drawn, this drawing is used as a base layer upon which each of the other plan views are drawn. For example, the floor plan may be drawn with black lines and labeled as the FLOOR PLAN layer, then the layer is changed and named FOUNDATION PLAN and the color is changed to red. All of the drawing information on the red layer is for the foundation plan. This process of changing layer names and colors continues until all of the drawings are complete. One advantage is that each layer may be turned on or off at the drafter's command. It is the same as tracing a drawing in manual drafting. Time is saved, and the drawings are all extremely accurate. When the drawings are to be plotted, each drawing may be converted to an individual file, or the layer to be plotted is turned on while the others are off. Any combination of plots may be reproduced, with select layers turned on.

There is now available structural CADD software that quickly and accurately draws structural steel shapes to exact specifications in plan, section, or elevation views. These structural steel packages also provide for beam details, integrating cutouts, framing angles, hole groups, welded or bolted connections, and complete dimensioning. An advantage is the capability of such programs to perform tedious standards specification sizing and calculations automatically. You do not have to spend time dimensioning hole patterns, because the program draws bolt or rivet hole patterns to your specifications, automatically displaying on center dimensions, hole diameters, and flange thicknesses. When framing angles are used, the package automatically draws, sizes, positions, dimensions, and notes the angle specifications along with bolt holes and dimensions. These CADD structural packages are often available with a variety of template menu overlays and symbol libraries. Detail views and symbology are often available from manufacturers of structural systems and components.

CADD APPLICATIONS

contain the same number of sheets or type of information, some of the representative drawings include: a floor plan; a foundation plan and details; a concrete slab plan and details; a roof framing plan and details; a roof drainage plan; building section(s); exterior elevations; a panel plan, elevations, and wall details.

Floor Plan

The *floor plan* is generally designed and drawn by the architect. This floor plan drawing is then used as the layout for the other associated drawing from consulting engineers—for example, the mechanical (HVAC), plumbing, electrical, and structural engineers. In some situations and depending on the structure, the structural engineering firm draws the floor plan as part of the set of drawings. (See Figure 23.58.)

Foundation Plan and Details

The purpose of the *foundation plan* is to show the supporting system for the walls, floor, and roof. The foundation may consist of continuous perimeter footings and walls that support the exterior and main bearing walls of the building. The footings in these loca-tions are rectangular in shape and are continuous concrete supports centered at exterior and interior bearing walls. (See Figure 23.59.) There are also pedestal footings that support the concentrated loads of particular elements of the structure. Interior support for the columns that support upper floor and roof loads are provided by pedestal footings, as shown in Figure 23.60. Anchor bolts and other metal connectors are shown and located on the foundation plan, as shown in Figure 23.61. A typical foundation plan is shown in Figure 23.62.

The foundation details are drawn to provide information on the concrete foundation at the perimeter walls, and the foundation and pedestals at the center support columns. Possible details include retaining walls, typical exterior foundation and rebar schedule, typical interior bearing wall, and typical pedestal and rebar schedule. Detail drawings are keyed to the plan view using detail markers. Detail markers are usually drawn as a circle of about $3/8"$ to $1/2"$ in diameter on the plan view and a coordinating circle of about $3/4"$ in diameter under the associated detail. Each detail marker is divided in half. The top half contains the detail number and the bottom identifies the page number on which the detail is drawn. Notice the detail markers associated with the example drawings in this chapter. The rebar schedules are charts placed adjacent to the detail that

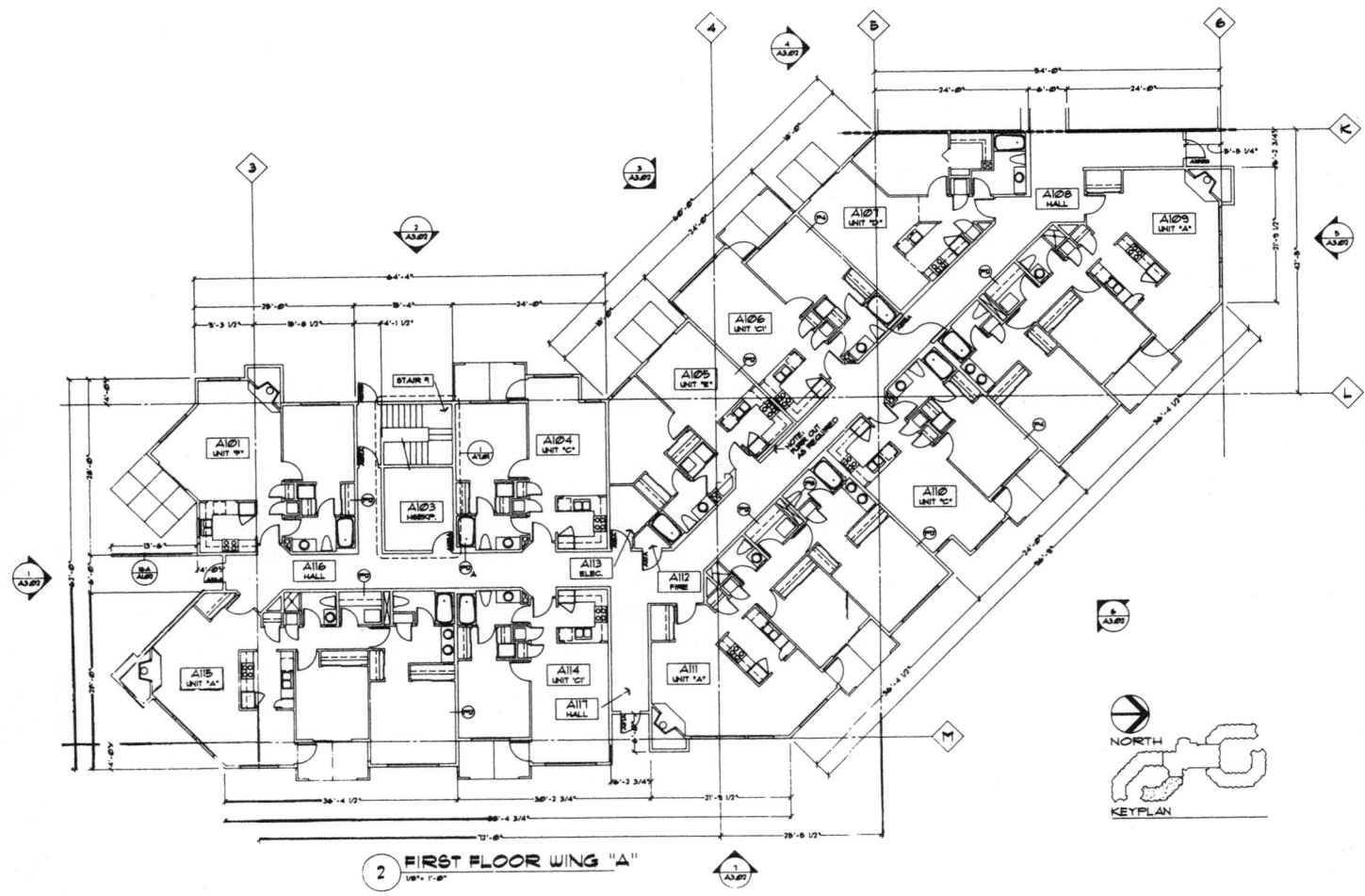

FIGURE 23.58 ■ CADD floor plan for a set of construction drawings. *Courtesy Soderstrom Architects, PC.*

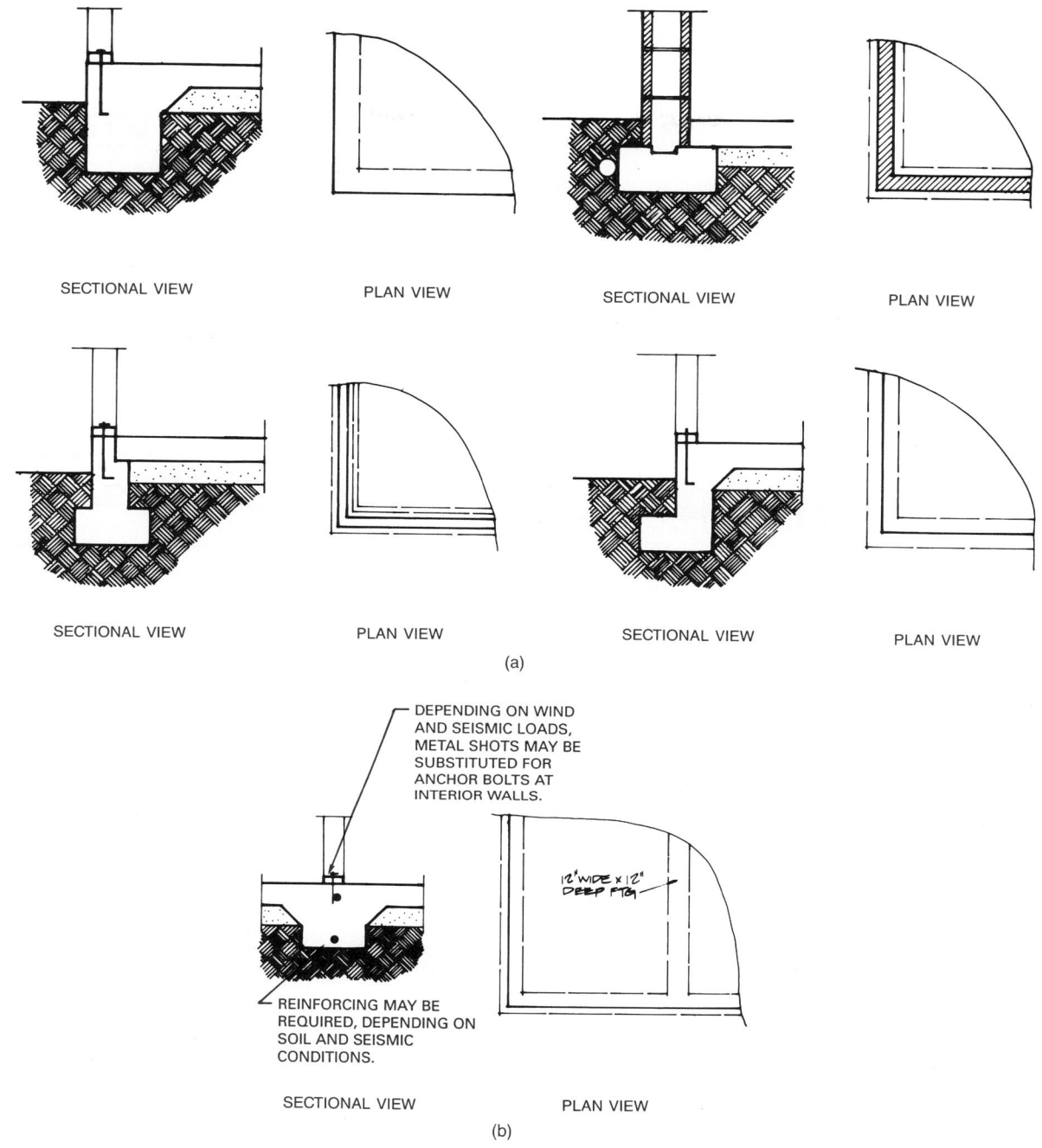

FIGURE 23.59 ■ (a) Common foundation and slab intersections; (b) footing and floor intersections at interior load-bearing wall.

key information about the rebar used to the drawing, as shown in Figure 23.60.

Concrete Slab Plan and Details

The *concrete slab plan* is drawn to outline the concrete that will become the floor(s). Items often found in slab plans include floor slabs, slab reinforcing, expansion joints, pedestal footings, metal connectors, anchor bolts, and any foundation cuts for doors or other openings. The openings and other items are located and labeled on the plan, as shown in Figure 23.63. In situations where tilt-up construction is used, the opening locations are dimensioned on the elevations.

The concrete slab details are drawn to provide information of the intersections of the concrete slabs. These details include interior and perimeter slab joints as in Figure 23.64, and other slab details. The drawings include such information as slab thicknesses, reinforcing sizes and locations, and slab elevations (heights), as shown in Figure 23.65.

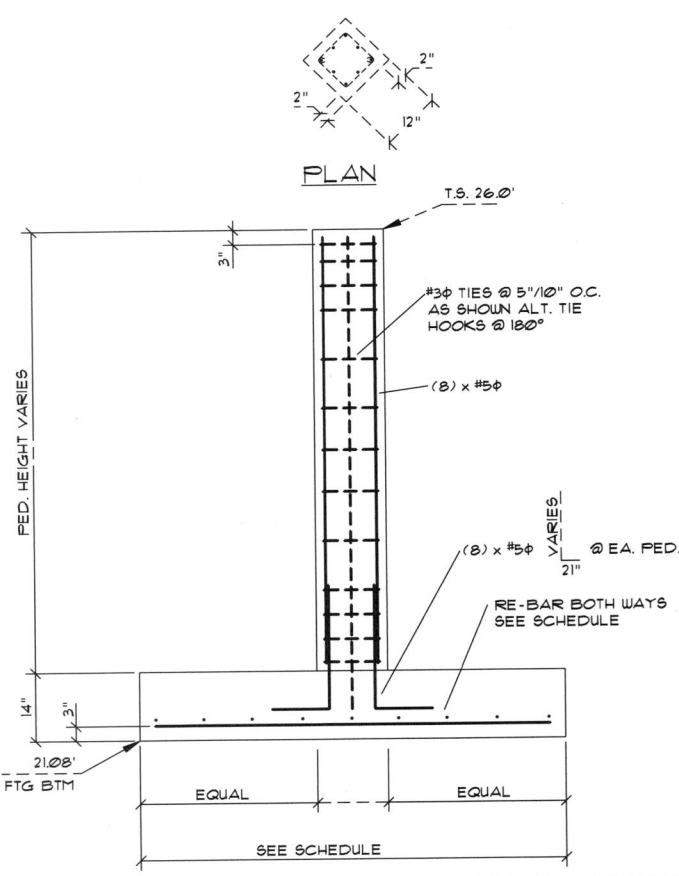

PLAN

FIGURE 23.60 ■ Pedestal footing detail.

FTG	PED HEIGHT	SIZE	REINFORCING STEEL
B-2	7.50	7'-0" ⊡	#6φ @10" O.C.
B-3	6.75	6'-6" ⊡	#6φ @12" O.C.
B-4	5.91	6'-6" ⊡	#6φ @12" O.C.
B-5	5.00	7'-0" ⊡	#6φ @10" O.C.

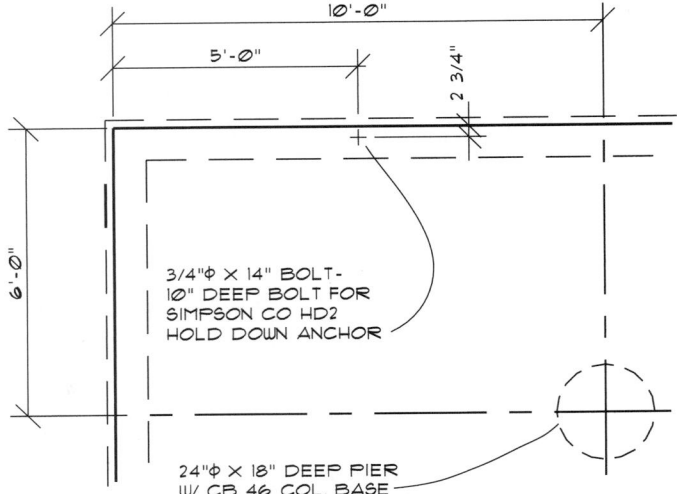

FIGURE 23.61 ■ Foundation plan showing location of metal connectors.

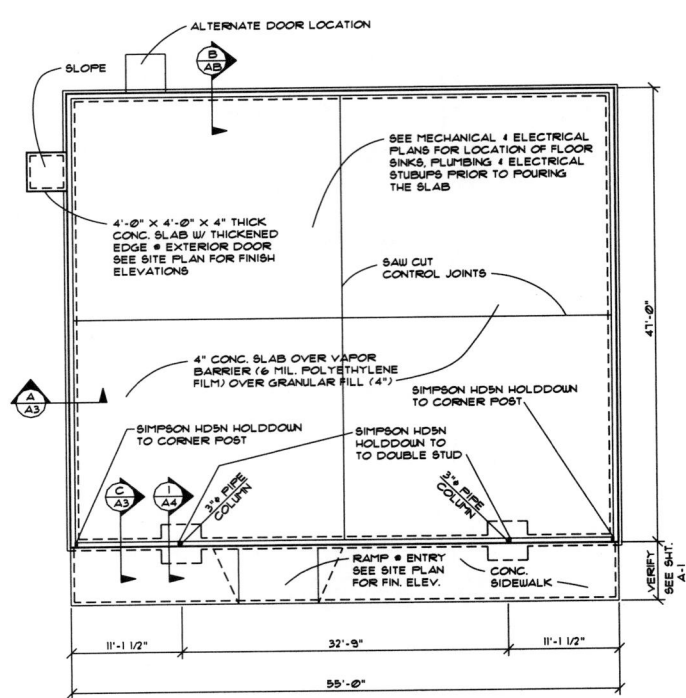

FIGURE 23.62 ■ A concrete slab foundation is commonly used for commercial structures. *Courtesy The Southland Corporation.*

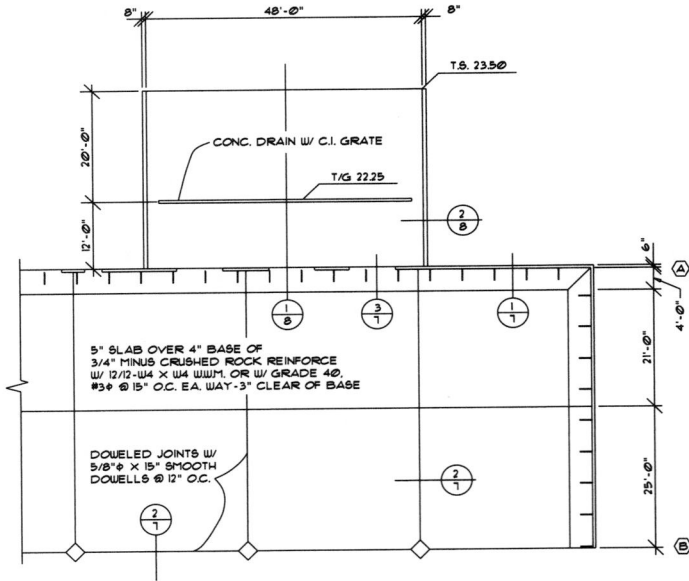

FIGURE 23.63 ■ A slab-on-grade plan shows the size and location of all concrete pours plus reinforcing specifications. *Courtesy Structureform Masters, Inc.*

Roof Framing Plan and Details

The purpose of the *roof framing plan* is to show the major structural components in plan view that occur at the roof level. A roof framing plan using truss framing is shown in Figure 23.3b (see

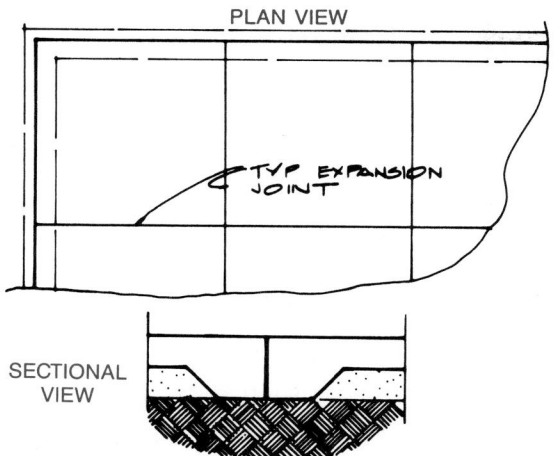

FIGURE 23.64 ■ Control joints are often placed in large slabs to resist cracking and allow the construction crews manageable areas to pour.

page 767). A panelized roof framing plan is shown in Figures 23.3b and 23.33 (see page 784).

Roof framing details are required to show the construction methods used at various member intersections in the building. Details may include the following intersections: wall to beam, beam to column, beam splices and connections, truss details, bottom chord bracing plan and details, purlin clips, cantilever locations, and roof drains. (See Figure 23.66).

Notice the elevation symbol shown in Figure 23.66e. This symbol is commonly used on structural drawings to give the elevation of locations from a known zero elevation. The zero elevation might be at the first floor or other good reference point such as the top of a foundation wall. These elevation symbols are used together with standard dimensioning practice as needed.

Roof Drainage Plan

The *roof drainage plan* may be part of a set of structural drawings for some buildings, although it may be considered part of the plumbing or piping drawings depending on the particular company's use and interpretation. The purpose of this drawing is to show the elevations of the roof and provide for adequate water drainage generally associated with low slope and flat roofs. (See Figure 23.67.) Some of the terminology associated with roof drainage plans includes: *roof drain* (RD)—a screened opening to allow for drainage; *overflow drain* (OD)—a backup in case the roof drains fail; *down spout* (DS)—usually a vertical pipe used to transport water from the roof; and *scupper* or *gutter*—a water collector usually on the outside of a wall at the roof level to funnel water from the roof drains to the down spouts.

Building Sections

The *building section* is used to show the relationship between the plans and details previously drawn. This drawing is considered a general arrangement or construction reference, as it is often drawn

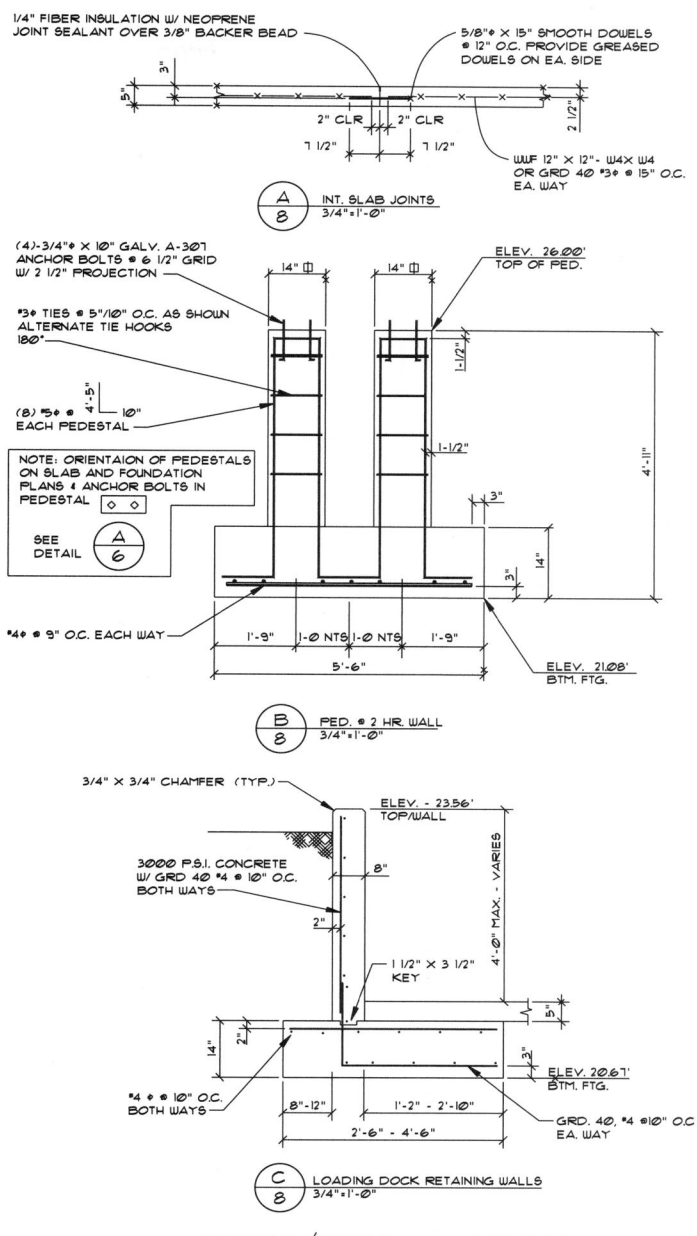

FIGURE 23.65 ■ Typical slab and footing details.

FOOTING	PEDESTAL HEIGHT	SIZE	REINFORCING STEEL
B-2	7.50	7'-0"⌷	#6∅ ∅10" O.C.
B-3	6.75	6'-6"⌷	#6∅ ∅12" O.C.
B-4	5.91	6'-6"⌷	#6∅ ∅12" O.C.
B-5	5.00	7'-0"⌷	#6∅ ∅10" O.C.
B-6	4.50	6'-0"⌷	#5∅ ∅10" O.C.
B-7	4.00	7'-0"⌷	#6∅ ∅10" O.C.

at a small scale. While some detailed information is provided with regard to building elevations (heights) and general dimensions, the overall section is not intended to provide explanation of building materials. The details more clearly serve this purpose, as they are drawn at a larger scale. Some less complex buildings, however, may show a great deal of construction information on the overall section and use fewer details. These overall sections, commonly

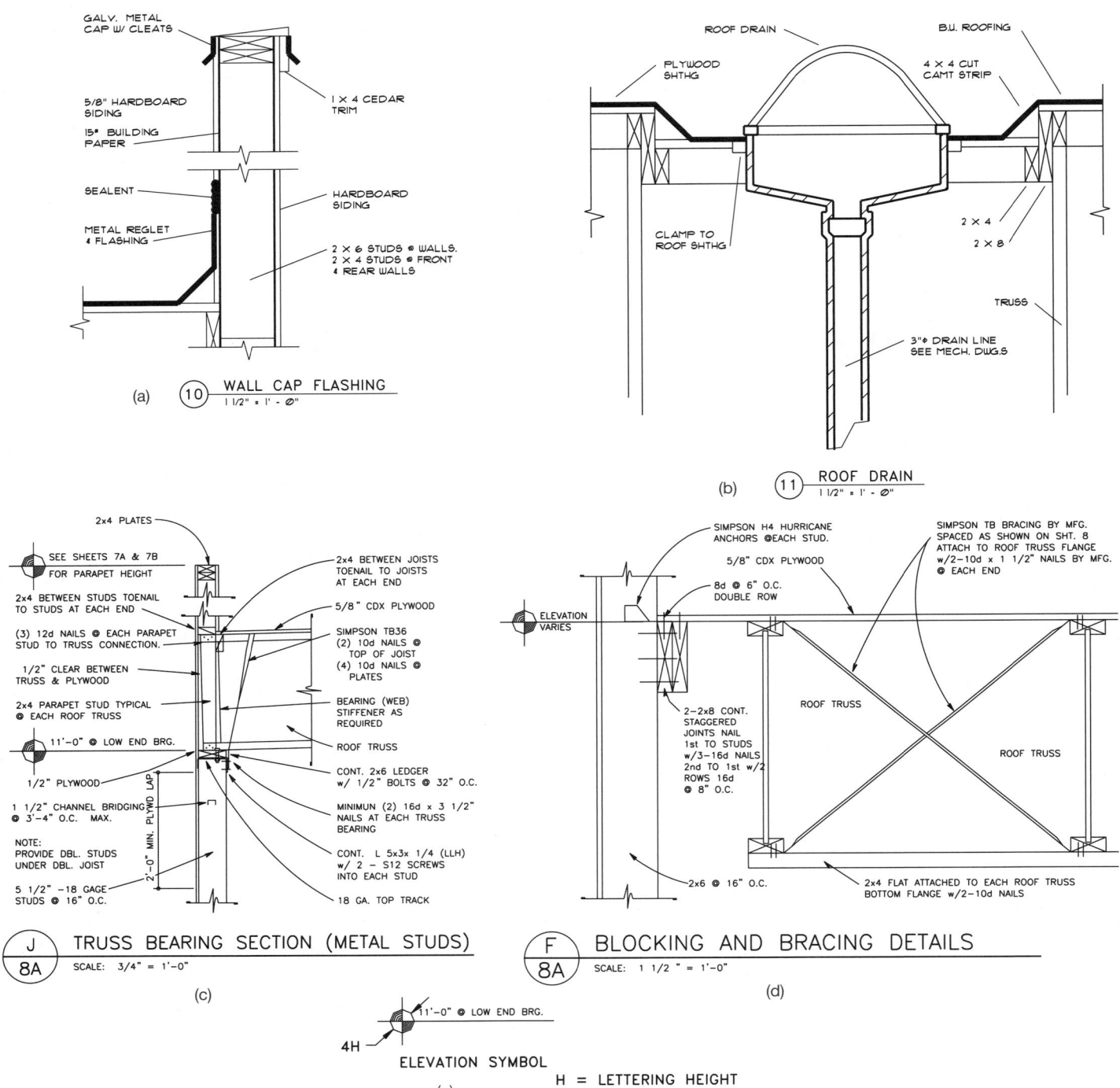

FIGURE 23.66 ■ Roof construction details. (**a and b**) *Courtesy The Southland Corporation.* (**c and d**) *Courtesy Wendy's International Inc.* (**e**) Elevation symbol is used to designate the elevation of a specific location. The elevation is above a known zero elevation on the structure.

called typical cross sections, show the general arrangement of the construction and often have detail coorelated to them. (See Figure 23.68.) In some situations, partial sections may be useful in describing portions of the construction that may not be effectively handled with the building section and may be larger areas than normally identified with a detailed section, as shown in Figure 23.69.

Exterior Elevations

The *exterior elevations* are drawings that show the external appearance of the building. An elevation is drawn at each side of the building to show the relationship of the building to the final grade, location of openings, wall heights, roof slopes, exterior building materials, and other exterior features. A front elevation is generally

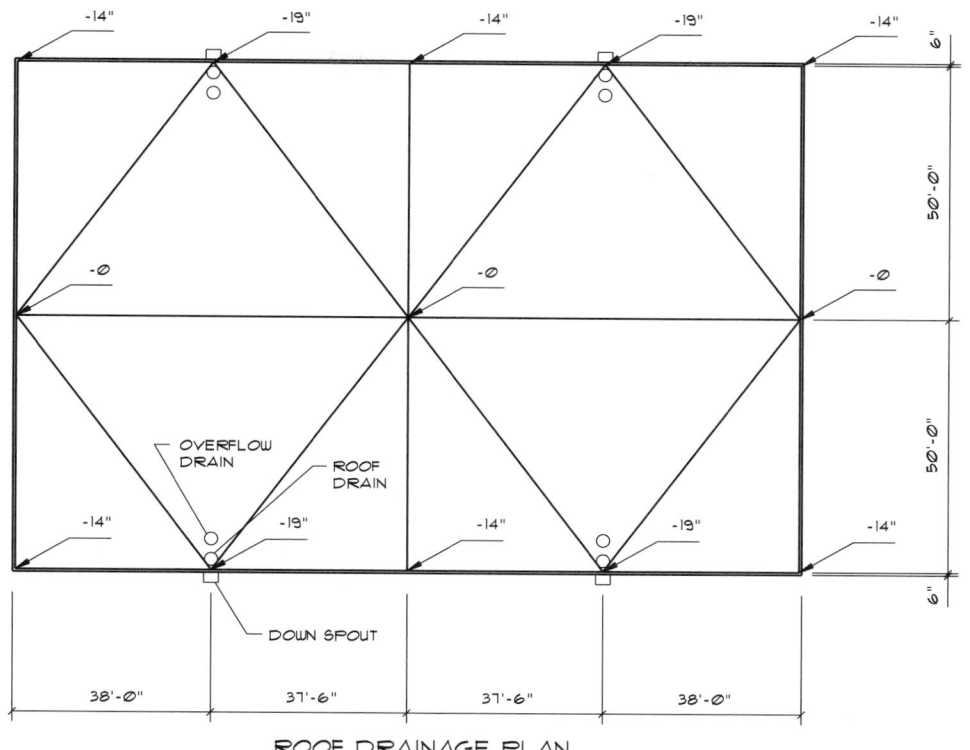

ROOF DRAINAGE PLAN

SCALE: 1/16" ====== 1'-0"

GENERAL NOTES:

1. ROOF AND OVERFLOW DRAINS SHALL BE LARGE GENERAL PURPOSE TYPE W/ NON-FERROUS DOMES AND 4"φ OUTLETS
2. OVERFLOW DRAINS SHALL BE SET W/ INLET 2" ABOVE ROOF DRAIN INLET AND SHALL BE CONNECTED TO DRAINS LINES INDEPENDENT FROM ROOF DRAINS.
3. USE A 4"H × 7W SCUPPER W/5" NOMINAL (3 3/4 × 5) RECTANGULAR CORRUGATED DOWNSPOUT.
   PROVIDE A 6" × 9" CONDUCTOR HEAD @ TOP OF DOWNSPOUT.

FIGURE 23.67 ■ Roof drainage plan.

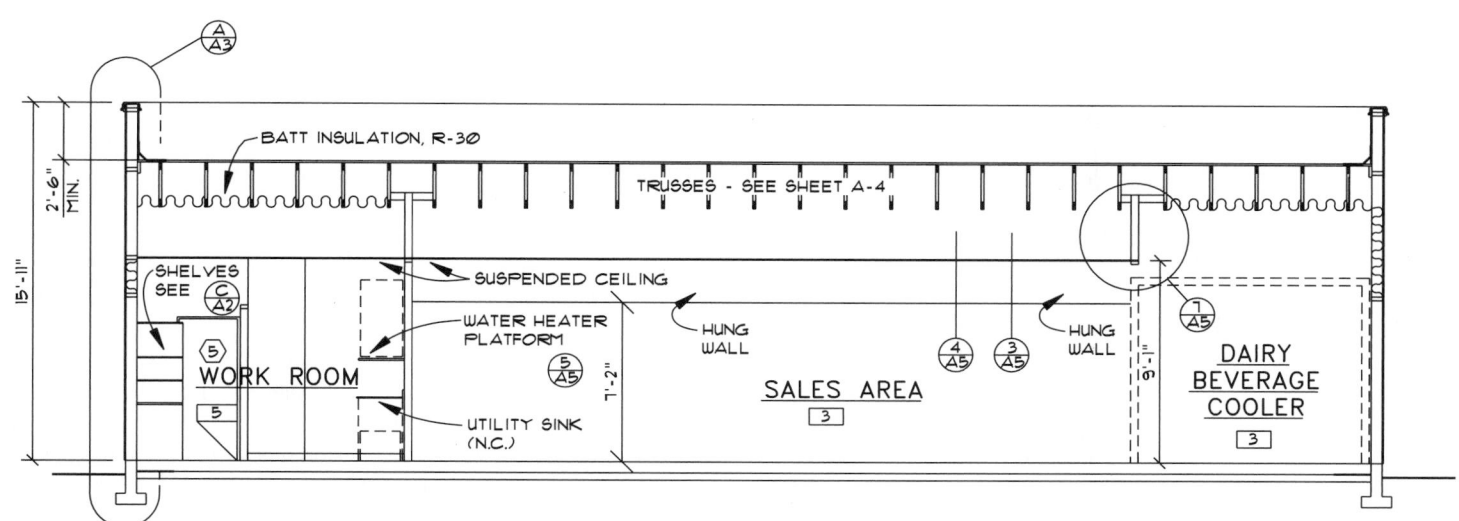

FIGURE 23.68 ■ Typical building sections for commercial construction are often drawn at a small scale to show major types of construction with specific information shown in details. *Courtesy The Southland Corporation.*

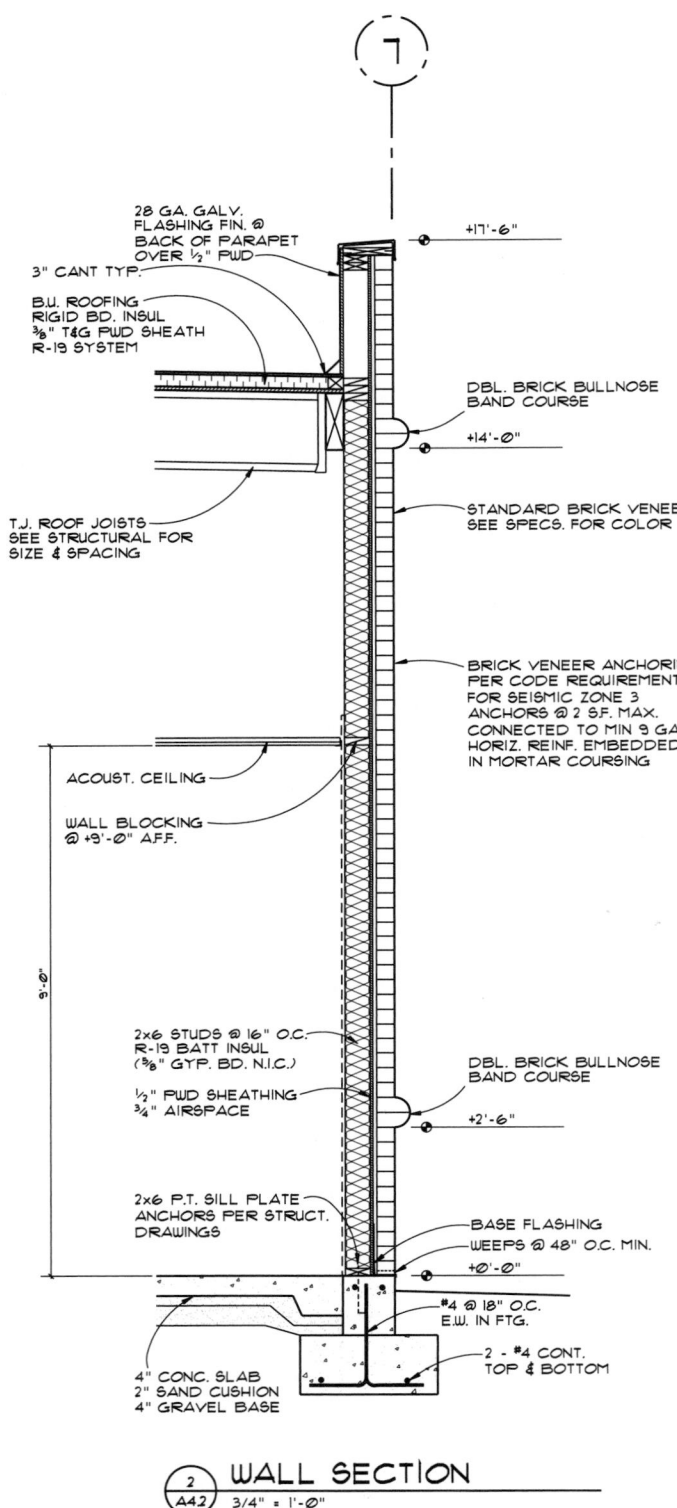

28 GA. GALV. FLASHING FIN. @ BACK OF PARAPET OVER ½" PWD

3" CANT TYP.

B.U. ROOFING RIGID BD. INSUL ⅜" T&G PWD SHEATH R-19 SYSTEM

T.J. ROOF JOISTS SEE STRUCTURAL FOR SIZE & SPACING

ACOUST. CEILING

WALL BLOCKING @ +9'-0" A.F.F.

2x6 STUDS @ 16" O.C. R-19 BATT INSUL (⅝" GYP. BD. N.I.C.)

½" PWD SHEATHING ¾" AIRSPACE

2x6 P.T. SILL PLATE ANCHORS PER STRUCT. DRAWINGS

4" CONC. SLAB 2" SAND CUSHION 4" GRAVEL BASE

+17'-6"

DBL. BRICK BULLNOSE BAND COURSE

+14'-0"

STANDARD BRICK VENEER SEE SPECS. FOR COLOR

BRICK VENEER ANCHORING PER CODE REQUIREMENTS FOR SEISMIC ZONE 3 ANCHORS @ 2 S.F. MAX. CONNECTED TO MIN 9 GA. HORIZ. REINF. EMBEDDED IN MORTAR COURSING

DBL. BRICK BULLNOSE BAND COURSE

+2'-6"

BASE FLASHING
WEEPS @ 48" O.C. MIN.
+0'-0"

#4 @ 18" O.C. E.W. IN FTG.

2 - #4 CONT. TOP & BOTTOM

9'-0"

WALL SECTION
2/A42 3/4" = 1'-0"

FIGURE 23.69 ■ Partial sections are used to clarify construction information through various portions of the structure. *Courtesy Ankrom Moisan Architects.*

the main entry view and is drawn at a ¼" = 1'–0" scale, depending on the size of the structure. Other elevations may be drawn at a smaller scale. The elevation's scale depends upon the size of the building, the amount of detail shown, and the sheet size used.

Many companies prefer to draw all elevations at the same scale showing an equal amount of detail in all views. An elevation may be omitted if it is the same as another. When this happens, the elevation may be labeled as RIGHT AND LEFT ELEVATION. Elevations are also often labeled by compass orientation, such as SOUTH ELEVATION. The elevations may be drawn showing a great deal of detail, as shown in Figure 23.70. Often, elevations for commercial buildings may be drawn at a small scale representing very little detail, as shown in Figure 23.71. In many situations the drafter may be required to draw interior elevations or details such as the interior finish elevation detail for Farrell's, as shown in Figure 23.72.

Panel Plan, Elevations, and Wall Details

The *panel plan* is used in tilt-up construction to show the location of the panels, as shown in Figure 23.18 (see page 773).

Panel elevations are used when the exterior elevations do not clearly show information about items located on or in the walls. Similarly to exterior elevations, the panel elevations will show door locations and reinforcing within walls and around openings. Dimensions associated with panel elevations provide both horizontal and vertical dimensions for openings and other features. (See Figure 23.19, page 774.)

Wall details are used to show the connection points of the concrete panels used in tilt-up construction, and connection details at the walls for other types of structures. (See Figure 23.73.)

DRAWING REVISIONS

Drawing revisions are common in the architectural, structural, and construction industry. Revisions can be caused for a number of reasons; for example, changes requested by the owner, job site corrections, correcting errors, or code changes. Changes are done in a formal manner by submitting an *addendum* to the contract, which is a written notification of the change or changes and is accompanied by a drawing that represents the change.

Revision Clouds

A revision cloud is placed around the area that is changed. The revision cloud is a cloudlike circle around the change as shown in Figure 23.74. CADD programs that are commonly used for architectural and structural drafting have commands that allow you to easily draw the revision cloud. There is also a triangle with a revision number inside that is placed next to the revision cloud or along the revision cloud line as shown in Figure 23.75, page 808. The triangle is commonly called a delta. The number is then correlated to a revision note placed somewhere on the drawing, or in the title block as shown in Figure 23.76, page 808. Each company has a desired location for revision notes, although common places are in the corners of the drawing, in a revision block or table, or in the title block. This practice is not as clearly defined as in ANSI/ASME standard drawings. The note is used to explain the change. If a reference is given in the title block, then detailed information about the revision is normally provided in the revision document that is filed with the project information. The revision document is typically filled

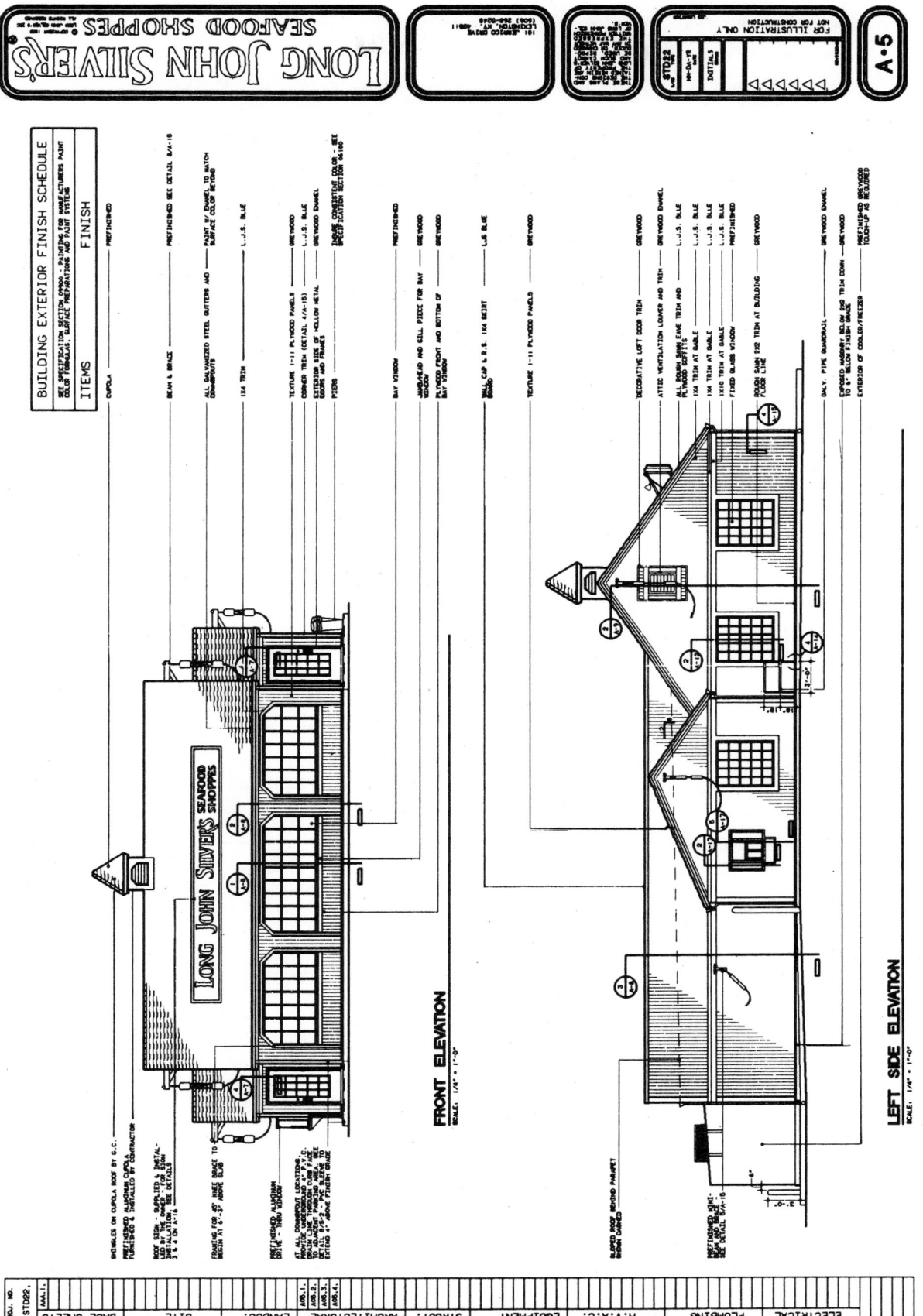

FIGURE 23.70 ■ CADD elevations of Long John Silver's Seafood Shoppes. *Courtesy Jerrico, Inc.*

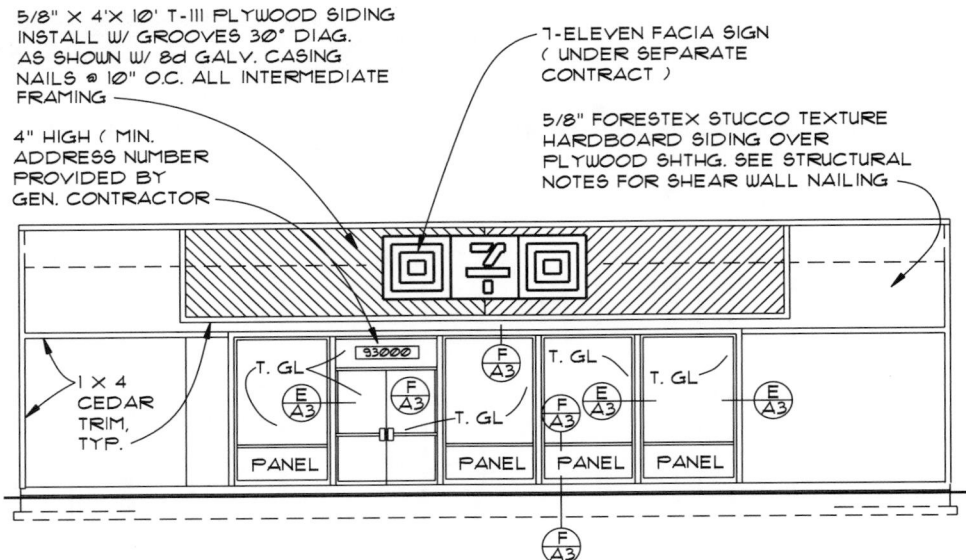

5/8" X 4'X 10' T-III PLYWOOD SIDING
INSTALL W/ GROOVES 30° DIAG.
AS SHOWN W/ 8d GALV. CASING
NAILS @ 10" O.C. ALL INTERMEDIATE
FRAMING

7-ELEVEN FACIA SIGN
(UNDER SEPARATE
CONTRACT)

5/8" FORESTEX STUCCO TEXTURE
HARDBOARD SIDING OVER
PLYWOOD SHTHG. SEE STRUCTURAL
NOTES FOR SHEAR WALL NAILING

4" HIGH (MIN.
ADDRESS NUMBER
PROVIDED BY
GEN. CONTRACTOR

1 X 4
CEDAR
TRIM,
TYP.

33000

T. GL

PANEL

FIGURE 23.71 ■ Elevations for commercial structures, such as this 7-11 store, often require little detail and may be drawn at a small scale. *Courtesy The Southland Corporation.*

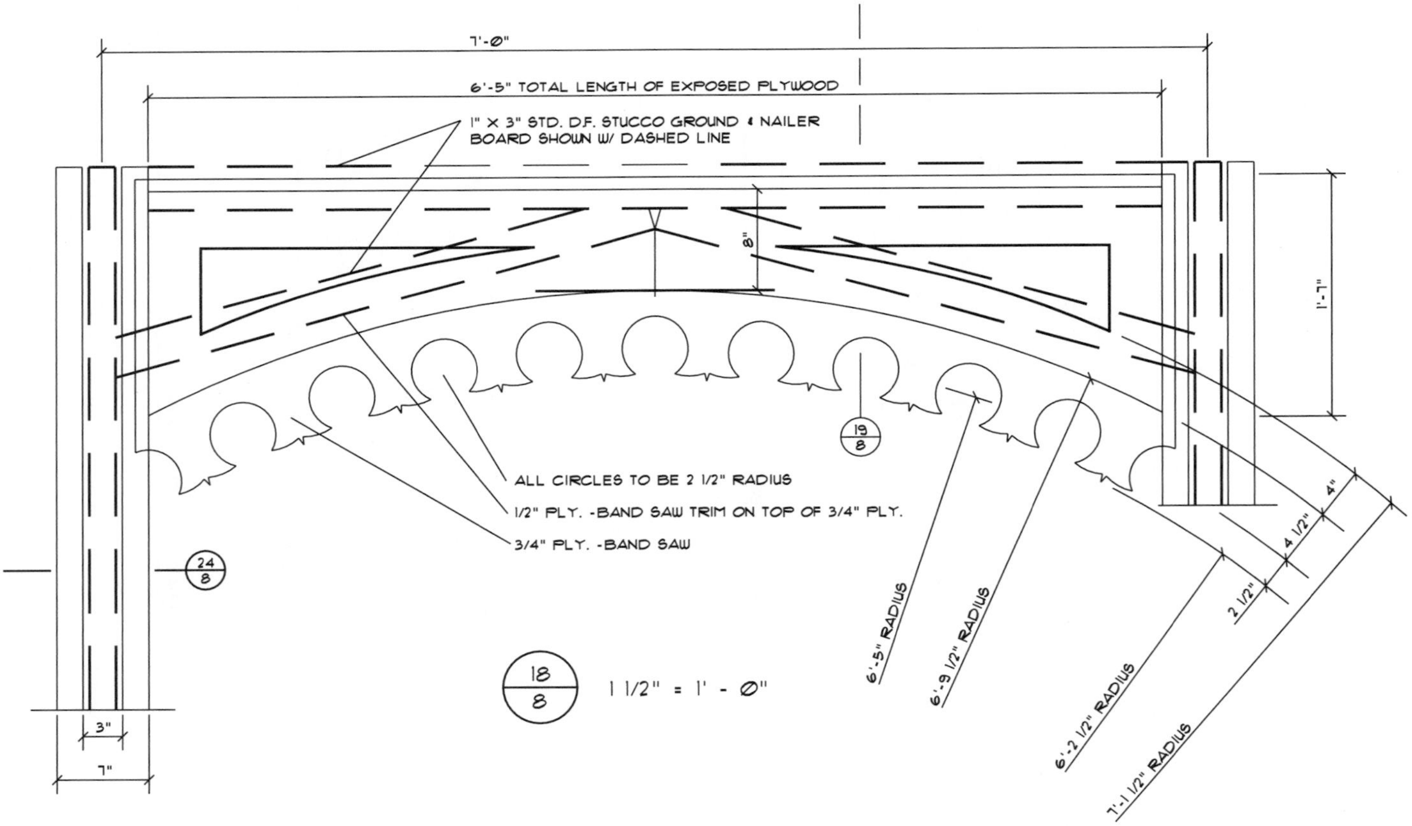

7'-0"

6'-5" TOTAL LENGTH OF EXPOSED PLYWOOD

1" X 3" STD. D.F. STUCCO GROUND & NAILER
BOARD SHOWN W/ DASHED LINE

8"

1'-7"

19
8

ALL CIRCLES TO BE 2 1/2" RADIUS

1/2" PLY. -BAND SAW TRIM ON TOP OF 3/4" PLY.

3/4" PLY. -BAND SAW

24
8

18
8

1 1/2" = 1' - 0"

6'-5" RADIUS

6'-9 1/2" RADIUS

6'-2 1/2" RADIUS

7'-1 1/2" RADIUS

2 1/2"

4 1/2"

4"

3"

7"

FIGURE 23.72 ■ In addition to structural details, the drafter may be required to draw interior elevations and details such as this finish detail for Farrell's Restaurant. *Courtesy Structureform Masters, Inc.*

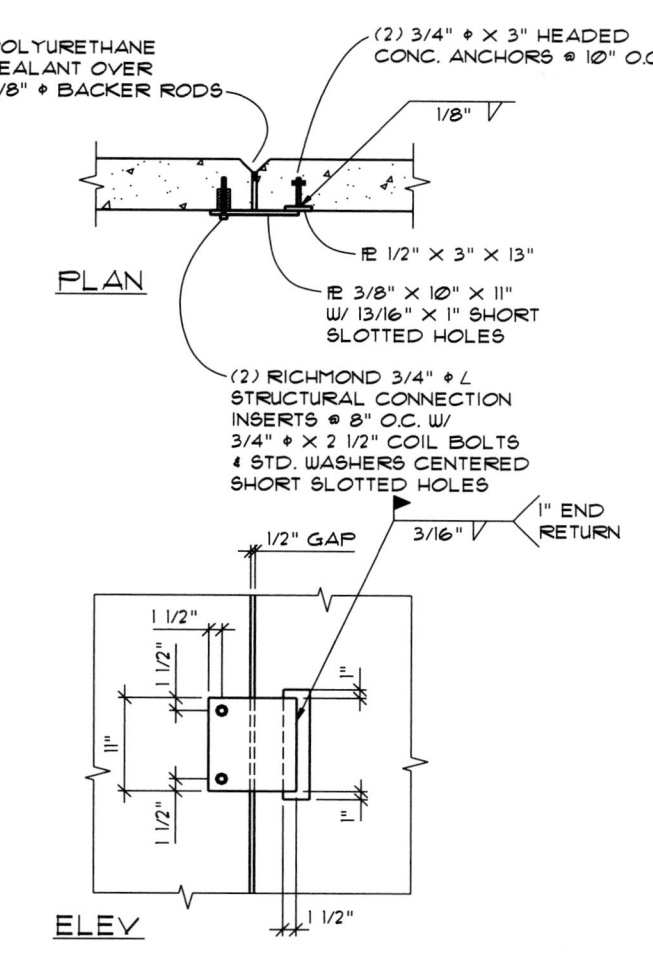

PLAN

POLYURETHANE
SEALANT OVER
5/8" ⌀ BACKER RODS

(2) 3/4" ⌀ X 3" HEADED
CONC. ANCHORS ⌀ 10" O.C.

1/8"

℗ 1/2" X 3" X 13"

℗ 3/8" X 10" X 11"
W/ 13/16" X 1" SHORT
SLOTTED HOLES

(2) RICHMOND 3/4" ⌀ ∠
STRUCTURAL CONNECTION
INSERTS ⌀ 8" O.C. W/
3/4" ⌀ X 2 1/2" COIL BOLTS
⅋ STD. WASHERS CENTERED
SHORT SLOTTED HOLES

1" END
RETURN

3/16"

1/2" GAP

1 1/2"

1 1/2"

1"

1 1/2"

1 1/2"

ELEV

CONNECTIONS TO BE SPACED
⌀ 5'-6" O.C. BEGINNING 12"
FROM THE SLAB.

1 / 3   PANEL CONNECTOR
1" = 1'-0"

FIGURE 23.73 ■ A typical wall connection detail.

out and filed for reference. Changes can cause increased costs in the project.

CADD Application

It is easy to draw revision clouds with CADD. AutoCAD, for example, has a REVCLOUD command that allows you to specify the revision cloud arc length and draw a revision cloud around any desired area. The command works by picking a start point and then moving the cursor in the desired direction to create the revision cloud, as shown in Figure 23.77a. Create the revision cloud in a pattern around the desired area while moving the cursor back toward the start point. Press the pick button when the cursor is back at the start point to complete the revision cloud as shown in Figure 23.77b.

BASIC DRAWING LAYOUT STEPS

The following gives you some basic steps that you can use when laying out a set of architectural/structural drawings. Not all elements of a set of drawings are demonstrated, but the steps used in the examples can be applied to any of the components in the set of working drawings.

Laying Out the Plan Views

The structural plan views are normally the drawings that begin the complete set. These include the types of drawings already discussed, such as the floor plan, foundation plan, and roof framing plan. After the plan views have been drawn, the sections and details may be properly correlated.

The Floor Plan, Step 1

Begin by laying out the floor plan at a scale that will provide the clearest representation on the sheet size to be used. Normally a

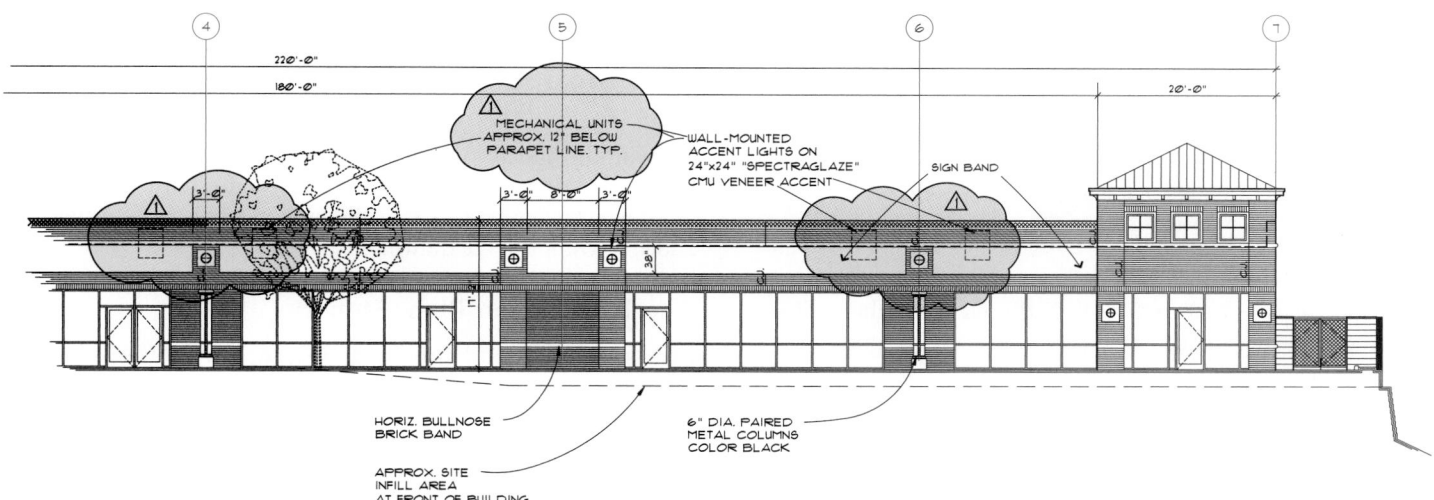

FIGURE 23.74 ■ A typical revision cloud and delta reference. *Portion of drawing courtesy Ankrom Moisan Architects.*

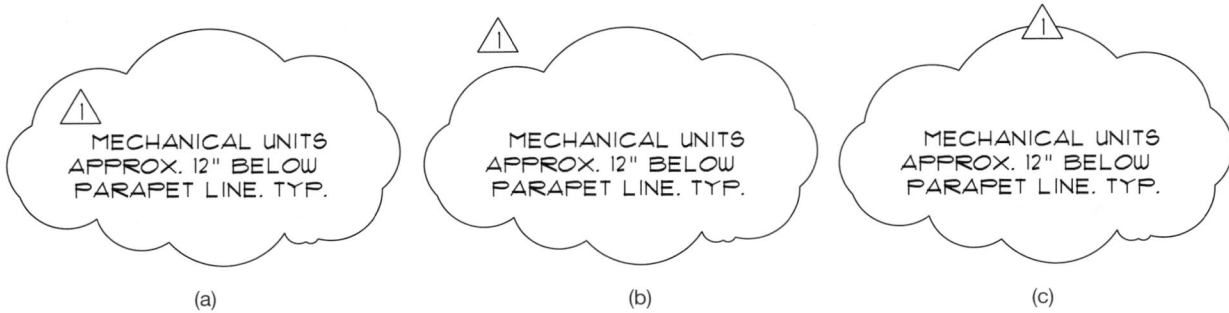

FIGURE 23.75 ■ Placement of the delta reference with the revision cloud: (**a**) delta inside of the revision cloud; (**b**) delta outside of the revision cloud; (**c**) delta inserted in the revision cloud line.

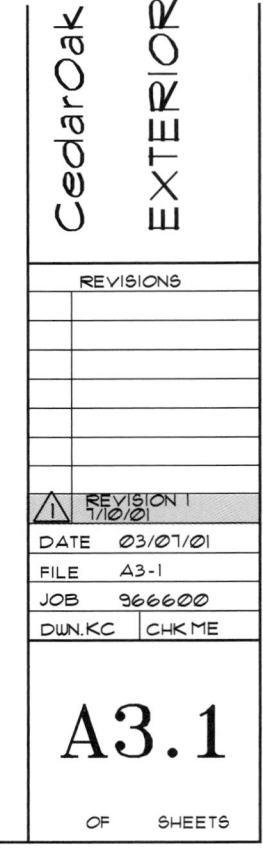

FIGURE 23.76 ■ Revision reference in the title block. The specific information about the revision is found in the job file. *Portion of title block courtesy Ankrom Moisan Architects.*

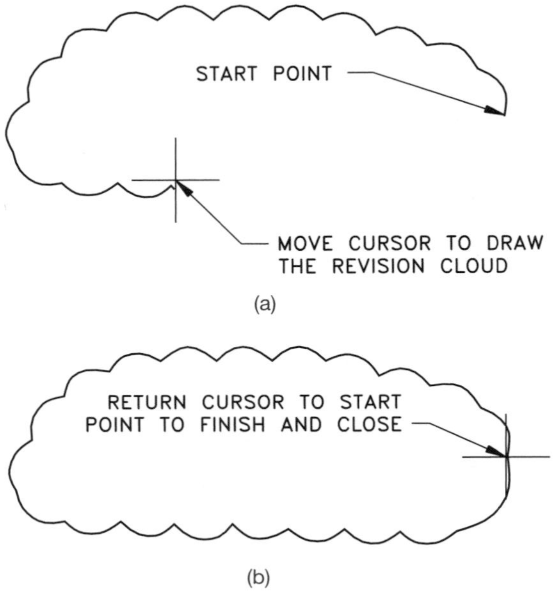

FIGURE 23.77 ■ The AutoCAD REVCLOUD command automatically draws arc segments along the path that you move the cursor.

scale range of $1/8" = 1'-0'$ to $1/4" = 1'-0"$ is used. Draw the outline of the building, leaving adequate space for dimensions and notes, as shown in Figure 23.78.

The Floor Plan, Step 2

Draw all internal walls, partitions, posts, and fixtures. Draw all dimensions. Structural drafting normally uses aligned dimension-

ing with the dimension numeral placed above the dimension line, and slashes placed where the dimension and extension lines meet, as shown in Figure 23.79. Finally, notes, symbols, and titles are added to complete the drawing.

The Foundation Plan, Step 1

After the floor plan has been drawn, turn off or freeze the layers that are not used to lay out the foundation such as partitions and fixtures. Leave the perimeter wall, bearing wall, and post layers turned on. The floor plan makes an excellent guide for the other plans because it is the base for all other construction.

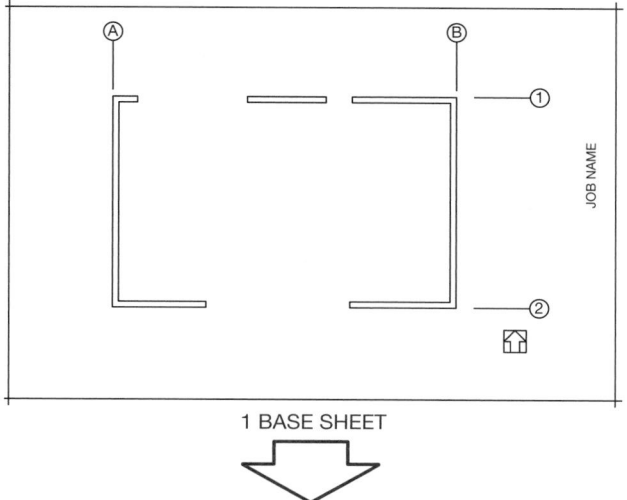

FIGURE 23.78 ■ Outline the floor plan. *Courtesy Structureform Masters, Inc.*

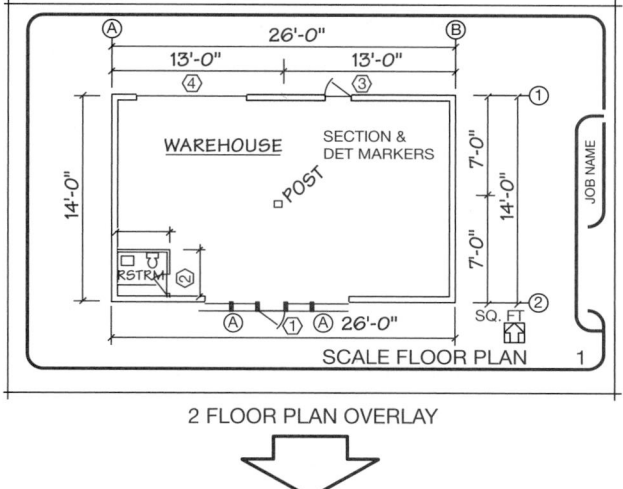

FIGURE 23.79 ■ Draw all interior features, dimensions, notes, and titles to complete the floor plan. *Courtesy Structureform Masters, Inc.*

The Foundation Plan, Step 2

Begin by drawing the foundation walls and footings. Next, add all dimensions, notes, and symbols. Finally, letter the title and scale, as shown in Figure 23.80.

The Roof Framing Plan

The layout of the roof framing plan works in much the same way as the foundation plan. Again, use the floor plan as a basis for the roof plan drawing. This gives you all of the layout features to quickly begin the new drawing. First, place all beams and purlins. Secondly, completely dimension the drawing and add all necessary notes and symbols. Finally, add all schedules and titles, as shown in Figure 23.81.

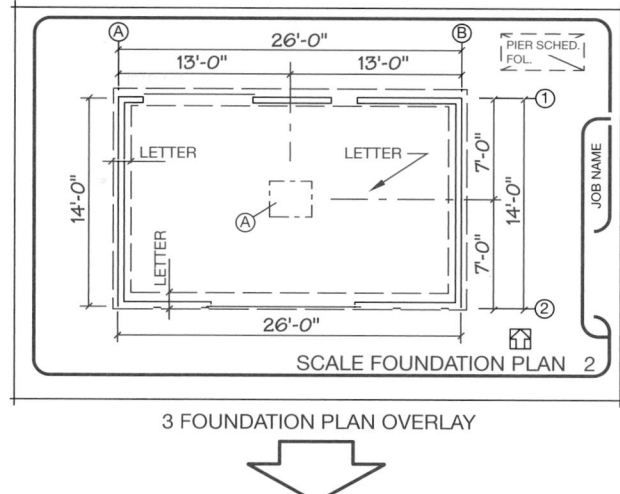

FIGURE 23.80 ■ Lay out the foundation with the floor plan as a guide. *Courtesy Structureform Masters, Inc.*

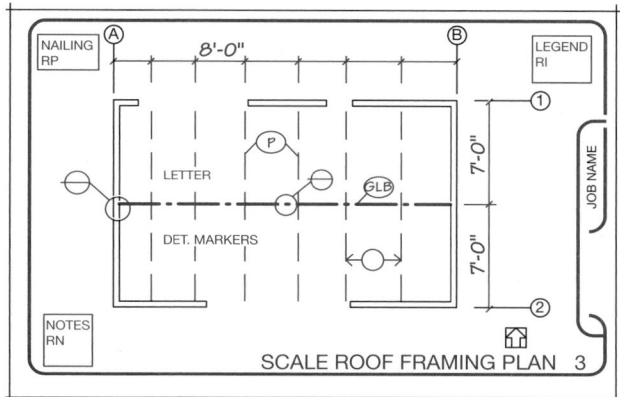

FIGURE 23.81 ■ Lay out the roof framing plan with the floor plan as a guide. *Courtesy Structureform Masters, Inc.*

The Details, Step 1

The detail drawings may be placed with the plan views or on separate sheets. In either case, the details and sections must be clearly correlated to the plans. The scale of the detail drawings depends on the size and complexity of the structure to be drawn. Select a scale that clearly shows all construction details without oversizing the drawing, thus wasting space and drawing time. Common scales for detail drawing range from $1/2" = 1'-0"$ to $3" = 1'-0"$. All the details may be the same scale, or the scale may change depending on the complexity of the drawing. In many cases it is necessary to arrange many details on one sheet or in an area on a sheet. Do this by laying out the details, beginning at the top left corner of the working area. Proceed with additional details placed in rows from left to right. The engineering sketch in Figure 23.2a (see page 766) is used as a guide for the following layout. Start the detail drawing by blocking out the major components of the structure. As you can see in the engineer's sketch, the beam on the left is 24" high and the

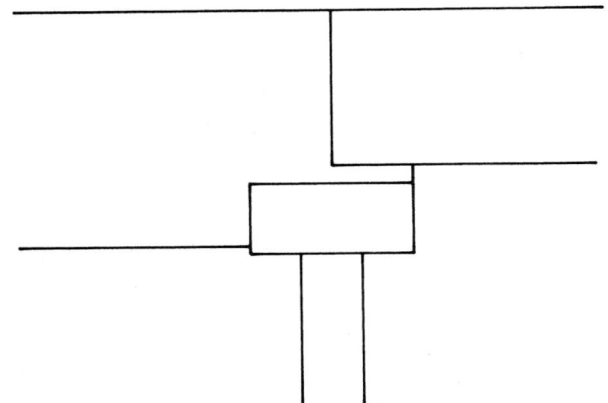

FIGURE 23.82 ■ Step 1—Block out components of the detail.

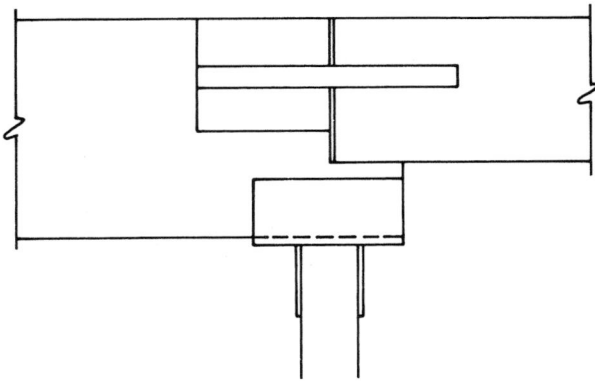

FIGURE 23.83 ■ Step 2—Draw all construction members.

beam on the right is 16" high. Thus the maximum beam height is 24". If you add another 12" for the post and 12 more inches for the notes, at a scale of 1" = 1'–0" you need a total drawing height of 48", or an actual height of 4" at the selected scale. The width of the detail is up to you. All you need is enough space to show all of the construction members. So, a width of about 4" should work. (See Figure 23.82.)

The Details, Step 2

Proceed by drawing in all of the construction members to scale. Refer to a vendor's catalog (Figure 23.38, page 787) to determine the actual sizes of the prefabricated connectors. Lay out the detail, as shown in Figure 23.83.

The Details, Step 3

Place all dimensions and notes on the drawing, followed by the title and scale, or the detail identification symbol and scale, as shown in Figure 23.2b (see page 766).

PICTORIAL DRAWINGS

Pictorial drawings such as isometrics are sometimes used in the set of architectural/structural drawings when it is determined to be necessary to represent something more clearly than in a two-dimensional drawing. Pictorial drawings are not required, but are done when the specific office feels that they aid in the visualization of specific construction applications. Figure 23.84 shows the use of pictorial drawings. Pictorial drawings are covered in detail in Chapter 19 of this text.

WEB SITE RESEARCH

The following Web sites can assist you in doing additional research on subjects such as manufacturing materials, manufacturing practices, tooling, and related areas:

www.aluminum.org—Aluminum Association.
www.asme.org—American Society of Mechanical Engineers.
www.aise.org—Association of Iron and Steel Engineers.
www.industrialpress.com—*Machinery's Handbook.* Materials and processes.
www.worldsteel.org—International Iron and Steel Institute.
www.csinet.org—Construction Specifications Institute.
www.aisc.org—American Institute for Steel Construction.
www.bcewp.com—Boise Cascade Engineered Wood Products Division.
www.wii.com—Willamette Industries.
www.awc.org—American Wood Council.
www.aci-int.org—American Concrete Institute.
www.tilt-up.org—The Tilt-Up Concrete Association.
www.concrete.com—The industry portal for concrete services and information.
www.precast.org—National Precast Concrete.
www.steel-sci.org—The Steel Construction Institute.
www.strongtie.com—Simpson Strong Tie information and catalog.
www.pci.org—Precast/Prestressed Concrete Institute.
www.cpci.ca—Canadian Precast/Prestressed Concrete Institute.
www.buildingteam.com—Building and construction industry links.
www.arcat.com—Building materials information, specifications, and CAD details.

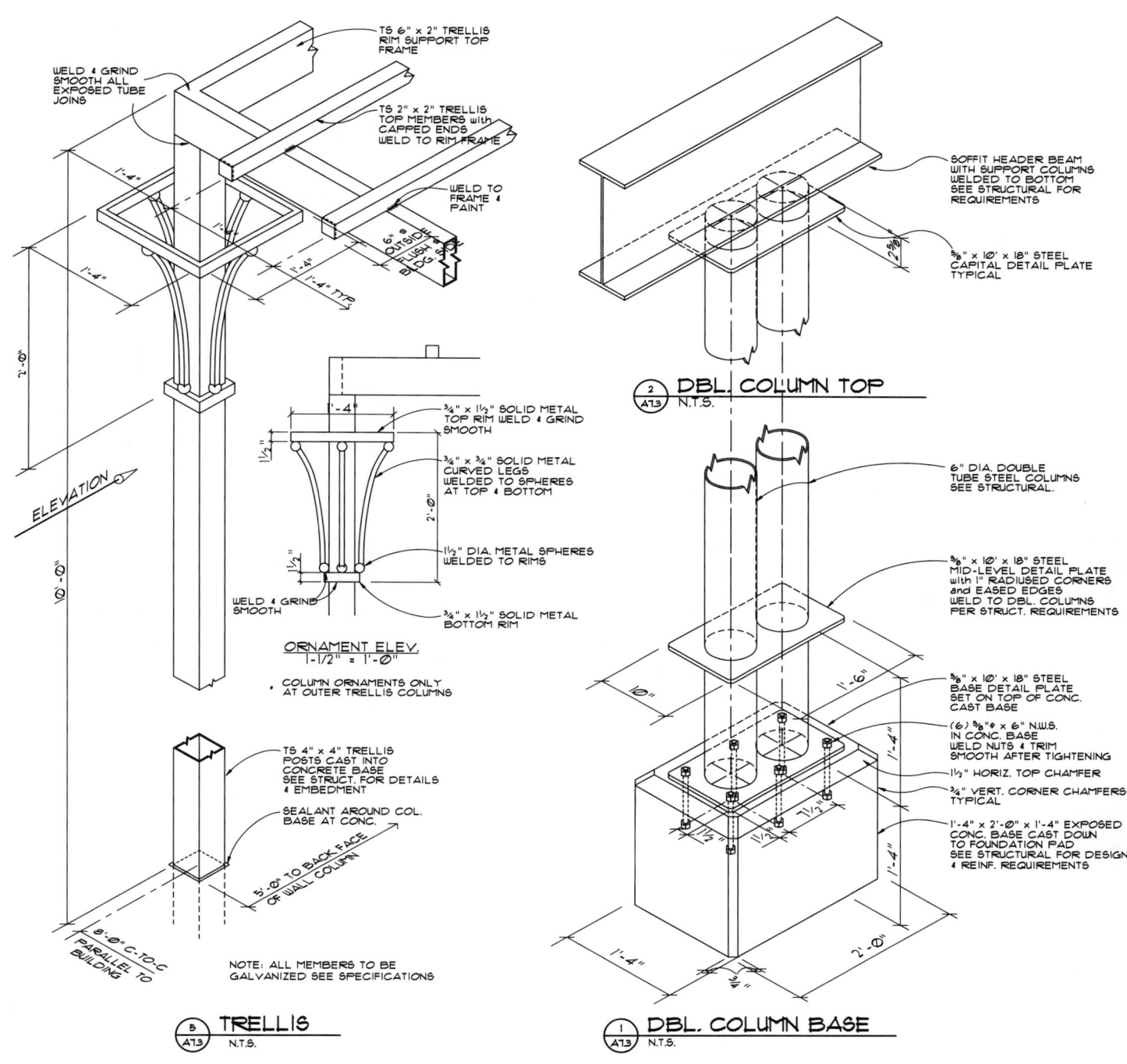

FIGURE 23.84 ■ Pictorial detail drawings. *Courtesy Ankrom Moisan Architects.*

CADD APPLICATIONS

Structural engineering software programs are available that integrate structural analysis, design, and drafting. A 3-D model is made during the design and drafting process as shown in Figure 23.85. The model may be updated at any time from new design information. The model may be used to automatically create framing plans and elevations as shown in Figure 23.86. The software package also creates construction details for a variety of materials. The construction details may be drawn at any scale and stored for later use. Parametric design of welding symbols and specifications allows the designer to change a variable and automatically update the entire drawing. Detail models as shown in Figure 23.87 may be converted to traditional 2-D views complete with dimensions and specifications.

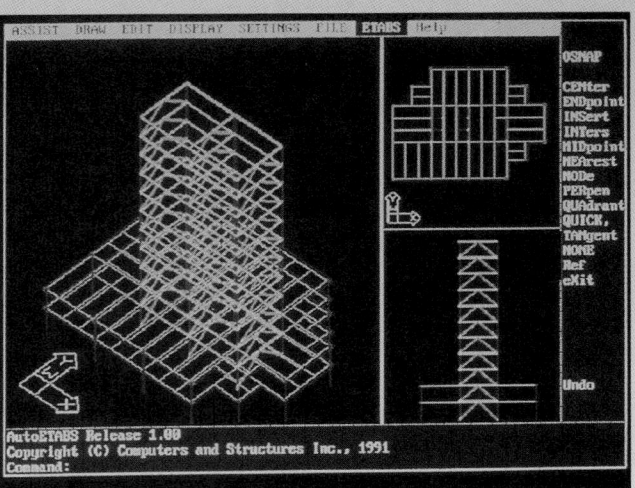

FIGURE 23.86 ■ Creating framing plans and elevations with the 3-D model. *Courtesy Computers and Structures, Inc.*

FIGURE 23.85 ■ A 3-D model is made during the CADD design and drafting process. *Photo courtesy of Softdesk.*

FIGURE 23.87 ■ A 3-D CADD model of a structural detail. *Photo courtesy of Softdesk.*

PROFESSIONAL PERSPECTIVE

The drafting procedure for the preparation of structural drawings is often complicated by the amount of information to be contained in these drawings. Structural drawings are normally done on large sheets with as much information as possible placed on each sheet. The first task is to determine the sheet size. This may already be decided as a standard within the company: for example, 22 × 34. The factors that should be considered when selecting sheet size or when drawing on predetermined sheet sizes are:

■ The size of the plan.

■ The number of dimensions and notes needed.

■ The scale used.

■ The number of details to be placed on the sheet.

■ The amount of free space required for possible future revisions.

All of these things must be considered, because you do not want to end up with some sheets that are overcrowded and others that have little information. Use CADD layers such as FL1, FL2, FDN, ROOF, ELEV, and NOTES.

MATH APPLICATION

DESIGNING AN OBSERVATION PLATFORM

Problem: Suppose a building code states that the rise between stair treads cannot exceed 8" for a 12" tread. Find the minimum distance for dimension L and a corresponding angle A in Figure 23.88 of an observation platform.

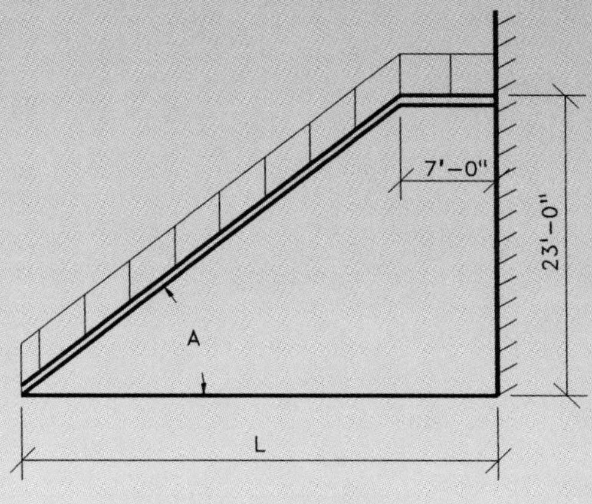

FIGURE 23.88 ■ Observation platform.

Solution: Geometrical figures are *similar* when they have the same shape. They are *congruent* when they have the same shape *and* size. When two triangles are similar, proportional equations can be written. In this design problem, we have similar triangles: the large staircase and the smaller single tread. (See Figure 23.89.)

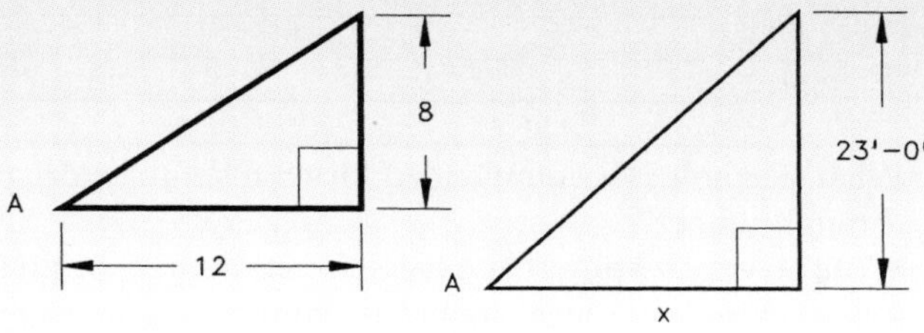

FIGURE 23.89 ■ Two similar triangles.

Set up a proportion

$$\frac{23}{x} = \frac{8}{12}$$

which gives

$$x = 34.5'$$

Then dimension L is an additional 7'.

$$L = 34.5 + 7 = 41.5' \text{ or } \mathbf{41'\text{–}6"}$$

Also, being right triangles, angle A can be found by the application of one trig function:

$$A = \text{Inv tan } \frac{8}{12} = 33.7°$$

It is often necessary for the drafter to calculate the weight and cubic yard of concrete for bills of materials, cost estimates, and construction purposes. The following formulas can be used to make these calculations:

(Length × Width × Height) × 150 = Total weight
(Length × Width × Height) ÷ 27 = Cubic yard

Problem: Given a precast concrete panel with the dimensions 18'–4" long, 1'–4" wide, and 2'–0" high, calculate the weight in pounds, and the volume in yards.

Solution:

18'–4" × 1'–4" × 2'–0" = 48.88889 cubic feet
48.88889 × 150 = 7333.3333 lb
48.88889 ÷ 27 = 1.81 cubic yards

If there are holes and/or cutouts, it is necessary to calculate the combined volume and weight of these and subtract them from the total.

CHAPTER 23

Structural Drafting Test

DIRECTIONS

Answer the questions with short complete statements or drawings as needed.

1. Describe situations in structural drafting when object lines may be drawn thicker than standard width and when they may be drawn thinner than standard width.
2. Complete the statement: The best structural lettering is _____.
3. Identify the primary components of concrete.
4. Identify the three basic methods of using concrete for building construction.
5. What is the diameter of a #5 steel rebar?
6. What do the parts of the callout 6 × 6 - #10 WWF mean?

a. 6 × 6 _____

b. #10 _____

c. WWF _____

7. Describe the purpose of placing reinforcing steel in poured-in-place concrete.

8. Identify three poured-in-place concrete floor systems.

9. Define *precast concrete.*

10. Define *precast-prestressed concrete.*

11. Define *tilt-up construction.*

12. Give a sample anchor bolt callout.

13. Give a sample rebar callout.

14. How should concrete rebar clear distances be dimensioned?

15. Give a sample slab-on-grade callout.

16. Define *structural engineering drawings.*

17. Define *shop drawings.*

18. What is another name for shop drawings?

19. Describe the modular principles of concrete block construction.

20. How is a bond beam constructed?

21. Define *pilaster.*

22. Describe a panelized roof system.

23. Give a sample timber callout.

24. List the three common types of laminated beams.

25. Give a sample glu-lam callout.

26. Give a sample callout that may be used to label lumber floor joists.

27. How should all lumber and timber be dimensioned on a drawing?

28. Put the following plywood information together into a proper callout:
   a. Exterior APA plywood
   b. 3/4" thick
   c. Group 1
   d. 48/24
   e. A grade veneer one side and C grade other side
   f. Nail w/ 12d 4" OC edges and 12" OC field

29. List at least three advantages of using steel studs.

30. Name at least three types of structural systems commonly used for prefabricated steel structure systems.

31. Define *girts* and *purlins.*

32. Define *plates* as associated with steel materials and give the symbol for plate steel.

33. Describe bars as associated with steel materials.

34. What do the components of the steel shape callout W24 × 55 mean?
   a. W _____
   b. 24 _____
   c. 55 _____

35. Identify the callout symbol for the three I-shaped structural materials.

36. Give a sample callout for a C shape.

37. Give a sample callout for flat bar.

38. Give a sample plate callout.

39. Give a sample rectangular tubing callout.

40. Give a sample pipe callout.

41. Give a sample equal leg angle callout.

42. Give a sample bolt callout.

43. What type of bolt head is assumed unless otherwise specified on a drawing?

44. Give the "standard" (steel-to-steel) recommended hole diameter for a 5/8"∅ bolt.

45. Interpret the parts of this note: #5 @ 12" OC EACH WAY.
   a. #5 _____
   b. @ _____
   c. 12" OC _____
   d. EACH WAY _____

46. If lengths are required for structural members, how should they be given?

47. How should location dimensions be established to the following members?
   a. Wide flange and other "I" shapes _____

   b. Channels _____

   c. Angles _____

   d. Plates _____

48. What size hole is recommended for #8 rebar to be driven through timber?

49. Write an explanation that covers the basics of coordinating working drawings, including numbering of pages, coordinating details and sections, and laying out details and sections.

50. Explain how the revision process works in structural drafting, including the use of revision clouds and the related revision note.

CHAPTER

23 *Structural Drafting Problems*

DIRECTIONS

1. Please read all related instructions before you begin working.

2. Use manual or computer-aided drafting as required by your course guidelines. Use an architectural lettering style or architectural CADD font style unless otherwise specified by your instructor.

3. Use the engineering sketches to prepare a complete set of drawings for the given problems.

4. Do all drawings on 22" × 34" or 24" × 36" sheets unless otherwise specified by your instructor.

5. Use proper sectioning and detailing techniques as correlated to the engineering sketches. Some rec-

ommended scales are provided. You may increase or decrease the scale depending on the available space and the complexity of the section or detail. Sections and details should be drawn at a minimum scale of 3/8" = 1'–0". Judgment should be used if a section or detail requires a lot of information; a larger scale ranging from 3/4" to 1-1/2" = 1'–0" should then be used.

6. It is recommended that you evaluate the entire set of sketches for each problem before beginning to draw. Information needed to draw one problem may be found on the sketch or data for another. The problems are real drafting situations, so you will interpret rough engineering sketches and prepare formal drawings using the best drafting techniques and standards that you have learned.

7. It is suggested that drawing sheets be organized with as much information as possible without crowding or reducing clarity. If additional drawing area is required, it would be better to add another sheet than to overcrowd the drawing.

8. Some problems in this chapter may contain errors, missing information, or slight inaccuracies. This is intentional and is meant to encourage you to apply appropriate problem-solving methods, engineering, and drafting standards in order to solve the problems. This is meant to force you to think about each part and how parts fit together in the structure. As in "real-world" projects, the engineering problem should be considered as a basis for your preliminary layouts. Always question inaccuracies in project designs and consult with the proper standards and other sources. In some cases, an error might be the source of engineering changes provided by your instructor; however, this is determined by your specific course objectives. Other situations may require that corrections be made during the development of the original design drawings. This is not intended as a source of frustration, but is considered to be part of the engineering drafter's daily responsibility in project development.

PROBLEM 23-1   Exterior pole light, sign footing, curb details, and sidewalk and paving details

Draw the following details on one sheet unless otherwise specified by your instructor. *Drawings courtesy of Wendy's International, Inc.*

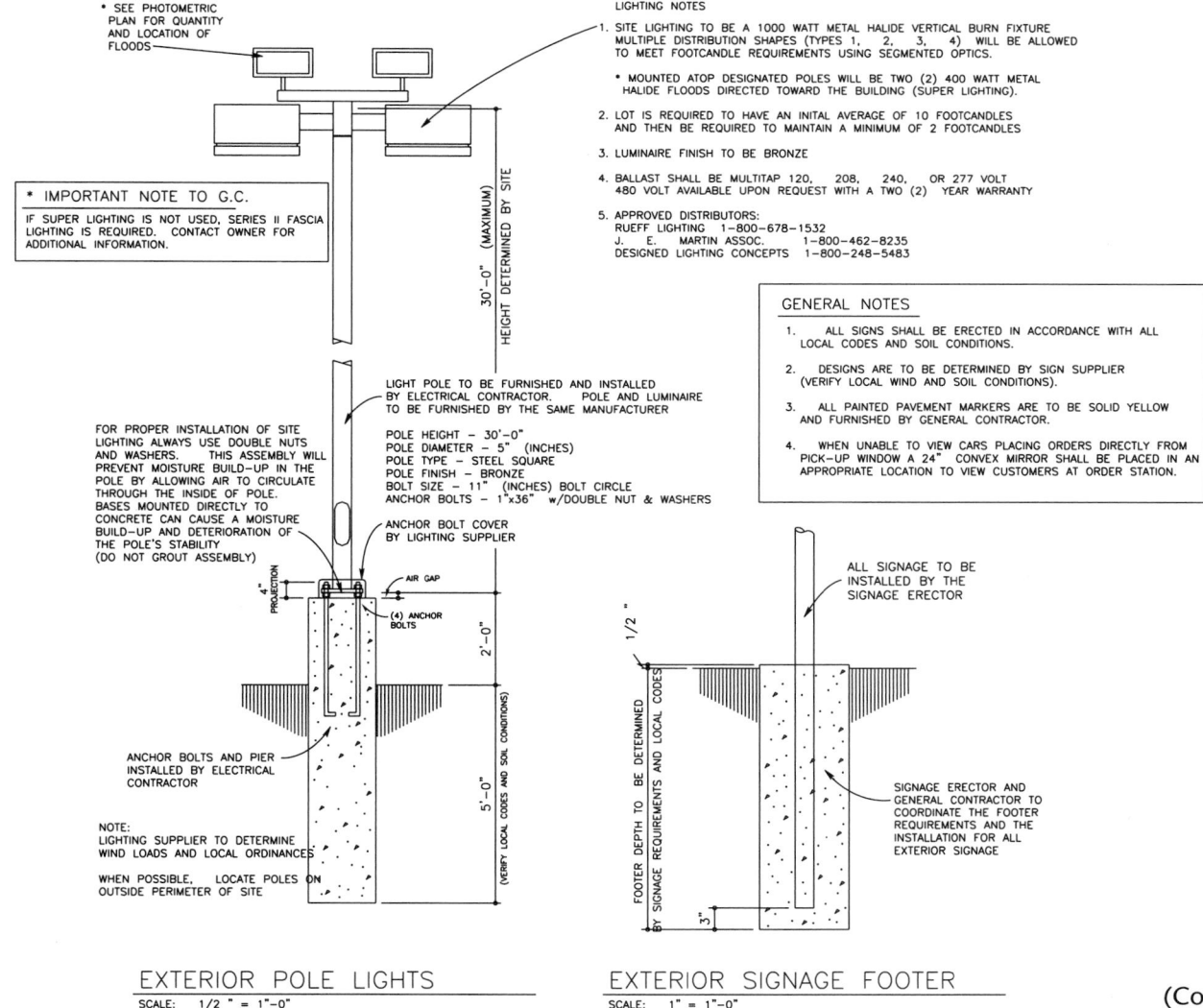

EXTERIOR POLE LIGHTS
SCALE: 1/2 " = 1'–0"

EXTERIOR SIGNAGE FOOTER
SCALE: 1" = 1'–0"

(Continued)

PROBLEM 23-1 (continued)

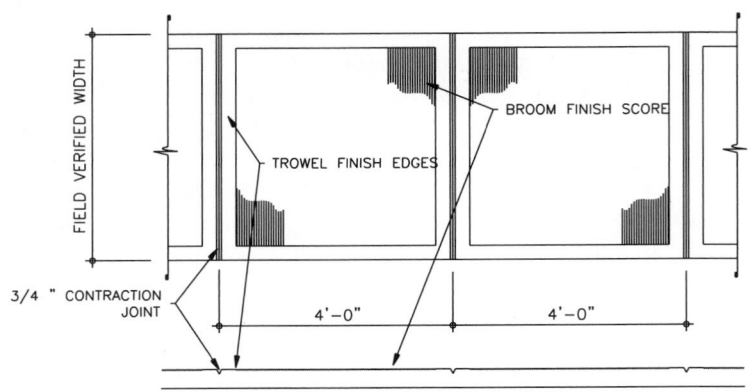

SIDEWALK FINISH DETAIL

NO SCALE

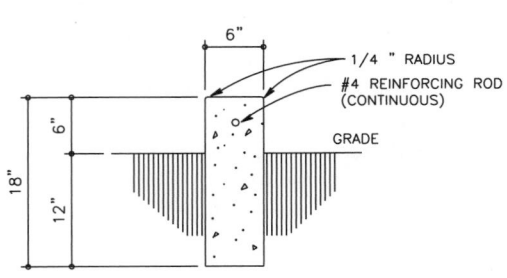

NOTE: ALL CONCRETE CURBS TO HAVE EXPANSION JOINTS
OR SAW CUTS NOT MORE THAN 20'-0" APART

CONCRETE CURB DETAIL

SCALE: 1" = 1"-0"

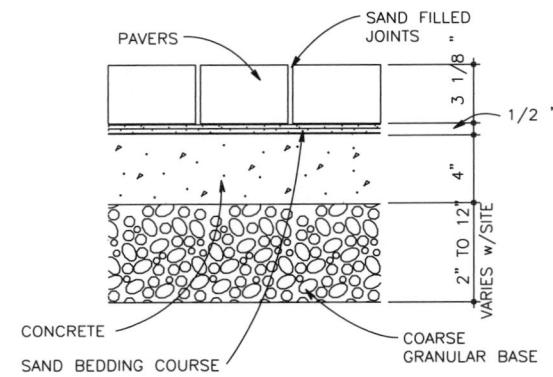

PAVER DETAIL

NO SCALE:

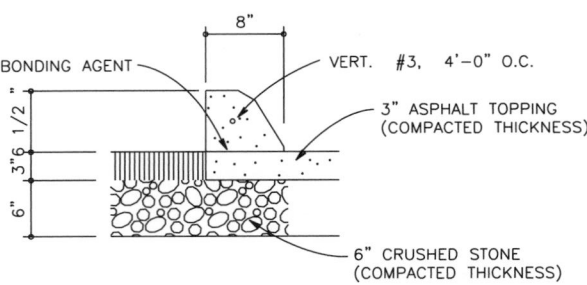

EXTRUDED CURB DETAIL

SCALE: 1" = 1"-0"

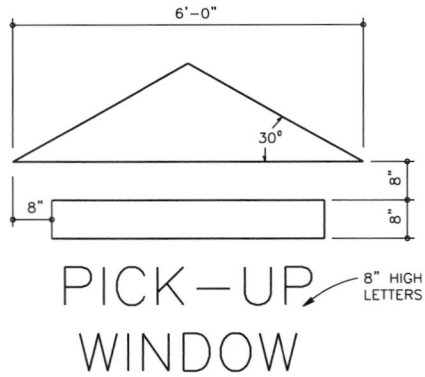

PICK—UP
WINDOW

PAVEMENT MARKER DETAILS

SCALE: 1/2 " = 1'-0"

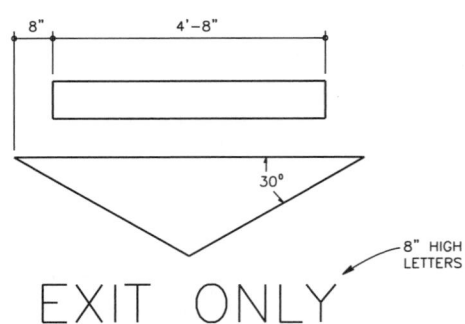

EXIT ONLY

PAVEMENT MARKER DETAILS

SCALE: 1/2 " = 1'-0"

PROBLEM 23-2 Trash enclosure drawings

Given the following layouts, draw the trash enclosure plan, elevation, sections, and details on one sheet, unless otherwise specified by your instructor.

Drawings courtesy of Wendy's International, Inc.

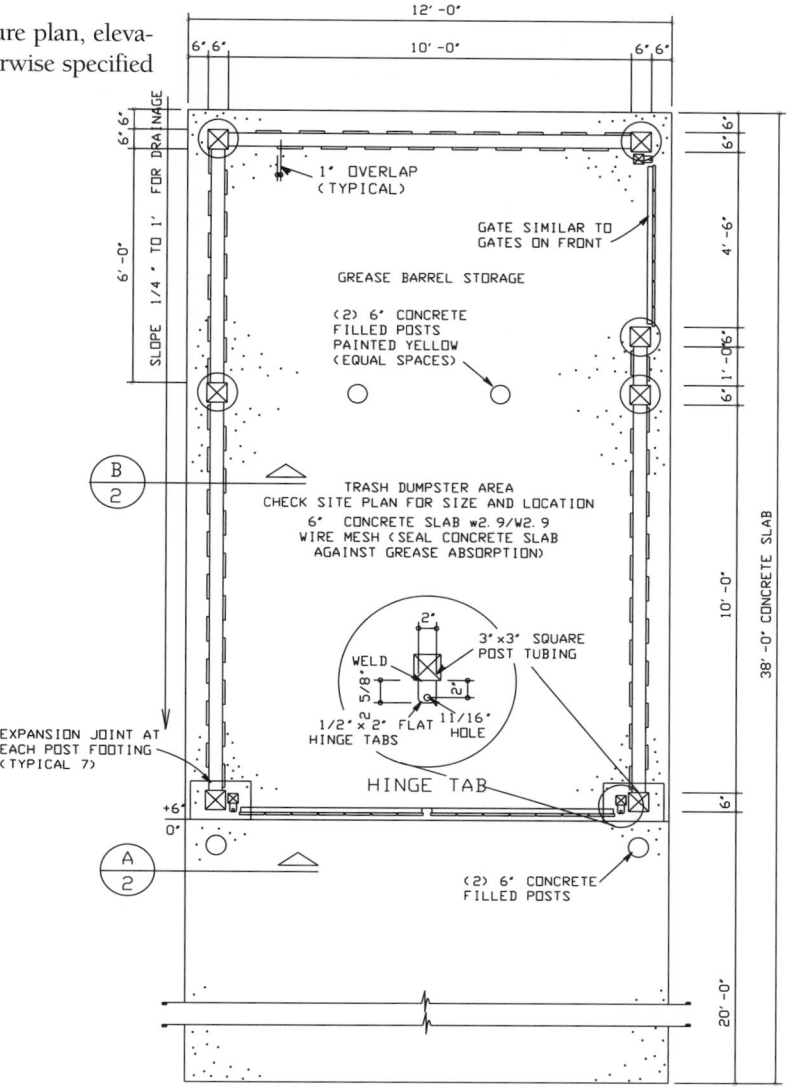

STANDARD TRASH ENCLOSURE PLAN
SCALE: 3/8 " = 1'-0"

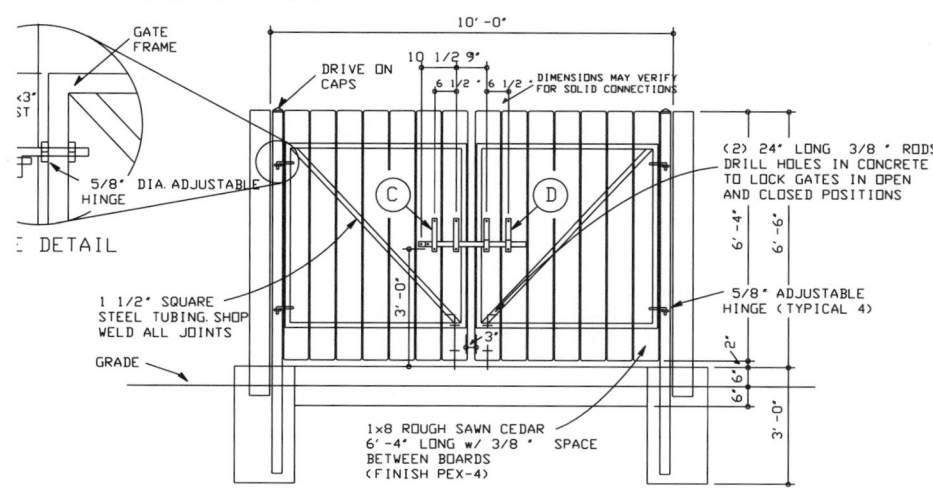

TYPICAL FRONT ELEVATION
SCALE: 3/8 " = 1'-0"

(Continued)

PROBLEM 23-2 continued

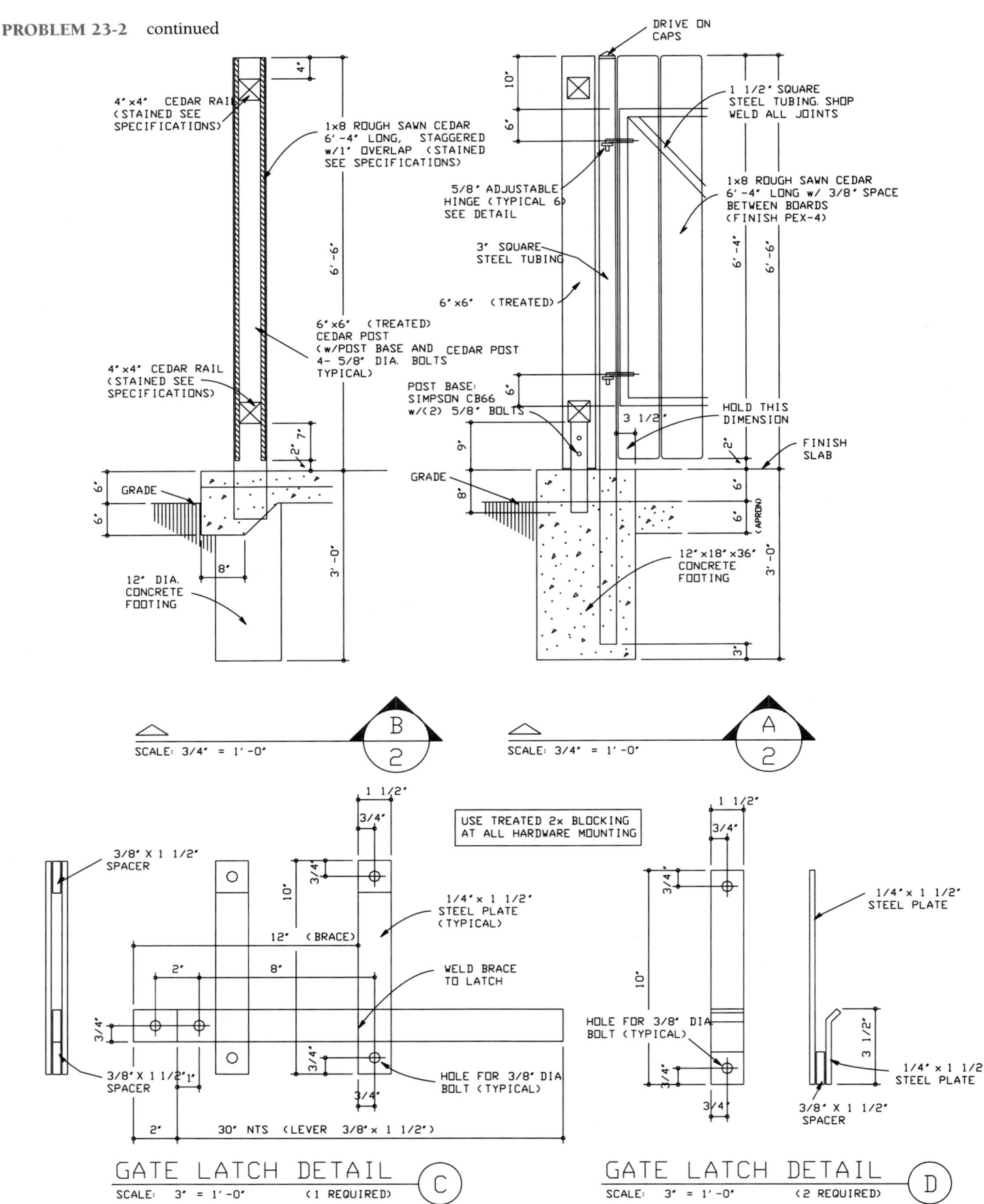

Problems 23-3 through 23-19 are the drawings for a 2400-square-foot storage building.

PROBLEM 23-3   Storage building floor plan

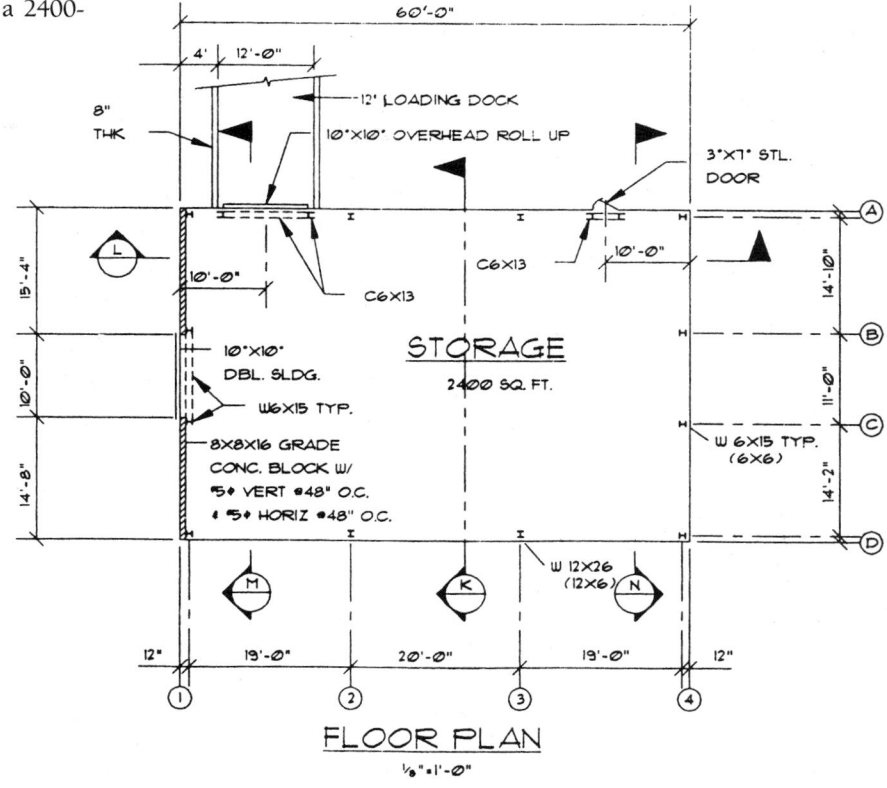

FLOOR PLAN
⅛"=1'-0"

PROBLEM 23-4   Foundation plan

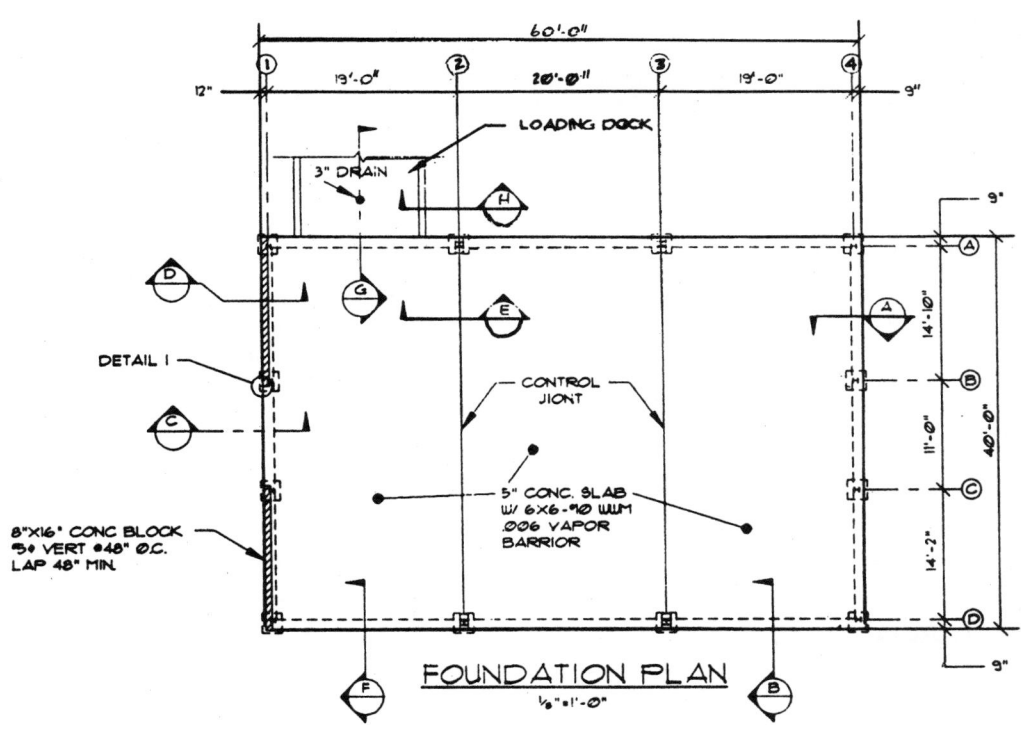

FOUNDATION PLAN
⅛"=1'-0"

GENERAL NOTES:
1. ALL CONC. TO BE 2500 PSI @28 DAYS MIN.
   COMP. STRENGTH.
2. ASSUME SOIL BEAR PRESSURE IS 2000 PSF
3. LAP ALL STEEL 40 DIA. MIN.

PROBLEM 23-5 Section A, Footing section

PROBLEM 23-6 Section B, Footing section

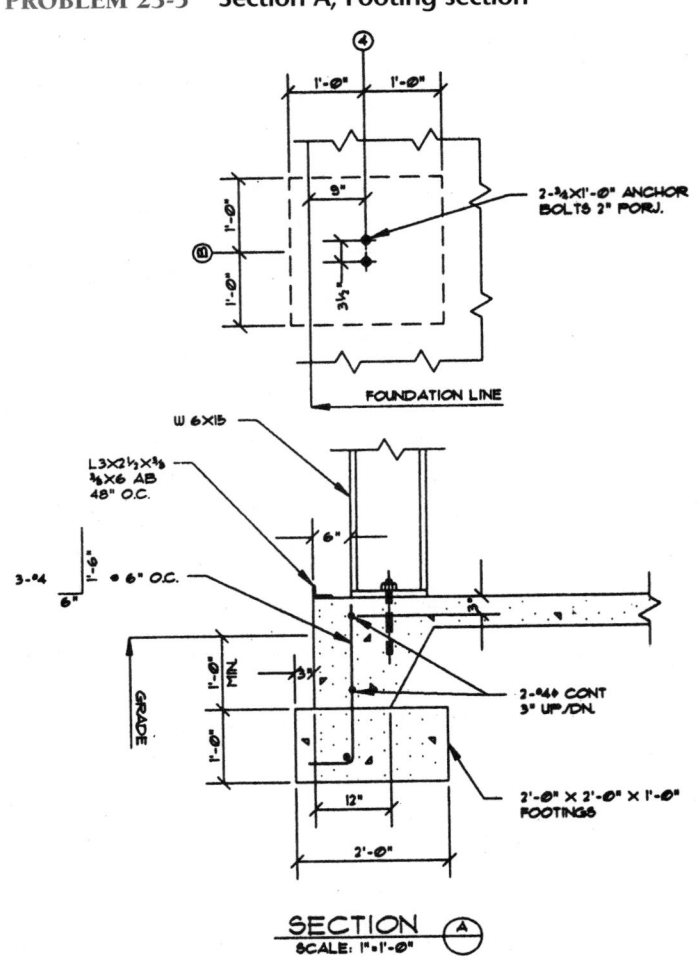

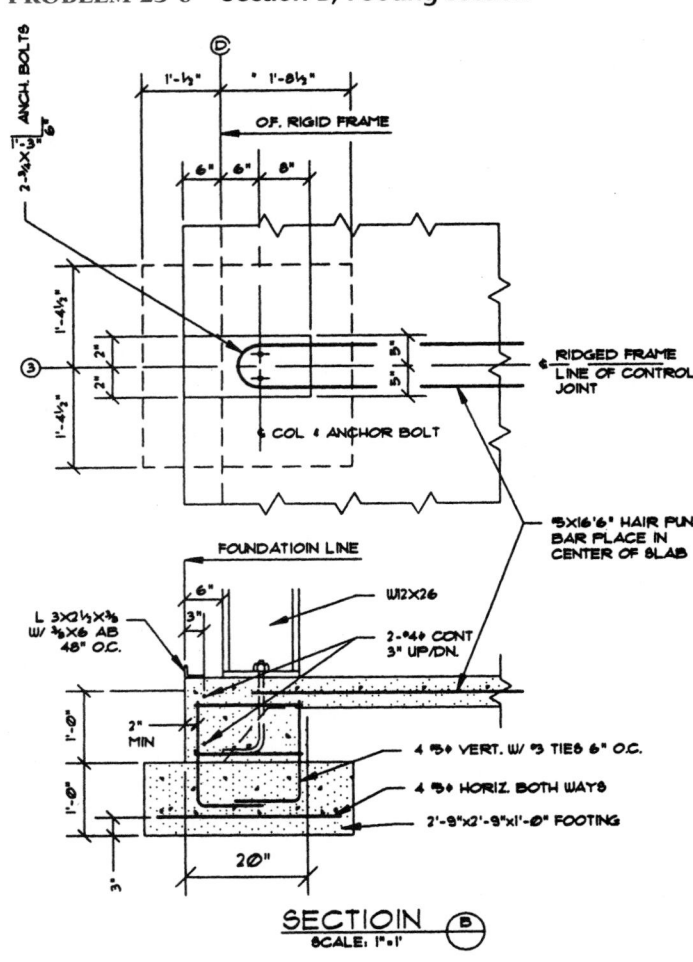

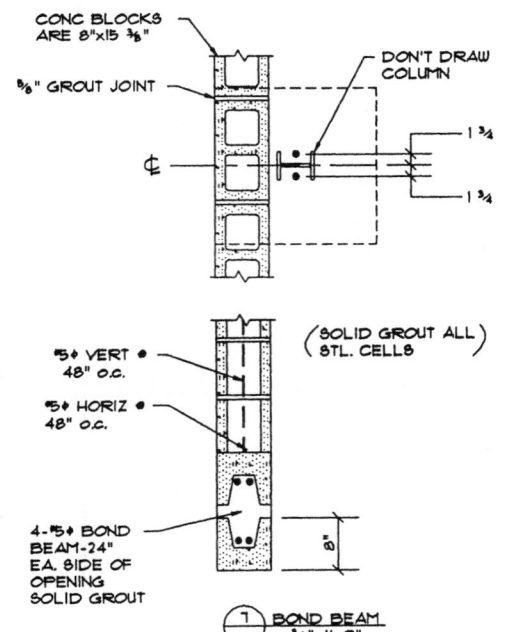

PROBLEM 23-7 Section C, Footing and concrete block
section; Section D, Footing section; Detail 1, Bond
beam detail

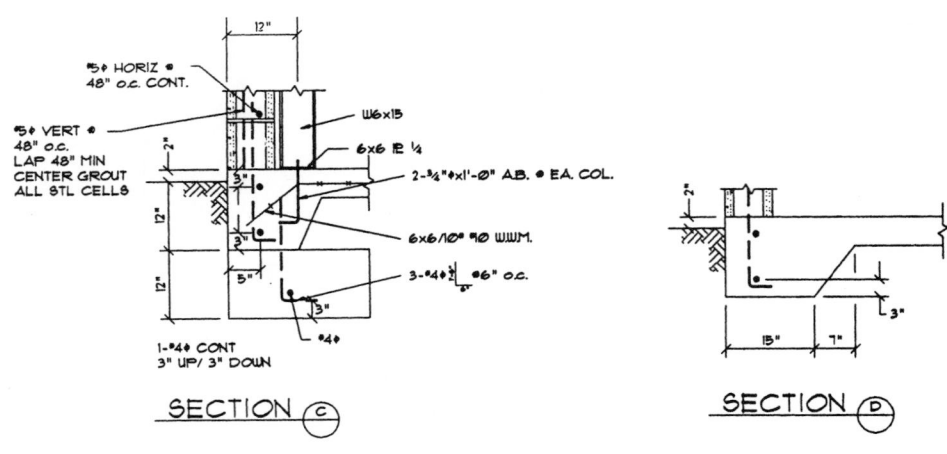

PROBLEM 23-8   Section E, Slab footing section; Section F, Slab joint section

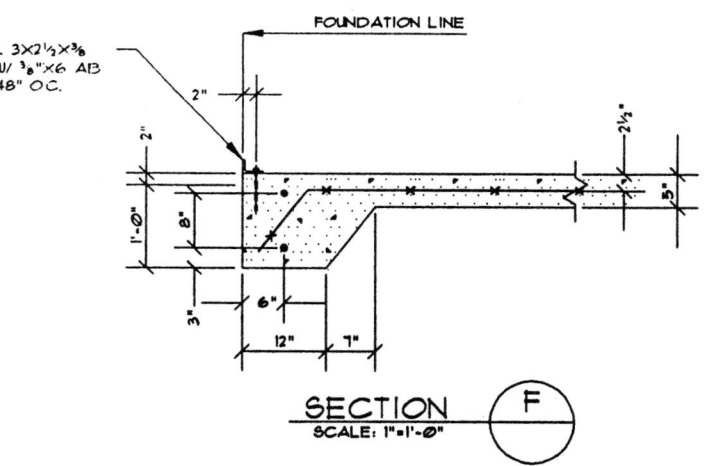

SECTION (F)
SCALE: 1"=1'-0"

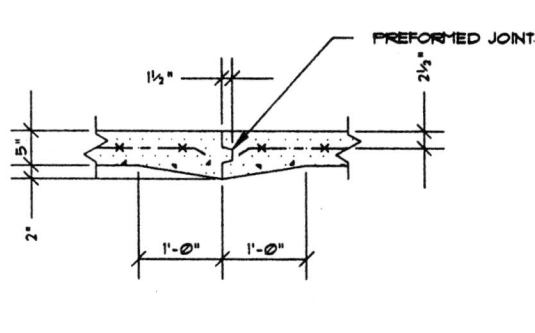

SECTION (E)
SCALE: 1"=1'-0"

PROBLEM 23-10   Sections H and J, Loading dock wall sections

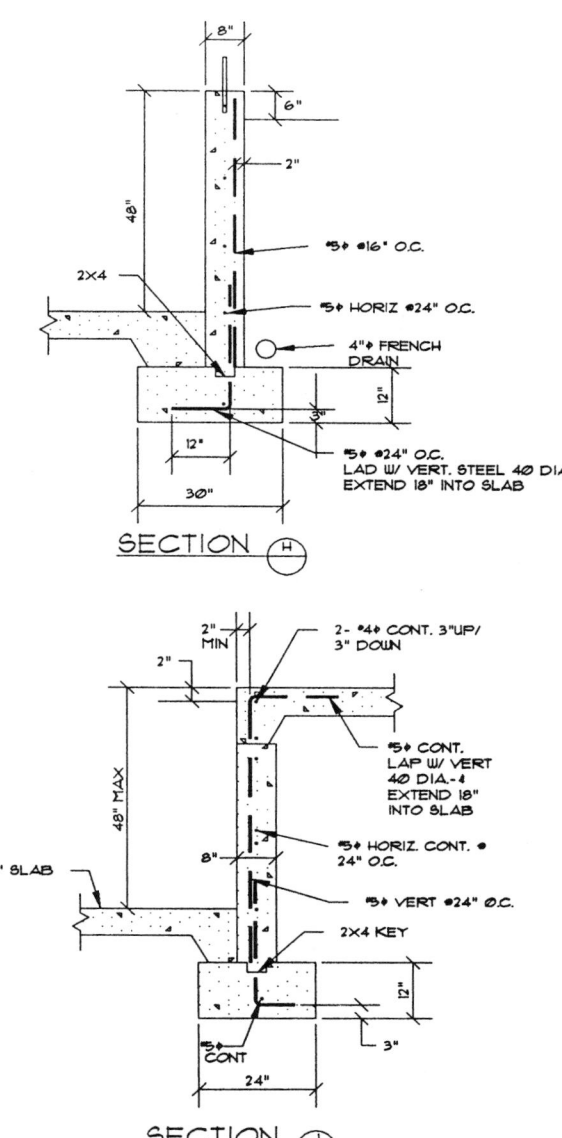

SECTION (H)

SECTION (J)

PROBLEM 23-9   Section G, Loading dock ramp section

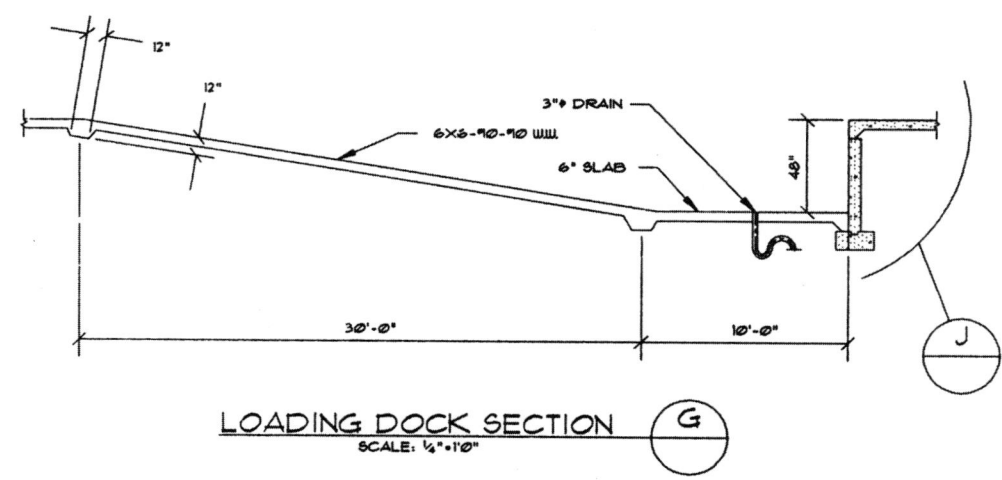

LOADING DOCK SECTION (G)
SCALE: ½"=1'0"

PROBLEM 23-11　Roof framing plan

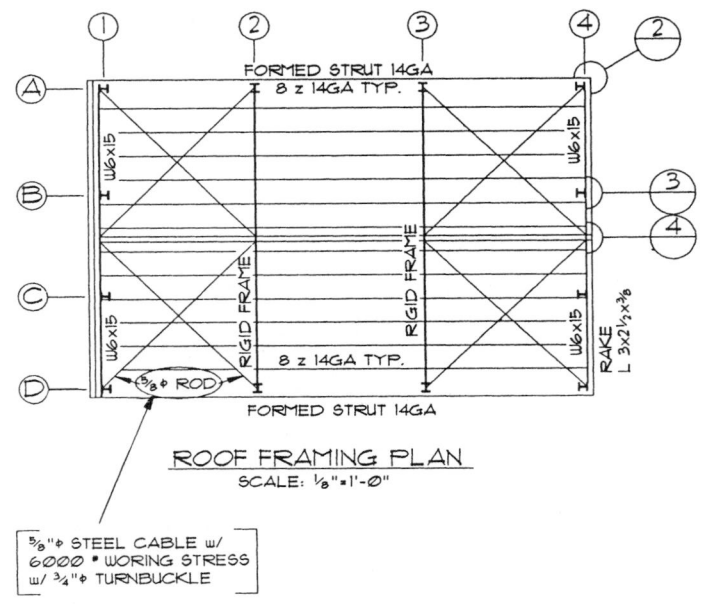

ROOF FRAMING PLAN
SCALE: 1/8"=1'-0"

5/8"Ø STEEL CABLE w/
6000 # WORING STRESS
w/ 3/4"Ø TURNBUCKLE

PROBLEM 23-13　Flange detail and connection section

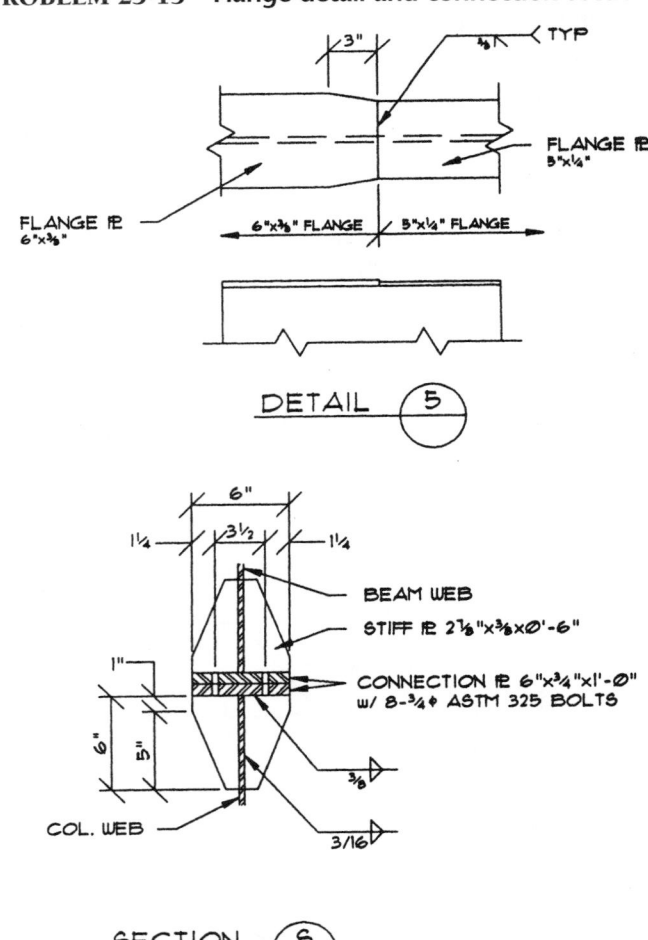

DETAIL　5

SECTION　S

PROBLEM 23-12　Section K, Typical building section with detail

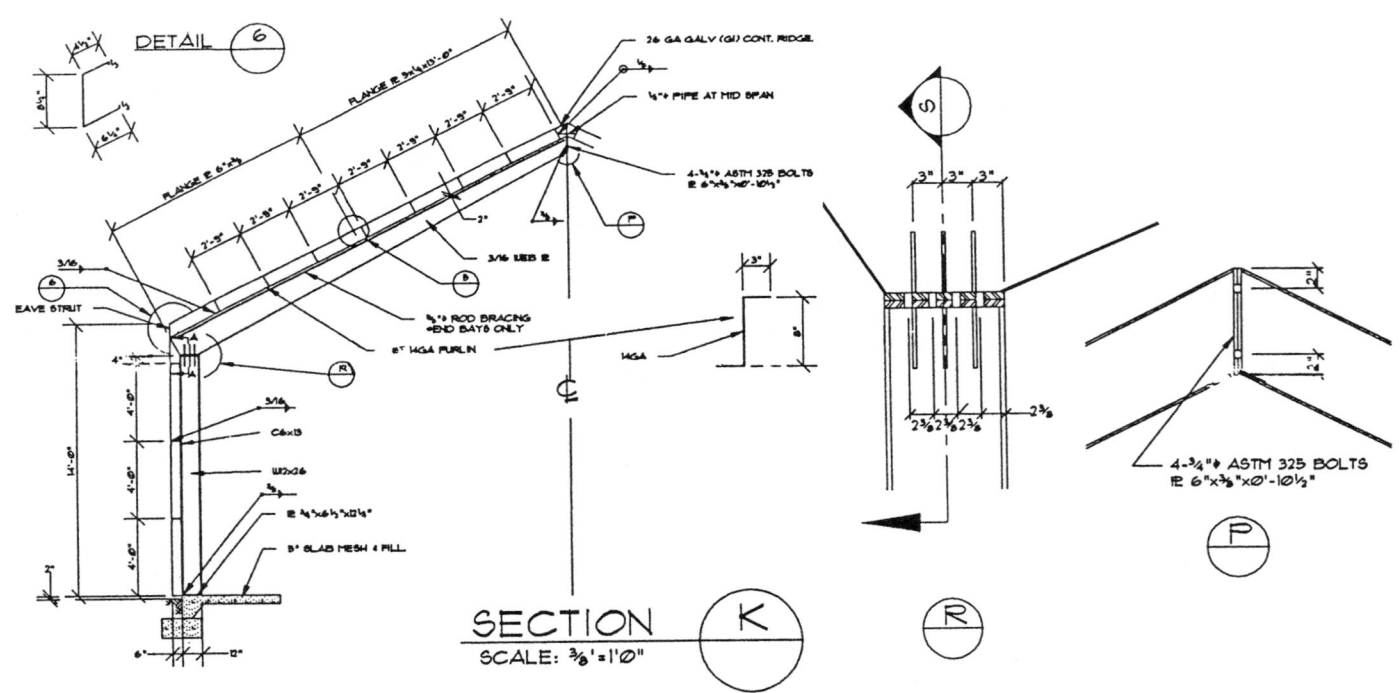

SECTION　K
SCALE: 3/8'=1'0"

PROBLEM 23-14   End wall section

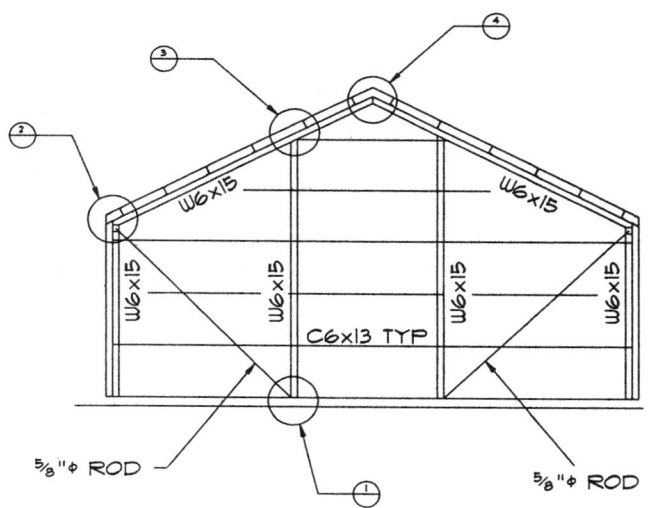

PROBLEM 23-15   Fire wall section

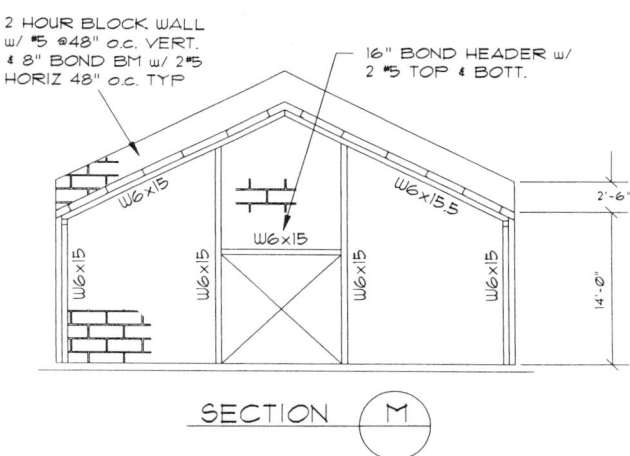

NOTE: TYP BLOCK WALL CONN TO BMS &
COLS USE ¾"φ ANCHOR BOLTS @ 4'-0" o.c.

PROBLEM 23-16   Section side wall framing

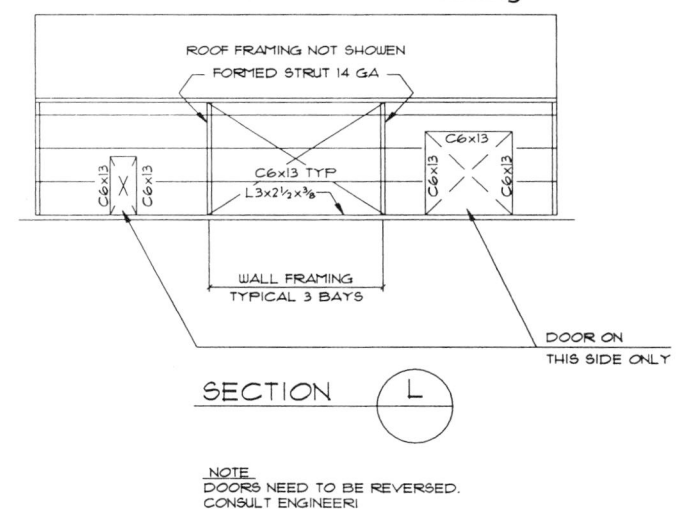

NOTE
DOORS NEED TO BE REVERSED.
CONSULT ENGINEER!

PROBLEM 23-17   Column base and cap connection details

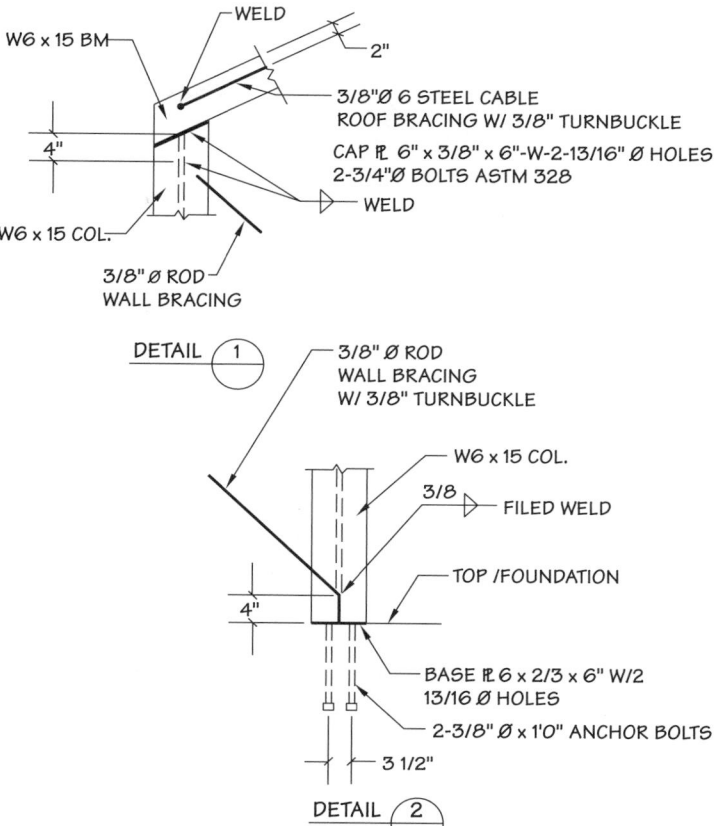

PROBLEM 23-18   Column cap and ridge connection details

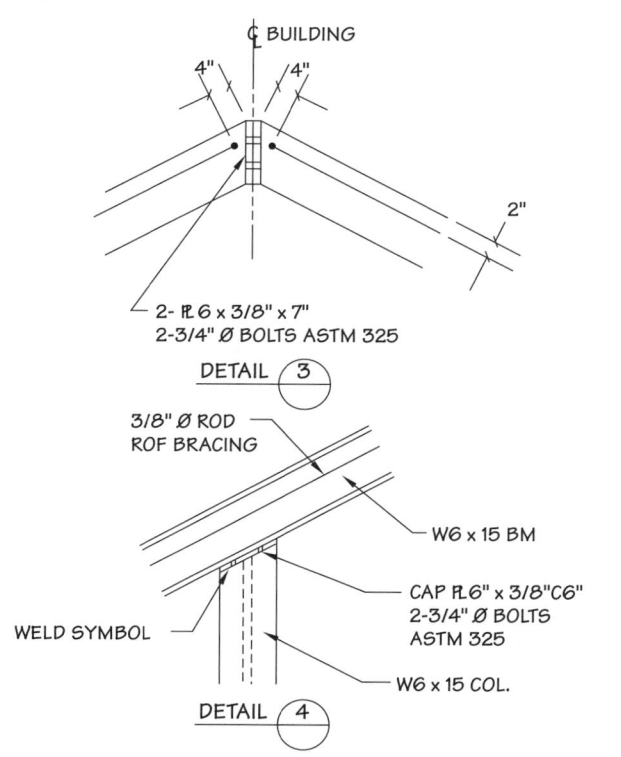

PROBLEM 23-19 **Elevations**

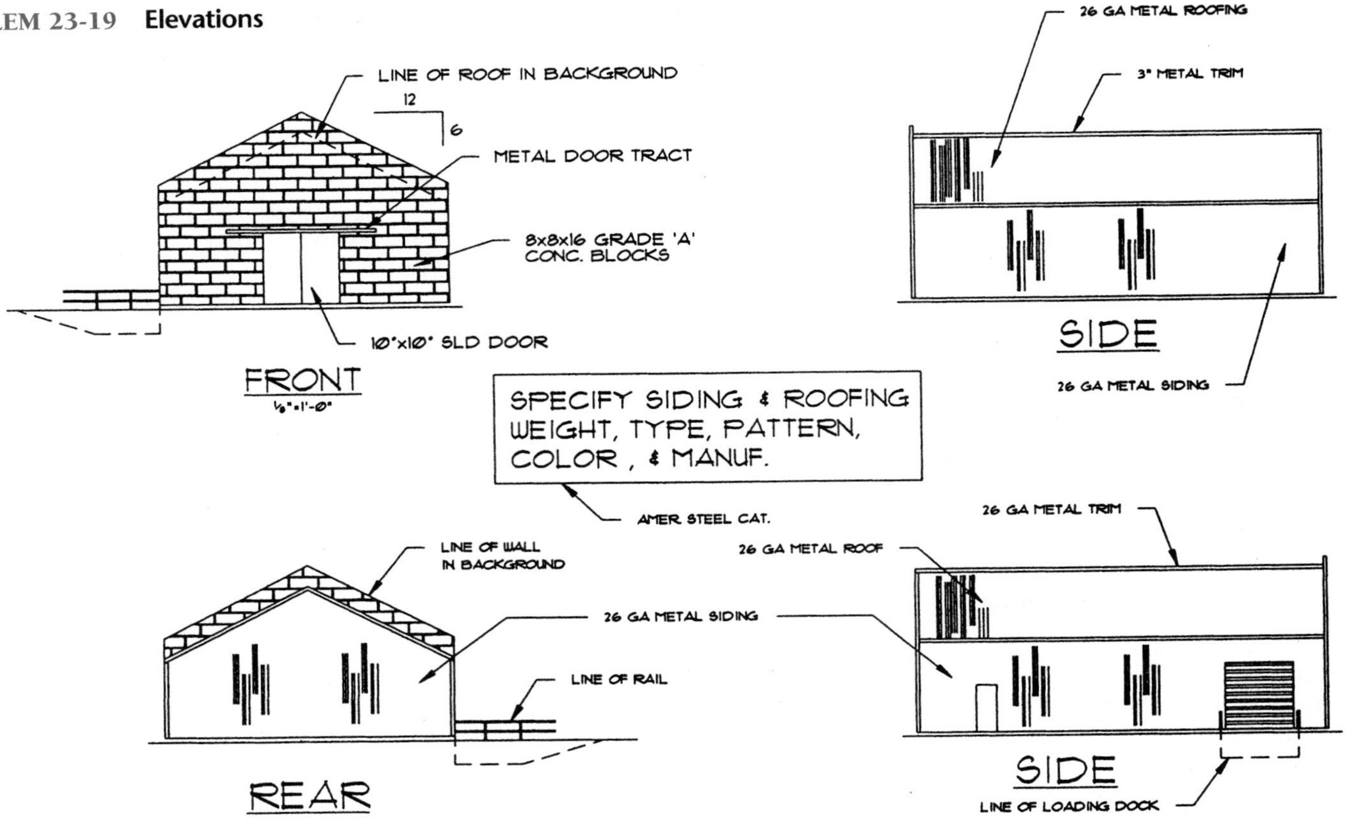

LINE OF ROOF IN BACKGROUND
12
6
METAL DOOR TRACT
8x8x16 GRADE 'A' CONC. BLOCKS
10'x10' SLD DOOR
FRONT
1/8"=1'-0"
26 GA METAL ROOFING
3" METAL TRIM
SIDE
26 GA METAL SIDING
SPECIFY SIDING & ROOFING WEIGHT, TYPE, PATTERN, COLOR, & MANUF.
AMER STEEL CAT.
LINE OF WALL IN BACKGROUND
26 GA METAL TRIM
26 GA METAL ROOF
26 GA METAL SIDING
LINE OF RAIL
REAR
SIDE
LINE OF LOADING DOCK

Problems 23-20 and 23-21 are for a warehouse facility.

PROBLEM 23-20 **Floor plan and schedules, 3/32" = 1'-0"**

NOTE: Alternate problems may be developed using plumbing, electrical, and HVAC overlays after studying Chapters 22, 24, and 25. Other drawings may be designed to complete a set of plans based on the content of this chapter if appropriate with course guidelines.

49'-4"
49'-4"
32'
50'-6"
101'
UNIT 1
2HR SEPARATION WALL
UNIT 2
BATH 1
BATH 2
OFFICE 1
OFFICE 2
28'-6" 30' 30' 30' 30' 30' 30' 30' 30' 28'-6"
7.5
11 10 9 8 7 6 5 4 3 2 1
10'
3'
BATH
15'
BATH ENLARGED

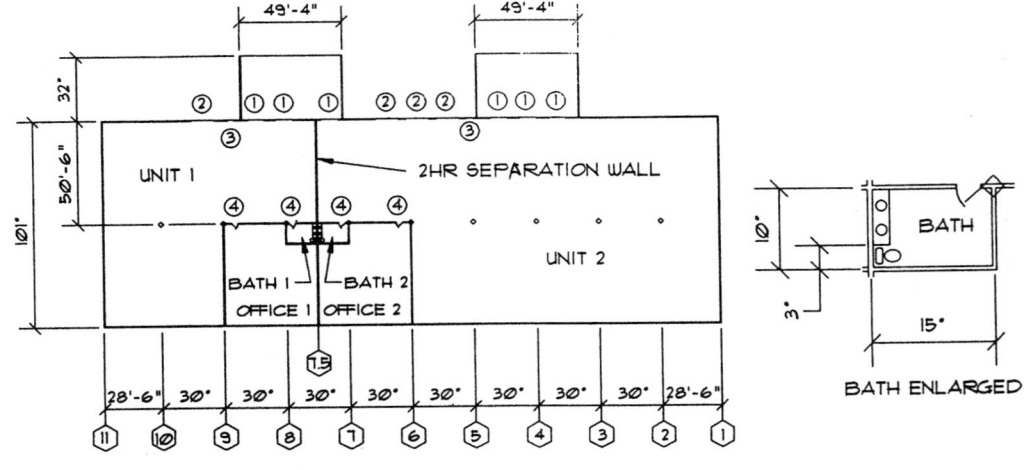

GAR DR OPENERS
OSCO LDM COMMERCIAL OR EQUIV.
1/2HP CHAIN DRIVE w/ INSTANT REVERSE MOTOR, QUICK DISCONNECT ARM, REVERSING DOOR EDGE, PULL CORD AND 3 BUTTON COMMAND STATION.

DOOR SCHEDULE

KEY	SIZE	DESCRIPTION
1	8'x8'	KINNEAY COMMERCIAL STEEL INSULATE OVER-HEAD, WEATHER STRIP ALL AROUND, STD 14" R TRACK
2	11'x12'	SAME
3	3'x7'	MESKER FRAME DOUBLE LAMINATED WIRE MESH LITE, ONE HR FIRE RATED, SCHLAGE A53PD HDWARE
4	3'x7'	THERMA-TRU 105M, SCHLAGE A53PD HDWARE

PROBLEM 23-21   Foundation plan, 3/32" = 1'–0"

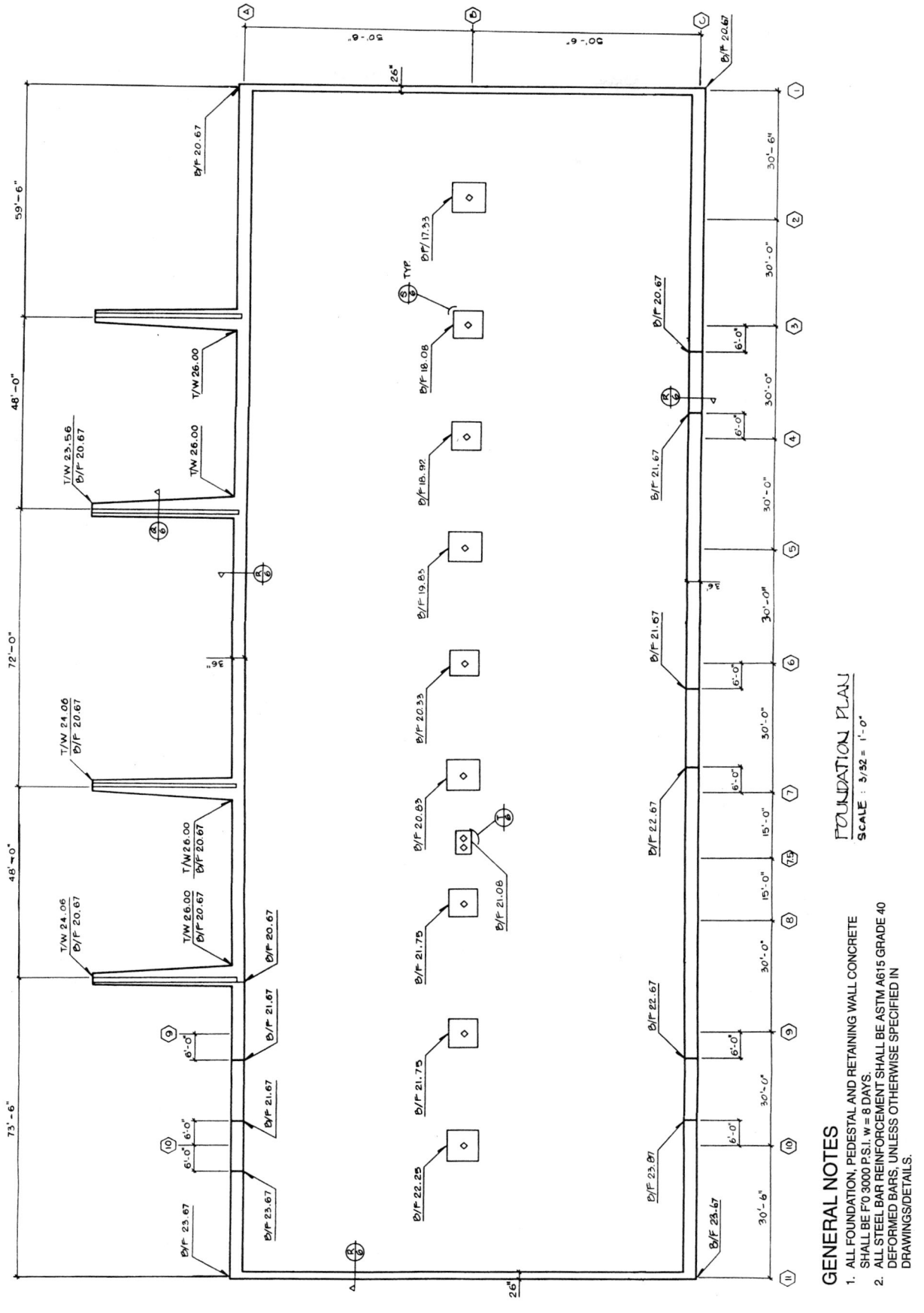

FOUNDATION PLAN
SCALE : 3/32 = 1'-0"

GENERAL NOTES

1. ALL FOUNDATION, PEDESTAL AND RETAINING WALL CONCRETE
    SHALL BE F'0 3000 P.S.I. w = 8 DAYS.

2. ALL STEEL BAR REINFORCEMENT SHALL BE ASTM A615 GRADE 40
    DEFORMED BARS, UNLESS OTHERWISE SPECIFIED IN
    DRAWINGS/DETAILS.

PROBLEM 23-22   Decorative interior elevations

Use a scale of ³⁄₃₂" = 1'–0" to measure the layout shown below. Convert to a ¼" = 1'–0" scale on appropriately sized vellum.
Courtesy Structureform Masters, Inc.

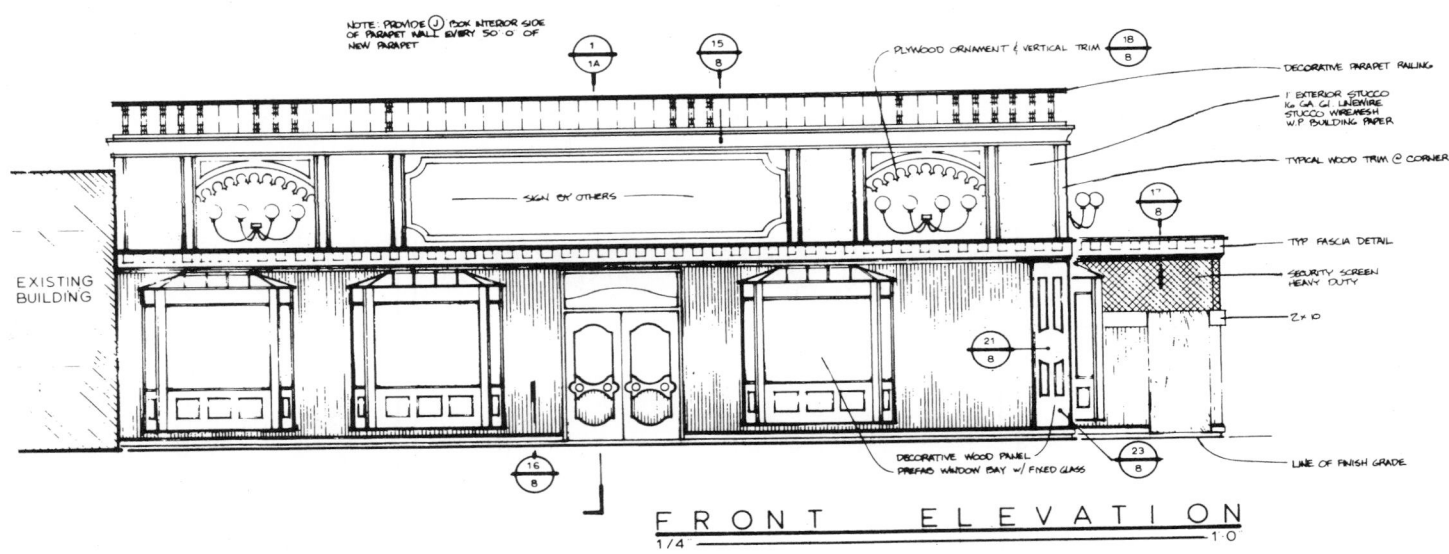

Problems 23-23 through 23-25 are pictorial drawings. Create isometric drawings from the given problems.

PROBLEM 23-23   Header detail

Courtesy Wendy's International, Inc.

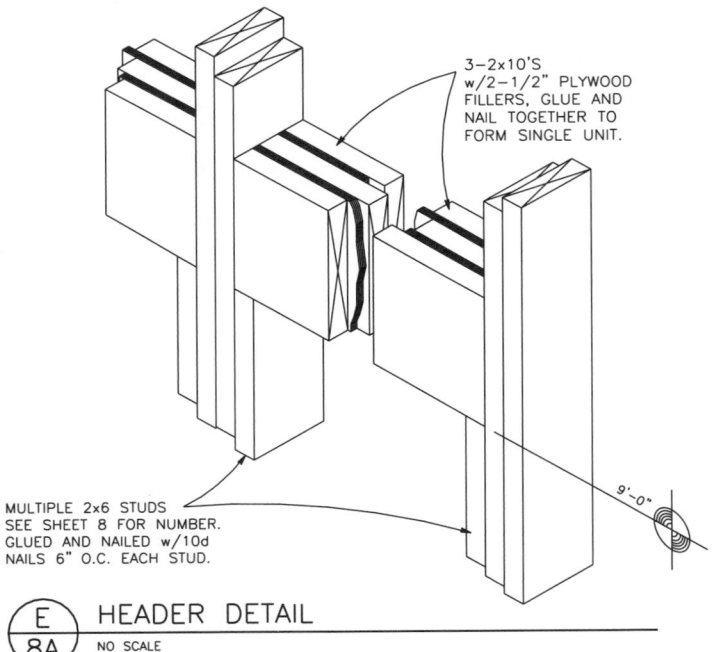

PROBLEM 23-24   Base detail

Courtesy Wendy's International, Inc.

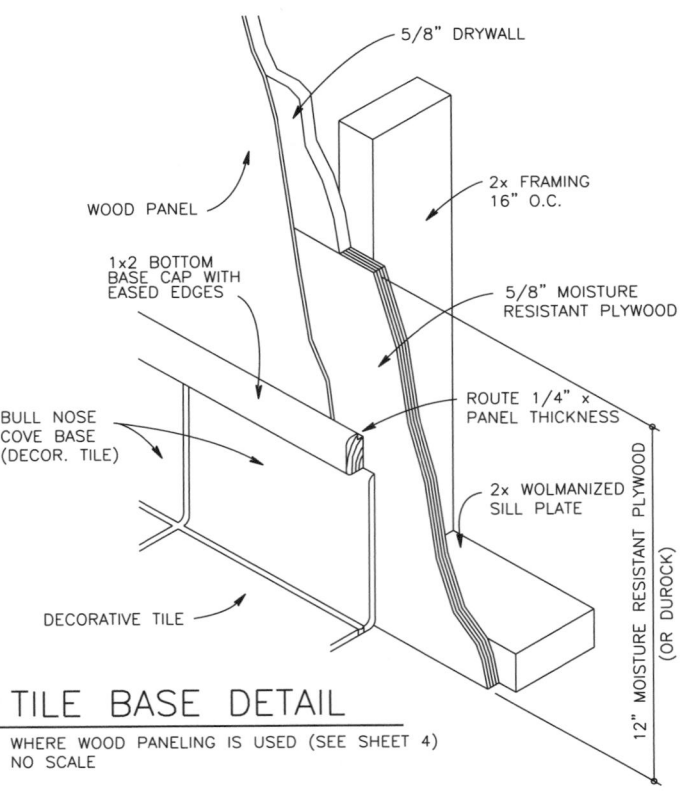

PROBLEM 23-25   Trellis detail

Courtesy Ankrom Moisan Architects

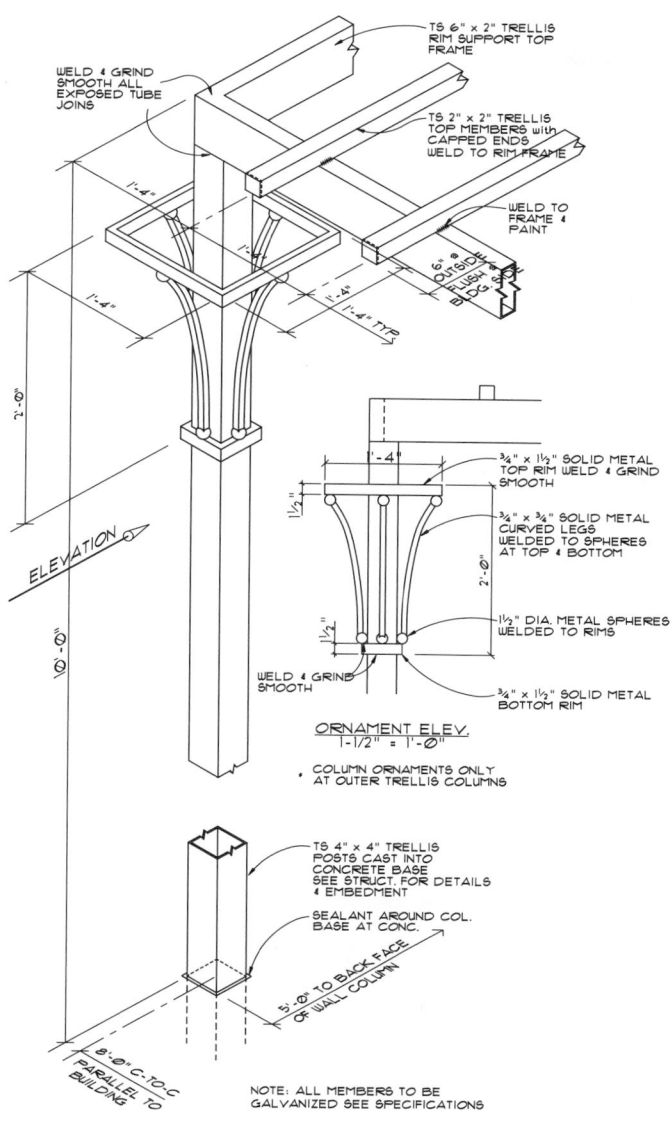

WELD & GRIND SMOOTH ALL EXPOSED TUBE JOINS

TS 6" x 2" TRELLIS RIM SUPPORT TOP FRAME

TS 2" x 2" TRELLIS TOP MEMBERS with CAPPED ENDS WELD TO RIM FRAME

WELD TO FRAME & PAINT

1'-4"

1'-4"

1'-4" TYP.

2'-0"

ELEVATION

10'-0"

¾" x 1½" SOLID METAL TOP RIM WELD & GRIND SMOOTH

¾" x ¾" SOLID METAL CURVED LEGS WELDED TO SPHERES AT TOP & BOTTOM

1½" DIA. METAL SPHERES WELDED TO RIMS

WELD & GRIND SMOOTH

¾" x 1½" SOLID METAL BOTTOM RIM

1'-4"

2'-0"

ORNAMENT ELEV.
1-1/2" = 1'-0"

• COLUMN ORNAMENTS ONLY AT OUTER TRELLIS COLUMNS

TS 4" x 4" TRELLIS POSTS CAST INTO CONCRETE BASE SEE STRUCT. FOR DETAILS & EMBEDMENT

SEALANT AROUND COL. BASE AT CONC.

5'-0" TO BACK FACE OF WALL-COLUMN

8'-0" C-TO-C PARALLEL TO BUILDING

NOTE: ALL MEMBERS TO BE GALVANIZED SEE SPECIFICATIONS

TRELLIS
B / A13   N.T.S.

MATH PROBLEMS

PROBLEM 23-26   Determine the values of the angle made by the cross member and the length of the cross member for the pier support shown below.

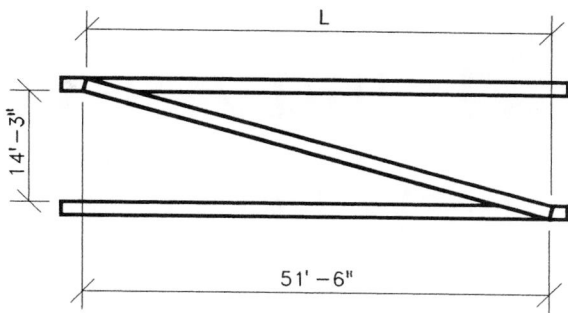

L

14'-3"

51'-6"

PROBLEM 23-27   Two straight roads intersect to form an angle of 80°. Find the shortest distance from one road to a gas station on the other road 1500 ft. from the junction.

PROBLEM 23-28   Two buildings with flat roofs are 160' apart. From the roof of the shorter building, 50' in height, the angle of elevation to the edge of the roof of the taller building is 17°. How high is the taller building?

PROBLEM 23-29   A ladder, with its pivoting foot in the street, makes an angle of 38° with the street when its top rests on a building on one side of the street and makes an angle of 45° with the street when its top rests on a building on the other side of the street. If the ladder is 40' long, how wide is the street?

PROBLEM 23-30   What is the angle of elevation of the sun if a telephone pole that is 60.0' long casts a shadow 72.3' long?

PROBLEM 23-31   Calculate the weight in pounds, and the volume in yards, for a precast concrete panel with the dimensions 20'-0" long, 4'-0" wide, and 1'-0" high.

PROBLEM 23-32   How tall is a tree located 205 ft. from a house if the angle of elevation to the top of the tree is 14° (as measured by a 5' 6" person at the house).

PROBLEM 23-33   A steel beam has its ends sitting on vertical columns projecting upward from level ground. One column is 42' high and the other is 47' high. The distance along the ground between the columns is 82'. How long is the steel beam and what is its angle of elevation?

24 CHAPTER

Heating, Ventilating, and Air-Conditioning (HVAC)/Pattern Development and Precision Sheet Metal Drafting

OBJECTIVES

After completing this chapter, you will:

- Discuss the purpose and function of HVAC systems.
- Prepare complete HVAC drawings, including plans, schedules, and details.
- Draw sheet metal pattern developments and intersections.
- Calculate and apply bend allowances to sheet metal components.
- Draw and completely dimension precision sheet metal fabrication drawings.
- Use an engineering problem as an example for HVAC and sheet metal drawing solutions.

THE ENGINEERING DESIGN APPLICATION

Your company produces a wide range of sheet metal products for various customers, ranging from HVAC ductwork to precision housing structures for electronic equipment. Your current drawing project is a chassis for an electronic testing device.

With HVAC drawings, it is not usually necessary to provide bend allowances. In this project, however, the final development requires tighter tolerances for all of the components to mount properly within the chassis. Calculating bend allowances enables you to determine the required size of the flat pattern.

While there are a number of methods for determining bend allowance, your company standards reference the tables listed in the *Machinery's Handbook*. By checking the table for the type of metal you are using plus the bend radius and thickness of material, you are able to determine the appropriate bend allowances.

Applying the bend allowance to each bend in the flat pattern ensures that when the material is formed into its final shape, the desired final dimensions can be achieved. Figure 24.1 shows an example flat pattern layout.

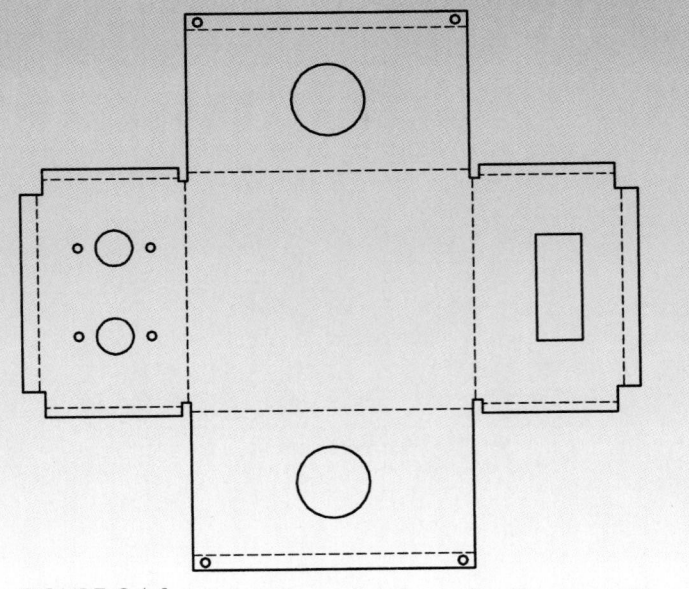

FIGURE 24.1 ■ Sample flat pattern layout for the engineering problem.

Most residential and commercial structures have *heating, ventilating, and air-conditioning (HVAC)* systems. These systems are also commonly known as *mechanical systems.* The purpose of most heating and air-conditioning systems is to help maintain a normal comfort zone for the occupants living or working in the structure. In other applications, such as meat lockers, the product may be the reason for environmental controls. The function of ventilating systems is to provide air movement or exchange within the structure. Ventilation is required when excess heat, fumes, moisture, odor, or pollutants must be removed and fresh air replaced. Most residential structures do not use HVAC plans unless the system is complex, or required by the lending or code enforcement agency. The architect of a commercial or public building often consults with an HVAC engineer for the design, drafting, and installation supervision of the HVAC system. The consulting engineer prepares all of the designs and specifications for the project and the drafter either works directly from written specifications or from engineering sketches. In many cases the engineer prepares rough sketches directly on a copy of the floor plans. The HVAC drafter then creates a new drawing using proper lines, symbols, dimensions, and notes. Some companies use an HVAC overlay where the base sheet is the floor plan and another sheet, known as the overlay, is used to draw the HVAC plan.

HVAC DRAFTING AND PATTERN DEVELOPMENT

HVAC systems are made up of mechanical equipment such as the furnace, air conditioner or ventilator, and ductwork. The *duct* is generally sheet metal pipe designed as the passageway for conveying air from the HVAC equipment to the source. Types of duct shapes include cylindrical, oval, and rectangular. Connectors such as elbows and transition shapes that allow conversion from round to square or rectangular are also used. Any duct shape is possible depending on the application. In most residential applications, standard premanufactured ductwork is used. In commercial structures, the ductwork may be premanufactured or custom-built in a sheet metal shop. Flat patterns are made when custom sheet metal shapes are required. These patterns are made by a sheet metal layout person or a sheet metal drafter. The flat pattern drawings, known as pattern developments, are transferred to thick paper. The paper pattern is then transferred to flat sheet metal to be formed into the desired duct shape. CADD systems are being used to develop the flat pattern and transfer the pattern to the metal.

Other industries also make sheet metal patterns. Whenever flat metal is bent into shape, a sheet metal pattern is required. Examples are auto body parts, storage facilities, or electronics chassis components.

HVAC SYSTEMS
Central Forced-air Systems

Central forced-air systems, among the most common systems for climate heating and air-conditioning, circulate the air from the living spaces through or around heating or cooling devices. A thermostat starts the cycle as a fan forces the air into ducts. These ducts connect to openings called diffusers, or air supply registers, which put warm air (WA) or cold air (CA) in the room. The air enters the room and either heats or cools as needed. Air then flows from the room through another opening called a return-air register (RA) and into the return duct. The return duct directs the air from the room over a heating or cooling device, depending on which is needed. If cool air is required, the return air is passed over the surface of a cooling coil. If warm air is required, the return air is passed over either the surface of a combustion chamber (the part of a furnace where fuel is burned) or a heating coil. The conditioned air is picked up again by the fan and the cycle is repeated. Figure 24.2 shows the air cycle in a forced-air system.

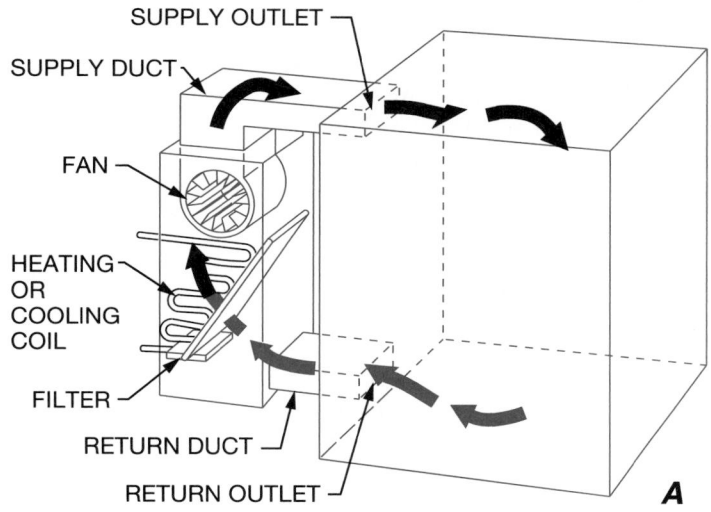

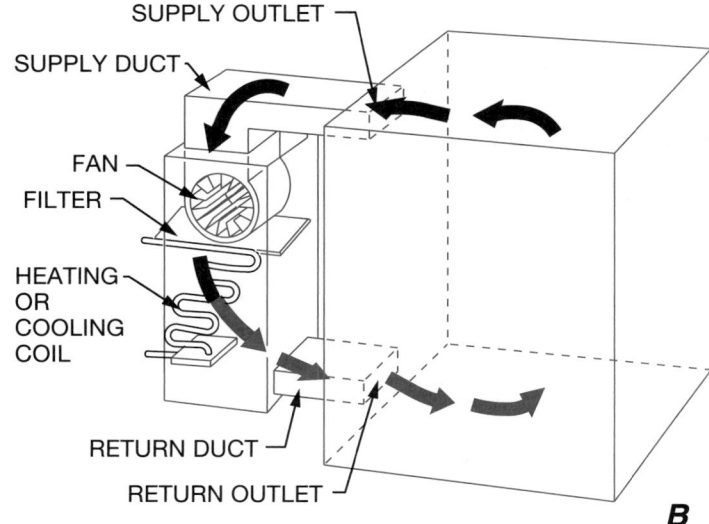

FIGURE 24.2 ■ **(a)** Down draft forced air system air cycle. **(b)** Up draft forced air system air cycle.

Refrigeration

Refrigeration, the most common type of cooling system, is based on the principle that as a liquid changes to vapor it absorbs large amounts of heat. The boiling point of a liquid can be altered by changing the pressure applied to the liquid so that when a gas is changed to a liquid, it will give up the heat. The basic parts of a refrigeration system are the cooling coil (evaporator), compressor (air pump), condenser (where vaporized refrigerant is liquified), and expansion valve. Common refrigerants boil at low temperatures. Figure 24.3 shows a pictorial diagram of the refrigeration cycle.

Hot Water System

In a hot water system, water is heated as it circulates around the combustion chamber of a fuel-fired boiler. The water is then circulated through pipes to radiators or convectors in the rooms. In a one-pipe system, one pipe leaves the boiler and runs through the building and back to the boiler as shown in Figure 24.4. In a two-pipe system, one pipe supplies heated water to all the outlets, while the other is a return pipe, which carries the water back to the boiler for reheating as shown in Figure 24.5. Hot water systems use a pump, called a circulator, to move the water through the system. The water is kept at a temperature of 150°–180°F in the boiler. When heat is needed, a thermostat starts the circulator pump.

Zoned Control System

A *zoned* system allows for one or more heaters and one thermostat per room. No ductwork is required, and only the heaters in occu-

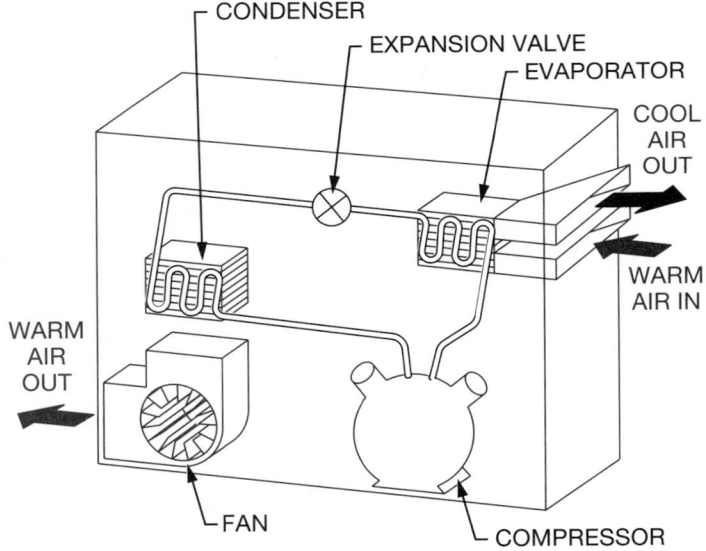

FIGURE 24.3 ■ Pictorial diagram of refrigeration cycle. *From Lang. Principles of Air Conditioning, 3d ed., Delmar Publishers Inc.*

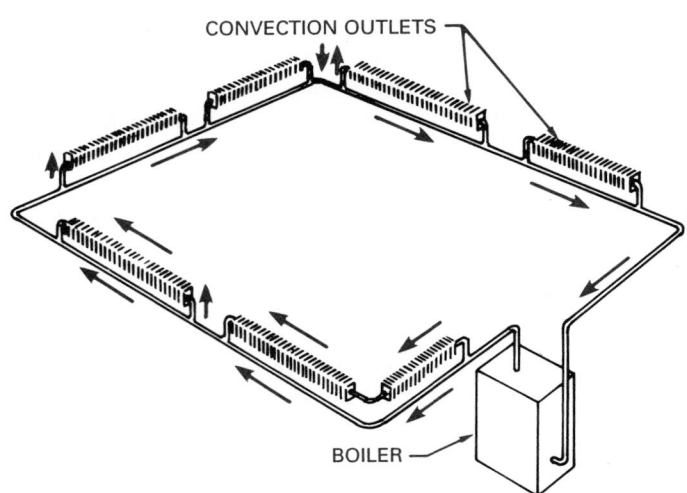

FIGURE 24.4 ■ One-pipe hot water system.

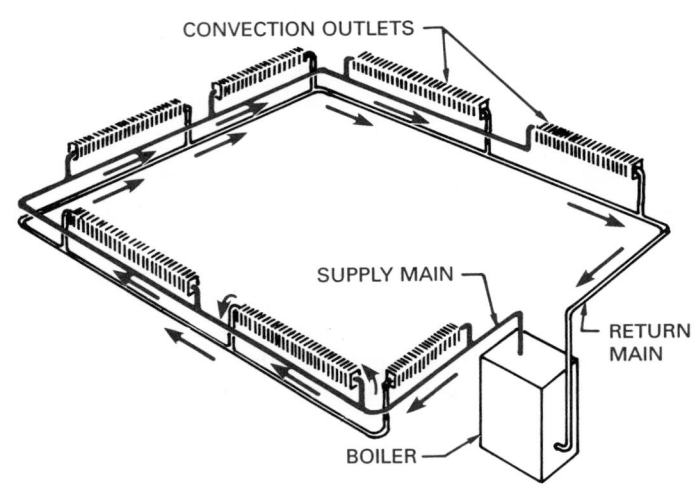

FIGURE 24.5 ■ Two-pipe hot water system.

pied rooms need to be turned on. One of the major differences between a zoned and central system is flexibility. A zoned system allows the occupant to determine how many areas are heated and how much energy is used.

Radiant Heat

Radiant heating and cooling systems function on the basis of providing a comfortable environment by means of controlling surface temperatures and minimizing excessive air movement within the space. Surface-mounted radiant panels provide comfort at a lower thermostat temperature than other systems. Radiant systems vary from oil or gas hot water piping in the ceiling or floor, to electric coils, wiring, or elements at the ceiling.

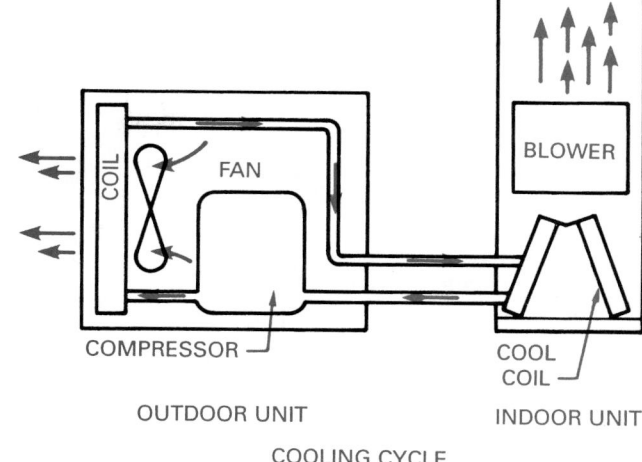

FIGURE 24.6 ■ Heat pump heating and cooling cycle. *Courtesy Lennox Industries, Inc.*

Heat Pump System

The heat pump is a forced-air central heating and cooling system that operates using a compressor and a circulating liquid gas refrigerant. Heat is extracted from the outside air and pumped inside the structure. The heat pump supplies up to three times as much heat per year for the same amount of electrical consumption. A standard electrical forced-air heating system works best when outside air is above 20° F. In the summer the cycle is reversed and the unit operates as an air conditioner. In this mode the heat is extracted from the inside air and pumped outside. On the cooling cycle the heat pump also acts as a dehumidifier. Figure 24.6 shows how a heat pump works.

VENTILATION

There are a number of reasons why ventilation of a structure or area is necessary. Residential applications include bath, kitchen, and laundry exhaust fans. In commercial applications ventilation may be necessary to exhaust fumes, pollutants, or moisture.

Sources of Pollutants

There are a number of sources of pollutants that make it necessary to plan ventilation systems. Moisture in the form of relative humidity can cause structural damage and health problems such as respiratory troubles and microbial growth. Each individual can produce up to one gallon of water vapor per day.

Indoor combustion from such items as gas-fired or wood-burning appliances, or fireplaces, can generate a variety of pollutants, including carbon monoxide and nitrogen oxides.

Humans and pets can transmit diseases through the air by exhaling a variety of bacterial and viral contaminants.

Tobacco smoke may contribute chemical compounds to the air environment. The pollution can affect smokers and nonsmokers.

Formaldehyde in glues used in construction materials such as plywood, particleboard, or even carpet and furniture causes pollution, as do the components of certain insulations. Formaldehyde has been considered a factor in certain diseases, eye irritation, and respiratory problems.

Radon is a naturally occurring radioactive gas that breaks down into compounds that may cause cancer when large quantities are inhaled over a long period of time. Radon may be more apparent in a structure that contains a great deal of concrete or is located in certain geographic areas of the country. Radon can be scientifically monitored at a nominal cost. Barriers can be built that help reduce the concern of radon contamination.

Household products such as aerosols and crafts materials such as glues and paints can contribute a number of toxic pollutants.

Air-to-air Heat Exchangers

The air-to-air heat exchanger is a heat-recovery ventilation device that pulls stale, polluted, warm air from the living or working space through a duct system and transfers the heat in that air to the fresh cold air being pulled into the structure. Heat exchangers do not produce heat; they only move heat from one airstream to the other. The heat transfers to the fresh airstream in the core of the heat exchanger. The core is usually designed to avoid mixing of the two airstreams to ensure that indoor pollutants are removed. Moisture in the stale air condenses in the core and is drained from the unit. Figure 24.7 shows the basic components and function of an air-to-air heat exchanger. Knowledgeable HVAC engineers, designers, and contractors are able to implement air-to-air heat exchanger technology by the selection of the proper size of a ducted system.

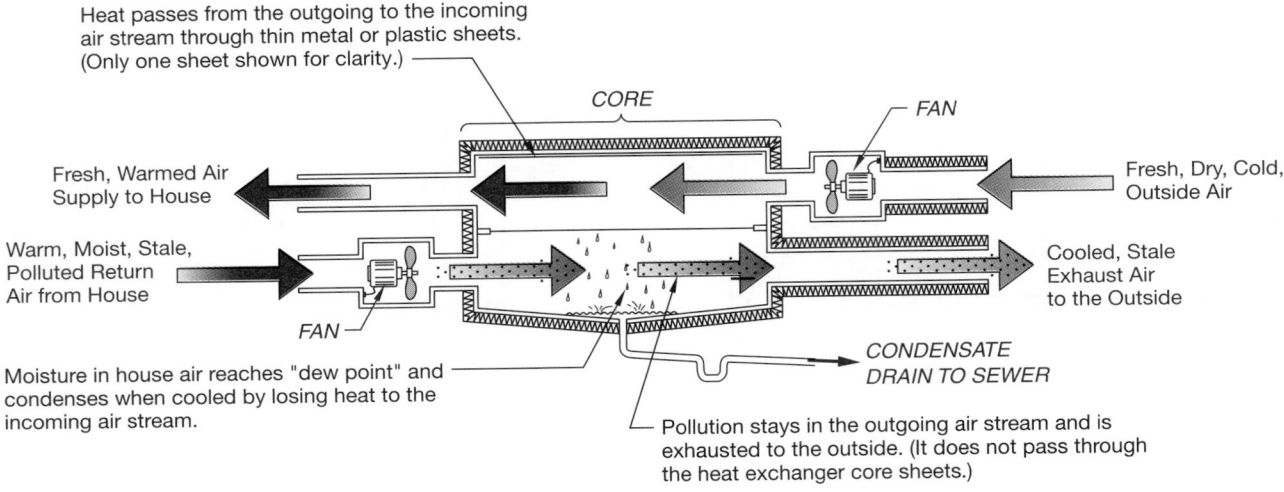

Heat passes from the outgoing to the incoming air stream through thin metal or plastic sheets. (Only one sheet shown for clarity.)

CORE

FAN

Fresh, Warmed Air Supply to House

Fresh, Dry, Cold, Outside Air

Warm, Moist, Stale, Polluted Return Air from House

Cooled, Stale Exhaust Air to the Outside

FAN

Moisture in house air reaches "dew point" and condenses when cooled by losing heat to the incoming air stream.

CONDENSATE DRAIN TO SEWER

Pollution stays in the outgoing air stream and is exhausted to the outside. (It does not pass through the heat exchanger core sheets.)

FIGURE 24.7 ■ Components and function of an air-to-air heat exchanger. *Courtesy U.S. Department of Energy.*

THERMOSTAT

The thermostat is an automatic mechanism for controlling the amount of heating or cooling given by a central or zoned heating or cooling system. The thermostat symbol is shown in Figure 24.8.

The location of the thermostat is an important consideration to the proper functioning of the system. For zoned units there may be thermostats placed in each room or a central panel placed in a convenient location. For central systems there may be one or more thermostats, depending on the layout of the system or the number of units required to service the structure. For example, an office complex may have a split system where the structure is divided into two or more zones. Each individual segment of the system has its own thermostat.

Several factors contribute to the effective placement of the thermostat for a central system. A good common location is near the center of the structure and close to a return-air duct. The air entering the return-air duct is usually temperate, thus causing little variation in temperature. This is a location where an average temperature reading can be achieved. There should be no drafts. Avoid locations where sunlight or a heat register could cause an unreliable reading. Avoid placement near an exterior opening or on an outside wall. Avoid placement near stairs or a similar traffic area where significant

bouncing or shaking could cause the mechanism to alter the actual reading.

HVAC SYMBOLS

There are over a hundred HVAC symbols that may be used in residential and commercial heating plans. Only a few of the possible symbols are typically used in residential HVAC drawings. The symbols are divided into heating, ventilating, and air-conditioning symbols. Figure 24.9 shows some common HVAC symbols. Line symbols are also commonly used in HVAC drafting to represent related piping. These line symbols are commonly solid or dashed lines that are broken periodically with an abbreviation inserted in the break. The abbreviation identifies the pipe application as shown in Figure 24.9. Sheet metal conduit (duct) and air-conditioning templates are available to help speed manual drafting, and CADD menus and symbols libraries may be used to customize drafting practices.

ANSI Standard symbols adopted by ANSI are available in the document *Graphical Symbols for Heating, Ventilating, and Air Conditioning*, ANSI Y32.2.4. Standard symbols are often modified to fit individual needs, with a legend placed on the drawing for interpretation.

HVAC DRAWINGS

Drawings for the HVAC system show the size and location of all equipment, ductwork, and components with accurate symbols, specifications, notes, and schedules that form the basis of contract requirements for construction. Specifications are documents that accompany the drawings and contain all pertinent written information related to the HVAC system.

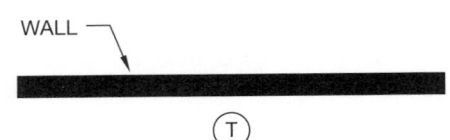

WALL

T

FIGURE 24.8 ■ Thermostat floor plan symbol.

EQUIPMENT SYMBOLS

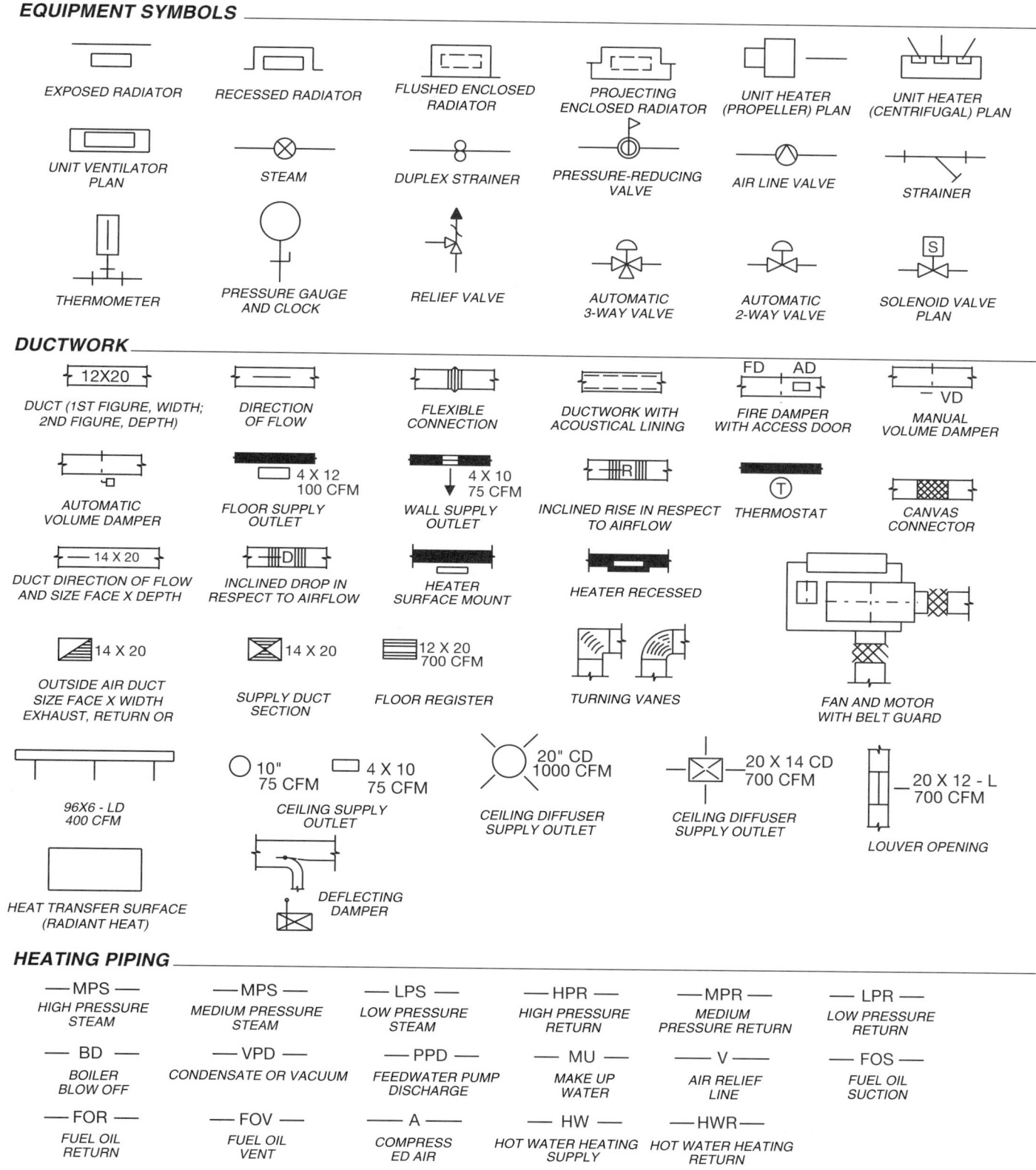

EXPOSED RADIATOR

RECESSED RADIATOR

FLUSHED ENCLOSED RADIATOR

PROJECTING ENCLOSED RADIATOR

UNIT HEATER (PROPELLER) PLAN

UNIT HEATER (CENTRIFUGAL) PLAN

UNIT VENTILATOR PLAN

STEAM

DUPLEX STRAINER

PRESSURE-REDUCING VALVE

AIR LINE VALVE

STRAINER

THERMOMETER

PRESSURE GAUGE AND CLOCK

RELIEF VALVE

AUTOMATIC 3-WAY VALVE

AUTOMATIC 2-WAY VALVE

SOLENOID VALVE PLAN

DUCTWORK

12X20

DUCT (1ST FIGURE, WIDTH; 2ND FIGURE, DEPTH)

DIRECTION OF FLOW

FLEXIBLE CONNECTION

DUCTWORK WITH ACOUSTICAL LINING

FD AD

FIRE DAMPER WITH ACCESS DOOR

VD

MANUAL VOLUME DAMPER

AUTOMATIC VOLUME DAMPER

4 X 12 100 CFM

FLOOR SUPPLY OUTLET

4 X 10 75 CFM

WALL SUPPLY OUTLET

R

INCLINED RISE IN RESPECT TO AIRFLOW

T

THERMOSTAT

CANVAS CONNECTOR

14 X 20

DUCT DIRECTION OF FLOW AND SIZE FACE X DEPTH

D

INCLINED DROP IN RESPECT TO AIRFLOW

HEATER SURFACE MOUNT

HEATER RECESSED

14 X 20

OUTSIDE AIR DUCT SIZE FACE X WIDTH EXHAUST, RETURN OR

14 X 20

SUPPLY DUCT SECTION

12 X 20 700 CFM

FLOOR REGISTER

TURNING VANES

FAN AND MOTOR WITH BELT GUARD

96X6 - LD 400 CFM

10" 75 CFM

4 X 10 75 CFM

CEILING SUPPLY OUTLET

20" CD 1000 CFM

CEILING DIFFUSER SUPPLY OUTLET

20 X 14 CD 700 CFM

CEILING DIFFUSER SUPPLY OUTLET

20 X 12 - L 700 CFM

LOUVER OPENING

HEAT TRANSFER SURFACE (RADIANT HEAT)

DEFLECTING DAMPER

HEATING PIPING

— MPS —
HIGH PRESSURE STEAM

— MPS —
MEDIUM PRESSURE STEAM

— LPS —
LOW PRESSURE STEAM

— HPR —
HIGH PRESSURE RETURN

— MPR —
MEDIUM PRESSURE RETURN

— LPR —
LOW PRESSURE RETURN

— BD —
BOILER BLOW OFF

— VPD —
CONDENSATE OR VACUUM

— PPD —
FEEDWATER PUMP DISCHARGE

— MU —
MAKE UP WATER

— V —
AIR RELIEF LINE

— FOS —
FUEL OIL SUCTION

— FOR —
FUEL OIL RETURN

— FOV —
FUEL OIL VENT

— A —
COMPRESS ED AIR

— HW —
HOT WATER HEATING SUPPLY

—HWR—
HOT WATER HEATING RETURN

FIGURE 24.9 ■ Common HVAC symbols.

(Continued)

AIR-CONDITIONING PIPING

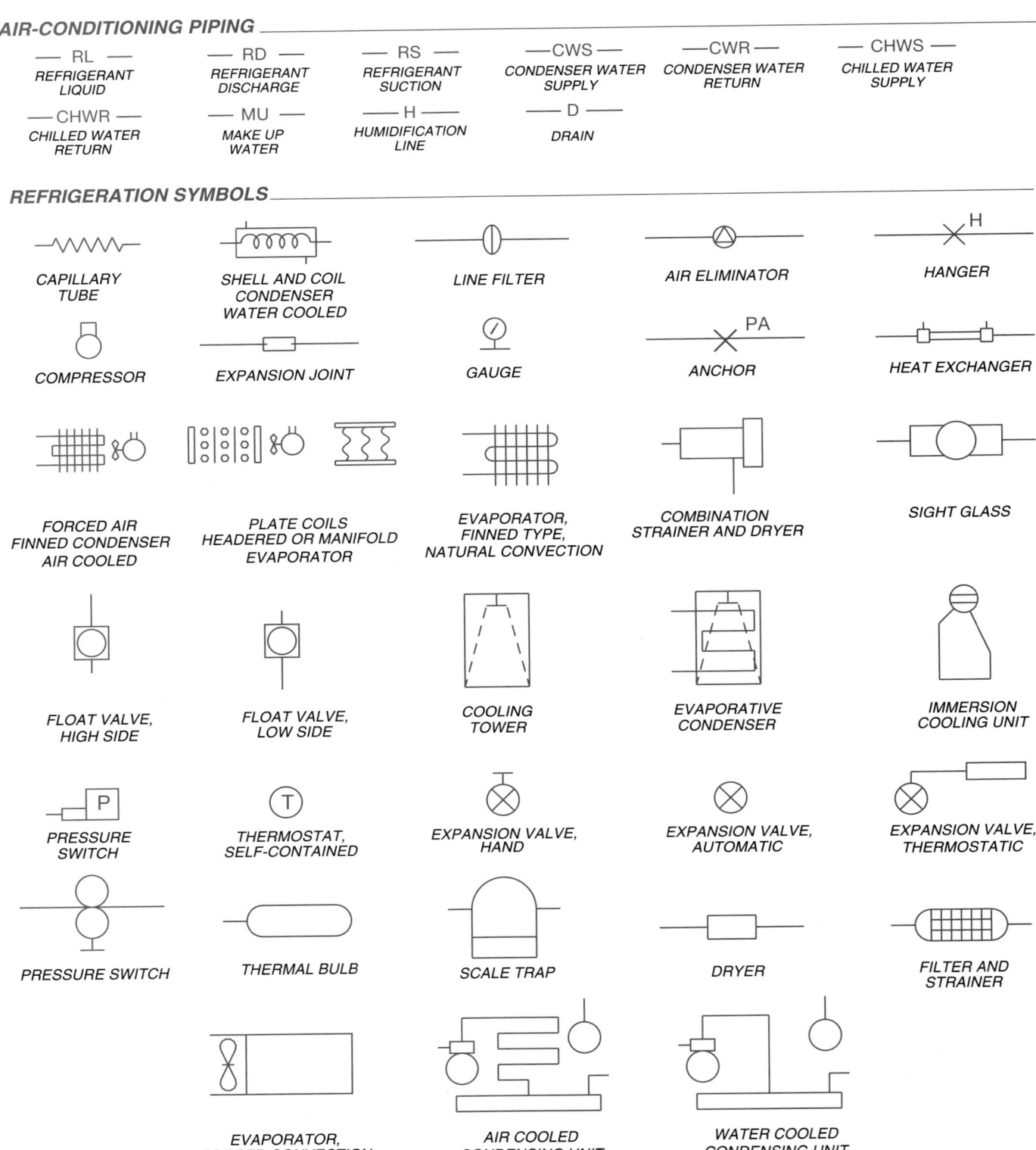

— RL —
REFRIGERANT
LIQUID

— RD —
REFRIGERANT
DISCHARGE

— RS —
REFRIGERANT
SUCTION

—CWS—
CONDENSER WATER
SUPPLY

—CWR—
CONDENSER WATER
RETURN

— CHWS —
CHILLED WATER
SUPPLY

— CHWR —
CHILLED WATER
RETURN

— MU —
MAKE UP
WATER

— H —
HUMIDIFICATION
LINE

— D —
DRAIN

REFRIGERATION SYMBOLS

CAPILLARY
TUBE

SHELL AND COIL
CONDENSER
WATER COOLED

LINE FILTER

AIR ELIMINATOR

HANGER

COMPRESSOR

EXPANSION JOINT

GAUGE

ANCHOR

HEAT EXCHANGER

FORCED AIR
FINNED CONDENSER
AIR COOLED

PLATE COILS
HEADERED OR MANIFOLD
EVAPORATOR

EVAPORATOR,
FINNED TYPE,
NATURAL CONVECTION

COMBINATION
STRAINER AND DRYER

SIGHT GLASS

FLOAT VALVE,
HIGH SIDE

FLOAT VALVE,
LOW SIDE

COOLING
TOWER

EVAPORATIVE
CONDENSER

IMMERSION
COOLING UNIT

PRESSURE
SWITCH

THERMOSTAT,
SELF-CONTAINED

EXPANSION VALVE,
HAND

EXPANSION VALVE,
AUTOMATIC

EXPANSION VALVE,
THERMOSTATIC

PRESSURE SWITCH

THERMAL BULB

SCALE TRAP

DRYER

FILTER AND
STRAINER

EVAPORATOR,
FORCED CONVECTION

AIR COOLED
CONDENSING UNIT

WATER COOLED
CONDENSING UNIT

FIGURE 24.9 ■ continued

HVAC PICTORIALS

CADD applications often make pictorial representations much easier to implement, especially when the HVAC program allows direct conversion from the plan view to the pictorial. The CADD pictorials, known as graphic models, may be used to view the HVAC system from any angle or orientation. Some CADD programs automatically analyze the layout for obstacles where an error in design may result in a duct that does not have a clear path. One of the biggest advantages of the CADD system is that when changes are made in the HVAC plan, these changes are simultaneously corrected on all drawings, schedules, and lists of materials. Figure 24.10 shows a CADD-generated perspective of HVAC duct routing.

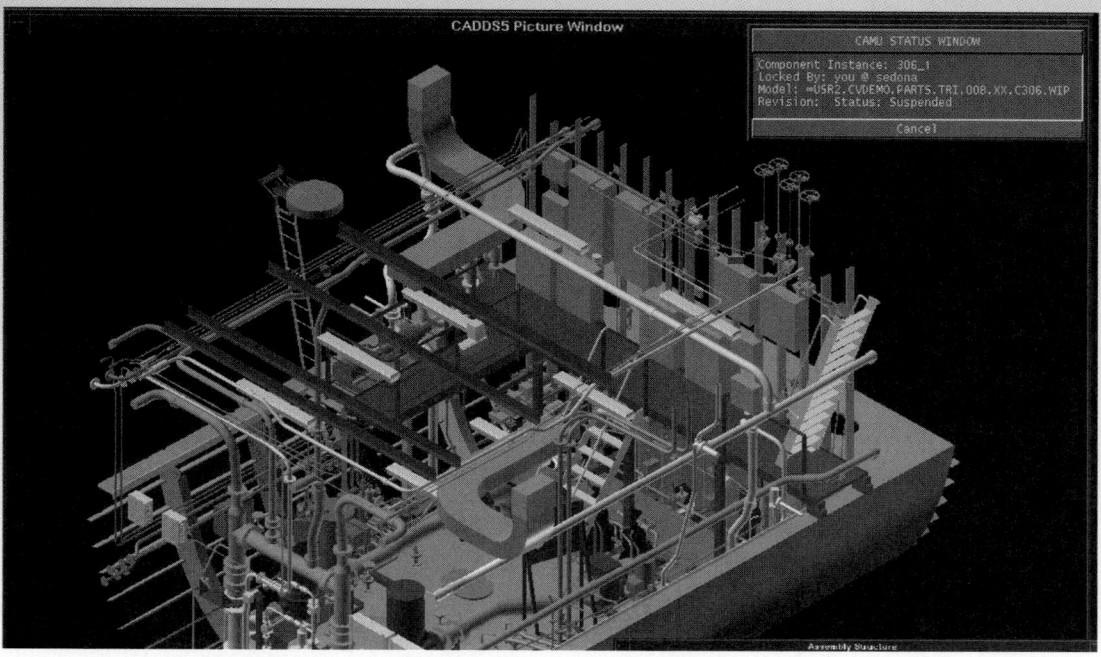

FIGURE 24.10 ■ CADD-generated perspective of HVAC duct routing. *Courtesy Computervision Corporation.*

Drawings may be prepared by the architect, architectural drafter, or the heating contractor when a complete HVAC layout is necessary. Figure 24.11 shows a heating plan for a residential structure.

For commercial structures, the HVAC plan may be prepared by an HVAC engineer as a consultant for the architect. The consulting engineer is responsible for the HVAC design and installation. The engineer determines the placement of all equipment and the location of all duct runs and components. He or she also determines all of the specifications for unit and duct size based on calculations of structure volume, exterior surface areas and construction materials, rate of airflow, and pressure. The engineer may prepare single-line sketches or submit data and calculations to a design drafter who prepares design sketches or final drawings. Drafters without design experience work from engineering or design sketches to prepare formal drawings. A single-line engineer's sketch is shown in Figure 24.12. The next step in the HVAC design is for the drafter to convert the rough sketch into a preliminary drawing. This preliminary drawing goes back to the engineer and architect for verification and corrections or changes. The final step in the design process is for the drafter to implement the design changes on the preliminary drawing to establish the final HVAC drawing. The final HVAC drawing is shown in Figure 24.13.

Convert an engineering sketch to a formal drawing using manual or CADD in this manner:

1. Draw duct runs using thick (0.7 or 0.9 mm) line widths.

2. Label duct sizes within the duct when appropriate, or use a note with a leader to the duct in other situations.

3. Duct sizes may be noted as 22 × 12 or 22/12, where the first number, 22, is the duct width and the second numeral, 12, indicates the duct depth.

4. Place notes on the drawing to avoid crowding. Aligned techniques may be used where horizontal notes read from the bottom of the sheet and vertical notes read from the right side of the sheet. Make notes clear and concise.

5. Refer to schedules to get specific drawing information that may not otherwise be available on the sketch.

6. Label equipment to clearly stand out from other information on the drawing either blocked out or bold.

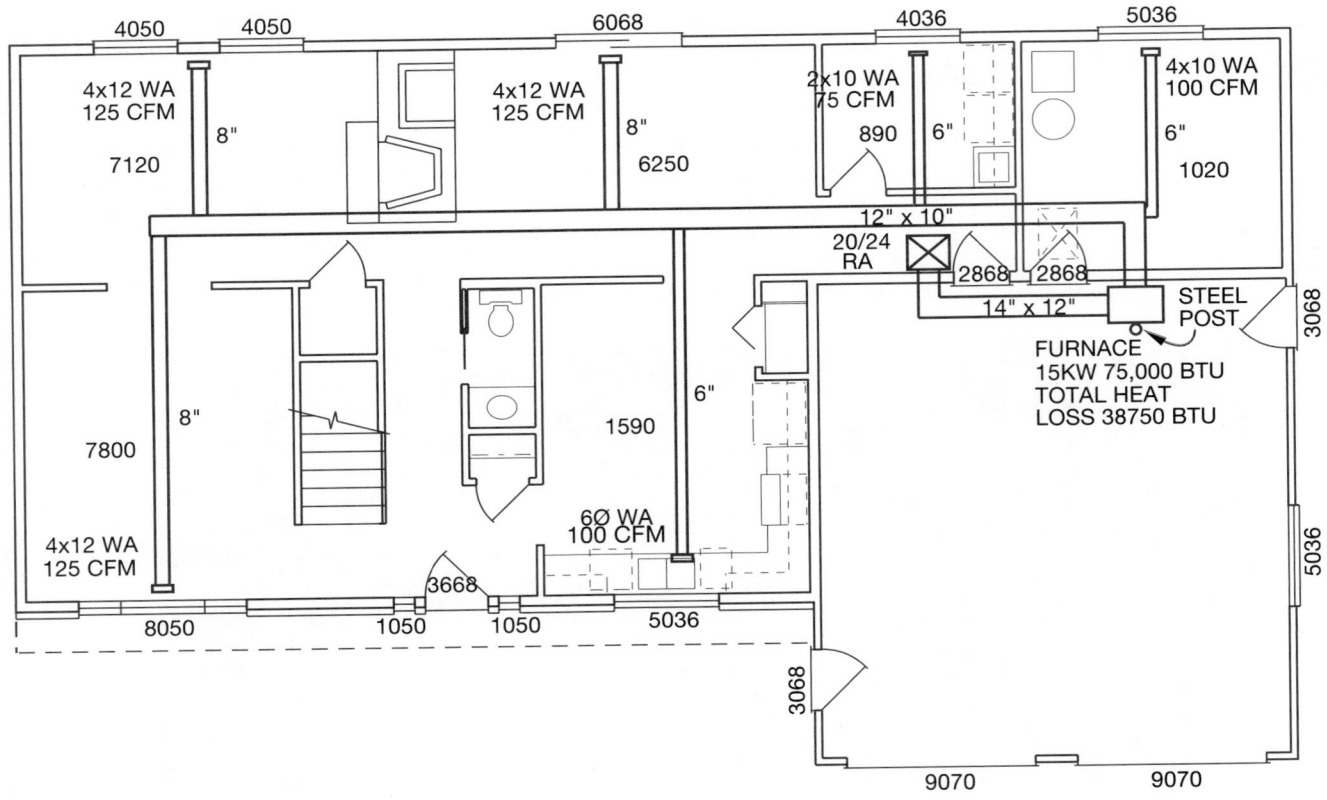

FIGURE 24.11 ■ A detailed forced-air plan for a residential structure.

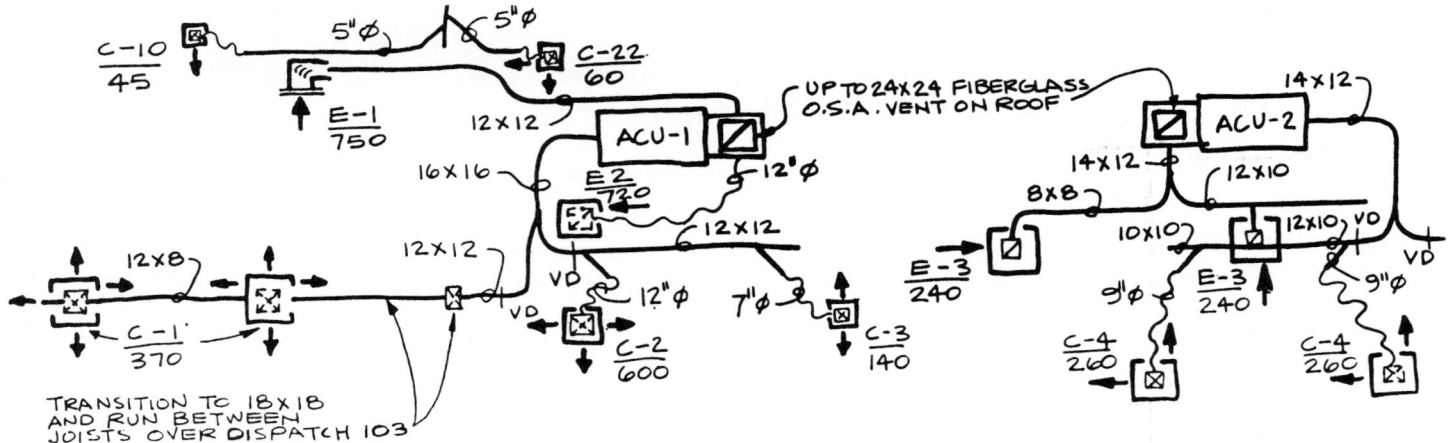

FIGURE 24.12 ■ Single-line HVAC engineer's sketch.

Several examples of duct system elements are shown comparing the engineering sketch and formal drawing in Figure 24.14.

Single- and Double-line HVAC Plans

HVAC plans are drawn over the outline of the floor plan or as an overlay. The floor plan layout is drawn first using thin lines as a base sheet for the HVAC layout and other overlays. The HVAC plan is then drawn using thick lines and notes for contrast with the floor plan. The HVAC plan shows the placement of equipment and

duckwork. The size (in inches) and shape (with symbols, $\varnothing$ = round, $\square$ = square or rectangular) of ductwork and system component labeling is placed on the drawing or keyed to schedules. Drawings may be either single-line or double-line, depending on the needs of the client or how much detail must be shown. Single-line drawings are easier and faster to draw. In many situations they are adequate to provide the equipment placement and duct routing as shown in Figure 24.15. Double-line drawings take up more space and are more time-consuming to draw than single-line, but they are often necessary when complex systems require more detail as shown in Figure 24.13.

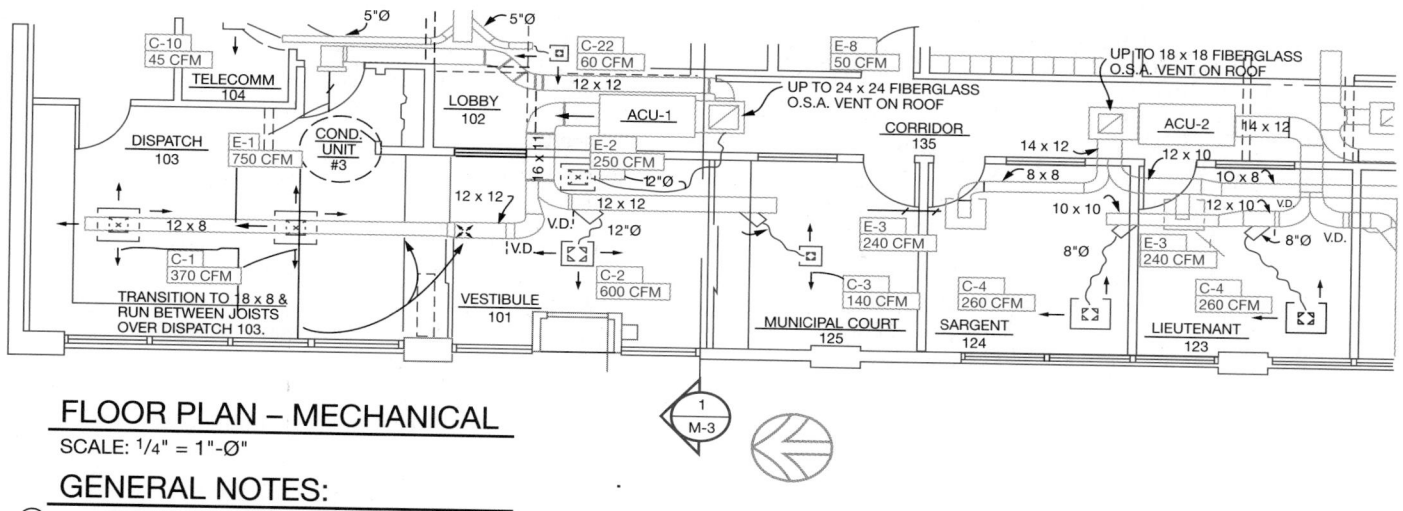

FLOOR PLAN – MECHANICAL

SCALE: 1/4" = 1"-Ø"

GENERAL NOTES:

1. VERIFY ALL EXISTING CONDITIONS AT SITE.

2. SEE ARCHITECTURAL FLOOR PLAN FOR 1-HR RATED AREAS. PROVIDE FIRE DAMPERS AS SCHEDULED AND AS REQ'D BY CODE.

FIGURE 24.13 ▪ HVAC plan for the engineer's sketch shown in Figure 24.12. *Courtesy W. Alan Gold Consulting Mechanical Engineer and Robert Evenson Associates AIA Architects.*

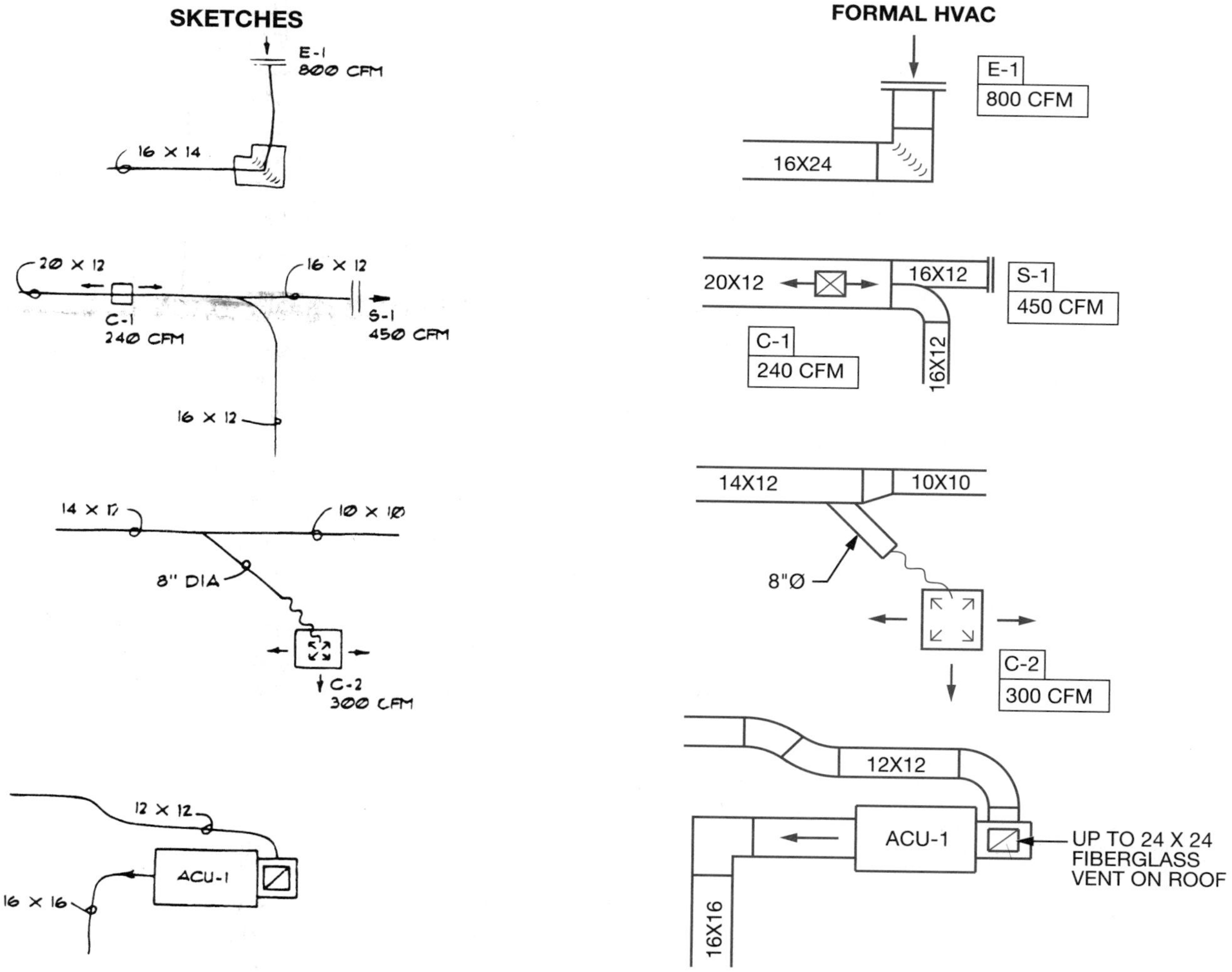

FIGURE 24.14 ▪ Examples showing engineering sketches converted to formal HVAC drawings.

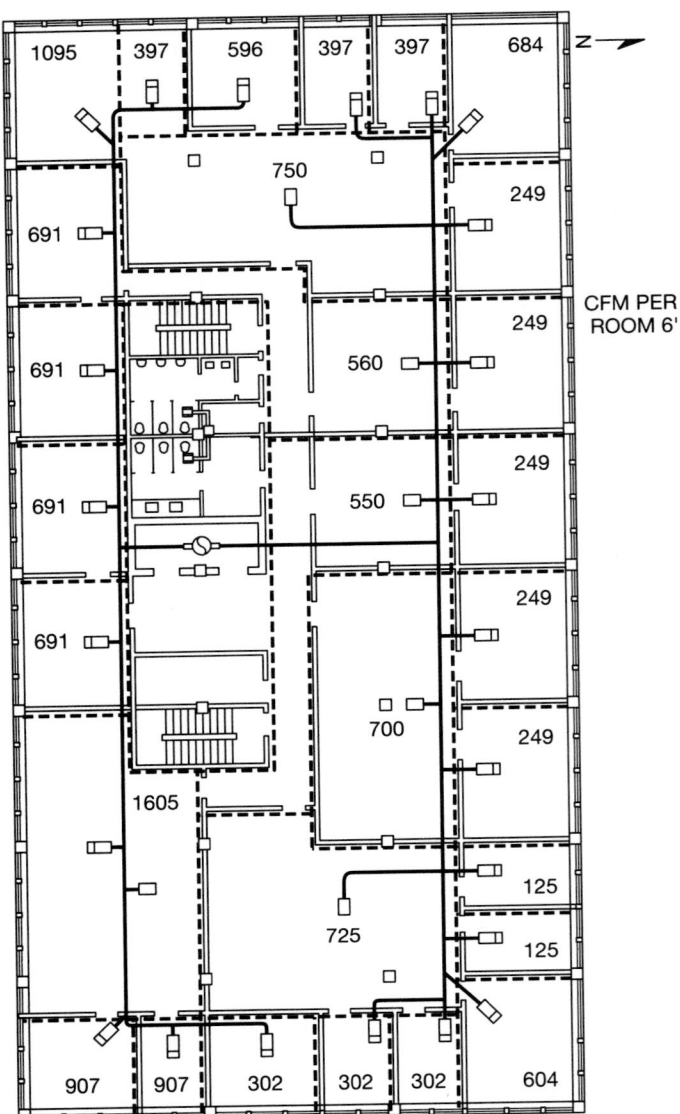

FIGURE 24.15 ■ Single-line ducted HVAC system showing a layout of the proposed trunk and runout ductwork. *Courtesy The Trane Company, La Crosse, WI.*

HVAC Symbol Specifications

There are a large variety of HVAC symbols that can be placed on a drawing. Symbols and notes are used to show and label floor, ceiling, and wall ducts, diffusers, and grills. Duct runs can be shown in plan view and in section. Ducts and components can be drawn with double-line or single-line representations. The size of double-line ducts is represented by the width of the duct in plan view. The outline of the duct runs is drawn with thick lines so that the ducts and related connections and equipment contrast with the rest of the drawing. The sectional view displays both the width and height. Single-line drawings are commonly created with a very thick line representing the duct run and with symbols to show the related connections and equipment. Common symbol specifications for

double-line and single-line HVAC drawings are shown in Figure 24.16a. Many companies include an HVAC legend on their drawings. This legend also provides information about the standards that are used by the company. Figure 24.16b shows an example of an HVAC legend. This legend is another good place to find information about how to create HVAC drawings.

CADD Applications

HVAC drafting standards are important whether the drawings are created manually or with CADD. Many companies have a standards manual or instructions that correlate with the way the HVAC drawings are created. The numbered standards instructions in Figure 24.17a correlate to the numbers found on the sample drawing in Figure 24.17b. Follow these instructions as you create your own HVAC drawings. This is a common example of HVAC standards. Keep in mind that other slight variations are found in industry.

Detail Drawings

Detail drawings are used to clarify specific features of the HVAC plan. Single- and double-line drawings are intended to establish the general arrangement of the system; they do not always provide enough information to fabricate specific components. When further clarification of features is required, detail drawings are made. A detail drawing is an enlarged view(s) of equipment, equipment installations, duct components, or any feature that is not defined on the plan. Detail drawings may be scaled or unscaled and provide adequate views and dimensions for sheet metal shops to prepare fabrication patterns as shown in Figure 24.18, page 842.

Section Drawings

Sections or sectional views are used to show and describe the interior portions of an object or structure that would otherwise be difficult to visualize. Section drawings may be used to provide a clear representation of construction details or a profile of the HVAC plan as taken through one or more locations in the building. There are two basic types of section drawings used in HVAC. One method is used to show the construction of the HVAC system in relationship to the structure. In this case the building is sectioned and the duct system is shown unsectioned. This drawing provides a profile of the HVAC system. There may be one or more sections taken through the structure, depending on the complexity of the project. The building structure may be drawn using thin lines as shown in Figure 24.19, page 843. Figure 24.19 is a section through the HVAC plan shown in Figure 24.13. The other sectioning method is used to show detail of equipment, or to show how parts of an assembly fit together. (See Figure 24.20, page 843.)

Schedules

Numbered symbols that are used on the HVAC plan to key specific items to charts are known as schedules. These schedules are used to describe items such as ceiling outlets, supply and exhaust grills,

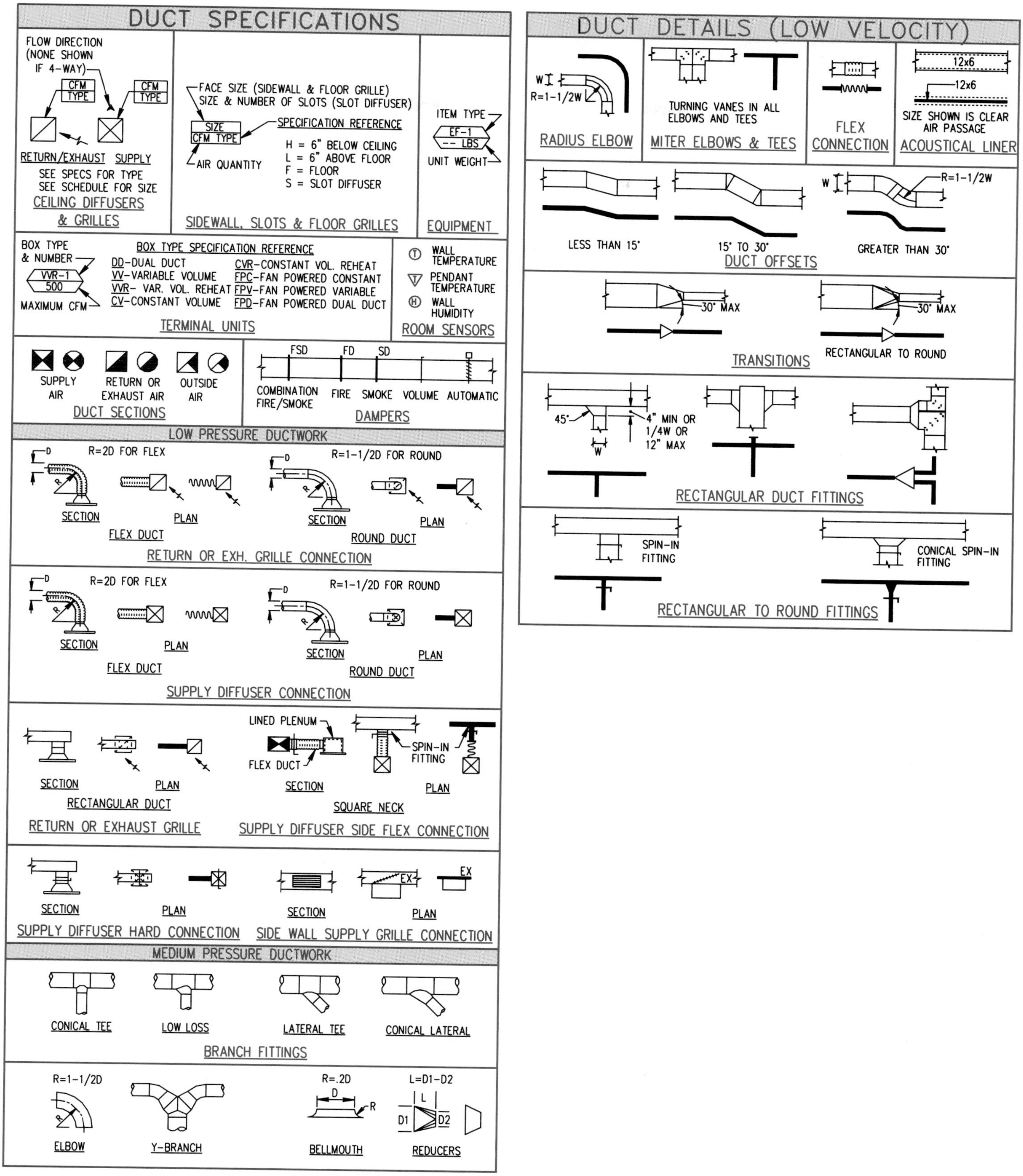

DUCT SPECIFICATIONS

FLOW DIRECTION
(NONE SHOWN
IF 4-WAY)

CFM
TYPE

CFM
TYPE

RETURN/EXHAUST   SUPPLY
SEE SPECS FOR TYPE
SEE SCHEDULE FOR SIZE
**CEILING DIFFUSERS
& GRILLES**

FACE SIZE (SIDEWALL & FLOOR GRILLE)
SIZE & NUMBER OF SLOTS (SLOT DIFFUSER)

SIZE
CFM TYPE

SPECIFICATION REFERENCE

H = 6" BELOW CEILING
L = 6" ABOVE FLOOR
F = FLOOR
S = SLOT DIFFUSER

AIR QUANTITY

SIDEWALL, SLOTS & FLOOR GRILLES

ITEM TYPE
EF-1
-- LBS
UNIT WEIGHT

EQUIPMENT

BOX TYPE
& NUMBER

VVR-1
500

MAXIMUM CFM

BOX TYPE SPECIFICATION REFERENCE
DD-DUAL DUCT   CVR-CONSTANT VOL. REHEAT
VV-VARIABLE VOLUME   FPC-FAN POWERED CONSTANT
VVR-VAR. VOL. REHEAT FPV-FAN POWERED VARIABLE
CV-CONSTANT VOLUME   FPD-FAN POWERED DUAL DUCT

TERMINAL UNITS

T WALL TEMPERATURE
V PENDANT TEMPERATURE
H WALL HUMIDITY

ROOM SENSORS

SUPPLY   RETURN OR   OUTSIDE
AIR   EXHAUST AIR   AIR
DUCT SECTIONS

FSD   FD   SD

COMBINATION   FIRE SMOKE VOLUME AUTOMATIC
FIRE/SMOKE
DAMPERS

LOW PRESSURE DUCTWORK

D   R=2D FOR FLEX   D   R=1-1/2D FOR ROUND

SECTION   PLAN   SECTION   PLAN
FLEX DUCT   **ROUND DUCT**
RETURN OR EXH. GRILLE CONNECTION

D   R=2D FOR FLEX   D   R=1-1/2D FOR ROUND

SECTION   PLAN   SECTION   PLAN
FLEX DUCT   **ROUND DUCT**
SUPPLY DIFFUSER CONNECTION

SECTION   PLAN   LINED PLENUM
                             SPIN-IN FITTING
SECTION   PLAN   FLEX DUCT
RECTANGULAR DUCT   SECTION   PLAN
RETURN OR EXHAUST GRILLE **SQUARE NECK**
SUPPLY DIFFUSER SIDE FLEX CONNECTION

SECTION   PLAN   SECTION   EX   PLAN   EX
SUPPLY DIFFUSER HARD CONNECTION **SIDE WALL SUPPLY GRILLE CONNECTION**

MEDIUM PRESSURE DUCTWORK

CONICAL TEE   LOW LOSS   LATERAL TEE   CONICAL LATERAL
BRANCH FITTINGS

R=1-1/2D   R=.2D   L=D1-D2
                        D   R   L
                                       D1 D2

ELBOW   Y-BRANCH   BELLMOUTH   REDUCERS

DUCT DETAILS (LOW VELOCITY)

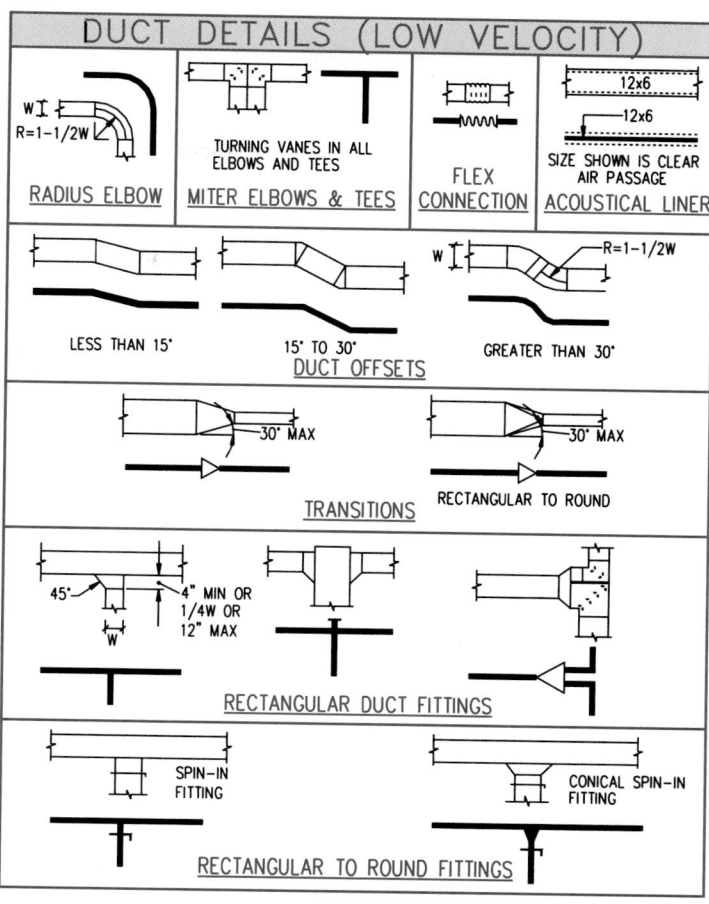

W
R=1-1/2W
RADIUS ELBOW

TURNING VANES IN ALL
ELBOWS AND TEES
MITER ELBOWS & TEES

FLEX CONNECTION

12x6
12x6
SIZE SHOWN IS CLEAR
AIR PASSAGE
ACOUSTICAL LINER

LESS THAN 15°   15° TO 30°   W R=1-1/2W   GREATER THAN 30°
DUCT OFFSETS

30° MAX   30° MAX
TRANSITIONS   **RECTANGULAR TO ROUND**

45°   4" MIN OR
       1/4W OR
W   12" MAX
RECTANGULAR DUCT FITTINGS

SPIN-IN
FITTING

CONICAL SPIN-IN
FITTING

RECTANGULAR TO ROUND FITTINGS

(a)

(Continued)

FIGURE 24.16 ■ (a) Common HVAC symbol specifications. *Courtesy PAE Consulting Engineers.* (b) An HVAC legend is commonly found on drawing to help communicate drafting standards. *Courtesy Interface Engineering.*

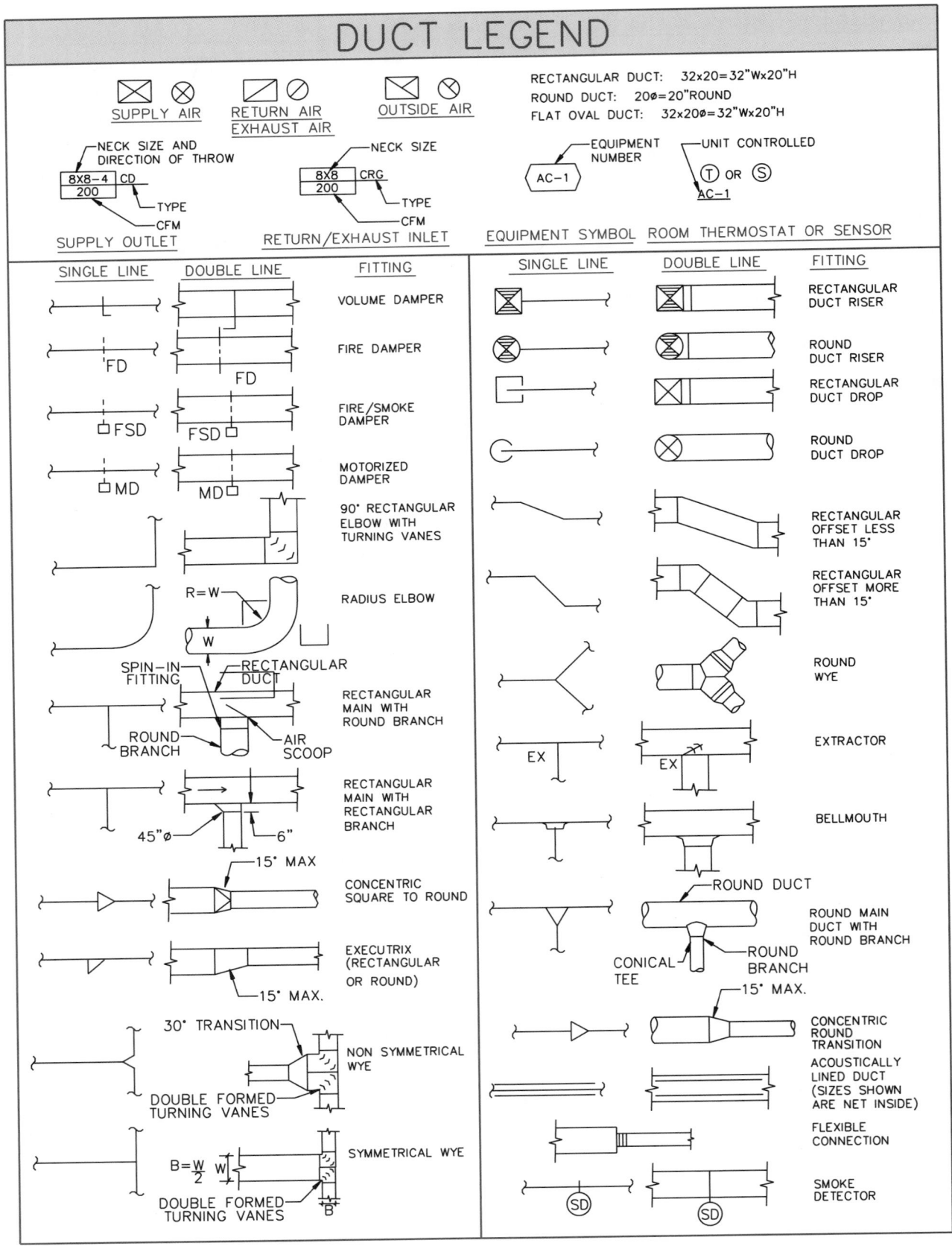

(b)

FIGURE 24.16 ■ continued

INSTRUCTIONS FOR NOTES:

1. DO NOT PUT TEXT ON/UNDER DUCT LINES. MOVE TEXT IF NEW DUCT GOES OVER OLD TEXT.

2. INSERT REVISION CLOUD DELTA IN CLOUD LINE.
   Example

3. IF POSSIBLE STRETCH OR EXTEND LINES INSTEAD OF ADDING SHORT SEGMENTS TO EXISTING LINES.

4. BREAK PIPE LINES FOR ALL UNIONS, VALVES AND OTHER PLUMBING FITTINGS. Example ⊢⊣⊳⊲◻

5. USE THICK LINES FOR WASTE AND EXTRA THICK LINES FOR SINGLE LINE DUCT WORK.

6. HVAC VENT LAYER SHOULD BE USED FOR STEAM OR FUEL OIL VENTING ONLY, USE PLUMBING VENT IN ALL OTHER VENT CASES.

7. DO NOT BREAK LEADER LINES FOR OTHER LEADERS UNLESS INSTRUCTED OTHERWISE BY ENGINEER. TRY TO MOVE THINGS UNTIL ROOM FOR CORRECT LEADER AND NOTES ARE FOUND.

8. TO LABEL TERMINAL UNITS:
   USE LARGE EQUIPMENT BOX TO LABEL ALL TERMINAL UNITS.
   Example

   | °R-12 |
   | 1000 |

9. TO LABEL EQUIPMENT:
   USE LARGE EQUIPMENT BOX TO LABEL ALL EQUIPMENT.
   Example

   | AHU-2 |
   | XX LBS |

10. DIFFUSER BOXES:
    USE THE SMALL DIFFUSER BOX FOR RETURN/EXHAUST AND SUPPLY CEILING DIFFUSERS AND GRILLES.
    Example ⊠

    | 500 |
    | C-1 |

    USE THE LARGE DIFFUSER BOX FOR SIDEWALL, SLOTS AND FLOOR GRILLES.
    Example

    | 6x48 |
    | 400 H-12 |

11. DO NOT PUT DIMENSIONS OR LABELS FOR DUCTS OR OTHER EQUIPMENT IN A DUCT THAT IS NOT THE ONE BEING LABELED.

12. READ AND USE THE MECHANICAL LEGENDS FOR EXAMPLES AND OTHER INFORMATION NOT SHOWN ON THIS SHEET.

13. WHEN LABELING A DUCT UP OR DOWN, WITH AN ARROW, THE FIRST NUMBER OF THE LABEL IS THE HORIZONTAL DIMENSION AND THE SECOND IS THE VERTICAL DIMENSION REGARDLESS OF WHICH SIDE THE ARROW POINTS.

    Example
    40x24
    40x24

DRAWING INSTRUCTIONS

1. TERMINAL UNITS ("R'S & "S) GO ON DUCT LAYER.

2. IF LABEL IS °R (VARIABLE VOLUME REHEAT) THEN USE TERMINAL UNIT W/REHEAT COIL.

3. LEADERS ON MULTI-LINE TEXT COME FROM THE UPPER LEFT OR THE LOWER RIGHT OF THE TEXT BLOCK

4. LEFT JUSTIFY ALL MULTI-LINE NOTES.

5. BREAKS FOR HIDDEN LINES ON DUCT & DIFFUSERS OR BREAKS IN PIPE LINES THAT CROSS NEED GAPS IN THE LINES. USE:
   'SB' BREAK COMMAND FOR BREAKING LINES IN DIFFUSERS AND SPECIAL CASES ONLY.
   'B' BREAK COMMAND FOR BREAKING LINES OVER PIPING AND DOUBLE LINE DUCT LINES.
   'BB' BREAK COMMAND FOR BREAKING A LINE OVER SINGLE LINE DUCT AND WASTE PIPE LINES.

6. TYPICAL OF #'s ARE WRITTEN (TYP. #), NOT (TYP #) OR (TYP. OF #).

7. THERMOSTATS - USE CIRCLE T, THERMDOT & ARC - 3 FROM DOUBLE DUCT MENU TO KEEP ON CORRECT LAYERS.

8. USE STRAIGHT SPIN IN FITTINGS FOR ROUND DUCT OFF OF SQUARE DUCT (LOW PRESSURE DUCT ONLY).

9. USE CONICAL FITTINGS FOR ROUND TAKE-OFFS FROM ROUND DUCT (MEDIUM PRESSURE ONLY).

10. USE TICKS AND NOT ARROWS FOR ANY DIMENSIONING ON DRAWINGS.

11. ALL LETTERED ('SYMBOLS') HEX NOTES ARE TO BE PRIOR (LEFT SIDE) TO WRITTEN NOTES DISREGARDING WHICH SIDE LEADER COMES OFF. ALL NUMBER HEX NOTES ARE TO BE AFTER (RIGHT SIDE) WRITTEN NOTES DISREGARDING WHICH SIDE LEADER COMES OFF.

12. ALL SUPPLY AND RETURN BRANCHES WILL HAVE DAMPERS. UNLESS OTHERWISE NOTED BY ENGINEER.

13. USE LOOP LEADERS TO LABEL SINGLE LINE PIPE AND ARROWS TO LABEL DROPS, RISERS AND VALVES ETC.

14. USE THE ARC ARROW (<-C-) TO LABEL RELOCATE EXISTING AND PUT HEX WITH R IN LINE. MOVE DIFFUSER LABEL TO LABEL NEW LOCATION AND PUT HEX R WITH LABEL. DIFFUSER IN NEW LOCATION IS ON NEW, NOT EXISTING LAYER.

15. DEMO LAYER FOR HVAC & DOUBLE LINE PIPE USES AN L3 LINE (MAGENTA) AND HIDDENA LINETYPE. EACH ENTITY TYPE SHOULD HAVE ITS OWN DEMO LAYER (DUCT, DIFFUSERS, DOUBLE LINE PIPES ETC.).

16. TO DEMO SINGLE LINE PIPE USE "DEMO X's" ROUTINE FROM THE LISP DROP-DOWN MENU. EACH ROUTINE CREATES A LAYER AND USES CORRECT COLOR FOR X's. (HVAC PIPING TYPES SUCH AS CHWS, HTWS, CDS ETC. CAN BE FOUND IN HVAC PIPING LEGEND.)

17. BREAK DUCT (OR PIPE) BETWEEN CAP AND DUCT (OR PIPE).

18. PIPES TO °R's DO NOT HAVE DROPS UNLESS REQUESTED BY ENGINEER.

19. USE SQUARE DUCT BREAK FROM HVAC DROP-DOWN MENU AND PUT BREAK ON DUCT LAYER OF TYPE BEING BROKEN.

20. INTERNAL SOUND INSULATION IS ON THE DUCT LAYER OFFSET 3" USING THE HIDDENA LINETYPE AND COLOR 3 (GREEN).

(Continued)

(a)

FIGURE 24.17 ■ The (a) numbered standards instructions used for creating standard HVAC symbols and drafting applications correlate to the circled numbers on (b) the sample drawing. *Courtesy PAE Consulting Engineers.*

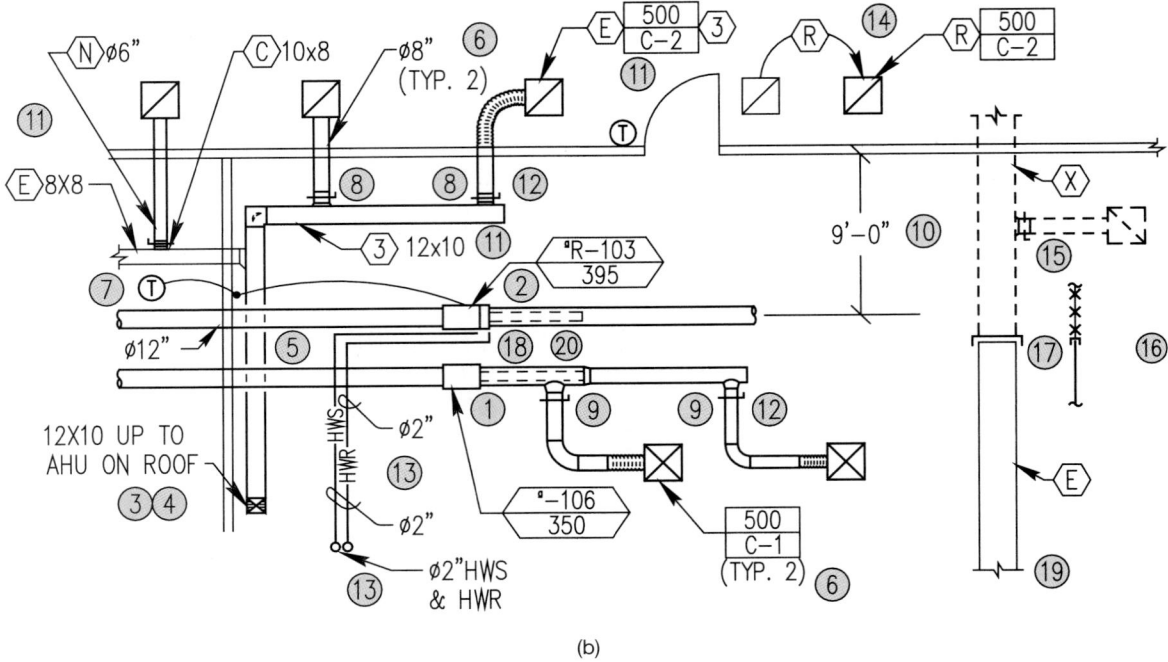

(b)

FIGURE 24.17 ■ continued

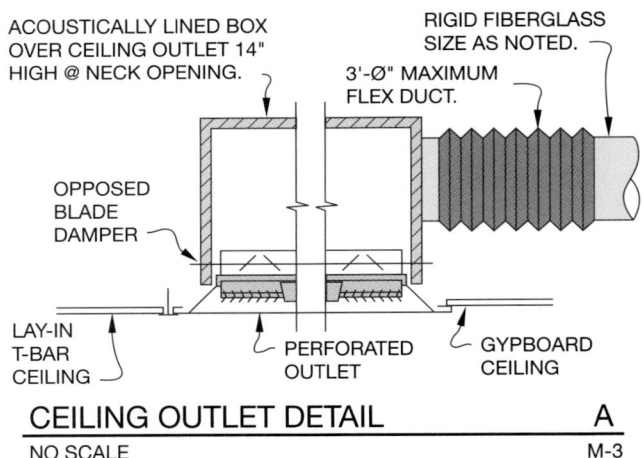

CEILING OUTLET DETAIL   A

NO SCALE   M-3

FIGURE 24.18 ■ Sample detail drawing. *Courtesy W. Alan Gold Consulting Mechanical Engineer and Robert Evenson Associates AIA Architects.*

hardware, and equipment. Schedules are charts of materials or products that include size, description, quantity used, capacity, location, vendor's specification, and any other information needed to construct or finish the system. Schedules aid the drawing by keeping it clear of unnecessary notes. They are generally placed in any convenient area of the drawing field or on a separate sheet. Items on the plan may be keyed to schedules by using a letter and number combination such as C-1 for CEILING OUTLET NO. 1, E-1 for EXHAUST GRILL NO. 1, or ACU-1 for EQUIPMENT UNIT

NO. 1. The exhaust grill schedule keyed to the HVAC plan in Figure 24.13 may be set up as a chart as shown in Figure 24.21.

Pictorial Drawings

Pictorial drawings may be isometric or oblique as shown in Figure 24.22. Isometric, oblique, and perspective techniques are discussed in Chapter 19. Pictorial drawings are usually not drawn to scale. They are used in HVAC for a number of applications, such as assisting in visualization of the duct system, and when the plan and sectional views are not adequate to show difficult duct routing. Manually drawn pictorials are time-consuming and generally not used unless necessary.

SHEET METAL DRAFTING

Sheet metal drafting is used in any industry where flat material is used for fabrication into desired shapes. One of the most common applications is the HVAC industry, although sheet metal shapes are common in the automotive, electronics, and other related industries. Sheet metal drafting is also called pattern development.

DRAWING REVISIONS

Drawing revisions are common on HVAC projects. Revisions can be caused for a number of reasons, for example, changes requested by the owner, job site corrections, correcting errors, or code changes. Changes are done in a formal manner by submitting an *addendum* to the contract, which is a written notification of the change that is accompanied by a drawing that represents the change.

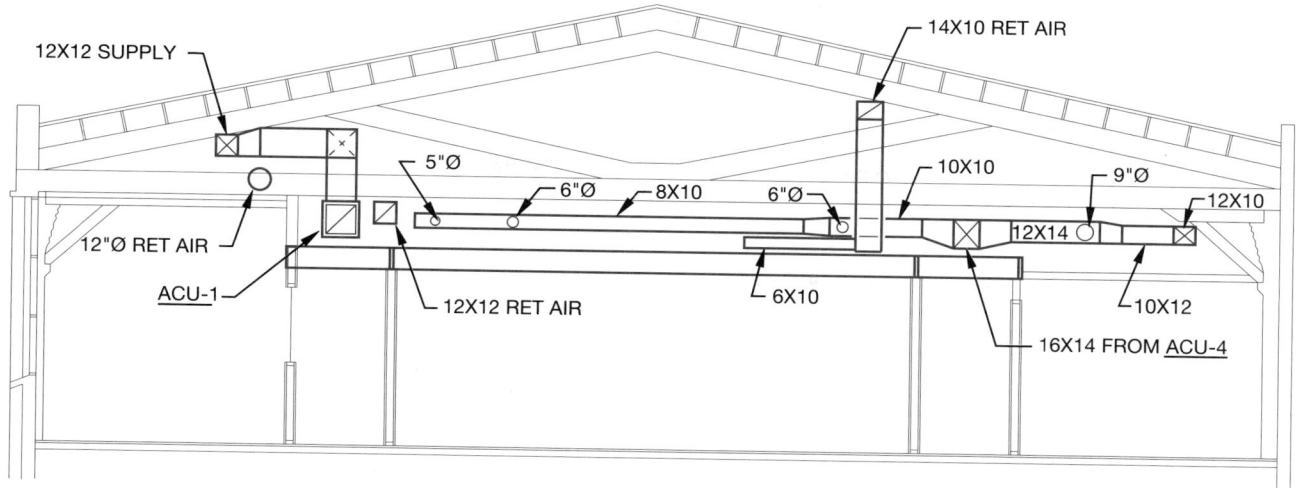

FIGURE 24.19 ■ Section drawing. *Courtesy W. Alan Gold Consulting Mechanical Engineer and Robert Evenson Associates AIA Architects.*

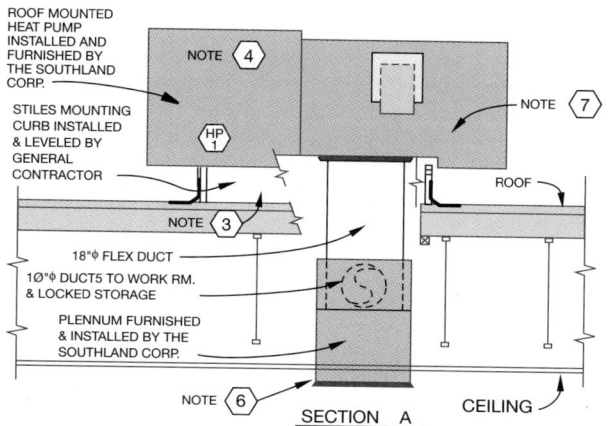

FIGURE 24.20 ■ Detailed section showing HVAC equipment installation. *Courtesy The Southland Corporation.*

Revision Clouds

A revision cloud is placed around the area that is changed. The revision cloud is a cloudlike circle around the change. A triangle with a revision number inside it is placed next to the revision cloud or along the revision cloud line. The triangle is commonly called a delta. The number is then correlated to a revision note placed somewhere on the drawing or in the title block. A complete discussion covering revisions related to architectural drawings is found in the Structural Drafting chapter.

EXHAUST GRILL SCHEDULE

| SYMBOL | SIZE | CFM | LOCATION | DAMPER TYPE | | | | | REMARKS |
				FIRE DPR.	KEY OP. OPPBLD	KEY OP. EXTR	NO DPR.	TYPE	
E – 1	24x12	75Ø	HIGH WALL	✕	✕			4	
E – 2	18x18	72Ø	CEILING	✕	✕			2	
E – 3	1Øx1Ø	24Ø	CEILING		✕			1	24x24 PANEL
E – 4	1Øx1Ø	35Ø	CEILING		✕			1	
E – 5	12x12	28Ø	CEILING		✕			1	
E – 6	1Øx1Ø	5ØØ	CEILING	✕	✕			1	
E – 7	6x6	35Ø			✕			2	
E – 8	12x6	5Ø			✕			3	
E – 9	12x6	2ØØ			✕			3	
E – 10	12x8	15Ø			✕			3	
E – 11	1Øx1Ø	29Ø			✕			3	
E – 12	9x4	16Ø			✕			1	24x24 PANEL
E – 13	9x4	75	HIGH WALL	✕	✕			4	

TYPE 1: KRUGER 119Ø SERIES STEEL PERFORATED FRAME 23 FOR LAY-IN TILE

TYPE 2: KRUGER 119Ø SERIES STEEL PERFORATED FRAME 22 FOR SURFACE MOUNT

TYPE 3: KRUGER EGC-5: 1/2"x1/2"x1/2" ALUMINUM GRID.

TYPE 4: KRUGER S8ØH: 35° HORIZ. BLADES 3/4" O.C.

FIGURE 24.21 ■ Exhaust grill schedule.

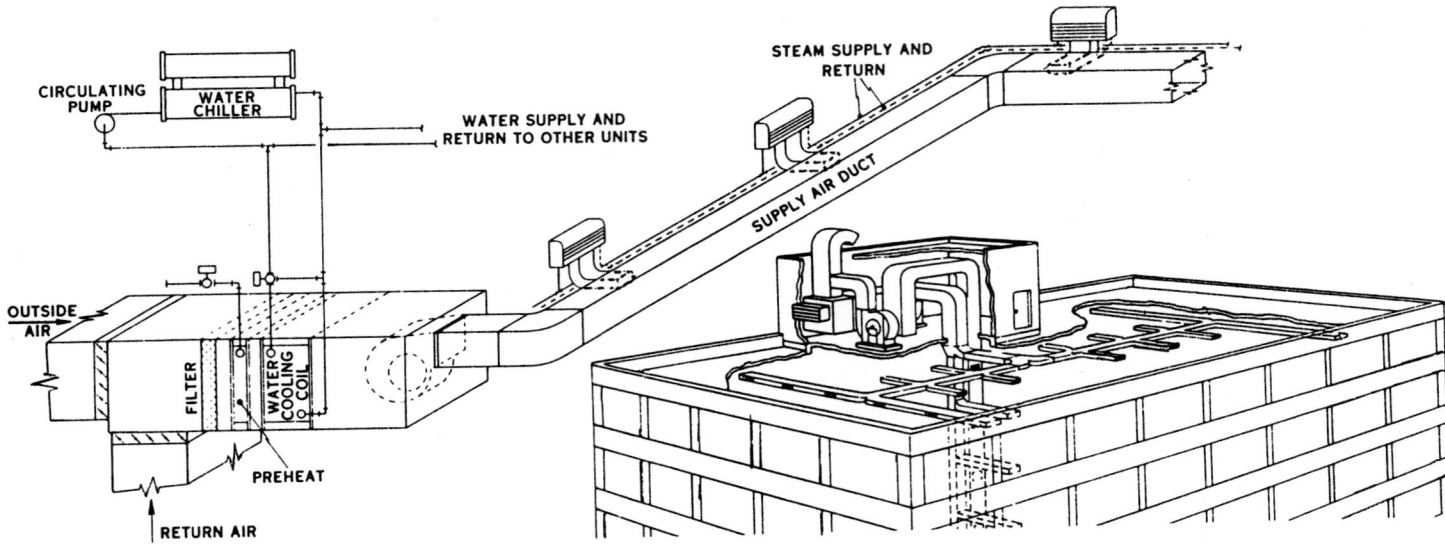

FIGURE 24.22 ■ Pictorial drawings. *Courtesy The Trane Company, La Crosse, WI.*

HVAC INDUSTRY

CADD APPLICATIONS

HVAC CADD software is available that allows you to place duct fittings and then automatically size ducts in accordance with common mechanical equipment suppliers' specifications. The floor plan is commonly used as a reference layer and the HVAC plan is a separate layer. The CADD drafter follows these simple steps:

Step 1. The drawing begins with the preliminary layout drawn as the duct centerlines. (See Figure 24.23.)

Step 2. Select supply and return registers from the template menu symbols library and add the symbols to the

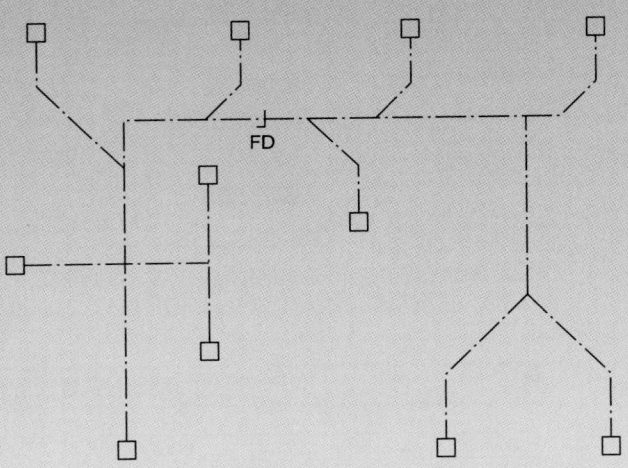

FIGURE 24.24 ■ Select supply and return registers from a tablet menu.

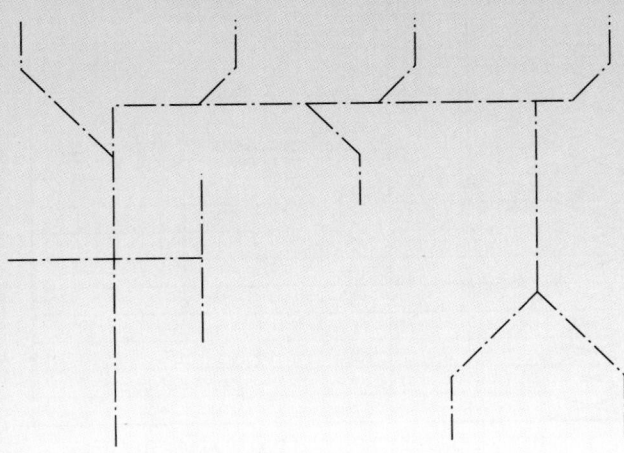

FIGURE 24.23 ■ The drawing begins with the preliminary layout drawn as duct centerlines.

end of the centerlines where appropriate. (See Figure 24.24)

Step 3. The program then automatically identifies and records the lengths of individual duct runs, and tags each run. Fittings are located and identified by the type of intersection. (See Figure 24.25.) While all of this drawing information is added to the layout, the computer automatically gathers design information into a file for duct sizing based on a

CADD APPLICATIONS *(continued)*

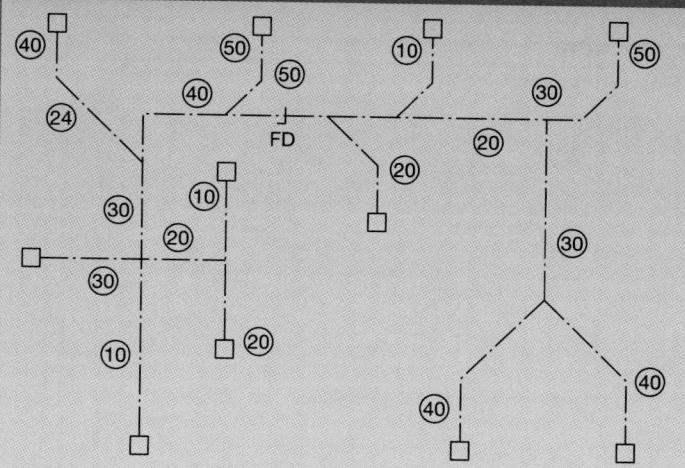

FIGURE 24.25 ■ Fittings are located and identified by the type of intersection.

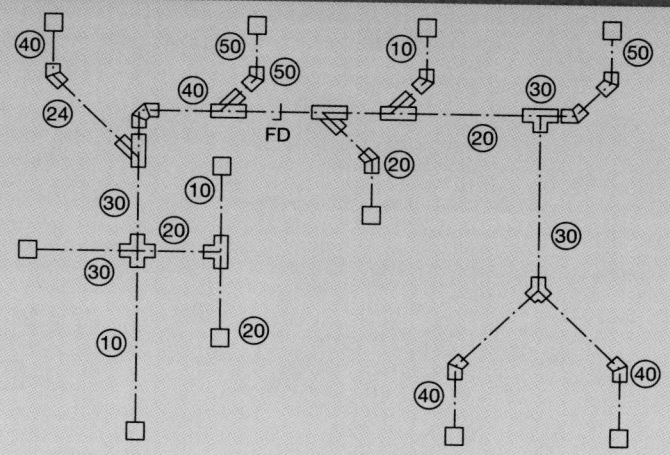

FIGURE 24.26 ■ HVAC symbols are drawn as accurate double-line symbols exactly to ANSI Y32.2.4 standards.

specific mechanical manufacturer's specifications that you select.

Step 4. After the fitting location and sizes are determined, the program transforms each fitting into accurate double-line symbols exactly to the ANSI Y32.2.4 standard. (See Figure 24.26.)

Step 5. When the fittings are in place, the program calculates and draws the connecting ducts, adding couplings automatically at the maximum duct lengths. If a transition is needed in a duct run, the program recommends the location and all you have to do is pick a transition fitting from the menu library. See Figure 24.27 for the complete HVAC layout.

An added advantage to using HVAC CADD software is that the program automatically records information, while you draw, to generate a complete bill of materials. The systems that offer you the greatest flexibility and productivity are designed as a parametric package. This type of program allows you to set the design parameters that you want and then the computer automatically draws and details according to these settings. As you draw, information such as the type of fitting, CFM, and gauge is placed with each fitting. A complete HVAC plan and related schedules are shown in Figure 24.28.

CADD HVAC packages provide a variety of tablet menu overlays that assist in the rapid selection of symbols.

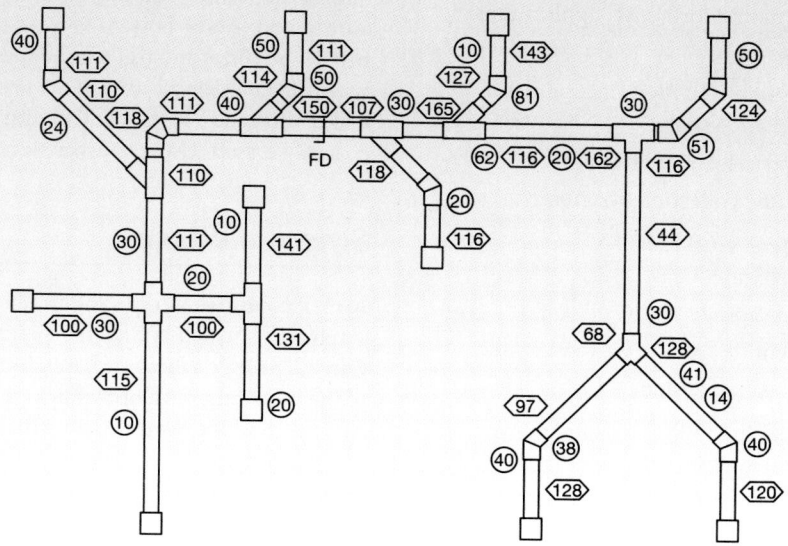

FIGURE 24.27 ■ The finished HVAC plan.

(Continued)

CADD APPLICATIONS (continued)

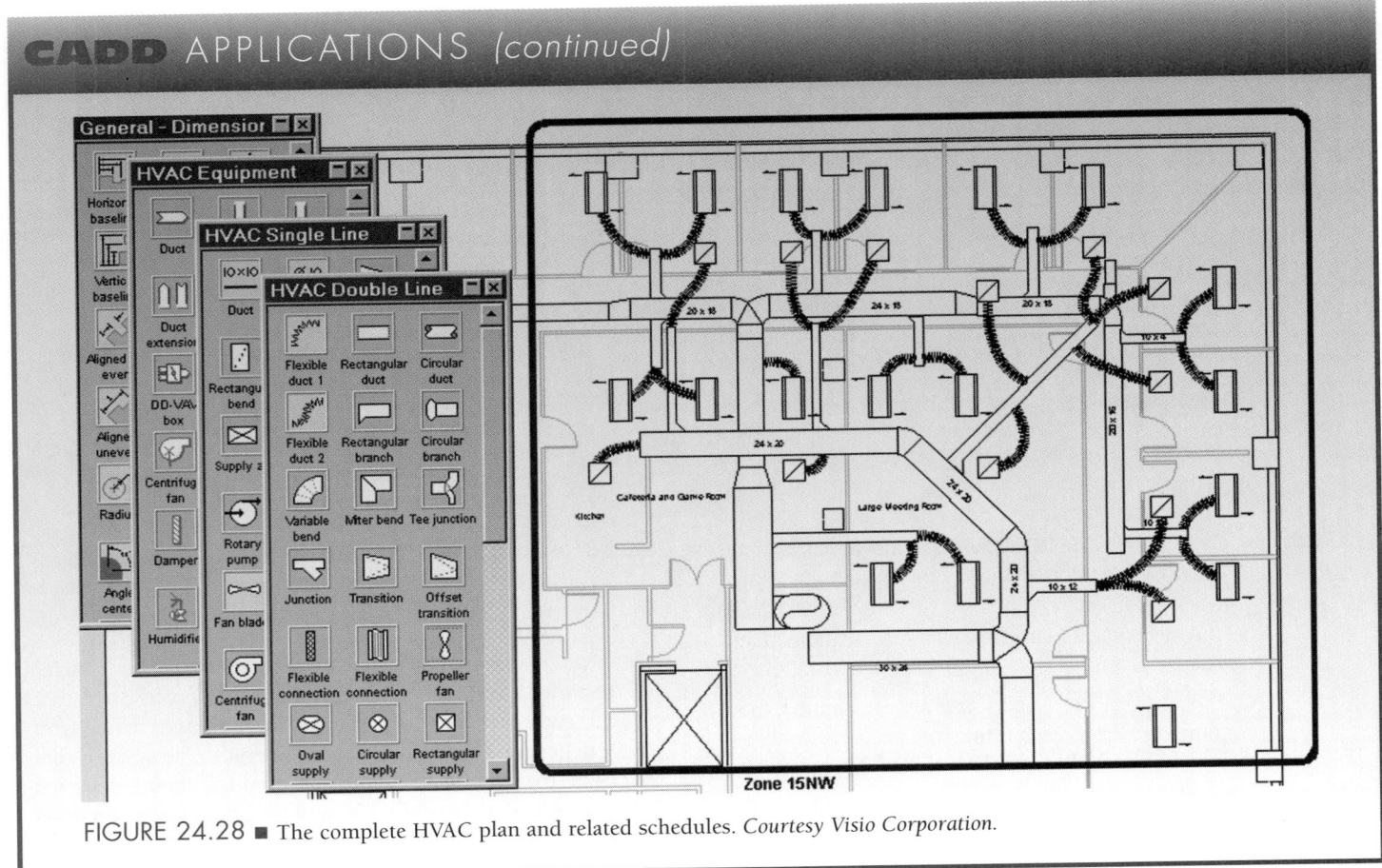

FIGURE 24.28 ■ The complete HVAC plan and related schedules. *Courtesy Visio Corporation.*

CADD Layers for HVAC Drawings

The American Institute of Architects (AIA) CADD Layer Guidelines establish the heading "Mechanical" as the major group identification for HVAC-related CADD layers. Following are some of the recommended CADD layer names for HVAC applications:

Layer Name	Description
M-CHIM	Prefabricated chimneys
M-CMPA	Compressed air systems
M-CONT	Controls and instrumentation
M-DUST	Dust and fume collection systems
M-ENER	Energy management systems
M-EXHS	Exhaust systems
M-FUEL	Fuel system piping
M-HVAC	HVAC system
M-HOTW	Hot water heating system
M-CWTR	Chilled water system
M-REFG	Refrigeration system
M-STEM	Steam system
M-ELEV	Elevations
M-SECT	Sections
M-DETL	Details
M-SCHD	Schedules and title block sheets

PATTERN DEVELOPMENT

The principle of pattern development is based on laying out geometric shapes in true size and shape flat patterns. The fundamental concepts involved in making patterns for basic geometric shapes may be used in the development of any pattern. In most situations a front and top or bottom view should be drawn to help establish true-length lines and true size shapes. *The key to pattern development is any line or element used in the development must be in true length.* Use construction lines for all preliminary layout work so errors may be easily erased. Use a construction layer when using CADD.

Stretch-out Line

A *stretch-out line* is typically the beginning line upon which measurements are made and the pattern development is established. Carefully observe how the stretch-out lines are established for each of the following developments as this is the first process in making a layout. Specific developments that do not begin with a stretch-out line will be identified. The instruction provided for the following pattern development is broken down into the most basic procedures. Individual shortcuts may be taken after adequate experience has been gained. It is extremely important to maintain a high degree of accuracy. Scale drawings and transfer dimensions very carefully. Another important consideration is to accurately label the elements of the views with numbers and/or letters and then transfer these labels to the pattern. This procedure may seem unnecessary on simple developments, but on complex patterns it is absolutely necessary. It is recommended that beginners label all developments as suggested in the given procedures.

Seams

When sheet metal parts are bent and formed into the desired shape, a seam results where the ends of the pattern come together. The fastening method depends on the kind and thickness of material, on the fabrication processes available, and on the end use of the part. Sheet metal components that must hold gases or liquid, or are pressurized, may require soldering or welding. Other applications may use mechanical seams, which hold the parts together by pressure lapped metal, metal clips, or pop rivets. Some of the most common seams used in the sheet metal fabrication industry are shown in Figure 24.29. Extra material may be required on the pattern to allow for seaming. For the purpose of discussion, in the following procedures and problems, either a single- or double-lap seam is used. For a single-lap seam, add the given amount to one side of the pattern. For double-lap seams, add the given amount to both sides of the pattern. The corner of a seam may be cut off at an angle (usually 45°) if it interferes with adjacent parts during fastening.

Hemmed edges are necessary when an exposed edge of a pattern must be strengthened. When hems are used, extra material must be added to the pattern on the side of the hem. Figure 24.30 shows some common hems.

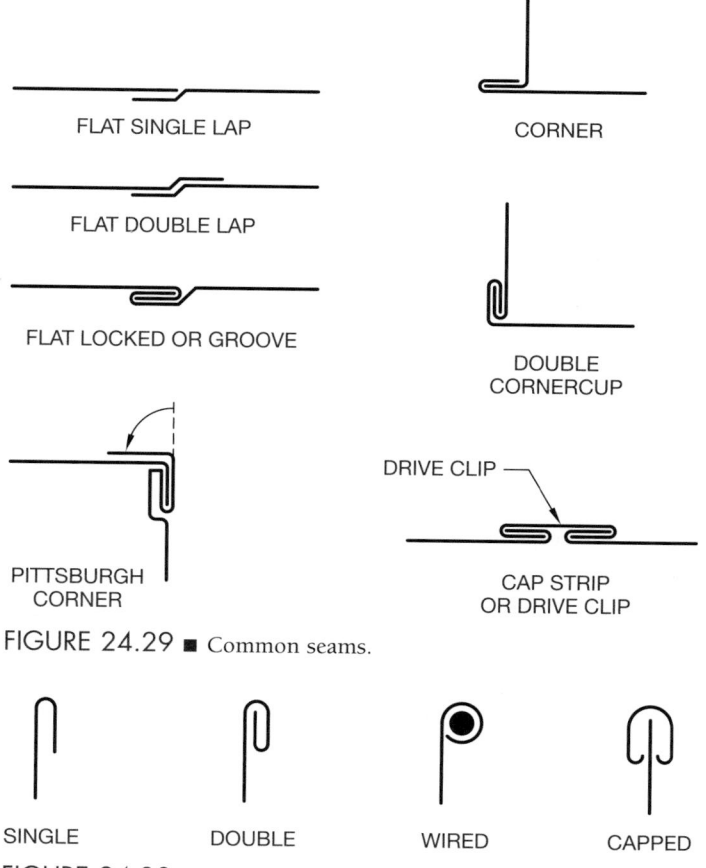

FIGURE 24.29 ■ Common seams.

FIGURE 24.30 ■ Common hems.

Rectangular (Right) or Square Prism

Commonly referred to as a box (open ended in this example), the rectangular or right prism may be developed as follows:

Step 1. Draw the front and top view, label the corners, and establish the stretch-out line off the base of the front view and perpendicular to the height line. (See Figure 24.31.)

Step 2. Beginning at 1 in the top view use dividers to measure the true-length distance from 1 to 2. Transfer this dimension to the stretch-out line, starting at any point near the front view. Continue this process by transferring the distance from 2 to 3, 3 to 4, and 4 to 1 to the stretch-out line. (See Figure 24.32.) You must end at the point you began; point 1 in this example.

Step 3. From each of the points established in step 2, draw vertical construction lines to meet a horizontal line drawn from the true-length (TL) height in the front view. (See Figure 24.33.)

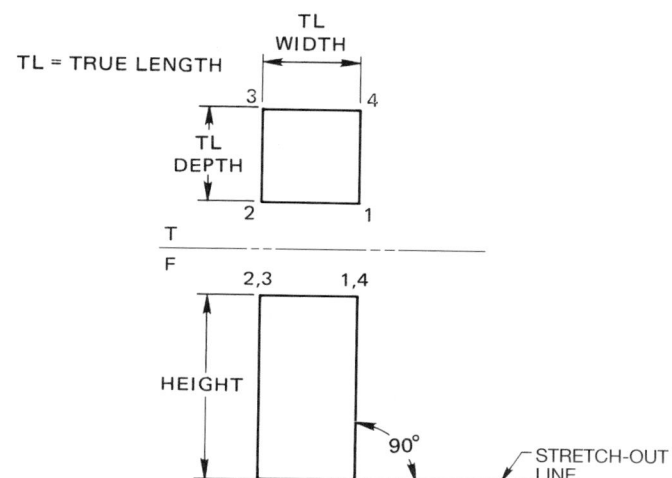

FIGURE 24.31 ■ Step 1—right prism development.

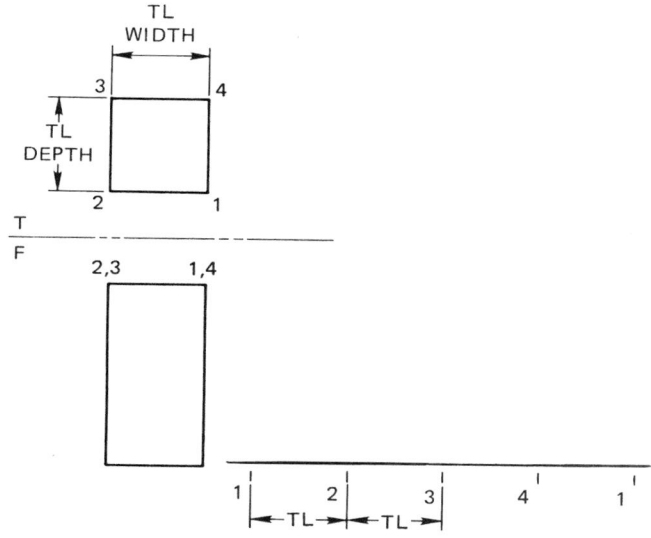

FIGURE 24.32 ■ Step 2—right prism development.

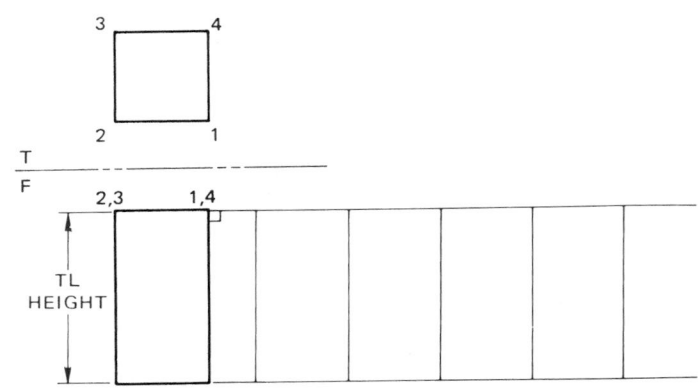

FIGURE 24.33 ■ Step 3—right prism development.

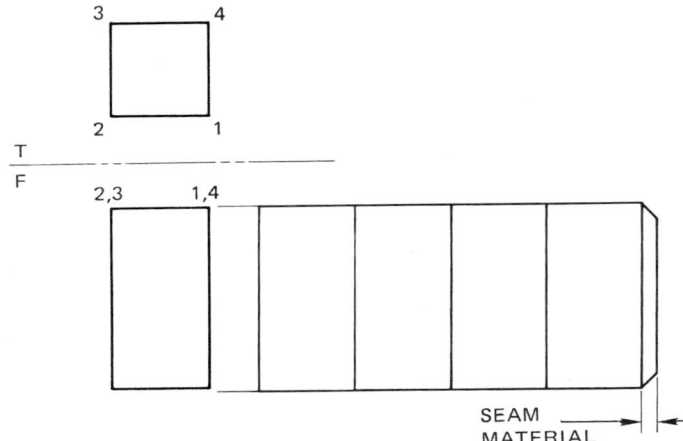

FIGURE 24.34 ■ Step 4—right prism development.

Step 4. Darken in object lines. Notice that thick lines are used where the pattern sides make a bend. Add any required seam material. (See Figure 24.34.) The pattern is now ready to be cut out and formed into the given shape or transferred to sheet metal for fabrication.

Truncated Prism

A sheet metal part such as a prism, pyramid, or cone is considered to be *truncated* if a portion is cut off, generally at an angle.

Step 1. Proceed as described in steps 1 through 3 for a right prism. Begin with the shortest element as the seam, when possible. The shortest seam is stronger, easier to fabricate, and requires less materials. This is true for all pattern development. (See Figure 24.35.)

Step 2. Darken all object lines by connecting the ends of the true-height elements to form the outline of the object. Darken in all bend lines and add seam material as shown in Figure 24.36.

Regular (Right) Cylinder

One of the most common shapes is the cylinder. The following procedure may be used for the development of any right cylindrical shape:

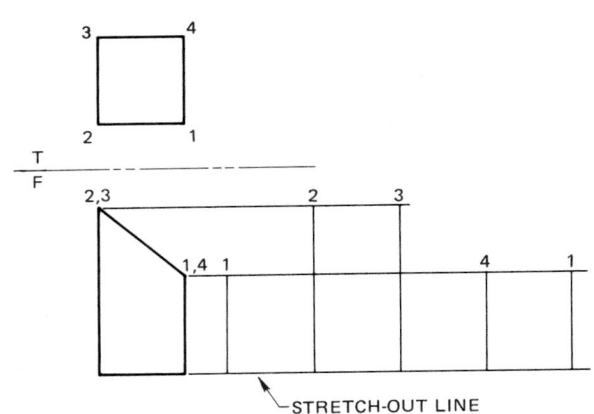

FIGURE 24.35 ■ Step 1—truncated prism development.

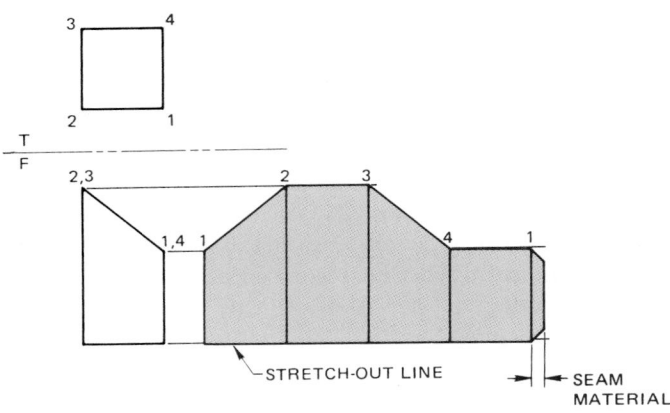

FIGURE 24.36 ■ Step 2—truncated prism development.

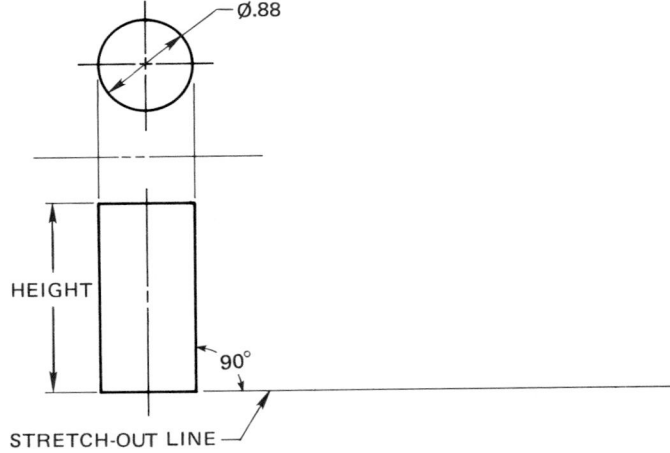

FIGURE 24.37 ■ Step 1—right cylinder development.

Step 1. Draw the top and front views. Establish the stretch-out line perpendicular to the side of the front view. The stretch-out line may be placed anywhere next to the front view, but it must be perpendicular to the side. (See Figure 24.37.)

Step 2. Establish the length of the stretch-out line equal to the circumference of the circle with the formula $C = \pi D$. The diameter is .88 in., so $C = 3.14(\pi) \times .88 = 2.76$ in. Now,

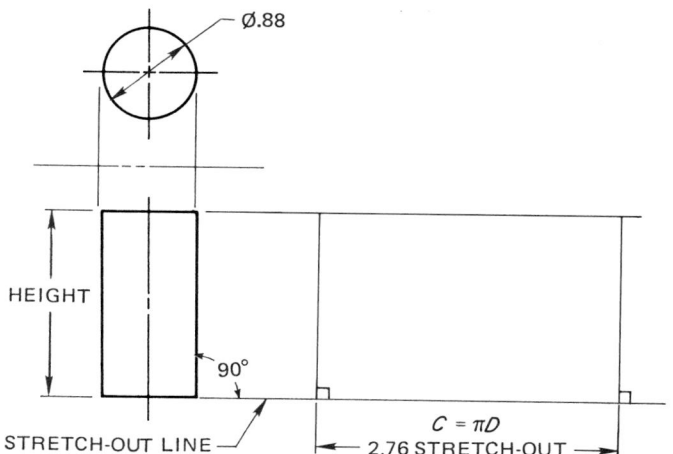

FIGURE 24.38 ■ Step 2—right cylinder development.

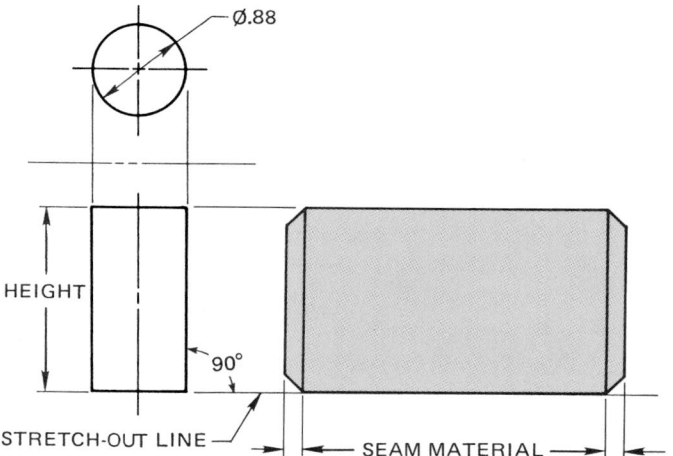

FIGURE 24.39 ■ Step 3—right cylinder development.

from the ends of the stretch-out line, draw perpendicular lines that meet a line projected from the true height in the front view as shown in Figure 24.38.

Step 3. Darken the outline of the pattern and add seam material as shown in Figure 24.39, which shows a double-lap seam.

Truncated Cylinder

Truncated cylinders have many HVAC applications with development procedures similar to the regular cylinder.

Step 1. Draw the top and front views. Divide the top view into twelve equal parts. More divisions establish better accuracy; less give less accuracy. Twelve divisions have been selected because of ease and effectiveness. Work carefully with a sharp pencil to maintain the best accuracy possible. Number each element in the top view and extend the numbering system into the front view. (See Figure 24.40.)

Step 2. Establish the stretch-out line perpendicular to the side of the front view and the length with the formula $C = \pi D$ as

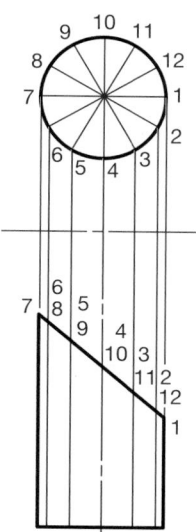

FIGURE 24.40 ■ Step 1—truncated cylinder development.

previously discussed. Divide the stretch-out into twelve equal parts (refer to dividing a line into equal parts, Chapter 6). Number each part from 1 through 12, ending at 1 (the seam). From each part on the stretch-out line, draw a perpendicular construction line equal to the total height of the cylinder. (See Figure 24.41.)

Step 3. Project the true height of each line segment in the front view to the correspondingly numbered line in the pattern layout. Where these lines intersect, a pattern of points is established. Use an irregular curve to connect the points forming the curved outline of the truncated cylinder pattern. Darken the outline of the pattern and add seams.

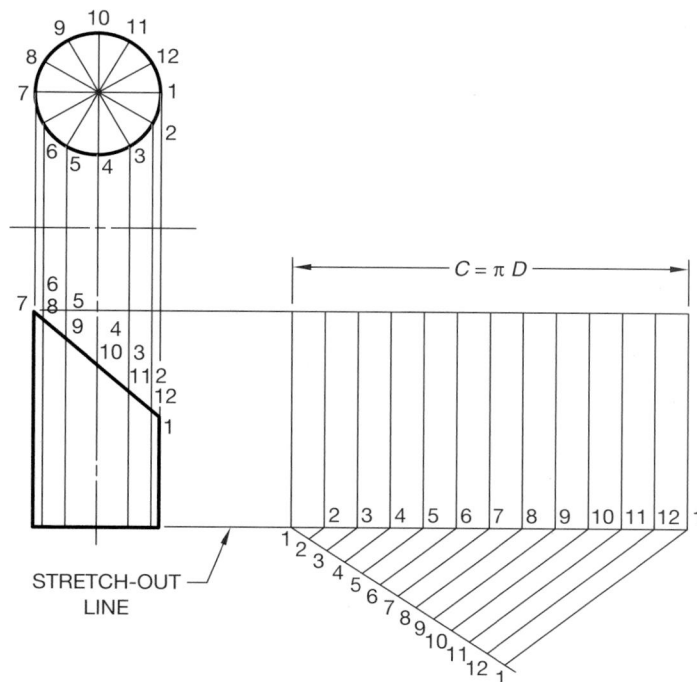

FIGURE 24.41 ■ Step 2—truncated cylinder development.

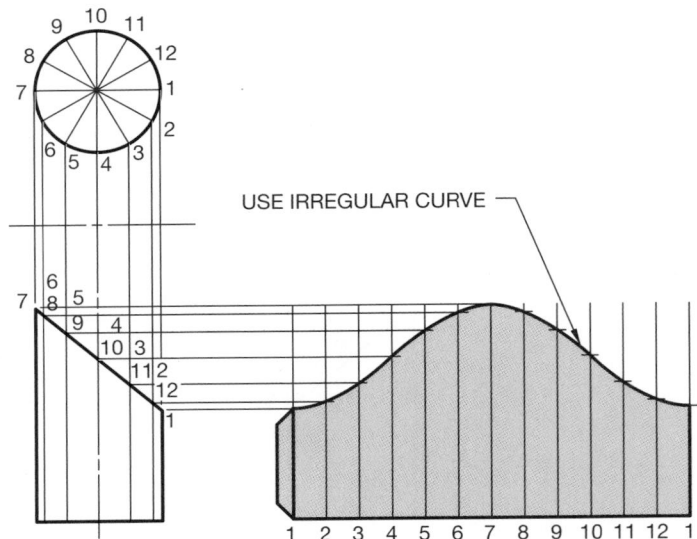

FIGURE 24.42 ■ Step 3—truncated cylinder development.

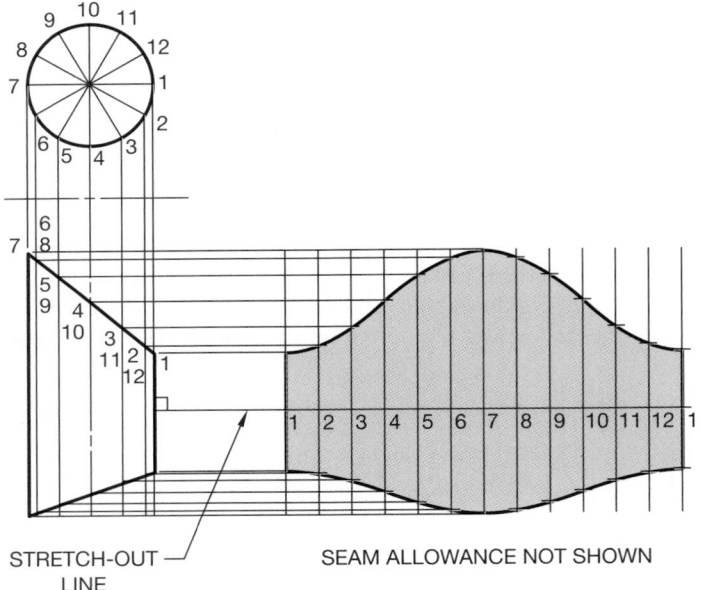

FIGURE 24.43 ■ Development of cylinder truncated on both ends.

The division lines may be left as thin lines to represent a curved object. Construction lines should not have to be erased if properly drawn. (See Figure 24.42.)

When the cylinder is truncated on both ends, the process is the same, with the stretch-out line established perpendicular to the side at any desired location. (See Figure 24.43.)

Cylindrical Elbow

Cylindrical elbows are used to make turns or corners in ductwork. The turns may be any number of degrees. Standard cylindrical elbows are made up of any given number of truncated cylinders. Each piece of the elbow may be developed as demonstrated in the

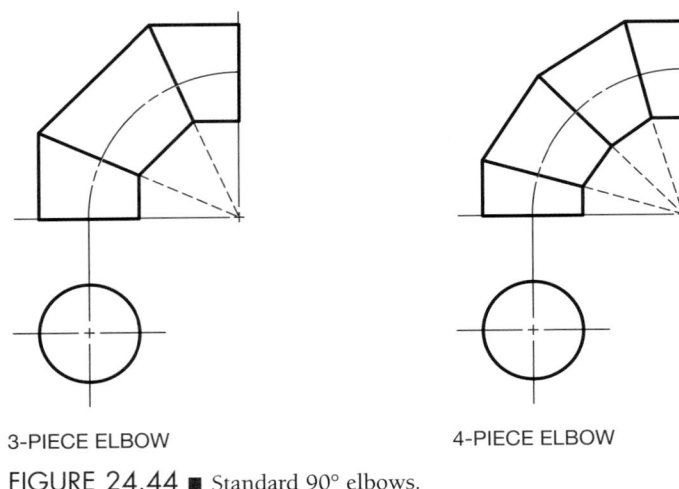

3-PIECE ELBOW   4-PIECE ELBOW

FIGURE 24.44 ■ Standard 90° elbows.

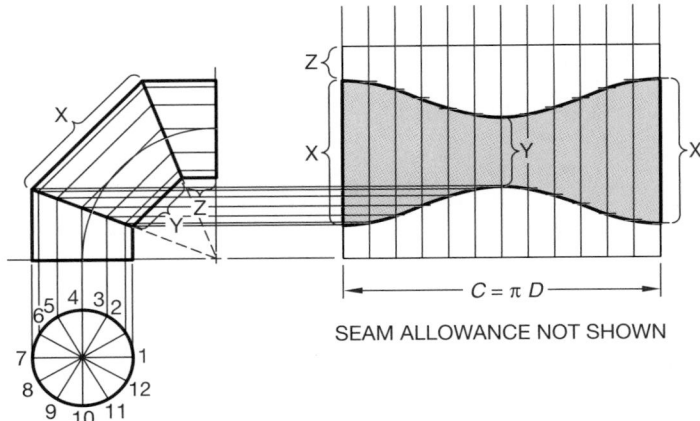

SEAM ALLOWANCE NOT SHOWN

FIGURE 24.45 ■ Development of a three-piece cylindrical elbow.

previous discussion. Figure 24.44 shows standard 90° three-piece and four-piece elbows. Each piece of the three-piece elbow is developed as a truncated cylinder, with the patterns developed separately or together by alternating starting elements as shown in Figure 24.45.

Cone

The size of the cone is established by the base diameter, and the height and may be developed as follows:

Step 1. Given the height and base diameter, draw the front and top (or bottom) views. Divide the top (bottom) view into twelve equal parts. Number each part and project these to the cone base in the front view. At the base in the front view, extend each point to the vertex of the cone. (See Figure 24.46.)

Step 2. The stretch-out line for a right circular cone development is a true circular arc. The radius of the arc is taken from the true-length element measured from zero to one in the front view. (See Figure 24.46.) At any convenient location where you have adequate space, draw an arc with the true length radius. (See Figure 24.47.)

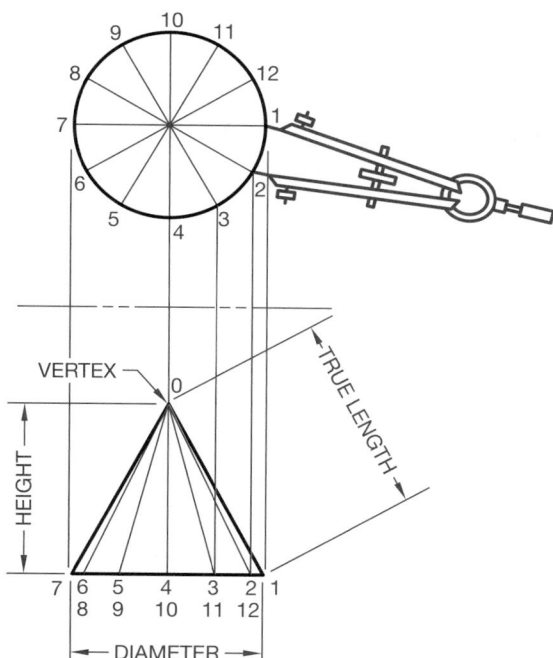

FIGURE 24.46 ■ Step 1—cone development.

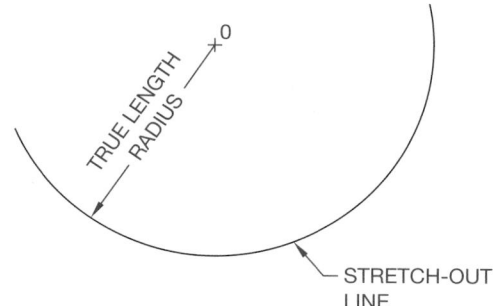

FIGURE 24.47 ■ Step 2—cone development.

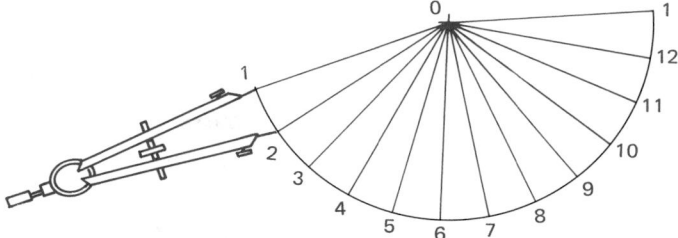

FIGURE 24.48 ■ Step 3—cone development.

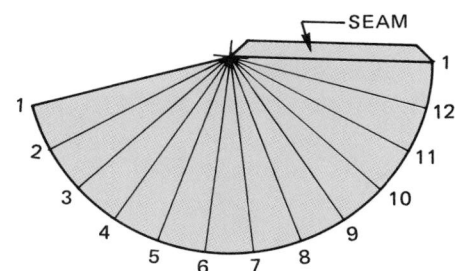

FIGURE 24.49 ■ Step 4—cone development.

Truncated Cone

The procedure for developing a truncated cone is the same as the regular cone previously described. Always begin by developing a pattern for a full regular cone. After step 3, the true length of each line segment from the vertex to the truncated line in the front view must be established. Project each point to the true-length line. Then measure the true length of each element from the vertex along the true-length line as shown in Figure 24.50. Transfer the true length of each individual line to the corresponding line in the development. (See Figure 24.50.) Finally, darken the outline using an irregular curve, where necessary, and add seam material as shown in Figure 24.50.

Offset Cone

The offset cone has the vertex offset from the center of the base. The procedure is slightly different from the development of a regular cone as the true length of each line element must be found. Refer to Figure 24.51 as you read the following discussion:

Step 1. Draw the top and front views. Divide the top view base circle into twelve equal parts. Project each point to the vertex and establish the same elements in the front view.

Step 2. The true lengths of elements 0–1 and 0–7 are true length in the front view. Establish the true-length diagram for the other elements using the revolution method as shown in color in Figure 24.51.

Step 3. There will be no stretch-out line in this development. Begin the development by laying out the true length of element 0–7 in a convenient location. Take a compass and lay out the true length of 0–6 and 0–8 on each side of 0–7. Set another compass with the radius equal to the distance X in the top view. This measurement is used several

Step 3. Go to the top view and use dividers to establish the increment from point 1 to point 2. Rotate the dividers around the circle, adjusting the dividers until you have accurately established the increment at least half way around the circle. (See Figure 24.46.) Beginning at any point on the stretch-out line, use the established increment to locate twelve equal spaces. Remember, if you begin at 1, you must end at 1. Now, connect each point along the stretch-out line to the vertex 0 with construction lines. (See Figure 24.48.) When using CADD, this procedure may be done quickly and accurately. The AutoCAD MEASURE command may be used to establish the circumference on the stretch-out line followed by the DIVIDE command to find the twelve equal spaces.

Step 4. Darken the outline of the cone pattern and add seam material as necessary. (See Figure 24.49.)

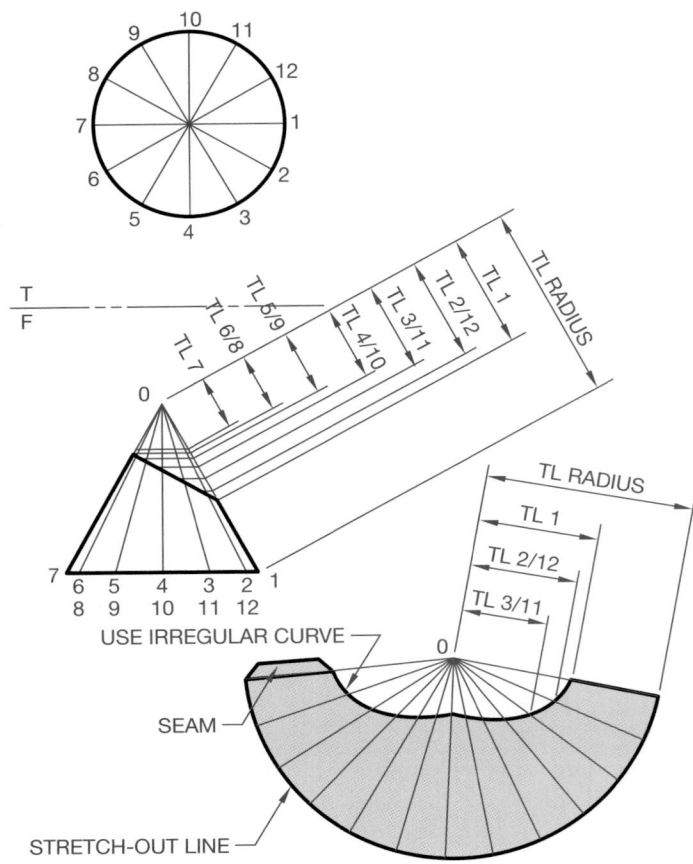

FIGURE 24.50 ■ Truncated cone development.

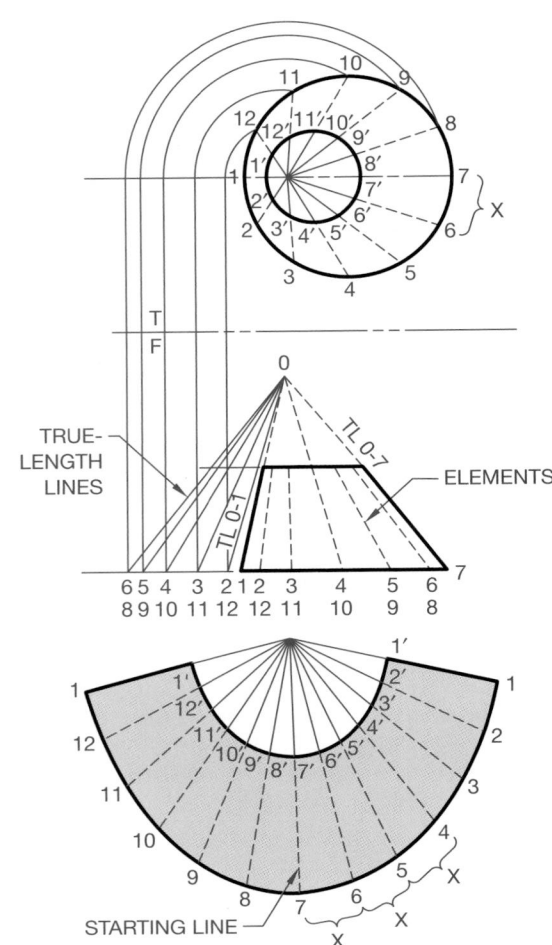

FIGURE 24.51 ■ Offset cone development.

times, so keep your compass at this setting. Using the distance *X*, set the compass point at 7 and draw an arc that intersects the arcs drawn for 0–6 and 0–8. This procedure locates the exact position of elements 0–6 and 0–8. Follow this same method for each of the elements, 0–5 and 0–9, 0–4 and 0–10, 0–3 and 0–11, 0–2 and 0–12, and 0–1 at each end. You now have a series of points along the base that may be connected with an irregular curve.

Step 4. The true lengths of elements 0–1 and 0–7 are in the front view. Use the true-length diagram to establish the true lengths of the other elements, 0–2 through 0–12. Transfer these true lengths to the corresponding elements in the development and connect the points with an irregular curve. Darken the outline and add seam material, if required. (See Figure 24.51.)

Pyramid

Pyramids are developed in a manner similar to a cone as described in the following steps:

Step 1. Draw the top and front views and establish the true length of the edge of a side or lateral edge. Set the compass with a radius equal to the true length and draw an arc in a convenient location. This is the stretch-out line. (See Figure 24.52.)

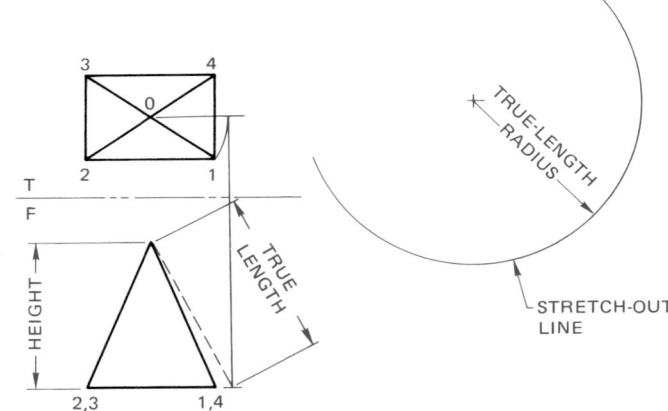

FIGURE 24.52 ■ Right pyramid development.

Step 2. Use dividers or a compass to lay out the true lengths from points 1–2, 2–3, 3–4, and 4–1 in the top view and transfer these distances consecutively to the stretch-out line. Connect points 1, 2, 3, 4, and 1 in the development and also connect these points to the vertex 0. Darken all lines and add seam allowances as shown in Figure 24.53.

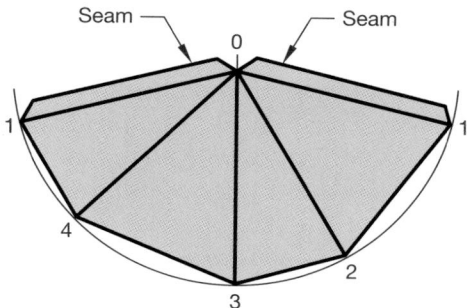

FIGURE 24.53 ■ Pyramid development.

Truncated Pyramid

The procedure for the development of a truncated pyramid is the same as for a regular pyramid except that the true length of each element from the vertex to the truncated line must be established and transferred to the pattern as shown in Figure 24.54.

Transition Piece

Also known as a *square to round,* this *transition piece* is a duct component that provides a change in shape from square or rectangular to round. Transition pieces may be designed to fit any given situation, but the pattern development technique is always as follows: (Note:

Accuracy and the use of a number system is very important. Do all layout work using construction lines or a CADD construction layer. There may be other techniques that work in some unique situations, but the method described here works with any transition piece configuration.)

Step 1. Draw the top and front views. Divide the circle in the top view into twelve equal parts. Connect the points in each quarter of the circle to the adjacent corner of the square. Project the same corresponding system to the front view. Number and letter each point in each view as shown in Figure 24-55. The transition piece shown in Figure 24.55 is made up of a series of triangles; for example, 1,A,D; 1,2,A; and 2,3,A. The development progresses by attaching the true size and shape of each triangle together in order. This technique is known as *triangulation.* There is no stretch-out line.

Step 2. Each element of the development must be in true length. Establish the true-length diagram using revolution as shown in Figure 24.56. Only one set of true-length elements is required because the given problem is a right transition piece (symmetrical about both axes).

Step 3. Begin the pattern development on a sheet in an area where there is a lot of space. Start with the true size and shape of triangle 1,A,D. The true lengths of 1–A and 1–D are the same and may be found in the true-length diagram. The true length of A–D is in the top view. (See Figure 24.57.) A review of Chapter 6, Geometric Construc-

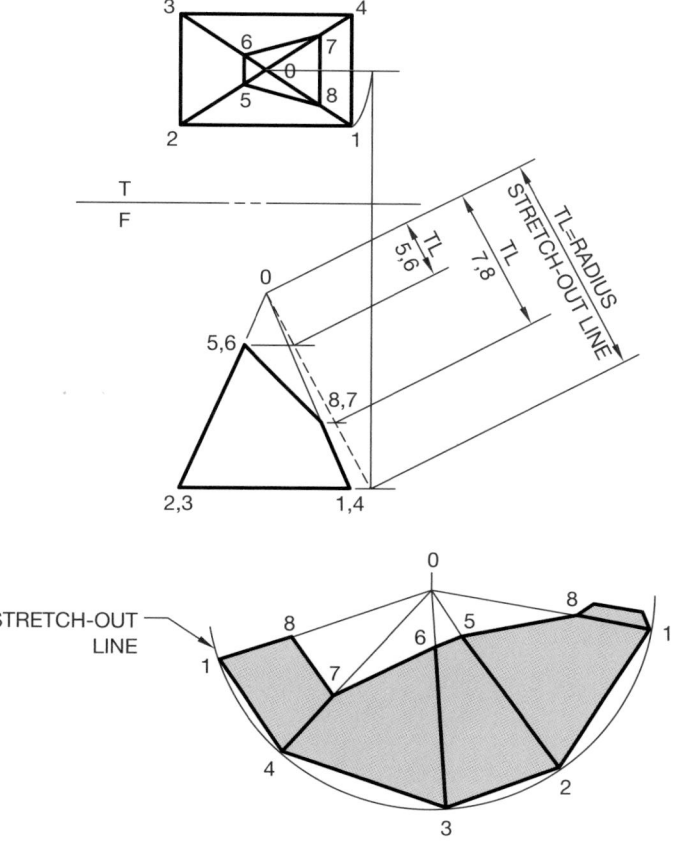

FIGURE 24.54 ■ Truncated pyramid development.

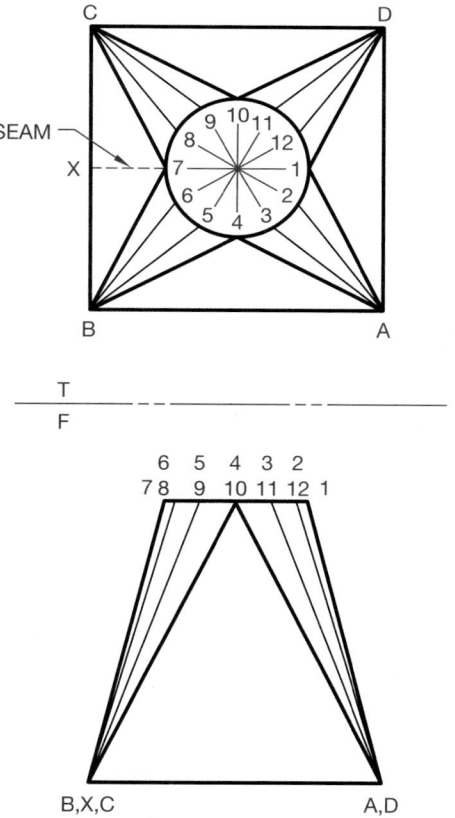

FIGURE 24.55 ■ Step 1—right transition piece view setup.

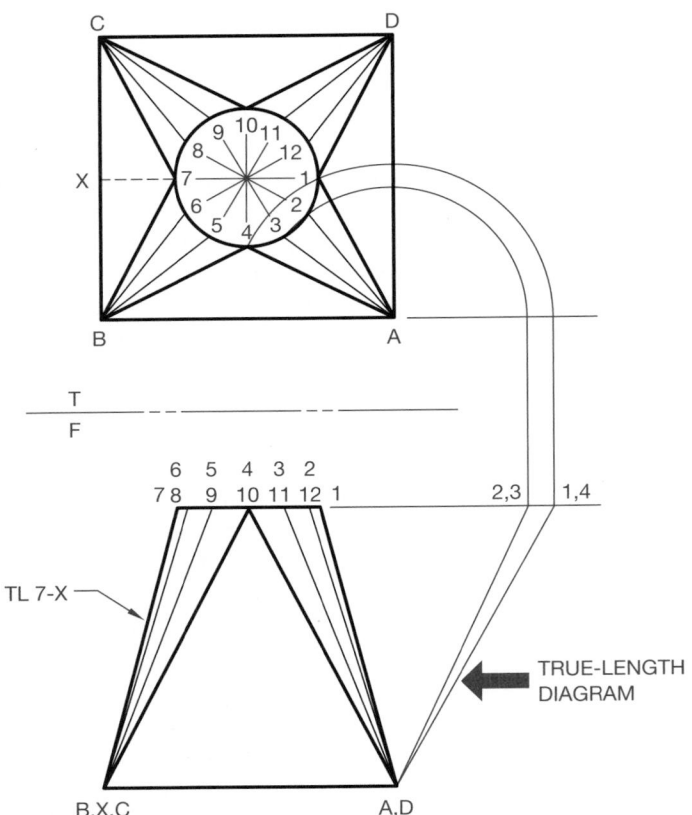

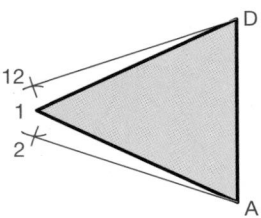

FIGURE 24.58 ■ Step 4—transition piece development; continue the pattern development both ways from the starting triangle.

for the distance from 1–2. Keep this compass setting, as it is used several times. With this setting, scribe two arcs from point 1 that intersect the previously drawn arcs. These intersections are points 2 and 12. Connect points 2–A and 12–D. Do not connect points 1,2 and 12 yet. (See Figure 24.58.) Do all work with construction lines until complete.

Step 5. Continue the same procedure, working both ways around the transition piece until the entire development is complete. Every triangle must be included and the pattern will end on both sides with the element 7–X, which is true length in the front view. The two final triangles are right triangles, because each is half of the full side that is an equilateral triangle. Remember that every line transferred from the views to the pattern must be true length. It is recommended that the numbering system be used to avoid errors. When all points have been established, the outline may be darkened. A series of points form the inside curve. Connect these points with an irregular curve. If any point is out of alignment with the others, you have made an error at this location. Notice in Figure 24.59 that thick

THE FOLLOWING TRUE LENGTHS ARE THE SAME FOR THIS RIGHT TRANSITION PIECE:

GROUP I	GROUP II
1-A	2-A
4-A	3-A
4-B	5-B
7-B	6-B
7-C	8-C
10-C	9-C
10-D	11-D
1-D	12-D

FIGURE 24.56 ■ Step 2—transition piece development; true-length diagram.

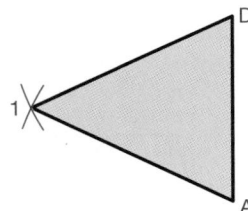

FIGURE 24.57 ■ Step 3—transition piece development; triangulation, true size and shape of starting triangle.

tion, and Chapters 9 and 10, Descriptive Geometry I and II, may be helpful here.

Step 4. Work both ways from triangle 1,A,D in Figure 24.57 to develop the adjacent triangles. From points A and D draw arcs equal to A–2 and D–12, respectively. Set a compass

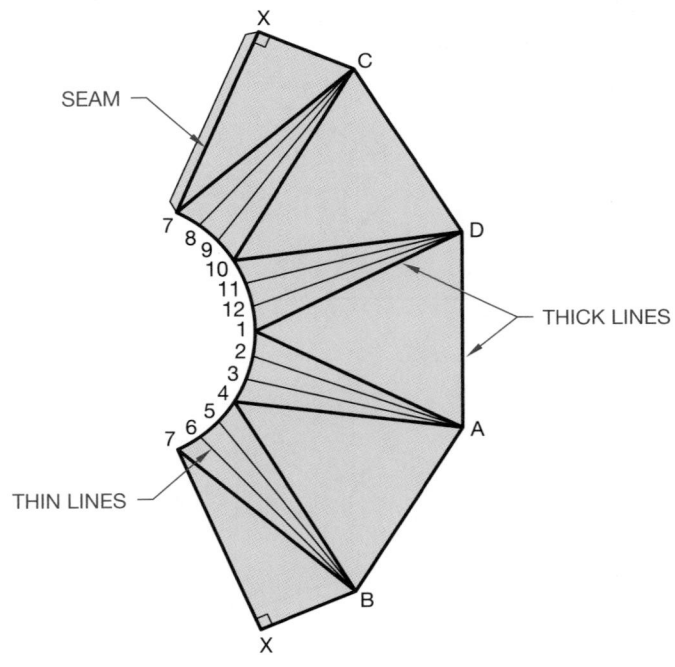

FIGURE 24.59 ■ Step 5—complete the transition piece development.

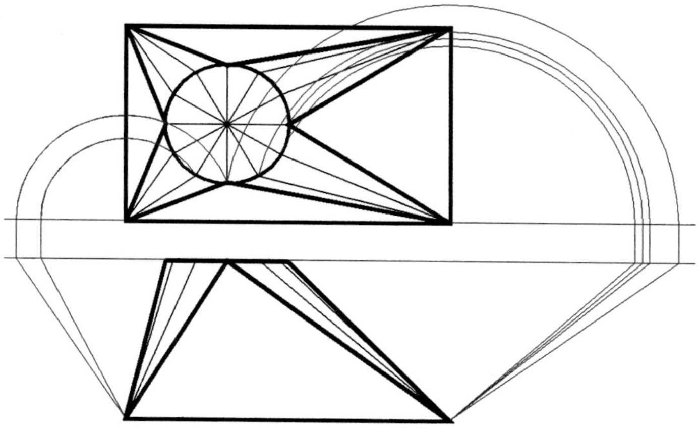

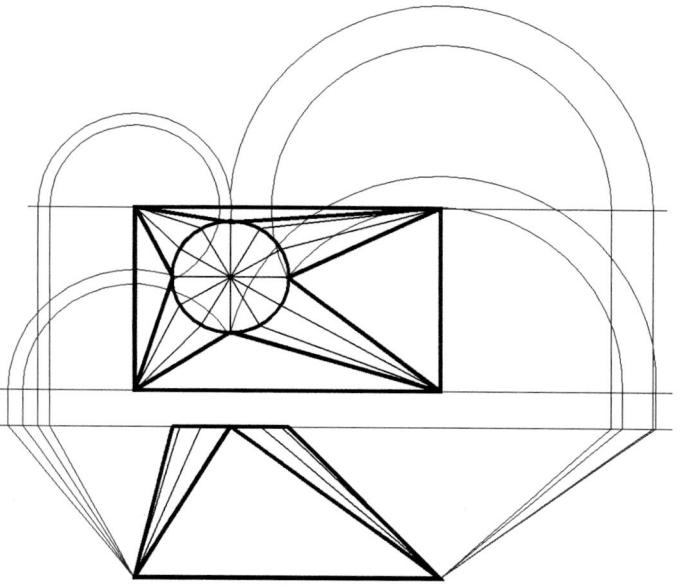

FIGURE 24.60 ■ Setting up a symmetrical offset transition piece; two true-length diagrams are required.

ONLY THE TRUE LENGTHS OF QUADRANT
ELEMENTS ARE SHOWN FOR CLARITY

FIGURE 24.61 ■ Setting up an offset transition piece; four true-length diagrams are required.

lines in the pattern are either outlines or bend lines, while thin lines form a smooth curve or contour. Cut out your pattern and see for yourself how it fits together.

The right transition piece used in the process previously described is the easiest transition to develop because it is symmetrical. If the round is offset as shown in Figure 24.60, then true-length diagrams are required for both halves.

When the round is offset as in Figure 24.61, true-length diagrams are necessary for the elements on each quarter of the transition. In these more complex transitions, a numbering system and accurate work are very important. Colored lines may be used to help keep true-length diagrams clearly separate.

FIGURE 24.62 ■ Curve-to-curve transition.

Curve-to-curve Triangulation

Triangulation is a technique that is used for the development of transition pieces or any other pattern where a series of triangles is used to form the desired shape. Figure 24.62 shows a part that makes a transition from one curved shape to another. In this situation the adjacent curves must be divided into a series of triangles before the pattern can be developed. The triangles help define one shape in relationship to the other. (See Figure 24.63.) After the views have been drawn, the following procedure is similar to the development of a transition piece:

Step 1. The part is symmetrical, so one set of true-length lines must be established. (See Figure 24.64.) Be careful when laying out and numbering the true-length lines. Some drafters even use colored lines to keep elements coordinated. A numbering system and accuracy are critical for this type of problem.

Step 2. After the true-length elements have been established, the development technique is the same as the transition piece. (See Figure 24.65.)

INTERSECTIONS

When one geometric shape meets another, the line of intersection between the shapes must be determined before the pattern of each piece can be developed. The key to determining the intersection between parts is setting up views with numbered true-length elements that correlate between views. The points of intersection of line elements at planes may then be projected between views.

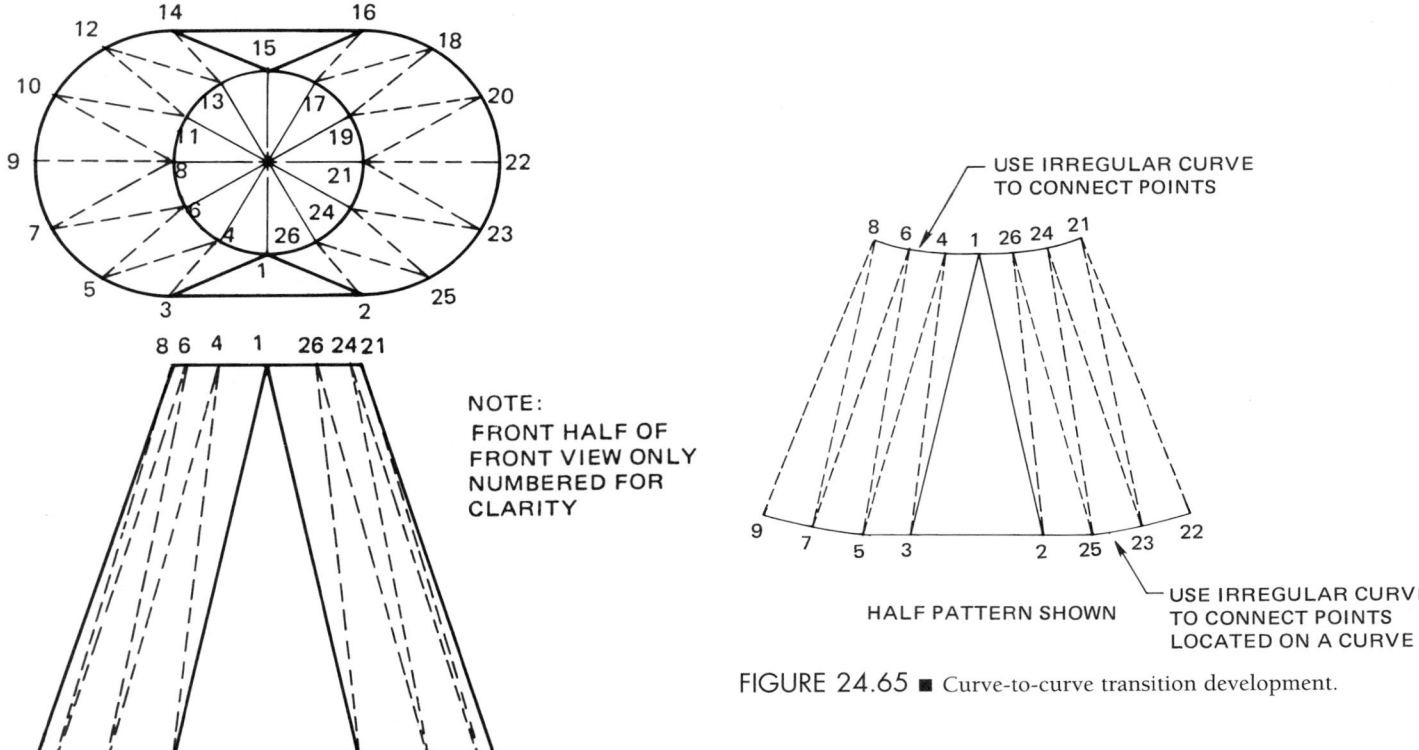

FIGURE 24.63 ■ Setting up the triangulation on a curve-to-curve transition.

NOTE:
FRONT HALF OF
FRONT VIEW ONLY
NUMBERED FOR
CLARITY

USE IRREGULAR CURVE
TO CONNECT POINTS

HALF PATTERN SHOWN

USE IRREGULAR CURVE
TO CONNECT POINTS
LOCATED ON A CURVE

FIGURE 24.65 ■ Curve-to-curve transition development.

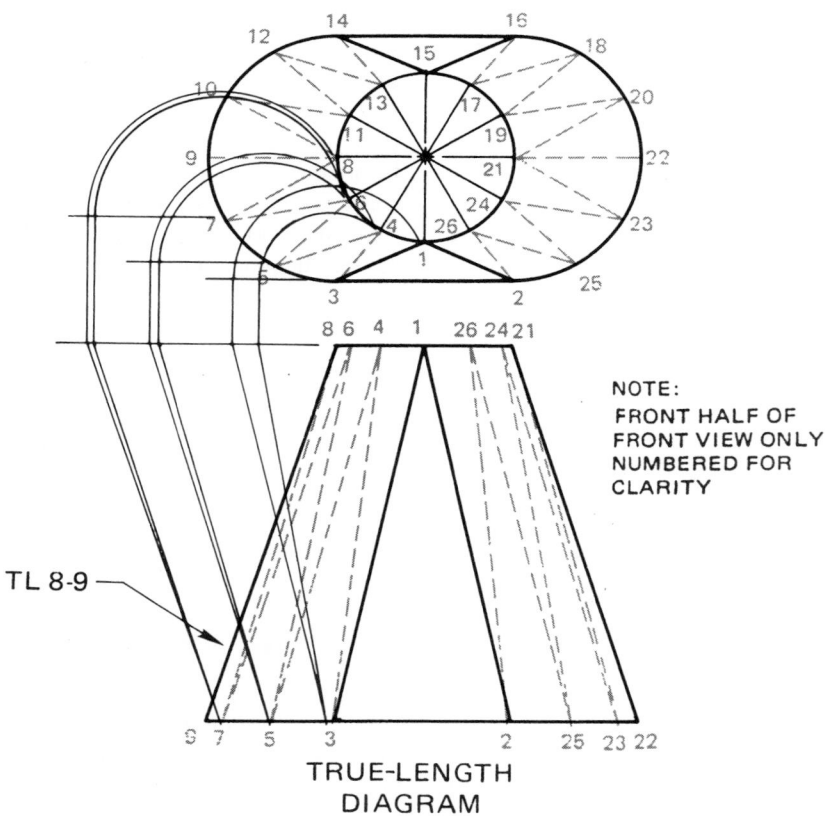

TL 8-9

NOTE:
FRONT HALF OF
FRONT VIEW ONLY
NUMBERED FOR
CLARITY

TRUE-LENGTH
DIAGRAM

FIGURE 24.64 ■ Curve-to-curve transition, true-length diagram.

CADD APPLICATIONS

SHEET METAL PATTERN DESIGN

CADD software is available that automatically produces nearly any sheet metal flat pattern layout you need. Figure 24.66 shows a variety of available shapes and patterns. Full-size plots can also be generated on the computer screen or the information can be transferred to computer numerical control burners, lasers, punch presses, or arc machines so the pattern can be cut or punched, as necessary. The programs can automatically calculate sheet layout for the best material conservation possible. Parametric design of sheet metal fabrication packages allows you to enter specific information about the design; for example, the variables H, Y, X, OX, and D shown on the transition piece in Figure 24.67. After you input the variable information, the program automatically draws the flat pattern on the screen and provides a print file of all the development information for your reference as shown in Figure 24.68.

SAMPLE SHAPES:

PATTERN SHAPES:

FIGURE 24.66 ■ A variety of sheet metal shapes and patterns may be drawn on a CADD system. *Courtesy C.A.M. Systems, Inc.*

USER INPUT SAMPLE:

Easy to operate comprehensive, self-prompting. Simply enter dimensions of desired shape.

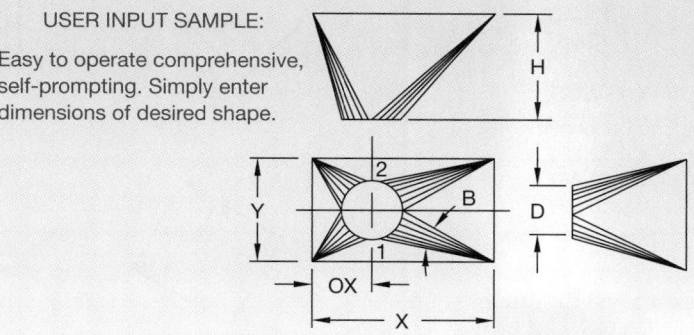

RECTANGLE TO ROUND
ONE OFFSET — NO ANGLES

Rectangle to Length X Direction	X	(12.125)
Rectangle to Length Y Direction	Y	(10.500)
Height	D	(5.375)
Number of Breaks per Corner	H	(12.562)
Select X in Y Offset	B	(5.0)
Offset in X Direction	OX	(12.125)

(Offset may be in X or Y direction from lower left corner of the rectangle)

FIGURE 24.67 ■ Parametric CADD packages are available to design duct fittings by giving variable information. *Courtesy C.A.M. Systems, Inc.*

(Continued)

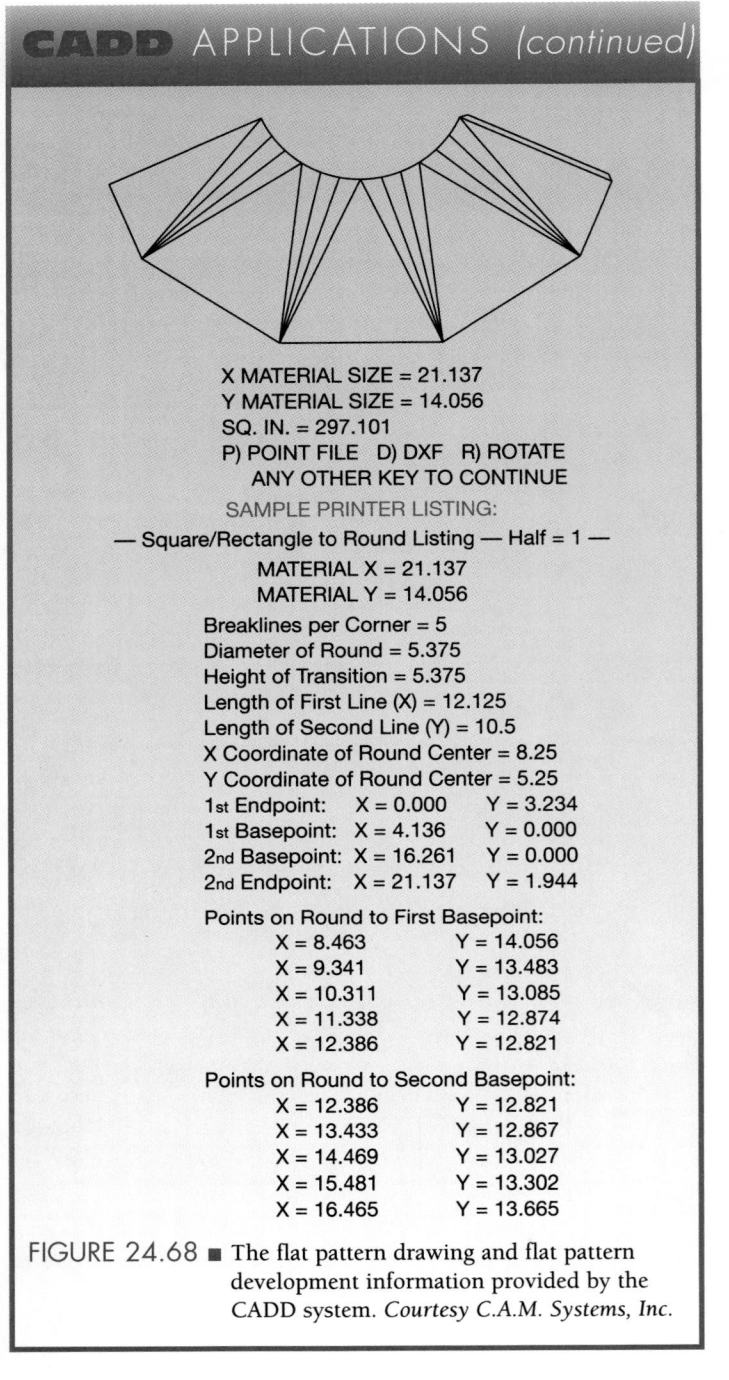

X MATERIAL SIZE = 21.137
Y MATERIAL SIZE = 14.056
SQ. IN. = 297.101
P) POINT FILE   D) DXF   R) ROTATE
ANY OTHER KEY TO CONTINUE
SAMPLE PRINTER LISTING:
— Square/Rectangle to Round Listing — Half = 1 —
MATERIAL X = 21.137
MATERIAL Y = 14.056

Breaklines per Corner = 5
Diameter of Round = 5.375
Height of Transition = 5.375
Length of First Line (X) = 12.125
Length of Second Line (Y) = 10.5
X Coordinate of Round Center = 8.25
Y Coordinate of Round Center = 5.25
1st Endpoint: X = 0.000   Y = 3.234
1st Basepoint: X = 4.136   Y = 0.000
2nd Basepoint: X = 16.261   Y = 0.000
2nd Endpoint: X = 21.137   Y = 1.944

Points on Round to First Basepoint:
X = 8.463   Y = 14.056
X = 9.341   Y = 13.483
X = 10.311   Y = 13.085
X = 11.338   Y = 12.874
X = 12.386   Y = 12.821

Points on Round to Second Basepoint:
X = 12.386   Y = 12.821
X = 13.433   Y = 12.867
X = 14.469   Y = 13.027
X = 15.481   Y = 13.302
X = 16.465   Y = 13.665

FIGURE 24.68 ■ The flat pattern drawing and flat pattern development information provided by the CADD system. *Courtesy C.A.M. Systems, Inc.*

Intersecting Prisms

The line of intersection between intersecting prisms may be found by projecting the true lengths of individual line elements between views. The corresponding points of intersection where lines intersect planes establish a series of points that are connected to form the lines of intersection. Several examples are shown in Figure 24.69. After the lines of intersection are determined, the pattern development may be made of the resulting shape.

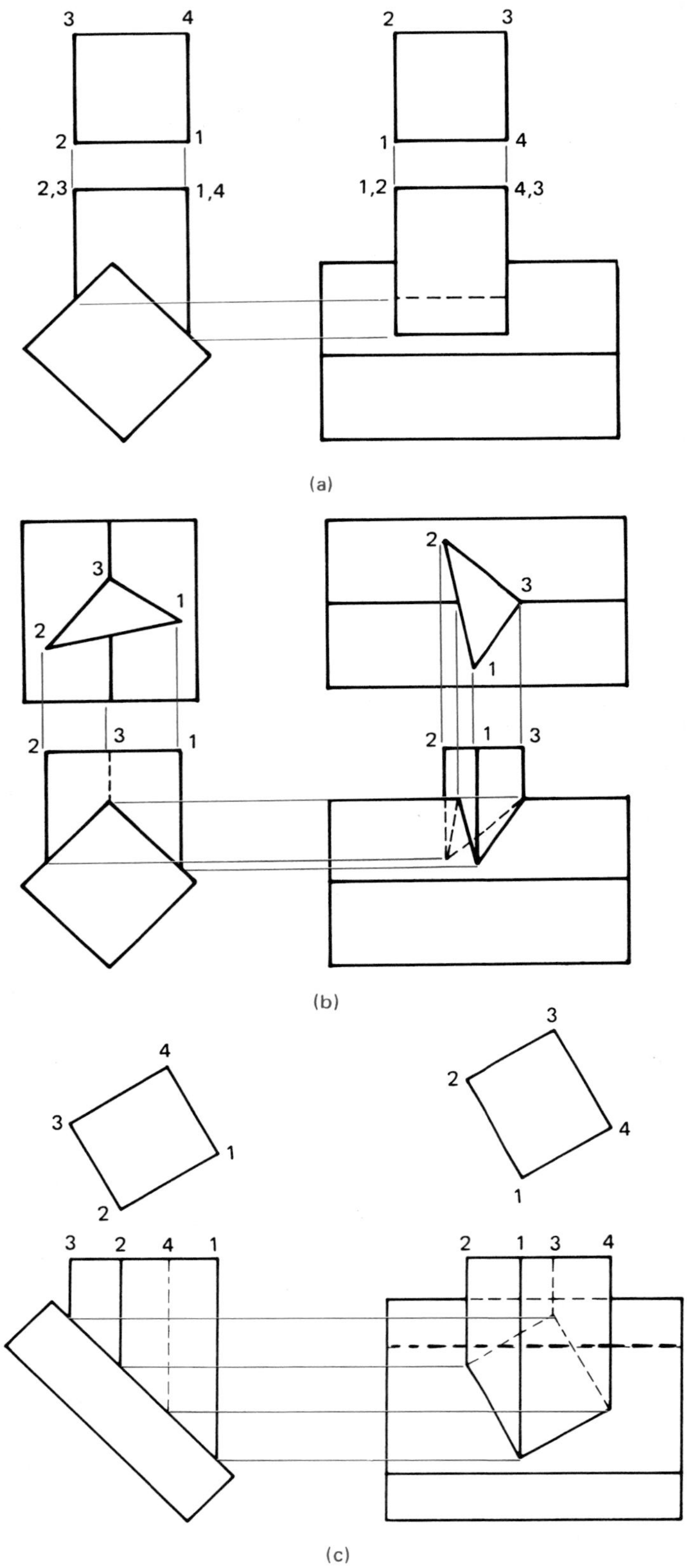

FIGURE 24.69 ■ Intersecting prisms.

Intersecting Cylinders

The line of intersection between intersecting cylinders is determined in a manner that is similar to that for intersecting prisms. A series of line elements is established on the intersecting cylinder. Points of intersection of these line elements are then plotted. When the points are connected, the line of intersection is formed. The following procedure provides a typical example:

Step 1. Establish two adjacent views and show the circular view of the intersecting cylinder in both views. Divide the circular views of the intersecting cylinder into twelve equal parts in each view and number the points correspondingly. Be sure that your numbering system correlates between views as shown in Figure 24.70.

Step 2. Project point 1 in the front view down until it meets the intersecting circle. Then project to the side view until it meets the same corresponding point 1 projected down in the side view. This is the point of intersection for line element 1. Follow the same procedure for each of the other eleven points. When complete, you have a series of points of intersection as shown in Figure 24.71.

Step 3. Connect the points of intersection in the side view to establish the line of intersection as shown in Figure 24.72. Now pattern developments of each cylinder can be made as discussed earlier in this chapter and as shown in Figure 24.72.

No matter how the intersecting cylinders are arranged, the technique is the same. The circular views are always drawn looking into the intersecting cylinders even when the cylinders intersect at an angle as in Figure 24.73. When the cylinders are offset, the back half of the intersecting cylinder appears hidden as shown in Figure 24.73. This example also shows how the patterns for the cylinders are drawn.

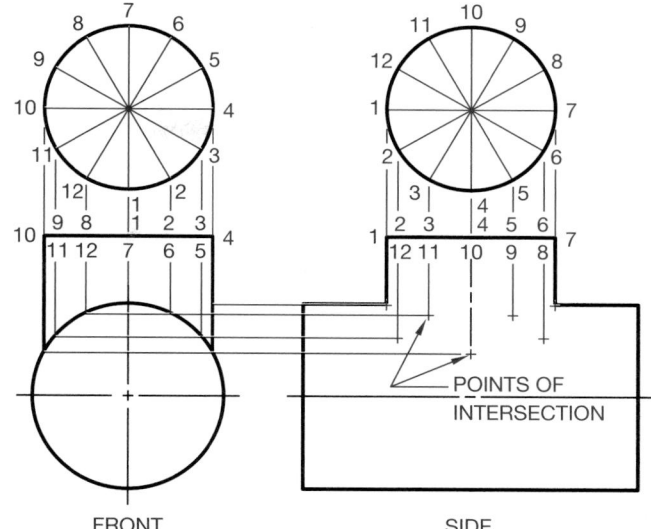

FIGURE 24.71 ∎ Establishing points of intersection for intersecting cylinders.

Cylinder Intersecting Cone

The procedure used to get the line of intersection between a cylinder and cone is a little more complex, but may be done given any situation if the following steps are used:

Step 1. Draw the front and top views. Draw the circular view of the intersecting cylinder in both views and divide these circles into twelve equal parts. Number the parts of each circle so the numbering system correlates between views. (See Figure 24.74, page 862.)

Step 2. Beginning with point number 1 in the front view, project this point until it intersects the true-length element of the cone, 0–X. This is the point of intersection of point 1 in the front view. From this point of intersection, project to the top view until the projection meets the same corresponding point projected from the circle in the top view. This is the point of intersection of point 1 in the top view. (See Figure 24.75, page 862.)

Step 3. Now the process becomes a little more difficult. Project point 2 from the front view to the true-length element of the cone, 0–X (points 2 and 12 are on the same projection line if the object is symmetrical). From this point, project point 2 into the top view until intersecting 0–X. Set a compass with a radius from 0 to this point of intersection and draw an arc. Project points 2 and 12 in the top view until they intersect this arc. These are the points of intersection of points 2 and 12 in the top view. Now, project points 2 and 12 from the top view until they intersect the projection line for 2 and 12 in the front view. This is the point of intersection for 2 and 12 in the front view. (See Figure 24.75.)

Step 4. Continue this process until each point has been located in each view. (See Figure 24.76, page 862.) Darken the line of intersection by connecting the points of intersection

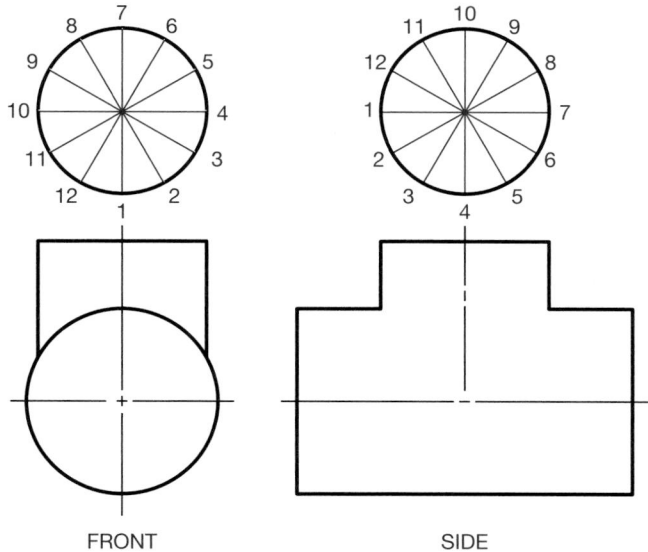

FIGURE 24.70 ∎ Setting up intersecting cylinders.

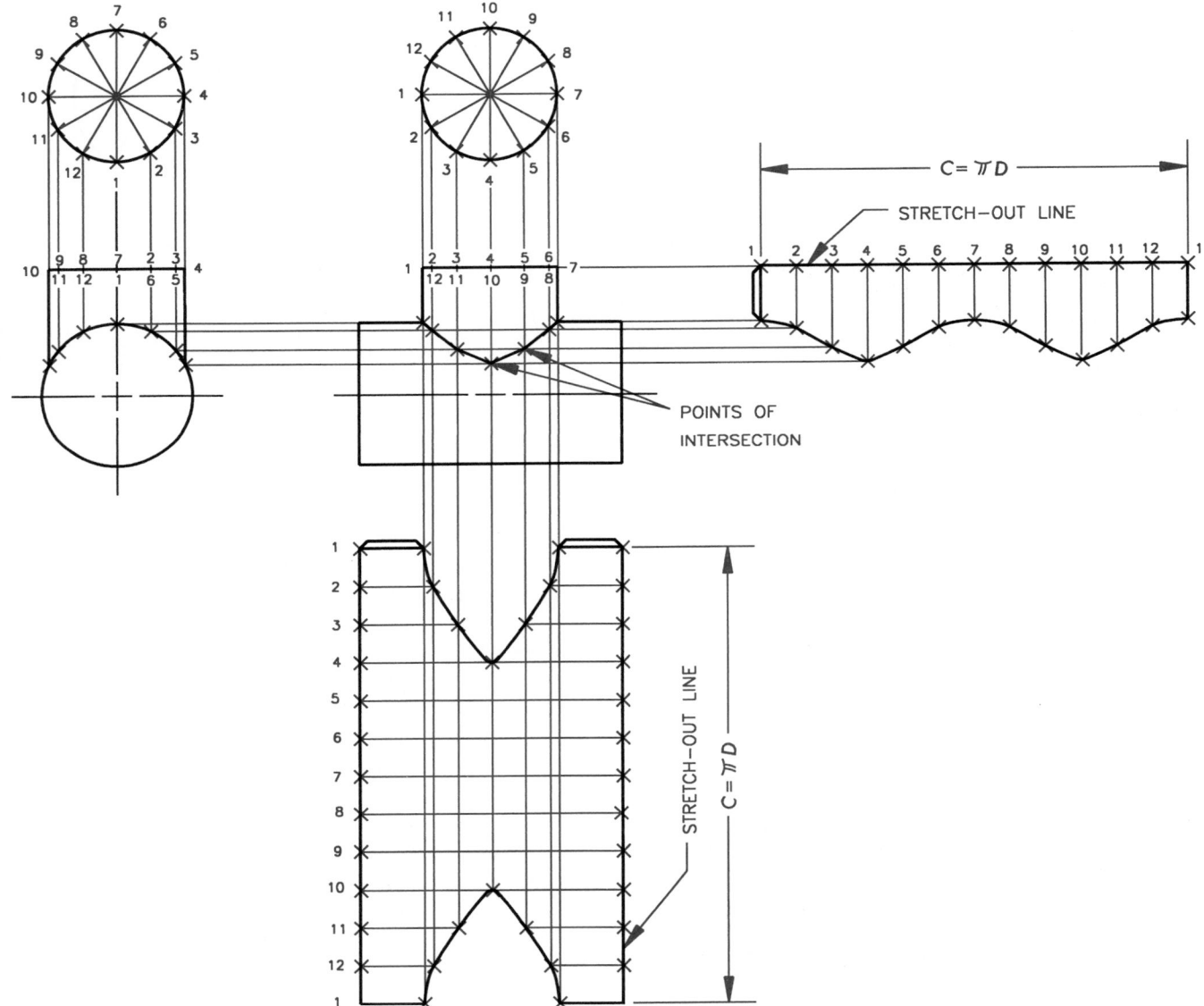

FIGURE 24.72 ■ Line of intersection for intersecting cylinders and the pattern for both cylinders.

with an irregular curve. Be sure hidden lines are properly represented. (See Figure 24.77.)

Step 5. The pattern development for the cone is started by using the true-length side of the cone to draw the stretch-out line arc. (This is the edge view.) Look at Figure 24.77 as you follow these steps: Lay out the circumference of the cone base along the stretch-out line, as you learned earlier in this chapter. Then, establish element 0–X in the middle of the pattern; this line coincides with 0–X in the front and top views. In the front view, project a radius from the vertex and begin where the projection lines from the circle next intersect the true-length line (0–X); the points for elements 1 and 7 are where the arcs from the front view cross line 0–X in the pattern. Repeat this for each intersection, and extend these radii past the 0–X line in the pattern.

In the top view, take measurements from the 0–X line at the point where the arcs passing through points 2 and 12 intersect. Measure from this intersection to point 2. The distance from 0–X to point 2 should be the same as from 0–X to point 12 because this object is symmetrical. Take this measurement and transfer it to the pattern by measuring from the intersection of the 2,12 radius on the 0–X line, and place a point on each side of the 0–X line on this radius. Continue this process for each pair of points (3,11; 4,10; 5,9; and 6,8) until each point has been transferred into the pattern. Connect the points in the pattern to establish the cutout where the cylinder intersects. Make a pattern development for the cylinder using the same procedure that you have used for other cylinder developments.

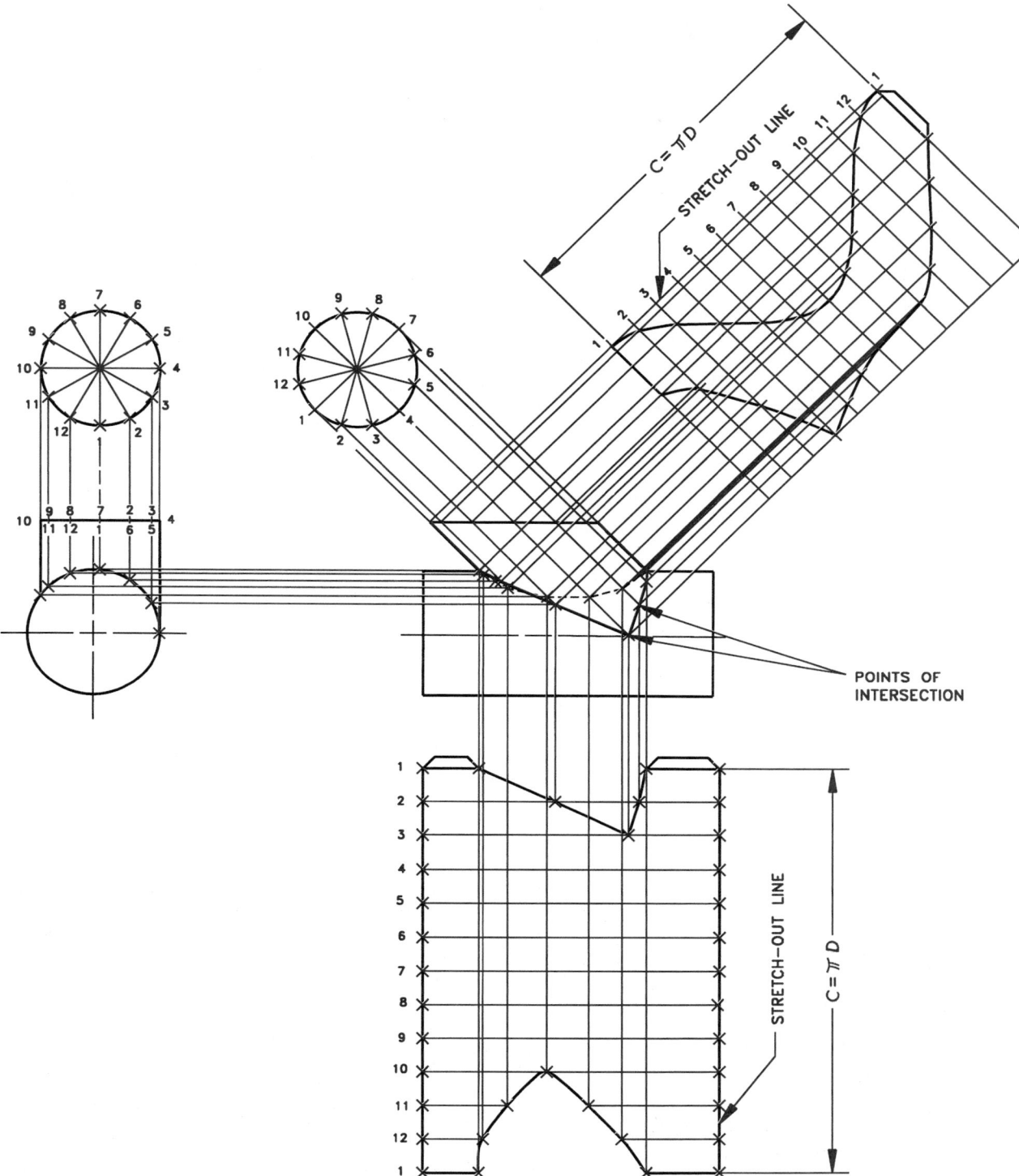

FIGURE 24.73 ■ Offset intersecting cylinders and the pattern for both cylinders.

MATERIAL BENDING AND CHASSIS LAYOUT

In precision sheet metal applications such as fabrication for electronics chassis components or sheet metal appliance body parts, the condition of material when bent must be taken into consideration.

Bend Allowance

Bend allowance is the amount of extra material needed for a bend to compensate for the compression during the bending process. This consideration is often not critical to HVAC sheet metal bending due to the larger tolerances applied to these components.

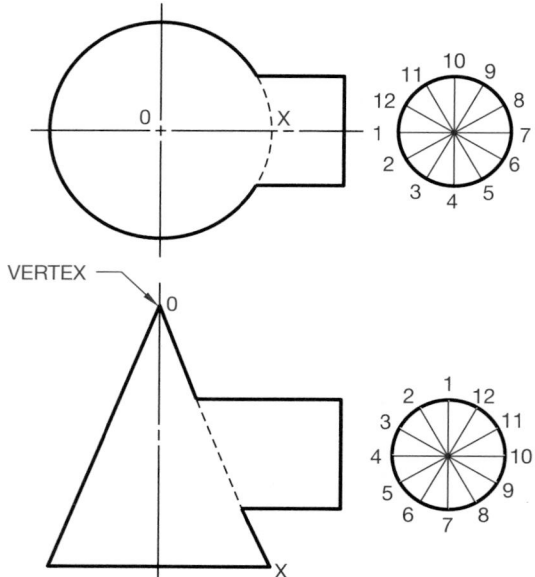

FIGURE 24.74 ■ Set up intersection between a cone and cylinder.

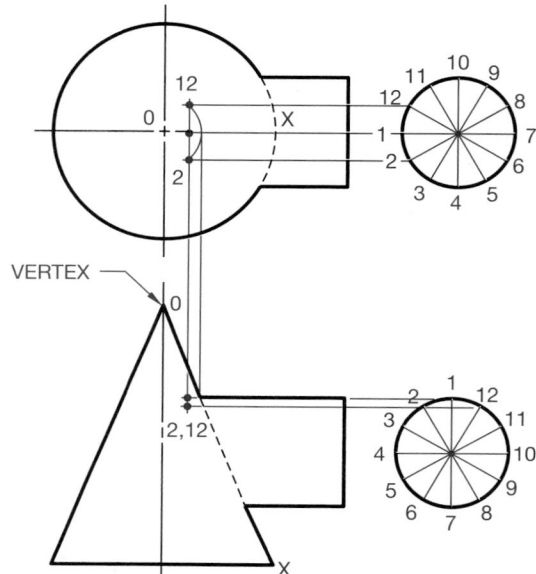

FIGURE 24.75 ■ Determine points of intersection between cone and cylinder.

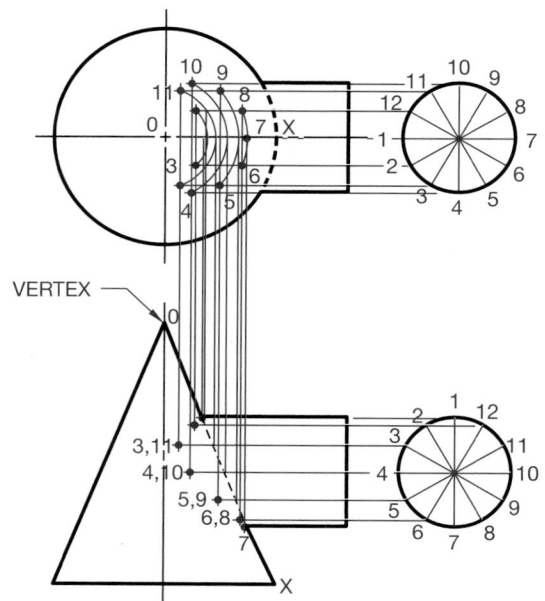

FIGURE 24.76 ■ Determine points of intersection between cone and cylinder.

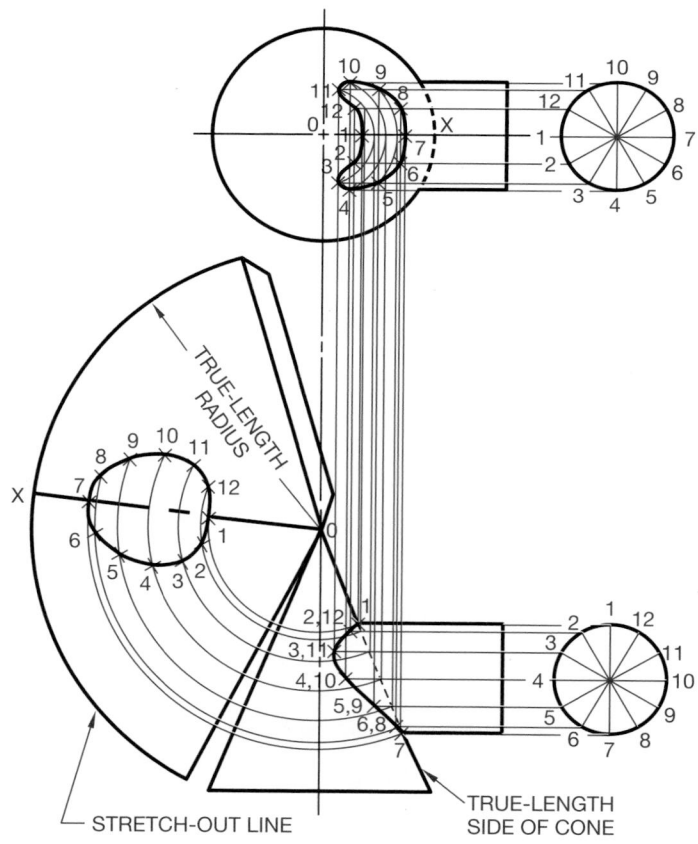

FIGURE 24.77 ■ Connect points of intersection to establish line of intersection between cone and cylinder and the pattern development for the cone.

Bend allowance is important when close tolerances must be held or when thick material must be bent or formed into desired shapes. The purpose of a bend allowance calculation is to determine the overall dimension of the flat pattern, so when bent the desired final dimension is achieved. There are a number of slightly different methods used to calculate bend allowance. There are different formulas in the *Machinery's Handbook, ASME Handbook,* and in most textbooks. Many companies use formulas derived from a proven method or by individual experimentation. The amount of bend allowance depends on the material thickness, type of material, bending process, degree of bend, and bend radius. The grain of material, surface condition, and amount of

PRECISION SHEET METAL FABRICATION

CADD applications are available for developing flat pattern layouts and converting existing drawings into flat layouts. These packages can automatically calculate bend allowances and notch locations. All you need to enter in the computer for bend allowance calculations is material thickness and type, and bend radius in degrees. The added advantage of some systems is the ability to interface with CAM applications software, which allows information that is generated at the CADD workstation to be sent to the fabrication shop to run a computer numerical control metal break or punch press. A drawing generated by a CADD precision sheet metal program is shown in Figure 24.78.

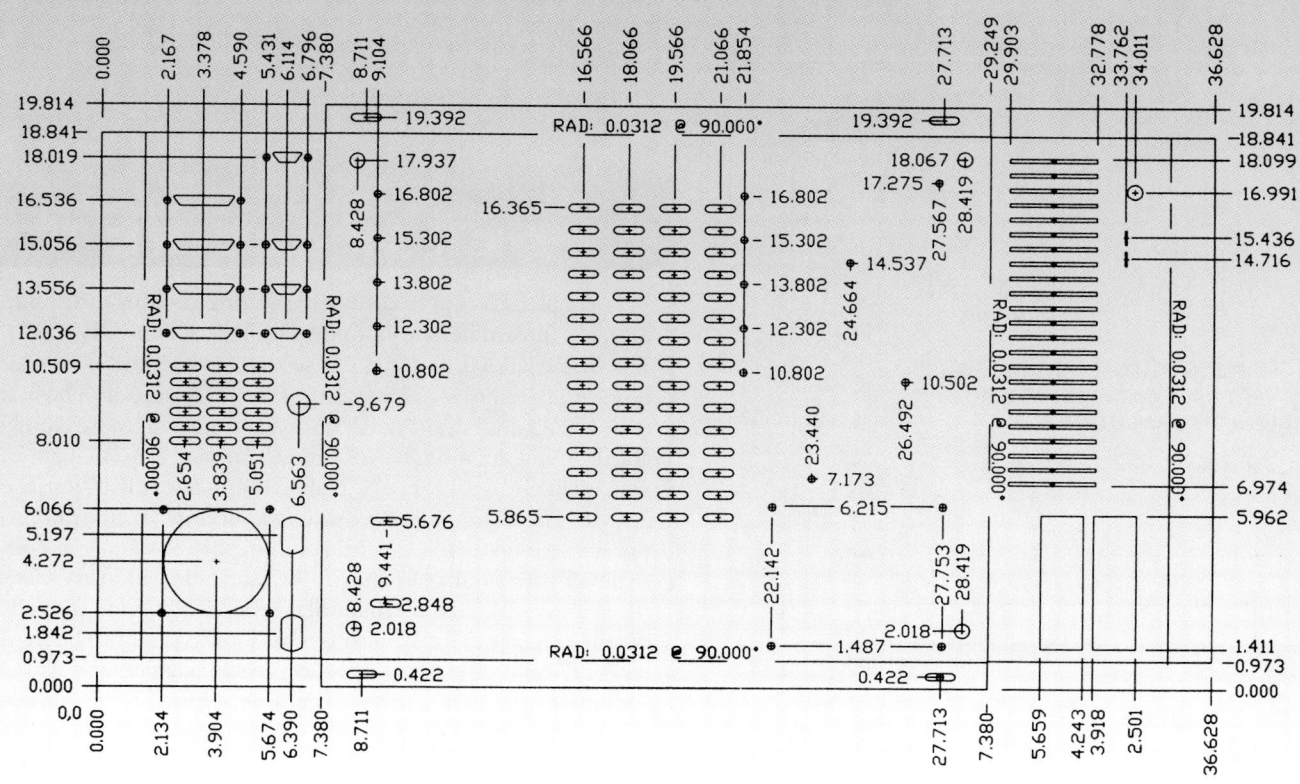

FIGURE 24.78 ■ ■ A CADD-generated precision sheet metal fabrication drawing. *Courtesy C.A.M. Systems, Inc.*

lubrication used also influence bend allowance. The only exact way to establish bend allowance for a specific application is to experiment with the equipment and material to be used. Some companies have made these tests and have developed charts that give the bend allowance for the type of material, material thickness, and bend radius.

When material bends, there is compression on the inside of the bend and stretching on the outside. Somewhere between, there is a neutral zone where neither stretching nor compression occurs; this is called the *neutral axis*. The neutral axis is approximately four-tenths of the thickness from inside of the bend, but this depends on the material and other factors. Information related to calculating the bend allowance and length of the flat pattern is shown in Figure 24.79. The following formula is used to calculate the length of the flat pattern (straight stock before bending):

Length of Flat Pattern = X + Y + Z
X = Horizontal Dimension to Bend = B − R − C
Y = Vertical Dimension to Bend = A − R − C
Z = Length of Neutral Axis
C = Material Thickness
R = Bend Radius
Z = 2(R + .4C) × π ÷ 4(90° bend)

Given the sheet metal bend shown in Figure 24.80, determine the length of the flat pattern.

X = B − R − C = 12.250 − .125 − .125 = 12.000
Y = A − R − C = 4.875 − .125 − .125 = 4.625
Z = 2(R + .4C) × π ÷ 4 = 2(.125 + .4 × .125)
    × 3.14 ÷ 4 = .275
Length of Flat Pattern = X + Y + Z = 12.000 + 4.625 + .275 = 16.900

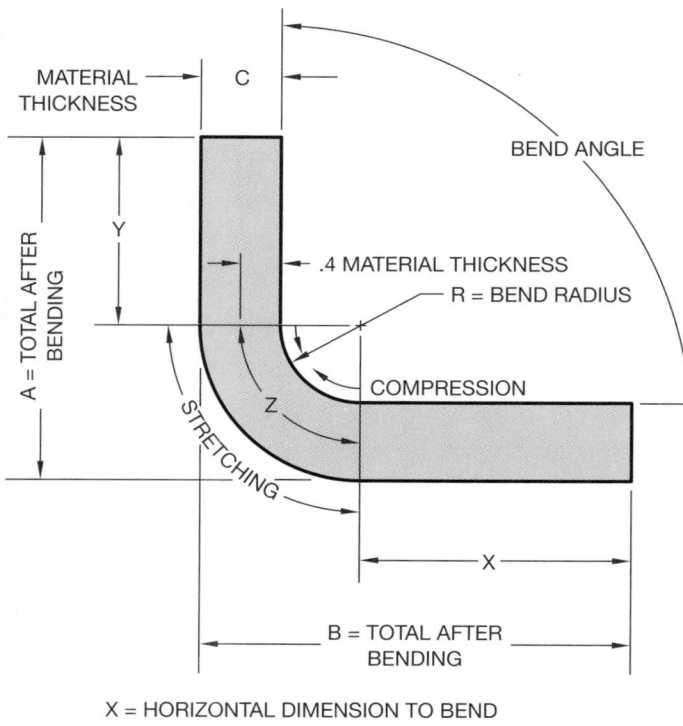

X = HORIZONTAL DIMENSION TO BEND
Y = VERTICAL DIMENSION TO BEND
Z = LENGTH OF NEUTRAL AXIS

FIGURE 24.79 ■ Bend allowance variables.

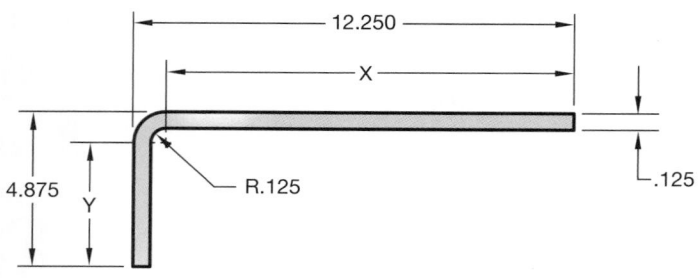

FIGURE 24.80 ■ Sample bend allowance problem.

Bend Relief

A corner with two edges bent in the same direction has internal stresses that may cause a crack at the corner. When necessary, some material at the corner may be cut away to help relieve this stress. This cutout at the corner is called *bend relief*. (See Figure 24.81.)

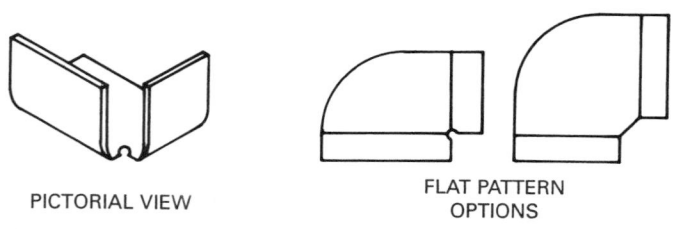

PICTORIAL VIEW

FLAT PATTERN
OPTIONS

FIGURE 24.81 ■ Bend relief.

Chassis and Precision Sheet Metal Layout

Electronic assemblies often require shields, frames, panels, and chassis to be fabricated. Layout drawings are required for the fabrication of these items. Preparing chassis layouts and other precision sheet metal drawings requires close tolerances and bend allowance calculations. The method of dimensioning these parts includes standard unidirectional, arrowless, or tabular dimensioning systems. (Refer to Chapter 12, Dimensioning.) A precision sheet metal pattern may be drawn as a flat pattern, as a finished product, or with the finished part shown and the flat pattern drawn using phantom lines.

PROFESSIONAL
PERSPECTIVE

This chapter has covered the engineering drafting that takes place in three different, but closely allied, engineering fields. The first is HVAC drafting, the second is sheet metal fabrication, and the third is precision sheet metal fabrication. The first two, HVAC and sheet metal fabrication, are the most closely allied. If you are a drafter for a mechanical (HVAC) engineer, you are likely to draw HVAC plans, details, and schedules. The drawings from your office are sent to a sheet metal fabricator so the duct components can be built and delivered to the construction site. Your colleague in the fabrication industry takes the drawing from the mechanical engineer's office and converts the duct shapes to flat pattern layouts for fabrication. The third type of drafting deals with sheet metal shapes and parts. Although these components are not used in the HVAC industry, they are used in the precision sheet metal fabrication industry. The difference is that this industry normally deals with closer tolerances. The types of items made in the precision sheet metal industry are electronics chassis and automotive (cars, trucks, and tractors) sheet metal components. In many cases, HVAC fabrication drawings do not even have dimensions; the pattern itself is used to fabricate the duct shape. On the other hand, the dimensioning and tolerancing for precision sheet metal parts are critical; and arrowless, datum, and tabular dimensioning is often used with close tolerances to achieve an accurate layout.

If you enter the HVAC industry, you should become very familiar with HVAC vendors' catalogs and manuals, and duct design applications engineering information available from mechanical suppliers such as Trane, Apec, and Elite. If you enter the precision sheet metal fabrication industry, you need to have a good understanding of tolerancing, material bending, and precision sheet metal fabrication methods.

SHEET METAL BEND ANGLES

It is required to find bend angle A for a sheet of metal having the cross section shown in Figure 24.82. The solution to this problem requires working with *two* right triangles because thickness of the material must be taken into account. Figure 24.83 is a drawing of the solution. OML stands for outside mold line. The steps that were taken are the following:

1. Find the angle of the large triangle with Inv tan $\left(\dfrac{.75}{1.25}\right) = 31.0°$.

2. Find the hypotenuse of both triangles with $\sqrt{1.25^2 + .75^2} = 1.4578$.

3. Find the angle of the slender, shaded triangle with Inv sin $\left(\dfrac{.125}{1.4578}\right) = 4.9°$

4. Subtract the two angles to find the bend angle.

$$A = 31.0 - 4.9 = \textbf{26.1°}$$

If the true distance along the bend (OML to OML) is known, then the problem is much simpler and requires working with only one right triangle.

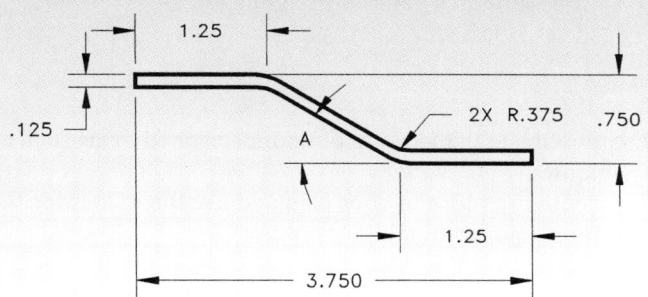

FIGURE 24.82 ▪ Sheet metal with two identical bends.

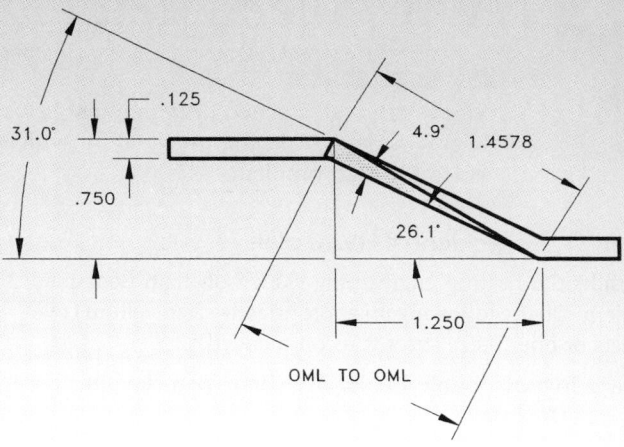

FIGURE 24.83 ▪ Triangles drawn to solve the problem.

MATH APPLICATION

CHAPTER 24
HVAC/Sheet Metal Drafting Test

DIRECTIONS

Answer the questions with short complete statements or drawings as needed.

PART 1 • HVAC DRAFTING

1. HVAC systems are also commonly known as _____ systems.
2. What is the general purpose of an HVAC system?
3. Under what general conditions is ventilation required?
4. Define *duct*.
5. Name two types of forced-air systems.
6. What is the basic principle of refrigeration?
7. Name two types of hot water systems.
8. How does a zoned control system differ from a central forced-air system?
9. Describe the basic function of a heat pump system.
10. Identify six sources of pollutants in a structure.
11. Describe the basic function of an air-to-air heat exchanger.
12. Define *thermostat*.
13. List three factors that contribute to the proper placement of a thermostat.
14. Name two types of HVAC plans (based on line technique).
15. What are detail drawings used for in HVAC drafting?
16. What are section drawings used for in HVAC drafting?
17. Describe two types of sectioning practices used in HVAC drafting.
18. When are pictorial drawings used in HVAC drafting?
19. Describe HVAC schedules and their importance.
20. How are schedules keyed to the HVAC plan?

PART 2 · SHEET METAL DRAFTING

21. Describe the basic principle of pattern development.
22. What is the key to pattern development?
23. Define *stretch-out line*.
24. Name and sketch four types of seams.
25. Name and sketch three types of hemmed edges.
26. Define *truncated*.
27. A transition piece is also known as a _____.
28. Why is it important to have a working knowledge of descriptive geometry for pattern development?
29. Define *triangulation*.
30. Define *bend allowance*.
31. Define *bend relief*.
32. Describe three different methods that may be used to draw precision sheet metal patterns.

CHAPTER 24

HVAC/Sheet Metal Drafting Problems

DIRECTIONS

1. Please read all related instructions before you begin working. Specific information will be provided for each problem.

2. Use manual or computer-aided drafting as required by your course guidelines.

PROBLEM 24-1   Residential HVAC plan

Given: Residential heating engineering sketch of main floor plan and basement. Do the following on appropriately sized vellum (two B-size sheets or one C-size sheet is recommended):

1. Make a formal double-line HVAC floor plan layout at a 1/4" = 1'–0" scale.

2. Approximate the location of undimensioned items such as windows.

3. Use thin lines for the floor plan and use thick lines for the heating equipment and duct runs.

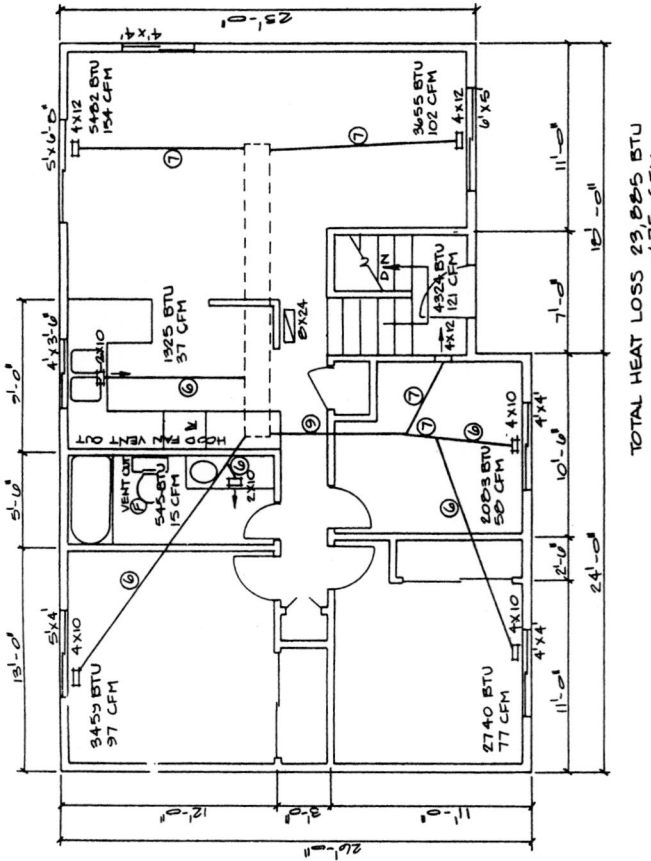

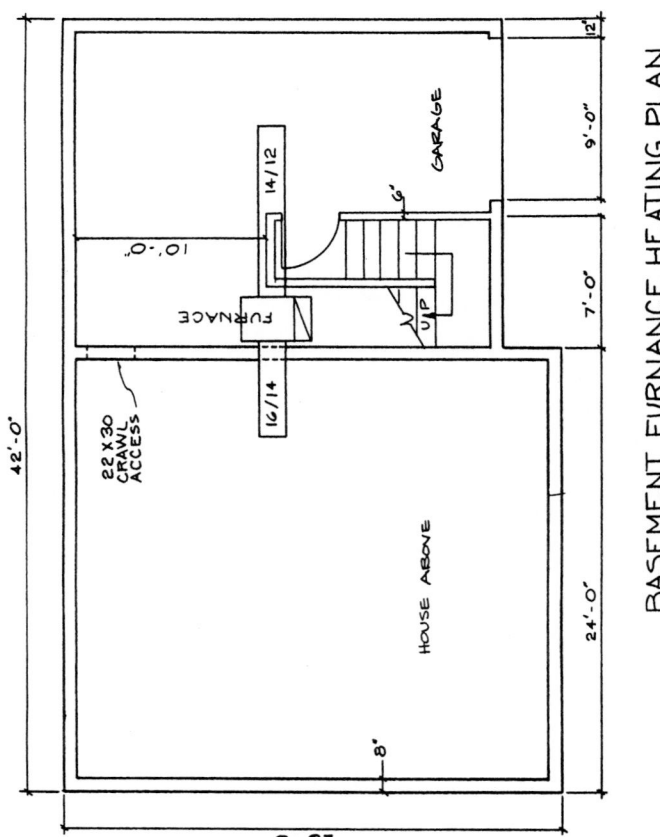

BASEMENT FURNACE HEATING PLAN

PROBLEM 24-2   Residential air-to-air heat exchanger plan

Given: Residential air-to-air heat exchanger ducting engineering sketch of basement floor plan. Do the following on appropriately sized vellum (one B- or C-size sheet is recommended):

1. Make a formal single-line air-to-air heat exchanger floor plan layout at a 1/4" = 1'–0" scale.

2. Approximate the location of undimensioned items such as doors.

3. Use thin lines for the floor plan and use thick lines for the air-to-air heat exchanger equipment and duct runs.

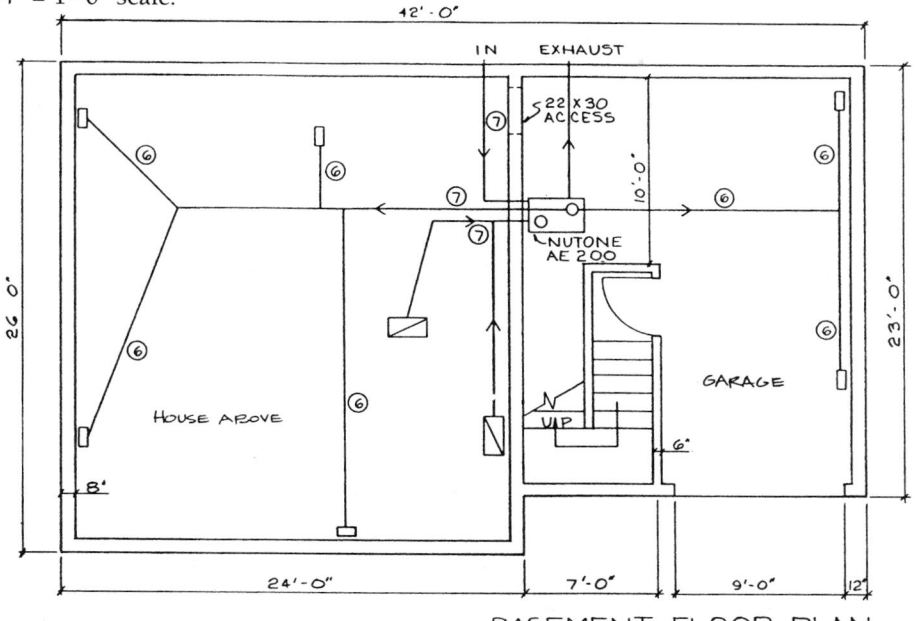

BASEMENT FLOOR PLAN
AIR-TO-AIR HEAT EXCHANGER PLAN

PROBLEM 24-3   Commercial HVAC plan

Given the following:

1. HVAC floor plan engineering layout at approximately 1/16" = 1'–0". The engineer's layout is rough, so round off dimensions to the nearest convenient units at 6" intervals. For example, if the dimension you scale reads 24 ft. 3 in., round off to 24 ft. 0 in. The floor plan will not require dimensioning; therefore, the representation is more important than the specific dimensions.
2. Related schedules.
3. Engineer's sketch for exhaust hood.

SCHEDULES				
CEILING OUTLET SCHEDULE				
Symbol	Size	CFM	Damper Type	Panel Size
C-10	9 × 9	230	Key Operated	12 × 12
C-11	8 × 8	185	Key Operated	12 × 12
C-12	6 × 6	40	Key Operated	12 × 12
C-13	6 × 6	45	Key Operated	12 × 12
C-14	6 × 18	300	Fire Damper	24 × 24
SUPPLY GRILL SCHEDULE				
Symbol	Size	CFM	Location	Damper Type
S-1	20 × 8	450	High Wall	Key Operation
S-2	12 × 12	450	High Wall	External Operation

EXHAUST GRILL SCHEDULE				
Symbol	Size	CFM	Location	Damper Type
E-5	18 × 24	1000	Low Wall	No Damper

ROOF EXHAUST FAN SCHEDULE			
Symbol	Area Served	CFM	Fan Specifications
REF-1	Solvent Tank	900	1/4 HP, 12 in. Nonspark Wheel, 1050 Max. Outlet Velocity

Do the following on appropriately sized sheet (D size is recommended; all required items will fit on one sheet with careful planning):

1. Make a formal double-line HVAC floor plan layout at a 1/4" = 1'–0" scale. (Note: You measured the given engineer's sketch at 1/16" = 1'–0".) Now convert the established dimensions to a formal drawing at 1/4" = 1'–0". Approximate the location of the HVAC duct runs and equipment in proportion to the presentation on the sketch. Assume that the single-line sketch is the centerline of the ducts.
2. Prepare correlated schedules in the space available. Set up the schedules in a manner *similar* to the examples in Figure 24.21, page 843, for layout.
3. Make a detail drawing of the exhaust hood either scaled or unscaled. Make the detail large enough to clearly show the features. Refer to Figure 24.18, page 842, for an example of a detail drawing.
4. Approximate the location of doors, windows, and fixtures.

(continued)

PROBLEM 24-3 continued

5. Draw the floor plan using thin lines and the HVAC components with thick lines for contrast. Use appropriate CADD layers.

Note: Do not include notes and dimensions for wall thickness, door sizes, and tangent. Top view of exhaust hood detail is drawn as a transition piece, similar to Figure 24.55.

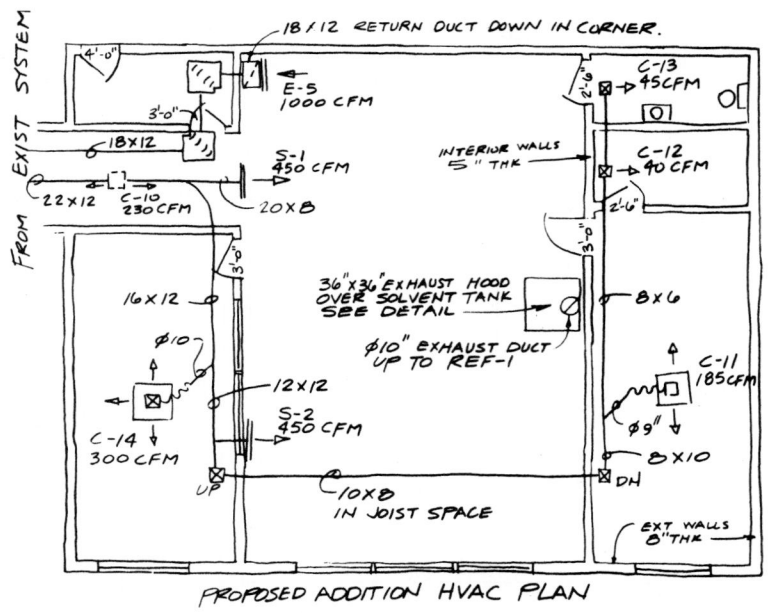

PROPOSED ADDITION HVAC PLAN

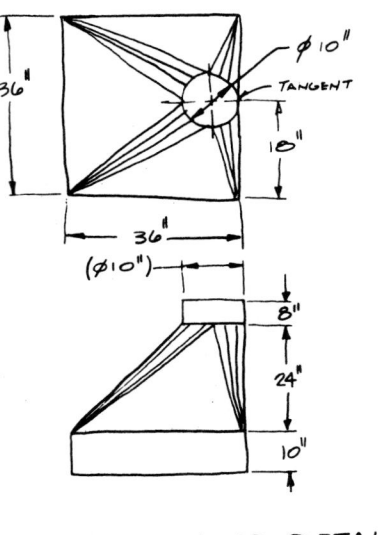

EXHAUST HOOD DETAIL

PROBLEM 24-4 Commercial HVAC plan

Problem courtesy Wendy's International, Inc.
   Given the following:

1. HVAC floor plan single-line engineering layout at approximately 1/8" = 1'–0". The provided problem is rough, so round off to the nearest convenient units at 6" intervals. For example, if a dimension reads 24'–3", round off too 24'–0" or 24'–6" as you prefer. The floor plan does not require dimensioning; therefore, the representation is more important than the actual dimensions.
2. Related general notes, schedules, schematics, and details.

Do the following on an appropriately sized sheet or sheets:

1. Make a formal double-line HVAV floor plan at a scale of 1/4" = 1'–0". (Note: You measured the given plan using a 1/8" = 1'–0" scale. Use the established dimensions to create the formal drawing at a scale of 1/4" = 1'–0".) Approximate the location of the HVAC duct runs and equipment in proportion to the presentation given on the engineering layout. Assume that the single-line layout is the centerline of the ducts.
2. Approximate the location and size of doors, windows, and fixtures.

PROBLEM 24-4 Continued

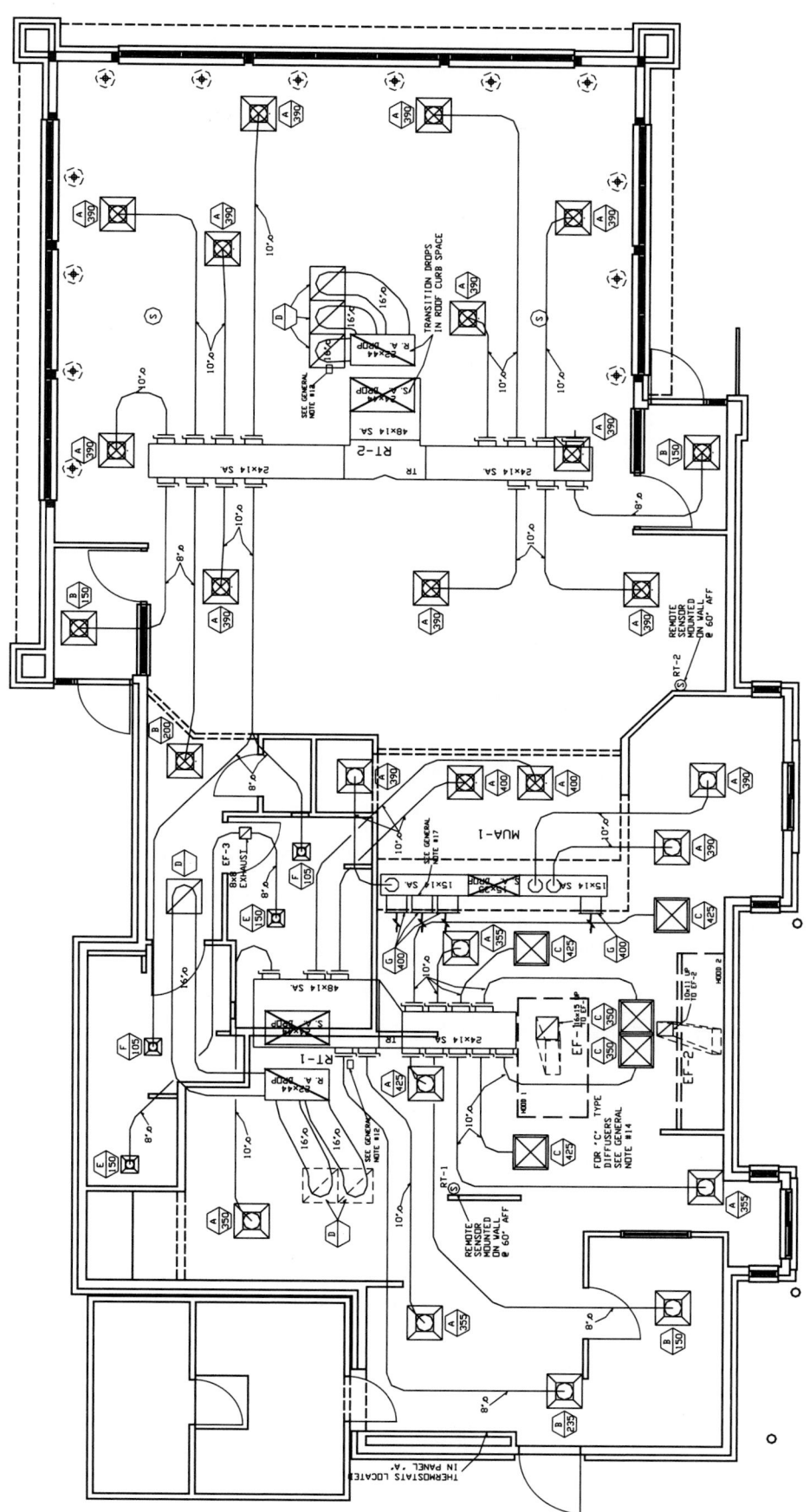

PROBLEM 24-4 continued

3. Prepare correlated general notes and schedules.
4. Create the necessary schematic and detail drawings with their correlated notes and specifications.

5. Draw the floor plan using thin lines and the HVAV components with thick lines for contrast. Use appropriate CADD layers.

GENERAL NOTES

1) HVAC SYSTEM DESIGNED TO MEET OUTSIDE TEMP OF 93 DEGREES F SUMMER AND 0 DEGREES F WINTER. CONTACT 'TRANE NATIONAL ACCOUNTS' FOR REQUIREMENTS AT OTHER TEMPERATURE CONDITIONS.

2) PROVIDE 1 YEAR WARRANTY ON WORKMANSHIP AND ON EQUIPMENT BY MECHANICAL CONTRACTOR. 1 YEAR PARTS, 5 YEAR COMPRESSOR AND 10 YEAR HEAT EXCHANGER WARRANTIES ON THE ROOFTOP UNITS BY MANUFACTURER

3) INSULATED FLEXIBLE DUCT MAY BE USED IN MAXIMUM LENGTHS OF 7'–0' PER BRANCH RUN. THE BALANCE OF THE RUN SHALL BE HARD PIPE W/1" INSULATED SLEEVE. FLEXIBLE DUCT SHALL BE OWENS CORNING INL. 25 OR EQUAL.

4) SPIN-IN FITTING W/ DAMPER SHALL BE FLEXAIRE RF OF METALAIRE MBSD. TYPICAL OF ALL BRANCH DUCT RUNS.

5) ALL DUCTWORK TO BE RUN ABOVE THE SUSPENDED CEILING.

6) HVAC CONTRACTOR TO RECEIVE AND SET ROOF MOUNTED CONDENSING UNITS AND PROVIDE MOUNTING RAILS. REFER TO STRUCTURAL PLAN 8 FOR LOCATIONS OF ROOF MOUNTED EQUIPMENT, CURB, AND RAIL DETAILS.

7) DUCTWORK SHOWN IS FINISHED O. D. DIMENSION AND TO BE INSULATED W/ 1 1/2" DUCT WRAP, INCLUDING DIFFUSERS.

8) RETURN DROPS FROM RT-1 AND RT-2 TO BE LINED, PINNED AND GLUED W/1" INSULATION PER INDUSTRY STANDARDS.

9) COMPLETED INSTALLATIONS SHALL CONFORM TO ALL APPLICABLE LOCAL, STATE, AND FEDERAL CODES AND ORDINANCES, INCLUDING, BUT NOT LIMITED TO THE LATEST EDITIONS OF THE FOLLOWING STATE BUILDING CODE, NFPA-90A, NFPA-96 AND NFPA-101.

10) SUPPLY AND RETURN AIR DUCT DROPS FROM ROOF TOP UNITS TO BE ISOLATED FROM UNIT VIBRATION WITH FLEXIBLE DUCT CONNECTORS.

11) ALL SUPPLY & RETURN AIR DUCTS AND DIFFUSERS ARE TO BE INSULATED.

12) RUN 18/8 THERMOSTAT WIRE FROM FAN CONTROL PANEL TO ROOFTOP UNITS AND 18/2 THERMOSTAT WIRE TO OUTSIDE AIR DAMPER.

13) HVAC CONTRACTOR TO TEST CHECK AND BALANCE ALL ROOFTOP UNITS, EXHAUST HOODS. EXHAUST FANS AND DIFFUSERS TO SPECIFIED CFM'S AND SUBMIT A WRITTEN REPORT TO THE OWNER.

14) **"C" DIFFUSERS TO HAVE 22" X 22" BOX FIELD FABRICATED W/12" COLLAR INSTALLED IN SIDE OF BOX. TO BE FIELD MEASURED FOR 12" HEIGHT**

15) THERMOSTATS ARE PROVIDED BY ELECTRICIAN IN FAN CONTROL PANEL (SEE SHEET #14). HVAC CONTRACTOR TO MOUNT REMOTE SENSORS AND RUN CONTROL WIRES FOR STATS AND SENSORS OUTSIDE OF THE FAN CONTROL PANEL.

16) ALL CURBS TO BE ONE PIECE WELDED AND INSULATED 18 GA STEEL. 1 1/2" RIGID INSULATION GLUED TO INSIDE OF CURB. ALL CURBS TO HAVE 1 1/2" WOOD NAILER ATTACHED TO TOP OF CURB.

17) FRAMING IN THIS AREA TO BE DOUBLE 2X4 @17 1/2" D.C. TO ACCOMMODATE AIR DEVICE PENETRATIONS.

18) ALL 90 DEGREE BENDS IN SUPPLY AIR DUCTS TO HAVE TURNING VANES. THE DISMANTLING OF ANY DUCT WILL BE REQUIRED TO VERIFY VANES.

19) ALL DUCT WORK IS TO BE INSTALLED PER ANSI SPECIFICATIONS.

20) PIEZO ALERTS FOR SMOKE DETECTORS INSTALLED BY HVAC CONTRACTOR MOUNTING LOCATION SHOWN ON ELECTRICAL PLAN.

21) HVAC CONTRACTOR TO SUPPLY AND INSTALL WIRE FOR THERMOSTAT AND SMOKE DETECTOR CIRCUITS.

22) HVAC CONTRACTOR TO CONNECT MUA-1 LOW VOLTAGE CONTROL TERMINALS TO CONTROL PANEL TERMINALS.

PROBLEM 24-4 continued

HVAC EQUIPMENT SCHEDULE

TRANE MODEL	MARK	TONS	S/A FAN			VOLT	HEATING INPUT	HEATING OUTPUT	TOTAL COOLING	MAX FLA @ 208V.	O.F.A. C.F.M.	NOTES
			HP	CFM	SP							
GAS HEAT												
YFD150C3H0	RTU-1	12.5	3	5000	.75	208/3/60	250 MBH	203 MBH	151 MBH	82	1250	①②
YFD150C3H0	RTU-2	12.5	3	5000	.75	208/3/60	250 MBH	203 MBH	151 MBH	82	1250	①②
AIR WISE MUA												
TBA-250/DX-C7.	MUA-1	7.5	2	2350	.5	208/3/60	250 MBH	200 MBH	90 MBH	43.8	100%	①③

NOTES: ① MOTORIZED OUTSIDE AIR DAMPERS INTERLOCKED WITH EXHAUST FANS IN FAN CONTROL PANEL.

② ELECTRICAL DISCONNECT SWITCH, SMOKE DETECTOR, AND CONVENIENCE OUTLET PROVIDED BY ROOFTOP MFG.

③ ELECTRICAL DISCONNECT SWITCH AND SMOKE DETECTOR PROVIDED BY ROOFTOP MANUFACTURER.

AIR DEVICE SCHEDULE

MARK	QUANTITY	NECK SIZE	TYPE			C.F.M.	MOUNTING		PATTERN	DUTY			METAL AIRE MODEL No.	TRANE MODEL No.	NOTES
			DIFFUSER	REGISTER	GRILL		LAY-IN	SURFACE		SUPPLY	RETURN	EXHAUST			
A	21	10"	o			355/390/425	o		4W	o			7500	BNASDIF821B	①②
B	5	8"	o			150/235	o		4W	o			7500	BNASDIF823B	①②
C	5	22"x22"	o			425		o	DUMP	o			7000R-6	BNARGRL002B	SEE NOTE #14
D	6	20"x20"			o			o	LOUVERED		o		7000R-6	BNARGRL002B	
E	2	8"		o		150		o				o	7000-1R	BNARGRL058B	①②
F	2	8"	o			105		o	4W	o			7000-1	BNASDIF102B	①②
G	4	14"x10"			o	400		o	DBL. DEFL	o			61DH014x10		① OPPOSED BLADE DAMPER

Notes: ① WHITE ② ROUND NECK

NOTE: EQUAL UNITS BY CARNES SERIES 4300-4, AND ANEMOSTAT SERIES PREO ARE ACCEPTABLE
NO SUBSTITUTIONS PERMITTED ON CORPORATE OWNED STORES.

FAN SCHEDULE

MARK	C.F.M.	S.P.	RPM	HP	VOLT	ACCESSORIES					PENN MODEL NUMBER	NOTES
						BDD	BS	DISC.	V-BELT	D-DRIVE		
EF-1	3000	1"	1300	1 1/2	208/3/60			●	●		FX13BFTH-W	①②
EF-2	1350	1"	1000	1	208/3/60			o	o		FX13BFTH-W	①②
EF-3	300	3/8 "	1550	1/8	115/1/60			o		o	XR-94	①②③

① FANS WILL BE SUPPLIED AS PART OF THE EXHAUST HOOD PACKAGE

② FOR U.L. APPROVED GREASE RATING USE PENN FX13BFTH-W FANS.

③ RPM'S ARE FOR INITIAL SET-UP. VERIFY CFM DURING STORE BALANCE.

AIR BALANCE SCHEDULE

MARK	SUPPLY AIR	RETURN AIR	OUTSIDE AIR	EXHAUST AIR	RESULTING PRESSURE
RT-1	5000	3750	+1250		+1250
RT-2	5000	3750	+1250		+1250
MUA-1			+2350		+2350
EF-1				-3000	-3000
EF-2				-1350	-1350
EF-3				-300	-300
TOTALS			+4850	-4650	+200

EXHAUST HOOD SCHEDULE

MARK	SIZE	SUPPLY OPENING	EXHAUST OPENING	ROOF OPENING	DUCT AT CURB
HOOD-1	48"x84"	10"x24"	30"x30"	16"x15"
HOOD-2	32"x126"	6"x18"	30"x30"	10"x11"

① ALL DUCTWORK SHALL BE FABRICATED ACCORDING TO NFPA 96 & LOCAL CODES.

② FINAL BALANCING OF THE SYSTEM SHALL BE PERFORMED AS OUTLINED IN INSTRUCTION SHEET FURNISHED BY HOOD SUPPLIER BY CERTIFIED TEST & BALANCE CO.

③ EXHAUST HOOD DUCT IS A 16 GA. BLACK STEEL W/CONT'S. WELDED SEAMS, PAINTED TO OWNER'S SPEC'S.

④ EXHAUST AIR OPENINGS TO BE CUT IN FACTORY ALL HOODS TO BE UL CERTIFIED AND LABELED.

HVAC EQUIPMENT & MATERIAL PACKAGE

FOR QUOTATION ON THE FOLLOWING EQUIPMENT CONTACT WENDYS ACCOUNT COORDINATOR, TRANE NATIONAL ACCOUNTS PHONE: 1-800-872-6330

EQUIPMENT PACKAGE:
RT-1, RT-2 AND MUA-1 AS PER SCHEDULE THIS SHEET. IF DESIGN CONDITIONS DIFFER FROM THOSE LISTED, CONTACT TRANE (LISTED ABOVE) FOR APPROPRIATE EQUIPMENT.

ROOF CURB PACKAGE:
INCLUDES CURBS FOR RT-1, RT-2, MUA-1, EF-1, EF-2, EF-3 AND 3 SETS OF ROOF RAILS. ALL CURBS ARE INSULATED AND COMPLY W/ NFPA 96 REQUIREMENTS FOR HEIGHT. CURBS PROVIDED FOR EF-1 AND EF-2 MUST BE HINGED AND CHAINED.

AIR DEVICE PACKAGE:
TO INCLUDE A THRU G AND ALL SQUARE TO ROUND TRANSITIONS. ALL DIFFUSERS TO BE ALUMINUM FACE.

EQUIPMENT AND MATERIAL PACKAGES ARE IN STOCK AND SHOULD BE ORDERED IMMEDIATELY UPON RECEIPT OF HVAC CONTRACT TO AVOID CONSTRUCTION DELAY.

EXHAUST HOOD PACKAGE

HOOD & FAN PACKAGE SUPPLIED BY WENDY'S ON ALL COMPANY OWNED STORES.

VERIFY WITH OWNER'S REPRESENTITIVE WHO IS TO SUPPLY HOOD & FAN PACKAGE ON FRANCHISE STORES.

INCLUDES KITCHEN HOODS AND EXHAUST FANS EF-1 THRU EF-3.

CURBS FOR EXHAUST FANS ARE SUPPLIED WITH HVAC MATERIALS PACKAGE.

COORDINATE THE DELIVERY OF THIS PACKAGE WITH THE WENDY'S JOB SUPERVISOR. INSTALLATION BY HVAC CONTRACTOR.

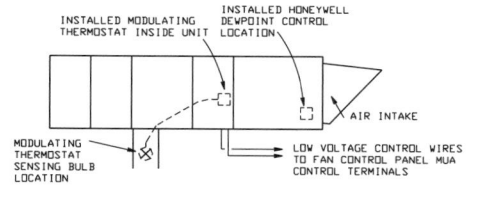

1. RELAY (   ) TO BE OMRON LY1F-24AC OR EQUAL W/AMP DRAW NO GREATER THAN 50mA WITH RELAY MOUNTING BASE. RELAY TO BE MOUNTED IN CONTROL CABINET IN ROOF TOP UNITS.

2. DUCT STAT DIFFERENTIAL THUMBWHEEL TO BE SET TO 5°F DIFFERENTIAL. SET POINT OF 65° F.

3. ALL ITEMS ARE FIELD FURNISHED AND INSTALLED.

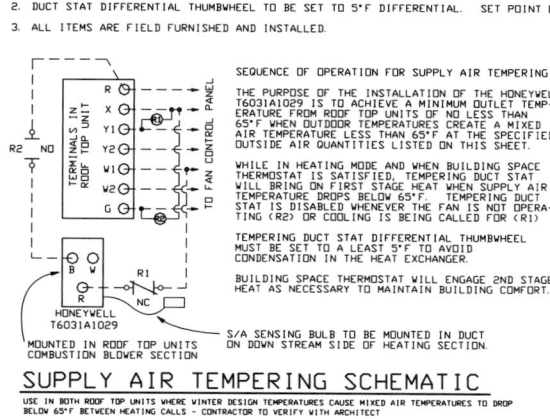

SEQUENCE OF OPERATION FOR SUPPLY AIR TEMPERING

THE PURPOSE OF THE INSTALLATION OF THE HONEYWELL T6031A1029 IS TO ACHIEVE A MINIMUM OUTLET TEMPERATURE FROM ROOF TOP UNITS OF NO LESS THAN 65°F WHEN OUTDOOR TEMPERATURES CREATE A MIXED AIR TEMPERATURE LESS THAN 65°F AT THE SPECIFIED OUTSIDE AIR QUANTITIES LISTED ON THIS SHEET.

WHILE IN HEATING MODE AND WHEN BUILDING SPACE THERMOSTAT IS SATISFIED, TEMPERING DUCT STAT WILL BRING ON FIRST STAGE HEAT WHEN SUPPLY AIR TEMPERATURE DROPS BELOW 65°F. TEMPERING DUCT STAT IS DISABLED WHENEVER THE FAN IS NOT OPERATING (R2) OR COOLING IS BEING CALLED FOR (R1)

TEMPERING DUCT STAT DIFFERENTIAL THUMBWHEEL MUST BE SET TO A LEAST 5°F TO AVOID CONDENSATION IN THE HEAT EXCHANGER.

BUILDING SPACE THERMOSTAT WILL ENGAGE 2ND STAGE HEAT AS NECESSARY TO MAINTAIN BUILDING COMFORT.

SUPPLY AIR TEMPERING SCHEMATIC

USE IN BOTH ROOF TOP UNITS WHERE WINTER DESIGN TEMPERATURES CAUSE MIXED AIR TEMPERATURES TO DROP BELOW 65°F BETWEEN HEATING CALLS - CONTRACTOR TO VERIFY WITH ARCHITECT

CONTROL PANEL. THESE CONTROLS ARE INTERLOCKED WITH HOODS TO PROVIDE REPLACEMENT AIR WHENEVER HOODS ARE OPERATING. HEATING OR COOLING OF MAKE UP AIR STREAM IS ACTIVATED BY DEW POINT SENSOR AND DUCT MOUNTED THERMOSTAT AS REQUIRED.

MAKE UP AIR CONTROLS

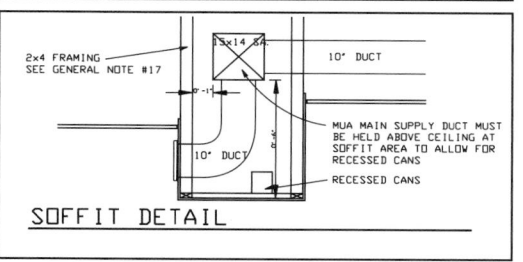

SOFFIT DETAIL

PROBLEM 24-5 Commercial HVAC plan

Problem courtesy Interface Engineering.

Given the following:

1. HVAC floor plan single-line engineering layout at approximately 1/8" = 1'–0". The provided problem is rough, so round off to the nearest convenient units at 6" intervals. For example, if a dimension reads 24'–3", round off to 24'–0" or 24'–6" as you prefer. The floor plan does not require dimensioning; therefore, the representation is more important than the actual dimensions.

2. Related general notes, schedules, schematics, and details.

Do the following on an appropriately sized sheet or sheets:

1. Make a formal double-line HVAV floor plan at a scale of 1/4" = 1'–0". (Note: You measured the given plan using a 1/8" = 1'–0" scale. Use the established dimensions to create the formal drawing at a scale of 1/4" = 1'–0".) Approximate the location of the HVAC duct runs and equipment in proportion to the presentation given on the engineering layout. Assume that the single-line layout is the centerline of the ducts.

2. Approximate the location and size of doors, windows, and fixtures.

3. Prepare correlated general notes and schedules.

4. Create the necessary schematic and detail drawings with their correlated notes and specifications.

5. Draw the floor plan using thin lines and the HVAV components with thick lines for contrast. Use appropriate CADD layers.

GENERAL NOTES

A PROVIDE SYSTEMS WHICH CONFORM TO NATIONAL CODES, LOCAL CODES, AND OTHER APPLICABLE CODES AND REGULATIONS, INCLUDING SEISMIC REQUIREMENTS.

B COORDINATE FINAL LOCATION OF EQUIPMENT, DUCTS, DIFFUSERS AND GRILLES WITH STRUCTURE, REFLECTED CEILING PLAN, LIGHTING LAY–OUT AND FIRE SPRINKLER SYSTEM.

C PROVIDE FIRE DAMPERS AT FIRE RATED WALL, FLOOR, AND CEILING PENETRATIONS.

D PROVIDE EQUIPMENT WHICH CONFORMS TO APPLICABLE ENERGY EFFICIENCY CODES.

E INSTALL EQUIPMENT TO PROVIDE SERVICE CLEARANCES AS RECOMMENDED BY THE MANUFACTURER, AND AS REQUIRED BY CODE AND LOCAL INSPECTOR. PROVIDE CLEAR LABELING OF FILTER PANELS. VERIFY ADEQUATE ACCESS FOR ROUTINE MAINTENANCE.

F PROVIDE ROOF CURBS FOR EQUIPMENT REQUIRING A ROOF PENETRATION (i.e., FANS, INTAKE HOODS, ETC) PROVIDE EQUIPMENT SUPPORTS FOR ROOF MOUNTED EQUIPMENT NOT REQUIRING A PENETRATION (i.e. CONDENSING UNITS). COORDINATE ROOF CURBS AND SUPPORTS WITH ROOFING SYSTEM. SECURELY ATTACH EQUIPMENT TO CURB AND SUPPORTS.

G PROVIDE CONDENSATE TRAPS ON A.C. EQUIPMENT. ROUTE DRAIN PIPE TO CODE APPROVED LOCATION.

H MOUNT ALL THERMOSTATS AT 4'–0" A.F.F.

I ROUTE DUCTWORK THROUGH OPEN WEB JOIST. COORDINATE EXACT DUCT ROUTING WITH STRUCTURE IN FEILD. VERIFY MAXIMUM SIZE OF DUCT TO ROUTE THROUGH JOISTS WITH STRUCTURAL.

HVAC NOTES

(1) ROUTE 6"ø DUCT FROM CEILING EXHAUST FAN TO SIDEWALL VENT CAP. TRANSITION AS REQUIRED AT FAN AND EXTERIOR SIDEWALL VENT CAP. PROVIDE SIDEWALL VENT CAP WITH BACKDRAFT DAMPER. COORDINATE EXACT LOCATION OF SIDEWALL VENT CAP WITH ARCHITECT.

(2) PROVIDE TRANSITIONS FROM FULL SIZE INLET AND OUTLET OF UNIT TO DUCTWORK AS REQUIRED. LOCATE UNIT SO THAT DUCT DROPS ARE LOCATED BETWEEN STRUCTURAL MEMBERS. SEE DETAIL 1/M3.1.

(3) ROUTE EXHAUST DUCT UP THROUGH ROOF TO FAN INLET. TRANSITION EXHAUST DUCTWORK TO FULLSIZE OPENING OF FAN INLET. CENTER FAN BETWEEN ROOF TRUSSES. SEE DETAIL 2/M3.1.

(4) ROUTE 6X6 EXHAUST DUCT FROM CEF-1/2, UP TO ROOF CAP. TRANSITION EXHAUST DUCT TO FULL SIZE OPENING OF CEF-1/2 AND ROOF CAP. ROOF CAP TERMINATION TO BE MINIMUM 3 FEET FROM PROPERTY LINE. SEE DETAIL 4/M3.1.

(5) PROVIDE SMOKE DETECTOR IN MAIN SUPPLY AIR DUCT DROP FROM UNIT.

(6) COORDINATE CROSSOVER OF DUCTWORK WITH STRUCTURE AND CEILING. ROUTE DUCTWORK THROUGH OPEN WEB JOISTS TO AVOID DUCTWORK CONFLICTS. SEE GENERAL NOTE 'i'.

(7) CONNECT 6"ø FLUE VENT TO HOT WATER HEATER CONNECTION. ROUTE 6"ø TYPE 'B' FLUE VENT UP THROUGH ROOF TO AGA APPROVED ROOF CAP. SEE DETAIL 5/M3.1.

(8) PROVIDE TWO 12X12 COMBUSTION AIR LOUVERS WITH MINIMUM 0.30 SQUARE FEET OF FREE AREA EACH LOUVER. INSTALL ONE COMBUSTION AIR LOUVER THROUGH WALL 12 INCHES AFF. INSTALL SECOND COMBUSTION AIR LOUVER THROUGH WALL 12 INCHES BELOW CEILING. COVER INSIDE LOUVER OPENING WITH 1/4 INCH GALVANIZED SCREEN.

(9) PROVIDE TRANSITIONS FROM FULL SIZE OUTLET OF MAU-1 TO DUCTWORK AS REQUIRED. LOCATE UNIT SO THAT SUPPLY DUCT DROP IS LOCATED BETWEEN STRUCTURAL MEMBERS. SEE DETAIL 1/M3.1.

(10) GREASE EXHAUST DUCT AND EXHAUST FAN PROVIDED BY KITCHEN EQUIPMENT CONTRACTOR. COORDINATE INSTALLATION OF GREASE EXHAUST DUCT AND EXHAUST FAN WITH GENERAL CONTRACTOR AND KITCHEN EQUIPMENT CONTRACTOR. COORDINATE CONDITIONED AIR DUCT CROSSOVER WITH GREASE EXHAUST DUCT.

PROBLEM 24-5 continued

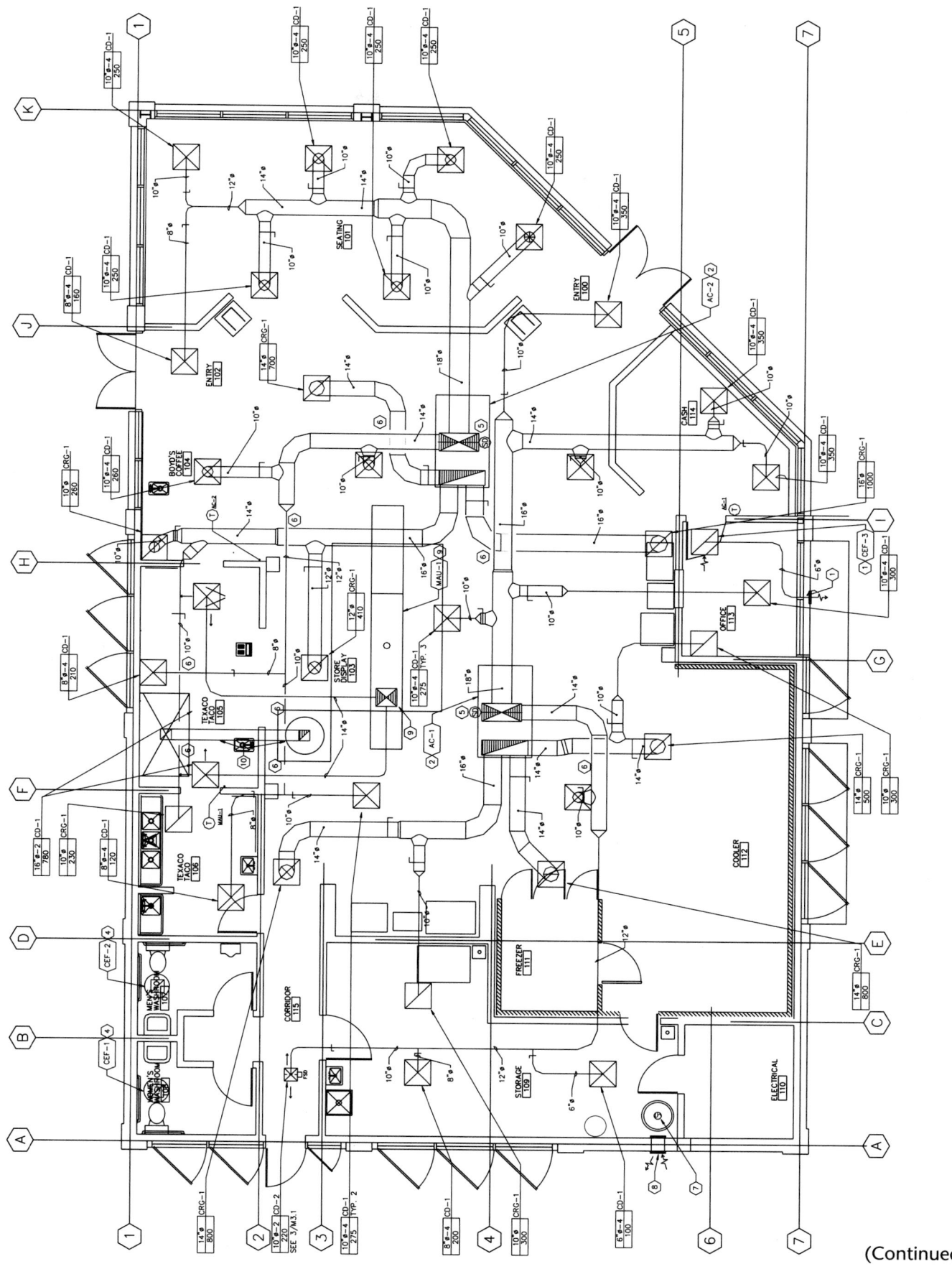

(Continued)

PROBLEM 24-5   continued

EXHAUST FAN SCHEDULE

SYMBOL	SERVING	TYPE	CFM	T.S.P. (IN)	MOTOR WATTS	MOTOR V/PH	MAX RPM	MODEL NO.	REMARKS
CEF-1	WOMENS RESTROOM	CEILING EXH. FAN	85	0.15	37	120/1	1450	GREENHECK SP 108	5.0 SONES MAX.
CEF-2	MENS RESTROOM	CEILING EXH. FAN	100	0.15	37	120/1	1450	GREENHECK SP 108	5.0 SONES MAX.
CEF-3	OFFICE	CEILING EXH. FAN	150	0.25	75	120/1	1690	GREENHECK SP 117	5.0 SONES MAX.

EQUIPMENT SCHEDULE

SYMBOL	DESCRIPTION	ELECTRICAL
AC-1	PACKAGED ROOFTOP A/C UNIT WITH GAS HEAT: 2700 CFM AT .8 IN. E.S.P. 345 CFM OUTSIDE AIR. 9.00 EER MIN. COOLING: 89.4 MBH TOTAL 66.6 SENSIBLE AT 76F EDB, 63F EWB, 84F AMBIENT MIN. HEATING: 120 MBH INPUT; 97 MBH OUTPUT WITH ROOF CURB AND ECONOMIZER. APPROXIMATE OPERATING WEIGHT: 1100 LBS. TRANE YCD090C3	208V/3 46.0 MCA. 2HP MOTOR
AC-2	PACKAGED ROOFTOP A/C UNIT WITH GAS HEAT: 2800 CFM AT .8 IN. E.S.P. 615 CFM OUTSIDE AIR. 9.00 EER MIN. COOLING: 90.6 MBH TOTAL 69.1 SENSIBLE AT 78F EDB, 64F EWB, 87F AMBIENT MIN. HEATING: 120 MBH INPUT; 97 MBH OUTPUT WITH ROOF CURB AND ECONOMIZER. APPROXIMATE OPERATING WEIGHT: 1100 LBS. TRANE YCD090C3	208V/3 46.0 MCA. 2HP MOTOR
MAU-1	EVAPORATIVE COOLING/INDIRECT GAS-FIRED MAKE-UP AIR UNIT: 1,560 CFM AT 0.8 IN ESP, 90F EDB, 68F EWB, 90% MIN. EVAP EFFICIENCY AT 600 MAXIMUM FPM FACE VELOCITY. HEATING: 75 MBH INPUT, 60 MBH OUTPUT. 1,000 LB OPERATING WEIGHT. PROVIDE WITH 12" MEDIA FILTER, ROOF CURB, INTAKE HOOD, DOWNTURN PLENUM CABINET. REZNOR MODEL HCRGB-75 HEATING UNIT WITH OPTIONAL EVAPORATIVE COOLER. NOTE: PROVIDE FOIL FACED INSULATION OF R-5 OR GREATER BELOW MAU-1 IN ROOF CURB SPACE. FOIL FACE TO FACE BOTTOM OF MAU-1. COORDINATE WITH ARCHITECT.	208V/3Ø 1/2 HP FAN MOTOR 120V/1Ø 1/70 HP PUMP MOTOR WIRED INTERNALLY.

DIFFUSER, REGISTER AND GRILLE SCHEDULE

SYMBOL	TYPE	FACE	FRAME	DAMPER	FINISH	MODEL NO.	REMARKS
CD-1	CEILING DIFFUSER	PERFORATED	CONCEALLED SPLINE	NONE	WHITE	CARNES SPFC	
CD-2	CEILING DIFFUSER	LOUVERED	SURFACE	NONE	WHITE	CARNES SKFA	
CEG-1	CEILING EXHAUST GRILLE	PERFORATED	CONCEALLED SPLINE	NONE	WHITE	CARNES SPJB	
CRG-1	CEILING RETURN GRILLE	PERFORATED	CONCEALLED SPLINE	NONE	WHITE	CARNES SPJB	

PROBLEM 24-5 continued

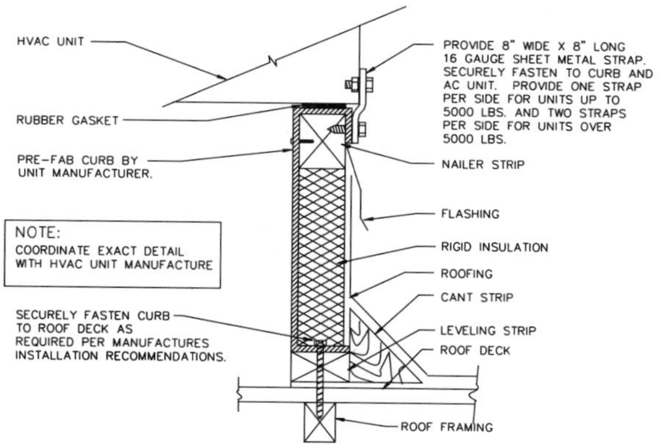

HVAC UNIT

RUBBER GASKET

PRE-FAB CURB BY
UNIT MANUFACTURER.

NOTE:
COORDINATE EXACT DETAIL
WITH HVAC UNIT MANUFACTURE

SECURELY FASTEN CURB
TO ROOF DECK AS
REQUIRED PER MANUFACTURES
INSTALLATION RECOMMENDATIONS.

PROVIDE 8" WIDE X 8" LONG
16 GAUGE SHEET METAL STRAP.
SECURELY FASTEN TO CURB AND
AC UNIT. PROVIDE ONE STRAP
PER SIDE FOR UNITS UP TO
5000 LBS. AND TWO STRAPS
PER SIDE FOR UNITS OVER
5000 LBS.

NAILER STRIP

FLASHING

RIGID INSULATION

ROOFING

CANT STRIP

LEVELING STRIP

ROOF DECK

ROOF FRAMING

1 AC UNIT CURB DETAIL
M3.1 NOT TO SCALE

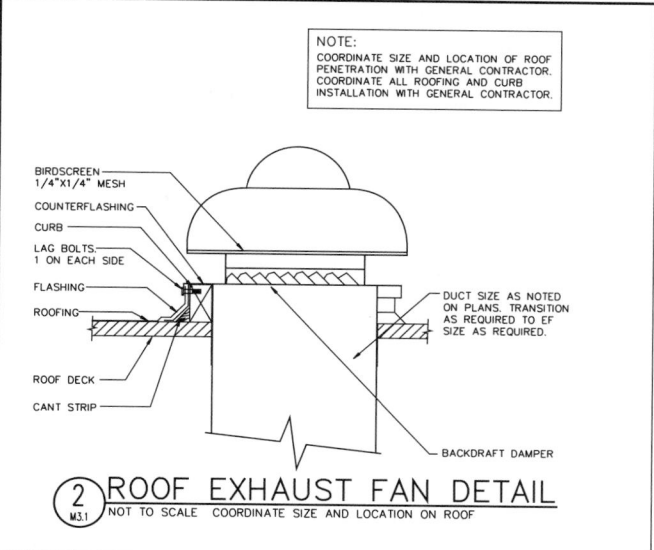

NOTE:
COORDINATE SIZE AND LOCATION OF ROOF
PENETRATION WITH GENERAL CONTRACTOR.
COORDINATE ALL ROOFING AND CURB
INSTALLATION WITH GENERAL CONTRACTOR.

BIRDSCREEN
1/4"X1/4" MESH

COUNTERFLASHING

CURB

LAG BOLTS.
1 ON EACH SIDE

FLASHING

ROOFING

ROOF DECK

CANT STRIP

DUCT SIZE AS NOTED
ON PLANS. TRANSITION
AS REQUIRED TO EF
SIZE AS REQUIRED.

BACKDRAFT DAMPER

2 ROOF EXHAUST FAN DETAIL
M3.1 NOT TO SCALE COORDINATE SIZE AND LOCATION ON ROOF

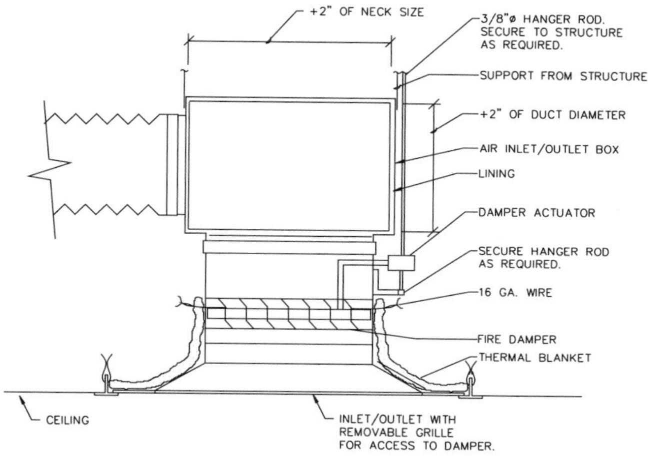

+2" OF NECK SIZE

3/8"ø HANGER ROD.
SECURE TO STRUCTURE
AS REQUIRED.

SUPPORT FROM STRUCTURE

+2" OF DUCT DIAMETER

AIR INLET/OUTLET BOX

LINING

DAMPER ACTUATOR

SECURE HANGER ROD
AS REQUIRED.

16 GA. WIRE

FIRE DAMPER

THERMAL BLANKET

CEILING

INLET/OUTLET WITH
REMOVABLE GRILLE
FOR ACCESS TO DAMPER.

THERMAL BLANKET TO BE 1 1/4"U.S. GYPSUM THERMOFIBER TYPE MINERAL WOOL.

**FIRE/SMOKE DAMPER
3 IN CEILING AIR INLET/OUTLET**
M3.1 NOT TO SCALE

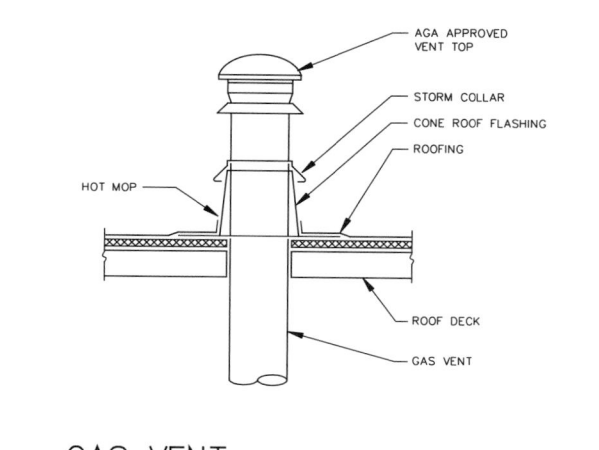

AGA APPROVED
VENT TOP

STORM COLLAR

CONE ROOF FLASHING

ROOFING

HOT MOP

ROOF DECK

GAS VENT

**GAS VENT
5 BUILT-UP ROOF PENETRATION**
M3.1 NOT TO SCALE

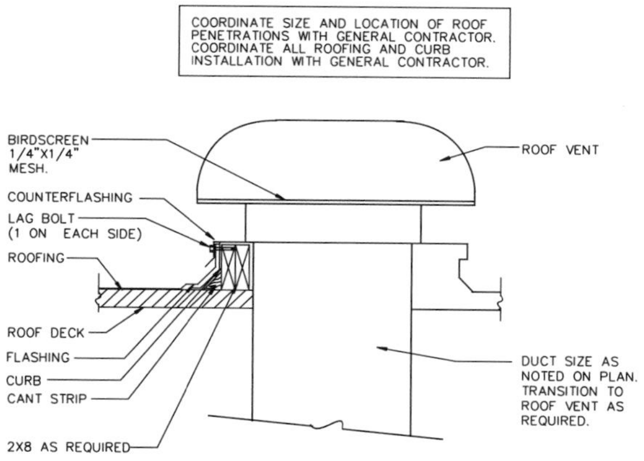

COORDINATE SIZE AND LOCATION OF ROOF
PENETRATIONS WITH GENERAL CONTRACTOR.
COORDINATE ALL ROOFING AND CURB
INSTALLATION WITH GENERAL CONTRACTOR.

BIRDSCREEN
1/4"X1/4"
MESH

COUNTERFLASHING

LAG BOLT
(1 ON EACH SIDE)

ROOFING

ROOF DECK

FLASHING

CURB

CANT STRIP

2X8 AS REQUIRED

ROOF VENT

DUCT SIZE AS
NOTED ON PLAN.
TRANSITION TO
ROOF VENT AS
REQUIRED.

4 ROOF VENT DETAIL
M3.1 NOT TO SCALE

PROBLEM 24-6   Exhaust duct system

Given: The engineer's sketch and specifications for an exhaust duct system. The sketch displays the top, front, and partial left side views of an exhaust duct system that could be found in any commercial solid-fuel exhaust. The exhaust pickup is rectangular in shape and the discharge throat is cylindrical. The directional path of the system is often obstructed and closely confined for reasons of design and operation of the system.

Do the following on appropriately sized paper:

1. Make a pattern development for each of the five exhaust duct components: (There will be five individual pattern development drawings.)
   A. Truncated cylinder
   B. Truncated cone
   C. Three-piece elbow
   D. Square-to-round transition piece
   E. Rectangular transitional elbow.
2. Use full scale unless otherwise specified by your instructor. A beam compass is necessary for manual drafting or use CADD.
3. Use a 3/8-in. single-lap seam on individual parts and between adjacent parts.
4. Show all layout and construction. Do not erase your construction lines.

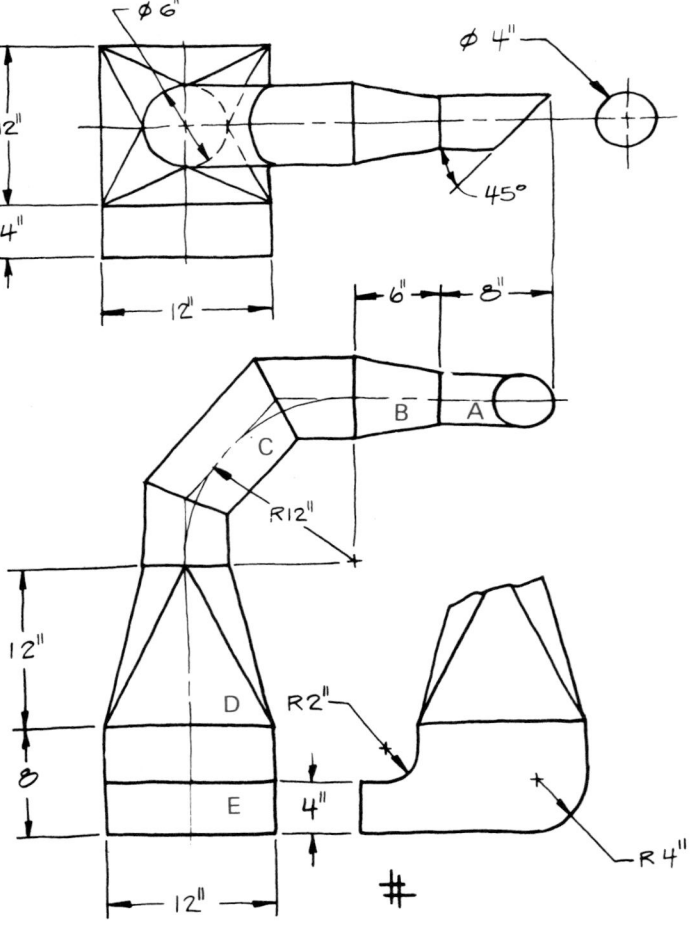

PROBLEM 24-7   Welding booth hood

Given: The engineer's sketch of a fabrication shop's welding booth hood.

Do the following:

1. Make a pattern development drawing of the pyramid-shaped hood and the shroud base at a scale of 1 1/2" = 1'-0" or full scale for CADD.
2. Provide the cutout for the window to be added later. Be careful to find the true location, and true size and shape of the cutout in the pattern.
3. No seam material allowance is required as the seams will be welded.
4. Show all layout and construction.

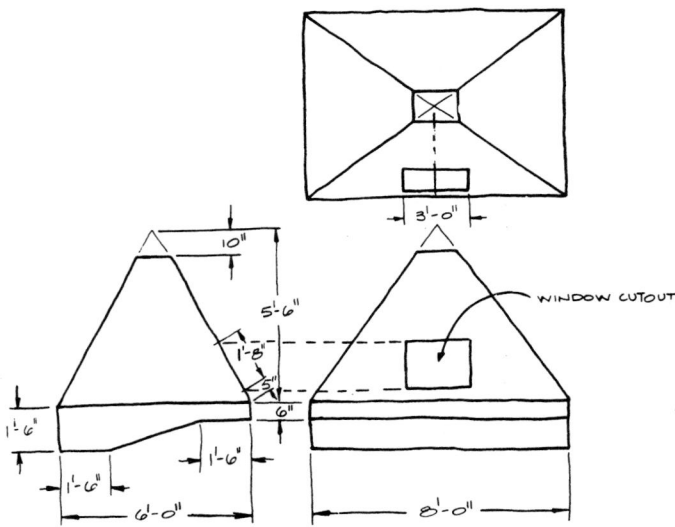

PROBLEM 24-8   Exhaust hood

Given: The exhaust hood detail from Problem 24-3.

Do the following:

1. Make a pattern development on appropriately sized layout for the transition piece, the top collar, and the base collar.
2. Use a half scale or full scale in CADD.
3. Provide a 1-in. single-lap seam for each part and between adjacent parts.
4. Show all layout and construction.

PROBLEM 24-9   Chemistry laboratory hood

Given: The engineer's rough sketch of the chemistry laboratory hood.

Do the following:

1. Make a pattern development drawing of the chemistry laboratory hood, including the top and bottom collars.
2. Use a 3" = 1'–0" scale, or full scale for CADD.
3. No seams required.
4. Show all layout and construction.

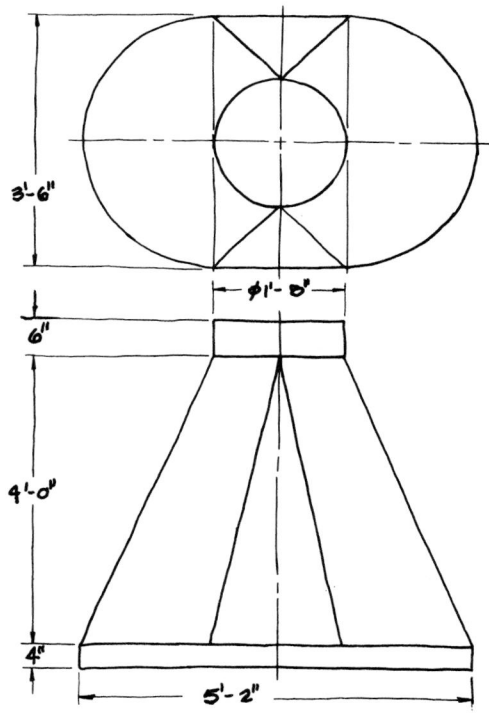

PROBLEM 24-10   Cylindrical duct intersection

Given: The engineer's computer sketch of intersecting cylindrical ducts.

Do the following:

1. Use a 1" = 1'–0" scale, or full scale for CADD.
2. Find the intersection between the cylindrical ducts.
3. Make a pattern development for each cylinder.
4. Use a 1-in. (scale) double-lap seam.
5. Show all layout and construction.

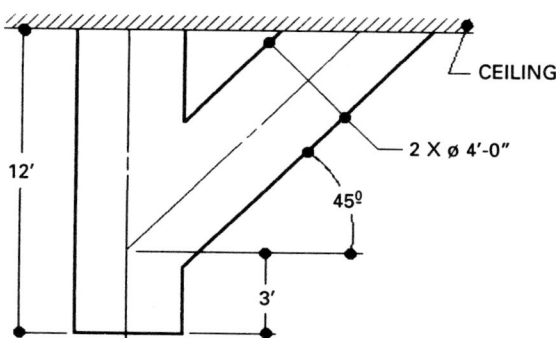

PROBLEM 24-11   Grain hopper

Given the following:

1. Use a 3/4" = 1'–0" scale, or full scale for CADD.
2. The engineer's sketch of the grain hopper.

Do the following:

1. Determine the line of intersection between the cylinder and the cone in both views.
2. Make the resulting pattern development of the cone and side intersecting cylinder.
3. No seam material allowance required.
4. Show all layout and construction.

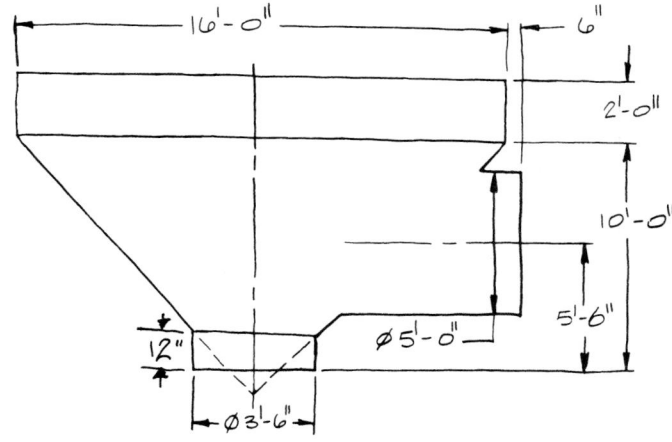

PROBLEM 24-12   Intersection

Given the following drawing, use your scale to double the size shown and draw the object. Determine the line of intersection between the parts.

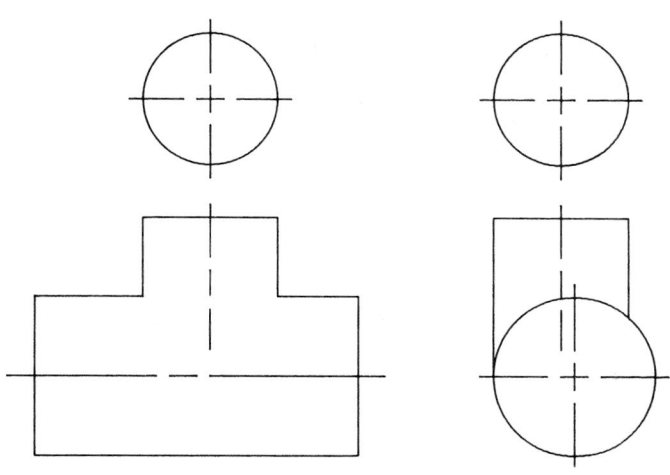

PROBLEM 24-13 Intersection

Given the following drawing, use your scale to double the size shown and draw the object. Determine the line of intersection between the parts.

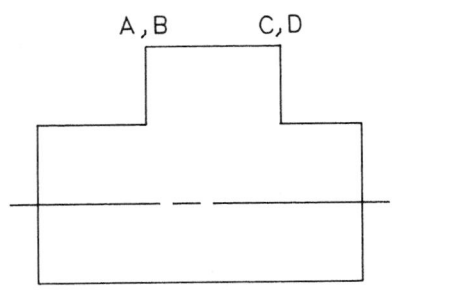
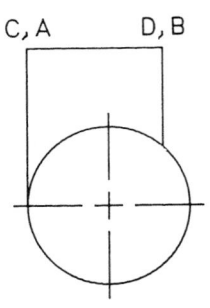

PROBLEM 24-14 Intersection

Given the following drawing, use your scale to double the size shown and draw the object. Complete the left-side view and determine the line of intersection between the parts in both views.

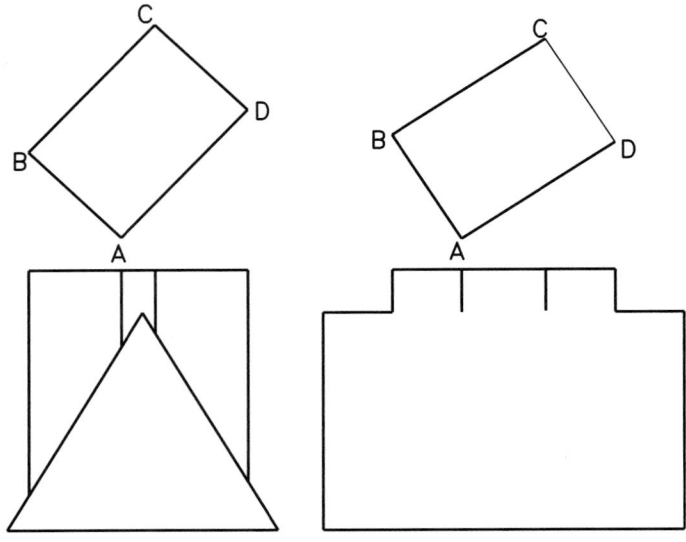

PROBLEM 24-15 Drawing displayed in form view and flat pattern

Draw the following object as shown:

Part Name: Mounting Bracket

Material: .25 THK 5086-H32

Courtesy TEMCO.

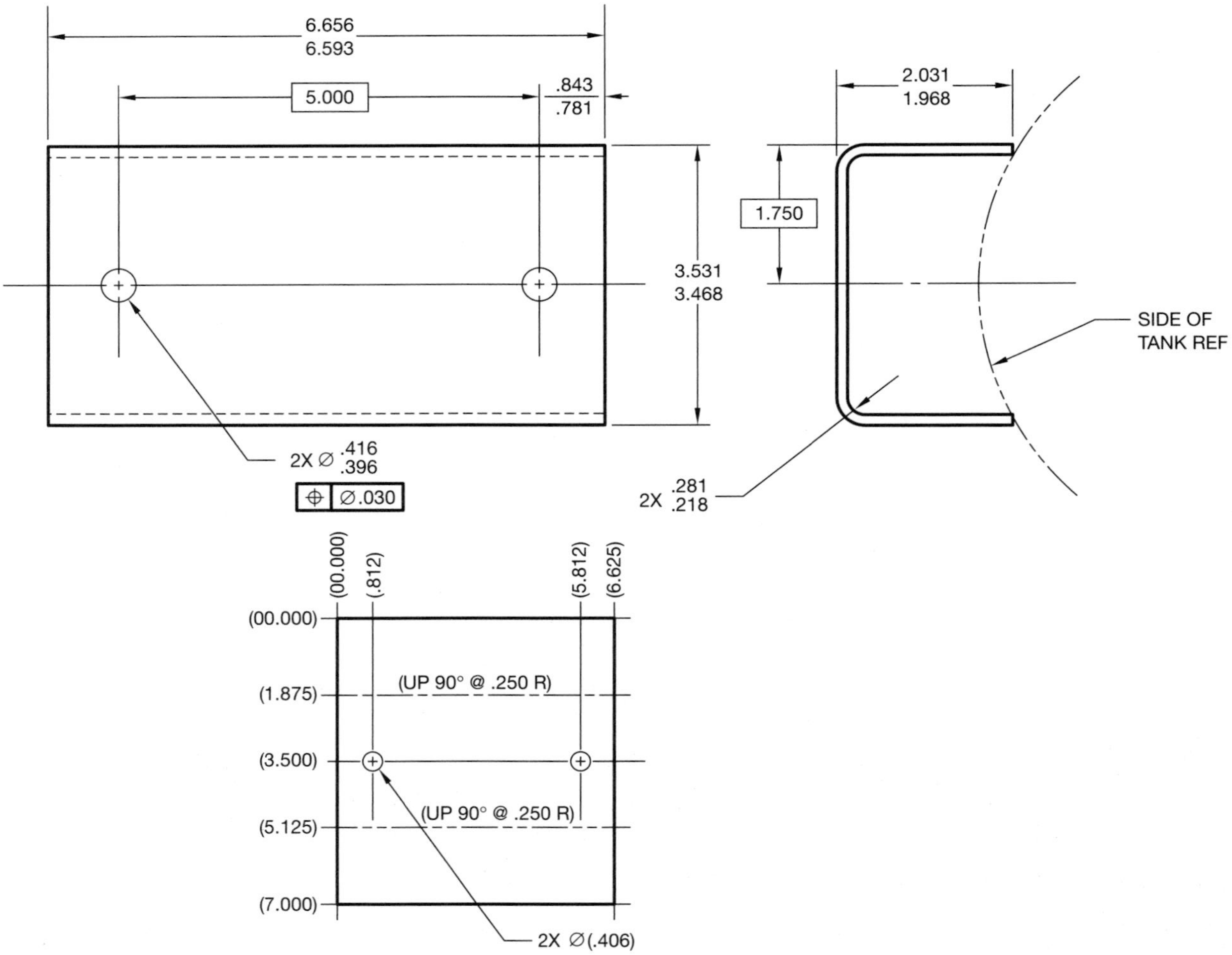

6.656
6.593

5.000

.843
.781

2.031
1.968

1.750

3.531
3.468

SIDE OF
TANK REF

2X Ø .416
.396

⊕ Ø.030

2X .281
.218

(00.000) (.812) (5.812) (6.625)

(00.000)

(1.875)   (UP 90° @ .250 R)

(3.500)

(5.125)   (UP 90° @ .250 R)

(7.000)

2X Ø(.406)

FLAT PATTERN
(SCALE .50X)

PROBLEM 24-16   Drawing displayed in form view and flat pattern *Courtesy TEMCO.*

Draw the following object as shown:

Part Name: Bracket

Material: 14 GA GALV CRS

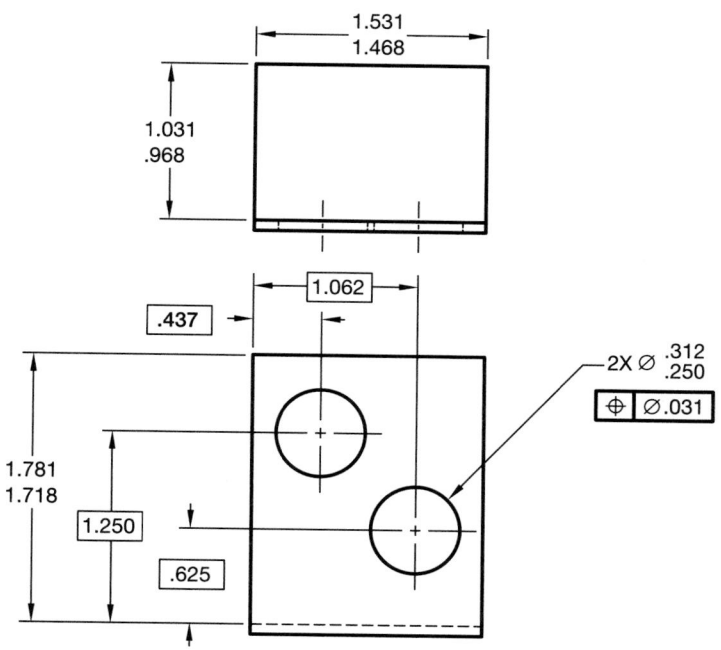

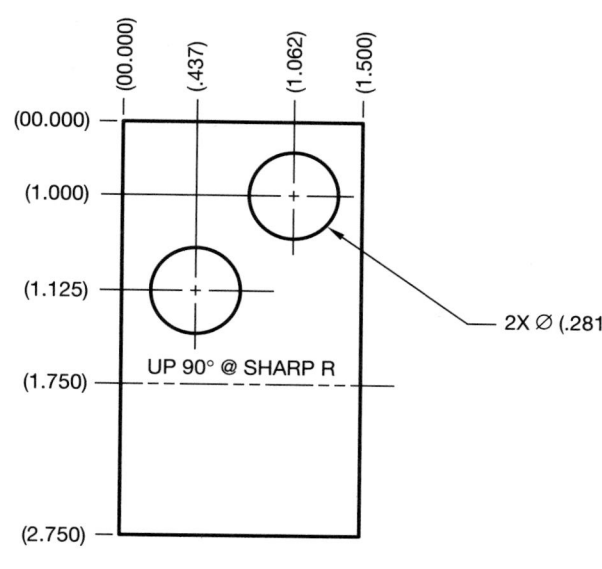

PROBLEM 24-17   Chassis layout

Part Name: Chassis

Material: Aluminum

Given: The engineer's rough sketch of a computer component chassis.

Do the following:

1. Make a flat pattern drawing of the given chassis on properly sized vellum. Full scale is recommended.
2. Use arrowless tabular dimensioning from the given datums.

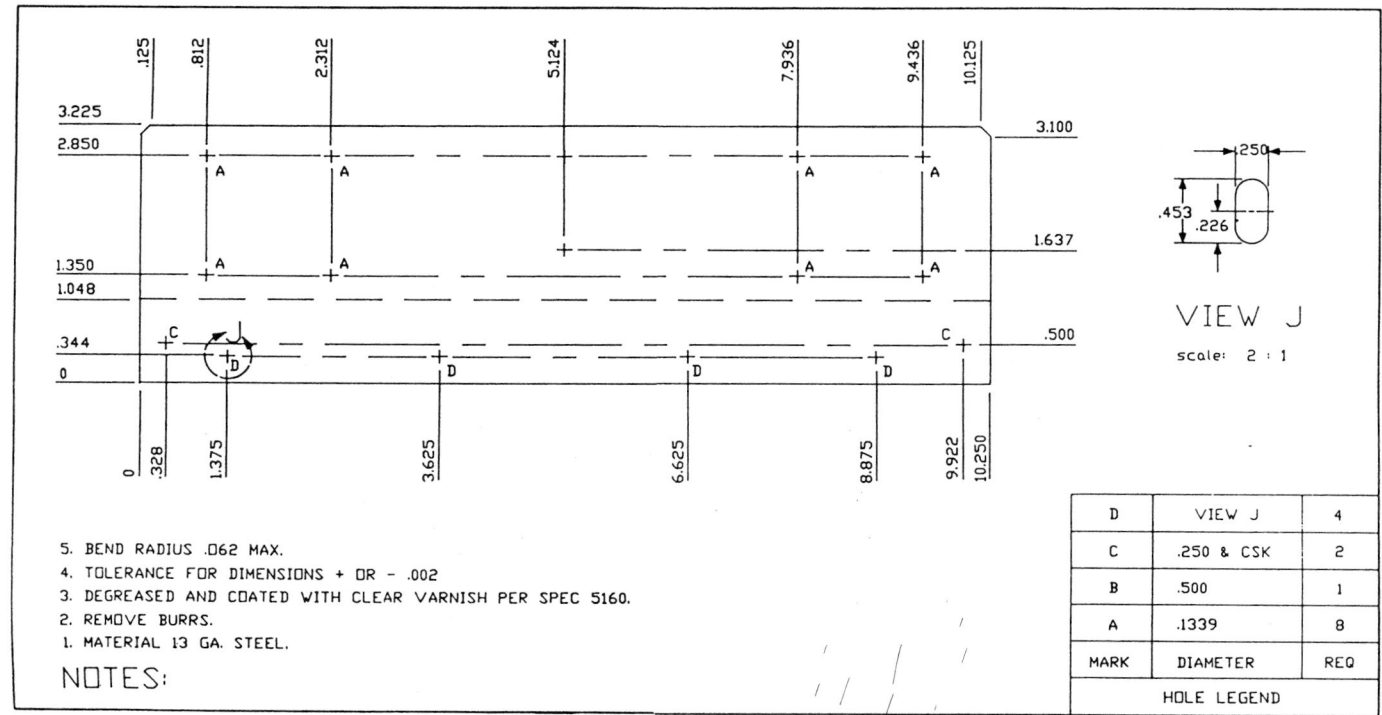

5. BEND RADIUS .062 MAX.
4. TOLERANCE FOR DIMENSIONS + OR - .002
3. DEGREASED AND COATED WITH CLEAR VARNISH PER SPEC 5160.
2. REMOVE BURRS.
1. MATERIAL 13 GA. STEEL.

NOTES:

MARK	DIAMETER	REQ
D	VIEW J	4
C	.250 & CSK	2
B	.500	1
A	.1339	8
MARK	DIAMETER	REQ

HOLE LEGEND

PROBLEM 24-18   Chassis layout

Given: The engineer's rough sketch of a computer component chassis.

Do the following:

1. Make a flat pattern drawing of the given chassis on properly sized vellum. Full scale is recommended. Establish the flat pattern layout dimension by bend allowance calculations. Show all math formulas and calculations on another paper.

2. Use arrowless tabular dimensioning from the given datums.

DIMENSIONING TABLE			
Hole Symbol	Hole Diameter	Depth	Quantity
A	.125	Thru	9
B	.250	Thru	2
C	.469	.030	2
D	See Detail D	Thru	1

3. Standard dimensioning is needed for the total length of flat pattern and a dimension from one datum to the bend line in the flat pattern.

BEND RADIUS = .062

DETAIL D

PROBLEM 24-19   Display of part in form view with flat pattern shown as phantom line

Draw the following object as shown:

Part Name: Plate-formed

Material: HC-112 6 mm THK

Problem based on original art courtesy Hyster Company.

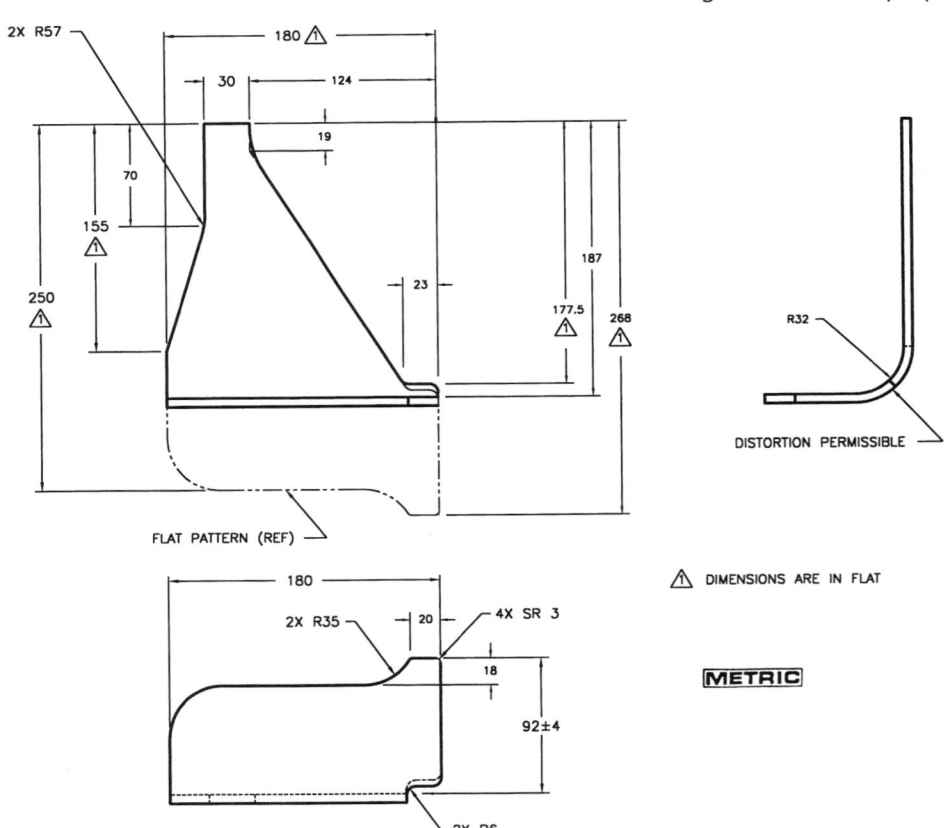

FLAT PATTERN (REF)

DISTORTION PERMISSIBLE

⚠ DIMENSIONS ARE IN FLAT

METRIC

PROBLEM 24-20   Display of part in form view and in flat pattern

Given the following engineer's layout, draw the flat pattern, and the formed view. Use geometric dimensioning and tolerancing as shown.

Part Name: Mounting Bracket
Material: 11 GA A569
Courtesy TEMCO.

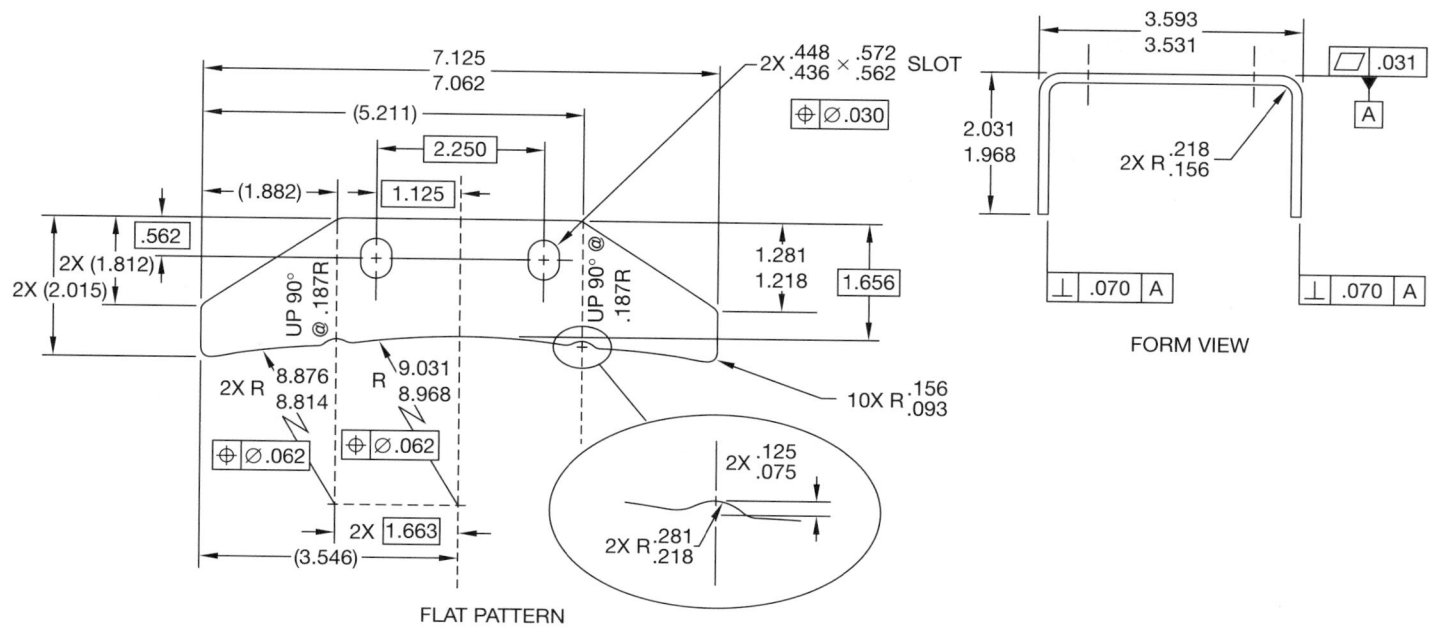

FLAT PATTERN

FORM VIEW

MATH PROBLEMS

Find the bend angle for each of the following figures:

PROBLEM 24-21

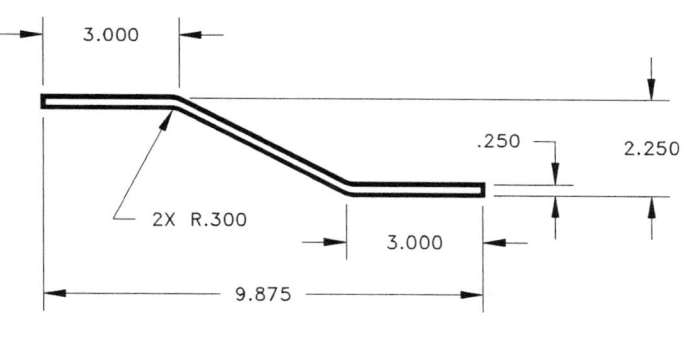

PROBLEM 24-22

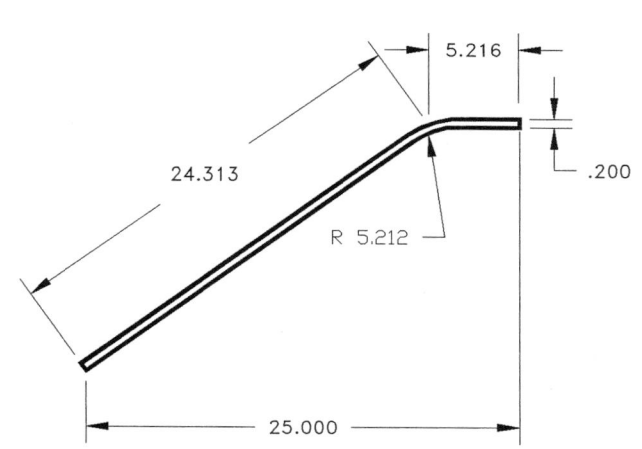

PROBLEM 24-23

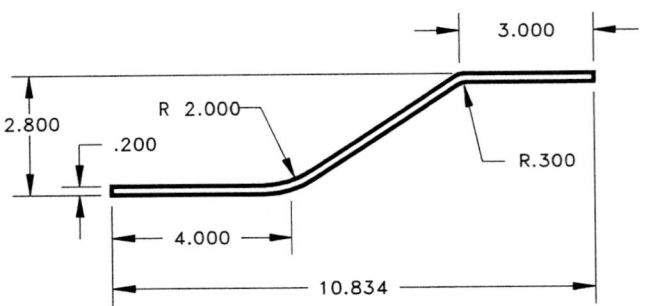

PROBLEM 24-25

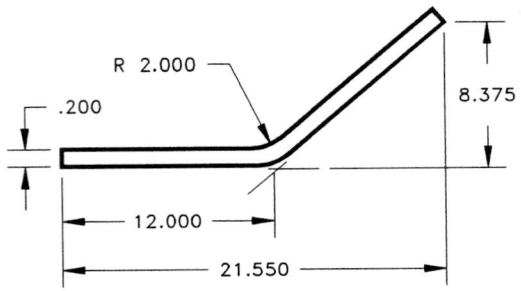

PROBLEM 24-24

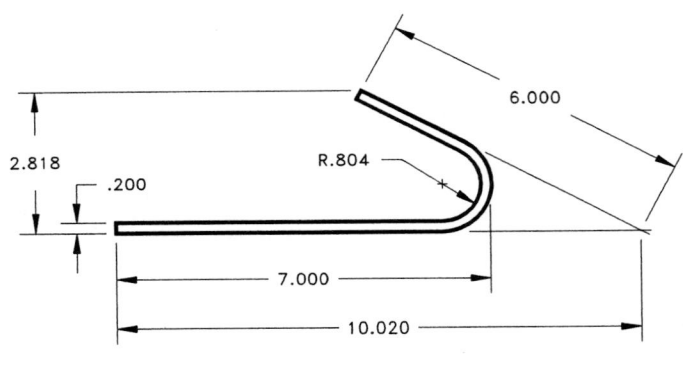

PROBLEM 24-26

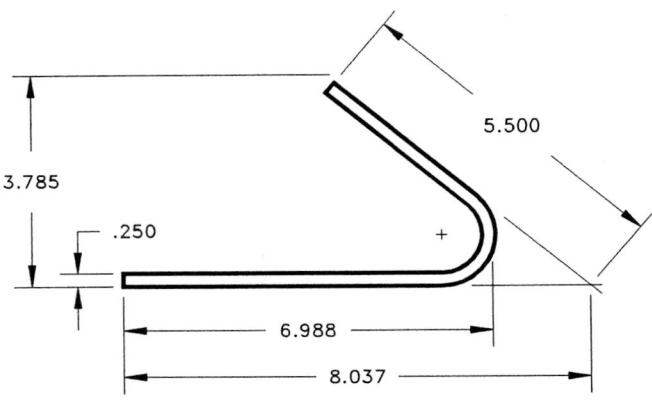

Electrical and Electronic Schematic Drafting

OBJECTIVES

After completing this chapter, you will:

■ Draw the following electrical diagrams: block, elementary, highway, wireless, and cable.

■ Draw a cable assembly.

■ Draw a set of electrical power system substation plans, including plot plan, bus plan and elevation, grounding layout and details, conduit installation details, electrical equipment layout, power panel detail, electrical floor plan, lighting plan, and security lighting plan.

■ Draw an industrial electrical schematic.

■ Prepare electrical drawings from engineering sketches.

■ Draw electronic block and schematic diagrams.

■ Create logic diagrams.

■ Prepare a printed circuit board layout.

■ Complete a marking and drilling drawing.

■ Prepare electronics pictorial drawings.

■ Make electronics drawings from given engineering sketches.

THE ENGINEERING DESIGN APPLICATION

In both electronic and electrical drafting, much of your work involves the use of symbols to show the components of your schematic or diagram. Having the proper tools available can greatly simplify your work in this area. If you are using manual drafting techniques, drawing templates are very useful for quick construction of standard symbols. CADD drafting offers additional advantages, however, because once a symbol has been drawn it can be used over and over again very simply. Addition-ally, CADD symbols can be created that automatically prompt for and place the required notational data in your drawing to properly define the symbol.

In this example, you have been provided a rough sketch of a circuit diagram. (See Figure 25.1.) By utilizing your CADD symbol library and adding the required notation, you are able to create the finished drawing. (Figure 25.2) with a minimum of time and effort.

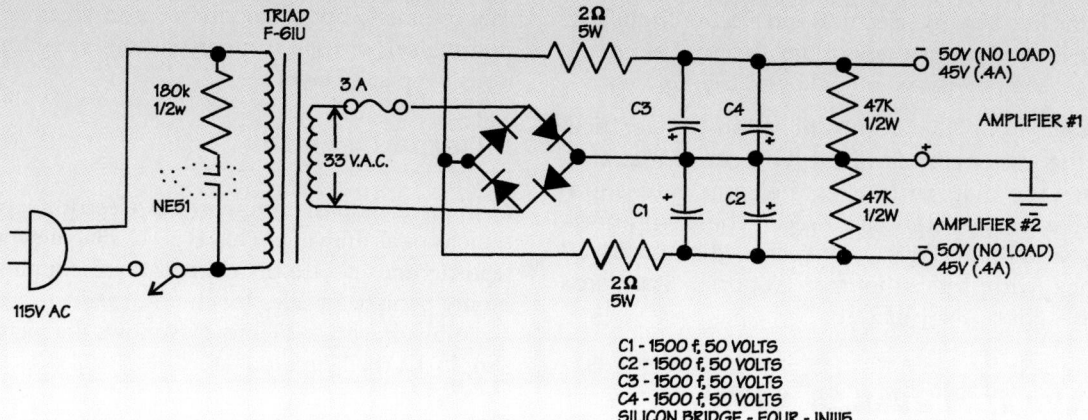

FIGURE 25.1 ■ Engineer's rough computer sketch of a circuit diagram.

(Continued)

THE ENGINEERING **DESIGN** APPLICATION *(continued)*

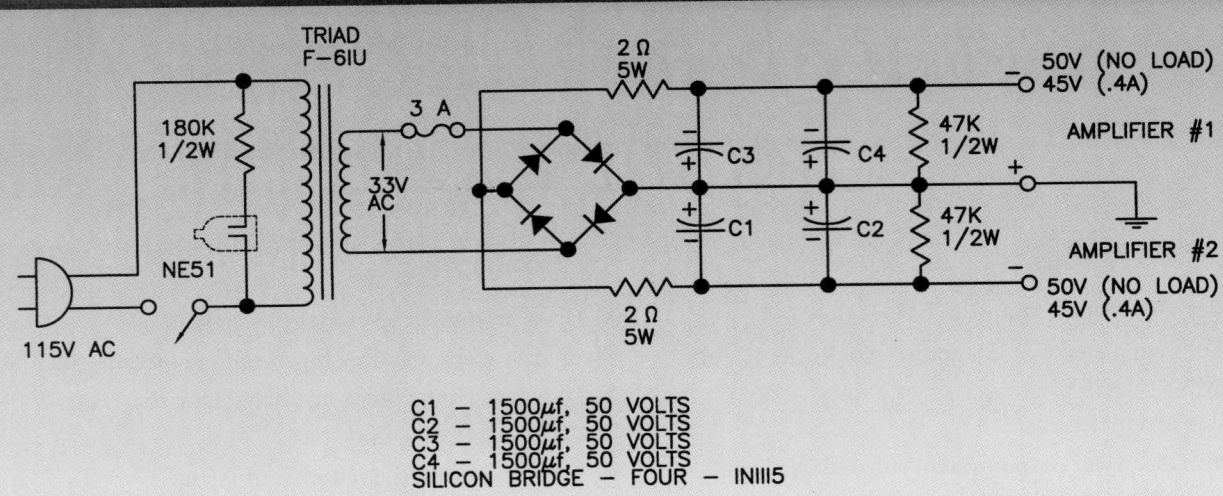

FIGURE 25.2 ■ The finished drawing from the engineer's sketch.

In the fully automated electronics industry, design begins with an engineer's sketch, which goes to an electronics technician who inputs the design into a computer. The computer system automatically evaluates the design and produces a prototype fabrication. The engineer then evaluates the system and makes modifications. After modifications are made and the system is completely tested and approved, the computer layout is sent to the drafter for engineering documentation. The drafter uses the CADD system to check design and drafting standards, evaluate engineering information for documentation accuracy, and correct component values. The drafter adds device tables that relate exactly how each component is attached to the board. The entire process is computerized. The drawing then progresses through the system to the automatic generation of the circuit boards. The drafter works closely with the circuit board computer program to ensure the accuracy of the product. This system may vary from one industry to the next, but as you can see, the electronics drafter is an important link between the engineering department and the final product.

Electrical drafting deals with concepts and symbols that relate to high-voltage applications from the production of electricity in power plants through distribution to industry and homes. While there is often a fine line between electrical and electronic drafting, *electronic drafting* is more oriented toward the design of electronic circuitry for radios, computers, and other low-voltage equipment.

ANSI The key to effective communication on electrical drawings is the use of standardized symbols so that anyone who uses the diagram makes the same interpretation. To ensure proper standardization, these engineering drawings and related documents should be prepared in accordance with the American National Standards Institute publication ANSI Y32.2.

FUNDAMENTALS OF ELECTRICAL DIAGRAMS

The purpose of electrical diagrams is to communicate information about the electrical system or *circuit* in a simple, easy-to-understand format of lines and symbols. *Electrical circuits* pro-vide the path for electrical flow from the source of electricity through system components and connections and back to the source. Electrical diagrams are generally not drawn to scale. The responsibility of the drafter is to organize the information in a logical, orderly manner without crowding or large variations in spacing layout.

Pictorial Diagram

Pictorial diagrams represent the electrical circuit as a three-dimensional drawing. This type of diagram provides the most realistic and easy-to-understand representation and may commonly be used in sales brochures, catalogs, service manuals, or assembly drawings. Figure 25.3 shows the pictorial drawing of a simple doorbell circuit.

Schematic Diagram

Schematic diagrams are drawn as a series of lines and symbols that represent the electrical current path and the components of the circuit. Figure 25.4 shows a schematic diagram of the doorbell circuit.

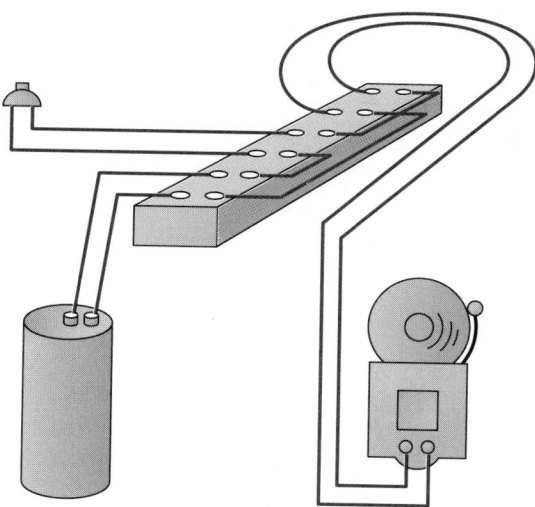

FIGURE 25.3 ■ Pictorial drawing of a simple doorbell circuit.

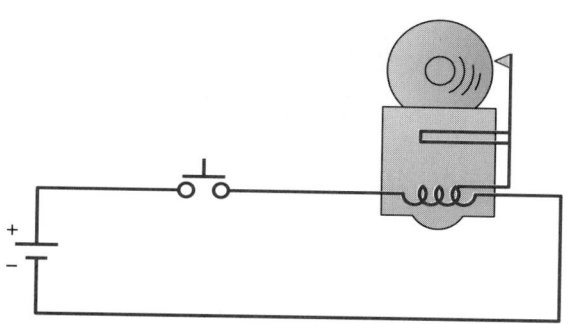

FIGURE 25.4 ■ Schematic diagram of doorbell circuit shown in Figure 25.3.

Block Diagram

The *block diagram* is a simplified version of the schematic diagram. Simplified symbols exhibit a minimum of detail of the component and generally no connections at individual *terminals* as shown in Figure 25.5.

Wiring Diagram

The *wiring diagram* is a type of schematic that shows all of the interconnections of the system components. These diagrams are often referred to as point-to-point interconnecting wiring diagrams. The wiring diagram is much more detailed than a standard schematic diagram because it shows the layout of individual wire runs. Figure 25.6 shows a wiring diagram of the doorbell circuit.

FIGURE 25.5 ■ Block diagram of the doorbell circuit shown in Figure 25.3.

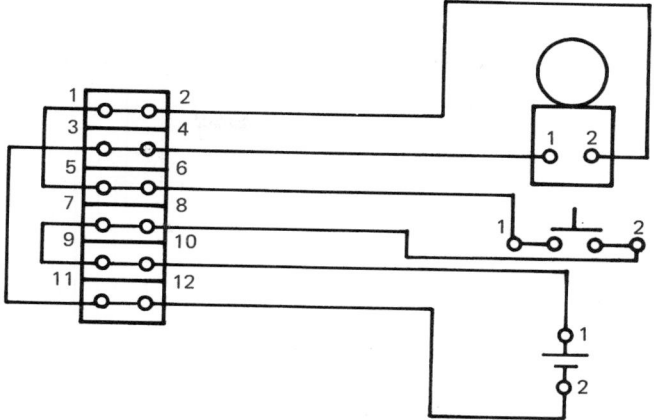

FIGURE 25.6 ■ Wiring diagram of the doorbell circuit shown in Figure 25.3.

Schematic Wiring Diagram

A *schematic wiring diagram* combines the simplicity of a schematic diagram and the completeness of a wiring diagram. The complete circuit is drawn as a series of lines and symbols that represent the electrical current path and the components of the circuit, plus the connection terminals are shown in their proper locations along the circuit. (See Figure 25.7.)

Highway Wiring Diagram

Also known as *highway diagrams*, these drawings are used for fabrication, quality control, and troubleshooting of the wiring of electrical circuits and systems. A highway wiring diagram is a simplified or condensed representation of a point-to-point interconnecting wiring diagram. Highway wiring diagrams may be used when it becomes difficult to draw individual connect lines because of diagram complexity or when it is not necessary to show all of the wires between terminal blocks.

The wiring lines are merged at convenient locations into main *trunk lines*, called *highways*, that run horizontally or vertically

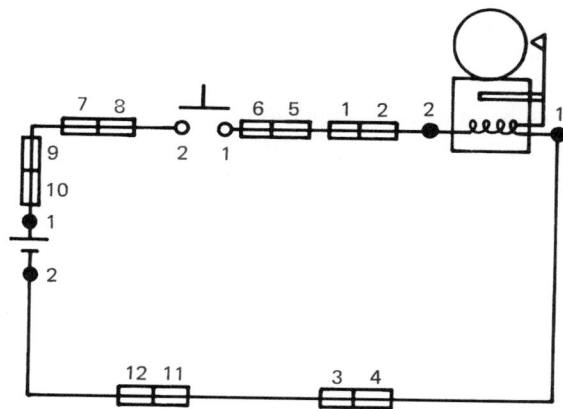

FIGURE 25.7 ■ Schematic wiring diagram of the doorbell circuit shown in Figure 25.3.

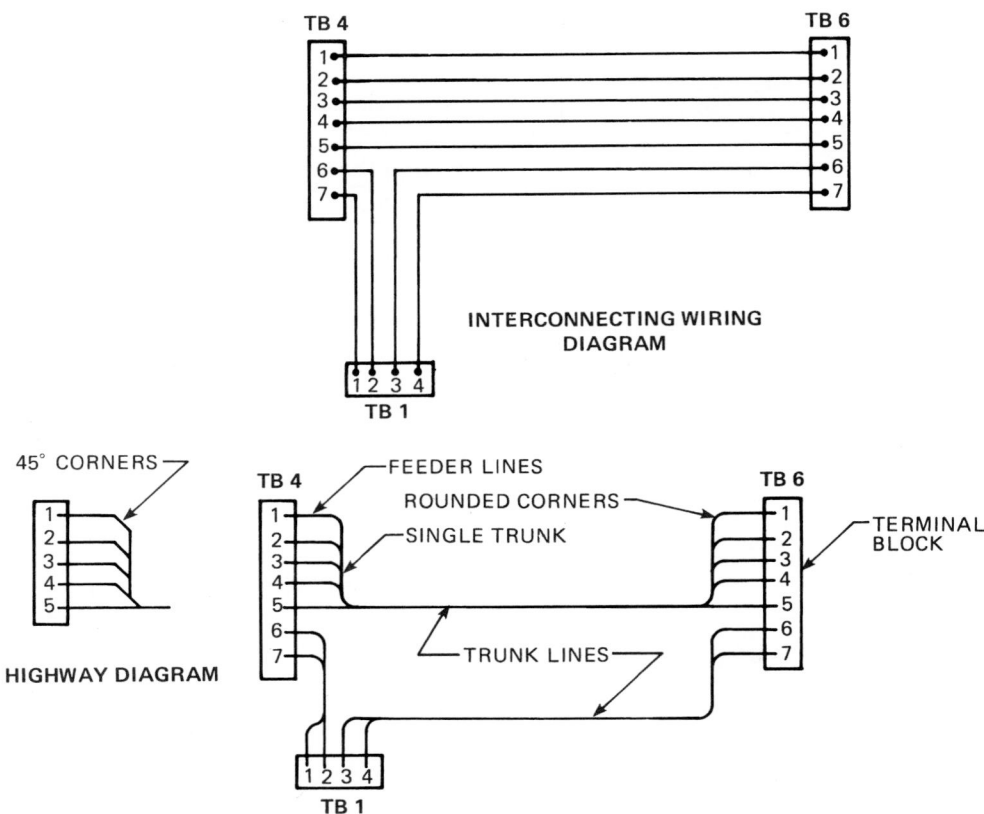

FIGURE 25.8 ■ An interconnecting wiring diagram and the same electrical system converted to a highway diagram.

between component symbols. The lines that run from the component to the trunk lines are called *feed lines*. The feed lines are identified by code letter, number, or a combination letter/number at the point where each line leaves the component. By reducing the number of lines drawn, the wiring diagram becomes easier to draw and interpret. Figure 25.8 shows an interconnecting wiring diagram and the same electricity system converted to a highway diagram.

A complete highway diagram has an identification system that guides the reader through the system, with a code at one terminal to the same corresponding terminal on another component. An identification system recommended by the American Standards Association (ASA) is a code made up of the wire destination, terminal number at the destination, wire size, and wire-covering color. Look at Figure 25.9 as you interpret the following code:

Component—M3	**Component—TB6**
TB6/1-B2	M3/3-B2
TB6 = Destination	M3 = Destination
1 = Terminal at Destination	3 = Terminal at Destination
B = Wire Size	B = Wire Size
2 = Color of Wire Covering	2 = Color of Wire Covering

Wireless Diagram

Wireless diagrams are similar to highway diagrams except that interconnecting lines are omitted as shown in Figure 25.10.

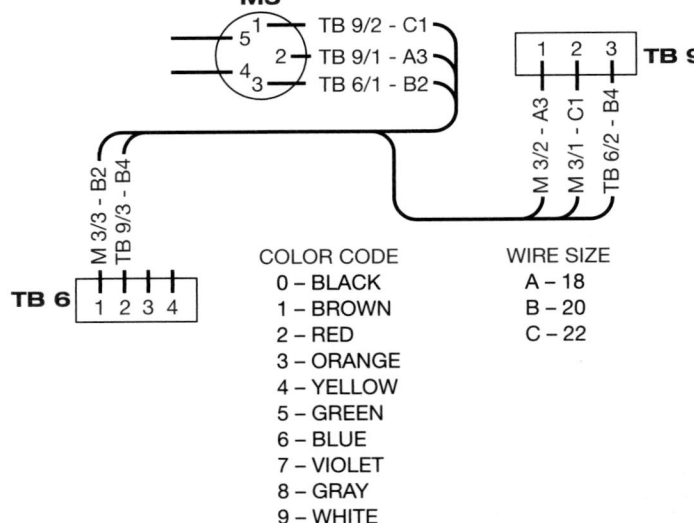

COLOR CODE	WIRE SIZE
0 – BLACK	A – 18
1 – BROWN	B – 20
2 – RED	C – 22
3 – ORANGE	
4 – YELLOW	
5 – GREEN	
6 – BLUE	
7 – VIOLET	
8 – GRAY	
9 – WHITE	

FIGURE 25.9 ■ Highway wiring diagram.

Cable Diagram and Assemblies

Cable diagrams are associated with *multiconductor* systems. A multiconductor is a *cable* or group of insulated wires put together in one sealed assembly. For example, the trunk line shown in the

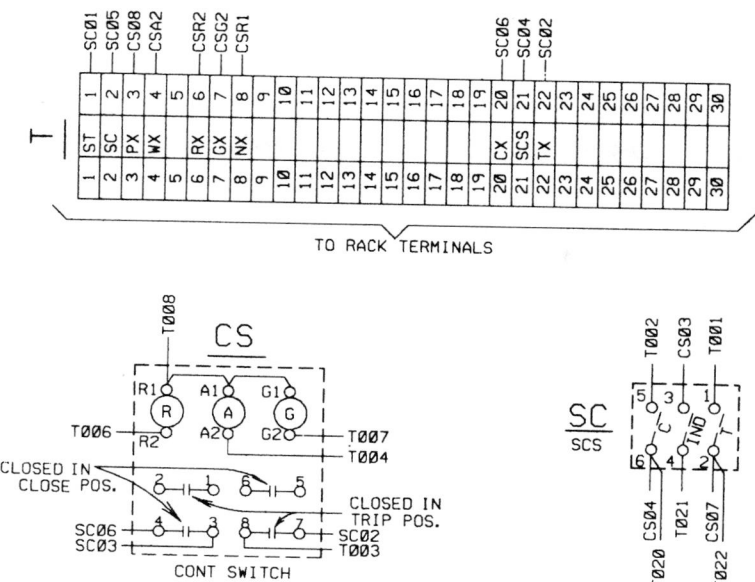

FIGURE 25.10 ■ CADD-generated wireless diagram. *Courtesy Bonneville Power Administration.*

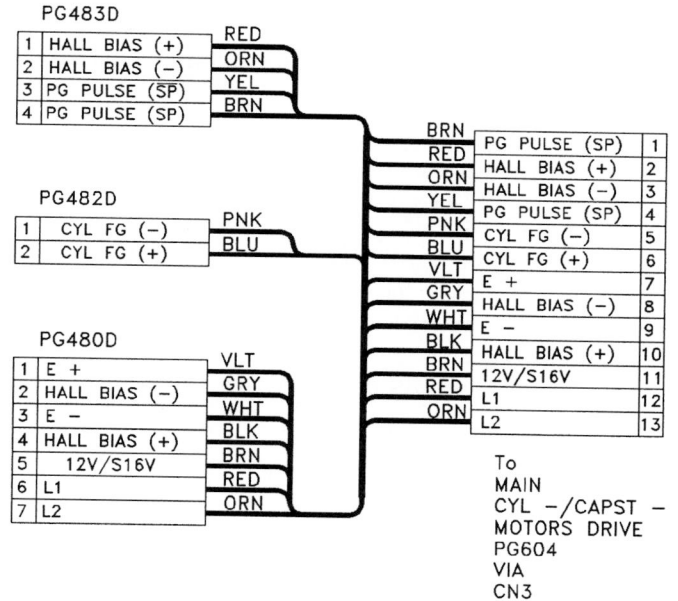

FIGURE 25.11 ■ Cable diagram. *Reproduced by permission of RCA Consumer Electronics.*

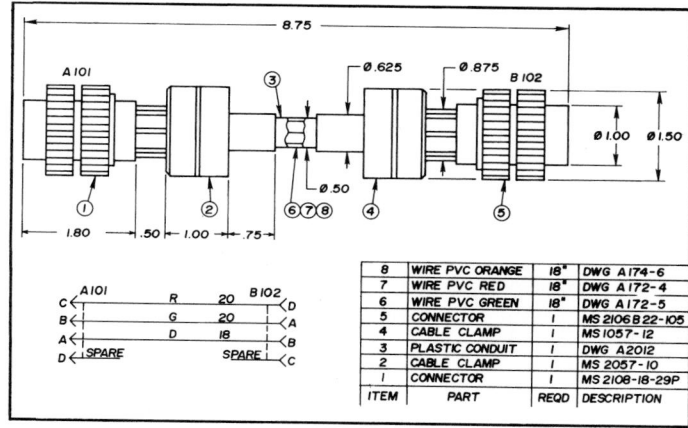

FIGURE 25.12 ■ Cable assembly.

highway diagram in Figure 25.8 or 25.9 could be a multiconductor. Cable systems are made up of insulated conductor wires, protective outer jacket or some other means of holding the wires together, and connectors at one end or both ends. Cables are used to connect components, equipment assemblies, or systems together. Cable diagrams usually provide circuit destination, conductor size, number of leads, conductor type, and power rating. A cable diagram is shown in Figure 25.11.

Also known as cable harness diagrams, *cable assemblies* are drawn to scale with dimensions and include a parts list that is coordinated with the drawing by identification balloons as shown in Figure 25.12.

GENERATION, TRANSMISSION, AND DISTRIBUTION OF ELECTRICITY

Electricity is generated around the world in hydroelectric, coal burning, and nuclear power plants. The electricity created at the generating station is increased in voltage at a step-up transformer and sent through high-voltage lines to a switching station where the electricity is retransmitted to various locations. Before the electricity can be used, it goes to a substation where a step-down transformer converts the high voltage to a lower voltage for heavy industry or transmission through lines to a distribution substation for further voltage reduction and distribution to light commercial and residential users.

THE ELECTRIC POWER SUBSTATION DESIGN DRAWINGS

A *substation* is the part of the electrical transmission system where electricity is switched or transformed from a very high voltage to a conveniently usable form for distribution to homes or businesses. Substation design drawings are an important part of any power supply system.

One-line Diagrams

The *one-line diagram* is a simple way for electrical engineers and drafters to communicate the design of an electrical power substation as shown in Figure 25.13.

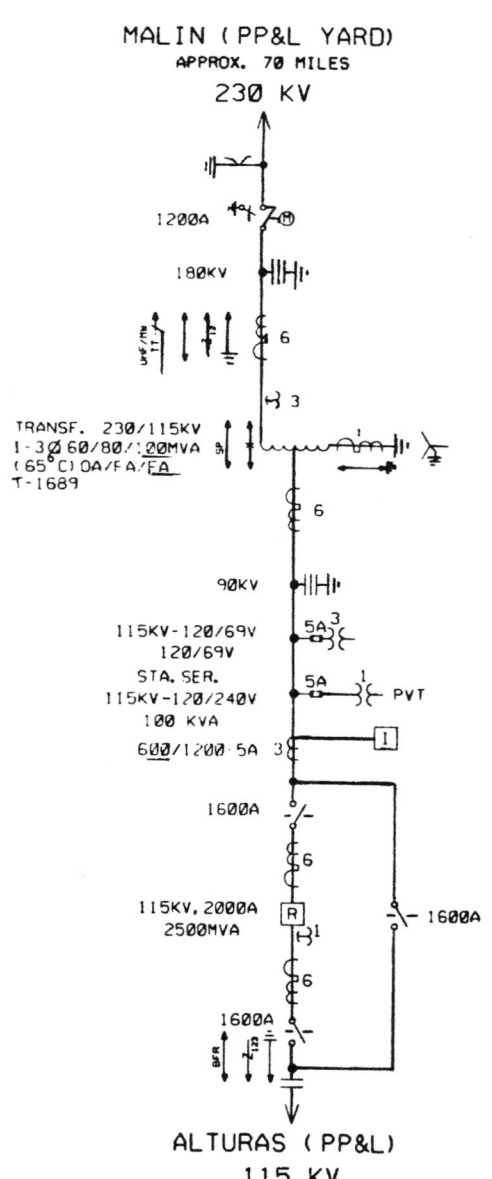

FIGURE 25.13 ■ CADD-generated one-line diagram. *Courtesy Bonneville Power Administration.*

Elementary Diagrams

Elementary diagrams provide the detail necessary for engineering analysis and operation of the substation equipment by operators or maintenance people.

Elementary diagram, DC (direct current) circuits, shows the direct current circuits that operate the relaying and controls for the substation equipment. Elementary diagram, AC (alternating current) circuits, depicts the circuits that provide information for both the protective relays and the instruments and meters used by people who work in the substation.

Schematic Diagrams

The *schematic diagram* is used to show the relationship of the components and equipment located in the substation. The schematic diagram is more detailed than the one-line diagram.

Electric Power Systems Schematic Symbols

ANSI Most standards for the preparation of electrical diagrams use symbols in accordance with the ANSI Y32.2 document. Any special variation to symbols should be shown in a legend or described in a general note. Commonly used electrical equipment symbols are shown in Figure 25.14.

Electrical relays are magnetic switching devices.

Nondirectional relays are relays that operate when current flows in either direction. *Directional relays* operate only when current flows in one direction.

Differential relays provide a switching connection between a circuit with two different voltage values. *Pilot relay* systems are controlled by communication devices and are designated by the type of communication circuit or the function of the relay system.

Supervisory control relays are used to check, monitor, or control other devices.

Plot Plan

The entire layout of a substation is shown by a plot-plan drawing. The *plot plan* is drawn similar to a map, showing the relationship between the elements of the substation in correct orientation to compass direction. The scale used for the plot plan generally ranges from 1" = 20' to 1" = 50'. Typical plot-plan symbols are shown in Figure 25.15. The items that are commonly found on a plot plan include:

1. Property boundaries and developed-yard boundary with chain link fence or block wall.

2. Primary bearing lines. *Bearing* is direction in relation to the Northwest or Northeast and Southwest or Southeast quadrants of a compass. For example, N 50°30'15"W is a line that is located at an angle of 50°30'15" toward West from North.

3. Service and access roads.

4. Buildings and other nonelectric structures.

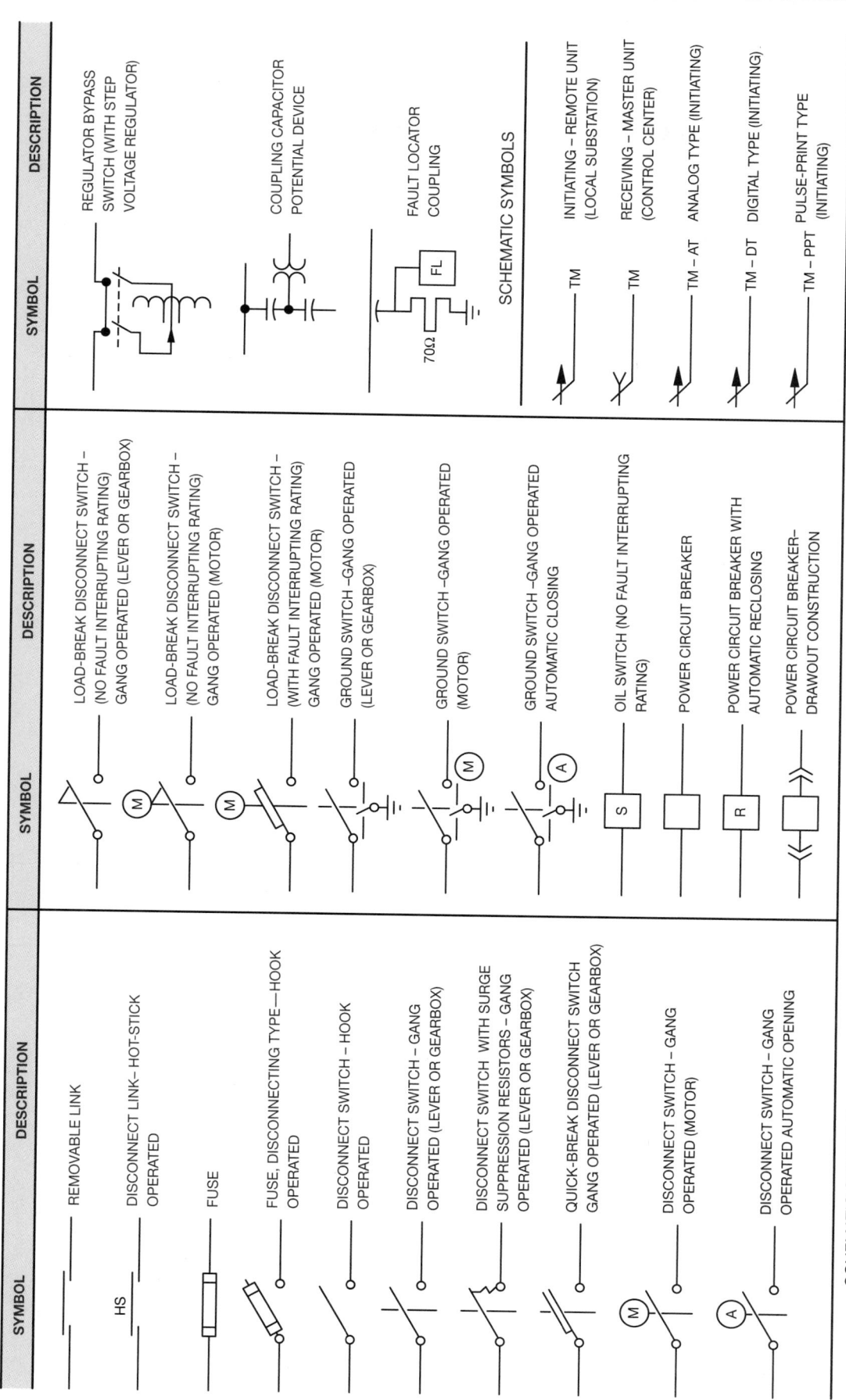

FIGURE 25.14 ■ Electric power systems schematic symbols.

RELAY SYMBOL	RELAY DESCRIPTION
	OVER, NONDIRECTIONAL
	UNDER, NONDIRECTIONAL
	RECEIVING, UNDER, DIRECTIONAL
	OVER, DIRECTIONAL, INITIATING (ARROWS INDICATE TRIPPING DIRECTION)
	RELAY WITH TRIP ATTACHMENT
	GROUND
T	TEMPERATURE
X	DIFFERENTIAL
Z	DISTANCE
C	CURRENT (ASSUMED IF LETTER C OMITTED)
V	VOLTAGE
W	POWER
F	FREQUENCY
	BALANCE
	COMMUNICATION CHANNEL (BASIC)
PW	PILOT WIRE
CC	POWER LINE CARRIER CURRENT
MW	MICROWAVE
VHF	RADIO
UHF	RADIO

FUNCTION SYMBOLS

RELAY SYMBOL	RELAY DESCRIPTION
	PHASE OVERCURRENT
	GROUND OVERCURRENT
	PHASE OVERCURRENT (INSTANTANEOUS)
	GROUND OVERCURRENT (INSTANTANEOUS)
FD	PHASE OVERCURRENT FAULT DETECTOR
BFR	PHASE OVERCURRENT – BREAKER FAILURE
MFD	PHASE OVERCURRENT – MULTIPHASE FAULT DETECTOR
DLF	PHASE OVERCURRENT – DEAD-LINE FAULT
V	OVERVOLTAGE
V	UNDERVOLTAGE
VC	VOLTAGE CONTROL (AUTOMATIC)
HLC	HOT-LINE CHECK (VOLTAGE)
F	OVERFREQUENCY
F	UNDERFREQUENCY
S	SYNCHRONISM-CHECK (FREQUENCY)

NONDIRECTIONAL SYMBOLS

RELAY SYMBOL	RELAY DESCRIPTION
	PHASE OVERCURRENT
	GROUND OVERCURRENT
	PHASE OVERCURRENT (INSTANTANEOUS)
	GROUND OVERCURRENT (INSTANTANEOUS)
G_0	ZERO-SEQUENCE OR POLARIZING OVERCURRENT
G_2	NEGATIVE SEQUENCE OVERCURRENT
W	POWER
Z_1	DISTANCE PHASE – ZONE 1 ONLY
Z_{12}	DISTANCE PHASE – ZONES 1 AND 2
Z_{123}	DISTANCE PHASE – ZONES 1, 2, AND 3
Z_2	DISTANCE PHASE – ZONE 2 ONLY
Z_{23}	DISTANCE PHASE – ZONES 2 AND 3
Z_3	DISTANCE PHASE – ZONE 3 ONLY
Z_{12}	DISTANCE GROUND – (ZONES INDICATED AS ABOVE)
OSB – Z	OUT-OF-STEP BLOCKING
OST – Z	OUT-OF-STEP TRIPPING

DIRECTIONAL SYMBOLS

APPLICATIONS – DIFFERENTIAL RELAYS

RELAY SYMBOL	RELAY DESCRIPTION
X	CURRENT DIFFERENTIAL (BASIC)
VX	VOLTAGE DIFFERENTIAL

APPLICATIONS – MISCELLANEOUS RELAYS

RELAY SYMBOL	RELAY DESCRIPTION
SP	SUDDEN PRESSURE
GP	HIGH GAS PRESSURE
GP	LOW GAS PRESSURE
T	HIGH TEMPERATURE
T	LOW TEMPERATURE
R_2	AUTOMATIC RECLOSING (R IN POWER CIRCUIT BREAKER SYMBOL. NUMBERS INDICATE 1-, 2-, 3-, OR 4- SHOT.)
	CONNECTION
	NONCONNECTION

DIFFERENTIAL AND MISCELLANEOUS SYMBOLS

RELAY SYMBOL	RELAY DESCRIPTION
PW	PILOT WIRE RELAYS
PW-TT	PILOT WIRE TRANSFER TRIP (INITIATING)
PW-TT	PILOT WIRE TRANSFER TRIP (RECEIVING)
CC-DC	CARRIER CURRENT DIRECTIONAL COMPARISON
CC-PC	CARRIER CURRENT PHASE COMPARISON
CC-TT	CARRIER CURRENT TRANSFER TRIP (INITIATING)
CC-TT	CARRIER CURRENT TRANSFER TRIP (RECEIVING)
CC-TT	CARRIER CURRENT TRANSFER TRIP (INITIATING AND RECEIVING)
MW-PC	MICROWAVE PHASE COMPARISON
MW-TT	MICROWAVE TRANSFER TRIP (INITIATING)
MW-TT	MICROWAVE TRANSFER TRIP (RECEIVING)
MW-TT	MICROWAVE TRANSFER TRIP (INITIATING AND RECEIVING)
VHF-TT	VHF RADIO TRANSFER TRIP (INITIATING)
VHF-TT	VHF RADIO TRANSFER TRIP (RECEIVING)
VHF-TT	VHF RADIO TRANSFER TRIP (INITIATING AND RECEIVING)
UHF-TT	UHF RADIO TRANSFER TRIP (INITIATING)
UHF-TT	UHF RADIO TRANSFER TRIP (RECEIVING)
UHF-TT	UHF RADIO TRANSFER TRIP (INITIATING AND RECEIVING)
MW-GD	MICROWAVE GENERATOR DROPPING
MW-LD	MICROWAVE LOAD DROPPING

PILOT RELAYING SCHEMES

RELAY SYMBOL	RELAY DESCRIPTION
SC	INITIATING – MASTER UNIT (CONTROL CENTER)
SC	RECEIVING – REMOTE UNIT (LOCAL SUBSTATION)
	NOTE: SUPERVISORY CONTROL OF EQUIPMENT IS SHOWN WITH A LETTER DENOTING THE MASTER UNIT LOCATION, AND AN EXPLANATORY NOTE IS MADE ON THE DRAWING. AN EXAMPLE FOLLOWS.
SC-D	RECEIVING – FROM DITMTER CENTER (D)
	THE LAST LETTER OR ABBREVIATION DENOTES THE CENTER LOCATION.

SUPERVISORY CONTROL SYMBOLS

FIGURE 25.14 ■ continued

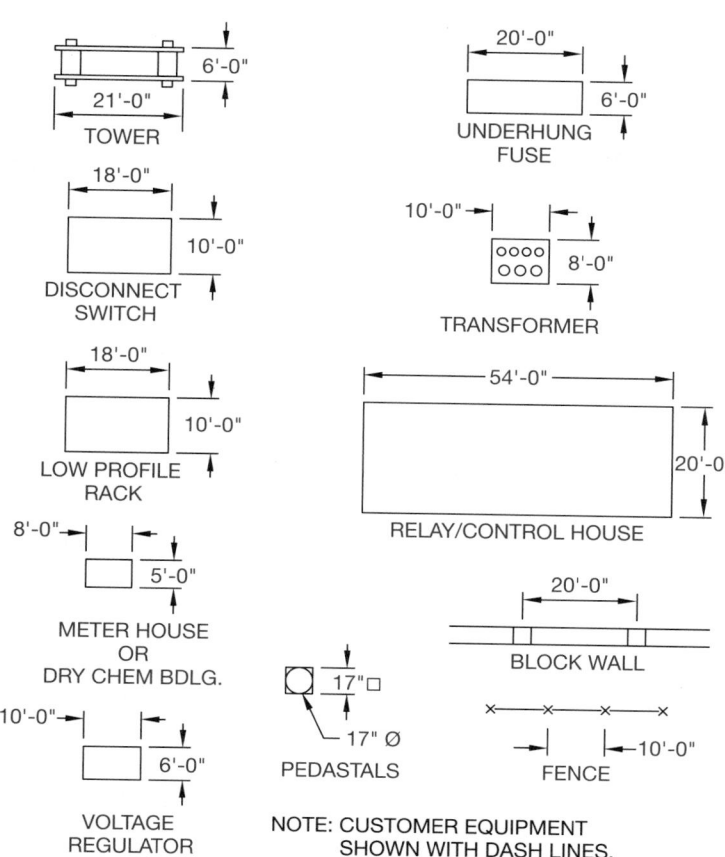

FIGURE 25.15 ■ Plot-plan symbols.

5. System electrical components such as switches, fuses, transformers, and racks.
6. Complete dimensioning.

Bus Layout

A *bus* is an aluminum or copper plate or tubing that carries the electrical current. The *bus layout* drawing is used by construction crews for the construction of the current-carrying portion of the substation. Bus layout drawings show the system in plan (top) view, as shown in Figure 25.16, and elevation (side) view as shown in Figure 25.17.

The components are identified with numbered balloons, which key each item to a bill of materials. Some common bus layout symbols are shown in Figure 25.18.

Grounding Layout and Details

There is a tremendous hazard in substations due to the possibility of high voltage occurring on pieces of equipment during fault conditions. The voltage at different pieces of equipment varies with the location in the yard and the fault current available. The amount of potential electrical shock depends on the voltages available during a fault condition. A *fault condition* is a short cir-

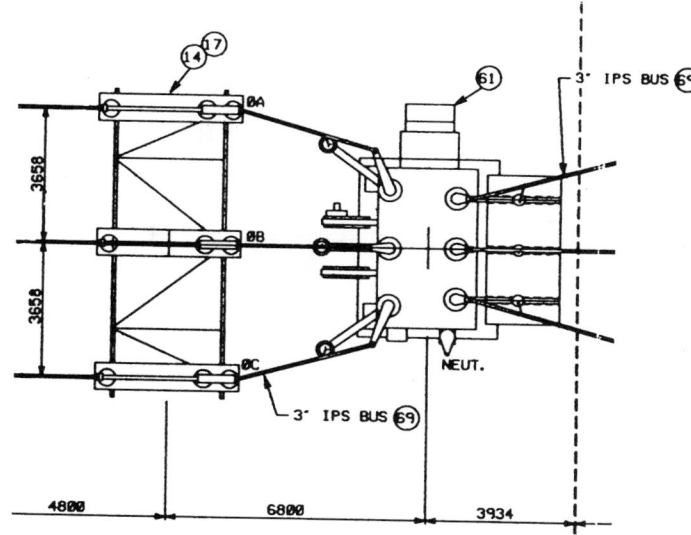

FIGURE 25.16 ■ CADD-generated bus layout plan. *Courtesy Bonneville Power Administration.*

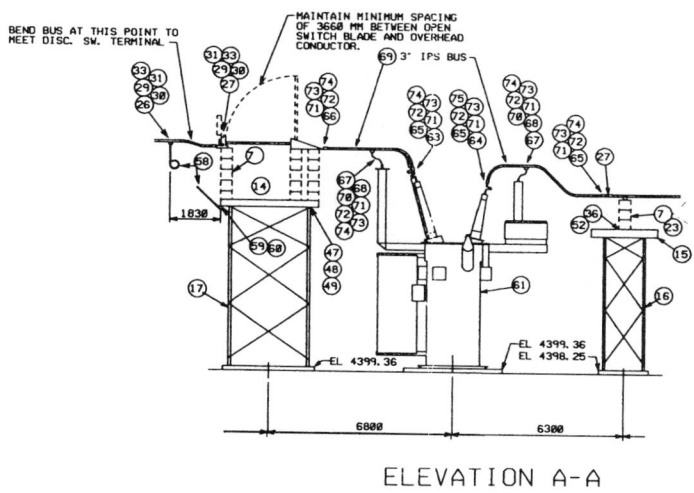

FIGURE 25.17 ■ CADD-generated bus elevations. *Courtesy Bonneville Power Administration.*

cuit, which is a zero resistance path for current flow. The voltage difference between the metal parts of the equipment and the ground surface on which a person stands must be maintained at a very low level. This is determined by the amount and location of the *ground grid,* which is the grounding system. Grounding layouts are often drawn at a scale of 1" = 10'–0" (1:50 metric) as shown in Figure 25.19.

Grounding details provide an enlarged view of how a structure or piece of equipment is grounded. Details are keyed to the grounding layout with letters; for example, "DETAIL H1," as shown in Figure 25.20.

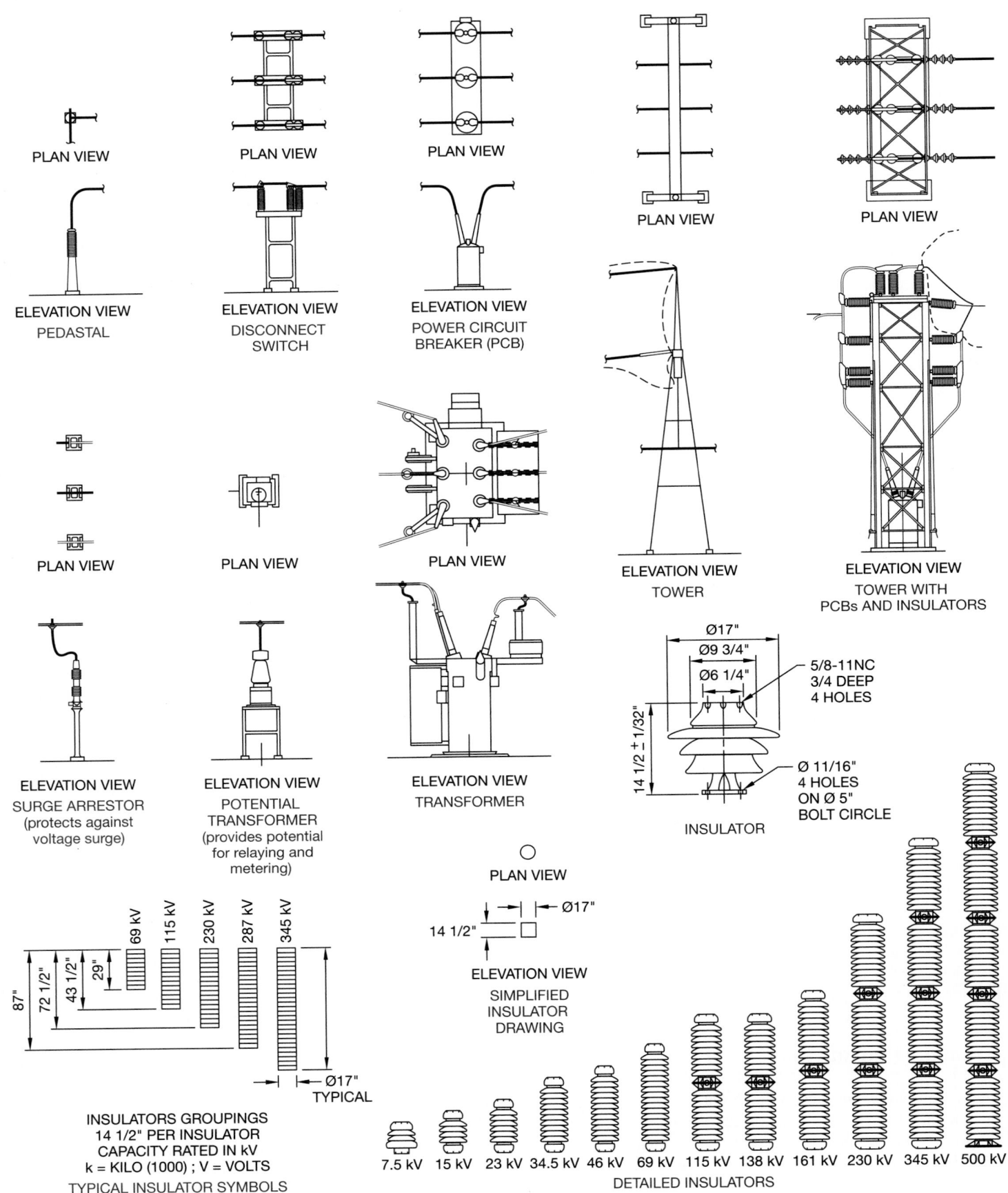

FIGURE 25.18 ■ Common bus layout symbols.

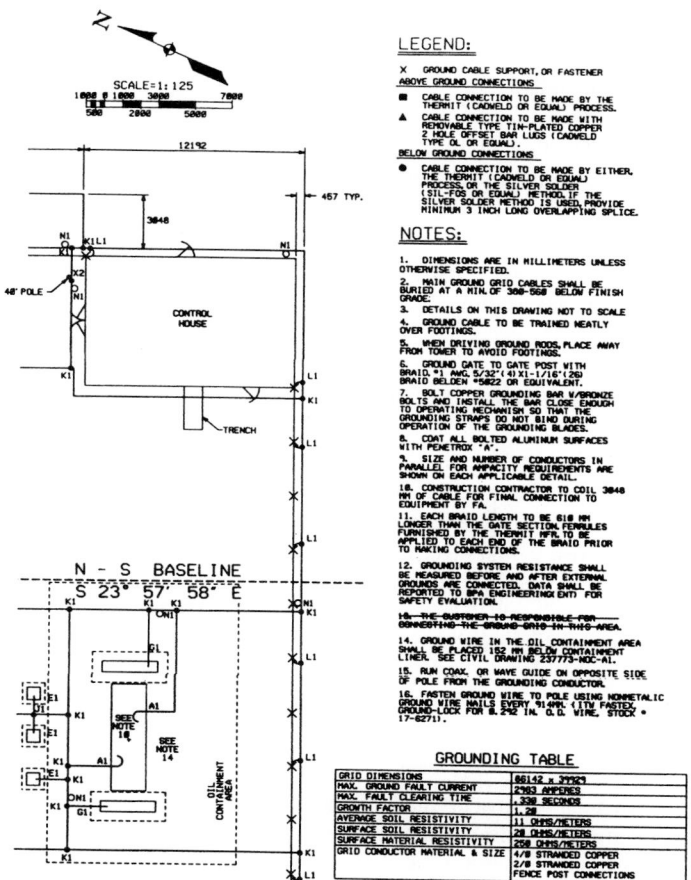

FIGURE 25.19 ■ Grounding layout. *Courtesy Bonneville Power Administration.*

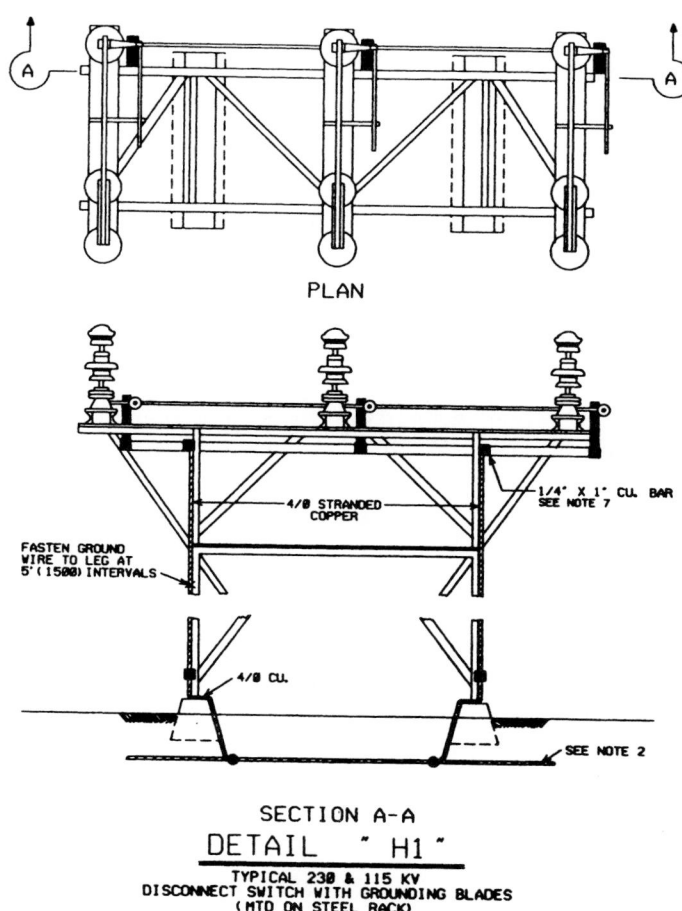

FIGURE 25.20 ■ Grounding details. *Courtesy Bonneville Power Administration.*

Conduit Installation Layout and Details

Each component in a substation has a specific function and is a different type of electrical equipment. The relay control house is a building that interconnects each piece of equipment using multiple conductor cables. (See Figure 25.21.)

Conduit detail drawings coordinate with the layout by providing construction details and the locations of various fittings, junction boxes, and brackets. Details correlate to the layout with callouts that give the detail identification and the page where the detail is located; for example, DETAIL N/2. These details are either drawn at a scale of 1/2" = 1'–0" or may be drawn without scale. (See Figure 25.22.) Various components in the drawing are keyed by balloons to a bill of materials.

RESIDENTIAL AND COMMERCIAL ELECTRICAL PLANS

The design of the electrical system is an important part of the total livability of a home and the function and safety of a commercial or industrial facility. Careful analysis should be given to the design placement of equipment and furniture, and the planned use of each

room. Local and national electrical codes provide specifications for installations and layout. Layout planning should play an important role in conjunction with code guidelines. An evaluation of need and code requirements should be closely compared so the electrical layout is not over- or underdesigned.

Architectural Electrical Symbols

Symbols and lines are used to show the electrical layout in a structure. In residential applications the electrical layout is often part of the floor plan. Commercial electrical plans are commonly an overlay of the base floor-plan sheet. Electrical symbols are generally 1/8 in. (3 mm) in height, where applicable. The electrical plan should be drawn in a clear, concise manner so the layout remains uncluttered. All lettering for switches and other notes should be 1/8 in. high, although a 5/32-in. (4 mm) height is used by some companies. Common electrical symbols are shown in Figure 25.23.

Switch symbols are drawn perpendicular to the wall and are placed to read from the right side or bottom of the sheet. The switch relay dashed line should intersect the symbol at right angles to the wall or the relay may begin next to the symbol. Do not mix

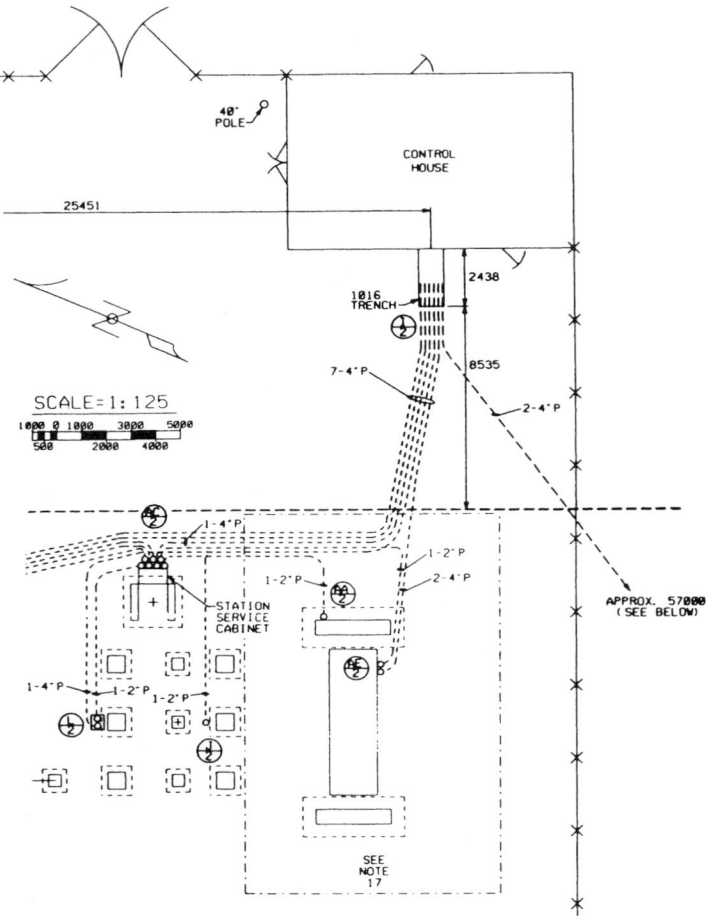

FIGURE 25.21 ■ Conduit installation layout drawing. *Courtesy Bonneville Power Administration.*

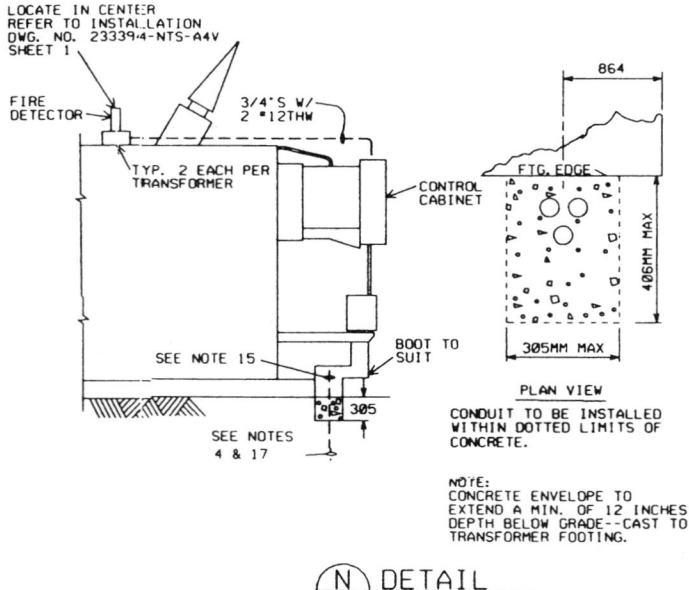

FIGURE 25.22 ■ Conduit detail. *Courtesy Bonneville Power Administration.*

methods. Verify the preferred procedure with your instructor or employer. (See Figure 25.24.) Figure 25.25 shows several typical electrical installations with switches to light outlets.

When special characteristics are required, such as a specific size fixture, a location requirement, or any other specification, a local note may be applied next to the electrical symbol to briefly describe the situation. (See Figure 25.26.) Where a specification affects electrical installations on the entire layout, general notes may be used. Some common errors related to the placement and practice of electrical floor-plan layout are shown in Figure 25.27.

Residential Electrical Plan Examples

Residential electrical plans are generally less complex than commercial plans. Figure 25.28 shows some maximum spacing recommendations for installation of wall outlets in a residential structure. A typical residential electrical layout is shown in Figure 25.29. A typical kitchen layout is shown in Figure 25.30. Refer to *Architectural Drafting and Design* by Jefferis and Madsen, Delmar Publishers, for complete architectural applications.

Commercial Electrical Plan Examples

Commercial electrical plans often follow much more detailed installation guidelines than residential applications. Electrical plan overlays are commonly used so the only information provided is that of electrical installations.

The electrical circuit switch legs for commercial applications are generally drawn as solid lines rather than dashed, as in residential electrical plans. The electrical circuit lines that continue from an installation to the service distribution panel are terminated next to the fixture and capped with an arrowhead, meaning that the circuit continues to the distribution panel. When multiple arrowheads are shown, this indicates the number of circuits in the electrical run. (See Figure 25.31.) For many installations where a number of circuit wires are used, the number of wires is indicated by slash marks placed in the circuit run. The number of slash marks equals the number of wires. (See Figure 25.32.)

There may be more than one commercial electrical overlay; for example, floor-plan lighting, electrical-plan power supplies, or reflected ceiling plan. The *floor-plan lighting layout* provides the location and identification of lighting fixtures and circuits. It is usually coordinated with a *lighting fixture schedule* as shown in Figure 25.33. In some applications a *power supply plan* is used to show all electrical outlets, junction boxes, and related circuits. (See Figure 25.34.) The *reflected ceiling plan* is used to show the layout for the suspended ceiling system as shown in Figure 25.35 (see page 898). Electrical plans for equipment installations may also be needed to supplement the power supply and lighting plans. Figure 25.36 (see page 898) shows the plan for roof installation of equipment. The drafter is also often required to draw schematic diagrams for specific electrical installations as shown in Figure 25.37 (see page 898).

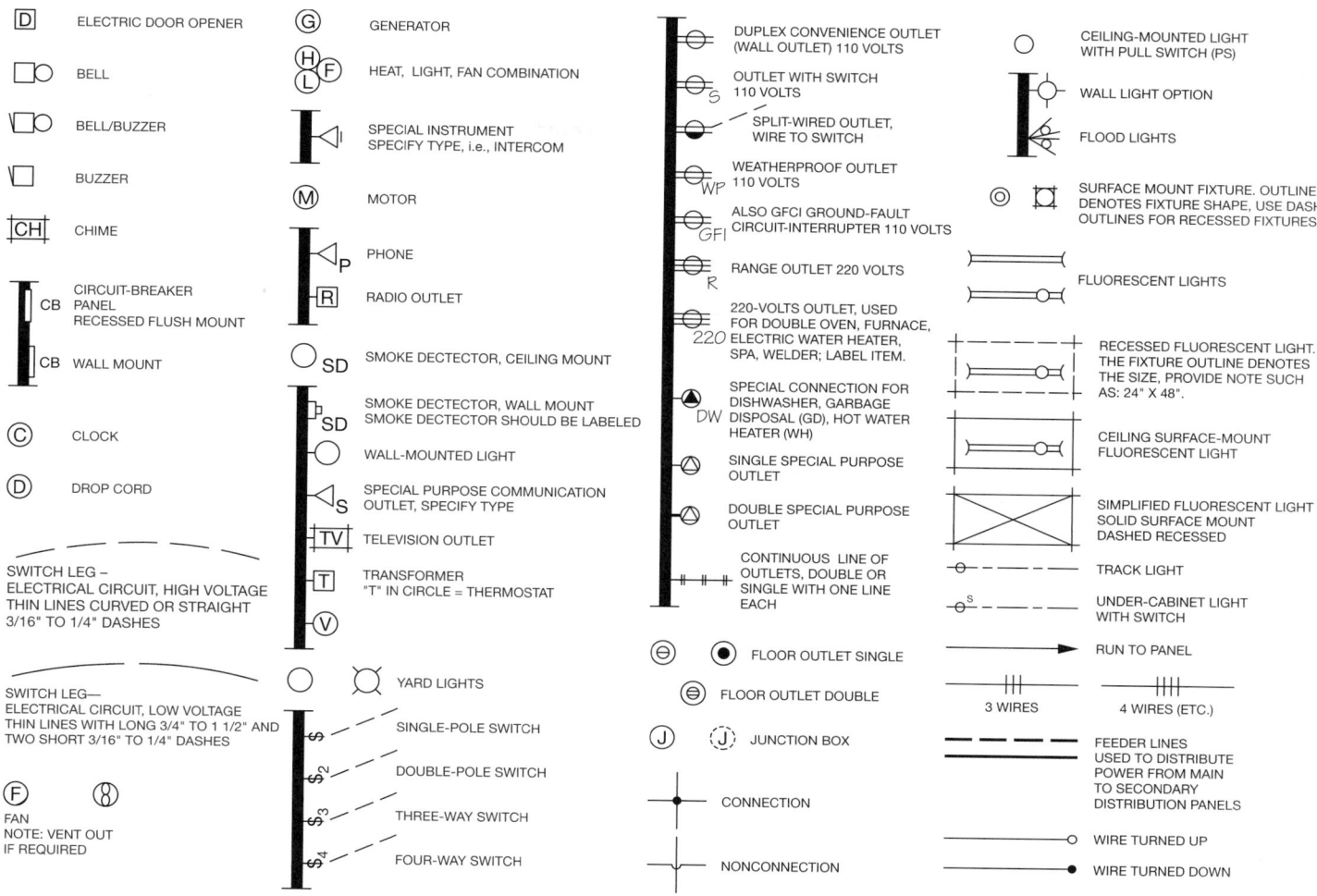

FIGURE 25.23 ■ Common floor-plan electrical symbols.

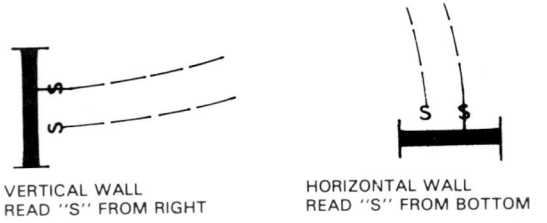

VERTICAL WALL
READ "S" FROM RIGHT

HORIZONTAL WALL
READ "S" FROM BOTTOM

FIGURE 25.24 ■ Switch symbol placement.

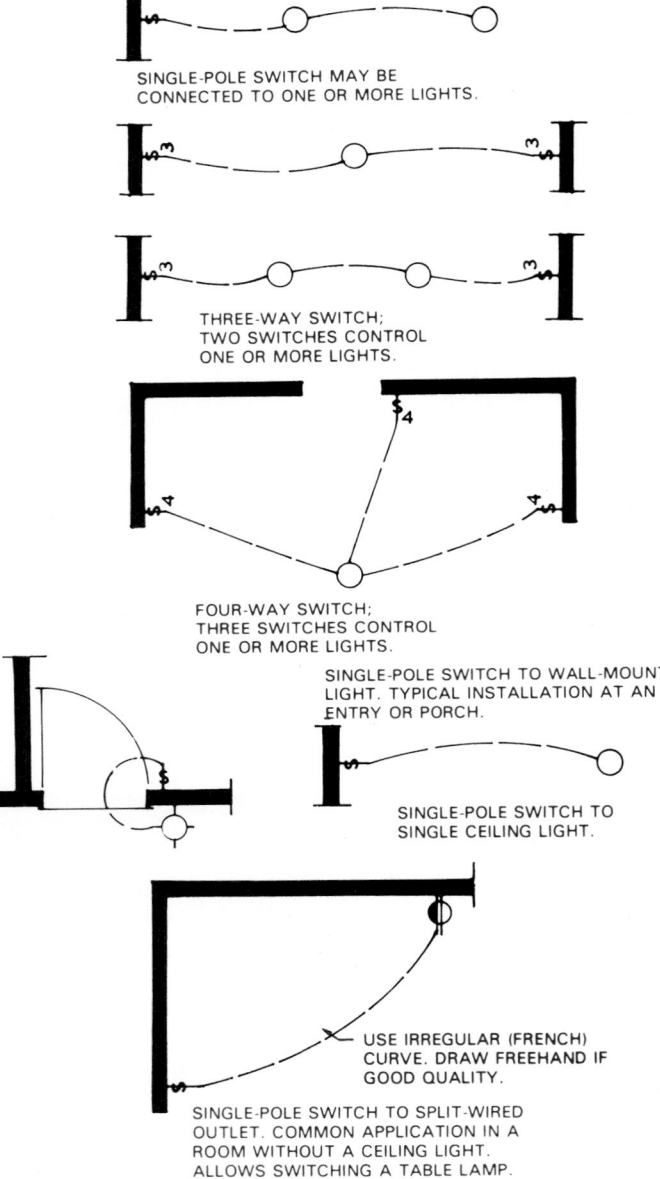

SINGLE-POLE SWITCH MAY BE
CONNECTED TO ONE OR MORE LIGHTS.

THREE-WAY SWITCH;
TWO SWITCHES CONTROL
ONE OR MORE LIGHTS.

FOUR-WAY SWITCH;
THREE SWITCHES CONTROL
ONE OR MORE LIGHTS.

SINGLE-POLE SWITCH TO WALL-MOUNTED
LIGHT. TYPICAL INSTALLATION AT AN
ENTRY OR PORCH.

SINGLE-POLE SWITCH TO
SINGLE CEILING LIGHT.

USE IRREGULAR (FRENCH)
CURVE. DRAW FREEHAND IF
GOOD QUALITY.

SINGLE-POLE SWITCH TO SPLIT-WIRED
OUTLET. COMMON APPLICATION IN A
ROOM WITHOUT A CEILING LIGHT.
ALLOWS SWITCHING A TABLE LAMP.

FIGURE 25.25 ■ Typical electrical installations.

FIXTURE SIZE   OUTLET HEIGHT   VENT TO OUTSIDE AIR / TIMED

FIGURE 25.26 ■ Special notes for electrical fixtures.

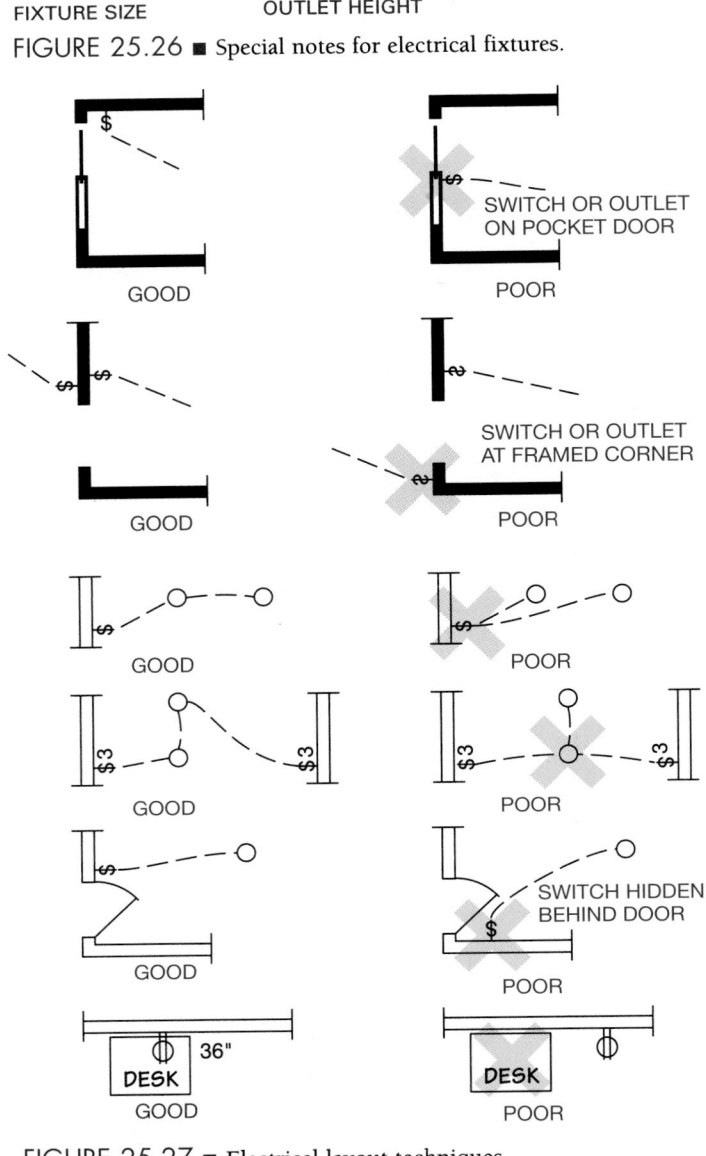

GOOD   SWITCH OR OUTLET ON POCKET DOOR / POOR

GOOD   SWITCH OR OUTLET AT FRAMED CORNER / POOR

GOOD   POOR

GOOD   POOR

GOOD   SWITCH HIDDEN BEHIND DOOR / POOR

DESK 36"
GOOD   DESK / POOR

FIGURE 25.27 ■ Electrical layout techniques.

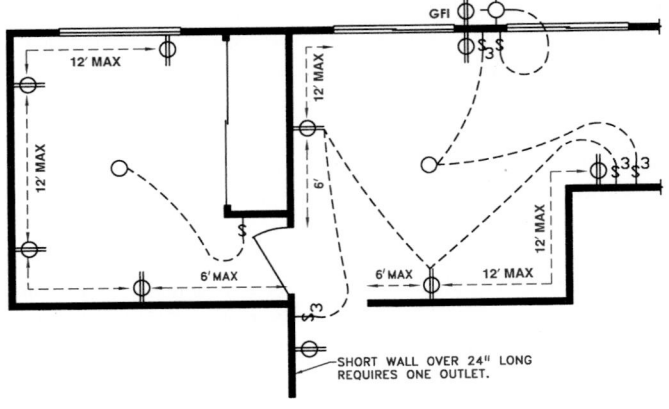

SHORT WALL OVER 24" LONG
REQUIRES ONE OUTLET.

FIGURE 25.28 ■ Maximum spacing requirements.

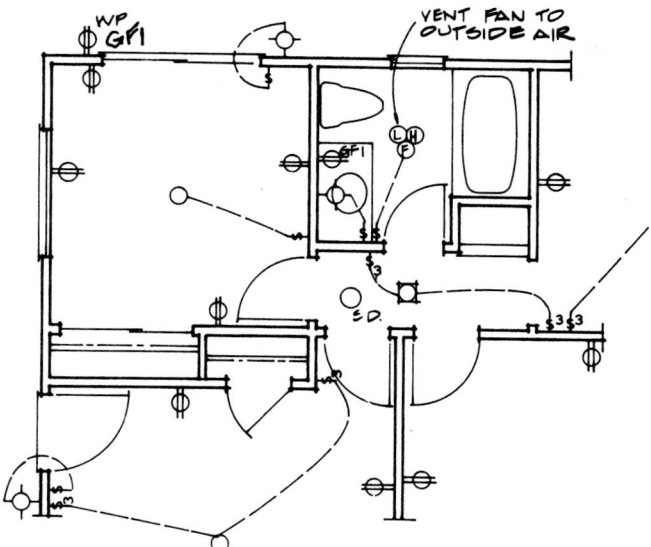

FIGURE 25.29 ■ Typical electrical layout.

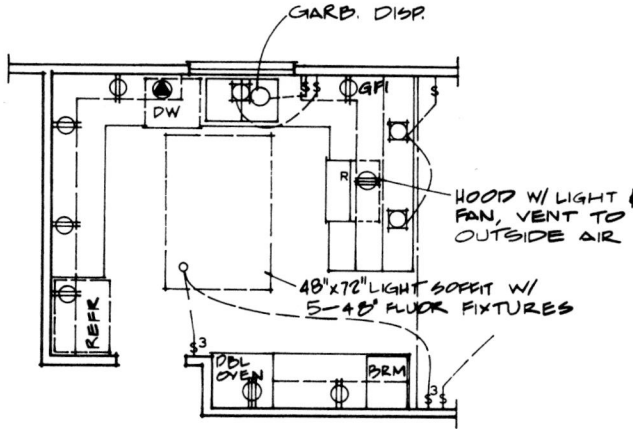

FIGURE 25.30 ■ Kitchen electrical layout.

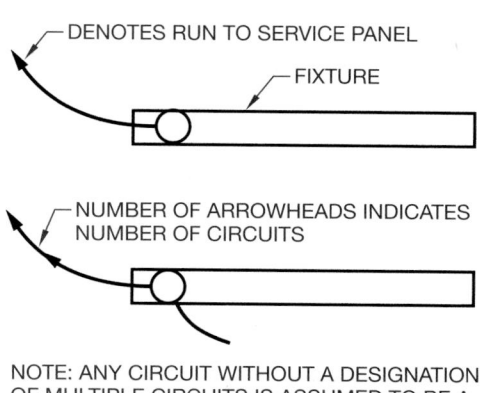

NOTE: ANY CIRCUIT WITHOUT A DESIGNATION OF MULTIPLE CIRCUITS IS ASSUMED TO BE A TWO-WIRE CIRCUIT.

FIGURE 25.31 ■ Electrical circuit designations.

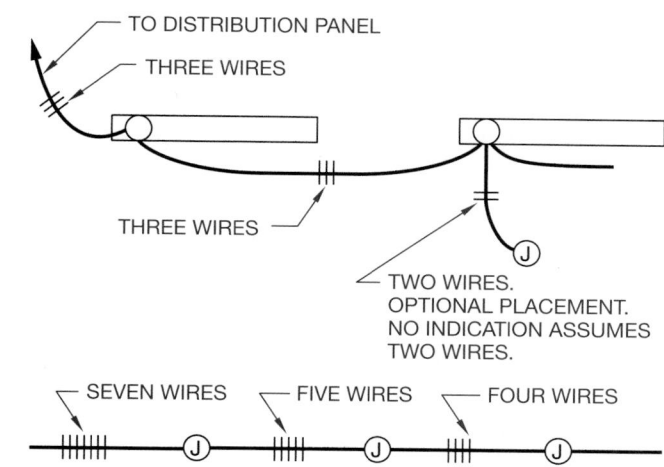

FIGURE 25.32 ■ Number of wires designated in circuit runs.

LIGHTING FIXTURE SCHEDULE

F1 – Surface mounted 8' open strip fluorescent. Lamps: (1) F96T12/LW/WM (75 watt). Manufacturer: Lithonia PUN 196 – 120V

F2 – Surface mounted 8' open strip fluorescent with damp location label and low temperature ballast. Lamps: (1) F96T12/LW/WM (75 watt). Manufacturer: Lithonia PUN 196 – DL – 120V

F3 – Surface mounted 4' open strip fluorescent. Lamps: (2) F48T12/LW/WM (30 watt). Manufacturer: Lithonia PUN 248 – 120V

F4 – Surface ceiling mounted vapor-tight incandescent with cast guard. Lamp: (1) 100W A19 Manufacturer: Steneo QVCXL—11GC

F5 – Surface mounted incandescent with prismatic lexan cylinder and dump location label. Lamp: (1) 100W A19 Manufacturer: Marco QB5NP – SA

F6 – Surface well mounted sodium vapor security flood light. Lexan lens and weather tight. Lamp: 70w Manufacturer: Crousshinds Sc – 711 – 70W HPS

F7 – Recessed ceiling mounted incandescent fan/light combination. Lamp: 10w Manufacturer: Broan Q678

F8 – 16' pole mounted sodium vapor flood area luminaire. Type III distribution flat lens. Bronze finish. Pole to be 16' straight square steel. Coated with paint to match fixture. See detail. Lamp: LU150 – 55 Manufacturer: ELSCO ZCHL – 150 – MPS – 16 – DP – 120 – B2A Alternate: Nu-Art QULT – III – MPS – 150

FIGURE 25.33 ■ Lighting needs are shown using an overlay of the floor plan and a lighting fixture schedule. *Courtesy The Southland Corporation.*

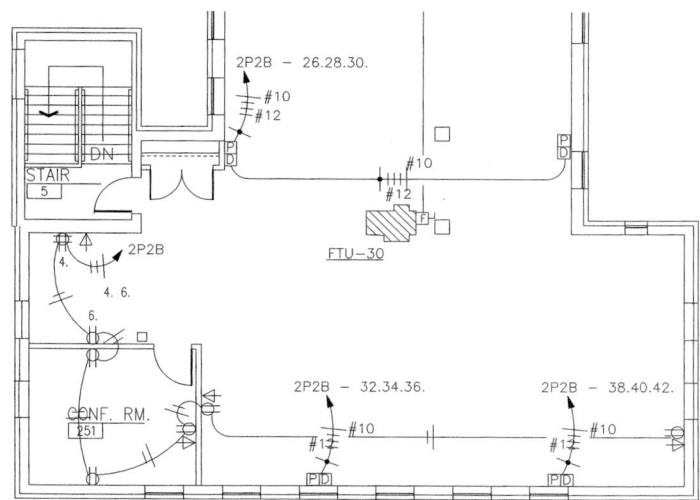

FIGURE 25.34 ■ Power supply plan. *Courtesy System Design Consultants.*

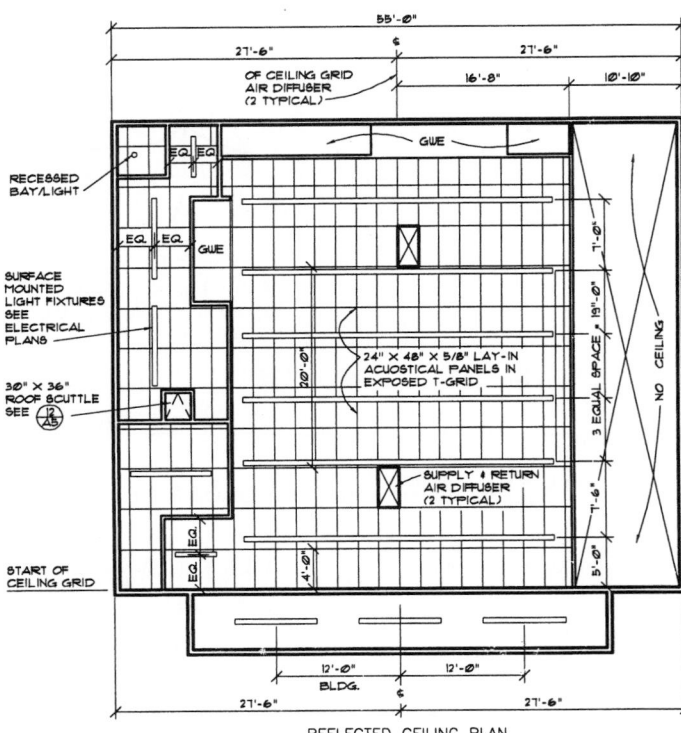

FIGURE 25.35 ■ A reflected ceiling plan to show the suspended ceiling layout. *Courtesy The Southland Corporation.*

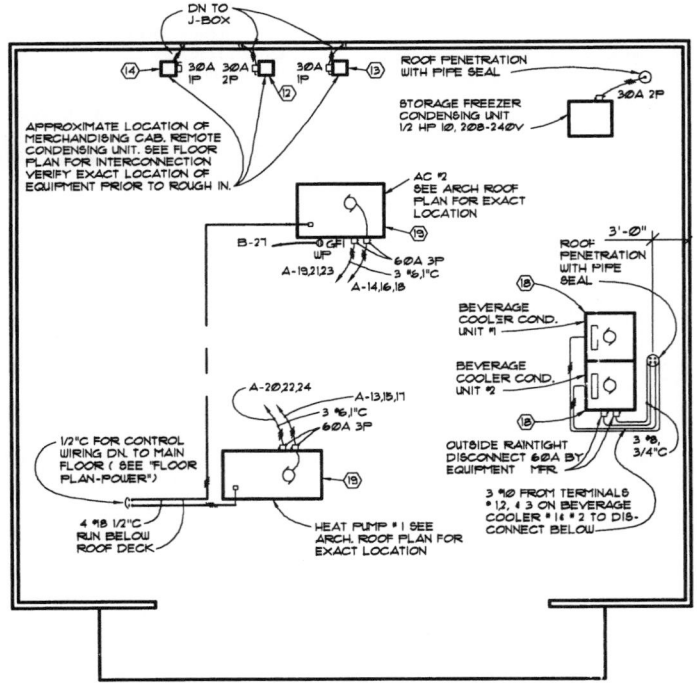

FIGURE 25.36 ■ To supplement the power supply and lighting plans, a plan showing the electrical needs of equipment on the roof is also drawn. *Courtesy The Southland Corporation.*

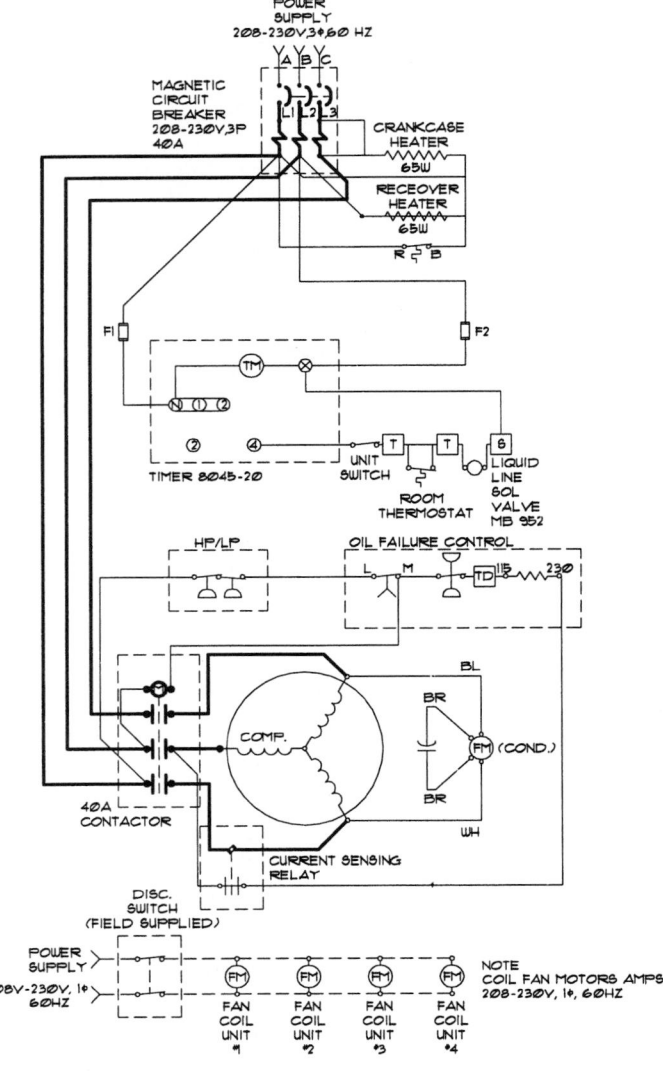

FIGURE 25.37 ■ Schematic wiring diagrams for specific applications. *Courtesy The Southland Corporation.*

ELECTRONIC SCHEMATIC DRAFTING

Electronics is the control of electrons for use in devices that are dependent on low voltage, amperage, and signal paths. Electronic drafters are typically referred to as technicians in the electronics industry. The technician's responsibility is to convert engineering sketches or instructions to formal drawings, or to revise existing drawings. This task dictates that drawings be prepared in a neat, organized manner using proper symbols. It is advisable for technicians to become familiar with the fundamentals of electronic technology, and as experience is gained in the industry, in-depth knowledge of product function is important for advancement into design and engineering.

ELECTRONIC DIAGRAMS

Many of the same types of diagrams used in electrical drafting are used in electronic drafting. There is only a slight difference between the appearance of electrical and electronic drawings. This text separates the two areas due to the specialized difference that exists between electrical power transmission systems and intricate electronic systems.

Block Diagrams

A *block diagram* outlines the path of a signal through a series of "steps" or "operations." The purpose of these drawings is to provide a quick interpretation of the relationship between the different components in electronic equipment. The steps or operations are shown in "blocks," which summarize a schematic drawing by omitting the details. A square or rectangular block is the usual basic symbol used in block diagrams. The size of the block is determined by the largest title going into the block, such as CONVERTER or AMPLIFIER. The block diagram should start in the upper left corner of the drawing and the components read from left to right, as shown in Figure 25.38. Graphic symbols are often used to demonstrate input and output devices, such as speakers, microphones, or antennas, as shown in Figure 25.39.

Technicians may be given a schematic diagram where the engineer has clearly defined the components to be placed in the block diagram. Designers in this field are able to determine the block diagram elements directly from a given schematic diagram. Figure 25.40 shows a block diagram prepared from a schematic diagram. In some situations, the drafter may prepare a block diagram from a pictorial diagram, as shown in Figure 25.41. Separate or optional components may be represented with a dashed line.

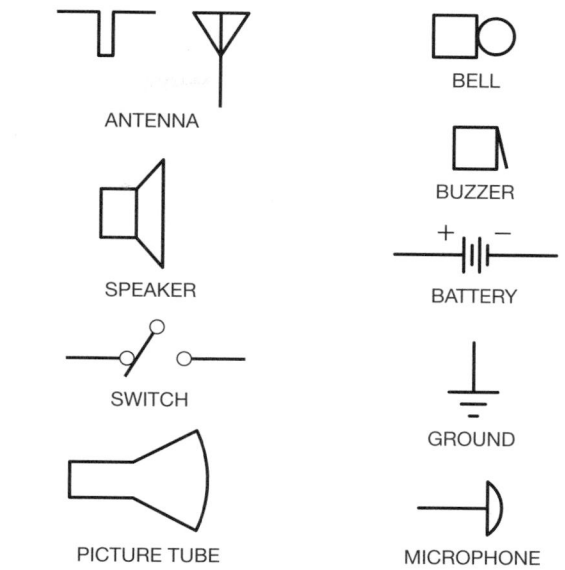

FIGURE 25.39 ■ Graphic symbols for block diagrams.

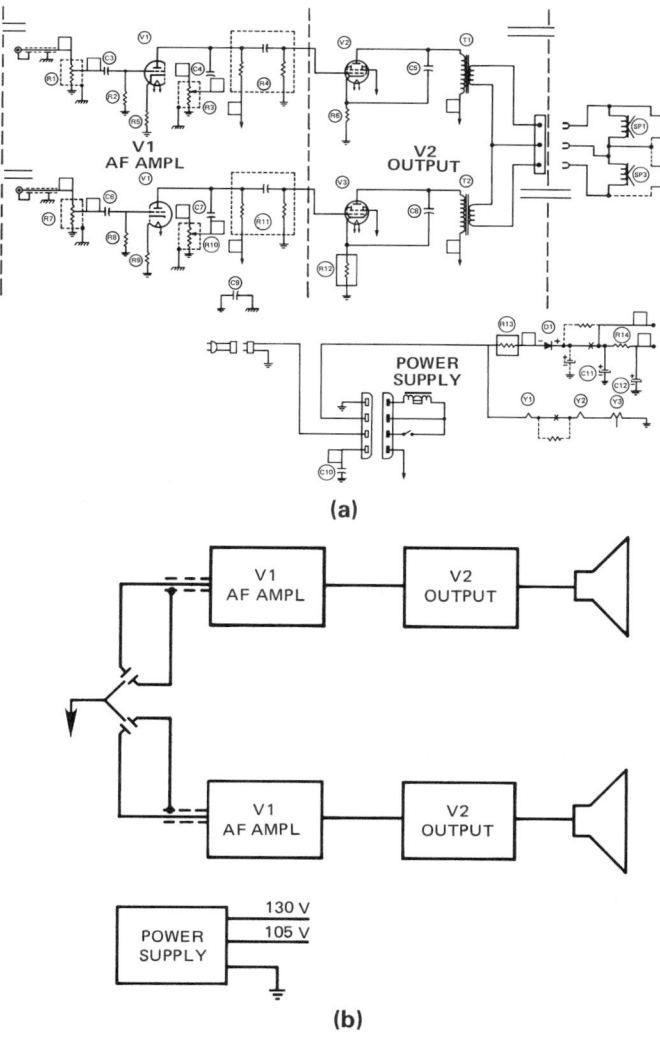

(a)

(b)

FIGURE 25.40 ■ (a) Schematic diagram is converted into (b) a block diagram.

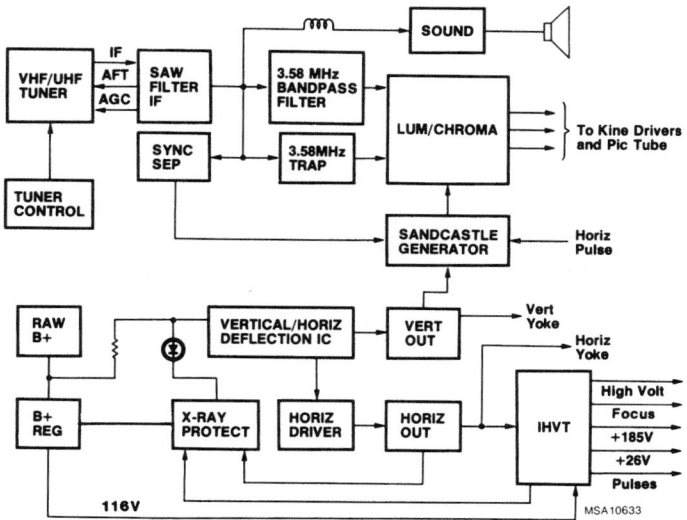

FIGURE 25.38 ■ Block diagram. *Reproduced by permission of RCA Consumer Electronics.*

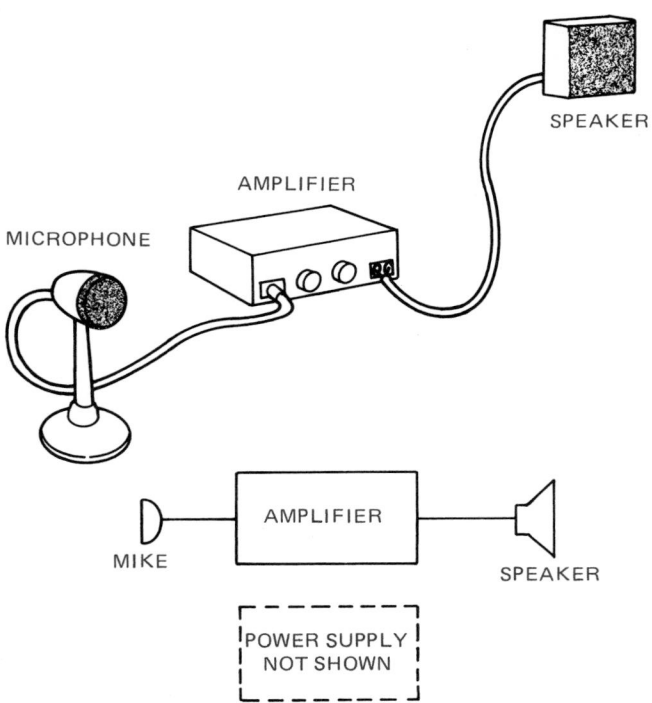

FIGURE 25.41 ■ Block diagram from a given pictorial drawing and optional components shown with dashed lines.

Schematic Diagrams

Preparation of a schematic diagram is typically the first stage of product development. *Schematic diagrams* provide the basic circuit connection information for electronic products. Components of the electronic system are often drawn in stages that represent the function of the device. Individual components are labeled with "reference designations," values, and suppliers identification. Reference designations tie the component directly with the schematic drawing. A reference designation is a letter (or letters) identifying a component, as follows:

Component	Reference Designation
Capacitor	C
Inductor	L
Resistor	R
Diode	D
Transistor	Q
Transformer	T
Switch	S
Semiconductor	CR
Integrated circuit	U

The drafter should know the product well enough to prepare schematics that are uncluttered and easy to read, follow standards, fit the size and shape of the product, and are technically correct. The following discussion provides enough electronic information to help the beginning drafter prepare a schematic.

Basic Electronic Symbols

The symbol and a brief definition of the following components are provided only for general information. A full understanding of function requires technical training.

Capacitor—C. A simple *capacitor* consists of two metal plates with wire connectors separated by an insulator. A capacitor in an electronic circuit opposes a change in voltage and stores electronic charge. (See Figure 25.42.)

Coil or Inductor—L. A *coil* or *inductor* is a conductor wound on a form or in a spiral that has inductive properties. *Inductance* is the property in an electronic circuit that opposes a change in current flow or where energy may be stored in a magnetic field, as in a transformer. (See Figure 25.43.)

Resistor—R. *Resistors* are components that resist the flow of electricity. They are used in applications where the circuit requires protection by reducing the flow of current. Resistors are available in fixed or variable values. Variable resistors allow the user to adjust the resistance. An example of a variable resistor is the volume control of a radio. The symbol is shown in Figure 25.44.

Semiconductor—Q, D, or CR. A simplified definition of a *semiconductor* is a device that provides a degree of resistance in an electronic circuit. There are various kinds of semiconductors used in electronics. Specific types of semiconductors are diodes, transistors, and others that are discussed later. Under certain conditions, these devices allow the current to pass through them freely, and under other conditions, they block the flow of current. The simplest of these devices is the *diode*. (See Figure 25.45.) The symbols picture the basic characteristics of the device. The arrow points in the direction in which conventional current flows. The arrow and bar are present in some form in almost all semiconductor symbols. The bar end of the diode is referred to as the *cathode* or negative side. The direction the symbol arrow points is important to the drafter and may be the only difference in the symbol for two different components. Figure 25.45 is not a complete illustration of semiconductor devices, but it does show how slight changes are used to indicate different components.

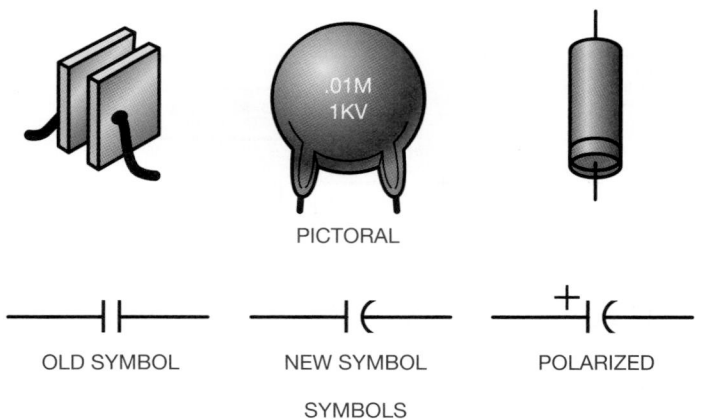

FIGURE 25.42 ■ Capacitor symbols.

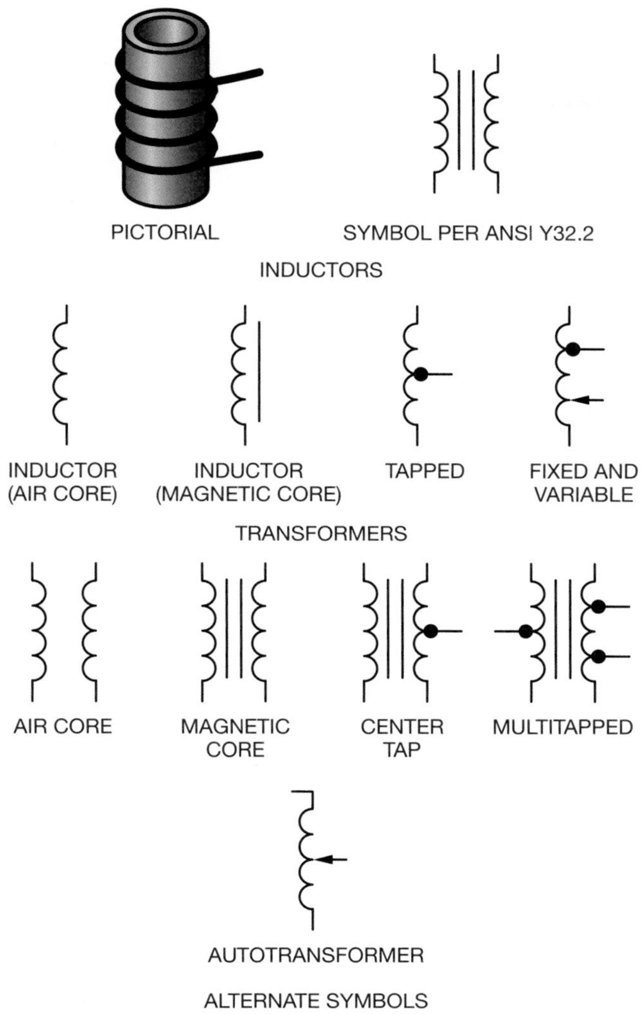

PICTORIAL   SYMBOL PER ANSI Y32.2

INDUCTORS

INDUCTOR   INDUCTOR   TAPPED   FIXED AND
(AIR CORE)   (MAGNETIC CORE)   VARIABLE

TRANSFORMERS

AIR CORE   MAGNETIC   CENTER   MULTITAPPED
                    CORE   TAP

AUTOTRANSFORMER

ALTERNATE SYMBOLS

FIGURE 25.43 ■ Coil (inductor) symbols.

PICTORIALS

NPN TRANSISTOR   PNP TRANSISTOR

JFET   JFET   UNIJUNCTION
UNIJUNCTION   FIELD-EFFECT   TRANSISTOR
(P-TYPE BASE)   (N-TYPE BASE)   (N-TYPE)

TRANSISTOR SYMBOLS

DIODE   DIODE   SILICON-CONTROLLED
SYMBOL   RECTIFIER

BRIDGE   INTEGRATED
RECTIFIER   CIRCUIT

LOGIC IC SYMBOLS

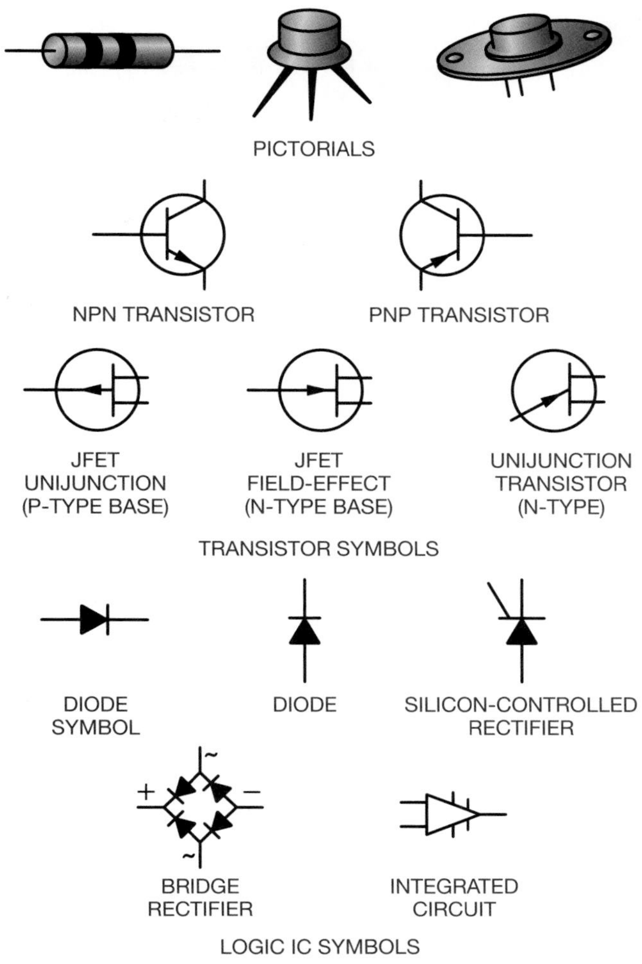

FIGURE 25.45 ■ Semiconductor symbols.

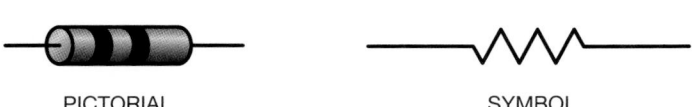

PICTORIAL   SYMBOL

FIGURE 25.44 ■ Resistor symbol.

Symbol Variations

An arrow through or to a symbol changes the component from a *fixed-value* device to a *variable-value* device. (See Figure 25.46.) A screwdriver symbol added to a schematic tells the technician that the component has a variable value by service adjustment, as shown in Figure 25.47.

Another group of symbol variations exists in coils, as shown in Figure 25.48.

Symbol Sizes

The actual size of symbols may vary slightly from one company or computer-aided design software to another. The uniformity of sym-

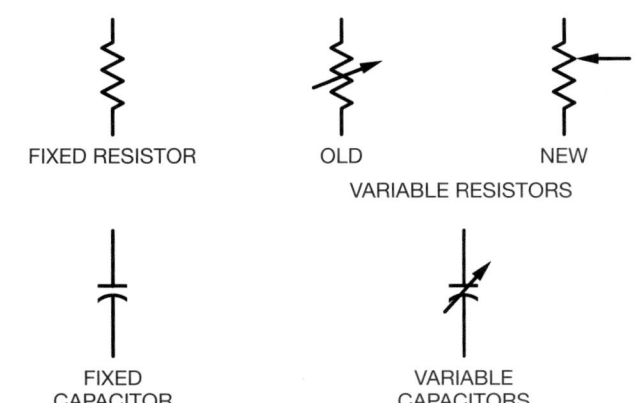

FIXED RESISTOR   OLD   NEW

VARIABLE RESISTORS

FIXED   VARIABLE
CAPACITOR   CAPACITORS

FIGURE 25.46 ■ Fixed and variable component symbols.

bols within a drawing is important. Figure 25.49 shows the recommended sizes of commonly used electronics symbols based on the drawing text height.

As an electronic drafter you may be working from rough engineering sketches.

REDUCE A SCREWDRIVER FROM

THIS: TO THIS:

AND ADD IT TO A BASIC SYMBOL SUCH AS:

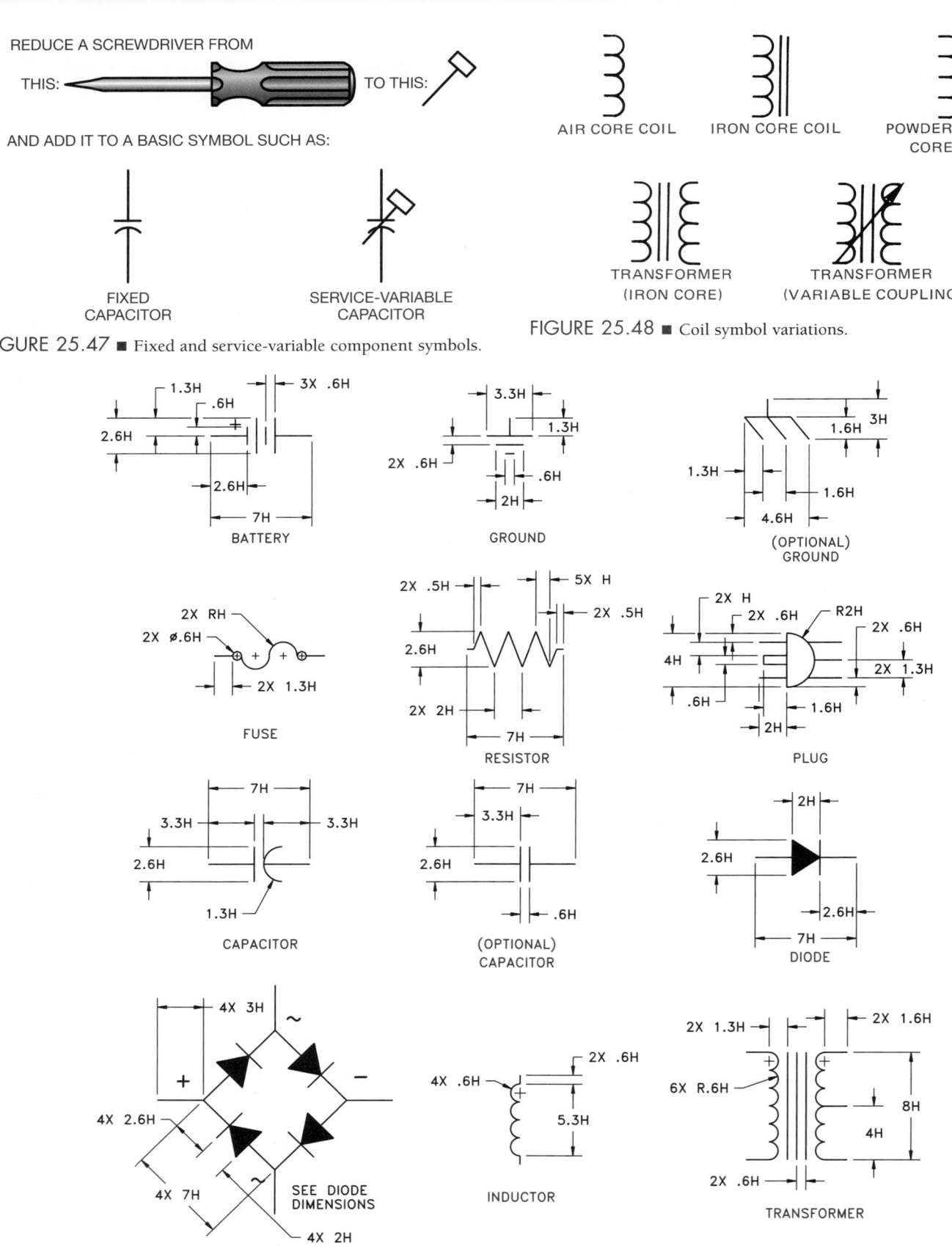

FIXED
CAPACITOR

SERVICE-VARIABLE
CAPACITOR

FIGURE 25.47 ■ Fixed and service-variable component symbols.

AIR CORE COIL   IRON CORE COIL   POWDERED IRON
CORE COIL

TRANSFORMER
(IRON CORE)

TRANSFORMER
(VARIABLE COUPLING)

FIGURE 25.48 ■ Coil symbol variations.

BATTERY

GROUND

(OPTIONAL)
GROUND

FUSE

RESISTOR

PLUG

CAPACITOR

(OPTIONAL)
CAPACITOR

DIODE

FULL BRIDGE RECTIFIER

INDUCTOR

H = LETTERING HEIGHT

TRANSFORMER

FIGURE 25.49 ■ Recommended sizes of commonly used electronics symbols based on drawing text height.

ANSI You need to be very familiar with standard symbols because engineers usually have operational problems on their minds and not the quality of drawings and sketches. Keep a copy of ANSI Y32.2 or the company standards handy for reference. When CADD is used, a symbol library or a template designed to ANSI standards is an excellent drafting aid. Standards are commonly maintained by the company symbols librarian.

Here are some sample potential communication problems:

1. Know current standards so if the symbol —||— is on a sketch, you know it is an old symbol of an electrical application for a capacitor.

2. Question symbols with odd details, such as the one in Figure 25.50. Does the line across the bottom of the resistor mean something special or was the engineer's pencil resting on the paper? In this case it could mean either, but be careful, as it could symbolize a variable resistor with a stop on the adjustment that will not allow it to go all the way to zero. If in doubt, ask the engineer.

Correct and Incorrect Applications for Schematic Diagrams

There are a variety of electronic drafting techniques that are more acceptable than others. While there are some methods that the beginning drafter should learn, there may be some situations when common applications may change due to the given situation or a difference between national and company standards. The examples shown in Figure 25.51 are some samples of recommended correct and incorrect applications.

Active Devices

An *active device* is an electronic component that contains voltage or current sources, such as a transistor and integrated circuit. *Transistors* are semiconductor devices in that they are conductors of electricity with resistance to electron flow, and are used to transfer or amplify an electronic signal. The key to the schematic layout is the location of the active devices. When schematics are designed, the active devices are often established first. Follow the engineering sketch as a guide for overall size and layout balance, and begin the schematic diagram by positioning the active devices. Inputs are usually on the left, and outputs on the right. Signal flow is across the device from left to right.

Bias Circuitry

Bias is the voltage applied to a circuit element to control the mode of operation. Bias may be fixed, forward, or reversed. The basic cir-

FIGURE 25.50 ▪ Confusing symbol sketch.

cuitry that makes any electronic device function is known as the *DC biasing*. The configuration of this basic circuit is resistors that are connected to an active device. Resistors are components that maintain resistance to the flow of an electric current. The resistors should be placed as close as possible to the active device without crowding or otherwise destroying the balance of the drawing. Resistors are normally placed vertically in the schematic, as shown in Figure 25.52. Transistors may be drawn one of two basic ways, depending on the bias direction by changing the arrow direction, as shown in Figure 25.53. After the active devices have been established on the drawing sheet, the next step is to add the resistors. There are a few common bias arrangements for transistor circuits, as shown in Figure 25.54 (see page 814).

Coupling Circuitry

After the bias circuitry has been established, the next step is generally the layout of the *coupling* circuitry. This is the point where each schematic becomes unique. The drafter's goal is to provide adequate space for all of the information while keeping the schematic compact and readable.

Making a Schematic Layout Sketch

Now that some fundamental guidelines have been provided for the schematic diagram layout, it is time to make a layout sketch. Making this sketch on graph paper helps in estimating the space required for component labels and values. The drafter often works from an engineer's design sketch. This sketch may be rough, but gives the layout characteristics, as shown in Figure 25.55. The engineering sketch may be difficult to read and interpret. For example, notice in Figure 25.55 the screened area where the lines cross. These crossing lines represent a question that you should answer. Is there a connection at this point or do the lines cross without making a connection? In this case, the resistors are part of the universal bias pattern as previously discussed, so you can conclude that a connection is intended. The standard representations for connections and nonconnections are shown in Figure 25.56. If there is any doubt about how to interpret the engineer's sketch, ask the engineer.

Now, make a layout sketch on graph paper, as shown in Figure 25.57. The blocked out areas next to the symbols represent intended component labeling.

There is more than one correct way to lay out a schematic. The same schematic with an optional layout is shown in Figure 25.58. The basic difference between this layout and the previous one is that the components attached to the top side of the active devices have been "folded down," as shown in Figure 25.59, page 906. The result of this folded-down form is a long, narrow schematic. This layout can be used to an advantage where the height of the schematic is fixed but the length can vary, as in fold-out pages in instruction manuals. This form gives the advantage of providing a clear space above the active device for labeling or circuit notations.

Labeling and Circuit Notations

There are two parts to the schematic layout—the symbols and the labeling. Without labeling, the symbols mean nothing more than

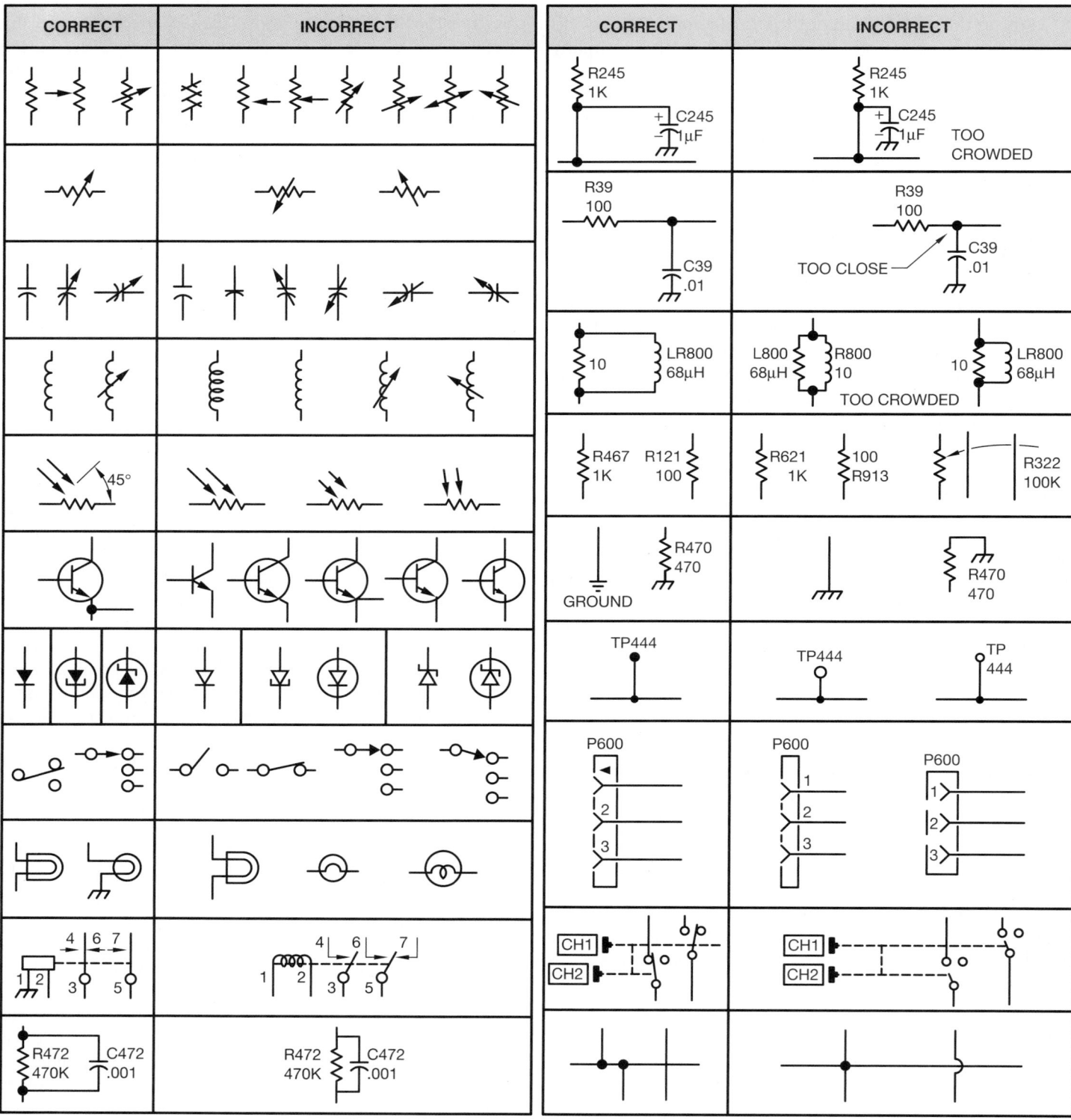

FIGURE 25.51 ■ Schematic do's and don'ts.

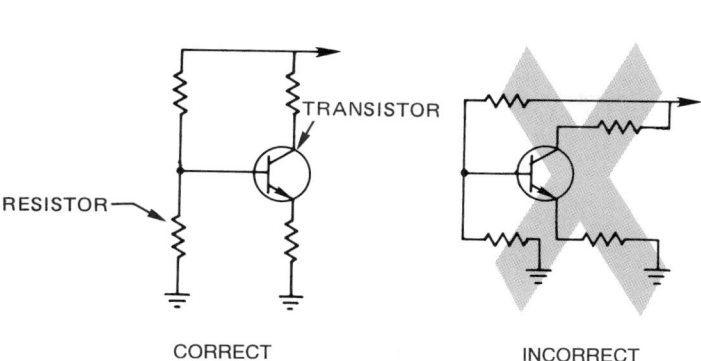

FIGURE 25.52 ■ Resistors placed in relation to the active device.

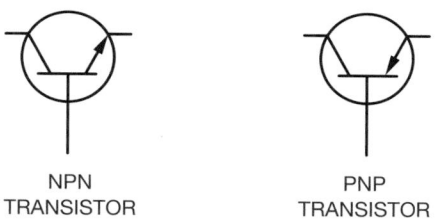

FIGURE 25.53 ■ Changing the bias direction of a transistor.

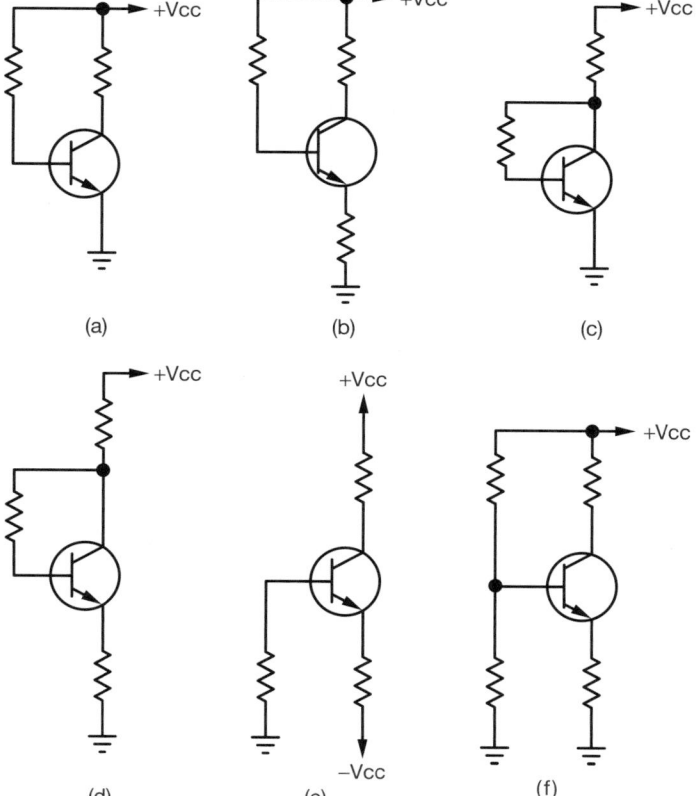

FIGURE 25.54 ■ Common transistor bias circuit arrangements.

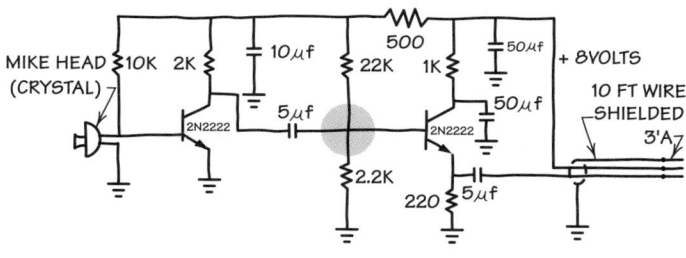

FIGURE 25.55 ■ Engineer's rough sketch.

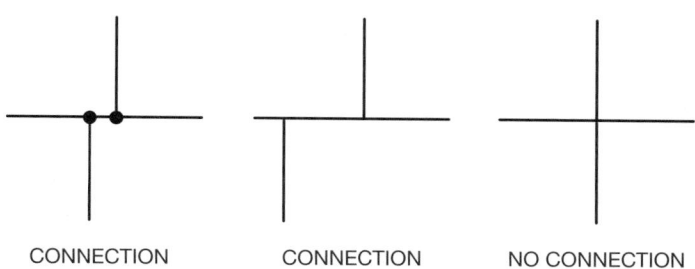

FIGURE 25.56 ■ Standard connection and nonconnection details.

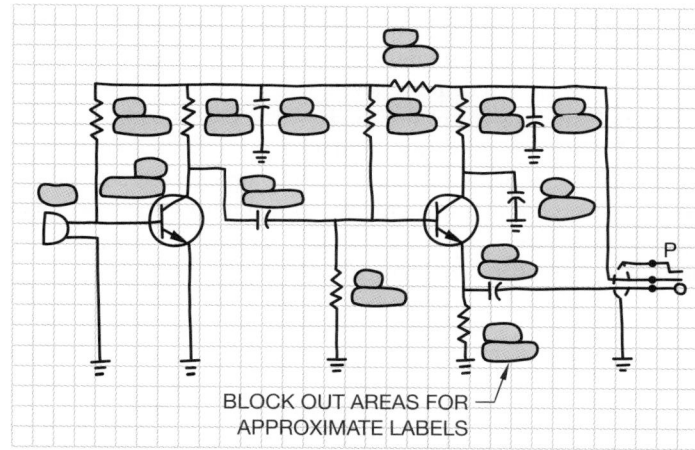

NOTE: SOMETIMES THE SKETCH COMES FROM THE ENGINEER WITHOUT CONNECTION DOTS AS SHOWN HERE.

FIGURE 25.57 ■ Schematic layout sketch.

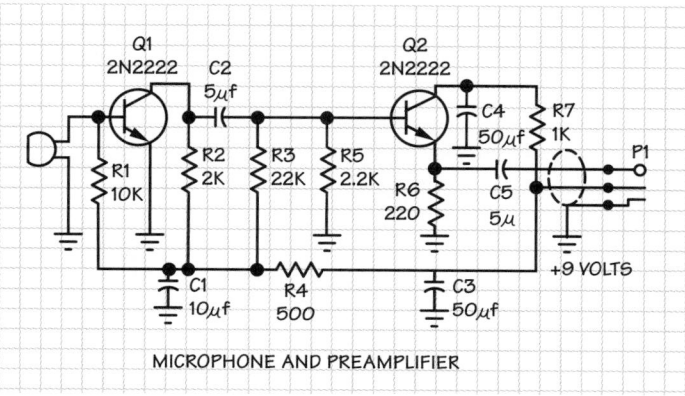

FIGURE 25.58 ■ Optional layout sketch.

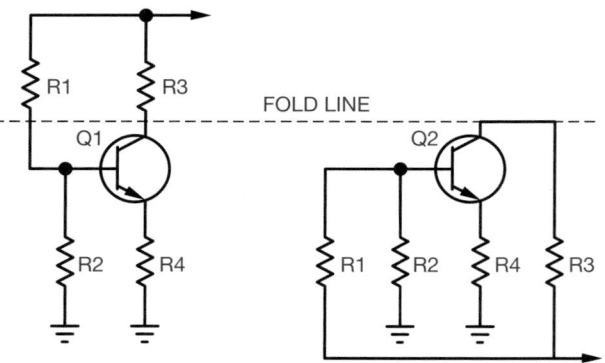

FIGURE 25.59 ■ Folding down the schematic.

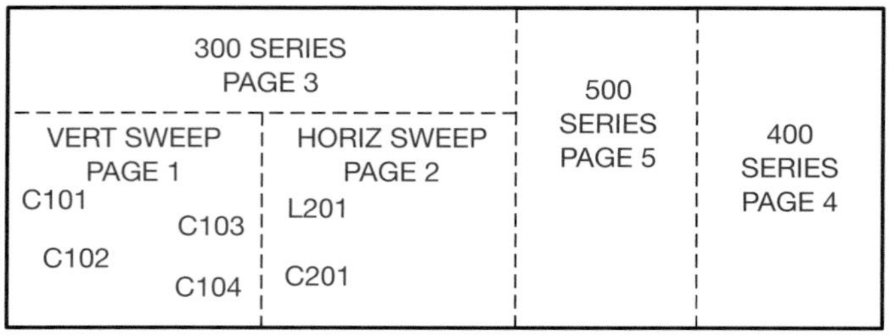

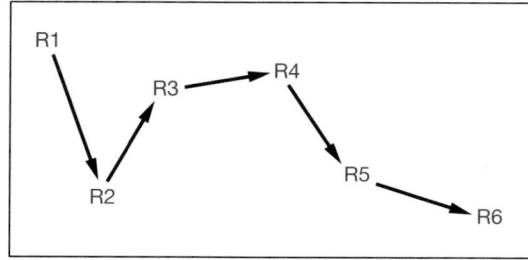

FIGURE 25.60 ■ Left-to-right component numbering sequence.

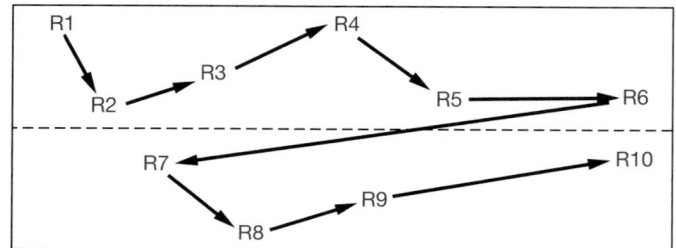

FIGURE 25.61 ■ Component numbering sequence applied in layers.

300 SERIES PAGE 3		500 SERIES PAGE 5	400 SERIES PAGE 4
VERT SWEEP PAGE 1 C101 C103 C102 C104	HORIZ SWEEP PAGE 2 L201 C201		

FIGURE 25.62 ■ Numbering components in groups.

the component they represent. The designating system has developed around a letter indicating the type of component followed by a number that gives the sequence of the part in the circuit. For example, the first resistor in the upper left of the schematic is designated R1. The next resistor to the right is R2, and so on across the schematic. The same holds true for capacitors, for example, C1, C2, and C3. Some of the other common component reference designators are: Transistor = Q, Inductor = L, Diode = D, Jack = J, and Plug = P. On simple circuits, the number sequence flows from left to right, as shown in Figure 25.60. In more complex schematics there can be two or more levels with each level flowing from left to right and top to bottom, as shown in Figure 25.61. Another method, used in units that contain subassemblies (individual groups of components that make up a complete system) or modular construction (component groups made up of equal components), is to assign the components a set of three- or four-digit numbers. For example, Figure 25.62 shows the vertical sweep module containing numbers from 100 to 199, the horizontal sweep module containing numbers from 200 to 299, the 300 series numbers ranging from 300 to 399, the 400 series numbers ranging from 400 to 499, and the 500 series numbers ranging from 500 to 599. The sequence of number series in schematics may be scrambled, established on the basis of physical layout, or arranged from left to right and top to bottom. This is also sometimes used for multiple sheet schematics. Notice the page numbers labeled in Figure 25.62. The annotation is usually done automatically by a computer program that generates numbers from left to right and top to bottom. The term *annotation* refers to numbers and text.

Units Used for Parts Values

Electronic component values run from extremely small values (0.000000000005) to extremely large values (2,500,000,000,000). In either case, numbers in this form take up too much space on the schematic. The numbers 1500 or 0.0003 would probably be acceptable. A zero should precede the decimal point for values less than one. It has become common practice to move the decimal point in groups of three and then modify the numerical name to indicate the number of places the decimal point has been moved. For example, if you have a 2,200-ohm resistor, you can move the decimal point three places to the left, drop the zeros, and change the numerical name to *kilohms,* which means 1000 ohms. The result is the designation 2.2 kilohms. The word *kilohms* may be abbreviated "k," thus reducing the notation to 2.2 k for a savings of more drafting space and time. The word *ohms* may be omitted because all resistors are rated in ohms. The following terms are used to rate or size common components:

■ Resistors = *ohms* = the unit of measurement of resistance. Symbol = Ω. Typical ranges are 0.1 ohm to 10m ohm.

■ Capacitor = *farad* = the unit of measure of capacitance. *Capacitance* is the property of an electric circuit to oppose a change in voltage. Symbol = *f.* Typical range is 10 pf to 10,000 μf.

■ Inductor or coil = *henry* = the unit of measurement of inductance. *Inductance* is the property of a component in an electric circuit that opposes a change in current flow. Symbol = H. Typical range is 1nH to 10H.

The following chart shows the values achieved by moving decimal points in given numbers:

Number	Move Decimal	Add Prefix	Abbreviation
1,000,000,000,000	Left 12 places	1 tera	1 T
1,000,000,000	Left 9 places	1 giga	1 G
1,000,000	Left 6 places	1 mega	1 M
1,000	Left 3 places	1 kilo	1 k
1	No change	1	1
0.001	Right 3 places	1 milli	1 m
0.000001	Right 6 places	1 micro	1 μ
0.000000001	Right 9 places	1 nano	1 n
0.000000000001	Right 12 places	1 pico	1 p

(TYPICAL RESISTOR RANGE OF VALUES; TYPICAL CAPACITOR RANGE OF VALUES; TYPICAL INDUCTOR RANGE OF VALUES)

There are some gray areas in this system. For example, there are no firm rules on which conversion for 500,000 is best—0.5 M or 500 k. However, 500 k is normally preferred. The number 1200 may be left as is or reduced to 1.2 k.

Military Identification Systems

Schematic diagrams that are drawn to meet military specifications provide a more elaborate component identification system. For example, a conventional part identification may be R304. In the military system, this part is labeled something like 14A2A4R304. This means that component R304 is located in subassembly 4 of subassembly 2A of assembly 14A. Military specifications are available covering every element used in the design, construction, and packaging of electronic equipment.

MIL-STD Some of the Military Standard Specifications (MIL-STD] include: MIL-STD-454, Standard General Requirements for Electronic Equipment, and MIL-STD-681, Identification Coding for Hook-up Wire.

Placement of Identification Numbers and Values

The placement of identification numbers is standard. The reference designator and value numbers are typically placed above horizontal components. (See Figure 25.63.) Other arrangements could cause confusion when components are not adequately spaced as shown in Figure 25.64.

For vertically drawn components the reference designator and the value are commonly placed on the right side of the symbol. (See Figure 25.65.)

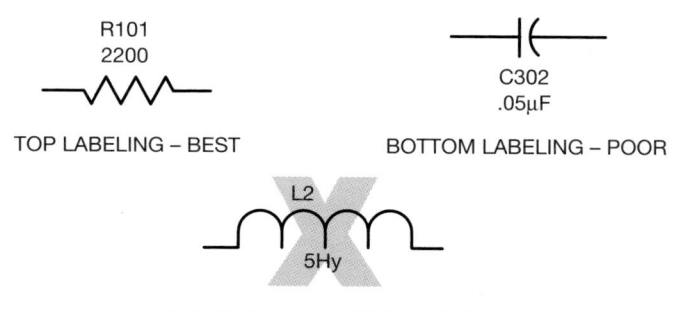

FIGURE 25.63 ■ Horizontal component labeling.

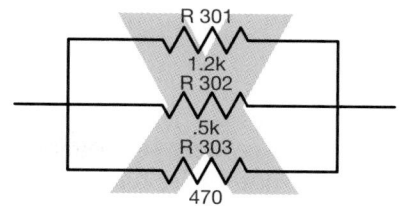

VERY CROWDED AND IMPROPERLY LABELED
(SPLIT LABELING CAUSES CONFUSION)

FIGURE 25.64 ■ An overcrowded schematic using split labeling. Avoid this practice.

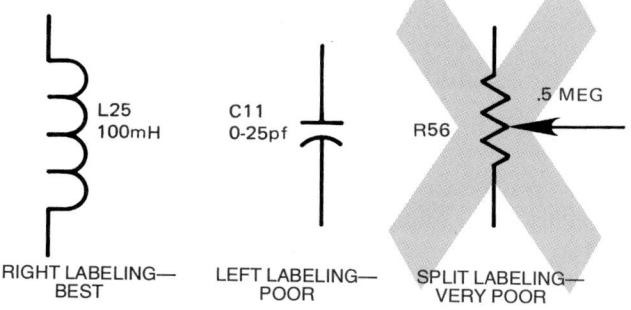

FIGURE 25.65 ■ Vertical component labeling.

The preferred technique in component labeling is placement of part identification at the top of all horizontal symbols and to the right of all vertical symbols, as shown in Figure 25.66.

A final drawing of the schematic used in the sketching layout examples is shown in Figure 25.67.

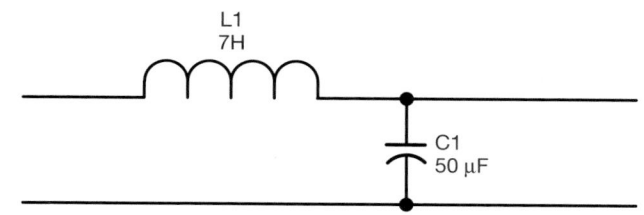

FIGURE 25.66 ■ Preferred labeling at the top of horizontal components and to the right of vertical components.

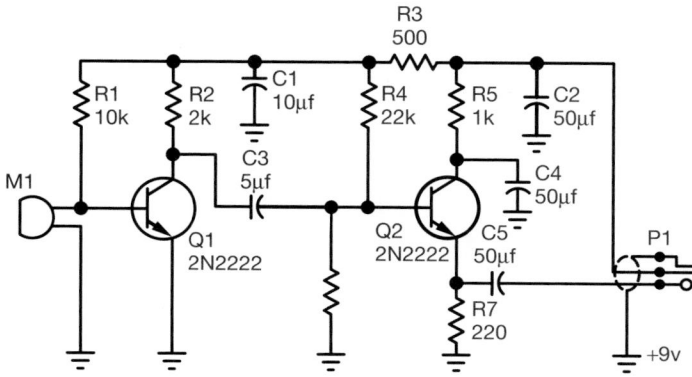

FIGURE 25.67 ■ Final drawing of schematic used in layout sketches (Figures 25.57 through 25.59).

Operational Amplifiers and Integrated Circuit Schematics

An *amplifier* (AMP) is a device that allows an input signal to control power and is capable of an output signal greater than the input signal. *An operational amplifier* (OP AMP) is a high-gain amplifier created from an integrated circuit. An *integrated circuit* (IC) is an electronic circuit that has been fabricated with extremely small components as an inseparable assembly known as a *chip*. You can become involved with integrated circuit schematics by drawing the schematic of the internal circuitry involved, or drawing the schematic for a system in which integrated circuits or operational amplifiers are used.

Internal Integrated Circuit Schematics

The internal circuitry of an integrated circuit is shown in Figure 25.68. The only exception is that in the IC diagram, the transistor symbols are not circled. The diagram is also sometimes enclosed by dashed lines in the form of a box. This box represents the package within which the IC is contained. Points are shown and labeled with *pin numbers,* which are external connectors for attachment to other circuitry. This usually means that all of the components shown in the schematic are made up of one piece of semiconductor material. Thus the definition of integrated circuit—all of the components of the circuit have been put together on one small piece of material. The schematic for an IC can be very simple or so complex that it contains several million components and hundreds of pins.

IC Symbols and Logic Circuits

In *logic circuits,* which are computer-oriented circuits, the schematics become a cross between a flow diagram and a schematic diagram. The internal components of an IC are self-contained for a specific function and may not be altered. Therefore, symbols are used that identify the IC. The shape of the symbol tells the technician what the device is expected to do. The numbers next to the input and output lines or pins tell the reader how to connect the IC to

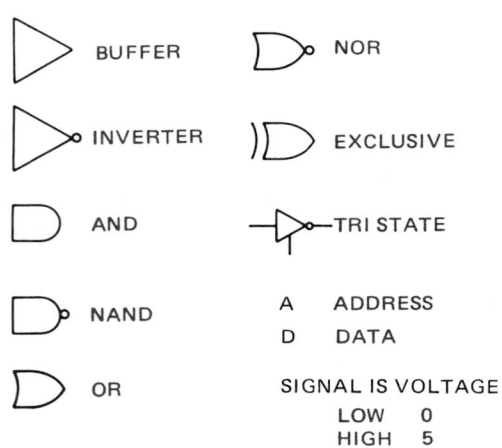

FIGURE 25.69 ■ The basic format of IC symbols.

make it work. The basic format of IC symbols is shown in Figure 25.69. The IC symbol that represents the internal IC shown in Figure 25.68 is shown in Figure 25.70. This symbol is called a NAND gate. The *gate* is the part of the system that makes the electronic circuit operate and permits an output only when a predetermined set of input conditions are met. The code NAND is the predetermined function of the gates. There are also AND, OR, and NOR functions. Simply put, the functions relate to a condition or pattern of conditions whereby an electrical pulse can be received at one pin when the input at another pin or pins is triggered at specific times.

The precise operation of each logic function is required knowledge for electronic technicians and design drafters, but premature for drafting fundamentals and beyond the scope of this study. Your concern at this time is to prepare a well-balanced, accurate drawing from engineering sketches. Notice how the IC schematic diagram shown in Figure 25.71 compares to previous schematics. The difference is in the IC symbols. While the integrated circuits may be complex in nature, the drafting task becomes easier when ICs are used. For example, the IC labeled TAA-500 in Figure 25.72 is an

FIGURE 25.70 ■ The IC symbol of the internal IC schematic shown in Figure 25.68.

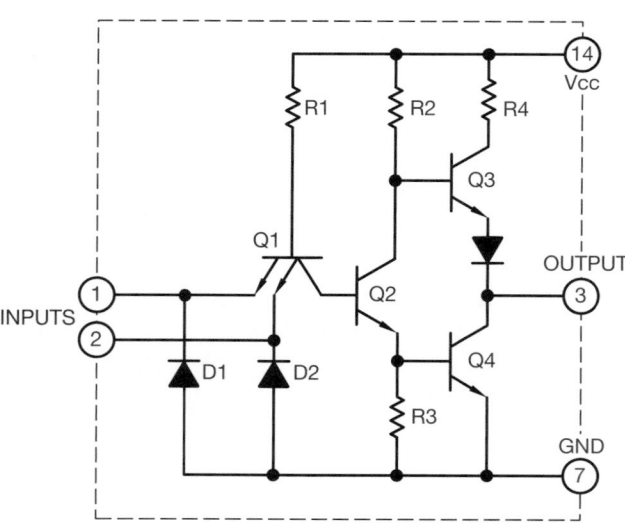

FIGURE 25.68 ■ The internal components of a logic integrated circuit.

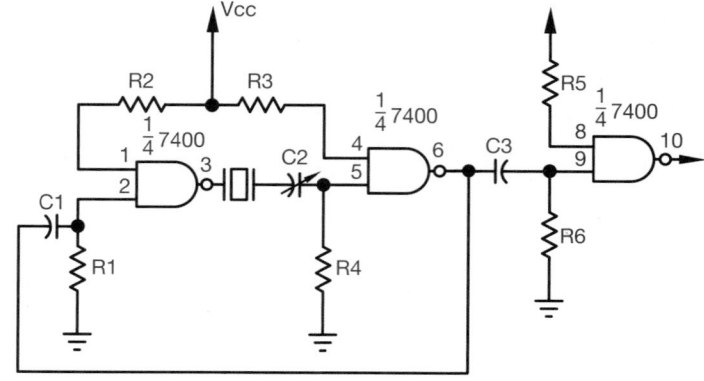

FIGURE 25.71 ■ A schematic diagram with an IC included.

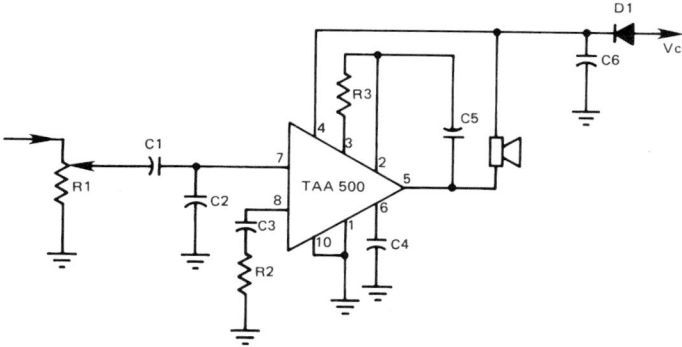

FIGURE 25.72 ■ Amplifier IC.

operational amplifier that contains 11 transistors, 5 diodes, 13 resistors, and 1 capacitor.

Logic diagrams are a type of schematic that are used to show the logical sequence of events in an electrical or electronic system. The basic electronic logic symbols and their functions are shown in Figure 25.73. Part of an electronic logic diagram is shown in Figure 25.74. These symbols are currently used, but may become obsolete.

TITLE	SYMBOL	
AND	AND	ALL INPUTS MUST BE PRESENT FOR OUTPUT.
OR	OR	ANY INPUT WILL ALLOW OUTPUT.
NOT	NOT	AN INPUT WILL RESULT IN NO OUTPUT.
TIME DELAY	TD 0-5	AN INPUT WILL ALLOW AN OUTPUT AFTER A TIME DELAY.
ON OFF	ON OFF	INPUT "A" WILL ALLOW AN OUTPUT EXCEPT WHEN THERE IS AN INPUT AT "B."

FIGURE 25.73 ■ Basic electronic logic symbols are currently being used, but are becomming obsolete.

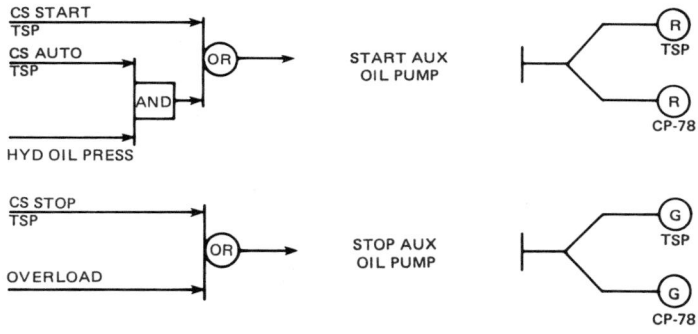

FIGURE 25.74 ■ Part of an electronic logic diagram.

Large–scale Integration

The next generation beyond the specific symbols (such as the gates) came about because of the rapid advancement of the electronic industry's ability to put more and more circuits on a single, small IC chip. Many calculators contain only one IC with hundreds of gates. The trend in drawing schematics for this *large-scale integration* (LSI) is simply to show the chip as a box. This type of schematic becomes like a block diagram with the name or function identified within the rectangle for each LSI. The LSI pin terminals are shown connected to the rest of the schematic. An LSI schematic diagram is shown in Figure 25.75. Keep in mind that each LSI block in the schematic may contain hundreds or thousands of components.

Symbol Sizes

The actual size of symbols may vary slightly from one company or computer-aided design software to another. The uniformity of symbols within a drawing is important. Figure 25.76 shows the recommended sizes of commonly used electronics symbols based on the drawing text height.

PRINTED CIRCUIT TECHNOLOGY

Schematic diagrams are designed and drawn to show the location of electronic components using symbols and lines to represent circuit paths or connections. In the production of the actual electronic product, the symbols shown on the schematic become electronic devices and the lines become wires that connect these devices. At one time, wires were soldered to component terminals and used as the circuit path between components. Today, wires are used as the connection cables that provide the source of current between pieces of electronic equipment. As electronic equipment designs have become increasingly smaller, the internal connection between electronic devices must also take up less space, be easier to install, and be extremely accurate. Printed circuit technology is the answer to these needs.

Schematic and Printed Circuit Board Accuracy

Accuracy and close attention to detail in the preparation of the schematic and printed circuit board are absolutely essential. Everything depends on the accuracy of these two items. The schematic and the printed circuit board cannot have a single mistake, because any mistake affects the master artwork, the drilling drawing, and the assembly and bill of materials.

Printed Circuit Design and Layout

Printed circuits (PC) form the interconnection between electronic devices. The base materials for circuit boards are special paper, plastic, glass, or Teflon. The quality of the *printed circuit board* (PCB) begins with the quality of the base material. Depending on the complexity of the electronics, printed circuits are prepared on one side or both sides of the board. In many applications, there are multiple boards or layers for one piece of electronic equipment.

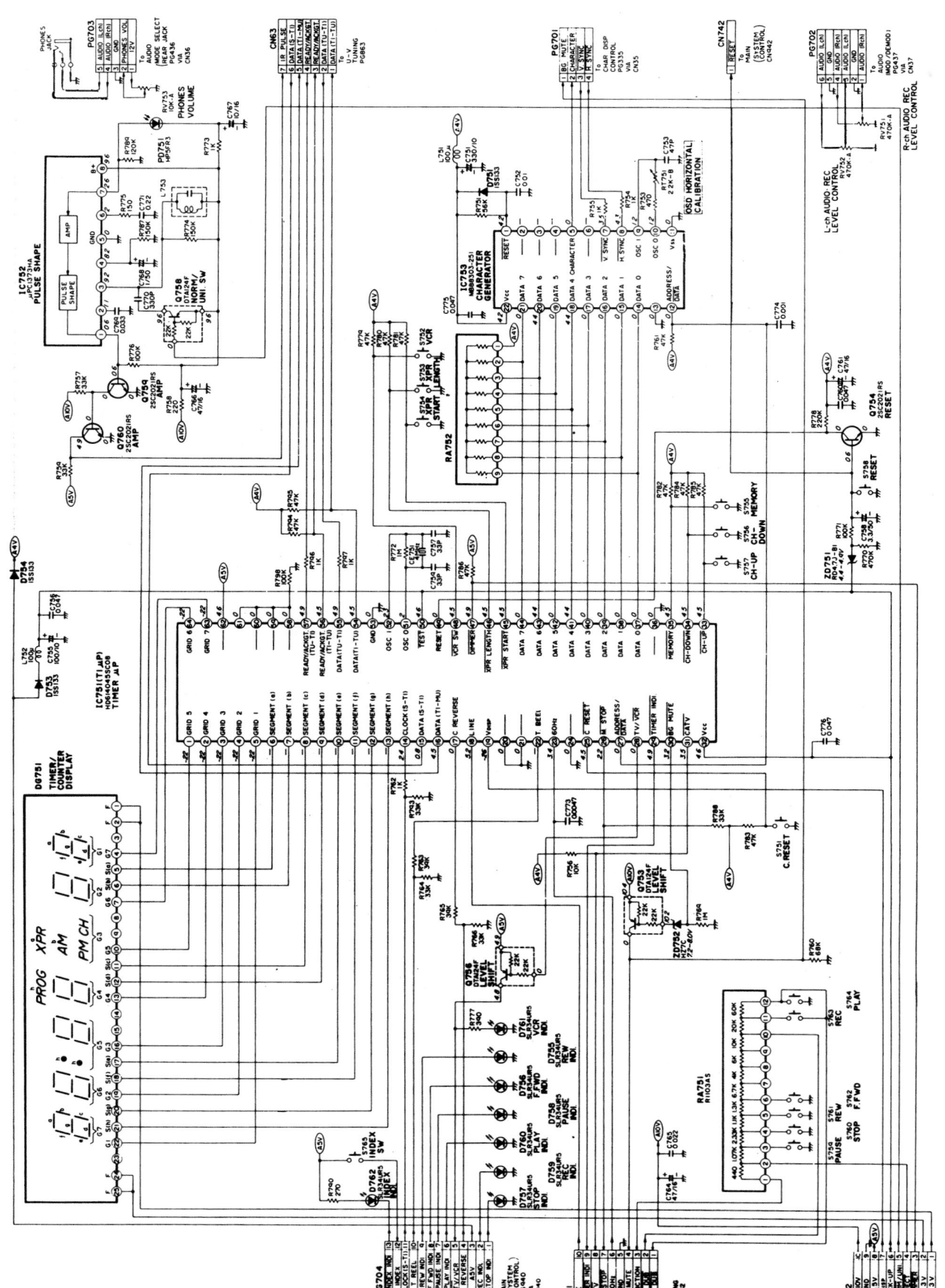

FIGURE 25.75 ■ A large-scale integration (LSI) schematic. Reproduced by permission of RCA Consumer Electronics.

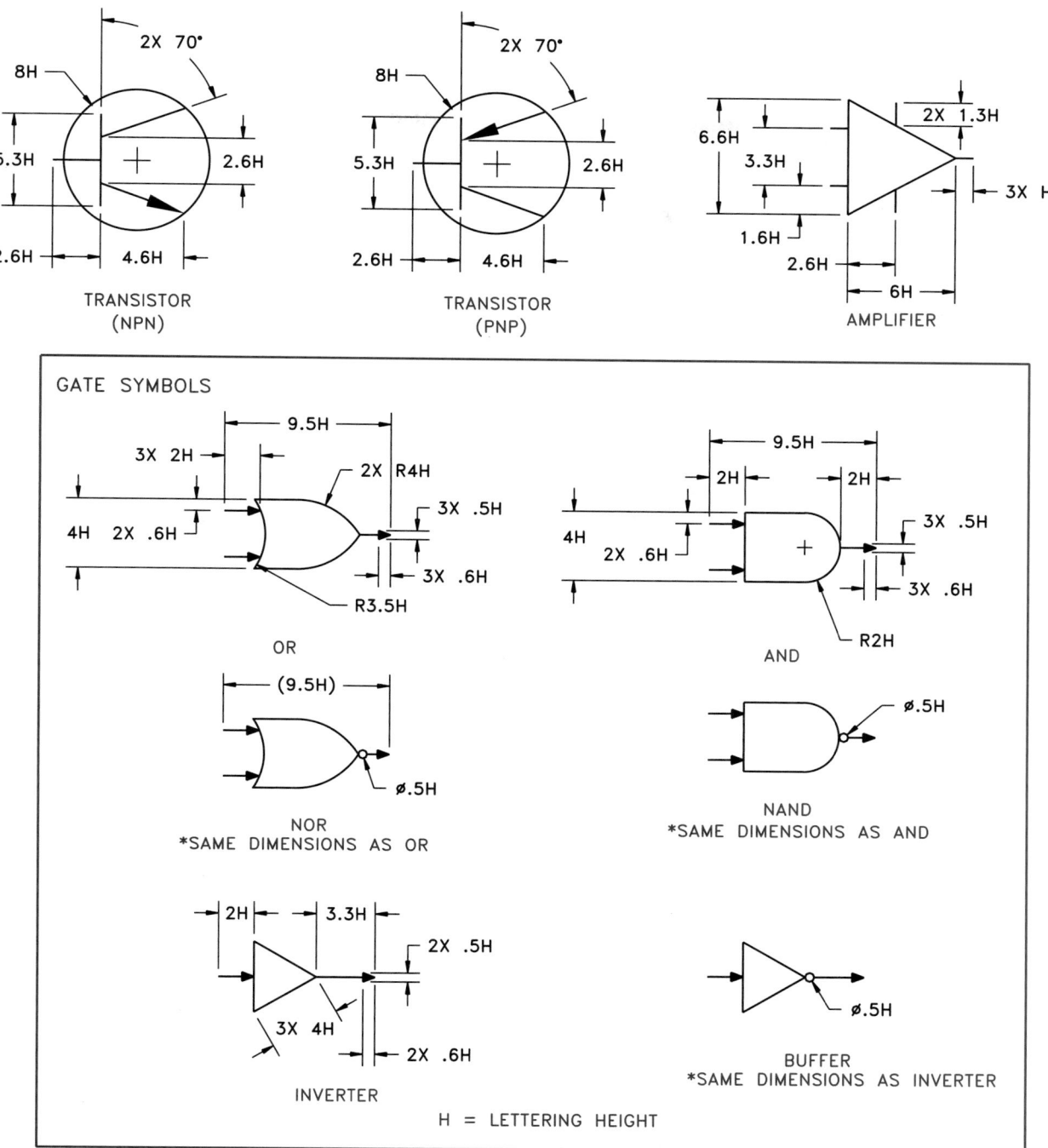

FIGURE 25.76 ■ MRecommended sizes of commonly used electronics symbols based on the drawing text height.

The printed circuits consist of pads and conductor lines that are made of thin conductive material such as copper on a base sheet or board. The *pads* or *lands* are the circuit termination locations where the electronic devices are attached. The pad normally has a location where a hole is to be drilled in the circuit board for mounting the device. Pads may be placed individually or in patterns depending on the connection characteristics of the device. (See Figure 25.77.) *Conductor traces* (lines) connect the pads to complete the circuit design. Design characteristics of conductors are shown in Figure 25.78. The size depends on the amount of current, the temperature, and the type of board specified. Conductors are .008 in. (0.2 mm) wide with a .007 in. (0.18 mm) space for many applications, but may be designed for applications with a width of .50 in. (12.7 mm) or more for large

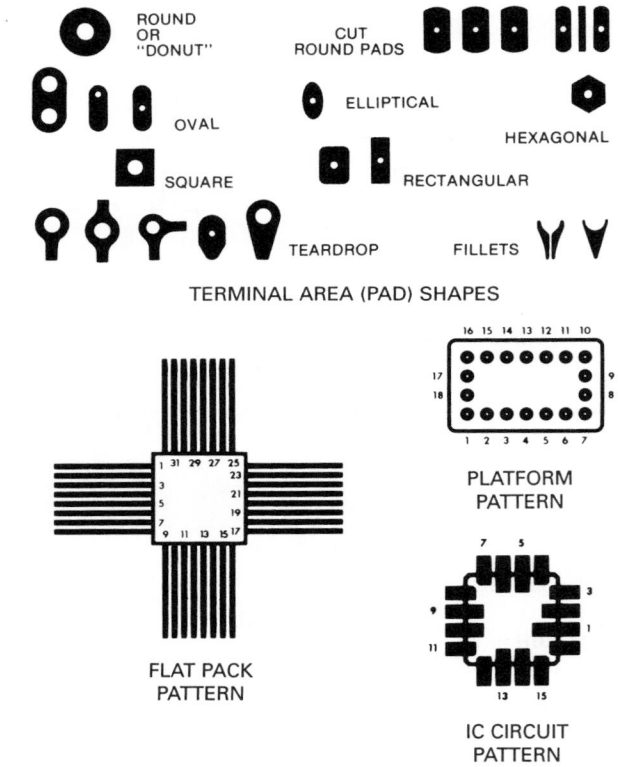

FIGURE 25.77 ■ Mounting pads. *Courtesy Bishop Graphics, Inc.*

current-carrying requirements. Factors that influence conductor spacing include product requirements and current specifications. For general uses, conductor spacing should be .007 in. (0.18 mm) minimum.

Printed Circuit Board Artwork

The *printed circuit artwork* is an accurate, scaled, undimensioned drawing used to produce the master pattern from which the actual board is manufactured. The artwork should be prepared from a design layout at an enlarged scale on polyester film using computer-aided design drafting. In addition to all electrical circuitry, the board should contain board edge identification marks; registration, datum, and indexing marks; photo reduction dimensions; scale designation; an identification number; and a revision number. Letters to be etched on the board should be .062 in. (1.57 mm) in height (minimum) and .015 in. (0.38 mm) in thickness (minimum).

Master Pattern Artwork

The *master pattern* is a one-to-one scale circuit pattern that is used to produce a printed circuit board. The master pattern artwork is normally prepared at an enlarged scale so that when it is reduced to a 1:1 ratio, its quality is enhanced.

RECOMMENDED	NOT RECOMMENDED
AVOID SHARP EXTERNAL ANGLES WHICH CAN CAUSE FOIL DELAMINATION. GOOD GOOD OK	
AVOID ACUTE INTERNAL ANGLES.	
ALWAYS USE THE SHORTEST PRACTICAL CIRCUIT ROUTING.	
MAINTAIN EQUAL SPACING WHERE CONDUCTORS PASS BETWEEN TERMINAL AREAS.	
AVOID LARGE MULTIPLE HOLE TERMINAL AREAS, WHICH MAY CAUSE THERMAL SOLDERING PROBLEMS AND NON-SYMMETRICAL SOLDER FILLETS.	
MAINTAIN UNIFORM PATTERN AROUND HOLE TO PRODUCE SYMMETRICAL SOLDER FILLETS.	
AVOID USING CONDUCTORS THE SAME SIZE AS TERMINALS, WHICH WILL CAUSE SOLDER TO FLOW AWAY FROM TERMINAL.	
PLATING BAR SHOULD EXTEND OUT FROM AND BEYOND BOARD EDGE TO FACILITATE FABRICATION.	

FIGURE 25.78 ■ Artwork pattern configuration, conductor traces, and taping techniques. *Courtesy Bishop Graphics, Inc.*

Grid System

The use of a grid system is essential in laying out and preparing the master pattern artwork. The grid system aids in the placement of pads and conductor traces. A *grid* is a network of equally spaced parallel lines running vertically and horizontally on a glass

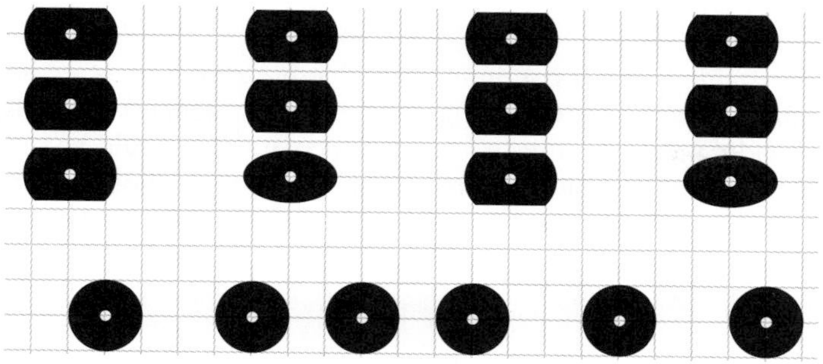

FIGURE 25.79 ■ Grid system. *Courtesy Bishop Graphics, Inc.*

or polyester film sheet, as shown in Figure 25.79. Standard spacing is .100 in. (2.54 mm), .050 in. (1.27 mm), or .025 in. (0.635 mm).

MIL-STD MIL-STD 275D specifies any multiple of .005 in. (0.127 mm).

IEC The International Electrotechnical Commission (IEC) specifies 0.5-mm and 0.1-mm increments.

Printed Circuit Scale

The printed circuit board layout should be prepared at an enlarged scale of 2:1 or 4:1 so that possible drafting errors or slight flaws are minimized when the layout is reduced to actual size. The accuracy of the PC layout is also critical. Sometimes it is necessary to prepare the PC layout at a 100:1 scale to ensure clarity and accuracy.

Board Size and Number of Layers

Printed circuit boards may be designed in three basic configurations: single layer, multilayer, and multilayer sandwich. *Single-layer* boards contain all printed wiring on one side with the components on the opposite side. *Multilayer* boards have printed circuits on both sides with most of the components on one side and the circuitry on the other. *Multilayer sandwich* boards consist of many thin boards laminated together, with the components on one or both sides of the external layers. As boards increase in complexity, they also cost more. However, the cost of a single multilayer board may be less than several single-layer boards for the same system.

The circuit board should be just large enough to contain the components and interconnections and remain economical to manufacture. One way to estimate the board size is to randomly place scaled component cutouts, allowing enough space for interconnections. Allow extra space for IC interconnections.

Solder Masks

Wave solder has become a common method of attaching components to printed circuit boards. A wave of molten solder is passed over the noncomponent side of the board, making all solder connections. A polymer coating, called a *solder mask* or *solder resist mask*, is applied to the board, covering all conductors except pads,

connector lands, and test points. Solder masks are used to prevent the bridging of solder between pads or conductor traces, to cut down on the amount of solder used, and to reduce the weight of the board.

Conductor Width and Spacing

Careful consideration must be given to conductor trace width and spacing during printed circuit design. Width and spacing that are too small can cause service problems in the circuitry. Width and spacing that are too great waste space and increase costs. Verify minimum width and spacing requirements with product and company specifications. Standard charts and specifications are available; but in general, conductor widths of .050 in. (1.27 mm) or .062 in. (1.57 mm), and minimum spacing between conductors of from .031 in. (0.79 mm) to .050 in. (1.27 mm) are recommended for low-voltage applications.

Component Terminal Holes

A printed circuit board should have a separate mounting hole for each component lead or terminal. Unsupported holes contain no conductive material. To determine the diameter of unsupported holes use the formula:

$$\text{Minimum hole diameter} = \text{Maximum lead diameter} + \text{Minimum drill tolerance}$$

Plated-through holes have conductive material plated on the inside wall to form a conductive connection between layers of the circuit board, as shown in Figure 25.80. Plated-through holes are drilled prior to plating. A general note "ALL HOLES TO BE PLATED THROUGH" should be specified. Usually plating thickness is specified as a minimum with an accepted tolerance of $-0 +100\%$. For plated holes, the minimum diameter should be greater than the minimum lead size plus .028 in. (0.71 mm). A second drilling process is required if the predrilled holes do not take into account the plating thickness. Drilling twice adds cost to the board. The number of different hole sizes should be kept to a minimum to help save manufacturing costs. Plating is specified in ounces (oz)

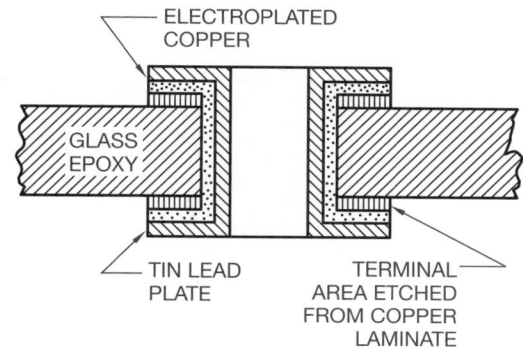

FIGURE 25.80 ■ Cross section of plated-through hole. *Courtesy Bishop Graphics, Inc.*

of copper. The base is typically ½ oz plating but can be up to 1–2 oz. Additionally, .0014" thick is specified as 1 oz of copper, which equals 1 oz per square foot.

Terminal Pads

There should be a separate terminal pad for each component lead or wire attachment. Terminal pads vary with designer preference and component characteristics. (See Figure 25.77 and 25.81.) The *minimum required annular ring* is the smallest part of the circular strip of conductive material surrounding a mounting hole that meets design requirements.

MIL-STD MIL-STD 275D specifies .015 in. (0.38 mm) minimum for unsupported holes, .005 in. (0.13 mm) minimum for plated-through holes on external layers, and .002 in. (0.05 mm) for internal layers of multilayer sandwich boards.

Ground Planes

A *ground plane* is a continuous conductive area used as a common reference point for circuit returns, signal potentials, shielding, or heat sinks. Ground plane patterns should be broken up so the conductive area is equal to about half of the nonconductive area. (See Figure 25.82.) Clearance should be provided between terminal pads and the ground plane, as shown in Figure 25.83.

Printed Circuit Board Layout

Before starting a printed circuit board layout, you should have a schematic or logic diagram, a parts list, design specifications, component sizes, lead-and-trace pattern and spacing, hole and terminal sizes, grid system, scale, board size and number of layers, and know the manufacturing processes to be used.

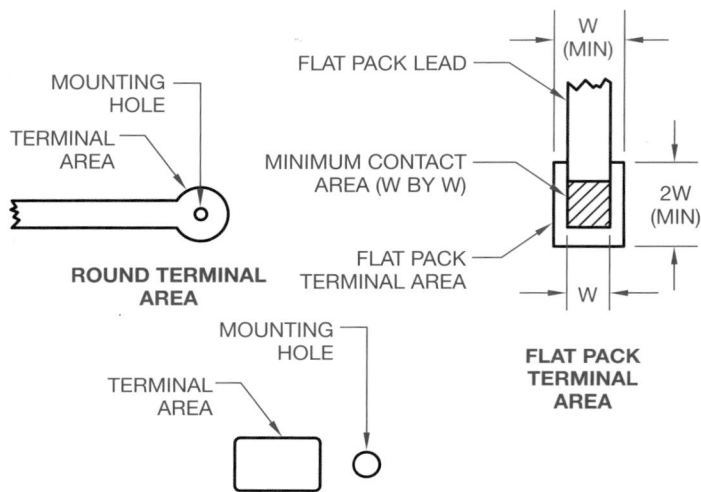

FIGURE 25.81 ■ Terminal areas. *Courtesy Bishop Graphics, Inc.*

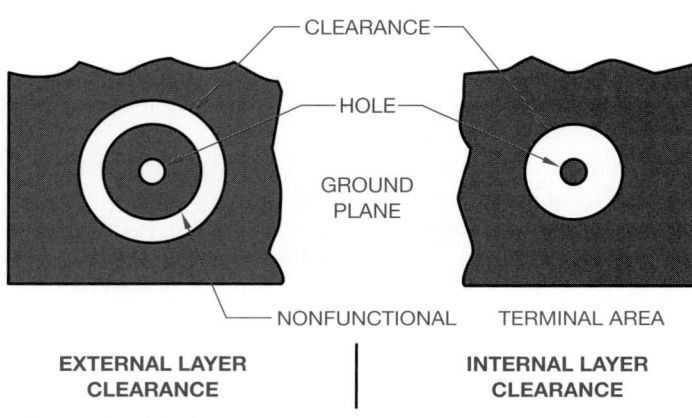

FIGURE 25.83 ■ Layer clearance at ground planes. *Courtesy Bishop Graphics, Inc.*

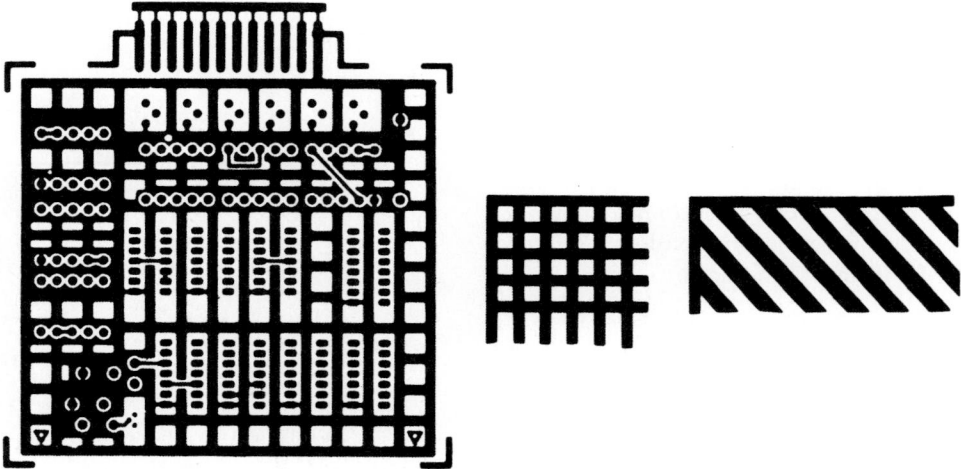

FIGURE 25.82 ■ Ground planes. *Courtesy Bishop Graphics, Inc.*

Silk Screen Artwork

The circuit board artwork may have a separate sheet called *silk screen* artwork containing component outlines and orientation symbols. This pattern is printed on the component side of the board after etching and plating. The reference designations should be placed so they are visible after component assembly. The component outlines should be located in the same position that the actual components occupy after assembly. The silk screen artwork should include registration marks that align with the pattern. (See Figure 25.84.)

Drilling Drawings

Drilling drawings are prepared after the master layout is complete. The *drilling* drawing is used to provide size and location dimensions for component and chassis mounting holes and the final dimensions for trimming the board. Drilling drawings are often set up using datum arrowless tabular dimensioning in accordance with computer numerical control manufacturing methods. (See Figure 25.85.) Refer to the previous discussion on hole sizes and to Chapter 12 for a more in-depth discussion of dimensioning practices.

Assembly Drawings

The printed circuit board *assembly* drawing is a complete engineering drawing, including components, assembly and fastening or soldering specifications, and a parts list or bill of materials. The parts list should include the: part number, item identification keyed to the assembly, quantity of each part, electronics designations for components, and Federal Code Identification, if required by contract. (See Figure 25.86.)

Printed Circuit Preparation

There are several methods used to prepare the final printed circuit board from a printed circuit layout; however, the two fundamental methods are the additive and the etching processes.

Additive Process

The *additive process* takes place on a board that is covered with a chemically etched material that will accept copper. The PCB layout design image is then transferred to the board using a silk screen or photo printing process. Copper conductor is then deposited on the image using electroless. *Electroless* is the depositing of a metal on another material through the action of an electric current. A solder resist mask is then applied to the copper deposit. This process

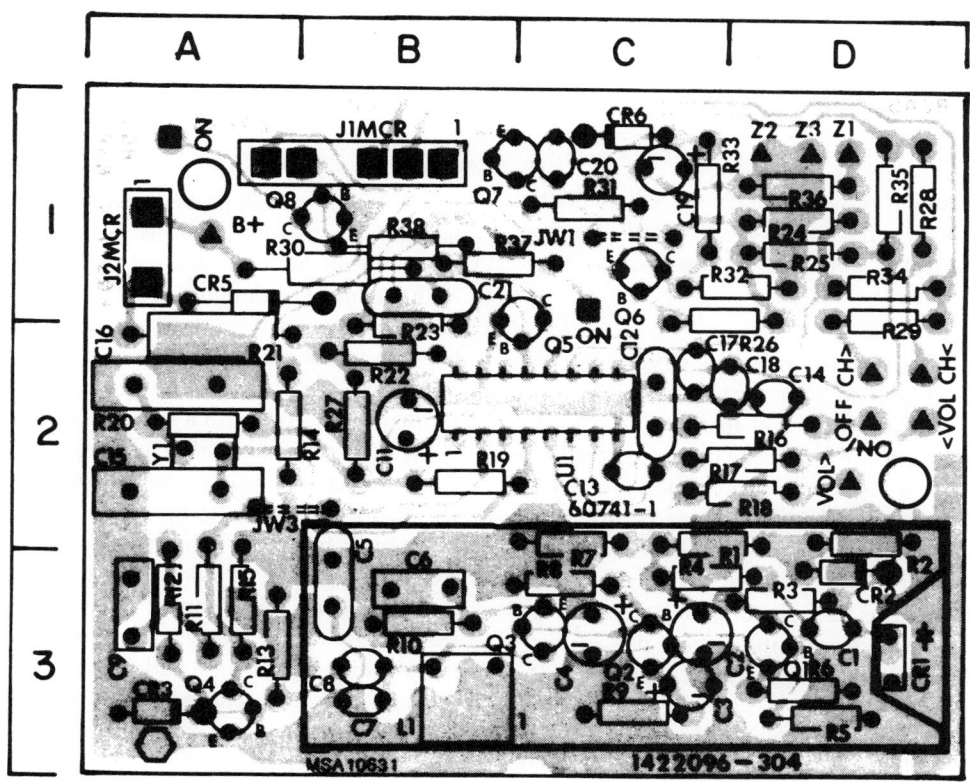

FIGURE 25.84 ■ Silk screen artwork. Reproduced by permission of RCA Consumer Electronics.

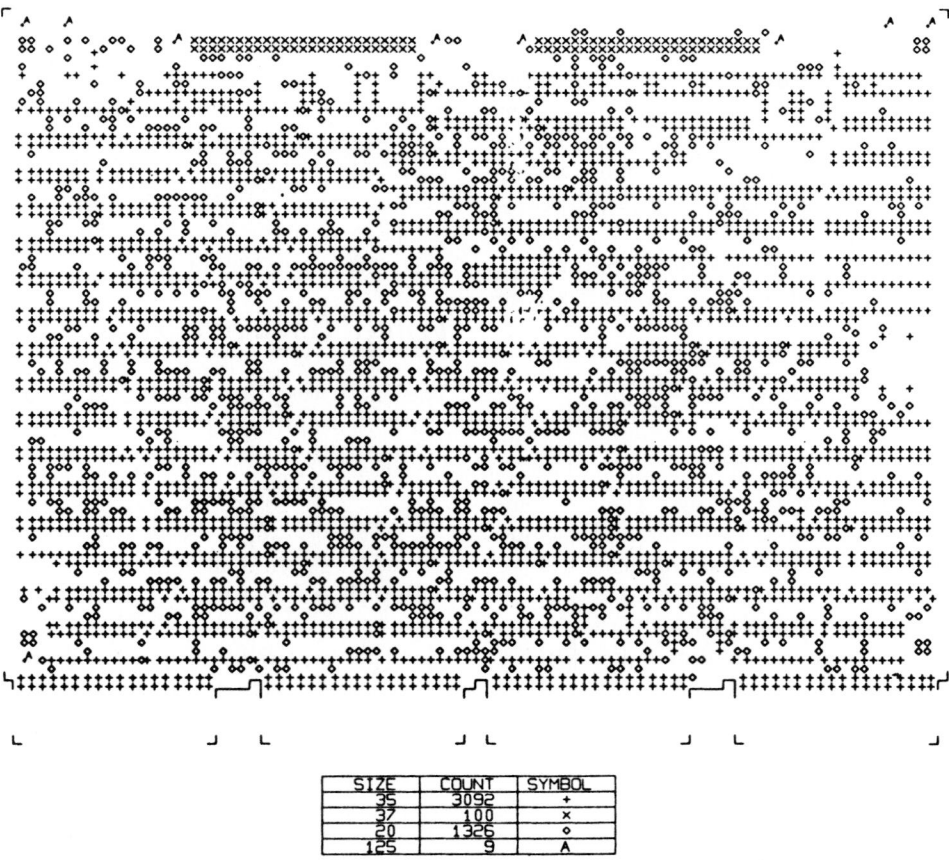

SIZE	COUNT	SYMBOL
35	3092	+
37	100	×
20	1326	o
125	9	^

FIGURE 25.85 ■ Drilling drawing. *Courtesy Floating Point Systems, Inc.*

prevents solder from bridging between closely spaced conductors when components are added to the printed circuit board. The board is then coated with a lacquer to protect the surface from corrosion.

Etched Process

A copper-covered base is used in the *etched process*. (Other conductive material may also be used.) The printed circuit layout is normally prepared at 2:1 or 4:1 scale and then photographically reduced to a photo negative at full scale. The copper-covered base is cleaned and treated with a light-sensitive emulsion. The base is then exposed to light through the negative. During this process, the emulsion hardens and an acid resist is formed at the printed circuit areas. The board is then placed in a chemical solution that etches away the copper coating in areas that are not protected by the acid resist. The board is then rinsed to stop the etching process and a protective coating is added. The resulting printed circuit is ready for components to be added.

Surface-mount Technology (SMT)

Surface-mounted devices (SMD) have emerged from the increasing need to get more electronic components into smaller places. Surface-mounted devices commonly take up to less than one-third the space of feed-through printed circuits. Conventional printed circuit board connectors require plated-through holes for each pin or connection point. Surface-mounted devices save the area, and are especially useful when designing very complex boards. The only drilled holes that are required are for chassis mounting or feed-through between layers.

In surface-mount technology, the traditional component lead-through is replaced with a solder paste to hold or *glue* the components in place on the surface of the printed circuit board. A solder paste template is made that allows for the placement of the paste at required areas. Then the components are automatically or manually placed into the solder paste, although most companies use robotic fabrication. A baking process allows the solder to liquify and then, upon solidification, the components remain soldered to the printed circuit. Some solderless connections have also been designed for surface-mounted devices.

Surface-mount technology is responsible for new types of electrical components and materials used in electronic circuits; for example, leadless plastic and ceramic IC chip housings. There are some problems, however, with surface-mounted devices; they are so dense that solder connections often become difficult, and testing electrical circuits becomes a problem because testing equipment may have difficulty connecting to very small fine-pitch parts. In

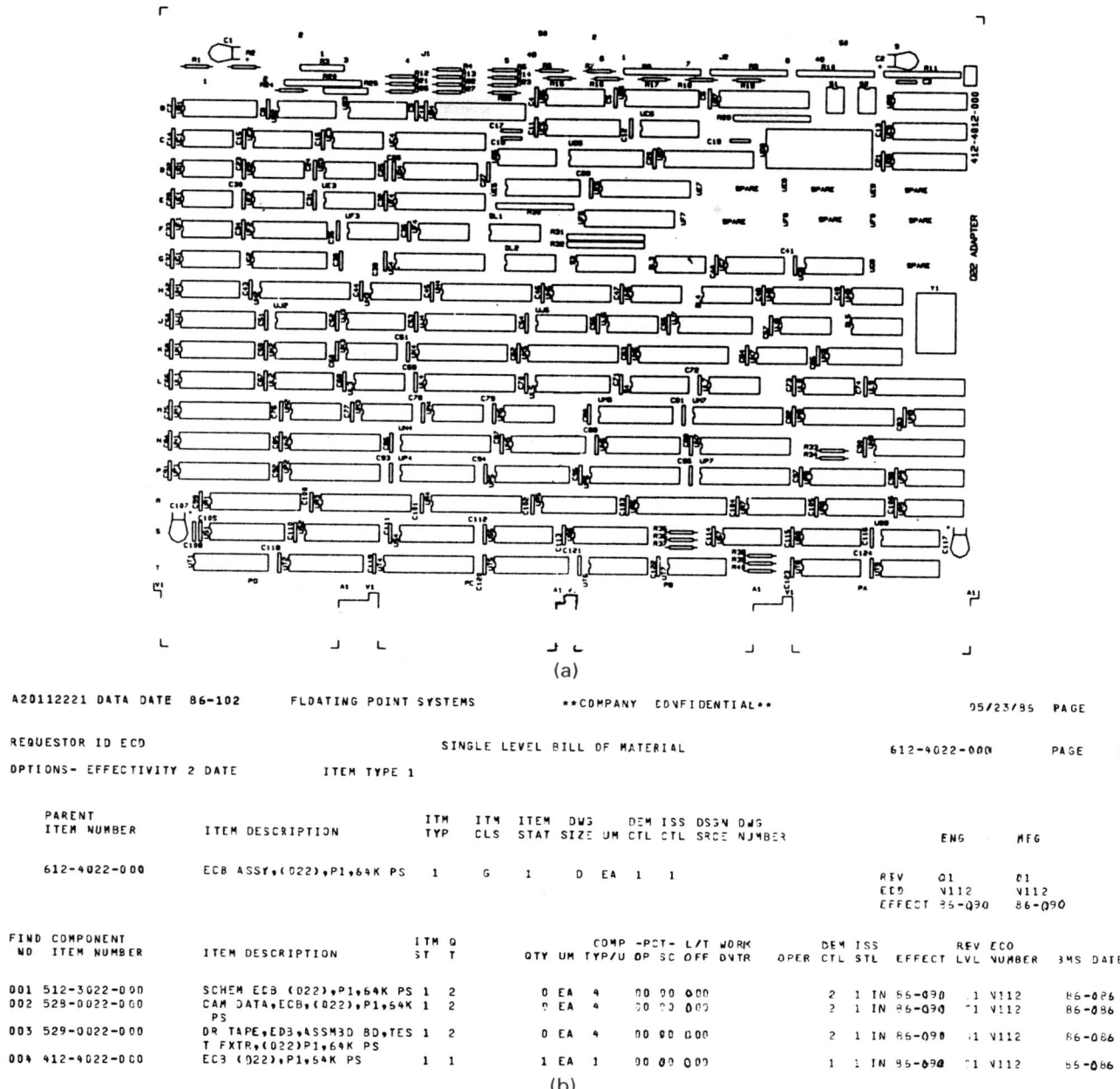

(a)

A20112221 DATA DATE 86-102   FLOATING POINT SYSTEMS   **COMPANY CONFIDENTIAL**   05/23/86 PAGE   1

REQUESTOR ID ECO   SINGLE LEVEL BILL OF MATERIAL   612-4022-000   PAGE   1

OPTIONS- EFFECTIVITY 2 DATE   ITEM TYPE 1

PARENT ITEM NUMBER	ITEM DESCRIPTION	ITM TYP	ITM CLS	ITEM STAT	DWG SIZE	UM	DEM CTL	ISS CTL	DSGN SRCE	DWG NUMBER	ENG	MFG
612-4022-000	ECB ASSY,(022),P1,64K PS	1	G	1	D	EA	1	1			REV 01	01
											ECO V112	V112
											EFFECT 86-090	86-090

FIND NO	COMPONENT ITEM NUMBER	ITEM DESCRIPTION	ITM ST	Q T	QTY	UM	COMP TYP/U	-PCT- OP	SC	L/T OFF	WORK DVTR	DEM OPER	ISS CTL	REV STL	ECO EFFECT	LVL	REV ECO NUMBER	BMS DATE
001	512-3022-000	SCHEM ECB (022),P1,64K PS	1	2	0	EA	4	00	00	000		2	1	IN	86-090	.1	V112	86-086
002	528-0022-000	CAM DATA,ECB,(022),P1,64K PS	1	2	0	EA	4	00	00	000		2	1	IN	86-090	.1	V112	86-086
003	529-0022-000	DR TAPE,EB3,ASSMBD BD,TEST FXTR,(022)P1,64K PS	1	2	0	EA	4	00	00	000		2	1	IN	86-090	.1	V112	86-086
004	412-4022-000	ECB (022),P1,64K PS	1	1	1	EA	1	00	00	000		1	1	IN	86-090	.1	V112	86-086

(b)

FIGURE 25.86 ■ **(a)** Assembly drawing. **(b)** Partial computerized parts list. *Courtesy Floating Point Systems, Inc.*

most cases, surface-mounted boards are replaced, rather than serviced, when maintenance is required.

PICTORIAL DRAWINGS

Three-dimensional drawings are frequently made to show pictorial representations of products for display in vendors' catalogs or instruction manuals. Pictorial assembly drawings are often used to show the physical arrangement of components so components are properly placed by factory workers during assembly. These drawings are also helpful for maintenance as they provide a realistic representation of the product.

One of the most effective uses of photodrafting is found in the electronics field. This is where a photo is taken of a complete assembly or product and drafting is used to label individual components. A common method of pictorial representation is the exploded technical illustration often used for parts identification and location, as shown in Figure 25.87.

Another type of pictorial diagram, known as a semipictorial wiring diagram, uses two-dimensional images of components in a diagram form for use in electrical or electronic applications. The idea is to show components as features that are recognizable by laypersons. This type of pictorial schematic is commonly used in automobile operations manuals, as shown in Figure 25.88.

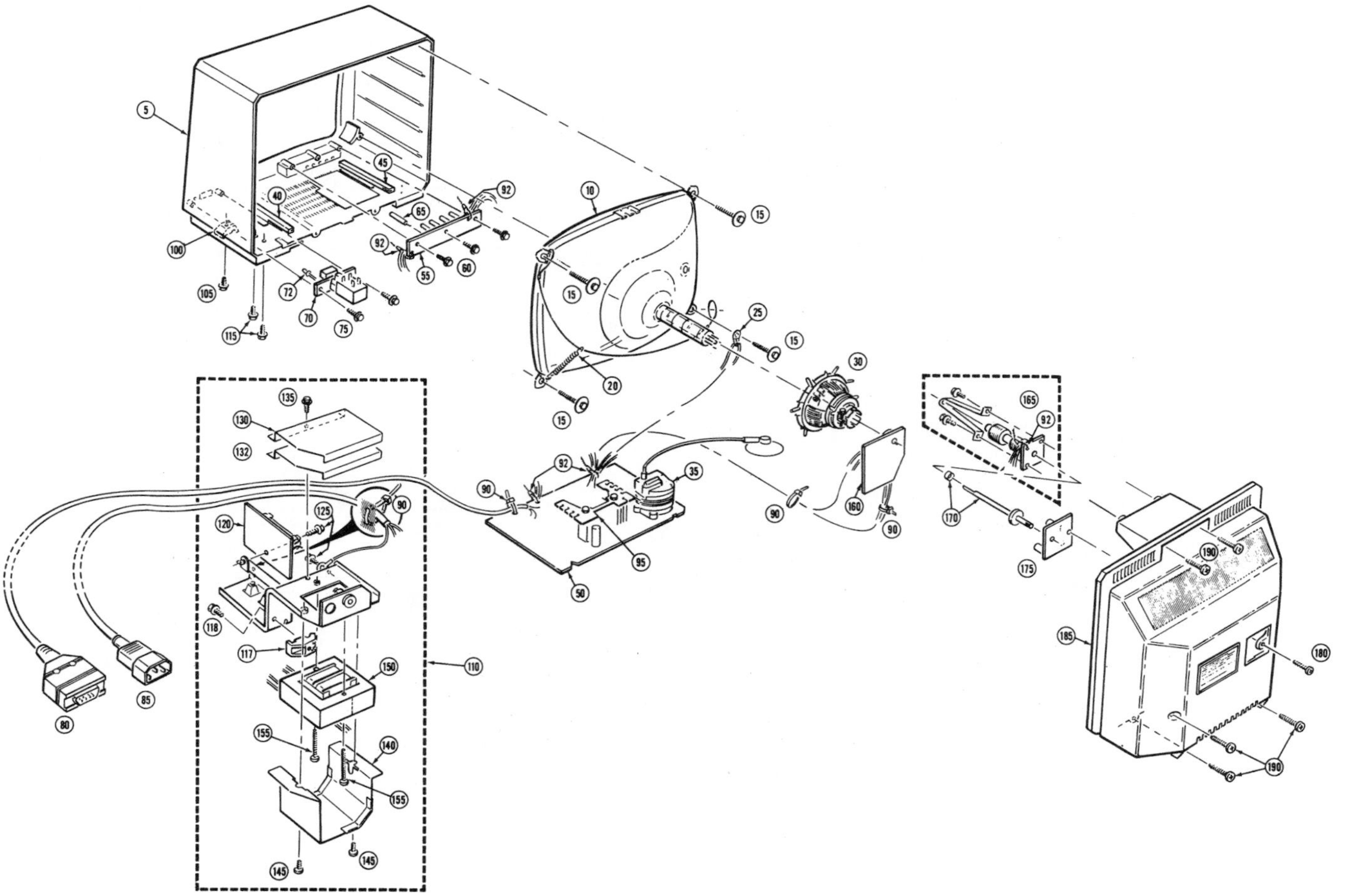

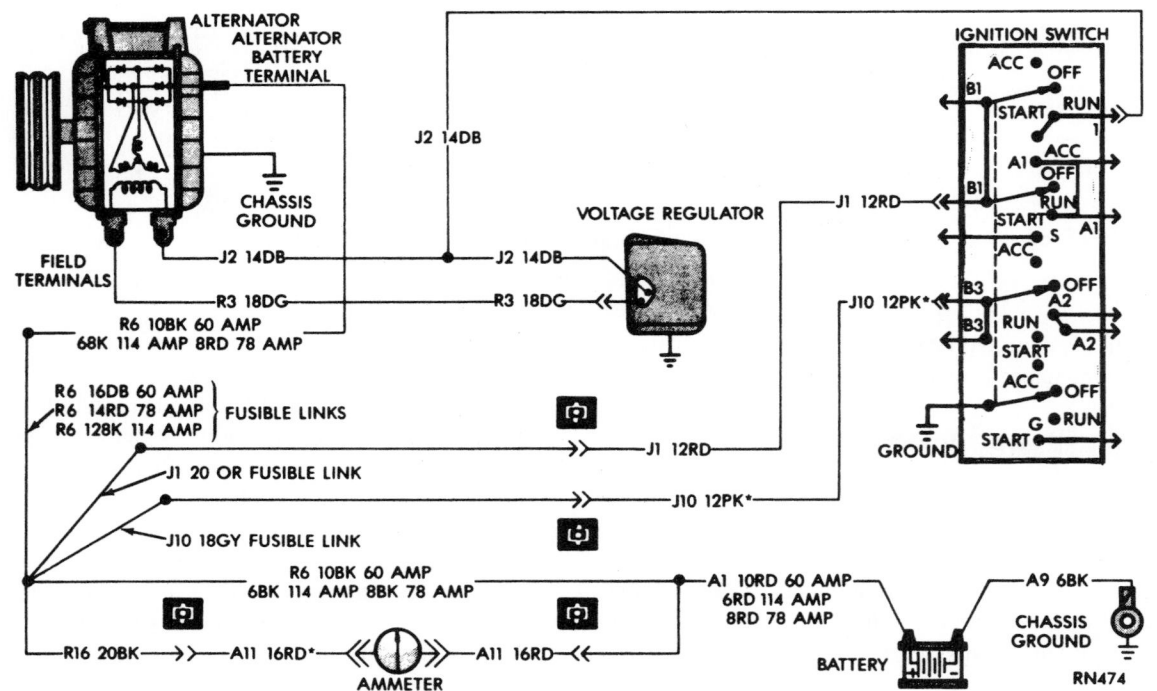

FIGURE 25.87 ■ Parts pictorial, exploded technical illustration. *Reprinted by permission of HEATH COMPANY.*

FIGURE 25.88 ■ Semipictorial wiring diagram. *Courtesy Chrysler Corporation.*

CADD APPLICATIONS

COMPUTER-AIDED ENGINEERING (CAE)

With electronic symbols established in a CADD template or symbols library, the design and layout of block diagrams, schematic diagrams, and other electronics drawings are greatly enhanced. The implementation of computers in this industry has advanced into what is known as *computer-aided engineering* (CAE), where the CADD function is taken further into the complete design, engineering, and functional analysis of the product. Beginning with the development of a computer-aided schematic design, the CAE program can perform the following functions:

1. It will stimulate the circuit operation.
2. It will test the system for possible problems.
3. It can identify and analyze design complexity to help reduce manufacturing costs.
4. Thermal characteristics of the circuitry are evaluated to identify possible overheating situations where heat sinks may be needed or mechanical cooling is required.
5. Arced or 45° corners, rather than sharp corners, are automatically placed in the conductor traces during the routing process.
6. Artwork masters are generated at 1:1 scale for all circuit levels, eliminating costly and error-prone photoreduction, as shown in Figure 25.89.
7. The solder resist mask is generated from the same database as the artwork master. This improves the manufacturability of the board and reduces external changes. (See Figure 25.90.)
8. The component layout that is created as part of the symbols can be extracted to create a marking drawing silk screen master, as shown in Figure 25.91.

9. The drill template can be generated from the same database as the artwork master and is used to determine plating requirements: generate a drill chart with sizes, quantities, and X-Y coordinate location dimensions; and create a drill pattern or provide direct communication to a computer numerical control (CNC) drilling machine to achieve repeatable board-to-board registration.
10. A description and reference designator may be placed on each symbol and may be used to simultaneously complete assembly drawings, parts lists, and other documentation.
11. A large-format digitizing table may be used to enable the designer to go from rough layout or existing works to tight, on-screen design.
12. Laser-directed photoplotters provide PCB artwork with accuracy tolerances up to 0.4 mil (0.000010 mm) along with image quality and speed that exceed the highest industry standards.
13. Printed circuit conductor routing may be performed automatically with completion on success on most boards of 97–100%. In complex board design, the incomplete part of the trace routing, if any, may be quickly engineered in the manual mode.
14. The designer or engineer may stop the routing process at any time to make component or connection changes, and restart the automatic routing process at the new location.
15. Component locations are designed along with trace routing.

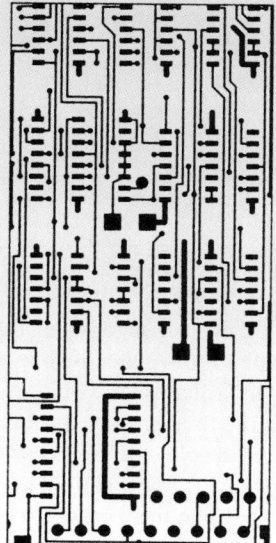

FIGURE 25.89 ■ Full scale CAE artwork eliminating the need to photoreduce. *Courtesy The Gerber Scientific Instrument Company.*

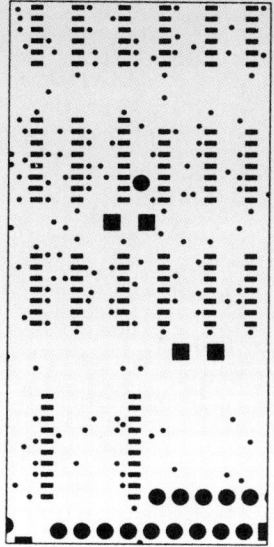

FIGURE 25.90 ■ The solder resist mask is generated from the same database as the artwork master shown in Figure 25.87. *Courtesy The Gerber Scientific Instrument Company.*

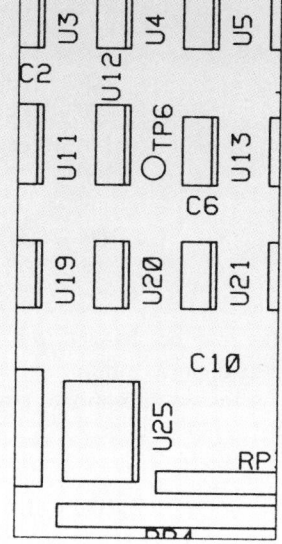

FIGURE 25.91 ■ Marking drawing created from symbols library used to design the artwork master. *Courtesy The Gerber Scientific Instrument Company.*

PROFESSIONAL PERSPECTIVE

As an entry-level drafter with training in electronics schematic drafting, you are able to make engineering drawing changes and prepare drawings from engineering sketches without specific knowledge of how electronic components and the systems function. However, in order to advance in skill level to designer or engineering technician, you need a thorough understanding of company requirements, the ability to communicate electrical and electronic terminology, and the knowledge of how the components go together. It is important for an engineering drafter in this business to visualize the systems and understand electrical clearance requirements. A lot of this can be learned on the job because the engineers are responsible for the design, but as you become experienced, you can assist in identifying engineering problems and troubleshooting systems. Schooling in communication skills and problem-solving techniques, along with electrical and electronic theory and practices, is essential. Helpful college courses in addition to your electronic drafting could include:

■ Computer basics—Windows.

■ Basic AC/DC electronics.

■ Solid state devices.

■ Digital logic.

■ Basic microprocessors.

■ Electrical physics.

When you go to work in this field, you will find that it is extremely important for you to check your work completely, making sure that electrical connections made are necessary. Nearly all drafting is done on a CADD system, so it is important for you to fully understand the computer operating systems. When you are bringing electronic symbols from a menu library, be sure that you leave adequate space to display the symbol, labeling, and circuit notations. Be sure that you set up the schematic in a logical layout sequence so the components, test points, and other items are identified from left to right and top to bottom. You need to know and understand the numbering systems relating to the schematic and the PCB design. This is all part of the knowledge you will gain with experience in the industry. An engineering drafter at Intel warns, "Keep an open mind on new CADD systems and better ways to do things. The hardest part of the job is to constantly learn new systems, because the industry changes so fast. To sum it up, be flexible."

MATH APPLICATION

WORKING WITH POWERS OF TEN

Powers of ten (scientific notation) is used to express the very large and small numbers found in electronics compactly, making multiplying or dividing them much easier. Here are some of the powers of ten.

$$10^4 = 10,000$$
$$10^3 = 1,000$$
$$10^2 = 100$$
$$10^1 = 10$$
$$10^0 = 1$$
$$10^{-1} = .1$$
$$10^{-2} = .01$$
$$10^{-3} = .001$$

A number may be changed to a power of ten form by first writing the digits as a number between 1 and 10, then multiplying it by the appropriate power of ten. Here are some examples to illustrate this:

$$3,400 = 3.4 \times 10^3$$
$$53,000,000 = 5.3 \times 10^7$$
$$.000051 = 5.1 \times 10^{-5}$$
$$.02 = 2 \times 10^{-2}$$

To Multiply: Multiply the numbers out front and add the exponents.

To Divide: Divide the numbers out front and subtract the exponent in the bottom from the exponent on top. Here are examples to illustrate:

$$(2 \times 10^3)(7 \times 10^8) = 14 \times 10^{11} = 1.4 \times 10^{12}$$
$$(3 \times 10^{-6})(2 \times 10^5) = 6 \times 10^{-6+5} = 6 \times 10^{-1}$$
$$(5 \times 10^{-20})(7 \times 10^{-3}) = 35 \times 10^{-23} = 3.5 \times 10^{-22}$$

These results can be obtained on the calculator using the EE, EXP, or EEX buttons, depending on the make of your calculator. For the last example, many calculators would use this sequence of buttons to obtain the final answer: 1.4 EE 15 ÷ 2 EE 5 +/– =. The calculator may display the answer without the "× 10" like this: 7^{19}. However, the answer should never be written down and communicated like this because it would be confused with 7 raised to the 19th power, which is an entirely different number! Always include the "× 10" when writing down a power of ten number.

CHAPTER
25

Electrical and Electronic Schematic Drafting Test

DIRECTIONS
Answer the questions with short complete statements or drawings as needed.

PART 1 • QUESTIONS

1. Explain the basic difference between electrical and electronic drafting.
2. Define *electrical circuit*.
3. Describe a pictorial diagram.
4. Draw a simple schematic diagram.
5. Draw a simple block diagram.
6. Draw a simple wiring diagram.
7. Draw a simple schematic wiring diagram.
8. Define *terminals*.
9. Briefly describe the transmission and distribution of electricity.
10. Describe the function of a substation.
11. Explain the purpose of a block diagram in an electrical power system design.
12. What is the document that provides most electrical symbol standards?
13. Describe the purpose of a substation plot plan.
14. What are the scales most commonly used to draw plot plans?
15. Define *bearing*.
16. Explain the purpose of a bus layout drawing.
17. What is the purpose of a ground layout?
18. What do grounding details show and at what scale are they drawn?
19. Define *conduit*.
20. Describe the purpose of conduit installation details.
21. What is the standard lettering height used on electrical floor plans?
22. How should switch symbols be drawn in relation to the wall upon which they are located?
23. Draw the following electrical symbols:
    Duplex convenience outlet
    Range outlet
    Recessed circuit breaker panel
    Phone
    Light
    Wall-mounted light
    Single-pole switch
    Simplified fluorescent light fixture
    Fan
24. What method is used to describe special characteristics in an electrical plan layout?

PART 2

1. Name the American National Standards Institute document that governs the use of electrical and electronic drafting symbols.

2. Describe the purpose of a block diagram.
3. What is the function of a capacitor?
4. Define *inductance*.
5. Describe the function of a resistor.
6. Define *semiconductor*.
7. Name two components that are considered semiconductors.
8. Draw and label a fixed and variable resistor.
9. Draw and label a circuit connection and a crossover.
10. Define *active devices*.
11. Define *bias*.
12. How should resistors be placed in relationship to the active devices?
13. What is the advantage of a folded-down schematic?
14. Describe the common sequence standard for component identification.
15. Give the standard letter identification for the following components: resistor, capacitor, transistor, tube, diode, inductor, plug, and jack.
16. Reduce the value 4,000 ohms to its simplest form.
17. Reduce the value 0.000000000001 farads to its simplest form.
18. Reduce the value 1,000,000 henrys to its simplest form.
19. Draw an example of a horizontal capacitor (C1, 5μf) and a vertical resistor (R1, 1,000 ohms) connected, and use the most popular labeling technique to show the identification and value of each.
20. Define *amplifier*.
21. Define *integrated circuit*.
22. What does the dashed line box around an internal IC schematic diagram denote?
23. Define *gate*.
24. Why is it possible that an IC schematic may be much easier to lay out and draw than a conventional schematic?
25. What does the abbreviation *LSI* denote?
26. How are electronic pictorials used in industry?
27. What is the main advantage of surface-mounted devices?
28. Printed circuits are used for what purpose?
29. Describe pads in PCB layout.
30. Why are holes drilled at the pad location?
31. Identify three factors that influence the size of conductor traces.
32. Make a general statement describing printed circuit artwork.
33. What scale is the master pattern?

34. What is the function of a grid system in PCB design?
35. Why is manual PCB layout usually prepared at an enlarged scale?
36. Describe the three basic PCB configurations: single layer, multilayer, and multilayer sandwich.
37. Describe a solder mask.
38. Describe unsupported and plated-through holes.
39. What general note should be provided when holes are to be plated prior to drilling?
40. Define *ground plane*.
41. What does the term *dolls* refer to in PCB layout?
42. Describe a silk screen artwork.
43. Describe a drilling drawing.
44. Identify the components that normally make up a complete assembly drawing.
45. Describe computer-aided engineering as related to electronics engineering.

CHAPTER 25

Electrical and Electronic Schematic Drafting Problems

DIRECTIONS

1. Please read all instructions before you begin working, unless otherwise specified by your instructor.
2. Use manual or computer-aided drafting as required by your course guidelines.
3. Use the attached engineering layouts to prepare each drawing. Refer to chapter coverage for symbol and component dimensions and also refer to previous drawings for information as you progress to each drawing.
4. Prepare well-balanced, easy-to-read drawings. Most drawings have no scale.
5. Many of the following problems contain symbols that are duplicated multiple times. These electrical problems demonstrate the power of CADD when you create one symbol and have the ability to use it several times on one or more drawings.
6. Additional information is provided for some problems.
7. Estimate unknown dimensions.

PART 1

PROBLEM 25-1   Switch panel

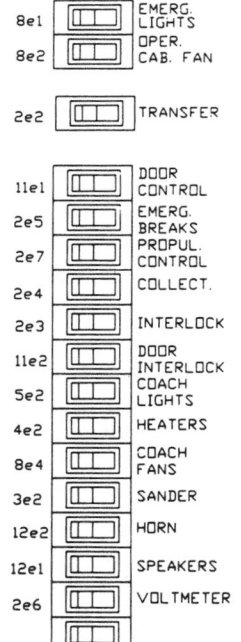

PROBLEM 25-2   Schematic

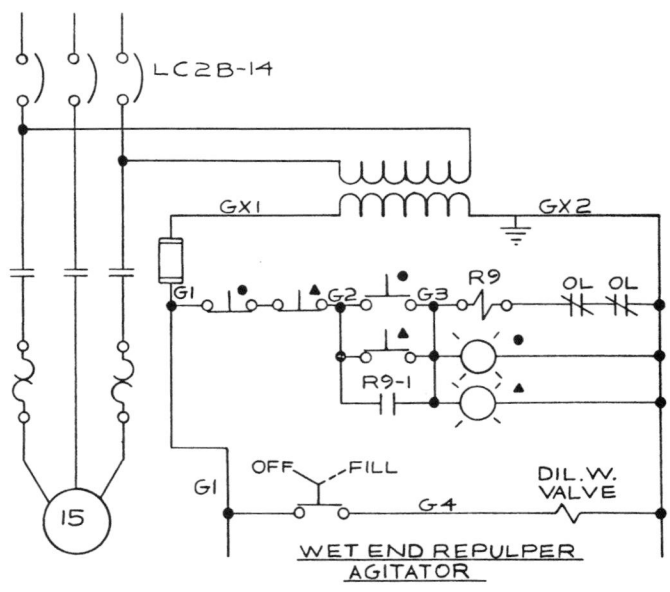

PROBLEM 25-3   Schematic

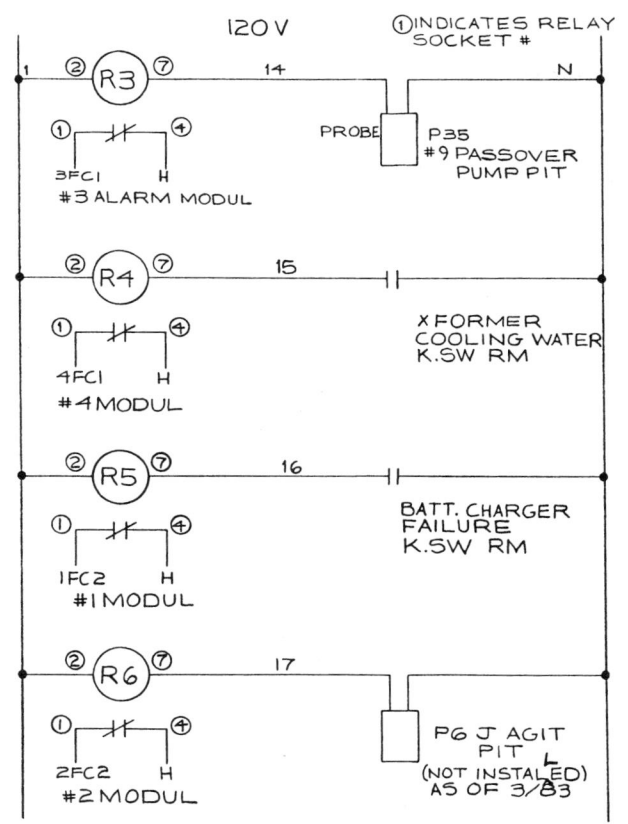

PROBLEM 25-5   Schematic

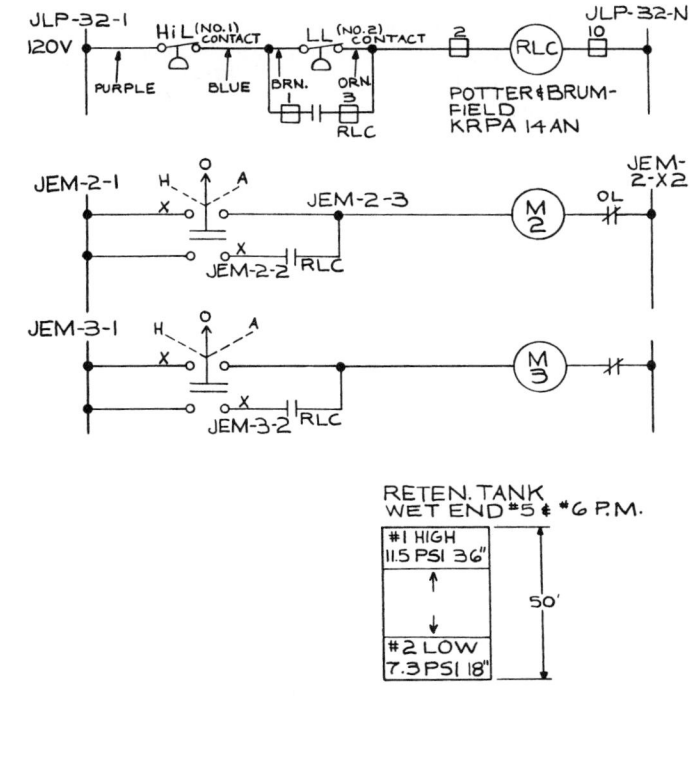

PROBLEM 25-4   Schematic

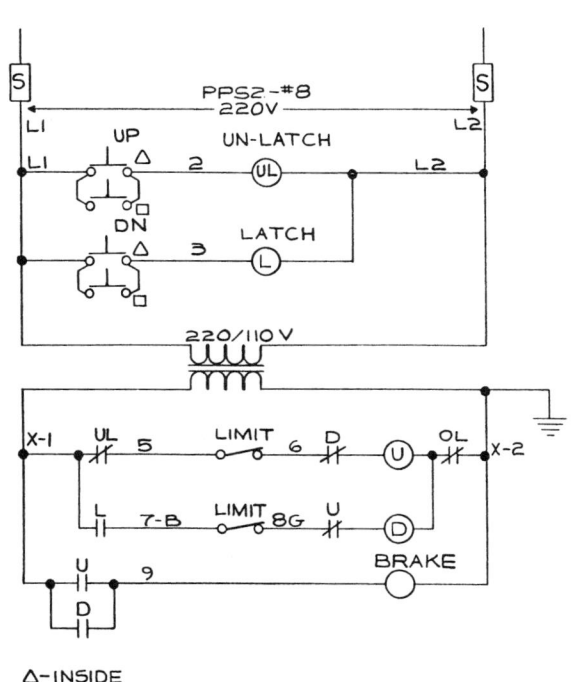

Δ—INSIDE
□—OUTSIDE

PROBLEM 25-6   Schematic

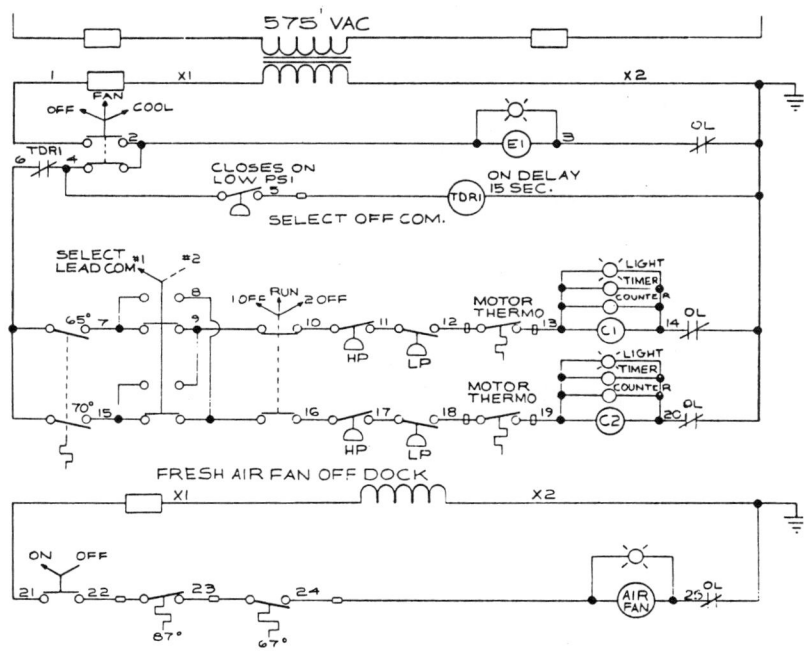

PROBLEM 25-7 Block diagram

Given the following engineering layout of the Potlatch-Pearl Substation, draw the block diagram on 11 × 17 polyester film or vellum unless otherwise specified by your instructor. There is no scale. The layout should neatly fill the sheet without crowding symbols or notes. Make all connection points Ø3/32 in. Provide the general note: INTERPRET PER ANSI Y32.2. *Courtesy Bonneville Power Administration.*

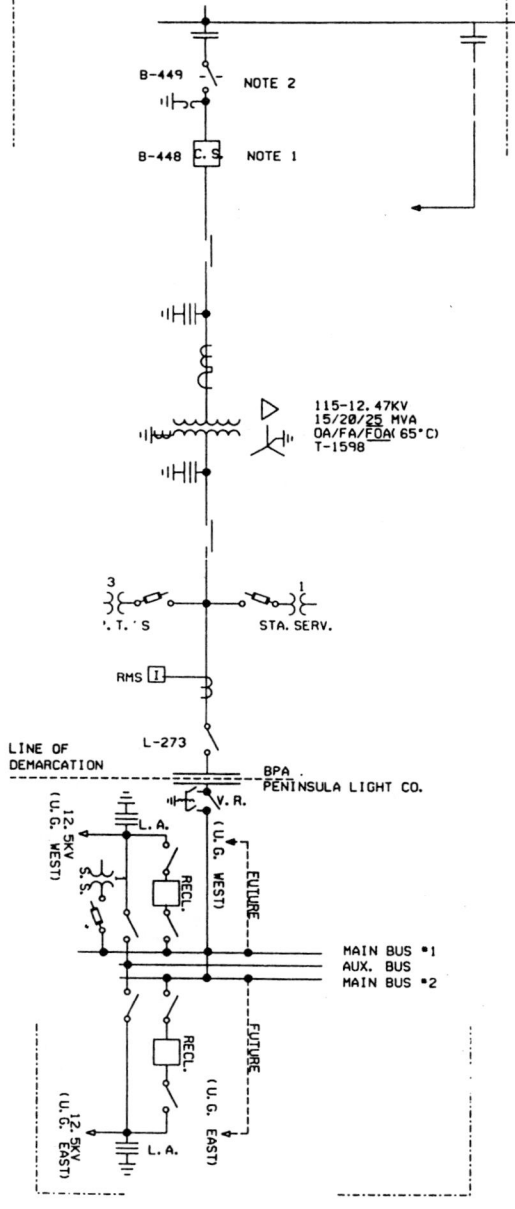

PROBLEM 25-8 Elementary diagram

Metering System.
Courtesy Bonneville Power Administration.

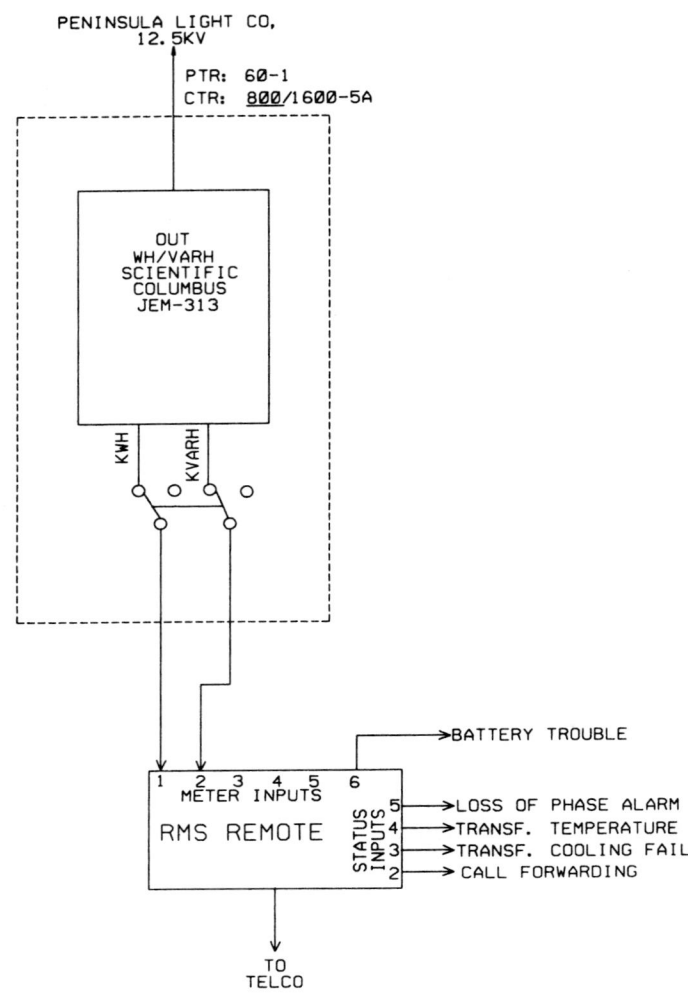

PROBLEM 25-9 Highway diagram

Draw all component outlines and feeder lines 0.5 mm wide and trunk lines 0.9 mm wide.

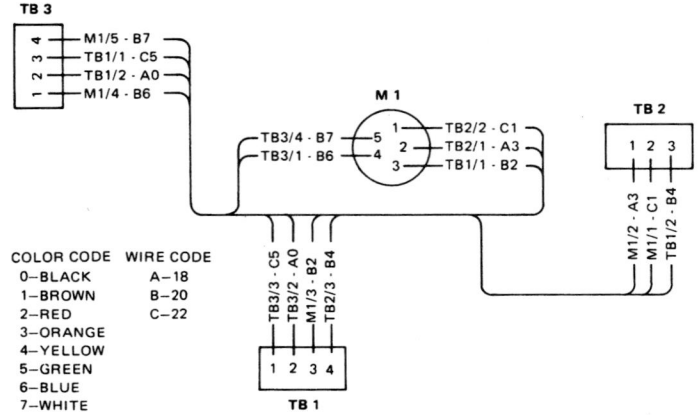

PROBLEM 25-10 Wireless diagram

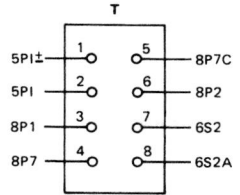

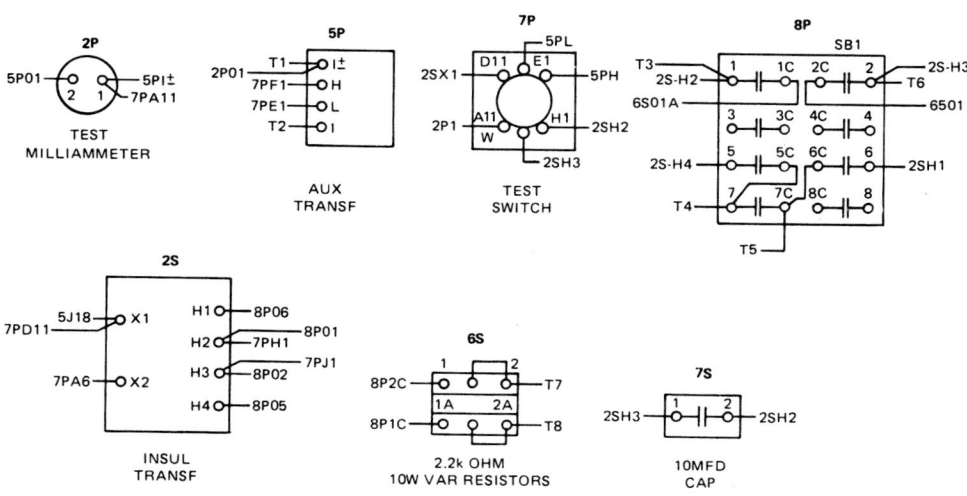

PROBLEM 25-11 Cable diagram

Display Panel-CVS-3 COND. Make all panel outlines and feeder lines 0.5 mm wide and all trunk lines 0.9 mm wide. Do all lettering ⅛ in. high, with component labels ³⁄₁₆ in. high and titles ¼ in. high.

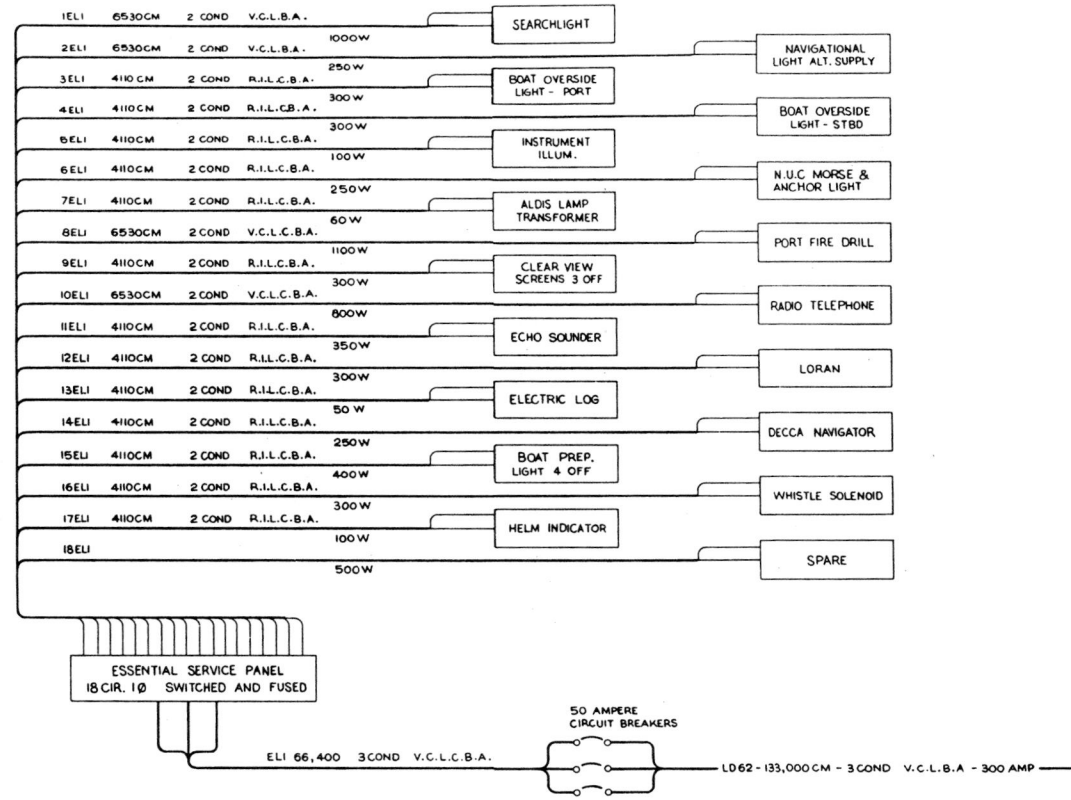

CABLE DIAGRAM FOR INSTRUMENT DISPLAY PANEL - CVS - 3 COND.

PROBLEM 25-12   Wiring diagram

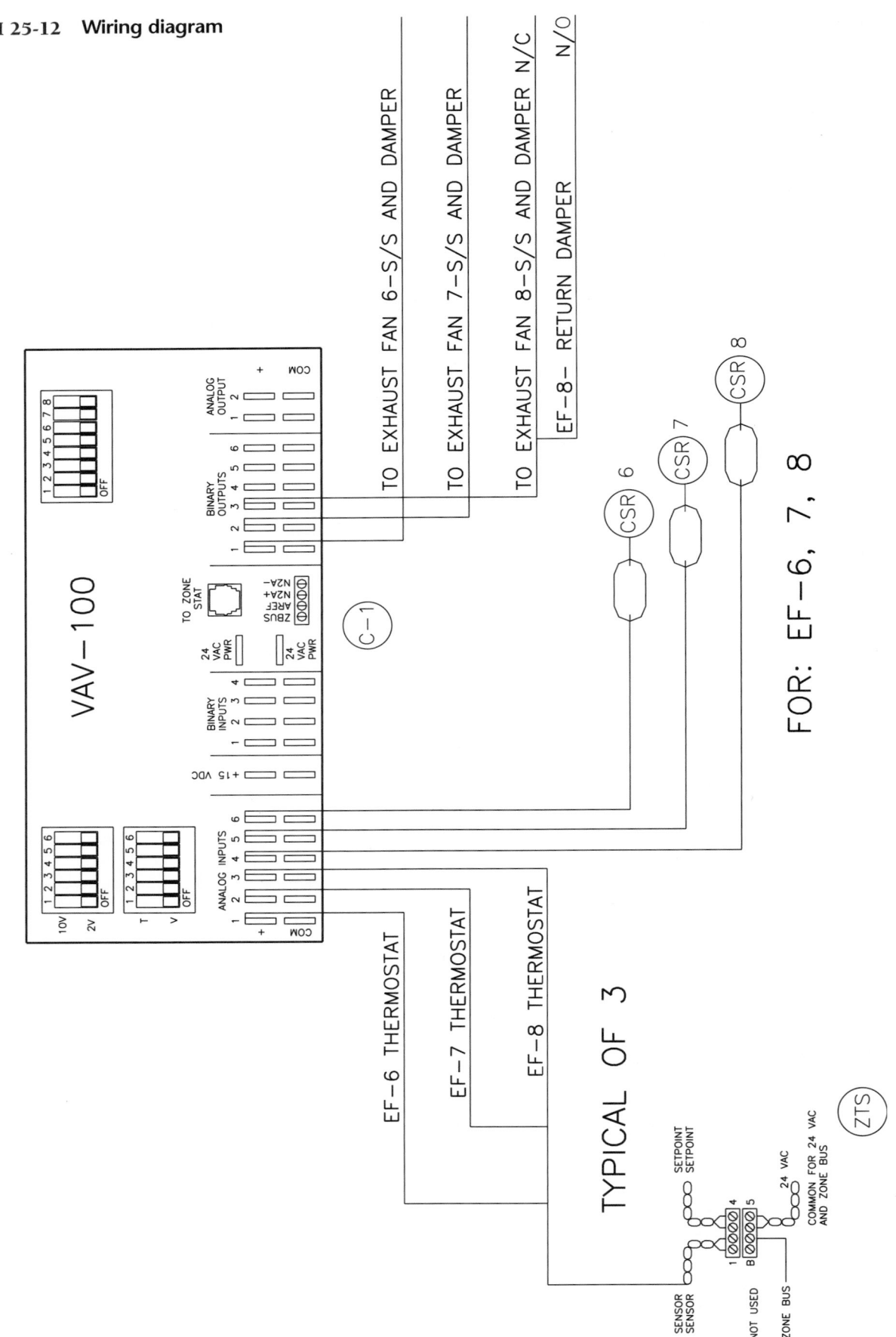

PROBLEM 25-13 Wireless diagram

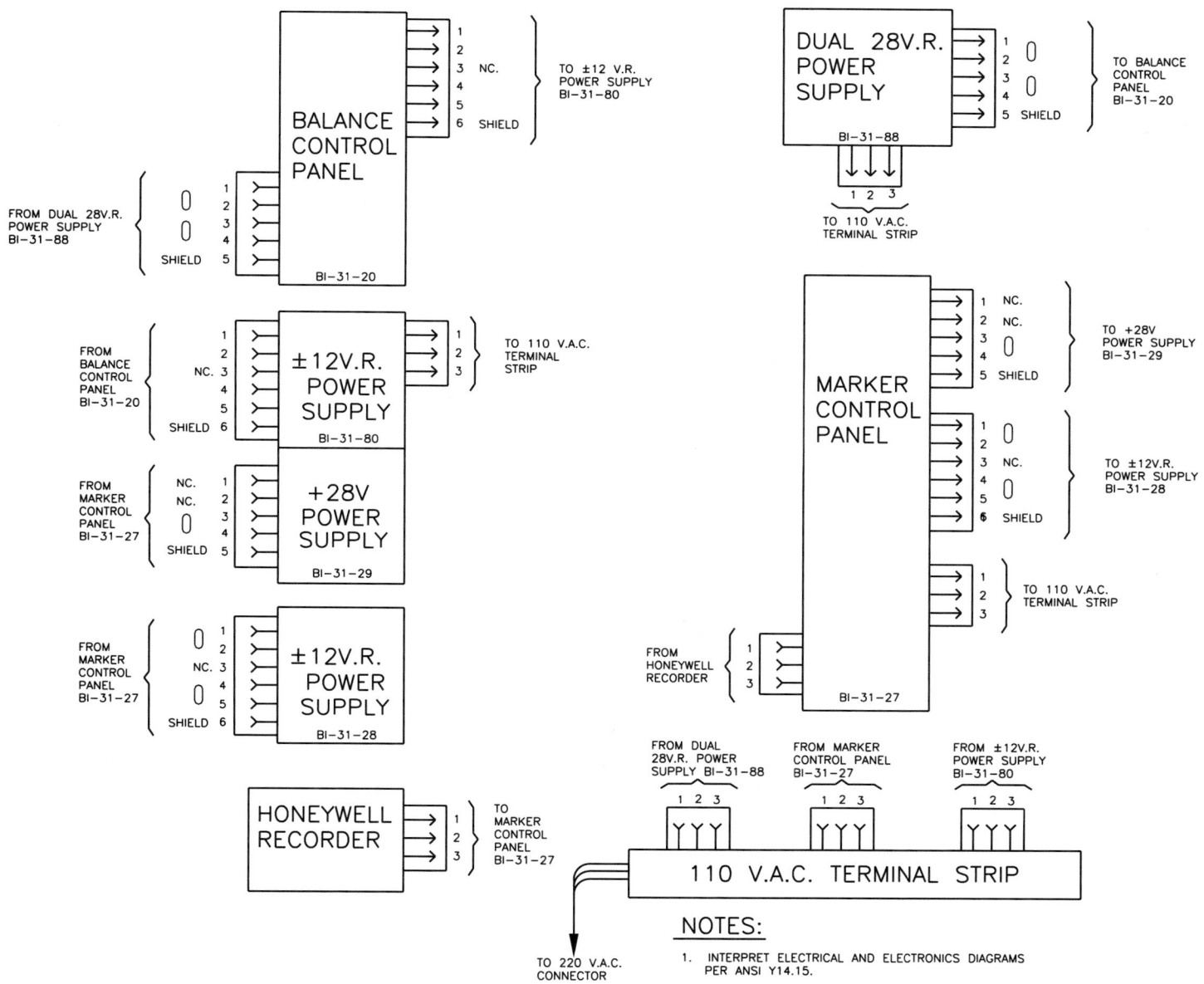

PROBLEM 25-14 Block diagram

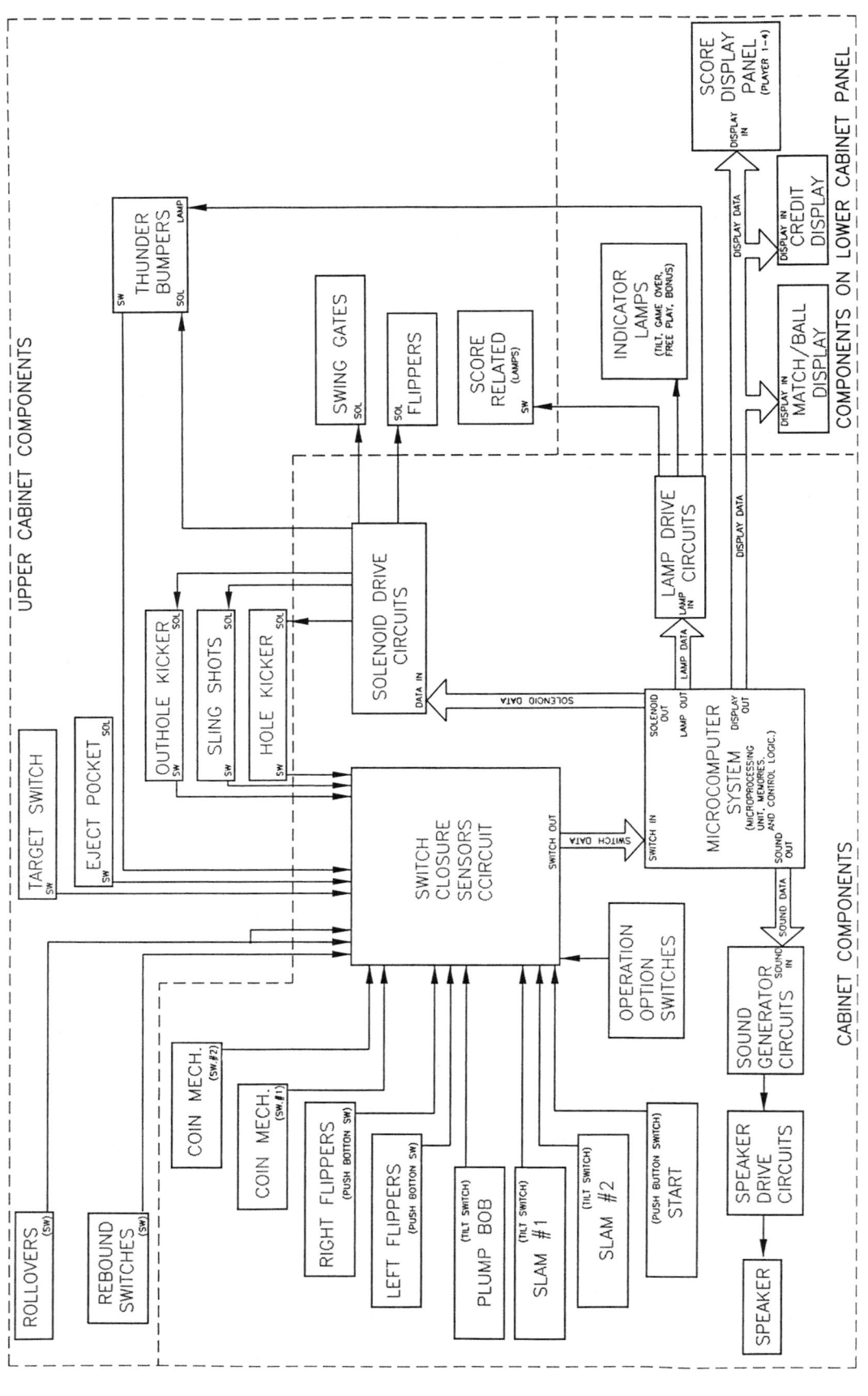

PROBLEM 25-15 Block diagram

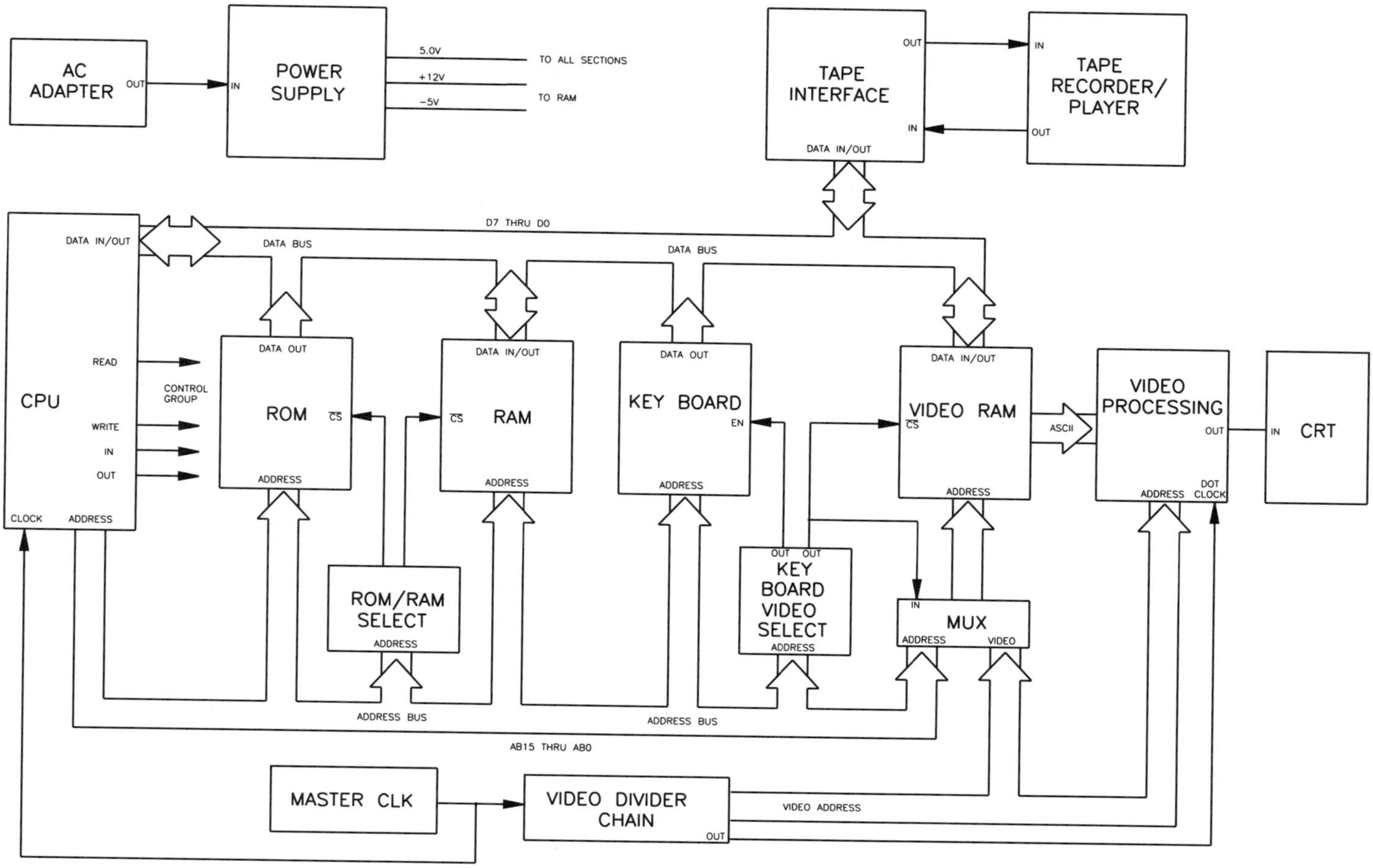

```
┌──────────┐         ┌──────────┐                              ┌──────────┐         ┌──────────┐
│    AC    │         │  POWER   │  5.0V ── TO ALL SECTIONS      │   TAPE   │ OUT ──▶ │   TAPE   │
│ ADAPTER  │ OUT ──▶ │  SUPPLY  │  +12V                        │INTERFACE │         │ RECORDER/│
│          │     IN  │          │  -5V  ── TO RAM              │          │ IN  ◀── OUT│ PLAYER  │
└──────────┘         └──────────┘                              │          │         └──────────┘
                                                               │ DATA IN/OUT│
```

(Block diagram: CPU, ROM, RAM, KEY BOARD, VIDEO RAM, VIDEO PROCESSING, CRT, ROM/RAM SELECT, KEY BOARD VIDEO SELECT, MUX, MASTER CLK, VIDEO DIVIDER CHAIN, with DATA BUS (D7 THRU D0) and ADDRESS BUS (AB15 THRU AB0) connections.)

PROBLEM 25-16 Wiring harness

Courtesy Flir Systems, Inc.

1. INTERPRET DRAWING IAW MIL-STD-100, CLASSIFICATION PER MIL-T-31000, PARA 3.6.4.

2. INTERPRET DIMENSIONS AND TOLERANCES PER ASME Y14.5M-1994.

△3. COVER SOLDER JOINTS USING ITEM 8 (TUBING-SHRINK 3/16").

△4. PREP WIRES AND ASSEMBLE CONTACTS PER XXXXXXXX.

△5. MARK REFERENCE DESIGNATORS ON HOUSING PER MIL-M-13231.

6. BAG ITEM AND IDENTIFY IAW MIL-STD-130, INCLUDE CURRENT REV LEVEL: 64869ASSYXXXXXXXX REV_.

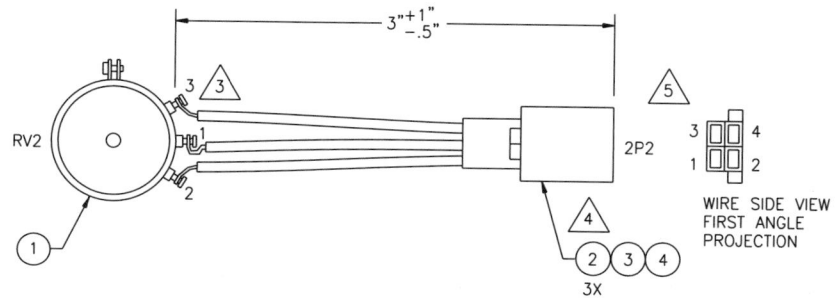

WIRE LIST				
WIRE #	GA/COLOR/ITEM	FROM	TO	SIGNAL
1	22/ORN/6	RV2-1	2P2-2	AZ +12V
2	22/GRN/7	RV2-2	2P2-3	AZ REF
3	22/BLU/5	RV2-3	2P2-4	AZ -12V
UNUSED POSITIONS: J3-D, E, F				

8		TUBING - SHRINK 3/16" BLK	.75"
7		WIRE - #22 STRD, TFE, GRN	4"
6		WIRE - #22 STRD, TFE, ORN	4"
5		WIRE - #22 STRD, TFE, BLU	4"
4		SHROUD - LATCHING DOUBLE ROW 4 POS	1
3		HOUSING - .025 SQ DUAL 4 POS LATCH	1
2		CONTACT - CRIMP PIN .025 SQ 22-26 GA	3
1		RESISTOR - VARI, 10K OHM SERVO MOUNT	1

PROBLEM 25-17   Wiring harness

Courtesy Flir Systems, Inc.

1. INTERPRET DRAWING IAW MIL-STD-100. CLASSIFICATION PER MIL-T-31000, PARA 3.6.4.

2. INTERPRET DIMENSIONS AND TOLERANCES PER ASME Y14.5M-1994.

3. OVERALL FINISHED LENGTH:
   W1 = 15.5"±1"
   W2 = 14"±1"
   W3 = 12.5"±1"
   W4 = 11"±1"

4. ASSEMBLE PER STD XXXXXXXX.

5. ALIGN P1, P2, P3, & P4. INSTALL FIRST CABLE TIE (ITEM 5) 2" MINIMUM FROM P1. EQUALLY SPACE REMAINING CABLE TIES.

6. DISCARD SOLDER TAIL.

7. INSTALL 1/2" OF ITEM 6 (TUBING – SHRINK 1/8") OVER EACH SOLDER TERMINATION ON ITEM 4 (CONNECTOR).

8. PREP AND ASSEMBLE ITEM 3 (CONNECTOR – COAX) PER PREP STD XXXXXXXX

9. BAG ITEMS AND IDENTIFY IAW MIL-STD-130, INCLUDE CURRENT REV LEVEL: 64869ASSYXXXXXXXX REV___.

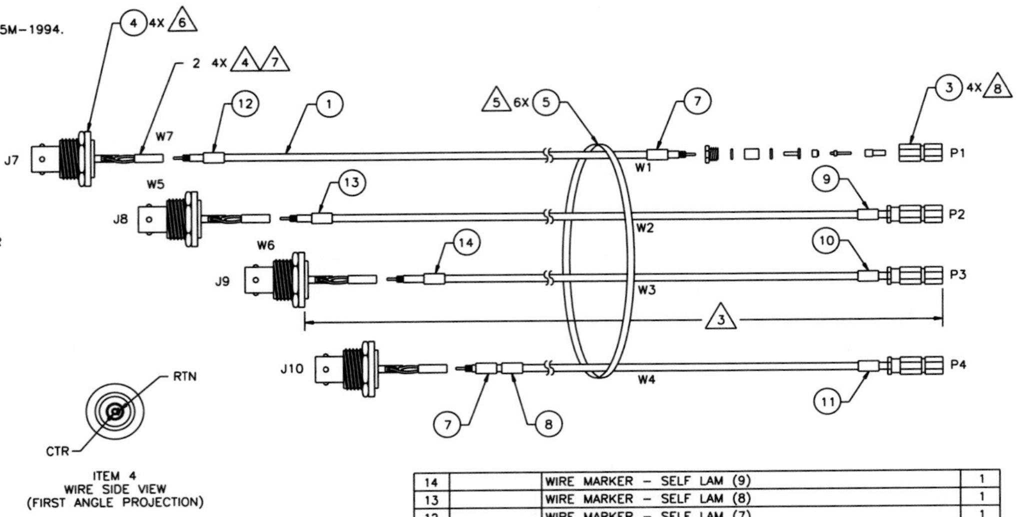

ITEM 4
WIRE SIDE VIEW
(FIRST ANGLE PROJECTION)

WIRE LIST						
WIRE	COLOR	FROM	TO	SIGNAL	CUT LENGTH	
W1	WHT	J7-CTR	P1	CENTER	15.5"	
W1	WHT/BLK	J7-RTN	P1	SHIELD	15.5"	
W2	WHT	J8-CTR	P2	CENTER	14"	
W2	WHT/BLK	J8-RTN	P2	SHIELD	14"	
W3	WHT	J9-CTR	P3	CENTER	12.5"	
W3	WHT/BLK	J9-RTN	P3	SHIELD	12.5"	
W4	WHT	J10-CTR	P4	CENTER	11"	
W4	WHT/BLU	J10-RTN	P4	SHIELD	11"	

14	WIRE MARKER – SELF LAM (9)	1
13	WIRE MARKER – SELF LAM (8)	1
12	WIRE MARKER – SELF LAM (7)	1
11	WIRE MARKER – SELF LAM (4)	1
10	WIRE MARKER – SELF LAM (3)	1
9	WIRE MARKER – SELF LAM (2)	1
8	WIRE MARKER – SELF LAM (0)	1
7	WIRE MARKER – SELF LAM (1)	2
6	TUBING – SHRINK 1/8"	4"
5	CABLE TIE – SELF LOCK .094 W X 3.62 L	6
4	CONNECTOR – TRIAX BULKHEAD INSULATED SOLDER TAIL	4
3	CONNECTOR – COAX (RG179/U), SCREW-ON	4
2	TERMINATION SLEEVE – COAX, RG316/U	4
1	CABLE – COAX, 75 OHM, RG179	56"

PROBLEM 25-18   Photocell wiring diagram

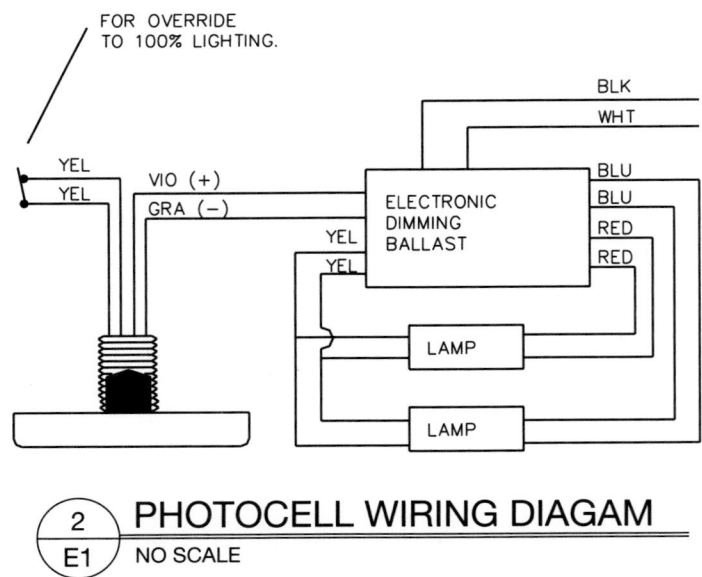

FOR OVERRIDE TO 100% LIGHTING.

2 / E1   PHOTOCELL WIRING DIAGAM   NO SCALE

PROBLEM 25-19 Block diagram

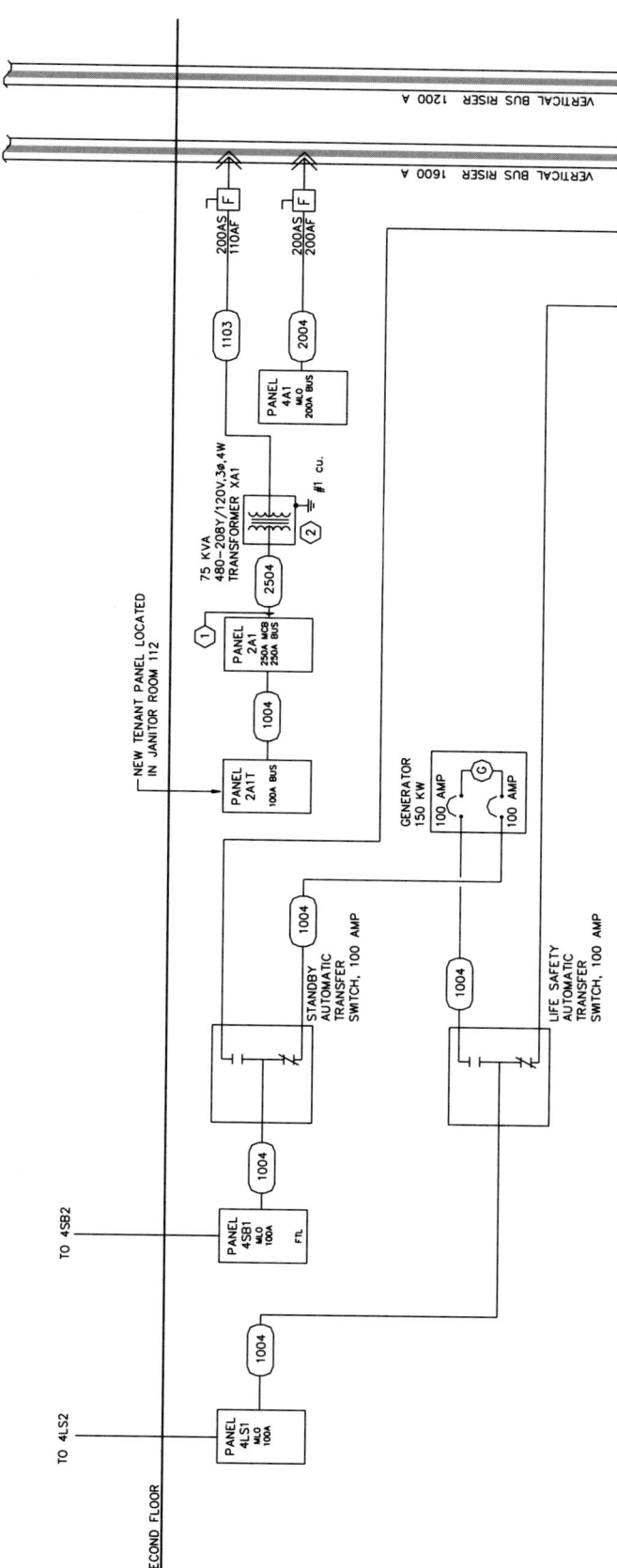

PROBLEM 25-20 Cable assembly

Given the cable assembly engineering layout in Figure 25.12, do the following (approximate dimensions not given):

1. Draw the cable assembly, wiring diagram, and bill of materials.
2. Use ANSI standard line widths for cable assembly and 0.5 mm lines for wiring diagram.
3. Use 1/8-in. lettering with 1/4-in. high letters in balloons and titles.
4. Make balloons ∅1/2 in.
5. Letter the following general notes:
1. INTERPRET DIMENSIONING AND TOLERANCING PER ANSI Y14.5M—1994.
2. INTERPRET DRAWING PER ANSI Y32.2.
3. DIMENSIONS ARE IN INCHES WITH TOLERANCES: FRACTIONS = ± 1/64, .XX = ±.010, .XXX = ±.005.
4. 32 MICRO IN FINISH ON METAL PARTS.
5. MATERIAL BRONZE.
6. USE SPEC 6712 FOR WIRE END PREPARATION.

PROBLEM 25-21 Bus layout plan view

Given the engineering layout shown in Figure 25.16 for Narrows Substation, draw the bus layout using a 3/16" = 1'–0" scale. The following line widths are recommended:

0.35 mm	Extension, dimension, balloon lines (make balloons ∅9/32 in.), all connections
0.50 mm	Component and bus lines, and lettering
0.60 mm	Titles

Draw north arrow S45° W. Bus lines are to be drawn from given dimensions and filled in solid except for connections. Dimensions not given are to be estimated.

PROBLEM 25-22 Bus elevations

Given the partial engineering layout shown in Figure 25.17 for Narrows Substation, draw the bus elevation. Follow all other instructions given in Problem 25–21.

PROBLEM 25-23 Grounding layout

Given the partial engineering layout shown in Figure 25.19 of the Narrows Substation, draw the grounding layout at a scale of 1" = 10'–0" Include notes and grounding table.

PROBLEM 25-24 Ground details

Given the partial engineering layout shown in Figure 25.20 of the Narrows Substation, draw the grounding detail.

PROBLEM 25-25 Conduit installation layout

Given the partial engineering layout shown in Figure 25.21 of the Narrows Substation, draw the conduit installation layout.

PROBLEM 25-26   Conduit installation details

Given the partial engineering layout of the conduit installation detail shown in Figure 25.22 for the Narrows Substation, draw the given views.

PROBLEM 25-27   Residential electrical

Given the typical electrical layouts shown in Figures 25.29 and 25.30, draw each at a scale of 1/4″ = 1′-0″. Place one next to the other. Estimate dimensions making your drawing about twice the size of the given drawing.

PROBLEM 25-28   Power panel detail (shown below)

Narrows Substation.
Courtesy Bonneville Power Administration.

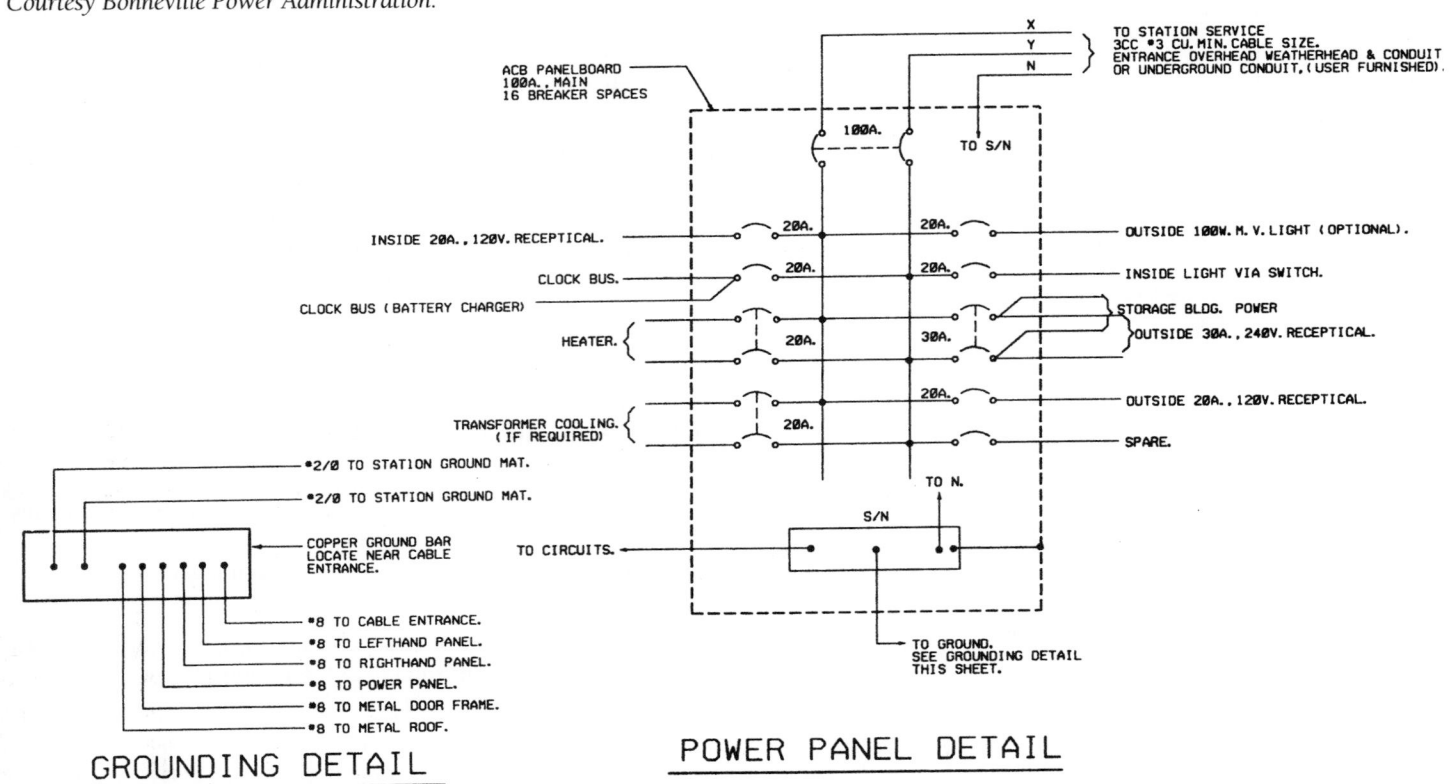

PROBLEM 25-29   Electrical floor plan

Given the partial engineering upper floor powder layout shown in Figure 25.34, draw the plan using a 1/4″ = 1′-0″ scale. Estimate dimensions, making your drawing about three times the size of the given drawing.

PROBLEM 25-30   Reflected ceiling plan

Given the engineering layout shown in Figure 25.35, draw the reflected ceiling plan at a 1/4" = 1'-0" scale. Estimate unknown dimensions.

PROBLEM 25-31   Roof plan—electrical

Given the partial engineering layout shown in Figure 25.36, draw the roof plan—electrical at a 1/4" = 1'-0" scale. Estimate unknown dimensions.

PROBLEM 25-32   Commercial schematic wiring diagram

Given the engineering layout shown in Figure 25.37, draw the beverage cooler condenser unit wiring diagram.

PROBLEM 25-33   First-floor lighting plan

Use the given problem layout to create the lighting plan shown. Use a 1/4" =1'-0" scale to draw the floor plan. Make the dimensions to your own specifications, but proportional to the given engineering drawing. Use appropriate layers when using CADD.
Courtesy Interface Engineering.

LUMINAIRE SCHEDULE

TYPE 'A':   RECESSED FLUORESCENT PARABOLIC LUMINAIRE. 2' BY 4'
STEEL HOUSING. LUMINAIRE PROVIDED BY LANDLORD.
THREE 32W T8 LAMPS.
NOMINAL INPUT WATTS: 93.

TYPE 'A1':   SAME AS TYPE 'A' EXCEPT CONTINUOUS DIMMING BALLAST.

LUMINAIRE SCHEDULE GENERAL NOTES

1   THIS LUMINAIRE SCHEDULE IS NOT COMPLETE WITHOUT A
COPY OF THE PROJECT MANUAL CONTAINING THE ELECTRICAL
SPECIFICATIONS.

2   T8 FLUORESCENT LAMPS TO BE 3500K WITH A MINIMUM CRI
OF 75.

NOTES THIS SHEET

1   2' X 4' PARABOLIC LUMINAIRES ARE TO BE PROVIDED BY LANDLORD

2   TYPE 'B' LUMINAIRES LOCATED BELOW CABINETS.

3   PHOTO CELL TO CONTROL ALL AMBIENT SECTORS AS SHOWN ON DRAWINGS.
ROUTE LOW VOLTAGE CONTROL CIRCUIT VIA PHOTOCELL. SEE DIAGRAM 2/E2.
VERIFY AMOUNT AND LOCATIONS OF PHOTO CELLS WITH MANUFACTURES REP.
PRIOR TO ROUGH-IN. PROVIDE WATTSTOPPER OR EQUAL.

4   CONNECT TO EMERGENCY LIGHTING CIRCUIT IN CORRIDOR.

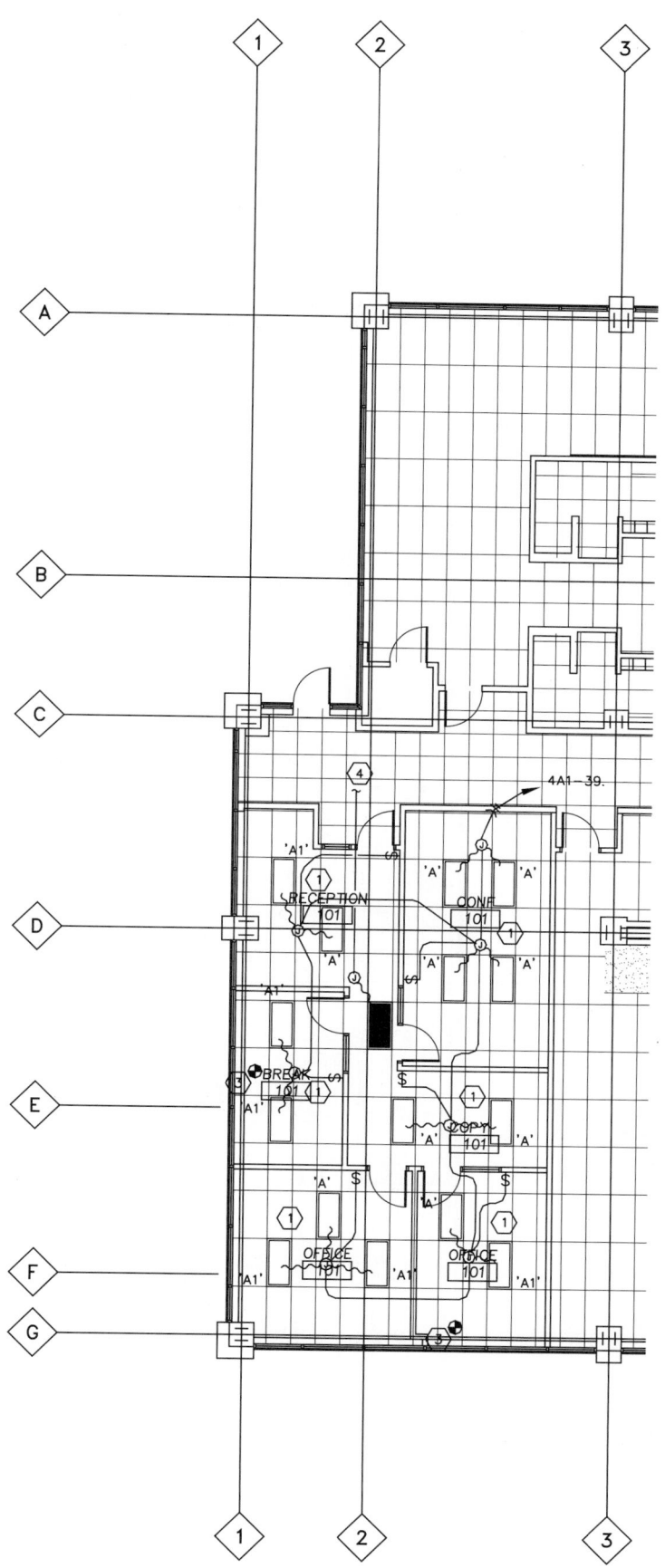

PROBLEM 25-34 **First-floor power plan**

Use the floor plan that you drew in Problem 25–33 as a guide to create the power plan shown in the given engineering drawing. Use appropriate layers when using CADD.

Courtesy Interface Engineering.

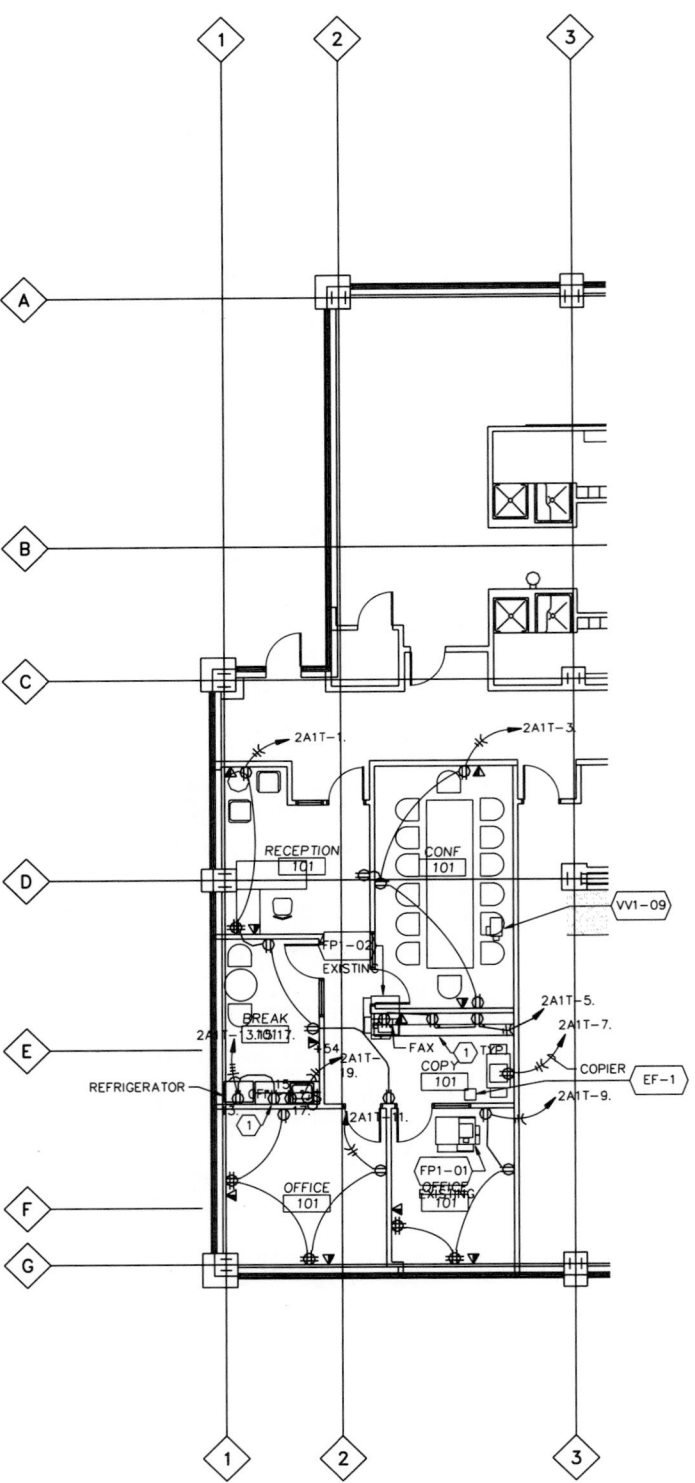

PART 2

DIRECTIONS

1. Follow previous instructions.
2. Use the selected engineering layouts and sketches to prepare each drawing. Keep in mind that engineering sketches may contain slight errors in format and symbol accuracy. Verify proper representation before drawing each symbol.

BLOCK AND SCHEMATIC DIAGRAMS

PROBLEM 25-35   Block diagram

Given the schematic diagram sketch that has been divided into stages, prepare a block diagram using 11 × 17 size. unless otherwise specified by your instructor. Make all lines 0.5 mm wide. Do all lettering 1/8" high with titles 1/4" high. Sketch does not show connection dots.

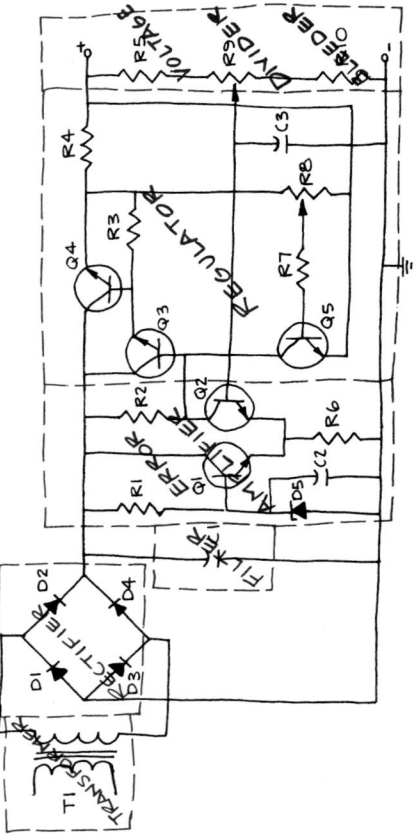

PROBLEM 25-36 Television receiver block diagram

Given the block diagram engineering layout, draw the block diagram using 11 × 17 size. unless otherwise specified by your instructor. Do all lettering 1/8" high with titles 1/4" high. Use 0.5 mm thick lines unless otherwise specified on the engineering sketch. *Courtesy RCA Consumer Electronics.*

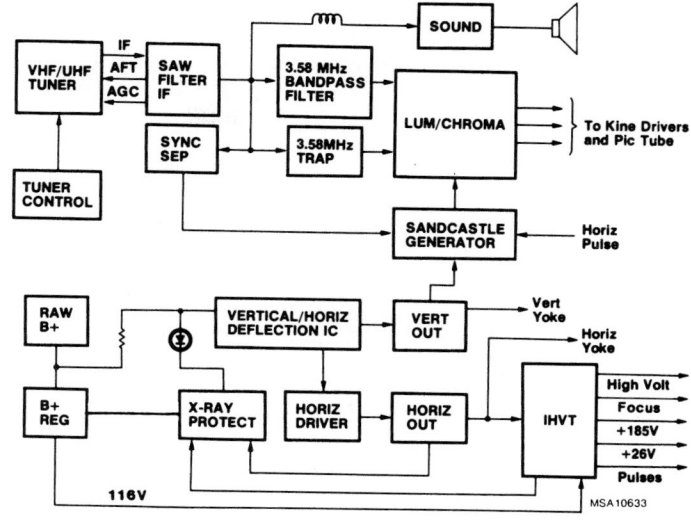

PROBLEM 25-37 Cylinder speed block diagram

Given the block diagram engineering layout, draw the block diagram using 17 × 22 size, unless otherwise specified by your instructor. Do all lettering 1/8" high with titles 1/4" high. Use 0.5-mm thick lines unless otherwise specified on the engineering sketch. *Courtesy RCA Consumer Electronics.*

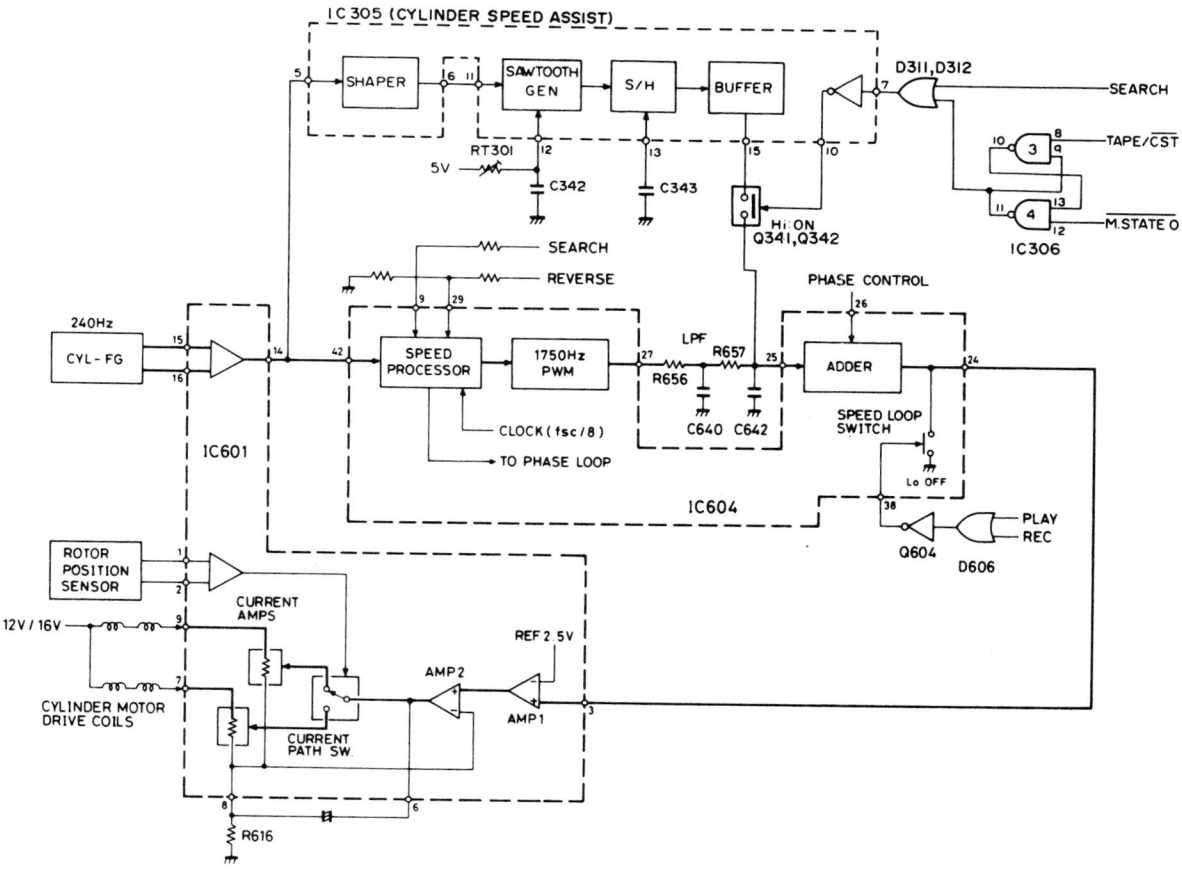

PROBLEM 25-38 Schematic diagram

Given the schematic engineering sketch, make a schematic diagram using 11 × 17 size. There is no scale, and the drawing must be balanced, uncluttered, and easy to read. Use the following instructions unless otherwise specified by your instructor:

1. Draw the entire schematic with 0.5-mm wide lines.

2. Do all lettering 1/8" high with titles 1/4" high.
3. Follow the left-to-right and top-to-bottom labeling system using coding as described in this chapter: R1, R2, R3 . . ., C1, C2. . . .
4. Label horizontal components above and vertical components to the right.

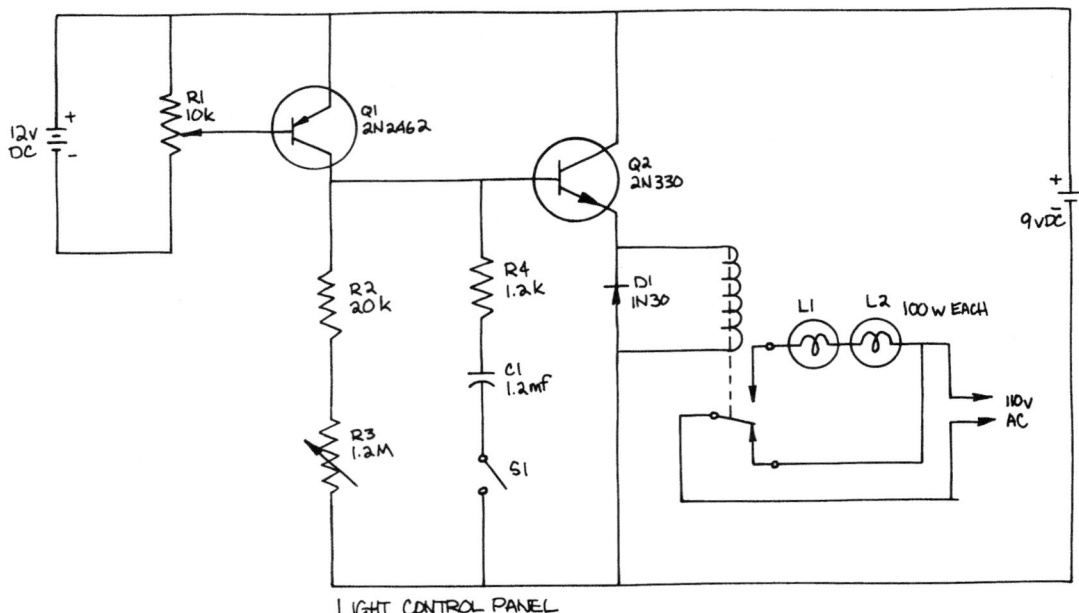

PROBLEM 25-39 Tuner schematic diagram

Given the schematic engineering layout, make a schematic diagram using 11 × 17 size. There is no scale, and the drawing must be balanced, uncluttered, and easy to read. Use the following instructions unless otherwise specified by your instructor.

1. Draw the entire schematic with 0.5-mm wide lines, and dashed stage lines 0.7 mm wide.
2. Do all lettering 1/8" high with titles 1/4" high.

3. Reference designators are not shown; use the left-to-right and top-to-bottom labeling system, using coding as described in this chapter: R1, R2, R3 . . ., C1, C2. . . .
4. Label horizontal components above and vertical components to the right.
5. Place connection dots as necessary.

Courtesy RCA Consumer Electronics.

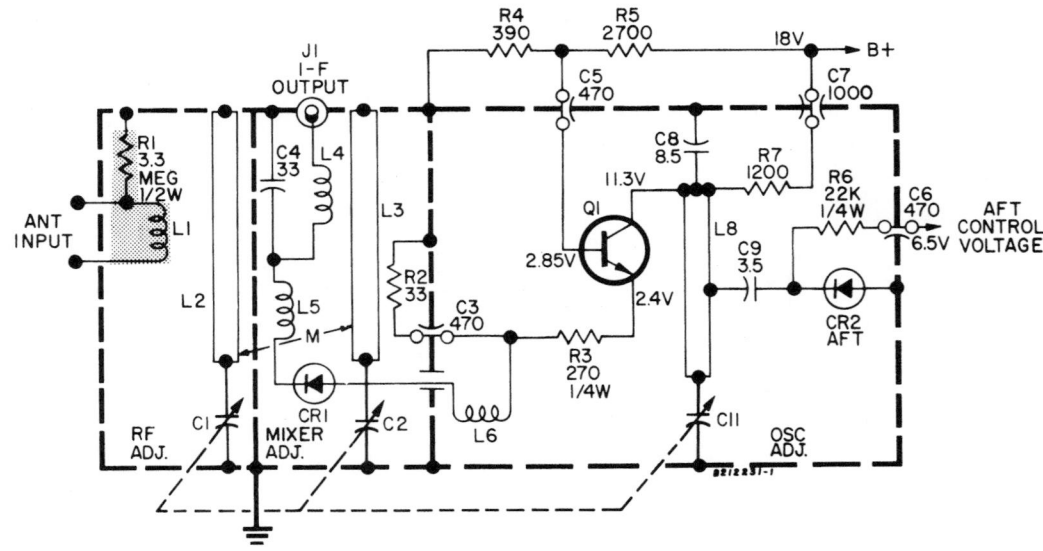

PROBLEM 25-40 Television receiver tuner schematic diagram

Given the schematic engineering sketch, make a schematic diagram using 17 × 22 size. There is no scale, and the drawing must be balanced, uncluttered, and easy to read. Use the following instructions unless otherwise specified by your instructor:

1. Draw the entire schematic with 0.5-mm wide lines, and dashed stage lines 0.7 mm wide.

2. Do all lettering 1/8" high with titles 1/4" high.

3. Reference designators are not shown; use the left-to-right and top-to-bottom labeling system, using coding as described in this chapter: R1, R2, R3 . . ., C1, C2. . . .

4. Label horizontal components above and vertical components to the right.

5. Place connection dots as necessary.

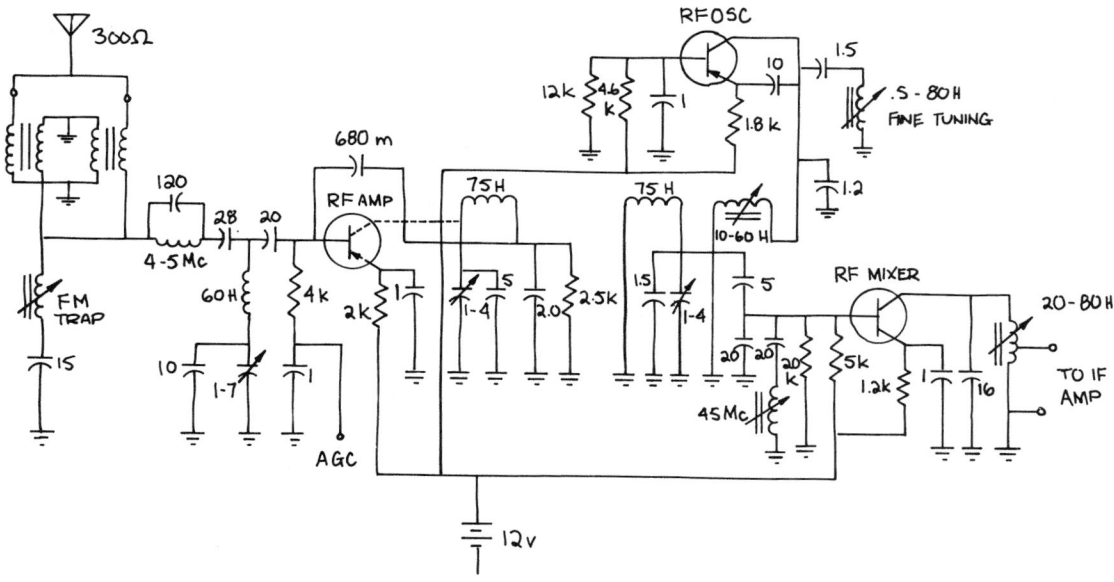

PROBLEM 25-41 Electronics schematic

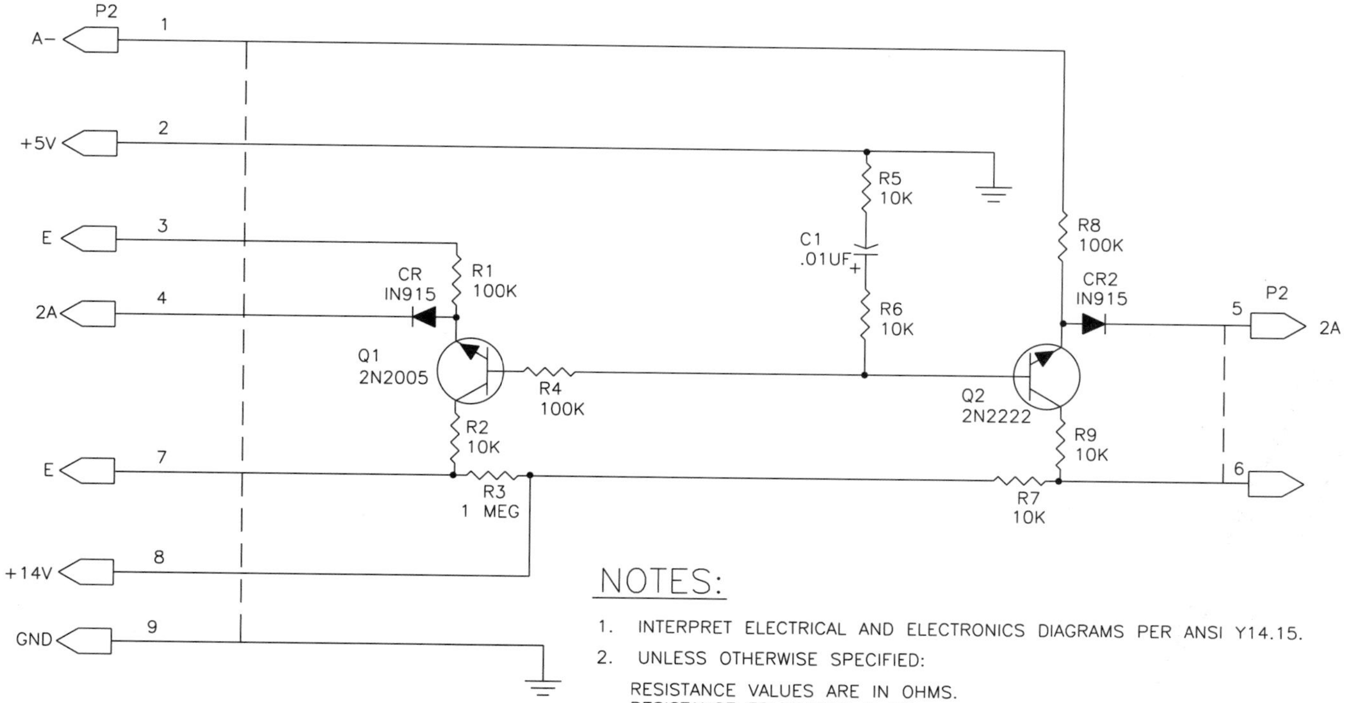

NOTES:

1.   INTERPRET ELECTRICAL AND ELECTRONICS DIAGRAMS PER ANSI Y14.15.
2.   UNLESS OTHERWISE SPECIFIED:

      RESISTANCE VALUES ARE IN OHMS.
      RESISTANCE TOLERANCE IS 5%.
      RESISTORS ARE 1/4 WATT.
      CAPACITANCE VALUES ARE IN MICROFARADS.
      CAPACITANCE TOLERANCE IS 10%.
      CAPACITOR VOLTAGE RATING IS 20V.
      INDUCTANCE VALUES ARE IN MICROHENRIES.

PROBLEM 25-42 **FM tuner schematic**

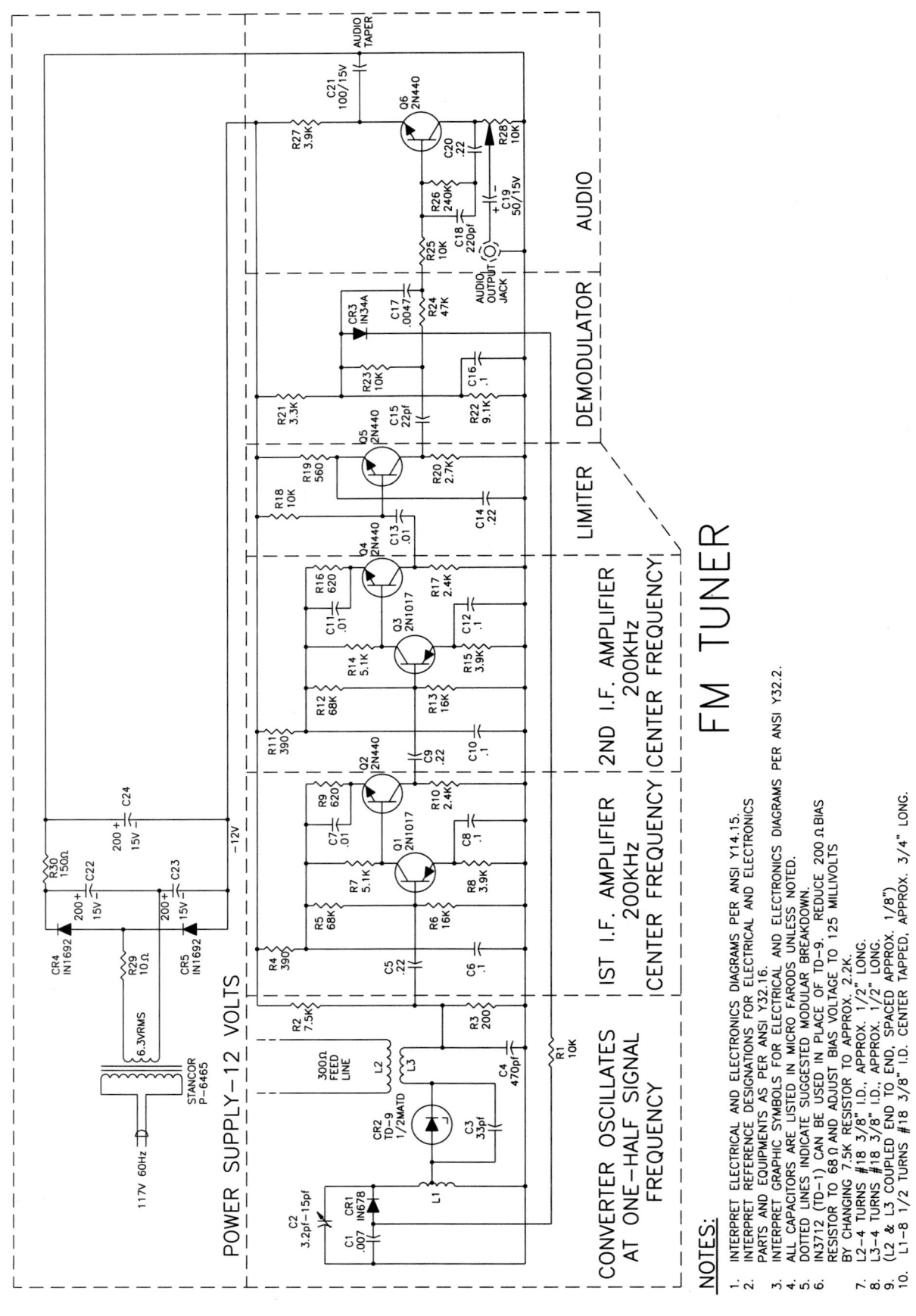

NOTES:

1. INTERPRET ELECTRICAL AND ELECTRONICS DIAGRAMS PER ANSI Y14.15.
2. INTERPRET REFERENCE DESIGNATIONS FOR ELECTRICAL AND ELECTRONICS
   PARTS AND EQUIPMENTS AS PER ANSI Y32.16.
3. INTERPRET GRAPHIC SYMBOLS FOR ELECTRICAL AND ELECTRONICS DIAGRAMS PER ANSI Y32.2.
4. ALL CAPACITORS ARE LISTED IN MICRO FARODS UNLESS NOTED.
5. DOTTED LINES INDICATE SUGGESTED MODULAR BREAKDOWN.
6. IN3712 (TD–1) CAN BE USED IN PLACE OF TD–9. REDUCE 200 Ω BIAS
   RESISTOR TO 68 Ω AND ADJUST BIAS VOLTAGE TO 125 MILLIVOLTS
   BY CHANGING 7.5K RESISTOR TO APPROX. 2.2K.
7. L2–4 TURNS #18 3/8" I.D., APPROX. 1/2" LONG.
8. L3–4 TURNS #18 3/8" I.D., APPROX. 1/2" LONG.
9. (L2 & L3 COUPLED END TO END, SPACED APPROX. 1/8")
10. L1–8 1/2 TURNS #18 3/8" I.D. CENTER TAPPED, APPROX. 3/4" LONG.

PROBLEM 25-43   Logic diagram

Given the IC schematic engineering layout for the logic diagram, make a schematic drawing using 17 × 22 size. Use the following instructions unless otherwise specified by your instructor:

1. Do all line work and lettering using 0.5-mm width.

2. Do all lettering 1/8" high with titles 1/4" high.
3. All connection points will be Ø3/32" filled in.
4. Provide the general note: INTERPRET PER ANSI Y 32.2.

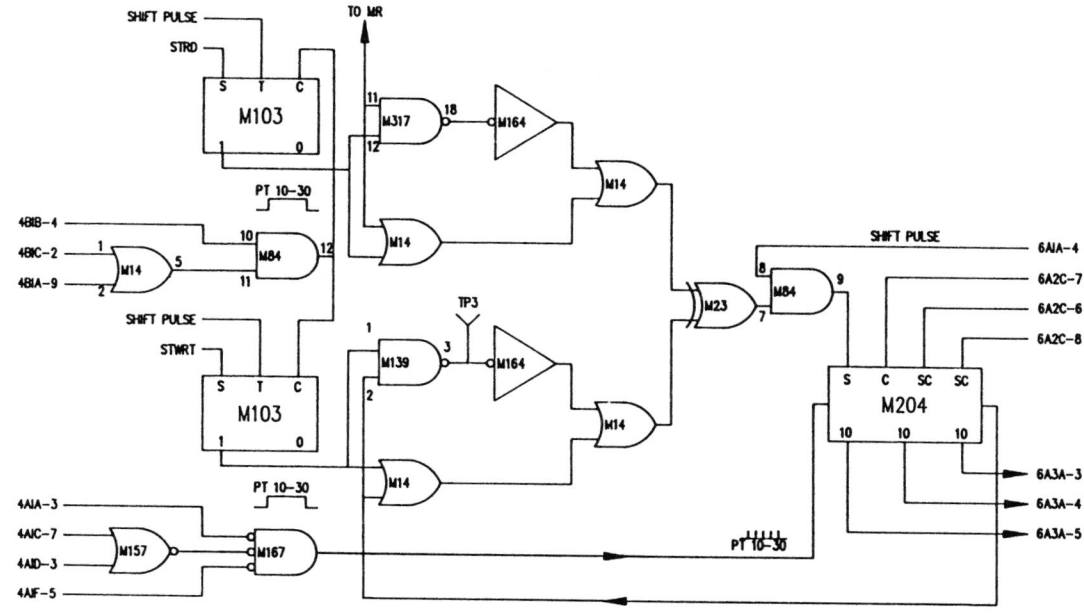

PROBLEM 25-44   Logic diagram

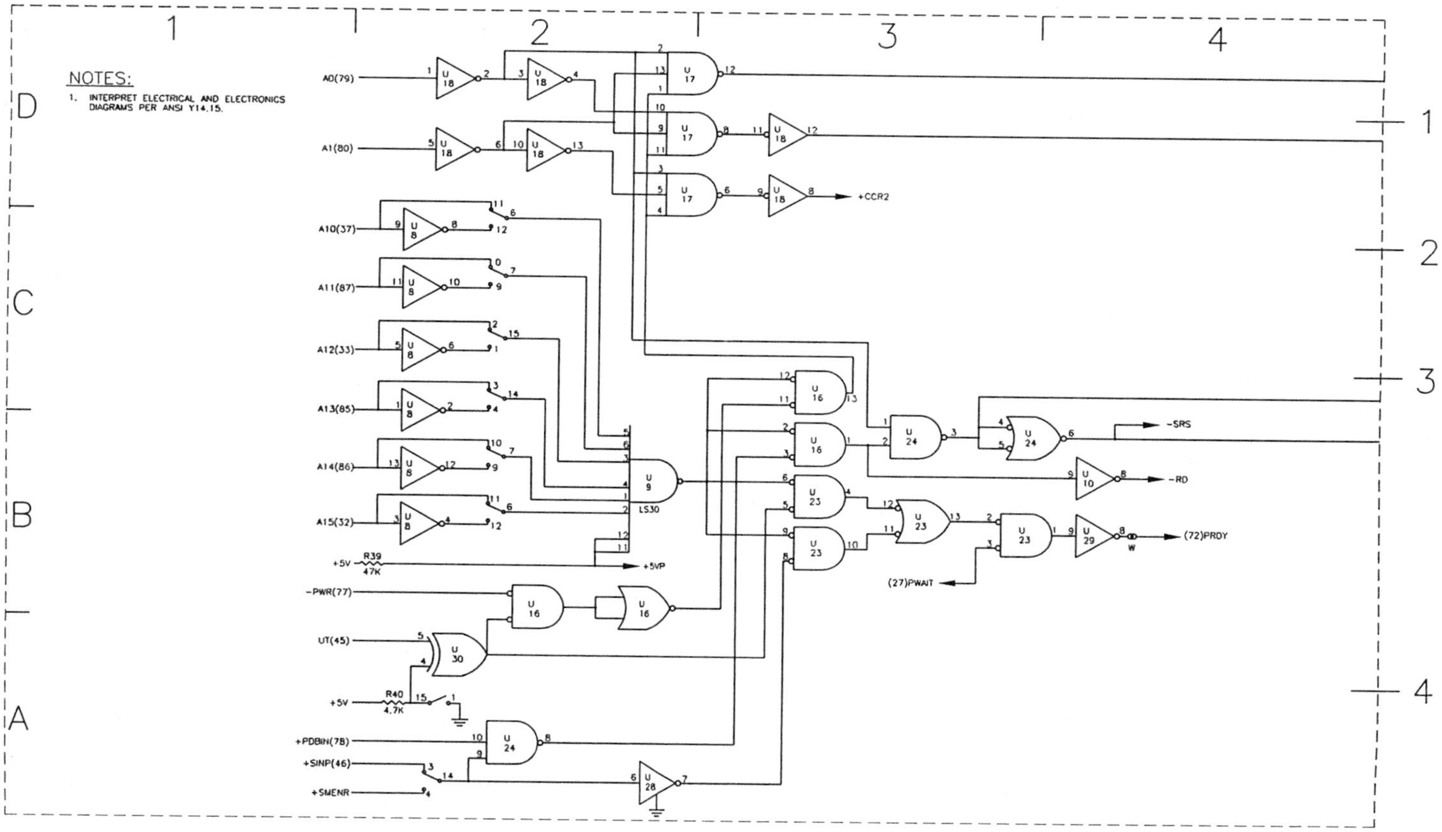

PROBLEM 25-45 Schematic diagram

Given the schematic engineering sketch, make a schematic diagram using 17 × 22 size. There is no scale, and the drawing must be balanced, uncluttered, and easy to read. Use the following instructions unless otherwise specified by your instructor:

1. Draw the entire schematic with 0.5-mm wide lines, and dashed stage lines 0.7 mm wide.
2. Do all lettering 1/8" high with titles 1/4" high.
3. Reference designators are not shown; use the left-to-right and top-to-bottom labeling system, using coding as described in this chapter: R1, R2, R3 . . ., C1, C2. . . .

4. Label horizontal components above and vertical components to the right.
5. Place connection dots as necessary.
   Notes: UNLESS OTHERWISE SPECIFIED.

1. Reference designators are for reference only and may not appear on part.
2. Resistor values are in DHMs.
3. For PWB/component assembly, see dwg. 100/01.
4. For parts, see SPL 100/02.
5. For printed wiring board, see dwg. 100/03.
   Courtesy RCA Consumer Electronics.

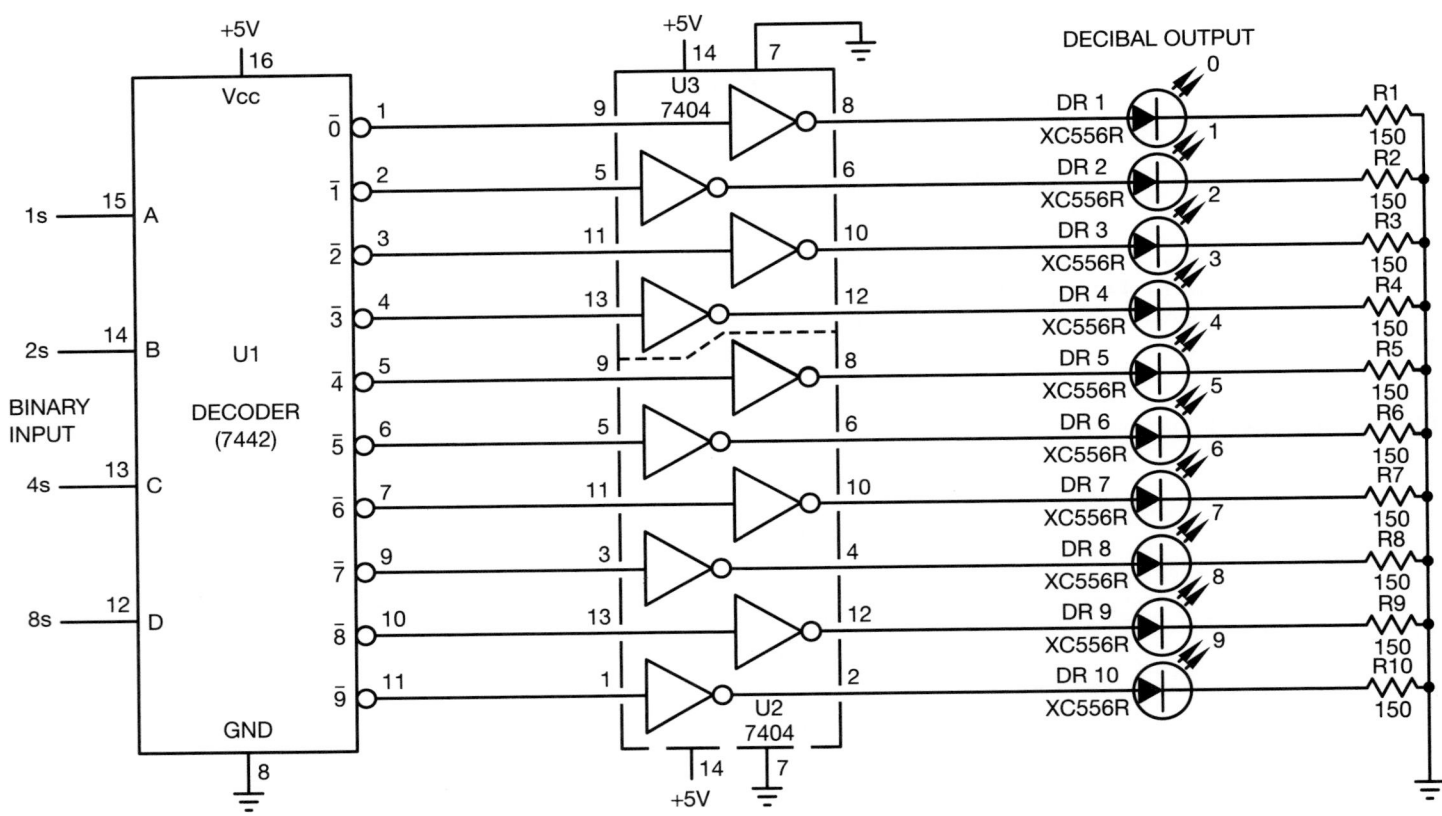

PROBLEM 25-46   Schematic diagram

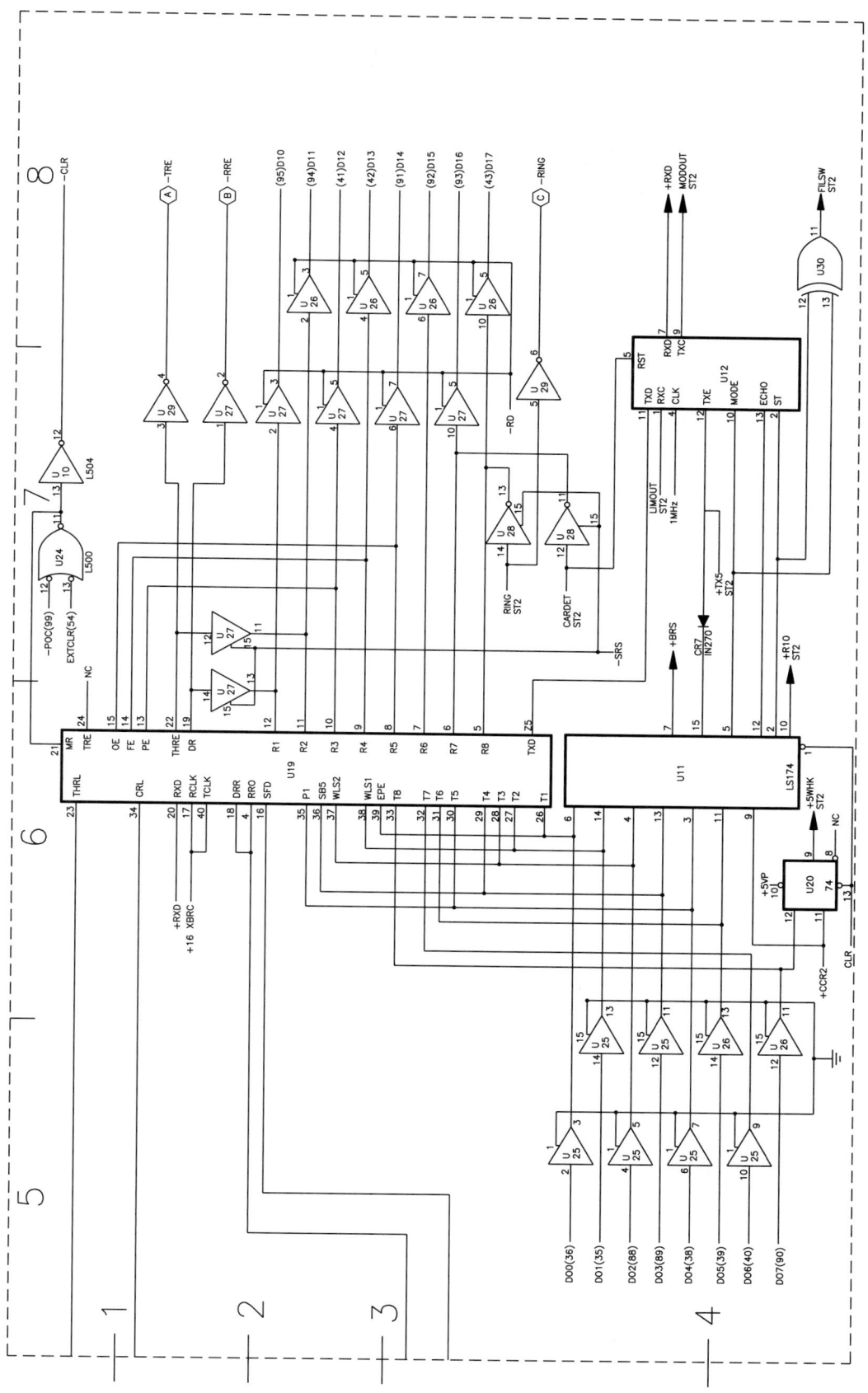

PROBLEM 25-47 Continuity tester

Given a schematic layout, create a schematic drawing, silk screen artwork, drilling drawing, and circuit side and component side board. Use 5/32" high lettering. All through hole components have AWB 24 size. There are four #35 mount holes that are .125 in. from each side at each corner. The finished size of the board will be 1-1/2" × 2-1/16" dimensions.

Reprinted from TECH DIRECTIONS, April 1999, Ann Arbor, MI. Courtesy Harry M. Hawkins, Professor Emeritus, Oswego State University of New York.

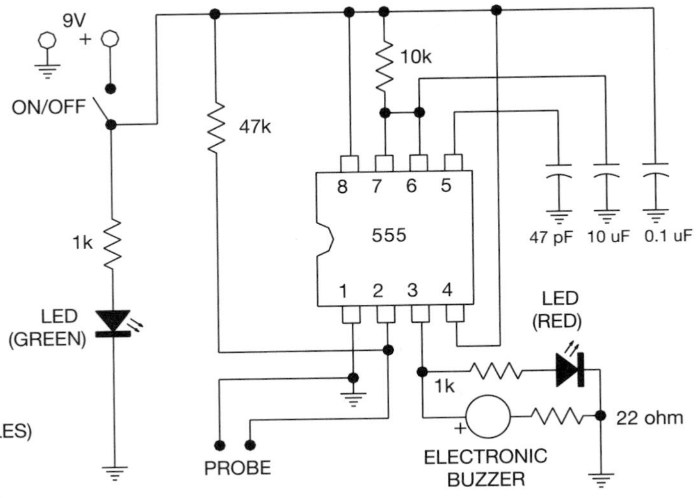

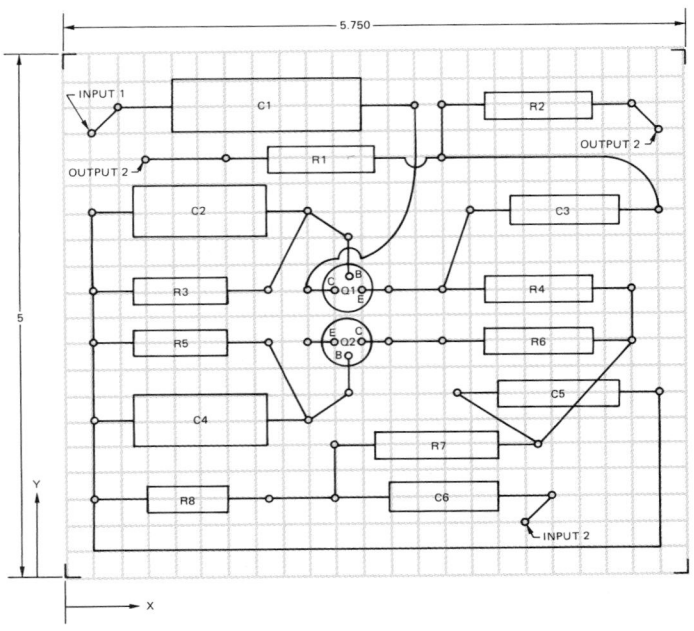

CONTINUITY TESTER

(a)

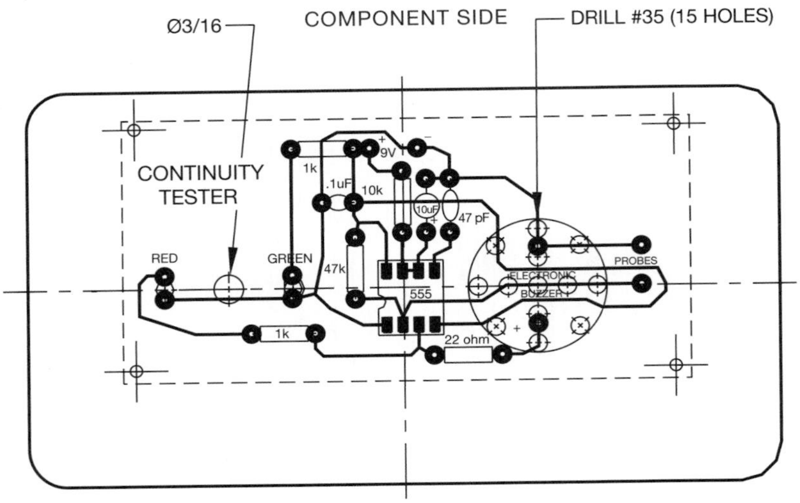

FOIL SIDE

(b)

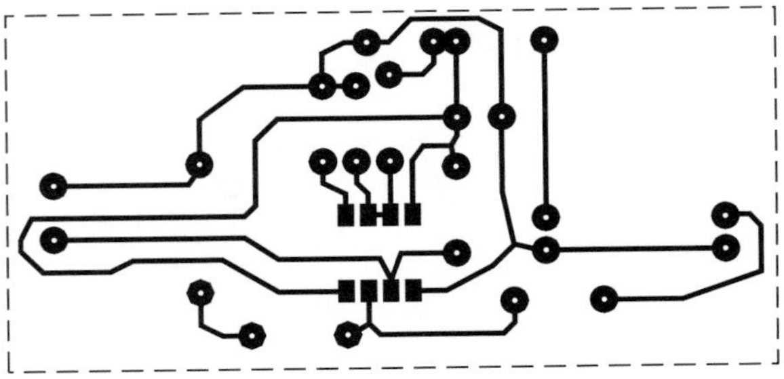

(c)

PROBLEM 25-48 Printed circuit board layout

Given the "component side" engineer's sketch of a printed circuit board, make a printed circuit board layout on the "back side" at 2:1 scale. Use the following instructions unless otherwise specified by your instructor:

1. Use PCB graphics or CADD software.
2. Use teardrop pads with fillet radius, .200" OD and .032" ID.
3. Use .080 trace line width.

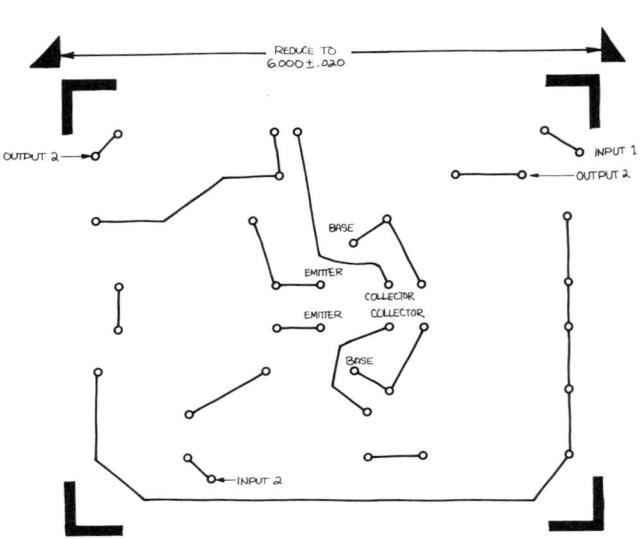

PROBLEM 25-49   Silk screen artwork

Given the PCB layout sketch from Problem 25–48, make a silk screen artwork at 1:1 scale. Use 5/32" high lettering with ink and mechanical lettering device or rub-on letters, unless otherwise specified by your instructor.

PROBLEM 25-50   Drilling drawing

Given the PCB layout from Problem 25–48 and the attached dimension table, prepare a drilling drawing using tabular dimensioning. Refer to Chapter 12, Dimensioning and Tolerancing. The holes may be shown using centerline "cross" at the center locations without drawing the actual circles. The 0,0 location is at the lower left corner of the PCB. Number the holes from top to bottom and left to right as coordinated with the given table. NOTE: Dimensions are given from component side.

HOLE NO.	X	Y	DIAMETER
1	.500	4.500	.032
2	3.250	4.500	.032
3	3.500	4.500	.032
4	5.250	4.500	.032
5	5.500	4.250	.032
6	.750	4.000	.032
7	1.500	4.000	.032
8	3.500	4.000	.032
9	.250	3.500	.032
10	2.250	3.500	.032
11	3.750	3.500	.032
12	5.500	3.500	.032
13	2.625	3.250	.032
14	.250	2.750	.032
15	1.875	2.750	.032
16	2.250	2.750	.032
17	3.000	2.750	.032
18	3.500	2.750	.032
19	5.250	2.750	.032
20	.250	2.250	.032
21	1.875	2.250	.032
22	2.250	2.250	.032
23	3.000	2.250	.032
24	3.500	2.250	.032
25	5.250	2.250	.032
26	2.625	1.750	.032
27	3.625	1.750	.032
28	5.500	1.750	.032
29	.250	1.500	.032
30	2.250	1.500	.032
31	2.500	1.250	.032
32	4.375	1.250	.032
33	.250	.750	.032
34	1.875	.750	.032
35	2.500	.750	.032
36	4.500	.750	.032
37	4.250	.500	.032

PICTORIAL DRAWING

Given the engineer's layout, make a pictorial drawing using any one of the specific shading techniques discussed in Chapter 19, Pictorial Drawings and Technical Illustrations. Scale the engineer's layout directly to determine dimensions. Increase the size of your final drawing. Place all labels on the final drawing in ink using mechanical lettering devices, or prepare a CADD ink plot if available. Use polyester film or vellum.

PROBLEM 25-51   Exploded technical illustration

Given the exploded technical illustration shown in Figure 25.87, select and draw the pictorial of any one subassembly unless otherwise specified by your instructor. Make your drawing larger but proportional to the given drawing.

PROBLEM 25-52   Pictorial schematic

Given the semipictorial schematic engineering layout shown in Figure 25.88, page 918, make a formal drawing.

MATH PROBLEMS

Convert these numbers to power of ten form:

PROBLEM 25-53   243

PROBLEM 25-54   14,000

PROBLEM 25-55   .004

Convert these numbers to regular form:

PROBLEM 25-56 6.7×10^3

PROBLEM 25-57 9×10^6

PROBLEM 25-58 6.2×10^{-3}

Perform these calculations.

PROBLEM 25-59 $(4 \times 10^6) (3 \times 10^{13})$

PROBLEM 25-60 $25 \times 10^{17}/5 \times 10^{11}$

PROBLEM 25-61 $(7 \times 10^{-6}) (5 \times 10^{-2})$

PROBLEM 25-62 $\dfrac{2.8 \times 10^{-6}}{8.1 \times 10^5}$

Access the *Online Companion for Engineering Drawing and Design*, 3e at *www.delmar.com* to view the following appendices:

American National Standards of Interest to Designers, Architects, and Drafters

Dimensioning and Tolerancing Symbols

Designation of Welding and Allied Processes by Letters

Symbols of Pipe Fittings and Valves

Math Instruction

APPENDIX A
ABBREVIATIONS

The following are standard abbreviations commonly used on working drawings:

Above finished floor . . . . .AFF
Accessory . . . . .ACCESS
Accumulate . . . . .ACCUM
Adaption . . . . .ADAPT
Addendum .ADD
Addition . . . . .ADD
Airfoil. . . . . .AF
Air pressure drop . . . . .APD
Alteration . . . . .ALT
Alternate . . . . .ALT
Alternating current . . . . .AC
Alternative . . . . .ALT
Altitude . . . . .ALT
Aluminum . . . . .AL
American Iron and
    Steel Institute . . . . .AISI
American Society for
    Testing Materials . . . . .ASTM
American Society of
    Mechanical Engineers . . . . .ASME
American Wire Gage . . . . .AWG
Ampere . . . . .AMP
Apparatus housing plenum . . . . .AHP
Approved .APPD
Approximate, Approximately . . . .APPROX
Architect, Architectural . . . . .ARCH
Assembly . . . . .ASM
Attach . . . . .ATT
Authorize .AUTH
Automatic .AUTO
Auxiliary . . . . .AUX
Average . . . . .AVG
Backdraft damper . . . . .BDD
Backward inclined . . . . .BI
Balance . . . . .BAL
Basement .BSMT
Battery . . . . .BAT
Bearing . . . . .BRG
Bill of materials . . . . .B/M
Bracket . . . . .BRKT
Brass . . . . .BRS
British thermal unit . . . . .BTU
British thermal units
    per hour . . . . .BTUH
Bronze . . . . .BRZ
Brown and Sharpe . . . . .B&S
Building . . . . .BLDG
Bushing . . . . .BUSH
Cadmium . . . . .CAD
Capacity . . . . .CAP

Carbon steel .CS
Cast iron CI
Cast steel . . . . .CS
Casting . . . . .CST
Ceiling or cooling . . . . .CLG
Center . . . . .CTR
Center of gravity . . . . .CG
Centerline . . . .(C̶) or CL
Centigrade C
Centimeter . . . . .CM
Chamfer .CHAM
Change . . . . .CHG
Check . . . . .CHK
Chief . . . . .CH
Chromium .CHR
Circular pitch . . . . .CP
Circumference . . . . .CIRC
Clockwise . . . . .CW
Cold drawn steel . . . . .CDS
Cold rolled steel . . . . .CRS
Company . . . . .CO
Composition . . . . .COMP
Compression, Compressor . . . . . . .COMP
Concentric . . . . .CONC
Concrete .CONC
Condition . . . . .COND
Conductor . . . . .COND
Conduit . . . . .CDT
Connect, Connection,
    Connector . . . . .CONN
Continue, Continued,
    Continuation . . . . .CONT
Control . . . . .CONT
Copper . . . . .CU
Corporation . . . . .CORP
Correspond . . . . .CORRES
Corrosion resistant steel . . . . .CRES
Counterbalance . . . . .CBAL
Counterbore . . . . .CBORE
Counterclockwise . . . . .CCW
Counterdrill . . . . .CDRILL
Countersink . . . . .CSK
Counterweight . . . . .CTWT
Cubic . . . . .CU
Cubic centimeter CC
Cubic feet per hour . . . . .CFH
Cubic feet per minute . . . . .CFM
Cubic feet per second . . . . .CFS
Cubic foot . . . . .CU FT
Cubic inch . . . . .CU IN
Cycle . . . . .CY
Dedendum . . . . .DED
Deflection .DEFL
Degree . . . .(°) or DEG
Department . . . . .DEPT
Detail . . . . .DET

Develop . . . . .DEV
Deviation .DEV
Dew point . . . . .DP
Diagonal .DIAG
Diameter . . . . .DIA, ∅
Diametral pitch . . . . .DP
Dimension . . . . .DIM.
Direct current . . . . .DC
Direct digital control . . . . .DDC
Distance . . . . .DIST
Division . . . . .DIV
Double width double inlet . . . . .DWDI
Down . . . . .DN
Drawing .DWG
Each . . . . .EA
Eccentric . . . . .ECC
Effective . . . . .EFF
Efficiency . . . . .EFF
Electric, Electrical . . . . .ELEC
Elevation .ELEV
Engineer .ENGR
Engineering . . . . .ENGRG
Entering air temperature . . . . .EAT
Entering dry bulb . . . . .EDB
Entering fluid temperature . . . . .EFT
Entering water temperature . . . . . . .EWT
Entering wet bulb . . . . .EWB
Equal . . . . .EQ
Equipment . . . . .EQUIP
Equivalent . . . . .EQUIV
Estimate . . . . .EST
Etcetera . . . . .ETC
Exhaust . . . . .EXH
Exhaust air .EA
Exhaust air damper . . . . .EAD
Existing . . . . .EXIST
Expansion . . . . .EXP
Extension .EXT
External . . . . .EXT
External static pressure . . . . .ESP
Extractor . . . . .EX
Extrusion .EXTR
Face velocity . . . . .FV
Fahrenheit . . . . .F
Feet, Foot . . . . .FT
Feet per minute . . . . .FPM
Feet per second . . . . .FPS
Figure . . . . .FIG
Fillister head . . . . .FIL HD
Filter . . . . .FILT
Finish . . . . .FIN
Fitting . . . . .FTG
Fixture unit .FU
Flat . . . . .FL
Flat head .FHD
Flexible . . . . .FLEX

Fluid pressure drop . . . . .FPD
Foot . . . .(') or FT
Foot pounds . . . . .FT LB
Forging .FORG
Forward .FWD
Forward curved . . . . .FC
Front . . . . .FRT
Future . . . . .FUT
Gage (Gauge) . . . . .GA
Gallon . . . . .GAL
Gallons per hour . . . . .GPH
Gallons per minute . . . . .GPM
Galvanize, Galvanized . . . . .GALV
Gasket . . . . .GSKT
Generator . . . . .GEN
Glycol . . . . .GLY
Grain . . . . .GRN
Gravity . . . . .G
Grind . . . . .GRD
Harden . . . . .HDN
Hardware . . . . .HDW
Head . . . . .HD
Heating . . . . .HTG
Heat treat . . . . .HT TR
Height . . . . .H or HGT
Hexagon .HEX
Horizontal . . . . .HORIZ
Horsepower . . . . .HP
Hot rolled steel . . . . .HRS
Hour . . . . .HR
Hydraulic . . . . .HYD
Illustration . . . . .ILLUS
Inch, Inches . . . .(") or IN
Inch ounce . . . . .IN. OZ
Inch pounds . . . . .IN. LB
Inclusive .INCL
Information . . . . .INFO
Inside diameter, Inside
  dimension . . . . .ID
Inside radius . . . . .IR
Inspection . . . . .INSP
Installation . . . . .INSTL
Instrument . . . . .INST
Insulation . . . . .INSL or INSUL
Interchangeable . . . . .INTCHG
Intermediate . . . . .INTER
Internal . . . . .INT
Invert elevation . . . . .IE
Isolator, Isolation . . . . .ISOL
Joint . . . . .JT
Kilometer .KM
Kilowatt . . . . .KW
Kilowatt hour . . . . .KWH
Knock out .KO
Laboratory . . . . .LAB
Leaving . . . . .LVG

Leaving air temp . . . . .LAT
Leaving dry bulb . . . . .LDB
Leaving fluid temperature . . . . .LFT
Leaving water
  temperature . . . . .LWT
Leaving wet bulb . . . . .LWB
Left hand .LH
Length . . . . .L or LG
Linear feet . . . . .LF
Lock washer . . . . .LWASH
Longitude, Longitudinal . . . . .LONG
Lower . . . . .LWR
Lubricate .LUB
Machine .MACH
Magnesium . . . . .MG
Maintenance . . . . .MAINT
Malleable .MAL
Manufacture, Manufacturer . . . . . . .MFR
Material .MATL
Maximum . . . . .MAX
Maximum material
  condition . . . . .MMC
Mechanical . . . . .MECH
Medium .MED
Memorandum . . . . .MEMO
Mercury . . . . .HG
Mile . . . . .MI
Miles per hour . . . . .MPH
Millimeter . . . . .MM
Minimum . . . . .MIN
Minute . . . .(′) or MIN
Miscellaneous . . . . .MISC
Modify . . . . .MOD
Molding .MLDG
Mounted .MTD
Mounting . . . . .MTG
National .NATL
National Electrical Mfg.
  Association . . . . .NEMA
National Machine Tool
  Builders Association . . . . .NMTBA
Negative . . . .(−) or NEG
Nickel . . . . .NI
No drawing . . . . .ND
Nominal .NOM
Nonstandard . . . . .NONSTD
Normally closed . . . . .NC
Normally open . . . . .NO
Not in contract . . . . .NIC
Number . . . . .NO
Obsolete .OBS
Of true position . . . . .OTP
On Center, On center
  distance . . . . .OC
Operate .OPER
Opposite .OPP

Optional .OPT
Original .ORIG
Ounce . . . . .OZ
Outside air . . . . .OSA
Outside air damper . . . . .OAD
Outside diameter . . . . .OD
Outside radius . . . . .OR
Oval head . . . . .OV HD
Overall . . . . .OA
Oxygen .OXY
Package, Packing . . . . .PKG
Page . . . . .P
Part . . . . .PT
Parting line . . . . .PL
Patent . . . . .PAT
Pattern .PATT
Perpendicular . . . . .PERP
Phase . . . . .PH
Pint . . . . .PT
Pitch . . . . .P
Pitch circle . . . . .PC
Pitch diameter . . . . .PD
Plan view .PV
Point on line . . . . .POL
Polypropylene . . . . .PP
Polyvinyl chloride . . . . .PVC
Position .POSN
Positive . . . .(+) or POS
Pound . . . . .LB
Pounds per square inch . . . . .PSI
Preliminary . . . . .PRELIM
Pressure . . . . .PRESS
Process .PROC
Product, Production . . . . .PROD
PVC coated steel . . . . .PVS
Quality .QUAL
Quantity .QTY
Quart . . . . .QT
Quarter .QTR
Radius . . . . .R (RAD)
Rear view .RV
Rectangular . . . . .RECT
Reduction . . . . .RED
Reference . . . . .REF
Regardless of feature size . . . . .RFS
Regular .REG
Reinforce . . . . .REINF
Relative humidity . . . . .RH
Remove .REM
Require . . . . .REQ
Required . . . . .REQD
Resistor .RES
Return air .RA
Return air damper . . . . .RAD
Reverse .REV
Revision .REV

Revolution . . . . .REV
Revolutions per minute . . . . .RPM
Right hand . . . . .RH
Root diameter . . . . .RD
Round . . . . .RD
Round head . . . . .RD HD
Rubber . . . . .RUB
Screw . . . . .SCR
Screw threads
  American National coarse . . . . .NC
  American National fine . . . . .NF
  American National extra fine . . . . .NEF
  American National 8 pitch 8N
  American National 12 pitch 12N
  American National 16 pitch 16N
  American Standard
    straight pipe coupling . . . . . . .NPSC
  American Standard taper pipe . . . . .NPT
  American Standard taper . . . . .NPTF
  Unified screw thread coarse . . . . . .UNC
  Unified screw thread fine . . . . .UNF
  Unified screw thread extra fine . .UNEF
  Unified screw thread 8 thread 8UN
  Unified screw thread 12 thread . . .12UN
  Unified screw thread 16 thread . . .16UN
  Unified screw thread special . . . . .UNS
Second . . . .(″) or SEC
Section . . . . .SECT
Sensible . . . . .SENS
Serial, Series . . . . .SER
Sheet . . . . .SH
Side view . . . . .SV
Similar . . . . .SIM
Single wall plenum . . . . .SWP
Single width single inlet . . . . .SWSI
Sketch . . . . .SK
Smoke damper . . . . .SD

Society of Automotive Engineers . . . . .SAE
Special . . . . .SPL
Specification . . . . .SPEC
Specific gravity . . . . .SP GR
Spherical . . . . .SPHER
Spot face . . . . .SF
Spring . . . . .SPR
Square . . . . .SQ
Square foot (feet) . . . . .SF
Square inch(es) . . . . .SQ IN
Stainless steel . . . . .SS
Standard . . . . .STD
Standard cubic feet per minute . . .SCFM
Static pressure . . . . .SP
Steel . . . . .STL
Structure, Structural . . . . .STRUCT
Supply air . . . . .SA
Support . . . . .SUPT
Switch . . . . .SW
Symbol . . . . .SYM
Symmetrical . . . . .SYM
Synthetic . . . . .SYN
Tangent . . . . .TAN
Technical .TECH
Teeth . . . . .T
Temperature . . . . .TEMP
Tensile strength . . . . .TS
Terminal .TERM
Theoretical . . . . .THEO
Thickness .THK
Thousand BTU per hour . . . . .MBH
Thread . . . . .THD
Through . . . . .THRU
Tolerance .TOL
Total static pressure . . . . .TSP
Tracer . . . . .TCR
Trademark .TM

Transformer . . . . .TRANS
Transmission . . . . .TRANS
Transverse . . . . .TRANSV
True length .TL
True position . . . . .TP
True view . . . . .TV
Turnbuckle . . . . .TRNBKL
Typical . . . . .TYP
Ultimate . . . . .ULT
United States . . . . .US
United States of America
  Standards Institute . . . . .USASI
United States gage . . . . .USG
Universal . . . . .UNIV
Unless otherwise specified . . . . .UOS
Upper . . . . .UPR
Vacuum . . . . .VAC
Velocity . . . . .V or VEL
Vent through roof . . . . .VTR
Versus . . . . .VS
Vertical . . . . .VERT
Volt(s) V
Volume . . . . .VOL
Volume damper . . . . .VD
Water gauge . . . . .WG
Water line .WL
Water pressure drop . . . . .WPD
Watt . . . . .W
Weight . . . . .WT
Width . . . . .W
With . . . . .W/
Without . . . . .W/O
Wood . . . . .WD
Wood screw . . . . .WD SCR
Wrought iron . . . . .WI
Yard . . . . .YD
Year . . . . .YR

APPENDIX B TABLES

TABLE 1 INCHES TO MILLIMETERS

in.	mm	in.	mm	in.	mm	in.	mm
1	25.4	26	660.4	51	1295.4	76	1930.4
2	50.8	27	685.8	52	1320.8	77	1955.8
3	76.2	28	711.2	53	1346.2	78	1981.2
4	101.6	29	736.6	54	1371.6	79	2006.6
5	127.0	30	762.0	55	1397.0	80	2032.0
6	152.4	31	787.4	56	1422.4	81	2057.4
7	177.8	32	812.8	57	1447.8	82	2082.8
8	203.2	33	838.2	58	1473.2	83	2108.2
9	228.6	34	863.6	59	1498.6	84	2133.6
10	254.0	35	889.0	60	1524.0	85	2159.0
11	279.4	36	914.4	61	1549.4	86	2184.4
12	304.8	37	939.8	62	1574.8	87	2209.8
13	330.2	38	965.2	63	1600.2	88	2235.2
14	355.6	39	990.6	64	1625.6	89	2260.6
15	381.0	40	1016.0	65	1651.0	90	2286.0
16	406.4	41	1041.4	66	1676.4	91	2311.4
17	431.8	42	1066.8	67	1701.8	92	2336.8
18	457.2	43	1092.2	68	1727.2	93	2362.2
19	482.6	44	1117.6	69	1752.6	94	2387.6
20	508.0	45	1143.0	70	1778.0	95	2413.0
21	533.4	46	1168.4	71	1803.4	96	2438.4
22	558.8	47	1193.8	72	1828.8	97	2463.8
23	584.2	48	1219.2	73	1854.2	98	2489.2
24	609.6	49	1244.6	74	1879.6	99	2514.6
25	635.0	50	1270.0	75	1905.0	100	2540.0

The above table is exact on the basis: 1 in. = 25.4 mm

TABLE 2 MILLIMETERS TO INCHES

in.	mm	in.	mm	in.	mm	in.	mm
1	0.039370	26	1.023622	51	2.007874	76	2.992126
2	0.078740	27	1.062992	52	2.047244	77	3.031496
3	0.118110	28	1.102362	53	2.086614	78	3.070866
4	0.157480	29	1.141732	54	2.125984	79	3.110236
5	0.196850	30	1.181102	55	2.165354	80	3.149606
6	0.236220	31	1.220472	56	2.204724	81	3.188976
7	0.275591	32	1.259843	57	2.244094	82	3.228346
8	0.314961	33	1.299213	58	2.283465	83	3.267717
9	0.354331	34	1.338583	59	2.322835	84	3.307087
10	0.393701	35	1.377953	60	2.362205	85	3.346457
11	0.433071	36	1.417323	61	2.401575	86	3.385827
12	0.472441	37	1.456693	62	2.440945	87	3.425197
13	0.511811	38	1.496063	63	2.480315	88	3.464567
14	0.551181	39	1.535433	64	2.519685	89	3.503937
15	0.590551	40	1.574803	65	2.559055	90	3.543307
16	0.629921	41	1.614173	66	2.598425	91	3.582677
17	0.669291	42	1.653543	67	2.637795	92	3.622047
18	0.708661	43	1.692913	68	2.677165	93	3.661417
19	0.748031	44	1.732283	69	2.716535	94	3.700787
20	0.787402	45	1.771654	70	2.755906	95	3.740157
21	0.826772	46	1.811024	71	2.795276	96	3.779528
22	0.866142	47	1.850394	72	2.834646	97	3.818898
23	0.905512	48	1.889764	73	2.874016	98	3.858268
24	0.944882	49	1.929134	74	2.913386	99	3.897638
25	0.984252	50	1.968504	75	2.952756	100	3.937008

The above table is approximate on the basis: 1 in. = 25.4 mm, 1/25.4 = 0.039370078740+

TABLE 3 INCHE/METRIC EQUIVALENTS

Fraction			Decimal Equivalent		Fraction			Decimal Equivalent	
			Customary (in.)	Metric (mm)				Customary (in.)	Metric (mm)
		1/64	.015625	0.3969			33/64	.515625	13.0969
	1/32		.03125	0.7938		17/32		.53125	13.4938
		3/64	.046875	1.1906			35/64	.546875	13.8906
1/16			.0625	1.5875	9/16			.5625	14.2875
		5/64	.078125	1.9844			37/64	.578125	14.6844
	3/32		.09375	2.3813		19/32		.59375	15.0813
		7/64	.109375	2.7781			39/64	.609375	15.4781
1/8			.1250	3.1750	5/8			.6250	15.8750
		9/64	.140625	3.5719			41/64	.640625	16.2719
	5/32		.15625	3.9688		21/32		.65625	16.6688
		11/64	.171875	4.3656			43/64	.671875	17.0656
3/16			.1875	4.7625	11/16			.6875	17.4625
		13/64	.203125	5.1594			45/64	.703125	17.8594
	7/32		.21875	5.5563		23/32		.71875	18.2563
		15/64	.234375	5.9531			47/64	.734375	18.6531
1/4			.250	6.3500	3/4			.750	19.0500
		17/64	.265625	6.7469			49/64	.765625	19.4469
	9/32		.28125	7.1438		25/32		.78125	19.8438
		19/64	.296875	7.5406			51/64	.796875	20.2406
5/16			.3125	7.9375	13/16			.8125	20.6375
		21/64	.328125	8.3384			53/64	.828125	21.0344
	11/32		.34375	8.7313		27/32		.84375	21.4313
		23/64	.359375	9.1281			55/64	.859375	21.8281
3/8			.3750	9.5250	7/8			.8750	22.2250
		25/64	.390625	9.9219			57/64	.890625	22.6219
	13/32		.40625	10.3188		29/32		.90625	23.0188
		27/64	.421875	10.7156			59/64	.921875	23.4156
7/16			.4375	11.1125	15/16			.9375	23.8125
		29/64	.453125	11.5094			61/64	.953125	24.2094
	15/32		.46875	11.9063		31/32		.96875	24.6063
		31/64	.484375	12.3031			63/64	.984375	25.0031
1/2			.500	12.7000	1			1.000	25.4000

TABLE 4 INCH/METRIC—CONVERSION

Measures of Length

1 millimeter (mm) = 0.03937 inch
1 centimeter (cm) = 0.39370 inch
1 meter (m) = 39.37008 inches
= 3.2808 feet
= 1.0936 yards
1 kilometer (km) = 0.6214 mile
1 inch = 25.4 millimeters (mm)
= 2.54 centimeters (cm)
1 foot = 304.8 millimeters (mm)
= 0.3048 meter (m)
1 yard = 0.9144 meter (m)
1 mile = 1.609 kilometers (km)

Measures of Area

1 square millimeter = 0.00155 square inch
1 square centimeter = 0.155 square inch
1 square meter = 10.764 square feet
= 1.196 square yards
1 square kilometer = 0.3861 square mile
1 square inch = 645.2 square millimeters
= 6.452 square centimeters
1 square foot = 929 square centimeters
= 0.0929 square meter
1 square yard = 0.836 square meter
1 square mile = 2.5899 square kilometers

Measures of Capacity (Dry)

1 cubic centimeter (cm^3) = 0.061 cubic inch
1 liter = 0.0353 cubic foot
= 61.023 cubic inches
1 cubic meter (m^3) = 35.315 cubic feet
= 1.308 cubic yards
1 cubic inch = 16.38706 cubic centimeters (cm^3)
1 cubic foot = 0.02832 cubic meter (m^3)
= 28.317 liters
1 cubic yard = 0.7646 cubic meter (m^3)

Measures of Capacity (Liquid)

1 liter = 1.0567 U.S. quarts
= 0.2642 U.S. gallon
= 0.2200 Imperial gallon
1 cubic meter (m^3) = 264.2 U.S. gallons
= 219.969 Imperial gallons
1 U.S. quart = 0.946 liter
1 Imperial quart = 1.136 liters
1 U.S. gallon = 3.785 liters
1 Imperial gallon = 4.546 liters

Measures of Weight

1 gram (g) = 15.432 grains
= 0.03215 ounce troy
= 0.03527 ounce avoirdupois
1 kilogram (kg) = 35.274 ounces avoirdupois
= 2.2046 pounds
1000 kilograms (kg) = 1 metric ton (t)
= 1.1023 tons of 2000 pounds
= 0.9842 ton of 2240 pounds
1 ounce avoirdupois = 28.35 grams (g)
1 ounce troy = 31.103 grams (g)
1 pound = 453.6 grams
= 0.4536 kilogram (kg)
1 ton of 2240 pounds = 1016 kilograms (kg)
= 1.016 metric tons
1 grain = 0.0648 gram (g)
1 metric ton = 0.9842 ton of 2240 pounds
= 2204.6 pounds

TABLE 5 RULES RELATIVE TO THE CIRCLE

To Find Circumference—
Multiply diameter by 3.1416 .... Or divide diameter by 0.3183

To Find Diameter—
Multiply circumference by 0.3183 .... Or divide circumference by 3.1416

To Find Radius—
Multiply circumference by 0.15915 .... Or divide circumference by 6.28318

To Find Side of an Inscribed Square—
Multiply diameter by 0.7071
Or multiply circumference by 0.2251 .... Or divide circumference by 4.4428

To Find Side of an Equal Square—
Multiply diameter by 0.8862 .... Or divide diameter by 1.1284
Or multiply circumference by 0.2821 .... Or divide circumference by 3.545

Square—
A side multiplied by 1.4142 equals diameter of its circumscribing circle.
A side multiplied by 4.443 equals circumference of its circumscribing circle.
A side multiplied by 1.128 equals diameter of an equal circle.
A side multiplied by 3.547 equals circumference of an equal circle.

To Find the Area of a Circle—
Multiply circumference by one-quarter of the diameter.
Or multiply the square of diameter by 0.7854
Or multiply the square of circumference by .07958
Or multiply the square of 1/2 diameter by 3.1416

To Find the Surface of a Sphere or Globe—
Multiply the diameter by the circumference.
Or multiply the square of diameter by 3.1416
Or multiply four times the square of radius by 3.1416

TABLE 6 STANDARD LINE TYPES

TABLE 7 ASTM AND SAE GRADE MARKINGS FOR STEEL BOLTS AND SCREWS

Grade Marking	Specification	Material
NO MARK	SAE—Grade 1	Low or Medium Carbon Steel
	ASTM—A307	Low Carbon Steel
	SAE—Grade 2	Low or Medium Carbon Steel
	SAE—Grade 5	Medium Carbon Steel, Quenched and Tempered
	ASTM—A 449	
	SAE—Grade 5.2	Low Carbon Martensite Steel, Quenched and Tempered
A 325	ASTM—A 325 Type 1	Medium Carbon Steel, Quenched and Tempered Radial dashes optional
A 325	ASTM—A 325 Type 2	Low Carbon Martensite Steel, Quenched and Tempered
A 325	ASTM—A 325 Type 3	Atmospheric Corrosion (Weathering) Steel, Quenched and Tempered
BC	ASTM—A 354 Grade BC	Alloy Steel, Quenched and Tempered
	SAE—Grade 7	Medium Carbon Alloy Steel, Quenched and Tempered, Roll Threaded After Heat Treatment
	SAE—Grade 8	Medium Carbon Alloy Steel, Quenched and Tempered
	ASTM—A 354 Grade BD	Alloy Steel, Quenched and Tempered
	SAE—Grade 8.2	Low Carbon Martensite Steel, Quenched and Tempered
A 490	ASTM—A 490 Type 1	Alloy Steel, Quenched and Tempered
A 490	ASTM—A 490 Type 3	Atmospheric Corrosion (Weathering) Steel, Quenched and Tempered

(Reprinted from The American Society of Mechanical Engineers—ANSI B18.2.1–1981 (R1992).)

TABLE 8 UNIFIED STANDARD SCREW THREAD SERIES

Sizes		Basic Major Diameter	THREADS PER INCH											Sizes
			Series with graded pitches			Series with constant pitches								
Primary	Secondary		Coarse UNC	Fine UNF	Extra fine UNEF	4UN	6UN	8UN	12UN	16UN	20UN	28UN	32UN	
0		0.0600	–	80	–	–	–	–	–	–	–	–	–	0
	1	0.0730	64	72	–	–	–	–	–	–	–	–	–	1
2		0.0860	56	64	–	–	–	–	–	–	–	–	–	2
	3	0.0990	48	56	–	–	–	–	–	–	–	–	–	3
4		0.1120	40	48	–	–	–	–	–	–	–	–	–	4
5		0.1250	40	44	–	–	–	–	–	–	–	–	–	5
6		0.1380	32	40	–	–	–	–	–	–	–	–	UNC	6
8		0.1640	32	36	–	–	–	–	–	–	–	–	UNC	8
10		0.1900	24	32	–	–	–	–	–	–	–	–	UNF	10
	12	0.2160	24	28	32	–	–	–	–	–	–	UNF	UNEF	12
1/4		0.2500	20	28	32	–	–	–	–	–	UNC	UNF	UNEF	1/4
5/16		0.3125	18	24	32	–	–	–	–	–	20	28	UNEF	5/16
3/8		0.3750	16	24	32	–	–	–	–	UNC	20	28	UNEF	3/8
7/16		0.4375	14	20	28	–	–	–	–	16	UNF	UNEF	32	7/16
1/2		0.5000	13	20	28	–	–	–	–	16	UNF	UNEF	32	1/2
9/16		0.5625	12	18	24	–	–	–	UNC	16	20	28	32	9/16
5/8		0.6250	11	18	24	–	–	–	12	16	20	28	32	5/8
	11/16	0.6875	–	–	24	–	–	–	12	16	20	28	32	11/16
3/4		0.7500	10	16	20	–	–	–	12	UNF	UNEF	28	32	3/4
	13/16	0.8125	–	–	20	–	–	–	12	16	UNEF	28	32	13/16
7/8		0.8750	9	14	20	–	–	–	12	16	UNEF	28	32	7/8
	15/16	0.9375	–	–	20	–	–	–	12	16	UNEF	28	32	15/16
1		1.0000	8	12	20	–	–	UNC	UNF	16	UNEF	28	32	1
	1 1/16	1.0625	–	–	18	–	–	8	12	16	20	28	–	1 1/16
1 1/8		1.1250	7	12	18	–	–	8	UNF	16	20	28	–	1 1/8
	1 3/16	1.1875	–	–	18	–	–	8	12	16	20	28	–	1 3/16
1 1/4		1.2500	7	12	18	–	–	8	UNF	16	20	28	–	1 1/4
	1 5/16	1.3125	–	–	18	–	–	8	12	16	20	28	–	1 5/16
1 3/8		1.3750	6	12	18	–	UNC	8	UNF	16	20	28	–	1 3/8
	1 7/16	1.4375	–	–	18	–	6	8	12	16	20	28	–	1 7/16
1 1/2		1.5000	6	12	18	–	UNC	8	UNF	16	20	28	–	1 1/2
	1 9/16	1.5625	–	–	18	–	6	8	12	16	20	–	–	1 9/16
1 5/8		1.6250	–	–	18	–	6	8	12	16	20	–	–	1 5/8
	1 11/16	1.6875	–	–	18	–	6	8	12	16	20	–	–	1 11/16
1 3/4		1.7500	5	–	–	–	6	8	12	16	20	–	–	1 3/4
	1 13/16	1.8125	–	–	–	–	6	8	12	16	20	–	–	1 13/16
1 7/8		1.8750	–	–	–	–	6	8	12	16	20	–	–	1 7/8
	1 15/16	1.9375	–	–	–	–	6	8	12	16	20	–	–	1 15/16
2		2.0000	4 1/2	–	–	–	6	8	12	16	20	–	–	2
	2 1/8	2.1250	–	–	–	–	6	8	12	16	20	–	–	2 1/8
2 1/4		2.2500	4 1/2	–	–	–	6	8	12	16	20	–	–	2 1/4
	2 3/8	2.3750	–	–	–	–	6	8	12	16	20	–	–	2 3/8
2 1/2		2.5000	4	–	–	UNC	6	8	12	16	20	–	–	2 1/2
	2 5/8	2.6250	–	–	–	4	6	8	12	16	20	–	–	2 5/8
2 3/4		2.7500	4	–	–	UNC	6	8	12	16	20	–	–	2 3/4
	2 7/8	2.8750	–	–	–	4	6	8	12	16	20	–	–	2 7/8
3		3.0000	4	–	–	UNC	6	8	12	16	20	–	–	3
	3 1/8	3.1250	–	–	–	4	6	8	12	16	–	–	–	3 1/8
3 1/4		3.2500	4	–	–	UNC	6	8	12	16	–	–	–	3 1/4
	3 3/8	3.3750	–	–	–	4	6	8	12	16	–	–	–	3 3/8
3 1/2		3.5000	4	–	–	UNC	6	8	12	16	–	–	–	3 1/2
	3 5/8	3.6250	–	–	–	4	6	8	12	16	–	–	–	3 5/8
3 3/4		3.7500	4	–	–	UNC	6	8	12	16	–	–	–	3 3/4
	3 7/8	3.8750	–	–	–	4	6	8	12	16	–	–	–	3 7/8
4		4.0000	4	–	–	UNC	6	8	12	16	–	–	–	4
	4 1/8	4.1250	–	–	–	4	6	8	12	16	–	–	–	4 1/8
4 1/4		4.2500	–	–	–	4	6	8	12	16	–	–	–	4 1/4
	4 3/8	4.3750	–	–	–	4	6	8	12	16	–	–	–	4 3/8
4 1/2		4.5000	–	–	–	4	6	8	12	16	–	–	–	4 1/2
	4 5/8	4.6250	–	–	–	4	6	8	12	16	–	–	–	4 5/8
4 3/4		4.7500	–	–	–	4	6	8	12	16	–	–	–	4 3/4
	4 7/8	4.8750	–	–	–	4	6	8	12	16	–	–	–	4 7/8
5		5.0000	–	–	–	4	6	8	12	16	–	–	–	5
	5 1/8	5.1250	–	–	–	4	6	8	12	16	–	–	–	5 1/8
5 1/4		5.2500	–	–	–	4	6	8	12	16	–	–	–	5 1/4
	5 3/8	5.3750	–	–	–	4	6	8	12	16	–	–	–	5 3/8
5 1/2		5.5000	–	–	–	4	6	8	12	16	–	–	–	5 1/2
	5 5/8	5.6250	–	–	–	4	6	8	12	16	–	–	–	5 5/8
5 3/4		5.7500	–	–	–	4	6	8	12	16	–	–	–	5 3/4
	5 7/8	5.8750	–	–	–	4	6	8	12	16	–	–	–	5 7/8
6		6.0000	–	–	–	4	6	8	12	16	–	–	–	6

TABLE 9 DIMENSIONS OF HEX CAP SCREWS (FINISHED HEX BOLTS)

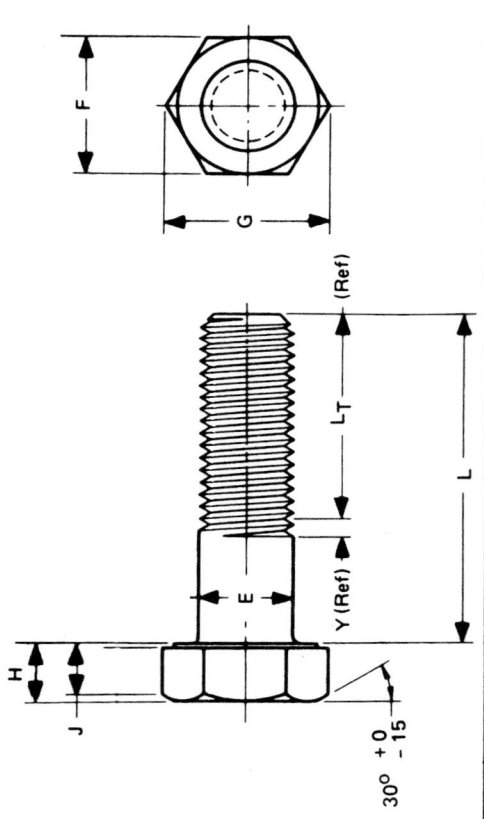

Nominal Size or Basic Product Dia (18)		E Body Dia (8)		F Width Across Flats			G Width Across Corners (4)		H Height			J Wrenching Height (4)	L_T Thread Length For Screw Lengths (10)		Y Transition Thread Length (10)	Runout of Bearing Surface FIM (5)
		Max	Min	Basic	Max	Min	Max	Min	Basic	Max	Min	Min	6 in. and Shorter Basic	Over 6 in. Basic	Max	Max
1/4	0.2500	0.2500	0.2450	7/16	0.438	0.428	0.505	0.488	5/32	0.163	0.150	0.106	0.750	1.000	0.250	0.010
5/16	0.3125	0.3125	0.3065	1/2	0.500	0.489	0.577	0.557	13/64	0.211	0.195	0.140	0.875	1.125	0.278	0.011
3/8	0.3750	0.3750	0.3690	9/16	0.562	0.551	0.650	0.628	15/64	0.243	0.226	0.160	1.000	1.250	0.312	0.012
7/16	0.4375	0.4375	0.4305	5/8	0.625	0.612	0.722	0.698	9/32	0.291	0.272	0.195	1.125	1.375	0.357	0.013
1/2	0.5000	0.5000	0.4930	3/4	0.750	0.736	0.866	0.840	5/16	0.323	0.302	0.215	1.250	1.500	0.385	0.014
9/16	0.5625	0.5625	0.5545	13/16	0.812	0.798	0.938	0.910	23/64	0.371	0.348	0.250	1.375	1.625	0.417	0.015
5/8	0.6250	0.6250	0.6170	15/16	0.938	0.922	1.083	1.051	25/64	0.403	0.378	0.269	1.500	1.750	0.455	0.017
3/4	0.7500	0.7500	0.7410	1 1/8	1.125	1.100	1.299	1.254	15/32	0.483	0.455	0.324	1.750	2.000	0.500	0.020
7/8	0.8750	0.8750	0.8660	1 5/16	1.312	1.285	1.516	1.465	35/64	0.563	0.531	0.378	2.000	2.250	0.556	0.023
1	1.0000	1.0000	0.9900	1 1/2	1.500	1.469	1.732	1.675	39/64	0.627	0.591	0.416	2.250	2.500	0.625	0.026
1 1/8	1.1250	1.1250	1.1140	1 11/16	1.688	1.631	1.949	1.859	11/16	0.718	0.658	0.461	2.500	2.750	0.714	0.029
1 1/4	1.2500	1.2500	1.2390	1 7/8	1.875	1.812	2.165	2.066	25/32	0.813	0.749	0.530	2.750	3.000	0.714	0.033
1 3/8	1.3750	1.3750	1.3630	2 1/16	2.062	1.994	2.382	2.273	27/32	0.878	0.810	0.569	3.000	3.250	0.833	0.036
1 1/2	1.5000	1.5000	1.4880	2 1/4	2.230	2.175	2.598	2.480	1 5/16	0.974	0.902	0.640	3.250	3.500	0.833	0.039
1 3/4	1.7500	1.7500	1.7380	2 5/8	2.625	2.538	3.031	2.893	1 3/32	1.134	1.054	0.748	3.750	4.000	1.000	0.046
2	2.0000	2.0000	1.9880	3	3.000	2.900	3.464	3.306	1 7/32	1.263	1.175	0.825	4.250	4.500	1.111	0.052
2 1/4	2.2500	2.2500	2.2380	3 3/8	3.375	3.262	3.897	3.719	1 3/8	1.423	1.327	0.933	4.750	5.000	1.111	0.059
2 1/2	2.5000	2.5000	2.4880	3 3/4	3.750	3.625	4.330	4.133	1 17/32	1.583	1.479	1.042	5.250	5.500	1.250	0.065
2 3/4	2.7500	2.7500	2.7380	4 1/8	4.125	3.988	4.763	4.546	1 11/16	1.744	1.632	1.151	5.750	6.000	1.250	0.072
3	3.0000	3.0000	2.9880	4 1/2	4.500	4.350	5.196	4.959	1 7/8	1.935	1.815	1.290	6.250	6.500	1.250	0.079

30° +0 / −15

TABLE 10 DIMENSIONS OF HEXAGON AND SPLINE SOCKET HEAD CAP SCREWS (1960 SERIES)

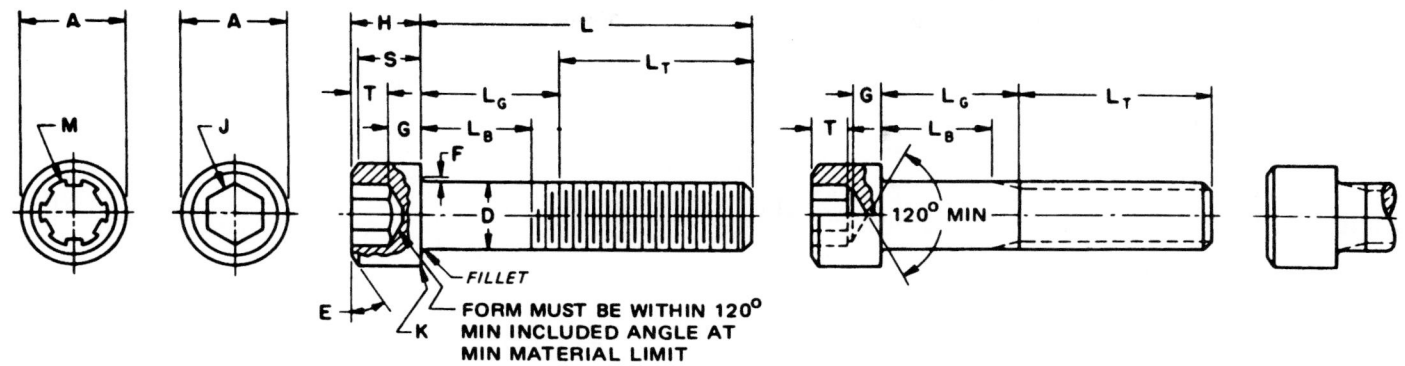

Nominal Size or Basic Screw Diameter		D Body Diameter		A Head Diameter		H Head Height		S Head Side Height	M Spline Socket Size	J Hexagon Socket Size		T Key Engagement	G Wall Thickness	K Chamfer or Radius
		Max	Min	Max	Min	Max	Min	Min	Nom	Nom		Min	Min	Max
0	0.0600	0.0600	0.0568	0.096	0.091	0.060	0.057	0.054	0.060	0.050		0.025	0.020	0.003
1	0.0730	0.0730	0.0695	0.118	0.112	0.073	0.070	0.066	0.072	1/16	0.062	0.031	0.025	0.003
2	0.0860	0.0860	0.0822	0.140	0.134	0.086	0.083	0.077	0.096	5/64	0.078	0.038	0.029	0.003
3	0.0990	0.0990	0.0949	0.161	0.154	0.099	0.095	0.089	0.096	5/64	0.078	0.044	0.034	0.003
4	0.1120	0.1120	0.1075	0.183	0.176	0.112	0.108	0.101	0.111	3/32	0.094	0.051	0.038	0.005
5	0.1250	0.1250	0.1202	0.205	0.198	0.125	0.121	0.112	0.111	3/32	0.094	0.057	0.043	0.005
6	0.1380	0.1380	0.1329	0.226	0.218	0.138	0.134	0.124	0.133	7/64	0.109	0.064	0.047	0.005
8	0.1640	0.1640	0.1585	0.270	0.262	0.164	0.159	0.148	0.168	9/64	0.141	0.077	0.056	0.005
10	0.1900	0.1900	0.1840	0.312	0.303	0.190	0.185	0.171	0.183	5/52	0.156	0.090	0.065	0.005
1/4	0.2500	0.2500	0.2435	0.375	0.365	0.250	0.244	0.225	0.216	3/16	0.188	0.120	0.095	0.008
5/16	0.3125	0.3125	0.3053	0.469	0.457	0.312	0.306	0.281	0.291	1/4	0.250	0.151	0.119	0.008
3/8	0.3750	0.3750	0.3678	0.562	0.550	0.375	0.368	0.337	0.372	5/16	0.312	0.182	0.143	0.008
7/16	0.4375	0.4375	0.4294	0.656	0.642	0.438	0.430	0.394	0.454	3/8	0.375	0.213	0.166	0.010
1/2	0.5000	0.5000	0.4919	0.750	0.735	0.500	0.492	0.450	0.454	3/8	0.375	0.245	0.190	0.010
5/8	0.6250	0.6250	0.6163	0.938	0.921	0.625	0.616	0.562	0.595	1/2	0.500	0.307	0.238	0.010
3/4	0.7500	0.7500	0.7406	1.125	1.107	0.750	0.740	0.675	0.620	5/8	0.625	0.370	0.285	0.010
7/8	0.8750	0.8750	0.8647	1.312	1.293	0.875	0.864	0.787	0.698	3/4	0.750	0.432	0.333	0.015
1	1.0000	1.0000	0.9886	1.500	1.479	1.000	0.988	0.900	0.790	3/4	0.750	0.495	0.380	0.015
1 1/8	1.1250	1.1250	1.1086	1.688	1.665	1.125	1.111	1.012	. . . .	7/8	0.875	0.557	0.428	0.015
1 1/4	1.2500	1.2500	1.2336	1.875	1.852	1.250	1.236	1.125	. . . .	7/8	0.875	0.620	0.475	0.015
1 3/8	1.3750	1.3750	1.3568	2.062	2.038	1.375	1.360	1.237	. . . .	1	1.000	0.682	0.523	0.015
1 1/2	1.5000	1.5000	1.4818	2.250	2.224	1.500	1.485	1.350	. . . .	1	1.000	0.745	0.570	0.015
1 3/4	1.7500	1.7500	1.7295	2.625	2.597	1.750	1.734	1.575	. . . .	1 1/4	1.250	0.870	0.665	0.015
2	2.0000	2.0000	1.9780	3.000	2.970	2.000	1.983	1.800	. . . .	1 1/2	1.500	0.995	0.760	0.015
2 1/4	2.2500	2.2500	2.2280	3.375	3.344	2.250	2.232	2.025	. . . .	1 3/4	1.750	1.120	0.855	0.031
2 1/2	2.5000	2.5000	2.4762	3.750	3.717	2.500	2.481	2.250	. . . .	1 3/4	1.750	1.245	0.950	0.031
2 3/4	2.7500	2.7500	2.7262	4.125	4.090	2.750	2.730	2.475	. . . .	2	2.000	1.370	1.045	0.031
3	3.0000	3.0000	2.9762	4.500	4.464	3.000	2.979	2.700	. . . .	2 1/4	2.250	1.495	1.140	0.031
3 1/4	3.2500	3.2500	3.2262	4.875	4.837	3.250	3.228	2.925	. . . .	2 1/4	2.250	1.620	1.235	0.031
3 1/2	3.5000	3.5000	3.4762	5.250	5.211	3.500	3.478	3.150	. . . .	2 3/4	2.750	1.745	1.330	0.031
3 3/4	3.7500	3.7500	3.7262	5.625	5.584	3.750	3.727	3.375	. . . .	2 3/4	2.750	1.870	1.425	0.031
4	4.0000	4.0000	3.9762	6.000	5.958	4.000	3.976	3.600	. . . .	3	3.000	1.995	1.520	0.031

TABLE 11 DIMENSIONS OF HEXAGON AND SPLINE SOCKET FLAT COUNTERSUNK HEAD CAP SCREWS

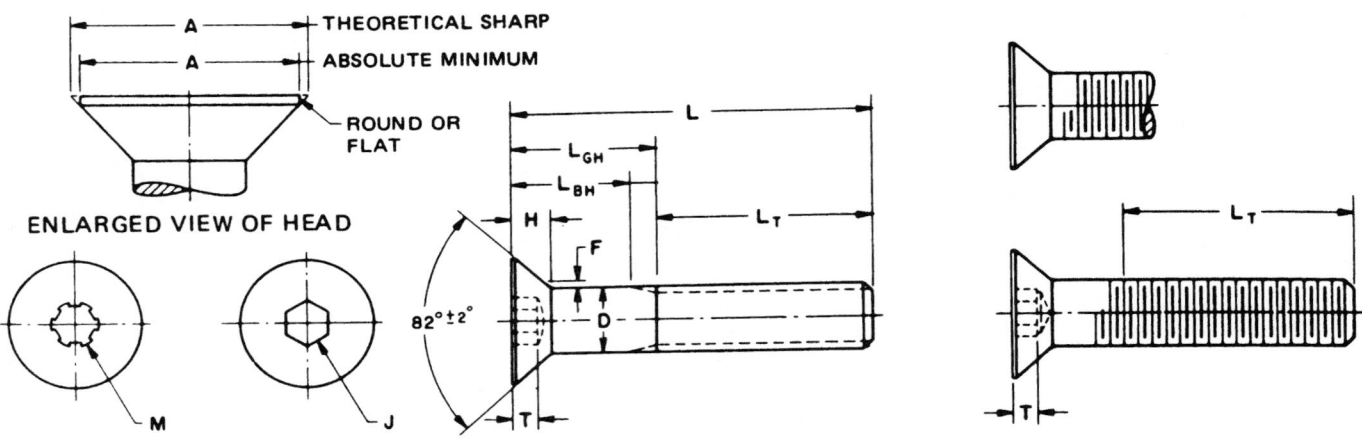

ENLARGED VIEW OF HEAD

Nominal Size or Basic Screw Diameter		D Body Diameter		A Head Diameter		H Heat Height		M Spline Socket Size	J Hexagon Socket Size	T Key Engagement	F Fillet Extension Above D Max
				Theoretical			Flushness				
		Max	Min	Sharp Max	Abs. Min	Reference	Tolerance		Nom	Min	Max
0	0.0600	0.0600	0.0568	0.138	0.117	0.044	0.006	0.048	0.035	0.025	0.006
1	0.0730	0.0730	0.0695	0.168	0.143	0.054	0.007	0.060	0.050	0.031	0.008
2	0.0860	0.0860	0.0822	0.197	0.168	0.064	0.008	0.060	0.050	0.038	0.010
3	0.0990	0.0990	0.0949	0.226	0.193	0.073	0.010	0.072	1/16 0.062	0.044	0.010
4	0.1120	0.1120	0.1075	0.255	0.218	0.083	0.011	0.072	1/16 0.062	0.055	0.012
5	0.1250	0.1250	0.1202	0.281	0.240	0.090	0.012	0.096	5/64 0.078	0.061	0.014
6	0.1380	0.1380	0.1329	0.307	0.263	0.097	0.013	0.096	5/64 0.078	0.066	0.015
8	0.1640	0.1640	0.1585	0.359	0.311	0.112	0.014	0.111	3/32 0.094	0.076	0.015
10	0.1900	0.1900	0.1840	0.411	0.359	0.127	0.015	0.145	1/8 0.125	0.087	0.015
1/4	0.2500	0.2500	0.2435	0.531	0.480	0.161	0.016	0.183	5/32 0.156	0.111	0.015
5/16	0.3125	0.3125	0.3053	0.656	0.600	0.198	0.017	0.216	3/16 0.188	0.135	0.015
3/8	0.3750	0.3750	0.3678	0.781	0.720	0.234	0.018	0.251	7/32 0.219	0.159	0.015
7/16	0.4375	0.4375	0.4294	0.844	0.781	0.234	0.018	0.291	1/4 0.250	0.159	0.015
1/2	0.5000	0.5000	0.4919	0.938	0.872	0.251	0.018	0.372	5/16 0.312	0.172	0.015
5/8	0.6250	0.6250	0.6163	1.188	1.112	0.324	0.022	0.454	3/8 0.375	0.220	0.015
3/4	0.7500	0.7500	0.7406	1.438	1.355	0.396	0.024	0.454	1/2 0.500	0.220	0.015
7/8	0.8750	0.8750	0.8647	1.688	1.604	0.468	0.025	. . .	9/16 0.562	0.248	0.015
1	1.0000	1.0000	0.9886	1.938	1.841	0.540	0.028	. . .	5/8 0.625	0.297	0.015
1 1/8	1.1250	1.1250	1.1086	2.188	2.079	0.611	0.031	. . .	3/4 0.750	0.325	0.031
1 1/4	1.2500	1.2500	1.2336	2.438	2.316	0.683	0.035	. . .	7/8 0.875	0.358	0.031
1 3/8	1.3750	1.3750	1.3568	2.688	2.553	0.755	0.038	. . .	7/8 0.875	0.402	0.031
1 1/2	1.5000	1.5000	1.4818	2.938	2.791	0.827	0.042	. . .	1 1.000	0.435	0.031

(Reprinted from The American Society of Mechanical Engineers—ANSI/ASME B18.3–1986 (R1993).)

TABLE 12 DIMENSIONS OF HEXAGON AND SPLINE SOCKET SET SCREWS

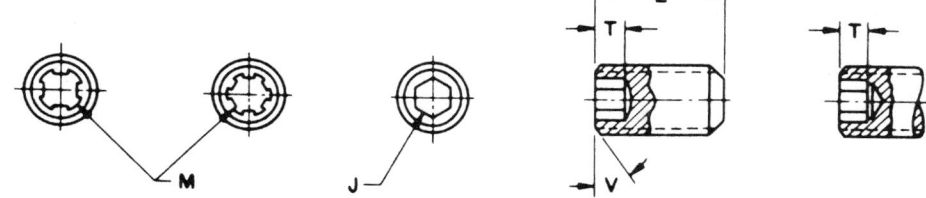

Nominal Size or Basic Screw Diameter		J Hexagon Socket Size		M Spline Socket Size	T Min Key Engagement to Develop Functional Capability of Key Hex Socket T_H Min	Spline Socket T_S Min	C Cup and Flat Point Diameters Max	Min	R Oval Point Radius Basic	Y Cone Point Angle 90° ±2° for These Nominal Lengths or Longer; 118° ±2° for Shorter Nominal Lengths
		Nom		Nom						
0	0.0600	0.028		0.033	0.050	0.026	0.033	0.027	0.045	5/64
1	0.0730	0.035		0.033	0.060	0.035	0.040	0.033	0.055	3/32
2	0.0860	0.035		0.048	0.060	0.040	0.047	0.039	0.064	7/64
3	0.0990	0.050		0.048	0.070	0.040	0.054	0.045	0.074	1/8
4	0.1120	0.050		0.060	0.070	0.045	0.061	0.051	0.084	5/32
5	0.1250	1/16	0.062	0.072	0.080	0.055	0.067	0.057	0.094	3/16
6	0.1380	1/16	0.062	0.072	0.080	0.055	0.074	0.064	0.104	3/16
8	0.1640	5/64	0.078	0.096	0.090	0.080	0.087	0.076	0.123	1/4
10	0.1900	3/32	0.094	0.111	0.100	0.080	0.102	0.088	0.142	1/4
1/4	0.2500	1/8	0.125	0.145	0.125	0.125	0.132	0.118	0.188	5/16
5/16	0.3125	5/32	0.156	0.183	0.156	0.156	0.172	0.156	0.234	3/8
3/8	0.3750	3/16	0.188	0.216	0.188	0.188	0.212	0.194	0.281	7/16
7/16	0.4375	7/32	0.219	0.251	0.219	0.219	0.252	0.232	0.328	1/2
1/2	0.5000	1/4	0.250	0.291	0.250	0.250	0.291	0.270	0.375	9/16
5/8	0.6250	5/16	0.312	0.372	0.312	0.312	0.371	0.347	0.469	3/4
3/4	0.7500	3/8	0.375	0.454	0.375	0.375	0.450	0.425	0.562	7/8
7/8	0.8750	1/2	0.500	0.595	0.500	0.500	0.530	0.502	0.656	1
1	1.0000	9/16	0.562	. . .	0.562	. . .	0.609	0.579	0.750	1 1/8
1 1/8	1.1250	9/16	0.562	. . .	0.562	. . .	0.689	0.655	0.844	1 1/4
1 1/4	1.2500	5/8	0.625	. . .	0.625	. . .	0.767	0.733	0.938	1 1/2
1 3/8	1.3750	5/8	0.625	. . .	0.625	. . .	0.848	0.808	1.031	1 5/8
1 1/2	1.5000	3/4	0.750	. . .	0.750	. . .	0.926	0.886	1.125	1 3/4
1 3/4	1.7500	1	1.000	. . .	1.000	. . .	1.086	1.039	1.312	2
2	2.0000	1	1.000	. . .	1.000	. . .	1.244	1.193	1.500	2 1/4

TABLE 12 (CONTINUED)

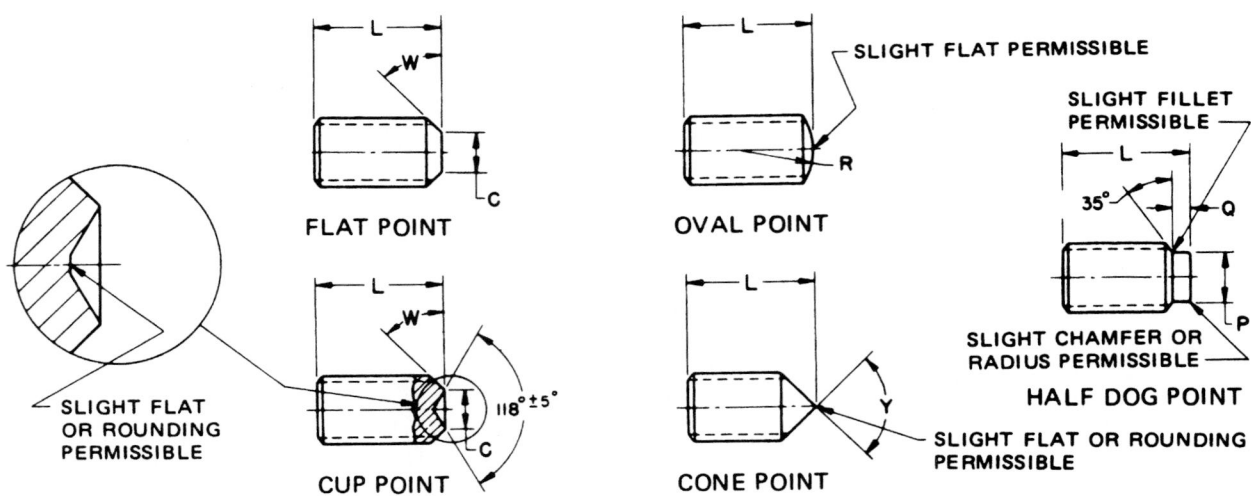

FLAT POINT

OVAL POINT

SLIGHT FLAT PERMISSIBLE

SLIGHT FILLET PERMISSIBLE

35°

SLIGHT CHAMFER OR RADIUS PERMISSIBLE

HALF DOG POINT

SLIGHT FLAT OR ROUNDING PERMISSIBLE

CUP POINT

118°±5°

CONE POINT

SLIGHT FLAT OR ROUNDING PERMISSIBLE

Nominal Size or Basic Screw Diameter		P		Q		B			B_1		
		Half Dog Point				Shortest Optimum Nominal Length To Which Column T_H Applies			Shortest Optimum Nominal Length To Which Column T_S Applies		
		Diameter		Length		Cup and Flat Points	90° Cone and Oval Points	Half Dog Points	Cup and Flat Points	90° Cone and Oval Points	Half Dog Point
		Max	Min	Max	Min						
0	0.0600	0.040	0.037	0.017	0.013	7/64	1/8	7/64	1/16	1/8	7/64
1	0.0730	0.049	0.045	0.021	0.017	1/8	9/64	1/8	3/32	9/64	1/8
2	0.0860	0.057	0.053	0.024	0.020	1/8	9/64	9/64	3/32	9/64	9/64
3	0.0990	0.066	0.062	0.027	0.023	9/64	5/32	5/32	3/32	5/32	5/32
4	0.1120	0.075	0.070	0.030	0.026	9/64	11/64	5/32	3/32	11/64	5/32
5	0.1250	0.083	0.078	0.033	0.027	3/16	3/16	11/64	1/8	3/16	11/64
6	0.1380	0.092	0.087	0.038	0.032	11/64	13/64	3/16	1/8	13/64	3/16
8	0.1640	0.109	0.103	0.043	0.037	3/16	7/32	13/64	3/16	7/32	13/64
10	0.1900	0.127	0.120	0.049	0.041	3/16	1/4	15/64	3/16	1/4	15/64
1/4	0.2500	0.156	0.149	0.067	0.059	1/4	5/16	19/64	1/4	5/16	19/64
5/16	0.3125	0.203	0.195	0.082	0.074	5/16	25/64	23/64	5/16	25/64	23/64
3/8	0.3750	0.250	0.241	0.099	0.089	3/8	7/16	7/16	3/8	7/16	7/16
7/16	0.4375	0.297	0.287	0.114	0.104	7/16	35/64	31/64	7/16	35/64	31/64
1/2	0.5000	0.344	0.334	0.130	0.120	1/2	39/64	35/64	1/2	39/64	35/64
5/8	0.6250	0.469	0.456	0.164	0.148	5/8	49/64	43/64	5/8	49/64	43/64
3/4	0.7500	0.562	0.549	0.196	0.180	3/4	29/32	51/64	3/4	29/32	51/64
7/8	0.8750	0.656	0.642	0.227	0.211	7/8	1 1/8	63/64	7/8	1 1/8	63/64
1	1.0000	0.750	0.734	0.260	0.240	1	1 17/64	1 1/8	. . .	. . .	. . .
1 1/8	1.1250	0.844	0.826	0.291	0.271	1 1/8	1 25/64	1 3/16	. . .	. . .	. . .
1 1/4	1.2500	0.938	0.920	0.323	0.303	1 1/4	1 1/2	1 5/16	. . .	. . .	. . .
1 3/8	1.3750	1.031	1.011	0.354	0.334	1 3/8	1 21/32	1 7/16	. . .	. . .	. . .
1 1/2	1.5000	1.125	1.105	0.385	0.365	1 1/2	1 51/64	1 9/16	. . .	. . .	. . .
1 3/4	1.7500	1.312	1.289	0.448	0.428	1 3/4	2 7/32	1 61/64	. . .	. . .	. . .
2	2.0000	1.500	1.474	0.510	0.490	2	2 25/64	2 5/64	. . .	. . .	. . .

(Reprinted from The American Society of Mechanical Engineers—ANSI/ASME B18.3–1986 (R1993).)

TABLE 13 DIMENSIONS OF SLOTTED FLAT COUNTERSUNK HEAD CAP SCREWS

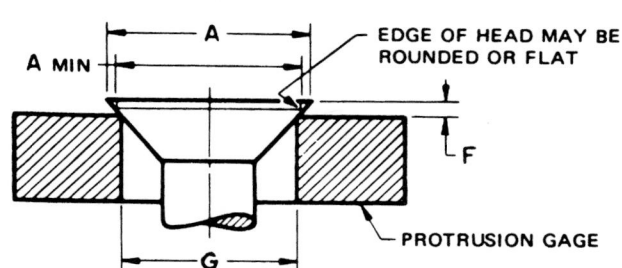

CAP SCREWS

FLAT

Type of Head

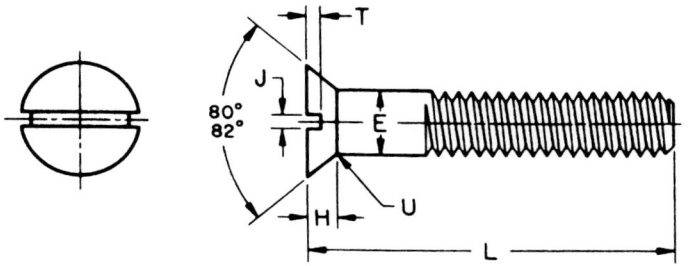

Nominal Size[1] or Basic Screw Diameter		E Body Diameter		A Head Diameter		H[2] Head Height	J Slot Width		T Slot Depth		U Fillet Radius	F[3] Protrusion Above Gaging Diameter		G[3] Gaging Diameter
		Max	Min	Max, Edge Sharp	Min, Edge Rounded or Flat	Ref	Max	Min	Max	Min	Max	Max	Min	
1/4	0.2500	0.2500	0.2450	0.500	0.452	0.140	0.075	0.064	0.068	0.045	0.100	0.046	0.030	0.424
5/16	0.3125	0.3125	0.3070	0.625	0.567	0.177	0.084	0.072	0.086	0.057	0.125	0.053	0.035	0.538
3/8	0.3750	0.3750	0.3690	0.750	0.682	0.210	0.094	0.081	0.103	0.068	0.150	0.060	0.040	0.651
7/16	0.4375	0.4375	0.4310	0.812	0.736	0.210	0.094	0.081	0.103	0.068	0.175	0.065	0.044	0.703
1/2	0.5000	0.5000	0.4930	0.875	0.791	0.210	0.106	0.091	0.103	0.068	0.200	0.071	0.049	0.756
9/16	0.5625	0.5625	0.5550	1.000	0.906	0.244	0.118	0.102	0.120	0.080	0.225	0.078	0.054	0.869
5/8	0.6250	0.6250	0.6170	1.125	1.020	0.281	0.133	0.116	0.137	0.091	0.250	0.085	0.058	0.982
3/4	0.7500	0.7500	0.7420	1.375	1.251	0.352	0.149	0.131	0.171	0.115	0.300	0.099	0.068	1.208
7/8	0.8750	0.8750	0.8660	1.625	1.480	0.423	0.167	0.147	0.206	0.138	0.350	0.113	0.077	1.435
1	1.0000	1.0000	0.9900	1.875	1.711	0.494	0.188	0.166	0.240	0.162	0.400	0.127	0.087	1.661
1 1/8	1.1250	1.1250	1.1140	2.062	1.880	0.529	0.196	0.178	0.257	0.173	0.450	0.141	0.096	1.826
1 1/4	1.2500	1.2500	1.2390	2.312	2.110	0.600	0.211	0.193	0.291	0.197	0.500	0.155	0.105	2.052
1 3/8	1.3750	1.3750	1.3630	2.562	2.340	0.665	0.226	0.208	0.326	0.220	0.550	0.169	0.115	2.279
1 1/2	1.5000	1.5000	1.4880	2.812	2.570	0.742	0.258	0.240	0.360	0.244	0.600	0.183	0.124	2.505

[1] Where specifying nominal size in decimals, zeros preceding decimal and in the fourth decimal place shall be omitted.
[2] Tabulated values determined from formula for maximum H, Appendix III.
[3] No tolerance for gaging diameter is given. If the gaging diameter of the gage used differs from tabulated value, the protrusion will be affected accordingly and the proper protrusion values must be recalculated using the formulas shown in Appendix II.
FOOTNOTES REFER TO ANSI B18.6.2–1972 (R1993).
(Reprinted from The American Society of Mechanical Engineers—ANSI B18.6.2–1972 (R1993).)

TABLE 14 DIMENSIONS OF SLOTTED ROUND HEAD CAP SCREWS

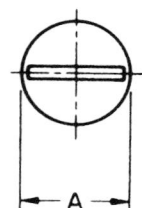

CAP SCREWS

ROUND

Type of Head

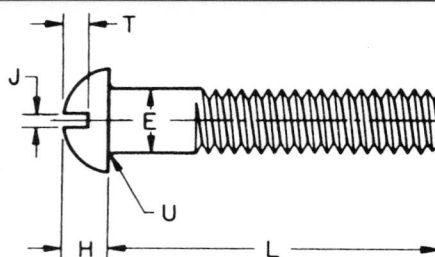

Nominal Size[1] or Basic Screw Diameter		E Body Diameter		A Head Diameter		H Head Height		J Slot Width		T Slot Depth		U Fillet Radius	
		Max	Min	Max	Min	Max	Min	Max	Min	Max	Min	Max	Min
1/4	0.2500	0.2500	0.2450	0.437	0.418	0.191	0.175	0.075	0.064	0.117	0.097	0.031	0.016
5/16	0.3125	0.3125	0.3070	0.562	0.540	0.245	0.226	0.084	0.072	0.151	0.126	0.031	0.016
3/8	0.3750	0.3750	0.3690	0.625	0.603	0.273	0.252	0.094	0.081	0.168	0.138	0.031	0.016
7/16	0.4375	0.4375	0.4310	0.750	0.725	0.328	0.302	0.094	0.081	0.202	0.167	0.047	0.016
1/2	0.5000	0.5000	0.4930	0.812	0.786	0.354	0.327	0.106	0.091	0.218	0.178	0.047	0.016
9/16	0.5625	0.5625	0.5550	0.937	0.909	0.409	0.378	0.118	0.102	0.252	0.207	0.047	0.016
5/8	0.6250	0.6250	0.6170	1.000	0.970	0.437	0.405	0.133	0.116	0.270	0.220	0.062	0.031
3/4	0.7500	0.7500	0.7420	1.250	1.215	0.546	0.507	0.149	0.131	0.338	0.278	0.062	0.031

[1]Where specifying nominal size in decimals, zeros preceding decimal and in the fourth decimal place shall be omitted.

(Reprinted from The American Society of Mechanical Engineers—ANSI B18.6.2–1972.)

TABLE 15 DIMENSIONS OF SLOTTED FILISTER HEAD CAP SCREWS

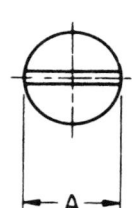

CAP SCREWS

FILLISTER

Type of Head

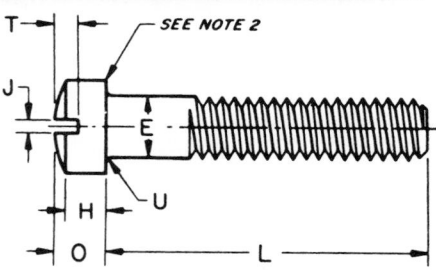

Nominal Size[1] or Basic Screw Diameter		E Body Diameter		A Head Diameter		H Head Side Height		O Total Head Height		J Slot Width		T Slot Depth		U Fillet Radius	
		Max	Min	Max	Min	Max	Min	Max	Min	Max	Min	Max	Min	Max	Min
1/4	0.2500	0.2500	0.2450	0.375	0.363	0.172	0.157	0.216	0.194	0.075	0.064	0.097	0.077	0.031	0.016
5/16	0.3125	0.3125	0.3070	0.437	0.424	0.203	0.186	0.253	0.230	0.084	0.072	0.115	0.090	0.031	0.016
3/8	0.3750	0.3750	0.3690	0.562	0.547	0.250	0.229	0.314	0.284	0.094	0.081	0.142	0.112	0.031	0.016
7/16	0.4375	0.4375	0.4310	0.625	0.608	0.297	0.274	0.368	0.336	0.094	0.081	0.168	0.133	0.047	0.016
1/2	0.5000	0.5000	0.4930	0.750	0.731	0.328	0.301	0.413	0.376	0.106	0.091	0.193	0.153	0.047	0.016
9/16	0.5625	0.5625	0.5550	0.812	0.792	0.375	0.346	0.467	0.427	0.118	0.102	0.213	0.168	0.047	0.016
5/8	0.6250	0.6250	0.6170	0.875	0.853	0.422	0.391	0.521	0.478	0.133	0.116	0.239	0.189	0.062	0.031
3/4	0.7500	0.7500	0.7420	1.000	0.976	0.500	0.466	0.612	0.566	0.149	0.131	0.283	0.223	0.062	0.031
7/8	0.8750	0.8750	0.8660	1.125	1.098	0.594	0.556	0.720	0.668	0.167	0.147	0.334	0.264	0.062	0.031
1	1.0000	1.0000	0.9900	1.312	1.282	0.656	0.612	0.803	0.743	0.188	0.166	0.371	0.291	0.062	0.031

[1]Where specifying nominal size in decimals, zeros preceding decimal and in the fourth decimal place shall be omitted.

[2]A slight rounding of the edges at periphery of head shall be permissible provided the diameter of the bearing circle is equal to no less than 90 percent of the specified minimum head diameter.

(Reprinted from The American Society of Mechanical Engineers—ANSI B18.6.2–1972.)

TABLE 16 DIMENSIONS OF SLOTTED FLAT COUNTERSUNK HEAD MACHINE SCREWS

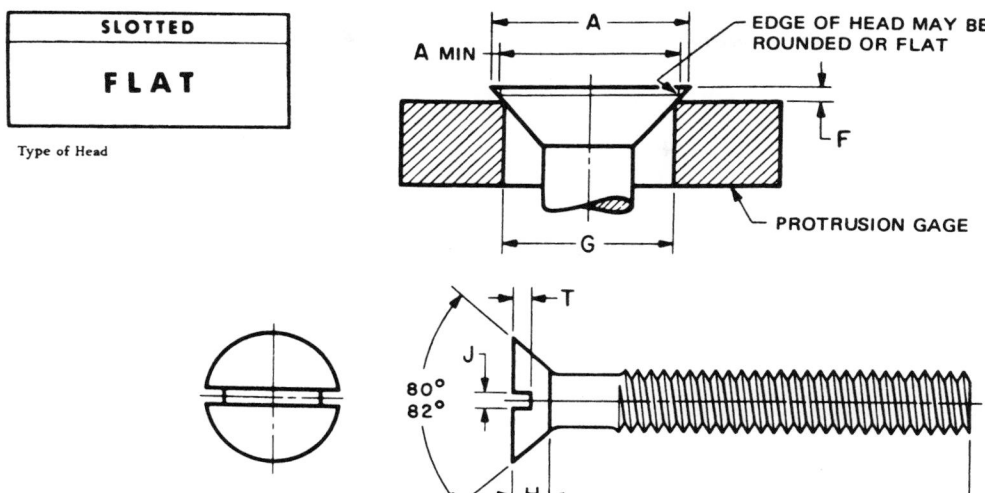

Nominal Size[1] or Basic Screw Diameter		L[2] These Lengths or Shorter are Undercut	A Head Diameter		H[3] Head Height	J Slot Width		T Slot Depth		F[4] Protrusion Above Gaging Diameter		G[4] Gaging Diameter
			Max. Edge Sharp	Min, Edge Rounded or Flat	Ref	Max	Min	Max	Min	Max	Min	
0000	0.0210	—	0.043	0.037	0.011	0.008	0.004	0.007	0.003	*	*	*
000	0.0340	—	0.064	0.058	0.016	0.011	0.007	0.009	0.005	*	*	*
00	0.0470	—	0.093	0.085	0.028	0.017	0.010	0.014	0.009	*	*	*
0	0.0600	1/8	0.119	0.099	0.035	0.023	0.016	0.015	0.010	0.026	0.016	0.078
1	0.0730	1/8	0.146	0.123	0.043	0.026	0.019	0.019	0.012	0.028	0.016	0.101
2	0.0860	1/8	0.172	0.147	0.051	0.031	0.023	0.023	0.015	0.029	0.017	0.124
3	0.0990	1/8	0.199	0.171	0.059	0.035	0.027	0.027	0.017	0.031	0.018	0.148
4	0.1120	3/16	0.225	0.195	0.067	0.039	0.031	0.030	0.020	0.032	0.019	0.172
5	0.1250	3/16	0.252	0.220	0.075	0.043	0.035	0.034	0.022	0.034	0.020	0.196
6	0.1380	3/16	0.279	0.244	0.083	0.048	0.039	0.038	0.024	0.036	0.021	0.220
8	0.1640	1/4	0.332	0.292	0.100	0.054	0.045	0.045	0.029	0.039	0.023	0.267
10	0.1900	5/16	0.385	0.340	0.116	0.060	0.050	0.053	0.034	0.042	0.025	0.313
12	0.2160	3/8	0.438	0.389	0.132	0.067	0.056	0.060	0.039	0.045	0.027	0.362
1/4	0.2500	7/16	0.507	0.452	0.153	0.075	0.064	0.070	0.046	0.050	0.029	0.424
5/16	0.3125	1/2	0.635	0.568	0.191	0.084	0.072	0.088	0.058	0.057	0.034	0.539
3/8	0.3750	9/16	0.762	0.685	0.230	0.094	0.081	0.106	0.070	0.065	0.039	0.653
7/16	0.4375	5/8	0.812	0.723	0.223	0.094	0.081	0.103	0.066	0.073	0.044	0.690
1/2	0.5000	3/4	0.875	0.775	0.223	0.106	0.091	0.103	0.065	0.081	0.049	0.739
9/16	0.5625	—	1.000	0.889	0.260	0.118	0.102	0.120	0.077	0.089	0.053	0.851
5/8	0.6250	—	1.125	1.002	0.298	0.133	0.116	0.137	0.088	0.097	0.058	0.962
3/4	0.7500	—	1.375	1.230	0.372	0.149	0.131	0.171	0.111	0.112	0.067	1.186

[1] Where specifying nominal size in decimals, zeros preceding decimal and in the fourth decimal place shall be omitted.
[2] Screws of these lengths and shorter shall have undercut heads as shown in Table 5.
[3] Tabulated values determined from formula for maximum H, Appendix V.
[4] No tolerance for gaging diameter is given. If the gaging diameter of the gage used differs from tabulated value, the protrusion will be affected accordingly and the proper protrusion values must be recalculated using the formulas shown in Appendix I.
* Not practical to gage.

For additional requirements refer to General Data on Pages 3, 4 and 5.

FOOTNOTES REFER TO ANSI B18.6.3—1972 (R1991).

(Reprinted from The American Society of Mechanical Engineers—ANSI B18.6.3–1972 (R1991).)

TABLE 17 DIMENSIONS OF HEX NUTS AND HEX JAM NUTS

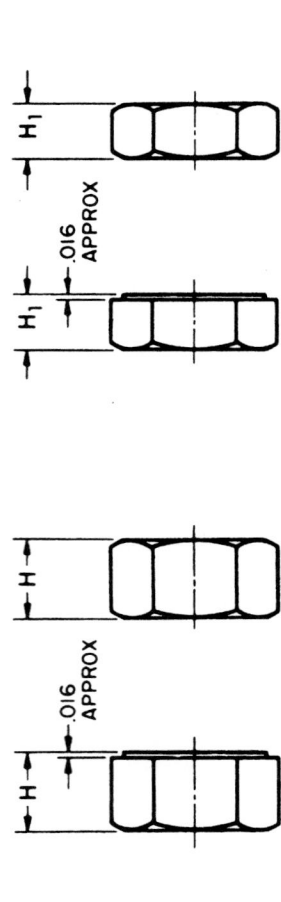

Nominal Size or Basic Major Dia of Thread		F Width Across Flats			G Width Across Corners		H Thickness Hex Nuts			H₁ Thickness Hex Jam Nuts			Hex Nuts Specified Proof Load — Runout of Bearing Face, FIR Max		Jam Nuts All Strength Levels
		Basic	Max	Min	Max	Min	Basic	Max	Min	Basic	Max	Min	Up to 150,000 psi	150,000 psi and Greater	
1/4	0.2500	7/16	0.438	0.428	0.505	0.488	7/32	0.226	0.212	5/32	0.163	0.150	0.015	0.010	0.015
5/16	0.3125	1/2	0.500	0.489	0.577	0.557	17/64	0.273	0.258	3/16	0.195	0.180	0.016	0.011	0.016
3/8	0.3750	9/16	0.562	0.551	0.650	0.628	21/64	0.337	0.320	7/32	0.227	0.210	0.017	0.012	0.017
7/16	0.4375	11/16	0.688	0.675	0.794	0.768	3/8	0.385	0.365	1/4	0.260	0.240	0.018	0.013	0.018
1/2	0.5000	3/4	0.750	0.736	0.866	0.840	7/16	0.448	0.427	5/16	0.323	0.302	0.019	0.014	0.019
9/16	0.5625	7/8	0.875	0.861	1.010	0.982	31/64	0.496	0.473	5/16	0.324	0.301	0.020	0.015	0.020
5/8	0.6250	15/16	0.938	0.922	1.083	1.051	35/64	0.559	0.535	3/8	0.387	0.363	0.021	0.016	0.021
3/4	0.7500	1 1/8	1.125	1.088	1.299	1.240	41/64	0.665	0.617	27/64	0.446	0.398	0.023	0.018	0.023
7/8	0.8750	1 5/16	1.312	1.269	1.516	1.447	3/4	0.776	0.724	31/64	0.510	0.458	0.025	0.020	0.025
1	1.0000	1 1/2	1.500	1.450	1.732	1.653	55/64	0.887	0.831	35/64	0.575	0.519	0.027	0.022	0.027
1 1/8	1.1250	1 11/16	1.688	1.631	1.949	1.859	31/32	0.999	0.939	39/64	0.639	0.579	0.030	0.025	0.030
1 1/4	1.2500	1 7/8	1.875	1.812	2.165	2.066	1 1/16	1.094	1.030	23/32	0.751	0.687	0.033	0.028	0.033
1 3/8	1.3750	2 1/16	2.062	1.994	2.382	2.273	1 11/64	1.206	1.138	25/32	0.815	0.747	0.036	0.031	0.036
1 1/2	1.5000	2 1/4	2.250	2.175	2.598	2.480	1 9/32	1.317	1.245	27/32	0.880	0.808	0.039	0.034	0.039

(Reprinted from The American Society of Mechanical Engineers—ANSI B18.2.2-1987 (R1993).)

TABLE 18 WOODRUFF KEY DIMENSIONS

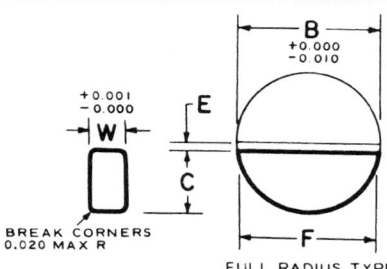

FULL RADIUS TYPE

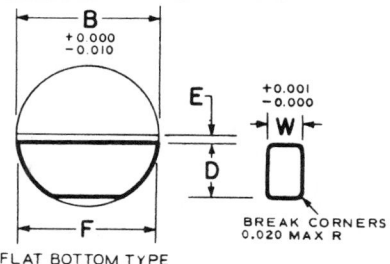

FLAT BOTTOM TYPE

Key No.	Nominal Key Size W × B	Actual Length F +0.000–0.010	Height of Key				Distance Below Center E
			C		D		
			Max	Min	Max	Min	
202	1/16 × 1/4	0.248	0.109	0.104	0.109	0.104	1/64
202.5	1/16 × 5/16	0.311	0.140	0.135	0.140	0.135	1/64
302.5	3/32 × 5/16	0.311	0.140	0.135	0.140	0.135	1/64
203	1/16 × 3/8	0.374	0.172	0.167	0.172	0.167	1/64
303	3/32 × 3/8	0.374	0.172	0.167	0.172	0.167	1/64
403	1/8 × 3/8	0.374	0.172	0.167	0.172	0.167	1/64
204	1/16 × 1/2	0.491	0.203	0.198	0.194	0.188	3/64
304	3/32 × 1/2	0.491	0.203	0.198	0.194	0.188	3/64
404	1/8 × 1/2	0.491	0.203	0.198	0.194	0.188	3/64
305	3/32 × 5/8	0.612	0.250	0.245	0.240	0.234	1/16
405	1/8 × 5/8	0.612	0.250	0.245	0.240	0.234	1/16
505	5/32 × 5/8	0.612	0.250	0.245	0.240	0.234	1/16
605	3/16 × 5/8	0.612	0.250	0.245	0.240	0.234	1/16
406	1/8 × 3/4	0.740	0.313	0.308	0.303	0.297	1/16
506	5/32 × 3/4	0.740	0.313	0.308	0.303	0.297	1/16
606	3/16 × 3/4	0.740	0.313	0.308	0.303	0.297	1/16
806	1/4 × 3/4	0.740	0.313	0.308	0.303	0.297	1/16
507	5/32 × 7/8	0.866	0.375	0.370	0.365	0.359	1/16
607	3/16 × 7/8	0.866	0.375	0.370	0.365	0.359	1/16
707	7/32 × 7/8	0.866	0.375	0.370	0.365	0.359	1/16
807	1/4 × 7/8	0.866	0.375	0.370	0.365	0.359	1/16
608	3/16 × 1	0.992	0.438	0.433	0.428	0.422	1/16
708	7/32 × 1	0.992	0.438	0.433	0.428	0.422	1/16
808	1/4 × 1	0.992	0.438	0.433	0.428	0.422	1/16
1008	5/16 × 1	0.992	0.438	0.433	0.428	0.422	1/16
1208	3/8 × 1	0.992	0.438	0.433	0.428	0.422	1/16
609	3/16 × 1 1/8	1.114	0.484	0.479	0.475	0.469	5/64
709	7/32 × 1 1/8	1.114	0.484	0.479	0.475	0.469	5/64
809	1/4 × 1 1/8	1.114	0.484	0.479	0.475	0.469	5/64
1009	5/16 × 1 1/8	1.114	0.484	0.479	0.475	0.469	5/64
610	3/16 × 1 1/4	1.240	0.547	0.542	0.537	0.531	5/64
710	7/32 × 1 1/4	1.240	0.547	0.542	0.537	0.531	5/64
810	1/4 × 1 1/4	1.240	0.547	0.542	0.537	0.531	5/64
1010	5/16 × 1 1/4	1.240	0.547	0.542	0.537	0.531	5/64
1210	3/8 × 1 1/4	1.240	0.547	0.542	0.537	0.531	5/64
811	1/4 × 1 3/8	1.362	0.594	0.589	0.584	0.578	3/32
1011	5/16 × 1 3/8	1.362	0.594	0.589	0.584	0.578	3/32
1211	3/8 × 1 3/8	1.362	0.594	0.589	0.584	0.578	3/32
812	1/4 × 1 1/2	1.484	0.641	0.636	0.631	0.625	7/64
1012	5/16 × 1 1/2	1.484	0.641	0.636	0.631	0.625	7/64
1212	3/8 × 1 1/2	1.484	0.641	0.636	0.631	0.625	7/64

All dimensions given are in inches.

The key numbers indicate nominal key dimensions. The last two digits give the nominal diameter B in eighths of an inch and the digits preceding the last two give the nominal width W in thirty-seconds of an inch.

Example: No. 204 indicates a key 2/32 × 4/8 or 1/16 × 1/2.
No. 808 indicates a key 8/32 × 8/8 or 1/4 × 1.
No. 1212 indicates a key 12/32 × 12/8 or 3/8 × 1 1/2.

(Continued)

TABLE 18 (CONTINUED)

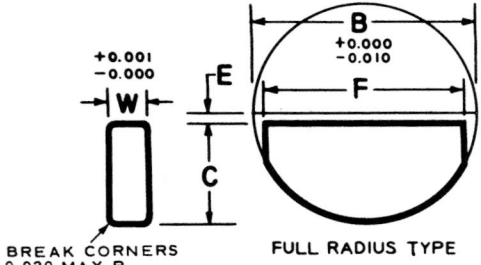

FULL RADIUS TYPE

BREAK CORNERS 0.020 MAX R

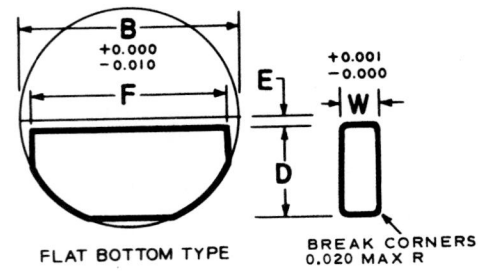

FLAT BOTTOM TYPE

BREAK CORNERS 0.020 MAX R

Key No.	Nominal Key Size W × B	Actual Length F +0.000–0.010	Height of Key				Distance Below Center E
			C		D		
			Max	Min	Max	Min	
617-1	3/16 × 2 1/8	1.380	0.406	0.401	0.396	0.390	21/32
817-1	1/4 × 2 1/8	1.380	0.406	0.401	0.396	0.390	21/32
1017-1	5/16 × 2 1/8	1.380	0.406	0.401	0.396	0.390	21/32
1217-1	3/8 × 2 1/8	1.380	0.406	0.401	0.396	0.390	21/32
617	3/16 × 2 1/8	1.723	0.531	0.526	0.521	0.515	17/32
817	1/4 × 2 1/8	1.723	0.531	0.526	0.521	0.515	17/32
1017	5/16 × 2 1/8	1.723	0.531	0.526	0.521	0.515	17/32
1217	3/8 × 2 1/8	1.723	0.531	0.526	0.521	0.515	17/32
822-1	1/4 × 2 3/4	2.000	0.594	0.589	0.584	0.578	25/32
1022-1	5/16 × 2 3/4	2.000	0.594	0.589	0.584	0.578	25/32
1222-1	3/8 × 2 3/4	2.000	0.594	0.589	0.584	0.578	25/32
1422-1	7/16 × 2 3/4	2.000	0.594	0.589	0.584	0.578	25/32
1622-1	1/2 × 2 3/4	2.000	0.594	0.589	0.584	0.578	25/32
822	1/4 × 2 3/4	2.317	0.750	0.745	0.740	0.734	5/8
1022	5/16 × 2 3/4	2.317	0.750	0.745	0.740	0.734	5/8
1222	3/8 × 2 3/4	2.317	0.750	0.745	0.740	0.734	5/8
1422	7/16 × 2 3/4	2.317	0.750	0.745	0.740	0.734	5/8
1622	1/2 × 2 3/4	2.317	0.750	0.745	0.740	0.734	5/8
1228	3/8 × 3 1/2	2.880	0.938	0.933	0.928	0.922	13/16
1428	7/16 × 3 1/2	2.880	0.938	0.933	0.928	0.922	13/16
1628	1/2 × 3 1/2	2.880	0.938	0.933	0.928	0.922	13/16
1828	9/16 × 3 1/2	2.880	0.938	0.933	0.928	0.922	13/16
2028	5/8 × 3 1/2	2.880	0.938	0.933	0.928	0.922	13/16
2228	11/16 × 3 1/2	2.880	0.938	0.933	0.928	0.922	13/16
2428	3/4 × 3 1/2	2.880	0.938	0.933	0.928	0.922	13/16

All dimensions given are in inches.

The key numbers indicate nominal key dimensions. The last two digits give the nominal diameter B in eighths of an inch and the digits preceding the last two give the nominal width W in thirty-seconds of an inch.

Example:

No. 617 indicates a key 6/32 × 17/8 or 3/16 × 2 1/8
No. 822 indicates a key 8/32 × 22/8 or 1/4 × 2 1/4
No. 1228 indicates a key 12/32 × 28/8 or 3/8 × 3 1/2

The key numbers with the -1 designation, while representing the nominal key size have a shorter length F and due to a greater distance below center E are less in height than the keys of the same number without the -1 designation.

(Reprinted from The American Society of Mechanical Engineers—ANSI B17.2–1967 (R1990).)

TABLE 19 WOODRUFF KEYSEAT DIMENSIONS

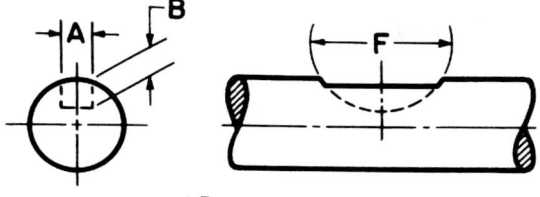

KEYSEAT-SHAFT

KEY ABOVE SHAFT

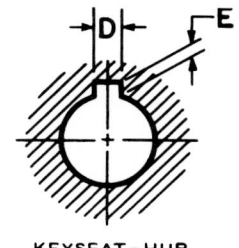

KEYSEAT-HUB

Key Number	Nominal Size Key	Keyseat – Shaft					Key Above Shaft	Keyseat – Hub	
		Width A*		Depth B	Diameter F		Height C	Width D	Depth E
		Min	Max	+0.005 −0.000	Min	Max	+0.005 −0.005	+0.002 −0.000	+0.005 −0.000
202	1/16 × 1/4	0.0615	0.0630	0.0728	0.250	0.268	0.0312	0.0635	0.0372
202.5	1/16 × 5/16	0.0615	0.0630	0.1038	0.312	0.330	0.0312	0.0635	0.0372
302.5	3/32 × 5/16	0.0928	0.0943	0.0882	0.312	0.330	0.0469	0.0948	0.0529
203	1/16 × 3/8	0.0615	0.0630	0.1358	0.375	0.393	0.0312	0.0635	0.0372
303	3/32 × 3/8	0.0928	0.0943	0.1202	0.375	0.393	0.0469	0.0948	0.0529
403	1/8 × 3/8	0.1240	0.1255	0.1045	0.375	0.393	0.0625	0.1260	0.0685
204	1/16 × 1/2	0.0615	0.0630	0.1668	0.500	0.518	0.0312	0.0635	0.0372
304	3/32 × 1/2	0.0928	0.0943	0.1511	0.500	0.518	0.0469	0.0948	0.0529
404	1/8 × 1/2	0.1240	0.1255	0.1355	0.500	0.518	0.0625	0.1260	0.0685
305	3/32 × 5/8	0.0928	0.0943	0.1981	0.625	0.643	0.0469	0.0948	0.0529
405	1/8 × 5/8	0.1240	0.1255	0.1825	0.625	0.643	0.0625	0.1260	0.0685
505	5/32 × 5/8	0.1553	0.1568	0.1669	0.625	0.643	0.0781	0.1573	0.0841
605	3/16 × 5/8	0.1863	0.1880	0.1513	0.625	0.643	0.0937	0.1885	0.0997
406	1/8 × 3/4	0.1240	0.1255	0.2455	0.750	0.768	0.0625	0.1260	0.0685
506	5/32 × 3/4	0.1553	0.1568	0.2299	0.750	0.768	0.0781	0.1573	0.0841
606	3/16 × 3/4	0.1863	0.1880	0.2143	0.750	0.768	0.0937	0.1885	0.0997
806	1/4 × 3/4	0.2487	0.2505	0.1830	0.750	0.768	0.1250	0.2510	0.1310
507	5/32 × 7/8	0.1553	0.1568	0.2919	0.875	0.895	0.0781	0.1573	0.0841
607	3/16 × 7/8	0.1863	0.1880	0.2763	0.875	0.895	0.0937	0.1885	0.0997
707	7/32 × 7/8	0.2175	0.2193	0.2607	0.875	0.895	0.1093	0.2198	0.1153
807	1/4 × 7/8	0.2487	0.2505	0.2450	0.875	0.895	0.1250	0.2510	0.1310
608	3/16 × 1	0.1863	0.1880	0.3393	1.000	1.020	0.0937	0.1885	0.0997
708	7/32 × 1	0.2175	0.2193	0.3237	1.000	1.020	0.1093	0.2198	0.1153
808	1/4 × 1	0.2487	0.2505	0.3080	1.000	1.020	0.1250	0.2510	0.1310
1008	5/16 × 1	0.3111	0.3130	0.2768	1.000	1.020	0.1562	0.3135	0.1622
1208	3/8 × 1	0.3735	0.3755	0.2455	1.000	1.020	0.1875	0.3760	0.1935
609	3/16 × 1 1/8	0.1863	0.1880	0.3853	1.125	1.145	0.0937	0.1885	0.0997
709	7/32 × 1 1/8	0.2175	0.2193	0.3697	1.125	1.145	0.1093	0.2198	0.1153
809	1/4 × 1 1/8	0.2487	0.2505	0.3540	1.125	1.145	0.1250	0.2510	0.1310
1009	5/16 × 1 1/8	0.3111	0.3130	0.3228	1.125	1.145	0.1562	0.3135	0.1622

TABLE 19 (CONTINUED)

Key Number	Nominal Size Key	Keyset – Shaft					Key Above Shaft	Keyseat – Hub	
		Width A*		Depth B	Diameter F		Height C	Width D	Depth E
		Min	Max	+0.005 −0.000	Min	Max	+0.005 −0.005	+0.002 −0.000	+0.005 −0.000
610	3/16 × 1 1/4	0.1863	0.1880	0.4483	1.250	1.273	0.0937	0.1885	0.0997
710	7/32 × 1 1/4	0.2175	0.2193	0.4327	1.250	1.273	0.1093	0.2198	0.1153
810	1/4 × 1 1/4	0.2487	0.2505	0.4170	1.250	1.273	0.1250	0.2510	0.1310
1010	5/16 × 1 1/4	0.3111	0.3130	0.3858	1.250	1.273	0.1562	0.3135	0.1622
1210	3/8 × 1 1/4	0.3735	0.3755	0.3545	1.250	1.273	0.1875	0.3760	0.1935
811	1/4 × 1 3/8	0.2487	0.2505	0.4640	1.375	1.398	0.1250	0.2510	0.1310
1011	5/16 × 1 3/8	0.3111	0.3130	0.4328	1.375	1.398	0.1562	0.3135	0.1622
1211	3/8 × 1 3/8	0.3735	0.3755	0.4015	1.375	1.398	0.1875	0.3760	0.1935
812	1/4 × 1 1/2	0.2487	0.2505	0.5110	1.500	1.523	0.1250	0.2510	0.1310
1012	5/16 × 1 1/2	0.3111	0.3130	0.4798	1.500	1.523	0.1562	0.3135	0.1622
1212	3/8 × 1 1/2	0.3735	0.3755	0.4485	1.500	1.523	0.1875	0.3760	0.1935
617-1	3/16 × 2 1/8	0.1863	0.1880	0.3073	2.125	2.160	0.0937	0.1885	0.0997
817-1	1/4 × 2 1/8	0.2487	0.2505	0.2760	2.125	2.160	0.1250	0.2510	0.1310
1017-1	5/16 × 2 1/8	0.3111	0.3130	0.2448	2.125	2.160	0.1562	0.3135	0.1622
1217-1	3/8 × 2 1/8	0.3735	0.3755	0.2135	2.125	2.160	0.1875	0.3760	0.1935
617	3/16 × 2 1/8	0.1863	0.1880	0.4323	2.125	2.160	0.0937	0.1885	0.0997
817	1/4 × 2 1/8	0.2487	0.2505	0.4010	2.125	2.160	0.1250	0.2510	0.1310
1017	5/16 × 2 1/8	0.3111	0.3130	0.3698	2.125	2.160	0.1562	0.3135	0.1622
1217	3/8 × 2 1/8	0.3735	0.3755	0.3385	2.125	2.160	0.1875	0.3760	0.1935
822-1	1/4 × 2 3/4	0.2487	0.2505	0.4640	2.750	2.785	0.1250	0.2510	0.1310
1022-1	5/16 × 2 3/4	0.3111	0.3130	0.4328	2.750	2.785	0.1562	0.3135	0.1622
1222-1	3/8 × 2 3/4	0.3735	0.3755	0.4015	2.750	2.785	0.1875	0.3760	0.1935
1422-1	7/16 × 2 3/4	0.4360	0.4380	0.3703	2.750	2.785	0.2187	0.4385	0.2247
1622-1	1/2 × 2 3/4	0.4985	0.5005	0.3390	2.750	2.785	0.2500	0.5010	0.2560
822	1/4 × 2 3/4	0.2487	0.2505	0.6200	2.750	2.785	0.1250	0.2510	0.1310
1022	5/16 × 2 3/4	0.3111	0.3130	0.5888	2.750	2.785	0.1562	0.3135	0.1622
1222	3/8 × 2 3/4	0.3735	0.3755	0.5575	2.750	2.785	0.1875	0.3760	0.1935
1422	7/16 × 2 3/4	0.4360	0.4380	0.5263	2.750	2.785	0.2187	0.4385	0.2247
1622	1/2 × 2 3/4	0.4985	0.5005	0.4950	2.750	2.785	0.2500	0.5010	0.2560
1228	3/8 × 3 1/2	0.3735	0.3755	0.7455	3.500	3.535	0.1875	0.3760	0.1935
1428	7/16 × 3 1/2	0.4360	0.4380	0.7143	3.500	3.535	0.2187	0.4385	0.2247
1628	1/2 × 3 1/2	0.4985	0.5005	0.6830	3.500	3.535	0.2500	0.5010	0.2560
1828	9/16 × 3 1/2	0.5610	0.5630	0.6518	3.500	3.535	0.2812	0.5635	0.2872
2028	5/8 × 3 1/2	0.6235	0.6255	0.6205	3.500	3.535	0.3125	0.6260	0.3185
2228	11/16 × 3 1/2	0.6860	0.6880	0.5893	3.500	3.535	0.3437	0.6885	0.3497
2428	3/4 × 3 1/2	0.7485	0.7505	0.5580	3.500	3.535	0.3750	0.7510	0.3810

Width A values were set with the maximum keyseat (shaft) width as that figure which will receive a key with the greatest amount of looseness consistent with assuring the key's sticking in the keyseat (shaft). Minimum keyseat width is that figure permitting the largest shaft distortion acceptable when assembling maximum key in minimum keyseat.

Dimensions A, B, C, D are taken at side intersection.

(Reprinted from The American Society of Mechanical Engineers—ANSI B17.2–1967 (R1990).)

TABLE 20 KEY SIZE VERSUS SHAFT DIAMETER

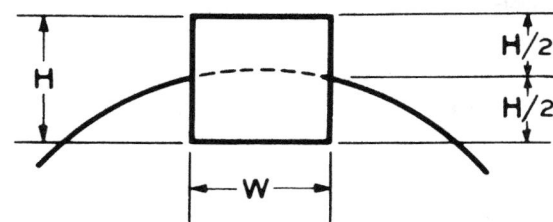

Nominal Shaft Diameter		Nominal Key Size			Nominal Keyset Depth	
		Width, W	Height, H		H/2	
Over	To (Incl)		Square	Rectangular	Square	Rectangular
5/16	7/16	3/32	3/32		3/64	
7/16	9/16	1/8	1/8	3/32	1/16	3/64
9/16	7/8	3/16	3/16	1/8	3/32	1/16
7/8	1-1/4	1/4	1/4	3/16	1/8	3/32
1-1/4	1-3/8	5/16	5/16	1/4	5/32	1/8
1-3/8	1-3/4	3/8	3/8	1/4	3/16	1/8
1-3/4	2-1/4	1/2	1/2	3/8	1/4	3/16
2-1/4	2-3/4	5/8	5/8	7/16	5/16	7/32
2-3/4	3-1/4	3/4	3/4	1/2	3/8	1/4
3-1/4	3-3/4	7/8	7/8	5/8	7/16	5/16
3-3/4	4-1/2	1	1	3/4	1/2	3/8
4-1/2	5-1/2	1-1/4	1-1/4	7/8	5/8	7/16
5-1/2	6-1/2	1-1/2	1-1/2	1	3/4	1/2
6-1/2	7-1/2	1-3/4	1-3/4	1-1/2*	7/8	3/4
7-1/2	9	2	2	1-1/2	1	3/4
9	11	2-1/2	2-1/2	1-3/4	1-1/4	7/8
11	13	3	3	2	1-1/2	1
13	15	3-1/2	3-1/2	2-1/2	1-3/4	1-1/4
15	18	4		3		1-1/2
18	22	5		3-1/2		1-3/4
22	26	6		4		2
26	30	7		5		2-1/2

*Some key standards show 1-1/4 in. Preferred size is 1-1/2 in. All dimensions given in inches.

Shaded areas:

For a stepped shaft, the size of a key is determined by the diameter of the shaft at the point of location of the key, regardless of the number of different diameters on the shaft.

Square-keys are preferred through 6 1/2-inch diameter shafts and rectangular keys for larger shafts. Sizes and dimensions in unshaded area are preferred.

If special considerations dictate the use of a keyseat in the hub shallower than the preferred nominal depth shown in Table 1, it is recommended that the tabulated preferred nominal standard keyseat be used in the shaft in all cases.

(Reprinted from The American Society of Mechanical Engineers—ANSI B17.1–1967 (R1989).)

TABLE 21 KEY DIMENSIONS AND TOLERANCES

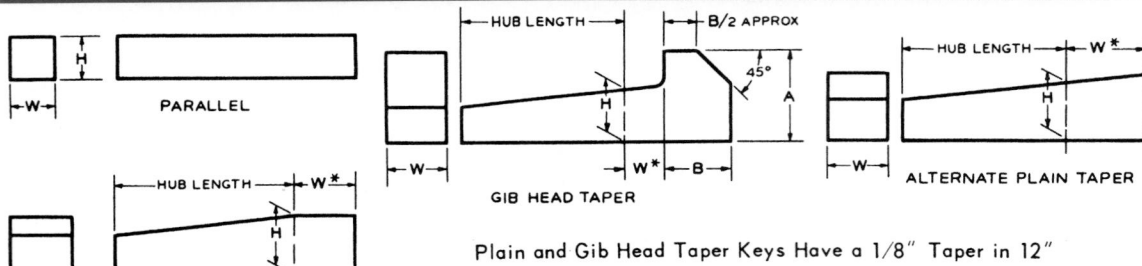

PARALLEL

PLAIN TAPER

GIB HEAD TAPER

ALTERNATE PLAIN TAPER

Plain and Gib Head Taper Keys Have a 1/8″ Taper in 12″

Key			Nominal Key Size Width, W		Tolerance			
					Width, W		Height, H	
			Over	To (Incl)				
Parallel	Square	Bar Stock	—	3/4	+0.000	−0.002	+0.000	−0.002
			3/4	1-1/2	+0.000	−0.003	+0.000	−0.003
			1-1/2	2-1/2	+0.000	−0.004	+0.000	−0.004
			2-1/2	3-1/2	+0.000	−0.006	+0.000	−0.006
		Keystock	—	1-1/4	+0.001	−0.000	+0.001	−0.000
			1-1/4	3	+0.002	−0.000	+0.002	−0.000
			3	3-1/2	+0.003	−0.000	+0.003	−0.000
	Rectangular	Bar Stock	—	3/4	+0.000	−0.003	+0.000	−0.003
			3/4	1-1/2	+0.000	−0.004	+0.000	−0.004
			1-1/2	3	+0.000	−0.005	+0.000	−0.005
			3	4	+0.000	−0.006	+0.000	−0.006
			4	6	+0.000	−0.008	+0.000	−0.008
			6	7	+0.000	−0.013	+0.000	−0.013
		Keystock	—	1-1/4	+0.001	−0.000	+0.005	−0.005
			1-1/4	3	+0.002	−0.000	+0.005	−0.005
			3	7	+0.003	−0.000	+0.005	−0.005
Taper	Plain or Gib Head Square or Rectangle		—	1-1/4	+0.001	−0.000	+0.005	−0.000
			1-1/4	3	+0.002	−0.000	+0.005	−0.000
			3	7	+0.003	−0.000	+0.005	−0.000

* For locating position of dimension *H*. Tolerance does not apply. All dimensions given in inches.

(Reprinted from The American Society of Mechanical Engineers—ANSI B17.1–1967 (R1989).)

TABLE 22 GIB HEAD NOMINAL DIMENSIONS

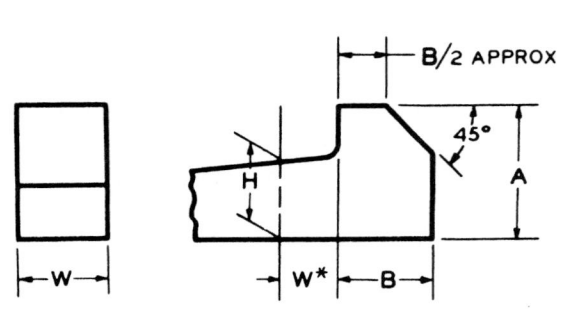

Nominal Key Size Width, W	Square			Rectangular		
	H	A	B	H	A	B
1/8	1/8	1/4	1/4	3/32	3/16	1/8
3/16	3/16	5/16	5/16	1/8	1/4	1/4
1/4	1/4	7/16	3/8	3/16	5/16	5/16
5/16	5/16	1/2	7/16	1/4	7/16	3/8
3/8	3/8	5/8	1/2	1/4	7/16	3/8
1/2	1/2	7/8	5/8	3/8	5/8	1/2
5/8	5/8	1	3/4	7/16	3/4	9/16
3/4	3/4	1-1/4	7/8	1/2	7/8	5/8
7/8	7/8	1-3/8	1	5/8	1	3/4
1	1	1-5/8	1-1/8	3/4	1-1/4	7/8
1-1/4	1-1/4	2	1-7/16	7/8	1-3/8	1
1-1/2	1-1/2	2-3/8	1-3/4	1	1-5/8	1-1/8
1-3/4	1-3/4	2-3/4	2	1-1/2	2-3/8	1-3/4
2	2	3-1/2	2-1/4	1-1/2	2-3/8	1-3/4
2-1/2	2-1/2	4	3	1-3/4	2-3/4	2
3	3	5	3-1/2	2	3-1/2	2-1/4
3-1/2	3-1/2	6	4	2-1/2	4	3

*For locating position of dimension *H*.

For larger sizes the following relationships are suggested as guides for establishing *A* and *B*.

$$A = 1.8 H \qquad B = 1.2 H$$

All dimensions given in inches.

(Reprinted from The American Society of Mechanical Engineers—ANSI B17.1–1967 (R1989).)

TABLE 23 CLASS 2 FIT FOR PARALLEL AND TAPER KEYS

Type of Key	Key Width		Side Fit			Top and Bottom Fit			
	Over	To (Incl)	Width Tolerance		Fit Range*	Depth Tolerance			Fit Range*
			Key	Keyset		Key	Shaft Keyseat	Hub Keyseat	
Parallel Square	—	1-1/4	+0.001 −0.000	+0.002 −0.000	0.002 CL 0.001 INT	+0.001 −0.000	+0.000 −0.015	+0.010 −0.000	0.030 CL 0.004 CL
	1-1/4	3	+0.002 −0.000	+0.002 −0.000	0.002 CL 0.002 INT	+0.002 −0.000	+0.000 −0.015	+0.010 −0.000	0.030 CL 0.003 CL
	3	3-1/2	+0.003 −0.000	+0.002 −0.000	0.002 CL 0.003 INT	+0.003 −0.000	+0.000 −0.015	+0.010 −0.000	0.030 CL 0.002 CL
Parallel Retangular	—	1-1/4	+0.001 −0.000	+0.002 −0.000	0.002 CL 0.001 INT	+0.005 −0.005	+0.000 −0.015	+0.010 −0.000	0.035 CL 0.000 CL
	1-1/4	3	+0.002 −0.000	+0.002 −0.000	0.002 CL 0.002 INT	+0.005 −0.005	+0.000 −0.015	+0.010 −0.000	0.035 CL 0.000 CL
	3	7	+0.003 −0.000	+0.002 −0.000	0.002 CL 0.003 INT	+0.005 −0.005	+0.000 −0.015	+0.010 −0.000	0.035 CL 0.000 CL
Taper	—	1-1/4	+0.001 −0.000	+0.002 −0.000	0.002 CL 0.001 INT	+0.005 −0.005	+0.000 −0.015	+0.010 −0.000	0.005 CL 0.025 INT
	1-1/4	3	+0.002 −0.000	+0.002 −0.000	0.002 CL 0.002 INT	+0.005 −0.005	+0.000 −0.015	+0.010 −0.000	0.005 CL 0.025 INT
	3	Δ	+0.003 −0.000	+0.002 −0.000	0.002 CL 0.003 INT	+0.005 −0.005	+0.000 −0.015	+0.010 −0.000	0.005 CL 0.025 INT

*Limits of variation. CL = Clearance; INT = Interference

Δ To (Incl) 3-1/2 Square and 7 Rectangular key widths.
All dimensions given in inches.

(Reprinted from The American Society of Mechanical Engineers—ANSI B17.1–1967 (R1989).)

TABLE 24 DECIMAL EQUIVALENTS AND TAP DRILL SIZES (LETTER AND NUMBER DRILL SIZES)

Fraction or Drill Size	Decimal Equivalent	Tap Size	Fraction or Drill Size	Decimal Equivalent	Tap Size	Fraction or Drill Size	Decimal Equivalent	Tap Size	Fraction or Drill Size	Decimal Equivalent	
Number Size Drills 80	.0135		39	.0995		15/64	.2344		19/32	.5938	16–11
79	.0145		38	.1015	5–40	Letter Size Drills B	.2380		39/64	.6094	
1/64	.0156		37	.1040	5–44	C	.2420		5/8	.6250	16–16
78	.0160		36	.1065	6–32	D	.2460		41/64	.6406	
77	.0180		7/64	.1094		1/4 E	.2500		21/32	.6562	4–10
76	.0200		35	.1100		F	.2570	5/16–18	43/64	.6719	
75	.0210		34	.1110	6–36	G	.2610		11/16	.6875	4–16
74	.0225		33	.1130	6–40	17/64	.2656		45/64	.7031	
73	.0240		32	.1160		H	.2660		23/32	.7188	
72	.0250		31	.1200		I	.2720	5/16–24	47/64	.7344	
71	.0260		1/8	.1250		J	.2770		3/4	.7500	
70	.0280		30	.1285		K	.2810		49/64	.7656	7/8–9
69	.0292		29	.1360	8–32,36	9/32	.2812		25/32	.7812	
68	.0310		28	.1405	8–40	L	.2900		51/64	.7969	
1/32	.0312		9/64	.1406		M	.2950		13/16	.8125	7/8–14
67	.0320		27	.1440		19/64	.2969		53/64	.8281	
66	.0330		26	.1470		N	.3020		27/32	.8438	
65	.0350		25	.1495	10–24	5/16	.3125	3/8–16	55/64	.8594	
64	.0360		24	.1520		O	.3160		7/8	.8750	1–8
63	.0370		23	.1540		P	.3230		57/64	.8906	
62	.0380		5/32	.1562		21/64	.3281		29/32	.9062	
61	.0390		22	.1570	10–30	Q	.3320	3/8–24	59/64	.9219	
60	.0400		21	.1590	10–32	R	.3390		15/16	.9375	1–12, 14
59	.0410		20	.1610		11/32	.3438		61/64	.9531	
58	.0420		19	.1660		S	.3480		31/32	.9688	
57	.0430		18	.1695		T	.3580		63/64	.9844	1 1/8–7
56	.0465		11/64	.1719		23/64	.3594		1	1.0000	
3/64	.0469	0–80	17	.1730		U	.3680	7/16–14	1 3/64	1.0469	1 1/8–12
55	.0520		16	.1770	12–24	3/8	.3750		1 7/64	1.1094	1 1/4–7
54	.0550	1–56	15	.1800		V	.3770		1 1/8	1.1250	
53	.0595	1–64, 72	14	.1820	12–28	W	.3860		1 11/64	1.1719	1 1/4–12
1/16	.0625		13	.1850	12–32	25/64	.3906	7/16–20	1 7/32	1.2188	1 3/8–6
52	.0635		3/16	.1875		X	.3970		1 1/4	1.2500	
51	.0670		12	.1890		Y	.4040		1 19/64	1.2969	1 3/8–12
50	.0700	2–56, 64	11	.1910		13/32	.4062		1 11/32	1.3438	1 1/2–6
49	.0730		10	.1935		Z	.4130		1 3/8	1.3750	
48	.0760		9	.1960		27/64	.4219	1/2–13	1 27/64	1.4219	1 1/2–12
5/64	.0781		8	.1990		7/16	.4375	1/2–20	1 1/2	1.5000	
47	.0785	3–48	7	.2010	1/4–20	29/64	.4531				
46	.0810		13/64	.2031		15/32	.4688				
45	.0820	3–56, 4–32	6	.2040		31/64	.4844	9/16–12			
44	.0860	4–36	5	.2055		1/2	.5000				
43	.0890	4–40	4	.2090		33/64	.5156	9/16–18			
42	.0935	4–48	3	.2130	1/4–28	17/32	.5312	5/8–11			
3/32	.0938		7/32	.2188		35/64	.5469				
41	.0960		2	.2210		9/16	.5625				
40	.0980		Letter Size Drills 1	.2280		37/64	.5781	5/8–18			
			A	.2340							

Pipe Thread Sizes

Thread	Drill	Thread	Drill
1/8–27		1 1/2–11 1/2	1 47/64
1/4–18	7/16	2–11 1/2	2 7/32
3/8–18	37/64	2 1/2–8	2 5/8
1/2–14	23/32	3–8	3 1/4
3/4–14	59/64	3 1/2–8	3 3/4
1–11 1/2	1 5/32	4–8	4 1/4
1 1/4–11 1/2	1 1/2		

Courtesy The L. S. Starrett Company, Athol, Massachusetts.

TABLE 25 GENERAL APPLICATIONS OF SAE STEELS

Application	SAE No.	Application	SAE No.
Adapters	1145	Chain pins, transmission	4320
Agricultural steel	1070	" " "	4815
" "	1080	" " "	4820
Aircraft forgings	4140	Chains, transmission	3135
Axles, front or rear	1040	" "	3140
" " "	4140	Clutch disks	1060
Axle shafts	1045	" "	1070
" "	2340	" "	1085
" "	2345	Clutch springs	1060
" "	3135	Coil springs	4063
" "	3140	Cold-headed bolts	4042
" "	3141	Cold-heading steel	30905
" "	4063	Cold-heading wire or rod	rimmed*
" "	4340	" " " "	1035
Ball-bearing races	52100	Cold-rolled steel	1070
Balls for ball bearings	52100	Connecting-rods	1040
Body stock for cars	rimmed*	" "	3141
Bolts, anchor	1040	Connecting-rod bolts	3130
Bolts and screws	1035	Corrosion resisting	51710
Bolts, cold-headed	4042	" "	30805
Bolts, connecting-rod	3130	Covers, transmission	rimmed*
Bolts, heat-treated	2330	Crankshafts	1045
Bolts, heavy-duty	4815	" "	1145
" "	4820	" "	3135
Bolts, steering-arm	3130	" "	3140
Brake levers	1030	" "	3141
" "	1040	Crankshafts, Diesel engine	4340
Bumper bars	1085	Cushion, springs	1060
Cams, free-wheeling	4615	Cutlery, stainless	51335
" "	4620	Cylinder studs	3130
Camshafts	1020	Deep-drawing steel	rimmed*
" "	1040	" " "	30905
Carburized parts	1020	Differential gears	4023
" "	1022	Disks, clutch	1070
" "	1024	" "	1060
" "	1320	Ductile steel	30905
" "	2317	Fan blades	1020
" "	2515	Fatigue resisting	4340
" "	3310	" "	4640
" "	3115	Fender stock for cars	rimmed*
" "	3120	Forgings, aircraft	4140
" "	4023	Forgings, carbon steel	1040
" "	4032	" " "	1045
" "	1117	Forgings, heat-treated	3240
" "	1118	" " "	5140

Application	SAE No.	Application	SAE No.
Forgings, heat-treated	6150	Key stock	1030
Forgings, high-duty	6150	" "	2330
Forgings, small or medium	1035	" "	3130
Forgings, large	1036	Leaf springs	1085
Free-cutting carbon steel	1111	" "	9260
" "	1113	Levers, brake	1030
Free-cutting chro.-ni.steel	30615	" "	1040
Free-cutting mang. steel	1132	Levers, gear shift	1030
" " "	1137	Levers, heat-treated	2330
Gears, carburized	1320	Lock-washers	1060
" "	2317	Mower knives	1085
" "	3115	Mower sections	1070
" "	3120	Music wire	1085
" "	3310	Nuts	3130
" "	4119	Nuts, heat-treated	2330
" "	4125	Oil-pans, automobile	rimmed*
" "	4320	Pinions, carburized	3115
" "	4615	" "	3120
" "	4620	" "	4320
" "	4815	Piston-pins	3115
" "	4820	" "	3120
Gears, heat-treated	2345	Plow beams	1070
Gears, car and truck	4027	Plow disks	1080
" "	4032	Plow shares	1080
Gears, cyanide-hardening	5140	Propeller shafts	2340
Gears, differential	4023	" "	2345
Gears, high duty	4640	" "	4140
" "	6150	Races, ball-bearing	52100
Gears, oil-hardening	3145	Ring gears	3115
" "	3150	" "	3120
" "	4340	" "	4119
" "	5150	Rings, snap	1060
Gears, ring	1045	Rivets	rimmed*
" "	3115	Rod and wire	killed*
" "	3120	Rod, cold-heading	1035
" "	4119	Roller bearings	4815
Gears, transmission	3115	Rollers for bearings	52100
" "	3120	Screws and bolts	1035
" "	4119	Screw stock, Bessemer	1111
Gears, truck and bus	3310	" " "	1112
" "	4320	" " "	1113
Gear shift levers	1030	Screw stock, open hearth	1115
Harrow disks	1080	Screws, heat-treated	2330
" "	1095	Seat springs	1095
Hay-rake teeth	1095	Shafts, axle	1045

Application	SAE No.	Application	SAE No.
Shafts, cyanide-hardening	5140	Steel, cold-heading	30905
Shafts, heavy-duty	4340	Steel, free-cutting carbon	11111
" "	6150	" "	1113
" "	4615	Steel, free-cutting chro.-ni.	30615
" "	4620	Steel, free-cutting mang.	1132
Shafts, oil-hardening	5150	Steel, minimum distortion	4615
Shafts, propeller	2340	" "	4620
" "	2345	" "	4640
" "	4140	Steel, soft ductile	30905
Shafts, transmission	4140	Steering arms	4042
Sheets and strips	rimmed*	Steering-arm bolts	3130
Snap rings	1060	Steering knuckles	3141
Spline shafts	1045	Steering-knuckle pins	4815
" "	1320	" "	4820
" "	2340	Studs	1040
" "	2345	" "	1111
" "	3115	Studs, cold-headed	4042
" "	3120	Studs, cylinder	3130
" "	3135	Studs, heat-treated	2330
" "	3140	Studs, heavy-duty	4815
" "	4023	" "	4820
" "	1060	Tacks	rimmed*
Spring clips	1095	Thrust washers	1060
Springs, coil	4063	Thrust washers, oil-harden	5150
" "	6150	Transmission shafts	4140
Springs, clutch	1060	Tubing	1040
Springs, cushion	1060	Tubing, front axle	4140
Springs, leaf	1085	Tubing, seamless	1030
" "	1095	Tubing, welded	1020
" "	4063	Universal joints	1145
" "	4068	Valve springs	1060
" "	9260	Washers, lock	1060
" "	6150	Welded structures	30705
Springs, hard-drawn coiled	1066	Wire and rod	killed*
Springs, oil-hardening	5150	Wire, cold-heading	rimmed*
Springs, oil-tempered wire	1066	" "	1035
Springs, seat	1095	Wire, hard-drawn spring	1045
Springs, valve	1060	" " "	1055
Spring wire	1045	Wire, music	1085
Spring wire, hard-drawn	1055	Wire, oil-tempered spring	1055
Spring wire, oil-tempered	1055	Wrist-pins, automobile	1020
Stainless irons	51210	Yokes	1145
" "	51710		
Steel, cold-rolled	1070		

*The "rimmed" and "killed" steels listed are in the SAE 1008, 1010, and 1015 group. See general description of these steels.

Reprinted by permission from Oberg, Jones, and Horton, *Machinery's Handbook,* 24th ed. (New York: Industrial Press, Inc., 1992), table 6, pp. 382–84.

TABLE 26 SURFACE ROUGHNESS PRODUCED BY COMMON PRODUCTION METHODS

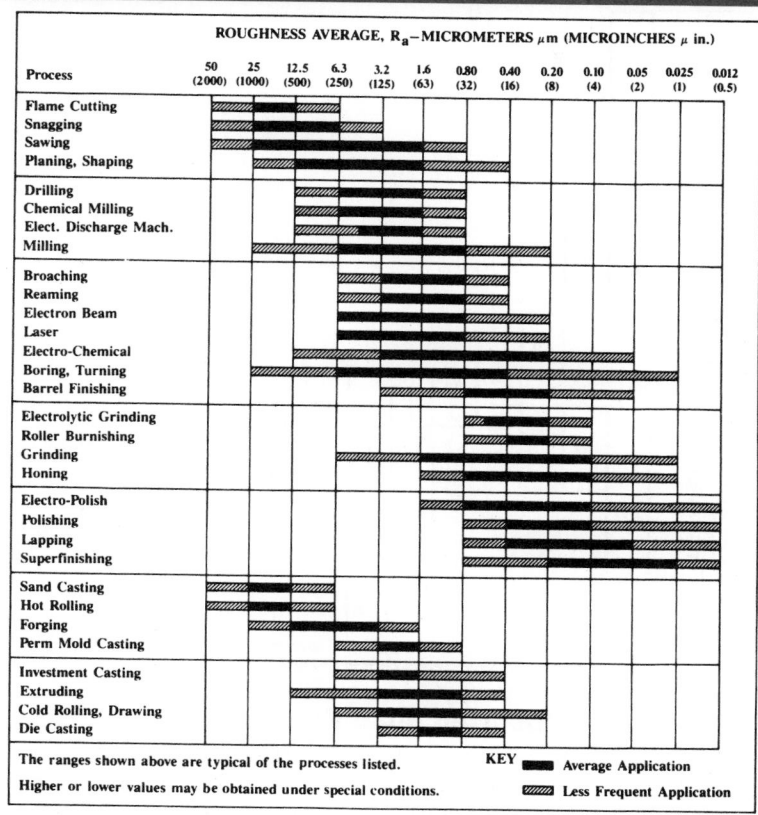

Reprinted by permission from Oberg, Jones, and Horton, *Machinery's Handbook*, 24th ed. (New York: Industrial Press, Inc., 1992), figure 5, p. 672.

TABLE 27 ALLOWANCES AND TOLERANCES

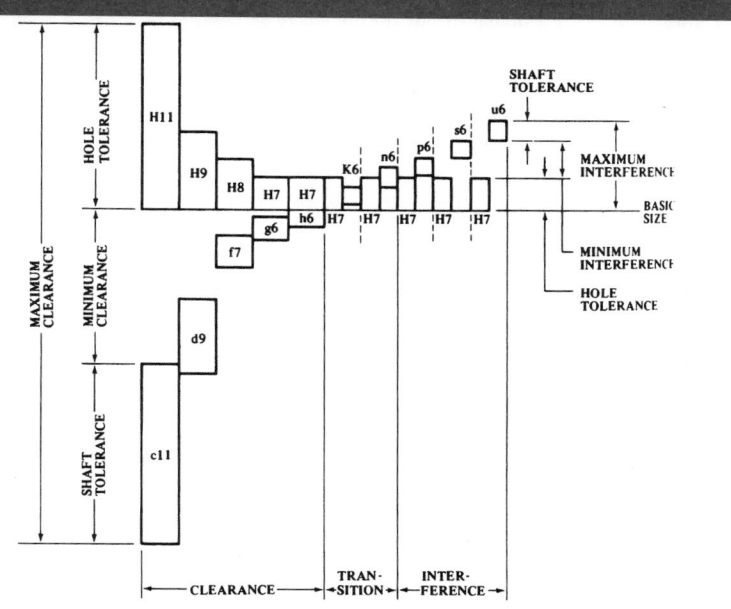

Preferred Hole Basis Fits

Reprinted by permission from Oberg, Jones, and Horton, *Machinery's Handbook*, 24th ed. (New York: Industrial Press, Inc., 1992), figure 2, p. 662.

TABLE 27 (CONTINUED)

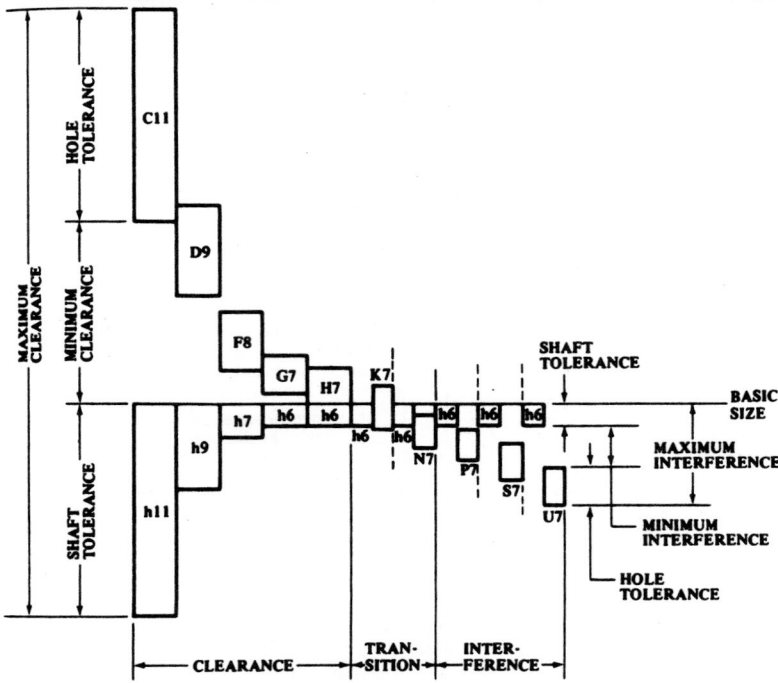

Preferred Shaft Basis Fits

Reprinted by permission from Oberg, Jones, and Horton, *Machinery's Handbook*, 24th ed. (New York: Industrial Press, Inc., 1992), figure 3, p. 623.

ISO SYMBOL		DESCRIPTION
Hole Basis	**Shaft Basis**	
H11/c11	C11/h11	*Loose running* fit for wide commercial tolerances or allowances on external members.
H9/d9	D9/h9	*Free running* fit not for use where accuracy is essential, but good for large temperature variations, high running speeds, or heavy journal pressures.
H8/f7	F8/h7	*Close running* fit for running on accurate machines and for accurate location at moderate speeds and journal pressures.
H7/g6	G7/h6	*Sliding* fit not intended to run freely, but to move and turn freely and locate accurately.
H7/h6	H7/h6	*Locational clearance* fit provides snug fit for locating stationary parts; but can be freely assembled and disassembled.
H7/k6	K7/h6	*Locational transition* fit for accurate location, a compromise between clearance and interference.
H7/n6	N7/h6	*Locational transition* fit for more accurate location where greater interference is permissible.
H7/p6[1]	P7/h6	*Locational interference* fit for parts requiring rigidity and alignment with prime accuracy of location but without special bore pressure requirements.
H7/s6	S7/h6	*Medium drive* fit for ordinary steel parts or shrink fits on light sections, the tightest fit usable with cast iron.
H7/u6	U7/h6	*Force* fit suitable for parts which can be highly stressed or for shrink fits where the heavy pressing forces required are impractical.

[1] Transition fit for basic sizes in range from 0 through 3 mm.

Description of Preferred Fits

Reprinted by permission from Oberg, Jones, and Horton, *Machinery's Handbook*, 24th ed. (New York: Industrial Press, Inc., 1992), figure 4, p. 624.

TABLE 28 AMERICAN NATIONAL STANDARD FITS

RUNNING AND SLIDING FITS

Limits are in thousandths of an inch.

Limits for hole and shaft are applied algebraically to the basic size to obtain the limits of size for the parts.

Data in **boldface** are in accordance with ABC agreements.

Symbols H5, g5, etc., are Hole and Shaft designations used in ABC System (Appendix I).

Nominal Size Range Inches Over To	Class RC 1 Limits of Clearance	Class RC 1 Standard Limits Hole H5	Class RC 1 Standard Limits Shaft g4	Class RC 2 Limits of Clearance	Class RC 2 Standard Limits Hole H6	Class RC 2 Standard Limits Shaft g5	Class RC 3 Limits of Clearance	Class RC 3 Standard Limits Hole H7	Class RC 3 Standard Limits Shaft f6	Class RC 4 Limits of Clearance	Class RC 4 Standard Limits Hole H8	Class RC 4 Standard Limits Shaft f7
0 – 0.12	0.1 0.45	+ 0.2 0	– 0.1 – 0.25	0.1 0.55	+ 0.25 0	– 0.1 – 0.3	0.3 0.95	+ 0.4 0	– 0.3 – 0.55	0.3 1.3	+ 0.6 0	– 0.3 – 0.7
0.12 – 0.24	0.15 0.5	+ 0.2 0	– 0.15 – 0.3	0.15 0.65	+ 0.3 0	– 0.15 – 0.35	0.4 1.12	+ 0.5 0	– 0.4 – 0.7	0.4 1.6	+ 0.7 0	– 0.4 – 0.9
0.24 – 0.40	0.2 0.6	0.25 0	– 0.2 – 0.35	0.2 0.85	+ 0.4 0	– 0.2 – 0.45	0.5 1.5	+ 0.6 0	– 0.5 – 0.9	0.5 2.0	+ 0.9 0	– 0.5 – 1.1
0.40 – 0.71	0.25 0.75	+ 0.3 0	– 0.25 – 0.45	0.25 0.95	+ 0.4 0	– 0.25 – 0.55	0.6 1.7	+ 0.7 0	– 0.6 – 1.0	0.6 2.3	+ 1.0 0	– 0.6 – 1.3
0.71 – 1.19	0.3 0.95	+ 0.4 0	– 0.3 – 0.55	0.3 1.2	+ 0.5 0	– 0.3 – 0.7	0.8 2.1	+ 0.8 0	– 0.8 – 1.3	0.8 2.8	+ 1.2 0	– 0.8 – 1.6
1.19 – 1.97	0.4 1.1	+ 0.4 0	– 0.4 – 0.7	0.4 1.4	+ 0.6 0	– 0.4 – 0.8	1.0 2.6	+ 1.0 0	– 1.0 – 1.6	1.0 3.6	+ 1.6 0	– 1.0 – 2.0
1.97 – 3.15	0.4 1.2	+ 0.5 0	– 0.4 – 0.7	0.4 1.6	+ 0.7 0	– 0.4 – 0.9	1.2 3.1	+ 1.2 0	– 1.2 – 1.9	1.2 4.2	+ 1.8 0	– 1.2 – 2.4
3.15 – 4.73	0.5 1.5	+ 0.6 0	– 0.5 – 0.9	0.5 2.0	+ 0.9 0	– 0.5 – 1.1	1.4 3.7	+ 1.4 0	– 1.4 – 2.3	1.4 5.0	+ 2.2 0	– 1.4 – 2.8
4.73 – 7.09	0.6 1.8	+ 0.7 0	– 0.6 – 1.1	0.6 2.3	+ 1.0 0	– 0.6 – 1.3	1.6 4.2	+ 1.6 0	– 1.6 – 2.6	1.6 5.7	+ 2.5 0	– 1.6 – 3.2
7.09 – 9.85	0.6 2.0	+ 0.8 0	– 0.6 – 1.2	0.6 2.6	+ 1.2 0	– 0.6 – 1.4	2.0 5.0	+ 1.8 0	– 2.0 – 3.2	2.0 6.6	+ 2.8 0	– 2.0 – 3.8
9.85 – 12.41	0.8 2.3	+ 0.9 0	– 0.8 – 1.4	0.8 2.9	+ 1.2 0	– 0.8 – 1.7	2.5 5.7	+ 2.0 0	– 2.5 – 3.7	2.5 7.5	+ 3.0 0	– 2.5 – 4.5
12.41 – 15.75	1.0 2.7	+ 1.0 0	– 1.0 – 1.7	1.0 3.4	+ 1.4 0	– 1.0 – 2.0	3.0 6.6	+ 0	– 3.0 – 4.4	3.0 8.7	+ 3.5 0	– 3.0 – 5.2
15.75 – 19.69	1.2 3.0	+ 1.0 0	– 1.2 – 2.0	1.2 3.8	+ 1.6 0	– 1.2 – 2.2	4.0 8.1	+ 1.6 0	– 4.0 – 5.6	4.0 10.5	+ 4.0 0	– 4.0 – 6.5
19.69 – 30.09	1.6 3.7	+ 1.2 0	– 1.6 – 2.5	1.6 4.8	+ 2.0 0	– 1.6 – 2.8	5.0 10.0	+ 3.0 0	– 5.0 – 7.0	5.0 13.0	+ 5.0 0	– 5.0 – 8.0
30.09 – 41.49	2.0 4.6	+ 1.6 0	– 2.0 – 3.0	2.0 6.1	+ 2.5 0	– 2.0 – 3.6	6.0 12.5	+ 4.0 0	– 6.0 – 8.5	6.0 16.0	+ 6.0 0	– 6.0 –10.0
41.49 – 56.19	2.5 5.7	+ 2.0 0	– 2.5 – 3.7	2.5 7.5	+ 3.0 0	– 2.5 – 4.5	8.0 16.0	+ 5.0 0	– 8.0 –11.0	8.0 21.0	+ 8.0 0	– 8.0 –13.0
56.19 – 76.39	3.0 7.1	+ 2.5 0	– 3.0 – 4.6	3.0 9.5	+ 4.0 0	– 3.0 – 5.5	10.0 20.0	+ 6.0 0	–10.0 –14.0	10.0 26.0	+10.0 0	–10.0 –16.0
76.39 –100.9	4.0 9.0	+ 3.0 0	– 4.0 – 6.0	4.0 12.0	+ 5.0 0	– 4.0 – 7.0	12.0 25.0	+ 8.0 0	–12.0 –17.0	12.0 32.0	+12.0 0	–12.0 –20.0
100.9 –131.9	5.0 11.5	+ 4.0 0	– 5.0 – 7.5	5.0 15.0	+ 6.0 0	– 5.0 – 9.0	16.0 32.0	+10.0 0	–16.0 –22.0	16.0 36.0	+16.0 0	–16.0 –26.0
131.9 –171.9	6.0 14.0	+ 5.0 0	– 6.0 – 9.0	6.0 19.0	+ 8.0 0	– 6.0 –11.0	18.0 38.0	+ 8.0 0	–18.0 –26.0	18.0 50.0	+20.0 0	–18.0 –30.0
171.9 –200	8.0 18.0	+ 6.0 0	– 8.0 –12.0	8.0 22.0	+10.0 0	– 8.0 –12.0	22.0 48.0	+16.0 0	–22.0 –32.0	22.0 63.0	+25.0 0	–22.0 –38.0

TABLE 28 (CONTINUED)

RUNNING AND SLIDING FITS *(continued)*

Limits are in thousandths of an inch.

Limits for hole and shaft are applied algebraically to the basic size to obtain the limits of size for the parts.

Data in **boldface** are in accordance with ABC agreements.

Symbols H8, e7, etc., are Hole and Shaft designations used in ABC System (Appendix I).

Class RC 5 Limits of Clearance	Class RC 5 Hole H8	Class RC 5 Shaft c7	Class RC 6 Limits of Clearance	Class RC 6 Hole H9	Class RC 6 Shaft e8	Class RC 7 Limits of Clearance	Class RC 7 Hole H9	Class RC 7 Shaft d8	Class RC 8 Limits of Clearance	Class RC 8 Hole H10	Class RC 8 Shaft c9	Class RC 9 Limits of Clearance	Class RC 9 Hole H11	Class RC 9 Shaft	Nominal Size Range Over	Nominal Size Range To
0.6 / 1.6	+0.6 / −0	−0.6 / −1.0	0.6 / 2.2	+1.0 / 0	−0.6 / −1.2	1.0 / 2.6	+1.0 / 0	−1.0 / −1.6	2.5 / 5.1	+1.6 / 0	−2.5 / −3.5	4.0 / 8.1	+2.5 / 0	−4.0 / −5.6	0	0.12
0.8 / 2.0	+0.7 / −0	−0.8 / −1.3	0.8 / 2.7	+1.2 / 0	−0.8 / −1.5	1.2 / 3.1	+1.2 / 0	−1.2 / −1.9	2.8 / 5.8	+1.8 / 0	−2.8 / −4.0	4.5 / 9.0	+3.0 / 0	−4.5 / −6.0	0.12	0.24
1.0 / 2.5	+0.9 / −0	−1.0 / −1.6	1.0 / 3.3	+1.4 / 0	−1.0 / −1.9	1.6 / 3.9	+1.4 / 0	−1.6 / −2.5	3.0 / 6.6	+2.2 / 0	−3.0 / −4.4	5.0 / 10.7	+3.5 / 0	−5.0 / −7.2	0.24	0.40
1.2 / 2.9	+1.0 / −0	−1.2 / −1.9	1.2 / 3.8	+1.6 / 0	−1.2 / −2.2	2.0 / 4.6	+1.6 / 0	−2.0 / −3.0	3.5 / 7.9	+2.8 / 0	−3.5 / −5.1	6.0 / 12.8	+4.0 / 0	−6.0 / −8.8	0.40	0.71
1.6 / 3.6	+1.2 / −0	−1.6 / −2.4	1.6 / 4.8	+2.0 / 0	−1.6 / −2.8	2.5 / 5.7	+2.0 / 0	−2.5 / −3.7	4.5 / 10.0	+3.5 / 0	−4.5 / −6.5	7.0 / 15.5	+5.0 / 0	−7.0 / −10.5	0.71	1.19
2.0 / 4.6	+1.6 / −0	−2.0 / −3.0	2.0 / 6.1	+2.5 / 0	−2.0 / −3.6	3.0 / 7.1	+2.5 / 0	−3.0 / −4.6	5.0 / 11.5	+4.0 / 0	−5.0 / −7.5	8.0 / 18.0	+6.0 / 0	−8.0 / −12.0	1.19	1.97
2.5 / 5.5	+1.8 / −0	−2.5 / −3.7	2.5 / 7.3	+3.0 / 0	−2.5 / −4.3	4.0 / 8.8	+3.0 / 0	−4.0 / −5.8	6.0 / 13.5	+4.5 / 0	−6.0 / −9.0	9.0 / 20.5	+7.0 / 0	−9.0 / −13.5	1.97	3.15
3.0 / 6.6	+2.2 / −0	−3.0 / −4.4	3.0 / 8.7	+3.5 / 0	−3.0 / −5.2	5.0 / 10.7	+3.5 / 0	−5.0 / −7.2	7.0 / 15.5	+5.0 / 0	−7.0 / −10.5	10.0 / 24.0	+9.0 / 0	−10.0 / −15.0	3.15	4.73
3.5 / 7.6	+2.5 / −0	−3.5 / −5.1	3.5 / 10.0	+4.0 / 0	−3.5 / −6.0	6.0 / 12.5	+4.0 / 0	−6.0 / −8.5	8.0 / 18.0	+6.0 / 0	−8.0 / −12.0	12.0 / 28.0	+10.0 / 0	−12.0 / −18.0	4.73	7.09
4.0 / 8.6	+2.8 / −0	−4.0 / −5.8	4.0 / 11.3	+4.5 / 0	−4.0 / −6.8	7.0 / 14.3	+4.5 / 0	−7.0 / −9.8	10.0 / 21.5	+7.0 / 0	−10.0 / −14.5	15.0 / 34.0	+12.0 / 0	−15.0 / −22.0	7.09	9.85
5.0 / 10.0	+3.0 / 0	−5.0 / −7.0	5.0 / 13.0	+5.0 / 0	−5.0 / −8.0	8.0 / 16.0	+5.0 / 0	−8.0 / −11.0	**12.0 / 25.0**	+8.0 / 0	−12.0 / −17.0	18.0 / 38.0	+12.0 / 0	−18.0 / −26.0	9.85	12.41
6.0 / 11.7	+3.5 / 0	−6.0 / −8.2	6.0 / 15.5	+6.0 / 0	−6.0 / −9.5	10.0 / 19.5	+6.0 / 0	−10.0 / −13.5	**14.0 / 29.0**	+9.0 / 0	−14.0 / −20.0	22.0 / 45.0	+14.0 / 0	−22.0 / −31.0	12.41	15.75
8.0 / 14.5	+4.0 / 0	−8.0 / −10.5	8.0 / 18.0	+6.0 / 0	−8.0 / −12.0	12.0 / 22.0	+6.0 / 0	−12.0 / −16.0	**16.0 / 32.0**	+10.0 / 0	−16.0 / −22.0	25.0 / 51.0	+16.0 / 0	−25.0 / −35.0	15.75	19.69
10.0 / 18.0	+5.0 / 0	−10.0 / −13.0	10.0 / 23.0	+8.0 / 0	−10.0 / −15.0	16.0 / 29.0	+8.0 / 0	−16.0 / −21.0	20.0 / 40.0	+12.0 / 0	−20.0 / −28.0	30.0 / 62.0	+20.0 / 0	−30.0 / −42.0	19.69	30.09
12.0 / 22.0	+6.0 / 0	−12.0 / −16.0	12.0 / 28.0	+10.0 / 0	−12.0 / −18.0	20.0 / 36.0	+10.0 / 0	−20.0 / −26.0	25.0 / 51.0	+16.0 / 0	−25.0 / −35.0	40.0 / 81.0	+25.0 / 0	−40.0 / −56.0	30.09	41.49
16.0 / 29.0	+8.0 / 0	−16.0 / −21.0	16.0 / 36.0	+12.0 / 0	−16.0 / −24.0	25.0 / 45.0	+12.0 / 0	−25.0 / −33.0	30.0 / 62.0	+20.0 / 0	−30.0 / −42.0	50.0 / 100	+30.0 / 0	−50.0 / −70.0	41.49	56.19
20.0 / 36.0	+10.0 / 0	−20.0 / −26.0	20.0 / 46.0	+16.0 / 0	−20.0 / −30.0	30.0 / 56.0	+16.0 / 0	−30.0 / −40.0	40.0 / 81.0	+25.0 / 0	−40.0 / −56.0	60.0 / 125	+40.0 / 0	−60.0 / −85.0	56.19	76.39
25.0 / 45.0	+12.0 / 0	−25.0 / −33.0	25.0 / 57.0	+20.0 / 0	−25.0 / −37.0	40.0 / 72.0	+20.0 / 0	−40.0 / −52.0	50.0 / 100	+30.0 / 0	−50.0 / −70.0	80.0 / 160	+50.0 / 0	−80.0 / −110	76.39	100.9
30.0 / 56.0	+16.0 / 0	−30.0 / −40.0	30.0 / 71.0	+25.0 / 0	−30.0 / −46.0	50.0 / 91.0	+25.0 / 0	−50.0 / −66.0	60.0 / 125	+40.0 / 0	−60.0 / −85.0	100 / 200	+60.0 / 0	−100 / −140	100.9	131.9
35.0 / 67.0	+20.0 / 0	−35.0 / −47.0	35.0 / 85.0	+30.0 / 0	−35.0 / −55.0	60.0 / 110.0	+30.0 / 0	−60.0 / −80.0	80.0 / 160	+50.0 / 0	−80.0 / −110	130 / 260	+80.0 / 0	−130 / −180	131.9	171.9
45.0 / 86.0	+25.0 / 0	−45.0 / −61.0	45.0 / 110.0	+40.0 / 0	−45.0 / −70.0	80.0 / 145.0	+40.0 / 0	−80.0 / −105.0	100 / 200	+60.0 / 0	−100 / −140	150 / 310	+100 / 0	−150 / −210	171.9	200

(Reprinted from The American Society of Mechanical Engineers—ANSI B4.1–1967 (R1987).)

AMERICAN NATIONAL STANDARD
PREFERRED METRIC LIMITS AND FITS

ANSI B4.2–1978

TABLE 28 (CONTINUED)

Dimensions in mm.

TABLE 2 PREFERRED HOLE BASIS CLEARANCE FITS

BASIC SIZE		LOOSE RUNNING			FREE RUNNING			CLOSE RUNNING			SLIDING			LOCATIONAL CLEARANCE		
		Hole H11	Shaft c11	Fit	Hole H9	Shaft d9	Fit	Hole H8	Shaft f7	Fit	Hole H7	Shaft g6	Fit	Hole H7	Shaft h6	Fit
1	MAX	1.060	0.940	0.180	1.025	0.980	0.070	1.014	0.994	0.030	1.010	0.998	0.018	1.010	1.000	0.016
	MIN	1.000	0.880	0.060	1.000	0.955	0.020	1.000	0.984	0.006	1.000	0.992	0.002	1.000	0.994	0.000
1.2	MAX	1.260	1.140	0.180	1.225	1.180	0.070	1.214	1.194	0.030	1.210	1.198	0.018	1.210	1.200	0.016
	MIN	1.200	1.080	0.060	1.200	1.155	0.020	1.200	1.184	0.006	1.200	1.192	0.002	1.200	1.194	0.000
1.6	MAX	1.660	1.540	0.180	1.625	1.580	0.070	1.614	1.594	0.030	1.610	1.598	0.018	1.610	1.600	0.016
	MIN	1.600	1.480	0.060	1.600	1.555	0.020	1.600	1.584	0.006	1.600	1.592	0.002	1.600	1.594	0.000
2	MAX	2.060	1.940	0.180	2.025	1.980	0.070	2.014	1.994	0.030	2.010	1.998	0.018	2.010	2.000	0.016
	MIN	2.000	1.880	0.060	2.000	1.955	0.020	2.000	1.984	0.006	2.000	1.992	0.002	2.000	1.994	0.000
2.5	MAX	2.560	2.440	0.180	2.525	2.480	0.070	2.514	2.494	0.030	2.510	2.498	0.018	2.510	2.500	0.016
	MIN	2.500	2.380	0.060	2.500	2.455	0.020	2.500	2.484	0.006	2.500	2.492	0.002	2.500	2.494	0.000
3	MAX	3.060	2.940	0.180	3.025	2.980	0.070	3.014	2.994	0.030	3.010	2.998	0.018	3.010	3.000	0.016
	MIN	3.000	2.880	0.060	3.000	2.955	0.020	3.000	2.984	0.006	3.000	2.992	0.002	3.000	2.994	0.000
4	MAX	4.075	3.930	0.220	4.030	3.970	0.090	4.018	3.990	0.040	4.012	3.996	0.024	4.012	4.000	0.020
	MIN	4.000	3.855	0.070	4.000	3.940	0.030	4.000	3.978	0.010	4.000	3.988	0.004	4.000	3.992	0.000
5	MAX	5.075	4.930	0.220	5.030	4.970	0.090	5.018	4.990	0.040	5.012	4.996	0.024	5.012	5.000	0.020
	MIN	5.000	4.855	0.070	5.000	4.940	0.030	5.000	4.978	0.010	5.000	4.988	0.004	5.000	4.992	0.000
6	MAX	6.075	5.930	0.220	6.030	5.970	0.090	6.018	5.990	0.040	6.012	5.996	0.024	6.012	6.000	0.020
	MIN	6.000	5.855	0.070	6.000	5.940	0.030	6.000	5.978	0.010	6.000	5.988	0.004	6.000	5.992	0.000
8	MAX	8.090	7.920	0.260	8.036	7.960	0.112	8.022	7.987	0.050	8.015	7.995	0.029	8.015	8.000	0.024
	MIN	8.000	7.830	0.080	8.000	7.924	0.040	8.000	7.972	0.013	8.000	7.986	0.005	8.000	7.991	0.000
10	MAX	10.090	9.920	0.260	10.036	9.960	0.112	10.022	9.987	0.050	10.015	9.995	0.029	10.015	10.000	0.024
	MIN	10.000	9.830	0.080	10.000	9.924	0.040	10.000	9.972	0.013	10.000	9.986	0.005	10.000	9.991	0.000
12	MAX	12.110	11.905	0.315	12.043	11.950	0.136	12.027	11.984	0.061	12.018	11.994	0.035	12.018	12.000	0.029
	MIN	12.000	11.795	0.095	12.000	11.907	0.050	12.000	11.966	0.016	12.000	11.983	0.006	12.000	11.989	0.000
16	MAX	16.110	15.905	0.315	16.043	15.950	0.136	16.027	15.984	0.061	16.018	15.994	0.035	16.018	16.000	0.029
	MIN	16.000	15.795	0.095	16.000	15.907	0.050	16.000	15.966	0.016	16.000	15.983	0.006	16.000	15.989	0.000
20	MAX	20.130	19.890	0.370	20.052	19.935	0.169	20.033	19.980	0.074	20.021	19.993	0.041	20.021	20.000	0.034
	MIN	20.000	19.760	0.110	20.000	19.883	0.065	20.000	19.959	0.020	20.000	19.980	0.007	20.000	19.987	0.000
25	MAX	25.130	24.890	0.370	25.052	24.935	0.169	25.033	24.980	0.074	25.021	24.993	0.041	25.021	25.000	0.034
	MIN	25.000	24.760	0.110	25.000	24.883	0.065	25.000	24.959	0.020	25.000	24.980	0.007	25.000	24.987	0.000
30	MAX	30.130	29.890	0.370	30.052	29.935	0.169	30.033	29.980	0.074	30.021	29.993	0.041	30.021	30.000	0.034
	MIN	30.000	29.760	0.110	30.000	29.883	0.065	30.000	29.959	0.020	30.000	29.980	0.007	30.000	29.987	0.000

**AMERICAN NATIONAL STANDARD
PREFERRED METRIC LIMITS AND FITS**

ANSI B4.2–1978

TABLE 28 (CONTINUED)

Dimensions in mm.

TABLE 2   PREFERRED HOLE BASIS CLEARANCE FITS (Continued)

BASIC SIZE		LOOSE RUNNING Hole H11	Shaft c11	Fit	FREE RUNNING Hole H9	Shaft d9	Fit	CLOSE RUNNING Hole H8	Shaft f7	Fit	SLIDING Hole H7	Shaft g6	Fit	LOCATIONAL CLEARANCE Hole H7	Shaft h6	Fit
40	MAX	40.160	39.880	0.440	40.062	39.920	0.204	40.039	39.975	0.089	40.025	39.991	0.050	40.025	40.000	0.041
	MIN	40.000	39.720	0.120	40.000	39.858	0.080	40.000	39.950	0.025	40.000	39.975	0.009	40.000	39.984	0.000
50	MAX	50.160	49.870	0.450	50.062	49.920	0.204	50.039	49.975	0.089	50.025	49.991	0.050	50.025	50.000	0.041
	MIN	50.000	49.710	0.130	50.000	49.858	0.080	50.000	49.950	0.025	50.000	49.975	0.009	50.000	49.984	0.000
60	MAX	60.190	59.860	0.520	60.074	59.900	0.248	60.046	59.970	0.106	60.030	59.990	0.059	60.030	60.000	0.049
	MIN	60.000	59.670	0.140	60.000	59.826	0.100	60.000	59.940	0.030	60.000	59.971	0.010	60.000	59.981	0.000
80	MAX	80.190	79.850	0.530	80.074	79.900	0.248	80.046	79.970	0.106	80.030	79.990	0.059	80.030	80.000	0.049
	MIN	80.000	79.660	0.150	80.000	79.826	0.100	80.000	79.940	0.030	80.000	79.971	0.010	80.000	75.981	0.000
100	MAX	100.220	99.830	0.610	100.087	99.880	0.294	100.054	99.964	0.125	100.035	99.988	0.069	100.035	100.000	0.057
	MIN	100.000	99.610	0.170	100.000	99.793	0.120	100.000	99.929	0.036	100.000	99.966	0.012	100.000	99.978	0.000
120	MAX	120.220	119.820	0.620	120.087	119.880	0.294	120.054	119.964	0.125	120.035	119.988	0.069	120.035	120.000	0.057
	MIN	120.000	119.600	0.180	120.000	119.793	0.120	120.000	119.929	0.036	120.000	119.966	0.012	120.000	119.978	0.000
160	MAX	160.250	159.790	0.710	160.100	159.855	0.345	160.063	159.957	0.146	160.040	159.986	0.079	160.040	160.000	0.065
	MIN	160.000	159.540	0.210	160.000	159.755	0.145	160.000	159.917	0.043	160.000	159.961	0.014	160.000	159.975	0.000
200	MAX	200.290	199.760	0.820	200.115	199.830	0.400	200.072	199.950	0.168	200.046	199.985	0.090	200.046	200.000	0.075
	MIN	200.000	199.470	0.240	200.000	199.715	0.170	200.000	199.904	0.050	200.000	199.956	0.015	200.000	199.971	0.000
250	MAX	250.290	249.720	0.860	250.115	249.830	0.400	250.072	249.950	0.168	250.046	249.985	0.090	250.046	250.000	0.075
	MIN	250.000	249.430	0.280	250.000	249.715	0.170	250.000	249.904	0.050	250.000	249.956	0.015	250.000	249.971	0.000
300	MAX	300.320	299.670	0.970	300.130	299.810	0.450	300.081	299.944	0.189	300.052	299.983	0.101	300.052	300.000	0.084
	MIN	300.000	299.350	0.330	300.000	299.680	0.190	300.000	299.892	0.056	300.000	299.951	0.017	300.000	299.968	0.000
400	MAX	400.360	399.600	1.120	400.140	399.790	0.490	400.089	399.938	0.208	400.057	399.982	0.111	400.057	400.000	0.093
	MIN	400.000	399.240	0.400	400.000	399.650	0.210	400.000	399.881	0.062	400.000	399.946	0.018	400.000	399.964	0.000
500	MAX	500.400	499.520	1.280	500.155	499.770	0.540	500.097	499.932	0.228	500.063	499.980	0.123	500.063	500.000	0.103
	MIN	500.000	499.120	0.480	500.000	499.615	0.230	500.000	499.869	0.068	500.000	499.940	0.020	500.000	499.960	0.000

(Reprinted from The American Society of Mechanical Engineers—ANSI B4.2–1978 (R1994)

AMERICAN NATIONAL STANDARD
PREFERRED METRIC LIMITS AND FITS

ANSI B4.2–1978

TABLE 28 (CONTINUED)

TABLE 4 PREFERRED SHAFT BASIS CLEARANCE FITS

Dimensions in mm.

BASIC SIZE		LOOSE RUNNING			FREE RUNNING			CLOSE RUNNING			SLIDING			LOCATIONAL CLEARANCE		
		Hole C11	Shaft h11	Fit	Hole D9	Shaft h9	Fit	Hole F8	Shaft h7	Fit	Hole G7	Shaft h6	Fit	Hole H7	Shaft h6	Fit
1	MAX	1•120	1•000	0•180	1•045	1•000	0•070	1•020	1•000	0•030	1•012	1•000	0•018	1•010	1•000	0•016
	MIN	1•060	0•940	0•060	1•020	0•975	0•020	1•006	0•990	0•006	1•002	0•994	0•002	1•000	0•994	0•000
1•2	MAX	1•320	1•200	0•180	1•245	1•200	0•070	1•220	1•200	0•030	1•212	1•200	0•018	1•210	1•200	0•016
	MIN	1•260	1•140	0•060	1•220	1•175	0•020	1•206	1•190	0•006	1•202	1•194	0•002	1•200	1•194	0•000
1•6	MAX	1•720	1•600	0•180	1•645	1•600	0•070	1•620	1•600	0•030	1•612	1•600	0•018	1•610	1•600	0•016
	MIN	1•660	1•540	0•060	1•620	1•575	0•020	1•606	1•590	0•006	1•602	1•594	0•002	1•600	1•594	0•000
2	MAX	2•120	2•000	0•180	2•045	2•000	0•070	2•020	2•000	0•030	2•012	2•000	0•018	2•010	2•000	0•016
	MIN	2•060	1•940	0•060	2•020	1•975	0•020	2•006	1•990	0•006	2•002	1•994	0•002	2•000	1•994	0•000
2•5	MAX	2•620	2•500	0•180	2•545	2•500	0•070	2•520	2•500	0•030	2•512	2•500	0•018	2•510	2•500	0•016
	MIN	2•560	2•440	0•060	2•520	2•475	0•020	2•506	2•490	0•006	2•502	2•494	0•002	2•500	2•494	0•000
3	MAX	3•120	3•000	0•180	3•045	3•000	0•070	3•020	3•000	0•030	3•012	3•000	0•018	3•010	3•000	0•016
	MIN	3•060	2•940	0•060	3•020	2•975	0•020	3•006	2•990	0•006	3•002	2•994	0•002	3•000	2•994	0•000
4	MAX	4•145	4•000	0•220	4•060	4•000	0•090	4•028	4•000	0•040	4•016	4•000	0•024	4•012	4•000	0•020
	MIN	4•070	3•925	0•070	4•030	3•970	0•030	4•010	3•988	0•010	4•004	3•992	0•004	4•000	3•992	0•000
5	MAX	5•145	5•000	0•220	5•060	5•000	0•090	5•028	5•000	0•040	5•016	5•000	0•024	5•012	5•000	0•020
	MIN	5•070	4•925	0•070	5•030	4•970	0•030	5•010	4•988	0•010	5•004	4•992	0•004	5•000	4•992	0•000
6	MAX	6•145	6•000	0•220	6•060	6•000	0•090	6•028	6•000	0•040	6•016	6•000	0•024	6•012	6•000	0•020
	MIN	6•070	5•925	0•070	6•030	5•970	0•030	6•010	5•988	0•010	6•004	5•992	0•004	6•000	5•992	0•000
8	MAX	8•170	8•000	0•260	8•076	8•000	0•112	8•035	8•000	0•050	8•020	8•000	0•029	8•015	8•000	0•024
	MIN	8•080	7•910	0•080	8•040	7•964	0•040	8•013	7•985	0•013	8•005	7•991	0•005	8•000	7•991	0•000
10	MAX	10•170	10•000	0•260	10•076	10•000	0•112	10•035	10•000	0•050	10•020	10•000	0•029	10•015	10•000	0•024
	MIN	10•080	9•910	0•080	10•040	9•964	0•040	10•013	9•985	0•013	10•005	9•991	0•005	10•000	9•991	0•000
12	MAX	12•205	12•000	0•315	12•093	12•000	0•136	12•043	12•000	0•061	12•024	12•000	0•035	12•018	12•000	0•029
	MIN	12•095	11•890	0•095	12•050	11•957	0•050	12•016	11•982	0•016	12•006	11•989	0•006	12•000	11•989	0•000
16	MAX	16•205	16•000	0•315	16•093	16•000	0•136	16•043	16•000	0•061	16•024	16•000	0•035	16•018	16•000	0•029
	MIN	16•095	15•890	0•095	16•050	15•957	0•050	16•016	15•982	0•016	16•006	15•989	0•006	16•000	15•989	0•000
20	MAX	20•240	20•000	0•370	20•117	20•000	0•169	20•053	20•000	0•074	20•028	20•000	0•041	20•021	20•000	0•034
	MIN	20•110	19•870	0•110	20•065	19•948	0•065	20•020	19•979	0•020	20•007	19•987	0•007	20•000	19•987	0•000
25	MAX	25•240	25•000	0•370	25•117	25•000	0•169	25•053	25•000	0•074	25•028	25•000	0•041	25•021	25•000	0•034
	MIN	25•110	24•870	0•110	25•065	24•948	0•065	25•020	24•979	0•020	25•007	24•987	0•007	25•000	24•987	0•000
30	MAX	30•240	30•000	0•370	30•117	30•000	0•169	30•053	30•000	0•074	30•028	30•000	0•041	30•021	30•000	0•034
	MIN	30•110	29•870	0•110	30•065	29•948	0•065	30•020	29•979	0•020	30•007	29•987	0•007	30•000	29•987	0•000

AMERICAN NATIONAL STANDARD
PREFERRED METRIC LIMITS AND FITS

ANSI B4.2–1978

TABLE 28 (CONTINUED)

TABLE 4   PREFERRED SHAFT BASIS CLEARANCE FITS (Continued)

Dimensions in mm.

BASIC SIZE		LOOSE RUNNING			FREE RUNNING			CLOSE RUNNING			SLIDING			LOCATIONAL CLEARANCE		
		Hole C11	Shaft h11	Fit	Hole D9	Shaft h9	Fit	Hole F8	Shaft h7	Fit	Hole G7	Shaft h6	Fit	Hole H7	Shaft h6	Fit
40	MAX	40.280	40.000	0.440	40.142	40.000	0.204	40.064	40.000	0.089	40.034	40.000	0.050	40.025	40.000	0.041
	MIN	40.120	39.840	0.120	40.080	39.938	0.080	40.025	39.975	0.025	40.009	39.984	0.009	40.000	39.984	0.000
50	MAX	50.290	50.000	0.450	50.142	50.000	0.204	50.064	50.000	0.089	50.034	50.000	0.050	50.025	50.000	0.041
	MIN	50.130	49.840	0.130	50.080	49.938	0.080	50.025	49.975	0.025	50.009	49.984	0.009	50.000	49.984	0.000
60	MAX	60.330	60.000	0.520	60.174	60.000	0.248	60.076	60.000	0.106	60.040	60.000	0.059	60.030	60.000	0.049
	MIN	60.140	59.810	0.140	60.100	59.926	0.100	60.030	59.970	0.030	60.010	59.981	0.010	60.000	59.981	0.000
80	MAX	80.340	80.000	0.530	80.174	80.000	0.248	80.076	80.000	0.106	80.040	80.000	0.059	80.030	80.000	0.049
	MIN	80.150	79.810	0.150	80.100	79.926	0.100	80.030	79.970	0.030	80.010	79.981	0.010	80.000	75.981	0.000
100	MAX	100.390	100.000	0.610	100.207	100.000	0.294	100.090	100.000	0.125	100.047	100.000	0.069	100.035	100.000	0.057
	MIN	100.170	99.780	0.170	100.120	99.913	0.120	100.036	99.965	0.036	100.012	99.978	0.012	100.000	99.978	0.000
120	MAX	120.400	120.000	0.620	120.207	120.000	0.294	120.090	120.000	0.125	120.047	120.000	0.069	120.035	120.000	0.057
	MIN	120.180	119.780	0.180	120.120	119.913	0.120	120.036	119.965	0.036	120.012	119.978	0.012	120.000	119.978	0.000
160	MAX	160.460	160.000	0.710	160.245	160.000	0.345	160.106	160.000	0.146	160.054	160.000	0.079	160.040	160.000	0.065
	MIN	160.210	159.750	0.210	160.145	159.900	0.145	160.043	159.960	0.043	160.014	159.975	0.014	160.000	159.975	0.000
200	MAX	200.530	200.000	0.820	200.285	200.000	0.400	200.122	200.000	0.168	200.061	200.000	0.090	200.046	200.000	0.075
	MIN	200.240	199.710	0.240	200.170	199.885	0.170	200.050	199.954	0.050	200.015	199.971	0.015	200.000	199.971	0.000
250	MAX	250.570	250.000	0.860	250.285	250.000	0.400	250.122	250.000	0.168	250.061	250.000	0.090	250.046	250.000	0.075
	MIN	250.280	249.710	0.280	250.170	249.885	0.170	250.050	249.954	0.050	250.015	249.971	0.015	250.000	249.971	0.000
300	MAX	300.650	300.000	0.970	300.320	300.000	0.450	300.137	300.000	0.189	300.069	300.000	0.101	300.052	300.000	0.084
	MIN	300.330	299.680	0.330	300.190	299.870	0.190	300.056	299.948	0.056	300.017	299.968	0.017	300.000	299.968	0.000
400	MAX	400.760	400.000	1.120	400.350	400.000	0.490	400.151	400.000	0.208	400.075	400.000	0.111	400.057	400.000	0.093
	MIN	400.400	399.640	0.400	400.210	399.860	0.210	400.062	399.943	0.062	400.018	399.964	0.018	400.000	399.964	0.000
500	MAX	500.880	500.000	1.280	500.385	500.000	0.540	500.165	500.000	0.228	500.083	500.000	0.123	500.063	500.000	0.103
	MIN	500.480	499.600	0.480	500.230	499.845	0.230	500.068	499.937	0.068	500.020	499.960	0.020	500.000	499.960	0.000

(Reprinted from The American Society of Mechanical Engineers—ANSI B4.2–1978 (R1994)

TABLE 28 (CONTINUED)

Table A1   Tolerance Zones for Internal (Hole) Dimensions (A14 through A9 and B14 through B9)

Dimensions in mm.

BASIC SIZE		A14	A13	A12	A11	A10	A9	B14	B13	B12	B11	B10	B9
OVER TO	0 3	+0•520 +0•270	+0•410 +0•270	+0•370 +0•270	+0•330 +0•270	+0•310 +0•270	+0•295 +0•270	+0•390 +0•140	+0•280 +0•140	+0•240 +0•140	+0•200 +0•140	+0•180 +0•140	+0•165 +0•140
OVER TO	3 6	+0•570 +0•270	+0•450 +0•270	+0•390 +0•270	+0•345 +0•270	+0•318 +0•270	+0•300 +0•270	+0•440 +0•140	+0•320 +0•140	+0•260 +0•140	+0•215 +0•140	+0•188 +0•140	+0•170 +0•140
OVER TO	6 10	+0•640 +0•280	+0•500 +0•280	+0•430 +0•280	+0•370 +0•280	+0•338 +0•280	+0•316 +0•280	+0•510 +0•150	+0•370 +0•150	+0•300 +0•150	+0•240 +0•150	+0•208 +0•150	+0•186 +0•150
OVER TO	10 14	+0•720 +0•290	+0•560 +0•290	+0•470 +0•290	+0•400 +0•290	+0•360 +0•290	+0•333 +0•290	+0•580 +0•150	+0•420 +0•150	+0•330 +0•150	+0•260 +0•150	+0•220 +0•150	+0•193 +0•150
OVER TO	14 18	+0•720 +0•290	+0•560 +0•290	+0•470 +0•290	+0•400 +0•290	+0•360 +0•290	+0•333 +0•290	+0•580 +0•150	+0•420 +0•150	+0•330 +0•150	+0•260 +0•150	+0•220 +0•150	+0•193 +0•150
OVER TO	18 24	+0•820 +0•300	+0•630 +0•300	+0•510 +0•300	+0•430 +0•300	+0•384 +0•300	+0•352 +0•300	+0•680 +0•160	+0•490 +0•160	+0•370 +0•160	+0•290 +0•160	+0•244 +0•160	+0•212 +0•160
OVER TO	24 30	+0•820 +0•300	+0•630 +0•300	+0•510 +0•300	+0•430 +0•300	+0•384 +0•300	+0•352 +0•300	+0•680 +0•160	+0•490 +0•160	+0•370 +0•160	+0•290 +0•160	+0•244 +0•160	+0•212 +0•160
OVER TO	30 40	+0•930 +0•310	+0•700 +0•310	+0•560 +0•310	+0•470 +0•310	+0•410 +0•310	+0•372 +0•310	+0•790 +0•170	+0•560 +0•170	+0•420 +0•170	+0•330 +0•170	+0•270 +0•170	+0•232 +0•170
OVER TO	40 50	+0•940 +0•320	+0•710 +0•320	+0•570 +0•320	+0•480 +0•320	+0•420 +0•320	+0•382 +0•320	+0•800 +0•180	+0•570 +0•180	+0•430 +0•180	+0•340 +0•180	+0•280 +0•180	+0•242 +0•180
OVER TO	50 65	+1•080 +0•340	+0•800 +0•340	+0•640 +0•340	+0•530 +0•340	+0•460 +0•340	+0•414 +0•340	+0•930 +0•190	+0•650 +0•190	+0•490 +0•190	+0•380 +0•190	+0•310 +0•190	+0•264 +0•190
OVER TO	65 80	+1•100 +0•360	+0•820 +0•360	+0•660 +0•360	+0•550 +0•360	+0•480 +0•360	+0•434 +0•360	+0•940 +0•200	+0•660 +0•200	+0•500 +0•200	+0•390 +0•200	+0•320 +0•200	+0•274 +0•200
OVER TO	80 100	+1•250 +0•380	+0•920 +0•380	+0•730 +0•380	+0•600 +0•380	+0•520 +0•380	+0•467 +0•380	+1•090 +0•220	+0•760 +0•220	+0•570 +0•220	+0•440 +0•220	+0•360 +0•220	+0•307 +0•220
OVER TO	100 120	+1•280 +0•410	+0•950 +0•410	+0•760 +0•410	+0•630 +0•410	+0•550 +0•410	+0•497 +0•410	+1•110 +0•240	+0•780 +0•240	+0•590 +0•240	+0•460 +0•240	+0•380 +0•240	+0•327 +0•240
OVER TO	120 140	+1•460 +0•460	+1•090 +0•460	+0•860 +0•460	+0•710 +0•460	+0•620 +0•460	+0•560 +0•460	+1•260 +0•260	+0•890 +0•260	+0•660 +0•260	+0•510 +0•260	+0•420 +0•260	+0•360 +0•260
OVER TO	140 160	+1•520 +0•520	+1•150 +0•520	+0•920 +0•520	+0•770 +0•520	+0•680 +0•520	+0•620 +0•520	+1•280 +0•280	+0•910 +0•280	+0•680 +0•280	+0•530 +0•280	+0•440 +0•280	+0•380 +0•280
OVER TO	160 180	+1•580 +0•580	+1•210 +0•580	+0•980 +0•580	+0•830 +0•580	+0•740 +0•580	+0•680 +0•580	+1•310 +0•310	+0•940 +0•310	+0•710 +0•310	+0•560 +0•310	+0•470 +0•310	+0•410 +0•310
OVER TO	180 200	+1•810 +0•660	+1•380 +0•660	+1•120 +0•660	+0•950 +0•660	+0•845 +0•660	+0•775 +0•660	+1•490 +0•340	+1•060 +0•340	+0•800 +0•340	+0•630 +0•340	+0•525 +0•340	+0•455 +0•340
OVER TO	200 225	+1•890 +0•740	+1•460 +0•740	+1•200 +0•740	+1•030 +0•740	+0•925 +0•740	+0•855 +0•740	+1•530 +0•380	+1•100 +0•380	+0•840 +0•380	+0•670 +0•380	+0•565 +0•380	+0•495 +0•380
OVER TO	225 250	+1•970 +0•820	+1•540 +0•820	+1•280 +0•820	+1•110 +0•820	+1•005 +0•820	+0•935 +0•820	+1•570 +0•420	+1•140 +0•420	+0•880 +0•420	+0•710 +0•420	+0•605 +0•420	+0•535 +0•420
OVER TO	250 280	+2•220 +0•920	+1•730 +0•920	+1•440 +0•920	+1•240 +0•920	+1•130 +0•920	+1•050 +0•920	+1•780 +0•480	+1•290 +0•480	+1•000 +0•480	+0•800 +0•480	+0•690 +0•480	+0•610 +0•480
OVER TO	280 315	+2•350 +1•050	+1•860 +1•050	+1•570 +1•050	+1•370 +1•050	+1•260 +1•050	+1•180 +1•050	+1•840 +0•540	+1•350 +0•540	+1•060 +0•540	+0•860 +0•540	+0•750 +0•540	+0•670 +0•540
OVER TO	315 355	+2•600 +1•200	+2•090 +1•200	+1•770 +1•200	+1•560 +1•200	+1•430 +1•200	+1•340 +1•200	+2•000 +0•600	+1•490 +0•600	+1•170 +0•600	+0•960 +0•600	+0•830 +0•600	+0•740 +0•600
OVER TO	355 400	+2•750 +1•350	+2•240 +1•350	+1•920 +1•350	+1•710 +1•350	+1•580 +1•350	+1•490 +1•350	+2•080 +0•680	+1•570 +0•680	+1•250 +0•680	+1•040 +0•680	+0•910 +0•680	+0•820 +0•680
OVER TO	400 450	+3•050 +1•500	+2•470 +1•500	+2•130 +1•500	+1•900 +1•500	+1•750 +1•500	+1•655 +1•500	+2•310 +0•760	+1•730 +0•760	+1•390 +0•760	+1•160 +0•760	+1•010 +0•760	+0•915 +0•760
OVER TO	450 500	+3•200 +1•650	+2•620 +1•650	+2•280 +1•650	+2•050 +1•650	+1•900 +1•650	+1•805 +1•650	+2•390 +0•840	+1•810 +0•840	+1•470 +0•840	+1•240 +0•840	+1•090 +0•840	+0•995 +0•840

(Reprinted from The American Society of Mechanical Engineers—ANSI B4.2–1978 (R1994).)

TABLE 28 (CONTINUED)

Table A13   Tolerance Zones for External (Shaft) Dimensions (a14 through a9 and b14 through b9)   Dimensions in mm.

BASIC SIZE		a14	a13	a12	a11	a10	a9	b14	b13	b12	b11	b10	b9
OVER	0	−0•270	−0•270	−0•270	−0•270	−0•270	−0•270	−0•140	−0•140	−0•140	−0•140	−0•140	−0•140
TO	3	−0•520	−0•410	−0•370	−0•330	−0•310	−0•295	−0•390	−0•280	−0•240	−0•200	−0•180	−0•165
OVER	3	−0•270	−0•270	−0•270	−0•270	−0•270	−0•270	−0•140	−0•140	−0•140	−0•140	−0•140	−0•140
TO	6	−0•570	−0•450	−0•390	−0•345	−0•318	−0•300	−0•440	−0•320	−0•260	−0•215	−0•188	−0•170
OVER	6	−0•280	−0•280	−0•280	−0•280	−0•280	−0•280	−0•150	−0•150	−0•150	−0•150	−0•150	−0•150
TO	10	−0•640	−0•500	−0•430	−0•370	−0•338	−0•316	−0•510	−0•370	−0•300	−0•240	−0•208	−0•186
OVER	10	−0•290	−0•290	−0•290	−0•290	−0•290	−0•290	−0•150	−0•150	−0•150	−0•150	−0•150	−0•150
TO	14	−0•720	−0•560	−0•470	−0•400	−0•360	−0•333	−0•580	−0•420	−0•330	−0•260	−0•220	−0•193
OVER	14	−0•290	−0•290	−0•290	−0•290	−0•290	−0•290	−0•150	−0•150	−0•150	−0•150	−0•150	−0•150
TO	18	−0•720	−0•560	−0•470	−0•400	−0•360	−0•333	−0•580	−0•420	−0•330	−0•260	−0•220	−0•193
OVER	18	−0•300	−0•300	−0•300	−0•300	−0•300	−0•300	−0•160	−0•160	−0•160	−0•160	−0•160	−0•160
TO	24	−0•820	−0•630	−0•510	−0•430	−0•384	−0•352	−0•680	−0•490	−0•370	−0•290	−0•244	−0•212
OVER	24	−0•300	−0•300	−0•300	−0•300	−0•300	−0•300	−0•160	−0•160	−0•160	−0•160	−0•160	−0•160
TO	30	−0•820	−0•630	−0•510	−0•430	−0•384	−0•352	−0•680	−0•490	−0•370	−0•290	−0•244	−0•212
OVER	30	−0•310	−0•310	−0•310	−0•310	−0•310	−0•310	−0•170	−0•170	−0•170	−0•170	−0•170	−0•170
TO	40	−0•930	−0•700	−0•560	−0•470	−0•410	−0•372	−0•790	−0•560	−0•420	−0•330	−0•270	−0•232
OVER	40	−0•320	−0•320	−0•320	−0•320	−0•320	−0•320	−0•180	−0•180	−0•180	−0•180	−0•180	−0•180
TO	50	−0•940	−0•710	−0•570	−0•480	−0•420	−0•382	−0•800	−0•570	−0•430	−0•340	−0•280	−0•242
OVER	50	−0•340	−0•340	−0•340	−0•340	−0•340	−0•340	−0•190	−0•190	−0•190	−0•190	−0•190	−0•190
TO	65	−1•080	−0•800	−0•640	−0•530	−0•460	−0•414	−0•930	−0•650	−0•490	−0•380	−0•310	−0•264
OVER	65	−0•360	−0•360	−0•360	−0•360	−0•360	−0•360	−0•200	−0•200	−0•200	−0•200	−0•200	−0•200
TO	80	−1•100	−0•820	−0•660	−0•550	−0•480	−0•434	−0•940	−0•660	−0•500	−0•390	−0•320	−0•274
OVER	80	−0•380	−0•380	−0•380	−0•380	−0•380	−0•380	−0•220	−0•220	−0•220	−0•220	−0•220	−0•220
TO	100	−1•250	−0•920	−0•730	−0•600	−0•520	−0•467	−1•090	−0•760	−0•570	−0•440	−0•360	−0•307
OVER	100	−0•410	−0•410	−0•410	−0•410	−0•410	−0•410	−0•240	−0•240	−0•240	−0•240	−0•240	−0•240
TO	120	−1•280	−0•950	−0•760	−0•630	−0•550	−0•497	−1•110	−0•780	−0•590	−0•460	−0•380	−0•327
OVER	120	−0•460	−0•460	−0•460	−0•460	−0•460	−0•460	−0•260	−0•260	−0•260	−0•260	−0•260	−0•260
TO	140	−1•460	−1•090	−0•860	−0•710	−0•620	−0•560	−1•260	−0•890	−0•660	−0•510	−0•420	−0•360
OVER	140	−0•520	−0•520	−0•520	−0•520	−0•520	−0•520	−0•280	−0•280	−0•280	−0•280	−0•280	−0•280
TO	160	−1•520	−1•150	−0•920	−0•770	−0•680	−0•620	−1•280	−0•910	−0•680	−0•530	−0•440	−0•380
OVER	160	−0•580	−0•580	−0•580	−0•580	−0•580	−0•580	−0•310	−0•310	−0•310	−0•310	−0•310	−0•310
TO	180	−1•580	−1•210	−0•980	−0•830	−0•740	−0•680	−1•310	−0•940	−0•710	−0•560	−0•470	−0•410
OVER	160	−0•580	−0•580	−0•580	−0•580	−0•580	~0•580	−0•310	−0•310	−0•310	−0•310	−0•310	−0•310
TO	180	−1•580	−1•210	−0•980	−0•830	−0•740	~0•680	−1•310	−0•940	−0•710	−0•560	−0•470	−0•410
OVER	180	−0•660	−0•660	−0•660	−0•660	−0•660	−0•660	−0•340	−0•340	−0•340	−0•340	−0•340	−0•340
TO	200	−1•810	−1•380	−1•120	−0•950	−0•845	−0•775	−1•490	−1•060	−0•800	−0•630	−0•525	−0•455
OVER	200	−0•740	−0•740	−0•740	−0•740	−0•740	−0•740	−0•380	−0•380	−0•380	−0•380	−0•380	−0•380
TO	225	−1•890	−1•460	−1•200	−1•030	−0•925	−0•855	−1•530	−1•100	−0•840	−0•670	−0•565	−0•495
OVER	225	−0•820	−0•820	−0•820	−0•820	−0•820	−0•820	−0•420	−0•420	−0•420	−0•420	−0•420	−0•420
TO	250	−1•970	−1•540	−1•280	−1•110	−1•005	−0•935	−1•570	−1•140	−0•880	−0•710	−0•605	−0•535
OVER	250	−0•920	−0•920	−0•920	−0•920	−0•920	−0•920	−0•480	−0•480	−0•480	−0•480	−0•480	−0•480
TO	280	−2•220	−1•730	−1•440	−1•240	−1•130	−1•050	−1•780	−1•290	−1•000	−0•800	−0•690	−0•610
OVER	280	−1•050	−1•050	−1•050	−1•050	−1•050	−1•050	−0•540	−0•540	−0•540	−0•540	−0•540	−0•540
TO	315	−2•350	−1•860	−1•570	−1•370	−1•260	−1•180	−1•840	−1•350	−1•060	−0•860	−0•750	−0•670
OVER	315	−1•200	−1•200	−1•200	−1•200	−1•200	−1•200	−0•600	−0•600	−0•600	−0•600	−0•600	−0•600
TO	355	−2•600	−2•090	−1•770	−1•560	−1•430	−1•340	−2•000	−1•490	−1•170	−0•960	−0•830	−0•740
OVER	355	−1•350	−1•350	−1•350	−1•350	−1•350	−1•350	−0•680	−0•680	−0•680	−0•680	−0•680	−0•680
TO	400	−2•750	−2•240	−1•920	−1•710	−1•580	−1•490	−2•080	−1•570	−1•250	−1•040	−0•910	−0•820
OVER	400	−1•500	−1•500	−1•500	−1•500	−1•500	−1•500	−0•760	−0•760	−0•760	−0•760	−0•760	−0•760
TO	450	−3•050	−2•470	−2•130	−1•900	−1•750	−1•655	−2•310	−1•730	−1•390	−1•160	−1•010	−0•915
OVER	450	−1•650	−1•650	−1•650	−1•650	−1•650	−1•650	−0•840	−0•840	−0•840	−0•840	−0•840	−0•840
TO	500	−3•200	−2•620	−2•280	−2•050	−1•900	−1•805	+2•390	−1•810	−1•470	−1•240	−1•090	−0•995

(Reprinted from The American Society of Mechanical Engineers—ANSI B4.2–1978 (R1994).)

APPENDIX C
UNIFIED SCREW THREAD VARIATIONS

THE FOLLOWING list is a ready reference of available standard series and selected combinations of Unified screw threads. Each thread is given as part of a proper thread note including major diameter, threads per inch, and series identification.

0–80 UNF	7/16–27 UNS	7/8–28 UN	1-3/8–18 UNEF	1-15/16–6 UN	2-7/8–12 UN
1–64 UNC	7/16–28 UNEF	7/8–32 UN	1-3/8–20 UN	1-15/16–8 UN	2-7/8–16 UN
1–72 UNF	7/16–32 UN	15/16–12 UN	1-3/8–28 UN	1-15/16–12 UN	2-7/8–20 UN
2–56 UNC	1/2–12 UNS	15/16–16 UN	1-7/16–6 UN	1-15/16–16 UN	3–4 UNC
2–64 UNF	1/2–13 UNC	15/16–20 UNEF	1-7/16–8 UN	1-15/16–20 UN	3–6 UN
3–48 UNC	1/2–14 UNS	15/16–28 UN	1-7/16–12 UN	2-4 1/2 UNC	3–8 UN
3–56 UNF	1/2–16 UN	15/16–32 UN	1-7/16–16 UN	2–6 UN	3–10 UNS
4–40 UNC	1/2–18 UNS	1–8 UNC	1-7/16–18 UNEF	2–8 UN	3–12 UN
4–48 UNF	1/2–20 UNF	1–10 UNS	1-7/16–20 UN	2–10 UNS	3–14 UNS
5–40 UNC	1/2–24 UNS	1–12 UNF	1-7/16–28 UN	2–12 UN	3–16 UN
5–44 UNF	1/2–27 UNS	1–14 UNS	1-1/2–6 UNC	2–14 UNS	3–18 UNS
6–32 UNC	1/2–28 UNEF	1–16 UN	1-1/2–8 UN	2–16 UN	3–20 UN
6–40 UNF	1/2–32 UN	1–18 UNS	1-1/2–10 UNS	2–18 UNS	3-1/8–6 UN
8–32 UNC	9/16–12 UNC	1–20 UNEF	1-1/2–12 UNF	2–20 UN	3-1/8–8 UN
8–36 UNF	9/16–14 UNS	1–24 UNS	1-1/2–14 UNS	2-1/16–16 UNS	3-1/8–12 UN
10–24 UNC	9/16–16 UN	1–27 UNS	1-1/2–16 UN	2-1/8–6 UN	3-1/8–16 UN
10–28 UNS	9/16–18 UNF	1–28 UN	1-1/2–18 UNEF	2-1/8–8 UN	3-1/4–4 UNC
10–32 UNF	9/16–20 UN	1–32 UN	1-1/2–20 UN	2-1/8–12 UN	3-1/4–6 UN
10–36 UNS	9/16–24 UNEF	1-1/16–8 UN	1-1/2–24 UNS	2-1/8–16 UN	3-1/4–8 UN
10–40 UNS	9/16–27 UNS	1-1/16–12 UN	1-1/2–28 UN	2-1/8–20 UN	3-1/4–10 UNS
10–48 UNS	9/16–28 UN	1-1/16–16 UN	1-9/16–6 UN	2-3/16–16 UNS	3 1/4–12 UN
10–56 UNS	9/16–32 UN	1-1/16–18 UNEF	1-9/16–8 UN	2-1/4-4-1/2 UNC	3 1/4–14 UNS
12–24 UNC	5/8–11 UNC	1-1/16–20 UN	1-9/16–12 UN	2-1/4–6 UN	3 1/4–16 UN
12–28 UNF	5/8–12 UN	1-1/16–28 UN	1-9/16–16 UN	2-1/4–8 UN	3-1/4–18 UNS
12–32 UNEF	5/8–14 UNS	1-1/8–7 UNC	1-9/16–18 UNEF	2-1/4–10 UNS	3-3/8–6 UN
12–36 UNS	5/8–16 UN	1-1/8–8 UN	1-9/16–20 UN	2-1/4–12 UN	3-3/8–8 UN
12–40 UNS	5/8–18 UNF	1-1/8–10 UNS	1-5/8–6 UN	2-1/4–14 UNS	3-3/8–12 UN
12–48 UNS	5/8–20 UN	1-1/8–12 UNF	1-5/8–8 UN	2-1/4–16 UN	3-3/8–16 UN
12–56 UNS	5/8–24 UNEF	1-1/8–14 UNS	1-5/8–10 UNS	2-1/4–18 UNS	3-1/2–4 UNC
1/4–20 UNC	5/8–27 UNS	1-1/8–16 UN	1-5/8–12 UN	2-1/4–20 UN	3-1/2–6 UN
1/4–24 UNS	5/8–28 UN	1-1/8–18 UNEF	1-5/8–14 UNS	2-5/16–16 UNS	3-1/2–8 UN
1/4–27 UNS	5/8–32 UN	1-1/8–20 UN	1-5/8–16 UN	2-3/8–6 UN	3-1/2–10 UNS
1/4–28 UNF	11/16–12 UN	1-1/8–24 UNS	1-5/8–18 UNEF	2-3/8–8 UN	3-1/2–12 UN
1/4–32 UNEF	11/16–16 UN	1-1/8–28 UN	1-5/8–20 UN	2-3/8–12 UN	3-1/2–14 UNS
1/4–36 UNS	11/16–20 UN	1-3/16–8 UN	1-5/8–24 UNS	2-3/8–16 UN	3-1/2–16 UN
1/4–40 UNS	11/16–24 UNEF	1-3/16–12 UN	1-11/16–6 UN	2-3/8–20 UN	3-1/2–18 UNS
1/4–48 UNS	11/16–28 UN	1-3/16–16 UN	1-11/16–8 UN	2-7/16–16 UNS	3-5/8–6 UN
1/4–56 UNS	11/16–32 UN	1-3/16–18 UNEF	1-11/16–12 UN	2-1/2–4 UNC	3-5/8–8 UN
5/16–18 UNC	3/4–10 UNC	1-3/16–20 UN	1-11/16–16 UN	2-1/2–6 UN	3-5/8–12 UN
5/16–20 UN	3/4–12 UN	1-3/16–28 UN	1-11/16–18 UNEF	2-1/2–8 UN	3-5/8–16 UN
5/16–24 UNF	3/4–14 UNC	1-1/4–7 UNC	1-11/16–20 UN	2-1/2–10 UNS	3-3/4–4 UNC
5/16–27 UNS	3/4–16 UNF	1-1/4–8 UN	1-3/4–5 UNC	2-1/2–12 UN	3-3/4–6 UN
5/16–28 UN	3/4–18 UNS	1-1/4–10 UNS	1-3/4–6 UN	2-1/2–14 UNS	3-3/4–8 UN
5/16–32 UNEF	3/4–20 UNEF	1-1/4–12 UNF	1-3/4–8 UN	2-1/2–16 UN	3-3/4–10 UNS
5/16–36 UNS	3/4–24 UNS	1-1/4–14 UNS	1-3/4–10 UNS	1-1/2–18 UNS	3-3/4–12 UN
5/16–40 UNS	3/4–27 UNS	1-1/4–16 UN	1-3/4–12 UN	2-1/2–20 UN	3-3/4–14 UNS
5/16–48 UNS	3/4–28 UN	1-1/4–18 UNEF	1-3/4–14 UNS	2-5/8–6 UN	3-3/4–16 UN
3/8–16 UNC	3/4–32 UN	1-1/4–20 UN	1-3/4–16 UN	2-5/8–8 UN	3-3/4–18 UNS
3/8–18 UNS	13/16–12 UN	1-1/4–24 UNS	1-3/4–18 UNS	2-5/8–12 UN	3-7/8–6 UN
3/8–20 UN	13/16–16 UN	1-1/4–28 UN	1-3/4–20 UN	2-5/8–16 UN	3-7/8–8 UN
3/8–24 UNF	13/16–20 UNEF	1-5/16–8 UN	1-13/16–6 UN	2-5/8–20 UN	3-7/8–12 UN
3/8–27 UNS	13/16–28 UN	1-5/16–12 UN	1-13/16–8 UN	2-3/4–4 UNC	3-7/8–16 UN
3/8–28 UN	13/16–32 UN	1-5/16–16 UN	1-13/16–12 UN	2-3/4–6 UN	4–4 UNC
3/8–32 UNEF	7/8–9 UNC	1-5/16–18 UNEF	1-13/16–20 UN	2-3/4–8 UN	4–6 UN
3/8–36 UNS	7/8–10 UNS	1-5/16–20 UN	1-7/8–6 UN	2-3/4–10 UNS	4–8 UN
3/8–40 UNS	7/8–12 UN	1-5/16–28 UN	1-7/8–8 UN	2-3/4–12 UN	4–10 UNS
.390–27 UNS	7/8–14 UNF	1-3/8–6 UNC	1-7/8–10 UNS	2-3/4–14 UNS	4–12 UN
7/16–14 UNC	7/8–16 UN	1-3/8–8 UN	1-7/8–12 UN	2-3/4–16 UN	4–14 UNS
7/16–16 UN	7/8–18 UNS	1-3/8–10 UNS	1-7/8–14 UNS	2-3/4–18 UNS	4–16 UN
7/16–18 UNS	7/8–20 UNEF	1-3/8–12 UNF	1-7/8–16 UN	2-3/4–20 UN	
7/16–20 UNF	7/8–24 UNS	1-3/8–14 UNS	1-7/8–18 UNS	2-7/8–6 UN	
7/16–24 UNS	7/8–27 UNS	1-3/8–16 UN	1-7/8–20 UN	2-7/8–8 UN	

APPENDIX D
METRIC SCREW THREAD VARIATIONS

THE FOLLOWING list is a ready reference of available standard coarse pitch series ISO metric screw threads. Each thread is given as part of a proper thread note, including metric symbol, major diameter, and thread pitch.

M1 × 0.25	M2.2 × 0.45	M6 × 1	M14 × 2	M30 × 3.5	M52 × 5
M1.1 × 0.25	M2.5 × 0.45	M7 × 1	M16 × 2	M33 × 3.5	M56 × 5.5
M1.2 × 0.25	M3 × 0.5	M8 × 1.25	M18 × 2.5	M36 × 4	M60 × 5.5
M1.4 × 0.3	M3.5 × 0.6	M9 × 1.25	M20 × 2.5	M39 × 4	M64 × 6
M1.6 × 0.35	M4 × 0.7	M10 × 1.5	M22 × 2.5	M42 × 4.5	M68 × 6
M1.8 × 0.35	M4.5 × 0.75	M11 × 1.5	M24 × 3	M45 × 4.5	
M2 × 0.4	M5 × 0.8	M12 × 1.75	M27 × 3	M48 × 5	

APPENDIX E
STRUCTURAL METAL SHAPE DESIGNATIONS

W SHAPES — Dimensions

Designation	Area A (In.²)	Depth d (In.)		Web Thickness t_w (In.)		$t_w/2$ (In.)	Flange Width b_f (In.)		Flange Thickness t_f (In.)		Distance T (In.)	Distance k (In.)	Distance k_1 (In.)
W 10×112	32.9	11.36	11 3/8	0.755	3/4	3/8	10.415	10 3/8	1.250	1 1/4	7 5/8	1 7/8	15/16
×100	29.4	11.10	11 1/8	0.680	11/16	3/8	10.340	10 3/8	1.120	1 1/8	7 5/8	1 3/4	7/8
× 88	25.9	10.84	10 7/8	0.605	5/8	5/16	10.265	10 1/4	0.990	1	7 5/8	1 5/8	13/16
× 77	22.6	10.60	10 5/8	0.530	1/2	1/4	10.190	10 1/4	0.870	7/8	7 5/8	1 1/2	13/16
× 68	20.0	10.40	10 3/8	0.470	1/2	1/4	10.130	10 1/8	0.770	3/4	7 5/8	1 3/8	3/4
× 60	17.6	10.22	10 1/4	0.420	7/16	1/4	10.080	10 1/8	0.680	11/16	7 5/8	1 5/16	3/4
× 54	15.8	10.09	10 1/8	0.370	3/8	3/16	10.030	10	0.615	5/8	7 5/8	1 1/4	11/16
× 49	14.4	9.98	10	0.340	5/16	3/16	10.000	10	0.560	9/16	7 5/8	1 3/16	11/16
W 10× 45	13.3	10.10	10 1/8	0.350	3/8	3/16	8.020	8	0.620	5/8	7 7/8	1 1/4	11/16
× 39	11.5	9.92	9 7/8	0.315	5/16	3/16	7.985	8	0.530	1/2	7 7/8	1 1/8	11/16
× 33	9.71	9.73	9 3/4	0.290	5/16	3/16	7.960	8	0.435	7/16	7 7/8	1 1/16	11/16
W 10× 30	8.84	10.47	10 1/2	0.300	5/16	3/16	5.810	5 3/4	0.510	1/2	8 5/8	15/16	1/2
× 26	7.61	10.33	10 3/8	0.260	1/4	1/8	5.770	5 3/4	0.440	7/16	8 5/8	7/8	1/2
× 22	6.49	10.17	10 1/8	0.240	1/4	1/8	5.750	5 3/4	0.360	3/8	8 5/8	3/4	1/2
W 10× 19	5.62	10.24	10 1/4	0.250	1/4	1/8	4.020	4	0.395	3/8	8 5/8	13/16	1/2
× 17	4.99	10.11	10 1/8	0.240	1/4	1/8	4.010	4	0.330	5/16	8 5/8	3/4	1/2
× 15	4.41	9.99	10	0.230	1/4	1/8	4.000	4	0.270	1/4	8 5/8	11/16	7/16
× 12	3.54	9.87	9 7/8	0.190	3/16	1/8	3.960	4	0.210	3/16	8 5/8	5/8	7/16

AMERICAN INSTITUTE OF STEEL CONSTRUCTION

W SHAPES — Dimensions

Designation	Area A (In.²)	Depth d (In.)		Web Thickness t_w (In.)		$t_w/2$ (In.)	Flange Width b_f (In.)		Flange Thickness t_f (In.)		Distance T (In.)	Distance k (In.)	Distance k_1 (In.)
W 8×67	19.7	9.00	9	0.570	9/16	5/16	8.280	8 1/4	0.935	15/16	6 1/8	1 7/16	11/16
×58	17.1	8.75	8 3/4	0.510	1/2	1/4	8.220	8 1/4	0.810	13/16	6 1/8	1 5/16	11/16
×48	14.1	8.50	8 1/2	0.400	3/8	3/16	8.110	8 1/8	0.685	11/16	6 1/8	1 3/16	5/8
×40	11.7	8.25	8 1/4	0.360	3/8	3/16	8.070	8 1/8	0.560	9/16	6 1/8	1 1/16	5/8
×35	10.3	8.12	8 1/8	0.310	5/16	3/16	8.020	8	0.495	1/2	6 1/8	1	9/16
×31	9.13	8.00	8	0.285	5/16	3/16	7.995	8	0.435	7/16	6 1/8	15/16	9/16
W 8×28	8.25	8.06	8	0.285	5/16	3/16	6.535	6 1/2	0.465	7/16	6 1/8	15/16	9/16
×24	7.08	7.93	7 7/8	0.245	1/4	1/8	6.495	6 1/2	0.400	3/8	6 1/8	7/8	9/16
W 8×21	6.16	8.28	8 1/4	0.250	1/4	1/8	5.270	5 1/4	0.400	3/8	6 5/8	13/16	1/2
×18	5.26	8.14	8 1/8	0.230	1/4	1/8	5.250	5 1/4	0.330	5/16	6 5/8	3/4	7/16
W 8×15	4.44	8.11	8 1/8	0.245	1/4	1/8	4.015	4	0.315	5/16	6 5/8	3/4	1/2
×13	3.84	7.99	8	0.230	1/4	1/8	4.000	4	0.255	1/4	6 5/8	11/16	7/16
×10	2.96	7.89	7 7/8	0.170	3/16	1/8	3.940	4	0.205	3/16	6 5/8	5/8	7/16
W 6×25	7.34	6.38	6 3/8	0.320	5/16	3/16	6.080	6 1/8	0.455	7/16	4 3/4	13/16	7/16
×20	5.87	6.20	6 1/4	0.260	1/4	1/8	6.020	6	0.365	3/8	4 3/4	3/4	7/16
×15	4.43	5.99	6	0.230	1/4	1/8	5.990	6	0.260	1/4	4 3/4	5/8	3/8
W 6×16	4.74	6.28	6 1/4	0.260	1/4	1/8	4.030	4	0.405	3/8	4 3/4	3/4	7/16
×12	3.55	6.03	6	0.230	1/4	1/8	4.000	4	0.280	1/4	4 3/4	5/8	3/8
× 9	2.68	5.90	5 7/8	0.170	3/16	1/8	3.940	4	0.215	3/16	4 3/4	9/16	3/8
W 5×19	5.54	5.15	5 1/8	0.270	1/4	1/8	5.030	5	0.430	7/16	3 1/2	13/16	7/16
×16	4.68	5.01	5	0.240	1/4	1/8	5.000	5	0.360	3/8	3 1/2	3/4	7/16
W 4×13	3.83	4.16	4 1/8	0.280	1/4	1/8	4.060	4	0.345	3/8	2 3/4	11/16	7/16

AMERICAN INSTITUTE OF STEEL CONSTRUCTION

W SHAPES — Dimensions

Designation	Area A (In²)	Depth d (In)		Web Thickness t_w (In)		Web $t_w/2$ (In)	Flange Width b_f (In)		Flange Thickness t_f (In)		Distance T (In)	Distance k (In)	Distance k_1 (In)
W 14×132	38.8	14.66	14 5/8	0.645	5/8	5/16	14.725	14 3/4	1.030	1	11 1/4	1 11/16	15/16
×120	35.3	14.48	14 1/2	0.590	9/16	5/16	14.670	14 5/8	0.940	15/16	11 1/4	1 5/8	15/16
×109	32.0	14.32	14 3/8	0.525	1/2	1/4	14.605	14 5/8	0.860	7/8	11 1/4	1 9/16	7/8
×99	29.1	14.16	14 1/8	0.485	1/2	1/4	14.565	14 5/8	0.780	3/4	11 1/4	1 7/16	7/8
×90	26.5	14.02	14	0.440	7/16	1/4	14.520	14 1/2	0.710	11/16	11 1/4	1 3/8	7/8
W 14×82	24.1	14.31	14 1/4	0.510	1/2	1/4	10.130	10 1/8	0.855	7/8	11	1 5/8	1
×74	21.8	14.17	14 1/8	0.450	7/16	1/4	10.070	10 1/8	0.785	13/16	11	1 9/16	15/16
×68	20.0	14.04	14	0.415	7/16	1/4	10.035	10	0.720	3/4	11	1 1/2	15/16
×61	17.9	13.89	13 7/8	0.375	3/8	3/16	9.995	10	0.645	5/8	11	1 7/16	15/16
W 14×53	15.6	13.92	13 7/8	0.370	3/8	3/16	8.060	8	0.660	11/16	11	1 7/16	15/16
×48	14.1	13.79	13 3/4	0.340	5/16	3/16	8.030	8	0.595	5/8	11	1 3/8	7/8
×43	12.6	13.66	13 5/8	0.305	5/16	3/16	7.995	8	0.530	1/2	11	1 5/16	7/8
W 14×38	11.2	14.10	14 1/8	0.310	5/16	3/16	6.770	6 3/4	0.515	1/2	12	1 1/16	5/8
×34	10.0	13.98	14	0.285	5/16	3/16	6.745	6 3/4	0.455	7/16	12	1	5/8
×30	8.85	13.84	13 7/8	0.270	1/4	1/8	6.730	6 3/4	0.385	3/8	12	15/16	5/8
W 14×26	7.69	13.91	13 7/8	0.255	1/4	1/8	5.025	5	0.420	7/16	12	15/16	9/16
×22	6.49	13.74	13 3/4	0.230	1/4	1/8	5.000	5	0.335	5/16	12	7/8	9/16

AMERICAN INSTITUTE OF STEEL CONSTRUCTION

W SHAPES — Dimensions

Designation	Area A (In²)	Depth d (In)		Web Thickness t_w (In)		Web $t_w/2$ (In)	Flange Width b_f (In)		Flange Thickness t_f (In)		Distance T (In)	Distance k (In)	Distance k_1 (In)
W 12×336	98.8	16.82	16 7/8	1.775	1 3/4	7/8	13.385	13 3/8	2.955	2 15/16	9 1/2	3 11/16	1 1/2
×305	89.6	16.32	16 3/8	1.625	1 5/8	13/16	13.235	13 1/4	2.705	2 11/16	9 1/2	3 7/16	1 7/16
×279	81.9	15.85	15 7/8	1.530	1 1/2	3/4	13.140	13 1/8	2.470	2 1/2	9 1/2	3 3/16	1 3/8
×252	74.1	15.41	15 3/8	1.395	1 3/8	11/16	13.005	13	2.250	2 1/4	9 1/2	2 15/16	1 5/16
×230	67.7	15.05	15	1.285	1 5/16	11/16	12.895	12 7/8	2.070	2 1/16	9 1/2	2 3/4	1 1/4
×210	61.8	14.71	14 3/4	1.180	1 3/16	5/8	12.790	12 3/4	1.900	1 7/8	9 1/2	2 5/8	1 1/4
×190	55.8	14.38	14 3/8	1.060	1 1/16	9/16	12.670	12 5/8	1.735	1 3/4	9 1/2	2 7/16	1 3/16
×170	50.0	14.03	14	0.960	15/16	1/2	12.570	12 5/8	1.560	1 9/16	9 1/2	2 1/4	1 1/8
×152	44.7	13.71	13 3/4	0.870	7/8	7/16	12.480	12 1/2	1.400	1 3/8	9 1/2	2 1/8	1 1/16
×136	39.9	13.41	13 3/8	0.790	13/16	7/16	12.400	12 3/8	1.250	1 1/4	9 1/2	1 15/16	1
×120	35.3	13.12	13 1/8	0.710	11/16	3/8	12.320	12 3/8	1.105	1 1/8	9 1/2	1 13/16	1
×106	31.2	12.89	12 7/8	0.610	5/8	5/16	12.220	12 1/4	0.990	1	9 1/2	1 11/16	15/16
×96	28.2	12.71	12 3/4	0.550	9/16	5/16	12.160	12 1/8	0.900	7/8	9 1/2	1 5/8	7/8
×87	25.6	12.53	12 1/2	0.515	1/2	1/4	12.125	12 1/8	0.810	13/16	9 1/2	1 1/2	7/8
×79	23.2	12.38	12 3/8	0.470	1/2	1/4	12.080	12 1/8	0.735	3/4	9 1/2	1 7/16	7/8
×72	21.1	12.25	12 1/4	0.430	7/16	1/4	12.040	12	0.670	11/16	9 1/2	1 3/8	7/8
×65	19.1	12.12	12 1/8	0.390	3/8	3/16	12.000	12	0.605	5/8	9 1/2	1 5/16	13/16
W 12×58	17.0	12.19	12 1/4	0.360	3/8	3/16	10.010	10	0.640	5/8	9 1/2	1 3/8	13/16
×53	15.6	12.06	12	0.345	3/8	3/16	9.995	10	0.575	9/16	9 1/2	1 1/4	13/16
W 12×50	14.7	12.19	12 1/4	0.370	3/8	3/16	8.080	8 1/8	0.640	5/8	9 1/2	1 3/8	13/16
×45	13.2	12.06	12	0.335	5/16	3/16	8.045	8	0.575	9/16	9 1/2	1 1/4	13/16
×40	11.8	11.94	12	0.295	5/16	3/16	8.005	8	0.515	1/2	9 1/2	1 1/4	3/4
W 12×35	10.3	12.50	12 1/2	0.300	5/16	3/16	6.560	6 1/2	0.520	1/2	10 1/2	1	9/16
×30	8.79	12.34	12 3/8	0.260	1/4	1/8	6.520	6 1/2	0.440	7/16	10 1/2	15/16	1/2
×26	7.65	12.22	12 1/4	0.230	1/4	1/8	6.490	6 1/2	0.380	3/8	10 1/2	7/8	1/2
W 12×22	6.48	12.31	12 1/4	0.260	1/4	1/8	4.030	4	0.425	7/16	10 1/2	7/8	1/2
×19	5.57	12.16	12 1/8	0.235	1/4	1/8	4.005	4	0.350	3/8	10 1/2	13/16	1/2
×16	4.71	11.99	12	0.220	1/4	1/8	3.990	4	0.265	1/4	10 1/2	3/4	1/2
×14	4.16	11.91	11 7/8	0.200	3/16	1/8	3.970	4	0.225	1/4	10 1/2	11/16	1/2

AMERICAN INSTITUTE OF STEEL CONSTRUCTION

W SHAPES
Dimensions

Designation	Area A (In.²)	Depth d (In.)	(frac)	Web Thickness t_w (In.)	(frac)	Web $t_w/2$ (In.)	Flange Width b_f (In.)	(frac)	Flange Thickness t_f (In.)	(frac)	Distance T (In.)	Distance k (In.)	Distance k_1 (In.)
W 18x119	35.1	18.97	19	0.655	5/8	5/16	11.265	11 1/4	1.060	1 1/16	15 1/2	1 3/4	15/16
x106	31.1	18.73	18 3/4	0.590	9/16	5/16	11.200	11 1/4	0.940	15/16	15 1/2	1 5/8	15/16
x 97	28.5	18.59	18 5/8	0.535	9/16	5/16	11.145	11 1/8	0.870	7/8	15 1/2	1 9/16	7/8
x 86	25.3	18.39	18 3/8	0.480	1/2	1/4	11.090	11 1/8	0.770	3/4	15 1/2	1 7/16	7/8
x 76	22.3	18.21	18 1/4	0.425	7/16	1/4	11.035	11	0.680	11/16	15 1/2	1 3/8	13/16
W 18x 71	20.8	18.47	18 1/2	0.495	1/2	1/4	7.635	7 5/8	0.810	13/16	15 1/2	1 1/2	7/8
x 65	19.1	18.35	18 3/8	0.450	7/16	1/4	7.590	7 5/8	0.750	3/4	15 1/2	1 7/16	13/16
x 60	17.6	18.24	18 1/4	0.415	7/16	1/4	7.555	7 1/2	0.695	11/16	15 1/2	1 3/8	13/16
x 55	16.2	18.11	18 1/8	0.390	3/8	3/16	7.530	7 1/2	0.630	5/8	15 1/2	1 5/16	13/16
x 50	14.7	17.99	18	0.355	3/8	3/16	7.495	7 1/2	0.570	9/16	15 1/2	1 1/4	13/16
W 18x 46	13.5	18.06	18	0.360	3/8	3/16	6.060	6	0.605	5/8	15 1/2	1 1/4	13/16
x 40	11.8	17.90	17 7/8	0.315	5/16	3/16	6.015	6	0.525	1/2	15 1/2	1 3/16	13/16
x 35	10.3	17.70	17 3/4	0.300	5/16	3/16	6.000	6	0.425	7/16	15 1/2	1 1/8	3/4
W 16x100	29.4	16.97	17	0.585	9/16	5/16	10.425	10 3/8	0.985	1	13 5/8	1 11/16	15/16
x 89	26.2	16.75	16 3/4	0.525	1/2	1/4	10.365	10 3/8	0.875	7/8	13 5/8	1 9/16	7/8
x 77	22.6	16.52	16 1/2	0.455	7/16	1/4	10.295	10 1/4	0.760	3/4	13 5/8	1 7/16	7/8
x 67	19.7	16.33	16 3/8	0.395	3/8	3/16	10.235	10 1/4	0.665	11/16	13 5/8	1 3/8	13/16
W 16x 57	16.8	16.43	16 3/8	0.430	7/16	1/4	7.120	7 1/8	0.715	11/16	13 5/8	1 3/8	7/8
x 50	14.7	16.26	16 1/4	0.380	3/8	3/16	7.070	7 1/8	0.630	5/8	13 5/8	1 5/16	13/16
x 45	13.3	16.13	16 1/8	0.345	3/8	3/16	7.035	7	0.565	9/16	13 5/8	1 1/4	13/16
x 40	11.8	16.01	16	0.305	5/16	3/16	6.995	7	0.505	1/2	13 5/8	1 3/16	13/16
x 36	10.6	15.86	15 7/8	0.295	5/16	3/16	6.985	7	0.430	7/16	13 5/8	1 1/8	3/4
W 16x 31	9.12	15.88	15 7/8	0.275	1/4	1/8	5.525	5 1/2	0.440	7/16	13 5/8	1 1/8	3/4
x 26	7.68	15.69	15 3/4	0.250	1/4	1/8	5.500	5 1/2	0.345	3/8	13 5/8	1 1/16	3/4

AMERICAN INSTITUTE OF STEEL CONSTRUCTION

W SHAPES
Dimensions

Designation	Area A (In.²)	Depth d (In.)	(frac)	Web Thickness t_w (In.)	(frac)	Web $t_w/2$ (In.)	Flange Width b_f (In.)	(frac)	Flange Thickness t_f (In.)	(frac)	Distance T (In.)	Distance k (In.)	Distance k_1 (In.)
W 14x730	215.0	22.42	22 3/8	3.070	3 1/16	1 9/16	17.890	17 7/8	4.910	4 15/16	11 1/4	5 9/16	2 3/16
x665	196.0	21.64	21 5/8	2.830	2 13/16	1 7/16	17.650	17 5/8	4.520	4 1/2	11 1/4	5 3/16	2 1/16
x605	178.0	20.92	20 7/8	2.595	2 5/8	1 5/16	17.415	17 3/8	4.160	4 3/16	11 1/4	4 13/16	1 15/16
x550	162.0	20.24	20 1/4	2.380	2 3/8	1 3/16	17.200	17 1/4	3.820	3 13/16	11 1/4	4 1/2	1 13/16
x500	147.0	19.60	19 5/8	2.190	2 3/16	1 1/8	17.010	17	3.500	3 1/2	11 1/4	4 3/16	1 3/4
x455	134.0	19.02	19	2.015	2	1	16.835	16 7/8	3.210	3 3/16	11 1/4	3 7/8	1 5/8
W 14x426	125.0	18.67	18 5/8	1.875	1 7/8	15/16	16.695	16 3/4	3.035	3 1/16	11 1/4	3 11/16	1 9/16
x398	117.0	18.29	18 1/4	1.770	1 3/4	7/8	16.590	16 5/8	2.845	2 7/8	11 1/4	3 1/2	1 1/2
x370	109.0	17.92	17 7/8	1.655	1 5/8	13/16	16.475	16 1/2	2.660	2 11/16	11 1/4	3 5/16	1 7/16
x342	101.0	17.54	17 1/2	1.540	1 9/16	13/16	16.360	16 3/8	2.470	2 1/2	11 1/4	3 1/8	1 3/8
x311	91.4	17.12	17 1/8	1.410	1 7/16	3/4	16.230	16 1/4	2.260	2 1/4	11 1/4	2 15/16	1 5/16
x283	83.3	16.74	16 3/4	1.290	1 5/16	11/16	16.110	16 1/8	2.070	2 1/16	11 1/4	2 3/4	1 1/4
x257	75.6	16.38	16 3/8	1.175	1 3/16	5/8	15.995	16	1.890	1 7/8	11 1/4	2 9/16	1 3/16
x233	68.5	16.04	16	1.070	1 1/16	9/16	15.890	15 7/8	1.720	1 3/4	11 1/4	2 3/8	1 3/16
x211	62.0	15.72	15 3/4	0.980	1	1/2	15.800	15 3/4	1.560	1 9/16	11 1/4	2 1/4	1 1/8
x193	56.8	15.48	15 1/2	0.890	7/8	7/16	15.710	15 3/4	1.440	1 7/16	11 1/4	2 1/8	1 1/16
x176	51.8	15.22	15 1/4	0.830	13/16	7/16	15.650	15 5/8	1.310	1 5/16	11 1/4	2	1 1/16
x159	46.7	14.98	15	0.745	3/4	3/8	15.565	15 5/8	1.190	1 3/16	11 1/4	1 7/8	1
x145	42.7	14.78	14 3/4	0.680	11/16	3/8	15.500	15 1/2	1.090	1 1/16	11 1/4	1 3/4	1

AMERICAN INSTITUTE OF STEEL CONSTRUCTION

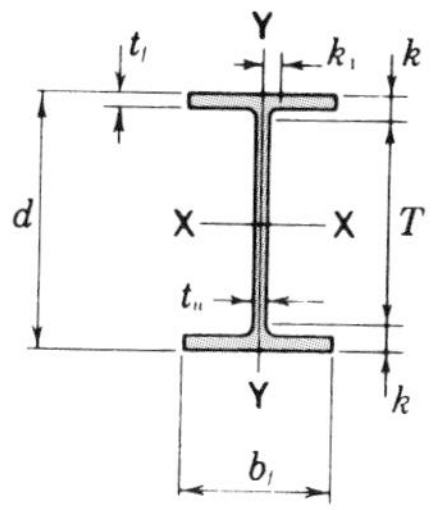

W SHAPES
Dimensions

Designation	Area A	Depth d		Web Thickness t_w		Web $\frac{t_w}{2}$	Flange Width b_f		Flange Thickness t_f		Distance T	Distance k	Distance k_1
	In.^{2}	In.		In.		In.	In.		In.		In.	In.	In.
W 36x300	88.3	36.74	36 3/4	0.945	15/16	1/2	16.655	16 5/8	1.680	1 11/16	31 1/8	2 13/16	1 1/2
x280	82.4	36.52	36 1/2	0.885	7/8	7/16	16.595	16 5/8	1.570	1 9/16	31 1/8	2 11/16	1 1/2
x260	76.5	36.26	36 1/4	0.840	13/16	7/16	16.550	16 1/2	1.440	1 7/16	31 1/8	2 9/16	1 1/2
x245	72.1	36.08	36 1/8	0.800	13/16	7/16	16.510	16 1/2	1.350	1 3/8	31 1/8	2 1/2	1 7/16
x230	67.6	35.90	35 7/8	0.760	3/4	3/8	16.470	16 1/2	1.260	1 1/4	31 1/8	2 3/8	1 7/16
W 36x210	61.8	36.69	36 3/4	0.830	13/16	7/16	12.180	12 1/8	1.360	1 3/8	32 1/8	2 5/16	1 1/4
x194	57.0	36.49	36 1/2	0.765	3/4	3/8	12.115	12 1/8	1.260	1 1/4	32 1/8	2 3/16	1 3/16
x182	53.6	36.33	36 3/8	0.725	3/4	3/8	12.075	12 1/8	1.180	1 3/16	32 1/8	2 1/8	1 3/16
x170	50.0	36.17	36 1/8	0.680	11/16	3/8	12.030	12	1.100	1 1/8	32 1/8	2	1 3/16
x160	47.0	36.01	36	0.650	5/8	5/16	12.000	12	1.020	1	32 1/8	1 15/16	1 1/8
x150	44.2	35.85	35 7/8	0.625	5/8	5/16	11.975	12	0.940	15/16	32 1/8	1 7/8	1 1/8
x135	39.7	35.55	35 1/2	0.600	5/8	5/16	11.950	12	0.790	13/16	32 1/8	1 11/16	1 1/8
W 33x241	70.9	34.18	34 1/8	0.830	13/16	7/16	15.860	15 7/8	1.400	1 3/8	29 3/4	2 3/16	1 3/16
x221	65.0	33.93	33 7/8	0.775	3/4	3/8	15.805	15 3/4	1.275	1 1/4	29 3/4	2 1/16	1 3/16
x201	59.1	33.68	33 5/8	0.715	11/16	3/8	15.745	15 3/4	1.150	1 1/8	29 3/4	1 15/16	1 1/8
W 33x152	44.7	33.49	33 1/2	0.635	5/8	5/16	11.565	11 5/8	1.055	1 1/16	29 3/4	1 7/8	1 1/8
x141	41.6	33.30	33 1/4	0.605	5/8	5/16	11.535	11 1/2	0.960	15/16	29 3/4	1 3/4	1 1/16
x130	38.3	33.09	33 1/8	0.580	9/16	5/16	11.510	11 1/2	0.855	7/8	29 3/4	1 11/16	1 1/16
x118	34.7	32.86	32 7/8	0.550	9/16	5/16	11.480	11 1/2	0.740	3/4	29 3/4	1 9/16	1 1/16
W 30x211	62.0	30.94	31	0.775	3/4	3/8	15.105	15 1/8	1.315	1 5/16	26 3/4	2 1/8	1 1/8
x191	56.1	30.68	30 5/8	0.710	11/16	3/8	15.040	15	1.185	1 3/16	26 3/4	1 15/16	1 1/16
x173	50.8	30.44	30 1/2	0.655	5/8	5/16	14.985	15	1.065	1 1/16	26 3/4	1 7/8	1 1/16
W 30x132	38.9	30.31	30 1/4	0.615	5/8	5/16	10.545	10 1/2	1.000	1	26 3/4	1 3/4	1 1/16
x124	36.5	30.17	30 1/8	0.585	9/16	5/16	10.515	10 1/2	0.930	15/16	26 3/4	1 11/16	1
x116	34.2	30.01	30	0.565	9/16	5/16	10.495	10 1/2	0.850	7/8	26 3/4	1 5/8	1
x108	31.7	29.83	29 7/8	0.545	9/16	5/16	10.475	10 1/2	0.760	3/4	26 3/4	1 9/16	1
x 99	29.1	29.65	29 5/8	0.520	1/2	1/4	10.450	10 1/2	0.670	11/16	26 3/4	1 7/16	1

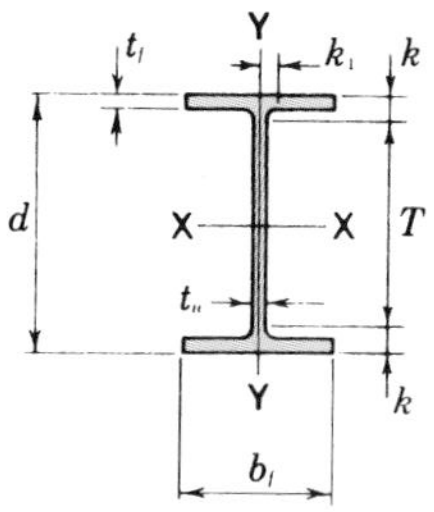

W SHAPES
Dimensions

Designation	Area A	Depth d		Web Thickness t_w		Web $\frac{t_w}{2}$	Flange Width b_f		Flange Thickness t_f		Distance T	Distance k	Distance k_1
	In.^{2}	In.		In.		In.	In.		In.		In.	In.	In.
W 27x178	52.3	27.81	27 3/4	0.725	3/4	3/8	14.085	14 1/8	1.190	1 3/16	24	1 7/8	1 1/16
x161	47.4	27.59	27 5/8	0.660	11/16	3/8	14.020	14	1.080	1 1/16	24	1 13/16	1
x146	42.9	27.38	27 3/8	0.605	5/8	5/16	13.965	14	0.975	1	24	1 11/16	1
W 27x114	33.5	27.29	27 1/4	0.570	9/16	5/16	10.070	10 1/8	0.930	15/16	24	1 5/8	15/16
x102	30.0	27.09	27 1/8	0.515	1/2	1/4	10.015	10	0.830	13/16	24	1 9/16	15/16
x 94	27.7	26.92	26 7/8	0.490	1/2	1/4	9.990	10	0.745	3/4	24	1 7/16	15/16
x 84	24.8	26.71	26 3/4	0.460	7/16	1/4	9.960	10	0.640	5/8	24	1 3/8	15/16
W 24x162	47.7	25.00	25	0.705	11/16	3/8	12.955	13	1.220	1 1/4	21	2	1 1/16
x146	43.0	24.74	24 3/4	0.650	5/8	5/16	12.900	12 7/8	1.090	1 1/16	21	1 7/8	1 1/16
x131	38.5	24.48	24 1/2	0.605	5/8	5/16	12.855	12 7/8	0.960	15/16	21	1 3/4	1 1/16
x117	34.4	24.26	24 1/4	0.550	9/16	5/16	12.800	12 3/4	0.850	7/8	21	1 5/8	1
x104	30.6	24.06	24	0.500	1/2	1/4	12.750	12 3/4	0.750	3/4	21	1 1/2	1
W 24x 94	27.7	24.31	24 1/4	0.515	1/2	1/4	9.065	9 1/8	0.875	7/8	21	1 5/8	1
x 84	24.7	24.10	24 1/8	0.470	1/2	1/4	9.020	9	0.770	3/4	21	1 9/16	15/16
x 76	22.4	23.92	23 7/8	0.440	7/16	1/4	8.990	9	0.680	11/16	21	1 7/16	15/16
x 68	20.1	23.73	23 3/4	0.415	7/16	1/4	8.965	9	0.585	9/16	21	1 3/8	15/16
W 24x 62	18.2	23.74	23 3/4	0.430	7/16	1/4	7.040	7	0.590	9/16	21	1 3/8	15/16
x 55	16.2	23.57	23 5/8	0.395	3/8	3/16	7.005	7	0.505	1/2	21	1 5/16	15/16
W 21x147	43.2	22.06	22	0.720	3/4	3/8	12.510	12 1/2	1.150	1 1/8	18 1/4	1 7/8	1 1/16
x132	38.8	21.83	21 7/8	0.650	5/8	5/16	12.440	12 1/2	1.035	1 1/16	18 1/4	1 13/16	1
x122	35.9	21.68	21 5/8	0.600	5/8	5/16	12.390	12 3/8	0.960	15/16	18 1/4	1 11/16	1
x111	32.7	21.51	21 1/2	0.550	9/16	5/16	12.340	12 3/8	0.875	7/8	18 1/4	1 5/8	15/16
x101	29.8	21.36	21 3/8	0.500	1/2	1/4	12.290	12 1/4	0.800	13/16	18 1/4	1 9/16	15/16
W 21x 93	27.3	21.62	21 5/8	0.580	9/16	5/16	8.420	8 3/8	0.930	15/16	18 1/4	1 11/16	1
x 83	24.3	21.43	21 3/8	0.515	1/2	1/4	8.355	8 3/8	0.835	13/16	18 1/4	1 9/16	15/16
x 73	21.5	21.24	21 1/4	0.455	7/16	1/4	8.295	8 1/4	0.740	3/4	18 1/4	1 1/2	15/16
x 68	20.0	21.13	21 1/8	0.430	7/16	1/4	8.270	8 1/4	0.685	11/16	18 1/4	1 7/16	7/8
x 62	18.3	20.99	21	0.400	3/8	3/16	8.240	8 1/4	0.615	5/8	18 1/4	1 3/8	7/8
W 21x 57	16.7	21.06	21	0.405	3/8	3/16	6.555	6 1/2	0.650	5/8	18 1/4	1 3/8	7/8
x 50	14.7	20.83	20 7/8	0.380	3/8	3/16	6.530	6 1/2	0.535	9/16	18 1/4	1 5/16	7/8
x 44	13.0	20.66	20 5/8	0.350	3/8	3/16	6.500	6 1/2	0.450	7/16	18 1/4	1 3/16	7/8

S SHAPES
Dimensions

Designation	Area A (In²)	Depth d (In.)	Depth (nom.)	Web Thickness t_w (In.)	t_w/2 (In.)	Flange Width b_f (In.)	b_f (nom.)	Flange Thickness t_f (In.)	t_f (nom.)	Distance T (In.)	Distance k (In.)	Grip (In.)	Max. Flge. Fastener (In.)
S 24x121	35.6	24.50	24-1/2	0.800	7/16	8.050	8	1.090	1-1/16	20-1/2	2	1-1/8	1
x106	31.2	24.50	24-1/2	0.620	5/16	7.870	7-7/8	1.090	1-1/16	20-1/2	2	1-1/8	1
S 24x100	29.3	24.00	24	0.745	3/8	7.245	7-1/4	0.870	7/8	20-1/2	1-3/4	7/8	1
x90	26.5	24.00	24	0.625	5/16	7.125	7-1/8	0.870	7/8	20-1/2	1-3/4	7/8	1
x80	23.5	24.00	24	0.500	1/4	7.000	7	0.870	7/8	20-1/2	1-3/4	7/8	1
S 20x96	28.2	20.30	20-1/4	0.800	7/16	7.200	7-1/4	0.920	15/16	16-3/4	1-3/4	15/16	1
x86	25.3	20.30	20-1/4	0.660	3/8	7.060	7	0.920	15/16	16-3/4	1-3/4	15/16	1
S 20x75	22.0	20.00	20	0.635	5/16	6.385	6-3/8	0.795	13/16	16-3/4	1-5/8	13/16	7/8
x66	19.4	20.00	20	0.505	1/4	6.255	6-1/4	0.795	13/16	16-3/4	1-5/8	13/16	7/8
S 18x70	20.6	18.00	18	0.711	3/8	6.251	6-1/4	0.691	11/16	15	1-1/2	11/16	7/8
x54.7	16.1	18.00	18	0.461	1/4	6.001	6	0.691	11/16	15	1-1/2	11/16	7/8
S 15x50	14.7	15.00	15	0.550	5/16	5.640	5-5/8	0.622	5/8	12-1/4	1-3/8	9/16	3/4
x42.9	12.6	15.00	15	0.411	1/4	5.501	5-1/2	0.622	5/8	12-1/4	1-3/8	9/16	3/4
S 12x50	14.7	12.00	12	0.687	3/8	5.477	5-1/2	0.659	11/16	9-1/2	1-7/16	11/16	3/4
x40.8	12.0	12.00	12	0.462	1/4	5.252	5-1/4	0.659	11/16	9-1/2	1-7/16	5/8	3/4
S 12x35	10.3	12.00	12	0.428	1/4	5.078	5-1/8	0.544	9/16	9-5/8	1-5/16	1/2	3/4
x31.8	9.35	12.00	12	0.350	3/16	5.000	5	0.544	9/16	9-5/8	1-5/16	1/2	3/4
S 10x35	10.3	10.00	10	0.594	5/16	4.944	5	0.491	1/2	7-3/4	1-1/8	1/2	3/4
x25.4	7.46	10.00	10	0.311	3/16	4.661	4-5/8	0.491	1/2	7-3/4	1-1/8	1/2	3/4
S 8x23	6.77	8.00	8	0.441	1/4	4.171	4-1/8	0.426	7/16	6	1	7/16	3/4
x18.4	5.41	8.00	8	0.271	1/8	4.001	4	0.426	7/16	6	1	7/16	3/4
S 7x20	5.88	7.00	7	0.450	1/4	3.860	3-7/8	0.392	3/8	5-1/8	15/16	3/8	3/4
x15.3	4.50	7.00	7	0.252	1/8	3.662	3-5/8	0.392	3/8	5-1/8	15/16	3/8	3/4
S 6x17.25	5.07	6.00	6	0.465	1/4	3.565	3-5/8	0.359	3/8	4-1/4	7/8	3/8	5/8
x12.5	3.67	6.00	6	0.232	1/8	3.332	3-3/8	0.359	3/8	4-1/4	7/8	3/8	5/8
S 5x14.75	4.34	5.00	5	0.494	1/4	3.284	3-1/4	0.326	5/16	3-3/8	13/16	5/16	5/8
x10	2.94	5.00	5	0.214	1/8	3.004	3	0.326	5/16	3-3/8	13/16	5/16	—
S 4x9.5	2.79	4.00	4	0.326	3/16	2.796	2-3/4	0.293	5/16	2-1/2	3/4	5/16	—
x7.7	2.26	4.00	4	0.193	1/8	2.663	2-5/8	0.293	5/16	2-1/2	3/4	5/16	—
S 3x7.5	2.21	3.00	3	0.349	3/16	2.509	2-1/2	0.260	1/4	1-5/8	11/16	1/4	—
x5.7	1.67	3.00	3	0.170	1/8	2.330	2-3/8	0.260	1/4	1-5/8	11/16	1/4	—

AMERICAN INSTITUTE OF STEEL CONSTRUCTION

M SHAPES
Dimensions

Designation	Area A (In²)	Depth d (In.)	Depth (nom.)	Web Thickness t_w (In.)	t_w/2 (In.)	Flange Width b_f (In.)	b_f (nom.)	Flange Thickness t_f (In.)	t_f (nom.)	Distance T (In.)	Distance k (In.)	Grip (In.)	Max. Flge. Fastener (In.)
M 14x18	5.10	14.00	14	0.215	1/8	4.000	4	0.270	1/4	12-3/4	5/8	1/4	3/4
M 12x11.8	3.47	12.00	12	0.177	1/8	3.065	3-1/8	0.225	1/4	10-7/8	9/16	1/4	—
M 10x9	2.65	10.00	10	0.157	1/8	2.690	2-3/4	0.206	3/16	8-7/8	9/16	3/16	—
M 8x6.5	1.92	8.00	8	0.135	1/16	2.281	2-1/4	0.189	3/16	7	1/2	3/16	—
M 6x20	5.89	6.00	6	0.250	1/8	5.938	6	0.379	3/8	4-1/4	7/8	3/8	7/8
M 6x4.4	1.29	6.00	6	0.114	1/16	1.844	1-7/8	0.171	3/16	5-1/8	7/16	3/16	—
M 5x18.9	5.55	5.00	5	0.316	3/16	5.003	5	0.416	7/16	3-1/4	7/8	7/16	7/8
M 4x13	3.81	4.00	4	0.254	1/8	3.940	4	0.371	3/8	2-3/8	13/16	3/8	3/4

AMERICAN INSTITUTE OF STEEL CONSTRUCTION

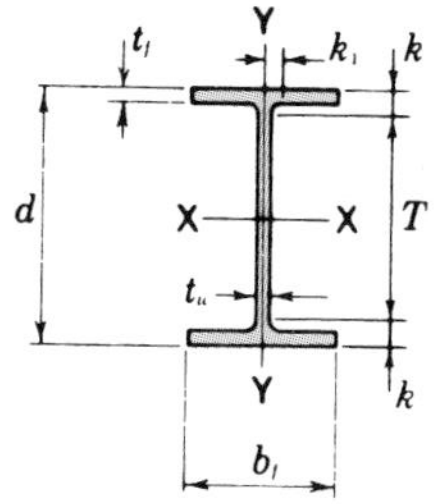

HP SHAPES
Dimensions

Designation	Area A	Depth d		Web Thickness t_w		$\frac{t_w}{2}$	Flange Width b_f		Flange Thickness t_f		Distance T	Distance k	Distance k_1
	In.²	In.		In.		In.	In.		In.		In.	In.	In.
HP 14x117	34.4	14.21	14 1/4	0.805	13/16	7/16	14.885	14 7/8	0.805	13/16	11 1/4	1 1/2	1 1/16
x102	30.0	14.01	14	0.705	11/16	3/8	14.785	14 3/4	0.705	11/16	11 1/4	1 3/8	1
x 89	26.1	13.83	13 7/8	0.615	5/8	5/16	14.695	14 3/4	0.615	5/8	11 1/4	1 5/16	15/16
x 73	21.4	13.61	13 5/8	0.505	1/2	1/4	14.585	14 5/8	0.505	1/2	11 1/4	1 3/16	7/8
HP 13x100	29.4	13.15	13 1/8	0.765	3/4	3/8	13.205	13 1/4	0.765	3/4	10 1/4	1 7/16	1
x 87	25.5	12.95	13	0.665	11/16	3/8	13.105	13 1/8	0.665	11/16	10 1/4	1 3/8	15/16
x 73	21.6	12.75	12 3/4	0.565	9/16	5/16	13.005	13	0.565	9/16	10 1/4	1 1/4	15/16
x 60	17.5	12.54	12 1/2	0.460	7/16	1/4	12.900	12 7/8	0.460	7/16	10 1/4	1 1/8	7/8
HP 12x 84	24.6	12.28	12 1/4	0.685	11/16	3/8	12.295	12 1/4	0.685	11/16	9 1/2	1 3/8	1
x 74	21.8	12.13	12 1/8	0.605	5/8	5/16	12.215	12 1/4	0.610	5/8	9 1/2	1 5/16	15/16
x 63	18.4	11.94	12	0.515	1/2	1/4	12.125	12 1/8	0.515	1/2	9 1/2	1 1/4	7/8
x 53	15.5	11.78	11 3/4	0.435	7/16	1/4	12.045	12	0.435	7/16	9 1/2	1 1/8	7/8
HP 10x 57	16.8	9.99	10	0.565	9/16	5/16	10.225	10 1/4	0.565	9/16	7 5/8	1 3/16	13/16
x 42	12.4	9.70	9 3/4	0.415	7/16	1/4	10.075	10 1/8	0.420	7/16	7 5/8	1 1/16	3/4
HP 8x 36	10.6	8.02	8	0.445	7/16	1/4	8.155	8 1/8	0.445	7/16	6 1/8	15/16	5/8

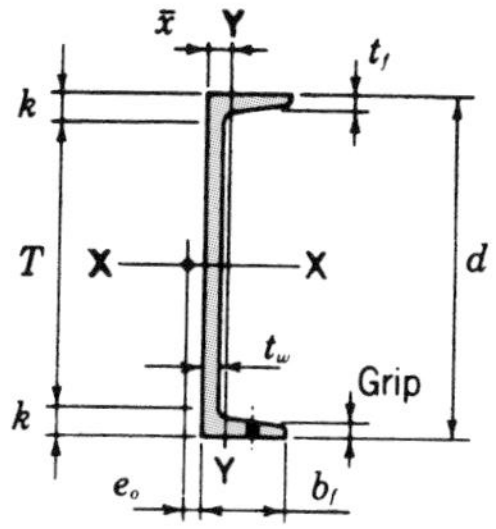

CHANNELS
AMERICAN STANDARD
Dimensions

Designation	Area A	Depth d	Web Thickness t_w		$\frac{t_w}{2}$	Flange Width b_f		Flange Average thickness t_f		Distance T	Distance k	Grip	Max. Flge. Fastener
	In.²	In.	In.		In.	In.		In.		In.	In.	In.	In.
C 15x50	14.7	15.00	0.716	11/16	3/8	3.716	3 3/4	0.650	5/8	12 1/8	1 7/16	5/8	1
x40	11.8	15.00	0.520	1/2	1/4	3.520	3 1/2	0.650	5/8	12 1/8	1 7/16	5/8	1
x33.9	9.96	15.00	0.400	3/8	3/16	3.400	3 3/8	0.650	5/8	12 1/8	1 7/16	5/8	1
C 12x30	8.82	12.00	0.510	1/2	1/4	3.170	3 1/8	0.501	1/2	9 3/4	1 1/8	1/2	7/8
x25	7.35	12.00	0.387	3/8	3/16	3.047	3	0.501	1/2	9 3/4	1 1/8	1/2	7/8
x20.7	6.09	12.00	0.282	5/16	1/8	2.942	3	0.501	1/2	9 3/4	1 1/8	1/2	7/8
C 10x30	8.82	10.00	0.673	11/16	5/16	3.033	3	0.436	7/16	8	1	7/16	3/4
x25	7.35	10.00	0.526	1/2	1/4	2.886	2 7/8	0.436	7/16	8	1	7/16	3/4
x20	5.88	10.00	0.379	3/8	3/16	2.739	2 3/4	0.436	7/16	8	1	7/16	3/4
x15.3	4.49	10.00	0.240	1/4	1/8	2.600	2 5/8	0.436	7/16	8	1	7/16	3/4
C 9x20	5.88	9.00	0.448	7/16	1/4	2.648	2 5/8	0.413	7/16	7 1/8	15/16	7/16	3/4
x15	4.41	9.00	0.285	5/16	1/8	2.485	2 1/2	0.413	7/16	7 1/8	15/16	7/16	3/4
x13.4	3.94	9.00	0.233	1/4	1/8	2.433	2 3/8	0.413	7/16	7 1/8	15/16	7/16	3/4
C 8x18.75	5.51	8.00	0.487	1/2	1/4	2.527	2 1/2	0.390	3/8	6 1/8	15/16	3/8	3/4
x13.75	4.04	8.00	0.303	5/16	1/8	2.343	2 3/8	0.390	3/8	6 1/8	15/16	3/8	3/4
x11.5	3.38	8.00	0.220	1/4	1/8	2.260	2 1/4	0.390	3/8	6 1/8	15/16	3/8	3/4
C 7x14.75	4.33	7.00	0.419	7/16	3/16	2.299	2 1/4	0.366	3/8	5 1/4	7/8	3/8	5/8
x12.25	3.60	7.00	0.314	5/16	3/16	2.194	2 1/4	0.366	3/8	5 1/4	7/8	3/8	5/8
x 9.8	2.87	7.00	0.210	3/16	1/8	2.090	2 1/8	0.366	3/8	5 1/4	7/8	3/8	5/8
C 6x13	3.83	6.00	0.437	7/16	3/16	2.157	2 1/8	0.343	5/16	4 3/8	13/16	5/16	5/8
x10.5	3.09	6.00	0.314	5/16	3/16	2.034	2	0.343	5/16	4 3/8	13/16	3/8	5/8
x 8.2	2.40	6.00	0.200	3/16	1/8	1.920	1 7/8	0.343	5/16	4 3/8	13/16	5/16	5/8
C 5x 9	2.64	5.00	0.325	5/16	3/16	1.885	1 7/8	0.320	5/16	3 1/2	3/4	5/16	5/8
x 6.7	1.97	5.00	0.190	3/16	1/8	1.750	1 3/4	0.320	5/16	3 1/2	3/4	—	—
C 4x 7.25	2.13	4.00	0.321	5/16	3/16	1.721	1 3/4	0.296	5/16	2 5/8	11/16	5/16	5/8
x 5.4	1.59	4.00	0.184	3/16	1/16	1.584	1 5/8	0.296	5/16	2 5/8	11/16	—	—
C 3x 6	1.76	3.00	0.356	3/8	3/16	1.596	1 5/8	0.273	1/4	1 5/8	11/16	—	—
x 5	1.47	3.00	0.258	1/4	1/8	1.498	1 1/2	0.273	1/4	1 5/8	11/16	—	—
x 4.1	1.21	3.00	0.170	3/16	1/16	1.410	1 3/8	0.273	1/4	1 5/8	11/16	—	—

CHANNELS MISCELLANEOUS Dimensions

Designation	Area A (In²)	Depth d (In.)	Web Thickness t_w (In.)	t_w	$\frac{t_w}{2}$ (In.)	Flange Width b_f (In.)	b_f	Flange Avg. thickness t_f (In.)	t_f	Distance T (In.)	Distance k (In.)	Grip (In.)	Max. Flge. Fastener (In.)
MC 18x58	17.1	18.00	0.700	11/16	3/8	4.200	4 1/4	0.625	5/8	15 1/4	1 3/8	5/8	1
x51.9	15.3	18.00	0.600	5/8	5/16	4.100	4 1/8	0.625	5/8	15 1/4	1 3/8	5/8	1
x45.8	13.5	18.00	0.500	1/2	1/4	4.000	4	0.625	5/8	15 1/4	1 3/8	5/8	1
x42.7	12.6	18.00	0.450	7/16	1/4	3.950	4	0.625	5/8	15 1/4	1 3/8	5/8	1
MC 13x50	14.7	13.00	0.787	13/16	3/8	4.412	4 3/8	0.610	5/8	10 1/4	1 3/8	5/8	1
x40	11.8	13.00	0.560	9/16	1/4	4.185	4 1/8	0.610	5/8	10 1/4	1 3/8	9/16	1
x35	10.3	13.00	0.447	7/16	1/4	4.072	4 1/8	0.610	5/8	10 1/4	1 3/8	9/16	1
x31.8	9.35	13.00	0.375	3/8	3/16	4.000	4	0.610	5/8	10 1/4	1 3/8	9/16	1
MC 12x50	14.7	12.00	0.835	13/16	7/16	4.135	4 1/8	0.700	11/16	9 3/8	1 5/16	11/16	1
x45	13.2	12.00	0.712	11/16	3/8	4.012	4	0.700	11/16	9 3/8	1 5/16	11/16	1
x40	11.8	12.00	0.590	9/16	5/16	3.890	3 7/8	0.700	11/16	9 3/8	1 5/16	11/16	1
x35	10.3	12.00	0.467	7/16	1/4	3.767	3 3/4	0.700	11/16	9 3/8	1 5/16	11/16	1
MC 12x37	10.9	12.00	0.600	5/8	5/16	3.600	3 5/8	0.600	5/8	9 3/8	1 5/16	5/8	7/8
x32.9	9.67	12.00	0.500	1/2	1/4	3.500	3 1/2	0.600	5/8	9 3/8	1 5/16	9/16	7/8
x30.9	9.07	12.00	0.450	7/16	1/4	3.450	3 1/2	0.600	5/8	9 3/8	1 5/16	9/16	7/8
MC 12x10.6	3.10	12.00	0.190	3/16	1/8	1.500	1 1/2	0.309	5/16	10 5/8	11/16	—	—
MC 10x41.1	12.1	10.00	0.796	13/16	3/8	4.321	4 3/8	0.575	9/16	7 1/2	1 1/4	5/8	7/8
x33.6	9.87	10.00	0.575	9/16	5/16	4.100	4 1/8	0.575	9/16	7 1/2	1 1/4	9/16	7/8
x28.5	8.37	10.00	0.425	7/16	3/16	3.950	4	0.575	9/16	7 1/2	1 1/4	9/16	7/8
MC 10x28.3	8.32	10.00	0.477	1/2	1/4	3.502	3 1/2	0.575	9/16	7 1/2	1 1/4	9/16	7/8
x25.3	7.43	10.00	0.425	7/16	3/16	3.550	3 1/2	0.500	1/2	7 3/4	1 1/8	1/2	7/8
x24.9	7.32	10.00	0.377	3/8	3/16	3.402	3 3/8	0.575	9/16	7 1/2	1 1/4	9/16	7/8
x21.9	6.43	10.00	0.325	5/16	3/16	3.450	3 1/2	0.500	1/2	7 3/4	1 1/8	1/2	7/8
MC 10x 8.4	2.46	10.00	0.170	3/16	1/16	1.500	1 1/2	0.280	1/4	8 5/8	11/16	—	—
MC 10x 6.5	1.91	10.00	0.152	1/8	1/16	1.127	1 1/8	0.202	3/16	9 1/8	7/16	—	—

AMERICAN INSTITUTE OF STEEL CONSTRUCTION

CHANNELS MISCELLANEOUS Dimensions

Designation	Area A (In²)	Depth d (In.)	Web Thickness t_w (In.)	t_w	$\frac{t_w}{2}$ (In.)	Flange Width b_f (In.)	b_f	Flange Avg. thickness t_f (In.)	t_f	Distance T (In.)	Distance k (In.)	Grip (In.)	Max. Flge. Fastener (In.)
MC 9x25.4	7.47	9.00	0.450	7/16	1/4	3.500	3 1/2	0.550	9/16	6 5/8	13/16	9/16	7/8
x23.9	7.02	9.00	0.400	3/8	3/16	3.450	3 1/2	0.550	9/16	6 5/8	13/16	9/16	7/8
MC 8x22.8	6.70	8.00	0.427	7/16	3/16	3.502	3 1/2	0.525	1/2	5 5/8	13/16	1/2	7/8
x21.4	6.28	8.00	0.375	3/8	3/16	3.450	3 1/2	0.525	1/2	5 5/8	13/16	1/2	7/8
MC 8x20	5.88	8.00	0.400	3/8	3/16	3.025	3	0.500	1/2	5 3/4	1 1/8	1/2	7/8
x18.7	5.50	8.00	0.353	3/8	3/16	2.978	3	0.500	1/2	5 3/4	1 1/8	1/2	7/8
MC 8x 8.5	2.50	8.00	0.179	3/16	1/16	1.874	1 7/8	0.311	5/16	6 1/2	3/4	5/16	5/8
MC 7x22.7	6.67	7.00	0.503	1/2	1/4	3.603	3 5/8	0.500	1/2	4 3/4	1 1/8	1/2	7/8
x19.1	5.61	7.00	0.352	3/8	3/16	3.452	3 1/2	0.500	1/2	4 3/4	1 1/8	1/2	7/8
MC 7x17.6	5.17	7.00	0.375	3/8	3/16	3.000	3	0.475	1/2	4 7/8	1 1/16	1/2	3/4
MC 6x18	5.29	6.00	0.379	3/8	3/16	3.504	3 1/2	0.475	1/2	3 7/8	1 1/16	1/2	7/8
x15.3	4.50	6.00	0.340	5/16	3/16	3.500	3 1/2	0.385	3/8	4 1/4	7/8	3/8	7/8
MC 6x16.3	4.79	6.00	0.375	3/8	3/16	3.000	3	0.475	1/2	3 7/8	1 1/16	1/2	3/4
x15.1	4.44	6.00	0.316	5/16	3/16	2.941	3	0.475	1/2	3 7/8	1 1/16	1/2	3/4
MC 6x12	3.53	6.00	0.310	5/16	1/8	2.497	2 1/2	0.375	3/8	4 3/8	13/16	3/8	5/8

AMERICAN INSTITUTE OF STEEL CONSTRUCTION

APPENDIX F
CORROSION-RESISTANT FLANGES AND FITTINGS

WELDING FITTINGS – DIMENSIONS

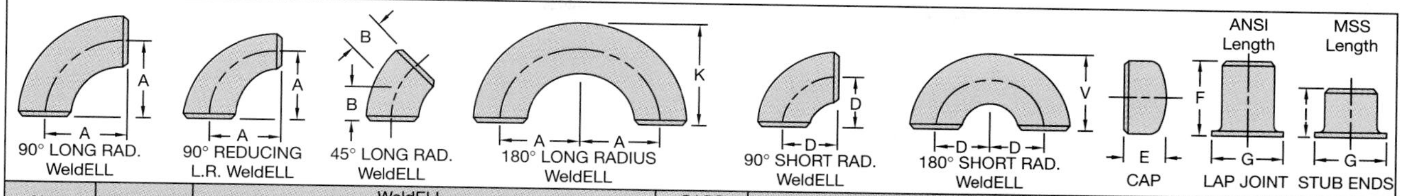
A
90° LONG RAD. WeldELL
A
90° REDUCING L.R. WeldELL
B
45° LONG RAD. WeldELL
K
A A
180° LONG RADIUS WeldELL
D
90° SHORT RAD. WeldELL
V
D D
180° SHORT RAD. WeldELL
E
CAP
ANSI Length
F
G
LAP JOINT
MSS Length
G
STUB ENDS

Nom. Pipe Size	Pipe O.D.	WeldELL A	WeldELL B	WeldELL D	WeldELL K	WeldELL V	CAPS E	STUB ENDS (see note below) G O.D. of Lap	Type A F (Length) ANSI Std.	Type A I (Length) MSS Std.	Type A Corner Radius	Type B,C▲ I (Length) MSS Std.	Type B,C▲ Corner Radius	Nom. Pipe Size
1/2	.840	1 1/2	5/8	–	1 7/8	–	1	1 3/8	3	2	1/8	2	1/32	1/2
3/4	1.050	1 1/8	–	–	1 11/16	–	1	1 11/16	3	2	1/8	2	1/32	3/4
1	1.315	1 1/2	7/8	1	2 3/16	1 5/8	1 1/2	2	4	2	1/8	2	1/32	1
1 1/4	1.660	1 7/8	1	1 1/4	2 3/4	2 1/16	1 1/2	2 1/2	4	2	3/16	2	1/32	1 1/4
1 1/2	1.900	2 1/4	1 1/8	1 1/2	3 1/4	2 7/16	1 1/2	2 7/8	4	2	1/4	2	1/32	1 1/2
2	2.375	3	1 3/8	2	4 3/16	3 3/16	1 1/2	3 5/8	6	2 1/2	5/16	2 1/2	1/32	2
2 1/2	2.875	3 3/4	1 3/4	2 1/2	5 3/16	3 15/16	1 1/2	4 1/8	6	2 1/2	5/16	2 1/2	1/32	2 1/2
3	3.500	4 1/2	2	3	6 1/4	4 3/4	2	5	6	2 1/2	3/8	2 1/2	1/32	3
3 1/2	4.000	5 1/4	2 1/4	3 1/2	7 1/4	5 1/2	2 1/2	5 1/2	6	3	3/8	3	1/32	3 1/2
4	4.500	6	2 1/2	4	8 1/4	6 1/4	2 1/2	6 3/16	6	3	7/16	3	1/32	4
5	5.563	7 1/2	3 1/8	5	10 5/16	7 3/4	3	7 5/16	8	3	7/16	3	1/16	5
6	6.625	9	3 3/4	6	12 5/16	9 5/16	3 1/2	8 1/2	8	3 1/2	1/2	3 1/2	1/16	6
8	8.625	12	5	8	16 5/16	12 5/16	4	10 5/8	8	4	1/2	4	1/16	8
10	10.750	15	6 1/4	10	20 3/8	15 3/8	5	12 3/4	10	5	1/2	5	1/16	10
12	12.750	18	7 1/2	12	24 3/8	18 3/8	6	15	10	6	1/2	6	1/16	12
14	14.000	21	8 3/4	14	28	21	6 1/2	16 1/4	12	–	1/2	–	–	14
16	16.000	24	10	16	32	24	7	18 1/2	12	–	1/2	–	–	16
18	18.000	27	11 1/4	18	36	27	8	21	12	–	1/2	–	–	18
20	20.000	30	12 1/2	20	40	30	9	23	12	–	1/2	–	–	20
24	24.000	36	15	24	48	26	10 1/2	27 1/2	12	–	1/2	–	–	24
30	30.000	45	18 1/2	30	60	45	10 1/2	–	–	–	–	–	–	30

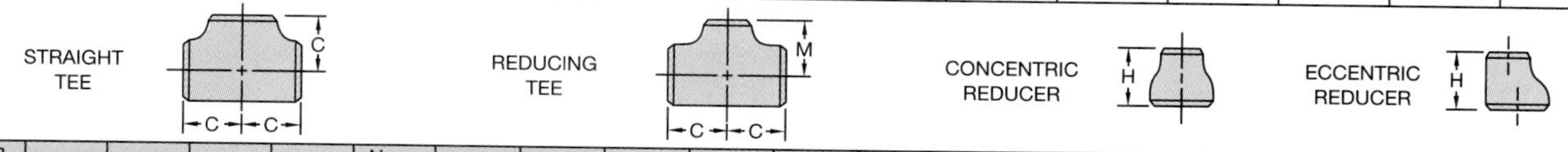
STRAIGHT TEE
C
C C
REDUCING TEE
M
C C
CONCENTRIC REDUCER
H
ECCENTRIC REDUCER
H

Nom. Pipe Size	Outlet	C	M	H
3/4	3/4	1 1/8	...	...
	1/2	1 1/8	1 1/8	1 1/2
1	1	1 1/2	...	...
	3/4	1 1/2	1 1/2	2
	1/2	1 1/2	1 1/2	2
1 1/4	1 1/4	1 7/8	...	...
	1	1 7/8	1 7/8	2
	3/4	1 7/8	1 7/8	2
	1/2	1 7/8	1 7/8	2
1 1/2	1 1/2	2 1/4	...	...
	1 1/4	2 1/4	2 1/4	2 1/2
	1	2 1/4	2 1/4	2 1/2
	3/4	2 1/4	2 1/4	2 1/2
	1/2	2 1/4	2 1/4	2 1/2
2	2	2 1/2	...	...
	1 1/2	2 1/2	2 3/8	3
	1 1/4	2 1/2	2 1/4	3
	1	2 1/2	2	3
	3/4	2 1/2	1 3/4	3
2 1/2	2 1/2	3	...	...
	2	3	2 3/4	3 1/2
	1 1/2	3	2 5/8	3 1/2
	1 1/4	3	2 1/2	3 1/2
	1	3	2 1/4	3 1/2
3	3	3 3/8	...	...
	2 1/2	3 3/8	3 1/4	3 1/2
	2	3 3/8	3	3 1/2
	1 1/2	3 3/8	2 7/8	3 1/2
	1 1/4	3 3/8	2 3/4	3 1/2
3 1/2	3 1/2	3 3/4	...	...
	3	3 3/4	3 5/8	4
	2 1/2	3 3/4	3 1/2	4
	2	3 3/4	3 1/4	4
	1 1/2	3 3/4	3 1/8	4
4	4	4 1/8	...	...
	3 1/2	4 1/8	4	4
	3	4 1/8	3 7/8	4
	2 1/2	4 1/8	3 3/4	4
	2	4 1/8	3 1/2	4
	1 1/2	4 1/8	3 3/8	4
5	5	4 7/8	...	...
	4	4 7/8	4 5/8	5
	3 1/2	4 7/8	4 1/2	5
	3	4 7/8	4 3/8	5
	2 1/2	4 7/8	4 1/4	5
	2	4 7/8	4 1/8	5
6	6	5 5/8	...	...
	5	5 5/8	5 3/8	5 1/2
	4	5 5/8	5 1/8	5 1/2
	3 1/2	5 5/8	5	5 1/2
	3	5 5/8	4 7/8	5 1/2
	2 1/2	5 5/8	4 3/4	5 1/2
8	8	7	...	...
	6	7	6 5/8	6
	5	7	6 3/8	6
	4	7	6 1/8	6
	3 1/2	7	6	6
10	10	8 1/2	...	...
	8	8 1/2	8	7
	6	8 1/2	7 5/8	7
	5	8 1/2	7 1/2	7
	4	8 1/2	7 1/4	7
12	12	10	...	...
	10	10	9 1/2	8
	8	10	9	8
	6	10	8 5/8	8
	5	10	8 1/2	8
14	14	11	...	...
	12	11	10 5/8	13
	10	11	10 1/8	13
	8	11	9 3/4	13
	6	11	9 3/8	13
16	16	12	...	...
	14	12	12	14
	12	12	11 5/8	14
	10	12	11 1/8	14
	8	12	10 3/4	14
	6	12	10 3/8	14
18	18	13 1/2	...	...
	16	13 1/2	13	15
	14	13 1/2	13	15
	12	13 1/2	12 5/8	15
	10	13 1/2	12 1/8	15
	8	13 1/2	11 3/4	15
20	20	15	...	...
	18	15	14 1/2	20
	16	15	14	20
	14	15	14	20
	12	15	13 5/8	20
	10	15	13 1/8	20
	8	15	12 3/4	20
24	24	17	...	...
	20	17	17	20
	18	17	16 1/2	20
	16	17	16	20
	14	17	16	20
	12	17	15 5/8	20
	10	17	15 1/8	20
30	30	22	...	...
	24	22	21	24
	20	22	20	24
	18	22	19 1/2	24
	16	22	19	24
	14	22	19	24
36	36	26 1/2	...	...
	30	26 1/2	25	24
	24	26 1/2	24	24
	20	26 1/2	23	24
	18	26 1/2	22 1/2	24
	16	26 1/2	22	24
42	42	30	...	...
	36	30	28	24
	30	30	28	24
	24	30	26	24
	20	30	26	24

NOTES:

Fittings having wall thicknesses of 0.065" are furnished with square ends unless otherwise specified.

STUB ENDS:

Type A for use with Lap Joint Flanges.

Types B and C for use with Slip-On-Flanges.

Types A and B are furnished with square inside corner, gasket surfaces machined and minimum lap thickness equal to nominal wall of barrel.

▲Dimensions of Type C are same as tabulated for Type B in 5S and 10S thicknesses. Type C is not available in 40S thicknesses. Type C has no fixed corner radius, lap face is not machined. Type C is only available in MSS length. Type B, usually purchased in the MSS length is also available in the ANSI length in Schedules 10S and 40S

FORGED FLANGES – DIMENSIONS

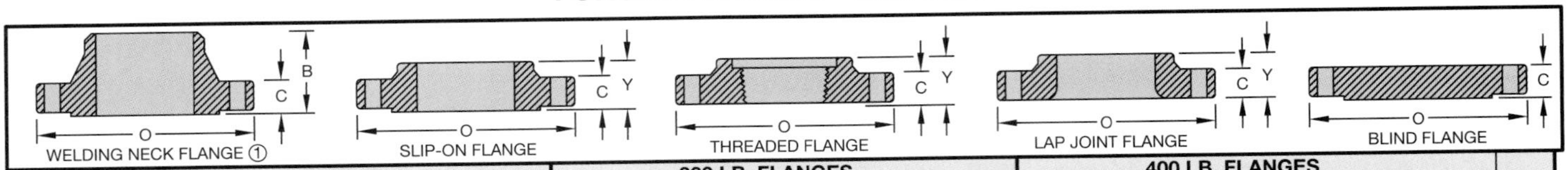
B
C
O
WELDING NECK FLANGE ①
C
Y
O
SLIP-ON FLANGE
C
Y
O
THREADED FLANGE
C
Y
O
LAP JOINT FLANGE
C
O
BLIND FLANGE

150 LB. FLANGES

Nom. Pipe Size	O	C②	Y② Weld Neck	Y② Slip On Thrd.	Y② Lap Joint	Bolt Circle	No.* and Size of Holes
1/2	3 1/2	7/16	1 7/8	5/8	5/8	2 3/8	4-5/8
3/4	3 7/8	1/2	2 1/16	5/8	5/8	2 3/4	4-5/8
1	4 1/4	9/16	2 3/16	11/16	11/16	3 1/8	4-5/8
1 1/4	4 5/8	5/8	2 1/4	13/16	13/16	3 1/2	4-5/8
1 1/2	5	11/16	2 7/16	7/8	7/8	3 7/8	4-5/8
2	6	3/4	2 1/2	1	1	4 3/4	4-3/4
2 1/2	7	7/8	2 3/4	1 1/8	1 1/8	5 1/2	4-3/4
3	7 1/2	15/16	2 3/4	1 3/16	1 3/16	6	4-3/4
3 1/2	8 1/2	15/16	2 13/16	1 1/4	1 1/4	7	8-3/4
4	9	15/16	3	1 5/16	1 5/16	7 1/2	8-3/4
5	10	15/16	3 1/2	1 7/16	1 7/16	8 1/2	8-7/8
6	11	1	3 1/2	1 9/16	1 9/16	9 1/2	8-7/8
8	13 1/2	1 1/8	4	1 3/4	1 3/4	11 3/4	8-7/8
10	16	1 3/16	4	1 15/16	1 15/16	14 1/4	12-1
12	19	1 1/4	4 1/2	2 3/16	2 3/16	17	12-1
14	21	1 3/8	5	2 1/4	3 1/8	18 3/4	12-1 1/8
16	23 1/2	1 7/16	5	2 1/2	3 7/16	21 1/4	16-1 1/8
18	25	1 9/16	5 1/2	2 11/16	3 13/16	22 3/4	16-1 1/4
20	27 1/2	1 11/16	5 11/16	2 7/8	4 1/16	25	20-1 1/4
24	32	1 7/8	6	3 1/4	4 3/8	29 1/2	20-1 3/8

300 LB. FLANGES

Nom. Pipe Size	O	C②	Y② Weld Neck	Y② Slip On Thrd.	Y② Lap Joint	Bolt Circle	No.* and Size of Holes
1/2	3 3/4	9/16	2 1/16	7/8	7/8	2 5/8	4-5/8
3/4	4 5/8	5/8	2 1/4	1	1	3 1/4	4-3/4
1	4 7/8	11/16	2 7/16	1 1/16	1 1/16	3 1/2	4-3/4
1 1/4	5 1/4	3/4	2 9/16	1 1/16	1 1/16	3 7/8	4-3/4
1 1/2	6 1/8	13/16	2 11/16	1 3/16	1 3/16	4 1/2	4-7/8
2	6 1/2	7/8	2 3/4	1 5/16	1 5/16	5	8-3/4
2 1/2	7 1/2	1	3	1 1/2	1 1/2	5 7/8	8-7/8
3	8 1/4	1 1/8	3 1/8	1 11/16	1 11/16	6 5/8	8-7/8
3 1/2	9	1 3/16	3 3/16	1 3/4	1 3/4	7 1/4	8-7/8
4	10	1 1/4	3 3/8	1 7/8	1 7/8	7 7/8	8-7/8
5	11	1 3/8	3 7/8	2	2	9 1/4	8-7/8
6	12 1/2	1 7/16	3 7/8	2 1/16	2 1/16	10 5/8	12-7/8
8	15	1 5/8	4 3/8	2 7/16	2 7/16	13	12-1
10	17 1/2	1 7/8	4 5/8	2 5/8	3 3/4	15 1/4	16-1 1/8
12	20 1/2	2	5 1/8	2 7/8	4	17 3/4	16-1 1/4
14	23	2 1/8	5 5/8	3	4 3/8	20 1/4	20-1 1/4
16	25 1/2	2 1/4	5 3/4	3 1/4	4 3/4	22 1/2	20-1 3/8
18	28	2 3/8	6 1/4	3 1/2	5 1/8	24 3/4	24-1 3/8
20	30 1/2	2 1/2	6 3/8	3 3/4	5 1/2	27	24-1 3/8
24	36	2 3/4	6 5/8	4 3/16	6	32	24-1 5/8

400 LB. FLANGES

Nom. Pipe Size	O	C②	Y② Weld Neck	Y② Slip On Thrd.	Y② Lap Joint	Bolt Circle	No.* and Size of Holes
1/2	3 3/4	9/16	2 1/16	7/8	7/8	2 5/8	4-5/8
3/4	4 5/8	5/8	2 1/4	1	1	3 1/4	4-3/4
1	4 7/8	11/16	2 7/16	1 1/16	1 1/16	3 1/2	4-3/4
1 1/4	5 1/4	13/16	2 5/8	1 1/8	1 1/8	3 7/8	4-3/4
1 1/2	6 1/8	7/8	2 3/4	1 1/4	1 1/4	4 1/2	4-7/8
2	6 1/2	1	2 7/8	1 7/16	1 7/16	5	8-3/4
2 1/2	7 1/2	1 1/8	3 1/8	1 5/8	1 5/8	5 7/8	8-7/8
3	8 1/4	1 1/4	3 1/4	1 13/16	1 13/16	6 5/8	8-7/8
3 1/2	9	1 3/8	3 3/8	1 15/16	1 15/16	7 1/4	8-1
4	10	1 3/8	3 1/2	2	2	7 7/8	8-1
5	11	1 1/2	4	2 1/8	2 1/8	9 1/4	8-1
6	12 1/2	1 5/8	4 1/16	2 1/4	2 1/4	10 5/8	12-1
8	15	1 7/8	4 5/8	2 11/16	2 11/16	13	12-1 1/8
10	17 1/2	2 1/8	4 7/8	2 7/8	4	15 1/4	16-1 1/4
12	20 1/2	2 1/4	5 3/8	3 1/8	4 1/4	17 3/4	16-1 3/8
14	23	2 3/8	5 7/8	3 5/16	4 5/8	20 1/4	20-1 3/8
16	25 1/2	2 1/2	6	3 11/16	5	22 1/2	20-1 1/2
18	28	2 5/8	6 1/2	3 7/8	5 3/8	24 3/4	24-1 1/2
20	30 1/2	2 3/4	6 5/8	4	5 3/4	27	24-1 5/8
24	36	3	6 7/8	4 1/2	6 1/4	32	24-1 7/8

600 LB. FLANGES

Nom. Pipe Size	O	C②	Y② Weld Neck	Y② Slip On Thrd.	Y② Lap Joint	Bolt Circle	No.* and Size of Holes
1/2	3 3/4	9/16	2 1/16	7/8	7/8	2 5/8	4-5/8
3/4	4 5/8	5/8	2 1/4	1	1	3 1/4	4-3/4
1	4 7/8	11/16	2 7/16	1 1/16	1 1/16	3 1/2	4-3/4
1 1/4	5 1/4	13/16	2 5/8	1 1/8	1 1/8	3 7/8	4-3/4
1 1/2	6 1/8	7/8	2 3/4	1 1/4	1 1/4	4 1/2	4-7/8
2	6 1/2	1	2 7/8	1 7/16	1 7/16	5	8-3/4
2 1/2	7 1/2	1 1/8	3 1/8	1 5/8	1 5/8	5 7/8	8-7/8
3	8 1/4	1 1/4	3 1/4	1 13/16	1 13/16	6 5/8	8-7/8
3 1/2	9	1 3/8	3 3/8	1 15/16	1 15/16	7 1/4	8-1
4	10 3/4	1 1/2	4	2 1/8	2 1/8	8 1/2	8-1
5	13	1 3/4	4 1/2	2 3/8	2 3/8	10 1/2	8-1 1/8
6	14	1 7/8	4 5/8	2 5/8	2 5/8	11 1/2	12-1 1/8
8	16 1/2	2 3/16	5 1/4	3	3	13 3/4	12-1 1/4
10	20	2 1/2	6	3 3/8	4 3/8	17	16-1 3/8
12	22	2 5/8	6 1/8	3 5/8	4 5/8	19 1/4	20-1 3/8
14	23 3/4	2 3/4	6 1/2	3 11/16	5	20 3/4	20-1 1/2
16	27	3	7	4 3/16	5 1/2	23 3/4	20-1 5/8
18	29 1/4	3 1/4	7 1/4	4 5/8	6	25 3/4	20-1 3/4
20	32	3 1/2	7 1/2	5	6 1/2	28 1/2	24-1 3/4
24	37	4	8	5 1/2	7 1/4	33	24-2

900 LB. FLANGES

Nom. Pipe Size	O	C②	Y② Weld Neck	Y② Slip On Thrd.	Y② Lap Joint	Bolt Circle	No.* and Size of Holes
1/2	4 3/4	7/8	2 3/8	1 1/4	1 1/4	3 1/4	4-7/8
3/4	5 1/8	1	2 3/4	1 3/8	1 3/8	3 1/2	4-7/8
1	5 7/8	1 1/8	2 7/8	1 5/8	1 5/8	4	4-1
1 1/4	6 1/4	1 1/8	2 7/8	1 5/8	1 5/8	4 3/8	4-1
1 1/2	7	1 1/4	3 1/4	1 3/4	1 3/4	4 7/8	4-1 1/8
2	8 1/2	1 1/2	4	2 1/4	2 1/4	6 1/2	8-1
2 1/2	9 5/8	1 5/8	4 1/8	2 1/2	2 1/2	7 1/2	8-1 1/8
3	9 1/2	1 1/2	4	2 1/8	2 1/8	7 1/2	8-1
3 1/2	...	...	...	...	...	...	...
4	11 1/2	1 3/4	4 1/2	2 3/4	2 3/4	9 1/4	8-1 1/4
5	13 3/4	2	5	3 1/8	3 1/8	11	8-1 3/8
6	15	2 3/16	5 1/2	3 3/8	3 3/8	12 1/2	12-1 1/4
8	18 1/2	2 1/2	6 3/8	4	4 1/2	15 1/2	12-1 1/2
10	21 1/2	2 3/4	7 1/4	4 1/4	5	18 1/2	16-1 1/2
12	24	3 1/8	7 7/8	4 5/8	5 5/8	21	20-1 1/2
14	25 1/4	3 3/8	8 3/8	5 1/8	6 1/8	22	20-1 5/8
16	27 3/4	3 1/2	8 1/2	5 1/4	6 1/2	24 1/4	20-1 3/4
18	31	4	9	6	7 1/2	27	20-2
20	33 3/4	4 1/4	9 3/4	6 1/4	8 1/4	29 1/2	20-2 1/8
24	41	5 1/2	11 1/2	8	10 1/2	35 1/2	20-2 5/8

1500 LB. FLANGES

Nom. Pipe Size	O	C②	Y② Weld Neck	Y② Slip On Thrd.	Y② Lap Joint	Bolt Circle	No.* and Size of Holes
1/2	4 3/4	7/8	2 3/8	1 1/4	1 1/4	3 1/4	4-7/8
3/4	5 1/8	1 7/8	2 3/4	1 3/8	1 3/8	3 1/2	4-7/8
1	5 7/8	1 1/8	2 7/8	1 5/8	1 5/8	4	4-1
1 1/4	6 1/4	1 1/8	2 7/8	1 5/8	1 5/8	4 3/8	4-1
1 1/2	7	1 1/4	3 1/4	1 3/4	1 3/4	4 7/8	4-1 1/8
2	8 1/2	1 1/2	4	2 1/4	2 1/4	6 1/2	8-1
2 1/2	9 5/8	1 5/8	4 1/8	2 1/2	2 1/2	7 1/2	8-1 1/8
3	10 1/2	1 7/8	4 5/8	2 7/8	2 7/8	8	8-1 1/4
3 1/2	...	...	...	...	...	...	...
4	12 1/4	2 1/8	4 7/8	3 9/16	3 9/16	9 1/2	8-1 3/8
5	14 3/4	2 7/8	6 1/8	4 1/8	4 1/8	11 1/2	8-1 5/8
6	15 1/2	3 1/4	6 3/4	4 11/16	4 11/16	12 1/2	12-1 1/2
8	19	3 5/8	8 3/8	5 5/8	5 5/8	15 1/2	12-1 3/4
10	23	4 1/4	10	6 1/4	7	19	12-2
12	26 1/2	4 7/8	11 1/8	7 1/8	8 5/8	22 1/2	16-2 1/8
14	29 1/2	5 1/4	11 3/4	...	9 1/2	25	16-2 3/8
16	32 1/2	5 3/4	12 1/4	...	10 1/4	27 3/4	16-2 5/8
18	36	6 3/8	12 7/8	...	10 7/8	30 1/2	16-2 7/8
20	38 3/4	7	14	...	11 1/2	32 3/4	16-3 1/8
24	46	8	16	...	13	39	16-3 5/8

2500 LB. FLANGES

Nom. Pipe Size	O	C②	Y② Weld Neck	Y② Slip On Thrd.	Y② Lap Joint	Bolt Circle	No.* and Size of Holes
1/2	5 1/4	1 3/16	2 7/8	1 9/16	1 9/16	3 1/2	4-7/8
3/4	5 1/2	1 1/4	3 1/8	1 11/16	1 11/16	3 3/4	4-7/8
1	6 1/4	1 3/8	3 1/2	1 7/8	1 7/8	4 1/4	4-1
1 1/4	7 1/4	1 1/2	3 3/4	2 1/16	2 1/16	5 1/8	4-1 1/8
1 1/2	8	1 3/4	4 3/8	2 3/8	2 3/8	5 3/4	4-1 1/4
2	9 1/4	2	5	2 3/4	2 3/4	6 3/4	8-1 1/8
2 1/2	10 1/2	2 1/4	5 5/8	3 1/8	3 1/8	7 3/4	8-1 1/4
3	12	2 5/8	6 5/8	3 5/8	3 5/8	9	8-1 3/8
4	14	3	7 1/2	4 1/4	4 1/4	10 3/4	8-1 5/8
5	16 1/2	3 5/8	9	5 1/8	5 1/8	12 3/4	8-1 7/8
6	19	4 1/4	10 3/4	6	6	14 1/2	8-2 1/8
8	21 3/4	5	12 1/2	7	7	17 1/4	12-2 1/8
10	26 1/2	6 1/2	16 1/2	9	9	21 1/4	12-2 5/8
12	30	7 1/4	18 1/4	10	10	24 3/8	12-2 7/8

150 LB. FLANGES ① ⑤

Nom. Pipe Size	O	C②	Y② Weld Neck	Y② Slip On Thrd.	Y② Lap Joint	Bolt Circle	No.* and Size of Holes
1/2	3 1/2	3/8	7/8	9/16	9/16	2 3/8	4-5/8
3/4	3 7/8	3/8	7/8	9/16	9/16	2 3/4	4-5/8
1	4 1/4	3/8	7/8	9/16	9/16	3 1/8	4-5/8
1 1/4	5 5/8	3/8	7/8	5/8	5/8	3 1/2	4-5/8
1 1/2	5	3/8	7/8	5/8	5/8	3 7/8	4-5/8
2	6	7/16	1	3/4	3/4	4 3/4	4-3/4
2 1/2	7	7/16	1	3/4	3/4	5 1/2	4-3/4
3	7 1/2	1/2	1 1/8	7/8	7/8	6	4-3/4
4	9	1/2	1 1/8	7/8	7/8	7 1/2	8-3/4
5	10	9/16	1 1/4	7/8	7/8	8 1/2	8-7/8
6	11	9/16	1 1/4	1 1/4	1 1/4	9 1/2	8-7/8
8	13 1/2	9/16	1 1/4	1 1/4	1 1/4	11 3/4	8-7/8
10	16	11/16	1 3/8	1 1/4	1 1/4	14 1/4	12-1
12	19	11/16	1 3/8	1 1/4	1 1/4	17	12-1

MSS 150 LB. FLANGES ⑤ ⑥

Nom. Pipe Size	O	C②	Y② Weld Neck	Y② Slip On Thrd.	Y② Lap Joint	Bolt Circle	No.* and Size of Holes
1/2	3 1/2	5/16	1 7/8	5/8	5/8	2 3/8	4-5/8
3/4	3 7/8	11/32	2 1/16	5/8	5/8	2 3/4	4-5/8
1	4 1/4	3/8	2 3/16	11/16	11/16	3 1/8	4-5/8
1 1/4	4 5/8	13/32	2 1/4	13/16	13/16	3 1/2	4-5/8
1 1/2	5	7/16	2 7/16	7/8	7/8	3 7/8	4-5/8
2	6	1/2	2 1/2	1	1	4 3/4	4-3/4
2 1/2	7	9/16	2 3/4	1 1/8	1 1/8	5 1/2	4-3/4
3	7 1/2	5/8	2 3/4	1 3/16	1 3/16	6	4-3/4
4	9	11/16	3	1 5/16	1 5/16	7 1/2	8-3/4
5	10	3/4	3 1/2	1 7/16	1 7/16	8 1/2	8-7/8
6	11	13/16	3 1/2	1 9/16	1 9/16	9 1/2	8-7/8
8	13 1/2	15/16	4	1 3/4	1 3/4	11 3/4	8-7/8
10	16	1	4	1 15/16	1 15/16	14 1/4	12-1
12	19	1 1/16	4 1/2	2 3/16	2 3/16	17	12-1

NOTES:
1. Always specify bore when ordering.
2. Includes 1/16" raised face in 150 lb. and 300 lb. Standard. Does not include 1/4" raised face in 400 lb. and heavier standards.
3. Other types, sizes, and facings on application.
4. For low pressure service up to 150 PSI at 500°F, 225 PSI at 150°F.
5. Drilling and OD match ANSI B16.5 150 lb. steel flange standard, MSS SP-42 150 lb. Corrosion resistant value standards and ANSI B16. 1 125 lb. cast iron flange standard. This class flange has flat face.
6. Thicknesses conform to MSS Standards.

*Bolt holes are 1/8" larger than recommended bolt.

APPENDIX G
THREADED FITTINGS AND THREADED COUPLINGS, REDUCERS, AND CAPS

Threaded Fittings—Class 2000, 3000 and 6000

DIMENSIONS (Inches)

Class	Dim	1/8	1/4	3/8	1/2	3/4	1	1 1/4	1 1/2	2	2 1/2	3	4
Class 2000	A		7/8	31/32	1 1/8	1 5/16	1 1/2	1 3/4	2	2 3/8	3	3 3/8	4 3/16
	B		29/32	1 1/16	1 5/16	1 9/16	1 27/32	2 7/32	2 1/2	3 1/32	3 11/16	4 5/16	5 3/4
	F		3/4	3/4	1	1 1/8	1 1/4	1 5/16	1 3/8	1 11/16	2 1/16	2 1/2	3 1/8
Class 3000	A	7/8	31/32	1 1/8	1 5/16	1 1/2	1 3/4	2	2 3/8	2 1/2	3 1/4	3 3/4	4 1/2
	B	29/32	1 1/16	1 5/16	1 9/16	1 27/32	2 7/32	2 1/2	3 1/32	3 11/32	4	4 3/4	6
	F	3/4	3/4	1	1 1/8	1 1/4	1 5/16	1 3/8	1 11/16	1 3/4	2 1/16	2 1/2	3 1/8
	H	1 1/4	1 1/4	1 1/2	1 5/8	1 5/8	1 7/8	2 1/4	2 5/8	2 15/16	3 5/16		
	J	7/8	7/8	1	1 1/8	1 3/8	1 3/4	2	2 1/8	2 1/2			
	K		1 7/8	2 1/8	2 9/16	3	3 1/2	3 15/16	4 1/4	5			
	L		2 11/16	3	3 9/16	4 1/8	4 13/16	5 3/8	6 7/16	6 5/8			
Class 6000	A	31/32	1 1/8	1 5/16	1 1/2	1 3/4	2	2 3/8	2 1/2	3 1/4	3 3/4	4 3/16	4 1/2
	B	1 1/16	1 5/16	1 9/16	1 27/32	2 7/32	2 1/2	3 1/32	3 11/32	4	4 3/4	5 3/4	6
	F	3/4	1	1 1/8	1 1/4	1 5/16	1 3/8	1 11/16	1 3/4	2 1/16	2 1/2	3 1/8	3 1/8
	H	1 1/4	1 1/2	1 5/8	1 7/8	2 1/4	2 5/8	2 15/16	3 5/16				
	J		7/8	1	1 1/8	1 1/8	1 3/4	2	2 1/8	2 1/2			
	K				2 9/16	3	3 1/2	3 15/16	4 3/4	5			
	L				3 9/16	4 1/8	4 13/16	5 3/8	6 7/16	6 5/8			

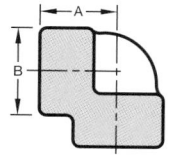
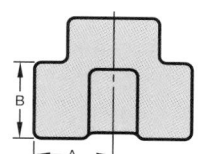
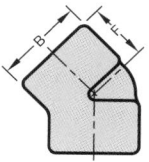
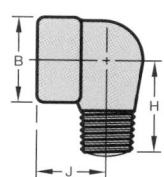
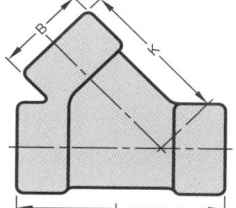
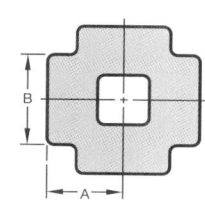

Threaded Couplings, Reducers and Caps—Class 3000 and 6000

DIMENSIONS (Inches)

Class	Dim	1/8	1/4	3/8	1/2	3/4	1	1 1/4	1 1/2	2	2 1/2	3	4
Class 3000	A	1 1/4	1 3/8	1 1/2	1 7/8	2	2 3/8	2 5/8	3 1/8	3 3/8	3 5/8	4 1/4	4 3/4
	B	3/4	3/4	7/8	1 1/8	1 3/8	1 3/4	2 1/4	2 1/2	3	3 5/8	4 1/4	5 1/2
	C	5/8	11/16	3/4	15/16	1	1 3/16	1 5/16	1 9/16	1 11/16	1 13/16	2 1/8	2 3/8
	D	15/16	1	1	1 1/4	1 7/16	1 5/8	1 3/4	1 3/4	1 7/8	2 3/8	2 9/16	2 11/16
Class 6000	A	1 1/4	1 3/8	1 1/2	1 7/8	2	2 3/8	2 5/8	3 1/8	3 3/8	3 5/8	4 1/4	4 3/4
	B	7/8	1	1 1/4	1 1/2	1 3/4	2 1/4	2 1/2	3	3 5/8	4 1/4	5	6 1/4
	C	5/8	11/16	3/4	15/16	1	1 3/16	1 5/16	1 9/16	1 11/16	1 13/16	2 1/8	2 3/8
	D	1	1 1/16	1 1/16	1 5/16	1 1/2	1 11/16	1 13/16	1 7/8	2	2 1/2	2 11/16	2 15/16

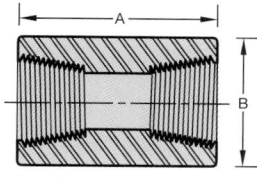
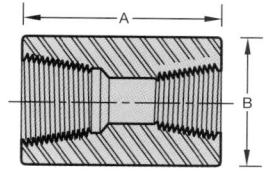
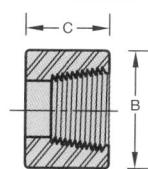
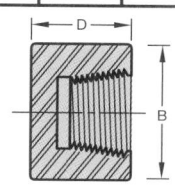

Courtesy Bonney Forge.

APPENDIX H
VALVE SPECIFICATIONS

CRANE

No. 438

GATE VALVES CLASS 125
¼″ to 3″

**Non-Rising Stem
Screwed Bonnet
Solid Wedge Disc**

No. 438, Threaded

RATINGS

Temp. F.	Psi Non-Shock
–20 to 150°	200
200	185
250	170
300	155
350	140
406	125
450	120

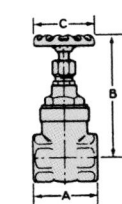
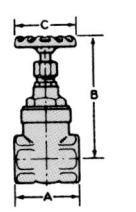

Weights and Dimensions

Valve N.P.S.	Weight—Pounds	Dimensions—Inches		
		A	B	C
¼	.6	1.64	3.44	1.75
⅜	.6	1.64	3.44	1.75
½	1.0	1.90	3.75	2.06
¾	1.4	2.14	4.38	2.75
1	2.1	2.47	4.88	2.75
1¼	3.1	3.08	5.63	3.06
1½	4.2	3.11	6.44	3.63
2	6.2	3.39	7.50	4.50
2½	11.3	4.25	9.06	5.00
3	16.0	4.59	9.69	5.00

"B" dimension is with valve open

CRANE

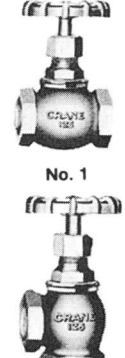

No. 1

No. 2

GLOBE AND ANGLE VALVES CLASS 125
⅛″ to 3″

Screwed Bonnet

**Globe
No. 1, Threaded**

**Angle
No. 2, Threaded**

RATINGS

Temp. F.	Psi Non-Shock
–20 to 150°	200
200	185
250	170
300	155
350	140
406	125
450	120

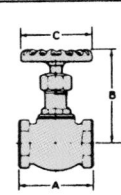

Weights and Dimensions

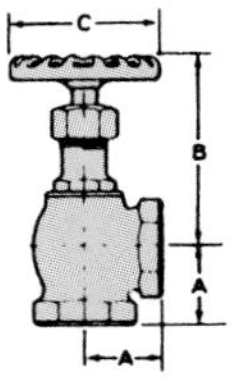

Valve N.P.S.	Weight—Pounds		Dimensions—Inches			
	No. 1	No. 2	A		B	C
			No. 1	No. 2		
⅛	.3	.3	1.44	.75	2.75	1.75
¼	.4	.4	1.63	.81	3.00	1.75
⅜	.5	.4	1.88	.94	3.25	2.03
½	1.0	.8	2.19	1.13	3.75	2.75
¾	1.4	1.4	2.69	1.38	4.25	2.75
1	2.2	2.0	3.19	1.63	5.00	3.00
1¼	3.3	3.0	3.69	1.81	5.50	3.72
1½	4.7	4.5	4.19	2.06	6.25	4.50
2	7.6	7.6	5.13	2.50	7.50	5.00
2½	11.6	—	6.06	—	7.50	5.00
3	19.5	—	7.06	—	9.25	6.00

"B" dimension is with valve open

CRANE

No. 366E

LIFT CHECK VALVES CLASS 300
¼″ to 3″

No. 366E Threaded

RATINGS Non-Shock

Temp. F.	Psi ¼″-2″	Psi 2½″-3″
-20-150°	1000	600
200	920	560
250	830	525
300	740	490
350	650	450
400	560	410
450	480	375
500	390	340
550	300	300

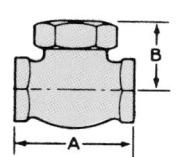
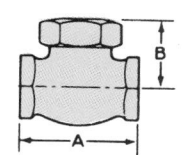

Weights and Dimensions

Valve N.P.S.	Weight—Pounds	Dimensions—Inches A	Dimensions—Inches B
¼	0.4	1.82	1.00
⅜	0.6	2.00	1.12
½	0.9	2.50	1.38
¾	1.5	2.94	1.88
1	2.6	3.50	2.00
1¼	4.2	4.06	2.38
1½	5.4	4.62	2.62
2	10.8	5.75	3.25
2½	15.6	6.88	3.88
3	24.0	8.00	4.50

CRANE

WEDGE GATE VALVES CLASS 125
2″ to 48″

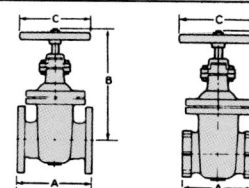

Non-Rising Stem   No. 461

Bronze Trim
No. 460, Threaded
No. 461, Flanged
All Iron
No. 473, Flanged

RATINGS

Temp. F.	Psi, Non-Shock 2 to 12″	Psi, Non-Shock 14 to 24″	Psi, Non-Shock 30 to 48″
-20 to 150°	200	150	150
200	190	135	115
225	180	130	100
250	175	125	85
275	170	120	65
300	165	110	50
325	155	105	
350	150	100	
375	145	*	
400	140	*	
425	130	*	
450	125	*	

*Use Crane 150-pound steel valves.

Weights and Dimensions

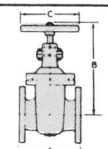

Valve N.P.S.	Weight—Pounds 460	Weight—Pounds 461	Weight—Pounds 473	Threaded A	Flanged A	Flanged B	Flanged C
2	25	30	30	5.38	7.00	11.31	8.0
2½	31	40	40	6.62	7.50	12.40	8.0
3	44	56	56	7.00	8.00	13.25	8.0
4	71	90	90	8.00	9.00	16.31	10.0
5	—	126	—	—	10.00	18.00	10.0
6	—	152	152	—	10.50	20.69	12.0
8	—	260	260	—	11.50	24.12	14.0

Valve N.P.S.	Weight—Pounds 461	Dimensions—Inches A	Dimensions—Inches B	Dimensions—Inches C
10	475	13.00	33.00	20.0
12	680	14.00	36.50	20.0
14	850	15.00	40.50	20.0
16	1280	16.00	47.25	22.0
18	1480	17.00	50.25	24.0
20	1840	18.00	54.75	24.0
24	2860	20.00	65.00	30.0
30	On Request	24.00	76.00	36.0
36	On Request	28.00	85.00	42.0
42	On Request	33.00	106.00	42.0
48	On Request	36.00	111.25	42.0

CRANE

GLOBE AND ANGLE VALVES CLASS 125 2″ to 10″

No. 353

Outside Screw & Yoke Bronze Trim

Globe No. 351, Flanged

Angle No. 353, Flanged

RATINGS

Temp. F.	Psi, Non-Shock	Temp. F.	Psi Non-Shock
–20 to 150°	200	**325**	155
200	190	**350**	150
225	180	**375**	145
250	175	**400**	140
275	170	**425**	130
300	165	**450**	125

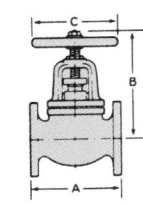 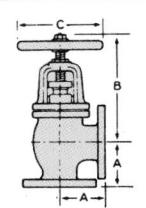

Weights and Dimensions

Valve N.P.S.	Weight—Pounds		Dimensions—Inches				
	351	353	351 A	353 A	351 B	353 B	C
2	34	32	8.00	4.00	11.12	11.00	8.00
2½	40	38	8.50	4.25	11.50	11.50	8.00
3	57	54	9.50	4.75	13.25	12.75	9.00
4	95	88	11.50	5.75	15.50	15.00	10.00
5	126	—	13.00	—	17.50	—	10.00
6	176	158	14.00	7.00	19.50	19.50	12.00
8	344	—	19.50	—	25.00	—	16.00
10	570	—	24.50	—	30.50	—	18.00

"B" dimension is with valve open.

CRANE

SWING CHECK VALVES CLASS 125 2″ to 24″

No. 373

Bolted Cap

Bronze Trim No. 372, Threaded No. 373, Flanged

All Iron No. 373½, Flanged

RATINGS

Temp. F.	Psi, Non-Shock	
	Sizes 2″-12″	Sizes 14″-24″
–20 to 150°	200	150
200	190	135
225	180	130
250	175	125
275	170	120
300	165	110
325	155	105
350	150	100
375	145	—
400	140	—
425	130	—
450	125	—

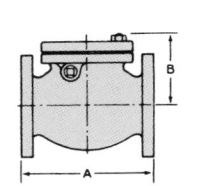

Weights and Dimensions

Valve N.P.S.	Weight—Pounds			Dimensions—Inches		
	372	373	373½	Threaded A	Flanged A	B
2	18	25	25	6.12	8.00	4.50
2½	22	34	34	7.25	8.50	5.38
3	29	44	44	8.00	9.50	5.88
4	54	75	75	9.25	11.50	6.62
5	—	103	—	—	13.00	7.75
6	—	127	127	—	14.00	8.25
8	—	230	230	—	19.50	10.25
10	—	510	510	—	24.50	11.60
12	—	695	695	—	27.50	13.60
14	—	875	—	—	31.00	15.75
16	—	1410	—	—	34.00	17.00
18	—	1540	—	—	38.50	17.40
20	—	1940	—	—	38.50	19.50
24	—	3000	—	—	51.00	20.50

APPENDIX I
SPUR AND HELICAL GEAR DATA

	Spur Gear Data	Suggested Number of Decimal Places
BASIC SPECIFICATIONS	NUMBER OF TEETH	
	DIAMETRAL PITCH	XX.XXXX
	PRESSURE ANGLE	XX°
	STANDARD PITCH DIAMETER	X.XXXX
	TOOTH FORM	
	ADDENDUM	.XXXX
	WHOLE DEPTH	.XXXX
	CALC. CIR. TOOTH THICKNESS ON STD. PITCH CIRCLE	.XXXX MAX. .XXXX MIN.
MANUFACTURING AND INSPECTION	GEAR TESTING RADIUS	X.XXXX MAX. X.XXXX MIN.
	AGMA QUALITY NUMBER	
	MAX. TOTAL COMPOSITE TOLERANCE	.XXXX
	MAX. TOOTH-TO-TOOTH COMPOSITE TOLERANCE	.XXXX
	MASTER GEAR SPECIFICATIONS	
	TESTING PRESSURE (OUNCES)	XX
	DIAMETER OF MEASURING PIN	.XXXX
	MEASUREMENT OVER TWO PINS (FOR SETUP ONLY)	X.XXXX MAX. X.XXXX MIN.
	OUTSIDE DIAMETER	+.000 X.XXX − .00X
	MAX. ROOT DIAMETER	X.XXX
ENGINEERING REFERENCES	MATING GEAR PART NUMBER	
	NUMBER OF TEETH IN MATING GEAR	
	OPERATING CENTER DISTANCE	X.XXXX MAX. X.XXXX MIN.

	Helical Data	Suggested Number of Decimal Places
BASIC SPECIFICATIONS	NUMBER OF TEETH	
	DIAMETRAL PITCH	XX.XXXX
	NORMAL DIAMETRAL PITCH	XX.XXXX
	NORMAL PRESSURE ANGLE	XX°
	HELIX ANGLE	XX.XXXX°
	HAND OF HELIX	L.H. OR R.H.
	STANDARD PITCH DIAMETER	X.XXXX
	TOOTH FORM	
	ADDENDUM	.XXXX
	WHOLE DEPTH	.XXXX
	CALC. NORMAL CIR. TOOTH THICKNESS ON STD. PITCH CIRCLE	.XXXX MAX. .XXXX MIN.
MANUFACTURING AND INSPECTION	MANUFACTURING AND INSPECTION	
	GEAR TESTING RADIUS	X.XXXX MAX. X.XXXX MIN.
	AGMA QUALITY NUMBER	
	MAX. TOTAL COMPOSITE TOLERANCE	.XXXX
	MAX. TOOTH-TO-TOOTH COMPOSITE TOLERANCE	.XXXX
	MASTER GEAR SPECIFICATIONS	
	TESTING PRESSURE (OUNCES)	XX
	DIAMETER OF MEASURING PIN	.XXXX
	MEASUREMENT OVER TWO PINS (FOR SETUP ONLY)	X.XXXX MAX. X.XXXX MIN.
	LEAD	
	OUTSIDE DIAMETER	+.000 X.XXX − .00X
	MAX. ROOT DIAMETER	X.XXXX
ENGINEERING REFERENCES	ENGINEERING REFERENCES	
	MATING GEAR PART NUMBER	
	NUMBER OF TEETH IN MATING GEAR	
	OPERATING CENTER DISTANCE	X.XXXX MAX. X.XXXX MIN.

GLOSSARY

Acme A thread system used especially for feed mechanisms.

Addendum (Spur Gear) The radial distance from the pitch circle to the top of the tooth.

Addendum Angle (Bevel Gear) The angle subtended by the addendum.

Aligned Section The cutting plane is staggered to pass through offset features of an object.

Alignment Charts Designed to graphically solve mathematical equation values using three or more scaled lines.

Allowance The tightest possible fit between two mating parts.

Alloys A mixture of two or more metals.

Amplifier (AMP) A device that allows an input signal to control power; capable of having an output signal greater than the input signal.

Angle of Repose The run-to-rise ratio of highway cut and fill.

Annealing Under certain heating and cooling conditions and techniques, steel may be softened.

Apparent Intersection This is a condition where lines or planes *look* like they may be intersecting, but in reality they may not be intersecting.

Auxiliary View A view that is required when a surface is not parallel to one of the principal planes of projection; the auxiliary projection plane is parallel to the inclined surface so that the surface may be viewed in its true size and shape.

Axis The centerline of a cylindrical feature.

Azimuth The clockwise measurement of the angle of a line, measured from the north of its reference meridian.

Backsight In surveying, the rod reading behind the level toward the point of beginning.

Ball Bearing A friction-reducer where balls roll in two grooved rings.

Base Circle (CAM) The smallest circle tangent to the CAM follower at the bottom of displacement.

Base Circle Diameter (Spur Gear) The diameter of a circle from which the involute tooth is generated.

Basic Dimension A numerical value used to describe the theoretically exact size, profile, orientation, or location of a feature or datum target. It is the basis from which permissible variations are established by tolerances on other dimensions, in notes, or in feature control frames.

Bearing (Civil) The measurement of the angle of a line, measured from either the north or the south meridian, whichever is nearer.

Bearing (Mechanical) A mechanical device that reduces friction between two surfaces.

Bearing Angle The bearing angle of a line is always 90 degrees or less and is identified either from the north or the south.

Bearing of a Line The angular relationship of the horizontal projection of the line relative to the compass, expressed in degrees.

Bearing Seal A rubber, felt, or plastic seal on the outer and inner ring of a bearing. Generally, it is filled with a special lubricant by the manufacturer.

Bearing Shield A metal plate on one or both sides of the bearing; serves to retain the lubricant and keep the bearing clean.

Bell and Spigot A pipe connection in which one end of a piece of pipe has a bell-shaped opening and the other end is tapered or notched to fit into the bell.

Bellcrank A link, pivoted near the center, that oscillates through an angle.

Bench Mark The name for a known point with a known elevation that is part of the geodetic control system. Another name for bench mark is *monument*.

Bend Allowance The amount of extra material needed for a bend to compensate for compression during the bending process.

Bend Relief Cutting away material at a corner to help relieve stress.

Bevel The term used to denote the slope of beams, as in structural engineering.

Bevel Gear Used to transmit power between intersecting shafts; takes the shape of a frustum of a cone.

Bias The voltage applied to a circuit element to control the mode of operation.

Bilateral Tolerance A tolerance in which variation is permitted in both directions from the specified dimension.

Bit Binary digit.

Bolt Circle Holes located in a circular pattern.

Bore To enlarge a hole with a single pointed machine tool in a lathe, drill press, or boring mill.

Boss   A cylindrical projection on the surface of a casting for forging.

Bow's Notation   A system of notation used to label a vector system. A letter is given to the space on each side of the vector, and each vector is then identified by the two letters on either side of it, read in a clockwise direction.

Broken Out Section   A portion of a part is broken away to clarify an interior feature; there is no associated cutting-plane line.

Bus   An aluminum or copper plate or tubing that carries the electrical current.

Bushing   A replaceable lining or sleeve used as a bearing surface.

Butt Weld   A form of pipe manufacture in which the seam of the pipe is a welded flat-faced joint. Also, a form of welding in which two pieces of material are "butted" against each other and welded.

Cabinet Oblique Drawing   A form of oblique drawing in which the receding lines are drawn at half scale, and usually at a 45-degree angle from horizontal.

CAD   Computer-aided design.

CAD/CAM   Computer-aided design/computer-aided manufacturing.

CADD   Computer-aided design and drafting.

CAE   Computer-aided engineering.

Cam   A machine part used to convert constant rotary motion into timed irregular motion.

Cam Motion   The base point from which to begin cam design. There are four basic types of motion: simple harmonic, constant velocity, uniform accelerated, and cycloidal.

Capacitor   An electronic component that opposes a change in voltage and storage of electronic energy.

Carburization   A process where carbon is introduced into the metal by heating to a specified temperature range while in contact with a solid, liquid, or gas material consisting of carbon.

Cartesian Coordinate System   A measurement system based on rectangular grids to measure width, height, and depth (X, Y, and Z).

Casting   An object or part produced by pouring molten metal into a mold.

Cavalier Oblique Drawing   A form of oblique drawing in which the receding lines are drawn true size, or full scale. Usually drawn at an angle of 45 horizontal degrees.

Central Processing Unit (CPU)   The processor and main memory chips in a computer. Specifically, the CPU is just the processor, but generally it refers to the computer.

Chain Dimensioning   Also known as point-to-point dimensioning when dimensions are established from one point to the next.

Chamfer   A slight surface angle used to relieve a sharp corner.

Chordal Addendum (Spur Gear)   The height from the top of the tooth to the line of the chordal thickness.

Chordal Thickness (Spur Gear)   The straight line thickness of a gear tooth on the pitch circle.

CIM   Computer-integrated manufacturing that combines CADD, CAM, and CAE into a controlled system.

Circular Pitch (Spur Gear)   The distance from a point on one tooth to the corresponding point on the adjacent tooth, measured on the pitch circle.

Circular Thickness (Spur Gear)   The length of an arc between the two sides of a gear tooth on the pitch circle.

Clearance (Spur Gear)   The radial distance between the top of a tooth and the bottom of the mating tooth space.

Coil or Inductor   A conductor wound on a form or in a spiral; contains inductance.

Cold Rolled Steel (CRS)   The additional cold forming of steel after initial hot rolling; cleans up hot formed steel.

Command   A specific instruction issued to the computer by the operator. The computer performs a function or task in response to a command.

Compressive   Pushing toward the point of currency, as in forces that are compressed.

Concentric   Two or more circles sharing the same center.

Concurrent Forces   Forces acting on a common point.

Cone Distance (Bevel Gear)   The slant height of the pitch cone.

Construction Lines   Very lightly drawn, nonreproducing lines used for the layout of a drawing.

Contour Interval   The distance in elevation between contour lines.

Contour Line   Denotes a series of connected points at a particular elevation.

Coplanar Forces   All lie in the same plane.

Counterbore   To cylindrically enlarge a hole; generally to allow the head of a screw or bolt to be recessed below the surface of an object.

Counterdrill   A machined hole that looks similar to a countersink/counterbore combination.

Countersink   Used to recess the tapered head of a fastener below the surface of an object.

Crank   A link, usually a rod or bar, that makes a complete revolution about a fixed point.

Crown Backing (Bevel Gear)   The distance between the cone apex and the outer tip of the gear teeth.

CRT   Cathode ray tube.

Cursor   A small rectangle, underline, or set of crosshairs that indicates present location on a video display screen. Also, a handheld input device used in conjunction with a digitizer.

Datum   A theoretically exact point, axis, or plane derived from the true geometric counterpart of a specified datum feature. The origin from which the location or geometric characteristics of features of a part are established.

Datum Dimensioning   A dimensioning system where each dimension originates from a common surface, plane, or axis.

Declination   A line that goes downward from its origin; assigned negative values.

Dedendum (Spur Gear)   The radial distance from the pitch circle to the bottom of the tooth.

Dedendum Angle (Bevel Gear)   The angle subtended by the dedendum.

Default   An action taken by computer software unless the operator specifies differently.

Detail   A drawing of an individual part that contains all of the views, dimensions, and specifications necessary to manufacture the part.

Diametral Pitch A ratio equal to the number of teeth on a gear per inch of pitch diameter.

Diazo A printing process that produces blue, black, or brown lines on various media (other resultant colors are also produced with certain special products). The print process is a combination of exposing an original in contact with a sensitized material exposed to an ultraviolet light, and then running the exposed material through an ammonia chamber to activate the remaining sensitized image to form the desired print. This is a fast and economical method of making prints commonly used in drafting.

Digitize The act of locating points and selecting commands using an input device (puck or stylus, used with a digitizer tablet).

Digitizer An electronically sensitized flat board or tablet that serves as a drawing surface for the input of graphics data. Images can be drawn or traced, and commands and symbols can be selected from a menu attached to the digitizer.

Dihedral Angle The angle that is formed by two intersecting planes.

Dimetric Drawing A pictorial drawing in which two axes form equal angles with the plane of projection. These can be greater than 90 but less than 180 and cannot have an angle of 120 degrees. The third axis may have an angle less or greater than the two equal axes.

Dip The slope of a stratum.

Displacement Diagram A graph; the curve on the diagram is a graph of the path of the cam follower. In the case of a drum cam displacement diagram, the diagram is actually the developed cylindrical surface of the cam.

Documentation Instruction manuals, guides, and tutorials provided with any computer hardware/software system.

Dowel Pin A cylindrical fastener used to retain parts in a fixed position or to keep parts aligned.

Draft The taper on the surface of a pattern for castings of the die for forgings, designed to help facilitate removal of the pattern from the mold or the part from the die. Draft is often 7–10 degrees but depends on the material and the process.

Drilling Drawing Used to provide size and location dimensions for trimming the printed circuit board.

Drum Cam A drum, or cylindrical, cam is a cylinder with a groove in its surface. As the cam rotates, the follower moves through the groove, producing a reciprocating motion parallel to the axis o the camshaft.

Drum Plotter A graphics pen plotter that can accommodate continuous feed paper, or in which sheet paper or film is attached to a sheet of flexible material mounted to a drum. The pen moves in one direction and the drum in the other.

Duct Sheet metal, plastic, or other material pipe designed as the passageway for conveying air from the HVAC equipment to the source.

Ductility The ability to be stretched, drawn, or hammered without breaking.

Eccentric Circle Not having the same center.

Electrical Relays Magnetic switching devices.

Electro-Discharge Machining (EDM) A process where material to be machined and an electrode are submerged in a fluid that does not conduct electricity, forming a barrier between the part and the electrode. A high-current, short-duration electrical charge is then used to remove material.

Electroless The depositing of metal on another material through the action of an electric current.

Electron Beam (EB) Generated by a heated tungsten filament used to cut or machine very accurate features in a part.

Element Any line, group of lines, shape, or group of shapes and text that is so defined by the computer operator.

Elementary Diagrams Diagrams that provide the detail necessary for engineering analysis and operation or maintenance of substation equipment.

Engineering Change Documents Documents used to initiate and implement a change to a production drawing; engineering change request (ECR) and engineering change notice (ECN) are examples.

Entity *See* element.

Equilibrant A vector that is equal in magnitude to the resultant and has the opposite direction and sense.

Equilibrium When a vector system has a resultant of zero, the system is said to be in equilibrium.

Exploded Assembly A pictorial assembly showing all parts removed from each other and aligned along axis lines.

Face Angle (Bevel Angle) The angle between the top of the teeth and the gear axis.

Fault Condition A short circuit that is a zero resistance path for electrical current flow.

Fillet A curve formed at the interior intersection between two or more surfaces.

Fixture A device for holding work in a machine tool.

Flange A thin rim around a part.

Flat-Bed Plotter A pen plotter where the drawing surface (bed) is oriented horizontally, and paper or film is attached to the surface by a vacuum or an electrostatic charge. The pen moves in both the x and y directions.

Floppy Disk A thin, circular, magnetic storage medium encased in a cover. It comes in 8", 5¼", and 3½" sizes.

Flowcharts Used to show organizational structure, steps, or progression in a process or system.

Flow Diagram A chart-type drawing that illustrates the organization of a system in a symbolic format.

Fold Lines The reference line of intersection between two reference planes in orthographic projection.

Follower The cam follower is a reciprocating device whose motion is produced by contact with the cam surface.

Font A specific type face, such as Helvetica or Gothic.

Foreshortened Line A line that appears shorter than its actual length, because it is at an angle to the line of sight.

Four-Bar Linkage The most commonly used linkage mechanism. It contains four links: a fixed link called the ground link, a pivoting link called a driver, another pivoted link called a follower, and a link between the driver and follower, called a coupler.

Free-Body Diagram A diagram that isolates and studies a part of the system of forces in an entire structure.

Full Section The cutting plane extends completely through the object.

Function Keys Extra keys on an alphanumeric keyboard that can be utilized in a computer program to represent different commands and functions. The active commands for function keys may change several times in a program.

Gate   The part of an electronic system that makes the electronic circuit operate; permits an output only when a predetermined set of input conditions are met.

Gear   A cylinder or cone with teeth on its contact surface; used to transmit motion and power from one shaft to another.

Gear Ratio   Any two mating gears have a relationship to each other called a gear ratio. This relationship is the same between any of the following: RPMs, number of teeth, and pitch diameters of the gears.

Gear Train   Formed when two or more gears are in contact.

Grade of a Line   A way to describe the inclination of a line in relation to the horizontal plane. The percent grade is the vertical rise divided by the horizontal run multiplied by 100.

Grade Slope   The percentage given to show amount of slope.

Graphical Kinematic Analysis   The process of drawing a particular mechanism in several phases of a full cycle to determine various characteristics of the mechanism.

Graphics Tablet   *See* digitizer.

Great Circle   One of an infinite number of circles from any point on the earth that is described by longitude and latitude.

Gunter's Chain   A 66-foot-long chain that Edmund Gunter invented; it is made up of 100 links. It is used in surveying.

Half Section   Used typically for symmetrical objects; the cutting-plane line actually cuts through one quarter of the part. The sectional view shows half of the interior and half of the exterior at the same time.

Hard Copy   A paper copy.

Hardware   The physical computer equipment.

HI   Height of instrument. In surveying, the calculation of the level, which is one factor needed to determine elevation.

Highway Diagram   A simplified or condensed representation of a point-to-point, interconnecting wiring diagram for an electrical circuit.

Hone   A method of finishing a hole or other surface to a desired close tolerance and fine surface finish using an abrasive.

Inclination   A line that goes upward from its origin; assigned positive values.

Inductance   The property in an electronic circuit that opposes a change in current flow or where energy may be stored in a magnetic field, as in a transformer.

Integrated Circuit (IC)   All of the components in a schematic are made up of one piece of semiconductor material.

Intersecting Lines   When lines are intersecting, the point of intersection is a point that lies on both lines.

Isogonic Chart   A chart showing isogonic lines.

Isogonic Lines   Shows how many degrees to the east or west the magnetic north or south pole is from the true North or South Pole.

Isometric Drawing   A form of pictorial drawing in which all three drawing axes form equal angles (120 degrees) with the plane of projection.

Jig   A device used for guiding a machine tool in the machining of a part or feature.

Joint   The connection point between two links.

Joystick   A graphics input device composed of a lever mounted in a small box that allows the user to control the movement of the cursor on the video display screen.

Kerf   A groove created by the cut of a saw.

Kinematics   The study of motion without regard to the forces causing the motion.

Lap Weld   A form of pipe manufacture in which the seam of the pipe is an angular "lap."

Large-Scale Integration (LSI)   More circuits on a single small IC chip.

Laser (Light Amplification by Stimulated Emission of Radiation)   A device that amplifies focused light waves and concentrates them in a narrow, very intense beam.

Latitude   The parallels around the earth that do not intersect. The equator is the longest line of latitude.

Lay   Describes the basic direction or configuration of the predominant surface pattern in a surface finish.

Layer   An individual aspect of a CADD drawing that makes a complete drawing when combined.

Lead (Worm Thread)   The distance that the thread advances axially in one revolution of the worm or thread.

Leveling   The process of determining elevation using backsight and foresight.

Lever   A link that moves back and forth through an angle; also known as a rocker.

Light Pen   A video display screen input device. It is a light-sensitive stylus connected to the terminal by a wire; enables the user to draw or select menu options directly on the screen.

Line of Sight   An imaginary straight line from the eye of the observer to a point on the object being observed. All lines of sight for a particular view are assumed to be parallel and are perpendicular to the projection plane involved.

Logic Diagrams   A type of schematic that is used to show the logical sequence in an electronic system.

Longitude   The meridians of the earth, which run from the North Pole to the South Pole. The lines of longitude are basically the same length.

Magnetic Declination   The degree difference between magnetic azimuth and true azimuth.

Malleable   The ability to be hammered or pressed into shape without breaking.

Master Pattern   A one-to-one scale circuit pattern that is used to produce a printed circuit board.

Maxwell Diagram   A combination vector diagram used to analyze the forces acting in a truss.

Mechanical Joint   A pipe connection that is a modification of the "bell and spigot" in which flanges and bolts are used with gaskets, packing rings, or grooved pipe ends providing a seal.

Mechanism   A combination of two or more machine members that work together to perform a specific motion.

Metes and Bounds   The system of describing portions of land by using lengths and boundaries.

Mouse   A handheld input device connected to the terminal by a wire. It is moved across a flat control the movement of the cursor on the screen. It rolls on a small ball that sends directional signals to the computer, and may have one or more buttons that serve as function keys.

Multiview Projection   The views of an object as projected upon two or more picture planes in orthographic projection.

Neck   A groove around a cylindrical part.

Nominal Size   The designation of the size for a commerical product.

Nomographs   A graphic representation of the relationship between two or more variables of a mathematical equation.

Nondestructive Testing (NDS)   Tests for potential defects in welds; they do not destroy or damage the weld or the part.

Normal Plane   A plane surface that is parallel to any of the primary projection planes.

Normalizing   A process of heating steel to a specific temperature and then allowing the material to cool slowly by air, bringing the steel to a normal state.

Numerical Control (NC)   A system of controlling a machine tool by means of numeric codes that direct the commands for the machine movements; computer numerical control (CNC) is a computer command control of the machine movement.

Oblique Drawing   A form of pictorial drawing in which the plane of projection is parallel to the front surface of the object and the receding angle is normally 45°.

Oblique Line   A straight line that is not parallel to any of the six principal planes.

Oblique Plane   Inclined to all of the principal projection planes.

Offset Section   The cutting plane is offset through staggered interior features of an object to show those features in section as if they were in the same plane.

Operational Amplifier (OPAMP)   A high-gain amplifier created from an integrated circuit.

Outside Diameter (Spur Gear)   The overall diameter of the gear; equal to the pitch diameter plus two addendum.

Pads   Or lands; the circuit-termination locations where the electronic devices are attached.

Pattern Development   Based on laying out geometric forms in true size and shape flat patterns.

Perspective Drawing   A form of pictorial drawing in which vanishing points are used to provide the depth and distortion that is seen with the human eye. Perspective drawings can be drawn using one, two, and three vanishing points.

Philadelphia Rod   A long pole with numbers and graduated sections that is used in surveying to determine elevation or distance.

Photodrafting   A combination of a photograph(s) with line work and lettering on a drawing.

Pictorial Drawing   A form of drawing that shows an object's depth. Three sides of the object can be seen in one view.

Pie Charts   Used for presentation purposes where portions of a circle represent quantity.

Piercing Point   A point where a particular line intersects a plane.

Pinion Gear   When two gears are mating, the pinion gear is the smaller, usually the driving gear.

Pitch   A distance of uniform measure determined at a point on one unit to the same corresponding point on the next unit; used in threads, springs, and other machine parts.

Pitch (Worm)   The distance from one tooth to the corresponding point on the next tooth measured parallel to the worm axis; equal to the circular pitch on the worm gear.

Pitch Angle (Bevel Gear)   The angle between an element of a pitch cone and its axis.

Pitch Diameter (Bevel Gear)   The diameter of the base of the pitch cone.

Pitch Diameter (Spur Gear)   The diameter of an imaginary pitch circle on which a gear tooth is designed. Pitch circles of two spur gears are tangent.

Plain Bearing   Based on a sliding action between the mating parts; also called sleeve or journal bearings.

Plane   A surface that is not curved or warped. It is a surface in which any two points may be connected by a straight line, and the straight line will always lie completely within the surface.

Plat   A tract of land showing building lots.

Plate Cam   A cam in the shape of a plate or disk. The motion of the follower is in a plane perpendicular to the axis of the camshaft.

POB   Point of beginning. Any point that has been determined to be the beginning of a survey.

Polar Charts   Designed by establishing polar coordinate scales where points are determined by an angle and distance from a center or pole.

Polyester Drafting Film   A high-quality drafting material with excellent reproduction, durability, and dimensional stability; also known by the trade name Mylar®.

Polygons   Enclosed figures such as triangles, squares, rectangles, parallelograms, and hexagons.

Pressure Angle   The direction of pressure between contacting gear teeth. It determines the size of the base circle and the shape of the involute spur gear tooth, commonly 20°.

Prime Circle (CAM)   A circle with a radius equal to the sum of the base circle radius and the roller follower radius.

Prime Meridian   The line of longitude that is given the 0 degree designation, and from which all other lines of longitude are measured.

Printed Circuits (PC)   Electronic circuits printed on a board that form the interconnection between electronic devices.

Printer   A device that receives data from the computer and converts it into alphanumeric or graphic printed images.

Profile   Shows what a section of land (or utility pipe, etc.) looks like in elevation.

Profile Line   A profile line is one that is parallel to the profile projection plane; its projection appears in true length in the profile view.

Projection Line   A projection line is a straight line at 90° to the fold line, which connects the projection of a point in a view to the projection of the same point in the adjacent view.

Projection Plane   A projection plane is an imaginary surface on which the view of the object is projected and drawn. This surface is imagined to exist between the object and the observer.

Public Land System   The land that was divided by a rectangular system of surveys, in which the main subdivisions are townships and sections.

Quench   To cool suddenly by plunging into water, oil, or other liquid.

Rack   Basically a straight bar with teeth on it. Theoretically, it is a spur gear with an infinite pitch diameter.

Radial Motion   Exists when the path of the motion forms a circle, the diameter of which is perpendicular to the center of the shaft; also known as rotational motion.

Ratio Scales   Special scales that are referred to as logarithmic and semilogarithmic.

Ream   To enlarge a hole slightly with a machine tool called a reamer to produce greater accuracy.

Rectangular System of Surveys   A system of describing the land that is part of a public land survey. Each one of these public land surveys uses townships, sections, quarter-sections, etc. to describe a particular piece of land.

Rectilinear Charts   Charts that are set up on a horizontal and vertical grid where the vertical axis identifies the quantities or values related to the horizontal values; also known as line charts.

Relief   A slight groove between perpendicular surfaces to provide clearance between the surfaces for machining.

Removed Section   A sectional view taken from the location of the section cutting plane and placed in any convenient location of the drawing, generally labeled in relation to the cutting plane.

Resistors   Components that contain resistance to the flow of electric current.

Revolution   An alternate method for solving descriptive geometry problems in which the observer remains stationary and the object is rotated to obtain various views.

Revolved Section   A sectional view established by revolving 90 degrees, in place, into a plane perpendicular to the line of sight, generally used to show the cross section of a part or feature that has consistent shape throughout the length.

Rib   A thin metal section between parts to reinforce while reducing weight in a part.

Right Angle   An angle of 90 degrees.

Rocker   A link that moves back and forth through an angle; also known as a lever.

Roller Bearings   A bearing composed of two grooved rings and a set of rollers. The rollers are the friction-reducing element.

Root Diameter (Spur Gear)   The diameter of a circle coinciding with the bottom of the tooth spaces.

Round   Two or more exterior surfaces rounded at their intersection.

RPM   Revolutions per minute.

Runouts   Characteristics of intersecting features, determined by locating the line of intersection between the mating parts.

Schematic Diagrams   Drawn as a series of lines and symbols that represent the electrical current path and the components of the circuit. Provides the basic circuit connection information for electronic products.

Semiconductors   Devices that provide a degree of resistance in an electronic circuit; types include diodes and transistors.

Skew Lines   Lines that are neither parallel nor intersecting.

Slider   A link that moves back and forth in a straight line.

Slope Angle   The angle in degrees that the line makes with the horizontal plane.

Socket Weld   A form of pipe connection in which a plain-end pipe is slipped into a larger opening or "socket" of a fitting. One exterior weld is required; thus, no weld material protrudes into the pipe.

Software   Computer programs stored on magnetic tape or disk that enable a computer to perform specific functions to accomplish a task.

Solder   An alloy of tin and lead.

Solder Mask   A polymer coating to prevent the bridging of solder between pads or conductor traces on a printed circuit board.

Solids Modeling   A design and engineering process in which a 3-D model of the actual part is created on the screen as a solid part showing no hidden features.

Space Diagram   A drawing of a vector system showing the correct direction and sense but not drawn to scale.

Spline   One of a series of keyways cut around a shaft and mating hole; generally used to transfer power from a shaft to a hub while allowing a sliding action between the parts.

Spur Gear   The simplest, most common type of gear used for transmitting motion between parallel shafts. Its teeth are straight and parallel to the shaft axis.

Stadia   Technique of measuring distance using a Philadelphia rod and level.

Station Point   In surveying, a fixed point from which measurements are made.

Stretch-Out Line   Typically, the beginning line upon which measurements are made and the pattern development is established.

Surface Charts   Designed to show values represented by the extent of a shaded area; also known as area charts.

Surface Finish   Refers to the roughness, waviness, lay, and flaws of a machine surface.

Surface Mount Technology (SMT)   The traditional component lead through is replaced with a solder paste to hold the electronic components in place on the surface of the printed circuit board and take up to less than one-third of the space of conventional PC boards.

Tangent   A straight or curved line that intersects a circle or arc at one point only; is always 90° relative to the center.

Taper   A conical shape on a shaft or hole, or the slope of a plane surface.

Tempering   A process of reheating normalized or hardened steel to a specified temperature, followed by cooling at a predetermined rate to achieve certain hardening characteristics.

Tensile Forces   Forces that pull away.

Tensile Strength   Ability to be stretched.

Thermoplastic   Plastic material may be heated and formed by pressure. Upon reheating, the shape can be changed.

Thermoset   Plastics are formed into permanent shape by heat and pressure and may not be altered after curing.

Thermostat   An automatic mechanism for controlling the amount of heating or cooling given by a central or zoned heating or cooling system.

Tilt-Up Construction   Method using formed wall panels that are lifted or tilted into place.

Tolerance   The total permissible variation in a size or location dimension.

Trackball   An input device consisting of a smooth ball mounted in a small box. A portion of the ball protrudes above the tope of the box and is rotated with the hand to move the cursor on the screen.

Transistors   Semiconductor devices in that they are conductors of electricity with resistance to electron flow applied and are used to transfer or amplify an electronic signal.

Transition Piece   A duct component that provides a change from square or rectangular to round; also known as a square to round.

Translational Motion   Linear motion.

Traverse   In surveying, a series of lines with directions and lengths that are connected at station points.

Triangulation   A technique used to lay out the true size and shape of a triangle with the true lengths of the sides; used in pattern development on objects such as the transition piece.

Trilinear Chart   Designed in the shape of an equilateral triangle; used to show the interrelationship between three variables on a three-dimensional diagram.

Trimetric Drawing   A type of pictorial drawing in which all three of the principal axes do not make equal angles with the plane of projection.

True Azimuth   The azimuth measured from the actual North or South Pole.

True Length or True Size and Shape   When the line of sight is perpendicular to a line, surface, or feature.

True Position   The theoretically exact location of a feature established by basic dimensions.

Turning Point   In surveying, this is each point at which the Philadelphia rod is placed and measured.

Ultrasonic Machining   A process where a high-frequency mechanical vibration is maintained in a tool designed to a desired shape; also known as impact grinding.

Undercut   A groove cut on the inside of a cylindrical hole.

Unilateral Tolerance   A tolerance in which variation is permitted in only one direction from the specified dimension.

Upset   A forging metal used to form a head or enlarged end on a shaft by pressure or hammering between dies.

Valve   Any mechanism, such as a gate, ball, flapper, or diaphragm, used to regulate the flow of fluids through a pipe.

Vector Analysis   A branch of mathematics that includes the manipulation of vectors.

Vector Diagram   A drawing of the vector system in which the vectors are drawn with the correct magnitude and sense, and to scale.

Vector Quantity   A quantity that requires both magnitude and direction for its complete description.

Vellum   A drafting paper with translucent properties.

Viewing-Plane Line   Represents the location of where a view is established.

Visualization   The process of recreating a three-dimensional image of an object in a person's mind.

Web   *See* rib.

Whole Depth (Spur Gear)   The full height of the tooth. It is equal to the sum of the addendum and the dedendum.

Wire Form   A three-dimensional form in which all edges and features show as lines, thus appearing to be constructed of wire.

Wireless Diagram   Similar to highway diagrams except that interconnecting lines are omitted. The interconnection of terminals is provided by coding.

Wiring Diagram   A type of schematic that shows all of the interconnections of the system components, also referred to as a point-to-point interconnecting wiring diagram.

Working Depth (Spur Gear)   The distance that a tooth occupies in the mating space. It is equal to two times the addendum.

Worm Gears   Used to transmit power between nonintersecting shafts. The worm is like a screw and has teeth similar to the teeth on a rack. The teeth on the worm gear are similar to the spur gear teeth, but they are curved to form the teeth on the worm.

Zero Declination   Places on the earth where the compass points exactly toward the true North or South Pole.

Zoning   A system of numbers along the top and bottom and letters along the left and right margins of a drawing used for ease of reading and locating items.

INDEX